P9-DGM-578

BIG IDEAS MATH®

A Common Core Curriculum

TEACHING EDITION

ALGEBRA 1

Ron Larson
Laurie Boswell

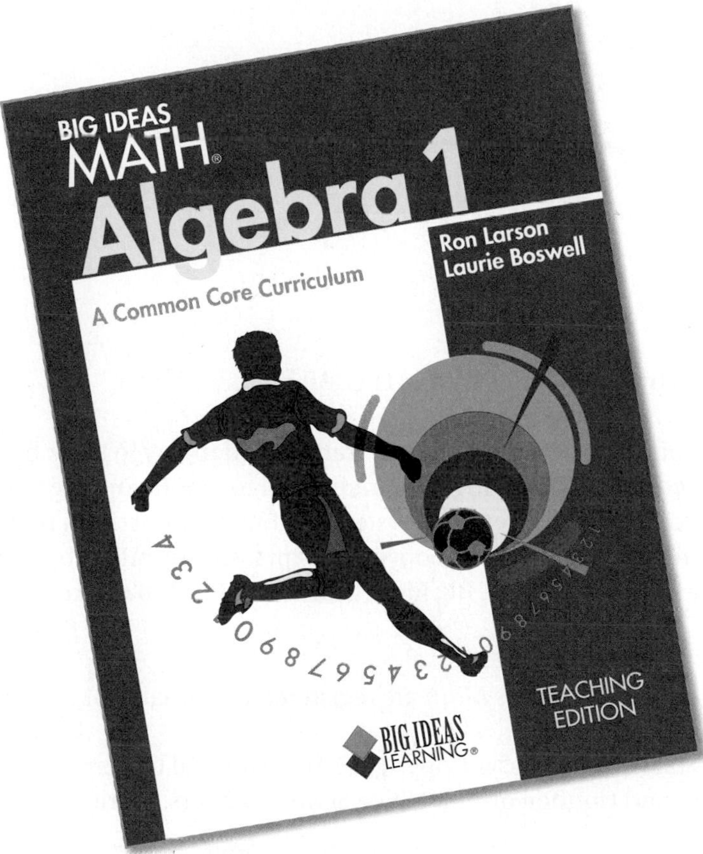

Erie, Pennsylvania
BigIdeasLearning.com

Big Ideas Learning, LLC
1762 Norcross Road
Erie, PA 16510-3838
USA

For product information and customer support, contact Big Ideas Learning at **1-877-552-7766** or visit us at ***BigIdeasLearning.com***.

Printed in the U.S.A.

ISBN 13: 978-1-60840-310-3
ISBN 10: 1-60840-310-6

2 3 4 5 6 7 8 9 10 WEB 16 15 14 13 12

AUTHORS

Ron Larson is a professor of mathematics at Penn State Erie, The Behrend College, where he has taught since receiving his Ph.D. in mathematics from the University of Colorado. Dr. Larson is well known as the lead author of a comprehensive program for mathematics that spans middle school, high school, and college courses. His high school and Advanced Placement books are published by Holt McDougal. Ron's numerous professional activities keep him in constant touch with the needs of students, teachers, and supervisors. Ron and Laurie Boswell began writing together in 1992. Since that time, they have authored over two dozen textbooks. In their collaboration, Ron is primarily responsible for the pupil edition and Laurie is primarily responsible for the teaching edition of the text.

Laurie Boswell is the Head of School and a mathematics teacher at the Riverside School in Lyndonville, Vermont. Dr. Boswell received her Ed.D. from the University of Vermont in 2010. She is a recipient of the Presidential Award for Excellence in Mathematics Teaching. Laurie has taught math to students at all levels, elementary through college. In addition, Laurie was a Tandy Technology Scholar, and served on the NCTM Board of Directors from 2002 to 2005. She currently serves on the board of NCSM, and is a popular national speaker. Along with Ron, Laurie has co-authored numerous math programs.

ABOUT THE BOOK

The Big Ideas Math Algebra 1 book is the newest book in the Big Ideas Math series. The program uses the same research-based strategy of a balanced approach to instruction that made the Big Ideas Math series so successful. This approach opens doors to abstract thought, reasoning, and inquiry as students persevere to answer the Essential Questions that drive instruction. The foundation of the program is the Common Core Standards for Mathematical Content and Standards for Mathematical Practice. This series exposes students to highly motivating and relevant problems that offer the depth and rigor needed to prepare them for Calculus and other college-level courses that they will study during their senior year in high school. The Big Ideas Math Algebra 1 book, along with the Red Accelerated book, completes the compacted pathway for middle school students.

Ron Larson Laurie Boswell

TEACHER REVIEWERS

- Aaron Eisberg
  Napa Valley Unified School District
  Napa, CA
- Gail Englert
  Norfolk Public Schools
  Norfolk, VA
- Alexis Kaplan
  Lindenwold Public Schools
  Lindenwold, NJ
- Lou Kwiatkowski
  Millcreek Township School District
  Erie, PA
- Marcela Mansur
  Broward County Public Schools
  Fort Lauderdale, FL
- Bonnie Pendergast
  Tolleson Union High School District
  Tolleson, AZ
- Tammy Rush
  Hillsborough County Public Schools
  Tampa, FL
- Patricia D. Seger
  Polk County Public Schools
  Bartow, FL
- Denise Walston
  Norfolk Public Schools
  Norfolk, VA

STUDENT REVIEWERS

- Ashley Benovic
- Vanessa Bowser
- Sara Chinsky
- Kaitlyn Grimm
- Lakota Noble
- Norhan Omar
- Jack Puckett
- Abby Quinn
- Victoria Royal
- Madeline Su
- Lance Williams

CONSULTANTS

Patsy Davis
Educational Consultant
Knoxville, Tennessee

Bob Fulenwider
Mathematics Consultant
Bakersfield, California

Deb Johnson
Differentiated Instruction Consultant
Missoula, Montana

Mark Johnson
Mathematics Assessment Consultant
Raymond, New Hampshire

Ryan Keating
Special Education Advisor
Gilbert, Arizona

Michael McDowell
Project-Based Instruction Specialist
Tahoe City, California

Sean McKeighan
Interdisciplinary Advisor
Norman, Oklahoma

Bonnie Spence
Differentiated Instruction Consultant
Missoula, Montana

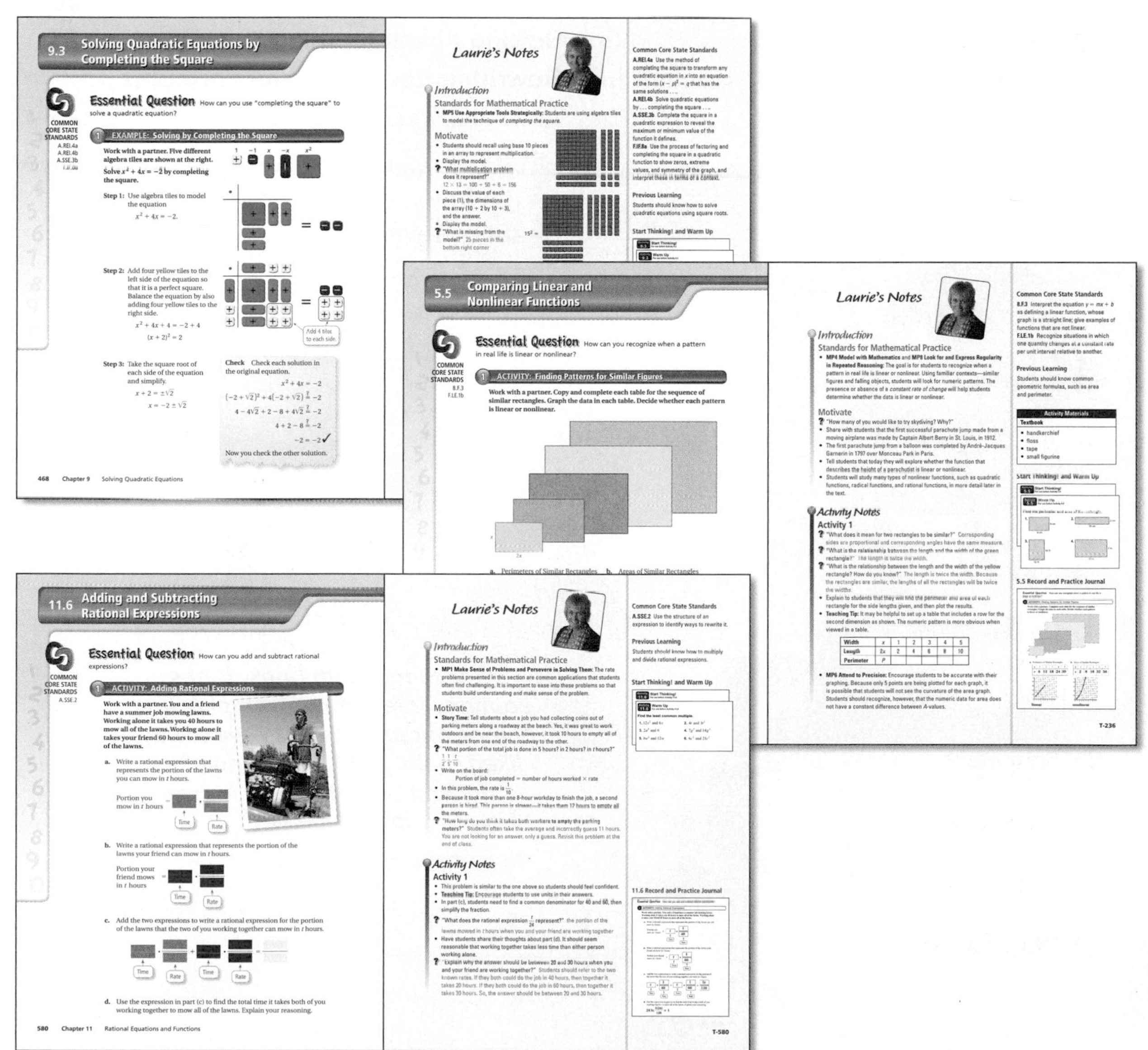

Solving Linear Equations

"I love my math book. It has so many interesting examples and homework problems. I have always liked math, but I didn't know how it could be used. Now I have lots of ideas."

Graphing and Writing Linear Equations

"I like starting each new lesson with a partner activity. I just moved to this school and the activities helped me make friends."

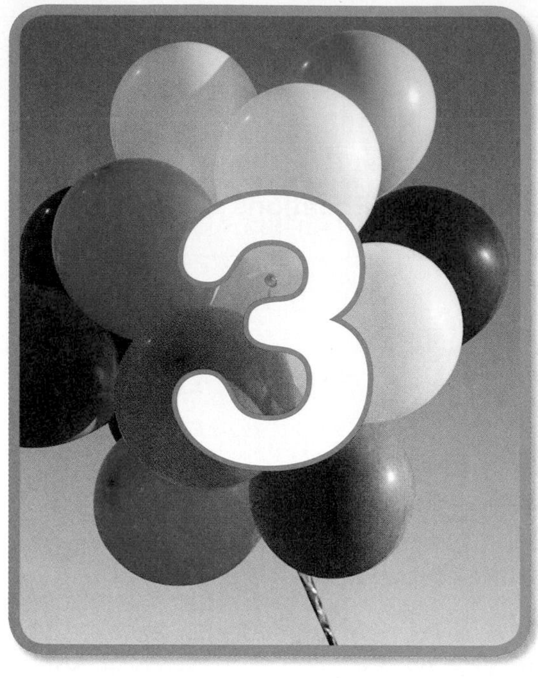

Solving Linear Inequalities

"I like having the book on the Internet. The online tutorials help me with my homework when I get stuck on a problem."

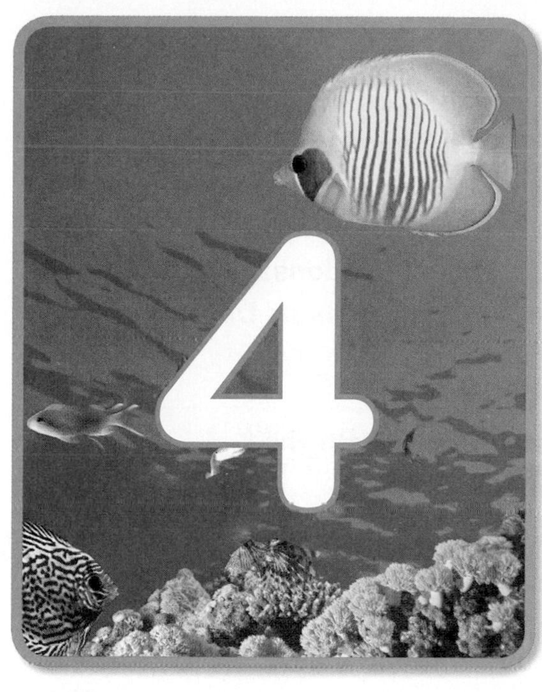

Solving Systems of Linear Equations

"I love the cartoons. They are funny and they help me remember the math. I want to be a cartoonist some day."

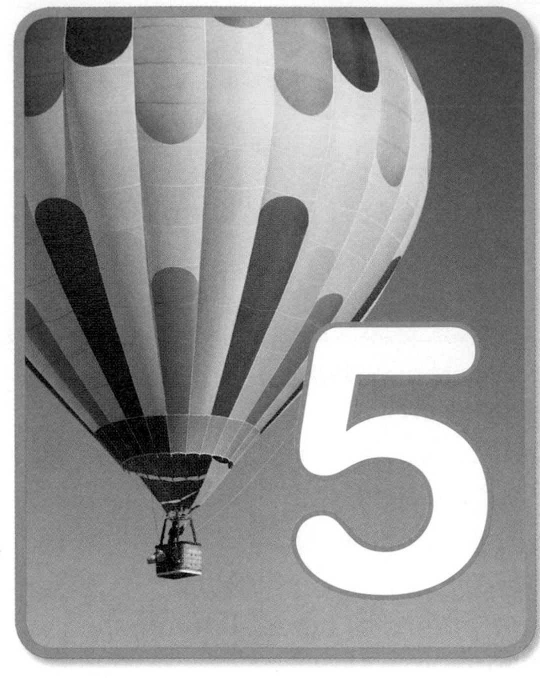

Linear Functions

"I like how I can click on the words in the book that is online and hear them read to me. I like to pronounce words correctly, but sometimes I don't know how to do that by just reading the words."

Exponential Equations and Functions

"I really liked the projects at the end of the book. The art project on symmetry in photos was my favorite. Someday I would like to study photography."

Polynomial Equations and Factoring

"I like how the glossary in the book is part of the index. When I couldn't remember how a vocabulary word was defined, I could go to the index and find where the word was defined in the book."

Graphing Quadratic Functions

"I like the practice tests in the book and online. I get really nervous on tests. So, having a practice test to work on at home helped me to chill out when the real test came."

Solving Quadratic Equations

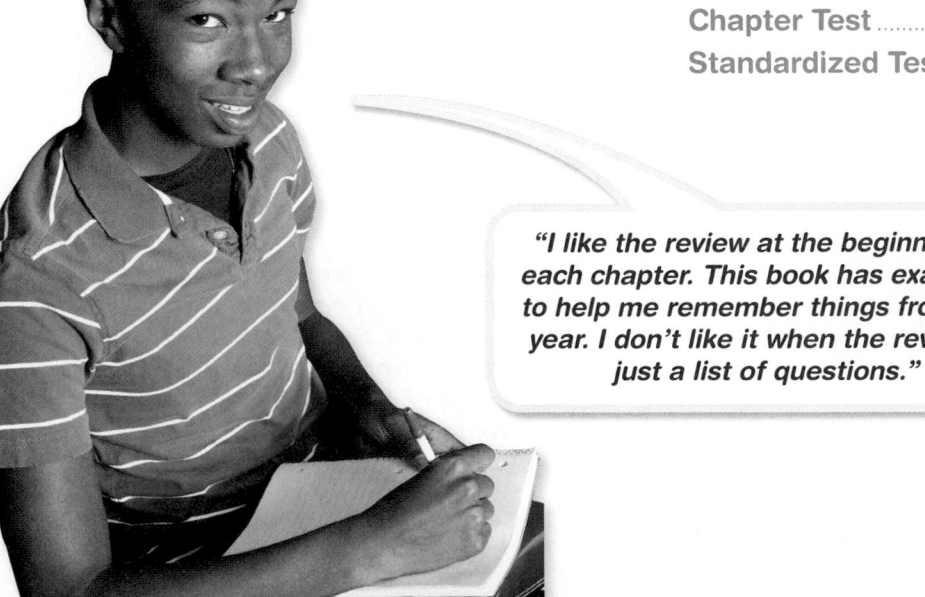

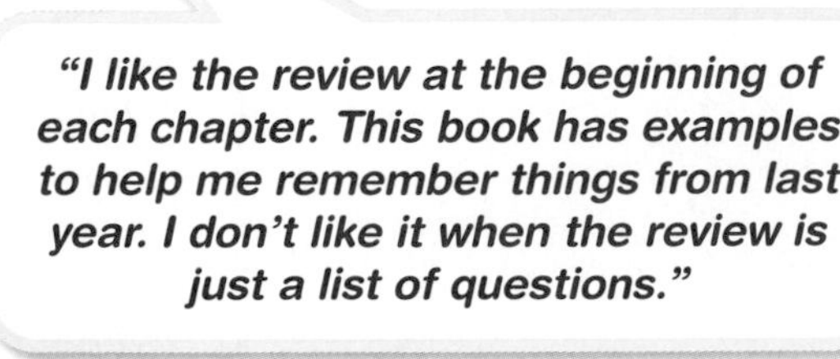

"I like the review at the beginning of each chapter. This book has examples to help me remember things from last year. I don't like it when the review is just a list of questions."

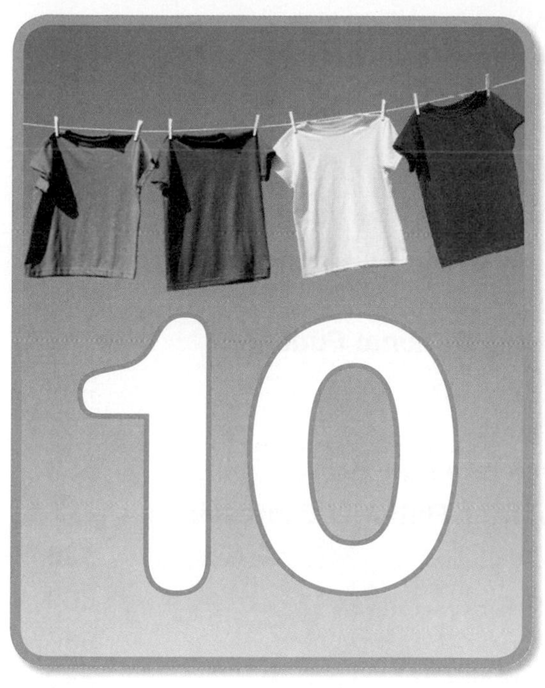

Square Root Functions and Geometry

"I like that the student book is available on my tablet. Now, I always have my book!"

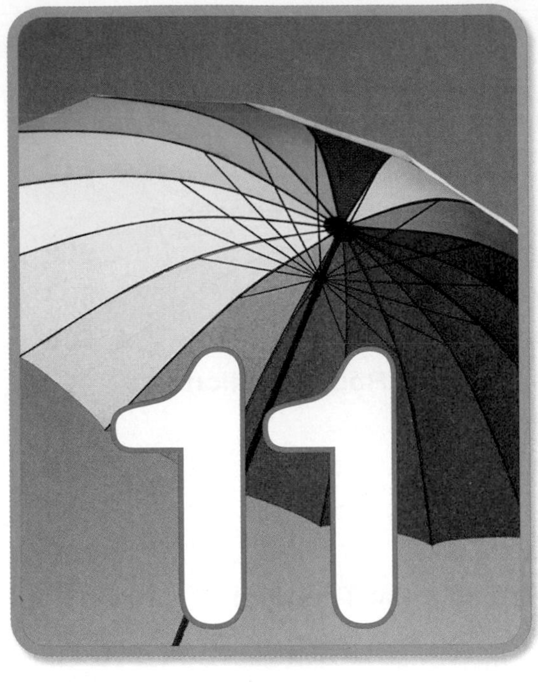

Rational Equations and Functions

"I like that the book teaches me to think. At first I just wanted someone to tell me how. Now I discover it!"

12

Data Analysis and Displays

Appendix A: My Big Ideas Projects

"I like the workbook (Record and Practice Journal). It saved me a lot of work to not have to copy all the questions and graphs."

PROGRAM OVERVIEW

Print

Available in print, online, and in digital format

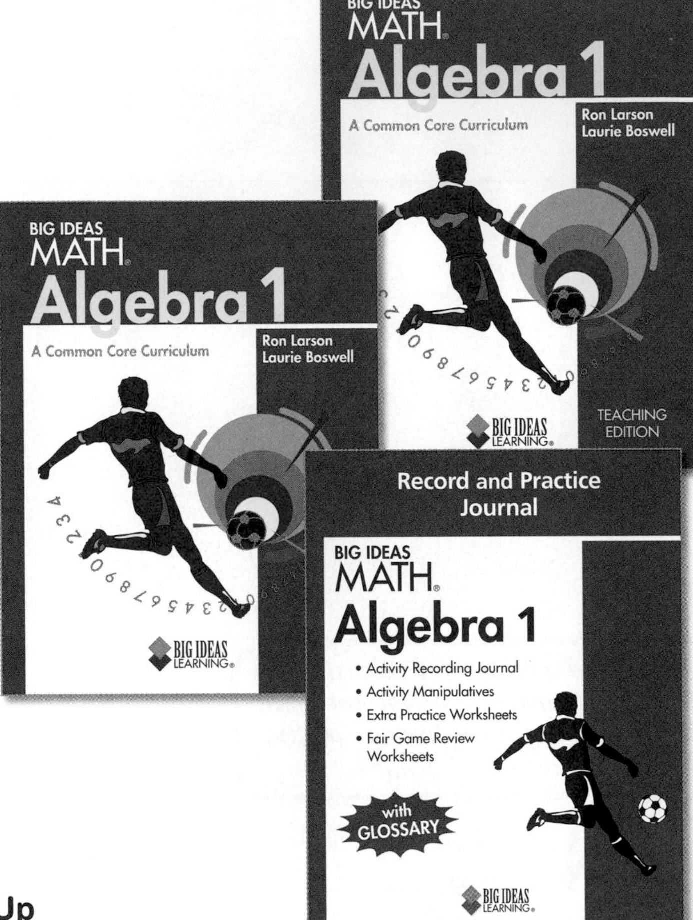

- **Pupil Edition**
- **Teaching Edition**
- **Record and Practice Journal**
- **Assessment Book**
  - **Pre-Course Test**
  - **Quizzes**
  - **Chapter Tests**
  - **Standardized Test Practice**
  - **Alternative Assessment**
  - **End-of-Course Tests**
- **Resources by Chapter**
  - **Start Thinking! and Warm Up**
  - **Family and Community Involvement: English and Spanish**
  - **School-to-Work**
  - **Graphic Organizers/Study Help**
  - **Financial Literacy**
  - **Technology Connection**
  - **Life Connections**
  - **Stories in History**
  - **Extra Practice**
  - **Enrichment and Extension**
  - **Puzzle Time**
  - **Projects with Rubrics**
  - **Cumulative Practice**

- Differentiating the Lesson
- Skills Review Handbook
- Basic Skills Handbook
- Worked-Out Solutions
- Lesson Plans
- Math Tool Paper

Skills Review Handbook

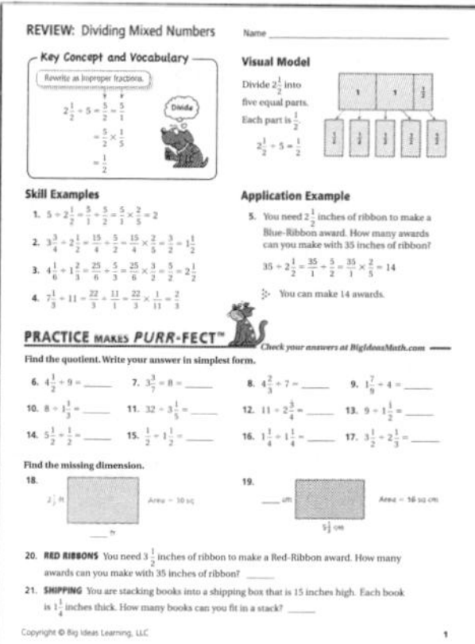
REVIEW: Dividing Mixed Numbers

Key Concept and Vocabulary

Visual Model

Skill Examples

Application Example

PRACTICE MAKES PURR-FECT™

Basic Skills Handbook

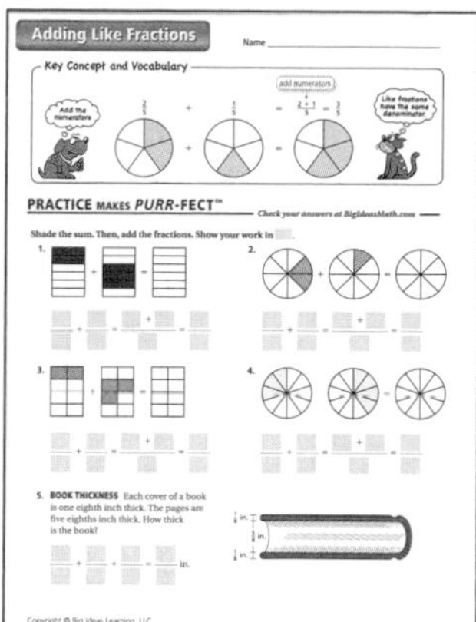
Adding Like Fractions

Key Concept and Vocabulary

PRACTICE MAKES PURR-FECT™

Technology

- Dynamic Assessment Resources
  - ExamView® Assessment Suite

- Dynamic Teaching Resources
  - Answer Presentation Tool
  - Dynamic Classroom
    - Interactive Whiteboard Lessons
    - Support for Mathematical Practices
  - Editable Ancillaries
  - Teaching Edition
- Dynamic Student Edition
  - Interactive Pupil Edition
  - Tutorials
  - Interactive Multi-Language Glossary: English and Spanish
  - Record and Practice Journal
  - Basic Skills Handbook
  - Skills Review Handbook
- *BigIdeasMath.com*
  - Dynamic Student Materials
  - Dynamic Teaching Materials

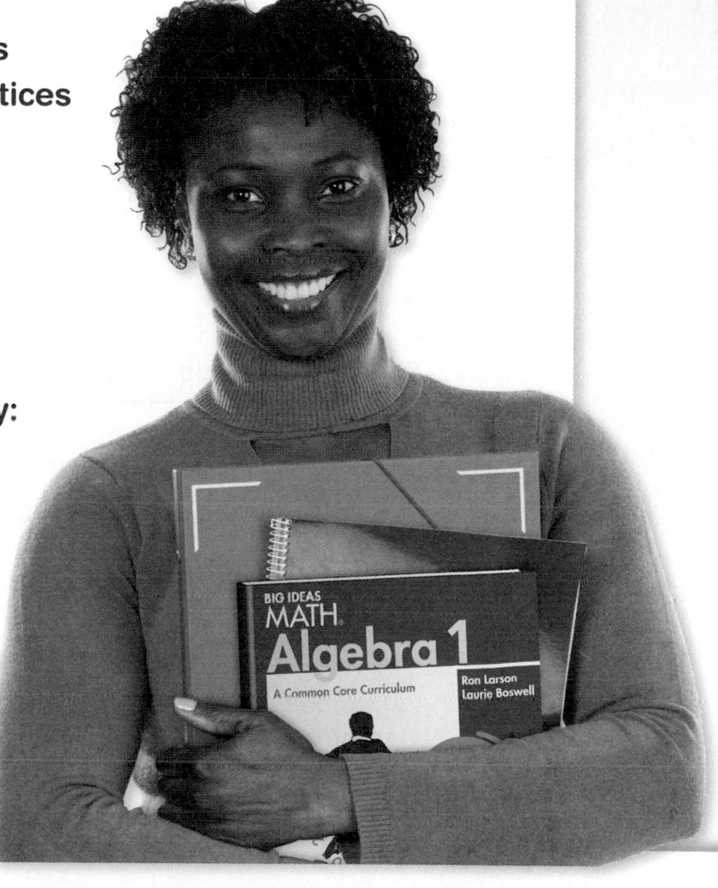

SCOPE AND

Regular Pathway

Grade 6

Ratios and Proportional Relationships	– Understand Ratio Concepts; Use Ratio Reasoning
The Number System	– Perform Fraction and Decimal Operations; Understand Rational Numbers
Expressions and Equations	– Write, Interpret, and Use Expressions and Equations
Geometry	– Solve Problems Involving Area, Surface Area, and Volume
Statistics and Probability	– Summarize Data Sets; Understand Variability

Grade 7

Ratios and Proportional Relationships	– Analyze Proportional Relationships
The Number System	– Perform Rational Number Operations
Expressions and Equations	– Generate Equivalent Expressions; Solve Problems Using Linear Equations
Geometry	– Understand Geometric Relationships; Solve Problems Involving Angles, Surface Area, and Volume
Statistics and Probability	– Analyze Populations; Find Probabilities of Events

Grade 8

The Number System	– Approximate Real Numbers; Perform Real Number Operations
Expressions and Equations	– Use Radicals and Integer Exponents; Connect Proportional Relationships and Lines; Solve Systems of Equations
Functions	– Define, Evaluate, and Compare Functions; Model Relationships
Geometry	– Understand Similarity; Apply the Pythagorean Theorem; Apply Volume Formulas
Statistics and Probability	– Analyze Bivariate Data

SEQUENCE

Compacted Pathway

Grade 6		
	Ratios and Proportional Relationships	– Understand Ratio Concepts; Use Ratio Reasoning
	The Number System	– Perform Fraction and Decimal Operations; Understand Rational Numbers
	Expressions and Equations	– Write, Interpret, and Use Expressions and Equations
	Geometry	– Solve Problems Involving Area, Surface Area, and Volume
	Statistics and Probability	– Summarize Data Sets; Understand Variability

Grade 7 Accelerated		
	Number and Quantity	– Analyze Proportional Relationships; Approximate Real Numbers; Perform Real Number Operations; Use Radicals and Integer Exponents
	Algebra	– Generate Equivalent Expressions; Connect Proportional Relationships and Lines; Solve Problems Using Linear Equations
	Geometry	– Understand Geometric Relationships; Understand Similarity; Solve Problems Involving Angles, Surface Area, and Volume
	Statistics and Probability	– Analyze Populations; Find Probabilities of Events

Algebra 1		
	Number and Quantity	– Use Rational Exponents; Perform Real Number Operations
	Algebra	– Solve Linear and Quadratic Equations; Solve Inequalities and Systems of Equations
	Functions	– Define, Evaluate, and Compare Functions; Write Sequences; Model Relationships
	Geometry	– Apply the Pythagorean Theorem
	Statistics and Probability	– Represent and Interpret Data; Analyze Bivariate Data

COMMON CORE STATE STANDARDS TO BOOK CORRELATION

After a standard is introduced, it is revisited many times in subsequent activities, lessons, and exercises.

Conceptual Category: Number and Quantity

Domain: The Real Number System

N.RN.1 Explain how the definition of the meaning of rational exponents follows from extending the properties of integer exponents to those values, allowing for a notation for radicals in terms of rational exponents.

- **Section 6.3** Radicals and Rational Exponents

N.RN.2 Rewrite expressions involving radicals and rational exponents using the properties of exponents.

- **Section 6.2** Properties of Exponents
- **Section 6.3** Radicals and Rational Exponents
- **Section 10.2** Solving Square Root Equations

N.RN.3 Explain why the sum or product of two rational numbers is rational; that the sum of a rational number and an irrational number is irrational; and that the product of a nonzero rational number and an irrational number is irrational.

- **Section 6.1** Properties of Square Roots

Domain: Quantities

N.Q.1 Use units as a way to understand problems and to guide the solution of multi-step problems; choose and interpret units consistently in formulas; choose and interpret the scale and the origin in graphs and data displays.
Found throughout. For example:

- **Section 1.4** Rewriting Equations and Formulas
- **Section 2.1** Graphing Linear Equations
- **Section 6.4** Exponential Functions

N.Q.2 Define appropriate quantities for the purpose of descriptive modeling.
Found throughout. For example:

- **Section 1.1** Solving Simple Equations
- **Section 3.2** Solving Inequalities Using Addition or Subtraction
- **Section 4.5** Systems of Linear Inequalities

N.Q.3 Choose a level of accuracy appropriate to limitations on measurement when reporting quantities.
Found throughout. For example:

- **Section 6.1** Properties of Square Roots
- **Section 10.2** Solving Square Root Equations
- **Section 12.5** Scatter Plots and Lines of Fit

Conceptual Category: Algebra

Domain: Seeing Structure in Expressions

A.SSE.1 Interpret expressions that represent a quantity in terms of its context.

a. Interpret parts of an expression, such as terms, factors, and coefficients.
- **Section 6.5** Exponential Growth
- **Section 6.6** Exponential Decay
- **Section 7.1** Polynomials

b. Interpret complicated expressions by viewing one or more of their parts as a single entity.
- **Section 6.5** Exponential Growth
- **Section 6.6** Exponential Decay

A.SSE.2 Use the structure of an expression to identify ways to rewrite it.
- **Section 7.6** Factoring Polynomials Using the GCF
- **Section 7.7** Factoring $x^2 + bx + c$
- **Section 7.8** Factoring $ax^2 + bx + c$
- **Section 7.9** Factoring Special Products
- **Section 11.3** Simplifying Rational Expressions
- **Section 11.4** Multiplying and Dividing Rational Expressions
- **Section 11.5** Dividing Polynomials
- **Section 11.6** Adding and Subtracting Rational Expressions

A.SSE.3 Choose and produce an equivalent form of an expression to reveal and explain properties of the quantity represented by the expression.

a. Factor a quadratic expression to reveal the zeros of the function it defines.
- **Section 7.6** Factoring Polynomials Using the GCF
- **Section 7.7** Factoring $x^2 + bx + c$
- **Section 7.8** Factoring $ax^2 + bx + c$
- **Section 7.9** Factoring Special Products

b. Complete the square in a quadratic expression to reveal the maximum or minimum value of the function it defines.
- **Section 9.3** Solving Quadratic Equations by Completing the Square

c. Use the properties of exponents to transform expressions for exponential functions.
- **Section 6.4** Exponential Functions
- **Section 6.5** Exponential Growth
- **Section 6.6** Exponential Decay

Domain: Arithmetic with Polynomials and Rational Expressions

A.APR.1 Understand that polynomials form a system analogous to the integers, namely, they are closed under the operations of addition, subtraction, and multiplication; add, subtract, and multiply polynomials.
- **Section 7.2** Adding and Subtracting Polynomials
- **Section 7.3** Multiplying Polynomials
- **Section 7.4** Special Products of Polynomials

Domain: Creating Equations

A.CED.1 Create equations and inequalities in one variable and use them to solve problems. *Include equations arising from linear and quadratic functions, and simple rational and exponential functions.*

- **Section 1.1** Solving Simple Equations
- **Section 1.2** Solving Multi-Step Equations
- **Section 1.3** Solving Equations with Variables on Both Sides
- **Section 3.1** Writing and Graphing Inequalities
- **Section 3.2** Solving Inequalities Using Addition or Subtraction
- **Section 3.3** Solving Inequalities Using Multiplication or Division
- **Section 3.4** Solving Multi-Step Inequalities
- **Section 6.4** Exponential Functions
- **Section 9.1** Solving Quadratic Equations by Graphing
- **Section 11.7** Solving Rational Equations

A.CED.2 Create equations in two or more variables to represent relationships between quantities; graph equations on coordinate axes with labels and scales.

- **Section 2.1** Graphing Linear Equations
- **Section 2.3** Graphing Linear Equations in Slope-Intercept Form
- **Section 2.4** Graphing Linear Equations in Standard Form
- **Section 2.5** Writing Equations in Slope-Intercept Form
- **Section 2.6** Writing Equations in Point-Slope Form
- **Section 2.7** Solving Real-Life Problems
- **Section 6.4** Exponential Functions
- **Section 8.1** Graphing $y = ax^2$
- **Section 10.1** Graphing Square Root Functions
- **Section 11.2** Graphing Rational Functions

A.CED.3 Represent constraints by equations or inequalities, and by systems of equations and/or inequalities, and interpret solutions as viable or non-viable options in a modeling context.

- **Section 2.5** Writing Equations in Slope-Intercept Form
- **Section 3.1** Writing and Graphing Inequalities
- **Section 3.2** Solving Inequalities Using Addition or Subtraction
- **Section 3.3** Solving Inequalities Using Multiplication or Division
- **Section 3.4** Solving Multi-Step Inequalities
- **Section 4.1** Solving Systems of Linear Equations by Graphing
- **Section 4.2** Solving Systems of Linear Equations by Substitution
- **Section 4.3** Solving Systems of Linear Equations by Elimination
- **Section 4.4** Solving Special Systems of Linear Equations
- **Section 4.5** Systems of Linear Inequalities

A.CED.4 Rearrange formulas to highlight a quantity of interest, using the same reasoning as in solving equations.

- **Section 1.4** Rewriting Equations and Formulas

Domain: Reasoning with Equations and Inequalities

8.EE.8 Analyze and solve pairs of simultaneous linear equations.

a. Understand that solutions to a system of two linear equations in two variables correspond to points of intersection of their graphs, because points of intersection satisfy both equations simultaneously.

- **Section 4.1** Solving Systems of Linear Equations by Graphing
- **Section 4.4** Solving Special Systems of Linear Equations

b. Solve systems of two linear equations in two variables algebraically, and estimate solutions by graphing the equations. Solve simple cases by inspection.

- **Section 4.1** Solving Systems of Linear Equations by Graphing
- **Section 4.2** Solving Systems of Linear Equations by Substitution
- **Section 4.3** Solving Systems of Linear Equations by Elimination
- **Section 4.4** Solving Special Systems of Linear Equations

c. Solve real-world and mathematical problems leading to two linear equations in two variables.

- **Section 4.1** Solving Systems of Linear Equations by Graphing
- **Section 4.2** Solving Systems of Linear Equations by Substitution
- **Section 4.3** Solving Systems of Linear Equations by Elimination
- **Section 4.4** Solving Special Systems of Linear Equations

A.REI.1 Explain each step in solving a simple equation as following from the equality of numbers asserted at the previous step, starting from the assumption that the original equation has a solution. Construct a viable argument to justify a solution method.

- **Section 1.1** Solving Simple Equations
- **Section 1.2** Solving Multi-Step Equations
- **Section 1.3** Solving Equations with Variables on Both Sides

A.REI.3 Solve linear equations and inequalities in one variable, including equations with coefficients represented by letters.

- **Section 1.1** Solving Simple Equations
- **Section 1.2** Solving Multi-Step Equations
- **Section 1.3** Solving Equations with Variables on Both Sides
- **Section 3.2** Solving Inequalities Using Addition or Subtraction
- **Section 3.3** Solving Inequalities Using Multiplication or Division
- **Section 3.4** Solving Multi-Step Inequalities
- **Section 6.4** Exponential Functions

A.REI.4 Solve quadratic equations in one variable.

a. Use the method of completing the square to transform any quadratic equation in x into an equation of the form $(x - p)^2 = q$ that has the same solutions. Derive the quadratic formula from this form.

- **Section 9.3** Solving Quadratic Equations by Completing the Square
- **Section 9.4** Solving Quadratic Equations Using the Quadratic Formula

b. Solve quadratic equations by inspection (e.g., for $x^2 = 49$), taking square roots, completing the square, the quadratic formula and factoring, as appropriate to the initial form of the equation. Recognize when the quadratic formula gives complex solutions and write them as $a \pm bi$ for real numbers a and b.

- **Section 7.5** Solving Polynomial Equations in Factored Form
- **Section 7.6** Factoring Polynomials Using the GCF
- **Section 7.7** Factoring $x^2 + bx + c$

- **Section 7.8** Factoring $ax^2 + bx + c$
- **Section 7.9** Factoring Special Products
- **Section 9.1** Solving Quadratic Equations by Graphing
- **Section 9.2** Solving Quadratic Equations Using Square Roots
- **Section 9.3** Solving Quadratic Equations by Completing the Square
- **Section 9.4** Solving Quadratic Equations Using the Quadratic Formula

A.REI.5 Prove that, given a system of two equations in two variables, replacing one equation by the sum of that equation and a multiple of the other produces a system with the same solutions.

- **Section 4.3** Solving Systems of Linear Equations by Elimination

A.REI.6 Solve systems of linear equations exactly and approximately (e.g., with graphs), focusing on pairs of linear equations in two variables.

- **Section 4.1** Solving Systems of Linear Equations by Graphing
- **Section 4.2** Solving Systems of Linear Equations by Substitution
- **Section 4.3** Solving Systems of Linear Equations by Elimination
- **Section 4.4** Solving Special Systems of Linear Equations

A.REI.7 Solve a simple system consisting of a linear equation and a quadratic equation in two variables algebraically and graphically.

- **Section 9.5** Solving Systems of Linear and Quadratic Equations

A.REI.10 Understand that the graph of an equation in two variables is the set of all its solutions plotted in the coordinate plane, often forming a curve (which could be a line).

- **Section 2.1** Graphing Linear Equations
- **Section 2.3** Graphing Linear Equations in Slope-Intercept Form
- **Section 2.4** Graphing Linear Equations in Standard Form
- **Section 2.6** Writing Equations in Point-Slope Form
- **Section 6.4** Exponential Functions
- **Section 8.1** Graphing $y = ax^2$
- **Section 10.1** Graphing Square Root Functions
- **Section 11.1** Direct and Inverse Variation
- **Section 11.2** Graphing Rational Functions

A.REI.11 Explain why the x-coordinates of the points where the graphs of the equations $y = f(x)$ and $y = g(x)$ intersect are the solutions of the equation $f(x) = g(x)$; find the solutions approximately, e.g., using technology to graph the functions, make tables of values, or find successive approximations. Include cases where $f(x)$ and/or $g(x)$ are linear, polynomial, rational, absolute value, exponential, and logarithmic functions.

- **Section 4.4** Solving Special Systems of Linear Equations
- **Section 6.4** Exponential Functions
- **Section 9.1** Solving Quadratic Equations by Graphing
- **Section 9.5** Solving Systems of Linear and Quadratic Equations

A.REI.12 Graph the solutions to a linear inequality in two variables as a half-plane (excluding the boundary in the case of a strict inequality), and graph the solution set to a system of linear inequalities in two variables as the intersection of the corresponding half-planes.

- **Section 3.5** Graphing Linear Inequalities in Two Variables
- **Section 4.5** Systems of Linear Inequalities

Conceptual Category: Functions

Domain: Interpreting Functions

8.F.1 Understand that a function is a rule that assigns to each input exactly one output. The graph of a function is the set of ordered pairs consisting of an input and the corresponding output.

- **Section 5.1** Domain and Range of a Function
- **Section 5.2** Discrete and Continuous Domains

8.F.2 Compare properties of two functions each represented in a different way (algebraically, graphically, numerically in tables, or by verbal descriptions).

- **Section 5.4** Function Notation
- **Section 6.5** Exponential Growth

8.F.3 Interpret the equation $y = mx + b$ as defining a linear function, whose graph is a straight line; give examples of functions that are not linear.

- **Section 2.3** Graphing Linear Equations in Slope-Intercept Form
- **Section 2.5** Writing Equations in Slope-Intercept Form
- **Section 5.3** Linear Function Patterns
- **Section 5.5** Comparing Linear and Nonlinear Functions

8.F.4 Construct a function to model a linear relationship between two quantities. Determine the rate of change and initial value of the function from a description of a relationship or from two (x, y) values, including reading these from a table or from a graph. Interpret the rate of change and initial value of a linear function in terms of the situation it models, and in terms of its graph or a table of values.

- **Section 2.7** Solving Real-Life Problems
- **Section 5.3** Linear Function Patterns

8.F.5 Describe qualitatively the functional relationship between two quantities by analyzing a graph (e.g., where the function is increasing or decreasing, linear or nonlinear). Sketch a graph that exhibits the qualitative features of a function that has been described verbally. *Found throughout. For example:*

- **Section 5.5** Comparing Linear and Nonlinear Functions
- **Section 8.1** Graphing $y = ax^2$
- **Section 8.3** Graphing $y = ax^2 + c$

F.IF.1 Understand that a function from one set (called the domain) to another set (called the range) assigns to each element of the domain exactly one element of the range. If f is a function and x is an element of its domain, then $f(x)$ denotes the output of f corresponding to the input x. The graph of f is the graph of the equation $y = f(x)$.

- **Section 5.1** Domain and Range of a Function
- **Section 5.2** Discrete and Continuous Domains
- **Section 5.4** Function Notation

F.IF.2 Use function notation, evaluate functions for inputs in their domains, and interpret statements that use function notation in terms of a context.

- **Section 5.4** Function Notation

F.IF.3 Recognize that sequences are functions, sometimes defined recursively, whose domain is a subset of the integers.

- **Section 5.6** Arithmetic Sequences
- **Section 6.7** Geometric Sequences

F.IF.4 For a function that models a relationship between two quantities, interpret key features of graphs and tables in terms of the quantities, and sketch graphs showing key features given a verbal description of the relationship.
Key features include: intercepts; intervals where the function is increasing, decreasing, positive, or negative; relative maximums and minimums; symmetries; end behavior; and periodicity.

- **Section 2.2** Slope of a Line
- **Section 2.3** Graphing Linear Equations in Slope-Intercept Form
- **Section 2.4** Graphing Linear Equations in Standard Form
- **Section 2.6** Writing Equations in Point-Slope Form
- **Section 2.7** Solving Real-Life Problems
- **Section 8.2** Focus of a Parabola
- **Section 8.4** Graphing $y = ax^2 + bx + c$
- **Section 8.5** Comparing Linear, Exponential, and Quadratic Functions
- **Section 10.1** Graphing Square Root Functions

F.IF.5 Relate the domain of a function to its graph and, where applicable, to the quantitative relationship it describes.

- **Section 5.1** Domain and Range of a Function
- **Section 5.2** Discrete and Continuous Domains

F.IF.6 Calculate and interpret the average rate of change of a function (presented symbolically or as a table) over a specified interval. Estimate the rate of change from a graph.

- **Section 2.2** Slope of a Line
- **Section 2.6** Writing Equations in Point-Slope Form
- **Section 8.5** Comparing Linear, Exponential, and Quadratic Functions

F.IF.7 Graph functions expressed symbolically and show key features of the graph, by hand in simple cases and using technology for more complicated cases.

a. Graph linear and quadratic functions and show intercepts, maxima, and minima.

- **Section 2.3** Graphing Linear Equations in Slope-Intercept Form
- **Section 8.4** Graphing $y = ax^2 + bx + c$
- **Section 8.5** Comparing Linear, Exponential, and Quadratic Functions

b. Graph square root, cube root, and piecewise-defined functions, including step functions and absolute value functions.

- **Section 5.4** Function Notation
- **Section 10.1** Graphing Square Root Functions

e. Graph exponential and logarithmic functions, showing intercepts and end behavior, and trigonometric functions, showing period, midline, and amplitude.

- **Section 6.4** Exponential Functions
- **Section 6.5** Exponential Growth
- **Section 6.6** Exponential Decay

F.IF.8 Write a function defined by an expression in different but equivalent forms to reveal and explain different properties of the function.

a. Use the process of factoring and completing the square in a quadratic function to show zeros, extreme values, and symmetry of the graph, and interpret these in terms of a context.

- **Section 9.3** Solving Quadratic Equations by Completing the Square

b. Use the properties of exponents to interpret expressions for exponential functions.

- **Section 6.4** Exponential Functions
- **Section 6.5** Exponential Growth
- **Section 6.6** Exponential Decay

F.IF.9 Compare properties of two functions each represented in a different way (algebraically, graphically, numerically in tables, or by verbal descriptions).

- **Section 5.4** Function Notation
- **Section 6.5** Exponential Growth

Domain: Building Functions

F.BF.1 Write a function that describes a relationship between two quantities.

a. Determine an explicit expression, a recursive process, or steps for calculation from a context. *Found throughout. For example:*

- **Section 2.5** Writing Equations in Slope-Intercept Form
- **Section 5.3** Linear Function Patterns
- **Section 6.5** Exponential Growth
- **Section 6.6** Exponential Decay

b. Combine standard function types using arithmetic operations.

- **Section 6.5** Exponential Growth

F.BF.2 Write arithmetic and geometric sequences both recursively and with an explicit formula, use them to model situations, and translate between the two forms.

- **Section 5.6** Arithmetic Sequences
- **Section 6.7** Geometric Sequences

F.BF.3 Identify the effect on the graph of replacing $f(x)$ by $f(x) + k$, $k\,f(x)$, $f(kx)$, and $f(x + k)$ for specific values of k (both positive and negative); find the value of k given the graphs. Experiment with cases and illustrate an explanation of the effects on the graph using technology. *Include recognizing even and odd functions from their graphs and algebraic expressions for them.*

- **Section 5.4** Function Notation
- **Section 6.4** Exponential Functions
- **Section 8.1** Graphing $y = ax^2$
- **Section 8.3** Graphing $y = ax^2 + c$
- **Section 8.4** Graphing $y = ax^2 + bx + c$

F.BF.4 Find inverse functions.

a. Solve an equation of the form $f(x) = c$ for a simple function f that has an inverse and write an expression for the inverse.

- **Section 11.2** Graphing Rational Functions

Domain: Linear, Quadratic, and Exponential Models

F.LE.1 Distinguish between situations that can be modeled with linear functions and with exponential functions.

a. Prove that linear functions grow by equal differences over equal intervals; and that exponential functions grow by equal factors over equal intervals.

- **Section 6.4** Exponential Functions

b. Recognize situations in which one quantity changes at a constant rate per unit interval relative to another.

- **Section 5.5** Comparing Linear and Nonlinear Functions

c. Recognize situations in which a quantity grows or decays by a constant percent rate per unit interval relative to another.
   - **Section 6.5** Exponential Growth
   - **Section 6.6** Exponential Decay

F.LE.2 Construct linear and exponential functions, including arithmetic and geometric sequences, given a graph, a description of a relationship, or two input-output pairs (include reading these from a table).
- **Section 5.3** Linear Function Patterns
- **Section 5.6** Arithmetic Sequences
- **Section 6.4** Exponential Functions
- **Section 6.7** Geometric Sequences

F.LE.3 Observe using graphs and tables that a quantity increasing exponentially eventually exceeds a quantity increasing linearly, quadratically, or (more generally) as a polynomial function.
- **Section 8.5** Comparing Linear, Exponential, and Quadratic Functions

F.LE.5 Interpret the parameters in a linear or exponential function in terms of a context.
- **Section 2.3** Graphing Linear Equations in Slope-Intercept Form
- **Section 6.4** Exponential Functions

Conceptual Category: Geometry

Domain: Geometric Measurement and Dimension

8.G.6 Explain a proof of the Pythagorean theorem and its converse.
- **Section 10.3** The Pythagorean Theorem
- **Section 10.4** Using the Pythagorean Theorem

8.G.7 Apply the Pythagorean theorem to determine unknown side lengths in right triangles in real-world and mathematical problems in two and three dimensions.
- **Section 10.3** The Pythagorean Theorem
- **Section 10.4** Using the Pythagorean Theorem

8.G.8 Apply the Pythagorean theorem to find the distance between two points in a coordinate system.
- **Section 10.4** Using the Pythagorean Theorem

Conceptual Category: Statistics and Probability

Domain: Interpreting Categorical and Quantitative Data

8.SP.1 Construct and interpret scatter plots for bivariate measurement data to investigate patterns of association between two quantities. Describe patterns such as clustering, outliers, positive or negative association, linear association, and nonlinear association.
- **Section 12.5** Scatter Plots and Lines of Fit
- **Section 12.6** Analyzing Lines of Fit

8.SP.2 Know that straight lines are widely used to model relationships between two quantitative variables. For scatter plots that suggest a linear association, informally fit a straight line, and informally assess the model fit by judging the closeness of the data points to the line.
- **Section 12.5** Scatter Plots and Lines of Fit

8.SP.3 Use the equation of a linear model to solve problems in the context of bivariate measurement data, interpreting the slope and intercept.

- **Section 12.5** Scatter Plots and Lines of Fit

8.SP.4 Understand that patterns of association can also be seen in bivariate categorical data by displaying frequencies and relative frequencies in a two-way table. Construct and interpret a two-way table summarizing data on two categorical variables collected from the same subjects. Use relative frequencies calculated for rows or columns to describe possible association between the two variables.

- **Section 12.7** Two-Way Tables

S.ID.1 Represent data with plots on the real number line (dot plots, histograms, and box plots).

- **Section 12.1** Measures of Central Tendency
- **Section 12.3** Box-and-Whisker Plots
- **Section 12.4** Shapes of Distributions
- **Section 12.8** Choosing a Data Display

S.ID.2 Use statistics appropriate to the shape of the data distribution to compare center (median, mean) and spread (interquartile range, standard deviation) of two or more different data sets.

- **Section 12.1** Measures of Central Tendency
- **Section 12.2** Measures of Dispersion
- **Section 12.3** Box-and-Whisker Plots
- **Section 12.4** Shapes of Distributions

S.ID.3 Interpret differences in shape, center, and spread in the context of the data sets, accounting for possible effects of extreme data points (outliers).

- **Section 12.1** Measures of Central Tendency
- **Section 12.2** Measures of Dispersion
- **Section 12.3** Box-and-Whisker Plots
- **Section 12.4** Shapes of Distributions

S.ID.5 Summarize categorical data for two categories in two-way frequency tables. Interpret relative frequencies in the context of the data (including joint, marginal, and conditional relative frequencies). Recognize possible associations and trends in the data.

- **Section 12.7** Two-Way Tables

S.ID.6 Represent data on two quantitative variables on a scatter plot, and describe how the variables are related.

a. Fit a function to the data; use functions fitted to data to solve problems in the context of the data. *Use given functions or choose a function suggested by the context. Emphasize linear and exponential models.*

- **Section 12.5** Scatter Plots and Lines of Fit

b. Informally assess the fit of a function by plotting and analyzing residuals.

- **Section 12.6** Analyzing Lines of Fit

c. Fit a linear function for a scatter plot that suggests a linear association.

- **Section 12.5** Scatter Plots and Lines of Fit

S.ID.7 Interpret the slope (rate of change) and the intercept (constant term) of a linear model in the context of the data.

- **Section 12.5** Scatter Plots and Lines of Fit

S.ID.8 Compute (using technology) and interpret the correlation coefficient of a linear fit.

- **Section 12.6** Analyzing Lines of Fit

S.ID.9 Distinguish between correlation and causation.

- **Section 12.6** Analyzing Lines of Fit

BOOK TO COMMON CORE STATE STANDARDS CORRELATION

Chapter 1

Solving Linear Equations

Number and Quantity

N.Q.1
N.Q.2

Algebra

A.CED.1
A.CED.4
A.REI.1
A.REI.3

Chapter 2

Graphing and Writing Linear Equations

Number and Quantity

N.Q.1

Algebra

A.CED.2
A.CED.3
A.REI.10

Functions

8.F.3
8.F.4
F.IF.4
F.IF.6
F.IF.7a
F.BF.1a
F.LE.5

Chapter 3

Solving Linear Inequalities

Number and Quantity

N.Q.2

Algebra

A.CED.1
A.CED.3
A.REI.3
A.REI.12

Chapter 4

Solving Systems of Linear Equations

Number and Quantity

N.Q.2

Algebra

A.CED.3
8.EE.8a–c
A.REI.5
A.REI.6
A.REI.11
A.REI.12

Chapter 5

Linear Functions

Functions

8.F.1
8.F.2
8.F.3
8.F.4
8.F.5
F.IF.1
F.IF.2
F.IF.3
F.IF.5
F.IF.7b
F.IF.9
F.BF.1a
F.BF.2
F.BF.3
F.LE.1b
F.LE.2

Chapter 6

Exponential Equations and Functions

Number and Quantity

N.RN.1
N.RN.2
N.RN.3
N.Q.1
N.Q.3

Algebra

A.SSE.1a–b
A.SSE.3c
A.CED.1
A.CED.2
A.REI.3
A.REI.10
A.REI.11

Functions

8.F.2
F.IF.3
F.IF.7e
F.IF.8b
F.IF.9
F.BF.1a–b
F.BF.2
F.BF.3
F.LE.1a, c
F.LE.2
F.LE.5

Chapter 7

Polynomial Equations and Factoring

Algebra

A.SSE.1a
A.SSE.2
A.SSE.3a
A.APR.1
A.REI.4b

Chapter 8

Graphing Quadratic Functions

Algebra

A.CED.2
A.REI.10

Functions

8.F.5
F.IF.4
F.IF.7a
F.BF.3
F.LE.3

Chapter 9

Solving Quadratic Equations

Algebra

A.SSE.3b
A.CED.1
A.REI.4a–b
A.REI.7
A.REI.11

Functions

F.IF.8a

Chapter 10

Square Root Functions and Geometry

Number and Quantity

N.RN.2
N.Q.3

Functions

F.IF.4
F.IF.7b

Algebra

A.CED.2
A.REI.10

Geometry

8.G.6
8.G.7
8.G.8

Chapter 11

Rational Equations and Functions

Algebra

A.SSE.2
A.CED.1
A.CED.2
A.REI.10

Functions

F.BF.4a

Chapter 12

Data Analysis and Displays

Number and Quantity

N.Q.3

Statistic and Probability

8.SP.1
8.SP.2
8.SP.3
8.SP.4
S.ID.1
S.ID.2
S.ID.3
S.ID.5
S.ID.6a–c
S.ID.7
S.ID.8
S.ID.9

A BALANCED APPROACH

The Common Core State Standards require students to do more than memorize how to solve problems. They define skills and knowledge that young people need to succeed academically in credit-bearing, college entry courses and in workforce training programs. Mastering the skills reflected in the Common Core State Standards prepares students for college and career.

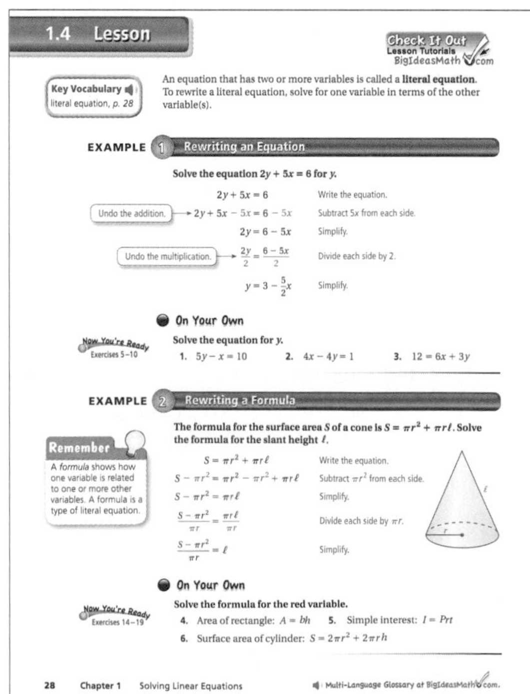

1.4 Lesson

Check It Out
Lesson Tutorials
BigIdeasMath.com

Key Vocabulary
literal equation, *p. 28*

An equation that has two or more variables is called a **literal equation**. To rewrite a literal equation, solve for one variable in terms of the other variable(s).

EXAMPLE 1 Rewriting an Equation

Solve the equation $2y + 5x = 6$ for y.

$2y + 5x = 6$	Write the equation.
Undo the addition. → $2y + 5x - 5x = 6 - 5x$	Subtract $5x$ from each side.
$2y = 6 - 5x$	Simplify.
Undo the multiplication. → $\frac{2y}{2} = \frac{6 - 5x}{2}$	Divide each side by 2.
$y = 3 - \frac{5}{2}x$	Simplify.

On Your Own

Now You're Ready
Exercises 5–10

Solve the equation for y.

1. $5y - x = 10$
2. $4x - 4y = 1$
3. $12 = 6x + 3y$

EXAMPLE 2 Rewriting a Formula

The formula for the surface area S of a cone is $S = \pi r^2 + \pi r\ell$. Solve the formula for the slant height ℓ.

Remember
A *formula* shows how one variable is related to one or more other variables. A formula is a type of literal equation.

$S = \pi r^2 + \pi r\ell$	Write the equation.
$S - \pi r^2 = \pi r^2 - \pi r^2 + \pi r\ell$	Subtract πr^2 from each side.
$S - \pi r^2 = \pi r\ell$	Simplify.
$\frac{S - \pi r^2}{\pi r} = \frac{\pi r\ell}{\pi r}$	Divide each side by πr.
$\frac{S - \pi r^2}{\pi r} = \ell$	Simplify.

On Your Own

Now You're Ready
Exercises 14–19

Solve the formula for the red variable.

4. Area of rectangle: $A = bh$
5. Simple interest: $I = Prt$
6. Surface area of cylinder: $S = 2\pi r^2 + 2\pi rh$

28 Chapter 1 Solving Linear Equations — Multi-Language Glossary at BigIdeasMath.com.

Essential Question How can you use a formula for one measurement to write a formula for a different measurement?

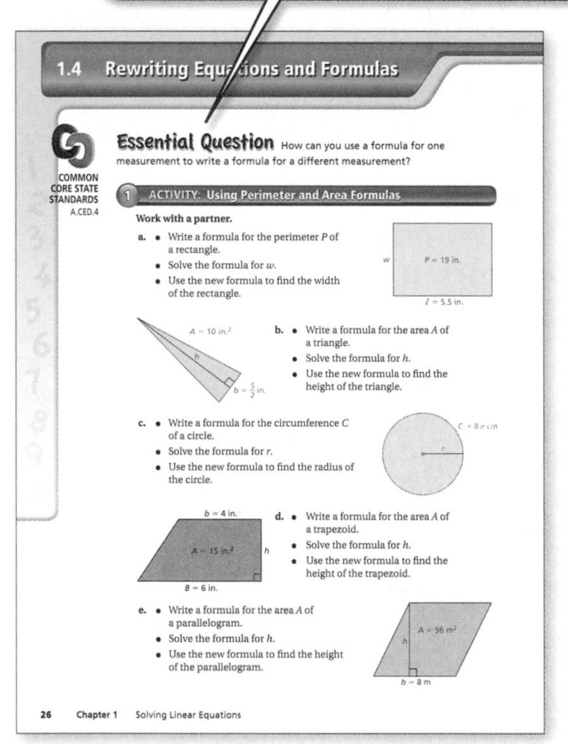

1.4 Rewriting Equations and Formulas

Essential Question How can you use a formula for one measurement to write a formula for a different measurement?

COMMON CORE STATE STANDARDS
A.CED.4

1 ACTIVITY: Using Perimeter and Area Formulas

Work with a partner.

a. • Write a formula for the perimeter P of a rectangle.
 • Solve the formula for w.
 • Use the new formula to find the width of the rectangle.

b. • Write a formula for the area A of a triangle.
 • Solve the formula for h.
 • Use the new formula to find the height of the triangle.

c. • Write a formula for the circumference C of a circle.
 • Solve the formula for r.
 • Use the new formula to find the radius of the circle.

d. • Write a formula for the area A of a trapezoid.
 • Solve the formula for h.
 • Use the new formula to find the height of the trapezoid.

e. • Write a formula for the area A of a parallelogram.
 • Solve the formula for h.
 • Use the new formula to find the height of the parallelogram.

26 Chapter 1 Solving Linear Equations

Research shows that students benefit from a program that includes equal exposure to discovery and direct instruction. By beginning each lesson with an inquiry-based activity, *Big Ideas Math* allows students to explore, question, explain, and persevere as they seek to answer Essential Questions that encourage abstract thought.

These rich activities are followed by direct instruction lessons, allowing for procedural fluency, modeling, and the opportunity to use clear precise mathematical language.

UNIQUE TEACHING EDITION

Standards for Mathematical Practice

- **MP1a Make Sense of Problems** and **MP3a Construct Viable Arguments:** The goal of this activity is for students to discuss, explain, and demonstrate how a literal equation may be solved using multiple approaches. For example, when solving $A = \frac{1}{2}bh$ for h, the first step could be dividing both sides by $\frac{1}{2}$, multiplying both sides by 2, or dividing both sides by $\frac{1}{2}b$. It is important to spend time discussing the multiple approaches.
- **MP7 Look for and Make Use of Structure:** Students should look at literal equations such as $A = \frac{1}{2}h(b + B)$ and recognize that there are 3 factors on the right side of the equation. Solving for h involves dividing both sides of the equation by the two remaining factors, $\frac{1}{2}$ and $(b + B)$.

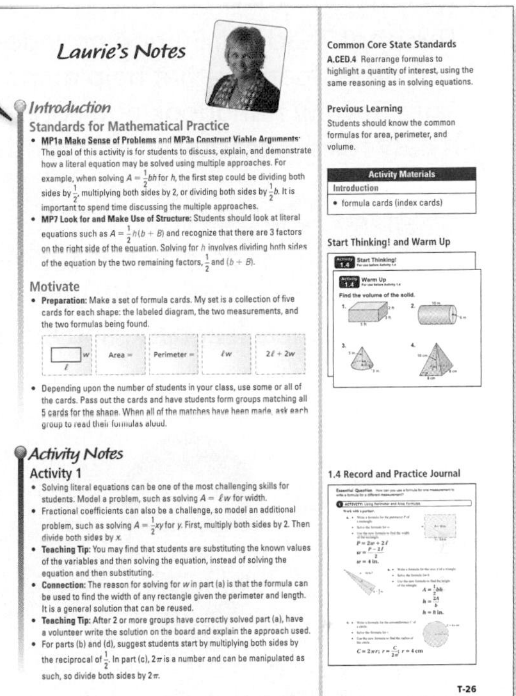

Laurie's Notes

Introduction

Standards for Mathematical Practice

- **MP1a Make Sense of Problems** and **MP3a Construct Viable Arguments:** The goal of this activity is for students to discuss, explain, and demonstrate how a literal equation may be solved using multiple approaches. For example, when solving $A = \frac{1}{2}bh$ for h, the first step could be dividing both sides by $\frac{1}{2}$, multiplying both sides by 2, or dividing both sides by $\frac{1}{2}b$. It is important to spend time discussing the multiple approaches.
- **MP7 Look for and Make Use of Structure:** Students should look at literal equations such as $A = \frac{1}{2}h(b + B)$ and recognize that there are 3 factors on the right side of the equation. Solving for h involves dividing both sides of the equation by the two remaining factors, $\frac{1}{2}$ and $(b + B)$.

Motivate

- **Preparation:** Make a set of formula cards. My set is a collection of five cards for each shape: the labeled diagram, the two measurements, and the two formulas being found.

w / ℓ	Area =	Perimeter =	ℓw	$2\ell + 2w$

- Depending upon the number of students in your class, use some or all of the cards. Pass out the cards and have students form groups matching all 5 cards for the shape. When all of the matches have been made, ask each group to read their formulas aloud.

Activity Notes

Activity 1

- Solving literal equations can be one of the most challenging skills for students. Model a problem, such as solving $A = \ell w$ for width.
- Fractional coefficients can also be a challenge, so model an additional problem, such as solving $A = \frac{1}{2}xy$ for y. First, multiply both sides by 2. Then divide both sides by x.
- **Teaching Tip:** You may find that students are substituting the known values of the variables and then solving the equation, instead of solving the equation and then substituting.
- **Connection:** The reason for solving for w in part (a) is that the formula can be used to find the width of any rectangle given the perimeter and length. It is a general solution that can be reused.
- **Teaching Tip:** After 2 or more groups have correctly solved part (a), have a volunteer write the solution on the board and explain the approach used.
- For parts (b) and (d), suggest students start by multiplying both sides by the reciprocal of $\frac{1}{2}$. In part (c), 2π is a number and can be manipulated as such, so divide both sides by 2π.

Common Core State Standards

A.CED.4 Rearrange formulas to highlight a quantity of interest, using the same reasoning as in solving equations.

Previous Learning

Students should know the common formulas for area, perimeter, and volume.

Activity Materials
Introduction
• formula cards (index cards)

Start Thinking! and Warm Up

Start Thinking!

Warm Up

Find the volume of the solid.

1.4 Record and Practice Journal

$P = 2w + 2\ell$

$w = \frac{P - 2\ell}{2}$

$w = 4$ in.

$A = \frac{1}{2}bh$

$h = \frac{2A}{b}$

$h = 8$ in.

$C = 2\pi r$; $r = \frac{C}{2\pi}$; $r = 4$ cm

T-26

The *Big Ideas Math* Teaching Edition is unique in its organization. It provides teachers with complete support as they teach the program. Using side-by-side pages, teachers have access to the full student page as they teach. Throughout the book Laurie Boswell shares insights that she has gained through years of teaching experience.

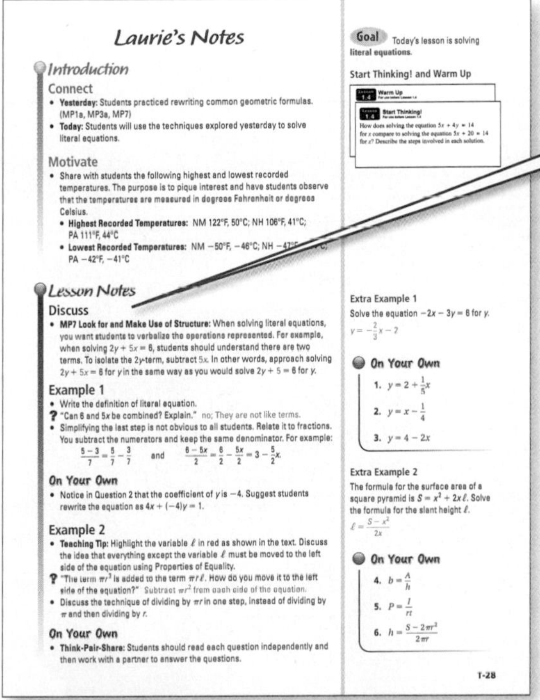

Laurie's Notes

Introduction

Connect

- **Yesterday:** Students practiced rewriting common geometric formulas. (MP1a, MP3a, MP7)
- **Today:** Students will use the techniques explored yesterday to solve literal equations.

Motivate

- Share with students the following highest and lowest recorded temperatures. The purpose is to pique interest and have students observe that the temperatures are measured in degrees Fahrenheit or degrees Celsius.
  - **Highest Recorded Temperatures:** NM 122°F, 50°C; NH 106°F, 41°C; PA 111°F, 44°C
  - **Lowest Recorded Temperatures:** NM −50°F, −46°C; NH −47°F, ...°C; PA −42°F, −41°C

Lesson Notes

Discuss

- **MP7 Look for and Make Use of Structure:** When solving literal equations, you want students to verbalize the operations represented. For example, when solving $2y + 5x = 6$, students should understand there are two terms. To isolate the $2y$-term, subtract $5x$. In other words, approach solving $2y + 5x = 6$ for y in the same way as you would solve $2y + 5 = 6$ for y.

Example 1

- Write the definition of literal equation.
- ? "Can 6 and $5x$ be combined? Explain." no; They are not like terms.
- Simplifying the last step is not obvious to all students. Relate it to fractions. You subtract the numerators and keep the same denominator. For example: $\frac{5-3}{7} = \frac{5}{7} - \frac{3}{7}$ and $\frac{6-5x}{2} = \frac{6}{2} - \frac{5x}{2} = 3 - \frac{5}{2}x$.

On Your Own

- Notice in Question 2 that the coefficient of y is −4. Suggest students rewrite the equation as $4x + (-4)y = 1$.

Example 2

- **Teaching Tip:** Highlight the variable ℓ in red as shown in the text. Discuss the idea that everything except the variable ℓ must be moved to the left side of the equation using Properties of Equality.
- ? "The term πr^2 is added to the term $\pi r\ell$. How do you move it to the left side of the equation?" Subtract πr^2 from each side of the equation.
- Discuss the technique of dividing by πr in one step, instead of dividing by π and then dividing by r.

On Your Own

- **Think-Pair-Share:** Students should read each question independently and then work with a partner to answer the questions.

Goal Today's lesson is solving **literal equations.**

Start Thinking! and Warm Up

Warm Up

Start Thinking!

How does solving the equation $5x + 4y = 14$ for x compare to solving the equation $5x + 20 = 14$ for x? Describe the steps involved in each solution.

Extra Example 1

Solve the equation $-2x - 3y = 6$ for y.

$y = -\frac{2}{3}x - 2$

On Your Own

1. $y = 2 + \frac{1}{5}x$
2. $y = x - \frac{1}{4}$
3. $y = 4 - 2x$

Extra Example 2

The formula for the surface area of a square pyramid is $S = x^2 + 2x\ell$. Solve the formula for the slant height ℓ.

$\ell = \frac{S - x^2}{2x}$

On Your Own

4. $b = \frac{A}{h}$
5. $P = \frac{I}{rt}$
6. $h = \frac{S - 2\pi r^2}{2\pi r}$

T-28

Discuss

- **MP7 Look for and Make Use of Structure:** When solving literal equations, you want students to verbalize the operations represented. For example, when solving $2y + 5x = 6$, students should understand there are two terms. To isolate the $2y$-term, subtract $5x$. In other words, approach solving $2y + 5x = 6$ for y in the same way as you would solve $2y + 5 = 6$ for y.

Each student page is accompanied by a support page. Laurie includes motivation suggestions, questioning strategies, and closure opportunities.

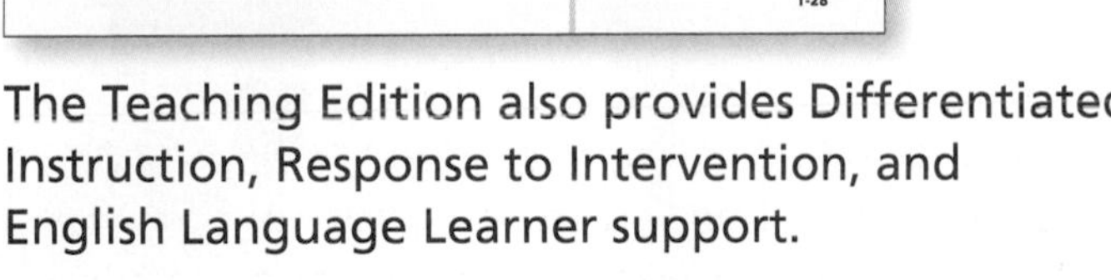

The Teaching Edition also provides Differentiated Instruction, Response to Intervention, and English Language Learner support.

CUSTOMIZED INSTRUCTION

Print Option

The print Teaching Edition provides teachers with help through Laurie's Notes and other features that help manage the classroom. The Chapter Resource Book, Skills Review Handbook and Assessment Book complete a teaching array that makes it easy for the teacher to differentiate, assess and teach.

Digital Option

Teachers can use 21st century technology tools found throughout the program to provide exciting ways to stimulate learning. These tools provide innovative electronic activities, timely feedback, and measures for accurate assessment.

Blended Option

Teachers will find that using the blended option provides them a multitude of ways to teach, differentiate, and assess. Teachers and students can customize their teaching and learning by blending the power of creative technology tools with the accessibility of print resources. Rich content and the combination of creative print and online resources allows for an engaging and challenging approach to teaching the Common Core State Standards.

The Dynamic Classroom

Regardless of the option you choose for your students, *The Dynamic Classroom* will be one of your most valued tools in the *Big Ideas Math* program. This powerful tool can be used with interactive whiteboards, and includes the following:

- Chapter Openers
- Math in History
- Start Thinking!
- Warm Ups
- Record and Practice Journal pages
- Virtual Manipulatives
- *On Your Own* Exercises
- Extra Examples
- Mini Assessments
- Closure Activities
- Graphic Organizers

PERSONALIZED LEARNING

The *Big Ideas Math* program offers teachers and students many ways to personalize and enrich the learning experience of all levels of learners.

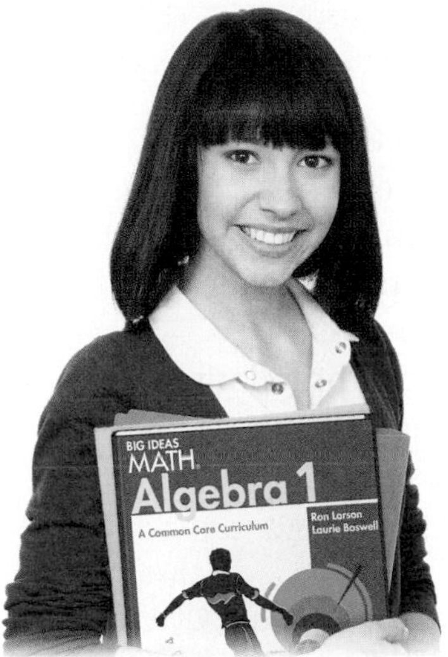

Lesson Tutorials Online

Two- to three-minute lesson tutorials provide colorful visuals and audio support for every example in the textbook. The Lesson Tutorials are valuable for students who miss a class, need a second explanation, or just need some help with a homework assignment. Parents can also use the tutorials to stay connected or to provide additional help at home.

The Dynamic Student Edition

This unique tool provides students with 21st century learning tools making it easier to directly interact with the underlying mathematics. From the dynamic student textbook to engaging tutorials, students use electronic manipulatives, flashcards and games to enhance their learning and understanding of math.

Differentiated Instruction

Through print and digital resources, the *Big Ideas Math* program completely supports the 3-Tier Response to Intervention model. Using research-based instructional strategies, teachers can reach, challenge and motivate each student with germane, high quality instruction targeted to individual needs.

Tier 3: Customized Learning Intervention

- Intensive Intervention Lessons
- Activities

Tier 2: Strategic Intervention

- Lesson Tutorials
- Basic Skills Handbook
- Skills Review Handbook
- Differentiated Instruction
- Game Closet

Tier 1: Daily Intervention

- Record and Practice Journal
- Fair Game Review
- Graphic Organizers
- Vocabulary Support
- Mini Assessments
- Game Closet
- Lesson Tutorials
- On Your Own

DIFFERENTIATED INSTRUCTION

Opening Doors to Learning

Two primary concerns while developing the *Big Ideas Math* program were the diversity of the student population and their different learning profiles. The authors developed a curriculum that helps teachers create classrooms that concentrate on learner needs by using the Universal Design for Learning model (UDL). The curriculum is designed to incorporate a wide variety of options, offering materials, methods, and assessments, so it is flexible and accommodating of individual student needs. By using Differentiated Instruction, teachers can open doors to learning that students are unable to open themselves.

English Language Learners

The *Big Ideas Math* program recognizes that English Language Learners (ELL) are a highly heterogeneous and complex group of students with diverse gifts, educational needs, backgrounds, languages, and goals. The writers used researched-based recommendations to develop a program specifically to assist these learners. In addition to global support, such as curriculum organized around Essential Questions involving both reading and writing, the program includes at-point-of-use ELL notes for the teacher; Family and Community Letters; ebooks with audio (English and Spanish); and a visual glossary.

Differentiating the Lesson Ancillary

Differentiating the Lesson, an online ancillary available at *BigIdeasMath.com,* provides complete teaching notes and worksheets that address the needs of the diverse learners in the classroom. The lessons engage students in activities that often incorporate visual learning and kinesthetic learning. Some lessons present an alternative approach to teaching the content while other lessons extend the concepts of the text in a challenging way for advanced students. Each chapter of *Differentiating the Lesson* begins with an overview of the differentiated lessons in the chapter, and describes the students who would most benefit from the approach used in each lesson.

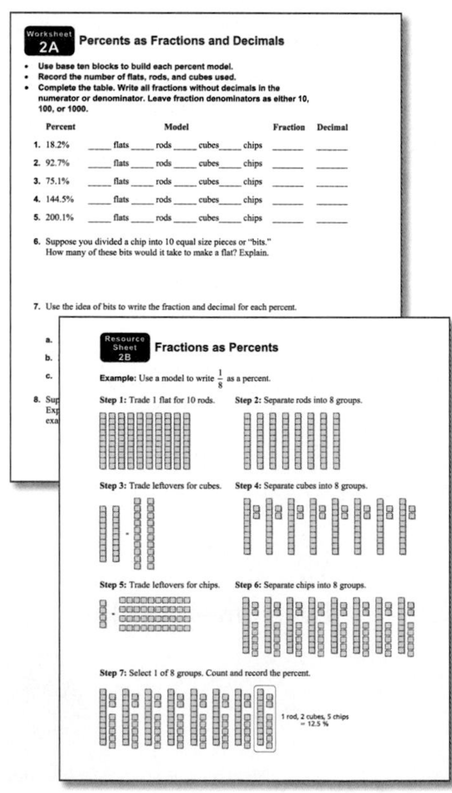

Worksheet 2A **Percents as Fractions and Decimals**

- **Use base ten blocks to build each percent model.**
- **Record the number of flats, rods, and cubes used.**
- **Complete the table. Write all fractions without decimals in the numerator or denominator. Leave fraction denominators as either 10, 100, or 1000.**

	Percent	Model	Fraction	Decimal
1.	18.2%	____ flats ____ rods ____ cubes ____ chips	____	____
2.	92.7%	____ flats ____ rods ____ cubes ____ chips	____	____
3.	75.1%	____ flats ____ rods ____ cubes ____ chips	____	____
4.	144.5%	____ flats ____ rods ____ cubes ____ chips	____	____
5.	200.1%	____ flats ____ rods ____ cubes ____ chips	____	____

6. Suppose you divided a chip into 10 equal size pieces or "bits." How many of these bits would it take to make a flat? Explain.

7. Use the idea of bits to write the fraction and decimal for each percent.

a.

b.

c.

8. Sup
Exp
exa

Resource Sheet 2B **Fractions as Percents**

Example: Use a model to write $\frac{1}{8}$ as a percent.

Step 1: Trade 1 flat for 10 rods.

Step 2: Separate rods into 8 groups.

Step 3: Trade leftovers for cubes.

Step 4: Separate cubes into 8 groups.

Step 5: Trade leftovers for chips.

Step 6: Separate chips into 8 groups.

Step 7: Select 1 of 8 groups. Count and record the percent.

1 rod, 2 cubes, 5 chips = 12.5 %

RESPONSE TO INTERVENTION

Through print and digital resources, the *Big Ideas Math* program completely supports the 3-Tier RTI model. Opportunities for daily assessment help identify areas of needs and easy-to-use resources are provided to support the education of all students.

Tier 1: Daily Intervention

The *Big Ideas Math* program uses research-based instructional strategies to ensure quality instruction. Vocabulary support, cooperative learning opportunities, and graphic organizers are included in the Pupil Edition, with additional strategies throughout the program. Daily student reviews and assessment guarantee that every student is making regular progress. Complete support helps teachers personalize instruction for every student.

- Record and Practice Journal
- Fair Game Reviews
- Graphic Organizers
- Vocabulary Support
- Mini Assessments
- Game Closet
- Lesson Tutorials
- On Your Own

Tier 2: Strategic Intervention

The *Big Ideas Math* program facilitates increased time and focus on instruction for students who are not responding effectively to Tier 1 intervention. The program's ancillary materials include additional support to assist teachers with the needs of these struggling learners. Extra Examples, Fair Game Reviews, Graphic Organizers, Study Tips, and Real-Life Applications enhance learning and engage the diverse students within today's math classrooms. Using the classroom and online resources provided, teachers can reach, challenge, and motivate each student with instruction targeted to their individual needs.

- Lesson Tutorials
- Basic Skills Handbook
- Skills Review Handbook
- Differentiated Instruction
- Game Closet

Tier 3: Customized Learning Intervention

Support for students working below grade level is also available.

- Intensive Intervention Lessons
- Activities

Skills Review Handbook

Basic Skills Handbook

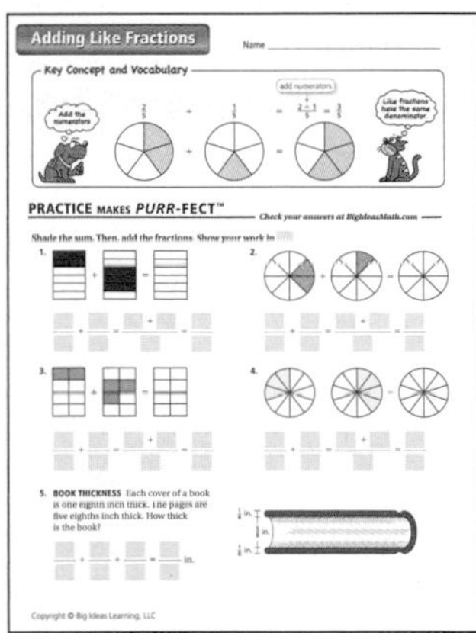

PACING GUIDE

Chapters 1–12: 163 Days

Chapter 1 — 8 Days

Chapter Opener	**1 Day**
Section 1.1	1 Day
Section 1.2	1 Day
Section 1.3	2 Days
Section 1.4	1 Day
Chapter Review/Chapter Tests	**2 Days**

Chapter 2 — 16 Days

Chapter Opener	**1 Day**
Section 2.1	1 Day
Section 2.2	2 Days
Section 2.3	1 Day
Section 2.4	2 Days
Study Help/Quiz	**1 Day**
Section 2.5	2 Days
Section 2.6	3 Days
Section 2.7	1 Day
Chapter Review/Chapter Tests	**2 Days**

Chapter 3 — 11 Days

Chapter Opener	**1 Day**
Section 3.1	2 Days
Section 3.2	0.5 Day
Section 3.3	0.5 Day
Study Help/Quiz	**1 Day**
Section 3.4	3 Days
Section 3.5	1 Day
Chapter Review/Chapter Tests	**2 Days**

Chapter 4 — 11 Days

Chapter Opener	**1 Day**
Section 4.1	2 Days
Section 4.2	1 Day
Study Help/Quiz	**1 Day**
Section 4.3	1 Day
Section 4.4	2 Days
Section 4.5	1 Day
Chapter Review/Chapter Tests	**2 Days**

Chapter 5 — 16 Days

Chapter Opener	**1 Day**
Section 5.1	2 Days
Section 5.2	1 Day
Section 5.3	2 Days
Study Help/Quiz	**1 Day**
Section 5.4	4 Days
Section 5.5	2 Days
Section 5.6	1 Day
Chapter Review/Chapter Tests	**2 Days**

Chapter 6 — 18 Days

Chapter Opener	**1 Day**
Section 6.1	3 Days
Section 6.2	2 Days
Section 6.3	1 Day
Study Help/Quiz	**1 Day**
Section 6.4	3 Days
Section 6.5	1 Day
Section 6.6	1 Day
Section 6.7	3 Days
Chapter Review/Chapter Tests	**2 Days**

Chapter 7 — 18 Days

Chapter Opener	1 Day
Section 7.1	1 Day
Section 7.2	2 Days
Section 7.3	2 Days
Section 7.4	2 Days
Study Help/Quiz	1 Day
Section 7.5	1 Day
Section 7.6	1 Day
Section 7.7	1 Day
Section 7.8	1 Day
Section 7.9	3 Days
Chapter Review/Chapter Tests	2 Days

Chapter 8 — 14 Days

Chapter Opener	1 Day
Section 8.1	2 Days
Section 8.2	1 Day
Section 8.3	2 Days
Study Help/Quiz	1 Day
Section 8.4	2 Days
Section 8.5	3 Days
Chapter Review/Chapter Tests	2 Days

Chapter 9 — 13 Days

Chapter Opener	1 Day
Section 9.1	2 Days
Section 9.2	1 Day
Section 9.3	1 Day
Study Help/Quiz	1 Day
Section 9.4	3 Days
Section 9.5	2 Days
Chapter Review/Chapter Tests	2 Days

Chapter 10 — 11 Days

Chapter Opener	1 Day
Section 10.1	2 Days
Section 10.2	2 Days
Study Help/Quiz	1 Day
Section 10.3	2 Days
Section 10.4	1 Day
Chapter Review/Chapter Tests	2 Days

Chapter 11 — 14 Days

Chapter Opener	1 Day
Section 11.1	2 Days
Section 11.2	3 Days
Section 11.3	1 Day
Section 11.4	1 Day
Study Help/Quiz	1 Day
Section 11.5	1 Day
Section 11.6	1 Day
Section 11.7	1 Day
Chapter Review/Chapter Tests	2 Days

Chapter 12 — 13 Days

Chapter Opener	1 Day
Section 12.1	1 Day
Section 12.2	1 Day
Section 12.3	1 Day
Section 12.4	1 Day
Study Help/Quiz	1 Day
Section 12.5	1 Day
Section 12.6	2 Days
Section 12.7	1 Day
Section 12.8	1 Day
Chapter Review/Chapter Tests	2 Days

Common Core State Standards for Mathematical Practice

Make sense of problems and persevere in solving them.

- Multiple representations are presented to help students move from concrete to representative and into abstract thinking
- *Essential Questions* help students focus and analyze
- *In Your Own Words* provide opportunities for students to look for meaning and entry points to a problem

Reason abstractly and quantitatively.

- Visual problem solving models help students create a coherent representation of the problem
- Opportunities for students to decontextualize and contextualize problems are presented in every lesson

Construct viable arguments and critique the reasoning of others.

- *Error Analysis*; *Different Words, Same Question*; and *Which One Doesn't Belong* features provide students the opportunity to construct arguments and critique the reasoning of others
- *Inductive Reasoning* activities help students make conjectures and build a logical progression of statements to explore their conjecture

Model with mathematics.

- Real-life situations are translated into diagrams, tables, equations, and graphs to help students analyze relations and to draw conclusions
- Real-life problems are provided to help students learn to apply the mathematics that they are learning to everyday life

Use appropriate tools strategically.

- *Graphic Organizers* support the thought process of what, when, and how to solve problems
- A variety of tool papers, such as graph paper, number lines, and manipulatives, are available as students consider how to approach a problem
- Opportunities to use the web, graphing calculators, and spreadsheets support student learning

Attend to precision.

- *On Your Own* questions encourage students to formulate consistent and appropriate reasoning
- Cooperative learning opportunities support precise communication

Look for and make use of structure.

- *Inductive Reasoning* activities provide students the opportunity to see patterns and structure in mathematics
- Real-world problems help students use the structure of mathematics to break down and solve more difficult problems

Look for and express regularity in repeated reasoning.

- Opportunities are provided to help students make generalizations
- Students are continually encouraged to check for reasonableness in their solutions

Go to *BigIdeasMath.com* for more information on the Common Core State Standards for Mathematical Practice.

Common Core State Standards for Mathematical Content for Algebra 1

Chapter Coverage for Standards

1 - 2 - 3 - 4 - 5 - 6 - 7 - 8 - 9 - 10 - 11 - 12

Conceptual Category Number and Quantity

- The Real Number System
- Quantities

1 - 2 - 3 - 4 - 5 - 6 - 7 - 8 - 9 - 10 - 11 - 12

Conceptual Category Algebra

- Seeing Structure in Expressions
- Arithmetic with Polynomials and Rational Expressions
- Creating Equations
- Reasoning with Equations and Inequalities

1 - 2 - 3 - 4 - 5 - 6 - 7 - 8 - 9 - 10 - 11 - 12

Conceptual Category Functions

- Interpreting Functions
- Building Functions
- Linear, Quadratic, and Exponential Models

1 - 2 - 3 - 4 - 5 - 6 - 7 - 8 - 9 - 10 - 11 - 12

Conceptual Category Geometry

- Geometric Measurement and Dimension

1 - 2 - 3 - 4 - 5 - 6 - 7 - 8 - 9 - 10 - 11 - 12

Conceptual Category Statistics and Probability

- Interpreting Categorical and Quantitative Data
- Making Inferences and Justifying Conclusions

Go to *BigIdeasMath.com* for more information on the Common Core State Standards for Mathematical Content.

How to Use Your Math Book

- Read the **Essential Question** in the activity.

  Work with a partner to decide **What Is Your Answer?**

  Now you are ready to do the **Practice** problems.

- Find the **Key Vocabulary** words, **highlighted in yellow**.

  Read their definitions. Study the concepts in each **Key Idea**.
  If you forget a definition, you can look it up online in the

  Multi-Language Glossary at BigIdeasMath.com.

- After you study each **EXAMPLE**, do the exercises in the **On Your Own**.

  Now You're Ready to do the exercises that correspond to the example.

  As you study, look for a **Study Tip** or a **Common Error**.

- The exercises are divided into 3 parts.

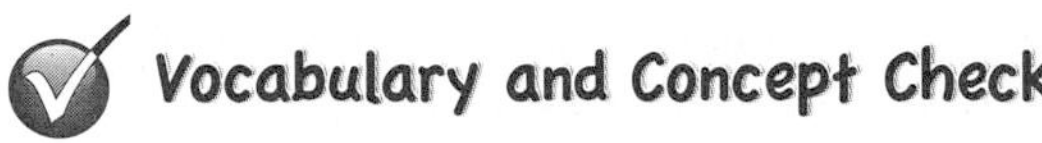

If an exercise has a 1 next to it, look back at Example 1 for help with that exercise.

More help is available at .

- To help study for your test, use the following.

Chapter Review **Chapter Test**

SCAVENGER HUNT

Use this *Scavenger Hunt* to find where things are in **Chapter 1**.

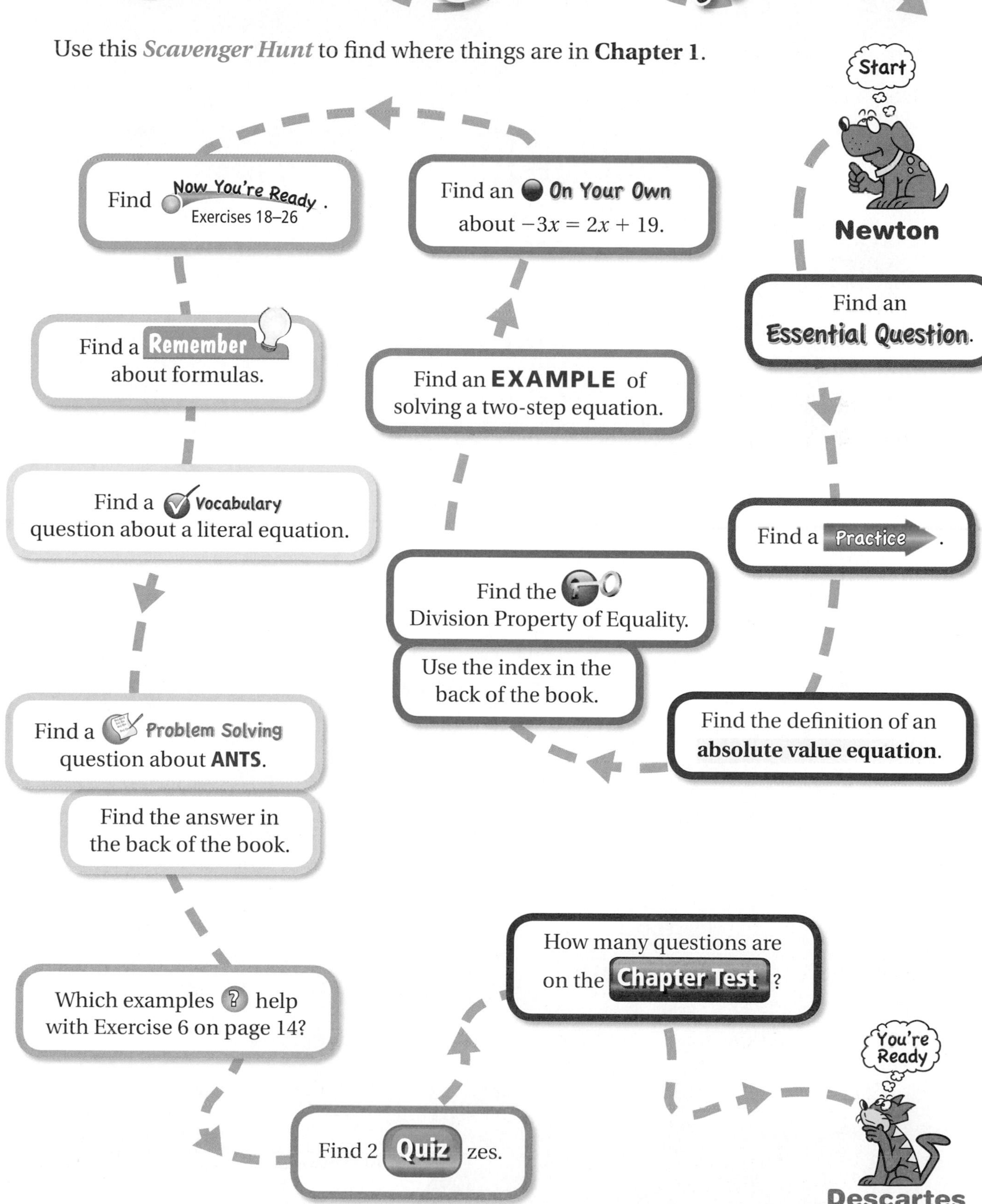

Start
Newton
Find an Essential Question.
Find a Practice.
Find the definition of an **absolute value equation**.
Use the index in the back of the book.
Find the Division Property of Equality.
Find an EXAMPLE of solving a two-step equation.
Find an On Your Own about $-3x = 2x + 19$.
Find Now You're Ready Exercises 18–26.
Find a Remember about formulas.
Find a Vocabulary question about a literal equation.
Find a Problem Solving question about **ANTS**.
Find the answer in the back of the book.
Which examples help with Exercise 6 on page 14?
Find 2 Quizzes.
How many questions are on the Chapter Test?
You're Ready
Descartes

1 Solving Linear Equations

"Dear Sir: Here is my suggestion for a good math problem."

"A box contains a total of 30 dog and cat treats. There are 5 times more dog treats than cat treats."

I need to learn to type so that I can write the story problems.

"How many of each type of treat are there?"

I think $D=RT$ stands for Descartes is Really Tired.

"Push faster, Descartes! According to the formula $R = D \div T$, the time needs to be 10 minutes or less to break our all-time speed record!"

Connections to Previous Learning

- Write, solve, and graph one-step linear equations.
- Solve problems using a formula.
- Find the absolute value of a number.

- Formulate and use different strategies to solve one-step and multi-step linear equations.
- Use properties of equality to rewrite an equation and to show two equations are equivalent.
- Use absolute value to add and subtract rational numbers.

- Create models to represent, analyze, and solve problems related to linear equations.
- Solve absolute value equations.
- Solve literal equations for a variable.

Math in History

There are two uses of the number 0 in mathematics.

★ Zero can be used as a place holder in a number system. For instance, the numbers 27 and 207 are different. The Mayans used zero in this way.

★ Zero can also be used to represent a number on the number line. The properties of 0, such as "the sum of zero and a number is that number" were described by Indian mathematicians over 3000 years ago.

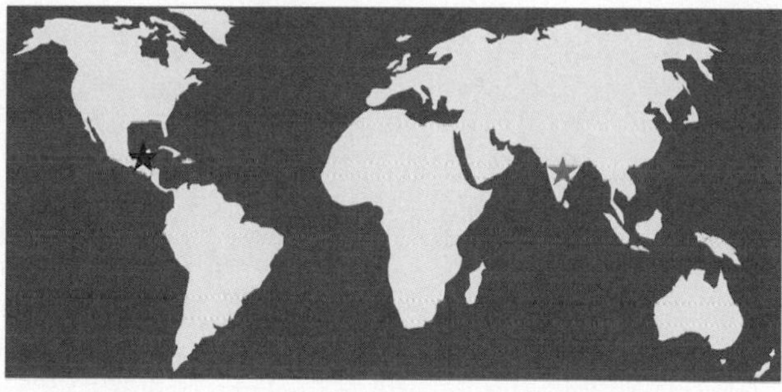

Pacing Guide for Chapter 1

Chapter Opener	1 Day
Section 1	1 Day
Section 2	1 Day
Section 3	2 Days
Section 4	1 Day
Chapter Review / Chapter Tests	2 Days
Total Chapter 1	8 Days
Year-to-Date	8 Days

Check Your Resources

- Record and Practice Journal
- Resources by Chapter
- Skills Review Handbook
- Assessment Book
- Worked-Out Solutions

Technology For the Teacher

The Dynamic Planning Tool
Editable Teacher's Resources at
BigIdeasMath.com

Common Core State Standards

7.NS.1d Apply properties of operations as strategies to add and subtract rational numbers.
7.NS.2c Apply properties of operations as strategies to multiply and divide rational numbers.

Additional Topics for Review

- Using operations with decimals and fractions
- Using order of operations to evaluate expressions
- Finding the absolute value of a number

Try It Yourself

1. -7	**2.** -13
3. 8	**4.** 32
5. -7	**6.** 2
7. -24	**8.** 63
9. -28	**10.** 4
11. -8	**12.** -4

Record and Practice Journal

1. -4	**2.** -12
3. -8	**4.** -8
5. 7	**6.** -7
7. 6	**8.** -12
9. 58°F	**10.** 2 floors
11. -10	**12.** -15
13. 30	**14.** 16
15. 6	**16.** 8
17. -3	**18.** -8
19. $60	**20.** 7 groups

Math Background Notes

Vocabulary Review

- Integer
- Absolute Value

Adding and Subtracting Integers

- Students should know how to add and subtract integers.
- Remind students how to add integers with the same sign. They should add the absolute values of the integers and then use the common sign.
- Remind students how to add integers with different signs. They should subtract the lesser absolute value from the greater absolute value and then use the sign of the integer with the greater absolute value.
- **Common Error:** Students may ignore the signs and just add the integers. Remind them of the meaning of absolute value. Make sure they understand that they should use the sign of the number that is farther from zero.
- Subtraction problems can be rewritten as addition problems. When subtracting an integer, add its opposite.
- **Common Error:** Students may change the sign of the first number, or forget to change the problem from subtraction to addition when changing the sign of the second number. Remind them that the first number is a starting point and will never change. Also remind students that the sign of the second number and the operation change.

Multiplying and Dividing Integers

- Students should know how to multiply and divide integers.
- When multiplying (or dividing) two integers with the same sign, the product (or quotient) is positive. When multiplying (or dividing) two integers with different signs, the product (or quotient) is negative.

Reteaching and Enrichment Strategies

If students need help. . .	If students got it. . .
Record and Practice Journal • Fair Game Review Skills Review Handbook Lesson Tutorials	Game Closet at *BigIdeasMath.com* Start the next section

What You Learned Before

"Once upon a time, there lived the most handsome dog who just happened to be a genius at math. He..."

Adding and Subtracting Integers (7.NS.1d)

Example 1 **Find 4 + (−12).**

$4 + (-12) = -8$

$|-12| > |4|$. So, subtract $|4|$ from $|-12|$.

Use the sign of −12.

Example 2 **Find −7 − (−16).**

$-7 - (-16) = -7 + 16$ Add the opposite of −16.

$= 9$ Add.

Try It Yourself

Add or subtract.

1. $-5 + (-2)$ **2.** $0 + (-13)$ **3.** $-6 + 14$

4. $19 - (-13)$ **5.** $-1 - 6$ **6.** $-5 - (-7)$

Multiplying and Dividing Integers (7.NS.2c)

Example 3 **Find −3 • (−5).**

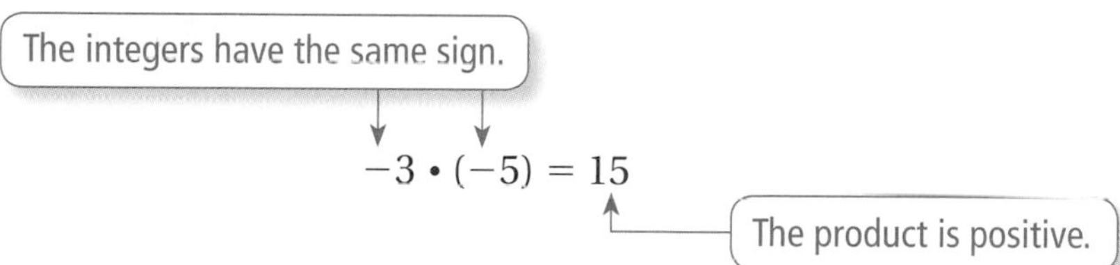
The integers have the same sign.
−3 • (−5) = 15
The product is positive.

Example 4 **Find 15 ÷ (−3).**

The integers have different signs.

$15 \div (-3) = -5$

The quotient is negative.

Try It Yourself

Multiply or divide.

7. $-3(8)$ **8.** $-7 \cdot (-9)$ **9.** $4 \cdot (-7)$

10. $-24 \div (-6)$ **11.** $-16 \div 2$ **12.** $12 \div (-3)$

1.1 Solving Simple Equations

COMMON CORE STATE STANDARDS
A.CED.1
A.REI.1
A.REI.3

Essential Question How can you use inductive reasoning to discover rules in mathematics? How can you test a rule?

1 ACTIVITY: Sum of the Angles of a Triangle

Work with a partner. Copy the triangles. Use a protractor to measure the angles of each triangle. Copy and complete the table to organize your results.

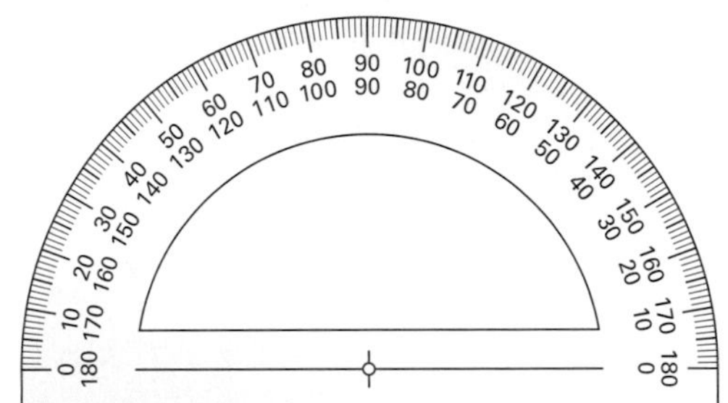

a.

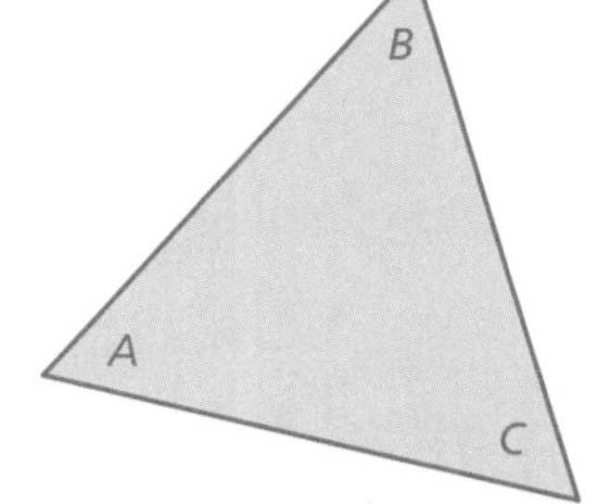

b.

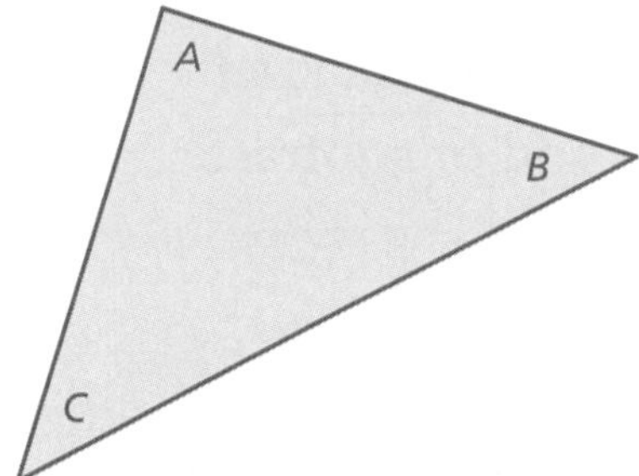

c.

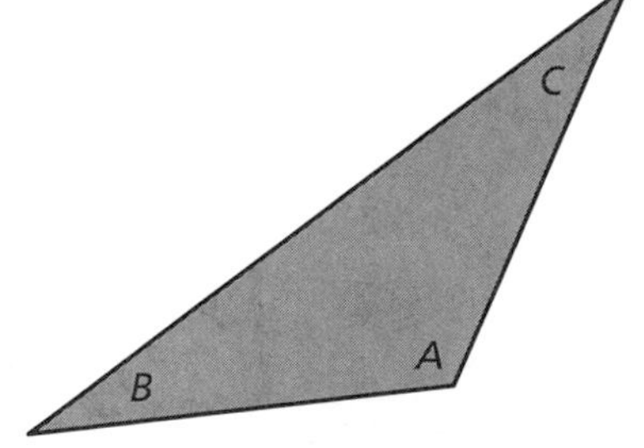

d.

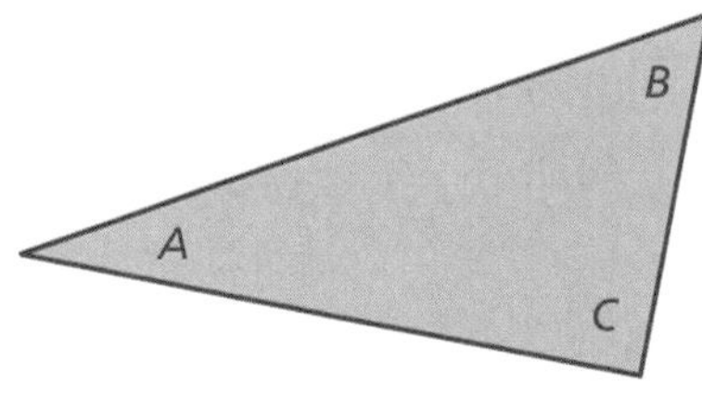

Triangle	Angle *A* (degrees)	Angle *B* (degrees)	Angle *C* (degrees)	*A* + *B* + *C*
a.				
b.				
c.				
d.				

Laurie's Notes

Introduction

Standards for Mathematical Practice

- **MP3a Construct Viable Arguments:** The goal of the activity is for students to observe, using inductive reasoning, that the sum of the angles of a triangle equals 180°. This can be expressed by writing a rule that is useful in solving other problems. In addition to making a conjecture, students test their conjecture (rule).
- **MP5 Use Appropriate Tools Strategically:** Students will be using protractors. Using a protractor can introduce the human error factor. Discuss measuring accurately with students. Ask if the results would be the same if they used dynamic geometry software.

Motivate

? "What do Tony Hawk, Shaun White, and Rodney Mullen have in common?" All are famous skateboarders. Shaun White is also a snowboarder.

- Today's activity is about angle measures. Boarders know a lot about angle measure, in particular the multiples of 180°, because of the different tricks they perform.

Activity Notes

Words of Wisdom

? "What does it mean to measure an angle?" Listen for an understanding of the rotation from one ray to a second ray. Both angles shown have the same measure, although some students would say the angle on the left is greater.

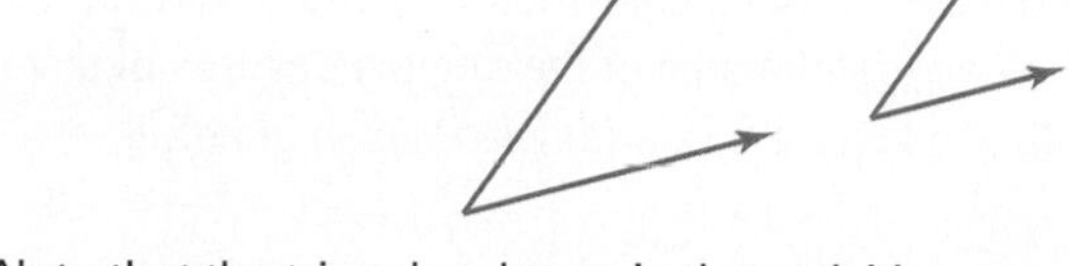

- Note that the triangles drawn in the activities are not shown in the standard orientation, with a side parallel to the horizontal edge. You do not want students to believe that triangles must have a horizontal base.
- Review with students how to place the protractor on the angle, and how to read the protractor. Instruct students to pay attention to precision and measure to the nearest degree.
- **MP6 Attend to Precision:** Notice that 0° does not always align with the bottom edge of some protractors, nor does the vertex of the angle always align with the bottom edge. It is common for students to align the bottom edge of the protractor with one ray of the angle, producing an error of more than 5°.

Activity 1

? "Do you see any pattern(s) in the table? Describe the pattern(s)." The sum of the angle measures is 180°, or close to 180°. Students might also mention that in part (a), all of the angles are congruent (same measure) and in parts (b) and (c), two of the angles are congruent.

- **FYI:** If a sum is significantly different from 180°, the student may have read the protractor incorrectly (i.e., they recorded 150° instead of 30°).

Common Core State Standards

A.CED.1 Create equations and inequalities in one variable and use them to solve problems.
A.REI.1 Explain each step in solving a simple equation . . .. Construct a viable argument to justify a solution method.
A.REI.3 Solve linear equations . . . in one variable, including equations with coefficients represented by letters.

Previous Learning

Students should know the vocabulary of angles, such as ray, vertex, acute, obtuse, right, and straight.

Activity Materials
Textbook
• protractors

Start Thinking! and Warm Up

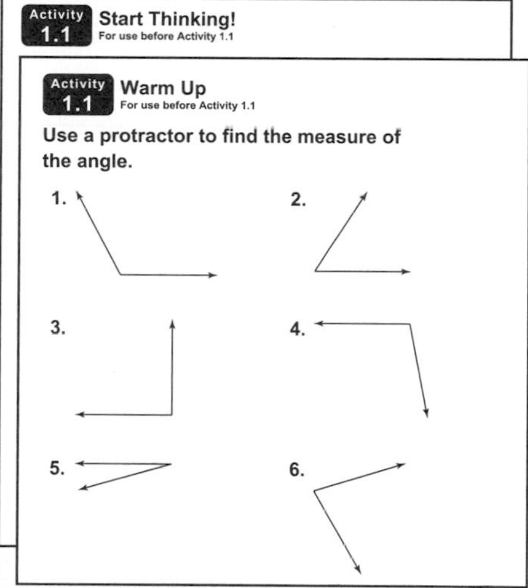

Activity 1.1 Start Thinking!
For use before Activity 1.1

Activity 1.1 Warm Up
For use before Activity 1.1

Use a protractor to find the measure of the angle.

1. 2. 3. 4. 5. 6.

1.1 Record and Practice Journal

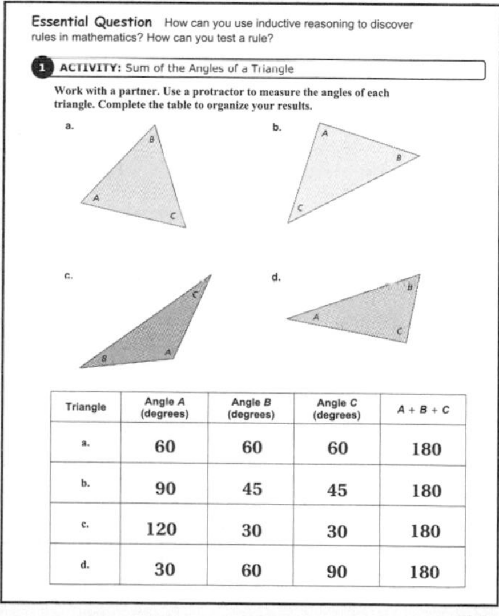

Essential Question How can you use inductive reasoning to discover rules in mathematics? How can you test a rule?

1 ACTIVITY: Sum of the Angles of a Triangle

Work with a partner. Use a protractor to measure the angles of each triangle. Complete the table to organize your results.

a. b. c. d.

Triangle	Angle A (degrees)	Angle B (degrees)	Angle C (degrees)	$A + B + C$
a.	60	60	60	180
b.	90	45	45	180
c.	120	30	30	180
d.	30	60	90	180

Differentiated Instruction

Kinesthetic

Ask two students to assist you at the board or overhead when solving equations. Assign one student to the left side of the equation and the other student to the right side. Each student is responsible for performing the operations on his or her side. Emphasize that to keep the equality, both students must perform the same operation to solve the equation.

1.1 Record and Practice Journal

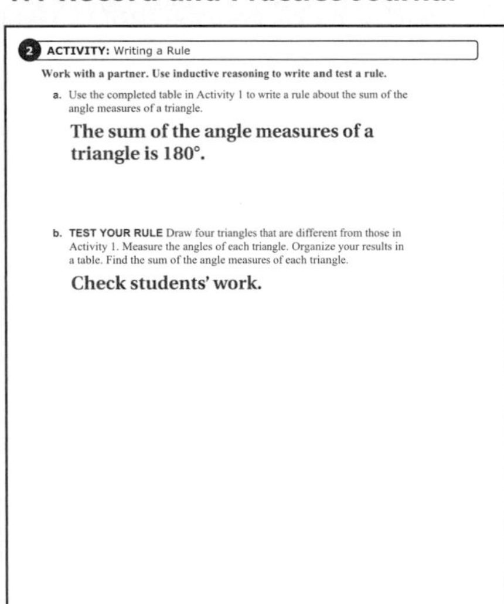

2 **ACTIVITY:** Writing a Rule

Work with a partner. Use inductive reasoning to write and test a rule.

a. Use the completed table in Activity 1 to write a rule about the sum of the angle measures of a triangle.

The sum of the angle measures of a triangle is 180°.

b. **TEST YOUR RULE** Draw four triangles that are different from those in Activity 1. Measure the angles of each triangle. Organize your results in a table. Find the sum of the angle measures of each triangle.

Check students' work.

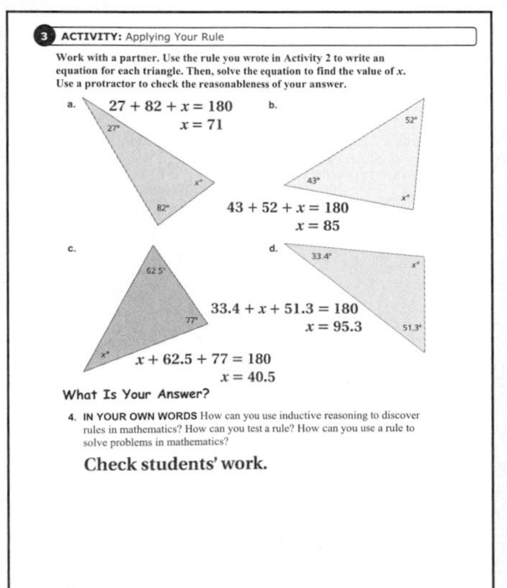

3 **ACTIVITY:** Applying Your Rule

Work with a partner. Use the rule you wrote in Activity 2 to write an equation for each triangle. Then, solve the equation to find the value of x. Use a protractor to check the reasonableness of your answer.

a. $27 + 82 + x = 180$
$x = 71$

b. $43 + 52 + x = 180$
$x = 85$

c. $x + 62.5 + 77 = 180$
$x = 40.5$

d. $33.4 + x + 51.3 = 180$
$x = 95.3$

What Is Your Answer?

4. **IN YOUR OWN WORDS** How can you use inductive reasoning to discover rules in mathematics? How can you test a rule? How can you use a rule to solve problems in mathematics?

Check students' work.

Laurie's Notes

Activity 2

- ? "What rule did you write for the sum of the angle measures of a triangle?" The sum of the angle measures of a triangle equals 180°.
- Suggest to students that they should use a straight edge to make their triangles larger so that it is easier to measure the angles.
- ? "Did you measure all three angles for each triangle?" Some students will only measure two angles and do a quick computation to find the third.
- **MP8 Look for and Express Regularity in Repeated Reasoning:** Mathematically proficient students understand the purpose of using a rule. Using the rule is more efficient when finding the third angle. The problem, however, asks students to test their rule, which is part of the process of inductive reasoning.

Activity 3

- Using a protractor to check for reasonableness should not be overlooked. Checking for reasonableness is part of the problem solving process.
- You should model the first problem and write the equation. Otherwise, students will do a computation to find the missing angle.
- **Write:** $27 + 82 + x = 180$. Students have solved equations previously and may simply write the answer as the second step: $x = 71$. Focus on the representation of equation solving instead of the intuitive sense of how to solve this addition equation. Model the second step by showing 109 subtracted from each side of the equation.
- Note that parts (c) and (d) integrate decimal review. Their answers should be exact.

What Is Your Answer?

? "What is inductive reasoning and how was it used in the activities today?" *Sample answer:* Inductive reasoning is writing a general rule based on examples. Today I found that the sum of the angle measures of several triangles equals 180°, so I wrote a rule for triangles in general.

Closure

- **Exit Ticket:** Two angles of a triangle measure 48.2° and 63.8°. Make a reasonable sketch of the triangle. Write and solve an equation to find the measure of the third angle. $48.2 + 63.8 + x = 180$; $x = 68$

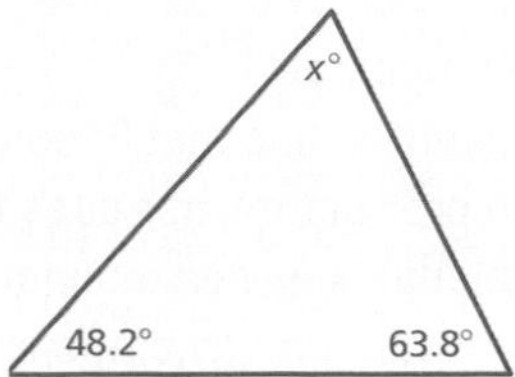

Technology For the Teacher

The Dynamic Planning Tool
Editable Teacher's Resources at *BigIdeasMath.com*

2 ACTIVITY: Writing a Rule

Work with a partner. Use inductive reasoning to write and test a rule.

a. Use the completed table in Activity 1 to write a rule about the sum of the angle measures of a triangle.

b. **TEST YOUR RULE** Draw four triangles that are different from those in Activity 1. Measure the angles of each triangle. Organize your results in a table. Find the sum of the angle measures of each triangle.

3 ACTIVITY: Applying Your Rule

Work with a partner. Use the rule you wrote in Activity 2 to write an equation for each triangle. Then, solve the equation to find the value of *x*. Use a protractor to check the reasonableness of your answer.

a.

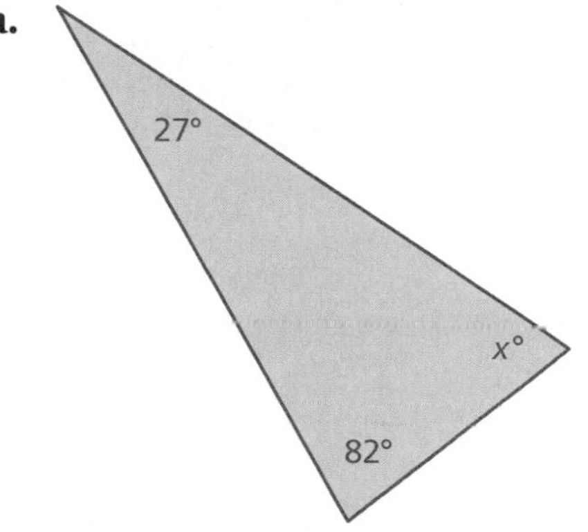

b.

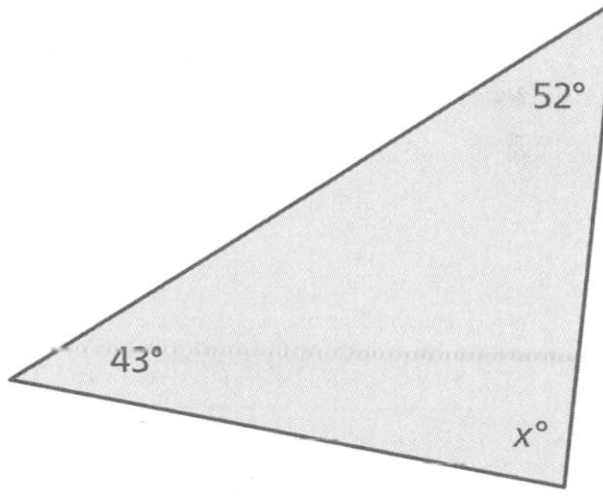

c.

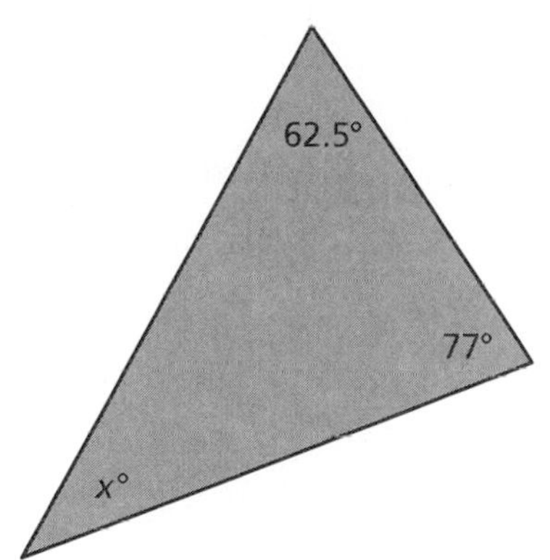

d.

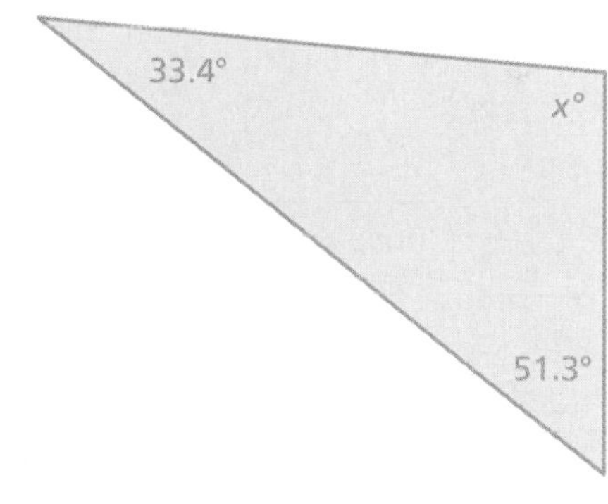

What Is Your Answer?

4. **IN YOUR OWN WORDS** How can you use inductive reasoning to discover rules in mathematics? How can you test a rule? How can you use a rule to solve problems in mathematics?

Practice Use what you learned about solving simple equations to complete Exercises 4–6 on page 7.

1.1 Lesson

Remember

Addition and subtraction are inverse operations.

Key Ideas

Addition Property of Equality

Words Adding the same number to each side of an equation produces an equivalent equation.

Algebra If $a = b$, then $a + c = b + c$.

Subtraction Property of Equality

Words Subtracting the same number from each side of an equation produces an equivalent equation.

Algebra If $a = b$, then $a - c = b - c$.

EXAMPLE 1 Solving Equations Using Addition or Subtraction

a. Solve $x - 7 = -6$.

$x - 7 = -6$ Write the equation.

Undo the subtraction. → $+7 \quad +7$ Add 7 to each side.

$x = 1$ Simplify.

∴ The solution is $x = 1$.

Check

$x - 7 = -6$

$1 - 7 \stackrel{?}{=} -6$

$-6 = -6$ ✓

b. Solve $y + 3.4 = 0.5$.

$y + 3.4 = 0.5$ Write the equation.

Undo the addition. → $-3.4 \quad -3.4$ Subtract 3.4 from each side.

$y = -2.9$ Simplify.

∴ The solution is $y = -2.9$.

Check

$y + 3.4 = 0.5$

$-2.9 + 3.4 \stackrel{?}{=} 0.5$

$0.5 = 0.5$ ✓

c. Solve $h + 2\pi = 3\pi$.

$h + 2\pi = 3\pi$ Write the equation.

Undo the addition. → $-2\pi \quad -2\pi$ Subtract 2π from each side.

$h = \pi$ Simplify.

∴ The solution is $h = \pi$.

Laurie's Notes

Introduction

Connect

- **Yesterday:** Students used the sum of the angle measures of a triangle to explore simple equation solving. (MP3a, MP5, MP6, MP8)
- **Today:** Students will use Properties of Equality to solve one-step equations.

Motivate

- Tell students that you are going to play a quick game of *REVERSO*. The directions are simple: you give a command to a student and your opponent must give the reverse (inverse) command to undo your command. For example, you say, "take 3 steps forward" and your opponent would say "take 3 steps backward." The goal is for students to think about inverse operations.
- Sample commands: turn lights on; step up on a chair; turn to your right; fold 2 sheets of paper; draw a square; open the door

Lesson Notes

Discuss

- When students begin to solve equations, *representation* is an important part in building their understanding. The three equations shown are equivalent, though for many students they *seem* different.

  $x - 3 = -7 \qquad -3 + x = -7 \qquad -7 = x - 3$
- **MP1 Make Sense of Problems and Persevere in Solving Them:** You are trying to help the students make sense of the problem. It is important that they pay attention to how equations are represented and how you present the operations that are performed on each side of the equation.
- **MP6 Attend to Precision:** The philosophy of this text is to address mathematical properties by definition. Example problems shown in this text focus on the process for finding solutions. You should explain to students that the properties are being used, even though they may not be mentioned specifically within the example.

Key Ideas

- Write the Key Ideas.
- Redefine *equivalent equations*. Two equations that have the same solution are *equivalent equations*.
- **Teaching Tip:** Use an alternate color to show adding (subtracting) c to (from) each side of the equation.
- Remind students of the big idea. Whatever you do to one side of the equation, you must do to the other side of the equation.

Example 1

- Work through each part. Note that the number being added to or subtracted from each side of the equation is written vertically below the number with which it will be combined.
- **MP6:** In part (a), the phrase "Add 7 to each side" is another way of saying "Use the Addition Property of Equality." Similarly, explain that parts (b) and (c) use the Subtraction Property of Equality.

Goal Today's lesson is solving one-step equations.

Start Thinking! and Warm Up

Lesson 1.1
Warm Up
For use before Lesson 1.1

Lesson 1.1 **Start Thinking!** For use before Lesson 1.1

The Addition Property of Equality states that adding the same number to each side of an equation produces an equivalent equation. What do you think the Subtraction, Multiplication, and Division Properties of Equality state?

Describe a real-life situation that you can relate to one of the properties of equality.

Extra Example 1

a. Solve $d - \frac{1}{4} = -\frac{1}{2}$. $-\frac{1}{4}$

b. Solve $m + 4.8 = 9.2$. 4.4

c. Solve $r - 6\pi = 2\pi$. 8π

On Your Own

1. $b = -7$ 2. $g = 0.8$
3. $k = -6$ 4. $r = 2\pi$
5. $t = -\frac{1}{2}$ 6. $z = -13.6$

Extra Example 2

a. Solve $\frac{2}{5}m = -4$. -10

b. Solve $3p = -\frac{2}{3}$. $-\frac{2}{9}$

On Your Own

7. $y = -28$ 8. $x = 6$
9. $w = 20$

English Language Learners

Vocabulary

In this section, students will learn to use inverse (or opposite) operations to solve equations. Students will use addition to solve a subtraction equation and use subtraction to solve an addition equation. Review these pairs of words that are essential to understanding mathematics. Give students one word of a pair and ask them to provide the opposite.

Examples:

odd, even	positive, negative
add, subtract	sum, difference
multiply, divide	product, quotient
plus, minus	

Laurie's Notes

Example 1 (continued)

? "What is the approximate value of π?" 3.14 "of 2π?" 6.28

- Remind students that 2π and 3π are (irrational) numbers, so you can treat these numbers as you would integers. It is common for students to think of π as a variable and they will say that there are two variables in $h + 2\pi$.

On Your Own

- Circulate as students work on these six questions. Remind students that it is the practice of *representing* their work that is important in these questions.

Key Ideas

- Write the Key Ideas.
- **Representation:** Review different ways in which multiplication is represented.

  $a(c) = b(c)$ $\quad ac = bc$ $\quad a \times c = b \times c$ $\quad a \cdot c = b \cdot c$
- Generally, when there are variables in equations, you do not want to use $\times$ to represent multiplication because it can be mistaken for a variable.
- **Representation**: Review different ways in which division is represented.

  $a \div c = b \div c$ $\quad \frac{a}{c} = \frac{b}{c}$ $\quad a/c = b/c$

Example 2

- Work through each part.
- Remind students that the goal is to solve for the variable so that it has a coefficient of 1.

? "What is the coefficient of n?" $-\frac{3}{4}$ "What operation is represented?" multiplication "How do you undo multiplication?" divide

? "What is equivalent to dividing by $-\frac{3}{4}$?" multiplying by $-\frac{4}{3}$

- When the coefficient is a fraction, it is more efficient to multiply by its multiplicative inverse (reciprocal).
- **MP6:** Part (a) uses the Multiplication Property of Equality and part (b) uses the Division Property of Equality. Remind students of the property names when explaining this example.
- The purpose of part (b) is to practice working with π in an algebraic expression.

On Your Own

- Circulate as students work on these three questions. Stress representation with students. Do not let them shortcut the process and simply record the answer. They should show what operation is being performed on each side of the equation.
- **Common Error:** Students may try to subtract 6π from πx or subtract πx from 6π. Remind them that the variable they are solving for is x.

On Your Own

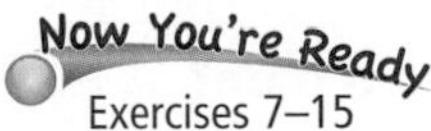

Exercises 7–15

Solve the equation. Check your solution.

1. $b + 2 = -5$ **2.** $g - 1.7 = -0.9$ **3.** $-3 = k + 3$

4. $r - \pi = \pi$ **5.** $t - \frac{1}{4} = -\frac{3}{4}$ **6.** $5.6 + z = -8$

Key Ideas

Remember

Multiplication and division are inverse operations.

Multiplication Property of Equality

Words Multiplying each side of an equation by the same number produces an equivalent equation.

Algebra If $a = b$, then $a \cdot c = b \cdot c$.

Division Property of Equality

Words Dividing each side of an equation by the same number produces an equivalent equation.

Algebra If $a = b$, then $a \div c = b \div c$, $c \neq 0$.

EXAMPLE 2 Solving Equations Using Multiplication or Division

a. Solve $-\frac{3}{4}n = -2$.

$-\frac{3}{4}n = -2$ Write the equation.

Use the reciprocal. → $-\frac{4}{3} \cdot \left(-\frac{3}{4}n\right) = -\frac{4}{3} \cdot (-2)$ Multiply each side by $-\frac{4}{3}$, the reciprocal of $-\frac{3}{4}$.

$n = \frac{8}{3}$ Simplify.

∴ Thc solution is $n = \frac{8}{3}$.

b. Solve $\pi x = 3\pi$.

$\pi x = 3\pi$ Write the equation.

Undo the multiplication. → $\frac{\pi x}{\pi} = \frac{3\pi}{\pi}$ Divide each side by π.

$x = 3$ Simplify.

∴ The solution is $x = 3$.

Check

$\pi x = 3\pi$

$\pi(3) \stackrel{?}{=} 3\pi$

$3\pi = 3\pi$ ✓

On Your Own

Now You're Ready
Exercises 18–26

Solve the equation. Check your solution.

7. $\frac{y}{4} = -7$ **8.** $6\pi = \pi x$ **9.** $0.09w = 1.8$

EXAMPLE 3 Standardized Test Practice

What value of k makes the equation $k + 4 \div 0.2 = 5$ true?

Ⓐ -15 Ⓑ -5 Ⓒ -3 Ⓓ 1.5

$$k + 4 \div 0.2 = 5$$ Write the equation.

$$k + 20 = 5$$ Divide 4 by 0.2.

$$-20 \quad -20$$ Subtract 20 from each side.

$$k = -15$$ Simplify.

∴ The correct answer is Ⓐ.

EXAMPLE 4 Real-Life Application

The melting point of bromine is −7°C.

The *melting point* of a solid is the temperature at which the solid becomes a liquid. The melting point of bromine is $\frac{1}{30}$ of the melting point of nitrogen. Write and solve an equation to find the melting point of nitrogen.

Words The melting point of bromine is $\frac{1}{30}$ of the melting point of nitrogen.

Variable Let n be the melting point of nitrogen.

Equation -7 $=$ $\frac{1}{30}$  n

$$-7 = \frac{1}{30}n$$ Write the equation.

$$30 \cdot (-7) = 30 \cdot \left(\frac{1}{30}n\right)$$ Multiply each side by 30.

$$-210 = n$$ Simplify.

∴ The melting point of nitrogen is −210°C.

On Your Own

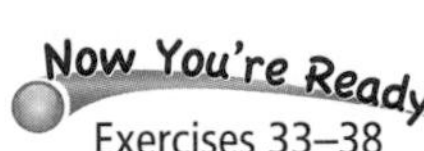

Exercises 33–38

10. Solve $p - 8 \div \frac{1}{2} = -3$.

11. Solve $q + |-10| = 2$.

12. The melting point of mercury is about $\frac{1}{4}$ of the melting point of krypton. The melting point of mercury is −39°C. Write and solve an equation to find the melting point of krypton.

Laurie's Notes

Lesson Notes

Example 3

? "What is $10 + 4 \div 2$?" listen for order of operations; Answer is 12, *not* 7.

- Students could use *Guess, Check, and Revise*. However, it is more efficient to use order of operations and then solve the equation.

Example 4

- **MP4 Model with Mathematics:** Throughout this text real-life applications are presented. Often, a mathematical model is written to represent the situation. Modeling with mathematics involves representing a situation algebraically, solving, and checking to make sure that the results make sense.
- Note the color-coding of the words and symbols. Discuss this feature with students. Students find it difficult to read a word problem and translate it into symbols. This skill is practiced throughout the text.
- **Representation:** The term $\frac{1}{30}n$ could also have been written as $\frac{n}{30}$. Make sure students understand why. It is how a fraction and number are multiplied.
- **FYI:** The final answer $-210 = n$ can also be written as $n = -210$.

On Your Own

- Remind students to perform the operations following the order of operations.
- **Common Error:** $8 \div \frac{1}{2} \neq 4$; $8 \div \frac{1}{2} = 16$

? "For Question 11, what does $|-10|$ mean?" absolute value of -10, which equals 10

Closure

- Describe in words how to solve a one-step equation.
- Write and solve a one-step equation.

Technology For the Teacher

Dynamic Classroom

The Dynamic Planning Tool
Editable Teacher's Resources at *BigIdeasMath.com*

Extra Example 3

Solve $w - 4 \div \frac{1}{2} = 5$. $w = 13$

Extra Example 4

The melting point of ice is $\frac{2}{9}$ of the melting point of candle wax. The melting point of ice is 32°F. Write and solve an equation to find the melting point of candle wax. $\frac{2}{9}x = 32$; 144°F

On Your Own

10. $p = 1$

11. $q = -8$

12. $-39 = \frac{1}{4}k$; -156°C

Vocabulary and Concept Check

1. $+$ and $-$ are inverses. $\times$ and $\div$ are inverses.
2. yes; The solution of each equation is $x = -3$.
3. $x - 3 = 6$; It is the only equation that does not have $x = 6$ as a solution.

Practice and Problem Solving

4. $x = 32$
5. $x = 57$
6. $x = 111$
7. $x = -5$
8. $g = 24$
9. $p = 21$
10. $y = -2.04$
11. $x = 9\pi$
12. $w = 10\pi$
13. $d = \frac{1}{2}$
14. $r = -\frac{7}{24}$
15. $n = -4.9$
16. $p - 14.50 = 53$; \$67.50
17. **a.** $105 = x + 14$; $x = 91$

    b. no; Because $82 + 9 = 91$, you did not knock down the last pin with the second ball of the frame.

Assignment Guide and Homework Check

Level	Assignment	Homework Check
Average	1–3, 10–14, 17, 21–24, 27–37 odd, 41, 45–48	7, 11, 16, 19, 21, 33
Advanced	1–3, 17, 24–27, 30–38 even, 39–48	17, 24, 30, 34, 39, 42

For Your Information

- **Exercise 17** Students may not know what a spare is in bowling. A spare means that all of the pins were knocked down after the second ball of a frame was thrown. To calculate the score after a spare, you add the number of pins knocked down on your next ball to 10. For example, if you got a spare in the first frame and then knocked down 6 pins on your next ball, your score for the first frame would be 16.

Common Errors

- **Exercises 4–6** Encourage students who choose the protractor to trace the triangle and extend the sides so they can get a more accurate reading.
- **Exercises 7–15** Students may perform the same operation on both sides instead of the opposite operation. Remind them that to solve for the variable, they must *undo* the operation by using the opposite (or inverse) operation.
- **Exercise 16** Students may write the wrong equation for the problem. Encourage them to rewrite the problem so that it is clear what equation they should write. Remind them that subtraction is not commutative.

1.1 Record and Practice Journal

Solve the equation. Check your solution.

1. $x + 5 = 16$ — $x = 11$
2. $11 = w - 12$ — $w = 23$
3. $\frac{3}{4} + z = \frac{5}{6}$ — $z = \frac{1}{12}$
4. $3y = 18$ — $y = 6$
5. $\frac{k}{7} = 10$ — $k = 70$
6. $\frac{4}{5}n = \frac{9}{10}$ — $n = \frac{9}{8}$
7. $x - 12 + 6 = 9$ — $x = 11$
8. $h + |-8| = 15$ — $h = 7$
9. $1.3(2) + p = 7.9$ — $p = 5.3$
10. A coupon subtracts \$5.16 from the price p of a shirt. You pay \$15.48 for the shirt after using the coupon. Write and solve an equation to find the original price of the shirt.

    $p - 5.16 = 15.48$; $p = \$20.64$
11. After a party, you have $\frac{1}{6}$ of the cookies you made left over. There are a dozen cookies left. How many cookies did you make for the party?

    $c = 72$

Technology For the Teacher
Answer Presentation Tool

1.1 Exercises

Vocabulary and Concept Check

1. **VOCABULARY** Which of the operations $+$, $-$, $\times$, and $\div$ are inverses of each other?
2. **VOCABULARY** Are the equations $3x = -9$ and $4x = -12$ equivalent? Explain.
3. **WHICH ONE DOESN'T BELONG?** Which equation does *not* belong with the other three? Explain your reasoning.

   $x - 2 = 4$ $x - 3 = 6$ $x - 5 = 1$ $x - 6 = 0$

Practice and Problem Solving

CHOOSE TOOLS **Find the value of x. Check the reasonableness of your answer.**

4. 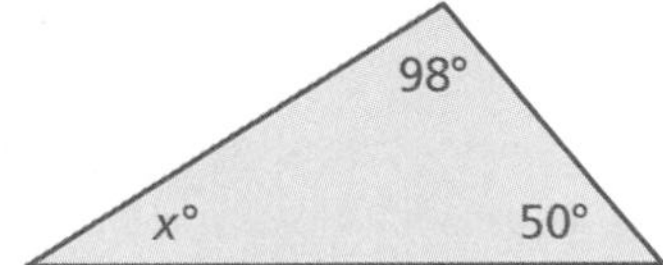

5. 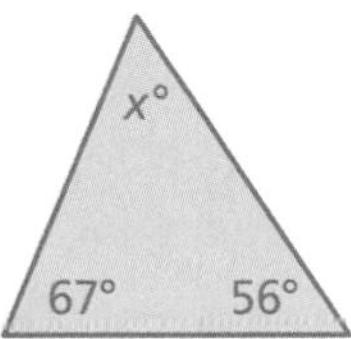

6. 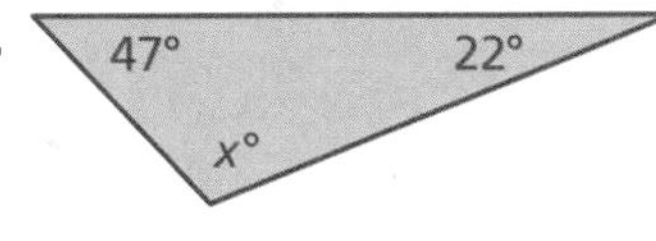

Solve the equation. Check your solution.

1
7. $x + 12 = 7$
8. $g - 16 = 8$
9. $-9 + p = 12$
10. $0.7 + y = -1.34$
11. $x - 8\pi = \pi$
12. $4\pi = w - 6\pi$
13. $\frac{5}{6} = \frac{1}{3} + d$
14. $\frac{3}{8} = r + \frac{2}{3}$
15. $n - 1.4 = -6.3$

16. **CONCERT** A discounted concert ticket is \$14.50 less than the original price p. You pay \$53 for a discounted ticket. Write and solve an equation to find the original price.

17. **BOWLING** Your friend's final bowling score is 105. Your final bowling score is 14 pins less than your friend's final score.

    a. Write and solve an equation to find your final score.

    b. Your friend made a spare in the tenth frame. Did you? Explain.

9	10	FINAL SCORE
8 – 89	7 / 6 105	105
6 3 82	9	?

Solve the equation. Check your solution.

② **18.** $7x = 35$ **19.** $4 = -0.8n$ **20.** $6 = -\frac{w}{8}$

21. $\frac{m}{\pi} = 7.3$ **22.** $-4.3g = 25.8$ **23.** $\frac{3}{2} = \frac{9}{10}k$

24. $-7.8x = -1.56$ **25.** $-2 = \frac{6}{7}p$ **26.** $3\pi d = 12\pi$

27. ERROR ANALYSIS Describe and correct the error in solving the equation.

✗
$$-1.5 + k = 8.2$$
$$k = 8.2 + (-1.5)$$
$$k = 6.7$$

28. TENNIS A gym teacher orders 42 tennis balls. Each package contains 3 tennis balls. Which of the following equations represents the number x of packages?

$x + 3 = 42$ $3x = 42$ $\frac{x}{3} = 42$ $x = \frac{3}{42}$

MODELING In Exercises 29–32, write and solve an equation to answer the question.

29. PARK You clean a community park for 6.5 hours. You earn $42.25. How much do you earn per hour?

30. SPACE SHUTTLE A space shuttle is scheduled to launch from Kennedy Space Center in 3.75 hours. What time is it now?

Launch Time
11:20 A.M.

31. BANKING After earning interest, the balance of an account is $420. The new balance is $\frac{7}{6}$ of the original balance. How much interest was earned?

Tallest Coasters at Cedar Point	
Roller Coaster	**Height (feet)**
Top Thrill Dragster	420
Millennium Force	310
Magnum XL-200	205
Mantis	?

32. ROLLER COASTER Cedar Point amusement park has some of the tallest roller coasters in the United States. The Mantis is 165 feet shorter than the Millennium Force. What is the height of the Mantis?

Common Errors

- **Exercises 18–26** Students may use the same operation instead of the opposite operation to get the variable by itself. Remind them that to *undo* the operation, they must use the opposite (or inverse) operation. Demonstrate that using the same operation will not work. For example:

Incorrect	Correct
$7x = 35$	$7x = 35$
$7 \cdot 7x = 35 \cdot 7$	$\frac{7x}{7} = \frac{35}{7}$
$49x = 245$	$x = 5$

- **Exercise 32** Students may skip the step of writing the equation and just subtract the difference in height from the height of the Millennium Force. Encourage them to develop the problem solving technique of writing the equation before solving. This skill will be useful later in mathematics.
- **Exercises 33–38** Students may forget to use the order of operations when solving for the variable. Remind them of the order of operations and encourage them to simplify both sides of the equation before solving.

English Language Learners

Vocabulary

Have students create a table in their notebooks of the common words used to indicate addition, subtraction, multiplication, and division.

For instance,

Addition	Subtraction
added to	subtracted from
plus	minus
sum of	difference of
more than	less than
increased by	decreased by
total of	fewer than
and	take away
Multiplication	**Division**
multiplied by	divided by
times	quotient of
product of	
twice	
of	

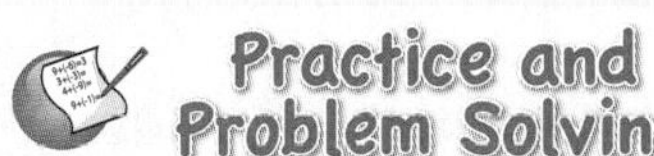

18. $x = 5$ **19.** $n = -5$

20. $w = -48$ **21.** $m = 7.3\pi$

22. $g = -6$ **23.** $k = 1\frac{2}{3}$

24. $x = 0.2$ **25.** $p = -2\frac{1}{3}$

26. $d = 4$

27. They should have added 1.5 to each side.

$$-1.5 + k = 8.2$$
$$k = 8.2 + 1.5$$
$$k = 9.7$$

28. $3x = 42$

29. $6.5x = 42.25$; \$6.50 per hour

30. $x + 3\frac{3}{4} = 11\frac{1}{3}$; 7:35 A.M.

31. $420 = \frac{7}{6}b$, $b = 360$; \$60

32. $x + 165 = 310$; 145 ft

Practice and Problem Solving

33. $h = -7$ **34.** $w = 19$

35. $q = 3.2$ **36.** $d = 0$

37. $x = -1\frac{4}{9}$ **38.** $p = -\frac{1}{12}$

39. greater than; Because a negative number divided by a negative number is a positive number.

40. *Sample answer:* $x - 2 = -4$, $\frac{x}{2} = -1$

41. 3 mg

42. See *Taking Math Deeper*.

43. 8 in.

44. **a.** \$18, \$27, \$45

b. *Sample answer:* Everyone did not do an equal amount of painting.

Fair Game Review

45. $7x - 4$ **46.** $1.6b - 3.2$

47. $\frac{25}{4}g - \frac{2}{3}$

48. A

Mini-Assessment

Solve the equation.

1. $t + 17 = 3$ $t = -14$
2. $-2\pi + d = -3\pi$ $d = -\pi$
3. $-13.5 = 2.7s$ $s = -5$
4. $\frac{2}{3}j = 8$ $j = 12$
5. You earn \$9.65 per hour. This week, you earned \$308.80 before taxes. Write and solve an equation to find the number of hours you worked this week. $9.65x = 308.8$; You worked 32 hours this week.

Taking Math Deeper

Exercise 42

A nice way to organize the given information is to put it into a table.

Use a table to organize the information.

	Total	Retake
Girls	x	$\frac{1}{4}x = 16$
Boys	y	$\frac{1}{8}y = 7$

Use the equations to solve for x and y.

Girls: $\frac{1}{4}x = 16$

$x = 64$

Boys: $\frac{1}{8}y = 7$

$y = 56$

Answer the question.

There are $64 + 56 = 120$ students in the class.

Project

Find out how many retakes were done at your school last year. Do the given ratios work for your school? What do you think are some of the reasons students have retakes?

Reteaching and Enrichment Strategies

If students need help. . .	If students got it. . .
Resources by Chapter • Practice A and Practice B • Puzzle Time Record and Practice Journal Practice Differentiating the Lesson Lesson Tutorials Skills Review Handbook	Resources by Chapter • Enrichment and Extension Start the next section

Solve the equation. Check your solution.

③ **33.** $-3 = h + 8 \div 2$ **34.** $12 = w - |-7|$ **35.** $q + |6.4| = 9.6$

36. $d - 2.8 \div 0.2 = -14$ **37.** $\frac{8}{9} = x + \frac{1}{3}(7)$ **38.** $p - \frac{1}{4} \cdot 3 = -\frac{5}{6}$

39. LOGIC Without solving, is the solution of $-2x = -15$ *greater than* or *less than* -15? Explain.

40. OPEN-ENDED Write a subtraction equation and a division equation that each has a solution of -2.

41. ANTS Some ant species can carry 50 times their body weight. It takes 32 ants to carry the cherry. About how much does each ant weigh?

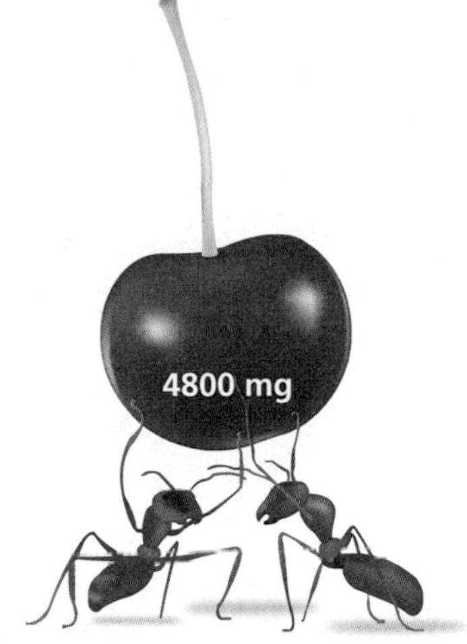

42. REASONING One-fourth of the girls and one-eighth of the boys in a class retake their school pictures. The photographer retakes pictures for 16 girls and 7 boys. How many students are in the class?

43. VOLUME The volume V of the cylinder is 72π cubic inches. Use the formula $V = Bh$ to find the height h of the cylinder.

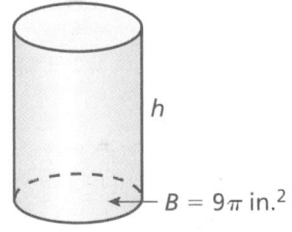

h

$B = 9\pi$ in.^{2}

44. Critical Thinking A neighbor pays you and two friends \$90 to paint her garage. The money is divided three ways in the ratio 2 : 3 : 5.

a. How much does each person receive?

b. What is one possible reason the money is not divided evenly?

Fair Game Review What you learned in previous grades & lessons

Simplify the expression. *(Skills Review Handbook)*

45. $2(x - 2) + 5x$ **46.** $0.4b - 3.2 + 1.2b$ **47.** $\frac{1}{4}g + 6g - \frac{2}{3}$

48. MULTIPLE CHOICE The temperature at 4 P.M. was -12°C. By 11 P.M. the temperature had dropped 14 degrees. What was the temperature at 11 P.M.? *(Skills Review Handbook)*

Ⓐ -26°C Ⓑ -2°C Ⓒ 2°C Ⓓ 26°C

1.2 Solving Multi-Step Equations

COMMON CORE STATE STANDARDS
A.CED.1
A.REI.1
A.REI.3

Essential Question How can you solve a multi-step equation? How can you check the reasonableness of your solution?

1 ACTIVITY: Solving for the Angles of a Triangle

Work with a partner. Write an equation for each triangle. Solve the equation to find the value of the variable. Then find the angle measures of each triangle. Use a protractor to check the reasonableness of your answer.

a.

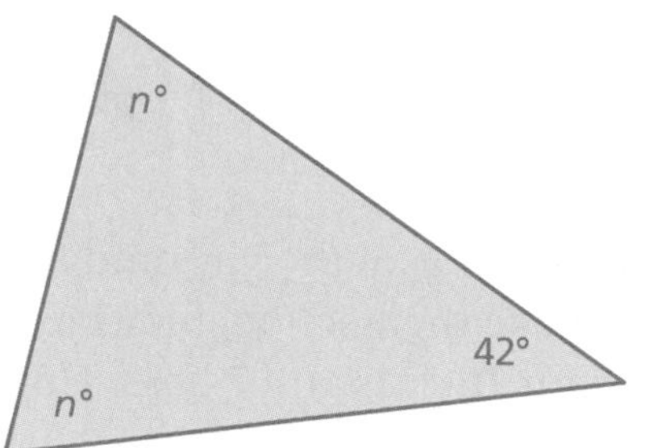

b.

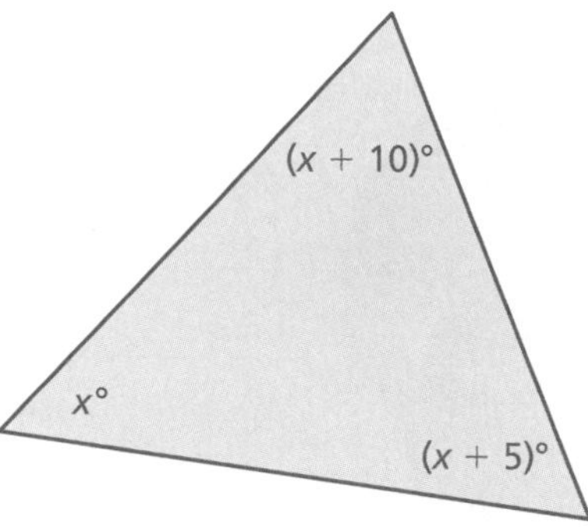

c.

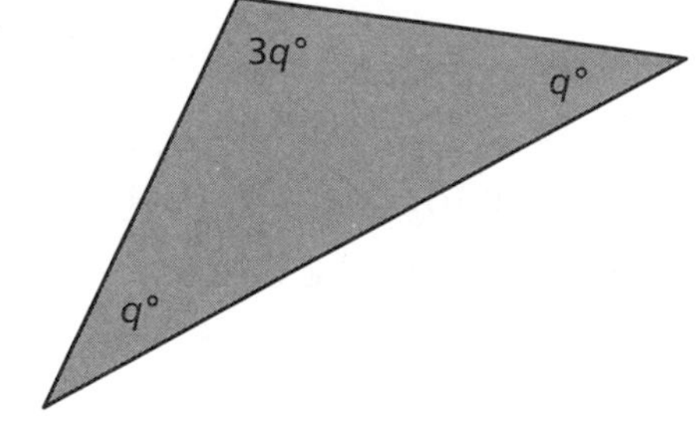

d.

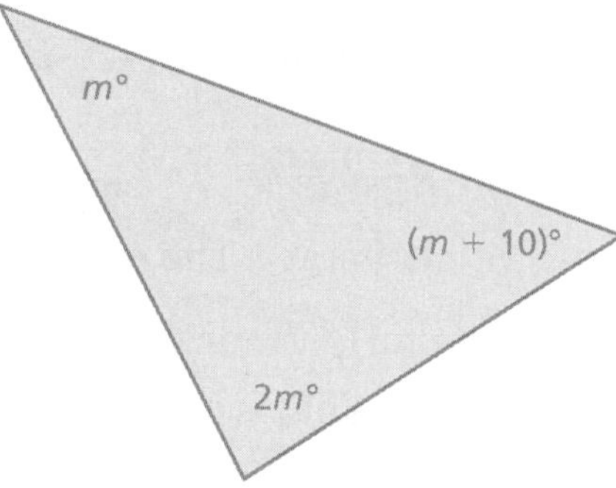

e.

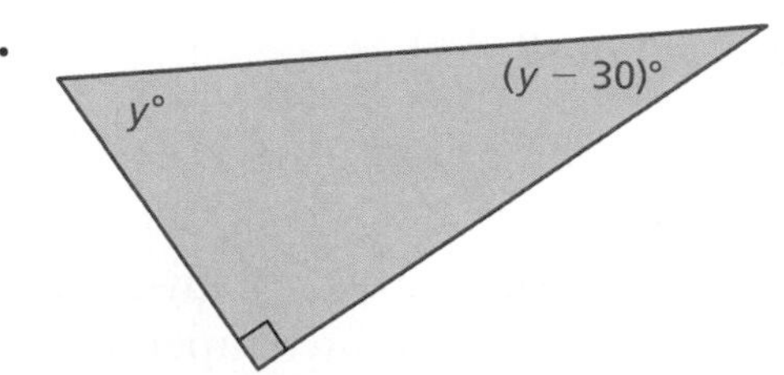

f.

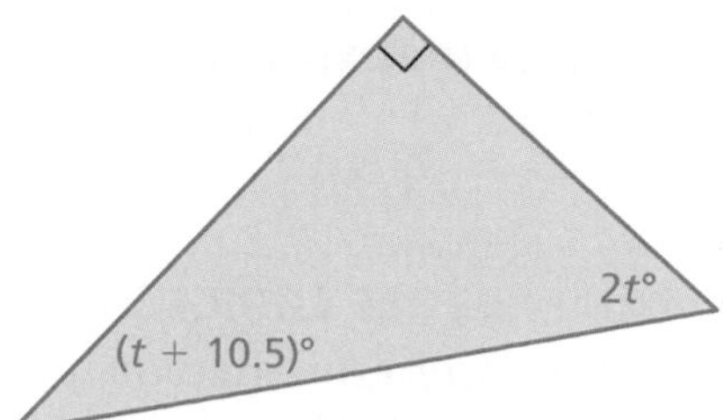

Laurie's Notes

Introduction

Standards for Mathematical Practice

- **MP1a Make Sense of Problems** and **MP2 Reason Abstractly and Quantitatively:** To make sense of solving multi-step equations, students must understand what the symbols and operations represent. When they evaluate expressions, they follow the order of operations. When they solve equations, they undo that process by performing the inverse operations in a reverse order. For example:

Evaluate $3x - 4$ when $x = 5$.		Solve $3x - 4 = 11$.	
$3(5) - 4$		$3x = 15$	$+ 4$
$15 - 4$	$\times 3$	$x = 5$	$\div 3$
11	$- 4$		

Mathematically proficient students make sense of this relationship and perform symbolic manipulations to solve for the variable.

Motivate

- Make a card for each student in your class. Write a variable term on each card. Students will walk around to find others with a card containing a *like term* to the one they are holding.

  Samples: $5x$, $-13x$, $5y$, $6xy$, x, $3.8x$, $\frac{1}{2}y$, $-3.8y$

- Ask students to explain what it means for terms to be *like* terms.

Activity Notes

Activity 1

? Ask a few questions to prepare students for the activity.

- "In the previous lesson, what did you conclude about the sum of the angle measures of a triangle?" sum = 180°
- "So if two angles measure 65° and 75°, what does the third angle measure?" 40°
- "If the angles of a triangle measure $x°$, $2x°$, and $3x°$, could you determine the measure of each angle?" Students should say yes.

- Model how to write and solve the equation $x + 2x + 3x = 180$. Be sure to mention like terms when solving. Ask about the coefficient of x.
- **Common Error:** After solving the equation, you still need to substitute the value into each angle expression to solve for each angle measure. Students sometimes forget this step.
- **FYI:** The triangles are drawn to scale, so the angle measures can be checked using a protractor.
- Ask for volunteers to show a few of the solutions at the board.

? "Why are there only two angles with variable expressions written in parts (e) and (f)?" The third angle in each is a right angle.

Common Core State Standards

A.CED.1 Create equations and inequalities in one variable and use them to solve problems.

A.REI.1 Explain each step in solving a simple equation . . .. Construct a viable argument to justify a solution method.

A.REI.3 Solve linear equations . . . in one variable, including equations with coefficients represented by letters.

Previous Learning

Students should know how to use inverse operations to solve one-step equations.

Activity Materials
Introduction
• index cards

Start Thinking! and Warm Up

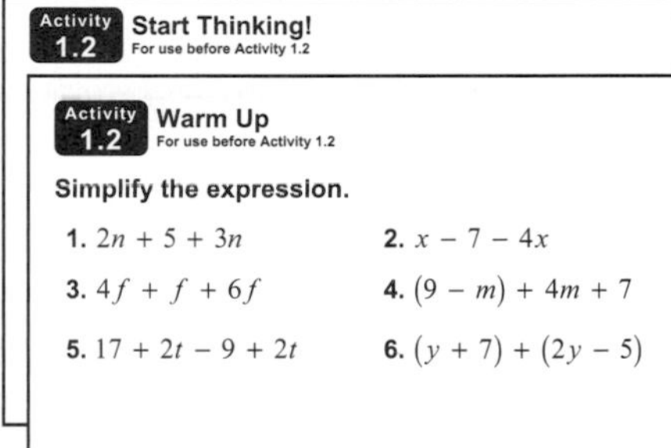

Activity 1.2 Start Thinking!
For use before Activity 1.2
Activity 1.2 Warm Up
For use before Activity 1.2
Simplify the expression.
1. 2n + 5 + 3n
2. x − 7 − 4x
3. 4f + f + 6f
4. (9 − m) + 4m + 7
5. 17 + 2t − 9 + 2t
6. (y + 7) + (2y − 5)

1.2 Record and Practice Journal

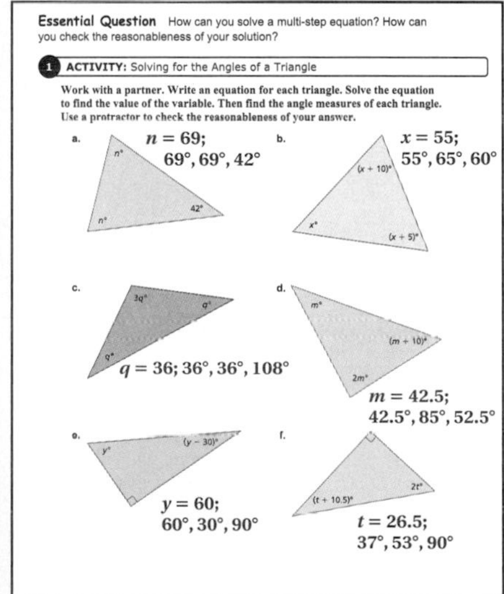

Differentiated Instruction

Auditory

Remind students that in order to solve an equation, the variable must be isolated on one side of the equation. The operations on the same side as the variable are those that need to be undone.

1.2 Record and Practice Journal

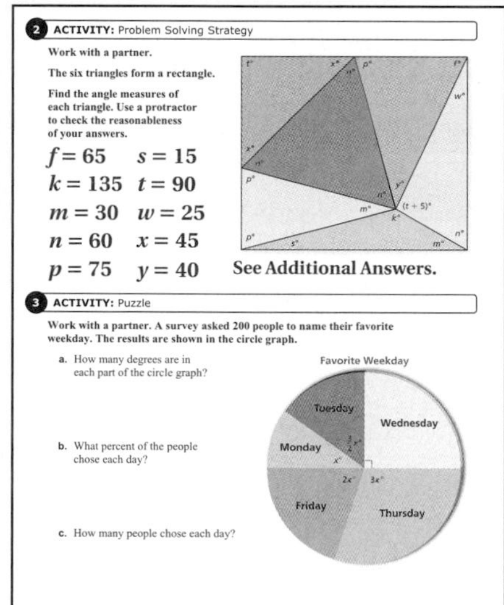

2 ACTIVITY: Problem Solving Strategy

Work with a partner.

The six triangles form a rectangle.

Find the angle measures of each triangle. Use a protractor to check the reasonableness of your answers.

$f = 65$ $s = 15$

$k = 135$ $t = 90$

$m = 30$ $w = 25$

$n = 60$ $x = 45$

$p = 75$ $y = 40$

See Additional Answers.

3 ACTIVITY: Puzzle

Work with a partner. A survey asked 200 people to name their favorite weekday. The results are shown in the circle graph.

a. How many degrees are in each part of the circle graph?

b. What percent of the people chose each day?

c. How many people chose each day?

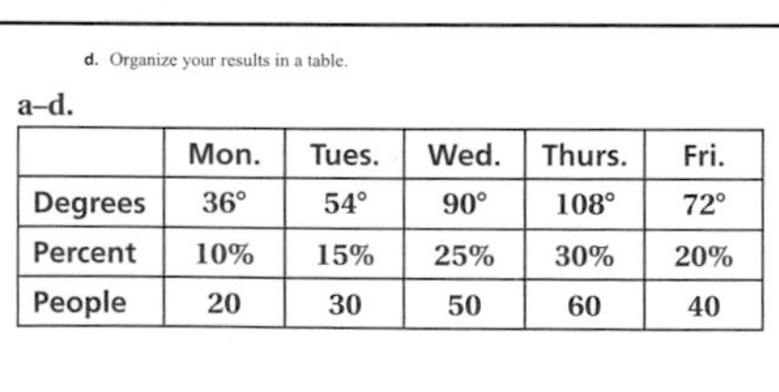

d. Organize your results in a table.

a–d.

	Mon.	Tues.	Wed.	Thurs.	Fri.
Degrees	36°	54°	90°	108°	72°
Percent	10%	15%	25%	30%	20%
People	20	30	50	60	40

What Is Your Answer?

4. IN YOUR OWN WORDS How can you solve a multi-step equation? How can you check the reasonableness of your solution?

Use inverse operations.

Check by substituting solution back into original equation.

Laurie's Notes

Activity 2

- Students will have different strategies for solving this puzzle.
- ? "Define a straight angle." An angle that measures 180°.
- Remind students to look for a variety of ways to check their answers.
- **MP1a:** It is important to hear about different entry points into the problem and strategies for solving. Discuss results and strategies for finding the angle measures. Listen for: right angles at the vertices of the rectangle; sum of angle measures forming a straight angle equals 180°; sum of angle measures about a point equals 360°; sum of the angle measures of a triangle equals 180°.
- Most students will not write a formal equation, but the thinking involved is an equation. For example, $k + m + s = 180$. If you know k and m, you can use mental math to solve for s.

Activity 3

- This example reviews fraction addition, mixed numbers, fraction division, and percents.
- ? Ask a few questions to help students begin the activity.
  - "How many people were surveyed?" 200
  - "What is the sum of the five central angle measures?" 360°
  - "What is the angle measure of the sector labeled Wednesday?" 90°
- Some students may use all five angles and set the expression equal to 360, while other students may only consider the four angles represented by a variable expression and set it equal to 270.
- ? "How do you find the percent each angle measure represents?"

  Convert $\frac{\text{angle measure}}{360}$ to a percent.

Words of Wisdom

- **MP1b Persevere in Solving Problems:** There are many steps in Activity 3, but it is possible to solve. This problem takes time and students will feel a sense of accomplishment when they finish.

Closure

- Find the angle measures in the right triangle. $x = 60$, $y = 30$, $z = 120$

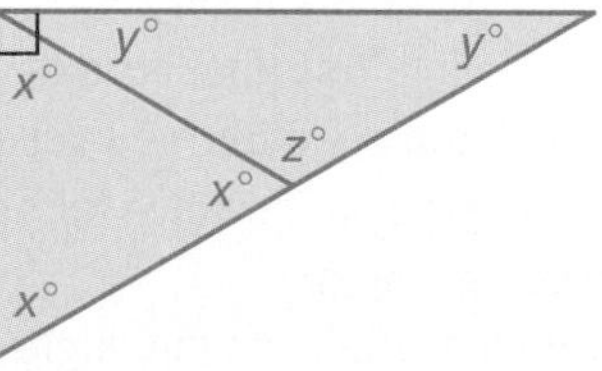

Technology For the Teacher

The Dynamic Planning Tool

Editable Teacher's Resources at *BigIdeasMath.com*

2 ACTIVITY: Problem-Solving Strategy

Work with a partner.

The six triangles form a rectangle.

Find the angle measures of each triangle. Use a protractor to check the reasonableness of your answers.

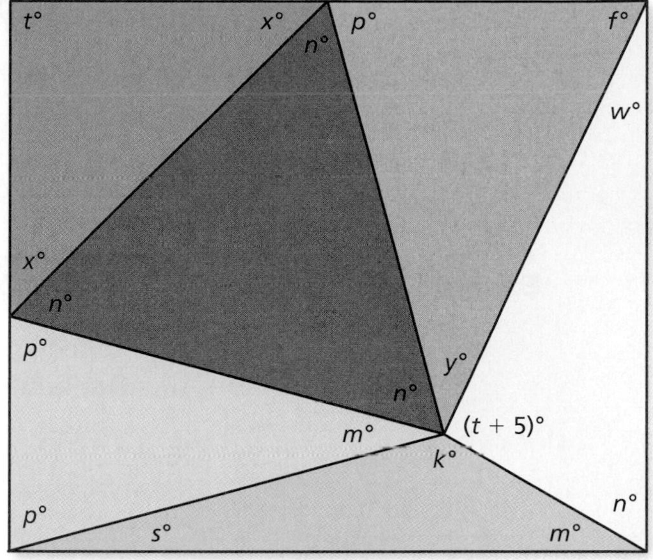

3 ACTIVITY: Puzzle

Work with a partner. A survey asked 200 people to name their favorite weekday. The results are shown in the circle graph.

a. How many degrees are in each part of the circle graph?

b. What percent of the people chose each day?

c. How many people chose each day?

d. Organize your results in a table.

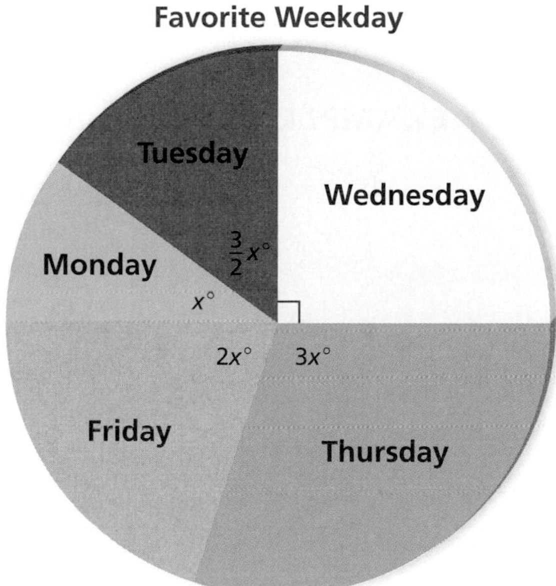

What Is Your Answer?

4. **IN YOUR OWN WORDS** How can you solve a multi-step equation? How can you check the reasonableness of your solution?

Practice Use what you learned about solving multi-step equations to complete Exercises 3–5 on page 14.

1.2 Lesson

Key Idea

Solving Multi-Step Equations

To solve multi-step equations, use inverse operations to isolate the variable.

EXAMPLE 1 Solving a Two-Step Equation

The height (in feet) of a tree after x years is $1.5x + 15$. After how many years is the tree 24 feet tall?

	$1.5x + 15 = 24$	Write an equation.
Undo the addition. →	$-15 \quad -15$	Subtract 15 from each side.
	$1.5x = 9$	Simplify.
Undo the multiplication. →	$\dfrac{1.5x}{1.5} = \dfrac{9}{1.5}$	Divide each side by 1.5.
	$x = 6$	Simplify.

∴ The tree is 24 feet tall after 6 years.

EXAMPLE 2 Combining Like Terms to Solve an Equation

Solve $8x - 6x - 25 = -35$.

	$8x - 6x - 25 = -35$	Write the equation.
	$2x - 25 = -35$	Combine like terms.
Undo the subtraction. →	$+25 \quad +25$	Add 25 to each side.
	$2x = -10$	Simplify.
Undo the multiplication. →	$\dfrac{2x}{2} = \dfrac{-10}{2}$	Divide each side by 2.
	$x = -5$	Simplify.

∴ The solution is $x = -5$.

On Your Own

Now You're Ready
Exercises 6–9

Solve the equation. Check your solution.

1. $-3z + 1 = 7$
2. $\frac{1}{2}x - 9 = -25$
3. $-4n - 8n + 17 = 23$

Laurie's Notes

Introduction

Connect

- **Yesterday:** Students developed an intuitive understanding about solving multi-step equations. (MP1, MP2)
- **Today:** Students will solve multi-step equations by using inverse operations to isolate the variable.

Motivate

- **Story Time:** Share with your students that you took a taxi recently from an airport to a hotel. The taxi driver charged \$18.90.

TAXI FARE	
\$2.50	INITIAL CHARGE
\$0.40	PER 1/5 MILE

? "How far was the hotel from the airport?" $\frac{41}{5}$, or 8.2 miles

- Students will probably try \$18.90 − \$2.50 and then divide by \$0.40. That will give the number of 1/5 miles.
- Explain that today they will solve equations that involve two operations just like the taxi problem.

Lesson Notes

Key Idea

- **Connection:** When you evaluate an expression, you follow the order of operations. Solving an equation undoes the evaluating, in reverse order. The goal is to isolate the variable term and then solve for the variable.

Example 1

- One way to explain the equation is to think of the tree as being 15 feet tall when being planted. It then grows 1.5 feet each year.
- **MP1a Make Sense of Problems:** Representing the problem in more than one way helps students make sense of the problem. Make a table to show the height of the tree from the first year to the sixth year. The table may help students see the connection between the coefficient 1.5, and how the height of the tree is changing.
- **MP6 Attend to Precision:** The two properties used to solve this equation are the Subtraction Property of Equality and the Division Property of Equality. Continue to use property names as you explain the examples in this section.

Example 2

- **MP6:** Remind students that they are using the Addition Property of Equality and the Division Property of Equality to solve this equation.

? "Why is $8x - 6x = 2x$?" Use the Distributive Property to subtract the terms; $8x - 6x = (8 - 6)x = 2x$.

On Your Own

- In Question 2, students may incorrectly divide both sides by $\frac{1}{2}$ and get $x = -8$. Remind students that dividing by $\frac{1}{2}$ is the same as multiplying by 2.

Goal Today's lesson is solving multi-step equations.

Start Thinking! and Warm Up

Lesson 1.2 **Start Thinking!** For use before Lesson 1.2

A multi-step equation requires two or more operations to solve the equation. Explain why the following situation can be modeled by a multi-step equation.

A plumber charges \$80 per hour for labor plus \$60 for parts.

Come up with your own scenario that can be modeled by a multi-step equation.

Extra Example 1

The height (in inches) of a plant after t days is $\frac{1}{2}t + 6$. After how many days is the plant 21 inches tall? 30 days

Extra Example 2

Solve $-2m + 4m + 5 = -3$. $m = -4$

On Your Own

1. $z = -2$
2. $x = -32$
3. $n = -0.5$

Extra Example 3

Solve $-4(3g - 5) + 10g = 19$. 0.5

Extra Example 4

You have scored 7, 10, 8, and 9 on four quizzes. Write and solve an equation to find the score you need on the fifth quiz so that your mean score is 8.

$\frac{x + 7 + 10 + 8 + 9}{5} = 8$; 6

On Your Own

4. $x = -1.5$ 5. $d = -1$

6. $\frac{88 + 92 + 87 + x}{4} = 90$; $x = 93$

English Language Learners

Vocabulary

English learners will benefit from understanding that a *term* is a number, a variable, or the product of a number and variable. *Like terms* are terms that have identical variable parts.

3 and 16 are like terms because they contain no variable.

$4x$ and $7x$ are like terms because they have the same variable x.

$5a$ and $5b$ are *not* like terms because they have different variables.

Laurie's Notes

Example 3

- Ask students to identify the operations involved in this equation. from left to right: multiplication (by 2), subtraction, multiplication ($5x$), addition
- **Note:** Combining like terms in the fourth step is not obvious to students. When the like terms are not adjacent, students are unsure of how to combine them. Rewrite the left side of the equation as $2 + (-10x) + 4$.
- **MP6:** Remind students that they are using the Subtraction Property of Equality and the Division Property of Equality to solve this equation.

Words of Wisdom

- Refer to the Study Tip. Instead of using the Distributive Property, both sides of the equation are divided by 2 in the third step. Explain to students that the left side of the equation is 2 times an expression. When the expression $2(1 - 5x)$ is divided by 2, it leaves the expression $1 - 5x$. In the next step, students want to add 1 to each side because of the subtraction operation shown. Again, it is helpful to write $1 - 5x$ as $1 + (-5x)$ so that it makes sense to students why 1 is subtracted from each side.
- **MP1 Make Sense of Problems and Persevere in Solving Them:** Sufficient time spent helping students make sense of multi-step equations is essential. Symbolic manipulation occurs in many contexts in mathematics and careful development of this proficiency is important.

Example 4

- You may need to review *mean* with the students.
- Discuss the information displayed in the table and write the equation.

? "Is it equivalent to write $\frac{x + 3.5}{5} = 1.5$ instead of $\frac{3.5 + x}{5} = 1.5$? Explain." yes; Commutative Property of Addition

- **FYI:** It may be helpful to write the third step with parentheses: $5\left(\frac{3.5 + x}{5}\right)$.
- **Note:** This is a classic question. When all of the data are known except for one, what is needed in order to achieve a particular average? Students often ask this in the context of wanting to know what they have to score on a test in order to achieve a certain average.

On Your Own

- Encourage students to work with a partner. Students need to be careful with multi-step equations and it is helpful to have a partner check each step.

Closure

- **Exit Ticket:** Solve $8x + 9 - 4x = 25$. Check your solution. $x = 4$

Technology For the Teacher

The Dynamic Planning Tool

Editable Teacher's Resources at *BigIdeasMath.com*

EXAMPLE 3 Using the Distributive Property to Solve an Equation

Solve $2(1 - 5x) + 4 = -8$.

$2(1 - 5x) + 4 = -8$	Write the equation.
$2(1) - 2(5x) + 4 = -8$	Use Distributive Property.
$2 - 10x + 4 = -8$	Multiply.
$-10x + 6 = -8$	Combine like terms.
$\underline{-6} \quad \underline{-6}$	Subtract 6 from each side.
$-10x = -14$	Simplify.
$\dfrac{-10x}{-10} = \dfrac{-14}{-10}$	Divide each side by -10.
$x = 1.4$	Simplify.

Study Tip

Here is another way to solve the equation in Example 3.

$$2(1 - 5x) + 4 = -8$$
$$2(1 - 5x) = -12$$
$$1 - 5x = -6$$
$$-5x = -7$$
$$x = 1.4$$

EXAMPLE 4 Real-Life Application

Use the table to find the number of miles x you need to run on Friday so that the mean number of miles run per day is 1.5.

Day	Miles
Monday	2
Tuesday	0
Wednesday	1.5
Thursday	0
Friday	x

Write an equation using the definition of mean.

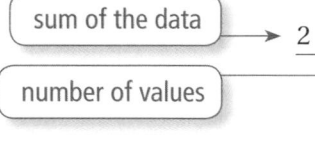

sum of the data / number of values	$\dfrac{2 + 0 + 1.5 + 0 + x}{5} = 1.5$	Write the equation.
	$\dfrac{3.5 + x}{5} = 1.5$	Combine like terms.
Undo the division.	$5 \cdot \dfrac{3.5 + x}{5} = 5 \cdot 1.5$	Multiply each side by 5.
	$3.5 + x = 7.5$	Simplify.
Undo the addition.	$\underline{-3.5} \qquad \underline{-3.5}$	Subtract 3.5 from each side.
	$x = 4$	Simplify.

∴ You need to run 4 miles on Friday.

On Your Own

Now You're Ready
Exercises 10 and 11

Solve the equation. Check your solution.

4. $-3(x + 2) + 5x = -9$

5. $5 + 1.5(2d - 1) = 0.5$

6. You scored 88, 92, and 87 on three tests. Write and solve an equation to find the score you need on the fourth test so that your mean test score is 90.

1.2 Exercises

Vocabulary and Concept Check

1. **WRITING** Write the verbal statement as an equation. Then solve.

   2 more than 3 times a number is 17.

2. **OPEN-ENDED** Explain how to solve the equation $2(4x - 11) + 9 = 19$.

Practice and Problem Solving

CHOOSE TOOLS Find the value of the variable. Then find the angle measures of the polygon. Check the reasonableness of your answer.

3. 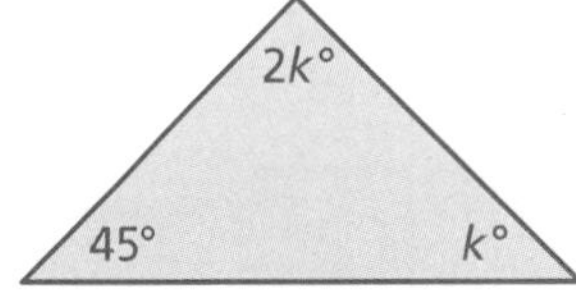

Sum of angle measures: 180°

4. 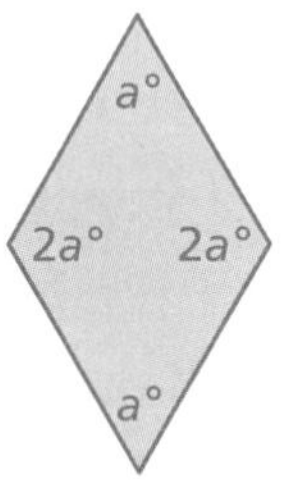

Sum of angle measures: 360°

5. 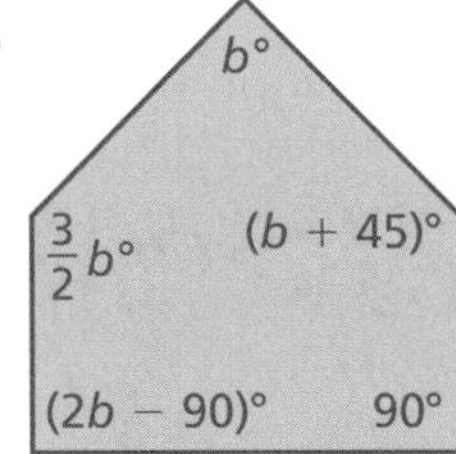

Sum of angle measures: 540°

Solve the equation. Check your solution.

① ② 6. $10x + 2 = 32$

7. $19 - 4c = 17$

8. $1.1x + 1.2x - 5.4 = -10$

9. $\frac{2}{3}h - \frac{1}{3}h + 11 = 8$

③ 10. $6(5 - 8v) + 12 = -54$

11. $21(2 - x) + 12x = 44$

12. **ERROR ANALYSIS** Describe and correct the error in solving the equation.

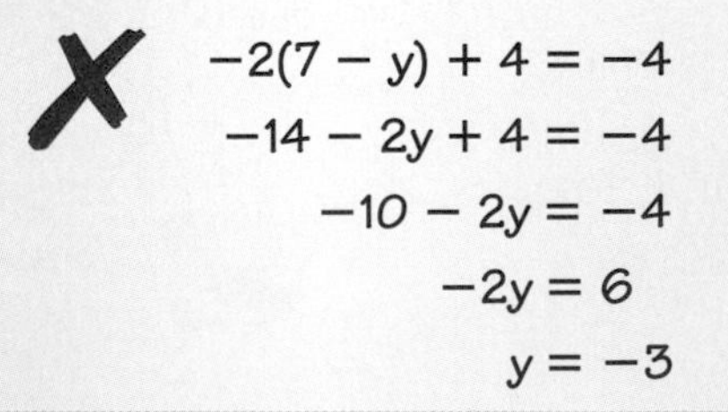

$$-2(7 - y) + 4 = -4$$
$$-14 - 2y + 4 = -4$$
$$-10 - 2y = -4$$
$$-2y = 6$$
$$y = -3$$

13. **WATCHES** The cost C (in dollars) of making n watches is represented by $C = 15n + 85$. How many watches are made when the cost is $385?

14. **HOUSE** The height of the house is 26 feet. What is the height x of each story?

Assignment Guide and Homework Check

Level	Assignment	Homework Check
Average	1–5, 7, 9, 11, 12–17, 19, 21, 23, 25–28	7, 11, 14, 16, 21
Advanced	1–5, 6–14 even, 15–28	10, 12, 14, 16, 21

Common Errors

- **Exercises 8 and 9** When combining like terms, students may square the variable. Remind them that $x^2 = x \cdot x$, and in these exercises they are not multiplying the variables. Remind them that when adding and subtracting variables, they perform the addition or subtraction on the coefficient of the variable.
- **Exercises 10 and 11** When using the Distributive Property, students may forget to distribute to all the values within the parentheses. Remind them that they need to distribute to all the values and encourage them to draw arrows showing the distribution, if needed.
- **Exercise 16** Students may struggle with writing the equation for this problem because of the tip that is added to the total. Encourage them to write an expression for the cost of the food and then add on the tip.
- **Exercises 18–23** Students may solve the equations for *a*, *b*, or *c*. Remind them that they are solving for *x*. Encourage them to make *x* a different color when solving so that it's easy to remember that they are solving for *x*.

1.2 Record and Practice Journal

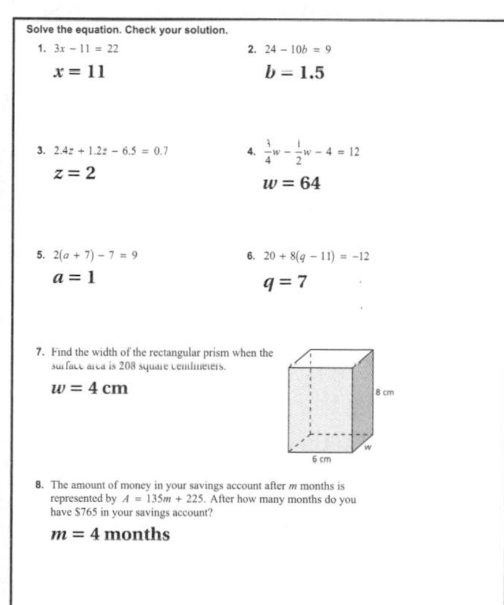

Solve the equation. Check your solution.

1. $3x - 11 = 22$
   $x = 11$
2. $24 - 10b = 9$
   $b = 1.5$
3. $2.4z + 1.2z - 6.5 = 0.7$
   $z = 2$
4. $\frac{3}{4}w - \frac{1}{2}w - 4 = 12$
   $w = 64$
5. $2(a + 7) - 7 = 9$
   $a = 1$
6. $20 + 8(q - 11) = -12$
   $q = 7$
7. Find the width of the rectangular prism when the surface area is 208 square centimeters.
   $w = 4$ cm

8. The amount of money in your savings account after *m* months is represented by $A = 135m + 225$. After how many months do you have \$765 in your savings account?
   $m = 4$ months

Technology For the Teacher
Answer Presentation Tool

Vocabulary and Concept Check

1. $2 + 3x = 17; x = 5$
2. *Sample answer:* Subtract 9 from each side. Divide each side by 2. Add 11 to each side. Divide each side by 4.

Practice and Problem Solving

3. $k = 45; 45°, 45°, 90°$
4. $a = 60; 60°, 120°, 60°, 120°$
5. $b = 90; 90°, 135°, 90°, 90°, 135°$
6. $x = 3$
7. $c = 0.5$
8. $x = -2$
9. $h = -9$
10. $v = 2$
11. $x = -\frac{2}{9}$
12. They did not distribute the −2 properly.

$$\begin{aligned} -2(7 - y) + 4 &= -4 \\ -14 + 2y + 4 &= -4 \\ 2y - 10 &= -4 \\ 2y &= 6 \\ y &= 3 \end{aligned}$$

13. 20 watches
14. 10 ft

Practice and Problem Solving

15. $4(b + 3) = 24$; 3 in.

16. $1.15(2p + 1.5) = 11.5$; \$4.25

17. $\dfrac{2580 + 2920 + x}{3} = 3000$; 3500 people

18. $x = \dfrac{3}{4} - a$

19. $x = \dfrac{-7}{b}$

20. $x = \dfrac{-8}{b}$

21. $x = \dfrac{3b}{2c}$

22. $x = \dfrac{12.5 + b}{a}$

23. $x = \dfrac{c - b}{a}$

24. See *Taking Math Deeper.*

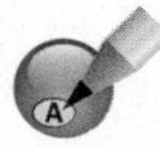

Fair Game Review

25. <	**26.** =
27. >	**28.** D

Mini-Assessment

Solve the equation.

1. $18 = 5a - 2a + 3$ $a = 5$
2. $2(4 - 2w) - 8 = -4$ $w = 1$
3. $2.3y + 4.4y - 3.7 = 16.4$ $y = 3$
4. $\frac{3}{4}z + \frac{1}{4}z - 6 = -5$ $z = 1$
5. The perimeter of the picture is 36 inches. What is the height of the picture? 10 in.

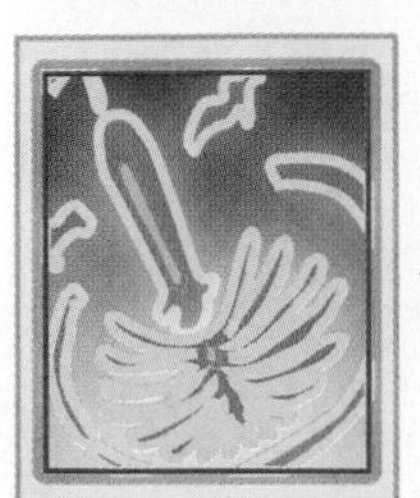

Taking Math Deeper

Exercise 24

This problem points out that mathematics and algebra are used in many different fields. This is a nice example using scoring at a diving competition.

 Begin by translating the scoring system into a mathematical formula.

minus the highest and lowest

Score = 0.6(degree of difficulty)(sum of countries' scores)

 Substitute the given information.

Let x = the degree of difficulty.

$77.7 = 0.6(x)(7.5 + 8.0 + 7.0 + 7.5 + 7.0)$

$77.7 = 0.6x(37)$

$77.7 = 22.2x$

$3.5 = x$

a. The degree of difficulty is 3.5.

 This question has many answers.

Let x = sum of the five countries' scores.

$97.2 = 0.6(4)(x)$

$97.2 = 2.4x$

$40.5 = x$

One possibility is the following:

b. 8.0 + 8.0 + 8.0 + 8.0 + 8.5 with a low score of 7.5 and a high score of 9.0

Project

Use the Internet or school library to find all the different dives that are scored in a diving competition. Find the degree of difficulty that goes with each dive.

Reteaching and Enrichment Strategies

If students need help. . .	If students got it. . .
Resources by Chapter • Practice A and Practice B • Puzzle Time Record and Practice Journal Practice Differentiating the Lesson Lesson Tutorials Skills Review Handbook	Resources by Chapter • Enrichment and Extension Start the next section

In Exercises 15–17, write and solve an equation to answer the question.

15. **POSTCARD** The area of the postcard is 24 square inches. What is the width b of the message (in inches)?

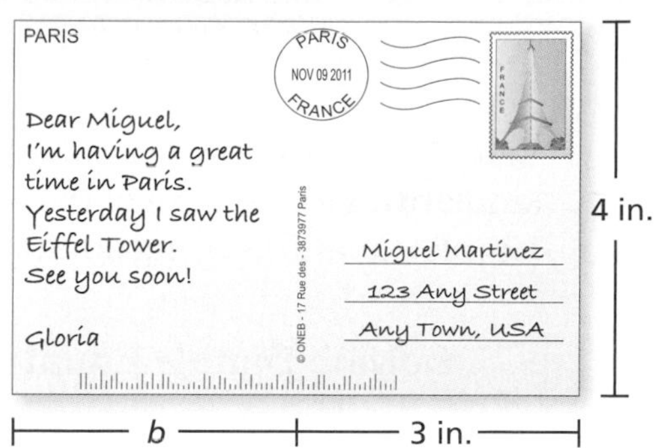

16. **BREAKFAST** You order two servings of pancakes and a fruit cup. The cost of the fruit cup is \$1.50. You leave a 15% tip. Your total bill is \$11.50. How much does one serving of pancakes cost?

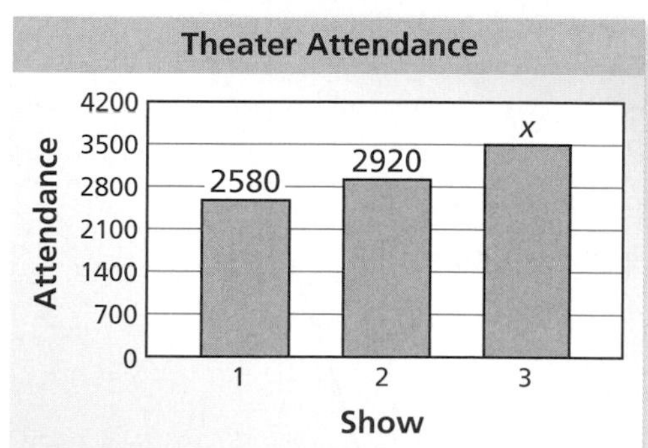

17. **PROBLEM SOLVING** How many people must attend the third show so that the average attendance for the three shows is 3000?

The letters *a*, *b*, and *c* represent constants. Solve the equation for *x*.

18. $x + a = \frac{3}{4}$

19. $bx = -7$

20. $2bx - bx = -8$

21. $4cx - b = 5b$

22. $ax - b = 12.5$

23. $ax + b = c$

24. **DIVING** Divers in a competition are scored by an international panel of judges. The highest and lowest scores are dropped. The total of the remaining scores is multiplied by the degree of difficulty of the dive. This product is multiplied by 0.6 to determine the final score.

 a. A diver's final score is 77.7. What is the degree of difficulty of the dive?

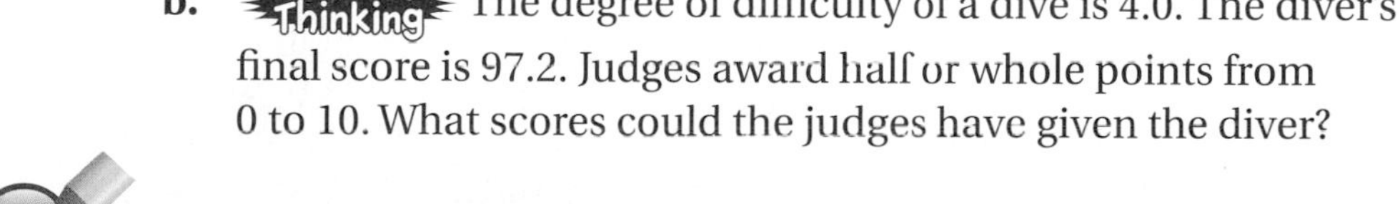

Judge	Russia	China	Mexico	Germany	Italy	Japan	Brazil
Score	7.5	8.0	6.5	8.5	7.0	7.5	7.0

 b. 

The degree of difficulty of a dive is 4.0. The diver's final score is 97.2. Judges award half or whole points from 0 to 10. What scores could the judges have given the diver?

© Paul Slaughter, www.slaughterphoto.com
Greg Louganis diving at the 1984 Olympics

Fair Game Review What you learned in previous grades & lessons

Let $a = 3$ and $b = -2$. Copy and complete the statement using <, >, or =.
(Skills Review Handbook)

25. $-5a$ ☐ 4

26. 5 ☐ $b + 7$

27. $a - 4$ ☐ $10b + 8$

28. **MULTIPLE CHOICE** What value of x makes the equation $x + 5 = 2x$ true? *(Skills Review Handbook)*

 Ⓐ -1 Ⓑ 0 Ⓒ 3 Ⓓ 5

1 Study Help

You can use a **Y chart** to compare two topics. List differences in the branches and similarities in the base of the Y. Here is an example of a Y chart that compares solving simple equations using addition to solving simple equations using subtraction.

Solving Simple Equations Using Addition

- Add the same number to each side of the equation.

Solving Simple Equations Using Subtraction

- Subtract the same number from each side of the equation.

Both:

- You can solve the equation in one step.
- You produce an equivalent equation.
- The variable can be on either side of the equation.
- It is always a good idea to check your solution.

On Your Own

Make Y charts to help you study and compare these topics.

1. solving simple equations using multiplication and solving simple equations using division
2. solving simple equations and solving multi-step equations

After you complete this chapter, make Y charts for the following topics.

3. solving equations with the variable on one side and solving equations with variables on both sides
4. solving multi-step equations and solving equations with variables on both sides
5. solving multi-step equations and rewriting literal equations
6. solving multi-step equations and solving absolute value equations

"I made a Y chart to compare and contrast Fluffy's characteristics with yours."

Sample Answers

1.

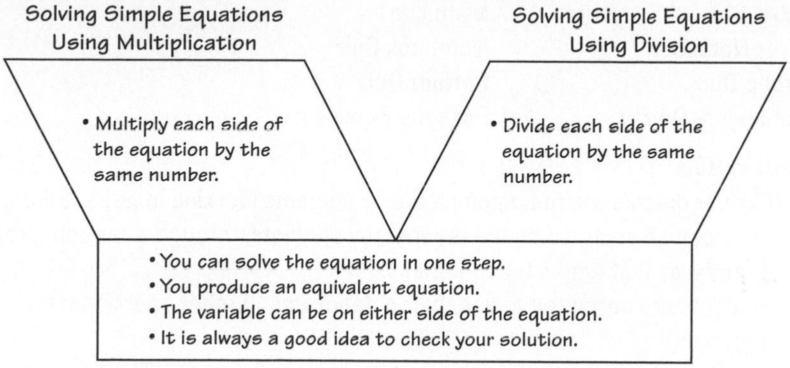

Solving Simple Equations Using Multiplication

- Multiply each side of the equation by the same number.

Solving Simple Equations Using Division

- Divide each side of the equation by the same number.

Both:

- You can solve the equation in one step.
- You produce an equivalent equation.
- The variable can be on either side of the equation.
- It is always a good idea to check your solution.

2.

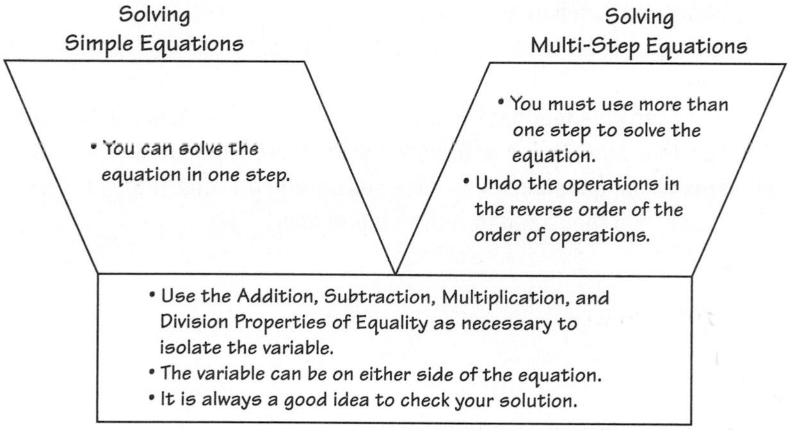

Solving Simple Equations

- You can solve the equation in one step.

Solving Multi-Step Equations

- You must use more than one step to solve the equation.
- Undo the operations in the reverse order of the order of operations.

Both:

- Use the Addition, Subtraction, Multiplication, and Division Properties of Equality as necessary to isolate the variable.
- The variable can be on either side of the equation.
- It is always a good idea to check your solution.

List of Organizers

Available at *BigIdeasMath.com*

Comparison Chart
Concept Circle
Definition (Idea) and Example Chart
Example and Non-Example Chart
Formula Triangle
Four Square
Information Frame
Information Wheel
Notetaking Organizer
Process Diagram
Summary Triangle
Word Magnet
Y Chart

About this Organizer

A **Y Chart** can be used to compare two topics. Students list differences between the two topics in the branches of the Y and similarities in the base of the Y. A Y chart serves as a good tool for assessing students' knowledge of a pair of topics that have subtle but important differences. You can include blank Y charts on tests or quizzes for this purpose.

Technology For the Teacher
Vocabulary Puzzle Builder

Answers

1. $y = \frac{1}{2}$
2. $w = 5\pi$
3. $m = 0.5$
4. $q = -3.6$
5. $k = 4$
6. $z = 16$
7. $n = \frac{3}{2}$
8. $t = 8$
9. $x = 60$; 55°, 60°, 65°
10. $x = 126$; 63°, 80°, 126°, 91°
11. $32
12. 50 ft, 150 ft, 75 ft, 180 ft
13. $230x = 1265$; 5.5 hours
14. $\frac{25 + 15 + 18 + p}{4} = 20$;
    $p = 22$ points

Assessment Book

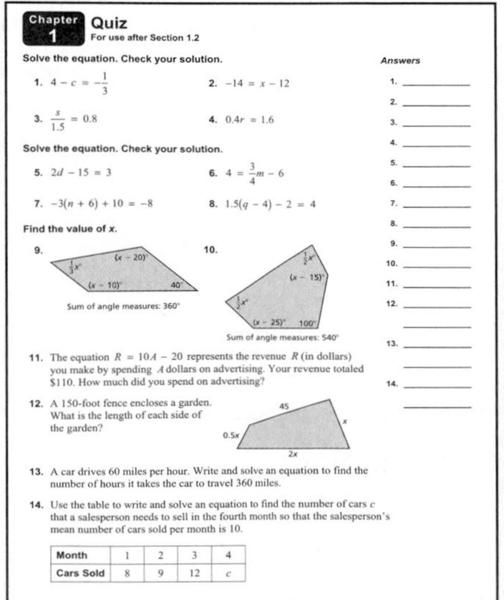

Chapter 1 Quiz
For use after Section 1.2

Solve the equation. Check your solution.

1. $4 - c = -\frac{1}{3}$
2. $-14 = x - 12$
3. $\frac{s}{1.5} = 0.8$
4. $0.4r = 1.6$

Solve the equation. Check your solution.

5. $2d - 15 = 3$
6. $4 = \frac{3}{4}m - 6$
7. $-3(n + 6) + 10 = -8$
8. $1.5(q - 4) - 2 = 4$

Find the value of *x*.

9. Sum of angle measures: 360°
10. Sum of angle measures: 540°

11. The equation $R = 10A - 20$ represents the revenue R (in dollars) you make by spending A dollars on advertising. Your revenue totaled $110. How much did you spend on advertising?
12. A 150-foot fence encloses a garden. What is the length of each side of the garden?

13. A car drives 60 miles per hour. Write and solve an equation to find the number of hours it takes the car to travel 360 miles.
14. Use the table to write and solve an equation to find the number of cars c that a salesperson needs to sell in the fourth month so that the salesperson's mean number of cars sold per month is 10.

Month	1	2	3	4
Cars Sold	8	9	12	c

Answers

1. ______
2. ______
3. ______
4. ______
5. ______
6. ______
7. ______
8. ______
9. ______
10. ______
11. ______
12. ______
13. ______
14. ______

Alternative Quiz Ideas

100% Quiz
Error Notebook
Group Quiz
Homework Quiz
Math Log
Notebook Quiz
Partner Quiz
Pass the Paper

Partner Quiz

- Partner quizzes are to be completed by students working in pairs. Student pairs can be selected by the teacher, by students, through a random process, or any way that works for your class.
- Students are permitted to use their notebooks and other appropriate materials.
- Each pair submits a draft of the quiz for teacher feedback. Then they revise their work and turn it in for a grade.
- When the pair is finished they can submit one paper, or each can submit their own.
- Teachers can give feedback in a variety of ways. It is important that the teacher does not reteach or provide the solution. The teacher can tell students which questions they have answered correctly, if they are on the right track, or if they need to rethink a problem.

Reteaching and Enrichment Strategies

If students need help. . .	If students got it. . .
Resources by Chapter • Study Help • Practice A and Practice B • Puzzle Time Lesson Tutorials *BigIdeasMath.com* Practice Quiz Practice from the Test Generator	Resources by Chapter • Enrichment and Extension • School-to-Work Game Closet at *BigIdeasMath.com* Start the next section

Technology For the Teacher

Answer Presentation Tool
Big Ideas Test Generator

Solve the equation. Check your solution. *(Section 1.1)*

1. $-\frac{1}{2} = y - 1$

2. $-3\pi + w = 2\pi$

3. $1.2m = 0.6$

4. $q + 2.7 = -0.9$

Solve the equation. Check your solution. *(Section 1.2)*

5. $-4k + 17 = 1$

6. $\frac{1}{4}z + 8 = 12$

7. $-3(2n + 1) + 7 = -5$

8. $2.5(t - 2) - 6 = 9$

Find the value of x. Then find the angle measures of the polygon. *(Section 1.2)*

9.

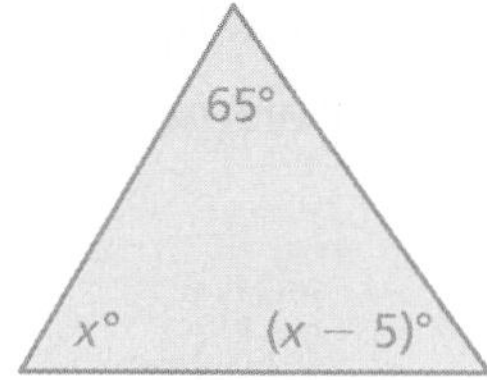

Sum of angle measures: 180°

10.

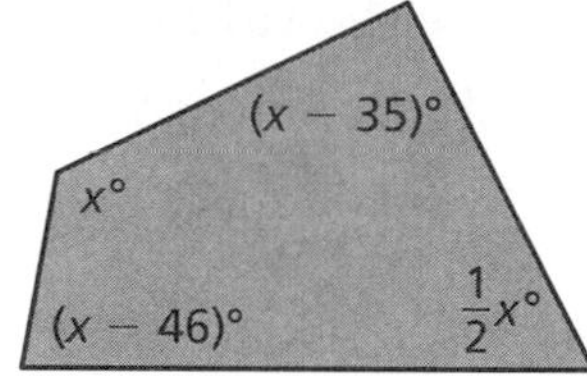

Sum of angle measures: 360°

11. JEWELER The equation $P = 2.5m + 35$ represents the price P (in dollars) of a bracelet, where m is the cost of the materials (in dollars). The price of a bracelet is \$115. What is the cost of the materials? *(Section 1.2)*

12. PASTURE A 455-foot fence encloses a pasture. What is the length of each side of the pasture? *(Section 1.2)*

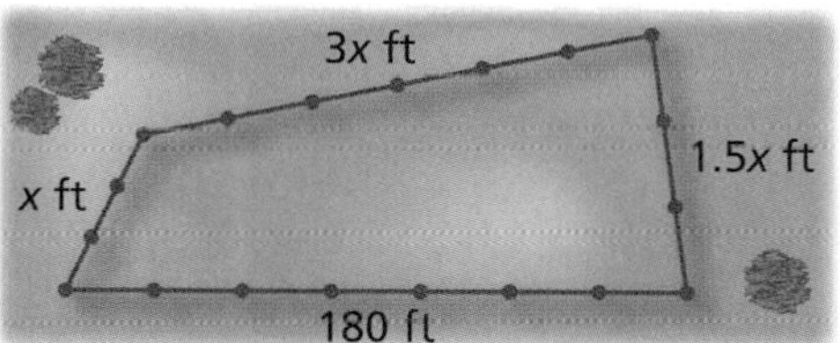

13. POSTERS A machine prints 230 movie posters each hour. Write and solve an equation to find the number of hours it takes the machine to print 1265 posters. *(Section 1.1)*

14. BASKETBALL Use the table to write and solve an equation to find the number of points p you need to score in the fourth game so that the mean number of points is 20? *(Section 1.2)*

Game	Points
1	25
2	15
3	18
4	p

1.3 Solving Equations with Variables on Both Sides

COMMON CORE STATE STANDARDS
A.CED.1
A.REI.1
A.REI.3

Essential Question How can you solve an equation that has variables on both sides?

1 ACTIVITY: Perimeter and Area

Work with a partner. Each figure has the unusual property that the value of its perimeter (in feet) is equal to the value of its area (in square feet).

- **Write an equation (value of perimeter = value of area) for each figure.**
- **Solve each equation for x.**
- **Use the value of x to find the perimeter and area of each figure.**
- **Check your solution by comparing the value of the perimeter and the value of the area of each figure.**

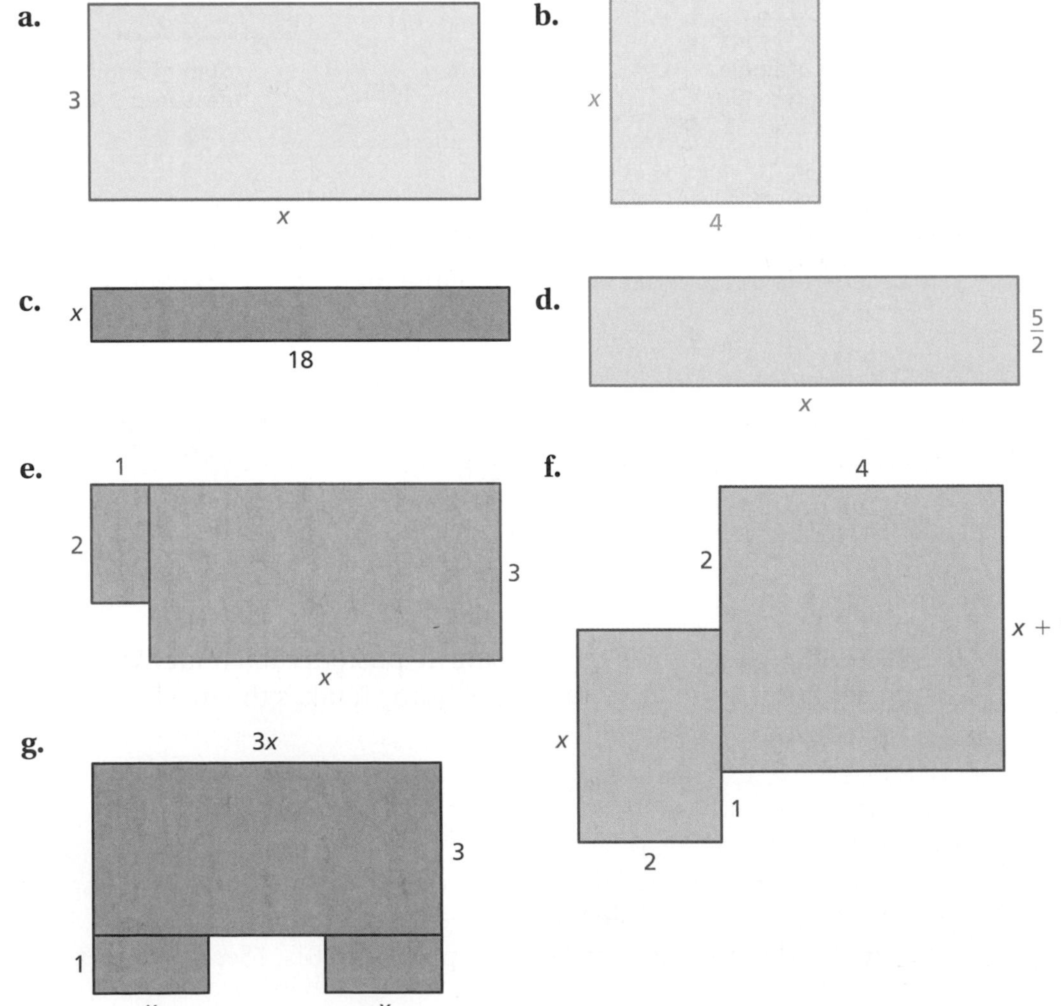

Laurie's Notes

Introduction

Standards for Mathematical Practice

- **MP2 Reason Abstractly and Quantitatively:** In these activities students have to create a symbolic representation of the problem using prior knowledge of particular geometric formulas. Once the problem has been represented, students reason abstractly as they manipulate the symbols. When finished, students check the reasonableness of their answers and attend to the units of measure in their answers.

Motivate

? "What balances with the cylinder? Explain."

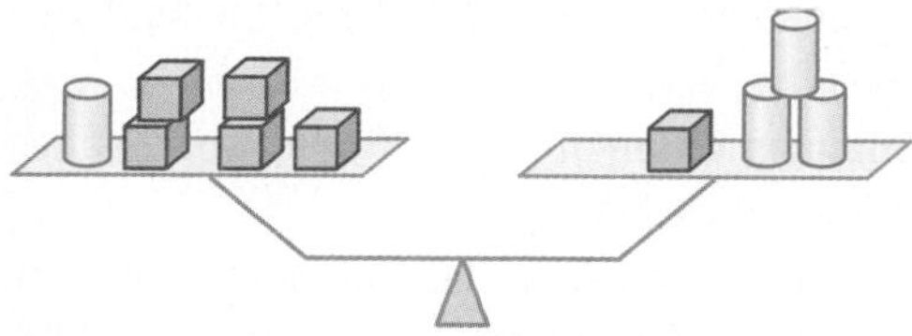

2 cubes; Remove one cube and one cylinder from each side. 2 cylinders balance with 4 cubes, so 1 cylinder would balance with 2 cubes.

- The balance problem is equivalent to $x + 5 = 3x + 1$, where x is a cylinder and the whole numbers represent cubes. This is an example of an equation with variables on both sides, the type students will solve today. Return to this equation at the end of class.
- If students are familiar with algebra tiles, you can model the problem using the tiles. The cylinder is replaced with an x-tile and the cubes are replaced with unit tiles.

Activity Notes

Activity 1

- Discuss with students the general concept of what it means to measure the attributes of a two-dimensional figure. In other words, what is the difference between a rectangle's perimeter and a rectangle's area? What type of units are used to measure each? linear units for perimeter and square units for area
- **FYI:** Be sure to make it clear that the directions are saying that perimeter and area are not the same, but their values are equal. For example, a square that measures 4 centimeters on each edge has a perimeter of 16 centimeters and an area of 16 square centimeters. The value (16) is the same, but the units of measure are not.

? Before students begin, ask a few review questions.
  - "How do you find the perimeter and the area of a rectangle?" $P = 2\ell + 2w$ and $A = \ell w$
  - "How do you find the perimeter and the area of a composite figure?" Listen for students' understanding that perimeter is the sum of all of the sides. The area is found in parts and then added together.
- Have a few groups share their work at the board, particularly for part (d), fractions, and part (g), algebraic expressions.

Common Core State Standards

A.CED.1 Create equations and inequalities in one variable and use them to solve problems.
A.REI.1 Explain each step in solving a simple equation . . .. Construct a viable argument to justify a solution method.
A.REI.3 Solve linear equations . . . in one variable, including equations with coefficients represented by letters.

Previous Learning

Students should know common formulas for perimeter, area, surface area, and volume.

Start Thinking! and Warm Up

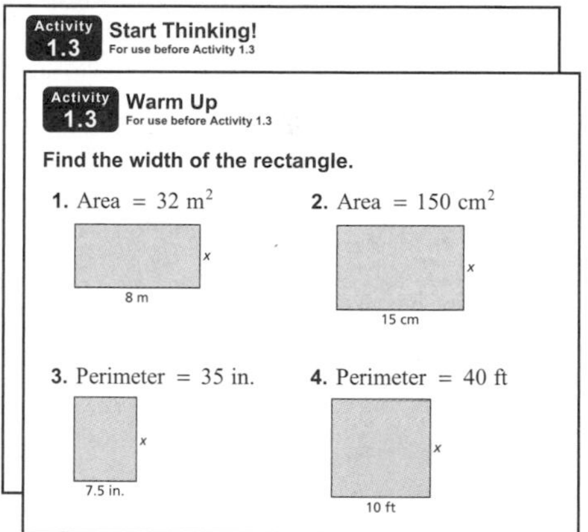

Activity 1.3 Start Thinking! For use before Activity 1.3

Activity 1.3 Warm Up For use before Activity 1.3

Find the width of the rectangle.

1. Area $= 32\text{ m}^2$ (8 m by x)
2. Area $= 150\text{ cm}^2$ (15 cm by x)
3. Perimeter $= 35$ in. (7.5 in. by x)
4. Perimeter $= 40$ ft (10 ft by x)

1.3 Record and Practice Journal

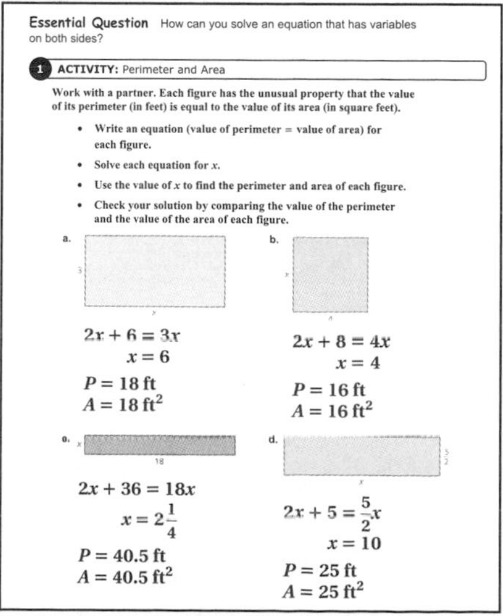

Essential Question How can you solve an equation that has variables on both sides?

1 ACTIVITY: Perimeter and Area

Work with a partner. Each figure has the unusual property that the value of its perimeter (in feet) is equal to the value of its area (in square feet).

- Write an equation (value of perimeter = value of area) for each figure.
- Solve each equation for x.
- Use the value of x to find the perimeter and area of each figure.
- Check your solution by comparing the value of the perimeter and the value of the area of each figure.

a. $2x + 6 = 3x$, $x = 6$, $P = 18$ ft, $A = 18\text{ ft}^2$

b. $2x + 8 = 4x$, $x = 4$, $P = 16$ ft, $A = 16\text{ ft}^2$

c. $2x + 36 = 18x$, $x = 2\frac{1}{4}$, $P = 40.5$ ft, $A = 40.5\text{ ft}^2$

d. $2x + 5 = \frac{5}{2}x$, $x = 10$, $P = 25$ ft, $A = 25\text{ ft}^2$

Differentiated Instruction

Auditory

Point out to students that skills used to solve equations in this lesson are the same skills they have used before. The goal is to isolate the variable on one side of the equation. Just as they used the Addition Property of Equality to remove a constant term from one side of the equation, they will use the same property to remove the variable term from one side of the equation.

1.3 Record and Practice Journal

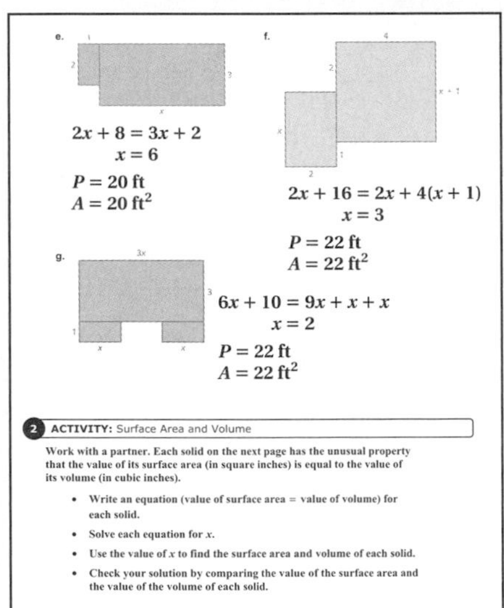

e. $2x + 8 = 3x + 2$
$x = 6$
$P = 20$ ft
$A = 20$ ft^2

f. $2x + 16 = 2x + 4(x + 1)$
$x = 3$
$P = 22$ ft
$A = 22$ ft^2

g. $6x + 10 = 9x + x + x$
$x = 2$
$P = 22$ ft
$A = 22$ ft^2

2 ACTIVITY: Surface Area and Volume

Work with a partner. Each solid on the next page has the unusual property that the value of its surface area (in square inches) is equal to the value of its volume (in cubic inches).

- Write an equation (value of surface area = value of volume) for each solid.
- Solve each equation for x.
- Use the value of x to find the surface area and volume of each solid.
- Check your solution by comparing the value of the surface area and the value of the volume of each solid.

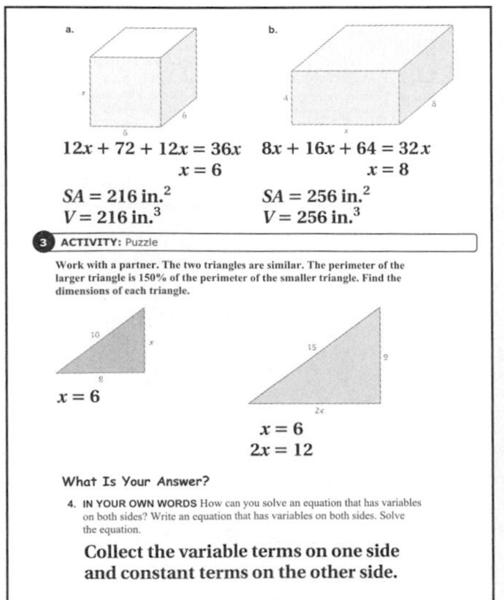

a. $12x + 72 + 12x = 36x$
$x = 6$
$SA = 216$ in.^{2}
$V = 216$ in.^{3}

b. $8x + 16x + 64 = 32x$
$x = 8$
$SA = 256$ in.^{2}
$V = 256$ in.^{3}

3 ACTIVITY: Puzzle

Work with a partner. The two triangles are similar. The perimeter of the larger triangle is 150% of the perimeter of the smaller triangle. Find the dimensions of each triangle.

$x = 6$

$x = 6$
$2x = 12$

What Is Your Answer?

4. IN YOUR OWN WORDS How can you solve an equation that has variables on both sides? Write an equation that has variables on both sides. Solve the equation.

Collect the variable terms on one side and constant terms on the other side.

Laurie's Notes

Activity 2

- **MP2:** This activity is similar to Activity 1. Students are asked to reason abstractly and quantitatively. Discuss the difference in units of measure (square and cubic units), and take time for students to look back at the process.

? "How do you find the surface area and volume of a rectangular prism?" $S = 2\ell w + 2\ell h + 2wh$ and $V = \ell wh$

- Students may guess that part (a) is a cube, suggesting $x = 6$. Ask students to verify their guess.

Activity 3

? "What are similar triangles?" Listen for an informal definition: same shape but not necessarily the same size; formally, corresponding sides are proportional and corresponding angles have the same measure (congruent).

- **MP1a Make Sense of Problems:** There are two different approaches students may take in solving this problem.
  - The first method is to solve the equation: 150% of the smaller triangle's perimeter is equal to the perimeter of the larger triangle.

    $$150\% \text{ of } (18 + x) = 24 + 2x$$
    $$1.5(18 + x) = 24 + 2x$$

    This method reviews decimal multiplication.
  - A second method is to use the definition of similar triangles to set up a proportion to solve for the missing sides.

    $$\frac{10}{15} = \frac{x}{9}$$

    Solve for x. Use this value to find the side labeled $2x$ in the larger triangle.

What Is Your Answer?

- **Neighbor Check:** Have students work independently and then have their neighbor check their work. Have students discuss any discrepancies.

Closure

- Describe how to solve $x + 5 = 3x + 1$. *Sample answer:* Subtract x from both sides, subtract 1 from both sides, and then divide both sides by 2.

Technology For the Teacher

The Dynamic Planning Tool
Editable Teacher's Resources at *BigIdeasMath.com*

2 ACTIVITY: Surface Area and Volume

Work with a partner. Each solid has the unusual property that the value of its surface area (in square inches) is equal to the value of its volume (in cubic inches).

- **Write an equation (value of surface area = value of volume) for each solid.**
- **Solve each equation for x.**
- **Use the value of x to find the surface area and volume of each solid.**
- **Check your solution by comparing the value of the surface area and the value of the volume of each solid.**

a.

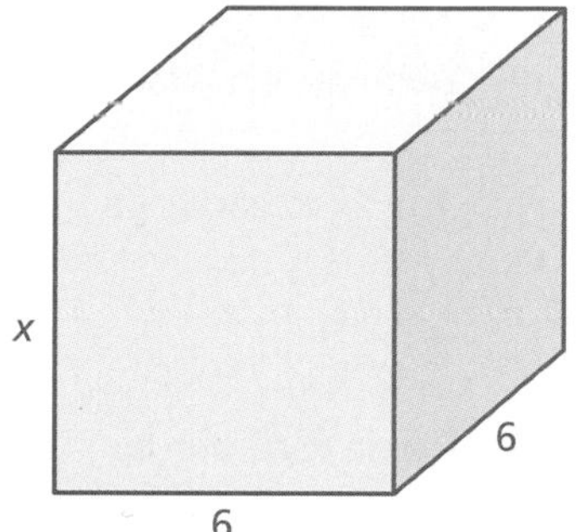

b.

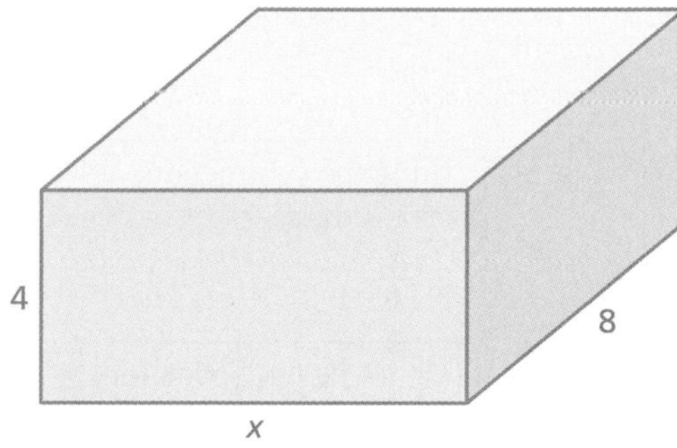

3 ACTIVITY: Puzzle

Work with a partner. The two triangles are similar. The perimeter of the larger triangle is 150% of the perimeter of the smaller triangle. Find the dimensions of each triangle.

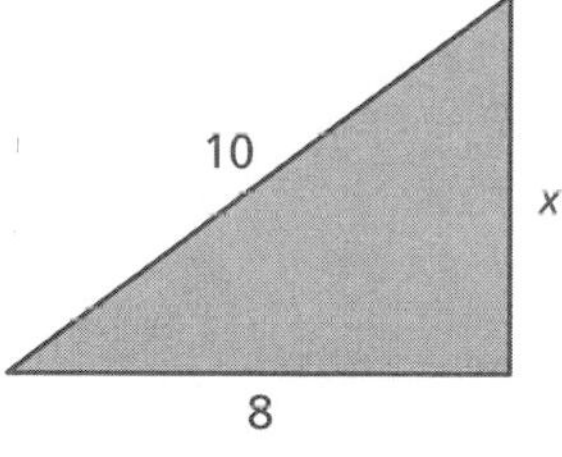

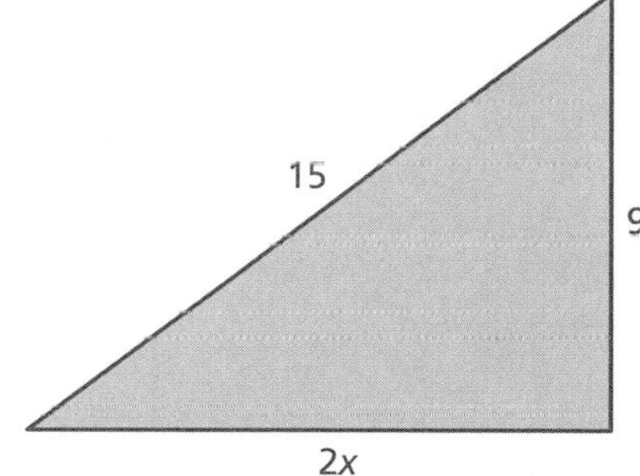

What Is Your Answer?

4. **IN YOUR OWN WORDS** How can you solve an equation that has variables on both sides? Write an equation that has variables on both sides. Solve the equation.

Practice Use what you learned about solving equations with variables on both sides to complete Exercises 3–5 on page 22.

1.3 Lesson

Solving Equations with Variables on Both Sides

To solve equations with variables on both sides, collect the variable terms on one side and the constant terms on the other side.

EXAMPLE 1 Solving an Equation with Variables on Both Sides

Solve $15 - 2x = -7x$. Check your solution.

$15 - 2x = -7x$ Write the equation.

Undo the subtraction. → $+ 2x \quad + 2x$ Add $2x$ to each side.

$15 = -5x$ Simplify.

Undo the multiplication. → $\frac{15}{-5} = \frac{-5x}{-5}$ Divide each side by -5.

$-3 = x$ Simplify.

The solution is $x = -3$.

Check

$15 - 2x = -7x$

$15 - 2(-3) \stackrel{?}{=} -7(-3)$

$21 = 21$ ✓

EXAMPLE 2 Solving Equations with Variables on Both Sides

a. Solve $3(5x + 2) = 15x$.

$3(5x + 2) = 15x$

$15x + 6 = 15x$

$- 15x \quad - 15x$

$6 = 0$ ✗

The equation $6 = 0$ is never true. So, the equation has no solution.

b. Solve $-2(4y + 1) = -8y - 2$.

$-2(4y + 1) = -8y - 2$

$-8y - 2 = -8y - 2$

$+ 8y \quad + 8y$

$-2 = -2$

The equation $-2 = -2$ is always true. So, the equation has infinitely many solutions.

Remember

When solving a linear equation that has no solution, you will obtain an equivalent equation that is not true for any value of x. When the equation has infinitely many solutions, you will obtain an equivalent equation that is true for all values of x.

On Your Own

Solve the equation. Check your solution, if possible.

1. $-3x = 2x + 19$
2. $4(1 - p) = -4p + 4$
3. $6m - m = \frac{5}{6}(6m - 10)$
4. $10k + 7 = -3 - 10k$

Now You're Ready
Exercises 6–14

Laurie's Notes

Introduction

Connect

- **Yesterday:** Students developed an intuitive understanding of solving equations with variables on both sides. (MP1a, MP2)
- **Today:** Students will solve equations with variables on both sides by using Properties of Equality and collecting variable terms on one side.

Motivate

? "How many of you have driven across the Mississippi River?"

- Share some information about the Mississippi River. It is the third longest river in North America, flowing 2350 miles from Lake Itasca in Minnesota to the Gulf of Mexico. The width of the river ranges from 20–30 feet at its narrowest to more than 11 miles at its widest.

Lesson Notes

Key Idea

- Discuss with students the vocabulary *variable term* and *constant term.*

? "In the expression $5x - 2 - 9x + y + 4$, what are the variable terms? Constant terms?" $5x$, $-9x$, and y; -2 and 4

- **Common Error:** Students forget to include the sign of the variable term.

Example 1

- **MP6 Attend to Precision:** The two properties used to solve this equation are the Addition Property of Equality and the Division Property of Equality. Continue to use property names as you explain the examples in this section.

Example 2

? "What operations are involved in part (a)?" Listen for 3 operations on the left and one operation on the right.

- As you work through part (a), remind students that the goal is to have the variables on one side and constants on the other side.
- **MP1a Make Sense of Problems** and **MP7 Look for and Make Use of Structure:** If time permits, solve part (a) differently. Begin by dividing both sides by 3. You want students to recognize that the left side has two factors, 3 and $(5x + 2)$, and dividing the left side by 3 yields the factor $5x + 2$. Dividing the right side by 3 yields $5x$. Finish solving the problem. Solving the problem in more than one way focuses attention on the structure of the equation and helps students make sense of the symbols.
- **Common Error:** When students distribute the -2 in part (b), they often forget to change the sign of the second term.
- Encourage students to check their solutions to part (b) by substituting any value of y into the equation. Try this with at least two different values of y to illustrate that the equation has more than one solution.

On Your Own

- **Neighbor Check:** Have students work independently and then have their neighbor check their work. Have students discuss any discrepancies.

Goal Today's lesson is solving equations with variables on both sides.

Start Thinking! and Warm Up

Lesson 1.3 Warm Up
For use before Lesson 1.3

Lesson 1.3 Start Thinking!
For use before Lesson 1.3

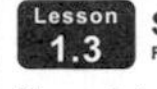

Try solving the equation $2x + 20 = 12x$ by first subtracting 20 from each side. What property allows you to do so? Does it help you get closer to finding a solution?

What is a better first step? Explain.

Extra Example 1

Solve $r = -5r + 18$. 3

Extra Example 2

Solve $6\left(1 + \frac{1}{2}x\right) = 2(x + 1)$. -4

On Your Own

1. $x = -3.8$
2. infinitely many solutions
3. no solution
4. -0.5

Extra Example 3

The legs of the right triangle have the same length. What is the area of the triangle? $1\frac{1}{8}$ square units

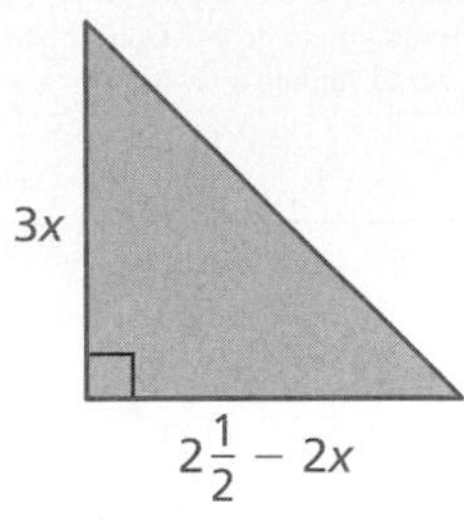

Extra Example 4

A boat travels 3 hours downstream at r miles per hour. On the return trip, the boat travels 5 miles per hour slower and takes 4 hours. What is the distance the boat travels each way? 60 mi

On Your Own

5. 36π
6. 50 mi

English Language Learners

Vocabulary

Remind English learners that *like terms* are terms with the same variables raised to the same power. As the number of terms in an equation increases, an important skill is to identify and combine like terms.

Laurie's Notes

Example 3

? "Define radius and diameter of a circle." Diameter is the distance across a circle through its center. Radius is the distance from a circle's center to any point on the circle and is equal to half the diameter.

- Solve the equation as shown.
- This example reviews the formula for the area of a circle. You might also ask about the formula for the circumference of a circle. $C = 2\pi r$

Example 4

- **MP1a Make Sense of Problems** and **MP4 Model with Mathematics:** In solving this real-life application students must explain the meaning of the problem to themselves and write a mathematical model that represents the situation algebraically. Students then solve and check their work to make sure that the results make sense.
- **MP6:** Remind students that they are using the Subtraction Property of Equality and the Division Property of Equality to solve this equation.

? "How do you find the distance traveled when you know the rate and time?" multiply; $d = rt$

- Discuss with students that the distance both ways is the same, a simple but not obvious fact.

? "If you travel 40 miles per hour for 2 hours, how far will you go? How about 40 miles per hour for a half hour?" 80 mi; 20 mi

? "How far do you travel at x miles per hour for 3 hours?" $3x$ mi

- Students need to read the time from the illustration. The rates for each direction are x and $(x + 2)$.
- Write and solve the equation as shown.

On Your Own

- **Think-Pair-Share:** Students should read each question independently and then work with a partner to answer the questions. When they have answered the questions, the pair should compare their answers with another group and discuss any discrepancies.

Closure

- **Exit Ticket:** Solve $6 - 2x = 4x - 9$. Check your solution. $x = 2.5$

Technology For the Teacher

The Dynamic Planning Tool
Editable Teacher's Resources at *BigIdeasMath.com*

EXAMPLE 3 Standardized Test Practice

The circles are identical. What is the area of each circle?

Ⓐ 2 Ⓑ 4 Ⓒ 16π Ⓓ 64π

The circles are identical, so the radius of each circle is the same.

$x + 2 = 2x$	Write an equation. The radius of the purple circle is $2x$.
$-x \quad -x$	Subtract x from each side.
$2 = x$	Simplify.

The area of each circle is $\pi r^2 = \pi(4)^2 = 16\pi$. So, the correct answer is Ⓒ.

EXAMPLE 4 Real-Life Application

A boat travels x miles per hour upstream on the Mississippi River. On the return trip, the boat travels 2 miles per hour faster. How far does the boat travel upstream?

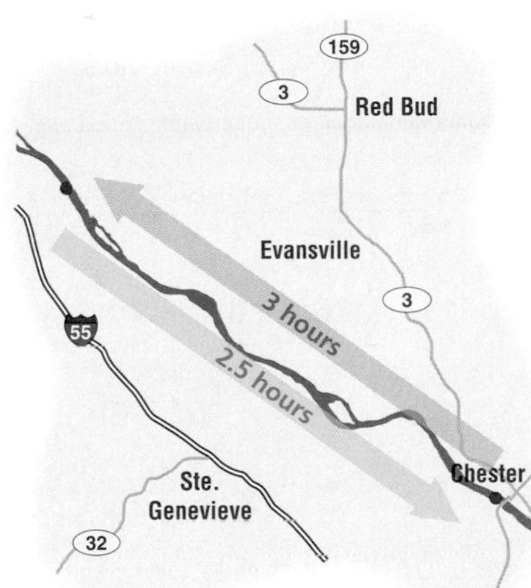

The speed of the boat on the return trip is $(x + 2)$ miles per hour.

Distance upstream = Distance of return trip

$3x = 2.5(x + 2)$	Write an equation.
$3x = 2.5x + 5$	Use Distributive Property.
$-2.5x \quad -2.5x$	Subtract $2.5x$ from each side.
$0.5x = 5$	Simplify.
$\frac{0.5x}{0.5} = \frac{5}{0.5}$	Divide each side by 0.5.
$x - 10$	Simplify.

The boat travels 10 miles per hour for 3 hours upstream. So, it travels 30 miles upstream.

On Your Own

5. **WHAT IF?** In Example 3, the diameter of the purple circle is $3x$. What is the area of each circle?
6. A boat travels x miles per hour from one island to another island in 2.5 hours. The boat travels 5 miles per hour faster on the return trip of 2 hours. What is the distance between the islands?

1.3 Exercises

Vocabulary and Concept Check

1. **WRITING** Is $x = 3$ a solution of the equation $3x - 5 = 4x - 9$? Explain.
2. **OPEN-ENDED** Write an equation that has variables on both sides and has a solution of -3.

Practice and Problem Solving

The value of the solid's surface area is equal to the value of the solid's volume. Find the value of x.

3. 

4. 

5. 

Solve the equation. Check your solution, if possible.

6. $m - 4 = 2m$
7. $3k - 1 = 7k + 2$
8. $-2x + 10 = -2(x + 5)$
9. $-24 - \frac{1}{8}p = \frac{3}{8}p$
10. $5(4w - 20) = 4(5w - 25)$
11. $\frac{3}{2}(16n + 3) = 24n$
12. $3(4z - 7) = -21 + 12z$
13. $0.1x = 0.2(x + 2)$
14. $\frac{1}{6}d + \frac{2}{3} = \frac{1}{4}(d - 2)$

15. **ERROR ANALYSIS** Describe and correct the error in solving the equation.

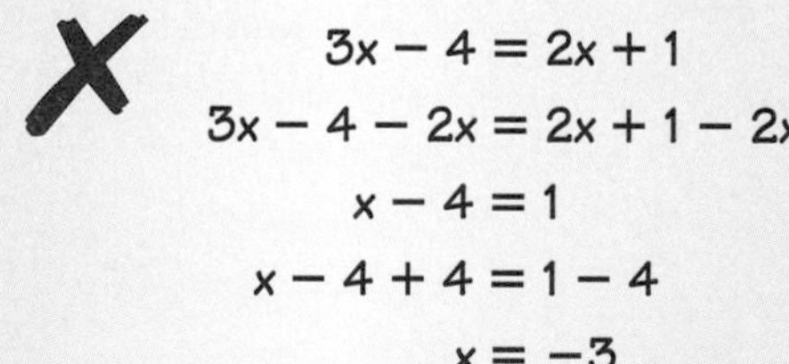

$$3x - 4 = 2x + 1$$
$$3x - 4 - 2x = 2x + 1 - 2x$$
$$x - 4 = 1$$
$$x - 4 + 4 = 1 - 4$$
$$x = -3$$

16. **TRAIL MIX** The equation $4.05p + 14.40 = 4.50(p + 3)$ represents the number p of pounds of peanuts you need to make trail mix. How many pounds of peanuts do you need for the trail mix?

17. **CARS** Write and solve an equation to find the number of miles you must drive to have the same cost for each of the car rentals.

$15 plus $0.50 per mile

$25 plus $0.25 per mile

Assignment Guide and Homework Check

Level	Assignment	Homework Check
Average	1–5, 9–19, 22, 27–30	9, 10, 12, 16, 18, 22
Advanced	1, 2, 12–15, 18–30	12, 18, 23, 25

For Your Information

- **Exercise 16** The equation represents a mixture problem in which peanuts are added to other ingredients, making trail mix. The equation shows that p pounds of peanuts that cost $4.05 per pound are added to other ingredients that cost a total of $14.40. This mixture creates $(p + 3)$ pounds of trail mix that costs $4.50 per pound.

Common Errors

- **Exercises 6–14** Students may perform the same operation instead of the opposite operation when trying to get the variable terms on the same side. Remind them that whenever a variable or number is moved from one side of the equal sign to the other, the opposite operation is used.
- **Exercises 6–14** Students may use the opposite operation when combining like terms on the same side of the equal sign. Remind them that the opposite operation is used only when moving the variable or number to the other side of the equation.
- **Exercises 16 and 17** Students may forget to write the units in their answers. Remind them that when units are given, the units need to be included in the answer.

1.3 Record and Practice Journal

Solve the equation. Check your solution.

1. $x + 16 = 9x$

   $x = 2$

2. $4y - 70 = 12y + 2$

   $y = -9$

3. $5(p + 6) = 8p$

   $p = 10$

4. $3(g - 7) = 2(10 + g)$

   $g = 41$

5. $1.8 + 7n = 9.5 - 4n$

   $n = 0.7$

6. $\frac{3}{7}w - 11 = -\frac{4}{7}w$

   $w = 11$

7. One movie club charges a $100 membership fee and $10 for each movie. Another club charges no membership fee but movies cost $15 each. Write and solve an equation to find the number of movies you need to buy for the cost of each movie club to be the same.

   $100 + 10x = 15x;\ x = 20$

8. Thirty percent of all the students in a school are in a play. All students except for 140 are in the play. How many students are in the school?

   200 students

Answer Presentation Tool

Vocabulary and Concept Check

1. no; When 3 is substituted for x, the left side simplifies to 4 and the right side simplifies to 3.
2. *Sample answer:* $4x + 1 = 3x - 2$

Practice and Problem Solving

3. $x = 13.2$ in.
4. $x = 10$ cm
5. $x = 7.5$ in.
6. $m = -4$
7. $k = -0.75$
8. no solution
9. $p = -48$
10. infinitely many solutions
11. no solution
12. infinitely many solutions
13. $x = -4$
14. $d = 14$
15. The 4 should have been added to the right side.

$$\begin{aligned} 3x - 4 &= 2x + 1 \\ 3x - 2x - 4 &= 2x + 1 - 2x \\ x - 4 &= 1 \\ x - 4 + 4 &= 1 + 4 \\ x &= 5 \end{aligned}$$

16. 2 lb
17. $15 + 0.5m = 25 + 0.25m$; 40 mi

Practice and Problem Solving

18. 3 units **19.** 7.5 units

20. 232 units

21. *Sample answer:*

a. $5x = 5x - 3$

b. $x + 4 = 2x - x + 4$

22. See *Taking Math Deeper.*

23. fractions; Because $\frac{1}{3}$ is hard to perform operations with when written as a decimal.

24. 10 mL **25.** 25 grams

26. square: 12 units
triangle: 10 units, 19 units, 19 units

Fair Game Review

27. 15.75 cm^3

28. 24 in.^{3}

29. about 153.86 ft^3

30. C

Mini-Assessment

Solve the equation.

1. $n - 4 = 3n + 6$ $n = -5$
2. $0.3(2w + 10) = 0.6w + 3$ infinitely many solutions
3. $-12p = 4(-3p + 6)$ no solution
4. $\frac{1}{3}v = -\frac{2}{3}\left(\frac{1}{2}v - 1\right)$ $v = 1$
5. The perimeter of the rectangle is equal to the perimeter of the square. What are the side lengths of each figure? rectangle: 4 units by 10 units; square: 7 units by 7 units

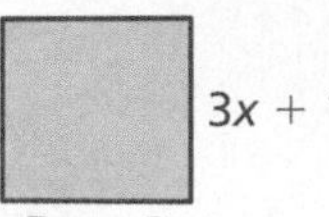

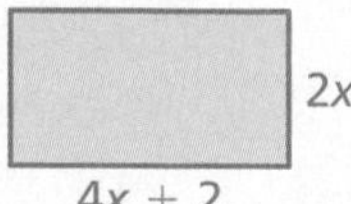

Taking Math Deeper

Exercise 22

This problem seems like it is easy, but it can actually be quite challenging.

Identify the key information in the table.

	Packing Material	Priority	Express
Box	\$2.25	\$2.50/lb	\$8.50/lb
Envelope	\$1.10	\$2.50/lb	\$8.50/lb

Write and solve an equation.

Let x = the weight of the DVD and packing material.

Cost of Mailing Box: $2.25 + 2.5x$

Cost of Mailing Envelope: $1.10 + 8.5x$

$$2.25 + 2.5x = 1.10 + 8.5x$$
$$1.15 = 6x$$
$$0.19 \approx x$$

Answer the question.

The weight of the DVD and packing material is about 0.19 pound, or about 3 ounces.

Project

Postage for special types of mail, such as priority mail, is determined by the weight of the package and the distance it needs to travel. Find the cost of sending a 15-ounce package from your house to Los Angeles, Washington, D.C., and Albuquerque.

Reteaching and Enrichment Strategies

If students need help. . .	If students got it. . .
Resources by Chapter • Practice A and Practice B • Puzzle Time Record and Practice Journal Practice Differentiating the Lesson Lesson Tutorials Skills Review Handbook	Resources by Chapter • Enrichment and Extension • School-to-Work Start the next section

A polygon is *regular* if each of its sides has the same length. Find the perimeter of the regular polygon.

18.

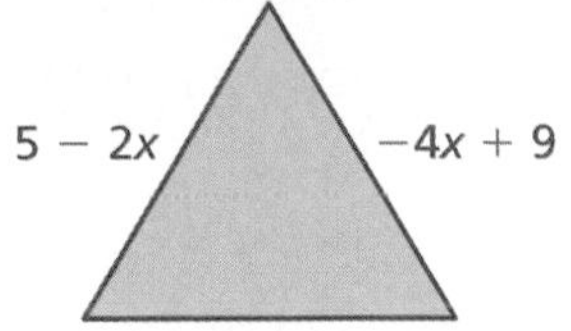

19.

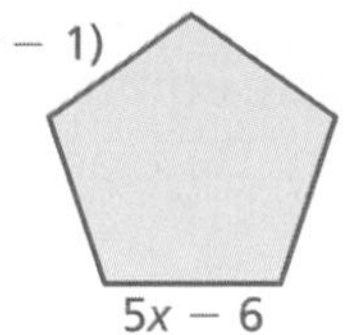

20. $x + 7$; $\frac{4}{3}x - \frac{1}{3}$

21. WRITING Write a linear equation that has (a) no solution and (b) infinitely many solutions. Justify your answers.

22. PRECISION The cost of mailing a DVD in an envelope by express mail is equal to the cost of mailing a DVD in a box by priority mail. What is the weight of the DVD with its packing material?

	Packing Material	Priority Mail	Express Mail
Box	\$2.25	\$2.50 per lb	\$8.50 per lb
Envelope	\$1.10	\$2.50 per lb	\$8.50 per lb

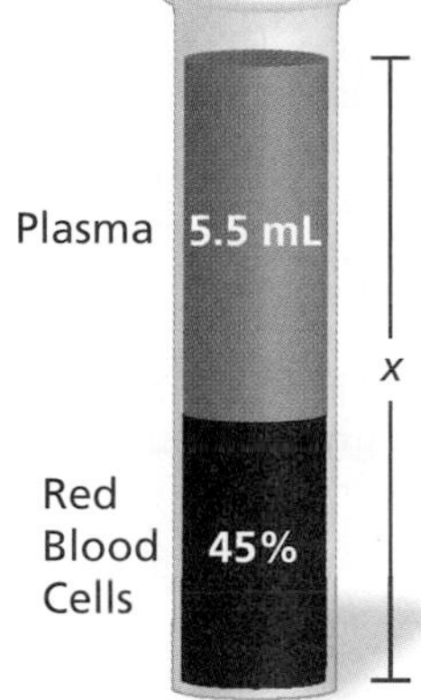

23. STRUCTURE Would you solve the equation $0.25x + 7 = \frac{1}{3}x - 8$ using fractions or decimals? Explain.

24. BLOOD SAMPLE The amount of red blood cells in a blood sample is equal to the total amount in the sample minus the amount of plasma. What is the total amount x of blood drawn?

25. NUTRITION One serving of oatmeal provides 16% of the fiber you need daily. You must get the remaining 21 grams of fiber from other sources. How many grams of fiber should you consume daily?

26. Geometry The perimeter of the square is equal to the perimeter of the triangle. What are the side lengths of each figure?

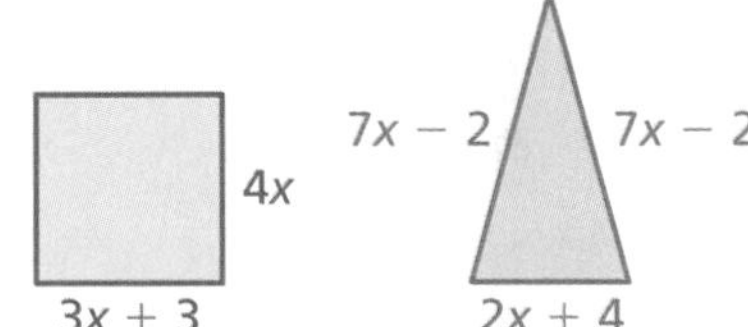

Fair Game Review What you learned in previous grades & lessons

Find the volume of the figure. Use 3.14 for π. *(Skills Review Handbook)*

27.

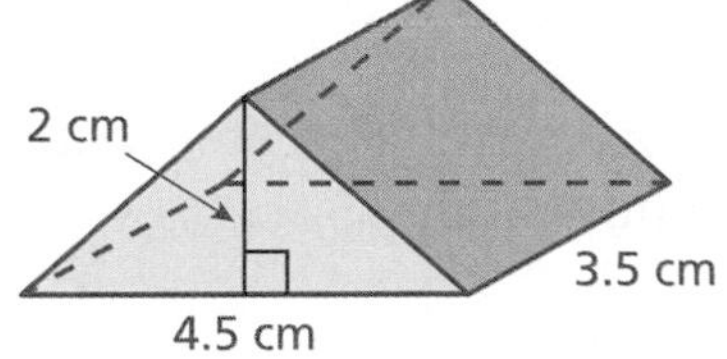

28.

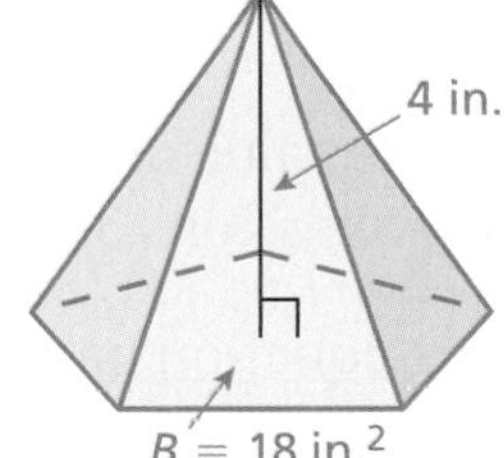

29.

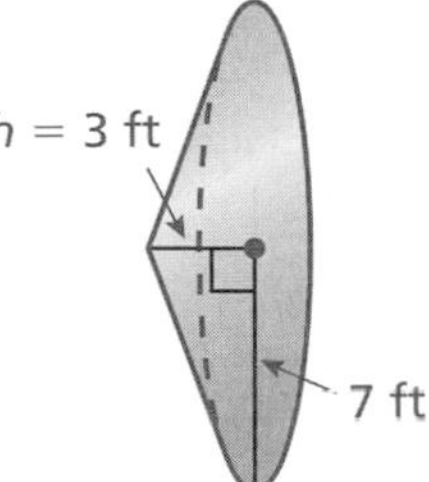

30. MULTIPLE CHOICE A car travels 480 miles on 15 gallons of gasoline. How many miles does the car travel per gallon? *(Section 1.1)*

(A) 28 mi/gal (B) 30 mi/gal (C) 32 mi/gal (D) 35 mi/gal

1.3b Solving Absolute Value Equations

Key Vocabulary
absolute value equation, *p. 24*

An **absolute value equation** is an equation that contains an absolute value expression. Here are three examples.

$$|x| = 2 \qquad |x + 1| = 5 \qquad 3|2x + 1| = 6$$

You can solve these types of equations by solving two related linear equations.

Key Idea

Solving Absolute Value Equations

To solve $|ax + b| = c$ for $c \geq 0$, solve the related linear equations

$$ax + b = c \quad \text{or} \quad ax + b = -c.$$

EXAMPLE 1 Solving Absolute Value Equations

a. Solve $|x - 4| = 6$. Graph the solutions.

Write two related linear equations for $|x - 4| = 6$. Then solve.

$x - 4 = 6$	*or*	$x - 4 = -6$	Write related linear equations.
$\underline{+4 \quad +4}$		$\underline{+4 \quad +4}$	Add 4 to each side.
$x = 10$	*or*	$x = -2$	Simplify.

The solutions are $x = -2$ and $x = 10$.

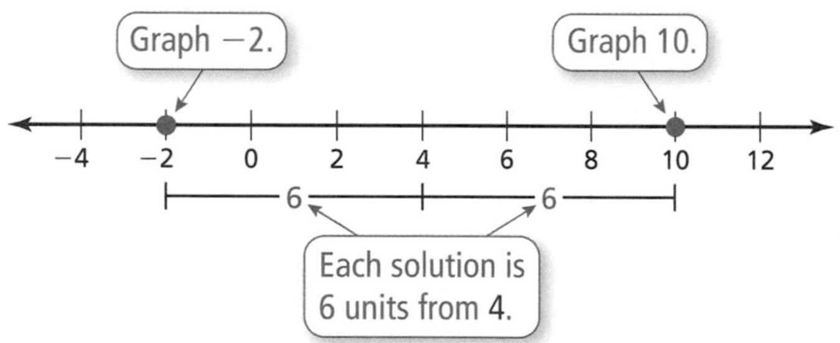

Check

$|x - 4| = 6$

$|-2 - 4| \stackrel{?}{=} 6$

$|-6| \stackrel{?}{=} 6$

$6 = 6$ ✓

$|x - 4| = 6$

$|10 - 4| \stackrel{?}{=} 6$

$|6| \stackrel{?}{=} 6$

$6 = 6$ ✓

b. Solve $|3x + 1| = -5$.

The absolute value of an expression must be greater than or equal to 0. The expression $|3x + 1|$ cannot equal -5.

So, the equation has no solution.

Practice

Solve the equation. Graph the solutions, if possible.

1. $|x| = 10$
2. $|x - 1| = 4$
3. $|3 + x| = -3$
4. $|4x - 5| = 8$

Laurie's Notes

Introduction

Connect

- **Yesterday:** Students solved equations with variables on both sides by using Properties of Equality and collecting variable terms on one side. (MP1a, MP4, MP6, MP7)
- **Today:** Students will solve absolute value equations.

Motivate

? "Can you think of an equation that has more than one answer?" Students may suggest something like $x + 4 = 4 + x$, which has infinitely many solutions.

? "What does $|5|$ mean?" absolute value of 5 "What does $|-5|$ mean?" absolute value of -5

? "What is the solution of $|x| = 5$?" 5 or -5

Lesson Notes

Discuss

- **MP1a Make Sense of Problems** and **MP4 Model with Mathematics:** Students used absolute value last year when they began their work with integers. This lesson extends the concept of absolute value to include a simple expression inside the absolute value bars, such as $|x + 1| = 4$ or $|2x + 1| = 4$. The approach to all of these problems is the same. Remind them of the number line model. For example:

Solve $|x| = 4$.

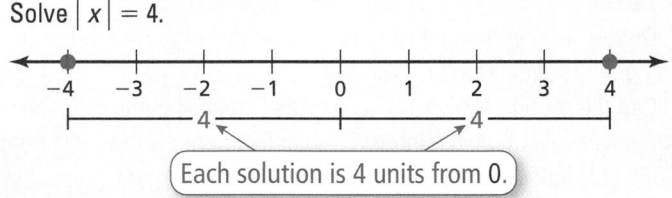

Each solution is 4 units from 0.

Key Idea

- Write the Key Idea on the board.
- **Teaching Tip:** When solving an absolute value equation, such as $|2x + 1| = 4$, place your fingers over the expression inside the absolute value bars and say, "We are looking for a quantity whose absolute value is 4. What is under my fingers is either equal to 4 or -4." That leads directly into writing the two equations that are solved. Covering up the expression helps focus student attention on the process, without them getting overwhelmed by symbols.

Example 1

- Write part (a) and say, "Either $x - 4 = 6$ or $x - 4 = -6$."
- **Common Error:** Students may quickly assume that because 10 is one solution, -10 is the second instead of -2. It is always important to remind students to check their work.

? Write part (b) and ask, "We are looking for a quantity whose absolute value is -5. What could the quantity be?" There is no value whose absolute value is -5.

Today's lesson is solving **absolute value equations.**

Start Thinking! and Warm Up

Lesson 1.3b Warm Up
For use before Lesson 1.3b

Lesson 1.3b Start Thinking!
For use before Lesson 1.3b

Recall that absolute value is the distance of a number from 0. The absolute value of -5 is $0 - (-5) = 5$. The absolute value of 5 is $5 - 0 = 5$.

What is the distance of 2 from 3? What is the distance of 4 from 3? Explain how these distances relate to the concept of absolute value.

Extra Example 1

a. Solve $|x + 6| = 2$. Graph the solutions. $x = -4$ or $x = -8$

b. Solve $|2x - 7| = -3$. The equation has no solution.

Practice

1. $x = 10$ or $x = -10$

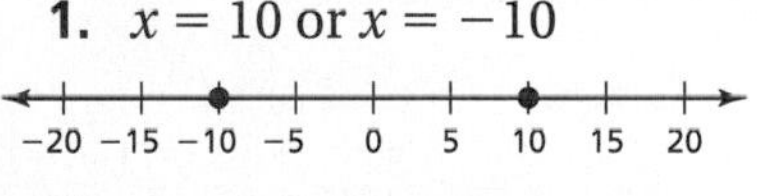

2. $x = 5$ or $x = -3$

3. no solution

4. $x = \frac{13}{4}$ or $x = -\frac{3}{4}$

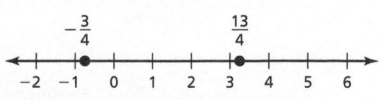

Record and Practice Journal Practice

See Additional Answers.

Extra Example 2

Solve $3|x - 2| + 9 = 15$.
$x = 0$ or $x = 4$

Extra Example 3

For a homework assignment, the minimum length of an essay is 2 pages. The maximum length is 6 pages. Write an absolute value equation that has these minimum and maximum lengths as its solutions. $|x - 4| = 2$

Practice

5. $x = 6$ or $x = -2$
6. $x = -\frac{3}{2}$ or $x = -\frac{11}{2}$
7. $x = 1$ or $x = -\frac{3}{5}$
8. *Sample answer:* $|x - 10| = 5$
9. *Sample answer:* $|x - 24| = 8$

Laurie's Notes

Example 1 (continued)

- Point out that in the Key Idea, a requirement for there to be a solution of an absolute value equation is $c \geq 0$, because the absolute value of a number cannot be negative. In part (b), $c < 0$. So, there is no solution.

Practice

- **Neighbor Check:** Have students work independently and then have their neighbor check their work. Have students discuss any discrepancies.

Example 2

- **MP7 Look for and Make Use of Structure:** You want students to recognize the similarity in solving a linear equation such as $x - 10 = -4$ and an absolute value equation such as $|3x + 9| - 10 = -4$. The first step is the same.
- **Common Error:** Students may assume there is no solution because of the negative quantity on the right side of the equation, like Example 1(b). However, when the absolute value expression is isolated, the right side of the equation becomes positive.
- **MP6 Attend to Precision:** The three properties used to solve this equation are the Addition Property of Equality, the Subtraction Property of Equality, and the Division Property of Equality. Remind students of the property names as you explain the steps in solving this equation.

Example 3

- Read through the problem and draw a number line.
- **?** "On the number line, where are the solutions located that represent how long the routine can be?" between 4 and 5
- **?** "Can the routine be exactly 4 minutes?" yes; 4 minutes is the minimum. "exactly 5 minutes?" yes; 5 minutes is the maximum.
- **?** "What is halfway between 4 and 5?" 4.5
- Continue to work through the problem as shown.
- Remind students to always check solutions.

Practice

- **Neighbor Check:** Have students work independently and then have their neighbor check their work. Have students discuss any discrepancies.

Closure

- Your friend says that the solutions of $|x - 2| = 9$ are $x = 11$ and $x = -11$. Explain your friend's error.

Technology For the Teacher

The Dynamic Planning Tool
Editable Teacher's Resources at *BigIdeasMath.com*

EXAMPLE 2 Solving an Absolute Value Equation

Solve $|3x + 9| - 10 = -4$.

$\lvert 3x + 9\rvert - 10 = -4$	Write the equation.
$+10 \quad +10$	Add 10 to each side.
$\lvert 3x + 9\rvert = 6$	Simplify.

Write two related linear equations for $|3x + 9| = 6$. Then solve.

$3x + 9 = 6$	*or*	$3x + 9 = -6$	Write related linear equations.
$-9 \quad -9$		$-9 \quad -9$	Subtract 9 from each side.
$3x = -3$	*or*	$3x = -15$	Simplify.
$\frac{3x}{3} = \frac{-3}{3}$		$\frac{3x}{3} = \frac{-15}{3}$	Divide each side by 3.
$x = -1$	*or*	$x = -5$	Simplify.

EXAMPLE 3 Real-Life Application

In a cheerleading competition, the minimum length of a routine is 4 minutes. The maximum length of a routine is 5 minutes. Write an absolute value equation that has these minimum and maximum lengths as its solutions.

Step 1: Graph the minimum and maximum lengths on a number line. Then find the point that is halfway between the lengths.

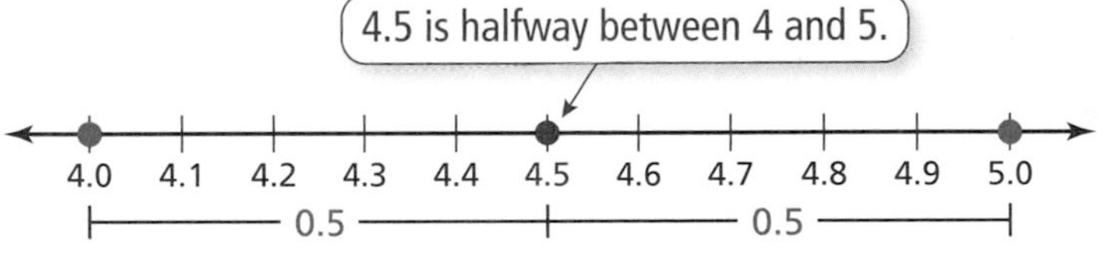

Step 2: Write the equation. Each solution is 0.5 unit from 4.5.

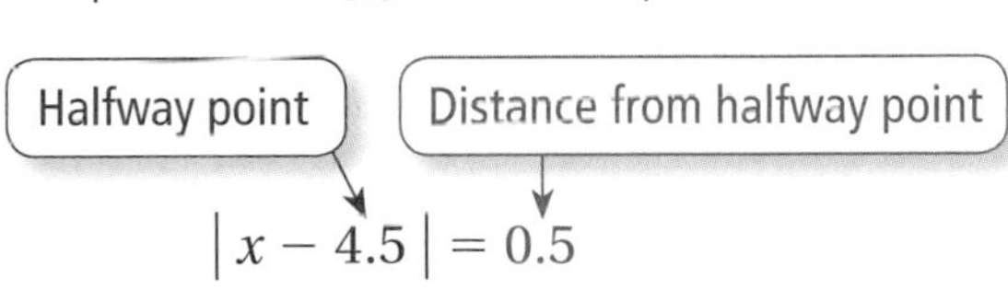

∴ The equation is $|x - 4.5| = 0.5$.

Practice

Solve the equation. Check your solutions.

5. $|x - 2| + 5 = 9$ **6.** $4|2x + 7| = 16$ **7.** $-2|5x - 1| - 3 = -11$

8. WRITING Write an absolute value equation that has 5 and 15 as its solutions.

9. POEM CONTEST For a poem contest, the minimum length of a poem is 16 lines. The maximum length is 32 lines. Write an absolute value equation that has these minimum and maximum lengths as its solutions.

1.4 Rewriting Equations and Formulas

COMMON CORE STATE STANDARDS

A.CED.4

Essential Question How can you use a formula for one measurement to write a formula for a different measurement?

1 ACTIVITY: Using Perimeter and Area Formulas

Work with a partner.

a.
- Write a formula for the perimeter P of a rectangle.
- Solve the formula for w.
- Use the new formula to find the width of the rectangle.

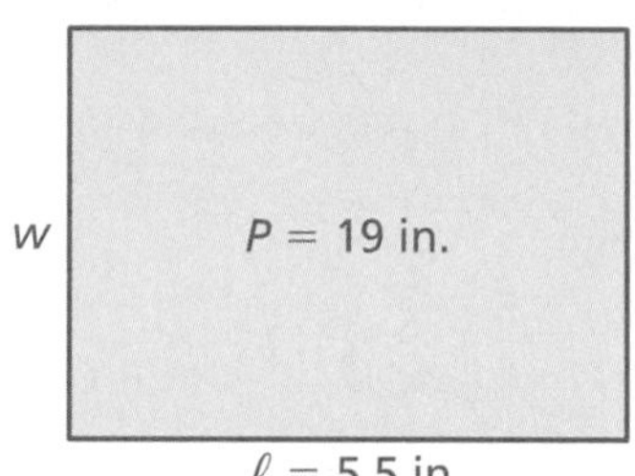

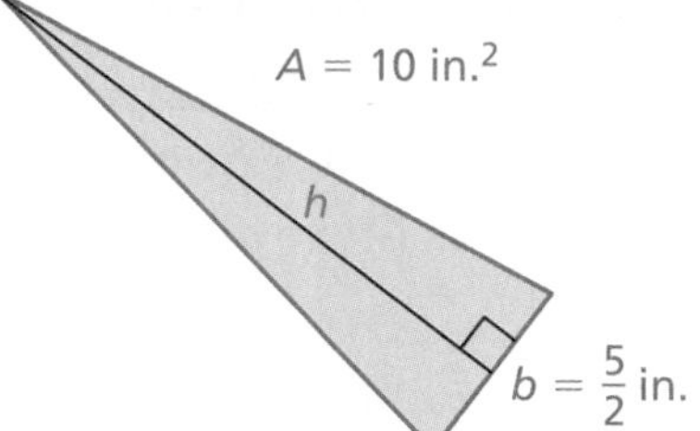

b.
- Write a formula for the area A of a triangle.
- Solve the formula for h.
- Use the new formula to find the height of the triangle.

c.
- Write a formula for the circumference C of a circle.
- Solve the formula for r.
- Use the new formula to find the radius of the circle.

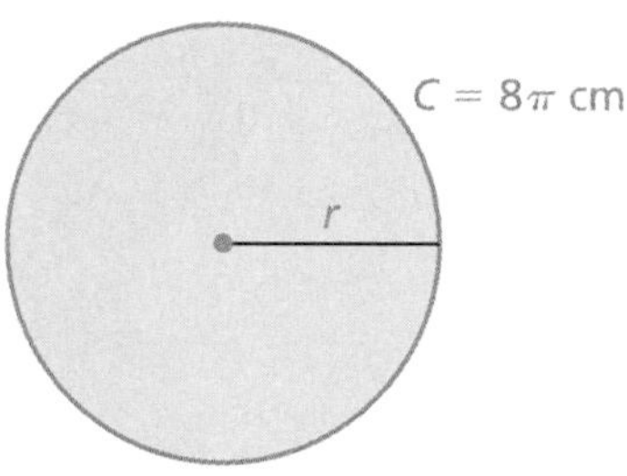

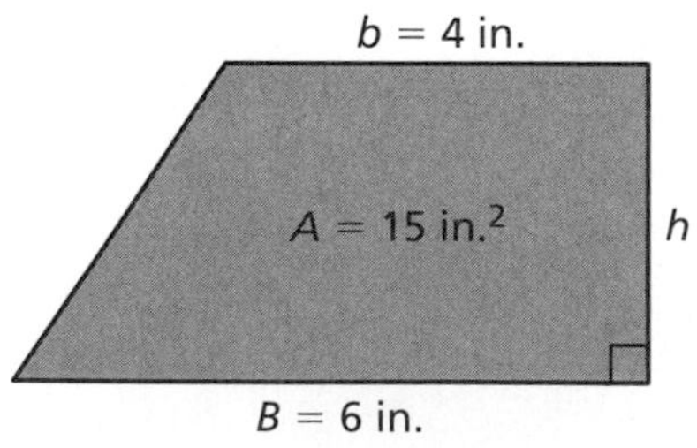

d.
- Write a formula for the area A of a trapezoid.
- Solve the formula for h.
- Use the new formula to find the height of the trapezoid.

e.
- Write a formula for the area A of a parallelogram.
- Solve the formula for h.
- Use the new formula to find the height of the parallelogram.

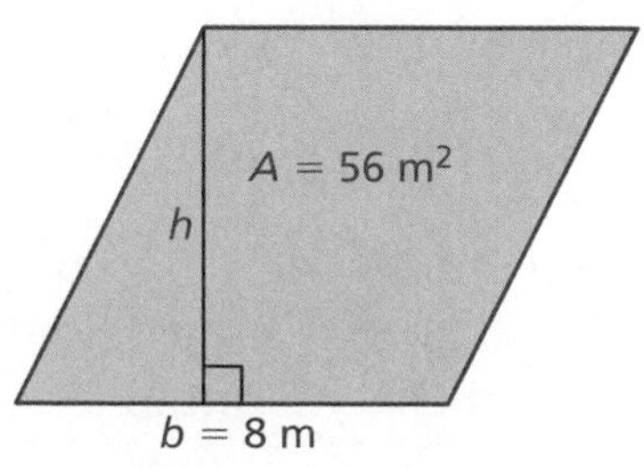

Laurie's Notes

Introduction

Standards for Mathematical Practice

- **MP1a Make Sense of Problems** and **MP3a Construct Viable Arguments:** The goal of this activity is for students to discuss, explain, and demonstrate how a literal equation may be solved using multiple approaches. For example, when solving $A = \frac{1}{2}bh$ for h, the first step could be dividing both sides by $\frac{1}{2}$, multiplying both sides by 2, or dividing both sides by $\frac{1}{2}b$. It is important to spend time discussing the multiple approaches.
- **MP7 Look for and Make Use of Structure:** Students should look at literal equations such as $A = \frac{1}{2}h(b + B)$ and recognize that there are 3 factors on the right side of the equation. Solving for h involves dividing both sides of the equation by the two remaining factors, $\frac{1}{2}$ and $(b + B)$.

Motivate

- **Preparation:** Make a set of formula cards. My set is a collection of five cards for each shape: the labeled diagram, the two measurements, and the two formulas being found.

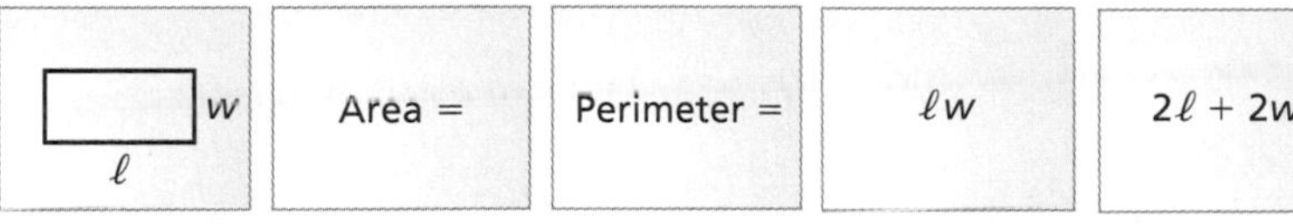

- Depending upon the number of students in your class, use some or all of the cards. Pass out the cards and have students form groups matching all 5 cards for the shape. When all of the matches have been made, ask each group to read their formulas aloud.

Activity Notes

Activity 1

- Solving literal equations can be one of the most challenging skills for students. Model a problem, such as solving $A = \ell w$ for width.
- Fractional coefficients can also be a challenge, so model an additional problem, such as solving $A = \frac{1}{2}xy$ for y. First, multiply both sides by 2. Then divide both sides by x.
- **Teaching Tip:** You may find that students are substituting the known values of the variables and then solving the equation, instead of solving the equation and then substituting.
- **Connection:** The reason for solving for w in part (a) is that the formula can be used to find the width of any rectangle given the perimeter and length. It is a general solution that can be reused.
- **Teaching Tip:** After 2 or more groups have correctly solved part (a), have a volunteer write the solution on the board and explain the approach used.
- For parts (b) and (d), suggest students start by multiplying both sides by the reciprocal of $\frac{1}{2}$. In part (c), 2π is a number and can be manipulated as such, so divide both sides by 2π.

Common Core State Standards

A.CED.4 Rearrange formulas to highlight a quantity of interest, using the same reasoning as in solving equations.

Previous Learning

Students should know the common formulas for area, perimeter, and volume.

Activity Materials
Introduction
• formula cards (index cards)

Start Thinking! and Warm Up

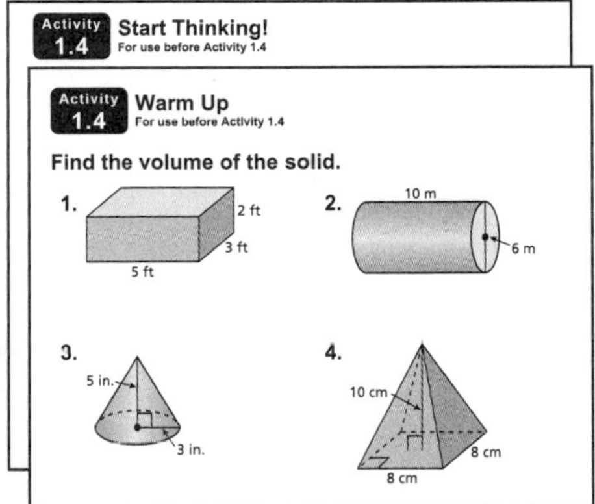

Activity 1.4 Start Thinking! For use before Activity 1.4

Activity 1.4 Warm Up For use before Activity 1.4

Find the volume of the solid.

1.
2.
3.
4.

1.4 Record and Practice Journal

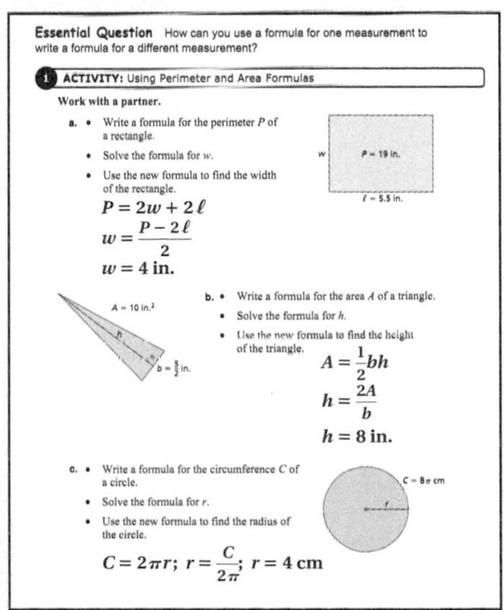

Essential Question How can you use a formula for one measurement to write a formula for a different measurement?

1 ACTIVITY: Using Perimeter and Area Formulas

Work with a partner.

a. • Write a formula for the perimeter P of a rectangle.
• Solve the formula for w.
• Use the new formula to find the width of the rectangle.

$P = 2w + 2\ell$

$w = \frac{P - 2\ell}{2}$

$w = 4$ in.

b. • Write a formula for the area A of a triangle.
• Solve the formula for h.
• Use the new formula to find the height of the triangle.

$A = \frac{1}{2}bh$

$h = \frac{2A}{b}$

$h = 8$ in.

c. • Write a formula for the circumference C of a circle.
• Solve the formula for r.
• Use the new formula to find the radius of the circle.

$C = 2\pi r$; $r = \frac{C}{2\pi}$; $r = 4$ cm

Differentiated Instruction

Kinesthetic

Have kinesthetic learners model the areas of the polygons on grid paper. Then compare their answers with the answers found using the area formulas.

1.4 Record and Practice Journal

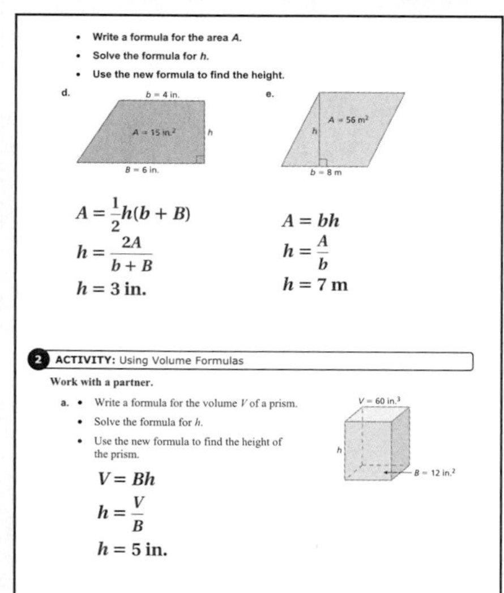

- Write a formula for the area A.
- Solve the formula for h.
- Use the new formula to find the height.

d. $b = 4$ in., $A = 15$ in.², $B = 6$ in.

$A = \frac{1}{2}h(b + B)$

$h = \frac{2A}{b + B}$

$h = 3$ in.

e. $A = 56$ m², $b = 8$ m

$A = bh$

$h = \frac{A}{b}$

$h = 7$ m

2 ACTIVITY: Using Volume Formulas

Work with a partner.

a. • Write a formula for the volume V of a prism.
• Solve the formula for h.
• Use the new formula to find the height of the prism.

$V = 60$ in.³, $B = 12$ in.²

$V = Bh$

$h = \frac{V}{B}$

$h = 5$ in.

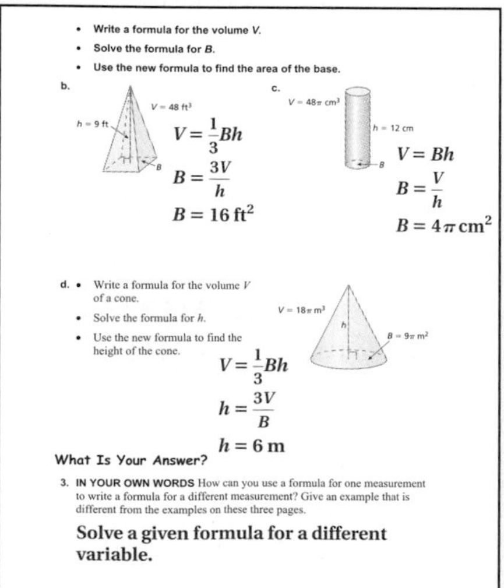

- Write a formula for the volume V.
- Solve the formula for B.
- Use the new formula to find the area of the base.

b. $V = 48$ ft³, $h = 9$ ft

$V = \frac{1}{3}Bh$

$B = \frac{3V}{h}$

$B = 16$ ft²

c. $V = 48\pi$ cm³, $h = 12$ cm

$V = Bh$

$B = \frac{V}{h}$

$B = 4\pi$ cm²

d. • Write a formula for the volume V of a cone.
• Solve the formula for h.
• Use the new formula to find the height of the cone.

$V = 18\pi$ m³, $B = 9\pi$ m²

$V = \frac{1}{3}Bh$

$h = \frac{3V}{B}$

$h = 6$ m

What Is Your Answer?

3. IN YOUR OWN WORDS How can you use a formula for one measurement to write a formula for a different measurement? Give an example that is different from the examples on these three pages.

Solve a given formula for a different variable.

Laurie's Notes

Activity 2

- This activity is similar to Activity 1, where students worked with perimeter and area formulas. In Activity 2, students will work with volume formulas.
- **MP7:** Note that all of the diagrams use B for the area of the base instead of having students use specific area formulas. Using this approach, the volume formulas for parts (a) and (c) are the same ($V = Bh$) and the volume formulas for parts (b) and (d) are the same $\left(V = \frac{1}{3}Bh\right)$. This helps students to recall that structurally the prism and the cylinder are similar, and that structurally the pyramid and the cone are similar.
- Use the *Teaching Tips* from Activity 1. Have students work in groups of 3 or 4 and post a correct solution on the board after 2 or more groups have been successful.
- **MP7:** For parts (b) and (d), suggest that students start by multiplying both sides by the reciprocal of $\frac{1}{3}$. Point out that multiplying by 3 and dividing by $\frac{1}{3}$ are equivalent.

What Is Your Answer?

- **Neighbor Check:** Have students work independently and then have their neighbor check their work. Have students discuss any discrepancies.

Closure

- Describe how to solve $d = rt$ for t. *Sample answer:* Divide both sides of the equation by r.

Technology For the Teacher

The Dynamic Planning Tool

Editable Teacher's Resources at *BigIdeasMath.com*

2 ACTIVITY: Using Volume Formulas

Work with a partner.

a. • Write a formula for the volume V of a prism.
• Solve the formula for h.
• Use the new formula to find the height of the prism.

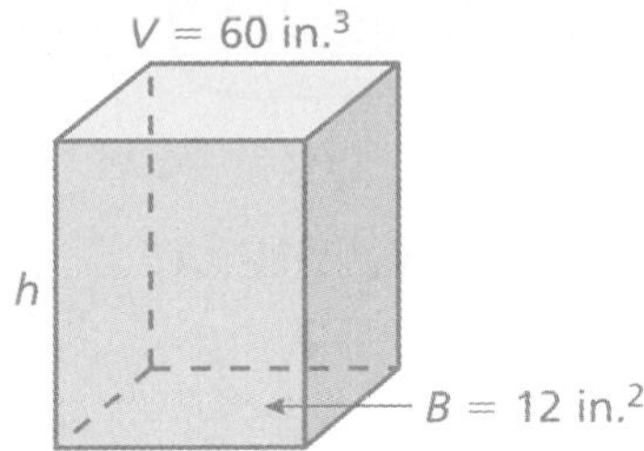

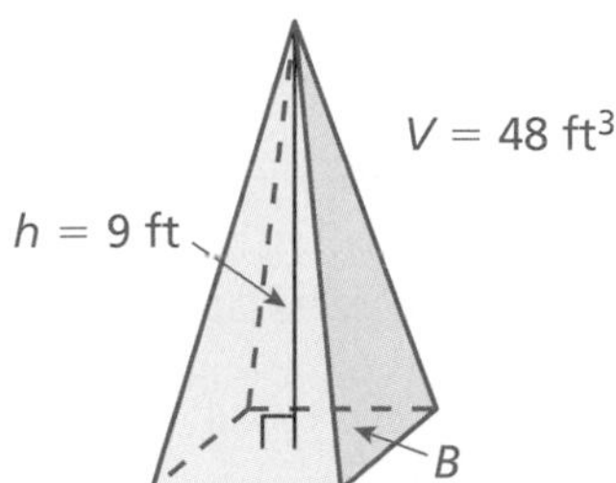

b. • Write a formula for the volume V of a pyramid.
• Solve the formula for B.
• Use the new formula to find the area of the base of the pyramid.

c. • Write a formula for the volume V of a cylinder.
• Solve the formula for B.
• Use the new formula to find the area of the base of the cylinder.

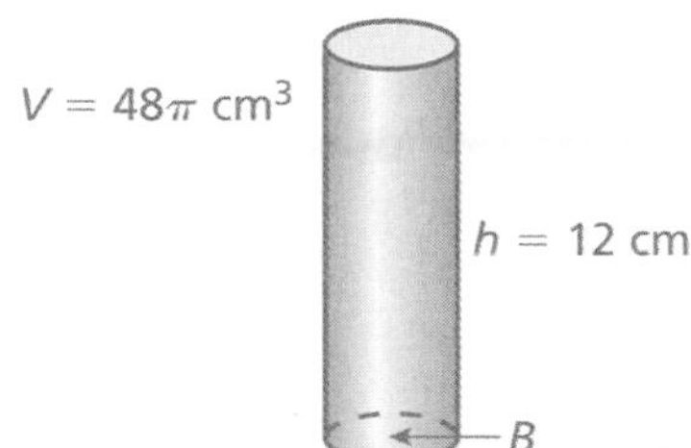

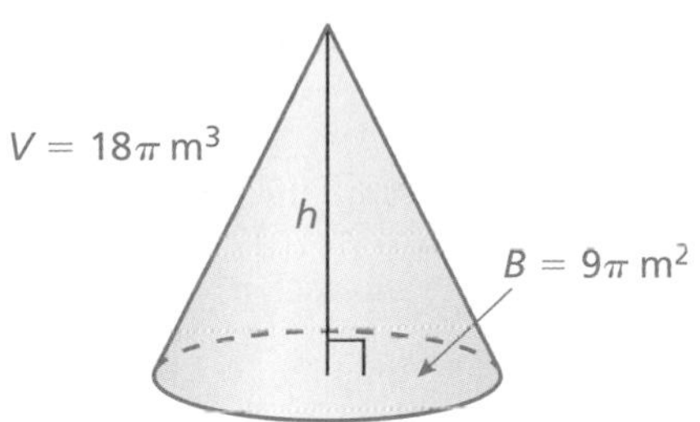

d. • Write a formula for the volume V of a cone.
• Solve the formula for h.
• Use the new formula to find the height of the cone.

What Is Your Answer?

3. **IN YOUR OWN WORDS** How can you use a formula for one measurement to write a formula for a different measurement? Give an example that is different from the examples on these two pages.

Practice Use what you learned about rewriting equations and formulas to complete Exercises 3 and 4 on page 30.

1.4 Lesson

Key Vocabulary
literal equation, *p. 28*

An equation that has two or more variables is called a **literal equation**. To rewrite a literal equation, solve for one variable in terms of the other variable(s).

EXAMPLE 1 Rewriting an Equation

Solve the equation $2y + 5x = 6$ for y.

	$2y + 5x = 6$	Write the equation.
Undo the addition. →	$2y + 5x - 5x = 6 - 5x$	Subtract $5x$ from each side.
	$2y = 6 - 5x$	Simplify.
Undo the multiplication. →	$\frac{2y}{2} = \frac{6 - 5x}{2}$	Divide each side by 2.
	$y = 3 - \frac{5}{2}x$	Simplify.

On Your Own

Solve the equation for y.

1. $5y - x = 10$
2. $4x - 4y = 1$
3. $12 = 6x + 3y$

EXAMPLE 2 Rewriting a Formula

Remember

A *formula* shows how one variable is related to one or more other variables. A formula is a type of literal equation.

The formula for the surface area S of a cone is $S = \pi r^2 + \pi r\ell$. Solve the formula for the slant height ℓ.

$S = \pi r^2 + \pi r\ell$	Write the equation.
$S - \pi r^2 = \pi r^2 - \pi r^2 + \pi r\ell$	Subtract πr^2 from each side.
$S - \pi r^2 = \pi r\ell$	Simplify.
$\frac{S - \pi r^2}{\pi r} = \frac{\pi r\ell}{\pi r}$	Divide each side by πr.
$\frac{S - \pi r^2}{\pi r} = \ell$	Simplify.

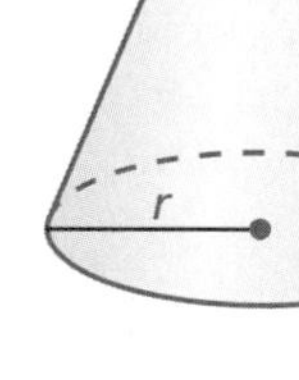

On Your Own

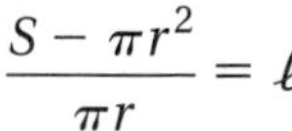

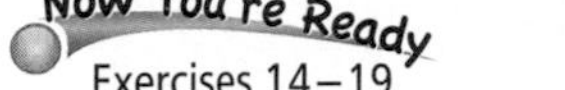

Solve the formula for the red variable.

4. Area of rectangle: $A = bh$
5. Simple interest: $I = Prt$
6. Surface area of cylinder: $S = 2\pi r^2 + 2\pi rh$

Laurie's Notes

Introduction

Connect

- **Yesterday:** Students practiced rewriting common geometric formulas. (MP1a, MP3a, MP7)
- **Today:** Students will use the techniques explored yesterday to solve literal equations.

Motivate

- Share with students the following highest and lowest recorded temperatures. The purpose is to pique interest and have students observe that the temperatures are measured in degrees Fahrenheit or degrees Celsius.
  - **Highest Recorded Temperatures:** NM 122°F, 50°C; NH 106°F, 41°C; PA 111°F, 44°C
  - **Lowest Recorded Temperatures:** NM −50°F, −46°C; NH −47°F, −44°C; PA −42°F, −41°C

Lesson Notes

Discuss

- **MP7 Look for and Make Use of Structure:** When solving literal equations, you want students to verbalize the operations represented. For example, when solving $2y + 5x = 6$, students should understand there are two terms. To isolate the $2y$-term, subtract $5x$. In other words, approach solving $2y + 5x = 6$ for y in the same way as you would solve $2y + 5 = 6$ for y.

Example 1

- Write the definition of literal equation.

? "Can 6 and $5x$ be combined? Explain." no; They are not like terms.

- Simplifying the last step is not obvious to all students. Relate it to fractions. You subtract the numerators and keep the same denominator. For example:

$$\frac{5-3}{7} = \frac{5}{7} - \frac{3}{7} \qquad \text{and} \qquad \frac{6-5x}{2} = \frac{6}{2} - \frac{5x}{2} = 3 - \frac{5}{2}x.$$

On Your Own

- Notice in Question 2 that the coefficient of y is −4. Suggest students rewrite the equation as $4x + (-4)y = 1$.

Example 2

- **Teaching Tip:** Highlight the variable ℓ in red as shown in the text. Discuss the idea that everything except the variable ℓ must be moved to the left side of the equation using Properties of Equality.

? "The term πr^2 is added to the term $\pi r \ell$. How do you move it to the left side of the equation?" Subtract πr^2 from each side of the equation.

- Discuss the technique of dividing by πr in one step, instead of dividing by π and then dividing by r.

On Your Own

- **Think-Pair-Share:** Students should read each question independently and then work with a partner to answer the questions.

Goal Today's lesson is solving **literal equations.**

Start Thinking! and Warm Up

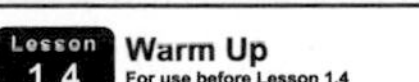

Lesson 1.4 **Start Thinking!** For use before Lesson 1.4

How does solving the equation $5x + 4y = 14$ for x compare to solving the equation $5x + 20 = 14$ for x? Describe the steps involved in each solution.

Extra Example 1

Solve the equation $-2x - 3y = 6$ for y.

$y = -\frac{2}{3}x - 2$

On Your Own

1. $y = 2 + \frac{1}{5}x$
2. $y = x - \frac{1}{4}$
3. $y = 4 - 2x$

Extra Example 2

The formula for the surface area of a square pyramid is $S = x^2 + 2x\ell$. Solve the formula for the slant height ℓ.

$\ell = \frac{S - x^2}{2x}$

On Your Own

4. $b = \frac{A}{h}$
5. $P = \frac{I}{rt}$
6. $h = \frac{S - 2\pi r^2}{2\pi r}$

English Language Learners

Vocabulary

Have students start a *Formula* page in their notebooks with the formulas used in this section. Each formula should be accompanied by a description of what each of the variables represents and an example. In the case of area formulas, units of measure should be included with the description (e.g., units and square units). As students progress throughout the year, additional formulas can be added to the *Formula* notebook page.

Extra Example 3

Solve the temperature formula $F = \frac{9}{5}C + 32$ for C. $C = \frac{5}{9}(F - 32)$

Extra Example 4

Which temperature is greater, 400°F or 200°C? 400°F

On Your Own

7. greater than

Laurie's Notes

Key Idea

- Write the formula for converting from degrees Fahrenheit to degrees Celsius.
- Use this formula if you know the temperature in degrees Fahrenheit and you want to find the temperature in degrees Celsius.

? "You are traveling abroad and the temperature is always stated in degrees Celsius. How can you figure out the temperature in degrees Fahrenheit, with which you are more familiar?" Students may recognize that you will want to have a different conversion formula that allows you to substitute for C and calculate F.

Example 3

? "What is the reciprocal of $\frac{5}{9}$?" $\frac{9}{5}$

- Remind students that multiplying by the reciprocal $\frac{9}{5}$ is more efficient than dividing by the fraction $\frac{5}{9}$.

Example 4

- **FYI:** The graphic on the left provides information about the temperature of a lightning bolt and the temperature of the surface of the sun. The two temperatures use different scales.

? "How can you compare two temperatures that are in different scales?" Listen for understanding that one of the temperatures must be converted.

? "How do you multiply $\frac{9}{5}$ times 30,000?" Students may recall that you can simplify before multiplying. Five divides into 30,000 six thousand times, so $6000 \times 9 = 54{,}000$.

? "Approximately how many times hotter is a lightning bolt than the surface of the sun?" 5 times This is a *cool* fact for students to know!

On Your Own

- **Neighbor Check:** Have students work independently and then have their neighbor check their work. Have students discuss any discrepancies.

Closure

- **Exit Ticket:** Solve $2x + 4y = 11$ for y. Check your solution. $y = -\frac{1}{2}x + \frac{11}{4}$

Technology For the Teacher

The Dynamic Planning Tool
Editable Teacher's Resources at *BigIdeasMath.com*

Temperature Conversion

A formula for converting from degrees Fahrenheit F to degrees Celsius C is

$$C = \frac{5}{9}(F - 32).$$

EXAMPLE 3 Rewriting the Temperature Formula

Solve the temperature formula for F.

$C = \frac{5}{9}(F - 32)$ Write the temperature formula.

Use the reciprocal. → $\frac{9}{5} \cdot C = \frac{9}{5} \cdot \frac{5}{9}(F - 32)$ Multiply each side by $\frac{9}{5}$, the reciprocal of $\frac{5}{9}$.

$\frac{9}{5}C = F - 32$ Simplify.

Undo the subtraction. → $\frac{9}{5}C + 32 = F - 32 + 32$ Add 32 to each side.

$\frac{9}{5}C + 32 = F$ Simplify.

∴ The rewritten formula is $F = \frac{9}{5}C + 32$.

EXAMPLE 4 Real-Life Application

Which has the greater temperature?

Convert the Celsius temperature of lightning to Fahrenheit.

$F = \frac{9}{5}C + 32$ Write the rewritten formula from Example 3.

$= \frac{9}{5}(30{,}000) + 32$ Substitute 30,000 for C.

$= 54{,}032$ Simplify.

∴ Because 54,032 °F is greater than 11,000 °F, lightning has the greater temperature.

On Your Own

7. Room temperature is considered to be 70 °F. Suppose the temperature is 23 °C. Is this greater than or less than room temperature?

1.4 Exercises

Vocabulary and Concept Check

1. **VOCABULARY** Is $-2x = \frac{3}{8}$ a literal equation? Explain.
2. **DIFFERENT WORDS, SAME QUESTION** Which is different? Find "both" answers.

Solve $4x - 2y = 6$ for y.	Solve $6 = 4x - 2y$ for y.
Solve $4x - 2y = 6$ for y in terms of x.	Solve $4x - 2y = 6$ for x in terms of y.

Practice and Problem Solving

3. **a.** Write a formula for the area A of a triangle.
   b. Solve the formula for b.
   c. Use the new formula to find the base of the triangle.

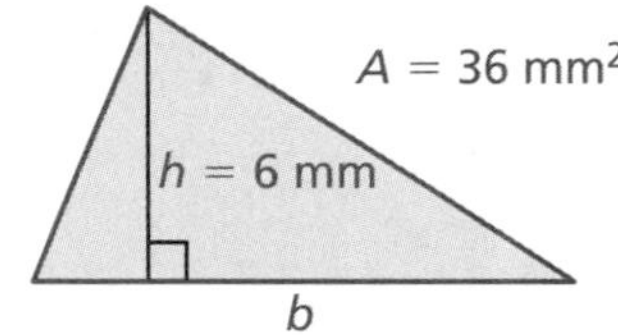

4. **a.** Write a formula for the volume V of a prism.
   b. Solve the formula for B.
   c. Use the new formula to find the area of the base of the prism.

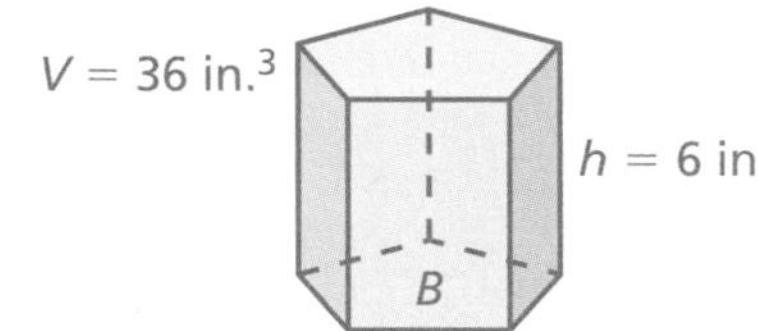

Solve the equation for *y*.

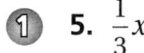

5. $\frac{1}{3}x + y = 4$
6. $3x + \frac{1}{5}y = 7$
7. $6 = 4x + 9y$
8. $\pi = 7x - 2y$
9. $4.2x - 1.4y = 2.1$
10. $6y - 1.5x = 8$
11. **ERROR ANALYSIS** Describe and correct the error in rewriting the equation.

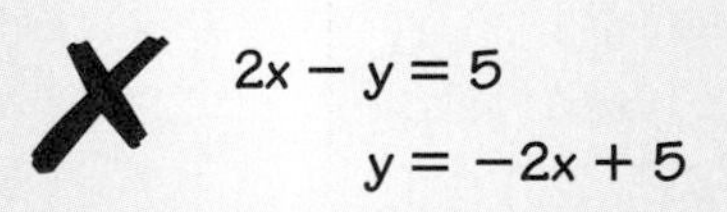

12. **TEMPERATURE** The formula $K = C + 273.15$ converts temperatures from Celsius C to Kelvin K.
    a. Solve the formula for C.
    b. Convert 300 K to Celsius.
13. **INTEREST** The formula for simple interest is $I = Prt$.
    a. Solve the formula for t.
    b. Use the new formula to find the value of t in the table.

I	\$75
P	\$500
r	5%
t	

Assignment Guide and Homework Check

Level	Assignment	Homework Check
Average	1–4, 8–11, 15–21 odd, 20, 24–28	8, 10, 17, 20
Advanced	1, 2, 10, 11, 17–28	10, 17, 20, 22

For Your Information

- **Exercise 2** *Different Words, Same Question* is a new type of exercise. Three of the four choices pose the same question using different words. The remaining choice poses a different question, so there are two answers.

Common Errors

- **Exercises 5–10** Students may solve the equation for the wrong variable. Remind them that they are solving the equation for *y*. Encourage them to make *y* a different color when solving so that it is easy for them to remember that they are solving for *y*.
- **Exercises 14–19** Each equation has a different step that could confuse students. Remind them to take their time when solving for the red variable. Remind them of the process of solving for a variable. They should start away from the variable and move toward it.

1.4 Record and Practice Journal

Solve the equation for *y*.

1. $2x + y = -9$

   $y = -2x - 9$

2. $4x - 10y = 12$

   $y = \frac{2}{5}x - \frac{6}{5}$

3. $13 = \frac{1}{6}y + 2x$

   $y = -12x + 78$

Solve the equation for the bold variable.

4. $V = \ell wh$

   $w = \frac{V}{\ell h}$

5. $f = \frac{1}{2}(r + 6.5)$

   $r = 2f - 6.5$

6. $S = 2\pi r^2 + 2\pi rh$

   $h = \frac{S - 2\pi r^2}{2\pi r}$

7. The formula for the area of a triangle is $A = \frac{1}{2}bh$.

   a. Solve the formula for h.

   $h = \frac{2A}{b}$

   b. Use the new formula to find the value of h.

   $h = 9$ in.

   $A = 54$ in.^{2}; 12 in.

Technology For the Teacher

Answer Presentation Tool

Vocabulary and Concept Check

1. no; The equation only contains one variable.
2. Solve $4x - 2y = 6$ for x in terms of y.; $x = \frac{3}{2} + \frac{1}{2}y$; $y = -3 + 2x$

Practice and Problem Solving

3. **a.** $A = \frac{1}{2}bh$

   b. $b = \frac{2A}{h}$

   c. $b = 12$ mm
4. **a.** $V = Bh$

   b. $B = \frac{V}{h}$

   c. $B = 6$ in.^{2}
5. $y = 4 - \frac{1}{3}x$
6. $y = 35 - 15x$
7. $y = \frac{2}{3} - \frac{4}{9}x$
8. $y = \frac{7}{2}x - \frac{\pi}{2}$
9. $y = 3x - 1.5$
10. $y = \frac{4}{3} + \frac{1}{4}x$
11. The y should have a negative sign in front of it.

    $2x - y = 5$

    $-y = -2x + 5$

    $y = 2x - 5$
12. **a.** $C = K - 273.15$

    b. 26.85°C
13. **a.** $t = \frac{I}{Pr}$

    b. $t = 3$ yr

Practice and Problem Solving

14. $t = \frac{d}{r}$

15. $m = \frac{e}{c^2}$

16. $C = R - P$

17. $\ell = \frac{A - \frac{1}{2}\pi w^2}{2w}$

18 $V = \frac{Bh}{3}$

19. $w = 6g - 40$

20. The rewritten formula is a general solution that can be reused.

21. **a.** $F = 32 + \frac{9}{5}(K - 273.15)$

b. 32°F

c. liquid nitrogen

22. See *Taking Math Deeper.*

23. $r^3 = \frac{3V}{4\pi}$; $r = 4.5$ in.

Fair Game Review

24. $3\frac{3}{4}$

25. $6\frac{2}{5}$

26. $\frac{1}{3}$

27. $1\frac{1}{4}$

28. D

Mini-Assessment

Solve the formula for the red variable.

1. Distance Formula: $d = rt$ $r = \frac{d}{t}$
2. Area of a triangle: $A = \frac{1}{2}bh$ $h = \frac{2A}{b}$
3. Circumference of a circle: $C = 2\pi r$ $r = \frac{C}{2\pi}$
4. The temperature in Portland, Oregon is 37°F. The temperature in Mobile, Alabama is 22°C. In which city is the temperature higher? Mobile, Alabama

Taking Math Deeper

Exercise 22

This problem is a nice review of circles and percents, as well as distance, rate, and time. It also has a bit of history related to George Ferris, who designed the first Ferris wheel for the 1893 World's Fair in Chicago.

Organize the given information.
Circumference (Navy Pier Ferris Wheel): $C = 439.6$ ft
Circumference (first Ferris wheel): x ft
Relationship: $439.6 = 0.56x$

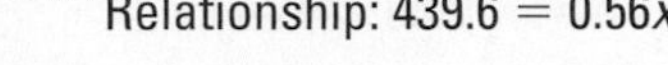

Find the radius of each wheel.

a. Radius (Navy Pier Ferris Wheel):

$$C = 2\pi r$$
$$439.6 \approx 2(3.14)r$$
$$70 = r$$

Circumference (first Ferris wheel):

$$439.6 = 0.56x$$
$$785 = x$$

b. Radius (first Ferris wheel):

$$785 \approx 2(3.14)R$$
$$125 = R$$

c. The first Ferris wheel made 1 revolution in 9 minutes. How fast was the wheel moving?

$$\text{rate} = \frac{785 \text{ ft}}{9 \text{ min}} \approx 87.2 \text{ ft per min}$$

It might be interesting for students to know that the first Ferris wheel had 36 cars, each of which held 60 people!

Project

Use your school's library or the Internet to find how long one revolution takes for the Ferris wheel on the Navy Pier in Chicago and the one in London, England. Which one has the greater circumference? Which one travels faster? How do you know?

Reteaching and Enrichment Strategies

If students need help. . .	If students got it. . .
Resources by Chapter • Practice A and Practice B • Puzzle Time Record and Practice Journal Practice Differentiating the Lesson Lesson Tutorials Skills Review Handbook	Resources by Chapter • Enrichment and Extension • School-to-Work Start the next section

Solve the equation for the red variable.

2 **14.** $d = rt$

15. $e = mc^2$

16. $R - C = P$

17. $A = \frac{1}{2}\pi w^2 + 2\ell w$

18. $B = 3\frac{V}{h}$

19. $g = \frac{1}{6}(w + 40)$

20. LOGIC Why is it useful to rewrite a formula in terms of another variable?

21. REASONING The formula $K = \frac{5}{9}(F - 32) + 273.15$ converts temperatures from Fahrenheit F to Kelvin K.

a. Solve the formula for F.

b. The freezing point of water is 273.15 Kelvin. What is this temperature in Fahrenheit?

c. The temperature of dry ice is −78.5 °C. Which is colder, dry ice or liquid nitrogen?

Navy Pier Ferris Wheel

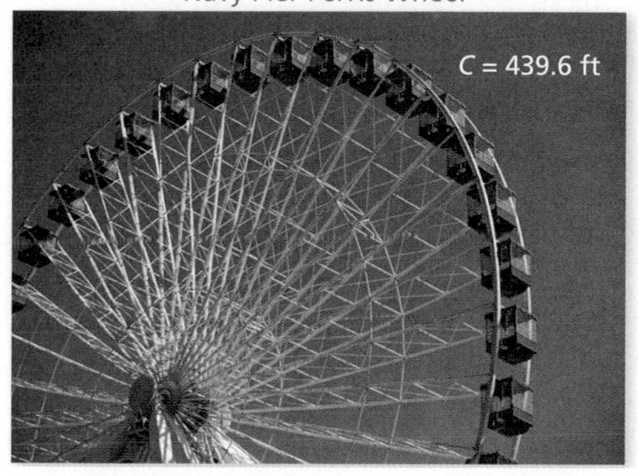

22. FERRIS WHEEL The Navy Pier Ferris Wheel in Chicago has a circumference that is 56% of the circumference of the first Ferris wheel built in 1893.

a. What is the radius of the Navy Pier Ferris Wheel?

b. What was the radius of the first Ferris wheel?

c. The first Ferris wheel took 9 minutes to make a complete revolution. How fast was the wheel moving?

23. Repeated Reasoning The formula for the volume of a sphere is $V = \frac{4}{3}\pi r^3$. Solve the formula for r^3. Use guess, check, and revise to find the radius of the sphere.

$V = 381.51$ in.^{3} r

Fair Game Review What you learned in previous grades & lessons

Multiply. *(Skills Review Handbook)*

24. $5 \times \frac{3}{4}$

25. $2.4 \times \frac{8}{3}$

26. $\frac{1}{4} \times \frac{3}{2} \times \frac{8}{9}$

27. $25 \times \frac{3}{5} \times \frac{1}{12}$

28. MULTIPLE CHOICE Which of the following is not equivalent to $\frac{3}{4}$? *(Skills Review Handbook)*

Ⓐ 0.75 Ⓑ 3 : 4 Ⓒ 75% Ⓓ 4 : 3

1.3–1.4 Quiz

Solve the equation. Check your solution. *(Section 1.3)*

1. $2(x + 4) = -5x + 1$

2. $\frac{1}{2}s = 4s - 21$

3. $8.3z = 4.1z + 10.5$

4. $3(b + 5) = 4(2b - 5)$

Solve the equation. Graph the solutions, if possible. *(Section 1.3)*

5. $|d + 10| = 6$

6. $-4|w - 1| = -8$

Solve the equation for *y*. *(Section 1.4)*

7. $6x - 3y = 9$

8. $8 = 2y - 10x$

Solve the formula for the red variable. *(Section 1.4)*

9. Volume of a cylinder: $V = \pi r^2 h$

10. Area of a trapezoid: $A = \frac{1}{2}h(b + B)$

11. TEMPERATURE In which city is the water temperature higher? *(Section 1.4)*

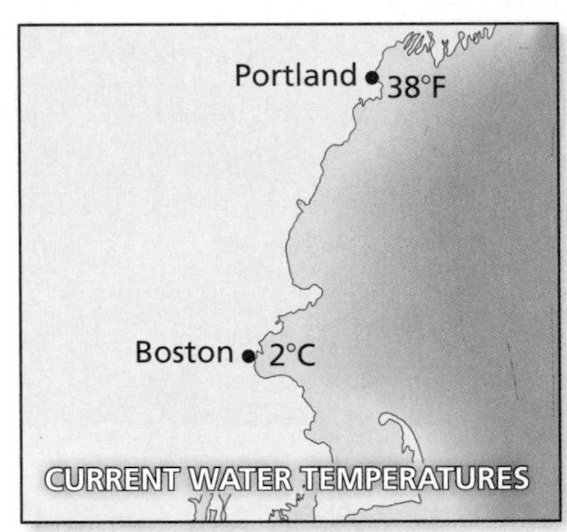

12. INTEREST The formula for simple interest I is $I = Prt$. Solve the formula for the interest rate r. What is the interest rate r if the principal P is \$1500, the time t is 2 years, and the interest earned I is \$90? *(Section 1.4)*

13. ROUTES From your home, the route to the store that passes the beach is 2 miles shorter than the route to the store that passes the park. What is the length of each route? *(Section 1.3)*

14. PERIMETER Use the triangle shown. *(Section 1.4)*

a. Write a formula for the perimeter P of the triangle.

b. Solve the formula for b.

c. Use the new formula to find b when a is 10 feet and c is 17 feet.

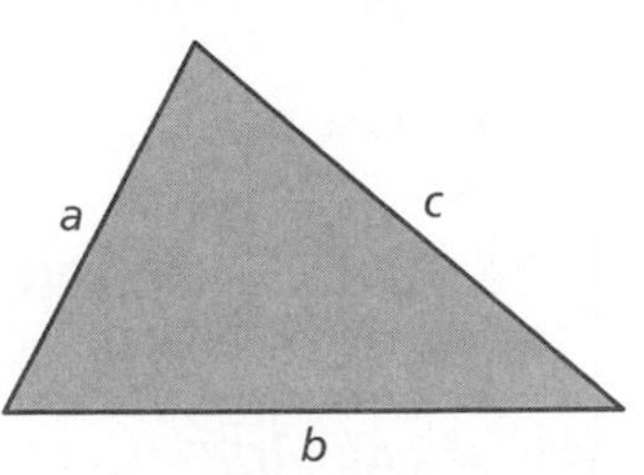

Perimeter = 42 feet

Alternative Assessment Options

Math Chat | Student Reflective Focus Question
Structured Interview | Writing Prompt

Math Chat

- Have students work in pairs. One student describes how to rewrite equations and formulas, giving examples. The other student probes for more information.
- The teacher should walk around the classroom listening to the pairs and asking questions to ensure understanding.

Study Help Sample Answers

Remind students to complete Graphic Organizers for the rest of the chapter.

3.

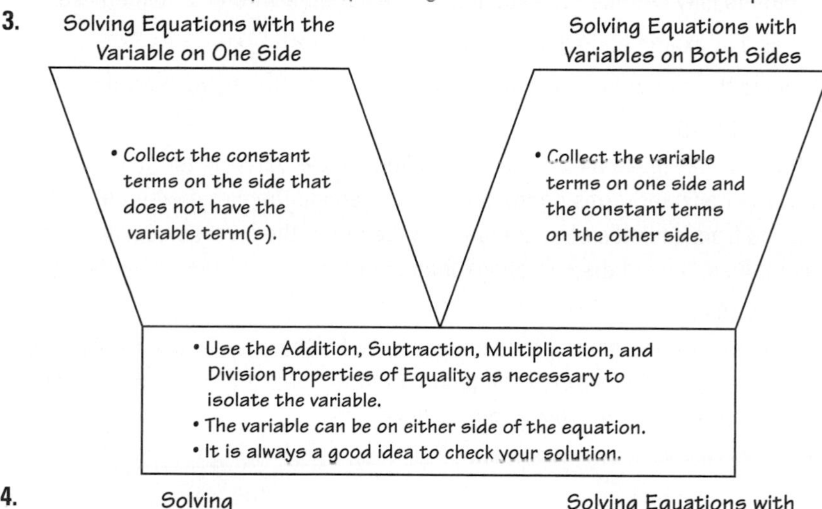

Solving Equations with the Variable on One Side
Solving Equations with Variables on Both Sides
• Collect the constant terms on the side that does not have the variable term(s).
• Collect the variable terms on one side and the constant terms on the other side.
• Use the Addition, Subtraction, Multiplication, and Division Properties of Equality as necessary to isolate the variable.
• The variable can be on either side of the equation.
• It is always a good idea to check your solution.

4.

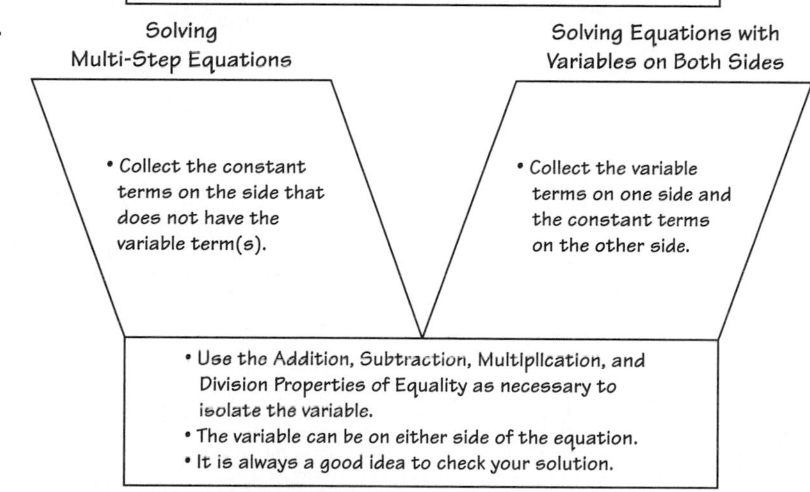

Solving Multi-Step Equations
Solving Equations with Variables on Both Sides
• Collect the constant terms on the side that does not have the variable term(s).
• Collect the variable terms on one side and the constant terms on the other side.
• Use the Addition, Subtraction, Multiplication, and Division Properties of Equality as necessary to isolate the variable.
• The variable can be on either side of the equation.
• It is always a good idea to check your solution.

5–6. Available at *BigIdeasMath.com*.

Reteaching and Enrichment Strategies

If students need help. . .	If students got it. . .
Resources by Chapter • Study Help • Practice A and Practice B • Puzzle Time Lesson Tutorials *BigIdeasMath.com* Practice Quiz Practice from the Test Generator	Resources by Chapter • Enrichment and Extension • School-to-Work Game Closet at *BigIdeasMath.com* Start the Chapter Review

Answers

1. $x = -1$
2. $s = 6$
3. $z = 2.5$
4. $b = 7$
5. $d = -4$ or $d = -16$
6. $w = 3$ or $w = -1$
7. $y = 2x - 3$
8. $y = 5x + 4$
9. $h = \dfrac{V}{\pi r^2}$
10. $b = \dfrac{2A}{h} - B$
11. Portland
12. 3%
13. passing beach: 13 miles
passing park: 15 miles
14. a. $P = a + b + c$
b. $b = P - a - c$
c. $b = 15$ feet

Technology For the Teacher

Answer Presentation Tool

Assessment Book

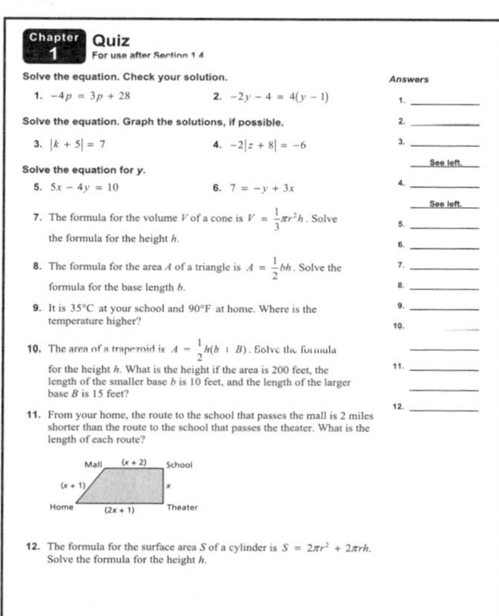

Chapter 1 Quiz
For use after Section 1.4

Solve the equation. Check your solution.

1. $-4p = 3p + 28$
2. $-2y - 4 = 4(y - 1)$

Solve the equation. Graph the solutions, if possible.

3. $|k + 5| = 7$
4. $-2|z + 8| = -6$

Solve the equation for y.

5. $5x - 4y = 10$
6. $7 = -y + 3x$

7. The formula for the volume V of a cone is $V = \frac{1}{3}\pi r^2 h$. Solve the formula for the height h.
8. The formula for the area A of a triangle is $A = \frac{1}{2}bh$. Solve the formula for the base length b.
9. It is 35°C at your school and 90°F at home. Where is the temperature higher?
10. The area of a trapezoid is $A = \frac{1}{2}h(b + B)$. Solve the formula for the height h. What is the height if the area is 200 feet, the length of the smaller base b is 10 feet, and the length of the larger base B is 15 feet?
11. From your home, the route to the school that passes the mall is 2 miles shorter than the route to the school that passes the theater. What is the length of each route?

Mall $(x + 2)$ School
$(x + 1)$ x
Home $(2x + 1)$ Theater

12. The formula for the surface area S of a cylinder is $S = 2\pi r^2 + 2\pi rh$. Solve the formula for the height h.

Answers
1. ____
2. ____
3. ____ See left.
4. ____ See left.
5. ____
6. ____
7. ____
8. ____
9. ____
10. ____
11. ____
12. ____

For the Teacher
Additional Review Options

- Big Ideas Test Generator
- Game Closet at *BigIdeasMath.com*
- Vocabulary Puzzle Builder
- Resources by Chapter
  - Puzzle Time
  - Study Help

Answers

1. $y = -19$
2. $n = -8$
3. $t = 12\pi$

Review of Common Errors

Exercises 1–5

- Students may perform the same operation that is in the equation instead of the inverse operation. Remind them that they must use an inverse operation to undo an operation. Also, remind them to check their solution in the original equation.

Exercises 6, 8, 14 and 15

- Students may multiply only one of the terms in parentheses by the factor outside the parentheses. Remind them how to correctly use the Distributive Property.

Exercises 7 and 9–12

- Students may change the exponent of the variable when combining like terms. For example, they may write the sum $x + x + \frac{1}{2}x + \frac{1}{2}x$ as $3x^4$. Remind them how to correctly combine like terms that have variables.

Exercises 13–15

- Students may make mistakes when collecting the variable terms on one side and the constant terms on the other side. Remind them that when a term is moved from one side of an equation to the other, the inverse operation is used. Also, remind them to check their solution in the original equation.

Exercises 19–23

- Students may be unsure about how to solve for the specified variable. Point out that they should work through the order of operations *backwards,* using inverse operations to isolate the variable.

1 Chapter Review

Review Key Vocabulary

absolute value equation, *p. 24*

literal equation, *p. 28*

Review Examples and Exercises

1.1 Solving Simple Equations *(pp. 2–9)*

The *boiling point* of a liquid is the temperature at which the liquid becomes a gas. The boiling point of mercury is about $\frac{41}{200}$ of the boiling point of lead. Write and solve an equation to find the boiling point of lead.

Let x be the boiling point of lead.

$\frac{41}{200}x = 357$ Write the equation.

$\frac{200}{41} \cdot \left(\frac{41}{200}x\right) = \frac{200}{41} \cdot 357$ Multiply each side by $\frac{200}{41}$.

$x \approx 1741$ Simplify.

The boiling point of lead is about 1741°C.

Exercises

Solve the equation. Check your solution.

1. $y + 8 = -11$
2. $3.2 = -0.4n$
3. $-\frac{t}{4} = -3\pi$

1.2 Solving Multi-Step Equations *(pp. 10–15)*

a. Solve $-4p - 9 = 3$.

$$\begin{aligned} -4p - 9 &= 3 \\ +9 \quad &\quad +9 \\ -4p &= 12 \\ \frac{-4p}{-4} &= \frac{12}{-4} \\ p &= -3 \end{aligned}$$

The solution is $p = -3$.

b. Solve $-14x + 28 + 6x = -44$.

$$\begin{aligned} -14x + 28 + 6x &= -44 \\ -8x + 28 &= -44 \\ -28 \quad &\quad -28 \\ -8x &= -72 \\ \frac{-8x}{-8} &= \frac{-72}{-8} \\ x &= 9 \end{aligned}$$

The solution is $x = 9$.

Exercises

Solve the equation. Check your solution.

4. $7y + 15 = -27$

5. $8 - \frac{3}{2}b = 11$

6. $-2(3z + 1) - 10 = 4$

7. $-3n - 2n + 9 = 29$

8. $2.5(4x - 6) - 5 = 10$

9. $\frac{2}{5}w + \frac{4}{5}w - 4 = 1$

Find the value of *x*. Then find the angle measures of the polygon.

10.

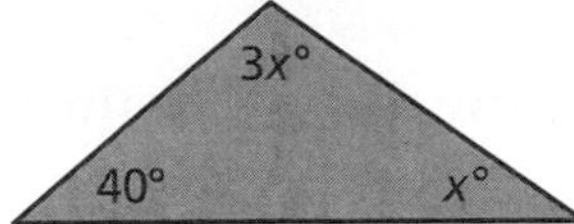

Sum of angle measures: 180°

11.

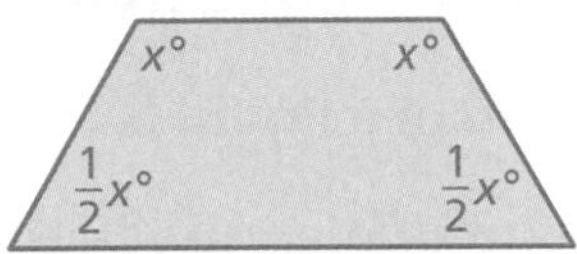

Sum of angle measures: 360°

12.

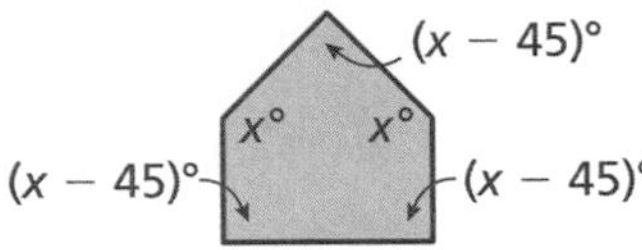

Sum of angle measures: 540°

1.3 Solving Equations with Variables on Both Sides (pp. 18–25)

a. Solve $3n - 2 = 11n + 18$.

$3n - 2 = 11n + 18$	Write the equation.
$\underline{-11n} \quad \underline{-11n}$	Subtract $11n$ from each side.
$-8n - 2 = 18$	Simplify.
$\underline{+2} \quad \underline{+2}$	Add 2 to each side.
$-8n = 20$	Simplify.
$\frac{-8n}{-8} = \frac{20}{-8}$	Divide each side by -8.
$n = -\frac{5}{2}$	Simplify.

∴ The solution is $n = -\frac{5}{2}$.

b. Solve $|x - 7| = 3$.

$\lvert x - 7 \rvert = 3$			Write the equation.
$x - 7 = 3$	*or*	$x - 7 = -3$	Write two related linear equations.
$\underline{+7} \quad \underline{+7}$		$\underline{+7} \quad \underline{+7}$	Add 7 to each side.
$x = 10$	*or*	$x = 4$	Simplify.

∴ The solutions are $x = 4$ and $x = 10$.

Review Game

Equation Puzzle

Materials per Group:

- envelope with 9 puzzle pieces

Directions:

Divide the class into small groups. Each group gets an envelope containing 9 equilateral triangle puzzle pieces. To put the puzzle pieces together, students must match the equation with its solution. When the puzzle is complete, it will also form an equilateral triangle.

Who Wins?

The first group to correctly put the puzzle together wins.

Tip:

If you are planning to use the puzzles again, laminate them before you cut them up. Also, copying the puzzle on different colors of paper or cardstock can be a time-saver when pieces fall on the floor.

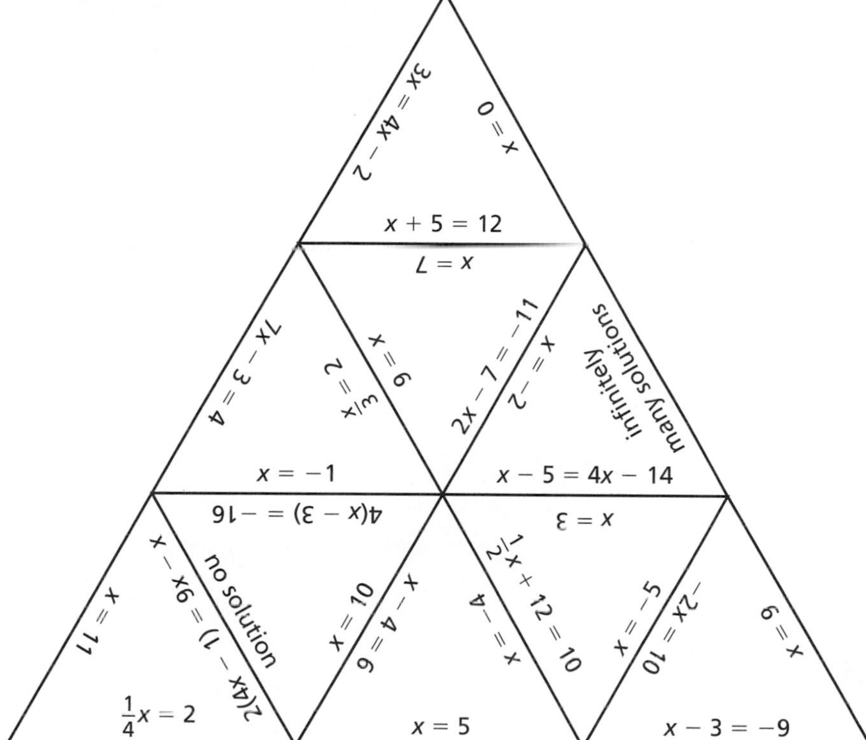

For the Student Additional Practice

- Lesson Tutorials
- Study Help (textbook)
- Student Website
  - Multi-Language Glossary
  - Practice Assessments

Answers

4. $y = -6$
5. $b = -2$
6. $z = -\frac{8}{3}$
7. $n = -4$
8. $x = 3$
9. $w = \frac{25}{6}$
10. $x = 35$; $40°, 105°, 35°$
11. $x = 120$; $60°, 120°, 120°, 60°$
12. $x = 135$; $90°, 135°, 90°, 135°, 90°$
13. $m = 6$
14. $p = 0.4$
15. $n = -19$
16. $x = 12$ or $x = -22$
17. $w = 5$ or $w = 4$
18. $y = \frac{13}{6}$ or $y = \frac{1}{6}$
19. $y = x - 6$
20. $y = 7 - 4x$
21. $y = \frac{1}{2}x + \frac{1}{2}$
22. a. $K = \frac{5}{9}(F - 32) + 273.15$

    b. about 388.71 K
23. a. $A = \frac{1}{2}h(b + B)$

    b. $h = \frac{2A}{b + B}$

    c. $h = 6$ cm

My Thoughts on the Chapter

Teacher Tip

Not allowed to write in your teaching edition? Use sticky notes to record your thoughts.

What worked. . .

What did not work. . .

What I would do differently. . .

Exercises

Solve the equation. Check your solution, if possible.

13. $5m - 1 = 4m + 5$

14. $3(5p - 3) = 5(p - 1)$

15. $\frac{2}{5}n + \frac{1}{10} = \frac{1}{2}(n + 4)$

Solve the equation. Check your solutions, if possible.

16. $|x + 5| = 17$

17. $|2w - 9| = 1$

18. $-3|6y - 7| + 10 = -8$

1.4 Rewriting Equations and Formulas *(pp. 26–31)*

The equation for a line in slope-intercept form is $y = mx + b$. Solve the equation for x.

$y = mx + b$ Write the equation.

$y - b = mx + b - b$ Subtract b from each side.

$y - b = mx$ Simplify.

$\frac{y - b}{m} = \frac{mx}{m}$ Divide each side by m.

$\frac{y - b}{m} = x$ Simplify.

∴ So, $x = \frac{y - b}{m}$.

Exercises

Solve the equation for y.

19. $5x - 5y = 30$

20. $14 = 8x + 2y$

21. $1 - 2y = -x$

22. **a.** The formula $F = \frac{9}{5}(K - 273.15) + 32$ converts a temperature from Kelvin K to Fahrenheit F. Solve the formula for K.

b. Convert 240 °F to Kelvin K. Round your answer to the nearest hundredth.

23. **a.** Write the formula for the area A of a trapezoid.

b. Solve the formula for h.

c. Use the new formula to find the height h of the trapezoid.

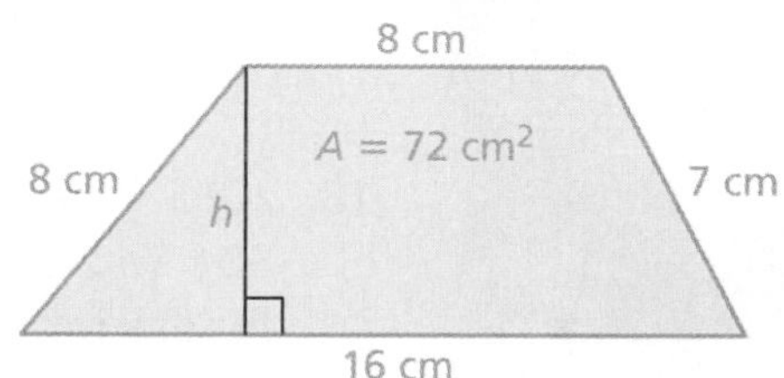

1 Chapter Test

Solve the equation. Check your solution.

1. $4 + y = 9.5$
2. $-\frac{x}{9} = -8$
3. $z - \frac{2}{3} = \frac{1}{8}$
4. $r - |-4| = 11$
5. $3.8n - 13 = 1.4n + 5$
6. $9(8d - 5) + 13 = 12d - 2$

Find the value of x. Then find the angle measures of the polygon.

7.

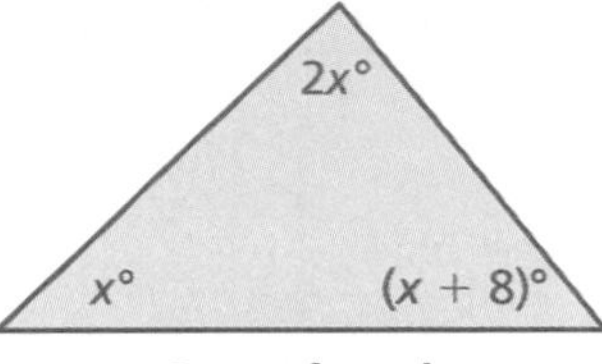

Sum of angle measures: 180°

8.

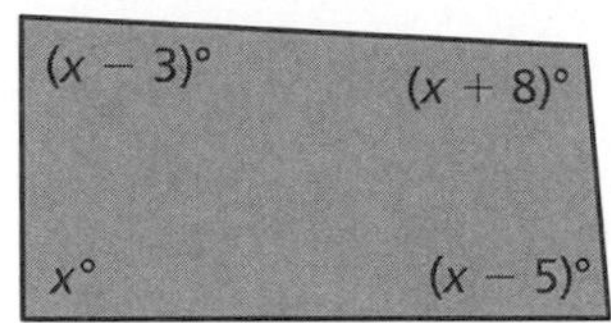

Sum of angle measures: 360°

Solve the equation. Graph the solutions, if possible.

9. $|2p - 3| = 7$
10. $5|3v - 8| = -10$

Solve the equation for y.

11. $1.2x - 4y = 28$
12. $0.5 = 0.4y - 0.25x$

Solve the formula for the red variable.

13. Perimeter of a rectangle: $P = 2\ell + 2w$
14. Distance formula: $d = rt$

15. **BASKETBALL** Your basketball team wins a game by 13 points. The opposing team scores 72 points. Explain how to find your team's score.

16. **CYCLING** You are biking at a speed of 18 miles per hour. You are 3 miles behind your friend who is biking at a speed of 12 miles per hour. Write and solve an equation to find the amount of time it takes for you to catch up to your friend.

17. **VOLCANOES** Two scientists are measuring lava temperatures. One scientist records a temperature of 1725 °F. The other scientist records a temperature of 950 °C. Which is the greater temperature? $\left(\text{Use } C = \frac{5}{9}(F - 32).\right)$

18. **JOBS** Your profit for mowing lawns this week is \$24. You are paid \$8 per hour and you paid \$40 for gas for the lawnmower. How many hours did you work this week?

Test Item References

Chapter Test Questions	Section to Review	Common Core State Standards
1–4, 15	1.1	A.CED.1, A.REI.1, A.REI.3
7, 8, 18	1.2	A.CED.1, A.REI.1, A.REI.3
5, 6, 9, 10, 16	1.3	A.CED.1, A.REI.1, A.REI.3
11–14, 17	1.4	A.CED.4

Test-Taking Strategies

Remind students to quickly look over the entire test before they start so that they can budget their time. When working with equations, students need to write all numbers and variables clearly, line up terms in each step, and not crowd their work. Have students use the **Stop** and **Think** strategy.

Common Assessment Errors

- **Exercises 1–4** Students may perform the same operation that is in the equation instead of the inverse operation. Remind them that they must use an inverse operation to undo an operation. Also, remind students to check their solution in the original equation.
- **Exercise 5** Students may make mistakes when collecting the variable terms on one side and the constant terms on the other side. Remind them that when a term is moved from one side of an equation to the other, the inverse operation is used. Also, remind them to check their solution in the original equation.
- **Exercise 6** Students may multiply only one of the terms in parentheses by the factor outside the parentheses. Remind them how to correctly use the Distributive Property.
- **Exercises 7 and 8** Students may change the exponent of the variable when combining like terms that have variables. For example, they may write the sum $2x + x + (x + 8)$ as $4x^3 + 8$. Remind them how to correctly combine like terms that have variables.
- **Exercises 11–14 and 17** Students may be unsure about how to solve for the specified variable. Point out that they should work through the order of operations *backwards,* using inverse operations to isolate the variable.

Reteaching and Enrichment Strategies

If students need help. . .	If students got it. . .
Resources by Chapter • Practice A and Practice B • Puzzle Time Record and Practice Journal Practice Differentiating the Lesson Lesson Tutorials Practice from the Test Generator Skills Review Handbook	Resources by Chapter • Enrichment and Extension • School-to-Work • Financial Literacy Game Closet at *BigIdeasMath.com* Start Standardized Test Practice

Answers

1. $y = 5.5$
2. $x = 72$
3. $z = \frac{19}{24}$
4. $r = 15$
5. $n = 7.5$
6. $d = 0.5$
7. $x = 43$; 43°, 86°, 51°
8. $x = 90$; 90°, 87°, 98°, 85°
9. $p = 5$ or $p = -2$
10. no solution
11. $y = 0.3x - 7$
12. $y = 0.625x + 1.25$
13. $w = \frac{P}{2} - \ell$
14. $r = \frac{d}{t}$
15. *Sample answer:* Write and solve the equation $x - 13 = 72$; $x = 85$ points
16. $18x = 12x + 3$; $\frac{1}{2}$ hour
17. 950° C
18. 8 hours

Assessment Book

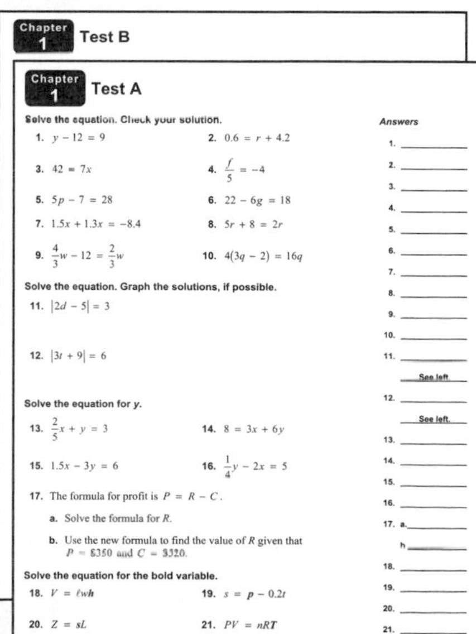

Chapter 1 Test B

Chapter 1 Test A

Solve the equation. Check your solution.

1. $y - 12 = 9$
2. $0.6 = r + 4.2$
3. $42 = 7x$
4. $\frac{f}{5} = -4$
5. $5p - 7 = 28$
6. $22 - 6g = 18$
7. $1.5x + 1.3x = -8.4$
8. $5r + 8 = 2r$
9. $\frac{4}{3}w - 12 = \frac{2}{3}w$
10. $4(3q - 2) = 16q$

Solve the equation. Graph the solutions, if possible.

11. $|2d - 5| = 3$
12. $|3t + 9| = 6$

Solve the equation for *y*.

13. $\frac{2}{5}x + y = 3$
14. $8 = 3x + 6y$
15. $1.5x - 3y = 6$
16. $\frac{1}{4}y - 2x = 5$
17. The formula for profit is $P = R - C$.
    a. Solve the formula for R.
    b. Use the new formula to find the value of R given that $P = \$350$ and $C = \$320$.

Solve the equation for the bold variable.

18. $V = \ell w\mathbf{h}$
19. $s = \mathbf{p} - 0.2t$
20. $Z = s\mathbf{L}$
21. $PV = nR\mathbf{T}$

Answers

1. ____
2. ____
3. ____
4. ____
5. ____
6. ____
7. ____
8. ____
9. ____
10. ____
11. See left.
12. See left.
13. ____
14. ____
15. ____
16. ____
17. a. ____ b. ____
18. ____
19. ____
20. ____
21. ____

Test-Taking Strategies
Available at *BigIdeasMath.com*

After Answering Easy Questions, Relax
Answer Easy Questions First
Estimate the Answer
Read All Choices before Answering
Read Question before Answering
Solve Directly or Eliminate Choices
Solve Problem before Looking at Choices
Use Intelligent Guessing
Work Backwards

About this Strategy

When taking a multiple choice test, be sure to read each question carefully and thoroughly. Before answering a question, determine exactly what is being asked, then eliminate the wrong answers and select the best choice.

Answers

1. A
2. I
3. B
4. 104 units
5. G

Technology For the Teacher

Big Ideas Test Generator

Item Analysis

1. **A.** Correct answer
   B. The student subtracts 4 from 32 instead of dividing 32 by 4.
   C. The student adds 4 to 32 instead of dividing 32 by 4.
   D. The student multiplies 4 by 32 instead of dividing 32 by 4.

2. **F.** The student correctly subtracts 3 from 39 but then multiplies instead of dividing.
   G. The student correctly subtracts 3 from 39 but then subtracts 2 instead of dividing.
   H. The student incorrectly adds 3 and 39 instead of subtracting and then performs division correctly.
   I. Correct answer

3. **A.** The student adds 5 to $65t$ instead of $55t$.
   B. Correct answer
   C. The student finds the time it takes for both vehicles to go 5 miles together instead of the time it takes one to catch up to the other.
   D. The student incorrectly subtracts 5 miles from t hours before multiplying.

4. **Gridded Response:** Correct answer: 104 units

   Common Error: The student correctly solves for x, but does not find the perimeter.

5. **F.** The student does not understand inverse operations.
   G. Correct answer
   H. The student subtracts r instead of dividing by r.
   I. The student does not understand inverse operations.

6. **A.** The student misunderstands that $3x$ and 5 are not like terms.
   B. Correct answer
   C. The student does not realize that the Distributive Property must first be used to multiply 2 by $x + 7$.
   D. The student does not realize that the Distributive Property must first be used to multiply 2 by $x + 7$.

Standardized Test Practice Icons

Gridded Response

Short Response (2-point rubric)

Extended Response (4-point rubric)

1 Standardized Test Practice

Test-Taking Strategy
Solve Directly or Eliminate Choices

When a cat wakes up, it's grumpy for x hours, where $2x-5x=x-4$. What's x?
Ⓐ 0 Ⓑ 1 Ⓒ 2 Ⓓ -3

Don't talk to me until I've had my morning milk.

"You can eliminate A and D. Then, solve directly to determine that the correct answer is B."

1. Which value of x makes the equation true? *(A.REI.3)*

$$4x = 32$$

A. 8 **C.** 36

B. 28 **D.** 128

2. A taxi ride costs \$3 plus \$2 for each mile driven. When you rode in a taxi, the total cost was \$39. This can be modeled by the equation below, where m represents the number of miles driven.

$$2m + 3 = 39$$

How long was your taxi ride? *(A.REI.3)*

F. 72 mi **H.** 21 mi

G. 34 mi **I.** 18 mi

3. A car traveling at a speed of 65 miles per hour is 5 miles behind a truck traveling at a speed of 55 miles per hour. Which equation can be used to find the amount of time t it takes for the car to catch up to the truck? *(A.CED.1)*

A. $65t + 5 = 55t$ **C.** $65t + 55t = 5$

B. $65t = 5 + 55t$ **D.** $65(t - 5) = 55t$

4. What is the perimeter of the square? *(A.CED.3, A.REI.3)*

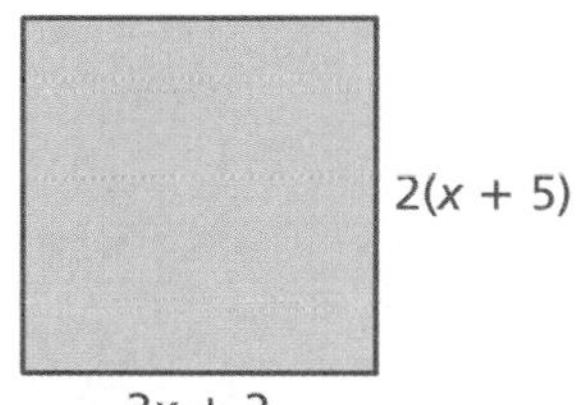

5. The formula below relates distance, rate, and time.

$$d = rt$$

Solve this formula for t. *(A.CED.4)*

F. $t = dr$ **H.** $t = d - r$

G. $t = \frac{d}{r}$ **I.** $t = \frac{r}{d}$

6. What could be the first step to solve the equation shown below? *(A.REI.1)*

$$3x + 5 = 2(x + 7)$$

A. Combine $3x$ and 5.

B. Multiply x by 2 and 7 by 2.

C. Subtract x from $3x$.

D. Subtract 5 from 7.

7. You work as a sales representative. You earn \$400 per week plus 5% of your total sales for the week. *(A.CED.1, A.REI.3)*

Part A Last week, you had total sales of \$5000. Find your total earnings. Show your work.

Part B One week, you earned \$1350. Let s represent your total sales that week. Write an equation that could be used to find s.

Part C Using your equation from Part B, find s. Explain all steps.

8. In 10 years, Maria will be 39 years old. Let m represent Maria's age today. Which equation can be used to find m? *(A.CED.1)*

F. $m = 39 + 10$

G. $m - 10 = 39$

H. $m + 10 = 39$

I. $10m = 39$

9. Which value of y makes the equation below true? *(A.REI.3)*

$$3y + 8 = 7y + 11$$

A. -4.75

B. -0.75

C. 0.75

D. 4.75

10. The equation below is used to convert a Celsius temperature C to its equivalent Fahrenheit temperature F.

$$F = \frac{9}{5}C + 32$$

Which formula can be used to convert a Fahrenheit temperature to its equivalent Celsius temperature? *(A.CED.4)*

F. $C = \frac{9}{5}(F - 32)$

G. $C = \frac{5}{9}F - 32$

H. $C = \frac{5}{9}(F + 32)$

I. $C = \frac{5}{9}(F - 32)$

Item Analysis (continued)

7. **4 points** The student demonstrates a thorough understanding of evaluating expressions, writing equations, and solving equations, and presents his or her steps clearly. The following answers should be obtained: Part A: \$650; Part B: $0.05s + 400 = 1350$; Part C: \$19,000.

   3 points The student demonstrates an essential but less than thorough understanding. In particular, the correct equation or its equivalent should be given in Part B, but an arithmetic error may have been performed in Part C.

   2 points The student demonstrates a partial understanding of the processes of writing and solving equations. Part A should be correctly completed, but the equation in Part B may be written incorrectly. Alternatively, the correct equation could be written in Part B, but Part C might display misunderstanding of how to proceed.

   1 point The student demonstrates a limited understanding of equation writing and solving, as well as working with percents. The student's response is incomplete and exhibits many flaws.

   0 points The student provided no response, a completely incorrect or incomprehensible response, or a response that demonstrates insufficient understanding of percents and equations.

8. **F.** The student misunderstands the problem and decides to add the two numbers together.

   G. The student understands that m and 39 are 10 apart but chooses subtraction instead of addition to relate them.

   H. Correct answer

   I. The student mistakes $10m$ for $10 + m$.

9. **A.** The student incorrectly simplifies the equation to $3y = 7y + 19$ and then solves for y.

   B. Correct answer

   C. The student incorrectly simplifies the equation to $-4y = -3$ or $4y = 3$ and then solves for y.

   D. The student incorrectly simplifies the equation to $3y + 19 = 7y$ and then solves for y.

10. **F.** The student correctly subtracts 32 from both sides of the equation but incorrectly multiplies by $\frac{9}{5}$.

    G. The student multiplies F by $\frac{5}{9}$ and then subtracts 32 instead of subtracting 32 from F and then multiplying by $\frac{5}{9}$.

    H. The student adds 32 to F instead of subtracting 32 from F.

    I. Correct answer

11. **Gridded Response:** Correct answer: 14 weeks

    Common Error: The student adds 35 and 175 to get 210 and then divides by 10 to get 21 as an answer.

Answers

6. B
7. *Part A* \$650

   Part B $0.05s + 400 = 1350$

   Part C \$19,000
8. H
9. B
10. I

Answers

11. 14 weeks
12. C
13. F
14. D
15. F

Answer for Extra Example

1. **A.** The student does not understand inverse operations.

   B. Correct answer

   C. The student does not understand inverse operations.

   D. The student subtracts $\frac{1}{3}B$ instead of dividing by $\frac{1}{3}B$.

Item Analysis (continued)

12. **A.** The student either subtracts variable terms incorrectly or assumes that if $0 = 0$, then $x = 0$.

B. The student incorrectly assumes that if $5 = 5$, then $x = 5$.

C. Correct answer

D. The student mixes up the rules for "no solution" and "infinitely many solutions."

13. **F.** Correct answer

G. The student incorrectly sets up one equation as $2x - 1 = 4$ instead of $2x + 1 = -4$.

H. The student incorrectly sets up one equation as $2x - 1 = -4$ instead of $2x + 1 = 4$.

I. The student does not understand that "no solution" is only used when absolute value expressions are set equal to a number less than 0.

14. **A.** The student distributes correctly but then makes a mistake combining the constant terms, yielding $2x = -11$.

B. The student does not distribute the left side of the equation correctly, yielding $6x - 3$.

C. The student combines $6x$ and $4x$ incorrectly, yielding $10x$ instead of $2x$.

D. Correct answer

15. **F.** Correct answer

G. The student incorrectly uses the fact that there are 4 items on one side and 2 on the other to get the ratio $\frac{2}{4} = \frac{1}{2}$.

H. The student incorrectly uses the fact that there are 4 items on one side and 2 on the other to get the ratio $\frac{2}{4} = \frac{1}{2}$ and then misuses the order of the ratio.

I. The student gets the correct ratio of $\frac{1}{3}$ but misuses it.

Extra Example

1. The formula for the volume V of a pyramid is $V = \frac{1}{3}Bh$. Solve the formula for the height h. *(A.CED.4)*

   A. $h = \frac{1}{3}VB$

   B. $h = \frac{3V}{B}$

   C. $h = \frac{V}{3B}$

   D. $h = V - \frac{1}{3}B$

11. You have already saved \$35 for a new cell phone. You need \$175 in all. You think you can save \$10 per week. At this rate, how many more weeks will you need to save money before you can buy the new cell phone? *(A.CED.1, A.REI.3)*

12. Solve $-8x - x + 5 = -6x + 5 - 3x$. *(A.REI.3)*

A. $x = 0$

B. $x = 5$

C. Infinitely many solutions

D. No solution

13. Solve $3|2x + 1| = 12$. *(A.REI.3)*

F. The solutions are $-\frac{5}{2}$ and $\frac{3}{2}$.

G. The solutions are $\frac{5}{2}$ and $\frac{3}{2}$.

H. The solutions are $-\frac{5}{2}$ and $-\frac{3}{2}$.

I. No solution

14. Which value of x makes the equation below true? *(A.REI.3)*

$$6(x - 3) = 4x - 7$$

A. -5.5

B. -2

C. 1.1

D. 5.5

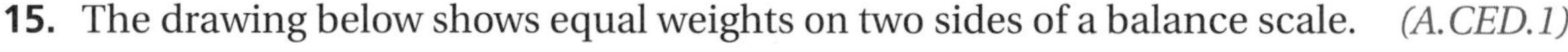

15. The drawing below shows equal weights on two sides of a balance scale. *(A.CED.1)*

What can you conclude from the drawing?

F. A mug weighs one-third as much as a trophy.

G. A mug weighs one-half as much as a trophy.

H. A mug weighs twice as much as a trophy.

I. A mug weighs three times as much as a trophy.

2 Graphing and Writing Linear Equations

"Okay Descartes, stand on the y-axis and try to intercept the pass when I throw."

"Here's an easy example of a line with a slope of 1."

"You eat one mouse treat the first day. Two treats the second day. And so on. Get it?"

Connections to Previous Learning

- Write, solve, and graph one-step linear equations.
- Construct and analyze tables, graphs, and equations to describe linear relationships and other simple relations.

- Write, solve, and graph multi-step equations.
- Graph proportional relationships and identify the unit rate as slope of linear functions.
- Use properties of equality to rewrite an equation and to show two equations are equivalent.

- Construct and analyze tables, graphs, and models to describe linear equations.
- Interpret slope and x- and y-intercepts when graphing a linear equation for a real world problem.
- Use tables, graphs, and models to represent, analyze, and solve real-life problems related to linear equations.

Math in History

The concept of writing all numbers by using only ten different symbols appears to have originated in India.

★ Here are some of the symbols that were used in the Brahmi system in India until around 400 A.D.

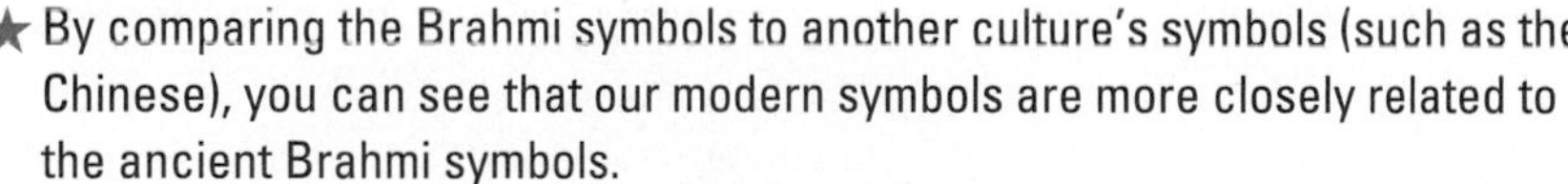

Notice that there is no symbol for 0. That concept was not yet devised. Also notice that the symbols for 2 and 3 are related to our modern symbols for 2 and 3.

Draw two horizontal bars quickly | Draw three horizontal bars quickly

★ By comparing the Brahmi symbols to another culture's symbols (such as the Chinese), you can see that our modern symbols are more closely related to the ancient Brahmi symbols.

1	2	3	4	5	6	7	8	9
一	二	三	四	五	六	七	八	九

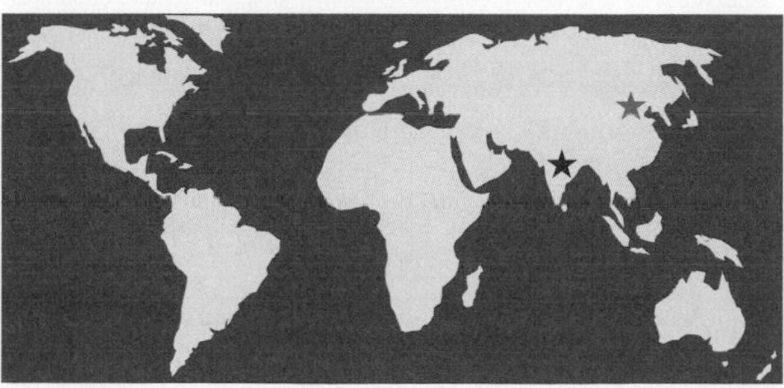

Pacing Guide for Chapter 2

Chapter Opener	1 Day
Section 1	1 Day
Section 2	2 Days
Section 3	1 Day
Section 4	2 Days
Study Help / Quiz	1 Day
Section 5	2 Days
Section 6	3 Days
Section 7	1 Day
Chapter Review / Chapter Tests	2 Days
Total Chapter 1	16 Days
Year-to-Date	24 Days

Check Your Resources

- Record and Practice Journal
- Resources by Chapter
- Skills Review Handbook
- Assessment Book
- Worked-Out Solutions

Technology For the Teacher

The Dynamic Planning Tool
Editable Teacher's Resources at *BigIdeasMath.com*

Common Core State Standards

6.EE.2c Evaluate expressions at specific values of their variables . . ..

6.NS.6c Find and position integers and other rational numbers on a horizontal or vertical number line diagram; find and position pairs of integers and other rational numbers on a coordinate plane.

Additional Topics for Review

- Order of Operations
- Exponents
- Plotting points in Quadrant I

Try It Yourself

1. -12
2. -23
3. 15
4. $4\frac{3}{4}$
5. $(0, 4)$
6. $(4, 2)$
7. Point R
8. Point N

Record and Practice Journal

1. 5
2. 16
3. -5
4. $-38\frac{1}{2}$
5. 108
6. 65
7. $-3\frac{7}{19}$
8. 262
9. \$50.00
10. $(-5, 0)$
11. $(3, -5)$
12. Point F
13. Point G
14. Point B, Point H
15. Point C, Point E

16–20.

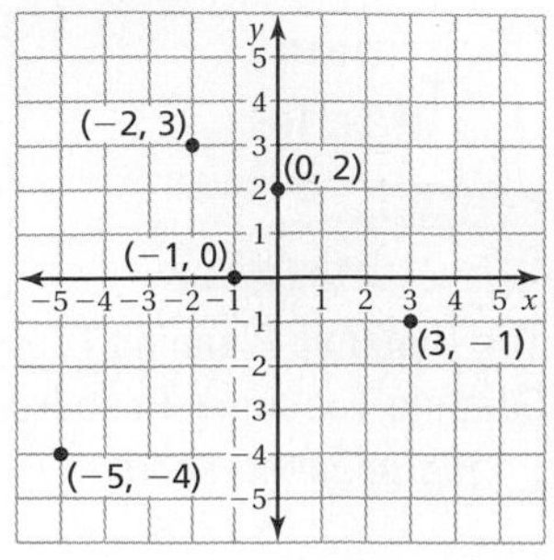

Math Background Notes

Vocabulary Review

- Evaluate
- Expression
- Order of Operations
- Substitute
- Coordinates

Evaluating Expressions Using Order of Operations

- Students should know how to substitute values into algebraic expressions and evaluate the results using order of operations.
- **Teaching Tip:** Sometimes color coding substitutions can help students to evaluate expressions. Each time you want to substitute a number in place of a variable, you must substitute your lead pencil for a colored pencil.
- Remind students that after they substitute values in for *x* and *y*, they must use the correct order of operations to continue simplifying the expression.
- **Common Error:** Encourage students to use a set of parentheses whenever they do a substitution. This will help students distinguish between subtracting 7 and multiplying by -7.

Plotting Points

- Students should know how to plot points in all four quadrants.
- **Common Error:** Students may write the coordinates backwards. Remind them that coordinates are written in alphabetical order with the *x* move (horizontal) written before the *y* move (vertical).
- **Common Error:** Students may also have difficulty with the negative numbers associated with plotting outside Quadrant I. Remind them that the negatives are directional. A negative *x*-value communicates a move to the left of the origin and a negative *y*-coordinate communicates a move downward from the origin.

Reteaching and Enrichment Strategies

If students need help. . .	If students got it. . .
Record and Practice Journal • Fair Game Review Skills Review Handbook Lesson Tutorials	Game Closet at *BigIdeasMath.com* Start the next section

What You Learned Before

"I estimate that we are on a slope of about −0.625. What do you think?"

Evaluating Expressions Using Order of Operations (6.EE.2c)

Example 1 **Evaluate $2xy + 3(x + y)$ when $x = 4$ and $y = 7$.**

$2xy + 3(x + y) = 2(4)(7) + 3(4 + 7)$	Substitute 4 for x and 7 for y.
$= 8(7) + 3(4 + 7)$	Use order of operations.
$= 56 + 3(11)$	Simplify.
$= 56 + 33$	Multiply.
$= 89$	Add.

Try It Yourself

Evaluate the expression when $a = \frac{1}{4}$ and $b = 6$.

1. $-8ab$

2. $16a^2 - 4b$

3. $\dfrac{5b}{32a^2}$

4. $12a + (b - a - 4)$

Plotting Points (6.NS.6c)

Example 2 **Write the ordered pair that corresponds to Point U.**

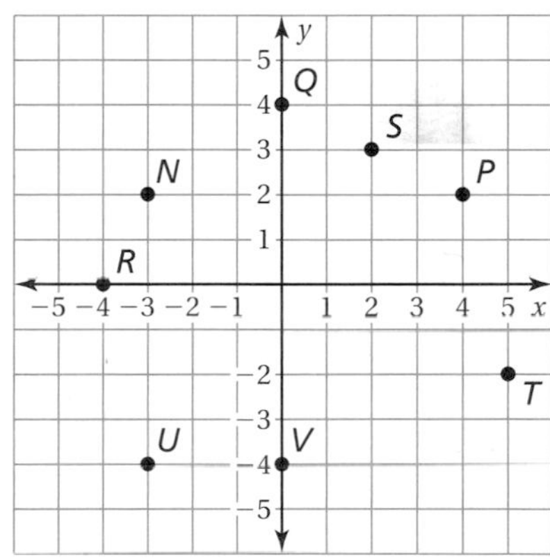

Point U is 3 units to the left of the origin and 4 units down. So, the x-coordinate is −3 and the y-coordinate is −4.

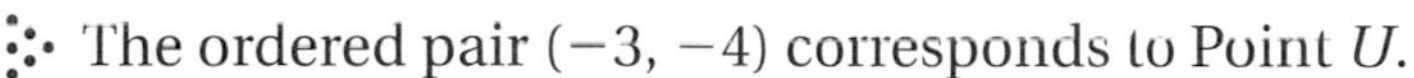

∴ The ordered pair (−3, −4) corresponds to Point U.

Example 3 **Which point is located at (5, −2)?**

Start at the origin. Move 5 units right and 2 units down.

∴ Point T is located at (5, −2).

Try It Yourself

Use the graph to answer the question.

5. Write the ordered pair that corresponds to Point Q.

6. Write the ordered pair that corresponds to Point P.

7. Which point is located at (−4, 0)?

8. Which point is located in Quadrant II?

2.1 Graphing Linear Equations

Essential Question How can you recognize a linear equation? How can you draw its graph?

COMMON CORE STATE STANDARDS
A.CED.2
A.REI.10

1 ACTIVITY: Graphing a Linear Equation

Work with a partner.

a. Use the equation $y = \frac{1}{2}x + 1$ to complete the table. (Choose any two x-values and find the y-values.)

	Solution Points	
x		
$y = \frac{1}{2}x + 1$		

b. Write the two ordered pairs given by the table. These are called **solution points** of the equation.

c. **PRECISION** Plot the two solution points. Draw a line *exactly* through the two points.

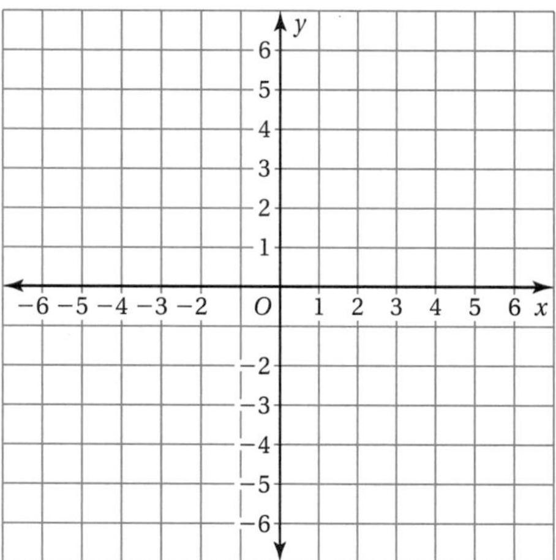

d. Find a different point on the line. Check that this point is a solution point of the equation $y = \frac{1}{2}x + 1$.

e. **LOGIC** Do you think it is true that *any* point on the line is a solution point of the equation $y = \frac{1}{2}x + 1$? Explain.

f. Choose five additional x-values for the table. (Choose positive and negative x-values.) Plot the five corresponding solution points. Does each point lie on the line?

	Solution Points				
x					
$y = \frac{1}{2}x + 1$					

g. **LOGIC** Do you think it is true that *any* solution point of the equation $y = \frac{1}{2}x + 1$ is a point on the line? Explain.

h. **THE MEANING OF A WORD** Why is $y = ax + b$ called a *linear equation*?

Laurie's Notes

Introduction

Standards for Mathematical Practice

- **MP4 Model with Mathematics:** The goal is for students to use a table of values and a graph to model a linear equation. Using multiple representations of linear equations deepens students' understanding and supports learning.
- Throughout this chapter, you may encounter applications that show a graph of discrete data with a line through the points. At this point in the text, we do not think it is necessary for students to distinguish between discrete data (plotting points only) and continuous data (plotting points along with the line). Students can draw a line through discrete points to help them solve an exercise. They will learn more about discrete and continuous data at a later time.

Motivate

- Play a game of coordinate BINGO.
- Distribute small coordinate grids to students. They should plot ten ordered pairs, where the x- and y-coordinates are integers between -4 and 4.
- Generate a random ordered pair in the grid. Write the integers from -4 to 4 on slips of paper and place them in a bag. Draw and replace an integer twice to generate the ordered pair, then write it on the board.
- Each time you record a new ordered pair, the students check to see if it is one of their 10 ordered pairs. If it is, they put an X there. The goal is to be the first person with three X's. If a student thinks they have won, they read their ordered pairs for you to check against the master list.

? "Are there ordered pairs that are not on lattice points, meaning the x- or y-coordinate is not an integer? Explain." yes; It's possible for the ordered pair to be $\left(3.5, \frac{1}{2}\right)$. Plot whatever example students give.

- Remind students that the ordered pairs are always (x, y), where x is the horizontal direction and y is the vertical direction.

Activity Notes

Activity 1

- Some students will recognize right away that if they substitute an even number for x, the y-coordinate will not be a fraction. It is likely that students will only try positive x-values. Encourage them to try negative values for x.
- In part (d), suggest that students consider only those ordered pairs that appear to be lattice points.
- **MP3a Construct Viable Arguments:** Listen and discuss student responses to the generalizations in parts (e) and (g).
- **Big Idea:** The goal of this activity is for students to recognize and understand two related, but different, ideas. 1) *All* solution points of a linear equation lie on the same line. 2) *All* points on the line are solution points of the equation.

Common Core State Standards

A.CED.2 Create equations in two or more variables to represent relationships between quantities; graph equations on coordinate axes with labels and scales.

A.REI.10 Understand that the graph of an equation in two variables is the set of all its solutions plotted in the coordinate plane, often forming a curve (which could be a line).

Previous Learning

Students should know about slope as a ratio. Students should know how to plot ordered pairs.

Activity Materials	
Introduction	**Textbook**
• small coordinate grid	• straightedge

Start Thinking! and Warm Up

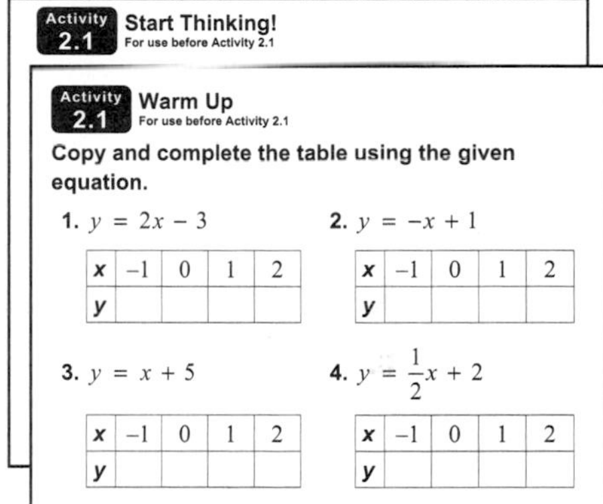

Activity 2.1 Start Thinking! For use before Activity 2.1

Activity 2.1 Warm Up For use before Activity 2.1

Copy and complete the table using the given equation.

1. $y = 2x - 3$

x	-1	0	1	2
y				

2. $y = -x + 1$

x	-1	0	1	2
y				

3. $y = x + 5$

x	-1	0	1	2
y				

4. $y = \frac{1}{2}x + 2$

x	-1	0	1	2
y				

2.1 Record and Practice Journal

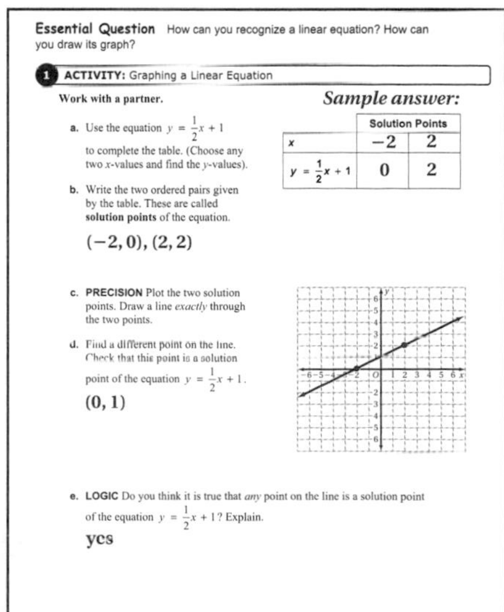

Essential Question How can you recognize a linear equation? How can you draw its graph?

1 ACTIVITY: Graphing a Linear Equation

Work with a partner.

Sample answer:

a. Use the equation $y = \frac{1}{2}x + 1$ to complete the table. (Choose any two x-values and find the y-values).

	Solution Points	
x	-2	2
$y = \frac{1}{2}x + 1$	0	2

b. Write the two ordered pairs given by the table. These are called **solution points** of the equation.

$(-2, 0), (2, 2)$

c. **PRECISION** Plot the two solution points. Draw a line *exactly* through the two points.

d. Find a different point on the line. Check that this point is a solution point of the equation $y = \frac{1}{2}x + 1$.

$(0, 1)$

e. **LOGIC** Do you think it is true that *any* point on the line is a solution point of the equation $y = \frac{1}{2}x + 1$? Explain.

yes

Differentiated Instruction

Kinesthetic

For students that are kinesthetic learners and have difficulty in plotting points in the coordinate plane, suggest they use a finger for tracing. Have the student place their finger at the origin and trace left or right along the x-axis to the first coordinate, then trace up or down to the second coordinate. Students should also practice writing the ordered pair of a plotted point. Guide students with questions such as, "Should you move left or right? How far? Should you move up or down? How far?"

2.1 Record and Practice Journal

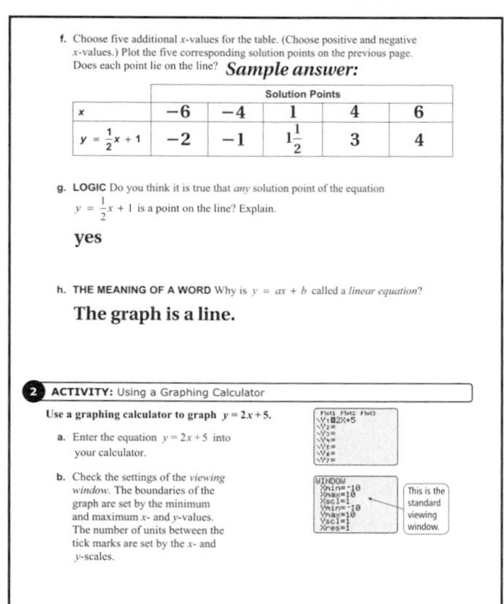

f. Choose five additional x-values for the table. (Choose positive and negative x-values.) Plot the five corresponding solution points on the previous page. Does each point lie on the line? *Sample answer:*

	Solution Points				
x	-6	-4	1	4	6
$y = \frac{1}{2}x + 1$	-2	-1	$1\frac{1}{2}$	3	4

g. **LOGIC** Do you think it is true that *any* solution point of the equation $y = \frac{1}{2}x + 1$ is a point on the line? Explain.

yes

h. **THE MEANING OF A WORD** Why is $y = ax + b$ called a *linear equation*?

The graph is a line.

2 ACTIVITY: Using a Graphing Calculator

Use a graphing calculator to graph $y = 2x + 5$.

a. Enter the equation $y = 2x + 5$ into your calculator.

b. Check the settings of the *viewing window*. The boundaries of the graph are set by the minimum and maximum x- and y-values. The number of units between the tick marks are set by the x- and y-scales.

This is the standard viewing window.

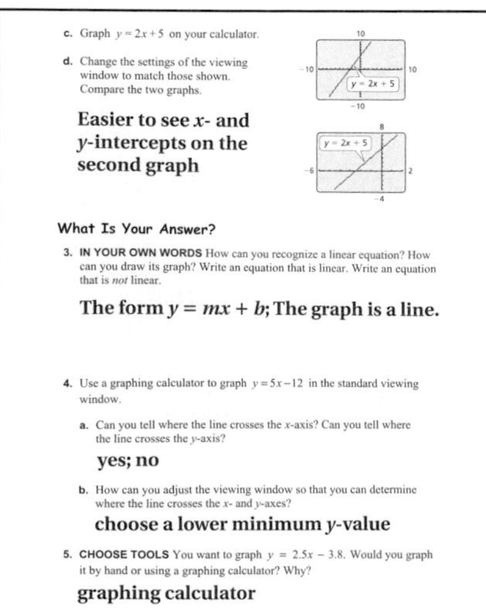

c. Graph $y = 2x + 5$ on your calculator.

d. Change the settings of the viewing window to match those shown. Compare the two graphs.

Easier to see x- and y-intercepts on the second graph

What Is Your Answer?

3. **IN YOUR OWN WORDS** How can you recognize a linear equation? How can you draw its graph? Write an equation that is linear. Write an equation that is *not* linear.

The form $y = mx + b$; The graph is a line.

4. Use a graphing calculator to graph $y = 5x - 12$ in the standard viewing window.

a. Can you tell where the line crosses the x-axis? Can you tell where the line crosses the y-axis?

yes; no

b. How can you adjust the viewing window so that you can determine where the line crosses the x- and y-axes?

choose a lower minimum y-value

5. **CHOOSE TOOLS** You want to graph $y = 2.5x - 3.8$. Would you graph it by hand or using a graphing calculator? Why?

graphing calculator

Laurie's Notes

Activity 2

- This is likely students' first experience with using a graphing calculator to graph a linear equation. Explain that the calculator can graph equations that are entered in the equation editor.
- Explain how to set the *standard viewing window*, or *standard viewing rectangle*.
- Because the viewing screen is a rectangle, one unit in the x-direction appears longer than one unit in the y-direction. When graphing by hand, you generally are using a square grid. It is important to point out this distinction to students because they will eventually graph $y = x$, which is a 45° line. It will not appear this way in the standard viewing rectangle.
- ? After the graph of $y = 2x + 5$ appears, ask, "Can you name a solution of this equation from looking at the graph?" Students may name solutions such as (0, 5) or (1, 7).
- ? After the viewing rectangle changes, ask, "Did the solutions of the equation change?" The graph appears less steep, however it is the same graph in a new view. The solutions have not changed.

What Is Your Answer?

- Discuss students' responses to the first question.
- Have students share the viewing window they selected for Question 4b.
- **MP5 Use Appropriate Tools Strategically:** This year students will be asked to graph many linear equations. In certain contexts they will need an accurate graph, while in other contexts a rough sketch will be sufficient. Students should use appropriate tools (paper and pencil versus technology) strategically.

Closure

- Find three ordered pairs that are solutions of the equation $y = 2x - 3$. Draw the graph. *Sample answer:* $(-1, -5)$, $(0, -3)$, and $(1, -1)$

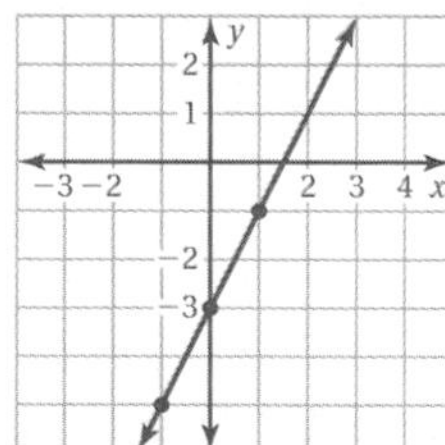

Technology For the Teacher

The Dynamic Planning Tool

Editable Teacher's Resources at *BigIdeasMath.com*

2 ACTIVITY: Using a Graphing Calculator

Use a graphing calculator to graph $y = 2x + 5$.

a. Enter the equation $y = 2x + 5$ into your calculator.

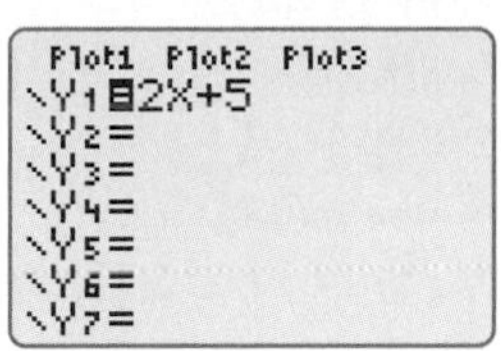

b. Check the settings of the *viewing window*. The boundaries of the graph are set by the minimum and maximum x- and y-values. The number of units between the tick marks are set by the x- and y-scales.

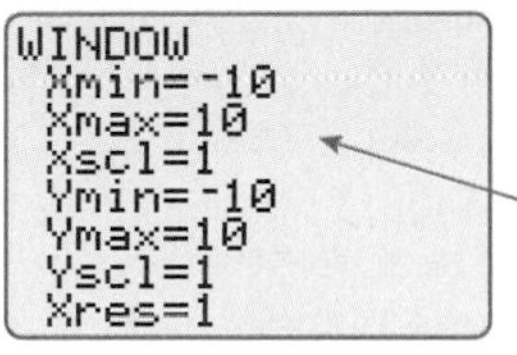

This is the standard viewing window.

c. Graph $y = 2x + 5$ on your calculator.

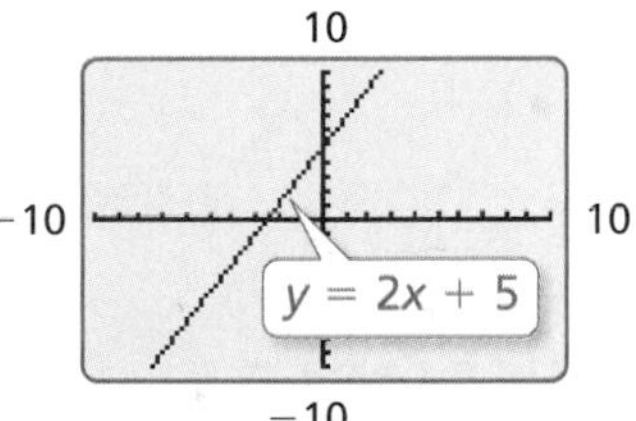

d. Change the settings of the viewing window to match those shown.

Compare the two graphs.

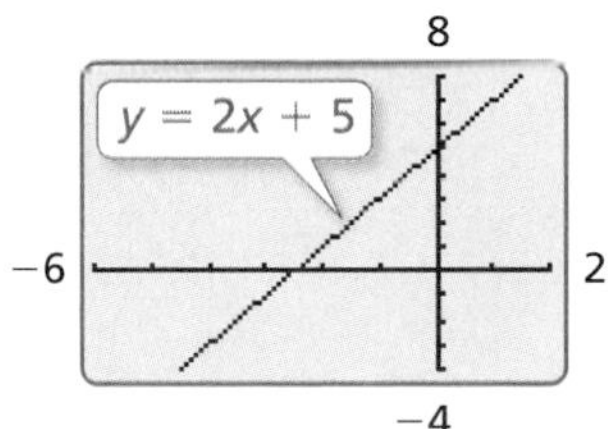

What Is Your Answer?

3. **IN YOUR OWN WORDS** How can you recognize a linear equation? How can you draw its graph? Write an equation that is linear. Write an equation that is *not* linear.

4. Use a graphing calculator to graph $y = 5x - 12$ in the standard viewing window.

   a. Can you tell where the line crosses the x-axis? Can you tell where the line crosses the y-axis?

   b. How can you adjust the viewing window so that you can determine where the line crosses the x- and y-axes?

5. **CHOOSE TOOLS** You want to graph $y = 2.5x - 3.8$. Would you graph it by hand or using a graphing calculator? Why?

Practice Use what you learned about graphing linear equations to complete Exercises 3 and 4 on page 46.

2.1 Lesson

Key Vocabulary
linear equation, *p. 44*
solution of a linear equation, *p. 44*

Key Idea

Linear Equations

A **linear equation** is an equation whose graph is a line. The points on the line are **solutions** of the equation.

You can use a graph to show the solutions of a linear equation. The graph below is for the equation $y = x + 1$.

x	y	(x, y)
-1	0	$(-1, 0)$
0	1	$(0, 1)$
2	3	$(2, 3)$

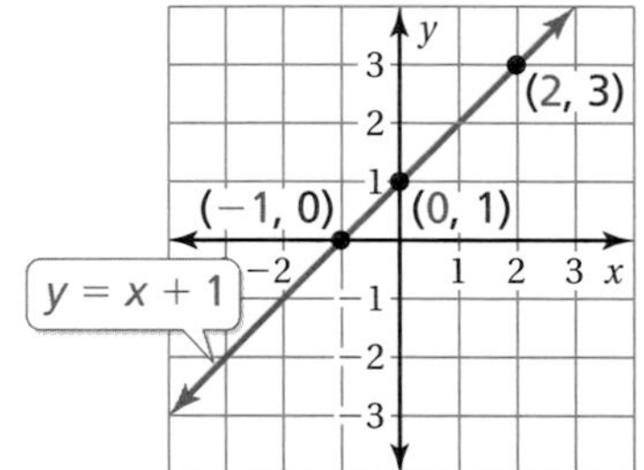

Remember

An ordered pair (x, y) is used to locate a point in a coordinate plane.

EXAMPLE 1 Graphing a Linear Equation

Graph $y = -2x + 1$.

Step 1: Make a table of values.

x	$y = -2x + 1$	y	(x, y)
-1	$y = -2(-1) + 1$	3	$(-1, 3)$
0	$y = -2(0) + 1$	1	$(0, 1)$
2	$y = -2(2) + 1$	-3	$(2, -3)$

Step 2: Plot the ordered pairs.

Step 3: Draw a line through the points.

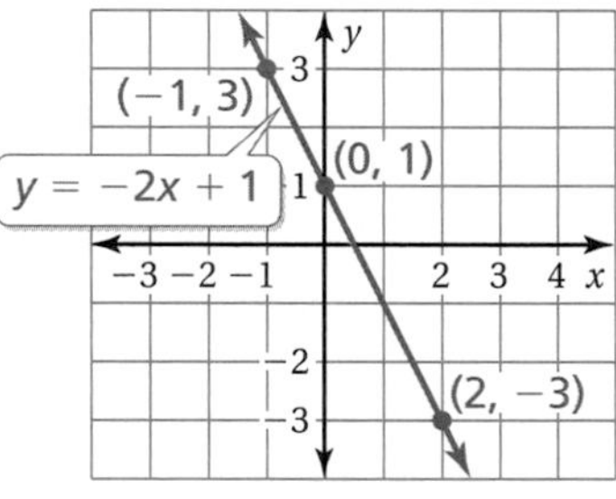

Check

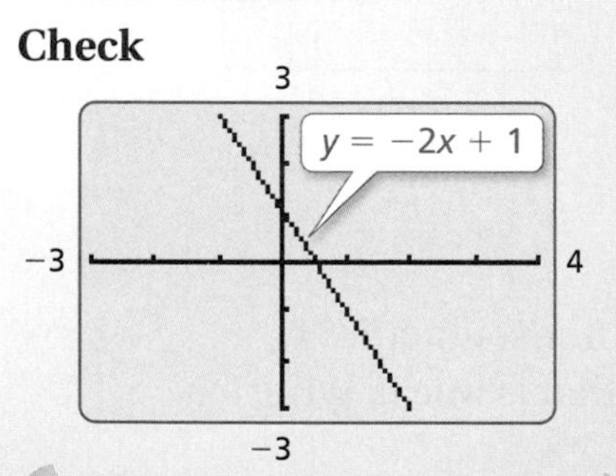

Key Idea

Graphing Horizontal and Vertical Lines

The graph of $y = b$ is a horizontal line passing through $(0, b)$.

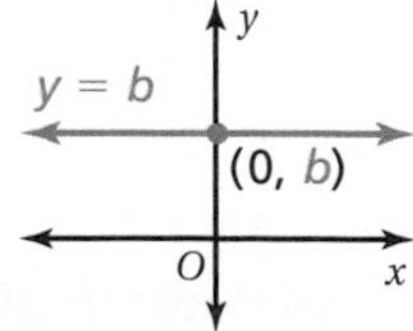

The graph of $x = a$ is a vertical line passing through $(a, 0)$.

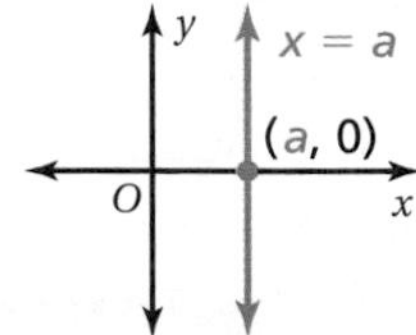

Laurie's Notes

Introduction

Connect

- **Yesterday:** Students explored the graphs of linear equations. (MP3a, MP4, MP5)
- **Today:** Students will graph linear equations using a table of values.

Motivate

- Discuss a fact about wind speeds related to Example 3. During a wild April storm in 1934, a wind gust of 231 miles per hour (372 kilometers per hour) pushed across the summit of Mt. Washington in New Hampshire. This wind speed still stands as the all-time surface wind speed record.

Lesson Notes

Key Idea

- Define *linear equation* and *solutions* of the equation.
- Note the use of color in the input-output table. The equation used is a simple equation that helps students focus on the representation of the solutions as ordered pairs. The y-coordinate is always 1 greater than the x-coordinate, just as the equation states.

Example 1

? As a quick review, ask a volunteer to review the rules for integer multiplication. If the factors have the same sign, the product is positive. If the factors have different signs, the product is negative.

- Write the 4-column table. Take the time to show how the x-coordinate is being substituted in the second column. The number in blue is the only quantity that varies (variable); the other quantities are always the same (constant). Values from the first and third columns form the ordered pair.

? "From the graph, can you estimate the solution when $x = \frac{1}{2}$? Verify your answer by solving the equation when $x = \frac{1}{2}$." yes; $\left(\frac{1}{2}, 0\right)$

- Students can use x- and y-intercepts, general slope of a line, and the table feature to check their graph on a graphing calculator.

Key Idea

- Students are sometimes confused by the equations $x = a$ and $y = b$. Explain to students that a and b are variables. They can equal any number.
- **Teaching Tip:** Another way to discuss the equation $y = b$ is to say that "y always equals a certain number, while x can equal anything." For example, if $y = -4$, the table of values will look like this:

x	-1	0	1	2
y	-4	-4	-4	-4

- **Teaching Tip:** Another way to discuss the equation $x = a$ is to say that "x always equals a certain number, while y can equal anything." For example, if $x = -2$, the table of values will look like this:

x	-2	-2	-2	-2
y	-1	0	1	2

Goal Today's lesson is graphing **linear equations.**

Start Thinking! and Warm Up

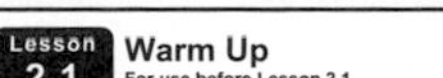

Lesson 2.1 Start Thinking!
For use before Lesson 2.1

Think about how much energy you have on an average day.

Graph your energy level (on a scale of 0 to 10) throughout an average day.

Are any sections of your graph linear?

Extra Example 1

Graph $y = \frac{1}{2}x - 3$.

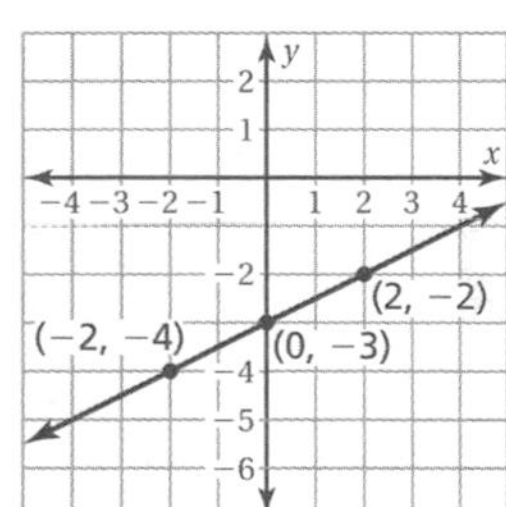

Extra Example 2

a. Graph $y = 4$.

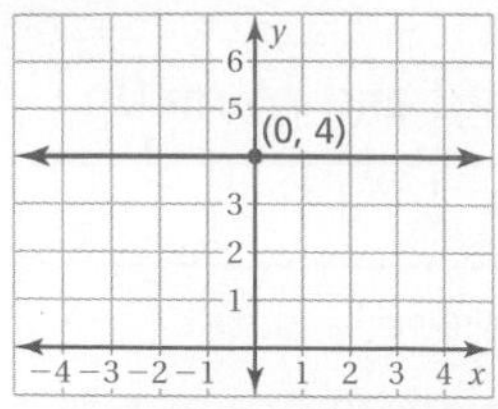

b. Graph $x = -1$.

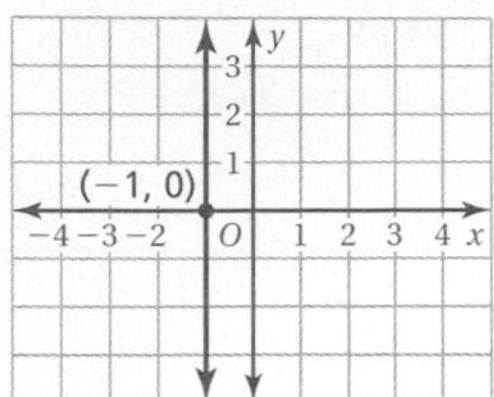

On Your Own

1–4. See Additional Answers.

Extra Example 3

The cost y (in dollars) for making friendship bracelets is $y = 0.5x + 2$, where x is the number of bracelets.

a. Graph the equation.

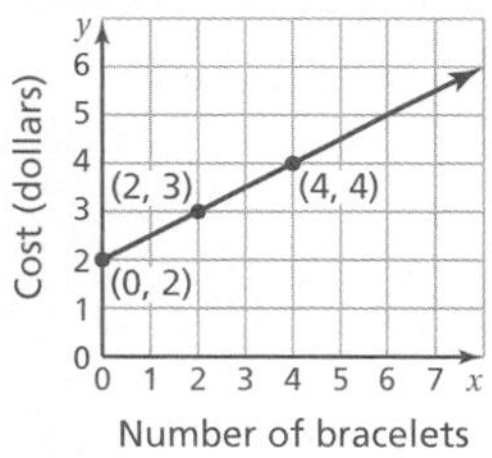

b. How many bracelets can be made for $10? 16

On Your Own

5. 8 hours after it enters the Gulf of Mexico

English Language Learners

Vocabulary

Make sure students understand that the graph of a *linear* equation is a *line*. Only two points are needed to graph a line, but if one of the points is incorrect the wrong line will be graphed. Plotting three points for a line in the coordinate plane and making sure that the points form a line provides students with a check when graphing.

Laurie's Notes

Example 2

? "What are other points on the line $y = -3$?" *Sample answer:* (5, −3), or anything of the form $(x, -3)$

? "What are other points on the line $x = 2$?" *Sample answer:* (2, −3), or anything of the form $(2, y)$

On Your Own

- Ask volunteers to share their graphs at the board.
- Students may ask how to graph $x = -4$ on their calculators. To create a graph using a graphing calculator, the equation must begin with "$y =$." Any equation not containing a y cannot be graphed on a calculator.

Example 3

? "What does x represent in the problem? What does y represent?" x = number of hours after the storm enters the Gulf of Mexico; y = wind speed

- Work through the problem using the 4-column table to generate solutions of the equation.
- Note that the y-coordinate is much greater than the x-coordinate. For this reason, a broken vertical axis is used. Students should *not* scale the y-axis beginning at 0.

? "Why are only non-negative numbers substituted for x?" Because x equals the number of hours after the storm enters the Gulf of Mexico, you do not know if the equation makes sense for x-values before that.

- Note that the ordered pairs are all located in Quadrant I because x is a non-negative number. Even though this restriction was not stated explicitly, you know from reading the description of x that it needs to be non-negative.
- In part (b), help students read the graph. Starting with a y-value of 74 on the y-axis, trace horizontally until you reach the graph of the line, and then trace straight down (vertically) to the x-axis. The x-coordinate is 4.

On Your Own

- **Neighbor Check:** Have students work independently and then have their neighbor check their work. Have students discuss any discrepancies.

Closure

- Explain how you know if an equation is linear. *Sample answer:* The graph of the equation is a line.

Technology For the Teacher

The Dynamic Planning Tool
Editable Teacher's Resources at *BigIdeasMath.com*

EXAMPLE 2 Graphing a Horizontal Line and a Vertical Line

a. Graph $y = -3$.

The graph of $y = -3$ is a horizontal line passing through $(0, -3)$. Draw a horizontal line through this point.

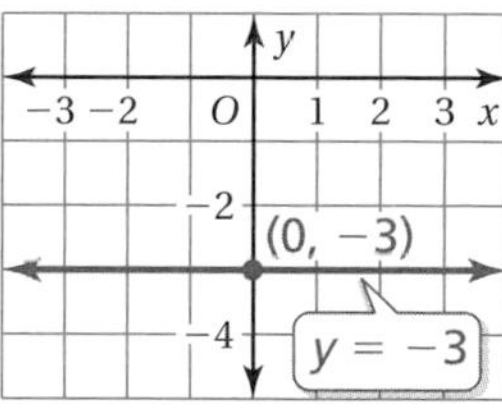

b. Graph $x = 2$.

The graph of $x = 2$ is a vertical line passing through $(2, 0)$. Draw a vertical line through this point.

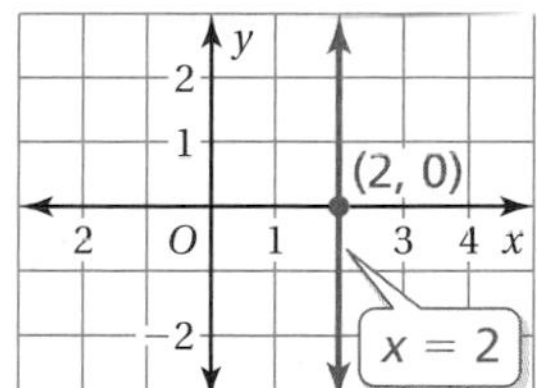

On Your Own

Now You're Ready Exercises 5–16

Graph the linear equation. Use a graphing calculator to check your graph, if possible.

1. $y = 3x$ 2. $y = -\frac{1}{2}x + 2$ 3. $x = -4$ 4. $y = -1.5$

EXAMPLE 3 Real-Life Application

The wind speed y (in miles per hour) of a tropical storm is $y = 2x + 66$, where x is the number of hours after the storm enters the Gulf of Mexico.

a. Graph the equation.

b. When does the storm become a hurricane?

A tropical storm becomes a hurricane when wind speeds are at least 74 miles per hour.

a. Make a table of values.

x	y = 2x + 66	y	(x, y)
0	$y = 2(0) + 66$	66	(0, 66)
1	$y = 2(1) + 66$	68	(1, 68)
2	$y = 2(2) + 66$	70	(2, 70)
3	$y = 2(3) + 66$	72	(3, 72)

Plot the ordered pairs and draw a line through the points.

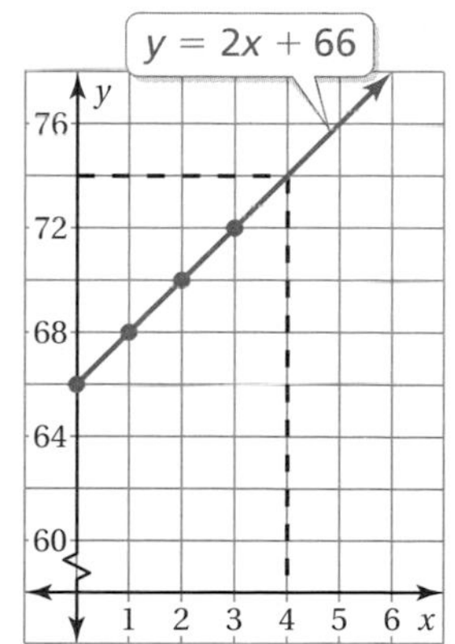

b. From the graph, you can see that $y = 74$ when $x = 4$. So, the storm becomes a hurricane 4 hours after it enters the Gulf of Mexico.

On Your Own

5. **WHAT IF?** In Example 3, the wind speed of the storm is $y = 1.5x + 62$. When does the storm become a hurricane?

2.1 Exercises

Vocabulary and Concept Check

1. **VOCABULARY** What type of graph represents the solutions of the equation $y = 2x + 3$?

2. **WHICH ONE DOESN'T BELONG?** Which equation does *not* belong with the other three? Explain your reasoning.

$y = 0.5x - 0.2$	$4x + 3 = y$	$y = x^2 + 6$	$\frac{3}{4}x + \frac{1}{3} = y$

Practice and Problem Solving

PRECISION Copy and complete the table. Plot the two solution points and draw a line *exactly* through the two points. Find a different solution point on the line.

3.

x		
$y = 3x - 1$		

4.

x		
$y = \frac{1}{3}x + 2$		

Graph the linear equation. Use a graphing calculator to check your graph, if possible.

5. $y = -5x$
6. $y = \frac{1}{4}x$
7. $y = 5$
8. $x = -6$
9. $y = x - 3$
10. $y = -7x - 1$
11. $y = -\frac{x}{3} + 4$
12. $y = \frac{3}{4}x - \frac{1}{2}$
13. $y = -\frac{2}{3}$
14. $y = 6.75$
15. $x = -0.5$
16. $x = \frac{1}{4}$

17. **ERROR ANALYSIS** Describe and correct the error in graphing the equation.

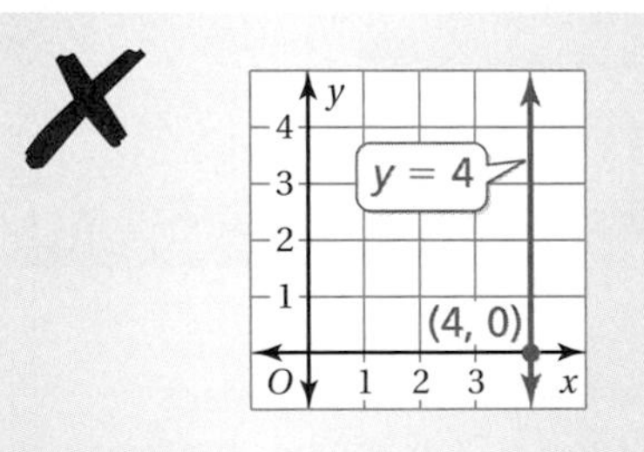

18. **MESSAGING** You sign up for an unlimited text messaging plan for your cell phone. The equation $y = 20$ represents the cost y (in dollars) for sending x text messages. Graph the equation. What does the graph tell you?

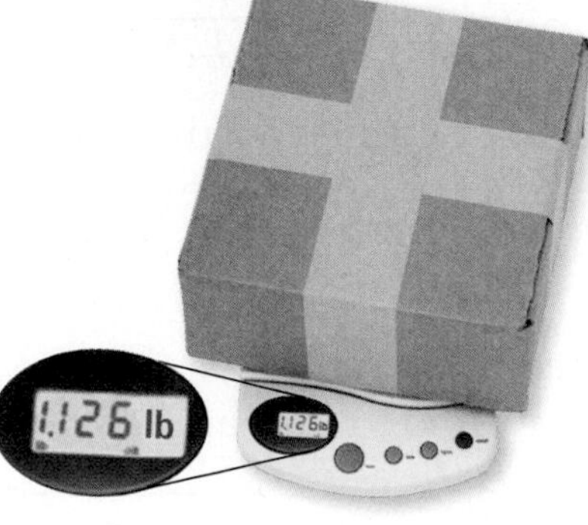

19. **MAIL** The equation $y = 2x + 3$ represents the cost y (in dollars) of mailing a package that weighs x pounds.
    a. Graph the equation.
    b. Use the graph to estimate how much it costs to mail the package.
    c. Use the equation to find exactly how much it costs to mail the package.

Assignment Guide and Homework Check

Level	Assignment	Homework Check
Average	1–4, 9–17, 21, 23–25, 28–32	11, 13, 21, 25
Advanced	1, 2, 10–26 even, 25, 27, 28–32	12, 14, 20, 26

For Your Information

- **Exercises 5–16** Students can use *x*- and *y*-intercepts, general slope of a line, and the table feature to check their graph on a graphing calculator.

Common Errors

- **Exercises 5–16** Students may make a calculation error for one of the ordered pairs in a table of values. If they only find two ordered pairs for the graph, they may not recognize their mistake. Encourage them to find at least three ordered pairs when drawing a graph.
- **Exercises 7, 13, and 14** Students may draw a vertical line through a point on the *x*-axis instead of through the corresponding point on the *y*-axis. Remind them that the equation is a horizontal line. Ask them to identify the *y*-coordinate for several *x*-coordinates. For example, what is the *y*-coordinate for $x = 5$? $x = 6$? $x = -4$? Students should answer with the same *y*-coordinate each time.
- **Exercises 8, 15, and 16** Students may draw a horizontal line through a point on the *y*-axis instead of through the corresponding point on the *x*-axis. Remind them that the equation is a vertical line. Ask them to identify the *x*-coordinate for several *y*-coordinates. For example, what is the *x*-coordinate for $y = 3$? $y = -1$? $y = 0$? Students should answer with the same *x*-coordinate each time.
- **Exercises 20–23** Students may make a mistake in solving for *y*, such as using the same operation instead of the opposite operation.

2.1 Record and Practice Journal

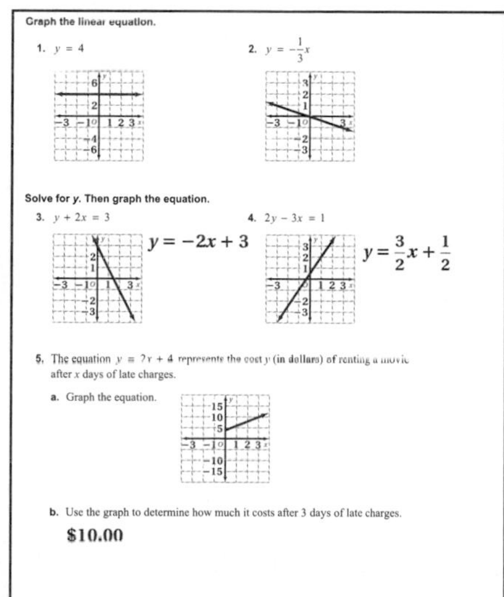

Graph the linear equation.

1. $y = 4$
2. $y = -\frac{1}{3}x$

Solve for *y*. Then graph the equation.

3. $y + 2x = 3$ — $y = -2x + 3$
4. $2y - 3x = 1$ — $y = \frac{3}{2}x + \frac{1}{2}$

5. The equation $y = 2x + 4$ represents the cost *y* (in dollars) of renting a movie after *x* days of late charges.
   a. Graph the equation.
   b. Use the graph to determine how much it costs after 3 days of late charges. **$10.00**

Technology For the Teacher
Answer Presentation Tool

Vocabulary and Concept Check

1. a line
2. $y = x^2 + 6$ does not belong because it is not a linear equation.

Practice and Problem Solving

3. *Sample answer:*

x	0	1
$y = 3x - 1$	−1	2

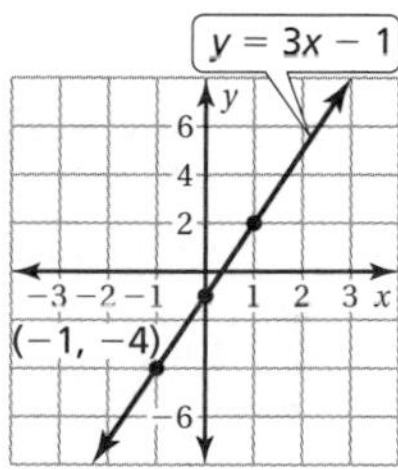

4. *Sample answer:*

x	0	3
$y = \frac{1}{3}x + 2$	2	3

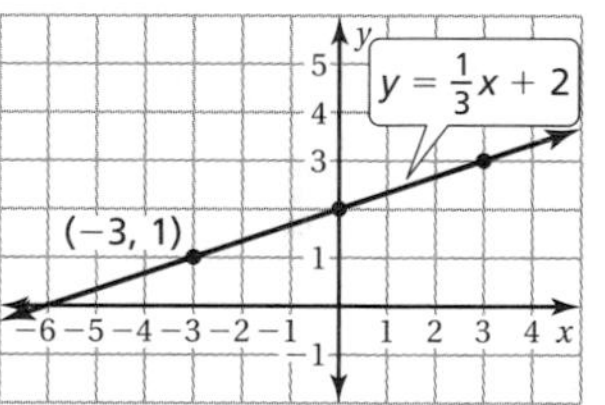

5.

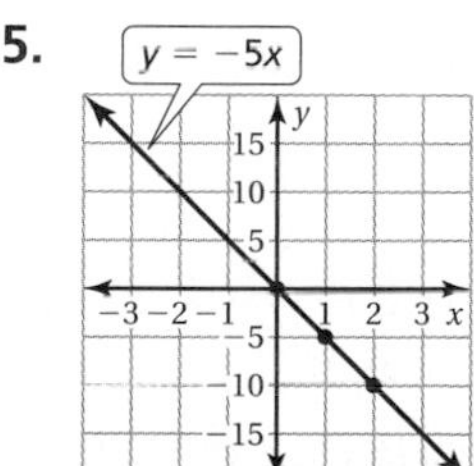

6.

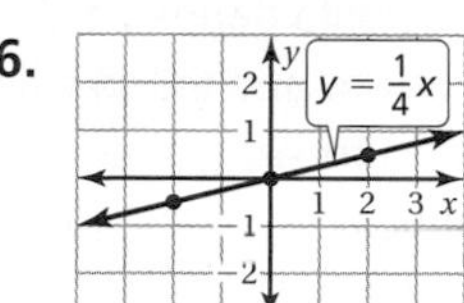

7–19. See Additional Answers.

Practice and Problem Solving

20. $y = 3x + 1$

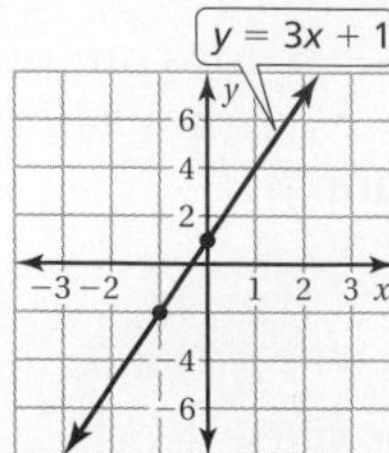

21–24. See Additional Answers.

25. See *Taking Math Deeper.*

26–27. See Additional Answers.

Fair Game Review

28. (5, 3) **29.** (−6, 6)

30. (2, −2) **31.** (−4, −3)

32. B

Mini-Assessment

Graph the linear equation.

1. Graph $y = -\frac{1}{2}x + 2$.

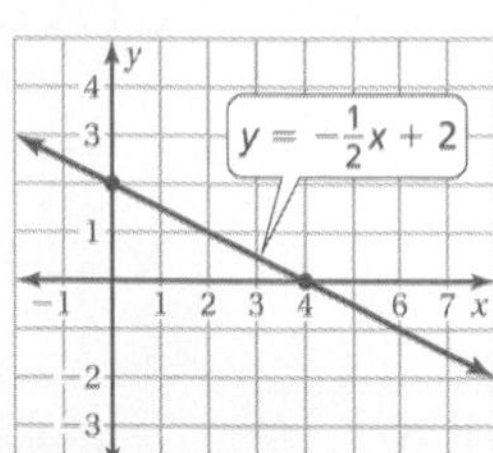

2. You have \$100 in your savings account and plan to deposit \$20 each month. Write and graph a linear equation that represents the balance in your account. $y = 20x + 100$

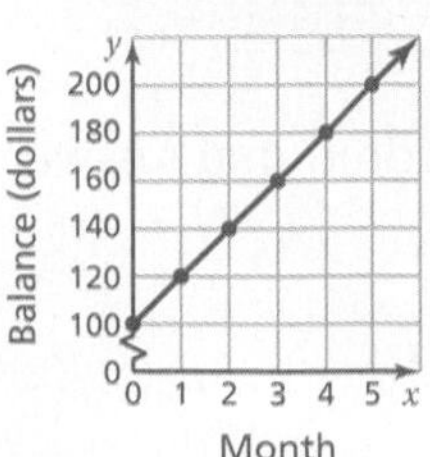

Taking Math Deeper

Exercise 25

Some of the information for this exercise is given in the photo and some is given in the text. It is a good idea to start by listing all of the given information.

List the given information.

- The camera can store 250 pictures.
- 1 second of video = 2 pictures
- Video time used = 90 seconds
- Let y = number of pictures
- Let x = number of seconds of video

a. Write and graph an equation for x and y.

$$y + 2x = 250$$
$$y = -2x + 250$$

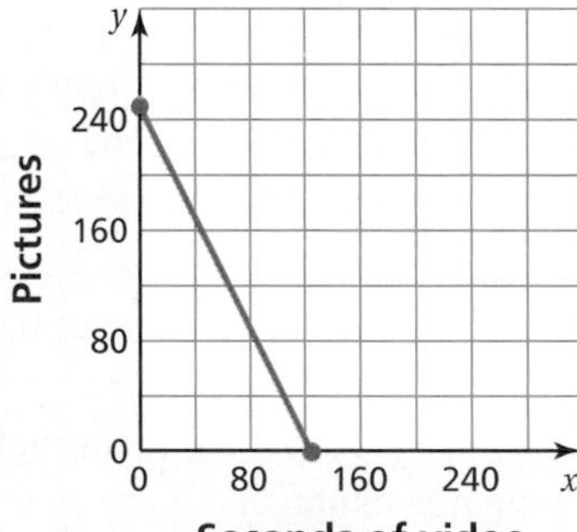

3 **b.** Answer the question.
When $x = 90$, the value of y is as follows.

$$y = -2x + 250$$
$$= -2(90) + 250$$
$$= 70$$

Your camera can store 70 pictures.

Project

Research digital cameras. Find the number of pictures that can be stored on five different cameras. Compare the prices of the cameras. What do you consider to be the better buy? Why?

Reteaching and Enrichment Strategies

If students need help. . .	If students got it. . .
Resources by Chapter • Practice A and Practice B • Puzzle Time Record and Practice Journal Practice Differentiating the Lesson Lesson Tutorials Skills Review Handbook	Resources by Chapter • Enrichment and Extension Start the next section

Solve for y. Then graph the equation. Use a graphing calculator to check your graph.

20. $y - 3x = 1$

21. $5x + 2y = 4$

22. $-\frac{1}{3}y + 4x = 3$

23. $x + 0.5y = 1.5$

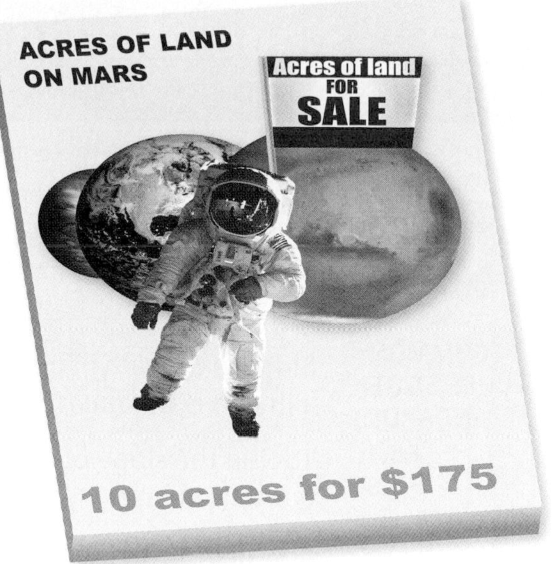

24. SAVINGS You have $100 in your savings account and plan to deposit $12.50 each month.

a. Write and graph a linear equation that represents the balance in your account.

b. How many months will it take you to save enough money to buy 10 acres of land on Mars?

25. CAMERA One second of video on your digital camera uses the same amount of memory as two pictures. Your camera can store 250 pictures.

a. Write and graph a linear equation that represents the number y of pictures your camera can store if you take x seconds of video.

b. How many pictures can your camera store after you take the video shown?

26. PROBLEM SOLVING Along the U.S. Atlantic Coast, the sea level is rising about 2 millimeters per year. How many millimeters has sea level risen since you were born? How do you know? Use a linear equation and a graph to justify your answer.

27.  The sum S of the measures of the angles of a polygon is $S = (n - 2) \cdot 180°$, where n is the number of sides of the polygon.

a. Plot four points (n, S) that satisfy the equation. Do the points lie on a line? Explain your reasoning.

b. Does the value $n = 3.5$ make sense in the context of the problem? Explain your reasoning.

Fair Game Review What you learned in previous grades & lessons

Write the ordered pair corresponding to the point. *(Skills Review Handbook)*

28. Point A

29. Point B

30. Point C

31. Point D

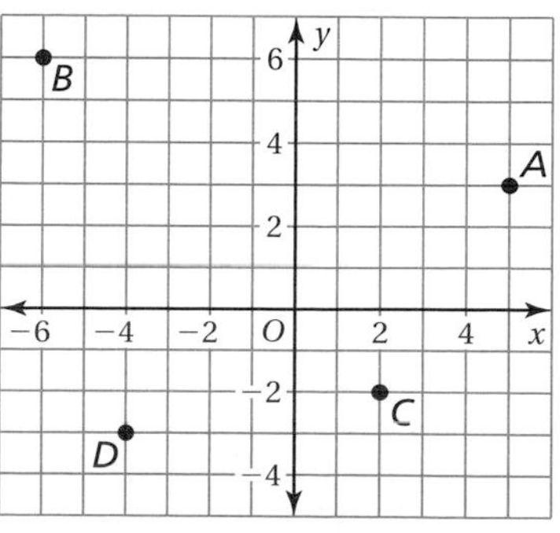

32. MULTIPLE CHOICE A debate team has 15 female members. The ratio of females to males is 3 : 2. How many males are on the debate team? *(Skills Review Handbook)*

 6 (B) 10  22 (D) 25

2.2 Slope of a Line

COMMON CORE STATE STANDARDS
F.IF.4
F.IF.6

Essential Question How can the slope of a line be used to describe the line?

Slope is the rate of change between any two points on a line. It is the measure of the *steepness* of the line.

To find the slope of a line, find the ratio of the change in y (vertical change) to the change in x (horizontal change).

$$\text{slope} = \frac{\text{change in } y}{\text{change in } x}$$

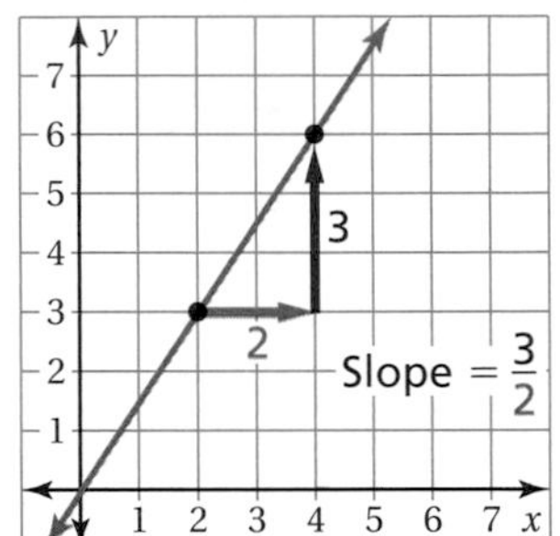

1 ACTIVITY: Finding the Slope of a Line

Work with a partner. Find the slope of each line using two methods.

Method 1: Use the two black points. •

Method 2: Use the two pink points. •

Do you get the same slope using each method? Why do you think this happens?

a.

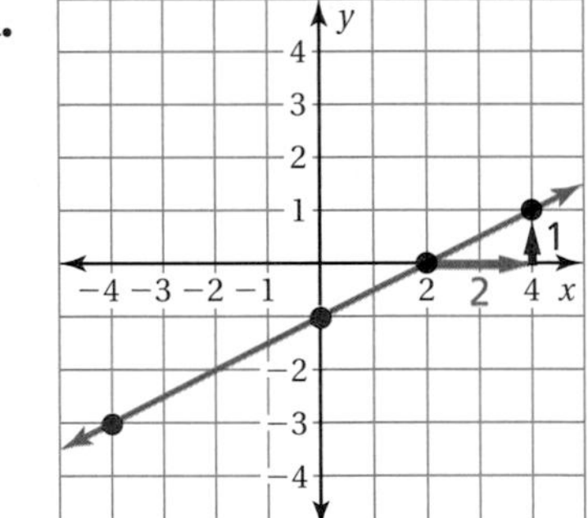

b.

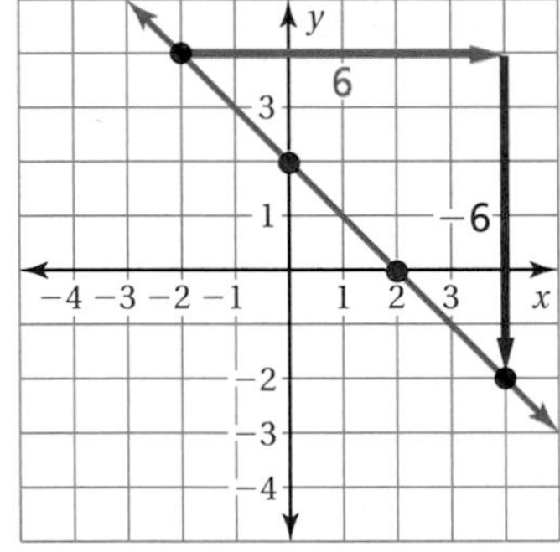

c.

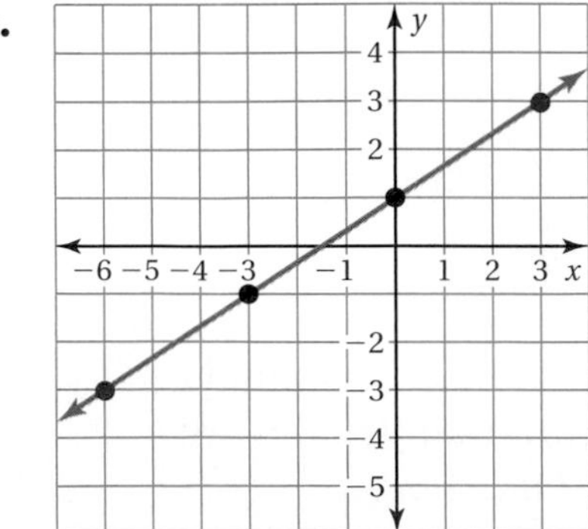

d.

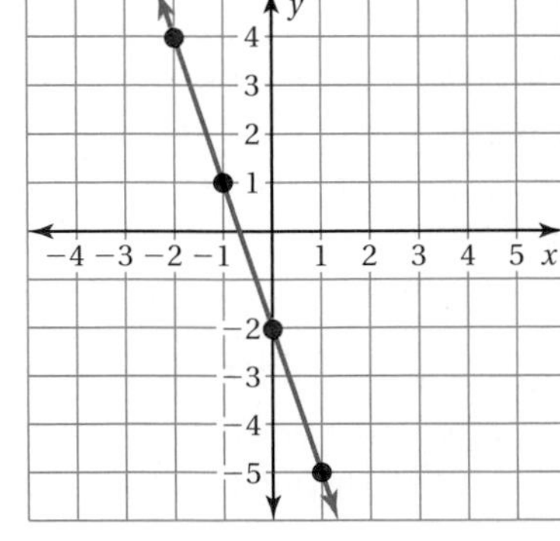

Laurie's Notes

Introduction

Standards for Mathematical Practice

- **MP1a Make Sense of Problems:** The goal is for students to use two different pairs of points on a line to find the line's slope. Drawing the arrow diagrams will help students to visualize the *slope triangle*. Students may recognize the triangles are similar and proportions can be formed.

Motivate

? "How many of you have been on a roller coaster?"

- Discuss with students what makes one roller coaster more thrilling than another. Students will usually describe how quickly the coaster drops or the steepness of the hill. This is similar to the *change in y* of a line when finding the slope.

Activity Notes

Discuss

? "Does anyone remember what is meant by slope of a line?" At least one student should recall that it measures the steepness of a line.

- Write the definition for slope. Sketch the graph shown to demonstrate what is meant by change in y (red vertical arrow) and change in x (blue horizontal arrow).
- Remind students that slope is always the change in y in the numerator and the change in x in the denominator. This can be confusing for students because graphs are read from left to right, and we have a tendency to move in the x-direction first. For this reason, students want to write the change in x in the numerator.

? "Can the change in x be negative? Explain." Yes; moving to the left horizontally is negative.

? "Can the change in y be negative? Explain." Yes; moving down vertically is negative.

Activity 1

- Encourage students to draw the change arrows for each pair of points. Label the change in x (or y) next to the arrow.
- **Big Idea:** The slope of a line is always the same regardless of what two ordered pairs are selected.
- **Common Error:** Students may forget to make the change negative when moving downward in the y-direction.
- **MP2 Reason Abstractly and Quantitatively** and **MP3b Critique the Reasoning of Others:** Students are asked to reason why the slopes are the same. They will form their own conjecture and listen to reasons offered by other students.

Common Core State Standards

F.IF.4 For a function that models a relationship between two quantities, interpret key features of graphs and tables in terms of the quantities, and sketch graphs showing key features given a verbal description of the relationship.

F.IF.6 Calculate and interpret the average rate of change of a function (presented symbolically or as a table) over a specified interval. Estimate the rate of change from a graph.

Previous Learning

Students should know that slope is the rate of change between two points on a line.

Start Thinking! and Warm Up

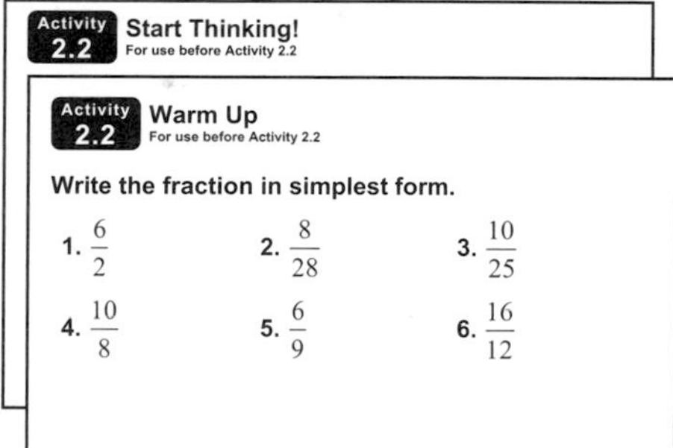

Activity 2.2 Start Thinking!
For use before Activity 2.2

Activity 2.2 Warm Up
For use before Activity 2.2

Write the fraction in simplest form.

1. $\frac{6}{2}$ 2. $\frac{8}{28}$ 3. $\frac{10}{25}$
4. $\frac{10}{8}$ 5. $\frac{6}{9}$ 6. $\frac{16}{12}$

2.2 Record and Practice Journal

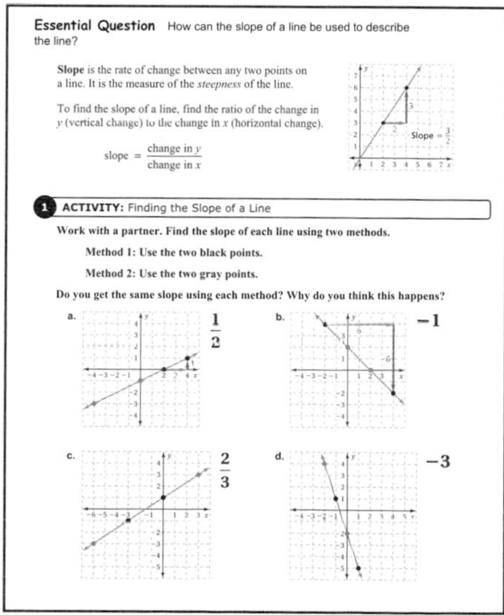

Essential Question How can the slope of a line be used to describe the line?

Slope is the rate of change between any two points on a line. It is the measure of the *steepness* of the line.

To find the slope of a line, find the ratio of the change in y (vertical change) to the change in x (horizontal change).

$$\text{slope} = \frac{\text{change in } y}{\text{change in } x}$$

1 ACTIVITY: Finding the Slope of a Line

Work with a partner. Find the slope of each line using two methods.

Method 1: Use the two black points.

Method 2: Use the two gray points.

Do you get the same slope using each method? Why do you think this happens?

a. $\frac{1}{2}$ b. -1

c. $\frac{2}{3}$ d. -3

Differentiated Instruction

Kinesthetic

Help students develop number sense about slope. Have them draw lines in the coordinate plane through the following pairs of points.

(0, 0) and (3, 5) (0, 0) and (3, 4)
(0, 0) and (3, 3) (0, 0) and (3, 2)
(0, 0) and (3, 1)

Next have students find the slope of each line. Point out that the line passing through (3, 3) has a slope of 1. The lines with y-coordinates greater than 3 have a slope greater than 1. The lines with y-coordinates less than 3 have a slope less than 1. For positive slopes, the steeper lines will have a greater slope.

2.2 Record and Practice Journal

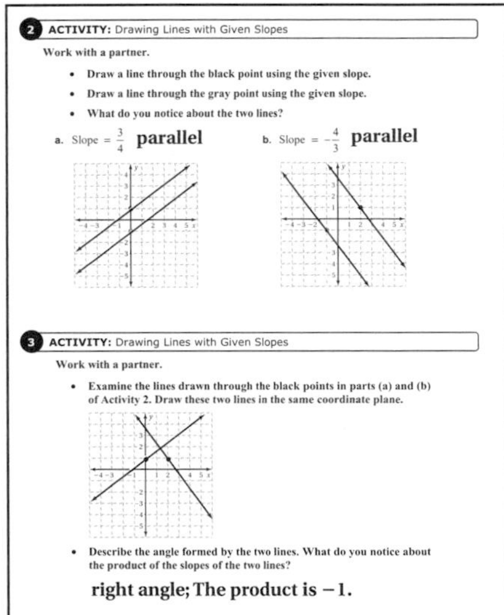

2 ACTIVITY: Drawing Lines with Given Slopes

Work with a partner.

- Draw a line through the black point using the given slope.
- Draw a line through the gray point using the given slope.
- What do you notice about the two lines?

a. Slope $= \frac{3}{4}$ **parallel** b. Slope $= -\frac{4}{3}$ **parallel**

3 ACTIVITY: Drawing Lines with Given Slopes

Work with a partner.

- Examine the lines drawn through the black points in parts (a) and (b) of Activity 2. Draw these two lines in the same coordinate plane.
- Describe the angle formed by the two lines. What do you notice about the product of the slopes of the two lines?

right angle; The product is -1.

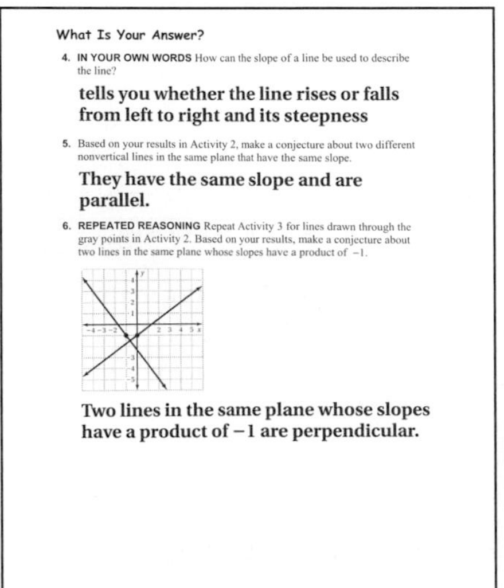

What Is Your Answer?

4. IN YOUR OWN WORDS How can the slope of a line be used to describe the line?

tells you whether the line rises or falls from left to right and its steepness

5. Based on your results in Activity 2, make a conjecture about two different nonvertical lines in the same plane that have the same slope.

They have the same slope and are parallel.

6. REPEATED REASONING Repeat Activity 3 for lines drawn through the gray points in Activity 2. Based on your results, make a conjecture about two lines in the same plane whose slopes have a product of -1.

Two lines in the same plane whose slopes have a product of -1 are perpendicular.

Laurie's Notes

Activity 2

- In addition to being able to determine the slope of a line that has been graphed, you want students to be able to draw a line that has a particular slope. This is the goal of Activity 2.

? "What does it mean for a line to have a slope of $\frac{3}{4}$?" For every 3 units of change in the y-direction, there is a change of 4 units in the x-direction.

? "What does it mean for a line to have a slope of $-\frac{4}{3}$?" For every -4 units of change in the y-direction, there is a change of 3 units in the x-direction. This is the same as 4 units in the y-direction and -3 units in the x-direction.

- **Teaching Tip:** If possible, give each pair of students two colored pencils. Have them draw the first line (through the black point) in one color and draw the second line (through the pink point) in the other color.

? "What do you notice about the two lines you have drawn?" Students should observe that the lines are parallel. Students may also observe that the positive slopes rise (from left to right) and the negative slopes fall (from left to right).

- **Big Idea:** Slope is a measure of the steepness of a line. Two different lines with the same slope are parallel (they have the same steepness).

Activity 3

- The goal of this activity is to have students observe that in Activity 2 the pairs of lines drawn in part (a) are perpendicular to the pairs of lines drawn in part (b). All students may not know this vocabulary word so they may say that the lines meet at right angles.
- **MP8 Look for and Express Regularity in Repeated Reasoning:** Students are asked to make an observation. If time permits students should test their conjecture with several examples (see Question 6).

What Is Your Answer?

- Discuss students' response to the first question.
- In the discussion of the slopes having a product of -1, review the definition of reciprocals.

Closure

- Plot the point (0, 3). Draw the line through this point that has a slope of $\frac{1}{3}$. Name two points on the line.

Sample answer: (3, 4), (−3, 2)

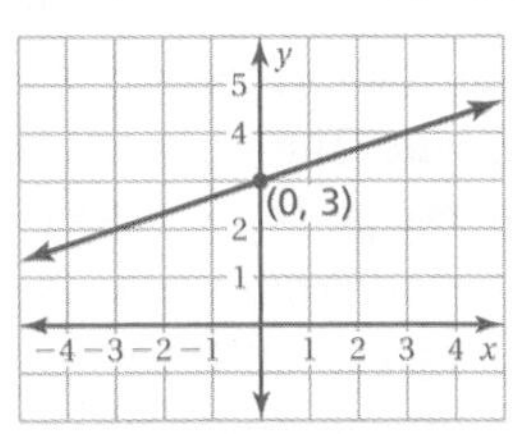

Technology For the Teacher

The Dynamic Planning Tool
Editable Teacher's Resources at *BigIdeasMath.com*

2 ACTIVITY: Drawing Lines with Given Slopes

Work with a partner.

- **Draw a line through the black point using the given slope.**
- **Draw a line through the pink point using the given slope.**
- **What do you notice about the two lines?**

a. Slope $= \frac{3}{4}$

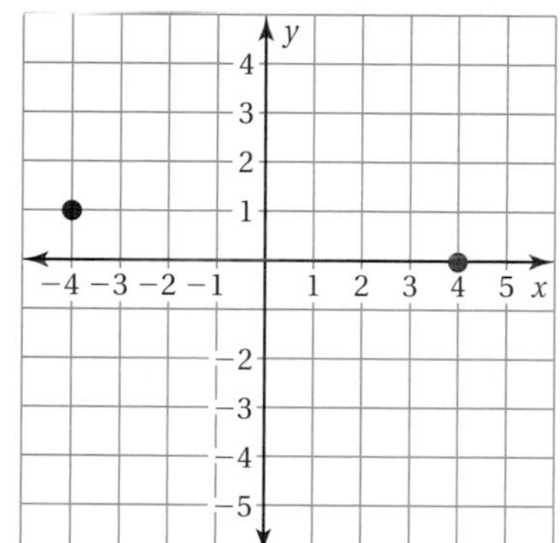

b. Slope $= -\frac{4}{3}$

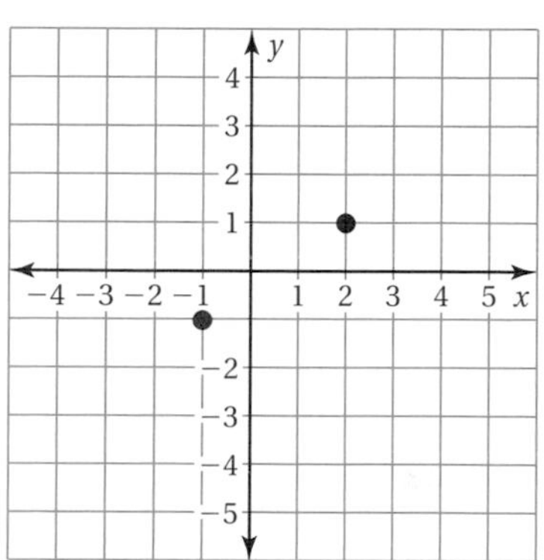

3 ACTIVITY: Drawing Lines with Given Slopes

Work with a partner.

- **Examine the lines drawn through the black points in parts (a) and (b) of Activity 2. Draw these two lines in the same coordinate plane.**
- **Describe the angle formed by the two lines. What do you notice about the product of the slopes of the two lines?**

What Is Your Answer?

4. **IN YOUR OWN WORDS** How can the slope of a line be used to describe the line?
5. Based on your results in Activity 2, make a conjecture about two different nonvertical lines in the same plane that have the same slope.
6. **REPEATED REASONING** Repeat Activity 3 for the lines drawn through the pink points in Activity 2. Based on your results, make a conjecture about two lines in the same plane whose slopes have a product of -1.

Practice Use what you learned about the slope of a line to complete Exercises 4–6 on page 53.

2.2 Lesson

Key Vocabulary
slope, *p. 50*
rise, *p. 50*
run, *p. 50*

Slope

The **slope** of a line is a ratio of the change in y (the **rise**) to the change in x (the **run**) between any two points, (x_1, y_1) and (x_2, y_2), on the line.

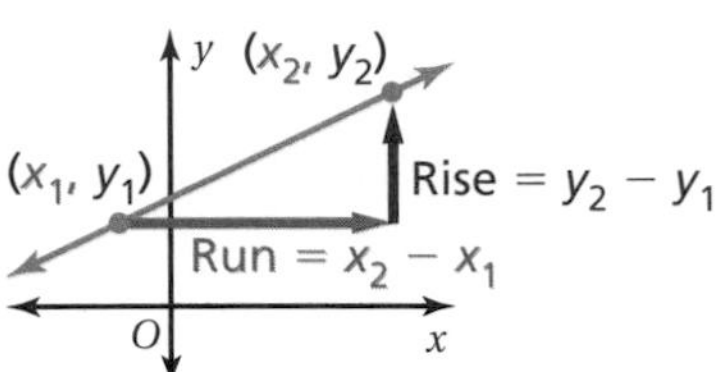

$$\text{slope} = \frac{\text{rise}}{\text{run}} = \frac{\text{change in } y}{\text{change in } x} = \frac{y_2 - y_1}{x_2 - x_1}$$

Positive slope

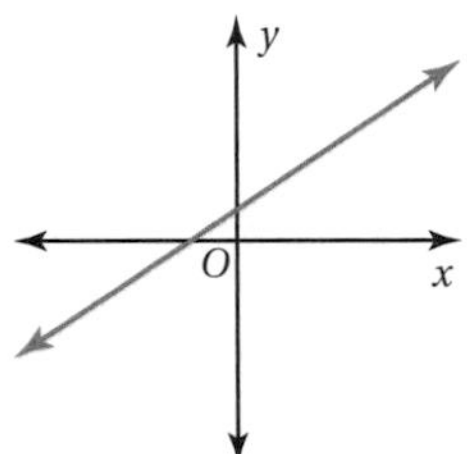

The line rises from left to right.

Negative slope

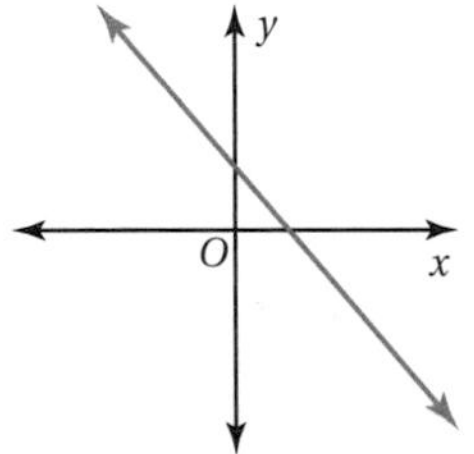

The line falls from left to right.

Reading

In the slope formula, x_1 is read as "x sub one" and y_2 is read as "y sub two". The numbers 1 and 2 in x_1 and y_2 are called *subscripts*.

EXAMPLE 1 Finding the Slope of a Line

Describe the slope of the line. Then find the slope.

a.

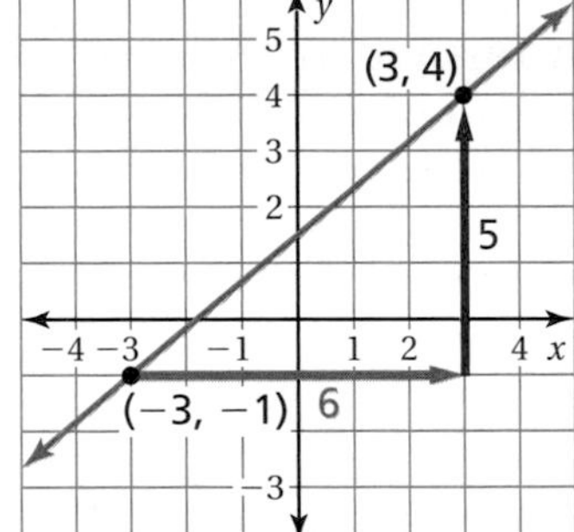

The line rises from left to right. So, the slope is positive. Let $(x_1, y_1) = (-3, -1)$ and $(x_2, y_2) = (3, 4)$.

$$\text{slope} = \frac{y_2 - y_1}{x_2 - x_1}$$

$$= \frac{4 - (-1)}{3 - (-3)}$$

$$= \frac{5}{6}$$

b.

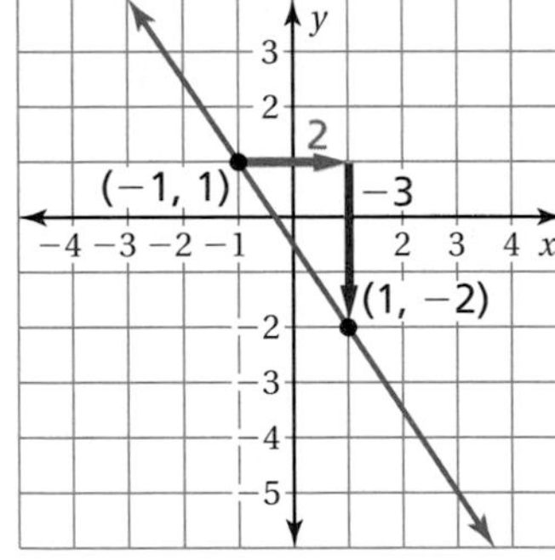

The line falls from left to right. So, the slope is negative. Let $(x_1, y_1) = (-1, 1)$ and $(x_2, y_2) = (1, -2)$.

$$\text{slope} = \frac{y_2 - y_1}{x_2 - x_1}$$

$$= \frac{-2 - 1}{1 - (-1)}$$

$$= \frac{-3}{2}, \text{ or } -\frac{3}{2}$$

Study Tip

When finding slope, you can label either point as (x_1, y_1) and the other point as (x_2, y_2).

Multi-Language Glossary at BigIdeasMath.com.

Laurie's Notes

Introduction

Connect

- **Yesterday:** Students explored slopes of lines. (MP1a, MP2, MP3b, MP8)
- **Today:** Students will find the slopes of lines in a variety of contexts.

Motivate

- Have students plot four points: $A(5, 0)$, $B(0, 5)$, $C(-5, 0)$, and $D(0, -5)$. Connect the points to form the quadrilateral *ABCD*.

? "What type of quadrilateral is *ABCD*?" Without proof, students should say square.

? "What is the slope of each side, meaning the slopes of the lines through *AB*, *BC*, *CD*, and *DA*?" Slopes of *AB* and *CD* are both −1. Slopes of *BC* and *DA* are both 1.

Lesson Notes

Key Idea

- Write the Key Idea. Define slope of a line.
- Note the use of color in the definition and on the graph. The *change in y* and the *vertical change arrow* are both red. The *change in x* and the *horizontal change arrow* are both blue.
- Discuss the difference in positive and negative slopes, a concept students explored yesterday.
- Remind students that graphs are read from left to right.
- Explain to students that you can also subtract coordinates to find the rise and run in addition to finding rise and run graphically.

Example 1

- **MP1a Make Sense of Problems:** Drawing the arrow diagrams will help students visualize the *slope triangle*.
- Students often ask if they can move in the *y*-direction first, followed by the *x*-direction. The answer is yes. Demonstrate this on either graph.
  - In part (a), start at $(-3, -1)$ and move up 5 units in the *y*-direction and then to the right 6 units in the *x*-direction. You will end at $(3, 4)$.
  - In part (b), start at $(-1, 1)$ and move down 3 units in the *y*-direction and then to the right 2 units in the *x*-direction. You will end at $(1, -2)$.
- Discuss the Study Tip. You can move in either direction first, and the labeling of the ordered pairs is arbitrary. Either point can be (x_1, y_1).

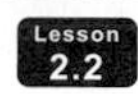

Goal Today's lesson is finding the **slope** of a line.

Start Thinking! and Warm Up

Lesson 2.2 Warm Up
For use before Lesson 2.2

Lesson 2.2 Start Thinking!
For use before Lesson 2.2

1. Each student must choose an ordered pair.
2. Choose a partner and work together to find the slope of the line joining your two points. Use a graph to help you.
3. Repeat the process several times with different partners.

Were any of the slopes positive? negative? zero?

Extra Example 1

Tell whether the slope of the line is *positive* or *negative*. Then find the slope.

a.

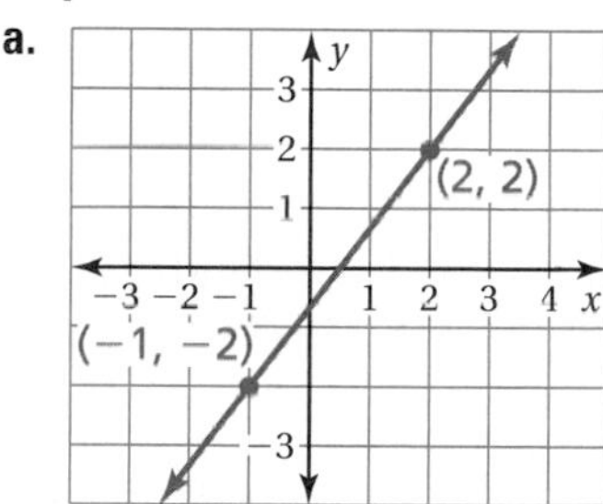

positive; $\frac{4}{3}$

b.

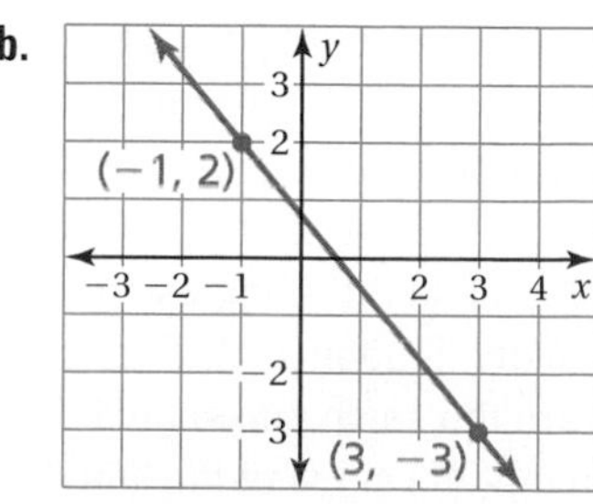

negative; $-\frac{5}{4}$

On Your Own

1. $-\frac{1}{5}$
2. $\frac{1}{3}$
3. $\frac{5}{2}$

Extra Example 2

Find the slope of the line. 0

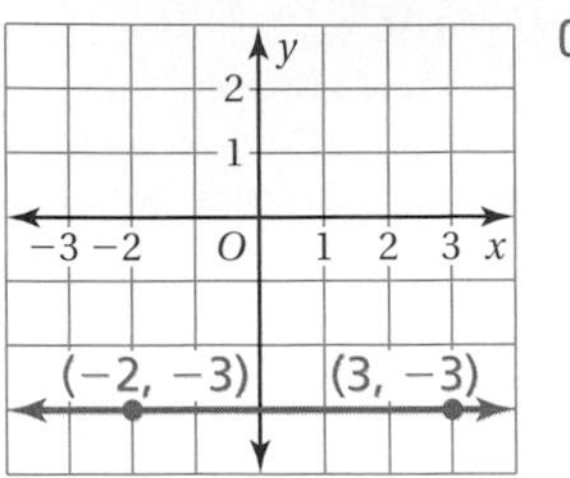

Extra Example 3

Find the slope of the line. undefined

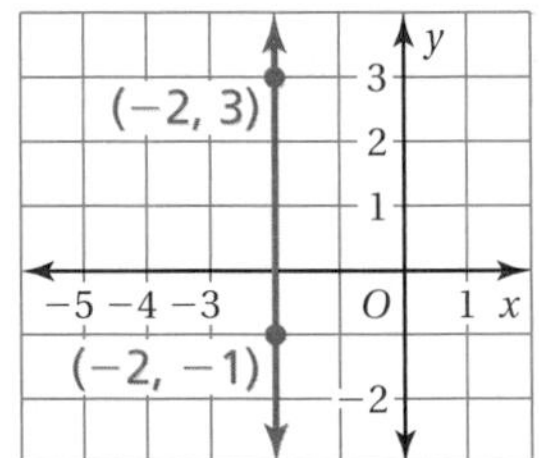

On Your Own

4. 0
5. undefined
6. undefined
7. because the change in y is zero; because the change in x is zero

English Language Learners

Comprehension

The Key Idea box states, "The slope of a line is a ratio of the change in y to the change in x between any two points on the line." Have students choose four points on a line. Use two points to find the slope. Then find the slope using the other two points. Students will find that the slopes are the same and should understand that the slope of the line is the same for the entire infinite length of the line.

Laurie's Notes

Example 2

? "How does a slope of $\frac{1}{2}$ compare to a slope of $\frac{1}{5}$? Describe the lines." A slope of $\frac{1}{2}$ runs 2 units for every 1 unit it rises. A slope of $\frac{1}{5}$ runs 5 units for each 1 unit it rises. A slope of $\frac{1}{5}$ is not as steep.

? "What would a slope of $\frac{1}{10}$ look like?" A slope of $\frac{1}{10}$ is less steep than a slope of $\frac{1}{5}$, so it is almost flat.

? "How steep do you think a horizontal line is?" Listen for students to describe a horizontal line as having no rise. In this example, they will see it has a slope of 0.

- This progression of questions is to help students visualize that as slopes of lines get less steep, the lines become horizontal.
- Work through the example.

Example 3

- Ask a series of questions similar to Example 2.

? "How does a slope of $\frac{9}{2}$ compare to a slope of $\frac{3}{2}$? Describe the lines." A slope of $\frac{9}{2}$ rises 9 units for every 2 units it runs. A slope of $\frac{3}{2}$ rises 3 units for every 2 units it runs. A slope of $\frac{9}{2}$ is steeper.

? "What would a slope of 10 look like?" A slope of 10 is steeper than a slope of $\frac{9}{2}$, so it is almost vertical.

? "How steep do you think a vertical line is?" Listen for students to describe a vertical line as having a slope of infinity.

- Work through the example.

On Your Own

? Listen to students' explanations of Question 7. If necessary, you might ask,

- "What is true about every y-coordinate for points on a horizontal line?" y-values are all the same
- "When you compute the rise, what will you get?" 0
- "What is true about every x-coordinate for points on a vertical line?" x-values are all the same
- "When you compute the run, what will you get?" 0

- Students should recall that 0 can be divided by a non-zero number and the quotient is 0. Division by 0 is undefined.

On Your Own

Now You're Ready
Exercises 7–9

Find the slope of the line.

1.

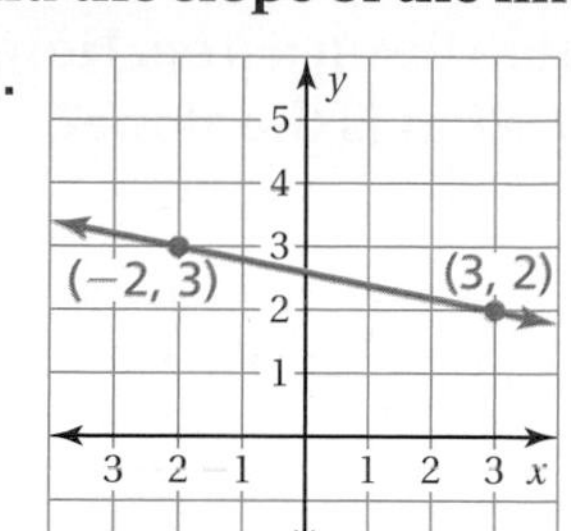

2.

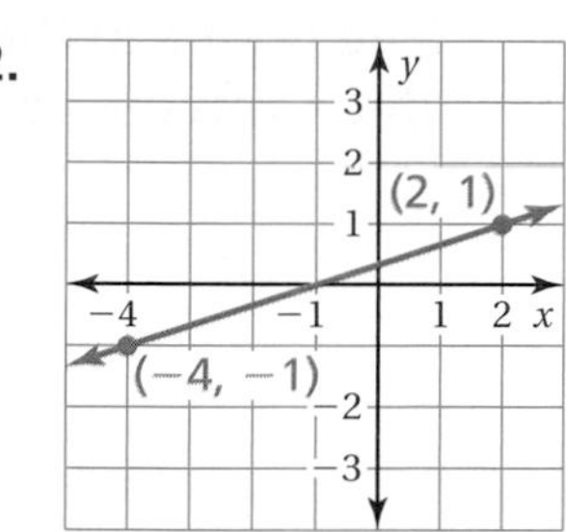

3.

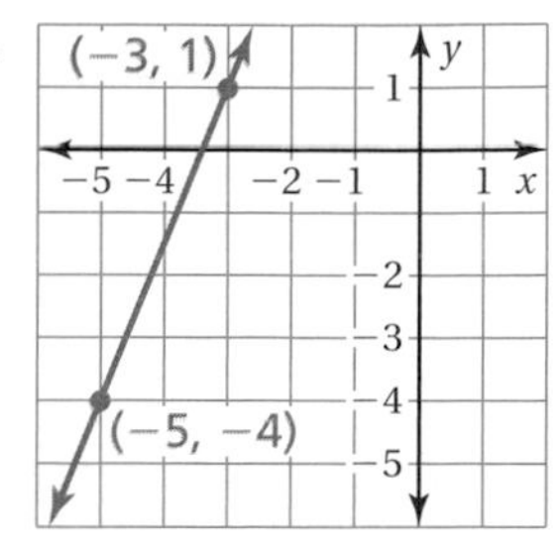

EXAMPLE 2 Finding the Slope of a Horizontal Line

Find the slope of the line.

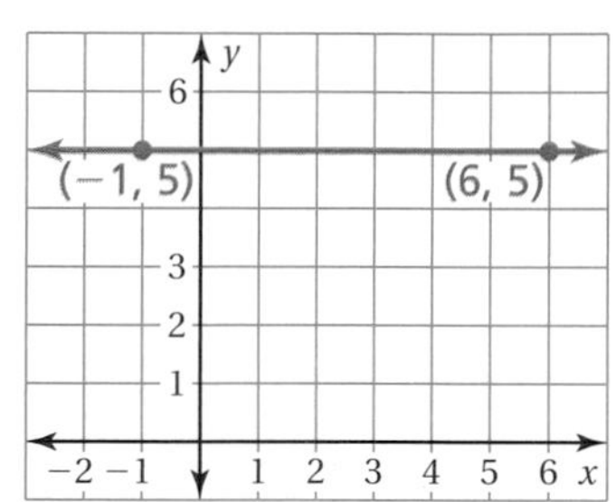

There is no change in y. So, the change in y is 0.

$$\text{slope} = \frac{y_2 - y_1}{x_2 - x_1}$$

$$= \frac{5 - 5}{6 - (-1)}$$

$$= \frac{0}{7}, \text{ or } 0$$

∴ The slope is 0.

EXAMPLE 3 Finding the Slope of a Vertical Line

Find the slope of the line.

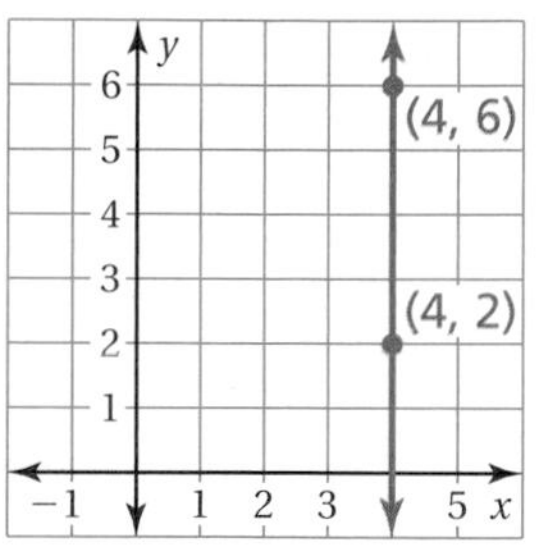

There is no change in x. So, the change in x is 0.

$$\text{slope} = \frac{y_2 - y_1}{x_2 - x_1}$$

$$= \frac{6 - 2}{4 - 4}$$

$$= \frac{4}{0} \quad ✗$$

∴ Because division by zero is undefined, the slope of the line is undefined.

On Your Own

Now You're Ready
Exercises 13–15

Find the slope of the line through the given points.

4. (1, −2), (7, −2) **5.** (−3, −3), (−3, −5) **6.** (0, 8), (0, 0)

7. How do you know that the slope of every horizontal line is 0? How do you know that the slope of every vertical line is undefined?

EXAMPLE 4 Finding Slope from a Table

The points in the table lie on a line. How can you find the slope of the line from the table? What is the slope?

x	1	4	7	10
y	8	6	4	2

Choose any two points from the table and use the slope formula.

Use the points $(x_1, y_1) = (1, 8)$ and $(x_2, y_2) = (4, 6)$.

$$\text{slope} = \frac{y_2 - y_1}{x_2 - x_1}$$

$$= \frac{6 - 8}{4 - 1}$$

$$= \frac{-2}{3}$$

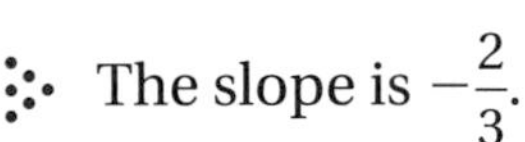

∴ The slope is $-\frac{2}{3}$.

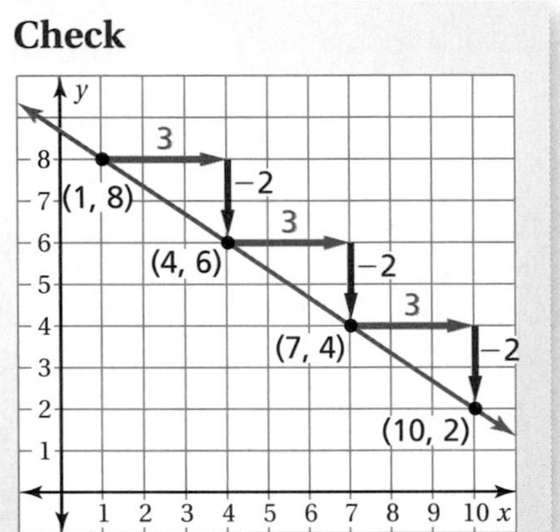

On Your Own

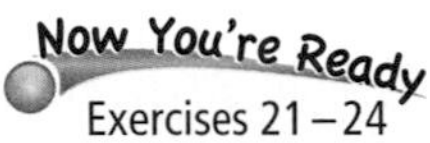

The points in the table lie on a line. How can you find the slope of the line from the table? What is the slope?

8.

x	1	3	5	7
y	2	5	8	11

9.

x	−3	−2	−1	0
y	6	4	2	0

Summary

Slope

Positive slope

The line rises from left to right.

Negative slope

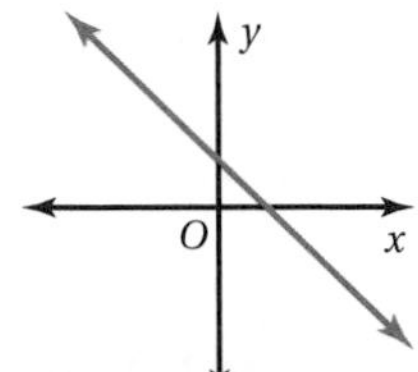

The line falls from left to right.

Slope of 0

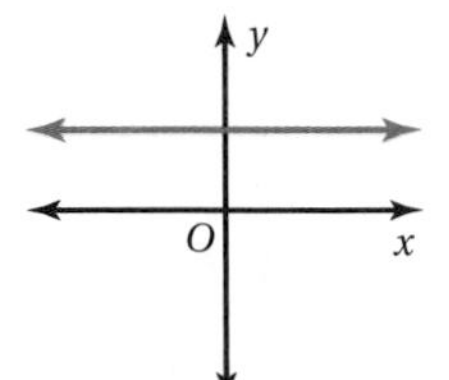

The line is horizontal.

Undefined slope

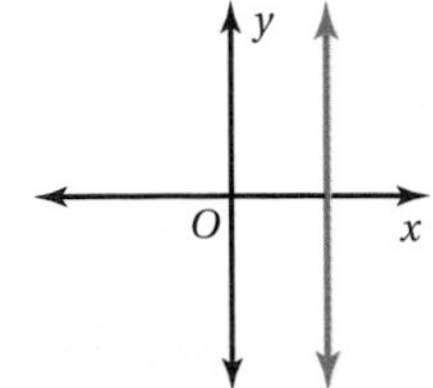

The line is vertical.

Laurie's Notes

Example 4

- ? "What do you notice about the x-values and the y-values?" The x-values are increasing by 3 and the y-values are decreasing by 2.
- Compute the slope between any two points in the table.
- Using (1, 8) and (4, 6): slope $= \frac{y_2 - y_1}{x_2 - x_1} = \frac{6 - 8}{4 - 1} = -\frac{2}{3}$
- Using (1, 8) and (7, 4): slope $= \frac{y_2 - y_1}{x_2 - x_1} = \frac{4 - 8}{7 - 1} = -\frac{4}{6} = -\frac{2}{3}$
- **Connection:** The slope is the same regardless of which two points are selected. The slope triangles that are formed are similar. (Note: A slope triangle is the triangle formed by the line and the change in x and change in y arrows.)
- ? "The line has a negative slope. What do you notice about the line?" The line falls from left to right.

On Your Own

- **Neighbor Check:** Have students work independently and then have their neighbor check their work. Have students discuss any discrepancies.
- **Connection:** In Question 8, students may recognize from the table that both x and y are increasing and the slope is positive. In Question 9, as x increases, the y-values are decreasing and the slope is negative.
- Students should observe by inspection the change in x and the change in y. This is *reading* the table.

Summary

- Students have computed the slopes of many lines today. Discuss the Summary by referring to previous examples.

Closure

- The points in the table lie on a line. How can you find the slope of the line from the table? What is the slope?

x	−1	0	1	2
y	−3	−1	1	3

Choose any two points from the table and use the slope formula.
slope = 2

The Dynamic Planning Tool
Editable Teacher's Resources at *BigIdeasMath.com*

Extra Example 4

The points in the table lie on a line. How can you find the slope of the line from the table? What is the slope?

x	−2	−1	0	1
y	−8	−5	−2	1

Choose any two points from the table and use the slope formula; 3

On Your Own

8. Choose any two points from the table and use the slope formula.

slope $= \frac{3}{2}$

9. Choose any two points from the table and use the slope formula.

slope $= -2$

Vocabulary and Concept Check

1. **a.** B and C

   b. A

   c. no; None of the lines are vertical.

2. *Sample answer:* When constructing a wheelchair ramp, you need to know the slope.

3. The line is horizontal.

Practice and Problem Solving

4.

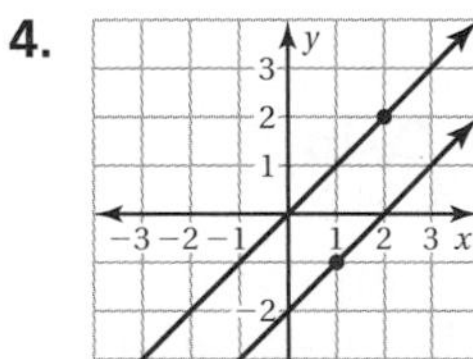

The lines are parallel.

5.

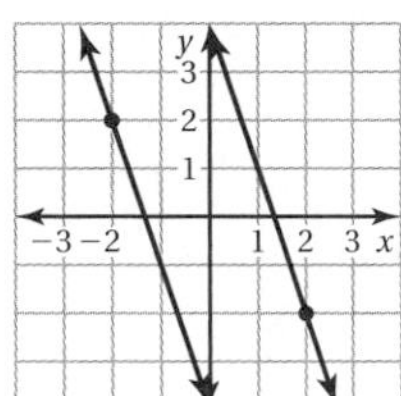

The lines are parallel.

6.

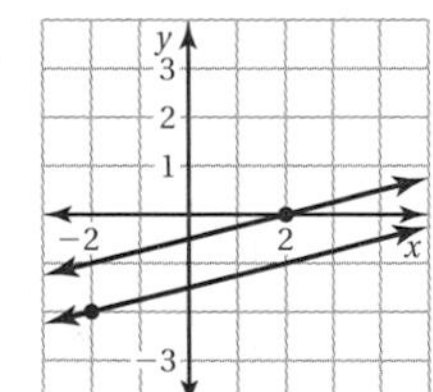

The lines are parallel.

7. $\frac{3}{4}$ **8.** $-\frac{5}{4}$

9. $-\frac{3}{5}$ **10.** $\frac{1}{6}$

11. 0 **12.** undefined

13. 0 **14.** undefined

15. undefined **16.** 2

17. $-\frac{11}{6}$ **18.** 0

Assignment Guide and Homework Check

Level	Assignment	Homework Check
Average	1–12, 13–35 odd, 20, 37–40	7, 11, 15, 21, 33
Advanced	1–3, 10–19, 22, 24, 27–36, 37–40	10, 22, 28, 32, 36

Common Errors

- **Exercises 7–12** Students may forget negatives, or include them when they are not needed. Remind them that if the line rises from left to right the slope is positive, and if the line falls from left to right the slope is negative.
- **Exercises 7–18** Students may find the reciprocal of the slope because they mix up rise and run. Remind them that the change in y is the numerator and the change in x is the denominator.

2.2 Record and Practice Journal

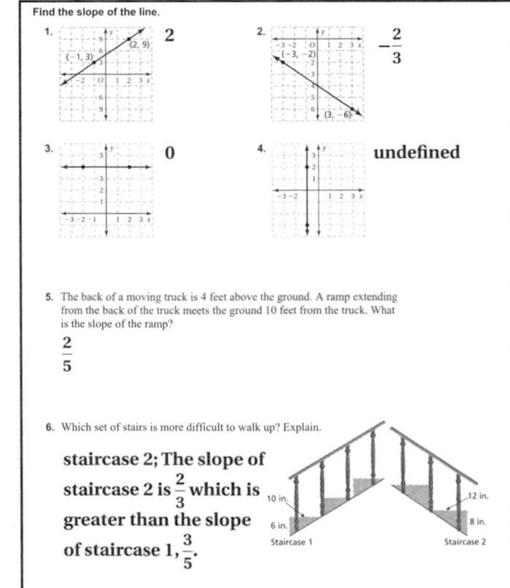

Find the slope of the line.

1. 2

2. $-\frac{2}{3}$

3. 0

4. undefined

5. The back of a moving truck is 4 feet above the ground. A ramp extending from the back of the truck meets the ground 10 feet from the truck. What is the slope of the ramp?

$\frac{2}{5}$

6. Which set of stairs is more difficult to walk up? Explain.

staircase 2; The slope of staircase 2 is $\frac{2}{3}$ which is greater than the slope of staircase 1, $\frac{3}{5}$.

Technology For the Teacher

Answer Presentation Tool

2.2 Exercises

Vocabulary and Concept Check

1. **CRITICAL THINKING** Refer to the graph.

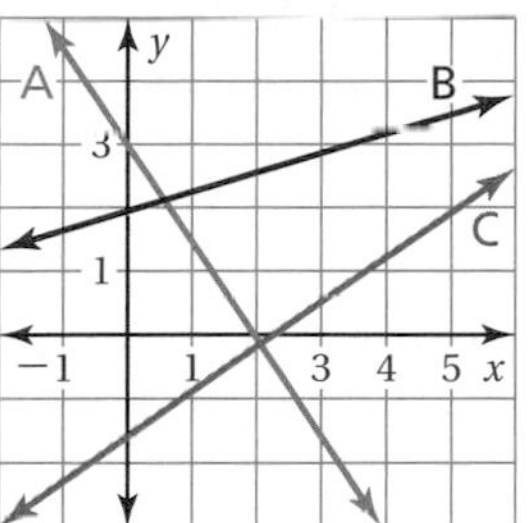

   a. Which lines have positive slopes?

   b. Which line has the steepest slope?

   c. Do any of the lines have undefined slope? Explain.

2. **OPEN-ENDED** Describe a real-life situation in which you need to know the slope.

3. **REASONING** The slope of a line is 0. What do you know about the line?

Practice and Problem Solving

Draw a line through each point using the given slope. What do you notice about the two lines?

4. Slope = 1

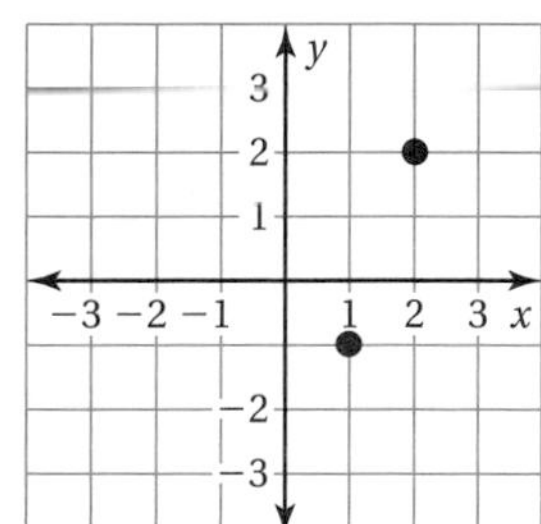

5. Slope = −3

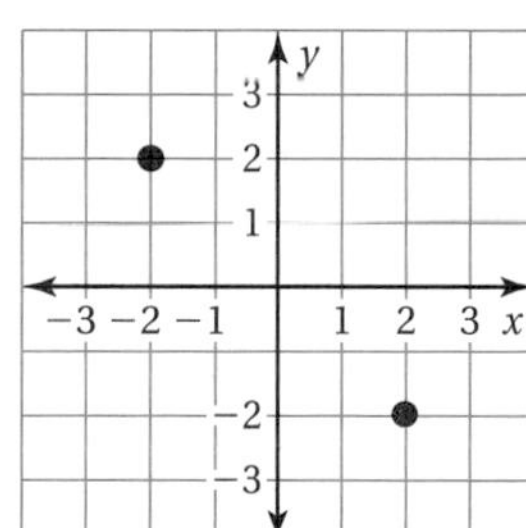

6. Slope = $\frac{1}{4}$

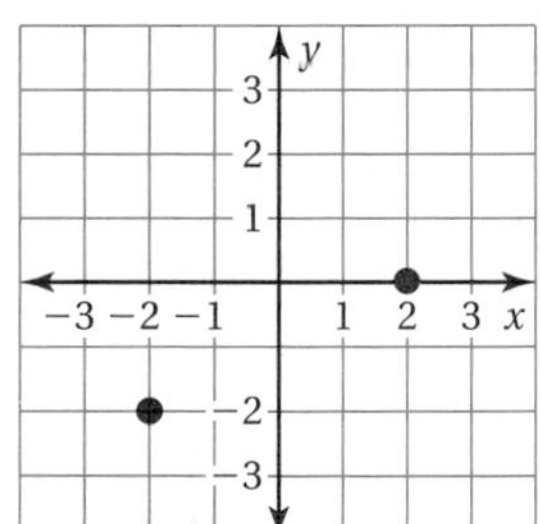

Find the slope of the line.

7.

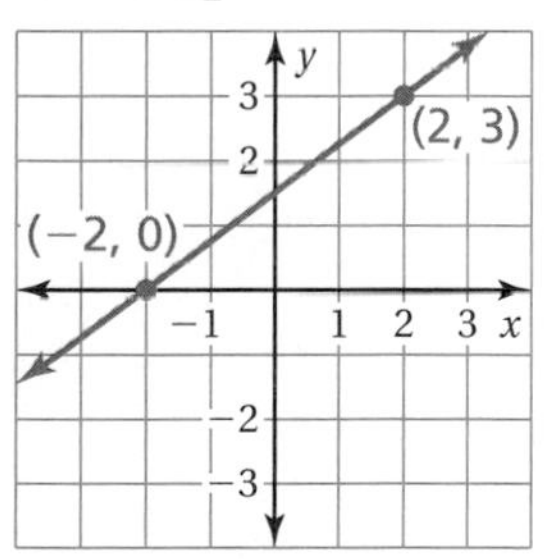

8.

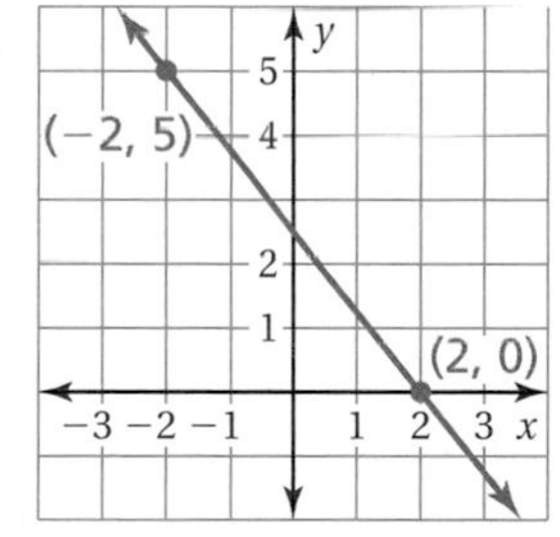

9.

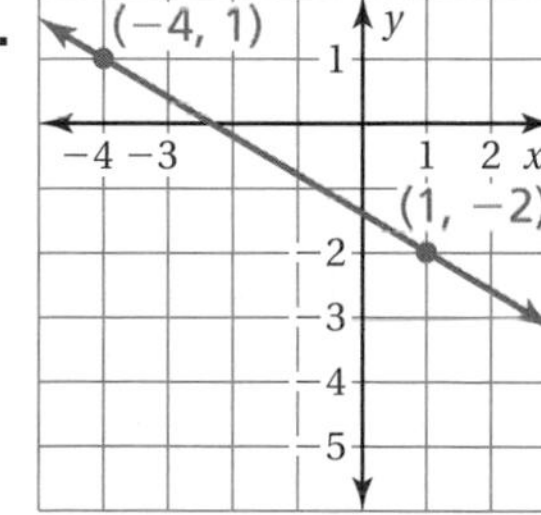

10.

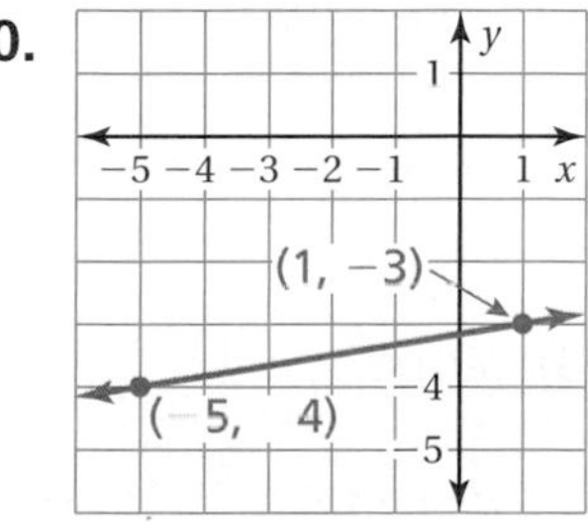

11.

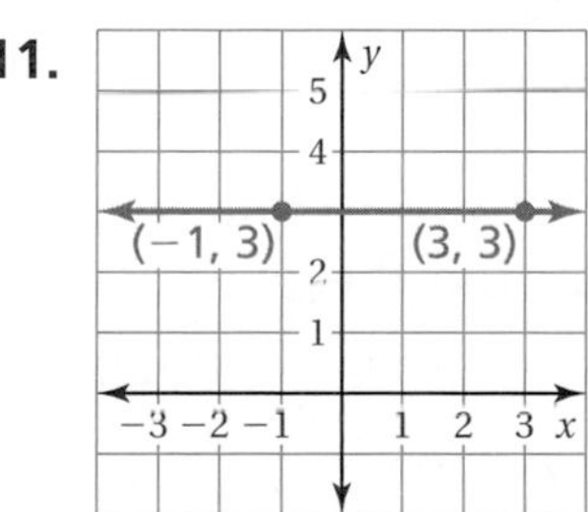

12.

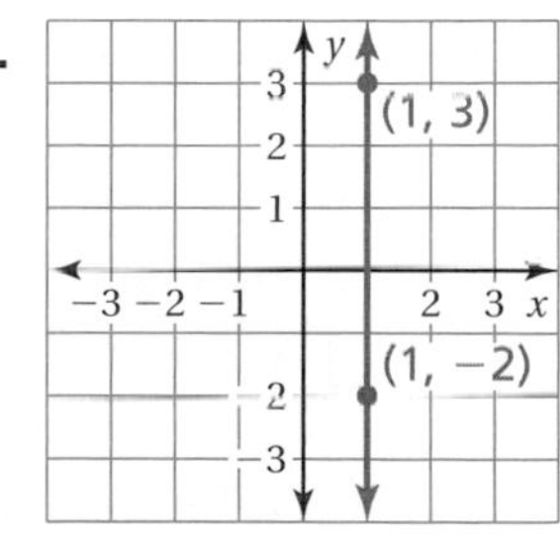

Find the slope of the line through the given points.

2 3 **13.** $(4, -1), (-2, -1)$ **14.** $(5, -3), (5, 8)$ **15.** $(-7, 0), (-7, -6)$

16. $(-3, 1), (-1, 5)$ **17.** $(10, 4), (4, 15)$ **18.** $(-3, 6), (2, 6)$

19. ERROR ANALYSIS Describe and correct the error in finding the slope of the line.

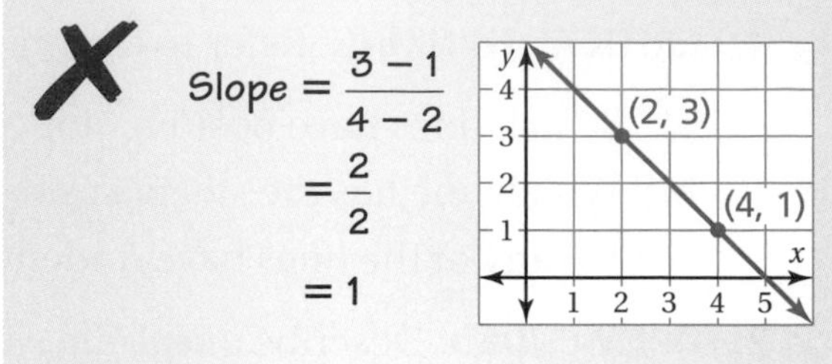

20. CRITICAL THINKING Is it more difficult to walk up the ramp or the hill? Explain.

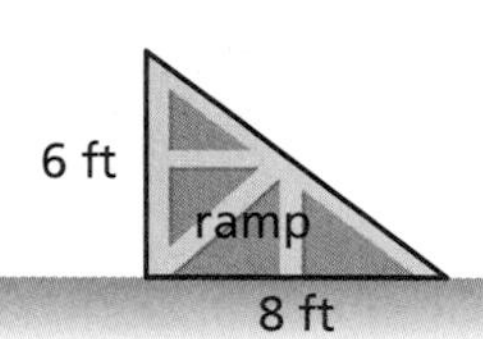

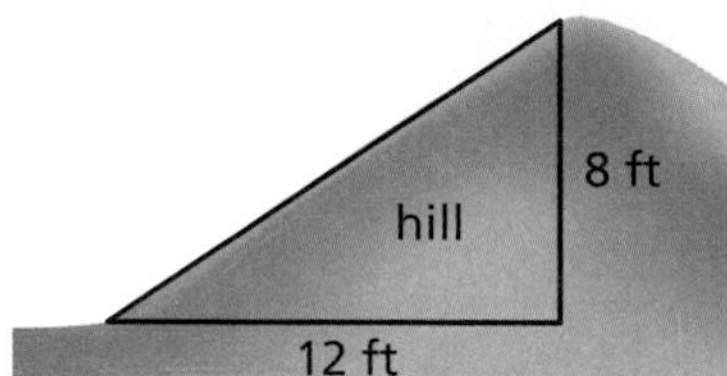

The points in the table lie on a line. How can you find the slope of the line from the table? What is the slope?

4 **21.**

x	1	3	5	7
y	2	10	18	26

22.

x	−3	2	7	12
y	0	2	4	6

23.

x	−6	−2	2	6
y	8	5	2	−1

24.

x	−8	−2	4	10
y	8	1	−6	−13

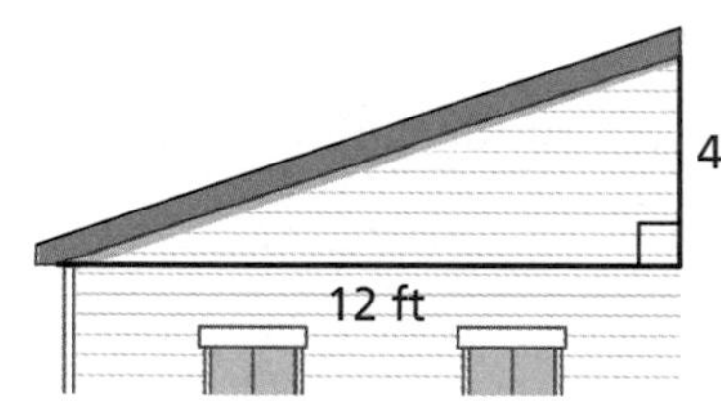

25. PITCH Carpenters refer to the slope of a roof as the *pitch* of the roof. Find the pitch of the roof.

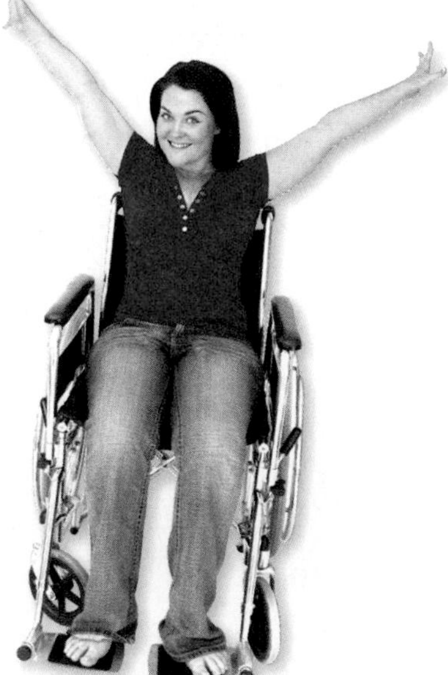

26. PROJECT The guidelines for a wheelchair ramp suggest that the ratio of the rise to the run be no greater than 1 : 12.

a. CHOOSE TOOLS Find a wheelchair ramp in your school or neighborhood. Measure its slope. Does the ramp follow the guidelines?

b. Design a wheelchair ramp that provides access to a building with a front door that is 2.5 feet higher than the sidewalk. Illustrate your design.

Use an equation to find the value of *k* so that the line that passes through the given points has the given slope.

27. $(1, 3), (5, k)$; slope $= 2$ **28.** $(-2, k), (2, 0)$; slope $= -1$

29. $(-4, k), (6, -7)$; slope $= -\frac{1}{5}$ **30.** $(4, -4), (k, -1)$; slope $= \frac{3}{4}$

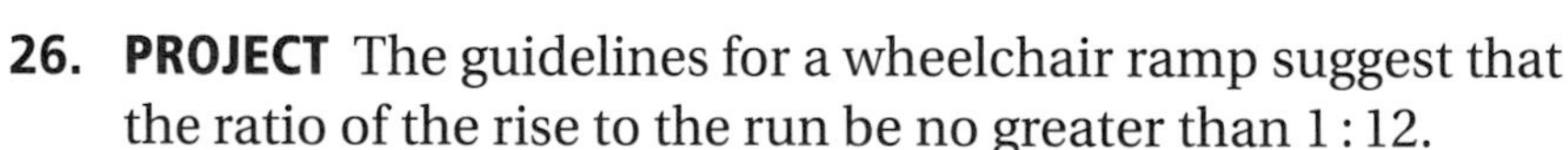

Common Errors

- **Exercise 20** Students may get confused because one of the slopes is negative and the other is positive. Tell them to think of the absolute values of the slopes when comparing. Encourage them to graph the slopes on a number line to check their answer.
- **Exercises 21–24** Students may find the change in *x* over the change in *y*. Remind them that slope is the change in *y* over the change in *x*.

Practice and Problem Solving

19. The denominator should be $2 - 4$.

slope $= -1$

20. The ramp because its slope is steeper.

21. Choose any two points from the table and use the slope formula.

slope $= 4$

22. Choose any two points from the table and use the slope formula.

slope $= \frac{2}{5}$

23. Choose any two points from the table and use the slope formula.

slope $= -\frac{3}{4}$

24. Choose any two points from the table and use the slope formula.

slope $= -\frac{7}{6}$

25. $\frac{1}{3}$

26. See Additional Answers.

27. $k = 11$ **28.** $k = 4$

29. $k = -5$ **30.** $k = 8$

Differentiated Instruction

Auditory

Discuss how the rate of change in a rate problem is related to slope. For example, the cost to travel on a turnpike (cost per mile) can be expressed as $\frac{\text{cost (in dollars)}}{\text{miles driven}}$, where the cost is the change in *y*-values and the miles driven is the change in *x*-values.

Practice and Problem Solving

31–35. See Additional Answers.

36. See *Taking Math Deeper.*

Fair Game Review

37.

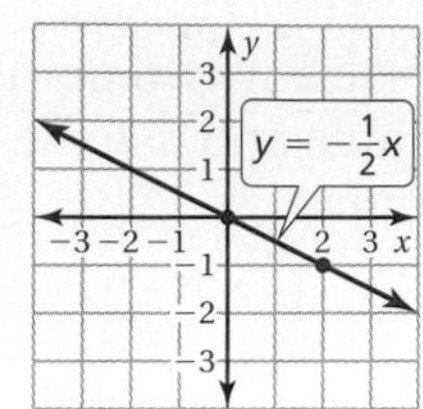

38.

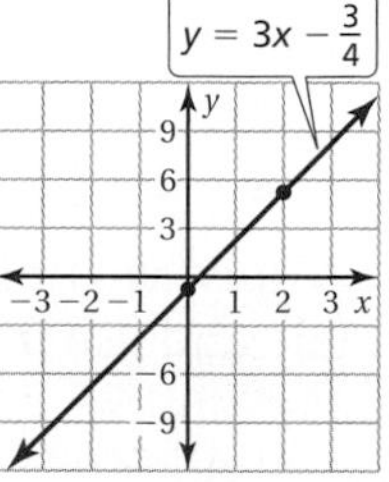

39.

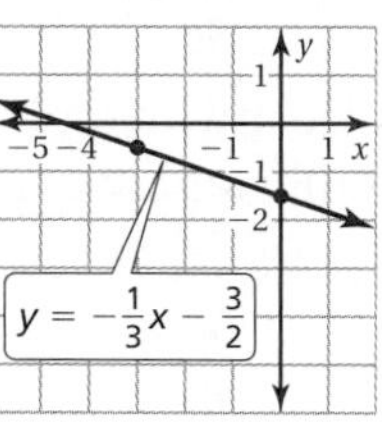

40. B

Mini-Assessment

Find the slope of the line.

1.

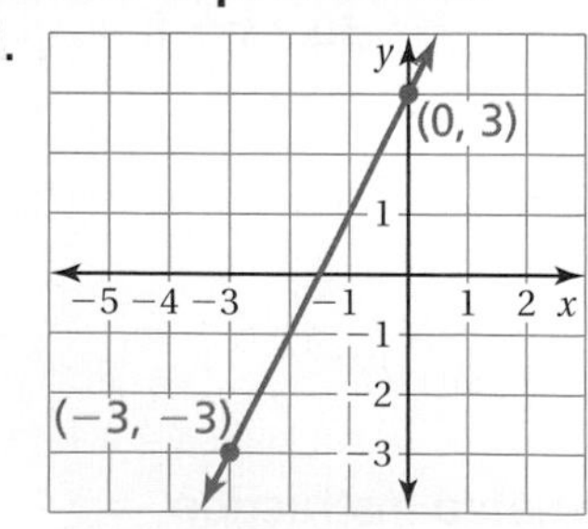

slope = 2

2.

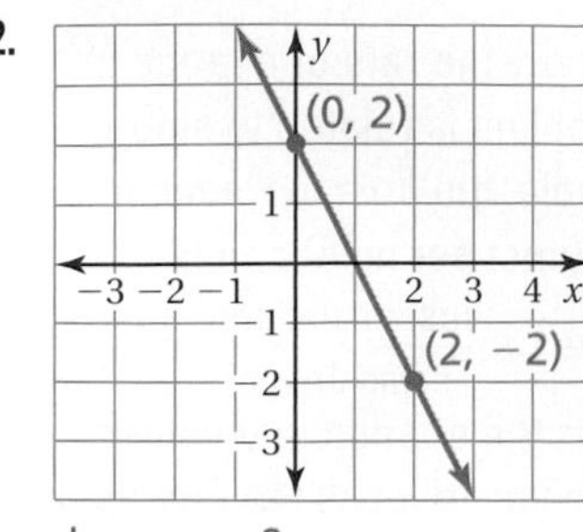

slope = −2

Taking Math Deeper

Exercise 36

This exercise is a nice example of the power of a diagram. Instead of using the drawing of the slide, encourage students to draw the slide in a coordinate plane. Once that is done, the question is easier to answer.

Draw a diagram.

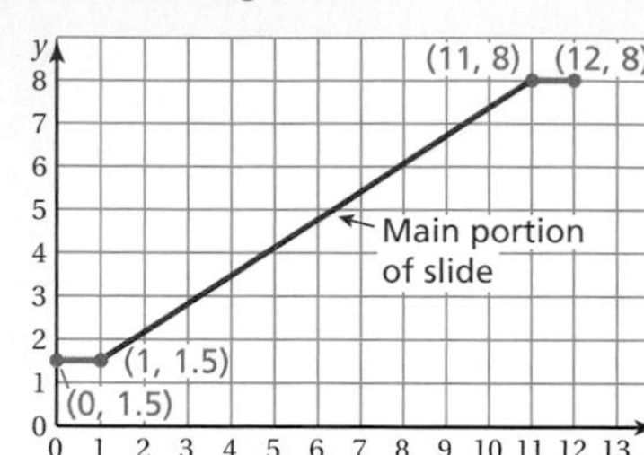

2 **a.** Find the slope of the slide.

$$\text{slope} = \frac{y_2 - y_1}{x_2 - x_1} = \frac{8 - 1.5}{11 - 1} = \frac{6.5}{10} = 0.65$$

Compare the slopes.

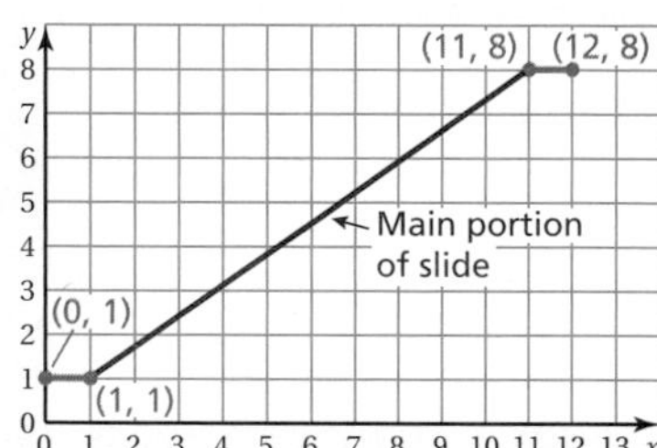

Because 0.7 > 0.65, the slide is steeper.

b.

$$\text{slope} = \frac{y_2 - y_1}{x_2 - x_1} = \frac{8 - 1}{11 - 1} = \frac{7}{10} = 0.7$$

Project

Many water parks and amusement parks have water slides. Find the height of a slide and calculate the slope of the main part of the slide.

Reteaching and Enrichment Strategies

If students need help...	If students got it...
Resources by Chapter • Practice A and Practice B • Puzzle Time Record and Practice Journal Practice Differentiating the Lesson Lesson Tutorials Skills Review Handbook	Resources by Chapter • Enrichment and Extension Start the next section

31. **TURNPIKE TRAVEL** The graph shows the cost of traveling by car on a turnpike.

   a. Find the slope of the line.

   b. Explain the meaning of the slope as a rate of change.

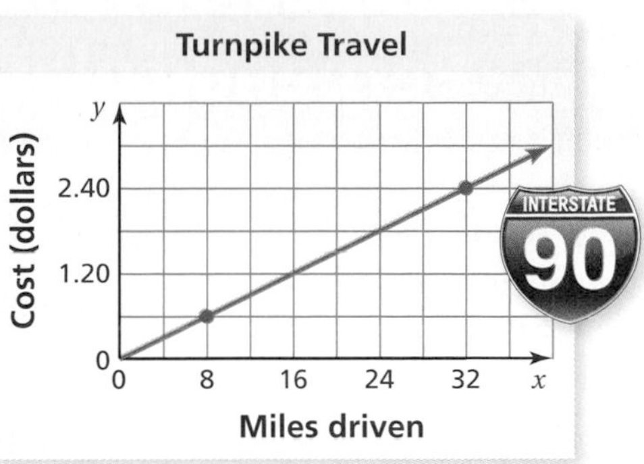

32. **BOAT RAMP** Which is steeper: the boat ramp or a road with a 12% grade? Explain. (*Note:* Road grade is the vertical increase divided by the horizontal distance.)

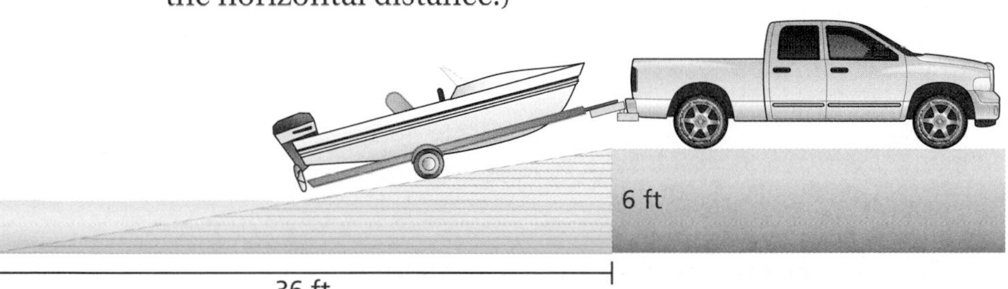

33. **REASONING** Do the points $A(-2, -1)$, $B(1, 5)$, and $C(4, 11)$ lie on the same line? Without using a graph, how do you know?

34. **BUSINESS** A small business earns a profit of \$6500 in January and \$17,500 in May. What is the rate of change in profit for this time period?

35. **STRUCTURE** Choose two points in the coordinate plane. Use the slope formula to find the slope of the line that passes through the two points. Then find the slope using the formula $\dfrac{y_1 - y_2}{x_1 - x_2}$. Are your results the same? Explain.

36. **Critical Thinking** The top and bottom of the slide are level with the ground, which has a slope of 0.

   a. What is the slope of the main portion of the slide?

   b. How does the slope change if the bottom of the slide is only 12 inches above the ground? Is the slide steeper? Explain.

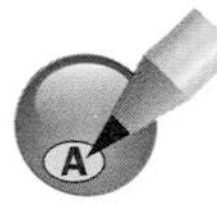

Fair Game Review What you learned in previous grades & lessons

Graph the linear equation. *(Section 2.1)*

37. $y = -\frac{1}{2}x$

38. $y = 3x \quad \frac{3}{4}$

39. $y = -\frac{x}{3} - \frac{3}{2}$

40. **MULTIPLE CHOICE** What is the prime factorization of 84? *(Skills Review Handbook)*

   Ⓐ $2 \times 3 \times 7$ Ⓑ $2^2 \times 3 \times 7$ Ⓒ $2 \times 3^2 \times 7$ Ⓓ $2^2 \times 21$

2.2b Slopes of Parallel and Perpendicular Lines

Key Vocabulary
perpendicular lines, *p. 57*

Study Tip

Vertical lines have undefined slopes.

Key Idea

Parallel Lines and Slopes

Two different lines in the same plane that never intersect are parallel lines. Nonvertical parallel lines have the same slope.

All vertical lines are parallel.

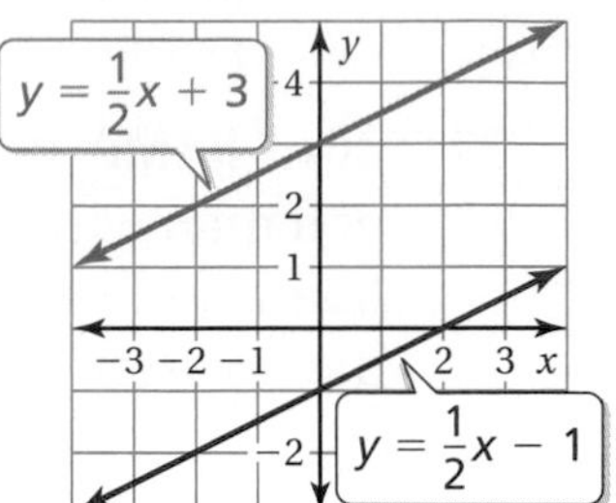

EXAMPLE 1 Identifying Parallel Lines

Which two lines are parallel? How do you know?

Find the slope of each line.

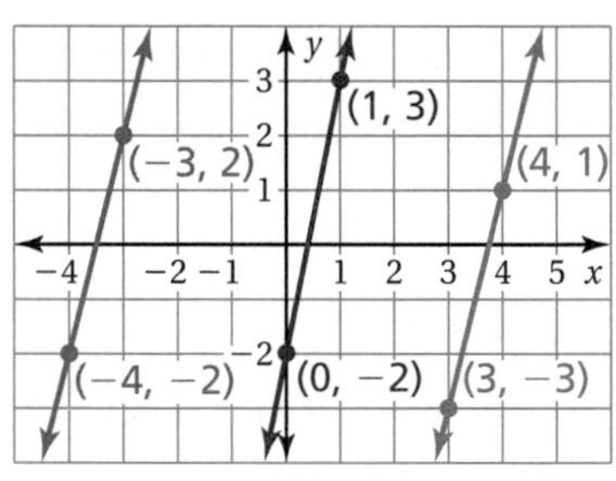

Blue Line	*Red Line*	*Green Line*
$\text{slope} = \frac{y_2 - y_1}{x_2 - x_1}$	$\text{slope} = \frac{y_2 - y_1}{x_2 - x_1}$	$\text{slope} = \frac{y_2 - y_1}{x_2 - x_1}$
$= \frac{-2 - 2}{-4 - (-3)}$	$= \frac{-2 - 3}{0 - 1}$	$= \frac{-3 - 1}{3 - 4}$
$= \frac{-4}{-1}$, or 4	$= \frac{-5}{-1}$, or 5	$= \frac{-4}{-1}$, or 4

The slope of the blue and green lines is 4. The slope of the red line is 5.

∴ The blue and green lines have the same slope, so they are parallel.

Practice

Which lines are parallel? How do you know?

1. 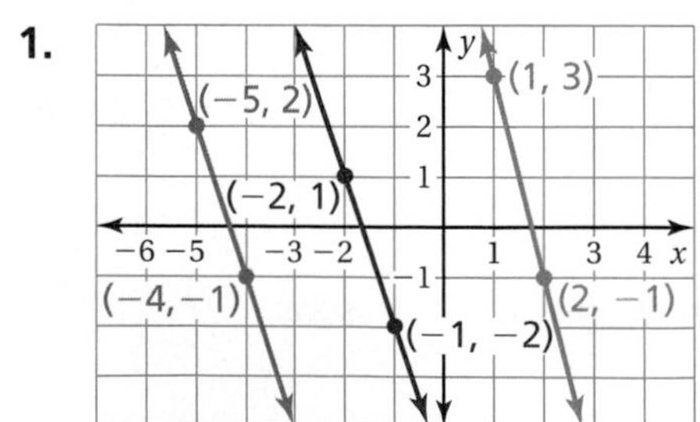

2. 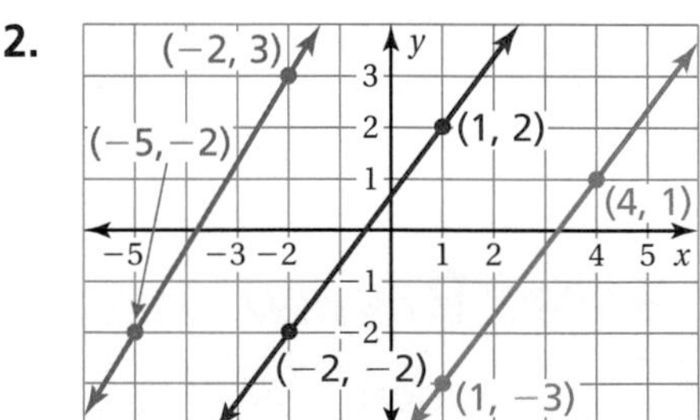

Are the given lines parallel? Explain your reasoning.

3. $y = -5, y = 3$
4. $y = 0, x = 0$
5. $x = -4, x = 1$

6. **GEOMETRY** The vertices of a quadrilateral are $A(-5, 3)$, $B(2, 2)$, $C(4, -3)$, and $D(-2, -2)$. How can you use slope to determine whether the quadrilateral is a parallelogram? Is it a parallelogram? Justify your answer.

Multi-Language Glossary at BigIdeasMath.com.

Laurie's Notes

Introduction

Connect

- **Yesterday:** Students found slopes of lines. (MP1a)
- **Today:** Students will use slope to determine if lines are parallel or perpendicular.

Motivate

- The words *parallel* and *perpendicular* can be confusing for students and difficult to remember.
- The word parallel comes from para-, "beside," and allelois, "each other," so parallel lines are lines that are beside each other. For example, parallel bars in gymnastics have bars that are beside each other.
- The word perpendicular comes from perpendicularis, meaning "vertical, as a plumb line." A plumb line is a cord with a weight attached that is used to determine perpendicularity.

Lesson Notes

Discuss

- Discuss the investigation for this section. Students graphed lines with a given slope through different ordered pairs.
- Refer to conjectures made regarding parallel and perpendicular lines.

Key Idea

- Write the Key Idea on the board.
- Model what a slope of $\frac{1}{2}$ means. Start at a point on the line and run 2 units for each unit you rise. Repeat for each line.

Example 1

? "How do you compute the slope for each line?" Students should use rise over run language, along with the formal definition, $\frac{y_2 - y_1}{x_2 - x_1}$.

- Work through the example. Students may look quickly and believe that all of the lines are parallel. They should compute the slope of each line to prove which lines are parallel.

Practice

- **MP3 Construct Viable Arguments and Critique the Reasoning of Others:** In Exercise 6, students are asked to make a conjecture and then justify their answer. Students need opportunities to construct viable arguments and share their thinking with other students.

Goal Today's lesson is determining if lines are parallel or perpendicular using slope.

Start Thinking! and Warm Up

Lesson 2.2b **Start Thinking!**
For use before Lesson 2.2b

Graph the linear equations $y = 3x + 2$ and $y = 3x - 4$ in a coordinate plane. What do you notice?

Graph the linear equations $y = -\frac{1}{2}x + 1$ and $y = 2x - 2$ in a coordinate plane. What do you notice?

Extra Example 1

Which lines are parallel? How do you know?

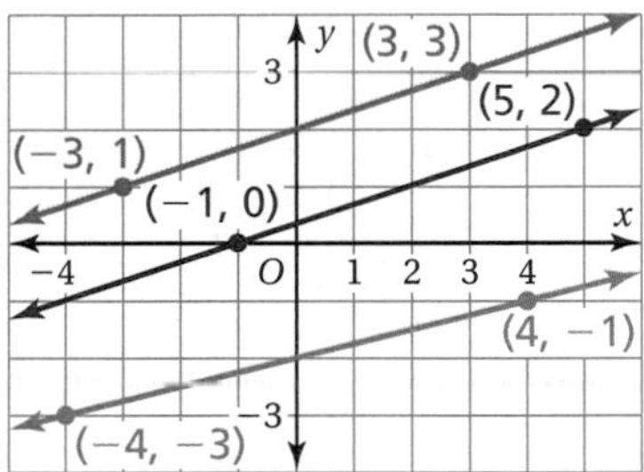

blue and red; They both have a slope of $\frac{1}{3}$.

Practice

1. blue and red; They both have a slope of -3.
2. red and green; They both have a slope of $\frac{4}{3}$.
3. yes; Both lines are horizontal and have a slope of 0.
4. no; $y = 0$ has a slope of 0 and $x = 0$ has an undefined slope.
5. yes; Both lines are vertical and have undefined slopes.
6. See Additional Answers.

Record and Practice Journal Practice

See Additional Answers.

Extra Example 2

Which lines are perpendicular? How do you know?

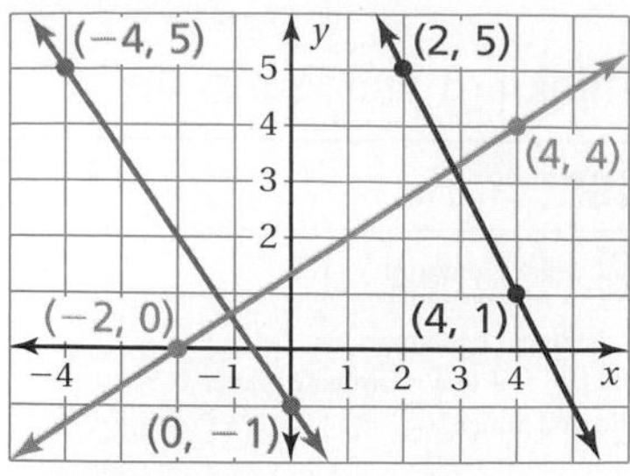

blue and green; The blue line has a slope of $-\frac{3}{2}$. The green line has a slope of $\frac{2}{3}$. The product of their slopes is $-\frac{3}{2} \cdot \frac{2}{3} = -1$.

Practice

7. blue and green; The blue line has a slope of 6. The green line has a slope of $-\frac{1}{6}$. The product of their slopes is $6 \cdot \left(-\frac{1}{6}\right) = -1$.
8. blue and green, red and green; The blue and red lines both have a slope of $-\frac{1}{2}$. The green line has a slope of 2. The product of their slopes is $2 \cdot \left(-\frac{1}{2}\right) = -1$.
9. yes; The line $x = -2$ is vertical. The line $y = 8$ is horizontal. A vertical line is perpendicular to a horizontal line.
10. no; Both lines are vertical and have undefined slopes.
11. yes; The line $x = 0$ is vertical. The line $y = 0$ is horizontal. A vertical line is perpendicular to a horizontal line.
12. See Additional Answers.

Laurie's Notes

Key Idea

- Write the Key Idea on the board.
- **Teaching Tip:** Use the corner of a piece of paper, placed at the point of intersection, to provide a visual model of what perpendicular means.

Example 2

- Work through the example as shown.
- Students may also refer to the slopes of perpendicular lines as being opposite reciprocals.

Practice

- Because vertical lines have undefined slope, students cannot multiply slopes to determine if the lines are perpendicular. In Exercises 9 and 11, students should give reasoning that states horizontal and vertical lines are perpendicular. In Exercise 10, students should give reasoning that states that the lines are not perpendicular because they are both vertical lines.
- **MP3:** In Exercise 12, students are asked to make a conjecture and then justify their answer. Students need opportunities to construct viable arguments and share their thinking with other students.

Closure

- **Exit Ticket:** Which lines are parallel? Which lines are perpendicular? How do you know?

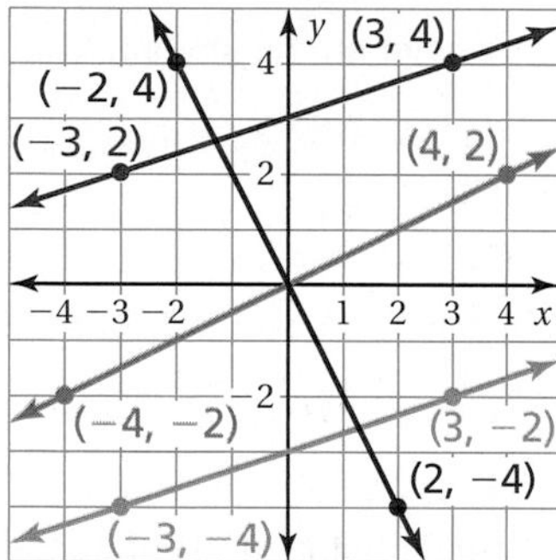

The red and green lines are parallel. They both have a slope of $\frac{1}{3}$. The black and blue lines are perpendicular. The product of their slopes is −1.

Technology For the Teacher

The Dynamic Planning Tool
Editable Teacher's Resources at *BigIdeasMath.com*

Key Idea

Perpendicular Lines and Slope

Two lines in the same plane that intersect to form right angles are **perpendicular lines**. Two nonvertical lines are perpendicular if and only if the product of their slopes is -1.

Vertical lines are perpendicular to horizontal lines.

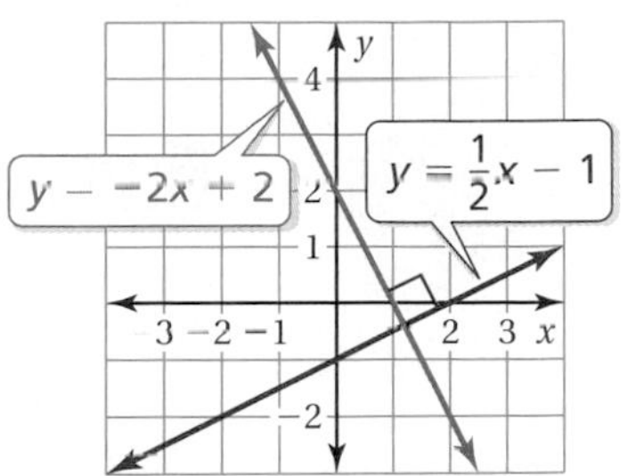

EXAMPLE 2 Identifying Perpendicular Lines

Which two lines are perpendicular? How do you know?

Find the slope of each line.

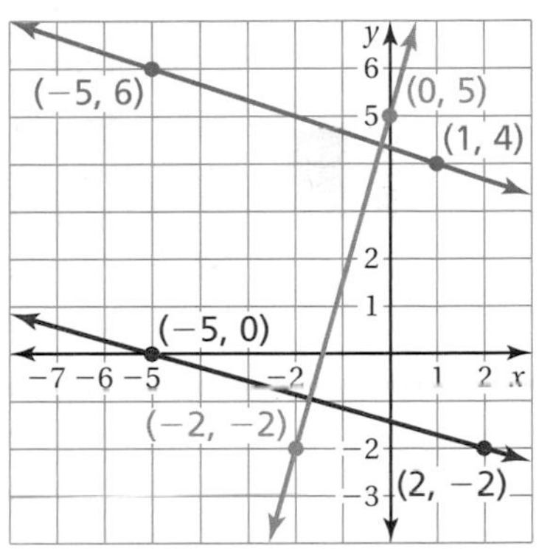

Blue Line	***Red Line***	***Green Line***
$\text{slope} = \dfrac{y_2 - y_1}{x_2 - x_1}$	$\text{slope} = \dfrac{y_2 - y_1}{x_2 - x_1}$	$\text{slope} = \dfrac{y_2 - y_1}{x_2 - x_1}$
$= \dfrac{4 - 6}{1 - (-5)}$	$= \dfrac{-2 - 0}{2 - (-5)}$	$= \dfrac{5 - (-2)}{0 - (-2)}$
$= \dfrac{-2}{6}$, or $-\dfrac{1}{3}$	$= -\dfrac{2}{7}$	$= \dfrac{7}{2}$

The slope of the red line is $-\frac{2}{7}$. The slope of the green line is $\frac{7}{2}$.

Because $-\frac{2}{7} \cdot \frac{7}{2} = -1$, the red and green lines are perpendicular.

Practice

Which lines are perpendicular? How do you know?

7.

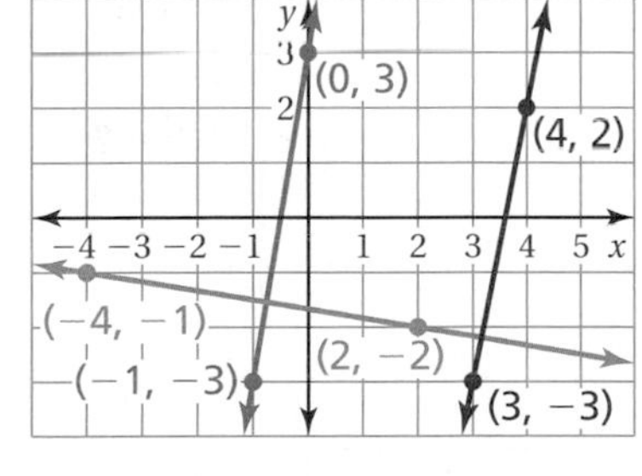

8.

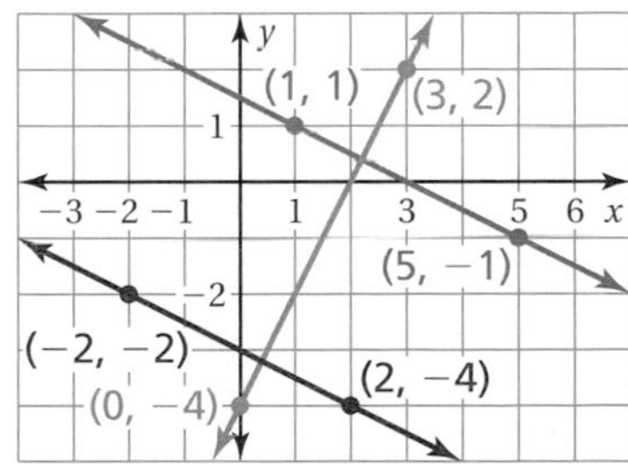

Are the given lines perpendicular? Explain your reasoning.

9. $x = -2, y = 8$

10. $x = -8, x = 7$

11. $y = 0, x = 0$

12. **GEOMETRY** The vertices of a parallelogram are $J(-5, 0)$, $K(1, 4)$, $L(3, 1)$, and $M(-3, -3)$. How can you use slope to determine whether the parallelogram is a rectangle? Is it a rectangle? Justify your answer.

2.3 Graphing Linear Equations in Slope-Intercept Form

COMMON CORE STATE STANDARDS
A.CED.2
A.REI.10
F.IF.4

Essential Question How can you describe the graph of the equation $y = mx + b$?

1 ACTIVITY: Finding Slopes and y-Intercepts

Work with a partner.

- **Graph the equation.**
- **Find the slope of the line.**
- **Find the point where the line crosses the y-axis.**

a. $y = -\frac{1}{2}x + 1$

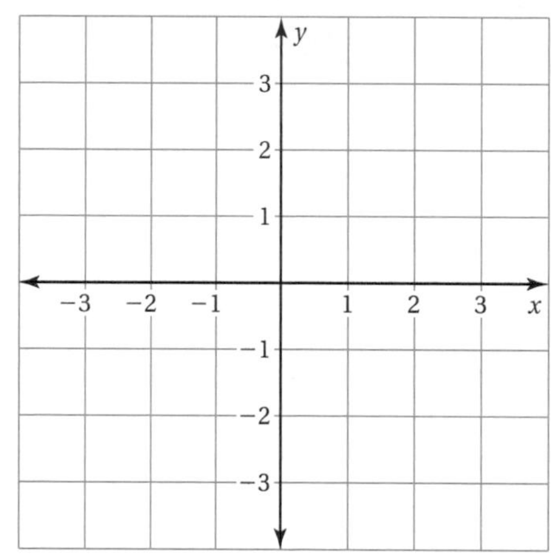

b. $y = -x + 2$

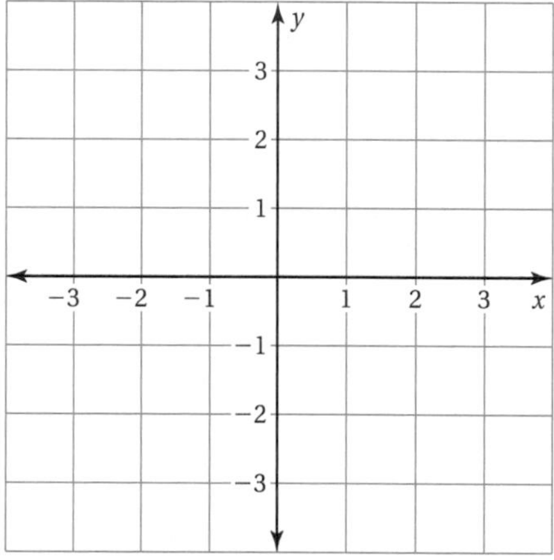

c. $y = -x - 2$

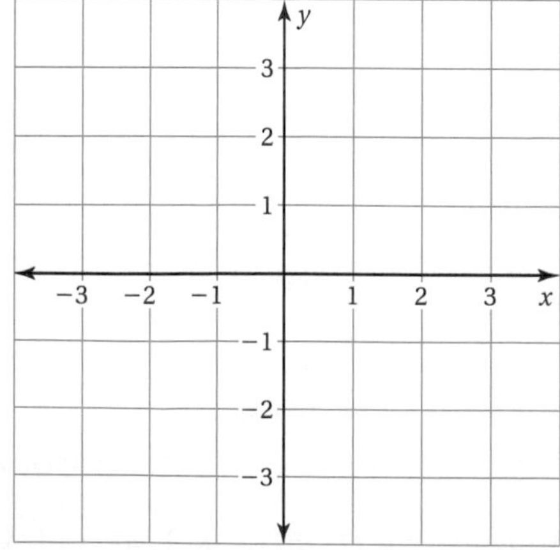

d. $y = \frac{1}{2}x + 1$

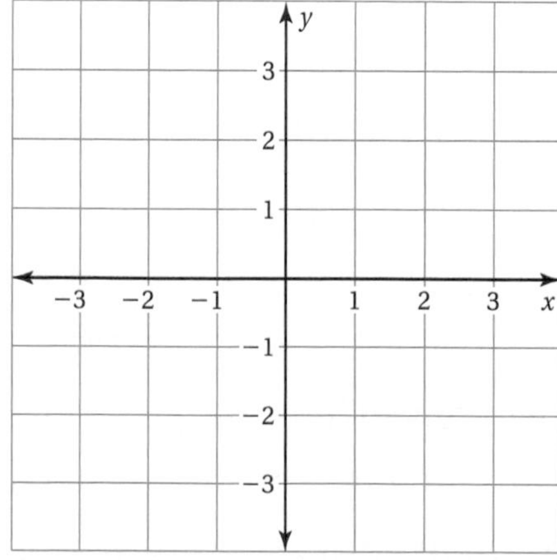

Laurie's Notes

Introduction

Standards for Mathematical Practice

- **MP3a Construct Viable Arguments** and **MP8 Look for and Express Regularity in Repeated Reasoning:** The goal is for students to discover that when equations are written in slope-intercept form, the coefficient of the x-term is the slope and the constant is the y-intercept. Students graph the lines using a table of values, a good review of this skill.

Motivate

- **Preparation:** Make three demonstration cards on 8.5"x 11" paper. The x-axis is labeled "time" and the y-axis is labeled "distance from home."
- Sample cards A, B, and C are shown.

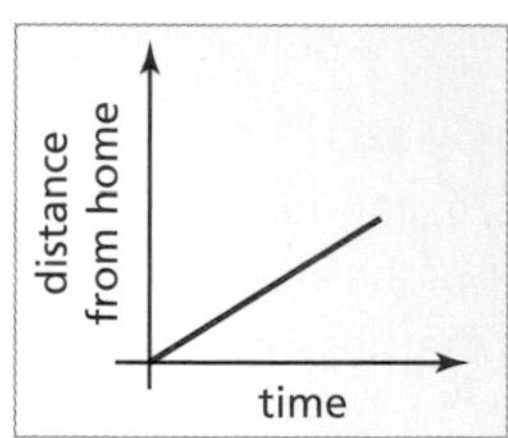

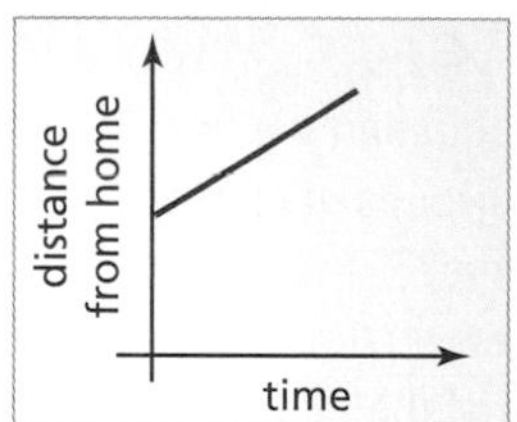

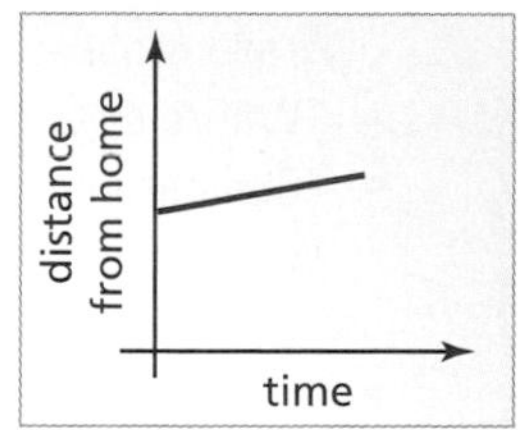

- Ask 3 students to hold the cards for the class to see.
- **?** "Consider how the axes are labeled. What does the slope of the line represent?" $\frac{\text{distance}}{\text{time}} = \text{rate}$
- **?** "What story does each card tell? How are the stories similar and different?" A: you begin at home; B: you travel at the same rate, but you start away from home; C: you start away from home, but you travel at a slower rate
- **Management Tip:** If you plan to use the demonstration cards again next year, laminate them.

Activity Notes

Activity 1

- **?** "How do you graph an equation?" Plot several points, then connect the points with a line.
- **?** "Is there a way to organize the points you need to plot?" Use an input-output table.
- Review with students how input-output tables are set up and what values of x they should substitute. When the coefficient of x is a fraction, it is wise to select x-values that are multiples of the denominator. This may help eliminate fractional values.
- **?** "How many points do you need in order to graph the equation?" Minimum is 2. Plot 3 to be safe.
- Remind students to solve the equation when $x = 0$. This will ensure that they find the point where the graph crosses the y-axis.
- The slope and the point where the line crosses the y-axis will be recorded in the table on the next page.
- Check students' work before going on to the Inductive Reasoning.

Common Core State Standards

A.CED.2 Create equations in two or more variables to represent relationships between quantities; graph equations on coordinate axes with labels and scales.

A.REI.10 Understand that the graph of an equation in two variables is the set of all its solutions plotted in the coordinate plane, often forming a curve (which could be a line).

F.IF.4 For a function that models a relationship between two quantities, interpret key features of graphs and tables in terms of the quantities, and sketch graphs showing key features given a verbal description of the relationship.

Previous Learning

Students should know how to find the slopes of lines.

Start Thinking! and Warm Up

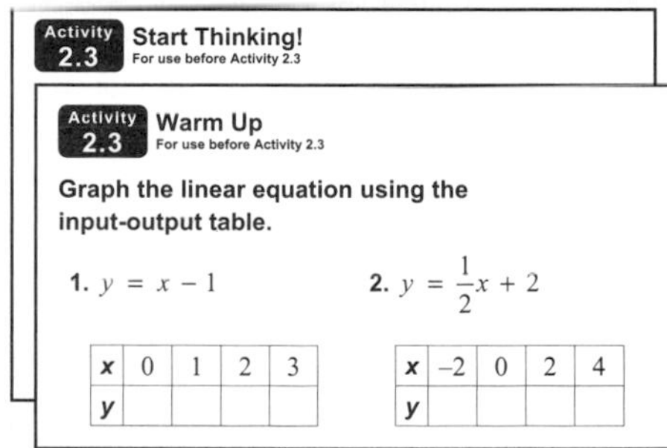

Activity 2.3 **Start Thinking!** For use before Activity 2.3

Activity 2.3 **Warm Up** For use before Activity 2.3

Graph the linear equation using the input-output table.

1. $y = x - 1$

x	0	1	2	3
y				

2. $y = \frac{1}{2}x + 2$

x	−2	0	2	4
y				

2.3 Record and Practice Journal

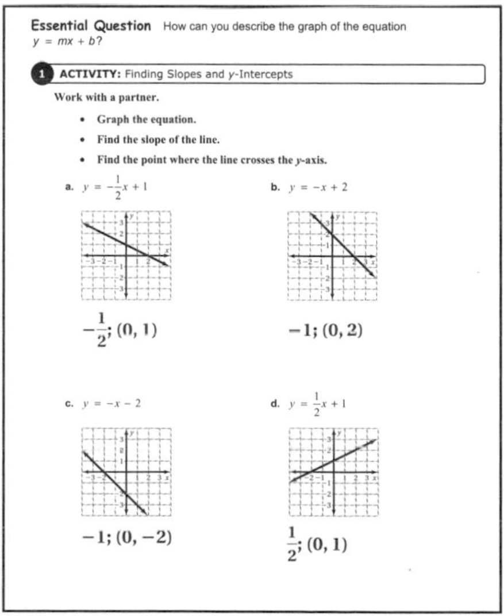

Essential Question How can you describe the graph of the equation $y = mx + b$?

1 ACTIVITY: Finding Slopes and y-Intercepts

Work with a partner.

- Graph the equation.
- Find the slope of the line.
- Find the point where the line crosses the y-axis.

a. $y = -\frac{1}{2}x + 1$ — $-\frac{1}{2}$; (0, 1)

b. $y = -x + 2$ — -1; (0, 2)

c. $y = -x - 2$ — -1; (0, −2)

d. $y = \frac{1}{2}x + 1$ — $\frac{1}{2}$; (0, 1)

English Language Learners

Build on Past Knowledge

Remind students from their study of rational numbers that the slope -2 can be written as the fraction $\frac{-2}{1}$. By writing the integer as a fraction, students can see that the slope has a run of 1 and a rise of -2. This will help in graphing linear equations.

2.3 Record and Practice Journal

Inductive Reasoning

Work with a partner. Graph each equation. Then complete the table.

	Equation	Description of Graph	Slope of Graph	Point of Intersection with y-axis
1a	2. $y = -\frac{1}{2}x + 1$	Line	$-\frac{1}{2}$	$(0, 1)$
1b	3. $y = -x + 2$	Line	-1	$(0, 2)$
1c	4. $y = -x - 2$	Line	-1	$(0, -2)$
1d	5. $y = \frac{1}{2}x + 1$	Line	$\frac{1}{2}$	$(0, 1)$
	6. $y = x + 2$	Line	1	$(0, 2)$
	7. $y = x - 2$	Line	1	$(0, -2)$
	8. $y = \frac{1}{2}x - 1$	Line	$\frac{1}{2}$	$(0, -1)$
	9. $y = -\frac{1}{2}x - 1$	Line	$-\frac{1}{2}$	$(0, -1)$
	10. $y = 3x + 2$	Line	3	$(0, 2)$
	11. $y = 3x - 2$	Line	3	$(0, -2)$
	12. $y = -2x + 3$	Line	-2	$(0, 3)$

What Is Your Answer?

13. **IN YOUR OWN WORDS** How can you describe the graph of the equation $y = mx + b$?

    a line with slope m and crosses the y-axis at $(0, b)$

    a. How does the value of m affect the graph of the equation?

    steepness of line

    b. How does the value of b affect the graph of the equation?

    Moves graph up and down.

    c. Check your answers to parts (a) and (b) with three equations that are not in the table.

    Check students' work.

14. Why do you think $y = mx + b$ is called the "slope-intercept" form of the equation of a line? Use drawings or diagrams to support your answer.

    m is the slope and b is the y-intercept.

Laurie's Notes

Inductive Reasoning

- Students begin by copying their results of the 4 graphs from Activity 1. Some students may be prepared to make a conjecture at this point. They can test their conjecture by trying the remaining questions. Give students sufficient time to complete the table. Provide grid paper for Questions 6–12.
- Students should begin to observe patterns as they complete the table.
- Circulate to ensure that graphs are drawn correctly.
- Encourage students to draw the directed arrows in order to help them find the slope of the line.
- Have students put a few graphs on the board to help facilitate discussion.

? When students have finished, ask a series of summary questions.

  - "Compare certain pairs of graphs such as 6 and 7; or 10 and 11. What do you observe?" They have the same steepness (slope) and the number at the end of the equation is the y-coordinate of where the graph crosses the y-axis.
  - "Where does the equation $y = x + 7$ cross the y-axis?" at $(0, 7)$
  - "Compare certain groups of graphs such as 3, 6, and 10; or 4, 7, and 11. What do you observe?" They cross the y-axis at the same point, but the slopes are different; the coefficient of x is the slope of the line.
  - "What is the slope of the equation $y = 7x + 2$?" slope $= 7$

Words of Wisdom

- Students may not use mathematical language to describe their observations. Listen for the concept, the vocabulary will come later.
- In equations such as $y = x - 2$, students do not always think of the subtraction operation as making the constant negative. You may need to remind students that this is the same as *adding the opposite*. So, $y = x - 2$ is equivalent to $y = x + (-2)$.

What Is Your Answer?

- These answers should follow immediately from discussing student observations.

Closure

- Refer back to the demonstration cards A, B, and C. Have students describe how the equations would be similar and how they would be different. *Sample answer:* A and B have the same slope, but different y-intercepts. B and C have different slopes, but the same y-intercept.

Technology For the Teacher

The Dynamic Planning Tool

Editable Teacher's Resources at *BigIdeasMath.com*

Inductive Reasoning

Work with a partner. Graph each equation. Then copy and complete the table.

Equation	Description of Graph	Slope of Graph	Point of Intersection with y-axis
1a **2.** $y = -\frac{1}{2}x + 1$	Line	$-\frac{1}{2}$	(0, 1)
1b **3.** $y = -x + 2$			
1c **4.** $y = -x - 2$			
1d **5.** $y = \frac{1}{2}x + 1$			
6. $y = x + 2$			
7. $y = x - 2$			
8. $y = \frac{1}{2}x - 1$			
9. $y = -\frac{1}{2}x - 1$			
10. $y = 3x + 2$			
11. $y = 3x - 2$			
12. $y = -2x + 3$			

What Is Your Answer?

13. **IN YOUR OWN WORDS** How can you describe the graph of the equation $y = mx + b$?

    a. How does the value of m affect the graph of the equation?

    b. How does the value of b affect the graph of the equation?

    c. Check your answers to parts (a) and (b) with three equations that are not in the table.

14. **LOGIC** Why do you think $y = mx + b$ is called the "slope-intercept" form of the equation of a line? Use drawings or diagrams to support your answer.

Practice Use what you learned about graphing linear equations in slope-intercept form to complete Exercises 4–6 on page 62.

Key Vocabulary

x-intercept, *p. 60*
y-intercept, *p. 60*
slope-intercept form, *p. 60*

Intercepts

The ***x*-intercept** of a line is the x-coordinate of the point where the line crosses the x-axis. It occurs when $y = 0$.

The ***y*-intercept** of a line is the y-coordinate of the point where the line crosses the y-axis. It occurs when $x = 0$.

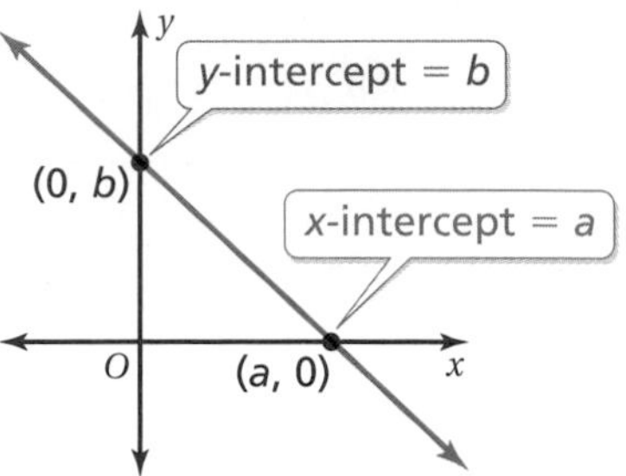

Slope-Intercept Form

Words A linear equation written in the form $y = mx + b$ is in **slope-intercept form**. The slope of the line is m and the y-intercept of the line is b.

Algebra $y = mx + b$ (slope: m; y-intercept: b)

EXAMPLE 1 Identifying Slopes and *y*-Intercepts

Find the slope and *y*-intercept of the graph of each linear equation.

a. $y = -4x - 2$

$y = -4x + (-2)$ Write in slope-intercept form.

∴ The slope is -4 and the y-intercept is -2.

b. $y - 5 = \frac{3}{2}x$

$y = \frac{3}{2}x + 5$ Add 5 to each side.

∴ The slope is $\frac{3}{2}$ and the y-intercept is 5.

On Your Own

Now You're Ready
Exercises 7–15

Find the slope and *y*-intercept of the graph of the linear equation.

1. $y = 3x - 7$

2. $y - 1 = -\frac{2}{3}x$

Multi-Language Glossary at BigIdeasMath.com.

Laurie's Notes

Introduction

Connect

- **Yesterday:** Students explored the connection between the equation of a line and its graph. (MP3a, MP8)
- **Today:** Students will use the slope-intercept form of a line to graph the line.

Motivate

- Share the following taxi information. All trips start at a convention center.

Destination	Distance	Taxi Fare
Football stadium	18.7 mi	\$39 approx.
Airport	12 mi	\$32 flat fee
Shopping district	9.5 mi	\$20 approx.

? "How do you think taxi fares are determined?" Answers will vary; listen for distance, number of passengers, tolls.

- Discuss why some locations, often involving airports, have flat fees associated with them.

Lesson Notes

Key Ideas

- Write the Key Ideas on the board. Draw the graph and discuss the vocabulary of this lesson: *x*-intercept, *y*-intercept, and slope-intercept form.
- Explain to students that the equation must be written with *y* as a function of *x*. This means that the equation must be solved for *y*.
- **FYI:** Students may ask why the letters *m* and *b* are used. Historically, there is no definitive answer. I tell my students that mathematicians, much older than myself, have used *m* for slope for centuries. Using *b* for the *y*-intercept appears to be an American phenomenon.

Example 1

? "What is a linear equation?" an equation whose graph is a line

- Write part (a). This is written in the form $y = mx + b$, enabling students to quickly identify the slope and *y*-intercept.
- Write part (b).

? "Is $y - 5 = \frac{3}{2}x$ in slope-intercept form?" no "Can you rewrite it so that it is?" yes; Add 5 to each side of the equation.

On Your Own

- **Think-Pair-Share:** Students should read each question independently and then work with a partner to answer the questions. When they have answered the questions, the pair should compare their answers with another group and discuss any discrepancies.

Goal Today's lesson is graphing the equation of a line written in **slope-intercept form**.

Lesson Materials
Textbook
• straightedge

Start Thinking! and Warm Up

Warm Up
For use before Lesson 2.3

Lesson 2.3 Start Thinking!
For use before Lesson 2.3

Describe a situation involving online shopping that can be modeled with a linear equation.

What is the slope?

What is the *y*-intercept?

Extra Example 1

Find the slope and *y*-intercept of the graph of each linear equation.

a. $y = \frac{3}{4}x - 5$

slope: $\frac{3}{4}$; *y*-intercept: -5

b. $y + \frac{1}{2} = -6x$

slope: -6; *y*-intercept: $-\frac{1}{2}$

On Your Own

1. slope: 3; *y*-intercept: -7
2. slope: $-\frac{2}{3}$; *y*-intercept: 1

Extra Example 2

Graph $y = -\frac{2}{3}x - 2$. Identify the x-intercept.

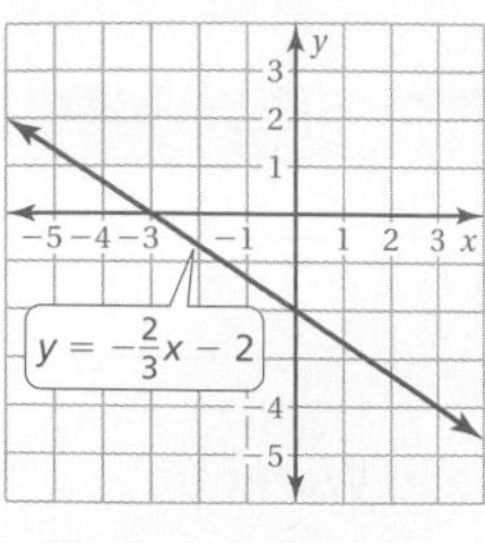

-3

Extra Example 3

The cost y (in dollars) for making friendship bracelets is $y = 0.5x + 2$, where x is the number of bracelets.

a. Graph the equation.

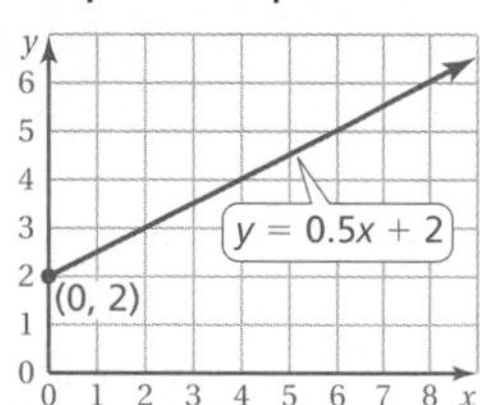

b. Interpret the slope and y-intercept.

The slope is 0.5. So, the cost per bracelet is \$0.50. The y-intercept is 2. So, there is an initial cost of \$2 to make the bracelets.

On Your Own

3–5. See Additional Answers.

Differentiated Instruction

Kinesthetic

When graphing a linear equation using the slope-intercept form, students must apply the slope correctly after plotting the point for the y-intercept. Have students plot (0, 3) in the coordinate plane. Then graph the lines $y = 4x + 3$, $y = \frac{1}{4}x + 3$, $y = -4x + 3$, and $y = -\frac{1}{4}x + 3$ in the same coordinate plane using (0, 3) as the starting point. Make sure students identify the correct rise and run for each line.

Laurie's Notes

Example 2

- ? "How can knowing the slope and the y-intercept help you graph a line?" Listen for student understanding of what slope and y-intercept mean.
- Remind students that a slope of -3 can be interpreted as $\frac{-3}{1} = \frac{3}{-1}$. Starting at the y-intercept, you can move to the right 1 unit and down 3 units, or to the left 1 unit and up 3 units. In both cases, you land on a point which satisfies the equation.
- ? **FYI:** "In this problem, you found the x-intercept by interpreting the slope and it coincidentally landed on the x-axis. How would you find the x-intercept without using a graph?" Set $y = 0$ and solve for x.

Example 3

- Write the equation $y = 2.5x + 2$ on the board.
- ? "What is the slope for this equation and what does it mean in the context of this problem?" 2.5; It costs \$2.50 for each mile you travel in the taxi. "What is the y-intercept and what does it mean in the context of this problem?" 2; The initial fee is \$2 when you sit down in the taxi.
- **MP2 Reason Abstractly and Quantitatively:** Mathematically proficient students make sense of the quantities and their relationships in problem situations. To develop this proficiency, students must be asked to interpret the meaning of the symbols.
- **MP6 Attend to Precision:** Suggest to students that because the slope is 2.5, any ratio equivalent to 2.5 can also be used, such as $\frac{2.5}{1} = \frac{5}{2}$. Using whole numbers instead of decimals improves the accuracy of graphing.
- Explain that the graph of this equation will only be in Quadrant I because it does not make sense to have a negative number of miles or a negative cost.
- ? "What is the cost for a 2-mile taxi ride? a 10-mile taxi ride?" \$7; \$27

On Your Own

- Students are asked to check their answers using a graphing calculator. It is helpful to build proficiency with using the graphing calculator so that the calculator becomes a useful tool in problem solving.

Closure

- **Exit Ticket:** Graph $y - 4 = 2x$ and identify the slope and y-intercept.

  slope = 2; y-intercept = 4

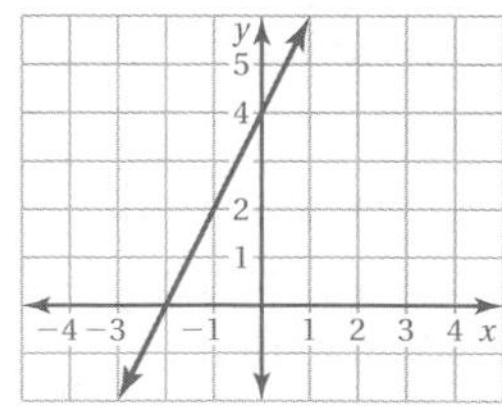

Technology For the Teacher

The Dynamic Planning Tool
Editable Teacher's Resources at *BigIdeasMath.com*

EXAMPLE 2 Graphing a Linear Equation in Slope-Intercept Form

Graph $y = -3x + 3$. Identify the x-intercept.

Step 1: Find the slope and y-intercept.

$$y = -3x + 3$$

slope → -3; y-intercept → 3

Check

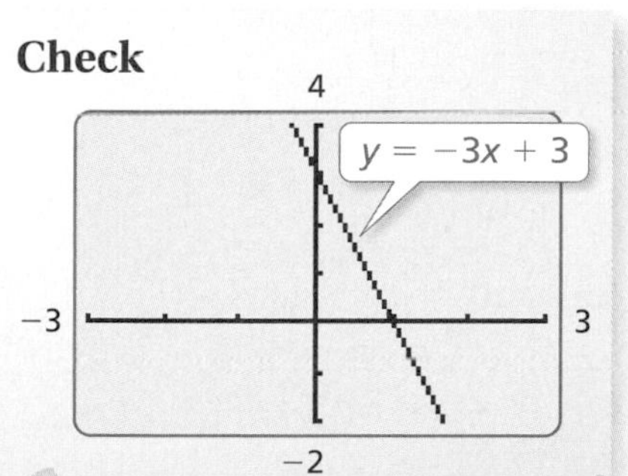

Step 2: The y-intercept is 3. So, plot (0, 3).

Step 3: Use the slope to find another point and draw the line.

$$\text{slope} = \frac{\text{rise}}{\text{run}} = \frac{-3}{1}$$

Plot the point that is 1 unit right and 3 units down from (0, 3). Draw a line through the two points.

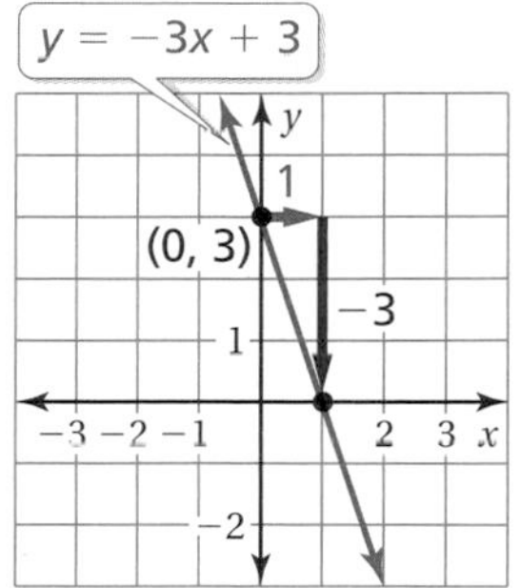

⁘ The line crosses the x-axis at (1, 0). So, the x-intercept is 1.

EXAMPLE 3 Real-Life Application

The cost y (in dollars) of taking a taxi x miles is $y = 2.5x + 2$. (a) Graph the equation. (b) Interpret the y-intercept and slope.

a. The slope of the line is $2.5 = \frac{5}{2}$. Use the slope and y-intercept to graph the equation.

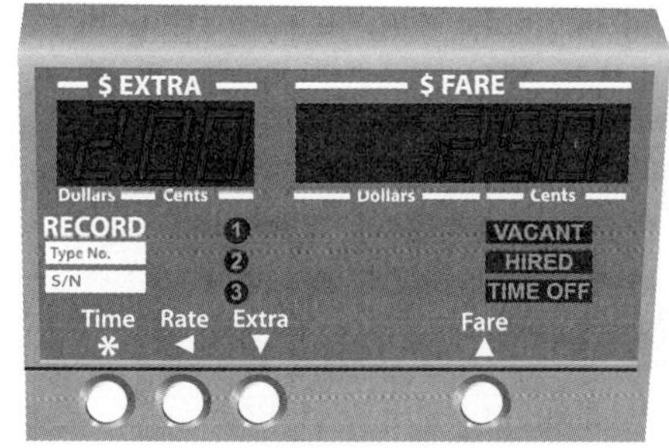

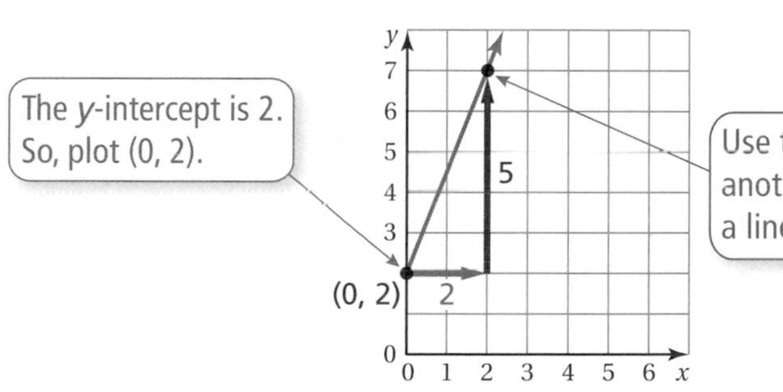

b. The slope is 2.5. So, the cost per mile is \$2.50. The y-intercept is 2. So, there is an initial fee of \$2 to take the taxi.

On Your Own

Graph the linear equation. Identify the x-intercept. Use a graphing calculator to check your answer.

3. $y = x - 4$

4. $y = -\frac{1}{2}x + 1$

5. In Example 3, the cost y (in dollars) of taking a different taxi x miles is $y = 2x + 1.5$. Interpret the y-intercept and slope.

2.3 Exercises

Vocabulary and Concept Check

1. **VOCABULARY** How can you find the x-intercept of the graph of $2x + 3y = 6$?
2. **CRITICAL THINKING** Is the equation $y = 3x$ in slope-intercept form? Explain.
3. **OPEN-ENDED** Describe a real-life situation that can be modeled by a linear equation. Write the equation. Interpret the y-intercept and slope.

Practice and Problem Solving

Match the equation with its graph. Identify the slope and y-intercept.

4. $y = 2x + 1$
5. $y = \frac{1}{3}x - 2$
6. $y = -\frac{2}{3}x + 1$

A. 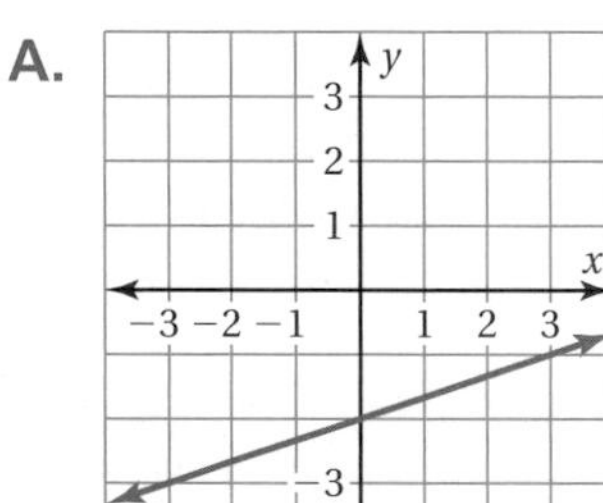

B. 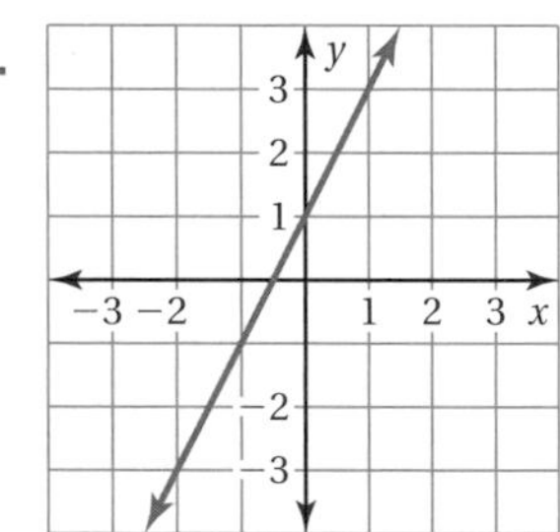

C. 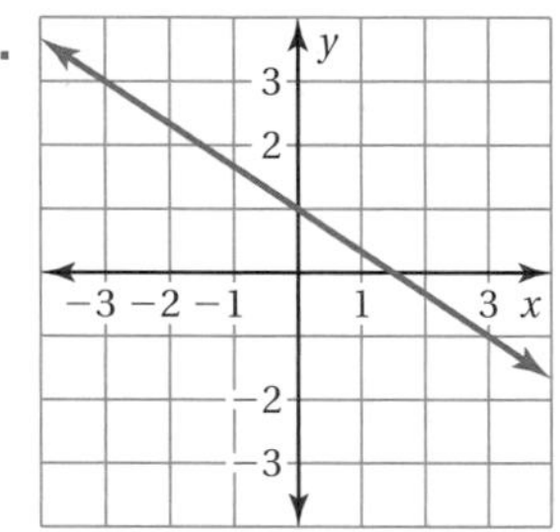

Find the slope and y-intercept of the graph of the linear equation.

7. $y = 4x - 5$
8. $y = -7x + 12$
9. $y = -\frac{4}{5}x - 2$
10. $y = 2.25x + 3$
11. $y + 1 = \frac{4}{3}x$
12. $y - 6 = \frac{3}{8}x$
13. $y - 3.5 = -2x$
14. $y + 5 = -\frac{1}{2}x$
15. $y = 1.5x + 11$

16. **ERROR ANALYSIS** Describe and correct the error in finding the slope and y-intercept of the graph of the linear equation.

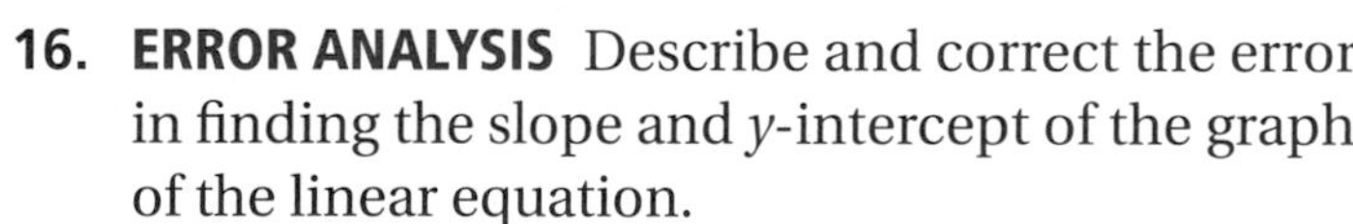

$y = 4x - 3$

The slope is 4 and the y-intercept is 3.

17. **SKYDIVING** A skydiver parachutes to the ground. The height y (in feet) of the skydiver after x seconds is $y = -10x + 3000$.

    a. Graph the equation.

    b. Interpret the x-intercept and slope.

Assignment Guide and Homework Check

Level	Assignment	Homework Check
Average	1–6, 13–16, 19–27 odd, 24, 26, 29–33	14, 16, 19, 23, 24
Advanced	1–6, 13–16, 18–28 even, 27, 29–33	14, 16, 18, 22, 24

Common Errors

- **Exercises 7–15** Students may forget to include negatives with the slope and/or y-intercept. Remind them to look at the sign in front of the slope and the y-intercept. Also remind students that the equation is $y = mx + b$. This means that if the linear equation has "minus b," then the y-intercept is negative.
- **Exercises 11–14** Students may identify the opposite y-intercept because they forget to solve for y. Remind them that slope-intercept form has y by itself, so they must solve for y before identifying the slope and y-intercept.
- **Exercises 18–23** Students may use the reciprocal of the slope when graphing and may find an incorrect x-intercept. Remind them that slope is *rise* over *run*, so the numerator represents vertical change, not horizontal.

2.3 Record and Practice Journal

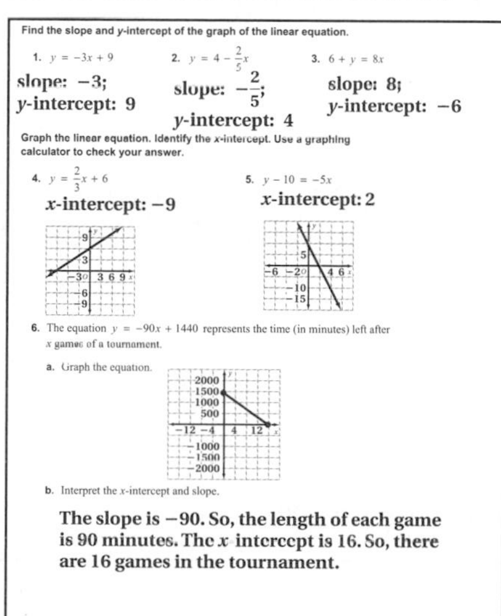
Find the slope and y-intercept of the graph of the linear equation.

1. $y = -3x + 9$
slope: -3; y-intercept: 9

2. $y = 4 - \frac{2}{5}x$
slope: $-\frac{2}{5}$; y-intercept: 4

3. $6 + y = 8x$
slope: 8; y-intercept: -6

Graph the linear equation. Identify the x-intercept. Use a graphing calculator to check your answer.

4. $y = \frac{2}{3}x + 6$
x-intercept: -9

5. $y - 10 = -5x$
x-intercept: 2

6. The equation $y = -90x + 1440$ represents the time (in minutes) left after x games of a tournament.

a. Graph the equation.

b. Interpret the x-intercept and slope.

The slope is -90. So, the length of each game is 90 minutes. The x-intercept is 16. So, there are 16 games in the tournament.

Answer Presentation Tool

Vocabulary and Concept Check

1. Find the x-coordinate of the point where the graph crosses the x-axis.
2. yes; The slope is 3 and the y-intercept is 0.
3. *Sample answer:* The amount of gasoline y (in gallons) left in your tank after you travel x miles is $y = -\frac{1}{20}x + 20$. The slope of $-\frac{1}{20}$ means the car uses 1 gallon of gas for every 20 miles driven. The y-intercept of 20 means there is originally 20 gallons of gas in the tank.

Practice and Problem Solving

4. B; slope: 2; y-intercept: 1
5. A; slope: $\frac{1}{3}$; y-intercept: -2
6. C; slope: $-\frac{2}{3}$; y-intercept: 1
7. slope: 4; y-intercept: -5
8. slope: -7; y-intercept: 12
9. slope: $-\frac{4}{5}$; y-intercept: -2
10. slope: 2.25; y-intercept: 3
11. slope: $\frac{4}{3}$; y-intercept: -1
12. slope: $\frac{3}{8}$; y-intercept: 6
13. slope: -2; y-intercept: 3.5
14. slope: $-\frac{1}{2}$; y-intercept: -5
15. slope: 1.5; y-intercept: 11

16–17. See Additional Answers.

Practice and Problem Solving

18.

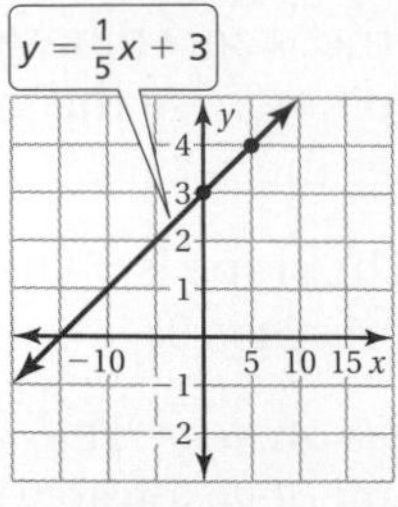

x-intercept: -15

19–27. See Additional Answers.

28. See *Taking Math Deeper.*

Fair Game Review

29. $y = 2x + 3$

30. $y = -\frac{4}{5}x + \frac{13}{5}$

31. $y = \frac{2}{3}x - 2$

32. $y = -\frac{7}{4}x + 2$

33. B

Mini-Assessment

Find the slope and y-intercept of the graph of the equation. Then graph the equation.

1. $y = -5x + 3$

slope $= -5$, y-intercept $= 3$

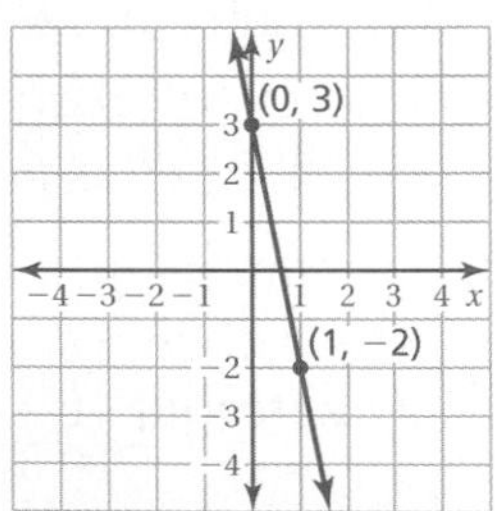

2. $y - 4 = \frac{1}{2}x$

slope $= \frac{1}{2}$, y-intercept $= 4$

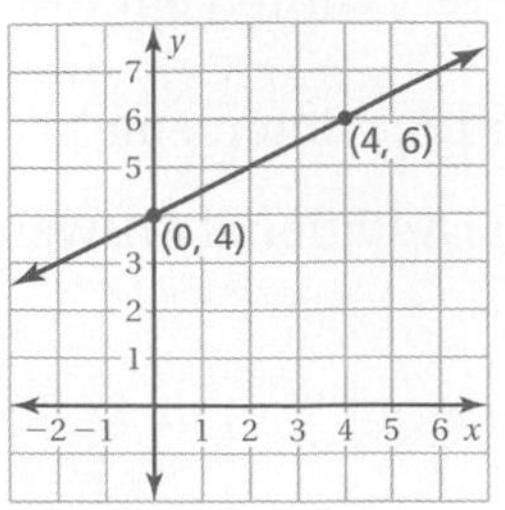

Taking Math Deeper

Exercise 28

This is a classic business problem. You have monthly costs for your business. The question is how much do you have to sell to cover your costs and start making a profit.

Organize the given information.

- The site sells 5 banner ads.
- Monthly income is $0.005 per click.
- It costs $120 per month to run the site.
- Let y be the monthly income.
- Let x be the number of clicks per month.

a. Write an equation for the income.

$y = 0.005x$

b. Graph the equation.

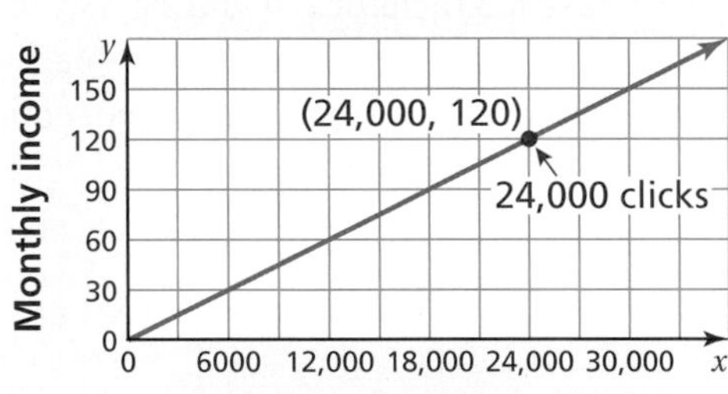

When the ads start to get 24,000 clicks a month, the income will be $120 per month. Each banner ad needs to average $\frac{24{,}000}{5} = 4800$ clicks. Any additional clicks per month will start earning a profit.

Project

Use the Internet or the school library to research methods for determining the number of clicks on a website.

Reteaching and Enrichment Strategies

If students need help...	If students got it...
Resources by Chapter • Practice A and Practice B • Puzzle Time Record and Practice Journal Practice Differentiating the Lesson Lesson Tutorials Skills Review Handbook	Resources by Chapter • Enrichment and Extension • School-to-Work Start the next section

Graph the linear equation. Identify the x-intercept. Use a graphing calculator to check your answer.

2 **18.** $y = \frac{1}{5}x + 3$ **19.** $y = 6x - 7$ **20.** $y = -\frac{8}{3}x + 9$

21. $y = -1.4x - 1$ **22.** $y + 9 = -3x$ **23.** $y - 4 = -\frac{3}{5}x$

24. PHONES The cost y (in dollars) of making a long distance phone call for x minutes is $y = 0.25x + 2$.

a. Graph the equation.

b. Interpret the slope and y-intercept.

25. APPLES Write a linear equation that models the cost y of picking x pounds of apples. Graph the equation.

26. ELEVATOR The basement of a building is 40 feet below ground level. The elevator rises at a rate of 5 feet per second. You enter the elevator in the basement. Write an equation that represents the height y (in feet) of the elevator after x seconds. Graph the equation.

27. REASONING You work in an electronics store. You earn a fixed amount of $35 per day, plus a 15% bonus on the merchandise you sell. Write an equation that models the amount y (in dollars) you earn for selling x dollars of merchandise in one day. Graph the equation.

28. Critical Thinking Six friends create a website. The website earns money by selling banner ads. The site has five banner ads. It costs $120 a month to operate the website.

a. A banner ad earns $0.005 per click. Write a linear equation that represents the monthly income y (in dollars) for x clicks.

b. Draw a graph of the equation in part (a). On the graph, label the number of clicks needed for the friends to start making a profit.

Fair Game Review What you learned in previous grades & lessons

Solve the equation for y. *(Section 1.4)*

29. $y - 2x = 3$ **30.** $4x + 5y = 13$ **31.** $2x - 3y = 6$ **32.** $7x + 4y = 8$

33. MULTIPLE CHOICE Which point is a solution of the equation $3x - 8y = 11$? *(Section 2.1)*

Ⓐ $(1, 1)$ Ⓑ $(1, -1)$ Ⓒ $(-1, 1)$ Ⓓ $(-1, -1)$

2.4 Graphing Linear Equations in Standard Form

COMMON CORE STATE STANDARDS
A.CED.2
A.REI.10
F.IF.4

Essential Question How can you describe the graph of the equation $ax + by = c$?

1 ACTIVITY: Using a Table to Plot Points

Work with a partner. You sold a total of $16 worth of tickets to a school concert. You lost track of how many of each type of ticket you sold.

$$\frac{\$4}{\text{Adult}} \cdot \text{Number of Adult Tickets} + \frac{\$2}{\text{Child}} \cdot \text{Number of Child Tickets} = \$16$$

a. Let x represent the number of adult tickets.
Let y represent the number of child tickets.
Write an equation that relates x and y.

b. Copy and complete the table showing the different combinations of tickets you might have sold.

Number of Adult Tickets, x					
Number of Child Tickets, y					

c. Plot the points from the table. Describe the pattern formed by the points.

d. If you remember how many adult tickets you sold, can you determine how many child tickets you sold? Explain your reasoning.

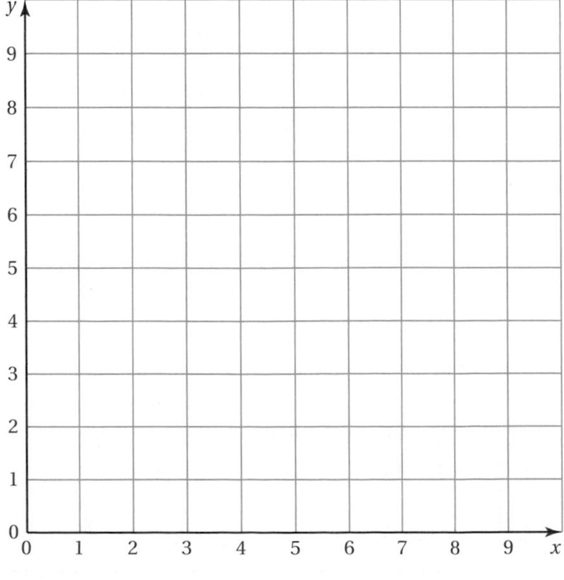

Laurie's Notes

Introduction

Standards for Mathematical Practice

- **MP7 Look for and Make Use of Structure:** In this lesson students will graph a linear equation in a new form. Mathematically proficient students discern a pattern or structure. Recognizing the equivalence of equations written in different forms requires that students be able to manipulate equations.

Motivate

- **Preparation:** Make a set of equation cards on strips of paper. The equations are all the same when simplified and need to be written large enough to be read by students sitting at the back of the classroom.
- Here is a sample set of equations: $y = 2x + 1$, $-2x + y = 1$, $2x - y = -1$, $4x - 2y = -2$
- Ask 4 students to stand at the front of the room and hold the cards so only they can see the equations.
- As you state an ordered pair, the students holding the cards determine if it is a solution of the equation they are holding. If it is, they raise their hand. If not, they do nothing. State several ordered pairs, four that are solutions and two that are not. Plot all of the points that you state. The four ordered pairs that are solutions will be in a line.

? "How many lines can pass through any two points?" one "How many lines pass through the four solutions points?" Students will say 4; now is the time to discuss the idea of one line written in different forms.

- Have each student reveal their equation to the class and read it aloud. Write each of the equations on the board.
- Explain to students that equations can be written in different forms. Today they will explore a new form of a linear equation.

Activity Notes

Activity 1

- Read the problem aloud. Discuss what the variables x and y represent.
- Note that a verbal model is shown for the equation $4x + 2y = 16$.

? "Could you have sold 5 adult tickets? Explain." No; 5 adult tickets would be \$20, which is too much.

- Students may say that they do not know how to figure out x and y. Students may not realize that there is more than one solution. Remind students that *Guess, Check, and Revise* would be an appropriate strategy to use.
- Discuss part (c). The points lie on a line.
- Discuss part (d). Students may not recognize that in knowing x, they can substitute and solve for y. This is not an obvious step for students.

? "Could $x = 1.5$? Explain." No, you cannot sell 1.5 tickets.

? "What are the different numbers of adult tickets that are possible to sell?" 0, 1, 2, 3, 4

- **Note:** This is an example of a discrete domain; there are only 5 possible values for the variable x. This will be taught at a later time.

Common Core State Standards

A.CED.2 Create equations in two or more variables to represent relationships between quantities; graph equations on coordinate axes with labels and scales.

A.REI.10 Understand that the graph of an equation in two variables is the set of all its solutions plotted in the coordinate plane, often forming a curve (which could be a line).

F.IF.4 For a function that models a relationship between two quantities, interpret key features of graphs and tables in terms of the quantities, and sketch graphs showing key features given a verbal description of the relationship.

Previous Learning

Students should know how to graph lines in slope-intercept form.

Start Thinking! and Warm Up

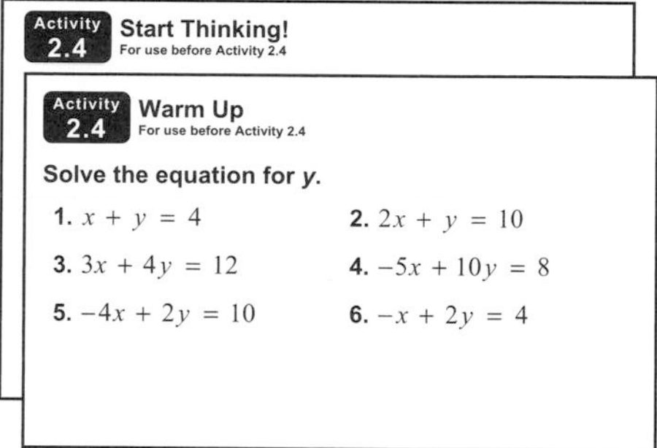

Activity 2.4 Start Thinking! For use before Activity 2.4

Activity 2.4 Warm Up For use before Activity 2.4

Solve the equation for y.

1. $x + y = 4$
2. $2x + y = 10$
3. $3x + 4y = 12$
4. $-5x + 10y = 8$
5. $-4x + 2y = 10$
6. $-x + 2y = 4$

2.4 Record and Practice Journal

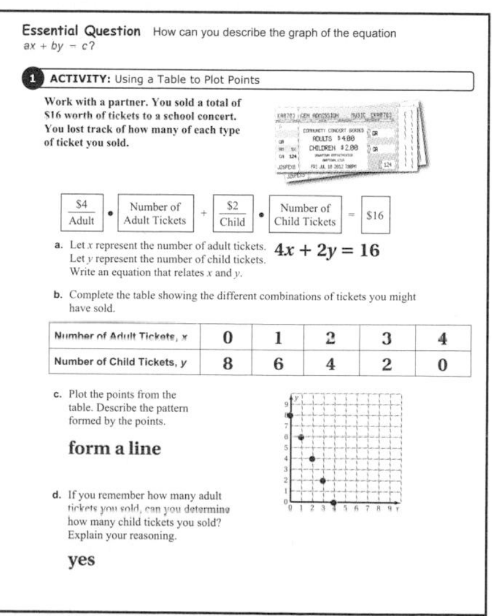

Essential Question How can you describe the graph of the equation $ax + by = c$?

1 ACTIVITY: Using a Table to Plot Points

Work with a partner. You sold a total of \$16 worth of tickets to a school concert. You lost track of how many of each type of ticket you sold.

\$4 Adult • Number of Adult Tickets + \$2 Child • Number of Child Tickets = \$16

a. Let x represent the number of adult tickets. Let y represent the number of child tickets. Write an equation that relates x and y. $4x + 2y = 16$

b. Complete the table showing the different combinations of tickets you might have sold.

Number of Adult Tickets, x	0	1	2	3	4
Number of Child Tickets, y	8	6	4	2	0

c. Plot the points from the table. Describe the pattern formed by the points.

form a line

d. If you remember how many adult tickets you sold, can you determine how many child tickets you sold? Explain your reasoning.

yes

Differentiated Instruction

Visual

Have students create a chart in their notebooks of the equation forms and how to graph them.

Slope-intercept form $y = mx + b$	• Plot $(0, b)$. • Use the slope m to plot a second point. • Draw a line through the two points.
Horizontal line $y = c$	• Draw a horizontal line through $(0, c)$.
Vertical line $x = c$	• Draw a vertical line through $(c, 0)$.
Standard form $ax + by = c$	• Find the y-intercept. • Find the x-intercept. • Plot the associated points. Draw a line through the two points.

2.4 Record and Practice Journal

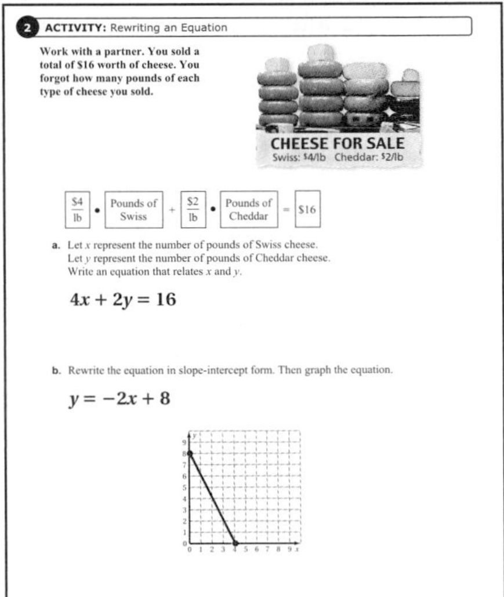

2 ACTIVITY: Rewriting an Equation

Work with a partner. You sold a total of $16 worth of cheese. You forgot how many pounds of each type of cheese you sold.

$\frac{\$4}{\text{lb}}$ • Pounds of Swiss + $\frac{\$2}{\text{lb}}$ • Pounds of Cheddar = $16

a. Let x represent the number of pounds of Swiss cheese. Let y represent the number of pounds of Cheddar cheese. Write an equation that relates x and y.

$4x + 2y = 16$

b. Rewrite the equation in slope-intercept form. Then graph the equation.

$y = -2x + 8$

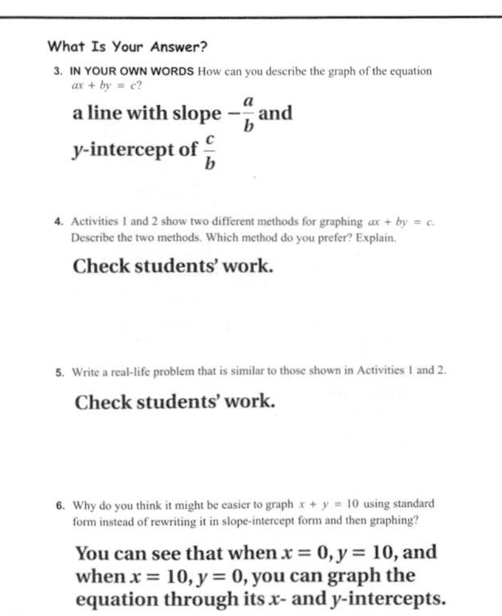

What Is Your Answer?

3. **IN YOUR OWN WORDS** How can you describe the graph of the equation $ax + by = c$?

a line with slope $-\frac{a}{b}$ and y-intercept of $\frac{c}{b}$

4. Activities 1 and 2 show two different methods for graphing $ax + by = c$. Describe the two methods. Which method do you prefer? Explain.

Check students' work.

5. Write a real-life problem that is similar to those shown in Activities 1 and 2.

Check students' work.

6. Why do you think it might be easier to graph $x + y = 10$ using standard form instead of rewriting it in slope-intercept form and then graphing?

You can see that when $x = 0, y = 10$, and when $x = 10, y = 0$, you can graph the equation through its x- and y-intercepts.

Laurie's Notes

Activity 2

- Read the problem aloud. Discuss what the variables x and y represent.
- Note that a verbal model is shown for the equation $4x + 2y = 16$.

? "Could you have sold 5 pounds of Swiss cheese? Explain." No; 5 pounds of Swiss cheese would be $20, which is too much.

- Give time for students to work with their partner. While this may be the same equation as Activity 1, the approach is different. Students are asked to write the equation in slope-intercept form. After the equation is in slope-intercept form, students can substitute a value for x, and find y. This is generally not the case for equations written in standard form.

? "Could $x = 1.5$? Explain." yes; You can buy a portion of a pound.

- **Note:** This is an example of a continuous domain; all numbers $0 \leq x \leq 4$ are possible. This will be taught at a later time.
- Students might observe that both examples have graphs in the first quadrant. This is common for real-life examples.

What Is Your Answer?

- **Question 3:** Students may guess that the graph is linear from Activity 1. However, some students may not be secure with this knowledge yet.
- Question 6 asks students to think about the process of graphing a line. Students consider the structure of the equation $x + y = 10$. When sharing their answers, listen for students to translate the equation as "the sum of two number is 10."

Closure

- Refer back to the equation cards. Rewrite the last three equations in slope-intercept form. $y = 2x + 1$

Technology For the Teacher

The Dynamic Planning Tool
Editable Teacher's Resources at *BigIdeasMath.com*

2 ACTIVITY: Rewriting an Equation

Work with a partner. You sold a total of $16 worth of cheese. You forgot how many pounds of each type of cheese you sold.

$$\frac{\$4}{\text{lb}} \cdot \text{Pounds of Swiss} + \frac{\$2}{\text{lb}} \cdot \text{Pounds of Cheddar} = \$16$$

a. Let x represent the number of pounds of Swiss cheese.

Let y represent the number of pounds of Cheddar cheese.

Write an equation that relates x and y.

b. Rewrite the equation in slope-intercept form. Then graph the equation.

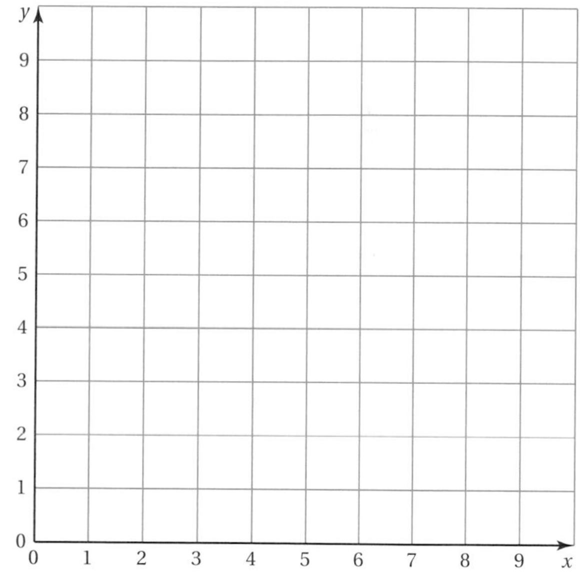

What Is Your Answer?

3. **IN YOUR OWN WORDS** How can you describe the graph of the equation $ax + by = c$?
4. Activities 1 and 2 show two different methods for graphing $ax + by = c$. Describe the two methods. Which method do you prefer? Explain.
5. Write a real-life problem that is similar to those shown in Activities 1 and 2.
6. Why do you think it might be easier to graph $x + y = 10$ using standard form instead of rewriting it in slope-intercept form and then graphing?

Practice Use what you learned about graphing linear equations in standard form to complete Exercises 3 and 4 on page 68.

2.4 Lesson

Study Tip

Any linear equation can be written in standard form.

Key Idea

Standard Form of a Linear Equation

The **standard form** of a linear equation is

$$ax + by = c$$

where a and b are not both zero.

EXAMPLE 1 Graphing a Linear Equation in Standard Form

Graph $-2x + 3y = -6$.

Step 1: Write the equation in slope-intercept form.

$-2x + 3y = -6$ Write the equation.

$3y = 2x - 6$ Add $2x$ to each side.

$y = \frac{2}{3}x - 2$ Divide each side by 3.

Step 2: Use the slope and y-intercept to graph the equation.

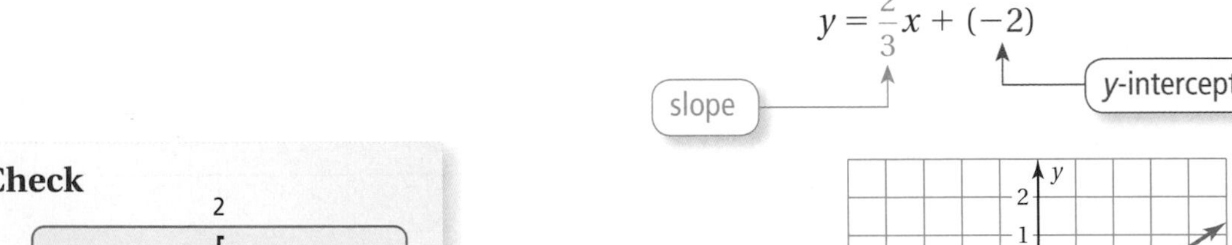

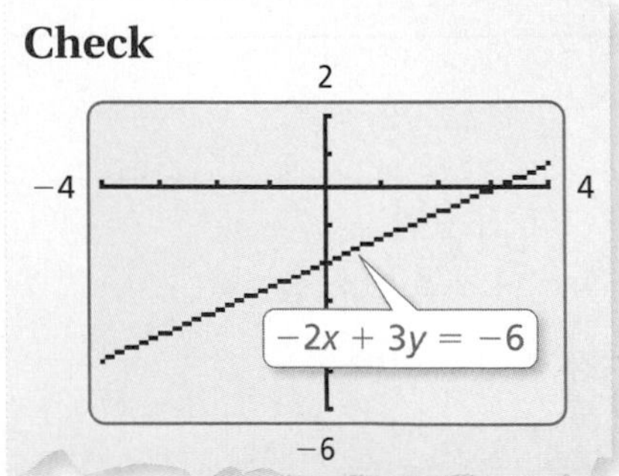

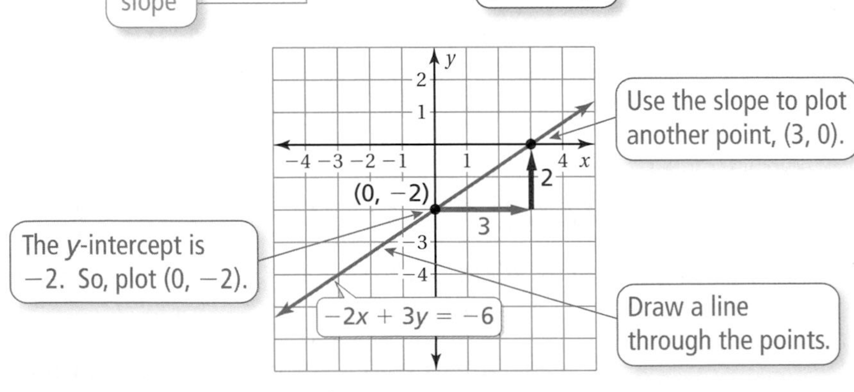

On Your Own

Now You're Ready
Exercises 5–10

Graph the linear equation. Use a graphing calculator to check your graph.

1. $x + y = -2$
2. $-\frac{1}{2}x + 2y = 6$
3. $-\frac{2}{3}x + y = 0$
4. $2x + y = 5$

Multi-Language Glossary at BigIdeasMath.com.

Laurie's Notes

Introduction

Connect

- **Yesterday:** Students explored the graph of an equation written in standard form. (MP7)
- **Today:** Students will graph equations written in standard form.

Motivate

? "How many pairs of numbers can you think of that have a sum of 5?" Encourage students to write their numbers on paper as ordered pairs. Example: (2, 3)

? "Did any of you include numbers that are not whole numbers?" Check to see if anyone had negative numbers or rational numbers.

- Ask one student to name an *x*-coordinate and another student to provide the *y*-coordinate. Plot the ordered pairs in a coordinate plane.

? "What do you think the equation of this line would be?" $x + y = 5$

Lesson Notes

Key Idea

- Define the standard form of a linear equation.
- Students may ask why both *a* and *b* cannot be zero. Explain that if $a = 0$ and $b = 0$, you would not have the equation of a line.
- **Teaching Tip:** Students are often confused when the standard form is written with parameters *a*, *b*, and *c*. Students see 5 variables. Show examples of equations written in standard form and identify *a*, *b*, and *c*.
- Ask students why they think $ax + by = c$ is called *standard* form. Students might suggest that the variables are on the left and a constant on the right.

Example 1

- Have students identify *a*, *b*, and *c*. $a = -2$, $b = 3$, and $c = -6$

? "How do you solve for *y*?" Add $2x$ to each side, then divide both sides by 3.

- Explain that the reason for rewriting the equation in slope-intercept form is so that the slope and the *y*-intercept can be used to graph the equation.
- **Common Error:** Students only divide one of the two terms on the right side of the equation by 3. Relate this to fraction operations. You are separating the expression into two terms and then simplifying.

? "Now that the equation is in slope-intercept form, explain how to graph the equation." Plot the ordered pair for the *y*-intercept. To plot another point, start at $(0, -2)$ and move to the right 3 units and up 2 units. Note that you can also move 3 units to the left and down 2 units. Connect these points with a line.

- Substitute the additional ordered pairs into the original equation to verify that they are solutions of the equation.

On Your Own

- In Questions 2 and 3, the fractional coefficients may present a problem.
- Remind students that equations must be solved for *y* in order to enter them in the equation editor of the graphing calculator.

Goal Today's lesson is graphing a line written in **standard form**.

Lesson Materials
Textbook
• straightedge

Start Thinking! and Warm Up

Lesson 2.4 Warm Up For use before Lesson 2.4

Lesson 2.4 Start Thinking! For use before Lesson 2.4

You have $40 to spend on turkey and cheese for a party. At the deli, turkey is $10 per pound and cheese is $6 per pound.

Is it easier to write an equation to represent the situation in *slope-intercept form* or *standard form*? Why?

Extra Example 1

Graph $3x - 2y = 2$.

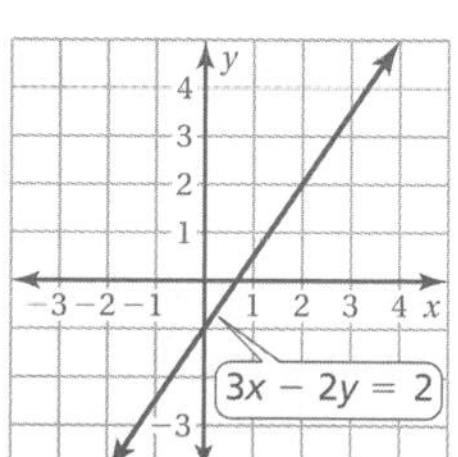

On Your Own

1. $x + y = -2$

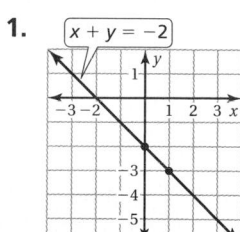

2. 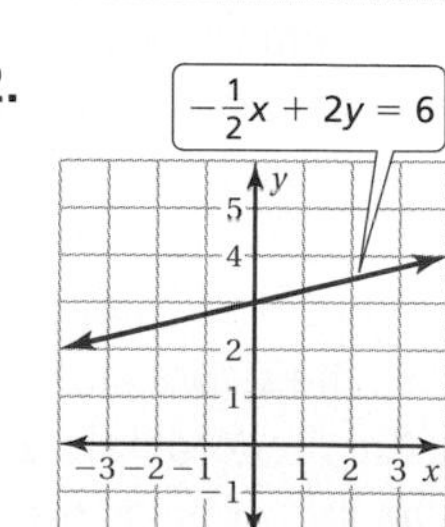

3–4. See Additional Answers.

Extra Example 2

Graph $5x - y = -5$ using intercepts.

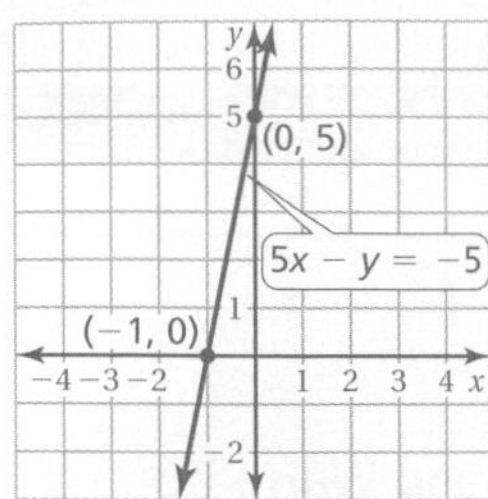

Extra Example 3

You have $2.40 to spend on grapes and bananas.

a. Graph the equation $1.2x + 0.6y = 2.4$, where x is the number of pounds of grapes and y is the number of pounds of bananas.

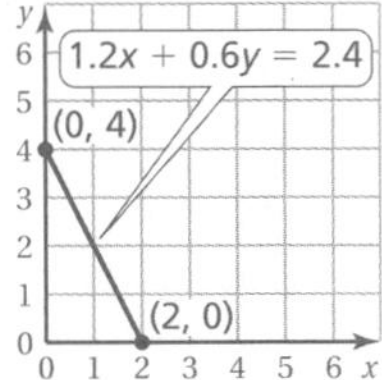

b. Interpret the intercepts. The x-intercept shows that you can buy 2 pounds of grapes, if you do not buy any bananas. The y-intercept shows that you can buy 4 pound of bananas, if you do not buy any grapes.

On Your Own

5–7. See Additional Answers.

English Language Learners

Vocabulary

For English learners, relate the word *intercept* with the football term *interception*. A defensive player on a football team crosses the path of the football to catch it and make an interception. Similarly, the y-intercept is the y-coordinate of the point where the line crosses the y-axis and the x-intercept is the x-coordinate of the point where the line crosses the x-axis.

Laurie's Notes

Example 2

- Start with a simple equation in standard form, such as $x + y = 4$. In this example, $a = 1$, $b = 1$, and $c = 4$. Explain to students that this could be solved for y by subtracting x from each side of the equation. Instead, you want to leave the equation as it was written.

? "Another way to think of this equation is *the sum of two numbers is 4*. Can you name some ordered pairs that would satisfy the equation?" Students should give many, including (0, 4) and (4, 0).

- Explain to students that sometimes an equation in standard form is graphed by using the two intercepts, instead of rewriting the equation in slope-intercept form.
- Write the equation shown: $x + 3y = -3$.

? "To find the x-intercept, what is the value of y? To find the y-intercept, what is the value of x?" 0; 0

- Finish the problem as shown.
- **Big Idea:** When the equation is in standard form, you can plot the points for the two intercepts and then draw the line through them.

Example 3

- Read the problem. Write the equation $1.5x + 0.6y = 6$ on the board.

? "What are the intercepts for this equation?" The x-intercept is 4 and the y-intercept is 10.

- Interpreting the intercepts in part (b) is an important step, particularly for real-life applications.
- Explain to students that negative values of x and y are not included in the graph because it does not make sense to have negative pounds of apples and bananas.

? "What is the cost of 2 pounds of apples and 5 pounds of bananas?" $6

? **MP1a Make Sense of Problems:** "What can you buy for $6?" *Sample answer:* 4 pounds of apples, or 10 pounds of bananas, or some other combination that is a solution. You are helping students make sense of the problem by asking them to interpret the symbolic representation.

On Your Own

- Students should work with a partner.

Closure

- **Writing Prompt:** To graph the equation $2x + y = 4$ … *Sample answer:* Find and plot the points for the x- and y-intercepts, then draw a line through these two points.

Technology For the Teacher

The Dynamic Planning Tool
Editable Teacher's Resources at *BigIdeasMath.com*

EXAMPLE 2 Graphing a Linear Equation in Standard Form

Graph $x + 3y = -3$ using intercepts.

Step 1: To find the x-intercept, substitute 0 for y.

$$x + 3y = -3$$
$$x + 3(0) = -3$$
$$x = -3$$

To find the y-intercept, substitute 0 for x.

$$x + 3y = -3$$
$$0 + 3y = -3$$
$$y = -1$$

Step 2: Graph the equation.

Check

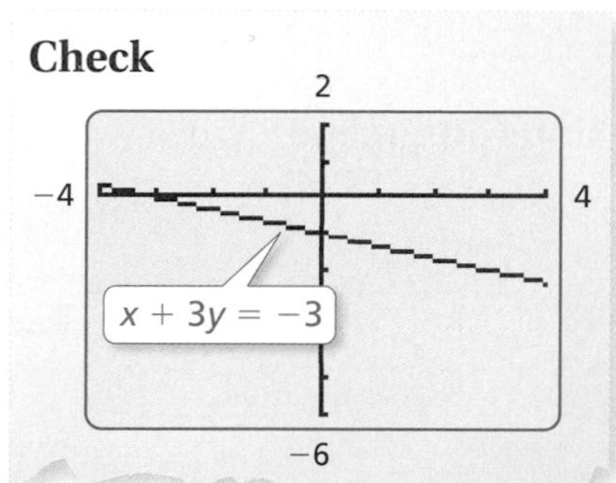

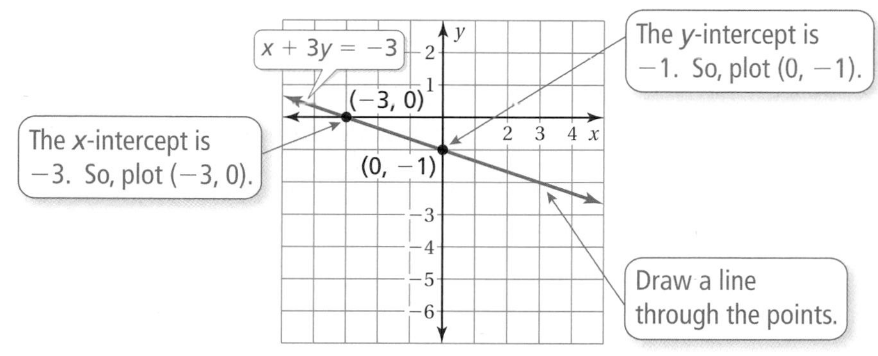

EXAMPLE 3 Real-Life Application

Bananas $0.60/pound

Apples $1.50/pound

You have \$6 to spend on apples and bananas. (a) Graph the equation $1.5x + 0.6y = 6$, where x is the number of pounds of apples and y is the number of pounds of bananas. (b) Interpret the intercepts.

a. Find the intercepts and graph the equation.

***x*-intercept**

$$1.5x + 0.6y = 6$$
$$1.5x + 0.6(0) = 6$$
$$x = 4$$

***y*-intercept**

$$1.5x + 0.6y = 6$$
$$1.5(0) + 0.6y = 6$$
$$y = 10$$

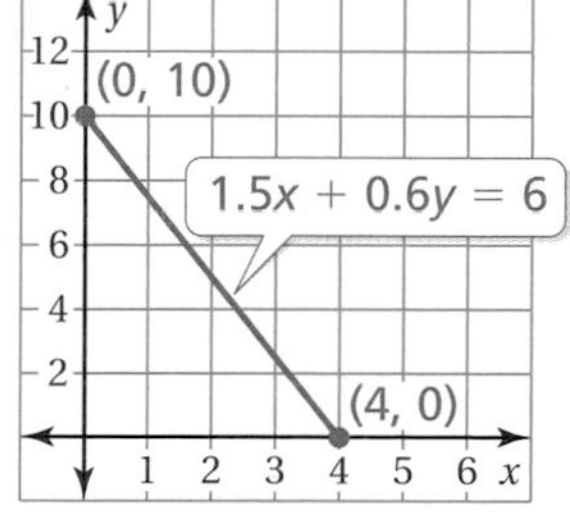

b. The x-intercept shows that you can buy 4 pounds of apples if you don't buy any bananas. The y-intercept shows that you can buy 10 pounds of bananas if you don't buy any apples.

On Your Own

Now You're Ready
Exercises 16–18

Graph the linear equation using intercepts. Use a graphing calculator to check your graph.

5. $2x - y = 8$

6. $x + 3y = 6$

7. **WHAT IF?** In Example 3, you buy y pounds of oranges instead of bananas. Oranges cost \$1.20 per pound. Graph the equation $1.5x + 1.2y = 6$. Interpret the intercepts.

2.4 Exercises

Vocabulary and Concept Check

1. **VOCABULARY** Is the equation $y = -2x + 5$ in standard form? Explain.

2. **REASONING** Does the graph represent a linear equation? Explain.

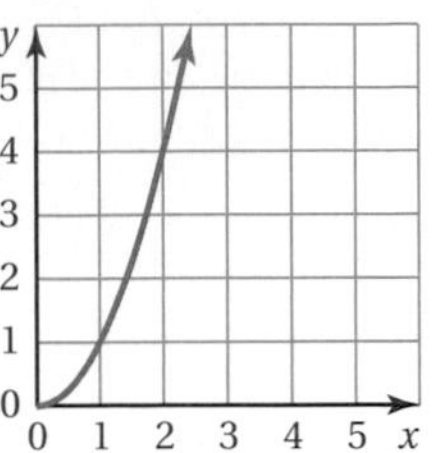

Practice and Problem Solving

Define two variables for the verbal model. Write an equation in slope-intercept form that relates the variables. Graph the equation.

3. $\frac{\$2.00}{\text{pound}}$ • Pounds of peaches + $\frac{\$1.50}{\text{pound}}$ • Pounds of apples = \$15

4. $\frac{16 \text{ miles}}{\text{hour}}$ • Hours biked + $\frac{2 \text{ miles}}{\text{hour}}$ • Hours walked = 32 miles

Write the linear equation in slope-intercept form.

1 5. $2x + y = 17$

6. $5x - y = \frac{1}{4}$

7. $-\frac{1}{2}x + y = 10$

Graph the linear equation. Use a graphing calculator to check your graph.

8. $-18x + 9y = 72$

9. $16x - 4y = 2$

10. $\frac{1}{4}x + \frac{3}{4}y = 1$

Use the graph to find the *x*- and *y*-intercepts.

11.

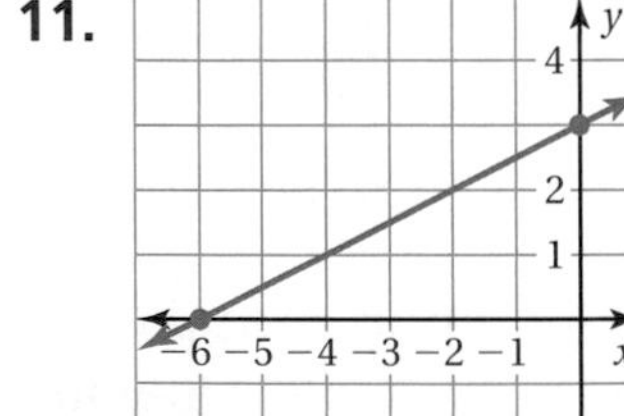

12.

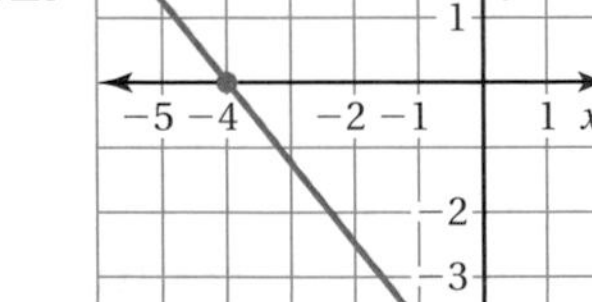

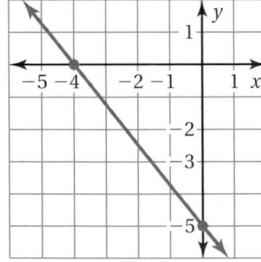

13.

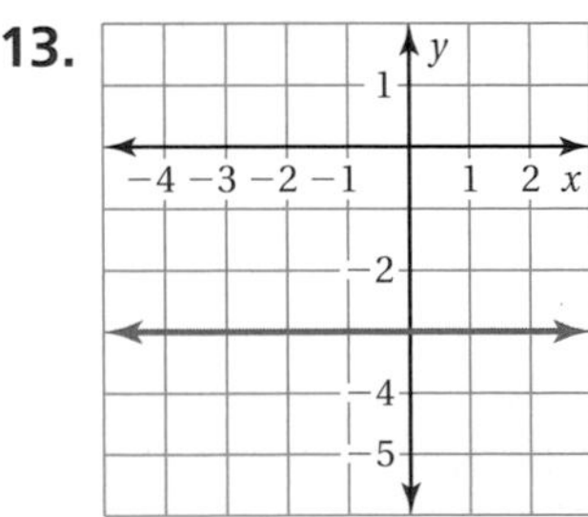

14. **ERROR ANALYSIS** Describe and correct the error in finding the x-intercept.

15. **BRACELET** A charm bracelet costs \$65, plus \$25 for each charm.

 a. Write an equation in standard form that represents the total cost of the bracelet.

 b. How much does the bracelet shown cost?

Assignment Guide and Homework Check

Level	Assignment	Homework Check
Average	1–4, 5–13 odd, 14–21, 24–26	7, 9, 13, 14, 16, 20
Advanced	1, 2, 8–14 even, 13, 16–23, 24–26	10, 13, 14, 16, 20, 22

Common Errors

- **Exercises 5–10** Students may use the same operation instead of the opposite operation when rewriting the equation in slope-intercept form.
- **Exercises 11 and 12, 16–18** Students may mix up the *x*- and *y*-intercepts. Remind them that the *x*-intercept is the *x*-coordinate of where the line crosses the *x*-axis and the *y*-intercept is the *y*-coordinate of where the line crosses the *y*-axis.
- **Exercise 13** Because the line is horizontal and there is no *x*-intercept, students may say that the *x*-intercept is zero. Remind them that this would mean that the *x*-intercept is at the origin; however, there is no *x*-intercept.

2.4 Record and Practice Journal

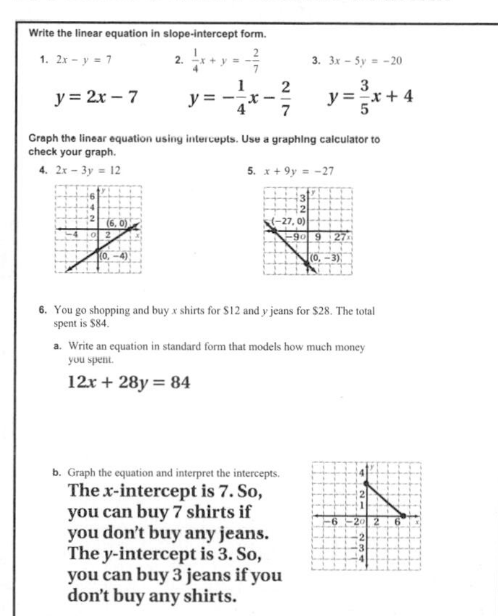

Write the linear equation in slope-intercept form.

1. $2x - y = 7$ $\quad y = 2x - 7$
2. $\frac{1}{4}x + y = -\frac{2}{7}$ $\quad y = -\frac{1}{4}x - \frac{2}{7}$
3. $3x - 5y = -20$ $\quad y = \frac{3}{5}x + 4$

Graph the linear equation using intercepts. Use a graphing calculator to check your graph.

4. $2x - 3y = 12$ (6, 0), (0, −4)
5. $x + 9y = -27$ (−27, 0), (0, −3)
6. You go shopping and buy *x* shirts for \$12 and *y* jeans for \$28. The total spent is \$84.
   a. Write an equation in standard form that models how much money you spent.
   $12x + 28y = 84$
   b. Graph the equation and interpret the intercepts.
   The *x*-intercept is 7. So, you can buy 7 shirts if you don't buy any jeans. The *y*-intercept is 3. So, you can buy 3 jeans if you don't buy any shirts.

Technology For the Teacher
Answer Presentation Tool

Vocabulary and Concept Check

1. no; The equation is in slope-intercept form.
2. no; The graph is not a line.

Practice and Problem Solving

3. x = pounds of peaches
   y = pounds of apples
   $y = -\frac{4}{3}x + 10$

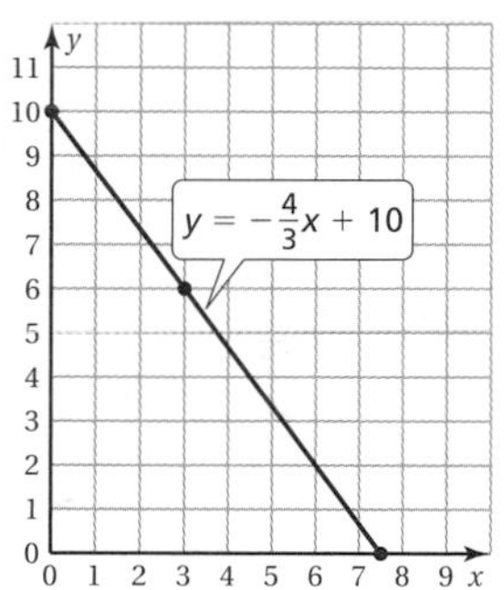

4. x = hours biked
   y = hours walked
   $y = -8x + 16$

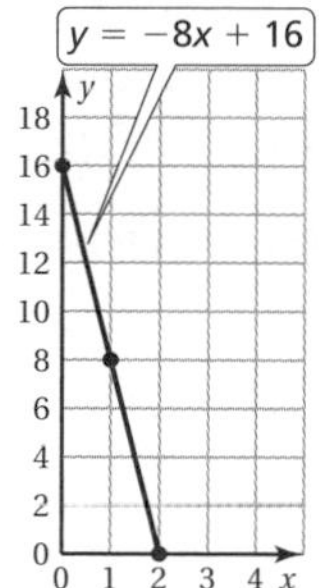

5. $y = -2x + 17$
6. $y = 5x - \frac{1}{4}$
7. $y = \frac{1}{2}x + 10$
8.

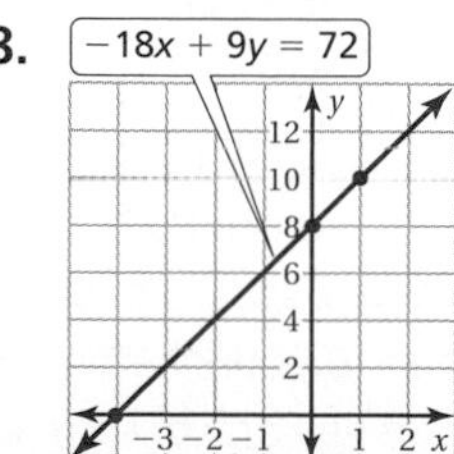

9–15. See Additional Answers.

Practice and Problem Solving

16–19. See Additional Answers.

20. See *Taking Math Deeper.*

21–23. See Additional Answers.

Fair Game Review

24. 1; 3; 5; 7; 9

25. 1; −2; −5; −8; −11

26. D

Mini-Assessment

1. Graph $-2x + 4y = 16$ using intercepts.

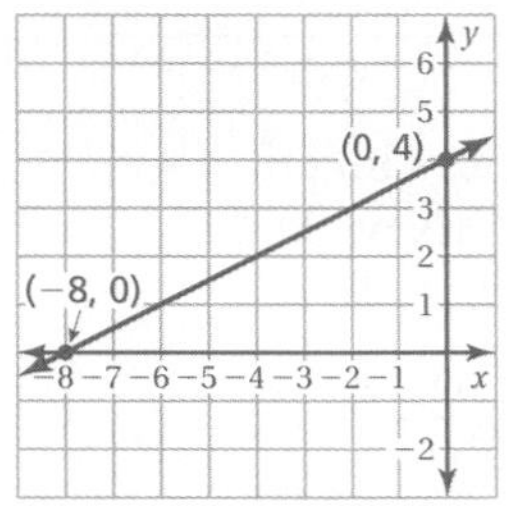

2. **You have $12 to spend on pears and oranges.**

   a. Graph the equation $1.2x + 0.8y = 12$, where x is the number of pounds of pears and y is the number of pounds of oranges.

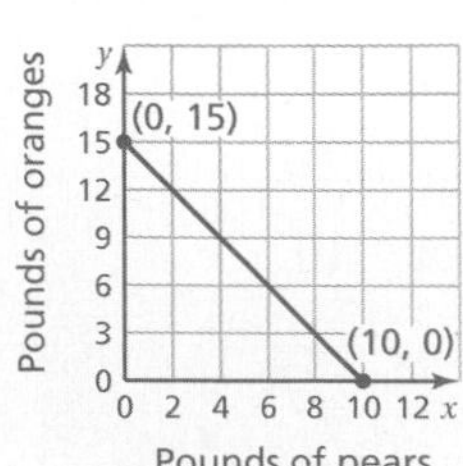

   b. Interpret the intercepts.
   The x-intercept shows that you can buy 10 pounds of pears if you do not buy any oranges. The y-intercept shows that you can buy 15 pounds of oranges if you do not buy any pears.

Taking Math Deeper

Exercise 20

As with many real-life problems, it helps to start by summarizing the given information.

Summarize the given information.

- Let x = days for renting boat.
- Let y = days for renting scuba gear.
- Cost of boat = \$250 per day.
- Cost of scuba gear = \$50 per day.
- Total spent = \$1000.

a. Write an equation.

$$250x + 50y = 1000$$

b. Graph the equation and interpret the intercepts.

$$y = -5x + 20$$

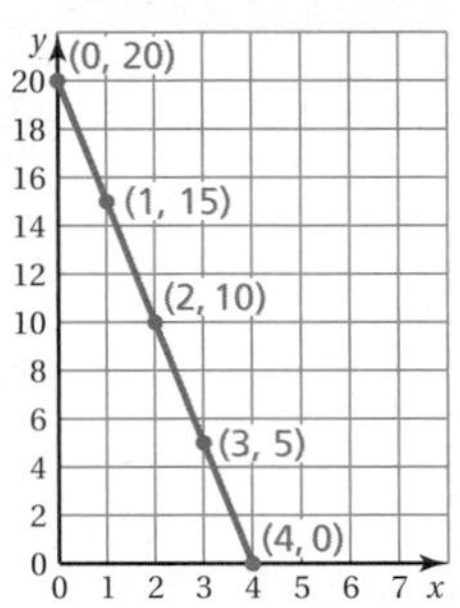

If $x = 0$, the group rented only the scuba gear for 20 days.
If $y = 0$, the group rented only the boat for 4 days.

Project

To go on a professional scuba diving tour, you need to be a certified diver. Use the school library or the Internet to research the requirements to become certified in scuba diving.

Reteaching and Enrichment Strategies

If students need help. . .	If students got it. . .
Resources by Chapter • Practice A and Practice B • Puzzle Time Record and Practice Journal Practice Differentiating the Lesson Lesson Tutorials Skills Review Handbook	Resources by Chapter • Enrichment and Extension • School-to-Work Start the next section

Graph the linear equation using intercepts. Use a graphing calculator to check your graph.

(2) **16.** $3x - 4y = -12$ **17.** $2x + y = 8$ **18.** $\frac{1}{3}x - \frac{1}{6}y = -\frac{2}{3}$

19. SHOPPING The amount of money you spend on x CDs and y DVDs is given by the equation $14x + 18y = 126$. Find the intercepts and graph the equation.

20. SCUBA Five friends go scuba diving. They rent a boat for x days and scuba gear for y days. The total spent is $1000.

a. Write an equation in standard form that represents the situation.

b. Graph the equation and interpret the intercepts.

21. MODELING You work at a restaurant as a host and a server. You earn $9.45 for each hour you work as a host and $7.65 for each hour you work as a server.

a. Write an equation in standard form that models your earnings.

b. Graph the equation.

Basic Information
Pay to the Order of: .... John Doe
\# of hours worked as .... host: x
\# of hours worked as .... server: y
Earnings for this pay .... period: $160.65

22. LOGIC Does the graph of every linear equation have an x-intercept? Explain your reasoning. Include an example.

23. Critical Thinking For a house call, a veterinarian charges $70, plus $40 an hour.

a. Write an equation that represents the total fee y charged by the veterinarian for a visit lasting x hours.

b. Find the x-intercept. Will this point appear on the graph of the equation? Explain your reasoning.

c. Graph the equation.

Fair Game Review What you learned in previous grades & lessons

Copy and complete the table of values. *(Skills Review Handbook)*

24.

x	−2	−1	0	1	2
$2x + 5$					

25.

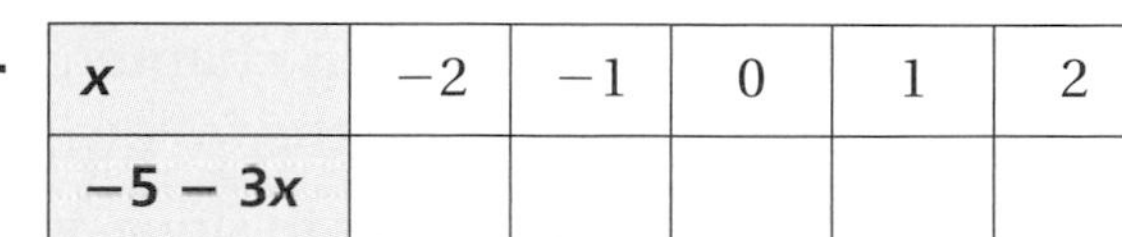

x	−2	−1	0	1	2
$-5 - 3x$					

26. MULTIPLE CHOICE Which value of x makes the equation $4x - 12 = 3x - 9$ true? *(Section 1.3)*

Ⓐ −1 Ⓑ 0 Ⓒ 1 Ⓓ 3

2 Study Help

You can use a **process diagram** to show the steps involved in a procedure. Here is an example of a process diagram for graphing a linear equation.

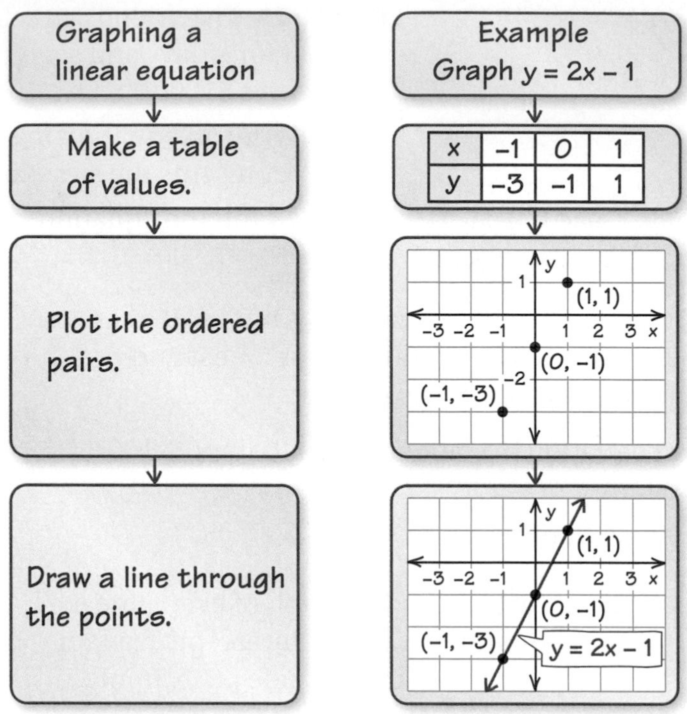

On Your Own

Make process diagrams with examples to help you study these topics.

1. finding the slope of a line
2. graphing a linear equation using
   a. slope and y-intercept
   b. x- and y-intercepts

After you complete this chapter, make process diagrams for the following topics.

3. writing equations in slope-intercept form
4. writing equations in point-slope form
5. writing equations of parallel lines
6. writing equations of perpendicular lines

"Here is a process diagram with suggestions for what to do if a hyena knocks on your door."

Sample Answers

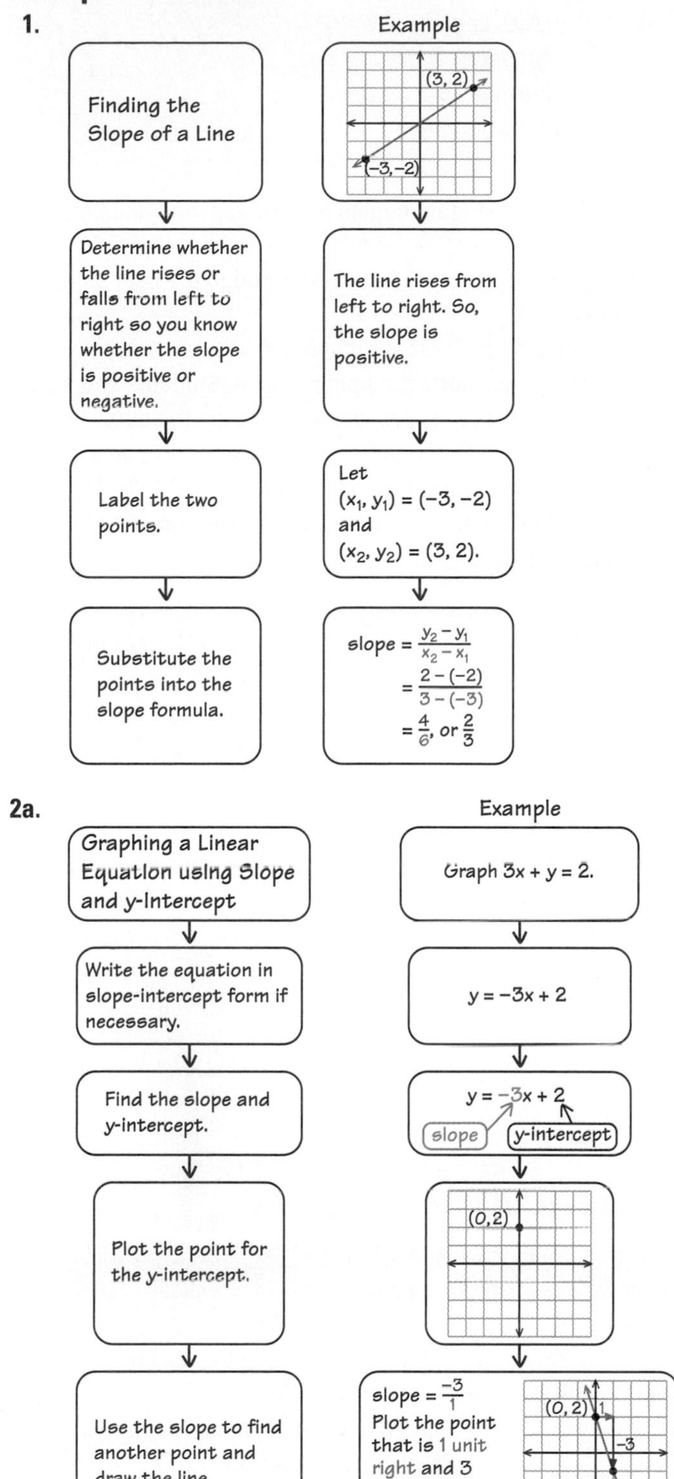

1.
Example
Finding the Slope of a Line
(3, 2)
(−3, −2)
Determine whether the line rises or falls from left to right so you know whether the slope is positive or negative.
The line rises from left to right. So, the slope is positive.
Label the two points.
Let $(x_1, y_1) = (-3, -2)$ and $(x_2, y_2) = (3, 2)$.
Substitute the points into the slope formula.
slope $= \frac{y_2 - y_1}{x_2 - x_1} = \frac{2-(-2)}{3-(-3)} = \frac{4}{6}$, or $\frac{2}{3}$
2a.
Example
Graphing a Linear Equation using Slope and y-Intercept
Graph $3x + y = 2$.
Write the equation in slope-intercept form if necessary.
$y = -3x + 2$
Find the slope and y-intercept.
$y = -3x + 2$
slope
y-intercept
Plot the point for the y-intercept.
(0, 2)
Use the slope to find another point and draw the line.
slope $= \frac{-3}{1}$
Plot the point that is 1 unit right and 3 units down from (0, 2).
(0, 2)
1
−3
$3x + y = 2$

2b. Available at *BigIdeasMath.com*.

List of Organizers

Available at *BigIdeasMath.com*

Comparison Chart
Concept Circle
Definition (Idea) and Example Chart
Example and Non-Example Chart
Formula Triangle
Four Square
Information Frame
Information Wheel
Notetaking Organizer
Process Diagram
Summary Triangle
Word Magnet
Y Chart

About this Organizer

A **Process Diagram** can be used to show the steps involved in a procedure. Process diagrams are particularly useful for illustrating procedures with two or more steps, and they can have one or more branches. As shown, students' process diagrams can have two parallel parts, in which the procedure is stepped out in one part and an example illustrating each step is shown in the other part. Or, the diagram can be made up of just one part, with example(s) included in the last "bubble" to illustrate the steps that precede it.

Technology For the Teacher
Vocabulary Puzzle Builder

Answers

1–4. See Additional Answers.

5. $-\frac{1}{2}$

6. 2

7. undefined

8. parallel slope: $-\frac{1}{2}$
perpendicular slope: 2

9. no; yes; The line $x = 1$ is vertical. The line $y = -1$ is horizontal. A vertical line is perpendicular to a horizontal line.

10. slope: $\frac{1}{4}$
y-intercept: -8

11. slope: -1
y-intercept: 3

12. x-intercept: 4
y-intercept: -6

13. x-intercept: 15
y-intercept: 3

14–16. See Additional Answers.

Assessment Book

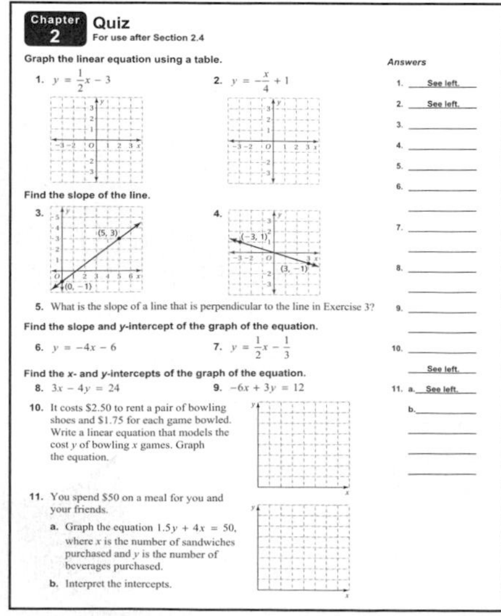

Chapter 2 **Quiz**
For use after Section 2.4

Graph the linear equation using a table.

1. $y = \frac{1}{2}x - 3$
2. $y = -\frac{x}{4} + 1$

Find the slope of the line.

3. (5, 3); (0, −1)
4. (−3, 1); (3, −1)
5. What is the slope of a line that is perpendicular to the line in Exercise 3?

Find the slope and y-intercept of the graph of the equation.

6. $y = -4x - 6$
7. $y = \frac{1}{2}x - \frac{1}{3}$

Find the x- and y-intercepts of the graph of the equation.

8. $3x - 4y = 24$
9. $-6x + 3y = 12$
10. It costs $2.50 to rent a pair of bowling shoes and $1.75 for each game bowled. Write a linear equation that models the cost y of bowling x games. Graph the equation.
11. You spend $50 on a meal for you and your friends.
    a. Graph the equation $1.5y + 4x = 50$, where x is the number of sandwiches purchased and y is the number of beverages purchased.
    b. Interpret the intercepts.

Answers

1. See left.
2. See left.
3.
4.
5.
6.
7.
8.
9.
10. See left.
11. a. See left.
    b.

Alternative Quiz Ideas

100% Quiz	Math Log
Error Notebook	Notebook Quiz
Group Quiz	Partner Quiz
Homework Quiz	**Pass the Paper**

Pass the Paper

- Work in groups of four. The first student copies the problem and does a step, explaining his or her work.
- The paper is passed and the second student works through the next step, also explaining his or her work.
- This process continues until the problem is completed.
- The second member of the group starts the next problem. Students should be allowed to question and debate as they are working through the quiz.
- Student groups can be selected by the teacher, by students, through a random process, or any way that works for your class.
- The teacher walks around the classroom listening to the groups and asks questions to ensure understanding.

Reteaching and Enrichment Strategies

If students need help. . .	If students got it. . .
Resources by Chapter • Study Help • Practice A and Practice B • Puzzle Time Lesson Tutorials *BigIdeasMath.com* Practice Quiz Practice from the Test Generator	Resources by Chapter • Enrichment and Extension • School-to-Work Game Closet at *BigIdeasMath.com* Start the next section

Technology For the Teacher

Answer Presentation Tool
Big Ideas Test Generator

2.1–2.4 Quiz

Check It Out
Progress Check
BigIdeasMath.com

Graph the linear equation using a table. *(Section 2.1)*

1. $y = -x + 8$
2. $y = \frac{x}{3} - 4$
3. $x = -1$
4. $y = 3.5$

Find the slope of the line. *(Section 2.2)*

5. 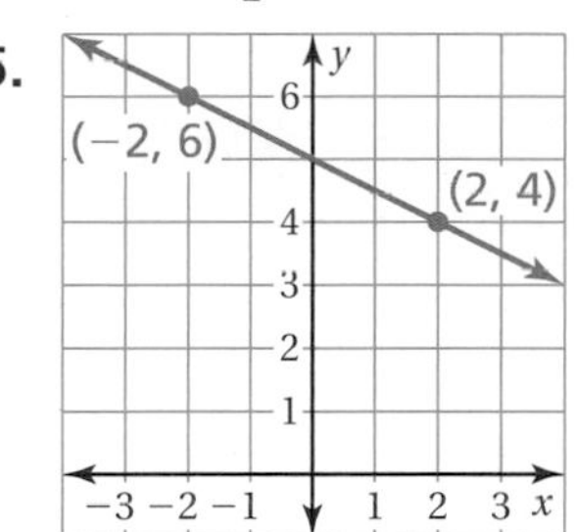

6. 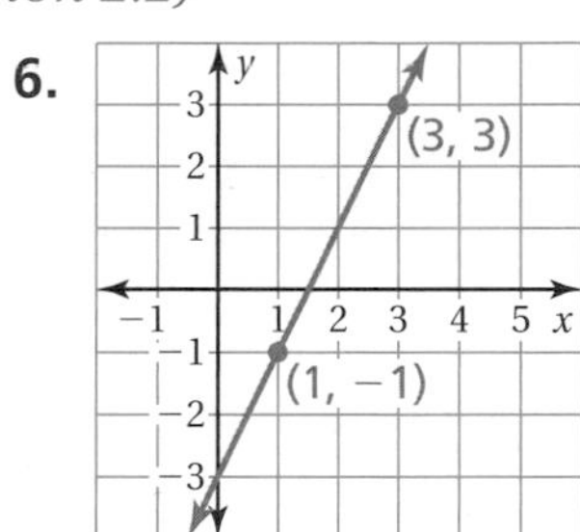

7. 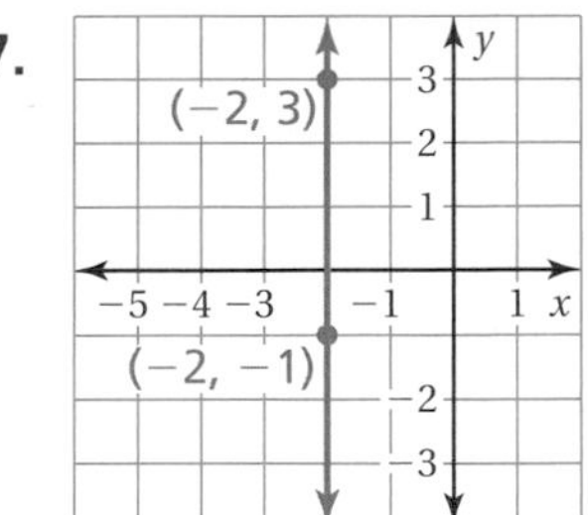

8. What is the slope of a line that is parallel to the line in Exercise 5? What is the slope of a line that is perpendicular to the line in Exercise 5? *(Section 2.2)*

9. Are the lines $y = -1$ and $x = 1$ parallel? Are they perpendicular? Justify your answer. *(Section 2.2)*

Find the slope and y-intercept of the graph of the linear equation. *(Section 2.3)*

10. $y = \frac{1}{4}x - 8$
11. $y = -x + 3$

Find the x- and y-intercepts of the graph of the equation. *(Section 2.4)*

12. $3x - 2y = 12$
13. $x + 5y = 15$

14. **BANKING** A bank charges \$3 each time you use an out-of-network ATM. At the beginning of the month, you have \$1500 in your bank account. You withdraw \$60 from your bank account each time you use an out-of-network ATM. Write and graph a linear equation that represents the balance in your account after you use an out-of-network ATM x times. *(Section 2.1)*

15. **STATE FAIR** Write a linear equation that models the cost y of one person going on x rides at the fair. Graph the equation. *(Section 2.3)*

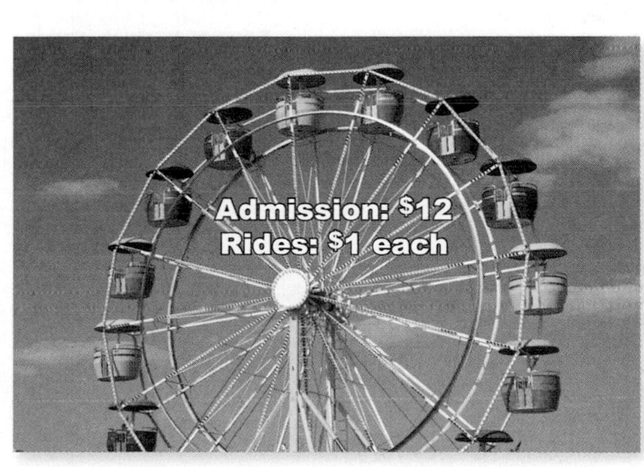

16. **PAINTING** You used \$90 worth of paint for a school float. *(Section 2.4)*

    a. Graph the equation $18x + 15y = 90$, where x is the number of gallons of blue paint and y is the number of gallons of white paint.

    b. Interpret the intercepts.

2.5 Writing Equations in Slope-Intercept Form

COMMON CORE STATE STANDARDS
8.F.3
A.CED.2
A.CED.3

Essential Question How can you write an equation of a line when you are given the slope and *y*-intercept of the line?

1 ACTIVITY: Writing Equations of Lines

Work with a partner.

- **Find the slope of each line.**
- **Find the *y*-intercept of each line.**
- **Write an equation for each line.**
- **What do the three lines have in common?**

a.

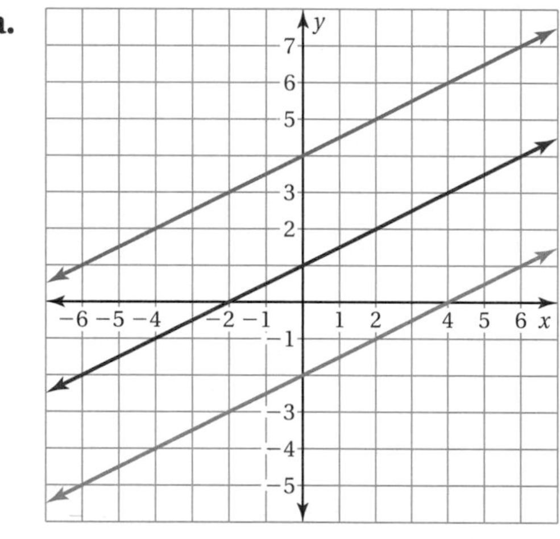

b.

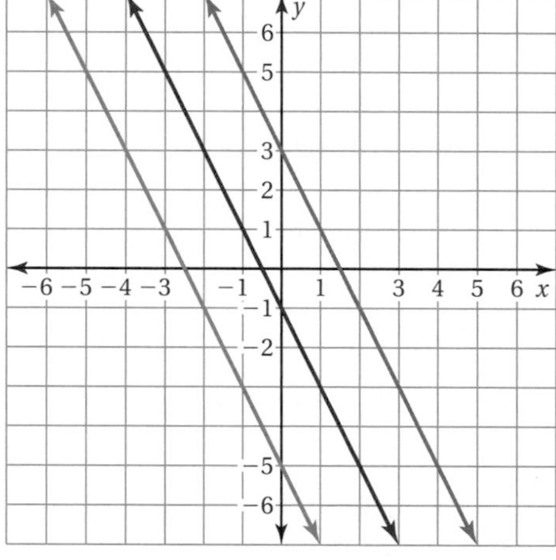

c.

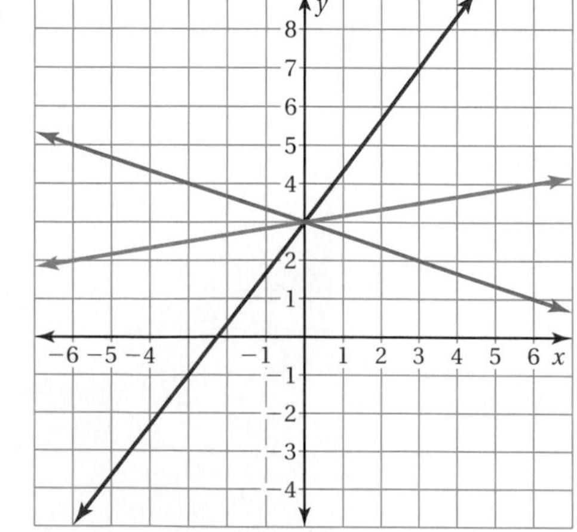

d.

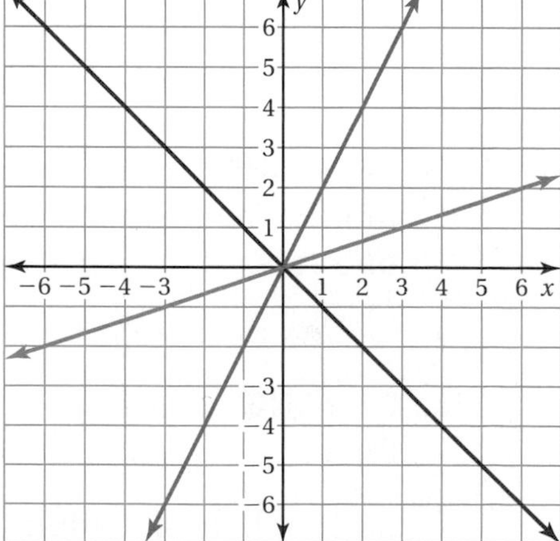

Laurie's Notes

Introduction

Standards for Mathematical Practice

- **MP1 Make Sense of Problems and Persevere in Solving Them** and **MP4 Model with Mathematics:** The goal of this lesson is for students to write equations of lines by first determining the slope and y-intercept from a graph. The visual model helps students make sense of and solve the problem. The graphical representation (model) helps students identify the important features of the line.

Motivate

- If there is sufficient space in your classroom, hallway, or school foyer, make coordinate axes using masking tape. Use a marker to scale each axis with integers -5 through 5.
- Take turns having two students be the *rope anchors* who then will make a line on the coordinate axes while other students observe.
- Here are a series of directions you can give and some follow-up questions. Remind students that slope is rise over run and that the equation of a line in slope-intercept form is $y = mx + b$.
  - ? Make the line $y = x$. "What is the slope?" 1 "What is the y-intercept?" 0
  - ? Keep the same slope, but make the y-intercept 2. "What is the equation of this line?" $y = x + 2$
  - ? Use the y-intercept 2, but make the slope steeper. "What is the slope of this line?" Answers will vary.
  - ? Keep the same y-intercept, but make the slope $\frac{1}{2}$. "What is the equation?" $y = \frac{1}{2}x + 2$
- **Management Tip:** This activity can also be done by drawing the axes on the board and having the students hold the rope against the board.

Activity Notes

Activity 1

- ? "How do you determine the slope of a line drawn in a coordinate plane?" Use two points that you are sure are on the graph and find the rise and run between the points.
- ? "Does it matter whether you move left-to-right or right-to-left when you're finding the rise and run? Explain." No; Either way the slope will be the same.
- Students may have difficulty writing the equation in slope-intercept form. They think it should be harder to do!
- **FYI:** When the y-intercept is negative, students may leave their equation as $y = 3x + (-4)$ instead of $y = 3x - 4$. Remind students that it is more common to represent the equation as $y = 3x - 4$.
- **Teaching Tip:** If you have a student that is color blind, refer to the lines by a number or letter scheme (1, 2, 3 or A, B, C).
- Ask students to share what they found in common for each trio of lines.

Common Core State Standards

8.F.3 Interpret the equation $y = mx + b$ as defining a linear function, whose graph is a straight line . . ..

A.CED.2 Create equatons in two or more variables to represent relationships between quantities; graph equations on coordinate axes with labels and scales.

A.CED.3 Represent constraints by equations . . . and interpret solutions as viable or nonviable options in a modeling context.

Previous Learning

Students should know how to find the slope of a line. Students should know about parallel lines.

Activity Materials
Introduction
• masking tape • rope or yarn

Start Thinking! and Warm Up

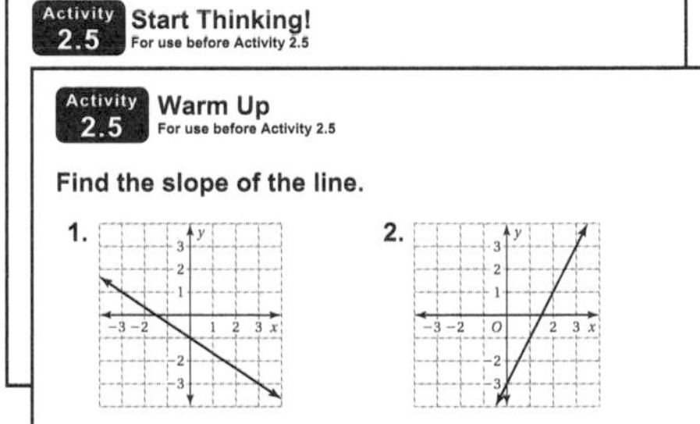

2.5 Record and Practice Journal

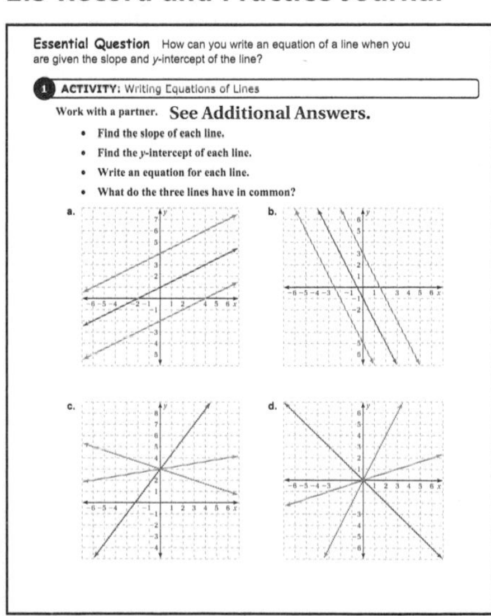

Differentiated Instruction

Visual

To avoid mistakes when substituting the variables, have students color code the slope and y-intercept of an equation.

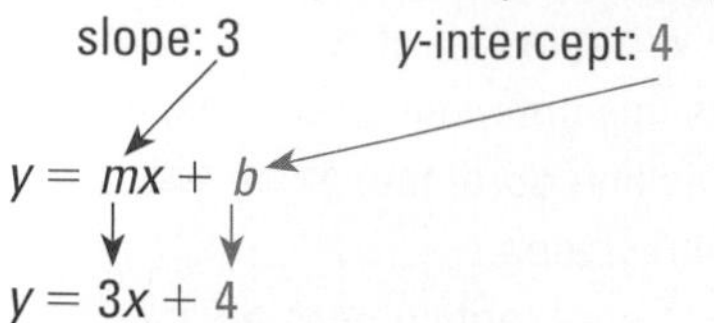

2.5 Record and Practice Journal

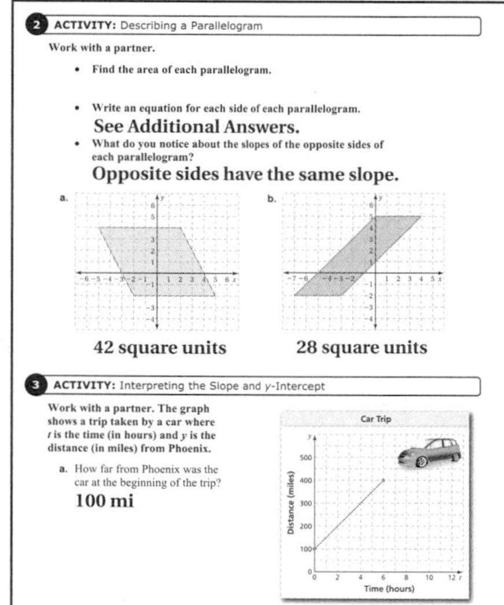

2 ACTIVITY: Describing a Parallelogram

Work with a partner.

- Find the area of each parallelogram.
- Write an equation for each side of each parallelogram.
  See Additional Answers.
- What do you notice about the slopes of the opposite sides of each parallelogram?
  Opposite sides have the same slope.

a. **42 square units**

b. **28 square units**

3 ACTIVITY: Interpreting the Slope and y-Intercept

Work with a partner. The graph shows a trip taken by a car where t is the time (in hours) and y is the distance (in miles) from Phoenix.

a. How far from Phoenix was the car at the beginning of the trip?
100 mi

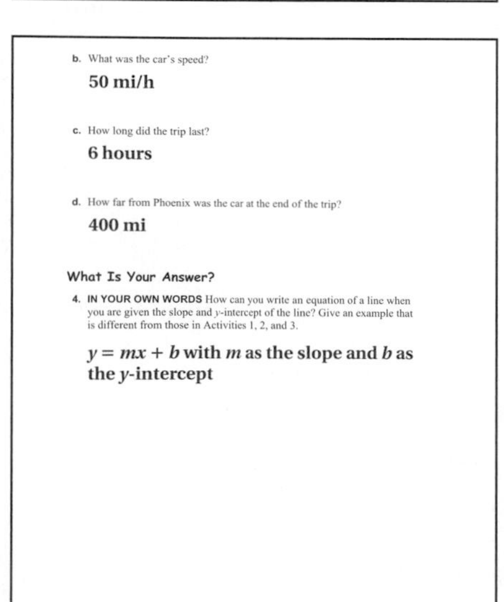

b. What was the car's speed?
50 mi/h

c. How long did the trip last?
6 hours

d. How far from Phoenix was the car at the end of the trip?
400 mi

What Is Your Answer?

4. **IN YOUR OWN WORDS** How can you write an equation of a line when you are given the slope and y-intercept of the line? Give an example that is different from those in Activities 1, 2, and 3.
$y = mx + b$ with m as the slope and b as the y-intercept

Laurie's Notes

Activity 2

? "How do you find the area of a parallelogram?" area = base × height

? "Are the base and height the sides of the parallelogram?" They could be if it's a rectangle. Otherwise, height is the perpendicular distance between the two bases.

- **MP1:** This is a good example of where students have the necessary skills but they will need to make sense of the problem and then persevere in working through the problem.
- To find the base and height, students simply count the units on the diagram. Note that the height for the parallelogram in part (b) is outside the parallelogram.
- **Common Error:** The slope of the horizontal sides is zero. Students may say that you cannot find the slope *for a flat line.*

? "What is the equation of a horizontal line?" $y = b$

- The challenge in this activity is writing the equations for the diagonal sides of the figure in part (a). Suggest that by extending the sides using the slope, the students should be able to determine the y-intercept.
- This activity reviews positive, negative, and zero slope. Area of a parallelogram is also reviewed.

Activity 3

- The graph in this problem represents a real-life context. Mathematically proficient students are able to interpret the mathematical results in the context of the situation or problem.
- If students have difficulty getting started with this activity, remind them to read the labels on the axes and interpret the y-intercept. The car was 100 miles from Phoenix at the beginning of the trip.
- Discuss answers to each part of the problem as a class.

? **Extension:** Draw the segment from (6, 400) to (12, 0) and explain that this represents the return trip. Ask the following questions.

  - "What is the slope of this line segment? What does the slope mean in the context of the problem?" slope ≈ -67; returning at a rate of about 67 mi/h
  - "What does the point (12, 0) mean in the context of the problem?" You have arrived in Phoenix and drove 12 hours.
  - "What would the graph look like if the car had stopped for 1 hour?" horizontal segment of length 1 unit

Closure

- **Exit Ticket:** What is the slope and y-intercept of the equation $y = 2x + 4$? slope = 2, y-intercept = 4 Write an equation of a line with a slope of 3 and a y-intercept of 1. $y = 3x + 1$

Technology For the Teacher

The Dynamic Planning Tool
Editable Teacher's Resources at *BigIdeasMath.com*

2 ACTIVITY: Describing a Parallelogram

Work with a partner.

- **Find the area of each parallelogram.**
- **Write an equation for each side of each parallelogram.**
- **What do you notice about the slopes of the opposite sides of each parallelogram?**

a.

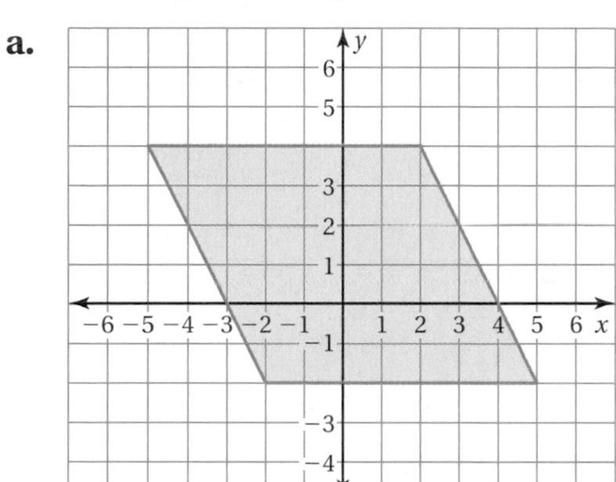

b.

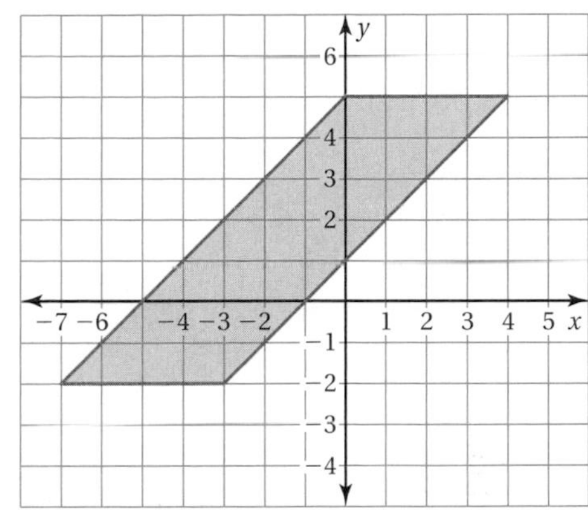

3 ACTIVITY: Interpreting the Slope and y-Intercept

Work with a partner. The graph shows a trip taken by a car where *t* is the time (in hours) and *y* is the distance (in miles) from Phoenix.

a. How far from Phoenix was the car at the beginning of the trip?

b. What was the car's speed?

c. How long did the trip last?

d. How far from Phoenix was the car at the end of the trip?

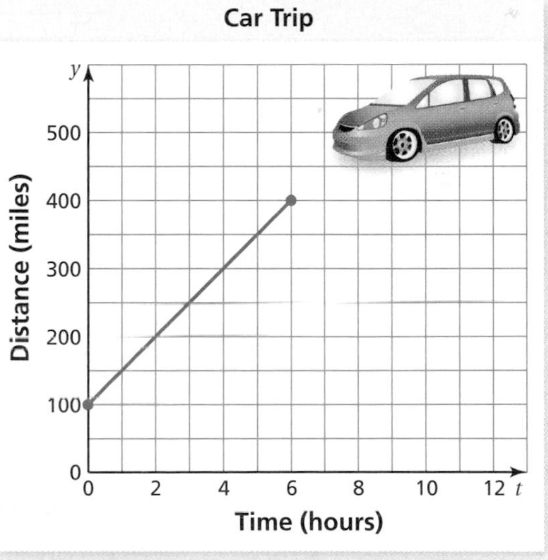

What Is Your Answer?

4. IN YOUR OWN WORDS How can you write an equation of a line when you are given the slope and y-intercept of the line? Give an example that is different from those in Activities 1, 2, and 3.

Practice Use what you learned about writing equations in slope-intercept form to complete Exercises 3 and 4 on page 76.

2.5 Lesson

EXAMPLE 1 Writing Equations in Slope-Intercept Form

Write an equation of the line in slope-intercept form.

a.

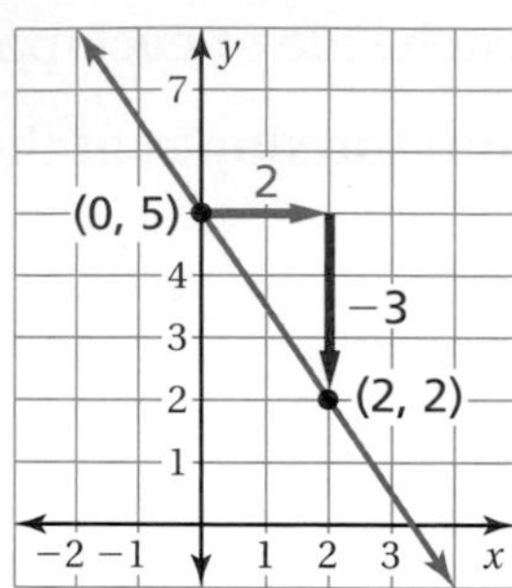

Find the slope and y-intercept.

$$\text{slope} = \frac{y_2 - y_1}{x_2 - x_1}$$

$$= \frac{2 - 5}{2 - 0}$$

$$= \frac{-3}{2}, \text{ or } -\frac{3}{2}$$

Because the line crosses the y-axis at $(0, 5)$, the y-intercept is 5.

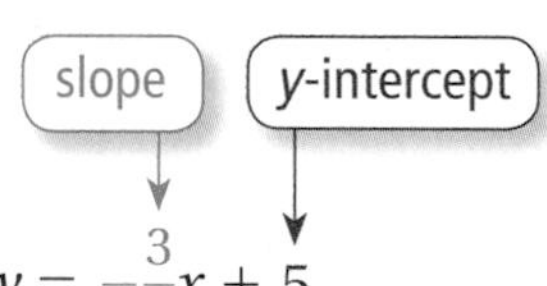

∴ So, the equation is $y = -\frac{3}{2}x + 5$.

Study Tip

After writing an equation, check that the given points are solutions of the equation.

b.

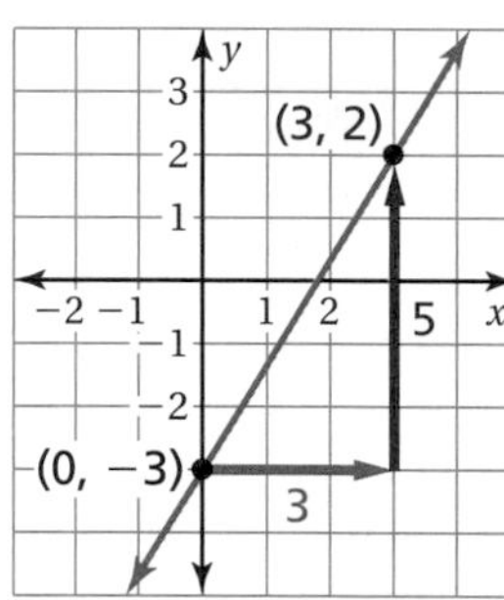

Find the slope and y-intercept.

$$\text{slope} = \frac{y_2 - y_1}{x_2 - x_1}$$

$$= \frac{-3 - 2}{0 - 3}$$

$$= \frac{-5}{-3}, \text{ or } \frac{5}{3}$$

Because the line crosses the y-axis at $(0, -3)$, the y-intercept is -3.

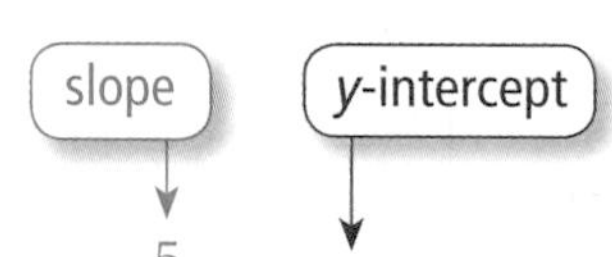

∴ So, the equation is $y = \frac{5}{3}x + (-3)$, or $y = \frac{5}{3}x - 3$.

On Your Own

Now You're Ready
Exercises 5–10

Write an equation of the line in slope-intercept form.

1. (graph with points (0, 2) and (1, 4); run 1, rise 2)

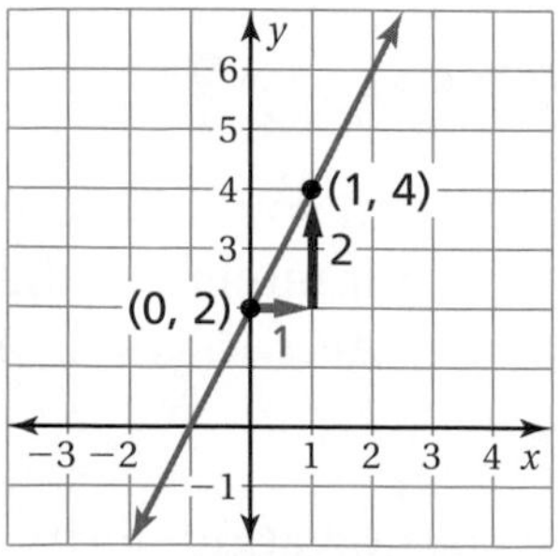

2.

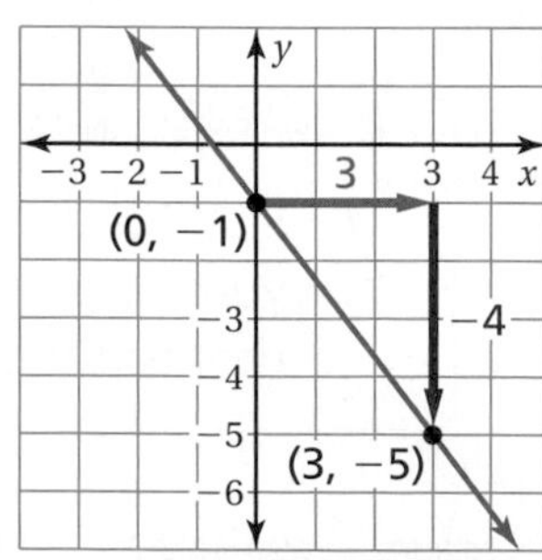

Laurie's Notes

Introduction

Connect

- **Yesterday:** Students developed an intuitive understanding about how to write the equation of a line when you know its slope and y-intercept. (MP1, MP4)
- **Today:** Students will write an equation of a line given its slope and y-intercept.

Motivate

- **Story Time:** Tell students that as a child you loved to dig tunnels in the sand. Ask if any of them like to dig tunnels or if they have traveled through tunnels. Hold a paper towel tube or other similar model to pique student interest. Share some facts about tunnels.
  - The world's longest overland tunnel is a 21-mile-long rail link under the Alps in Switzerland. The tunnel took eight years to build and cost \$3.5 billion. It reduces the time trains need to cross between Germany and Italy from 3.5 hours to just under 2 hours.
  - The world's longest underwater tunnel is Seikan Tunnel in Japan. It is 33.49 miles long and runs under the Tsugaru Strait. It opened in 1988 and took 17 years to construct.
  - The Channel Tunnel (Chunnel) connects England and France. It is 31 miles long and travels under the English Channel.

Lesson Notes

Discuss

- **MP1a Make Sense of Problems** and **MP4 Model with Mathematics:** In this lesson students will make quick visual inspection of linear graphs to approximate the slope and the y-intercept. This approximation is a helpful check when the slope and y-intercept are computed.

Example 1

- Write the slope-intercept form of an equation, $y = mx + b$. Review with students that the coefficient of the x-term is the slope, and the constant b is the y-intercept.
- Review with students how to compute slope.

? "What do you know about the slope of the line in part (a) by inspection? Explain." Slope is negative because the graph falls left to right.

? "What are the coordinates of the point where the line crosses the y-axis?" (0, 5)

- Use the slope and the y-intercept to write the equation.
- Work through part (b). Remind students that you want the more simplified equation $y = \frac{5}{3}x - 3$ instead of $y = \frac{5}{3}x + (-3)$. Stress that while both forms are correct, the simplified version is preferred.

On Your Own

- Before students begin these two problems, they should do a visual inspection. They should make a note of the sign of the slope and y-intercept. It is very easy to have the wrong sign(s) when the equation is written.

Goal Today's lesson is writing an equation of a line in slope-intercept form.

Lesson Materials
Introduction
• paper towel tube

Start Thinking! and Warm Up

Lesson 2.5 Warm Up
For use before Lesson 2.5

Lesson 2.5 Start Thinking!
For use before Lesson 2.5

A gym membership has a \$20 enrollment fee and costs \$40 per month.

Write an equation in slope-intercept form that represents the cost y after x months of joining the gym.

What does the slope represent?

What does the y-intercept represent?

Extra Example 1

Write an equation of the line in slope-intercept form.

a.

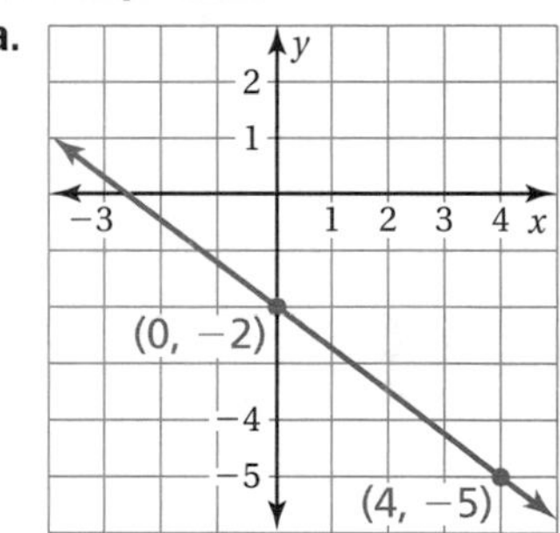

$y = -\frac{3}{4}x - 2$

b.

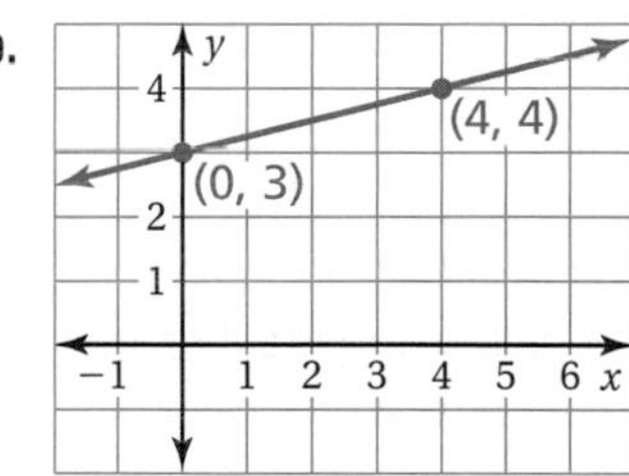

$y = \frac{1}{4}x + 3$

On Your Own

1. $y = 2x + 2$
2. $y = -\frac{4}{3}x - 1$

Extra Example 2

Write an equation of the line that passes through the points $(0, -1)$ and $(4, -1)$. $y = -1$

Extra Example 3

In Example 3, the points are (0, 3500) and (5, 1750).

a. Write an equation that represents the distance y (in feet) remaining after x months. $y = -350x + 3500$

b. How much time does it take to complete the tunnel? 10 months

On Your Own

3. $y = 5$

4. $8\frac{3}{4}$ mo

English Language Learners

Organization

Students will benefit by writing down the steps for writing an equation in slope-intercept form when given a graph. Have students write the steps in their notebooks. A poster with the steps could be posted in the classroom.

Step 1: Write the slope-intercept form of an equation.
Step 2: Determine the slope of the line.
Step 3: Determine the y-intercept of the line.
Step 4: Write the equation in slope-intercept form.

Laurie's Notes

Example 2

- Make a quick sketch of the graph to reference as you work the problem.
- When finding the slope, students are unsure of how to simplify $\frac{0}{3}$. This is a good time to review the difference between $\frac{0}{3}$ and $\frac{3}{0}$.
- **Teaching Tip:** To explain why $\frac{3}{0}$ is undefined, first write the problem $8 \div 4 = 2$ on the board. Then rewrite it as $4\overline{)8}$. To check, multiply the quotient (2) times the divisor (4) and you get the dividend (8). In other words, 2 multiplied by 4 is 8. Do the same thing with $\frac{3}{0}$. Rewrite it using long division, $0\overline{)3}$. What do you multiply 0 by to get 3? There is no quotient, so you say $\frac{3}{0}$ is undefined. You cannot divide by 0.
- **MP7 Look for and Make Use of Structure:** Students don't always recognize that $y = -4$ is a linear equation written in slope-intercept form. It helps to write the extra step of $y = (0)x + (-4)$ so students can see that the slope is 0. Students should recognize that $y = -4$ and $y = (0)x + (-4)$ are equivalent.

Example 3

- Ask a volunteer to read the problem. Discuss information that can be *read* from the graph.

? "By visual inspection, what do you know about the sign of the slope and the y-intercept in this problem?" The slope is negative. The y-intercept is positive.

? "What does a slope of -500 mean in the context of this problem?" A slope of -500 means that for each additional month of work, the distance left to complete is 500 feet less.

- The x-intercept for this graph is 7.
- Note that the graph is in Quadrant I. In the context of this problem, it doesn't make sense for time or distance to be negative.

On Your Own

- **MP6 Attend to Precision:** For Question 3, encourage students to sketch a graph of the line through the two points to give them a clue as to how to begin. The visual model is an approximation that can be used to check their final answer. This technique will help students start Question 4.

Closure

- **Writing Prompt:** For a line that has been graphed in a coordinate plane, you can write the equation by … finding the slope and y-intercept

Technology For the Teacher

The Dynamic Planning Tool
Editable Teacher's Resources at *BigIdeasMath.com*

EXAMPLE 2 Standardized Test Practice

Which equation is shown in the graph?

(A) $y = -4$ (B) $y = -3$

(C) $y = 0$ (D) $y = -3x$

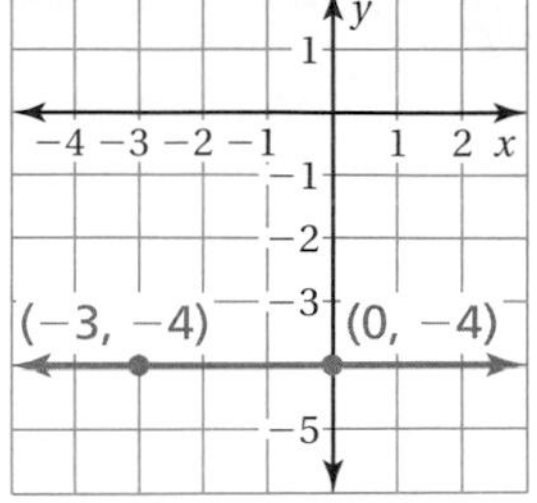

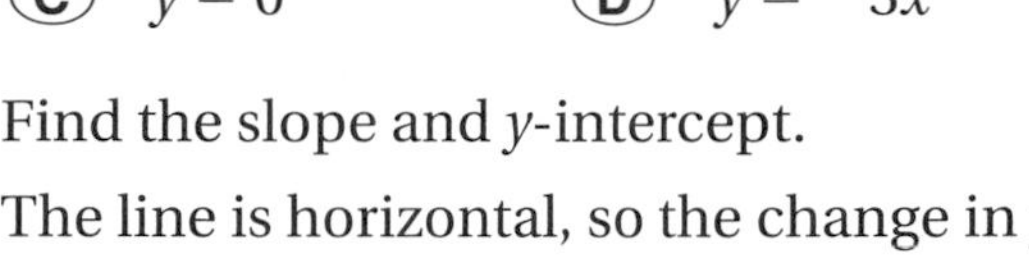

Find the slope and y-intercept.

The line is horizontal, so the change in y is 0.

$$\text{slope} = \frac{\text{change in } y}{\text{change in } x} = \frac{0}{3} = 0$$

Because the line crosses the y-axis at $(0, -4)$, the y-intercept is -4.

∴ So, the equation is $y = 0x + (-4)$, or $y = -4$. The correct answer is (A).

Remember

The graph of $y = a$ is a horizontal line that passes through $(0, a)$.

EXAMPLE 3 Real-Life Application

The graph shows the distance remaining to complete a tunnel. (a) Write an equation that represents the distance y (in feet) remaining after x months. (b) How much time does it take to complete the tunnel?

Engineers used tunnel boring machines like the ones shown above to dig an extension of the Metro Gold Line in Los Angeles. The new tunnels are 1.7 miles long and 21 feet wide.

Tunnel Digging

a. Find the slope and y-intercept.

$$\text{slope} = \frac{\text{change in } y}{\text{change in } x} = \frac{-2000}{4} = -500$$

Because the line crosses the y-axis at $(0, 3500)$, the y-intercept is 3500.

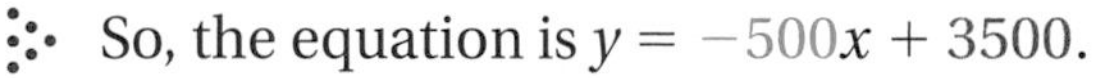

∴ So, the equation is $y = -500x + 3500$.

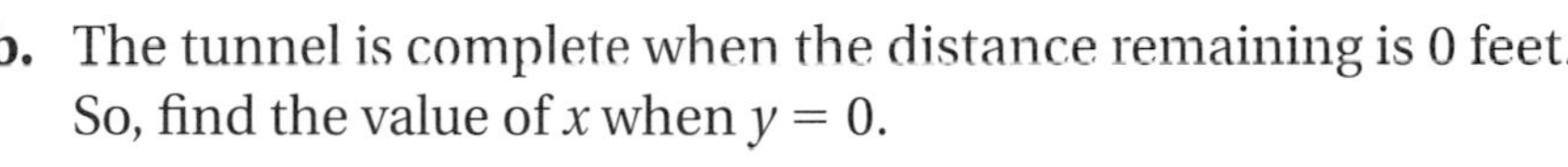

b. The tunnel is complete when the distance remaining is 0 feet. So, find the value of x when $y = 0$.

$y = -500x + 3500$	Write the equation.
$0 = -500x + 3500$	Substitute 0 for y.
$-3500 = -500x$	Subtract 3500 from each side.
$7 = x$	Solve for x.

∴ It takes 7 months to complete the tunnel.

On Your Own

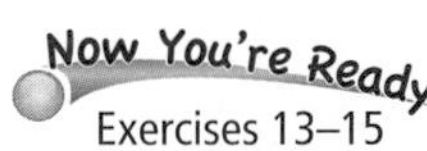

Now You're Ready
Exercises 13–15

3. Write an equation of the line that passes through $(0, 5)$ and $(4, 5)$.
4. **WHAT IF?** In Example 3, the points are $(0, 3500)$ and $(5, 1500)$. How long does it take to complete the tunnel?

2.5 Exercises

Vocabulary and Concept Check

1. **PRECISION** Explain how to find the slope of a line given the intercepts of the line.
2. **WRITING** Explain how to write an equation of a line using its graph.

Practice and Problem Solving

Write an equation for each side of the figure.

3. 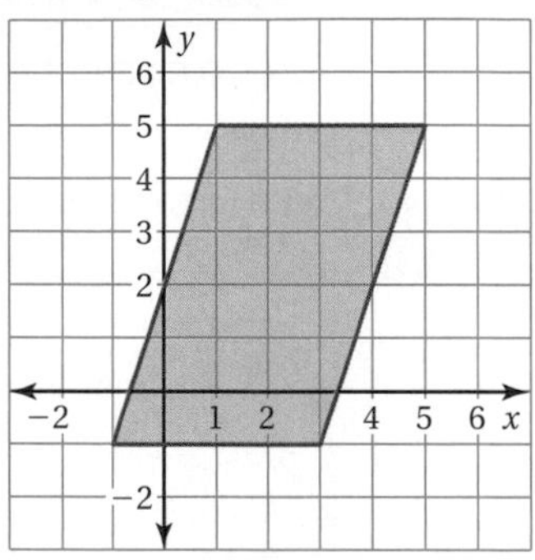

4. 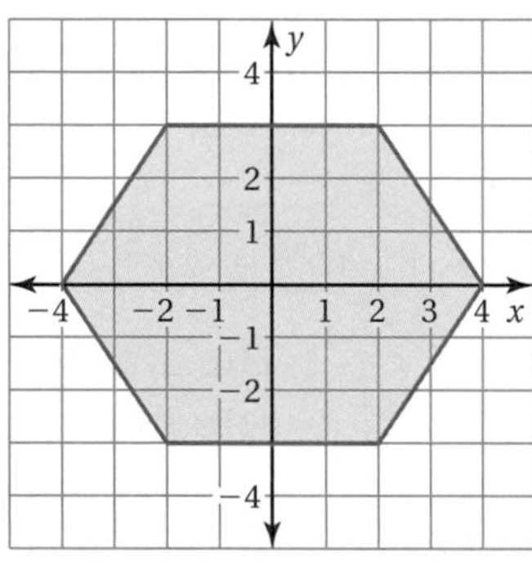

Write an equation of the line in slope-intercept form.

1 5. 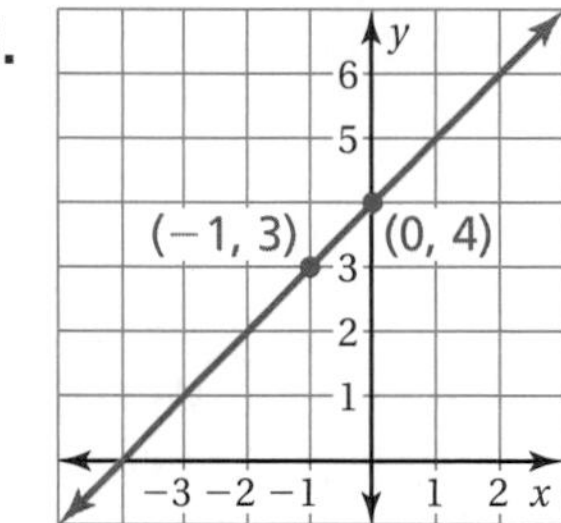

6. 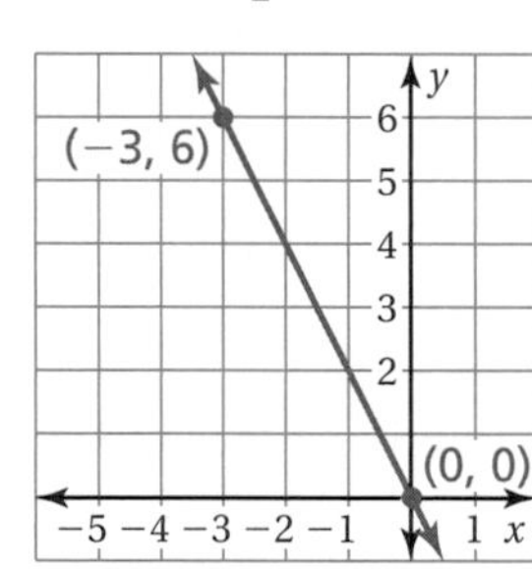

7. 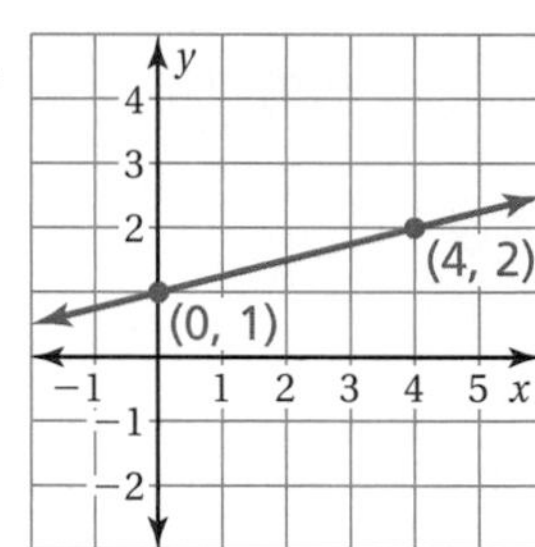

8. 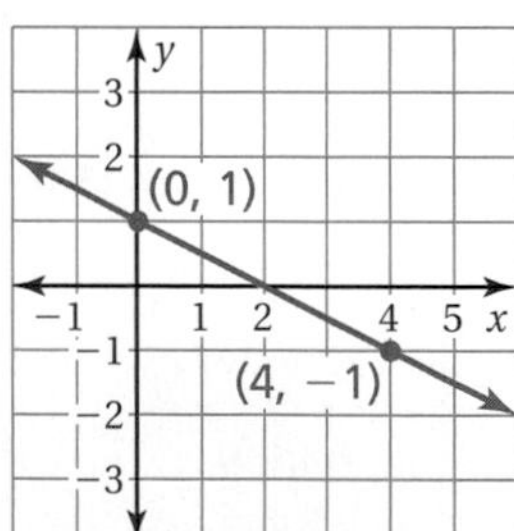

9. 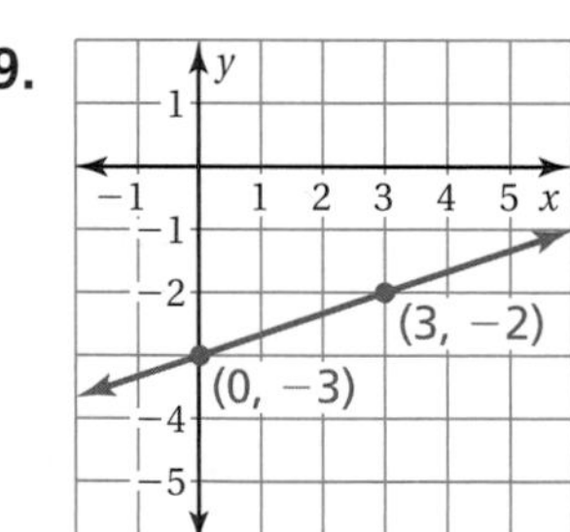

10. 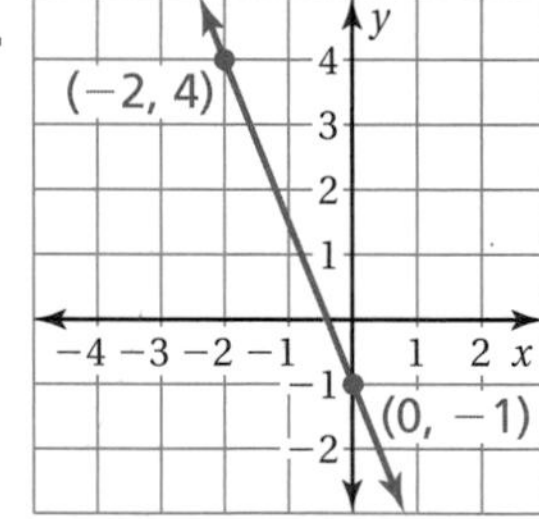

11. **ERROR ANALYSIS** Describe and correct the error in writing the equation of the line.

$y = \frac{1}{2}x + 4$

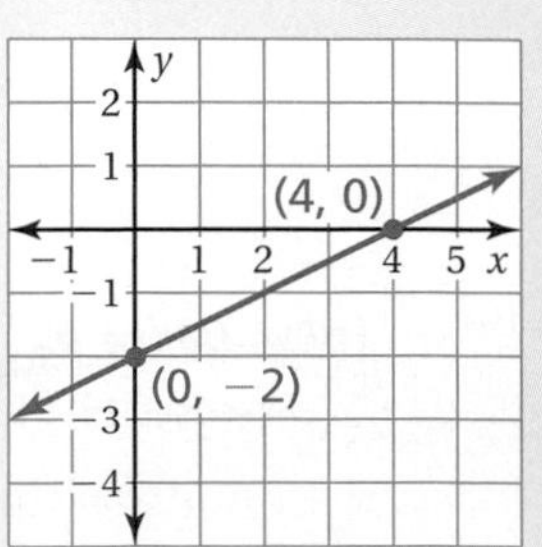

12. **BOA** A boa constrictor is 18 inches long at birth and grows 8 inches per year. Write an equation that represents the length y (in feet) of a boa constrictor that is x years old.

Assignment Guide and Homework Check

Level	Assignment	Homework Check
Average	1–4, 6–17, 20–24	6, 9, 12, 14, 16
Advanced	1, 2, 8–19, 20–24	8, 14, 16, 18

Common Errors

- **Exercises 5–10** Students may write the reciprocal of the slope or forget a negative sign. Remind them of the definition of slope. Ask students to predict the sign of the slope based on the rise or fall of the line.
- **Exercises 13–15** Students may write the wrong equation when the slope is zero. For example, instead of $y = 5$, students may write $x = 5$. Ask them what is the rise of the graph (zero) and write this in slope-intercept form with the y-intercept as well, such as $y = 0x + 5$. Then ask students what happens when a variable (or any number) is multiplied by zero. Rewrite the equation as $y = 5$.

2.5 Record and Practice Journal

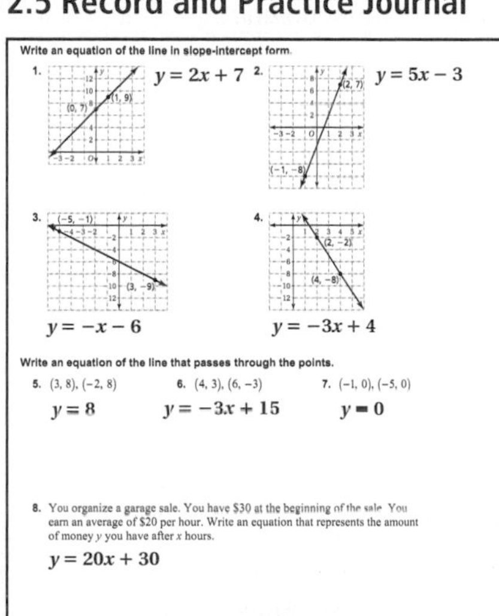

Write an equation of the line in slope-intercept form.

1. $y = 2x + 7$
2. $y = 5x - 3$
3. $y = -x - 6$
4. $y = -3x + 4$

Write an equation of the line that passes through the points.

5. (3, 8), (−2, 8) $y = 8$
6. (4, 3), (6, −3) $y = -3x + 15$
7. (−1, 0), (−5, 0) $y = 0$

8. You organize a garage sale. You have \$30 at the beginning of the sale. You earn an average of \$20 per hour. Write an equation that represents the amount of money y you have after x hours.

$y = 20x + 30$

Technology For the Teacher
Answer Presentation Tool

Vocabulary and Concept Check

1. *Sample answer:* Find the ratio of the rise to the run between the intercepts.
2. *Sample answer:* Find the slope of the line between any two points. Then find the y-intercept. The equation of the line is $y = mx + b$, where m is the slope and b is the y-intercept.

Practice and Problem Solving

3. $y = 3x + 2$;
   $y = 3x - 10$;
   $y = 5$;
   $y = -1$
4. $y = \frac{3}{2}x + 6$;
   $y = 3$;
   $y = -\frac{3}{2}x + 6$;
   $y = \frac{3}{2}x - 6$;
   $y = -3$;
   $y = -\frac{3}{2}x - 6$
5. $y = x + 4$
6. $y = -2x$
7. $y = \frac{1}{4}x + 1$
8. $y = -\frac{1}{2}x + 1$
9. $y = \frac{1}{3}x - 3$
10. $y = -\frac{5}{2}x - 1$
11. The x-intercept was used instead of the y-intercept.
    $y = \frac{1}{2}x - 2$
12. $y = \frac{2}{3}x + \frac{3}{2}$

Practice and Problem Solving

13. $y = 5$ **14.** $y = 0$

15. $y = -2$ **16.** $y = 0.7x + 10$

17. See Additional Answers.

18. $y = -140x + 500$

19. See *Taking Math Deeper*.

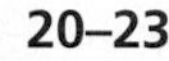

Fair Game Review

20–23.

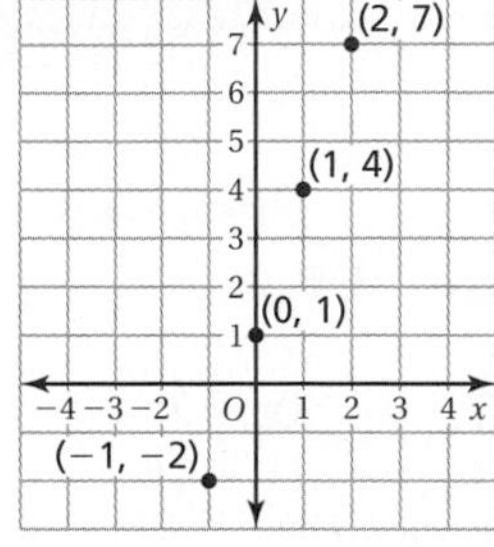

24. C

Mini-Assessment

Write an equation of the line in slope-intercept form.

1. $y = x + 2$

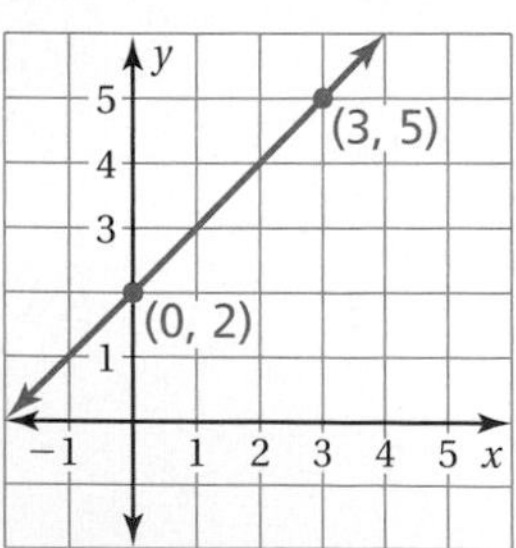

2. $y = -2x - 1$

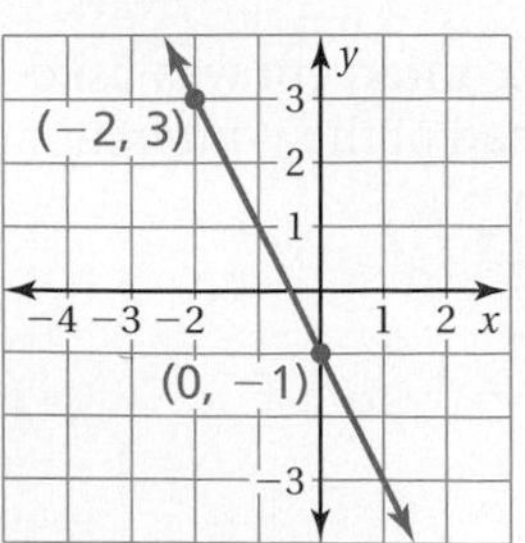

Taking Math Deeper

Exercise 19

This is a nice real-life problem using estimation. For this problem, remember that you are not looking for exact solutions. You want to know *about* how much the trees grow each year so that you can predict their approximate heights.

Estimate the heights in the photograph.

a. Height of 10-year-old tree: about 18 ft
Height of 8-year-old tree: about 14 ft

b. Plot the heights of the two trees.

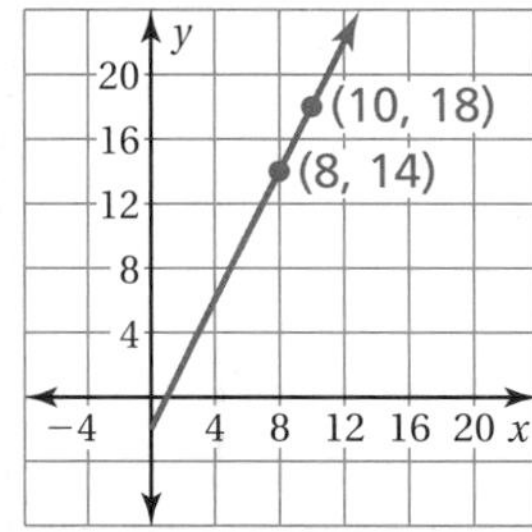

c. The trees are growing at a rate of about 2 feet per year. Because this would put the height of a 0-year-old tree at -2, it is better to adjust the rate of growth to be about 1.8 feet per year.

d. A possible equation for the growth rate is $y = 1.8x$.

Project

Research information about the palm tree. Pick any kind of palm tree in which you are interested. How old is the longest living palm tree?

Reteaching and Enrichment Strategies

If students need help...	If students got it...
Resources by Chapter • Practice A and Practice B • Puzzle Time Record and Practice Journal Practice Differentiating the Lesson Lesson Tutorials Skills Review Handbook	Resources by Chapter • Enrichment and Extension • School-to-Work • Financial Literacy Start the next section

Write an equation of the line that passes through the points.

② **13.** (2, 5), (0, 5) **14.** (−3, 0), (0, 0) **15.** (0, −2), (4, −2)

16. WALKATHON One of your friends gives you \$10 for a charity walkathon. Another friend gives you an amount per mile. After 5 miles, you have raised \$13.50 total. Write an equation that represents the amount y of money you have raised after x miles.

17. BRAKING TIME During each second of braking, an automobile slows by about 10 miles per hour.

a. Plot the points (0, 60) and (6, 0). What do the points represent?

b. Draw a line through the points. What does the line represent?

c. Write an equation of the line.

18. PAPER You have 500 sheets of notebook paper. After 1 week, you have 72% of the sheets left. You use the same number of sheets each week. Write an equation that represents the number y of pages remaining after x weeks.

19. Critical Thinking The palm tree on the left is 10 years old. The palm tree on the right is 8 years old. The trees grow at the same rate.

a. Estimate the height y (in feet) of each tree.

b. Plot the two points (x, y), where x is the age of each tree and y is the height of each tree.

c. What is the rate of growth of the trees?

d. Write an equation that represents the height of a palm tree in terms of its age.

Fair Game Review What you learned in previous grades & lessons

Plot the ordered pair in a coordinate plane. *(Skills Review Handbook)*

20. (1, 4) **21.** (−1, −2) **22.** (0, 1) **23.** (2, 7)

24. MULTIPLE CHOICE Which of the following statements is true? *(Section 2.3)*

Ⓐ The x-intercept is 5.

Ⓑ The x-intercept is −2.

Ⓒ The y-intercept is 5.

Ⓓ The y-intercept is −2.

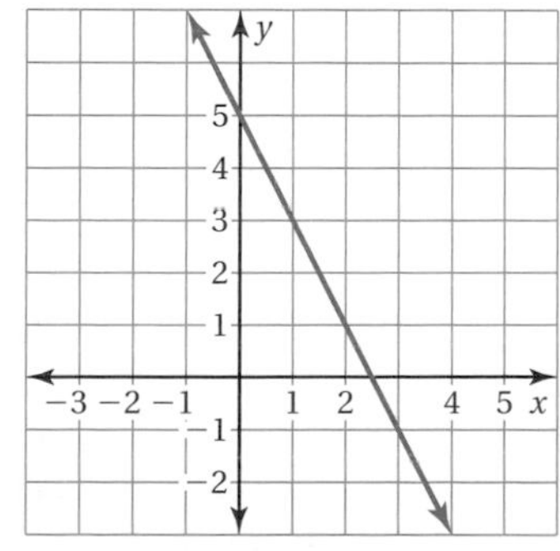

2.6 Writing Equations in Point-Slope Form

COMMON CORE STATE STANDARDS
A.CED.2
A.REI.10
F.IF.4
F.IF.6

Essential Question How can you write an equation of a line when you are given the slope and a point on the line?

1 ACTIVITY: Writing Equations of Lines

Work with a partner.

- **Sketch the line that has the given slope and passes through the given point.**
- **Find the *y*-intercept of the line.**
- **Write an equation of the line.**

a. $m = -2$

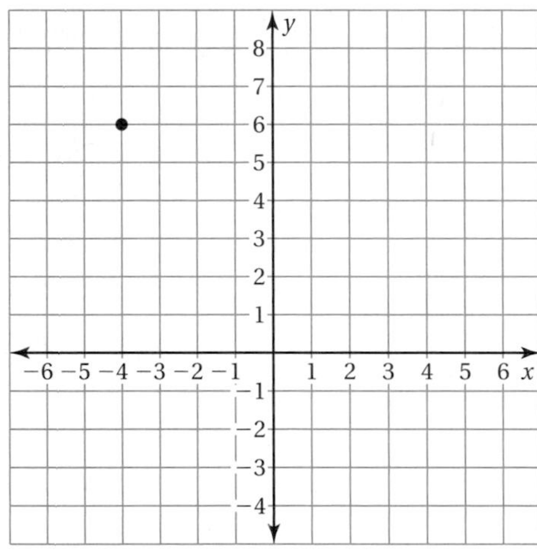

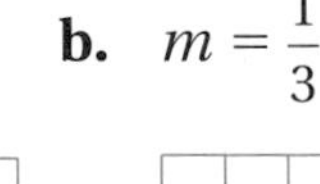

b. $m = \frac{1}{3}$

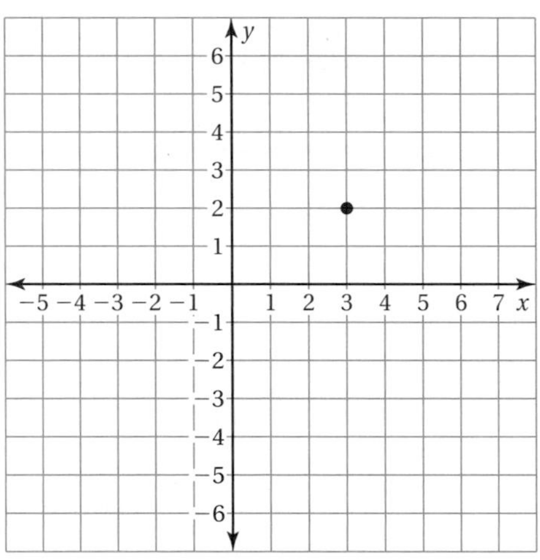

c. $m = -\frac{2}{3}$

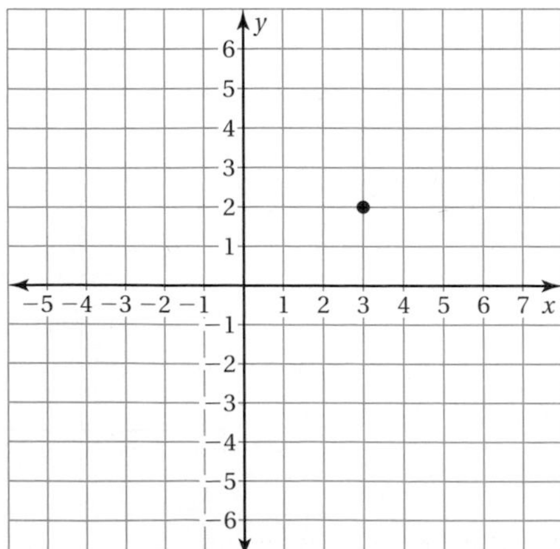

d. $m = \frac{5}{2}$

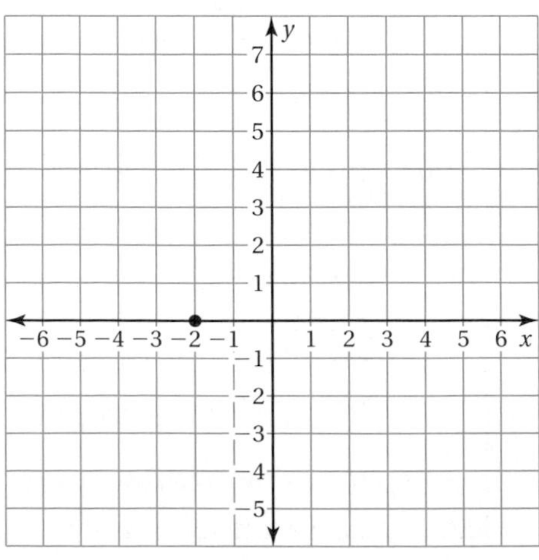

Laurie's Notes

Introduction

Standards for Mathematical Practice

- **MP1 Make Sense of Problems and Persevere in Solving Them:** The goal is for students to write equations of lines given a slope and a point. The slope may be stated explicitly or determined from a contextual setting. Students use the slope to graph the line and work backwards to find the *y*-intercept. There are different approaches students may use as they make sense of the problem.
- **MP3 Construct Viable Arguments and Critique the Reasoning of Others:** Take time for discussions and explanations so that students' reasoning is revealed.

Motivate

- Hold a piece of ribbon and a pair of scissors in your hands. Snip a one-foot piece of ribbon off. Repeat once or twice more.

? "Do you know how long my ribbon was when I first started?" no

- Your question should prompt students to ask two obvious questions: "How much are you cutting off each time?" and "How many times have you made a cut?" How much you cut off is the slope (-1). How many times you cut the ribbon helps students work backwards to find the length before any cuts were made, which is the *y*-intercept.

Activity Notes

Activity 1

? "What does it mean for a line to have a slope of -2? A slope of $\frac{1}{3}$?"

For every unit it runs, it falls 2 units. For every 3 units it runs, it rises 1.

- Students may also answer the last question by saying "over 1, down 2" and "over 3, up 1." These geometric answers are fine. Students will need this level of understanding to locate additional points on a line, in order to find the *y*-intercept.
- You cannot sketch the line immediately. You must first find additional points on the line. Students should start at the given point and use the slope to find additional points on the line. One of the points will give the *y*-intercept.
- For part (b), it might be helpful to think of the slope of $\frac{1}{3}$ as $\frac{-1}{-3}$. So, start at the point given and move left 3 units and then down 1 unit.
- **Common Error:** Students may interchange the rise and run. Have students look back at their graph to see if the slope looks correct to them.
- **Teaching Tip:** Encourage students to use a pencil and lightly trace the rise and run direction arrows as they locate additional points.
- To share student work, have transparency grids available for the overhead.

? "What made it possible to write the equation of the line?" The slope was given and by using the slope, it was possible to find the *y*-intercept. Then substitute into the formula $y = mx + b$.

- **MP1:** Asking the last question about slope and having students state their understanding helps them make sense of the problem.

Common Core State Standards

A.CED.2 Create equations in two or more variables . . .; graph equations on coordinate axes with labels and scales.
A.REI.10 Understand that the graph of an equation in two variables is the set of all its solutions plotted in the coordinate plane, ...
F.IF.4 For a function that models a relationship between two quantities, interpret key features of graphs and tables in terms of the quantities, and sketch graphs showing key features given a verbal description of the relationship.
F.IF.6 Calculate and interpret the average rate of change of a function . . . over a specified interval. Estimate the rate of change from a graph.

Previous Learning

Students should know how to plot ordered pairs and apply the definition of slope.

Start Thinking! and Warm Up

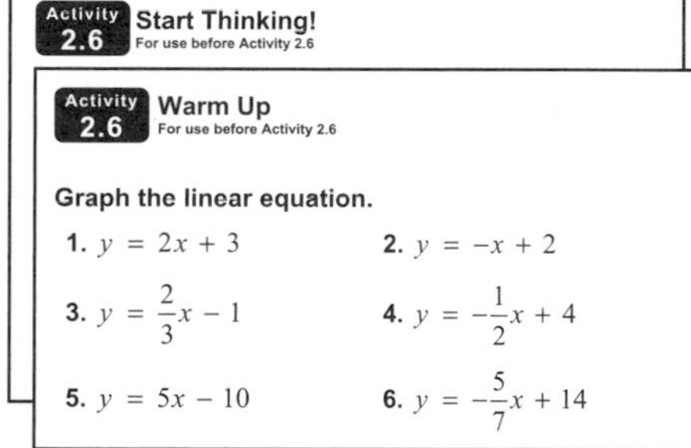

Activity 2.6 Start Thinking! For use before Activity 2.6

Activity 2.6 Warm Up For use before Activity 2.6

Graph the linear equation.

1. $y = 2x + 3$
2. $y = -x + 2$
3. $y = \frac{2}{3}x - 1$
4. $y = -\frac{1}{2}x + 4$
5. $y = 5x - 10$
6. $y = -\frac{5}{7}x + 14$

2.6 Record and Practice Journal

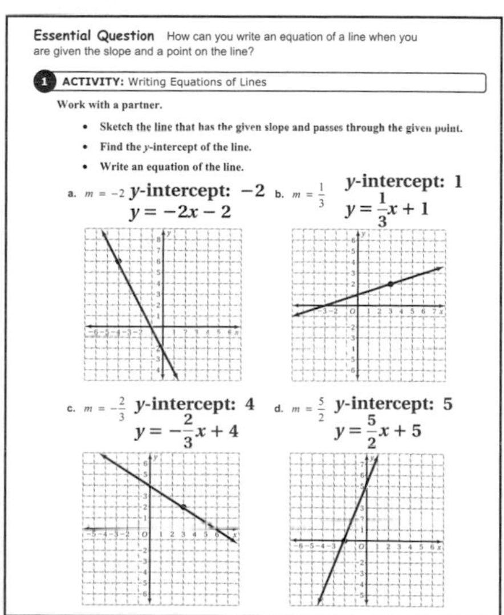

Essential Question How can you write an equation of a line when you are given the slope and a point on the line?

1 ACTIVITY: Writing Equations of Lines

Work with a partner.

- Sketch the line that has the given slope and passes through the given point.
- Find the *y*-intercept of the line.
- Write an equation of the line.

a. $m = -2$ *y*-intercept: -2 $y = -2x - 2$

b. $m = \frac{1}{3}$ *y*-intercept: 1 $y = \frac{1}{3}x + 1$

c. $m = -\frac{2}{3}$ *y*-intercept: 4 $y = -\frac{2}{3}x + 4$

d. $m = \frac{5}{2}$ *y*-intercept: 5 $y = \frac{5}{2}x + 5$

Differentiated Instruction

Kinesthetic

Write a list of linear equations on the board or overhead. Have students copy the equations onto index cards. On the back of each card students are to write the slope and y-intercept of the line. After the cards are completed, students can work in pairs to check each other's work. Finally, students can quiz each other with the flash cards they made.

2.6 Record and Practice Journal

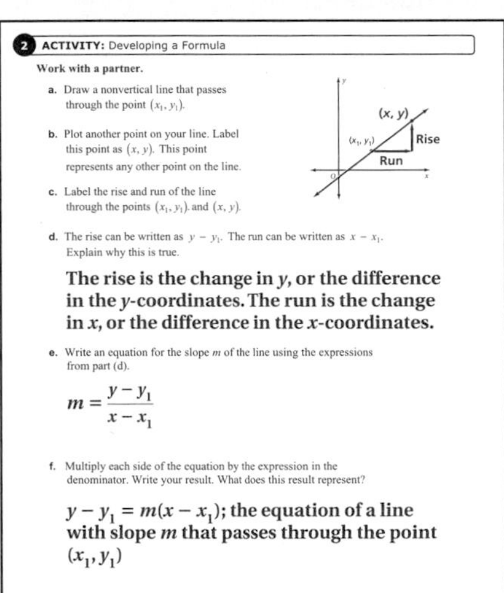

2 ACTIVITY: Developing a Formula

Work with a partner.

a. Draw a nonvertical line that passes through the point (x_1, y_1).

b. Plot another point on your line. Label this point as (x, y). This point represents any other point on the line.

c. Label the rise and run of the line through the points (x_1, y_1) and (x, y).

d. The rise can be written as $y - y_1$. The run can be written as $x - x_1$. Explain why this is true.

The rise is the change in y, or the difference in the y-coordinates. The run is the change in x, or the difference in the x-coordinates.

e. Write an equation for the slope m of the line using the expressions from part (d).

$$m = \frac{y - y_1}{x - x_1}$$

f. Multiply each side of the equation by the expression in the denominator. Write your result. What does this result represent?

$y - y_1 = m(x - x_1)$; the equation of a line with slope m that passes through the point (x_1, y_1)

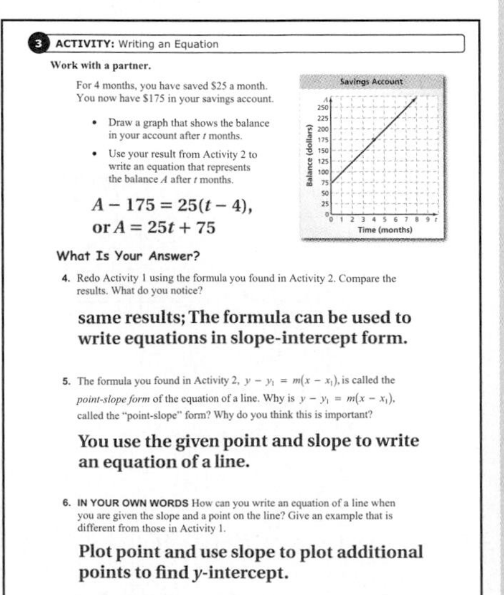

3 ACTIVITY: Writing an Equation

Work with a partner.

For 4 months, you have saved \$25 a month. You now have \$175 in your savings account.

- Draw a graph that shows the balance in your account after t months.
- Use your result from Activity 2 to write an equation that represents the balance A after t months.

$A - 175 = 25(t - 4)$, or $A = 25t + 75$

What Is Your Answer?

4. Redo Activity 1 using the formula you found in Activity 2. Compare the results. What do you notice?

same results; The formula can be used to write equations in slope-intercept form.

5. The formula you found in Activity 2, $y - y_1 = m(x - x_1)$, is called the *point-slope form* of the equation of a line. Why is $y - y_1 = m(x - x_1)$, called the "point-slope" form? Why do you think this is important?

You use the given point and slope to write an equation of a line.

6. **IN YOUR OWN WORDS** How can you write an equation of a line when you are given the slope and a point on the line? Give an example that is different from those in Activity 1.

Plot point and use slope to plot additional points to find y-intercept.

Laurie's Notes

Activity 2

- **Big Idea:** This activity helps develop the formula for finding an equation of a line given its slope and a point on the line. Students should see the relationship between the slope formula and the point-slope form.
- **MP1:** The steps of the derivation are provided. Encourage students to read carefully, discuss with their partner, and think about the process. Do not jump in too quickly to rescue students!
- Allow sufficient time before having a class discussion of this activity. Ask volunteers to share their work on the board. Sharing their process aloud helps students become more confident in their reasoning.

Activity 3

- If students do not understand what the slope is, suggest they work backwards and make a table of values.

Month, t	0	1	2	3	4
Balance in Account, A	\$75	\$100	\$125	\$150	\$175

- Students can use the table to draw the graph.

? Ask a few questions to guide students' understanding:
  - "What is the slope for this problem? What point is given?" slope = \$25; given point is (4, 175)
  - "Do you have enough information to write the equation?" yes; $A - 175 = 25(t - 4)$, or $A = 25t + 75$
  - "Explain why the slope is positive." You are putting money in the bank. Your account is growing.
- **MP1:** Take time for students to transform the equation into slope-intercept form. Students should interpret the slope and y-intercept in the context of the problem.

What Is Your Answer?

- **Neighbor Check:** Have students work independently and then have their neighbor check their work. Have students discuss any discrepancies.
- **MP7 Look for and Make Use of Structure:** In Question 4, the equations in point-slope form and slope-intercept form are equivalent, but structurally they look different. Take time for students to appreciate what information is known about the line from the form in which it is written.

Closure

- Refer back to the ribbon and scissors. If the ribbon is now 7 feet and you made 4 equal cuts of 1-foot length, write the equation that gives the length of the ribbon R after n cuts. $R = 11 - n$ or $R = -n + 11$

Technology For the Teacher

Dynamic Classroom

The Dynamic Planning Tool

Editable Teacher's Resources at *BigIdeasMath.com*

2 ACTIVITY: Developing a Formula

Work with a partner.

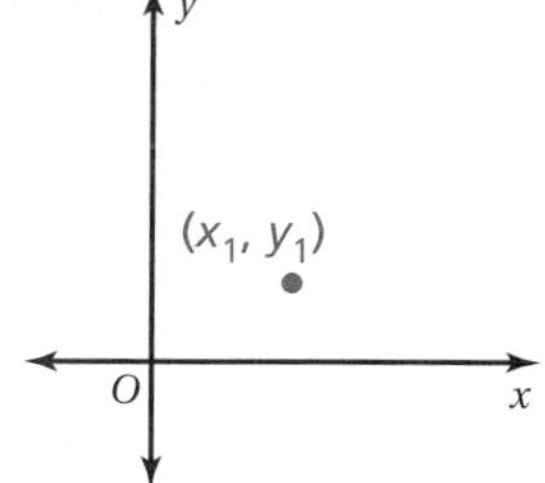

a. Draw a nonvertical line that passes through the point (x_1, y_1).

b. Plot another point on your line. Label this point as (x, y). This point represents any other point on the line.

c. Label the rise and run of the line through the points (x_1, y_1) and (x, y).

d. The rise can be written as $y - y_1$. The run can be written as $x - x_1$. Explain why this is true.

e. Write an equation for the slope m of the line using the expressions from part (d).

f. Multiply each side of the equation by the expression in the denominator. Write your result. What does this result represent?

3 ACTIVITY: Writing an Equation

Work with a partner.

For 4 months, you have saved $25 a month. You now have $175 in your savings account.

- Draw a graph that shows the balance in your account after t months.
- Use your result from Activity 2 to write an equation that represents the balance A after t months.

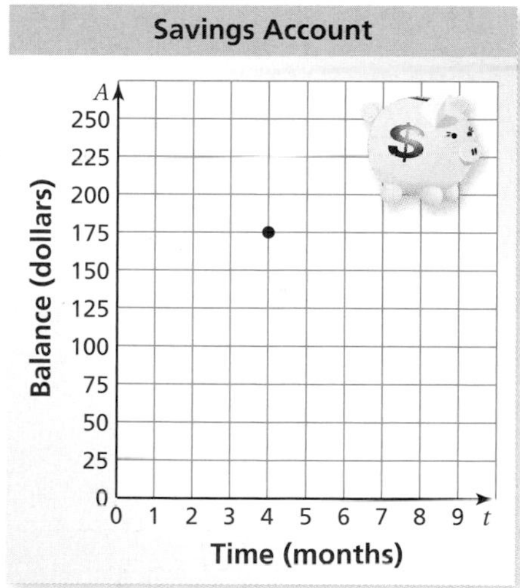

What Is Your Answer?

4. Redo Activity 1 using the formula you found in Activity 2. Compare the results. What do you notice?

5. The formula you found in Activity 2, $y - y_1 = m(x - x_1)$, is called the *point-slope form* of the equation of a line. Why is $y - y_1 = m(x - x_1)$ called the "point-slope" form? Why do you think it is important?

6. **IN YOUR OWN WORDS** How can you write an equation of a line when you are given the slope and a point on the line? Give an example that is different from those in Activity 1.

Practice Use what you learned about writing equations using a slope and a point to complete Exercises 3–5 on page 82.

2.6 Lesson

Key Vocabulary
point-slope form, *p. 80*

Point-Slope Form

Words A linear equation written in the form $y - y_1 = m(x - x_1)$ is in **point-slope form.** The line passes through the point (x_1, y_1) and the slope of the line is m.

Algebra $y - y_1 = m(x - x_1)$

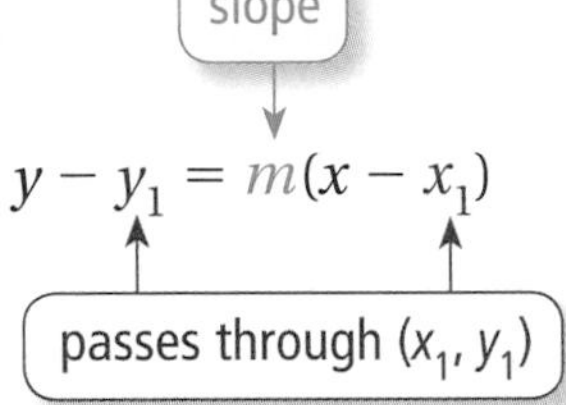

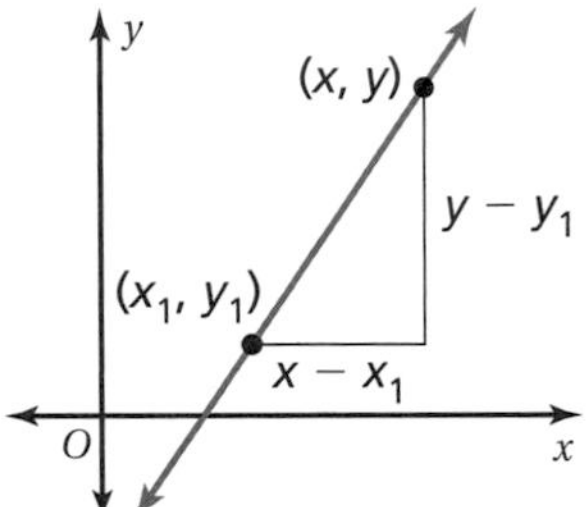

EXAMPLE 1 Writing an Equation Using a Slope and a Point

Write in point-slope form an equation of the line that passes through the point (−6, 1) with slope $\frac{2}{3}$.

$y - y_1 = m(x - x_1)$ Write the point-slope form.

$y - 1 = \frac{2}{3}[x - (-6)]$ Substitute $\frac{2}{3}$ for m, −6 for x_1, and 1 for y_1.

$y - 1 = \frac{2}{3}(x + 6)$ Simplify.

∴ So, the equation is $y - 1 = \frac{2}{3}(x + 6)$.

Check Check that (−6, 1) is a solution of the equation.

$y - 1 = \frac{2}{3}(x + 6)$ Write the equation.

$1 - 1 \stackrel{?}{=} \frac{2}{3}(-6 + 6)$ Substitute.

$0 = 0$ ✓ Simplify.

On Your Own

Exercises 6–11

Write in point-slope form an equation of the line that passes through the given point and has the given slope.

1. $(1, 2);\ m = -4$ **2.** $(7, 0);\ m = 1$ **3.** $(-8, -5);\ m = -\frac{3}{4}$

Multi-Language Glossary at BigIdeasMath.com.

Laurie's Notes

Introduction

Connect

- **Yesterday:** Students developed an intuitive understanding of how to write the equation of a line given the slope and a point. (MP1, MP3, MP7)
- **Today:** Students will write the equation of a line given the slope and a point.

Motivate

- ? "Have you seen an airplane come in for a landing either in real life, on the television, or in movies?" Most will answer yes.
- ? "Can you describe in words or with a picture what it looks like?" Listen for a smooth approach, meaning a constant rate of descent.
- ? "If the plane descends 200 feet per second, what is its height 5 seconds before it lands?" 1000 ft
- Make a sketch of this scenario and ask if it's possible to write an equation that models the height h of the airplane, t seconds before it lands.

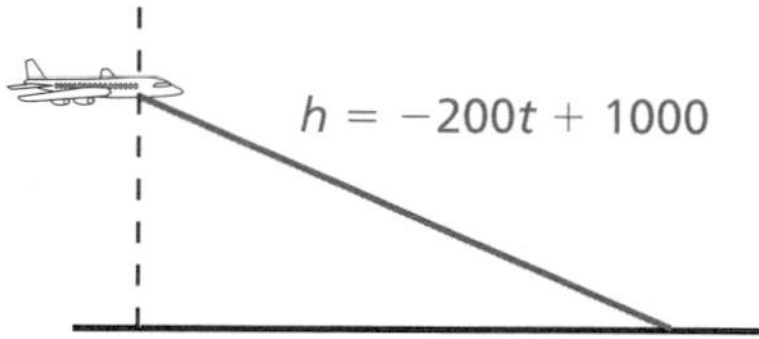

Lesson Notes

Key Idea

- ? Draw a coordinate plane and graph a point. "How many lines go through this point with a slope of $\frac{1}{2}$?" only one line
- Explain that the *point-slope form* of the equation of a line is equivalent to the slope-intercept form and is the equation of a unique line.
- Write the Key Idea on the board. Use of color is helpful.
- **Teaching Tip:** On a side board, write the formula for slope as $\frac{y - y_1}{x - x_1} = m$ so students are reminded of how this form of the equation was derived.

Discuss

- **MP1a Make Sense of Problems:** Although students derived the point-slope formula in the activity, they will have lingering questions about the use of subscripts. They might ask why the first point was not labeled (x, y) and the second point (x_1, y_1). The labels are arbitrary, the line could be sloping downward, and the points could be located in any quadrant.

Example 1

- Write the point-slope form of a linear equation.
- ? "What is the slope of the line?" $\frac{2}{3}$ "What point do we know the line passes through?" $(-6, 1)$
- Substitute the known information. Remind students that they are subtracting a negative, so they have $x + 6$ inside the parentheses.
- ? "How can we check if our equation is reasonable?" Students might suggest a quick sketch or rewriting the equation in slope-intercept form to see if the y-intercept makes sense.

Goal Today's lesson is writing the equation of a line given a slope and a point or two points.

Start Thinking! and Warm Up

Lesson 2.6 Warm Up
For use before Lesson 2.6

Lesson 2.6 Start Thinking!
For use before Lesson 2.6

How is writing the equation of a line given the slope and a point on the line similar to writing the equation of a line given the slope and y-intercept? How is it different?

Extra Example 1

Write in point-slope form an equation of the line that passes through the given point and has the given slope.

a. $(2, 2)$; $m = \frac{5}{2}$ $y - 2 = \frac{5}{2}(x - 2)$

b. $(3, -6)$; $m = -\frac{4}{3}$ $y + 6 = -\frac{4}{3}(x - 3)$

On Your Own

1. $y - 2 = -4(x - 1)$
2. $y - 0 = 1(x - 7)$
3. $y + 5 = -\frac{3}{4}(x + 8)$

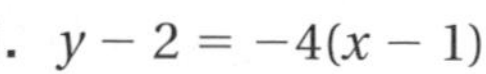

Extra Example 2

Write in slope-intercept form an equation of line that passes through the points $(-3, 0)$ and $(6, 3)$. $y = \frac{1}{3}x + 1$

Extra Example 3

You are pulling down your kite at a rate of 2 feet per second. After 3 seconds, your kite is 54 feet above you.

a. Write and graph an equation that represents the height y (in feet) of the kite above you after x seconds.
$y = -2x + 60$

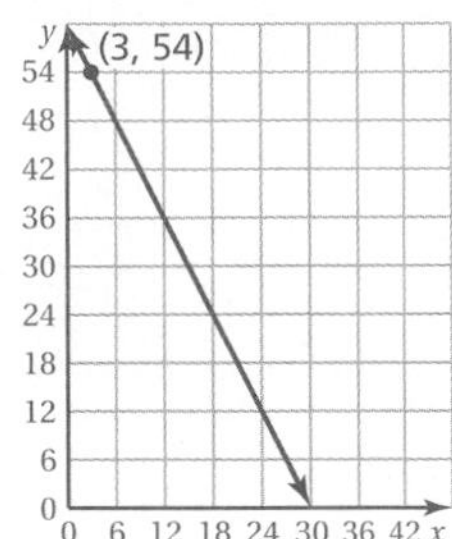

b. At what height was the kite flying? 60 ft

On Your Own

4. $y = -x - 1$
5. $y = 4x + 15$
6. $y = \frac{1}{2}x + 10$
7. $y = -10x + 55$

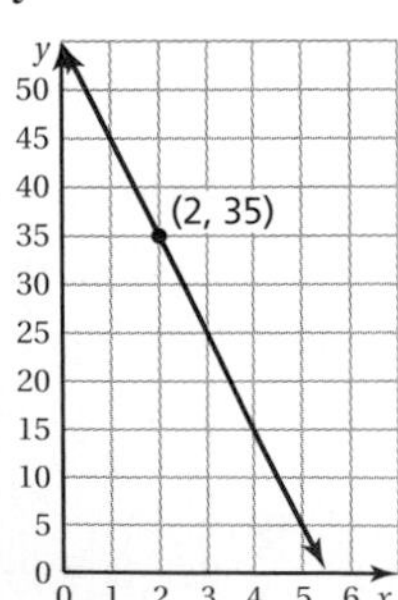

English Language Learners

Visual

Encourage English learners to plot the given point in the coordinate plane and then use the slope. The graph will give them a visual reference they can use when writing the equation.

Laurie's Notes

On Your Own

- **Neighbor Check:** Have students work independently and then have their neighbor check their work. Have students discuss any discrepancies.

Example 2

- Plot both points. Draw the line through the two points.
- ? "Is the slope positive or negative?" negative
- ? "How can you find the slope exactly?" Use the slope formula.
- ? "Can you estimate the *y*-intercept?" Listen for a positive number greater than 4.
- Continue to work through the problem as shown.
- ? "Do you think we would get the same equation if we had used $(5, -2)$ instead of $(2, 4)$? yes; Students may be unsure.
- **MP1a:** Work the problem again using $(5, -2)$ as shown in the Study Tip. Mathematically proficient students can make sense of why either point will result in the same equation.

Example 3

- Ask a volunteer to read the problem. Discuss information that can be *read* from the illustration.
- ? "Have any of you parasailed?" Wait for students to respond. Explain that you want a smooth descent, like an airplane.
- ? "What is the slope for this problem? How did you know?" Slope is -10. The arrow pointing down means the slope is negative.
- ? "Do we know a point that satisfies the equation?" yes, (2, 25)
- Write the point-slope formula. Substitute the known information.
- **Extension:** Have students determine when you reach the boat, meaning $y = 0$.

On Your Own

- Discuss student solutions. Check that signs of numbers are correct for Questions 4–6.

Closure

- **Exit Ticket:** Write an equation of the line with a slope of 2 that passes through the point $(-1, 4)$ in point-slope form and slope-intercept form.
$y - 4 = 2(x + 1)$; $y = 2x + 6$

Technology For the Teacher

The Dynamic Planning Tool
Editable Teacher's Resources at *BigIdeasMath.com*

EXAMPLE 2 Writing an Equation Using Two Points

Study Tip

You can use either of the given points to write the equation of the line.

Use $m = -2$ and $(5, -2)$.

$y - (-2) = -2(x - 5)$

$y + 2 = -2x + 10$

$y = -2x + 8$ ✓

Write in slope-intercept form an equation of the line that passes through the points (2, 4) and (5, −2).

Find the slope: $m = \frac{y_2 - y_1}{x_2 - x_1} = \frac{-2 - 4}{5 - 2} = \frac{-6}{3} = -2$

Then use the slope $m = -2$ and the point (2, 4) to write an equation of the line.

$y - y_1 = m(x - x_1)$	Write the point-slope form.
$y - 4 = -2(x - 2)$	Substitute −2 for m, 2 for x_1, and 4 for y_1.
$y - 4 = -2x + 4$	Use Distributive Property.
$y = -2x + 8$	Write in slope-intercept form.

EXAMPLE 3 Real-Life Application

10 feet per second

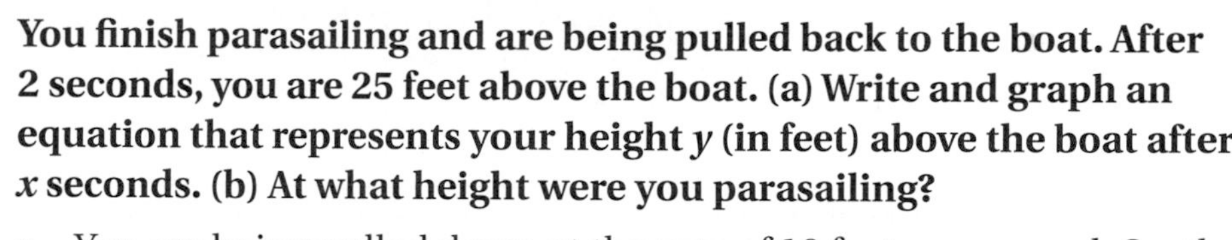

You finish parasailing and are being pulled back to the boat. After 2 seconds, you are 25 feet above the boat. (a) Write and graph an equation that represents your height *y* (in feet) above the boat after *x* seconds. (b) At what height were you parasailing?

a. You are being pulled down at the rate of 10 feet per second. So, the slope is −10. You are 25 feet above the boat after 2 seconds. So, the line passes through (2, 25). Use the point-slope form.

$y - 25 = -10(x - 2)$	Substitute for m, x_1, and y_1.
$y - 25 = -10x + 20$	Use Distributive Property.
$y = -10x + 45$	Write in slope-intercept form.

∴ So, the equation is $y = -10x + 45$.

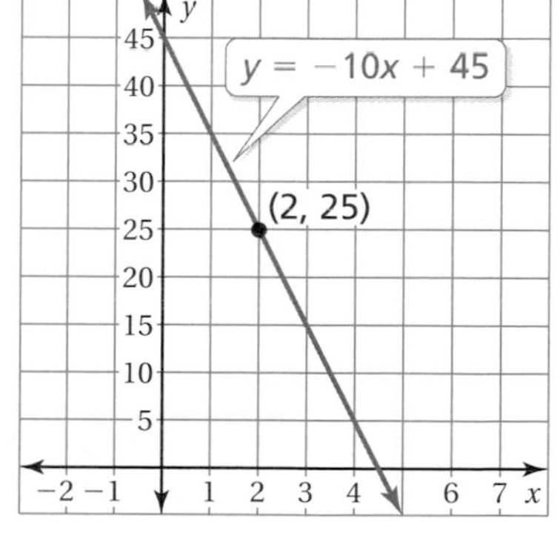

b. You start descending when $x = 0$. The y-intercept is 45. So, you were parasailing at a height of 45 feet.

On Your Own

Now You're Ready
Exercises 12–17

Write in slope-intercept form an equation of the line that passes through the given points.

4. (−2, 1), (3, −4) **5.** (−5, −5), (−3, 3) **6.** (−8, 6), (−2, 9)

7. WHAT IF? In Example 3, you are 35 feet above the boat after 2 seconds. Write and graph an equation that represents your height *y* (in feet) above the boat after *x* seconds.

2.6 Exercises

Vocabulary and Concept Check

1. **VOCABULARY** From the equation $y - 3 = -2(x + 1)$, identify the slope and a point on the line.
2. **WRITING** Describe how to write an equation of a line using (a) its slope and a point on the line, and (b) two points on the line.

Practice and Problem Solving

Use the point-slope form to write an equation of the line with the given slope that passes through the given point.

3. $m = \frac{1}{2}$

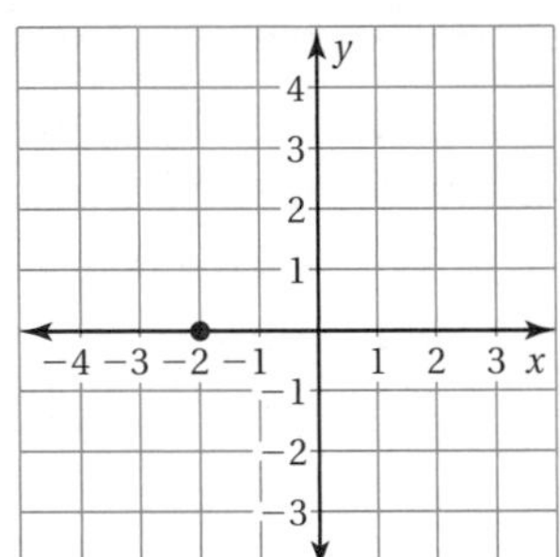

4. $m = -\frac{3}{4}$

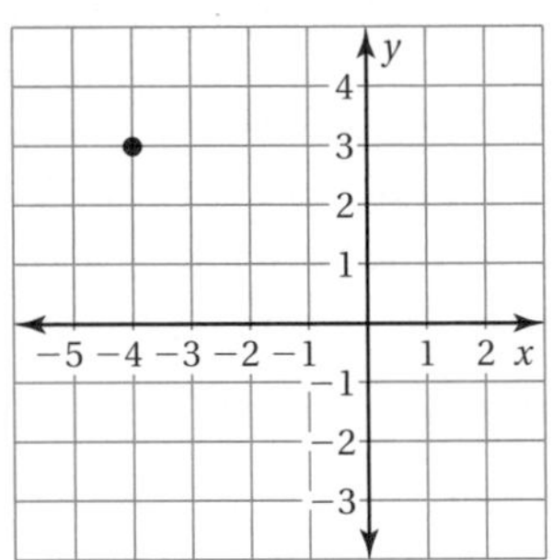

5. $m = -3$

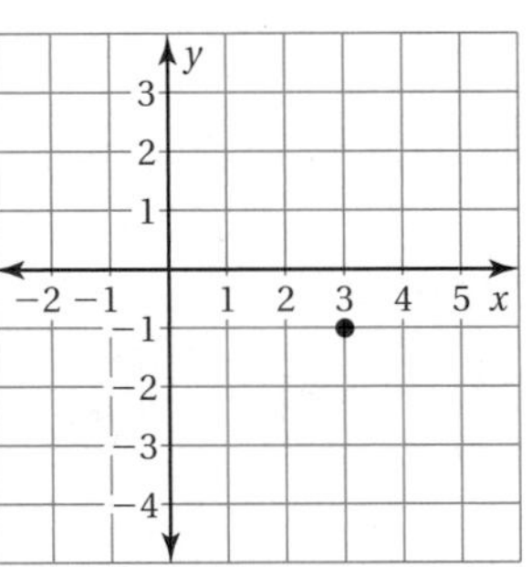

Write in point-slope form an equation of the line that passes through the given point and has the given slope.

1 6. $(3, 0)$; $m = -\frac{2}{3}$
7. $(4, 8)$; $m = \frac{3}{4}$
8. $(1, -3)$; $m = 4$
9. $(7, -5)$; $m = -\frac{1}{7}$
10. $(3, 3)$; $m = \frac{5}{3}$
11. $(-1, -4)$; $m = -2$

Write in slope-intercept form an equation of the line that passes through the given points.

2 12. $(-1, -1), (1, 5)$
13. $(2, 4), (3, 6)$
14. $(-2, 3), (2, 7)$
15. $(4, 1), (8, 2)$
16. $(-9, 5), (-3, 3)$
17. $(1, 2), (-2, -1)$

18. **CHEMISTRY** At 0 °C, the volume of a gas is 22 liters. For each degree the temperature T (in degrees Celsius) increases, the volume V (in liters) of the gas increases by $\frac{2}{25}$. Write an equation that represents the volume of the gas in terms of the temperature.

Assignment Guide and Homework Check

Level	Assignment	Homework Check
Average	1–5, 7, 9, 13, 15, 18, 19, 21, 24–27	7, 13, 18, 19
Advanced	1, 2, 10, 16, 19–23, 24–27	10, 16, 19, 22

Common Errors

- **Exercises 6–17** Students may forget to include negatives with the slope and coordinates, or they may apply them incorrectly. Remind them that when the coordinates are negative, they will be subtracting a negative after substituting in point-slope form, which is a positive.
- **Exercises 12–17** Students may use the reciprocal of the slope when writing the equation. Remind them that slope is the change in *y* over the change in *x*.
- **Exercise 18** Students might have trouble knowing which variable can be compared with *x* and *y* and may write the given point backwards. Review what the words "in terms of" mean when writing an equation. In this problem, *V* could be replaced by *y* and *T* could be replaced by *x*. Remind students to check their equation by substituting the given point and checking that it is a solution of the equation.
- **Exercise 20** Students may struggle when plotting the given points because π is used. Encourage them to scale the *y*-axis by increments of π.

2.6 Record and Practice Journal

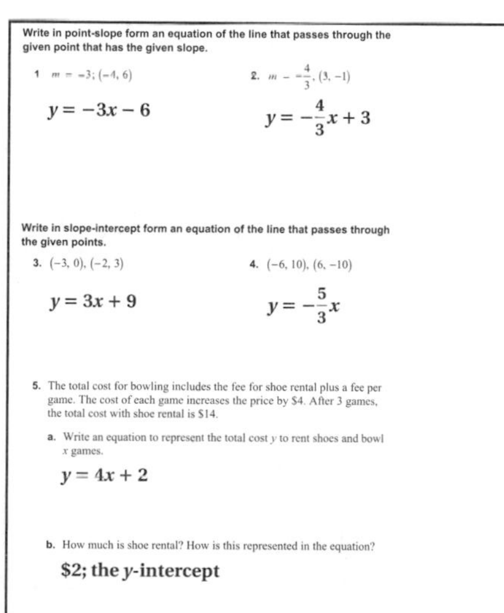

Write in point-slope form an equation of the line that passes through the given point that has the given slope.

1. $m = -3$; $(-4, 6)$

   $y = -3x - 6$

2. $m = -\frac{4}{3}$; $(3, -1)$

   $y = -\frac{4}{3}x + 3$

Write in slope-intercept form an equation of the line that passes through the given points.

3. $(-3, 0)$, $(-2, 3)$

   $y = 3x + 9$

4. $(-6, 10)$, $(6, -10)$

   $y = -\frac{5}{3}x$

5. The total cost for bowling includes the fee for shoe rental plus a fee per game. The cost of each game increases the price by $4. After 3 games, the total cost with shoe rental is $14.

   a. Write an equation to represent the total cost *y* to rent shoes and bowl *x* games.

   $y = 4x + 2$

   b. How much is shoe rental? How is this represented in the equation?

   $2; the *y*-intercept

Answer Presentation Tool

Vocabulary and Concept Check

1. slope = −2; $(-1, 3)$
2. a. Write the point-slope form. Substitute the slope for m and the point for (x_1, y_1). Simplify and check your work.

   b. First use the two points to find the slope. Then write the point-slope form. Substitute the slope for m and one of the points for (x_1, y_1). Simplify and check your work.

Practice and Problem Solving

3. $y - 0 = \frac{1}{2}(x + 2)$
4. $y - 3 = -\frac{3}{4}(x + 4)$
5. $y + 1 = -3(x - 3)$
6. $y - 0 = -\frac{2}{3}(x - 3)$
7. $y - 8 = \frac{3}{4}(x - 4)$
8. $y + 3 = 4(x - 1)$
9. $y + 5 = -\frac{1}{7}(x - 7)$
10. $y - 3 = \frac{5}{3}(x - 3)$
11. $y + 4 = -2(x + 1)$
12. $y = 3x + 2$
13. $y = 2x$
14. $y = x + 5$
15. $y = \frac{1}{4}x$
16. $y = -\frac{1}{3}x + 2$
17. $y = x + 1$
18. $V = \frac{2}{25}T + 22$

Practice and Problem Solving

19. a. $V = -4000x + 30{,}000$

b. \$30,000

20. a.

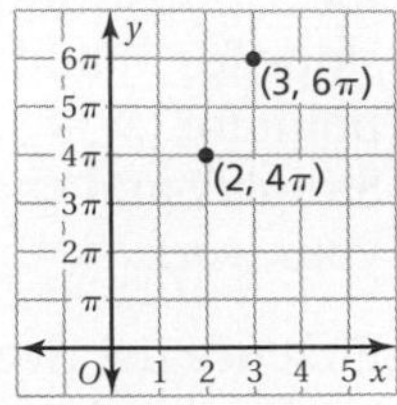

b. $y = 2\pi x$

21. See *Taking Math Deeper.*

22. a. $y = -2x + 68$

b. 68 ounces

c. after 34 seconds

23. a. $y = 14x - 108.5$

b. 4 meters

Fair Game Review

24. 45 **25.** 175

26. -4.5 **27.** D

Mini-Assessment

Write in point-slope an equation of the line that passes through the given point and has the given slope.

1. $(1, 4)$; $m = 3$ $y - 4 = 3(x + 1)$
2. $(-2, 1)$; $m = -2$ $y - 1 = -2(x + 2)$
3. $(3, 5)$; $m = 1$ $y - 5 = 1(x - 3)$
4. $(2, 1)$; $m = \frac{1}{2}$ $y - 1 = \frac{1}{2}(x - 2)$
5. **You rent a floor sander for \$24 per day. You pay \$82 for 3 days.**
   a. Write an equation that represents your total cost y (in dollars) after x days. $y = 24x + 10$
   b. Interpret the y-intercept. The y-intercept is 10. This means you paid a deposit fee of \$10 to rent the sander.

Taking Math Deeper

Exercise 21

The challenge in this biology problem is to interpret the given information as a rate of change (or slope) and as an ordered pair.

Translate the given information into math.

T = temperature (°F)
x = chirps per minute
Rate of change = 0.25 degree per chirp

Write an equation.

Given point: $(x, T) = (40, 50)$

With a slope of 0.25, you can determine that the T-intercept of the line is 40. So, the equation is

a. $T = 0.25x + 40$.

Use the equation.
If $x = 100$ chirps per minute, then

$T = 0.25(100) + 40$
b. $= 65°F$.

If $T = 96$, then you can find the number of chirps per minute as follows.

$96 = 0.25x + 40$
$56 = 0.25x$
$224 = x$

c. So, you would expect the cricket to make 224 chirps in one minute.

This relationship between temperature and cricket chirps was first published by Amos Dolbear in 1897 in an article called *The Cricket as a Thermometer.*

Project

Research other plants or animals that predict the temperature or weather.

Reteaching and Enrichment Strategies

If students need help. . .	If students got it. . .
Resources by Chapter • Practice A and Practice B • Puzzle Time Record and Practice Journal Practice Differentiating the Lesson Lesson Tutorials Skills Review Handbook	Resources by Chapter • Enrichment and Extension • School-to-Work • Financial Literacy • Technology Connection Start the next section

19. **CARS** After it is purchased, the value of a new car decreases \$4000 each year. After 3 years, the car is worth \$18,000.

   a. Write an equation that represents the value V (in dollars) of the car x years after it is purchased.

   b. What was the original value of the car?

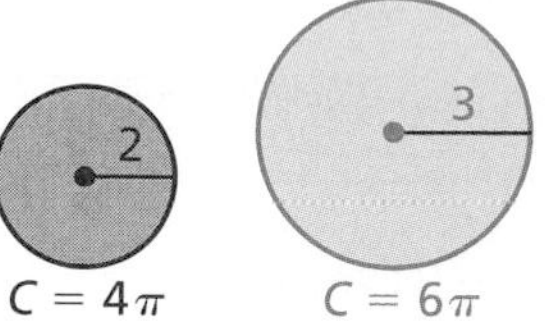

20. **CIRCUMFERENCE** Consider the circles shown.

   a. Plot the points $(2, 4\pi)$ and $(3, 6\pi)$.

   b. Write an equation of the line that passes through the two points.

21. **CRICKETS** According to Dolbear's Law, you can predict the temperature T (in degrees Fahrenheit) by counting the number x of chirps made by a snowy tree cricket in 1 minute. For each rise in temperature of 0.25°F, the cricket makes an additional chirp each minute.

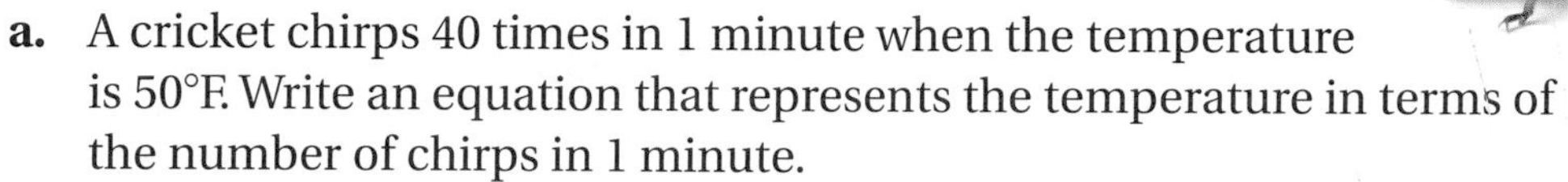

   a. A cricket chirps 40 times in 1 minute when the temperature is 50°F. Write an equation that represents the temperature in terms of the number of chirps in 1 minute.

   b. You count 100 chirps in 1 minute. What is the temperature?

   c. The temperature is 96 °F. How many chirps would you expect the cricket to make?

Leaning Tower of Pisa

22. **WATERING CAN** You water the plants in your classroom at a constant rate. After 5 seconds, your watering can contains 58 ounces of water. Fifteen seconds later, the can contains 28 ounces of water.

   a. Write an equation that represents the amount y (in ounces) of water in the can after x seconds.

   b. How much water was in the can when you started watering the plants?

   c. When is the watering can empty?

23. **Problem Solving** The Leaning Tower of Pisa in Italy was built between 1173 and 1350.

   a. Write an equation for the yellow line.

   b. The tower is 56 meters tall. How far off center is the top of the tower?

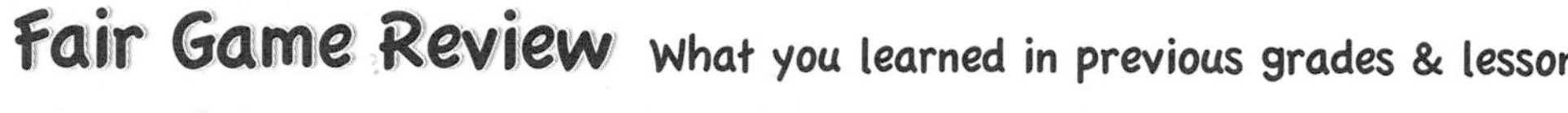

Fair Game Review What you learned in previous grades & lessons

Find the percent of the number. *(Skills Review Handbook)*

24. 15% of 300
25. 140% of 125
26. 6% of -75

27. **MULTIPLE CHOICE** What is the x-intercept of the equation $3x + 5y = 30$? *(Section 2.4)*

   Ⓐ -10　　Ⓑ -6　　Ⓒ 6　　Ⓓ 10

2.6b Writing Equations of Parallel and Perpendicular Lines

You can use the slope-intercept form or the point-slope form to write equations of parallel and perpendicular lines.

EXAMPLE 1 Writing an Equation of a Parallel Line

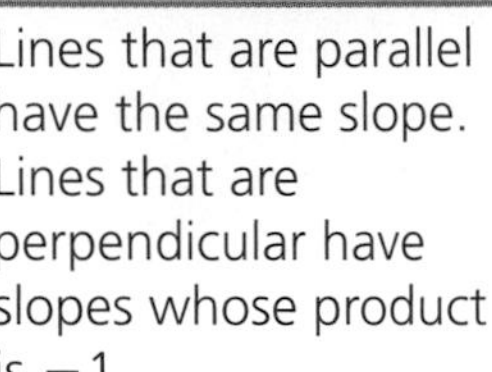

Lines that are parallel have the same slope. Lines that are perpendicular have slopes whose product is -1.

Write an equation of the line that passes through $(6, -2)$ and is parallel to the line $y = \frac{1}{2}x + 3$.

Step 1: Find the slope of the parallel line.

The slope of $y = \frac{1}{2}x + 3$ is $\frac{1}{2}$. So, the parallel line that passes through $(6, -2)$ has the same slope, $\frac{1}{2}$.

Step 2: Use the slope $\frac{1}{2}$ and the slope-intercept form to find the y-intercept of the parallel line that passes through $(6, -2)$.

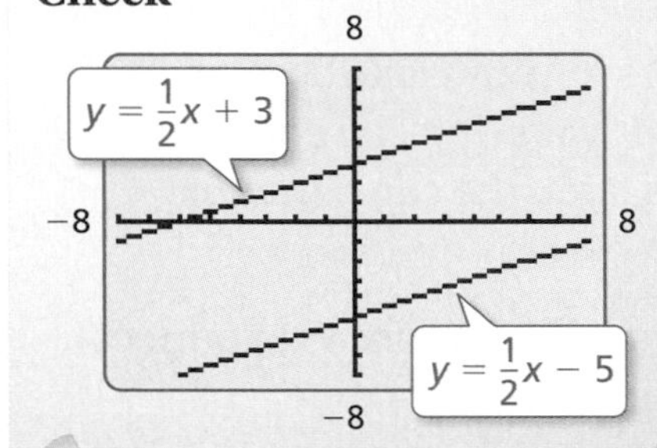

$y = mx + b$	Write the slope-intercept form.
$-2 = \frac{1}{2}(6) + b$	Substitute $\frac{1}{2}$ for m, 6 for x, and -2 for y.
$-2 = 3 + b$	Multiply.
$-5 = b$	Subtract 3 from each side.

The parallel line has a slope of $\frac{1}{2}$ and a y-intercept of -5.

So, an equation of the parallel line is $y = \frac{1}{2}x + (-5)$, or $y = \frac{1}{2}x - 5$.

Practice

Write an equation of the line that passes through the given point and is parallel to the given line. Use a graphing calculator to check your answer.

1. $(-2, 1)$; $y = 3x - 4$
2. $(6, -3)$; $y = -\frac{2}{3}x + 5$
3. $(-4, -5)$; $y = -4x - 1$

Write an equation of the line that passes through the given point and is parallel to the line shown in the graph.

4. $(2, -3)$
5. $(0, 0)$
6. $(-5, 8)$
7. $(-1, -7)$

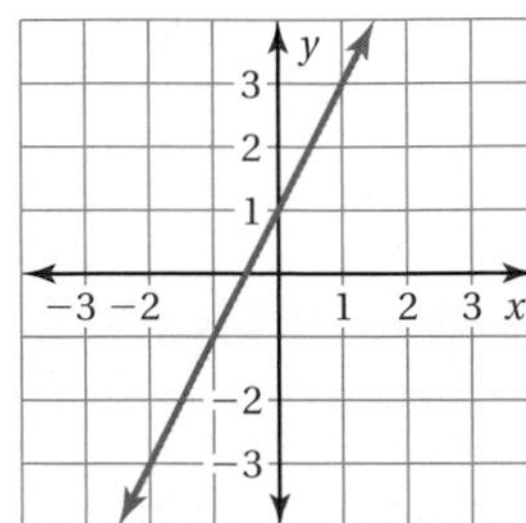

Laurie's Notes

Introduction

Connect

- **Yesterday:** Students wrote the equation of a line given the slope and a point or two points. (MP1a)
- **Today:** Students will write the equation of a line that is either parallel or perpendicular to a given line.

Motivate

- Ask if anyone has visited Salt Lake City. If they have, ask if they know anything about the manner in which the streets are named.
- Salt Lake City's method of numbering streets is similar to the idea of latitude and longitude. The city is set up like a grid with each address having coordinates indicating the distance west or east and north or south from Temple Square.
- Share that today students will write equations of lines that are parallel and perpendicular, like a grid.

Lesson Notes

Discuss

- **MP7 Look for and Make Use of Structure:** Students know two forms for writing linear equations. In this lesson, they will use one of the forms and apply what they know about slopes of parallel and perpendicular lines. Mathematically proficient students see the equation $y = mx + b$ and recognize that a slope and a point can be substituted in order to solve for the y-intercept, b.

? "What do you know about the slopes of parallel lines?" They are equal.

? "What do you know about the slopes of perpendicular lines?" The product of the slopes equals -1. They are negative reciprocals.

Example 1

- Sketch a coordinate plane and label $(6, -2)$.

? "How many lines pass through this point?" infinitely many

? "How many lines pass through this point with a specific slope?" one

? "We want to write the equation of the line parallel to $y = \frac{1}{2}x + 3$ that passes through $(6, -2)$. What will the slope be?" $\frac{1}{2}$

? "Do you know the y-intercept?" Some students may try to approximate the y-intercept, but in general they need to solve for b.

- Write the slope-intercept form of a line. Explain that they know the slope and a point on the line. Underline y, m, and x.

? "How can we find b?" Substitute what we know in the equation and solve for b.

- Check using a graphing calculator. Enter both equations in the equation editor. Verify that the equation found passes through $(6, -2)$.
- **MP1 Make Sense of Problems and Persevere in Solving Them:** The equation could also be found using the point-slope form. When it makes sense, take time to solve problems using more than one strategy. Ask students to identify what is the same/different about the approaches.

Goal Today's lesson is writing equations of parallel and perpendicular lines.

Start Thinking! and Warm Up

Lesson 2.6b **Warm Up** For use before Lesson 2.6b

Lesson 2.6b **Start Thinking!** For use before Lesson 2.6b

A line passes through the point $(2, 3)$ and is parallel to the line $y = 3x$. How can you find another point on the line?

A line passes through the point $(1, 1)$ and is perpendicular to the line $y = \frac{1}{4}x - 2$. How can you find another point on the line?

Extra Example 1

Write an equation of the line that passes through $(1, 2)$ and is parallel to the line $y = -3x - 4$.

$y = -3x + 5$

Practice

1. $y = 3x + 7$
2. $y = -\frac{2}{3}x + 1$
3. $y = -4x - 21$
4. $y = 2x - 7$
5. $y = 2x$
6. $y = 2x + 18$
7. $y = 2x - 5$

Record and Practice Journal Practice

See Additional Answers.

Laurie's Notes

Example 1 (continued)

- **MP5 Use Appropriate Tools Strategically:** If time permits, look at the table of values on the graphing calculator. Point out that for both equations, as x increases by 1, the y-values increase by $\frac{1}{2}$. By using technology as a tool, students see equations represented symbolically, graphically, and as a table of values. This will deepen their understanding of the concept of slope.

Practice

- Note that Exercises 4–7 require students to determine the slope of the line by *reading* the graph shown.

Example 2

- This example is similar to Example 1. What differs is how information about the problem is given. Students must determine the slope of the line by *reading* the graph shown.

? "What is the slope of the line shown in the graph? How do you know?" −4; found the rise and the run

? "What is the slope of the line perpendicular to this line?" $\frac{1}{4}$

- Write the point-slope form of a line. Explain that they know the slope and a point on the line. Underline y_1, m, and x_1.
- Substitute and write the equation in slope-intercept form.
- Check using a graphing calculator. Enter both equations in the equation editor. Verify that the equation found passes through (−3, 1).

? "Do the lines look perpendicular?" no

  - **Note:** For the lines to look perpendicular, select the *square* viewing rectangle. Show students how this is done.
- **MP1:** The equation could also be found using slope-intercept form as in Example 1.

Practice

- Note that Exercises 11–14 require students to determine the slope of the line by *reading* the graph shown.

Closure

- **Writing Prompt:** To write an equation of the line that passes through (−2, 4) and is parallel to $y = -3x + 4$, . . .

Technology For the Teacher

The Dynamic Planning Tool
Editable Teacher's Resources at *BigIdeasMath.com*

Extra Example 2

Write an equation of the line that passes through (1, −1) and is perpendicular to the line shown in the graph.

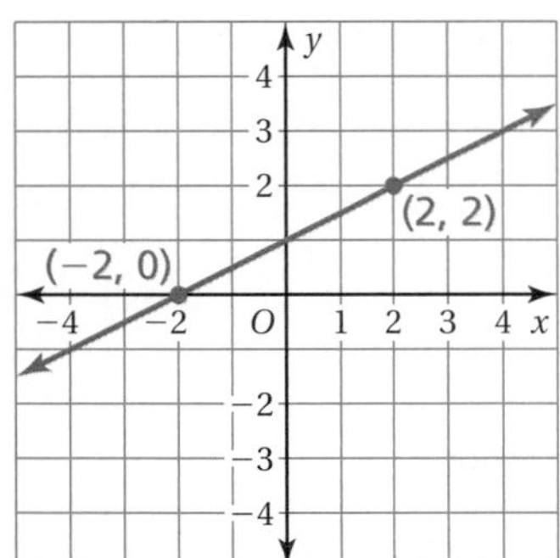

$y = -2x + 1$

Practice

8. $y = 3x - 10$
9. $y = -x - 1$
10. $y = \frac{1}{2}x + \frac{17}{2}$
11. $y = \frac{1}{5}x + \frac{19}{5}$
12. $y = \frac{1}{5}x + 6$
13. $y = \frac{1}{5}x - 1$
14. $y = \frac{1}{5}x - \frac{18}{5}$
15. See Additional Answers.

EXAMPLE 2 Writing an Equation of a Perpendicular Line

Write an equation of the line that passes through (−3, 1) and is perpendicular to the line shown in the graph.

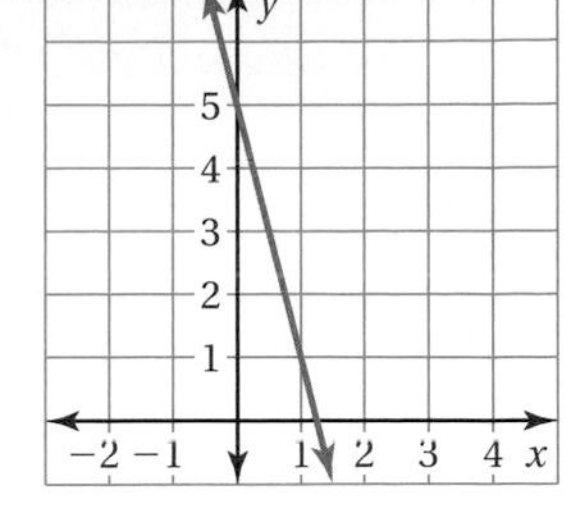

Step 1: Find the slope of the line in the graph.

(0, 5), 1, −4, (1, 1)

$$\text{slope} = \frac{\text{change in } y}{\text{change in } x} = \frac{-4}{1} = -4$$

Step 2: Find the slope of the perpendicular line.

The slope of the line in the graph is -4. Because $-4 \cdot \frac{1}{4} = -1$, the slope of the perpendicular line is $\frac{1}{4}$.

Step 3: Use the slope $m = \frac{1}{4}$ and the point-slope form to write an equation of the perpendicular line that passes through $(-3, 1)$.

$y - y_1 = m(x - x_1)$	Write the point-slope form.
$y - 1 = \frac{1}{4}[x - (-3)]$	Substitute $\frac{1}{4}$ for m, -3 for x_1, and 1 for y_1.
$y - 1 = \frac{1}{4}x + \frac{3}{4}$	Simplify.
$y = \frac{1}{4}x + \frac{7}{4}$	Add 1 to each side.

∴ So, an equation of the perpendicular line is $y = \frac{1}{4}x + \frac{7}{4}$.

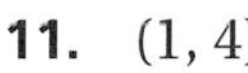

Practice

Write an equation of the line that passes through the given point and is perpendicular to the given line.

8. $(4, 2); y = -\frac{1}{3}x + 1$ **9.** $(0, -1); y = x - 6$ **10.** $(-3, 7); y = -2x - 5$

Write an equation of the line that passes through the given point and is perpendicular to the line shown in the graph.

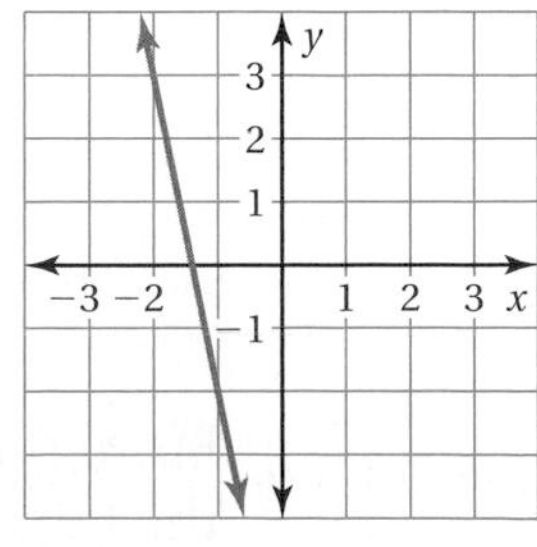

11. $(1, 4)$ **12.** $(0, 6)$

13. $(-5, -2)$ **14.** $(3, -3)$

15. REASONING Rework Example 1 using the point-slope form. Rework Example 2 using the slope-intercept form. Which method do you prefer? Explain your reasoning.

2.7 Solving Real-Life Problems

COMMON CORE STATE STANDARDS
8.F.4
A.CED.2
F.IF.4

Essential Question How can you use a linear equation in two variables to model and solve a real-life problem?

1 EXAMPLE: Writing a Story

Write a story that uses the graph at the right.

- **In your story, interpret the slope of the line, the *y*-intercept, and the *x*-intercept.**
- **Make a table that shows data from the graph.**
- **Label the axes of the graph with units.**
- **Draw pictures for your story.**

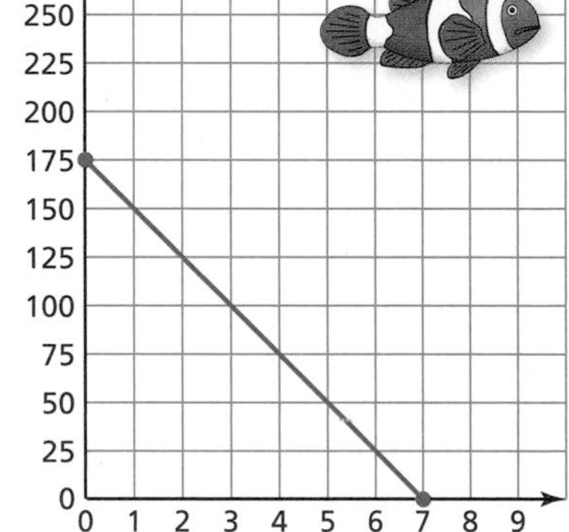

There are many possible stories. Here is one about a reef tank.

Tom works at an aquarium shop on Saturdays. One Saturday, when Tom gets to work, he is asked to clean a 175-gallon reef tank.

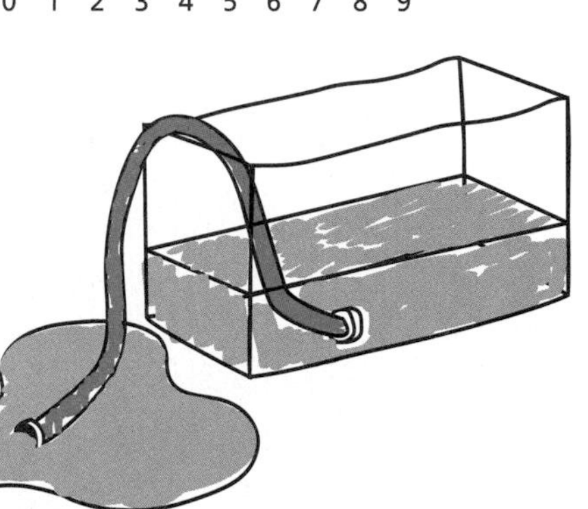

His first job is to drain the tank. He puts a hose into the tank and starts a siphon. Tom wonders if the tank will finish draining before he leaves work.

He measures the amount of water that is draining out and finds that 12.5 gallons drain out in 30 minutes. So, he figures that the rate is 25 gallons per hour. To see when the tank will be empty, Tom makes a table and draws a graph.

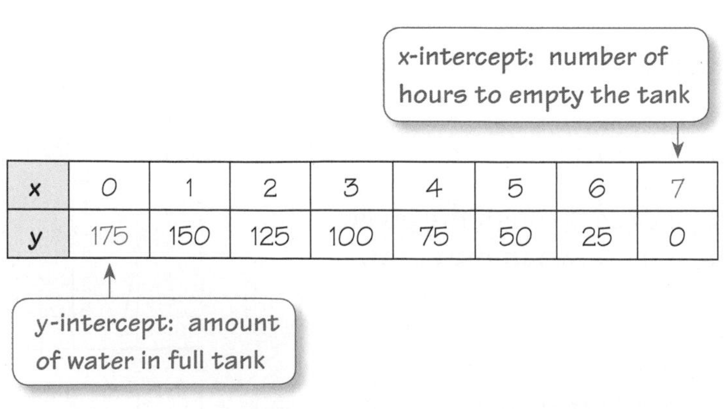

x	0	1	2	3	4	5	6	7
y	175	150	125	100	75	50	25	0

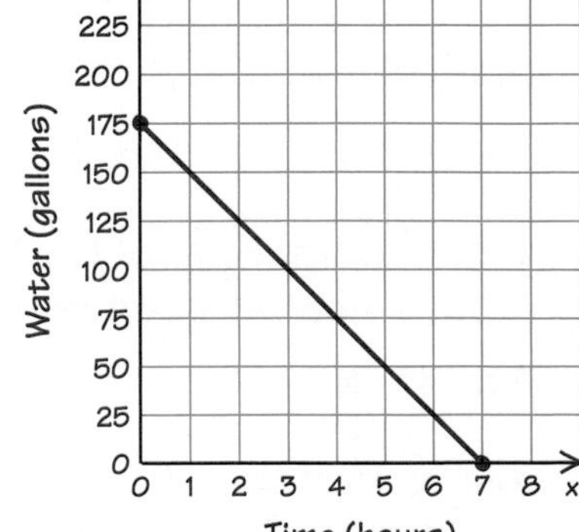

From the table and also from the graph, Tom sees that the tank will be empty after 7 hours. This will give him 1 hour to wash the tank before going home.

Laurie's Notes

Introduction

Standards for Mathematical Practice

- **MP4 Model with Mathematics:** The goal is for students to use a linear equation to model and solve a real-life problem. Modeling with mathematics involves representing a situation algebraically, solving, and checking to make sure that the results make sense.

Motivate

- **Time for a Story!** Tell students that you want to give your friend 10 pounds of chocolate for his birthday. The fancy bars of chocolate come in one-pound and half-pound bars. Hold up two rectangular blocks (or real bars).

? "What are some different ways I could give my friend the 10 pounds of chocolate?" Answers will vary.

- Students should eventually give the two *simple* answers of 10 one-pound bars and no half-pound bars, or 20 half-pound bars and no one-pound bars. If you let x equal the number of one-pound bars and y equal the number of half-pound bars, these two solutions are (10, 0) and (0, 20). Plot the two points.

Discuss

? Depending upon time, you may wish to ask a series of questions and have students answer them now or at the end of the lesson.

- "Are there other combinations besides these two that will equal 10 pounds?" Make a table to show additional solutions.
- "What is the slope of the line that goes through the two points?" -2
- "Could you write an equation for the line through the two points?" $y = -2x + 20$
- "If you bought 5 one-pound bars, how many half-pound bars would you need to purchase?" 10
- "If a one-pound bar costs twice as much as a half-pound bar, what do you know about all of the possible gift combinations?" all cost the same

Activity Notes

Example 1

- Ask one or more students to read through the sample provided.
- The problem does not have to be about fish tanks, but it does need to use the ordered pairs (0, 175) and (7, 0). The table of values will be the same regardless of the context selected. Notice that the equation is not determined, although this could be an extension for some or all students.
- Here are some suggestions for ordered pairs: (# of weeks, $ in the bank); (# of hours biking, kilometers left); (# of hours reading, pages to read)

Common Core State Standards

8.F.4 Construct a function to model a linear relationship between two quantities. Determine the rate of change and initial value of the function . . .. Interpret the rate of change and initial value of a linear function in terms of the situation it models, and in terms of its graph or a table of values.

A.CED.2 Create equations in two or more variables . . .; graph equations on coordinate axes with labels and scales.

F.IF.4 For a function that models a relationship between two quantities, interpret key features of graphs and tables in terms of the quantities, and sketch graphs showing key features given a verbal description of the relationship.

Previous Learning

Students should know how to write equations in slope-intercept form.

Start Thinking! and Warm Up

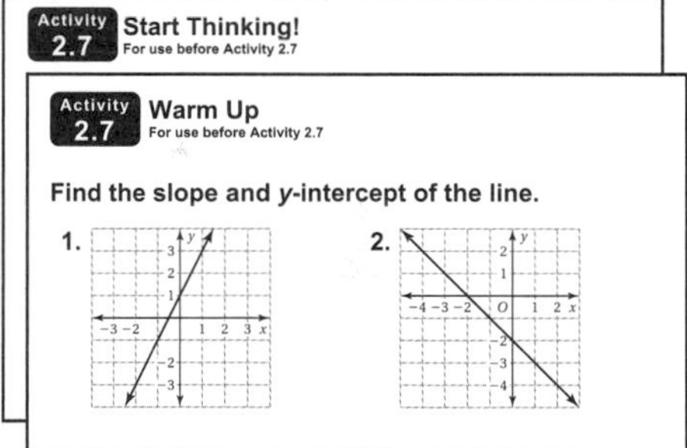

Activity 2.7 Start Thinking! For use before Activity 2.7

Activity 2.7 Warm Up For use before Activity 2.7

Find the slope and y-intercept of the line.

1.

2.

2.7 Record and Practice Journal

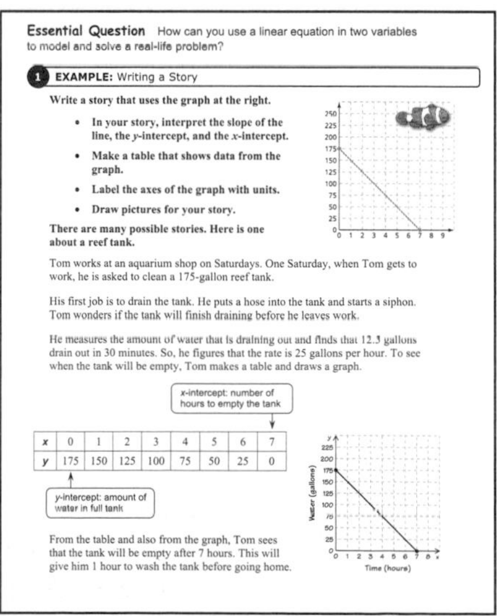

Essential Question How can you use a linear equation in two variables to model and solve a real-life problem?

1 EXAMPLE: Writing a Story

Write a story that uses the graph at the right.

- **In your story, interpret the slope of the line, the y-intercept, and the x-intercept.**
- **Make a table that shows data from the graph.**
- **Label the axes of the graph with units.**
- **Draw pictures for your story.**

There are many possible stories. Here is one about a reef tank.

Tom works at an aquarium shop on Saturdays. One Saturday, when Tom gets to work, he is asked to clean a 175-gallon reef tank.

His first job is to drain the tank. He puts a hose into the tank and starts a siphon. Tom wonders if the tank will finish draining before he leaves work.

He measures the amount of water that is draining out and finds that 12.5 gallons drain out in 30 minutes. So, he figures that the rate is 25 gallons per hour. To see when the tank will be empty, Tom makes a table and draws a graph.

x	0	1	2	3	4	5	6	7
y	175	150	125	100	75	50	25	0

From the table and also from the graph, Tom sees that the tank will be empty after 7 hours. This will give him 1 hour to wash the tank before going home.

English Language Learners

English Language Learners
For Activity 2, pair English learners with English speakers to write a story based on the English learners' culture. Encourage students to share how they incorporated the math into the story.

2.7 Record and Practice Journal

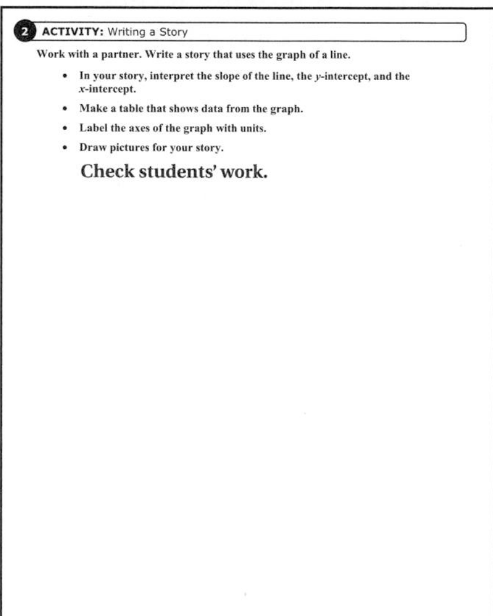

2 ACTIVITY: Writing a Story

Work with a partner. Write a story that uses the graph of a line.

- **In your story, interpret the slope of the line, the y-intercept, and the x-intercept.**
- **Make a table that shows data from the graph.**
- **Label the axes of the graph with units.**
- **Draw pictures for your story.**

Check students' work.

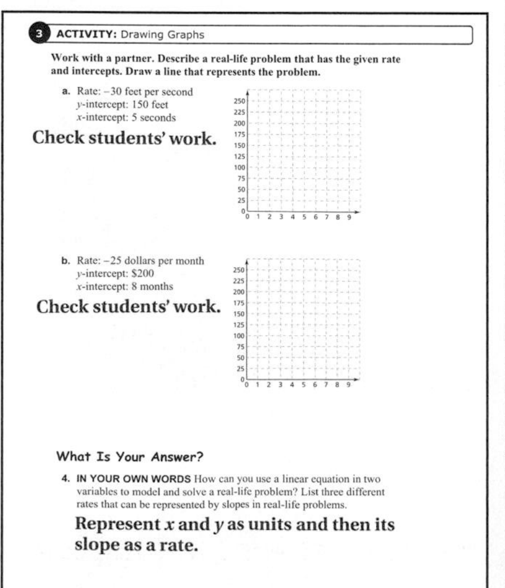

3 ACTIVITY: Drawing Graphs

Work with a partner. Describe a real-life problem that has the given rate and intercepts. Draw a line that represents the problem.

a. Rate: −30 feet per second
y-intercept: 150 feet
x-intercept: 5 seconds

Check students' work.

b. Rate: −25 dollars per month
y-intercept: $200
x-intercept: 8 months

Check students' work.

What Is Your Answer?

4. IN YOUR OWN WORDS How can you use a linear equation in two variables to model and solve a real-life problem? List three different rates that can be represented by slopes in real-life problems.

Represent x and y as units and then its slope as a rate.

Laurie's Notes

Activity 2

- **Interdisciplinary:** This activity is open-ended. The goal is to integrate language arts skills while practicing mathematical skills.
- You modeled how this problem can be done with the chocolate bars at the beginning of the lesson.
- You may wish to return to this problem *after* students have brainstormed three different rates in Question 4.
- **MP3 Construct Viable Arguments and Critique the Reasoning of Others:** Students should share their stories with the whole class, drawing a sketch of the graph with the axes labeled. When students are sharing their stories, you should attend to their reasoning, which includes use of stated assumptions, thought process, and justification of their conclusions.

Activity 3

- **MP6 Attend to Precision:** In this activity students are expected to determine a context, explain their reasoning, and attend to precision.

? "What are the ordered pairs for the intercepts?" (0, 150) and (5, 0)

? "Explain how the rate is found." The change is 150 feet in 5 seconds, which simplifies to 30 feet per 1 second.

? "Why is the rate negative?" The amount of feet is decreasing as time increases.

- Students may need to see the graph to understand why the rate is negative.
- **Extension:** Ask students to sketch a graph that represents a rate of 30 feet per second.

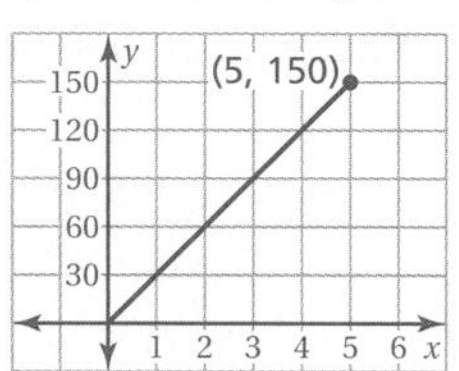

- **Connection:** For many contextual problems, the x- and y-intercepts will be positive numbers, and so the slope of the line between them is negative.

What Is Your Answer?

- **Whole Class Activity:** Have the class brainstorm 1 or 2 rates, and then ask students to think about two or three more on their own. This question will help students think about the variety of contexts in which linear equations can arise. Possible Rates: miles per hour, feet per second, miles per gallon, outs per inning, points per quarter, people per team, tiles per foot

Closure

- Refer back to the chocolate question given at the beginning of the lesson and ask students to write an equation for this problem. The rate of -2 means that for every 2 half-pound bars that are bought, there is one less one-pound bar. $y = -2x + 20$

Technology For the Teacher

The Dynamic Planning Tool
Editable Teacher's Resources at *BigIdeasMath.com*

2 ACTIVITY: Writing a Story

Work with a partner. Write a story that uses the graph of a line.

- **In your story, interpret the slope of the line, the y-intercept, and the x-intercept.**
- **Make a table that shows data from the graph.**
- **Label the axes of the graph with units.**
- **Draw pictures for your story.**

3 ACTIVITY: Drawing Graphs

Work with a partner. Describe a real-life problem that has the given rate and intercepts. Draw a line that represents the problem.

a. Rate: -30 feet per second
y-intercept: 150 feet
x-intercept: 5 seconds

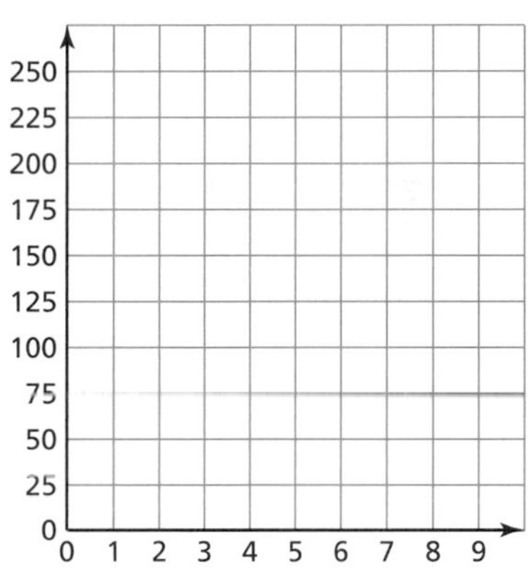

b. Rate: -25 dollars per month
y-intercept: \$200
x-intercept: 8 months

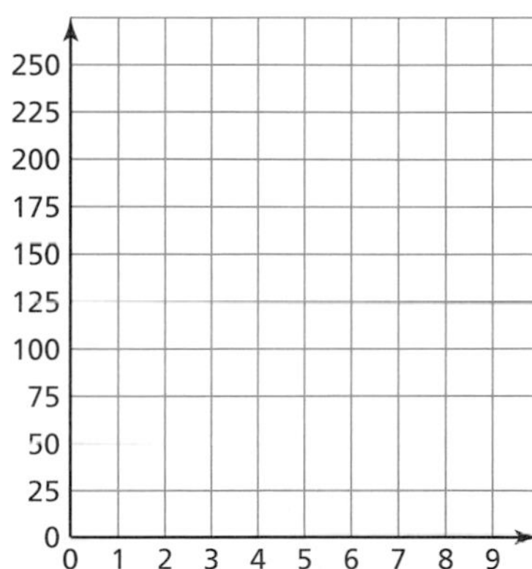

What Is Your Answer?

4. **IN YOUR OWN WORDS** How can you use a linear equation in two variables to model and solve a real-life problem? List three different rates that can be represented by slopes in real-life problems.

Practice Use what you learned about solving real-life problems to complete Exercises 4 and 5 on page 90.

2.7 Lesson

EXAMPLE 1 Real-Life Application

The percent y (in decimal form) of battery power remaining x hours after you turn on a laptop computer is $y = -0.2x + 1$. (a) Graph the equation. (b) Interpret the x- and y-intercepts. (c) After how many hours is the battery power at 75%?

a. Use the slope and the y-intercept to graph the equation.

$$y = -0.2x + 1$$

slope: -0.2; y-intercept: 1

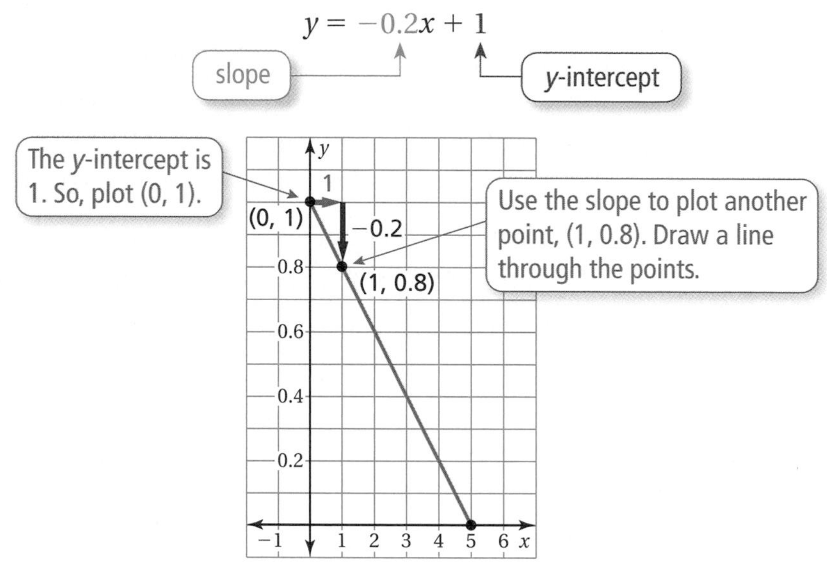

b. To find the x-intercept, substitute 0 for y in the equation.

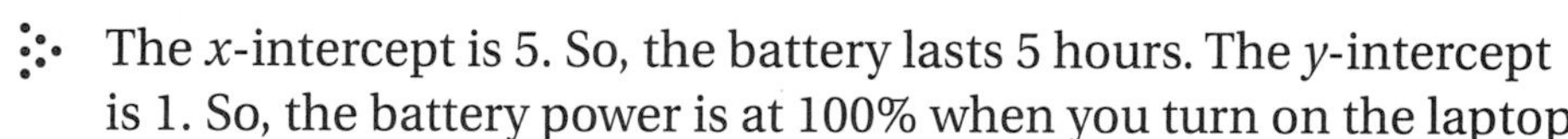

$y = -0.2x + 1$	Write the equation.
$0 = -0.2x + 1$	Substitute 0 for y.
$5 = x$	Solve for x.

∴ The x-intercept is 5. So, the battery lasts 5 hours. The y-intercept is 1. So, the battery power is at 100% when you turn on the laptop.

75% Remaining

c. Find the value of x when $y = 0.75$.

$y = -0.2x + 1$	Write the equation.
$0.75 = -0.2x + 1$	Substitute 0.75 for y.
$1.25 = x$	Solve for x.

∴ The battery power is at 75% after 1.25 hours.

On Your Own

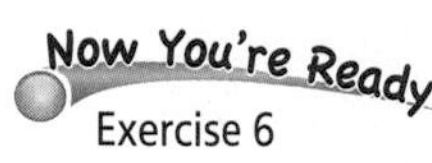

1. The amount y (in gallons) of gasoline remaining in a gas tank after driving x hours is $y = -2x + 12$. (a) Graph the equation. (b) Interpret the x- and y-intercepts. (c) After how many hours are there 5 gallons left?

Laurie's Notes

Introduction

Connect

- **Yesterday:** Students explored real-life problems involving rates, where the x- and y-intercepts were each positive. (MP3, MP4, MP6)
- **Today:** Students will solve real-life problems using equations, graphs, and intercepts.
- **FYI:** The goal is to bring the concepts related to linear equations, graphing, and writing, together into one lesson with more focus on the interpretation of slope and intercepts.

Motivate

- Hold a digital camera and take some pictures of your students. You could also pretend to be taking pictures. Continue to take pictures until someone finally asks how many you're going to take.
- Say, "Until my memory card is full! But don't worry, it's only a 64 megabyte (64 MB) card." If someone asks, tell them that every picture uses about 4 MB.

Discuss

- Discuss how many pictures can be stored on the card. Every time a picture is taken, the megabytes remaining decrease.

Lesson Notes

Example 1

- Discuss the use of a laptop computer and what happens when the computer is running off the battery versus using the AC adapter.
- Read the problem. Write what an ordered pair represents in words: (# of hours computer runs on battery, % of battery remaining as a decimal)

? "If you have just turned your fully charged computer on, how much battery power do you have?" 100%

? "What is the ordered pair associated with turning your computer on?" (0, 1)

- **Common Error:** Students may say (0, 100), but remind them that the percent needs to be in decimal form.
- Refer to the equation and ask about the slope and y-intercept.
- Plot the point for the y-intercept and use the slope to plot additional points on the graph.

? "The y-intercept means you just turned your computer on. What does the x-intercept mean?" The computer ran for 5 hours before the battery died.

? **MP1a Make Sense of Problems** and **MP2 Reason Abstractly and Quantitatively:** "What does a slope of 0.2 mean in the context of this problem?" If you lose 20% per hour, it will last 5 hours.

? **Big Idea:** Ask students why the graph is contained in Quadrant I only. It does not make sense for the battery power remaining to be greater than 1 (100%) or less than 0.

On Your Own

- This question is modeled after Example 1.

Goal Today's lesson is solving real-life problems.

Lesson Materials	
Introduction	**Textbook**
• digital camera • memory card	• straightedge

Start Thinking! and Warm Up

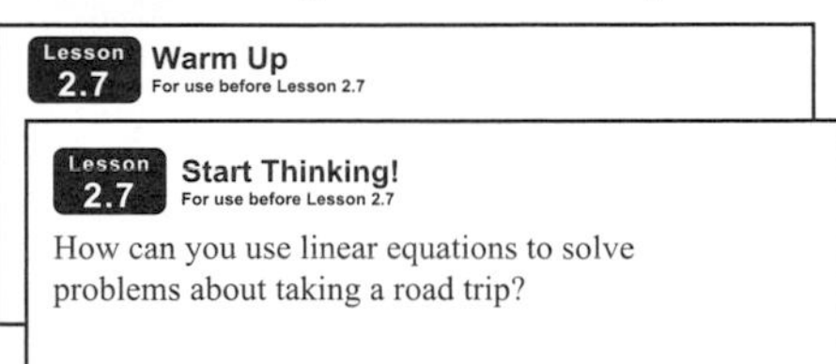

Lesson 2.7 Warm Up
For use before Lesson 2.7
Lesson 2.7 Start Thinking!
For use before Lesson 2.7
How can you use linear equations to solve problems about taking a road trip?

Extra Example 1

The percent y (in decimal form) of battery power remaining x hours after you turn on your handheld video game is $y = -0.3x + 1$.

a. Graph the equation.

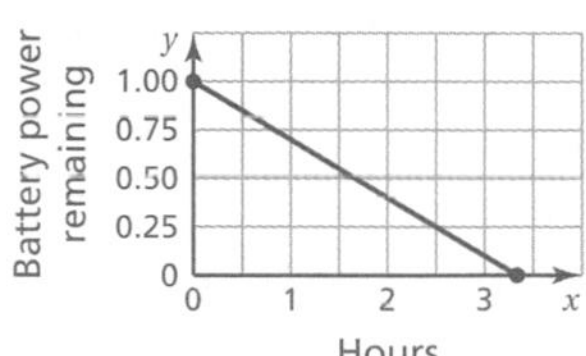

b. Interpret the x- and y-intercepts.

The x-intercept is $3\frac{1}{3}$. So, the battery lasts $3\frac{1}{3}$ hours. The y-intercept is 1. So, the battery power is at 100% when you turn it on.

c. After how many hours is the battery power at 40%? after 2 hours

On Your Own

1. See Additional Answers.

Laurie's Notes

Extra Example 2

The graph shows the cost y (in dollars) of a BMX (Bicycle Motocross) track membership and entry fees for x races at the track.

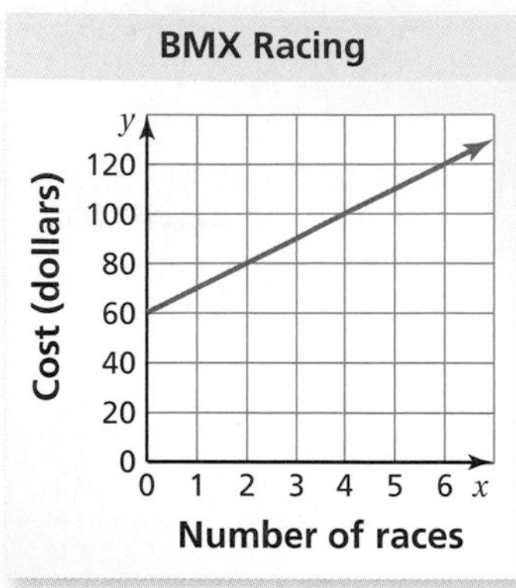

a. Find the slope and y-intercept. slope: 10; y-intercept: 60

b. Write an equation of the line. $y = 10x + 60$

c. How much does it cost to be a member and enter 4 races? $100

On Your Own

2. **a.** The slope is $\frac{3}{2}$. So, the flag is raised at a rate of $\frac{3}{2}$ feet per second.

   b. $y = \frac{3}{2}x + 3$

   c. 16.5 or $16\frac{1}{2}$ ft

Differentiated Instruction

Visual

Encourage students to write down notes when solving word problems, or to underline relevant information and cross out irrelevant information. Allow students to do this on handouts and tests.

Example 2

- Ask a student to read the problem. Write in words what the ordered pairs represent: (°C, °F)
- ? "Explain in words what the two ordered pairs represent." 0°C is the same as 32°F. 30°C is the same as 86°F.
- Draw a sketch of the graph. In order to find the slope, it is helpful for students to see the arrows representing the change in x (30) and the change in y (54).
- ? "The slope is $\frac{9}{5}$. What other information is needed to write the equation of this line?" y-intercept
- The y-intercept is shown in the graph. Write the equation.
- When students evaluate the equation for $C = 15$, they may make an error multiplying the fraction and whole number. Remind students of how multiplication of fractions is performed, and to divide out the common factor of 5 before multiplying.
- **Note:** Students have seen the conversion formula $F = \frac{9}{5}C + 32$ before. Because you want to reference the y-intercept, the equation is written in terms of x and y.
- **FYI:** This is a real-life application where it makes sense for the graph to be found in Quadrants I, II and III.
- ? "What do you think *mean temperature of Earth* means?" Answers will vary.

On Your Own

- In interpreting the slope, the graph is read from left to right, so every 2 seconds the flag's height increases 3 feet.
- ? "Why does it make sense that the y-intercept is not 0?" Flag's height does not start on the ground.

Closure

- Draw a graph of the memory card problem. Find and interpret the slope. slope $= -4$; Megabytes of free space decrease by 4 for each picture taken. Write the equation of the line. $y = -4x + 64$ Find and interpret the x- and y-intercepts. 16 and 64; When 16 pictures have been taken, the camera has 0 MB of free space. When 0 pictures have been taken, the camera has 64 MB of free space.

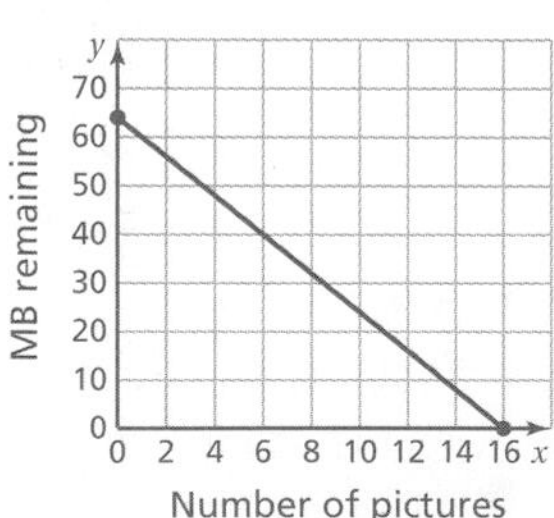

Technology For the Teacher

The Dynamic Planning Tool
Editable Teacher's Resources at *BigIdeasMath.com*

EXAMPLE 2 Real-Life Application

The graph relates temperatures y (in degrees Fahrenheit) to temperatures x (in degrees Celsius). (a) Find the slope and y-intercept. (b) Write an equation of the line. (c) What is the mean temperature of Earth in degrees Fahrenheit?

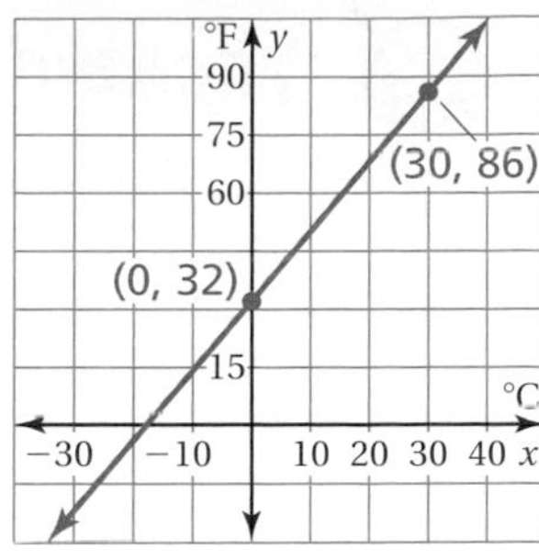

Mean Temperature: 15°C

a. slope $= \dfrac{\text{change in } y}{\text{change in } x} = \dfrac{54}{30} = \dfrac{9}{5}$

The line crosses the y-axis at (0, 32). So, the y-intercept is 32.

∴ The slope is $\frac{9}{5}$ and the y-intercept is 32.

b. Use the slope and y-intercept to write an equation.

slope → $\frac{9}{5}$; y-intercept → 32

∴ The equation is $y = \frac{9}{5}x + 32$.

c. In degrees Celsius, the mean temperature of Earth is 15°. To find the mean temperature in degrees Fahrenheit, find the value of y when $x = 15$.

$y = \frac{9}{5}x + 32$ Write the equation.

$= \frac{9}{5}(15) + 32$ Substitute 15 for x.

$= 59$ Simplify.

∴ The mean temperature of Earth is 59°F.

On Your Own

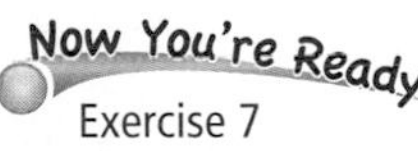

2. The graph shows the height y (in feet) of a flag x seconds after you start raising it up a flagpole.

   a. Find and interpret the slope.

   b. Write an equation of the line.

   c. What is the height of the flag after 9 seconds?

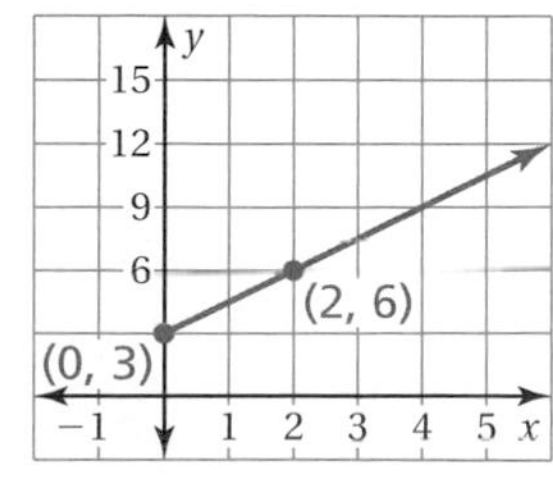

2.7 Exercises

Vocabulary and Concept Check

1. **REASONING** Explain how to find the slope, y-intercept, and x-intercept of the line shown.
2. **OPEN-ENDED** Describe a real-life situation that uses a negative slope.
3. **REASONING** In a real-life situation, what does the slope of a line represent?

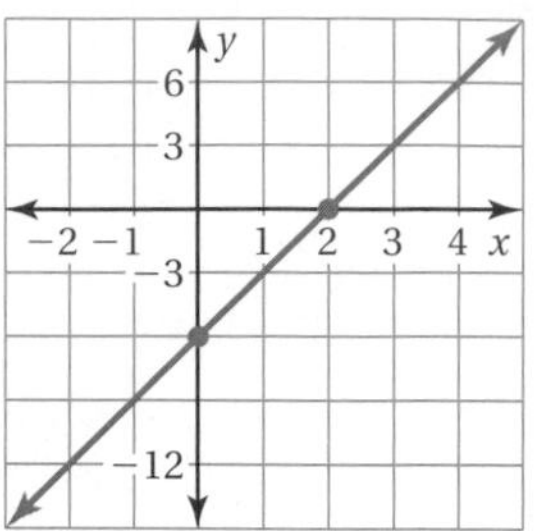

Practice and Problem Solving

Describe a real-life problem that has the given rate and intercepts. Draw a line that represents the problem.

4. Rate: -1.6 gallons per hour
   y-intercept: 16 gallons
   x-intercept: 10 hours

5. Rate: -3°F per hour
   y-intercept: 21°F
   x-intercept: 7 hours

1 6. **DOWNLOAD** You are downloading a song. The percent y (in decimal form) of megabytes remaining to download after x seconds is $y = -0.1x + 1$.

   a. Graph the equation.
   b. Interpret the x- and y-intercepts.
   c. After how many seconds is the download 50% complete?

2 7. **HIKING** The graph relates temperature y (in degrees Fahrenheit) to altitude x (in thousands of feet).

   a. Find the slope and y-intercept.
   b. Write an equation of the line.
   c. What is the temperature at sea level?

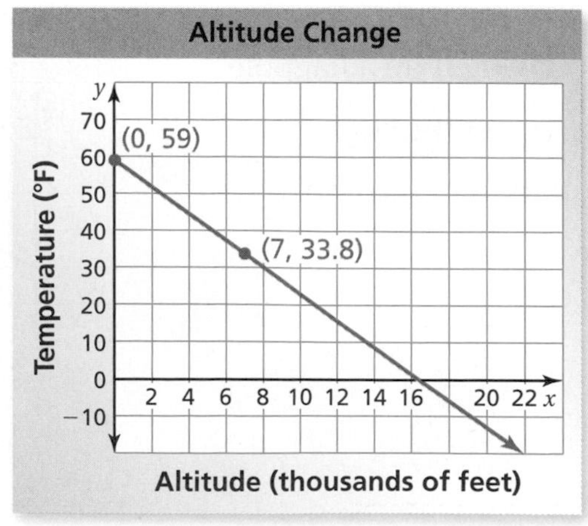

Assignment Guide and Homework Check

Level	Assignment	Homework Check
Average	1–9, 11–14	2, 3, 6, 7, 8
Advanced	1–3, 7–10, 11–14	2, 3, 7, 8

For Your Information

- **Exercise 8c** If you were to travel in a straight line, the speed would remain the same but the distance traveled would be less. So, it would take less time to make the trip.
- **Exercise 9** Ask students if their school lies on the line drawn between Denver and Beijing.

Common Errors

- **Exercise 6** Students may forget to convert the percent in part (c) to a decimal before substituting into the equation. Remind them that they need to convert percents to decimals before substituting.
- **Exercise 7** Students may struggle to find the change in *y* for the slope. Encourage them to focus on the *y*-coordinates and to write an expression that represents the change in temperature, 33.8 − 59, to help find the change in *y*.

2.7 Record and Practice Journal

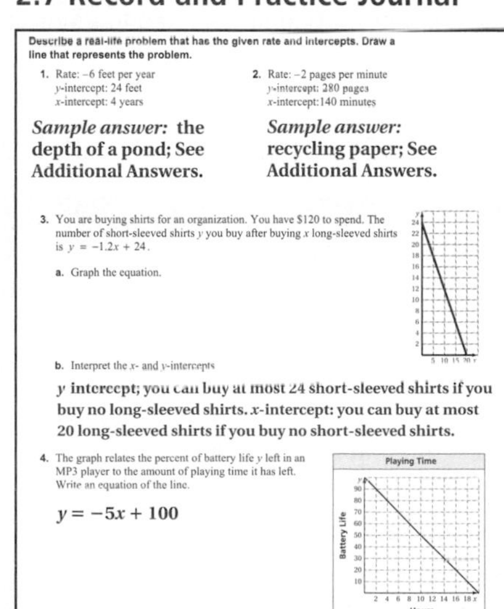

Describe a real-life problem that has the given rate and intercepts. Draw a line that represents the problem.

1. Rate: −6 feet per year
   y-intercept: 24 feet
   x-intercept: 4 years

   Sample answer: **the depth of a pond; See Additional Answers.**

2. Rate: −2 pages per minute
   y-intercept: 280 pages
   x-intercept: 140 minutes

   Sample answer: **recycling paper; See Additional Answers.**

3. You are buying shirts for an organization. You have $120 to spend. The number of short-sleeved shirts y you buy after buying x long-sleeved shirts is $y = -1.2x + 24$.

   a. Graph the equation.

   b. Interpret the x- and y-intercepts

   y intercept; you can buy at most 24 short-sleeved shirts if you buy no long-sleeved shirts. x-intercept: you can buy at most 20 long-sleeved shirts if you buy no short-sleeved shirts.

4. The graph relates the percent of battery life y left in an MP3 player to the amount of playing time it has left. Write an equation of the line.

   $y = -5x + 100$

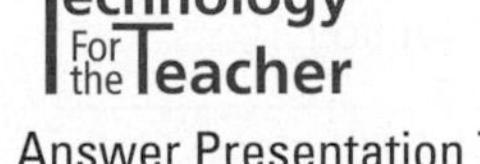

Vocabulary and Concept Check

1. The y-intercept is −6 because the line crosses the y-axis at the point (0, −6). The x-intercept is 2 because the line crosses the x-axis at the point (2, 0). You can use these two points to find the slope.

   $$\text{slope} = \frac{\text{change in } y}{\text{change in } x} = \frac{6}{2} = 3$$

2. *Sample answer:* a balloon descending toward the ground
3. *Sample answer:* the rate at which something is happening

Practice and Problem Solving

4–5. See Additional Answers.

6. a.

b. The x-intercept is 10. So, it takes 10 seconds to download the song. The y-intercept is 1. So, 100% of the song needs to be downloaded.

c. 5 sec

7. a. slope: −3.6
y-intercept: 59

b. $y = -3.6x + 59$

c. 59°F

Practice and Problem Solving

8. See Additional Answers.

9. a. Antananarivo: 19°S, 47°E
Denver: 39°N, 105°W
Brasilia: 16°S, 48°W
London: 51°N, 0°W
Beijing: 40°N, 116°E

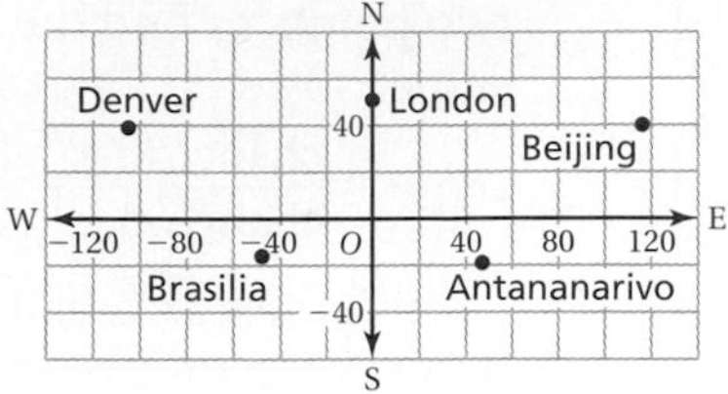

b. $y = \frac{1}{221}x + \frac{8724}{221}$

c. a place that is on the prime meridian

10. See *Taking Math Deeper*.

Fair Game Review

11. $h = \frac{5}{4}$

12. $k = 14$

13. $q = -2.3$

14. B

Mini-Assessment

1. You need $125 to buy an MP3 player. Your allowance per week is $5 and you earn $20 per lawn mowed.

a. Write an equation that represents your weekly income y (in dollars) for x lawns mowed. $y = 20x + 5$

b. Interpret the y-intercept. The y-intercept is 5. This is your allowance, the amount you started with before mowing lawns.

c. How many lawns do you need to mow to earn enough money in one week to buy the MP3 player? 6 lawns

Taking Math Deeper

Exercise 10

This is a classic "break-even" type of business problem. The band wants to invest $5000 in new equipment and is trying to project how many tickets need to be sold to pay for the equipment.

 Organize the given information.

R = band's revenue (income)
x = number of tickets sold
Income = $1500 + 30% of ticket sales
Price of each ticket = $20
Maximum capacity = 800

 Write an equation for the revenue.

$$R = 1500 + (30\% \text{ of } \$20 \text{ times } x)$$
$$= 1500 + 0.3(20)x$$
a. $$= 1500 + 6x$$

In other words, the band receives $6 per ticket. The organizers of the concert keep the remaining $14 to cover the expenses of auditorium rental, marketing, and salaries.

 Use the equation.
To find the number of tickets that need to be sold to earn a revenue of $5000, substitute 5000 for R and solve for x.

$$5000 = 1500 + 6x$$
$$3500 = 6x$$
$$583.3 \approx x$$

b. So, if the band can sell 584 tickets to the concert, it will earn enough to pay for the new equipment. The capacity of this auditorium is 800, so this is possible.

Project

Draw a poster that could be used to advertise a concert by your favorite band.

Reteaching and Enrichment Strategies

If students need help...	If students got it...
Resources by Chapter • Practice A and Practice B • Puzzle Time Record and Practice Journal Practice Differentiating the Lesson Lesson Tutorials Skills Review Handbook	Resources by Chapter • Enrichment and Extension • School-to-Work • Financial Literacy • Technology Connection • Life Connections Start the next section

8. **REASONING** Your family is driving from Cincinnati to St. Louis. The graph relates your distance from St. Louis y (in miles) and travel time x (in hours).

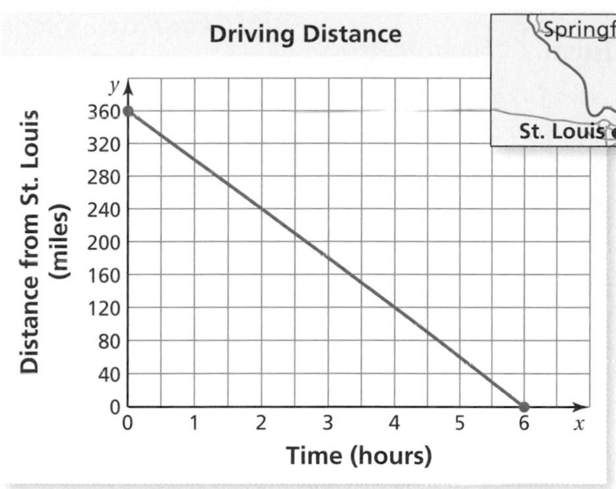

a. Interpret the x- and y-intercepts.

b. What is the slope? What does the slope represent in this situation?

c. Write an equation of the line. How would the graph and the equation change if you were able to travel in a straight line?

9. **PROJECT** Use a map or the Internet to find the latitude and longitude of your school to the nearest whole number. Then find the latitudes and longitudes of: Antananarivo, Madagascar; Denver, Colorado; Brasilia, Brazil; London, England; and Beijing, China.

a. Plot a point for each of the cities in the same coordinate plane. Let the positive y-axis represent north and the positive x-axis represent east.

b. Write an equation of the line that passes through Denver and Beijing.

c. In part (b), what geographic location does the y-intercept represent?

10. **Reasoning** A band is performing at an auditorium for a fee of \$1500. In addition to this fee, the band receives 30% of each \$20 ticket sold. The maximum capacity of the auditorium is 800 people.

a. Write an equation that represents the band's revenue R when x tickets are sold.

b. The band needs \$5000 for new equipment. How many tickets must be sold for the band to earn enough money to buy the new equipment?

Fair Game Review What you learned in previous grades & lessons

Solve the equation. Check your solution. *(Section 1.2)*

11. $-h - 7h + 13 = 3$

12. $4(k - 10) - 4 = 12$

13. $9 + 2.5(2q - 3) = -10$

14. **MULTIPLE CHOICE** Which equation is the slope-intercept form of $24x - 8y = 56$? *(Section 2.4)*

Ⓐ $y = -3x + 7$ Ⓑ $y = 3x - 7$ Ⓒ $y = -3x - 7$ Ⓓ $y = 3x + 7$

2.5–2.7 Quiz

Check It Out
Progress Check
BigIdeasMath.com

Write an equation of the line in slope-intercept form. *(Section 2.5)*

1.

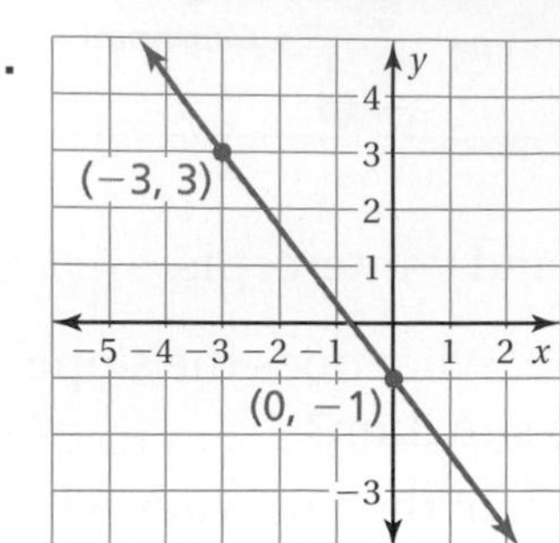

2.

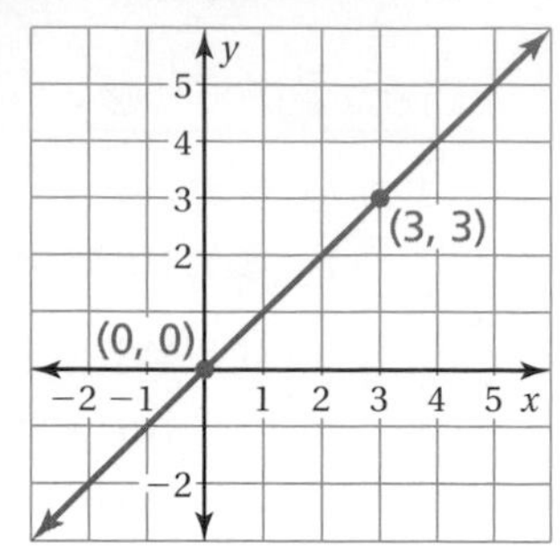

3.

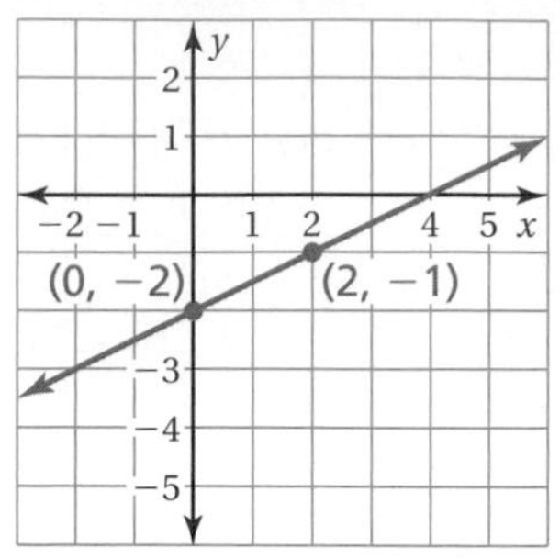

Write in point-slope form an equation of the line that passes through the given point and has the given slope. *(Section 2.6)*

4. $(1, 3)$; $m = 2$

5. $(-3, -2)$; $m = \frac{1}{3}$

6. $(-1, 4)$; $m = -1$

7. $(8, -5)$; $m = -\frac{1}{8}$

Write in slope-intercept form an equation of the line that passes through the given points. *(Section 2.6)*

8. $\left(0, -\frac{2}{3}\right)\left(-3, -\frac{2}{3}\right)$

9. $(4, 0)$, $(0, 4)$

10. Write an equation of the line that passes through $(2, -5)$ and is (a) parallel to and (b) perpendicular to the line $y = \frac{1}{3}x + 4$. *(Section 2.6)*

11. CONSTRUCTION A construction crew is extending a highway sound barrier that is 13 miles long. The crew builds $\frac{1}{2}$ mile per week. Write an equation for the length y (in miles) of the barrier after x weeks. *(Section 2.5)*

12. FISH POND You are draining a fish pond. The amount y (in liters) of water remaining after x hours is $y = -60x + 480$. (a) Graph the equation. (b) Interpret the x- and y-intercepts. *(Section 2.7)*

13. WATER A recreation department bought bottled water to sell at a fair. The graph shows the number y of bottles remaining after each hour x. *(Section 2.7)*

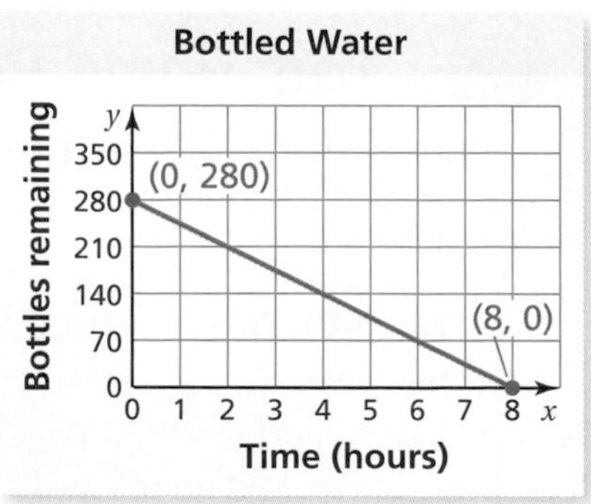

a. Find the slope and y-intercept.

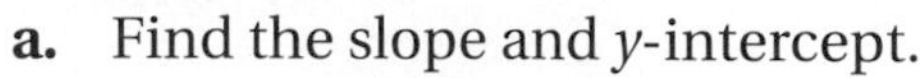

b. Write an equation of the line.

c. The fair started at 10 A.M. When did the recreation department run out of bottled water?

Alternative Assessment Options

Math Chat	Student Reflective Focus Question
Structured Interview	Writing Prompt

Math Chat

- Work in groups of four. Discuss the similarities and differences of writing equations of lines in slope-intercept form compared to writing equations of lines in point-slope form. When they are finished, they explain their findings to the other groups in the class.
- The teacher should walk around the classroom listening to the pairs and ask questions to ensure understanding.

Study Help Sample Answers

Remind students to complete Graphic Organizers for the rest of the chapter.

3.

	Example
Writing equations in slope-intercept form	Write an equation of the line in slope-intercept form. 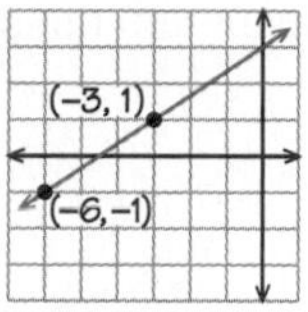
↓	↓
Find the slope *m*.	slope $= \frac{y_2 - y_1}{x_2 - x_1} = \frac{1-(-1)}{-3-(-6)} = \frac{2}{3}$
↓	↓
Find the y-intercept *b*.	Because the line crosses the y-axis at (0, 3), the y-intercept is 3.
↓	↓
Sustitute the slope and y-intercept into the form $y = mx + b$.	The equation is $y = \frac{2}{3}x + 3$.

4–6. Available at *BigIdeasMath.com*.

Reteaching and Enrichment Strategies

If students need help. . .	If students got it. . .
Resources by Chapter • Study Help • Practice A and Practice B • Puzzle Time Lesson Tutorials *BigIdeasMath.com* Practice Quiz Practice from the Test Generator	Resources by Chapter • Enrichment and Extension • School-to-Work Game Closet at *BigIdeasMath.com* Start the Chapter Review

Answers

1. $y = -\frac{4}{3}x - 1$
2. $y = x$
3. $y = \frac{1}{2}x - 2$
4. $y - 3 = 2(x - 1)$
5. $y + 2 = \frac{1}{3}(x + 3)$
6. $y - 4 = -1(x + 1)$
7. $y + 5 = -\frac{1}{8}(x - 8)$
8. $y = -\frac{2}{3}$
9. $y = -x + 4$
10. **a.** $y = \frac{1}{3}x - \frac{17}{3}$
    b. $y = -3x + 1$
11. $y = \frac{1}{2}x + 13$

12–13. See Additional Answers.

Technology For the Teacher
Answer Presentation Tool

Assessment Book

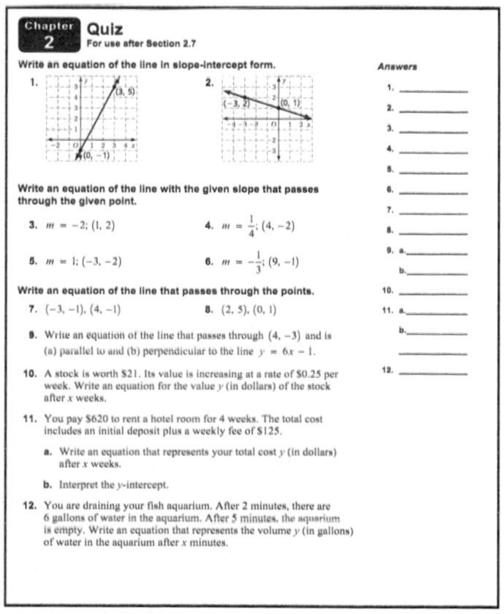

Chapter 2 Quiz
For use after Section 2.7

Write an equation of the line in slope-intercept form.

1. (3, 5), (0, −1)
2. (−3, 2), (0, 1)

Write an equation of the line with the given slope that passes through the given point.

3. $m = -2$; (1, 2)
4. $m = \frac{1}{4}$; (4, −2)
5. $m = 1$; (−3, −2)
6. $m = -\frac{1}{3}$; (9, −1)

Write an equation of the line that passes through the points.

7. (−3, −1), (4, −1)
8. (2, 5), (0, 1)
9. Write an equation of the line that passes through (4, −3) and is (a) parallel to and (b) perpendicular to the line $y = 6x - 1$.
10. A stock is worth $21. Its value is increasing at a rate of $0.25 per week. Write an equation for the value y (in dollars) of the stock after x weeks.
11. You pay $620 to rent a hotel room for 4 weeks. The total cost includes an initial deposit plus a weekly fee of $125.
    a. Write an equation that represents your total cost y (in dollars) after x weeks.
    b. Interpret the y-intercept.
12. You are draining your fish aquarium. After 2 minutes, there are 6 gallons of water in the aquarium. After 5 minutes, the aquarium is empty. Write an equation that represents the volume y (in gallons) of water in the aquarium after x minutes.

Answers
1. ______
2. ______
3. ______
4. ______
5. ______
6. ______
7. ______
8. ______
9. a. ______
b. ______
10. ______
11. a. ______
b. ______
______
12. ______

For the Teacher
Additional Review Options

- Big Ideas Test Generator
- Game Closet at *BigIdeasMath.com*
- Vocabulary Puzzle Builder
- Resources by Chapter
  - Puzzle Time
  - Study Help

Answers

1. 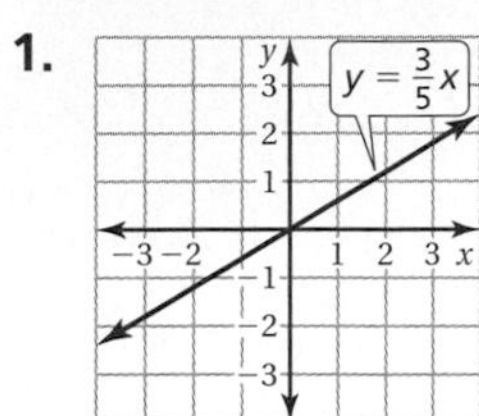

2. 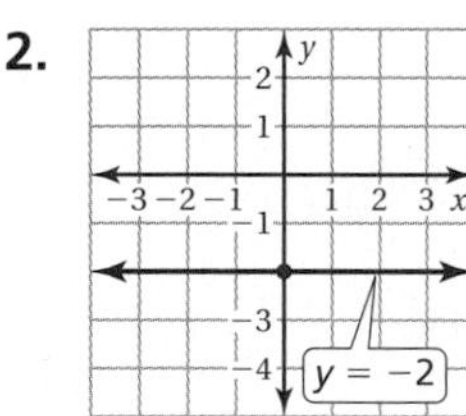

3. 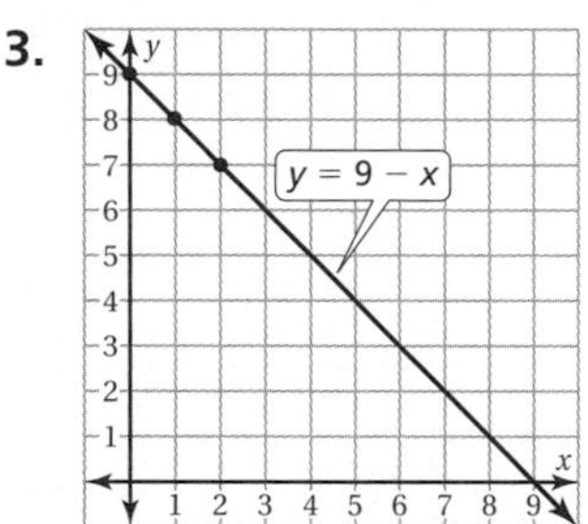

4. 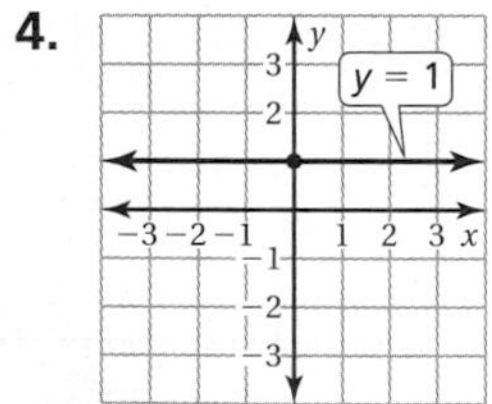

5. 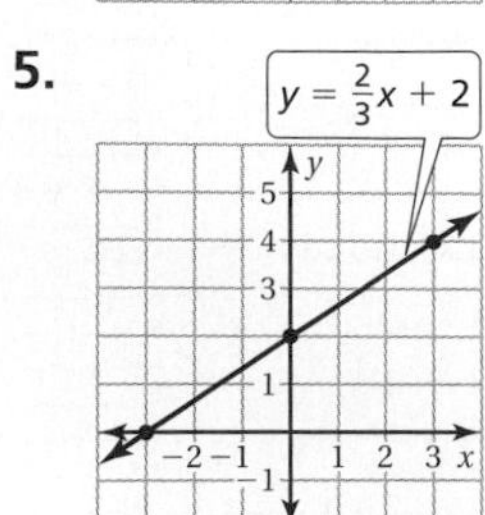

6. 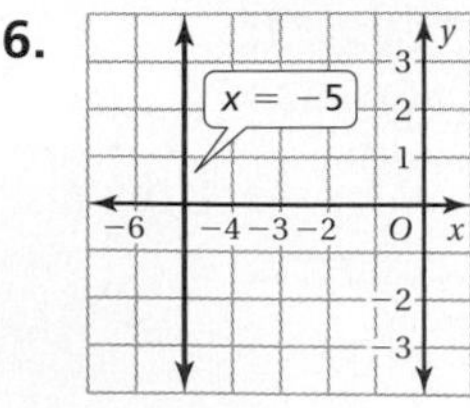

Review of Common Errors

Exercises 1–6

- Students may make a calculation error for one of the ordered pairs in a table of values. If they only find two ordered pairs for the graph, they may not recognize their mistake. Encourage them to find at least three ordered pairs when drawing a graph.

Exercises 2 and 4

- Students may draw a vertical line through a point on the *x*-axis instead of through the corresponding point on the *y*-axis. Remind them that the equation is a horizontal line. Ask them to identify the *y*-coordinate for several *x*-coordinates. For example, what is the *y*-coordinate for $x = 5$? $x = 6$? $x = -4$? Students should answer with the same *y*-coordinate each time.

Exercise 6

- Students may draw a horizontal line through a point on the *y*-axis instead of through the corresponding point on the *x*-axis. Remind them that the equation is a vertical line. Ask them to identify the *x*-coordinate for several *y*-coordinates. For example, what is the *x*-coordinate for $y = 2$? $y = -6$? $y = 0$? Students should answer with the same *x*-coordinate each time.

2 Chapter Review

Check It Out
Vocabulary Help
BigIdeasMath.com

Review Key Vocabulary

linear equation *p. 44*
solution of a linear equation, *p. 44*
slope, *p. 50*
rise, *p. 50*
run, *p. 50*
perpendicular lines, *p. 57*
x-intercept, *p. 60*
y-intercept, *p. 60*
slope-intercept form, *p. 60*
standard form, *p. 66*
point-slope form, *p. 80*

Review Examples and Exercises

2.1 Graphing Linear Equations *(pp. 42–47)*

Graph $y = 3x - 1$.

Step 1: Make a table of values.

x	$y = 3x - 1$	y	(x, y)
-2	$y = 3(-2) - 1$	-7	$(-2, -7)$
-1	$y = 3(-1) - 1$	-4	$(-1, -4)$
0	$y = 3(0) - 1$	-1	$(0, -1)$
1	$y = 3(1) - 1$	2	$(1, 2)$

Step 2: Plot the ordered pairs.

Step 3: Draw a line through the points.

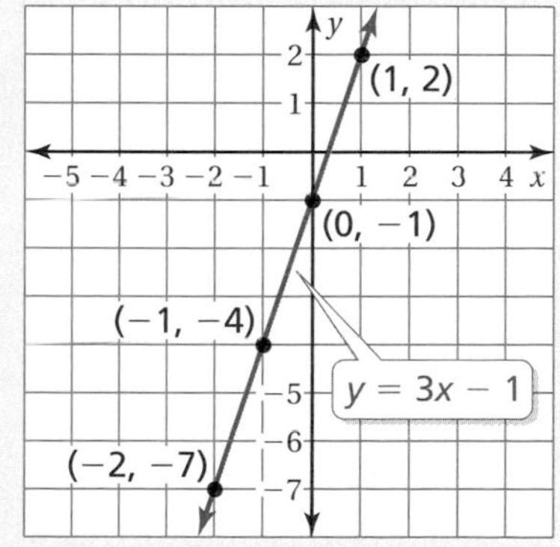

Exercises

Graph the linear equation.

1. $y = \frac{3}{5}x$

2. $y = -2$

3. $y = 9 - x$

4. $y = 1$

5. $y = \frac{2}{3}x + 2$

6. $x = -5$

2.2 Slope of a Line (pp. 48–57)

Find the slope of each line in the graph.

Red Line: slope $= \dfrac{y_2 - y_1}{x_2 - x_1} = \dfrac{5 - (-3)}{2 - 2} = \dfrac{8}{0}$

∴ The slope of the red line is undefined.

Blue Line: slope $= \dfrac{y_2 - y_1}{x_2 - x_1} = \dfrac{-1 - 2}{4 - (-3)} = \dfrac{-3}{7}$, or $-\dfrac{3}{7}$

Green Line: slope $= \dfrac{y_2 - y_1}{x_2 - x_1} = \dfrac{4 - 4}{5 - 0} = \dfrac{0}{5}$, or 0

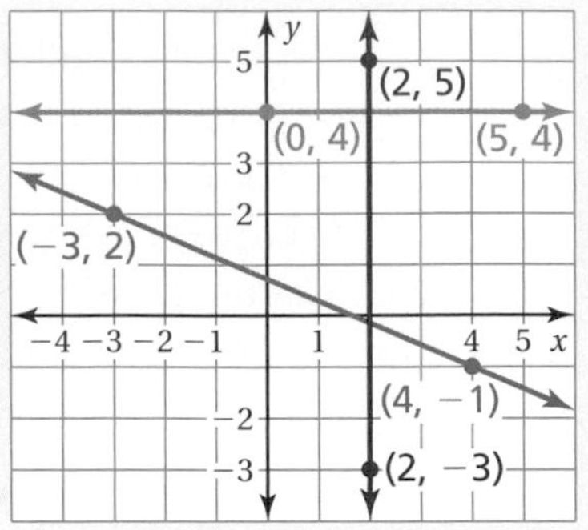

Exercises

The points in the table lie on a line. How can you find the slope of the line from the table? What is the slope?

7.

x	0	1	2	3
y	−1	0	1	2

8.

x	−2	0	2	4
y	3	4	5	6

9. Are the lines $x = 2$ and $y = 4$ parallel? Are they perpendicular? Explain.

2.3 Graphing Linear Equations in Slope-Intercept Form (pp. 58–63)

Graph $y = 0.5x - 3$. Identify the x-intercept.

Step 1: Find the slope and y-intercept.

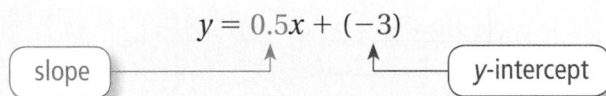

$$y = 0.5x + (-3)$$

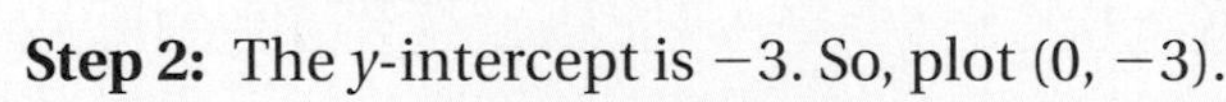

Step 2: The y-intercept is −3. So, plot (0, −3).

Step 3: Use the slope to find another point and draw the line.

$$\text{slope} = \frac{\text{rise}}{\text{run}} = \frac{1}{2}$$

Plot the point that is 2 units right and 1 unit up from (0, –3). Draw a line through the two points.

∴ The line crosses the x-axis at (6, 0). So, the x-intercept is 6.

Exercises

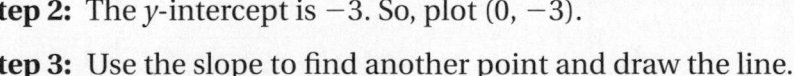

Graph the linear equation. Identify the x-intercept. Use a graphing calculator to check your answer.

10. $y = 2x - 6$ **11.** $y = -4x + 8$ **12.** $y = -x - 8$

Review of Common Errors (continued)

Exercises 7 and 8

- Students may find the reciprocal of the slope instead of the slope. Remind them that slope is change in *y* over change in *x*.

Exercises 10–12

- Students may forget to include negatives with the slope and/or *y*-intercept. Remind them to look at the sign in front of the slope and the *y*-intercept. Also remind students that the equation is $y = mx + b$. This means that if the linear equation has "minus *b*," then the *y*-intercept is negative.
- Students may use the reciprocal of the slope when graphing and may find an incorrect *x*-intercept. Remind them that slope is *rise* over *run*, so the numerator represents vertical change, not horizontal.

Answers

7. Choose any two points from the table and use the slope formula.

slope = 1

8. Choose any two points from the table and use the slope formula.

slope $= \dfrac{1}{2}$

9. no; yes; The line $x = 2$ is vertical. The line $y = 4$ is horizontal. A vertical line is perpendicular to a horizontal line.

10. 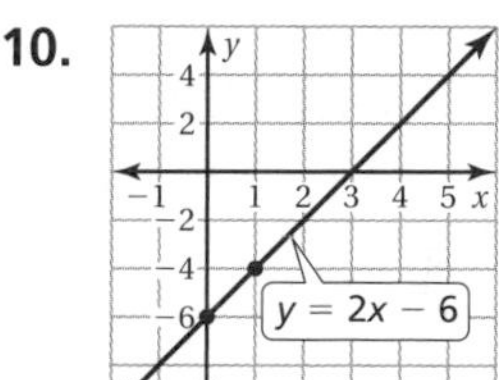

x-intercept: 3

11. 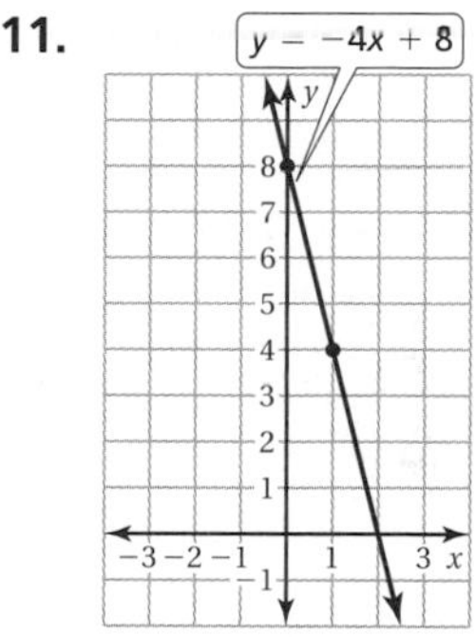

x-intercept: 2

12. 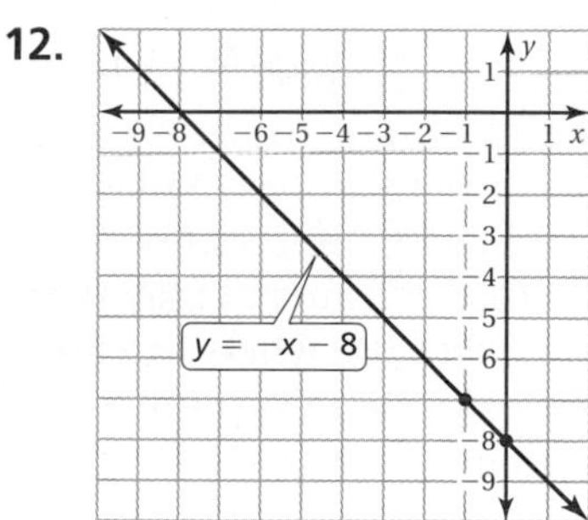

x-intercept: −8

Answers

13. 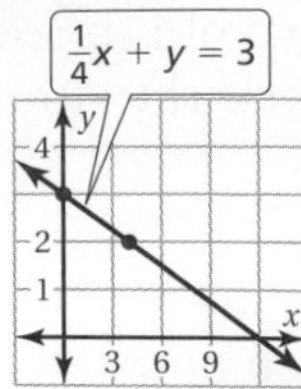

14. 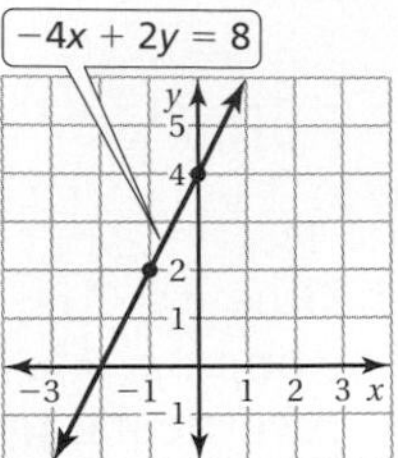

15. 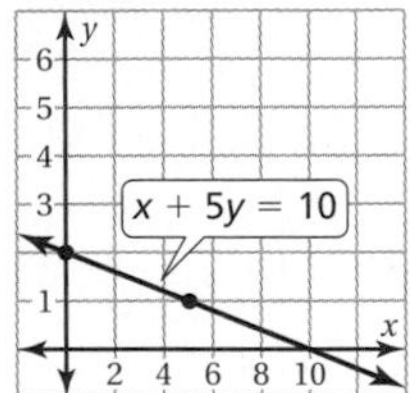

16. 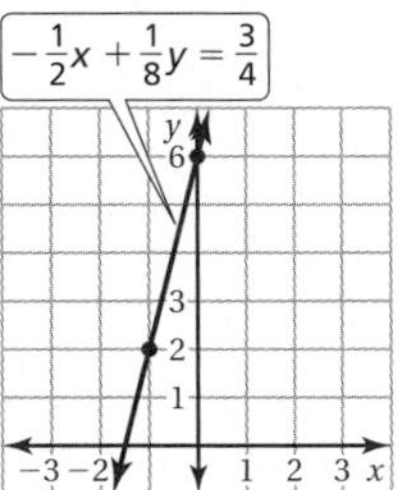

17. 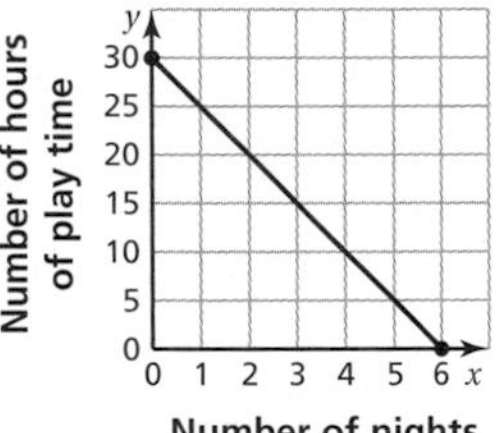

The x-intercept is the number of nights that cost \$180 when there are no hours of play time. The y-intercept is the number of hours of play time that cost \$180 when you do not leave your dog at the kennel for any nights.

Review of Common Errors (continued)

Exercises 13–17

- Students may use the same operation instead of the inverse operation when rewriting the equation in slope-intercept form. Remind them of the steps to rewrite an equation.
- Students may mix up the x- and y-intercepts. Remind them that the x-intercept is the x-coordinate of where the line crosses the x-axis and the y-intercept is the y-coordinate of where the line crosses the y-axis.

Exercises 18 and 19

- Students may write the reciprocal of the slope or forget a negative sign. Remind them of the definition of slope. Ask them to predict the sign of the slope based on the rise or fall of the line.

Exercises 20 and 21

- Students may write the wrong equation when the slope is zero. For example, instead of $y = 5$, students may write $x = 5$. Ask them what is the rise of the graph (zero) and write this in slope-intercept form with the y-intercept as well, such as $y = 0x + 5$. Then ask students what happens when a variable (or any number) is multiplied by zero. Rewrite the equation as $y = 5$.

Exercises 22 and 23

- Students may use the reciprocal of the slope when writing the equation. Remind them that slope is the change in y over the change in x.

Exercise 24

- Students may use the reciprocal of the slope when graphing and may find an incorrect x-intercept. Remind them that slope is *rise* over *run*, so the numerator represents vertical change, not horizontal.

2.4 Graphing Linear Equations in Standard Form *(pp. 64–69)*

Graph $8x + 4y = 16$.

Step 1: Write the equation in slope-intercept form.

$8x + 4y = 16$	Write the equation.
$4y = -8x + 16$	Subtract $8x$ from each side.
$y = -2x + 4$	Divide each side by 4.

Step 2: Use the slope and y-intercept to plot two points.

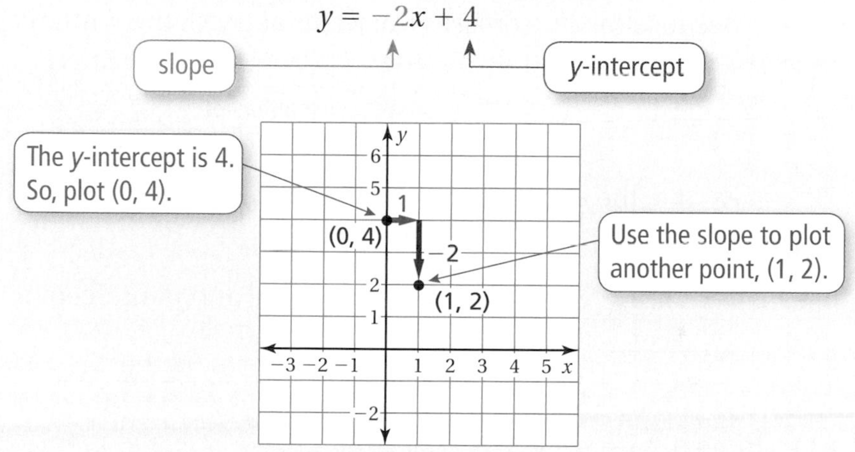

Step 3: Draw a line through the points.

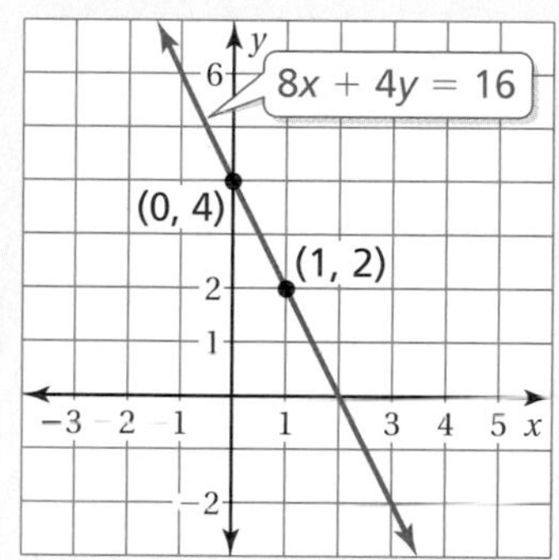

Exercises

Graph the linear equation.

13. $\frac{1}{4}x + y = 3$

14. $-4x + 2y = 8$

15. $x + 5y = 10$

16. $-\frac{1}{2}x + \frac{1}{8}y = \frac{3}{4}$

17. A dog kennel charges \$30 per night to board your dog and \$6 for each hour of play time. The amount of money you spend is given by $30x + 6y = 180$, where x is the number of nights and y is the number of hours of play time. Graph the equation and interpret the intercepts.

2.5 Writing Equations in Slope-Intercept Form (pp. 72–77)

Write an equation of the line in slope-intercept form.

a.

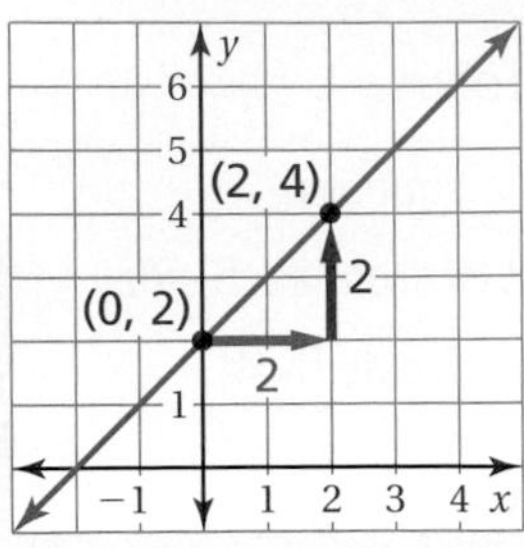

Find the slope and y-intercept.

$$\text{slope} = \frac{y_2 - y_1}{x_2 - x_1} = \frac{4 - 2}{2 - 0} = \frac{2}{2}, \text{ or } 1$$

Because the line crosses the y-axis at (0, 2), the y-intercept is 2.

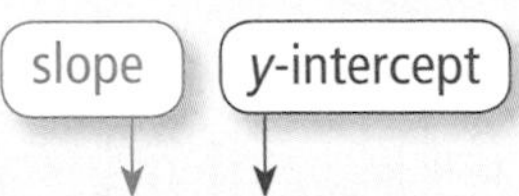

∴ So, the equation is $y = 1x + 2$, or $y = x + 2$.

b.

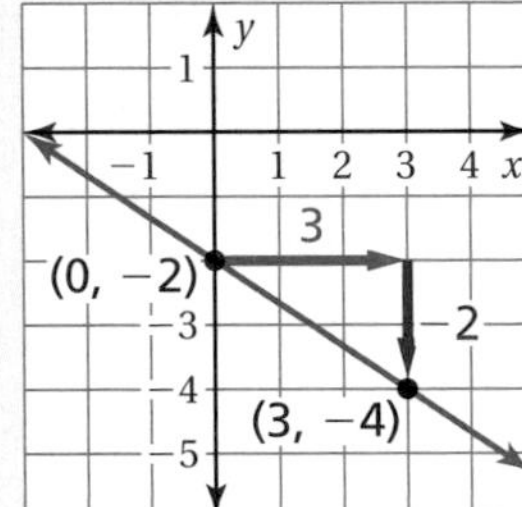

Find the slope and y-intercept.

$$\text{slope} = \frac{y_2 - y_1}{x_2 - x_1} = \frac{-4 - (-2)}{3 - 0} = \frac{-2}{3}, \text{ or } -\frac{2}{3}$$

Because the line crosses the y-axis at (0, −2), the y-intercept is −2.

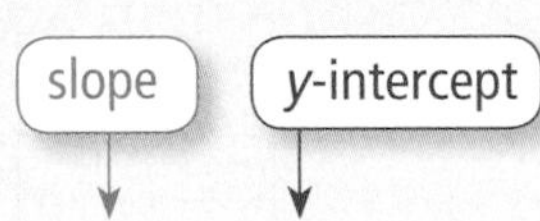

∴ So, the equation is $y = -\frac{2}{3}x + (-2)$, or $y = -\frac{2}{3}x - 2$.

Exercises

Write an equation of the line in slope-intercept form.

18.

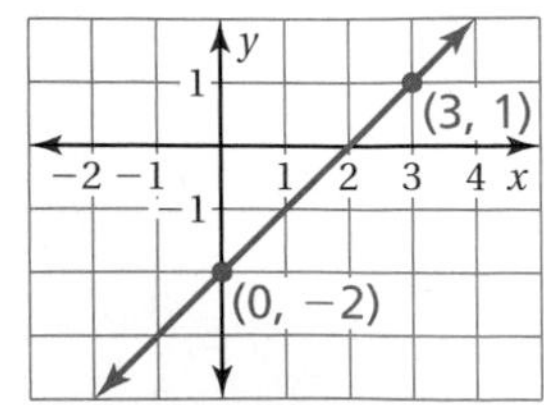

19.

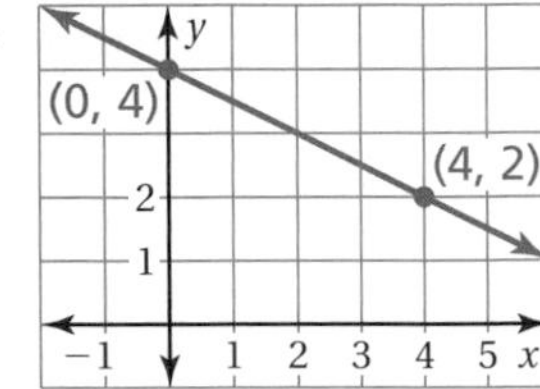

20. Write an equation of the line that passes through (0, 8) and (6, 8).

21. Write an equation of the line that passes through (0, −5) and (−5, −5).

Review Game

Graphing Linear Equations

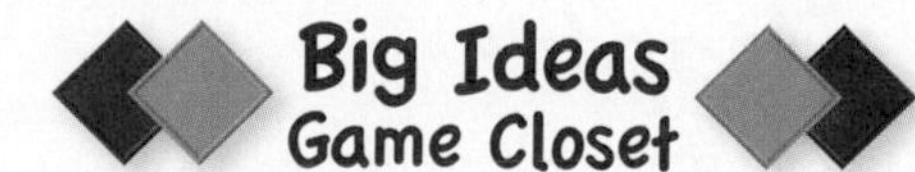

Materials per Group:

- map of the United States
- pencil
- straightedge

Directions:

On a map of the United States, students will place a coordinate plane with the origin located at Wichita, Kansas. The x-axis will go from -1700 miles to 1700 miles and the y-axis will go from -625 miles to 625 miles. These are roughly the dimensions of the United States.

The teacher will write equations and cities, in jumbled order, on the board. Students will work in groups and graph the equations to determine which line goes through which city.

Examples:

Dallas	$y = \frac{156}{50}x - 156$
Denver	$y = \frac{312}{625}x + 312$
Orlando	$y = \frac{100}{200}x + 100$
Chicago	$y = \frac{280}{600}x$
Las Vegas	$y = \frac{625}{1275}x - 625$

Who Wins?

The first group to correctly graph the lines and match the cities wins.

For the Student
Additional Practice

- Lesson Tutorials
- Study Help (textbook)
- Student Website
  - Multi-Language Glossary
  - Practice Assessments

Answers

18. $y = x - 2$

19. $y = -\frac{1}{2}x + 4$

20. $y = 8$

21. $y = -5$

22. $y - 4 = 3(x - 4)$

23. $y = -\frac{1}{2}x$

24. **a.** 

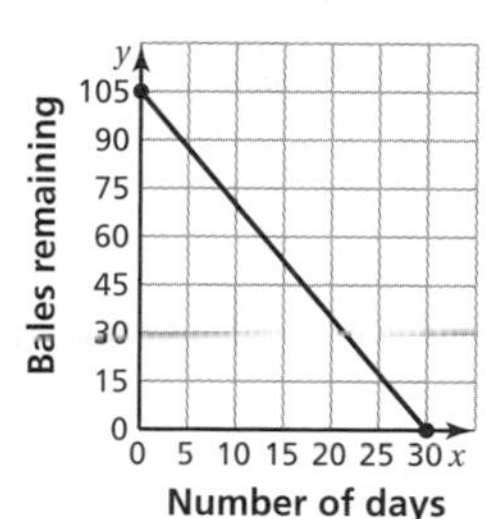

b. The x-intercept is the number of days it takes to feed all the hay to the cows. The y-intercept represents how many bales of hay there were originally.

c. 70 bales

My Thoughts on the Chapter

What worked. . .

Teacher Tip

Not allowed to write in your teaching edition? Use sticky notes to record your thoughts.

What did not work. . .

What I would do differently. . .

2.6 Writing Equations in Point-Slope Form (pp. 78–85)

Write in slope-intercept form an equation of the line that passes through the points (2, 1) and (3, 5).

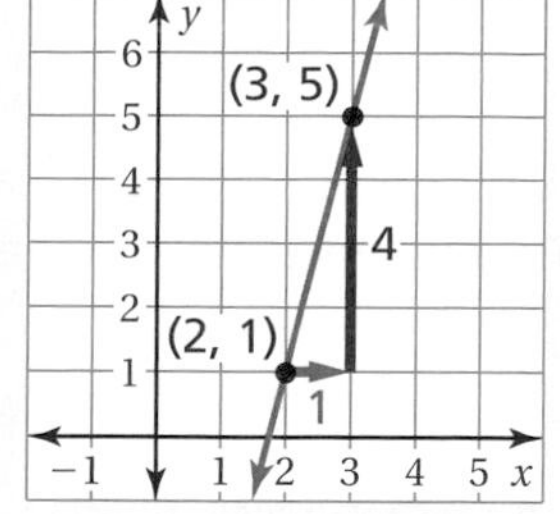

Find the slope.

$$m = \frac{y_2 - y_1}{x_2 - x_1} = \frac{5 - 1}{3 - 2} = \frac{4}{1}, \text{ or } 4$$

Then use the slope and one of the given points to write an equation of the line.

Use $m = 4$ and (2, 1).

$y - y_1 = m(x - x_1)$	Write the point-slope form.
$y - 1 = 4(x - 2)$	Substitute 4 for m, 2 for x_1, and 1 for y_1.
$y - 1 = 4x - 8$	Use Distributive Property.
$y = 4x - 7$	Write in slope-intercept form.

∴ So, the equation is $y = 4x - 7$.

Exercises

22. Write in point-slope form an equation of the line that passes through the point (4, 4) with slope 3.

23. Write in slope-intercept form an equation of the line that passes through the points (−4, 2) and (6, −3).

2.7 Solving Real-Life Problems (pp. 86–91)

The amount y (in dollars) of money you have left after playing x games at a carnival is $y = -0.75x + 10$. How much money do you have after playing eight games?

$y = -0.75x + 10$	Write the equation.
$= -0.75(8) + 10$	Substitute 8 for x.
$= 4$	Simplify.

∴ You have $4 left after playing 8 games.

Exercises

24. HAY The amount y (in bales) of hay remaining after feeding cows for x days is $y = -3.5x + 105$. (a) Graph the equation. (b) Interpret the x- and y-intercepts. (c) How many bales are left after 10 days?

2 Chapter Test

Find the slope and y-intercept of the graph of the linear equation.

1. $y = 6x - 5$

2. $y = 20x + 15$

3. $y = -5x - 16$

4. $y - 1 = 3x + 8.4$

5. $y + 4.3 = 0.1x$

6. $-\frac{1}{2}x + 2y = 7$

Graph the linear equation.

7. $y = 2x + 4$

8. $y = -\frac{1}{2}x - 5$

9. $-3x + 6y = 12$

10. Which lines are parallel? Which lines are perpendicular? Explain.

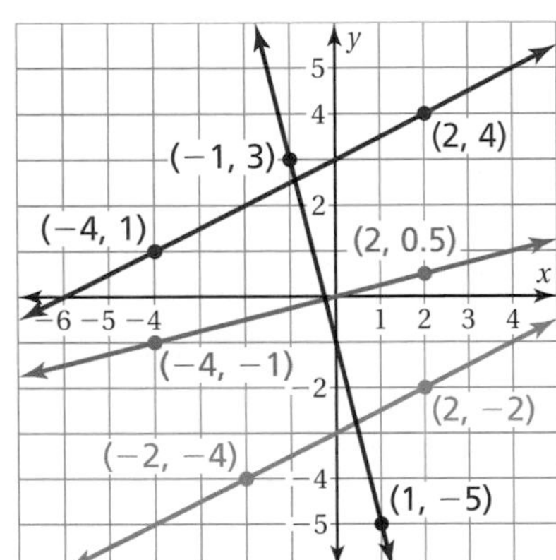

11. The points in the table lie on a line. How can you find the slope of the line from the table? What is the slope?

x	y
−1	−4
0	−1
1	2
2	5

Write an equation of the line in slope-intercept form.

12.

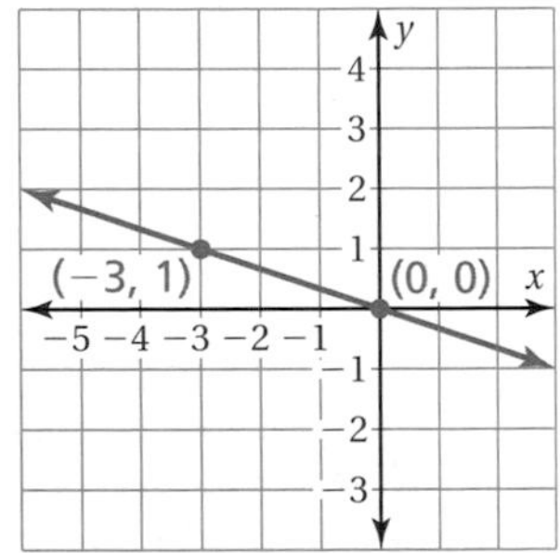

13.

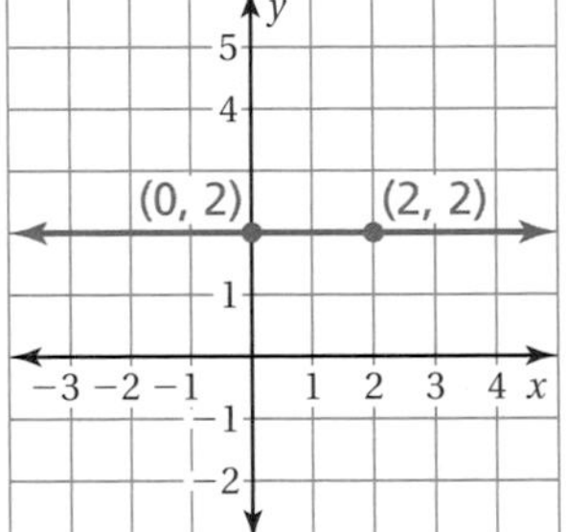

Write in slope-intercept form an equation of the line that passes through the given points.

14. $(-1, 5), (3, -3)$

15. $(-4, 1), (4, 3)$

16. $(-2, 5), (-1, 1)$

17. BRAILLE Because of its size and detail, Braille takes longer to read than text. A person reading Braille reads at 25% the rate of a person reading text.

a. Write and graph an equation that represents the average rate y of a Braille reader in terms of the average rate x of a text reader.

b. Interpret the solution (180, 45).

c. What happens to y as x increases? Explain.

Test Item References

Chapter Test Questions	Section to Review	Common Core State Standards
7, 8	2.1	A.CED.2, A.REI.10
10, 11	2.2	F.IF.4, F.IF.6
1–6	2.3	A.CED.2, A.REI.10, F.IF.4
9	2.4	A.CED.2, A.REI.10, F.IF.4
12, 13	2.5	8.F.3, A.CED.2, A.CED.3
14–16	2.6	A.CED.2, A.REI.10, F.IF.4, F.IF.6
17	2.7	8.F.4, A.CED.2, F.IF.4

Test-Taking Strategies

Remind students to quickly look over the entire test before they start so that they can budget their time. Students should jot down the formulas for slope-intercept form and point-slope form on the back of their test before they begin. Teach students to use the Stop and Think strategy before answering. **Stop** and carefully read the question, and **Think** about what the answer should look like.

Common Assessment Errors

- **Exercises 1–6** Students may use the reciprocal of the slope when graphing and may find an incorrect *x*-intercept. Remind them that slope is *rise* over *run*, so the numerator represents vertical change, not horizontal.
- **Exercises 7–9** Students may make a calculation error for one of the ordered pairs in a table of values. If they only find two ordered pairs for the graph, they may not recognize their mistake. Encourage them to find at least three ordered pairs when drawing a graph.
- **Exercises 12 and 13** Students may write the reciprocal of the slope and forget a negative sign. Ask them to predict the sign of the slope based on the rise or fall of the line.
- **Exercise 14–16** Students may use the reciprocal of the slope when writing the equation. Remind them that slope is the change in *y* over the change in *x*.
- **Exercise 17** Students may have difficulty using the given information to come up with an equation. Point out that the two rates are proportional.

Reteaching and Enrichment Strategies

If students need help. . .	If students got it. . .
Resources by Chapter • Practice A and Practice B • Puzzle Time Record and Practice Journal Practice Differentiating the Lesson Lesson Tutorials Practice from the Test Generator Skills Review Handbook	Resources by Chapter • Enrichment and Extension • School-to-Work • Financial Literacy • Technology Connection • Life Connections Game Closet at *BigIdeasMath.com* Start Standardized Test Practice

Answers

1. slope: 6; y-intercept: -5
2. slope: 20; y-intercept: 15
3. slope: -5; y-intercept: -16
4. slope: 3; y-intercept: 9.4
5. slope: 0.1; y-intercept: -4.3
6. slope: $\frac{1}{4}$; y-intercept: $\frac{7}{2}$

7–9. See Additional Answers.

10. The red and green lines are parallel. They both have a slope of $\frac{1}{2}$. The black and blue lines are perpendicular. The product of their slopes is -1.
11. Choose any two points from the table and use the slope formula.
    slope $= 3$
12. $y = -\frac{1}{3}x$
13. $y = 2$
14. $y = -4x - 3$
15. $y = \frac{1}{4}x + 2$
16. $y = -2x + 3$
17. See Additional Answers.

Assessment Book

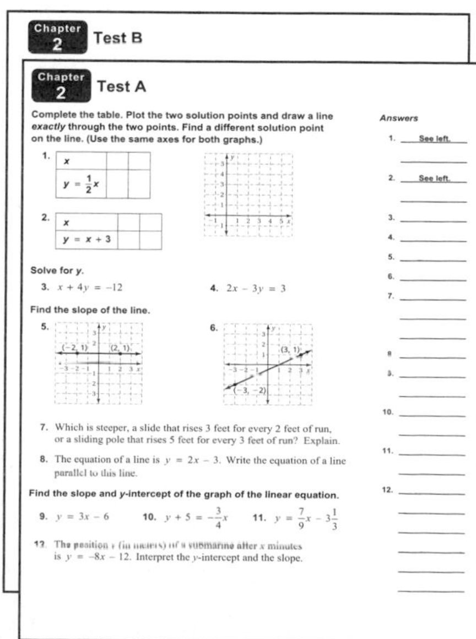

Test-Taking Strategies
Available at *BigIdeasMath.com*

After Answering Easy Questions, Relax
Answer Easy Questions First
Estimate the Answer
Read All Choices before Answering
Read Question before Answering
Solve Directly or Eliminate Choices
Solve Problem before Looking at Choices
Use Intelligent Guessing
Work Backwards

About this Strategy

When taking a multiple choice test, be sure to read each question carefully and thoroughly. After reading the question, estimate the answer before trying to solve.

Answers

1. A
2. H
3. A
4. H

Technology For the Teacher

Big Ideas Test Generator

Item Analysis

1. **A.** Correct answer
   B. The student reads the slope correctly, but uses the wrong point to identify the *y*-intercept.
   C. The student reads the *y*-intercept correctly, but miscalculates the slope.
   D. The student finds the slope and *y*-intercept incorrectly.

2. **F.** The student does not recognize the slope of 4 as $\frac{4}{1}$.
   G. The student recognizes that a slope of $\frac{4}{1}$ is equivalent to a slope of $\frac{8}{2}$, but the student incorrectly multiplies instead of adding.
   H. Correct answer
   I. The student incorrectly adds 1 and 4 to the coefficients of *a* and *b* respectively instead of adding 1 and 4 directly to *a* and *b*.

3. **A.** Correct answer
   B. The student accounts for the variable term correctly, but fails to realize that the starting point is three years prior to the point at which the value is 21,000.
   C. The student accounts for the variable term correctly, but is confused by the roles of 21,000 and three years time. The student decides that subtracting them makes sense because the car is depreciating.
   D. The student does not understand how to write the variable term or the constant term.

4. **F.** The student interchanges correct values for *x* and *y*.
   G. The student makes two errors: interchanging *x* and *y*, and assigning a negative sign incorrectly.
   H. Correct answer
   I. The student makes a mistake with a correct solution (4, 2) and assigns a negative value to 2, forgetting that there is already a minus sign in the equation.

2 Standardized Test Practice

Test-Taking Strategy
Estimate the Answer
In x days, your owner vacuums y = 107x hairs from the rug. How many in 3 days?
A 201 B 539 C 321 D 1,000,000
So, you prefer a hairless cat?

"Using estimation you can see that there are about 300 hairs. So, it's got to be C."

1. Which equation matches the line shown in the graph? *(A.REI.10)*

   A. $y = 2x - 2$

   B. $y = 2x + 1$

   C. $y = x - 2$

   D. $y = x + 1$

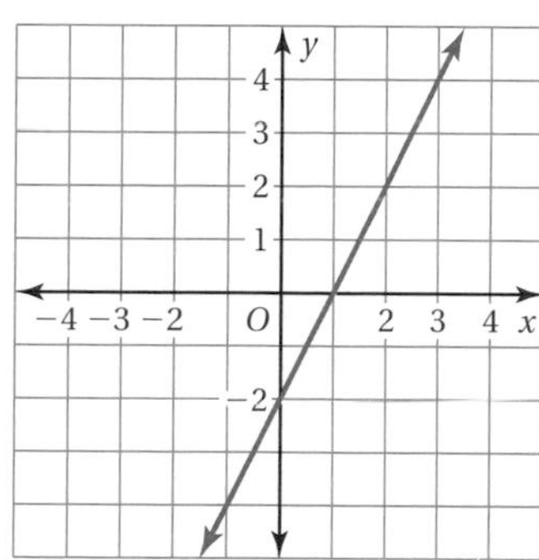

2. A line has a slope of 4 and passes through the point (a, b). Which point must also lie on this line? *(F.IF.6)*

   F. $(a, b + 4)$

   G. $(2a, 8b)$

   H. $(a + 1, b + 4)$

   I. $(2a, 5b)$

3. A car's value depreciates at a rate of \$2,500 per year. Three years after it was purchased, the car's value was \$21,000. Which equation can be used to find v, its value in dollars, n years after it was purchased? *(A.CED.2)*

   A. $v = 28{,}500 - 2{,}500n$

   B. $v = 21{,}000 - 2{,}500n$

   C. $v = 18{,}500 - 2{,}500n$

   D. $v = 18{,}500 - n$

4. The equation $6x - 5y = 14$ is written in standard form. Which point lies on the graph of this equation? *(A.REI.10)*

   F. $(-4, -1)$

   G. $(-2, 4)$

   H. $(-1, -4)$

   I. $(4, -2)$

5. The line shown in the graph below has a slope of -3. What is the equation of the line? *(A.REI.10)*

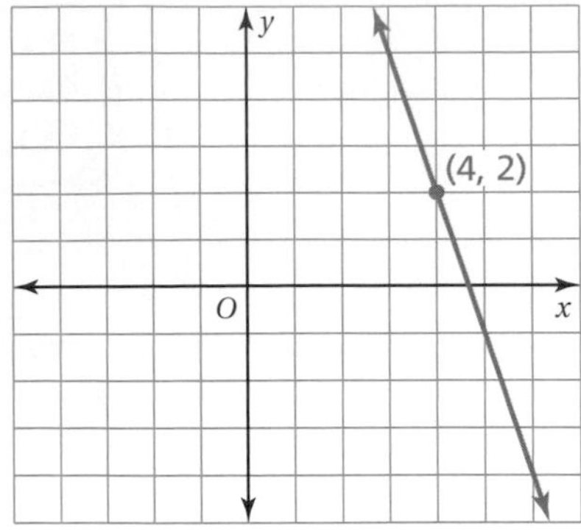

A. $y = 3x - 10$

B. $y = -3x + 10$

C. $y = -3x + 14$

D. $y = -3x - 14$

6. A cell phone plan costs \$10 per month plus \$0.10 for each minute used. Last month, you spent \$18.50 using this plan. This can be modeled by the equation below, where m represents the number of minutes used.

$$0.1m + 10 = 18.5$$

How many minutes did you use last month? *(A.REI.3)*

F. 8.4 min

G. 85 min

H. 185 min

I. 285 min

7. What is the slope of the line that passes through the points $(2, -2)$ and $(8, 1)$? *(F.IF.6)*

8. It costs \$40 to rent a car for one day. In addition, the rental agency charges you for each mile driven, as shown in the graph. *(F.IF.6)*

Part A Determine the slope of the line joining the points on the graph.

Part B Explain what the slope represents.

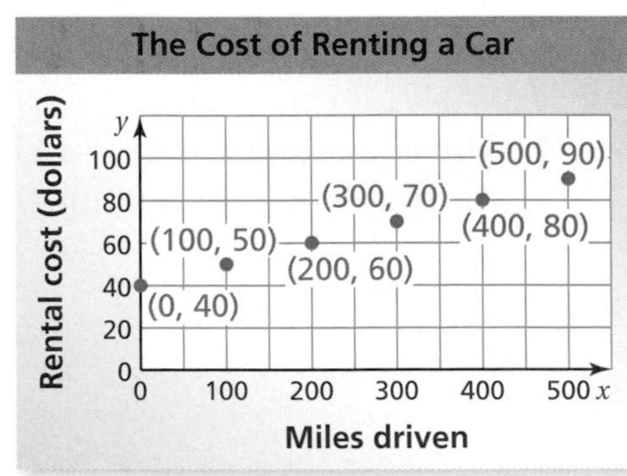

Item Analysis (continued)

5. **A.** The student uses the slope of 3 instead of -3.

   B. The student interchanges the coordinates when finding the equation.

   C. Correct answer

   D. The student works correctly, but uses the wrong sign on the term 14.

6. **F.** The student correctly subtracts 10 from both sides, but then subtracts 0.1 instead of dividing.

   G. Correct answer

   H. The student ignores 10 and simply divides by 0.1.

   I. The student adds 10 to both sides, and then divides by 0.1.

7. **Gridded Response:** Correct answer: 0.5, or $\frac{1}{2}$

   Common Error: The student performs subtraction incorrectly for the y-terms, yielding an answer of $\frac{1}{6}$ or $-\frac{1}{6}$.

8. **2 points** The student demonstrates a thorough understanding of the slope of a line and what it represents, explains the work fully, and calculates the slope accurately. The slope of the line is $\frac{50-40}{100-0} = \frac{10}{100} = \frac{1}{10} = 0.10$. The slope represents the rental cost per mile driven, \$0.10 per mile.

   1 point The student's work and explanations demonstrate a lack of essential understanding. The formula for the slope of a line is misstated, or the student incorrectly states what the slope of the line represents.

   0 points The student provides no response, a completely incorrect or incomprehensible response, or a response that demonstrates insufficient understanding of the slope of a line and what it represents.

Answers

5. C

6. G

7. $\frac{1}{2}$

8. *Part A* 0.10

   Part B \$0.10 per mile

Answers

9. D
10. I
11. B
12. H
13. A

Answer for Extra Example

1. **Gridded Response:**
   Correct answer: 1.25

   Common Error: The student does not properly distribute the factor 3 in the right hand side of the equation, yielding an incorrect answer of $x = 1$.

Item Analysis (continued)

9. **A.** The student correctly subtracts 7 from each side, but incorrectly adds $4x$ to $2x$ instead of subtracting.

B. The student correctly subtracts $2x$ from $4x$, but incorrectly subtracts 5 from each side instead of adding.

C. The student correctly adds 5 to each side, but incorrectly adds $2x$ to $4x$ instead of subtracting.

D. Correct answer

10. **F.** The student mistakes slope for meaning that a line passes through (0, 0).

G. The student mistakes a vertical line for zero slope.

H. The student mistakes slope for meaning that a line passes through (0, 0).

I. Correct answer

11. **A.** The student divides M by 3, but fails to divide $(K + 7)$ by 3.

B. Correct answer

C. The student divides K by 3, but fails to divide 7 by 3.

D. The student subtracts 7 instead of adding it to both sides.

12. **F.** The student picks the coefficient of x from the equation, failing to convert to slope-intercept form.

G. The student divides properly by 2, but attaches an incorrect sign since x initially has a positive coefficient.

H. Correct answer

I. The student correctly moves $5x$ to the opposite side of the equation, but fails to divide by 2.

13. **A.** Correct answer

B. The student divides 16 by 4 incorrectly. Also, the student does not divide each side by π.

C. The student divides 16 by 4 correctly, but does not divide each side by π.

D. The student incorrectly subtracts 4 from 16 instead of dividing. Also, the student does not divide each side by π.

Extra Example

1. What value of x makes the equation shown below true? *(A.REI.3)*

   $11x - 7 = 3(x + 1)$

9. Which value of x makes the equation below true? *(A.REI.3)*

$$7 + 2x = 4x - 5$$

A. -2

B. 1

C. 2

D. 6

10. Which line has a slope of 0? *(F.IF.6)*

F. 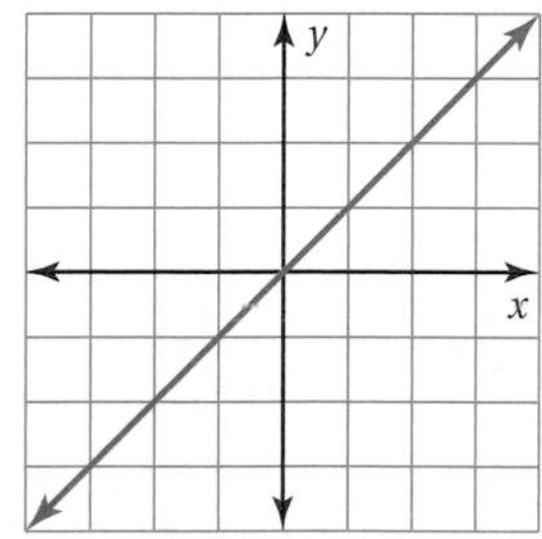

H. 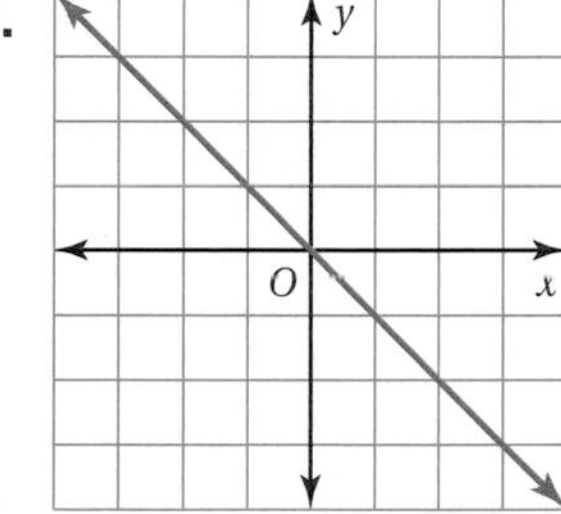

G. 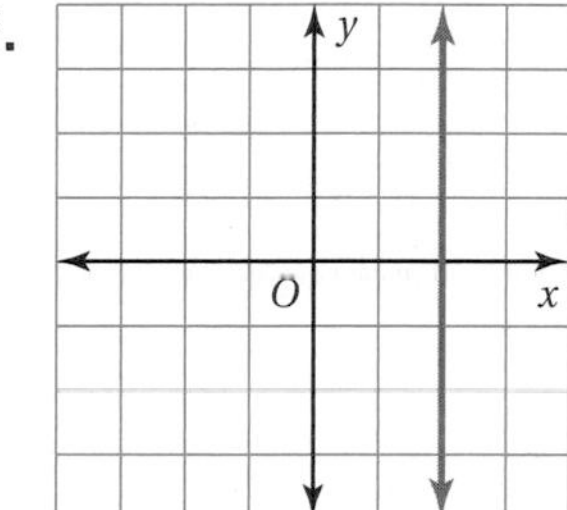

I. 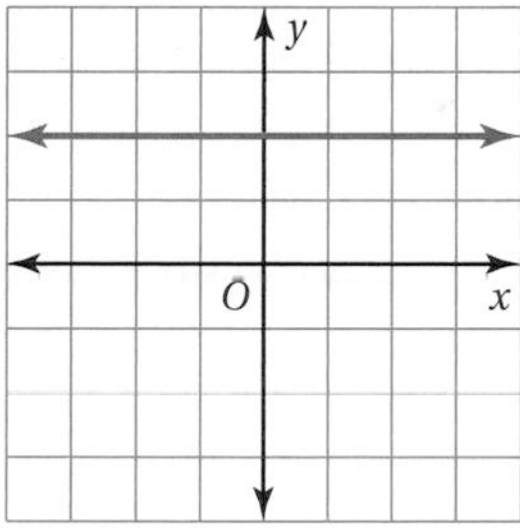

11. Solve the formula $K = 3M - 7$ for M. *(A.CED.4)*

A. $M = K + 7$

B. $M = \frac{K + 7}{3}$

C. $M = \frac{K}{3} + 7$

D. $M = \frac{K - 7}{3}$

12. The linear equation $5x + 2y = 10$ is written in standard form. What is the slope of the graph of this equation? *(F.IF.6)*

F. 5

G. 2.5

H. -2.5

I. -5

13. Solve $4\pi x = 16\pi$. *(A.REI.3)*

A. $x = 4$

B. $x = 2\pi$

C. $x = 4\pi$

D. $x = 12\pi$

3 Solving Linear Inequalities

3.1 Writing and Graphing Inequalities

3.2 Solving Inequalities Using Addition or Subtraction

3.3 Solving Inequalities Using Multiplication or Division

3.4 Solving Multi-Step Inequalities

3.5 Graphing Linear Inequalities in Two Variables

Connections to Previous Learning

- Write, solve, and graph one-step linear inequalities.

- Formulate and use different strategies to solve one-step and multi-step linear inequalities, including inequalities with rational coefficients.

- Write, solve, and graph one-step and multi-step inequalities in one and two variables.

Math in History

Long distance communication occurred in two basic ways in ancient cultures: by sight and by sound. Because the different sights (smoke signals) and sounds (drum beats) were limited, people had to restrict the communication to important topics, such as safety or danger.

★ The smoke signal messages that Native American tribes sent were simple, but important. Here are three of the signals used by the Apache Indian tribe.

- One puff: Something unusual is going on, but there's no cause for alarm or imminent danger.
- Two puffs: All is well. Camp is established and safe.
- Three puffs: This was an alarm signal, just as it is with the Boy Scouts today. A continuous column of smoke indicated great danger and a call for help.

★ In Africa, New Guinea, and South America, people used drum telegraphy to communicate with each other from great distances. For instance, when European expeditions came into the jungles to explore, they were often surprised to find that the message of their coming and their intention was carried through the woods in advance of their arrival.

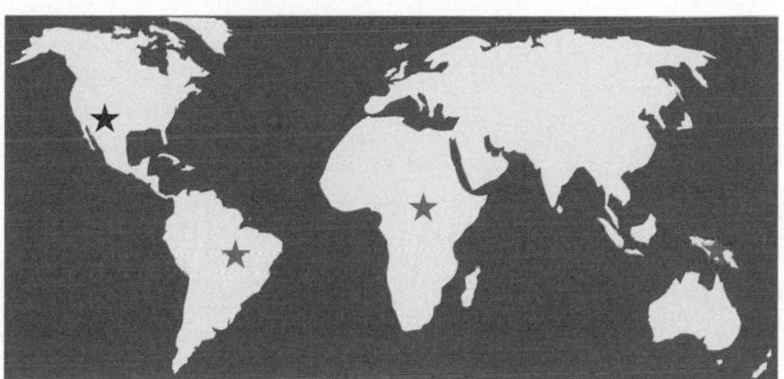

Pacing Guide for Chapter 3

Chapter Opener	1 Day
Section 1	2 Days
Section 2	0.5 Day
Section 3	0.5 Day
Study Help / Quiz	1 Day
Section 4	3 Days
Section 5	1 Day
Chapter Review / Chapter Tests	2 Days
Total Chapter 1	11 Days
Year-to-Date	35 Days

Check Your Resources

- Record and Practice Journal
- Resources by Chapter
- Skills Review Handbook
- Assessment Book
- Worked-Out Solutions

Technology For the Teacher

The Dynamic Planning Tool
Editable Teacher's Resources at
BigIdeasMath.com

Common Core State Standards

8.NS.1 Know that numbers that are not rational are called irrational. Understand informally that every number has a decimal expansion; for rational numbers show that the decimal expansion repeats eventually, and convert a decimal expansion which repeats eventually into a rational number.

6.EE.5 Understand solving an equation or inequality as a process of answering a question: which values from a specified set, if any, make the equation or inequality true? Use substitution to determine whether a given number in a specified set makes an equation or inequality true.

Additional Topics for Review

- Review square roots
- Converting between fractions and decimals
- Inequalities

Try It Yourself

1. $=$ **2.** $<$

3. $<$

4. −4 −2 0 2 4 6

5. −2 0 2 4 6 8

6. −2 0 2 4 6 8

7. 0 5 10 15 20

Record and Practice Journal

1. $>$ **2.** $=$

3. $<$ **4.** $<$

5. $>$ **6.** $<$

7. your friend; 5.6 ft is about 5 ft and 7 in.

8. 0 1 2 3 4 5 6

9. 2 3 4 5 6 7 8

10. −1 0 1 2 3 4 5

11–14. See Additional Answers.

Math Background Notes

Vocabulary Review

- Inequality
- Rational Number
- Solution Set
- Irrational Number

Comparing Real Numbers

- Students should be able to compare integers. Students have studied real numbers. Comparing irrational and rational numbers is a relatively new skill.
- Encourage students to convert the real numbers into a similar form before comparing them. In Example 1, it is helpful to convert the given numbers into fractions. In Example 2, it is helpful to convert to decimals.

Graphing Inequalities

- Students should be able to graph inequalities on a number line.
- Remind students that an equation usually produces a finite number of solutions, but an inequality produces an entire set of solutions. That is why an inequality requires you to shade the number line to describe the solutions.
- Remind students that inequalities containing $\leq$ or $\geq$ will require a closed circle. Inequalities containing $<$ or $>$ will require an open circle.
- **Teaching Tip:** Some students have difficulty deciding which side of the number line to shade. Encourage students to pick a test value on each side of the circle. Substitute each test value for *x*. Only one of the resulting inequalities will be true. Shade the number line on the side of the circle from which the valid test value was selected.

Reteaching and Enrichment Strategies

If students need help. . .	If students got it. . .
Record and Practice Journal • Fair Game Review Skills Review Handbook Lesson Tutorials	Game Closet at *BigIdeasMath.com* Start the next section

What You Learned Before

"Some people remember which is bigger by thinking that $<$ is the mouth of a hungry alligator who is trying to eat the LARGER number."

And this is supposed to help me sleep at night?

Comparing Real Numbers (8.NS.1)

Complete the number sentence with $<$, $>$, or $=$.

Example 1 $\frac{1}{3}$ $\square$ 0.3

$\frac{1}{3} = \frac{10}{30}$, $0.3 = \frac{3}{10} = \frac{9}{30}$

Because $\frac{10}{30}$ is greater than $\frac{9}{30}$,

$\frac{1}{3}$ is greater than 0.3.

∴ So, $\frac{1}{3} > 0.3$.

Example 2 $\sqrt{6}$ $\square$ 2.5

Use a calculator to estimate $\sqrt{6}$.

$\sqrt{6} \approx 2.45$

Because 2.45 is less than 2.5, $\sqrt{6}$ is less than 2.5.

∴ So, $\sqrt{6} < 2.5$.

Try It Yourself

Complete the number sentence with $<$, $>$, or $=$.

1. $\frac{1}{4}$ $\square$ 0.25 **2.** 0.1 $\square$ $\frac{1}{9}$ **3.** π $\square$ $\sqrt{10}$

Graphing Inequalities (6.EE.5)

Example 3 **Graph $x \geq 3$.**

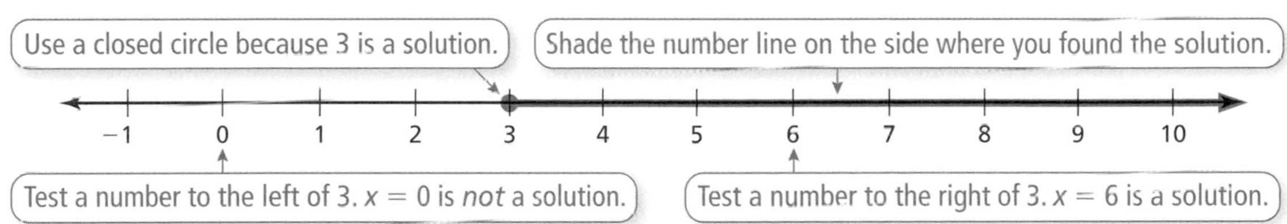
Use a closed circle because 3 is a solution.
Shade the number line on the side where you found the solution.
−1 0 1 2 3 4 5 6 7 8 9 10
Test a number to the left of 3. x = 0 is not a solution.
Test a number to the right of 3. x = 6 is a solution.

Example 4 **Graph $x < 2$.**

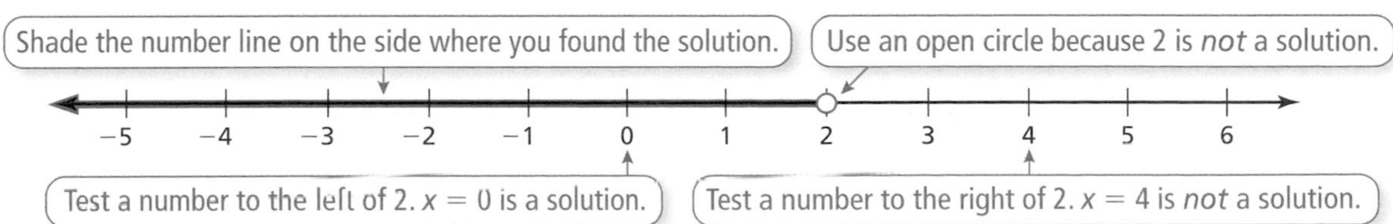
Shade the number line on the side where you found the solution.
Use an open circle because 2 is not a solution.
−5 −4 −3 −2 −1 0 1 2 3 4 5 6
Test a number to the left of 2. x = 0 is a solution.
Test a number to the right of 2. x = 4 is not a solution.

Try It Yourself

Graph the inequality.

4. $x \geq 0$ **5.** $x < 6$ **6.** $x \leq 4$ **7.** $x > 10$

3.1 Writing and Graphing Inequalities

COMMON CORE STATE STANDARDS
A.CED.1
A.CED.3

Essential Question How can you use an inequality to describe a real-life statement?

1 ACTIVITY: Writing and Graphing Inequalities

Work with a partner. Write an inequality for the statement. Then sketch the graph of all the numbers that make the inequality true.

a. **Statement:** The temperature t in Minot, North Dakota has never been below -36°F.

Inequality:

Graph:

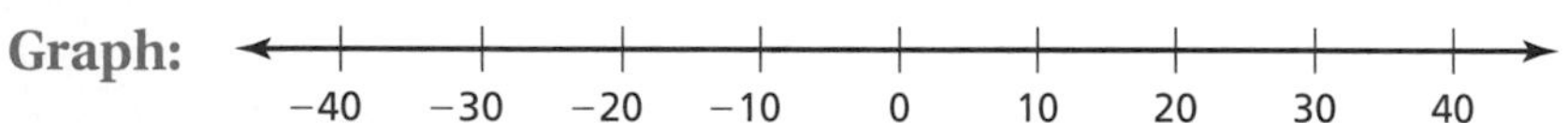

TIMM'S HILL
WISCONSIN'S HIGHEST
NATURAL POINT
ELEV. 1951.5 FT

b. **Statement:** The elevation e in Wisconsin is at most 1951.5 feet above sea level.

Inequality:

Graph:

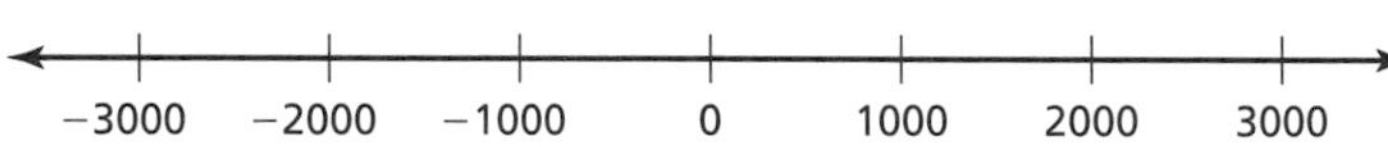

2 ACTIVITY: Writing and Graphing Inequalities

Work with a partner. Write an inequality for the graph. Then, in words, describe all the values of x that make the inequality true.

a.

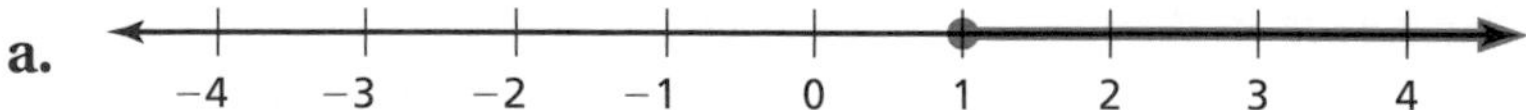

b.

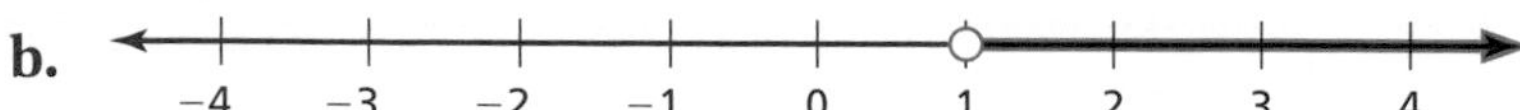

c.

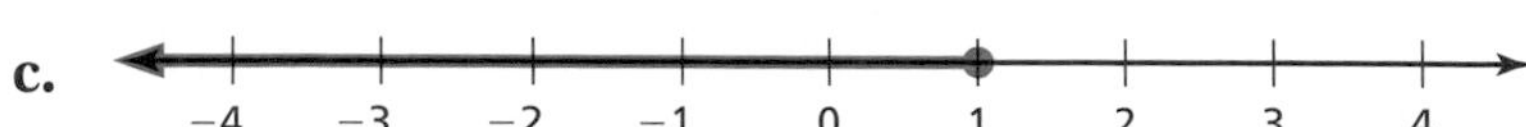

d.

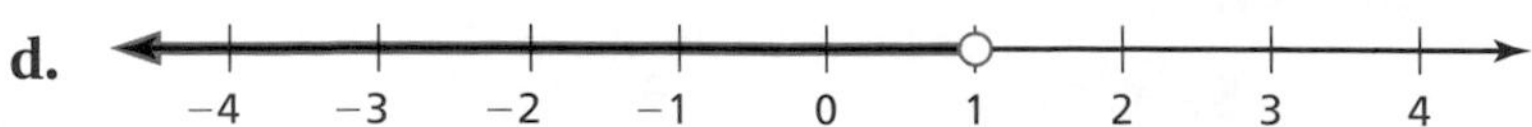

Laurie's Notes

Introduction

Standards for Mathematical Practice

- **MP1a Make Sense of Problems:** The goal is for students to discuss, explain, and demonstrate how to use an inequality to describe a real-life statement. Students are able to explain the correspondence between an inequality represented symbolically, as a verbal description, and as a graph. Remind students of the symbols used to express inequalities, the direction of an inequality, and the open circle/closed circle notation used when graphing an inequality.

Motivate

- **Preparation:** Write 8 inequalities on index cards. Draw the matching graphs on 8 strips of paper large enough to be seen by students across your room. Tape the 8 graphs in different locations around your room.
- Examples of inequalities to explore: $x > 4$; $x \leq -4$; $x > -4$; $x \leq 4$; $x < -2.5$; $x \leq -2.5$; $x > 3.5$; $x \geq 3.5$
- Explain to students that they are starting a new chapter today. Express your confidence in them, knowing that they will have little difficulty with graphing inequalities.
- Select 8 students at random and hand each an index card. Ask students to find their graphs and to go stand next to the graphs.

? After students have matched their cards to the graphs, ask each student to explain how they know their match is correct. What features of the graph did they look for? Listen for: open circle versus closed circle, shading the correct side of the number line.

- After all of the students have made their explanations, collect their cards. Next, ask 8 different students to go to one of the graphs and say aloud the inequality that is shown by the graph.

Activity Notes

Activity 1

- Students should work with their partner on this activity. Caution students to read carefully.
- In part (a), did students graph the temperatures that Minot experienced or did *not* experience? In part (b), did students translate "at most" correctly?

Activity 2

- Students should be familiar with the direction of the inequality and with the open/closed notation.
- This activity assesses a student's ability to distinguish between $x > 1$ and $x \geq 1$, and between $x \leq 1$ and $x < 1$.

Common Core State Standards

A.CED.1 Create equations and inequalities in one variable and use them to solve problems.

A.CED.3 Represent constraints by . . . inequalities . . . and interpret solutions as viable or non-viable options in a modeling context.

Previous Learning

Students should know how to graph numbers on a number line, solve single variable equations, and solve single variable inequalities using rational numbers.

Activity Materials	
Introduction	**Textbook**
• index cards • paper strips	• spaghetti • metric ruler

Start Thinking! and Warm Up

Activity 3.1 Start Thinking! For use before Activity 3.1

Activity 3.1 Warm Up For use before Activity 3.1

Measure the line segment to the nearest tenth of a centimeter.

1. ______
2. ______
3. ______
4. ______
5. ______
6. ______

3.1 Record and Practice Journal

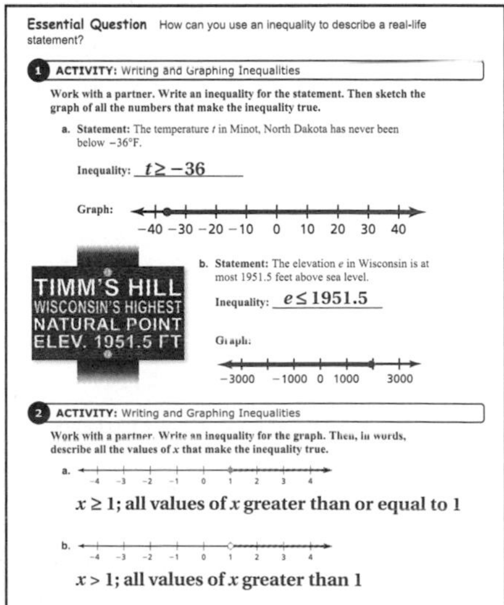

Essential Question How can you use an inequality to describe a real-life statement?

1 ACTIVITY: Writing and Graphing Inequalities

Work with a partner. Write an inequality for the statement. Then sketch the graph of all the numbers that make the inequality true.

a. **Statement:** The temperature t in Minot, North Dakota has never been below -36°F.

Inequality: $t \geq -36$

Graph: −40 −30 −20 −10 0 10 20 30 40

b. **Statement:** The elevation e in Wisconsin is at most 1951.5 feet above sea level.

Inequality: $e \leq 1951.5$

Graph: −3000 −1000 0 1000 3000

2 ACTIVITY: Writing and Graphing Inequalities

Work with a partner. Write an inequality for the graph. Then, in words, describe all the values of x that make the inequality true.

a. −4 −3 −2 −1 0 1 2 3 4

$x \geq 1$; all values of x greater than or equal to 1

b. −4 −3 −2 −1 0 1 2 3 4

$x > 1$; all values of x greater than 1

English Language Learners

Vocabulary and Symbols

Students should review the vocabulary and symbols for inequalities. Have students add a table of symbols and what the symbols mean to their notebooks. Students should add to the table as new phrases are used in the chapter.

Symbol	Phrase
$=$	is equal to
$\neq$	is not equal to
$<$	is less than
$\leq$	is less than or equal to
$>$	is greater than
$\geq$	is greater than or equal to

3.1 Record and Practice Journal

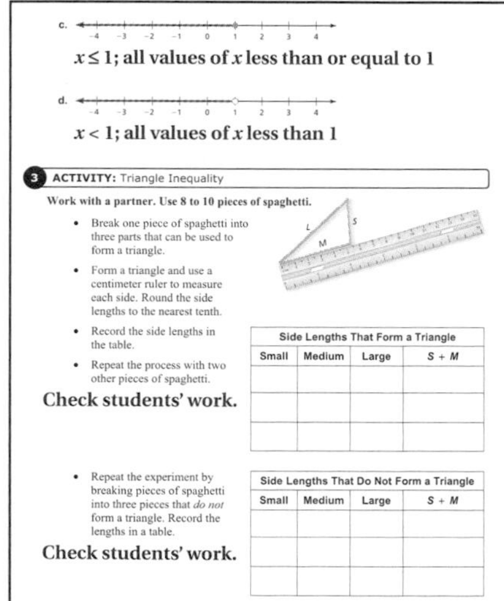

c. $x \leq 1$; all values of x less than or equal to 1

d. $x < 1$; all values of x less than 1

3 ACTIVITY: Triangle Inequality

Work with a partner. Use 8 to 10 pieces of spaghetti.

- Break one piece of spaghetti into three parts that can be used to form a triangle.
- Form a triangle and use a centimeter ruler to measure each side. Round the side lengths to the nearest tenth.
- Record the side lengths in the table.
- Repeat the process with two other pieces of spaghetti.

Check students' work.

Side Lengths That Form a Triangle			
Small	Medium	Large	$S + M$

- Repeat the experiment by breaking pieces of spaghetti into three pieces that *do not* form a triangle. Record the lengths in a table.

Check students' work.

Side Lengths That Do Not Form a Triangle			
Small	Medium	Large	$S + M$

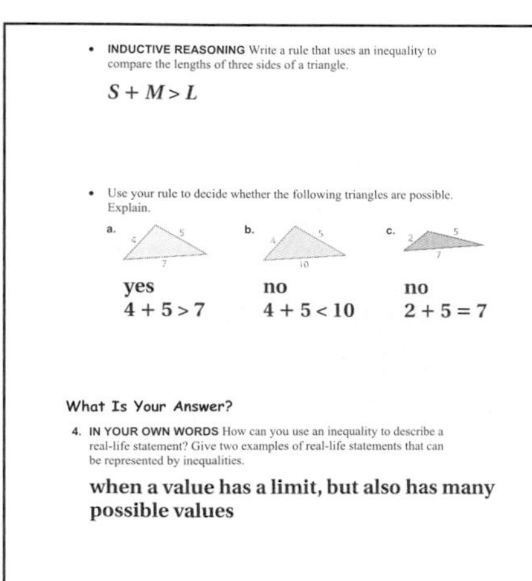

- **INDUCTIVE REASONING** Write a rule that uses an inequality to compare the lengths of three sides of a triangle.

$S + M > L$

- Use your rule to decide whether the following triangles are possible. Explain.

a. yes $4 + 5 > 7$

b. no $4 + 5 < 10$

c. no $2 + 5 = 7$

What Is Your Answer?

4. **IN YOUR OWN WORDS** How can you use an inequality to describe a real-life statement? Give two examples of real-life statements that can be represented by inequalities.

when a value has a limit, but also has many possible values

Laurie's Notes

Activity 3

- **MP4 Model with Mathematics:** In this activity, students will explore another property of triangles. Students will use pieces of spaghetti to model the construction of a triangle and draw conclusions about the side lengths that form a triangle. This concrete approach is very different from using a compass or dynamic software to model the same construction.
- **Management Tip:** Tell students your expectation is that the floor will remain spaghetti free.
- Distribute metric rulers and pieces of spaghetti to each pair of students.
- **MP6 Attend to Precision:** Circulate as students are working on the activity. Check to see that students are measuring to the nearest tenth of a centimeter.
- **MP3 Construct Viable Arguments and Critique the Reasoning of Others** and **MP8 Look for and Express Regularity in Repeated Reasoning:** Students are asked to use inductive reasoning to write a rule describing the pattern they observed when comparing the side lengths of a triangle. There are different, yet equivalent, ways of writing the rule. It is important for students to explain why their rule is plausible. It is also important for students to hear and critique the reasoning of their classmates.
- **Whole Class:** Discuss results with the class. Some students may not have observed a pattern for when the three lengths form a triangle.

? "Is there a group that would like to share their observation about when the lengths form a triangle and when they don't? Explain." Sum of the two shorter sides has to be greater than the longest side.

- **FYI:** Even though a triangle is shown for the last three parts of Activity 3, they may not be drawn to scale. In fact, parts (b) and (c) are *not* triangles.

What Is Your Answer?

- **Big Idea:** This is known as the *Triangle Inequality Theorem.* When the sum of the two shorter sides is less than the length of the longest side, a triangle is not formed, that is, the ends do *not* meet.

Closure

- **Exit Ticket:** Write a word description with a real-life context for each inequality. Then graph the inequality.

$x > 8$ *Sample answer:* You need to work more than 8 hours a day.

$x \leq -10$ *Sample answer:* The temperature will stay at or below -10°F.

Technology For the Teacher

The Dynamic Planning Tool
Editable Teacher's Resources at *BigIdeasMath.com*

3 ACTIVITY: Triangle Inequality

Work with a partner. Use 8 to 10 pieces of spaghetti.

- Break one piece of spaghetti into three parts that can be used to form a triangle.
- Form a triangle and use a centimeter ruler to measure each side. Round the side lengths to the nearest tenth.

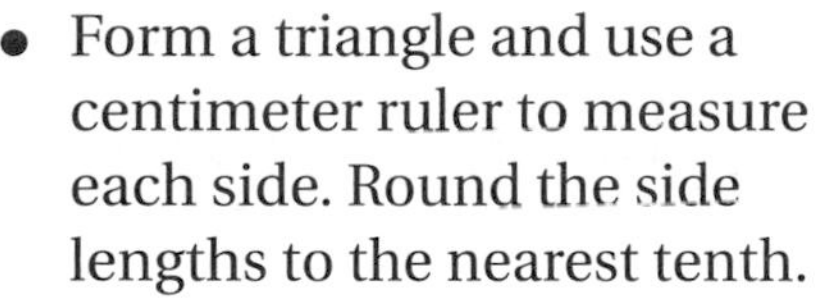

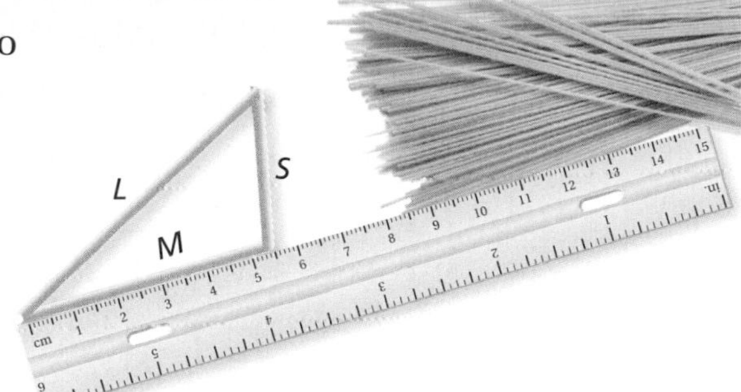

- Record the side lengths in a table.

Side Lengths That Form a Triangle			
Small	Medium	Large	$S + M$

- Repeat the process with two other pieces of spaghetti.
- Repeat the experiment by breaking pieces of spaghetti into three pieces that *do not* form a triangle. Record the lengths in a table.

Side Lengths That Do Not Form a Triangle			
Small	Medium	Large	$S + M$

- **INDUCTIVE REASONING** Write a rule that uses an inequality to compare the lengths of three sides of a triangle.
- Use your rule to decide whether the following triangles are possible. Explain.

a. 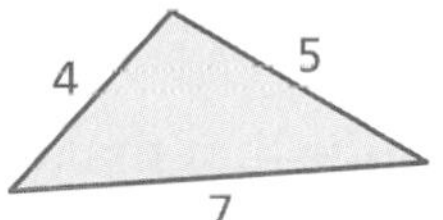

b. 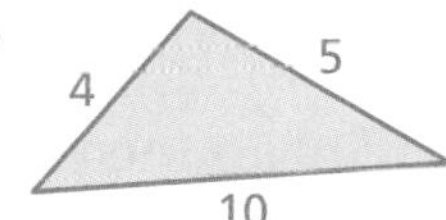

c. 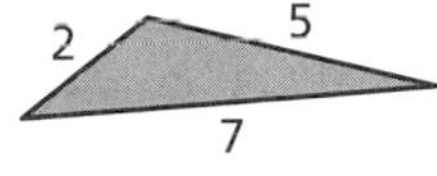

What Is Your Answer?

4. **IN YOUR OWN WORDS** How can you use an inequality to describe a real-life statement? Give two examples of real-life statements that can be represented by inequalities.

Practice Use what you learned about writing and graphing inequalities to complete Exercises 4 and 5 on page 108.

3.1 Lesson

Key Vocabulary
inequality, *p. 106*
solution of an inequality, *p. 106*
solution set, *p. 106*
graph of an inequality, *p. 107*

An **inequality** is a mathematical sentence that compares expressions. It contains the symbol $<$, $>$, $\leq$, or $\geq$. To write an inequality, look for the following phrases to determine where to place the inequality symbol.

Inequality Symbols				
Symbol	$<$	$>$	$\leq$	$\geq$
Key Phrases	• is less than • is fewer than	• is greater than • is more than	• is less than or equal to • is at most • is no more than	• is greater than or equal to • is at least • is no less than

EXAMPLE 1 Writing an Inequality

A number w minus 3.5 is less than or equal to -2. Write this sentence as an inequality.

A number w minus 3.5 is less than or equal to -2.

$$w - 3.5 \qquad \leq \qquad -2$$

∴ An inequality is $w - 3.5 \leq -2$.

On Your Own

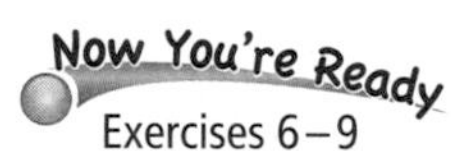

Exercises 6–9

Write the word sentence as an inequality.

1. A number b is fewer than 30.4.
2. Twice a number k is at least $-\frac{7}{10}$.

A **solution of an inequality** is a value that makes the inequality true. An inequality can have more than one solution. The set of all solutions of an inequality is called the **solution set**.

Value of x	$x + 5 \geq -2$	Is the inequality true?
-6	$-6 + 5 \overset{?}{\geq} -2$ $-1 \geq -2$ ✓	yes
-7	$-7 + 5 \overset{?}{\geq} -2$ $-2 \geq -2$ ✓	yes
-8	$-8 + 5 \overset{?}{\geq} -2$ $-3 \ngeq -2$ ✗	no

Reading

The symbol $\ngeq$ means "is not greater than or equal to."

Multi-Language Glossary at BigIdeasMath.com.

Laurie's Notes

Introduction

Connect

- **Yesterday:** Students reviewed how to graph and write an inequality. (MP1a, MP3, MP4, MP6, MP8)
- **Today:** Students will translate inequalities from words to symbols and check to see if a value is a solution of the inequality.

Motivate

- **Story Time:** You are planning to visit several theme parks and notice in doing your research that some of the rides have height restrictions.

Attraction	Restriction	Inequality
Dinosaur	Minimum is now 40 inches	$h \geq 40$
Primeval Whirl	Must be at least 48 inches	$h \geq 48$
Bay Slide	Must be under 60 inches	$h < 60$

- Ask students to write each as an inequality, where h is the rider's height.
- In today's lesson, they will be translating words to symbols.

Lesson Notes

Discuss

- Write the definition of an inequality.
- Review the four inequality symbols and key phrases or words that suggest each inequality.

Example 1

? "Would $3.5 - w \leq -2$ be equivalent to $w - 3.5 \leq -2$? Explain." no; Subtraction is *not* commutative.

? "Is there another way to say $w - 3.5$? Explain." yes; the difference of w and 3.5

On Your Own

- **Think-Pair-Share:** Students should read each question independently and then work with a partner to answer the questions. When they have answered the questions, the pair should compare their answers with another group and discuss any discrepancies.

Discuss

- Discuss what is meant by a solution of an inequality. Inequalities can, and generally do, have more than one solution. All of the solutions are collectively referred to as the **solution set**.
- It is helpful to write the inequality and substitute the value you are checking, as shown in the table.
- **Common Error:** Students will often make the mistake of thinking $-1 \leq -2$, forgetting that relationships are reversed on the negative side of 0; $-1 \geq -2$.

Goal Today's lesson is writing and graphing **inequalities**.

Start Thinking! and Warm Up

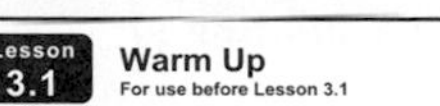

Lesson 3.1 **Start Thinking!**
For use before Lesson 3.1

Write a sentence involving a real-life situation that can be modeled using an inequality.

Which inequality symbol applies: <, ≤, >, or ≥?

Extra Example 1

A number b plus 2.7 is greater than or equal to 3. Write this sentence as an inequality. $b + 2.7 \geq 3$

On Your Own

1. $b < 30.4$
2. $2k \geq -\frac{7}{10}$

Laurie's Notes

Extra Example 2

Tell whether -2 is a solution of each inequality.

a. $x - 4 < -10$ no

b. $2.3x > -5$ yes

On Your Own

3. yes
4. no
5. yes

Extra Example 3

Graph $y \geq -5$.

On Your Own

6. 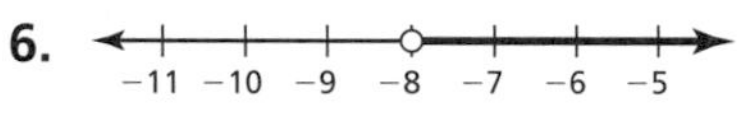

7. 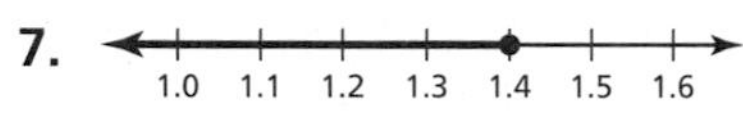

8. 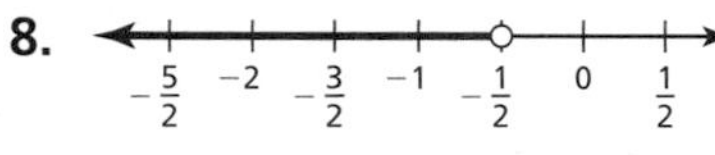

9. 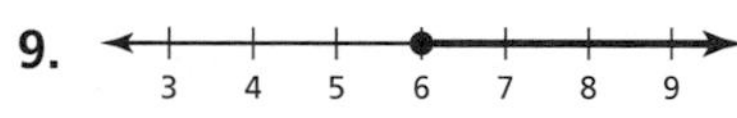

Differentiated Instruction

Auditory

Stress to students the importance of reading a statement and translating it into an expression, equation, or inequality. The word "is" plays an important role in the meaning of the statement. For instance, *six less than a number* translates to $x - 6$, while *six is less than a number* translates to $6 < x$.

Example 2

? "How do you determine if -4 is a solution of an inequality?" Substitute -4 for the variable, simplify, and decide if the inequality is true.

- Work through each example as shown. In part (b), students must recall that the product of two negatives is a positive.

On Your Own

- **Common Error:** In Question 4, when students substitute for *m*, the result is $5 - (-6)$ which is 11.
- Ask volunteers to share their work at the board.

Discuss

- Discuss what is meant by the graph of an inequality. Remind students of the difference between the open and closed circles.

Example 3

- A number is tested on each side of the boundary point. This is a technique that demonstrates what it means to have a boundary point. On one side of the boundary point are all of the values which satisfy the inequality, and on the other side are all of the values which do *not* satisfy the inequality.

On Your Own

- In Question 8, check to see that students locate $-\frac{1}{2}$ correctly.
- In Question 9, students must first evaluate $\sqrt{36}$.
- Ask students to share their graphs at the board.

Closure

- **Writing Prompt:** To decide if a number is a solution of the inequality, you . . .

Technology For the Teacher

The Dynamic Planning Tool
Editable Teacher's Resources at *BigIdeasMath.com*

EXAMPLE 2 Checking Solutions

Tell whether −4 is a solution of each inequality.

a. $x + 8 < -3$

$x + 8 < -3$ Write the inequality.

$-4 + 8 \stackrel{?}{<} -3$ Substitute −4 for x.

$4 \nless -3$ ✗ Simplify.

4 is *not* less than −3.

So, −4 is *not* a solution of the inequality.

b. $-4.5x > -21$

$-4.5x > -21$ Write the inequality.

$-4.5(-4) \stackrel{?}{>} -21$ Substitute −4 for x.

$18 > -21$ ✓ Simplify.

18 is greater than −21.

So, −4 is a solution of the inequality.

On Your Own

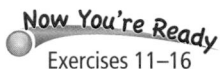

Now You're Ready
Exercises 11–16

Tell whether −6 is a solution of the inequality.

3. $c + 4 < -1$ **4.** $5 - m \leq 10$ **5.** $21 \div x \geq -3.5$

The **graph of an inequality** shows all of the solutions of the inequality on a number line. An open circle ○ is used when a number is *not* a solution. A closed circle ● is used when a number is a solution. An arrow to the left or right shows that the graph continues in that direction.

EXAMPLE 3 Graphing an Inequality

Graph $y \leq -3$.

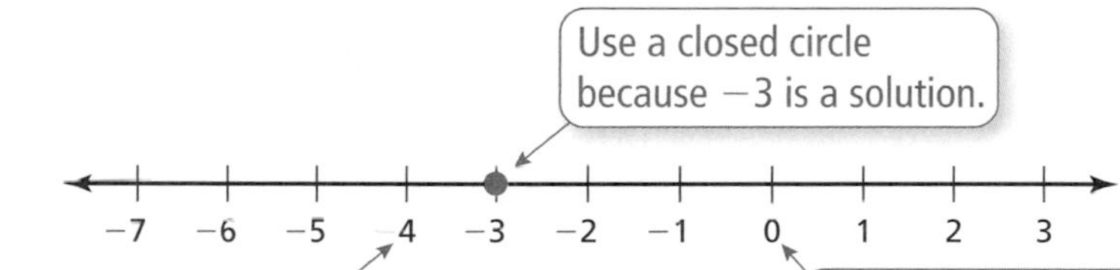
Use a closed circle because −3 is a solution.
−7 −6 −5 −4 −3 −2 −1 0 1 2 3
Test a number to the left of −3. y = −4 is a solution.
Test a number to the right of −3. y = 0 is not a solution.

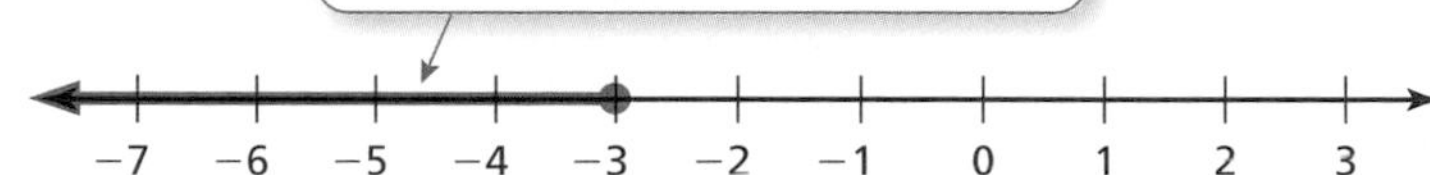
Shade the number line on the side where you found the solution.
−7 −6 −5 −4 −3 −2 −1 0 1 2 3

On Your Own

Now You're Ready
Exercises 17–20

Graph the inequality on a number line.

6. $b > -8$ **7.** $g \leq 1.4$ **8.** $r < -\frac{1}{2}$ **9.** $v \geq \sqrt{36}$

3.1 Exercises

Vocabulary and Concept Check

1. **VOCABULARY** Would an open circle or a closed circle be used in the graph of the inequality $k < 250$? Explain.

2. **DIFFERENT WORDS, SAME QUESTION** Which is different? Write "both" inequalities.

w is greater than or equal to -7.	w is no less than -7.
w is no more than -7.	w is at least -7.

3. **REASONING** Do $x \geq -9$ and $-9 \geq x$ represent the same inequality? Explain.

Practice and Problem Solving

Write an inequality for the graph. Then, in words, describe all the values of x that make the inequality true.

4. −3 0 3 6 9 12 15 18

5. 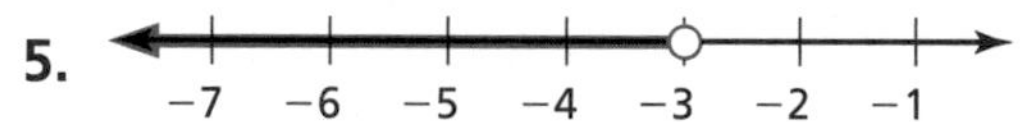

Write the word sentence as an inequality.

① 6. A number x is no less than -4.

7. A number y added to 5.2 is less than 23.

8. A number b multiplied by -5 is at most $-\frac{3}{4}$.

9. A number k minus 8.3 is greater than 48.

10. **ERROR ANALYSIS** Describe and correct the error in writing the word sentence as an inequality.

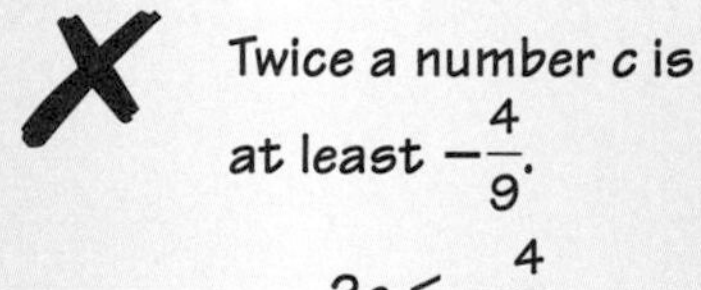

Twice a number c is at least $-\frac{4}{9}$.
$2c \leq -\frac{4}{9}$

Tell whether the given value is a solution of the inequality.

② 11. $s + 6 \leq 12;\ s = 4$

12. $15n > -3;\ n = -2$

13. $a - 2.5 \leq 1.6;\ a = 4.1$

14. $-3.3q > -13;\ q = 4.6$

15. $\frac{4}{5}h \geq -4;\ h = -15$

16. $\frac{1}{12} - p < \frac{1}{3};\ p = \frac{1}{6}$

Graph the inequality on a number line.

③ 17. $g \geq -6$

18. $q > 1.25$

19. $z < 11\frac{1}{4}$

20. $w \leq -\sqrt{64}$

21. **DRIVING** When you are driving with a learner's license, a licensed driver who is 21 years of age or older must be with you. Write an inequality that represents this situation.

Assignment Guide and Homework Check

Level	Assignment	Homework Check
Average	1–5, 7–19 odd, 10, 18, 23–25, 28–31	9, 10, 13, 17, 24
Advanced	1–5, 10–20 even, 22–31	10, 12, 20, 22, 24

Common Errors

- **Exercises 6–9** Students may struggle with knowing which inequality symbol to use. Encourage them to put the word sentence into a real-life context and to use the table in the lesson that explains what symbol matches each phrase.
- **Exercises 11–16** Students may try to solve for the variable instead of substituting the given value into the inequality and determining if that value is a solution of the inequality. Remind them that they are not solving inequalities yet, just checking a number to see if it is a solution.
- **Exercises 17–20** Students may use a closed circle instead of an open circle and vice versa. They may also shade the wrong side of the number line. Review how to graph inequalities and encourage students to test a value on each side of the circle.

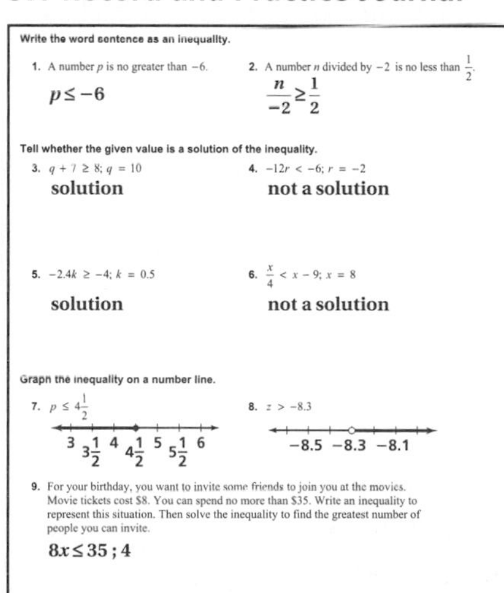
3.1 Record and Practice Journal

Write the word sentence as an inequality.

1. A number p is no greater than -6.
   $p \leq -6$
2. A number n divided by -2 is no less than $\frac{1}{2}$.
   $\frac{n}{-2} \geq \frac{1}{2}$

Tell whether the given value is a solution of the inequality.

3. $q + 7 \geq 8;\ q = 10$
   solution
4. $-12r < -6;\ r = -2$
   not a solution
5. $-2.4k \geq -4;\ k = 0.5$
   solution
6. $\frac{x}{4} < x - 9;\ x = 8$
   not a solution

Graph the inequality on a number line.

7. $p \leq 4\frac{1}{2}$
   3 $3\frac{1}{2}$ 4 $4\frac{1}{2}$ 5 $5\frac{1}{2}$ 6
8. $z > -8.3$
   −8.5 −8.3 −8.1
9. For your birthday, you want to invite some friends to join you at the movies. Movie tickets cost $8. You can spend no more than $35. Write an inequality to represent this situation. Then solve the inequality to find the greatest number of people you can invite.
   $8x \leq 35$; 4

Answer Presentation Tool

Vocabulary and Concept Check

1. An open circle would be used because 250 is not a solution.
2. w is no more than -7.; $w \leq -7$; $w \geq -7$
3. no; $x \geq -9$ is all values of x greater than or equal to -9. $-9 \geq x$ is all values of x less than or equal to -9.

Practice and Problem Solving

4. $x \geq 9$; all values of x greater than or equal to 9
5. $x < -3$; all values of x less than -3
6. $x \geq -4$
7. $y + 5.2 < 23$
8. $-5b \leq -\frac{3}{4}$
9. $k - 8.3 > 48$
10. The inequality symbol is reversed. $2c \geq -\frac{4}{9}$
11. yes
12. no
13. yes
14. no
15. no
16. yes
17. −7 −6 −5 −4 −3 −2 −1
18. 0 0.25 0.5 0.75 1 1.25 1.5
19. $10\frac{1}{2}$ $10\frac{3}{4}$ 11 $11\frac{1}{4}$ $11\frac{1}{2}$ $11\frac{3}{4}$ 12
20. −12 −11 −10 −9 −8 −7 −6
21. $x \geq 21$

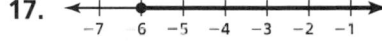

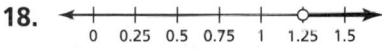

Practice and Problem Solving

22. yes **23.** yes

24. maybe; If your friend is 10, 11, or 12, then your friend can play "E 10+" games, but is not old enough for "T" games. If your friend is 13 or older, then your friend can play "T" games.

25. See Additional Answers.

26. See *Taking Math Deeper.*

27. a. $m < n; n \leq p$

b. $m < p$

c. no; Because n is no more than p and m is less than n, m cannot be equal to p.

Fair Game Review

28. 15 **29.** -1.7

30. 10π **31.** D

Mini-Assessment

Write the word sentence as an inequality.

1. A number m multiplied by -4.9 is at most 5. $-4.9m \leq 5$
2. A number p minus 1.1 is greater than or equal to $-\frac{2}{3}$. $p - 1.1 \geq -\frac{2}{3}$
3. A number h divided by 4 is less than -7.5. $\frac{h}{4} < -7.5$

Graph the inequality on a number line.

4. $x > -2.9$

−2.9

−4 −3 −2 −1 0 1

5. $a \leq -5.25$

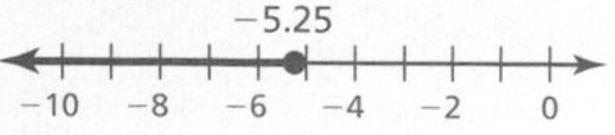

Taking Math Deeper

Exercise 26

This is a practical problem for anyone who is planning to fly. This size restriction applies only to carry-on luggage, not to luggage that is checked. For students who have not thought of the differences, it might be interesting for them to think about the advantages of carry-on luggage.

- No chance of luggage not arriving
- No waiting for luggage at destination
- No extra fees for luggage

Draw a diagram showing the length, width, and height of a carry-on bag.

Find some possible combinations for which $\ell + w + h \leq 45$.

Bag	ℓ	w	h
Standard size → A	22 in.	14 in.	9 in.
B	20 in.	14 in.	11 in.
C	18 in.	14 in.	13 in.

Students might find it interesting to discover which of their three choices has the greatest volume. In the three examples in the table, the volumes are 2272, 3080, and 3276 cubic inches.

In general, the more cube-like the luggage, the greater the volume. So, the maximum volume would be with luggage that is 15 inches by 15 inches by 15 inches, which has a volume of 3375 cubic inches.

Reteaching and Enrichment Strategies

If students need help...	If students got it...
Resources by Chapter • Practice A and Practice B • Puzzle Time Record and Practice Journal Practice Differentiating the Lesson Lesson Tutorials Skills Review Handbook	Resources by Chapter • Enrichment and Extension Start the next section

Tell whether the given value is a solution of the inequality.

22. $3p > 5 + p;\ p = 4$

23. $\frac{y}{2} \geq y - 11;\ y = 18$

24. LOGIC Each video game rating is matched with the inequality that represents the suggested ages of players. Your friend is old enough to play "E 10+" games. Is your friend old enough to play "T" games? Explain.

$x \geq 3$

$x \geq 6$

$x \geq 10$

$x \geq 13$

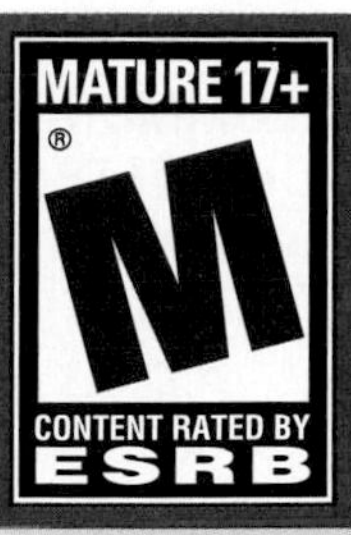

$x \geq 17$

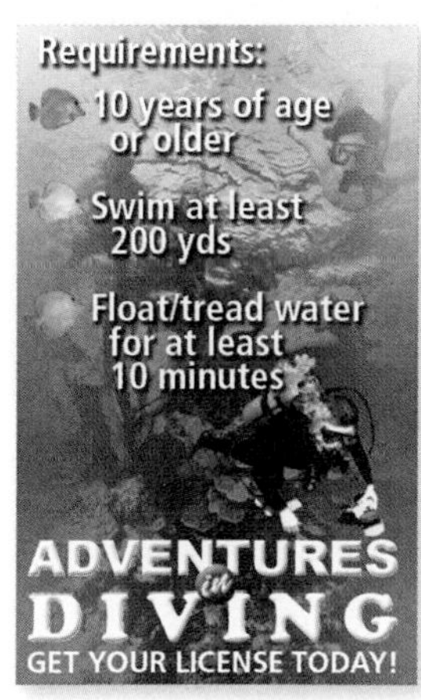

25. SCUBA DIVING Three requirements for a scuba diving training course are shown.

a. Write and graph three inequalities that represent the requirements.

b. You can swim 10 lengths of a 25-yard pool. Do you satisfy the swimming requirement of the course? Justify your answer.

26. REPEATED REASONING On an airplane, the maximum sum of the length, width, and height of a carry-on bag is 45 inches. Find three different sets of dimensions that are reasonable for a carry-on bag. Use a diagram to justify your answer.

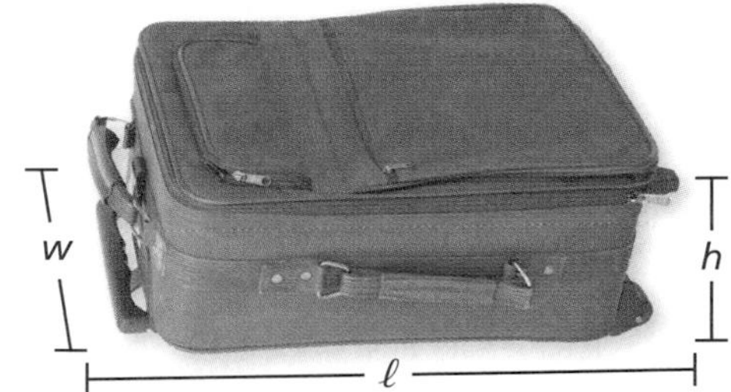

27. 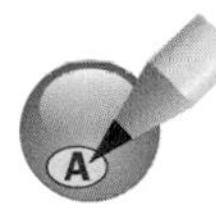

A number m is less than another number n. The number n is less than or equal to a third number p.

a. Write two inequalities representing these relationships.

b. Describe the relationship between m and p.

c. Can m be equal to p? Explain.

Fair Game Review What you learned in previous grades & lessons

Solve the equation. Check your solution. *(Section 1.1)*

28. $r - 12 = 3$ **29.** $4.2 + p = 2.5$ **30.** $n - 3\pi = 7\pi$

31. MULTIPLE CHOICE Which of the following is the equation of the line in slope-intercept form? *(Section 2.5)*

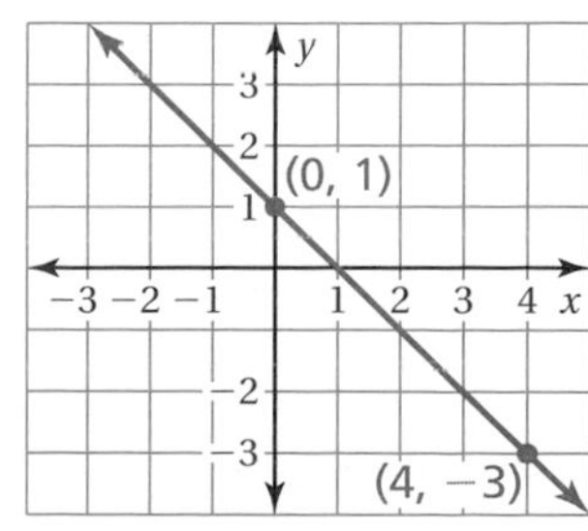

Ⓐ $y = -2x + 1$ Ⓑ $y = -x - 1$

Ⓒ $y = x + 1$ Ⓓ $y = -x + 1$

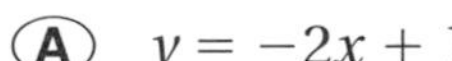

3.2 Solving Inequalities Using Addition or Subtraction

Essential Question How can you use addition or subtraction to solve an inequality?

COMMON CORE STATE STANDARDS
A.CED.1
A.CED.3
A.REI.3

1 ACTIVITY: Quarterback Passing Efficiency

Work with a partner. The National Collegiate Athletic Association (NCAA) uses the following formula to rank the passing efficiency P of quarterbacks.

$$P = \frac{8.4Y + 100C + 330T - 200N}{A}$$

Y = total length of all completed passes (in Yards)

C = Completed passes

T = passes resulting in a Touchdown

N = iNtercepted passes

A = Attempted passes

M = incoMplete passes

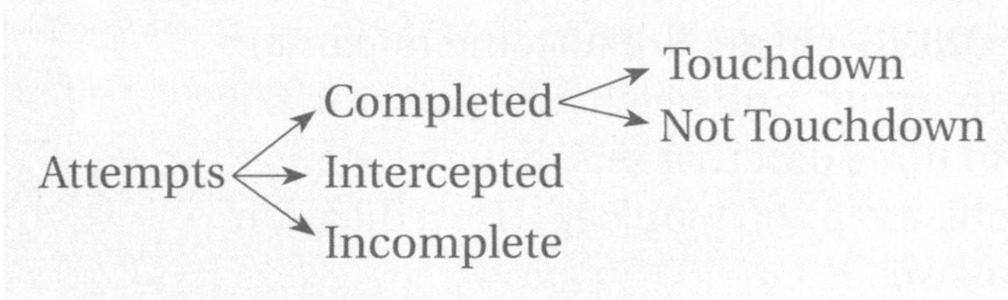

Which of the following equations or inequalities are true relationships among the variables? Explain your reasoning.

a. $C + N < A$ **b.** $C + N \leq A$ **c.** $T < C$ **d.** $T \leq C$

e. $N < A$ **f.** $A > T$ **g.** $A - C \geq M$ **h.** $A = C + N + M$

2 ACTIVITY: Quarterback Passing Efficiency

Work with a partner. Which of the following quarterbacks has a passing efficiency rating that satisfies the inequality $P > 100$? Show your work.

Player	Attempts	Completions	Yards	Touchdowns	Interceptions
A	149	88	1065	7	9
B	400	205	2000	10	3
C	426	244	3105	30	9
D	188	89	1167	6	15

Laurie's Notes

Introduction

Standards for Mathematical Practice

- **MP7 Look for and Make Use of Structure:** Students have solved equations using addition and subtraction. Structurally, inequalities involving these operations are solved in the same fashion. Mathematically proficient students recognize the similarity between solving $x + 9 = 17$ and $x + 9 < 17$.

Motivate and Discuss

- Wear a football related piece of clothing today, if you own one.
- Set the tone by tossing a few passes in class with a small foam football. Ask a statistician to record your efforts in a table at the board. Use 3 columns: **C**ompleted, **IN**tercepted, and Inco**M**plete.
- I recommend **A**ttempting 10 short passes to students nearby. You may need to give permission to have a pass intercepted.

? Ask the following questions.

- "How many passes did I attempt?" 10 Record this next to the table.
- "Can I complete more passes than I attempt?" no
- "Are *completed passes* + *incomplete passes* always *less than or equal to attempted passes*?" yes
- "Are *completed passes* + *incomplete passes* always *less than attempted passes*?" No, they could be equal.

Activity Notes

Activity 1

- The tree diagram should be a helpful aid to students.
- Discuss students' answers and their reasoning when they have finished.
- For parts (c) and (d), point out the need to pay attention to the inequality symbol. It is possible, though unlikely, that $T = C$. In that case, the inequality $T < C$ may *not* be true, while the inequality $T \le C$ is true.
- **MP2 Reason Abstractly and Quantitatively:** There will be some heated discussion about the inequalities, but remember to ask, is it *possible* versus is it *probable*. Take time for students to explain their reasoning.

Activity 2

- You may want to allow calculators to increase speed and accuracy, or you may want to use this as an opportunity to review computation skills.
- **Common Error:** Students may forget order of operations. The computation in the numerator must be completed before dividing by the denominator. On a calculator, this can be done by using parentheses, or simply by pressing the *Enter* key before dividing by the denominator.
- Suggest to students that they write the formula, and then rewrite it substituting the values for the variables.

? Ask the following questions.

- "Which player(s) were above average, meaning $P > 100$?" A and C
- "Which player(s) were average, meaning $P = 100$?" B
- "So, was player D below average, meaning $P < 100$?" yes

Common Core State Standards

A.CED.1 Create equations and inequalities in one variable and use them to solve problems.

A.CED.3 Represent constraints by . . . inequalities . . . and interpret solutions as viable or non-viable options in a modeling context.

A.REI.3 Solve linear . . . inequalities in one variable . . . .

Previous Learning

Students should know how to solve equations using addition and subtraction. Students should be able to evaluate expressions.

Activity Materials	
Introduction	**Textbook**
• foam football	• calculator

Start Thinking! and Warm Up

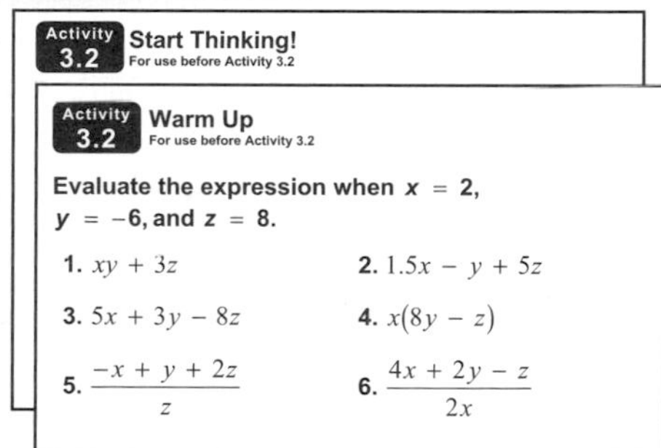

Activity 3.2 Start Thinking! For use before Activity 3.2

Activity 3.2 Warm Up For use before Activity 3.2

Evaluate the expression when $x = 2$, $y = -6$, and $z = 8$.

1. $xy + 3z$
2. $1.5x - y + 5z$
3. $5x + 3y - 8z$
4. $x(8y - z)$
5. $\frac{-x + y + 2z}{z}$
6. $\frac{4x + 2y - z}{2x}$

3.2 Record and Practice Journal

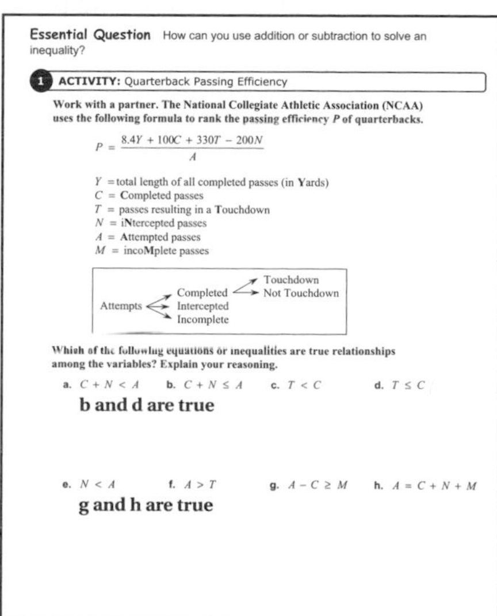

Essential Question How can you use addition or subtraction to solve an inequality?

1 **ACTIVITY:** Quarterback Passing Efficiency

Work with a partner. The National Collegiate Athletic Association (NCAA) uses the following formula to rank the passing efficiency P of quarterbacks.

$$P = \frac{8.4Y + 100C + 330T - 200N}{A}$$

Y = total length of all completed passes (in **Y**ards)
C = **C**ompleted passes
T = passes resulting in a **T**ouchdown
N = i**N**tercepted passes
A = **A**ttempted passes
M = inco**M**plete passes

Which of the following equations or inequalities are true relationships among the variables? Explain your reasoning.

a. $C + N < A$ b. $C + N \le A$ c. $T < C$ d. $T \le C$

b and d are true

e. $N < A$ f. $A > T$ g. $A - C \ge M$ h. $A = C + N + M$

g and h are true

English Language Learners

Vocabulary

It is important that English learners understand the difference between *is less than* and *is less than or equal to,* as well as *is greater than* and *is greater than or equal to.* Give each student a card with one of the numbers $-10, -9, -8, -7, -6, \ldots, 10$. (Include more numbers if your class is larger.) Tell students to stand up if their number *is less than* (say a number), and then *is less than or equal to* that same number. The class should discuss the difference between the two. This can be repeated with *is greater than* and *is greater than or equal to.*

3.2 Record and Practice Journal

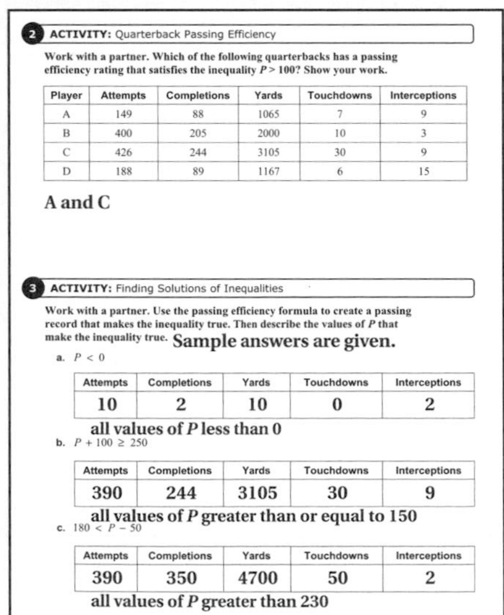

2 ACTIVITY: Quarterback Passing Efficiency

Work with a partner. Which of the following quarterbacks has a passing efficiency rating that satisfies the inequality $P > 100$? Show your work.

Player	Attempts	Completions	Yards	Touchdowns	Interceptions
A	149	88	1065	7	9
B	400	205	2000	10	3
C	426	244	3105	30	9
D	188	89	1167	6	15

A and C

3 ACTIVITY: Finding Solutions of Inequalities

Work with a partner. Use the passing efficiency formula to create a passing record that makes the inequality true. Then describe the values of P that make the inequality true. **Sample answers are given.**

a. $P < 0$

Attempts	Completions	Yards	Touchdowns	Interceptions
10	2	10	0	2

all values of P less than 0

b. $P + 100 \geq 250$

Attempts	Completions	Yards	Touchdowns	Interceptions
390	244	3105	30	9

all values of P greater than or equal to 150

c. $180 < P - 50$

Attempts	Completions	Yards	Touchdowns	Interceptions
390	350	4700	50	2

all values of P greater than 230

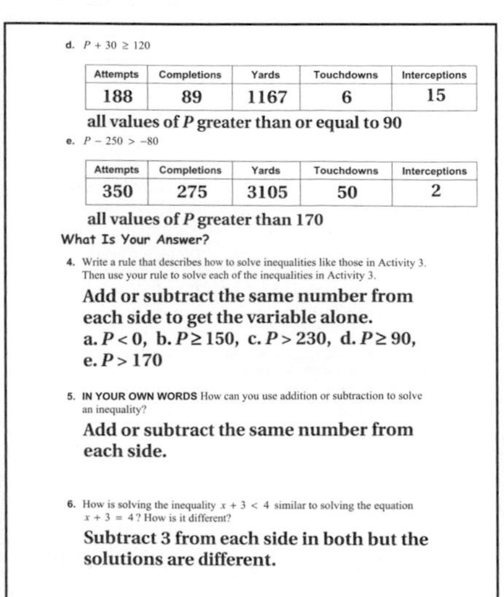

d. $P + 30 \geq 120$

Attempts	Completions	Yards	Touchdowns	Interceptions
188	89	1167	6	15

all values of P greater than or equal to 90

e. $P - 250 > -80$

Attempts	Completions	Yards	Touchdowns	Interceptions
350	275	3105	50	2

all values of P greater than 170

What Is Your Answer?

4. Write a rule that describes how to solve inequalities like those in Activity 3. Then use your rule to solve each of the inequalities in Activity 3.

Add or subtract the same number from each side to get the variable alone.
a. $P < 0$, b. $P \geq 150$, c. $P > 230$, d. $P \geq 90$, e. $P > 170$

5. **IN YOUR OWN WORDS** How can you use addition or subtraction to solve an inequality?

Add or subtract the same number from each side.

6. How is solving the inequality $x + 3 < 4$ similar to solving the equation $x + 3 = 4$? How is it different?

Subtract 3 from each side in both but the solutions are different.

Laurie's Notes

Activity 3

- **MP2:** Now that students have had the opportunity to work with the formula as stated, it is time to put a twist on the problem. Notice that four of the inequalities in this activity force students to think about solving the inequality before they begin. In the second problem, if $P + 100 \geq 250$, then it means $P \geq 150$. Students should decontextualize the situation and manipulate the symbols to solve the inequality.
- Remind students that yards do not count unless the pass is completed.
- Answers will vary for this activity, but suggest to students that they keep the numbers as simple as possible. For instance, one possible answer for the first question is (1, 0, 0, 0, 1), where only 1 pass is attempted and it is intercepted. The result is $P = -200$, and so $P < 0$.
- **MP2:** Students will need to do a little trial and error with these problems. They should start to ask themselves, "*What happens when I increase this variable, but decrease another variable?*" Students need the opportunity to practice this type of reasoning.

What Is Your Answer?

- **Think-Pair-Share:** Students should read each question independently and then work with a partner to answer the questions. When they have answered the questions, the pair should compare their answers with another group and discuss any discrepancies.

Closure

- **Exit Ticket:**
  If $a < b$ is true, is $a \leq b$ also true? Explain. yes; Because in both cases a is less than b.
  If $a \leq b$ is true, is $a < b$ also true? Explain. no; Because if a equals b, then a cannot be less than b.

Technology For the Teacher

The Dynamic Planning Tool
Editable Teacher's Resources at *BigIdeasMath.com*

3 ACTIVITY: Finding Solutions of Inequalities

Work with a partner. Use the passing efficiency formula to create a passing record that makes the inequality true. Then describe the values of *P* that make the inequality true.

a. $P < 0$

Attempts	Completions	Yards	Touchdowns	Interceptions

b. $P + 100 \geq 250$

Attempts	Completions	Yards	Touchdowns	Interceptions

c. $180 < P - 50$

Attempts	Completions	Yards	Touchdowns	Interceptions

d. $P + 30 \geq 120$

Attempts	Completions	Yards	Touchdowns	Interceptions

e. $P - 250 > -80$

Attempts	Completions	Yards	Touchdowns	Interceptions

What Is Your Answer?

4. Write a rule that describes how to solve inequalities like those in Activity 3. Then use your rule to solve each of the inequalities in Activity 3.
5. **IN YOUR OWN WORDS** How can you use addition or subtraction to solve an inequality?
6. How is solving the inequality $x + 3 < 4$ similar to solving the equation $x + 3 = 4$? How is it different?

Practice Use what you learned about solving inequalities using addition or subtraction to complete Exercises 3–5 on page 114.

3.2 Lesson

Study Tip

You can solve inequalities the same way you solve equations. Use inverse operations to get the variable by itself.

Addition Property of Inequality

Words If you add the same number to each side of an inequality, the inequality remains true.

Numbers

$$\begin{array}{rcl} -3 & < & 2 \\ +4 & & +4 \\ \hline 1 & < & 6 \end{array}$$

Algebra

$$\begin{array}{rcl} x-3 & > & -10 \\ +3 & & +3 \\ \hline x & > & -7 \end{array}$$

Subtraction Property of Inequality

Words If you subtract the same number from each side of an inequality, the inequality remains true.

Numbers

$$\begin{array}{rcl} -3 & < & 1 \\ -5 & & -5 \\ \hline -8 & < & -4 \end{array}$$

Algebra

$$\begin{array}{rcl} x+7 & > & -20 \\ -7 & & -7 \\ \hline x & > & -27 \end{array}$$

These properties are also true for $\leq$ and $\geq$.

EXAMPLE 1 Solving an Inequality Using Addition

Solve $x - 6 \geq -10$. Graph the solution.

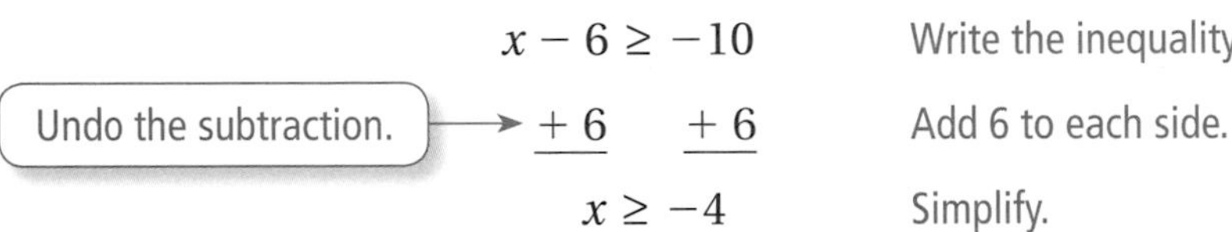

$$\begin{array}{rcll} x-6 & \geq & -10 & \text{Write the inequality.} \\ +6 & & +6 & \text{Add 6 to each side.} \\ \hline x & \geq & -4 & \text{Simplify.} \end{array}$$

Undo the subtraction.

∴ The solution is $x \geq -4$.

Study Tip

To check a solution, you check some numbers that are solutions and some that are not.

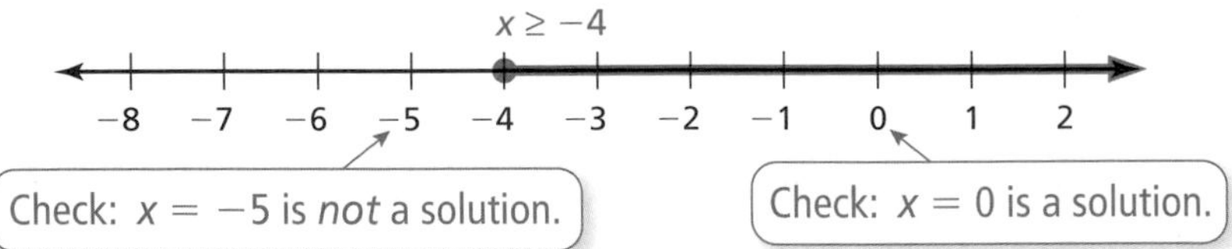

On Your Own

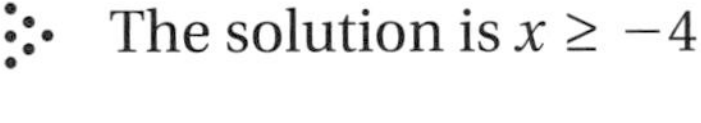

Solve the inequality. Graph the solution.

1. $b - 2 > -9$
2. $m - 3.8 \leq 5$
3. $\frac{1}{4} > y - \frac{1}{4}$

Laurie's Notes

Introduction

Connect

- **Yesterday:** Students explored inequalities and solved simple inequalities using mental math. (MP2, MP7)
- **Today:** Students will use the Addition and Subtraction Properties of Inequality to solve inequalities.

Motivate

- Airlines have guidelines for the maximum weight of luggage, depending upon whether it is carry-on or checked luggage.
- ? "If there is a maximum weight restriction of 50 pounds for a checked bag, what inequality does this suggest?" $w \le 50$
- Suggest different scenarios. If my bag weighs 40.5 pounds, how much more can I add? If my bag weighs 56.4 pounds, how much must I remove?
- Today's lesson involves solving inequalities of this type.

Lesson Notes

Key Ideas

- These properties should look very familiar, as they are similar to the Addition and Subtraction Properties of Equality used in solving equations.
- **Teaching Tip:** I summarize these two properties in the following way:
  I am older than you. In two years, I will still be older than you.
  Laurie's age > Student's age — if $a > b$
  Laurie's age + 2 > Student's age + 2 — then $a + c > b + c$
  Two years ago, I was older than you.
  Laurie's age > Student's age — if $a > b$
  Laurie's age − 2 > Student's age − 2 — then $a - c > b - c$

Example 1

- ? "How do you isolate the variable, meaning get x by itself?" Add 6 to each side of the inequality.
- **MP6 Attend to Precision:** Remind students that they are using the Addition Property of Inequality. Adding 6 is the inverse operation of subtracting 6.
- Solve, graph, and check. Note the *Study Tip*.

On Your Own

- **Think-Pair-Share:** Students should read each question independently and then work with a partner to answer the questions. When they have answered the questions, the pair should compare their answers with another group and discuss any discrepancies.
- I often have students who like to rewrite inequalities so that the variable is on the left. If they do, they must be extremely careful. The inequality symbol must be reversed. Note if $4 < x$, then $x > 4$.
- These problems integrate a review of fraction and decimal operations.

Today's lesson is solving inequalities using addition or subtraction.

Start Thinking! and Warm Up

Lesson 3.2 Warm Up
For use before Lesson 3.2

Lesson 3.2 Start Thinking!
For use before Lesson 3.2

Students are selling magazine subscriptions for a fundraiser. The student who sells the most magazines wins a prize. You have sold 26 subscriptions. Your friend is currently in first place, having sold 41. How many subscriptions do you have to sell in order to move into first place?

What does this have to do with solving inequalities using addition or subtraction?

Extra Example 1

Solve $f - 4 \le 1$. Graph the solution.
$f \le 5$

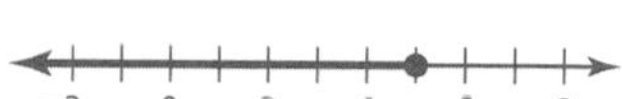

On Your Own

1. $b > -7$;

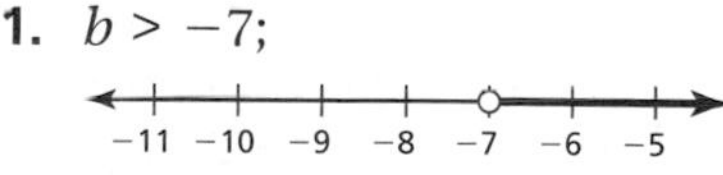

2. $m \le 8.8$;

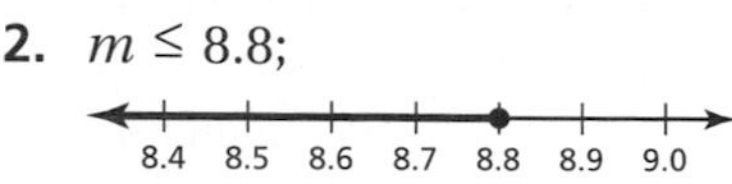

3. $\frac{1}{2} > y$;

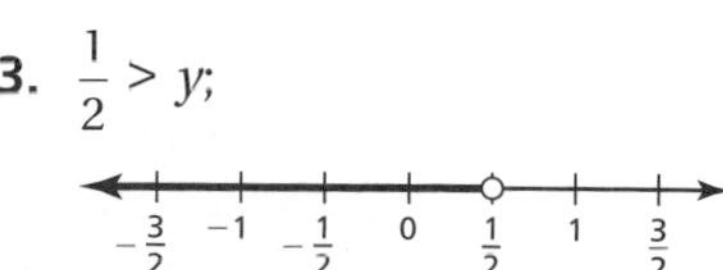

Extra Example 2

Solve $-7 \geq 2.3 + x$. Graph the solution.
$x \leq -9.3$

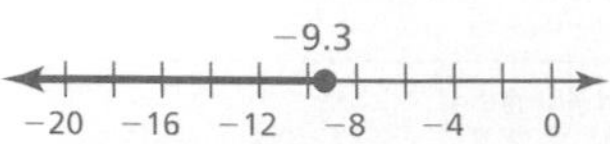

On Your Own

4. $k \leq -8$;

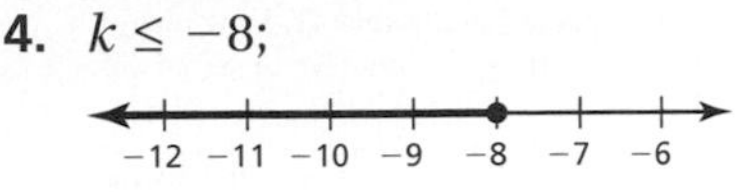

5. $\frac{1}{6} \leq z$;

$0 \quad \frac{1}{6} \quad \frac{2}{6} \quad \frac{3}{6} \quad \frac{4}{6} \quad \frac{5}{6} \quad 1$

6. $p > -3$;

$-5 \quad -4 \quad -3 \quad -2 \quad -1 \quad 0 \quad 1$

Extra Example 3

You have raised \$225 for a charity. Your goal is to raise at least \$600. Write and solve an inequality that represents the amount of money you need to raise to reach your goal. $225 + x \geq 600$; $x \geq \$375$

On Your Own

7. $32.5 + w \leq 50$;
$w \leq 17.5$ lb

Differentiated Instruction

Visual

To show students that the Addition and Subtraction Properties of Inequality are true, graph two numbers, -2 and 5, on the number line. Write the ordered relationship, $-2 < 5$, on the board. Now add 4 to each number. Both points move the same distance to the right, so their ordered positions remain the same, $2 < 9$. Now subtract 6 from each of the original numbers. Both points move 6 units to the left, so their ordered positions remain the same, $-8 < -1$.

Laurie's Notes

Example 2

? "What operation is being performed on the right side of the inequality?" addition

? "How do you undo an addition problem?" subtract

- In a problem such as this, point out that the inequality can be rewritten as $-8 > x + 1.4$. This is possible because of the Commutative Property of Addition. Some students feel more comfortable with the inequality written in this form.
- Solve, graph, and check.

On Your Own

- **Think-Pair-Share:** Students should read each question independently and then work with a partner to answer the questions. When they have answered the questions, the pair should compare their answers with another group and discuss any discrepancies.
- These problems integrate a review of fraction and decimal operations.

Example 3

- Ask a volunteer to read the problem.
- **MP1a Make Sense of Problems:** Take time to write each stage of the solution: words, variables, and inequality. Notice the use of color coding. Color can be a useful tool to help students make sense of a problem.
- Set up the inequality and solve as shown. Note that the constant terms are aligned vertically.

On Your Own

- **Neighbor Check:** Have students work independently and then have their neighbor check their work. Have students discuss any discrepancies.

Closure

- **Exit Ticket:** Solve and graph.

$x + 3.8 \leq -9$ $\quad x \leq -12.8$

−12.8
−20 −16 −12 −8 −4 0

$\frac{2}{5} > x - \frac{3}{4}$ $\quad x < 1\frac{3}{20}$

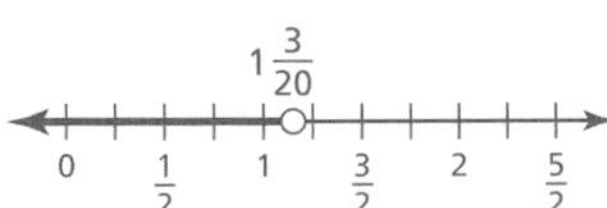

Technology For the Teacher

The Dynamic Planning Tool
Editable Teacher's Resources at *BigIdeasMath.com*

EXAMPLE 2 Solving an Inequality Using Subtraction

Solve $-8 > 1.4 + x$. Graph the solution.

Undo the addition.

$$\begin{aligned} -8 &> 1.4 + x && \text{Write the inequality.} \\ -1.4 & \quad\ -1.4 && \text{Subtract 1.4 from each side.} \\ -9.4 &> x && \text{Simplify.} \end{aligned}$$

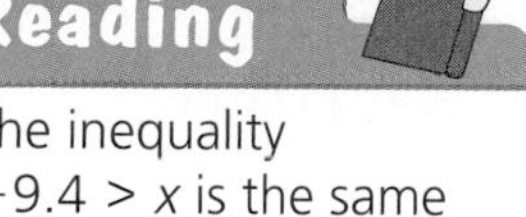

The inequality $-9.4 > x$ is the same as $x < -9.4$.

∴ The solution is $x < -9.4$.

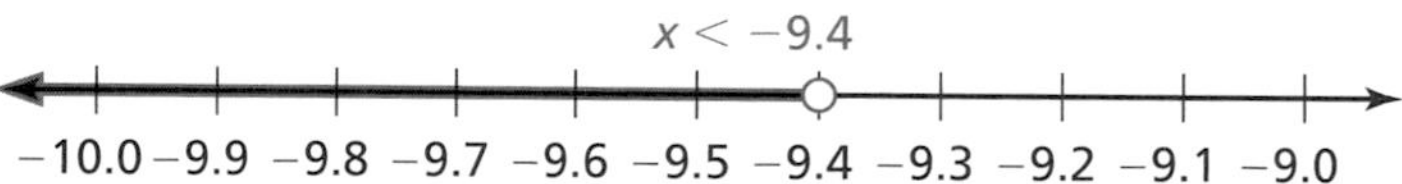

On Your Own

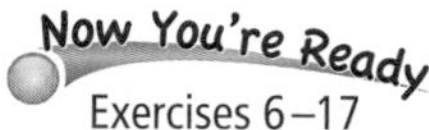

Exercises 6–17

Solve the inequality. Graph the solution.

4. $k + 5 \leq -3$

5. $\frac{5}{6} \leq z + \frac{2}{3}$

6. $p + 0.7 > -2.3$

EXAMPLE 3 Real-Life Application

On a train, carry-on bags can weigh no more than 50 pounds. Your bag weighs 24.8 pounds. Write and solve an inequality that represents the amount of weight you can add to your bag.

Words	Weight of your bag	plus	amount of weight you can add	is no more than	the weight limit.
Variable	Let w be the possible weight you can add.				
Inequality	24.8	+	w	$\leq$	50

$$\begin{aligned} 24.8 + w &\leq 50 && \text{Write the inequality.} \\ -24.8 \quad & \quad -24.8 && \text{Subtract 24.8 from each side.} \\ w &\leq 25.2 && \text{Simplify.} \end{aligned}$$

∴ You can add no more than 25.2 pounds to your bag.

On Your Own

7. **WHAT IF?** Your carry-on bag weighs 32.5 pounds. Write and solve an inequality that represents the possible weight you can add to your bag.

3.2 Exercises

Vocabulary and Concept Check

1. **REASONING** Is the inequality $r - 5 \leq 8$ the same as $8 \leq r - 5$? Explain.

2. **WHICH ONE DOESN'T BELONG?** Which inequality does *not* belong with the other three? Explain your reasoning.

$c + \frac{7}{2} \leq \frac{3}{2}$ $c + \frac{7}{2} \geq \frac{3}{2}$ $\frac{3}{2} \geq c + \frac{7}{2}$ $c - \frac{3}{2} \leq -\frac{7}{2}$

Practice and Problem Solving

Use the formula in Activity 1 to create a passing record that makes the inequality true.

3. $P \geq 180$
4. $P + 40 < 110$
5. $280 \leq P - 20$

Solve the inequality. Graph the solution.

6. $y - 3 \geq 7$
7. $t - 8 > -4$
8. $n + 11 \leq 20$
9. $a + 7 > -1$
10. $5 < v - \frac{1}{2}$
11. $\frac{1}{5} > d + \frac{4}{5}$
12. $-\frac{2}{3} \leq g - \frac{1}{3}$
13. $m + \frac{7}{4} \leq \frac{11}{4}$
14. $11.2 \leq k + 9.8$
15. $h - 1.7 < -3.2$
16. $0 > s + \pi$
17. $5 \geq u - 4.5$

18. **ERROR ANALYSIS** Describe and correct the error in graphing the solution of the inequality.

$5 \geq x - 5$

$10 \geq x$

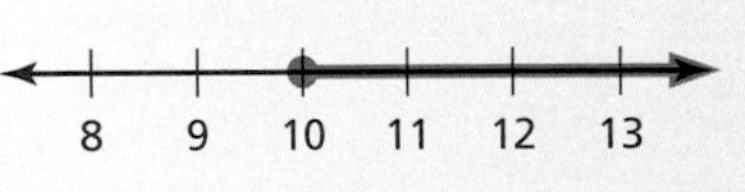

19. **PROBLEM SOLVING** The maximum volume of a great white pelican's bill is about 700 cubic inches.

    a. A pelican scoops up 100 cubic inches of water. Write and solve an inequality that represents the additional volume the pelican's bill can contain.

    b. A pelican's stomach can contain about one-third the maximum amount that its bill can contain. Write an inequality that represents the volume of the pelican's stomach.

Assignment Guide and Homework Check

Level	Assignment	Homework Check
Average	1–5, 9–17 odd, 18, 21, 24, 25, 28–32	11, 15, 21, 24
Advanced	1–5, 12–18 even, 21–24, 26–32	12, 22, 24, 26

Common Errors

- **Exercises 6–17** When solving the inequality, students may use the same operation instead of the opposite operation. Remind them that solving inequalities is similar to solving equations, so they should use the opposite operation to solve the inequality.
- **Exercises 6–17** Students may reverse the direction of the inequality symbol when adding or subtracting. Remind them that the inequality symbol does not change direction when adding to or subtracting from both sides. Review the *Numbers* part of the *Key Idea.*
- **Exercises 20–22** Students may write the wrong formula before solving the inequality. Encourage them to write a formula with variables and then substitute the values given in the figure to solve.

3.2 Record and Practice Journal

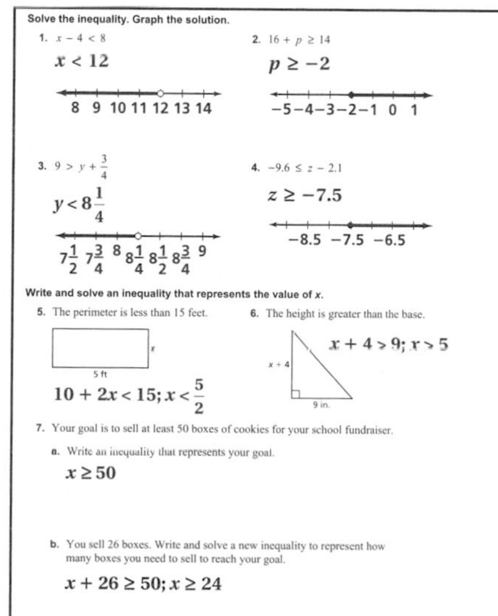

Solve the inequality. Graph the solution.

1. $x - 4 < 8$

   $x < 12$

   8 9 10 11 12 13 14

2. $16 + p \geq 14$

   $p \geq -2$

   −5 −4 −3 −2 −1 0 1

3. $9 > y + \frac{3}{4}$

   $y < 8\frac{1}{4}$

   $7\frac{1}{2}$ $7\frac{3}{4}$ 8 $8\frac{1}{4}$ $8\frac{1}{2}$ $8\frac{3}{4}$ 9

4. $-9.6 \leq z - 2.1$

   $z \geq -7.5$

   −8.5 −7.5 −6.5

Write and solve an inequality that represents the value of *x*.

5. The perimeter is less than 15 feet.

   x, 5 ft

   $10 + 2x < 15; x < \frac{5}{2}$

6. The height is greater than the base.

   x + 4, 9 in.

   $x + 4 > 9; x > 5$

7. Your goal is to sell at least 50 boxes of cookies for your school fundraiser.

   a. Write an inequality that represents your goal.

   $x \geq 50$

   b. You sell 26 boxes. Write and solve a new inequality to represent how many boxes you need to sell to reach your goal.

   $x + 26 \geq 50; x \geq 24$

Technology For the Teacher

Answer Presentation Tool

Vocabulary and Concept Check

1. no; The solution of $r - 5 \leq 8$ is $r \leq 13$ and the solution of $8 \leq r - 5$ is $r \geq 13$.
2. $c + \frac{7}{2} \geq \frac{3}{2}$; It is the only one whose solution is $c \geq -2$. The solution of the other three inequalities is $c \leq -2$.

Practice and Problem Solving

3. *Sample answer:* $A = 350, C = 275, Y = 3105, T = 50, N = 2$
4. *Sample answer:* $A = 500, C = 205, Y = 1700, T = 10, N = 17$
5. *Sample answer:* $A = 400, C = 380, Y = 6510, T = 83, N = 0$
6. $y \geq 10$;

   6 7 8 9 10 11 12

7. $t > 4$;

   0 1 2 3 4 5 6

8. $n \leq 9$;

   6 7 8 9 10 11 12

9. $a > -8$;

   −11 −10 −9 −8 −7 −6 −5

10. $5\frac{1}{2} < v$;

    4 $4\frac{1}{2}$ 5 $5\frac{1}{2}$ 6 $6\frac{1}{2}$ 7

11–17. See Additional Answers.

18. The wrong side of the number line is shaded.

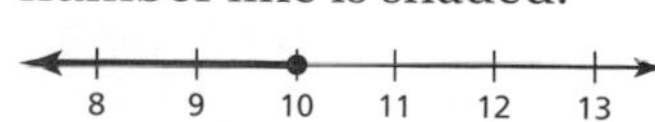

19. a. $100 + V \leq 700$; $V \leq 600 \text{ in.}^3$

    b. $V \leq \frac{700}{3} \text{ in.}^3$

Practice and Problem Solving

20. $4 + 4 + x < 16; x < 8$ ft

21. $x + 2 > 10; x > 8$ m

22. $10 + 10 + 12 + 12 + x \leq 60;$
$x \leq 16$ in.

23. 5

24. $x - 2 \geq 4; x \geq 6$ ft

25. **a.** $4500 + x \geq 12{,}000;$
$x \geq 7500$ points

b. This changes the number added to x by 60%, so the inequality becomes $7200 + x \geq 12{,}000$. So, you need less points to advance to the next level.

26. See *Taking Math Deeper.*

27. $2\pi h + 2\pi \leq 15\pi; h \leq 6.5$ mm

Fair Game Review

28. 2 **29.** 10

30. $\frac{15}{4}$ **31.** 12

32. 7 **33.** 0.5

34. $\frac{2}{3}$ **35.** $2\sqrt{3}$

Mini-Assessment

Solve the inequality. Graph the solution.

1. $-6 \leq u - 4$ $u \geq -2$

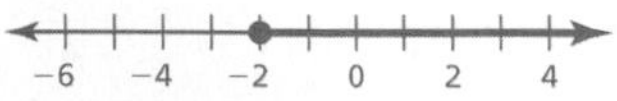

−6
−4
−2
0
2
4

2. $q - 2.5 \geq 6.3$ $q \geq 8.8$

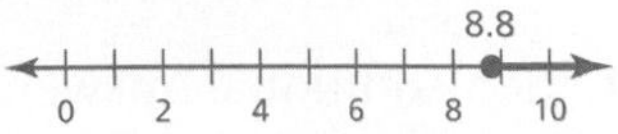

8.8
0
2
4
6
8
10

3. $s + 9 < 33$ $s < 24$

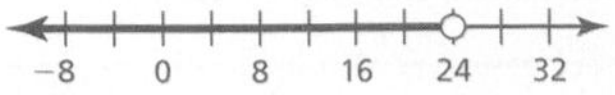

−8
0
8
16
24
32

4. $-\frac{2}{3} \geq f + \frac{1}{3}$ $f \leq -1$

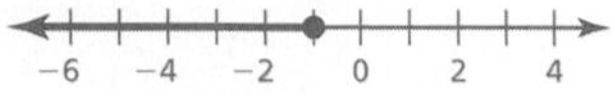

−6
−4
−2
0
2
4

Taking Math Deeper

Exercise 26

Some students may not have had experiences with an electrical circuit that overloads and triggers the circuit breaker. This problem is a nice opportunity to familiarize students with the fact that different appliances use different amounts of electricity.

1 Write an inequality.

Let x = amount of additional electricity used.

1100 = amount used by the microwave oven.

1800 = amount of electricity that overloads the circuit.

$1100 + x < 1800$

Don't overload.

2 Solve the inequality.

a. $1100 + x < 1800$

$x < 700$ watts

3 Answer the question.

You cannot plug in the hot plate or the toaster because each one uses more than 700 watts.

b. You *can* plug in the clock radio and the blender without overloading the circuit.

Appliance	Watts
Clock radio	50
Blender	300
Hot plate	1200
Toaster	800

Project

It might be interesting for students to research different types of appliances and how much electricity each one uses. In this example, you can see that appliances that generate heat use a lot of electricity. It would follow that a crockpot uses quite a bit less electricity than an oven or a range.

Reteaching and Enrichment Strategies

If students need help. . .	If students got it. . .
Resources by Chapter • Practice A and Practice B • Puzzle Time Record and Practice Journal Practice Differentiating the Lesson Lesson Tutorials Skills Review Handbook	Resources by Chapter • Enrichment and Extension Start the next section

Write and solve an inequality that represents the value of x.

20. The perimeter is less than 16 feet.

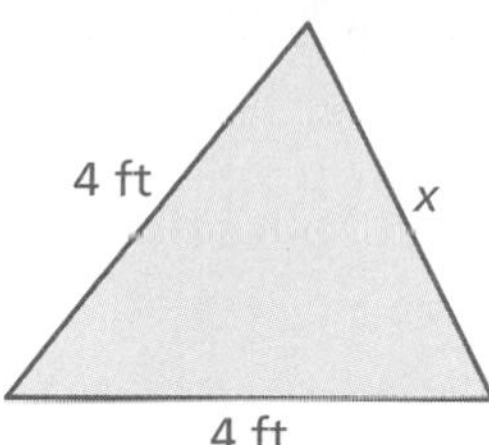

21. The base is greater than the height.

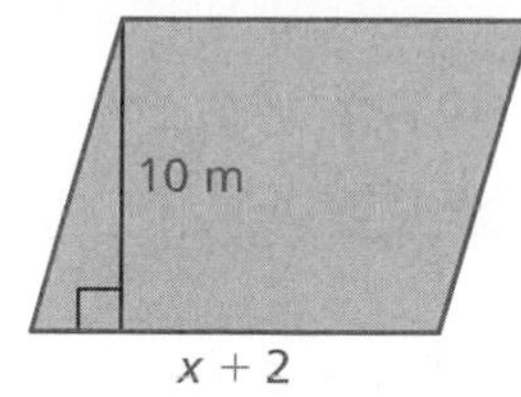

22. The perimeter is less than or equal to 5 feet.

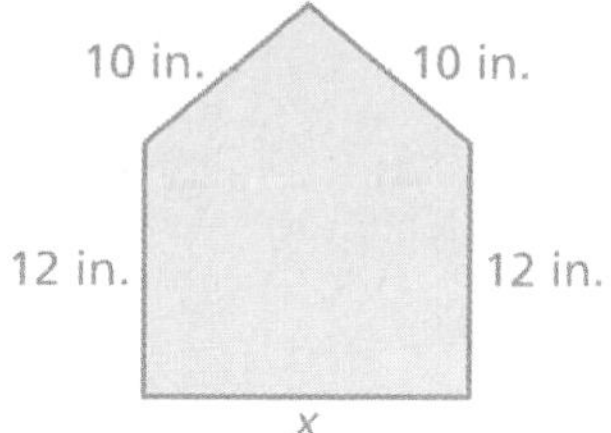

23. REASONING The solution of $w + c \leq 8$ is $w \leq 3$. What is the value of c?

24. FENCE The hole for a fence post is 2 feet deep. The top of the fence post needs to be at least 4 feet above the ground. Write and solve an inequality that represents the required length of the fence post.

25. VIDEO GAME You need at least 12,000 points to advance to the next level of a video game.

a. Write and solve an inequality that represents the number of points you need to advance.

b. You find a treasure chest that increases your score by 60%. Explain how this changes the inequality.

26. MODELING A circuit overloads at 1800 watts of electricity. A microwave that uses 1100 watts of electricity is plugged into the circuit.

a. Use a model to write and solve an inequality that represents the additional number of watts you can plug in without overloading the circuit.

b. In addition to the microwave, what two appliances in the table can you plug in without overloading the circuit? Explain.

Appliance	Watts
Clock radio	50
Blender	300
Hot plate	1200
Toaster	800

27. Critical Thinking The maximum surface area of the solid is 15π square millimeters. Write and solve an inequality that represents the height of the cylinder.

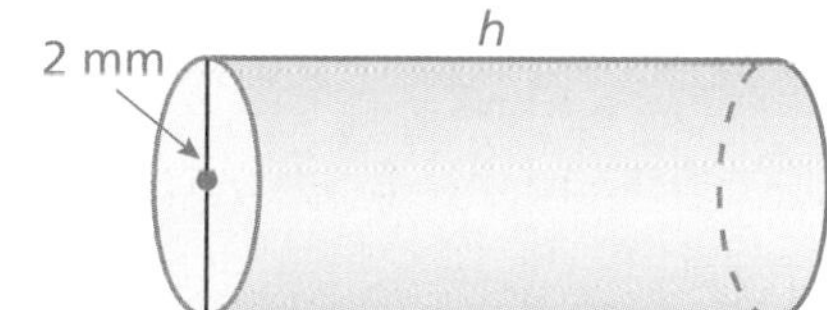

Fair Game Review What you learned in previous grades & lessons

Solve the equation. *(Section 1.1)*

28. $6 = 3x$ **29.** $\frac{r}{5} = 2$ **30.** $4c = 15$ **31.** $8 = \frac{2}{3}b$

Find the square root. *(Skills Review Handbook)*

32. $\sqrt{49}$ **33.** $\sqrt{0.25}$ **34.** $\sqrt{\frac{4}{9}}$ **35.** $\sqrt{12}$

3.3 Solving Inequalities Using Multiplication or Division

COMMON CORE STATE STANDARDS
A.CED.1
A.CED.3
A.REI.3

Essential Question How can you use multiplication or division to solve an inequality?

1 ACTIVITY: Using a Table to Solve an Inequality

Work with a partner.

- **Copy and complete the table.**
- **Decide which graph represents the solution of the inequality.**
- **Write the solution of the inequality.**

a. $3x \le 6$

x	-1	0	1	2	3	4	5
$3x$							
$3x \overset{?}{\le} 6$							

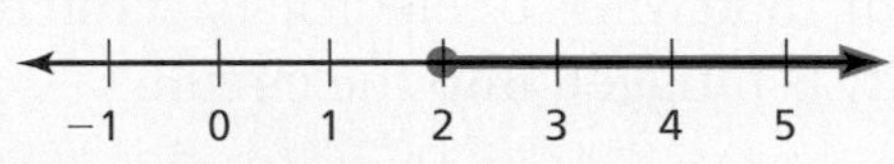

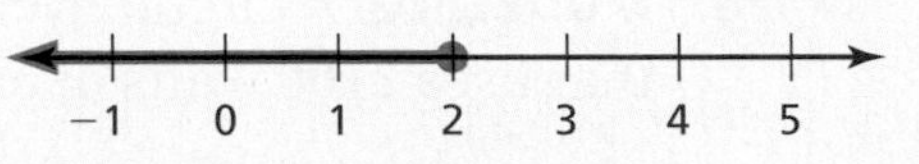

b. $-2x > 4$

x	-5	-4	-3	-2	-1	0	1
$-2x$							
$-2x \overset{?}{>} 4$							

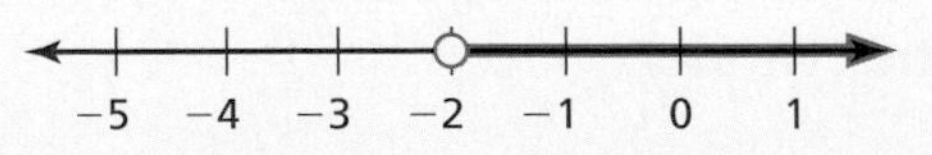

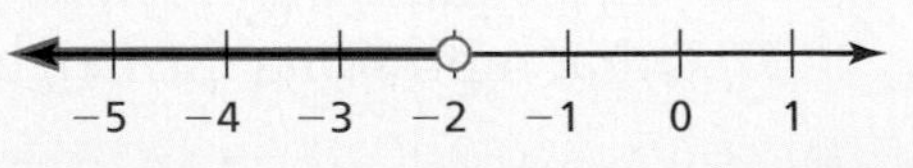

2 ACTIVITY: Writing a Rule

Work with a partner. Use a table to solve each inequality.

a. $3x > 3$ **b.** $4x \le 4$ **c.** $-2x \ge 6$ **d.** $-5x < 10$

Write a rule that describes how to solve inequalities like those in Activity 1. Then use your rule to solve each of the four inequalities above.

Laurie's Notes

Introduction

Standards for Mathematical Practice

- **MP7 Look For and Make Use of Structure:** Students have solved equations using multiplication and division. Structurally, inequalities involving these operations are solved in the same fashion. Mathematically proficient students recognize the similarity between solving $4x = 36$ and $4x < 36$.
- **MP8 Look For and Express Regularity in Repeated Reasoning:** The approach in this investigation is to use a table of values to see what numbers satisfy the inequality. The problems involving positive coefficients behave as expected. It is the problems involving negative coefficients that seem not to work as expected—from the student perspective. Through repeated trials students should recognize that when the coefficient is negative, the inequality symbol is *reversed.*

Motivate

- Ask a series of questions and record the students' solutions.
  - ? "What integers are solutions of $x > 4$?" 5, 6, 7, ...
  - ? "What integers are solutions of $-x > 4$, meaning what numbers have an opposite that is greater than 4?" $-5, -6, -7, \ldots$
  - ? "What integers are solutions of $x < -4$?" $-5, -6, -7, \ldots$
- Leave these 3 problems on the board and refer to them at the end of class.

Activity Notes

Activity 1

- Explain to students that for each inequality, they are to evaluate one side of the inequality and then decide if the inequality is satisfied. This means, *is the value of x a solution of the inequality?* Students will write *yes* or *no* in the third row of the table to indicate if the value is a solution or not.
- Using the information in the table, students decide which graph represents the solution. Finally, they write the solution.
- ? Discuss the results with your students.
  - "What did you find as the solution for part (a)?" $x \le 2$
  - "Is this what you would have expected?" Likely, they will say yes.
  - "What did you find as the solution for part (b)?" $x < -2$
  - "Is this what you would have expected?" Likely, they will say no.
- Do not tell students a rule at this point. Simply say that perhaps they need to try a few more problems to help figure out what is going on.

Activity 2

? "Did any of the inequalities have solutions that you expected?" Yes, the first two inequalities; Listen for students to say that for the last two inequalities, the number part of the solution was expected but the inequality symbol was switched.

? "Did you notice a difference in the first two inequalities versus the last two?" They noticed the positive coefficient in parts (a) and (b), versus the negative coefficient in parts (c) and (d).

- Students might not be quite ready to write a rule. Let the uncertainty be unresolved. You don't need to solve every problem immediately.

Common Core State Standards

A.CED.1 Create . . . inequalities in one variable and use them to solve problems.
A.CED.3 Represent constraints by . . . inequalities . . . and interpret solutions as viable or non-viable options in a modeling context.
A.REI.3 Solve linear . . . inequalities in one variable . . ..

Previous Learning

Students should know how to solve equations using multiplication and division. Students should be able to evaluate expressions and decide if a number is a solution of an inequality.

Start Thinking! and Warm Up

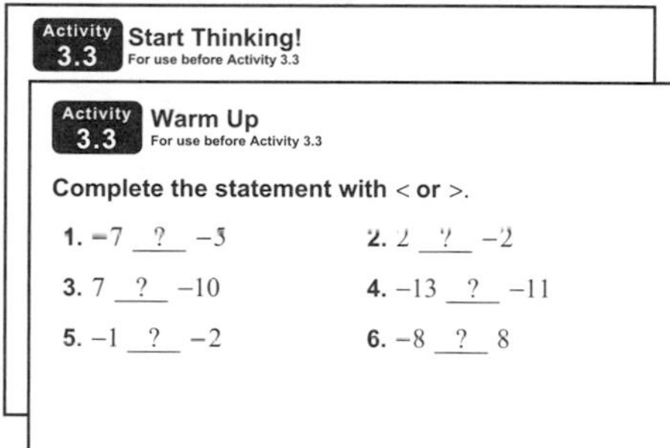

Activity 3.3 Start Thinking!
For use before Activity 3.3

Activity 3.3 Warm Up
For use before Activity 3.3

Complete the statement with $<$ or $>$.

1. -7 ___?___ -5
2. 2 ___?___ -2
3. 7 ___?___ -10
4. -13 ___?___ -11
5. -1 ___?___ -2
6. -8 ___?___ 8

3.3 Record and Practice Journal

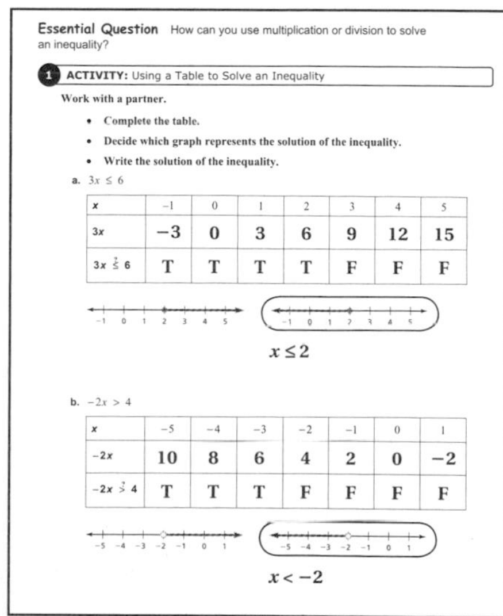

Essential Question How can you use multiplication or division to solve an inequality?

1 ACTIVITY: Using a Table to Solve an Inequality

Work with a partner.

- Complete the table.
- Decide which graph represents the solution of the inequality.
- Write the solution of the inequality.

a. $3x \le 6$

x	−1	0	1	2	3	4	5
$3x$	−3	0	3	6	9	12	15
$3x \overset{?}{\le} 6$	T	T	T	T	F	F	F

$x \le 2$

b. $-2x > 4$

x	−5	−4	−3	−2	−1	0	1
$-2x$	10	8	6	4	2	0	−2
$-2x \overset{?}{>} 4$	T	T	T	F	F	F	F

$x < -2$

English Language Learners

Pair Activity

Create index cards with problems similar to those in Activities 2 and 4. Pair English learners with English speakers and give 5 cards to each pair. Have students work together to solve the inequalities. When students have completed the problems, check their work and give them another set of cards.

3.3 Record and Practice Journal

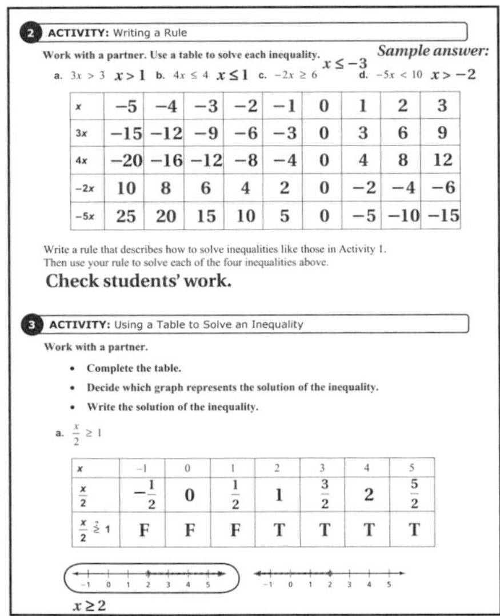

2 ACTIVITY: Writing a Rule

Work with a partner. Use a table to solve each inequality. *Sample answer:*

a. $3x > 3$ $x > 1$ b. $4x \le 4$ $x \le 1$ c. $-2x \ge 6$ $x \le -3$ d. $-5x < 10$ $x > -2$

x	−5	−4	−3	−2	−1	0	1	2	3
$3x$	−15	−12	−9	−6	−3	0	3	6	9
$4x$	−20	−16	−12	−8	−4	0	4	8	12
$-2x$	10	8	6	4	2	0	−2	−4	−6
$-5x$	25	20	15	10	5	0	−5	−10	−15

Write a rule that describes how to solve inequalities like those in Activity 1. Then use your rule to solve each of the four inequalities above.

Check students' work.

3 ACTIVITY: Using a Table to Solve an Inequality

Work with a partner.

- Complete the table.
- Decide which graph represents the solution of the inequality.
- Write the solution of the inequality.

a. $\frac{x}{2} \ge 1$

x	−1	0	1	2	3	4	5
$\frac{x}{2}$	$-\frac{1}{2}$	0	$\frac{1}{2}$	1	$\frac{3}{2}$	2	$\frac{5}{2}$
$\frac{x}{2} \overset{?}{\ge} 1$	F	F	F	T	T	T	T

$x \ge 2$

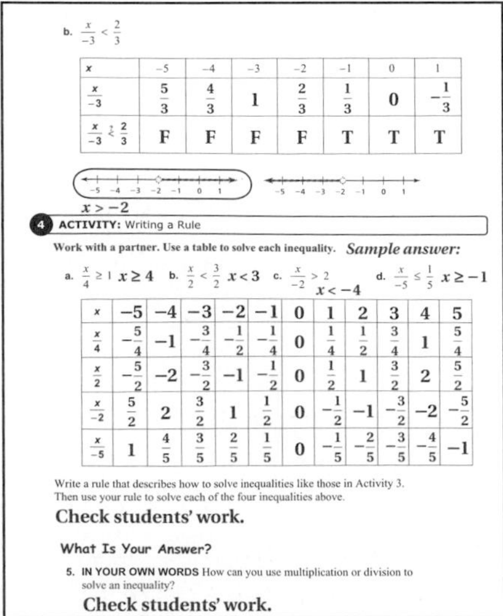

b. $\frac{x}{-3} < \frac{2}{3}$

x	−5	−4	−3	−2	−1	0	1
$\frac{x}{-3}$	$\frac{5}{3}$	$\frac{4}{3}$	1	$\frac{2}{3}$	$\frac{1}{3}$	0	$-\frac{1}{3}$
$\frac{x}{-3} \overset{?}{<} \frac{2}{3}$	F	F	F	F	T	T	T

$x > -2$

4 ACTIVITY: Writing a Rule

Work with a partner. Use a table to solve each inequality. *Sample answer:*

a. $\frac{x}{4} \ge 1$ $x \ge 4$ b. $\frac{x}{2} < \frac{3}{2}$ $x < 3$ c. $\frac{x}{-2} > 2$ $x < -4$ d. $\frac{x}{-5} \le \frac{1}{5}$ $x \ge -1$

x	−5	−4	−3	−2	−1	0	1	2	3	4	5
$\frac{x}{4}$	$-\frac{5}{4}$	−1	$-\frac{3}{4}$	$-\frac{1}{2}$	$-\frac{1}{4}$	0	$\frac{1}{4}$	$\frac{1}{2}$	$\frac{3}{4}$	1	$\frac{5}{4}$
$\frac{x}{2}$	$-\frac{5}{2}$	−2	$-\frac{3}{2}$	−1	$-\frac{1}{2}$	0	$\frac{1}{2}$	1	$\frac{3}{2}$	2	$\frac{5}{2}$
$\frac{x}{-2}$	$\frac{5}{2}$	2	$\frac{3}{2}$	1	$\frac{1}{2}$	0	$-\frac{1}{2}$	−1	$-\frac{3}{2}$	−2	$-\frac{5}{2}$
$\frac{x}{-5}$	1	$\frac{4}{5}$	$\frac{3}{5}$	$\frac{2}{5}$	$\frac{1}{5}$	0	$-\frac{1}{5}$	$-\frac{2}{5}$	$-\frac{3}{5}$	$-\frac{4}{5}$	−1

Write a rule that describes how to solve inequalities like those in Activity 3. Then use your rule to solve each of the four inequalities above.

Check students' work.

What Is Your Answer?

5. **IN YOUR OWN WORDS** How can you use multiplication or division to solve an inequality?

Check students' work.

Laurie's Notes

Activity 3

- Explain that the next two activities are similar to the first two except they involve using multiplication to solve the inequality instead of division.
- Give time for students to work through the two problems.

? Discuss the results with your students.

- "What did you find as the solution for part (a)?" $x \ge 2$
- "Is this what you would have expected?" Likely, they will say yes.
- "What did you find as the solution for part (b)?" $x > -2$
- "Is this what you would have expected?" Likely, they will say no.

- **MP1 Make Sense of Problems and Persevere in Solving Them:** Again, resist the temptation to simply tell students a rule. Suggest that trying additional problems might help them. Students will make sense of the problem as they persevere through the process of repeated trials.

Activity 4

- Give time for students to work through the four problems with their partner.

? "Did any of the inequalities have solutions that you expected?" Again, listen for the same comments from students as before. The first two problems have expected solutions. The last two problems had the number part of the solution expected, but the inequality symbol switched.

- After working through all four problems, students should have a sense that solving these inequalities is the same as solving equations except when the coefficient is negative.

What Is Your Answer?

- **Neighbor Check:** Have students work independently and then have their neighbor check their work. Have students discuss any discrepancies.

Closure

- Refer to the three inequalities written at the beginning of class.

$x > 4$ $\qquad -x > 4$ $\qquad x < -4$

? "Which inequalities have the same solution?" $-x > 4$ and $x < -4$

? "Is this consistent with what you discovered in the activities? Explain." yes; Listen for comments about the negative coefficient of x and the switching of the inequality symbol.

Technology For the Teacher

The Dynamic Planning Tool
Editable Teacher's Resources at *BigIdeasMath.com*

3 ACTIVITY: Using a Table to Solve an Inequality

Work with a partner.

- **Copy and complete the table.**
- **Decide which graph represents the solution of the inequality.**
- **Write the solution of the inequality.**

a. $\frac{x}{2} \geq 1$

x	-1	0	1	2	3	4	5
$\frac{x}{2}$							
$\frac{x}{2} \overset{?}{\geq} 1$							

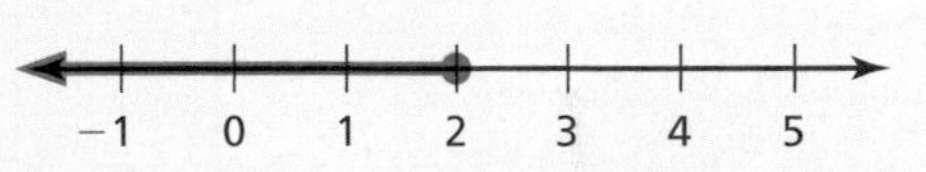

b. $\frac{x}{-3} < \frac{2}{3}$

x	-5	-4	-3	-2	-1	0	1
$\frac{x}{-3}$							
$\frac{x}{-3} \overset{?}{<} \frac{2}{3}$							

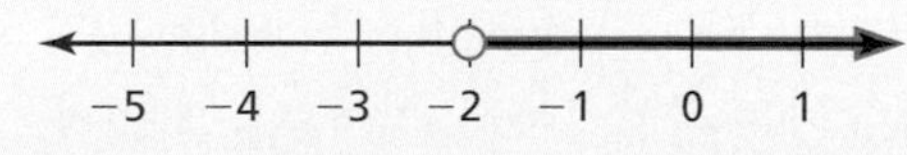

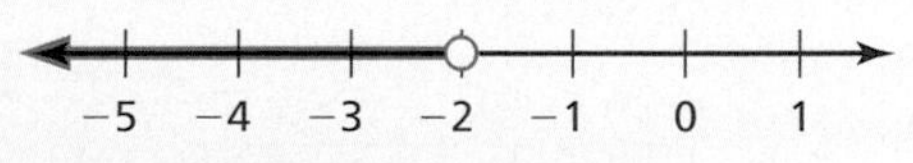

4 ACTIVITY: Writing a Rule

Work with a partner. Use a table to solve each inequality.

a. $\frac{x}{4} \geq 1$ **b.** $\frac{x}{2} < \frac{3}{2}$ **c.** $\frac{x}{-2} > 2$ **d.** $\frac{x}{-5} \leq \frac{1}{5}$

Write a rule that describes how to solve inequalities like those in Activity 3. Then use your rule to solve each of the four inequalities above.

What Is Your Answer?

5. **IN YOUR OWN WORDS** How can you use multiplication or division to solve an inequality?

Practice Use what you learned about solving inequalities using multiplication or division to complete Exercises 4–9 on page 121.

3.3 Lesson

Check It Out
Lesson Tutorials
BigIdeasMath.com

Remember

Multiplication and division are inverse operations.

Key Idea

Multiplication and Division Properties of Inequality (Case 1)

Words If you multiply or divide each side of an inequality by the same *positive* number, the inequality remains true.

Numbers

$$-6 < 8 \qquad\qquad 6 > -8$$

$$2 \cdot (-6) < 2 \cdot 8 \qquad\qquad \frac{6}{2} > \frac{-8}{2}$$

$$-12 < 16 \qquad\qquad 3 > -4$$

Algebra

$$\frac{x}{2} < -9 \qquad\qquad 4x > -12$$

$$2 \cdot \frac{x}{2} < 2 \cdot (-9) \qquad\qquad \frac{4x}{4} > \frac{-12}{4}$$

$$x < -18 \qquad\qquad x > -3$$

These properties are also true for ≤ and ≥.

EXAMPLE 1 Solving an Inequality Using Multiplication

Solve $\frac{x}{8} > -5$. Graph the solution.

$\frac{x}{8} > -5$	Write the inequality.
Undo the division. → $8 \cdot \frac{x}{8} > 8 \cdot (-5)$	Multiply each side by 8.
$x > -40$	Simplify.

The solution is $x > -40$.

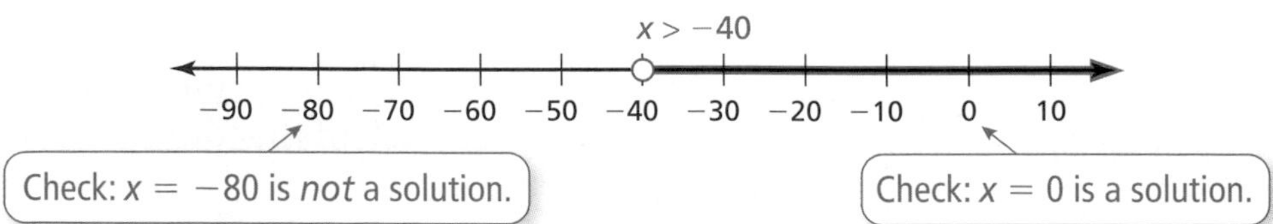

x > −40
−90 −80 −70 −60 −50 −40 −30 −20 −10 0 10
Check: x = −80 is *not* a solution.
Check: x = 0 is a solution.

On Your Own

Solve the inequality. Graph the solution.

1. $a \div 2 < 4$

2. $\frac{n}{7} \geq -1$

3. $-6.4 \geq \frac{w}{5}$

Laurie's Notes

Introduction

Connect

- **Yesterday:** Students gained an intuitive understanding of solving inequalities involving multiplication and division. (MP1, MP7, MP8)
- **Today:** Students will use the Multiplication and Division Properties of Inequality to solve inequalities.

Motivate

? "Have you heard of Ultimate?" Answers will vary.

- It is a sport played with a flying disc at colleges, high schools, and some middle schools. There are 10 simple rules, one of which is there aren't any officials! Pretty cool.
- The popularity of the sport has skyrocketed, but there is *at most* one-fifth the numbers of students playing Ultimate as there are playing lacrosse. If there are 26 students playing Ultimate, what is the minimum number playing lacrosse? $\frac{1}{5}x \geq 26$; $x \geq 130$ players

Lesson Notes

Key Idea

- These properties should look familiar, as they are similar to the Multiplication and Division Properties of Equality used in solving equations.
- Note that the properties are restricted to multiplying and dividing by a *positive* number. This is very important.

Example 1

? "How do you isolate the variable, meaning get x by itself?" Multiply by 8 on each side of the inequality.

- **MP6 Attend to Precision:** Remind students that they are using the Multiplication Property of Inequality. Multiplying by 8 is the inverse operation of dividing by 8.
- **Representation:** Multiplication is represented by the dot notation, and -5 is enclosed in parentheses for clarity only. Otherwise, students might become confused and think 5 is being subtracted.

On Your Own

- **Think-Pair-Share:** Students should read each question independently and then work with a partner to answer the questions. When they have answered the questions, the pair should compare their answers with another group and discuss any discrepancies.
- Division is represented in different ways in Questions 1 and 2. The second representation is more common in algebra.
- After solving the inequality in Question 3, the result will be $-32 \geq w$. Students can also rewrite this as $w \leq -32$. The direction of the inequality symbol is reversed *only* because the solution is being rewritten with the variable on the left side of the inequality statement.

Goal Today's lesson is solving inequalities using multiplication or division.

Start Thinking! and Warm Up

Lesson 3.3 Warm Up
For use before Lesson 3.3

Lesson 3.3 Start Thinking!
For use before Lesson 3.3

In football, a team has four attempts to advance the ball at least 10 yards, which earns them a first down.

Write an inequality to describe the average yards per down for a team who has earned a first down in a set of 4 downs.

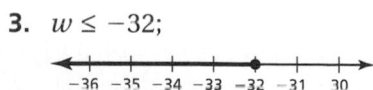

Extra Example 1

Solve $\frac{d}{5} < -7$. Graph the solution.

$d < -35$

−50 −40 −30 −20 −10 0

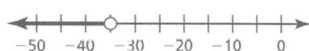

On Your Own

1. $a < 8$;

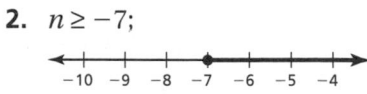

4 5 6 7 8 9 10

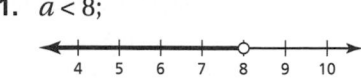

2. $n \geq -7$;

−10 −9 −8 −7 −6 −5 −4

3. $w \leq -32$;

−36 −35 −34 −33 −32 −31 30

Extra Example 2

Solve $2x \geq 12$. Graph the solution.
$x \geq 6$

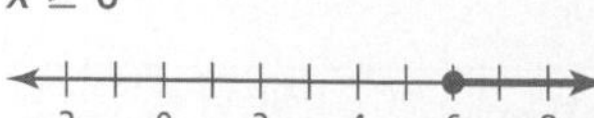

On Your Own

4. $b \geq 9$;

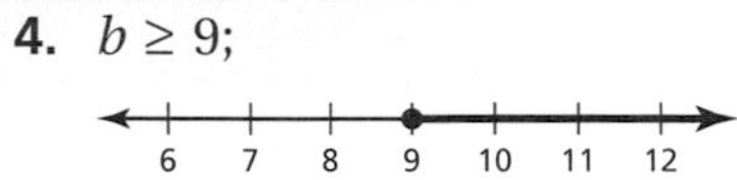

5. $k > -5$;

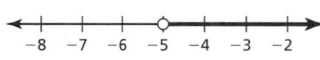

6. $q < -12$;

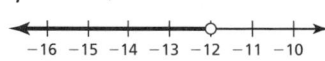

Differentiated Instruction

Visual

Use the inequality $6 < 9$ to show students why it is necessary to reverse the inequality symbol when multiplying or dividing by a negative number.

Add -3 to each side. The result is $3 < 6$, a true statement.

Subtract -3 from each side. The result is $9 < 12$, a true statement.

Multiply each side by -3. If the inequality is *not* reversed, the statement $-18 < -27$ is false. By reversing the inequality, the statement $-18 > -27$ is true.

Divide each side by -3. If the inequality is *not* reversed, the statement $-2 < -3$ is false. By reversing the inequality, the statement $-2 > -3$ is true.

Laurie's Notes

Example 2

? "What operation is being performed on the left side of the inequality?" multiplication

? "How do you undo a multiplication problem?" divide

- Solve, graph, and check.

On Your Own

- **Think-Pair-Share:** Students should read each question independently and then work with a partner to answer the questions. When they have answered the questions, the pair should compare their answers with another group and discuss any discrepancies.
- Notice that although all of the coefficients are positive, sometimes the constant is negative. This is important in helping students understand when the direction of the inequality symbol is going to be reversed. The focus is on the sign of the coefficient, not the sign of the constant.
- For Question 6, remind students that after solving this inequality, the result will be $-12 > q$. Students can also rewrite this as $q < -12$. The direction of the inequality symbol is reversed *only* because the solution is being rewritten with the variable on the left side of the inequality statement.
- These problems integrate review of decimal operations.

Key Idea

- These properties look identical to what they have been using in the lesson, *except* now the direction of the inequality symbol must be reversed for the inequality to remain true because they are multiplying or dividing by a *negative* quantity!
- The short version of the property: When you multiply or divide by a negative quantity, reverse the direction of the inequality symbol.
- **Common Error:** When students solve $2x < -4$, they sometimes reverse the inequality symbol because there's a negative number in the problem. The inequality symbol is reversed *only* when the coefficient is negative, not when the constant is negative.

EXAMPLE 2 Solving an Inequality Using Division

Solve $3x \leq -24$. Graph the solution.

	$3x \leq -24$	Write the inequality.
Undo the multiplication. →	$\dfrac{3x}{3} \leq \dfrac{-24}{3}$	Divide each side by 3.
	$x \leq -8$	Simplify.

∴ The solution is $x \leq -8$.

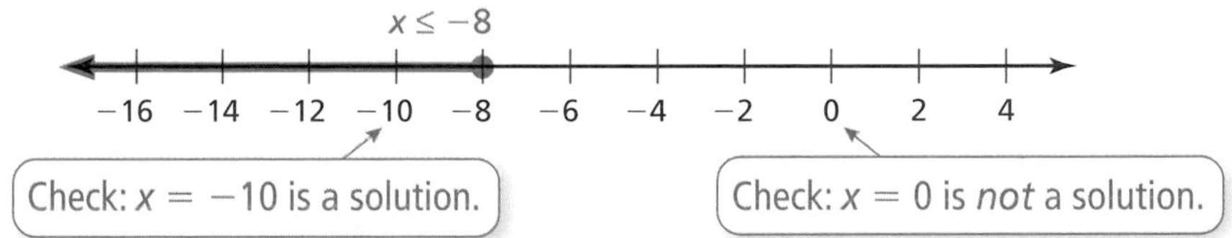

On Your Own

Now You're Ready
Exercises 10–18

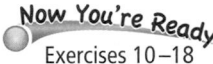

Solve the inequality. Graph the solution.

4. $4b \geq 36$ **5.** $2k > -10$ **6.** $-18 > 1.5q$

Key Idea

Multiplication and Division Properties of Inequality (Case 2)

Words If you multiply or divide each side of an inequality by the same *negative* number, the direction of the inequality symbol must be reversed for the inequality to remain true.

Numbers

$$-6 < 8 \qquad\qquad 6 > -8$$

$$(\ \ 2) \cdot (-6) > (-2) \cdot 8 \qquad\qquad \frac{6}{-2} < \frac{-8}{-2}$$

$$12 > -16 \qquad\qquad -3 < 4$$

Algebra

$$\frac{x}{-6} < 3 \qquad\qquad -5x > 30$$

$$-6 \cdot \frac{x}{-6} > -6 \cdot 3 \qquad\qquad \frac{-5x}{-5} < \frac{30}{-5}$$

$$x > -18 \qquad\qquad x < -6$$

These properties are also true for ≤ and ≥.

Common Error

A negative sign in an inequality does not necessarily mean you must reverse the inequality symbol.

Only reverse the inequality symbol when you multiply or divide both sides by a negative number.

EXAMPLE 3 Solving an Inequality Using Multiplication

Solve $\frac{y}{-3} > 2$. Graph the solution.

$\frac{y}{-3} > 2$ — Write the inequality.

Undo the division. → $-3 \cdot \frac{y}{-3} < -3 \cdot 2$ — Multiply each side by −3. Reverse the inequality symbol.

$y < -6$ — Simplify.

The solution is $y < -6$.

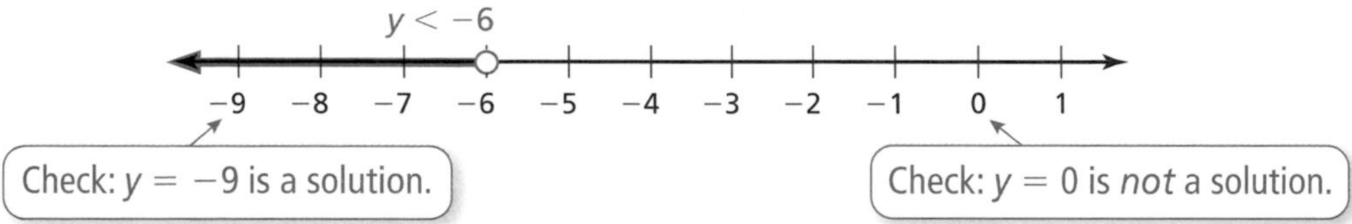

EXAMPLE 4 Solving an Inequality Using Division

Solve $-7y \leq -35$. Graph the solution.

$-7y \leq -35$ — Write the inequality.

Undo the multiplication. → $\frac{-7y}{-7} \geq \frac{-35}{-7}$ — Divide each side by −7. Reverse the inequality symbol.

$y \geq 5$ — Simplify.

The solution is $y \geq 5$.

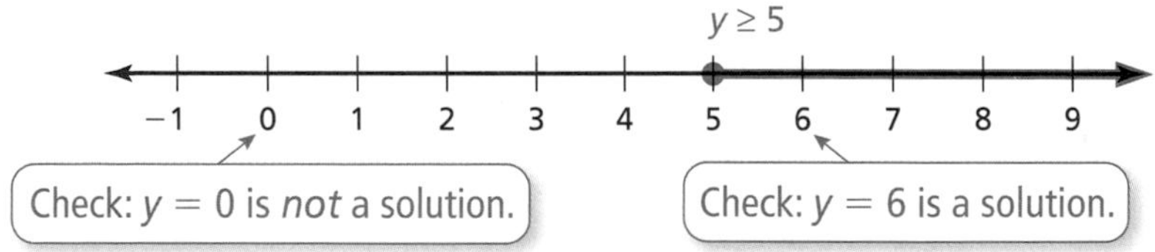

On Your Own

Now You're Ready
Exercises 27–35

Solve the inequality. Graph the solution.

7. $\frac{p}{-4} < 7$

8. $\frac{x}{-5} \leq -5$

9. $1 \geq -\frac{1}{10}z$

10. $-9m > 63$

11. $-2r \geq -22$

12. $-0.4y \geq -12$

Laurie's Notes

Lesson Notes

Example 3

- Write the example.
- ? "What operation is being performed?" division by -3
- ? "How do you undo dividing by -3?" multiply by -3
- Solve as usual, but remember to reverse the direction of the inequality symbol.
- When graphing, remember to use an open circle because the inequality is strictly *less than.*

Example 4

- Write the example.
- ? "What operation is being performed?" multiplication by -7
- ? "How do you undo multiplying by -7?" divide by -7
- Solve as usual, but remember to reverse the direction of the inequality symbol. Remember, the quotient of two negatives is positive.
- ? "Should you use an open or closed circle?" Use a closed circle because the inequality is greater than or equal to.

On Your Own

- **Neighbor Check:** Have students work independently and then have their neighbor check their work. Have students discuss any discrepancies.
- Have students share their work at the board.

Closure

- **Exit Ticket:** Solve and graph.

$\frac{x}{-3} \le -9$ $x \ge 27$

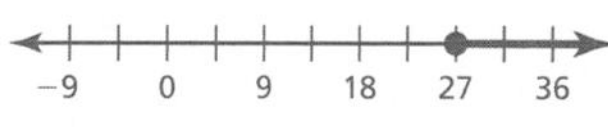

$-8 > 4x$ $x < -2$

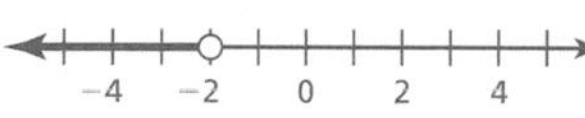

The Dynamic Planning Tool
Editable Teacher's Resources at *BigIdeasMath.com*

Extra Example 3

Solve $\frac{c}{-4} \le 3$. Graph the solution.

$c \ge -12$

Extra Example 4

Solve $-3j > -9$. Graph the solution.

$j < 3$

On Your Own

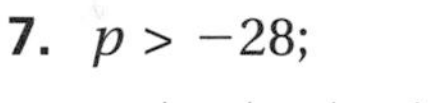

7. $p > -28$;

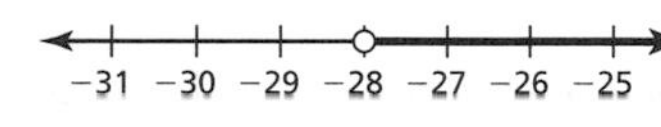

8. $x \ge 25$;

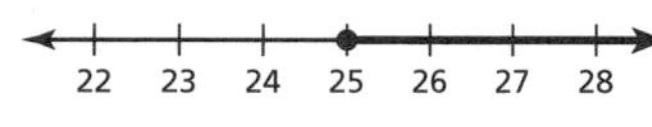

9. $z \ge -10$;

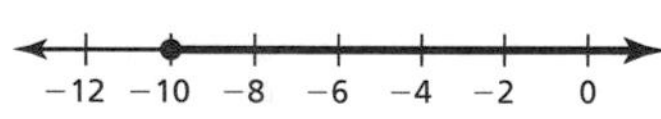

10. $m < -7$;

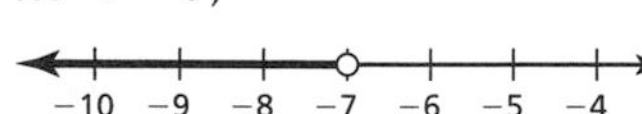

11. $r \le 11$;

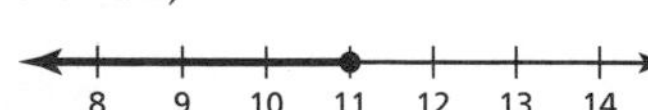

12. $y \le 30$;

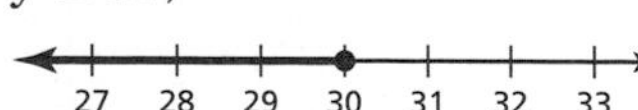

Vocabulary and Concept Check

1. Multiply each side of the inequality by 6.
2. The first inequality is divided by a positive number. The second inequality is divided by a negative number. Because this inequality is divided by a negative number, the direction of the inequality symbol must be reversed.
3. *Sample answer:* $-3x < 6$

Practice and Problem Solving

4. $x < 1$ **5.** $x \geq -1$

6. $x < -3$ **7.** $x \leq -3$

8. $x < -5$ **9.** $x \leq \frac{3}{2}$

10. $n > 6$;

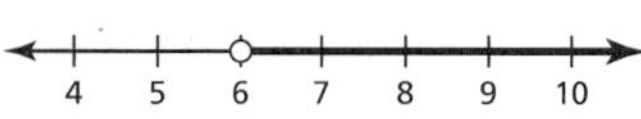

11. $c \leq -36$;

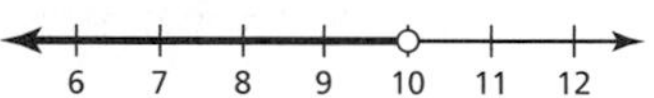

12. $m < 10$;

6 7 8 9 10 11 12

13–18. See Additional Answers.

19. The inequality sign should not have been reversed.

$$\frac{x}{2} < -5$$

$$2 \cdot \frac{x}{2} < 2 \cdot (-5)$$

$$x < -10$$

20. $\frac{x}{3} \leq 4; x \leq 12$

21. $\frac{x}{8} < -2; x < -16$

22. $4x \geq -12; x \geq -3$

23. $5x > 20; x > 4$

Assignment Guide and Homework Check

Level	Assignment	Homework Check
Average	1–9, 13–41 odd, 36, 38, 48–52	15, 21, 29, 36, 39
Advanced	1–9, 16–22 even, 19, 25, 32–42 even, 37, 44–52	16, 20, 34, 36, 42

Common Errors

- **Exercises 10–18** Students may perform the same operation on both sides instead of the opposite operation when solving the inequality. Remind them that solving inequalities is similar to solving equations.
- **Exercises 10–18** When there is a negative in the inequality, students may reverse the direction of the inequality symbol. Remind them that they only reverse the direction when they are multiplying or dividing by a negative number. All of these exercises keep the same inequality symbol.

3.3 Record and Practice Journal

Solve the inequality. Graph the solution.

1. $5n < 75$ — $n < 15$ (13 15 17)
2. $\frac{x}{6} \leq -12$ — $x \leq -72$ (−74 −72 −70)
3. $-15t > -60$ — $t < 4$ (1 2 3 4 5 6)
4. $-4q \geq 122$ — $q \leq -30.5$ (−30.5)
5. $-8p < \frac{4}{5}$ — $p > -\frac{1}{10}$ ($-\frac{1}{10}$ $\frac{1}{10}$ $\frac{3}{10}$)
6. $-9 \geq 2.4m$ — $m \leq -3.75$ (−3.75)
7. $-\frac{r}{2} \leq -11$ — $r \geq 22$ (20 22 24)
8. $-\frac{t}{6} > 1.2$ — $t < -7.2$ (−7.4 −7.2 −7)
9. $-4 \geq \frac{q}{-0.1}$ — $q \geq 0.4$ (0.2 0.4 0.6)
10. To win a trivia game, you need at least 60 points. Each question is worth 4 points. Write and solve an inequality that represents the number of questions you need to answer correctly to win the game.
$4x \geq 60; x \geq 15$

Technology For the Teacher
Answer Presentation Tool

3.3 Exercises

Vocabulary and Concept Check

1. **VOCABULARY** Explain how to solve $\frac{x}{6} < -5$.
2. **WRITING** Explain how solving $2x < -8$ is different from solving $-2x < 8$.
3. **OPEN-ENDED** Write an inequality that is solved using the Division Property of Inequality where the inequality symbol needs to be reversed.

Practice and Problem Solving

Use a table to solve the inequality.

4. $4x < 4$
5. $-2x \le 2$
6. $-5x > 15$
7. $\frac{x}{-3} \ge 1$
8. $\frac{x}{-2} > \frac{5}{2}$
9. $\frac{x}{4} \le \frac{3}{8}$

Solve the inequality. Graph the solution.

1 2 10. $3n > 18$
11. $\frac{c}{4} \le -9$
12. $1.2m < 12$
13. $-14 > x \div 2$
14. $\frac{w}{5} \ge -2.6$
15. $5 < 2.5k$
16. $4x \le -\frac{3}{2}$
17. $2.6y \le -10.4$
18. $10.2 > \frac{b}{3.4}$

19. **ERROR ANALYSIS** Describe and correct the error in solving the inequality.

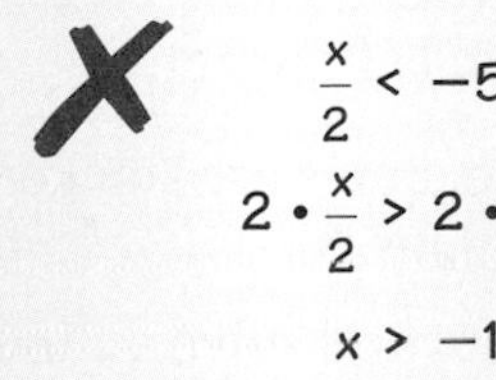

Write the word sentence as an inequality. Then solve the inequality.

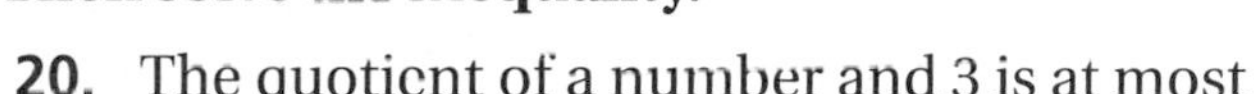

20. The quotient of a number and 3 is at most 4.
21. A number divided by 8 is less than -2.
22. Four times a number is at least -12.
23. The product of 5 and a number is greater than 20.

24. **CAMERA** You earn \$9.50 per hour at your summer job. Write and solve an inequality that represents the number of hours you need to work in order to buy a digital camera that costs \$247.

25. **COPIES** You have \$3.65 to make copies. Write and solve an inequality that represents the number of copies you can make.

26. **SPEED LIMIT** The maximum speed limit for a school bus is 55 miles per hour. Write and solve an inequality that represents the number of hours it takes to travel 165 miles in a school bus.

Solve the inequality. Graph the solution.

③ ④ 27. $-2n \le 10$

28. $-5w > 30$

29. $\frac{h}{-6} \ge 7$

30. $-8 < -\frac{1}{3}x$

31. $-2y < -11$

32. $-7d \ge 56$

33. $2.4 > -\frac{m}{5}$

34. $\frac{k}{-0.5} \le 18$

35. $-2.5 > \frac{b}{-1.6}$

36. **ERROR ANALYSIS** Describe and correct the error in solving the inequality.

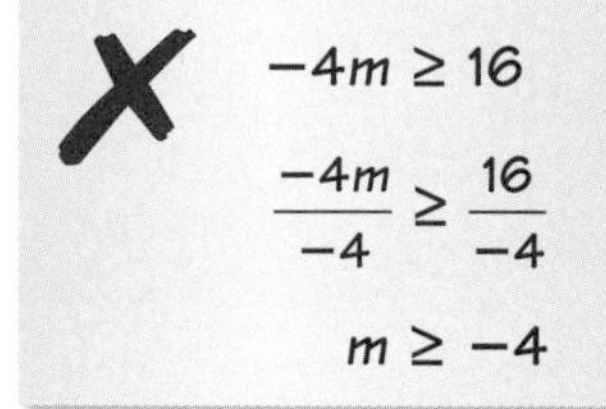

37. **CRITICAL THINKING** Are all numbers greater than zero solutions of $-x > 0$? Explain.

38. **TRUCKING** In many states, the maximum height (including freight) of a vehicle is 13.5 feet.

    a. Five crates are stacked vertically on the bed of the truck. Is this legal? Explain.

    b. Write and solve an inequality to justify your answer to part (a).

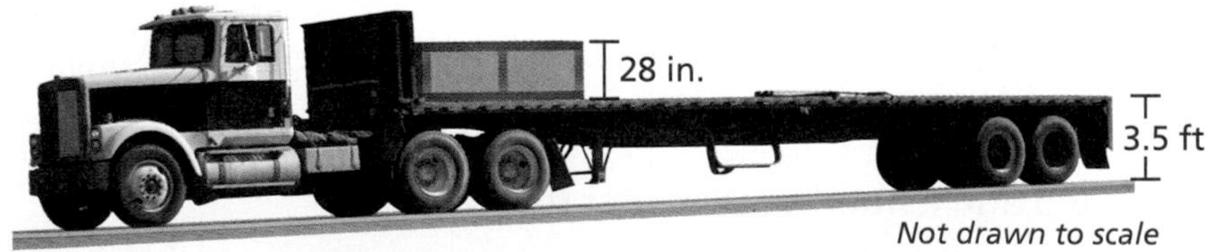

Write and solve an inequality that represents the value of x.

39. Area $\ge$ 102 cm^2

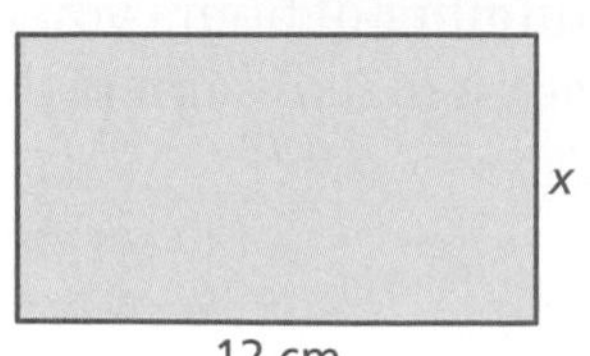

40. Area < 30 ft^2

Common Errors

- **Exercise 25** Students may forget to change the cost per copy to a decimal. In the photo, the cost is given as a whole number of cents. Some students will write $25x \le 3.65$. Point out to them that they need to write the cost per copy in dollars (0.25). Unit analysis can and should be used when working with rates.
- **Exercises 27–35** Students may forget to reverse the inequality symbol when multiplying or dividing by a negative number. Remind them of this rule. Encourage students to substitute values into the original inequality to check that the solution is correct.
- **Exercise 41** Students may write an incorrect inequality before solving. They may write $\frac{x}{3} < 80$ because there are three friends. However, the student is included in the trip as well, so there are 4 people going on the trip. The inequality should be $\frac{x}{4} < 80$.

Differentiated Instruction

Visual

Students may question why they are asked to graph the solution of an inequality, but not asked to graph the solution of an equation. The graph of an equation is just a point on the number line. The graph of an inequality provides more information, because of the infinite number of possible solutions.

Practice and Problem Solving

24. $9.5x \ge 247$; $x \ge 26$ h

25. $0.25x \le 3.65$; $x \le 14.6$; You can make at most 14 copies.

26. $55x \ge 165$; $x \ge 3$ h

27. $n \ge -5$;

−6 −5 −4 −3 −2 −1 0

28. $w < -6$;

−8 −7 −6 −5 −4 −3 −2

29. $h \le -42$;

−46 −45 −44 −43 −42 −41 −40

30–35. See Additional Answers.

36. They forgot to reverse the inequality symbol.

$$-4m \ge 16$$
$$\frac{-4m}{-4} \le \frac{16}{-4}$$
$$m \le -4$$

37. no; You need to solve the inequality for x. The solution is $x < 0$. Therefore, numbers greater than 0 are not solutions.

38. a. no; The maximum height allowed is 162 inches, and 5 crates on the truck has a height of 182 inches.

b. $\frac{28}{12}x \le 10$; $x \le \frac{30}{7}$, or $4\frac{2}{7}$ At most 4 crates can be stacked.

39. $12x \ge 102$; $x \ge 8.5$ cm

Practice and Problem Solving

40. $5x < 30; x < 6$ ft

41. $\frac{x}{4} < 80; x < \$320$

42. See Additional Answers.

43. *Answer should include, but is not limited to:* Make sure students use the correct number of months that the CD has been out.

44. See *Taking Math Deeper*.

45. $n \geq -6$ and $n \leq -4$;

46. $x \geq 2$;

47. $m < 20$;

Fair Game Review

48. 4 **49.** $8\frac{1}{4}$

50. 66 **51.** 84

52. A

Mini-Assessment

Solve the inequality. Graph the solution.

1. $2z < 4$ $z < 2$

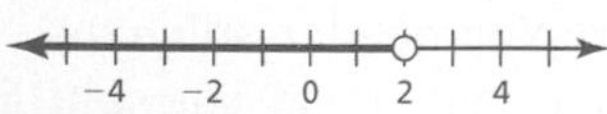

2. $15 \geq 5k$ $k \leq 3$

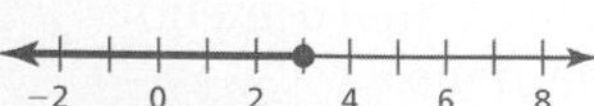

3. $\frac{m}{4} \geq -3$ $m \geq -12$

4. $3 < \frac{\ell}{-6}$ $\ell < -18$

5. $-6p < -36$ $p > 6$

Taking Math Deeper

Exercise 44

Double (or compound) inequalities, like those in Exercises 44–47, can often be written using a single inequality statement.

Begin by graphing each inequality.

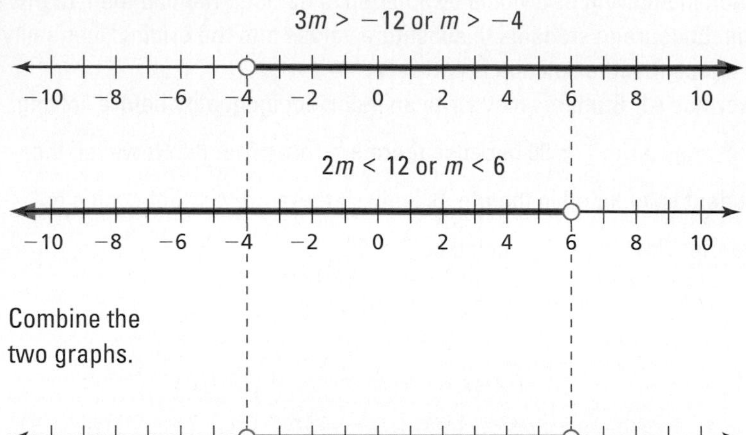

2 Combine the two graphs.

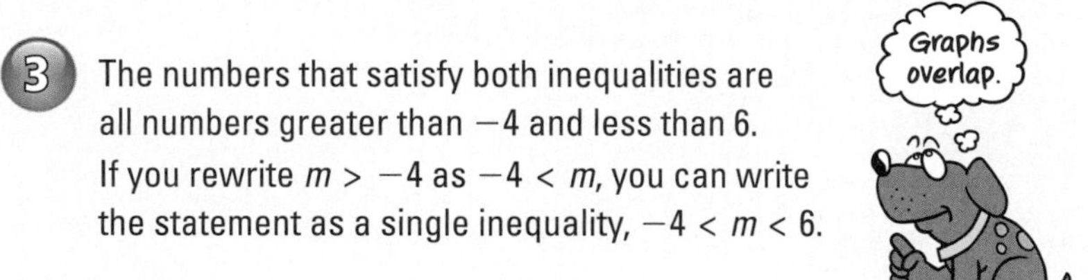

3 The numbers that satisfy both inequalities are all numbers greater than -4 and less than 6. If you rewrite $m > -4$ as $-4 < m$, you can write the statement as a single inequality, $-4 < m < 6$.

Project

Use the newspaper, Internet, TV, radio, or any other source to record and graph at least 10 different uses of inequalities.

Reteaching and Enrichment Strategies

If students need help. . .	If students got it. . .
Resources by Chapter • Practice A and Practice B • Puzzle Time Record and Practice Journal Practice Differentiating the Lesson Lesson Tutorials Skills Review Handbook	Resources by Chapter • Enrichment and Extension • School-to-Work Start the next section

41. **TRIP** You and three friends are planning a trip. You want to keep the cost below \$80 per person. Write and solve an inequality that represents the total cost of the trip.

42. **PRECISION** Explain why the direction of the inequality symbol must be reversed when multiplying or dividing by the same negative number.

43. **PROJECT** Choose two musical artists to research.

   a. Use the Internet or a magazine to complete the table.

   b. Find and compare the average number of copies sold per month for each CD. Which CD do you consider to be the most successful? Explain.

   c. Assume each CD continues to sell at the average rate. Write and solve an inequality that represents the number of months it will take for the total number of copies sold to exceed twice the current number sold.

Artist	Name of CD	Release Date	Current Number of Copies Sold
1.			
2.			

Structure **Describe all numbers that satisfy *both* inequalities. Include a graph with your description.**

44. $3m > -12$ and $2m < 12$

45. $\frac{n}{2} \geq -3$ and $\frac{n}{-4} \geq 1$

46. $2x \geq -4$ and $2x \geq 4$

47. $\frac{m}{-4} > -5$ and $\frac{m}{4} < 10$

Fair Game Review What you learned in previous grades & lessons

Solve the equation. *(Section 1.2)*

48. $-4w + 5 = -11$

49. $4(x - 3) = 21$

50. $\frac{v}{6} - 7 = 4$

51. $\frac{m + 300}{4} = 96$

52. **MULTIPLE CHOICE** Which of the following is *not* a solution of $p - 3.9 \geq 0.8$? *(Section 3.2)*

Ⓐ $p = -4.5$ Ⓑ $p = 4.7$ Ⓒ $p = 4.75$ Ⓓ $p = 5$

3 Study Help

You can use a **four square** to organize information about a topic. Each of the four squares can be a category, such as *definition, vocabulary, example, non-example, words, algebra, table, numbers, visual, graph,* or *equation*. Here is an example of a four square for an inequality.

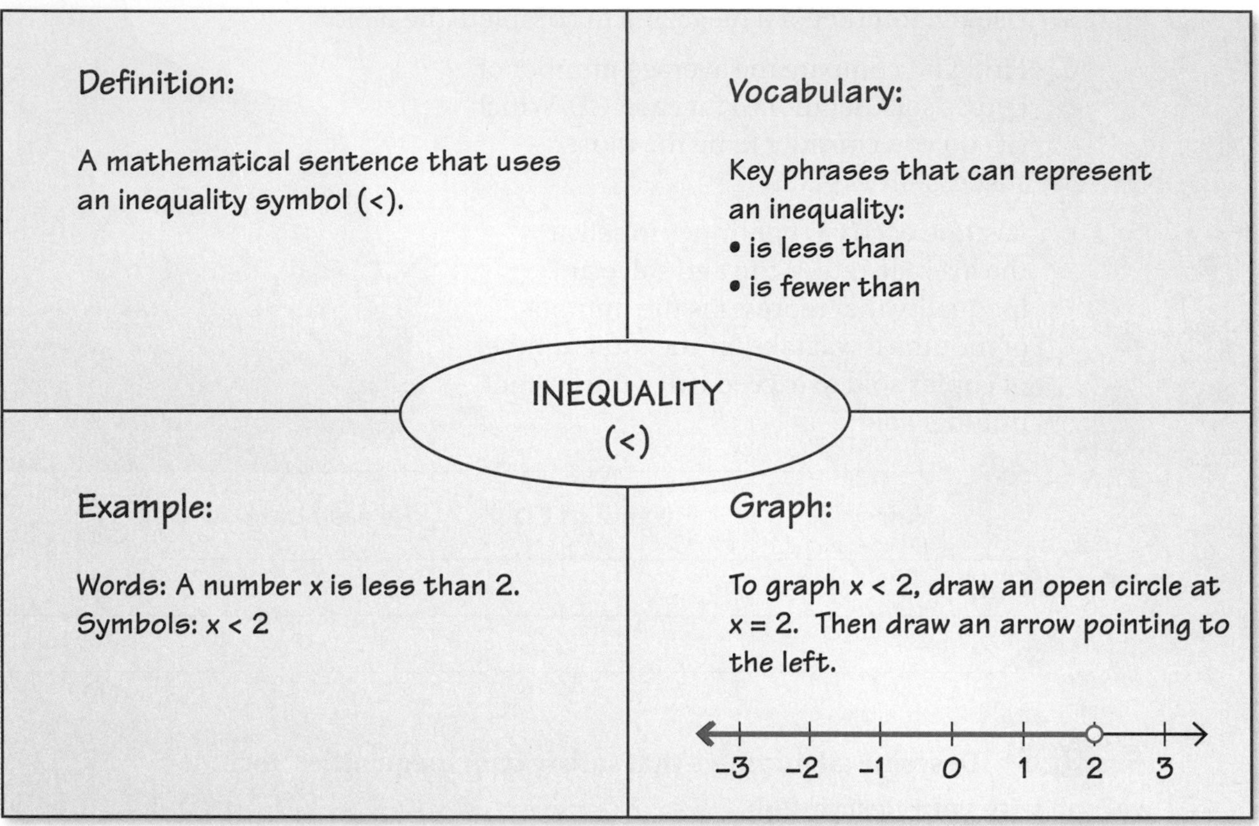

Definition:
A mathematical sentence that uses an inequality symbol (<).
Vocabulary:
Key phrases that can represent an inequality:
• is less than
• is fewer than
INEQUALITY (<)
Example:
Words: A number x is less than 2.
Symbols: x < 2
Graph:
To graph x < 2, draw an open circle at x = 2. Then draw an arrow pointing to the left.
−3 −2 −1 0 1 2 3

On Your Own

Make four squares to help you study these topics.

1. inequality (≥)
2. solving an inequality using addition
3. solving an inequality using subtraction
4. solving an inequality using multiplication
5. solving an inequality using division

After you complete this chapter, make four squares for the following topics.

6. solving a compound inequality
7. graphing an inequality in two variables

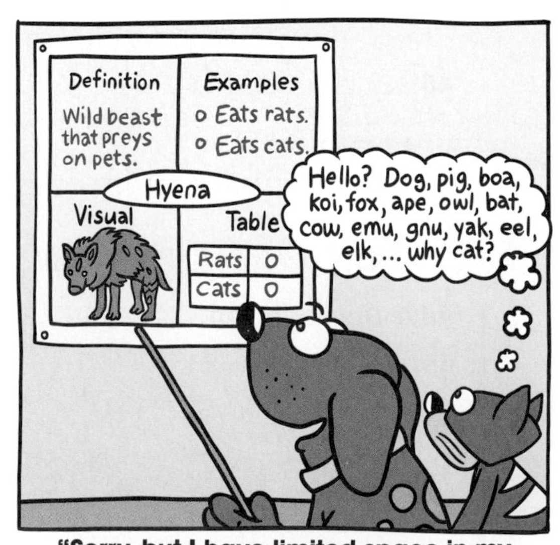

"Sorry, but I have limited space in my four square. I needed pet names with only three letters."

Sample Answers

1.

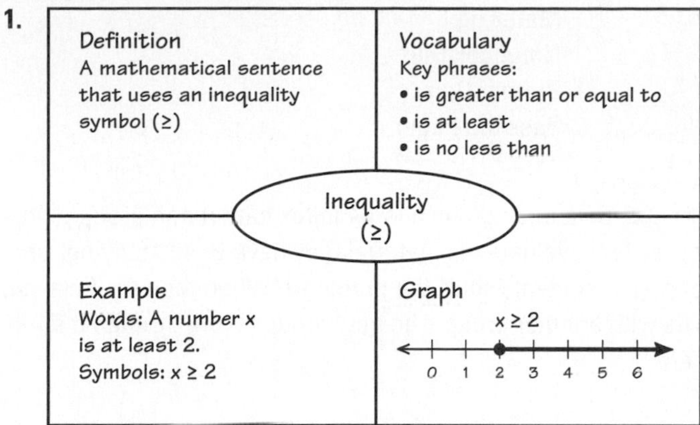

2.

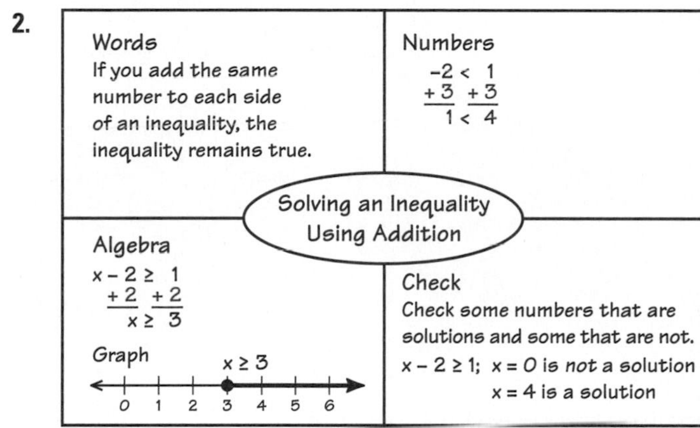

Words
If you add the same number to each side of an inequality, the inequality remains true.
Numbers
−2 < 1
+3 +3
1 < 4
Solving an Inequality Using Addition
Algebra
x − 2 ≥ 1
+2 +2
x ≥ 3
Graph
x ≥ 3
0 1 2 3 4 5 6
Check
Check some numbers that are solutions and some that are not.
x − 2 ≥ 1; x = 0 is not a solution
x = 4 is a solution

3.

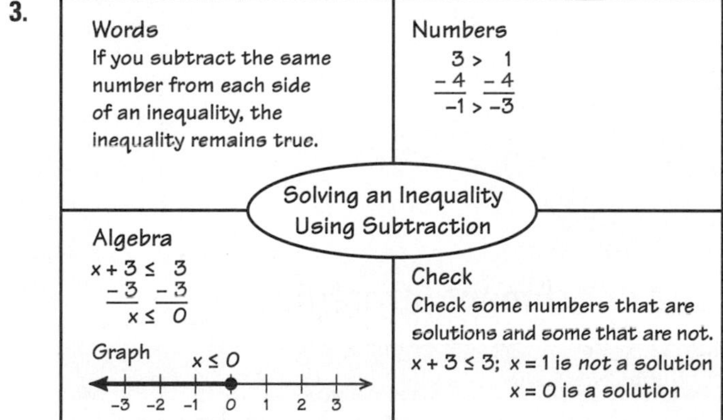

Words
If you subtract the same number from each side of an inequality, the inequality remains true.
Numbers
3 > 1
−4 −4
−1 > −3
Solving an Inequality Using Subtraction
Algebra
x + 3 ≤ 3
−3 −3
x ≤ 0
Graph
x ≤ 0
−3 −2 −1 0 1 2 3
Check
Check some numbers that are solutions and some that are not.
x + 3 ≤ 3; x = 1 is not a solution
x = 0 is a solution

4–5. Available at *BigIdeasMath.com*.

List of Organizers

Available at *BigIdeasMath.com*

Comparison Chart
Concept Circle
Definition (Idea) and Example Chart
Example and Non-Example Chart
Formula Triangle
Four Square
Information Frame
Information Wheel
Notetaking Organizer
Process Diagram
Summary Triangle
Word Magnet
Y Chart

About this Organizer

A **Four Square** can be used to organize information about a topic. Students write the topic in the "bubble" in the middle of the four square. Then students write concepts related to the topic in the four squares surrounding the bubble. Any concept related to the topic can be used. Encourage students to include concepts that will help them learn the topic. Students can place their four squares on note cards to use as a quick study reference.

Technology For the Teacher
Vocabulary Puzzle Builder

Answers

1. $x + 1 < -13$
2. $t - 1.6 \leq 9$
3. yes 4. no
5. −13 −12 −11 −10 −9 −8 −7
6. 6.4 6.5 6.6 6.7 6.8 6.9 7
7. $x < 6$

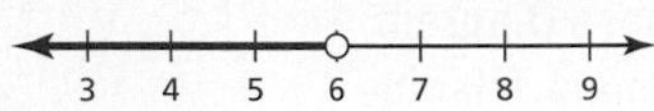

8. $g \geq 16$

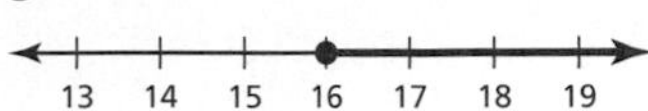

9. $h \leq -8$

10. $1 < p$

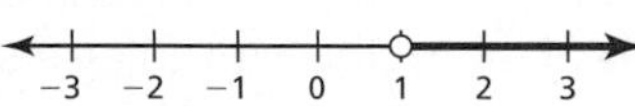

11. $n \leq 12$;

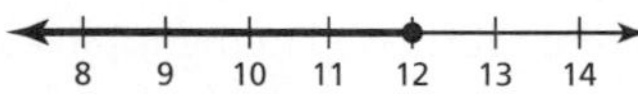

12. $y \leq -15$;
−18 −17 −16 −15 −14 −13 −12

13. $\frac{x}{6} > 9$; $x > 54$

14–15. See Additional Answers.

16. 7

17. $5b < 35$; $b < 7$ ft

Assessment Book

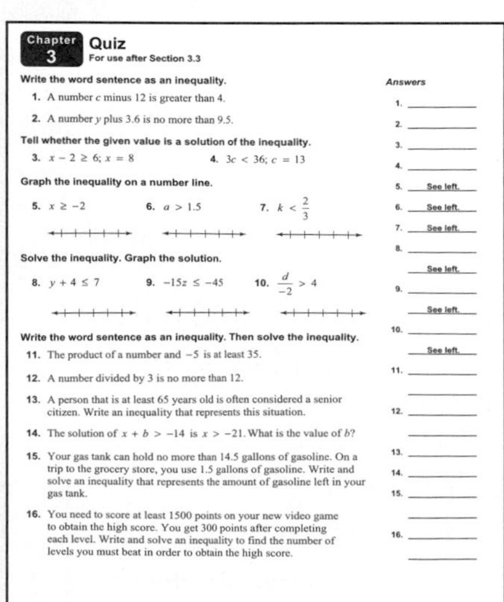

Chapter 3 **Quiz**
For use after Section 3.3

Write the word sentence as an inequality.

1. A number c minus 12 is greater than 4.
2. A number y plus 3.6 is no more than 9.5.

Tell whether the given value is a solution of the inequality.

3. $x - 2 \geq 6$; $x = 8$
4. $3c < 36$; $c = 13$

Graph the inequality on a number line.

5. $x \geq -2$
6. $a > 1.5$
7. $k < \frac{2}{3}$

Solve the inequality. Graph the solution.

8. $y + 4 \leq 7$
9. $-15z \leq -45$
10. $\frac{d}{-2} > 4$

Write the word sentence as an inequality. Then solve the inequality.

11. The product of a number and −5 is at least 35.
12. A number divided by 3 is no more than 12.
13. A person that is at least 65 years old is often considered a senior citizen. Write an inequality that represents this situation.
14. The solution of $x + b > -14$ is $x > -21$. What is the value of b?
15. Your gas tank can hold no more than 14.5 gallons of gasoline. On a trip to the grocery store, you use 1.5 gallons of gasoline. Write and solve an inequality that represents the amount of gasoline left in your gas tank.
16. You need to score at least 1500 points on your new video game to obtain the high score. You get 300 points after completing each level. Write and solve an inequality to find the number of levels you must beat in order to obtain the high score.

Answers
1. ______
2. ______
3. ______
4. ______
5. See left.
6. See left.
7. See left.
8. ______ See left.
9. ______ See left.
10. ______ See left.
11. ______
12. ______
13. ______
14. ______
15. ______
16. ______

Alternative Quiz Ideas

100% Quiz	Math Log
Error Notebook	Notebook Quiz
Group Quiz	Partner Quiz
Homework Quiz	Pass the Paper

Group Quiz

Students work in groups. Give each group a large index card. Each group writes five questions that they feel evaluate the material they have been studying. On a separate piece of paper, students solve the problems. When they are finished, they exchange cards with another group. The new groups work through the questions on the card.

Reteaching and Enrichment Strategies

If students need help. . .	If students got it. . .
Resources by Chapter • Study Help • Practice A and Practice B • Puzzle Time Lesson Tutorials *BigIdeasMath.com* Practice Quiz Practice from the Test Generator	Resources by Chapter • Enrichment and Extension • School-to-Work Game Closet at *BigIdeasMath.com* Start the next section

Technology For the Teacher

Answer Presentation Tool
Big Ideas Test Generator

3.1–3.3 Quiz

Check It Out
Progress Check
BigIdeasMath.com

Write the word sentence as an inequality. *(Section 3.1)*

1. A number x plus 1 is less than -13.

2. A number t minus 1.6 is at most 9.

Tell whether the given value is a solution of the inequality. *(Section 3.1)*

3. $12n < -2;\ n = -1$

4. $y + 4 < -3;\ y = -7$

Graph the inequality on a number line. *(Section 3.1)*

5. $x > -10$

6. $w < 6.8$

Solve the inequality. Graph the solution. *(Section 3.2 and Section 3.3)*

7. $x - 2 < 4$

8. $g + 14 \geq 30$

9. $h - 1 \leq -9$

10. $\frac{3}{2} < p + \frac{1}{2}$

11. $\frac{n}{-6} \geq -2$

12. $-4y \geq 60$

Write the word sentence as an inequality. Then solve the inequality. *(Section 3.3)*

13. The quotient of a number and 6 is more than 9.

14. Five times a number is at most -10.

LIFEGUARDS NEEDED
Take Our Training Course NOW!!!
Lifeguard Training Requirements
- Swim at least 100 yards
- Tread water for at least 5 minutes
- Swim 10 yards or more underwater without taking a breath

15. LIFEGUARD Three requirements for a lifeguard training course are shown. *(Section 3.1)*

a. Write and graph three inequalities that represent the requirements.

b. You can swim 350 feet. Do you satisfy the swimming requirement of the course? Explain.

16. REASONING The solution of $x - a > 4$ is $x > 11$. What is the value of a? *(Section 3.2)*

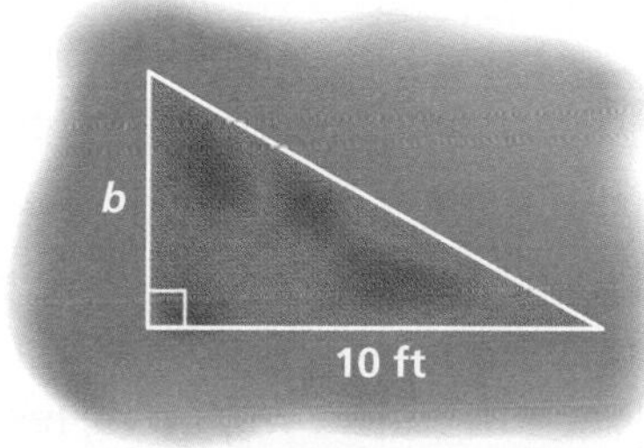

17. GARDEN The area of the triangular garden must be less than 35 square feet. Write and solve an inequality that represents the value of b. *(Section 3.3)*

3.4 Solving Multi-Step Inequalities

Essential Question How can you use an inequality to describe the area and perimeter of a composite figure?

COMMON CORE STATE STANDARDS
A.CED.1
A.CED.3
A.REI.3

1 ACTIVITY: Areas and Perimeters of Composite Figures

Work with a partner.

a. For what values of x will the area of the blue region be greater than 12 square units?

b. For what values of x will the sum of the inner and outer perimeters of the blue region be greater than 20 units?

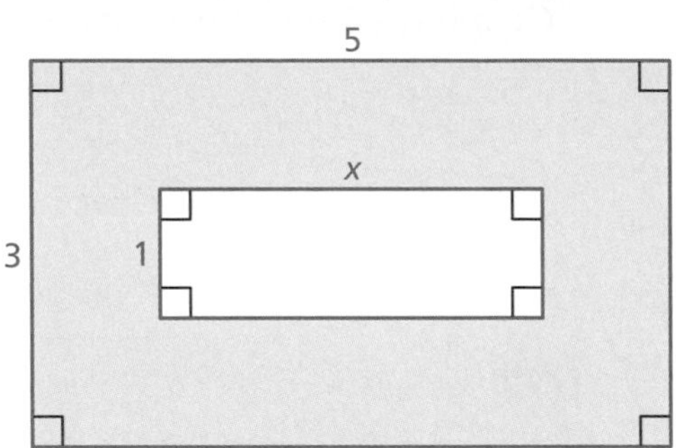

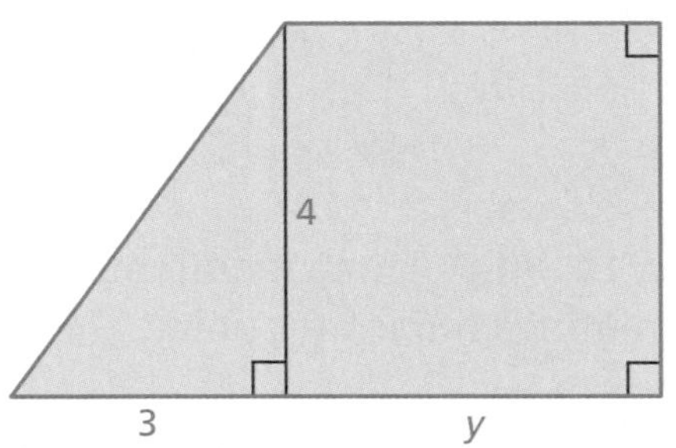

c. For what values of y will the area of the trapezoid be less than or equal to 10 square units?

d. For what values of y will the perimeter of the trapezoid be less than or equal to 16 units?

e. For what values of w will the area of the red region be greater than or equal to 36 square units?

f. For what values of w will the sum of the inner and outer perimeters of the red region be greater than 47 units?

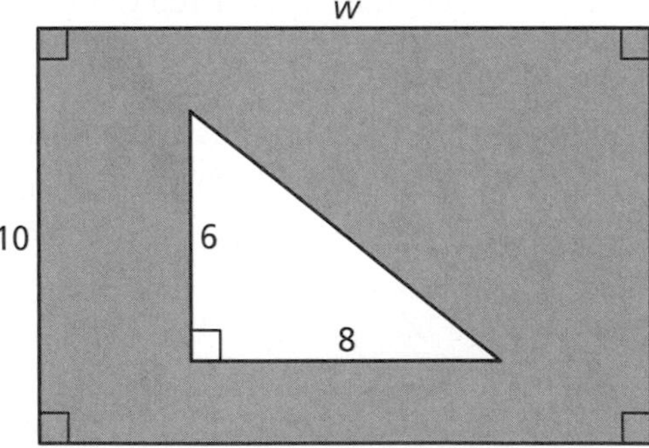

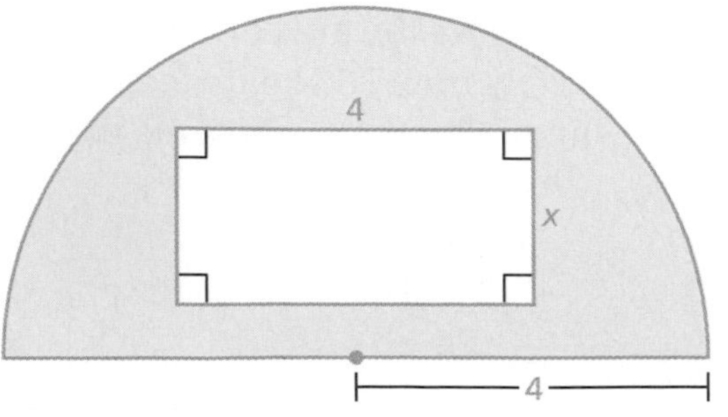

g. For what values of x will the area of the yellow region be less than 4π square units?

h. For what values of x will the sum of the inner and outer perimeters of the yellow region be less than $4\pi + 20$ units?

Laurie's Notes

Introduction

Standards for Mathematical Practice

- **MP7 Look For and Make Use of Structure:** Students have solved multi-step equations. Structurally, multi-step inequalities are solved in the same fashion. Mathematically proficient students recognize the similarity between solving $-4x - 5 = 41$ and $-4x - 5 < 41$. Students must recall that when the variable term has been isolated, if the coefficient is negative, the inequality symbol is reversed.
- **MP2 Reason Abstractly and Quantitatively:** A geometric context is used to generate the inequality. Once the inequality is written, the goal is for students to decontextualize the problem and manipulate the symbols. After solving the problem, students should look back at their solution and ask if it makes sense.

Motivate

- A regulation Ultimate field is 64 meters long and 37 meters wide, with two end zones of 18 meters each. Sometimes the length (64 meters) varies due to using existing fields. Let the length be x. Draw a sketch and find the area of the field in terms of x.

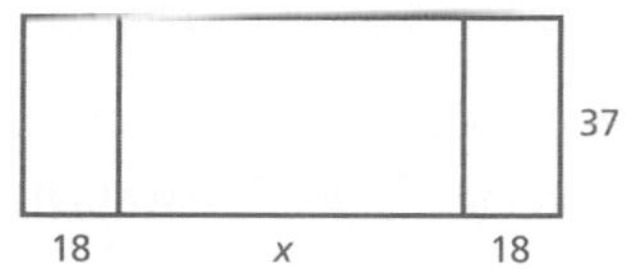

Area $= 37(x + 36) = (37x + 1332)\ \text{m}^2$

Activity Notes

Activity 1

- This activity uses composite figures. Sometimes instead of adding figures together, one figure is removed from another. For each problem, students will need to write a variable expression for the composite area or perimeter. This expression will then become part of an inequality to be solved.
- ? "How do you find the area of just the blue region?" Subtract the area of the inner rectangle from the area of the larger rectangle.
- ? "How do you find the total perimeter of the blue region?" Add the perimeter of the inner rectangle to the perimeter of the larger rectangle.
- Discuss with students that $x > 0$. At $x = 0$, the inner rectangle no longer exists. This is true for any geometric figure.
- These are neither trivial, nor simple problems. They require that students be familiar with the formulas and solve inequalities. My students always say that it feels like they're solving an equation and at the end they go back to make sure that they've paid attention to the inequality symbol.
- **MP2:** In parts (e) and (f), students should notice that the width of the red rectangle must be greater than 8, the base of the inner triangle.
- **MP2:** In part (g), the length of a segment from the midpoint of the base of the rectangle to an upper corner of the rectangle must be less than 4, the radius of the circle. So, $x^2 + 2^2 < 4^2$, or $x < 2\sqrt{3}$.
- As students are solving the problems, you may need to remind them about collecting like terms and the Distributive Property.

Common Core State Standards

A.CED.1 Create . . . inequalities in one variable and use them to solve problems.

A.CED.3 Represent constraints by . . . inequalities . . . and interpret solutions as viable or non-viable options in a modeling context.

A.REI.3 Solve linear . . . inequalities in one variable . . ..

Previous Learning

Students should know how to solve multi-step equations and one-step inequalities.

Start Thinking! and Warm Up

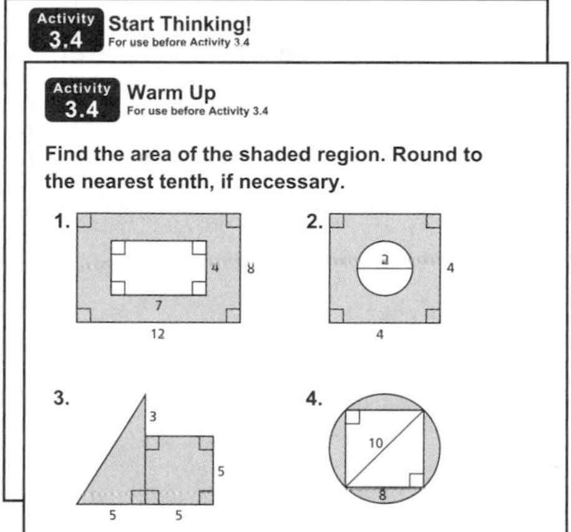

Activity 3.4 **Start Thinking!** For use before Activity 3.4

Activity 3.4 **Warm Up** For use before Activity 3.4

Find the area of the shaded region. Round to the nearest tenth, if necessary.

3.4 Record and Practice Journal

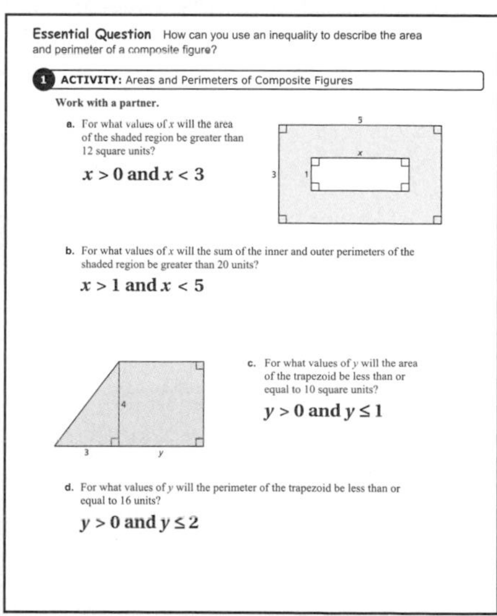

Essential Question How can you use an inequality to describe the area and perimeter of a composite figure?

1 **ACTIVITY:** Areas and Perimeters of Composite Figures

Work with a partner.

a. For what values of x will the area of the shaded region be greater than 12 square units?

$x > 0$ and $x < 3$

b. For what values of x will the sum of the inner and outer perimeters of the shaded region be greater than 20 units?

$x > 1$ and $x < 5$

c. For what values of y will the area of the trapezoid be less than or equal to 10 square units?

$y > 0$ and $y \le 1$

d. For what values of y will the perimeter of the trapezoid be less than or equal to 16 units?

$y > 0$ and $y \le 2$

Differentiated Instruction

Visual

Discuss with students the composite figure in Activity 1(a). Ask students if the number -1 makes sense as a solution. Ask students if the number 6 makes sense as a solution. Point out that in application problems, students need to determine whether the solution of an inequality makes sense as an answer.

3.4 Record and Practice Journal

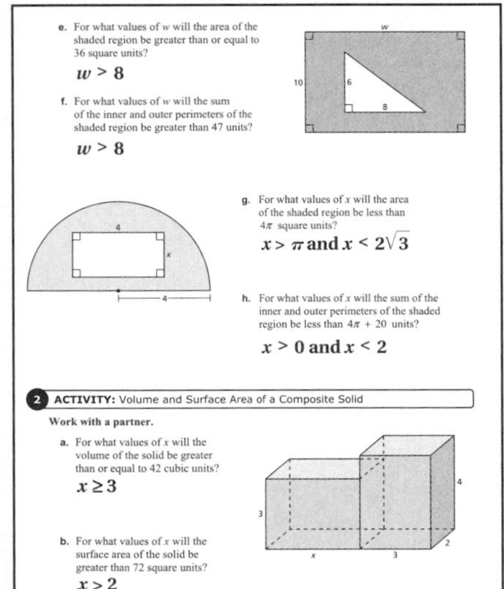

e. For what values of w will the area of the shaded region be greater than or equal to 36 square units?

$w > 8$

f. For what values of w will the sum of the inner and outer perimeters of the shaded region be greater than 47 units?

$w > 8$

g. For what values of x will the area of the shaded region be less than 4π square units?

$x > \pi$ and $x < 2\sqrt{3}$

h. For what values of x will the sum of the inner and outer perimeters of the shaded region be less than $4\pi + 20$ units?

$x > 0$ and $x < 2$

2 ACTIVITY: Volume and Surface Area of a Composite Solid

Work with a partner.

a. For what values of x will the volume of the solid be greater than or equal to 42 cubic units?

$x \geq 3$

b. For what values of x will the surface area of the solid be greater than 72 square units?

$x > 2$

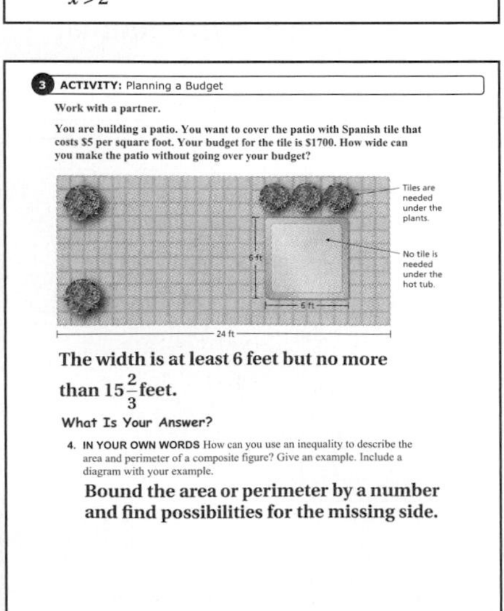

3 ACTIVITY: Planning a Budget

Work with a partner.

You are building a patio. You want to cover the patio with Spanish tile that costs $5 per square foot. Your budget for the tile is $1700. How wide can you make the patio without going over your budget?

The width is at least 6 feet but no more than $15\frac{2}{3}$ feet.

What Is Your Answer?

4. **IN YOUR OWN WORDS** How can you use an inequality to describe the area and perimeter of a composite figure? Give an example. Include a diagram with your example.

Bound the area or perimeter by a number and find possibilities for the missing side.

Laurie's Notes

Activity 2

- While eating in a restaurant, I noticed a cardboard model that had been folded from a net. It was a great example of a composite solid, a rectangular prism and a trapezoidal prism that looked like a fire truck! Of course, I asked the waiter if I could take it home.
- In part (a), students find the volume of each prism separately and add, $V = 6x + 24$. Then, they solve $6x + 24 \geq 42$.
- In part (b), students find the surface area of each prism and subtract the rectangular region that has been counted twice.

  Left prism: $S = 2(2x) + 2(6) + 2(3x) = 10x + 12$

  Right prism: $S = 2(12) + 2(6) + 2(8) = 52$

  Inside surface: $A = 6 \times 2 = 12$ (to be subtracted)

  Summary: Surface Area $= 10x + 12 + 52 - 12 = 10x + 52$

  Solve: $10x + 52 > 72$
- **MP3 Construct Viable Arguments and Critique the Reasoning of Others:** Ask two volunteers who have their work displayed in a neat and organized fashion to share their solutions at the board. You want to model good problem solving, as well as good mathematics. Hearing the reasoning process of classmates helps students improve their own problem solving strategies.

Activity 3

- Ask a student to read the problem.

? "Do you know the finished size of the patio?" no

- The width x is going to vary depending upon the budget. You want the patio as wide as possible.

? "How do you find the area of the patio if it is going to vary in size?" Use a variable to express the width.

- Area $= 24x - 36$; Cost: $5(24x - 36) \leq 1700$

$$120x - 180 \leq 1700$$
$$120x \leq 1880$$
$$x \leq 15\frac{2}{3}$$

Closure

- Refer to the Ultimate playing field from the beginning of class. How long can the field be so that the perimeter, including end zones, is less than 268 meters? less than 61 m

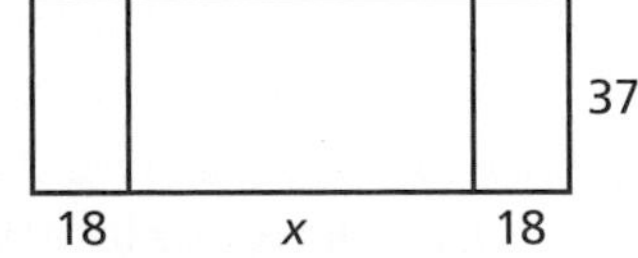

Technology For the Teacher

The Dynamic Planning Tool
Editable Teacher's Resources at *BigIdeasMath.com*

2 ACTIVITY: Volume and Surface Area of a Composite Solid

Work with a partner.

a. For what values of x will the volume of the solid be greater than or equal to 42 cubic units?

b. For what values of x will the surface area of the solid be greater than 72 square units?

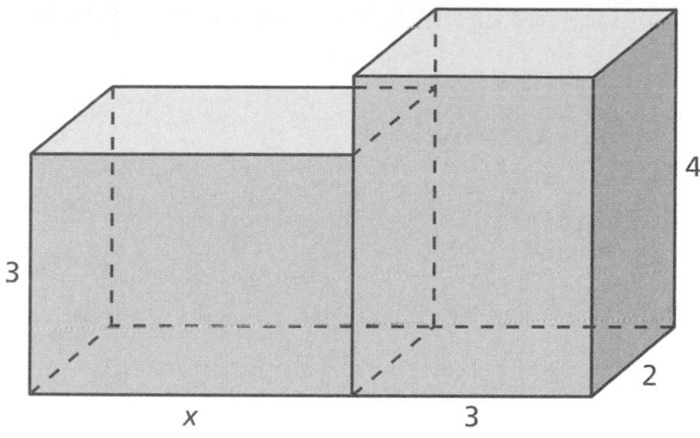

3 ACTIVITY: Planning a Budget

Work with a partner.

You are building a patio. You want to cover the patio with Spanish tile that costs \$5 per square foot. Your budget for the tile is \$1700. How wide can you make the patio without going over your budget?

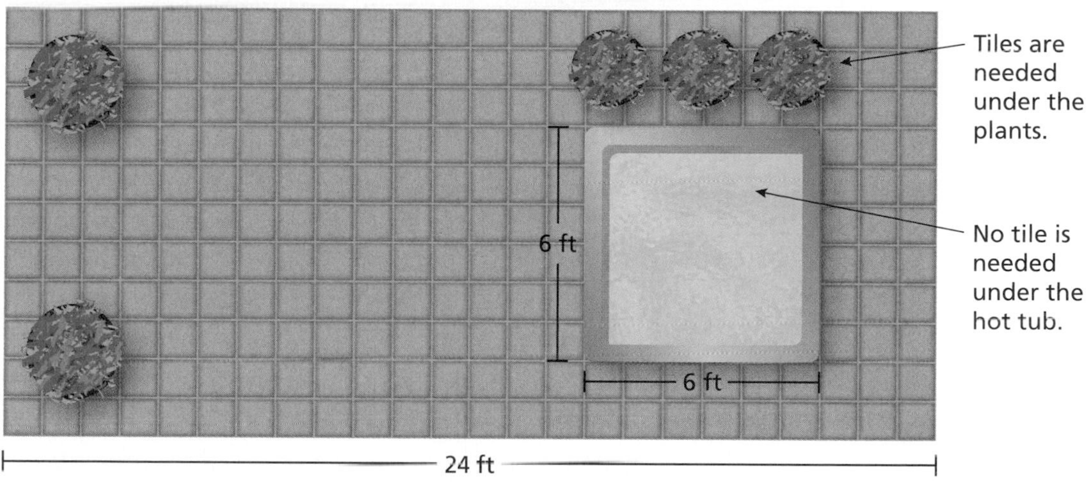

What Is Your Answer?

4. **IN YOUR OWN WORDS** How can you use an inequality to describe the area and perimeter of a composite figure? Give an example. Include a diagram with your example.

Practice Use what you learned about solving multi-step inequalities to complete Exercises 3 and 4 on page 130.

3.4 Lesson

You can use the properties of inequality to solve multi-step inequalities the same way you use the properties of equality to solve multi-step equations.

EXAMPLE 1 Solving a Multi-Step Inequality

Solve $\frac{y}{-6} + 7 < 9$. Graph the solution.

	$\frac{y}{-6} + 7 < 9$	Write the inequality.
Undo the addition.	$\underline{-7 \quad\; -7}$	Subtract 7 from each side.
	$\frac{y}{-6} < 2$	Simplify.
Undo the division.	$-6 \cdot \frac{y}{-6} > -6 \cdot 2$	Multiply each side by −6. Reverse the inequality symbol.
	$y > -12$	Simplify.

The solution is $y > -12$.

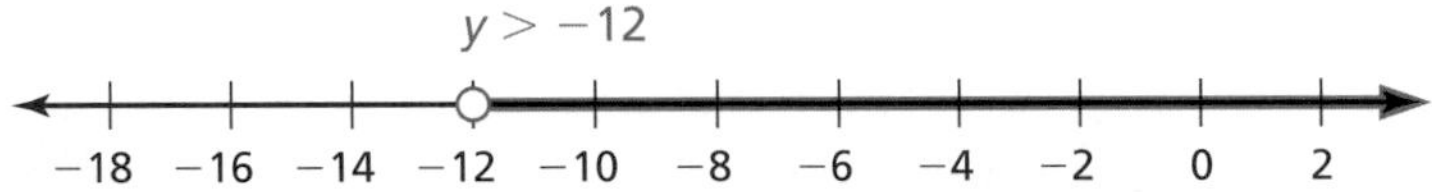

On Your Own

Now You're Ready
Exercises 5–10

Solve the inequality. Graph the solution.

1. $4b - 1 < 7$
2. $8 + 9c \geq -28$
3. $\frac{n}{-2} + 11 > 12$

When solving an inequality, if you obtain an inequality that is true, such as $-5 < 0$, then the solution is the set of *all real numbers*. If you obtain an inequality that is false, such as $3 \leq -2$, then the inequality has *no solutions*.

EXAMPLE 2 Solving an Inequality with No Solution

Solve $8x - 3 > 4(2x + 3)$.

$8x - 3 > 4(2x + 3)$	Write the inequality.
$8x - 3 > 8x + 12$	Distributive Property
$\underline{-8x} \qquad \underline{-8x}$	Subtract 8x from each side.
$-3 \not> 12$ ✗	Simplify.

The inequality $-3 > 12$ is false. So, there are no solutions.

Laurie's Notes

Introduction

Connect

- **Yesterday:** Students reviewed geometric formulas and gained an intuitive understanding of solving multi-step inequalities. (MP2, MP3, MP7)
- **Today:** Students will solve and graph multi-step inequalities.

Motivate

? "How many of you have played the game *Trivial Pursuit*® or a variation of it?" Answers will vary.

- This popular trivia game was created in 1979 by two friends who had lost some pieces to the game Scrabble®. They decided to create a new game. The rest is history. *Trivial Pursuit*® has sold more than 88 million copies in 26 countries, and in 17 languages. There are versions of the game that focus on sports, pop culture, and regional geography.

Lesson Notes

Discuss

- You solve multi-step inequalities the same way you solve multi-step equations. You only need to remember to change the direction of the inequality symbol if you multiply or divide by a negative quantity.
- Recall that solving an equation undoes the evaluating in reverse order. The goal is to isolate the variable.

Example 1

? "What operations are being performed on the left side of the inequality?" division and addition

? "What is the first step in isolating the variable, meaning getting the *y*-term by itself?" Subtract 7 from each side of the inequality.

? "To solve for *y*, what is the last step?" Multiply each side by −6 and change the direction of the inequality symbol.

- Graph and check. Remember to use an open circle because the variable cannot equal −12.

Example 2

- Once the students use the Distributive Property, they may be unsure about the next step. Students may choose to work with the variable terms or with the constant terms. Solve the inequality in more than one way so students recognize that the result is the same.

? "Does a solution of $-3 > 12$ make sense?" no

? "How can you interpret a result of $-3 > 12$?" There are no values of *x* that make the inequality true.

- **MP1a Make Sense of Problems:** Have students substitute different values for *x*. This does not prove that there is no solution, but it helps students make sense of the problem.

Goal Today's lesson is solving multi-step inequalities.

Start Thinking! and Warm Up

Lesson 3.4 Warm Up
For use before Lesson 3.4

Lesson 3.4 Start Thinking!
For use before Lesson 3.4

What are your math quiz grades so far this grading period?

What must you earn on the next quiz in order to have at least a C quiz average? a B? an A?

Extra Example 1

Solve $17 \le 3y - 4$. Graph the solution.
$y \ge 7$

On Your Own

1. $b < 2$;

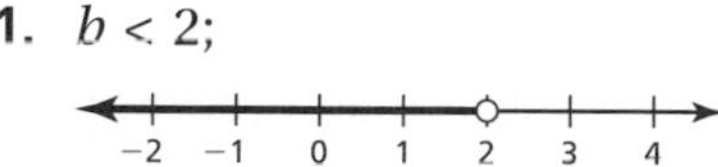

2. $c \ge -4$;

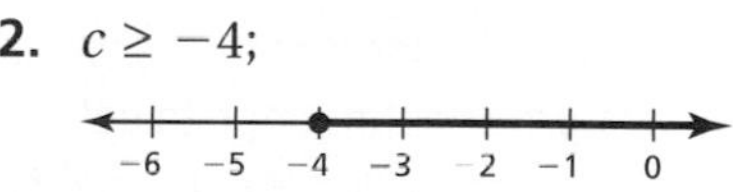

3. $n < -2$;

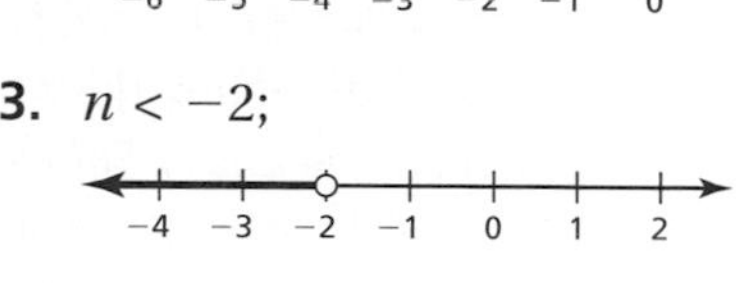

Extra Example 2

Solve $2(5x - 3) < 10x + 7$.
The solution is the set of all real numbers.

Extra Example 3

Solve $12 > -2(y - 4)$. Graph the solution. $y > -2$

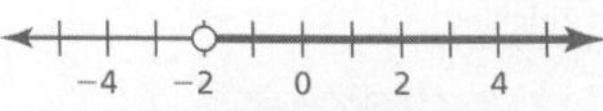

Extra Example 4

In Example 4, suppose you need a mean score of at least 85 to advance to the next round of the trivia game. What score do you need on the fifth game to advance? You need at least 73 points to advance to the next round.

On Your Own

4. all real numbers;

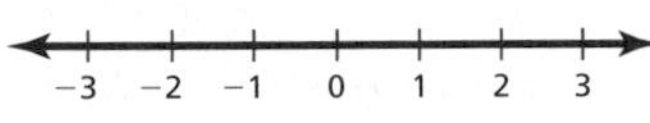

5. no solutions
6. at least 88 points

English Language Learners

Vocabulary

Give English learners the opportunity to use precise language to solve an inequality. Write the inequality $-2x + 4 > 8$ on the board. Have one student come to the board. For each step of the solution, call on another student to give the instruction for solving. The instructions should be given in complete sentences. The instructions for the inequality are:

(1) Subtract 4 from each side.
(2) Simplify.
(3) Divide each side by -2. Reverse the inequality symbol.
(4) Simplify.

Laurie's Notes

Example 3

- This example is similar to the last in that the variable terms are identical on each side of the inequality, so they subtract out. Unlike the last example, the remaining inequality is true.
- **?** "Does a solution of $-2 \leq 7$ make sense?" yes
- **?** "How can you interpret a result of $-2 \leq 7$?" All values of x make the inequality true.
- **?** "If all values of x make the inequality true, what should the graph look like?" All of the number line is shaded.
- **MP1a:** Have students substitute different values for x. This does not prove that all values of x make the inequality true, but it helps students make sense of the problem.

Example 4

- This is a classic problem. Students always want to know what they have to score on a test in order to have a (mean) average of ____. This is the same type of problem.
- **?** "How do you compute a mean?" Sum the data and divide by the number of data values.
- Set up the problem to compute the mean. Because you want your score to be a minimum of 90, you need to set the mean greater than or equal to 90.
- You need at least a 98 to advance to the next level. Hopefully this score is attainable. Often with my students, the score they need to achieve isn't possible on one test!

On Your Own

- **Common Error:** If students solve Question 5 by distributing the -4, it is very possible they will write $-12n - 4$ instead of $-12n + 4$. For the factor $3n - 1$, they need to remember to *add the opposite* so that the initial equation could be written as $-4[3n + (-1)] > -12n + 5.2$. Then distribute the -4.

Closure

- **Writing:** How are these problems alike? How are they different?

  $3n - 4 = -25$ $\qquad$ $3n - 4 > -25$ $\qquad$ $3n - 4 \leq -25$

  Sample answer: They are alike because they each use the expressions $(3n - 4)$ and (-25). They are different because of the way the expressions are related: equal to, greater than, and less than or equal to.

Technology For the Teacher

The Dynamic Planning Tool
Editable Teacher's Resources at *BigIdeasMath.com*

EXAMPLE 3 Standardized Test Practice

Which graph represents the solution of $2(5x - 1) \leq 7 + 10x$?

Ⓐ 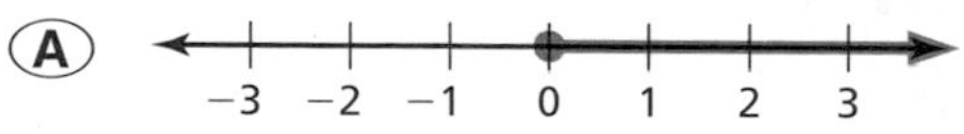

Ⓑ 

Ⓒ −3 −2 −1 0 1 2 3

Ⓓ 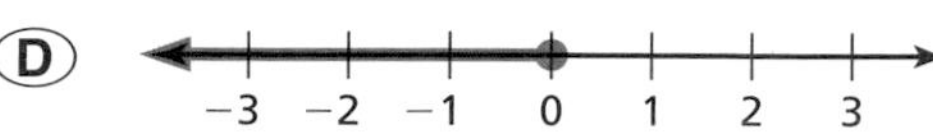

Study Tip

The graph of the set of all real numbers is the entire number line.

$2(5x - 1) \leq 7 + 10x$	Write the inequality.
$10x - 2 \leq 7 + 10x$	Distributive Property
$\underline{-10x} \qquad \underline{-10x}$	Subtract $10x$ from each side.
$-2 \leq 7$	Simplify.

∴ The inequality $-2 \leq 7$ is true. So, the solution is the set of all real numbers. The correct answer is Ⓑ.

EXAMPLE 4 Real-Life Application

Trivia Challenge

Your Scores

- 95 **Game 1:** Very impressive!
- 91 **Game 2:** Good job!
- 77 **Game 3:** You can do better!
- 89 **Game 4:** Nice work!

You need a mean score of at least 90 to advance to the next round of the trivia game. What score do you need on the fifth game to advance?

Use the definition of mean to write and solve an inequality. Let x be the score on the fifth game.

$$\frac{95 + 91 + 77 + 89 + x}{5} \geq 90$$

The meaning of the phrase "at least" is greater than or equal to.

$\dfrac{352 + x}{5} \geq 90$	Simplify.
$5 \cdot \dfrac{352 + x}{5} \geq 5 \cdot 90$	Multiply each side by 5.
$352 + x \geq 450$	Simplify.
$\underline{-352} \qquad \underline{-352}$	Subtract 352 from each side.
$x \geq 98$	Simplify.

Remember

The mean in Example 4 is equal to the sum of the game scores divided by the number of games.

∴ You need at least 98 points to advance to the next round.

On Your Own

Now You're Ready Exercises 12–20

Solve the inequality, if possible.

4. $2(k - 5) < 2k + 5$

5. $-4(3n - 1) > -12n + 5.2$

6. **WHAT IF?** In Example 4, you need a mean score of at least 88 to advance to the next round of the trivia game. What score do you need on the fifth game to advance?

3.4 Exercises

Vocabulary and Concept Check

1. **WRITING** Compare and contrast solving multi-step inequalities and solving multi-step equations.

2. **WRITING** How do you know when an inequality has no solutions? How do you know when the solution of an inequality is the set of all real numbers?

Practice and Problem Solving

3. For what values of k will the perimeter of the octagon be less than or equal to 64 units?

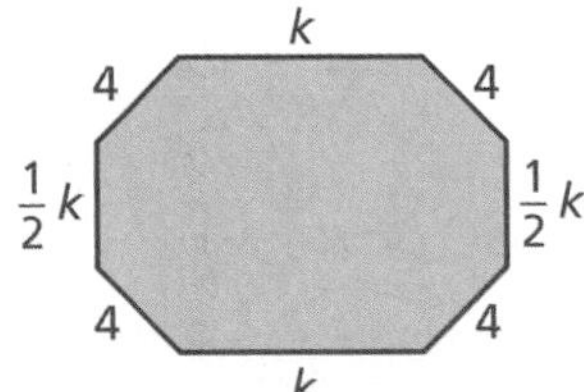

4. For what values of h will the surface area of the solid be greater than 46 square units?

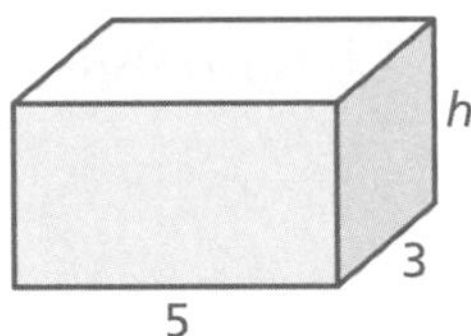

Solve the inequality. Graph the solution.

1 5. $7b + 4 \geq 11$

6. $2v - 4 < 8$

7. $1 - \frac{m}{3} \leq 6$

8. $\frac{4}{5} < 3w - \frac{11}{5}$

9. $1.8 < 0.5 - 1.3p$

10. $-2.4r + 9.6 \geq 4.8$

11. **ERROR ANALYSIS** Describe and correct the error in solving the inequality.

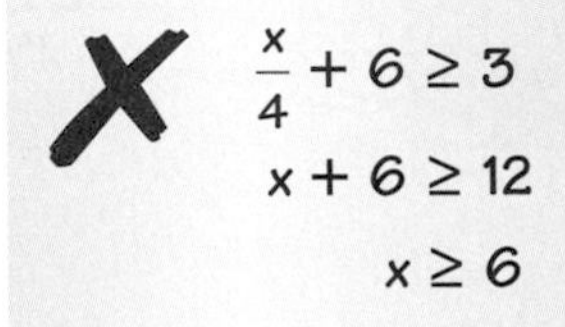

Solve the inequality, if possible.

2 3 12. $6(g + 2) \leq 18$

13. $4(y - 2) \geq 4y - 9$

14. $-10 \geq \frac{5}{3}(h - 3)$

15. $-\frac{1}{3}(u + 2) > 5$

16. $2.7 > 0.9(n - 1.7)$

17. $10 > -2.5(z - 3.1)$

18. $5(w + 4) \leq 5w + 20$

19. $-(6 - x) < x - 7.5$

20. $12c - 5 > 3(4c + 1)$

21. **ATM** Write and solve an inequality that represents the number of $20 bills you can withdraw from the account without going below the minimum balance.

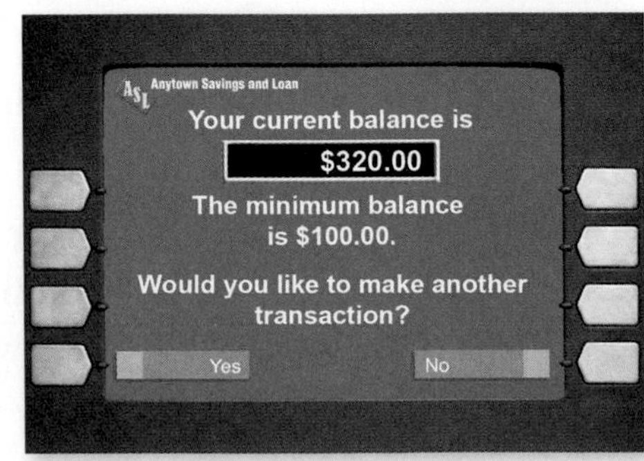

Assignment Guide and Homework Check

Level	Assignment	Homework Check
Average	1–4, 7–25 odd, 29–33	7, 15, 21, 23
Advanced	1–4, 11, 14, 18, 22–33	14, 23, 25, 26

For Your Information

- **Exercises 26 and 28** These exercises involve the Pythagorean Theorem. You may want to review this topic with students before they try solving these exercises.

Common Errors

- **Exercises 5–10** Students may incorrectly multiply or divide before adding to or subtracting from both sides. Remind them that they should work backward through the order of operations, or that they should start away from the variable and move toward it.
- **Exercises 5–10, 12–20** Students may forget to reverse the inequality symbol when multiplying or dividing by a negative number. Encourage them to write the inequality symbol that they should have in the solution before solving.
- **Exercises 12–20** If students distribute before solving, they may forget to distribute the number to the second term. Remind them that they need to distribute to everything within the parentheses. Encourage students to draw arrows to represent the multiplication.

3.4 Record and Practice Journal

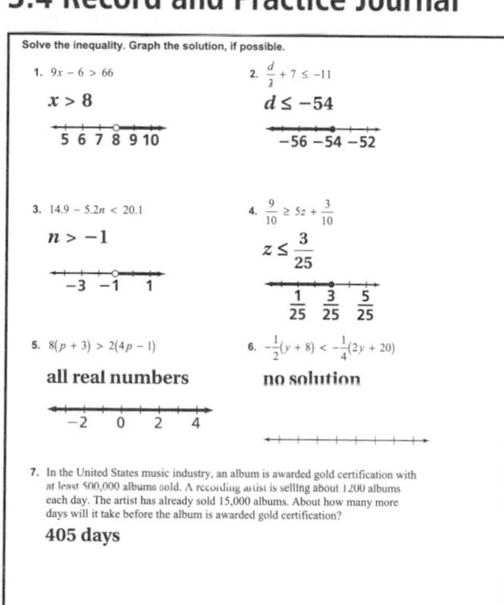

Solve the inequality. Graph the solution, if possible.

1. $9x - 6 > 66$

   $x > 8$

   5 6 7 8 9 10

2. $\frac{d}{3} + 7 \le -11$

   $d \le -54$

   −56 −54 −52

3. $14.9 - 5.2n < 20.1$

   $n > -1$

   −3 −1 1

4. $\frac{9}{10} \ge 5z + \frac{3}{10}$

   $z \le \frac{3}{25}$

   $\frac{1}{25}$ $\frac{3}{25}$ $\frac{5}{25}$

5. $8(p + 3) > 2(4p - 1)$

   all real numbers

   −2 0 2 4

6. $-\frac{1}{2}(y + 8) < -\frac{1}{4}(2y + 20)$

   no solution

7. In the United States music industry, an album is awarded gold certification with at least 500,000 albums sold. A recording artist is selling about 1200 albums each day. The artist has already sold 15,000 albums. About how many more days will it take before the album is awarded gold certification?

   405 days

Technology For the Teacher

Answer Presentation Tool

Vocabulary and Concept Check

1. *Sample answer:* They use the same techniques, but when solving an inequality, you must be careful to reverse the inequality symbol when you multiply or divide by a negative number.
2. *Sample answer:* When solving an inequality, if you obtain an inequality that is false, such as $3 \le -2$, then the inequality has *no solutions.* If you obtain an inequality that is true, such as $-5 < 0$, then the solution is the set of *all real numbers.*

Practice and Problem Solving

3. $k > 0$ and $k \le 16$ units
4. $h > 1$ unit
5. $b \ge 1$;

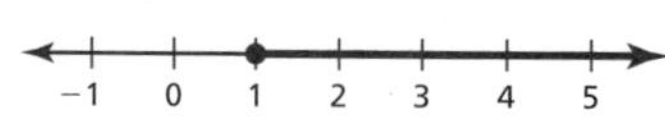

6. $v < 6$;

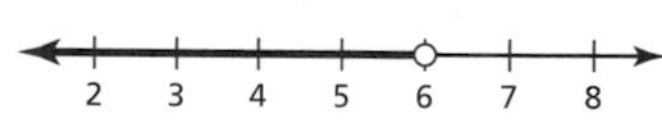

7. $m \ge -15$;

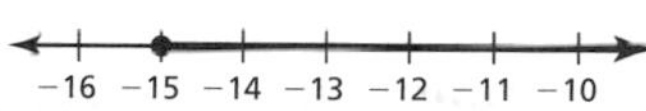

8. $w > 1$;

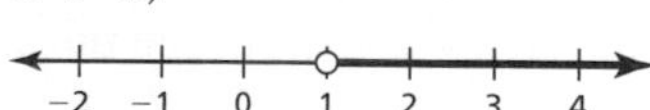

9. $p < -1$;

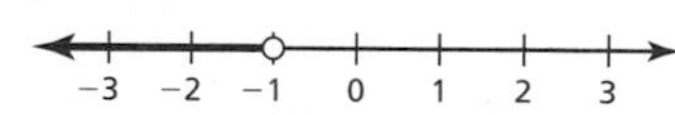

10. $r \le 2$;

11. See Additional Answers.
12. $g \le 1$
13. all real numbers

14–21. See Additional Answers.

Practice and Problem Solving

22. $x \le 6$;

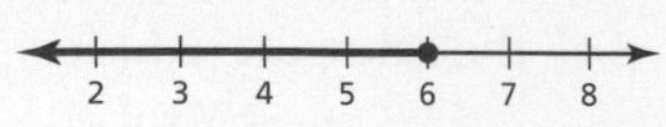

23. $b < 3$;

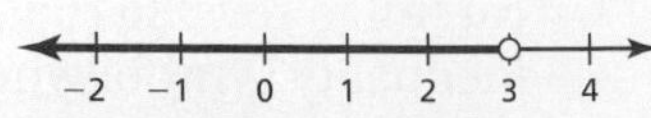

24. $\frac{3}{16}x + 2 \le 11$;
$x > 0$ and $x \le 48$ lines

25. $500 - 20x \ge 100$;
$x > \$0$ and $x \le \$20$ per hour

26. See *Taking Math Deeper.*

27. See Additional Answers.

28. $r \ge 3$ units

Fair Game Review

29.

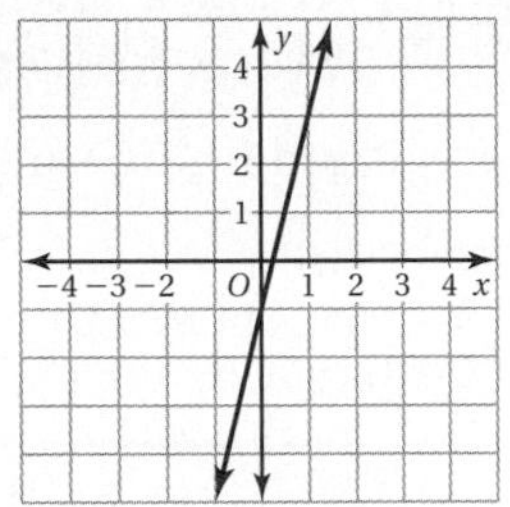

30–32. See Additional Answers.

33. A

Mini-Assessment

Solve the inequality. Graph the solution.

1. $2x + 4 < 10$ $x < 3$

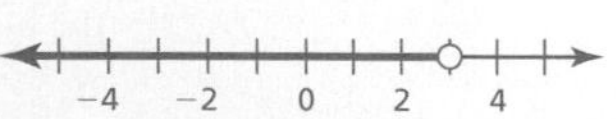

2. $3 \le \frac{y}{-5} + 7$ $y \le 20$

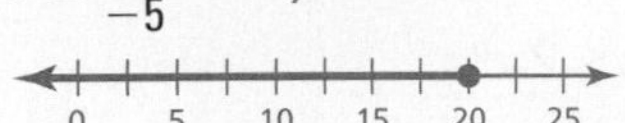

3. $-4.2 - 1.1b \le 2.4$ $b \ge -6$

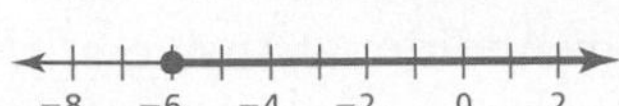

4. $\frac{2}{3}m + \frac{2}{3} \ge -\frac{1}{3}$ $m \ge -\frac{3}{2}$

−2 $-\frac{3}{2}$ −1 $-\frac{1}{2}$ 0 $\frac{1}{2}$

Taking Math Deeper

Exercise 26

Many inequality problems are easier to solve using equations. This is an example of such a problem.

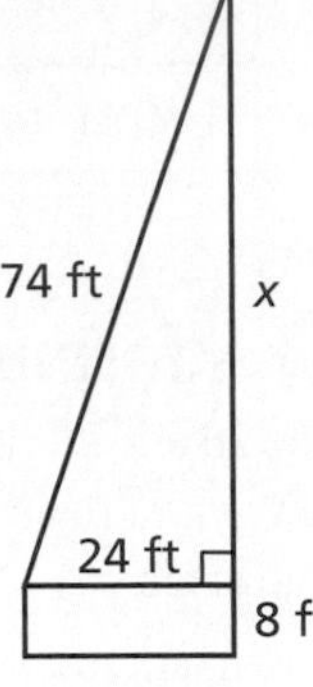

Draw a diagram and label the dimensions.

Use the Pythagorean Theorem to solve for x.

$$x^2 + 24^2 = 74^2$$
$$x^2 + 576 = 5476$$
$$x^2 = 4900$$
$$x = 70 \text{ ft}$$

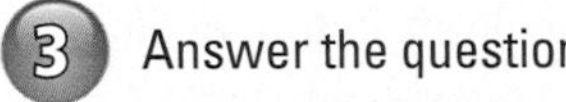
Answer the question.

When the fire truck is exactly 24 feet from the building, the ladder can reach a total height of 78 feet. If the fire truck is farther away from the building, the ladder will not reach as high.

Let S represent the number of stories.
An inequality for S is given by:

$$S \le \frac{\text{number of feet ladder can reach}}{\text{10 feet per story}}$$

$$S \le \frac{78}{10}$$

$$S \le 7.8$$

Project

Many buildings are much taller than the ladder on a fire truck can reach. Use the Internet to research what fire codes exist for these buildings. What special safety measures are required for skyscrapers?

Reteaching and Enrichment Strategies

If students need help...	If students got it...
Resources by Chapter • Practice A and Practice B • Puzzle Time Record and Practice Journal Practice Differentiating the Lesson Lesson Tutorials Skills Review Handbook	Resources by Chapter • Enrichment and Extension • School-to-Work Start the next section

Solve the inequality. Graph the solution.

22. $5x - 2x + 7 \leq 15 + 10$

23. $7b - 12b + 1.4 > 8.4 - 22$

24. **TYPING** One line of text on a page uses about $\frac{3}{16}$ of an inch. There are 1-inch margins at the top and bottom of a page. Write and solve an inequality to find the number of lines that can be typed on a page that is 11 inches long.

25. **WOODWORKING** A woodworker builds a cabinet in 20 hours. The cabinet is sold at a store for \$500. Write and solve an inequality that represents the hourly wage the store can pay the woodworker and still make a profit of at least \$100.

26. **FIRE TRUCK** The height of one story of a building is about 10 feet. The bottom of the ladder on the fire truck must be at least 24 feet away from the building. Write and solve an inequality to find the number of stories the ladder can reach.

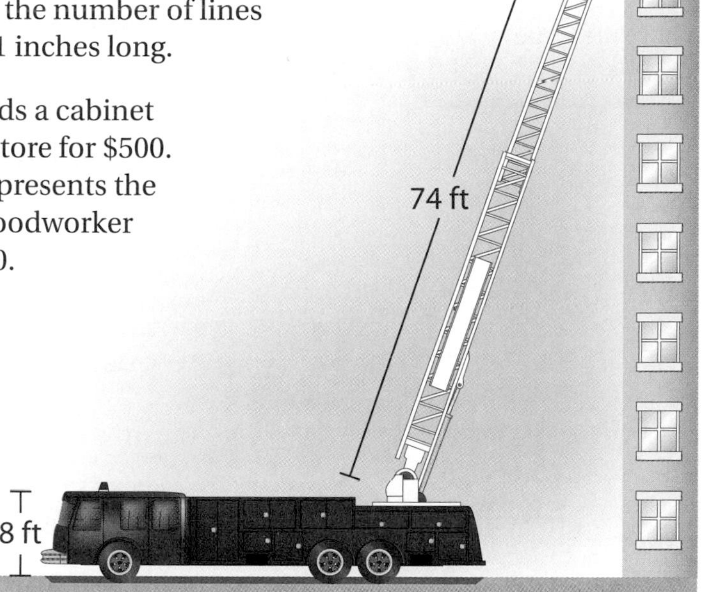

27. **REASONING** A drive-in movie theater charges \$3.50 per car.

a. The drive-in has already admitted 100 cars. Write and solve an inequality to find how many more cars the drive-in needs to admit to earn at least \$500.

b. The theater increases the price by \$1 per car. How does this affect the total number of cars needed to earn \$500? Explain.

28. Challenge For what values of r will the area of the shaded region be greater than or equal to $9(\pi - 2)$?

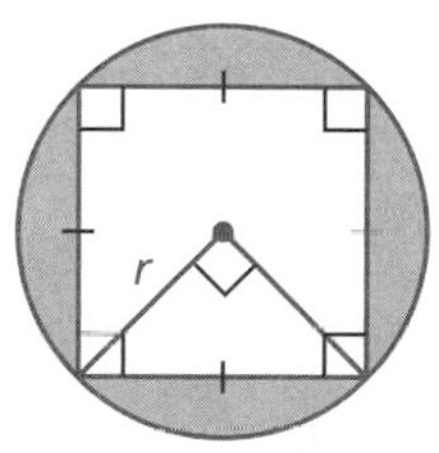

Fair Game Review What you learned in previous grades & lessons

Graph the linear equation. *(Section 2.1)*

29. $y = 4x - 1$

30. $y = -4$

31. $x = 5$

32. $y = -\frac{1}{2}x + 3$

33. **MULTIPLE CHOICE** Which of the following is shown in the graph? *(Section 2.4)*

Ⓐ $3x + 4y = -12$

Ⓑ $3x - 4y = -12$

Ⓒ $3x + 4y = 12$

Ⓓ $3x - 4y = 12$

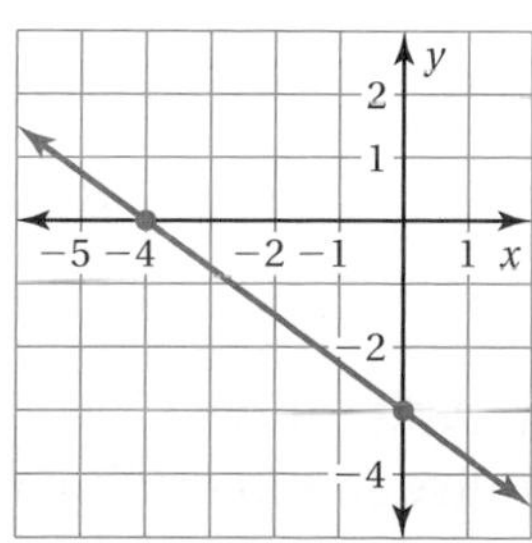

3.4b Solving Compound Inequalities

Key Vocabulary
compound inequality, p. 132
absolute value inequality, p. 134

A **compound inequality** is an inequality formed by joining two inequalities with the word "and" or the word "or."

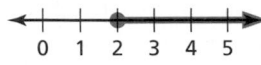

Solutions of a compound inequality with "and" consist of numbers that are solutions of both inequalities.

$x \geq 2$

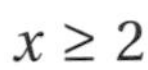

$x < 5$

$2 \leq x$ *and* $x < 5$
$2 \leq x < 5$

Solutions of a compound inequality with "or" consist of numbers that are solutions of at least one of the inequalities.

$y \leq -2$

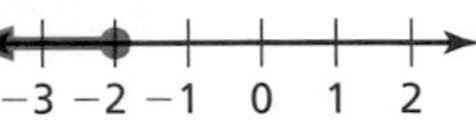

$y > 1$

$y \leq -2$ *or* $y > 1$

EXAMPLE 1 Writing and Graphing Compound Inequalities

Write each word sentence as an inequality. Graph the inequality.

a. A number x is greater than -8 and less than or equal to 4.

A number x is greater than -8 and less than or equal to 4.

$x > -8$ *and* $x \leq 4$

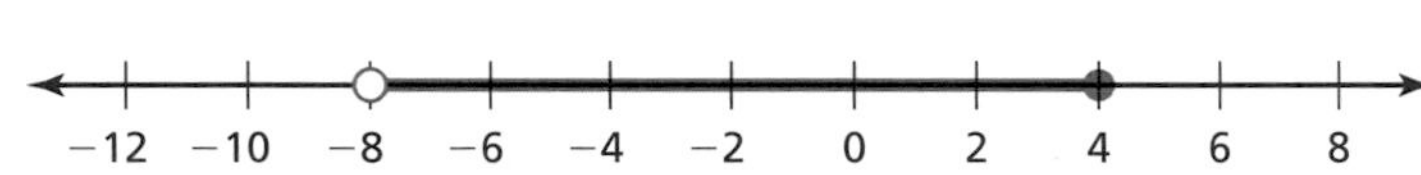

Study Tip

A compound inequality with "and" can be written as a single inequality. For example, you can write $x > -8$ and $x \leq 4$ as $-8 < x \leq 4$.

b. A number y is at most 0 or at least 7.

A number y is at most 0 or at least 7.

$y \leq 0$ *or* $y \geq 7$

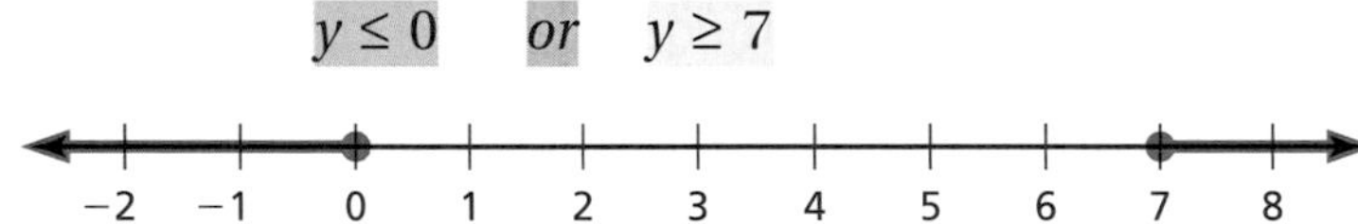

Practice

In Exercises 1–4, write the word sentence as an inequality. Graph the inequality.

1. A number k is more than 3 and less than 9.
2. A number n is greater than or equal to 6 and no more than 11.
3. A number w is fewer than -10 or no less than -6.
4. A number z is less than or equal to -5 or more than 4.
5. Write an inequality to describe the graph.

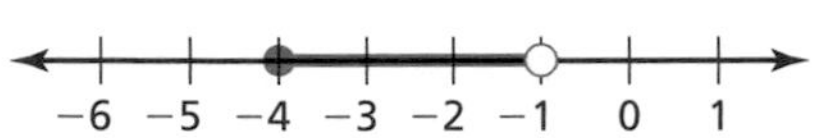

6. The world's longest human life span is 122 years. Write and graph a compound inequality that describes the ages of all humans.

Laurie's Notes

Introduction

Connect

- **Yesterday:** Students solved and graphed multi-step inequalities. (MP1a)
- **Today:** Students will solve and graph compound inequalities.

Motivate

- Share this image of the ride restrictions at an amusement park.

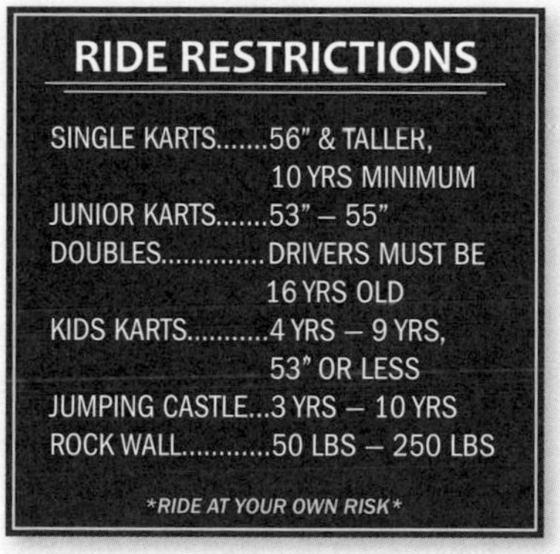

RIDE RESTRICTIONS

SINGLE KARTS.......56" & TALLER, 10 YRS MINIMUM
JUNIOR KARTS.......53" – 55"
DOUBLES.............DRIVERS MUST BE 16 YRS OLD
KIDS KARTS...........4 YRS – 9 YRS, 53" OR LESS
JUMPING CASTLE...3 YRS – 10 YRS
ROCK WALL...........50 LBS – 250 LBS

RIDE AT YOUR OWN RISK

- ? "How would you write an inequality for the height requirement for the single karts?" $h \geq 56$
- ? "How would you write an inequality for the height requirement for junior karts?" *Sample answer:* $h \geq 53$ and $h \leq 55$
- Explain to students that today they will solve inequalities with the words AND and OR.

Lesson Notes

Discuss

- **MP1a Make Sense of Problems:** As students work through this lesson, they should connect their thinking to prior work with solving equations and inequalities, graphing solutions, and absolute value.
- ? "What does it mean if your parent says, 'If you want to go to the movies this weekend, you have to clean your room and sweep the kitchen'?" You have to do two things; clean your room and sweep the kitchen.
- ? "What does it mean if your parent says, 'If you want to go to the movies this weekend, you have to clean your room or sweep the kitchen'?" You have to do one of two things; clean your room or sweep the kitchen.
- ? "Could you do both tasks even if your parent said OR?" yes
- Define compound inequality and give examples of each.

Example 1

- Write the compound inequality in words.
- Discuss the Study Tip. Be sure that students understand that when they encounter a compound inequality written in the form $a < x < b$, the conjunction is always AND.
- As you graph say, "We want to shade the values of x that are greater than -8 and less than or equal to 4."
- ? "What does the phrase 'at most' mean?" As an example, if you have at most 3 cookies, then you could have 3, 2, 1, or no cookies.
- Solve part (b) as shown.

Practice

- Have volunteers write and graph the inequality for 2 or 3 of the problems.

Goal Today's lesson is solving **compound inequalities.**

Start Thinking! and Warm Up

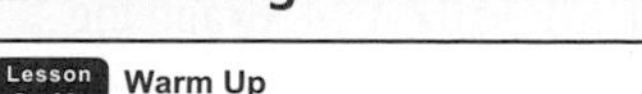

Lesson 3.4b Warm Up
For use before Lesson 3.4b

Lesson 3.4b Start Thinking!
For use before Lesson 3.4b

Use a blue colored pencil and a yellow colored pencil.

On a number line, graph the inequality $x \geq -2$ using the blue colored pencil.

On the same number line, graph the inequality $x \leq 4$ using the yellow colored pencil.

Describe the numbers in the segment of the number line where the two colors overlap.

Extra Example 1

Write each word sentence as an inequality. Graph the inequality.

a. A number x is less than 2 or greater than 5. $x < 2$ *or* $x > 5$

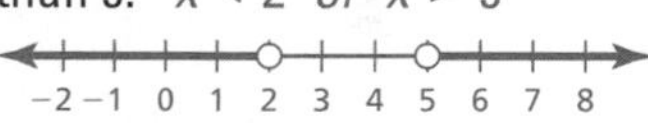

b. A number y is more than -4 and less than or equal to 3. $y > -4$ *and* $y \leq 3$

Practice

1. $3 < k < 9$;
2. $6 \leq n \leq 11$; 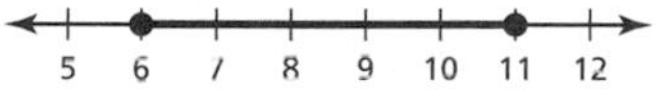
3. $w < -10$ or $w \geq -6$;
4. $z \leq -5$ or $z > 4$; 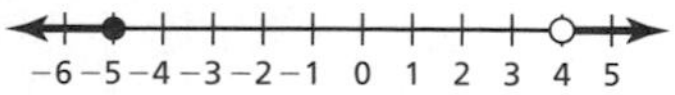
5. $-4 \leq x < -1$;
6. $0 < x \leq 122$;

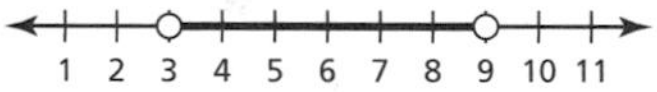

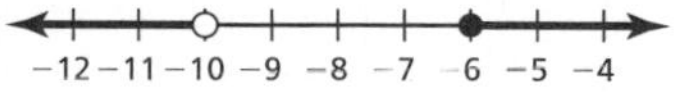

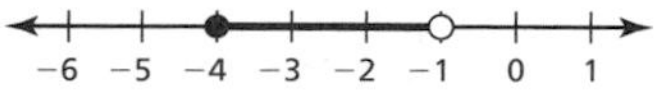

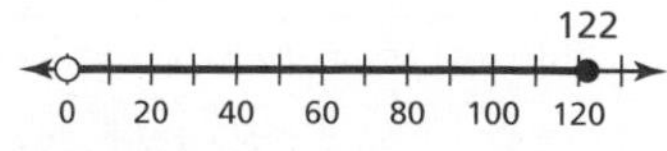

Record and Practice Journal Practice

See Additional Answers.

Extra Example 2

Solve $-1 \leq 2 - 3x < 14$. Graph the solution. $-4 < x \leq 1$

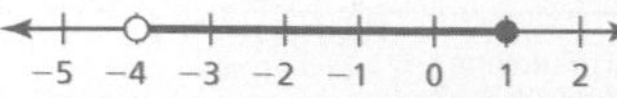

Extra Example 3

Solve $4x + 11 \leq 3$ or $2x - 8 > -6$. Graph the solution. $x \leq -2$ or $x > 1$

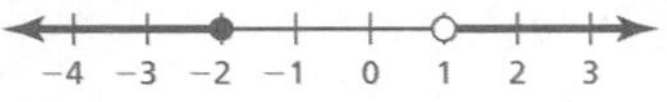

Practice

7. $9 < x < 12$;

8. $-2 \leq x < 2$;

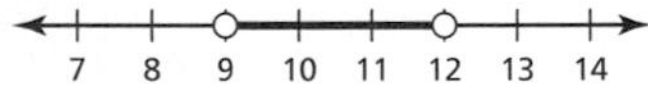

9. $-2 < x \leq 3$;

10. $x \leq -3$ or $x \geq 3$;

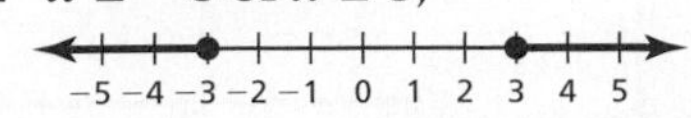

11. $x > -6$ or $x \leq -7$;

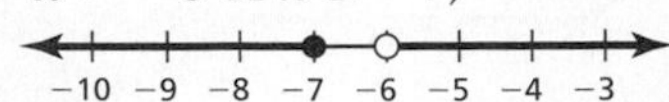

Laurie's Notes

Example 2

- Discuss the general process for solving compound inequalities. Explain that each inequality can be solved separately. If the compound inequality with AND is written as a single inequality statement, perform the same operation to each of the three parts (left, middle, and right).
- Write the inequality and ask a volunteer to read it.

? "This inequality is equivalent to what two inequalities?" $-3 < -2x + 1$ and $-2x + 1 \leq 9$

- Explain that to isolate the $-2x$ term, subtract 1 from each of the three parts.
- **Common Error:** When dividing each of the three parts by -2, students may forget to reverse the direction of the inequality symbols.
- Discuss why the solution $2 > x \geq -4$ is rewritten. By convention we read inequalities from the least value (-4) to the greatest value (2) just as we would read the graph of the solution from left to right.

? "How do you check your solution?" Substitute values between -4 and 2.

? "Are -4 and 2 part of the solution set?" -4 is a solution but 2 is not.

- Be sure to substitute values not in the solution set to show that they do not make the inequality true.
- **MP1a:** Take time to solve the inequality as shown in the example and as shown in the Study Tip. You want students to understand that both approaches result in the same solution.

Example 3

? Write the inequality and ask, "How would you solve an OR compound inequality?" *Sample answer:* Solve the two inequalities separately and use OR when writing the answer.

- Solve each inequality as shown.
- When checking the solution, students should recognize that a solution, say $x = -2$, satisfies one of the two inequalities but not both. Values that are not solutions, say $x = 0$, satisfy neither inequality.

Practice

- **Neighbor Check:** Have students work independently and then have their neighbor check their work. Have students discuss any discrepancies.
- Have students share their work at the board.

You can solve compound inequalities by solving two inequalities separately. When a compound inequality with "and" is written as a single inequality, you can solve the inequality by performing the same operation on each expression.

EXAMPLE 2 Solving a Compound Inequality with "And"

Solve $-3 < -2x + 1 \leq 9$. Graph the solution.

$-3 < -2x + 1 \leq 9$	Write the inequality.
$\underline{-1} \qquad \underline{-1} \quad \underline{-1}$	Subtract 1 from each expression.
$-4 < -2x \leq 8$	Simplify.
$\frac{-4}{-2} > \frac{-2x}{-2} \geq \frac{8}{-2}$	Divide each expression by -2. Reverse the inequality symbols.
$2 > x \geq -4$	Simplify.

∴ The solution is $-4 \leq x < 2$.

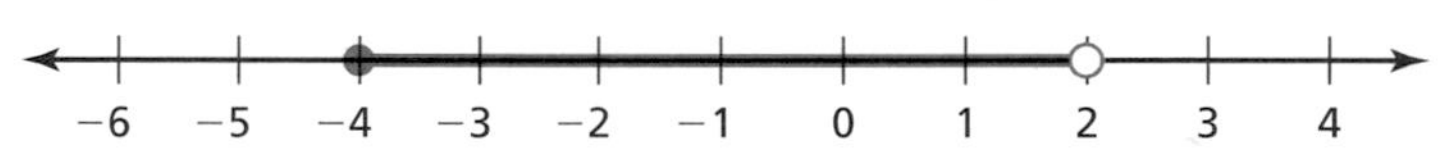

Study Tip

You can also solve the inequality in Example 2 by solving the inequalities

$-3 < -2x + 1$

and

$-2x + 1 \leq 9$

separately.

EXAMPLE 3 Solving a Compound Inequality with "Or"

Solve $3x - 5 < -8$ *or* $2x - 1 > 5$. Graph the solution.

$3x - 5 < -8$	*or*	$2x - 1 > 5$	Write the inequality.
$\underline{+5} \quad \underline{+5}$		$\underline{+1} \quad \underline{+1}$	Addition Property of Inequality
$3x < -3$	*or*	$2x > 6$	Simplify.
$\frac{3x}{3} < \frac{-3}{3}$	*or*	$\frac{2x}{2} > \frac{6}{2}$	Division Property of Inequality
$x < -1$	*or*	$x > 3$	Simplify.

∴ The solution is $x < -1$ *or* $x > 3$.

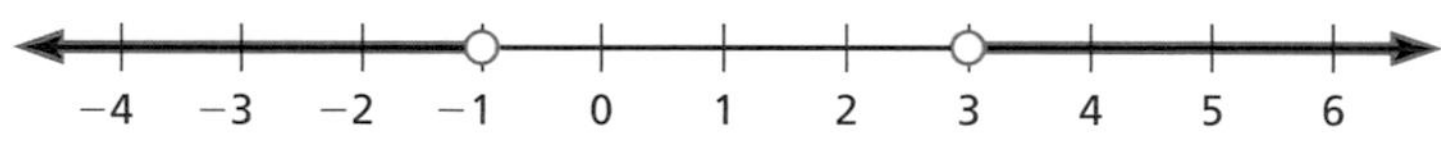

Practice

Solve the inequality. Graph the solution.

7. $4 < x - 5 < 7$

8. $-1 \leq 2x + 3 < 7$

9. $15 > -3x + 9 \geq 0$

10. $4x + 1 \leq -11$ *or* $3x - 4 \geq 5$

11. $-2x - 7 < 5$ *or* $-5x + 6 \geq 41$

Study Tip

When an absolute value expression is on the left side of an inequality, use an "and" statement for $<$ and $\leq$, and an "or" statement for $>$ and $\geq$.

An **absolute value inequality** is an inequality that contains an absolute value expression. For example, $|x| < 2$ and $|x| > 2$ are absolute value inequalities.

The distance between x and 0 is less than 2.

$|x| < 2$

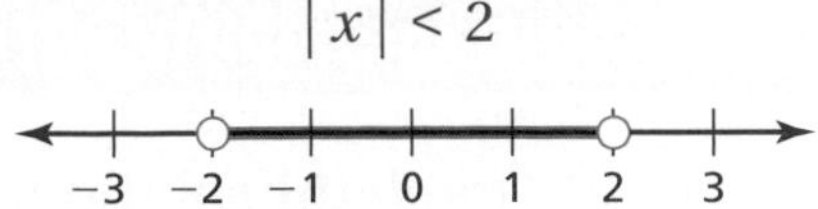

The graph of $|x| < 2$ is $x > -2$ *and* $x < 2$.

The distance between x and 0 is greater than 2.

$|x| > 2$

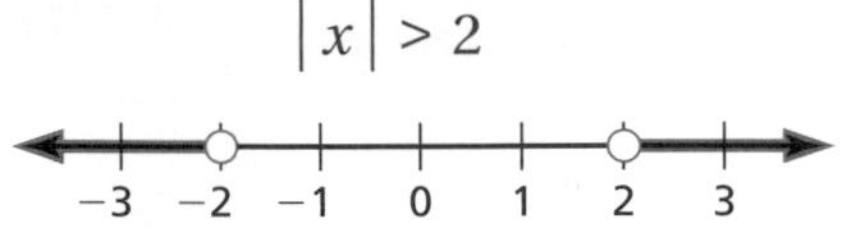

The graph of $|x| > 2$ is $x < -2$ *or* $x > 2$.

You can solve these types of inequalities by solving a compound inequality.

Key Idea

Solving Absolute Value Inequalities

To solve $|ax + b| < c$ for $c > 0$, solve the compound inequality

$$ax + b > -c \quad \textit{and} \quad ax + b < c.$$

To solve $|ax + b| > c$ for $c > 0$, solve the compound inequality

$$ax + b < -c \quad \textit{or} \quad ax + b > c.$$

In the inequalities above, you can replace $<$ with $\leq$ and $>$ with $\geq$.

EXAMPLE 4 Solving Absolute Value Inequalities

a. Solve $|x + 7| \leq 2$. Graph the solution.

Use $|x + 7| \leq 2$ to write a compound inequality. Then solve.

$x + 7 \geq -2$	*and*	$x + 7 \leq 2$	Write compound inequality.
$\underline{-7} \quad \underline{-7}$		$\underline{-7} \quad \underline{-7}$	Subtract 7 from each side.
$x \geq -9$	*and*	$x \leq -5$	Simplify.

∴ The solution is $x \geq -9$ *and* $x \leq -5$.

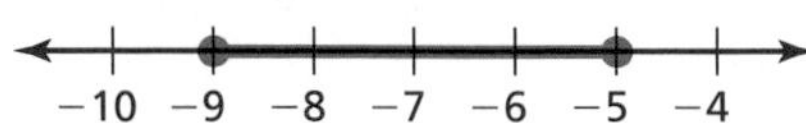

b. Solve $|8x - 11| < 0$.

The absolute value of an expression must be greater than or equal to 0. The expression $|8x - 11|$ cannot be less than 0.

∴ So, the inequality has no solution.

Laurie's Notes

Discuss

- Students solved absolute value equations in Section 1.3b. Now they are combining that skill with their understanding of compound inequalities to solve absolute value inequalities.

? "What does $|x| = 2$ mean geometrically?" all values that are 2 units from 0

- Graph the solutions for $|x| = 2$.

? "What does $|x| < 2$ mean geometrically?" all values that are less than 2 units from 0

- Graph the solutions for $|x| < 2$.

? "What does $|x| > 2$ mean geometrically?" all values that are more than 2 units from 0

- Graph the solutions for $|x| > 2$.
- Connect these two types of absolute value inequalities to student understanding of AND and OR compound inequalities.

Key Idea

- Write the Key Idea on the board.
- **Teaching Tip:** When solving an absolute value inequality such as $|2x + 1| < 4$, I place my fingers over the expression inside the absolute value bars and say, "We are looking for a quantity whose absolute value is less than 4 units from 0. What is under my fingers is between -4 AND 4." That leads directly into writing the compound inequality $-4 < 2x + 1 < 4$. By covering up the expression, I find it helps focus student attention on the process, without them getting overwhelmed by symbols. If the problem had been $|2x + 1| > 4$, I would have said, "We are looking for a quantity whose absolute value is more than 4 units from 0. What is under my fingers is less than -4 OR greater than 4."

Example 4

? Write the absolute value inequality and ask, "How could you write this as a compound inequality?" $-2 \leq x + 7 \leq 2$

? "How do you know it should be an AND and not an OR compound inequality?" The absolute value is within 2 units of 0.

- Check values that are solutions and verify that values that are *not* solutions do not satisfy the absolute value inequality.
- **Extension:** Students may observe that the solution is a segment of length 4 with a midpoint of -7. The solution might be described as values that are within 2 units of -7.

? Write the problem in part (b) and ask, "What value(s) of x will make the expression on the left side of the inequality less than 0? Explain." none; Any value substituted for x, then multiplied by 8 and decreased by 11 will not be negative once you find the absolute value of the result.

Extra Example 4

a. Solve $|2x + 7| > 3$. Graph the solution. $x < -5$ or $x > -2$

−7 −6 −5 −4 −3 −2 −1 0

b. Solve $|x - 6| < 0$. The inequality has no solution.

Laurie's Notes

Extra Example 5

Solve $5|x + 3| - 2 \le 18$.
$x \le 1$ *and* $x \ge -7$

Extra Example 6

In a poll, 52% of voters say they plan to reelect the mayor. The poll has a margin of error of ±4 percentage points. Write and solve an absolute value inequality to find the least and greatest percents of voters who plan to reelect the mayor.
$|x - 52| \le 4$; The least percent of voters is 48%. The greatest percent of voters is 56%.

Practice

12. $x \le -1$ *or* $x \ge 7$;

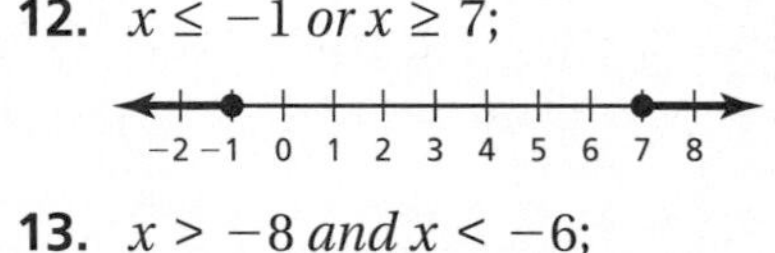

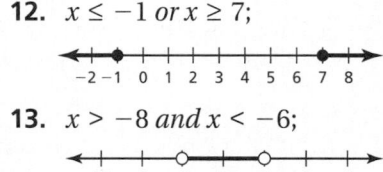

13. $x > -8$ *and* $x < -6$;

14–19. See Additional Answers.

Mini-Assessment

Write the word sentence as an inequality. Graph the inequality.

1. A number x is no more than 0 or no less than 5. $x \le 0$ *or* $x \ge 5$;

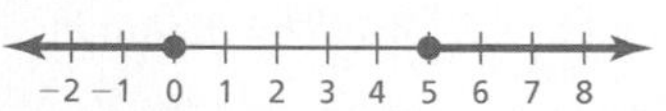

2. A number p is greater than -4 fewer than 2. $-4 < p < 2$;

Solve the inequality. Graph the solution.

3. $x + 2 < -1$ *or* $-2x - 3 \le -9$
$x < -3$ *or* $x \ge 3$;

4. $|x + 2| \le 3$
$-5 \le x \le 1$;

Example 5

- **MP7 Look For and Make Use of Structure:** When solving this inequality, students need to think of $|2x - 5|$ as a term that needs to be isolated first. Students should think solving $4|2x - 5| + 1 > 29$ is similar to solving $4x + 1 > 29$. Additionally, solving $4x + 1 > 29$ is similar to solving $4x + 1 = 29$. To isolate the variable (or the absolute value expression), subtract 1 and then divide by 4.
- **MP1a:** By connecting the process back to prior skills, students generally can follow and make sense of the problems. Performing the process on their own is still a skill that needs to be practiced.

Example 6

? "Have any of you heard of the expression *margin of error*? Can you explain it?" Students may know it has to do with surveys or polls.

- **FYI:** *Margin of error* is a statistical term relating to a *confidence interval*. If a poll reports with 95% confidence that 80% of the people surveyed would buy Brand A with a 3% margin of error, it means that if the poll were repeated 100 times, the percentage who say they would buy Brand A will range between 77% and 83% most (95%) of the time. The margin of error is related to the sample size, a concept beyond the scope of this lesson.
- Ask a volunteer to read the problem.
- My experience is that students will immediately *get it*, meaning they will say 45 to 49 percent of the voters will vote for the mayor. Take the time to write the model and talk about how the absolute value inequality can model this problem.

Practice

- **Neighbor Check:** Have students work independently and then have their neighbor check their work. Have students discuss any discrepancies.
- Have students share their work at the board.

Closure

- **Exit Ticket:** Solve and graph.

$|x - 7| < 13$ $-6 < x < 20$

$|x + 4| \ge 8$ $x \ge 4$ or $x \le -12$

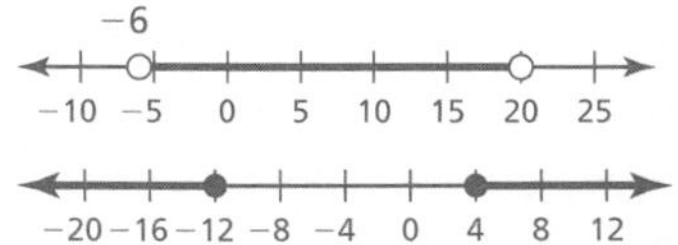

Technology For the Teacher

The Dynamic Planning Tool
Editable Teacher's Resources at *BigIdeasMath.com*

EXAMPLE 5 Solving an Absolute Value Inequality

Solve $4|2x-5|+1>29$.

$4|2x-5|+1>29$	Write the inequality.
$|2x-5|>7$	Isolate the absolute value expression.

Use $|2x-5|>7$ to write a compond inequality. Then solve.

$2x-5<-7$	*or*	$2x-5>7$	Write compound inequality.
$\underline{+5\quad +5}$		$\underline{+5\quad +5}$	Add 5 to each side.
$2x<-2$	*or*	$2x>12$	Simplify.
$\frac{2x}{2}<\frac{-2}{2}$	*or*	$\frac{2x}{2}>\frac{12}{2}$	Divide each side by 2.
$x<-1$	*or*	$x>6$	Simplify.

EXAMPLE 6 Real-Life Application

In a poll, 47% of voters say they plan to reelect the mayor. The poll has a margin of error of ±2 percentage points. Write and solve an absolute value inequality to find the least and greatest percents of voters who plan to reelect the mayor.

Words | Actual percent of voters | minus | percent of voters in poll | is less than or equal to | the margin of error.

Variable Let x represent the actual percent of voters who plan on reelecting the mayor.

Inequality $|x - 47| \leq 2$

$x-47\geq -2$	*and*	$x-47\leq 2$	Write compound inequality.
$\underline{+47\quad +47}$		$\underline{+47\quad +47}$	Add 47 to each side.
$x\geq 45$	*and*	$x\leq 49$	Simplify.

∴ The least percent of voters who plan to reelect the mayor is 45%. The greatest percent of voters who plan to reelect the mayor is 49%.

Practice

Solve the inequality. Graph the solution, if possible.

12. $|x-3|\geq 4$ **13.** $|x+7|<1$ **14.** $11\geq|4x-5|$

15. $|8x-9|<0$ **16.** $3|2x+5|-8\geq 19$ **17.** $-2|x-10|+1>-7$

18. NUMBER SENSE What is the solution of $|4x-2|\geq -6$? Explain.

19. MODELING In Example 6, 44% of the voters say they plan to reelect the mayor. The poll has a margin of error of ±3 percentage points. Use a model to write and solve an absolute value inequality to find the least and greatest percents of voters who plan to reelect the mayor.

3.5 Graphing Linear Inequalities in Two Variables

COMMON CORE STATE STANDARDS
A.REI.12

Essential Question How can you use a coordinate plane to solve problems involving linear inequalities?

1 ACTIVITY: Graphing Inequalities

Work with a partner.

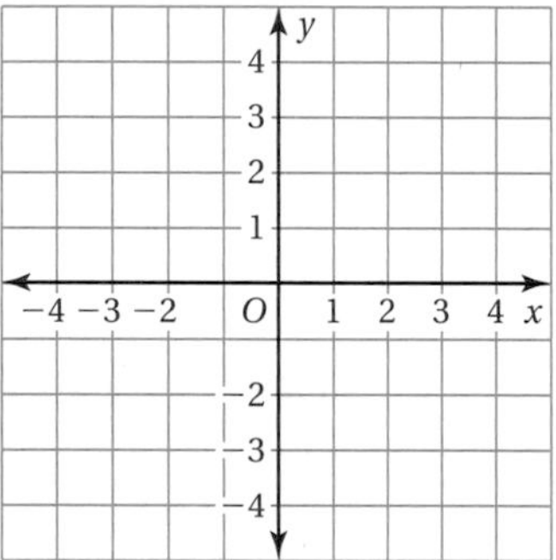

a. Graph $y = x + 1$ in a coordinate plane.

b. Choose three points that lie above the graph of $y = x + 1$. Substitute the values of x and y of each point in the inequality $y > x + 1$. If the substitutions result in true statements, plot the points on the graph.

c. Choose three points that lie below the graph of $y = x + 1$. Substitute the values of x and y of each point in the inequality $y > x + 1$. If the substitutions result in true statements, plot the points on the graph.

d. To graph $y > x + 1$, would you choose points above or below $y = x + 1$?

e. Choose a point that lies on the graph of $y = x + 1$. Substitute the values of x and y in the inequality $y > x + 1$. What do you notice? Do you think the graph of $y > x + 1$ includes the points that lie on the graph of $y = x + 1$? Explain your reasoning.

f. Explain how you could change the inequality so that it includes the points that lie on the graph of $y = x + 1$.

2 ACTIVITY: Writing and Graphing Inequalities

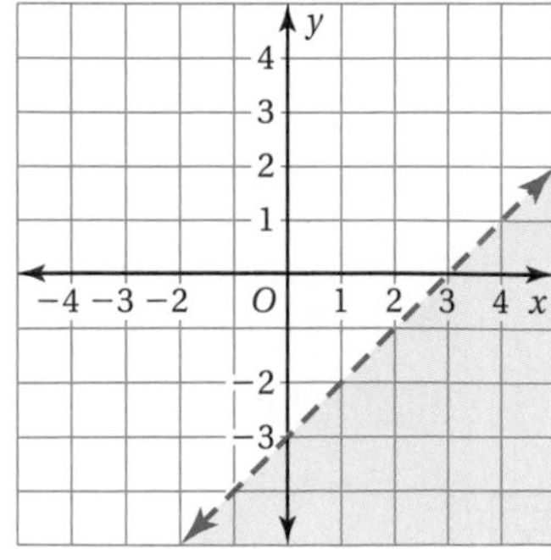

Work with a partner. The graph of a linear inequality in two variables shows all the solutions of the inequality in a coordinate plane. An ordered pair (x, y) is a solution of an inequality if the inequality is true when the values of x and y are substituted in the inequality.

a. Write an equation for the graph of the dashed blue line.

b. The solutions of an inequality are represented by the shaded region. In words, describe the solutions of the inequality.

c. Write an inequality for the graph. Which inequality symbol did you use? Explain your reasoning.

Laurie's Notes

Introduction

Standards for Mathematical Practice

- **MP6 Attend to Precision:** Students have graphed linear inequalities in one variable. Today they extend their understanding to linear inequalities in two variables. They should be able to explain the similarity of the processes, identifying why a particular symbol ($<, \leq, >, \geq$) was used and why a dashed or solid line was used.

Motivate

- Before class begins, place a piece of rope, yarn, or tape on the floor so that the classroom is divided into two *halves*. The two halves do not need to be the same size, but you do want some student desks in each half.
- Draw a line on the board and plot a point.
- Explain that the point divides the line into two parts called rays (half-lines). The point is the "boundary" between the two rays.

? "What would divide a plane into two half-planes?" a line

- A line is the "boundary" that divides the plane into two half-planes.
- Refer to the boundary line in the classroom. Students are on one side of the line or the other, or perhaps they are "on" the line.
- Explain that today students will graph linear inequalities in two variables. A line will divide the coordinate plane into two parts. The solution will be on one side of the line or the other.

Activity Notes

Activity 1

- In this activity, students will graph a line in the coordinate plane and then decide which side of the line contains points that satisfy the inequality.
- Students may find it helpful to make a table of values.

x	y	$y > x + 1$
1	4	$4 > 1 + 1$ (yes)
2	-3	$-3 > 2 + 1$ (no)

- When students have finished, ask a student to share their reasoning for parts (d), (e), and (f).
- Students may not pay close attention to the difference between graphing $y > x + 1$ and $y = x + 1$.

? "When you graphed inequalities in one variable how did you distinguish between $x > 2$ and $x \geq 2$?" The graph of $x > 2$ has an open circle at 2 and is shaded to the right of the circle. The graph of $x \geq 2$ has a closed circle at 2 and is shaded to the right of the circle.

- Explain that when graphing inequalities in two variables, a similar distinction will need to be made. Let the students derive that distinction on their own.

Common Core State Standards

A.REI.12 Graph the solutions to a linear inequality in two variables as a half-plane (excluding the boundary in the case of a strict inequality), ...

Previous Learning

Students should know how to graph a linear equation.

Activity Materials	
Introduction	**Textbook**
• rope, yarn, or tape	• calculator

Start Thinking! and Warm Up

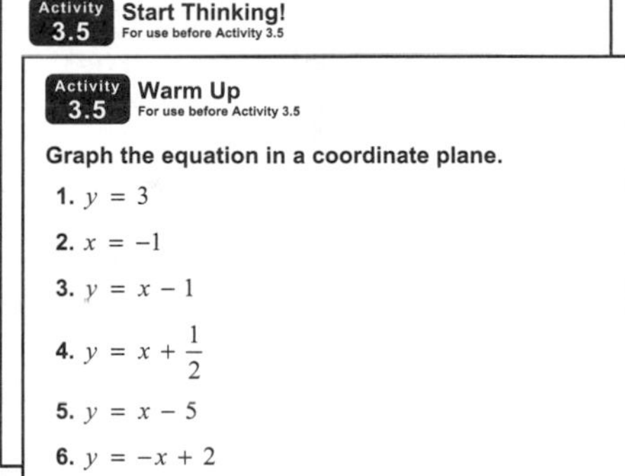

Activity 3.5 Start Thinking! For use before Activity 3.5

Activity 3.5 Warm Up For use before Activity 3.5

Graph the equation in a coordinate plane.

1. $y = 3$
2. $x = -1$
3. $y = x - 1$
4. $y = x + \frac{1}{2}$
5. $y = x - 5$
6. $y = -x + 2$

3.5 Record and Practice Journal

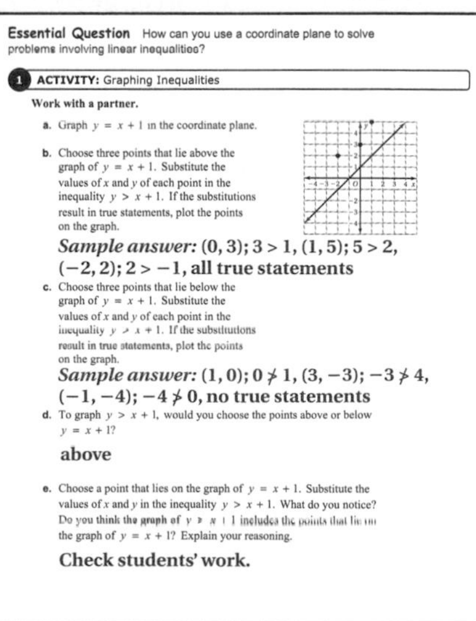

Essential Question How can you use a coordinate plane to solve problems involving linear inequalities?

1 ACTIVITY: Graphing Inequalities

Work with a partner.

a. Graph $y = x + 1$ in the coordinate plane.

b. Choose three points that lie above the graph of $y = x + 1$. Substitute the values of x and y of each point in the inequality $y > x + 1$. If the substitutions result in true statements, plot the points on the graph.

Sample answer: **(0, 3); 3 > 1, (1, 5); 5 > 2, (−2, 2); 2 > −1, all true statements**

c. Choose three points that lie below the graph of $y = x + 1$. Substitute the values of x and y of each point in the inequality $y > x + 1$. If the substitutions result in true statements, plot the points on the graph.

Sample answer: **(1, 0); $0 \not> 1$, (3, −3); $-3 \not> 4$, (−1, −4); $-4 \not> 0$, no true statements**

d. To graph $y > x + 1$, would you choose the points above or below $y = x + 1$?

above

e. Choose a point that lies on the graph of $y = x + 1$. Substitute the values of x and y in the inequality $y > x + 1$. What do you notice? Do you think the graph of $y > x + 1$ includes the points that lie on the graph of $y = x + 1$? Explain your reasoning.

Check students' work.

Differentiated Instruction

Kinesthetic

Some students benefit by having access to graphing calculators or the graphing software programs available on many computer operating systems. Students can enter the inequalities, view their graphs, and check solution points.

3.5 Record and Practice Journal

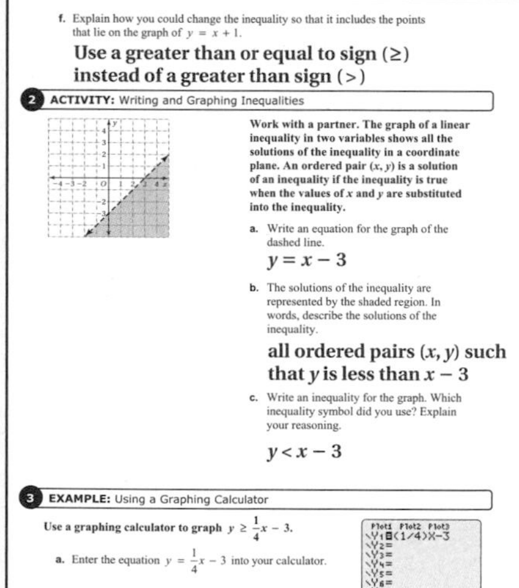

f. Explain how you could change the inequality so that it includes the points that lie on the graph of $y = x + 1$.

Use a greater than or equal to sign (≥) instead of a greater than sign (>)

2 ACTIVITY: Writing and Graphing Inequalities

Work with a partner. The graph of a linear inequality in two variables shows all the solutions of the inequality in a coordinate plane. An ordered pair (x, y) is a solution of an inequality if the inequality is true when the values of x and y are substituted into the inequality.

a. Write an equation for the graph of the dashed line.

$y = x - 3$

b. The solutions of the inequality are represented by the shaded region. In words, describe the solutions of the inequality.

all ordered pairs (x, y) such that y is less than $x - 3$

c. Write an inequality for the graph. Which inequality symbol did you use? Explain your reasoning.

$y < x - 3$

3 EXAMPLE: Using a Graphing Calculator

Use a graphing calculator to graph $y \geq \frac{1}{4}x - 3$.

a. Enter the equation $y = \frac{1}{4}x - 3$ into your calculator.

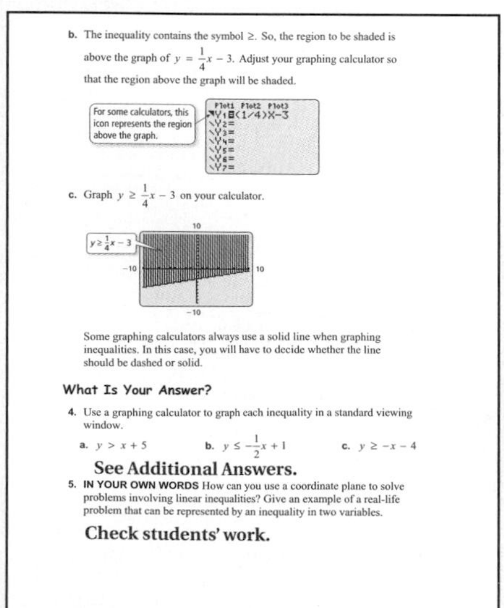

b. The inequality contains the symbol ≥. So, the region to be shaded is above the graph of $y = \frac{1}{4}x - 3$. Adjust your graphing calculator so that the region above the graph will be shaded.

For some calculators, this icon represents the region above the graph.

c. Graph $y \geq \frac{1}{4}x - 3$ on your calculator.

Some graphing calculators always use a solid line when graphing inequalities. In this case, you will have to decide whether the line should be dashed or solid.

What Is Your Answer?

4. Use a graphing calculator to graph each inequality in a standard viewing window.

a. $y > x + 5$ b. $y \leq -\frac{1}{2}x + 1$ c. $y \geq -x - 4$

See Additional Answers.

5. IN YOUR OWN WORDS How can you use a coordinate plane to solve problems involving linear inequalities? Give an example of a real-life problem that can be represented by an inequality in two variables.

Check students' work.

Laurie's Notes

Activity 2

- This is students' first encounter with a dashed line versus a solid line. My experience has been that students select the < symbol for the dashed line, reasoning that it is analogous to the open circle on a number line.
- Take time for students to select ordered pairs in the shaded portion to verify that they are solutions. Ordered pairs above and on the line are not solutions.

Example 3

- **Teacher Tip:** It is possible that more than one type of graphing calculator is used in your classroom. You may want to group students that are using the same type of calculator together to act as a support group.

? "How do you enter the fraction $\frac{1}{4}$ on your calculator?" Answers may vary depending on the calculator. Students should recognize that the decimal 0.25 could also be used.

- Explain the steps necessary to produce the shading for the inequality. Show students the icon for shading the region below the graph.
- Explain to students that some calculators always use the solid line. They will need to determine whether ordered pairs on the line are part of the solution set.

What Is Your Answer?

- Walk around to observe the graphing screens of calculators as students work Question 4.
- Have students share the real-life examples they wrote for Question 5.

Closure

- Which inequality represents the graph? B

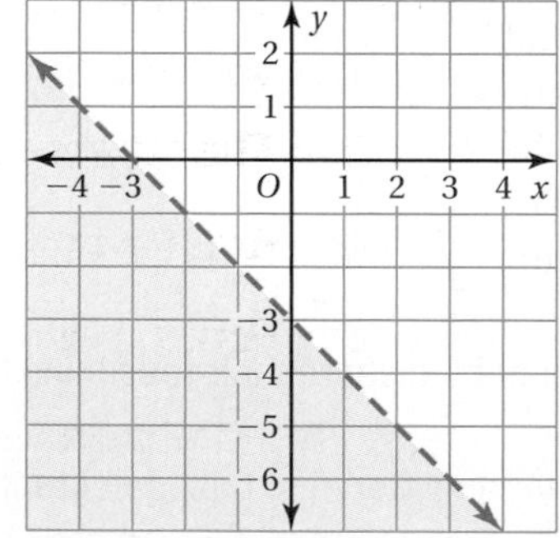

A. $y > x - 3$

B. $y < -x - 3$

C. $y \geq -x + 3$

D. $y \leq -x - 3$

Technology For the Teacher

The Dynamic Planning Tool

Editable Teacher's Resources at *BigIdeasMath.com*

3 EXAMPLE: Using a Graphing Calculator

Use a graphing calculator to graph $y \geq \frac{1}{4}x - 3$.

a. Enter the equation $y = \frac{1}{4}x - 3$ into your calculator.

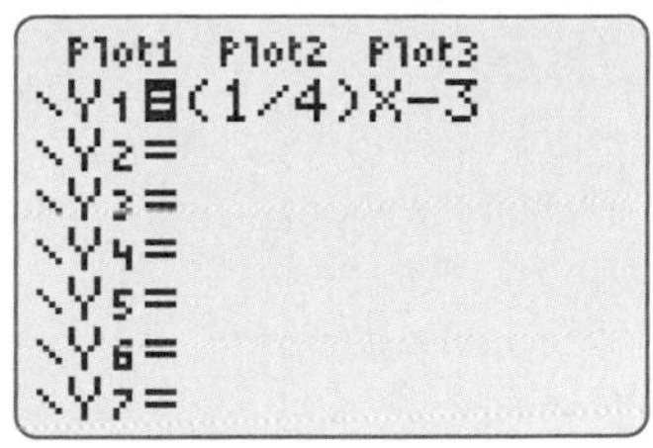

b. The inequality contains the symbol ≥. So, the region to be shaded is above the graph of $y = \frac{1}{4}x - 3$. Adjust your graphing calculator so that the region above the graph will be shaded.

For some calculators, this icon represents the region above the graph.

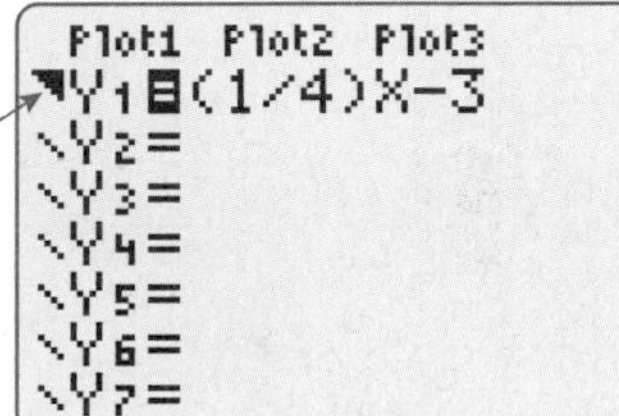

c. Graph $y \geq \frac{1}{4}x - 3$ on your calculator.

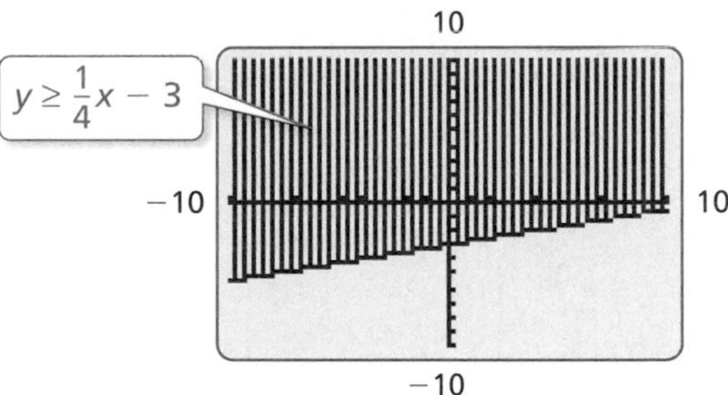

Some graphing calculators always use a solid line when graphing inequalities. In this case, you will have to decide whether the line should be dashed or solid.

What Is Your Answer?

4. Use a graphing calculator to graph each inequality in a standard viewing window.

   a. $y > x + 5$ **b.** $y \leq -\frac{1}{2}x + 1$ **c.** $y \geq -x - 4$

5. **IN YOUR OWN WORDS** How can you use a coordinate plane to solve problems involving linear inequalities? Give an example of a real-life problem that can be represented by an inequality in two variables.

Practice Use what you learned about writing and graphing inequalities to complete Exercises 8–10 on page 141.

3.5 Lesson

Key Vocabulary
linear inequality in two variables, *p. 138*
solution of a linear inequality, *p. 138*
graph of a linear inequality, *p. 138*
half-planes, *p. 138*

A **linear inequality in two variables** x and y can be written as

$$ax + by < c \qquad ax + by \leq c \qquad ax + by > c \qquad ax + by \geq c$$

where a, b, and c are real numbers. A **solution of a linear inequality** in two variables is an ordered pair (x, y) that makes the inequality true.

EXAMPLE 1 Checking Solutions of a Linear Inequality

Tell whether the ordered pair is a solution of the inequality.

a. $2x + y < -3$; $(-1, 9)$

$2x + y < -3$	Write the inequality.
$2(-1) + 9 \overset{?}{<} -3$	Substitute -1 for x and 9 for y.
$7 \nless -3$ ✗	Simplify. 7 is *not* less than -3.

So, $(-1, 9)$ is *not* a solution of the inequality.

b. $x - 3y \geq 8$; $(2, -2)$

$x - 3y \geq 8$	Write the inequality.
$2 - 3(-2) \overset{?}{\geq} 8$	Substitute 2 for x and -2 for y.
$8 \geq 8$ ✓	Simplify. 8 is equal to 8.

So, $(2, -2)$ is a solution of the inequality.

On Your Own

Now You're Ready
Exercises 11–18

Tell whether the ordered pair is a solution of the inequality.

1. $x + y > 0$; $(-2, 2)$

2. $4x - y \geq 5$; $(0, 0)$

3. $5x - 2y \leq -1$; $(-4, -1)$

4. $-2x - 3y < 15$; $(5, -7)$

A dashed boundary line means that points on the line are *not* solutions. A solid boundary line means that points on the line are solutions.

The **graph of a linear inequality** in two variables shows all of the solutions of the inequality in a coordinate plane.

All solutions of $y < 2x$ lie on one side of the *boundary line* $y = 2x$.

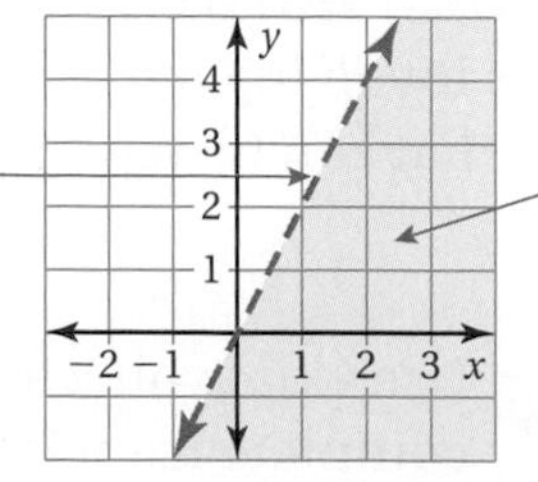

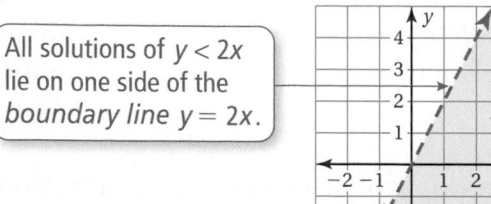

The boundary line divides the coordinate plane into two **half-planes**. The shaded half-plane is the graph of $y < 2x$.

Multi-Language Glossary at BigIdeasMath.com.

Laurie's Notes

Introduction

Connect

- **Yesterday:** Students explored the graphs of linear inequalities. (MP6)
- **Today:** Students will graph linear inequalities in two variables.

Motivate

- **Acting Time:** Wear an apron and/or bring a kitchen tool. Explain that you want to supplement your income before the holidays, so you are going to make your famous fudge brownies and nutty trail mix to sell at [name a school or town event]. Your profit is $1 on each brownie and $2 on each bag of trail mix.

? "My goal is $500. If I sell *x* brownies and *y* bags of trail mix, what equation represents a profit of $500?" $x + 2y = 500$

? "Name two ordered pairs that satisfy the equation and interpret what they mean." *Sample answer:* (500, 0) means you sold 500 brownies and no trail mix and (0, 250) means you sold 250 bags of trail mix and no brownies.

- Do a quick sketch of the line.

? "I changed my goal. I want to make *at least* $500. How do I represent my goal now?" $x + 2y \geq 500$

- Explain that today's lesson is about linear inequalities in two variables.

Lesson Notes

Discuss

- Define linear inequalities in two variables. Refer to the example used in the Motivate.
- Explain that while the examples are in a form that resembles standard form, linear inequalities can also be written in a form that resembles slope-intercept form.

Example 1

? "Recall my brownie and trail mix problem. Would selling 275 brownies and 120 bags of trail mix meet my goal? Explain." yes; $275 + 2 \times 120 = \$515$, which is greater than $500

- Substituting the ordered pair in the inequality is one way to check if you have a solution without graphing.
- Work through each problem as shown.

On Your Own

- Students may make a careless error in computation. Checking their results with a neighbor should catch errors.

Goal Today's lesson is graphing a **linear inequality in two variables.**

Start Thinking! and Warm Up

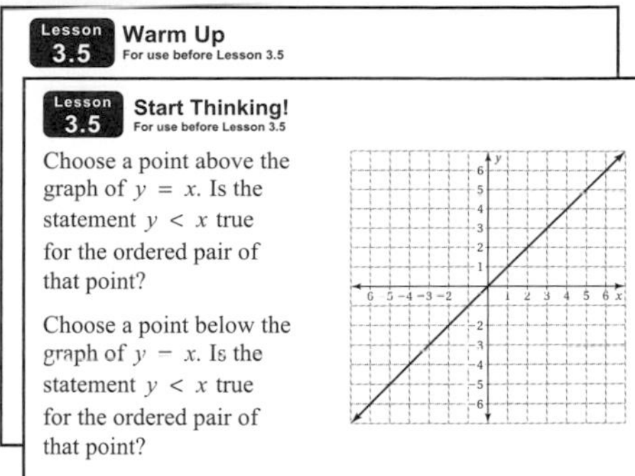
Lesson 3.5 Warm Up
For use before Lesson 3.5

Lesson 3.5 Start Thinking!
For use before Lesson 3.5

Choose a point above the graph of $y = x$. Is the statement $y < x$ true for the ordered pair of that point?

Choose a point below the graph of $y = x$. Is the statement $y < x$ true for the ordered pair of that point?

Extra Example 1

Tell whether the ordered pair is a solution of the inequality.

a. $3x - y > 5$; (3, 0) yes

b. $2x + 4y \leq -6$, (2, −2) no

On Your Own

1. no
2. no
3. yes
4. yes

Extra Example 2

a. Graph $y < -1$ in a coordinate plane.

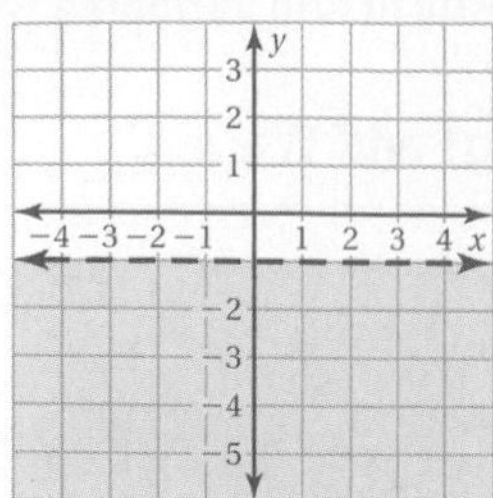

b. Graph $x \geq 2$ in a coordinate plane.

On Your Own

5.

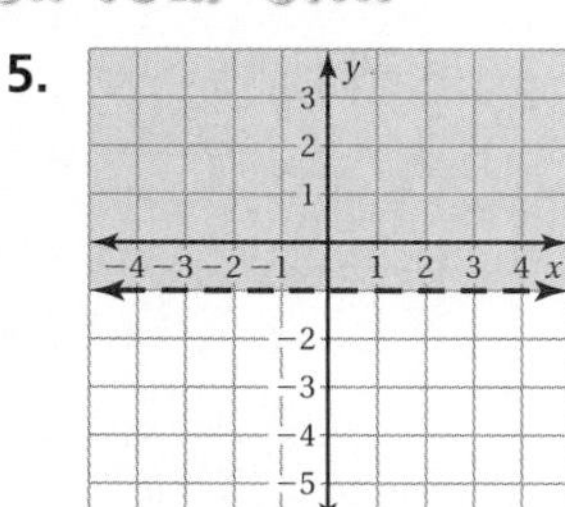

6.

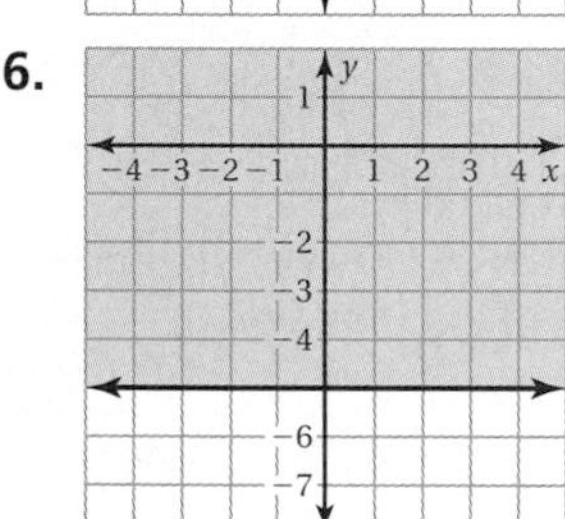

7–8. See Additional Answers.

English Language Learners

Pair Activity

Pair English learners with English speakers. Assign a problem-solving exercise from the exercise set. Have students make a poster illustrating the solution steps to the problem. Students should include writing the inequality, solving the inequality for y, graphing the inequality, and showing and interpreting solution points.

Laurie's Notes

Discuss

- Define the graph of a linear inequality in two variables. Relate the solid and dashed lines to the closed and open circles used when inequalities in one variable are graphed on a number line.
- The closed and open circles are boundaries, creating two half-lines. The solid and dashed lines are boundaries, creating two half-planes.

Key Idea

- Write and discuss the steps to graph a linear inequality in two variables.
- **?** "What test point would simplify the computation?" It is likely that at least one student will think of the origin as a good test point.
- Remind students that if the line passes through the origin, they need to select a different test point. Reasonable ordered pairs on the axes are typically good choices.

Example 2

- **?** "Describe the graph of $y = 2$." horizontal line through (0, 2)
- **?** "Describe the graph of $x = 1$." vertical line through (1, 0)
- **MP1a Make Sense of Problems:** After each example is completed, students should verbalize that the points in the solution set have y-coordinates that are less than or equal to 2 (part (a)) and x-coordinates that are greater than 1 (part (b)).

On Your Own

- Check students' work. In Question 8, be sure that students shaded the graph correctly. The inequality $3.5 > x$ is equivalent to $x < 3.5$.

Graphing a Linear Inequality in Two Variables

Step 1 Graph the boundary line for the inequality. Use a dashed line for $<$ or $>$. Use a solid line for $\leq$ or $\geq$.

Step 2 Test a point that is not on the boundary line to determine if it is a solution of the inequality.

Step 3 If the test point is a solution, shade the half-plane that contains the point. If the test point is *not* a solution, shade the half-plane that does *not* contain the point.

It is convenient to use the origin as a test point because it is easily substituted. However, you must choose a different test point if the origin is on the boundary line.

EXAMPLE 2

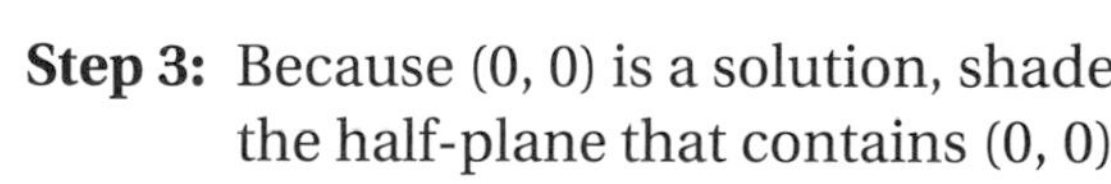

a. Graph $y \leq 2$ in a coordinate plane.

Step 1: Graph $y = 2$. Use a solid line because the inequality symbol is $\leq$.

Step 2: Test (0, 0).

$y \leq 2$ Write the inequality.

$0 \leq 2$ ✓ Substitute.

Step 3: Because (0, 0) is a solution, shade the half-plane that contains (0, 0).

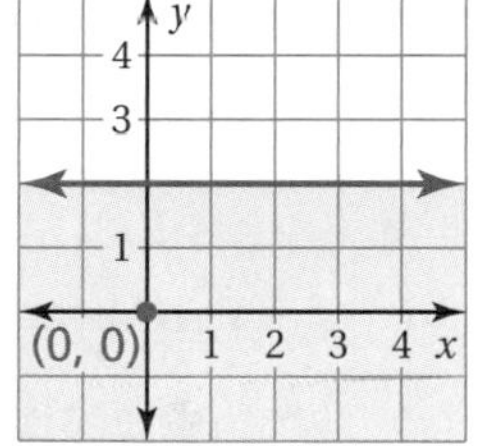

b. Graph $x > 1$ in a coordinate plane.

Step 1: Graph $x = 1$. Use a dashed line because the inequality symbol is $>$.

Step 2: Test (0, 0).

$x > 1$ Write the inequality.

$0 \ngtr 1$ ✗ Substitute.

Step 3: Because (0, 0) is *not* a solution, shade the half-plane that does *not* contain (0, 0).

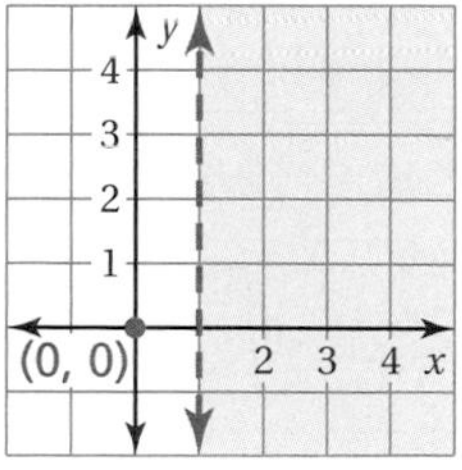

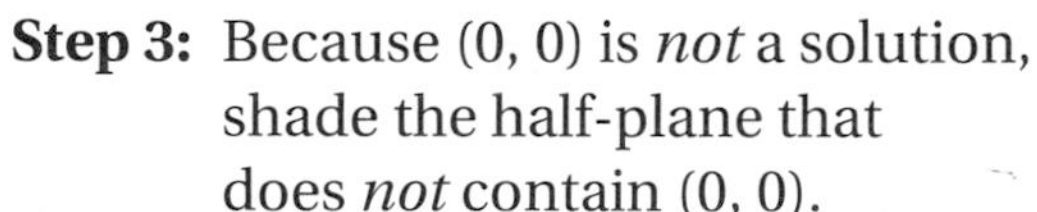

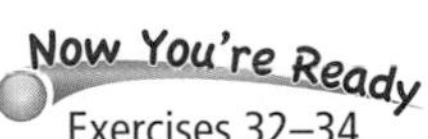

Exercises 32–34

Graph the inequality in a coordinate plane.

5. $y > -1$
6. $y \geq -5$
7. $x \leq -4$
8. $3.5 > x$

EXAMPLE 3 Graphing Linear Inequalities in Two Variables

Graph $-x + 2y > 2$ in a coordinate plane.

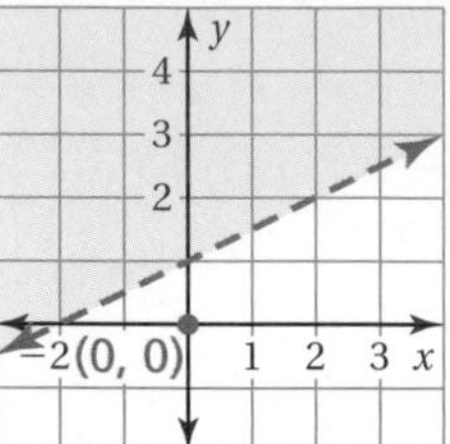

Step 1: Graph $-x + 2y = 2$, or $y = \frac{1}{2}x + 1$.

Use a dashed line because the inequality symbol is >.

Step 2: Test (0, 0).

$-x + 2y > 2$	Write the inequality.
$-(0) + 2(0) \overset{?}{>} 2$	Substitute.
$0 \not> 2$ ✗	Simplify.

Step 3: Because (0, 0) is *not* a solution, shade the half-plane that does *not* contain (0, 0).

Check

5, −2, 3, −1

EXAMPLE 4 Real-Life Application

You can spend at most \$10 on grapes and apples for a fruit salad. Grapes cost \$2.50 per pound and apples cost \$1 per pound. Write and graph an inequality for the amounts of grapes and apples you can buy. Identify and interpret two solutions of the inequality.

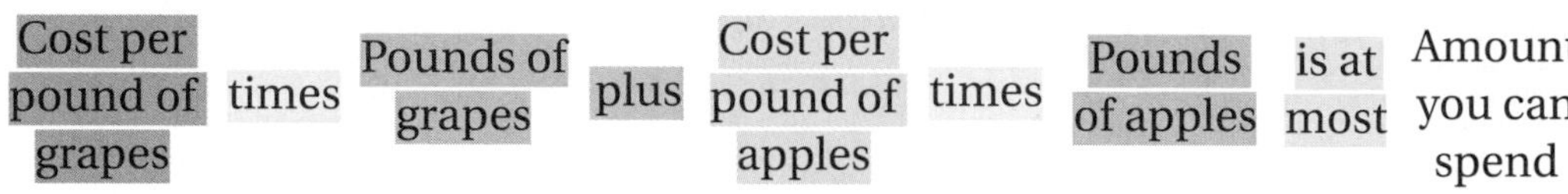

Words	Cost per pound of grapes	times	Pounds of grapes	plus	Cost per pound of apples	times	Pounds of apples	is at most	Amount you can spend

Variables Let *x* be pounds of grapes and *y* be pounds of apples.

Inequality	2.50	$\cdot$	x	+	1	$\cdot$	y	$\le$	10

Step 1: Graph $2.5x + y = 10$, or $y = -2.5x + 10$. Use a solid line because the inequality symbol is ≤.

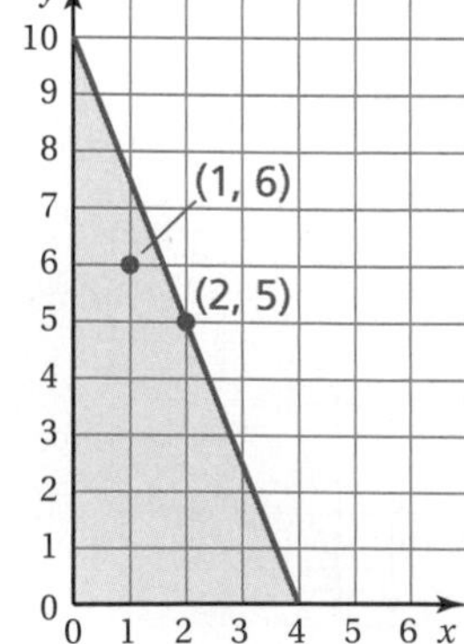

Step 2: Test (0, 0).

$2.5x + y \le 10$	Write the inequality.
$2.5(0) + 0 \overset{?}{\le} 10$	Substitute.
$0 \le 10$ ✓	Simplify.

Step 3: Because (0, 0) is a solution, shade the half-plane that contains (0, 0).

∴ Two possible solutions are (1, 6) and (2, 5). So, you can buy 1 pound of grapes and 6 pounds of apples, or 2 pounds of grapes and 5 pounds of apples.

On Your Own

Now You're Ready
Exercises 35–40

Graph the inequality in a coordinate plane.

9. $x + y \le -4$ **10.** $x - 2y < 0$ **11.** $2x + 2y \ge 3$

Laurie's Notes

Example 3

- Students used their calculators in the investigation, though the form of the equation was different.

? "How do you enter $-x + 2y = 2$ in your calculator?" You first have to solve the equation for y.

- **MP6 Attend to Precision:** Ask students how the process would change if the problem had been $-x - 2y > 2$. When solving for y, it would be necessary to change the direction of the inequality symbol. Mathematically proficient students can apply their knowledge of solving inequalities in one variable to solving inequalities in two variables.

Example 4

- This example is similar to the Motivate problem where there was a boundary of \$500 that you were trying to reach. In this example, there is a boundary of \$10 that you cannot exceed.
- Help students write the verbal model for this example. Students often think, "Cost of grapes plus cost of apples cannot exceed \$10." Defining the variables is a step students often omit.
- Note the use of colors to help students translate the model.

? "What are the intercepts for the boundary line? Interpret their meaning." (0, 10) means you bought 10 pounds of apples and no grapes and (4, 0) means you bought 4 pounds of grapes and no apples.

On Your Own

? "When equations are written in standard form, what technique might be helpful when graphing the line?" Determine the two intercepts.

- Have students share their work at the board.

Closure

- **Phone Message:** Write a brief phone message that you would leave for a friend that missed today's class. Explain how to graph a linear inequality in two variables.

  Sample answer: First, graph the boundary line. If the inequality symbol is $\le$ or $\ge$, use a solid line. Otherwise, use a dashed line. Test a point on one side of the line by substituting the x-value and y-value into the inequality. If the point makes the inequality true, shade the half of the plane that contains the point. Otherwise, shade the half of the plane on the other side of the line.

Technology For the Teacher

The Dynamic Planning Tool
Editable Teacher's Resources at *BigIdeasMath.com*

Extra Example 3

Graph $2x - y \ge 1$ in a coordinate plane.

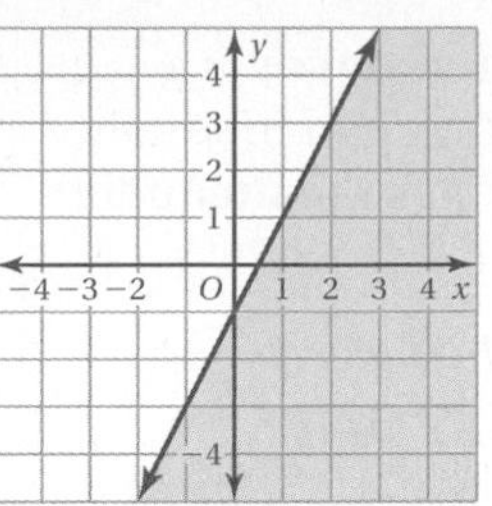

Extra Example 4

You can spend at most \$24 on party supplies. Party hats cost \$0.50 each and helium-filled balloons cost \$2 each. Write and graph an inequality for the amounts of party hats and balloons you can buy. Identify and interpret two solutions of the inequality.

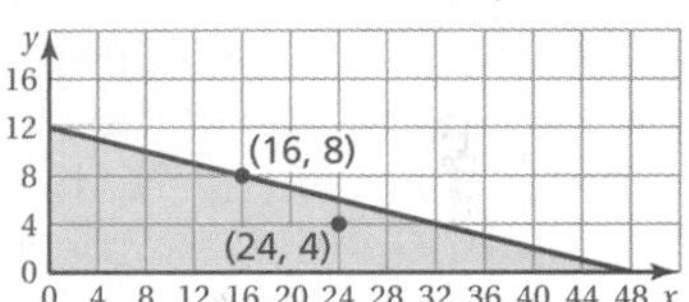

Two possible solutions are (16, 8) and (24, 4). So, you can buy 16 party hats and 8 balloons, or 24 party hats and 4 balloons.

On Your Own

9.

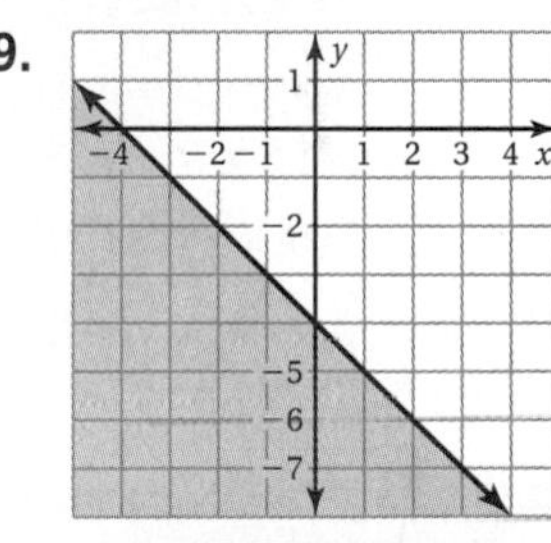

10.

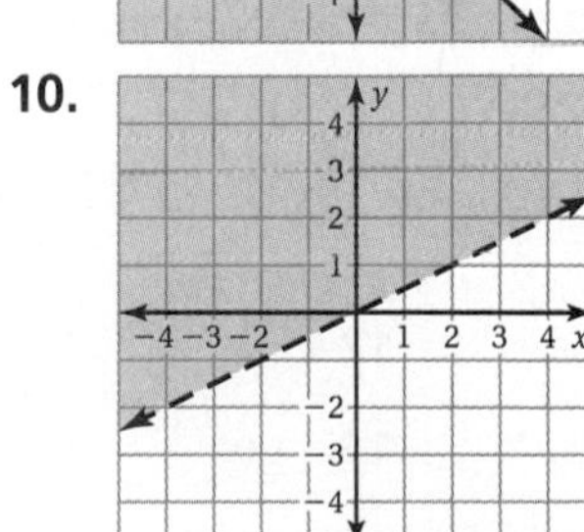

11.

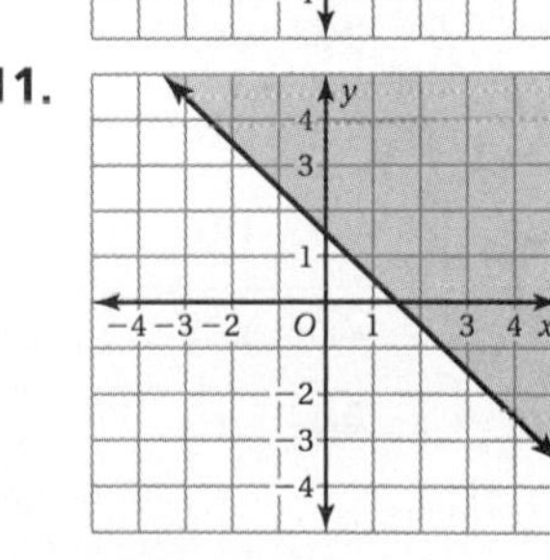

Vocabulary and Concept Check

1. An ordered pair is a solution of an inequality if it makes the inequality true.
2. *Sample answer:* $3x + 2y > 15$
3. The graph of a linear equation in two variables will be a solid line. The graph of a linear inequality in two variables could be a solid or dashed line, and half of the coordinate plane will be shaded.
4. If the test point (not on the boundary line) is a solution, then everything on the same side of the line as the point is also a solution. Similarly, if the test point is not a solution, then everything on the other side of the line is a solution.
5. C
6. A
7. B

Practice and Problem Solving

8. All the points above the line $y = x - 1$
9. All the points on or above the line $y = -x + 5$
10. All the points below the line $y = x - 2$
11. yes
12. no
13. no
14. yes
15. yes
16. yes
17. no
18. no
19. yes
20. no
21. no
22. yes
23. no
24. yes

Assignment Guide and Homework Check

Level	Assignment	Homework Check
Average	1–10, 11–25 odd, 32–44 even, 55–58	15, 25, 36, 42
Advanced	1–10, 32–54 even, 43, 53, 55–58	36, 42, 46, 54

Common Errors

- **Exercises 11–18** Students may make careless errors in computation. Remind them that subtracting a negative number is the same as adding its opposite.

3.5 Record and Practice Journal

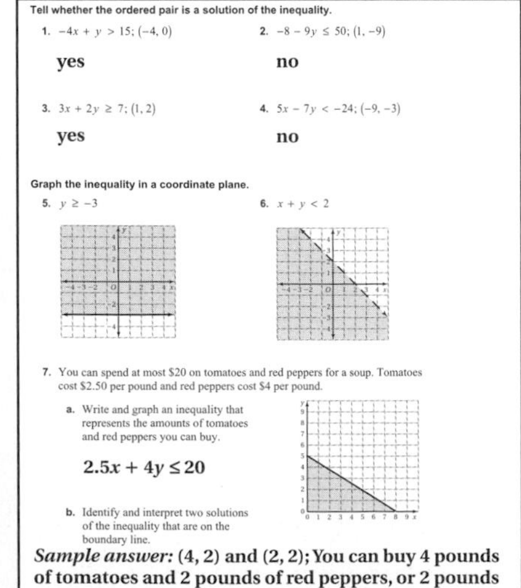

Tell whether the ordered pair is a solution of the inequality.

1. $-4x + y > 15$; $(-4, 0)$

   yes

2. $-8 - 9y \le 50$; $(1, -9)$

   no

3. $3x + 2y \ge 7$; $(1, 2)$

   yes

4. $5x - 7y < -24$; $(-9, -3)$

   no

Graph the inequality in a coordinate plane.

5. $y \ge -3$
6. $x + y < 2$

7. You can spend at most \$20 on tomatoes and red peppers for a soup. Tomatoes cost \$2.50 per pound and red peppers cost \$4 per pound.

   a. Write and graph an inequality that represents the amounts of tomatoes and red peppers you can buy.

   $2.5x + 4y \le 20$

   b. Identify and interpret two solutions of the inequality that are on the boundary line.

   ***Sample answer:* (4, 2) and (2, 2); You can buy 4 pounds of tomatoes and 2 pounds of red peppers, or 2 pounds of tomatoes and 2 pounds of red peppers.**

Technology For the Teacher

Answer Presentation Tool

3.5 Exercises

Vocabulary and Concept Check

1. **VOCABULARY** How can you tell whether an ordered pair is a solution of an inequality?
2. **OPEN-ENDED** Write an example of an inequality in two variables.
3. **WRITING** Compare the graph of a linear inequality in two variables with the graph of a linear equation in two variables.
4. **REASONING** Why do you only need to test one point when graphing a linear inequality?

Match the inequality with its graph.

5. $x > -1$
6. $y > 1$
7. $x < 1$

A.

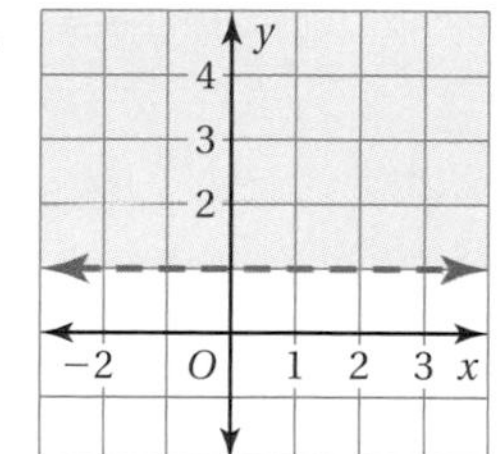

B.

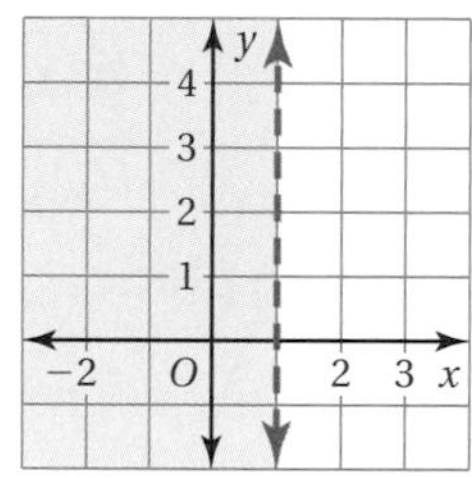

C.

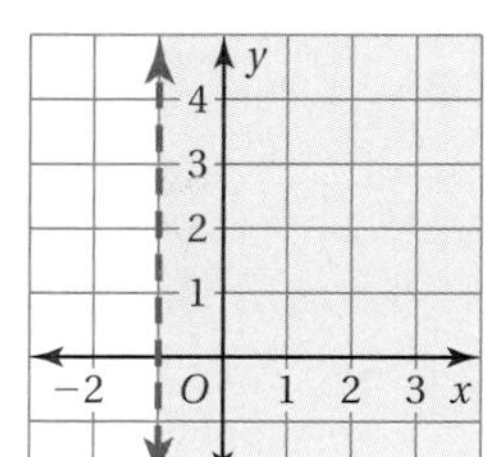

Practice and Problem Solving

In words, describe the solutions of the inequality.

8. $y > x - 1$
9. $y \geq -x + 5$
10. $y < x - 2$

Tell whether the ordered pair is a solution of the inequality.

1 11. $x + y < 7$; $(6, -1)$
12. $2x - y \leq 0$; $(-2, -5)$
13. $x + 3y \geq -2$; $(-4, -2)$
14. $3x + 2y > -6$; $(0, 0)$
15. $-6x + 4y \leq 5$; $(3, -5)$
16. $3x - 5y \geq -8$; $(-1, 1)$
17. $-x - 6y > 12$; $(-8, 2)$
18. $-4x - 8y < -15$; $(-6, 3)$

Tell whether the ordered pair is a solution of the inequality whose graph is shown.

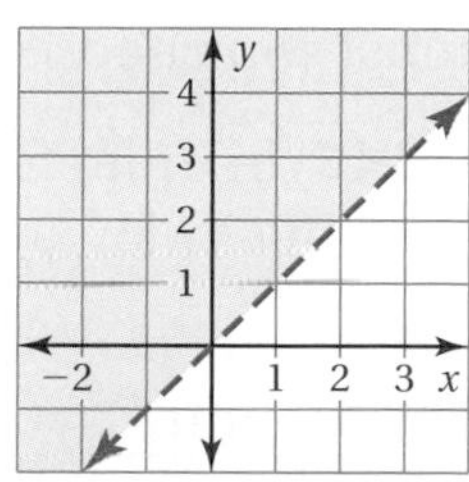

19. $(0, 4)$
20. $(0, 0)$
21. $(-1, -2)$
22. $(-1, 3)$
23. $(3, 3)$
24. $(-2, -1)$

25. **FABRIC** You can spend at most \$60 on lace. Cotton lace is \$2 per yard and linen lace is \$3 per yard. Write an inequality for the amounts of lace you can buy. Can you buy 12 yards of cotton lace and 15 yards of linen lace? Explain.

In Exercises 26–28, use the inequality $2x + y < -1$.

26. Write the equation of the boundary line in slope-intercept form.

27. Tell whether you would use a solid line or a dashed line to graph the boundary line. Then graph the boundary line.

28. Test the point (0, 0) in the inequality. Is the test point a solution? If so, shade the half-plane that contains the point. If not, shade the half-plane that does *not* contain the point.

Match the inequality with its graph.

29. $3x - 2y \le 6$ **30.** $3x - 2y < 6$ **31.** $3x - 2y \ge 6$

A.

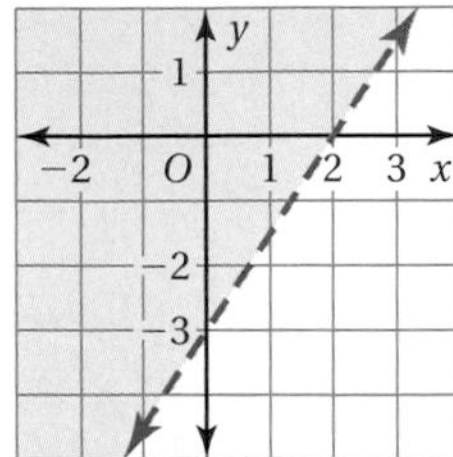

B.

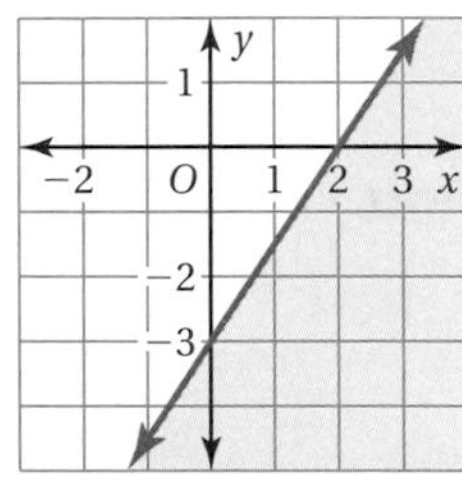

C.

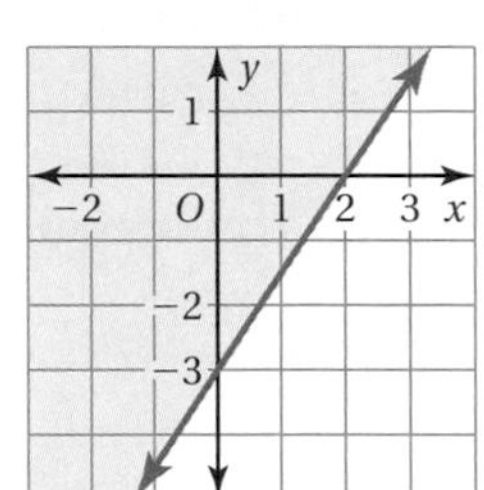

Graph the inequality in a coordinate plane.

2 **32.** $y < 5$ **33.** $x \ge -3$ **34.** $x < 2$

3 **35.** $y \le 3x - 1$ **36.** $-2x + y > -4$ **37.** $3x - 2y \ge 0$

38. $5x - 2y \le 6$ **39.** $2x - y < -3$ **40.** $-x + 4y > -2$

ERROR ANALYSIS Describe and correct the error in graphing the inequality.

41. $y < -x + 1$

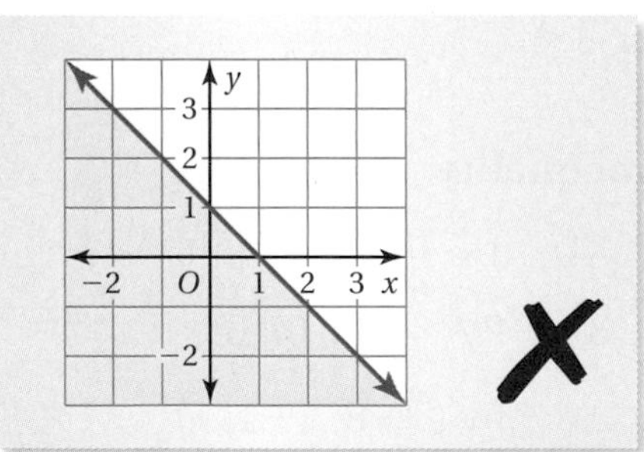

42. $y \le 3x - 2$

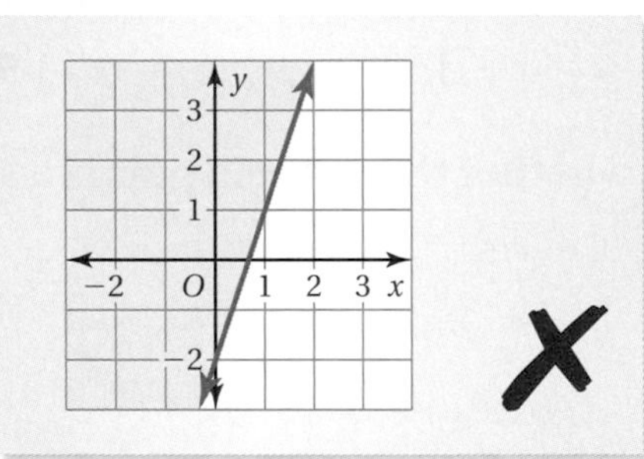

43. CRITICAL THINKING When graphing a linear inequality in two variables, why must you choose a test point that is *not* on the boundary line?

44. MODELING In order for the drama club to cover the expenses of producing a play, at least $1500 worth of tickets must be sold.

a. Use a model to write an inequality that represents this situation.

b. Graph the inequality.

c. Eighty adults and 110 students attend the play. Does the drama club cover its expenses? Explain.

Common Errors

- **Exercise 32** Students may draw a vertical line through a point on the x-axis instead of a horizontal line through the corresponding point on the y-axis. Ask them to identify the y-coordinate for several x-coordinates. For example, what is the y-coordinate for $x = 5$? $x = 2$? $x = -4$? Students should answer with the same y-coordinate each time.
- **Exercises 33 and 34** Students may draw a horizontal line through a point on the y-axis instead of a vertical line through the corresponding point on the x-axis. Ask them to identify the x-coordinate for several y-coordinates. For example, what is the x-coordinate for $y = 2$? $y = 0$? $y = -5$? Students should answer with the same x-coordinate each time.
- **Exercises 37–39** Students may forget to reverse the inequality symbol when solving for y. Remind them of the rule for multiplying or dividing by a negative number.
- **Exercises 32–40** Students may use the wrong boundary line and/or shade the wrong half-plane. Remind them to use a test point to check that their graph is correct.

English Language Learners

Simplified Language

Word problems pose an additional challenge for English learners. You can simplify the statement of word problems so that students can apply their problem-solving skills. Students benefit from comparing the original problem to the simplified statement and learn how to determine which information in the problem is critical to the solution.

Practice and Problem Solving

25. $2x + 3y \leq 60$; no, you cannot buy 12 yards of cotton lace and 15 yards of linen lace because (12, 15) is not a solution of the inequality.

26. $y < -2x - 1$

27. The boundary line will be dashed.

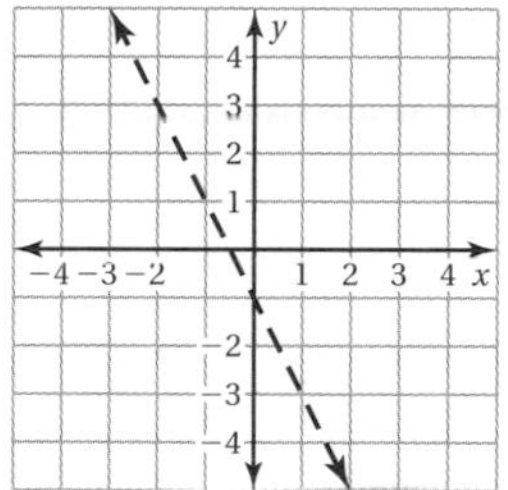

28. (0, 0) is not a solution of the inequality.

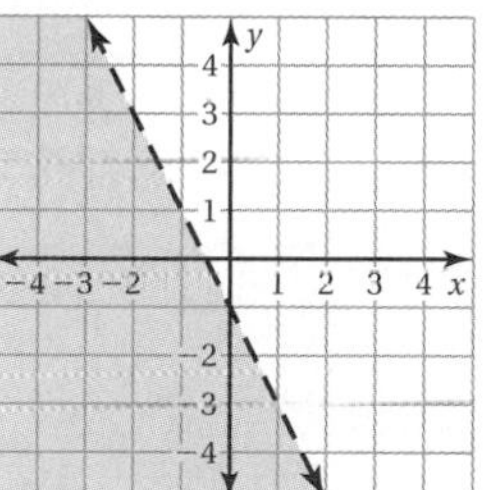

29. C

30. A

31. B

32.

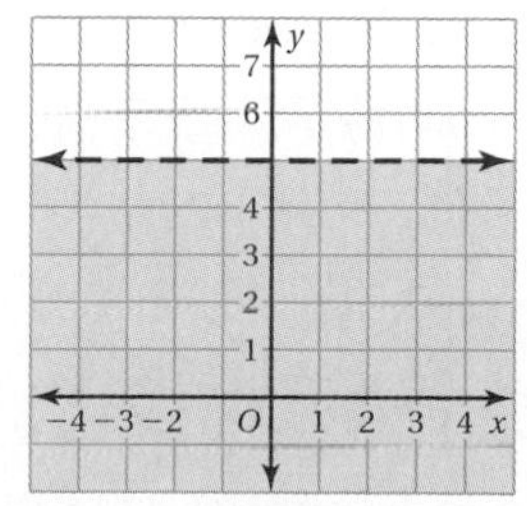

33–44. See Additional Answers.

Practice and Problem Solving

45. yes **46.** no

47. no **48.** yes

49. $y > 2x + 1$

50. $y > \frac{1}{2}x + 2$

51. $y \le -\frac{1}{2}x - 2$

52. infinitely many solutions

53. See Additional Answers.

54. See *Taking Math Deeper*.

Fair Game Review

55. 256 **56.** −8

57. 243 **58.** B

Mini-Assessment

Tell whether the ordered pair is a solution of the inequality.

1. $4x - y > 2$; (1, 3) no
2. $-2x + 5y \le -1$; (−2, −1) yes

Graph the inequality in a coordinate plane.

3. $y \ge -2$

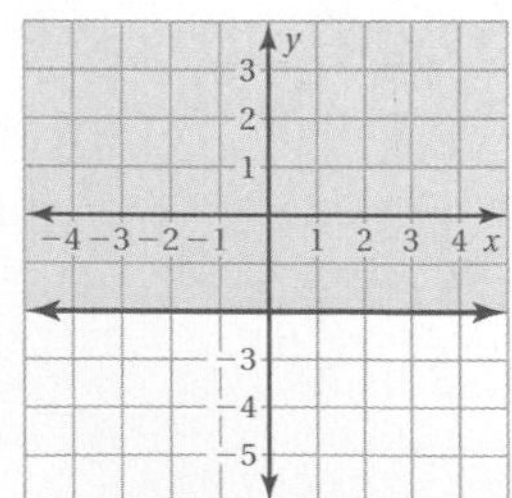

4. $x + 2y < 6$

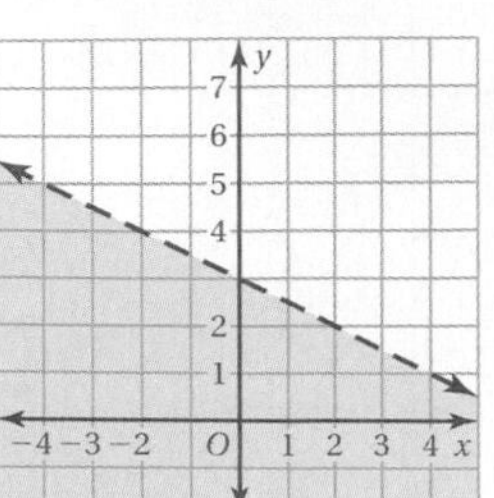

Taking Math Deeper

Exercise 54

The challenge in this problem is to incorporate the information about the delivery person's weight when writing the inequality.

1 Write the inequality.

Because the delivery person weighs 200 pounds, the most that the packages can weigh is $2000 - 200 = 1800$ pounds. Let x be the number of small boxes and y be the number of large boxes.

a. Weight of small boxes plus Weight of large boxes is at most 1800.

$$40x + 75y \le 1800$$

2 Graph the inequality using the intercepts.

No large boxes:
$40x + 75(0) = 1800$
$x = 45$

No small boxes:
$40(0) + 75y = 1800$
$y = 24$

Plot (45, 0) and (0, 24).

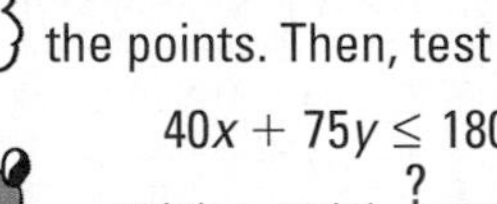

Draw a solid line through the points. Then, test (0, 0).

$$40x + 75y \le 1800$$
$$40(0) + 75(0) \overset{?}{\le} 1800$$
$$0 \le 1800$$

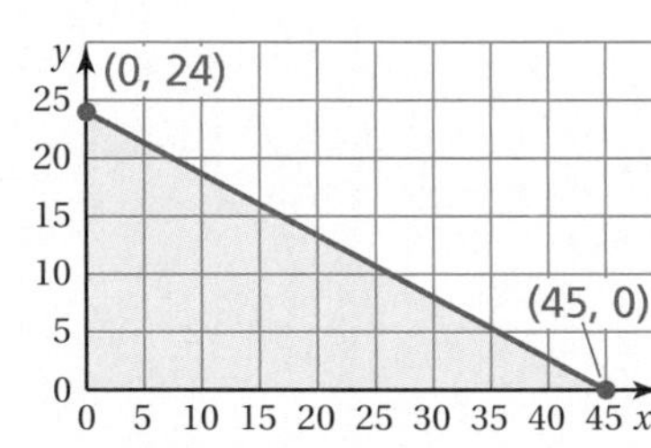

Shade the half-plane that contains (0, 0).

b. The points (45, 0) and (0, 24) are on the boundary line. So, the delivery person can take 45 small boxes and 0 large boxes or 0 small boxes and 24 large boxes.

3 Interpret solutions.

c. Solutions on the boundary line involve a significant number of boxes. It may not be practical for one delivery person to transport that many boxes in one trip.

Reteaching and Enrichment Strategies

If students need help. . .	If students got it. . .
Resources by Chapter • Practice A and Practice B • Puzzle Time Record and Practice Journal Practice Differentiating the Lesson Lesson Tutorials Skills Review Handbook	Resources by Chapter • Enrichment and Extension • School-to-Work • Financial Literacy Start the next section

Tell whether the ordered pair is a solution of the inequality.

45. $y < \frac{1}{3}x + \frac{1}{4}$; $(6, 2)$

46. $2.5 - y \leq 1.8x$; $(0.5, 1.5)$

47. $0.2x + 1.6y \geq -1$; $(10, -2.2)$

48. $2x - \frac{2}{3}y > -5$; $\left(\frac{3}{4}, 4\right)$

Write an inequality that represents the graph.

49. 

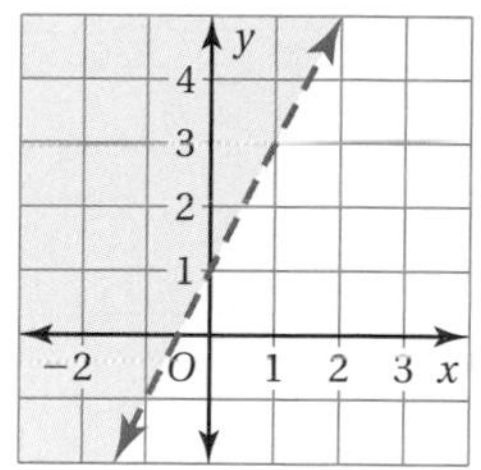

50. 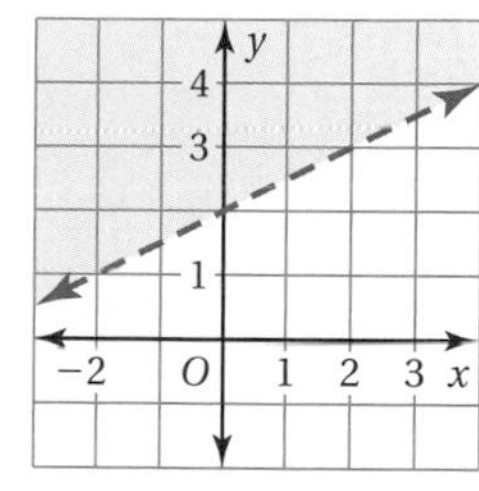

51. 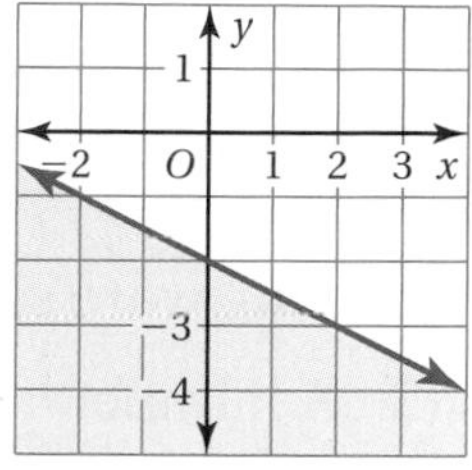

52. REASONING How many solutions does the inequality $2x + y \geq 5$ have?

53. PROBLEM SOLVING After buying your admission ticket, you have \$9 to spend at the movies. Arcade games cost \$0.75 per game and soft drinks cost \$2.25.

a. Write and graph an inequality that represents the numbers of arcade games you can play and soft drinks you can buy.

b. Identify and interpret two solutions of the inequality.

54. 

Large boxes weigh 75 pounds and small boxes weigh 40 pounds.

a. Write and graph an inequality that represents the numbers of large and small boxes a 200-pound delivery person can take on the elevator.

b. Identify and interpret two solutions of the inequality that are on the boundary line.

c. Explain why the solutions in part (b) might not be practical in real life.

Weight Limit
2000 lb

Fair Game Review What you learned in previous grades & lessons

Multiply. *(Skills Review Handbook)*

55. $4 \cdot 4 \cdot 4 \cdot 4$

56. $(-2) \cdot (-2) \cdot (-2)$

57. $3 \cdot 3 \cdot 3 \cdot 3 \cdot 3$

58. MULTIPLE CHOICE Which graph represents the solution of $-5(x - 9) \geq -35$? *(Section 3.4)*

Ⓐ 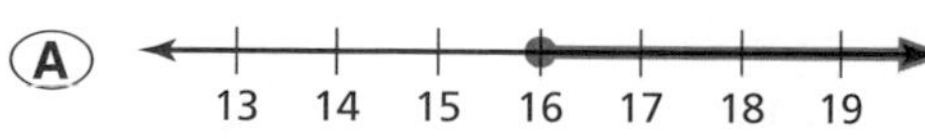

Ⓑ 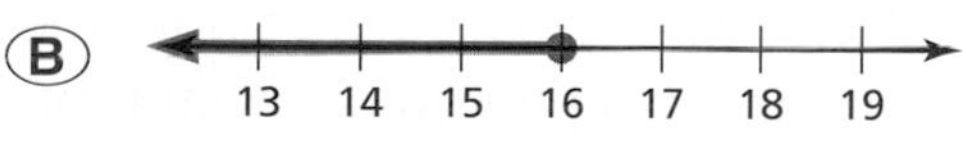

Ⓒ −19 −18 −17 −16 −15 −14 −13

Ⓓ 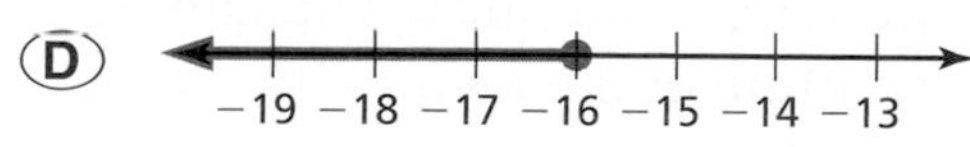

3.4–3.5 Quiz

Solve the inequality. Graph the solution. *(Section 3.4)*

1. $2m + 1 \geq 7$
2. $\frac{n}{6} - 8 \leq 2$
3. $2 - \frac{j}{5} > 7$
4. $\frac{5}{4} > -3w - \frac{7}{4}$

Write the word sentence as an inequality. Graph the inequality. *(Section 3.4)*

5. A number h is greater than 1 and less than 6.
6. A number q is less than or equal to -3 or at least 2.

Solve the inequality. Graph the solution, if possible. *(Section 3.4)*

7. $7 > -2y + 5 > -3$
8. $3z + 2 \leq -10$ *or* $z - 7 \geq -5$
9. $|2b - 1| \leq 3$
10. $-4|r - 1| + 7 < -9$

Graph the inequality in a coordinate plane. *(Section 3.5)*

11. $y \geq -8$
12. $x < 6$
13. $x + y > 5$
14. $4x - 4y \leq 8$

15. **PARTY** You buy lunch for guests at a party. You can spend no more than \$100. You will spend \$20 on beverages and \$10 per guest on sandwiches. Write and solve an inequality to find the number of guests you can invite to the party. *(Section 3.4)*

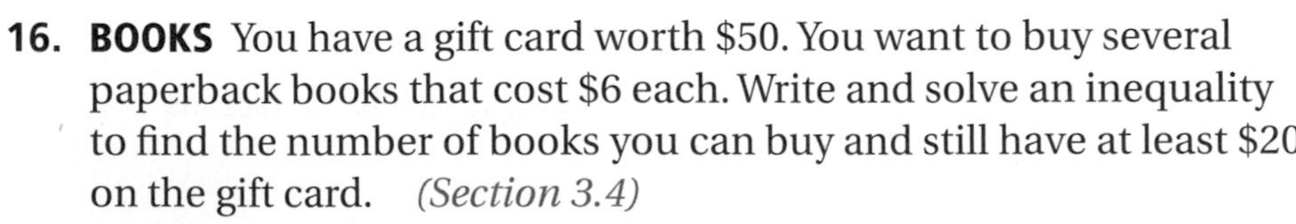

16. **BOOKS** You have a gift card worth \$50. You want to buy several paperback books that cost \$6 each. Write and solve an inequality to find the number of books you can buy and still have at least \$20 on the gift card. *(Section 3.4)*

17. **SUPPLIES** You have \$6 to spend on pens and notebooks. Pens cost \$0.75 each and notebooks cost \$1.50 each. Write and graph an inequality that represents the numbers of pens and notebooks you can buy. Identify and interpret a solution of the inequality. *(Section 3.5)*

Alternative Assessment Options

Math Chat
Structured Interview
Student Reflective Focus Question
Writing Prompt

Student Reflective Focus Question

Ask students to summarize the similarities and differences between solving multi-step inequalities and solving multi-step equations. Be sure that they include examples. Select students at random to present to the class.

Study Help Sample Answers

Remind students to complete Graphic Organizers for the rest of the chapter.

6.

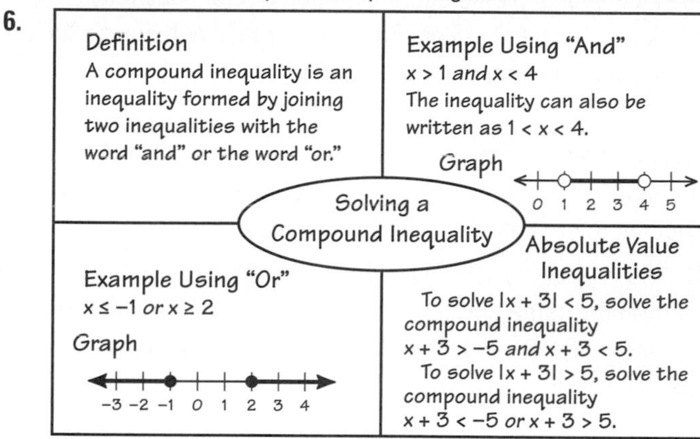

7.

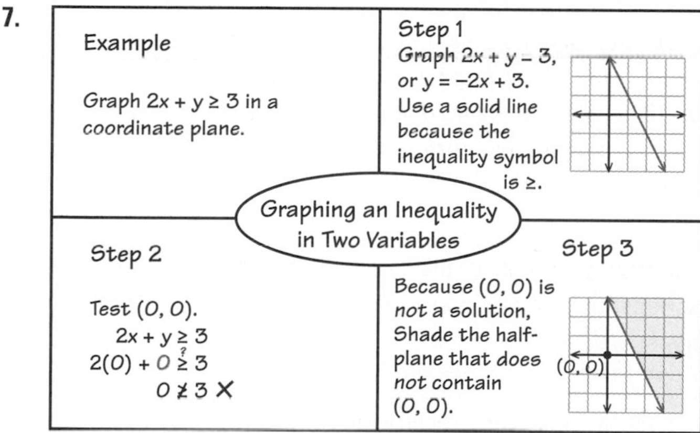

Reteaching and Enrichment Strategies

If students need help...	If students got it...
Resources by Chapter • Study Help • Practice A and Practice B • Puzzle Time Lesson Tutorials *BigIdeasMath.com* Practice Quiz Practice from the Test Generator	Resources by Chapter • Enrichment and Extension • School-to-Work Game Closet at *BigIdeasMath.com* Start the Chapter Review

Answers

1. $m \geq 3$;
2. $n \leq 60$;
3. $j < -25$;
4. $w > -1$;
5. $1 < h < 6$;
6. $q \leq -3$ *or* $q \geq 2$;
7. $-1 < y < 4$;

8–14. See Additional Answers.

15. $10x + 20 \leq 100$; $x \leq 8$ guests

16. $6x + 20 \leq 50$; $x \leq 5$ books

17. See Additional Answers.

Technology For the Teacher

Answer Presentation Tool

Assessment Book

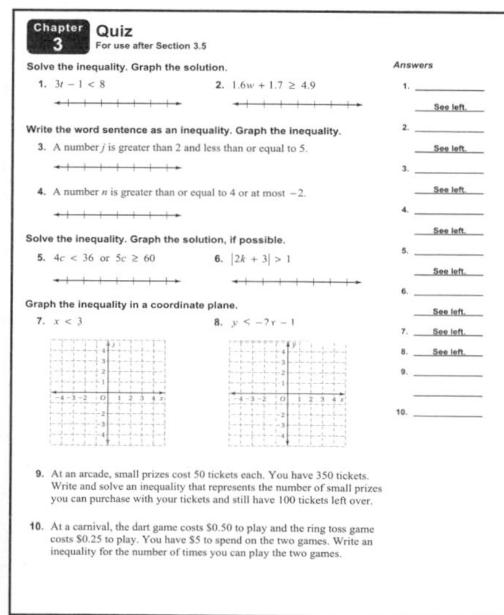

Chapter 3 Quiz
For use after Section 3.5

Solve the inequality. Graph the solution.

1. $3t - 1 < 8$
2. $1.6w + 1.7 \geq 4.9$

Write the word sentence as an inequality. Graph the inequality.

3. A number j is greater than 2 and less than or equal to 5.
4. A number n is greater than or equal to 4 or at most -2.

Solve the inequality. Graph the solution, if possible.

5. $4c < 36$ or $5c \geq 60$
6. $|2k + 3| > 1$

Graph the inequality in a coordinate plane.

7. $x < 3$
8. $y < -2x - 1$

9. At an arcade, small prizes cost 50 tickets each. You have 350 tickets. Write and solve an inequality that represents the number of small prizes you can purchase with your tickets and still have 100 tickets left over.

10. At a carnival, the dart game costs $0.50 to play and the ring toss game costs $0.25 to play. You have $5 to spend on the two games. Write an inequality for the number of times you can play the two games.

Answers
1. See left.
2. See left.
3. See left.
4. See left.
5. See left.
6. See left.
7. See left.
8. See left.
9.
10.

For the Teacher

Additional Review Options

- Big Ideas Test Generator
- Game Closet at *BigIdeasMath.com*
- Vocabulary Puzzle Builder
- Resources by Chapter
  - Puzzle Time
  - Study Help

Answers

1. $v < -2$
2. $x - \frac{1}{4} \le -\frac{3}{4}$
3. no
4. yes
5. 0.9 1.0 1.1 1.2 1.3 1.4 1.5
6. $9\frac{1}{2}$ $9\frac{3}{4}$ 10 $10\frac{1}{4}$ $10\frac{1}{2}$ $10\frac{3}{4}$ 11
7. $b < 5$;

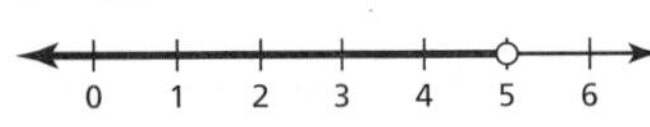

8. $x \le 13$;

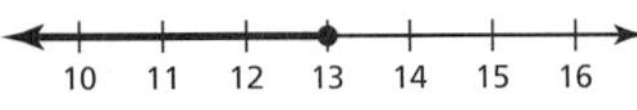

9. $y \ge -3$;

10. $x \ge 8$;

11. $z < -11$;

−15 −14 −13 −12 −11 −10 −9

12. $q \le 9$;

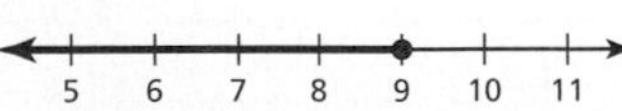

13. $x < 2$;

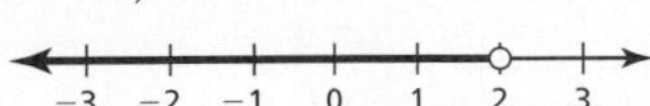

14. $z \ge -16$;

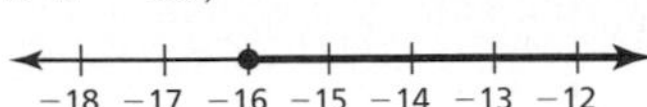

15. $w < -4$;

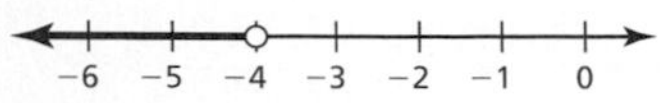

Review of Common Errors

Exercises 1 and 2

- Students may struggle knowing which inequality symbol to use. Encourage them to put the word sentence into a real-life context and to use the table in the lesson that explains what symbol matches each phrase.

Exercises 3 and 4

- Students may try to solve for the variable instead of substituting the given value into the inequality and determining if that value is a solution of the inequality.

Exercises 5 and 6

- Students may use a closed circle instead of an open circle. They may shade the wrong side of the number line. Review how to graph inequalities. Encourage them to test a value on each side of the circle.

Exercises 7–9

- Students may reverse the direction of the inequality symbol when adding or subtracting. Remind them that the inequality symbol does not change direction when adding to or subtracting from both sides.

Exercises 7–18

- Students may perform the same operation instead of the inverse operation when solving the inequality. Remind them that solving inequalities is similar to solving equations.

Exercises 10–18

- When there is a negative in the inequality, students may reverse the direction of the inequality symbol. Remind them that they only reverse the direction when they are multiplying or dividing by a negative number.
- Students may forget to reverse the inequality symbol when multiplying or dividing by a negative number. Encourage them to write the inequality symbol that they should have in the solution before solving.

Exercises 13–18

- Students may incorrectly multiply or divide before adding to or subtracting from both sides. Remind them that they should work backwards through the order of operations.

Exercises 19–21

- Students may use the wrong boundary line and/or shade the wrong half-plane. Remind them to use a test point to check that their graph is correct.

3 Chapter Review

Check It Out
Vocabulary Help
BigIdeasMath.com

Review Key Vocabulary

inequality, *p. 106*
solution of an inequality, *p. 106*
solution set, *p. 106*
graph of an inequality, *p. 107*
compound inequality, *p. 132*
absolute value inequality, *p. 134*
linear inequality in two variables, *p. 138*
solution of a linear inequality, *p. 138*
graph of a linear inequality, *p. 138*
half-planes, *p. 138*

Review Examples and Exercises

3.1 Writing and Graphing Inequalities (pp. 104–109)

a. Four plus a number w is at least $-\frac{1}{2}$. Write this sentence as an inequality.

Four plus a number w | is at least | $-\frac{1}{2}$.

$4 + w \quad \geq \quad -\frac{1}{2}$

∴ An inequality is $4 + w \geq -\frac{1}{2}$.

b. Graph $m > 4$.

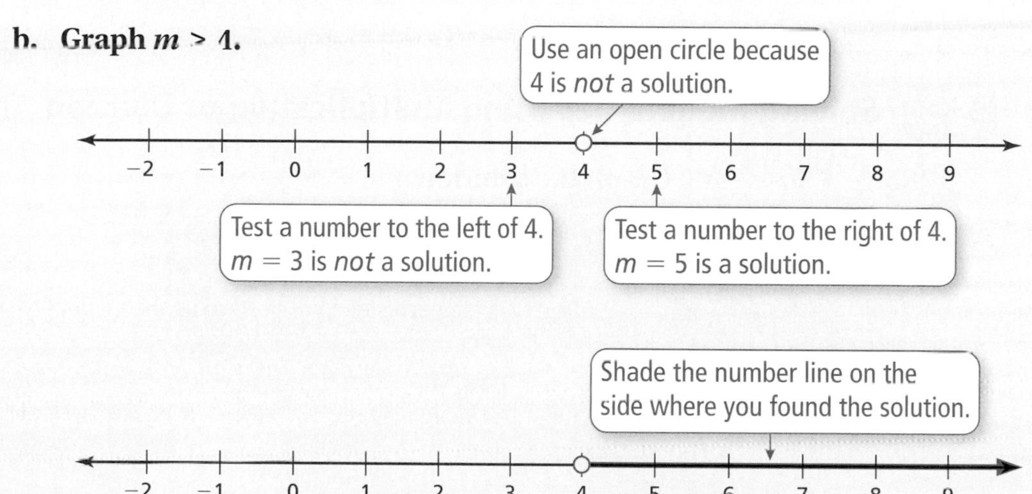

Use an open circle because 4 is *not* a solution.
−2
−1
0
1
2
3
4
5
6
7
8
9
Test a number to the left of 4. m = 3 is *not* a solution.
Test a number to the right of 4. m = 5 is a solution.
Shade the number line on the side where you found the solution.

Exercises

Write the word sentence as an inequality.

1. A number v is less than -2.
2. A number x minus $\frac{1}{4}$ is no more than $-\frac{3}{4}$.

Tell whether the given value is a solution of the inequality.

3. $10 - q < 3;\ q = 6$
4. $12 \div m \geq -4;\ m = -3$

Graph the inequality on a number line.

5. $p < 1.2$
6. $n > 10\frac{1}{4}$

3.2 Solving Inequalities Using Addition or Subtraction (pp. 110–115)

Solve $-4 < n - 3$. Graph the solution.

	$-4 < n - 3$	Write the inequality.
Undo the subtraction. →	$\underline{+3 \quad\quad +3}$	Add 3 to each side.
	$-1 < n$	Simplify.

The solution is $n > -1$.

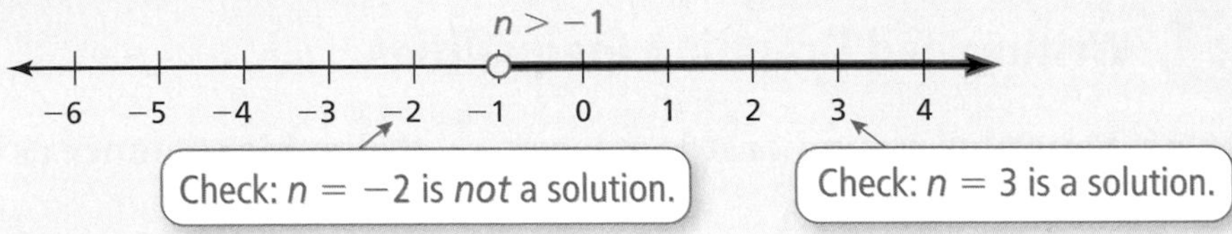

n > −1
−6 −5 −4 −3 −2 −1 0 1 2 3 4
Check: n = −2 is *not* a solution.
Check: n = 3 is a solution.

Exercises

Solve the inequality. Graph the solution.

7. $b + 13 < 18$ **8.** $x - 3 \leq 10$ **9.** $y + 1 \geq -2$

3.3 Solving Inequalities Using Multiplication or Division (pp. 116–123)

Solve $-8a \geq -48$. Graph the solution.

	$-8a \geq -48$	Write the inequality.
Undo the multiplication. →	$\frac{-8a}{-8} \leq \frac{-48}{-8}$	Divide each side by -8. Reverse the inequality symbol.
	$a \leq 6$	Simplify.

The solution is $a \leq 6$.

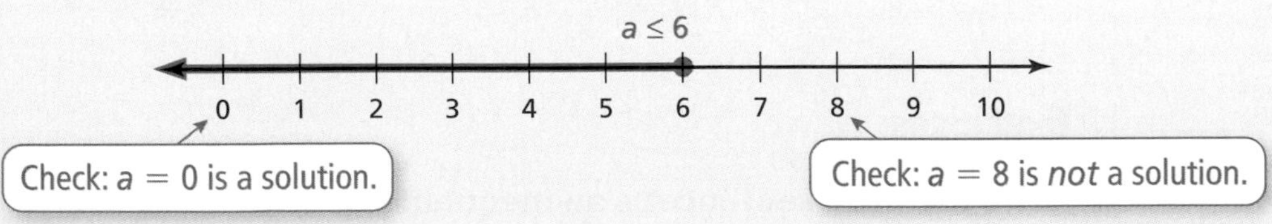

a ≤ 6
0 1 2 3 4 5 6 7 8 9 10
Check: a = 0 is a solution.
Check: a = 8 is *not* a solution.

Exercises

Solve the inequality. Graph the solution.

10. $\frac{x}{2} \geq 4$ **11.** $4z < -44$ **12.** $-2q \geq -18$

Review Game

Musical Toss

Materials

- soft object that can be tossed around
- a device to play music
- old homework, quiz, and test questions

Directions

Divide the class into pairs (groups of two).

Designate one pair of students to play the music and write the problems on the board. Pairs of students should be switched periodically.

The remaining members of the class will stand in a circle with each pair clearly identifiable.

When the music starts, the soft object is tossed to a pair of students and the problem is written on the board. That pair has to solve the problem and toss the object to another pair before the music stops.

Who wins?

The group holding the object when the music stops is eliminated. This will continue until there is one group remaining, the winner.

For the Student Additional Practice

- Lesson Tutorials
- Study Help (textbook)
- Student Website
  - Multi-Language Glossary
  - Practice Assessments

Answers

16. $1 < x < 11$;

17. $x \leq 1$ *or* $x \geq 3$;

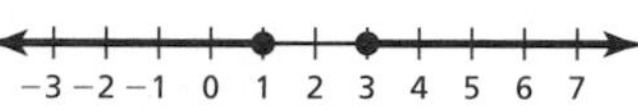

18. $x < 2$ *or* $x > 4$;

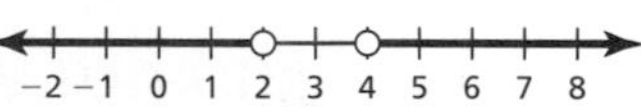

19.

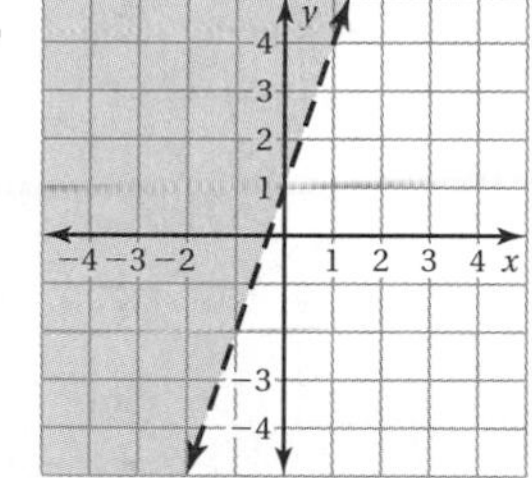

20.

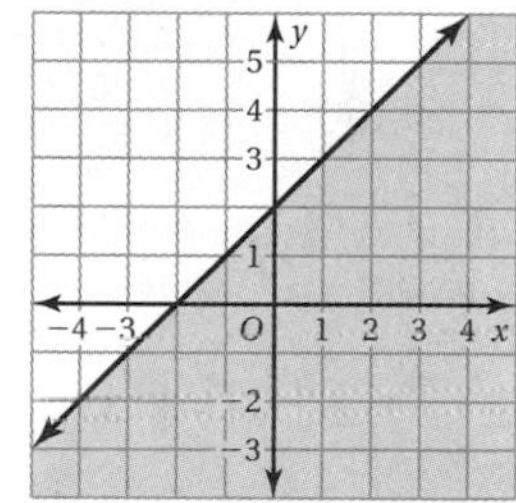

21.

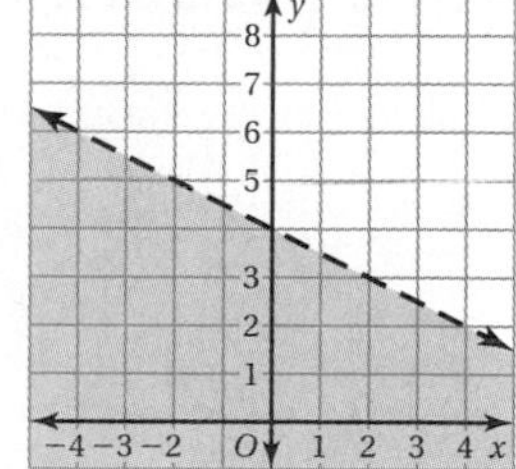

My Thoughts on the Chapter

Teacher Tip

Not allowed to write in your teaching edition? Use sticky notes to record your thoughts.

What worked. . .

What did not work. . .

What I would do differently. . .

3.4 Solving Multi-Step Inequalities (pp. 126–135)

Solve $2x - 3 \le -9$. Graph the solution.

	$2x - 3 \le -9$	Write the inequality.
Step 1: Undo the subtraction. →	$+3 \quad +3$	Add 3 to each side.
	$2x \le -6$	Simplify.
Step 2: Undo the multiplication. →	$\frac{2x}{2} \le \frac{-6}{2}$	Divide each side by 2.
	$x \le -3$	Simplify.

∴ The solution is $x \le -3$.

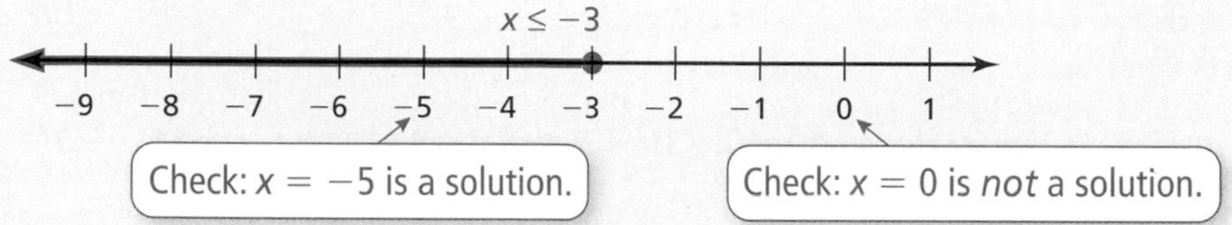

Exercises

Solve the inequality. Graph the solution.

13. $4x + 3 < 11$

14. $\frac{z}{-4} - 3 \le 1$

15. $-3w - 4 > 8$

16. $4 > x - 7 > -6$

17. $2x + 2 \le 4$ *or* $x + 2 \ge 5$

18. $|x - 3| > 1$

3.5 Graphing Linear Inequalities in Two Variables (pp. 136–143)

Graph $4x + 2y \ge -6$ in a coordinate plane.

Step 1: Graph $4x + 2y = -6$, or $y = -2x - 3$. Use a solid line because the inequality symbol is $\ge$.

Step 2: Test (0, 0).

$4x + 2y \ge -6$	Write the inequality.
$4(0) + 2(0) \overset{?}{\ge} -6$	Substitute.
$0 \ge -6$ ✓	Simplify.

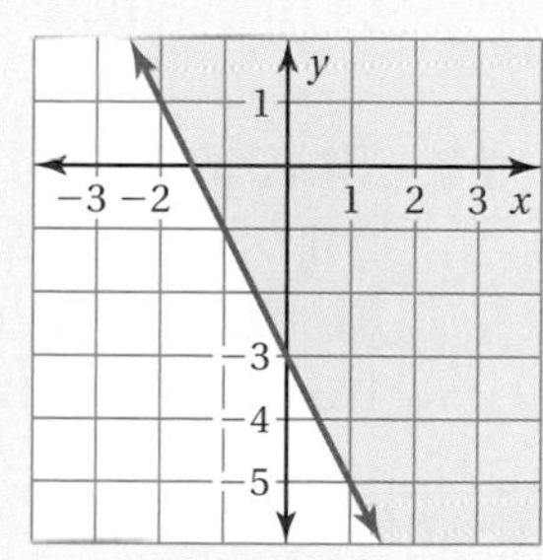

Step 3: Because (0, 0) is a solution, shade the half-plane that contains (0, 0).

Exercises

Graph the inequality in a coordinate plane.

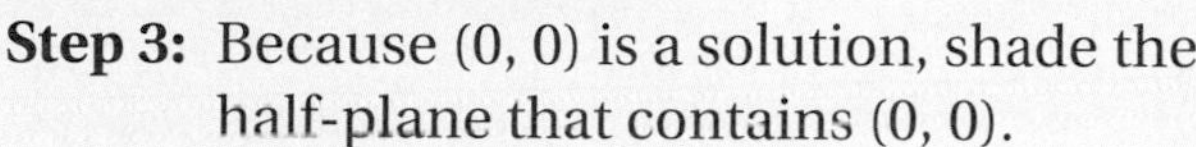

19. $-9x + 3y > 3$

20. $-2x + 2y \le 4$

21. $5x + 10y < 40$

3 Chapter Test

Check It Out
Test Practice
BigIdeasMath.com

Write the word sentence as an inequality.

1. A number j plus 20.5 is greater than or equal to 50.
2. A number r multiplied by $\frac{1}{7}$ is less than -14.

Tell whether the given value is a solution of the inequality.

3. $v - 2 \le 7;\ v = 9$
4. $\frac{3}{10}p < 0;\ p = 10$
5. $-3n \ge 6;\ n = -3$

Solve the inequality. Graph the solution.

6. $n - 3 > -3$
7. $x - \frac{7}{8} \le \frac{9}{8}$
8. $-6b \ge -30$
9. $\frac{y}{-4} \ge 13$
10. $3v - 7 \ge -13.3$
11. $-5(t + 11) < -60$
12. $3 \le x + 5 \le 9$
13. $3x - 2 \le 4$ *or* $x - 4 \ge 6$
14. $|x + 5| < 12$

Graph the inequality in a coordinate plane.

15. $x > -6$
16. $y < 2$
17. $x \ge -1$
18. $3x + y \ge 7$
19. $4x + 2y \le 8$
20. $3x - 9y > 18$

21. **VOTING** U.S. citizens must be at least 18 years of age on Election Day to vote. Write an inequality that represents this situation.

22. **GARAGE** The vertical clearance for a hotel parking garage is 10 feet. Write and solve an inequality that represents the height (in feet) of the vehicle.

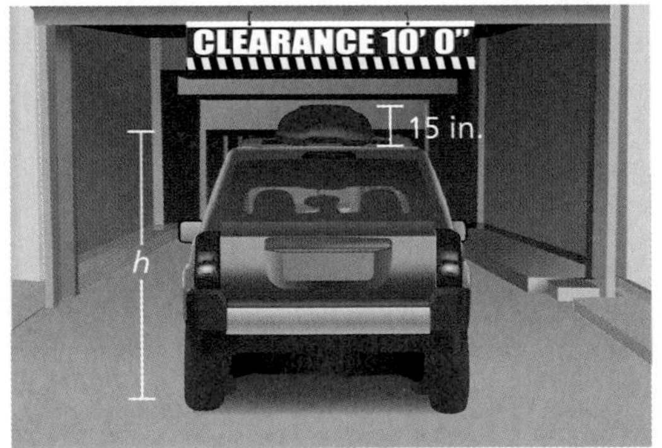

23. **TRADING CARDS** You have \$25 to buy trading cards online. Each pack of cards costs \$4.50. Shipping costs \$2.95. Write and solve an inequality to find the number of packs of trading cards you can buy.

24. **SCIENCE QUIZZES** The table shows your scores on four science quizzes. What score do you need on the fifth quiz to have a mean score of at least 80?

Quiz	1	2	3	4	5
Score (%)	76	87	73	72	?

Test Item References

Chapter Test Questions	Section to Review	Common Core State Standards
1–5, 21	3.1	A.CED.1, A.CED.3
6, 7, 22	3.2	A.CED.1, A.CED.3, A.REI.3
8 and 9	3.3	A.CED.1, A.CED.3 , A.REI.3
10–14, 23, 24	3.4	A.CED.1, A.CED.3 , A.REI.3
15–20	3.5	A.REI.12

Test-Taking Strategies

Remind students to quickly look over the entire test before they start so that they can budget their time. When writing word phrases as inequalities, students can get confused by the subtle differences in wording, such as "is no more than" and "is no less than." Encourage students to think very carefully about which inequality symbol is implied by the wording. Teach the students to use the Stop and Think strategy before answering. **Stop** and carefully read the question, and **Think** about what the answer should look like.

Common Assessment Errors

- **Exercises 1 and 2** Students may not use the correct inequality symbol. Remind them to put the word sentence into a real-life context.
- **Exercises 6–14** Remind them that they only reverse the direction of the inequality symbol when they are multiplying or dividing by a negative number.
- **Exercises 6–14** Students may perform the same operation instead of the inverse operation when solving the inequality. Remind them that solving inequalities is similar to solving equations.
- **Exercises 6–14** Students may use the wrong circle and/or shade the wrong side of the number line. Remind them to test a value on each side of the circle.
- **Exercises 15–20** Students may use the wrong boundary line and/or shade the wrong half-plane. Remind them to use a test point to check that their graph is correct.

Reteaching and Enrichment Strategies

If students need help...	If students got it...
Resources by Chapter • Practice A and Practice B • Puzzle Time Record and Practice Journal Practice Differentiating the Lesson Lesson Tutorials Practice from the Test Generator Skills Review Handbook	Resources by Chapter • Enrichment and Extension • School-to-Work • Financial Literacy Game Closet at *BigIdeasMath.com* Start Standardized Test Practice

Answers

1. $j + 20.5 \geq 50$
2. $\frac{1}{7}r < -14$
3. yes
4. no
5. yes
6. $n > 0$;

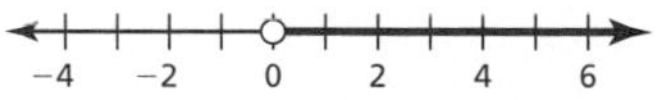

7. $x \leq 2$;

8. $b \leq 5$;

9. $y \leq -52$;

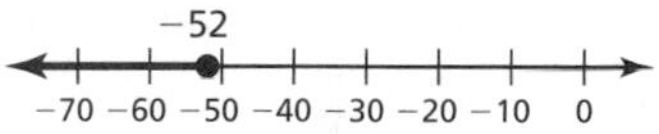

10. $v \geq -2.1$;

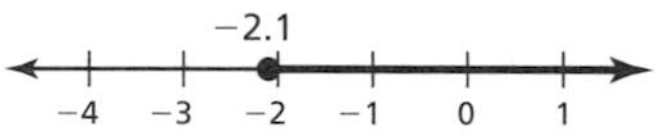

11–20. See Additional Answers.

21. $x \geq 18$
22. $h + 1.25 < 10$; $h < 8.75$ feet
23. $4.5x + 2.95 \leq 25$; $x \leq 4.9$; at most 4 packs of trading cards
24. at least 92%

Assessment Book

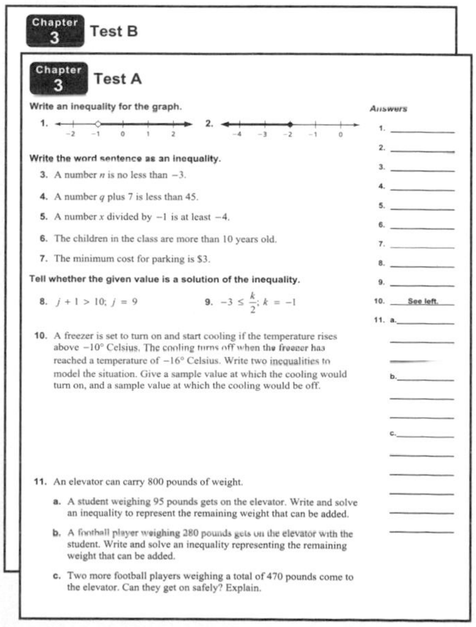

Chapter 3 Test B

Chapter 3 Test A

Write an inequality for the graph.

1. (graph) 2. (graph)

Write the word sentence as an inequality.

3. A number n is no less than −3.
4. A number q plus 7 is less than 45.
5. A number x divided by −1 is at least −4.
6. The children in the class are more than 10 years old.
7. The minimum cost for parking is $3.

Tell whether the given value is a solution of the inequality.

8. $j + 1 > 10$; $j = 9$
9. $-3 \leq \frac{k}{2}$; $k = -1$

10. A freezer is set to turn on and start cooling if the temperature rises above −10° Celsius. The cooling turns off when the freezer has reached a temperature of −16° Celsius. Write two inequalities to model the situation. Give a sample value at which the cooling would turn on, and a sample value at which the cooling would be off.

11. An elevator can carry 800 pounds of weight.
    a. A student weighing 95 pounds gets on the elevator. Write and solve an inequality to represent the remaining weight that can be added.
    b. A football player weighing 280 pounds gets on the elevator with the student. Write and solve an inequality representing the remaining weight that can be added.
    c. Two more football players weighing a total of 470 pounds come to the elevator. Can they get on safely? Explain.

Answers

1. ____
2. ____
3. ____
4. ____
5. ____
6. ____
7. ____
8. ____
9. ____
10. See left.
11. a. ____ b. ____ c. ____

Test-Taking Strategies
Available at *BigIdeasMath.com*

After Answering Easy Questions, Relax
Answer Easy Questions First
Estimate the Answer
Read All Choices before Answering
Read Question before Answering
Solve Directly or Eliminate Choices
Solve Problem before Looking at Choices
Use Intelligent Guessing
Work Backwards

About this Strategy
When taking a multiple choice test, be sure to read each question carefully and thoroughly. After skimming the test and answering the easy questions, stop for a few seconds, take a deep breath, and relax. Work through the remaining questions carefully, using your knowledge and test-taking strategies. Remember, you already completed many of the questions on the test!

Answers
1. B
2. G
3. 1.75
4. D
5. I

Technology For the Teacher
Big Ideas Test Generator

Item Analysis

1. **A.** The student finds the slope correctly but then fails to find the y-intercept.
   B. Correct answer
   C. The student finds the reciprocal of the slope and then uses the second point to find the y-intercept.
   D. The student finds the reciprocal of the slope and then uses the first point to find the y-intercept.

2. **F.** The student mistakes the roles of slope and intercept, thinking same intercept means what same slope means.
   G. Correct answer
   H. The student chooses a conclusion for lines that have the same intercept and the same slope, overlooking that these lines have different slopes.
   I. The student is confused by the problem.

3. **Gridded Response:** Correct answer: 1.75
   Common Error: The student subtracts 11 from -4 instead of adding, yielding an answer of -3.75.

4. **A.** The student confuses the area and perimeter formulas (or the concepts) and writes the less than symbol instead of the greater than symbol.
   B. The student writes the less than symbol instead of the greater than symbol.
   C. The student confuses the area and perimeter formulas (or the concepts).
   D. **Correct answer**

5. **F.** The student makes an error subtracting integers or is unsure about greater than or less than with negative integers.
   G. The student makes an error subtracting integers or is unsure about greater than or less than with negative integers.
   H. The student makes an error subtracting integers.
   I. Correct answer

3 Standardized Test Practice

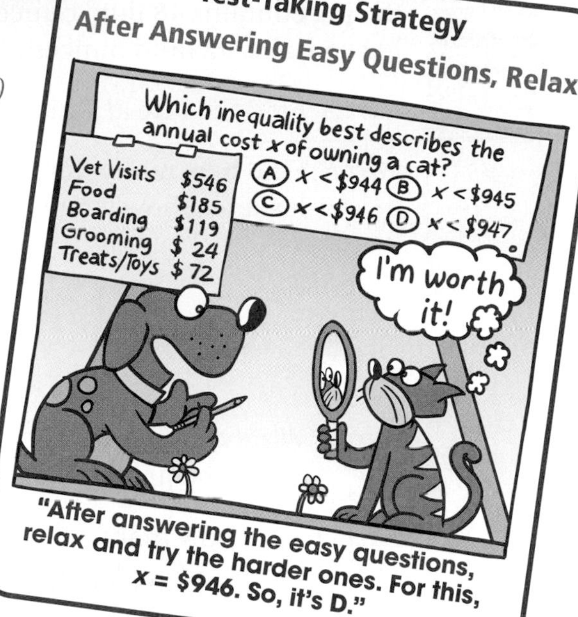

Test-Taking Strategy
After Answering Easy Questions, Relax
Which inequality best describes the annual cost x of owning a cat?
Vet Visits $546
Food $185
Boarding $119
Grooming $ 24
Treats/Toys $ 72
Ⓐ x < $944 Ⓑ x < $945
Ⓒ x < $946 Ⓓ x < $947
I'm worth it!
"After answering the easy questions, relax and try the harder ones. For this, x = $946. So, it's D."

1. A line contains the points $(-3, 5)$ and $(6, 8)$. What is the equation of the line? *(8.F.4, F.LE.2)*

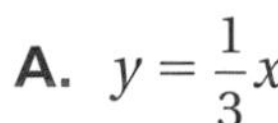

   A. $y = \frac{1}{3}x$ **C.** $y = 3x - 10$

   B. $y = \frac{1}{3}x + 6$ **D.** $y = 3x + 14$

2. Two lines have the same y-intercept. The slope of one line is 1 and the slope of the other line is -1. What can you conclude? *(F.IF.4)*

   F. The lines are parallel.

   G. The lines meet at exactly one point.

   H. The lines meet at more than one point.

   I. The situation described is impossible.

3. What value of x makes the equation below true? *(A.REI.3)*

$$4x - 11 = -4$$

4. The perimeter of the triangle shown below is greater than 50 centimeters. Which inequality represents this algebraically? *(A.CED.1)*

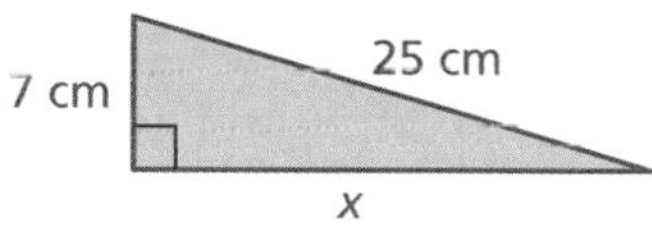

   A. $\frac{1}{2}(7x) < 50$ **C.** $\frac{1}{2}(7x) > 50$

   B. $x + 32 < 50$ **D.** $x + 32 > 50$

5. Which value is a solution of $x - 2 \geq -3$? *(A.REI.3)*

   F. -6 **H.** $-\frac{3}{2}$

   G. -5 **I.** -1

6. Water is leaking from a jug at a constant rate. After leaking for 2 hours, the jug contains 48 fluid ounces of water. After leaking for 5 hours, the jug contains 42 fluid ounces of water. *(8.F.4)*

Part A Find the rate at which water is leaking from the jug.

Part B Find how many fluid ounces of water were in the jug before it started leaking. Show your work and explain your reasoning.

Part C Write an equation that shows how many fluid ounces y of water are left in the jug after it has been leaking for h hours.

Part D Find how many hours it will take the jug to empty entirely. Show your work and explain your reasoning.

7. Which graph represents the inequality below? *(A.REI.12)*

$$3x + 6y > 6$$

A.

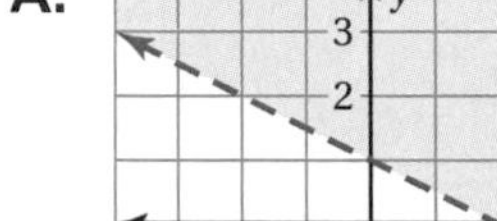

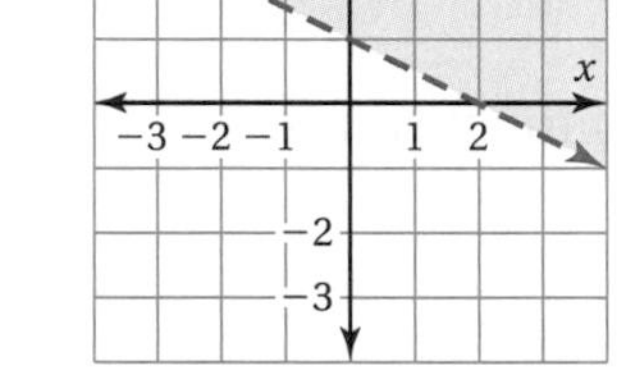

C.

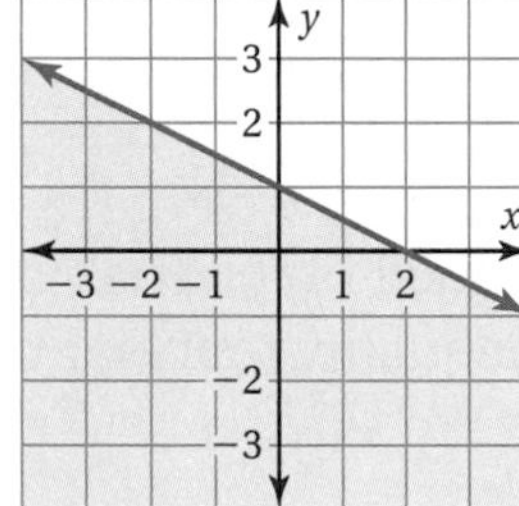

B.

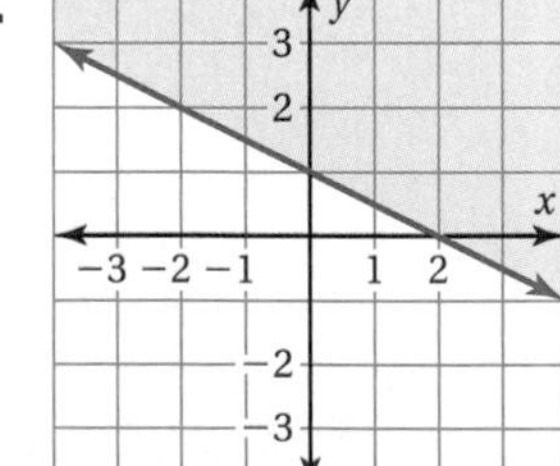

D.

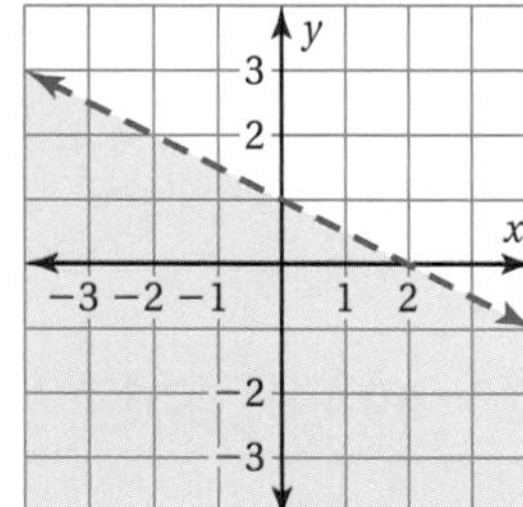

8. Solve $-5x - 2 \geq 8$. *(A.REI.3)*

F. $x < -2$

G. $x \leq -2$

H. $x \leq -\frac{6}{5}$

I. $x \geq -2$

Item Analysis (continued)

6. **4 points** The student demonstrates a thorough understanding of how to analyze a problem situation and translate it into a linear equation by building its components step-by-step. In addition, the student is able to set the equation equal to a given value and solve it in clear, complete steps. In Part A, the water is leaking at a rate of 2 fluid ounces per hour. In Part B, there were 52 fluid ounces in the jug before it started leaking. In Part C, the equation is $y = 52 - 2h$, or its equivalent. In Part D, the jug will be empty after 26 hours.

   3 points The student demonstrates an essential but less than thorough understanding of the problem situation and how to attack it. There may be a minor error made along the way, but subsequent work is consistent with the error.

   2 points The student demonstrates a partial understanding of how to interpret the problem and translate it algebraically. The student's work and explanations demonstrate a lack of essential understanding. For example, the student may calculate the rate correctly in Part A, but then overlooks the fact that the leaking process began two hours before the jug contained 48 fluid ounces. Additionally, the student will have trouble translating the work in Parts A and B into an equation.

   1 point The student demonstrates limited understanding. The student's response is incomplete and exhibits many flaws.

   0 points The student provides no response, a completely incorrect or incomprehensible response, or a response that demonstrates insufficient understanding of how to work with a verbal situation that must be translated into algebra.

7. **A.** Correct answer

   B. The student incorrectly makes the boundary line solid and shades the half-plane below the line instead of above.

   C. The student incorrectly makes the boundary line solid.

   D. The student incorrectly shades the half-plane below the boundary line instead of above.

8. **F.** The student solves the inequality correctly, but forgets to include the "equal to" part of the symbol.

   G. Correct answer

   H. The student subtracts 2 from each side instead of adding.

   I. The student solves the inequality correctly, but fails to reverse the direction of the inequality symbol when dividing by a negative coefficient.

Answers

6. *Part A* 2 fluid ounces per hour

   Part B 52 fluid ounces

   Part C $y = -2h + 52$

   Part D 26 hours

7. A

8. G

Answers

9. C
10. 225 calories
11. F
12. D

Item Analysis (continued)

9. **A.** The student makes an error subtracting or dividing integers.

   B. The student makes an error subtracting or dividing integers and then fails to reverse the direction of the inequality symbol when dividing by a negative coefficient.

   C. Correct answer

   D. The student fails to reverse the direction of the inequality symbol when dividing by a negative coefficient.

10. **Gridded Response:** Correct answer: 225 calories

    Common Error: The student misreads the graph.

11. **F.** Correct answer

    G. The student makes an arithmetic error relating *c* and *m*.

    H. The student picks an equation that fits the point (10, 90).

    I. The student interchanges the roles of *c* and *m*.

12. **A.** The student misunderstands the vocabulary. "Fewer than" does not include "equal to."

    B. The student switches the meaning of the symbols.

    D. The student does not include "equal to" in either part of the statement.

    D. Correct answer

Answer for Extra Example

1. **Gridded Response:**
   Correct answer: 5.6

   Common Error: The student fails to distribute 10 across the expression on the left side of the equation, yielding the incorrect solution of 0.2.

Extra Example

1. What value of x makes this equation true? *(A.REI.3)*

   $10(x - 3) = 5x - 2$

9. Which graph represents the inequality below? *(A.REI.3)*

$$-2x + 3 < 1$$

A. −4 −3 −2 −1 0 1 2 3 4 x

C. −4 −3 −2 −1 0 1 2 3 4 x

B. −4 −3 −2 −1 0 1 2 3 4 x

D. −4 −3 −2 −1 0 1 2 3 4 x

The graph below shows how many calories *c* are burned during *m* minutes of playing basketball. Use the graph for Exercises 10 and 11.

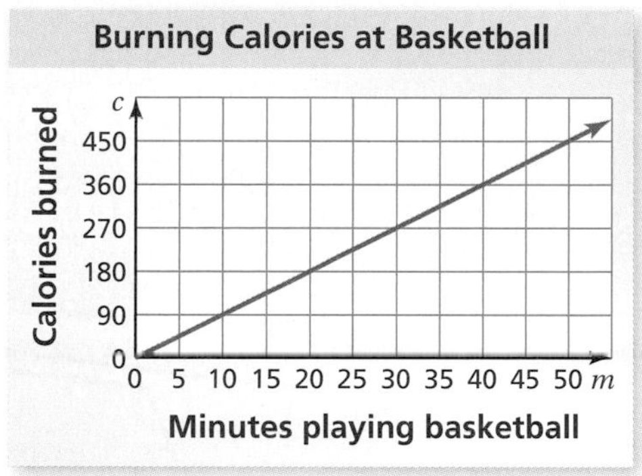

10. How many calories are burned in 25 minutes? *(A.CED.3)*

11. Which equation represents the graph? *(A.CED.2)*

F. $c - 9m$

H. $c = m + 80$

G. $c = 90m$

I. $m = 9c$

12. Which word sentence represents the inequality below? *(A.REI.3)*

$$3 \leq x \leq 8$$

A. A number x is fewer than 8 or no less than 3.

B. A number x is less than or equal to 3 or more than 8.

C. A number x is more than 3 and less than 8.

D. A number x is greater than or equal to 3 and no more than 8.

4 Solving Systems of Linear Equations

4.1 Solving Systems of Linear Equations by Graphing

4.2 Solving Systems of Linear Equations by Substitution

4.3 Solving Systems of Linear Equations by Elimination

4.4 Solving Special Systems of Linear Equations

4.5 Systems of Linear Inequalities

"Can you graph a system of linear equations that shows the number of biscuits and treats that I am going to share with you?"

"Hey look over here. Can you estimate the solution of the system of linear equations that I made with these cattails?"

Connections to Previous Learning

- Reason about and solve simple one-variable equations and inequalities.

- Write and solve one-step and two-step linear equations in one variable with one solution, no solution, or infinitely many solutions.

- Solve systems of two linear equations in two variables algebraically and graphically.
- Graph the solution set to a system of linear inequalities in two variables as the intersection of the corresponding half-planes.

Math in History

Many ancient cultures used a counting frame or abacus to perform calculations. Most of these cultures used a base ten system like the Roman or Chinese system. The abacus shown below is representing the number 2786.

★ Examples of the use of an abacus in Rome date back to around the first century A.D.

★ Examples of the use of an abacus in China date back to around the 14th century A.D.

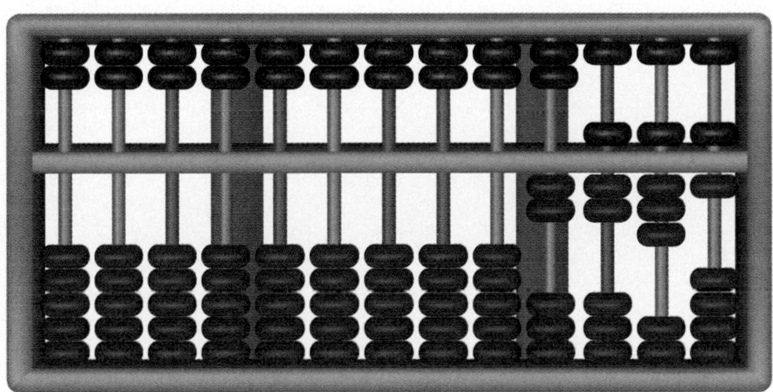

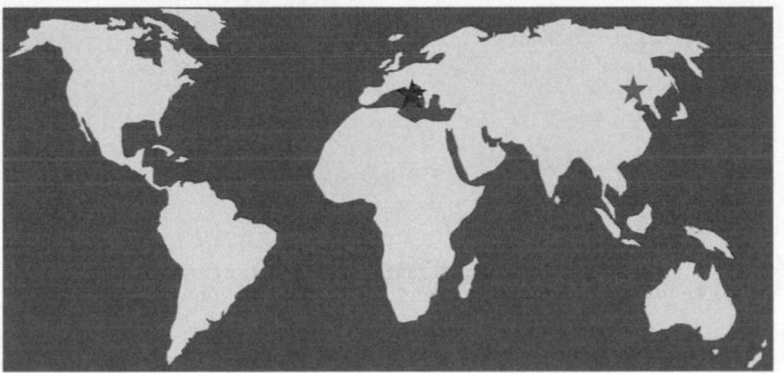

Pacing Guide for Chapter 4

Chapter Opener	1 Day
Section 1	2 Days
Section 2	1 Day
Study Help / Quiz	1 Day
Section 3	1 Day
Section 4	2 Days
Section 5	1 Day
Chapter Review/ Chapter Tests	2 Days
Total Chapter 4	11 Days
Year-to-Date	46 Days

Check Your Resources

- Record and Practice Journal
- Resources by Chapter
- Skills Review Handbook
- Assessment Book
- Worked-Out Solutions

Technology For the Teacher

The Dynamic Planning Tool
Editable Teacher's Resources at
BigIdeasMath.com

Common Core State Standards

A.REI.3 Solve linear inequalities in one variable.

A.REI.12 Graph the solutions to a linear inequality in two variables as a half-plane (excluding the boundary in the case of a strict inequality), . . .

Additional Topics for Review

- Solving simple equations
- Graphing linear equations

Try It Yourself

1. $x = 3$ **2.** $w = 4$

3. $z = 13$ **4.** $c = -17$

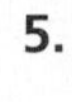

5.

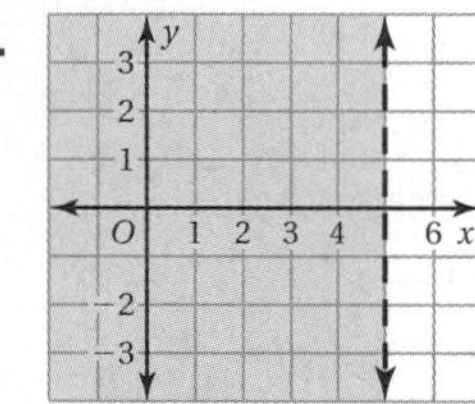

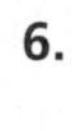

6.

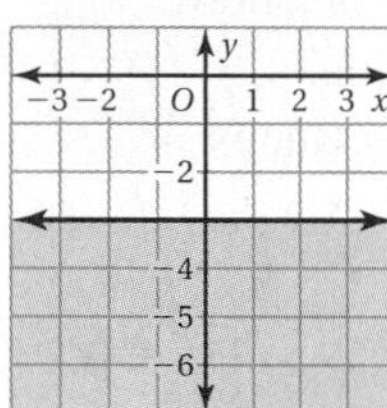

7–8. See Additional Answers.

Record and Practice Journal

1. $y = 2$ **2.** $a = -3$

3. $k = 5$ **4.** $m = 6$

5. $t = -4$ **6.** $h = 9$

7. 45 calculators

8.

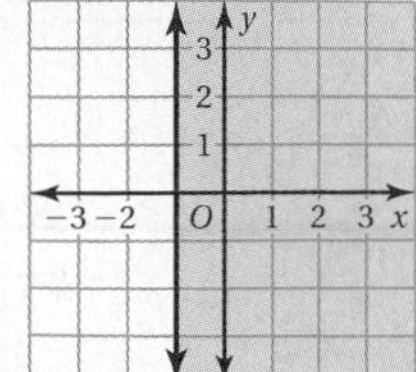

9–12. See Additional Answers.

Math Background Notes

Vocabulary Review

- Linear inequality in two variables
- Solution of a linear inequality in two variables
- Graph of a linear inequality in two variables

Solving Multi-Step Equations

- Students should know how to solve multi-step equations.
- To solve multi-step equations, use inverse operations to isolate the variable on one side of the equation.
- Remind students to isolate the variable term on one side of the equation before dividing each side by the coefficient of the variable term.
- **Common Error:** Students may use the Distributive Property incorrectly. Emphasize the second step in Example 1.

Graphing Linear Inequalities

- Students should know how to graph linear inequalities in two variables.
- Remind students to graph the boundary line using a dashed line when the inequality symbol is $<$ or $>$, and using a solid line when the inequality symbol is $\leq$ or $\geq$. Then test a point on one side of the boundary line to determine which half-plane to shade.
- **Common Error:** Students may forget to reverse the inequality symbol when they solve for y in Exercise 8. Remind them to reverse the inequality symbol when they multiply or divide each side by a negative number.

Reteaching and Enrichment Strategies

If students need help. . .	If students got it. . .
Record and Practice Journal • Fair Game Review Skills Review Handbook Lesson Tutorials	Game Closet at *BigIdeasMath.com* Start the next section

What You Learned Before

This hurts infinitely.

"Hold your tail a bit lower."

Solving Multi-Step Equations (A.REI.3)

Example 1 **Solve $4x - 2(3x + 1) = 16$.**

$4x - 2(3x + 1) = 16$	Write the equation.
$4x - 6x - 2 = 16$	Use Distributive Property.
$-2x - 2 = 16$	Combine like terms.
$-2x = 18$	Add 2 to each side.
$x = -9$	Divide each side by -2.

∴ The solution is $x = -9$.

Try It Yourself

Solve the equation. Check your solution.

1. $-5x + 8 = -7$

2. $7w + w - 15 = 17$

3. $-3(z - 8) + 10 = -5$

4. $2 = 10c - 4(2c - 9)$

Graphing Linear Inequalities (A.REI.12)

Example 2 **Graph $4x + y \geq -2$ in a coordinate plane.**

Step 1: Graph $4x + y = -2$. Use a solid line because the inequality symbol is $\geq$.

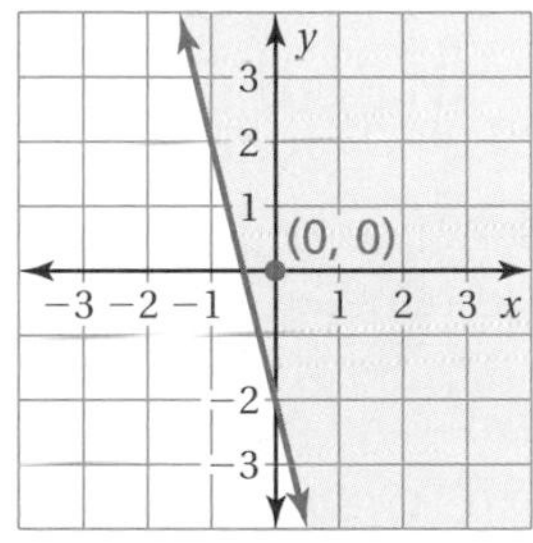

Step 2: Test (0, 0).

$4x + y \geq -2$	Write the inequality.
$4(0) + 0 \overset{?}{\geq} -2$	Substitute.
$0 \geq -2$ ✓	Simplify.

Step 3: Because (0, 0) is a solution, shade the half-plane that contains (0, 0).

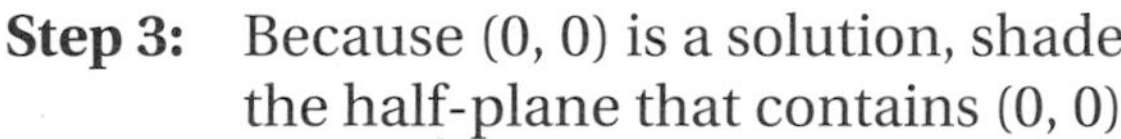

Try It Yourself

Graph the inequality in a coordinate plane.

5. $x < 5$

6. $y \leq -3$

7. $x + y > -8$

8. $x - 2y \geq 6$

4.1 Solving Systems of Linear Equations by Graphing

Essential Question How can you solve a system of linear equations?

COMMON CORE STATE STANDARDS
8.EE.8a
8.EE.8b
8.EE.8c
A.CED.3
A.REI.6

1 ACTIVITY: Writing a System of Linear Equations

Work with a partner.

Your family starts a bed-and-breakfast. They spend \$500 fixing up a bedroom to rent. The cost for food and utilities is \$10 per night. Your family charges \$60 per night to rent the bedroom.

a. Write an equation that represents the costs.

Cost, C (in dollars) = \$10 per night • Number of nights, x + \$500

b. Write an equation that represents the revenue (income).

Revenue, R (in dollars) = \$60 per night • Number of nights, x

c. A set of two (or more) linear equations is called a **system of linear equations.** Write the system of linear equations for this problem.

2 ACTIVITY: Using a Table to Solve a System

Use the cost and revenue equations from Activity 1 to find how many nights your family needs to rent the bedroom before recovering the cost of fixing up the bedroom. This is the *break-even point.*

a. Copy and complete the table.

x	0	1	2	3	4	5	6	7	8	9	10	11
C												
R												

b. How many nights does your family need to rent the bedroom before breaking even?

Laurie's Notes

Introduction

Standards for Mathematical Practice

- **MP1 Make Sense of Problems and Persevere in Solving Them:** Students will investigate a situation that can be represented by a system of equations today. They will write a system of equations that represents the situation. They will have the opportunity to use different approaches to solve the system and discover the form and meaning of the solution. Encourage them to persevere in the different solution approaches.

Motivate

- Write the "geometric equations" on the board. Explain that each square represents the same quantity, as does each triangle. Have students work with a partner to figure out what the square and triangle represent.

  $\square + \square + \square + \triangle = 47$

  $\square - \triangle = 1$
- Ask a volunteer to share their solution, and explain how they figured out the answer. $\triangle = 11, \square = 12$
- Share with students that this is the type of problem they will be working on in Chapter 4.

Activity Notes

Activity 1

- Discuss what is known about your costs and your income.
- **Financial Literacy:** Do not assume that students are knowledgeable about concepts such as *costs* (fixed and variable) and *revenue* (income). Explain these words as you use them.
- Point out to students that the units in the verbal model agree. This means that "dollars per night × nights" is equal to dollars. So, the units in the equation are dollars = dollars + dollars.
- Ask a pair of students to share the equations they wrote.
- Discuss the definition of a system of linear equations.

Activity 2

- Read the problem aloud. Define and discuss the break-even point.
- ? "Why would a business want to know the break-even point?" You want to know how many nights it will take before you start to make money.
- Give time for students to work with their partners to fill in the table. Make sure that all students are using correct equations.
- ? **Extension:** "What patterns do you observe in the table?" The two rows continue to get closer together until they are finally equal at $x = 10$. Then the revenue is greater than the cost.
- Make sure to interpret the answer to part (b). When you rent the room for 10 nights, the costs and the revenue both equal $600. A solution of each equation is (10, 600).

Common Core State Standards

8.EE.8a Understand that solutions to a system of two linear equations in two variables correspond to points of intersection of their graphs, because points of intersection satisfy both equations simultaneously.

8.EE.8b Solve systems of two linear equations in two variables algebraically, and estimate solutions by graphing the equations. Solve simple cases by inspection.

A.CED.3 Represent constraints by systems of equations and interpret solutions as viable or nonviable options in a modeling context.

Also **8.EE.8c** and **A.REI.6**

Previous Learning

Students should know how to graph linear equations.

Start Thinking! and Warm Up

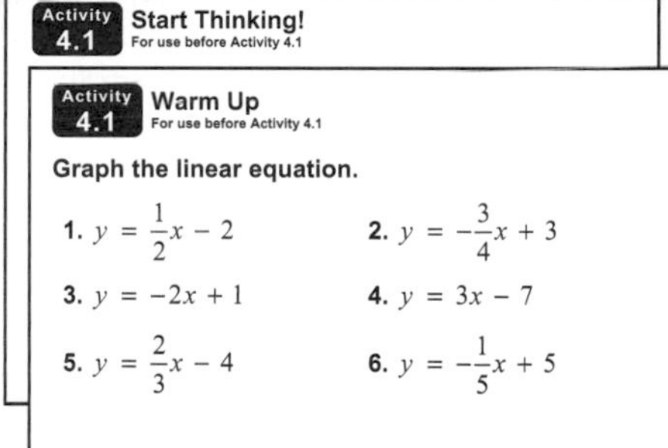

Activity 4.1 **Start Thinking!** For use before Activity 4.1

Activity 4.1 **Warm Up** For use before Activity 4.1

Graph the linear equation.

1. $y = \frac{1}{2}x - 2$
2. $y = -\frac{3}{4}x + 3$
3. $y = -2x + 1$
4. $y = 3x - 7$
5. $y = \frac{2}{3}x - 4$
6. $y = -\frac{1}{5}x + 5$

4.1 Record and Practice Journal

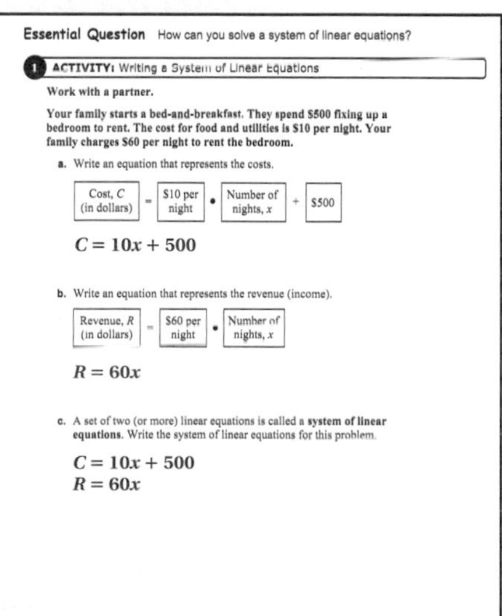

Essential Question How can you solve a system of linear equations?

1 ACTIVITY: Writing a System of Linear Equations

Work with a partner.

Your family starts a bed-and-breakfast. They spend $500 fixing up a bedroom to rent. The cost for food and utilities is $10 per night. Your family charges $60 per night to rent the bedroom.

a. Write an equation that represents the costs.

Cost, C (in dollars) = $10 per night • Number of nights, x + $500

$C = 10x + 500$

b. Write an equation that represents the revenue (income).

Revenue, R (in dollars) = $60 per night • Number of nights, x

$R = 60x$

c. A set of two (or more) linear equations is called a **system of linear equations**. Write the system of linear equations for this problem.

$C = 10x + 500$
$R = 60x$

English Language Learners

Pair Activity

Pair each English learner with an English speaker. Ask both students to solve the system graphically, but let one use a graphing utility while the other student makes the graphs by hand. Students then compare their answers. Partners should alternate solution methods as they continue to solve problems.

4.1 Record and Practice Journal

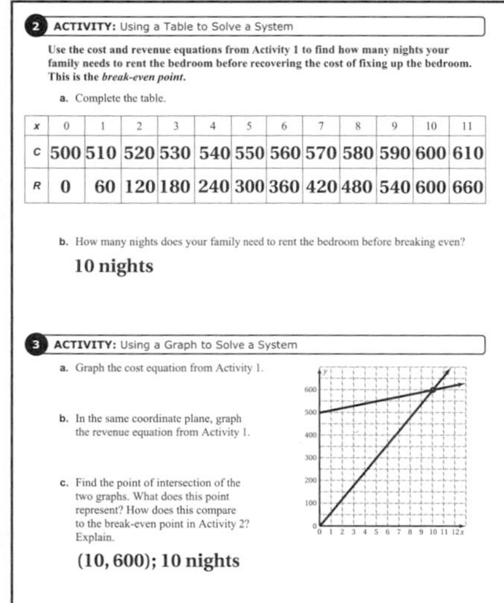

2 ACTIVITY: Using a Table to Solve a System

Use the cost and revenue equations from Activity 1 to find how many nights your family needs to rent the bedroom before recovering the cost of fixing up the bedroom. This is the *break-even point*.

a. Complete the table.

x	0	1	2	3	4	5	6	7	8	9	10	11
C	500	510	520	530	540	550	560	570	580	590	600	610
R	0	60	120	180	240	300	360	420	480	540	600	660

b. How many nights does your family need to rent the bedroom before breaking even?

10 nights

3 ACTIVITY: Using a Graph to Solve a System

a. Graph the cost equation from Activity 1.

b. In the same coordinate plane, graph the revenue equation from Activity 1.

c. Find the point of intersection of the two graphs. What does this point represent? How does this compare to the break-even point in Activity 2? Explain.

(10, 600); 10 nights

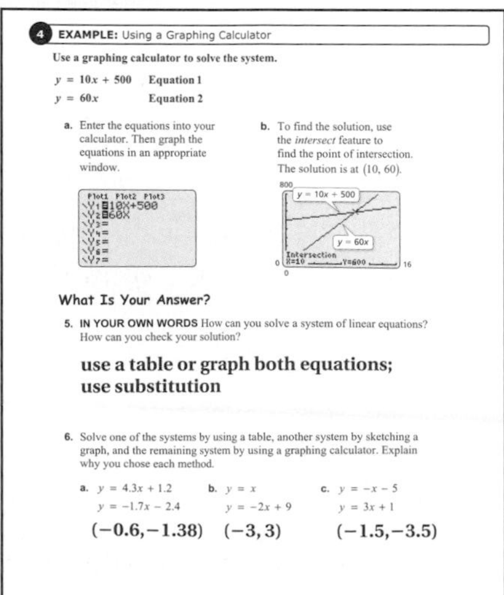

4 EXAMPLE: Using a Graphing Calculator

Use a graphing calculator to solve the system.

$y = 10x + 500$ Equation 1

$y = 60x$ Equation 2

a. Enter the equations into your calculator. Then graph the equations in an appropriate window.

b. To find the solution, use the *intersect* feature to find the point of intersection. The solution is at (10, 60).

What Is Your Answer?

5. **IN YOUR OWN WORDS** How can you solve a system of linear equations? How can you check your solution?

use a table or graph both equations; use substitution

6. Solve one of the systems by using a table, another system by sketching a graph, and the remaining system by using a graphing calculator. Explain why you chose each method.

a. $y = 4.3x + 1.2$, $y = -1.7x - 2.4$ — **(−0.6, −1.38)**

b. $y = x$, $y = -2x + 9$ — **(−3, 3)**

c. $y = -x - 5$, $y = 3x + 1$ — **(−1.5, −3.5)**

Laurie's Notes

Activity 3

- ? "Look at the scaling of the axes for this activity. What are the units that will be used to graph each of these equations?" number of nights, dollars
- The cost equation is $C = 10x + 500$. The revenue equation is $R = 60x$.
- ? "In what form is this Cost equation? What strategy can be used to graph the equation?" slope-intercept form; Students might say just plot the points from the table in Activity 2. Others might say plot the point for the y-intercept and then use a slope of 10 (which is equivalent to right 10 units, up 100 units).
- ? "In what form is the Revenue equation? What strategy can be used to graph the equation?" similar response to previous question
- Provided that students have graphed the equations carefully, the lines should intersect at (10, 600).
- Make sure to interpret the answer to part (c). When you rent the room for 10 nights, the costs and the revenue both equal $600. A point on each line is (10, 600). It is the point of intersection for the graphs.

Example 4

- Students have used the graphing calculator before to graph equations. They will enter both equations in the equation editor.
- ? "Do you think the standard viewing window allows us to see both graphs? Explain." no; Activity 3 shows that a much greater range is needed.
- Students can set the window manually using insight from Activity 3.
- ? "Name a solution of the first equation ($y = 10x + 500$) that is *not* a solution of the second equation ($y = 60x$)." *Sample answer:* (0, 500)
- ? "Name a solution of the second equation ($y = 60x$) that is *not* a solution of the first equation ($y = 10x + 500$)." *Sample answer:* (0, 0)
- ? "Name a solution of *both* equations." (10, 600)
- ? "Name an ordered pair that is *not* a solution of either equation." *Sample answer:* (10, 10)

What Is Your Answer?

- **MP6 Attend to Precision:** In Question 5, students should realize the importance of checking solutions of systems of equations.
- **MP5 Use Appropriate Tools Strategically:** In Question 6, students are choosing between a graphing utility and pencil and paper.

Closure

- **Phone call:** Write a brief script for a phone conversation with a friend that was not in class today. Explain what a system of linear equations is and how you solve a system of linear equations.

Technology For the Teacher

The Dynamic Planning Tool

Editable Teacher's Resources at *BigIdeasMath.com*

3 ACTIVITY: Using a Graph to Solve a System

a. Graph the cost equation from Activity 1.

b. In the same coordinate plane, graph the revenue equation from Activity 1.

c. Find the point of intersection of the two graphs. What does this point represent? How does this compare to the break-even point in Activity 2? Explain.

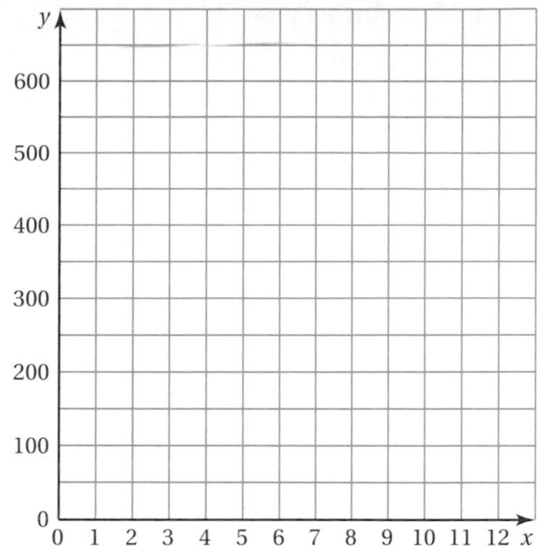

4 EXAMPLE: Using a Graphing Calculator

Use a graphing calculator to solve the system. $y = 10x + 500$ Equation 1

$y = 60x$ Equation 2

a. Enter the equations into your calculator. Then graph the equations in an appropriate window.

```
Plot1 Plot2 Plot3
\Y1■10X+500
\Y2■60X
\Y3=
\Y4=
\Y5=
\Y6=
\Y7=
```

b. To find the solution, use the *intersect* feature to find the point of intersection. The solution is (10, 600).

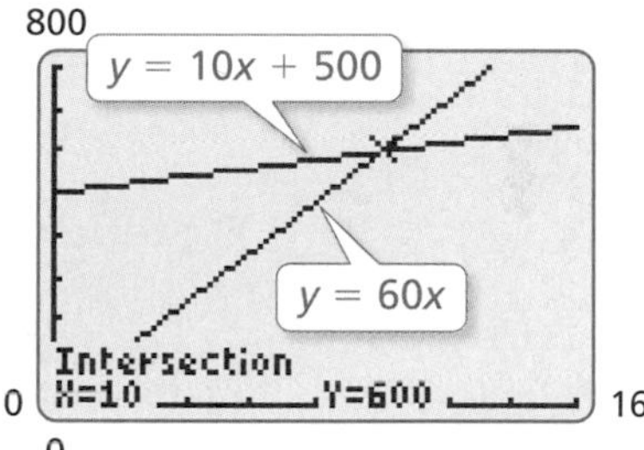

What Is Your Answer?

5. **IN YOUR OWN WORDS** How can you solve a system of linear equations? How can you check your solution?

6. Solve one of the systems by using a table, another system by sketching a graph, and the remaining system by using a graphing calculator. Explain why you chose each method.

   a. $y = 4.3x + 1.2$
   $y = -1.7x - 2.4$

   b. $y = x$
   $y = -2x + 9$

   c. $y = -x - 5$
   $y = 3x + 1$

Practice Use what you learned about systems of linear equations to complete Exercises 4–6 on page 158.

4.1 Lesson

Key Vocabulary

system of linear equations, *p. 156*

solution of a system of linear equations, *p. 156*

A **system of linear equations** is a set of two or more linear equations in the same variables. An example is shown below.

$y = x + 1$ Equation 1

$y = 2x - 7$ Equation 2

A **solution of a system of linear equations** in two variables is an ordered pair that is a solution of each equation in the system. The solution of a system of linear equations is the point of intersection of the graphs of the equations.

Reading

A system of linear equations is also called a *linear system*.

Key Idea

Solving a System of Linear Equations by Graphing

Step 1 Graph each equation in the same coordinate plane.

Step 2 Estimate the point of intersection.

Step 3 Check the point from Step 2 by substituting for x and y in each equation of the original system.

EXAMPLE 1 Solving a System of Linear Equations by Graphing

Solve the system by graphing. $\mathbf{y = 2x + 5}$ Equation 1

$\mathbf{y = -4x - 1}$ Equation 2

Step 1: Graph each equation.

Step 2: Estimate the point of intersection. The graphs appear to intersect at $(-1, 3)$.

Step 3: Check the point from Step 2.

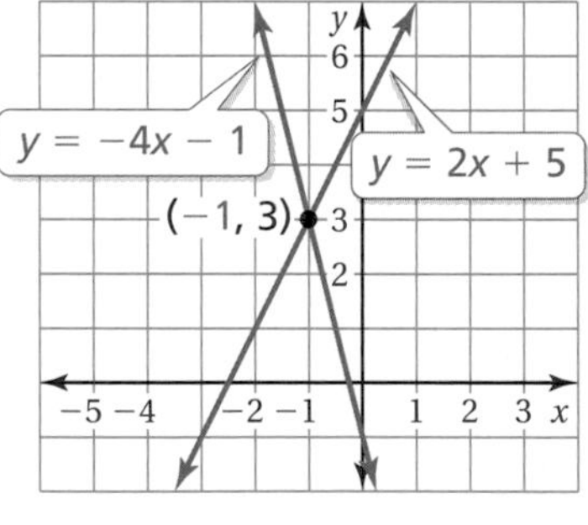

Equation 1	Equation 2
$y = 2x + 5$	$y = -4x - 1$
$3 \stackrel{?}{=} 2(-1) + 5$	$3 \stackrel{?}{=} -4(-1) - 1$
$3 = 3$ ✓	$3 = 3$ ✓

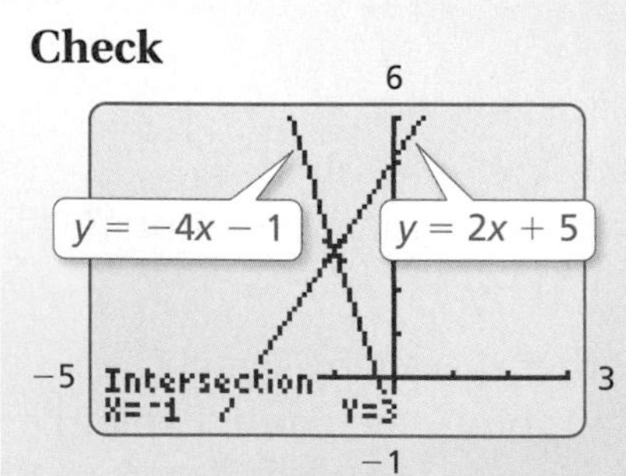

The solution is $(-1, 3)$.

On Your Own

Now You're Ready
Exercises 10–12

Solve the system of linear equations by graphing.

1. $y = x - 1$
   $y = -x + 3$

2. $y = -5x + 14$
   $y = x - 10$

3. $y = x$
   $y = 2x + 1$

Laurie's Notes

Introduction

Connect

- **Yesterday:** Students investigated different approaches to solving a system of equations. (MP1, MP5, MP6)
- **Today:** Students will solve systems of linear equations by graphing.

Motivate

- Share a story about a trip to Indianapolis, where Market Street intersects Meridian Street at Monument Circle. Draw a sketch.

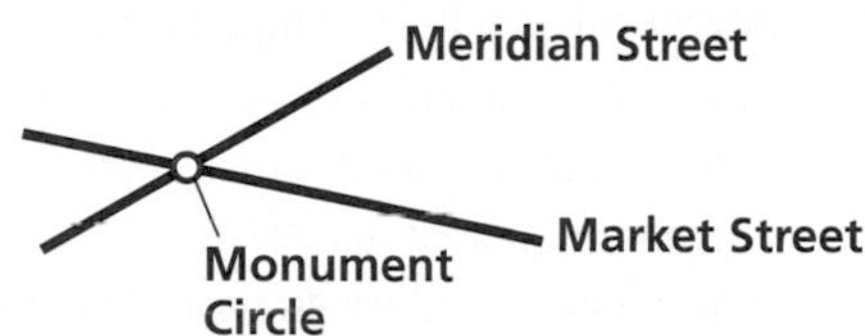

? Ask if students have ever visited a town or city where a monument was located in the middle of two streets.

- **Connection:** The monument is located on both streets. In other words, you will find the monument where the streets intersect.

Lesson Notes

Discuss

- Define a system of linear equations.

? "What is a solution of linear equation in two variables?" an ordered pair that satisfies the equation

- Define a solution of a system of linear equations.

? "How many solutions do you think a system of linear equations can have and why?" Students are likely to suggest that there can be only one solution where the lines intersect. Do not correct this response—they will discover the rest of the possibilities in time.

Key Idea

- Discuss the steps for solving a system of linear equations by graphing.
- Checking solutions is important, especially if the graphing is done by hand.

? "Why do you have to check your answer in both equations?" It is possible for an ordered pair to satisfy only one of the equations.

Example 1

? "Will the lines intersect? Explain." yes; The slopes are different.

- This is a great time to review graphing an equation in slope-intercept form. One method is to plot the point for the y-intercept then find a second point on the graph using the slope. Another method is to make a table of values and plot the ordered pairs.
- Work through the problem as shown.

? "Should we trust our eyes? What if the solution is actually $(-1.1, 3.2)$?" Checking the solution in the original equations will confirm our guess.

Goal Today's lesson is solving a **system of linear equations** by graphing.

Start Thinking! and Warm Up

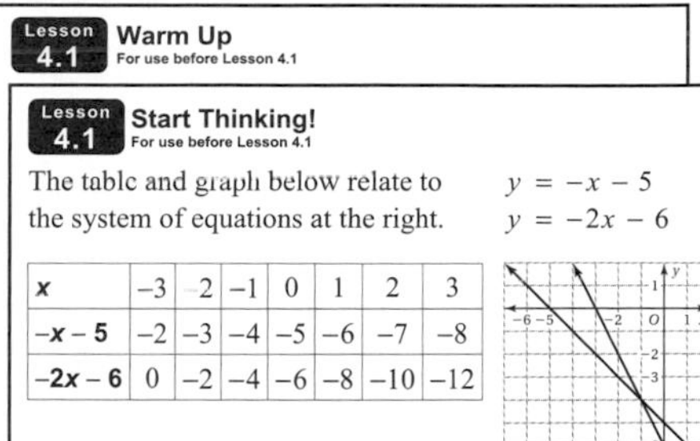

Lesson 4.1 **Warm Up**
For use before Lesson 4.1

Lesson 4.1 **Start Thinking!**
For use before Lesson 4.1

The table and graph below relate to the system of equations at the right.

$y = -x - 5$
$y = -2x - 6$

x	−3	−2	−1	0	1	2	3
$-x - 5$	−2	−3	−4	−5	−6	−7	−8
$-2x - 6$	0	−2	−4	−6	−8	−10	−12

What is the solution of the system? Is it easier to see the solution of the system using the table or the graph? In general, is using a table or graph a better method? Explain.

Extra Example 1

Solve the system by graphing.

$y = -2x + 2$

$y = 3x - 3$

(1, 0)

On Your Own

1. (2, 1)
2. (4, −6)
3. (−1, −1)

Extra Example 2

In Example 2, the kicker makes a total of 6 extra points and field goals and scores 12 points. Write and solve a system of equations to find the number of extra points and field goals. 3 extra points, 3 field goals

On Your Own

4. $(-3, 5)$
5. $(-2, -7)$
6. $(4, -8)$
7. $x + y = 7$
   $x + 3y = 17$
   $(2, 5)$; 2 extra points, 5 field goals

Differentiated Instruction

Visual

Remind students that the graph of an equation is a line in which all of the ordered pairs satisfy the equation. In a system of linear equations, the point of intersection of the two lines represents the ordered pair that satisfies both equations. Have students identify the x-value and the y-value of the solution and what the values represent in the problem.

Laurie's Notes

On Your Own

- **Neighbor Check:** Have students work independently and then have their neighbor check their work. Have students discuss any discrepancies.

Example 2

- **MP1 Make Sense of Problems and Persevere in Solving Them:** Ask a student to read the problem. Discuss vocabulary as needed. Ask questions to clarify the problem such as, "How many times did the kicker score points? How many of each type of kick was made? How many points were scored on field goals? What will the variables represent?"
- Students can be careless when defining variables. Be sure students use the definitions from the problem statement: x = the *number* of extra points made and y = the *number* of field goals made.
- This is a great time to review graphing an equation in standard form. One method is to graph the line using the intercepts $(0, y)$ and $(x, 0)$. A second method is to rewrite the equation in slope-intercept form.

? "What is your estimate for the point of intersection?" $(6, 2)$

- Remind students to check the solution in *both* equations.
- If time permits, use a graphing calculator to graph the equations.

On Your Own

- Check students' work.
- **Question 6:** Ask a volunteer to share the solution. Ask students whether they worked with the fractional coefficient of the x-term in the first equation or they multiplied through by 2 to eliminate the fraction. In the second equation, did they divide through by 2 to simplify the equation before graphing? Students should be comfortable with these different approaches.

Closure

- **Exit Ticket:** Solve by graphing.
  $y = 2x + 3$
  $y = -x + 6$
  $(1, 5)$

Technology For the Teacher

The Dynamic Planning Tool
Editable Teacher's Resources at *BigIdeasMath.com*

EXAMPLE 2 Real-Life Application

A kicker on a football team scores 1 point for making an extra point and 3 points for making a field goal. The kicker makes a total of 8 extra points and field goals in a game and scores 12 points. Write and solve a system of linear equations to find the number x of extra points and the number y of field goals.

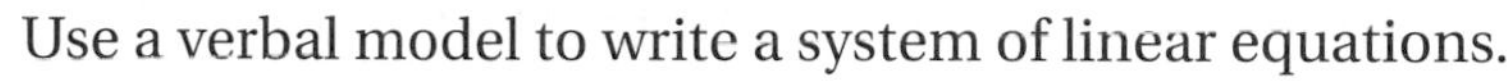

Use a verbal model to write a system of linear equations.

Number of extra points, x + Number of field goals, y = Total number of kicks

Points per extra point • Number of extra points, x + Points per field goal • Number of field goals, y = Total number of points

The system is: $x + y = 8$ Equation 1

$x + 3y = 12$ Equation 2

Step 1: Graph each equation.

Step 2: Estimate the point of intersection. The graphs appear to intersect at (6, 2).

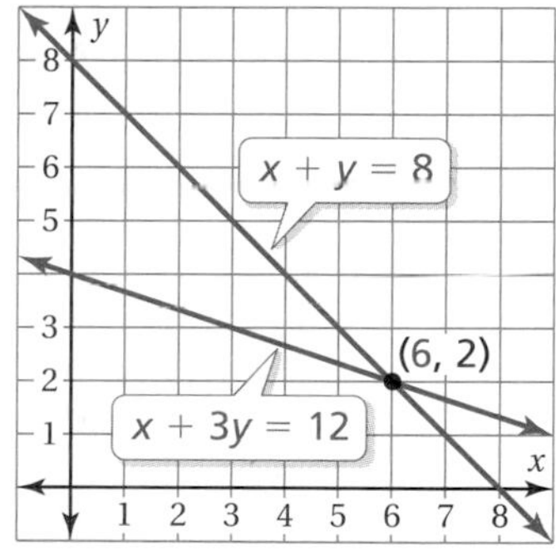

Step 3: Check your point from Step 2.

Equation 1	Equation 2
$x + y = 8$	$x + 3y = 12$
$6 + 2 \stackrel{?}{=} 8$	$6 + 3(2) \stackrel{?}{=} 12$
$8 = 8$ ✓	$12 = 12$ ✓

The solution is (6, 2). So, the kicker made 6 extra points and 2 field goals.

Check

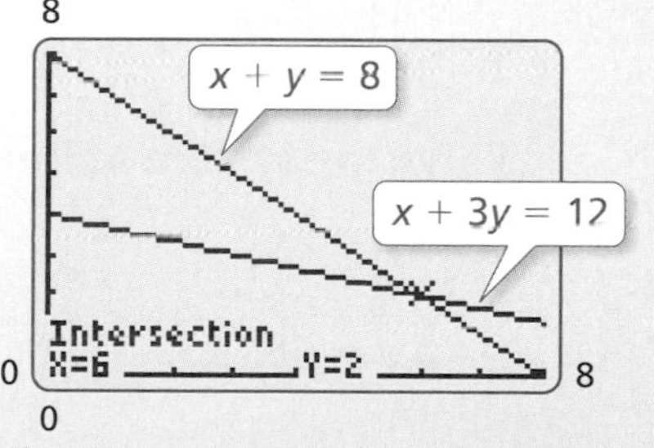

Study Tip

It may be easier to graph the equations in a system by rewriting the equations in slope-intercept form.

On Your Own

Now You're Ready
Exercises 13–15

Solve the system of linear equations by graphing.

4. $y = -4x - 7$
   $x + y = 2$

5. $x - y = 5$
   $-3x + y = -1$

6. $\frac{1}{2}x + y = -6$
   $6x + 2y = 8$

7. **WHAT IF?** In Example 2, the kicker makes a total of 7 extra points and field goals and scores 17 points. Write and solve a system of linear equations to find the numbers of extra points and field goals.

4.1 Exercises

Vocabulary and Concept Check

1. **VOCABULARY** Do the equations $4x - 3y = 5$ and $7y + 2x = -8$ form a system of linear equations? Explain.
2. **WRITING** What does it mean to solve a system of equations?
3. **WRITING** You graph a system of linear equations and the solution appears to be (3, 4). How can you verify that the solution is (3, 4)?

Practice and Problem Solving

Use a table to find the break-even point. Check your solution.

4. $C = 15x + 150$
   $R = 45x$
5. $C = 24x + 80$
   $R = 44x$
6. $C = 36x + 200$
   $R = 76x$

Match the system of linear equations with the corresponding graph. Use the graph to estimate the solution. Check your solution.

7. $y = 1.5x - 2$
   $y = -x + 13$
8. $y = x + 4$
   $y = 3x - 1$
9. $y = \frac{2}{3}x - 3$
   $y = -2x + 5$

A.
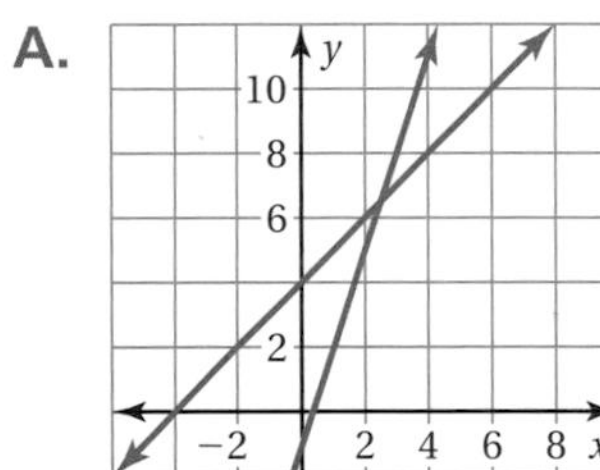

B.
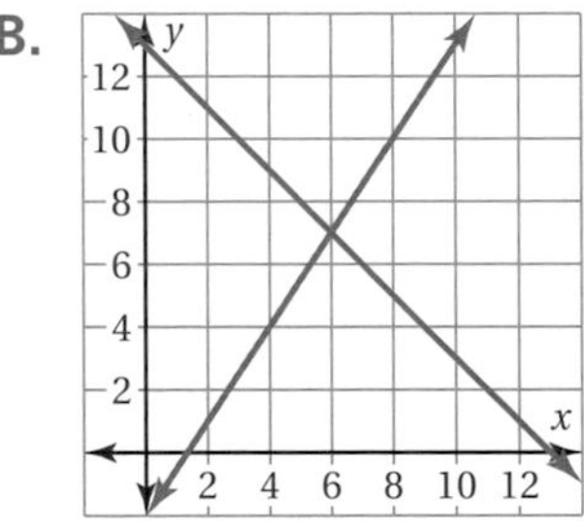

C.
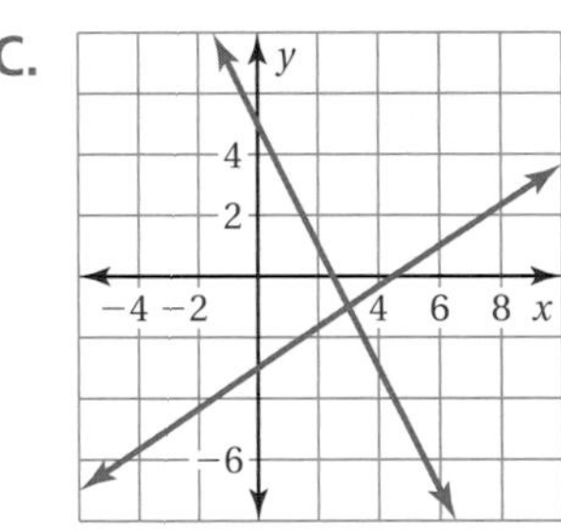

Solve the system of linear equations by graphing.

1 10. $y = 2x + 9$
   $y = 6 - x$

11. $y = -x - 4$
   $y = \frac{3}{5}x + 4$

12. $y = 2x + 5$
   $y = \frac{1}{2}x - 1$

2 13. $x + y = 27$
   $y = x + 3$

14. $y - x = 17$
   $y = 4x + 2$

15. $x - y = 7$
   $0.5x + y = 5$

16. **CARRIAGE RIDES** The cost C (in dollars) for the care and maintenance of a horse and carriage is $C = 15x + 2000$, where x is the number of rides.

   a. Write an equation for the revenue R in terms of the number of rides.

   b. How many rides are needed to break even?

Assignment Guide and Homework Check

Level	Assignment	Homework Check
Average	1–11, 13, 16, 17, 21, 25–28	3, 4, 10, 17
Advanced	1–15, 17, 18, 20–23, 25–28	5, 15, 17, 23

Common Errors

- **Exercises 7–9** Students may use the graph beneath each system to estimate its solution. Remind them to first use the *y*-intercepts and slopes of the equations in each system to find the correct graph.
- **Exercises 10–15** Students may not show enough of the graph, so the lines will not intersect. Encourage them to extend their lines until they intersect. All of the systems of linear equations in this section have a solution.
- **Exercises 17–19** Students may try to visually estimate the point of intersection of the graphs on their graphing utility screen. Remind them to use the *intersect* feature to estimate the point of intersection.

4.1 Record and Practice Journal

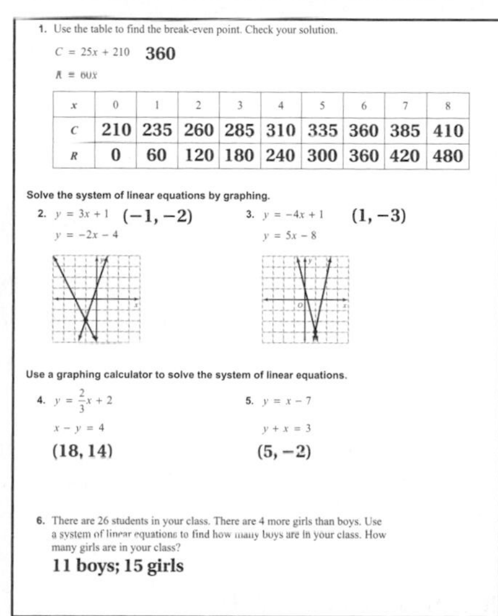

1. Use the table to find the break-even point. Check your solution.

$C = 25x + 210$ **360**

$R = 60x$

x	0	1	2	3	4	5	6	7	8
C	210	235	260	285	310	335	360	385	410
R	0	60	120	180	240	300	360	420	480

Solve the system of linear equations by graphing.

2. $y = 3x + 1$ **(−1, −2)**
$y = -2x - 4$

3. $y = -4x + 1$ **(1, −3)**
$y = 5x - 8$

Use a graphing calculator to solve the system of linear equations.

4. $y = \frac{2}{3}x + 2$
$x - y = 4$
(18, 14)

5. $y = x - 7$
$y + x = 3$
(5, −2)

6. There are 26 students in your class. There are 4 more girls than boys. Use a system of linear equations to find how many boys are in your class. How many girls are in your class?
11 boys; 15 girls

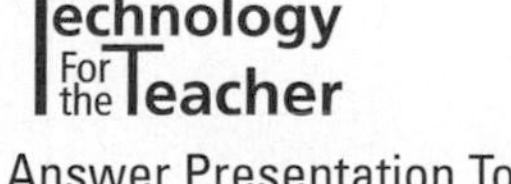

Answer Presentation Tool

Vocabulary and Concept Check

1. yes; The equations are linear and in the same variables.
2. Find the ordered pair (x, y) that represents the point of intersection of the graphs of the equations in the system.
3. Check whether (3, 4) is a solution of each equation.

Practice and Problem Solving

4. (5, 225)
5. (4, 176)
6. (5, 380)
7. B; (6, 7)
8. A; (2.5, 6.5)
9. C; (3, −1)
10. (−1, 7)
11. (−5, 1)
12. (−4, −3)
13. (12, 15)
14. (5, 22)
15. (8, 1)
16. **a.** $R = 35x$
    b. 100 rides

Practice and Problem Solving

17. (5, 1.5)

18. (1, 3)

19. (−6, 2)

20. Only the x-value is given; The solution is (4, 3).

21. no; Two lines cannot intersect in exactly two points.

22. 26 math problems, 16 science problems

23. See *Taking Math Deeper.*

24. **a.** $y = 0.5x + 2.5$
$y = 0.4x + 5.8$

b. yes; month 33

Fair Game Review

25. $c = 8$ **26.** $y = 4$

27. $x = 11$ **28.** B

Mini-Assessment

Solve the system of linear equations by graphing.

1. $y = 2x + 6$
$y = -2x - 2$ (−2, 2)

2. $y = 3x + 9$
$y = -\frac{1}{4}x - 4$ (−4, −3)

3. $2x + y = 4$
$y = x - 5$ (3, −2)

4. A wallet contains 23 bills. All the bills are \$1 bills and \$5 bills. There are 7 more \$1 bills than \$5 bills. How much money does the wallet contain? \$55

Taking Math Deeper

Exercise 23

One way to look at this problem is to make a table to compare your cumulative distances at half-hour intervals.

Time	Your Distance	Friend's Distance
0 hour	0 mile	0.5 mile
0.5 hour	1.7 miles	2 miles
1 hour	3.4 miles	3.5 miles
1.5 hours	5.1 miles	5 miles
2 hours	6.8 miles	6.5 miles
2.5 hours	8.5 miles	8 miles

Because you are a half mile behind your friend, your distance is 0 mile and your friend's distance is 0.5 mile.

You paddle at 3.4 miles per hour, so your distance after 0.5 hour is half of that, or 1.7 miles. Use the rate of 1.7 miles per half-hour to complete your distances in the table.

Your friend paddles at 3 miles per hour, so your friend's distance after 0.5 hour is 1.5 miles. Use the rate of 1.5 miles per half-hour to complete your friend's distances in the table.

a. The table shows that you are 0.1 mile behind your friend after one hour and 0.1 mile ahead of your friend after 1.5 hours. You can deduce that you pass your friend halfway between those two times. So, you will catch up to your friend after paddling for 1.25 hours.

How far have you traveled?

By similar reasoning, the distance you have traveled when you catch up to your friend is the mean of 3.4 miles and 5.1 miles. So you catch up to your friend after traveling $(3.4 + 5.1)/2 = 4.25$ miles.

b. The table shows that your friend is 0.5 mile behind you when you finish the race.

Reteaching and Enrichment Strategies

If students need help. . .	If students got it. . .
Resources by Chapter • Practice A and Practice B • Puzzle Time Record and Practice Journal Practice Differentiating the Lesson Lesson Tutorials Skills Review Handbook	Resources by Chapter • Enrichment and Extension Start the next section

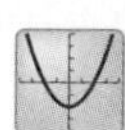 **Use a graphing calculator to solve the system of linear equations.**

17. $2.2x + y = 12.5$
$1.4x - 4y = 1$

18. $2.1x + 4.2y = 14.7$
$-5.7x - 1.9y = -11.4$

19. $-1.1x - 5.5y = -4.4$
$0.8x - 3.2y = -11.2$

20. ERROR ANALYSIS Describe and correct the error in solving the system of linear equations.

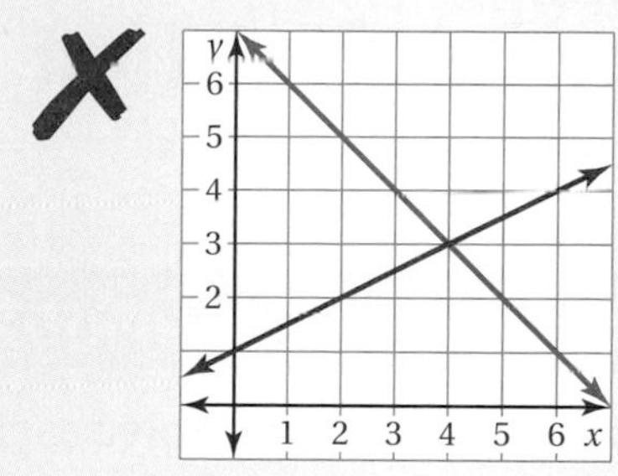

21. REASONING Is it possible for a system of two linear equations to have exactly two solutions? Explain your reasoning.

22. MODELING You have a total of 42 math and science problems for homework. You have 10 more math problems than science problems. How many problems do you have in each subject? Use a system of linear equations to justify your answer.

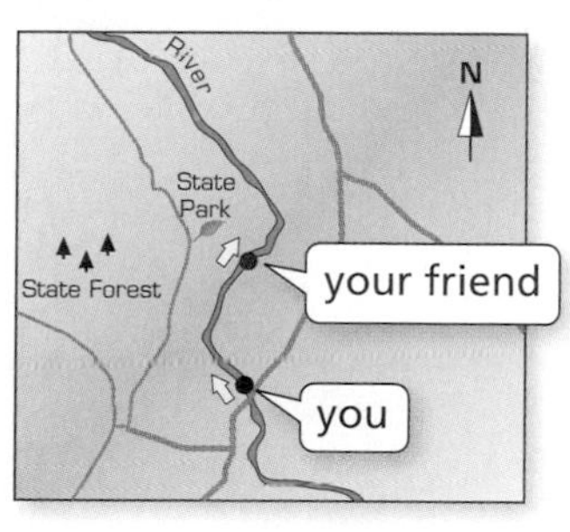

23. CANOE RACE You and your friend are in a canoe race. Your friend is a half mile in front of you and paddling 3 miles per hour. You are paddling 3.4 miles per hour.

a. You are 8.5 miles from the finish line. How long will it take you to catch up to your friend?

b. You both maintain your paddling rates for the remainder of the race. How far ahead of your friend will you be when you cross the finish line?

24. 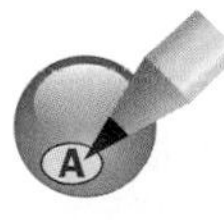 Your friend is trying to grow her hair as long as her cousin's hair. The table shows their hair lengths (in inches) in different months.

Month	Friend's Hair (in.)	Cousin's Hair (in.)
3	4	7
8	6.5	9

a. Write a system of linear equations that represents this situation.

b. Will your friend's hair ever be as long as her cousin's hair? If so, in what month?

Fair Game Review What you learned in previous grades & lessons

Solve the equation. Check your solution. *(Section 1.2)*

25. $\frac{3}{4}c - \frac{1}{4}c + 3 = 7$

26. $5(2 - y) + y = -6$

27. $6x - 3(x + 8) = 9$

28. MULTIPLE CHOICE The graph of which equation is perpendicular to the graph of $y = 2x + 1$? *(Section 2.2)*

Ⓐ $y = -2x - 1$ Ⓑ $y = -\frac{1}{2}x + 2$ Ⓒ $y = \frac{1}{2}x - 1$ Ⓓ $y = 2x + 2$

4.2 Solving Systems of Linear Equations by Substitution

Essential Question How can you use substitution to solve a system of linear equations?

COMMON CORE STATE STANDARDS
8.EE.8b
8.EE.8c
A.CED.3
A.REI.6

1 ACTIVITY: Using Substitution to Solve a System

Work with a partner. Solve each system of linear equations using two methods.

Method 1: Solve for x first.

Solve for x in one of the equations. Use the expression for x to find the solution of the system. Explain how you did it.

Method 2: Solve for y first.

Solve for y in one of the equations. Use the expression for y to find the solution of the system. Explain how you did it.

$y = 6x - 11$

Is the solution the same using both methods?

a. $6x - y = 11$
$2x + 3y = 7$

b. $2x - 3y = -1$
$x - y = 1$

c. $3x + y = 5$
$5x - 4y = -3$

d. $5x - y = 2$
$3x - 6y = 12$

e. $x + y = -1$
$5x + y = -13$

f. $2x - 6y = -6$
$7x - 8y = 5$

2 ACTIVITY: Writing and Solving a System of Equations

Work with a partner.

a. Roll a pair of number cubes that have different colors. Then write the ordered pair shown by the number cubes. The ordered pair at the right is (3, 4).

x-value

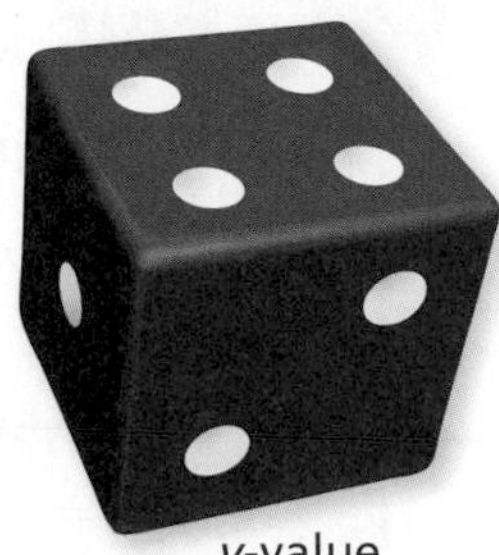

y-value

b. Write a system of linear equations that has this ordered pair as its solution.

c. Exchange systems with your partner and use one of the methods from Activity 1 to solve the system.

Laurie's Notes

Introduction

Standards for Mathematical Practice

- **MP1 Make Sense of Problems and Persevere in Solving Them:** Students will investigate another technique for solving a system of equations today. Students have knowledge of systems, solutions of systems, and symbolic manipulation skills. Students will have the opportunity to make different attempts at solving a system. Encourage them to persevere and try different approaches.

Motivate

- Play a game of "Zip, Zap, Zoop," which is a combination of several games.
- **Directions:** Stand in a circle. Count around the circle. When your number is a *multiple* of 4, say "**Zip**" instead of the number. When your number *contains* a 4, say "**Zap.**" When your number is *both* a multiple of 4 *and* contains a 4, say "**Zoop.**"
- The counting will go: 1, 2, 3, zoop, 5, 6, 7, zip, 9, 10, 11, zip, 13, zap, . . .
- They are *substituting* an expression for a number.
- The faster they count the funnier it becomes!

Activity Notes

Activity 1

- Given the equation $y = 6x - 11$, students are not familiar with the phrase, "use the expression for y." It may be helpful to highlight $6x - 11$ and identify it as "an expression for y."
- As a hint, tell students to think about the game they just played. They replaced certain numbers with words.
- **?** After one of the equations has been solved for a variable ask, "What does it mean to solve a system of linear equations?" Find an ordered pair that satisfies both equations.
- **?** "If there were only one equation with one variable, would you be able to solve it?" yes
- **?** "Can you see a way to use this system to get an equation with one variable?" Answers will vary.
- To save time have different groups do different pairs of problems.
- Discuss whether the results for each method are the same.

Activity 2

- Students can struggle with trying to write any equation that passes through a given point. For the example (3, 4), ask students to think about the relationship between the two numbers: x is one less than y, so adding 1 to x is equal to y. This suggests the equation $y = x + 1$. A second equation can be generated by thinking, "You can double the x-value and subtract 2 to get the y-value." This suggests $y = 2x - 2$.
- Take time for students to share their strategies for generating their systems and to solve each other's system.

Common Core State Standards

8.EE.8b Solve systems of two linear equations in two variables algebraically, and estimate solutions by graphing the equations. Solve simple cases by inspection.

8.EE.8c Solve real-world and mathematical problems leading to two linear equations in two variables.

A.CED.3 Represent constraints by systems of equations and interpret solutions as viable or nonviable options in a modeling context.

A.REI.6 Solve systems of linear equations exactly and approximately, focusing on pairs of linear equations in two variables.

Previous Learning

Students should know how to solve and evaluate linear equations.

Start Thinking! and Warm Up

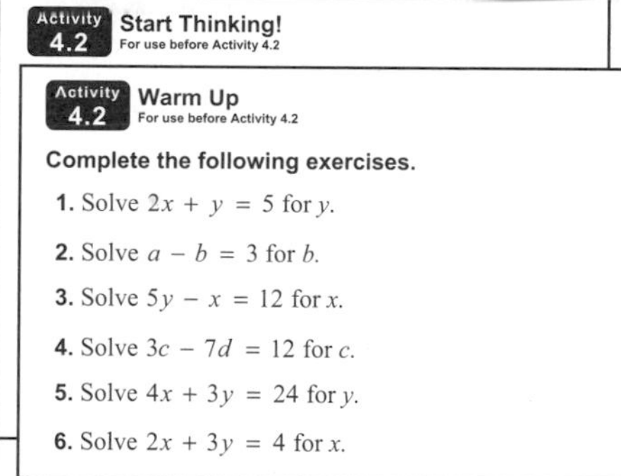

Activity 4.2 Start Thinking!
For use before Activity 4.2

Activity 4.2 Warm Up
For use before Activity 4.2

Complete the following exercises.

1. Solve $2x + y = 5$ for y.
2. Solve $a - b = 3$ for b.
3. Solve $5y - x = 12$ for x.
4. Solve $3c - 7d = 12$ for c.
5. Solve $4x + 3y = 24$ for y.
6. Solve $2x + 3y = 4$ for x.

4.2 Record and Practice Journal

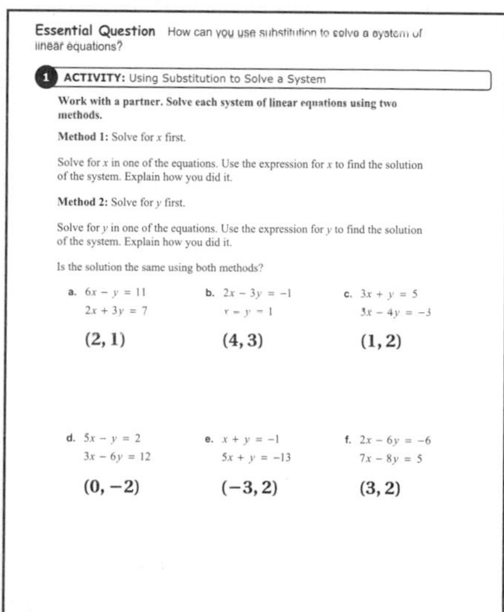

Essential Question How can you use substitution to solve a system of linear equations?

1 ACTIVITY: Using Substitution to Solve a System

Work with a partner. Solve each system of linear equations using two methods.

Method 1: Solve for x first.

Solve for x in one of the equations. Use the expression for x to find the solution of the system. Explain how you did it.

Method 2: Solve for y first.

Solve for y in one of the equations. Use the expression for y to find the solution of the system. Explain how you did it.

Is the solution the same using both methods?

a. $6x - y = 11$, $2x + 3y = 7$ **(2, 1)**

b. $2x - 3y = -1$, $x - y = 1$ **(4, 3)**

c. $3x + y = 5$, $5x - 4y = -3$ **(1, 2)**

d. $5x - y = 2$, $3x - 6y = 12$ **(0, −2)**

e. $x + y = -1$, $5x + y = -13$ **(−3, 2)**

f. $2x - 6y = -6$, $7x - 8y = 5$ **(3, 2)**

Differentiated Instruction

Pair Activity

Pair students. Each pair gets a set of two clue cards, one card to each student. Partners may not show each other their cards, but must communicate their clues verbally. Each pair must work together to answer the questions on their clue cards.

Prepare clue cards ahead of time.

Sample set:

Card 1: Three times Toni's age added to twice Sam's age is 88 years. How old is Toni?

Card 2: Sam's age reduced by half of Toni's age is 16 years. How old is Sam?

Answers: Toni: 14, Sam: 23

4.2 Record and Practice Journal

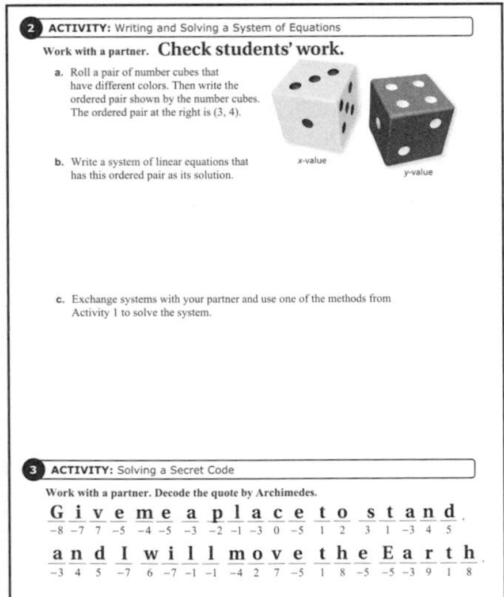

2 **ACTIVITY:** Writing and Solving a System of Equations

Work with a partner. **Check students' work.**

a. Roll a pair of number cubes that have different colors. Then write the ordered pair shown by the number cubes. The ordered pair at the right is (3, 4).

b. Write a system of linear equations that has this ordered pair as its solution.

c. Exchange systems with your partner and use one of the methods from Activity 1 to solve the system.

3 **ACTIVITY:** Solving a Secret Code

Work with a partner. Decode the quote by Archimedes.

G i v e m e a p l a c e t o s t a n d
−8 −7 7 −5 −4 −5 −3 −2 −1 −3 0 −5 1 2 3 1 −3 4 5

a n d I w i l l m o v e t h e E a r t h
−3 4 5 −7 6 −7 −1 −1 −4 2 7 −5 1 8 −5 −5 −3 9 1 8

(A, C) $x + y = -3$, $x - y = -3$ **(−3, 0)**

(D, E) $x + y = 0$, $x - y = 10$ **(5, −5)**

(G, H) $x + y = 0$, $x - y = -16$ **(−8, 8)**

(I, L) $x + 2y = -9$, $2x - y = -13$ **(−7, −1)**

(M, N) $x + 2y = 4$, $2x - y = -12$ **(−4, 4)**

(O, P) $x + 2y = -2$, $2x - y = 6$ **(2, −2)**

(R, S) $2x + y = 21$, $x - y = 6$ **(9, 3)**

(T, U) $2x + y = -7$, $x - y = 10$ **(1, −9)**

(V, W) $2x + y = 20$, $x - y = 1$ **(7, 6)**

What Is Your Answer?

4. IN YOUR OWN WORDS How can you use substitution to solve a system of linear equations?

Check students' work.

Laurie's Notes

Activity 3

- **FYI:** Archimedes is considered one of the greatest mathematicians in history. He was born in Syracuse, Greece in 287 B.C. He was killed during the siege of Syracuse in 212 B.C. by a Roman soldier who did not realize he was Archimedes. Some of his greatest contributions to mathematics were in the area of Geometry. He was also an accomplished engineer and an inventor. He invented the screw pump for raising water up an inclined plane and explained how levers work.
- In this activity students may actually use trial and error to guess at the solution versus using the substitution method. They may also solve a limited number of problems, start to decode the message, and guess at the remaining letters.
- Encourage students to check their solution by substituting the ordered pair into each of the equations.

What Is Your Answer?

- **MP7 Look for and Make Use of Structure:** Students discovered in Activity 1 that given a system of two equations in two variables, they can solve one equation for one of the variables and then substitute the expression for that variable in the other equation to find the value of the other variable.

Closure

- Write a system of linear equations that could be solved easily by substitution. Write a system of linear equations that could *not* be solved easily by substitution. Explain your reasoning.

Technology For the Teacher

Dynamic Classroom

The Dynamic Planning Tool
Editable Teacher's Resources at *BigIdeasMath.com*

3 ACTIVITY: Solving a Secret Code

Work with a partner. Decode the quote by Archimedes.

__ __ __ __   __ __   __   __ __ __ __ __   __ __   __ __ __ __ __,
-8 -7 7 -5 -4 -5 -3 -2 -1 -3 0 -5 1 2 3 1 -3 4 5

__ __ __   __   __ __ __ __   __ __ __ __   __ __ __   __ __ __ __ __.
-3 4 5 -7 6 -7 -1 -1 -4 2 7 -5 1 8 -5 -5 -3 9 1 8

(A, C) $x + y = -3$
$x - y = -3$

(D, E) $x + y = 0$
$x - y = 10$

(G, H) $x + y = 0$
$x - y = -16$

(I, L) $x + 2y = -9$
$2x - y = -13$

(M, N) $x + 2y = 4$
$2x - y = -12$

(O, P) $x + 2y = -2$
$2x - y = 6$

(R, S) $2x + y = 21$
$x - y = 6$

(T, U) $2x + y = -7$
$x - y = 10$

(V, W) $2x + y = 20$
$x - y = 1$

What Is Your Answer?

4. **IN YOUR OWN WORDS** How can you use substitution to solve a system of linear equations?

Practice Use what you learned about systems of linear equations to complete Exercises 4–6 on page 164.

4.2 Lesson

Another way to solve systems of linear equations is to use substitution.

Key Idea

Solving a System of Linear Equations by Substitution

Step 1 Solve one of the equations for one of the variables.

Step 2 Substitute the expression from Step 1 into the other equation and solve for the other variable.

Step 3 Substitute the value from Step 2 into one of the original equations and solve.

EXAMPLE 1 Solving a System of Linear Equations by Substitution

Solve the system by substitution. $\mathbf{y = 2x - 4}$ Equation 1

$\mathbf{7x - 2y = 5}$ Equation 2

Step 1: Equation 1 is already solved for y.

Step 2: Substitute $2x - 4$ for y in Equation 2.

$7x - 2y = 5$	Equation 2
$7x - 2(2x - 4) = 5$	Substitute $2x - 4$ for y.
$7x - 4x + 8 = 5$	Use the Distributive Property.
$3x + 8 = 5$	Combine like terms.
$3x = -3$	Subtract 8 from each side.
$x = -1$	Divide each side by 3.

Step 3: Substitute -1 for x in Equation 1 and solve for y.

$y = 2x - 4$	Equation 1
$= 2(-1) - 4$	Substitute -1 for x.
$= -2 - 4$	Multiply.
$= -6$	Subtract.

The solution is $(-1, -6)$.

Check

Equation 1

$y = 2x - 4$

$-6 \stackrel{?}{=} 2(-1) - 4$

$-6 = -6$ ✓

Equation 2

$7x - 2y = 5$

$7(-1) - 2(-6) \stackrel{?}{=} 5$

$5 = 5$ ✓

On Your Own

Now You're Ready
Exercises 10–15

Solve the system of linear equations by substitution. Check your solution.

1. $y = 2x + 3$
   $y = 5x$
2. $4x + 2y = 0$
   $y = \frac{1}{2}x - 5$
3. $x = 5y + 3$
   $2x + 4y = -1$

Laurie's Notes

Introduction

Connect

- **Yesterday:** Students discovered how to use substitution to solve a system of linear equations. (MP1, MP7)
- **Today:** Students will solve systems of linear equations by substitution.

Motivate

- Share a cooking story about salsa.

	Cilantro	Tomatoes	Onion
Summer Salsa	$\frac{1}{2}$ cup	3 cups	$\frac{3}{4}$ cup
Romero's Salsa	$1\frac{1}{4}$ cups	8 cups	2 cups

? "Do you think algebra can help a cook figure out how much salsa of each type can be made if you have 5 cups of cilantro and 40 cups of crushed tomatoes?" Comments will vary.

- Explain to students that the techniques they will study today are used to solve problems of this type.

Lesson Notes

Discuss

- Students have solved a system of linear equations by graphing. Substitution is a second way to solve a system of linear equations.

Key Idea

- Discuss the steps in solving a system of linear equations by substitution.
- At the end of Step 2, you can add "You now know either the *x*- or *y*-coordinate of the ordered pair that satisfies both equations. Next, you need to find the other coordinate."
- Remind students to check their solution in both equations.

Example 1

- Write Equation 1 and Equation 2.

? "What form is Equation 1 written in?" slope-intercept form

? "What form is Equation 2 written in?" standard form

- Say, "Equation 1 is already solved for *y*, so you can go right to Step 2. Substitute the expression $2x - 4$ for *y* in the second equation."
- **MP6 Attend to Precision:** Students may get sloppy and say they are "plugging in for *y*." "Plugging in" is not a mathematical operation or process. It is better to say that they are "substituting for *y*," so they become familiar with the math terminology they are expected to know.

? After the second step ask, "What does $x = -1$ mean in the context of this problem?" *Sample answer:* The *x*-coordinate of the solution is -1.

? "How do we determine the *y*-coordinate of the solution?" Substitute -1 for *x* in one of the original equations and solve for *y*.

- Check the solution.

Goal Today's lesson is solving a system of linear equations by substitution.

Start Thinking! and Warm Up

Lesson 4.2 Warm Up
For use before Lesson 4.2

Lesson 4.2 Start Thinking!
For use before Lesson 4.2

Solve each system first by graphing and then by substitution. Which system is easier to solve by graphing? Which system is easier to solve by substitution? Explain.

$2x + y = 5$
$3x + 5y = 18$

$y = \frac{1}{2}x + 3$
$y = -\frac{3}{4}x - 2$

Extra Example 1

Solve the system by substitution.

$y = 3x - 4$
$5x - 2y = 10$
$(-2, -10)$

On Your Own

1. $(1, 5)$
2. $(2, -4)$
3. $\left(\frac{1}{2}, -\frac{1}{2}\right)$

Laurie's Notes

On Your Own

- Note that in all of these systems, one of the variables has already been solved for explicitly.

Example 2

- ? **MP1 Make Sense of Problems and Persevere in Solving Them:** Ask a student to read the problem. Ask, "Is it possible that you bought only turkey burgers or only veggie burgers? Explain." no; You spent \$90, but 50 turkey burgers would cost \$100 and 50 veggie burgers would cost \$75.
- Students can be careless when defining variables. Be sure they use:
  $x =$ number of turkey burgers
  $y =$ number of veggie burgers
- ? "There are two equations, each with two variables. Can we solve for one of the variables in either equation?" yes
- ? "Is there a choice that might be easier than another? Explain." yes; It may be easiest to solve for x or y in the first equation because both coefficients are 1, and there are no decimals.
- Discuss the Study Tip.
- Work through the problem as shown.
- If time permits, solve the system again, solving for y in Step 1. Compare the answers.
- ? "Could this system be solved by graphing?" yes

On Your Own

- Observe student work. It is likely that some students will solve for x and some for y. When students have finished ask two volunteers to share their work at the board, demonstrating each method.

Closure

- **Exit Ticket:** Solve the system by substitution and by graphing.
  $y = 3x + 1$
  $y = x + 3$
  (1, 4)

Extra Example 2

A weightlifter uses a total of 12 plates to add 260 pounds to a bar. He uses 45-pound plates and 10-pound plates. Write and solve a system of equations to find the number x of 45-pound plates and the number y of 10-pound plates he uses.

$x + y = 12$
$45x + 10y = 260$
four 45-pound plates, eight 10-pound plates

On Your Own

4. $x + y = 100$
   $2x + 3y = 240$
   60 cups of lemonade, 40 cups of orange juice

English Language Learners

Simplifying the Language

To help your students understand and remember the steps for solving a system of linear equations by substitution, present a simplified version of the steps.

Step 1: Solve.

Step 2: Substitute and solve.

Step 3: Substitute and solve.

Ask students to explain what to solve and what to substitute in each step.

Technology For the Teacher

The Dynamic Planning Tool
Editable Teacher's Resources at *BigIdeasMath.com*

EXAMPLE 2 Real-Life Application

You buy a total of 50 turkey burgers and veggie burgers for \$90. You pay \$2 per turkey burger and \$1.50 per veggie burger. Write and solve a system of linear equations to find the number x of turkey burgers and the number y of veggie burgers you buy.

Use a verbal model to write a system of linear equations.

Number of turkey burgers, x	+	Number of veggie burgers, y	=	Total number of burgers

Cost per turkey burger	•	Number of turkey burgers, x	+	Cost per veggie burger	•	Number of veggie burgers, y	=	Total cost

The system is: $x + y = 50$ Equation 1

$2x + 1.5y = 90$ Equation 2

Study Tip

It is easiest to solve for a variable that has a coefficient of 1 or −1.

Step 1: Solve Equation 1 for x.

$x + y = 50$ Equation 1

$x = 50 - y$ Subtract y from each side.

Step 2: Substitute $50 - y$ for x in Equation 2.

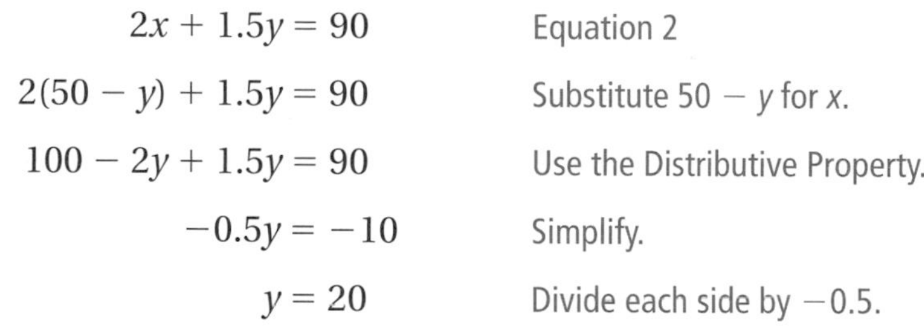

$2x + 1.5y = 90$ Equation 2

$2(50 - y) + 1.5y = 90$ Substitute $50 - y$ for x.

$100 - 2y + 1.5y = 90$ Use the Distributive Property.

$-0.5y = -10$ Simplify.

$y = 20$ Divide each side by −0.5.

Check

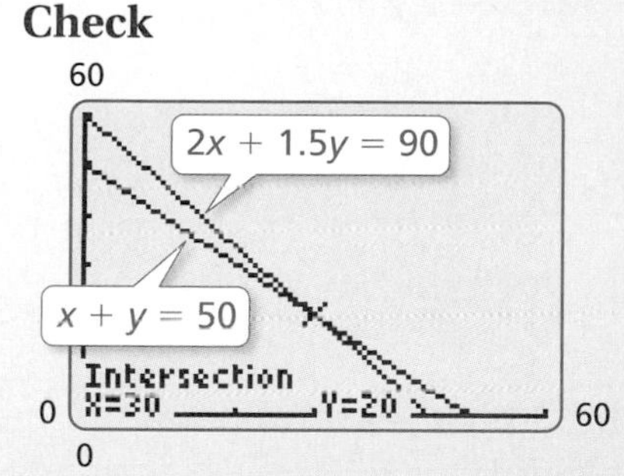

Step 3: Substitute 20 for y in Equation 1 and solve for x.

$x + y = 50$ Equation 1

$x + 20 = 50$ Substitute 20 for y.

$x = 30$ Subtract 20 from each side.

∴ You buy 30 turkey burgers and 20 veggie burgers.

On Your Own

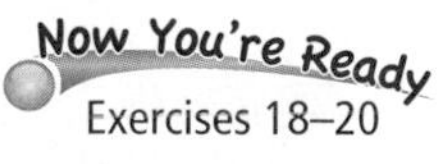

Exercises 18–20

4. A juice stand sells lemonade for \$2 per cup and orange juice for \$3 per cup. The juice stand sells a total of 100 cups of juice for \$240. Write and solve a system of linear equations to find the number of cups of lemonade and the number of cups of orange juice sold.

4.2 Exercises

Vocabulary and Concept Check

1. **WRITING** Describe how to solve a system of linear equations by substitution.
2. **NUMBER SENSE** When solving a system of linear equations by substitution, how do you decide which variable to solve for in Step 1?
3. **REASONING** Does solving a system of linear equations by graphing give the same solution as solving by substitution? Explain your reasoning.

Practice and Problem Solving

Write a system of linear equations that has the ordered pair as its solution. Use a method from Activity 1 to solve the system.

4. 

5. 

6. 

Tell which equation you would use in Step 1 when solving the system by substitution. Explain your reasoning.

7. $2x + 3y = 5$
   $4x - y = 3$

8. $\frac{2}{3}x + 5y = -1$
   $x + 6y = 0$

9. $2x + 10y = 14$
   $5x - 9y = 1$

Solve the system of linear equations by substitution. Check your solution.

1 10. $y = x - 4$
   $y = 4x - 10$

11. $y = 2x + 5$
   $y = 3x - 1$

12. $x = 2y + 7$
   $3x - 2y = 3$

13. $4x - 2y = 14$
   $y = \frac{1}{2}x - 1$

14. $2x = y - 10$
   $x + 7 = y$

15. $8x - \frac{1}{3}y = 0$
   $12x + 3 = y$

16. **SCHOOL CLUBS** There are a total of 64 students in a drama club and a yearbook club. The drama club has 10 more students than the yearbook club.
    a. Write a system of linear equations that represents this situation.
    b. How many students are in the drama club? the yearbook club?

17. **THEATER** A drama club earns $1040 from a production. A total of 64 adult tickets and 132 student tickets are sold. An adult ticket costs twice as much as a student ticket.
    a. Write a system of linear equations that represents this situation.
    b. What is the cost of each ticket?

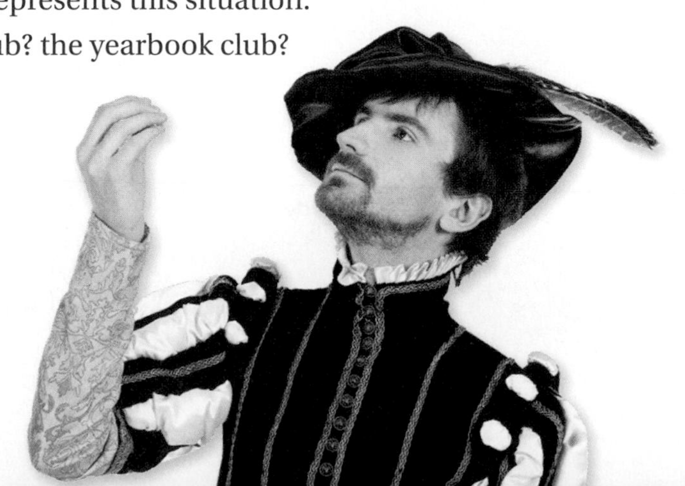

Assignment Guide and Homework Check

Level	Assignment	Homework Check
Average	1–7, 10, 11, 16, 19, 21, 22, 26–29	7, 10, 19, 21
Advanced	1–8, 13–16, 20–29	15, 16, 20, 23

Common Errors

- **Exercises 4–6** Students may write equations that do not give the correct solutions. For instance, in Exercise 4, a student might write one equation as $2x + 3y = 5$. Tell students to check each equation to make sure that the ordered pair represented by the dice is a solution.
- **Exercises 10–15, 18–20** Students may find one coordinate of the solution and stop there. Remind them that the answer is a coordinate pair representing both an x-value and a y-value.

4.2 Record and Practice Journal

Solve the system of linear equations by substitution. Check your solution.

1. $y = -2x + 4$
   $-x + 3y = -9$
   (3, −2)

2. $\frac{3}{4}x - 5y = 7$
   $x = -4y + 12$
   $\left(11, \frac{1}{4}\right)$

3. $5x - y = 4$
   $2x + 2y = 16$
   (2, 6)

4. $2x + 3y = 0$
   $8x + 9y = 18$
   (9, −6)

5. A gas station sells a total of 4500 gallons of regular gas and premium gas in one day. The ratio of gallons of regular gas sold to gallons of premium gas sold is 7 : 2.

   a. Write a system of linear equations that represents this situation.
   $x + y = 4500$
   $2x = 7y$

   b. How many gallons sold were regular gas? premium gas?
   Regular: 3500 gal; Premium 1000 gal

Technology For the Teacher
Answer Presentation Tool

Vocabulary and Concept Check

1. **Step 1:** Solve one of the equations for one of the variables.
   Step 2: Substitute the expression from Step 1 into the other equation and solve.
   Step 3: Substitute the value from Step 2 into one of the original equations and solve.
2. If possible, solve for a variable that has a coefficient of 1 or −1, or that is easy to solve.
3. sometimes; A solution obtained by graphing may not be exact.

Practice and Problem Solving

4. *Sample answer:* $x + y = 5$
   $x - y = -1$
5. *Sample answer:* $x + 2y = 6$
   $x - y = 3$
6. *Sample answer:*
   $x - y = 1$
   $2x - 3y = -3$
7. $4x - y = 3$; The coefficient of y is −1.
8. $x + 6y = 0$; The coefficient of x is 1, and there is no constant.
9. $2x + 10y = 14$; Dividing by 2 to solve for x yields integers.
10. (2, −2)
11. (6, 17)
12. (−1, −3)
13. (4, 1)
14. (−3, 4)
15. $\left(\frac{1}{4}, 6\right)$
16. **a.** $x + y = 64$
    $x = y + 10$
    b. 37 students; 27 students

Practice and Problem Solving

17. a. $x = 2y$

$64x + 132y = 1040$

b. adult tickets: $8; student tickets: $4

18. $(-3, -3)$

19. $(-2, 4)$ **20.** $(6, -3)$

21. The expression for y was substituted back into the same equation; Solution: $(2, 1)$

22. $y = 2.5x$

$2x + y = 180$

base angles: 40°, third angle: 100°

23. 30 cats, 35 dogs

24. 26

25. See *Taking Math Deeper.*

Fair Game Review

26. $3x - 7y = 9$

27. $2x - 5y = -8$

28. $6x - y = 3$

29. B

Mini-Assessment

Solve the system of linear equations by substitution.

1. $y = 3x - 2$

$y = -x + 6$ $(2, 4)$

2. $2y + 8 = x$

$8x + y = -21$ $(-2, -5)$

3. $4x + 3y = 26$

$2x - 3y = -14$ $(2, 6)$

4. You spent \$56 on food and clothes. You spent \$18 more on clothes than on food. Write and solve a system of equations to find how much you spent on each.

$x + y = 56$ clothes: \$37

$y - x = 18$ food: \$19

Taking Math Deeper

Exercise 25

One way to visualize the problem is to make a diagram.

Make a diagram.

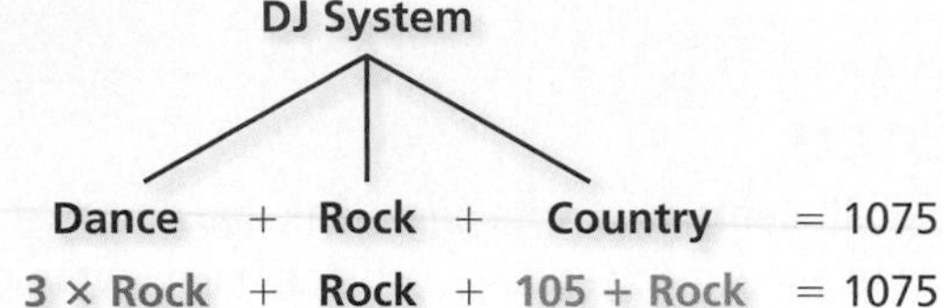

Dance + Rock + Country = 1075

3 × Rock + Rock + 105 + Rock = 1075

The DJ has 1075 songs on her system, so the sum of the numbers of dance, rock, and country songs is 1075.

The dance selection is 3 times the size of the rock selection, so you can substitute 3 × Rock for the number of dance songs.

There are 105 more country songs than rock songs, so you can substitute 105 + Rock for the number of country songs.

Now you can find the number r of rock songs.

Write and solve an equation.

$$3r + r + (105 + r) = 1075$$
$$5r + 105 = 1075$$
$$5r = 970$$
$$r = 194$$

There are 194 rock songs.

Substitute and solve.

Dance = 3 × Rock
= 3 × 194
= 582

Country = 105 + Rock
= 105 + 194
= 299

There are 582 dance, 194 rock, and 299 country songs on the system.

Reteaching and Enrichment Strategies

If students need help. . .	If students got it. . .
Resources by Chapter • Practice A and Practice B • Puzzle Time Record and Practice Journal Practice Differentiating the Lesson Lesson Tutorials Skills Review Handbook	Resources by Chapter • Enrichment and Extension Start the next section

Solve the system of linear equations by substitution. Check your solution.

2 **18.** $y - x = 0$
$2x - 5y = 9$

19. $x + 4y = 14$
$3x + 7y = 22$

20. $-2x - 5y = 3$
$3x + 8y = -6$

21. ERROR ANALYSIS Describe and correct the error in solving the system of linear equations.

✗		Step 1:	Step 2:
$2x + y = 5$	Equation 1	$2x + y = 5$	$2x + (-2x + 5) = 5$
$3x - 2y = 4$	Equation 2	$y = -2x + 5$	$2x - 2x + 5 = 5$
			$5 = 5$

22. STRUCTURE The measure of the obtuse angle in the isosceles triangle is two and a half times the measure of one base angle. Write and solve a system of linear equations to find the measures of all the angles.

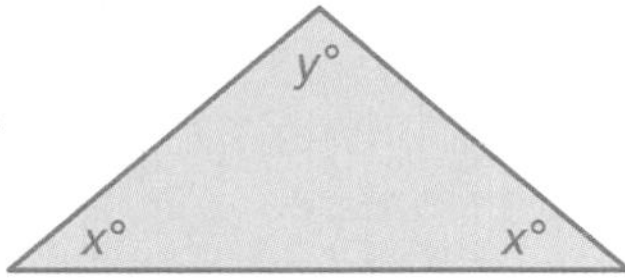

23. ANIMAL SHELTER An animal shelter has a total of 65 abandoned cats and dogs. The ratio of cats to dogs is 6 : 7. How many cats are in the shelter? How many dogs are in the shelter? Justify your answers.

24. NUMBER SENSE The sum of the digits of a two-digit number is 8. When the digits are reversed, the number increases by 36. Find the original number.

25. Repeated Reasoning A DJ has a total of 1075 dance, rock, and country songs on her system. The dance selection is three times the size of the rock selection. The country selection has 105 more songs than the rock selection. How many songs on the system are dance? rock? country?

Fair Game Review What you learned in previous grades & lessons

Write the equation in standard form. *(Section 2.4)*

26. $3x - 9 = 7y$

27. $8 - 5y = -2x$

28. $6x = y + 3$

29. MULTIPLE CHOICE Use the figure to find the measure of $\angle 2$. *(Skills Review Handbook)*

Ⓐ 17° Ⓑ 73°

Ⓒ 83° Ⓓ 107°

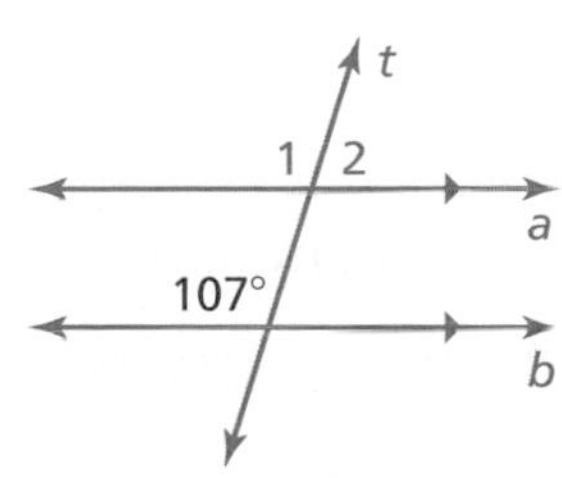

4 Study Help

You can use a **notetaking organizer** to write notes, vocabulary, and questions about a topic. Here is an example of a notetaking organizer for solving systems of linear equations by graphing.

Write important vocabulary or formulas in this space. →

system of linear equations solution of a system of linear equations	Solving systems of linear equations by graphing **Step 1:** Graph each equation. **Step 2:** Estimate the point of intersection. **Step 3:** Check the point from Step 2. Example: Solve the system. $y = x + 1$ $y = -2x - 2$

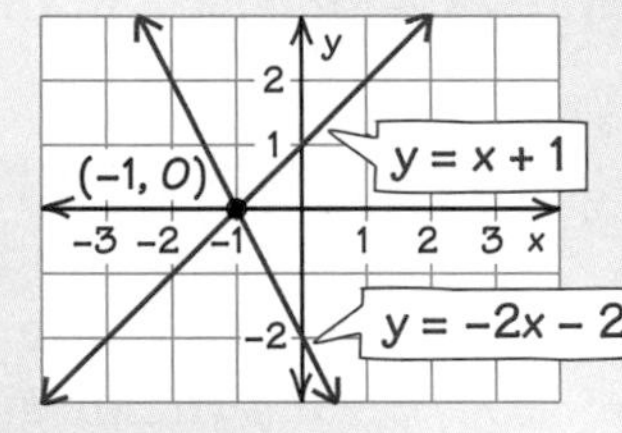

The solution is (−1, 0).

Will a system of linear equations always have a solution?

← Write your notes about the topic in this space.

← Write your questions about the topic in this space.

On Your Own

Make a notetaking organizer to help you study this topic.

1. solving systems of linear equations by substitution

After you complete this chapter, make notetaking organizers for the following topics.

2. solving systems of linear equations by elimination
3. graphing systems of linear inequalities

"My notetaking organizer has me thinking about retirement when I won't have to fetch sticks anymore."

Sample Answers

1.

	Solving systems of linear equations by substitution
system of linear equations solution of a system of linear equations	Step 1: Solve one of the equations for one of the variables. Step 2: Substitute the expression from Step 1 into the other equation and solve for the other variable. Step 3: Substitute the value from Step 2 into one of the original equations and solve. Example: Solve the system. $x + y = 5$ $2x = 3y$ Step 1: $y = 5 - x$ Step 2: $2x = 3(5 - x)$ $2x = 15 - 3x$ $5x = 15$ $x = 3$ Step 3: $3 + y = 5$ $y = 2$ The solution is (3, 2).
How do you decide which equation to use to solve for one of the variables in Step 1?	

List of Organizers

Available at *BigIdeasMath.com*

Comparison Chart
Concept Circle
Definition (Idea) and Example Chart
Example and Non-Example Chart
Formula Triangle
Four Square
Information Frame
Information Wheel
Notetaking Organizer
Process Diagram
Summary Triangle
Word Magnet
Y Chart

About this Organizer

A **Notetaking Organizer** can be used to write notes, vocabulary, and questions about a topic. In the space on the left, students write important vocabulary or formulas. In the space on the right, students write their notes about the topic. In the space at the bottom, students write their questions about the topic. A notetaking organizer can also be used as an assessment tool, in which blanks are left for students to complete.

Technology For the Teacher

Vocabulary Puzzle Builder

Answers

1. B; (1, −1)
2. C; (3, 0)
3. A; (−2, −3)
4. (4, 5)
5. (−1, 4)
6. (3, −5)
7. (6, −2)
8. (8, 3)
9. (−4, −1)
10. a. $y = 2x + 15$ Members

    $y = 3x$ Nonmembers

    b. It is beneficial when you rent more than 15 new release movies per year.
11. 23 and 15; (23, 15) is the solution of the system

    $x + y = 38$

    $x - y = 8$.
12. 60 ft by 30 ft
13. 63 nurses; 14 doctors

Assessment Book

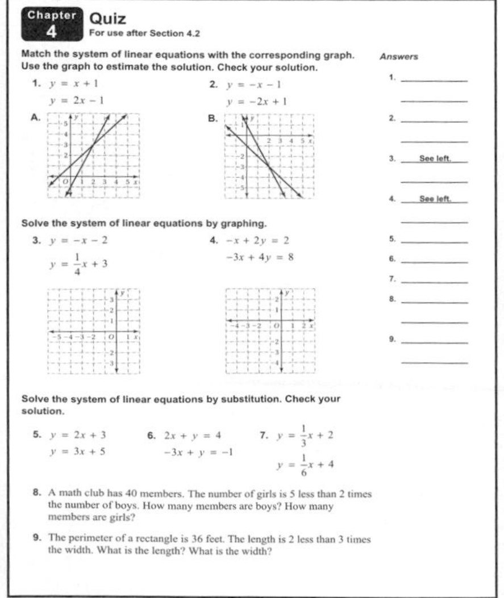

Chapter 4 **Quiz**
For use after Section 4.2

Match the system of linear equations with the corresponding graph. Use the graph to estimate the solution. Check your solution.

1. $y = x + 1$
   $y = 2x - 1$
2. $y = -x - 1$
   $y = -2x + 1$

A. B.

Solve the system of linear equations by graphing.

3. $y = -x - 2$
   $y = \frac{1}{4}x + 3$
4. $-x + 2y = 2$
   $-3x + 4y = 8$

Solve the system of linear equations by substitution. Check your solution.

5. $y = 2x + 3$
   $y = 3x + 5$
6. $2x + y = 4$
   $-3x + y = -1$
7. $y = \frac{1}{3}x + 2$
   $y = \frac{1}{6}x + 4$

8. A math club has 40 members. The number of girls is 5 less than 2 times the number of boys. How many members are boys? How many members are girls?

9. The perimeter of a rectangle is 36 feet. The length is 2 less than 3 times the width. What is the length? What is the width?

Answers

1. ______
2. ______
3. See left.
4. See left.
5. ______
6. ______
7. ______
8. ______
9. ______

Alternative Quiz Ideas

100% Quiz
Error Notebook
Group Quiz
Homework Quiz
Math Log
Notebook Quiz
Partner Quiz
Pass the Paper

Error Notebook

An error notebook provides an opportunity for students to analyze and learn from their errors. Have students make an error notebook for this chapter. They should work in their notebook a little each day. Give students the following directions.

- Use a notebook and divide the page into three columns.
- Label the first column *problem*, second column *error*, and third column *correction*.
- In the first column, write down the problem on which the errors were made. Record the source of the problem (homework, quiz, in-class assignment).
- The second column should show the exact error that was made. Include a statement of why you think the error was made. This is where the learning takes place, so it is helpful to use a different color ink for the work in this column.
- The last column contains the corrected problems and comments that will help with future work.
- Separate each problem with horizontal lines.

Reteaching and Enrichment Strategies

If students need help. . .	If students got it. . .
Resources by Chapter • Study Help • Practice A and Practice B • Puzzle Time Lesson Tutorials *BigIdeasMath.com* Practice Quiz Practice from the Test Generator	Resources by Chapter • Enrichment and Extension • School-to-Work Game Closet at *BigIdeasMath.com* Start the next section

Technology For the Teacher

Answer Presentation Tool
Big Ideas Test Generator

4.1–4.2 Quiz

Match the system of linear equations with the corresponding graph. Use the graph to estimate the solution. Check your solution. *(Section 4.1)*

1. $y = x - 2$
$y = -2x + 1$

2. $y = x - 3$
$y = -\frac{1}{3}x + 1$

3. $y = \frac{1}{2}x - 2$
$y = 4x + 5$

A.

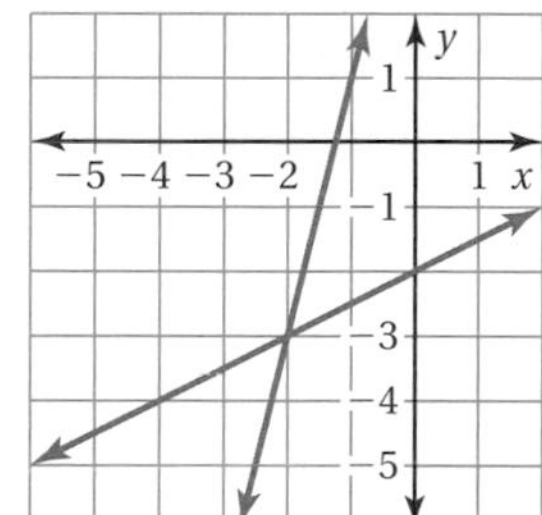

B.

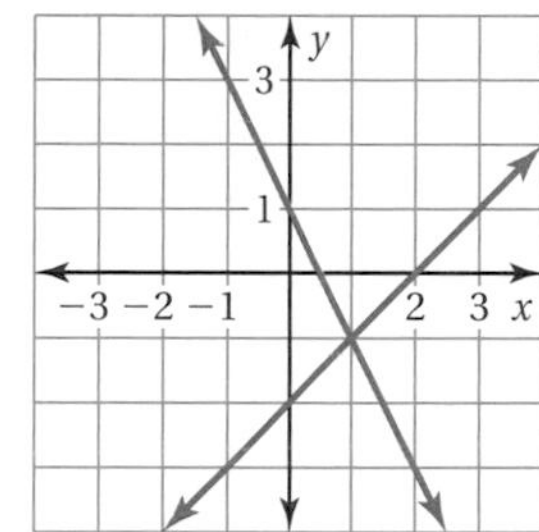

C.

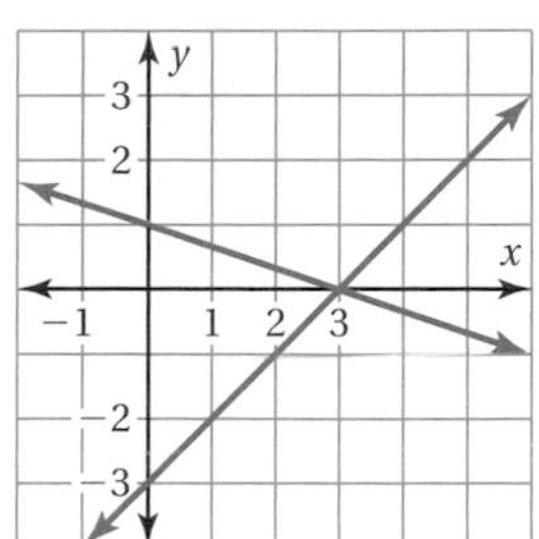

Solve the system of linear equations by graphing. *(Section 4.1)*

4. $y = 2x - 3$
$y = -x + 9$

5. $6x + y = -2$
$y = -3x + 1$

6. $4x + 2y = 2$
$3x = 4 - y$

Solve the system of linear equations by substitution. Check your solution. *(Section 4.2)*

7. $y = x - 8$
$y = 2x - 14$

8. $x = 2y + 2$
$2x - 5y = 1$

9. $x - 5y = 1$
$-2x + 9y = -1$

10. MOVIE CLUB Members of a movie rental club pay a \$15 annual membership fee and \$2 for new release movies. Nonmembers pay \$3 for new release movies. *(Section 4.1)*

a. Write a system of linear equations that represents this situation.

b. When is it beneficial to have a membership?

11. NUMBER SENSE The sum of two numbers is 38. The greater number is 8 more than the other number. Find each number. Use a system of linear equations to justify your answer. *(Section 4.1)*

12. VOLLEYBALL The length of a sand volleyball court is twice its width. The perimeter is 180 feet. Find the length and width of the sand volleyball court. *(Section 4.2)*

13. MEDICAL STAFF A hospital employs a total of 77 nurses and doctors. The ratio of nurses to doctors is 9 : 2. How many nurses are employed at the hospital? How many doctors are employed at the hospital? *(Section 4.2)*

4.3 Solving Systems of Linear Equations by Elimination

COMMON CORE STATE STANDARDS
8.EE.8b
8.EE.8c
A.CED.3
A.REI.5
A.REI.6

Essential Question How can you use elimination to solve a system of linear equations?

1 ACTIVITY: Using Elimination to Solve a System

Work with a partner. Solve each system of linear equations using two methods.

Method 1: Subtract.

Subtract Equation 2 from Equation 1. What is the result? Explain how you can use the result to solve the system of equations.

Method 2: Add.

Add the two equations. What is the result? Explain how you can use the result to solve the system of equations.

Is the solution the same using both methods?

a. $2x + y = 4$
$2x - y = 0$

b. $3x - y = 4$
$3x + y = 2$

c. $x + 2y = 7$
$x - 2y = -5$

2 ACTIVITY: Using Elimination to Solve a System

Work with a partner.

$2x + y = 2$ Equation 1
$x + 5y = 1$ Equation 2

a. Can you add or subtract the equations to solve the system of linear equations? Explain.

b. Explain what property you can apply to Equation 1 in the system so that the *y* coefficients are the same.

c. Explain what property you can apply to Equation 2 in the system so that the *x* coefficients are the same.

d. You solve the system in part (b). Your partner solves the system in part (c). Compare your solutions.

e. Use a graphing calculator to check your solution.

Laurie's Notes

Introduction

Standards for Mathematical Practice

- **MP1 Make Sense of Problems and Persevere in Solving Them:** Students will investigate a third technique for solving a system of linear equations today. Students have the prerequisite knowledge. Students will use more than one approach to solve a system. Encourage them to persevere and try to understand why these approaches work.
- There are multiple avenues for solving a system using the elimination technique. Students may solve the system using several different approaches, all of which lead to the same answer.

Motivate

- Draw the following sketch on the board. Make it clear to students that the right sides represent 13 pounds of weight and 5 pounds of weight.

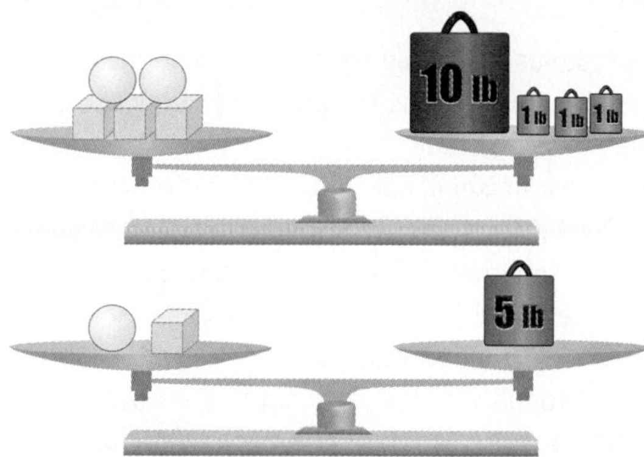

- Ask students to draw a sketch of a balance scale with 4 cubes and 3 balls at the left, 18 pounds at the right. Relate this to adding equations.
- Ask students to draw a sketch of a balance scale with 2 cubes and 1 ball at the left, 8 pounds at the right. Relate this to subtracting equations.
- Explain to students that today they will perform operations with equations similar to those they represented with the balance scales.

Activity Notes

Activity 1

- Students may not understand what it means to add or subtract equations. Tell them to think about the balance scales—they added the left sides of the scales and the right sides of the scales.
- As you walk around, you may need to remind students what it means to solve a system.

? After students find the value of one variable, ask "How can you find the value of the other variable?" Substitute the known value into one of the original equations to solve for the other variable.

- Discuss the solutions found using each method (addition and subtraction).

? "Can you eliminate one variable by adding the equations for any system of two linear equations? Explain." no; One of the variables must have coefficients that are opposites in the two equations for this to work.

Common Core State Standards

A.CED.3 Represent constraints by systems of equations and interpret solutions as viable or nonviable options in a modeling context.

A.REI.5 Prove that, given a system of two equations in two variables, replacing one equation by the sum of that equation and a multiple of the other produces a system with the same solutions.

A.REI.6 Solve systems of linear equations exactly and approximately, focusing on pairs of linear equations in two variables.

Also **8.EE.8b and 8.EE.8c**

Previous Learning

Students should know how to solve and evaluate linear equations.

Start Thinking! and Warm Up

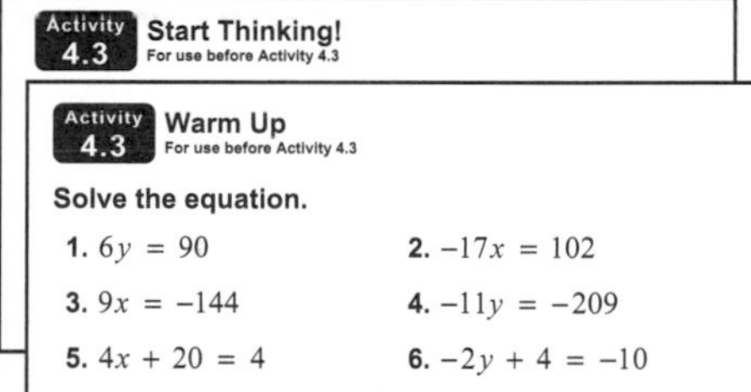

Activity 4.3 Start Thinking!
For use before Activity 4.3

Activity 4.3 Warm Up
For use before Activity 4.3

Solve the equation.

1. $6y = 90$
2. $-17x = 102$
3. $9x = -144$
4. $-11y = -209$
5. $4x + 20 = 4$
6. $-2y + 4 = -10$

4.3 Record and Practice Journal

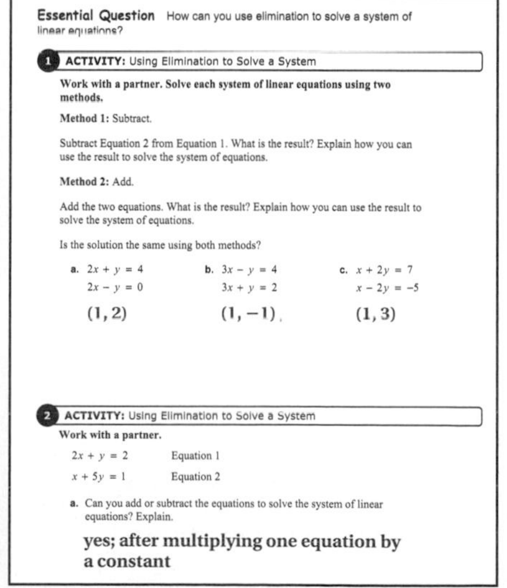

Essential Question How can you use elimination to solve a system of linear equations?

1 **ACTIVITY:** Using Elimination to Solve a System

Work with a partner. Solve each system of linear equations using two methods.

Method 1: Subtract.

Subtract Equation 2 from Equation 1. What is the result? Explain how you can use the result to solve the system of equations.

Method 2: Add.

Add the two equations. What is the result? Explain how you can use the result to solve the system of equations.

Is the solution the same using both methods?

a. $2x + y = 4$, $2x - y = 0$ $(1, 2)$

b. $3x - y = 4$, $3x + y = 2$ $(1, -1)$

c. $x + 2y = 7$, $x - 2y = -5$ $(1, 3)$

2 **ACTIVITY:** Using Elimination to Solve a System

Work with a partner.

$2x + y = 2$ Equation 1

$x + 5y = 1$ Equation 2

a. Can you add or subtract the equations to solve the system of linear equations? Explain.

yes; after multiplying one equation by a constant

Differentiated Instruction

Kinesthetic

It may help students to write each term of a system on a small slip of paper. When a variable is eliminated, the student removes the terms related to that variable. This helps them to grasp the concept.

4.3 Record and Practice Journal

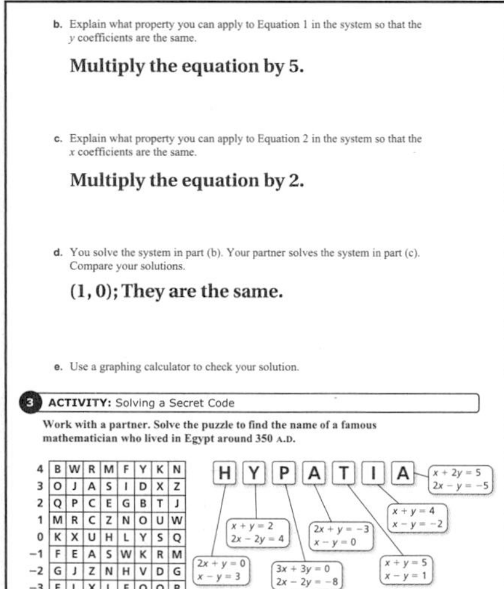

b. Explain what property you can apply to Equation 1 in the system so that the y coefficients are the same.

Multiply the equation by 5.

c. Explain what property you can apply to Equation 2 in the system so that the x coefficients are the same.

Multiply the equation by 2.

d. You solve the system in part (b). Your partner solves the system in part (c). Compare your solutions.

(1, 0); They are the same.

e. Use a graphing calculator to check your solution.

3 ACTIVITY: Solving a Secret Code

Work with a partner. Solve the puzzle to find the name of a famous mathematician who lived in Egypt around 350 A.D.

4	B	W	R	M	F	Y	K	N
3	O	J	A	S	I	D	X	Z
2	Q	P	C	E	G	B	T	J
1	M	R	C	Z	N	O	U	W
0	K	X	U	H	L	Y	S	Q
−1	F	E	A	S	W	K	R	M
−2	G	J	Z	N	H	V	D	G
−3	E	L	X	L	F	Q	O	B
	−3	−2	−1	0	1	2	3	4

H Y P A T I A

H: $2x + y = 0$, $x - y = 3$

Y: $x + y = 2$, $2x - 2y = 4$

P: $3x + 3y = 0$, $2x - 2y = -8$

A: $2x + y = -3$, $x - y = 0$

T: $x + y = 5$, $x - y = 1$

I: $x + y = 4$, $x - y = -2$

A: $x + 2y = 5$, $2x - y = -5$

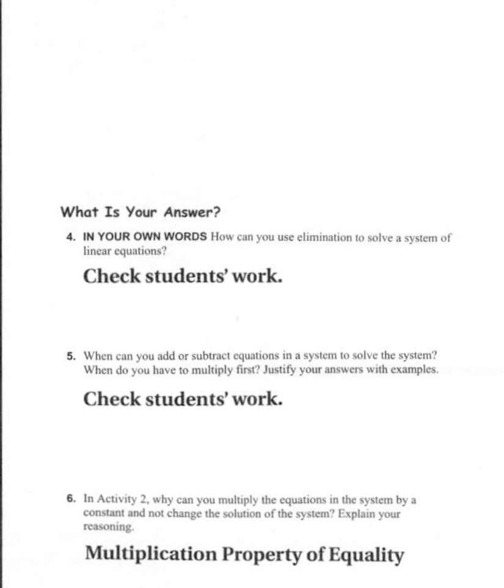

What Is Your Answer?

4. **IN YOUR OWN WORDS** How can you use elimination to solve a system of linear equations?

Check students' work.

5. When can you add or subtract equations in a system to solve the system? When do you have to multiply first? Justify your answers with examples.

Check students' work.

6. In Activity 2, why can you multiply the equations in the system by a constant and not change the solution of the system? Explain your reasoning.

Multiplication Property of Equality

Laurie's Notes

Activity 1 (continued)

- Discuss how elimination compares to the substitution method. With elimination, you add or subtract equations to eliminate a variable. With substitution, you write an equation in one variable by substituting an expression for the other variable. In both methods, you solve the resulting equation for the variable and use its value to find the other variable.

Activity 2

? **MP7 Look for and Make Use of Structure:** In parts (b) and (c), students may first think of properties such as the Commutative and Associative Properties. Refer to the first balance scale and ask, "If you double the number of cubes and balls on one side, and the weight on the other, will the scale balance?" yes "What will each side weigh?" 26 pounds

? "What property did you use?" Multiplication Property of Equality

- When students finish, discuss the process and the solutions.

Activity 3

- Several of the systems are easily solved by adding the equations to eliminate a variable. To solve the remaining systems, the students can multiply through by a constant.

? When students have finished, ask "How did you solve the system for the third letter?" Students may have multiplied the second equation by 3/2, or multiplied the first equation by 2 and the second equation by 3. Compare the two approaches.

- Well-known quote associated with Hypatia: "Reserve your right to think, for even to think wrongly is better than not to think at all."
- Hypatia (370–415) was the daughter of a mathematician and is the first known woman mathematician. Hypatia wrote many mathematical treatises, and was noted for her ability to explain complex mathematical ideas clearly.

What Is Your Answer?

- **MP7:** Students need to realize that one pair of like variable terms in a system must have the same or opposite coefficients before you can eliminate a variable by adding or subtracting the equations.

Closure

- Write a system of linear equations that can be solved easily by elimination. Write a system of linear equations that *cannot* be solved easily by elimination. Explain your reasoning.

Technology For the Teacher

The Dynamic Planning Tool

Editable Teacher's Resources at *BigIdeasMath.com*

3 ACTIVITY: Solving a Secret Code

Work with a partner. Solve the puzzle to find the name of a famous mathematician who lived in Egypt around 350 A.D.

4	B	W	R	M	F	Y	K	N
3	O	J	A	S	I	D	X	Z
2	Q	P	C	E	G	B	T	J
1	M	R	C	Z	N	O	U	W
0	K	X	U	H	L	Y	S	Q
−1	F	E	A	S	W	K	R	M
−2	G	J	Z	N	H	V	D	G
−3	E	L	X	L	F	Q	O	B
	−3	−2	−1	0	1	2	3	4

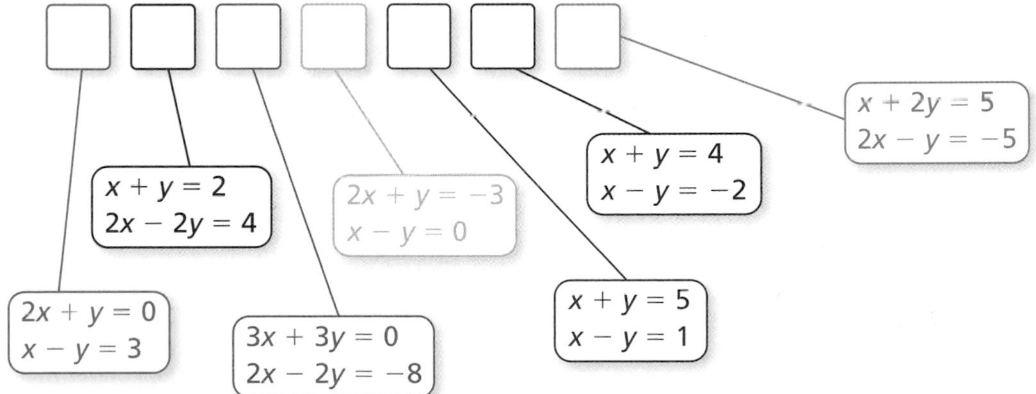

What Is Your Answer?

4. **IN YOUR OWN WORDS** How can you use elimination to solve a system of linear equations?
5. When can you add or subtract equations in a system to solve the system? When do you have to multiply first? Justify your answers with examples.
6. In Activity 2, why can you multiply equations in the system by a constant and not change the solution of the system? Explain your reasoning.

Practice Use what you learned about systems of linear equations to complete Exercises 4–6 on page 173.

4.3 Lesson

Solving a System of Linear Equations by Elimination

Step 1 Multiply, if necessary, one or both equations by a constant so at least one pair of like terms has the same or opposite coefficients.

Step 2 Add or subtract the equations to eliminate one of the variables.

Step 3 Solve the resulting equation for the remaining variable.

Step 4 Substitute the value from Step 3 into one of the original equations and solve.

EXAMPLE 1 Solving a System of Linear Equations by Elimination

Study Tip

Because the coefficients of x are the same, you can also solve the system by subtracting in Step 2.

$$\begin{aligned} x + 3y &= -2 \\ x - 3y &= 16 \\ \hline 6y &= -18 \end{aligned}$$

So, $y = -3$.

Solve the system by elimination. $\quad x + 3y = -2$ Equation 1; $\quad x - 3y = 16$ Equation 2

Step 1: The coefficients of the y-terms are already opposites.

Step 2: Add the equations.

$x + 3y = -2$ Equation 1

$x - 3y = 16$ Equation 2

$2x = 14$ Add the equations.

Step 3: Solve for x.

$2x = 14$ Equation from Step 2

$x = 7$ Divide each side by 2.

Step 4: Substitute 7 for x in one of the original equations and solve for y.

$x + 3y = -2$ Equation 1

$7 + 3y = -2$ Substitute 7 for x.

$3y = -9$ Subtract 7 from each side.

$y = -3$ Divide each side by 3.

The solution is $(7, -3)$.

Check

Equation 1

$x + 3y = -2$

$7 + 3(-3) \stackrel{?}{=} -2$

$-2 = -2$ ✓

Equation 2

$x - 3y = 16$

$7 - 3(-3) \stackrel{?}{=} 16$

$16 = 16$ ✓

On Your Own

Now You're Ready
Exercises 7–12

Solve the system of linear equations by elimination. Check your solution.

1. $2x - y = 9$; $4x + y = 21$
2. $-5x + 2y = 13$; $5x + y = -1$
3. $3x + 4y = -6$; $7x + 4y = -14$

Laurie's Notes

Introduction

Connect

- **Yesterday:** Students discovered how and when they can add or subtract the equations in a system to eliminate a variable. (MP1, MP7)
- **Today:** Students will solve systems of linear equations by elimination.

Motivate

- **Play the Opposites Game:** Pair students. Each student has a piece of paper and a pencil. Tell students that they will be given one minute to make a list of words that are opposites. One partner writes a word while the other partner writes the opposite of that word (example: hot and cold).
- Time the game for one minute.
- At the end of the game determine which pair has the longest list. Have them read their list of words, alternating between partners. Ask one or two others to share any new words from their lists.

? "What is the opposite of 4? -3.8? $7x$?" -4; 3.8; $-7x$

- Today students will use like terms whose coefficients are the same or opposite.

Lesson Notes

Discuss

- **MP3a Construct Viable Arguments:** Students have solved systems of linear equations by graphing and substitution. Elimination is a third method. Students should continue to focus on what method makes sense to use when solving a given system. They should choose a method based on the coefficients of the system and the form in which the equations are written. Students should be able to explain why they select a particular method.

Key Idea

- Discuss the steps for solving a system of linear equations by elimination. Students often find the steps very wordy, but the process is fairly simple and straightforward.
- As you work through the first example, refer back to the guidelines as you begin each step.

Example 1

- Write Equation 1 and Equation 2.

? "What are the coefficients of the x-terms?" Both are 1.

? "What are the coefficients of the y-terms?" 3 and -3

- **Teaching Tip:** Line up like terms vertically so that you can add the terms in each column.

? "When you add the equations, what is the sum?" $2x = 14$

? "How do you determine the y-coordinate of the solution?" Substitute 7 for x in one of the original equations.

- Check the solution.

Goal Today's lesson is solving a system of linear equations by elimination.

Start Thinking! and Warm Up

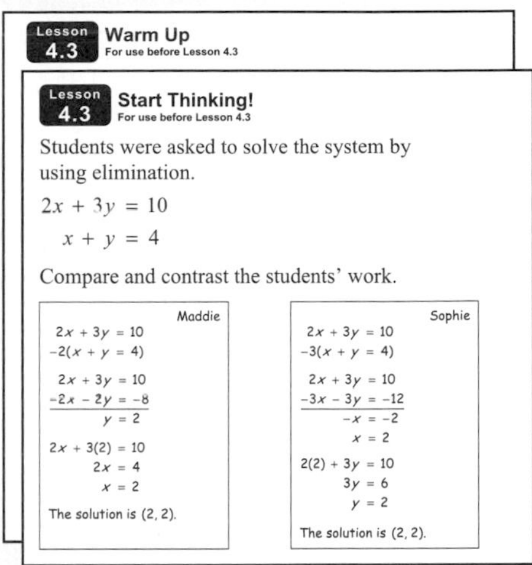

Lesson 4.3 Warm Up
For use before Lesson 4.3

Lesson 4.3 Start Thinking!
For use before Lesson 4.3

Students were asked to solve the system by using elimination.

$2x + 3y = 10$

$x + y = 4$

Compare and contrast the students' work.

Maddie	Sophie
$2x + 3y = 10$	$2x + 3y = 10$
$-2(x + y = 4)$	$-3(x + y = 4)$
$2x + 3y = 10$	$2x + 3y = 10$
$-2x - 2y = -8$	$-3x - 3y = -12$
$y = 2$	$-x = -2$
$2x + 3(2) = 10$	$x = 2$
$2x = 4$	$2(2) + 3y = 10$
$x = 2$	$3y = 6$
	$y = 2$
The solution is (2, 2).	The solution is (2, 2).

Extra Example 1

Solve the system by elimination.

$3x - y = 14$

$-3x + 4y = 16$

(8, 10)

On Your Own

1. (5, 1)
2. $(-1, 4)$
3. $(-2, 0)$

Laurie's Notes

Example 1 (continued)

- Discuss with students other methods they could use to solve the system and whether they would select that method. For instance, another way to use elimination is to subtract one equation from the other to eliminate the variable x. Graphing might not be as efficient because the equations are in standard form. Substitution could be used.

On Your Own

- **Common Error:** Question 3 does not have coefficients that are opposites. Students may subtract the equations, but neglect to handle the subtraction properly on the right side. Make sure they subtracted $-6 - (-14)$.
- Ask volunteers to share their work at the board.

Example 2

- Write the system of equations.

? "What do you notice about this system?" Listen for this answer: None of the coefficients of like variable terms are the same or opposite.

? "Can you think of a way to rewrite the equations so that either the x-terms or the y-terms have coefficients that are the same or opposite?" Multiply through one equation by a constant.

? "If we wanted to eliminate the x-term, what could we do? Multiply the second equation by -3 and add, or multiply the second equation by 3 and subtract.

- Discuss the two options: Multiplying the second equation by -3 and adding, or multiplying by 3 and subtracting.
- Work through the problem using both approaches.
- Although both approaches give you the same answers, sometimes fewer computation errors occur when equations are added rather than subtracted. Some students prefer adding equations.
- **Common Error:** When multiplying through by the constant, be sure that students multiply *every* term by the constant. It helps to use color:

$$(3)(-2x - 4y) = 14(3)$$

Remind students that they are using the Distributive Property.

? Before students try the problems on their own, ask "What if this system had been written this way?"

$$5y - 6x = 25$$
$$-2x - 4y = 14$$

Students should recognize that the like terms are not lined up in the same columns, so one of the equations should be rewritten.

On Your Own

- Have students check with their neighbors as they work through the problems. Remind students to use care when setting up the addition or subtraction. It is very frustrating to check your answer and find out that you made an error at the beginning of your solution process.

Extra Example 2

Solve the system by elimination.

$3x - 4y = 33$

$4x + 3y = 19$

$(7, -3)$

On Your Own

4. $(3, 2)$
5. $(-6, -1)$
6. $(0, 3)$

EXAMPLE 2 Solving a System of Linear Equations by Elimination

Solve the system by elimination. $-6x + 5y = 25$ Equation 1

$-2x - 4y = 14$ Equation 2

Step 1: Multiply Equation 2 by 3.

$-6x + 5y = 25$ → $-6x + 5y = 25$ Equation 1

$-2x - 4y = 14$ **Multiply by 3.** → $-6x - 12y = 42$ Revised Equation 2

Study Tip

In Example 2, notice that you can also multiply Equation 2 by −3 and then add the equations.

Step 2: Subtract the equations.

$-6x + 5y = 25$ Equation 1

$-6x - 12y = 42$ Revised Equation 2

$17y = -17$ Subtract the equations.

Step 3: Solve for y.

$17y = -17$ Equation from Step 2

$y = -1$ Divide each side by 17.

Step 4: Substitute −1 for y in one of the original equations and solve for x.

$-2x - 4y = 14$ Equation 2

$-2x - 4(-1) = 14$ Substitute −1 for y.

$-2x + 4 = 14$ Multiply.

$-2x = 10$ Subtract 4 from each side.

$x = -5$ Divide each side by −2.

∴ The solution is $(-5, -1)$.

Check

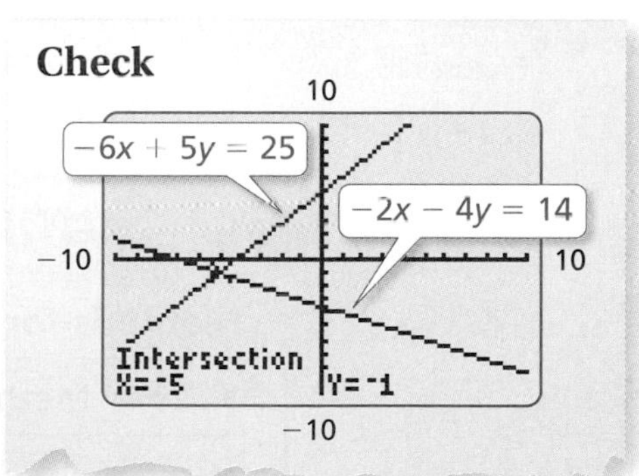

On Your Own

Solve the system of linear equations by elimination. Check your solution.

4. $3x + y = 11$
$6x + 3y = 24$

5. $4x - 5y = -19$
$-x - 2y = 8$

6. $5y = 15 - 5x$
$y = -2x + 3$

EXAMPLE 3 Real-Life Application

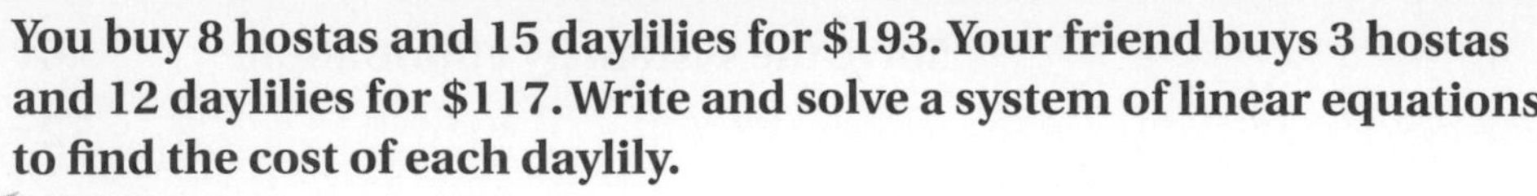

You buy 8 hostas and 15 daylilies for \$193. Your friend buys 3 hostas and 12 daylilies for \$117. Write and solve a system of linear equations to find the cost of each daylily.

Use a verbal model to write a system of linear equations.

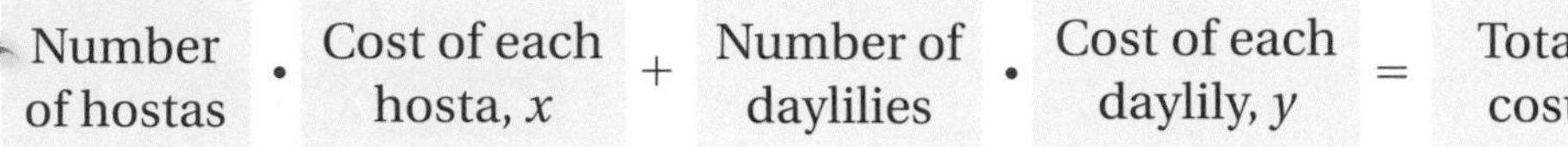

Number of hostas
•
Cost of each hosta, x
+
Number of daylilies
•
Cost of each daylily, y
=
Total cost

The system is: $8x + 15y = 193$ Equation 1 (You)

$3x + 12y = 117$ Equation 2 (Your friend)

Step 1: To find the cost y of each daylily, eliminate the x-terms. Multiply Equation 1 by 3. Multiply Equation 2 by 8.

$8x + 15y = 193$ **Multiply by 3.** → $24x + 45y = 579$ Revised Equation 1

$3x + 12y = 117$ **Multiply by 8.** → $24x + 96y = 936$ Revised Equation 2

Step 2: Subtract the revised equations.

$24x + 45y = 579$ Revised Equation 1

$\underline{24x + 96y = 936}$ Revised Equation 2

$-51y = -357$ Subtract the equations.

Step 3: Solving the equation $-51y = -357$ gives $y = 7$.

∴ Each daylily costs \$7.

On Your Own

7. A landscaper buys 4 peonies and 9 geraniums for \$190. Another landscaper buys 5 peonies and 6 geraniums for \$185. Write and solve a system of linear equations to find the cost of each peony.

Summary

Methods for Solving Systems of Linear Equations

Method	When to Use
Graphing *(Lesson 4.1)*	To estimate solutions
Substitution *(Lesson 4.2)*	When one of the variables in one of the equations has a coefficient of 1 or -1
Elimination *(Lesson 4.3)*	When at least one pair of like terms has the same or opposite coefficients
Elimination (Multiply First) *(Lesson 4.3)*	When one of the variables cannot be eliminated by adding or subtracting the equations

Laurie's Notes

Example 3

- Ask a volunteer to read and summarize the problem: Two people bought different numbers of two types of flowers and paid different amounts.
- Although the problem is to find the cost per daylily, you need a second variable for the cost per hosta.
- Discuss the verbal model used to generate each equation.

? "What does Equation 1 represent?" the amount of money you spent for your flowers

? "What does Equation 2 represent?" the amount of money your friend spent for the flowers

? "What do you notice about this system of equations that is different than the first two examples?" listen for this answer: Neither pair of like terms has a coefficient that is a multiple of the other.

- Students should recognize that 24 is the least common multiple of 8 and 3. Work through the problem as shown.
- Discuss alternative approaches to this problem.
  - Multiply Equation 2 by 8/3 and subtract to eliminate the x-terms.
  - Multiply Equation 1 by 4 and Equation 2 by 5 and subtract to eliminate the y-terms.
- **Extension:** Determine the cost of each hosta plant.

Summary

- **MP1a Make Sense of Problems** and **MP3a:** The summary box reviews the last three lessons. It is important for students to understand that the approach they use usually depends on what the system looks like. Students should be able to make wise decisions about the approach they choose and explain the reasoning behind their choice.

Closure

Write an example of a system for each condition relative to the solution by elimination.

- Adding or subtracting equations will eliminate one of the variables.
- You need to multiply one of the equations by an integer before you add or subtract.
- You need to multiply both of the equations by an integer (or one equation by a fraction) before you add or subtract.

Technology For the Teacher

The Dynamic Planning Tool
Editable Teacher's Resources at *BigIdeasMath.com*

Extra Example 3

There are 340 calories in 2 cups of cereal with 1 cup of milk. There are 570 calories in 3 cups of the cereal with 2 cups of milk. Write and solve a system of linear equations to find the number x of calories in 1 cup of the cereal without milk.

$2x + y = 340$
$3x + 2y = 570$;
110 calories

On Your Own

7. $4x + 9y = 190$
   $5x + 6y = 185$
   $x = 25$; $25 per peony

English Language Learners

Pair Activity

Pair English learners with English speakers. One partner uses an example covered in class to quiz the other about the process. The quizzer looks at the steps for solving a system by elimination in the Key Idea on page 170. Without revealing the step number, the quizzer reads one randomly chosen step at a time to the other student, and asks where the step was carried out in the example. This helps students get comfortable with the language used.

Vocabulary and Concept Check

1. **Step 1:** Multiply, if necessary, one or both equations by a constant so at least one pair of like terms has the same or opposite coefficients.
   Step 2: Add or subtract the equations to eliminate one of the variables.
   Step 3: Solve the resulting equation for the remaining variable.
   Step 4: Substitute the value from Step 3 into one of the original equations and solve.
2. Use multiplication when at least one pair of like terms are not the same or opposites.
3. $2x + 3y = 11$
   $3x - 2y = 10$;
   You have to use multiplication to solve the system by elimination.

Practice and Problem Solving

4. (2, 1)
5. (6, 2)
6. (−1, 3)
7. (2, 1)
8. (−1, 3)
9. (1, −3)
10. (4, −1)
11. (3, 2)
12. (−2, −5)
13. The student added y-terms, but subtracted x-terms and constants; solution (1, 2)
14. a. $x + y = 58$
    $x - y = 14$
    b. You sell 36 tickets, your friend sells 22 tickets.
15. a. $2x + y = 10$
    $2x + 3y = 22$
    b. 6 minutes

Assignment Guide and Homework Check

Level	Assignment	Homework Check
Average	1–9, 13, 17, 18, 25, 27, 32–35	7, 9, 18, 25, 27
Advanced	1–8, 10, 12, 13, 15, 17–20, 24, 25, 27–35	9, 15, 18, 27, 29

Common Errors

- **Exercises 4–12** Students may make careless errors. Remind students to line up like terms neatly in columns to avoid confusion about what terms to add or subtract. Also, take care with subtraction—it may help to change the sign of each term in the equation to be subtracted, so you can just add the terms.
- **Exercise 14** Students may struggle writing an equation to represent "You sell 14 more tickets than your friend." Explain that it helps to ask the question "How does your number of tickets compare to your friend's number of tickets?"

4.3 Record and Practice Journal

Solve the system of linear equations by elimination. Check your solution.

1. $x + y = 7$
   $3x - y = 1$
   (2, 5)
2. $-2x - 5y = -8$
   $-2x + y = 16$
   (−6, 4)
3. $8x - 9y = 7$
   $2x - 3y = -5$
   (11, 9)
4. $-5x + 3y = -6$
   $9x - 4y = 1$
   (−3, −7)
5. A high school has a total of 850 students. There are 60 more female students than there are male students.
   a. Write a system of linear equations that represents this situation.
   $x + y = 850$
   $x = y + 60$
   b. How many students are female? male?
   455 females; 395 males

Technology For the Teacher
Answer Presentation Tool

4.3 Exercises

Vocabulary and Concept Check

1. **WRITING** Describe how to solve a system of linear equations by elimination.
2. **NUMBER SENSE** When should you use multiplication to solve a system of linear equations by elimination?
3. **WHICH ONE DOESN'T BELONG?** Which system of equations does *not* belong with the other three? Explain your reasoning.

$3x + 3y = 3$	$-2x + y = 6$	$2x + 3y = 11$	$x + y = 5$
$2x - 3y = 7$	$2x - 3y = -10$	$3x - 2y = 10$	$3x - y = 3$

Practice and Problem Solving

Use a method from Activity 1 to solve the system.

4. $x + y = 3$
   $x - y = 1$
5. $-x + 3y = 0$
   $x + 3y = 12$
6. $3x + 2y = 3$
   $3x - 2y = -9$

Solve the system of linear equations by elimination. Check your solution.

1 7. $x + 3y = 5$
   $-x - y = -3$
8. $x - 2y = -7$
   $3x + 2y = 3$
9. $4x + 3y = -5$
   $-x + 3y = -10$
10. $2x + 7y = 1$
    $2x - 4y = 12$
11. $2x + 5y = 16$
    $3x - 5y = -1$
12. $3x - 2y = 4$
    $6x - 2y = -2$

13. **ERROR ANALYSIS** Describe and correct the error in solving the system of linear equations.

$5x + 2y = 9$ Equation 1
$3x - 2y = -1$ Equation 2
$2x \quad = 10$
$x = 5$
The solution is $(5, -8)$.

14. **RAFFLE TICKETS** You and your friend are selling raffle tickets for a new laptop. You sell 14 more tickets than your friend sells. Together, you and your friend sell 58 tickets.
    a. Write a system of linear equations that represents this situation.
    b. How many tickets do each of you sell?

15. **JOGGING** You can jog around your block twice and the park once in 10 minutes. You can jog around your block twice and the park 3 times in 22 minutes.
    a. Write a system of linear equations that represents this situation.
    b. How long does it take you to jog around the park?

Solve the system of linear equations by elimination. Check your solution.

2 3 **16.** $2x - y = 0$
$3x - 2y = -3$

17. $x + 4y = 1$
$3x + 5y = 10$

18. $-2x + 3y = 7$
$5x + 8y = -2$

19. $3x + 3 = 3y$
$2x - 6y = 2$

20. $2x - 6 = 4y$
$7y = -3x + 9$

21. $5x = 4y + 8$
$3y = 3x - 3$

22. ERROR ANALYSIS Describe and correct the error in solving the system of linear equations.

✗

$x + y = 1$ Equation 1 **Multiply by −5.** $-5x + 5y = -5$

$5x + 3y = -3$ Equation 2 $5x + 3y = -3$

$8y = -8$

$y = -1$

The solution is (2, −1).

23. REASONING For what values of a and b should you solve the system by elimination?

a. $4x - y = 3$
$ax + 10y = 6$

b. $x - 7y = 6$
$-6x + by = 9$

24. AIRPLANES Two airplanes are flying to the same airport. Their positions are shown in the graph. Write a system of linear equations that represents this situation. Solve the system by elimination to justify your answer.

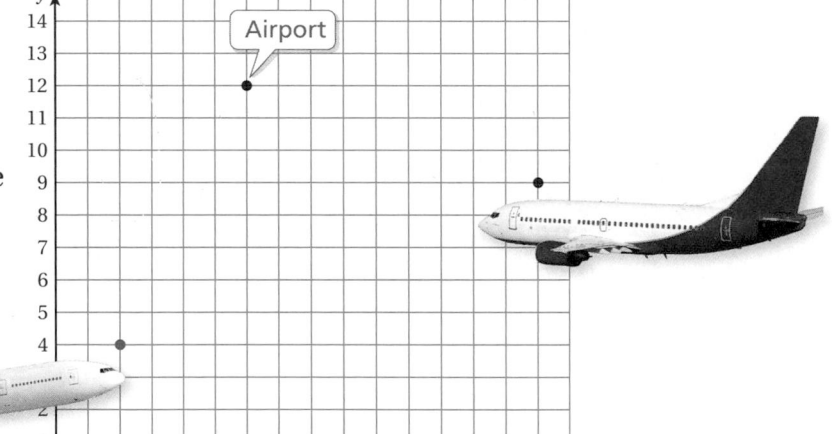

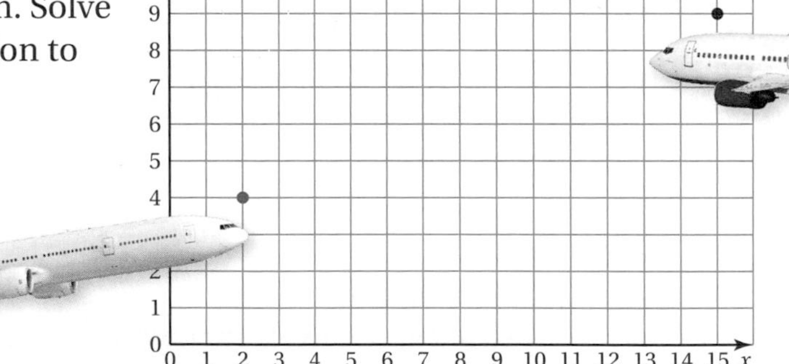

25. TEST PRACTICE The table shows the number of correct answers on a practice standardized test. You score 86 points on the test and your friend scores 76 points.

	You	Your Friend
Multiple Choice	23	28
Short Response	10	5

a. Write a system of linear equations that represents this situation.

b. How many points is each type of question worth?

Common Errors

- **Exercises 16–21** Students may make errors in multiplying through the whole equation. Tell them to clearly show what they are multiplying each equation by so they do not forget during the process. They can check the sign and coefficient of each term after multiplying.
- **Exercise 24** Students may not write a correct system of equations for the situation. Ask them to consider what line each airplane has to take to fly to the airport.
- **Exercise 25** Students may write a system of equations by reading across the table horizontally. Stress that your total number of points is equal to the sum of your points for multiple choice questions and for short response questions which are both listed in the column for "You."
- **Exercise 27** Students may use an incorrect problem-solving strategy. Make sure they realize that they first need to solve for the cost per hour for each activity, then use those rates to answer the question.
- **Exercise 29** Students may fail to write an equation for the situation using the gold percents. Explain that the total amount of gold contributed by each alloy is equal to the amount of gold in the final mixture. Each amount of gold is given by the percent times the number of ounces.
- **Exercise 30** Students may not write a system of equations that represents the problem correctly. Tell them this is an extension of the formula $d = r \times t$ where the rate is increased or decreased by the speed of the current.
- **Exercise 31** Students may use an incorrect approach. Tell them to look for a way to add a multiple of one equation to another equation to eliminate two of the variables. This will allow them to solve for the third variable. They can then substitute the value of this variable back into two of the equations to write a new system of two equations in two variables, which they know how to solve.

16. (3, 6) **17.** (5, −1)

18. (−2, 1) **19.** (−2, −1)

20. (3, 0) **21.** (4, 3)

22. The y-term was multiplied by 5 instead of −5; solution: (−3, 4)

23. **a.** ±4

b. ±7

24. $y = 2x$

$y = -\frac{1}{3}x + 14;$

Solution: (6, 12)

25. **a.** $23x + 10y = 86$

$28x + 5y = 76$

b. Multiple choice: 2 points each; Short response: 4 points each

English Language Learners

Auditory

Ask students to explain how they used the table to set up their system of equations in Exercise 25. Then ask students to explain how they used the table to set up their system of equations in Exercise 27. Discuss the general factors you need to take into account in using a table to write the equations for a system of equations.

Practice and Problem Solving

26. no; You cannot sell -6 tickets.

27. \$95

28. *Sample answer:* Find the line perpendicular to $2x + y = 0$ through $(2, -4)$; $x - 2y = 10$; Solution: $(2, -4)$

29. 5 grams of 90% gold alloy, 3 grams of 50% gold alloy

30. See *Taking Math Deeper*.

31. $(-1, 2, 1)$

Fair Game Review

32. yes **33.** yes

34. no **35.** D

Mini-Assessment

Solve the system of linear equations by elimination.

1. $5x - 2y = 18$
$-5x + 3y = -22$ $(2, -4)$

2. $2x + 4y = 20$
$-3x + 4y = 30$ $(-2, 6)$

3. $4x - 2y = 2$
$7x - 3y = 6$ $(3, 5)$

4. You have 33 quarters and dimes in a jar. The jar contains a total of \$4.95. Write and solve a system of equations to find the number x of dimes and the number y of quarters.
$x + y = 33$
$0.1x + 0.25y = 4.95$;
$(22, 11)$ or 22 dimes and 11 quarters

Taking Math Deeper

Exercise 30

This is a classic riverboat problem. Students may find the problem complicated, so it helps to organize the information in a table.

Make a table.

	Distance	Rate	Time
Downstream	10 mi	$(B + C)$ mi/h	1/2 h
Upstream	10 mi	$(B - C)$ mi/h	5/6 h

Distance: The boat travels 10 miles in each direction.

Time: The trip downstream takes 30 minutes and the trip upstream takes 50 minutes. Because speed is measured in miles per hour, think of 30 minutes as 1/2 hour and 50 minutes as 5/6 hour.

Rate: The boat's speed B relative to the water is increased or decreased by the rate C of the current. So, the boat's actual rate of speed is $B + C$ downstream and $B - C$ upstream.

Now use the formula $d = r \times t$ to write and solve a set of equations.

Write a set of equations.

$10 = \frac{1}{2}(B + C)$ **Multiply by 2.** $20 = B + C$ Downstream

$10 = \frac{5}{6}(B - C)$ **Multiply by 6/5.** $12 = B - C$ Upstream

$32 = 2B$ Add equations.

The speed of the boat relative to the water is $B = 32/2 = 16$ miles per hour.

Substitute.

Substitute 16 for B in $20 = B + C$ and solve for C.

$$20 = B + C$$
$$20 = 16 + C$$
$$4 = C$$

The speed of the current is 4 miles per hour.

Reteaching and Enrichment Strategies

If students need help...	If students got it...
Resources by Chapter • Practice A and Practice B • Puzzle Time Record and Practice Journal Practice Differentiating the Lesson Lesson Tutorials Skills Review Handbook	Resources by Chapter • Enrichment and Extension • School-to-Work Start the next section

26. LOGIC You solve a system of equations in which x represents the number of adult tickets sold and y represents the number of student tickets sold. Can $(-6, 24)$ be the solution of the system? Explain your reasoning.

27. VACATION The table shows the activities of two tourists at a vacation resort. You want to go parasailing for one hour and horseback riding for two hours. How much do you expect to pay?

	Parasailing	Horseback Riding	Total Cost
Tourist 1	2 hours	5 hours	\$205
Tourist 2	3 hours	3 hours	\$240

28. REASONING The solution of a system of linear equations is $(2, -4)$. One equation in the system is $2x + y = 0$. Explain how you could find a second equation for the system. Then find a second equation. Solve the system by elimination to justify your answer.

29. JEWELER A metal alloy is a mixture of two or more metals. A jeweler wants to make 8 grams of 18 carat gold, which is 75% gold. The jeweler has an alloy that is 90% gold and an alloy that is 50% gold. How much of each alloy should the jeweler use?

30. PROBLEM SOLVING A power boat takes 30 minutes to travel 10 miles downstream. The return trip takes 50 minutes. What is the speed of the current?

31. Critical Thinking Solve the system of equations by elimination.

$$2x - y + 3z = -1$$
$$x + 2y - 4z = -1$$
$$y - 2z = 0$$

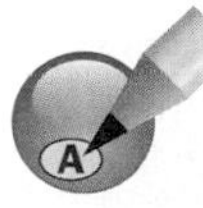

Fair Game Review What you learned in previous grades & lessons

Decide whether the two equations are equivalent. *(Section 1.2 and Section 1.3)*

32. $4n + 1 = n - 8$

$3n = -9$

33. $2a + 6 = 12$

$a + 3 = 6$

34. $7v - \frac{3}{2} = 5$

$14v - 3 = 15$

35. MULTIPLE CHOICE Which line has the same slope as $y = \frac{1}{2}x - 3$? *(Section 2.3)*

Ⓐ $y = -2x + 4$ Ⓑ $y = 2x + 3$ Ⓒ $y - 2x = 5$ Ⓓ $2y - x = 7$

4.4 Solving Special Systems of Linear Equations

Essential Question Can a system of linear equations have no solution? Can a system of linear equations have many solutions?

COMMON CORE STATE STANDARDS
8.EE.8a
8.EE.8b
8.EE.8c
A.CED.3
A.REI.6
A.REI.11

1 ACTIVITY: Writing a System of Linear Equations

Work with a partner. Your cousin is 3 years older than you. Your ages can be represented by two linear equations.

$y = t$ Your age

$y = t + 3$ Your cousin's age

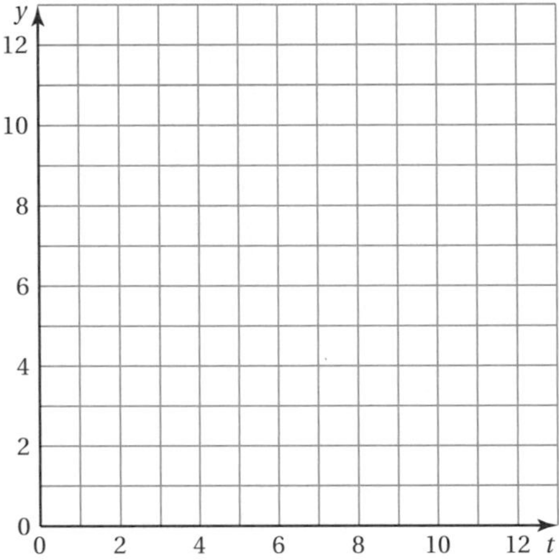

a. Graph both equations in the same coordinate plane.

b. What is the vertical distance between the two graphs? What does this distance represent?

c. Do the two graphs intersect? If not, what does this mean in terms of your age and your cousin's age?

2 ACTIVITY: Using a Table to Solve a System

Work with a partner. You invest \$500 for equipment to make dog backpacks. Each backpack costs you \$15 for materials. You sell each backpack for \$15.

a. Copy and complete the table for your cost *C* and your revenue *R*.

x	0	1	2	3	4	5	6	7	8	9	10
C											
R											

b. When will your company break even? What is wrong?

Laurie's Notes

Introduction

Standards for Mathematical Practice

- **MP1 Make Sense of Problems and Persevere in Solving Them:** Students have learned three techniques for solving a system of linear equations. In this lesson, they use more than one technique to solve the same system, and in doing so recognize that not all systems have a single solution. Using a variety of strategies helps students make sense of the new possible outcomes for a solution of a system of equations.

Motivate

- If you have an overhead projector or document camera, place 2 pieces of spaghetti on display. Say, "These represent two lines. Right now they are intersecting."

? "Is there any other relationship they could have?" Listen for parallel (non-intersecting) and a reasonable description of coinciding lines.

- Now place a transparency of a coordinate grid on top of the spaghetti.
- Suggest that lines, when graphed, do not always have to intersect.
- **Management Tip:** I use spaghetti for many models throughout the year, so I keep a box of spaghetti in my desk.

Activity Notes

Activity 1

? "How do you measure the vertical distance between two graphs?" Use a vertical line on the graph to count the units from one graph to the other.

- **Part (c):** Students will likely say that the two are not the same age. The solution actually says more: you and your cousin will never be the same age at the same time. Your cousin will always be 3 years older.

Activity 2

- Have students write the cost equation and the revenue equation. Students should turn back to the previous lesson as needed for help.
- Give time for students to work with their partner to fill in the table. Make sure that all students are using correct equations.

? **Extension:** "What patterns do you observe in the table?" Both rows of numbers are increasing by the same amount (15). The values are not getting closer together.

- Discuss student answers for part (b). Students should recognize that when you sell an item for the exact price that it costs, you will never make a profit, nor pay off the original investment of $500.

? **Extension:** "If you were to graph this system, what would you expect the graph to look like?" parallel lines; same slope

Common Core State Standards

A.CED.3 Represent constraints by systems of equations and interpret solutions as viable or nonviable options in a modeling context.

A.REI.6 Solve systems of linear equations exactly and approximately, focusing on pairs of linear equations in two variables.

A.REI.11 Explain why the x-coordinates of the points where the graphs of the equations $y = f(x)$ and $y = g(x)$ intersect are the solutions of the equation $f(x) = g(x)$; find the solutions approximately, using technology to graph the functions.

Also **8.EE.8a, b, and c**

Previous Learning

Students should know how to solve equations with variables on both sides.

Start Thinking! and Warm Up

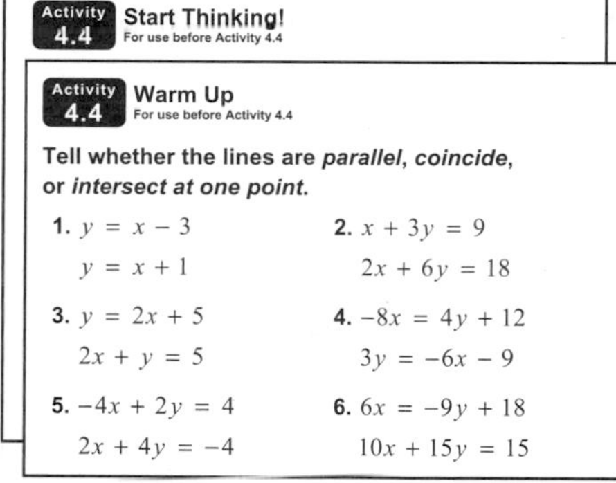

Activity 4.4 Start Thinking! For use before Activity 4.4

Activity 4.4 Warm Up For use before Activity 4.4

Tell whether the lines are *parallel*, *coincide*, or *intersect at one point*.

1. $y = x - 3$, $y = x + 1$
2. $x + 3y = 9$, $2x + 6y = 18$
3. $y = 2x + 5$, $2x + y = 5$
4. $-8x = 4y + 12$, $3y = -6x - 9$
5. $-4x + 2y = 4$, $2x + 4y = -4$
6. $6x = -9y + 18$, $10x + 15y = 15$

4.4 Record and Practice Journal

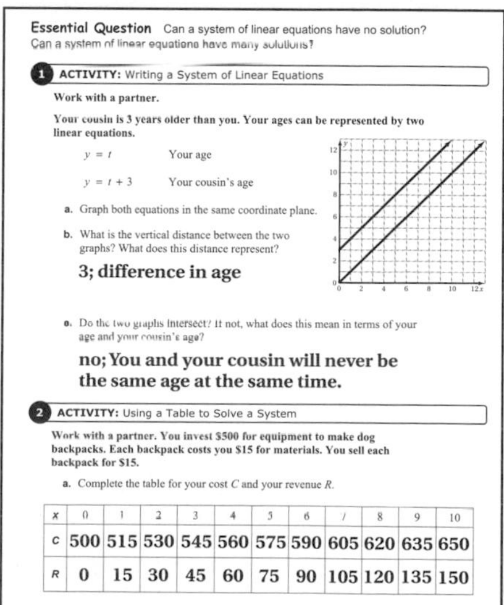

Essential Question Can a system of linear equations have no solution? Can a system of linear equations have many solutions?

1 ACTIVITY: Writing a System of Linear Equations

Work with a partner.

Your cousin is 3 years older than you. Your ages can be represented by two linear equations.

$y = t$ Your age

$y = t + 3$ Your cousin's age

a. Graph both equations in the same coordinate plane.

b. What is the vertical distance between the two graphs? What does this distance represent?

3; difference in age

c. Do the two graphs intersect? If not, what does this mean in terms of your age and your cousin's age?

no; You and your cousin will never be the same age at the same time.

2 ACTIVITY: Using a Table to Solve a System

Work with a partner. You invest $500 for equipment to make dog backpacks. Each backpack costs you $15 for materials. You sell each backpack for $15.

a. Complete the table for your cost C and your revenue R.

x	0	1	2	3	4	5	6	7	8	9	10
C	500	515	530	545	560	575	590	605	620	635	650
R	0	15	30	45	60	75	90	105	120	135	150

Differentiated Instruction

Visual

To help students remember that lines with the same slope are parallel, write "If the slopes on the board are =, then the lines are parallel." Relate that the segments that make up the equal sign are parallel segments.

4.4 Record and Practice Journal

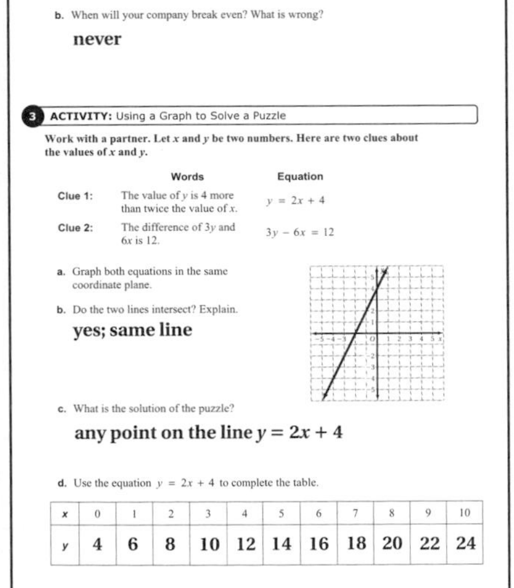

b. When will your company break even? What is wrong?

never

3 ACTIVITY: Using a Graph to Solve a Puzzle

Work with a partner. Let x and y be two numbers. Here are two clues about the values of x and y.

	Words	Equation
Clue 1:	The value of y is 4 more than twice the value of x.	$y = 2x + 4$
Clue 2:	The difference of $3y$ and $6x$ is 12.	$3y - 6x = 12$

a. Graph both equations in the same coordinate plane.

b. Do the two lines intersect? Explain.

yes; same line

c. What is the solution of the puzzle?

any point on the line $y = 2x + 4$

d. Use the equation $y = 2x + 4$ to complete the table.

x	0	1	2	3	4	5	6	7	8	9	10
y	4	6	8	10	12	14	16	18	20	22	24

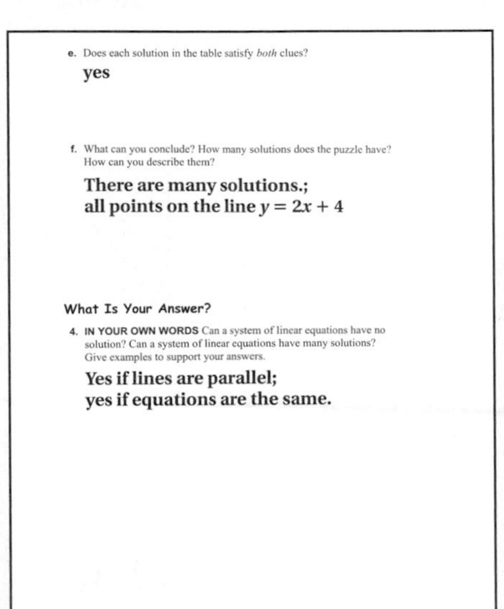

e. Does each solution in the table satisfy *both* clues?

yes

f. What can you conclude? How many solutions does the puzzle have? How can you describe them?

There are many solutions.; all points on the line $y = 2x + 4$

What Is Your Answer?

4. IN YOUR OWN WORDS Can a system of linear equations have no solution? Can a system of linear equations have many solutions? Give examples to support your answers.

Yes if lines are parallel; yes if equations are the same.

Laurie's Notes

Activity 3

- Students enjoy puzzles. Present the next activity in this context.
- ? "Look at the words for each clue and then look at the equation. Does the translation make sense? Explain." Listen to student explanations.
- ? "In what form is each equation written? What strategy can be used to graph each equation?" Listen to student explanations.
- Provided students have graphed the equations carefully, the lines should coincide. There is only one graph.
- Discuss answers to part (f). The fact that the two graphs coincide means the equations are equivalent and have the same graph. There are many solutions to the puzzle! Help students recognize this by selecting two or three ordered pairs from the table. Talk through the substitution of the ordered pair into each equation. All of the ordered pairs will be solutions of both equations.
- **Extension:** Students have heard of puzzles where you start with a number, do a few computations, and eventually end up with a number twice the original. Here's an example. (I have annotated the algebraic representation to the right.) Suggest to students that they start with a small number so that the computation is simple.

Step 1: Pick a number, perhaps your age.	x
Step 2: Add 10.	$x + 10$
Step 3: Multiply by 4.	$4(x + 10) = 4x + 40$
Step 4: Divide by 2.	$(4x + 40) \div 2 = 2x + 20$
Step 5: Subtract 20.	$2x + 20 - 20 = 2x$

Announce that you should now have a number that is twice what you started with. In relation to this lesson, you would have:

$$y = \frac{4(x + 10)}{2} - 20 \text{ and } y = 2x.$$

The graph of both lines is $y = 2x$.

What Is Your Answer?

- This question tries to help students summarize the two additional cases for the solution of a system of linear equations.

Closure

- Sketch a graph of a system of equations that has no solution.
- Sketch a graph of a system of equations that has many solutions.

Technology For the Teacher

The Dynamic Planning Tool

Editable Teacher's Resources at *BigIdeasMath.com*

3 ACTIVITY: Using a Graph to Solve a Puzzle

Work with a partner. Let x and y be two numbers. Here are two clues about the values of x and y.

	Words	Equation
Clue 1:	y is 4 more than twice the value of x.	$y = 2x + 4$
Clue 2:	The difference of $3y$ and $6x$ is 12.	$3y - 6x = 12$

a. Graph both equations in the same coordinate plane.

b. Do the two lines intersect? Explain.

c. What is the solution of the puzzle?

d. Use the equation $y = 2x + 4$ to complete the table.

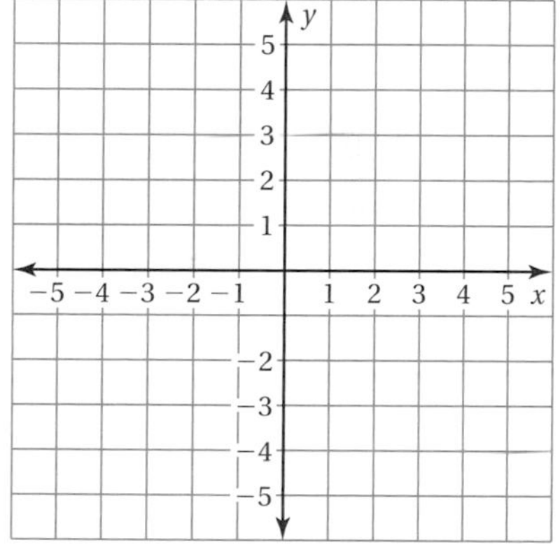

x	0	1	2	3	4	5	6	7	8	9	10
y											

e. Does each solution in the table satisfy *both* clues?

f. What can you conclude? How many solutions does the puzzle have? How can you describe them?

What Is Your Answer?

4. **IN YOUR OWN WORDS** Can a system of linear equations have no solution? Can a system of linear equations have many solutions? Give examples to support your answers.

Practice Use what you learned about special systems of linear equations to complete Exercises 3 and 4 on page 180.

4.4 Lesson

Solutions of Systems of Linear Equations

A system of linear equations can have *one solution, no solution,* or *infinitely many solutions.*

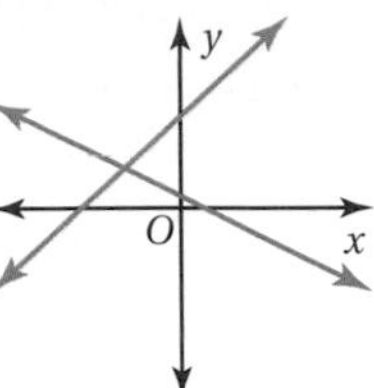

One solution

The lines intersect.

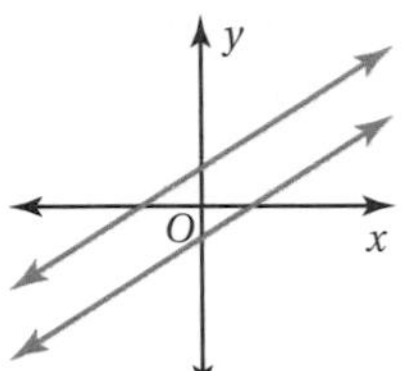

No solution

The lines are parallel.

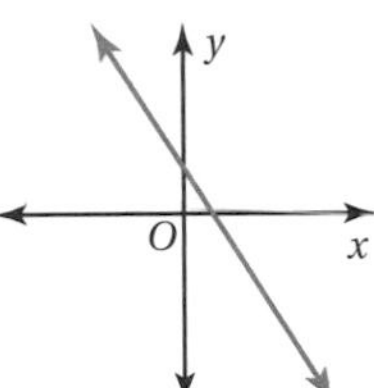

Infinitely many solutions

The lines are the same.

EXAMPLE 1 Solving a System: No Solution

Solve the system. $\quad y = 3x + 1 \quad$ Equation 1

$\qquad\qquad\qquad\quad y = 3x - 5 \quad$ Equation 2

Method 1: Solve by graphing.

Graph each equation.

The lines have the same slope and different y-intercepts. So, the lines are parallel.

Because parallel lines do not intersect, there is no point that is a solution of both equations.

So, the system of linear equations has no solution.

Method 2: Solve by substitution.

Substitute $3x - 5$ for y in Equation 1.

$y = 3x + 1$ Equation 1

$3x - 5 = 3x + 1$ Substitute $3x - 5$ for y.

$-5 \neq 1$ ✗ Subtract $3x$ from each side.

The equation $-5 = 1$ is never true. So, the system of linear equations has no solution.

On Your Own

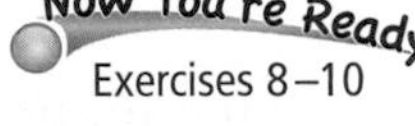

Solve the system of linear equations. Check your solution.

1. $y = -x + 3$
   $y = -x + 5$
2. $y = -5x - 2$
   $5x + y = 0$
3. $x = 2y + 10$
   $2x + 3y = -1$

Laurie's Notes

Introduction

Connect

- **Yesterday:** Students explored the graphs of two special systems of linear equations. (MP1)
- **Today:** Students will solve special systems of linear equations.

Motivate

- Make a set of cards in advance: equation card, slope card, and y-intercept card. Make enough cards so that everyone in the class has a card.

Examples:

$y = 3x - 4$	$m = 3$	$b = -4$
$3x + y = 4$	$m = -3$	$b = 4$
$y = 4x - 2$	$m = 4$	$b = -2$
$y = 2 - 4x$	$m = -4$	$b = 2$

? "What is the general form of an equation written in slope-intercept form?" $y = mx + b$

- Distribute all of the cards and have students form a matching set of 3.
- The goal of this quick activity is to review the slope-intercept form of an equation.

Lesson Notes

Key Idea

- Write the Key Idea. Connect this back to the spaghetti used yesterday to introduce special systems of linear equations.

Example 1

? "How do you graph equations in slope-intercept form?" Plot the point for the y-intercept and then use the slope to locate two additional points.

- Because both equations are in slope-intercept form, students will observe the same slope for each equation and will not be surprised that the lines are parallel. Graphing is a visual check that the lines are parallel.

? "What if you did not notice that the lines had the same slope and you tried to solve the system algebraically? What do you think will happen?" Students are not likely to have a sense for what will happen.

- **MP7 Look for and Make Use of Structure:** Because both equations are solved for y, substitution is an approach that makes sense.
- Work through Method 2. Point out that solving the system algebraically leads to a statement $-5 = 1$ that is never true. You can interpret this false statement to mean that the system has no solutions, because there are no ordered pairs that satisfy both equations. So, the lines must be parallel.

On Your Own

- Students may quickly observe that the first two systems have no solution because the slopes are the same, so the lines are parallel. Make sure they base their answers on sound reasoning.

Goal Today's lesson is solving a system of equations with no solution or infinitely many solutions.

Lesson Materials	
Introduction	**Textbook**
• cards	• straightedge

Start Thinking! and Warm Up

Lesson 4.4 **Start Thinking!** For use before Lesson 4.4

How is solving a linear equation with no solution similar to solving a system of linear equations with no solution? How is it different?

How is solving a linear equation whose solution is all real numbers similar to solving a system of linear equations with infinitely many solutions? How it is different?

Extra Example 1

Solve the system.

$y = -2x + 5$

$y = -2x + 1$

no solution

On Your Own

1. no solution
2. no solution
3. $(4, -3)$

Extra Example 2

The perimeter of the trapezoid is 10 units. The perimeter of the triangle is 5 units. Write and solve a linear system to find the values of x and y.

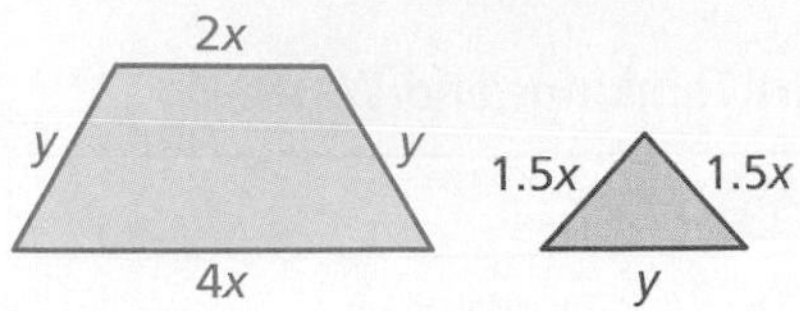

infinitely many solutions

On Your Own

4. (0, 3)
5. no solution
6. infinitely many solutions; all points on the line $y = \frac{1}{2}x - \frac{5}{2}$
7. There is no solution because the resulting equation $y = -\frac{1}{2}x + \frac{27}{4}$ is parallel to the equation of the perimeter of the triangle.

Laurie's Notes

Example 2

- Ask a volunteer to read the problem. Check to see that students are comfortable with finding the perimeter of a triangle and a rectangle.
- Write the equations. Simplify each.
- Students can rewrite each equation in slope-intercept form to graph. They will notice that the equations are the same.

? "What do you think will happen if we try to solve the system algebraically?" Answers will vary.

- **MP8 Look for and Express Regularity in Repeated Reasoning:** Work through Method 2. Point out that solving the system algebraically leads to a statement that is always true, 0 = 0. You can interpret this identity to mean that the system has infinitely many solutions, because there are infinitely many ordered pairs that satisfy both equations and the lines coincide.
- Note that there are infinitely many solutions of this system. However, they are limited to Quadrant I because the perimeter cannot be negative.

On Your Own

- **Think-Pair-Share:** Students should read each question independently and then work with a partner to answer the questions. When they have answered the questions, the pair should compare their answers with another group and discuss any discrepancies.

Closure

- **Exit Ticket:** Write a system of equations that will have no solution. *Sample answer:* $y = 2x + 4$ and $y = 2x - 6$

Differentiated Instruction

Visual

Make a poster of the Key Idea box on the number of solutions of systems of linear equations. Use color to help English learners make connections between concepts and language.

Technology For the Teacher

The Dynamic Planning Tool
Editable Teacher's Resources at *BigIdeasMath.com*

EXAMPLE 2 Solving a System: Infinitely Many Solutions

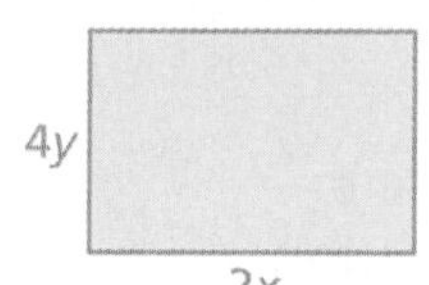

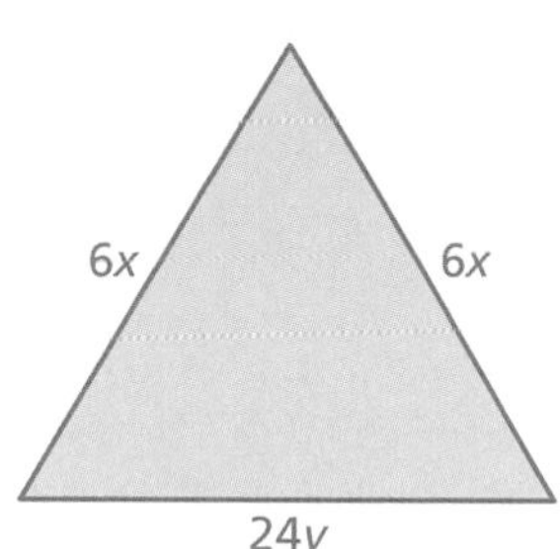

The perimeter of the rectangle is 36 units. The perimeter of the triangle is 108 units. Write and solve a system of linear equations to find the values of *x* and *y*.

Perimeter of rectangle		*Perimeter of triangle*	
$2(2x) + 2(4y) = 36$		$6x + 6x + 24y = 108$	
$4x + 8y = 36$	Equation 1	$12x + 24y = 108$	Equation 2

The system is: $4x + 8y = 36$ Equation 1

$12x + 24y = 108$ Equation 2

Method 1: Solve by graphing.

Graph each equation.

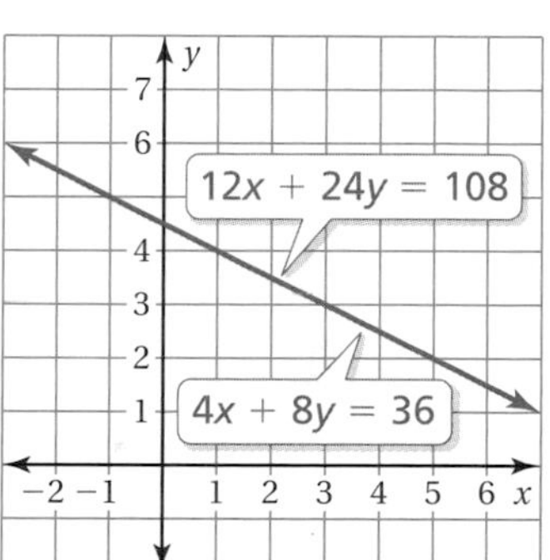

The lines have the same slope and the same *y*-intercept. So, the lines are the same.

Because the lines are the same, all the points on the line are solutions of both equations. So, the system of linear equations has infinitely many solutions.

Method 2: Solve by elimination.

Multiply Equation 1 by 3 and subtract the equations.

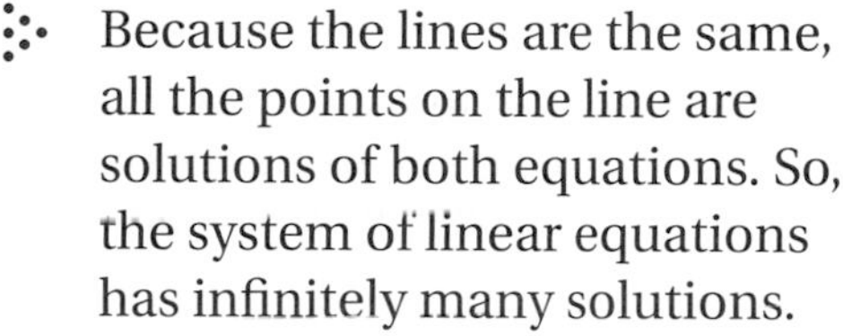

$4x + 8y = 36$ → $12x + 24y = 108$	Revised Equation 1
$12x + 24y = 108$ → $12x + 24y = 108$	Equation 2
$0 = 0$	Subtract.

The equation $0 = 0$ is always true. So, the solutions are all the points on the line $4x + 8y = 36$. The system of linear equations has infinitely many solutions.

On Your Own

Now You're Ready
Exercises 11–13

Solve the system of linear equations. Check your solution.

4. $x + y = 3$
$x - y = -3$

5. $2x + y = 5$
$4x + 2y = 0$

6. $2x - 4y = 10$
$-12x + 24y = -60$

7. **WHAT IF?** What happens to the solution in Example 2 if the perimeter of the rectangle is 54 units? Explain.

4.4 Exercises

Vocabulary and Concept Check

1. **WRITING** Describe the difference between the graph of a system of linear equations that has *no solution* and the graph of a system of linear equations that has *infinitely many solutions.*

2. **REASONING** When solving a system of linear equations algebraically, how do you know when the system has *no solution*? *infinitely many solutions*?

Practice and Problem Solving

Let x and y be two numbers. Find the solution of the puzzle.

3. y is $\frac{1}{3}$ more than 4 times the value of x.

   The difference of $3y$ and $12x$ is 1.

4. $\frac{1}{2}$ of x plus 3 is equal to y.

   x is 6 more than twice the value of y.

Without graphing, determine whether the system of linear equations has *one solution, infinitely many solutions,* or *no solution.* Explain your reasoning.

5. $y = 5x - 9$
   $y = 5x + 9$

6. $y = 6x + 2$
   $y = 3x + 1$

7. $y = 8x - 2$
   $y - 8x = -2$

Solve the system of linear equations. Check your solution.

1 8. $y = 2x - 2$
   $y = 2x + 9$

9. $y = 3x + 1$
   $-x + 2y = -3$

10. $y = \frac{\pi}{3}x + \pi$
    $-\pi x + 3y = -6\pi$

2 11. $y = -\frac{1}{6}x + 5$
    $x + 6y = 30$

12. $\frac{1}{3}x + y = 1$
    $2x + 6y = 6$

13. $-2x + y = 1.3$
    $2(0.5x - y) = 4.6$

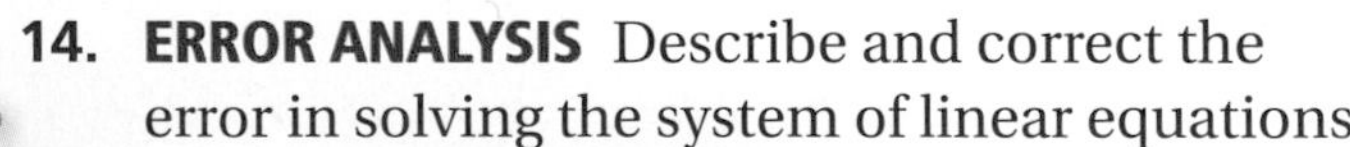

14. **ERROR ANALYSIS** Describe and correct the error in solving the system of linear equations.

$y = -2x + 4$
$y = -2x + 6$

The lines have the same slope so there are infinitely many solutions.

15. **PIG RACE** In a pig race, your pig gets a head start of 3 feet and is running at a rate of 2 feet per second. Your friend's pig is also running at a rate of 2 feet per second. A system of linear equations that represents this situation is $y = 2x + 3$ and $y = 2x$. Will your friend's pig catch up to your pig? Explain.

Assignment Guide and Homework Check

Level	Assignment	Homework Check
Average	1–9, 11, 12, 14, 15, 16, 21, 23–26	2, 3, 9, 21
Advanced	1–13, 17–26	4, 8, 17, 21

Common Errors

- **Exercises 5–13** Students may see that the slope is the same for both equations and immediately say that the system of linear equations has no solution. Remind them that they need to compare the slope *and* y-intercepts when determining the number of solutions.
- **Exercises 9–13** Students may make calculation errors when solving the equations for y. Encourage them to be careful when solving for y.

4.4 Record and Practice Journal

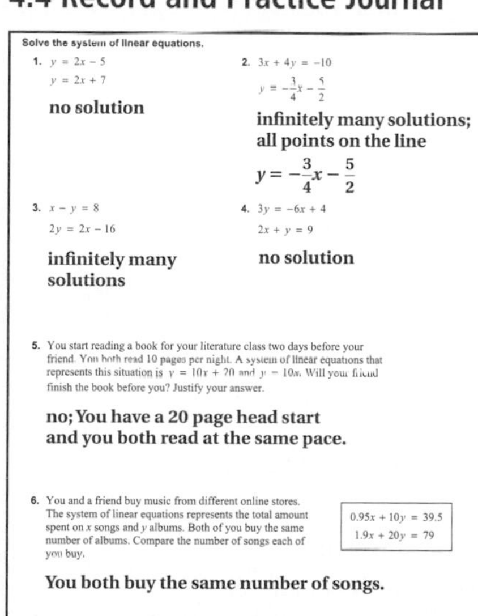

Solve the system of linear equations.

1. $y = 2x - 5$
   $y = 2x + 7$
   no solution

2. $3x + 4y = -10$
   $y = -\frac{3}{4}x - \frac{5}{2}$
   infinitely many solutions; all points on the line $y = -\frac{3}{4}x - \frac{5}{2}$

3. $x - y = 8$
   $2y = 2x - 16$
   infinitely many solutions

4. $3y = -6x + 4$
   $2x + y = 9$
   no solution

5. You start reading a book for your literature class two days before your friend. You both read 10 pages per night. A system of linear equations that represents this situation is $y = 10x + 20$ and $y = 10x$. Will your friend finish the book before you? Justify your answer.
   no; You have a 20 page head start and you both read at the same pace.

6. You and a friend buy music from different online stores. The system of linear equations represents the total amount spent on x songs and y albums. Both of you buy the same number of albums. Compare the number of songs each of you buy.
   $0.95x + 10y = 39.5$
   $1.9x + 20y = 79$
   You both buy the same number of songs.

Vocabulary and Concept Check

1. The graph of a system with no solution is two parallel lines, and the graph of a system with infinitely many solutions is one line.
2. When solving a system of linear equations algebraically, you know the system has no solution when you reach an invalid statement such as $-7 = 2$; infinitely many solutions: a valid statement such as $2 = 2$.

Practice and Problem Solving

3. infinitely many solutions; all points on the line $y = 4x + \frac{1}{3}$
4. no solution
5. no solution; The lines have the same slope and different y-intercepts.
6. one solution; The lines have different slopes.
7. infinitely many solutions; The lines are identical.
8. no solution
9. $(-1, -2)$
10. no solution
11. infinitely many solutions; all points on the line $y = -\frac{1}{6}x + 5$
12. infinitely many solutions; all points on the line $y = \frac{1}{3}x + 1$
13. $(-2.4, -3.5)$
14. There are different y-intercepts, so the lines are parallel; no solution
15. no; because they are running at the same speed and your pig had a head start

Practice and Problem Solving

16. one solution; Because the lines have different slopes, they will intersect in one point.

17. See Additional Answers.

18. a. 6 h

    b. You both work the same number of hours.

19. $y = 0.99x + 10$

    $y = 0.99x$

    no; Because you paid \$10 before buying the same number of songs at the same price, you spend \$10 more.

20. $a = b$: always; The lines are parallel; $a \geq b$: sometimes; The lines are parallel for $a = b$; $a < b$: never; The lines have different slopes and different y-intercepts.

21. See *Taking Math Deeper*.

22. $a = 2$, $b = 2$; yes; Both equations are the same.

Fair Game Review

23–25. See Additional Answers.

26. B

Mini-Assessment

Solve the system of linear equations.

1. $2x + 3y = 5$

   $2x + 3y = 7$

   no solution

2. $x + 2y = 12$

   $y = -\frac{1}{2}x + 6$

   infinitely many solutions; all points on the line $y = -\frac{1}{2}x + 6$

3. $-3x + 2y = 2$

   $4x + 3y = 20$

   one solution; (2, 4)

Taking Math Deeper

Exercise 21

You can approach this problem using the Guess-and-Test method, by finding a set of possible costs for Group 1 and testing them for Group 2.

Guess.

Let L = the price of a lift ticket and S = the price of a ski rental. Make a guess using $L = 10$. Find the corresponding value of S for Group 1.

$36L + 18S = 684$	Equation for Group 1
$36(10) + 18S = 684$	Substitute 10 for L.
$18S = 324$	Subtract 360 from each side.
$S = 18$	Divide each side by 18.

Group 1 could have paid \$10 per lift ticket and \$18 per ski rental.

Test.

Could Group 2 have paid \$10 per lift ticket and \$18 per ski rental?

$24L + 12S = 456$	Equation for Group 2
$24(10) + 12(18) \stackrel{?}{=} 456$	Substitute.
$456 = 456$ ✓	Simplify.

So, Group 2 could have paid these prices. Are other prices possible?

Guess-and-Test another set of prices.

Group 1: Guess $L = 15$. Find S.

$36L + 18S = 684$

$36(15) + 18S = 684$

$18S = 144$

$S = 8$

Group 2: Test $L = 15$ and $S = 8$.

$24(15) + 12(8) \stackrel{?}{=} 456$

$456 = 456$ ✓

Another possible set of prices is \$15 per lift ticket and \$8 per ski rental.

Because the given information leads to more than one set of possible prices, it is not possible to determine how much each lift ticket costs.

Reteaching and Enrichment Strategies

If students need help. . .	If students got it. . .
Resources by Chapter • Practice A and Practice B • Puzzle Time Record and Practice Journal Practice Differentiating the Lesson Lesson Tutorials Skills Review Handbook	Resources by Chapter • Enrichment and Extension • School-to-Work Start the next section

16. **REASONING** One equation in a system of linear equations has a slope of -3. The other equation has a slope of 4. How many solutions does the system have? Explain.

17. **LOGIC** How can you use the slopes and y-intercepts of equations in a system of linear equations to determine whether the system has *one solution, infinitely many solutions,* or *no solution*? Explain your reasoning.

$$4x + 8y = 64$$
$$8x + 16y = 128$$

18. **MONEY** You and a friend both work two different jobs. The system of linear equations represents the total earnings for x hours worked at the first job and y hours worked at the second job. Your friend earns twice as much as you.
    - **a.** One week, both of you work 4 hours at the first job. How many hours do you and your friend work at the second job?
    - **b.** Both of you work the same number of hours at the second job. Compare the number of hours you each work at the first job.

19. **DOWNLOADS** You download a digital album for \$10. Then you and your friend download the same number of individual songs for \$0.99 each. Write a system of linear equations that represents this situation. Will you and your friend spend the same amount of money? Explain.

20. **REASONING** Does the system shown *always, sometimes,* or *never* have no solution when $a = b$? $a \geq b$? $a < b$? Explain your reasoning.

$$y = ax + 1$$
$$y = bx + 4$$

21. **SKIING** The table shows the number of lift tickets and ski rentals sold to two different groups. Is it possible to determine how much each lift ticket costs? Justify your answer.

Group	1	2
Number of Lift Tickets	36	24
Number of Ski Rentals	18	12
Total Cost (dollars)	684	456

22. **Precision** Find the values of a and b so the system shown has the solution (2, 3). Does the system have any other solutions? Explain.

$$12x - 2by = 12$$
$$3ax - by = 6$$

Fair Game Review What you learned in previous grades & lessons

Graph the inequality in a coordinate plane. *(Section 3.5)*

23. $3x + y \geq 6$ **24.** $-3x - 4y \geq 4$ **25.** $-4x + 3y < -12$

26. **MULTIPLE CHOICE** What is the solution of $-2(y + 5) \leq 16$? *(Section 3.4)*

Ⓐ $y \leq -13$ Ⓑ $y \geq -13$ Ⓒ $y \leq -3$ Ⓓ $y \geq -3$

4.4b Solving Linear Equations by Graphing

Solving Equations Using Graphs

Step 1: To solve the equation $ax + b = cx + d$, write two linear equations.

$$ax + b = cx + d$$

$y = ax + b$ and $y = cx + d$

Step 2: Graph the system of linear equations. The x-value of the solution of the system of linear equations is the solution of the equation $ax + b = cx + d$.

EXAMPLE 1 Solving an Equation Using a Graph

Solve $x - 2 = -\frac{1}{2}x + 1$ using a graph. Check your solution.

Step 1: Write a system of linear equations using each side of the equation.

$$x - 2 = -\frac{1}{2}x + 1$$

$y = x - 2$ $\quad$ $y = -\frac{1}{2}x + 1$

Step 2: Graph the system.

$y = x - 2$

$y = -\frac{1}{2}x + 1$

The graphs intersect at (2, 0).

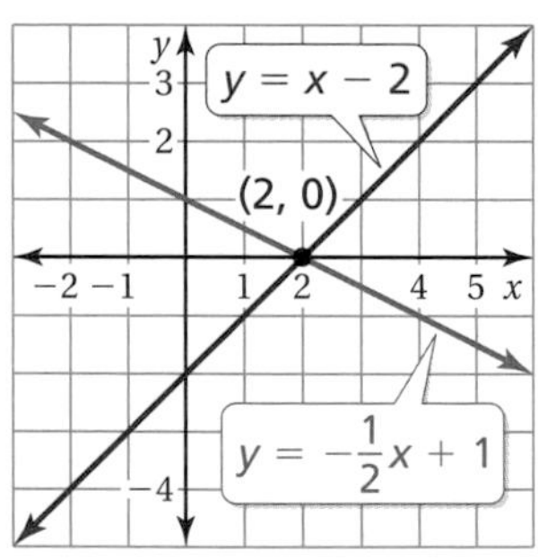

So, the solution of the equation is $x = 2$.

Check

$x - 2 = -\frac{1}{2}x + 1$

$2 - 2 \stackrel{?}{=} -\frac{1}{2}(2) + 1$

$0 = 0$ ✓

Practice

Use a graph to solve the equation. Check your solution.

1. $2x + 3 = 4$
2. $2x = x - 3$
3. $3x + 1 = 3x + 2$
4. $\frac{1}{3}x = x + 8$
5. $1.5x + 2 = 11 - 3x$
6. $3 - 2x = -2x + 3$
7. **STRUCTURE** Write an equation with variables on both sides that has no solution. How can you change the equation so that it has infinitely many solutions?

Laurie's Notes

Introduction

Connect

- **Yesterday:** Students solved special systems of linear equations both algebraically and by graphing. (MP7, MP8)
- **Today:** Students will solve an equation with variables on both sides by graphing a system of equations.

Motivate

- ? "Have any of you seen the play *Little Shop of Horrors*? There is a flytrap-like alien plant that lives on human blood and eventually grows large enough to swallow people whole."
- In reality, the Venus flytrap does not grow rapidly. However, the Thuja Giant, a fast growing evergreen tree, grows 3–5 feet per year.

Lesson Notes

Words of Wisdom

- Students often do not see the point of learning a new technique when the old technique worked well. The graphical approach helps to show the connection between the algebraic and geometric approaches. Some equations are easier to graph than they are to manipulate.

Example 1

- **MP5 Use Appropriate Tools Strategically** and **MP7 Look for and Make Use of Structure:** Students have solved systems by graphing previously. Today they will solve a linear equation with variables on both sides. The big idea is to think of each side of the equation as a linear equation by writing y = left side of equation, and y = right side of equation.
- Set each side of the equation equal to y to create the system of equations.
- The graph of each linear equation shows the value of the expression for different values of x. So, the intersection shows the value of x for which the expressions are equal.
- The check is algebraic. The correct value of x makes the original equation true.

Practice

- ? For Exercise 5 ask, "How do you graph a slope of 1.5?" Think of $\frac{rise}{run} = \frac{3}{2}$.
- Take time to have students share their thinking and work for Exercise 7. Students who found this problem difficult will benefit from hearing a variety of approaches.

Goal Today's lesson is solving linear equations by graphing.

Start Thinking! and Warm Up

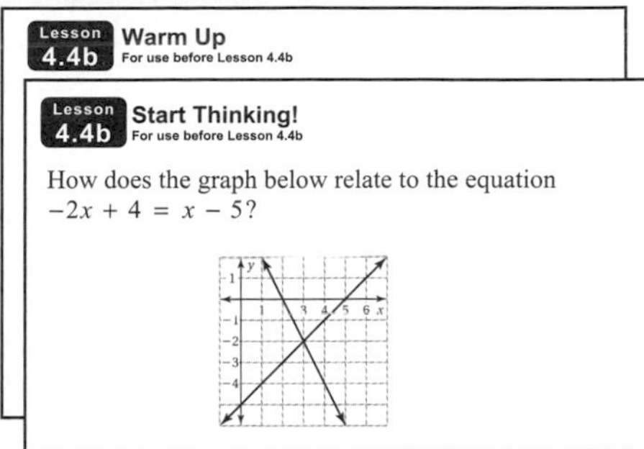

Lesson 4.4b Warm Up
For use before Lesson 4.4b

Lesson 4.4b Start Thinking!
For use before Lesson 4.4b

How does the graph below relate to the equation $-2x + 4 = x - 5$?

Extra Example 1

Solve $x + 3 = \frac{1}{2}x + 1$ using a graph. Check your solution. $x = -4$

Practice

1. $x = \frac{1}{2}$
2. $x = -3$
3. no solution
4. $x = -12$
5. $x = 2$
6. all real values of x
7. *Sample answer:* $6x - 3 = 6x$; Subtract 3 from the right side.

Record and Practice Journal Practice
See Additional Answers.

Laurie's Notes

Extra Example 2

In Example 2, Plant A grows 0.4 inch per month. Plant B grows three times faster per month.

a. Use the model to write an equation. $0.4x + 12 = 1.2x + 9$

b. After how many months x are the plants the same height? 3.75 mo

Practice

8. $x = 2.6$

9. $x = \frac{21}{2}$

10. $x = -20.5$

11. 6 mo

Mini-Assessment

Use a graph to solve the equation.

1. $-\frac{1}{3}x + 5 = 4x - 8$ $x = 3$

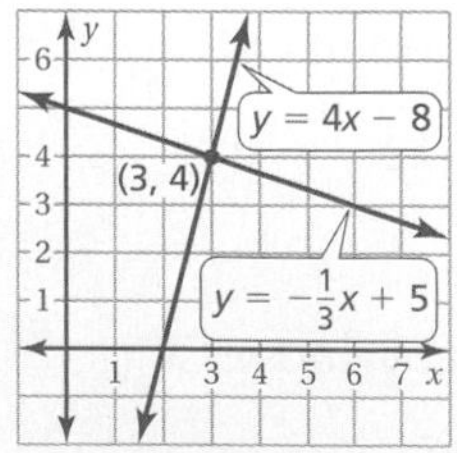

2. $-\frac{3}{2}x - 2 = -2x - 3$ $x = -2$

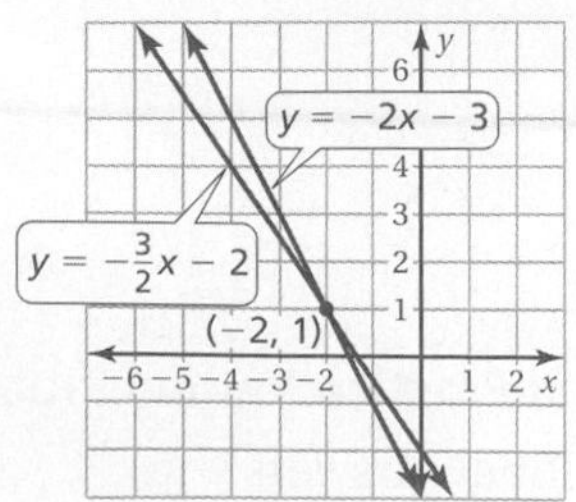

Example 2

- Ask a volunteer to read the problem. Students will also need to read information from the diagram.
- The verbal model helps to focus attention on the goal, which is to determine *when* the two *plants will be the same height.*
- The growth rate is stated in terms of inches per month and is multiplied by months. So, you are adding inches to inches on each side of the equation.

$$\frac{\text{inches}}{\cancel{\text{months}}} \times \cancel{\text{months}} + \text{inches}$$

- Finish working through the problem as shown.
- **?** "How tall are the plants after 5 months?" 15 inches
- **?** "How can we check our solution?" Substitute $x = 5$ into the original equation; both sides equal 15.
- Discuss the Study Tip. This is a second way to solve the equation.

Practice

- You may wish to have students use a graphing calculator to complete the problems.

Closure

- **Exit Ticket:** Solve the equation $3x - 4 = \frac{1}{2}x + 1$. $x = 2$

Technology For the Teacher

The Dynamic Planning Tool
Editable Teacher's Resources at *BigIdeasMath.com*

EXAMPLE 2 Real-Life Application

Plant A grows 0.6 inch per month. Plant B grows twice as fast.

a. Use the model to write an equation.

b. After how many months x are the plants the same height?

Plant A: 12 in. Plant B: 9 in.

Growth rate • Months, x + Original height = Growth rate • Months, x + Original height

a. The equation is $0.6x + 12 = 1.2x + 9$.

b. Write a system of linear equations using each side of the equation. Then use a graphing calculator to graph the system.

$$0.6x + 12 = 1.2x + 9$$

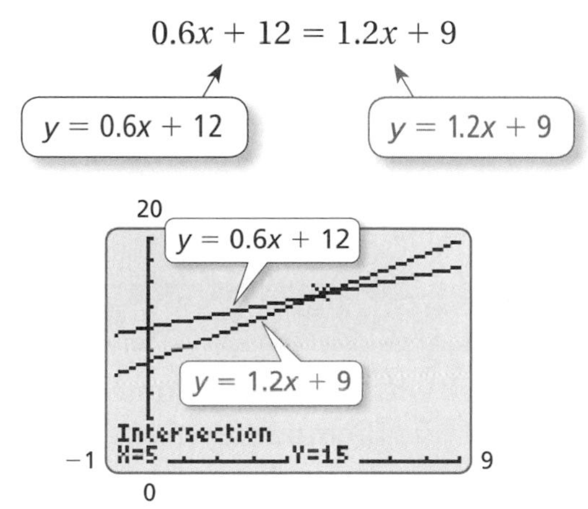

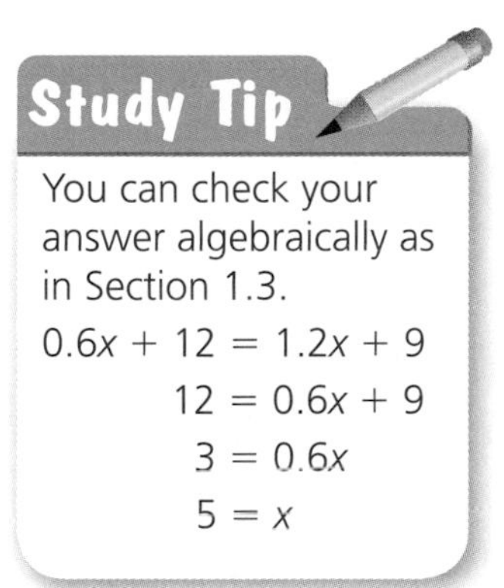

Study Tip

You can check your answer algebraically as in Section 1.3.

$$0.6x + 12 = 1.2x + 9$$
$$12 = 0.6x + 9$$
$$3 = 0.6x$$
$$5 = x$$

The solution of the system is (5, 15).

So, the plants are both 15 inches tall after 5 months.

Practice

Use a graph to solve the equation. Check your solution.

8. $6x - 2 = x + 11$

9. $\frac{4}{3}x - 1 = \frac{2}{3}x + 6$

10. $1.75x = 2.25x + 10.25$

11. **WHAT IF?** In Example 2, the growth rate of Plant A is 0.5 inch per month. After how many months x are the plants the same height?

4.5 Systems of Linear Inequalities

COMMON CORE STATE STANDARDS
A.CED.3
A.REI.12

Essential Question How can you sketch the graph of a system of linear inequalities?

1 ACTIVITY: Graphing Linear Inequalities

Work with a partner. Match the linear inequality with its graph.

$2x + y \leq 4$ Inequality 1

$2x - y \leq 0$ Inequality 2

a.

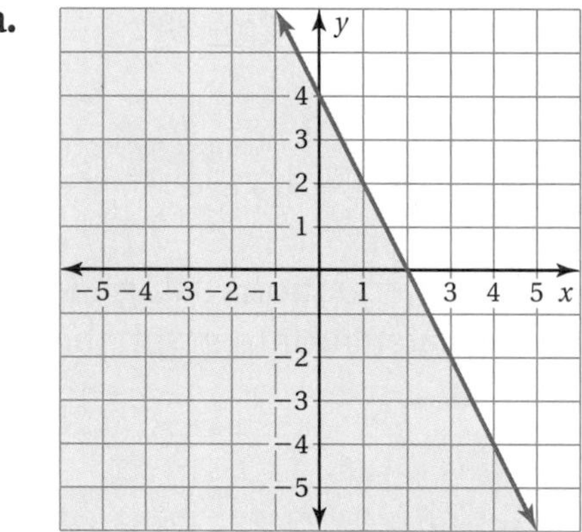

b.

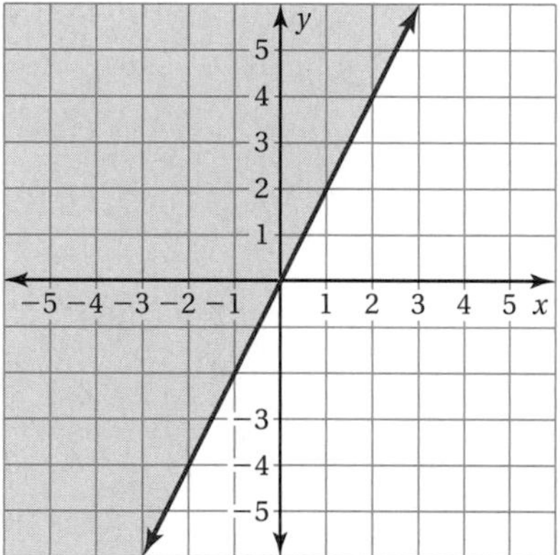

2 ACTIVITY: Graphing a System of Linear Inequalities

Work with a partner. Consider the system of linear inequalities given in Activity 1.

$2x + y \leq 4$ Inequality 1

$2x - y \leq 0$ Inequality 2

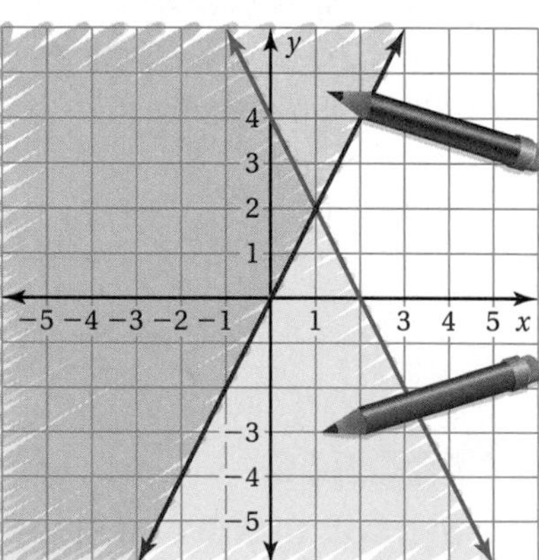

Use colored pencils to shade the solutions of the two linear inequalities. When you graph both inequalities in the same coordinate plane, what do you get?

Describe each of the shaded regions in the graph at the right. What does the unshaded region represent?

Laurie's Notes

Introduction

Standards for Mathematical Practice

- **MP6 Attend to Precision:** Students have graphed linear inequalities and systems of equations. In graphing and solving a system of linear inequalities, students must pay attention to the inequality symbol. They use it to determine which half-plane to shade and whether to use a solid line or a dashed line. Students must also pay attention to the definition of a solution of a system of linear inequalities. It is possible for an ordered pair to be a solution to one inequality, but not all of the inequalities.

Motivate

- In a coordinate plane, sketch a square region bounded by $x = -2$, $x = 2$, $y = -2$, and $y = 2$. Shade the interior.
- Ask students to write the equations for the lines forming the sides of the square.
- ? "The equations represent the sides of the square. How might you represent the interior of the square?" Listen for ideas about writing inequalities. "Name ordered pairs in the interior and in the exterior of the square." Answers vary.
- Explain that in a previous lesson they graphed linear inequalities. In this lesson they will graph 2 or more linear inequalities on the same coordinate grid.

Activity Notes

Activity 1

- Graphing a system of linear inequalities combines the skills of graphing linear inequalities and graphing linear systems.
- It should not take students long to match each linear inequality with its graph. Some may simply do a test point such as (0, 0). Others may solve the inequality for y and consider the slope of the line.
- ? "Which graph matches Inequality 1 and how did you decide?" graph (a); Explanations will vary.
- ? "How do the graphs of $2x + y \leq 4$ and $2x + y < 4$ differ?" The boundary line is solid for $2x + y \leq 4$ and dashed for $2x + y < 4$.
- ? "What does the shaded portion of a linear inequality represent?" ordered pairs that are *solutions* of the inequality

Activity 2

- Systems of linear inequalities are not yet defined. If students ask what a system of linear inequalities is, turn the question back to them and ask what they think it is. Some students will be able to give a description.
- ? When students finish their graphs ask, "What portion of your graph represents the solution of the system of linear inequalities?" the region where the two colors overlap, including the boundary lines
- **Extension:** Have students identify the 4 regions in their graph and interpret what each region represents.

Common Core State Standards

A.CED.3 Represent constraints by inequalities, and by systems of inequalities and interpret solutions as viable or nonviable options in a modeling context.

A.REI.12 Graph the solutions to a linear inequality in two variables as a half-plane (excluding the boundary in the case of a strict inequality), and graph the solution set to a system of linear inequalities in two variables as the intersection of the corresponding half-planes.

Previous Learning

Students should know how to solve and graph linear inequalities and how to solve systems of linear equations.

Start Thinking! and Warm Up

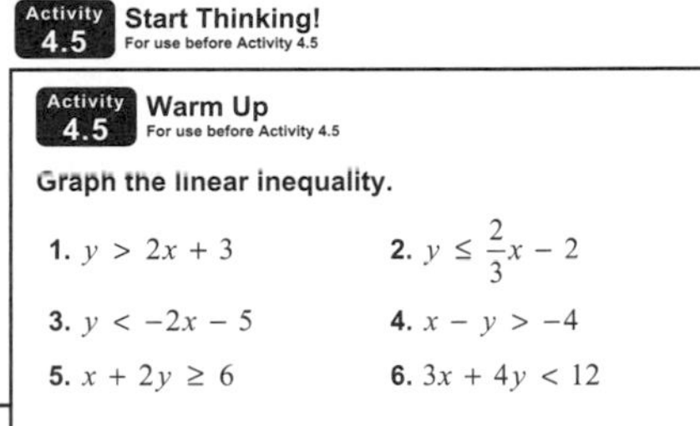

Activity 4.5 Start Thinking! For use before Activity 4.5

Activity 4.5 Warm Up For use before Activity 4.5

Graph the linear inequality.

1. $y > 2x + 3$
2. $y \leq \frac{2}{3}x - 2$
3. $y < -2x - 5$
4. $x - y > -4$
5. $x + 2y \geq 6$
6. $3x + 4y < 12$

4.5 Record and Practice Journal

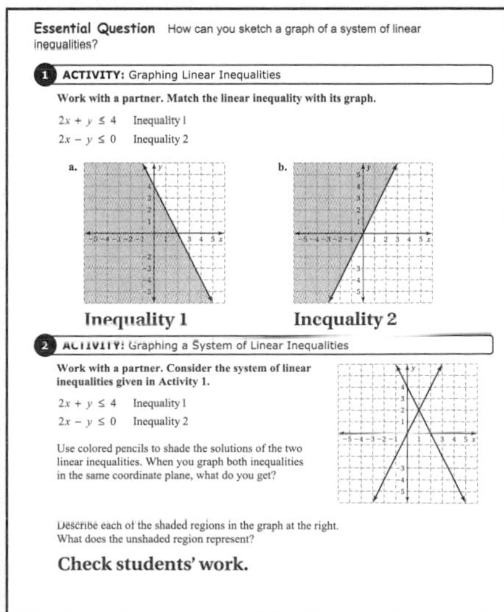

Essential Question How can you sketch a graph of a system of linear inequalities?

1 ACTIVITY: Graphing Linear Inequalities

Work with a partner. Match the linear inequality with its graph.

$2x + y \leq 4$ Inequality 1

$2x - y \leq 0$ Inequality 2

a. Inequality 1

b. Inequality 2

2 ACTIVITY: Graphing a System of Linear Inequalities

Work with a partner. Consider the system of linear inequalities given in Activity 1.

$2x + y \leq 4$ Inequality 1

$2x - y \leq 0$ Inequality 2

Use colored pencils to shade the solutions of the two linear inequalities. When you graph both inequalities in the same coordinate plane, what do you get?

Describe each of the shaded regions in the graph at the right. What does the unshaded region represent?

Check students' work.

Differentiated Instruction

Auditory

Students may not understand why some points in shaded regions are not solutions of the system. Explain that an ordered pair must satisfy *every* inequality in the system to be a solution of the system. Only points in the region where the shadings from all the inequalities overlap represent solutions of the system.

4.5 Record and Practice Journal

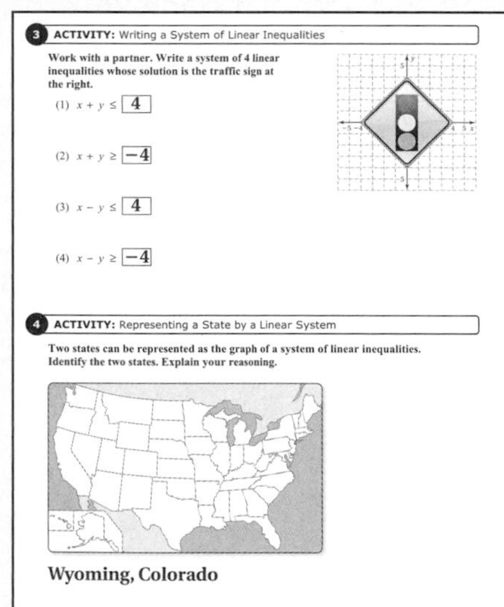

3 ACTIVITY: Writing a System of Linear Inequalities

Work with a partner. Write a system of 4 linear inequalities whose solution is the traffic sign at the right.

(1) $x + y \le$ **4**

(2) $x + y \ge$ **−4**

(3) $x - y \le$ **4**

(4) $x - y \ge$ **−4**

4 ACTIVITY: Representing a State by a Linear System

Two states can be represented as the graph of a system of linear inequalities. Identify the two states. Explain your reasoning.

Wyoming, Colorado

What Is Your Answer?

5. IN YOUR OWN WORDS How can you sketch the graph of a system of linear inequalities?

Graph each inequality in the same coordinate plane and shade the solution of each. The solution to the system is the region where the shading overlaps.

6. When graphing a system of linear inequalities, which region represents the solution of the system? Do you think all systems have a solution? Explain.

The region where the solutions overlap; no

Laurie's Notes

Activity 3

? "What shape is the traffic sign?" square; Some students may call it a diamond, however a diamond is not a geometric shape.

- The inequalities that students write are based on the lines containing the four sides of the sign.
- When students finish, have them identify (match) the inequality that corresponds to each side of the sign.
- Check to see that the solution is indeed the interior of the sign.
- **Extension:** A STOP sign is nested in the interior of the traffic sign. Ask what additional inequalities are needed to form the octagon. (2 horizontal and 2 vertical)

Activity 4

- This is not a trick question. It is simply to have students recognize that systems of linear inequalities can enclose a polygon. The states of Colorado and Wyoming have boundaries that are approximately straight segments.

What Is Your Answer?

- **MP7 Look for and Make Use of Structure:** Students discern the relationships between the systems of linear inequalities they worked with and their graphs to develop a general strategy for graphing a system of linear inequalities.

Closure

- Write a system of linear inequalities that would have a triangular region as a solution.

  Sample answer: $x > 0$

  $y > 0$

  $y < -x + 4$

Technology For the Teacher

The Dynamic Planning Tool

Editable Teacher's Resources at *BigIdeasMath.com*

3 ACTIVITY: Writing a System of Linear Inequalities

Work with a partner. Write a system of 4 linear inequalities whose solution is the traffic sign at the right.

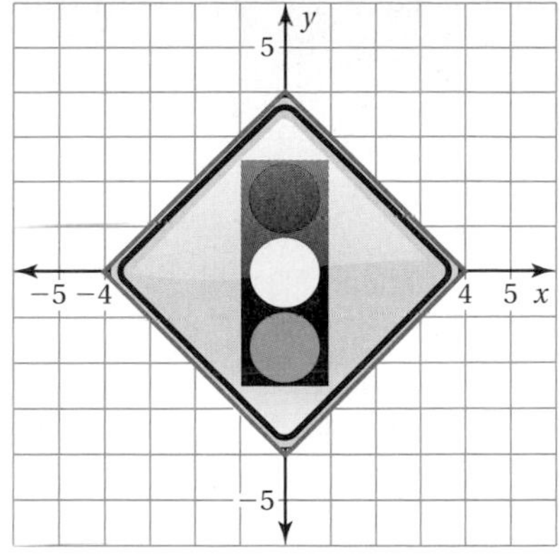

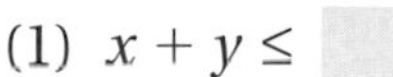

(1) $x + y \leq$ ▭

(2) $x + y \geq$ ▭

(3) $x - y \leq$ ▭

(4) $x - y \geq$ ▭

4 ACTIVITY: Representing a State by a Linear System

Two states can be represented as the graph of a system of linear inequalities. Identify the two states. Explain your reasoning.

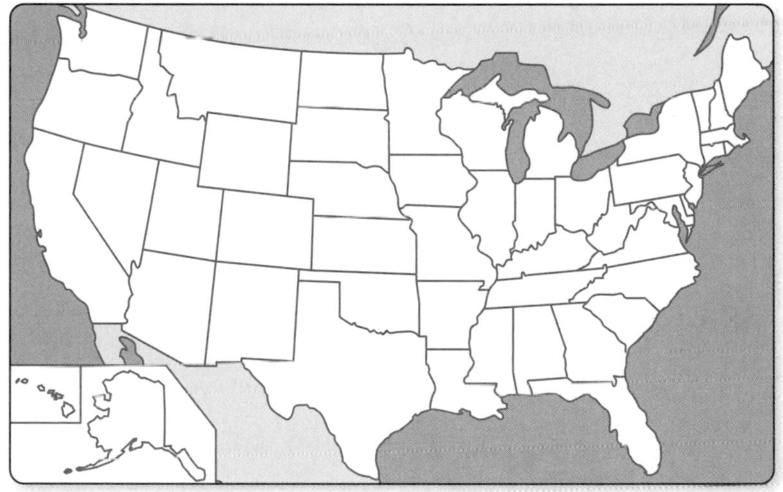

What Is Your Answer?

5. **IN YOUR OWN WORDS** How can you sketch the graph of a system of linear inequalities?

6. When graphing a system of linear inequalities, which region represents the solution of the system? Do you think all systems have a solution? Explain.

Use what you learned about systems of linear inequalities to complete Exercises 7–9 on page 189.

4.5 Lesson

Key Vocabulary

system of linear inequalities, *p. 186*

solution of a system of linear inequalities, *p. 186*

graph of a system of linear inequalities, *p. 186*

A **system of linear inequalities** is a set of two or more linear inequalities in the same variables. An example is shown below.

$y < x + 2$ Inequality 1

$y \geq 2x - 1$ Inequality 2

A **solution of a system of linear inequalities** in two variables is an ordered pair that is a solution of each inequality in the system.

EXAMPLE 1 Checking Solutions

Tell whether each ordered pair is a solution of the system.

$\boldsymbol{y < 2x}$ Inequality 1

$\boldsymbol{y \geq x + 1}$ Inequality 2

a. (3, 5)

Inequality 1	Inequality 2
$y < 2x$	$y \geq x + 1$
$5 \overset{?}{<} 2(3)$	$5 \overset{?}{\geq} 3 + 1$
$5 < 6$ ✓	$5 \geq 4$ ✓

∴ (3, 5) is a solution of both inequalities. So, it is a solution of the system.

b. (−2, 0)

Inequality 1	Inequality 2
$y < 2x$	$y \geq x + 1$
$0 \overset{?}{<} 2(-2)$	$0 \overset{?}{\geq} -2 + 1$
$0 \not< -4$ ✗	$0 \geq -1$ ✓

∴ (−2, 0) is not a solution of both inequalities. So, it is not a solution of the system.

On Your Own

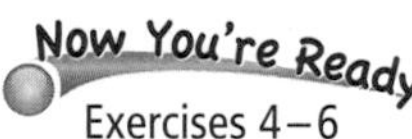

Exercises 4–6

Tell whether the ordered pair is a solution of the system of linear inequalities.

1. $y < 5$
   $y > x - 4$; $(-1, 5)$

2. $y \leq -2x + 5$
   $y < x + 3$; $(0, -1)$

The **graph of a system of linear inequalities** is the graph of all of the solutions of the system.

Graphing a System of Linear Inequalities

Step 1 Graph each inequality in the same coordinate plane.

Step 2 Find the intersection of the half-planes. This intersection is the graph of the system.

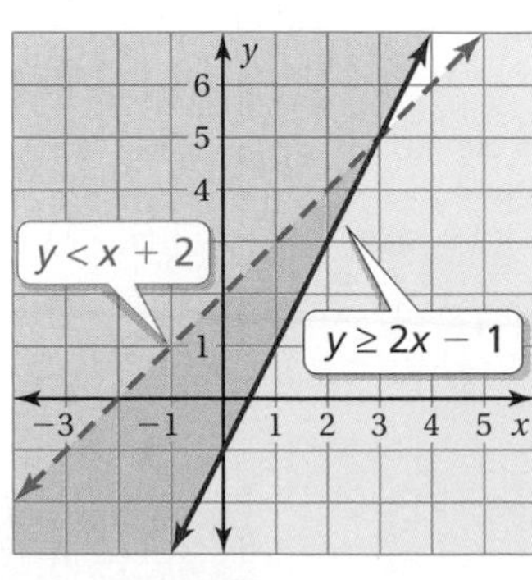

Multi-Language Glossary at BigIdeasMath.com.

Laurie's Notes

Introduction

Connect

- **Yesterday:** Students investigated graphs of systems of linear inequalities. (MP6, MP7)
- **Today:** Students will solve systems of linear inequalities by graphing.

Motivate

- Ask students to describe what a cubic foot looks like. They should use their hands and/or point to something in the classroom as a reference. Then ask about a cubic yard. A small baby's playpen is a reasonable model of a cubic yard.

? "About how much does a cubic foot of dry sand weigh?" Answers will vary, but a reasonable estimate is 100 pounds.

? "About how much does a cubic yard of dry sand weigh?" 27×100 pounds $= 2700$ pounds

- Ask students to visualize the last beach they were on and the amount of sand there.

Lesson Notes

Discuss

- Students worked with a system of linear inequalities yesterday without the formal definition.
- Write and discuss the two definitions.

Example 1

- Review the four inequality symbols: $<, \leq, >, \geq$. Remind students that $4 \leq 4$ is a true statement, but $4 < 4$ is not a true statement.
- To be a solution, an ordered pair must satisfy *all* of the inequalities. Check each ordered pair in both inequalities.

? "Is it possible for an ordered pair to satisfy one inequality and not the other? Explain." Yes; an ordered pair could be in the shaded region for inequality 1 but not in the shaded region for inequality 2.

On Your Own

- **Think-Pair-Share:** Students should read each question independently and then work with a partner to answer the questions. When they have answered the questions, the pair should compare their answers with another group and discuss any discrepancies.

Key Idea

- Discuss the graph of a linear system.
- Write and discuss the Key Idea.

? "Is it possible that you shade the half-planes and there is no intersection? Explain." yes; Two parallel lines may have half-planes that do not intersect.

Goal Today's lesson is solving **systems of linear inequalities.**

Start Thinking! and Warm Up

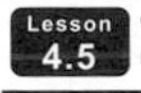

Warm Up
For use before Lesson 4.5

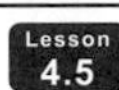

Start Thinking!
For use before Lesson 4.5

A system of linear equations can have one single solution, no solution, or infinitely many solutions. Is this the case for a system of two linear inequalities? Why or why not? Be sure to include sketches in your explanation.

Extra Example 1

Tell whether each ordered pair is a solution of the system.

$y \geq \frac{1}{2}x$

$y < x + 2$

a. (2, 1) yes

b. (0, 2) no

On Your Own

1. no
2. yes

Extra Example 2

Graph the system.

$x > 1$

$y \leq \frac{1}{2}x + 1$

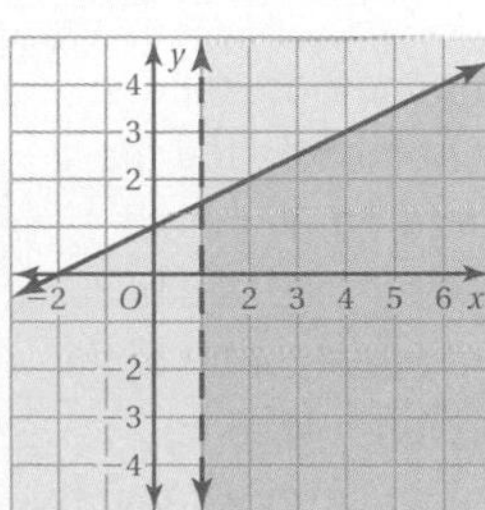

Extra Example 3

Graph the system.

$y < x + 2$

$-x + y \geq -2$

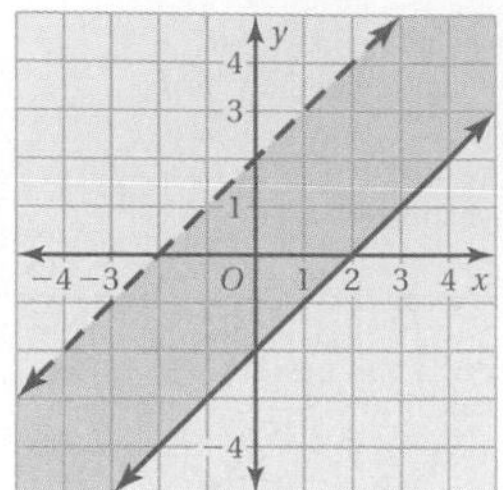

On Your Own

3.

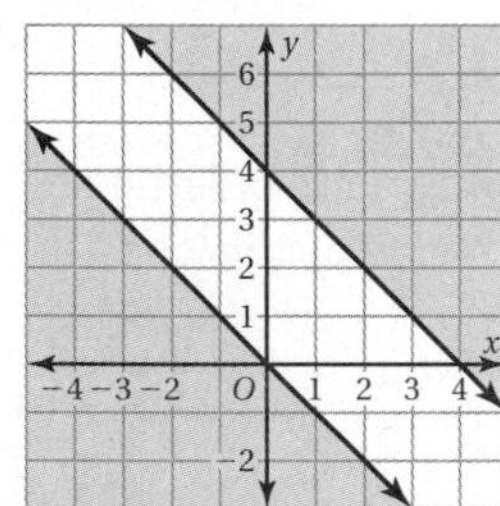

4.

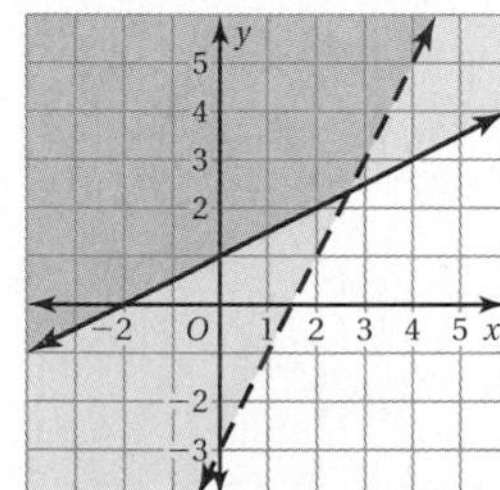

5.

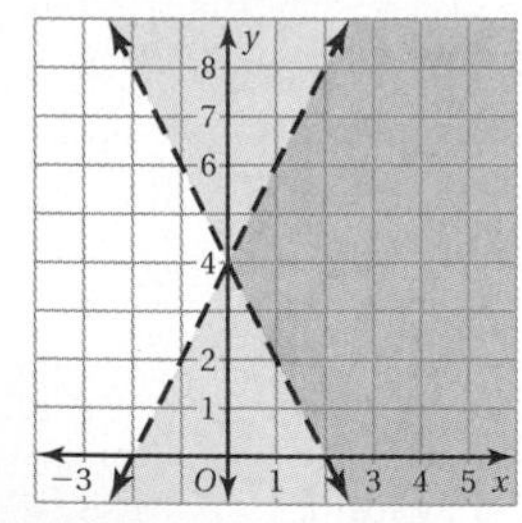

Differentiated Instruction

Auditory Learners

Tell students that the method used in Example 2 is sometimes called the graph-and-check method, because students *graph* the inequalities and then *check* points in the solution region. Students should get in the habit of checking their graphs using test points.

Laurie's Notes

Example 2

- Write the system of inequalities.
- ? "Describe the graph of $y = 3$." horizontal line through (0, 3)
- ? "Will the graph of $y \leq 3$ be the half-plane above or below the line $y = 3$?" below
- Explain to students that you can visually select a solution point from the region where the shading overlaps. You can also use the graph to select ordered pairs that are not part of the solution.

Example 3

- **MP7 Look for and Make Use of Structure:** The inequalities are written in standard form. Students may leave them in this form, or they may choose to rewrite them in slope-intercept form to sketch their graphs.
- ? "What is the slope of Inequality 1?" -2
- ? "What is the slope of Inequality 2?" -2
- ? "What is the relationship between the two lines?" They are parallel.
- ? **MP8 Look for and Express Regularity in Repeated Reasoning:** "How do you know which side of each boundary line to shade?" Use a test point such as (0, 0) in each inequality.
- Shading the wrong half-plane will completely change the solution region (it will no longer be correct), so testing a point is a very important step.
- **Connection:** When you solved systems of linear equations, parallel lines resulted in "no solution." Do not automatically conclude there is no solution when the graph of a system of inequalities involves parallel lines. It is possible that the shading will overlap.

On Your Own

- **Neighbor Check:** Have students work independently and then have their neighbor check their work. Have students discuss any discrepancies.
- Have volunteers share their work at the board.
- **Extension:** Have students graph inequalities using a graphing calculator. If time permits, you could have students use calculators to check their solutions.

EXAMPLE 2 Graphing a System of Linear Inequalities

Study Tip

For help with graphing linear inequalities, see Section 3.5.

Graph the system.

$y \leq 3$ Inequality 1

$y > x + 2$ Inequality 2

Step 1: Graph each inequality.

Step 2: Find the intersection of the half-planes. One solution is $(-3, 1)$.

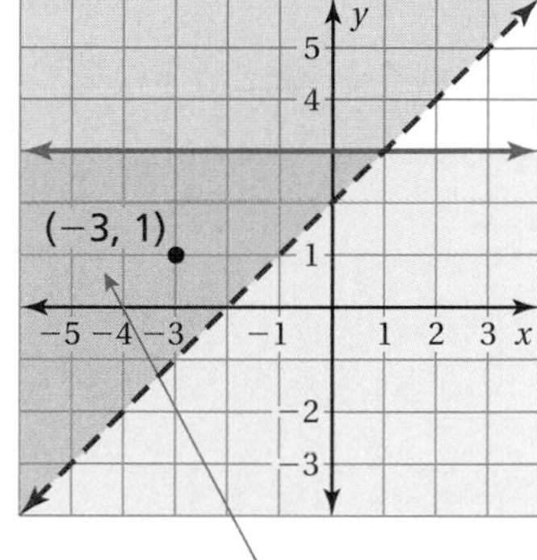

(−3, 1)
The solution is the purple shaded region.

Check

Verify that $(-3, 1)$ is a solution of each inequality.

Inequality 1	Inequality 2
$y \leq 3$	$y > x + 2$
$1 \leq 3$ ✓	$1 \overset{?}{>} -3 + 2$
	$1 > -1$ ✓

EXAMPLE 3 Graphing a System of Linear Inequalities: No Solution

Graph the system.

$2x + y < -1$ Inequality 1

$2x + y > 3$ Inequality 2

Step 1: Graph each inequality.

Step 2: Find the intersection of the half-planes.

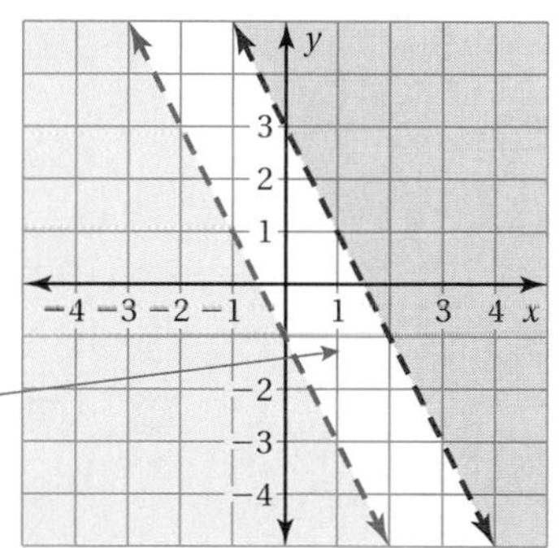

The lines are parallel and the half-planes do not intersect.

∴ So, the system has no solution.

On Your Own

Now You're Ready
Exercises 7–15

Graph the system of linear inequalities.

3. $y \geq -x + 4$
$x + y \leq 0$

4. $y > 2x - 3$
$y \geq \frac{1}{2}x + 1$

5. $-2x + y < 4$
$2x + y > 4$

EXAMPLE 4 Writing a System of Linear Inequalities

Write a system of linear inequalities represented by the graph.

The horizontal boundary line passes through $(0, -2)$. So, an equation of the line is $y = -2$.

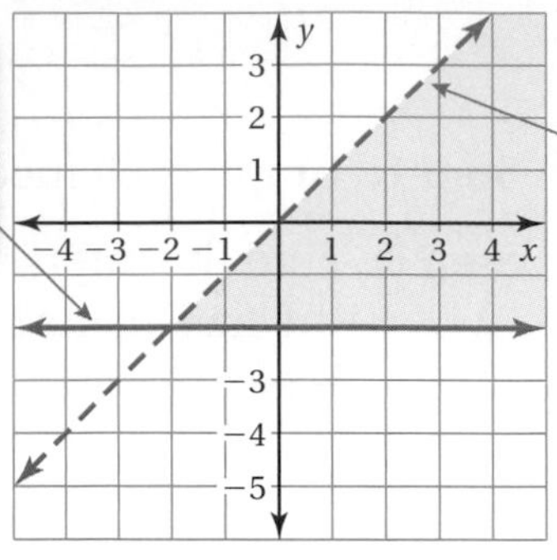

The slope of the other boundary line is 1 and the y-intercept is 0. So, an equation of the line is $y = x$.

The shaded region is *above* the *solid* boundary line, so the inequality is $y \geq -2$.

The shaded region is *below* the *dashed* boundary line, so the inequality is $y < x$.

∴ The system is $y \geq -2$ and $y < x$.

EXAMPLE 5 Real-Life Application

You have at most 8 hours to spend at the mall and at the beach. You want to spend at least 2 hours at the mall and more than 4 hours at the beach. Write and graph a system that represents the situation. How much time could you spend at each location?

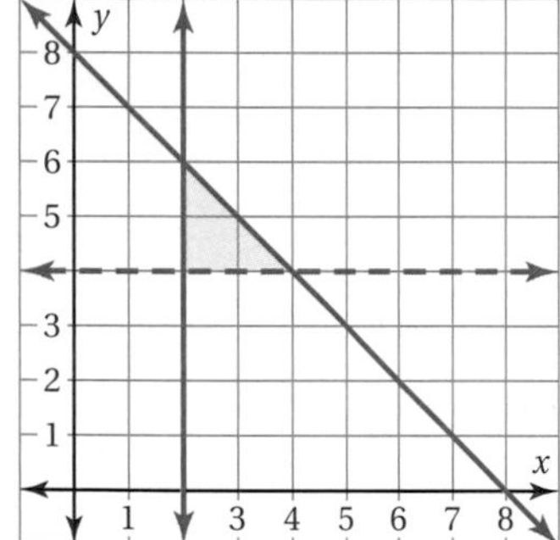

Use the constraints to write a system of linear inequalities. Let x be the number of hours at the mall and let y be the number of hours at the beach.

$x + y \leq 8$	at most 8 hours at the mall and at the beach
$x \geq 2$	at least 2 hours at the mall
$y > 4$	more than 4 hours at the beach

Graph the system. One ordered pair in the solution region is $(2.5, 5)$.

∴ So, you could spend 2.5 hours at the mall and 5 hours at the beach.

On Your Own

Now You're Ready
Exercises 24–26

Write a system of linear inequalities represented by the graph.

6.

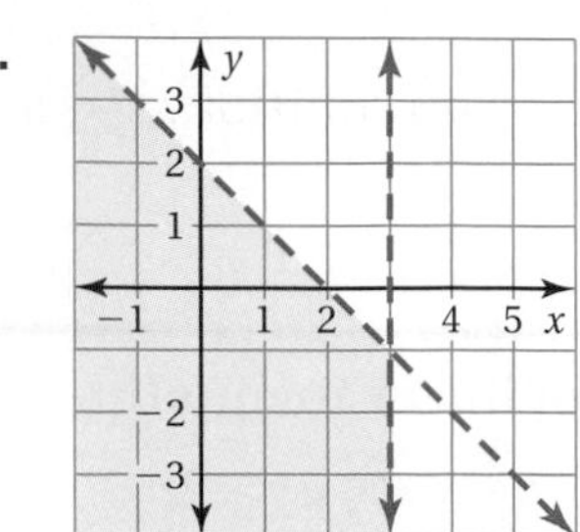

7.

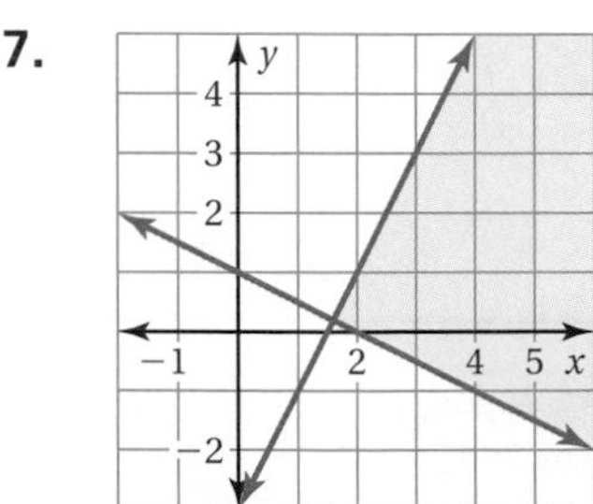

8. **WHAT IF?** In Example 5, you want to spend at least 3 hours at the mall. How does this change the system? Is $(2.5, 5)$ still a solution? Explain.

Laurie's Notes

Example 4

- Students are expected to write a system of linear inequalities represented by a graph.
- Graph the system of linear inequalities on the board.

? "What is the general equation of a horizontal line passing through (0, b)?" $y = b$

? "What is the slope of the diagonal line? Explain." 1; The line *rises* 1 unit for each *run* of 1 unit.

? "What inequality symbol should you use for the diagonal line? Why?" the "less than" symbol $<$; The shaded region is below the line and the line is dashed.

Example 5

? "What does the word *constraint* mean?" limitation or restriction

- Explain that in this problem, there are constraints on the variables. Each constraint can be represented by an inequality.
- Define the variables. This is a step that students often want to skip. Write "x = the number of hours at the mall" and "y = the number of hours at the beach."

? "How do you represent *at most 8 hours to spend at the mall and at the beach*?" $x + y \le 8$

? "How do you represent *at least 2 hours at the mall*?" $x \ge 2$

? "How do you represent *more than 4 hours at the beach*?" $y > 4$

- Check that students have the correct inequality symbols.
- Continue to work the problem as shown.

? Ask about different ordered pairs that are a solution of all, some, or none of the inequalities. Sample: "Is (1, 7) a solution? Explain." no; Satisfies 2 of the 3 inequalities, not all 3.

Closure

- Name an ordered pair that is a solution of the system of linear inequalities.
  Sample answer: (0, 0)
- Name an ordered pair that is *not* a solution of the system of linear inequalities.
  Sample answer: (3, 3)

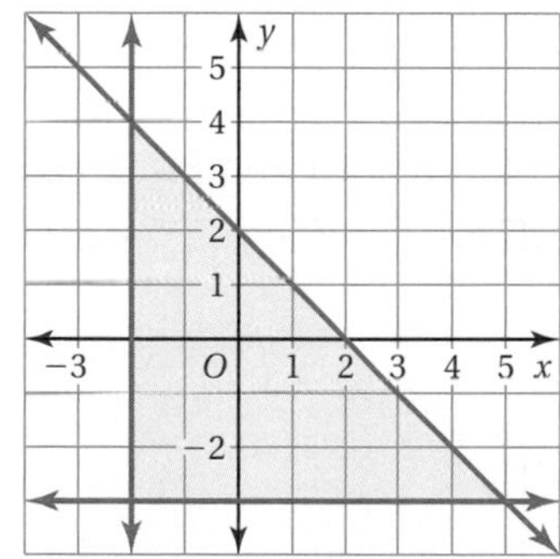

Technology For the Teacher

The Dynamic Planning Tool
Editable Teacher's Resources at *BigIdeasMath.com*

Extra Example 4

Write a system of linear inequalities represented by the graph.

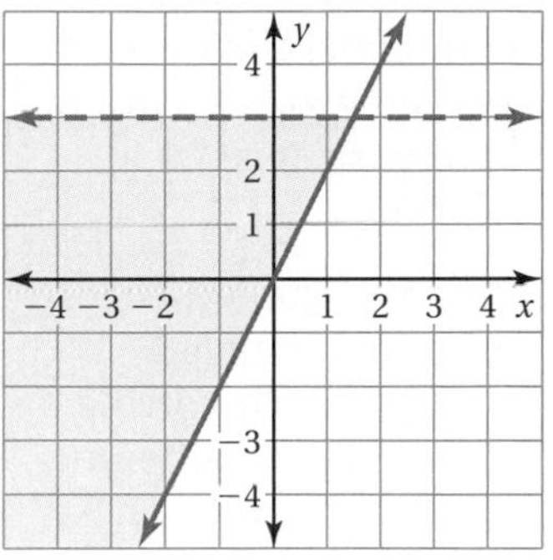

$y \ge 2x$

$y < 3$

Extra Example 5

In Example 5, you want to spend at most 7 hours at the mall and at the beach. How much time could you spend at each location?

Sample answer: 2 hours at the mall, 5 hours at the beach

On Your Own

6. $y < -x + 2$
   $x < 3$

7. $y \ge -\frac{1}{2}x + 1$
   $y \le 2x - 3$

8. New system:
   $x + y \le 8$
   $x \ge 3$
   $y > 4$
   No, (2.5, 5) is not a solution.

Vocabulary and Concept Check

1. Substitute its coordinates for x and y in each inequality of the system and simplify each side. When both resulting inequalities are true, the ordered pair is a solution. Otherwise, it is not a solution.
2. They are similar because both can be solved using graphs involving lines. They are different because the solution is an intersection of lines for equations, and an intersection of regions for inequalities.
3. no; The point is not part of the solution of the inequality bordered by the dashed line.

Practice and Problem Solving

4. yes **5.** no

6. no

7–15. See Additional Answers.

16. **a.** x = pounds of blueberries
y = pounds of strawberries

b. $x + y \geq 3$
$4x + 3y \leq 21$

c.

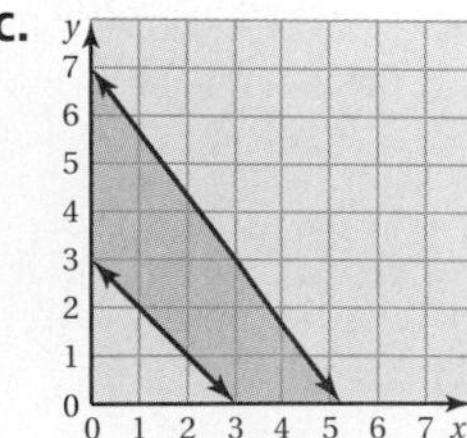

d. yes; This is more than 3 pounds and costs \$19, which is less than \$21.

Assignment Guide and Homework Check

Level	Assignment	Homework Check
Average	1–10, 18–21, 23, 24, 35–38	3, 7, 9, 23
Advanced	1–9, 13–22, 25–29 odd, 33–38	3, 8, 9, 16, 29

Common Errors

- **Exercises 7–15** Students may shade the wrong half-plane for one or both inequalities. Remind students to use a test point when they choose the half-plane to shade.
- **Exercises 7–15** Students may use the wrong style of boundary line for one or both inequalities. Remind students to use a solid line for ≤ and ≥ and a dashed line for < and >.

4.5 Record and Practice Journal

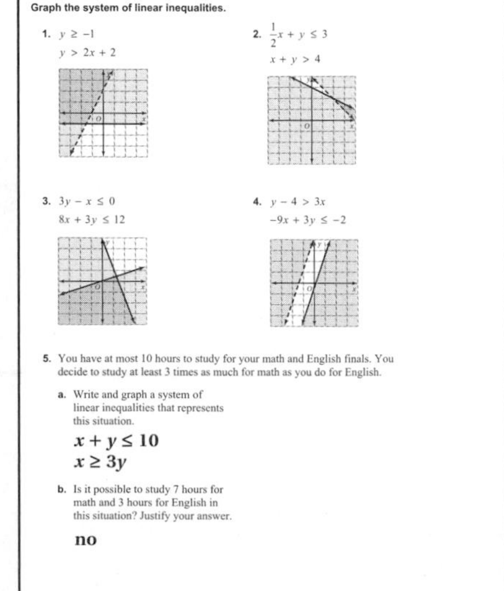

Graph the system of linear inequalities.

1. $y \geq -1$
   $y > 2x + 2$
2. $\frac{1}{2}x + y \leq 3$
   $x + y > 4$
3. $3y - x \leq 0$
   $8x + 3y \leq 12$
4. $y - 4 > 3x$
   $-9x + 3y \leq -2$
5. You have at most 10 hours to study for your math and English finals. You decide to study at least 3 times as much for math as you do for English.
   a. Write and graph a system of linear inequalities that represents this situation.
   $x + y \leq 10$
   $x \geq 3y$
   b. Is it possible to study 7 hours for math and 3 hours for English in this situation? Justify your answer.
   no

Technology For the Teacher
Answer Presentation Tool

4.5 Exercises

Vocabulary and Concept Check

1. **VOCABULARY** How can you verify that an ordered pair is a solution of a system of linear inequalities?

2. **WRITING** How are solving systems of linear inequalities and systems of linear equations similar? How are they different?

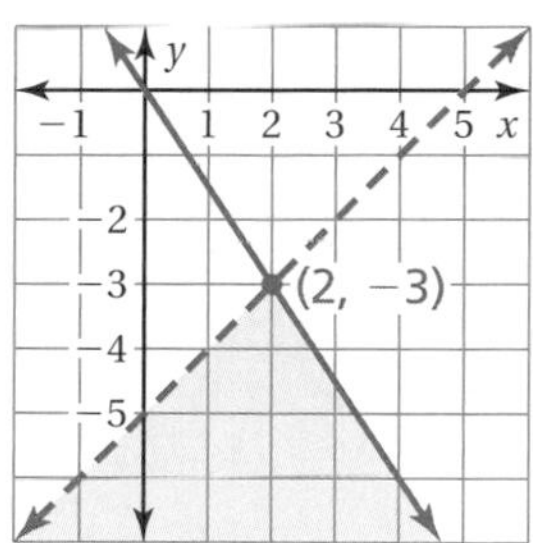

3. **REASONING** Is the point shown a solution of the system of linear inequalities? Explain.

Practice and Problem Solving

Tell whether the ordered pair is a solution of the system of linear inequalities.

4. $y < 4$
   $y > x + 3$; $(-5, 2)$

5. $y > -2$
   $y \le x - 5$; $(1, -1)$

6. $y \le x + 7$
   $y \ge 2x + 3$; $(0, 0)$

Graph the system of linear inequalities.

7. $y < -3$
   $y \ge 5x$

8. $y > -x + 3$
   $-2x + y \ge 0$

9. $x + y > 1$
   $-x - y < -3$

10. $y < -2$
    $y > 2$

11. $y \ge -5$
    $y < 3x + 1$

12. $x + y > 4$
    $y > \frac{3}{2}x - 9$

13. $-x + y < -1$
    $-x - 1 \ge -y$

14. $2x + y \le 5$
    $y + 2 \ge -2x$

15. $-2x - 5y < 15$
    $-4x > 10y + 60$

16. **MUFFINS** You can spend at most \$21 on fruit. Blueberries cost \$4 per pound and strawberries cost \$3 per pound. You need at least 3 pounds to make muffins.

    a. Define the variables.

    b. Write a system of linear inequalities that represents this situation.

    c. Graph the system of linear inequalities.

    d. Is it possible to buy 4 pounds of blueberries and 1 pound of strawberries in this situation? Justify your answer.

ERROR ANALYSIS **Describe and correct the error in graphing the system of linear inequalities.**

17. $y \geq x + 3$

$y < -x - 2$

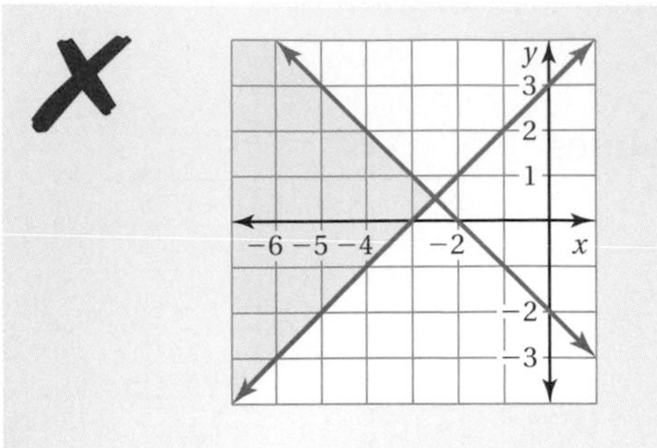

18. $y \leq 3x + 4$

$y > \frac{1}{2}x + 2$

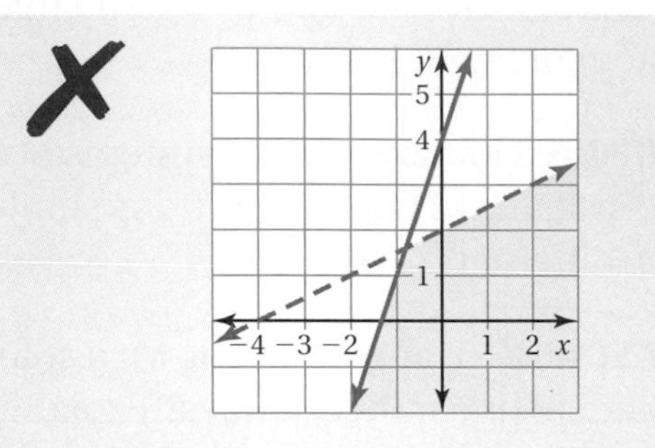

Match the graph with the corresponding system of linear inequalities.

19.

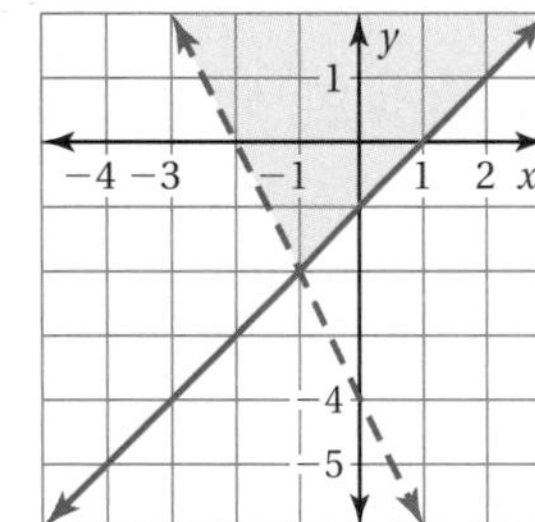

20.

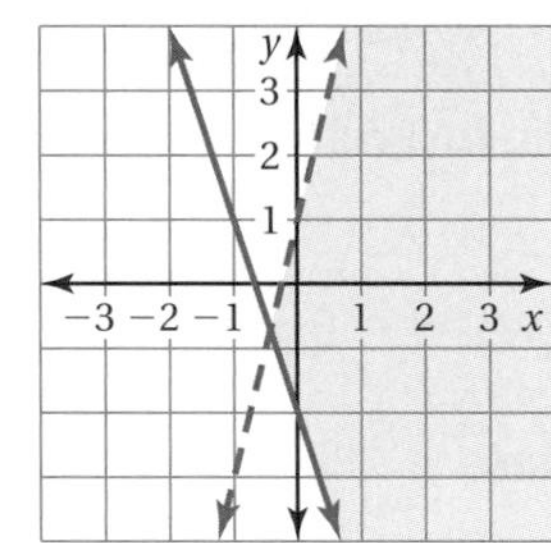

21.

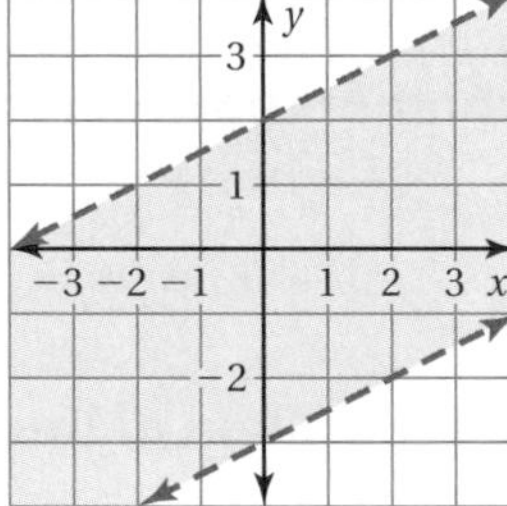

A. $y < 4x + 1$

$y \geq -3x - 2$

B. $-x + y \geq -1$

$2x + y > -4$

C. $-\frac{1}{2}x + y < 2$

$-2x + 4y > -12$

22. REASONING Describe the intersection of the half-planes of the system shown.

$$x - y \leq 4$$
$$x - y \geq 4$$

23. JOBS You earn \$12 per hour working as a manager at a grocery store. You also coach a soccer team for \$10 per hour. You need to earn at least \$110 per week, but you do not want to work more than 20 hours per week.

a. Write and graph a system of linear inequalities that represents this situation.

b. Identify and interpret one solution of the system.

Common Errors

- **Exercise 22** Students may say that there is no intersection or that the intersection is all the points in the plane. Encourage them to graph the system so they can see that two inequalities share the boundary line.
- **Exercise 23** Students may interpret their solution backwards, stating the number of hours at the grocery store as the number of hours coaching and vice versa. Tell them to make sure they define their variables and pay attention to the definitions as they set up and solve the system.
- **Exercises 24–26** Students may make careless errors. Remind them that they can write an equation of the boundary line in point-slope form. They can then determine the type of inequality symbol to use based on the style of the boundary line and whether the shading is above or below the line.
- **Exercises 27–29** Students may draw all three boundary lines first, then incorrectly shade the half-planes. Tell them to sketch the complete graph for each inequality, one at a time.
- **Exercise 30** Students may have incorrect shading or boundary lines. Explain that to find the boundary lines, they can rewrite the absolute value inequality as a compound inequality. Then they can use test points to determine the shading.

Practice and Problem Solving

17. The solid line is incorrect for the graph of $y < -x - 2$; Change it to a dashed line.

18. Incorrect half-plane shaded for $y > \frac{1}{2}x + 2$; Shade above the dashed line.

19. B **20.** A

21. C

22. The intersection of the half-planes is the line $y = x - 4$.

23. **a.** $x + y \leq 20$

$12x + 10y \geq 110$

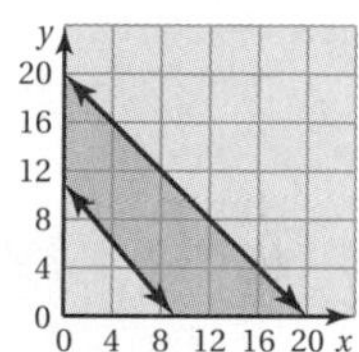

b. *Sample answer:* (10, 8); Work 10 hours at the grocery store and 8 hours as a coach.

24. $y \leq \frac{1}{2}x + 4$

$y > \frac{3}{2}x + 1$

25. $y > -2x - 1$

$y < -2x - 3$

26. $y \geq \frac{2}{3}x - 2$

$y \geq -3x + 2$

English Language Learners

Alternate Language

Word problems pose difficulties for English learners. Have a student read the problem, then discuss the intended meaning of each sentence. Students can often clear up misinterpretations this way.

Practice and Problem Solving

27–30. See Additional Answers.

31. **a.** *Sample answer:* $y < 2x - 3$

b. *Sample answer:* $y > 2x - 3$

32. See *Taking Math Deeper*.

33. *Sample answer:* You drive 6 hours and your friend drives 8 hours for a total of 14 hours and a distance of 900 miles each day.

34. **a.** $y \le 2x + 1$

$y \ge -3$

$y \le -2x + 9$

b. 32 square units

Fair Game Review

35. -5 **36.** -5

37. 1 **38.** C

Mini-Assessment

1. Tell whether $(-3, 5)$ is a solution of the system.

   $y \ge -x + 2$

   $y < 4$ no

2. Graph the system.

   $y > 2x$

   $y \le -2x + 4$

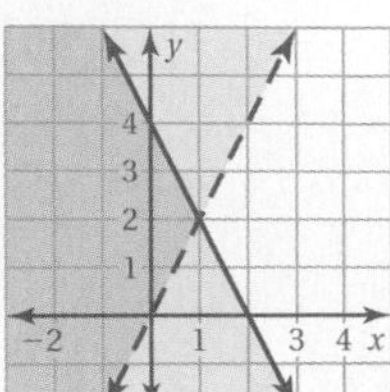

3. On a project, you spend more than twice as much time as your partner who spends at least 5 hours. The total of both of your hours is less than 20. Identify one solution of this situation. you: 13 h, your partner: 6 h

Taking Math Deeper

Exercise 32

One way to look at this problem is to consider the possible whole numbers of hours that you can spend playing games and on rides.

Make an array.

Let g = the number of hours playing games.

Let r = the number of hours on rides.

Make an 8×8 array of ordered pairs (g, r) representing up to 8 hours playing games and up to 8 hours on rides in whole numbers of hours.

(0, 8) (1, 8) (2, 8) (3, 8) (4, 8) (5, 8) (6, 8) (7, 8) (8, 8)
(0, 7) (1, 7) (2, 7) (3, 7) (4, 7) (5, 7) (6, 7) (7, 7) (8, 7)
(0, 6) (1, 6) (2, 6) (3, 6) (4, 6) (5, 6) (6, 6) (7, 6) (8, 6)
(0, 5) (1, 5) **(2, 5)** (3, 5) (4, 5) (5, 5) (6, 5) (7, 5) (8, 5)
(0, 4) (1, 4) (2, 4) (3, 4) (4, 4) (5, 4) (6, 4) (7, 4) (8, 4)
(0, 3) (1, 3) (2, 3) (3, 3) (4, 3) (5, 3) (6, 3) (7, 3) (8, 3)
(0, 2) (1, 2) (2, 2) (3, 2) (4, 2) (5, 2) (6, 2) (7, 2) (8, 2)
(0, 1) (1, 1) (2, 1) (3, 1) (4, 1) (5, 1) (6, 1) (7, 1) (8, 1)
(0, 0) (1, 0) (2, 0) (3, 0) (4, 0) (5, 0) (6, 0) (7, 0) (8, 0)

Use the problem constraints.

You can spend at most 8 hours at the park, so you cross out the ordered pairs that represent a total of more than 8 hours. (Use red.)

You want to spend less than 3 hours playing games, so cross out the ordered pairs with a g-coordinate of 3 or more. (Use blue.)

You want to spend at least 4 hours on rides, so cross out the ordered pairs with an r-coordinate of less than 4. (Use green.)

Any of the 12 remaining ordered pairs can be used to answer the question. One answer is 2 hours playing games and 5 hours on rides.

How do the possibilities change when you consider fractional portions of an hour?

Reteaching and Enrichment Strategies

If students need help. . .	If students got it. . .
Resources by Chapter • Practice A and Practice B • Puzzle Time Record and Practice Journal Practice Differentiating the Lesson Lesson Tutorials Skills Review Handbook	Resources by Chapter • Enrichment and Extension • School-to-Work • Financial Literacy Start the next section

4 **Write a system of linear inequalities represented by the graph.**

24. 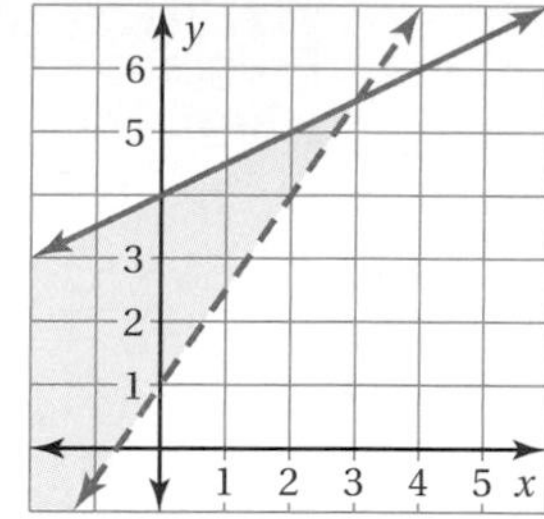

25. 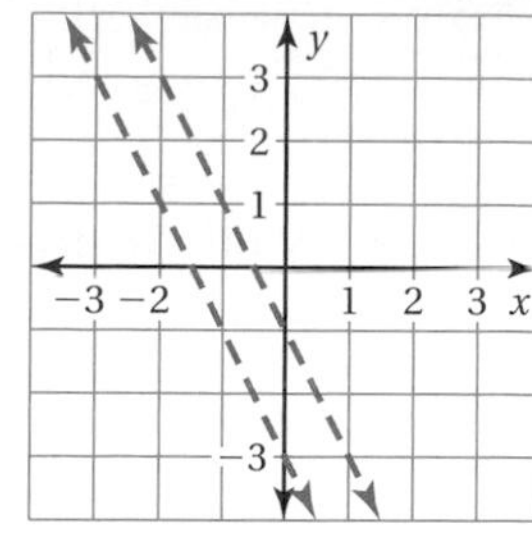

26. 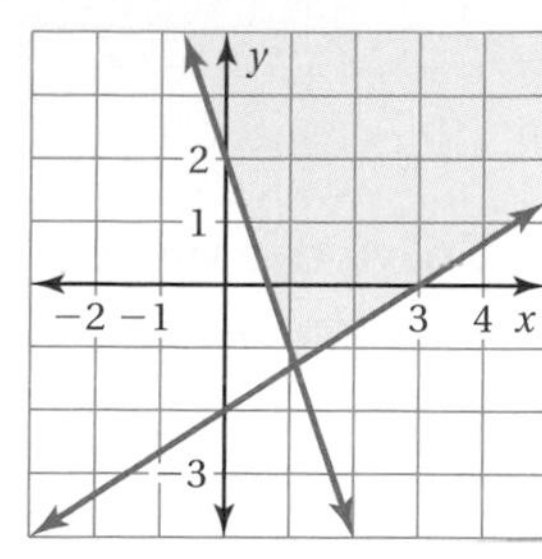

Graph the system of linear inequalities.

27. $y > 1$
$x \geq 2$
$y > x - 1$

28. $y \leq 5x - 6$
$y > 0.5x - 4$
$y < -x + 7$

29. $-4x + 2y < 12$
$6x + y \leq 9$
$-9x + 3y \geq -15$

30. **STRUCTURE** Write a system of linear inequalities that is equivalent to $|y| < x$ where $x > 0$. Graph the system.

31. **REPEATED REASONING** One inequality in a system is $-4x + 2y > 6$. Write another inequality so the system has (a) *no solution* and (b) *infinitely many solutions.*

32. **AMUSEMENT PARK** You have at most 8 hours to spend at an amusement park. You want to spend less than 3 hours playing games and at least 4 hours on rides. How much time can you spend on each activity?

33. **ROAD TRIP** On a road trip, you drive about 70 miles per hour and your friend drives about 60 miles per hour. The plan is to drive less than 15 hours and at least 600 miles each day. Your friend will drive more hours than you. Identify and interpret one solution of this situation.

34. **Geometry** The following points are the vertices of a triangle.

$$(2, 5), (6, -3), (-2, -3)$$

a. Write a system of linear inequalities that represents the triangle.

b. Find the area of the triangle.

Fair Game Review What you learned in previous grades & lessons

Evaluate the expression when $a = -2$, $b = 3$, and $c = -1$. *(Skills Review Handbook)*

35. $4a - bc$

36. $ab + c^2$

37. $-3c - ac$

38. **MULTIPLE CHOICE** What is the solution of $2(x - 4) = -(-x + 3)$? *(Section 1.3)*

Ⓐ $x = -5$ Ⓑ $x = 2$ Ⓒ $x = 5$ Ⓓ $x = 7$

4.3–4.5 Quiz

Solve the system of linear equations by elimination. Check your solution. *(Section 4.3)*

1. $x + 2y = 4$
$-x - y = 2$

2. $2x - y = 1$
$x + 3y - 4 = 0$

3. $3x = -4y + 10$
$4x + 3y = 11$

Solve the system of linear equations. Check your solution. *(Section 4.4)*

4. $3x - 2y = 16$
$6x - 4y = 32$

5. $4y = x - 8$
$-\frac{1}{4}x + y = -1$

6. $-2x + y = -2$
$3x + y = 3$

Use a graph to solve the equation. Check your solution. *(Section 4.4)*

7. $4x - 1 = 2x$

8. $-\frac{1}{2}x + 1 = -x + 1$

9. $1 - 3x = -3x + 2$

Graph the system of linear inequalities. *(Section 4.5)*

10. $y \le \frac{1}{2}x + 1$
$y > -x - 1$

11. $2x + y \ge -3$
$2x < -y - 4$

12. $-5x + y + 1 > 0$
$\frac{3}{4}x + y \ge -2$

Write a system of linear inequalities represented by the graph. *(Section 4.5)*

13.

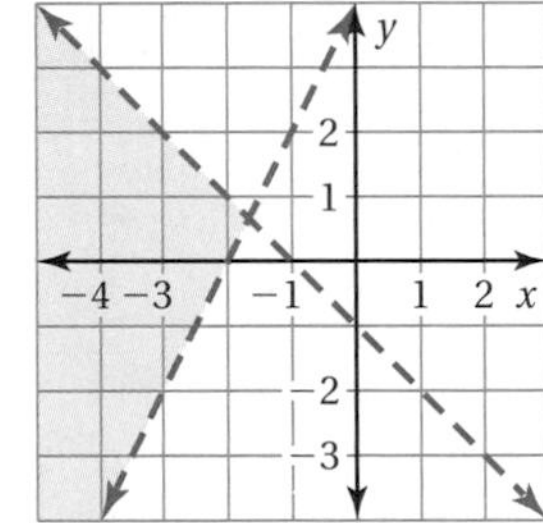

14.

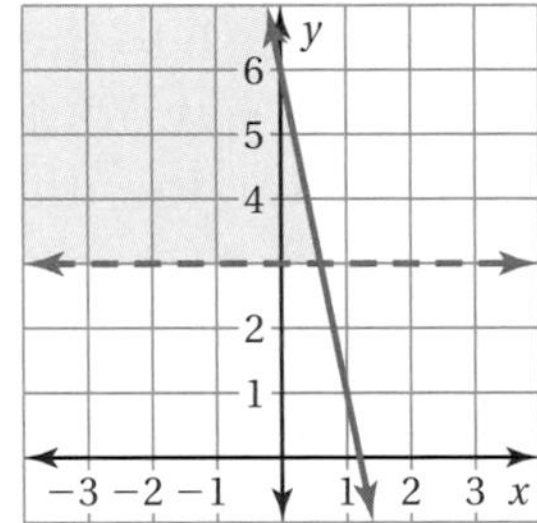

15. RENTALS A business rents bicycles and in-line skates. Bicycle rentals cost \$25 per day and in-line skate rentals cost \$20 per day. The business has 20 rentals today and makes \$455. *(Section 4.3)*

a. Write a system of linear equations that represents this situation.

b. How many bicycle rentals and in-line skate rentals did the business have today?

16. JOBS You earn \$11 per hour delivering pizzas. You also work part-time at a convenience store where you earn \$9 per hour. You want to earn at least \$150 per week, but you can only work 25 hours per week. How many hours can you work at each job? *(Section 4.5)*

Alternative Assessment Options

Math Chat	Student Reflective Focus Question
Structured Interview	Writing Prompt

Structured Interview

Interviews can occur formally or informally. Ask a student to perform a task and to explain it as they work. Have them describe their thought process. Probe the student for more information. Do not ask leading questions. Keep a rubric or notes.

Teacher Prompts	Student Answers	Teacher Notes
Tell me a story about earning money. Include: Job 1: \$9/h, Job 2: \$12/h Hours: at most 25 Earn: \$220/wk or more	Last week, I worked 12 hours at \$9 per hour and 10 hours at \$12 per hour for a total of \$228	Student understands the constraints of the system concerning hours and total earnings.

Study Help Sample Answers

2.

system of linear equations solution of a system of linear equations	Solving systems of linear equations by elimination Step 1: Rewrite equations as needed. Step 2: Combine equations to eliminate a variable. Step 3: Solve for the remaining variable. Step 4: Substitute the value from Step 3 into one of the original equations and solve. Example: Solve the system. $-x + y = 5$ $3x - y = -1$ Step 1: Don't need to rewrite – can add to eliminate y. Step 2: $-x + y = 5$ Add equations. $\underline{3x - y = -1}$ $2x \quad = 4$ Step 3: $x = 2$ Solve for x. Step 4: $-(2) + y = 5$ Substitute 2 for x in Eq 1. $y = 7$ Solve for y. The solution is (2, 7).
What happens when adding equations eliminates both variables?	

3. Available at *BigIdeasMath.com*

Reteaching and Enrichment Strategies

If students need help. . .	If students got it. . .
Resources by Chapter • Study Help • Practice A and Practice B • Puzzle Time Lesson Tutorials *BigIdeasMath.com* Practice Quiz Practice from the Test Generator	Resources by Chapter • Enrichment and Extension • School-to-Work Game Closet at *BigIdeasMath.com* Start the Chapter Review

Answers

1. $(-8, 6)$
2. $(1, 1)$
3. $(2, 1)$
4. infinitely many solutions
5. no solution
6. $(1, 0)$
7. $x = 0.5$
8. $x = 0$
9. no solution
10.

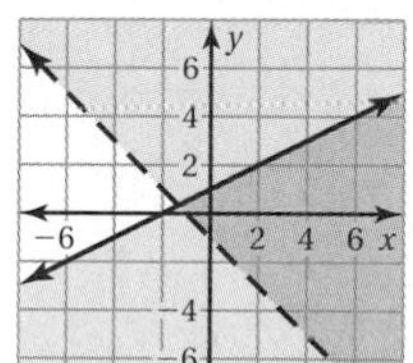

11–12. See Additional Answers.

13. $y < -x - 1$; $y > 2x + 4$
14. $y > 3$; $y \leq -5x + 6$
15. a. $x + y = 20$; $25x + 20y = 455$
    b. 11 bicycles, 9 in-line skates
16. *Sample answers:*
    pizza: 5 h, store: 12 h
    pizza: 12 h, store: 10 h

Technology For the Teacher

Answer Presentation Tool

Assessment Book

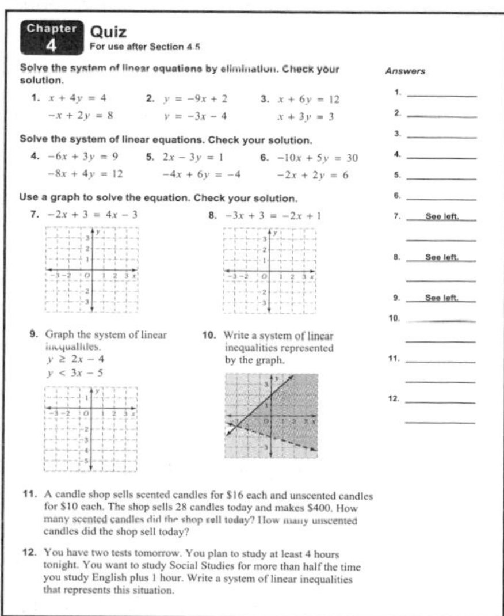

Chapter 4 Quiz
For use after Section 4.5

Solve the system of linear equations by elimination. Check your solution.

1. $x + 4y = 4$; $-x + 2y = 8$
2. $y = -9x + 2$; $y = -3x - 4$
3. $x + 6y = 12$; $x + 3y = 3$

Solve the system of linear equations. Check your solution.

4. $-6x + 3y = 9$; $-8x + 4y = 12$
5. $2x - 3y = 1$; $-4x + 6y = -4$
6. $-10x + 5y = 30$; $-2x + 2y = 6$

Use a graph to solve the equation. Check your solution.

7. $-2x + 3 = 4x - 3$
8. $-3x + 3 = -2x + 1$

9. Graph the system of linear inequalities.
   $y \geq 2x - 4$
   $y < 3x - 5$

10. Write a system of linear inequalities represented by the graph.

11. A candle shop sells scented candles for \$16 each and unscented candles for \$10 each. The shop sells 28 candles today and makes \$400. How many scented candles did the shop sell today? How many unscented candles did the shop sell today?

12. You have two tests tomorrow. You plan to study at least 4 hours tonight. You want to study Social Studies for more than half the time you study English plus 1 hour. Write a system of linear inequalities that represents this situation.

Answers

1. ____ 2. ____ 3. ____ 4. ____ 5. ____ 6. ____ 7. See left. 8. See left. 9. See left. 10. ____ 11. ____ 12. ____

For the Teacher

Additional Review Options

- Big Ideas Test Generator
- Game Closet at *BigIdeasMath.com*
- Vocabulary Puzzle Builder
- Resources by Chapter
  - Puzzle Time
  - Study Help

Answers

1. $(5, 7)$
2. $(6, -2)$
3. $(-4, -2)$

Review of Common Errors

Exercises 1–3

- Students may not show enough of the graph, so the lines will not intersect. Encourage them to extend their lines until they intersect.

Exercises 4–6

- Students may find one coordinate of the solution and stop there. Remind them that the answer is a coordinate pair representing both an x-value and a y-value.

Exercise 7

- Students may make careless errors. Remind students to line up like terms neatly in columns to avoid confusion about what terms to add or subtract. Also, take care with subtraction—it may help to change the sign of each term in the equation to be subtracted, so you can just add the terms.

Exercises 8–10

- Students may see that the slope is the same for both equations and immediately say that the system of linear equations has no solution. Remind them that they need to compare the slope *and* y-intercepts when determining the number of solutions. Encourage them to check algebraically and graphically that the system of linear equations has the number of solutions they found.

Exercise 11

- Students may forget the process. Remind them to write a system of two equations by setting y equal to each side of the equation.

Exercises 12–14

- Students may shade the wrong half-plane for one or both inequalities. Remind students to use a test point when they choose the half-plane to shade.

Exercises 12–14

- Students may use the wrong style of boundary line for one or both inequalities. Remind students to use a solid line for $\leq$ and $\geq$ and a dashed line for $<$ and $>$.

4 Chapter Review

Check It Out
Vocabulary Help
BigIdeasMath.com

Review Key Vocabulary

system of linear equations, *p. 156*
solution of a system of linear equations, *p. 156*
system of linear inequalities, *p. 186*
solution of a system of linear inequalities, *p. 186*
graph of a system of linear inequalities, *p. 186*

Review Examples and Exercises

4.1 Solving Systems of Linear Equations by Graphing *(pp. 154–159)*

Solve the system by graphing. $\quad \mathbf{y = -2x}$ Equation 1

$\mathbf{y = 3x + 5}$ Equation 2

Step 1: Graph each equation.

Step 2: Estimate the point of intersection. The graphs appear to intersect at $(-1, 2)$.

Step 3: Check the point from Step 2.

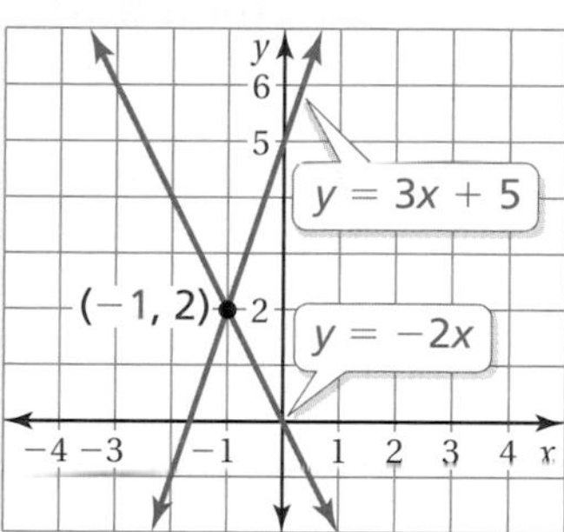

$y = -2x$	$y = 3x + 5$
$2 \stackrel{?}{=} -2(-1)$	$2 \stackrel{?}{=} 3(-1) + 5$
$2 = 2$ ✓	$2 = 2$ ✓

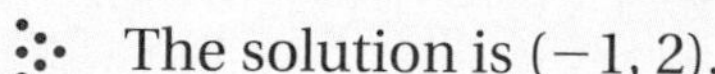

∴ The solution is $(-1, 2)$.

Exercises

Solve the system of linear equations by graphing.

1. $y = 2x - 3$
$y = x + 2$

2. $y = -x + 4$
$x + 3y = 0$

3. $x - y = -2$
$2x - 3y = -2$

4.2 Solving Systems of Linear Equations by Substitution *(pp. 160–165)*

Solve the system by substitution. $\quad \mathbf{x = 1 + y}$ Equation 1

$\mathbf{x + 3y = 13}$ Equation 2

Step 1: Equation 1 is already solved for x.

Step 2: Substitute $1 + y$ for x in Equation 2.

$1 + y + 3y = 13$ Substitute $1 + y$ for x.

$y = 3$ Solve for y.

Step 3: Substituting 3 for y in Equation 1 gives $x = 4$.

∴ The solution is $(4, 3)$.

Exercises

Solve the system of linear equations by substitution. Check your solution.

4. $y = -3x - 7$
   $y = x + 9$

5. $\frac{1}{2}x + y = -4$
   $y = 2x + 16$

6. $-x + 5y = 28$
   $x + 3y = 20$

4.3 Solving Systems of Linear Equations by Elimination *(pp. 168–175)*

You have a total of 5 quarters and dimes in your pocket. The value of the coins is \$0.80. Write and solve a system of linear equations to find the number *x* of dimes and the number *y* of quarters in your pocket.

Use a verbal model to write a system of linear equations.

Number of dimes, x + Number of quarters, y = Number of coins

Value of a dime • Number of dimes, x + Value of a quarter • Number of quarters, y = Total value

The system is $x + y = 5$ and $0.1x + 0.25y = 0.8$.

Step 1: Multiply Equation 2 by 10.

$x + y = 5$		$x + y = 5$	Equation 1
$0.1x + 0.25y = 0.8$	**Multiply by 10.**	$x + 2.5y = 8$	Revised Equation 2

Step 2: Subtract the equations.

$x + y = 5$	Equation 1
$x + 2.5y = 8$	Revised Equation 2
$-1.5y = -3$	Subtract the equations.

Step 3: Solving the equation $-1.5y = -3$ gives $y = 2$.

Step 4: Substitute 2 for y in one of the original equations and solve for x.

$x + y = 5$	Equation 1
$x + 2 = 5$	Substitute 2 for y.
$x = 3$	Subtract 2 from each side.

So, you have 3 dimes and 2 quarters in your pocket.

Exercises

7. **GIFT BASKET** A gift basket that contains jars of jam and packages of bread mix costs \$45. There are 8 items in the basket. Jars of jam cost \$6 each and packages of bread mix cost \$5 each. Write and solve a system of linear equations to find the number of jars of jam and the number of packages of bread mix in the gift basket.

Review Game

Systems of Inequalities

Materials per Group:

- 2 sheets of graph paper
- 2 pencils
- 2 books

Directions

- Pair students. The game is for 2 players.
- Each player writes and graphs a system of inequalities on graph paper. Players do not show each other their systems. They each stand up a book to hide their graphs.
- The object of the game is to be the first to determine your opponent's system.
- The players give each other the equations of the boundary lines for their systems. Each player graphs the opponent's boundary lines on their graph paper in a different coordinate plane from their own system.
- Player 1 takes the first turn by calling off the coordinates of a point to player 2. Player 2 acknowledges whether the point is or is *not* a solution of his or her system. Player 1 keeps track of "hits" and "misses" on their graph of the opponent's boundary lines. Plot a point to represent a hit, or an "X" to represent a miss.
- Players alternate turns asking ordered pairs.
- At the end of each turn, the player has the option of naming the opponent's system by writing the inequalities. If the inequalities are incorrect, however, the player misses a turn—the opponent gets to check two points before the player gets another turn.
- In essence, you are trying to determine whether each inequality of your opponent's system involves $<$, $\leq$, $>$, or $\geq$ by strategically choosing points. The game can be modified to systems of more than two inequalities.

Who wins?

The first player to correctly name the opponent's system of linear inequalities wins the game.

For the Student
Additional Practice

- Lesson Tutorials
- Study Help (textbook)
- Student Website
  Multi-Language Glossary
  Practice Assessments

Answers

4. $(-4, 5)$ **5.** $(-8, 0)$

6. $(2, 6)$

7. $x + y = 8$
$6x + 5y = 45$; 5 jars of jam, 3 packages of bread mix

8. $(-5, 0)$

9. infinitely many solutions; all points on the line $y = \frac{3}{2}x - \frac{1}{2}$

10. no solution

11. $x = -4$

12.

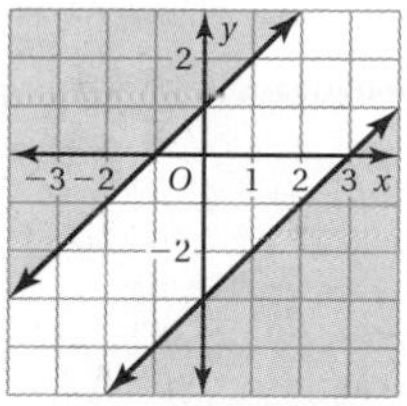

13.

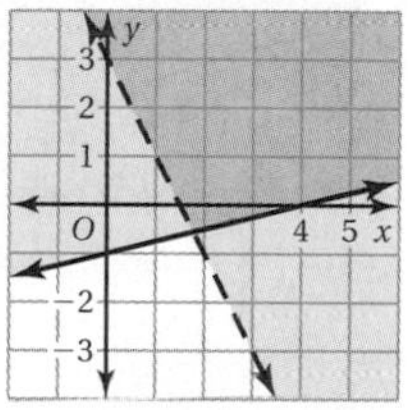

14.

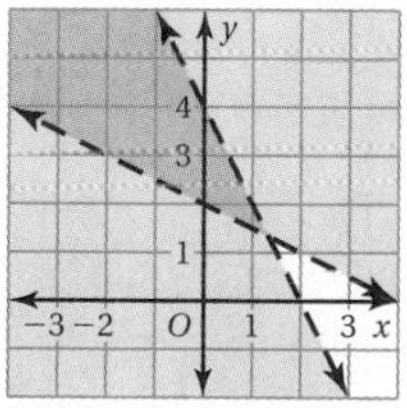

My Thoughts on the Chapter

Teacher Tip

Not allowed to write in your teaching edition? Use sticky notes to record your thoughts.

What worked. . .

What did not work. . .

What I would do differently. . .

4.4 Solving Special Systems of Linear Equations (pp. 176–183)

Solve the system. $y = -5x - 8$ Equation 1

$y = -5x + 4$ Equation 2

Solve by substitution. Substitute $-5x + 4$ for y in Equation 1.

$y = -5x - 8$ Equation 1

$-5x + 4 = -5x - 8$ Substitute $-5x + 4$ for y.

$4 \neq -8$ ✗ Add $5x$ to each side.

∴ The equation $4 = -8$ is never true. So, the system of linear equations has no solution.

Exercises

Solve the system of linear equations. Check your solution.

8. $x + 2y = -5$
   $x - 2y = -5$

9. $3x - 2y = 1$
   $9x - 6y = 3$

10. $8x - 2y = 16$
    $-4x + y = 8$

11. Use a graph to solve $2x - 9 = 7x + 11$. Check your solution.

4.5 Systems of Linear Inequalities (pp. 184–191)

Graph the system. $y < x - 2$ Inequality 1

$y \geq 2x - 4$ Inequality 2

Step 1: Graph each inequality.

Step 2: Find the intersection of the half-planes. One solution is $(0, -3)$.

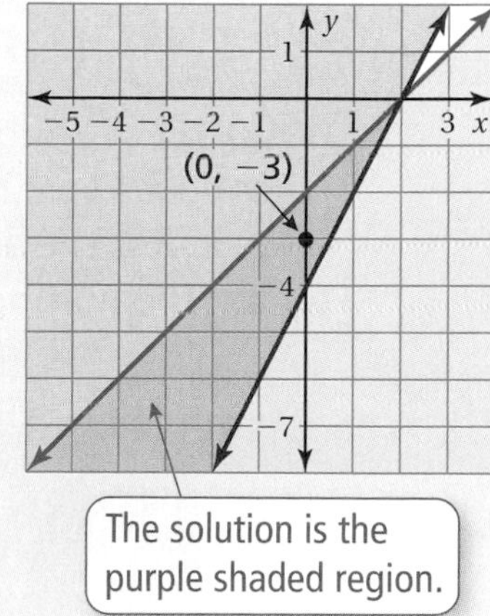

The solution is the purple shaded region.

Check

Verify that $(0, -3)$ is a solution of each inequality.

Inequality 1	Inequality 2
$y < x - 2$	$y \geq 2x - 4$
$-3 \overset{?}{<} 0 - 2$	$-3 \overset{?}{\geq} 2(0) - 4$
$-3 < -2$ ✓	$-3 \geq -4$ ✓

Exercises

Graph the system of linear inequalities.

12. $y \leq x - 3$
    $y \geq x + 1$

13. $y > -2x + 3$
    $y \geq \frac{1}{4}x - 1$

14. $x + 2y > 4$
    $2x + y < 4$

4 Chapter Test

Solve the system of linear equations by graphing.

1. $y = 4 - x$
   $y = x - 4$

2. $y = \frac{1}{2}x + 10$
   $y = 4x - 4$

3. $y + x = 0$
   $3y + 6x = -9$

Solve the system of linear equations by substitution. Check your solution.

4. $-3x + y = 2$
   $-x + y - 4 = 0$

5. $x + y = 20$
   $y = 2x - 1$

6. $x - y = 3$
   $x + 2y = -6$

Solve the system of linear equations by elimination. Check your solution.

7. $2x + y = 3$
   $x - y = 3$

8. $x + y = 12$
   $3x = 2y + 6$

9. $-2x + y + 3 = 0$
   $3x + 4y = -1$

Without graphing, determine whether the system of linear equations has *one solution*, *infinitely many solutions*, or *no solution*. Explain your reasoning.

10. $y = 4x + 8$
    $y = 5x + 1$

11. $2y = 16x - 2$
    $y = 8x - 1$

12. $y = -3x + 2$
    $6x + 2y = 10$

Use a graph to solve the equation. Check your solution.

13. $\frac{1}{4}x - 4 = \frac{3}{4}x + 2$

14. $8x - 14 = -2x - 4$

Graph the system of linear inequalities.

15. $y > \frac{1}{2}x + 4$
    $2y \leq x + 4$

16. $y \geq -\frac{2}{3}x + 1$
    $-3x + y > -2$

17. $x + y < 1$
    $5x + y > 4$

18. **BOUQUET** A bouquet of lilies and tulips has 12 flowers. Lilies cost \$3 each and tulips cost \$2 each. The bouquet costs \$32. Write and solve a system of linear equations to find the number of lilies and tulips in the bouquet.

19. **DINNER** How much does it cost for two specials and two glasses of milk?

20. **SHOPPING** You have \$110 to spend at the mall. You want to buy at most 6 articles of clothing. A clothing store sells shirts for \$12 and pairs of pants for \$18. You want to have at least \$20 left over for food.

    a. Write and graph a system of linear inequalities that represents this situation.

    b. How many shirts and pairs of pants can you buy at the store?

Test Item References

Chapter Test Questions	Section to Review	Common Core State Standards
1–3	4.1	A.CED.3, 8.EE.8, A.REI.6
4–6, 18	4.2	A.CED.3, 8.EE.8b, 8.EE.8c, A.REI.6
7–9, 19	4.3	A.CED.3, 8.EE.8b, 8.EE.8c, A.REI.5, A.REI.6
10–14	4.4	A.CED.3, 8.EE.8, A.REI.6, A.REI.11
15–17, 20	4.5	A.CED.3, A.REI.12

Test-Taking Strategies

Remind students to quickly look over the entire test before they start so that they can budget their time. This test involves solving systems of equations and inequalities, and the answers take on several different forms. So, it is important that students use the **Stop** and **Think** strategy before they answer a question.

Common Assessment Errors

- **Exercises 4–6** Students may find one coordinate of the solution and stop there. Remind them that the answer is a coordinate pair representing both an x-value and a y-value.
- **Exercises 10–12** Students may only compare the slopes. Remind them that they must also compare the y-intercepts to determine the number of solutions.
- **Exercises 13–14** Students may forget the process. Remind them to write a system of two equations by setting y equal to each side of the equation.
- **Exercises 15–17** Students may shade the wrong half-plane for one or both inequalities. Remind students to use a test point when they choose the half-plane to shade.
- **Exercises 15–17** Students may use the wrong style of boundary line for one or both inequalities. Remind students to use a solid line for $\leq$ and $\geq$ and a dashed line for $<$ and $>$.

Reteaching and Enrichment Strategies

If students need help...	If students got it...
Resources by Chapter • Practice A and Practice B • Puzzle Time Record and Practice Journal Practice Differentiating the Lesson Lesson Tutorials Practice from the Test Generator Skills Review Handbook	Resources by Chapter • Enrichment and Extension • School-to-Work • Financial Literacy Game Closet at *BigIdeasMath.com* Start Standardized Test Practice

Answers

1. (4, 0) **2.** (4, 12)

3. (−3, 3) **4.** (1, 5)

5. (7, 13) **6.** (0, −3)

7. (2, −1) **8.** (6, 6)

9. (1, −1)

10. one solution; The lines have different slopes.

11. infinitely many solutions; The equations represent the same line.

12. no solution; The lines have the same slope and different y-intercepts.

13. $x = -12$ **14.** $x = 1$

15–18. See Additional Answers.

19. $16.10

20. a. $x + y \leq 6$

$12x + 18y \leq 90$

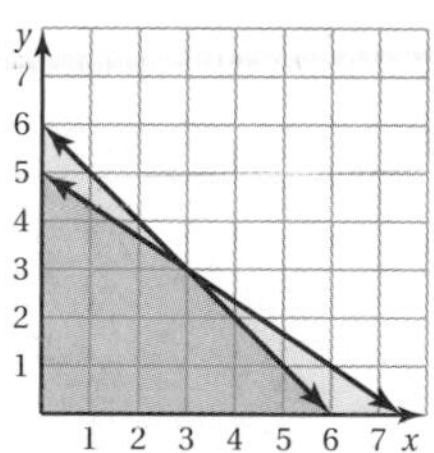

b. *Sample answer:* 2 shirts, 3 pants

Assessment Book

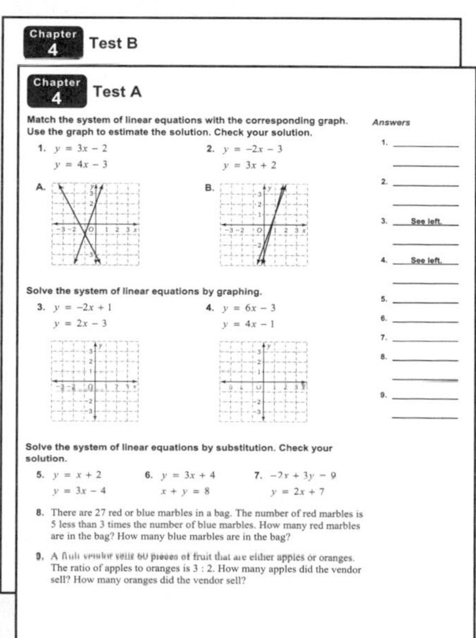

Chapter 4 Test B

Chapter 4 Test A

Match the system of linear equations with the corresponding graph. Use the graph to estimate the solution. Check your solution.

1. $y = 3x - 2$
   $y = 4x - 3$
2. $y = -2x - 3$
   $y = 3x + 2$

A. B.

Solve the system of linear equations by graphing.

3. $y = -2x + 1$
   $y = 2x - 3$
4. $y = 6x - 3$
   $y = 4x - 1$

Solve the system of linear equations by substitution. Check your solution.

5. $y = x + 2$
   $y = 3x - 4$
6. $y = 3x + 4$
   $x + y = 8$
7. $-2x + 3y = 9$
   $y = 2x + 7$

8. There are 27 red or blue marbles in a bag. The number of red marbles is 5 less than 3 times the number of blue marbles. How many red marbles are in the bag? How many blue marbles are in the bag?

9. A fruit vendor sells 60 pieces of fruit that are either apples or oranges. The ratio of apples to oranges is 3 : 2. How many apples did the vendor sell? How many oranges did the vendor sell?

Answers

1. ____
2. ____
3. See left.
4. See left.
5. ____
6. ____
7. ____
8. ____
9. ____

Test-Taking Strategies

Available at *BigIdeasMath.com*

After Answering Easy Questions, Relax
Answer Easy Questions First
Estimate the Answer
Read All Choices before Answering
Read Question before Answering
Solve Directly or Eliminate Choices
Solve Problem before Looking at Choices
Use Intelligent Guessing
Work Backwards

About this Strategy

When taking a multiple choice test, be sure to read each question carefully and thoroughly. Look closely for words that change the meaning of the question like not, never, all, every, and always.

Answers

1. D
2. F
3. 4
4. A

Technology For the Teacher

Big Ideas Test Generator

Item Analysis

1. **A.** The student finds a solution point for the first equation.

   B. The student finds a solution point for the first equation.

   C. The student makes an error rewriting the second equation in slope-intercept form.

   D. Correct answer

2. **F.** Correct answer

   G. The student uses the wrong symbol to represent shading above the boundary line.

   H. The student uses the wrong symbol to represent the dashed boundary line.

   I. The student uses the wrong symbol to represent shading above the boundary line and a dashed boundary line.

3. **Gridded Response:** Correct answer: 4

   Common error: The student uses the reciprocal of the slope instead of the opposite of the reciprocal of the slope.

4. **A.** Correct answer

   B. The student uses the wrong style of boundary line for the given symbol.

   C. The student shades the wrong half-plane.

   D. The student shades the wrong half-plane and uses the wrong style of boundary line for the given symbol.

5. **F.** Correct answer

   G. The student checks solutions by substituting for *x* and *y* in reverse order.

   H. The student finds a solution for the first equation.

   I. The student finds a solution for the second equation.

4 Standardized Test Practice

Test-Taking Strategy
Read Question Before Answering

A leash? Me? I don't think so!
I buy 2 toys and 1 new leash for \$7. You buy 3 toys and 1 new leash for \$9. How much is a new leash?
Ⓐ \$2 Ⓑ \$3 Ⓒ \$4 Ⓓ \$5

"Take your time and read the question carefully before choosing your answer."

1. What is the solution of the system of equations shown below? *(A.REI.6)*

$$y = -\frac{2}{3}x - 1$$

$$4x + 6y = -6$$

A. $\left(-\frac{3}{2}, 0\right)$ **C.** No solution

B. $(0, -1)$ **D.** Infinitely many solutions

2. Which inequality is shown in the coordinate plane? *(A.REI.12)*

F. $y > -5$

G. $y < -5$

H. $y \geq -5$

I. $y \leq -5$

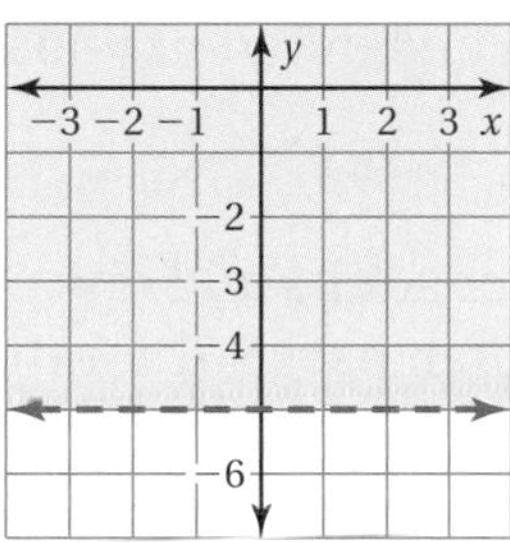

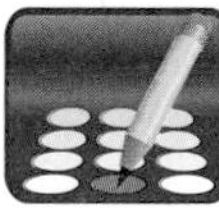

3. What is the slope of a line that is perpendicular to the line $y = -0.25x + 3$? *(F.IF.6)*

4. Which graph shows the solution of $-4x + y > -3$? *(A.REI.12)*

A.

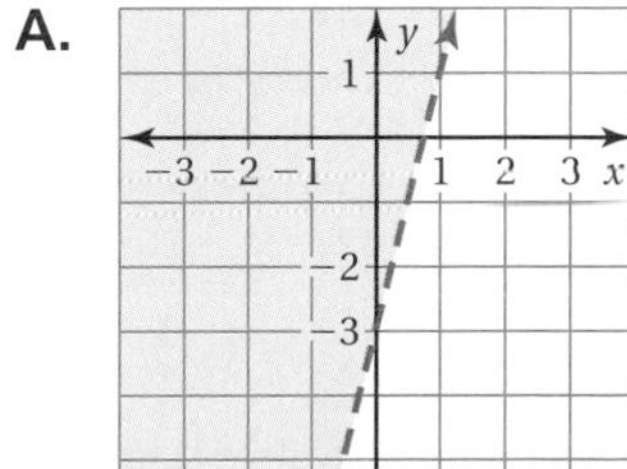

C.

y
1
−3 −2 −1 1 2 3 x
−2
−3

B.

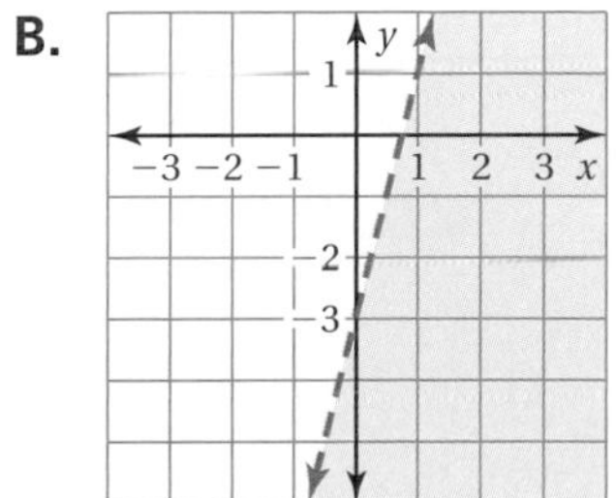

D.

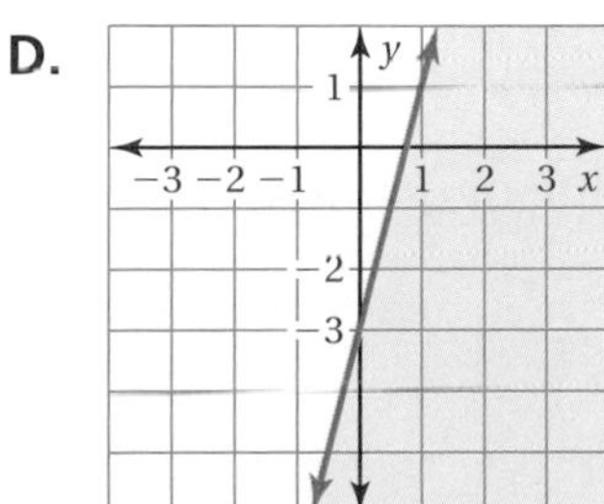

5. Which point is a solution of the system of equations shown below? *(A.REI.6)*

$$x + 3y = 10$$
$$x = 2y - 5$$

F. $(1, 3)$

G. $(3, 1)$

H. $(55, -15)$

I. $(-35, -15)$

6. A system of two linear equations has no solution. What can you conclude about the graphs of the two equations? *(8.EE.8b)*

A. The lines have the same slope and the same y-intercept.

B. The lines have the same slope and different y-intercepts.

C. The lines have different slopes and the same y-intercept.

D. The lines have different slopes and different y-intercepts.

7. A scenic train ride has one price for adults and one price for children. One family of two adults and two children pays \$62 for the train ride. Another family of one adult and four children pays \$70. Which system of linear equations can be used to find the price x for an adult and the price y for a child? *(8.EE.8c)*

F. $2x + 2y = 70$
$x + 4y = 62$

G. $x + y = 62$
$x + y = 70$

H. $2x + 2y = 62$
$4x + y = 70$

I. $2x + 2y = 62$
$x + 4y = 70$

8. Which graph shows the solution of $-\frac{x}{4} - 10 > -18$? *(A.REI.3)*

A. 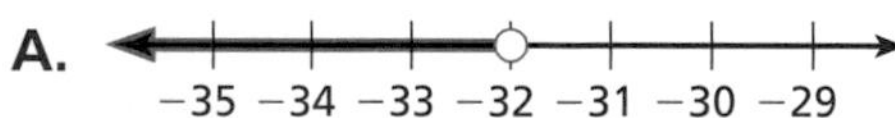

B. 29 30 31 32 33 34 35

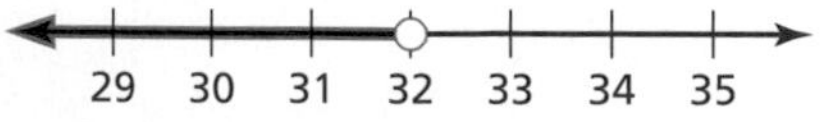

C. −35 −34 −33 −32 −31 −30 −29

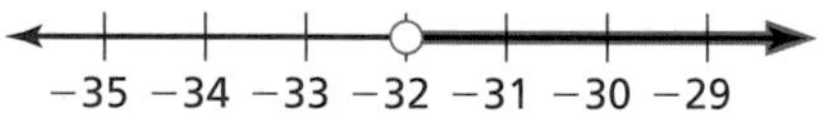

D. 29 30 31 32 33 34 35

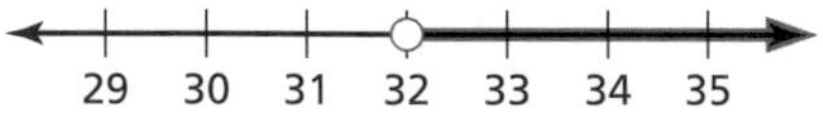

9. What value of w makes the equation below true? *(A.REI.3)*

$$7w - 3w = 2(3w + 11)$$

Item Analysis (continued)

6. **A.** The student thinks the lines must coincide.
B. Correct answer
C. The student reverses the idea that the slopes are the same and the y-intercepts are different.
D. The student incorrectly reasons that both the slopes and y-intercepts must be different for there to be no solution.

7. **F.** The student uses the wrong cost for each situation.
G. The student defines x as the amount spent on adults and y as the amount spent on children.
H. The student defines x as the amount per child and y as the amount per adult.
I. Correct answer

8. **A.** The student fails to reverse the sign on the right side when each side is multiplied by -4.
B. Correct answer
C. The student fails to reverse both the sign on the right side and the inequality symbol when each side is multiplied by -4.
D. The student fails to reverse the inequality symbol when each side is multiplied by -4.

9. **Gridded Response:** Correct answer: -11
Common error: $-\frac{11}{2}$; The student fails to use the distributive property correctly by multiplying 11 by 2.

10. **F.** The student forgets to reverse the sign of -1 when subtracting it from 4.
G. The student finds a perpendicular line instead of a parallel line.
H. Correct answer
I. The student determines the slope incorrectly as the change in x divided by the change in y.

11. **A.** The student considers the graph of the second inequality.
B. The student associates the negative number with Quadrant II.
C. Correct answer
D. The student thinks the system includes the graphs of both inequalities.

Answers

5. F
6. B
7. I
8. B
9. -11

Answers

10. H
11. C
12. $7.50
13. H
14. B
15. G

Answer for Extra Example

1. **A.** The student considers the graph of the individual inequalities rather than the solution.
   B. The student considers the graph of the individual inequalities rather than the solution.
   C. The student chooses the most familiar situation.
   D. Correct answer

Item Analysis (continued)

12. **2 points** The student demonstrates a thorough understanding of how to solve a system of linear equations, explains the work fully, and applies the solution correctly to the context of the problem. The solution is (7.5, 10). The x-value represents the cost of each T-shirt, $7.50.

1 point The student's work and explanations demonstrate a lack of essential understanding. The system of linear equations was set up or solved incorrectly, or, if the solution is correct, was applied incorrectly to the problem.

0 points The student provides no response, a completely incorrect or incomprehensible response, or a response that demonstrates insufficient understanding of how to solve a system of linear equations.

13. **F.** The student fails to distribute 12 correctly.
G. The student incorrectly isolates y by dividing each side by $12y$ instead of subtracting $12y$.
H. Correct answer
I. The student first divides each side by 32, then fails to distribute $\frac{3}{8}$ correctly.

14. **A.** The student adds $\frac{1}{3}$ instead of multiplying by $\frac{1}{3}$.
B. Correct answer
C. The student represents the sum of one-third of a number and ten as $\frac{1}{3}(n + 10)$
D. The student adds 10 to the wrong side of the equation.

15. **F.** The student adds $7y$ to the right side instead of subtracting.
G. Correct answer
H. The student inverts the fraction.
I. The student fails to divide the right side by 4.

Extra Example

1. The solution of a system of two linear inequalities is a half-plane. Which is true about graphs of the inequalities? *(A.REI.12)*
   A. The boundary lines are solid lines.
   B. The boundary lines are dashed lines.
   C. The boundary lines intersect.
   D. The boundary lines are parallel.

10. The graph of which equation is parallel to the line that passes through the points $(-1, 5)$ and $(4, 7)$? *(F.IF.6)*

F. $y = \frac{2}{3}x + 6$

G. $y = -\frac{5}{2}x + 4$

H. $y = \frac{2}{5}x + 1$

I. $y = \frac{5}{2}x - 1$

11. Which of the following is true for the system of inequalities shown below? *(A.REI.12)*

$$y < -2$$

$$y > 3x + 5$$

A. The graph of the system is located in Quadrants I, II, and III.

B. The graph of the system is located in Quadrant II only.

C. The graph of the system is located in Quadrant III only.

D. The graph of the system is located in Quadrants I, II, III, and IV.

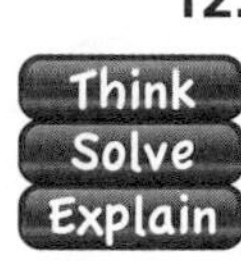

12. You buy 3 T-shirts and 2 pairs of shorts for \$42.50. Your friend buys 5 T-shirts and 3 pairs of shorts for \$67.50. Use a system of linear equations to find the cost of each T-shirt. Show your work and explain your reasoning. *(8.EE.8c)*

13. The two figures have the same area. What is the value of y? *(A.CED.1)*

F. $\frac{1}{4}$

G. $\frac{15}{8}$

H. 3

I. 8

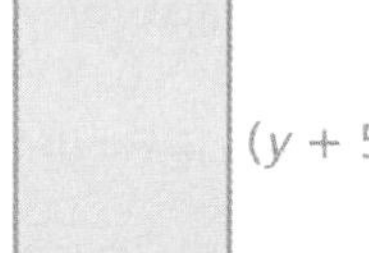

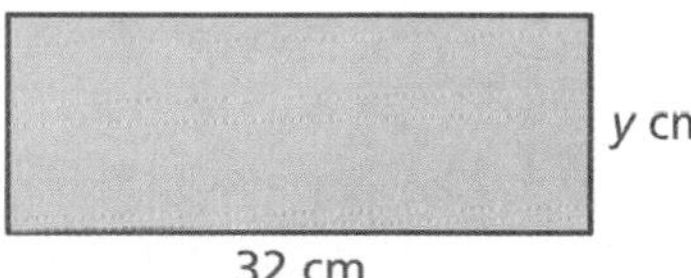

14. The sum of one-third of a number and 10 is equal to 13. What is the number? *(A.CED.1)*

A. $\frac{8}{3}$ **B.** 9 **C.** 29 **D.** 69

15. Solve the equation $4x + 7y = 16$ for x. *(A.CED.4)*

F. $x = 4 + \frac{7}{4}y$

G. $x = 4 - \frac{7}{4}y$

H. $x = 4 + \frac{4}{7}y$

I. $x = 16 - 7y$

5 Linear Functions

"Here's how I remember that the range is the *y*-values."

"I draw a cabin on the *y*-axis. Then, I hum 'Home, Home on the range'."

"It is my treat-converter function machine. However many cat treats I input, the machine outputs TWICE that many dog biscuits. Isn't that cool?"

Connections to Previous Learning

- Construct and analyze tables, graphs, and equations that represent linear relationships between dependent and independent variables.
- Identify and plot ordered pairs in all four quadrants.

- Understand the connections between proportional relationships, lines, and linear equations.

- Construct and analyze tables, graphs, and models to represent, analyze, and solve problems related to linear functions, including analysis of domain, range, and the difference between discrete and continuous data.
- Translate among representations of linear functions in words, tables, graphs, equations, and function notation.
- Compare graphs of linear and nonlinear functions.

Math in History

Formulas for the volumes of solids have been known and used in many cultures.

★ There are records of Japanese mathematicians using a Chinese text that was written around 200 B.C. The third section of the text contains methods for finding the volumes of prisms, cylinders, pyramids, and cones.

★ Around 628 A.D., a book called Brahmasphutasiddhanta was written by the Indian mathematician Brahmagupta. The book has 25 chapters. In one of them, he describes methods for calculating the volume of a prism and a cone.

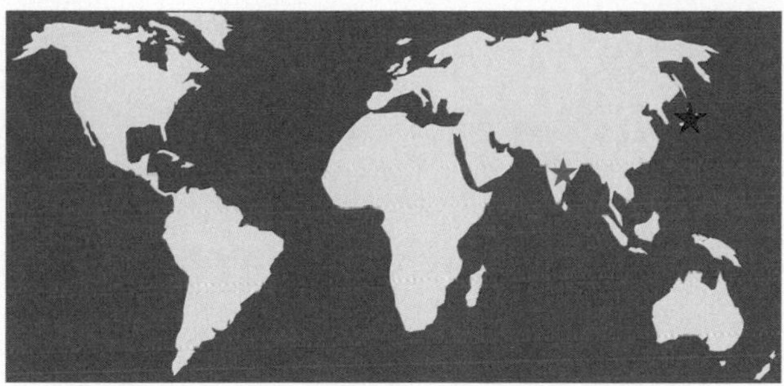

Pacing Guide for Chapter 5

Chapter Opener	1 Day
Section 1	2 Days
Section 2	1 Day
Section 3	2 Days
Study Help / Quiz	1 Day
Section 4	4 Days
Section 5	2 Days
Section 6	1 Day
Chapter Review / Chapter Tests	2 Days
Total Chapter 5	16 Days
Year-to-Date	62 Days

Check Your Resources

- Record and Practice Journal
- Resources by Chapter
- Skills Review Handbook
- Assessment Book
- Worked-Out Solutions

Technology For the Teacher

The Dynamic Planning Tool
Editable Teacher's Resources at
BigIdeasMath.com

Common Core State Standards

6.EE.9 Use variables to represent two quantities . . . that change in relationship to one another . . .. Analyze the relationship between the dependent and independent variables using graphs and tables . . ..

Additional Topics for Review

- Adding and subtracting decimals and fractions
- Multiplying fractions and decimals
- Plotting points and identifying coordinates in Quadrant I

Try It Yourself

1. As the input decreases by 2, the output increases by 3.
2. As the input increases by 2, the output increases by 1.
3. As the input decreases by 1, the output decreases by 3.5.

Record and Practice Journal

1. As the input increases by 1, the output increases by 2.
2. As the input increases by 2, the output increases by 5.
3. As the input increases by 4, the output increases by 3.
4. As the input increases by 1, the output decreases by 7.
5. As the hours increase by 1, the customers increase by 15.
6. Input Output

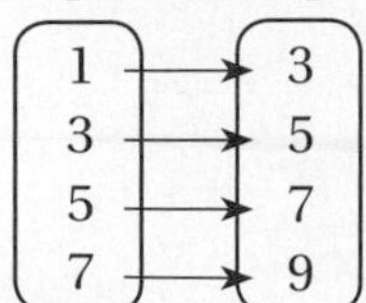

As the input increases by 2, the output increases by 2.

7–10. See Additional Answers.

Math Background Notes

Vocabulary Review

- Input
- Output
- Mapping Diagram

Recognizing Patterns

- Students have been working with patterns since elementary school. They should know how to use mapping diagrams and In and Out tables. They should know how to use graphs to represent and identify patterns.
- Remind students that mapping diagrams, In and Out tables, graphs, and words are four different ways that can be used to express or describe a pattern.
- **Common Error:** Some students may try to find a pattern between the input and output values. Remind them that they are not searching for how the input values relate to output values but rather, how the input values relate to one another and how the output values relate to one another.
- **Teaching Tip:** Some students may have difficulty identifying the pattern. Encourage these students to search for context clues first. For example, are the input values getting progressively greater? If so, the pattern will most likely involve addition or multiplication. If the numbers are all even, try adding or multiplying by even numbers first to try to find the pattern.
- **Teaching Tip:** Rather than trying to identify the pattern using the points on a graph, encourage students to transfer the information contained in the ordered pairs into a mapping diagram as in Example 3. Then use the mapping diagram to find the pattern.

Reteaching and Enrichment Strategies

If students need help. . .	If students got it. . .
Record and Practice Journal • Fair Game Review Skills Review Handbook Lesson Tutorials	Game Closet at *BigIdeasMath.com* Start the next section

What You Learned Before

STTRIIIKKKE three. You're out!

"Do you think the stripes in this shirt make me look too linear?"

Describe the pattern of inputs and outputs.

Example 1

Input	Output
2	0
4	3
6	6
8	9

∴ As the input increases by 2, the output increases by 3.

Example 2

Input, x	6	1	−4	−9	−14
Output, y	7	8	9	10	11

∴ As the input x decreases by 5, the output y increases by 1.

Example 3 **Draw a mapping diagram for the graph. Then describe the pattern of inputs and outputs.**

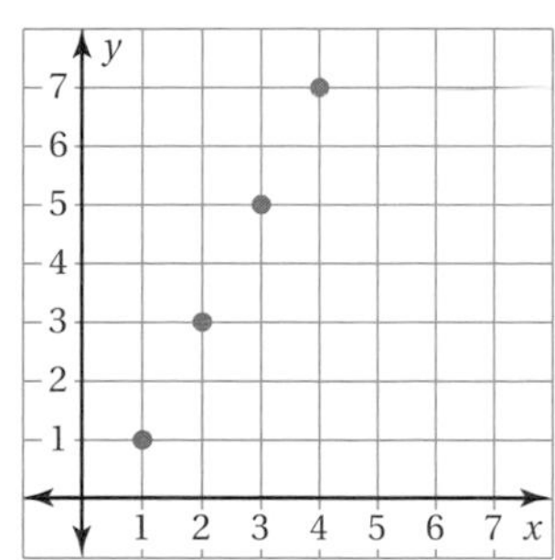

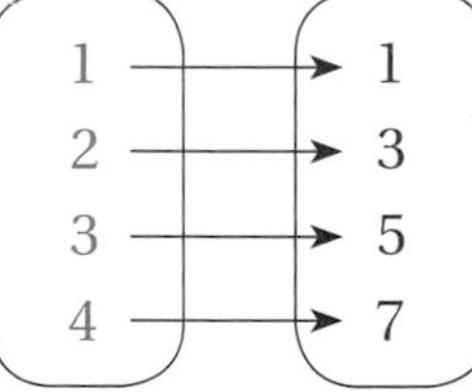

∴ As the input increases by 1, the output increases by 2.

Try It Yourself

Describe the pattern of inputs x and outputs y.

1.

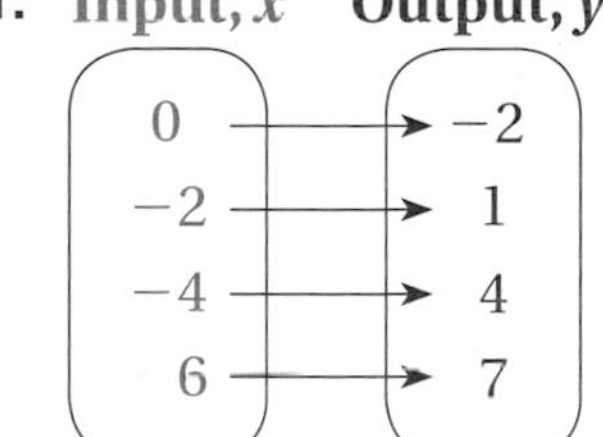

2.

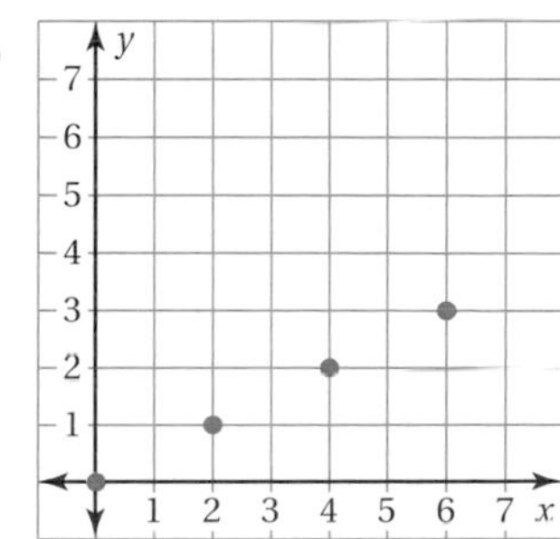

3.

Input, x	0	−1	−2	3	−4
Output, y	7	3.5	0	−3.5	−7

5.1 Domain and Range of a Function

Essential Question How can you find the domain and range of a function?

COMMON CORE STATE STANDARDS
8.F.1
F.IF.1
F.IF.5

1 ACTIVITY: The Domain and Range of a Function

Work with a partner. In Activity 1 in Section 2.4, you completed the table shown below. The table shows the number of adult and child tickets sold for a school concert.

input →	Number of Adult Tickets, x	0	1	2	3	4
output →	Number of Child Tickets, y	8	6	4	2	0

The variables x and y are related by the linear equation $4x + 2y = 16$.

a. Write the equation in *function form* by solving for y.

b. The **domain** of a function is the set of all input values. Find the domain of the function.

Domain =

Why is $x = 5$ not in the domain of the function?

Why is $x = \frac{1}{2}$ not in the domain of the function?

c. The **range** of a function is the set of all output values. Find the range of the function.

Range =

d. Functions can be described in many ways.

- by an equation
- by an input-output table
- in words
- by a graph
- as a set of ordered pairs

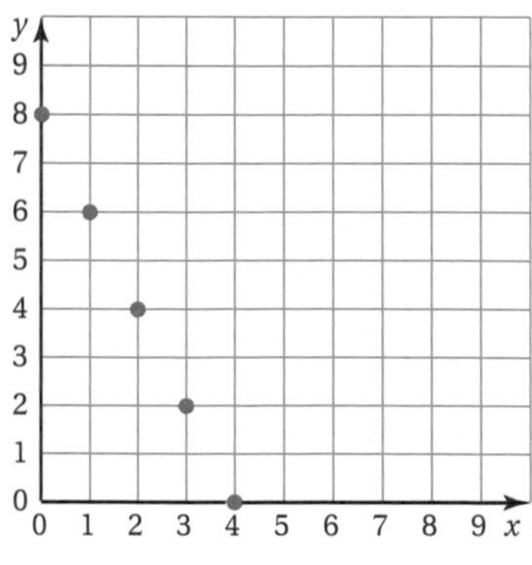

Use the graph to write the function as a set of ordered pairs.

Laurie's Notes

Introduction

Standards for Mathematical Practice

- **MP6 Attend to Precision:** The language of functions and the ways in which functions are represented are important in this investigation. The language of functions can be challenging for students, more so than the actual concepts. Students need to pay attention to the definitions of function, domain, and range.

Motivate

- Ask for a volunteer who will not mind you measuring his or her head for a hat. See the chart for sizes.

Hat Size	$6\frac{1}{2}$	$6\frac{5}{8}$	$6\frac{3}{4}$	$6\frac{7}{8}$	7	$7\frac{1}{8}$	$7\frac{1}{4}$	$7\frac{3}{8}$	$7\frac{1}{2}$	$7\frac{5}{8}$	$7\frac{3}{4}$	$7\frac{7}{8}$	8
Inches	$20\frac{1}{2}$	$20\frac{7}{8}$	$21\frac{1}{4}$	$21\frac{5}{8}$	22	$22\frac{1}{2}$	$22\frac{7}{8}$	$23\frac{1}{4}$	$23\frac{5}{8}$	24	$24\frac{3}{8}$	$24\frac{3}{4}$	$25\frac{1}{4}$
Centimeters	52	53	54	55	56	57	58	59	60	61	62	63	64
	X-SMALL		SMALL		MEDIUM	LARGE			X-LARGE		XX-LARGE		

? "What is the input for determining your hat size?" size of your head in inches or centimeters

? "What are the outputs?" hat size as a number $\left(6\frac{1}{2}, 6\frac{5}{8}, \ldots\right)$ or a category (X-small, small, ...)

- **Discuss:** Relate the input and output for determining the hat size to the new vocabulary, domain and range.

Activity Notes

Activity 1

- This problem involves reading the language of functions. Students should pay attention to the vocabulary.

? "Recall that the linear equation $4x + 2y = 16$ is written in standard form. What other common form have you used to write linear equations?" slope-intercept form

- Writing the equation in slope-intercept form results in an equation in function form. Students may write $y = 8 - 2x$, which is equivalent to $y = -2x + 8$.
- This problem focuses attention on the domain, the set of all input values. We also describe domain as the *permissible values* for x, meaning what numbers can be substituted for x. In the context of this problem, it makes sense that x can only be a whole number between 0 and 4, inclusive.
- Once the domain values are determined, a set of output values, the range, result.

? "Why are the ordered pairs of the graph not connected?" The function has whole number solutions. You cannot have fractional or negative numbers of tickets sold.

? "How many solutions are there for this function?" five

Common Core State Standards

8.F.1 Understand that a function is a rule that assigns to each input exactly one output. The graph of a function is the set of ordered pairs consisting of an input and the corresponding output.
F.IF.1 Understand that a function from one set (called the domain) to another set (called the range) assigns to each element of the domain exactly one element of the range . . ..
F.IF.5 Relate the domain of a function to its graph and, where applicable, to the quantitative relationship it describes.

Previous Learning

Students should know how to determine solutions of a linear equation.

Start Thinking! and Warm Up

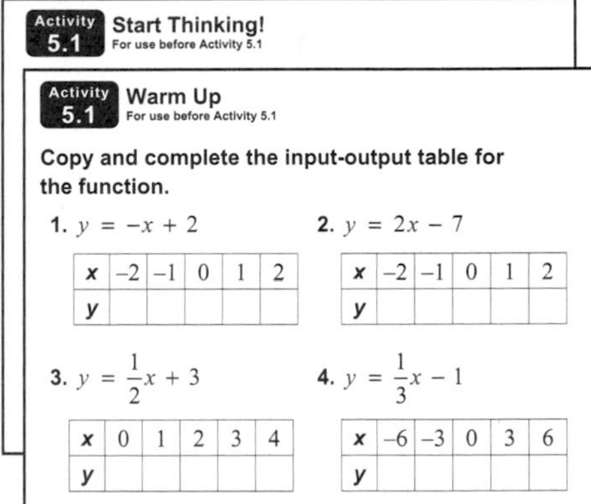

Activity 5.1 Start Thinking! For use before Activity 5.1

Activity 5.1 Warm Up For use before Activity 5.1

Copy and complete the input-output table for the function.

1. $y = -x + 2$

x	−2	−1	0	1	2
y					

2. $y = 2x - 7$

x	−2	−1	0	1	2
y					

3. $y = \frac{1}{2}x + 3$

x	0	1	2	3	4
y					

4. $y = \frac{1}{3}x - 1$

x	−6	−3	0	3	6
y					

5.1 Record and Practice Journal

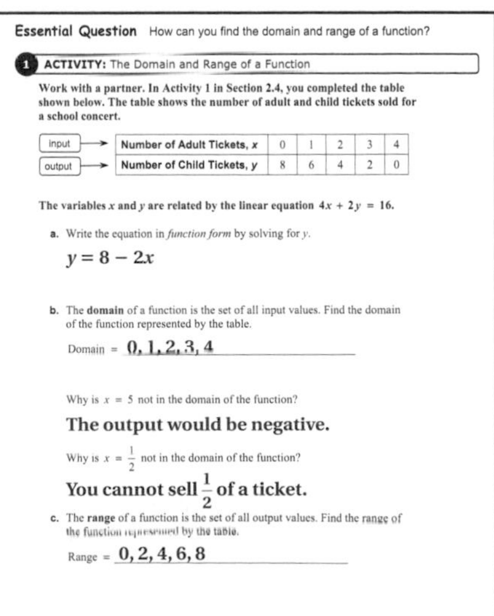

Essential Question How can you find the domain and range of a function?

1 ACTIVITY: The Domain and Range of a Function

Work with a partner. In Activity 1 in Section 2.4, you completed the table shown below. The table shows the number of adult and child tickets sold for a school concert.

input →	Number of Adult Tickets, x	0	1	2	3	4
output →	Number of Child Tickets, y	8	6	4	2	0

The variables x and y are related by the linear equation $4x + 2y = 16$.

a. Write the equation in *function form* by solving for y.

$y = 8 - 2x$

b. The **domain** of a function is the set of all input values. Find the domain of the function represented by the table.

Domain = 0, 1, 2, 3, 4

Why is $x = 5$ not in the domain of the function?

The output would be negative.

Why is $x = \frac{1}{2}$ not in the domain of the function?

You cannot sell $\frac{1}{2}$ of a ticket.

c. The **range** of a function is the set of all output values. Find the range of the function represented by the table.

Range = 0, 2, 4, 6, 8

English Language Learners

Vocabulary

Students will find it helpful to relate the words *input* and *output* with the prepositions *in* and *out*.

5.1 Record and Practice Journal

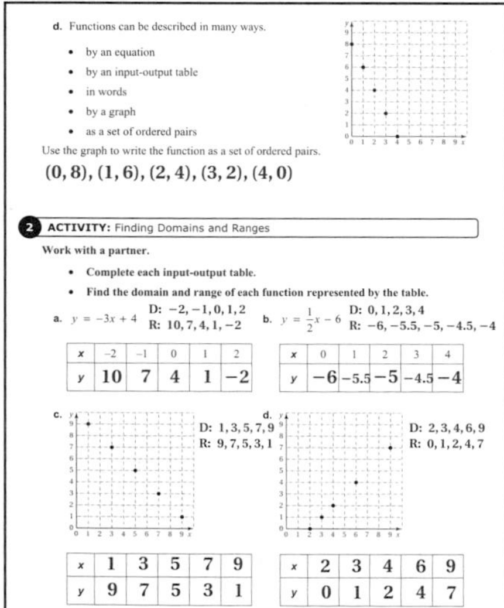

d. Functions can be described in many ways.

- by an equation
- by an input-output table
- in words
- by a graph
- as a set of ordered pairs

Use the graph to write the function as a set of ordered pairs.

(0, 8), (1, 6), (2, 4), (3, 2), (4, 0)

2 ACTIVITY: Finding Domains and Ranges

Work with a partner.

- **Complete each input-output table.**
- **Find the domain and range of each function represented by the table.**

a. $y = -3x + 4$ D: −2, −1, 0, 1, 2 R: 10, 7, 4, 1, −2

x	−2	−1	0	1	2
y	10	7	4	1	−2

b. $y = \frac{1}{2}x - 6$ D: 0, 1, 2, 3, 4 R: −6, −5.5, −5, −4.5, −4

x	0	1	2	3	4
y	−6	−5.5	−5	−4.5	−4

c. D: 1, 3, 5, 7, 9 R: 9, 7, 5, 3, 1

x	1	3	5	7	9
y	9	7	5	3	1

d. D: 2, 3, 4, 6, 9 R: 0, 1, 2, 4, 7

x	2	3	4	6	9
y	0	1	2	4	7

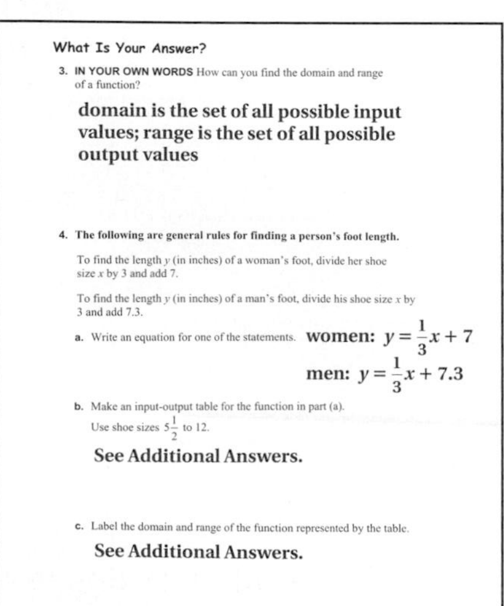

What Is Your Answer?

3. **IN YOUR OWN WORDS** How can you find the domain and range of a function?

domain is the set of all possible input values; range is the set of all possible output values

4. **The following are general rules for finding a person's foot length.**

To find the length y (in inches) of a woman's foot, divide her shoe size x by 3 and add 7.

To find the length y (in inches) of a man's foot, divide his shoe size x by 3 and add 7.3.

a. Write an equation for one of the statements. **women:** $y = \frac{1}{3}x + 7$

men: $y = \frac{1}{3}x + 7.3$

b. Make an input-output table for the function in part (a). Use shoe sizes $5\frac{1}{2}$ to 12.

See Additional Answers.

c. Label the domain and range of the function represented by the table.

See Additional Answers.

Laurie's Notes

Activity 2

- **Connection:** Students have graphed linear equations whose domain and range were the set of real numbers. In fact, students plotted 3 to 4 points that satisfied the equation and then connected the points to graph the linear equation. If the domain is restricted to only a finite set of values, the range becomes restricted to a finite set of values, and the number of solutions is finite.
- Students may ask why they are only using five values for x in parts (a) and (b), and why they are not connecting the ordered pairs in the graph in parts (c) and (d). Again, the focus is on domain and range. Remind students that if you only use certain domain values (inputs for x), the range values (outputs for y) are determined by using the function rule (equation).

? "What is the resulting range for the function $y = -3x + 4$ with a domain of $-2, -1, 0, 1, 2$?" 10, 7, 4, 1, −2

- Students often make these problems more difficult than they are. Remind them to use substitution to evaluate the functions (equations) for parts (a) and (b), then use their eyesight to read the ordered pair solutions for parts (c) and (d). Then students record their answers in the input-output table.

? To focus attention on the language of functions, ask students to describe the function rule for each problem. For instance, the function rule for part (a) is *multiply the input by −3 and then add 4.* Students will need to find the function rule for parts (c) and (d). $y = 10 - x$; $y = x - 2$

What Is Your Answer?

- **Question 4 Extension:** Gather data from students who are willing to share their shoe size and to have their foot measured. How well does the rule fit the data?
- **MP4 Model with Mathematics:** Students should start to recognize that the real-life applications in this book demonstrate how functions can model familiar phenomena.

Closure

- **Writing Prompt:** Describe what is meant by the domain and range of a function.

Technology For the Teacher

The Dynamic Planning Tool
Editable Teacher's Resources at *BigIdeasMath.com*

2 ACTIVITY: Finding Domains and Ranges

Work with a partner.

- **Copy and complete each input-output table.**
- **Find the domain and range of the function represented by the table.**

a. $y = -3x + 4$

x	-2	-1	0	1	2
y					

b. $y = \frac{1}{2}x - 6$

x	0	1	2	3	4
y					

c.

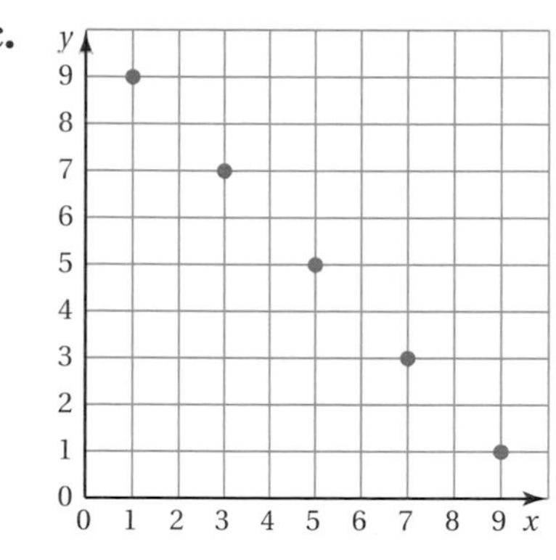

x					
y					

d.

x					
y					

What Is Your Answer?

3. **IN YOUR OWN WORDS** How can you find the domain and range of a function?
4. **The following are general rules for finding a person's foot length.**

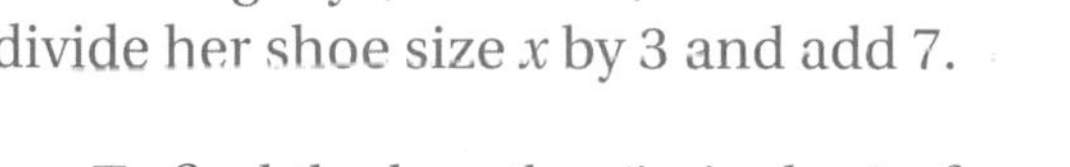
To find the length *y* (in inches) of a woman's foot, divide her shoe size *x* by 3 and add 7.

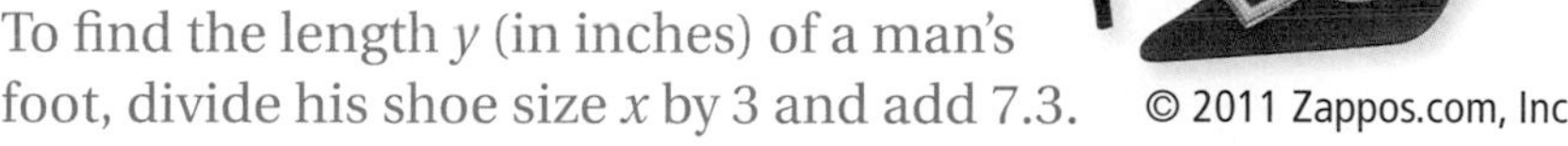
To find the length *y* (in inches) of a man's foot, divide his shoe size *x* by 3 and add 7.3.

a. Write an equation for one of the statements.

b. Make an input-output table for the function in part (a). Use shoe sizes $5\frac{1}{2}$ to 12.

c. Label the domain and range of the function on the table.

Use what you learned about the domain and range of a function to complete Exercise 3 on page 206.

5.1 Lesson

Key Vocabulary
function, *p. 204*
domain, *p. 204*
range, *p. 204*
independent variable, *p. 204*
dependent variable, *p. 204*

Key Idea

Functions

A **function** is a relationship that pairs each *input* with exactly one *output*. The **domain** is the set of all possible input values. The **range** is the set of all possible output values.

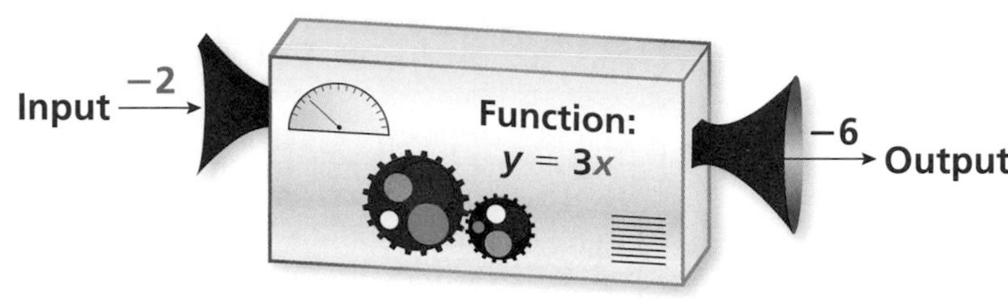

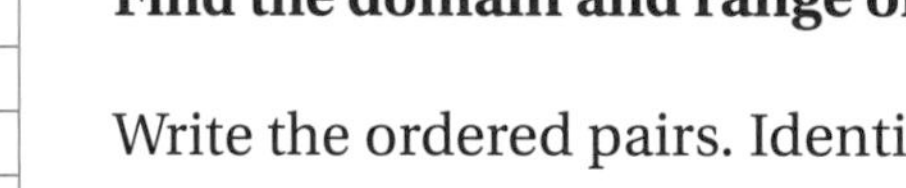

EXAMPLE 1 Finding Domain and Range from a Graph

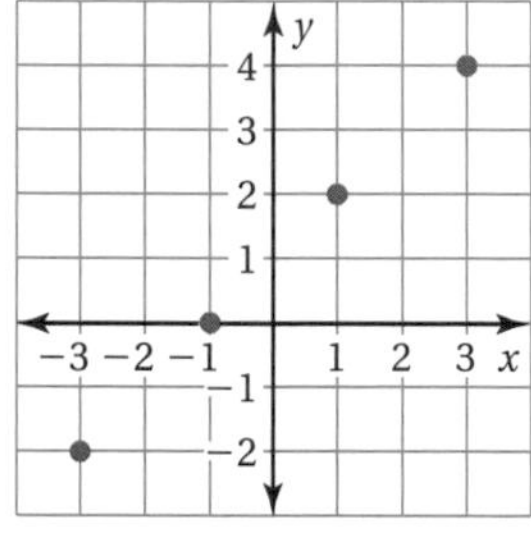

Find the domain and range of the function represented by the graph.

Write the ordered pairs. Identify the inputs and outputs.

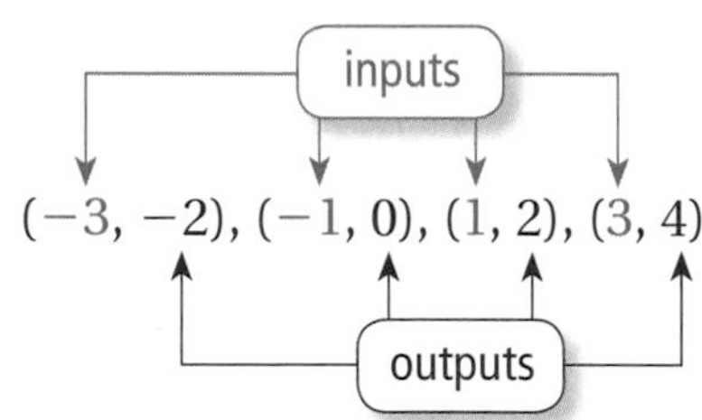

The domain is −3, −1, 1, and 3. The range is −2, 0, 2, and 4.

On Your Own

Now You're Ready
Exercises 4–6

Find the domain and range of the function represented by the graph.

1. 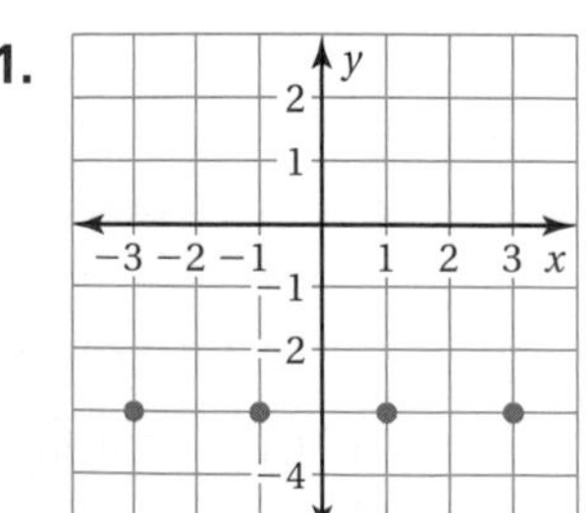

2. 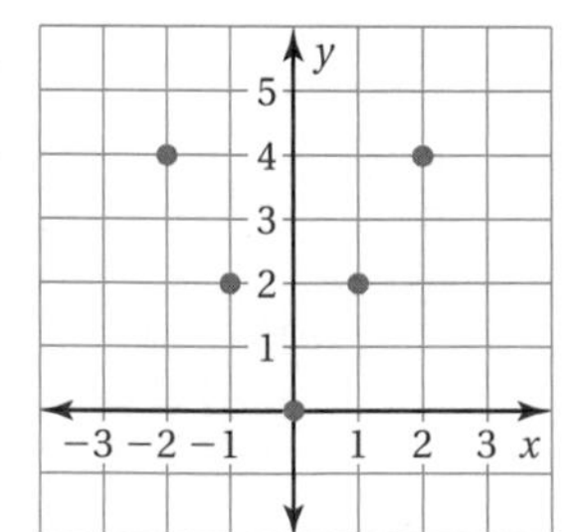

When an equation represents a function, the variable that represents input values is the **independent variable** because it can be *any* value in the domain. The variable that represents output values is the **dependent variable** because it *depends* on the value of the independent variable.

Multi-Language Glossary at BigIdeasMath.com.

Laurie's Notes

Introduction

Connect

- **Yesterday:** Students explored the domain and range of functions by revisiting familiar problems. (MP4, MP6)
- **Today:** Students will identify the domain and range of a function from a graph and table of values.

Motivate

? "Could someone describe how a vending machine works?" Listen for: put in money, make a selection, and item comes out.

- Explain that a vending machine is like a function. You make a selection (the input) and a specific item comes out (the output).
- Sometimes there are several inputs that give the same output (3 different buttons for the same bottled water), but there are never several different outputs of the same input (if the vending machine is working properly).

Lesson Notes

Key Idea

- Write the Key Idea. The graphic of a *function machine* should help students conceptualize the idea of entering an input value, applying a rule, and obtaining the output value.
- In discussing the definition of a function, describe what is meant by *each input is paired with exactly one output.* This means:
    - one unique input yields one unique output, or
    - two or more inputs yield the same output.
    - It is *not* a function if one unique input yields more than one output.

Example 1

- Point out to students that the inputs, or x-coordinates, are the domain, and the outputs, or y-coordinates, are the range.

? "Is each input paired with exactly one output?" yes

On Your Own

- **Question 1:** Students may incorrectly say this is not a function because there is only one number in the range (students should not list it 4 times). Every domain value is paired with the same range value, and that is okay. Remind students it is only the repeat of x-values they need to consider.

Write

- Write the definitions of independent variable and dependent variable.
- Use an example to help make the point about the dependent variable *depending* on the value of the independent variable.
- **Example:** In the equation $y = 0.99x$, the cost y in dollars of buying songs *depends* on the number x of songs. So, x is the independent variable (any number of songs may be purchased) and y is the dependent variable (the cost depends on the number of songs purchased).

Goal Today's lesson is identifying the **domain** and **range** of a **function**.

Start Thinking! and Warm Up

Start Thinking!
For use before Lesson 5.1

The installation and set-up fees for cable Internet come to \$150. The monthly cost for Internet access is \$40 per month.

Write a function for the cost y of x months of Internet service.

What are the domain and range of the function if you only have Internet service for 6 months?

Extra Example 1

Find the domain and range of the function represented by the graph.

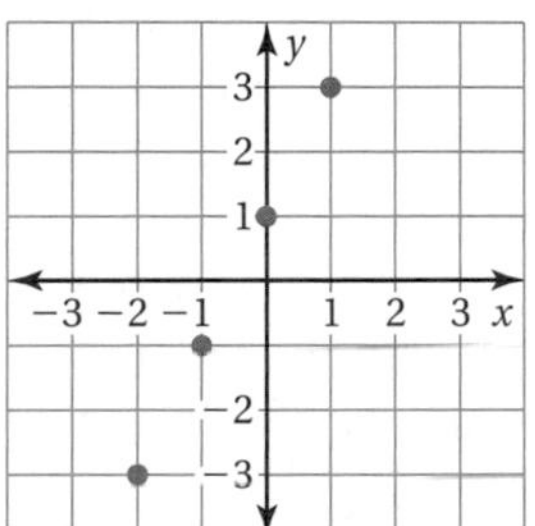

domain: −2, −1, 0, 1; range: −3, −1, 1, 3

On Your Own

1. domain: −3, −1, 1, 3
   range: −3
2. domain: −2, −1, 0, 1, 2
   range: 0, 2, 4

Extra Example 2

The function $y = 24x$ gives the number y bottles of water you have after buying x cases. (a) Identify the independent and dependent variables. independent variable: x; dependent variable: y
(b) The domain is 0, 1, 2, and 3. What is the range? 0, 24, 48, 72

Extra Example 3

x	0	1	2	3	4
y	0.11	0.18	0.27	0.37	0.47

a. The table shows the percent y (in decimal form) of the moon that was visible at midnight x days after August 10, 2013. Interpret the domain and range. The domain is 0, 1, 2, 3, and 4. So, the table shows the data for August 10, 11, 12, 13, and 14. The range is 0.11, 0.18, 0.27, 0.37, and 0.47. So, the table shows that the moon was more visible each day.

b. What percent of the moon was visible on August 12, 2013? 27%

On Your Own

3. a. independent variable: x; dependent variable: y

b. 14, 10, 6, 2

4. a. The domain is 0, 1, 2, 3, and 4. It represents March 24, 25, 26, 27, and 28

The range is 0.19, 0.29, 0.39, 0.49, and 0.59. These amounts are increasing, so the moon was more visible each day.

b. 59%

Differentiated Instruction

Kinesthetic

Have students build their own function machine using a shoe box, index cards, and sticky notes. Write the function on the sticky note and put it on the box. Write numbers on the index cards to use as the input and output values of the function.

Laurie's Notes

Example 2

- Write the function and discuss what the variables x and y represent.
- Discuss with students that equations written in function form have the independent variable and all computations on one side of the equation, and the dependent variable on the other side. When the value of the independent variable changes, and the arithmetic is performed, the dependent variable changes.

? "Why is the amount left in the bottle the dependent variable?" The amount left is the output, and it depends on how many gulps have been taken.

- **Extension:** Ask students to describe the patterns in the input-output table. The x-values increase by 1 and the y-values decrease by 3.

Example 3

- **FYI:** During the full moon, the moon's illuminated side is facing Earth and appears to be completely illuminated by direct sunlight. During a new moon, the moon's unilluminated side is facing Earth and the moon is not visible.

? MP1a Make Sense of Problems: Ask a few questions to help students make sense of the problem.

- "What is the independent variable in this problem?" number of days after May 19, 2014 "What is the dependent variable?" percent of the moon visible at midnight, written as a decimal
- "What day does $x = 3$ represent?" May 22, 2014
- Ask a volunteer to make a statement relating the independent and dependent variables. Listen for: the percent of the moon visible at midnight depends on how many days past May 19, 2014 it is.
- "Is the moon becoming more or less visible in the week following May 19, 2014? Explain." less; The percents are decreasing.

On Your Own

- **Question 4:** Ask students if the data suggests that they are moving towards a full moon or away from a full moon.

Closure

- **Exit Ticket:** Make an input-output table for the function $y = -2x + 3$ using the inputs $-2, 0, 2, 4$.

x	-2	0	2	4
y	7	3	-1	-5

Technology For the Teacher

The Dynamic Planning Tool
Editable Teacher's Resources at *BigIdeasMath.com*

EXAMPLE 2 Finding the Range of a Function

The function $y = -3x + 12$ gives the amount y (in fluid ounces) of juice remaining in a bottle after you take x gulps. (a) Identify the independent and dependent variables. (b) The domain is 0, 1, 2, 3, and 4. What is the range?

a. Because the amount y remaining depends on the number x of gulps, y is the dependent variable and x is the independent variable.

b. Make an input-output table to find the range.

Input, x	$-3x + 12$	Output, y
0	$-3(0) + 12$	12
1	$-3(1) + 12$	9
2	$-3(2) + 12$	6
3	$-3(3) + 12$	3
4	$-3(4) + 12$	0

∴ The range is 12, 9, 6, 3, and 0.

EXAMPLE 3 Real-Life Application

The table shows the percent y (in decimal form) of the moon that was visible at midnight x days after May 19, 2014. (a) Interpret the domain and range. (b) What percent of the moon was visible on May 21, 2014?

x	y
0	0.76
1	0.65
2	0.54
3	0.43
4	0.32

a. Zero days after May 19 is May 19. One day after May 19 is May 20. So, the domain of 0, 1, 2, 3, and 4 represents May 19, 20, 21, 22, and 23.

The range is 0.76, 0.65, 0.54, 0.43, and 0.32. These amounts are decreasing, so the moon was less visible each day.

b. May 21, 2014 corresponds to the input $x = 2$. When $x = 2$, $y = 0.54$. So, 0.54, or 54% of the moon was visible on May 21, 2014.

On Your Own

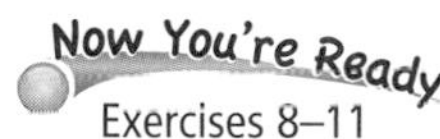

3. The function $y = -4x + 14$ gives the number y of avocados you have left after making x batches of guacamole.

   a. Identify the independent and dependent variables.

   b. The domain is 0, 1, 2, and 3. What is the range?

4. The table shows the percent y (in decimal form) of the moon that was visible at midnight x days after March 24, 2015.

x	0	1	2	3	4
y	0.19	0.29	0.39	0.49	0.59

   a. Interpret the domain and range.

   b. What percent of the moon was visible on March 28, 2015?

5.1 Exercises

Check It Out
Help with Homework
BigIdeasMath.com

Vocabulary and Concept Check

1. **VOCABULARY** How are independent variables and dependent variables different?
2. **DIFFERENT WORDS, SAME QUESTION** Which is different? Find "both" answers.

Find the range of the function represented by the table.

Find the inputs of the function represented by the table.

Find the x-values of the function represented by (2, 7), (4, 5), and (6, −1).

Find the domain of the function represented by (2, 7), (4, 5), and (6, −1).

x	2	4	6
y	7	5	−1

Practice and Problem Solving

3. The number of earrings and headbands you can buy with \$24 is represented by the equation $8x + 4y = 24$. The table shows the numbers of earrings and headbands.

Earrings, x	0	1	2	3
Headbands, y	6	4	2	0

   a. Write the equation in function form.
   b. Find the domain and range.
   c. Why is $x = 6$ not in the domain of the function?

Find the domain and range of the function represented by the graph.

1 4. 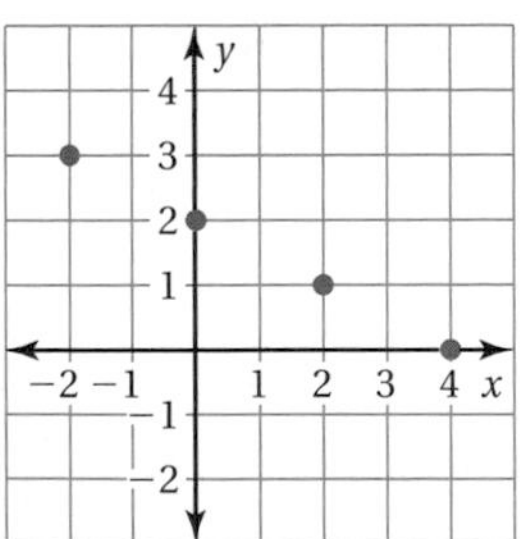

5. 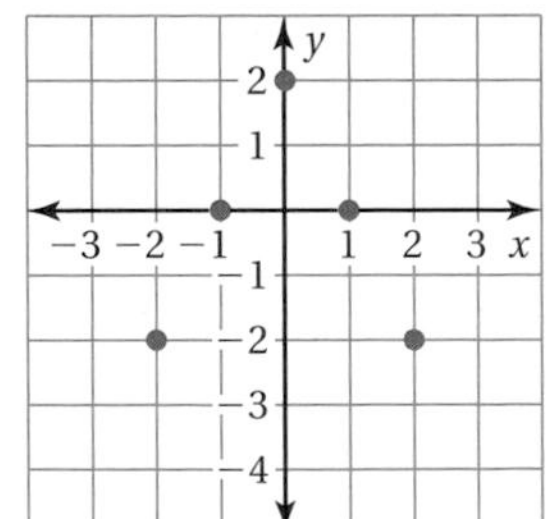

6. 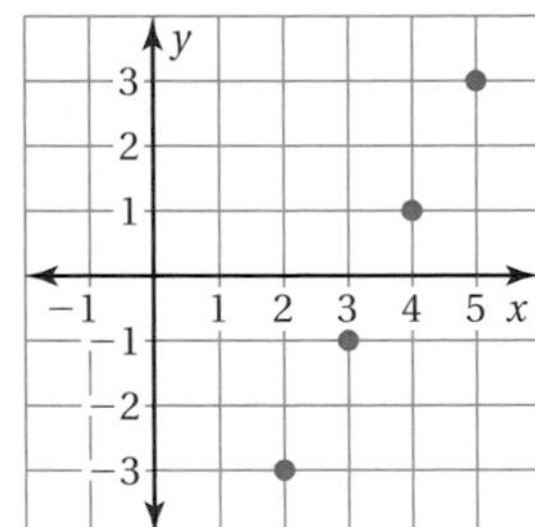

7. **ERROR ANALYSIS** Describe and correct the error in finding the domain and range of the function represented by the graph.

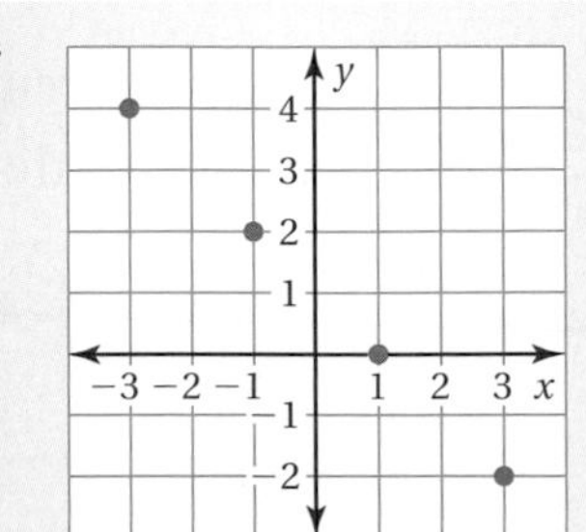
The domain is −2, 0, 2, and 4.
The range is −3, −1, 1, 3.

2 8. **PARKING METER** The number of quarters you put into a parking meter affects the amount of time on the meter. Identify the independent and dependent variables.

Assignment Guide and Homework Check

Level	Assignment	Homework Check
Average	1–7, 9–12, 15–19	2, 4, 10, 12
Advanced	1–3, 6, 7, 9–19	2, 6, 10, 12

For Your Information

- **Exercise 13** This problem is an example of a continuous domain. Discrete and continuous domains will be discussed in the next section.

Common Errors

- **Exercises 4–6** Students may mix up the domain and range. For example, a student may give all the *y*-values of the coordinates as the domain. This can happen with all the ordered pairs, or only one or two. Remind students that the *x*-coordinates are the domain and the *y*-coordinates are the range, as shown in Example 1.
- **Exercises 9–11** Students may make mistakes when substituting the values of *x* and solving for *y*. For example, a student may write $y = 6(2) + 2 = 6(4) = 24$ instead of $y = 6(2) + 2 = 12 + 2 = 14$. Remind them of the order of operations.

5.1 Record and Practice Journal

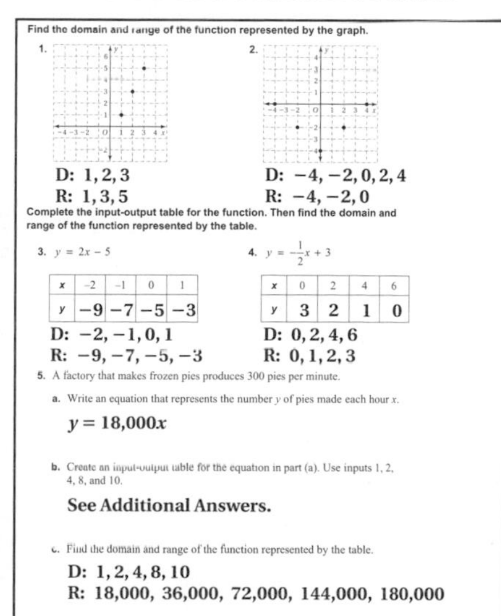

Find the domain and range of the function represented by the graph.

1. D: 1, 2, 3
   R: 1, 3, 5

2. D: −4, −2, 0, 2, 4
   R: −4, −2, 0

Complete the input-output table for the function. Then find the domain and range of the function represented by the table.

3. $y = 2x - 5$

x	−2	−1	0	1
y	−9	−7	−5	−3

D: −2, −1, 0, 1
R: −9, −7, −5, −3

4. $y = -\frac{1}{2}x + 3$

x	0	2	4	6
y	3	2	1	0

D: 0, 2, 4, 6
R: 0, 1, 2, 3

5. A factory that makes frozen pies produces 300 pies per minute.

   a. Write an equation that represents the number *y* of pies made each hour *x*.

   $y = 18{,}000x$

   b. Create an input-output table for the equation in part (a). Use inputs 1, 2, 4, 8, and 10.

   See Additional Answers.

   c. Find the domain and range of the function represented by the table.

   D: 1, 2, 4, 8, 10
   R: 18,000, 36,000, 72,000, 144,000, 180,000

Technology For the Teacher
Answer Presentation Tool

Vocabulary and Concept Check

1. *Sample answer:* An independent variable represents an input value, and a dependent variable represents an output value.
2. Find the range of the function represented by the table.; 7, 5, −1; 2, 4, 6

Practice and Problem Solving

3. **a.** $y = 6 - 2x$

   b. domain: 0, 1, 2, 3
   range: 6, 4, 2, 0

   c. $x = 6$ is not in the domain because it would make *y* negative, and it is not possible to buy a negative number of headbands.

4. domain: −2, 0, 2, 4
   range: 3, 2, 1, 0

5. domain: −2, −1, 0, 1, 2
   range: −2, 0, 2

6. domain: 2, 3, 4, 5
   range: −3, −1, 1, 3

7. The domain and range are switched. The domain is −3, −1, 1, and 3. The range is −2, 0, 2, and 4.

8. independent variable: the number of quarters; dependent variable: the amount of time on the meter

9.

x	−1	0	1	2
y	−4	2	8	14

domain: −1, 0, 1, 2
range: −4, 2, 8, 14

10.

x	0	4	8	12
y	−2	−3	−4	−5

domain: 0, 4, 8, 12
range: −2, −3, −4, −5

Practice and Problem Solving

11.

x	-1	0	1	2
y	1.5	3	4.5	6

domain: $-1, 0, 1, 2$
range: 1.5, 3, 4.5, 6

12. **a.** domain: 1, 2, 3
range: 6.856, 7.923, 8.135

b. The domain represents the round of competition. The range represents the scores received by the vaulter.

c. 7.638

13. See *Taking Math Deeper*.

14. See Additional Answers.

Fair Game Review

15–18. See Additional Answers.

19. D

Mini-Assessment

Find the domain and range of the function represented by the graph.

1.

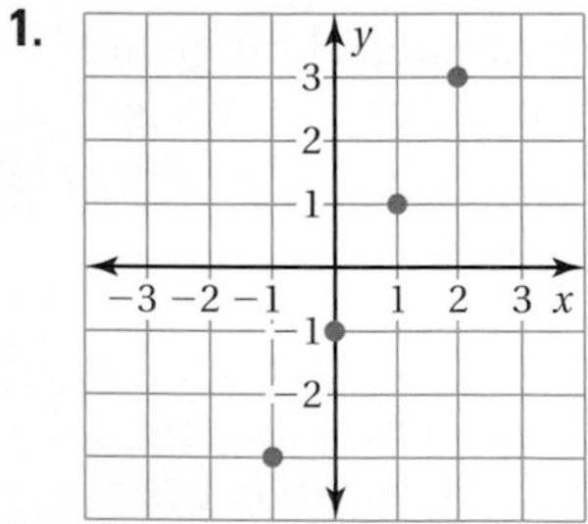

domain: $-1, 0, 1, 2$;
range: $-3, -1, 1, 3$

2.

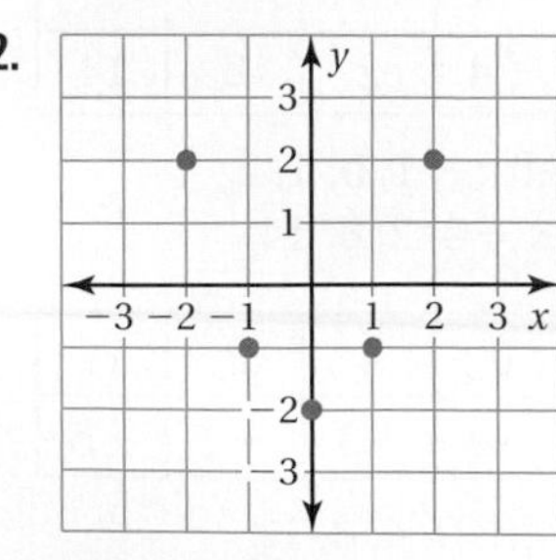

domain: $-2, -1, 0, 1, 2$;
range: $-2, -1, 2$

Taking Math Deeper

Exercise 13

The manatee is a gray, waterplant-eating mammal that reaches up to 15 feet in length and can weigh more than 2000 pounds. Humans (especially motor boats) are the cause of about half of all manatee deaths.

Find a function.

Let x = manatee's weight in pounds.
Let y = weight of food per day.

a. $y = 12\%$ of x
$= 0.12x$

x is the input, so it is the independent variable. y is the output, so it is the dependent variable.

Describe the function.
Create an input-output table for the function. Then, identify the domain and range.

b.

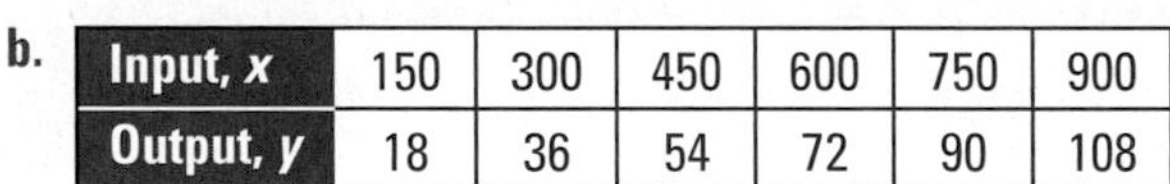

Input, x	150	300	450	600	750	900
Output, y	18	36	54	72	90	108

For the function, the domain is the set of positive numbers that are possible weights of manatees. An approximation would be the positive numbers up to 2000.
The range would be the positive numbers up to about 240.

c. *For the input-output table shown*, the domain is the x-values and the range is the y-values.

Use the function.
The manatees weigh a total of
$300 + 750 + 1050 = 2100$ pounds.

d. In a day, these manatees would eat
$y = 0.12(2100) = 252$ pounds of food.
In a week, they would eat
$7(252) = 1764$ pounds of food.

Reteaching and Enrichment Strategies

If students need help. . .	If students got it. . .
Resources by Chapter • Practice A and Practice B • Puzzle Time Record and Practice Journal Practice Differentiating the Lesson Lesson Tutorials Skills Review Handbook	Resources by Chapter • Enrichment and Extension Start the next section

Copy and complete the input-output table for the function. Then find the domain and range of the function represented by the table.

9. $y = 6x + 2$

x	-1	0	1	2
y				

10. $y = -\frac{1}{4}x - 2$

x	0	4	8	12
y				

11. $y = 1.5x + 3$

x	-1	0	1	2
y				

12. VAULTING In the sport of vaulting, a vaulter performs a routine while on a moving horse. For each round x of competition, the vaulter receives a score y from 1 to 10.

x	y
1	6.856
2	7.923
3	8.135

a. Find the domain and range of the function represented by the table.

b. Interpret the domain and range.

c. What is the mean score of the vaulter?

13. MANATEE A manatee eats the equivalent of about 12% of its body weight each day.

a. Write an equation that represents the amount y (in pounds) of food a manatee eats each day for its weight x. Identify the independent variable and the dependent variable.

b. Make an input-output table for the equation in part (a). Use the inputs 150, 300, 450, 600, 750, and 900.

c. Find the domain and range of the function represented by the table.

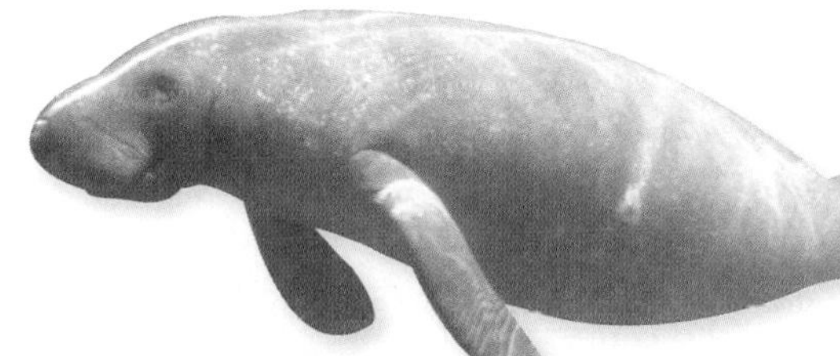

d. The weights of three manatees are 300 pounds, 750 pounds, and 1050 pounds. What is the total amount of food that these three manatees eat in a day? in a week?

14. Precision Describe the domain and range of the function.

a. $y = |x|$ **b.** $y = -|x|$ **c.** $y = |x| - 6$ **d.** $y = -|x| + 4$

Fair Game Review What you learned in previous grades & lessons

Graph the linear equation. *(Section 2.1)*

15. $y = 2x + 8$ **16.** $5x + 6y = 12$ **17.** $-x - 3y = 2$ **18.** $y = 7x - 5$

19. MULTIPLE CHOICE The minimum number of people needed for a group rate at an amusement park is 8. Which inequality represents the number of people needed to get the group rate? *(Section 3.1)*

Ⓐ $x \leq 8$ Ⓑ $x > 8$ Ⓒ $x < 8$ Ⓓ $x \geq 8$

5.1b Relations and Functions

Key Vocabulary
relation, *p. 208*
Vertical Line Test, *p. 209*

A **relation** pairs inputs with outputs. A relation that pairs each input with *exactly one* output is a function.

EXAMPLE 1 Determining Whether Relations are Functions

Determine whether each relation is a function.

a. (−2, 2), (−1, 2), (0, 2), (1, 0), (2, 0)

Every input has exactly one output.

∴ So, the relation is a function.

b.

Input	−2	−1	0	0	1	2
Output	3	4	5	6	7	8

The input 0 has two outputs, 5 and 6.

∴ So, the relation is *not* a function.

c.

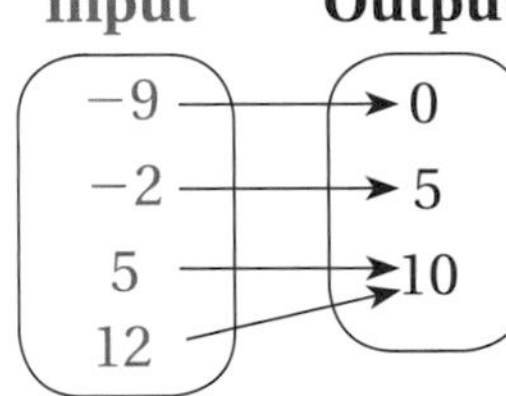

Every input has exactly one output.

∴ So, the relation is a function.

Practice

Determine whether the relation is a function.

1. (−5, 0), (0, 0), (5, 0)
(5, 10), (10, 10)

2.

Input	Output
2	2.6
4	5.2
6	7.8

3.

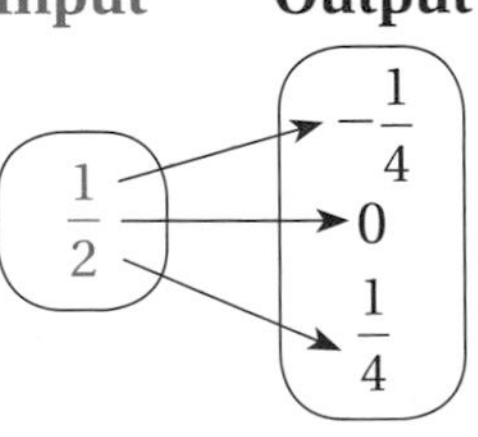

Determine whether the statement is *true* or *false*. Explain your reasoning.

4. Every function is a relation.

5. Every relation is a function.

6. When you switch the inputs and outputs of any function, the resulting relation is a function.

7. REASONING You record the number x of runs scored by the winning team and the number y of runs scored by the losing team for each softball game in a team's season. Does the relation necessarily represent a function? Explain.

Laurie's Notes

Introduction

Connect

- **Yesterday:** Students identified the domain and range of a function. (MP1a)
- **Today:** Students will determine whether a relation is a function.

Motivate

- Copy the following table without specifying what the data represents.

	Jan	Feb	Mar	Apr	May	Jun
Honolulu	81	81	82	83	85	87
Anchorage	22	26	34	44	56	63

? "What do you think the data represents?" the average monthly high temperatures

? "Can there be more than one average monthly high for any location?" no

? "Can different locations have the same average monthly high?" yes

- Discuss the definition of function again. The average monthly high temperature in a given location is a function of the month.

Lesson Notes

Discuss

- Write the definition of a relation. The definition is so simple that students can find it confusing. Simply stated, a relation is a set of ordered pairs. Functions take it one step further by specifying that each input is paired with *exactly one* output.

Example 1

- The relations are represented in three different ways. In part (c), make sure students can identify the four ordered pairs.
- Part (a) and part (c) show that a relation is a function when output values are repeated, as long as no input values are repeated.
- Have students do a quick graph for each of the three problems. You can refer back to these graphs after Example 2.

Practice

- **MP3a Construct Viable Arguments:** Have students share their reasoning for the true-false exercises. Having a counterexample to show that a statement is false is helpful in presenting an argument.

Goal Today's lesson is determining whether a **relation** is a function.

Start Thinking! and Warm Up

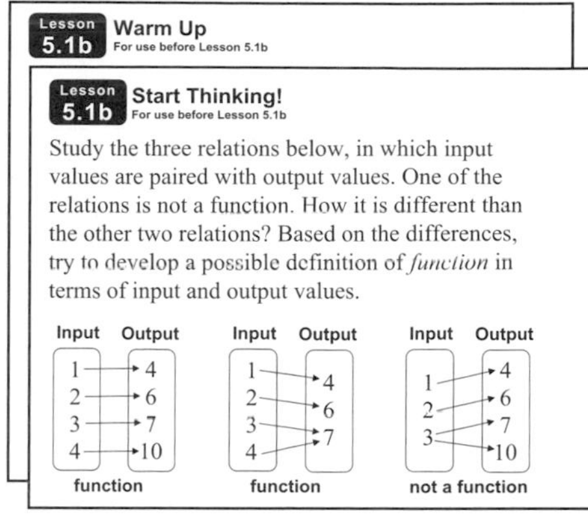

Lesson 5.1b **Warm Up** For use before Lesson 5.1b

Lesson 5.1b **Start Thinking!** For use before Lesson 5.1b

Study the three relations below, in which input values are paired with output values. One of the relations is not a function. How it is different than the other two relations? Based on the differences, try to develop a possible definition of *function* in terms of input and output values.

Extra Example 1

Determine whether each relation is a function.

a. (0, 2), (1, 4), (2, 6), (2, 7) not a function

b.

Input	−2	−1	0	1	2
Output	2	1	0	1	2

function

c. Input Output not a function

Input	Output
2	5
4	10
6	15

Practice

1. not a function
2. function
3. not a function
4. true; A function is a relation by definition.
5. false; A relation is not a function when an input has more than one output.

Record and Practice Journal Practice

See Additional Answers.

Extra Example 2

Determine whether each graph represents a function.

a.

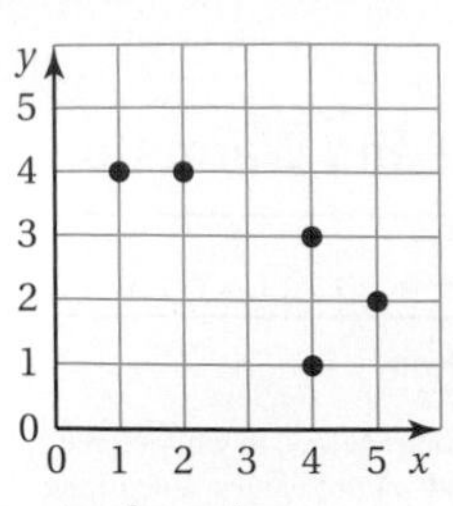

not a function

b.

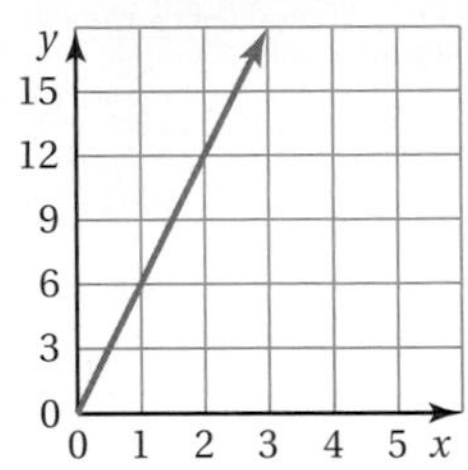

function

Practice

6. false; For a function in which two inputs have the same output, the switch leads to an input with two outputs.
7. no; In any two games, if the winning teams had the same numbers of runs while the losing teams had different numbers of runs, then the relation is not a function.
8. function
9. function
10. not a function
11. no; All linear equations represent functions except for those representing vertical lines.

Laurie's Notes

Key Idea

- Write the Key Idea.
- Explain that the Vertical Line Test is a visual way to look at a set of ordered pairs (a relation) and determine whether the relation is a function.
- In addition to the examples shown, look back at the graphs students made for Example 1. Use a pencil or piece of spaghetti held vertically to model the Vertical Line Test.

Example 2

- Have students use their pencils to model the Vertical Line Test for each problem.
- In part (a), point out that a vertical line also passes through (5, 5), and (5, 1). Stress that when *any* vertical line passes through more than one point of the graph of a relation, the relation is not a function.

Practice

- In Exercise 11, students should understand that all of the linear equations they wrote and graphed in Chapter 2 represented functions except for those representing vertical lines.

Closure

- Write a set of three ordered pairs that represent a function.
  Sample answer: (−2, −4), (0, 0), (2, 4)
- Write a set of three ordered pairs that do not represent a function.
  Sample answer: (−1, −3), (1, 1), (1, 3)

Technology For the Teacher

The Dynamic Planning Tool
Editable Teacher's Resources at *BigIdeasMath.com*

You can use a vertical line test to determine whether a graph represents a function.

Vertical Line Test

Words A graph represents a function when no vertical line passes through more than one point on the graph.

Examples **Function**

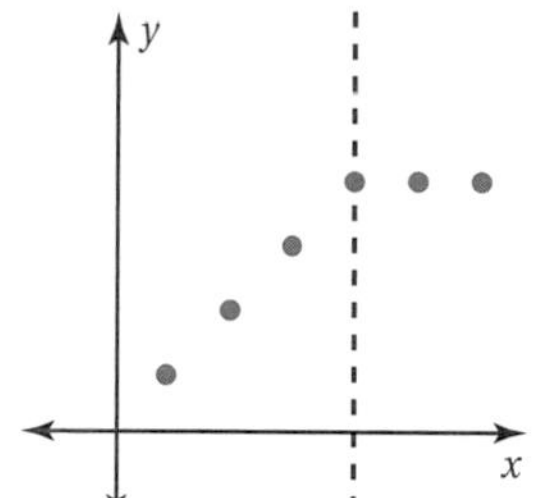

Not a function

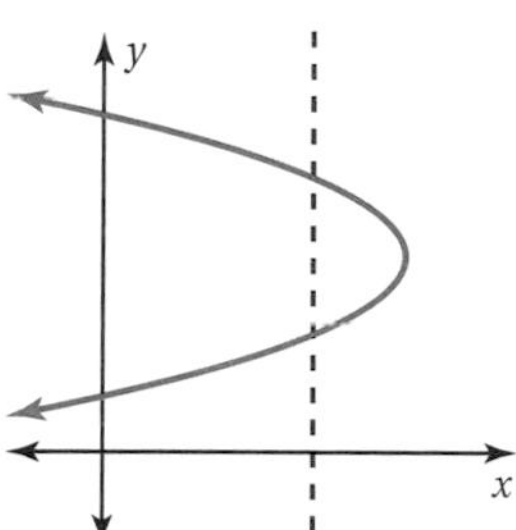

EXAMPLE 2 Using the Vertical Line Test

Determine whether each graph represents a function.

a.

You can draw a vertical line through (2, 2) and (2, 5).

∴ So, the graph does *not* represent a function.

b.

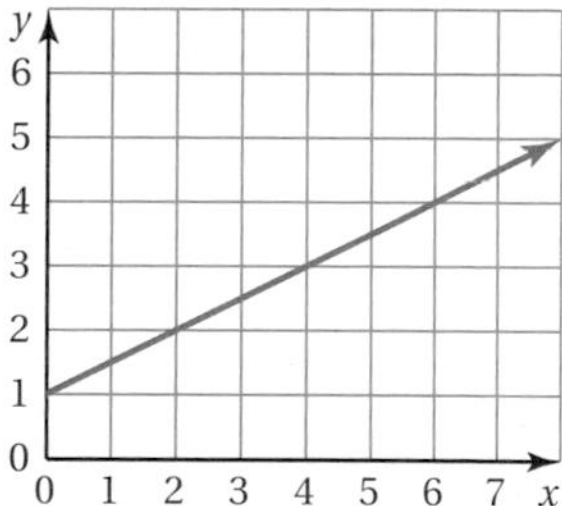

No vertical line can be drawn through two points on the graph.

∴ So, the graph represents a function.

Practice

Determine whether the graph represents a function.

8.

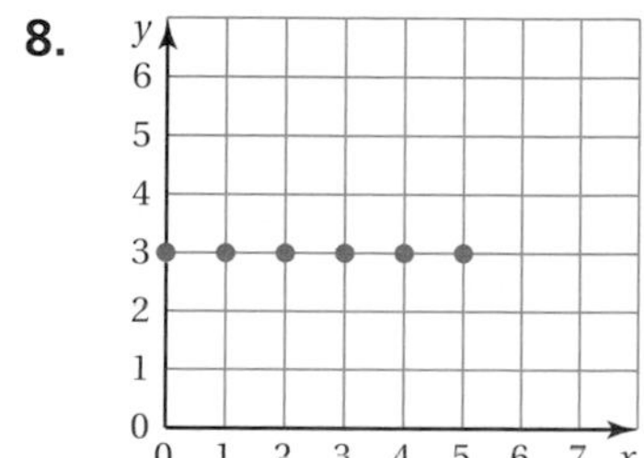

9.

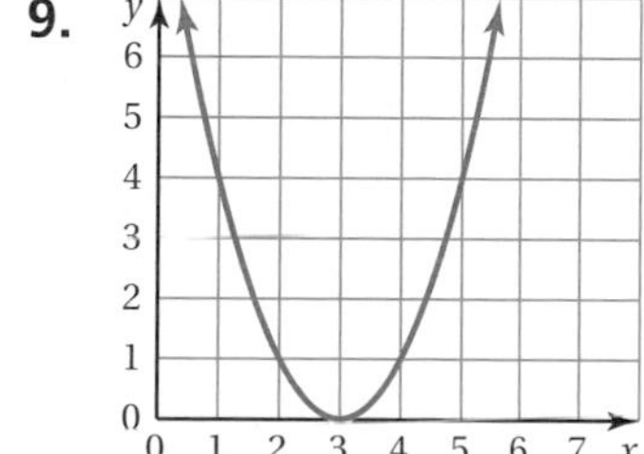

10.

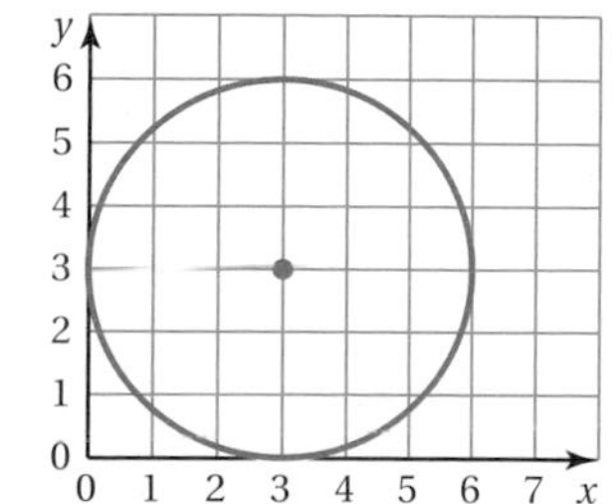

11. REASONING You studied linear equations in Chapter 2. Do all linear equations represent functions? Explain your reasoning.

5.2 Discrete and Continuous Domains

COMMON CORE STATE STANDARDS
8.F.1
F.IF.1
F.IF.5

Essential Question How can you decide whether the domain of a function is discrete or continuous?

1 EXAMPLE: Discrete and Continuous Domains

In Activities 1 and 2 in Section 2.4, you studied two real-life problems represented by the same equation.

$$4x + 2y = 16 \quad \text{or} \quad y = -2x + 8$$

a.

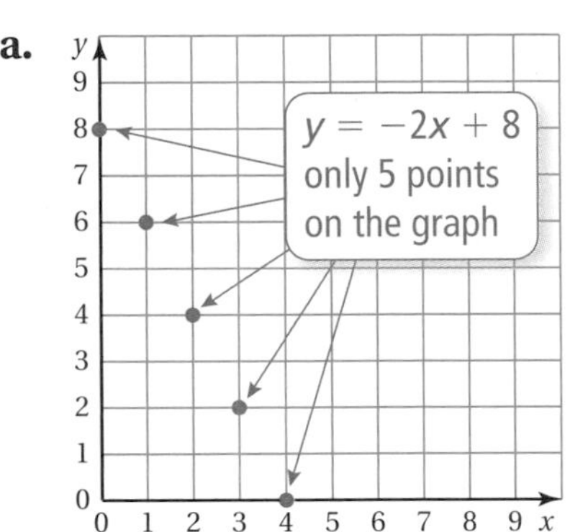

Domain (x-values): 0, 1, 2, 3, 4

Range (y-values): 8, 6, 4, 2, 0

The domain is **discrete** because it consists of only the numbers 0, 1, 2, 3, and 4.

b.

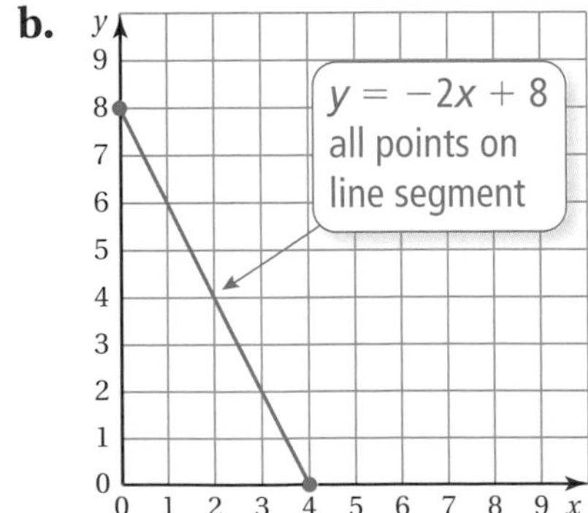

Domain (x-values): $0 \le x \le 4$

Range (y-values): $0 \le y \le 8$

The domain is **continuous** because it consists of all numbers from 0 to 4 on the number line.

Laurie's Notes

Introduction

Standards for Mathematical Practice

- **MP1a Make Sense of Problems:** The goal is for students to discuss, explain, and demonstrate the difference between discrete and continuous domains. Students have a sense of what contexts make sense for using certain numbers in an interval versus using all the numbers in an interval. In this lesson, they will learn the vocabulary of discrete and continuous domains.

Motivate

- Pass out strips of paper. Have students fold the paper in half.
- On one half of the paper, have students describe a variable that must be an integer (e.g., 4, -5, 20). On the other half, have them describe a variable that could be a fraction or decimal (e.g., $2\frac{1}{2}$, 4.8, -3.1). Students should write a description of the variable instead of giving a numerical example.

Samples:

People at a movie (130)	Hours you work (2.2)
Problems assigned (18)	Length of fingernails (1.6 cm)
Buses on the road (15)	Yards lost on the play $\left(8\frac{1}{2}\right)$

- Have students tear the paper in half and place in two piles.
- Read and discuss examples in each pile.

Activity Notes

Example 1

- Discuss the two problems on this page. Make note of the vocabulary, discrete and continuous, used to describe the domain of each problem.

? "Why are the ordered pairs of the graph not connected in part (a)?" The function has whole number solutions. You cannot have fractional or negative numbers of tickets sold.

? "What do the intercepts of each graph represent?" Part (a): (0, 8) represents 8 child tickets sold and no adult tickets sold; (4, 0) represents 4 adult tickets sold and no child tickets sold; Part (b): (0, 8) represents 8 pounds of Cheddar sold and no pounds of Swiss sold; (4, 0) represents 4 pounds of Swiss sold and no pounds of Cheddar sold.

- Review the inequality notation used to describe the domain and range for part (b). Students learned compound inequalities in Chapter 3. Write the inequalities and ask students to read them.

? "What does the compound inequality $0 \le x \le 4$ mean?" all the numbers from 0 to 4, including 0 and 4 "What does the compound inequality $0 \le y \le 8$ mean?" all the numbers from 0 to 8, including 0 and 8

- Refer to the variables described by students on the strips of paper. The integer variables described are examples of discrete domains. The fraction or decimal variables described are examples of continuous domains.
- **Common Error:** It is incorrect to say that discrete domains are finite and continuous domains are infinite. A domain can be discrete and infinite, such as the set of counting numbers.

Common Core State Standards

8.F.1 Understand that a function is a rule that assigns to each input exactly one output. The graph of a function is the set of ordered pairs consisting of an input and the corresponding output.

F.IF.1 Understand that a function from one set (called the domain) to another set (called the range) assigns to each element of the domain exactly one element of the range . . ..

F.IF.5 Relate the domain of a function to its graph and, where applicable, to the quantitative relationship it describes.

Previous Learning

Students should know how to find solutions of a linear equation.

Activity Materials
Introduction
• paper strips

Start Thinking! and Warm Up

Activity 5.2 Start Thinking!
For use before Activity 5.2

Activity 5.2 Warm Up
For use before Activity 5.2

Find the domain and range of the function represented by the graph.

1. 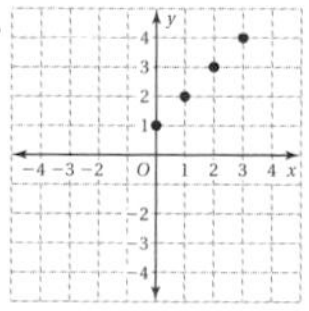

2. 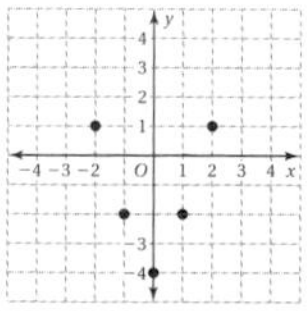

5.2 Record and Practice Journal

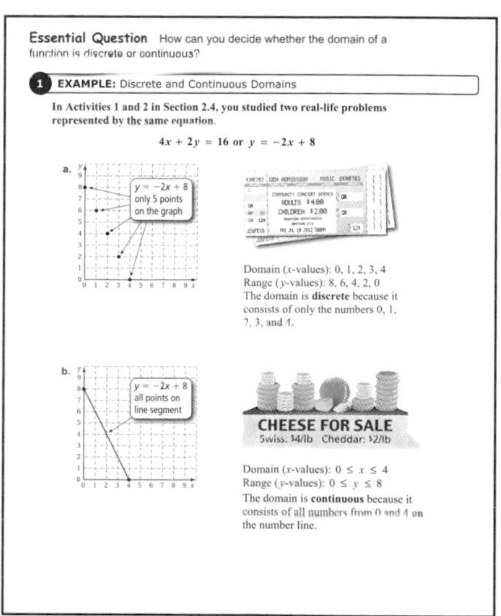

Essential Question How can you decide whether the domain of a function is discrete or continuous?

1 EXAMPLE: Discrete and Continuous Domains

In Activities 1 and 2 in Section 2.4, you studied two real-life problems represented by the same equation.

$$4x + 2y = 16 \text{ or } y = -2x + 8$$

a. 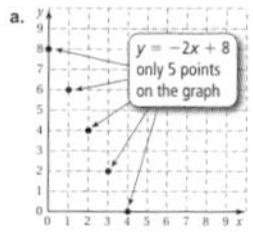

$y = -2x + 8$
only 5 points
on the graph

ADULTS $4.00
CHILDREN $2.00

Domain (x-values): 0, 1, 2, 3, 4
Range (y-values): 8, 6, 4, 2, 0
The domain is **discrete** because it consists of only the numbers 0, 1, 2, 3, and 4.

b. 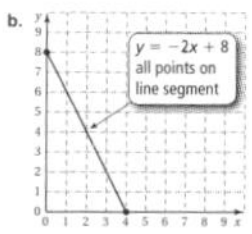

$y = -2x + 8$
all points on
line segment

CHEESE FOR SALE
Swiss: $4/lb Cheddar: $2/lb

Domain (x-values): $0 \le x \le 4$
Range (y-values): $0 \le y \le 8$
The domain is **continuous** because it consists of all numbers from 0 and 4 on the number line.

Differentiated Instruction

Auditory

Help students understand the concept of a function by showing them how it is used in everyday life. For example, the number of plates to set on the dinner table is a function of the number of people expected to eat. A person earning an hourly wage has an income that is a function of the number of hours worked. Discuss with students other instances of functions in life.

5.2 Record and Practice Journal

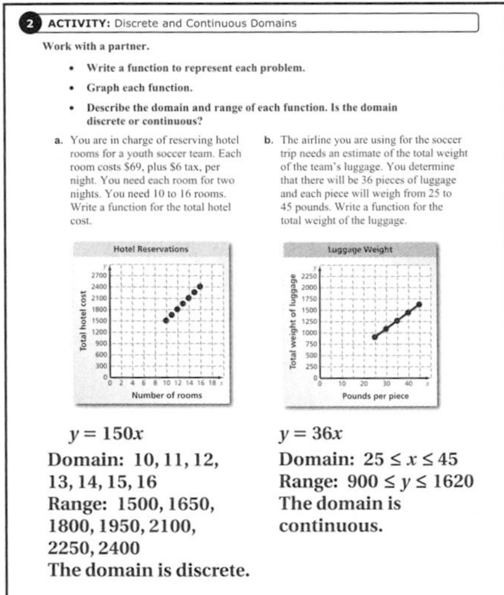

2 ACTIVITY: Discrete and Continuous Domains

Work with a partner.

- **Write a function to represent each problem.**
- **Graph each function.**
- **Describe the domain and range of each function. Is the domain discrete or continuous?**

a. You are in charge of reserving hotel rooms for a youth soccer team. Each room costs \$69, plus \$6 tax, per night. You need each room for two nights. You need 10 to 16 rooms. Write a function for the total hotel cost.

b. The airline you are using for the soccer trip needs an estimate of the total weight of the team's luggage. You determine that there will be 36 pieces of luggage and each piece will weigh from 25 to 45 pounds. Write a function for the total weight of the luggage.

a. $y = 150x$
Domain: 10, 11, 12, 13, 14, 15, 16
Range: 1500, 1650, 1800, 1950, 2100, 2250, 2400
The domain is discrete.

b. $y = 36x$
Domain: $25 \le x \le 45$
Range: $900 \le y \le 1620$
The domain is continuous.

What Is Your Answer?

3. **IN YOUR OWN WORDS** How can you decide whether the domain of a function is discrete or continuous? Describe two real-life examples of functions: one with a discrete domain and one with a continuous domain.

discrete consists of only certain numbers in an interval; continuous consists of all numbers in an interval

Laurie's Notes

Words of Wisdom

- **MP1 Make Sense of Problems and Persevere in Solving Them:** Students may ask you to explain each of the two problems, meaning, show them how to do the problems! Resist the tendency to jump in and solve the problems for them. Students should work with their partner and work through the problem. They need to read and think! Trust that they have the ability to make sense of the problem and expect them to persevere.

Activity 2

- **Part (a):** Rooms are \$75 a night (\$69 + \$6). If 10 rooms are rented for 2 nights, the cost is $\$75 \times 10 \times 2 = \$1500$. Write a verbal model to help you find an equation: Cost = cost per night × number of rooms × 2 nights; if x = the number of rooms rented, the equation is $y = (75)2x = 150x$.
- **Part (b):** There are 36 pieces of luggage that will vary in weight between 25 and 45 pounds. The total weight equals the number of pieces of luggage times the average weight of the pieces. If x = average weight of the pieces of luggage, the equation is $y = 36x$.

? "Is the domain of part (a) discrete or continuous? Explain." discrete; The number of rooms must be a whole number.

? "Is the domain of part (b) discrete or continuous? Explain." continuous; The weight of each piece of luggage is between 25 and 45 pounds, inclusive, but could be a fraction of a pound.

What Is Your Answer?

- **Think-Pair-Share:** Students should read the question independently and then work with a partner to answer the question. When they have answered the question, the pair should compare their answer with another group and discuss any discrepancies.

Closure

- **Exit Ticket:** Sketch a graph of the two examples suggested in Question 3. Students should scale the axes with reasonable numbers.

Technology For the Teacher

The Dynamic Planning Tool
Editable Teacher's Resources at *BigIdeasMath.com*

2 ACTIVITY: Discrete and Continuous Domains

Work with a partner.

- **Write a function to represent each problem.**
- **Graph each function.**
- **Describe the domain and range of each function. Is the domain discrete or continuous?**

a. You are in charge of reserving hotel rooms for a youth soccer team. Each room costs \$69, plus \$6 tax, per night. You need each room for two nights. You need 10 to 16 rooms. Write a function for the total hotel cost.

b. The airline you are using for the soccer trip needs an estimate of the total weight of the team's luggage. You determine that there will be 36 pieces of luggage and each piece will weigh from 25 to 45 pounds. Write a function for the total weight of the luggage.

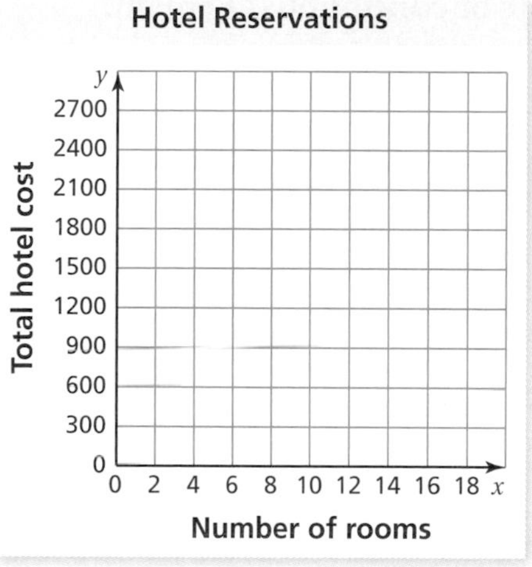

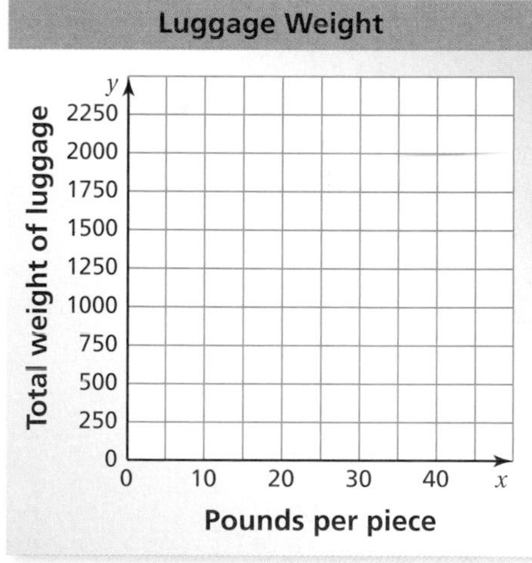

What Is Your Answer?

3. **IN YOUR OWN WORDS** How can you decide whether the domain of a function is discrete or continuous? Describe two real-life examples of functions: one with a discrete domain and one with a continuous domain.

Practice Use what you learned about discrete and continuous domains to complete Exercises 3 and 4 on page 214.

5.2 Lesson

Key Vocabulary
discrete domain, *p. 212*
continuous domain, *p. 212*

Key Idea

Discrete and Continuous Domains

A **discrete domain** is a set of input values that consists of only certain numbers in an interval.

Example: Integers from 1 to 5

A **continuous domain** is a set of input values that consists of all numbers in an interval.

Example: All numbers from 1 to 5

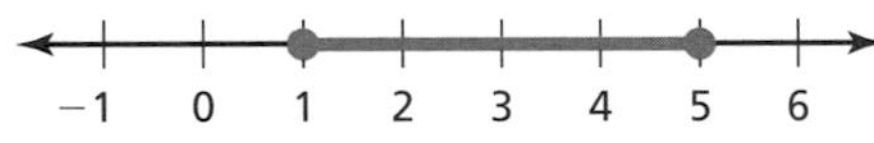

EXAMPLE 1 Graphing Discrete Data

The function $y = 15.95x$ represents the cost y (in dollars) of x tickets for a museum. Graph the function using a domain of 0, 1, 2, 3, and 4. Is the domain discrete or continuous? Explain.

Make an input-output table.

Input, x	$15.95x$	Output, y	Ordered Pair, (x, y)
0	15.95(0)	0	(0, 0)
1	15.95(1)	15.95	(1, 15.95)
2	15.95(2)	31.9	(2, 31.9)
3	15.95(3)	47.85	(3, 47.85)
4	15.95(4)	63.8	(4, 63.8)

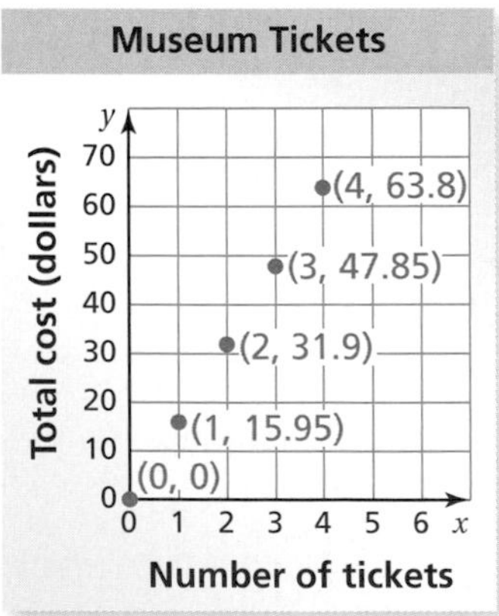

Plot the ordered pairs. Because you cannot buy part of a ticket, the graph consists of individual points.

∴ So, the domain is discrete.

On Your Own

1. The function $m = 50 - 9d$ represents the amount of money m (in dollars) you have after buying d DVDs. Graph the function. Is the domain discrete or continuous? Explain.

Laurie's Notes

Introduction

Connect

- **Yesterday:** Students explored problems with discrete and continuous domains. (MP1)
- **Today:** Students will graph functions and determine whether the domain is discrete or continuous.

Motivate

- **FYI:** Share some trivia about the words *continuous* and *discrete.*
- The word *continuous* derives from a Latin root meaning *to hang together* or *to cohere.* This same root gives us the noun *continent* (an expanse of land unbroken by sea).
- The word *discrete* derives from a Latin root meaning *to separate.* This same root yields the verb *discern* (to recognize as distinct or separate) and the cognate *discreet* (to show discernment).

Lesson Notes

Key Idea

- The number line graphs of each example should help students visualize the difference between these two types of domains.
- **Common Error:** Discrete functions do not need to exclude fractions or decimals. A discrete domain might be shoe sizes from 6 to 9, including the half sizes such as $6\frac{1}{2}$.

Example 1

- The table displayed is a good reminder of how equations are evaluated and how solutions can be recorded as ordered pairs.

? "Is this data set a function? Explain." yes; Each input is paired with exactly one output.

- **MP6 Attend to Precision:** Students may ask why the outputs are not written with the $ symbol and a digit in the hundredths position. If describing the answer to a contextual problem, such as "what is the cost for 2 people to visit the museum?", the answer would be stated with the units ($31.90). Otherwise, the *y*-coordinate is stated as a real number.

? "What do you notice about the range of this discrete function?" The range is discrete also.

On Your Own

? "Could you buy 0 DVDs? 1 DVD? 2 DVDs? What is the greatest number of DVDs you have money to purchase?" yes; yes; yes; 5

? "Is it possible to spend all of your money on DVDs? Explain." no; If you buy 5 DVDs, you have $5 remaining which is not enough to buy a sixth DVD.

Goal Today's lesson is determining whether a domain is **discrete** or **continuous**.

Start Thinking! and Warm Up

Lesson 5.2 Start Thinking!
For use before Lesson 5.2

Discuss whether the following functions have discrete or continuous domains.

A. The air temperature over the course of a day

B. The cost of hot dogs

C. The distance traveled on a road trip

D. The weight of a baby over his first month

E. The cost of parking for a certain number of hours at a parking garage

Extra Example 1

The function $y = 12.90x$ represents the cost y (in dollars) of x admission tickets for a museum. Graph the function using a domain of 0, 1, 2, 3, and 4. Is the domain discrete or continuous? Explain.

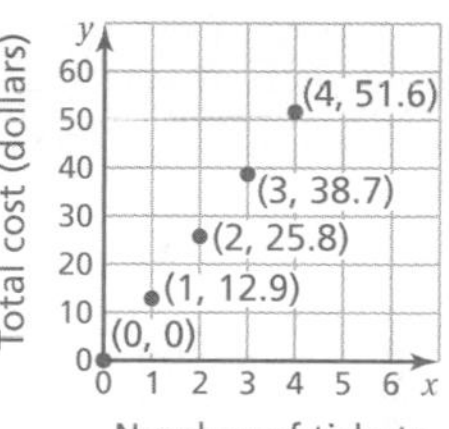

The domain is discrete because you cannot buy part of a ticket.

On Your Own

1. 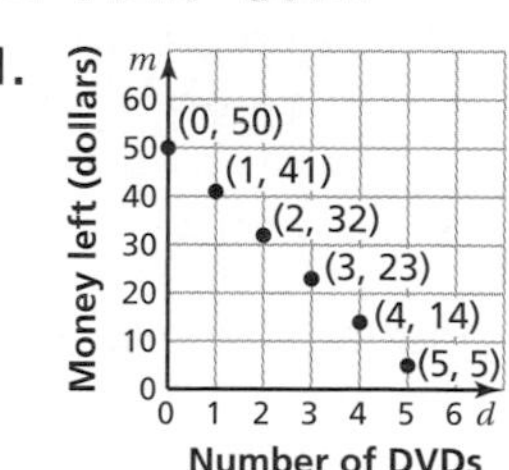

The domain is discrete because you cannot buy part of a DVD.

Extra Example 2

A cereal bar contains 155 calories. The number c of calories consumed is a function of the number b of bars eaten. Graph the function. Is the domain discrete or continuous?

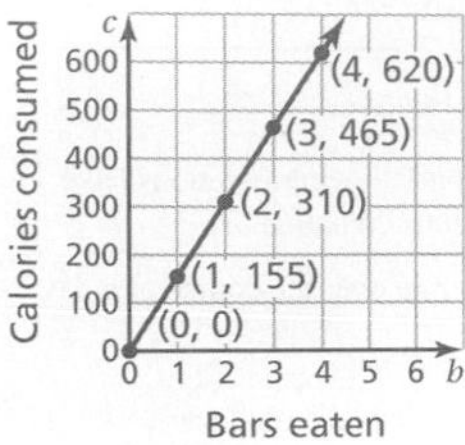

The domain is continuous because you can eat part of a cereal bar.

Extra Example 3

You conduct an experiment on the distance traveled at 55 miles per hour. (a) What is the domain of the function? $1 \le t \le 4$ (b) Is the domain discrete or continuous? The domain is continuous because the time can be any value between 1 and 4, inclusive.

Input Time, t (hours)	Output Distance, d (miles)
1	55
2	110
3	165
4	220

On Your Own

2. See Additional Answers.
3. The domain is discrete because you cannot have part of a story.

English Language Learners

Vocabulary

Have students add the key vocabulary words *function, domain, range, independent variable, dependent variable, relation, Vertical Line Test, discrete domain*, and *continuous domain* to their notebooks. Definitions, examples, and pictures should accompany the words.

Laurie's Notes

Example 2

- Read the problem. This context will be familiar to students. Students will recognize that it is possible to eat some portion of a cereal bar, meaning this is a continuous domain.
- Discuss the graph of this function. Although the table of values stops at the ordered pair (4, 520), it is possible to consume more than 4 cereal bars.
- The domain is *restricted* in the sense that the number of cereal bars must be non-negative.

? **Extension:** "What is the slope and c-intercept of this function?" 130; 0

Example 3

- Read the problem and discuss the table.

? "What are the values of the domain?" Listen for: time (in seconds) between 2 and 10, inclusive.

? "What are the values of the range?" Listen for: distance (in miles) between 0.434 and 2.17, inclusive.

On Your Own

- These are nice questions that provide another context for understanding the difference between continuous and discrete domains.

Closure

- **Exit Ticket:** Explain how to determine if a graph has a continuous or discrete domain. Listen for: if the points on the graph are connected then it is continuous, but if the points are separated then it is discrete.

The Dynamic Planning Tool
Editable Teacher's Resources at *BigIdeasMath.com*

EXAMPLE 2 Graphing Continuous Data

A cereal bar contains 130 calories. The number *c* of calories consumed is a function of the number *b* of bars eaten. Graph the function. Is the domain discrete or continuous?

Make an input-output table.

Input, *b*	Output, *c*	Ordered Pair, (*b*, *c*)
0	0	(0, 0)
1	130	(1, 130)
2	260	(2, 260)
3	390	(3, 390)
4	520	(4, 520)

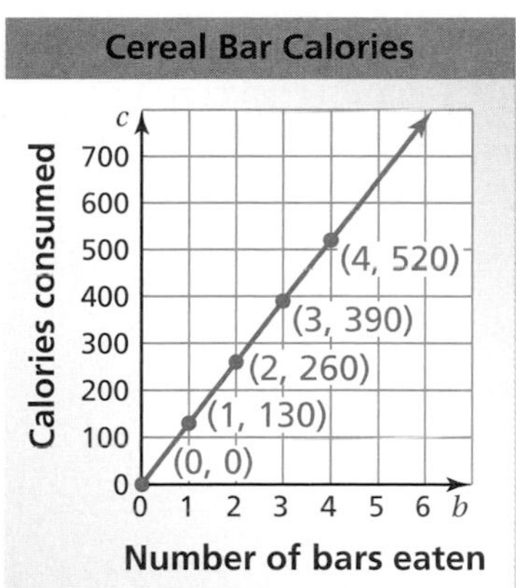

Plot the ordered pairs. Because you can eat part of a cereal bar, *b* can be any value greater than or equal to 0. Draw a line through the points.

So, the domain is continuous.

EXAMPLE 3 Standardized Test Practice

Input Time, *t* (seconds)	Output Distance, *d* (miles)
2	0.434
4	0.868
6	1.302
8	1.736
10	2.170

You conduct an experiment on the speed of sound waves in dry air at 86 °F. You record your data in a table. Which of the following is true?

Ⓐ The domain is $2 \le t \le 10$ and it is discrete.

Ⓑ The domain is $2 \le t \le 10$ and it is continuous.

Ⓒ The domain is $0.434 \le d \le 2.17$ and it is discrete.

Ⓓ The domain is $0.434 \le d \le 2.17$ and it is continuous.

The domain is the set of possible input values, or the time *t*. The time *t* can be any value from 2 to 10. So, the domain is continuous.

The correct answer is Ⓑ.

On Your Own

Now You're Ready Exercises 5–8

2. A 20-gallon bathtub is draining at a rate of 2.5 gallons per minute. The number *g* of gallons remaining is a function of the number *m* of minutes. Graph the function. Is the domain discrete or continuous?

3. Is the domain discrete or continuous? Explain.

Input Number of Stories	1	2	3
Output Height of Building (feet)	12	24	36

5.2 Exercises

Vocabulary and Concept Check

1. **VOCABULARY** Explain how continuous domains and discrete domains are different.

2. **WRITING** Describe how you can use a graph to determine whether a domain is discrete or continuous.

Practice and Problem Solving

Describe the domain and range of the function. Is the domain discrete or continuous?

3. 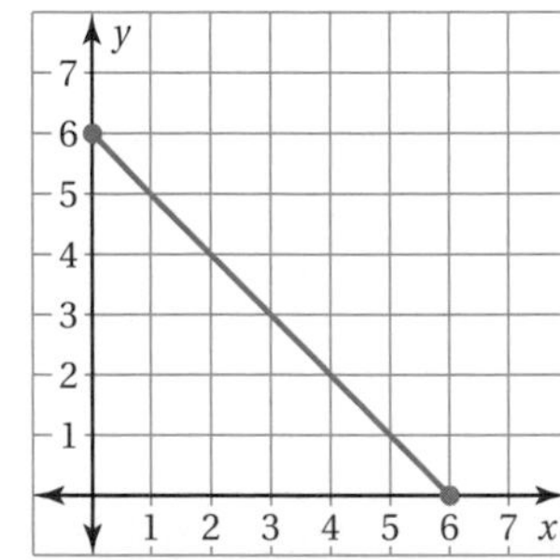

4. 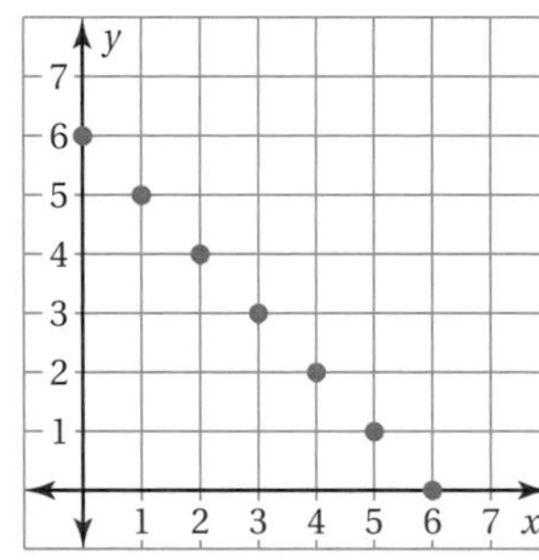

Graph the function. Is the domain discrete or continuous?

5.

Input Bags, x	Output Marbles, y
2	20
4	40
6	60

6.

Input Years, x	Output Height of a Tree, y (feet)
0	3
1	6
2	9

7.

Input Width, x (inches)	Output Volume, y (cubic inches)
5	50
10	100
15	150

8.

Input Hats, x	Output Cost, y (dollars)
0	0
1	8.45
2	16.9

9. **ERROR ANALYSIS** Describe and correct the error made in the statement about the domain.

2.5 is in the domain.

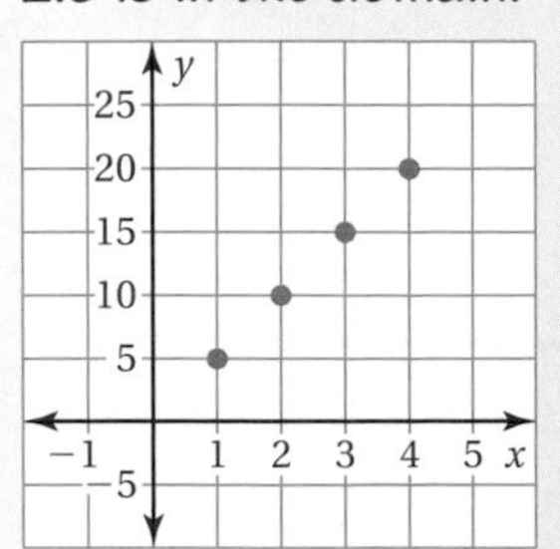

10. **YARN** The function $m = 40 - 8.5b$ represents the amount m of money (in dollars) that you have after buying b balls of yarn. Graph the function using a domain of 0, 1, 2, and 3. Is the domain discrete or continuous?

Assignment Guide and Homework Check

Level	Assignment	Homework Check
Average	1–4, 7–13, 16–19	2, 8, 10, 12
Advanced	1–4, 8–19	2, 8, 12, 14

Common Errors

- **Exercises 5–8** Students may mistake the output for the domain and say that the domain is continuous when it is actually discrete. Encourage them to think about the context of the problem. For example, you cannot buy part of a hat, so the domain is discrete.
- **Exercises 5–8** When graphing the function, students may connect the points without considering if the data are discrete or continuous. Remind them that the graph displays discrete and continuous data differently, so the graphs should be different.
- **Exercise 12** Students may say that the function has a discrete domain because length is often given as a whole number. Encourage them to think about a context for the function, for example, the length of a snake.

5.2 Record and Practice Journal

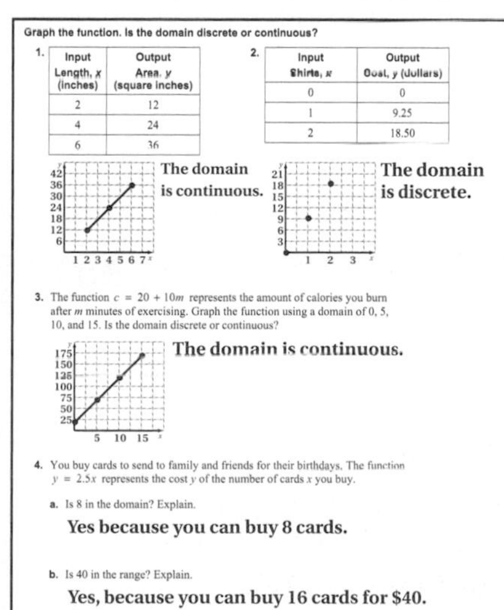
Graph the function. Is the domain discrete or continuous?

1.

Input Length, x (inches)	Output Area, y (square inches)
2	12
4	24
6	36

The domain is continuous.

2.

Input Shirts, x	Output Cost, y (dollars)
0	0
1	9.25
2	18.50

The domain is discrete.

3. The function $c = 20 + 10m$ represents the amount of calories you burn after m minutes of exercising. Graph the function using a domain of 0, 5, 10, and 15. Is the domain discrete or continuous?

The domain is continuous.

4. You buy cards to send to family and friends for their birthdays. The function $y = 2.5x$ represents the cost y of the number of cards x you buy.

a. Is 8 in the domain? Explain.

Yes because you can buy 8 cards.

b. Is 40 in the range? Explain.

Yes, because you can buy 16 cards for $40.

Technology For the Teacher

Answer Presentation Tool

Vocabulary and Concept Check

1. A discrete domain consists of only certain numbers in an interval, whereas a continuous domain consists of all numbers in an interval.
2. If the graph is a line covering all inputs on an interval, then it is a continuous domain. If a graph consists of just points, then it is a discrete domain.

Practice and Problem Solving

3. domain: $0 \le x \le 6$
range: $0 \le y \le 6$;
continuous
4. domain: 0, 1, 2, 3, 4, 5, 6
range: 0, 1, 2, 3, 4, 5, 6;
discrete
5. 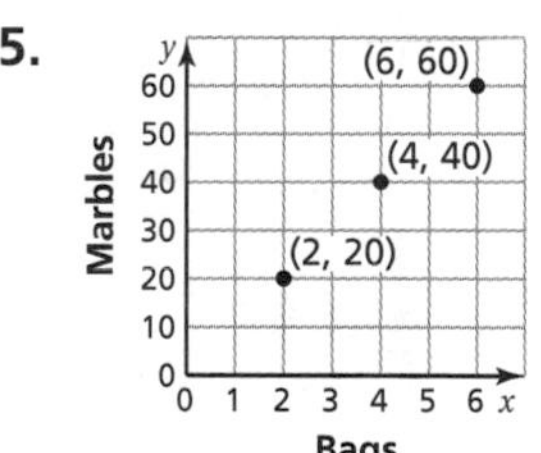

discrete

6–8. See Additional Answers.

9. 2.5 is *not* in the domain, because the domain is discrete and consists only of the integers 1, 2, 3, and 4.
10. 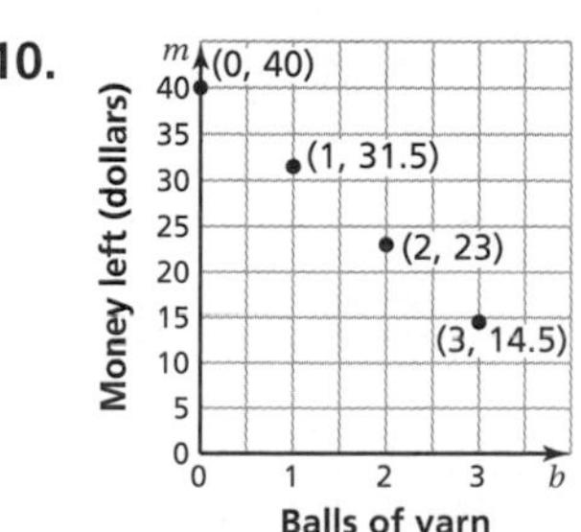

discrete

Practice and Problem Solving

11. a. independent variable: c; dependent variable: t

b. discrete

12. a. 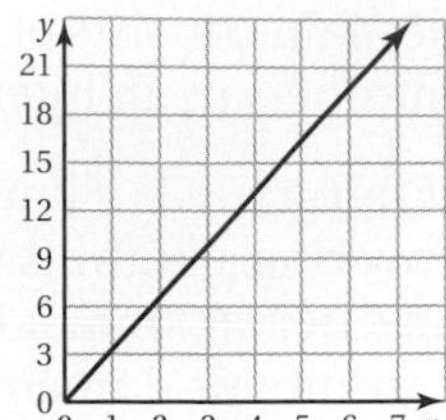

independent variable: x; dependent variable: y

b. continuous

13. no; A height can be any positive number.

14. See *Taking Math Deeper.*

15. See Additional Answers.

Fair Game Review

16. 1 **17.** $-\frac{5}{2}$

18. $\frac{1}{3}$ **19.** C

Mini-Assessment

Graph the function. Is the domain discrete or continuous?

1.

Cups, x	0	3	6	9
Cost, y ($)	0	6	12	18

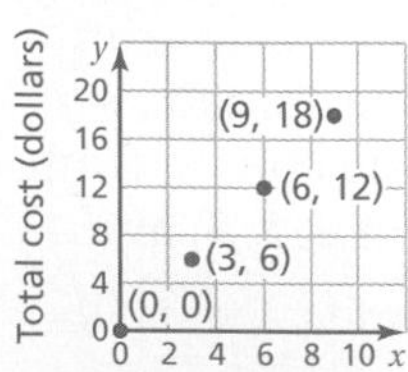

discrete

2.

Time, x (h)	0	1	2	3
Distance, y (mi)	0	55	110	165

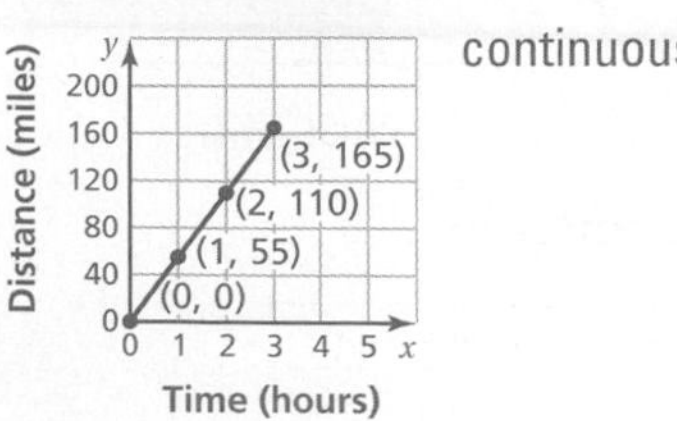

continuous

Taking Math Deeper

Exercise 14

You can make a table for this problem showing the possible numbers x of books and their corresponding weights y (in pounds).

1 Make a table.

Input, Books, x	Output, Weight, y (in pounds)
0	0.0
1	5.2
2	10.4
3	15.6
4	20.8
5	26.0

Input, Books, x	Output, Weight, y (in pounds)
6	31.2
7	36.4
8	41.6
9	46.8
10	52.0

2 Consider the domain and range.

The numbers x of books are the inputs that make up the domain. The corresponding weights y are the outputs that make up the range.

a. You can see that 52 is the greatest value in the range. The box holds up to 10 books, so the maximum weight is 52 pounds.

b. Because at most 10 books can fit into the box, the greatest value in the domain is 10. So, 15 is not in the domain.

3 Graph the function.

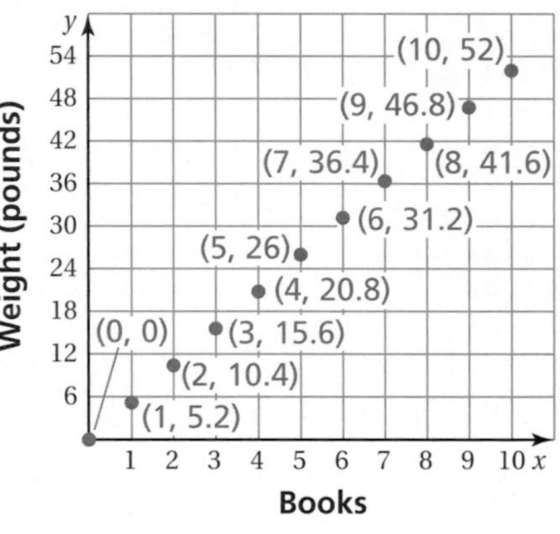

c. Because the domain of the function consists of the whole numbers 0 through 10, the domain is discrete.

Reteaching and Enrichment Strategies

If students need help. . .	If students got it. . .
Resources by Chapter • Practice A and Practice B • Puzzle Time Record and Practice Journal Practice Differentiating the Lesson Lesson Tutorials Skills Review Handbook	Resources by Chapter • Enrichment and Extension Start the next section

11. **TICKETS** The number t of tickets sold at a concert is a function of the ticket cost c.

   a. Which variable is independent? dependent?

   b. Is the domain discrete or continuous?

12. **DISTANCE** The function $y = 3.28x$ converts length from x meters to y feet.

   a. Graph the function. Which variable is independent? dependent?

   b. Is the domain discrete or continuous?

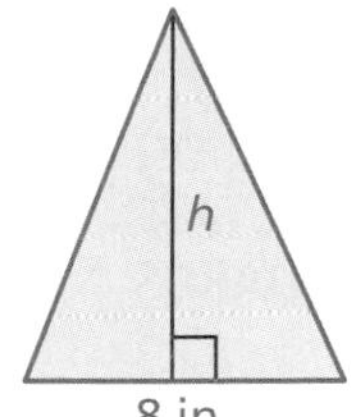

13. **LOGIC** The area A of the triangle is a function of the height h. Your friend says the domain is discrete. Is he correct? Explain.

14. **PACKING** You are packing books into a box. The box can hold at most 10 books. The function $y = 5.2x$ represents the weight y (in pounds) of x books.

   a. Is 52 in the range? Explain.

   b. Is 15 in the domain? Explain.

   c. Graph the function. Is the domain discrete or continuous?

15. **Reasoning** Describe a real-world situation for the given constraints.

   a. A negative number in the domain and the domain is continuous

   b. A negative number in the range and the domain is discrete

Fair Game Review What you learned in previous grades & lessons

Find the slope of the line. *(Section 2.2)*

16. 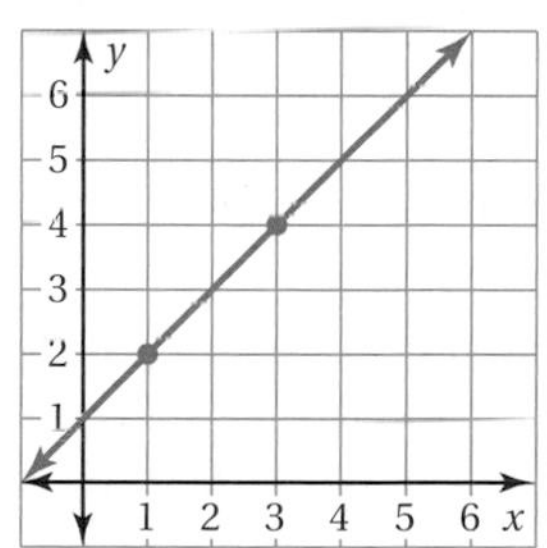

17. 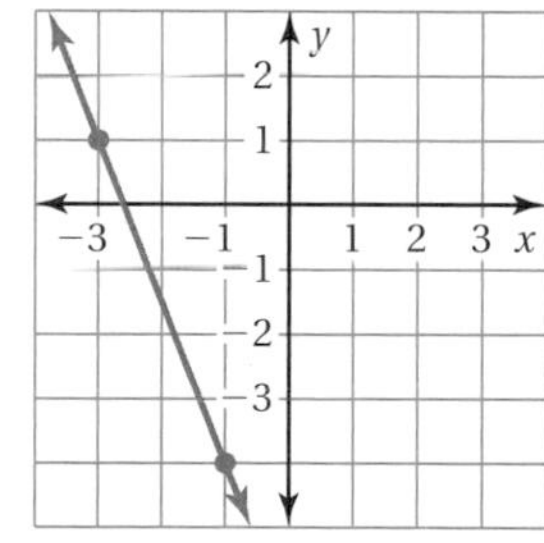

18. 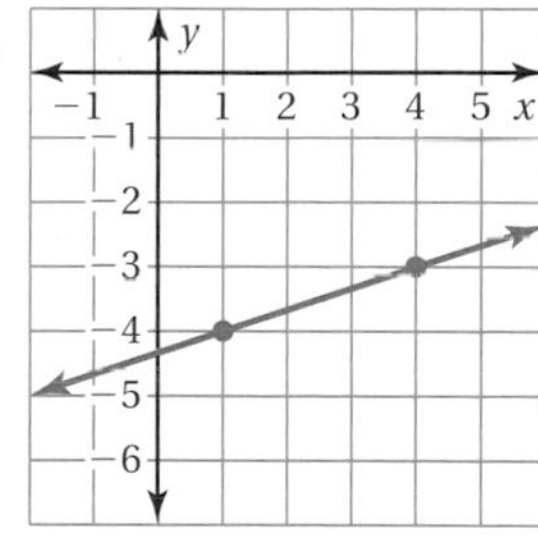

19. **MULTIPLE CHOICE** What is the y-intercept of the graph of the linear equation? *(Section 2.3)*

   (A) -4 (B) -2

   (C) 2 (D) 4

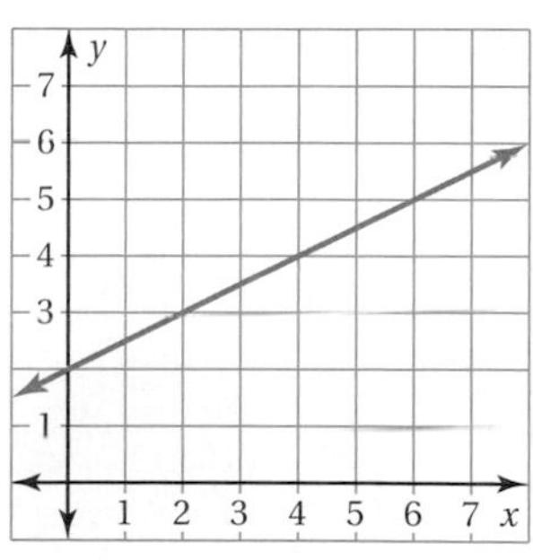

5.3 Linear Function Patterns

Essential Question How can you use a linear function to describe a linear pattern?

COMMON CORE STATE STANDARDS
8.F.3
8.F.4
F.BF.1a
F.LE.2

1 ACTIVITY: Finding Linear Patterns

Work with a partner.

- **Plot the points from the table in a coordinate plane.**
- **Write a linear equation for the function represented by the graph.**

a.

x	0	2	4	6	8
y	150	125	100	75	50

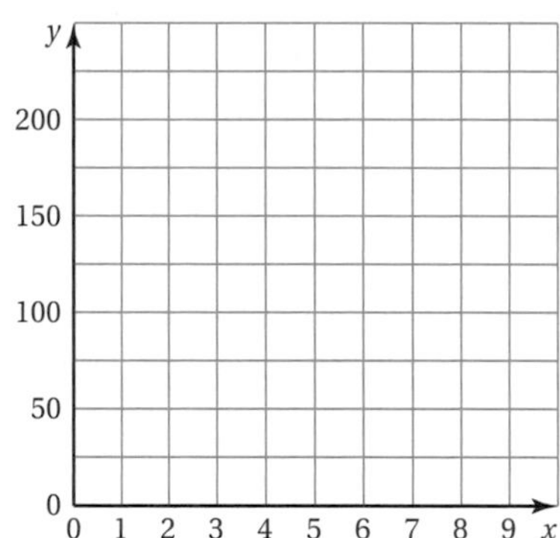

b.

x	4	6	8	10	12
y	15	20	25	30	35

y
40
30
20
10
0
0 4 8 12 16 x

c.

x	−4	−2	0	2	4
y	4	6	8	10	12

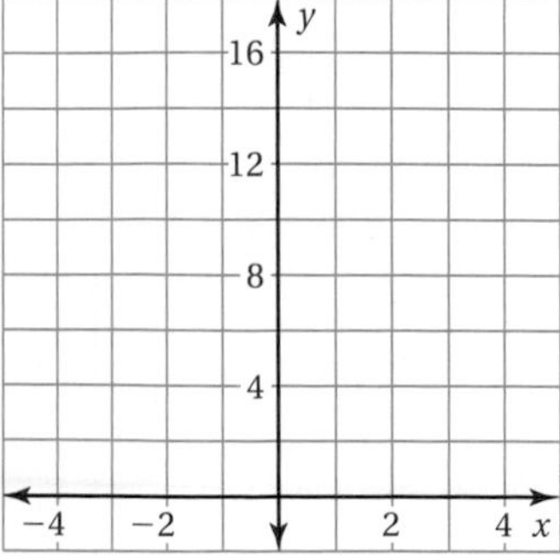

d.

x	−4	−2	0	2	4
y	1	0	−1	−2	−3

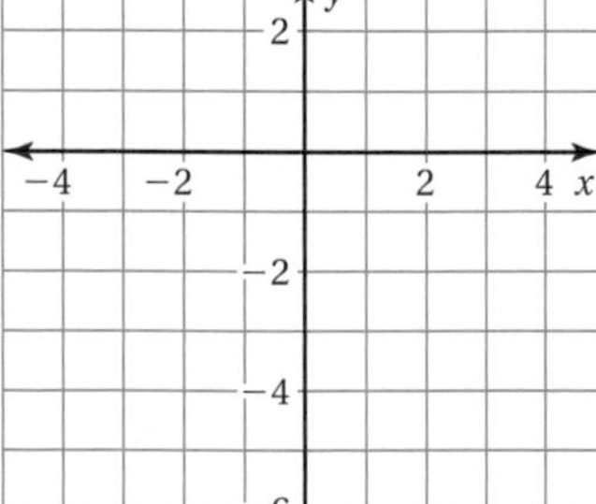

Laurie's Notes

Introduction

Standards for Mathematical Practice

- **MP8 Look for and Express Regularity in Repeated Reasoning:** The goal is for students to recognize that a linear pattern occurs when there is a constant rate of change in a table of values or in a graph. In Chapter 2, students wrote linear equations in slope-intercept form. What is new is the language—the linear equation is referred to as a linear function. Students will recognize a pattern (constant rate of change) in the data and write the function.

Motivate

- Do a quick matching game with students. Have 4–5 graphs on the board with slopes and *y*-intercepts that are different enough so that students can distinguish between them. Write the equations in a list. Have students work with a partner to match the correct equation with each graph.
- Make sure that students are still focusing on key information from the graph. Is it increasing or decreasing from left to right? Is the slope steeper than 1 or close to 0? Is the *y*-intercept positive or negative?

Activity Notes

Activity 1

? "What do you notice about the scaling on the axes for each problem?" Answers will vary. Students should recognize the difference of how the *x*- and *y*-axes are scaled in each problem.

? "For each problem, you are asked to write a linear equation for the function. How will you do this?" Find the slope and *y*-intercept.

- Give sufficient time for students to work through the four problems.
- From the graphs, students should be able to determine the slope. It is important that students pay attention to how the axes are scaled when they record values for rise and run.
- **MP1 Make Sense of Problems and Persevere in Solving Them** and **MP8 Look for and Express Regularity in Repeated Reasoning:** From the table of values, the *y*-intercept is given for 3 of the 4 problems. In part (a), the ordered pair (0, 150) gives the *y*-intercept, $b = 150$. To find the slope from the table, notice that every time *x* increases by 2, *y* decreases by 25. You can recognize this as a constant rate of change in which the run is 2 and the rise is -25. So, $m = \frac{\text{rise}}{\text{run}} = \frac{-25}{2} = -12.5$. Now write the equation in slope-intercept form, $y = -12.5x + 150$.
- When students have finished, check their equations.

? "What numeric patterns do you see in the table?" Listen for how the *x*- and *y*-values are changing.

- Make sure students recognize the connection between the numeric patterns in the table and the slope of the line.
- **FYI:** For students, recognizing a pattern in the table is the easy part. Helping students translate the pattern into a slope, and then into an equation, is the challenging part. This takes practice.

Common Core State Standards

8.F.3 Interpret the equation $y = mx + b$ as defining a linear function, whose graph is a straight line; . . ..

8.F.4 Construct a function to model a linear relationship between two quantities. Determine the rate of change and initial value of the function from a description of a relationship or from two (x, y) values, including reading these from a table or from a graph. Interpret the rate of change and initial value of a linear function in terms of the situation it models, and in terms of its graph or a table of values.

Also **F.BF.1a** and **F.LE.2**

Previous Learning

Students should know how to write a linear equation in slope-intercept form. Students should know common geometric formulas, such as area and perimeter.

Start Thinking! and Warm Up

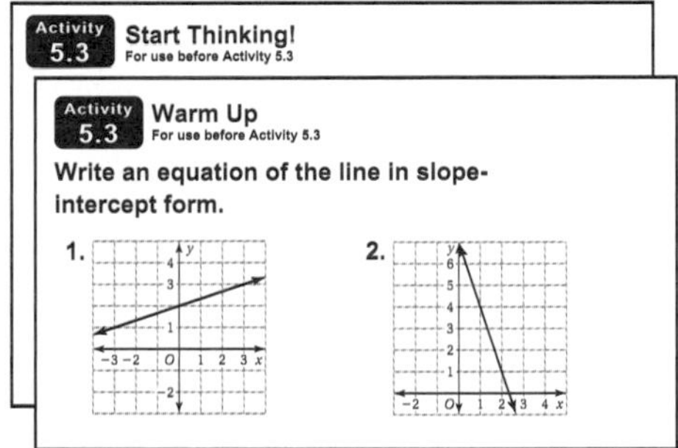

Activity 5.3 Start Thinking! For use before Activity 5.3

Activity 5.3 Warm Up For use before Activity 5.3

Write an equation of the line in slope-intercept form.

1.

2.

5.3 Record and Practice Journal

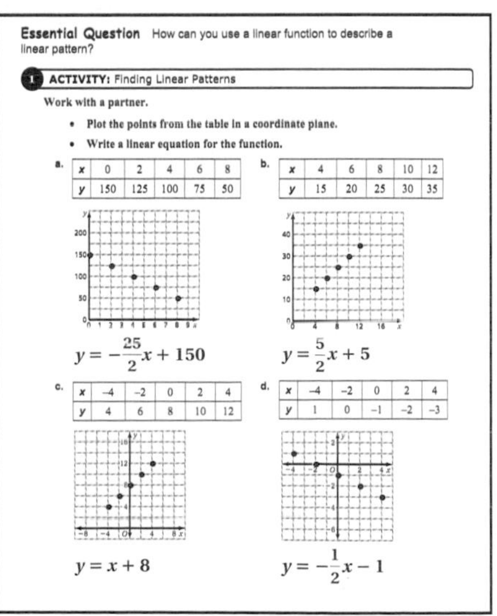

Essential Question How can you use a linear function to describe a linear pattern?

1 ACTIVITY: Finding Linear Patterns

Work with a partner.

- Plot the points from the table in a coordinate plane.
- Write a linear equation for the function.

a.

x	0	2	4	6	8
y	150	125	100	75	50

$y = -\frac{25}{2}x + 150$

b.

x	4	6	8	10	12
y	15	20	25	30	35

$y = \frac{5}{2}x + 5$

c.

x	−4	−2	0	2	4
y	4	6	8	10	12

$y = x + 8$

d.

x	−4	−2	0	2	4
y	1	0	−1	−2	−3

$y = -\frac{1}{2}x - 1$

Differentiated Instruction

Visual

Explain to students that representing a function table as a list of ordered pairs is for convenience. Once the function is represented by ordered pairs, it can be graphed in a coordinate plane. This is a visual representation of the function and is an excellent way to show students the connection between algebra and geometry.

5.3 Record and Practice Journal

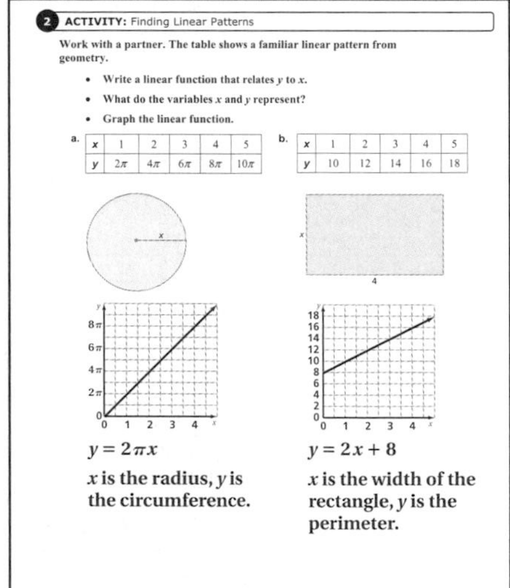

2 ACTIVITY: Finding Linear Patterns

Work with a partner. The table shows a familiar linear pattern from geometry.

- Write a linear function that relates y to x.
- What do the variables x and y represent?
- Graph the linear function.

a.

x	1	2	3	4	5
y	2π	4π	6π	8π	10π

$y = 2\pi x$

x is the radius, y is the circumference.

b.

x	1	2	3	4	5
y	10	12	14	16	18

$y = 2x + 8$

x is the width of the rectangle, y is the perimeter.

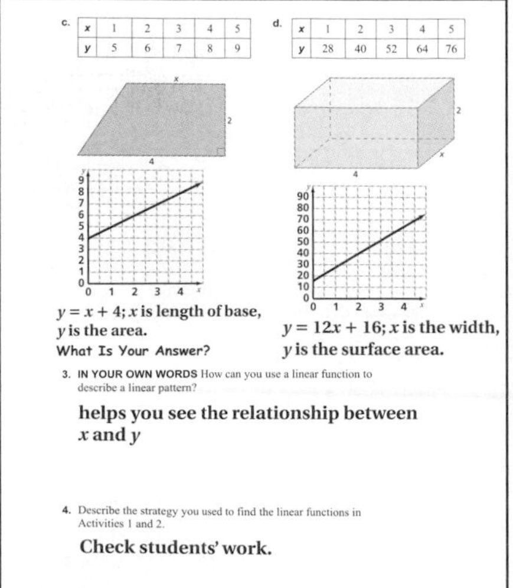

c.

x	1	2	3	4	5
y	5	6	7	8	9

$y = x + 4$; x is length of base, y is the area.

d.

x	1	2	3	4	5
y	28	40	52	64	76

$y = 12x + 16$; x is the width, y is the surface area.

What Is Your Answer?

3. IN YOUR OWN WORDS How can you use a linear function to describe a linear pattern?

helps you see the relationship between x and y

4. Describe the strategy you used to find the linear functions in Activities 1 and 2.

Check students' work.

Laurie's Notes

Activity 2

- The challenge in these problems is that the equation relates to a geometric formula. The figure shown for each problem should provide a hint as to what the variables x and y represent in the problem.
- **Part (a):** Two formulas involving π and circles are circumference ($C = 2\pi r$) and area ($A = \pi r^2$). Substitute the value of x for the radius in each formula. The value of the circumference will match the y-values in the table.
- **Part (b):** Two formulas involving rectangles are perimeter ($P = 2\ell + 2w$) and area ($A = \ell w$). Substitute 4 for the length and the value of x for the width in each formula. The value of the perimeter will match the y-values in the table.

? "Could y represent the perimeter for part (c)? Explain." no; You only know 3 of the 4 side lengths, and the sum of the three sides you know is greater than y.

- **Part (c):** The formula for the area of a trapezoid is $A = (b + B)h \div 2$. Substitute 4 for B, 2 for h, and the value of x for the length of the shorter base. The value of the area will match the y-values in the table.
- **Part (d):** Two formulas involving a rectangular prism are surface area ($S = 2\ell w + 2wh + 2\ell h$) and volume ($V = \ell wh$). Substitute 4 for the length, 2 for the height, and the value of x for the width in each formula. The value of the surface area will match the y-values in the table.

? Extension: "In part (c), how does the diagram of the trapezoid change as the value of x increases?" When $x = 4$, the trapezoid becomes a rectangle. When $x > 4$, the upper base becomes the longer of the two bases.

- Note that in each of these problems, there is a numeric pattern in the table. Have students describe the numeric pattern. Encourage them to use language such as "As x increases by 1, y increases by 2π."

What Is Your Answer?

- **Think-Pair-Share:** Students should read each question independently and then work with a partner to answer the questions. When they have answered the questions, the pair should compare their answers with another group and discuss any discrepancies.

Closure

- **Exit Ticket:** Plot the points given in the table and write a linear equation for the function.

x	−2	0	2	4	6
y	2	3	4	5	6

$y = \frac{1}{2}x + 3$

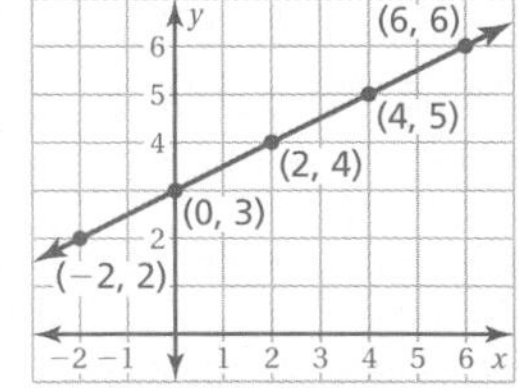

The Dynamic Planning Tool

Editable Teacher's Resources at *BigIdeasMath.com*

2 ACTIVITY: Finding Linear Patterns

Work with a partner. The table shows a familiar linear pattern from geometry.

- **Write a linear function that relates y to x.**
- **What do the variables x and y represent?**
- **Graph the linear function.**

a.

x	1	2	3	4	5
y	2π	4π	6π	8π	10π

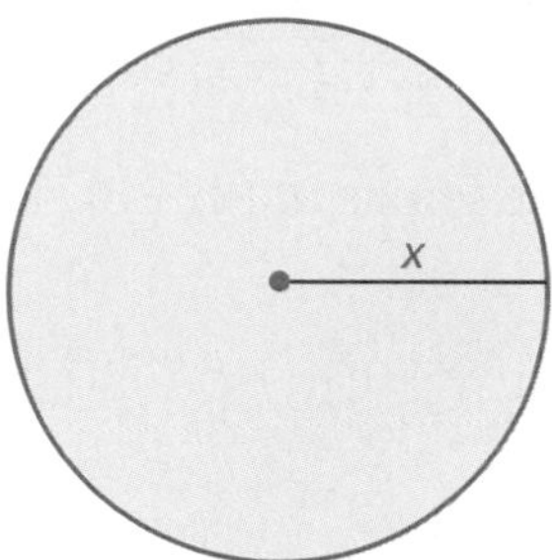

b.

x	1	2	3	4	5
y	10	12	14	16	18

x

4

c.

x	1	2	3	4	5
y	5	6	7	8	9

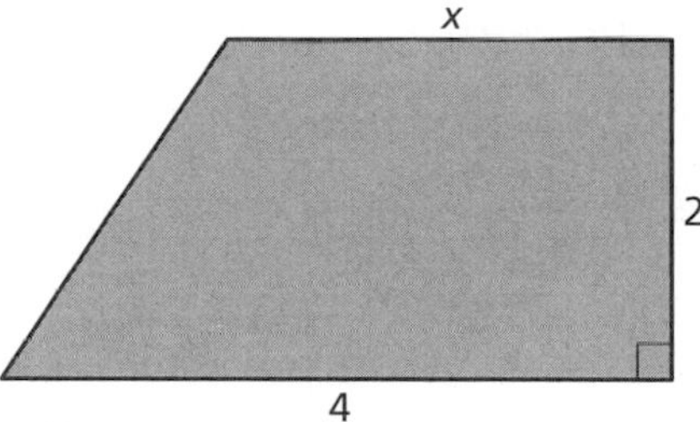

d.

x	1	2	3	4	5
y	28	40	52	64	76

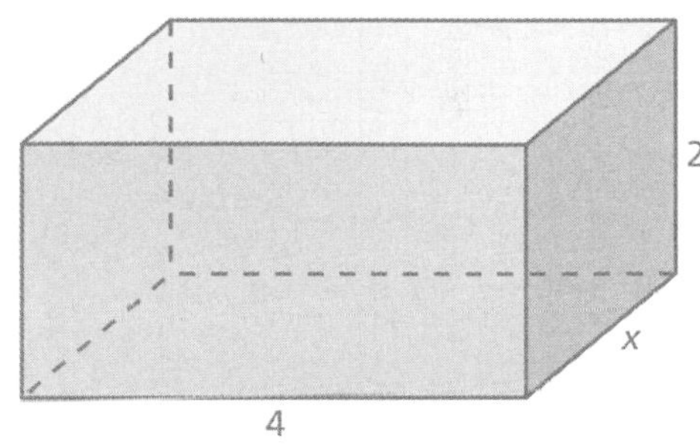

What Is Your Answer?

3. **IN YOUR OWN WORDS** How can you use a linear function to describe a linear pattern?
4. Describe the strategy you used to find the linear functions in Activities 1 and 2.

Practice Use what you learned about linear function patterns to complete Exercises 4 and 5 on page 220.

5.3 Lesson

Key Vocabulary
linear function, *p. 218*

A **linear function** is a function whose graph is a nonvertical line. A linear function can be written in the form $y = mx + b$.

EXAMPLE 1 Finding a Linear Function Using a Graph

Use the graph to write a linear function that relates y to x.

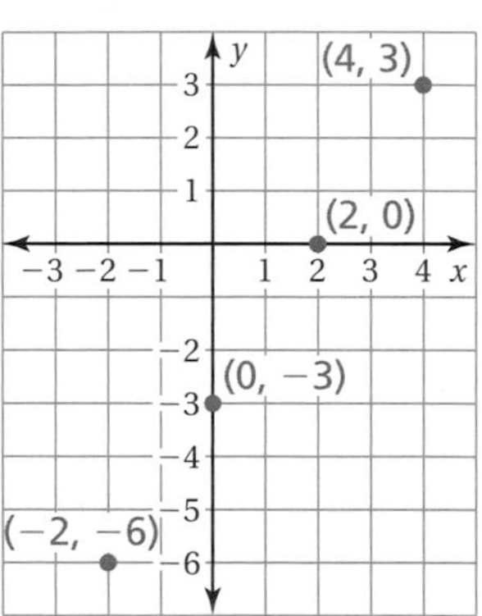

The points lie on a line. Find the slope and y-intercept of the line.

$$\text{slope} = \frac{\text{change in } y}{\text{change in } x} = \frac{3 - 0}{4 - 2} = \frac{3}{2}$$

Because the line crosses the y-axis at $(0, -3)$, the y-intercept is -3.

So, the linear function is $y = \frac{3}{2}x - 3$.

EXAMPLE 2 Finding a Linear Function Using a Table

Use the table to write a linear function that relates y to x.

x	-3	-2	-1	0
y	9	7	5	3

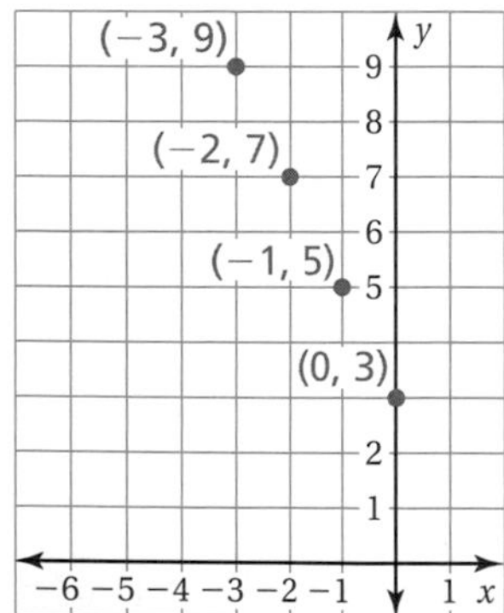

Plot the points in the table.

The points lie on a line. Find the slope and y-intercept of the line.

$$\text{slope} = \frac{\text{change in } y}{\text{change in } x} = \frac{9 - 7}{-3 - (-2)} = \frac{2}{-1} = -2$$

Because the line crosses the y-axis at $(0, 3)$, the y-intercept is 3.

So, the linear function is $y = -2x + 3$.

On Your Own

Now You're Ready
Exercises 6–11

Use the graph or table to write a linear function that relates y to x.

1.

2.

x	-2	-1	0	1
y	2	2	2	2

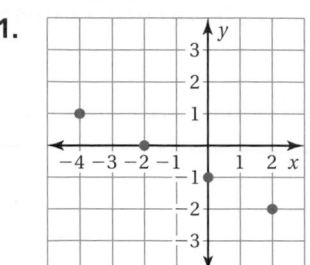

Laurie's Notes

Introduction

Connect

- **Yesterday:** Students gained additional practice in writing linear equations. (MP1, MP8)
- **Today:** Students will write linear functions by recognizing patterns in graphical and tabular information.

Motivate

- Tell the story of Amos Dolbear, who in 1898 noticed that warmer crickets seemed to chirp faster. Dolbear made a detailed study of cricket chirp rates based on the temperature of the crickets' environment and came up with the cricket chirping temperature formula known as Dolbear's Law. Remember that the formula is actually a linear function with a slope and y-intercept!

Lesson Notes

Example 1

- Students worked on similar problems in Chapter 2. The difference here is the function terminology.
- Review the definition of a function and the vocabulary associated with functions: domain and range.
- Write the definition of a linear function.
- **Connection:** Remind students of the Vertical Line Test. A vertical line cannot be a function.
- **Extension:** From the definition, students might guess that there are other types of functions besides linear functions. Draw a parabola or sine wave to make this connection.
- **Teaching Tip:** To find the slope, draw a right triangle with the hypotenuse between two of the points. Label the legs of the triangle to represent the rise and run. Then compute the slope.

? "Does it matter what two points you select to find the slope? Explain." no; The ratio of rise to run will be the same because the slope triangles are actually similar. It is unlikely students will say this; however, it is the case.

- **MP8 Look for and Express Regularity in Repeated Reasoning:** Demonstrate that it does not matter what two points are selected to compute the slope. The slope between (0, −3) and (2, 0) is $\frac{3}{2}$. The slope between (0, −3) and (4, 3) is $\frac{6}{4} = \frac{3}{2}$.

Example 2

- Plot the ordered pairs and repeat the steps from Example 1.

? "Can you tell anything about the slope without plotting the points? Explain." yes; As x increases by 1 (run), y decreases by 2 (rise).

On Your Own

- **Common Error:** Students may say the slope for Question 1 is −2 instead of $-\frac{1}{2}$. It is very easy to state the reciprocal of the slope.
- **Question 2:** Students may need to graph this function. Once graphed, they will recognize this as a horizontal line whose equation is $y = 2$.

Goal Today's lesson is writing a **linear function** from a graph or a table of values.

Start Thinking! and Warm Up

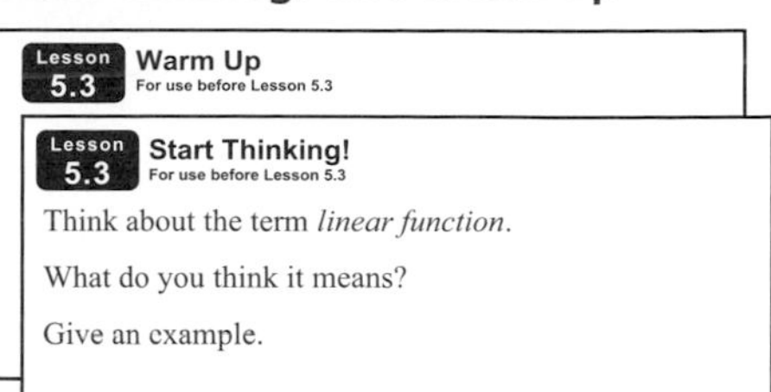
Lesson 5.3 Warm Up
For use before Lesson 5.3

Lesson 5.3 Start Thinking!
For use before Lesson 5.3

Think about the term *linear function.*

What do you think it means?

Give an example.

Extra Example 1

Use the graph to write a linear function that relates y to x.

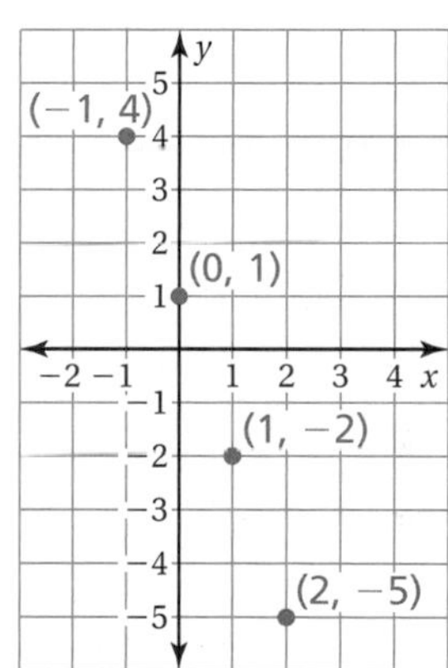

$y = -3x + 1$

Extra Example 2

Use the table to write a linear function that relates y to x.

x	−2	0	2	4
y	−2	−1	0	1

$y = \frac{1}{2}x - 1$

 **On Your Own**

1. $y = -\frac{1}{2}x - 1$
2. $y = 2$

Extra Example 3

Graph the data.

Hours Jogging, x	Calories Burned, y
2	800
4	1600
6	2400
8	3200

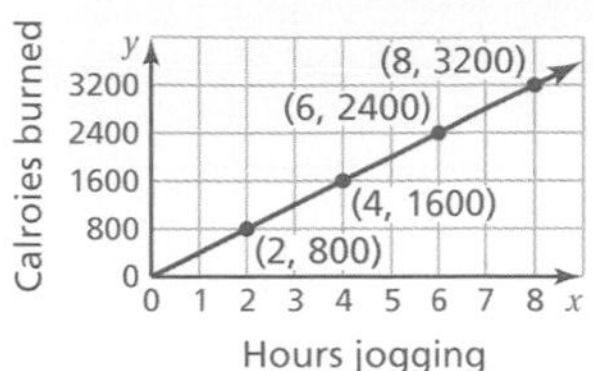

a. Is the domain discrete or continuous? continuous

b. Write a linear function that relates y to x. $y = 400x$

c. How many calories do you burn in 2.5 hours? 1000 calories

On Your Own

3.

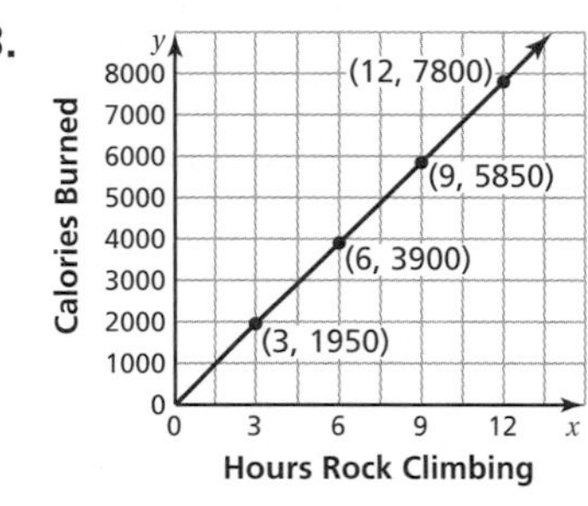

a. continuous

b. $y = 650x$

c. 3575 calories

English Language Learners

Classroom

This chapter gives English learners a chance to share with the rest of the class and the opportunity to build their confidence. Many examples and exercises use tables and graphs giving English learners a rest from interpreting sentences.

Laurie's Notes

Example 3

- Read the problem and discuss the ordered pairs in the table.
- **?** Ask questions to check understanding.
    - "Is the slope positive or negative? How do you know?" Listen for positive slope and for recognition that the x- and y-values in the table are both increasing.
    - "What is the domain? Is it continuous?" hours kayaking ≥ 0; yes
    - "What is the range?" calories burned ≥ 0
    - "Explain what a slope of 300 means in the context of this problem." The person is burning 300 calories per hour kayaking.
    - "Explain why a y-intercept of 0 makes sense." If you haven't kayaked yet, you haven't burned any calories.
- **Extension:** Ask students to determine how long the person would have to kayak in order to burn 1000 calories (i.e., given y, solve for x).

On Your Own

- Compare the slope of this line to the slope of the line in Example 3. Which slope is greater? Which line is steeper? Question 3 has the greater slope and the line is steeper.

Summary

- Discuss the Summary. Students should be able to describe how a table of values and a graph can represent a function. They should also be able to describe how to *read* information from the graph and table in order to write the linear function. This includes recognizing the y-intercept and the pattern of the constant rate of change (the slope).

Closure

- **Exit Ticket:** Write the table of values on the board and ask students to write the equation that relates the temperature to the number of chirps. Acknowledge that this is an approximation and not every Snowy Tree cricket will chirp exactly the same.

Chirps per minute	0	16	32	48	64
Temperature (°F)	40	44	48	52	56

$T = 0.25x + 40$ (Have students check this equation with the one they wrote for Exercise 21 in Section 2.6. It is the same equation.)

Technology For the Teacher

The Dynamic Planning Tool
Editable Teacher's Resources at *BigIdeasMath.com*

EXAMPLE 3 Real-Life Application

Hours Kayaking, x	Calories Burned, y
2	600
4	1200
6	1800
8	2400

Graph the data in the table. (a) Is the domain discrete or continuous? (b) Write a linear function that relates y to x. (c) How many calories do you burn in 4.5 hours?

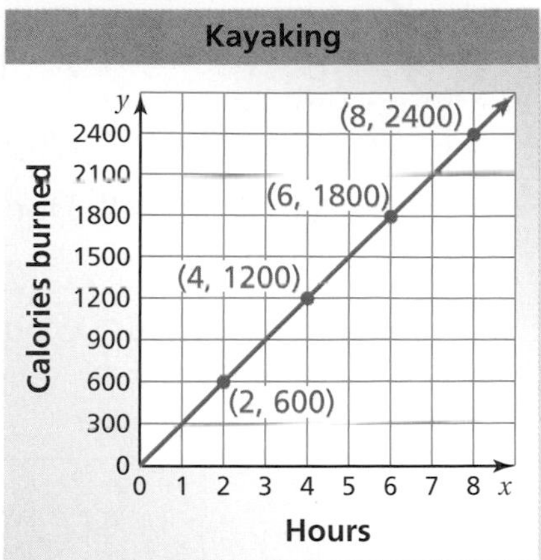

a. Plot the points. Time can represent any value greater than or equal to 0, so the domain is continuous. Draw a line through the points.

b. The y-intercept is 0 and the slope is $\frac{1200 - 600}{4 - 2} = \frac{600}{2} = 300$.

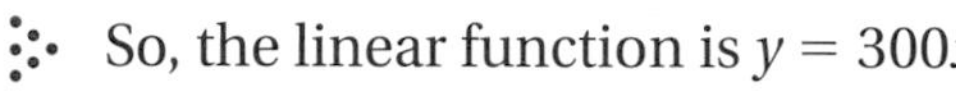

∴ So, the linear function is $y = 300x$.

c. Find the value of y when $x = 4.5$.

$y = 300x$ Write the equation.

$= 300(4.5)$ Substitute 4.5 for x.

$= 1350$ Multiply.

∴ You burn 1350 calories in 4.5 hours of kayaking.

On Your Own

Hours Rock Climbing, x	Calories Burned, y
3	1950
6	3900
9	5850
12	7800

3. Graph the data in the table.

a. Is the domain discrete or continuous?

b. Write a linear function that relates y to x.

c. How many calories do you burn in 5.5 hours?

Summary

Representing a Function

Words An output is 2 more than the input.

Equation $y = x + 2$

Input-Output Table

Input, x	-1	0	1	2
Output, y	1	2	3	4

Graph

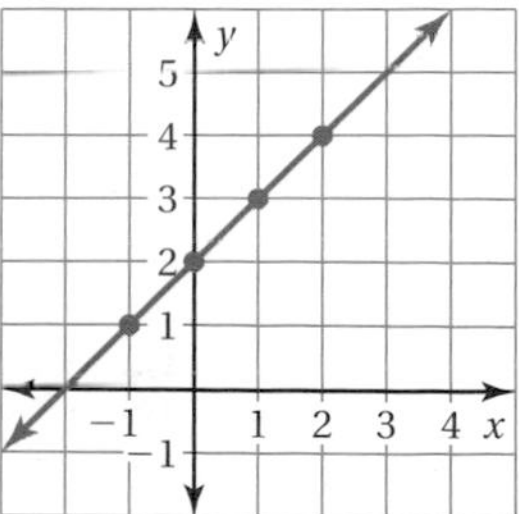

5.3 Exercises

Vocabulary and Concept Check

1. **VOCABULARY** Describe four ways to represent a function.
2. **VOCABULARY** Does the graph represent a linear function? Explain.
3. **REASONING** Do all linear functions have a y-intercept? Explain.

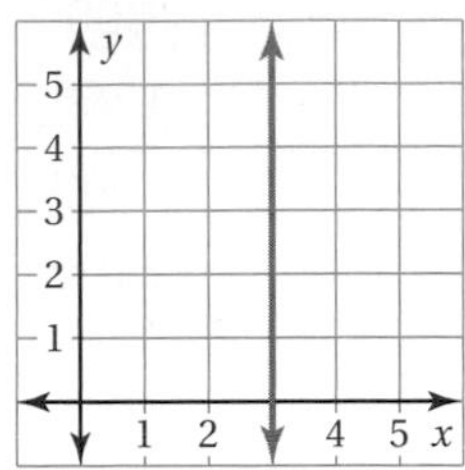

Practice and Problem Solving

The table shows a familiar linear pattern from geometry. Write a linear function that relates y to x. What do the variables x and y represent? Graph the linear function.

4.

x	1	2	3	4	5
y	π	2π	3π	4π	5π

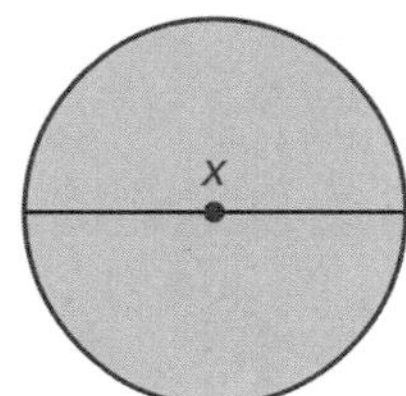

5.

x	1	2	3	4	5
y	2	4	6	8	10

4

x

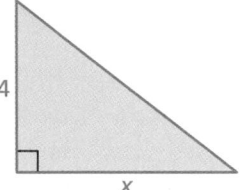

Use the graph or table to write a linear function that relates y to x.

6. 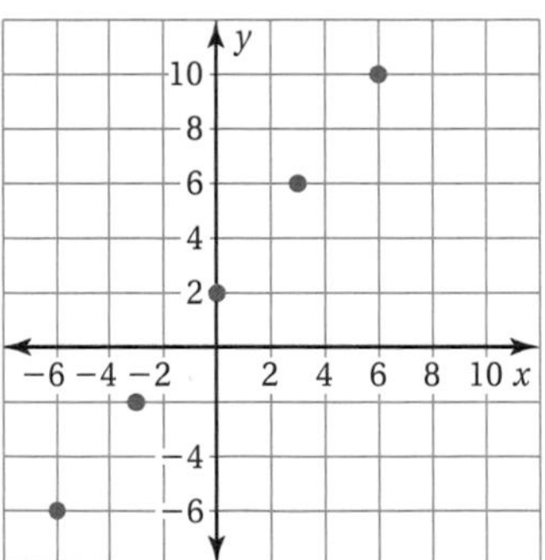

7. 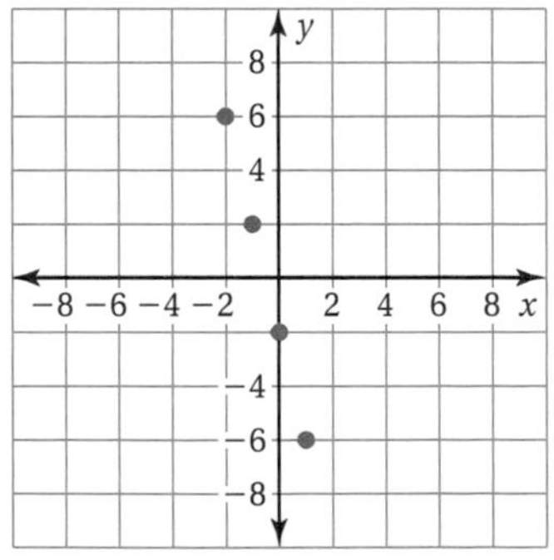

8. 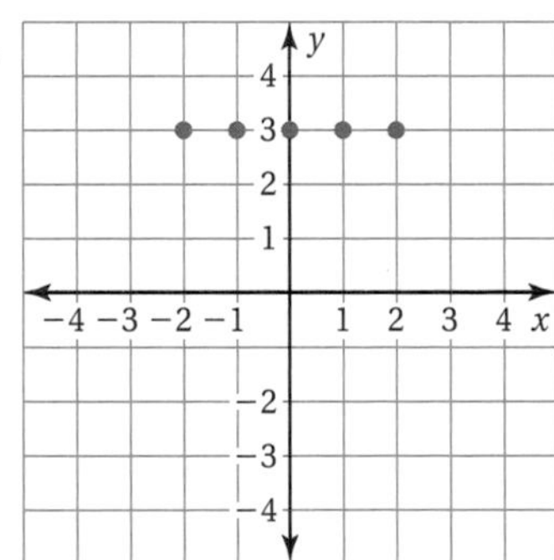

9.

x	−2	−1	0	1
y	−4	−2	0	2

10.

x	−8	−4	0	4
y	2	1	0	−1

11.

x	−3	0	3	6
y	3	5	7	9

12. **MOVIES** The table shows the cost y (in dollars) of renting x movies.
    a. Which variable is independent? dependent?
    b. Graph the data. Is the domain discrete or continuous?
    c. Write a function that relates y to x.
    d. How much does it cost to rent three movies?

Number of Movies, x	0	1	2	4
Cost, y	0	3	6	12

Assignment Guide and Homework Check

Level	Assignment	Homework Check
Average	1–11, 13, 17–20	2, 7, 9, 13
Advanced	1–5, 7–15 odd, 14, 16–20	2, 7, 9, 13

Common Errors

- **Exercises 6 and 7** Students may find the wrong slope because they may misread the scale on an axis. Encourage them to label the points and to use the points they know to write the slope.
- **Exercise 8** Students may not remember how to write the equation for a horizontal line. They may write $x = 3$ instead of $y = 3$. Encourage them to think about the slope-intercept form of an equation.
- **Exercises 9–11** Students may write the reciprocal of the slope when writing the equation from the table. Encourage them to substitute a point into the equation and check to make sure that the equation is true for that point.

5.3 Record and Practice Journal

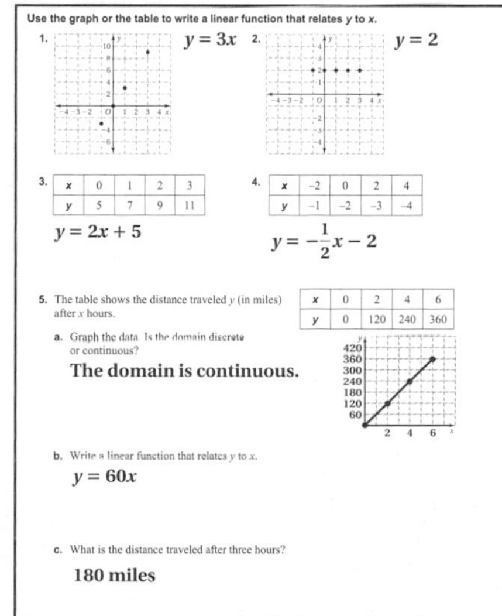

Use the graph or the table to write a linear function that relates y to x.

1. $y = 3x$
2. $y = 2$
3.

x	0	1	2	3
y	5	7	9	11

$y = 2x + 5$

4.

x	-2	0	2	4
y	-1	-2	-3	-4

$y = -\frac{1}{2}x - 2$

5. The table shows the distance traveled y (in miles) after x hours.

x	0	2	4	6
y	0	120	240	360

a. Graph the data. Is the domain discrete or continuous?
The domain is continuous.

b. Write a linear function that relates y to x.
$y = 60x$

c. What is the distance traveled after three hours?
180 miles

Technology For the Teacher
Answer Presentation Tool

Vocabulary and Concept Check

1. words, equation, table, graph
2. no; The vertical line has more than one output for the input $x = 3$.
3. yes; All nonvertical lines intersect the y-axis.

Practice and Problem Solving

4. $y = \pi x$; x is the diameter; y is the circumference.

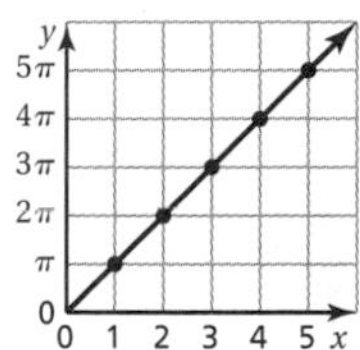

5. $y = 2x$; x is the base of the triangle; y is the area of the triangle.

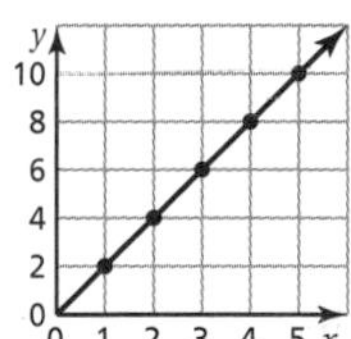

6. $y = \frac{4}{3}x + 2$
7. $y = -4x - 2$
8. $y = 3$
9. $y = 2x$
10. $y = -\frac{1}{4}x$
11. $y = \frac{2}{3}x + 5$
12. a. independent variable: x; dependent variable: y

b.

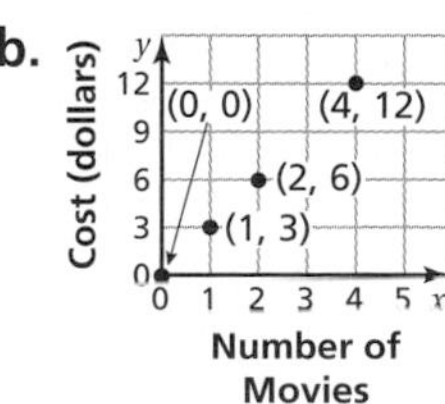

discrete

c. $y = 3x$

d. $9

Practice and Problem Solving

13. a. 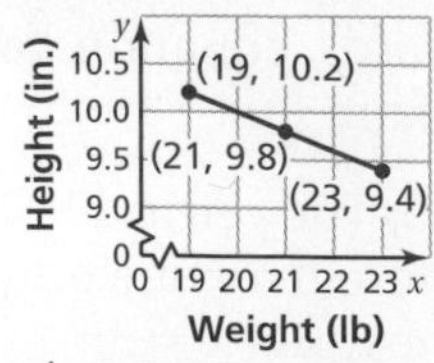

linear

b. $y = -0.2x + 14$

c. 9.7 in.

14. yes; A horizontal line passes the Vertical Line Test.

15. See *Taking Math Deeper.*

16. See Additional Answers.

Fair Game Review

17. -4; -2; 1

18. 8; 2; -7

19. -1.25; -0.25; 1.25

20. B

Mini-Assessment

Use the graph or table to write a linear function that relates *y* to *x*.

1. 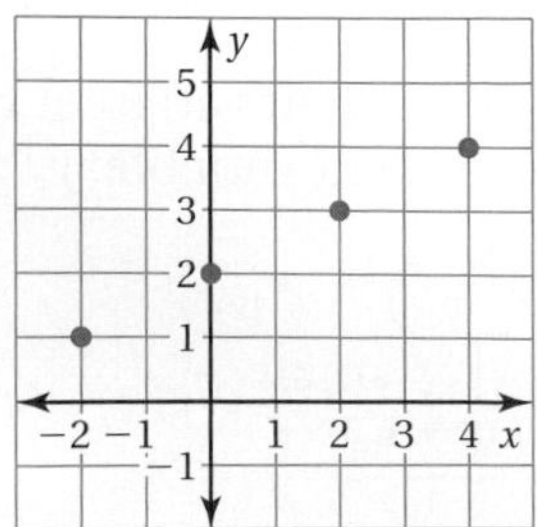

$y = \frac{1}{2}x + 2$

2.

x	−2	−1	0	1
y	9	4	−1	−6

$y = -5x - 1$

Taking Math Deeper

Exercise 15

Students might find it interesting to discover that there is a correlation between years of education and salary. Of course, the correlation only relates annual salaries. There are many examples of people with no years of education beyond high school who have big salaries.

 Graph the data. Describe the pattern.

a. 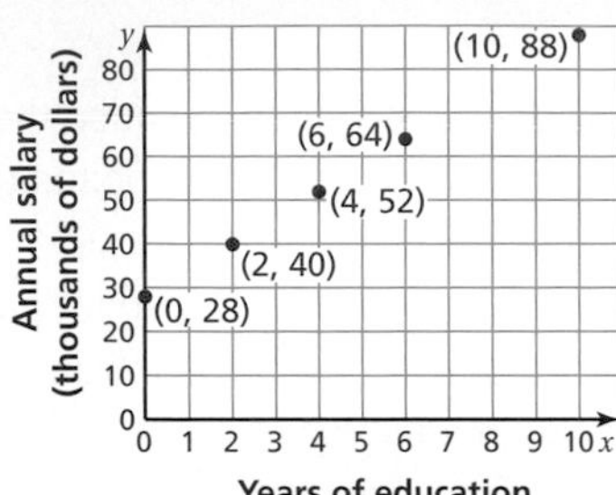

The pattern is that for every 2 years of additional education, the annual salary increases by $12,000.

 Write a function.

Let x = years of education beyond high school.
Let y = annual salary.

y-intercept = 28
slope = 6
$y = 6x + 28$

What about 8?

 Use the function.

For 8 years of education beyond high school, the annual salary is

b. $y = 6(8) + 28 = 76$, or $76,000.

Check this on the graph to see that it fits the pattern.

Project

Select four careers in which you might be interested. List the annual salary for each. Also list the amount of education required for each career.

Reteaching and Enrichment Strategies

If students need help. . .	If students got it. . .
Resources by Chapter • Practice A and Practice B • Puzzle Time Record and Practice Journal Practice Differentiating the Lesson Lesson Tutorials Skills Review Handbook	Resources by Chapter • Enrichment and Extension • School-to-Work Start the next section

13. **BIKE JUMPS** A bunny hop is a bike trick in which the rider brings both tires off the ground without using a ramp. The table shows the height y (in inches) of a bunny hop on a bike that weighs x pounds.

Weight, x	19	21	23
Height, y	10.2	9.8	9.4

a. Graph the data. Then describe the pattern.

b. Write a linear function that relates the height of a bunny hop to the weight of the bike.

c. What is the height of a bunny hop on a bike that weighs 21.5 pounds?

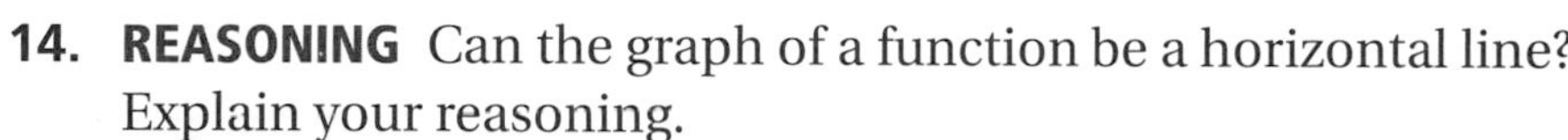

14. **REASONING** Can the graph of a function be a horizontal line? Explain your reasoning.

Years of Education, x	Annual Salary, y
0	28
2	40
4	52
6	64
10	88

15. **SALARY** The table shows a person's annual salary y (in thousands of dollars) after x years of education beyond high school.

a. Graph the data. Then describe the pattern.

b. What is the annual salary of the person after 8 years of education beyond high school?

16. Problem Solving The Heat Index is calculated using the relative humidity and the temperature. For every 1 degree increase in the temperature from 94°F to 98°F at 75% relative humidity, the Heat Index rises 4°F.

a. On a summer day, the relative humidity is 75%, the temperature is 94°F, and the Heat Index is 122°F. Construct a table that relates the temperature t to the Heat Index H. Start the table at 94°F and end it at 98°F.

b. Identify the independent and dependent variables.

c. Write a linear function that represents this situation.

d. Estimate the Heat Index when the temperature is 100°F.

Fair Game Review What you learned in previous grades & lessons

Evaluate the expression when $x = -2, 0,$ and 3. *(Skills Review Handbook)*

17. $x - 2$

18. $3x + 2$

19. $0.5x - 0.25$

20. **MULTIPLE CHOICE** Which expression has a value less than 1? *(Skills Review Handbook)*

Ⓐ $\frac{1}{5^{-2}}$ Ⓑ 5^{-2} Ⓒ 5^0 Ⓓ 5^2

5 Study Help

You can use a **comparison chart** to compare two topics. Here is an example of a comparison chart for domain and range.

	Domain	Range
Definition	the set of all possible input values	the set of all possible output values
Algebra Example: $y = mx + b$	x-values	corresponding y-values
Ordered pairs Example: (−4, 0), (−3, 1), (−2, 2), (−1, 3)	−4, −3, −2, −1	0, 1, 2, 3
Table Example: x: −1, 0, 2, 3 y: 1, 0, 4, 9	−1, 0, 2, 3	0, 1, 4, 9
Graph Example:	−3, −1, 2, 3	−1, 1, 2

On Your Own

Make comparison charts to help you study and compare these topics.

1. independent variable and dependent variable
2. discrete domain and continuous domain
3. linear functions with positive slopes and linear functions with negative slopes

After you complete this chapter, make a comparison chart for the following topics.

4. linear functions and nonlinear functions

"Creating a comparison chart causes canines to crystalize concepts."

Sample Answers

1.

	Independent Variable	Dependent Variable
Definition	The variable that represents the input values	The variable that represents the output values
Algebra Example: $y = mx + b$	x-values	y-values
Ordered pairs Example: (−2, 0), (−1, 5) (0, 10), (1, 15)	−2, −1, 0, 1	0, 5, 10, 15
Table Example: x: 3, 5, 7, 9 y: 1, −1, −1, 1	3, 5, 7, 9	−1, 1
Graph Example:	−2, −1, 0, 1, 2	−1, 0, 1

2.

	Discrete Domain	Continuous Domain
Definition	Domain that consists of only certain numbers in an interval	Domain that consists of all numbers in an interval
Words	• integers from 0 through 4 • the number of people attending a play	• all numbers from 0 through 4 • gallons of gasoline in a fuel tank
Table	Input Number in group, x / Output Total cost of tickets, y (dollars): 2 / 15 3 / 22.5 4 / 30 5 / 37.5	Input Years, x / Output Height of a tree, y (feet): 0 / 1 1 / 3 2 / 5 3 / 7
Graphs Number line Coordinate plane		

3. Available at *BigIdeasMath.com*.

List of Organizers

Available at *BigIdeasMath.com*

Comparison Chart
Concept Circle
Definition (Idea) and Example Chart
Example and Non-Example Chart
Formula Triangle
Four Square
Information Frame
Information Wheel
Notetaking Organizer
Process Diagram
Summary Triangle
Word Magnet
Y Chart

About this Organizer

A **Comparison Chart** can be used to compare two topics. Students list different aspects of the two topics in the left column. These can include *algebra, definition, description, equation(s), graph(s), table(s),* and *words*. Students write about or give examples illustrating these aspects in the other two columns for the topics being compared. Comparison charts are particularly useful with topics that are related but that have distinct differences. Students can place their comparison charts on note cards to use as a quick study reference.

Technology For the Teacher
Vocabulary Puzzle Builder

Answers

1. domain: −4, −1, 2, 5
   range: 2, 1, 0, −1
2. domain: −2, −1, 0, 1, 2
   range: −1, 1, 3
3. domain: −1, 1, 3, 5
   range: −1
4. 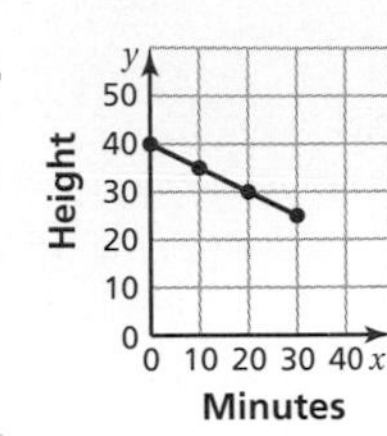

   continuous
5. 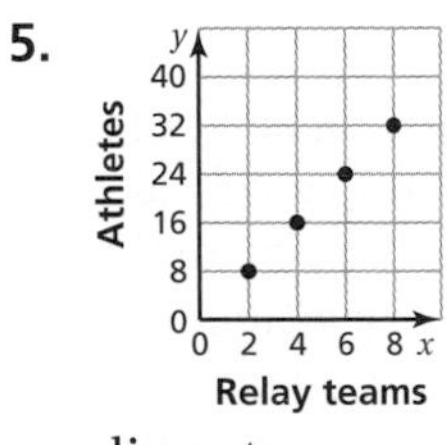

   discrete
6. $y = 2x - 4$
7. $y = \frac{2}{3}x - 1$
8. See Additional Answers.
9. **a.** $R = 2A + 2$

   b. \$22 million
10. See Additional Answers.

Assessment Book

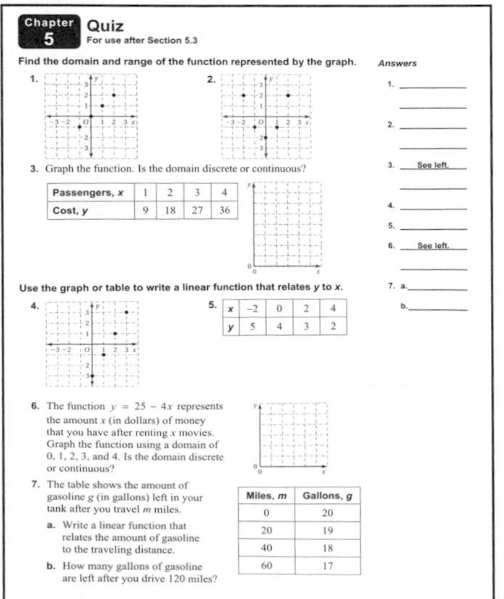

Chapter 5 Quiz
For use after Section 5.3

Find the domain and range of the function represented by the graph.

Answers

1.

2.

1. ________

2. ________

3. Graph the function. Is the domain discrete or continuous?

Passengers, x	1	2	3	4
Cost, y	9	18	27	36

3. See left.

Use the graph or table to write a linear function that relates y to x.

4.

5.

x	−2	0	2	4
y	5	4	3	2

4. ________
5. ________
6. See left.
7. a. ________
   b. ________

6. The function $y = 25 - 4x$ represents the amount x (in dollars) of money that you have after renting x movies. Graph the function using a domain of 0, 1, 2, 3, and 4. Is the domain discrete or continuous?

7. The table shows the amount of gasoline g (in gallons) left in your tank after you travel m miles.
   a. Write a linear function that relates the amount of gasoline to the traveling distance.
   b. How many gallons of gasoline are left after you drive 120 miles?

Miles, m	Gallons, g
0	20
20	19
40	18
60	17

Alternative Quiz Ideas

100% Quiz
Error Notebook
Group Quiz
Homework Quiz
Math Log
Notebook Quiz
Partner Quiz
Pass the Paper

Partner Quiz

- Students should work in pairs. Each pair should have a small white board.
- The teacher selects certain problems from the quiz and writes one on the board.
- The pairs work together to solve the problem and write their answer on the white board.
- Students show their answers and, as a class, discuss any differences.
- Repeat for as many problems as the teacher chooses.
- For the word problems, teachers may choose to have students read them out of the book.

Reteaching and Enrichment Strategies

If students need help. . .	If students got it. . .
Resources by Chapter • Study Help • Practice A and Practice B • Puzzle Time Lesson Tutorials *BigIdeasMath.com* Practice Quiz Practice from the Test Generator	Resources by Chapter • Enrichment and Extension • School-to-Work Game Closet at *BigIdeasMath.com* Start the next section

Technology For the Teacher

Answer Presentation Tool
Big Ideas Test Generator

5.1–5.3 Quiz

Check It Out
Progress Check
BigIdeasMath.com

Find the domain and range of the function represented by the graph. *(Section 5.1)*

1. 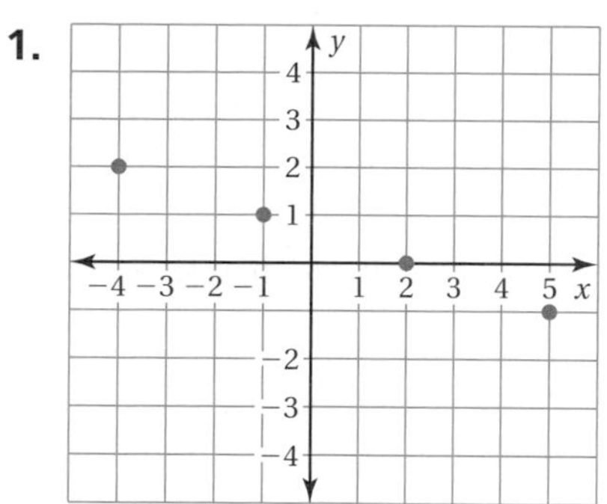

2. 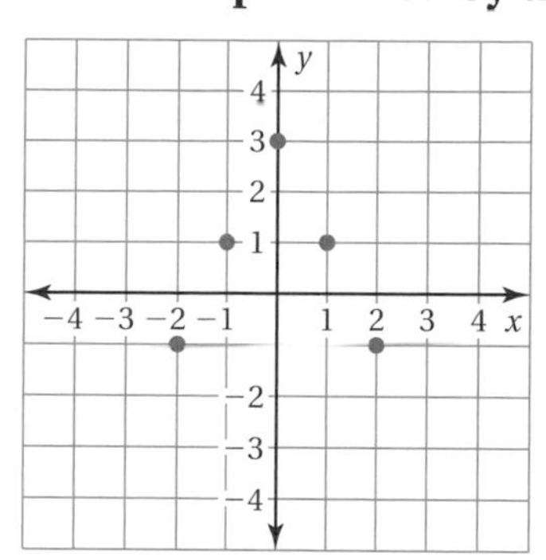

3. 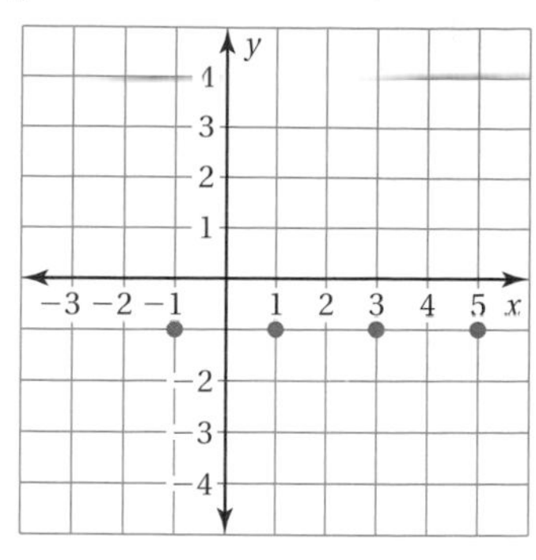

Graph the function. Is the domain discrete or continuous? *(Section 5.2)*

4.

Minutes, x	0	10	20	30
Height, y	40	35	30	25

5. 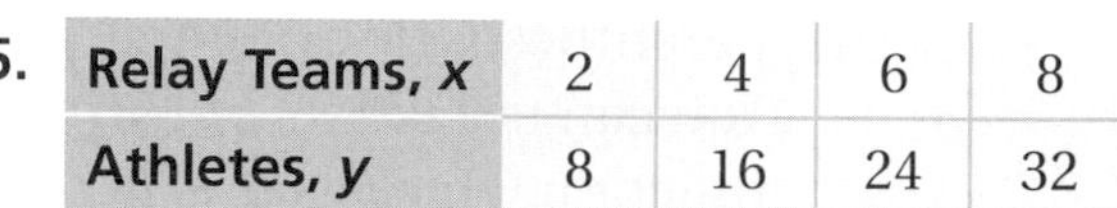

Relay Teams, x	2	4	6	8
Athletes, y	8	16	24	32

Use the graph or table to write a linear function that relates y to x. *(Section 5.3)*

6.

7.

x	y
−3	−3
0	−1
3	1
6	3

8. VIDEO GAME The function $m = 30 - 3r$ represents the amount m (in dollars) of money you have after renting r video games. Graph the function using a domain of 0, 1, 2, 3, and 4. Is the domain discrete or continuous? *(Section 5.2)*

9. ADVERTISING The table shows the revenue R (in millions of dollars) of a company when it spends A (in millions of dollars) on advertising. *(Section 5.3)*

a. Write a linear function that relates the revenue to the advertising cost.

b. What is the revenue of the company when it spends \$10 million on advertising?

Advertising, A	Revenue, R
0	2
2	6
4	10
6	14
8	18

10. WATER Water accounts for about 60% of a person's body weight. *(Section 5.1)*

a. Write an equation that represents the water weight y of a person who weighs x pounds. Identify the independent variable and the dependent variable.

b. Make an input-output table for the equation in part (a). Use the inputs 100, 120, 140, and 160.

c. Find the domain and range of the function represented by the table.

5.4 Function Notation

COMMON CORE STATE STANDARDS
F.BF.3
F.IF.1
F.IF.2
F.IF.7b

Essential Question How can you use function notation to represent a function?

By naming a function f, you can write the function using **function notation.**

$f(x) = 2x - 3$ Function notation

This is read as "f of x equals $2x$ minus 3." The notation $f(x)$ is another name for y. When function notation is used, the parentheses do not imply multiplication. You can use letters other than f to name a function. The letters g, h, j, and k are often used to name functions.

1 ACTIVITY: Matching Functions with Their Graphs

Work with a partner. Match each function with its graph.

a. $f(x) = 2x - 3$

b. $g(x) = -x + 2$

c. $h(x) = x^2 - 1$

d. $j(x) = 2x^2 - 3$

A.
4
−6 6
−4

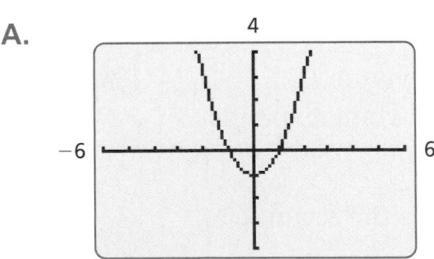

B.

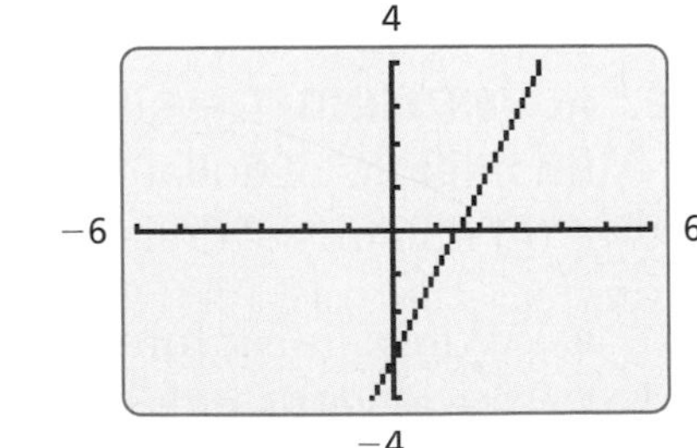

C.
4
−6 6
−4

D.

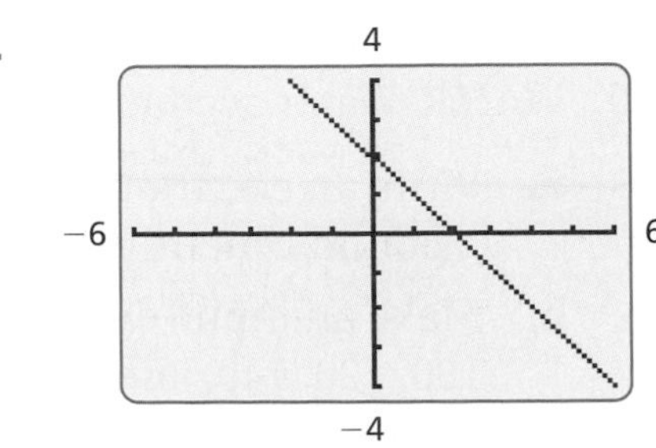

Laurie's Notes

Introduction

Standards for Mathematical Practice

- **MP6 Attend to Precision:** This lesson focuses on the notation associated with functions, and how the notation can be used to describe a vertical shift. The concepts are really not new, however, the notation can be confusing for students. Be careful with language and notation as you work through this lesson.

Motivate

- Make a coordinate grid on the classroom floor or foyer of the school using masking tape. Use a dark marker to scale the axes $[-6, 6]$ in each direction.
- Form two groups of about five students each. Call the groups f and g. Assign the x-values -1, 0, 1, 2, and 3 to each group, one x-value to each student. (Expand the set of x-values for a group of more than five.)
- Have the students in group f plot their respective points for the linear function $y = x - 1$. Have group g plot the points for $y = x - 3$. Ask questions about slopes, y-intercepts, and parallel lines. Students should be able to name the ordered pairs they represent.
- Say to each group, "Add 2 to your y-value and move to the coordinates of your new ordered pair." Repeat the questions asked earlier.
- **Big Idea:** The lines have translated 2 units up. The new functions are $y = x + 1$ for group f, and $y = x - 1$ for group g.

Activity Notes

Discuss

- Introduce function notation. Connect the notation to the groups who plotted themselves as ordered pairs. In function notation, the original equations for each group are $f(x) = x - 1$ and $g(x) = x - 3$.
- Discuss how to read function notation.

Activity 1

- Students should be able to match the linear functions using the slopes and y-intercepts as clues. To match the quadratic functions, they can consider the y-intercepts.

? "What strategies did you use to match the functions in parts (a) and (b)?" Listen for information about the slopes and/or y-intercepts.

? "How did you know that graphs A and C did not match $f(x)$ or $g(x)$?" The graphs were not lines.

? "What strategies did you use to match the functions in parts (c) and (d)?" Listen for information about the y-intercepts.

- If time permits, evaluate the function $j(x) = 2x^2 - 3$ for several values of x to confirm that the ordered pairs are on the graph.

Common Core State Standards

F.BF.3 Identify the effect on the graph of replacing $f(x)$ by $f(x) + k$ . . . for specific values of k (both positive and negative); find the value of k given the graphs. Experiment with cases . . . .

F.IF.1 . . . If f is a function and x is an element of its domain, then $f(x)$ denotes the output of f corresponding to the input x. The graph of f is the graph of the equation $y = f(x)$.

F.IF.2 Use function notation, evaluate functions for inputs in their domains, and interpret statements that use function notation in terms of a context.

F.IF.7b Graph . . . piecewise-defined functions, including step functions and absolute value functions.

Previous Learning

Students should know how to write, graph, and determine the domain and range of linear equations.

Start Thinking! and Warm Up

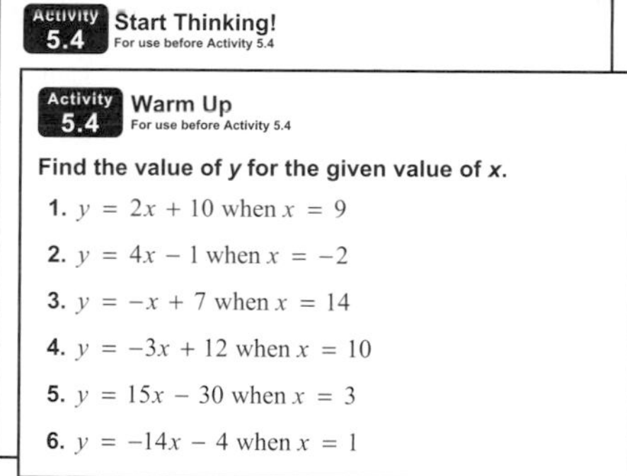

Activity 5.4 Start Thinking! For use before Activity 5.4

Activity 5.4 Warm Up For use before Activity 5.4

Find the value of y for the given value of x.

1. $y = 2x + 10$ when $x = 9$
2. $y = 4x - 1$ when $x = -2$
3. $y = -x + 7$ when $x = 14$
4. $y = -3x + 12$ when $x = 10$
5. $y = 15x - 30$ when $x = 3$
6. $y = -14x - 4$ when $x = 1$

5.4 Record and Practice Journal

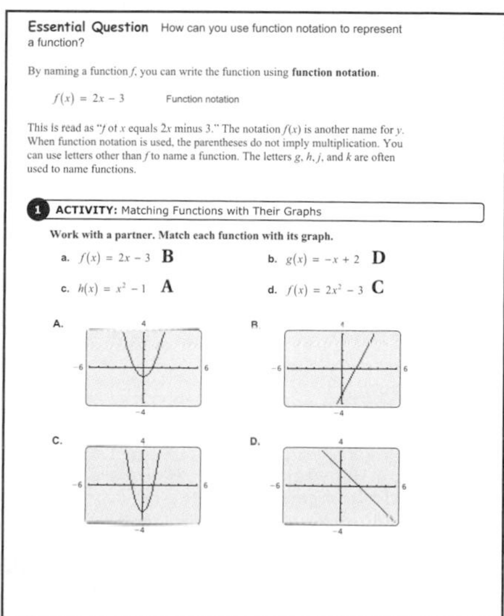

Essential Question How can you use function notation to represent a function?

By naming a function f, you can write the function using **function notation**.

$f(x) = 2x - 3$ Function notation

This is read as "f of x equals $2x$ minus 3." The notation $f(x)$ is another name for y. When function notation is used, the parentheses do not imply multiplication. You can use letters other than f to name a function. The letters g, h, j, and k are often used to name functions.

1 ACTIVITY: Matching Functions with Their Graphs

Work with a partner. Match each function with its graph.

a. $f(x) = 2x - 3$ **B**

b. $g(x) = -x + 2$ **D**

c. $h(x) = x^2 - 1$ **A**

d. $f(x) = 2x^2 - 3$ **C**

A. B. C. D.

Differentiated Instruction

Auditory and Kinesthetic

Emphasize that function notation is used as an alternative way to find y for a given value of x. Have your students write out and compare the problem statements below.

- Given $y = 2x - 3$, find the value of y when $x = 4$.
- Given $f(x) = 2x - 3$, find $f(4)$.

5.4 Record and Practice Journal

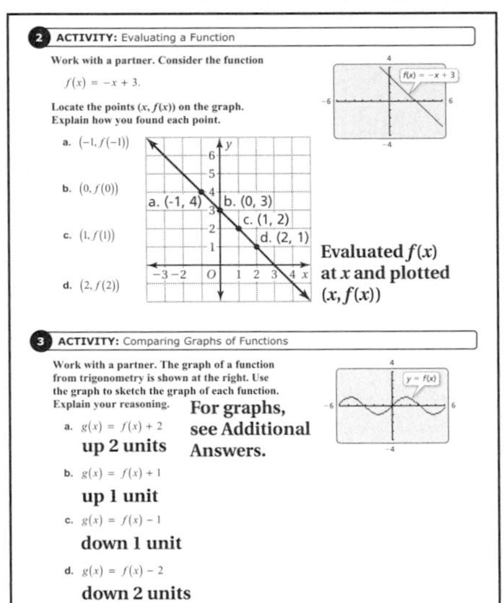

2 ACTIVITY: Evaluating a Function

Work with a partner. Consider the function

$f(x) = -x + 3.$

Locate the points $(x, f(x))$ on the graph. Explain how you found each point.

a. $(-1, f(-1))$

b. $(0, f(0))$

c. $(1, f(1))$

d. $(2, f(2))$

Evaluated $f(x)$ at x and plotted $(x, f(x))$

3 ACTIVITY: Comparing Graphs of Functions

Work with a partner. The graph of a function from trigonometry is shown at the right. Use the graph to sketch the graph of each function. Explain your reasoning.

For graphs, see Additional Answers.

a. $g(x) = f(x) + 2$ **up 2 units**

b. $g(x) = f(x) + 1$ **up 1 unit**

c. $g(x) = f(x) - 1$ **down 1 unit**

d. $g(x) = f(x) - 2$ **down 2 units**

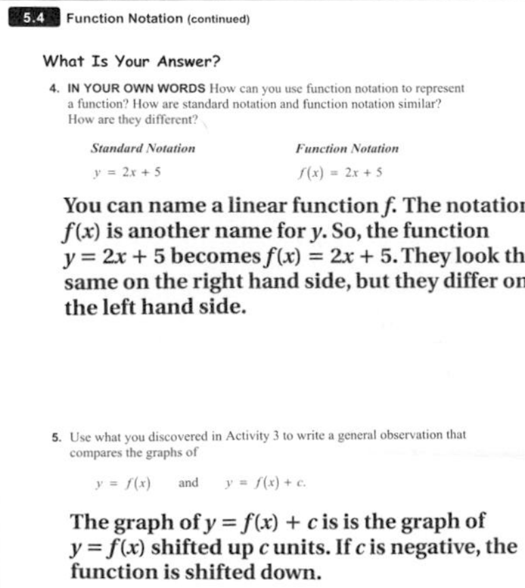

5.4 Function Notation (continued)

What Is Your Answer?

4. IN YOUR OWN WORDS How can you use function notation to represent a function? How are standard notation and function notation similar? How are they different?

Standard Notation	Function Notation
$y = 2x + 5$	$f(x) = 2x + 5$

You can name a linear function f. The notation $f(x)$ is another name for y. So, the function $y = 2x + 5$ becomes $f(x) = 2x + 5$. They look the same on the right hand side, but they differ on the left hand side.

5. Use what you discovered in Activity 3 to write a general observation that compares the graphs of

$y = f(x)$ and $y = f(x) + c.$

The graph of $y = f(x) + c$ is is the graph of $y = f(x)$ shifted up c units. If c is negative, the function is shifted down.

Laurie's Notes

Activity 2

- Discuss how the ordered pairs are related to the form (x, y). The ordered pair $(2, f(2))$ is the same as $(2, y)$, where y is the value of the function when $x = 2$. So, the notation $f(2)$ represents the value of the function f when $x = 2$. You can read $f(2)$ as "f of 2."
- Students may be confused by the notation. Continue to remind them of what $f(-1)$ means, what $f(0)$ means, and so on.

Activity 3

- Explain that trigonometry is the study of how the sides and angles of a triangle are related to each other. A trigonometric function is a function of an angle (independent variable). Many students will study trigonometry in high school.
- Relate this activity to the activity they did at the start of class when they used their bodies to plot points. The original function written as $y = x - 1$ or $f(x) = x - 1$, was changed to $f(x) + 2$. You could rename this new function $m(x)$ and write $m(x) = f(x) + 2$.

What Is Your Answer?

- In Question 5, discuss the meaning of $f(x)$ in the equations $y = f(x)$ and $y = f(x) + c$.
- In Question 5, check to see what observations students made about the graph of $y = f(x) + c$.

Closure

- At the beginning of class, group f plotted the function $y = x - 1$, also written $f(x) = x - 1$. Describe the graph of the function $g(x) = f(x) - 3$. The graph of g is a translation 3 units down of the graph of f.

Technology For the Teacher

The Dynamic Planning Tool
Editable Teacher's Resources at *BigIdeasMath.com*

2 ACTIVITY: Evaluating a Function

Work with a partner. Consider the function

$$f(x) = -x + 3.$$

Locate the points $(x, f(x))$ on the graph. Explain how you found each point.

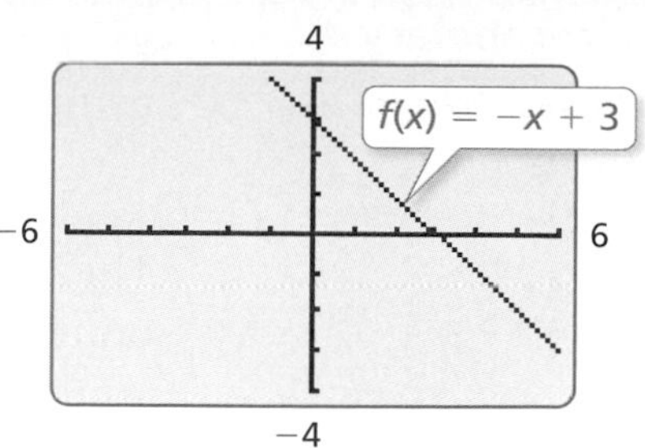

a. 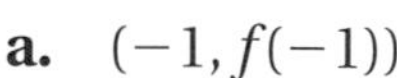 $(-1, f(-1))$

b. $(0, f(0))$

c. $(1, f(1))$

d. $(2, f(2))$

3 ACTIVITY: Comparing Graphs of Functions

Work with a partner. The graph of a function from trigonometry is shown at the right. Use the graph to sketch the graph of each function. Explain your reasoning.

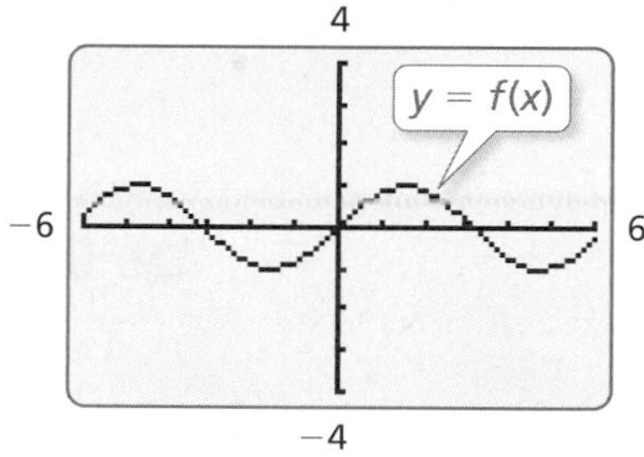

a. $g(x) = f(x) + 2$

b. $g(x) = f(x) + 1$

c. $g(x) = f(x) - 1$

d. $g(x) = f(x) - 2$

What Is Your Answer?

4. **IN YOUR OWN WORDS** How can you use function notation to represent a function? How are standard notation and function notation similar? How are they different?

Standard Notation	*Function Notation*
$y = 2x + 5$	$f(x) = 2x + 5$

5. Use what you discovered in Activity 3 to write a general observation that compares the graphs of

$$y = f(x) \quad \text{and} \quad y = f(x) + c.$$

Practice Use what you learned about function notation to complete Exercises 4–6 on page 229.

5.4 Lesson

Key Vocabulary
function notation, *p. 226*

In Section 5.3, you learned that you can write a linear function in the form $y = mx + b$. By naming a linear function f, you can also write the function using **function notation.**

$$f(x) = mx + b \qquad \text{Function notation}$$

The notation $f(x)$ is another name for y. If f is a function and x is in its domain, then $f(x)$ represents the output of f corresponding to the input x. You can use letters other than f to name a function, such as g or h.

EXAMPLE 1 Evaluating a Function

The notation $f(x)$ is read as "the value of f at x" or "f of x." It does not mean "f times x."

Evaluate $f(x) = -4x + 7$ when $x = 2$.

$f(x) = -4x + 7$	Write the function.
$f(2) = -4(2) + 7$	Substitute 2 for x.
$= -8 + 7$	Multiply.
$= -1$	Add.

∴ When $x = 2$, $f(x) = -1$.

On Your Own

Now You're Ready
Exercises 4–9

Evaluate the function when $x = -4$, 0, and 3.

1. $f(x) = 2x - 5$
2. $g(x) = -x - 1$

EXAMPLE 2 Solving for the Independent Variable

For $h(x) = \frac{2}{3}x - 5$, find the value of x for which $h(x) = -7$.

$h(x) = \frac{2}{3}x - 5$	Write the function.
$-7 = \frac{2}{3}x - 5$	Substitute -7 for $h(x)$.
$-2 = \frac{2}{3}x$	Add 5 to each side.
$-3 = x$	Multiply each side by $\frac{3}{2}$.

∴ When $x = -3$, $h(x) = -7$.

On Your Own

Now You're Ready
Exercises 11–16

Find the value of x so that the function has the given value.

3. $f(x) = 6x + 9$; $f(x) = 21$
4. $g(x) = -\frac{1}{2}x + 3$; $g(x) = -1$

Laurie's Notes

Introduction

Connect

- **Yesterday:** Students used graphs of functions and vertical shifts of graphs of functions to explore function notation. (MP6)
- **Today:** Students will use function notation to evaluate functions, graph functions, and vertically shift the graphs of functions.

Motivate

- **Story Time:** Share some of the history of functions and function notation with students.
- **17th Century:** Gottfried Wilhelm Leibniz was the first person to use the term "function" in a mathematical sense.
- **18th Century:** Leonard Euler introduced the notation $f(x)$ that students will learn about today.

Lesson Notes

Discuss

- Review function language and notation. Explain that the linear function $y = mx + b$ can also be represented as $f(x) = mx + b$. The notation $f(x)$ represents the value of the function for the input x.
- Remind students that $f(x)$ does not mean f times x.

Example 1

? "What does it mean to evaluate an expression?" Substitute a value for the variable and perform the computations to find the value of the expression.

? "What does it mean to evaluate an equation?" Substitute a value for the input variable and perform the computations to find the value of the output variable.

- **Teaching Tip:** Write $f(x) = -4x + 7$ and $y = -4x + 7$ separated by a vertical line. Evaluate each when $x = 2$. Students will feel comfortable with the "y" notation. Seeing the parallel work with function notation helps them understand that the output is the same. The language and notation have changed, the process has not.

Example 2

? "In terms of inputs and outputs, what is the problem asking us to do?" Find the input when the output is -7.

- It might be helpful to translate the problem by writing (?, -7).
- When solving for x in the last step, students often want to divide both sides of the equation by $\frac{2}{3}$. Dividing by $\frac{2}{3}$ is equivalent to multiplying by $\frac{3}{2}$.
- In addition to writing the final solution "When $x = -3$, $h(x) = -7$," write "The ordered pair $(-3, -7)$ is a solution."

On Your Own

- Remind students to look back and check their work.

Goal Today's lesson is using **function notation** to evaluate functions, graph functions, and perform vertical translations of graphs of functions.

Start Thinking! and Warm Up

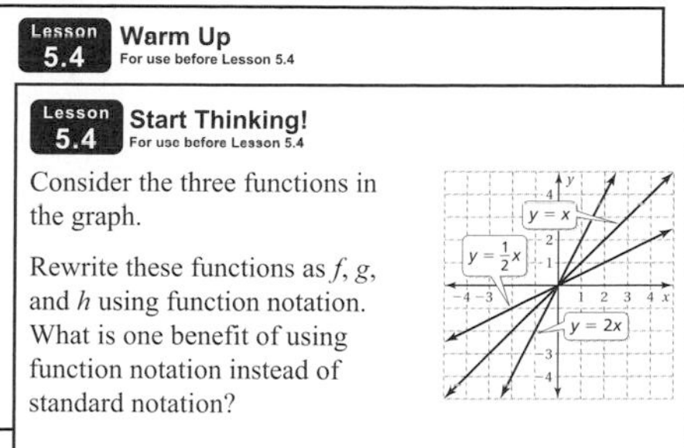
Lesson 5.4 Warm Up
For use before Lesson 5.4

Lesson 5.4 Start Thinking!
For use before Lesson 5.4

Consider the three functions in the graph.

Rewrite these functions as f, g, and h using function notation. What is one benefit of using function notation instead of standard notation?

Extra Example 1

Evaluate $f(x) = 3x - 8$ when $x = -2$. -14

On Your Own

1. $f(-4) = -13$; $f(0) = -5$; $f(3) = 1$
2. $g(-4) = 3$; $g(0) = -1$; $g(3) = -4$

Extra Example 2

For $g(x) = 3x - 8$, find the value of x for which $g(x) = -5$. $x = 1$

On Your Own

3. $x = 2$ 4. $x = 8$

Extra Example 3

Graph $g(x) = -\frac{1}{2}x + 2$.

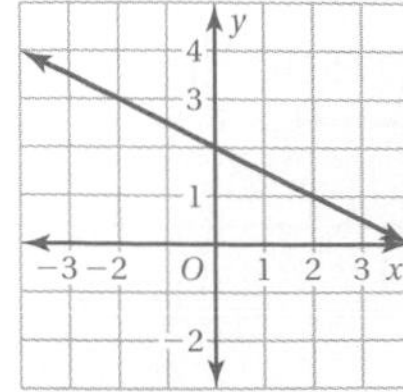

On Your Own

5. 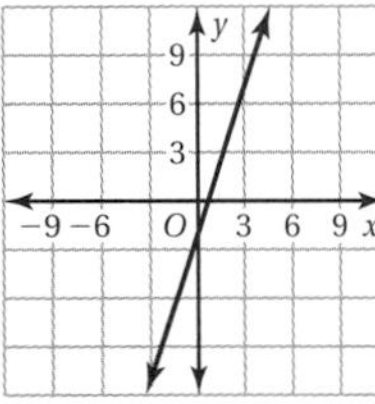

6. 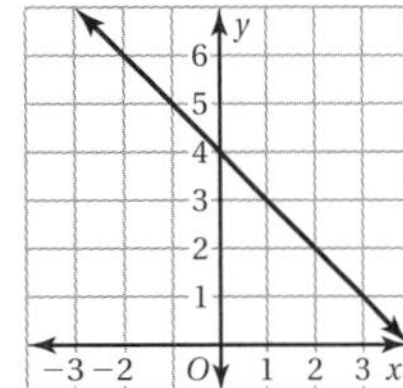

7. 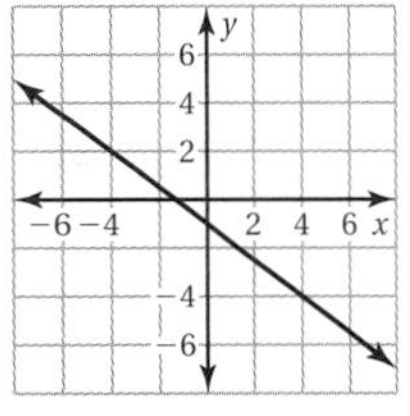

English Language Learners

Vocabulary

To help your students understand what translation means, explain that to translate a graph 5 units up, you simply "move" or "shift" the graph 5 units up.

Laurie's Notes

Example 3

- Write the function $f(x) = 2x + 5$.
- ? "How did we write this linear function in Chapter 2?" $y = 2x + 5$
- ? "What do you know about this linear function?" It has a positive slope of 2 and a y-intercept of 5.
- **MP8 Look for and Express Regularity in Repeated Reasoning:** Make the table of values. Ask students to describe the pattern in the data. As the input increases by 1, the output increases by 2.
- When the ordered pairs are plotted, reinforce this pattern by drawing the slope triangle.
- The vertical axis is labeled y, rather than $f(x)$. Either notation could be used. Because you want to make reference to the y-intercept, y is used.
- ? **MP6 Attend to Precision:** To practice the language of functions you can ask, "What is $f(1)$?" 7 "What is the value of the function at $x = -2$?" 1

On Your Own

- In Question 7, inputs that are multiples of 4 will result in outputs that are integers.
- **Extension:** Students could check their graphs using a graphing calculator. In the equation editor, Y1 replaces $f(x)$, Y2 replaces $g(x)$, and Y3 replaces $h(x)$.

Key Idea

- Write the Key Idea.
- Recall the motivator from yesterday's class. Students used their bodies to graph a linear function. Then they added 2 to their y-value and moved to the new position.
- A vertical translation can be up ($k > 0$) or down ($k < 0$).
- **Extension:** You could discuss and briefly show other transformations.
- **Extension:** A vertical translation can be modeled on a graphing calculator. The screenshot shows equations that will graph the original function, $f(x) = x - 3$, and a translation 2 units up of the graph of f.

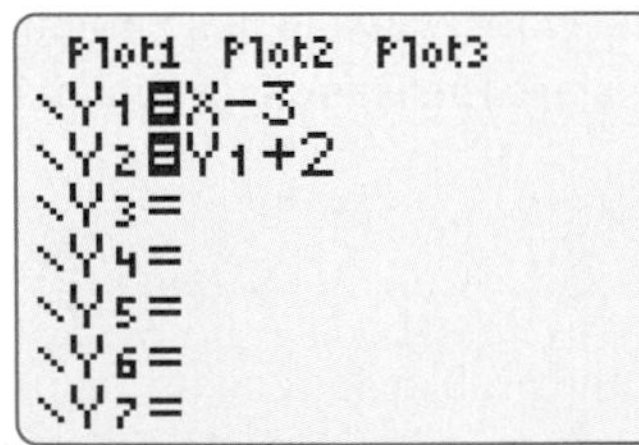

EXAMPLE 3 Graphing a Linear Function in Function Notation

Graph $f(x) = 2x + 5$.

Step 1: Make a table of values.

x	-2	-1	0	1	2
$f(x)$	1	3	5	7	9

Step 2: Plot the ordered pairs.

Step 3: Draw a line through the points.

Study Tip

The graph of $f(x)$ consists of the points $(x, f(x))$.

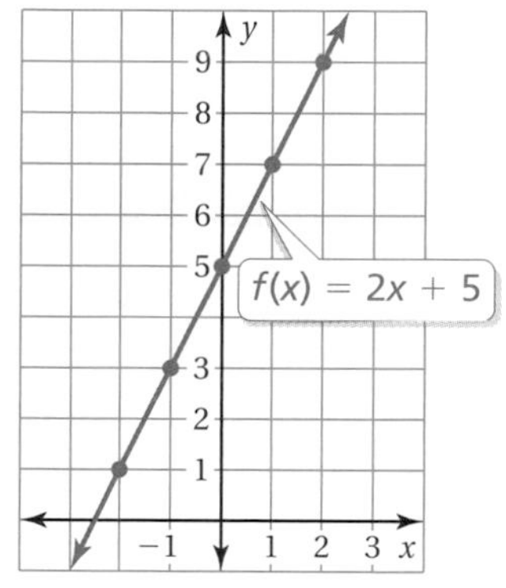

On Your Own

Now You're Ready
Exercises 22–27

Graph the linear function.

5. $f(x) = 3x - 2$ **6.** $g(x) = -x + 4$ **7.** $h(x) = -\frac{3}{4}x - 1$

Key Idea

Vertical Translations

The graph of $f(x) + k$ is a vertical translation of the graph of $f(x)$, where $k \neq 0$.

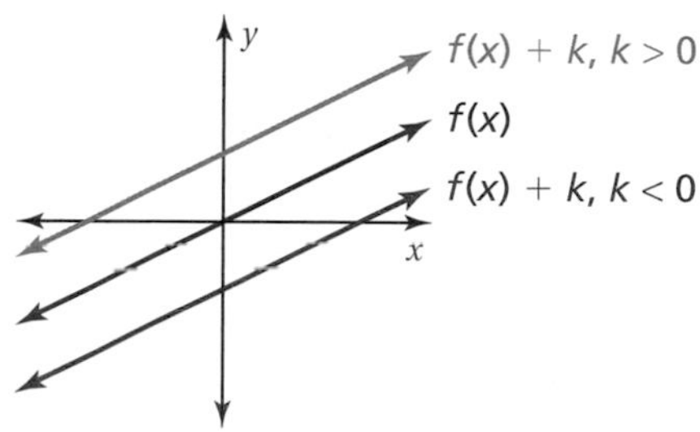

In vertical translations of graphs of linear functions, the graphs have the same slope but different y-intercepts.

EXAMPLE 4 Comparing Graphs of Linear Functions

Graph $g(x) = x - 3$. Compare the graph to the graph of $f(x) = x$.

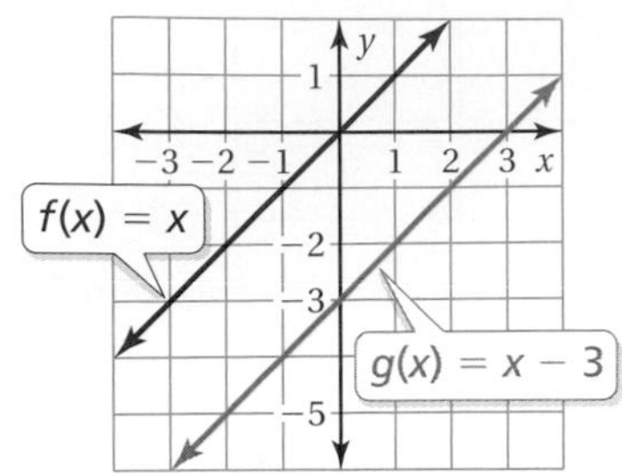

Use the slope and y-intercept to graph the equations.

$g(x) = x - 3$

$= 1x + (-3)$

slope y-intercept

$f(x) = x$

$= 1x + 0$

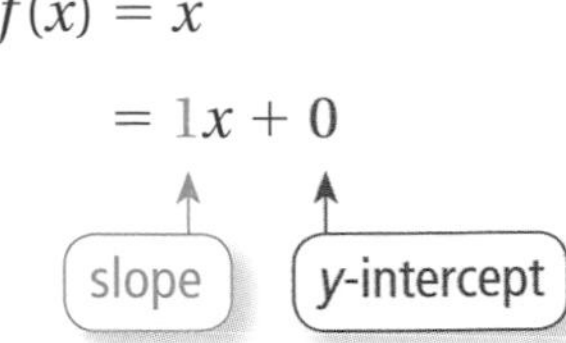

∴ The graphs have the same slope but different y-intercepts. The graph of g is a translation 3 units down of the graph of f.

EXAMPLE 5 Standardized Test Practice

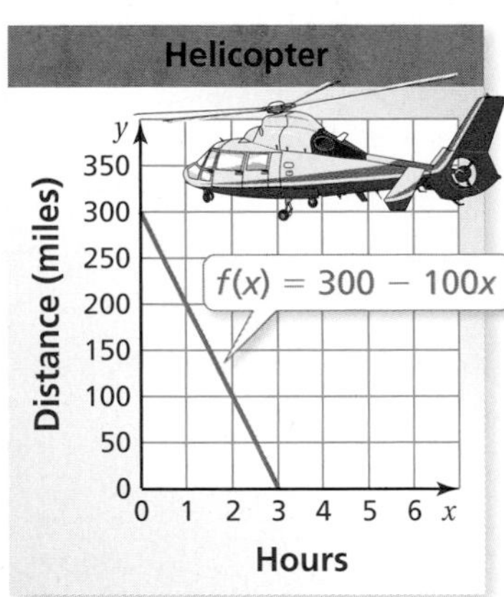

The graph shows the number y of miles a helicopter is from its destination after x hours on its first flight. On its second flight, the helicopter travels at the same speed but 50 miles farther. Which statement is true about the graph of the function that represents the second flight compared to the graph of the function that represents the first flight?

Ⓐ The slope decreases.

Ⓑ The slope increases.

Ⓒ The graph is a translation 50 units down.

Ⓓ The graph is a translation 50 units up.

The helicopter travels at the same speed on both flights. So, the graphs have the same slope. You can eliminate choices A and B.

Because the helicopter travels 50 miles farther on the second flight, it is 50 miles farther from its destination when $x = 0$. So, the graph of the function that represents the second flight is a vertical translation 50 units up of the graph of the function that represents the first flight.

∴ The correct answer is Ⓓ.

On Your Own

Now You're Ready
Exercises 29–31

Graph the function. Compare the graph to the graph of $f(x) = -2x$.

8. $g(x) = -2x + 3$

9. $h(x) = -2x - 5$

10. **WHAT IF?** In Example 5, the helicopter travels the same distance but 50 miles per hour faster on the second flight. How does the graph of the function that represents the second flight compare to the graph of the function that represents the first flight?

Laurie's Notes

Example 4

- This example is a nice review of using the slope and y-intercept to graph linear functions.
- ? "What is the slope of the graph of each function?" slope of graph of g: 1; slope of graph f: 1
- **Common Error:** Students may say that the slope is x. Remind students that the coefficient of x is 1, and you usually do not write the coefficient when it is 1.
- ? "What is the y-intercept of the graph of each function?" y-intercept of graph of g: -3; y-intercept of graph of f: 0

Example 5

- Ask a volunteer to read the problem.
- Take time to interpret the meaning of the original function $f(x) = 300 - 100x$. Generate a few ordered pairs for the function.

Hours, x	0	1	2
Miles from Destination, y	300	200	100

- ? "What is the slope and what does it mean in the context of the problem?" slope $= -100$; Each time x increases by 1 hour, the miles from destination decreases by 100 miles, so the speed of the helicopter is 100 mph.
- ? "How far is the helicopter from its destination to start?" 300 miles
- ? "What will change in the function representing the second flight? the y-intercept
- Finish working through the problem as shown.

On Your Own

- If white boards are available, use them for students to sketch their graphs.
- **Teaching Tip:** Graph $f(x) = -2x$ in one color and the other two functions using a second color.

Closure

- If $f(x) = 2x - 4$, what is the equation of a translation 3 units down of the graph of f? $g(x) = 2x - 7$

Technology For the Teacher

The Dynamic Planning Tool
Editable Teacher's Resources at *BigIdeasMath.com*

Extra Example 4

Graph $g(x) = -x + 2$. Compare the graph to the graph of $f(x) = -x$.

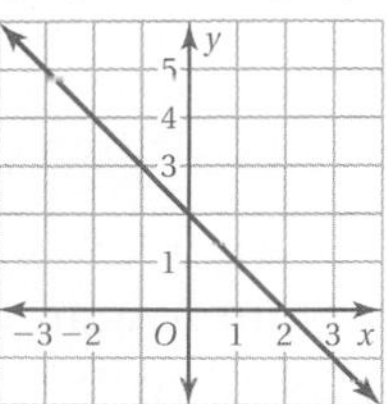

The graph of g is a translation 2 units up of the graph of f.

Extra Example 5

In Example 5, explain how to change the graph of f to represent the situation when the helicopter travels at the same speed on a trip that is 20 miles shorter than the original trip. Translate the graph of f down 20 units.

On Your Own

8. 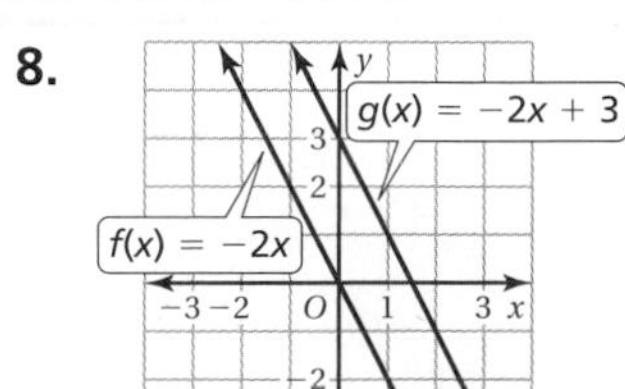

The graph of g is a translation 3 units up of the graph of f.

9. 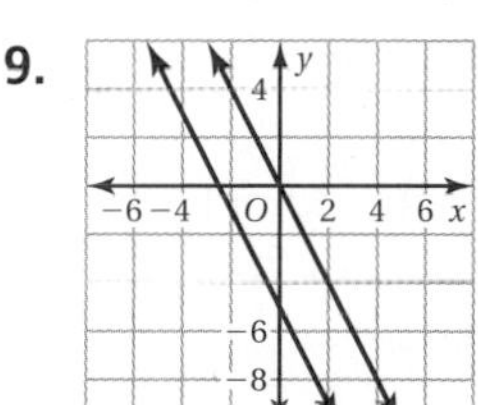

The graph of h is a translation 5 units down of the graph of f.

10. *Sample answer:* The graph has the same y-intercept but a steeper slope.

Vocabulary and Concept Check

1. Function notation assigns a name such as f to a function and $f(x)$ represents the value of the function at x. Example: $y = 3x + 1$ can be written as $f(x) = 3x + 1$.
2. It represents your height when you were 13 years old.
3. a line; Changing b translates the graph vertically.

Practice and Problem Solving

4. $f(-2) = 4; f(0) = 6; f(5) = 11$
5. $g(-2) = -8; g(0) = -2; g(5) = 13$
6. $h(-2) = 13; h(0) = 9; h(5) = -1$
7. $h(-2) = -5; h(0) = -7; h(5) = -12$
8. $g(-2) = -15; g(0) = -3; g(5) = 27$
9. $f(-2) = 12; f(0) = 2; f(5) = -23$
10. $g(-2)$ does not mean multiply g by -2, it means the value of g at -2.

    $g(-2) = 4(-2) + 6$

    $g(-2) = -8 + 6$

    $g(-2) = -2$

11. $x = 1$
12. $x = -3$
13. $x = -2$
14. $x = 5$
15. $x = -6$
16. $x = 15$
17. a. $198

    b. 25 hours
18. a. $77.50

    b. 8 tickets

Assignment Guide and Homework Check

Level	Assignment	Homework Check
Average	1–6, 10, 11–17 odd, 23, 25, 29, 43–46	4, 10, 17, 23, 29
Advanced	1–3, 4–18 even, 22–26 even, 29–32, 40–46	6, 10, 18, 24, 32

Common Errors

- **Exercises 4–9** Students may confuse the signs when they substitute for *x*, especially when they skip steps. Encourage students to take the time to carry out all the steps.
- **Exercises 11–16** Students may set up the problem incorrectly. Help them to recognize that a value is given for the dependent variable. They should substitute that value for the dependent variable in the function and then solve for *x*.
- **Exercises 17–18** Students may have difficulty relating the real-life quantities to the function. Encourage them to first specify the quantities represented by the dependent and independent variables. Then, in each part, substitute the value for the appropriate variable and solve for the other variable.

5.4 Record and Practice Journal

Find the value of *x* so that the function has the given value.

1. $g(x) = 8x - 11; g(x) = 5$

   $x = 2$
2. $v(x) = -2x - 2; v(x) = 12$

   $x = -7$
3. $k(x) = \frac{3}{2}x + 7; k(x) = -2$

   $x = -6$
4. $j(x) = -4x + 9; j(x) = 8$

   $x = \frac{1}{4}$

Graph the linear function.

5. $f(x) = -3x + 1$
6. $h(x) = \frac{1}{4}x - 2$
7. You have $124. You earn $22 each time you mow your neighbor's lawn. The function $s(x) = 22x + 124$ represents your total savings.

   a. What will your savings be after you mow 5 times?

   $234

   b. How many times do you have to mow to save a total of $300?

   8 times

Technology For the Teacher

Answer Presentation Tool

5.4 Exercises

Vocabulary and Concept Check

1. **VOCABULARY** What is function notation? Give an example.
2. **VOCABULARY** Your height can be represented by a function $h(x)$ where x is your age. What does $h(13)$ represent?
3. **WRITING** What type of graph is given by $y = mx + b$? How does changing the value of b affect the graph?

Practice and Problem Solving

Evaluate the function when $x = -2, 0,$ and 5.

① 4. $f(x) = x + 6$
5. $g(x) = 3x - 2$
6. $h(x) = -2x + 9$
7. $h(x) = -x - 7$
8. $g(x) = 6x - 3$
9. $f(x) = -5x + 2$

10. **ERROR ANALYSIS** Describe and correct the error in evaluating the function $g(x) = 4x + 6$ when $x = -2$.

✗
$$g(-2) = 4(-2) + 6$$
$$-2g = -8 + 6$$
$$-2g = -2$$
$$g = 1$$

Find the value of x so that the function has the given value.

② 11. $h(x) = -7x + 10;\ h(x) = 3$
12. $t(x) = -3x - 5;\ t(x) = 4$
13. $n(x) = 4x + 15;\ n(x) = 7$
14. $p(x) = 6x - 12;\ p(x) = 18$
15. $q(x) = \frac{1}{3}x - 2;\ q(x) = -4$
16. $r(x) = -\frac{4}{5}x + 7;\ r(x) = -5$

17. **SUMMER JOB** You earn \$11 per hour working at a grocery store during the summer. The function $p(x) = 11x$ represents the amount you earn for working x hours.
   a. You work 18 hours. How much do you earn?
   b. How many hours do you have to work to earn \$275?

18. **ORCHESTRA** A group of friends are buying tickets to the orchestra. Each ticket costs \$17.50 and one of the friends has a coupon for \$10. The function $C(x) = 17.5x - 10$ represents the total cost of buying x tickets.
   a. How much does it cost to buy 5 tickets?
   b. How many tickets can you buy with \$130.00?

Match the function with its graph.

19. $f(x) = -2x - 2$

20. $g(x) = \frac{1}{2}x + 2$

21. $h(x) = \frac{1}{2}x - 2$

A.

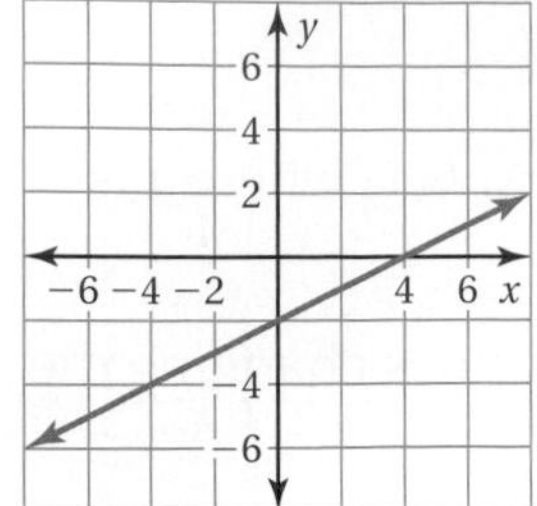

B.

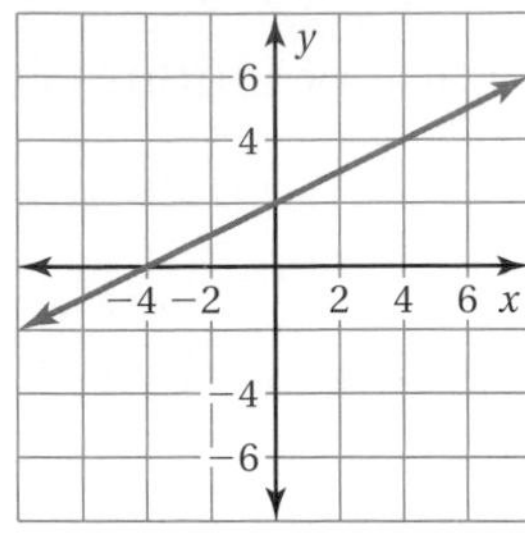

C.

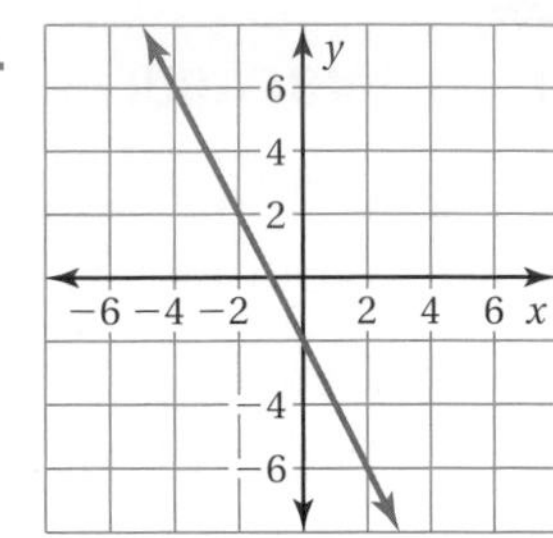

Graph the linear function.

3 **22.** $f(x) = 4x + 1$

23. $g(x) = -2x - 5$

24. $h(x) = -\frac{1}{2}x - 3$

25. $f(x) = \frac{3}{5}x + 2$

26. $g(x) = 7x - 4$

27. $h(x) = -6x + 3$

28. ATMOSPHERIC TEMPERATURE Under normal conditions, the atmospheric temperature drops 3.5°F per 1000 feet of altitude up to 40,000 feet. When the outside temperature is 80°F, the atmospheric temperature can be modeled by $t(x) = -3.5x + 80$, where x is the altitude in thousands of feet.

a. Graph the function and identify its domain and range.

b. Find and interpret the value of x so that $t(x) = -25$.

Graph the function. Compare the graph to the graph of $f(x) = 3x$.

4 **29.** $g(x) = 3x + 2$

30. $n(x) = 3x - 7$

31. $v(x) = 3x - \frac{7}{2}$

32. DECK The function $C(x) = 25x + 50$ represents the labor cost for Jones Remodeling to build a deck, where x is the number of hours. Sample labor costs from their main competitor, Premiere Remodeling, are shown in the table.

Hours	Cost
2	\$130
4	\$160
6	\$190

a. Which cost function has the greater rate of change? What does the rate of change represent?

b. The graph of which cost function has the greater y-intercept? Interpret the y-intercept.

c. The job is estimated to take 8 hours. Which company would you hire? Explain your reasoning.

Common Errors

- **Exercises 19–27** Incorrect graphs are the result of careless mistakes or a lack of understanding about either slope-intercept form or function notation. Look for a pattern of errors by a student to determine how to help.
- **Exercises 29–31** Students may describe the translation incorrectly. Make sure they realize that the slope is the same and only the y-intercept is different. The sign of the y-intercept determines whether the vertical translation is up or down.
- **Exercises 33–35** Students may have incorrect answers due to poor graph quality. Remind them to check their solution in each function.
- **Exercises 37–39** Students may have a poor strategy for solving the problem or may get confused along the way. Tell them to plan out their strategy and clearly label the steps of their procedure to avoid confusion.

19. C **20.** B

21. A

22.

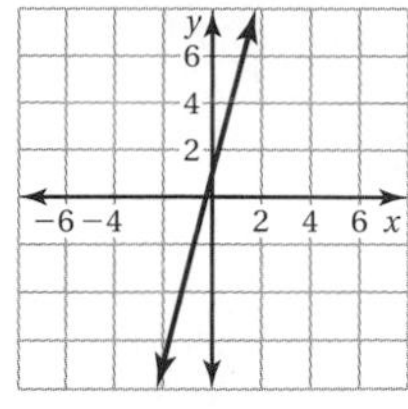

23.

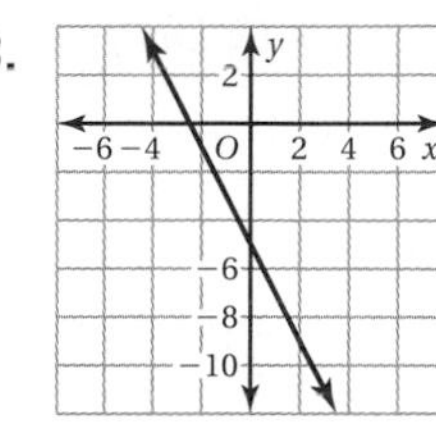

24.

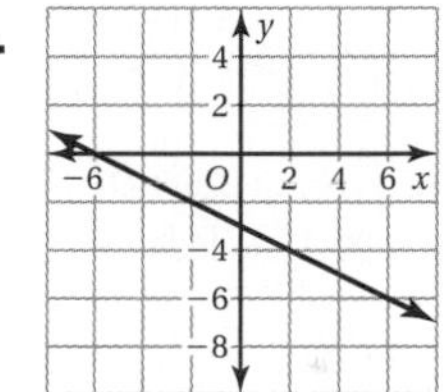

25–31. See Additional Answers.

32. **a.** Jones Remodeling: The rate of change represents the hourly labor cost.

b. Premiere Remodeling: The y-intercept represents the initial cost without labor charges.

c. Premiere Remodeling; For 8 hours of labor, Premiere Remodeling charges \$220 while Jones Remodeling charges \$250.

33.

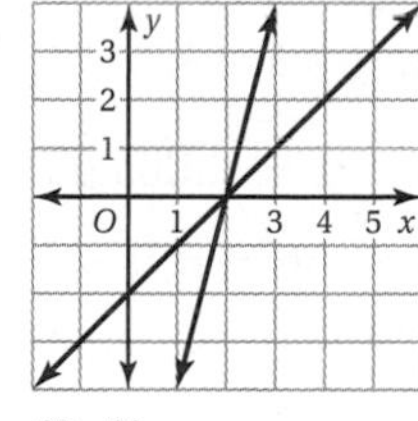

(2, 0)

34.

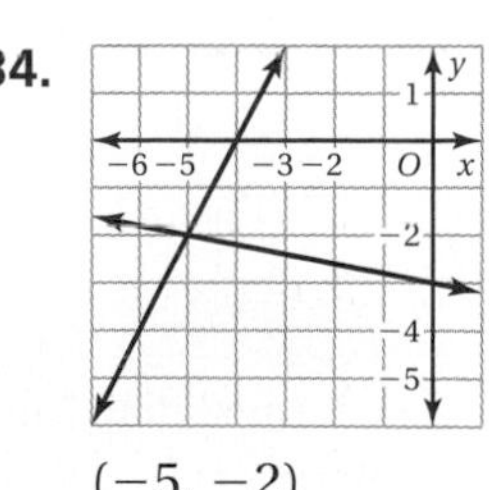

(−5, −2)

Practice and Problem Solving

35.

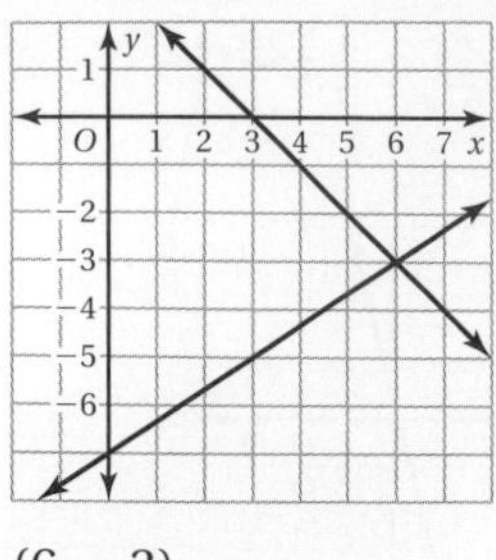

$(6, -3)$

36. See *Taking Math Deeper.*

37. $k = -1$ **38.** $k = 4$

39. $k = -4$

40. *Sample answer:* The second graph is a vertical translation 5 inches down of the first graph, where $0 < x < 15$.

41. Translate the graph of $y = x$ four units to the left.

42. $f(g(x)) = 12x - 5$; $g(f(x)) = 12x - 20$

Fair Game Review

43. $y = x$ **44.** $y = -2x + 1$

45. $y = 1$ **46.** B

Mini-Assessment

1. Evaluate $f(x) = 3x - 7$ when $x = -4, 0,$ and 6. $f(-4) = -19$; $f(0) = -7$; $f(6) = 11$
2. For $g(x) = -\frac{1}{4}x + 5$, find the value of x for which $g(x) = 8$. $x = -12$
3. Graph $g(x) = \frac{1}{2}x - 1$. Compare the graph to the graph of $f(x) = \frac{1}{2}x$.

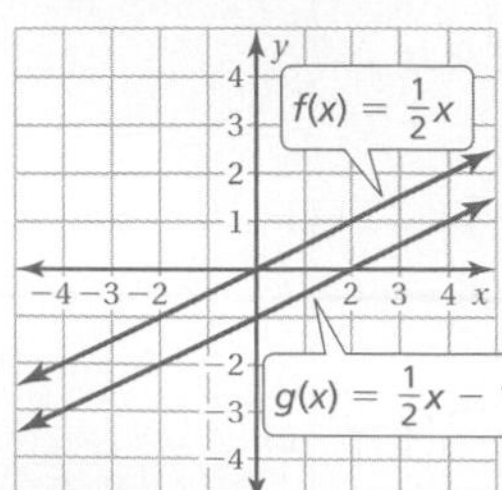

The graph of g is a translation 1 unit down of the graph of f.

Taking Math Deeper

Exercise 36

MP5 Use Appropriate Tools Strategically. There are several tools and methods that can be used to solve this problem.

Use a calculator!

Pencil and Paper: algebraically *(Lesson 1.3)* and graphically *(Lesson 4.4b)*

Graphing Calculator: graphically *(Lesson 4.4b)*

Because f and g are linear functions involving decimals, this provides a great opportunity to use a graphing calculator.

Graph the functions in the standard viewing window.

10

−10 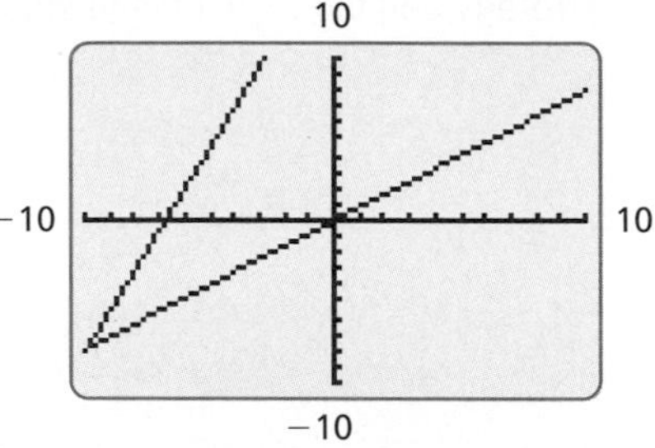 10

−10

In the standard viewing window, the intersection of the graphs is at the edge of the screen. Adjust the viewing window so that the intersection is visible.

10

−20 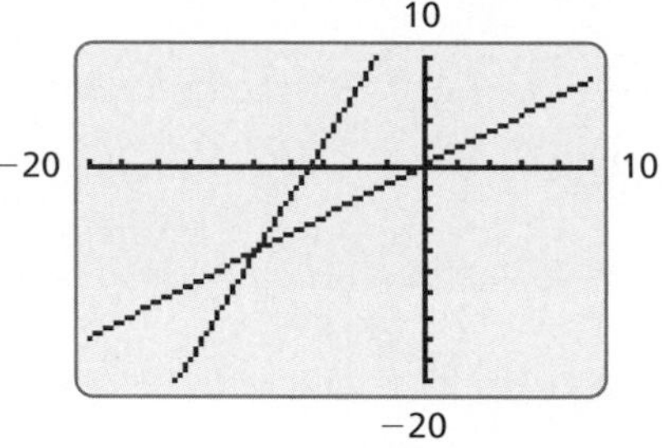 10

−20

Use the *intersect* feature of the graphing calculator to find the point of intersection. The solution is the x-coordinate of the intersection of the two lines. The point of intersection is $(-10, -8)$, so the solution is $x = -10$.

10

−20 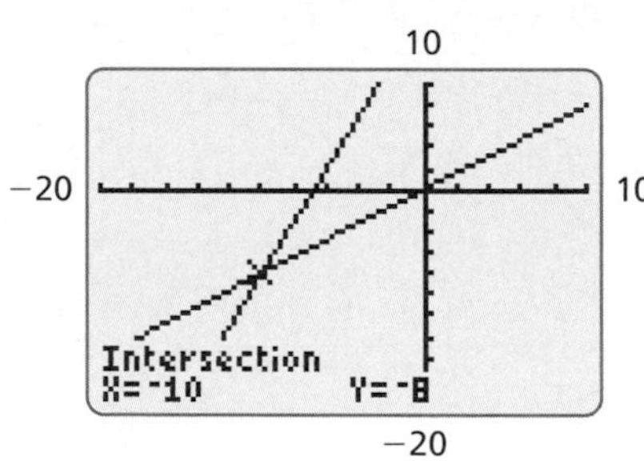 10

−20

Reteaching and Enrichment Strategies

If students need help. . .	If students got it. . .
Resources by Chapter • Practice A and Practice B • Puzzle Time Record and Practice Journal Practice Differentiating the Lesson Lesson Tutorials Skills Review Handbook	Resources by Chapter • Enrichment and Extension • School-to-Work Start the next section

Graph the functions $f(x)$ and $g(x)$ in the same coordinate plane. Use the graph to solve $f(x) = g(x)$.

33. $f(x) = x - 2$

$g(x) = 4x - 8$

34. $f(x) = -\frac{1}{5}x - 3$

$g(x) = 2x + 8$

35. $f(x) = \frac{2}{3}x - 7$

$g(x) = -x + 3$

36. CHOOSE TOOLS What tool would you use to solve $f(x) = g(x)$ when $f(x) = 2.5x + 17$ and $g(x) = 0.8x$? Explain. Then solve $f(x) = g(x)$.

Given $f(x) = 2x + 1$, find the value of k so that the graph is $f(x) + k$.

37.

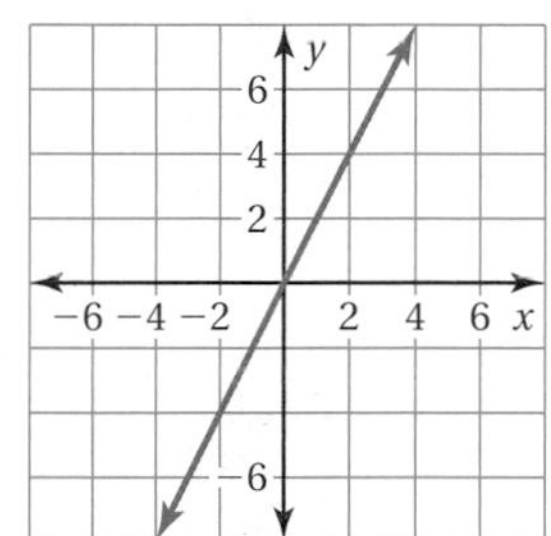

38.

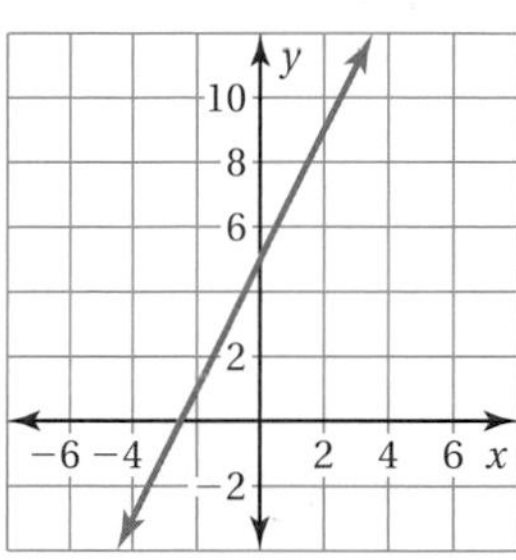

39.

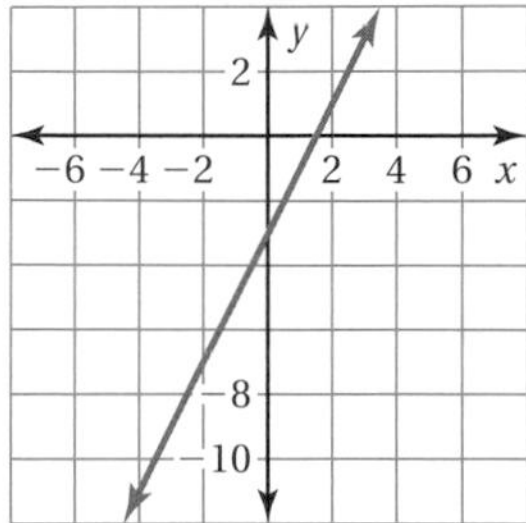

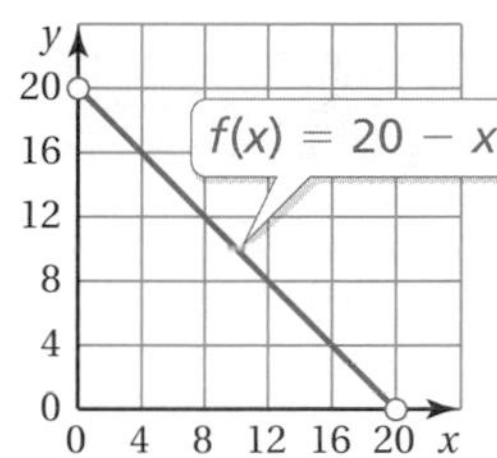

40. PERIMETER The graph shows the relationship between the width y and length x of a rectangle in inches. A second rectangle has a perimeter that is 10 inches less than the perimeter of the first rectangle. How does the graph relating the width and length of the second rectangle compare to the graph shown?

41. CRITICAL THINKING The graph of $y = x + 4$ is a translation 4 units up of the graph of $y = x$. How can you obtain the graph of $y = x + 4$ from the graph of $y = x$ using a horizontal translation?

42. Structure Given that $f(x) = 3x - 5$ and $g(x) = 4x$, write a function that represents $f(g(x))$ and a function that represents $g(f(x))$.

Fair Game Review What you learned in previous grades & lessons

Write in slope-intercept form an equation of the line that passes through the given points. *(Section 2.6)*

43. (0, 0), (4, 4)

44. (−4, 9), (1, −1)

45. (−2, 1), (3, 1)

46. MULTIPLE CHOICE You buy a pair of gardening gloves for \$2.25 and x packets of seeds for \$0.88 each. Which equation represents the total cost y? *(Skills Review Handbook)*

Ⓐ $y = 0.88x - 2.25$

Ⓑ $y = 0.88x + 2.25$

Ⓒ $y = 2.25x - 0.88$

Ⓓ $y = 2.25x + 0.88$

5.4b Special Functions

Key Vocabulary

piecewise function, *p. 232*
step function, *p. 233*
absolute value function, *p. 234*

Piecewise Function

A **piecewise function** is a function defined by two or more equations. Each "piece" of the function applies to a different part of its domain. An example is shown below.

$$y = \begin{cases} x - 2, & \text{if } x \le 0 \\ 2x + 1, & \text{if } x > 0 \end{cases}$$

- The expression $x - 2$ gives the value of y when x is less than or equal to 0.
- The expression $2x + 1$ gives the value of y when x is greater than 0.

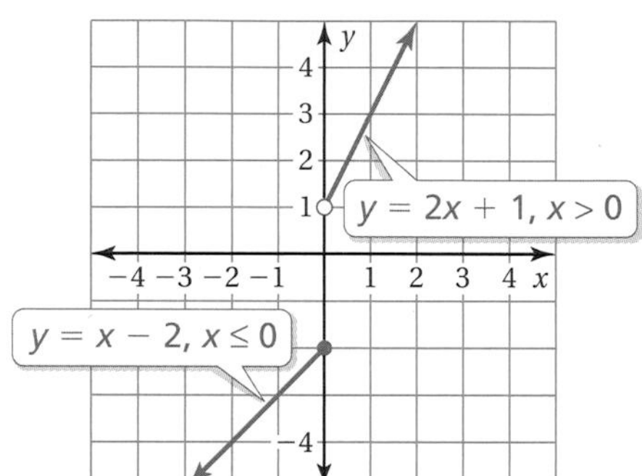

EXAMPLE 1 Graphing a Piecewise Function

Graph $y = \begin{cases} -x - 4, & \text{if } x < 0 \\ x, & \text{if } x \ge 0 \end{cases}$**. Describe the domain and range.**

Step 1: Graph $y = -x - 4$ for $x < 0$. Because x is not equal to 0, use an open circle at $(0, -4)$.

Step 2: Graph $y = x$ for $x \ge 0$. Because x is greater than or equal to 0, use a closed circle at $(0, 0)$.

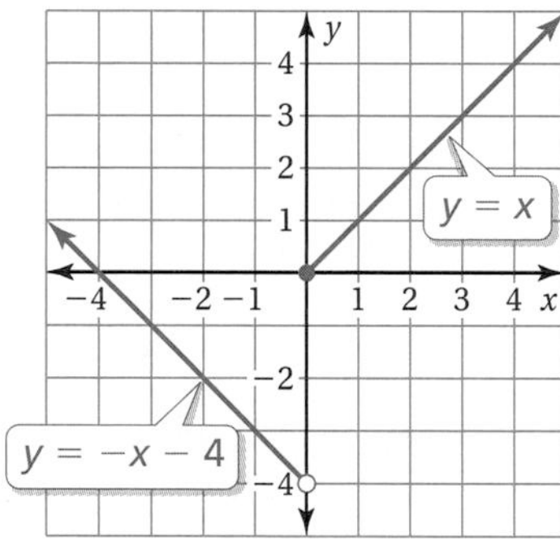

∴ The domain is all real numbers. The range is $y > -4$.

Practice

Graph the function. Describe the domain and range.

1. $y = \begin{cases} x + 3, & \text{if } x \le 0 \\ -x, & \text{if } x > 0 \end{cases}$

2. $y = \begin{cases} x - 2, & \text{if } x < 0 \\ 4x, & \text{if } x \ge 0 \end{cases}$

3. $y = \begin{cases} -3x - 2, & \text{if } x \le 1 \\ x + 1, & \text{if } x > 1 \end{cases}$

4. $y = \begin{cases} 2x, & \text{if } x < -1 \\ -2x, & \text{if } x \ge -1 \end{cases}$

5. $y = \begin{cases} 1, & \text{if } x < -3 \\ x - 1, & \text{if } -3 \le x \le 3 \\ -2, & \text{if } x > 3 \end{cases}$

6. $y = \begin{cases} -x + 2, & \text{if } x \le -2 \\ 5, & \text{if } -2 < x < 1 \\ 3x, & \text{if } x \ge 1 \end{cases}$

7. **REASONING** Does $y = \begin{cases} 1 - x, & \text{if } x \le 0 \\ x - 1, & \text{if } x \ge -2 \end{cases}$ represent a function? Explain your reasoning.

Laurie's Notes

Introduction

Connect

- **Yesterday:** Students used function notation to evaluate and graph functions and perform vertical translations of graphs of functions. (MP6, MP8)
- **Today:** Students will graph piecewise functions.

Motivate

- Ask students if they have seen a utility bill where the cost per kilowatt hour depended upon the amount used. For instance, the first 5000 KWH might be \$0.1782 per KWH, then \$0.1219 per KWH for the next 3000 KWH, and so on.
- Explain that this type of billing scheme is an example of the type of function they will learn about today.

Lesson Notes

Key Idea

- Write the definition and connect it to the Motivate.
- Explain the piecewise-function notation. Depending on the value of x, a different expression is used to evaluate and graph the function.

Example 1

- Write the function.
- ? "How do you know which expression to use?" It depends on what x is; if $x < 0$ you use $y = -x - 4$ and if $x \geq 0$ you use $y = x$.
- Graph the function as shown.
- **Alternate Method:** Lightly graph each line using the slope and y-intercept. Then use the domains to decide what portion of each line to keep, and what to erase. This method is sometimes easier for students when the change between the expressions does not occur at $x = 0$ and when there are more than two expressions.
- In stating the range, students should notice that all of the y-values are greater than -4. Sometimes students will find this difficult to answer until I move my hand in an upward motion starting at $y = -4$.
- **Extension:** "Find the domain and range of the function in the Key Idea." domain: all real numbers; range: $y \leq -2$, $y > 1$

Practice

- The exercises provide a good review of graphing linear functions.
- Exercises 5 and 6 have 3 expressions. Compound inequalities are used to define the domain of the second expressions.
- Students need to recall the definition of a function in Exercise 7. Here, x-values between -2 and 0 are defined two ways, providing two different y-values. So, this is not a function.

Goal Today's lesson is graphing **piecewise functions**, including **step functions** and **absolute value functions**.

Start Thinking! and Warm Up

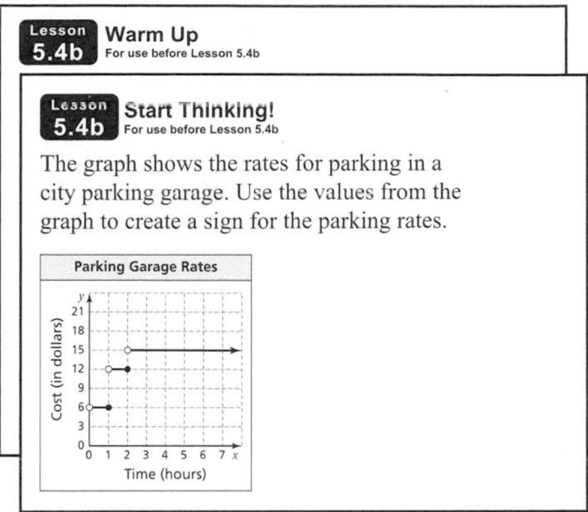

Extra Example 1

Graph $y = \begin{cases} -2, & \text{if } x \leq 0 \\ x, & \text{if } x > 0 \end{cases}$.

Describe the domain and range.

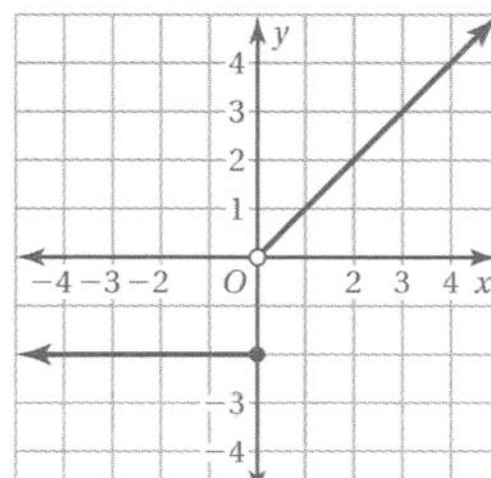

The domain is all real numbers. The range is $y = -2$ or $y > 0$.

Practice

1–6. See Additional Answers.

7. no; It fails the Vertical Line Test.

Record and Practice Journal Practice

See Additional Answers.

Extra Example 2

Write a piecewise function for the graph.

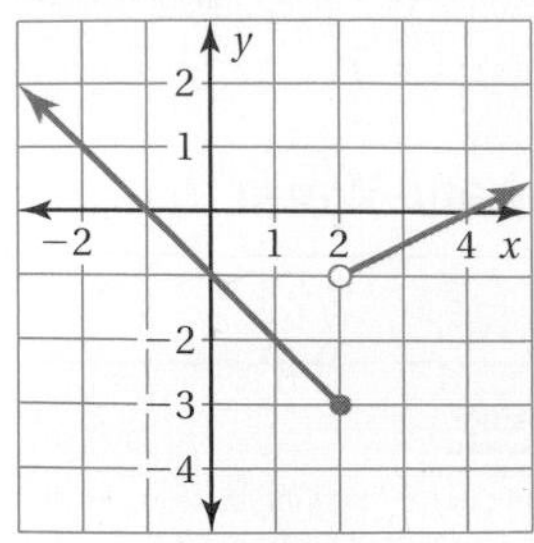

$$f(x) = \begin{cases} -x - 1, & \text{if } x \le 2 \\ \frac{1}{2}x - 2, & \text{if } x > 2 \end{cases}$$

Extra Example 3

A store charges \$3 to rent a DVD for one day and \$2 for each additional day. Write and graph a step function that represents the relationship between the number of days x and the total cost of renting the DVD for up to 5 days.

$$f(x) = \begin{cases} 3, & \text{if } 0 < x \le 1 \\ 5, & \text{if } 1 < x \le 2 \\ 7, & \text{if } 2 < x \le 3 \\ 9, & \text{if } 3 < x \le 4 \\ 11, & \text{if } 4 < x \le 5 \end{cases}$$

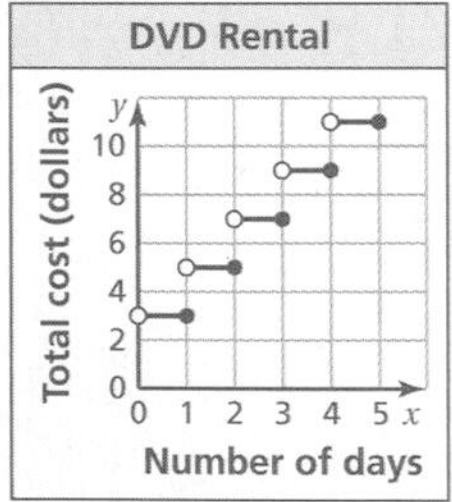

Practice

8. $f(x) = \begin{cases} -x - 1, & \text{if } x \le 0 \\ x + 2, & \text{if } x > 0 \end{cases}$

9. $f(x) = \begin{cases} -4, & \text{if } x < 0 \\ -x, & \text{if } x \ge 0 \end{cases}$

10. $f(x) = \begin{cases} -x - 2, & \text{if } x \le -2 \\ 2, & \text{if } -2 < x < 1 \\ 2x - 3, & \text{if } x \ge 1 \end{cases}$

11. See Additional Answers.

Laurie's Notes

Example 2

? "How can you write an equation of each line from looking at the graph?" Listen for a correct description of a method.

- Remind students that they know how to "read" a graph to determine the slope and y-intercept. If the slope is not obvious, students will need to calculate the slope from two points on the line.
- Students may find it easier to write the equation of each line first, and then write the values of x for which each expression is defined.

Discuss

- Define a step function, noting that it is a special type of piecewise function.
- The name *step* function comes from the graph of the function, which looks like a staircase.

Example 3

- The table helps organize the data. Compound inequalities are used to define the days.
- Because there are five days, the step function will have 5 expressions. Each expression is a constant.
- Point out that function notation, $f(x)$, is used.

? "Why do the line segments have an open circle on the left side and closed circle on the right?" Listen for students to explain what happens when you go beyond a whole number of days.

- When graphing, say, "It costs \$50 to rent a karaoke machine for any portion of the first day. Once you have it for a portion of the second day, the cost jumps to \$75."

? "What does it cost to rent the machine for 3 days?" \$100

? "What does it cost to rent the machine for 3 1/2 days?" \$125

- **Common Error:** Some students may say that the cost for 3 days is \$50 + \$75 + \$100, thinking you pay for each "step" in the process.

Practice

- First students need to determine the equation for the lines, and then think about the restriction for the domain.
- To find the function in Exercise 10, students may find it helpful to extend the lines so that the y-intercepts are located on the graph.
- Discuss the solutions as a class.

EXAMPLE 2 Writing a Piecewise Function

Write a piecewise function for the graph.

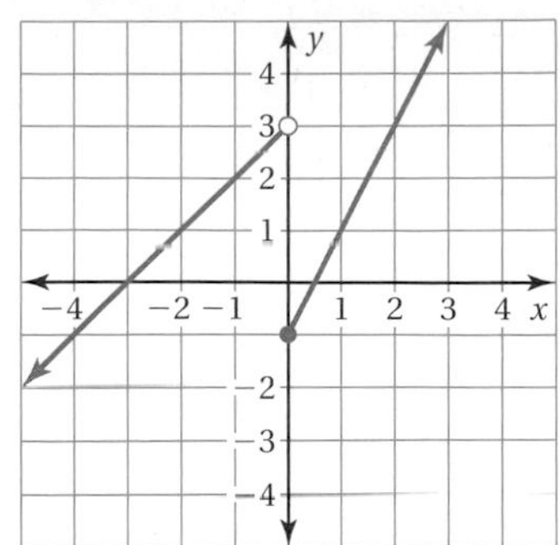

Each "piece" of the function is linear.

When $x < 0$, the graph is the line given by $y = x + 3$.

When $x \geq 0$, the graph is the line given by $y = 2x - 1$.

So, a piecewise function for the graph is $f(x) = \begin{cases} x + 3, & \text{if } x < 0 \\ 2x - 1, & \text{if } x \geq 0 \end{cases}$.

Study Tip

The graph of a step function can look like a staircase.

A **step function** is a piecewise function defined by constant values over its domain. The graph of a step function consists of a series of line segments.

EXAMPLE 3 Graphing a Step Function

You rent a karaoke machine for 5 days. The rental company charges \$50 for the first day and \$25 for each additional day. Write and graph a step function that represents the relationship between the number of days x and the total cost of renting the karaoke machine.

Use a table to organize the information.

Time (days)	Total Cost
$0 < x \leq 1$	50
$1 < x \leq 2$	75
$2 < x \leq 3$	100
$3 < x \leq 4$	125
$4 < x \leq 5$	150

$$f(x) = \begin{cases} 50, & \text{if } 0 < x \leq 1 \\ 75, & \text{if } 1 < x \leq 2 \\ 100, & \text{if } 2 < x \leq 3 \\ 125, & \text{if } 3 < x \leq 4 \\ 150, & \text{if } 4 < x \leq 5 \end{cases}$$

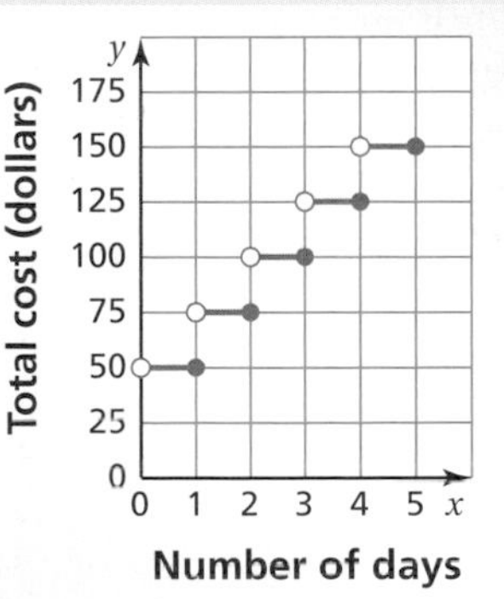

Practice

Write a piecewise function for the graph.

8.

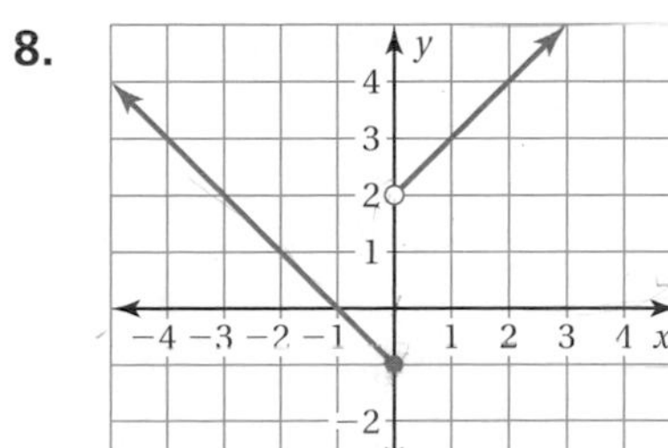

9.

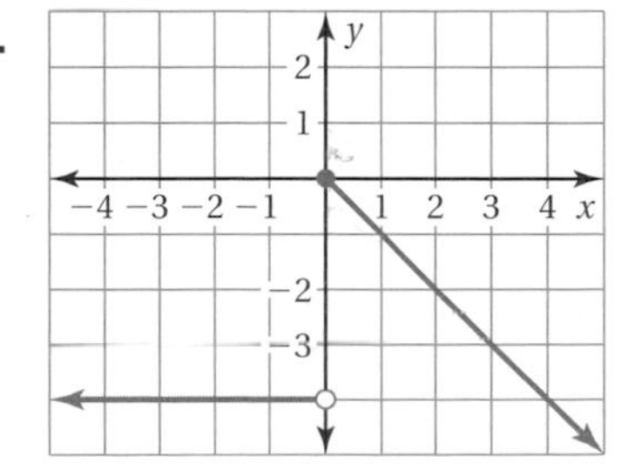

10.

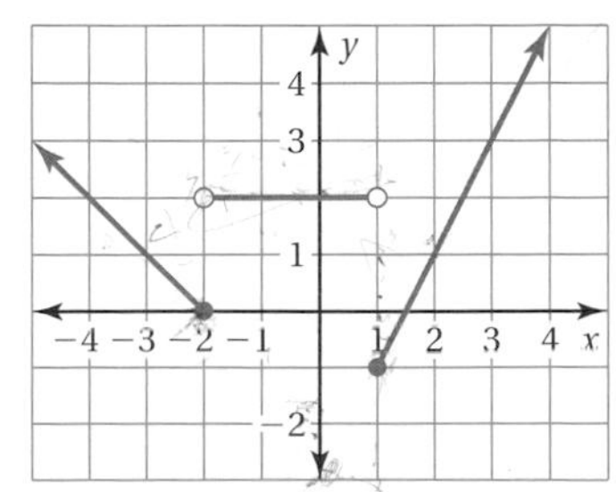

11. **LANDSCAPING** A landscaper rents a wood chipper for 4 days. The rental company charges \$100 for the first day and \$50 for each additional day. Write and graph a step function that represents the relationship between the number of days x and the total cost of renting the chipper.

Key Idea

Study Tip

The absolute value function $f(x) = |x|$ can be written as a piecewise function.

$$f(x) = \begin{cases} -x, & \text{if } x < 0 \\ 0, & \text{if } x = 0 \\ x, & \text{if } x > 0 \end{cases}$$

Absolute Value Function

An **absolute value function** has a V-shaped graph that opens up or down.

The most basic absolute value function is $f(x) = |x|$.

The absolute value of a number is always nonnegative. So, the range of $f(x) = |x|$ is $y \geq 0$.

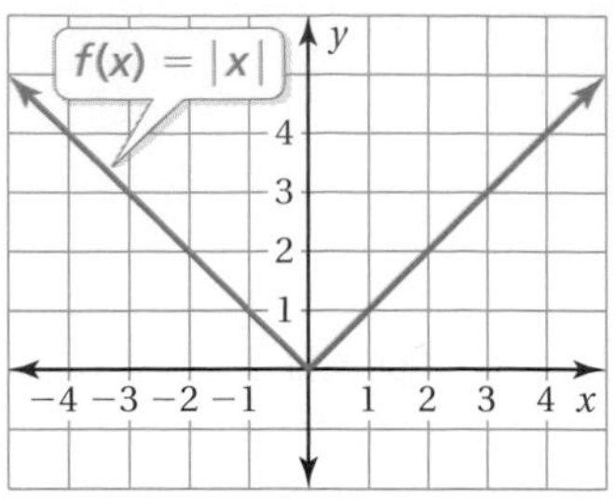

EXAMPLE 4 Graphing Absolute Value Functions

Graph each function. Compare the graph to the graph of $y = |x|$. Describe the domain and range.

a. $y = |x| + 3$

Step 1: Make a table of values.

x	−2	−1	0	1	2
y	5	4	3	4	5

Step 2: Plot the ordered pairs.

Step 3: Draw the V-shaped graph.

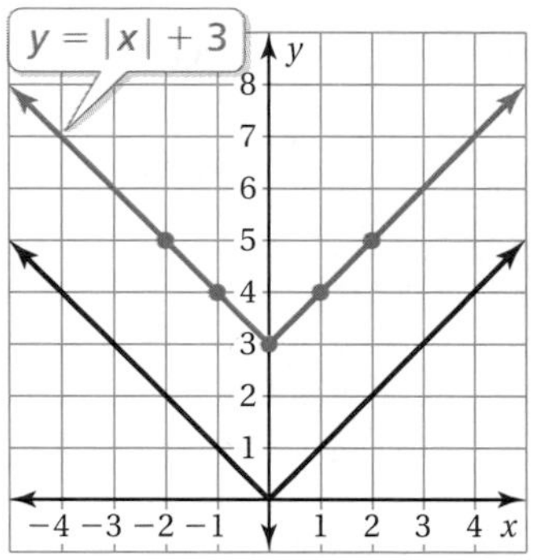

Study Tip

The function $y = |x| + 3$ can be written as a piecewise function.

$$f(x) = \begin{cases} -x + 3, & \text{if } x < 0 \\ x + 3, & \text{if } x \geq 0 \end{cases}$$

∴ The graph of $y = |x| + 3$ is a translation 3 units up of the graph of $y = |x|$. The domain is all real numbers. The range is $y \geq 3$.

b. $y = |x - 2|$

Step 1: Make a table of values.

x	0	1	2	3	4
y	2	1	0	1	2

Step 2: Plot the ordered pairs.

Step 3: Draw the V-shaped graph.

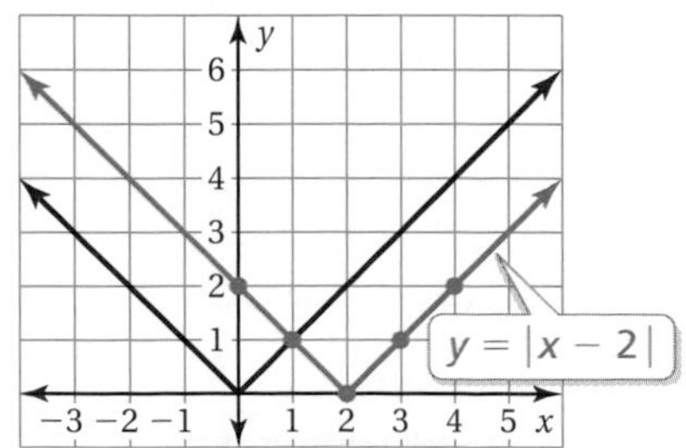

∴ The graph of $y = |x - 2|$ is a translation 2 units to the right of the graph of $y = |x|$. The domain is all real numbers. The range is $y \geq 0$.

Laurie's Notes

Key Idea

- ? "What does the absolute value of a number mean?" They may say if the number is positive, it stays the same, and if the number is negative, you make it positive. The language is not precise, but they understand that the result is different for positive and negative numbers.
- Make a table of values for $f(x) = |x|$ using integer domain values from -3 to 3. Plot the points and draw the V-shaped graph. Students should recognize the two parts of the graph can be described by the functions $f(x) = -x$ and $f(x) = x$.
- **Connection:** The absolute value function can be written as a piecewise function.
- Write the algebraic definition in the Study Tip. There are three expressions depending upon whether x is positive, negative, or zero.
- The piecewise function could be written with two expressions if $x > 0$ and $x = 0$ are combined to $x \geq 0$.
- The shape of the graph is always a V and the range of $f(x) = |x|$ is $y \geq 0$.

Discuss

- **MP7 Look for and Make Use of Structure:** Students are familiar with the graph of $y = x$ and how the slope and y-intercept are affected by transformations. When graphing absolute value equations of the form $y = a|x - h| + k$, the parameters a, h, and k affect the graph of $y = |x|$ in similar ways.
- Explain how adding k outside the absolute value affects the y-values, meaning it is a vertical shift. Similarly, explain how subtracting h inside the absolute value affects the x-values, meaning it is a horizontal shift.

Example 4

- ? "When you graph $y = x$ and $y = x + b$ on the same graph, what does b do to the graph of $y = x$?" shifts it up or down
- Work through the steps in part (a).
- ? "How does this graph compare to $y = |x|$?" shifted up 3 units "How has the range changed?" range increased by 3
- Point out that this is a *translation* 3 units up.
- Work through the steps in part (b).
- ? "How does this graph compare to $y = |x|$?" shifted 2 units to the right "How has the range changed?" range did not change
- Point out that this is a *translation* 2 units right.
- The Study Tip shows how to write the absolute value function in part (a) as a piecewise function with two expressions. Point out that 0 can be included in the domain of either expression.
- If time permits, write the function in part (b) as a piecewise function.

Extra Example 4

Graph each function. Compare the graph to the graph of $y = |x|$. Describe the domain and range.

a. $y = |x| - 2$

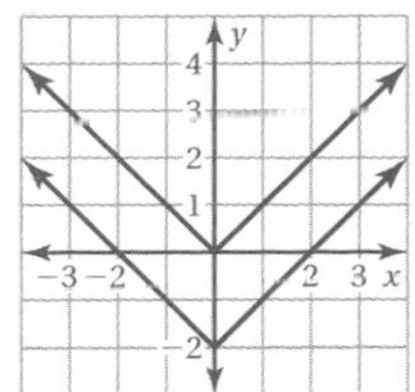

The graph of $y = |x| - 2$ is a translation 2 units down of the graph of $y = |x|$. The domain is all real numbers. The range is $y \geq -2$.

b. $y = |x + 3|$

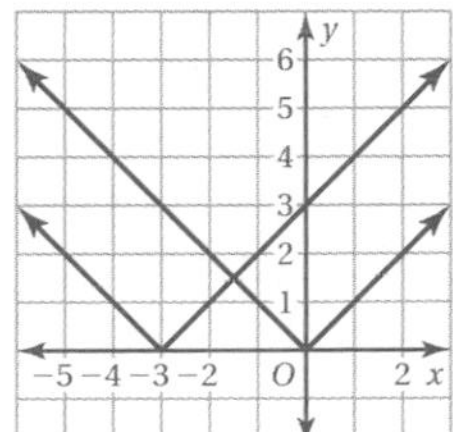

The graph of $y = |x + 3|$ is a translation 3 units to the left of the graph of $y = |x|$. The domain is all real numbers. The range is $y \geq 0$.

Extra Example 5

Graph $y = -2|x|$. Compare the graph to the graph of $y = |x|$. Describe the domain and range.

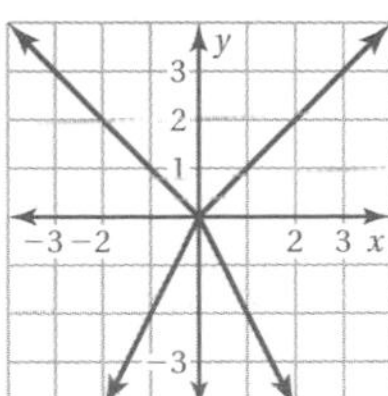

The graph of $y = -2|x|$ opens down and is wider than the graph of $y = |x|$. The domain is all real numbers. The range is $y \leq 0$.

Practice

12–20. See Additional Answers.

21. $y = |x| - 7$

22. $y = |x + 10|$

23. $y = |x - 5| - 1$

24. $y = |x + 6| + 4$

25. **a.** positive k: translation up; negative k: translation down

b. positive h: translation to the right; negative h: translation to the left

c. positive a: Graph is narrower when $a > 1$ and is wider when $a < 1$; negative a: Graph opens down, is narrower when $a < -1$, and is wider when $a > -1$.

26. $x = -2, 4$

27. $x = -7, 3$

28. $x = -9, -5$

29. *Sample answer:*

$$y = \begin{cases} -x - 4, & \text{if } x < -4 \\ x + 4, & \text{if } x \geq -4 \end{cases}$$

30. See Additional Answers.

Mini-Assessment

Graph the function. Describe the domain and range.

1. $y = \begin{cases} -x - 4, & \text{if } x \leq -2 \\ x - 2, & \text{if } x > -2 \end{cases}$

See Additional Answers.

2. $y = -\frac{1}{2}|x + 2| - 1$

See Additional Answers.

3. You rent inline skates at a rate of \$6 for the first hour and \$2 for each additional hour. Write and graph a step function that represents the relationship between the number of hours x and the total cost of the skates.

See Additional Answers.

Laurie's Notes

Example 5

? "When you graph $y = x$ and $y = mx$ on the same graph, what does m do to the graph of $y = x$?" It changes the steepness of the line.

- Work through the steps in the example.
- Discuss the transformations of the graph of $y = a|x|$.
- Because a is negative, the graph opens down.
- Because $|a| < 1$, the graph is wider than the graph of $y = |x|$.
- If time permits, investigate the graphs of absolute value functions using a graphing calculator.

Practice

- Students may describe the graphs of Exercises 16, 17, and 20 as being *wider* and *narrower* than the graph of $y = |x|$.
- Exercises 21–24 could be checked using a graphing calculator.
- Take time to discuss Exercise 25 completely for positive and negative values in each part.
- Exercises 26–28 connect the algebraic method and the graphical method of solving absolute value equations.

Closure

- Graph $y = |x + 2| - 3$ and compare it to the graph of $y = |x|$.

Technology For the Teacher

The Dynamic Planning Tool
Editable Teacher's Resources at *BigIdeasMath.com*

EXAMPLE 5 Graphing Absolute Value Functions

Graph $y = -\frac{1}{2}|x|$. Compare the graph to the graph of $y = |x|$. Describe the domain and range.

Step 1: Make a table of values.

x	-2	-1	0	1	2
y	-1	$-\frac{1}{2}$	0	$-\frac{1}{2}$	-1

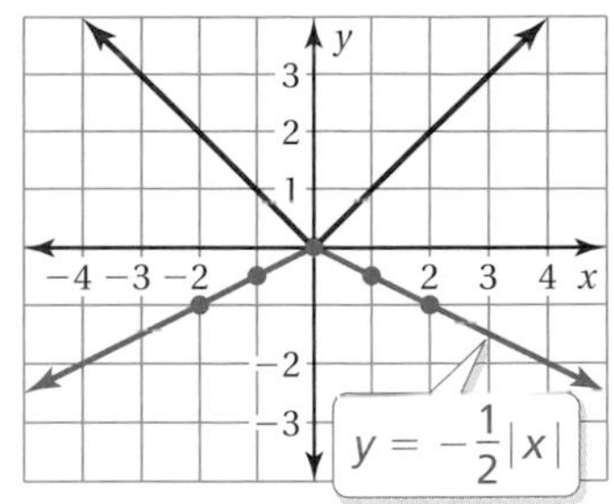

Step 2: Plot the ordered pairs.

Step 3: Draw the V-shaped graph.

∴ The graph of $y = -\frac{1}{2}|x|$ opens down and is wider than the graph of $y = |x|$. The domain is all real numbers. The range is $y \le 0$.

Practice

Graph the function. Compare the graph to the graph of $y = |x|$. Describe the domain and range.

12. $y = |x| - 1$

13. $y = |x| + 5$

14. $y = |x + 4|$

15. $y = |x - 3|$

16. $y = \frac{1}{4}|x|$

17. $y = -3|x|$

18. $y = |x + 1| - 2$

19. $y = -|x - 5| + 1$

20. $y = 4|x| - 4$

Write an equation for the given translation of $y = |x|$.

21. 7 units down

22. 10 units left

23. 1 unit down and 5 units right

24. 4 units up and 6 units left

25. REASONING Explain how the graph of each function compares to the graph of $y = |x|$ for positive and negative values of k, h, and a.

a. $y = |x| + k$ **b.** $y = |x - h|$ **c.** $y = a|x|$

Solve each equation using a graph. Check your solution.

26. $|x - 1| = 3$

27. $|x + 2| - 6 = -1$

28. $2|x + 7| = 4$

29. STRUCTURE Rewrite the function $y = |x + 4|$ using piecewise notation.

30. STRUCTURE Graph $y = \begin{cases} -x + 5, & \text{if } x \le 0 \\ |x|, & \text{if } x > 0 \end{cases}$. Describe the domain and range.

5.5 Comparing Linear and Nonlinear Functions

COMMON CORE STATE STANDARDS
8.F.3
F.LE.1b

Essential Question How can you recognize when a pattern in real life is linear or nonlinear?

1 ACTIVITY: Finding Patterns for Similar Figures

Work with a partner. Copy and complete each table for the sequence of similar rectangles. Graph the data in each table. Decide whether each pattern is linear or nonlinear.

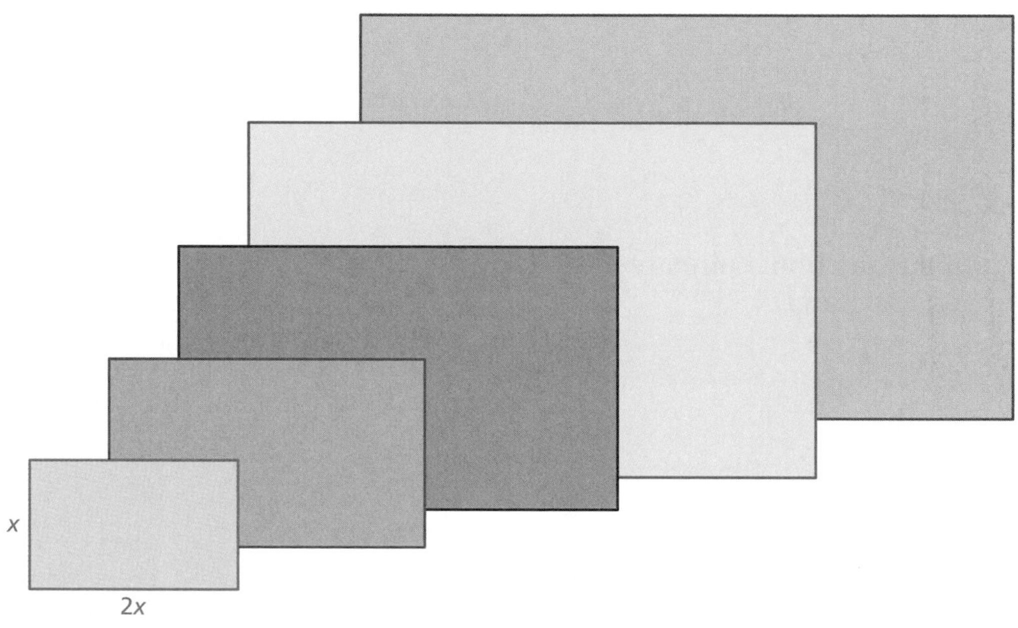

a. Perimeters of Similar Rectangles

x	1	2	3	4	5
P					

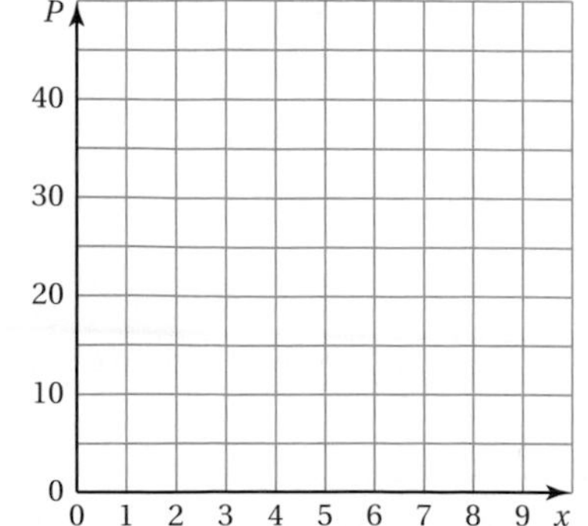

b. Areas of Similar Rectangles

x	1	2	3	4	5
A					

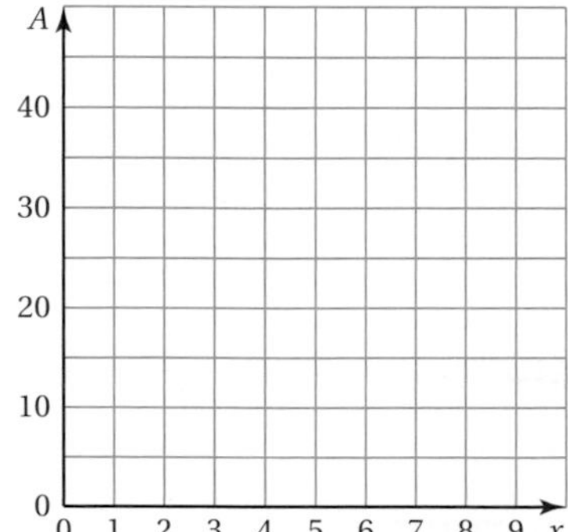

Laurie's Notes

Introduction

Standards for Mathematical Practice

- **MP4 Model with Mathematics** and **MP8 Look for and Express Regularity in Repeated Reasoning:** The goal is for students to recognize when a pattern in real life is linear or nonlinear. Using familiar contexts—similar figures and falling objects, students will look for numeric patterns. The presence or absence of a *constant rate of change* will help students determine whether the data is linear or nonlinear.

Motivate

? "How many of you would like to try skydiving? Why?"

- Share with students that the first successful parachute jump made from a moving airplane was made by Captain Albert Berry in St. Louis, in 1912.
- The first parachute jump from a balloon was completed by André-Jacques Garnerin in 1797 over Monceau Park in Paris.
- Tell students that today they will explore whether the function that describes the height of a parachutist is linear or nonlinear.
- Students will study many types of nonlinear functions, such as quadratic functions, radical functions, and rational functions, in more detail later in the text.

Activity Notes

Activity 1

? "What does it mean for two rectangles to be similar?" Corresponding sides are proportional and corresponding angles have the same measure.

? "What is the relationship between the length and the width of the green rectangle?" The length is twice the width.

? "What is the relationship between the length and the width of the yellow rectangle? How do you know?" The length is twice the width. Because the rectangles are similar, the lengths of all the rectangles will be twice the widths.

- Explain to students that they will find the perimeter and area of each rectangle for the side lengths given, and then plot the results.
- **Teaching Tip:** It may be helpful to set up a table that includes a row for the second dimension as shown. The numeric pattern is more obvious when viewed in a table.

Width	x	1	2	3	4	5
Length	$2x$	2	4	6	8	10
Perimeter	P					

- **MP6 Attend to Precision:** Encourage students to be accurate with their graphing. Because only 5 points are being plotted for each graph, it is possible that students will not see the curvature of the area graph. Students should recognize, however, that the numeric data for area does not have a constant difference between A-values.

Common Core State Standards

8.F.3 Interpret the equation $y = mx + b$ as defining a linear function, whose graph is a straight line; give examples of functions that are not linear.

F.LE.1b Recognize situations in which one quantity changes at a constant rate per unit interval relative to another.

Previous Learning

Students should know common geometric formulas, such as area and perimeter.

Activity Materials
Textbook
• handkerchief • floss • tape • small figurine

Start Thinking! and Warm Up

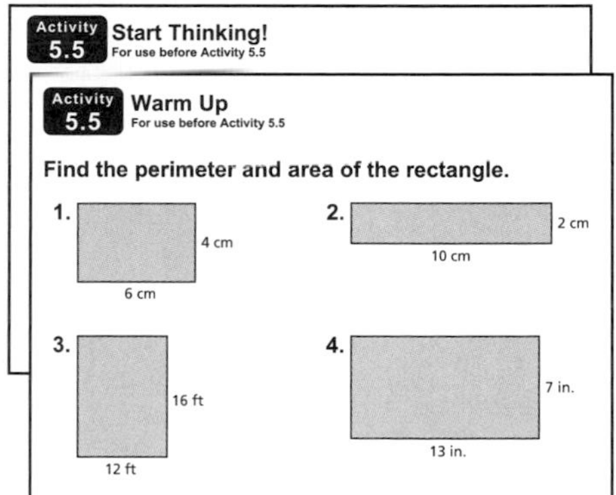

Activity 5.5 Start Thinking! For use before Activity 5.5

Activity 5.5 Warm Up For use before Activity 5.5

Find the perimeter and area of the rectangle.

5.5 Record and Practice Journal

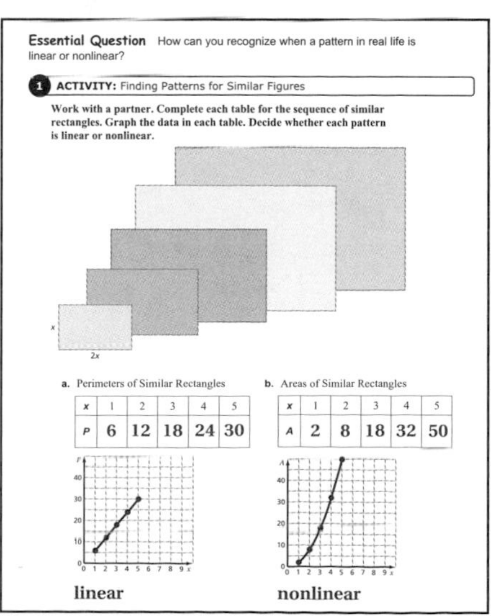

Essential Question How can you recognize when a pattern in real life is linear or nonlinear?

1 ACTIVITY: Finding Patterns for Similar Figures

Work with a partner. Complete each table for the sequence of similar rectangles. Graph the data in each table. Decide whether each pattern is linear or nonlinear.

a. Perimeters of Similar Rectangles

x	1	2	3	4	5
P	6	12	18	24	30

linear

b. Areas of Similar Rectangles

x	1	2	3	4	5
A	2	8	18	32	50

nonlinear

Differentiated Instruction

Visual

Students may be able to describe how the sequence of output numbers is changing, for example, *start with 2 and add 3*, but they may find it difficult to write a function rule for changing an input value to an output value. If students determine that the output increases or decreases by a constant value as the input increases, the function will have an ax term. Have students create function tables for equations such as $y = x + 3$, $y = 4x - 1$, and $y = 0.5x$ to see this pattern.

5.5 Record and Practice Journal

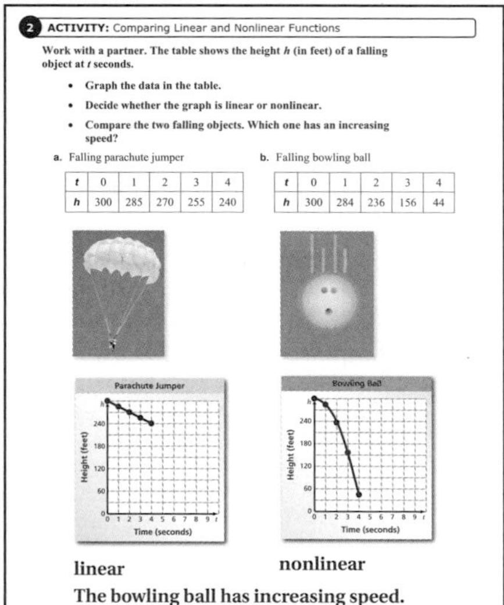

2 ACTIVITY: Comparing Linear and Nonlinear Functions

Work with a partner. The table shows the height h (in feet) of a falling object at t seconds.

- Graph the data in the table.
- Decide whether the graph is linear or nonlinear.
- Compare the two falling objects. Which one has an increasing speed?

a. Falling parachute jumper

t	0	1	2	3	4
h	300	285	270	255	240

b. Falling bowling ball

t	0	1	2	3	4
h	300	284	236	156	44

linear

nonlinear

The bowling ball has increasing speed.

What Is Your Answer?

3. IN YOUR OWN WORDS How can you recognize when a pattern in real life is linear or nonlinear? Describe two real-life patterns: one that is linear and one that is nonlinear. Use patterns that are different from those described in Activities 1 and 2.

If the rate of change is constant, the pattern is linear.

Laurie's Notes

Activity 2

- This activity is similar to Activity 1, except the ordered pairs are already given. Discuss the two falling objects—one with a parachute and one that is free falling.
- ? "Is there a difference in the rate at which two objects fall when one is attached to a parachute and the other is left to free fall? Explain." Listen for discussion of rate. It is unlikely students will bring up acceleration.
- You could make a small parachute using a handkerchief, tape, floss, and a small figurine to model a parachute-controlled fall and a free fall.
- **MP6:** Again, it is necessary for students to be accurate when plotting the ordered pairs given the scale on the y-axis.
- ? After students have plotted the points, ask about the two graphs. First note that the two graphs begin at the same height (y-intercept), 300 feet.
  - "How far has the jumper fallen after 4 seconds?" 60 ft "How far has the bowling ball fallen after 4 seconds?" 256 ft
  - "Describe the flight of the jumper." falling at a constant rate of 15 ft/sec
  - "Describe the flight of the bowling ball." Listen for students to describe that the bowling ball is picking up speed as it falls.
- **Extension:** Students could write a linear equation for the jumper, but not the bowling ball.

What Is Your Answer?

- Students may need help thinking of real-life patterns that are nonlinear. You might suggest area or volume relationships, or even simple story graphs about time and distance.

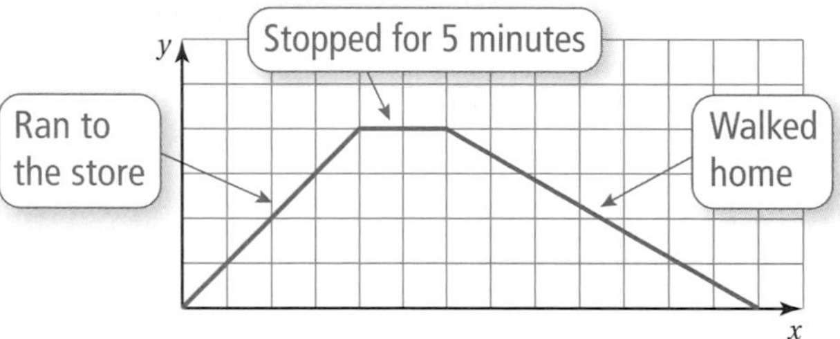

Closure

- Draw two functions with a domain of $x \geq 0$. Have one that is linear and one that is nonlinear. Describe how the graphs are alike and how they are different. Answers will vary.

Technology For the Teacher

The Dynamic Planning Tool

Editable Teacher's Resources at *BigIdeasMath.com*

2 ACTIVITY: Comparing Linear and Nonlinear Functions

Work with a partner. The table shows the height *h* (in feet) of a falling object at *t* seconds.

- **Graph the data in the table.**
- **Decide whether the graph is linear or nonlinear.**
- **Compare the two falling objects. Which one has an increasing speed?**

a. Falling parachute jumper

t	0	1	2	3	4
h	300	285	270	255	240

Parachute Jumper

Height (feet): 0, 60, 120, 180, 240, *h*

Time (seconds): 0, 1, 2, 3, 4, 5, 6, 7, 8, 9, *t*

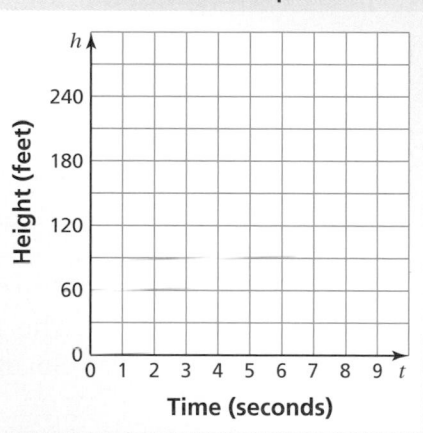

b. Falling bowling ball

t	0	1	2	3	4
h	300	284	236	156	44

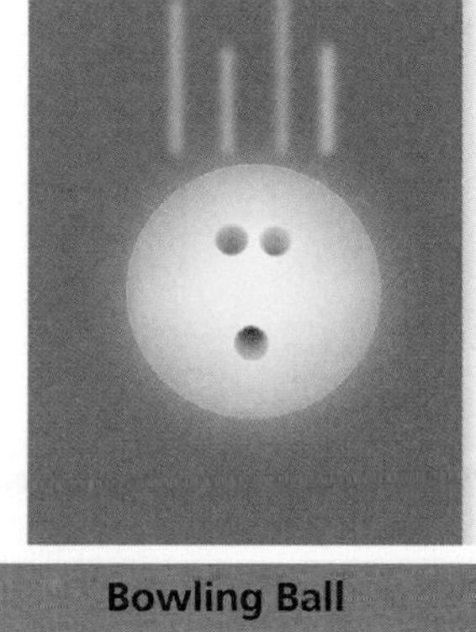

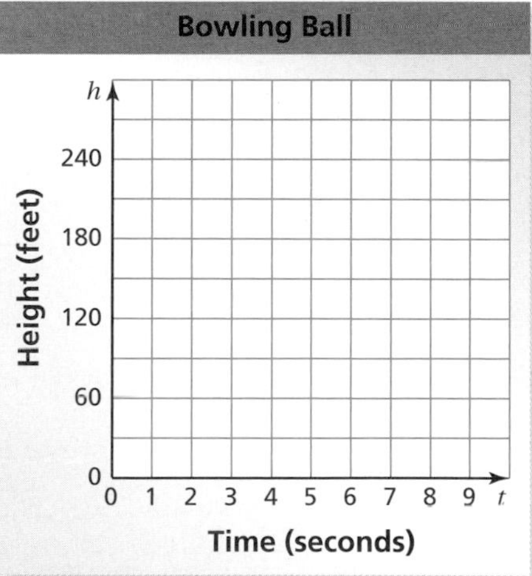

What Is Your Answer?

3. **IN YOUR OWN WORDS** How can you recognize when a pattern in real life is linear or nonlinear? Describe two real-life patterns: one that is linear and one that is nonlinear. Use patterns that are different from those described in Activities 1 and 2.

Use what you learned about comparing linear and nonlinear functions to complete Exercises 3–6 on page 240.

5.5 Lesson

Key Vocabulary
nonlinear function, *p. 238*

The graph of a linear function shows a constant rate of change. A **nonlinear function** does not have a constant rate of change. So, its graph is *not* a line.

EXAMPLE 1 Identifying Functions from Tables

Does the table represent a *linear* or *nonlinear* function? Explain.

Study Tip

A constant rate of change describes a quantity that changes by equal amounts over equal intervals.

a.

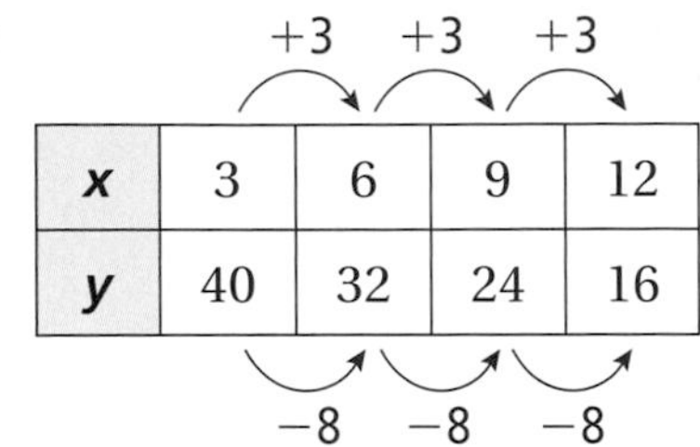

	+3	+3	+3	
x	3	6	9	12
y	40	32	24	16
	−8	−8	−8	

As x increases by 3, y decreases by 8. The rate of change is constant. So, the function is linear.

b.

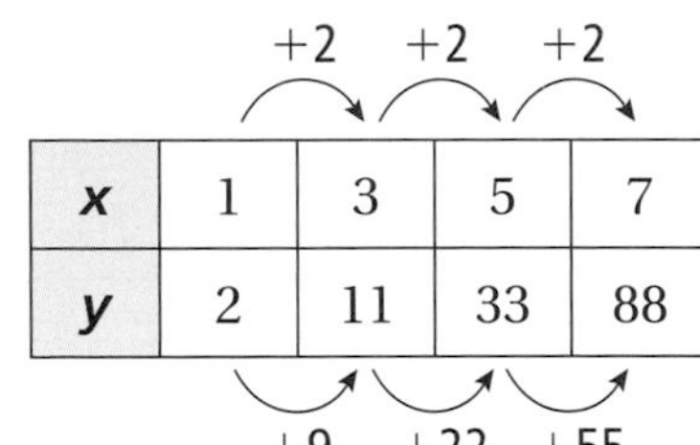

	+2	+2	+2	
x	1	3	5	7
y	2	11	33	88
	+9	+22	+55	

As x increases by 2, y increases by different amounts. The rate of change is *not* constant. So, the function is nonlinear.

EXAMPLE 2 Identifying Functions from Graphs

Does the graph represent a *linear* or *nonlinear* function? Explain.

a.

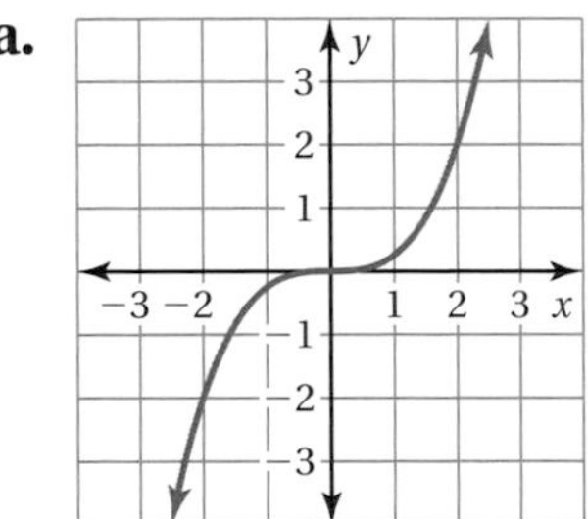

The graph is *not* a line.
So, the function is nonlinear.

b.

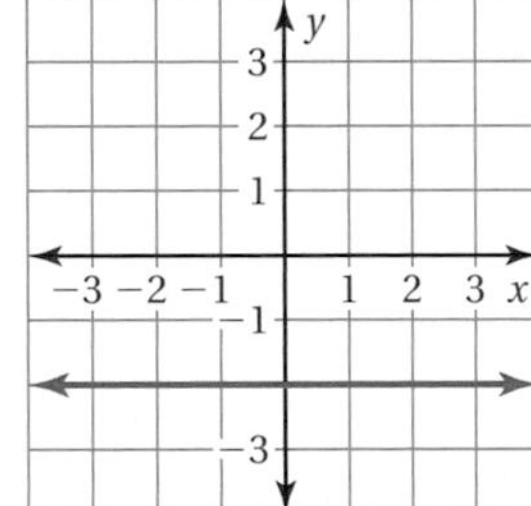

The graph is a line.
So, the function is linear.

On Your Own

Now You're Ready
Exercises 3–11

Does the table or graph represent a *linear* or *nonlinear* function? Explain.

1.

x	y
0	25
7	20
14	15
21	10

2.

x	y
2	8
4	4
6	0
8	−4

3.

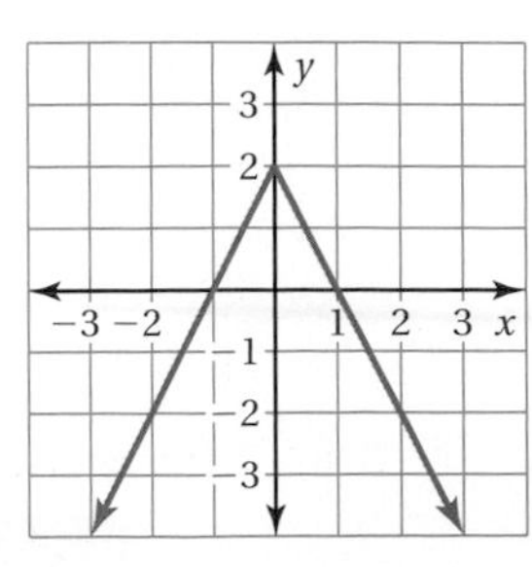

Multi-Language Glossary at BigIdeasMath.com.

Laurie's Notes

Introduction

Connect

- **Yesterday:** Students explored the graphs of functions that were linear and nonlinear. (MP4, MP6, MP8)
- **Today:** Students will compare linear and nonlinear functions.

Motivate

- Ask 5 students to complete a table of values, where the domain is the same for 5 functions.

	−3	−2	−1	0	1	2	3
$y = x + 2$	−1	0	1	2	3	4	5
$y = x - 2$	−5	−4	−3	−2	−1	0	1
$y = 2x$	−6	−4	−2	0	2	4	6
$y = \frac{x}{2}$	$-\frac{3}{2}$	−1	$-\frac{1}{2}$	0	$\frac{1}{2}$	1	$\frac{3}{2}$
$y = x^2$	9	4	1	0	1	4	9

- Spend time discussing the many patterns in the table. Discuss one function at a time. Ask students for their observations about patterns, changes in *y*-values, slope, and *y*-intercept.
- For $y = x^2$, students want it to have a constant slope. Draw a quick plot of the points and show it is not a linear function.

Lesson Notes

Example 1

- Copy the first table of values. Draw attention to the change in *x* (increasing by 3 each time) and the change in *y* (decreasing by 8 each time). Because the rate of change is constant, the function is linear.
- Copy the second table of values. Draw attention to the change in *x* (increasing by 2 each time) and the change in *y* (increasing by different amounts each time). This is a nonlinear function.

Example 2

- Part (b) may seem obvious, but the horizontal line seems like a special case to students. They may not be sure it is a linear function.

? "What is the slope of this line?" 0 "What is the constant rate of change?" Each time *x* increases by 1, *y* stays the same.

On Your Own

? "What are the constant rates of change for Questions 1 and 2?" $-\frac{5}{7}$; −2

? "Why is Question 3 not a linear function?" There are two parts of this function. The rate of change is positive, then negative.

Goal Today's lesson is comparing linear and **nonlinear functions**.

Start Thinking! and Warm Up

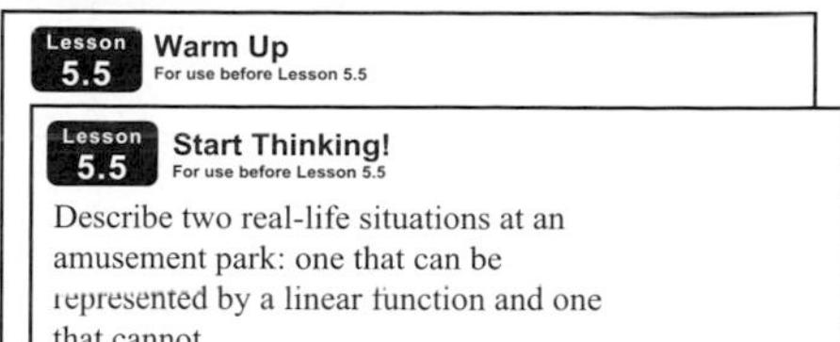
Lesson 5.5 Warm Up
For use before Lesson 5.5

Lesson 5.5 Start Thinking!
For use before Lesson 5.5

Describe two real-life situations at an amusement park: one that can be represented by a linear function and one that cannot.

Extra Example 1

Does the table represent a *linear* or *nonlinear* function? Explain.

x	3	4	5	6
y	1	2	3	4

linear; As *x* increases by 1, *y* increases by 1.

Extra Example 2

Does the graph represent a *linear* or *nonlinear* function? Explain.

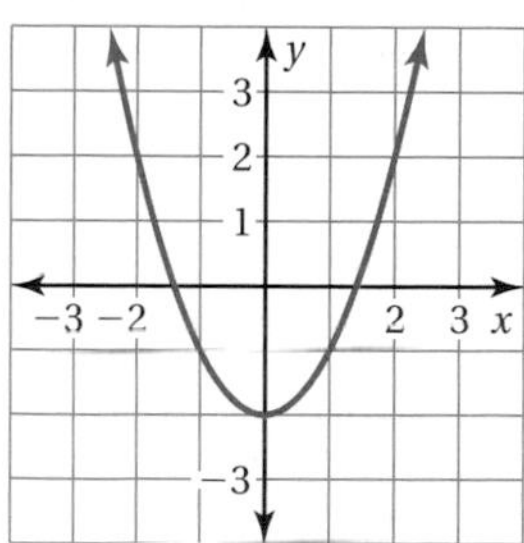

nonlinear; The graph is *not* a line.

On Your Own

1. linear; As *x* increases by 7, *y* decreases by 5.
2. linear; As *x* increases by 2, *y* decreases by 4.
3. nonlinear; The graph is not a line.

Extra Example 3

Does $y = 6x - 3$ represent a *linear* function? Yes, the equation is written in slope-intercept form.

Extra Example 4

Account A earns simple interest. Account B earns compound interest. The table shows the balances for 5 years. Graph the data and compare the graphs.

Year, t	Account A Balance	Account B Balance
0	\$50	\$50
1	\$55	\$55
2	\$60	\$60.50
3	\$65	\$66.55
4	\$70	\$73.21
5	\$75	\$80.53

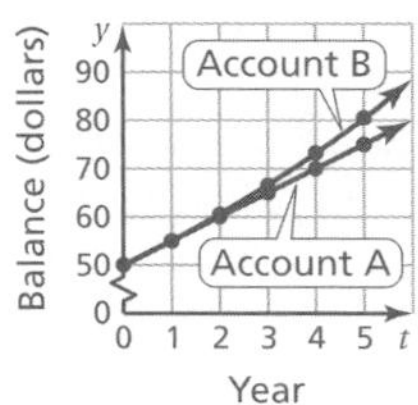

The function representing the balance of Account A is linear. The function representing the balance of Account B is nonlinear.

On Your Own

4. linear; The equation is in slope-intercept form.
5. linear; You can rewrite the equation in slope-intercept form.
6. nonlinear; You can rewrite the equation in slope-intercept form.

English Language Learners

Vocabulary

Begin the lesson by reviewing the terms *function* and *linear function.* Define *nonlinear function* and compare it to linear function.

Laurie's Notes

Example 3

- Discuss each equation. Remind students that all linear functions can be written in slope-intercept form.
- **MP7 Look for and Make Use of Structure:** Students often see $y = \frac{4}{x}$ and $y = \frac{x}{4}$ as *the same kind of function.* So, many students think this will be a linear function. Remind students of how fractions are multiplied, and use the examples $\frac{4}{x}$ and $\frac{x}{4}$.

$$\frac{4}{x} = \frac{4}{1} \cdot \frac{1}{x} = 4 \cdot \frac{1}{x} \qquad \frac{x}{4} = \frac{x}{1} \cdot \frac{1}{4} = x \cdot \frac{1}{4} = \frac{1}{4} \cdot x$$

  So, $y = \frac{x}{4}$ is linear with a slope of $\frac{1}{4}$. The equation $y = \frac{4}{x}$ cannot be written as a linear equation.
- **Note:** The equation $y = \frac{4}{x}$ shows inverse variation.

Example 4

- **Financial Literacy:** Ask a volunteer to read the problem. In addition to looking at linear and nonlinear functions, you also want to integrate financial literacy skills when appropriate. It is a good idea for students to become familiar with the simple interest formula $I = Prt$.
- Point out that both functions have the same initial value. Explain to students that they should interpret the initial value as the starting balance, or the principal, in the context of the problem.

? "Each time the year increases by 1, what happens to the balance of Account A?" It increases by \$10.

? "Each time the year increases by 1, what happens to the balance of Account B?" It increases by a greater amount each year.

- **Extension:** Show students how to calculate the values in the table. Account A's balance is found using $I = Prt$, where $P = 100$, $r = 0.1$, and $t =$ year. Account B's balance can also be found using $I = Prt$, but the principal is changing each year.

On Your Own

- Students should see the exponent of 2 in Question 6 and quickly decide that the function is nonlinear.

Closure

- **Exit Ticket:** Describe how to determine if a function is linear or nonlinear from (a) the equation, (b) a table of values, and (c) a graph.

Technology For the Teacher

The Dynamic Planning Tool
Editable Teacher's Resources at *BigIdeasMath.com*

EXAMPLE 3 Standardized Test Practice

Which equation represents a *nonlinear* function?

Ⓐ $y = 4.7$ Ⓑ $y = \pi x$

Ⓒ $y = \frac{4}{x}$ Ⓓ $y = 4(x - 1)$

You can rewrite the equations $y = 4.7$, $y = \pi x$, and $y = 4(x - 1)$ in slope-intercept form. So, they are linear functions.

You cannot rewrite the equation $y = \frac{4}{x}$ in slope-intercept form.

So, it is a nonlinear function.

∴ The correct answer is Ⓒ.

EXAMPLE 4 Real-Life Application

Study Tip

In Example 4, the *initial value* of each function is $100.

Account A earns simple interest. Account B earns compound interest. The table shows the balances for 5 years. Graph the data and compare the graphs.

Year, t	Account A Balance	Account B Balance
0	$100	$100
1	$110	$110
2	$120	$121
3	$130	$133.10
4	$140	$146.41
5	$150	$161.05

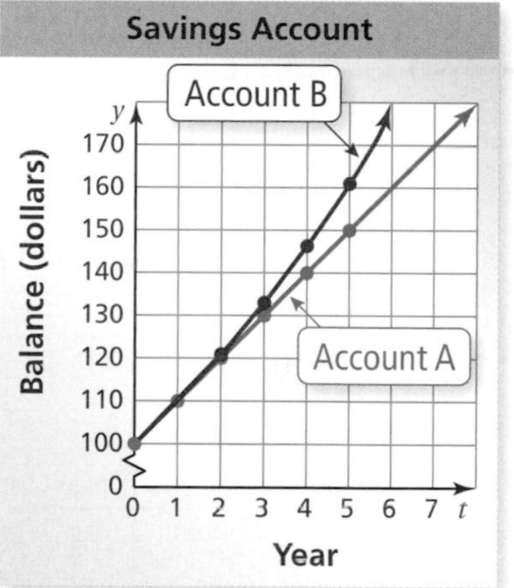

Both graphs show that the balances are positive and increasing.

The balance of Account A has a constant rate of change of $10. So, the function representing the balance of Account A is linear.

The balance of Account B increases by different amounts each year. Because the rate of change is not constant, the function representing the balance of Account B is nonlinear.

On Your Own

Now You're Ready
Exercises 12–14

Does the equation represent a *linear* or *nonlinear* function? Explain.

4. $y = x + 5$ 5. $y = \frac{4x}{3}$ 6. $y = 1 - x^2$

5.5 Exercises

Vocabulary and Concept Check

1. **VOCABULARY** Describe how linear functions and nonlinear functions are different.

2. **WHICH ONE DOESN'T BELONG?** Which equation does *not* belong with the other three? Explain your reasoning.

$5y = 2x$ $\quad$ $y = \frac{2}{5}x$ $\quad$ $10y = 4x$ $\quad$ $5xy = 2$

Practice and Problem Solving

Graph the data in the table. Decide whether the function is *linear* or *nonlinear*.

3.

x	0	1	2	3
y	4	8	12	16

4.

x	1	2	3	4
y	1	2	6	24

5.

x	6	5	4	3
y	21	15	10	6

6.

x	−1	0	1	2
y	−7	−3	1	5

Does the table or graph represent a *linear* or *nonlinear* function? Explain.

2 7.

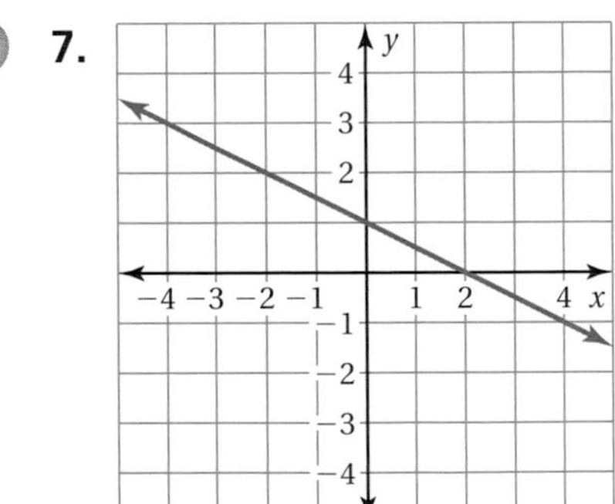

8.

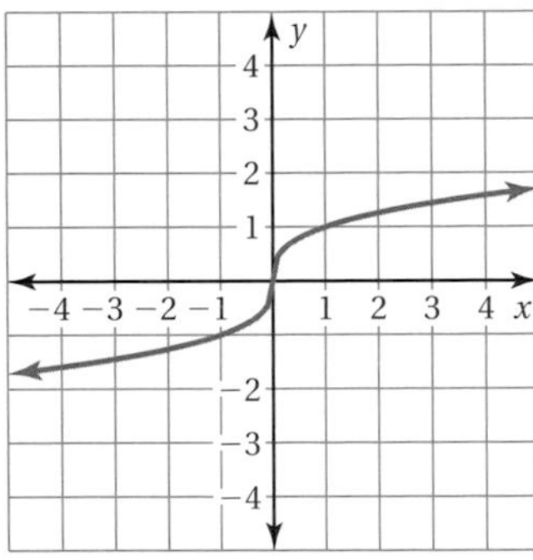

9.

x	5	11	17	23
y	7	11	15	19

10.

x	−3	−1	1	3
y	9	1	1	9

11. **VOLUME** The table shows the volume V (in cubic feet) of a cube with a side length of x feet. Does the table represent a linear or nonlinear function? Explain.

Side Length, x	1	2	3	4	5	6	7	8
Volume, V	1	8	27	64	125	216	343	512

Assignment Guide and Homework Check

Level	Assignment	Homework Check
Average	1–10, 13–17, 20–23	8, 9, 14, 16
Advanced	1–6, 8–14 even, 15–19, 20–23	8, 14, 16, 17

Common Errors

- **Exercises 3–6, 9, and 10** Students may say that the function is linear because the *x*-values are increasing or decreasing by the same amount each time. Encourage them to examine the *y*-values to see if the graph represents a line.
- **Exercises 12–14** Students may not rewrite the equation in slope-intercept form and will guess whether the equation is linear. Remind them to attempt to write the equation in slope-intercept form as a check.
- **Exercise 15** Students may try to graph the coordinate pairs to determine if the function is linear and make an incorrect assumption depending upon how they scale their axes. Encourage them to examine the change in *y* for each *x*-value.

5.5 Record and Practice Journal

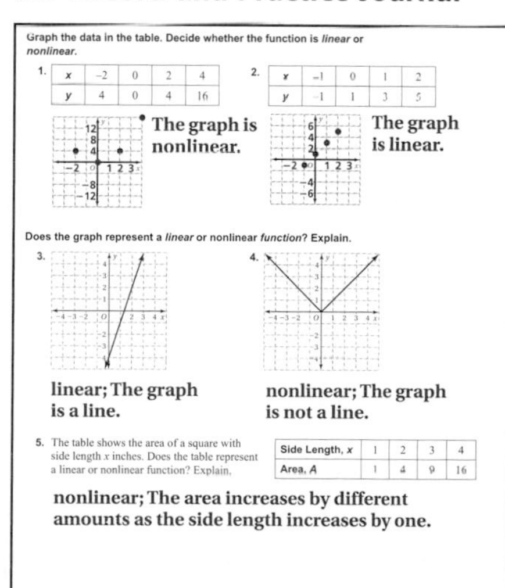
Graph the data in the table. Decide whether the function is *linear* or *nonlinear*.

1.

x	−2	0	2	4
y	4	0	4	16

The graph is nonlinear.

2.

x	−1	0	1	2
y	−1	1	3	5

The graph is linear.

Does the graph represent a *linear* or nonlinear *function*? Explain.

3. linear; The graph is a line.

4. nonlinear; The graph is not a line.

5. The table shows the area of a square with side length x inches. Does the table represent a linear or nonlinear function? Explain.

Side Length, x	1	2	3	4
Area, A	1	4	9	16

nonlinear; The area increases by different amounts as the side length increases by one.

Technology For the Teacher

Answer Presentation Tool

Vocabulary and Concept Check

1. A linear function has a constant rate of change. A nonlinear function does not have a constant rate of change.
2. $5xy = 2$; You cannot rewrite the equation in slope-intercept form.

Practice and Problem Solving

3. 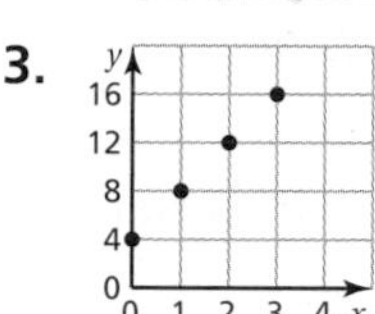

linear

4. 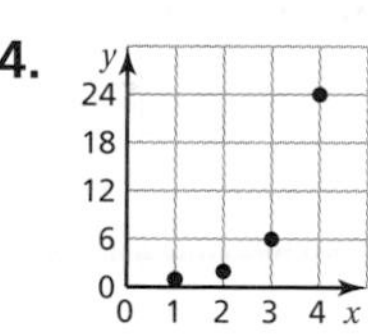

nonlinear

5. 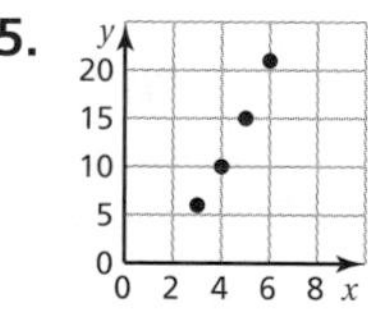

nonlinear

6. 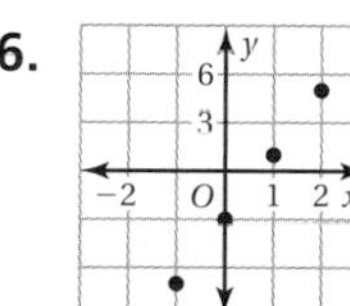

linear

7. linear; The graph is a line.
8. nonlinear; The graph is not a line.
9. linear; As x increases by 6, y increases by 4.
10. nonlinear; As x increases by 2, y changes by different amounts.
11. nonlinear; As x increases by 1, V increases by different amounts.

Practice and Problem Solving

12. linear; You can rewrite the equation in slope-intercept form.

13. linear; You can rewrite the equation in slope-intercept form.

14. nonlinear; You cannot rewrite the equation in slope-intercept form.

15. nonlinear; As x decreases by 65, y increases by different amounts.

16. See *Taking Math Deeper*.

17. **a.** linear; The equation that represents the function is $h = 1.6x$.

b. tree B; It is growing at a rate of 1.6 feet per year.

18. linear; As the height increases by 1, the volume increases by 9π.

19. See Additional Answers.

Fair Game Review

20. 7 **21.** −6

22. ±3 **23.** C

Mini-Assessment

Does the table or graph represent a *linear* or *nonlinear* function? Explain.

1.

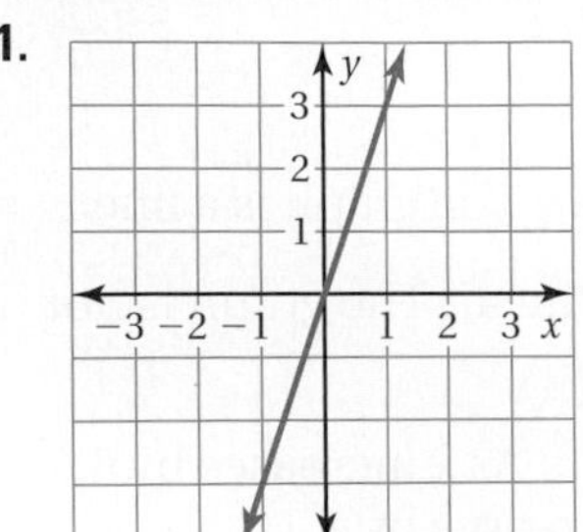

linear; The graph is a line.

2.

x	−2	0	2	4
y	8	0	8	64

nonlinear; The rate of change is not constant.

Taking Math Deeper

Exercise 16

Students can learn a valuable lesson about mathematics from this problem. Even though the problem does not specifically ask them to draw a graph, it is still a good idea. *Seeing* the relationship between pounds and cost is easier than simply finding the relationship using algebra.

1 Plot the two given points.

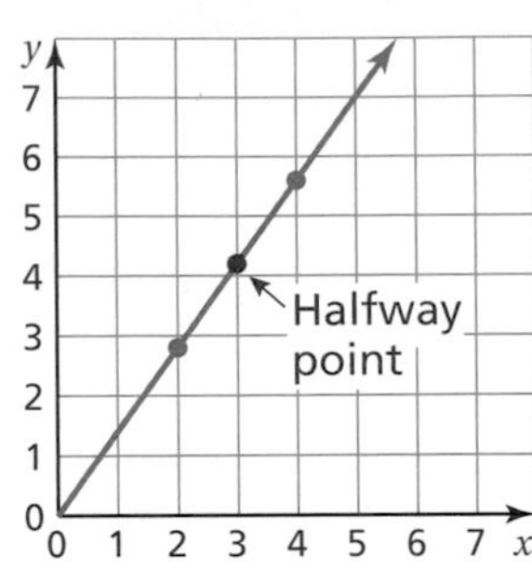

2 Find the halfway point.

Because you want the table to represent a linear function and 3 is halfway between 2 and 4, you need to find the number that is halfway between \$2.80 and \$5.60. This number is the mean of \$2.80 and \$5.60.

a. $\text{Mean} = \frac{2.80 + 5.60}{2} = 4.20$

3 Write a function.

Let x = pounds of seeds.

Let y = cost.

y-intercept = 0

$\text{slope} = \frac{5.60 - 2.80}{4 - 2} = 1.4$

b. $y = 1.4x$

c. The initial value is \$0.

d. The function does not have a maximum value because you can always increase the cost by increasing the amount of sunflower seeds purchased.

Project

Plant some sunflower seeds. Keep track of the progress of the plants until they bloom.

Reteaching and Enrichment Strategies

If students need help. . .	If students got it. . .
Resources by Chapter • Practice A and Practice B • Puzzle Time Record and Practice Journal Practice Differentiating the Lesson Lesson Tutorials Skills Review Handbook	Resources by Chapter • Enrichment and Extension • School-to-Work • Financial Literacy Start the next section

Does the equation represent a *linear* or *nonlinear* function? Explain.

3 **12.** $2x + 3y = 7$

13. $y + x = 4x + 5$

14. $y = \frac{8}{x^2}$

15. **LIGHT** The frequency y (in terahertz) of a light wave is a function of its wavelength x (in nanometers). Does the table represent a linear or nonlinear function? Explain.

Color	Red	Yellow	Green	Blue	Violet
Wavelength, x	660	595	530	465	400
Frequency, y	454	504	566	645	749

16. **MODELING** The table shows the cost y (in dollars) of x pounds of sunflower seeds.

Pounds, x	Cost, y
2	2.80
3	?
4	5.60

a. What is the missing y-value that makes the table represent a linear function?

b. Write a linear function that represents the cost y of x pounds of seeds.

c. What is the initial value of the function?

d. Does the function have a maximum value? Explain your reasoning.

17. **TREES** Tree A grows at a rate of 1.5 feet per year. The table shows the height h (in feet) of Tree B after x years.

Years, x	Height, h
0	0
2	3.2
5	8

a. Does the table represent a linear or nonlinear function? Explain.

b. Which tree is growing at a faster rate? Explain.

18. **PRECISION** The radius of the base of a cylinder is 3 feet. Is the volume of the cylinder a linear or nonlinear function of the height of the cylinder? Explain.

19. Number Sense The ordered pairs represent a function.

(0, 0), (1, 1), (2, 4), (3, 9), and (4, 16)

a. Graph the ordered pairs and describe the pattern. Is the function linear or nonlinear?

b. Write an equation that represents the function.

Fair Game Review What you learned in previous grades & lessons

Find the square root(s). *(Skills Review Handbook)*

20. $\sqrt{49}$

21. $-\sqrt{36}$

22. $\pm\sqrt{9}$

23. **MULTIPLE CHOICE** Which of the following equations has a slope of −2 and passes through the point (2, 3)? *(Section 2.6)*

Ⓐ $y = -2x + 6$ Ⓑ $y - 3 = -2(x + 2)$ Ⓒ $y = -2x + 7$ Ⓓ $y - 2 = -2(x - 3)$

5.6 Arithmetic Sequences

Essential Question How are arithmetic sequences used to describe patterns?

COMMON CORE STATE STANDARDS
F.BF.2
F.IF.3
F.LE.2

1 ACTIVITY: Describing a Pattern

Work with a partner.

- **Use the figures to complete the table.**
- **Plot the points in your completed table.**
- **Describe the pattern of the y-values.**

a. $n = 1$ $n = 2$ $n = 3$ $n = 4$ $n = 5$

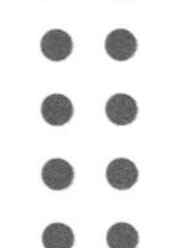

Number of Rows, n	1	2	3	4	5
Number of Dots, y					

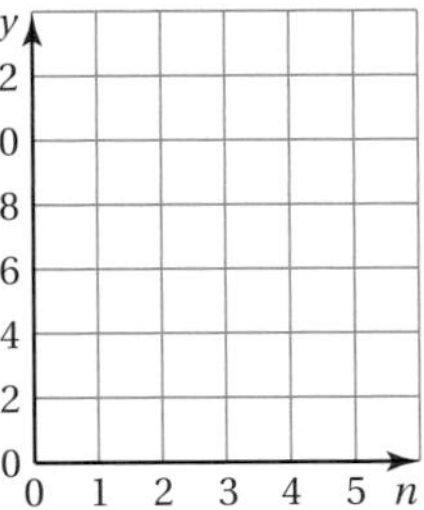

b. $n = 1$ $n = 2$ $n = 3$ $n = 4$ $n = 5$

Number of Stars, n	1	2	3	4	5
Number of Sides, y					

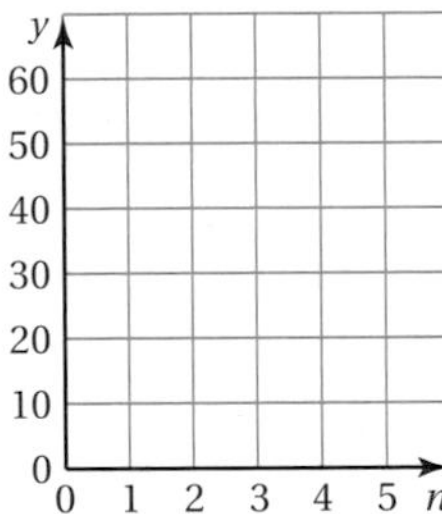

c. $n = 1$ $n = 2$ $n = 3$ $n = 4$ $n = 5$

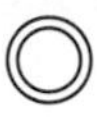

n	1	2	3	4	5
Number of Circles, y					

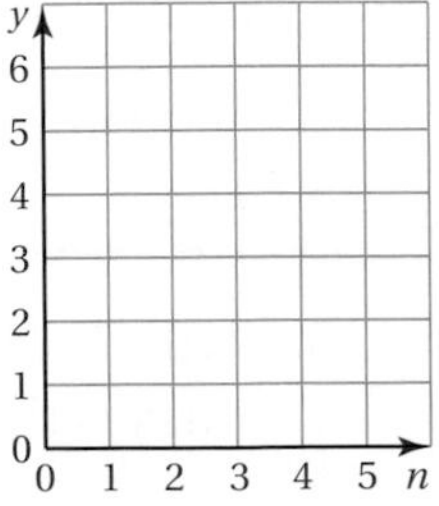

Laurie's Notes

Introduction

Standards for Mathematical Practice

- **MP4 Model with Mathematics** and **MP8 Look for and Express Regularity in Repeated Reasoning:** The goal is for students to recognize a numeric pattern in a sequence of numbers. Students have worked with ordered pairs using variables *x* and *y*. Today's investigation, however, uses the variables *n* and *y*, where *n* is the position of a term in a sequence and *y* is the value of the term. A sequence can have a pattern that uses one operation or a combination of operations. Students will use the position number and describe the numeric pattern modeled.

Motivate

- It's time to lead a class cheer! When students enter class, ask them to repeat after you, "Two, four, six, eight, we think math is really great!" Write on the board: 2, 4, 6, 8, . . .
- Now have them try a new cheer. "Three, five, seven, nine, we think math is really fine!" Write on the board: 3, 5, 7, 9, . . .
- Finally have them try one last cheer. "One, four, seven, ten, the year is about to end!" Write on the board: 1, 4, 7, 10, . . .

? "Are there any differences in the sequences I wrote?" One possible answer is that the first sequence is all evens, the second is all odds, and the third alternates odd then even. Students might also observe that the first two sequences are increasing by two and the third sequence is increasing by three.

? "Can all of the sequences be continued?" Yes

? "Name the next number in each sequence." 10; 11; 13

- Explain that today students will look at sequences of numbers and record the numbers in order.

Activity Notes

Activity 1

- Students are familiar with plotting ordered pairs. In this activity, they must first generate the ordered pairs. Students should recognize a pattern in their graphs, namely that the points lie on a line.
- **MP6 Attend to Precision:** Students should recognize that the scaling on the vertical axis of the graph is different than the scaling on the horizontal axis in parts (a) and (b).
- **Extension:** "What would the graph in part (a) look like if both axes were scaled by ones?" The graph would look steeper.

Common Core State Standards

F.BF.2 Write arithmetic sequences . . . with an explicit formula and use them to model situations . . ..

F.IF.3 Recognize that sequences are functions, sometimes defined recursively, whose domain is a subset of the integers.

F.LE.2 Construct linear functions, including arithmetic . . . sequences, given a graph, a description of a relationship, or two input-output pairs (include reading these from a table).

Previous Learning

Students should be familiar with collecting and organizing data.

Start Thinking! and Warm Up

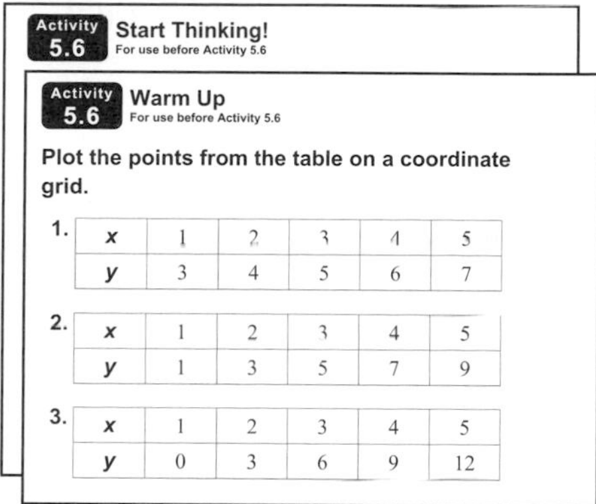

Activity 5.6 Start Thinking! For use before Activity 5.6

Activity 5.6 Warm Up For use before Activity 5.6

Plot the points from the table on a coordinate grid.

1.

x	1	2	3	4	5
y	3	4	5	6	7

2.

x	1	2	3	4	5
y	1	3	5	7	9

3.

x	1	2	3	4	5
y	0	3	6	9	12

5.6 Record and Practice Journal

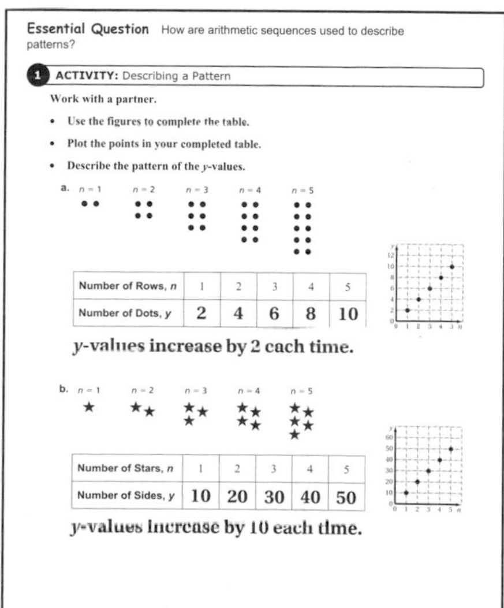

Essential Question How are arithmetic sequences used to describe patterns?

1 ACTIVITY: Describing a Pattern

Work with a partner.

- Use the figures to complete the table.
- Plot the points in your completed table.
- Describe the pattern of the *y*-values.

a.

Number of Rows, *n*	1	2	3	4	5
Number of Dots, *y*	2	4	6	8	10

y-values increase by 2 each time.

b.

Number of Stars, *n*	1	2	3	4	5
Number of Sides, *y*	10	20	30	40	50

y-values increase by 10 each time.

Differentiated Instruction

Kinesthetic

Use stacks of coins to explain the concept of a common difference to students. Model the arithmetic sequence 1, 3, 5, 7, . . . with four stacks of coins. Ask how many coins are needed to make the first stack as high as the second, the second stack as high as the third, and so on. Explain that the answer to each question is 2, and this is the common difference of the sequence.

5.6 Record and Practice Journal

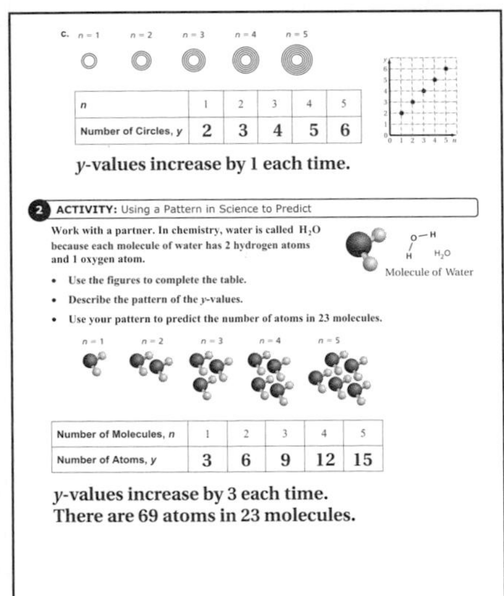

c. $n = 1$ $n = 2$ $n = 3$ $n = 4$ $n = 5$

n	1	2	3	4	5
Number of Circles, y	2	3	4	5	6

y-values increase by 1 each time.

2 ACTIVITY: Using a Pattern in Science to Predict

Work with a partner. In chemistry, water is called H_2O because each molecule of water has 2 hydrogen atoms and 1 oxygen atom.

Molecule of Water

- Use the figures to complete the table.
- Describe the pattern of the y-values.
- Use your pattern to predict the number of atoms in 23 molecules.

$n = 1$ $n = 2$ $n = 3$ $n = 4$ $n = 5$

Number of Molecules, n	1	2	3	4	5
Number of Atoms, y	3	6	9	12	15

y-values increase by 3 each time.
There are 69 atoms in 23 molecules.

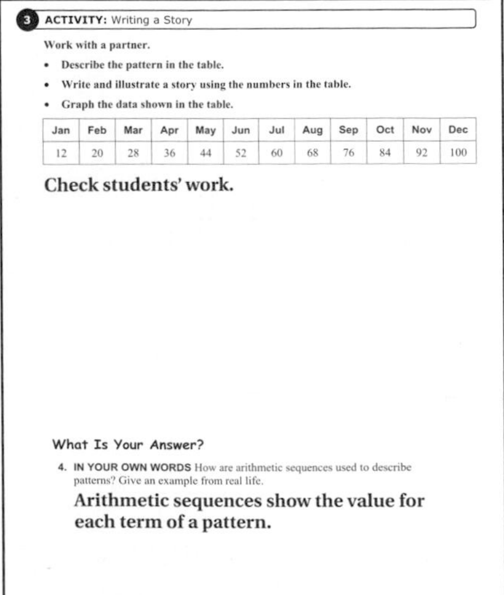

3 ACTIVITY: Writing a Story

Work with a partner.

- Describe the pattern in the table.
- Write and illustrate a story using the numbers in the table.
- Graph the data shown in the table.

Jan	Feb	Mar	Apr	May	Jun	Jul	Aug	Sep	Oct	Nov	Dec
12	20	28	36	44	52	60	68	76	84	92	100

Check students' work.

What Is Your Answer?

4. **IN YOUR OWN WORDS** How are arithmetic sequences used to describe patterns? Give an example from real life.

Arithmetic sequences show the value for each term of a pattern.

Laurie's Notes

Activity 2

- Discuss the molecular model. There are three atoms in the one molecule. There are two sizes and colors of spheres to denote the different atoms—hydrogen and oxygen.
- When students have finished, ask about their table of values.

? "What is the pattern of the y-values?" increasing by 3

? "How did recognizing the pattern help you determine the number of atoms in 23 molecules?" Listen for: multiply the number of molecules by 3

Activity 3

- Give time for partners to brainstorm with one another. Then give quiet time for each pair to write their stories.
- Ask several students to share their stories.
- **MP6:** Ask a volunteer to make a graph of the data. Be sure the student labels the axes with appropriate descriptions, i.e. months on the horizontal axis.

What Is Your Answer?

- Have students share their responses.

Closure

- Record and graph the first five values in one of the cheers from the start of class.

Technology For the Teacher

The Dynamic Planning Tool
Editable Teacher's Resources at *BigIdeasMath.com*

2 ACTIVITY: Using a Pattern in Science to Predict

Work with a partner. In chemistry, water is called H_2O because each molecule of water has 2 hydrogen atoms and 1 oxygen atom.

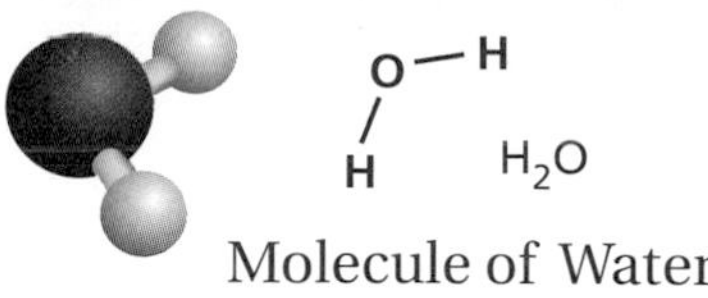

Molecule of Water

- **Use the figures to complete the table.**
- **Describe the pattern of the *y*-values.**
- **Use your pattern to predict the number of atoms in 23 molecules.**

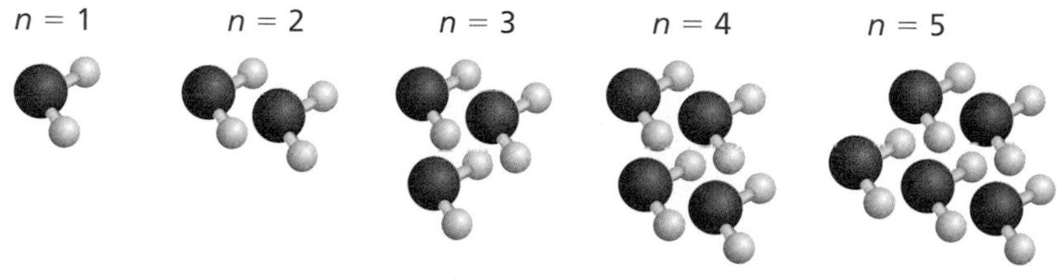

Number of Molecules, *n*	1	2	3	4	5
Number of Atoms, *y*					

3 ACTIVITY: Writing a Story

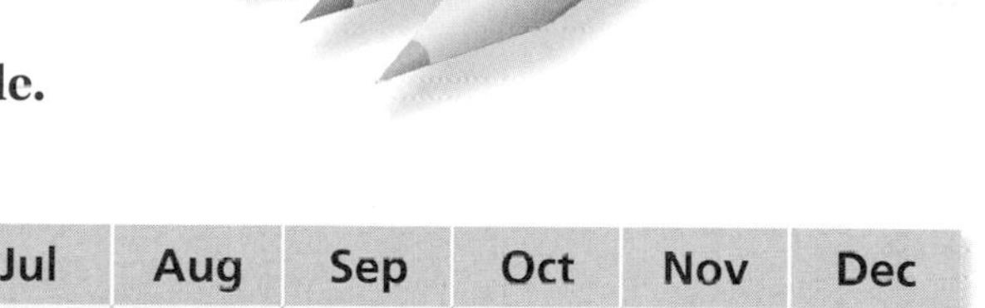

Work with a partner.

- **Describe the pattern in the table.**
- **Write and illustrate a story using the numbers in the table.**
- **Graph the data shown in the table.**

Jan	Feb	Mar	Apr	May	Jun	Jul	Aug	Sep	Oct	Nov	Dec
12	20	28	36	44	52	60	68	76	84	92	100

What Is Your Answer?

4. **IN YOUR OWN WORDS** How are arithmetic sequences used to describe patterns? Give an example from real life.

Practice Use what you learned about arithmetic sequences to complete Exercise 3 on page 247.

5.6 Lesson

Key Vocabulary
sequence, *p. 244*
term, *p. 244*
arithmetic sequence, *p. 244*
common difference, *p. 244*

A **sequence** is an ordered list of numbers. Each number in a sequence is called a **term.** Each term a_n has a specific position n in the sequence.

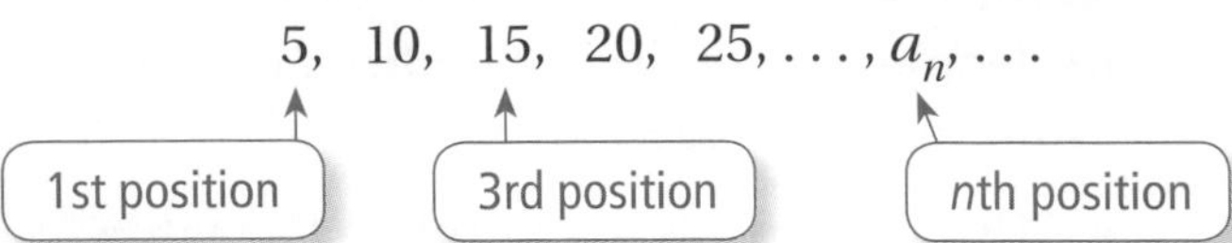

Key Idea

Arithmetic Sequence

In an **arithmetic sequence,** the difference between consecutive terms is the same. This difference is called the **common difference.** Each term is found by adding the common difference to the previous term.

5, 10, 15, 20, . . . Terms of an arithmetic sequence

+5 +5 +5 ← Common difference

EXAMPLE 1 Extending an Arithmetic Sequence

Write the next three terms of the arithmetic sequence −7, −14, −21, −28, ...

Use a table to organize the terms and find the pattern.

Position	1	2	3	4
Term	−7	−14	−21	−28

+(−7) +(−7) +(−7) ← Each term is 7 less than the previous term. So, the common difference is −7.

Add −7 to a term to find the next term.

Position	1	2	3	4	5	6	7
Term	−7	−14	−21	−28	−35	−42	−49

+(−7) +(−7) +(−7)

∴ The next three terms are −35, −42, and −49.

On Your Own

Write the next three terms of the arithmetic sequence.

1. −12, 0, 12, 24, . . . **2.** 0.2, 0.6, 1, 1.4, . . . **3.** 4, $3\frac{3}{4}$, $3\frac{1}{2}$, $3\frac{1}{4}$, . . .

Multi-Language Glossary at BigIdeasMath.com.

Laurie's Notes

Introduction

Connect

- **Yesterday:** Students explored arithmetic sequences. (MP4, MP6, MP8)
- **Today:** Students will describe and represent arithmetic sequences using an equation.

Motivate

- Before students arrive, write each number of an arithmetic sequence on separate index cards (Example: 13, 17, 21, 25, 29). Hand an index card to each of the first 5 students that enter the room.
- Tell the 5 students that they are holding numbers in an arithmetic sequence and to position themselves in order.
- Students may realize that the order could be increasing or decreasing.

  Version A: 13, 17, 21, 25, 29 **Version B:** 29, 25, 21, 17, 13

? "If 13 is in the 1st position, what number is in the 4th position?" 25
"What can you add to each number to get the next number?" 4

? "If 29 is in the 1st position, what number is in the 4th position?" 17
"What can you add to each number to get the next number?" −4

Discuss

- Discuss that arithmetic sequences can be increasing or decreasing. If the sequence is increasing, a positive number is added to each term. If the sequence is decreasing, a negative number is added to each term.
- **FYI:** Although subtracting 4 and adding −4 are equivalent, we will describe arithmetic sequences as an additive process; therefore this is an important concept to discuss with students.

Lesson Notes

Key Idea

- Write the Key Idea on the board, highlighting the vocabulary and the notation of using three dots (. . .) to indicate that the pattern continues.
- Connect the Key Idea to the index card motivation. The common difference was 4 in Version A and −4 in Version B.

Example 1

- Explain that a table is used to organize the terms in an arithmetic sequence because it helps in recognizing patterns.
- It is important for students to note that not all sequences of numbers are arithmetic. For example, the sequence 1, 2, 4, 7, 11, . . . has an observable pattern (add 1, add 2, add 3, add 4), but it is not arithmetic.

On Your Own

- These problems integrate review of integer, decimal, and fraction computation.

Goal Today's lesson is writing an equation to represent an **arithmetic sequence**.

Lesson Materials
Introduction
• index cards

Start Thinking! and Warm Up

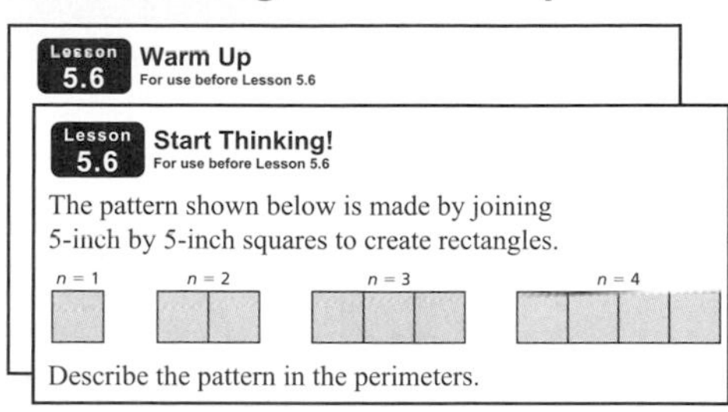

Extra Example 1

Write the next three terms of the arithmetic sequence 10, 7.5, 5, 2.5, ...

0, −2.5, −5

On Your Own

1. 36, 48, 60
2. 1.8, 2.2, 2.6
3. $3, 2\frac{3}{4}, 2\frac{1}{2}$

Extra Example 2

Graph the arithmetic sequence 5, 4, 3, 2, ... What do you notice?

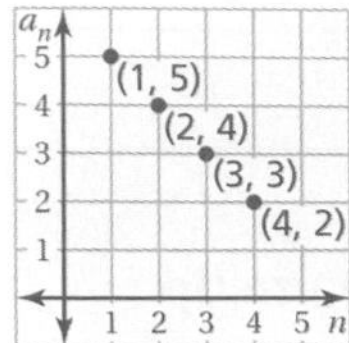

The points of the graph lie on a line.

On Your Own

4. 15, 18, 21

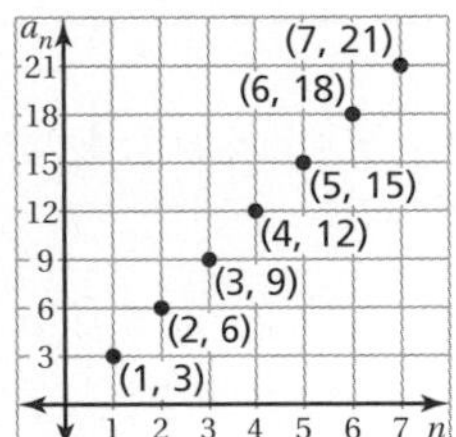

5. −4, −6, −8

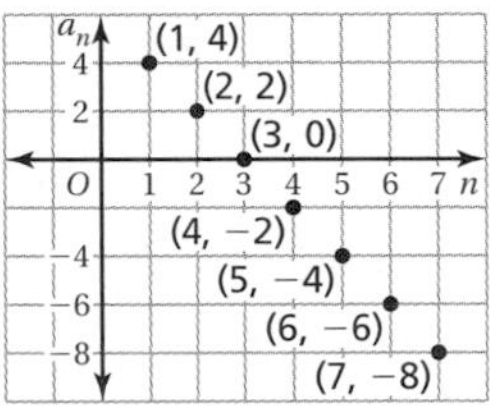

6. 0.2, 0, −0.2

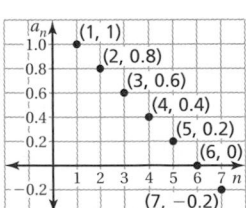

Laurie's Notes

Example 2

- **MP6 Attend to Precision:** Note that the axes are labeled n and a_n. These are the input and output variables. Point out that the scaling on the axes is different.
- **?** "What is the common difference in the terms?" 4
- **?** "How would you describe this sequence of numbers?" multiples of 4
- **?** "How would you describe the graph of the terms?" The points are on a line.
- **Big Idea:** The points on the graph of an arithmetic sequence lie on a line. The slope of the line is the common difference, or the constant rate of change.

On Your Own

- When students graph Question 6, encourage them to use a different scale for the vertical axis. It is okay to scale by tenths instead of ones.
- **?** "What is true about all three graphs?" They are all linear.

Discuss

- Students have extended an arithmetic sequence and graphed the terms. Now they are going to write an equation for the sequence.
- The common difference corresponds to a constant rate of change, or slope.
- One way for students to determine an equation for the sequence is to write consecutive terms in a horizontal fashion. Start by drawing 5 blanks for the first five terms of a sequence.

  _____, _____, _____, _____, _____

  The first term is a_1 so fill in the first blank. The common difference is d, so the second term will be $a_1 + d$.

  a_1, $a_1 + d$, _____, _____, _____

  Ask the students to fill in the next three blanks. The nth term will be $a_1 + (n - 1)d$. So, $a_n = a_1 + (n - 1)d$.
- An alternative approach is to use point-slope form using the point $(1, a_1)$ and the common difference d as the slope.

$$y - y_1 = m(x - x_1)$$
$$a_n - a_1 = d(n - 1)$$
$$a_n = a_1 + d(n - 1)$$
$$a_n = a_1 + (n - 1)d$$

Key Idea

- Now that the equation has been developed, describe it as the equation for an arithmetic sequence. Read the equation based on the meanings of the variables:
  "The nth term of the sequence is equal to the first term plus one less than the number of terms multiplied by the common difference."

EXAMPLE 2 Graphing an Arithmetic Sequence

Graph the arithmetic sequence 4, 8, 12, 16, ... What do you notice?

Make a table. Then plot the ordered pairs (n, a_n).

Position, n	Term, a_n
1	4
2	8
3	12
4	16

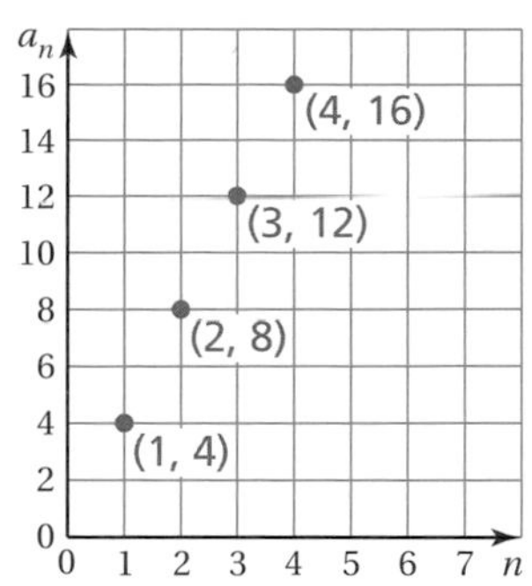

∴ The points of the graph lie on a line.

On Your Own

Now You're Ready
Exercises 25–28

Write the next three terms of the arithmetic sequence. Then graph the sequence.

4. 3, 6, 9, 12, . . .

5. 4, 2, 0, −2, . . .

6. 1, 0.8, 0.6, 0.4, . . .

Because consecutive terms of an arithmetic sequence have a common difference, the sequence has a constant rate of change. So, the points of any arithmetic sequence lie on a line. You can use the first term and the common difference to write a linear function that describes an arithmetic sequence.

Position, n	Term, a_n	Written using a_1 and d	Numbers
1	first term, a_1	a_1	4
2	second term, a_2	$a_1 + d$	$4 + 4 = 8$
3	third term, a_3	$a_1 + 2d$	$4 + 2(4) = 12$
4	fourth term, a_4	$a_1 + 3d$	$4 + 3(4) = 16$
⋮	⋮	⋮	⋮
n	nth term, a_n	$a_1 + (n-1)d$	$4 + (n-1)(4)$

Key Idea

Equation for an Arithmetic Sequence

Let a_n be the nth term of an arithmetic sequence with first term a_1 and common difference d. The nth term is given by

$$a_n = a_1 + (n-1)d.$$

EXAMPLE 3 Writing an Equation for an Arithmetic Sequence

Study Tip

Notice that the equation in Example 3 is of the form $y = mx + b$, where y is replaced by a_n and x is replaced by n.

Write an equation for the nth term of the arithmetic sequence 14, 11, 8, 5, ... Then find a_{50}.

The first term is 14 and the common difference is −3.

$a_n = a_1 + (n - 1)d$ — Equation for an arithmetic sequence

$a_n = 14 + (n - 1)(-3)$ — Substitute 14 for a_1 and −3 for d.

$a_n = -3n + 17$ — Simplify.

Use the equation to find the 50th term.

$a_n = -3n + 17$ — Write the equation.

$a_{50} = -3(50) + 17$ — Substitute 50 for n.

$= -133$ — Simplify.

EXAMPLE 4 Real-Life Application

Online bidding for a purse increases \$5 for each bid after the \$60 initial bid.

Bid Number	1	2	3
Bid Amount	\$60	\$65	\$70

a. Write an equation for the nth term of the arithmetic sequence.

The first term is 60 and the common difference is 5.

$a_n = a_1 + (n - 1)d$ — Equation for an arithmetic sequence

$a_n = 60 + (n - 1)5$ — Substitute 60 for a_1 and 5 for d.

$a_n = 5n + 55$ — Simplify.

b. The winning bid is \$90. How many bids were there?

Use the equation to find the value of n for which $a_n = 90$.

$a_n = 5n + 55$ — Write the equation.

$90 = 5n + 55$ — Substitute 90 for a_n.

$35 = 5n$ — Subtract 55 from each side.

$7 = n$ — Divide each side by 5.

∴ There were 7 bids.

Check

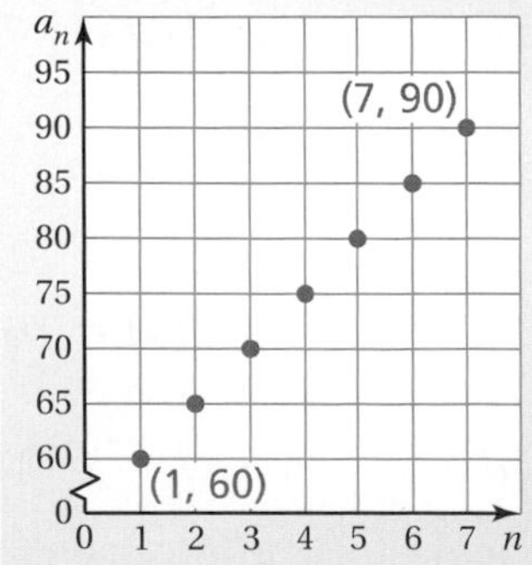

(Graph: a_n vs. n, points from (1, 60) to (7, 90))

On Your Own

Now You're Ready
Exercises 33–38

Write an equation for the nth term of the arithmetic sequence. Then find a_{25}.

7. 4, 5, 6, 7, . . .

8. 8, 16, 24, 32, . . .

9. −2, −1, 0, 1, . . .

10. **WHAT IF?** In Example 4, the winning bid is \$105. How many bids were there?

Laurie's Notes

Example 3

- Tell students to write the sequence 14, 11, 8, 5, ...
- ? "We want to write the equation for this sequence of numbers. What is the common difference?" −3
- ? "What is the first term?" 14
- Write an equation for the arithmetic sequence, make the substitutions, and simplify.
- Be sure to discuss the Study Tip.

Example 4

- In this example, students should be able to find the common difference and state the first term. Writing the equation should be familiar from their work in Chapter 2.
- Students need to be careful with the Distributive Property.
- Students generally find that the process makes sense. They need to work slowly and not get bogged down by the symbols. It is important to continually use the vocabulary "the first term" instead of "a_1" and "the common difference" instead of "d."
- **Extension:** Enter the sequence in a graphing calculator. Make a table of values and a graph.

On Your Own

- Ask students what part of the solution of Example 4 they will use to answer Question 10.

Closure

- Which of the following are arithmetic sequences? For those that are, identify the common difference and write an equation for the nth term of the arithmetic sequence.
  - **a.** −1, −2, −3, −4, . . . yes; −1; $a_n = -n$
  - **b.** 8, 10, 13, 17, . . . no
  - **c.** 8, 4, 2, 1, . . . no

Technology For the Teacher

The Dynamic Planning Tool
Editable Teacher's Resources at *BigIdeasMath.com*

English Language Learners

Group Activity
Give groups of students index cards with arithmetic sequences and instruct them to work as a group to write an equation for the nth term of the arithmetic sequence. This will give English language learners a chance to interact with classmates and communicate mathematically.

Extra Example 3

Write an equation for the nth term of the arithmetic sequence 96, 92, 88, 84, ... Then find a_{14}. $a_n = -4n + 100$; $a_{14} = 44$

Extra Example 4

In Example 4, the winning bid is 185. How many bids were there? 26

On Your Own

7. $a_n = n + 3$; $a_{25} = 28$
8. $a_n = 8n$; $a_{25} = 200$
9. $a_n = n - 3$; $a_{25} = 22$
10. 10 bids

Vocabulary and Concept Check

1. subtract a term from the following term
2. The graphs are similar in that they are both linear. The graphs are different in that the graph of a linear function is a line, whereas the graph of an arithmetic sequence consists of points.

Practice and Problem Solving

3.

n	1	2	3	4
y	25	50	75	100

As n increases by 1, y increases by 25.

4. 13, 24, 35
5. 21.5, 25, 28.5
6. $4\frac{1}{2}$, 9, $13\frac{1}{2}$
7. 5
8. -2
9. 25
10. $\frac{1}{2}$
11. -1.5
12. 150
13. 22, 25, 28
14. 45, 56, 67
15. 36, 41, 46
16. $-60, -90, -120$
17. $0.1, -0.2, -0.5$
18. $\frac{1}{6}, 0, -\frac{1}{6}$
19. 4, 6, 8, 10, . . .; yes; There is a common difference of 2.
20. yes; 13
21. no
22. no
23. yes; 6
24. The difference between a term and its previous term is -1, not 1. So, the common difference is -1.

Assignment Guide and Homework Check

Level	Assignment	Homework Check
Average	1, 2, 7–17 odd, 24, 25, 29, 33–39 odd, 46–49	13, 24, 29, 33
Advanced	1, 2, 14–30 even, 33–38, 41, 46–49	14, 24, 30, 41

Common Errors

- **Exercises 4–6** Students may subtract the common difference instead of adding to get the next term. Remind them that the common difference is the amount you add to one term to get the next term.
- **Exercises 7–12** Students may subtract a term from the previous term to find the common difference. Explain that because the common difference is the amount you add to one term to get the next term, subtract a term from the next term to find the common difference.

5.6 Record and Practice Journal

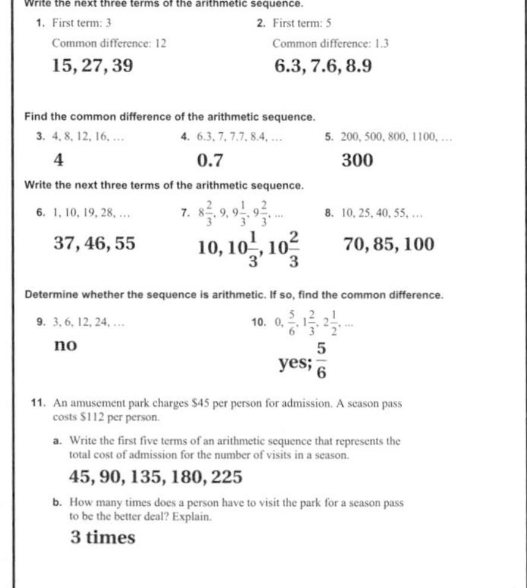

Write the next three terms of the arithmetic sequence.

1. First term: 3
   Common difference: 12
   15, 27, 39
2. First term: 5
   Common difference: 1.3
   6.3, 7.6, 8.9

Find the common difference of the arithmetic sequence.

3. 4, 8, 12, 16, … **4**
4. 6.3, 7, 7.7, 8.4, … **0.7**
5. 200, 500, 800, 1100, … **300**

Write the next three terms of the arithmetic sequence.

6. 1, 10, 19, 28, … **37, 46, 55**
7. $8\frac{2}{3}, 9, 9\frac{1}{3}, 9\frac{2}{3}, \ldots$ **$10, 10\frac{1}{3}, 10\frac{2}{3}$**
8. 10, 25, 40, 55, … **70, 85, 100**

Determine whether the sequence is arithmetic. If so, find the common difference.

9. 3, 6, 12, 24, … **no**
10. $0, \frac{5}{6}, 1\frac{2}{3}, 2\frac{1}{2}, \ldots$ **yes; $\frac{5}{6}$**

11. An amusement park charges $45 per person for admission. A season pass costs $112 per person.
    a. Write the first five terms of an arithmetic sequence that represents the total cost of admission for the number of visits in a season.
    45, 90, 135, 180, 225
    b. How many times does a person have to visit the park for a season pass to be the better deal? Explain.
    3 times

Technology For the Teacher
Answer Presentation Tool

5.6 Exercises

Vocabulary and Concept Check

1. **VOCABULARY** How do you find the common difference of an arithmetic sequence?
2. **WRITING** How are the graphs of arithmetic sequences and linear functions similar? How are they different?

Practice and Problem Solving

Use the figures to complete the table. Then describe the pattern of the y-values.

3. $n = 1$  $n = 2$ 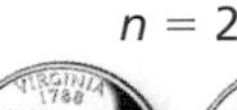  $n = 3$ 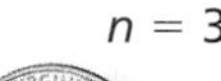  $n = 4$   

Number of Quarters, n	1	2	3	4
Number of Cents, y				

Write the next three terms of the arithmetic sequence.

4. First term: 2
   Common difference: 11
5. First term: 18
   Common difference: 3.5
6. First term: 0
   Common difference: $4\frac{1}{2}$

Find the common difference of the arithmetic sequence.

7. 5, 10, 15, 20, . . .
8. 16.1, 14.1, 12.1, 10.1, . . .
9. 100, 125, 150, 175, . . .
10. $3, 3\frac{1}{2}, 4, 4\frac{1}{2}, \ldots$
11. 6.5, 5, 3.5, 2, . . .
12. 350, 500, 650, 800, . . .

Write the next three terms of the arithmetic sequence.

13. 10, 13, 16, 19, . . .
14. 1, 12, 23, 34, . . .
15. 16, 21, 26, 31, . . .
16. 60, 30, 0, −30, . . .
17. 1.3, 1, 0.7, 0.4, . . .
18. $\frac{5}{6}, \frac{2}{3}, \frac{1}{2}, \frac{1}{3}, \ldots$

19. **PATTERN** Write a sequence to represent the number of smiley faces in each group. Is the sequence arithmetic? Explain.

Determine whether the sequence is arithmetic. If so, find the common difference.

20. 13, 26, 39, 52, . . .

21. 5, 9, 14, 20, . . .

22. 6, 12, 24, 48, . . .

23. 69, 75, 81, 87, . . .

24. ERROR ANALYSIS Describe and correct the error in finding the common difference of the arithmetic sequence.

> ✗ 2, 1, 0, −1, . . .
> +1 +1 +1
> The common difference is 1.

Write the next three terms of the arithmetic sequence. Then graph the sequence.

2 **25.** 7, 6.4, 5.8, 5.2, . . .

26. −15, 0, 15, 30, . . .

27. $\frac{1}{2}, \frac{5}{8}, \frac{3}{4}, \frac{7}{8}, \ldots$

28. −1, −3, −5, −7, . . .

29. NUMBER SENSE The first term of an arithmetic sequence is 3. The common difference of the sequence is 1.5 times the first term. Write the next three terms of the sequence. Then graph the sequence.

30. DOMINOES The first row of a dominoes display has 10 dominoes. Each row after the first has two more dominoes than the row before it. Write the first five terms of the sequence that represents the number of dominoes in each row. Then graph the sequence.

31. ZOO A zoo charges \$8 per person for admission.

a. Copy and complete the table.

b. Do the costs in your table show an arithmetic sequence? If so, graph the sequence.

c. What is the cost for one person to visit the zoo six times?

d. An annual family pass costs \$130. How many times does a family of five have to visit the zoo for the annual pass to be the better deal? Explain.

Number of Visits in One Year	Cost
1	\$8
2	
3	
4	

32. REPEATED REASONING Firewood is stacked in a pile. The bottom row has 20 logs and the top row has 14 logs. Each row has one more log than the row above it. How many logs are in the pile?

Write an equation for the *n*th term of the arithmetic sequence. Then find a_{10}.

3 **33.** −5, −4, −3, −2, . . .

34. −3, −6, −9, −12, . . .

35. $\frac{1}{2}, 1, 1\frac{1}{2}, 2, \ldots$

36. 10, 11, 12, 13, . . .

37. −10, −20, −30, −40, . . .

38. $\frac{1}{7}, \frac{2}{7}, \frac{3}{7}, \frac{4}{7}, \ldots$

Common Errors

- **Exercises 21 and 22** Students may call it an arithmetic sequence because they see a pattern. Remind them that only a sequence in which there is a common difference between terms is arithmetic.
- **Exercises 25–28** Students may use something other than the integers 1 through 7 as the domain values for the plotted points. Remind them that the domain consists of the positions of the terms, numbered from 1 to *n*, the number of terms.
- **Exercise 29** Students may use 1.5 as the common difference. Tell them to read the problem carefully.
- **Exercises 33–38** Students may use the form $a_n = a_1 + nd$. This omits the first term and puts each term in the wrong position. Explain that they must use the formula on page 245 and solve for a_n.
- **Exercise 43** Students may think that an unending sequence is continuous. Remind them that a continuous domain consists of all numbers in an interval.
- **Exercise 45** Students may think the domain is continuous because the units are minutes. Remind them that because the domain is for a sequence, they only use certain values.

Practice and Problem Solving

25–30. See Additional Answers for graphs.

25. 4.6, 4, 3.4

26. 45, 60, 75

27. $1, 1\frac{1}{8}, 1\frac{1}{4}$

28. −9, −11, −13

29. 7.5, 12, 16.5

30. 10, 12, 14, 16, 18

31. a.

Number of Visits in One Year	Cost
1	\$8
2	\$16
3	\$24
4	\$32

b. yes;

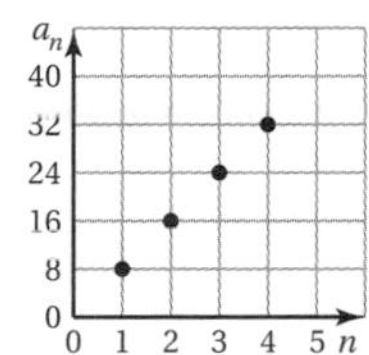

c. \$48

d. 4 times; The cost for the family to visit the zoo 3 times is \$120, which is less than the \$130 pass. The cost for the family to visit the zoo 4 times is \$160, which is more than the \$130 pass.

32. 119 logs

33. $a_n = n - 6; a_{10} = 4$

34. $a_n = -3n; a_{10} = -30$

35. $a_n = \frac{1}{2}n; a_{10} = 5$

36. $a_n - n + 9; a_{10} = 19$

37. $a_n = -10n; a_{10} = -100$

38. $a_n = \frac{1}{7}n; a_{10} = \frac{10}{7}$

Practice and Problem Solving

39. $a_n = -20n + 120$

40. dependent; Each term depends on its position.

41. See *Taking Math Deeper.*

42. *Sample answer:* $-7, -10, -13, -16, \ldots, a_n = -3n - 4$; $20, 17, 14, 11, \ldots, a_n = 23 - 3n$

43. discrete; A sequence of n terms has a domain of the integers 1 through n.

44. **a.** $a_n = 2n + 10$

b. They are equivalent.

c. 28 trees

45. **a.** $a_n = 5n$

b. discrete

c. Substitute the number of minutes in a day, 1440, for n.

Fair Game Review

46. $(-2, -4)$ **47.** $(4, -2)$

48. $(4, -4)$ **49.** D

Mini-Assessment

1. Write the next three terms of the arithmetic sequence 0, 1, 2, 3, ...
   4, 5, 6
2. Write an equation for the *n*th term of the arithmetic sequence $-3, -2, -1, 0, \ldots$.
   $y = n - 4$
3. Graph the arithmetic sequence 1.5, 3, 4.5, 6, ... What do you notice?

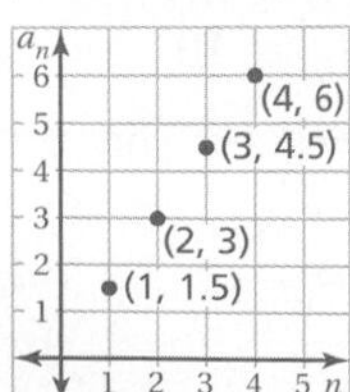

The points of the graph lie on a line.

Taking Math Deeper

Exercise 41

This problem can be used to introduce one of the most famous concepts in the history of mathematics: the difference between velocity (speed) and acceleration. This difference is a key concept in the study of calculus.

Understand the problem. Make a table.

a.

Position, n	Speed, a_n	Rule
1	32	$31 + 1(1) = 32$
2	33	$31 + 2(1) = 33$
3	34	$31 + 3(1) = 34$
4	35	$31 + 4(1) = 35$
n	a_n	$31 + n(1) = a_n$

Alternatively, you could use time instead of position. Time is the number of seconds *elapsed* from the present time. It is common (in precalculus and calculus) to represent the current time by $t = 0$. The entries in the first column would then be 0, 1, 2, 3, and t.

Write an equation for the *n*th term of the arithmetic sequence. Let n represent the position in the sequence and let a_n represent the speed.

b. $a_n = 31 + n$

c. Assuming you continue speeding up in the same pattern until you reach the speed limit, the 34th term is 65. So, the domain consists of the integers from 1 to 34.

The equation would be different using time instead of position. Let t represent the elapsed time. The equation would then be $a_t = 32 + t$.

It would be good for students to learn some terminology connected with this type of motion. Because the car's *velocity* is increasing, we say that the car is *accelerating.*

Reteaching and Enrichment Strategies

If students need help. . .	If students got it. . .
Resources by Chapter • Practice A and Practice B • Puzzle Time Record and Practice Journal Practice Differentiating the Lesson Lesson Tutorials Skills Review Handbook	Resources by Chapter • Enrichment and Extension • School-to-Work • Financial Literacy • Technology Connection Start the next section

39. **MOVIE REVENUE** A movie earns \$100 million the first week it is released. The movie earns \$20 million less each additional week. Write an equation for the nth term of the arithmetic sequence.

40. **REASONING** Are the terms of an arithmetic sequence independent or dependent? Explain your reasoning.

41. **SPEED** On a highway, you take 3 seconds to increase your speed from 32 to 35 miles per hour. Your speed increases the same amount each second.

    a. Write the first four terms of the sequence that represents your speed each second.
    b. Write an equation that describes the arithmetic sequence.
    c. The speed limit is 65 miles per hour. What is the domain of the function?

42. **OPEN-ENDED** Write the first four terms of two different arithmetic sequences with a common difference of -3. Write an equation for the nth term of each sequence.

43. **REASONING** Is the domain of an arithmetic sequence discrete or continuous? Describe the types of numbers in the domain.

44. **EARTH DAY** You and a group of friends take turns planting 2 trees each at a campsite. After the first person plants 2 trees, there are 12 trees at the campsite.

    a. Write an equation for the nth term of the sequence.
    b. What do you notice about the slope given by the equation and the common difference of the sequence?
    c. After 8 more people plant trees, how many trees are at the campsite?

45. 

The number of births in a country each minute after midnight January 1st can be estimated by the sequence in the table.

Minutes after Midnight January 1st	1	2	3	4
Babies Born	5	10	15	20

    a. Write an equation for the nth term of the sequence.
    b. Is the domain discrete or continuous?
    c. Explain how to use your function to estimate the number of births in a day.

Fair Game Review What you learned in previous grades & lessons

Solve the system of linear equations by graphing. *(Section 4.1)*

46. $y = 2x$

    $y = 3x + 2$

47. $y = -2x + 6$

    $y = \frac{1}{4}x - 3$

48. $y + x = 0$

    $y + 2 = -\frac{1}{2}x$

49. **MULTIPLE CHOICE** What expression is equivalent to 4^5? *(Skills Review Handbook)*

    (A) $4 \cdot 5$ (B) $4 \cdot 4 \cdot 4 \cdot 4$ (C) 5^4 (D) $4 \cdot 4 \cdot 4 \cdot 4 \cdot 4$

5.4–5.6 Quiz

Evaluate the function when $x = -4, 0,$ and 2. *(Section 5.4)*

1. $f(x) = x - 2$

2. $g(x) = 7x + 3$

3. $h(x) = -\frac{1}{4}x + 5$

Graph the function. Compare the graph to the graph of $f(x) = 4x$. *(Section 5.4)*

4. $g(x) = 4x + 1$

5. $h(x) = 4x - 2$

6. $n(x) = 4x - 6$

Graph the function. Compare the graph to the graph of $y = |x|$. Describe the domain and range. *(Section 5.4)*

7. $y = |x| + 2$

8. $y = |x - 6|$

9. $y = 2|x|$

Does the table or graph represent a *linear* or *nonlinear* function? Explain. *(Section 5.5)*

10.

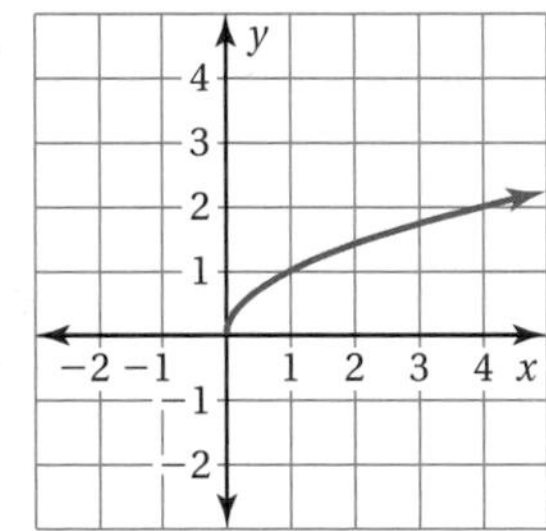

11.

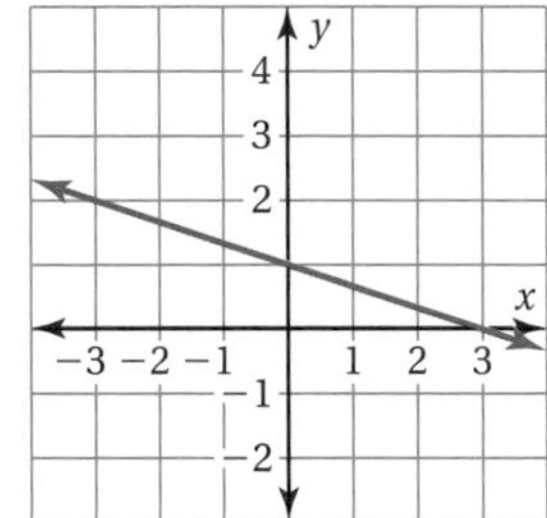

12.

x	y
0	3
3	0
6	3
9	6

Write an equation for the *n*th term of the arithmetic sequence. Then find a_{15}. *(Section 5.6)*

13. 5, 6, 7, 8, . . .

14. −3, −2, −1, 0, . . .

15. 4, 8, 12, 16, . . .

16. −1.5, −0.5, 0.5, 1.5, . . .

17. HIGH-SPEED RAIL A high-speed passenger train travels at 110 miles per hour. The function $d(x) = 1375 - 110x$ represents the distance (in miles) the train is from its destination after x hours. How far is the train from its destination after 8 hours? *(Section 5.4)*

18. CHICKEN SALAD The equation $y = 7.9x$ represents the cost y (in dollars) of buying x pounds of chicken salad. Does this equation represent a linear or nonlinear function? Explain. *(Section 5.5)*

19. PHONE BILL The table shows your phone bill for each minute over your plan limit. *(Section 5.6)*

Extra Minute	1	2	3
Phone Bill	\$40.40	\$40.85	\$41.30

a. Write an equation for the *n*th term of the arithmetic sequence.

b. Your phone bill is \$45.35. How many extra minutes were billed to your account?

Alternative Assessment Options

Math Chat
Student Reflective Focus Question
Structured Interview
Writing Prompt

Writing Prompt

Ask students to write two different stories. One story should involve data whose domain is continuous and the other story should involve data whose domain is discrete. Both sets of data should be linear. Students should graph their data and write linear functions that relate *y* to *x*. They should include a summary in each story describing how they know whether the domains are discrete or continuous. Students can share their stories and summaries with the class.

Study Help Sample Answers

4.

	Linear Functions	Nonlinear Functions
Definition/Description	A *linear function* is a function whose graph is a line. The graph of a linear function shows a constant rate of change.	A *nonlinear function* is a function whose graph is *not* a line. A nonlinear function does not have a constant rate of change.
Equations	$y = \frac{1}{2}x - 3$ $3x + 2y = 10$ $y = -x$	$y = x^2$ $y = \frac{1}{x}$ $y = \lvert x \rvert$
Tables	x: −2, −1, 0, 1, 2; y: −6, −3, 0, 3, 6 x: −2, −1, 0, 1, 2; y: 1, 1, 1, 1, 1	x: −2, −1, 0, 1, 2; y: 4, 1, 0, 1, 4 x: −2, −1, 0, 1, 2; y: 2, 1, 0, 1, 2
Graphs		

Reteaching and Enrichment Strategies

If students need help...	If students got it...
Resources by Chapter • Study Help • Practice A and Practice B • Puzzle Time Lesson Tutorials *BigIdeasMath.com* Practice Quiz Practice from the Test Generator	Resources by Chapter • Enrichment and Extension • School-to-Work Game Closet at *BigIdeasMath.com* Start the Chapter Review

Answers

1. $f(-4) = -6; f(0) = -2; f(2) = 0$
2. $g(-4) = -25; g(0) = 3; g(2) = 17$
3. $h(-4) = 6; h(0) = 5; h(2) = 4\frac{1}{2}$

4–9. See Additional Answers.

10. nonlinear; The graph is not a line.
11. linear; The graph is a line.
12. nonlinear; As *x* increases by 3, *y* changes by different amounts.
13. $a_n = n + 4; a_{15} = 19$
14. $a_n = n - 4; a_{15} = 11$
15. $a_n = 4n; a_{15} = 60$
16. $a_n = n - 2.5; a_{15} = 12.5$
17. 495 mi
18. linear; You can rewrite the equation in slope-intercept form.
19. **a.** $a_n = 0.45n + 39.95$

    b. 12 minutes

Technology For the Teacher
Answer Presentation Tool

Assessment Book

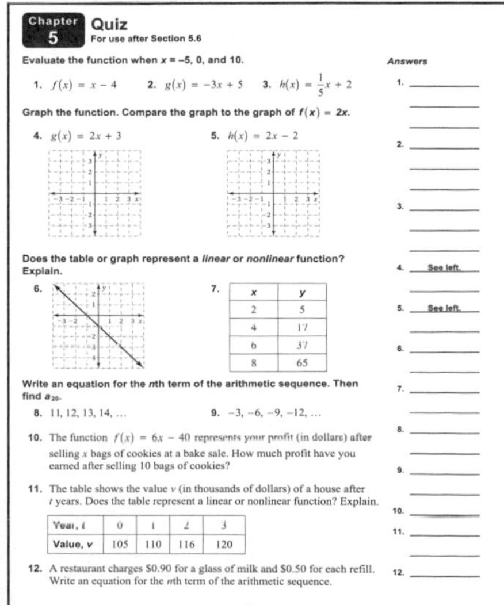

Chapter 5 Quiz
For use after Section 5.6

Evaluate the function when $x = -5$, 0, and 10.

1. $f(x) = x - 4$
2. $g(x) = -3x + 5$
3. $h(x) = \frac{1}{5}x + 2$

Graph the function. Compare the graph to the graph of $f(x) = 2x$.

4. $g(x) = 2x + 3$
5. $h(x) = 2x - 2$

Does the table or graph represent a *linear* or *nonlinear* function? Explain.

6.

7.

x	y
2	5
4	17
6	37
8	65

Write an equation for the *n*th term of the arithmetic sequence. Then find a_{20}.

8. 11, 12, 13, 14, ...
9. −3, −6, −9, −12, ...
10. The function $f(x) = 6x - 40$ represents your profit (in dollars) after selling *x* bags of cookies at a bake sale. How much profit have you earned after selling 10 bags of cookies?
11. The table shows the value *v* (in thousands of dollars) of a house after *t* years. Does the table represent a linear or nonlinear function? Explain.

Year, *t*	0	1	2	3
Value, *v*	105	110	116	120

12. A restaurant charges $0.90 for a glass of milk and $0.50 for each refill. Write an equation for the *n*th term of the arithmetic sequence.

Answers
1. ______
2. ______
3. ______
4. See left.
5. See left.
6. ______
7. ______
8. ______
9. ______
10. ______
11. ______
12. ______

For the Teacher
Additional Review Options

- Big Ideas Test Generator
- Game Closet at *BigIdeasMath.com*
- Vocabulary Puzzle Builder
- Resources by Chapter
  Puzzle Time
  Study Help

Answers

1. domain: $-5, -4, -3, -2$
   range: $3, 1, -1, -3$
2. domain: $-2, -1, 0, 1, 2$
   range: $-4, -2, 0$

Review of Common Errors

Exercises 1 and 2

- Students may confuse the domain and range. Remind them that the domain is the set of all possible input values (the *x*-coordinates) and the range is the set of all possible output values (the *y*-coordinates).

Exercises 3 and 4

- Students may see decimal numbers in a table and think that the function is continuous, or see whole numbers in a table and think that the function is discrete. Encourage them to think about the context of the problem. Point out that you can drive for part of an hour or part of a mile, but you cannot buy part of a stamp.
- Students may graph the function incorrectly. Remind them that discrete data points are not connected, but continuous data points are.

Exercise 5

- Students may try to write the linear function without first finding the slope and *y*-intercept, or they may use the reciprocal of the slope in their function. Encourage them to check their work by making sure that all of the given points are solutions.

Exercise 6

- Students may notice that all of the values of *y* are the same and not be able to write the linear function. Encourage them to plot the points, and if necessary, remind them how to write the equation of a horizontal line.

Exercises 7–9

- Students may confuse the signs when they substitute for *x*, especially when they skip steps. Encourage students to take the time to carry out all the steps.

Exercises 10 and 11

- Students may describe the translation incorrectly.

Exercises 12 and 13

- Students may guess their answer, or they may think that because the *x*-values are increasing by the same amount, the function is linear. Encourage them to examine the *y*-values or to plot the given points so that they can tell if the table represents a linear or nonlinear function.

Exercises 14–16

- Students may use the form $a_n = a_1 + nd$. This omits the first term and puts each term in the wrong position. Explain that they must use the formula on page 245 and solve for a_n.

5 Chapter Review

Review Key Vocabulary

function, *p. 204*
domain, *p. 204*
range, *p. 204*
independent variable, *p. 204*
dependent variable, *p. 204*
relation, *p. 208*
Vertical Line Test, *p. 209*
discrete domain, *p. 212*
continuous domain, *p. 212*
linear function, *p. 218*
function notation, *p. 226*
piecewise function, *p. 232*
step function, *p. 233*
absolute value function, *p. 234*
nonlinear function, *p. 238*
sequence, *p. 244*
term, *p. 244*
arithmetic sequence, *p. 244*
common difference, *p. 244*

Review Examples and Exercises

5.1 Domain and Range of a Function *(pp. 202–209)*

Find the domain and range of the function represented by the graph.

Write the ordered pairs. Identify the inputs and outputs.

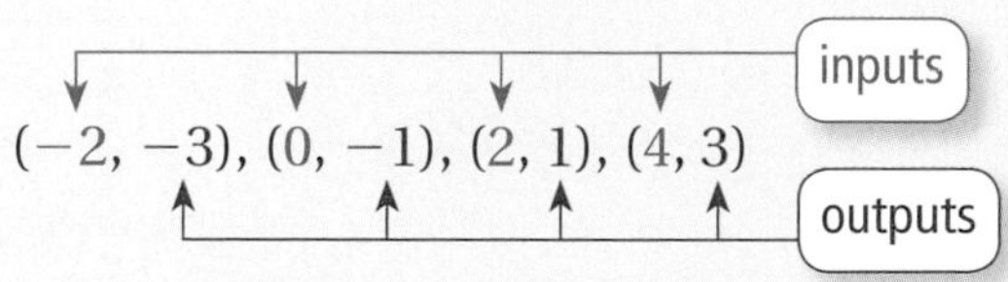

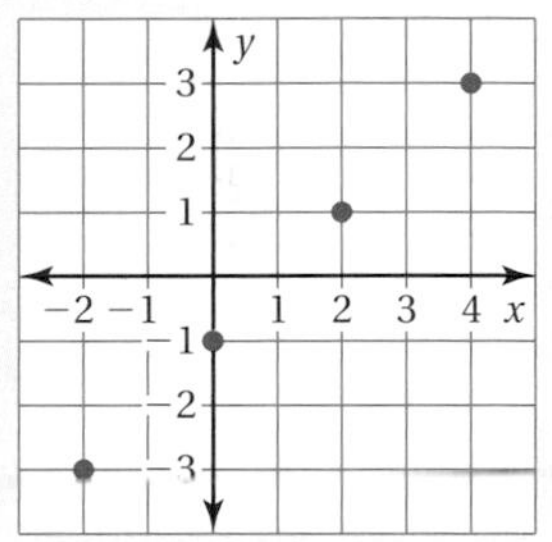

∴ The domain is −2, 0, 2, and 4.
The range is −3, −1, 1, and 3.

Exercises

Find the domain and range of the function represented by the graph.

1. 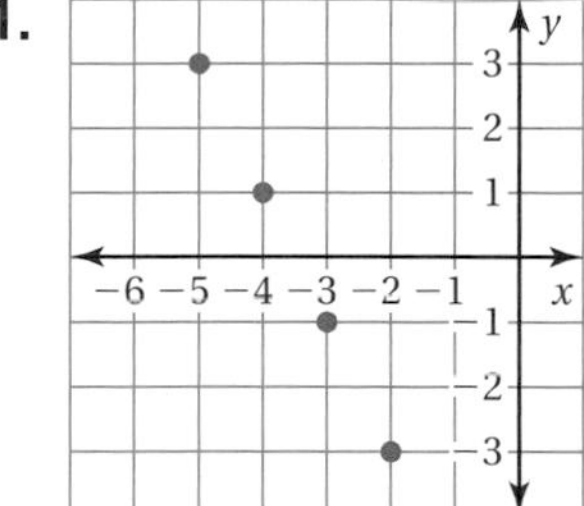

2. 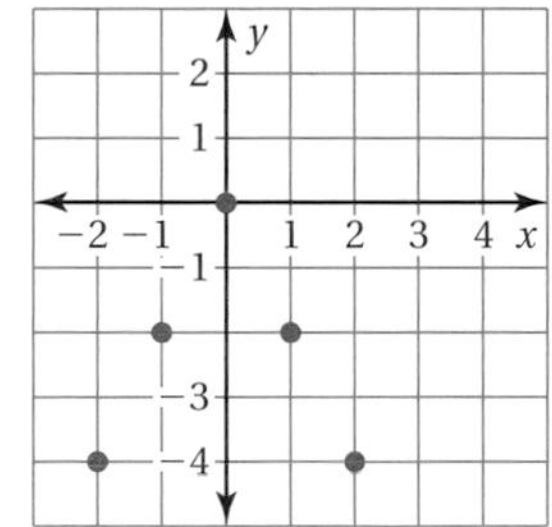

5.2 Discrete and Continuous Domains *(pp. 210–215)*

A yearbook costs \$19.50. The graph shows the cost *y* of *x* yearbooks. Is the domain discrete or continuous?

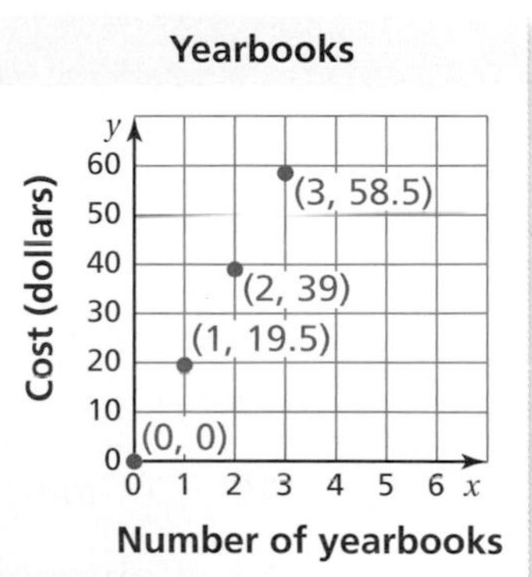

Because you cannot buy part of a yearbook, the graph consists of individual points.

∴ So, the domain is discrete.

Exercises

Graph the function. Is the domain discrete or continuous?

3.

Hours, x	0	1	2	3	4
Miles, y	0	4	8	12	16

4.

Stamps, x	20	40	60	80	100
Cost, y	8.4	16.8	25.2	33.6	42

5.3 Linear Function Patterns (pp. 216–221)

Use the graph to write a linear function that relates y to x.

The points lie on a line. Find the slope and y-intercept of the line.

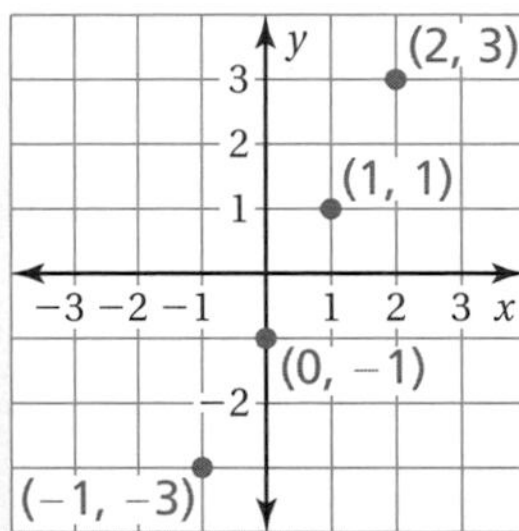

$$\text{slope} = \frac{\text{change in } y}{\text{change in } x} = \frac{3-1}{2-1} = \frac{2}{1} = 2$$

Because the line crosses the y-axis at $(0, -1)$, the y-intercept is -1.

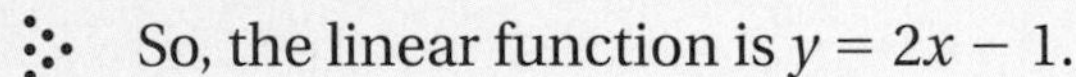

∴ So, the linear function is $y = 2x - 1$.

Exercises

Use the graph or table to write a linear function that relates y to x.

5.

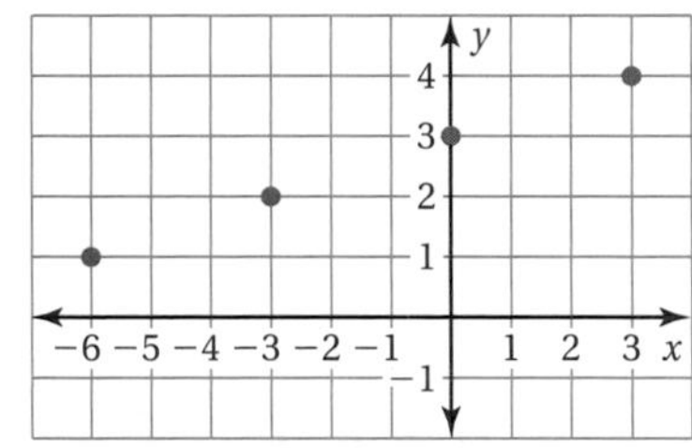

6.

x	−2	0	2	4
y	−7	−7	−7	−7

5.4 Function Notation (pp. 224–235)

Evaluate $f(x) = 3x - 20$ when $x = 4$.

$f(x) = 3x - 20$ — Write the function.

$f(4) = 3(4) - 20$ — Substitute 4 for x.

$= -8$ — Simplify.

Exercises

Evaluate the function when $x = -5$, 0, and 2.

7. $f(x) = 5x + 12$
8. $g(x) = -1.5x - 1$
9. $h(x) = 7 - 3x$
10. Compare the graph of $f(x) = -3x - 1$ to the graph of $g(x) = -3x$.
11. Compare the graph of $y = |x| + 1$ to the graph of $y = |x|$.

Review Game

Writing Linear Functions

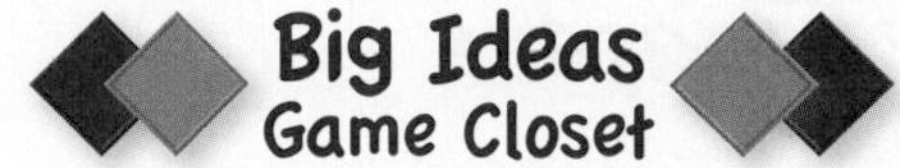

Materials per Group:

- paper
- two yard sticks
- pencils

Directions:

- Divide the class into an even number of groups.
- Groups pair up to compete against each other.
- Each pair of groups makes a paper football.
- Students in each pair of groups take turns flicking the football with their fingers as high and as far as they can. Students from the other group measure and record the length and height that the football travels. Both groups have to agree on the measurements. Length can be measured after the football has come to rest, but height must be measured while it is moving.
- Each student writes the domain and range of both the ascent and descent of the football when they took their turn. The domain is the length traveled and the range is the height traveled. (See figure below.) Students write one linear function to approximate the ascent and another linear function to approximate the descent.

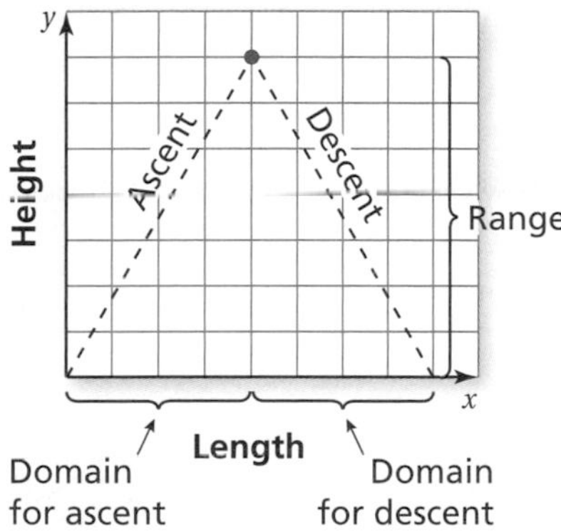

Who Wins?

Each student earns their group a point for each inch achieved in length and height. Points only count if the linear functions correctly model the motion of the football and the domains and ranges are clearly identified. The group with the most points wins.

For the Student Additional Practice

- Lesson Tutorials
- Study Help (textbook)
- Student Website
  - Multi-Language Glossary
  - Practice Assessments

Answers

3. 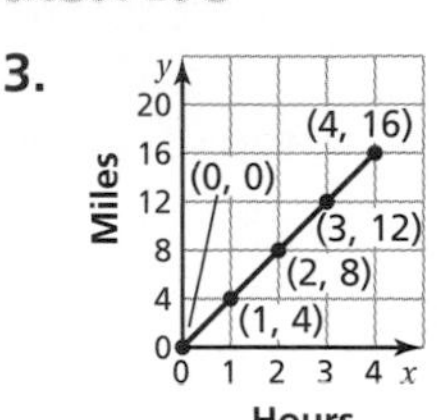

continuous

4. 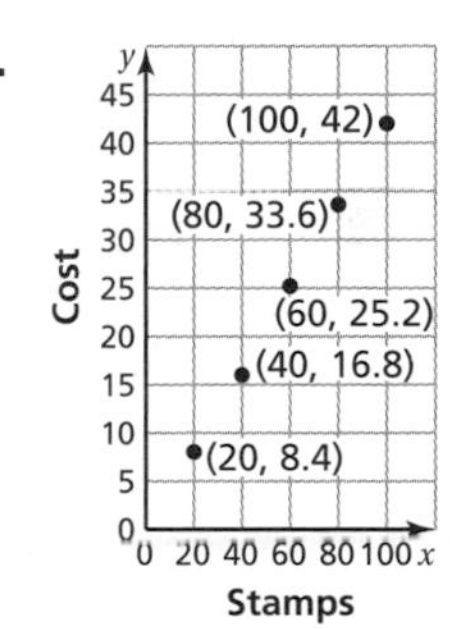

discrete

5. $y = \frac{1}{3}x + 3$
6. $y = -7$
7. $f(-5) = -13; f(0) = 12; f(2) = 22$
8. $g(-5) = 6.5; g(0) = -1; g(2) = -4$
9. $h(-5) = 22; h(0) = 7; h(2) = 1$
10. The graph of f is a translation 1 unit down of the graph of g.
11. The graph of $y = |x| + 1$ is a translation 1 unit up of the graph of $y = |x|$.
12. linear; As x increases by 3, y increases by 9.
13. nonlinear; As x increases by 2, y changes by different amounts.
14. $a_n = -n + 12; a_{30} = -18$
15. $a_n = 6n; a_{30} = 180$
16. $a_n = 2n - 11; a_{30} = 49$

My Thoughts on the Chapter

Teacher Tip

Not allowed to write in your teaching edition? Use sticky notes to record your thoughts.

What worked. . .

What did not work. . .

What I would do differently. . .

5.5 Comparing Linear and Nonlinear Functions (pp. 236–241)

Does the table represent a *linear* or *nonlinear* function? Explain.

a.

+2 +2 +2

x	0	2	4	6
y	0	1	4	9

+1 +3 +5

As *x* increases by 2, *y* increases by different amounts. The rate of change is *not* constant. So, the function is nonlinear.

b.

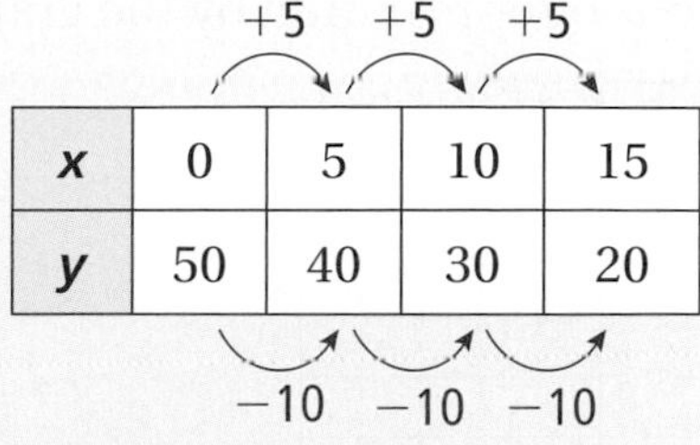

+5 +5 +5

x	0	5	10	15
y	50	40	30	20

−10 −10 −10

As *x* increases by 5, *y* decreases by 10. The rate of change is constant. So, the function is linear.

Exercises

Does the table represent a *linear* or *nonlinear* function? Explain.

12.

x	3	6	9	12
y	1	10	19	28

13.

x	1	3	5	7
y	3	1	1	3

5.6 Arithmetic Sequences (pp. 242–249)

Write an equation for the *n*th term of the arithmetic sequence −3, −5, −7, −9, ... Then find a_{20}.

The first term is −3 and the common difference is −2.

$a_n = a_1 + (n - 1)d$ — Equation for an arithmetic sequence

$a_n = -3 + (n - 1)(-2)$ — Substitute −3 for a_1 and −2 for *d*.

$a_n = -2n - 1$ — Simplify.

Use the equation to find the 20th term.

$a_{20} = -2(20) - 1$ — Substitute 20 for *n*.

$= -41$ — Simplify.

Exercises

Write an equation for the *n*th term of the arithmetic sequence. Then find a_{30}.

14. 11, 10, 9, 8, . . .

15. 6, 12, 18, 24, . . .

16. −9, −7, −5, −3, . . .

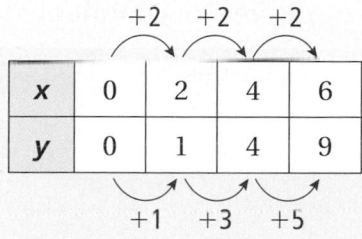

5 Chapter Test

Check It Out
Test Practice
BigIdeasMath.com

1. Find the domain and range of the function represented by the graph.

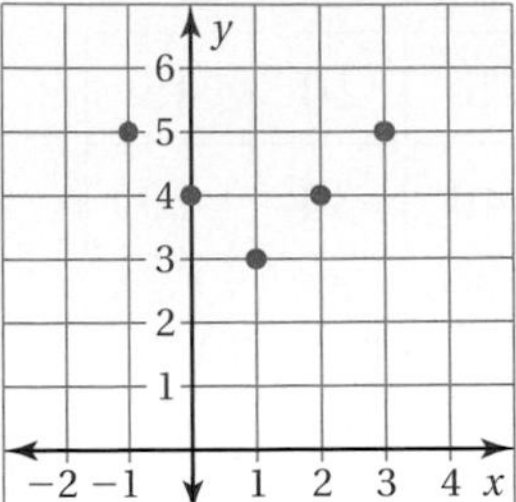

2. Graph the function. Is the domain discrete or continuous?

Minutes, x	Gallons, y
0	60
5	45
10	30
15	15

3. Use the graph to write a linear function that relates y to x.

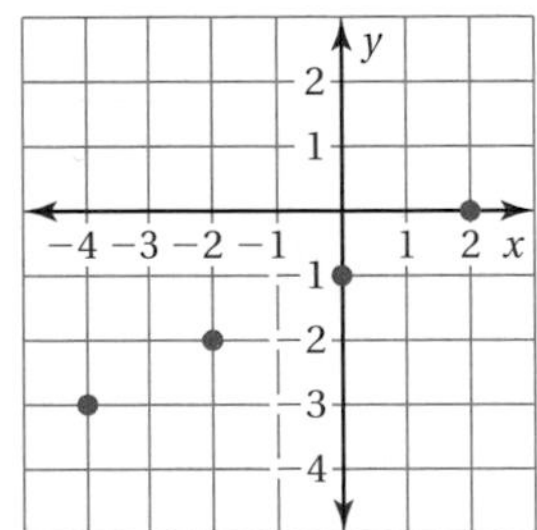

4. Does the table represent a *linear* or *nonlinear* function? Explain.

x	0	2	4	6
y	8	0	−8	−16

Evaluate the function when $x = -3, 0,$ and 6.

5. $f(x) = 9x - 10$

6. $g(x) = 2.5x + 5$

7. $h(x) = 15 - 3x$

8. Compare the graph of $h(x) = 5x + 2$ to the graph of $f(x) = 5x$.

9. Compare the graph of $y = |x + 3| - 2$ to the graph of $y = |x|$.

10. Graph $f(x) = \begin{cases} -x, & \text{if } x \le 0 \\ x + 5, & \text{if } x > 0 \end{cases}$. Describe the domain and range.

Write an equation for the nth term of the arithmetic sequence. Then find a_{25}.

11. 6, 12, 18, 24, . . .

12. −6, −5, −4, −3, . . .

13. 3, 1, −1, −3, . . .

14. **FOOD DRIVE** You are putting cans of food into boxes for a food drive. One box holds 30 cans of food. Write a linear function using function notation that represents the number of cans of food that will fit in x boxes. Is the domain discrete or continuous?

15. **SEATING** The first row of a theater has 20 seats. Each row after the first has two more seats than the row before it. Write an equation for the number of seats in the nth row. How many seats are in row 20?

16. **SURFACE AREA** A function relates the surface area S (in square inches) of a cube to the side length x (in inches) of the cube. Is the function linear or nonlinear? Explain.

Test Item References

Chapter Test Questions	Section to Review	Common Core State Standards
1	5.1	8.F.1, F.IF.1, F.IF.5
2	5.2	8.F.1, F.IF.1, F.IF.5
3	5.3	8.F.3, 8.F.4, F.BF.1a, F.LE.2
5–10, 14	5.4	F.IF.1, F.IF.2, F.IF.7b, F.BF.3
4, 16	5.5	8.F.3, F.LE.1b
11–13, 15	5.6	F.IF.3, F.BF.2, F.LE.2

Test-Taking Strategies

Remind students to quickly look over the entire test before they start so that they can budget their time. This test involves analyzing pairs of concepts that students can easily confuse, such as domain and range, input and output, rise and run, discrete and continuous, and linear and nonlinear. So, it is important that students use the **Stop** and **Think** strategy before they answer a question.

Common Assessment Errors

- **Exercise 1** Students may confuse the domain and range. Remind them that the domain is the set of all possible input values (the x-coordinates) and the range is the set of all possible output values (the y-coordinates).
- **Exercise 2** Students may see whole numbers in a table and think that the function is discrete. Encourage them to consider the context.
- **Exercise 3** Students may try to write the linear function without first finding the slope and y-intercept, or they may use the reciprocal of the slope in the function. Encourage them to check their work by making sure that all of the given points are solutions.
- **Exercises 5–7** Students may confuse the signs when they substitute for x, especially when they skip steps. Encourage students to take the time to carry out all the steps.
- **Exercises 11–13** Students may use the form $a_n = a_1 + nd$. This omits the first term and puts each term in the wrong position. Explain that they must use the formula on page 245 and solve for a_n.

Reteaching and Enrichment Strategies

If students need help. . .	If students got it. . .
Resources by Chapter • Practice A and Practice B • Puzzle Time Record and Practice Journal Practice Differentiating the Lesson Lesson Tutorials Practice from the Test Generator Skills Review Handbook	Resources by Chapter • Enrichment and Extension • School-to-Work • Financial Literacy • Technology Connection Game Closet at *BigIdeasMath.com* Start Standardized Test Practice

Answers

1. domain: $-1, 0, 1, 2, 3$
   range: $5, 4, 3$

2. 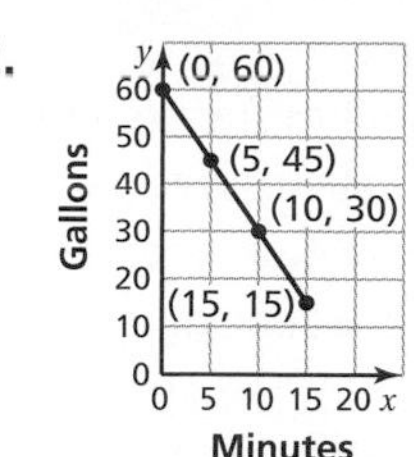

   continuous

3. $y = \frac{1}{2}x - 1$

4. linear; As x increases by 2, y decreases by 8.

5. $f(-3) = -37; f(0) = -10;$ $f(6) = 44$

6. $g(-3) = -2.5; g(0) = 5;$ $g(6) = 20$

7. $h(-3) = 24; h(0) = 15;$ $h(6) = -3$

8–10. See Additional Answers.

11. $a_n = 6n; a_{25} = 150$

12. $a_n = n - 7; a_{25} = 18$

13. $a_n = -2n + 5; a_{25} = -45$

14. $c(x) = 30x$; discrete

15. $a_n = 2n + 18$; 58 seats

16. See Additional Answers.

Assessment Book

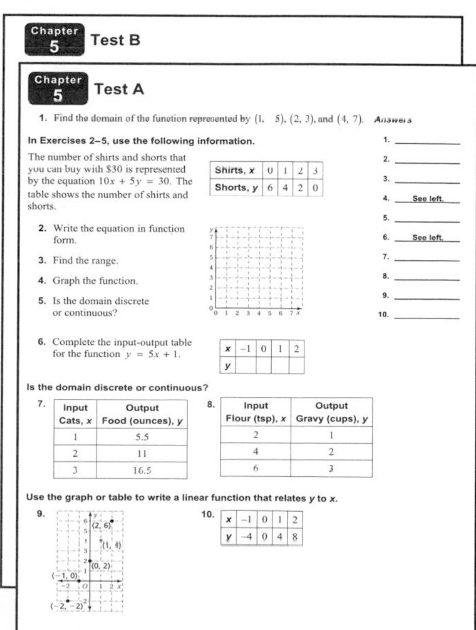

Chapter 5 Test B

Chapter 5 Test A

1. Find the domain of the function represented by (1, 5), (2, 3), and (4, 7).

In Exercises 2–5, use the following information.

The number of shirts and shorts that you can buy with \$30 is represented by the equation $10x + 5y = 30$. The table shows the number of shirts and shorts.

Shirts, x	0	1	2	3
Shorts, y	6	4	2	0

2. Write the equation in function form.
3. Find the range.
4. Graph the function.
5. Is the domain discrete or continuous?

6. Complete the input-output table for the function $y = 5x + 1$.

x	−1	0	1	2
y				

Is the domain discrete or continuous?

7.

Input Cats, x	Output Food (ounces), y
1	5.5
2	11
3	16.5

8.

Input Flour (tsp), x	Output Gravy (cups), y
2	1
4	2
6	3

Use the graph or table to write a linear function that relates y to x.

9. (2, 6), (1, 4), (0, 2), (−1, 0), (−2, −2)

10.

x	−1	0	1	2
y	−4	0	4	8

Answers
1. ____
2. ____
3. ____
4. See left.
5. ____
6. See left.
7. ____
8. ____
9. ____
10. ____

Test-Taking Strategies
Available at *BigIdeasMath.com*

After Answering Easy Questions, Relax
Answer Easy Questions First
Estimate the Answer
Read All Choices before Answering
Read Question before Answering
Solve Directly or Eliminate Choices
Solve Problem before Looking at Choices
Use Intelligent Guessing
Work Backwards

About this Strategy
When taking a multiple choice test, be sure to read each question carefully and thoroughly. One way to answer the question is to work backwards. Try putting the responses into the question, one at a time, and see if you get a correction solution.

Answers
1. D
2. F
3. \$200
4. B

Technology For the Teacher

Big Ideas Test Generator

Item Analysis

1. **A.** The student thinks that reversing the domain gets you the range.

   B. The student takes 5 away from each domain element, ignoring the coefficient 0.2.

   C. The student performs an arithmetic error subtracting integers.

   D. Correct answer

2. **F.** Correct answer

   G. The student picks the number 1 from the formula.

   H. The student uses the numbers in the formula: $1 - 0.25 = 0.75$.

   I. The student picks the number 0.25 from the formula.

3. **Gridded Response:** Correct answer: \$200

   Common Error: The student divides \$500 by 5, or \$800 by 10.

4. **A.** The student thinks that a line with a negative slope is nonlinear.

   B. Correct answer

   C. The student thinks that a horizontal line is nonlinear.

   D. The student thinks that a steep line with a positive slope is nonlinear.

5 Standardized Test Practice

1. The domain of the function $y = 0.2x - 5$ is 5, 10, 15, 20. What is the range of this function? *(8.F.1)*

   A. 20, 15, 10, 5 **C.** 4, 3, 2, 1

   B. 0, 5, 10, 15 **D.** $-4, -3, -2, -1$

Test-Taking Strategy
Work Backwards

For x cats, a litter box is changed $y=3x$ times per month. How many cats are there when $y=12$?
Ⓐ1 Ⓑ2 Ⓒ3 Ⓓ4

Share a litter box? Please!

KEEP OFF!

"Work backwards by trying 1, 2, 3, and 4. You will see that 3(4) = 12. So, D is correct."

2. A toy runs on a rechargeable battery. During use, the battery loses power at a constant rate. The percent P of total power left in the battery after x hours can be found using the equation shown below. When will the battery be fully discharged? *(A.REI.3)*

$$P = -0.25x + 1$$

   F. After 4 hours of use

   G. After 1 hour of use

   H. After 0.75 hour of use

   I. After 0.25 hour of use

3. A limousine company charges a fixed cost for a limousine and an hourly rate for its driver. It costs \$500 to rent the limousine for 5 hours and \$800 to rent the limousine for 10 hours. What is the fixed cost, in dollars, to rent the limousine? *(8.F.4)*

4. Which graph shows a nonlinear function? *(F.LE.1b)*

A.

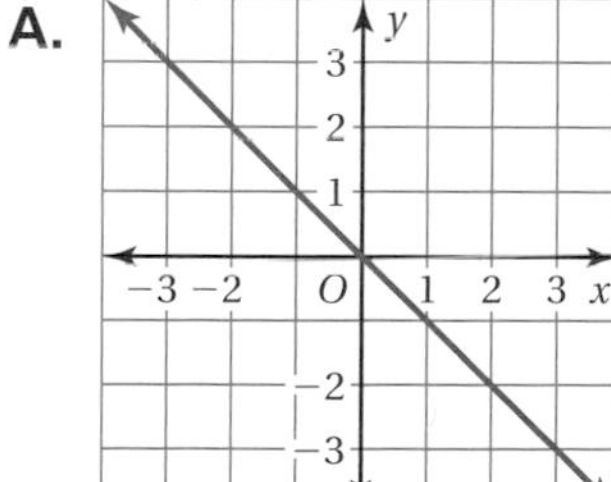

C.

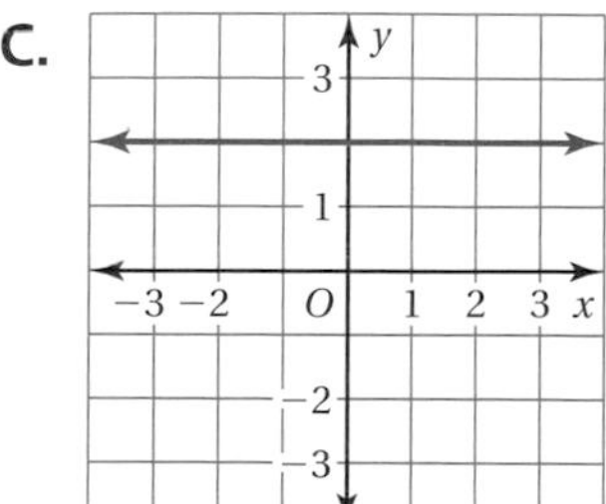

B.

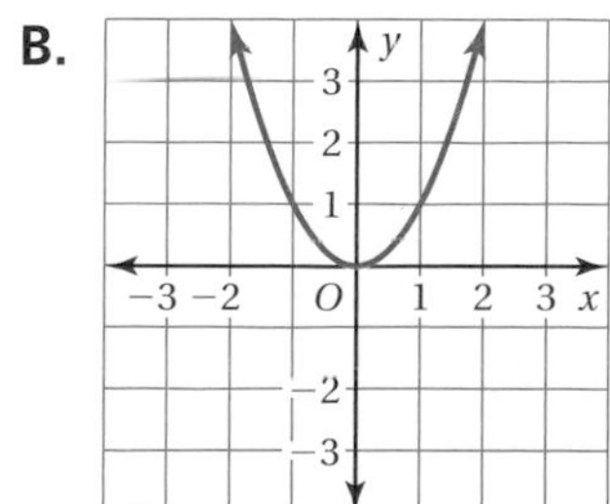

D.

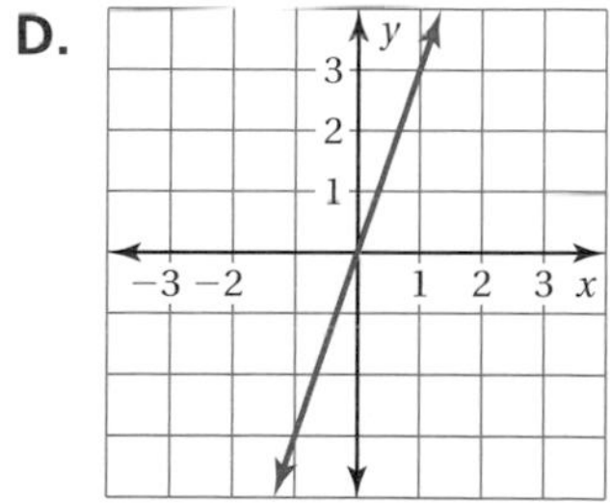

5. The equations $y = -x + 4$ and $y = \frac{1}{2}x - 8$ form a system of linear equations. The table below shows the y value for each equation at six different values of x. *(A.REI.6)*

x	0	2	4	6	8	10
$y = -x + 4$	4	2	0	−2	−4	−6
$y = \frac{1}{2}x - 8$	−8	−7	−6	−5	−4	−3

What can you conclude from the table?

F. The system has one solution, when $x = 0$.

G. The system has one solution, when $x = 4$.

H. The system has one solution, when $x = 8$.

I. The system has no solution.

6. The temperature fell from 54 degrees Fahrenheit to 36 degrees Fahrenheit over a six-hour period. The temperature fell by the same number of degrees each hour. How many degrees Fahrenheit did the temperature fall each hour? *(A.CED.1)*

7. What is the domain of the function graphed in the coordinate plane below? *(8.F.1)*

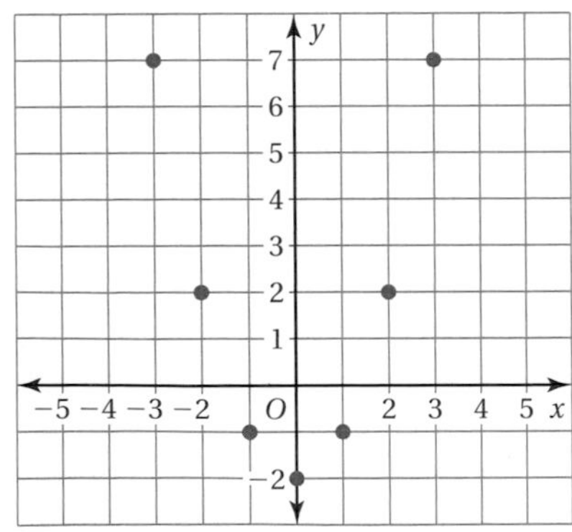

A. 0, 1, 2, 3

B. −2, −1, 2, 7

C. −3, −2, −1, 0, 1, 2, 3

D. −2, −1, 0, 1, 2, 3, 7

8. What value of w makes the equation below true? *(A.REI.3)*

$$\frac{w}{3} = 3(w - 1) - 1$$

F. $\frac{3}{2}$

G. $\frac{5}{4}$

H. $\frac{3}{4}$

I. $\frac{1}{2}$

Item Analysis (continued)

5. **F.** The student associates 0 with a solution.

   G. The student selects this choice because the first equation equals 0 when $x = 4$.

   H. Correct answer

   I. The student has no idea how to interpret the table, or fails to see that the functions are equal at $x = 8$.

6. **Gridded Response:** Correct answer: 3°F

   Common Error: The student subtracts 36 from 54, but fails to divide the result by 6.

7. **A.** The student only includes the domain values greater than or equal to 0.

   B. The student finds the range.

   C. Correct answer

   D. The student puts the domain and range together.

8. **F.** Correct answer

   G. The student multiplies both sides by 3, but fails to distribute the 3 correctly on the right side.

   H. The student starts by distributing the 3 on the right side incorrectly.

   I. The student multiplies both sides by 3, distributes the 3 correctly, but does not distribute the 9 correctly. Alternatively, the student starts by distributing the 3 on the right side correctly. But when multiplying both sides by 3, the student does not distribute it across the expression $3w - 4$ correctly.

9. **A.** The student subtracts 5 and -3 incorrectly.

   B. The student subtracts 5 and -3 incorrectly, and subtracts -4 and 1 incorrectly.

   C. Correct answer

   D. The student subtracts -4 and 1 incorrectly.

Answers

5. H
6. 3°F
7. C
8. F

Answers

9. C

10. H

11. *Part A* yes

    Part B no

12. C

Item Analysis (continued)

10. **F.** The student picks the slope of the problem and ignores the need to find a y-intercept.

    G. The student picks the slope of the problem and the y-value from the point (6, 1) for the y-intercept.

    H. Correct answer

    I. The student approaches the problem properly, but misplaces a negative sign for y.

11. **2 points** The student demonstrates a thorough understanding of how to determine whether data show a linear function or a nonlinear function, explains the work fully, and relates perimeter to a linear function and area to a nonlinear function. The first table shows a linear function. The second table shows a nonlinear function.

    1 point The student's work and explanations demonstrate a lack of essential understanding. The slope formula is used incorrectly or a graph of the data is incomplete.

    0 points The student provides no response, a completely incorrect or incomprehensible response, or a response that demonstrates insufficient understanding of linear functions and nonlinear functions.

12. **A.** The student fails to recognize that "at least 14" includes the value 14.

    B. The student reverses the inequality and fails to include the value 14.

    C. Correct answer

    D. The student reverses the inequality.

Answers for Extra Examples

1. **A.** The student ignores the starting amount of $300.

   B. The student combines the starting and weekly amounts incorrectly.

   C. The student reverses the roles of the starting and weekly amounts.

   D. Correct answer

2. **F.** The student identifies the domain incorrectly.

   G. The student identifies the domain incorrectly.

   H. Correct answer

   I. The student identifies the domain correctly but thinks it is discrete.

Extra Examples

1. Deanna started a savings account with $300 and added $20 per month to the account. Let n be the number of weeks that Deanna added money to the account and let a be the total amount in her account. Which equation describes the relationship between a and n? *(F.BF.1a)*

   A. $a = 20n$

   B. $a = 320n$

   C. $a = 300n + 20$

   D. $a = 20n + 300$

2. Julia is studying how attendance at an amusement park is related to temperature. To build her study, Julia uses the temperature each day as an input value and amusement park attendance as a daily output value. Which statement best describes this situation? *(F.IF.5)*

   F. The domain is daily attendance and it is continuous.

   G. The domain is daily attendance and it is discrete.

   H. The domain is temperature and it is continuous.

   I. The domain is temperature and it is discrete.

9. What is the slope of the line shown in the graph below? *(F.IF.6)*

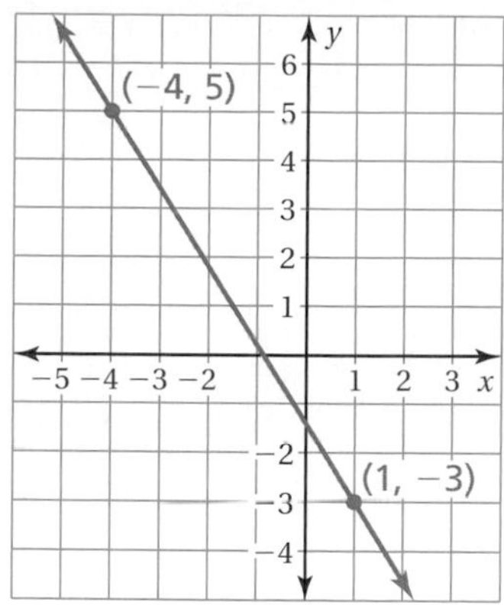

A. $-\frac{2}{5}$

B. $-\frac{2}{3}$

C. $-\frac{8}{5}$

D. $-\frac{8}{3}$

10. A line with a slope of $\frac{1}{3}$ passes through the point (6, 1). What is the equation of the line? *(A.CED.2)*

F. $y = \frac{1}{3}x$

G. $y = \frac{1}{3}x + 1$

H. $x - 3y = 3$

I. $x + 3y = 3$

11. Think Solve Explain The tables show how the perimeter and area of a square are related to its side length. Examine the data in the table. *(F.LE.1b)*

Side Length	1	2	3	4	5	6
Perimeter	4	8	12	16	20	24

Side Length	1	2	3	4	5	6
Area	1	4	9	16	25	36

Part A Does the first table show a linear function? Explain your reasoning.

Part B Does the second table show a linear function? Explain your reasoning.

12. In many states, you must be at least 14 years old to operate a personal watercraft. Which inequality represents this situation? *(A.CED.1)*

A. $y > 14$

B. $y < 14$

C. $y \geq 14$

D. $y \leq 14$

6 Exponential Equations and Functions

1 + 100 + 100²
FLEAS
"If one flea had 100 babies, and each baby grew up and had 100 babies, ..."

+ 100³
Help! I can't see the expiration date on my flea collar.
"... and each of those babies grew up and had 100 babies, you would have 1,010,101 fleas."

"Here's how I remember the square root of 2."

Can't I just use a calculator?
"February is the 2nd month. It has 28 days. Split 28 into 14 and 14. Move the decimal to get 1.414."

Connections to Previous Learning

- Extend understandings of numbers to the system of rational numbers.
- Evaluate expressions involving whole-number exponents.
- Reason about and solve one-variable equations.

- Use rational numbers to approximate irrational numbers.
- Work with radicals and integer exponents.
- Solve real-life and mathematical problems using numerical and algebraic expressions and equations.

- Use properties of rational and irrational numbers.
- Extend the properties of exponents to rational exponents.
- Write and graph exponential equations and functions to model, analyze, and solve real-world problems.

Math in History

Before the use of calculators, it was common for textbooks to have a table of square roots. These were calculated by various methods.

★ Isaac Newton (1643–1727) was a famous English mathematician who invented calculus. Here is his method for calculating a square root.

1. You want to find the square root of x.
2. Let G be your best guess.
3. To find the next G, call it G'. $G' = \dfrac{x + G^2}{2G}$
4. This G' becomes your next G. Go back to Step 2 and continue until you have reached your desired accuracy.

★ During the 17th century, many people invented calculating machines. Here is one by Frenchman Blaise Pascal (1623–1662).

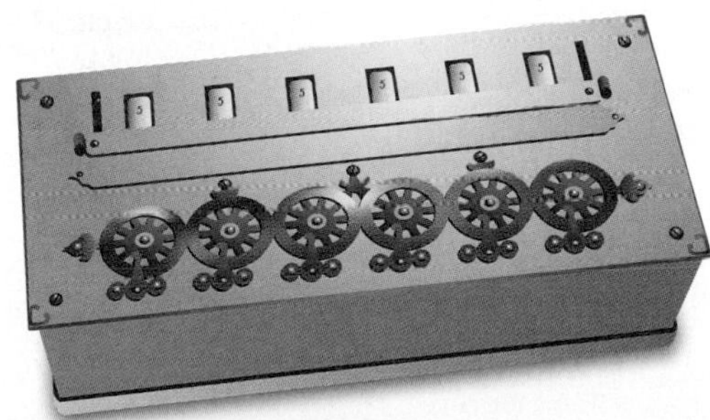

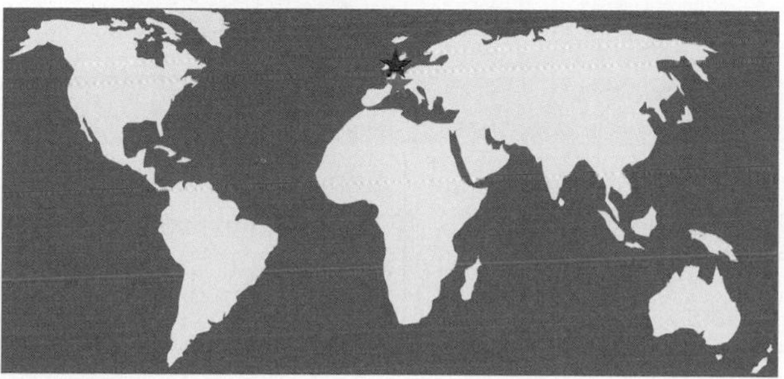

Pacing Guide for Chapter 6

Chapter Opener	1 Day
Section 1	3 Days
Section 2	2 Days
Section 3	1 Day
Study Help / Quiz	1 Day
Section 4	3 Days
Section 5	1 Day
Section 6	1 Day
Section 7	3 Days
Chapter Review / Chapter Tests	2 Days
Total Chapter 6	18 Days
Year-to-Date	80 Days

Check Your Resources

- Record and Practice Journal
- Resources by Chapter
- Skills Review Handbook
- Assessment Book
- Worked-Out Solutions

Technology For the Teacher

The Dynamic Planning Tool
Editable Teacher's Resources at
BigIdeasMath.com

Common Core State Standards

8.EE.1 Know and apply the properties of integer exponents to generate equivalent numerical expressions.
8.EE.2 Use square root and cube root symbols to represent solutions to equations of the form $x^2 = p$ and $x^3 = p$, where p is a positive rational number. Evaluate square roots of small perfect squares and cube roots of small perfect cubes. Know that $\sqrt{2}$ is irrational.
F.BF.2 Write arithmetic and geometric sequences both recursively and with an explicit formula, use them to model situations, and translate between the two forms.

Additional Topics for Review

- Exponents
- Writing scientific notation
- Factoring whole numbers
- Comparing linear and nonlinear functions
- Graphing functions

Try It Yourself

1. 71 **2.** 106

3. −12 **4.** 8

5. 11 and −11 **6.** −2

7. $a_n = 2n$

8. $a_n = -3n + 9$

9. $a_n = -7n + 29$

Record and Practice Journal

1. 3 **2.** −35

3. 23 **4.** −4

5. 9 **6.** −5

7. ± 4 **8.** $s = 7$ ft

9. $a_n = 4n - 3$

10. $a_n = -12n + 13$

11. $a_n = -6n + 24$

12. $a_n = 9n - 33$

13. a. $g_n = -2.5n + 22$

b. 2 gal

Math Background Notes

Vocabulary Review

- Order of Operations
- Perfect Square
- Radicand
- Common Difference
- Square Root
- Radical
- Arithmetic Sequence
- Equation for an Arithmetic Sequence

Using Order of Operations

- Students should know the order of operations, given by **PEMDAS.**
- Although Example 1 includes each of the four categories of operations, many problems do not. Students only need to apply the correct order of the operations in the problem they are solving.

Finding Square Roots

- Students should know how to find the square root of a perfect square.
- Remind students to check each answer by squaring it to see if they get the radicand of the original expression.
- **Common Error:** Students may not place the correct sign on the square root. Remind them that every positive number has a positive *and* a negative square root. The positive square root is denoted $\sqrt{n}$ and the negative square root is denoted $-\sqrt{n}$.

Writing an Equation for an Arithmetic Sequence

- Students should know how to write an equation for an arithmetic sequence.
- Remind students that the common difference is the amount you add to one term to get the next term.
- **Common Error:** Students may use the form $a_n = a_1 + nd$. This omits the first term and puts each term in the wrong position. Explain that they must use the formula $a_n = a_1 + (n - 1)d$ and solve for a_n.

Reteaching and Enrichment Strategies

If students need help. . .	If students got it. . .
Record and Practice Journal • Fair Game Review Skills Review Handbook Lesson Tutorials	Game Closet at *BigIdeasMath.com* Start the next section

What You Learned Before

"It's called the Power of Negative One, Descartes!"

Using Order of Operations (8.EE.1)

Example 1 **Evaluate $10^2 \div (30 \div 3) - 4(3 - 9) + 5^0$.**

First:	Parentheses	$10^2 \div (30 \div 3) - 4(3 - 9) + 5^0 = 10^2 \div 10 - 4(-6) + 5^0$
Second:	Exponents	$= 100 \div 10 - 4(-6) + 1$
Third:	Multiplication and Division (from left to right)	$= 10 + 24 + 1$
Fourth:	Addition and Subtraction (from left to right)	$= 35$

Try It Yourself

Evaluate the expression.

1. $12\left(\frac{14}{2}\right) - 3^3 + 15 - 2^0$ **2.** $5^2 \cdot 8 \div 2^2 + 20 \cdot 3 - 4$ **3.** $-7 + 16 \cdot 4^{-2} + (10 - 4^2)$

Finding Square Roots (8.EE.2)

Example 2 **Find $-\sqrt{81}$.**

$-\sqrt{81}$ represents the negative square root. Because $9^2 = 81$, $-\sqrt{81} = -\sqrt{9^2} = -9$.

Try It Yourself

4. Find $\sqrt{64}$. **5.** Find $\pm\sqrt{121}$. **6.** Find $-\sqrt{4}$.

Writing an Equation for an Arithmetic Sequence (F.BF.2)

Example 3 **Write an equation for the *n*th term of the arithmetic sequence 5, 15, 25, 35, ...**

The first term is 5 and the common difference is 10.

$a_n = a_1 + (n - 1)d$ Equation for an arithmetic sequence

$a_n = 5 + (n - 1)10$ Substitute 5 for a_1 and 10 for d.

$a_n = 10n - 5$ Simplify.

Try It Yourself

Write an equation for the *n*th term of the arithmetic sequence.

7. 2, 4, 6, 8, . . . **8.** 6, 3, 0, −3, . . . **9.** 22, 15, 8, 1, . . .

6.1 Properties of Square Roots

COMMON CORE STATE STANDARDS
N.RN.3

Essential Question How can you multiply and divide square roots?

Recall that when you multiply a number by itself, you square the number.

Symbol for squaring is 2nd power. → $4^2 = 4 \cdot 4$

$= 16$ 4 squared is 16.

To "undo" this, take the square root of the number.

Symbol for square root is a radical sign. → $\sqrt{16} = \sqrt{4^2} = 4$ The square root of 16 is 4.

1 ACTIVITY: Finding Square Roots

Work with a partner. Use a square root symbol to write the side length of the square. Then find the square root. Check your answer by multiplying.

a. Sample: $s = \sqrt{81} = 9$ ft

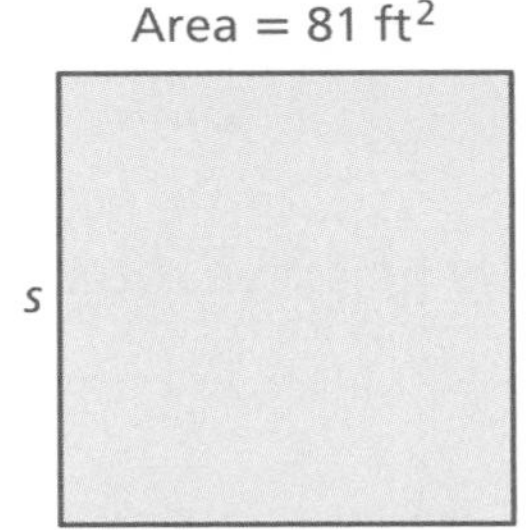

Check

$$\begin{array}{r} 9 \\ \times 9 \\ \hline 81 \end{array} \checkmark$$

∴ The side length of the square is 9 feet.

b. Area = 121 yd^2

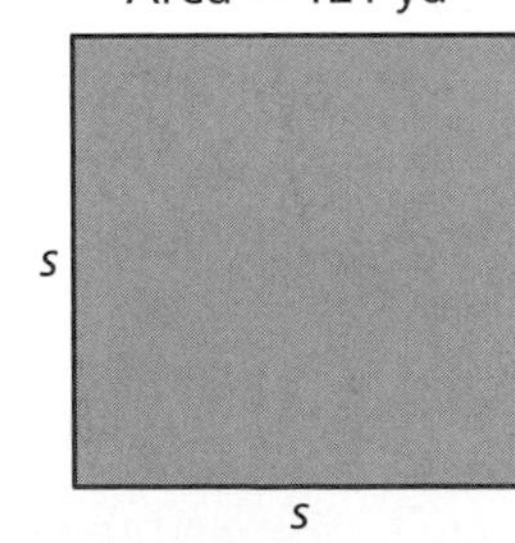

c. Area = 324 cm^2

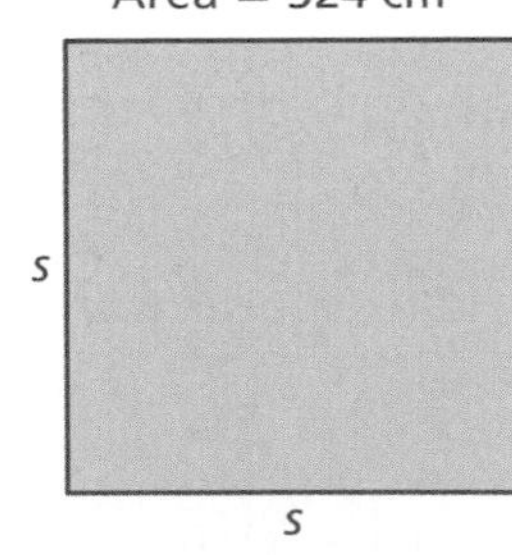

d. Area = 361 mi^2

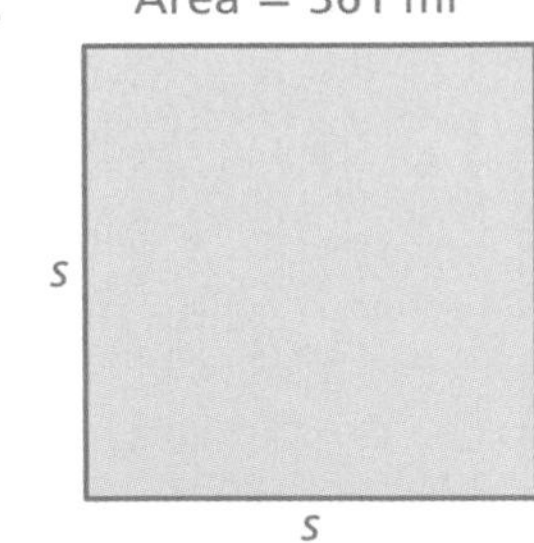

e. Area = 2.89 in.^{2}

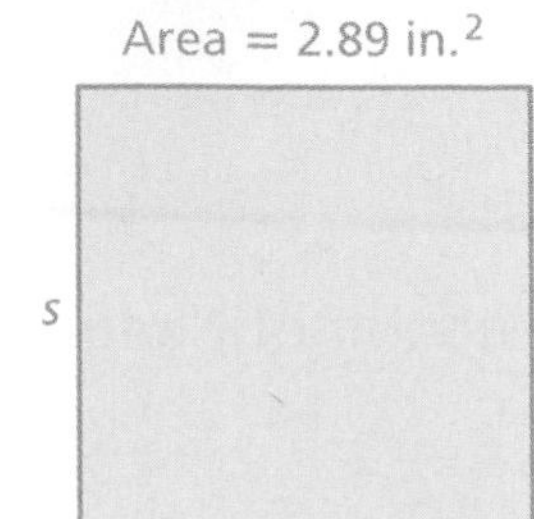

f. Area = 6.25 m^2

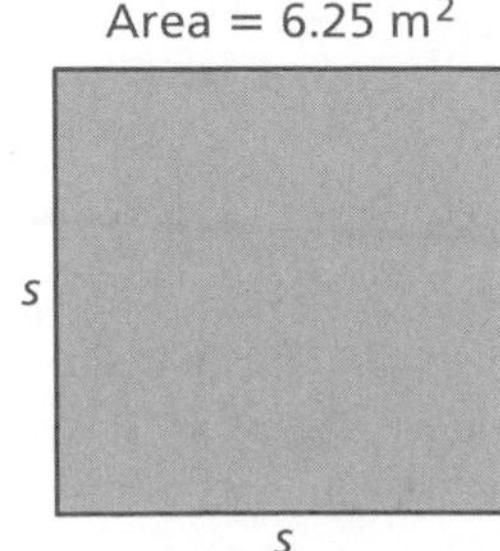

g. Area = $\frac{16}{25}$ ft^2

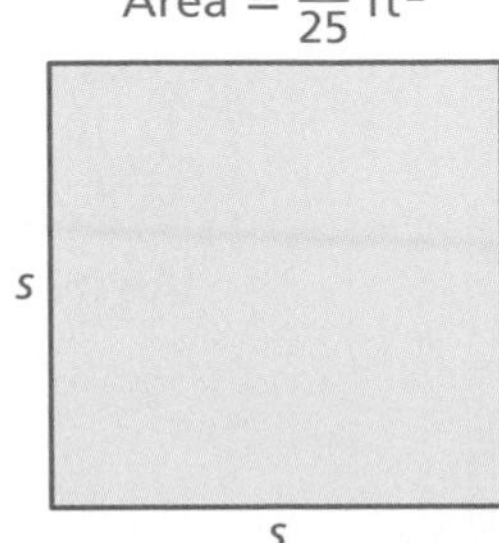

Laurie's Notes

Introduction

Standards for Mathematical Practice

- **MP7 Look for and Make Use of Structure:** Students will simplify and evaluate expressions involving square roots. They will see that a radicand with two terms is different than a radicand with two factors. Students will also work with quotients where the numerator is a binomial. Structurally, $\frac{a+b}{c}$ is equal to $\frac{a}{c} + \frac{b}{c}$. This is how fractions are added. Recognizing this structure will be helpful in simplifying certain radical expressions.

Motivate

- As a warm-up, write several numeric sequences and ask students to think-pair-share a description of the patterns. Examples:
  2, 5, 11, 23, . . . double the previous term and add 1
  2, 3, 5, 8, 13, . . . add consecutive terms
  2, 3, 5, 7, . . . prime numbers; Students may ask for more information.
  1, 4, 9, 16, . . . perfect squares
- Ask partners to share their descriptions.
- Be sure to include the last example. Perfect squares will be part of the investigation today.

Activity Notes

Discuss

- Discuss the vocabulary: exponent, power, square root.

Activity 1

? "When you know the dimensions of a square, how can you find the area of the square?" Square the dimensions.

? "When you know the area of a square, how can you find the dimensions of the square?" Find the square root of the area.

- Students should recognize the perfect squares, though they may be unsure when the area is not a whole number.
- As you circulate, you can stimulate reasoning by asking, "What number can you multiply by itself to get 289?" 17 "So what number can you multiply by itself to get 2.89?" 1.7
- If students are not using calculators, they will need time for trial and error on these problems.

Common Core State Standards

N.RN.3 Explain why the sum or product of two rational numbers is rational; that the sum of a rational number and an irrational number is irrational; and that the product of a nonzero rational number and an irrational number is irrational.

Previous Learning

Students should know how to find the square root of a perfect square and how to use the order of operations to evaluate expressions.

Start Thinking! and Warm Up

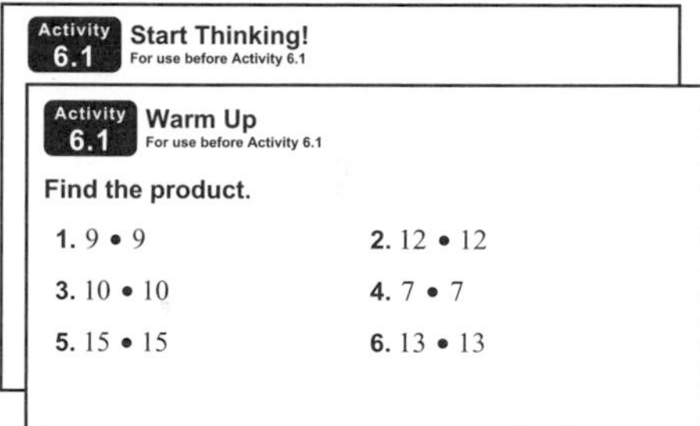

Activity 6.1 Start Thinking! For use before Activity 6.1

Activity 6.1 Warm Up For use before Activity 6.1

Find the product.

1. 9 • 9
2. 12 • 12
3. 10 • 10
4. 7 • 7
5. 15 • 15
6. 13 • 13

6.1 Record and Practice Journal

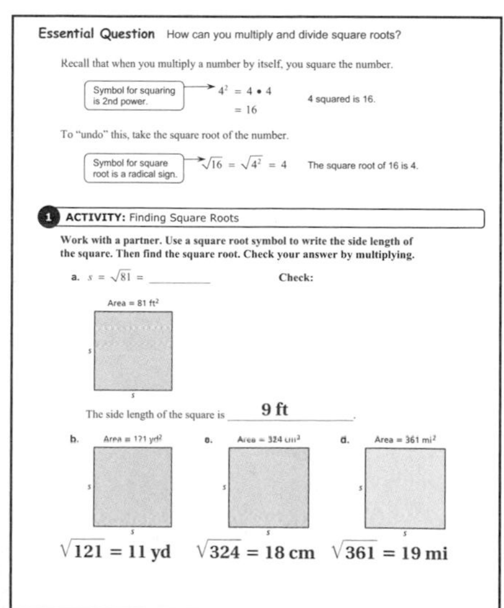

Essential Question How can you multiply and divide square roots?

Recall that when you multiply a number by itself, you square the number.

Symbol for squaring is 2nd power. $4^2 = 4 \cdot 4 = 16$ 4 squared is 16.

To "undo" this, take the square root of the number.

Symbol for square root is a radical sign. $\sqrt{16} = \sqrt{4^2} = 4$ The square root of 16 is 4.

1 ACTIVITY: Finding Square Roots

Work with a partner. Use a square root symbol to write the side length of the square. Then find the square root. Check your answer by multiplying.

a. $s = \sqrt{81} =$ ______ Check:

Area = 81 ft²

The side length of the square is 9 ft.

b. Area = 121 yd² c. Area = 324 cm² d. Area = 361 mi²

$\sqrt{121} = 11$ yd $\sqrt{324} = 18$ cm $\sqrt{361} = 19$ mi

English Language Learners

Build on Past Knowledge

Remind students of inverse operations. Addition and subtraction are inverse operations, as are multiplication and division. Taking the square root of a number is the inverse of squaring a number and squaring a number is the inverse of taking the square root of a number.

6.1 Record and Practice Journal

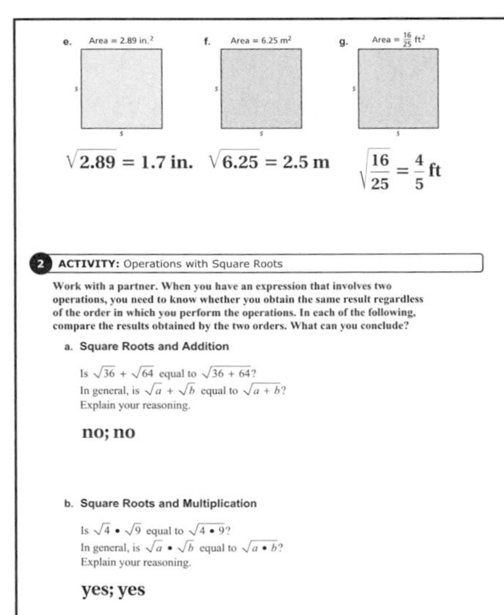

e. Area = 2.89 in.² f. Area = 6.25 m² g. Area = $\frac{16}{25}$ ft²

$\sqrt{2.89} = 1.7$ in. $\sqrt{6.25} = 2.5$ m $\sqrt{\frac{16}{25}} = \frac{4}{5}$ ft

2 ACTIVITY: Operations with Square Roots

Work with a partner. When you have an expression that involves two operations, you need to know whether you obtain the same result regardless of the order in which you perform the operations. In each of the following, compare the results obtained by the two orders. What can you conclude?

a. **Square Roots and Addition**

Is $\sqrt{36} + \sqrt{64}$ equal to $\sqrt{36 + 64}$?
In general, is $\sqrt{a} + \sqrt{b}$ equal to $\sqrt{a + b}$?
Explain your reasoning.

no; no

b. **Square Roots and Multiplication**

Is $\sqrt{4} \cdot \sqrt{9}$ equal to $\sqrt{4 \cdot 9}$?
In general, is $\sqrt{a} \cdot \sqrt{b}$ equal to $\sqrt{a \cdot b}$?
Explain your reasoning.

yes; yes

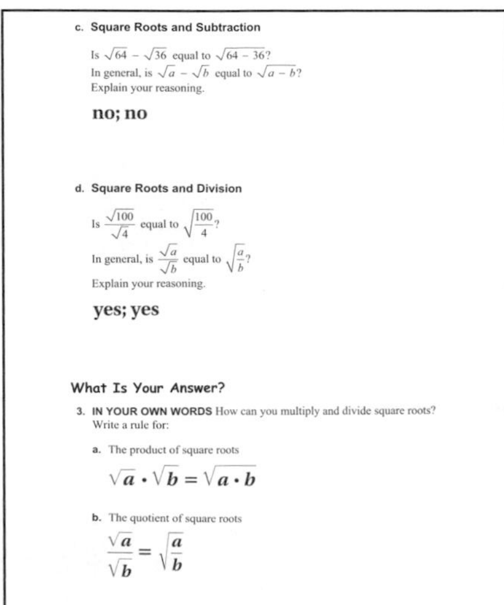

c. **Square Roots and Subtraction**

Is $\sqrt{64} - \sqrt{36}$ equal to $\sqrt{64 - 36}$?
In general, is $\sqrt{a} - \sqrt{b}$ equal to $\sqrt{a - b}$?
Explain your reasoning.

no; no

d. **Square Roots and Division**

Is $\frac{\sqrt{100}}{\sqrt{4}}$ equal to $\sqrt{\frac{100}{4}}$?
In general, is $\frac{\sqrt{a}}{\sqrt{b}}$ equal to $\sqrt{\frac{a}{b}}$?
Explain your reasoning.

yes; yes

What Is Your Answer?

3. **IN YOUR OWN WORDS** How can you multiply and divide square roots? Write a rule for:

a. The product of square roots

$\sqrt{a} \cdot \sqrt{b} = \sqrt{a \cdot b}$

b. The quotient of square roots

$\frac{\sqrt{a}}{\sqrt{b}} = \sqrt{\frac{a}{b}}$

Laurie's Notes

Activity 2

- Calculators are not necessary to perform these operations. The numbers involved are all perfect squares.
- Introduce the activity by posing a question about a numeric expression that involves two operations, such as addition and multiplication.

? "Does the order in which the operations are performed matter?" yes

- Explain that in each part of this activity, you will determine whether order matters when combining one of the four basic operations with taking square roots.
- **MP3 Construct Viable Arguments and Critique the Reasoning of Others:** Encourage students to build strong arguments to support their conclusions. They should understand that one counterexample is sufficient to say that order does matter. However, they should try other examples to support their reasoning when they feel that order does not matter. Suggest that working with perfect squares is helpful. For instance, they may compare $\sqrt{16} \cdot \sqrt{25}$ to $\sqrt{16 \cdot 25}$, and so on.
- Ask students to summarize their findings and discuss their arguments. They are often surprised by the results. They try to make the connection to the Commutative and Associative Properties where addition and multiplication work, but subtraction and division do not.

What Is Your Answer?

- Students may not be able to phrase the statement efficiently, but listen for "The product of the square roots of two numbers is equal to the square root of the product of the two numbers." A similar statement can be made for quotients.

Closure

- Simplify $\sqrt{25 \cdot 81}$. 45
- Simplify $\sqrt{\frac{25}{81}}$. $\frac{5}{9}$

Technology For the Teacher

The Dynamic Planning Tool
Editable Teacher's Resources at *BigIdeasMath.com*

2 ACTIVITY: Operations with Square Roots

Work with a partner. When you have an expression that involves two operations, you need to know whether you obtain the same result regardless of the order in which you perform the operations. In each of the following, compare the results obtained by the two orders. What can you conclude?

a. Square Roots and Addition

Is $\sqrt{36} + \sqrt{64}$ equal to $\sqrt{36 + 64}$?

In general, is $\sqrt{a} + \sqrt{b}$ equal to $\sqrt{a + b}$?

Explain your reasoning.

b. Square Roots and Multiplication

Is $\sqrt{4} \cdot \sqrt{9}$ equal to $\sqrt{4 \cdot 9}$?

In general, is $\sqrt{a} \cdot \sqrt{b}$ equal to $\sqrt{a \cdot b}$?

Explain your reasoning.

c. Square Roots and Subtraction

Is $\sqrt{64} - \sqrt{36}$ equal to $\sqrt{64 - 36}$?

In general, is $\sqrt{a} - \sqrt{b}$ equal to $\sqrt{a - b}$?

Explain your reasoning.

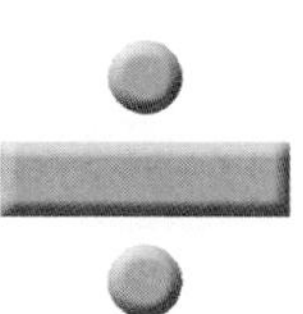

d. Square Roots and Division

Is $\frac{\sqrt{100}}{\sqrt{4}}$ equal to $\sqrt{\frac{100}{4}}$?

In general, is $\frac{\sqrt{a}}{\sqrt{b}}$ equal to $\sqrt{\frac{a}{b}}$?

Explain your reasoning.

What Is Your Answer?

3. **IN YOUR OWN WORDS** How can you multiply and divide square roots? Write a rule for:
   a. The product of square roots
   b. The quotient of square roots

Use what you learned about square roots to complete Exercises 3–5 on page 264.

6.1 Lesson

Product Property of Square Roots

Algebra	Numbers
$\sqrt{xy} = \sqrt{x} \cdot \sqrt{y}$, where $x, y \geq 0$	$\sqrt{9 \cdot 5} = \sqrt{9} \cdot \sqrt{5} = 3\sqrt{5}$

Quotient Property of Square Roots

Algebra	Numbers
$\sqrt{\frac{x}{y}} = \frac{\sqrt{x}}{\sqrt{y}}$, where $x \geq 0$ and $y > 0$	$\sqrt{\frac{3}{4}} = \frac{\sqrt{3}}{\sqrt{4}} = \frac{\sqrt{3}}{2}$

EXAMPLE 1 Simplifying Square Roots

a. $\sqrt{150} = \sqrt{25 \cdot 6}$ — Factor using the greatest perfect square factor.

$= \sqrt{25} \cdot \sqrt{6}$ — Product Property of Square Roots

$= 5\sqrt{6}$ — Simplify.

A square root is simplified when the radicand has no perfect square factors other than 1.

b. $\sqrt{\frac{15}{64}} = \frac{\sqrt{15}}{\sqrt{64}}$ — Quotient Property of Square Roots

$= \frac{\sqrt{15}}{8}$ — Simplify.

EXAMPLE 2 Evaluating Square Roots

Evaluate $\sqrt{b^2 - 4ac}$ when $a = 2$, $b = -8$, and $c = 4$.

$\sqrt{b^2 - 4ac} = \sqrt{(-8)^2 - 4(2)(4)}$ — Substitute.

$= \sqrt{32}$ — Simplify.

$= \sqrt{16 \cdot 2}$ — Factor.

$= \sqrt{16} \cdot \sqrt{2}$ — Product Property of Square Roots

$= 4\sqrt{2}$ — Simplify.

On Your Own

Now You're Ready
Exercises 6–17

Simplify the expression.

1. $\sqrt{\frac{23}{9}}$

2. $-\sqrt{80}$

3. $\sqrt{\frac{27}{100}}$

4. Evaluate $\sqrt{b^2 - 4ac}$ when $a = 2$, $b = -6$, and $c = -5$.

Laurie's Notes

Introduction

Connect

- **Yesterday:** Students explored the properties of operations with square roots. (MP3, MP7)
- **Today:** Students will use properties of square roots to simplify and evaluate expressions involving square roots.

Motivate

? "Have any of you heard of the Ellipse in Washington, DC?" It is an oval-shaped park located within President's Park South, just south of the White House. Laid out by the Army Corps of Engineers from 1877 to 1880, it has a major axis of 1058.26 feet, a minor axis of 902.85 feet, a perimeter (circumference) of 3086.87 feet, and an area of about 17 acres.

- In today's lesson, students will find the circumference of an elliptical room.

Lesson Notes

Key Ideas

- Write and discuss the two properties. Demonstrate them using numbers.
- Read and discuss the notation shown for numbers. Explain that "three square root five" or $3\sqrt{5}$ is in simplified form.
- Remind students that a square root is simplified when the radicand has no perfect square factors other than 1.

Example 1

? "How can you write 150 as a product of two factors?" Answers will vary. Write all the possibilities: 2×75, 3×50, 5×30, 6×25, and 10×15.

? "Which factor pair includes a perfect square?" 6×25

- By convention, $5\sqrt{6}$ is generally used rather than $\sqrt{6}(5)$. This is similar to the convention of writing $5x$ rather than $x5$.
- To simplify a square root expression, perform any operations and factor to remove any perfect square factors other than 1 from the radicand.

Example 2

- To evaluate square roots, substitute for the variables and then simplify the square root expression.
- The expression shown is the discriminant of the quadratic formula in Chapter 9. They don't need to know the name of it now, or how it is used.

? "How many operations are used to evaluate $(-8)^2 - (4)(2)(4)$? Explain." 4; squaring −8, two multiplications, and subtraction

? "Does the order in which the operations are performed matter? Explain." yes; The order of operations is needed to get the correct answer.

- Point out that because $b = -8$ you need parentheses in $(-8)^2$.

? "Which factor pair of 32 includes the greatest perfect square?" 16×2

- If time permits, have students use a graphing calculator to evaluate the original expression and the simplified expression.

Goal Today's lesson is simplifying expressions involving square roots.

Start Thinking! and Warm Up

Lesson 6.1 Warm Up
For use before Lesson 6.1

Lesson 6.1 Start Thinking!
For use before Lesson 6.1

Consider the following square root expressions.

$\sqrt{108}$ $\quad$ $6\sqrt{3}$

Are the expressions equal? How do you know?

The expression $6\sqrt{3}$ is *simplified,* but $\sqrt{108}$ is not. Using this example, what do you think it means for a square root to be simplified?

Extra Example 1

Simplify each expression.

a. $\sqrt{28}$ $2\sqrt{7}$

b. $\sqrt{\frac{11}{100}}$ $\frac{\sqrt{11}}{10}$

Extra Example 2

Evaluate $\sqrt{b^2 - 4ac}$ when $a = 3$, $b = 4$, and $c = -2$.
$2\sqrt{10}$

On Your Own

1. $\sqrt{\frac{23}{3}}$
2. $-4\sqrt{5}$
3. $\frac{3\sqrt{3}}{10}$
4. $2\sqrt{19}$

Extra Example 3

Simplify $\frac{18 + \sqrt{72}}{6}$. $3 + \sqrt{2}$

Extra Example 4

Use the formula in Example 4 to estimate the circumference of the ellipse to the nearest inch.

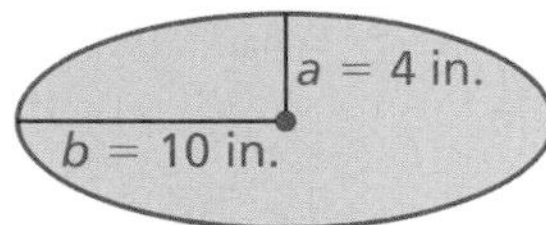

about 48 in.

On Your Own

5. $4 + 2\sqrt{2}$
6. $\frac{-1 - 3\sqrt{3}}{4}$
7. $\frac{1 - \sqrt{7}}{3}$
8. $4\pi\sqrt{29} \approx 68$ ft

Differentiated Instruction

Auditory

Sentences that are worded in a clever way help some students to learn concepts more easily. The sentence "The square root of a product is equal to the product of the square roots" has a structure which helps students remember the Product Property of Square Roots.

Laurie's Notes

Example 3

- **MP7 Look for and Make Use of Structure:** Discuss the operations involved: addition, division, and taking a square root. The fraction bar acts as a grouping symbol. So, the operations in the numerator are performed before the division.
- **?** "Can you write 8 as a product of two factors, one of which is a perfect square other than 1?" yes; 4×2
- Students may say that $\frac{6 + 2\sqrt{2}}{2}$ does not look simpler than $\frac{6 + \sqrt{8}}{2}$. They may also think there is nothing else that can be done to it, because they do not recognize the common factor of 2.
- You can show how to divide out the common factor of 2 in the last step either by using the Distributive Property or by adding fractions.
- **Extension:** Ask students to approximate $3 + \sqrt{2}$. (between 4 and 5)

Example 4

- Draw two concentric circles and label the radii as 15 feet and 20 feet.
- **?** "Can you find the circumference of each circle? Explain." yes; Use the formula $C = 2\pi r$ to obtain $C = 32\pi$ ft and $C = 40\pi$ ft.
- Draw an ellipse inside the concentric circles.
- **?** "How do you think the circumference of the ellipse compares to the circumferences of the two circles?" The circumference of the ellipse is between the circumferences of the two circles.
- Work through the example as shown. The factors of 328 will not be obvious. The divisibility rule for 4 is helpful, however.

16 ft
20 ft

On Your Own

- Ask students to show their solutions for Questions 5–7. Each radicand has a perfect square factor, but the solution methods vary.

Closure

- Which (if any) of the expressions $\sqrt{24}$, $\sqrt{25}$, and $\sqrt{26}$ cannot be simplified? Explain. $\sqrt{26}$ cannot be simplified because it does not have any perfect square factors except 1.

Technology For the Teacher

The Dynamic Planning Tool
Editable Teacher's Resources at *BigIdeasMath.com*

EXAMPLE 3 Simplifying Radical Expressions

Simplify $\frac{6+\sqrt{8}}{2}$.

$$\frac{6+\sqrt{8}}{2} = \frac{6+\sqrt{4 \cdot 2}}{2}$$ Factor the radicand.

$$= \frac{6+\sqrt{4} \cdot \sqrt{2}}{2}$$ Product Property of Square Roots

$$= \frac{6+2\sqrt{2}}{2}$$ Simplify.

$$= 3+\sqrt{2}$$ Divide.

EXAMPLE 4 Real-Life Application

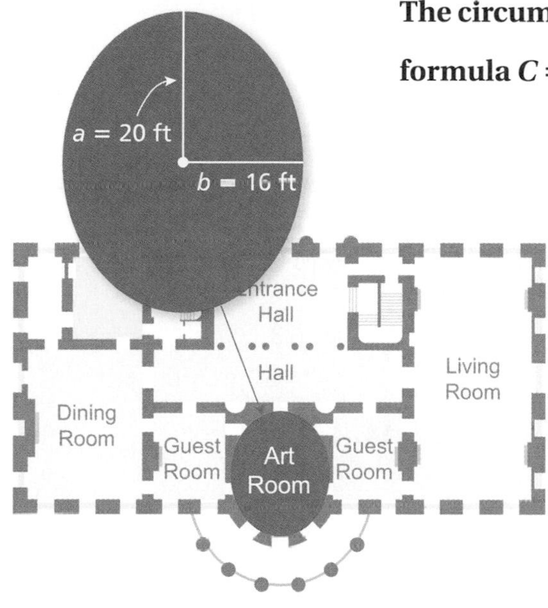

The circumference C of the art room in a mansion is given by the formula $C = 2\pi\sqrt{\frac{a^2+b^2}{2}}$. Find the circumference of the room.

$$C = 2\pi\sqrt{\frac{a^2+b^2}{2}}$$ Write formula.

$$= 2\pi\sqrt{\frac{20^2+16^2}{2}}$$ Substitute.

$$= 2\pi\sqrt{328}$$ Simplify.

$$= 2\pi\sqrt{4 \cdot 82}$$ Factor.

$$= 2\pi \cdot \sqrt{4} \cdot \sqrt{82}$$ Product Property of Square Roots

$$= 4\pi\sqrt{82}$$ Simplify.

The circumference of the room is $4\pi\sqrt{82}$, or about 114 feet.

On Your Own

Now You're Ready
Exercises 21–26

Simplify the expression.

5. $\frac{8+\sqrt{32}}{2}$

6. $\frac{-1-\sqrt{27}}{4}$

7. $\frac{2-\sqrt{28}}{2(3)}$

8. Use the formula in Example 4 to find the circumference of an ellipse in which $a = 14$ feet and $b = 6$ feet.

6.1 Exercises

Vocabulary and Concept Check

1. **WRITING** How do you know when the square root of a positive integer is simplified?
2. **WRITING** How is the Product Property of Square Roots similar to the Quotient Property of Square Roots?

Practice and Problem Solving

Find the dimensions of the square. Check your answer.

3. Area = 64 ft^2

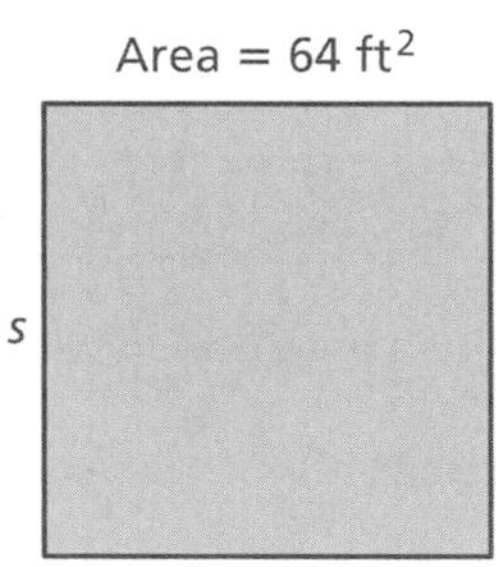

4. Area = 144 in.^{2}

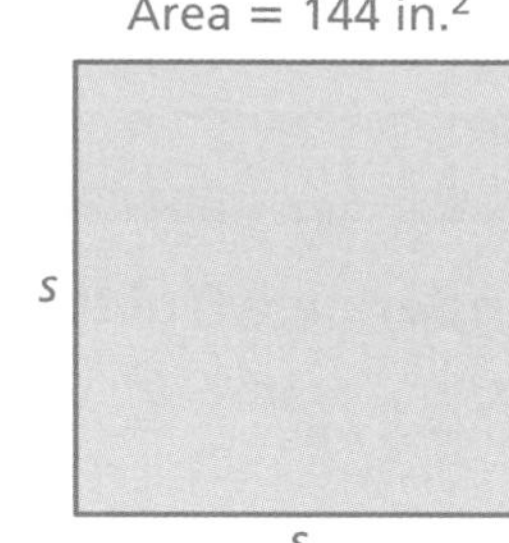

5. Area = $\frac{9}{16}$ cm^2

s, s

Simplify the expression.

1 6. $\sqrt{18}$ 7. $-\sqrt{200}$ 8. $\sqrt{12}$

9. $\sqrt{48}$ 10. $\sqrt{125}$ 11. $-\sqrt{\frac{23}{64}}$

12. $-\sqrt{\frac{65}{121}}$ 13. $\sqrt{\frac{18}{49}}$ 14. $\sqrt{\frac{25}{36}}$

Evaluate the expression when $x = -2$, $y = 8$, and $z = \frac{1}{2}$.

2 15. $\sqrt{x^2 + yz}$ 16. $\sqrt{2x^2 + y^2}$ 17. $\sqrt{y - 44xz}$

18. **ERROR ANALYSIS** Describe and correct the error in simplifying the expression.

$$\sqrt{\frac{20}{9}} = \frac{\sqrt{20}}{\sqrt{9}} = \frac{\sqrt{20}}{3}$$

19. **ELECTRICITY** The electric current I (in amperes) an appliance uses is given by the formula $I = \sqrt{\frac{P}{R}}$, where P is the power (in watts) and R is the resistance (in ohms). Find the current an appliance uses when the power is 147 watts and the resistance is 4 ohms.

Assignment Guide and Homework Check

Level	Assignment	Homework Check
Average	1–5, 6–28 even, 19, 29, 33–36	12, 16, 19, 22, 28
Advanced	1–5, 7–31 odd, 20, 28, 32, 33–36	11, 20, 23, 28, 32

For Your Information

- **Exercises 29–31** Without restrictions on the variables, absolute values are needed for the solution to be equivalent to the original expression. For example, if y can be negative in Exercise 30, then the solution would be $5|y|\sqrt{z}$. Students will learn more about this in a later course.

Common Errors

- **Exercises 6–14** Students may leave a perfect square factor in the radicand, or they may move the wrong factor outside the radical. Remind students to factor the radicand using its greatest perfect square factor, and bring the square root of that factor outside the radical.
- **Exercises 15–17** Students may rewrite the square root of a sum or difference as the sum or difference of square roots. Remind them that there are no properties for addition and subtraction corresponding to the product and quotient properties of square roots.
- **Exercises 21–26** Students may fail to divide out common factors from the numerator and denominator. Remind them to simplify the radicand first, then look for common factors in the numerator and denominator.
- **Exercises 29–31** Students may leave a perfect square factor in the radicand. Remind students to factor out the greatest perfect square factor.

6.1 Record and Practice Journal

1. $\sqrt{44}$
   $2\sqrt{11}$
2. $-\sqrt{175}$
   $-5\sqrt{7}$
3. $-\sqrt{\frac{10}{49}}$
   $-\frac{\sqrt{10}}{7}$
4. $\sqrt{\frac{54}{81}}$
   $\frac{\sqrt{6}}{3}$
5. $\frac{8-\sqrt{112}}{4}$
   $2-\sqrt{7}$
6. $\frac{-6+\sqrt{72}}{18}$
   $\frac{-1+\sqrt{2}}{3}$

Simplify the expression. Assume all variables are positive.

7. $\sqrt{36xy^2}$
   $6y\sqrt{x}$
8. $\sqrt{50x^2yz^3}$
   $5xz\sqrt{2yz}$
9. A trampoline has an area of 49π square feet. What is the diameter of the trampoline?
   14 ft

Answer Presentation Tool

Vocabulary and Concept Check

1. when the radicand has no perfect square factors other than 1
2. both allow you to rewrite and simplify square root expressions

Practice and Problem Solving

3. $s = 8$ ft
4. $s = 12$ in.
5. $s = \frac{3}{4}$ cm
6. $3\sqrt{2}$
7. $-10\sqrt{2}$
8. $2\sqrt{3}$
9. $4\sqrt{3}$
10. $5\sqrt{5}$
11. $-\frac{\sqrt{23}}{8}$
12. $-\frac{\sqrt{65}}{11}$
13. $\frac{3\sqrt{2}}{7}$
14. $\frac{5}{6}$
15. $2\sqrt{2}$
16. $6\sqrt{2}$
17. $2\sqrt{13}$
18. did not simplify the radical in the numerator;
    $\sqrt{\frac{20}{9}} = \frac{\sqrt{20}}{\sqrt{9}} = \frac{2\sqrt{5}}{3}$
19. $\frac{7\sqrt{3}}{2} \approx 6.1$ amperes

Practice and Problem Solving

20. $\frac{\sqrt{14}}{2} \approx 1.9$ sec
21. $3 + \sqrt{11}$
22. $-1 - \sqrt{2}$
23. $2 + 2\sqrt{3}$
24. $\frac{-3 - 4\sqrt{5}}{6}$
25. $\frac{1 + \sqrt{7}}{2}$
26. $\frac{2 - 2\sqrt{2}}{5}$
27. $3\sqrt{10} \approx 9.5$ ft^3
28. $280\sqrt{14} \approx 1047.7$ ft^2
29. $xy\sqrt{42}$
30. $5y\sqrt{z}$
31. $3xy\sqrt{2xz}$
32. See *Taking Math Deeper*.

Fair Game Review

33. 243 34. 16
35. 125 36. B

Mini-Assessment

Simplify the expression.

1. $\sqrt{75}$ $5\sqrt{3}$
2. $-\sqrt{\frac{27}{16}}$ $-\frac{3\sqrt{3}}{4}$
3. $\frac{4 - \sqrt{20}}{2}$ $2 - \sqrt{5}$
4. Evaluate $\sqrt{2x^2 + y^2z}$ when $x = 7$, $y = -4$, and $z = -3$. $5\sqrt{2}$

Taking Math Deeper

Exercise 32

This exercise provides great practice for standard A.CED.4. It will be more challenging for students that do not remember the formula for the surface area of a cube. These students will have to use the diagram to develop a formula.

Develop a formula for the surface area A of the cube.

Each face of the cube is a square with side length s and area s^2.

Cubes have 6 identical faces.

So, a formula for the surface area of the cube is $A = 6s^2$.

To write the side length s as a function of the surface area A, solve the formula for s.

Solve for s.

$A = 6s^2$ Write the formula.

$\frac{A}{6} = s^2$ Divide each side by 6.

$\sqrt{\frac{A}{6}} = s$ Take the positive square root of each side.

Use this function to find the side length of a cube with a surface area of 72 square feet.

$s = \sqrt{\frac{A}{6}}$ Write the function.

$= \sqrt{\frac{72}{6}}$ Substitute 72 for A.

$= \sqrt{12}$ Divide.

$= 2\sqrt{3}$ Simplify.

So, the side length s of the cube is $2\sqrt{3}$, or about 3.5 feet.

Reteaching and Enrichment Strategies

If students need help. . .	If students got it. . .
Resources by Chapter • Practice A and Practice B • Puzzle Time Record and Practice Journal Practice Differentiating the Lesson Lesson Tutorials Skills Review Handbook	Resources by Chapter • Enrichment and Extension Start the next section

20. **BASEBALL** You drop a baseball from a height of 56 feet. Use the expression $\sqrt{\frac{h}{16}}$, where h is the height (in feet), to find the time (in seconds) it takes the baseball to hit the ground.

Simplify the expression.

3 21. $\frac{6 + \sqrt{44}}{2}$

22. $\frac{-7 - \sqrt{98}}{7}$

23. $\frac{10 + \sqrt{300}}{5}$

24. $\frac{-3 - \sqrt{80}}{6}$

25. $\frac{2 + \sqrt{28}}{4}$

26. $\frac{-4 + \sqrt{32}}{-2(5)}$

27. **VOLUME** A pet store installs a new aquarium in your teacher's classroom. What is the volume of the aquarium?

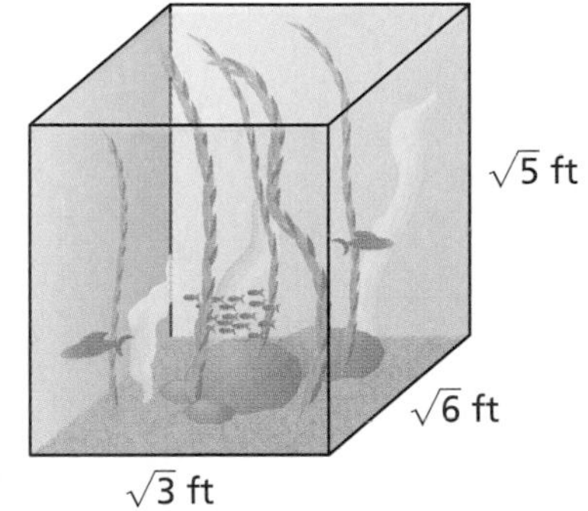

28. **BILLBOARD** What is the area of the rectangular billboard?

Simplify the expression. Assume all variables are positive.

29. $\sqrt{42x^2y^2}$

30. $\sqrt{25y^2z}$

31. $\sqrt{18x^3y^2z}$

32. **Modeling** Write an equation that represents the side length s of a cube as a function of the surface area A of the cube. Find the side length when the surface area is 72 square feet.

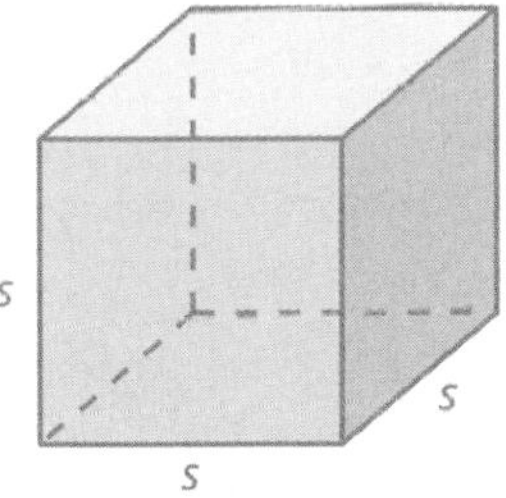

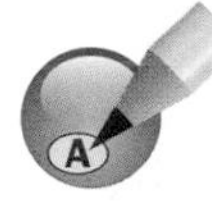

Fair Game Review What you learned in previous grades & lessons

Evaluate the expression. *(Skills Review Handbook)*

33. 3^5

34. 2^4

35. 5^3

36. **MULTIPLE CHOICE** Which value is equivalent to $6(0.2)^3$? *(Skills Review Handbook)*

(A) 0.008 (B) 0.048 (C) 1.728 (D) 3.6

6.1b Real Number Operations

Key Vocabulary
closed, *p. 266*

A set of numbers is **closed** under an operation when the operation performed on any two numbers in the set results in a number that is also in the set. For example, the set of integers is closed under addition, subtraction, and multiplication. This means that if a and b are two integers, then $a + b$, $a - b$, and ab are also integers.

ACTIVITY 1: Sums and Products of Rational Numbers

The table shows several sums and products of rational numbers. Complete the table.

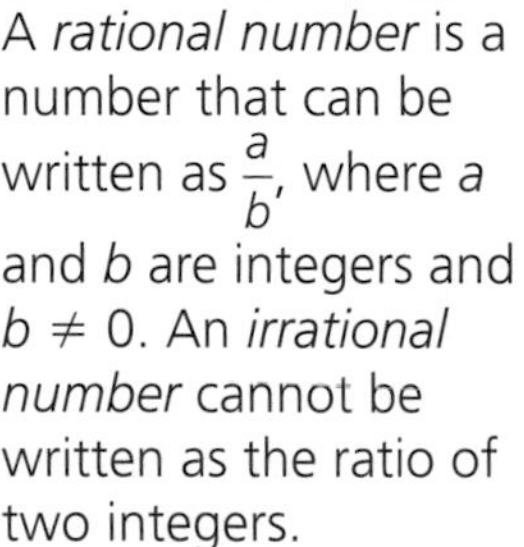

Remember

A *rational number* is a number that can be written as $\frac{a}{b}$, where a and b are integers and $b \neq 0$. An *irrational number* cannot be written as the ratio of two integers.

Sum or Product	Answer	Rational or Irrational?
$12 + 5$		
$-4 + 9$		
$\frac{4}{5} + \frac{2}{3}$		
$0.74 + 2.1$		
3×8		
-4×6		
3.1×0.6		
$\frac{3}{4} \times \frac{5}{7}$		

ACTIVITY 2: Sums of Rational and Irrational Numbers

The table shows several sums of rational and irrational numbers. Complete the table.

Sum	Answer	Rational or Irrational?
$1 + \sqrt{5}$		
$\sqrt{2} + \frac{5}{6}$		
$4 + \pi$		
$-8 + \sqrt{10}$		

Practice

1. Using the results in Activity 1, do you think the set of rational numbers is closed under addition? under multiplication? Explain your reasoning.
2. Using the results in Activity 2, what do you notice about the sum of a rational number and an irrational number?

Laurie's Notes

Introduction

Connect

- **Yesterday:** Students simplified expressions involving square roots. (MP7)
- **Today:** Students will explore closure under addition, subtraction, multiplication, and division for rational and irrational numbers.

Motivate

- Note: The following is an oddity.
- Write the following on the board: $\frac{16}{64} = \frac{1}{4}$ and then explain that you "simplified" $\frac{16}{64}$ by "crossing out" the 6s: $\frac{1\not{6}}{\not{6}4}$.

? "What do you think about this method of simplifying fractions?" Expect a range of comments.

- **Discuss:** Fractions are not simplified in this manner. Only factors can be divided out and the 6s are not factors!
- Explain that in today's lesson they will need to pay attention to the structure of the numbers and how the numbers are represented.

Lesson Notes

Discuss

- Write the definition of a set of numbers *closed* under an operation.
- Give an example: When you add two counting numbers, the sum is another counting number.

? "When you subtract two counting numbers, is the difference another counting number?" not always; $4 - 1$ is a counting number but $1 - 4$ is an integer.

- **MP3a Construct Viable Arguments:** The set of integers is closed under addition, subtraction, and multiplication. Have students give a counterexample to show that integers are not closed under division.

Activity 1

? "What is a rational number?" a number that can be written as $\frac{a}{b}$, where a and b are integers and $b \neq 0$

? "How can you show that 4, 0.4, and $0.\overline{4}$ are all rational numbers?" Rewrite them as $\frac{4}{1}, \frac{4}{10}, \frac{4}{9}$.

- These problems serve as a nice skill check for adding and multiplying rational numbers.

Activity 2

- Because the terms being added are from different sets (rational numbers and irrational numbers), this activity is not about closure.
- Often, you cannot simplify the sum of a rational number and an irrational number. The second sum, however, can be written in another form: $\sqrt{2} + \frac{5}{6} = \frac{6\sqrt{2}}{6} + \frac{5}{6} = \frac{6\sqrt{2} + 5}{6}$.

Goal Today's lesson is determining closure and related patterns for real number operations.

Start Thinking! and Warm Up

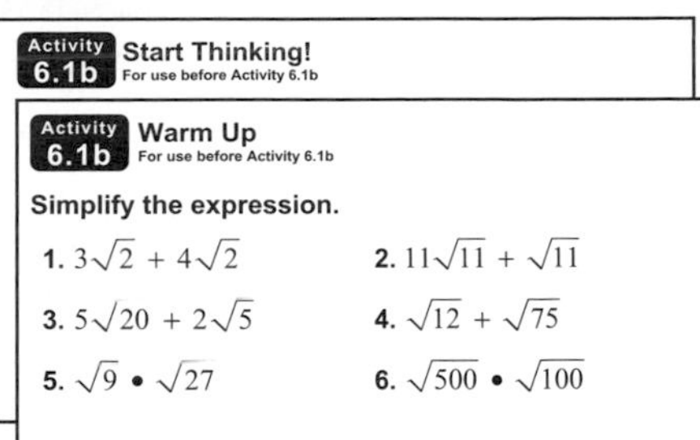

Activity 6.1b Start Thinking! For use before Activity 6.1b

Activity 6.1b Warm Up For use before Activity 6.1b

Simplify the expression.

1. $3\sqrt{2} + 4\sqrt{2}$
2. $11\sqrt{11} + \sqrt{11}$
3. $5\sqrt{20} + 2\sqrt{5}$
4. $\sqrt{12} + \sqrt{75}$
5. $\sqrt{9} \bullet \sqrt{27}$
6. $\sqrt{500} \bullet \sqrt{100}$

6.1b Record and Practice Journal

A set of numbers is **closed** under an operation when the operation performed on any two numbers in the set results in a number that is also in that set. For example, the set of integers is closed under addition, subtraction, and multiplication. This means that if a and b are two integers, then $a + b$, $a - b$, and ab are also integers.

1 ACTIVITY: Sums and Products of Rational Numbers

The table shows several sums and products of rational numbers. Complete the table.

Sum or Product	Answer	Rational or Irrational?
$12 + 5$	17	rational
$-4 + 9$	5	rational
$\frac{4}{5} + \frac{2}{3}$	$\frac{22}{15}$	rational
$0.74 + 2.1$	2.84	rational
3×8	24	rational
-4×6	-24	rational
3.1×0.6	1.86	rational
$\frac{3}{4} \times \frac{5}{7}$	$\frac{15}{28}$	rational

2 ACTIVITY: Sums of Rational and Irrational Numbers

The table shows several sums of rational and irrational numbers. Complete the table.

Sum	Answer	Rational or Irrational?
$1 + \sqrt{5}$	$1 + \sqrt{5}$	irrational
$\sqrt{2} + \frac{5}{6}$	$\frac{6\sqrt{2} + 5}{6}$	irrational
$4 + \pi$	$4 + \pi$	irrational
$-8 + \sqrt{10}$	$-8 + \sqrt{10}$	irrational

Practice

1. Using the results in Activity 1, do you think the set of rational numbers is closed under addition? under multiplication? Explain your reasoning.
   yes; yes
2. Using the results in Activity 2, what do you notice about the sum of a rational number and an irrational number?
   The sum is an irrational number.

Laurie's Notes

Activity 2 (continued)

- Students should recognize that the sum always includes an irrational number.

Practice

- **MP3a:** Students should conclude that the set of rational numbers is closed under addition and multiplication. The sets of whole numbers and integers are also closed under these operations.
- Students should conclude that the sum of a rational number and an irrational number is always irrational.

Activity 3

- This is similar to the last activity except that students are finding the product instead of the sum.
- Remind students that you do not leave a radicand with a perfect square factor greater than 1. The first problem can be simplified as $12\sqrt{3}$.

Activity 4

- Remind students how to solve $3\sqrt{2} + 5\sqrt{2}$. Structurally, this is similar to the problem $3x + 5x$. The underlying property that enables the terms to be added is the Distributive Property.
- Discuss each problem and how the sum or product is found.

Practice

- **MP3a:** In Exercise 5, check whether students can give a counterexample using two different irrational numbers. For instance, a counterexample could be $\sqrt{12} \div \sqrt{3} = \sqrt{4} = 2$ (rational).
- **MP3a:** In Exercise 6, students may need guidance. If students are secure with addition and subtraction of fractions, they should be able to follow the argument written for the general case (with variables).

Closure

- Give an example showing that the sum of two rational numbers can be a whole number.
- Give an example showing that the product of two rational numbers can be a whole number.

Technology For the Teacher

The Dynamic Planning Tool
Editable Teacher's Resources at *BigIdeasMath.com*

6.1b Record and Practice Journal

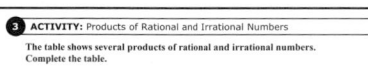

3 ACTIVITY: Products of Rational and Irrational Numbers

The table shows several products of rational and irrational numbers. Complete the table.

Product	Answer	Rational or Irrational?
$6 \cdot \sqrt{12}$	$12\sqrt{3}$	irrational
$-2 \cdot \pi$	-2π	irrational
$\frac{2}{5} \cdot \sqrt{3}$	$\frac{2\sqrt{3}}{5}$	irrational
$0 \times \sqrt{6}$	0	rational

4 ACTIVITY: Sums and Products of Irrational Numbers

The table shows several sums and products of irrational numbers. Complete the table.

Sum or Product	Answer	Rational or Irrational?
$3\sqrt{2} + 5\sqrt{2}$	$8\sqrt{2}$	irrational
$\sqrt{12} + \sqrt{27}$	$5\sqrt{3}$	irrational
$\sqrt{7} + \pi$	$\sqrt{7} + \pi$	irrational
$-\pi + \pi$	0	rational
$\pi \cdot \sqrt{7}$	$\pi\sqrt{7}$	irrational
$\sqrt{5} \times \sqrt{2}$	$\sqrt{10}$	irrational
$4\pi \cdot \sqrt{3}$	$4\pi\sqrt{3}$	irrational
$\sqrt{3} \cdot \sqrt{3}$	3	rational

Practice

3. Using the results in Activity 3, is the product of a rational number and an irrational number always irrational? Explain.

   no

4. Using the results in Activity 4, do you think the set of irrational numbers is closed under addition? under multiplication? Explain your reasoning.

   no; no

5. **CRITICAL THINKING** Is the set of irrational numbers closed under division? If not, find a counterexample. (A *counterexample* is an example that shows that a statement is false.)

   no; *Sample answer:* $\frac{\sqrt{3}}{\sqrt{3}} = 1$

6. **STRUCTURE** The set of integers is closed under addition and multiplication. Use this information to show that the sum and product of two rational numbers are always rational numbers.

   Check students' work.

ACTIVITY 3 Products of Rational and Irrational Numbers

The table shows several products of rational and irrational numbers. Complete the table.

Product	Answer	Rational or Irrational?
$6 \cdot \sqrt{12}$		
$-2 \cdot \pi$		
$\frac{2}{5} \cdot \sqrt{3}$		
$0 \times \sqrt{6}$		

ACTIVITY 4 Sums and Products of Irrational Numbers

The table shows several sums and products of irrational numbers. Complete the table.

Sum or Product	Answer	Rational or Irrational?
$3\sqrt{2} + 5\sqrt{2}$		
$\sqrt{12} + \sqrt{27}$		
$\sqrt{7} + \pi$		
$-\pi + \pi$		
$\pi \cdot \sqrt{7}$		
$\sqrt{5} \times \sqrt{2}$		
$4\pi \cdot \sqrt{3}$		
$\sqrt{3} \times \sqrt{3}$		

Practice

3. Using the results in Activity 3, is the product of a rational number and an irrational number always irrational? Explain.
4. Using the results in Activity 4, do you think the set of irrational numbers is closed under addition? under multiplication? Explain your reasoning.
5. **CRITICAL THINKING** Is the set of irrational numbers closed under division? If not, find a counterexample. (A *counterexample* is an example that shows that a statement is false.)
6. **STRUCTURE** The set of integers is closed under addition and multiplication. Use this information to show that the sum and product of two rational numbers are always rational numbers.

6.2 Properties of Exponents

COMMON CORE STATE STANDARDS
N.RN.2

Essential Question How can you use inductive reasoning to observe patterns and write general rules involving properties of exponents?

1 ACTIVITY: Writing a Rule for Products of Powers

Work with a partner. Write the product of the two powers as a single power. Then, write a *general rule* for finding the product of two powers with the same base.

a. Sample: $(3^4)(3^3) = (3 \cdot 3 \cdot 3 \cdot 3)(3 \cdot 3 \cdot 3) = 3^7$

b. $(2^2)(2^3) =$

c. $(4^1)(4^5) =$

d. $(5^3)(5^5) =$

e. $(x^2)(x^6) =$

2 ACTIVITY: Writing a Rule for Quotients of Powers

Work with a partner. Write the quotient of the two powers as a single power. Then, write a *general rule* for finding the quotient of two powers with the same base.

a. Sample: $\dfrac{3^4}{3^2} = \dfrac{3 \cdot 3 \cdot \cancel{3} \cdot \cancel{3}}{\cancel{3} \cdot \cancel{3}} = 3^2$

b. $\dfrac{4^3}{4^2} =$

c. $\dfrac{2^5}{2^2} =$

d. $\dfrac{x^6}{x^3} =$

e. $\dfrac{3^4}{3^4} =$

3 ACTIVITY: Writing a Rule for Powers of Powers

Work with a partner. Write the expression as a single power. Then, write a *general rule* for finding a power of a power.

a. Sample: $(3^2)^3 = (3 \cdot 3)(3 \cdot 3)(3 \cdot 3) = 3^6$

b. $(2^2)^4 =$

c. $(7^3)^2 =$

d. $(y^3)^3 =$

e. $(x^4)^2 =$

Laurie's Notes

Introduction

Standards for Mathematical Practice

- **MP8 Look for and Express Regularity in Repeated Reasoning:** Today, students will write powers in expanded form to recognize patterns. Students often confuse the five different properties they are discovering today. When this happens, encourage them to recall the process of expanding expressions to develop the properties.

Motivate

- **Story Time:** Tell students you have been offered a special salary for one month. On day 1 you will receive 1¢, on day 2 you will receive 2¢, day 3 is 4¢, and so on, with your salary doubling every school day for the month. There are 23 school days this month. Should you take the new salary?
- Give time for students to start the tabulation. Let them use a calculator for speed. The table below shows the daily pay.

1	$2 = 2^1$	$4 = 2^2$	$8 = 2^3$
$16 = 2^4$	$3^2 = 25$	$64 = 2^6$	$128 = 2^7$
$256 = 2^8$	$512 = 2^9$	$1024 = 2^{10}$	$2048 = 2^{11}$
$4096 = 2^{12}$	$8192 = 2^{13}$	$16{,}384 = 2^{14}$	$32{,}768 = 2^{15}$
$65{,}536 = 2^{16}$	$131{,}072 = 2^{17}$	$262{,}144 = 2^{18}$	$542{,}298 = 2^{19}$
$1{,}048{,}576 = 2^{20}$	$2{,}097{,}152 = 2^{21}$	$4{,}194{,}304 = 2^{22}$	

- In this penny doubling problem, each day you are paid a power of 2. Your salary is actually the *sum* of all of these amounts.

Activity Notes

Discuss

- Inductive reasoning is using specific examples to make a general statement. Encourage students to focus on the title of each activity and to include the steps for writing the expanded expressions as they copy each example.

Activity 1

? "What does 3^3 mean?" $3 \times 3 \times 3$ "What does the exponent of a power tell you?" how many times to use the base as a factor

- In writing the general rule, expect correct vocabulary, such as "To multiply two powers with the same base, add the exponents and keep the base."

Activity 2

- In this activity, you divide out common factors to simplify the expression.
- Expect correct vocabulary, such as "When you divide two powers with the same base, subtract the exponents and keep the same base."
- **FYI:** All the quotients result in nonnegative exponents. The formal lesson includes quotients that result in negative exponents.

Common Core State Standards

N.RN.2 Rewrite expressions involving radicals and rational exponents using the properties of exponents.

Previous Learning

Students should know how to evaluate expressions involving integer exponents.

Activity Materials
Introduction
• calculators

Start Thinking! and Warm Up

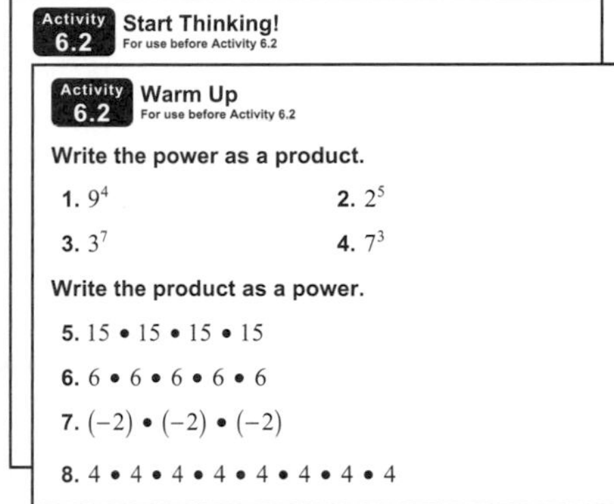

Activity 6.2 **Start Thinking!** For use before Activity 6.2

Activity 6.2 **Warm Up** For use before Activity 6.2

Write the power as a product.

1. 9^4
2. 2^5
3. 3^7
4. 7^3

Write the product as a power.

5. $15 \bullet 15 \bullet 15 \bullet 15$
6. $6 \bullet 6 \bullet 6 \bullet 6 \bullet 6$
7. $(-2) \bullet (-2) \bullet (-2)$
8. $4 \bullet 4 \bullet 4 \bullet 4 \bullet 4 \bullet 4 \bullet 4 \bullet 4$

6.2 Record and Practice Journal

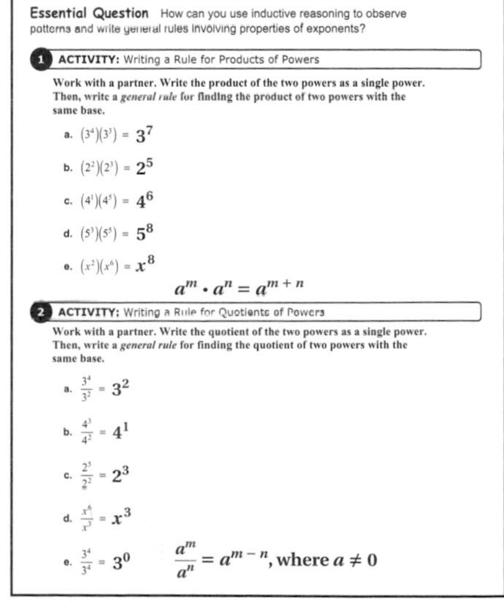

Essential Question How can you use inductive reasoning to observe patterns and write general rules involving properties of exponents?

1 ACTIVITY: Writing a Rule for Products of Powers

Work with a partner. Write the product of the two powers as a single power. Then, write a *general rule* for finding the product of two powers with the same base.

a. $(3^4)(3^3) = 3^7$

b. $(2^2)(2^3) = 2^5$

c. $(4^1)(4^5) = 4^6$

d. $(5^3)(5^5) = 5^8$

e. $(x^2)(x^6) = x^8$

$$a^m \cdot a^n = a^{m+n}$$

2 ACTIVITY: Writing a Rule for Quotients of Powers

Work with a partner. Write the quotient of the two powers as a single power. Then, write a *general rule* for finding the quotient of two powers with the same base.

a. $\frac{3^4}{3^2} = 3^2$

b. $\frac{4^3}{4^2} = 4^1$

c. $\frac{2^5}{2^2} = 2^3$

d. $\frac{x^6}{x^3} = x^3$

e. $\frac{3^4}{3^4} = 3^0$ $\quad \frac{a^m}{a^n} = a^{m-n}$, where $a \neq 0$

English Language Learners

Vocabulary

Remind English learners that when they see a negative exponent, they should think *reciprocal*. Review the meaning of the word reciprocal. Students often think that because the exponent is negative, the expression is negative. Remind them that a number of the form x^a cannot be negative unless the base is negative.

6.2 Record and Practice Journal

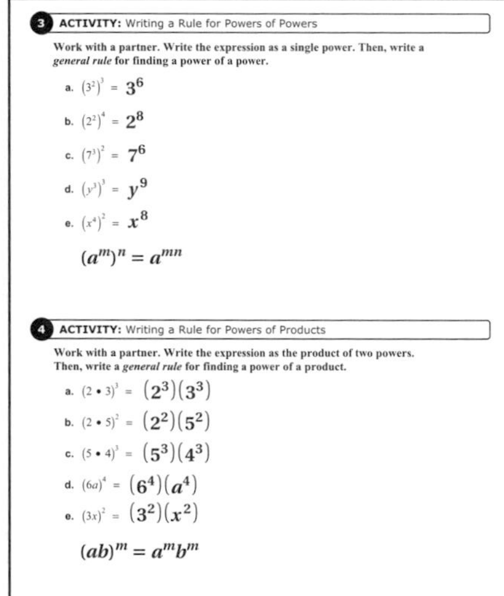

3 ACTIVITY: Writing a Rule for Powers of Powers

Work with a partner. Write the expression as a single power. Then, write a *general rule* for finding a power of a power.

a. $(3^2)^3 = 3^6$

b. $(2^2)^4 = 2^8$

c. $(7^3)^2 = 7^6$

d. $(y^3)^3 = y^9$

e. $(x^4)^2 = x^8$

$(a^m)^n = a^{mn}$

4 ACTIVITY: Writing a Rule for Powers of Products

Work with a partner. Write the expression as the product of two powers. Then, write a *general rule* for finding a power of a product.

a. $(2 \cdot 3)^3 = (2^3)(3^3)$

b. $(2 \cdot 5)^2 = (2^2)(5^2)$

c. $(5 \cdot 4)^3 = (5^3)(4^3)$

d. $(6a)^4 = (6^4)(a^4)$

e. $(3x)^2 = (3^2)(x^2)$

$(ab)^m = a^m b^m$

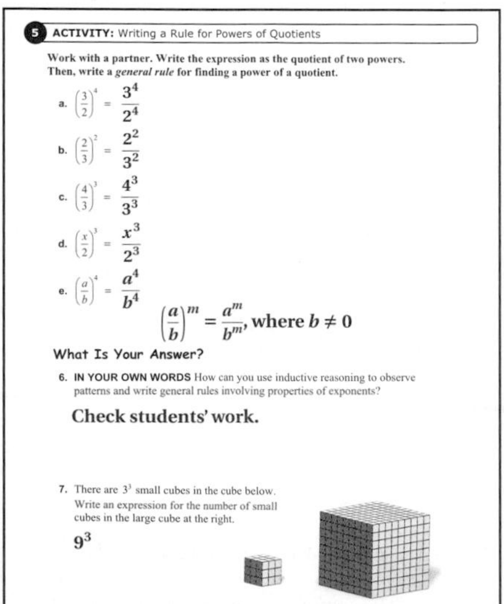

5 ACTIVITY: Writing a Rule for Powers of Quotients

Work with a partner. Write the expression as the quotient of two powers. Then, write a *general rule* for finding a power of a quotient.

a. $\left(\frac{3}{2}\right)^4 = \frac{3^4}{2^4}$

b. $\left(\frac{2}{3}\right)^2 = \frac{2^2}{3^2}$

c. $\left(\frac{4}{3}\right)^3 = \frac{4^3}{3^3}$

d. $\left(\frac{x}{2}\right)^3 = \frac{x^3}{2^3}$

e. $\left(\frac{a}{b}\right)^4 = \frac{a^4}{b^4}$

$\left(\frac{a}{b}\right)^m = \frac{a^m}{b^m}$, where $b \neq 0$

What Is Your Answer?

6. IN YOUR OWN WORDS How can you use inductive reasoning to observe patterns and write general rules involving properties of exponents?

Check students' work.

7. There are 3^3 small cubes in the cube below. Write an expression for the number of small cubes in the large cube at the right.

9^3

Laurie's Notes

Activity 3

- Students often confuse this property with the Product of Powers Property (Activity 1). Make sure students can explain how $4^3 \cdot 4^2$ is different from $(4^3)^2$. Expanding the expressions helps to demonstrate this.
- When students have difficulty recognizing a pattern, have them expand the power in stages: $(2^2)^4 = 2^2 \cdot 2^2 \cdot 2^2 \cdot 2^2 = (2 \cdot 2)(2 \cdot 2)(2 \cdot 2)(2 \cdot 2) = 2^8$.
- Expect correct vocabulary, such as "When you raise a power to a power, multiply the exponents and keep the same base."

Activity 4

- "Products of powers" and "powers of products" sound very similar to students. Ask a volunteer to describe the difference.
- In an expression such as $(6a)^4$, students are often unsure how to simplify $6a \cdot 6a \cdot 6a \cdot 6a$. The Commutative and Associative Properties allow you to rewrite this as $6 \cdot 6 \cdot 6 \cdot 6 \cdot a \cdot a \cdot a \cdot a$.
- Expect correct vocabulary, such as "When you raise a product to a power, raise each base to the power and multiply the two powers."

Activity 5

- "Quotients of powers" and "powers of quotients" sound very similar to students. Ask a volunteer to describe the difference.
- **Common Error:** Students may think that the exponent is "distributed" in the same way that the factor is distributed in the Distributive Property. Raising to a power and distributing are not the same.
- Expect correct vocabulary, such as "When you raise a quotient to a power, raise both the numerator and denominator to the power."

What Is Your Answer?

- **MP8** Both questions involve repeated reasoning.

Closure

- Write an equivalent expression for the following.

a. $(3^2)(3^3)$ 3^5 **b.** $\frac{3^5}{3^2}$ 3^3 **c.** $(3^2)^3$ 3^6

d. $(2 \cdot 3)^3$ $2^3 \cdot 3^3$ **e.** $\left(\frac{2}{3}\right)^3$ $\frac{2^3}{3^3}$

Technology For the Teacher

The Dynamic Planning Tool

Editable Teacher's Resources at *BigIdeasMath.com*

4 ACTIVITY: Writing a Rule for Powers of Products

Work with a partner. Write the expression as the product of two powers. Then, write a *general rule* for finding a power of a product.

a. Sample: $(2 \cdot 3)^3 = (2 \cdot 3)(2 \cdot 3)(2 \cdot 3) = (2^3)(3^3)$

b. $(2 \cdot 5)^2 =$

c. $(5 \cdot 4)^3 =$

d. $(6a)^4 =$

e. $(3x)^2 =$

5 ACTIVITY: Writing a Rule for Powers of Quotients

Work with a partner. Write the expression as the quotient of two powers. Then, write a *general rule* for finding a power of a quotient.

a. Sample: $\left(\frac{3}{2}\right)^4 = \frac{3}{2} \cdot \frac{3}{2} \cdot \frac{3}{2} \cdot \frac{3}{2} = \frac{3 \cdot 3 \cdot 3 \cdot 3}{2 \cdot 2 \cdot 2 \cdot 2} = \frac{3^4}{2^4}$

b. $\left(\frac{2}{3}\right)^2 =$

c. $\left(\frac{4}{3}\right)^3 =$

d. $\left(\frac{x}{2}\right)^3 =$

e. $\left(\frac{a}{b}\right)^4 =$

What Is Your Answer?

6. **IN YOUR OWN WORDS** How can you use inductive reasoning to observe patterns and write general rules involving properties of exponents?

7. There are 3^3 small cubes in the cube below. Write an expression for the number of small cubes in the large cube at the right.

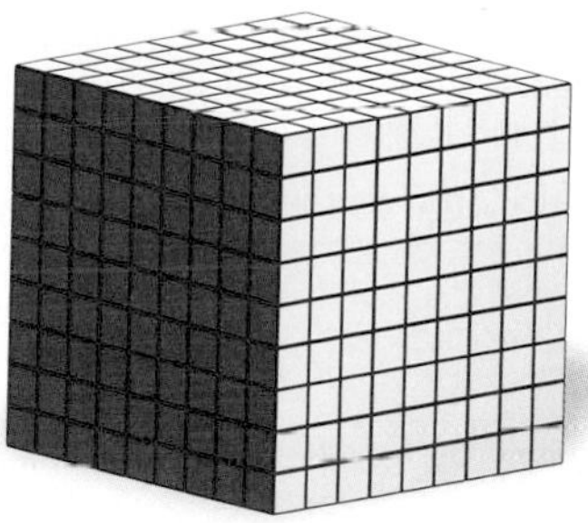

Practice Use what you learned about exponents to complete Exercises 6–11 on page 273.

6.2 Lesson

Product of Powers Property

Words To multiply powers with the same base, add their exponents.

Numbers $4^6 \cdot 4^3 = 4^{6+3} = 4^9$ **Algebra** $a^m \cdot a^n = a^{m+n}$

Quotient of Powers Property

Words To divide powers with the same base, subtract their exponents.

Numbers $\dfrac{4^6}{4^3} = 4^{6-3} = 4^3$ **Algebra** $\dfrac{a^m}{a^n} = a^{m-n}$, where $a \neq 0$

Power of a Power Property

Words To find a power of a power, multiply the exponents.

Numbers $(4^6)^3 = 4^{6 \cdot 3} = 4^{18}$ **Algebra** $(a^m)^n = a^{mn}$

Remember

For any integer n and any nonzero integer a, $a^0 = 1$ and $a^{-n} = \dfrac{1}{a^n}$.

EXAMPLE 1 Using Properties of Exponents

Simplify. Write your answer using only positive exponents.

a. $3^2 \cdot 3^6 = 3^{2+6}$ Product of Powers Property

$= 3^8$ Simplify.

The base is 3. Add the exponents.

b. $\dfrac{(-4)^2}{(-4)^7} = (-4)^{2-7}$ Quotient of Powers Property

$= (-4)^{-5}$ Simplify.

$= \dfrac{1}{(-4)^5}$ Definition of negative exponent

The base is -4. Subtract the exponents.

c. $(z^4)^{-3} = z^{4 \cdot (-3)}$ Power of a Power Property

$= z^{-12}$ Simplify.

$= \dfrac{1}{z^{12}}$ Definition of negative exponent

The base is z. Multiply the exponents.

On Your Own

Exercises 12–17

Simplify. Write your answer using only positive exponents.

1. $10^4 \cdot 10^{-6}$
2. $x^9 \cdot x^{-9}$
3. $\dfrac{-5^8}{-5^4}$
4. $\dfrac{y^6}{y^7}$
5. $(6^{-2})^{-5}$
6. $(w^{12})^5$

Laurie's Notes

Introduction

Connect

- **Yesterday:** Students discovered properties of exponents. (MP8)
- **Today:** Students will use properties of exponents to simplify expressions.

Motivate

- The $10,000 bill, which is no longer in circulation, would be much easier to carry than the same amount in pennies.

? Ask a few questions about money.

  - "How many pennies equal $10,000?" $100 \times 10{,}000 = 1{,}000{,}000$ or 10^6
  - "How many dimes equal $10,000?" $10 \times 10{,}000 = 100{,}000$ or 10^5
  - "How many $10 bills equal $10,000?" $\frac{1}{10} \times 10{,}000 = 1000$ or 10^3
  - "How many $100 bills equal $10,000?" $\frac{1}{100} \times 10{,}000 = 100$ or 10^2

Lesson Notes

Discuss

- **MP7 Look for and Make Use of Structure:** Before beginning the formal lesson, review zero and negative exponents: $a^0 = 1$ and $a^{-n} = \frac{1}{a^n}$. Also, remind students that 0^0 is undefined.

Key Ideas

- Write and discuss the three properties. These three properties are grouped together because they are about an operation being done to one or more powers (multiplying, dividing, raising to a power).
- If needed, write the expanded form of the powers in each example.
- Stress that in the first two properties the base is the same.

Example 1

? Write the first expression and ask, "What is the common base?" 3

? "We are multiplying two powers with the same base. What do we do with the exponents?" Add the exponents.

- The second expression is different from the quotients in the activity because the exponent in the denominator is greater than the exponent in the numerator. Use the Quotient of Powers Property to simplify. Then verify that the results are the same when the expressions are expanded and the common factors are divided out.
- **Common Error:** Students believe that $(-4)^{-5}$ is the final answer. Remind them that answers need to be written using only positive exponents.

On Your Own

- Ask what the base is for Question 3. It is not -5. If -5 were the base, then the problem would be written $\frac{(-5)^8}{(-5)^4}$.

Goal Today's lesson is simplifying expressions involving exponents.

Start Thinking! and Warm Up

Lesson 6.2 Warm Up
For use before Lesson 6.2

Lesson 6.2 Start Thinking!
For use before Lesson 6.2

In Activity 6.2, you found the rule for finding the product of powers: $a^m \bullet a^n = a^{m+n}$.

Which of the following expressions can be simplified using the product of powers rule?

$2^6 \bullet 2^7$ $\quad 3^2 \bullet 4^2$ $\quad (-2)^2 \bullet (-2)^{12}$

$c^3 \bullet c^4$ $\quad x^3 \bullet y^1$ $\quad a^2 \bullet 4^2$

Extra Example 1

Simplify. Write your answer using only positive exponents.

a. $(-5)^3(-5)^6$ $\quad (-5)^9$

b. $\frac{x^4}{x^6}$ $\quad \frac{1}{x^2}$

c. $(7^3)^{-5}$ $\quad \frac{1}{7^{15}}$

On Your Own

1. $\frac{1}{10^2}$
2. 1
3. 5^4
4. $\frac{1}{y}$
5. 6^{10}
6. w^{60}

Laurie's Notes

Key Ideas

- Write and discuss the two properties. These two properties involve two different bases.
- **Common Misconception:** Students often refer to both of these properties as "distributing" the exponent. They will say, "When you distribute the exponent . . ." This is not the Distributive Property.

Example 2

- **FYI:** In this text, expressions containing variables will be completely simplified, as in Example 2. Expressions without variables may not be completely simplified, as in Example 1.
- Write the first expression. There are two factors, -1.5 and y. The product is squared. Using the Power of a Product Property, each factor is squared.
- **Common Error:** In a problem such as $(-1.5y)^2$, students will often square the variable, but not the coefficient.
- The solution for the third expression has several steps. The Power of a Quotient Property is used in the first step. The next step is *not* obvious to students. A negative exponent is defined as $a^{-n} = \frac{1}{a^n}$. Students do not see this as the "reciprocal of a^n." Discuss the definition using the idea of a "reciprocal."
- **MP7:** Off to the side, do a simpler problem such as

  $$\left(\frac{3}{4}\right)^{-2} = \frac{3^{-2}}{4^{-2}} = 3^{-2} \cdot \frac{1}{4^{-2}} = \frac{1}{3^2} \cdot 4^2 = \frac{4^2}{3^2} = \left(\frac{4}{3}\right)^2.$$

  Discuss the result. $\frac{3}{4}$ and $\frac{4}{3}$ are reciprocals.

On Your Own

- In Question 8, students may ask whether to place the negative sign with the 4, n, or both when they rewrite the expression as a quotient of powers. Thinking of the expression as a negative quantity with an odd exponent means that the final answer is negative, so they can place the negative sign in front of the whole quotient.
- In Question 9, several properties must be used to simplify the expression. Have a student show the solution and talk through the steps as they are written.

Extra Example 2

Simplify. Write your answer using only positive exponents.

a. $(-2x)^3$ $-8x^3$

b. $\left(\frac{-3}{b}\right)^2$ $\frac{9}{b^2}$

c. $\left(\frac{4}{3y}\right)^{-3}$ $\frac{27y^3}{64}$

On Your Own

7. $\frac{1}{1000y^3}$
8. $-\frac{1024}{n^5}$
9. $\frac{1}{32k^{10}}$
10. $\frac{49}{36c^2}$

Differentiated Instruction

Visual

Remind students that the Product of Powers Property can only be applied to powers having the same base. Have students highlight each unique base with a different color. Then add the exponents.

$$\begin{aligned} 4^2 \cdot 4^3 - (3^2)^2 &= 4^2 \cdot 4^3 - 3^2 \cdot 3^2 \\ &= 4^{2+3} - 3^{2+2} \\ &= 4^5 - 3^4 \\ &= 1024 - 81 \\ &= 943 \end{aligned}$$

Key Ideas

Power of a Product Property

Words To find a power of a product, find the power of each factor and multiply.

Numbers $(3 \cdot 2)^5 = 3^5 \cdot 2^5$ **Algebra** $(ab)^m = a^m b^m$

Power of a Quotient Property

Words To find a power of a quotient, find the power of the numerator and the power of the denominator and divide.

Numbers $\left(\frac{3}{2}\right)^5 = \frac{3^5}{2^5}$ **Algebra** $\left(\frac{a}{b}\right)^m = \frac{a^m}{b^m}$, where $b \neq 0$

EXAMPLE 2 Using Properties of Exponents

Simplify. Write your answer using only positive exponents.

a. $(-1.5y)^2 = (-1.5)^2 \cdot y^2$ Power of a Product Property

$= 2.25y^2$ Simplify.

b. $\left(\frac{a}{-10}\right)^3 = \frac{a^3}{(-10)^3}$ Power of a Quotient Property

$= -\frac{a^3}{1000}$ Simplify.

c. $\left(\frac{2x}{3}\right)^{-5} = \frac{(2x)^{-5}}{3^{-5}}$ Power of a Quotient Property

$= \frac{3^5}{(2x)^5}$ Definition of negative exponent

$= \frac{3^5}{2^5 x^5}$ Power of a Product Property

$= \frac{243}{32x^5}$ Simplify.

On Your Own

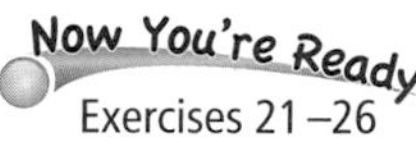
Now You're Ready
Exercises 21–26

Simplify. Write your answer using only positive exponents.

7. $(10y)^{-3}$

8. $\left(-\frac{4}{n}\right)^5$

9. $\left(\frac{1}{2k^2}\right)^5$

10. $\left(\frac{6c}{7}\right)^{-2}$

EXAMPLE 3 Standardized Test Practice

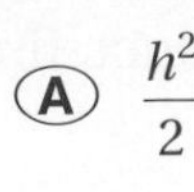

Which expression represents the volume of the cylinder?

Ⓐ $\dfrac{h^2}{2}$ Ⓑ $\dfrac{\pi h^2}{4}$ Ⓒ $\dfrac{\pi h^3}{2}$ Ⓓ $\dfrac{\pi h^3}{4}$

$V = \pi r^2 h$	Formula for volume of a cylinder
$= \pi\left(\dfrac{h}{2}\right)^2 (h)$	Substitute $\dfrac{h}{2}$ for r.
$= \pi\left(\dfrac{h^2}{2^2}\right)(h)$	Power of a Quotient Property
$= \dfrac{\pi h^3}{4}$	Simplify.

∴ The correct answer is Ⓓ.

EXAMPLE 4 Real-Life Application

A jellyfish emits about 1.25×10^8 particles of light, or photons, in 6.25×10^{-4} second. How many photons does the jellyfish emit each second? Write your answer in scientific notation and in standard form.

Divide to find the unit rate.

$\dfrac{1.25 \times 10^8}{6.25 \times 10^{-4}}$ ← photons / ← seconds	Write the rate.
$= \dfrac{1.25}{6.25} \times \dfrac{10^8}{10^{-4}}$	Rewrite.
$= 0.2 \times 10^{12}$	Simplify.
$= 2 \times 10^{11}$	Write in scientific notation.

Remember

A number is written in scientific notation when it is of the form $a \times 10^b$, where $1 \le a < 10$ and b is an integer.

∴ The jellyfish emits 2×10^{11}, or 200,000,000,000 photons per second.

On Your Own

11. In Example 3, which expression represents the area of a base of the cylinder?

12. It takes the Sun about 2.3×10^8 years to orbit the center of the Milky Way. It takes Pluto about 2.5×10^2 years to orbit the Sun. How many times does Pluto orbit the Sun while the Sun completes one orbit around the Milky Way? Write your answer in scientific notation.

Laurie's Notes

Example 3

? "How do you find the volume of a cylinder?" Multiply the area of the base and the height.

? "What is the area of the base?" $\pi r^2 = \pi\left(\frac{h}{2}\right)^2$

- In the third step, make sure students square both the numerator and denominator. When a variable and constant are involved, students sometimes think that only the variable should be squared.

Example 4

? "How do you say the number of photons given in the problem statement in standard form?" one hundred twenty-five million

? "How do you say the amount of time in standard form?" six hundred twenty-five millionths

- The question is asking about a rate: photons emitted per second.

? "What is the difference between a ratio and a rate?" A rate is a special ratio. A ratio is a comparison of two quantities using division, and a rate is a ratio of two quantities with different units.

- In the second step, the expression has been rewritten. To explain this step, use a simpler problem: $\frac{ab}{cd} = \frac{a}{c} \cdot \frac{b}{d}$. This is how fractions are multiplied.
- Simplify each expression.
- You may need to remind students of how to write a number in scientific notation. Refer to the Remember Box.

On Your Own

? "In Question 12, what is the dividend?" 2.3×10^8

? "In Question 12, what is the divisor?" 2.5×10^2

Closure

- Give the name of each property and ask students to write an example using the property.

Technology For the Teacher

The Dynamic Planning Tool
Editable Teacher's Resources at *BigIdeasMath.com*

Extra Example 3

Write an expression for the volume of the cylinder.

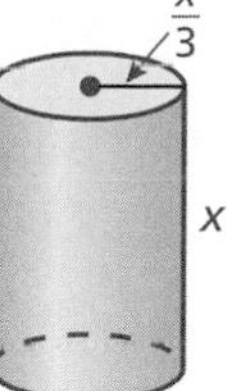

$\frac{\pi x^3}{9}$

Extra Example 4

A 100-watt light bulb emits about 8×10^8 photons in 8×10^{-10} second. How many photons does the light bulb emit each second? Write your answer in scientific notation and in standard form.

The light bulb emits about 1×10^{18}, or 1,000,000,000,000,000,000 photons each second.

On Your Own

11. B

12. 9.2×10^5 times

Vocabulary and Concept Check

1. D
2. A
3. B
4. C
5. Simplify 3^{6-3}; 3^3; 3^9

Practice and Problem Solving

6. n^7
7. x^2
8. c^{15}
9. $64b^3$
10. $\frac{k^5}{243}$
11. $64a^4$
12. 8^5
13. b^{11}
14. 12^5
15. $\frac{1}{d^3}$
16. 5^{20}
17. $\frac{1}{x^6}$
18. The exponents were multiplied instead of added; $x^5 \cdot x^{-2} = x^{5+(-2)} = x^3$
19. The exponents were added instead of multiplied; $(m^3)^4 = m^{3 \cdot 4} = m^{12}$
20. 10^7 nanometers

Assignment Guide and Homework Check

Level	Assignment	Homework Check
Average	1–5, 7–27 odd, 20, 31, 32, 33, 36, 40, 47–50	13, 19, 20, 23, 32, 36
Advanced	1–5, 12–32 even, 33, 36, 40, 42–46, 47–50	18, 26, 32, 36, 44, 46

For Your Information

- **Exercises 6–26** If students have trouble with the properties, encourage them to write the expressions as repeated multiplication and then simplify.

Common Errors

- **Exercises 6, 12, and 13** Students may multiply the exponents instead of adding the exponents. Remind them of the Product of Powers Property.
- **Exercises 9, 11, 21, and 24–26** Students may forget to apply the exponent to the coefficient. Remind them of the Power of a Product Property.
- **Exercises 7, 11, 14, and 15** Students may divide the exponents instead of subtracting the exponents. Remind them of the Quotient of Powers Property.
- **Exercises 10, 22, 23, and 26** Students may forget to apply the exponent to the denominator. Remind them of the Power of a Quotient Property.
- **Exercises 8, 16, and 17** Students may add the exponents instead of multiplying the exponents. Remind them of the Power of a Power Property.

6.2 Record and Practice Journal

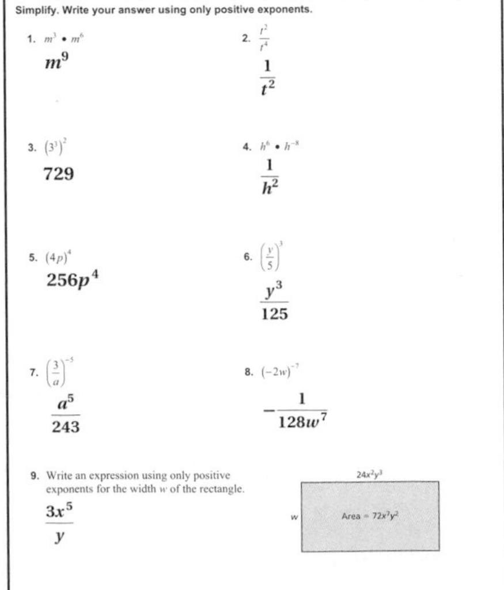

Simplify. Write your answer using only positive exponents.

1. $m^3 \cdot m^6$

   m^9

2. $\frac{t^2}{t^4}$

   $\frac{1}{t^2}$

3. $(3^3)^2$

   729

4. $h^6 \cdot h^{-8}$

   $\frac{1}{h^2}$

5. $(4p)^4$

   $256p^4$

6. $\left(\frac{y}{5}\right)^3$

   $\frac{y^3}{125}$

7. $\left(\frac{3}{a}\right)^{-5}$

   $\frac{a^5}{243}$

8. $(-2w)^{-7}$

   $-\frac{1}{128w^7}$

9. Write an expression using only positive exponents for the width w of the rectangle.

   $\frac{3x^5}{y}$

Technology For the Teacher

Answer Presentation Tool

6.2 Exercises

Vocabulary and Concept Check

MATCHING **Match the property with its example.**

1. Quotient of Powers Property
2. Power of a Power Property
3. Power of a Quotient Property
4. Power of a Product Property

A. $(4^5)^2 = 4^{5 \cdot 2}$ B. $\left(\frac{5}{2}\right)^4 = \frac{5^4}{2^4}$ C. $(5 \cdot 2)^4 - 5^4 \cdot 2^4$ D. $\frac{4^5}{4^2} = 4^{5-2}$

5. **DIFFERENT WORDS, SAME QUESTION** Which is different? Find "both" answers.

Simplify $3^3 \cdot 3^6$.	Simplify 3^{3+6}.	Simplify 3^{6-3}.	Simplify $3^6 \cdot 3^3$.

Practice and Problem Solving

Simplify the expression.

6. $(n^4)(n^3)$
7. $\frac{x^5}{x^3}$
8. $(c^5)^3$
9. $(4b)^3$
10. $\left(\frac{k}{3}\right)^5$
11. $\frac{(2a)^6}{a^2}$

Simplify. Write your answer using only positive exponents.

1 12. $8^{-2} \cdot 8^7$
13. $b^4 \cdot b^7$
14. $\frac{12^7}{12^2}$
15. $\frac{d^5}{d^8}$
16. $(5^5)^4$
17. $(x^3)^{-2}$

ERROR ANALYSIS **Describe and correct the error in simplifying the expression.**

18.

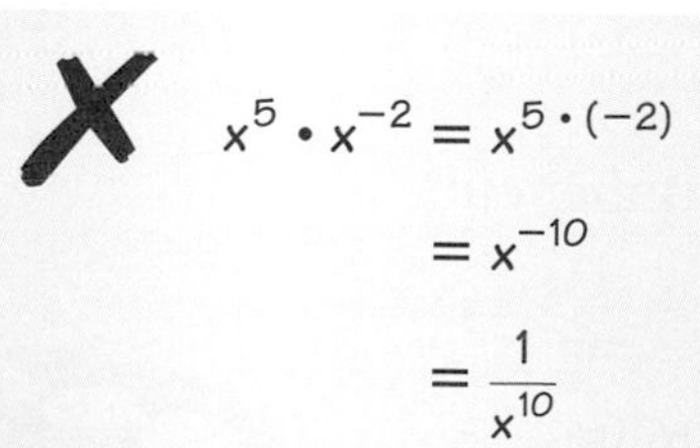

19.

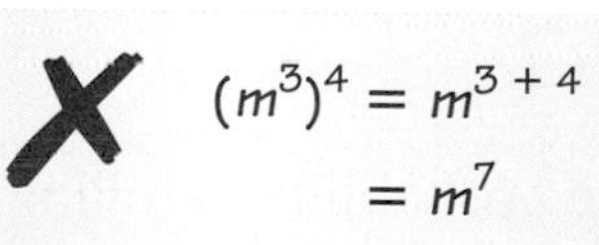

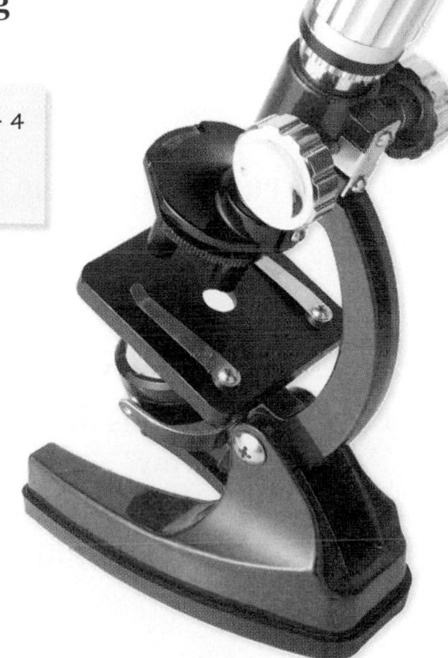

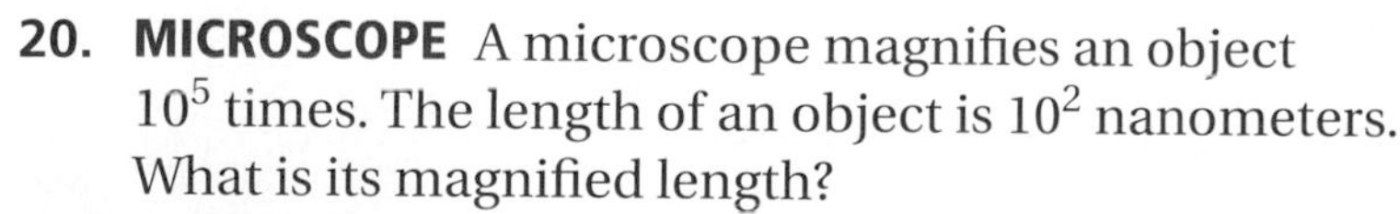

20. **MICROSCOPE** A microscope magnifies an object 10^5 times. The length of an object is 10^2 nanometers. What is its magnified length?

Simplify. Write your answer using only positive exponents.

2 3 **21.** $(6.2y)^2$ **22.** $\left(\frac{w}{4}\right)^4$ **23.** $\left(-\frac{6}{d}\right)^{-2}$

24. $(7p)^{-3}$ **25.** $(-5x)^5$ **26.** $\left(\frac{3n^3}{4}\right)^2$

27. ERROR ANALYSIS Describe and correct the error in simplifying the expression.

✗ $\left(\frac{x^3}{3}\right)^2 = \frac{(x^3)^2}{3} = \frac{x^6}{3}$

28. OPEN-ENDED Use the properties of exponents to write three expressions equivalent to x^8.

29. REASONING Are the expressions $(a^4)^2$ and a^{4^2} equivalent? Explain your reasoning.

30. GEOMETRY Consider Cube A and Cube B.

a. Which property of exponents should you use to find the volume of each cube?

b. How can you use the Power of a Quotient Property to find how many times greater the volume of Cube B is than the volume of Cube A?

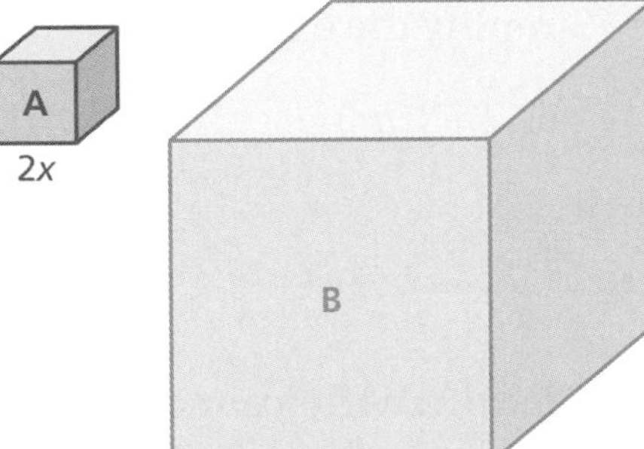

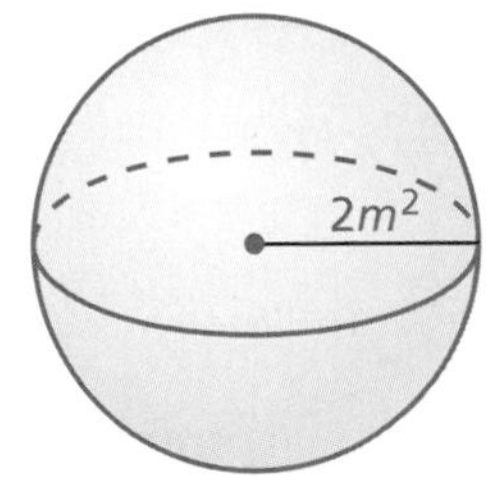

31. SPHERE The volume V of a sphere is $V = \frac{4}{3}\pi r^3$, where r is the radius. What is the volume of the sphere in terms of m and π?

32. PROBABILITY The probability of rolling a 6 on a number cube is $\frac{1}{6}$. The probability of rolling a 6 twice in a row is $\left(\frac{1}{6}\right)^2 = \frac{1}{36}$.

a. Write an expression that represents the probability of rolling a 6 n times in a row.

b. What is the probability of rolling a 6 five times in a row?

c. What is the probability of flipping heads on a coin five times in a row?

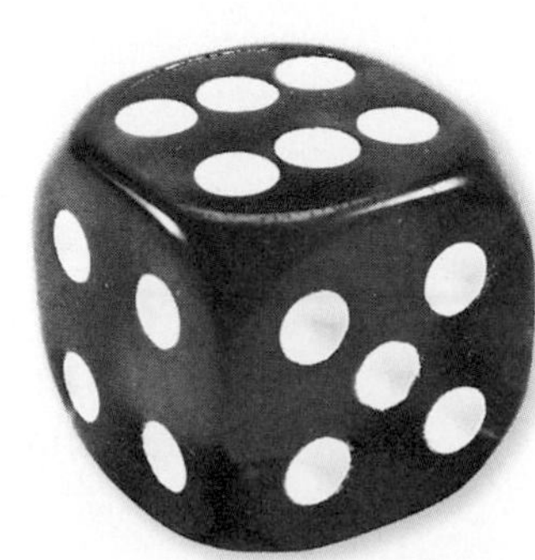

Common Errors

- **Exercises 33–38, and 44** Students may find the product or quotient and leave the decimal factor in the wrong form. Remind them that the factor in scientific notation must be greater than or equal to 1 and less than 10.
- **Exercises 39–41** Students may try to combine powers that do not have the same base. Remind them that the Product of Powers and Quotient of Powers Properties can only be used with powers that have the same base.

Practice and Problem Solving

21. $38.44y^2$ **22.** $\frac{w^4}{256}$

23. $\frac{d^2}{36}$ **24.** $\frac{1}{343p^3}$

25. $-3125x^5$ **26.** $\frac{9n^6}{16}$

27. The exponent was not applied to the denominator.

$\left(\frac{x^3}{3}\right)^2 = \frac{(x^3)^2}{3^2} = \frac{x^6}{9}$

28. *Sample answer:* $x^3 \cdot x^5$; $(x^2)^4$; $\frac{x^2}{x^{-6}}$

29. no; $(a^4)^2 = a^8$ by the Power of a Power Property. $a^{4^2} = a^{16}$ because $4^2 = 16$.

30. **a.** Power of a Product Property

b. Find the ratio of the volumes $\frac{(8x)^3}{(2x)^3}$. Using the Power of a Quotient Property, $\frac{(8x)^3}{(2x)^3} = \left(\frac{8x}{2x}\right)^3$. This simplifies to $4^3 = 64$. The volume of Cube B is 64 times the volume of Cube A.

31. $V = \frac{32\pi m^6}{3}$

32. **a.** $\left(\frac{1}{6}\right)^n$ **b.** $\frac{1}{7776}$

c. $\frac{1}{32}$

Differentiated Instruction

Kinesthetic

Have students use grid paper when converting numbers from scientific notation to standard form and from standard form to scientific notation. Write the number with one digit in each square. Place the decimal point on the line between the squares. Students may find it easier to count the number of squares than the number of digits.

Practice and Problem Solving

33. 5.1×10^{-3}

34. 4.88×10^{7}

35. 3.456×10^{-9}

36. 7.5×10^{-3}

37. 4×10^{-2}

38. 5×10^{2}

39. $\dfrac{y^{12}}{216x^6}$ **40.** $-\dfrac{n^8}{4m^6}$

41. $\dfrac{5b^5c^{14}}{12}$ **42.** $(2xy)^3$

43. $14a^2b$ microns

44. Earth: 500 sec, or 8 min, 20 sec; Jupiter: 2600 sec, or 43 min, 20 sec; Neptune: 15,000 sec, or 4 h, 10 min

45. 4.4 on the Richter Scale

46. *See Taking Math Deeper.*

Fair Game Review

47. $4\sqrt{3}$ **48.** $\dfrac{\sqrt{70}}{6}$

49. $\dfrac{6\sqrt{5}}{11}$ **50.** D

Mini-Assessment

Simplify. Write your answer using only positive exponents.

1. $5^{11} \cdot 5^{-5}$ 5^6

2. $\dfrac{(-6)^5}{(-6)^8}$ $\dfrac{1}{(-6)^3}$

3. $(7^2)^3$ 7^6

4. $(1.4x^3)^2$ $1.96x^6$

5. $\left(\dfrac{-10}{4x}\right)^{-3}$ $-\dfrac{8x^3}{125}$

Taking Math Deeper

Exercise 46

One way to solve this problem is to use substitution.

Begin by rewriting the first equation so that the right side is equivalent to the right side of the second equation.

$$\frac{k^{2x}}{k^y} = k^{13}$$

$$\frac{k^{2x}}{k^y} \cdot k^{15} = k^{13} \cdot k^{15}$$

$$k^{2x - y + 15} = k^{28}$$

2 Simplify the second equation.

$$(k^x k^{2y})^2 = k^{28}$$

$$k^{2x}k^{4y} = k^{28}$$

$$k^{2x + 4y} = k^{28}$$

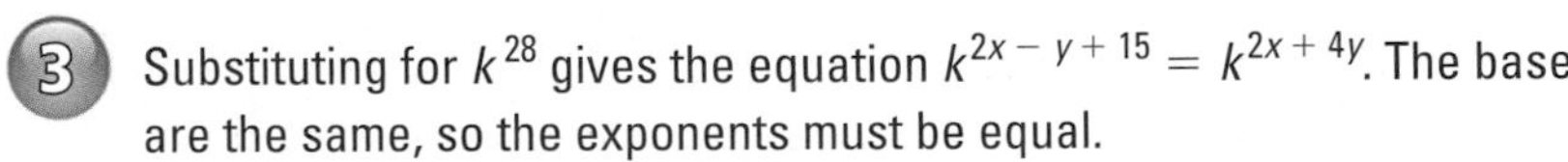

3 Substituting for k^{28} gives the equation $k^{2x - y + 15} = k^{2x + 4y}$. The bases are the same, so the exponents must be equal.

$$2x - y + 15 = 2x + 4y$$

$$15 = 5y$$

$$3 = y$$

By substituting 3 for y in one of the original equations, you will find $x = 8$.

So, the solution is $x = 8$, $y = 3$.

Reteaching and Enrichment Strategies

If students need help. . .	If students got it. . .
Resources by Chapter • Practice A and Practice B • Puzzle Time Record and Practice Journal Practice Differentiating the Lesson Lesson Tutorials Skills Review Handbook	Resources by Chapter • Enrichment and Extension Start the next section

Evaluate the expression. Write your answer in scientific notation.

4 33. $(3.4 \times 10^2)(1.5 \times 10^{-5})$
34. $(6.1 \times 10^{-3})(8 \times 10^9)$
35. $(4.8 \times 10^{-4})(7.2 \times 10^{-6})$

36. $\frac{(3 \times 10^3)}{(4 \times 10^5)}$
37. $\frac{(6.4 \times 10^{-7})}{(1.6 \times 10^{-5})}$
38. $\frac{(3.9 \times 10^{-5})}{(7.8 \times 10^{-8})}$

Simplify. Write your answer using only positive exponents.

39. $(6x^2y^{-4})^{-3}$
40. $\frac{(2m)^{-2}n^5}{-m^4n^{-3}}$
41. $\frac{15b^{-3}c^4}{(6b^{-4}c^{-5})^2}$

42. **REASONING** Write $8x^3y^3$ as the power of a product.

43. **COMPUTER CHIP** The area of a rectangular computer chip is $112a^3b^2$ square microns. The width is $8ab$ microns. What is the length?

44. **PROBLEM SOLVING** The speed of light is approximately 3×10^5 kilometers per second. The table shows the average distance each planet is from the Sun. How long does it take sunlight to reach Earth? Jupiter? Neptune?

Planet	Average Distance from the Sun (km)
Mercury	5.8×10^7
Venus	1.1×10^8
Earth	1.5×10^8
Mars	2.3×10^8
Jupiter	7.8×10^8
Saturn	1.4×10^9
Uranus	2.9×10^9
Neptune	4.5×10^9

45. **RICHTER SCALE** The Richter Scale is used to compare the intensities of earthquakes. An increase of 1 in magnitude on the Richter Scale represents a tenfold increase in intensity. An earthquake registers 7.4 on the Richter Scale and is followed by an aftershock that is 1000 times less intense. What is the magnitude of the aftershock?

46. **Precision** Find x and y when $\frac{k^{2x}}{k^y} = k^{13}$ and $(k^xk^{2y})^2 = k^{28}$. Explain how you found your answer.

Fair Game Review What you learned in previous grades & lessons

Simplify the expression. *(Section 6.1)*

47. $\sqrt{48}$
48. $\sqrt{\frac{70}{36}}$
49. $\sqrt{\frac{180}{121}}$

50. **MULTIPLE CHOICE** Which of the following is the solution of $\frac{x}{3} < -6$? *(Section 3.3)*

(A) $x > -2$ (B) $x < -2$ (C) $x > -18$ (D) $x < -18$

6.3 Radicals and Rational Exponents

COMMON CORE STATE STANDARDS
N.RN.1
N.RN.2

Essential Question How can you write and evaluate an *n*th root of a number?

Recall that you cube a number as follows.

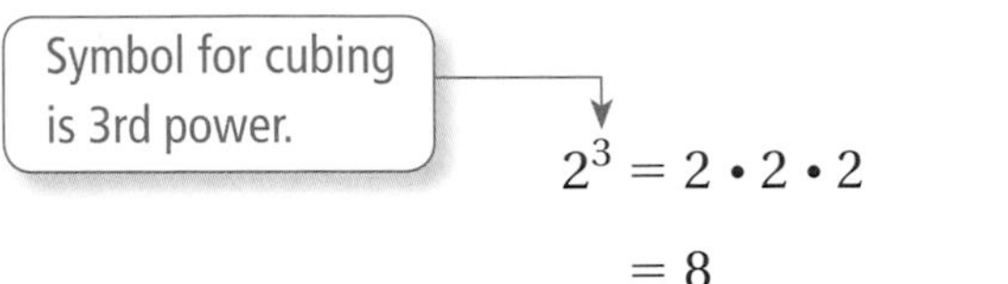

$$2^3 = 2 \cdot 2 \cdot 2$$
$$= 8$$ 2 cubed is 8.

To "undo" this, take the cube root of the number.

Symbol for cube root is $\sqrt[3]{}$.

$$\sqrt[3]{8} = \sqrt[3]{2^3} = 2$$ The cube root of 8 is 2.

1 ACTIVITY: Finding Cube Roots

Work with a partner. Use a cube root symbol to write the side length of the cube. Then find the cube root. Check your answer by multiplying. Which cube is the largest? Which two are the same size? Explain your reasoning.

a. Volume = 27 ft^3

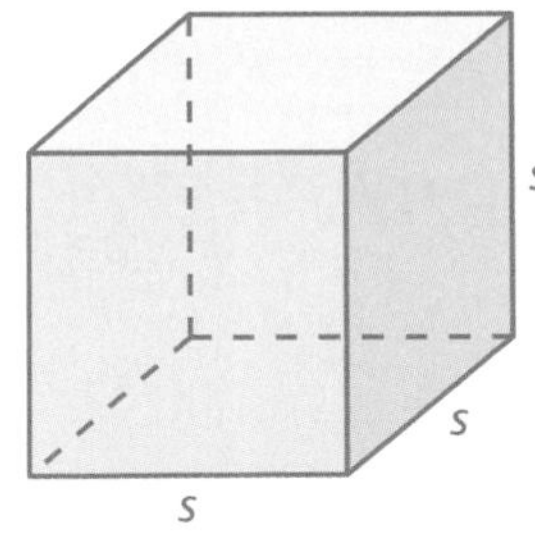

b. Volume = 125 cm^3

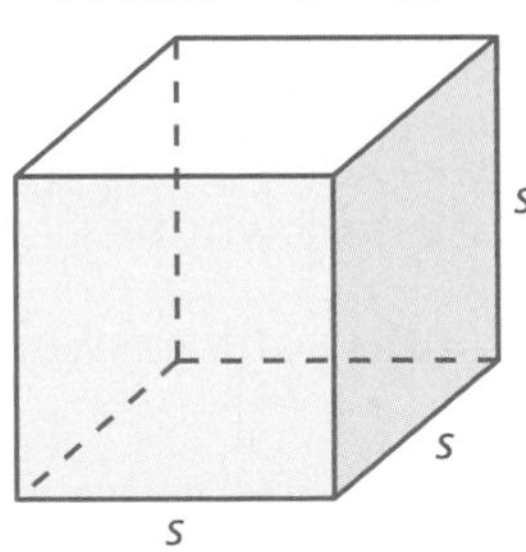

c. Volume = 3375 in.^3

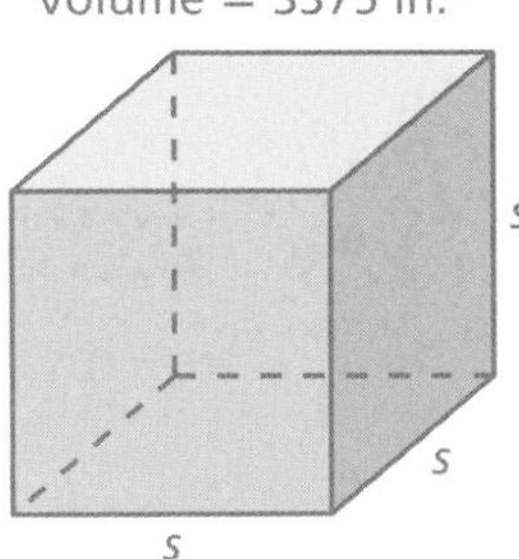

d. Volume = 3.375 m^3

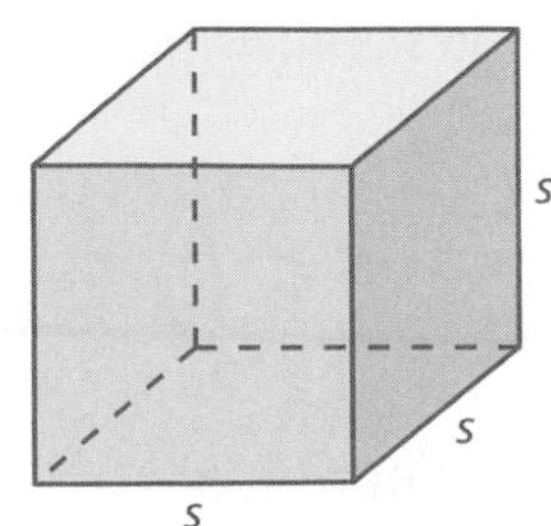

e. Volume = 1 yd^3

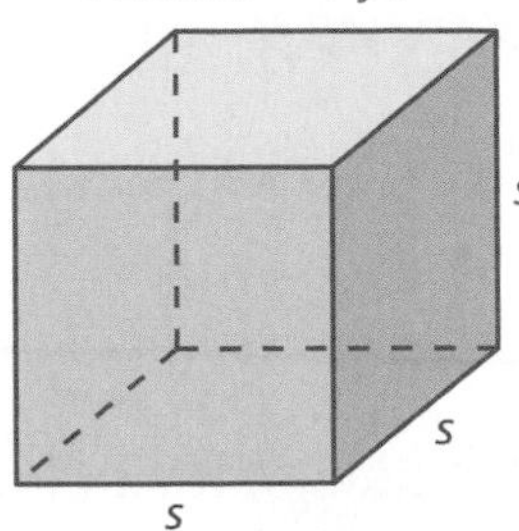

f. Volume = $\frac{125}{8}$ mm^3

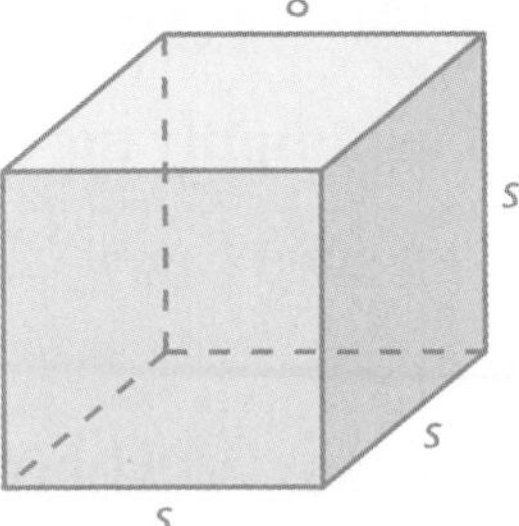

Cubes are not drawn to scale.

Laurie's Notes

Introduction

Standards for Mathematical Practice

- **MP2 Reason Abstractly and Quantitatively:** In this section, students will work with real numbers in forms that are very new to them. Real numbers will be represented in radical form and rational exponent form. Having a sense of how consecutive powers of a whole number grow will help students reason quantitatively about whole numbers with fractional exponents. Examples:

$3^1 = 3$	$3^2 = 9$	$3^3 = 27$	$3^4 = 81$
$81^1 = 81$	$81^{3/4} = 27$	$81^{2/4} = 9$	$81^{1/4} = 3$

Motivate

- **Story Time:** Tell students that you had "8 yards" of crushed stone delivered to your house for a new walkway. Ask, "What do you think 8 yards of gravel looks like?"
- Students may not know that crushed stone is generally sold by the cubic yard and the word "cubic" is rarely stated. So, "8 yards" refers to 8 cubic yards. Would the crushed stone fit inside your classroom? closet? locker room?

? "If the crushed stone were dumped inside a cube-shaped container, what would the dimensions of the cube be?" 2 yards on each edge

? "What would the dimensions be in feet?" 6 feet on each edge

? "What would the volume be in cubic feet?" 216 ft^3

- Make the connection that $8 \text{ yd}^3 = 216 \text{ ft}^3$.

Activity Notes

Discuss

- Explain that in the first activity they will find the dimensions of a cube. Introduce the cube root symbol.
- If you have a model of a cube in your room, have a student use it to describe the meaning of the volume of a cube and the side (edge) lengths of a cube.

Activity 1

- Students will find a calculator helpful in testing their guesses for side length. For instance, they probably do not know that $15^3 = 3375$, although the 5 in the ones place should be an important clue!
- When students have finished, spend time discussing the two additional questions posed.

? **MP2:** "How did you determine which cube was the largest?" Listen for something about converting units, and that 1 meter is longer than 1 yard.

- **Extension:** Put the cubes in order from greatest volume to least volume.

Common Core State Standards

N.RN.1 Explain how the definition of the meaning of rational exponents follows from extending the properties of integer exponents to those values, allowing for a notation for radicals in terms of rational exponents.

N.RN.2 Rewrite expressions involving radicals and rational exponents using the properties of exponents.

Previous Learning

Students should know how to evaluate powers and how to find the square roots of numbers.

Start Thinking! and Warm Up

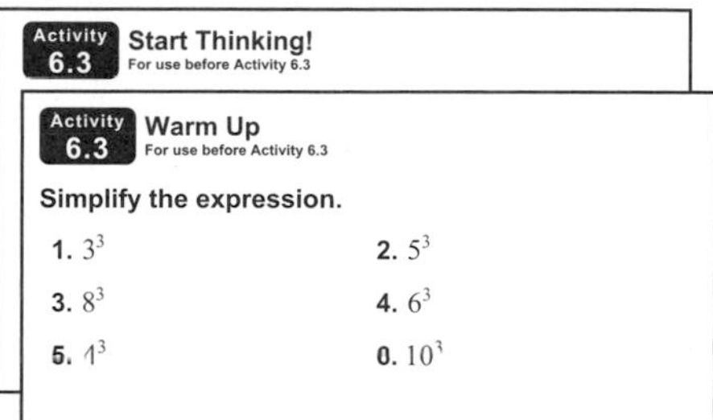

Activity 6.3 Start Thinking! For use before Activity 6.3

Activity 6.3 Warm Up For use before Activity 6.3

Simplify the expression.

1. 3^3
2. 5^3
3. 8^3
4. 6^3
5. 1^3
6. 10^3

6.3 Record and Practice Journal

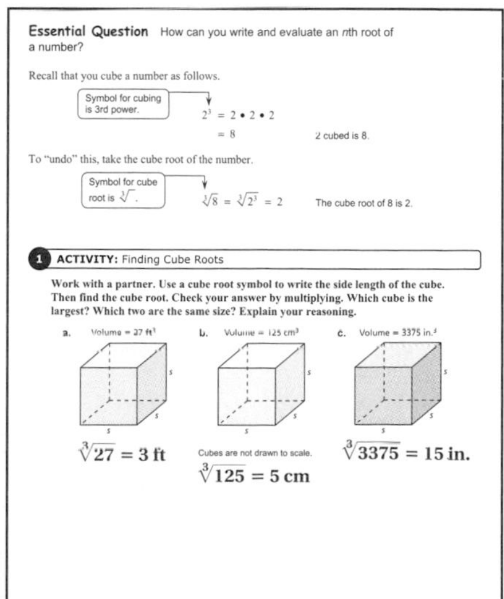

Essential Question How can you write and evaluate an *n*th root of a number?

Recall that you cube a number as follows.

Symbol for cubing is 3rd power.

$2^3 = 2 \cdot 2 \cdot 2$

$= 8$ 2 cubed is 8.

To "undo" this, take the cube root of the number.

Symbol for cube root is $\sqrt[3]{\ }$.

$\sqrt[3]{8} = \sqrt[3]{2^3} = 2$ The cube root of 8 is 2.

1 ACTIVITY: Finding Cube Roots

Work with a partner. Use a cube root symbol to write the side length of the cube. Then find the cube root. Check your answer by multiplying. Which cube is the largest? Which two are the same size? Explain your reasoning.

a. Volume = 27 ft^3

$\sqrt[3]{27} = 3 \text{ ft}$

b. Volume = 125 cm^3

Cubes are not drawn to scale.

$\sqrt[3]{125} = 5 \text{ cm}$

c. Volume = 3375 in.^3

$\sqrt[3]{3375} = 15 \text{ in.}$

Differentiated Instruction

Dependent Learners

Reinforce self-learning techniques by telling students to ask themselves questions like "What does $\sqrt[3]{8}$ mean? How can I state its meaning in words? Is there another way to represent this?"

Demonstrate how these types of questions lead to conceptual thinking, such as "The cube root of 8 is the number that you cube to get 8, or the number you multiply by itself three times to get 8. Another way to represent this is: $\sqrt[3]{8} \cdot \sqrt[3]{8} \cdot \sqrt[3]{8} = 8$."

6.3 Record and Practice Journal

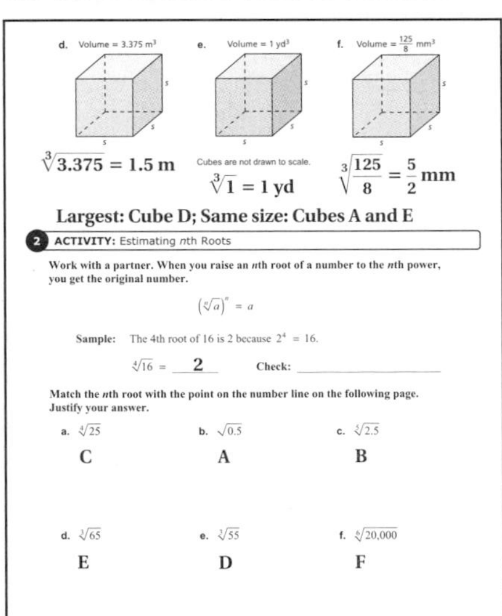

d. Volume = 3.375 m^3 e. Volume = 1 yd^3 f. Volume = $\frac{125}{8}$ mm^3

$\sqrt[3]{3.375} = 1.5$ m

Cubes are not drawn to scale.

$\sqrt[3]{1} = 1$ yd

$\sqrt[3]{\frac{125}{8}} = \frac{5}{2}$ mm

Largest: Cube D; Same size: Cubes A and E

2 ACTIVITY: Estimating *n*th Roots

Work with a partner. When you raise an *n*th root of a number to the *n*th power, you get the original number.

$$\left(\sqrt[n]{a}\right)^n = a$$

Sample: The 4th root of 16 is 2 because $2^4 = 16$.

$\sqrt[4]{16} =$ **2** **Check:** ____________

Match the *n*th root with the point on the number line on the following page. Justify your answer.

a. $\sqrt[4]{25}$ **C** b. $\sqrt{0.5}$ **A** c. $\sqrt[5]{2.5}$ **B**

d. $\sqrt[3]{65}$ **E** e. $\sqrt[3]{55}$ **D** f. $\sqrt[6]{20{,}000}$ **F**

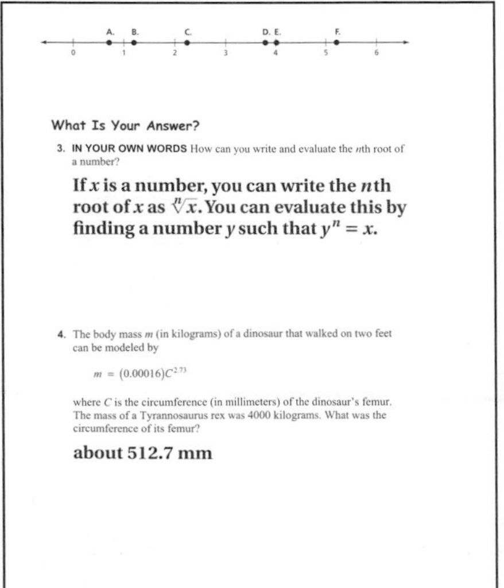

What Is Your Answer?

3. **IN YOUR OWN WORDS** How can you write and evaluate the *n*th root of a number?

If x is a number, you can write the nth root of x as $\sqrt[n]{x}$. You can evaluate this by finding a number y such that $y^n = x$.

4. The body mass m (in kilograms) of a dinosaur that walked on two feet can be modeled by

$$m = (0.00016)C^{2.73}$$

where C is the circumference (in millimeters) of the dinosaur's femur. The mass of a Tyrannosaurus rex was 4000 kilograms. What was the circumference of its femur?

about 512.7 mm

Laurie's Notes

Activity 1 (continued)

- After finishing this activity, students should be comfortable with the idea that cubing and finding a cube root are inverse operations. Have students write the results for each cube, then substitute to write an equivalent statement involving the cube. For example, in part (a) write $\sqrt[3]{27} = 3 \rightarrow \sqrt[3]{3^3} = 3$.

Activity 2

- Write on the board: "When you raise the *n*th root of a number to the *n*th power, you get the original number." Relate this to the solution statements in the first activity, and note the distinction. In Activity 1, the cube root of a cubed number is the number. In the statement above, the cube of the cube root of a number is the number.

  Activity 1: $\sqrt[3]{8^3} = 8$ Activity 2: $\left(\sqrt[3]{8}\right)^3 = 8$
- Use this example to talk about *n*th roots for values of *n* other than 3.
- Describe the sample problem: To find the fourth root of 16, ask yourself "What number can you use as a factor 4 times to get a product of 16?"
- **MP2:** Discuss the justifications for part (a). Listen for student reasoning, such as $\sqrt[4]{25}$ is between 2 and 3 because $2^4 = 16$ and $3^4 = 81$. Also, $\sqrt[4]{25}$ is closer to 2 than 3.
- **MP3 Construct Viable Arguments and Critique the Reasoning of Others:** Ask volunteers to share their answers, explaining how they made their matches.

What Is Your Answer?

? "Could you lift 4000 kilograms? About how many pounds is 4000 kilograms?" no; 8800 lb

? "Where is the femur on a human? Do you think the size of your femur is greater than that of a dinosaur?" thighbone; yes, for some dinosaurs

Closure

- Explain how to find the length of a cube with a volume of 300 cubic inches.

Technology For the Teacher

The Dynamic Planning Tool
Editable Teacher's Resources at *BigIdeasMath.com*

2 ACTIVITY: Estimating *n*th Roots

Work with a partner. When you raise an *n*th root of a number to the *n*th power, you get the original number.

$$(\sqrt[n]{a})^n = a$$

Sample: The 4th root of 16 is 2 because $2^4 = 16$.

$\sqrt[4]{16} = 2$ **Check:** $2^4 = 2 \cdot 2 \cdot 2 \cdot 2 = 16$ ✓

Match the *n*th root with the point on the number line. Justify your answer.

a. $\sqrt[4]{25}$ **b.** $\sqrt{0.5}$ **c.** $\sqrt[5]{2.5}$

d. $\sqrt[3]{65}$ **e.** $\sqrt[3]{55}$ **f.** $\sqrt[6]{20{,}000}$

What Is Your Answer?

3. **IN YOUR OWN WORDS** How can you write and evaluate the *n*th root of a number?

4. The body mass m (in kilograms) of a dinosaur that walked on two feet can be modeled by

$$m = (0.00016)C^{2.73}$$

where C is the circumference (in millimeters) of the dinosaur's femur. The mass of a Tyrannosaurus rex was 4000 kilograms. What was the circumference of its femur?

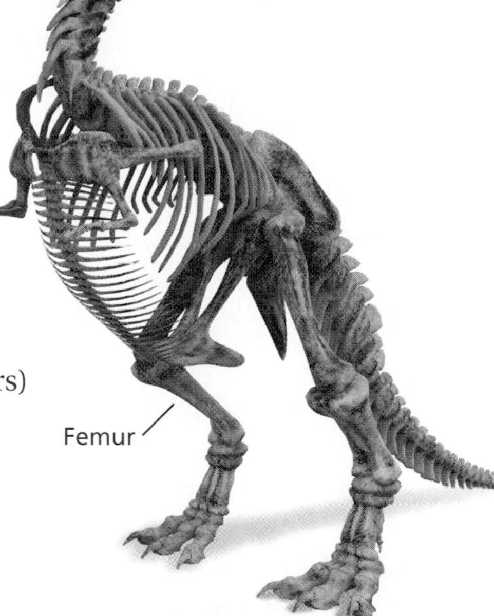

Practice Use what you learned about cube roots to complete Exercises 3–5 on page 280.

6.3 Lesson

Key Vocabulary
*n*th root, *p. 278*

When $b^n = a$ for an integer n greater than 1, b is an ***n*th root** of a.

$\sqrt[n]{a}$ *n*th root of *a*

The *n*th roots of a number may be real numbers or *imaginary numbers.* You will study imaginary numbers in a future course.

EXAMPLE 1 Finding *n*th Roots

Study Tip

In Example 1b, although $3^4 = 81$ and $(-3)^4 = 81$, $\sqrt[4]{81} = 3$ because the radical symbol indicates the positive root.

Simplify each expression.

a. $\sqrt[3]{64}$

$$\sqrt[3]{64} = \sqrt[3]{4 \cdot 4 \cdot 4}$$
$$= 4$$

b. $\sqrt[4]{81}$

$$\sqrt[4]{81} = \sqrt[4]{3 \cdot 3 \cdot 3 \cdot 3}$$
$$= 3$$

Key Idea

Rational Exponents

Words The *n*th root of a positive number a can be written as a power with base a and an exponent of $1/n$.

Numbers $\sqrt[4]{81} = 81^{1/4}$ **Algebra** $\sqrt[n]{a} = a^{1/n}$

EXAMPLE 2 Simplifying Expressions with Rational Exponents

Reading

When $n = 2$, the 2 is typically not written with the radical sign.

Simplify each expression.

a. $400^{1/2}$

$400^{1/2} = \sqrt{400}$	Write the expression in radical form.
$= \sqrt{20 \cdot 20}$	Rewrite.
$= 20$	Simplify.

b. $243^{1/5}$

$243^{1/5} = \sqrt[5]{243}$	Write the expression in radical form.
$= \sqrt[5]{3 \cdot 3 \cdot 3 \cdot 3 \cdot 3}$	Rewrite.
$= 3$	Simplify.

On Your Own

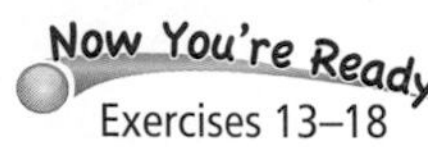

Simplify the expression.

1. $\sqrt[3]{216}$
2. $\sqrt[5]{32}$
3. $\sqrt[4]{625}$
4. $49^{1/2}$
5. $343^{1/3}$
6. $64^{1/6}$

Laurie's Notes

Introduction

Connect

- **Yesterday:** Students discovered what it means to find the *n*th root of a number. (MP2, MP3)
- **Today:** Students will evaluate expressions involving *n*th roots and rational exponents.

Motivate

- Discuss rubber band balls: how to make them, large examples, etc. You can measure to find the circumference of a rubber band ball.

? "How can you find the radius?" Solve the circumference formula for *r*.

- Discuss finding the square root to solve for *r*.

? "When you know the volume of a sphere, how can you find the radius?" Solve for *r* in the formula for the volume of a sphere, $V = \frac{4}{3}\pi r^3$.

Lesson Notes

Discuss

- Write a few problems like these on the board.

$10^2 = 100$, so $\sqrt{100} =$ ____ $\quad$ $4^3 = 64$, so $\sqrt[3]{64} =$ ____

$10^4 = 10{,}000$, so $\sqrt[4]{10{,}000} =$ ____ $\quad$ $2^5 = 32$, so $\sqrt[5]{32} =$ ____

- Write the definition of *n*th root and refer to the problems above.
- **FYI:** The square root of a negative number is not a real number. The cube root of a negative number, however, is a negative real number. For instance, $\sqrt[3]{-8} = -2$. Students may study *imaginary numbers* in a future course.

Example 1

? "What number can you multiply by itself three times to get 64?" 4

? "What number can you multiply by itself four times to get 81?" 3 or −3

- Reference the Study Tip. Because the radical symbol indicates the positive root, the only answer for $\sqrt[4]{81}$ is 3, not ±3.

Discuss

- Use $\sqrt{n} \cdot \sqrt{n} = n$ and $n^{1/2} \cdot n^{1/2} = n^1$ to deduce $\sqrt{n} = n^{1/2}$. Use properties of exponents with rational exponents.

$(a^{1/2})^2 = a^{1/2 \cdot 2} = a^1 = a$ $\quad$ The square of $a^{1/2}$ is *a*. So, $\sqrt{a} = a^{1/2}$.

$(a^{1/3})^3 = a^{1/3 \cdot 3} = a^1 = a$ $\quad$ The cube of $a^{1/3}$ is *a*. So, $\sqrt[3]{a} = a^{1/3}$.

$(a^{1/4})^4 = a^{1/4 \cdot 4} = a^1 = a$ $\quad$ So, $\sqrt[4]{a} = a^{1/4}$.

You can generalize this pattern to determine that $\sqrt[n]{a} = a^{1/n}$.

Key Idea

- Write the definition of rational exponents. Refer to previous examples.
- It may be helpful to demonstrate the *n*th root key on a calculator at this time and compare it to raising a number to a rational exponent.
- **Language:** $\sqrt[n]{a}$ is the radical form; $a^{1/n}$ is the rational exponent form.

Goal Today's lesson is evaluating expressions involving ***n*th roots** and rational exponents.

Lesson Materials
Textbook
• calculators

Start Thinking! and Warm Up

Lesson 6.3
Start Thinking!
For use before Lesson 6.3

What do you think *rational exponents* means?

Give three examples of expressions that contain rational exponents.

Extra Example 1

Simplify each expression.

a. $\sqrt[3]{1000}$ 10

b. $\sqrt[7]{128}$ 2

English Language Learners

Build on Past Knowledge

Remind students that taking the square root of a number is the inverse of squaring a number. Likewise, taking the *n*th root of a number is the inverse of multiplying the number by itself *n* times.

Extra Example 2

Simplify each expression.

a. $1600^{1/2}$ 40

b. $625^{1/4}$ 5

On Your Own

1. 6 **2.** 2

3. 5 **4.** 7

5. 7 **6.** 2

Extra Example 3

Simplify each expression.

a. $216^{2/3}$ 36

b. $25^{3/2}$ 125

On Your Own

7. 16 **8.** 243

9. 64

Extra Example 4

In Example 4, a beach ball has a volume of 33,510 cubic inches. Find the radius to the nearest inch. Use 3.14 for π.
about 20 in.

On Your Own

10. about 16 in.

Laurie's Notes

Example 2

? "How can you rewrite $400^{1/2}$?" $\sqrt{400}$ "What number can be multiplied by itself to get 400?" 20

- Explain that the 2 is typically not written with the radical sign in a square root.
- Students should recognize 243 as 3 to the fifth power.

On Your Own

- **MP2 Reason Abstractly and Quantitatively:** In Question 1, $5^3 = 125$ and $10^3 = 1000$, so $\sqrt[3]{216}$ is between 5 and 10, but closer to 5.

Discuss

- By rewriting a radical expression in rational exponent form, properties of exponents can be used to simplify the expression.

Example 3

- Begin by writing the expression. Explain that you can rewrite the exponent as $3 \cdot \left(\frac{1}{4}\right)$, but it is usually easier to evaluate the root first.
- Work through the example slowly, so students can understand the steps.
- Note: You can also write the expressions in radical form and then simplify.

Example 4

- Relate this problem to the Motivate. The volume of a sphere is $V = \frac{4}{3}\pi r^3$. If time permits, show the steps for solving this equation for r.

? "Can you use the photo to estimate the radius?" about 2 to 3 feet

- The solution calls for a calculator, but without a calculator you could simplify $339 \div 12.56 \approx 26.99$ and estimate that $\sqrt[3]{26.99} \approx 3.0$.

On Your Own

? "How many cubic inches are in a cubic foot?" 1728

Closure

- **Writing Prompt:** To evaluate $36^{5/2}$, you . . .

Technology For the Teacher

The Dynamic Planning Tool
Editable Teacher's Resources at *BigIdeasMath.com*

You can use properties of exponents to simplify expressions involving rational exponents.

EXAMPLE 3 Using Properties of Exponents

a. $16^{3/4} = 16^{(1/4) \cdot 3}$ Rewrite the exponent.

$= (16^{1/4})^3$ Power of a Power Property

$= 2^3$ Evaluate the fourth root of 16.

$= 8$ Evaluate power.

b. $27^{4/3} = 27^{1/3 \cdot 4}$ Rewrite the exponent.

$= (27^{1/3})^4$ Power of a Power Property

$= 3^4$ Evaluate the third root of 27.

$= 81$ Evaluate power.

On Your Own

Exercises 20–25

Simplify the expression.

7. $64^{2/3}$ **8.** $9^{5/2}$ **9.** $256^{3/4}$

EXAMPLE 4 Real-Life Application

Volume = 113 cubic feet

The radius r of a sphere is given by the equation $r = \left(\frac{3V}{4\pi}\right)^{1/3}$, where V is the volume of the sphere. Find the radius of the beach ball to the nearest foot. Use 3.14 for π.

$r = \left(\frac{3V}{4\pi}\right)^{1/3}$ Write the equation.

$= \left[\frac{3(113)}{4(3.14)}\right]^{1/3}$ Substitute 113 for V and 3.14 for π.

$= \left(\frac{339}{12.56}\right)^{1/3}$ Multiply.

≈ 3 Use a calculator.

The radius of the beach ball is about 3 feet.

On Your Own

10. **WHAT IF?** In Example 4, the volume of the beach ball is 17,000 cubic inches. Find the radius to the nearest inch. Use 3.14 for π.

6.3 Exercises

Vocabulary and Concept Check

1. **WRITING** Explain how to simplify $81^{1/4}$.
2. **WHICH ONE DOESN'T BELONG?** Which expression does *not* belong with the other three? Explain your reasoning.

$(\sqrt[3]{27})^2$ $27^{2/3}$ 3^2 $27^{3/2}$

Practice and Problem Solving

Find the dimensions of the cube. Check your answer.

3. Volume = 64 in.^{3}

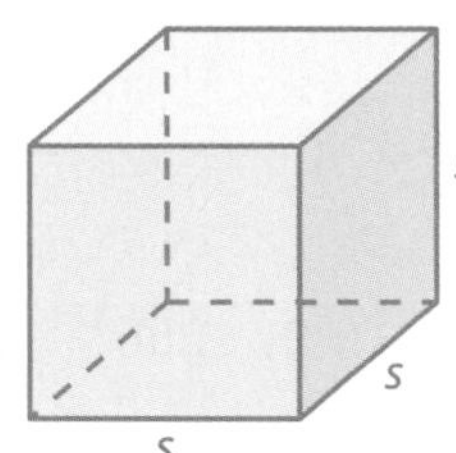

4. Volume = 216 cm^3

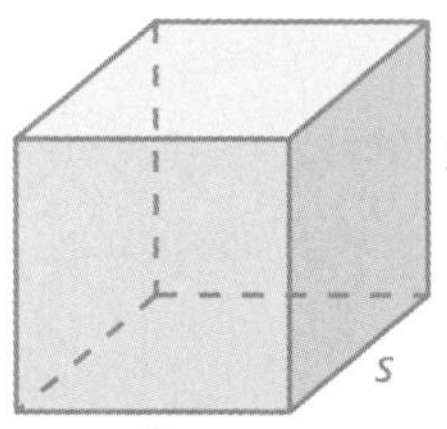

5. Volume = $\frac{343}{512}$ ft^3

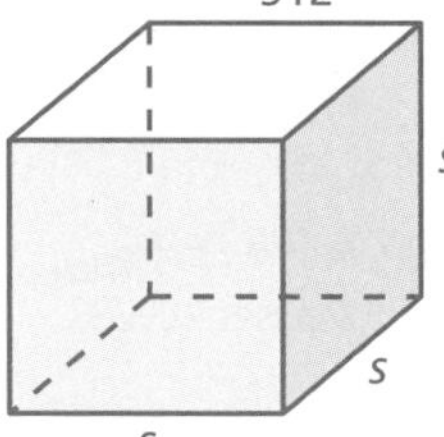

Write the expression in rational exponent form.

6. $\sqrt[7]{5}$
7. $(\sqrt[3]{4})^2$
8. $(\sqrt[5]{8})^4$

Write the expression in radical form.

9. $15^{1/3}$
10. $140^{1/7}$
11. $78^{2/5}$

12. **ERROR ANALYSIS** Describe and correct the error in writing the expression in rational exponent form.

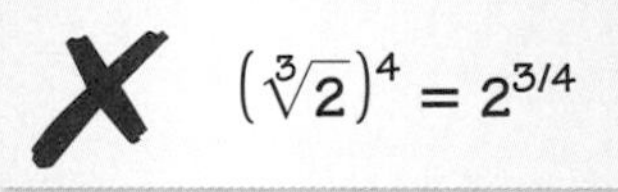

Simplify the expression.

1 13. $\sqrt[4]{256}$
14. $\sqrt[3]{125}$
15. $\sqrt[5]{1024}$

2 16. $128^{1/7}$
17. $1000^{1/3}$
18. $81^{1/2}$

19. **BAKE SALE** A math club is having a bake sale. Find the length and width of the bake sale sign.

π is SWEET!
Math Club Bake Sale this Saturday
$4^{1/2}$ ft
$\sqrt[6]{729}$ ft

Assignment Guide and Homework Check

Level	Assignment	Homework Check
Average	1–5, 6–28 even, 19, 30–33	14, 19, 20, 26, 28
Advanced	1–5, 12, 13–25 odd, 26–29, 30–33	19, 23, 26, 28, 29

Common Errors

- **Exercises 6–11** To write the rational exponent form, students may use the index as the numerator and the exponent as the denominator. To write the radical form, students may use the numerator as the index and the denominator as the exponent. Remind students that the index should be in the denominator and the exponent should be in the numerator.
- **Exercise 19** Students may answer that the sign is $4\frac{1}{2}$ feet wide. Make sure they realize that the $\frac{1}{2}$ is an exponent.
- **Exercises 20–25** Students may solve by rewriting $a^{b/n}$ as $\frac{a^b}{a^n}$. Tell them to rewrite the exponent b/n as $\frac{1}{n} \cdot b$. Then find the nth root of a and apply the exponent b to the result.

Technology For the Teacher

Answer Presentation Tool

6.3 Record and Practice Journal

Simplify the expression.

1. $\sqrt[5]{243}$ **3**
2. $\sqrt[3]{64}$ **4**
3. $144^{1/2}$ **12**
4. $729^{1/3}$ **9**
5. $1000^{2/3}$ **100**
6. $64^{3/2}$ **512**
7. $81^{3/4}$ **27**
8. $32^{7/5}$ **128**
9. The radius r of the base of a cylinder is given by the equation $r = \left(\frac{V}{\pi h}\right)^{1/2}$, where V is the volume of the cylinder and h is the height of the cylinder. Find the radius of the cylinder to the nearest inch. Use 3.14 for π.

   3 in.

6 in.

Volume = 170 in.^{3}

Vocabulary and Concept Check

1. 3; Find the fourth root of 81.
2. $27^{3/2}$; All other expressions are equal to 9.

Practice and Problem Solving

3. $s = 4$ in.
4. $s = 6$ cm
5. $s = \frac{7}{8}$ ft
6. $5^{1/7}$
7. $4^{2/3}$
8. $8^{4/5}$
9. $\sqrt[3]{15}$
10. $\sqrt[7]{140}$
11. $(\sqrt[5]{78})^2$
12. The root and power are switched; $2^{4/3}$
13. 4
14. 5
15. 4
16. 2
17. 10
18. 9
19. length: 3 ft, width: 2 ft

Practice and Problem Solving

20. 8

21. 25

22. 216

23. 9

24. 32

25. 2401

26. about 1 in.

27. $a^{m/n}$; The mth power of the nth root of a positive number a can be written as a power with base a and an exponent of m/n.

28. about 1.38 ft

29. always; $\left(x^{1/3}\right)^3 = x^{3/3} = x^1 = x$

30. sometimes; $x^{1/3} = x^{-3}$ when $x = 1$.

31. always; By definition, $x^{1/3} = \sqrt[3]{x}$.

32. See *Taking Math Deeper*.

33. always; $x^{(2/3) - (1/3)} = x^{1/3} = \sqrt[3]{x}$ by definition.

34. sometimes; $x = x^{1/3} \cdot x^3$ when $x = 0$ and $x = 1$.

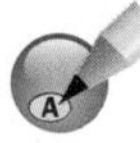

Fair Game Review

35–37. See Additional Answers.

38. A

Mini-Assessment

Simplify the expression.

1. $\sqrt[3]{729}$ 9
2. $1296^{1/4}$ 6
3. $64^{5/6}$ 32
4. $81^{5/4}$ 243
5. Use the formula $V \approx 7.66\ell^3$ to estimate the side length ℓ of a regular dodecahedron with a volume V of 490 cubic inches about 4 inches

Taking Math Deeper

Exercise 32

Students may be inclined to say *never*, because taking the cube root of a nonnegative number typically decreases the number, whereas cubing a nonnegative number typically increases the number. It is a good idea in questions like these to check special cases, such as $x = 0$ and $x = 1$. Taking a graphical approach, however, will account for all cases.

Graph the related functions $y = x^{1/3}$ and $y = x^3$ for nonnegative values of x.

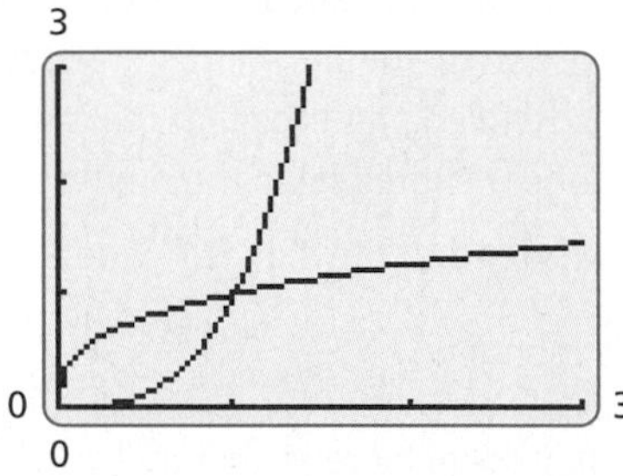

Interpret the graphs.

The graphs intersect when $x = 0$ and $x = 1$.

Depending on the viewing window, the graphs may not appear to intersect when $x = 0$. Use different viewing windows to convince students.

Check both values and then answer the question.

$x = 0$: $0^{1/3} = 0$ and $0^3 = 0$ **$x = 1$:** $1^{1/3} = 1$ and $1^3 = 1$

For a nonnegative real number x, $x^{1/3} = x^3$ when $x = 0$ and $x = 1$.

So, $x^{1/3}$ is *sometimes* equal to x^3.

Challenge students to think about a negative x-value that satisfies the equation. Have them verify their result using a graphing calculator.

Reteaching and Enrichment Strategies

If students need help...	If students got it...
Resources by Chapter • Practice A and Practice B • Puzzle Time Record and Practice Journal Practice Differentiating the Lesson Lesson Tutorials Skills Review Handbook	Resources by Chapter • Enrichment and Extension • School-to-Work Start the next section

Simplify the expression.

3 **20.** $32^{3/5}$ **21.** $125^{2/3}$ **22.** $36^{3/2}$

23. $243^{2/5}$ **24.** $128^{5/7}$ **25.** $343^{4/3}$

26. PAPER CUPS The radius r of the base of a cone is given by the equation $r = \left(\frac{3V}{\pi h}\right)^{1/2}$, where V is the volume of the cone and h is the height of the cone. Find the radius of the paper cup to the nearest inch. Use 3.14 for π.

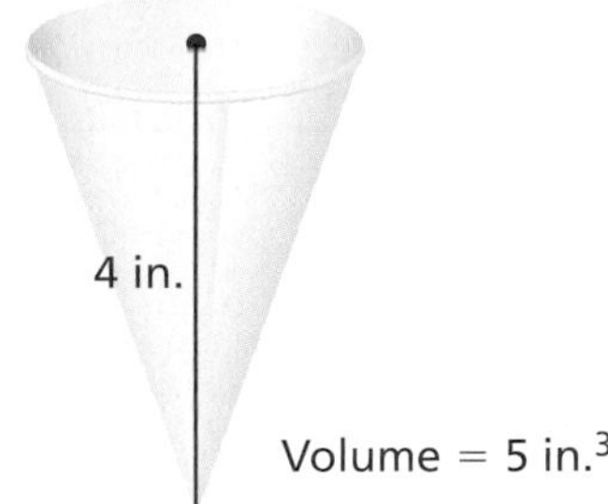

27. WRITING Explain how to write $(\sqrt[n]{a})^m$ in rational exponent form.

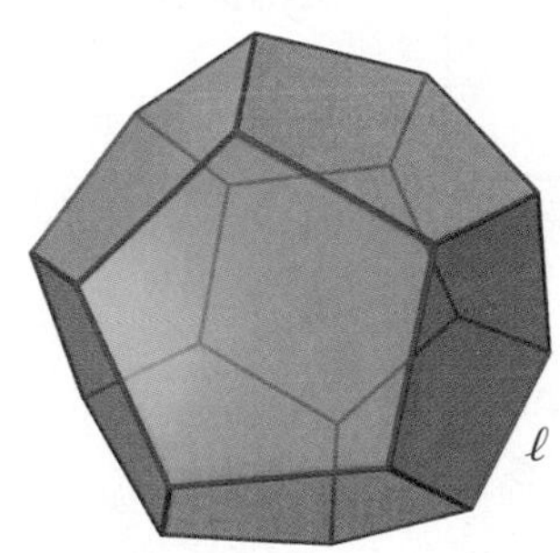

28. PROBLEM SOLVING The formula for the volume of a regular dodecahedron is $V \approx 7.66\,\ell^3$, where ℓ is the length of an edge. The volume of the dodecahedron is 20 cubic feet. Estimate the edge length.

Logic **Determine whether the statement is *always, sometimes,* or *never* true. Let x be a nonnegative real number. Justify your answer.**

29. $(x^{1/3})^3 = x$ **30.** $x^{1/3} = x^{-3}$ **31.** $x^{1/3} = \sqrt[3]{x}$

32. $x^{1/3} = x^3$ **33.** $\frac{x^{2/3}}{x^{1/3}} = \sqrt[3]{x}$ **34.** $x = x^{1/3} \cdot x^3$

Fair Game Review What you learned in previous grades & lessons

Graph the linear equation. *(Section 2.3 and Section 2.4)*

35. $y = -2x + 1$ **36.** $4x - 2y = 6$ **37.** $y = -\frac{1}{3}x - 5$

38. MULTIPLE CHOICE Which equation is shown in the graph? *(Section 2.1)*

Ⓐ $y = -\frac{1}{2}x + 1$ Ⓑ $y = -\frac{1}{2}x - 1$

Ⓒ $y = \frac{1}{2}x - 1$ Ⓓ $y = \frac{1}{2}x + 1$

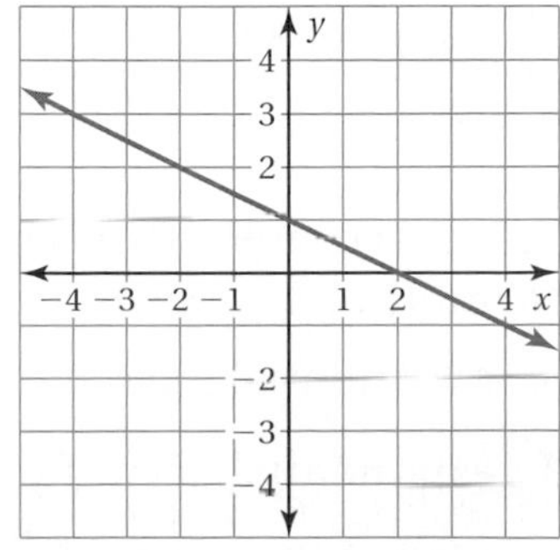

6 Study Help

You can use an **information frame** to help you organize and remember concepts. Here is an example of an information frame for the Product of Powers Property.

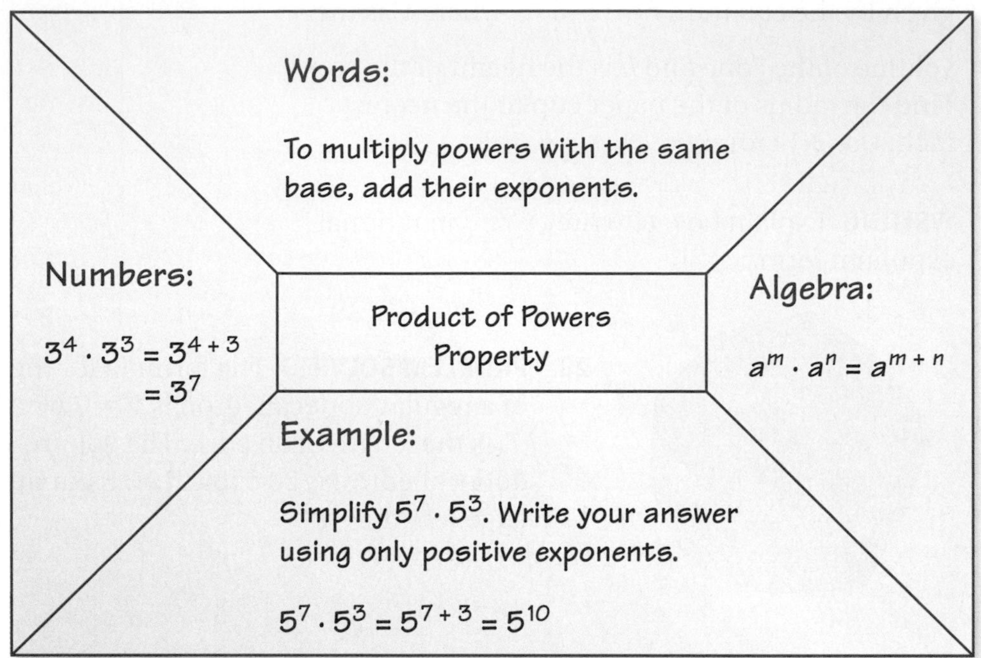

On Your Own

Make information frames to help you study these topics.

1. Product Property of Square Roots
2. Quotient Property of Square Roots
3. Quotient of Powers Property
4. Power of a Power Property
5. Power of a Product Property
6. Power of a Quotient Property
7. rational exponents

After you complete this chapter, make information frames for the following topics.

8. exponential growth functions
9. exponential decay functions
10. geometric sequences

"Dear Mom, I am sending you an information frame card for Mother's Day!"

Sample Answers

1.

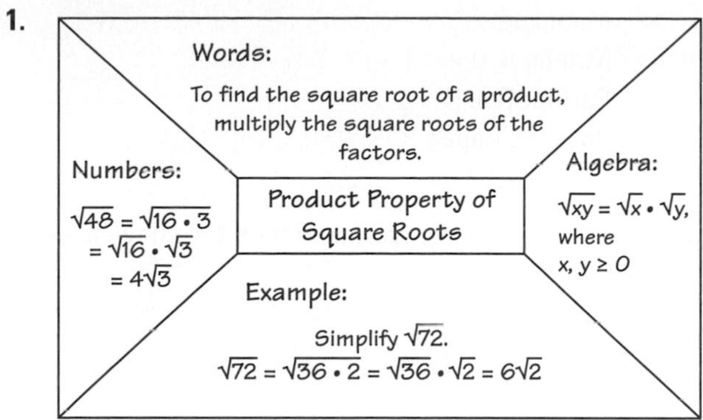

Words:
To find the square root of a product, multiply the square roots of the factors.
Numbers:
$\sqrt{48} = \sqrt{16 \cdot 3}$
$= \sqrt{16} \cdot \sqrt{3}$
$= 4\sqrt{3}$
Product Property of Square Roots
Algebra:
$\sqrt{xy} = \sqrt{x} \cdot \sqrt{y}$, where $x, y \geq 0$
Example:
Simplify $\sqrt{72}$.
$\sqrt{72} = \sqrt{36 \cdot 2} = \sqrt{36} \cdot \sqrt{2} = 6\sqrt{2}$

2.

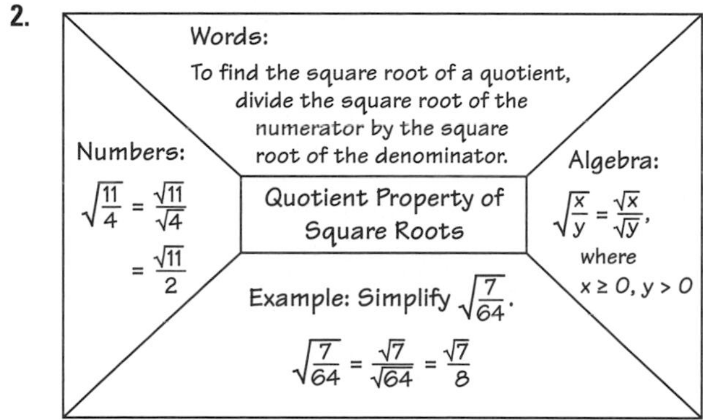

Words:
To find the square root of a quotient, divide the square root of the numerator by the square root of the denominator.
Numbers:
$\sqrt{\frac{11}{4}} = \frac{\sqrt{11}}{\sqrt{4}}$
$= \frac{\sqrt{11}}{2}$
Quotient Property of Square Roots
Algebra:
$\sqrt{\frac{x}{y}} = \frac{\sqrt{x}}{\sqrt{y}}$, where $x \geq 0, y > 0$
Example: Simplify $\sqrt{\frac{7}{64}}$.
$\sqrt{\frac{7}{64}} = \frac{\sqrt{7}}{\sqrt{64}} = \frac{\sqrt{7}}{8}$

3.

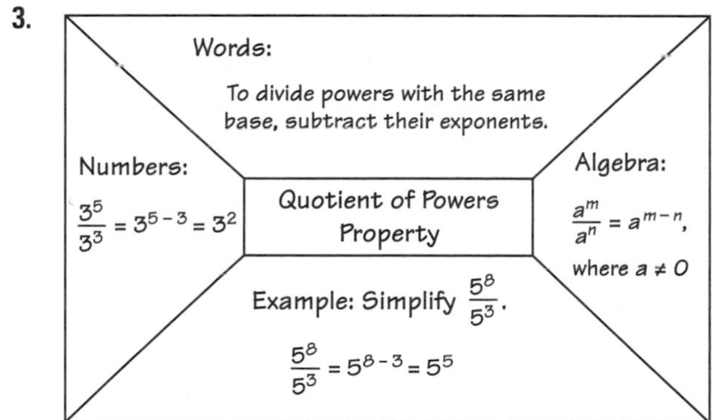

Words:
To divide powers with the same base, subtract their exponents.
Numbers:
$\frac{3^5}{3^3} = 3^{5-3} = 3^2$
Quotient of Powers Property
Algebra:
$\frac{a^m}{a^n} = a^{m-n}$, where $a \neq 0$
Example: Simplify $\frac{5^8}{5^3}$.
$\frac{5^8}{5^3} = 5^{8-3} = 5^5$

4.

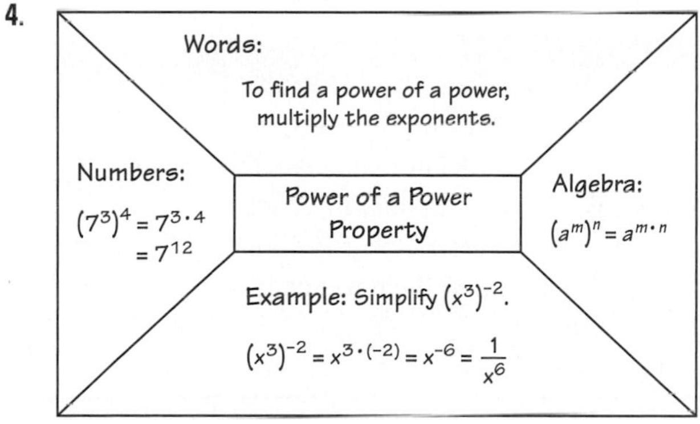

Words:
To find a power of a power, multiply the exponents.
Numbers:
$(7^3)^4 = 7^{3 \cdot 4}$
$= 7^{12}$
Power of a Power Property
Algebra:
$(a^m)^n = a^{m \cdot n}$
Example: Simplify $(x^3)^{-2}$.
$(x^3)^{-2} = x^{3 \cdot (-2)} = x^{-6} = \frac{1}{x^6}$

5–7. Available at *BigIdeasMath.com*

List of Organizers

Available at *BigIdeasMath.com*

Comparison Chart
Concept Circle
Definition (Idea) and Example Chart
Example and Non-Example Chart
Formula Triangle
Four Square
Information Frame
Information Wheel
Notetaking Organizer
Process Diagram
Summary Triangle
Word Magnet
Y Chart

About this Organizer

An **Information Frame** can be used to help students organize and remember concepts. Students write the topic in the middle rectangle. Then students write related concepts in the spaces around the rectangle. Related concepts can include *Words, Numbers, Algebra,* and *Example* as shown in the sample answers, but other related concepts can include *Definition, Non-Example, Visual, Procedure, Details,* and *Vocabulary*. Students can place their information frames on note cards to use as a quick study reference.

Technology For the Teacher
Vocabulary Puzzle Builder

Answers

1. $2\sqrt{5}$
2. $\dfrac{\sqrt{11}}{9}$
3. $2 - \sqrt{3}$
4. $-2 + \sqrt{5}$
5. $2\sqrt{14}$
6. $3\sqrt{3}$
7. 3^6
8. k^{12}
9. $\dfrac{1}{16y^2}$
10. $\dfrac{r^3}{8}$
11. 3
12. 2
13. 64
14. 32
15. $2\sqrt{15}\ \text{ft}^3$
16. no; $\pi - \pi = 0$
17. a. 10^{12}

    b. 10^5

    c. 1000 decigrams; 1000 decigrams $= 10^3 \cdot 10^{-1} = 10^2$ g

    10,000 milligrams $= 10^4 \cdot 10^{-3} = 10^1$ g

    $10^2 \text{ g} > 10^1 \text{ g}$

Assessment Book

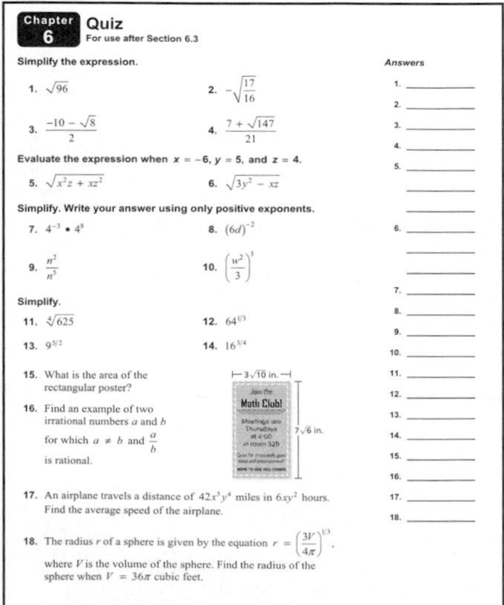

Chapter 6 Quiz
For use after Section 6.3

Simplify the expression.

1. $\sqrt{96}$
2. $-\sqrt{\dfrac{17}{16}}$
3. $\dfrac{-10 - \sqrt{8}}{2}$
4. $\dfrac{7 + \sqrt{147}}{21}$

Evaluate the expression when $x = -6$, $y = 5$, and $z = 4$.

5. $\sqrt{x^2z + xz^2}$
6. $\sqrt{3y^2 - xz}$

Simplify. Write your answer using only positive exponents.

7. $4^{-3} \cdot 4^8$
8. $(6d)^{-2}$
9. $\dfrac{n^2}{n^5}$
10. $\left(\dfrac{w^2}{3}\right)^3$

Simplify.

11. $\sqrt[4]{625}$
12. $64^{1/3}$
13. $9^{5/2}$
14. $16^{3/4}$
15. What is the area of the rectangular poster?
16. Find an example of two irrational numbers a and b for which $a \neq b$ and $\dfrac{a}{b}$ is rational.

17. An airplane travels a distance of $42x^5y^4$ miles in $6xy^2$ hours. Find the average speed of the airplane.
18. The radius r of a sphere is given by the equation $r = \left(\dfrac{3V}{4\pi}\right)^{1/3}$, where V is the volume of the sphere. Find the radius of the sphere when $V = 36\pi$ cubic feet.

Answers

1. ____ 2. ____ 3. ____ 4. ____ 5. ____ 6. ____ 7. ____ 8. ____ 9. ____ 10. ____ 11. ____ 12. ____ 13. ____ 14. ____ 15. ____ 16. ____ 17. ____ 18. ____

Alternative Quiz Ideas

100% Quiz	Math Log
Error Notebook	Notebook Quiz
Group Quiz	Partner Quiz
Homework Quiz	Pass the Paper

100% Quiz
This is a quiz where students are given the answers and then they have to explain and justify each answer.

Reteaching and Enrichment Strategies

If students need help. . .	If students got it. . .
Resources by Chapter • Study Help • Practice A and Practice B • Puzzle Time Lesson Tutorials *BigIdeasMath.com* Practice Quiz Practice from the Test Generator	Resources by Chapter • Enrichment and Extension • School-to-Work Game Closet at *BigIdeasMath.com* Start the next section

Technology For the Teacher

Answer Presentation Tool
Big Ideas Test Generator

6.1–6.3 Quiz

Simplify the expression. *(Section 6.1)*

1. $\sqrt{20}$

2. $\sqrt{\frac{11}{81}}$

3. $\frac{4 - \sqrt{12}}{2}$

4. $\frac{-6 + \sqrt{45}}{3}$

Evaluate the expression when $x = 2, y = -3$, and $z = 6$. *(Section 6.1)*

5. $\sqrt{x + y^2 z}$

6. $\sqrt{3xz - y^2}$

Simplify. Write your answer using only positive exponents. *(Section 6.2)*

7. $3^2 \cdot 3^4$

8. $(k^4)^3$

9. $(4y)^{-2}$

10. $\left(\frac{r}{2}\right)^3$

Simplify. *(Section 6.3)*

11. $\sqrt[3]{27}$

12. $16^{1/4}$

13. $512^{2/3}$

14. $4^{5/2}$

15. CEDAR CHEST You store blankets in a cedar chest. What is the volume of the cedar chest? *(Section 6.1)*

16. CRITICAL THINKING Is the set of irrational numbers closed under subtraction? If not, find a counterexample. *(Section 6.1)*

Unit of Mass	Mass
gigagram	10^9 grams
megagram	10^6 grams
kilogram	10^3 grams
hectogram	10^2 grams
dekagram	10^1 grams
decigram	10^{-1} gram
centigram	10^{-2} gram
milligram	10^{-3} gram
microgram	10^{-6} gram
nanogram	10^{-9} gram

17. METRIC UNITS The table shows several units of mass. *(Section 6.2)*

a. How many times larger is a kilogram than a nanogram? Write your answer using only positive exponents.

b. How many times smaller is a milligram than a hectogram? Write your answer using only positive exponents.

c. Which is greater, 10,000 milligrams or 1000 decigrams? Explain your reasoning.

6.4 Exponential Functions

Essential Question What are the characteristics of an exponential function?

COMMON CORE STATE STANDARDS
A.REI.3
A.REI.11
F.BF.3
F.IF.7e
F.LE.1a
F.LE.2

1 ACTIVITY: Describing an Exponential Function

Work with a partner. The graph below shows estimates of the population of Earth from 5000 B.C. through 1500 A.D. at 500-year intervals.

a. Describe the pattern.

b. Did Earth's population increase by the same *amount* or the same *percent* for each 500-year period? Explain.

c. Assume the pattern continued. Estimate Earth's population in 2000.

d. Use the Internet to find Earth's population in 2000. Did the pattern continue? If not, why did the pattern change?

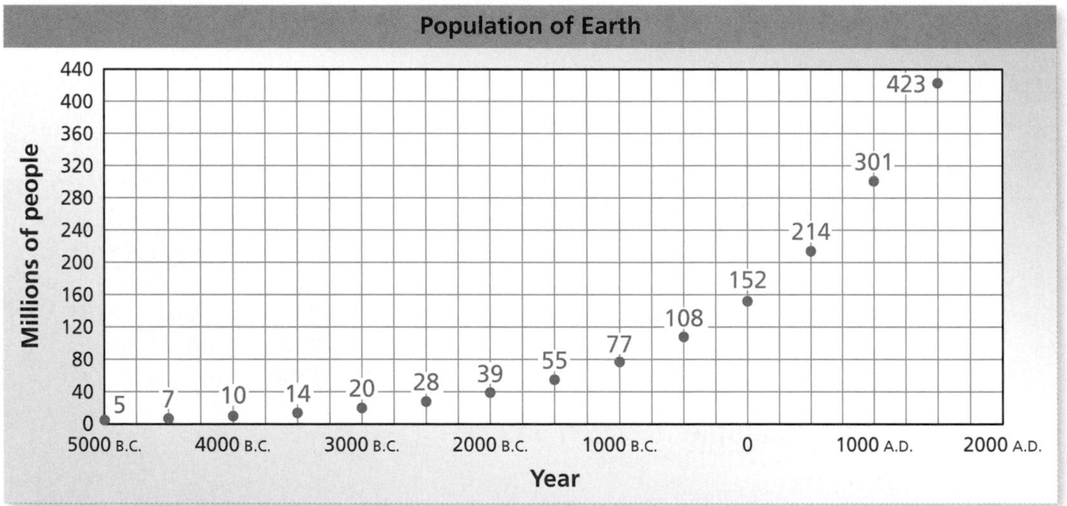

4000 B.C.
Civilization begins to develop in Mesopotamia.

3000 B.C.
Stonehenge is built in England.

2000 B.C.
Middle Kingdom in Egypt

1000 B.C.
Approximate beginning of Iron Age

Laurie's Notes

Introduction

Standards for Mathematical Practice

- **MP3a Construct Viable Arguments** and **MP7 Look for and Make Use of Structure:** Students are comfortable working with linear functions. To develop comfort with exponential functions, they must pay attention to the order of operations. The function $y = 3(5)^x$ represents 5 to the x power multiplied by 3. The function $y = 25(2)^x$ represents 2 to the x power multiplied by 25. The structure of the general equation helps students to recognize that the first function is increasing more rapidly because a base of 5 grows more quickly than a base of 2.

Motivate

- Pose this scenario. You give a test and scores range from 40% to 100%. To help your students, you decide to increase each score by 10% of the score (10% scaling).

? "Does everyone's score increase?" yes

? "Does everyone's score increase the same amount?" no

? "Which students receive the least benefit? the greatest benefit? Explain." A 40% test score receives the least benefit (4 points); A 100% test score receives the greatest benefit (10 points).

- Without asking about fairness, discuss that the *percent of change* is the same (10%) for all tests, but the *amount of change* is not. In a linear function, the amount of change is the same, so this function is not linear. Explain that this section will explore this idea further.

Activity Notes

Discuss

- In this activity, students will use a graph and a table to describe the patterns of exponential functions. Students do not yet have the language of exponential functions, but listen for descriptions that suggest a multiplicative pattern.

Activity 1

- In part (a), try to evoke a stronger description than "The graph curves." Discuss "Is there a way to describe the curve? Look at the numbers. Is there a numerical pattern? The scale of each axis is very different. How does that influence the graph?"
- In part (b), students should quickly recognize that the *amount of growth* was not constant. Students will have more difficulty in describing the percent of growth. Refer back to the 10% scaling (a score of 50 increased to 55). Students may estimate a change of a little less than 50%.

Common Core State Standards

F.BF.3 Identify the effect on the graph of replacing $f(x)$ by $f(x) + k$, $kf(x)$, $f(kx)$, and $f(x + k)$ for specific values of k (both positive and negative); find the value of k given the graphs. Experiment with cases . . . .

F.IF.7e Graph exponential . . . functions, showing intercepts and end behavior . . . .

F.LE.1a Prove that . . . exponential functions grow by equal factors over equal intervals.

F.LE.2 Construct . . . exponential functions, including . . . geometric sequences, given a graph, a description of a relationship, or two input-output pairs (include reading these from a table).

Also **A.REI.3, A.REI.11**

Previous Learning

Students should know how to evaluate and graph functions. Students should know how to find a rate of change and a percent of change.

Start Thinking! and Warm Up

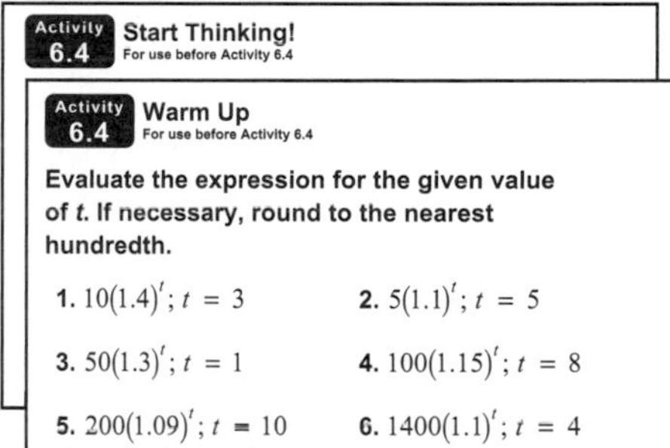

Activity 6.4 Start Thinking! For use before Activity 6.4

Activity 6.4 Warm Up For use before Activity 6.4

Evaluate the expression for the given value of *t*. If necessary, round to the nearest hundredth.

1. $10(1.4)^t;\ t = 3$
2. $5(1.1)^t;\ t = 5$
3. $50(1.3)^t;\ t = 1$
4. $100(1.15)^t;\ t = 8$
5. $200(1.09)^t;\ t = 10$
6. $1400(1.1)^t;\ t = 4$

6.4 Record and Practice Journal

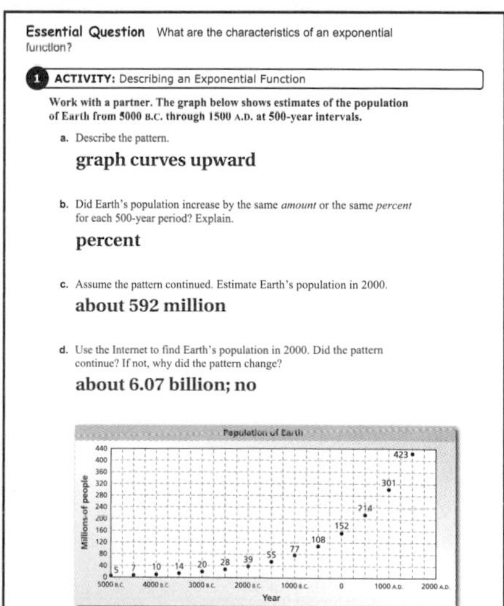

Essential Question What are the characteristics of an exponential function?

1 **ACTIVITY:** Describing an Exponential Function

Work with a partner. The graph below shows estimates of the population of Earth from 5000 B.C. through 1500 A.D. at 500-year intervals.

a. Describe the pattern.

graph curves upward

b. Did Earth's population increase by the same *amount* or the same *percent* for each 500-year period? Explain.

percent

c. Assume the pattern continued. Estimate Earth's population in 2000.

about 592 million

d. Use the Internet to find Earth's population in 2000. Did the pattern continue? If not, why did the pattern change?

about 6.07 billion; no

Differentiated Instruction

Inclusion

Students with fine-motor problems may find it difficult to graph exponential functions and get frustrated. Make sure these students understand the general shape of an exponential function and have them graph on a coordinate plane without grid lines. Also show how a spreadsheet program or graphing calculator can be used to generate a graph.

6.4 Record and Practice Journal

2 ACTIVITY: Modeling an Exponential Function

Work with a partner. Use the following exponential function to complete the table. Compare the results with the data in Activity 1.

$P = 152(1.406)^{t/500}$

Year	t	Population from Activity 1	P
5000 B.C.	−5000	5	5.03
4500 B.C.	−4500	7	7.08
4000 B.C.	−4000	10	9.95
3500 B.C.	−3500	14	13.99
3000 B.C.	−3000	20	19.68
2500 B.C.	−2500	28	27.66
2000 B.C.	−2000	39	38.90
1500 B.C.	−1500	55	54.69
1000 B.C.	−1000	77	76.89
500 B.C.	−500	108	108.11
1 B.C.	0	152	152.00
500 A.D.	500	214	213.71
1000 A.D.	1000	301	300.48
1500 A.D.	1500	423	422.47

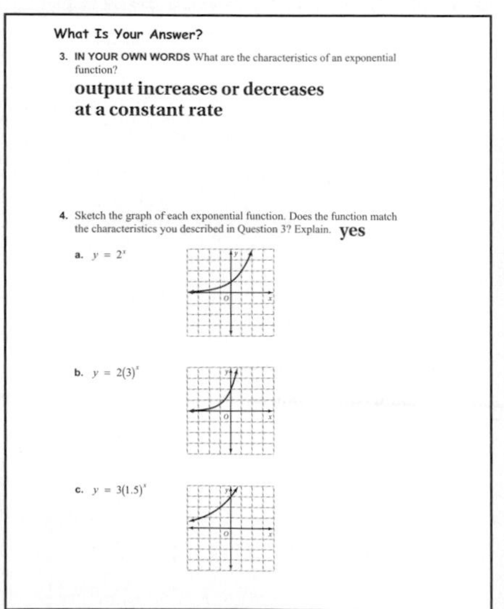

What Is Your Answer?

3. **IN YOUR OWN WORDS** What are the characteristics of an exponential function?
   output increases or decreases at a constant rate

4. Sketch the graph of each exponential function. Does the function match the characteristics you described in Question 3? Explain. **yes**

   a. $y = 2^x$

   b. $y = 2(3)^x$

   c. $y = 3(1.5)^x$

Laurie's Notes

Activity 1 (continued)

- **Connection:** Remind students that a percent of change is the amount of change divided by the original amount. For the data in the graph, the percent of change can be calculated as shown in this partial table.

Year, *t*	−4500	−4000	−3500	−3000	−2500	−2000
Change/Previous	2/5	3/7	4/10	6/14	8/20	11/28
Percent of Change	40%	43%	40%	43%	40%	39%

- Have students find the average percent of change to compare in Activity 2. about 41%

Activity 2

- The function can be used without a formal exponential function definition.
- ? "What is the independent variable of this function? What does it represent? Where is the independent variable located in the equation?" *t*, the time in years; in the numerator of the exponent
- ? "What is the dependent variable of this function? What does it represent?" *P*; the population in millions
- If students are not familiar with using the exponent key on their calculators, model the first year for them. Point out that the exponent $t/500$ is only applied to the base 1.406. It is not applied to the constant 152.
- **MP3a:** Discuss and compare the results to the data in Activity 1.
- ? "How does the average you found in Activity 1 relate to the function in Activity 2?" Students may or may not see a relationship.
- **MP7:** The base of the exponential function is 1.406, which equals 140.6%. You can think of 140.6% as about a 40% increase.
- **Extension:** Use the function to predict the population in 2000 A.D.

What Is Your Answer?

- **Common Error:** In Question 4, students may forget about order of operations. The exponent is applied to the base, not the constants.
- If students are using a graphing calculator, the window influences how they *see* the function. This can be explored as time permits.

Closure

- Describe how the range values are changing. 50% increase

x	1	2	3	4
y	4	6	9	13.5

Technology For the Teacher

The Dynamic Planning Tool
Editable Teacher's Resources at *BigIdeasMath.com*

2 ACTIVITY: Modeling an Exponential Function

Work with a partner. Use the following exponential function to complete the table. Compare the results with the data in Activity 1.

$$P = 152(1.406)^{t/500}$$

1 B.C.
Augustus Caesar controls most of the Mediterranean world. (Use $t = 0$ to approximate 1 B.C.)

1000 A.D.
Song Dynasty has about one-fifth of Earth's population.

Year	t	Population from Activity 1	P
5000 B.C.	−5000		
4500 B.C.	−4500		
4000 B.C.	−4000		
3500 B.C.	−3500		
3000 B.C.	−3000		
2500 B.C.	−2500		
2000 B.C.	−2000		
1500 B.C.	1500		
1000 B.C.	−1000		
500 B.C.	−500		
1 B.C.	0		
500 A.D.	500		
1000 A.D.	1000		
1500 A.D.	1500		

What Is Your Answer?

3. **IN YOUR OWN WORDS** What are the characteristics of an exponential function?

4. Sketch the graph of each exponential function. Does the function match the characteristics you described in Question 3? Explain.

   a. $y = 2^x$ **b.** $y = 2(3)^x$ **c.** $y = 3(1.5)^x$

Practice Use what you learned about exponential functions to complete Exercises 4 and 5 on page 289.

6.4 Lesson

Check It Out
Lesson Tutorials
BigIdeasMath.com

Key Vocabulary
exponential function, p. 286

A function of the form $y = ab^x$, where $a \neq 0$, $b \neq 1$, and $b > 0$ is an **exponential function.** The exponential function $y = ab^x$ is a nonlinear function that changes by equal factors over equal intervals.

EXAMPLE 1 Identifying Functions

Does each table represent a *linear* or an *exponential* function? Explain.

a.

+1 +1 +1

x	0	1	2	3
y	2	4	6	8

+2 +2 +2

As x increases by 1, y increases by 2. The rate of change is constant. So, the function is linear.

b.

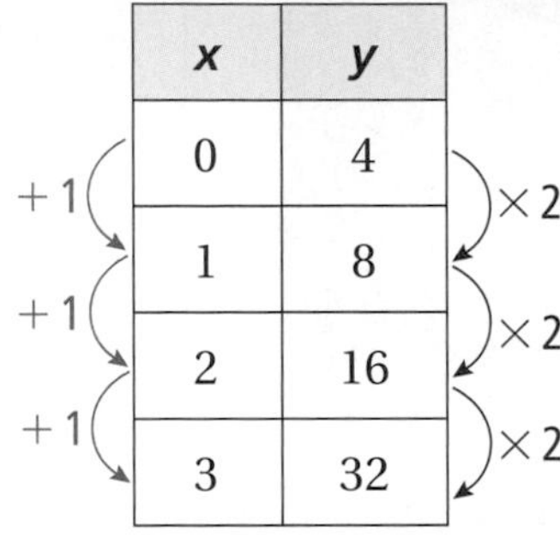

	x	y	
	0	4	
+1			×2
	1	8	
+1			×2
	2	16	
+1			×2
	3	32	

As x increases by 1, y is multiplied by 2. So, the function is exponential.

EXAMPLE 2 Evaluating Exponential Functions

Evaluate each function for the given value of x.

a. $y = -2(5)^x$; $x = 3$

$y = -2(5)^x$	Write the function.	$y = 3(0.5)^x$
$= -2(5)^3$	Substitute for x.	$= 3(0.5)^{-2}$
$= -2(125)$	Evaluate the power.	$= 3(4)$
$= -250$	Multiply.	$= 12$

b. $y = 3(0.5)^x$; $x = -2$

On Your Own

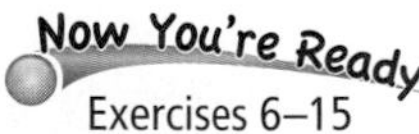

Now You're Ready
Exercises 6–15

Does the table represent a *linear* or an *exponential* function? Explain.

1.

x	0	1	2	3
y	8	4	2	1

2.

x	y
−4	1
0	0
4	−1
8	−2

Evaluate the function when $x = -2, 0,$ and $\frac{1}{2}$.

3. $y = 2(9)^x$

4. $y = 1.5(2)^x$

Multi-Language Glossary at BigIdeasMath.com.

Laurie's Notes

Introduction

Connect

- **Yesterday**: Students used data to explore growth for an exponential function. (MP3a, MP7)
- **Today**: Students will use the growth pattern of a set of data to determine whether the data represent a linear or an exponential function.

Motivate

- Search online for a short video about bacteria reproducing through binary fission. Show the video to your students.
- A single bacterium, splitting in two every 20 minutes, would produce nearly 5000 billion-billion bacteria in 1 day.
- ? "How many reproductive cycles (20-minute time periods) do the bacteria experience in a 24-hour day?" There are 3 per hour, so there are $3 \times 24 = 72$ in a 24-hour day.
- ? "How do you write 5000 billion-billion?" $5000 \times 10^9 \times 10^9 = 5 \times 10^{21}$, or in standard form, a 5 followed by 21 zeros
- Explain that today's lesson will look at a type of function that can be used to model bacteria population growth.

Lesson Notes

Discuss

- Write the definition of an exponential function. Do not dwell on the constraints ($a \neq 0$, $b \neq 1$, and $b > 0$) at this point. Instead, return to the constraints after students have a sense of how the function behaves. Point out that the independent variable x is the exponent of the function, b is the base, and a is a constant.
- Further explain the statement that an exponential function changes by equal factors over equal intervals. Compare this to a linear function, where the function changes by an equal amount over equal intervals.

Example 1

- ? "How can you determine whether the data represent a linear function?" A linear function has a constant rate of change.
- ? "Do the data in part (a) change by equal amounts over equal intervals? Explain." yes; The function increases by 2 for each 1-unit increase in x.
- ? "What is the equation of this function?" $y = 2x + 2$
- Point out the distinction in part (b). The function increases by a factor of 2 (it doubles) for each 1-unit increase in x.
- **FYI:** The equation of this function is $y = 4(2)^x$.
- **Connection:** Students should recognize that the y-values are powers of 2.

Example 2

- Work through the steps in part (a). Students may forget the order of operations and simplify $(-2)(5) = -10$ before cubing. Remind students to apply the exponent to the base 5 only.

Goal Today's lesson is writing and graphing **exponential functions**.

Lesson Materials
Introduction
• bacteria video

Start Thinking! and Warm Up

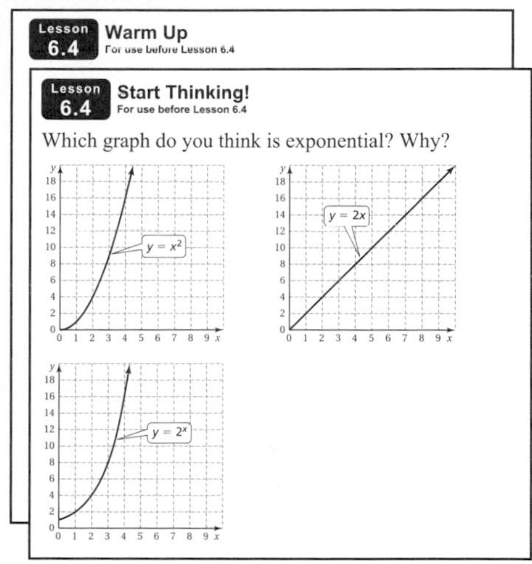

Lesson 6.4 Warm Up
For use before Lesson 6.4
Lesson 6.4 Start Thinking!
For use before Lesson 6.4
Which graph do you think is exponential? Why?
y = x²
y = 2x
y = 2ˣ

Extra Example 1

Does each table represent a *linear* or an *exponential* function? Explain.

a.

x	0	1	2	3
y	5	10	15	20

linear; As x increases by 1, y increases by 5.

b.

x	0	1	2	3
y	1	3	9	27

exponential; As x increases by 1, y is multiplied by 3.

Extra Example 2

Evaluate the function $y = -4(2)^x$ when $x = -1, 0,$ and 3. $-2, -4, -32$

On Your Own

1. exponential; As x increases by 1, y is multiplied by $\frac{1}{2}$.
2. linear; As x increases by 4, y decreases by 1.
3. $\frac{2}{81}$, 2, 6
4. 0.375, 1.5, $1.5\sqrt{2}$

English Language Learners

Illustrate

Help English language learners with vocabulary and concepts. Create posters of different exponential functions. Label the functions, domains, ranges, and intercepts. Display the posters in the classroom so that students can easily refer to them.

Extra Example 3

Graph $y = 5^x$. Describe the domain and range.

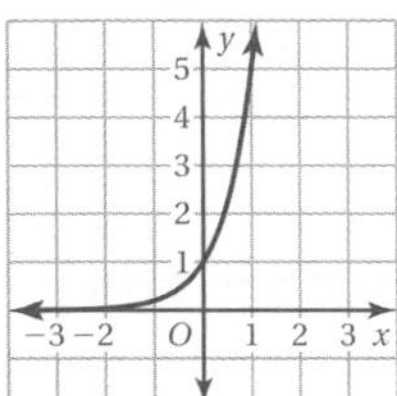

domain: all real numbers;
range: all positive real numbers

Extra Example 4

Graph $y = 5^x - 1$. Describe the domain and range. Compare the graph to the graph of $y = 5^x$.

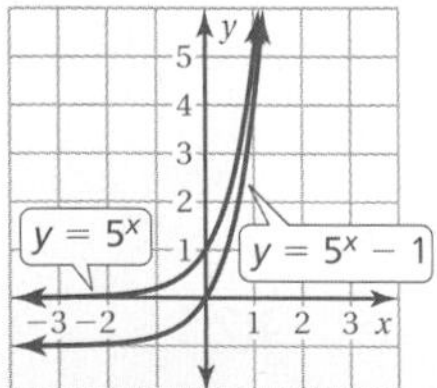

domain: all real numbers;
range: all real numbers greater than -1;
The graph is a translation 1 unit down of the graph of $y = 5^x$.

On Your Own

5–8. See Additional Answers.

Laurie's Notes

Example 2 (continued)

? "What property can you use in part (b) to raise 0.5 to a negative power?" $a^{-n} = \frac{1}{a^n}$.

- **MP8 Look for and Express Regularity in Repeated Reasoning:** Show your students another way involving rewriting the expression.

  $(0.5)^{-2} = (0.5)^{(2)(-1)} = \left[\left(\frac{1}{2}\right)^2\right]^{-1} = \left(\frac{1}{4}\right)^{-1} = 4$

  After a few examples, they learn the pattern and quit using all the steps.

On Your Own

- Question 1 is not obvious to all students because the function *decreases* at each interval by a factor of $\frac{1}{2}$. Question 2 is not obvious to all students because the equal intervals are 4 units long instead of 1.

Example 3

- **FYI:** The concepts of growth and decay come in later sections.
- Write the equation and ask a student to read it. Listen for "y equals 2 to the xth power." It is important for students to realize that they are graphing powers of 2.
- Complete the table of values and plot the points.
- You can draw a smooth curve through the plotted points because the domain in this problem is the set of all real numbers x. In some application problems, the domain is restricted to discrete values by context, like a 20-minute doubling period for bacteria.

? "What is the y-intercept? the x-intercept? 1; no x-intercept

Example 4

- Write the equation and ask a student to read it. Listen for "y equals 2 to the xth power plus 3." It is important for students to realize that they are graphing 3 more than the powers of 2.
- Complete the table of values and plot the points.

? "What is the y-intercept? the x-intercept?" 4; no x-intercept

? "How is this graph related to the previous graph?" It is the same graph shifted up 3 units.

- Notice how the range is affected by the vertical translation.
- **Connection:** When students graphed linear functions, they learned that $y = x$ and $y = x + 3$ are parallel lines and the second graph is a vertical shift up 3 units. In Example 4, the graph of $y = 2^x + 3$ is a vertical shift up 3 units of the graph of $y = 2^x$.

On Your Own

- In Question 6, students may be surprised that the graph is decreasing. They may describe the graph as "sloping downward" or refer to slope. Remind students that lines have a slope, exponential functions do not.

EXAMPLE 3 Graphing an Exponential Function

Graph $y = 2^x$. Describe the domain and range.

Step 1: Make a table of values.

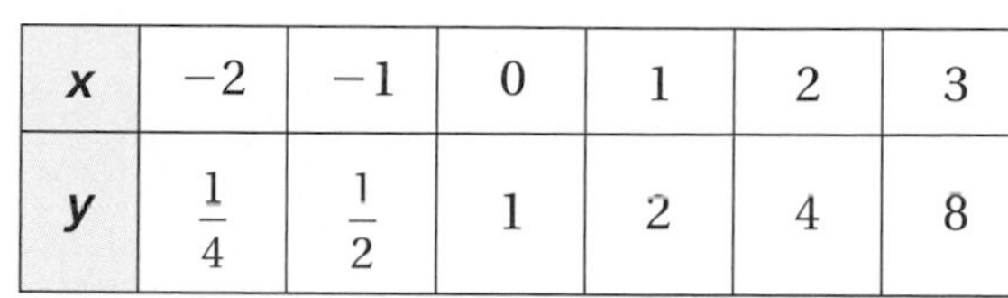

x	-2	-1	0	1	2	3
y	$\frac{1}{4}$	$\frac{1}{2}$	1	2	4	8

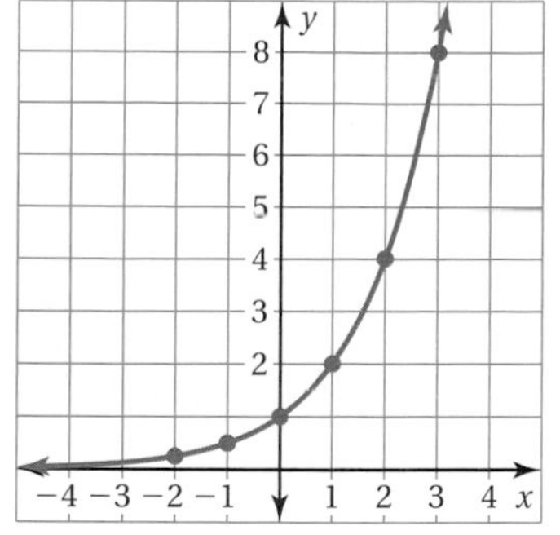

Step 2: Plot the ordered pairs.

Step 3: Draw a smooth curve through the points.

From the graph, you can see that the domain is all real numbers and the range is all positive real numbers.

Study Tip

In Example 3, you can substitute any value for x. So, the domain is all real numbers.

EXAMPLE 4 Graphing a Vertical Translation

Graph $y = 2^x + 3$. Describe the domain and range. Compare the graph to the graph of $y = 2^x$.

Step 1: Make a table of values.

x	-2	-1	0	1	2	3
y	$\frac{13}{4}$	$\frac{7}{2}$	4	5	7	11

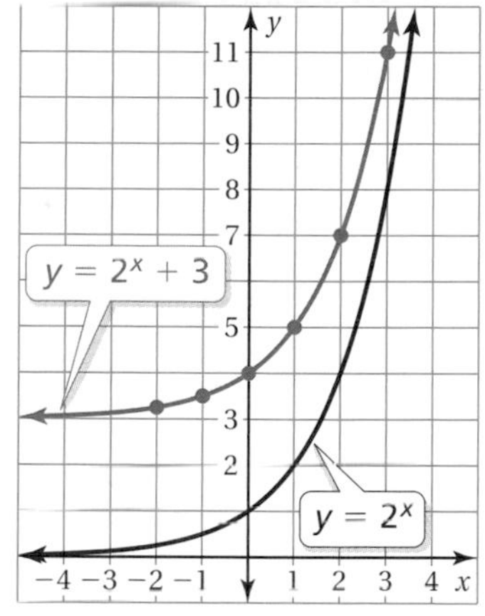

Step 2: Plot the ordered pairs.

Step 3: Draw a smooth curve through the points.

From the graph, you can see that the domain is all real numbers and the range is all real numbers greater than 3. The graph of $y = 2^x + 3$ is a translation 3 units up of the graph of $y = 2^x$.

Remember

In Section 5.4, you learned that the graph of $f(x) + k$ is a vertical translation of the graph of $f(x)$.

On Your Own

Now You're Ready Exercises 21–23 and 27–29

Graph the function. Describe the domain and range.

5. $y = 3^x$ **6.** $y = \left(\frac{1}{2}\right)^x$ **7.** $y = -2\left(\frac{1}{4}\right)^x$

8. Graph $y = \left(\frac{1}{2}\right)^x - 2$. Describe the domain and range. Compare the graph to the graph of $y = \left(\frac{1}{2}\right)^x$.

Study Tip

To find the y-intercept of the graph of $y = ab^x$, substitute 0 for x.

$y = ab^0$

$y = a(1)$

$y = a$

So, the y-intercept is a.

For an exponential function of the form $y = ab^x$, the y-values change by a factor of b as x increases by 1. Also notice that a is the y-intercept.

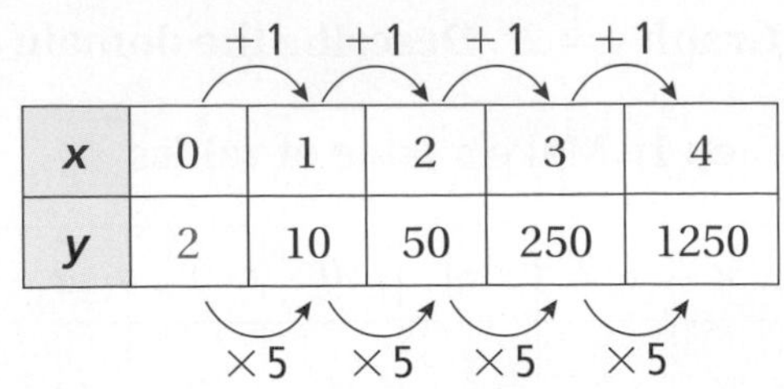

x	0	1	2	3	4
y	2	10	50	250	1250

$y = 2(5)^x$

EXAMPLE 5 Real-Life Application

The graph represents a bacteria population y after x days.

a. Write an exponential function that represents the population.

Use the graph to make a table of values.

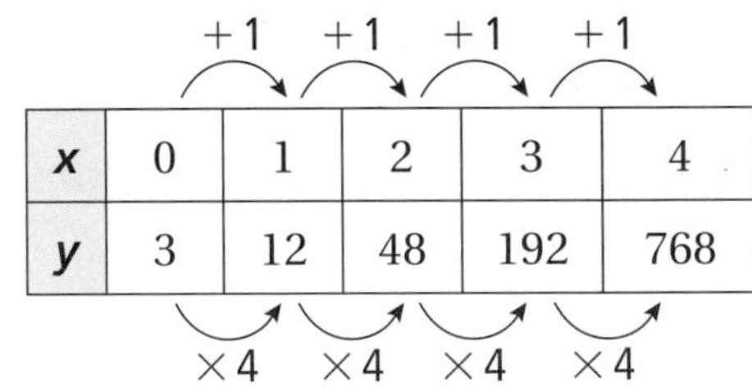

x	0	1	2	3	4
y	3	12	48	192	768

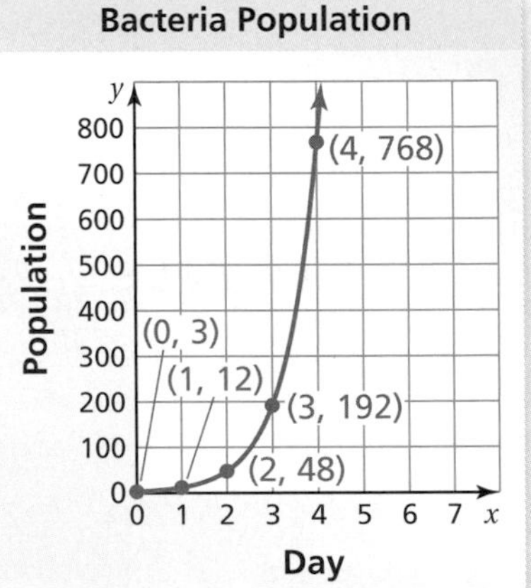

Study Tip

For help with rational exponents, see Section 6.3.

The y-intercept is 3 and the y-values increase by a factor of 4 as x increases by 1.

So, the population can be modeled by $y = 3(4)^x$.

b. Find the population after 12 hours and after 5 days.

12 hours $= \frac{1}{2}$ day

Population after 12 hours		*Population after 5 days*
$y = 3(4)^x$	Write the function.	$y = 3(4)^x$
$= 3(4)^{1/2}$	Substitute for x.	$= 3(4)^5$
$= 3(2)$	Evaluate the power.	$= 3(1024)$
$= 6$	Multiply.	$= 3072$

There are 6 bacteria after 12 hours and 3072 bacteria after 5 days.

On Your Own

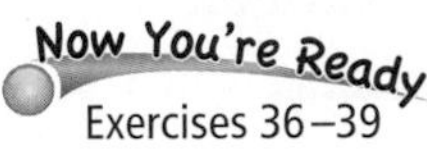

Exercises 36–39

9. A bacteria population y after x days can be represented by an exponential function whose graph passes through (0, 100) and (1, 200).

   a. Write a function that represents the population.

   b. Find the population after 6 days. Does this bacteria population grow faster than the bacteria population in Example 5? Explain.

Laurie's Notes

Discuss

- Before the next example, you can do some discovery. Have students use their graphing utilities to graph $y = 2^x$, $y = 3(2)^x$, and $y = 6(2)^x$.
- Have students use the *trace* or *table* feature to determine the y-intercepts of the graphs. The y-intercepts of the three graphs are 1, 3, and 6.

? "What is the y-intercept of the graph of $y = ab^x$?" a

- The Study Tip describes a way to answer this question algebraically.

Example 5

- Students use a graph to write an exponential function in Example 5.

? "What type of information do you need to write a linear function?" two points or the slope and a point

- You can write an exponential function of the form $y = ab^x$ when you know the y-intercept a and the base b of the function.

? "Look at the table of values. Each time x increases by 1 unit, by what factor does y change? What is the y-intercept?" 4; 3

- Make sure students know what x and y represent. x is the number of days and y is the number of bacteria.
- In part (b), students review rational exponents.

? "What symbol is equivalent to a rational exponent of $\frac{1}{2}$?" the square root symbol

On Your Own

- Students may need guidance. The ordered pairs can be used to determine that for each 1-unit increase in x, the y-value increases by a factor of 2, and the y-intercept is 100.

Closure

- Match the function with the graph.

1. $y = 2(3)^x$ B
2. $y = 5(2)^x$ C
3. $y = 3(5)^x$ A

A.

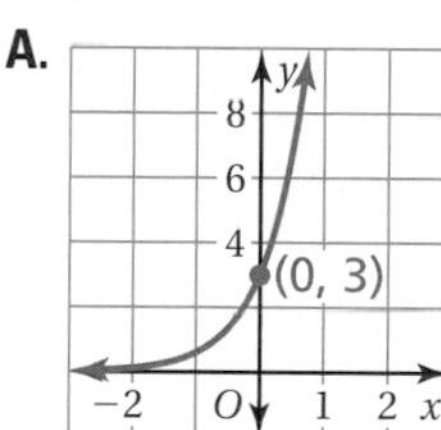

B.

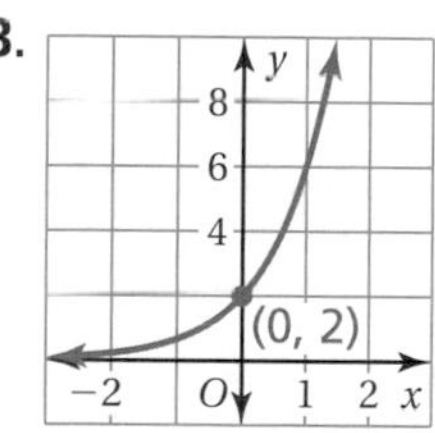

C.

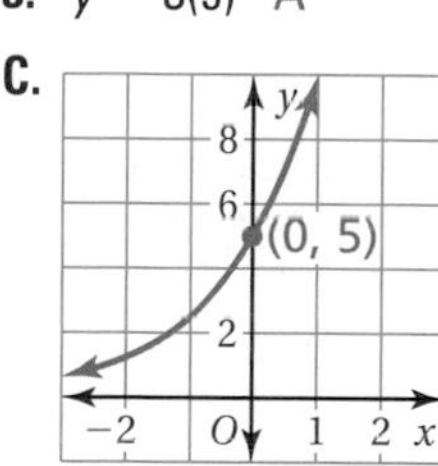

Extra Example 5

A bacteria population y after x hours can be represented by an exponential function whose graph passes through (0, 20) and (1, 60).

a. Write a function that represents the population. $y = 20(3)^x$

b. Find the population after 5 hours. There are 4860 bacteria after 5 hours.

On Your Own

9. a. $y = 100(2)^x$

   b. 6400; No, this population doubles every day, but the population in Example 5 quadruples every day.

Vocabulary and Concept Check

1. A linear function changes by a constant amount over equal intervals. An exponential function changes by equal factors over equal intervals.
2. *Sample answer:*

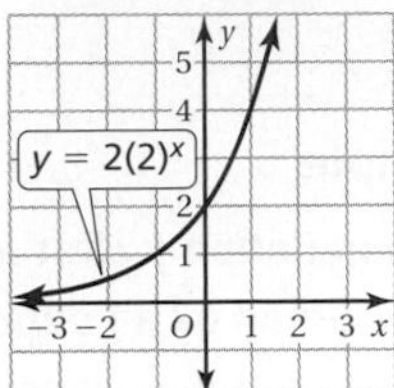

3. $f(x) = (-3)^x$; It is not an exponential function.

Practice and Problem Solving

4.

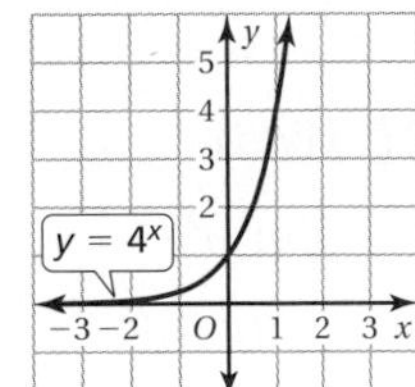

5.

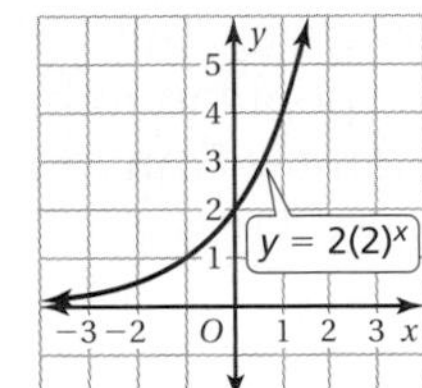

6. linear; As x increases by 1, y increases by 2.
7. exponential; As x increases by 1, y is multiplied by 2.
8. exponential; As x increases by 1, y is multiplied by 4.
9. linear; As x increases by 3, y decreases by 9.

10. 9
11. 1.5
12. -100
13. 8
14. 72
15. 2
16. $(0.5)^{-2}$ should have been evaluated first;

$$g(-2) = 6(0.5)^{-2}$$
$$= 6(4)$$
$$= 24$$

Assignment Guide and Homework Check

Level	Assignment	Homework Check
Average	1–3, 6–16 even, 17, 21–29 odd, 26, 36, 38, 40, 43–46	17, 23, 26, 27, 40
Advanced	1–3, 11–17 odd, 21–29 odd, 30, 36, 38, 40–42, 43–46	17, 23, 29, 40, 42

Common Errors

- **Exercise 9** Students may say that the function is exponential because the y-values are relatively far apart. Remind them that an exponential function changes by equal factors over equal intervals.
- **Exercises 10–15** Students may multiply before evaluating the power. Remind them of the order of operations.

6.4 Record and Practice Journal

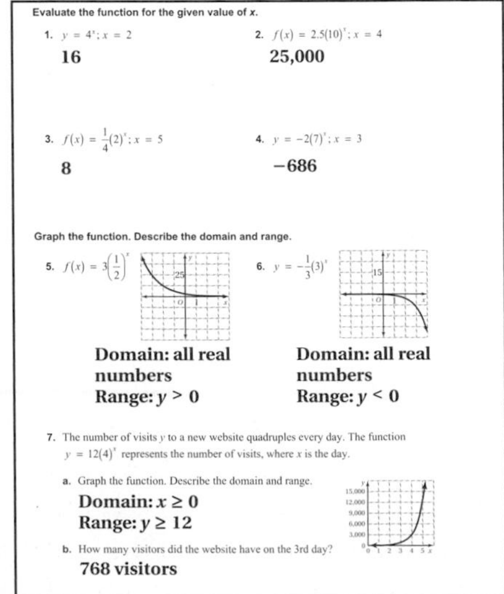

Evaluate the function for the given value of x.

1. $y = 4^x; x = 2$

16

2. $f(x) = 2.5(10)^x; x = 4$

25,000

3. $f(x) = \frac{1}{4}(2)^x; x = 5$

8

4. $y = -2(7)^x; x = 3$

−686

Graph the function. Describe the domain and range.

5. $f(x) = 3\left(\frac{1}{2}\right)^x$

Domain: all real numbers
Range: $y > 0$

6. $y = -\frac{1}{3}(3)^x$

Domain: all real numbers
Range: $y < 0$

7. The number of visits y to a new website quadruples every day. The function $y = 12(4)^x$ represents the number of visits, where x is the day.

a. Graph the function. Describe the domain and range.

Domain: $x \geq 0$
Range: $y \geq 12$

b. How many visitors did the website have on the 3rd day?

768 visitors

Technology For the Teacher

Answer Presentation Tool

6.4 Exercises

Vocabulary and Concept Check

1. **VOCABULARY** Describe how linear and exponential functions change over equal intervals.

2. **OPEN-ENDED** Sketch an increasing exponential function whose graph has a y-intercept of 2.

3. **WHICH ONE DOESN'T BELONG?** Which equation does *not* belong with the other three? Explain your reasoning.

$y = 3^x$ | $f(x) = 2(4)^x$ | $f(x) = (-3)^x$ | $y = 5(3)^x$

Practice and Problem Solving

Sketch the graph of the exponential function.

4. $y = 4^x$

5. $y = 2(2)^x$

Does the table represent a *linear* or an *exponential* function? Explain.

1 6.

x	y
0	−2
1	0
2	2
3	4

7.

x	y
1	6
2	12
3	24
4	48

8.

x	−1	0	1	2
y	0.25	1	4	16

9.

x	−3	0	3	6
y	10	1	−8	−17

Evaluate the function for the given value of x.

2 10. $y = 3^x;\ x = 2$

11. $f(x) = 3(2)^x;\ x = -1$

12. $y = -4(5)^x;\ x = 2$

13. $f(x) = 0.5^x;\ x = -3$

14. $f(x) = \frac{1}{3}(6)^x;\ x = 3$

15. $y = \frac{1}{4}(4)^x;\ x = \frac{3}{2}$

16. **ERROR ANALYSIS** Describe and correct the error in evaluating the function.

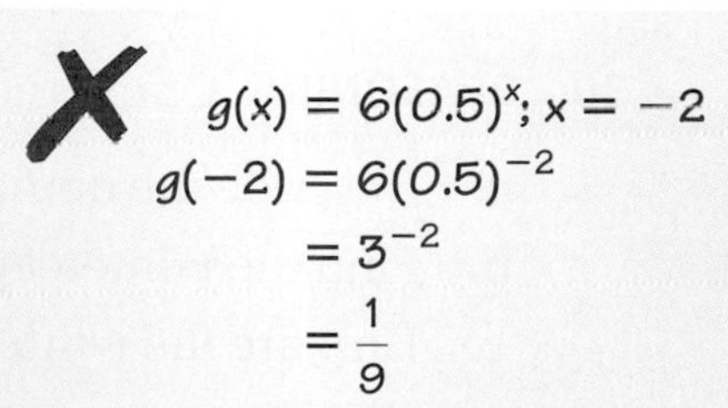

17. **CALCULATOR** You graph an exponential function on a calculator. You zoom in repeatedly at 25% of the screen size. The function $y = 0.25^x$ represents the percent (in decimal form) of the original screen display that you see, where x is the number of times you zoom in. You zoom in twice. What percent of the original screen do you see?

Match the function with its graph.

18. $f(x) = -3(4)^x$

19. $y = 2(0.5)^x$

20. $y = 4(1.5)^x$

A. 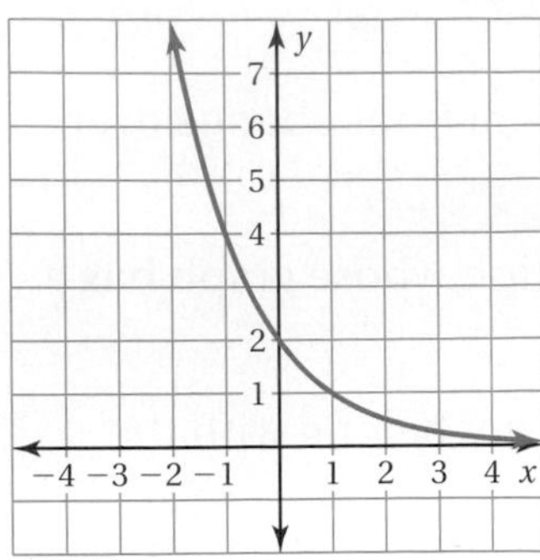

B. 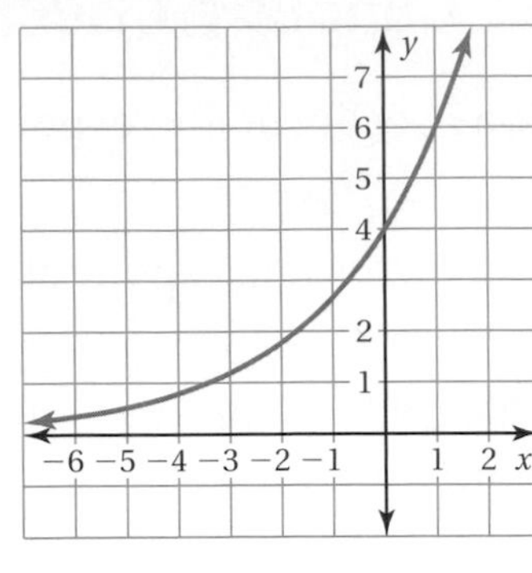

C. 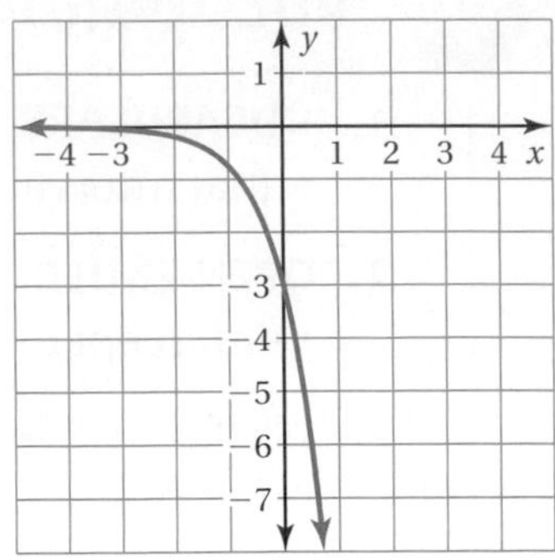

Graph the function. Describe the domain and range.

3 21. $y = 9^x$

22. $f(x) = -7^x$

23. $f(x) = 4\left(\frac{1}{4}\right)^x$

24. **LOGIC** Describe the graph of $y = a(2)^x$ when a is (a) positive and (b) negative. (c) How does the graph change as a changes?

25. **NUMBER SENSE** Consider the graph of $f(x) = 2(b)^x$. How do the graphs differ when $b > 1$ and $0 < b < 1$?

26. **COYOTES** A population y of coyotes in a national park triples every 20 years. The function $y = 15(3)^x$ represents the population, where x is the number of 20-year periods.

    a. Graph the function. Describe the domain and range.

    b. Find and interpret the y-intercept.

    c. How many coyotes are in the national park after 20 years?

Graph the function. Describe the domain and range. Compare the graph to the graph of $y = 3^x$.

4 27. $y = 3^x - 1$

28. $y = 3^x + 3$

29. $y = 3^x - \frac{1}{2}$

30. **REASONING** Graph the function $f(x) = -2^x$. Then graph $g(x) = -2^x - 3$.

    a. Describe the domain and range of each function.

    b. Find the y-intercept of the graph of each function.

    c. How are the y-intercept, domain, and range affected by the translation?

31. **REASONING** When does an exponential function intersect the x-axis? Give an example to justify your answer.

Common Errors

- **Exercise 22** Students may think of the expression as $(-7)^x$ rather than $-(7)^x$. Remind them of the order of operations.
- **Exercise 23** Students may multiply before evaluating the power. Remind them of the order of operations.
- **Exercises 32–34** Students may find the wrong value of k because they mentally picture the graph of $g(x)$ incorrectly. Encourage students to graph $g(x)$ so they can recognize how the given graph is translated from the graph of $g(x)$.
- **Exercises 36–39** Students may write incorrect equations. Remind them that in an exponential equation of the form $y = ab^x$, a is the value of y when $x = 0$, and b is the factor that y changes for each 1-unit increase in x.

17. 6.25% **18.** C

19. A **20.** B

21.

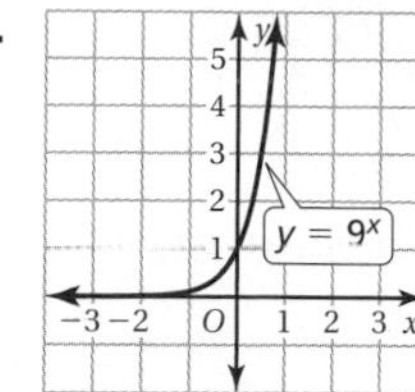

domain: all real numbers; range: all positive real numbers

22.

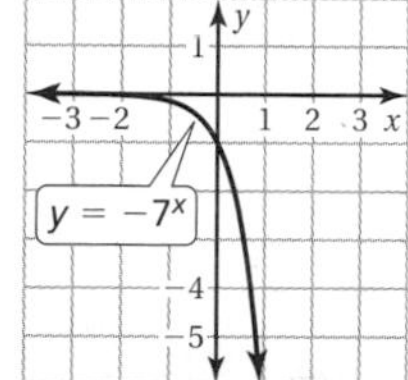

domain: all real numbers; range: all negative real numbers

23. See Additional Answers.

24. **a.** The graph is above the x-axis, and it increases from left to right.

b. The graph is below the x-axis, and it decreases from left to right.

Sample answer: As a changes, the y-intercept changes.

25. When $b > 1$, the graph rises from left to right. When $0 < b < 1$, the graph falls from left to right.

26–31. See Additional Answers.

32. $k = 1$ **33.** $k = -1$

34. $k = 2$

Differentiated Instruction

Inclusion

In the lessons on exponential functions, students are expected to make many tables and graphs. Provide a resource page with blank tables and enlarged coordinate grids for them to use.

Practice and Problem Solving

35. 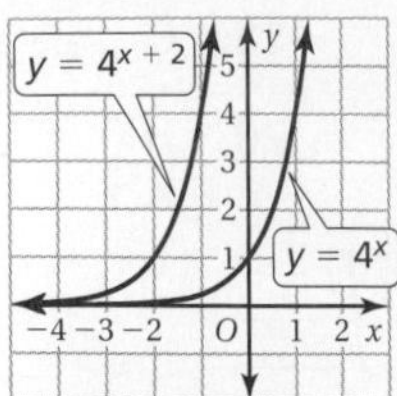

The graph is translated 2 units to the left.

36. $y = 0.5(2)^x$

37. $y = -8\left(\frac{1}{2}\right)^x$

38. $y = 2(4)^x$

39. $y = -3(5)^x$

40. a. $y = 40\left(\frac{3}{2}\right)^x$

b. about 304 visitors

41. See *Taking Math Deeper*.

42. $g(x) = 2^{x-3} + 4$

Fair Game Review

43. 0.23 44. 0.03

45. 1.5 46. C

Mini-Assessment

1. Does the table represent a *linear* or an *exponential* function? Explain.

x	1	2	3	4
y	2	4	8	16

exponential; As x increases by 1, y is multiplied by 2.

2. Evaluate the function $y = 2(4)^x$ when $x = 2$. 32

3. Graph $y = 2(2)^x$. Describe the domain and range.

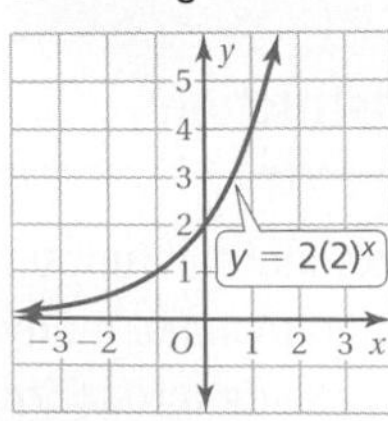

domain: all real numbers;
range: all positive real numbers

Taking Math Deeper

Exercise 41

Students can solve this exercise using repeated multiplication.

1. Use a table to examine how the sales increase exponentially over time.

Year	Increase in Sales	Yearly Sales
0		3300
1	$3300(0.06) = 198$	$3300 + 198 = 3498$
2	$3498(0.06) \approx 210$	$3498 + 210 = 3708$
3	$3708(0.06) \approx 222$	$3708 + 222 = 3930$
4	$3930(0.06) \approx 236$	$3930 + 236 = 4166$
5	$4166(0.06) \approx 250$	$4166 + 250 = 4416$
6	$4416(0.06) \approx 265$	$4416 + 265 = 4681$

2. Answer the question.
The store expects to sell about 4681 gas grills in year 6.

3. Use an equation to justify the answer. Let y be the number of gas grills sold in year x.

The table shows that the y-intercept is 3300. Each year the total sales are 1.06 times the sales in the previous year. So, the equation is $y = 3300(1.06)^x$.

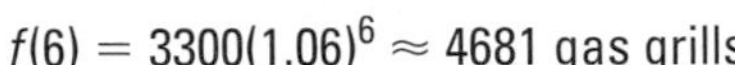

$f(6) = 3300(1.06)^6 \approx 4681$ gas grills

Project

Research a major company that produces consumer goods and review its most recent *annual report*. Why do you think it is important for companies to project future sales?

Reteaching and Enrichment Strategies

If students need help. . .	If students got it. . .
Resources by Chapter • Practice A and Practice B • Puzzle Time Record and Practice Journal Practice Differentiating the Lesson Lesson Tutorials Skills Review Handbook	Resources by Chapter • Enrichment and Extension • School-to-Work Start the next section

Given $g(x) = 0.25^x - 1$, find the value of k so that the graph is $g(x) + k$.

32. 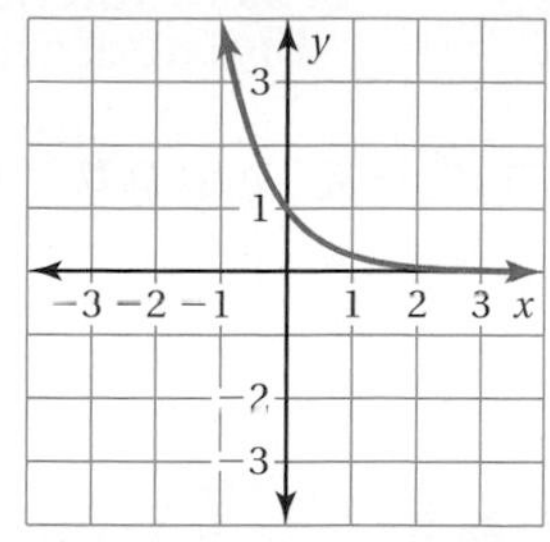

33. 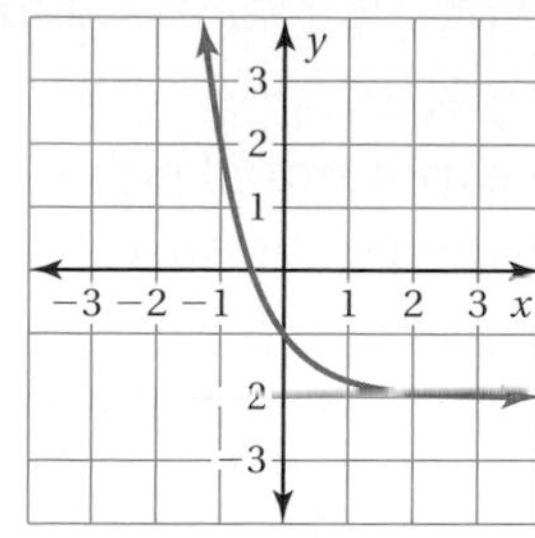

34. 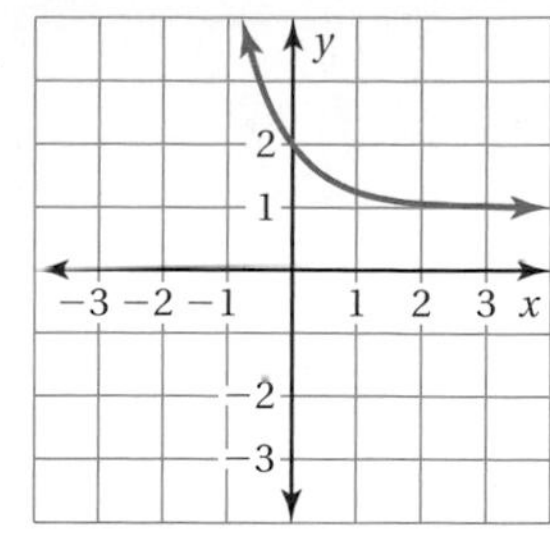

35. **REASONING** Graph $g(x) = 4^{x+2}$. Compare the graph to the graph of $f(x) = 4^x$.

Write an exponential function represented by the graph or table.

5 36. 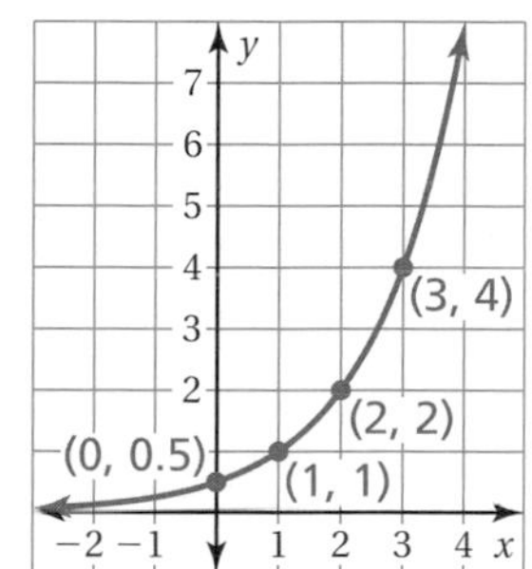

37. 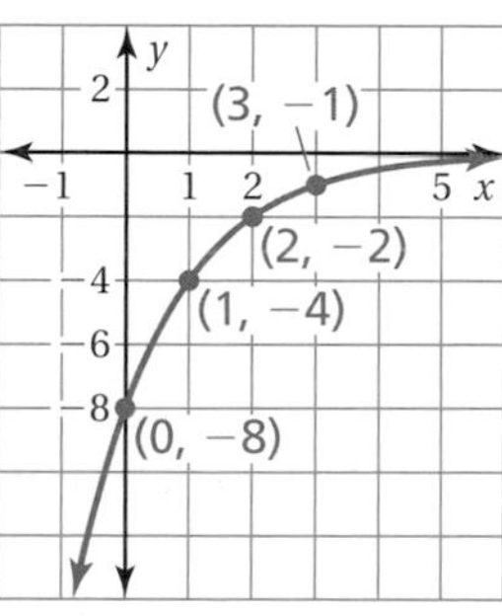

38.

x	0	1	2	3
y	2	8	32	128

39.

x	0	1	2	3
y	−3	−15	75	−375

40. **ART GALLERY** The graph represents the number y of visitors to a new art gallery after x months.
   a. Write an exponential function that represents this situation.
   b. Approximate the number of visitors after 5 months.

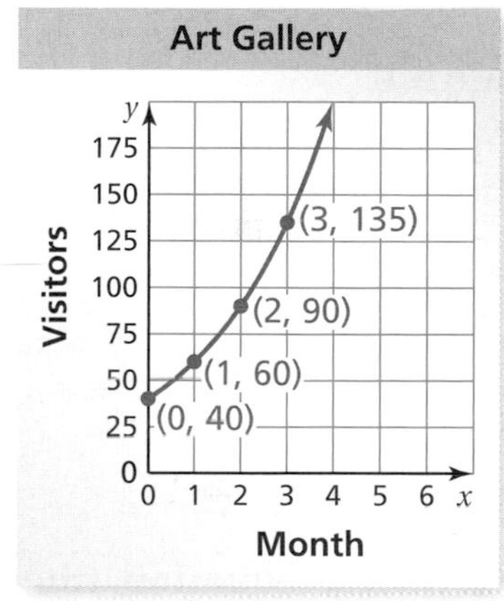

41. **SALES** A sales report shows that 3300 gas grills were purchased from a chain of hardware stores last year. The store expects grill sales to increase 6% each year. About how many grills does the store expect to sell in year 6? Use an equation to justify your answer.

42. **Structure** The graph of g is a translation 4 units up and 3 units right of the graph of $f(x) = 2^x$. Write an equation for g.

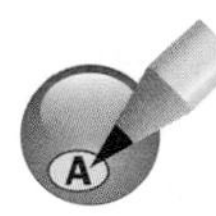

Fair Game Review What you learned in previous grades & lessons

Write the percent as a decimal. *(Skills Review Handbook)*

43. 23%
44. 3%
45. 150%

46. **MULTIPLE CHOICE** Which of the following is equivalent to 100(0.95)? *(Skills Review Handbook)*

(A) 0.95 (B) 9.5 (C) 95 (D) 950

6.4b Solving Exponential Equations

To solve an exponential equation of the form $b^x = b^y$ when $b > 0$ and $b \neq 1$, solve the equation $x = y$.

EXAMPLE 1 Solving Exponential Equations

a. Solve $5^x = 125$.

$5^x = 125$	Write the equation.
$5^x = 5^3$	Rewrite 125 as 5^3.
$x = 3$	Equate the exponents.

b. Solve $4^x = 2^{x-3}$.

$4^x = 2^{x-3}$	Write the equation.
$(2^2)^x = 2^{x-3}$	Rewrite 4 as 2^2.
$2^{2x} = 2^{x-3}$	Power of a Power Property
$2x = x - 3$	Equate the exponents.
$x = -3$	Solve for *x*.

Check

$$4^x = 2^{x-3}$$
$$4^{-3} \stackrel{?}{=} 2^{-3-3}$$
$$\frac{1}{4^3} \stackrel{?}{=} \frac{1}{2^6}$$
$$\frac{1}{64} = \frac{1}{64} \ \checkmark$$

c. Solve $9^{x+2} = 27^x$.

$9^{x+2} = 27^x$	Write the equation.
$(3^2)^{x+2} = (3^3)^x$	Rewrite 9 as 3^2 and 27 as 3^3.
$3^{2x+4} = 3^{3x}$	Power of a Power Property
$2x + 4 = 3x$	Equate the exponents.
$4 = x$	Solve for *x*.

Check

$$9^{x+2} = 27^x$$
$$9^{4+2} \stackrel{?}{=} 27^4$$
$$531{,}441 = 531{,}441 \ \checkmark$$

Practice

Solve the equation. Check your solution, if possible.

1. $3^x = 81$ **2.** $2^x = 32$ **3.** $\frac{1}{16} = 4^x$

4. $10^x = 10^{x+1}$ **5.** $\left(\frac{1}{5}\right)^x = \left(\frac{1}{5}\right)^{3x}$ **6.** $6^{x-5} = 36^x$

7. $100^{5x+2} = 1000^{4x-1}$ **8.** $32^{1-x} = 8^{2x-2}$ **9.** $\left(\frac{1}{8}\right)^{x-5} = 4^x$

10. NUMBER SENSE Explain how you can use mental math to solve the equation $8^{x-4} = 1$.

11. REASONING Why does this method for solving $b^x = b^y$ not work when $b = 1$? Give an example to justify your answer.

Laurie's Notes

Introduction

Connect

- **Yesterday:** Students wrote and graphed exponential functions. (MP8)
- **Today:** Students will solve exponential equations of the form $b^x = b^y$.

Motivate

- Before class, write the following values on index cards.

2^4	3^4	2^6	$9^{3/2}$	2^5
4^2	9^2	8^2	3^3	$4^{5/2}$

- Play a quick matching game with 10 students. Hand each student a card and ask them to find the student whose card has an equivalent value.
- When the students have found their matches, have each pair of students explain why the values are equivalent.

Lesson Notes

For Your Information

- **MP5 Use Appropriate Tools Strategically:** Today, students will solve exponential equations both by using properties of exponents and by using a graphing calculator.
- **MP7 Look for and Make Use of Structure:** To solve an exponential equation of the form $b^x - b^y$, set the exponents equal and solve. Other strategies for solving exponential functions will be studied in a future course.

Example 1

- In part (a), ask, "5 to what power is 125?" Students will quickly say 3, but work through the problem anyway, to show them the process.
- In part (b), students are unsure about equating exponents when they see the equation in the form $b^x = b^y$. Say, "If the two sides of the equation are equal and the bases are equal, then the exponents have to be equal."
- **?** "What common base can you use to rewrite each side of the equation in part (c)?" 3
- Rewrite each side of the equation using a base of 3. The properties of exponents must be used. When the Power of a Power Property is used, the Distributive Property is also used to multiply 2 and $x + 2$.
- Finish working through the problem as shown.

Practice

- The common base to use for each side of the equation should be fairly obvious to students. When negative exponents are involved, students must be careful with the signs.
- Exercise 4 may confuse students initially. There is no value of x that makes sense for this equation.
- Ask volunteers to share their work at the board for Exercises 1–9.
- In Exercise 10, students must think "8 raised to what power would be 1?"

Goal Today's lesson is solving exponential equations.

Start Thinking! and Warm Up

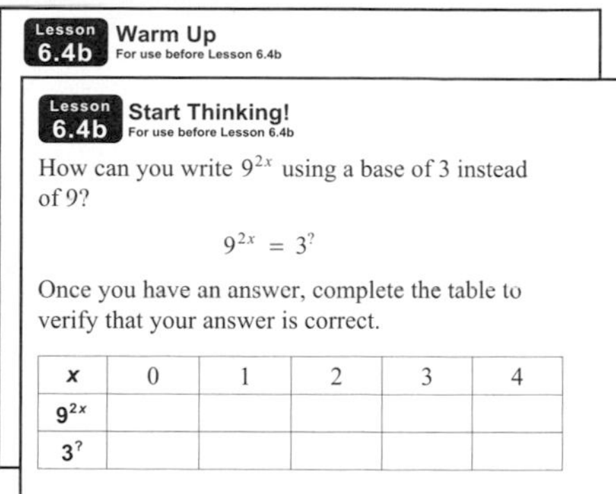

Lesson 6.4b Warm Up
For use before Lesson 6.4b

Lesson 6.4b Start Thinking!
For use before Lesson 6.4b

How can you write 9^{2x} using a base of 3 instead of 9?

$$9^{2x} = 3^?$$

Once you have an answer, complete the table to verify that your answer is correct.

x	0	1	2	3	4
9^{2x}					
$3^?$					

Extra Example 1

a. Solve $4^x = 256$. $x = 4$

b. Solve $3^x = 9^{x+1}$. $x = -2$

c. Solve $64^x = 16^{x+3}$. $x = 6$

Practice

1. $x = 4$

2. $x = 5$

3. $x = -2$

4. no solution

5. $x = 0$

6. $x = -5$

7. $x = \frac{7}{2}$

8. $x = 1$

9. $x = 3$

10. *Sample answer:* The exponent $x - 4$ must be equal to 0 for 8^{x-4} to be equal to 1. You can use mental math to solve the equation $x - 4 = 0$; $x = 4$.

11. *Sample answer:* Because 1 raised to any power is equal to 1, $1^x = 1^y$ is true for all real numbers x and y. The solution method fails in this case because it only gives the solution values for which $x = y$.

Record and Practice Journal Practice

See Additional Answers.

Extra Example 2

Use a graphing calculator to solve $\left(\frac{1}{3}\right)^{x-3} = 4$. $x \approx 1.74$

Practice

12. $x \approx -1.71$
13. $x \approx 0.85$
14. $x \approx 0.67$
15. $x \approx -10.00$
16. $x \approx -1.77$
17. no solution

Mini-Assessment

Solve the equation. Check your solution, if possible.

1. $8^x = 512$ $x = 3$
2. $27 = 81^x$ $x = \frac{3}{4}$
3. $8^{x+1} = 4^{2x+3}$ $x = -3$

Use a graphing calculator to solve the equation.

4. $6^{x-2} = 3$ $x \approx 2.61$
5. $3^{-x-4} = x + 7$ $x \approx -4.74$

Laurie's Notes

Discuss

- An exponential equation that cannot easily be rewritten in the form $b^x = b^y$ can be solved by graphing.
- Treat each side of the equation as a function and graph. Look for the intersection of the graphs. At the point of intersection, the two functions have the same function value for the same value of x.

Example 2

- **MP5:** Write the original equation. This is an exponential equation that you do not want to try to write in the form $b^x = b^y$. A graphing calculator is the appropriate tool.
- Write a system of equations using each side of the equation. Students should recall how to use the intersection feature on their calculators. The solution is not a rational number, so an approximation is used.
- Encourage students to check their solution algebraically.

Practice

- **Think-Pair-Share:** Students should read each question independently and then work with a partner to answer the questions. When they have answered the questions, the pair should compare their answers with another group and discuss any discrepancies.

Closure

- Solve $4^{2x} = 8^{x-1}$ using both methods. $x = -3$

Technology For the Teacher

The Dynamic Planning Tool
Editable Teacher's Resources at *BigIdeasMath.com*

EXAMPLE 2 Solving an Equation by Graphing

Use a graphing calculator to solve $\left(\frac{1}{2}\right)^{x-1} = 7$.

Step 1: Write a system of equations using each side of the equation.

$$y = \left(\frac{1}{2}\right)^{x-1} \qquad \text{Equation 1}$$

$$y = 7 \qquad \text{Equation 2}$$

Step 2: Enter the equations into your calculator. Then graph the equations in a standard viewing window.

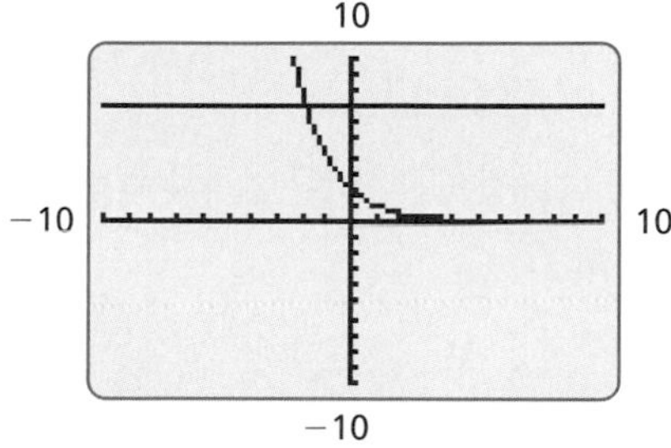

Step 3: Use the *intersect* feature to find the point of intersection. It is at about $(-1.81, 7)$.

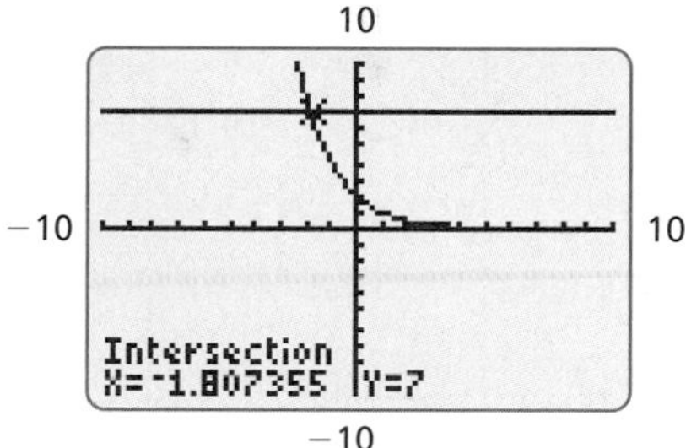

So, the solution is $x \approx -1.81$.

Check: Check the solution algebraically.

$$\left(\frac{1}{2}\right)^{x-1} = 7 \qquad \text{Write the equation.}$$

$$\left(\frac{1}{2}\right)^{-1.81-1} \stackrel{?}{=} 7 \qquad \text{Substitute } -1.81 \text{ for } x.$$

$$7.01 \approx 7 \checkmark \qquad \text{Use a calculator.}$$

Practice

Use a graphing calculator to solve the equation.

12. $4^{x+3} = 6$

13. $2^x = 1.8$

14. $4 = 8^x$

15. $\left(\frac{3}{4}\right)^{x+2} = 10$

16. $2^{-x-3} = 3^{x+1}$

17. $5^x = -4^{x+4}$

6.5 Exponential Growth

Essential Question What are the characteristics of exponential growth?

COMMON CORE STATE STANDARDS
A.SSE.1a
A.SSE.1b
F.IF.7e

1 ACTIVITY: Comparing Types of Growth

Work with a partner. Describe the pattern of growth for each sequence and graph. How many of the patterns represent exponential growth? Explain your reasoning.

a. 1, 4, 7, 10, 13, 16, 19, 22, 25, 28, 31

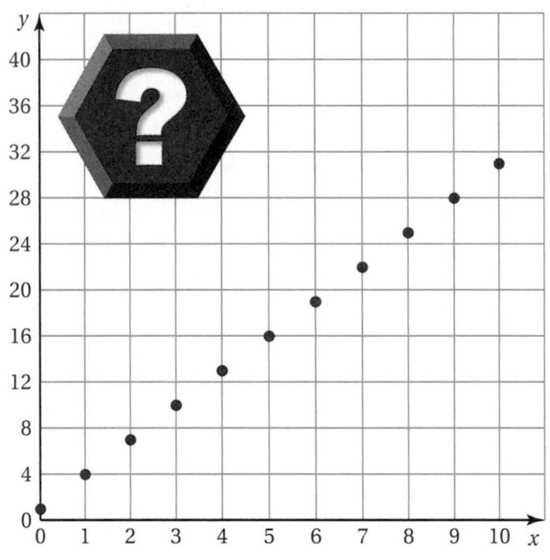

b. 1.0, 1.4, 2.0, 2.7, 3.8, 5.4, 7.5, 10.5, 14.8, 20.7, 28.9

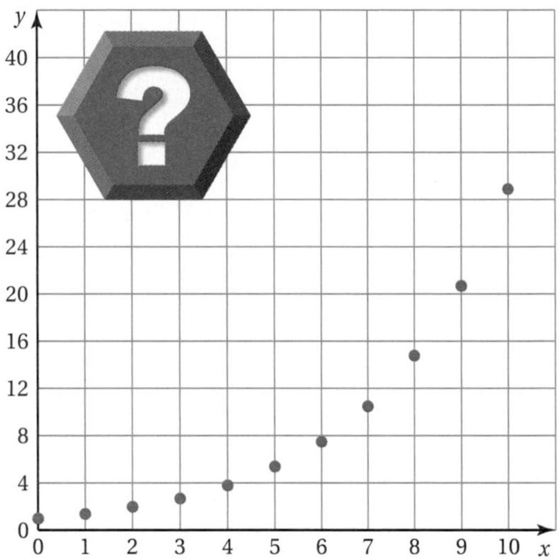

c. 1.0, 1.3, 2.3, 4.0, 6.3, 9.3, 13.0, 17.3, 22.3, 28.0, 34.3

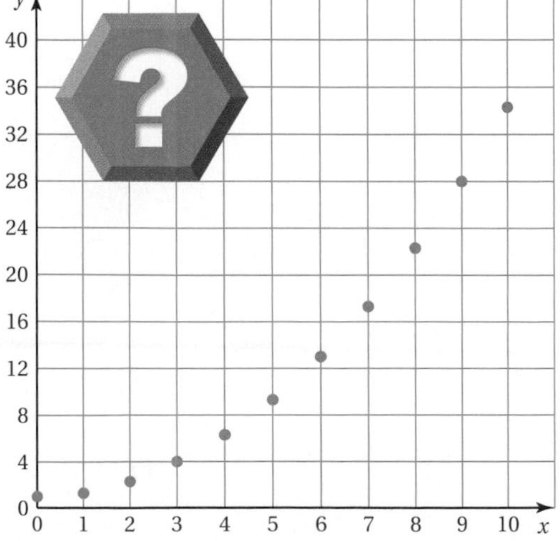

d. 1.0, 1.6, 2.4, 3.4, 4.7, 6.4, 8.7 11.5, 15.3, 20.2, 26.6

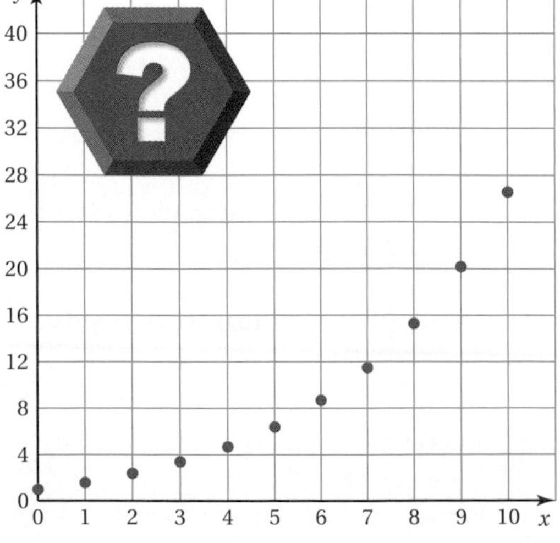

Laurie's Notes

Introduction

Standards for Mathematical Practice

- **MP3a Construct Viable Arguments:** Students will form arguments to support whether sets of data show exponential growth. Their arguments should connect to the definition and properties of exponential functions.

Motivate

- Play "What Am I?" with your students. Provide the following clues:
  - I have excellent eyesight.
  - I love fish and I am a good swimmer.
  - I have about 7000 feathers.
  - My wing span ranges from 72 to 90 inches.
  - I can fly about 30 miles per hour in level flight.
  - I became the national emblem in 1782.
- Explain that they will solve a problem about bald eagles today.

Activity Notes

Discuss

- In this activity, students are asked to describe the pattern of growth for each sequence and graph. Tell your students to base their decisions on more than a cursory look at the graphs—Don't just say "it looks exponential."

Activity 1

? "How are all four sequences alike?" Listen for ideas like: Each has 11 terms; Each is increasing; Each starts with 1.

? "What is your hunch about the types of graphs these are?" Likely responses are that the first is linear and last three are exponential.

? "How are the last three graphs different from each other?" Each set of points forms a curve of a different shape and other than the first point of each graph, all the points are different.

? "How can you decide whether each pattern represents exponential growth?" Students should describe a process to determine whether consecutive terms change by common factors.

- One way to determine whether each sequence is exponential is to substitute the coordinates of the first point, (0, 1), in $y = ab^x$ to find that $a = 1$. Then, use the second point to determine the value of b. Because $a = 1$, the base b is the common factor between terms. So, check whether each term is b times the previous term. If so, the sequence is exponential. The sequence in part (b) is exponential, the others are not.
- Students will wonder why the graphs in parts (c) and (d) *look* exponential but are not. Discuss how the end behavior of these graphs may be different from the graph of an exponential function.
- **Big Idea:** Exponential functions have a constant percent of change (common factor).
- **Extension:** Students should be able to write the linear function for part (a).

Common Core State Standards

A.SSE.1a Interpret parts of an expression, such as terms, factors, and coefficients.

A.SSE.1b Interpret complicated expressions by viewing one or more of their parts as a single entity.

F.IF.7e Graph exponential . . . functions, showing intercepts and end behavior, ...

Previous Learning

Students should know the characteristics of exponential functions. Students should know how to find a percent of change.

Start Thinking! and Warm Up

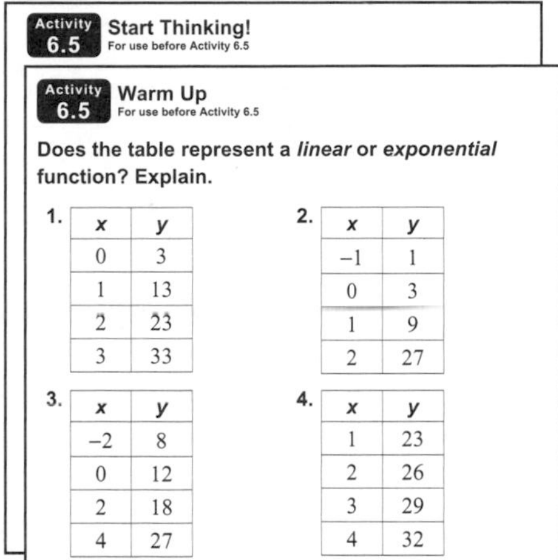

Activity 6.5 Start Thinking! For use before Activity 6.5

Activity 6.5 Warm Up For use before Activity 6.5

Does the table represent a *linear* or *exponential* function? Explain.

1.

x	y
0	3
1	13
2	23
3	33

2.

x	y
−1	1
0	3
1	9
2	27

3.

x	y
−2	8
0	12
2	18
4	27

4.

x	y
1	23
2	26
3	29
4	32

6.5 Record and Practice Journal

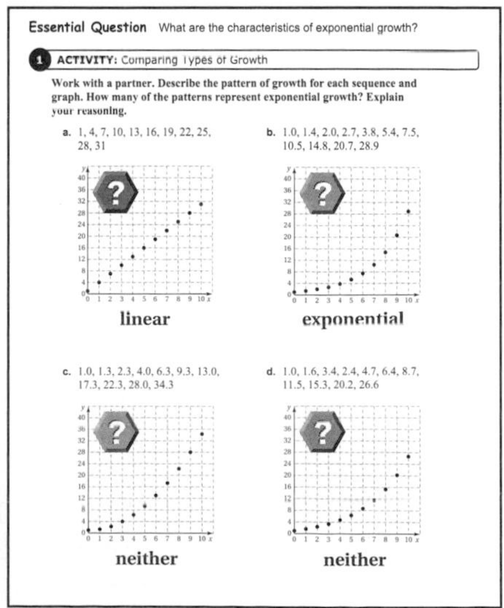

Essential Question What are the characteristics of exponential growth?

1 ACTIVITY: Comparing Types of Growth

Work with a partner. Describe the pattern of growth for each sequence and graph. How many of the patterns represent exponential growth? Explain your reasoning.

a. 1, 4, 7, 10, 13, 16, 19, 22, 25, 28, 31

b. 1.0, 1.4, 2.0, 2.7, 3.8, 5.4, 7.5, 10.5, 14.8, 20.7, 28.9

linear

exponential

c. 1.0, 1.3, 2.3, 4.0, 6.3, 9.3, 13.0, 17.3, 22.3, 28.0, 34.3

d. 1.0, 1.6, 3.4, 2.4, 4.7, 6.4, 8.7, 11.5, 15.3, 20.2, 26.6

neither

neither

Differentiated Instruction

Inclusion

Relate concepts of linear equations with the concepts of exponential equations.

	Linear	*Exponential*
Equation	$y = mx + b$	$y = ab^x$
Growth	additive	multiplicative
Growth represented by	m	b
Initial value when $x = 0$ (y-intercept)	b	a

6.5 Record and Practice Journal

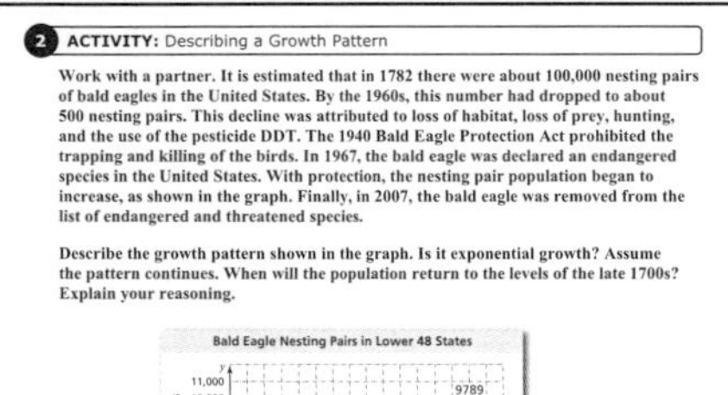

2 ACTIVITY: Describing a Growth Pattern

Work with a partner. It is estimated that in 1782 there were about 100,000 nesting pairs of bald eagles in the United States. By the 1960s, this number had dropped to about 500 nesting pairs. This decline was attributed to loss of habitat, loss of prey, hunting, and the use of the pesticide DDT. The 1940 Bald Eagle Protection Act prohibited the trapping and killing of the birds. In 1967, the bald eagle was declared an endangered species in the United States. With protection, the nesting pair population began to increase, as shown in the graph. Finally, in 2007, the bald eagle was removed from the list of endangered and threatened species.

Describe the growth pattern shown in the graph. Is it exponential growth? Assume the pattern continues. When will the population return to the levels of the late 1700s? Explain your reasoning.

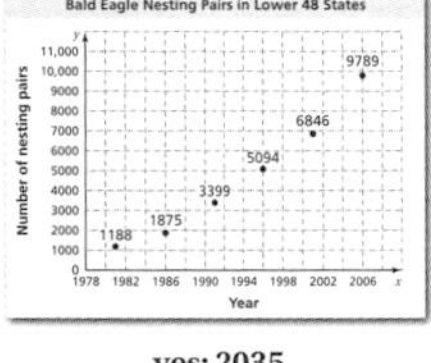

yes; 2035

What Is Your Answer?

3. IN YOUR OWN WORDS What are the characteristics of exponential growth? How can you distinguish exponential growth from other growth patterns?

increases by a constant factor over time

4. Which of the following are examples of exponential growth? Explain.

a. Growth of the balance of a savings account

exponential growth

b. Speed of the moon in orbit around Earth

not exponential growth

c. Height of a ball that is dropped from a height of 100 feet

not exponential growth

Laurie's Notes

Activity 2

- **MP3a:** This activity is similar to Activity 1. Students construct an argument about whether the data is exponential. Although the x-values are not stated explicitly, students should be able to judge that they occur at equal intervals (about 5 years apart). The y-values are labeled on the graph.
- Ask a student to read the introduction about bald eagles. Ask if any students have seen a bald eagle in its natural habitat.
- Give students time to discuss strategies and organize their work. Using a calculator will help students to focus on the process and concept rather than the computations.
- Expect students to find and compare the factors between consecutive terms.

$$1875 \div 1188 \approx 1.58$$
$$3399 \div 1875 \approx 1.81$$
$$5094 \div 3399 \approx 1.50$$
$$6846 \div 5094 \approx 1.34$$
$$9789 \div 6846 \approx 1.43$$

? "What is the approximate rate of growth of the population?" The population is growing by about 50% every 5 years.

- Students can solve the equation $100{,}000 = 1.5^x$ to find the number of 5-year time periods that will pass before the population catches up to that of the late 1700s.

What Is Your Answer?

- In Question 3, students should discuss the common ratio between equally spaced values.
- Students may have some idea about the contexts discussed in Question 4.

Closure

- Is either sequence of numbers exponential? Explain.

a. 3, 6, 12, 24, 48, . . . yes; The common factor is 2.

b. 3, 6, 9, 12, 15, . . . no; There is no common factor.

Technology For the Teacher

The Dynamic Planning Tool
Editable Teacher's Resources at *BigIdeasMath.com*

2 ACTIVITY: Predicting a Future Event

Work with a partner. It is estimated that in 1782 there were about 100,000 nesting pairs of bald eagles in the United States. By the 1960s, this number had dropped to about 500 nesting pairs. This decline was attributed to loss of habitat, loss of prey, hunting, and the use of the pesticide DDT.

The 1940 Bald Eagle Protection Act prohibited the trapping and killing of the birds. In 1967, the bald eagle was declared an endangered species in the United States. With protection, the nesting pair population began to increase, as shown in the graph. Finally, in 2007, the bald eagle was removed from the list of endangered and threatened species.

Describe the growth pattern shown in the graph. Is it exponential growth? Assume the pattern continues. When will the population return to the levels of the late 1700s? Explain your reasoning.

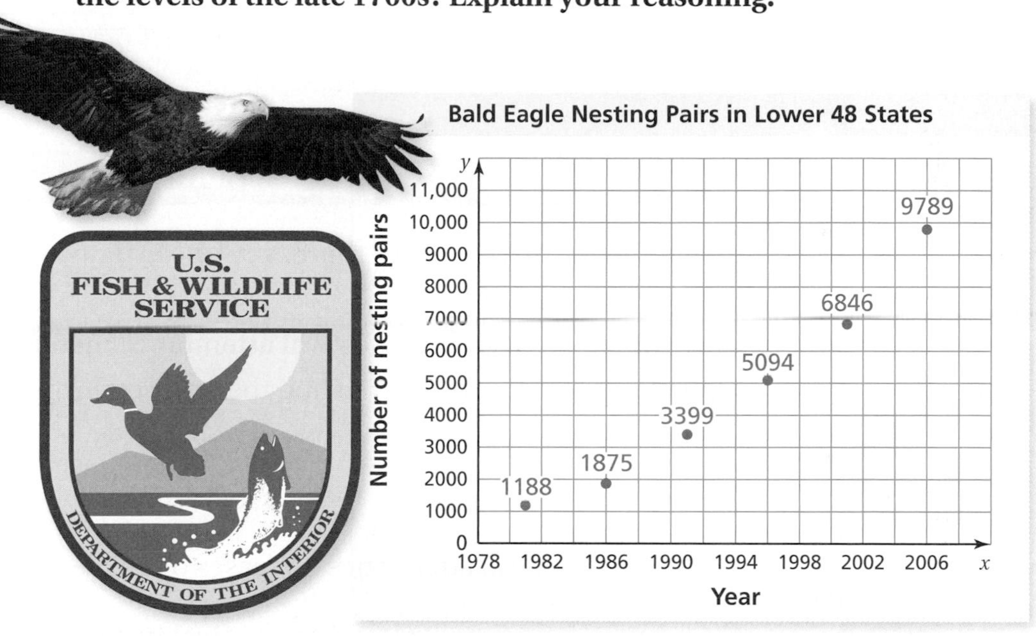

What Is Your Answer?

3. **IN YOUR OWN WORDS** What are the characteristics of exponential growth? How can you distinguish exponential growth from other growth patterns?

4. Which of the following are examples of exponential growth? Explain.
   a. Growth of the balance of a savings account
   b. Speed of the moon in orbit around Earth
   c. Height of a ball that is dropped from a height of 100 feet

Use what you learned about exponential growth to complete Exercises 3 and 4 on page 298.

6.5 Lesson

Key Vocabulary
exponential growth, *p. 296*
exponential growth function, *p. 296*
compound interest, *p. 297*

Exponential growth occurs when a quantity increases by the same factor over equal intervals of time.

Key Idea

Exponential Growth Functions

A function of the form $y = a(1 + r)^t$, where $a > 0$ and $r > 0$, is an **exponential growth function.**

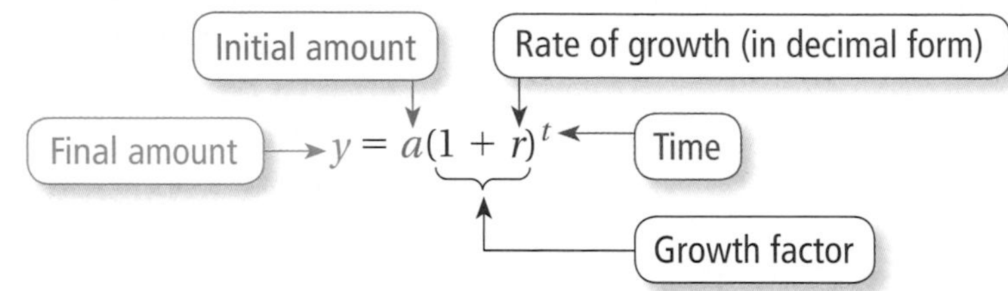

Study Tip

Notice that an exponential growth function is of the form $y = ab^x$, where b is replaced by $1 + r$ and x is replaced by t.

EXAMPLE 1 Using an Exponential Growth Function

The function $y = 150{,}000(1.1)^t$ represents the attendance y at a music festival t years after 2010.

a. By what percent does the festival attendance increase each year?

Use the growth factor $1 + r$ to find the rate of growth.

$1 + r = 1.1$ — Write an equation.

$r = 0.1$ — Subtract 1 from each side.

So, the festival attendance increases by 10% each year.

b. How many people will attend the festival in 2014? Round your answer to the nearest ten thousand.

The value $t = 4$ represents 2014.

$y = 150{,}000(1.1)^t$ — Write exponential growth function.

$= 150{,}000(1.1)^4$ — Substitute 4 for t.

$= 219{,}615$ — Use a calculator.

About 220,000 people will attend the festival in 2014.

On Your Own

1. The function $y = 500{,}000(1.15)^t$ represents the number y of members of a website t years after 2010.

   a. By what percent does the website membership increase each year?

   b. How many members will there be in 2016? Round your answer to the nearest hundred thousand.

Laurie's Notes

Introduction

Connect

- **Yesterday:** Students determined whether sequences show exponential growth. (MP3a)
- **Today:** Students will write and evaluate exponential growth functions.

Motivate

- Simulate the rumor game. You tell two students a rumor (no homework tonight). In 15 seconds, they each tell two students the rumor. This process continues.

? "How many students will hear the rumor after 1 minute?" 32

- In a small class, all the students will know the rumor in less than a minute.

Lesson Notes

Key Idea

- Write the definition of an exponential growth function. The number of variables in the general equation will create anxiety for some students. If possible, use different colors to highlight the different parts of the equation.
- **Connection:** Help students recognize that the base b in the general exponential equation $y = ab^x$ is replaced by $1 + r$ and the independent variable x is replaced by t.
- The initial amount a is the amount present before any growth occurs. It is still the y-intercept when time $t = 0$.
- The base is the common factor and it is called the growth factor for exponential growth. If $r = 0.5$ then $1 + r = 1.5 = 150\%$. The growth rate is 50% and the growth factor is 150%.

Example 1

- Ask a student to read the problem while you write the equation.

? "What was the initial amount and what does it mean?" 150,000; the number of people that attended the music festival in 2010 when $t = 0$

? "What is the growth factor? What is the growth rate?" 1.1; 0.1 or 10%

? "How can you predict the festival attendance in 2014?" Evaluate y for $t = 4$.

On Your Own

? **MP7 Look for and Make Use of Structure:** "What is the initial amount and what does it mean?" 500,000; the number of members in 2010 when $t = 0$

Discuss

? "What is the simple interest formula? Explain the variables." $I = Prt$; P is the principal (initial amount), r is the annual interest rate (in decimal form), t is the time in years, I is the interest.

- *Compound interest* is an example of exponential growth. The formula is an exponential growth function.

Goal Today's lesson is writing and evaluating **exponential growth functions.**

Lesson Materials
Textbook
• calculators

Start Thinking! and Warm Up

Lesson 6.5 **Warm Up**
For use before Lesson 6.5

Lesson 6.5 **Start Thinking!**
For use before Lesson 6.5

Britney deposits \$500 in a savings account that earns 3% annual interest compounded yearly.

Riley deposits \$400 in a savings account that earns 6% annual interest compounded yearly.

How much money does each person have in his or her account after 1 year? 2 years? 3 years?

After how many years will Riley have more money in his account than Britney?

Extra Example 1

The function $y = 40{,}000(1.05)^t$ represents the population of a town t years after 2010.

a. By what percent does the population increase each year? 5%

b. What will the population be in 2017? Round your answer to the nearest hundred. 56,300

On Your Own

1. a. 15%

b. 1,200,000

Extra Example 2

You deposit \$1000 in a savings account that earns 3% annual interest compounded yearly. Write a function for the balance after t years.

$y = 1000(1.03)^t$

Extra Example 3

You deposit \$300 in a money market account. Interest is compounded yearly. After 1 year, the account has a balance of \$321. Write and graph a function that represents the balance y (in dollars) after t years. $y = 300(1.07)^t$

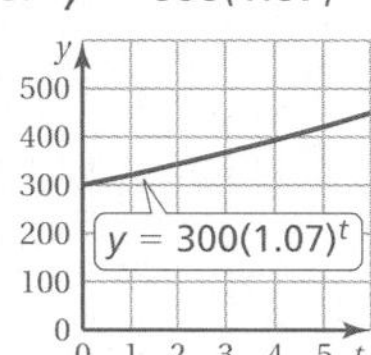

On Your Own

2. $y = 500(1.04)^t$

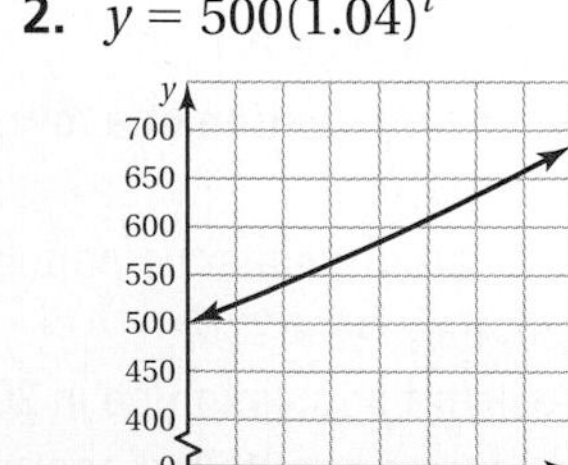

English Language Learners

Pair Activity

Engage English language learners in conversation. Assign each student a problem to solve. Then group students in pairs and have them discuss their problems and solutions.

Laurie's Notes

Key Idea

- Write the definition and formula for compound interest. Use different colors to highlight the different parts of the equation.
- Notice that n appears twice in the formula. The expression nt represents the number of times interest is compounded in t years and $\frac{r}{n}$ represents the interest rate used at each compounding.

Example 2

? "What information do you know?" principal: \$100, annual interest rate: 5%, number of compoundings per year: 1, number of years: unknown

- Because $n = 1$, the formula simplifies to an equation that looks like the standard exponential growth formula.

Example 3

- Ask a student to read the problem while you copy the table shown.
- **FYI:** The same table was used in Lesson 5.5.

? "What information do you know from the table? Explain your reasoning." The principal is \$100. The interest rate is 10%, because each year the balance is 10% greater than the last. Interest is compounded once a year, because the balance is exactly 10% greater each year.

- Substitute what is known and simplify. The formula simplifies to an equation that looks like the standard exponential growth formula.
- Examples 2 and 3 have the same principal, number of compoundings per year, and number of years. Example 3 has twice the interest rate.
- Graph the two equations to compare the earnings.

? "How much more does the 10% money market account earn than the 5% savings account after 5 years?" \$33.42 (\$61.05 − \$27.63)

On Your Own

- **Extension:** If time permits, explore the problem in Question 2 for a \$1000 initial deposit. How much interest does each amount (\$500 and \$1000) earn after 5 years? \$608.33; \$1216.65

Closure

- What do you know about the exponential function $y = 800(1.25)^t$? Initial amount: 800; growth factor: 1.25; growth rate 25%

Technology For the Teacher

The Dynamic Planning Tool
Editable Teacher's Resources at *BigIdeasMath.com*

Key Idea

Compound Interest

Compound interest is interest earned on the principal *and* on previously earned interest. The balance y of an account earning compound interest is

$$y = P\left(1 + \frac{r}{n}\right)^{nt}.$$

P = principal (initial amount)
r = annual interest rate (in decimal form)
t = time (in years)
n = number of times interest is compounded per year

EXAMPLE 2 Writing a Function

Study Tip

For interest compounded yearly, you can substitute 1 for n in the formula to get $y = P(1 + r)^t$.

You deposit $100 in a savings account that earns 5% annual interest compounded yearly. Write a function for the balance after *t* years.

$y = P\left(1 + \frac{r}{n}\right)^{nt}$	Write compound interest formula.
$y = 100\left(1 + \frac{0.05}{1}\right)^{(1)(t)}$	Substitute 100 for P, 0.05 for r, and 1 for n.
$y = 100(1.05)^t$	Simplify.

EXAMPLE 3 Real-Life Application

The table shows the balance of a money market account over time.

Year, t	Balance
0	$100
1	$110
2	$121
3	$133.10
4	$146.41
5	$161.05

a. Write a function for the balance after *t* years.

From the table, you know the balance increases 10% each year.

$y = a(1 + r)^t$	Write exponential growth function.
$y = 100(1 + 0.1)^t$	Substitute 100 for a and 0.1 for r.
$y = 100(1.1)^t$	Simplify.

Saving Money

Balance vs. Year: $y = 100(1.1)^t$; $y = 100(1.05)^t$ (y-axis 0–200, t-axis 0–7)

b. Graph the functions from part (a) and Example 2 in the same coordinate plane. Compare the account balances.

The money market account earns 10% interest each year and the savings account earns 5% interest each year. So, the balance of the money market account increases faster.

On Your Own

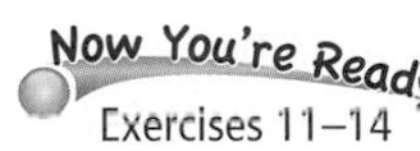

2. You deposit $500 in a savings account that earns 4% annual interest compounded yearly. Write and graph a function that represents the balance y (in dollars) after t years.

6.5 Exercises

Vocabulary and Concept Check

1. **VOCABULARY** When does the exponential function $y = a(1 + r)^t$ represent an exponential growth function?
2. **VOCABULARY** The population of a city grows by 3% each year. What is the growth factor?

Practice and Problem Solving

Describe the pattern of growth for the sequence.

3. 1.0, 1.2, 1.4, 1.7, 2.1, 2.5, 3.0, 3.6, 4.3, 5.2, 6.2
4. 1, 7, 13, 19, 25, 31, 37, 43, 49, 55, 61

Identify the initial amount a and the rate of growth r (as a percent) of the exponential function. Evaluate the function when $t = 5$. Round your answer to the nearest tenth.

1 5. $y = 25(1.2)^t$
6. $f(t) = 12(1.05)^t$
7. $d(t) = 1500(1.074)^t$
8. $y = 175(1.028)^t$
9. $g(t) = 6.7(2)^t$
10. $h(t) = 1.8^t$

Write and graph a function that represents the situation.

2 3 11. You deposit \$800 in an account that earns 7% annual interest compounded yearly.
12. Your \$35,000 annual salary increases by 4% each year.
13. A population of 210,000 increases by 12.5% each year.
14. Sales of \$10,000 increase by 70% each year.
15. **ERROR ANALYSIS** The growth rate of a bacteria culture is 150% each hour. Initially, there are 10 bacteria. Describe and correct the error in finding the number of bacteria in the culture after 8 hours.

$b(t) = 10(1.5)^t$

$b(8) = 10(1.5)^8 \approx 256.3$

After 8 hours, there are about 256 bacteria in the culture.

16. **INVESTMENT** The function $y = 7500(1.08)^t$ represents the value y of an investment after t years.
    a. What is the initial investment?
    b. What is the value of the investment after 6 years?
17. **POPULATION** The population of a city has been increasing by 2% annually. In 2000, the population was 315,000. Predict the population of the city in 2020. Round your answer to the nearest thousand.

Assignment Guide and Homework Check

Level	Assignment	Homework Check
Average	1–4, 5–17 odd, 16, 21, 25–28	5, 11, 15, 16, 21
Advanced	1–4, 6–16 even, 15, 18–28	8, 12, 15, 16, 21

Common Errors

- **Exercises 5–10** Students may confuse the rate of growth and the growth factor. Point out that the rate of growth determines the increase from one year to the next and the growth factor determines the new total amount.
- **Exercises 11–14** Students may forget to change the percents to decimals. Remind them to use the decimal form of the growth rate.
- **Exercise 20** Students may see the word "triples" and automatically answer 300%. Remind students that tripling means 100% becomes 300%, so the growth rate is 200%.

6.5 Record and Practice Journal

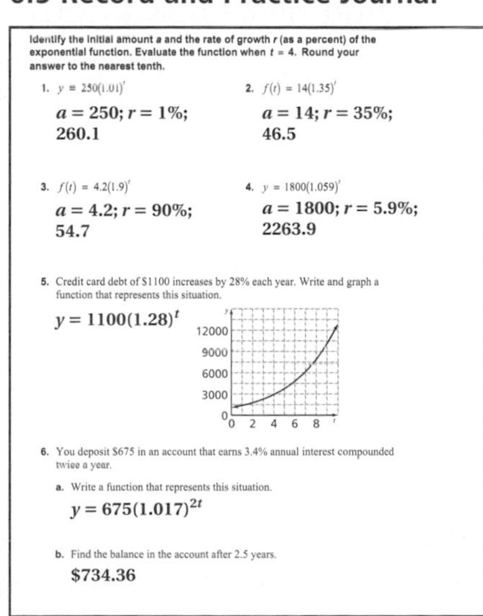
Identify the initial amount *a* and the rate of growth *r* (as a percent) of the exponential function. Evaluate the function when $t = 4$. Round your answer to the nearest tenth.

1. $y = 250(1.01)^t$

   $a = 250; r = 1\%;$ 260.1

2. $f(t) = 14(1.35)^t$

   $a = 14; r = 35\%;$ 46.5

3. $f(t) = 4.2(1.9)^t$

   $a = 4.2; r = 90\%;$ 54.7

4. $y = 1800(1.059)^t$

   $a = 1800; r = 5.9\%;$ 2263.9

5. Credit card debt of \$1100 increases by 28% each year. Write and graph a function that represents this situation.

   $y = 1100(1.28)^t$

6. You deposit \$675 in an account that earns 3.4% annual interest compounded twice a year.

   a. Write a function that represents this situation.

   $y = 675(1.017)^{2t}$

   b. Find the balance in the account after 2.5 years.

   \$734.36

Technology For the Teacher
Answer Presentation Tool

Vocabulary and Concept Check

1. when $a > 0$ and $r > 0$
2. 1.03

Practice and Problem Solving

3. The terms are multiplied by a factor of about 1.2, so the pattern shows exponential growth.
4. The terms increase by 6, so the pattern shows linear growth.
5. $a = 25, r = 20\%; 62.2$
6. $a = 12, r = 5\%; 15.3$
7. $a = 1500, r = 7.4\%; 2143.4$
8. $a = 175, r = 2.8\%; 200.9$
9. $a = 6.7, r = 100\%; 214.4$
10. $a = 1, r = 80\%; 18.9$
11. $y = 800(1.07)^t$

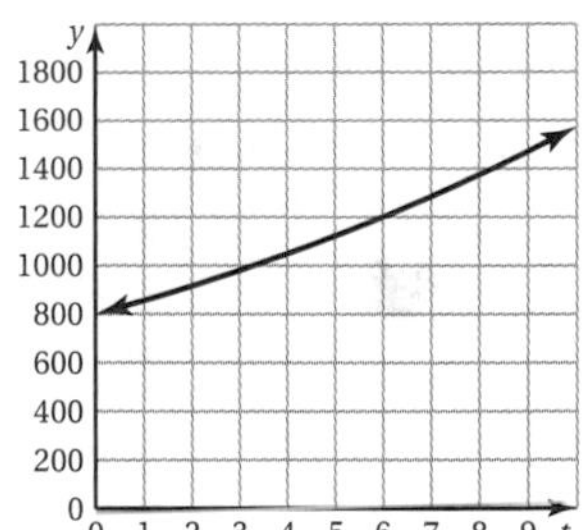

12–14. See Additional Answers.

15. Because the growth rate is 150%, the growth factor should be 2.5, not 1.5.

    $b(t) = 10(1 + 1.5)^t = 10(2.5)^t$

    $b(8) = 10(2.5)^8 \approx 15{,}258.8$

    After 8 hours, there are about 15,259 bacteria in the culture.

16. **a.** \$7500

    b. \$11,901.56

17. 468,000

Practice and Problem Solving

18. $A = 2000(1.0125)^{4t}$; \$2564.07

19. $A = 6200(1.007)^{12t}$; \$7029.46

20. 200%; Tripling means 100% becomes 300%, a growth of 200%.

21. $C(t) = 9000(1.003)^{12t} + 480t$; $C(t)$ represents the total amount of money you have saved after t years.

22. See *Taking Math Deeper*.

23. yes; You can use the properties of exponents to transform the yearly balance function to the monthly balance function. You can use the compound interest formula with an annual interest rate of 2.4% compounded monthly to verify this.

24. **a.** $c(t) = 29{,}000(\sqrt{2})^t$

b. about 10.75 billion

Fair Game Review

25. $\frac{4}{9}$ **26.** $\frac{1}{64}$

27. $\frac{81}{625}$ **28.** B

Mini-Assessment

Identify the initial amount *a* and the rate of growth *r* (as a percent) of the exponential function. Evaluate the function when *t* = 3. Round your answer to the nearest tenth.

1. $y = 5(1.6)^t$
$a = 5, r = 60\%; 20.5$

2. $f(t) = 17(1.25)^t$
$a = 17, r = 25\%; 33.2$

3. $g(t) = 2.9(2)^t$
$a = 2.9, r = 100\%; 23.2$

Taking Math Deeper

Exercise 22

Instead of attempting to create an equation to solve this problem, investigate what happens when a number is repeatedly doubled.

Make a table. Let x be the number of tickets sold during the first hour.

I bet x is a one- or two-digit integer.

Hour	1	2	3	4	5	6
Tickets	x	$2x$	$4x$	$8x$	$16x$	$32x$

Hour	7	8	9	10	11	12
Tickets	$64x$	$128x$	$256x$	$512x$	$1024x$	$2048x$

Interpret the table.

The table shows that a total of $4095x$ tickets were sold. You do not know the exact number of tickets sold unless you know the amount sold in the first hour.

You can find when a portion of the tickets were sold by *working backwards*. Notice that the number of tickets sold in the 12th hour is about one-half of the total number of tickets sold, because $\frac{2048x}{4095x} \approx \frac{1}{2}$.

Also, notice that the number of tickets sold in the 11th hour is about one-quarter of the total number of tickets sold, because $\frac{1024x}{4095x} \approx \frac{1}{4}$.

This means that the other one-quarter of the tickets were sold prior to the 11th hour.

Answer the question.

After 10 hours, about one-quarter of the tickets were sold.

Reteaching and Enrichment Strategies

If students need help. . .	If students got it. . .
Resources by Chapter • Practice A and Practice B • Puzzle Time Record and Practice Journal Practice Differentiating the Lesson Lesson Tutorials Skills Review Handbook	Resources by Chapter • Enrichment and Extension • School-to-Work • Financial Literacy Start the next section

Write a function that represents the situation. Find the balance in the account after the given time period.

18. \$2000 deposit that earns 5% annual interest compounded quarterly; 5 years

19. \$6200 deposit that earns 8.4% annual interest compounded monthly; 18 months

20. NUMBER SENSE During a flu epidemic, the number of sick people triples every week. What is the growth rate as a percent? Explain your reasoning.

21. SAVINGS You deposit \$9000 in a savings account that earns 3.6% annual interest compounded monthly. You also save \$40 per month in a safe at home. Write a function $C(t) = b(t) + h(t)$, where $b(t)$ represents the balance of your savings account and $h(t)$ represents the amount in your safe after t years. What does $C(t)$ represent?

22. REASONING The number of concert tickets sold doubles every hour. After 12 hours, all of the tickets are sold. After how many hours are about one-fourth of the tickets sold? Explain your reasoning.

23. YOU BE THE TEACHER The balance of a savings account can be modeled by the function $b(t) = 5000(1.024)^t$, where t is the time in years. To model the monthly balance, a student writes

$$b(t) = 5000(1.024)^t = 5000(1.024)^{\left(\frac{1}{12}\cdot 12\right)t} = 5000\left(1.024^{\frac{1}{12}}\right)^{12t} \approx 5000(1.002)^{12t}.$$

Is the student correct? Explain your reasoning.

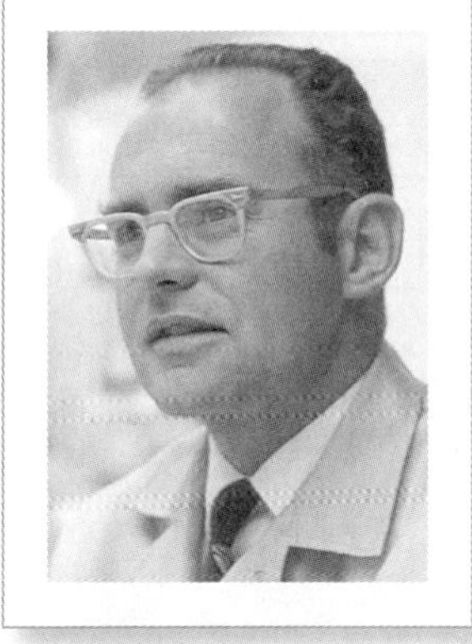

24. Critical Thinking Gordon Moore stated that the number of transistors that can be placed on an integrated circuit will double every 2 years. This trend is known as Moore's Law. In 1978, the Intel®8086 held 29,000 transistors on an integrated circuit.

a. Write a function that represents Moore's Law, where t is the number of years since 1978.

b. How many transistors could be placed on an integrated circuit in 2015?

Fair Game Review What you learned in previous grades & lessons

Simplify the expression. *(Section 6.2)*

25. $\left(\frac{2}{3}\right)^2$ **26.** $\left(\frac{1}{4}\right)^3$ **27.** $\left(\frac{3}{5}\right)^4$

28. MULTIPLE CHOICE The domain of the function $y = 4x - 3$ is 1, 4, 7, 10, and 13. Which number is *not* in the range of the function? *(Section 5.1)*

Ⓐ 1 Ⓑ 10 Ⓒ 13 Ⓓ 25

6.6 Exponential Decay

Essential Question What are the characteristics of exponential decay?

COMMON CORE STATE STANDARDS
A.SSE.1a
A.SSE.1b
F.IF.7e

1 ACTIVITY: Comparing Types of Decay

Work with a partner. Describe the pattern of decay for each sequence and graph. Which of the patterns represent exponential decay? Explain your reasoning.

a. 30.0, 24.3, 19.2, 14.7, 10.8, 7.5, 4.8, 2.7, 1.2, 0.3, 0.0

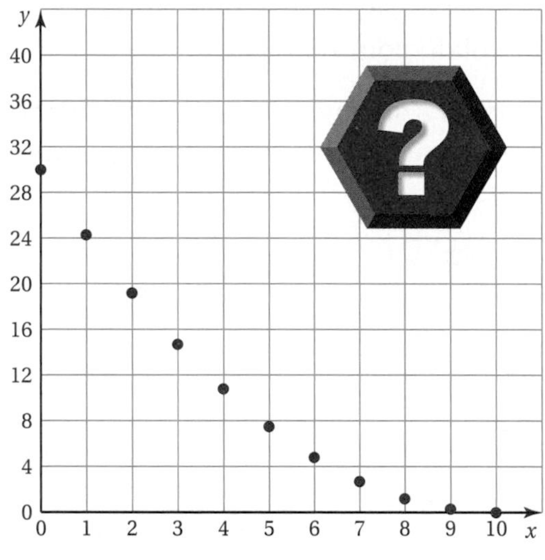

b. 30, 27, 24, 21, 18, 15, 12, 9, 6, 3, 0

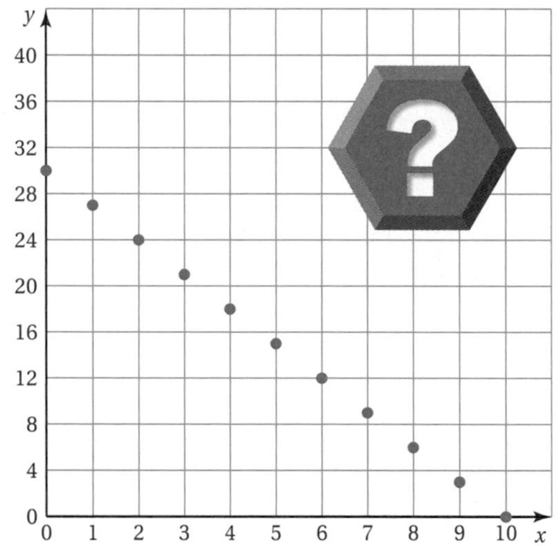

c. 30.0, 24.0, 19.2, 15.4, 12.3, 9.8, 7.9, 6.3, 5.0, 4.0, 3.2

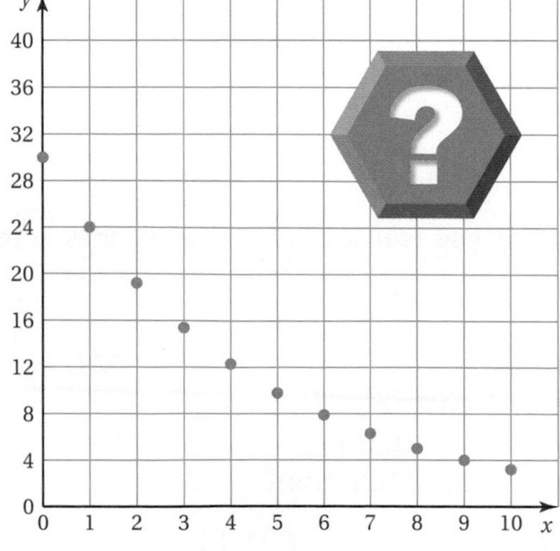

d. 30.0, 29.7, 28.8, 27.3, 25.2, 22.5, 19.2, 15.3, 10.8, 5.7, 0.0

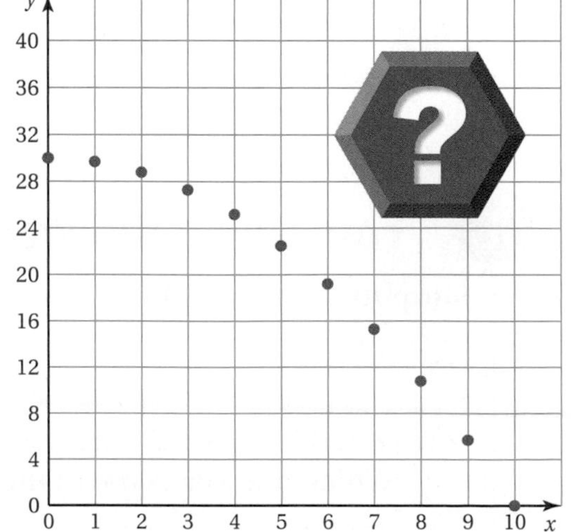

Laurie's Notes

Introduction

Standards for Mathematical Practice

- **MP3a Construct Viable Arguments:** Similar to the last lesson, students are asked to determine whether a data set is exponential. In this case, they are working with exponential decay. In making their decision, they will need to construct an argument that connects to the definition of an exponential function. Eyesight alone will not be enough. Listen carefully to the language that students use in forming their argument.

Motivate

? "How many of you have watched an episode of CSI on television?"

? "Why do you like the show?"

- Premiering in 2000, CSI is responsible for a marked increase in the study of forensic science. Today, students will explore a possible CSI scenario.

Activity Notes

Activity 1

- In this activity, students are asked to describe the pattern of decay for each sequence and graph.
- Based on their experience from the last activity, students should not immediately say that part (b) is linear and the other three are exponential.
- Students should be prepared to look for a common ratio between consecutive terms.
- What students may not have realized yet is that the common ratio must be between 0 and 1 for exponential decay.

? "How are all four graphs alike?" *Sample answers:* All have 11 ordered pairs plotted; All have the same window; All have the same y-intercept, (0, 30).

? "What is your hunch about the types of graphs these might be?" Likely they will say part (b) is linear and the other three *may* be exponential.

? "How are the three nonlinear graphs different?" They have different ending points; The graph in part (d) curves differently.

? "How will you decide if the pattern represents exponential decay?" Students should describe a process to determine the common factor (base of the exponential function) between consecutive terms.

- **Common Error:** Students will incorrectly divide the greater number by the lesser number. They should divide a y-value by the previous y-value. For instance, in part (a) $24.3 \div 30.0 = 0.81$ instead of $30.0 \div 24.3 \approx 1.23$.
- The graph in part (c) is the only one with exponential decay, because $24 \div 30 = 0.8$, $19.2 \div 24 = 0.8$, $15.4 \div 19.2 \approx 0.8$, etc.
- **Big Idea:** Exponential functions have a constant percent of change (common factor).
- **Extension:** Students should be able to write the equation for the linear function in part (b). $y = -3x + 30$

Common Core State Standards

A.SSE.1a Interpret parts of an expression, such as terms, factors, and coefficients.

A.SSE.1b Interpret complicated expressions by viewing one or more of their parts as a single entity.

F.IF.7e Graph exponential and logarithmic functions, showing intercepts and end behavior, and trigonometric functions, showing period, midline, and amplitude.

Previous Learning

Students should know the characteristics of exponential functions. Students should know how to find a percent of change.

Start Thinking! and Warm Up

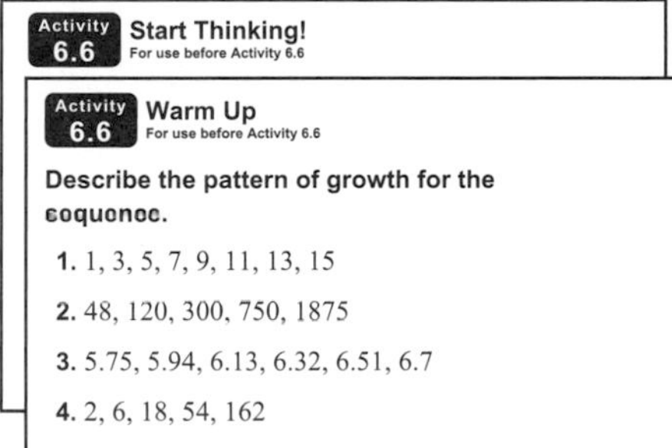

Activity 6.6 Start Thinking!
For use before Activity 6.6

Activity 6.6 Warm Up
For use before Activity 6.6

Describe the pattern of growth for the sequence.

1. 1, 3, 5, 7, 9, 11, 13, 15
2. 48, 120, 300, 750, 1875
3. 5.75, 5.94, 6.13, 6.32, 6.51, 6.7
4. 2, 6, 18, 54, 162

6.6 Record and Practice Journal

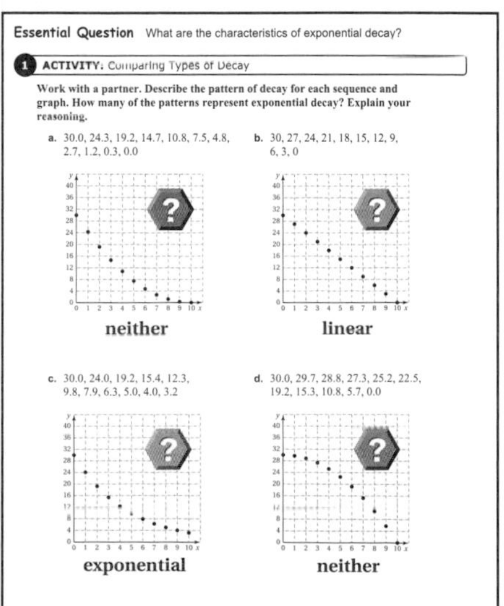

Essential Question What are the characteristics of exponential decay?

1 ACTIVITY: Comparing Types of Decay

Work with a partner. Describe the pattern of decay for each sequence and graph. How many of the patterns represent exponential decay? Explain your reasoning.

a. 30.0, 24.3, 19.2, 14.7, 10.8, 7.5, 4.8, 2.7, 1.2, 0.3, 0.0 — **neither**

b. 30, 27, 24, 21, 18, 15, 12, 9, 6, 3, 0 — **linear**

c. 30.0, 24.0, 19.2, 15.4, 12.3, 9.8, 7.9, 6.3, 5.0, 4.0, 3.2 — **exponential**

d. 30.0, 29.7, 28.8, 27.3, 25.2, 22.5, 19.2, 15.3, 10.8, 5.7, 0.0 — **neither**

English Language Learners

Connection

Help English language learners by making the connection to other fields of study. The exponential function lessons have many application problems that relate to other subjects, such as forensic science, savings, car depreciation, and population growth.

6.6 Record and Practice Journal

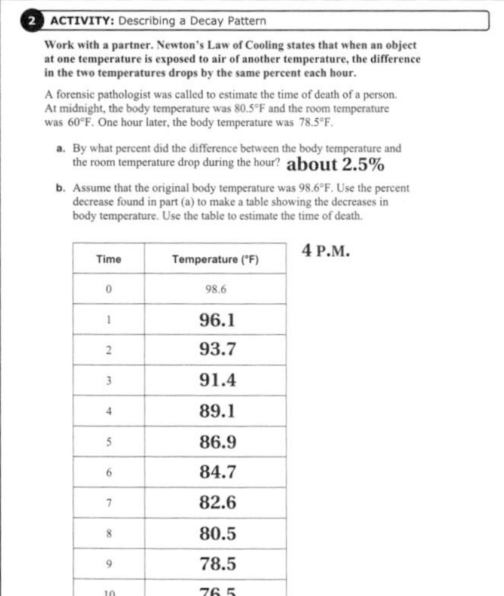

2 ACTIVITY: Describing a Decay Pattern

Work with a partner. Newton's Law of Cooling states that when an object at one temperature is exposed to air of another temperature, the difference in the two temperatures drops by the same percent each hour.

A forensic pathologist was called to estimate the time of death of a person. At midnight, the body temperature was 80.5°F and the room temperature was 60°F. One hour later, the body temperature was 78.5°F.

a. By what percent did the difference between the body temperature and the room temperature drop during the hour? **about 2.5%**

b. Assume that the original body temperature was 98.6°F. Use the percent decrease found in part (a) to make a table showing the decreases in body temperature. Use the table to estimate the time of death.

Time	Temperature (°F)
0	98.6
1	**96.1**
2	**93.7**
3	**91.4**
4	**89.1**
5	**86.9**
6	**84.7**
7	**82.6**
8	**80.5**
9	**78.5**
10	**76.5**

4 P.M.

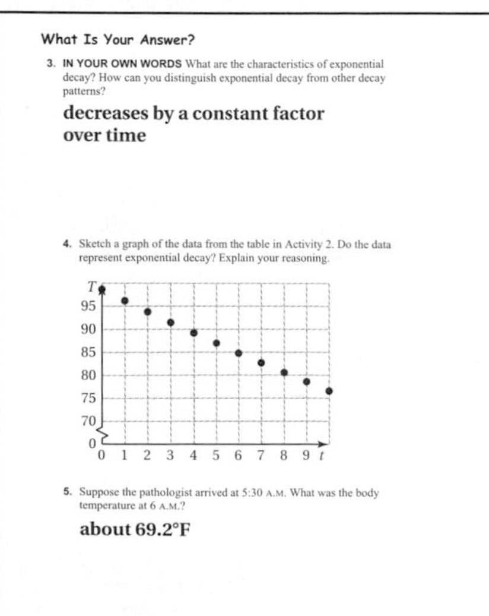

What Is Your Answer?

3. **IN YOUR OWN WORDS** What are the characteristics of exponential decay? How can you distinguish exponential decay from other decay patterns?

decreases by a constant factor over time

4. Sketch a graph of the data from the table in Activity 2. Do the data represent exponential decay? Explain your reasoning.

5. Suppose the pathologist arrived at 5:30 A.M. What was the body temperature at 6 A.M.?

about 69.2°F

Laurie's Notes

Activity 2

- **MP3a:** This activity is similar to the last with students trying to construct an argument as to why the context is exponential.
- Ask a student to read the introduction about Newton's Law of Cooling. Perhaps they have studied it in science. Discuss what the law states. The difference (subtraction) in the temperatures drops by the same percent (common ratio) each hour.
- Work through part (a) with students to insure they get started correctly.

Time	Body Temp.	Room Temp.	Difference
Midnight	80.5°	60°	20.5°
1 A.M.	78.5°	60°	18.5°

The percent decrease is $(20.5 - 18.5)/20.5 \approx 9.8\%$, or about 10%.

- The body temperature is dropping by 10% of the difference between the body temperature and 60° each hour.
- Students now start to fill in the table in part (b).

Time	Drop in Body Temperature	Body Temperature
0	10% of $(98.6° - 60°) = 3.86°$	$98.6° - 3.86° \approx 94.7°$
1	10% of $(94.7° - 60°) = 3.47°$	$94.7° - 3.47° \approx 91.2°$

- Students need to be organized with their data and careful with how they round answers.
- **Working Backwards:** Students will determine that the time of death was approximately 6 hours before the body was found.

What Is Your Answer?

- In Question 3, students should discuss the common ratio between equally spaced values.

Closure

- How are exponential growth and exponential decay the same? Quantities change by the same factor over equal intervals. How are they different? For exponential growth, the factor is greater than 1. For exponential decay, the factor is greater than 0 and less than 1.

Technology For the Teacher

The Dynamic Planning Tool

Editable Teacher's Resources at *BigIdeasMath.com*

2 ACTIVITY: Describing a Decay Pattern

Work with a partner. Newton's Law of Cooling states that when an object at one temperature is exposed to air of another temperature, the difference in the two temperatures drops by the same percent each hour.

A forensic pathologist was called to estimate the time of death of a person. At midnight, the body temperature was 80.5°F and the room temperature was 60°F. One hour later, the body temperature was 78.5°F.

a. By what percent did the difference between the body temperature and the room temperature drop during the hour?

b. Assume that the original body temperature was 98.6°F. Use the percent decrease found in part (a) to make a table showing the decreases in body temperature. Use the table to estimate the time of death.

Time	Temperature (°F)
0	98.6
1	
2	
3	
4	
5	
6	
7	
8	
9	
10	

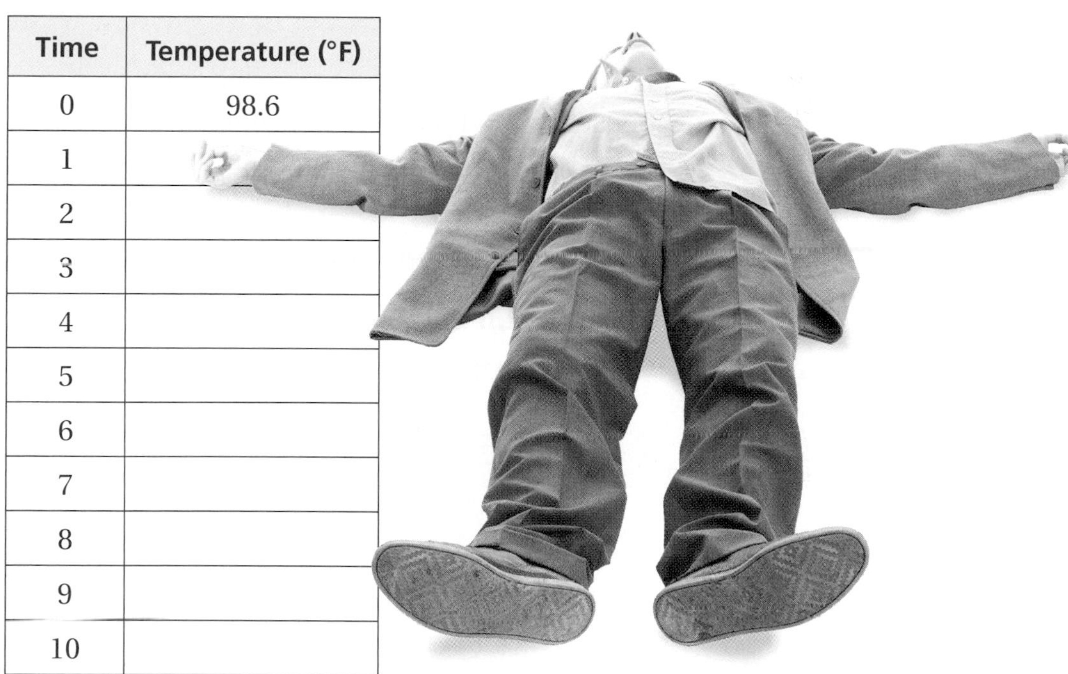

What Is Your Answer?

3. **IN YOUR OWN WORDS** What are the characteristics of exponential decay? How can you distinguish exponential decay from other decay patterns?
4. Sketch a graph of the data from the table in Activity 2. Do the data represent exponential decay? Explain your reasoning.
5. Suppose the pathologist arrived at 5:30 A.M. What was the body temperature at 6 A.M.?

Practice Use what you learned about exponential decay to complete Exercises 3 and 4 on page 304.

6.6 Lesson

Check It Out
Lesson Tutorials
BigIdeasMath.com

Key Vocabulary
exponential decay, *p. 302*
exponential decay function, *p. 302*

Exponential decay occurs when a quantity decreases by the same factor over equal intervals of time.

Key Idea

Exponential Decay Functions

A function of the form $y = a(1 - r)^t$, where $a > 0$ and $0 < r < 1$, is an **exponential decay function.**

$$\text{Final amount} \longrightarrow y = \overset{\text{Initial amount}}{a}(\underbrace{1 - \overset{\text{Rate of decay (in decimal form)}}{r}}_{\text{Decay factor}})^{t} \longleftarrow \text{Time}$$

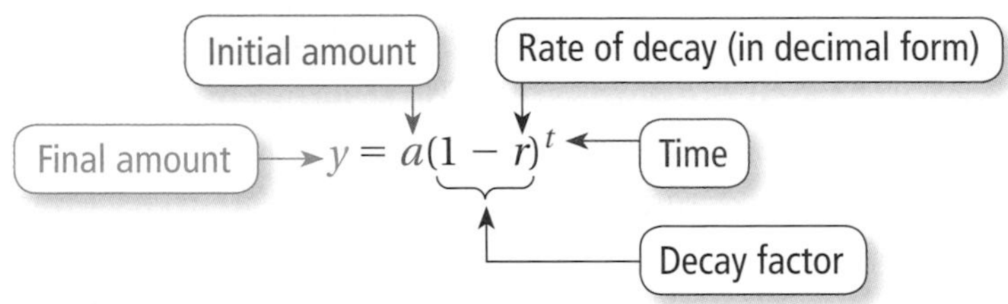

Study Tip

Notice that an exponential decay function is of the form $y = ab^x$, where b is replaced by $1 - r$ and x is replaced by t.

For exponential growth, the value inside the parentheses is greater than 1 because r is added to 1. For exponential decay, the value inside the parentheses is less than 1 because r is subtracted from 1.

EXAMPLE 1 Identifying Exponential Growth and Decay

Determine whether each table represents an *exponential growth function*, an *exponential decay function*, or *neither*.

a.

	x	y	
	0	270	
+1	1	90	$\times\frac{1}{3}$
+1	2	30	$\times\frac{1}{3}$
+1	3	10	$\times\frac{1}{3}$

As x increases by 1, y is multiplied by $\frac{1}{3}$.

So, the table represents an exponential decay function.

b.

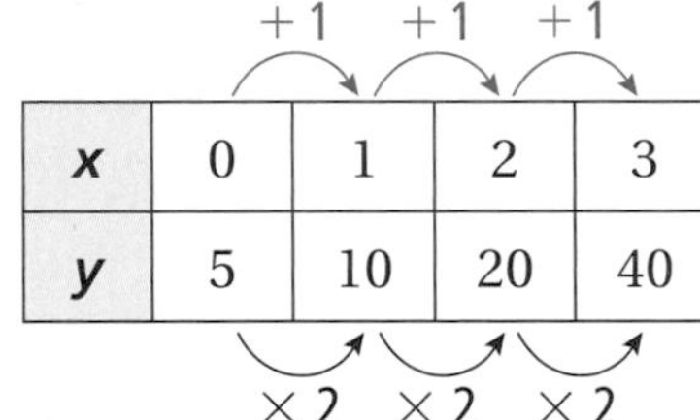

	+1	+1	+1	
x	0	1	2	3
y	5	10	20	40
	×2	×2	×2	

As x increases by 1, y is multiplied by 2.

So, the table represents an exponential growth function.

On Your Own

Now You're Ready
Exercises 8–13

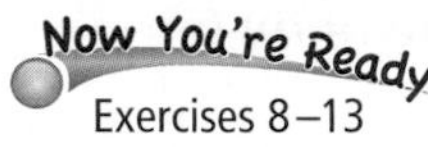

Determine whether the table represents an *exponential growth function*, an *exponential decay function*, or *neither*.

1.

x	0	1	2	3
y	64	16	4	1

2.

x	1	3	5	7
y	4	11	18	25

Laurie's Notes

Introduction

Connect

- **Yesterday:** Students determined whether sequences show exponential decay. (MP3a)
- **Today:** Students will write and evaluate exponential decay functions.

Motivate

- As a whole class or smaller groups, distribute items that will decay (dice, unit algebra tiles, or pennies) and a cup or bag.
- Students record the initial amount ($t = 0$) of items. Pour the items out of the cup or bag. Remove the even numbers (dice), the positive tiles, or the heads (pennies). Record the number of items that remain ($t = 1$).
- Repeat this process until two or fewer items remain.
- Have students keep this data and return to it at the end of class.

Lesson Notes

Key Idea

- The number of variables in the function will create tension for some students. The different colors highlight the different parts of the equation.
- **Connection:** The base b in the general exponential equation $y = ab^x$ is replaced by $1 - r$ and the independent variable x is replaced by t.
- The initial amount a is the amount present before any decay occurs. It is the y-intercept when time $t = 0$.
- The base is the common factor and is called the decay factor. If $r = 0.25$, then $1 - r = 0.75 = 75\%$. The decay rate is 25% and the decay factor is 75%.
- Students should think of the decay factor as *what remains*. If a quantity decreases 25%, then 75% of the quantity remains.

Example 1

- **?** "Is the data in part (a) decreasing by the same factor over equal time intervals? Explain." yes; The y-values are multiplied by the same factor, $\frac{1}{3}$.
- The decay factor is $\frac{1}{3}$. This is what remains of the previous amount.
- The rate of decay is $\frac{2}{3}$, meaning $\frac{2}{3}$ of the previous amount is lost.
- **?** "Is the data in part (b) decreasing by the same factor over equal time intervals? Explain." no; The y-values are multiplied by a factor greater than 1. This is exponential growth.

On Your Own

- Caution students to actually test the data and not assume that the tables show exponential decay and exponential growth.
- **Think-Pair-Share:** Students should read each question independently and then work with a partner to answer the questions. When they have answered the questions, the pair should compare their answers with another group and discuss any discrepancies.

Goal Today's lesson is writing and evaluating **exponential decay functions.**

Lesson Materials
Introduction
• dice, algebra tiles, or pennies

Start Thinking! and Warm Up

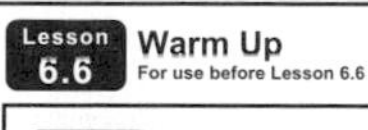
Lesson 6.6 Warm Up
For use before Lesson 6.6

Lesson 6.6 Start Thinking!
For use before Lesson 6.6

How are the graphs of exponential growth functions and exponential decay functions similar? How are they different?

Extra Example 1

Determine whether the table represents an *exponential growth function*, an *exponential decay function*, or *neither.*

x	1	2	3	4
y	4	12	36	108

exponential growth function

On Your Own

1. exponential decay function
2. neither

Differentiated Instruction

Mnemonics can help students remember what the variables stand for in exponential growth and decay functions.

$y = a(1 + r)^t$ Exponential growth

$y = a(1 - r)^t$ Exponential decay

a – The letter that *begins* the alphabet is the value you *begin* with.

y – The letter near the *end* of the alphabet is the value you *end* with.

r – the percentage *rate* at which the value increases or decreases

t – the *time* (in years)

Extra Example 2

The function $P = 5100(0.95)^t$ represents the population P of a town after t years. By what percent does the population decrease each year? 5%

On Your Own

3. 10%

Extra Example 3

In Example 3, the car loses 10% of its value every year.

a. Write a function that represents the value y (in dollars) of the car after t years. $y = 21{,}500(0.9)^t$

b. Graph the function from part (a). Use the graph to estimate the value of the car after 4 years.

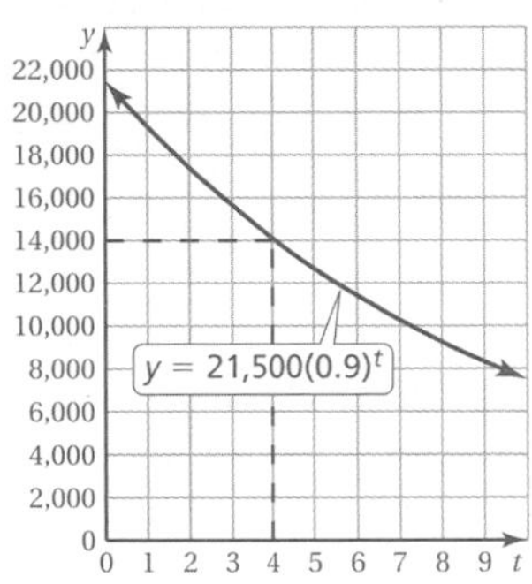

about $14,000

On Your Own

4. a. $y = 21{,}500(0.91)^t$

b.

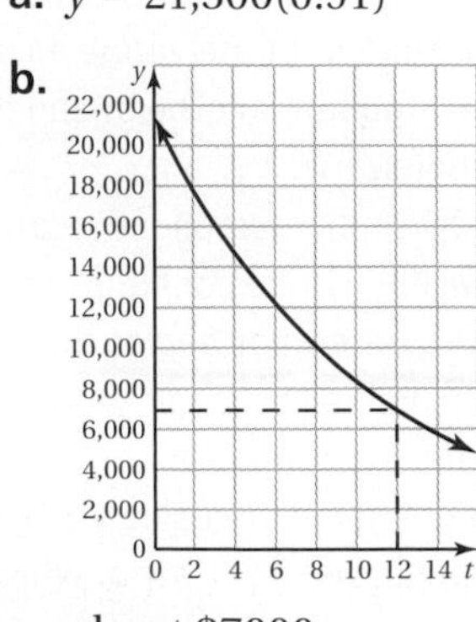

about $7000

Laurie's Notes

Example 2

- ? "Is P an exponential function? Explain." yes; It is of the general form $y = ab^x$. "Is P an exponential decay function? Explain." yes; It is of the general form $y = a(1 - r)^x$.
- ? **MP7: Look for and Make Use of Structure:** "What is the initial population of the town?" 4870
- ? "What is the rate of decay and what does it mean?" 6%; The population of the town is decreasing by 6% each year.
- **Extension:** Find the population of the town after 4 years. about 3802

On Your Own

- Ask students to interpret the constant 275 and to find the area of the coral reef after 2 years. initial area; about 223 square miles If time permits, students could graph the equation.

Example 3

- Work through the problem as shown.
- Point out to students that the car has lost more than 50% of its value over the 6-year period.
- ? "Is the amount the car loses in value the same each year? Explain." no; It loses more the first year than any other year. The percent of the value that it loses is the same.
- **Extension:** Graph the equation. Discuss the steepness of the graph over time and the fact that the car loses more of its value in the first year than any other year.

On Your Own

- ? "Do you think that all cars depreciate at the same rate?" Students should have some sense that many factors influence the rate at which cars depreciate.

Closure

- Write a model for the Motivate done at the beginning of class. The model should be $y = a(0.5)^t$ where a is the initial amount. Ask how well the model fits the data.

Technology For the Teacher

The Dynamic Planning Tool
Editable Teacher's Resources at *BigIdeasMath.com*

EXAMPLE 2 Interpreting an Exponential Decay Function

The function $P = 4870(0.94)^t$ represents the population P of a town after t years. By what percent does the population decrease each year?

Use the decay factor $1 - r$ to find the rate of decay.

$1 - r = 0.94$ — Write an equation.

$r = 0.06$ — Solve for r.

So, the population of the town decreases by 6% each year.

On Your Own

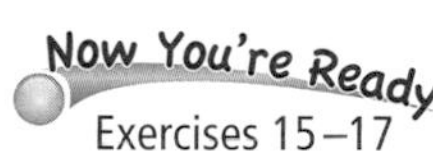

3. The function $A = 275\left(\frac{9}{10}\right)^t$ represents the area A (in square miles) of a coral reef after t years. By what percent does the area of the coral reef decrease each year?

EXAMPLE 3 Real-Life Application

The value of a car is \$21,500. It loses 12% of its value every year.

a. Write a function that represents the value y (in dollars) of the car after t years.

$y = a(1 - r)^t$ — Write exponential decay function.

$y = 21{,}500(1 - 0.12)^t$ — Substitute 21,500 for a and 0.12 for r.

$y = 21{,}500(0.88)^t$ — Simplify.

b. Graph the function from part (a). Use the graph to estimate the value of the car after 6 years.

From the graph, you can see that the y-value is about 10,000 when $t = 6$.

So, the value of the car is about \$10,000 after 6 years.

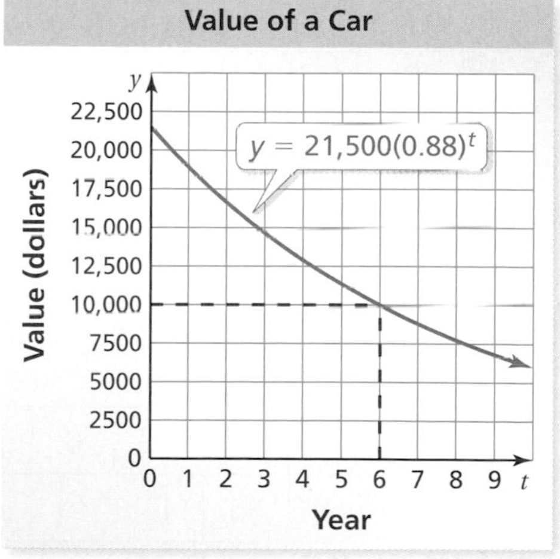

Check

$y = 21{,}500(0.88)^t$

$= 21{,}500(0.88)^6$

≈ 9985 ✓

On Your Own

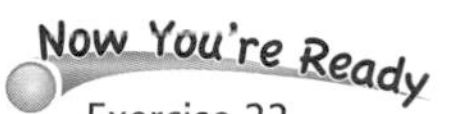

4. **WHAT IF?** The car loses 9% of its value every year.

   a. Write a function that represents the value y (in dollars) of the car after t years.

   b. Graph the function from part (a). Estimate the value of the car after 12 years. Round your answer to the nearest thousand.

6.6 Exercises

Vocabulary and Concept Check

1. **WRITING** When does the function $y = ab^x$ represent exponential growth? exponential decay?
2. **VOCABULARY** What is the decay factor in the function $y = a(1 - r)^t$?

Practice and Problem Solving

Describe the pattern of decay for the sequence.

3. 28, 26, 24, 22, 20, 18, 16, 14, 12, 10, 8
4. 256, 192, 144, 108, 81, 60.8, 45.6, 34.2, 25.6, 19.2, 14.4

Determine whether the graph represents an *exponential growth function*, an *exponential decay function*, or *neither*.

5.

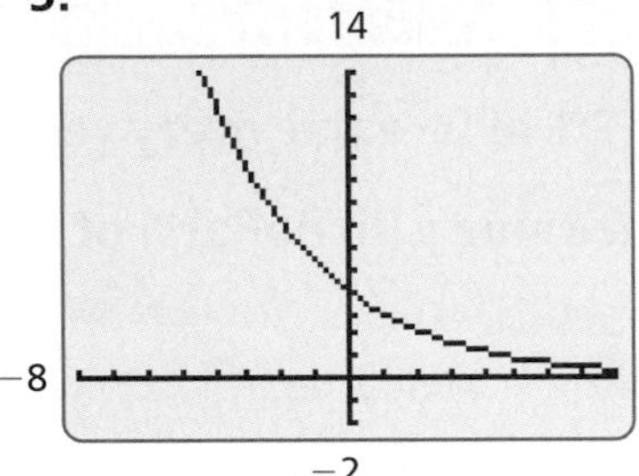

6.

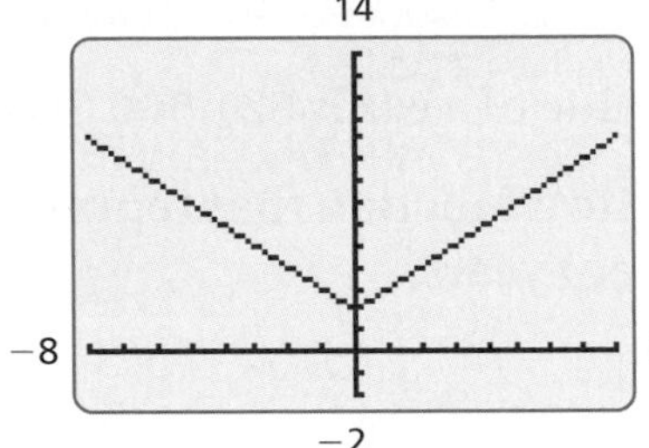

7.

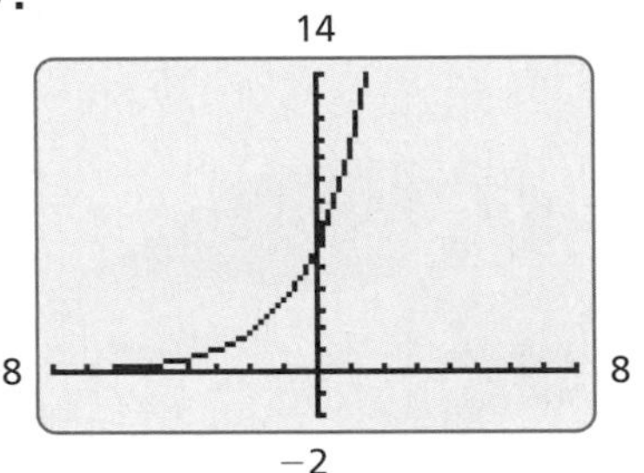

Determine whether the table represents an *exponential growth function*, an *exponential decay function*, or *neither*.

1 8.

x	0	1	2	3
y	17	51	153	459

9.

x	1	2	3	4
y	32	28	24	20

10.

x	1	2	3	4
y	625	125	25	5

11.

x	2	4	6	8
y	256	64	16	4

12.

x	2	4	6	8
y	35	42	49	42

13.

x	3	5	7	9
y	6	216	7776	279,936

14. **CAMPER** The table shows the value of a camper t years after it is purchased.
    a. Determine whether the table represents an *exponential growth function*, an *exponential decay function*, or *neither*.
    b. What is the value of the camper after 5 years?

t	Value
1	$24,000
2	$19,200
3	$15,360
4	$12,288

Assignment Guide and Homework Check

Level	Assignment	Homework Check
Average	1–4, 9–17 odd, 14, 22, 25–28	9, 11, 14, 15, 22
Advanced	1–4, 9–17 odd, 14, 21–28	9, 11, 14, 17, 22

Common Errors

- **Exercise 1** Students may incorrectly think that when $b > 1$, the function represents exponential decay and that when $0 < b < 1$, the function represents exponential growth. Remind them of the definitions.
- **Exercises 8–13** Students may have difficulty determining the factor used to increase or decrease the dependent variable. Encourage them to look for patterns.
- **Exercises 15–17** Students may write the decay factor as the rate of decay. Remind them of the definition.
- **Exercises 22–23** Students may forget to change the percents (rates of decay) to decimals before subtracting them from 1 to find the decay factor.

6.6 Record and Practice Journal

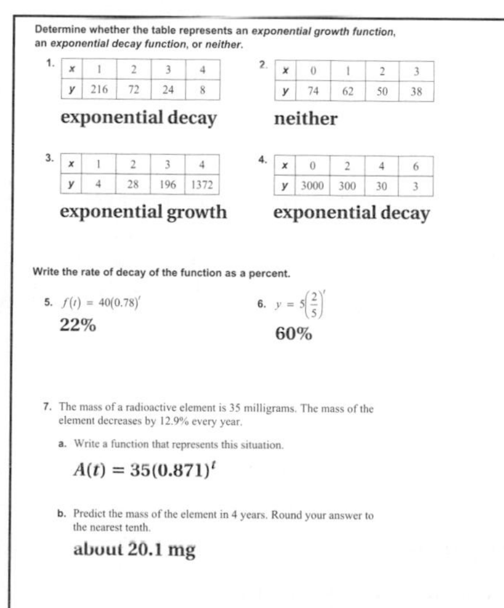

Determine whether the table represents an *exponential growth function*, an *exponential decay function*, or *neither*.

1.

x	1	2	3	4
y	216	72	24	8

exponential decay

2.

x	0	1	2	3
y	74	62	50	38

neither

3.

x	1	2	3	4
y	4	28	196	1372

exponential growth

4.

x	0	2	4	6
y	3000	300	30	3

exponential decay

Write the rate of decay of the function as a percent.

5. $f(t) = 40(0.78)^t$

22%

6. $y = 5\left(\frac{2}{5}\right)^t$

60%

7. The mass of a radioactive element is 35 milligrams. The mass of the element decreases by 12.9% every year.

a. Write a function that represents this situation.

$A(t) = 35(0.871)^t$

b. Predict the mass of the element in 4 years. Round your answer to the nearest tenth.

about 20.1 mg

Technology For the Teacher

Answer Presentation Tool

Vocabulary and Concept Check

1. When $b > 1$, the function represents exponential growth. When $0 < b < 1$, the function represents exponential decay.
2. $1 - r$

Practice and Problem Solving

3. The rate of change of the sequence is a constant -2, so it is a linear decay pattern.
4. The sequence changes by a factor of approximately 0.75, so it is an exponential decay pattern.
5. exponential decay function
6. neither
7. exponential growth function
8. exponential growth function
9. neither
10. exponential decay function
11. exponential decay function
12. neither
13. exponential growth function
14. a. exponential decay function

    b. $9830.40

Practice and Problem Solving

15. 20%　　**16.** 5%

17. 25%　　**18.** B

19. A　　**20.** C

21. See Additional Answers.

22. $p(t) = 250{,}000(0.985)^t$; 214,933

23. See *Taking Math Deeper.*

24. **a.** decrease; The value of the function at 4 is less than the value of the function at 2. Because the function value is decreasing as x increases and the function is exponential, it will decrease by the same percent over time. The function represents exponential decay.

b. 6

c. $y = 6(0.5)^x$

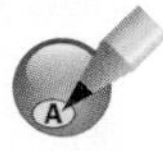

Fair Game Review

25. $a_n = 3n + 6$; 51

26. $a_n = -2n + 5$; -25

27. $a_n = -4n - 3$; -63

28. A

Mini-Assessment

Determine whether the table represents an *exponential growth function,* an *exponential decay function,* or *neither.*

1.

x	0	1	2	3
y	320	80	20	5

exponential decay function

2.

x	0	2	4	6
y	3	12	48	192

exponential growth function

Taking Math Deeper

Exercise 23

Students can look for an entry point to the solution by first using repetitive multiplication.

Use a table to examine how the air pressure decreases exponentially over time. The tire loses 8% of its air each day, so multiply the air pressure by $1 - 0.08 = 0.92$ until it is at or below 24 psi.

Time, x (days)	Air Pressure, y (psi)
0	32
1	$32(0.92) = 29.44$
2	$29.44(0.92) = 27.0848$
3	$27.0848(0.92) = 24.918016$
4	$24.918016(0.92) = 22.92457472$

Interpret the table.

The air pressure drops to 24 psi between 3 days (or 72 hours) and 4 days (or 98 hours) after the tire began losing air. The air pressure was 32 psi at **noon** on Monday, so the TPMS will alert the driver sometime between noon on Thursday and noon on Friday.

Use a graph to answer the question, focusing on x-values between 3 and 4. The situation can be modeled by $y = 32(0.92)^x$. Notice the change in variables from t to x, because graphing calculators use x.

The air pressure is about 24 psi after 3.4 days. This is slightly less than three and a half days, which occurs late Thursday night (before midnight).

So, the TPMS will alert the driver on Thursday.

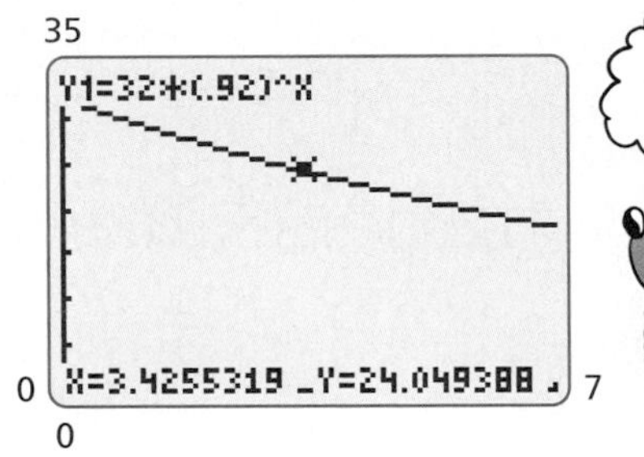

Project

Research tire pressure monitoring systems. How do these systems work?

Reteaching and Enrichment Strategies

If students need help. . .	If students got it. . .
Resources by Chapter • Practice A and Practice B • Puzzle Time Record and Practice Journal Practice Differentiating the Lesson Lesson Tutorials Skills Review Handbook	Resources by Chapter • Enrichment and Extension • School-to-Work • Financial Literacy • Technology Connection Start the next section

Write the rate of decay of the function as a percent.

② 15. $y = 4(0.8)^t$

16. $f(t) = 30(0.95)^t$

17. $g(t) = \left(\frac{3}{4}\right)^t$

Match the exponential function with its graph.

18. $y = 10(1.3)^t$

19. $h(t) = 6\left(\frac{7}{8}\right)^t$

20. $y = 2(0.6)^t$

A. 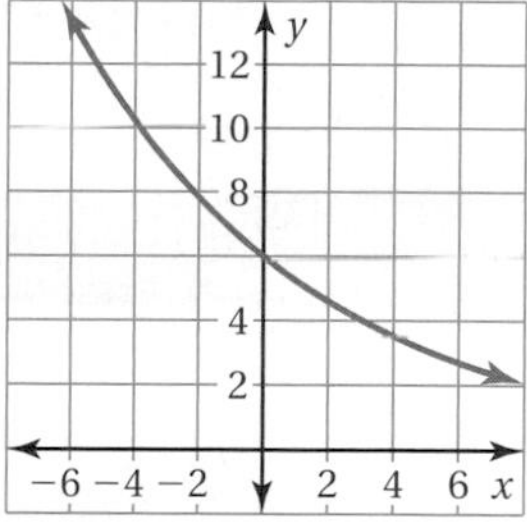

B. 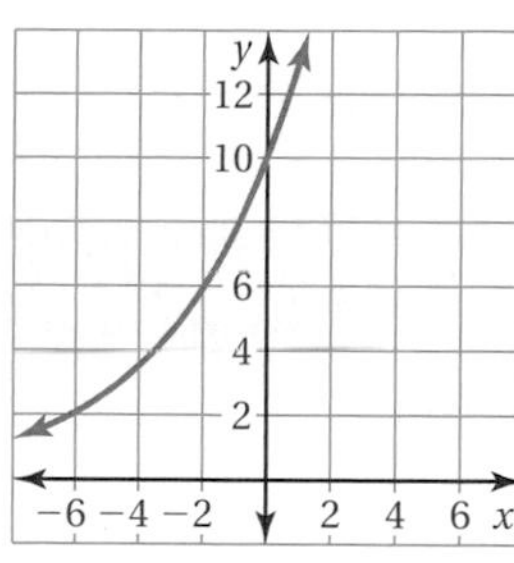

C. 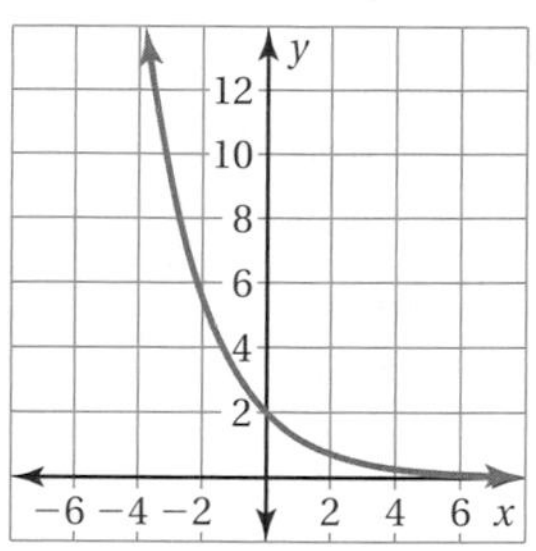

21. **CHOOSE TOOLS** When would you graph an exponential decay function by hand? When would you use a graphing calculator? Explain your reasoning.

③ 22. **POPULATION** A city has a population of 250,000. The population is expected to decrease by 1.5% annually for the next decade. Write a function that represents this situation. Then predict the population in 10 years.

23. **TIRE PRESSURE** At noon on Monday, the air pressure of a tire is 32 pounds per square inch (psi). The tire loses 8% of its air every day. The tire pressure monitoring system (TPMS) will alert the driver when the tire pressure is less than or equal to 24 psi. On what day of the week will the TPMS alert the driver? Use the *trace* feature of a graphing calculator to help find the answer.

24. **Structure** The graph of an exponential function passes through $\left(2, \frac{3}{2}\right)$ and $\left(4, \frac{3}{8}\right)$.

   a. Do the y-values increase or decrease as x increases? How do you know?

   b. Find the y-intercept of the graph.

   c. Write an exponential function that represents the graph.

Fair Game Review What you learned in previous grades & lessons

Write an equation for the nth term of the arithmetic sequence. Then find a_{15}. *(Section 5.6)*

25. 9, 12, 15, 18, . . .

26. 3, 1, −1, −3, . . .

27. −7, −11, −15, −19, . . .

28. **MULTIPLE CHOICE** What is the solution of the linear system? *(Section 4.3)*

$$2x - 5y = 11$$
$$5x - 3y = -1$$

Ⓐ (−2, −3) Ⓑ (−2, 3)

Ⓒ (2, −3) Ⓓ (2, 3)

6.7 Geometric Sequences

Essential Question How are geometric sequences used to describe patterns?

COMMON CORE STATE STANDARDS
F.BF.2
F.IF.3
F.LE.2

1 ACTIVITY: Describing Calculator Patterns

Work with a partner.

- **Enter the keystrokes on a calculator and record the results in the table.**
- **Describe the pattern.**

a. Step 1 

Step 2 

Step 3 × 2 =

Step 4 

Step 5  

Step	1	2	3	4	5
Calculator Display					

b. Step 1 

Step 2 

Step 3 

Step 4 

Step 5 × . 5 =

Step	1	2	3	4	5
Calculator Display					

c. Use a calculator to make your own sequence. Start with any number and multiply by 3 each time. Record your results in the table.

Step	1	2	3	4	5
Calculator Display					

Laurie's Notes

Introduction

Standards for Mathematical Practice

- **MP4 Model with Mathematics** and **MP5 Use Appropriate Tools Strategically:** Students have worked with exponential functions for three sections and should have a good sense now of how exponential functions model real-life applications. The calculator is a useful tool in performing the repeated computations so that the focus is on the process (repeated common factor) versus the computation skill.

Motivate

- Write on the board: KB, MB, GB.
- ? "Does anyone know what these letters stand for?"
- You may need to give a hint that they are abbreviations. They stand for kilobytes (2^{10} bytes), megabytes (2^{20} bytes), and gigabytes (2^{30} bytes).
- Explain that these represent large numbers of bytes and are often used to describe storage capacity for computers.

Activity Notes

Activity 1

- Students should be able to follow the keystrokes shown for the first two problems. Individual calculators may have a feature (such as ANS) that will appear on the screens.
- Part (a) multiplies by 2, therefore it is generating powers of 2 ($2^1 = 2, 2^2 = 4, 2^3 = 8, 2^4 = 16, 2^5 = 32$).
- ? "There are 2^{10} bytes in 1 kilobyte. Continue the pattern in part (a) to write the number of bytes in 1 kilobyte as a whole number." 1024
- Part (b) multiplies by $\frac{1}{2}$, starting with 64.
- ? When students have finished, ask them to describe the connection between parts (a) and (b).
- **Extension:** Ask students what would happen if they continue multiplying by $\frac{1}{2}$ in part (b). It is common for students to predict that the numbers will become negative. The numbers are decreasing, but they will always be positive. $64, 32, 16, 8, 4, 2, 1, \frac{1}{2}, \frac{1}{4}, \frac{1}{8}, \ldots$
- **Teaching Tip:** Students love to explore how many times they can repeat a particular task on a calculator, such as multiplying by 3. Give time for students to explore this question if it comes up. It doesn't take much time to reach an overflow error on the calculator, which is a good experience.

Common Core State Standards

F.BF.2 Write arithmetic and geometric sequences both recursively and with an explicit formula, use them to model situations, and translate between the two forms.

F.LE.2 Construct linear and exponential functions, including arithmetic and geometric sequences, given a graph, a description of a relationship, or two input-output pairs (include reading these from a table).

Also **F.IF.3**

Previous Learning

Students should know how to write an exponential function. Students should know how to write an equation for an arithmetic sequence.

Start Thinking! and Warm Up

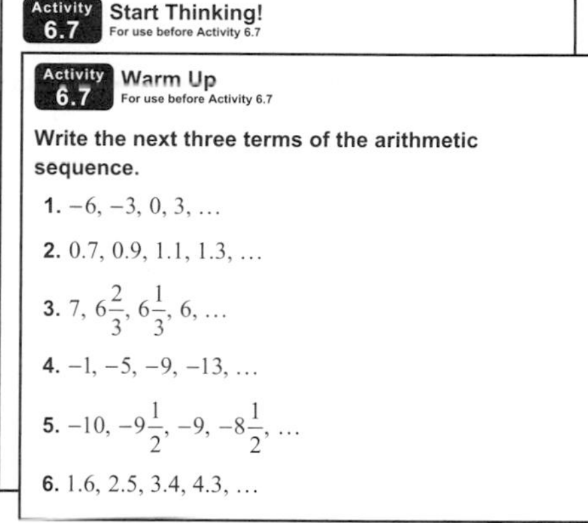

Activity 6.7 Start Thinking! For use before Activity 6.7

Activity 6.7 Warm Up For use before Activity 6.7

Write the next three terms of the arithmetic sequence.

1. $-6, -3, 0, 3, \ldots$
2. $0.7, 0.9, 1.1, 1.3, \ldots$
3. $7, 6\frac{2}{3}, 6\frac{1}{3}, 6, \ldots$
4. $-1, -5, -9, -13, \ldots$
5. $-10, -9\frac{1}{2}, -9, -8\frac{1}{2}, \ldots$
6. $1.6, 2.5, 3.4, 4.3, \ldots$

6.7 Record and Practice Journal

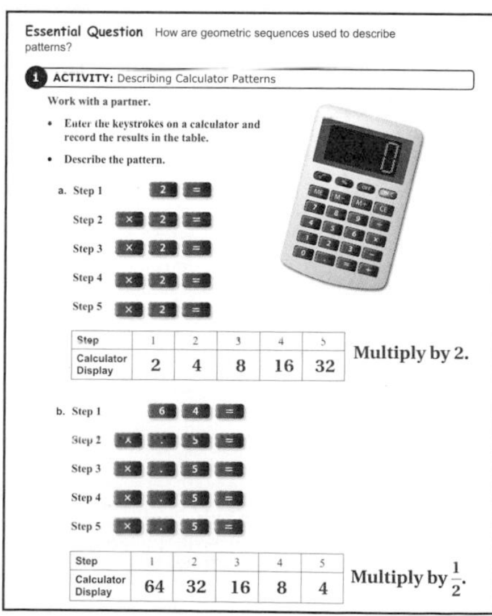

Essential Question How are geometric sequences used to describe patterns?

1 **ACTIVITY:** Describing Calculator Patterns

Work with a partner.

- Enter the keystrokes on a calculator and record the results in the table.
- Describe the pattern.

a. Step 1: 2 =
Step 2: × 2 =
Step 3: × 2 =
Step 4: × 2 =
Step 5: × 2 =

Step	1	2	3	4	5
Calculator Display	2	4	8	16	32

Multiply by 2.

b. Step 1: 6 4 =
Step 2: × . 5 =
Step 3: × . 5 =
Step 4: × . 5 =
Step 5: × . 5 =

Step	1	2	3	4	5
Calculator Display	64	32	16	8	4

Multiply by $\frac{1}{2}$.

Differentiated Instruction

Kinesthetic

Use stacks of coins to explain the concept of a common ratio to students. Model the geometric sequence 1, 3, 9, 27, . . . with four stacks of coins. State that the second stack is 3 times as high as the first stack. Ask how the height of the third stack compares to the height of the second stack. Ask the same question for the third and fourth stacks. Explain that each stack is 3 times taller than the previous stack, so 3 is the common ratio of the sequence.

6.7 Record and Practice Journal

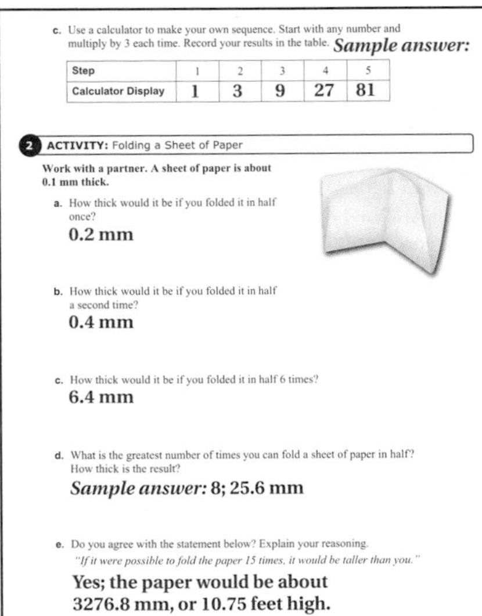

c. Use a calculator to make your own sequence. Start with any number and multiply by 3 each time. Record your results in the table. ***Sample answer:***

Step	1	2	3	4	5
Calculator Display	1	3	9	27	81

2 ACTIVITY: Folding a Sheet of Paper

Work with a partner. A sheet of paper is about 0.1 mm thick.

a. How thick would it be if you folded it in half once?

0.2 mm

b. How thick would it be if you folded it in half a second time?

0.4 mm

c. How thick would it be if you folded it in half 6 times?

6.4 mm

d. What is the greatest number of times you can fold a sheet of paper in half? How thick is the result?

Sample answer: **8; 25.6 mm**

e. Do you agree with the statement below? Explain your reasoning.

"If it were possible to fold the paper 15 times, it would be taller than you."

Yes; the paper would be about 3276.8 mm, or 10.75 feet high.

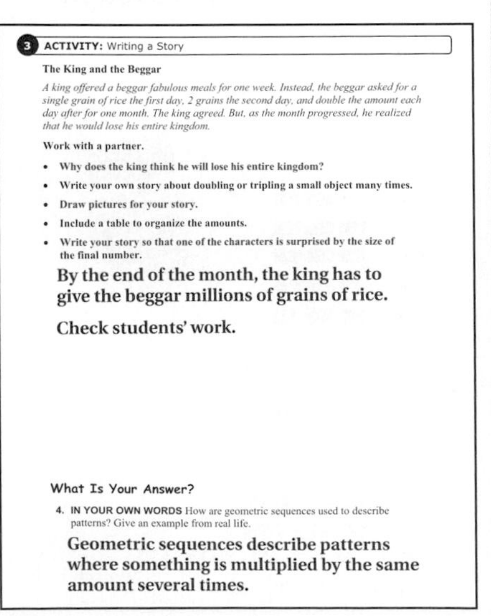

3 ACTIVITY: Writing a Story

The King and the Beggar

A king offered a beggar fabulous meals for one week. Instead, the beggar asked for a single grain of rice the first day, 2 grains the second day, and double the amount each day after for one month. The king agreed. But, as the month progressed, he realized that he would lose his entire kingdom.

Work with a partner.

- **Why does the king think he will lose his entire kingdom?**
- **Write your own story about doubling or tripling a small object many times.**
- **Draw pictures for your story.**
- **Include a table to organize the amounts.**
- **Write your story so that one of the characters is surprised by the size of the final number.**

By the end of the month, the king has to give the beggar millions of grains of rice.

Check students' work.

What Is Your Answer?

4. **IN YOUR OWN WORDS** How are geometric sequences used to describe patterns? Give an example from real life.

Geometric sequences describe patterns where something is multiplied by the same amount several times.

Laurie's Notes

Activity 2

- Provide paper for students to explore this activity. Students may be surprised at the results of this activity.
- For an interesting website about folding paper in half, go to *www.pomonahistorical.org/12times.htm.*
- Students may not be familiar with their heights in metric units. You could have them measure themselves with a meter stick. You may need to remind students that 10 mm = 1 cm, so 100 mm = 10 cm and 1000 mm = 100 cm.
- **FYI:** Robert Wadlow (1918–1940) was the tallest known human being in history. He was 8 feet 11 inches, or about 272 centimeters tall.
- **Extension:** Offer a challenge to students. "How many times would you need to fold the paper so that the stack would reach the moon (assuming that you could fold the paper that many times)?" Students will need to find the distance to the moon and then convert to millimeters. The answer of 42 folds will seem surprisingly low to students.

Activity 3

- Students may have heard a story similar to this before. After this activity, students should have a sense of how quickly something grows when it is doubled repeatedly.
- Ask several students to share their stories.
- **Extension:** Ask students to write final drafts of their stories, complete with illustrations and tables. Display stories on a bulletin board.

What Is Your Answer?

- Although no formal definition of geometric sequences is given in the activity, students should realize that they have been *multiplying* by a constant, rather than *adding* a constant as with arithmetic sequences.
- Have students share their examples.

Closure

- Complete the table by multiplying the result of the previous step by 2.

Step	1	2	3	4	5
Result	3	6	12	24	48

Technology For the Teacher

Dynamic Classroom

The Dynamic Planning Tool

Editable Teacher's Resources at *BigIdeasMath.com*

2 ACTIVITY: Folding a Sheet of Paper

Work with a partner. A sheet of paper is about 0.1 mm thick.

a. How thick would it be if you folded it in half once?

b. How thick would it be if you folded it in half a second time?

c. How thick would it be if you folded it in half 6 times?

d. What is the greatest number of times you can fold a sheet of paper in half? How thick is the result?

e. Do you agree with the statement below? Explain your reasoning.

"If it were possible to fold the paper 15 times, it would be taller than you."

3 ACTIVITY: Writing a Story

The King and the Beggar

A king offered a beggar fabulous meals for one week. Instead, the beggar asked for a single grain of rice the first day, 2 grains the second day, and double the amount each day after for one month. The king agreed. But, as the month progressed, he realized that he would lose his entire kingdom.

Work with a partner.

- **Why does the king think he will lose his entire kingdom?**
- **Write your own story about doubling or tripling a small object many times.**
- **Draw pictures for your story.**
- **Include a table to organize the amounts.**
- **Write your story so that one of the characters is surprised by the size of the final number.**

What Is Your Answer?

4. **IN YOUR OWN WORDS** How are geometric sequences used to describe patterns? Give an example from real life.

Use what you learned about geometric sequences to complete Exercise 4 on page 310.

6.7 Lesson

Key Vocabulary
geometric sequence, *p. 308*
common ratio, *p. 308*

Geometric Sequence

In a **geometric sequence,** the ratio between consecutive terms is the same. This ratio is called the **common ratio.** Each term is found by multiplying the previous term by the common ratio.

1, 5, 25, 125, . . . Terms of a geometric sequence

$\times 5$ $\times 5$ $\times 5$ ← Common ratio

EXAMPLE 1 Extending a Geometric Sequence

Write the next three terms of the geometric sequence 3, 6, 12, 24, ...

Use a table to organize the terms and extend the pattern.

Position	1	2	3	4	5	6	7
Term	3	6	12	24	48	96	192

$\times 2$ $\times 2$ $\times 2$ $\times 2$ $\times 2$ $\times 2$

Each term is twice the previous term. So, the common ratio is 2.

Multiply a term by 2 to find the next term.

∴ The next three terms are 48, 96, and 192.

EXAMPLE 2 Graphing a Geometric Sequence

Graph the geometric sequence 32, 16, 8, 4, 2, ... What do you notice?

Make a table. Then plot the ordered pairs (n, a_n).

Position, n	1	2	3	4	5
Term, a_n	32	16	8	4	2

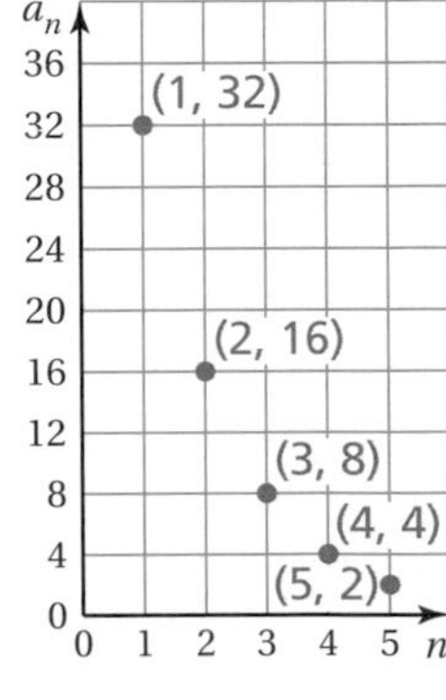

∴ The points of the graph appear to lie on an exponential curve.

On Your Own

Now You're Ready
Exercises 11–16

Write the next three terms of the geometric sequence. Then graph the sequence.

1. 1, 3, 9, 27, . . . **2.** 64, 16, 4, 1, . . . **3.** 80, −40, 20, −10, . . .

Laurie's Notes

Introduction

Connect

- **Yesterday:** Students explored geometric sequences. (MP4, MP5)
- **Today:** Students will extend and graph geometric sequences. Students will also write an equation for the *n*th term of a geometric sequence.

Motivate

- Before students arrive, write each number of a geometric sequence on separate index cards (Example: 6, 12, 24, 48, 96). Hand an index card to each of the first 5 students that enter the room.
- Tell the 5 students that they are holding numbers in a geometric sequence and to position themselves in order.
- Students may realize that the order could be increasing or decreasing.

Version A: 6, 12, 24, 48, 96 **Version B:** 96, 48, 24, 12, 6

? "If 6 is in the 1st position, how can you generate the terms that follow?" multiply by 2

? "If 96 is in the 1st position, how can you generate the terms that follow?" multiply by $\frac{1}{2}$

? "Is this the same number you would multiply 48 by to get 24?" yes

Lesson Notes

Discuss

- Discuss that geometric sequences can be increasing, decreasing, or neither. It would be neither when the common ratio is negative. The terms alternate between positive and negative.

Key Idea

- Write the Key Idea on the board, highlighting the vocabulary.
- Connect the Key Idea to the index card motivation.

Example 1

- Explain that a table is used to organize the terms in a geometric sequence because it helps in recognizing patterns.
- Remind students that they can find the common ratio by finding the ratio of a term to the previous term.

Example 2

- Discuss using the notation n and a_n instead of x and y. The position is n and the value of the term at the nth position is a_n.
- **MP5 Use Appropriate Tools Strategically:** Notice that the scales in the graph are not the same. Use a graphing utility to display the graph (a) as shown in the book and (b) by using the same scale for each axis.

On Your Own

- Student may believe that the sequence in Question 3 is not a geometric sequence. They will say that the terms decrease by half each time, but the change of sign may confuse them.

Goal Today's lesson is writing equations for **geometric sequences**.

Start Thinking! and Warm Up

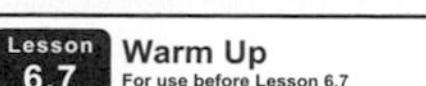

Lesson 6.7 Warm Up
For use before Lesson 6.7

Lesson 6.7 Start Thinking!
For use before Lesson 6.7

In an *arithmetic sequence*, each term is found by adding the same number to the previous term.

In a *geometric sequence*, each term is found by multiplying the previous term by the same number.

Come up with a way to remember how arithmetic and geometric sequences are different. Share your answer with your class.

Extra Example 1

Write the next three terms of the geometric sequence 3, 9, 27, 81, ...
243, 729, 2187

Extra Example 2

Graph the geometric sequence 80, 40, 20, 10, 5, ... What do you notice?

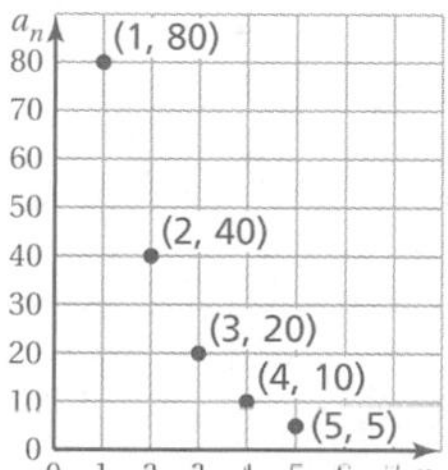

The points of the graph appear to lie on an exponential curve.

On Your Own

1. 81, 243, 729

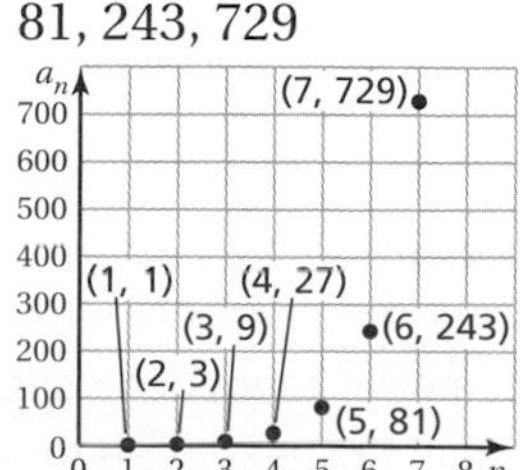

2–3. See Additional Answers

Extra Example 3

The table shows the results of clicking the *zoom-in* button on the mapping website in Example 3.

Zoom-in Clicks	1	2	3
Map Side Length (miles)	160	80	40

a. Write an equation for the nth term of the geometric sequence.
$a_n = 160(0.5)^{n-1}$

b. Find and interpret a_5. 10; The side length of the square map after 5 clicks is 10 miles.

On Your Own

4. 10 clicks

English Language Learners

Group Activity

Give groups of students index cards with geometric sequences and instruct them to find the next three terms as a group. This will give English language learners a chance to interact with classmates and communicate mathematically.

Laurie's Notes

Discuss

- It should not be a surprise to students that geometric sequences behave like exponential functions, except when the terms alternate signs as with Question 3 on the previous page. A geometric sequence is also a set of discrete data points, not a continuous curve.
- Write a table similar to what is shown to develop the connection between geometric sequences and exponential functions.
- Spend time talking about the nth term and why the number of times the common factor has been used is $n - 1$, not n. Using a simpler example may help.

  3 6 12 24 48

  The 5th term is 48 which is $3 \cdot 2 \cdot 2 \cdot 2 \cdot 2$. The common ratio is 2 and it has been used 4 times.

Key Idea

- Write the equation for a geometric sequence.
- **Connection:** Help students see the connection between this equation and the general form of an exponential equation.

Example 3

- Ask how many have used a mapping application on a computer or smartphone. Most students probably have so they will be familiar with the language used in the problem.
- Write the general equation for a geometric sequence.
- ? "What is the first term, meaning what is the side length of the map after 1 click?" 5 "What is the common ratio?" 2
- Substitute the known information to write the equation.
- Continue to work the problem as shown.

On Your Own

- **Think-Pair-Share:** Students should read the question independently and then work with a partner to answer the question. When they have answered the question, the pair should compare their answer with another group and discuss any discrepancies.

Closure

At the start of class two geometric sequences were formed.

Version A: 6, 12, 24, 48, 96 **Version B:** 96, 48, 24, 12, 6

Write an equation for each. $a_n = 6(2)^{n-1}$; $a_n = 96(0.5)^{n-1}$

Technology For the Teacher

The Dynamic Planning Tool
Editable Teacher's Resources at *BigIdeasMath.com*

Because consecutive terms of a geometric sequence change by equal factors, the points of any geometric sequence with a positive common ratio lie on an exponential curve. You can use the first term and the common ratio to write an exponential function that describes a geometric sequence.

Position, n	Term, a_n	Written using a_1 and r	Numbers
1	first term, a_1	a_1	1
2	second term, a_2	a_1r	$1 \cdot 5 = 5$
3	third term, a_3	a_1r^2	$1 \cdot 5^2 = 25$
4	fourth term, a_4	a_1r^3	$1 \cdot 5^3 = 125$
⋮	⋮	⋮	⋮
n	nth term, a_n	a_1r^{n-1}	$1 \cdot 5^{n-1}$

Key Idea

Study Tip

Notice that $a_n = a_1r^{n-1}$ is of the form $y = ab^x$.

Equation for a Geometric Sequence

Let a_n be the nth term of a geometric sequence with first term a_1 and common ratio r. The nth term is given by

$$a_n = a_1r^{n-1}.$$

EXAMPLE 3 Real-Life Application

Clicking the *zoom-out* button on a mapping website doubles the side length of the square map.

Zoom-out Clicks	1	2	3
Map Side Length (miles)	5	10	20

a. Write an equation for the *n*th term of the geometric sequence.

The first term is 5 and the common ratio is 2.

$a_n = a_1r^{n-1}$ Equation for a geometric sequence

$a_n = 5(2)^{n-1}$ Substitute 5 for a_1 and 2 for r.

b. Find and interpret a_8.

Use the equation to find the 8th term.

$a_n = 5(2)^{n-1}$ Write the equation.

$= 5(2)^{8-1}$ Substitute 8 for n.

$= 640$ Simplify.

The side length of the square map after 8 clicks is 640 miles.

On Your Own

Now You're Ready
Exercises 25–28

4. **WHAT IF?** After how many clicks on the *zoom-out* button is the side length of the map 2560 miles?

6.7 Exercises

Vocabulary and Concept Check

1. **WRITING** How are arithmetic sequences and geometric sequences different?
2. **REASONING** Compare and contrast the two sequences.

   2, 4, 6, 8, 10, . . .   2, 4, 8, 16, 32, . . .

3. **CRITICAL THINKING** Why do the points of a geometric sequence lie on an exponential curve only when the common ratio is positive?

Practice and Problem Solving

4. Enter 4 on a calculator. Multiply by 6 four times. Record your results in the table. Describe the pattern.

Step	1	2	3	4	5
Calculator Display					

Find the common ratio of the geometric sequence.

5. 3, −12, 48, −192, . . .
6. 200, 100, 50, 25, . . .
7. 7640, 764, 76.4, 7.64, . . .
8. 9, −18, 36, −72, . . .
9. 0.1, 0.9, 8.1, 72.9, . . .
10. 5, 1, $\frac{1}{5}$, $\frac{1}{25}$, . . .

Write the next three terms of the geometric sequence. Then graph the sequence.

1 2 11. 2, 10, 50, 250, . . .
12. −7, 14, −28, 56, . . .
13. 81, −27, 9, −3, . . .
14. −375, −75, −15, −3, . . .
15. 36, 6, 1, $\frac{1}{6}$, . . .
16. $\frac{1}{49}$, $\frac{1}{7}$, 1, 7, . . .

17. **ERROR ANALYSIS** Describe and correct the error in writing the next three terms of the geometric sequence.

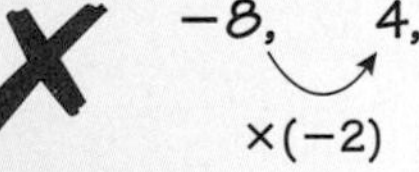
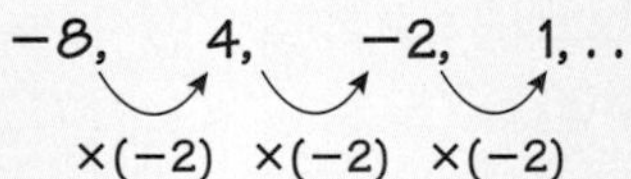

18. **BADMINTON** A badminton tournament begins with 128 teams. After the first round, 64 teams remain. After the second round, 32 teams remain. How many teams remain after the third, fourth, and fifth rounds?

Tell whether the sequence is *geometric, arithmetic,* or *neither*.

19. −8, 0, 8, 16, . . .
20. −1, 3, −5, 7, . . .
21. 1, 4, 9, 16, . . .
22. $\frac{3}{49}$, $\frac{3}{7}$, 3, 21, . . .
23. 192, 24, 3, $\frac{3}{8}$, . . .
24. −25, −18, −12, −7, . . .

Assignment Guide and Homework Check

Level	Assignment	Homework Check
Average	1–3, 5–27 odd, 18, 34–37	5, 11, 17, 18, 27
Advanced	1–3, 6–28 even, 17, 29–37	12, 17, 18, 28, 32

Common Errors

- **Exercises 5–16** Student may confuse geometric sequences with arithmetic sequences and try to find the common difference. Remind them that geometric sequences have a common ratio.
- **Exercises 11–16** Students may label the y-axis using terms of the sequence. Remind students that they must use equal intervals.
- **Exercises 19–24** Students may only check the relationship between two consecutive terms in a sequence to determine if it is arithmetic or geometric. Remind students to check the relationship for every term in the sequence.

6.7 Record and Practice Journal

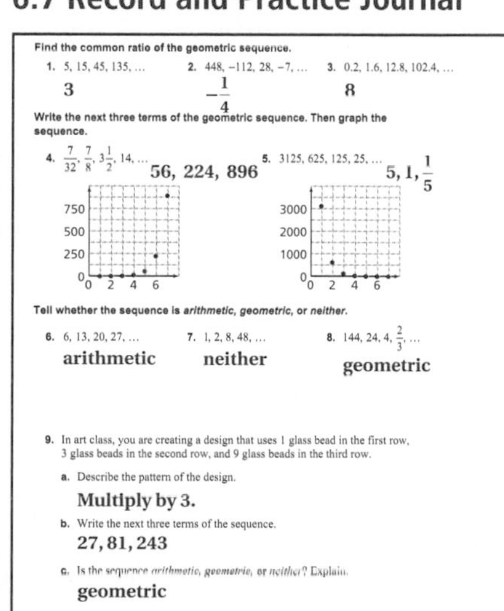
Find the common ratio of the geometric sequence.

1. 5, 15, 45, 135, … **3**
2. 448, −112, 28, −7, … $-\frac{1}{4}$
3. 0.2, 1.6, 12.8, 102.4, … **8**

Write the next three terms of the geometric sequence. Then graph the sequence.

4. $\frac{7}{32}, \frac{7}{8}, 3\frac{1}{2}, 14, \ldots$ **56, 224, 896**
5. 3125, 625, 125, 25, … $5, 1, \frac{1}{5}$

Tell whether the sequence is *arithmetic*, *geometric*, or *neither*.

6. 6, 13, 20, 27, … **arithmetic**
7. 1, 2, 8, 48, … **neither**
8. 144, 24, 4, $\frac{2}{3}$, … **geometric**

9. In art class, you are creating a design that uses 1 glass bead in the first row, 3 glass beads in the second row, and 9 glass beads in the third row.
   a. Describe the pattern of the design. **Multiply by 3.**
   b. Write the next three terms of the sequence. **27, 81, 243**
   c. Is the sequence *arithmetic*, *geometric*, or *neither*? Explain. **geometric**

Technology For the Teacher
Answer Presentation Tool

Vocabulary and Concept Check

1. *Sample answer:* You add a number to extend arithmetic sequences and you multiply by a number to extend geometric sequences.
2. *Sample answer:* Both have a first term of 2 and consist of even numbers. The first sequence is arithmetic with a common difference of 2, and the second sequence is geometric with a common ratio of 2.
3. When the common ratio is negative, a_n will alternate between being positive and negative. These points do not lie on an exponential curve.

Practice and Problem Solving

4.

Step	1	2	3	4	5
Calculator Display	4	24	144	864	5184

The numbers get larger. Each number is 6 times the previous number.

5. −4
6. 0.5
7. 0.1
8. −2
9. 9
10. $\frac{1}{5}$

11–18. See Additional Answers.

19. arithmetic
20. neither
21. neither
22. geometric
23. geometric
24. neither

Practice and Problem Solving

25. $a_n = (-5)^{n-1}$; 15,625
26. $a_n = 2(4)^{n-1}$; 8192
27. $a_n = 5(3)^{n-1}$; 3645
28. $a_n = 2(7)^{n-1}$; 235,298
29. **a.** $a_n = (6)^{n-1}$
    b. all positive integers; discrete
30. $a_n = 9(3)^{n-1}$; 59,049
31. dependent; the terms are the output of the function; the position is the independent variable
32. **a.** $a_n = 0.01(2)^{n-1}$
    b. \$167,772
    c. no; The student should have chosen to live on campus because \$167,772 is more like the cost of a house rather than the cost of room and board for less than 1 month.
33. See *Taking Math Deeper.*

Fair Game Review

34. $-4n$ **35.** $9x - 10$

36. $3y - 6$ **37.** C

Mini-Assessment

Write the next three terms of the geometric sequence.

1. −6, −12, −24, −48, . . . −96, −192, −384
2. 405, 135, 45, 15, . . . 5, $\frac{5}{3}$, $\frac{5}{9}$
3. 3, −9, 27, −81, . . . 243, −729, 2187

Taking Math Deeper

Exercise 33

This problem introduces important calculus concepts of *infinite sequences* and *limits.*

Understand the problem. Make a table. Note that 16 gallons is 16 × 128, or 2048 fluid ounces.

a.

Day, n	1	2	3	4	5
Expression	$\frac{3}{4}(2048)$	$\frac{3}{4}\left[\frac{3}{4}(2048)\right]$	$\frac{3}{4}\left[\frac{3}{4}\left(\frac{3}{4}\right)\right](2048)$	$\frac{3}{4}\left[\frac{3}{4}\left(\frac{3}{4}\right)\left(\frac{3}{4}\right)(2048)\right]$	$\frac{3}{4}\left[\frac{3}{4}\left(\frac{3}{4}\right)\left(\frac{3}{4}\right)\left(\frac{3}{4}\right)\right](2048)$
Amount, a_n (fluid ounces)	1536	1152	864	648	486

Note that at the end of each day, there is three-fourths the amount of soup there was at the end of the previous day. So, this is a geometric sequence with a common ration of $\frac{3}{4}$.

Write an equation.

The first term is 1536 and the common ratio is $\frac{3}{4}$.

$$a_n = a_1 r^{n-1}$$

b. $a_n = 1536\left(\frac{3}{4}\right)^{n-1}$

c. This question is challenging. The theoretical answer is that the soup will *never* be all gone. The "real-life" answer is that at a certain point, it is physically impossible to remove one-fourth of a quantity . . . so the process is not possible.

Reteaching and Enrichment Strategies

If students need help. . .	If students got it. . .
Resources by Chapter • Practice A and Practice B • Puzzle Time Record and Practice Journal Practice Differentiating the Lesson Lesson Tutorials Skills Review Handbook	Resources by Chapter • Enrichment and Extension • School-to-Work • Financial Literacy • Technology Connection • Life Connections Start the next section

Write an equation for the *n*th term of the geometric sequence. Then find a_7.

③ **25.** 1, −5, 25, −125, . . .

26. 2, 8, 32, 128, . . .

27.

n	1	2	3	4
a_n	5	15	45	135

28.

n	1	2	3	4
a_n	2	14	98	686

29. **CHAIN EMAIL** You start a chain email and send it to 6 friends. The process continues and each of your friends forwards the email to 6 people.

a. Write an equation for the *n*th term of the geometric sequence.

b. Describe the domain. Is the domain discrete or continuous?

30. **REASONING** What is the 9th term of a geometric sequence where $a_3 = 81$ and $r = 3$?

31. **PRECISION** Are the terms of a geometric sequence independent or dependent? Explain your reasoning.

32. **ROOM AND BOARD** A college student makes a deal with her parents to live at home instead of living on campus. She will pay her parents \$0.01 for the first day of the month, \$0.02 for the second day, \$0.04 for the third day, and so on.

a. Write an equation for the *n*th term of the geometric sequence.

b. What will she pay on the 25th day?

c. Did the student make a good choice or should she have chosen to live on campus? Explain.

33. Repeated Reasoning A soup kitchen makes 16 gallons of soup. Each day, a quarter of the soup is served and the rest is saved for the next day.

a. Write the first five terms of the sequence of the number of fluid ounces of soup left each day.

b. Write an equation to represent the sequence.

c. When is all the soup gone? Explain.

Fair Game Review What you learned in previous grades & lessons

Simplify the expression. *(Skills Review Handbook)*

34. $2n - 6n$

35. $2(4x - 5) + x$

36 $4(y - 1) - (y + 2)$

37. **MULTIPLE CHOICE** What is the solution of $6(3 - x) = -4x + 12$? *(Section 1.3)*

Ⓐ $x = -3$ Ⓑ $x = -2$ Ⓒ $x = 3$ Ⓓ $x = 6$

6.7b Recursively Defined Sequences

Key Vocabulary
recursive rule, *p. 312*

In Sections 5.6 and 6.7, you wrote *explicit* equations for sequences. Now, you will write *recursive* equations for sequences. A **recursive rule** gives the beginning term(s) of a sequence and an equation that indicates how any term a_n in the sequence relates to the previous term.

Key Idea

Recursive Equation for an Arithmetic Sequence

$a_n = a_{n-1} + d$, where d is the common difference.

Recursive Equation for a Geometric Sequence

$a_n = r \cdot a_{n-1}$, where r is the common ratio.

EXAMPLE 1 Writing Terms of Recursively Defined Sequences

Write the first six terms of each sequence. Then graph each sequence.

a. $\boldsymbol{a_1 = 2, a_n = a_{n-1} + 3}$

$a_1 = 2$

$a_2 = a_1 + 3 = 2 + 3 = 5$

$a_3 = a_2 + 3 = 5 + 3 = 8$

$a_4 = a_3 + 3 = 8 + 3 = 11$

$a_5 = a_4 + 3 = 11 + 3 = 14$

$a_6 = a_5 + 3 = 14 + 3 = 17$

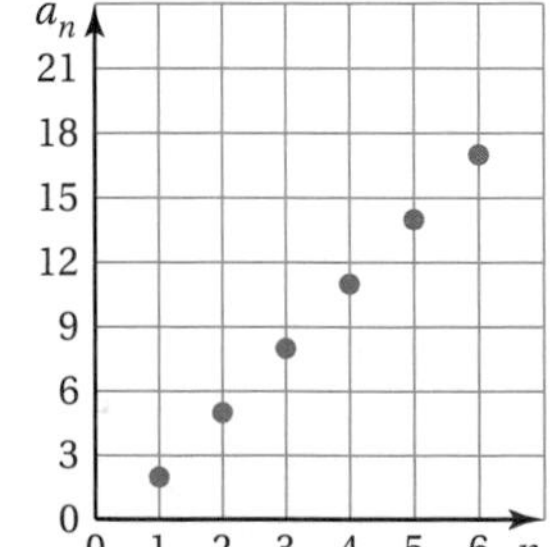

b. $\boldsymbol{a_1 = 1, a_n = 3a_{n-1}}$

$a_1 = 1$

$a_2 = 3a_1 = 3(1) = 3$

$a_3 = 3a_2 = 3(3) = 9$

$a_4 = 3a_3 = 3(9) = 27$

$a_5 = 3a_4 = 3(27) = 81$

$a_6 = 3a_5 = 3(81) = 243$

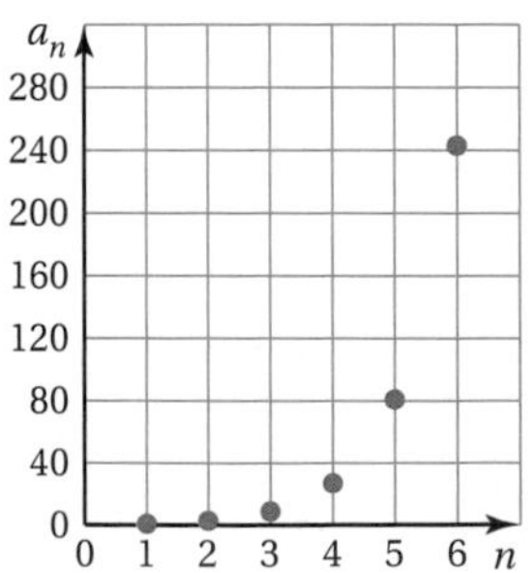

Practice

Write the first six terms of the sequence. Then graph the sequence.

1. $a_1 = 0, a_n = a_{n-1} - 8$

2. $a_1 = -7.5, a_n = a_{n-1} + 2.5$

3. $a_1 = -36, a_n = \frac{1}{2}a_{n-1}$

4. $a_1 = 0.7, a_n = 10a_{n-1}$

Multi-Language Glossary at BigIdeasMath.com.

Laurie's Notes

Introduction

Connect

- **Yesterday:** Students extended and graphed geometric sequences. Students also wrote an equation for the *n*th term of a geometric sequence. (MP5)
- **Today:** Students will write recursive rules for sequences.

Motivate

- Write the following sequence of numbers on the board: 0, 1, 1, 2, 3, 5, 8, . . . and ask students to continue the sequence. After many of the students are able to add to the sequence ask what the pattern is.
- These are the Fibonacci numbers. By definition, the first two Fibonacci numbers are 0 and 1 and each subsequent number is the sum of the previous two. This set of numbers is named after Leonardo of Pisa, who was known as Fibonacci. Fibonacci numbers are found in pine cones, pineapples, and leaf arrangements in plants.

? "How would you find the 20th Fibonacci number?" You need to know the 18th and 19th to find the 20th.

Lesson Notes

Discuss

- **MP1 Make Sense of Problems and Persevere in Solving Them** Recursive sequences are a new type of sequence for students to think about and work with. Conceptually, recursive rules can be simple to describe, however, the notation can be challenging for some students. The idea that you cannot determine the 12th term in the sequence without knowing the 11th term is also a new experience for students. The lesson is designed so that students can make connections to arithmetic and geometric sequences.

Key Idea

- Write each definition and share a few additional words to describe each. For the arithmetic sequence, the *n*th term is the previous term plus some common difference. For the geometric sequence, the *n*th term is the previous term times some common ratio.
- Explain the subscript notation. Students sometimes think that with a_{n-1} they are supposed to distribute and write $a_n - a$. Make sure they understand that a_{n-1} is the term before a_n and a_{n-3} would be the third term before a_n.

Goal Today's lesson is writing **recursive rules** for sequences.

Start Thinking! and Warm Up

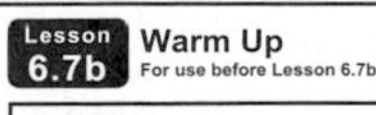

Lesson 6.7b **Start Thinking!** For use before Lesson 6.7b

Describe the sequence 5, 10, 20, 40, 80, 160, ...

You can write this as

$a_n = 2a_{n-1}$, where $a_1 = 5$.

When you write the equation, why does the value of a_1 need to be given?

Record and Practice Journal Practice

See Additional Answers.

Extra Example 1

Write the first six terms of each sequence. Then graph each sequence.

a. $a_1 = 5, a_n = a_{n-1} - 2$

5, 3, 1, −1, −3, −5

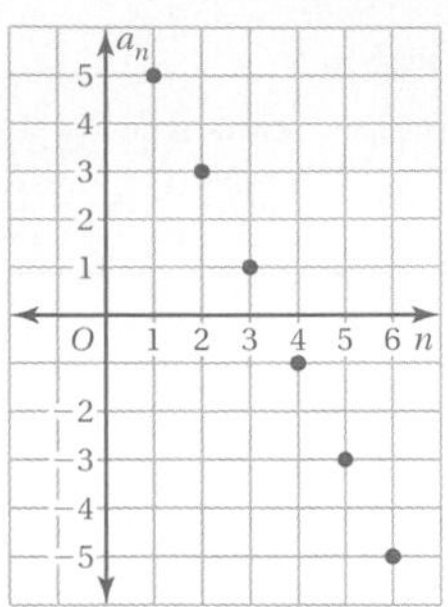

b. $a_1 = 0.5, a_n = 4a_{n-1}$

0.5, 2, 8, 32, 128, 512

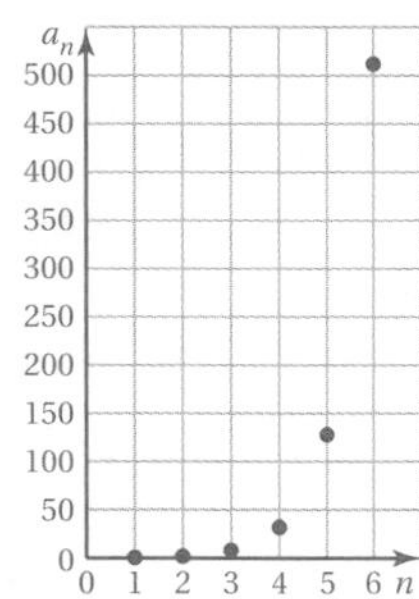

Extra Example 2

Write a recursive rule for each sequence.

a. −12, −5, 2, 9, 16, . . .

$a_1 = -12, a_n = a_{n-1} + 7$

b. 3, 18, 108, 648, 3888, . . .

$a_1 = 3, a_n = 6a_{n-1}$

Practice

1–4. See Additional Answers.

5. $a_1 = 8, a_n = a_{n-1} - 5$

6. $a_1 = 1.3, a_n = a_{n-1} - 1.3$

7. $a_1 = 4, a_n = 5a_{n-1}$

8. $a_1 = 1600,$
$a_n = -0.25a_{n-1}$

9. $a_1 = 2, a_n = a_{n-1} + 1.5$

Laurie's Notes

Example 1

- Write the first recursive rule. Tell the students that the first term is 2 and to find a term you add 3 to the previous term. Sometimes the language "seed" is used to refer to the first term. The first term is the seed and it gets us started.
- ? "If the first term is 2, how do we find the second term?" Add 3 to the first term. "How do we find the third term?" Add 3 more to the second term.
- After generating the sequence and plotting the points, anticipate that students will be confused why this linear graph is not just written as $y = 3x - 1$, for $x = 1, 2, 3, \ldots$. It could be, if we wanted to write an explicit equation, which we will do in Example 3.
- Write the second recursive rule. Tell students that the first term is 1 and to find a term you multiply the previous term by 3.
- ? "If the first term is 1, how do we find the second term?" Multiply the first term by 3. "How do we find the third term?" Multiply the second term by 3.
- The terms of this recursive sequence are the powers of 3 and is equivalent to the exponential equation $y = 3^x$ for $x = 1, 2, 3, \ldots$.
- **Connection**: If the recursive rule describes an arithmetic or geometric sequence, the explicit form would be a linear or exponential function with a discrete domain of $x = 1, 2, 3, \ldots$.

Practice

- Exercise 1 is a decreasing arithmetic sequence and Exercise 3 is a decreasing geometric sequence.

Discuss

- A second skill after generating the terms of a recursively defined sequence is to write the recursive rule when the terms are known.

Example 2

- Write the sequence and organize it in a table.
- ? "Is the sequence arithmetic or geometric?" arithmetic "How do you know?" There is a common difference of 12.
- Work the problem as shown. Be sure that the students understand what the first term is and what the recursive rule for the sequence is.
- Write the second sequence and organize it in a table.
- ? "Is the sequence arithmetic or geometric?" geometric "How do you know?" There is a common ratio of $\frac{1}{5}$.
- Work the problem as shown. Remind students to state the first term and the recursive rule.

Practice

- Have each student work with a partner.
- Discuss the solutions as a whole class.

EXAMPLE 2 Writing Recursive Rules

Write a recursive rule for each sequence.

a. −30, −18, −6, 6, 18, . . .

Use a table to organize the terms and find the pattern.

Position	1	2	3	4	5
Term	−30	−18	−6	6	18

+ 12 + 12 + 12 + 12

The sequence is arithmetic with first term −30 and common difference 12.

$a_n = a_{n-1} + d$ Recursive equation (arithmetic)

$a_n = a_{n-1} + 12$ Substitute 12 for *d*.

So, a recursive rule for the sequence is $a_1 = -30$, $a_n = a_{n-1} + 12$.

b. 500, 100, 20, 4, 0.8, . . .

Use a table to organize the terms and find the pattern.

Position	1	2	3	4	5
Term	500	100	20	4	0.8

$\times\frac{1}{5}$ $\times\frac{1}{5}$ $\times\frac{1}{5}$ $\times\frac{1}{5}$

The sequence is geometric with first term 500 and common ratio $\frac{1}{5}$.

$a_n = r \cdot a_{n-1}$ Recursive equation (geometric)

$a_n = \frac{1}{5}a_{n-1}$ Substitute $\frac{1}{5}$ for *r*.

So, a recursive rule for the sequence is $a_1 = 500$, $a_n = \frac{1}{5}a_{n-1}$.

Practice

Write a recursive rule for the sequence.

5. 8, 3, −2, −7, −12, . . .

6. 1.3, 2.6, 3.9, 5.2, 6.5, . . .

7. 4, 20, 100, 500, 2500, . . .

8. 1600, −400, 100, −25, 6.25, . . .

9. **SUNFLOWERS** Write a recursive rule for the height of the sunflower over time.

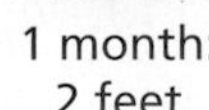

1 month: 2 feet
2 months: 3.5 feet
3 months: 5 feet
4 months: 6.5 feet

EXAMPLE 3 Translating Recursive Rules into Explicit Equations

Write an explicit equation for each recursive rule.

a. $a_1 = 25, a_n = a_{n-1} - 10$

The recursive rule represents an arithmetic sequence with first term 25 and common difference -10.

$a_n = a_1 + (n - 1)d$	Equation for an arithmetic sequence
$a_n = 25 + (n - 1)(-10)$	Substitute 25 for a_1 and -10 for d.
$a_n = -10n + 35$	Simplify.

b. $a_1 = 19.6, a_n = -0.5a_{n-1}$

The recursive rule represents a geometric sequence with first term 19.6 and common ratio -0.5.

$a_n = a_1r^{n-1}$	Equation for a geometric sequence
$a_n = 19.6(-0.5)^{n-1}$	Substitute 19.6 for a_1 and -0.5 for r.

EXAMPLE 4 Translating Explicit Equations into Recursive Rules

Write a recursive rule for each explicit equation.

a. $a_n = -2n + 3$

The explicit equation represents an arithmetic sequence with first term $-2(1) + 3 = 1$ and common difference -2.

$a_n = a_{n-1} + d$	Recursive equation (arithmetic)
$a_n = a_{n-1} + (-2)$	Substitute -2 for d.

So, a recursive rule for the sequence is $a_1 = 1, a_n = a_{n-1} - 2$.

b. $a_n = -3(2)^{n-1}$

The explicit equation represents a geometric sequence with first term -3 and common ratio 2.

$a_n = r \cdot a_{n-1}$	Recursive equation (geometric)
$a_n = 2a_{n-1}$	Substitute 2 for r.

So, a recursive rule for the sequence is $a_1 = -3, a_n = 2a_{n-1}$.

Practice

Write an explicit equation for the recursive rule.

10. $a_1 = -45, a_n = a_{n-1} + 20$

11. $a_1 = 13, a_n = -3a_{n-1}$

Write a recursive rule for the explicit equation.

12. $a_n = -n + 1$

13. $a_n = -2.5(2)^{n-1}$

Laurie's Notes

Discuss

- A third skill is to write explicit equations for the recursive rules. If students have made the connection between arithmetic sequences and linear functions, and between geometric sequences and exponential functions, these problems should make sense.

Example 3

- Take time to generate the first few terms of the recursively defined sequence.
- Write the recursive equation for an arithmetic sequence. Substitute the known information and simplify.
- Generate the first few terms of the arithmetic sequence to verify that the results are the same.
- Repeat the same process for part (b), the geometric sequence.
- Note that the sign of the common ratio (negative) causes the terms to alternate signs.

Example 4

? "Do you think it is possible to begin with an explicit equation and rewrite it into a recursive rule?" Students may or may not have a sense about this question.

? "In part (a), what are the first few terms of the sequence $a_n = -2n + 3$?" $1, -1, -3, -5, \ldots$

? "What is the first term?" 1 "Does the explicit equation represent an arithmetic or a geometric sequence?" arithmetic "What is the common difference?" -2

? "Do you have enough information to write a recursive rule?" yes

? "In part (b), what are the first few terms of the sequence $a_n = -3(2)^{n-1}$?" $-3, -6, -12, -24, \ldots$

? "What is the first term?" -3 "Does the explicit equation represent an arithmetic or a geometric sequence?" geometric "What is the common ratio?" 2

? "Do you have enough information to write a recursive rule?" yes

Practice

- **Think-Pair-Share:** Students should read each question independently and then work with a partner to answer the questions. When they have answered the questions, the pair should compare their answers with another group and discuss any discrepancies.

Extra Example 3

Write an explicit equation for each recursive rule.

a. $a_1 = 9, a_n = a_{n-1} + 8$
$a_n = 8n + 1$

b. $a_1 = 12.5, a_n = 18a_{n-1}$
$a_n = 12.5(18)^{n-1}$

Extra Example 4

Write a recursive rule for each explicit equation.

a. $a_n = 14n - 8$
$a_1 = 6, a_n = a_{n-1} + 14$

b. $a_n = 4(5)^{n-1}$
$a_1 = 4, a_n = 5a_{n-1}$

Practice

10. $a_n = 20n - 65$

11. $a_n = 13(-3)^{n-1}$

12. $a_1 = 0, a_n = a_{n-1} - 1$

13. $a_1 = -2.5, a_n = 2a_{n-1}$

14. $a_1 = 5, a_2 = 6,$
$a_n = a_{n-2} + a_{n-1}$;
45, 73, 118

15. $a_1 = -3, a_2 = -4,$
$a_n = a_{n-2} + a_{n-1}$;
$-29, -47, -76$

16. $a_1 = 1, a_2 = 1,$
$a_n = a_{n-1} - a_{n-2}$;
$0, -1, -1$

17. $a_1 = 4, a_2 = 3,$
$a_n = a_{n-2} - a_{n-1}$;
$7, -11, 18$

Extra Example 5

Write a recursive rule for the sequence 10, 12, −2, 14, −16, 30, ... Then write the next 3 terms of the sequence.

$a_1 = 10$, $a_2 = 12$, $a_n = a_{n-2} - a_{n-1}$; −46, 76, −122

Practice

18. $a_1 = 2, a_2 = 3,$
$a_n = a_{n-2} \cdot a_{n-1};$
1944, 209,952

19. $a_1 = -2, a_2 = 2.5,$
$a_n = a_{n-2} \cdot a_{n-1};$
−781.25, −48,828.125

20. See Additional Answers.

Mini-Assessment

1. Write the first six terms of the sequence $a_1 = -10$, $a_n = a_{n-1} + 3$. Then graph the sequence.

−10, −7, −4, −1, 2, 5

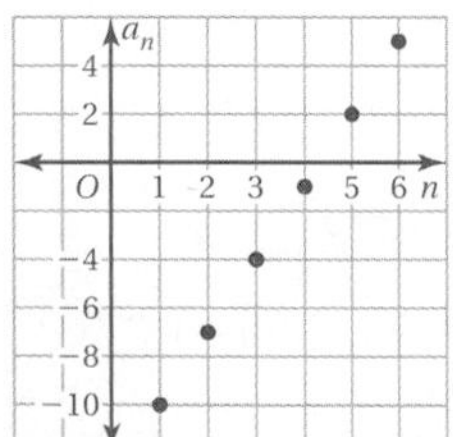

2. Write a recursive rule for the sequence 729, −243, 81, −27, 9, ...

$a_1 = 729$, $a_n = -\frac{1}{3}a_{n-1}$

3. Write an explicit equation for the recursive rule $a_1 = 22$, $a_n = 6a_{n-1}$. $a_n = 22(6)^{n-1}$

4. Write a recursive rule for the explicit equation $a_n = -11n - 2$.

$a_1 = -13$, $a_n = a_{n-1} - 11$

5. Write a recursive rule for the sequence 1, 3, 6, 10, 15, ... Then write the next 3 terms of the sequence.

$a_1 = 1$, $a_n = a_{n-1} + n$; 21, 28, 36

Laurie's Notes

Example 5

- The Fibonacci numbers were discussed at the beginning of class. Review how the sequence of numbers is generated: each number in the sequence, after the first two, can be found by adding the two previous terms.
- **FYI:** The photo shows an example of Fibonacci numbers in nature. There are 13 green spirals and 21 blue spirals in the flower shown.
- This is a different type of recursive rule because it is not arithmetic or geometric. The fact that it is recursive to begin with is obvious to students because they were able to extend the sequence by recognizing the pattern.
- Writing the recursive rule for this sequence should not be a problem for students at this point in the lesson.
- **Extension:** There are many different opportunities for students to explore Fibonacci numbers further. The connection of the Fibonacci numbers to the Golden Ratio is a popular topic.

Practice

- **MP1:** Exercises 14–17 give students a chance to make sense of problems and preserve in solving them. They may not immediately see a pattern in the numbers. After a period of time, you could offer one additional number in the sequence and that may be enough for students to make a connection.
- Exercises 18 and 19 involve products as stated in the direction line.
- Additional information about the Fibonacci spiral can be found on the Internet. Also, students may find the Fibonacci spiral and its connection to shells interesting.

Closure

- Create a recursive arithmetic sequence.
  $a_1 =$ ________ and $a_n =$ ________
- Create a recursive geometric sequence.
  $a_1 =$ ________ and $a_n =$ ________

Technology For the Teacher

The Dynamic Planning Tool
Editable Teacher's Resources at *BigIdeasMath.com*

You can write recursive rules for sequences that are neither arithmetic nor geometric. One way is to look for patterns in the sums of consecutive terms.

EXAMPLE 5 Writing Recursive Rules for Other Sequences

Write a recursive rule for the sequence 1, 1, 2, 3, 5, 8, ... Then write the next 3 terms of the sequence.

The sequence does not have a common difference or a common ratio. Find the sums of consecutive terms.

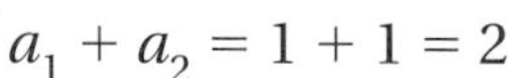

$a_1 + a_2 = 1 + 1 = 2$ 2 is the third term.

$a_2 + a_3 = 1 + 2 = 3$ 3 is the fourth term.

$a_3 + a_4 = 2 + 3 = 5$ 5 is the fifth term.

$a_4 + a_5 = 3 + 5 = 8$ 8 is the sixth term.

So, a recursive equation for the sequence is $a_n = a_{n-2} + a_{n-1}$. Use the equation to find the next three terms.

$a_7 = a_5 + a_6 = 5 + 8 = 13$ $a_8 = a_6 + a_7 = 8 + 13 = 21$ $a_9 = a_7 + a_8 = 13 + 21 = 34$

The sequence in Example 5 is called the *Fibonacci sequence.* This pattern is naturally occurring in many objects, such as flowers.

A recursive rule for the sequence is $a_1 = 1$, $a_2 = 1$, $a_n = a_{n-2} + a_{n-1}$. The next three terms are 13, 21, and 34.

Practice

Write a recursive rule for the sequence. Then write the next 3 terms of the sequence.

14. 5, 6, 11, 17, 28, . . .

15. −3, −4, −7, −11, −18, . . .

16. 1, 1, 0, −1, −1, 0, 1, 1, . . .

17. 4, 3, 1, 2, −1, 3, −4, . . .

Use a pattern in the products of consecutive terms to write a recursive rule for the sequence. Then write the next 2 terms of the sequence.

18. 2, 3, 6, 18, 108, . . .

19. −2, 2.5, −5, −12.5, 62.5, . . .

20. GEOMETRY Consider squares 1–6 in the diagram.

a. Write a sequence in which each term a_n is the side length of square n.

b. What is the name of this sequence? What is the next term of this sequence?

c. Use the term in part (b) to add another square to the diagram and extend the spiral.

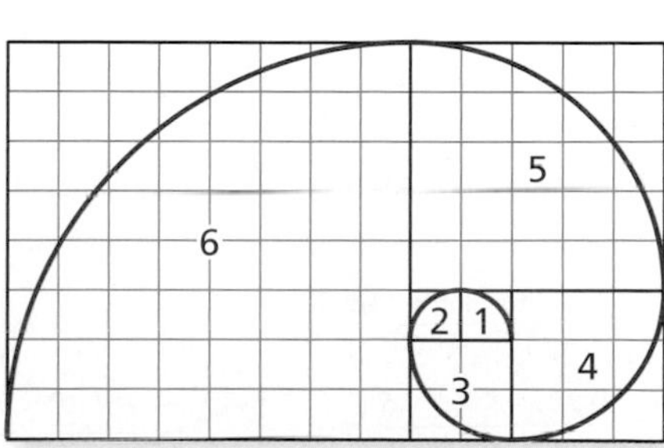

6.4–6.7 Quiz

Check It Out
Progress Check
BigIdeasMath.com

Does the table represent a *linear* or an *exponential* function? Explain. *(Section 6.4)*

1.

x	1	2	3	4
y	5	10	15	20

2.

x	2	4	6	8
y	5	10	20	40

Graph the function. Describe the domain and range. *(Section 6.4)*

3. $y = 5^x$

4. $y = -2\left(\frac{1}{6}\right)^x$

Solve the equation. Check your solution, if possible. *(Section 6.4)*

5. $8^{x+2} = 64^{4x+1}$

6. $7^{2x-6} = 49^{3x-11}$

Determine whether the table represents an *exponential growth function*, an *exponential decay function*, or *neither*. *(Section 6.6)*

7.

x	0	1	2	3
y	7	21	63	189

8.

x	1	2	3	4
y	14,641	1331	121	11

Write the next three terms of the geometric sequence. Then graph the sequence. *(Section 6.7)*

9. 15, −45, 135, −405, . . .

10. 768, 192, 48, 12, . . .

Write a recursive rule for the sequence. *(Section 6.7)*

11. 5, 11, 17, 23, . . .

12. −14, 28, −56, 112, . . .

13. **SAVINGS ACCOUNT** You deposit \$2500 in a savings account that earns 6% annual interest compounded yearly. *(Section 6.5)*

    a. Write and graph a function that represents the balance y (in dollars) after t years.

    b. What is the balance after 5 years?

14. **CURRENCY** A country's base unit of currency is valued at US\$2. The country's base unit of currency loses about 3.9% of its value every month. *(Section 6.6)*

    a. Write a function that represents the value y (in U.S. dollars) of the base unit of currency after t months.

    b. What is the value of the country's base unit of currency after 1.5 years?

Alternative Assessment Options

Math Chat	Student Reflective Focus Question
Structured Interview	Writing Prompt

Math Chat

- Have individual students work problems from the quiz on the board. The student explains the process used and justifies each step. Students in the class ask questions of the student presenting.
- The teacher explores the thought process of the student presenting, but does not teach or ask leading questions.

Study Help Sample Answers

Remind students to complete Graphic Organizers for the rest of the chapter.

8.

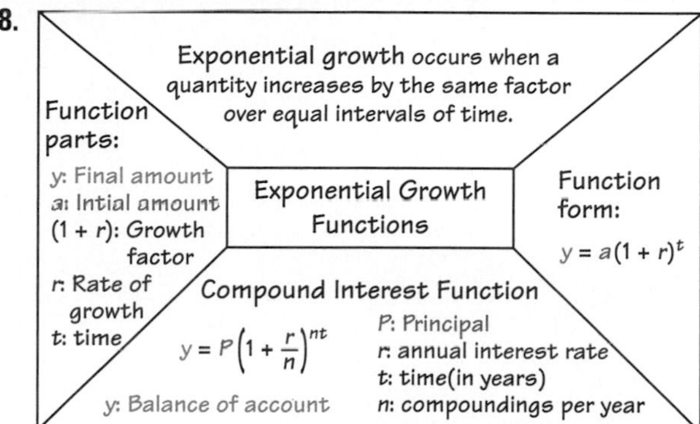

9.

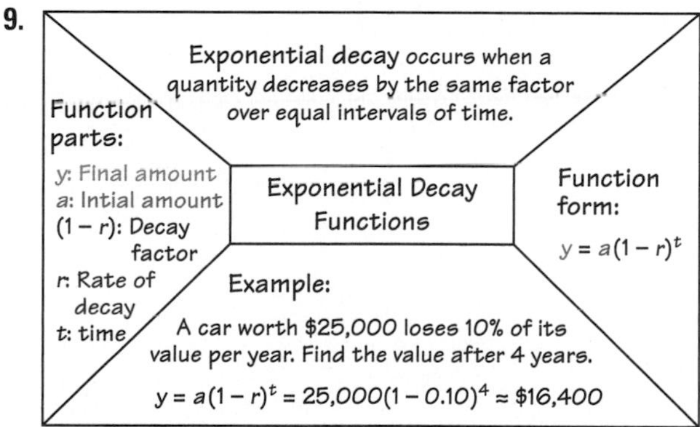

10. Available at *BigIdeasMath.com.*

Reteaching and Enrichment Strategies

If students need help. . .	If students got it. . .
Resources by Chapter • Study Help • Practice A and Practice B • Puzzle Time Lesson Tutorials *BigIdeasMath.com* Practice Quiz Practice from the Test Generator	Resources by Chapter • Enrichment and Extension • School-to-Work Game Closet at *BigIdeasMath.com* Start the Chapter Review

Answers

1. linear

2. exponential

3–4. See Additional Answers.

5. $x = 0$ **6.** $x = 4$

7. exponential growth function

8. exponential decay function

9–10. See Additional Answers.

11. $a_1 = 5, a_n = a_{n-1} + 6$

12. $a_1 = -14, a_n = (-2)a_{n-1}$

13. a. $y = 2500(1.06)^t$

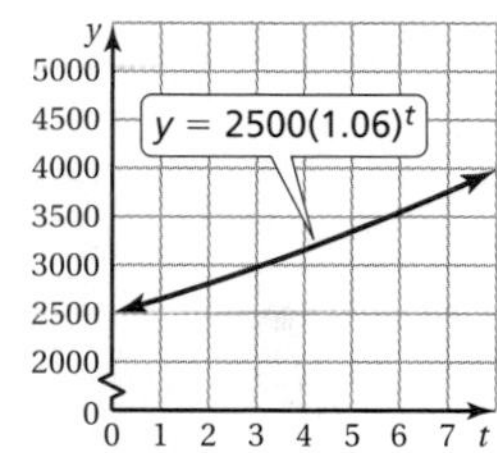

b. $3345.56

14. a. $y = 2(0.961)^t$

b. about US$0.98

Technology For the Teacher

Answer Presentation Tool

Assessment Book

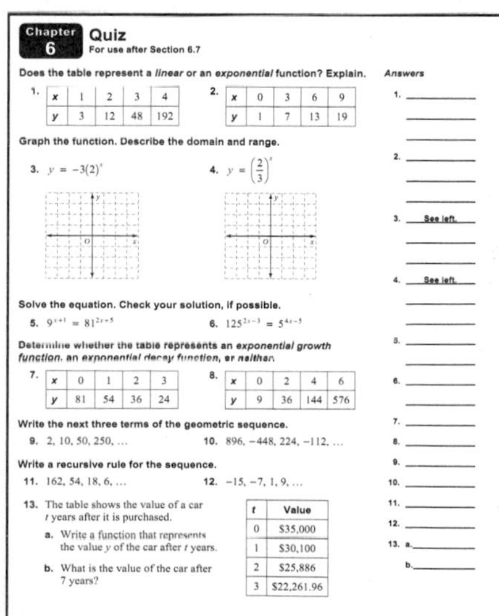

Chapter 6 **Quiz**
For use after Section 6.7

Does the table represent a *linear* or an *exponential* function? Explain.

1.

x	1	2	3	4
y	3	12	48	192

2.

x	0	3	6	9
y	1	7	13	19

Graph the function. Describe the domain and range.

3. $y = -3(2)^x$ 4. $y = \left(\frac{2}{3}\right)^x$

Solve the equation. Check your solution, if possible.

5. $9^{x+1} = 81^{2x+5}$ 6. $125^{2x-3} = 5^{4x-5}$

Determine whether the table represents an *exponential growth function*, an *exponential decay function*, or *neither*.

7.

x	0	1	2	3
y	81	54	36	24

8.

x	0	2	4	6
y	9	36	144	576

Write the next three terms of the geometric sequence.

9. 2, 10, 50, 250, … 10. 896, −448, 224, −112, …

Write a recursive rule for the sequence.

11. 162, 54, 18, 6, … 12. −15, −7, 1, 9, …

13. The table shows the value of a car t years after it is purchased.

a. Write a function that represents the value y of the car after t years.

b. What is the value of the car after 7 years?

t	Value
0	$35,000
1	$30,100
2	$25,886
3	$22,261.96

Answers

1. ____
2. ____
3. See left.
4. See left.
5. ____
6. ____
7. ____
8. ____
9. ____
10. ____
11. ____
12. ____
13. a. ____ b. ____

For the Teacher
Additional Review Options

- Big Ideas Test Generator
- Game Closet at *BigIdeasMath.com*
- Vocabulary Puzzle Builder
- Resources by Chapter
    Puzzle Time
    Study Help

Answers

1. $4\sqrt{6}$

2. $2\sqrt{2}$

3. $4 + \sqrt{3}$

Review of Common Errors

- **Exercises 1–3** Students may leave a perfect square factor in the radicand, or they may move the wrong factor outside the radical. Remind students to factor the radicand using its greatest perfect square factor, and bring the square root of that factor outside the radical.
- **Exercise 2** Students may rewrite the square root of a sum or difference as the sum or difference of square roots. Remind them that there are no properties for addition and subtraction corresponding to the product and quotient properties of square roots.
- **Exercise 3** Students may fail to divide out common factors from the numerator and denominator. Remind them to simplify the radicand first, then look for common factors in the numerator and denominator.

6 Chapter Review

Check It Out
Vocabulary Help
BigIdeasMath.com

Review Key Vocabulary

closed, *p. 266*
nth root, *p. 278*
exponential function, *p. 286*
exponential growth, *p. 296*
exponential growth function, *p. 296*
compound interest, *p. 297*
exponential decay, *p. 302*
exponential decay function, *p. 302*
geometric sequence, *p. 308*
common ratio, *p. 308*
recursive rule, *p. 312*

Review Examples and Exercises

6.1 Properties of Square Roots *(pp. 260–267)*

Evaluate $\sqrt{b^2 - 4ac}$ when $a = -2$, $b = 2$, and $c = 5$.

$$\sqrt{b^2-4ac} = \sqrt{2^2 - 4(-2)(5)} \quad \text{Substitute.}$$
$$= \sqrt{44} \quad \text{Simplify.}$$
$$= \sqrt{4 \cdot 11} \quad \text{Factor.}$$
$$= \sqrt{4} \cdot \sqrt{11} \quad \text{Product Property of Square Roots}$$
$$= 2\sqrt{11} \quad \text{Simplify.}$$

Exercises

Evaluate the expression when $x = 3$, $y = 4$, and $z = 2$.

1. $\sqrt{xy^2z}$

2. $\sqrt{2z + y}$

3. $\dfrac{8 + \sqrt{xy}}{z}$

6.2 Properties of Exponents *(pp. 268–275)*

Simplify $\left(\dfrac{3x}{4}\right)^{-4}$. Write your answer using only positive exponents.

$$\left(\frac{3x}{4}\right)^{-4} = \frac{(3x)^{-4}}{4^{-4}} \quad \text{Power of a Quotient Property}$$
$$= \frac{4^4}{(3x)^4} \quad \text{Definition of negative exponent}$$
$$= \frac{4^4}{3^4x^4} \quad \text{Power of a Product Property}$$
$$= \frac{256}{81x^4} \quad \text{Simplify.}$$

Exercises

Simplify. Write your answer using only positive exponents.

4. $y^3 \cdot y^{-3}$ 5. $\dfrac{x^4}{x^7}$ 6. $(xy^2)^3$ 7. $\left(\dfrac{2x}{5y}\right)^{-2}$

6.3 Radicals and Rational Exponents *(pp. 276–281)*

Simplify each expression.

a. $\sqrt[3]{512} = \sqrt[3]{8 \cdot 8 \cdot 8} = 8$ Rewrite and simplify.

b. $900^{1/2} = \sqrt{900}$ Write the expression in radical form.

$= \sqrt{30 \cdot 30}$ Rewrite.

$= 30$ Simplify.

Exercises

Simplify the expression.

8. $\sqrt[3]{8}$ 9. $64^{1/2}$ 10. $625^{3/4}$

6.4 Exponential Functions *(pp. 284–293)*

a. Graph $y = 4^x$.

Step 1: Make a table of values.

x	−1	0	1	2	3
y	0.25	1	4	16	64

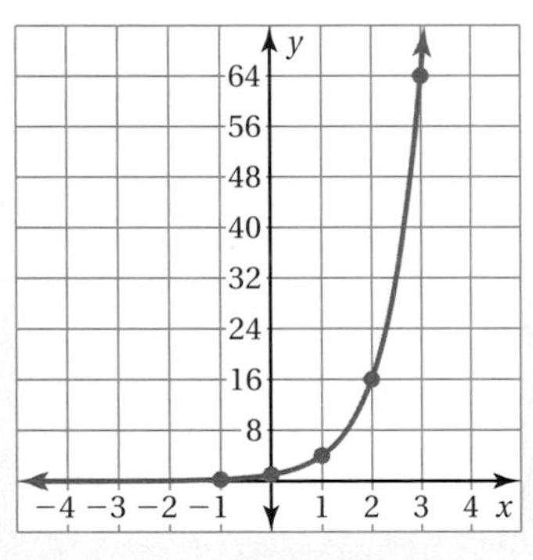

Step 2: Plot the ordered pairs.

Step 3: Draw a smooth curve through the points.

b. Write an exponential function represented by the graph.

Use the graph to make a table of values.

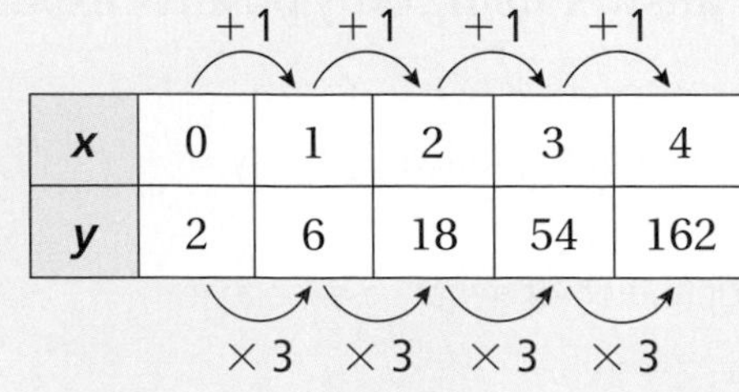

x	0	1	2	3	4
y	2	6	18	54	162

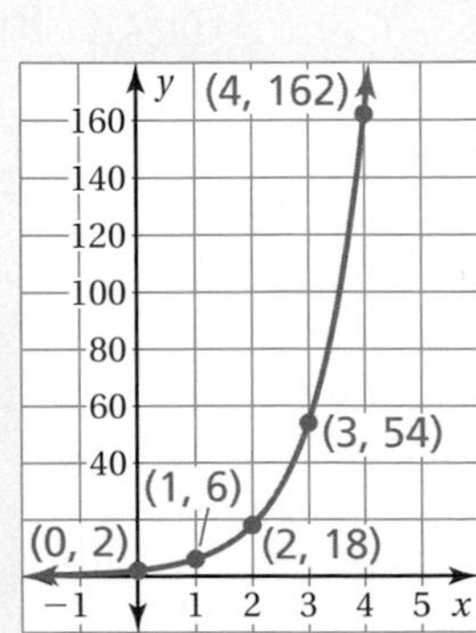

The y-intercept is 2 and the y-values increase by a factor of 3 as x increases by 1.

So, the exponential function is $y = 2(3)^x$.

Review of Common Errors (continued)

- **Exercise 4** Students may multiply the exponents instead of adding the exponents. Remind of the Product of Powers Property.
- **Exercise 5** Students may divide the exponents instead of subtracting the exponents. Remind them of the Quotient of Powers Property.
- **Exercise 6** Students may add the exponents instead of multiplying the exponents. Remind them of the Power of a Power Property.
- **Exercise 7** Students may forget to apply the exponent to the denominator. Remind them of the Power of a Quotient Property.
- **Exercise 10** Students may solve by rewriting $a^{b/n}$ as $\frac{a^b}{a^n}$. Tell them to rewrite the exponent b/n as $\frac{1}{n} \cdot b$. Then find the nth root of a and apply the exponent b to the result.

Answers

4. 1
5. $\frac{1}{x^3}$
6. x^3y^6
7. $\frac{25y^2}{4x^2}$
8. 2
9. 8
10. 125

Answers

11. 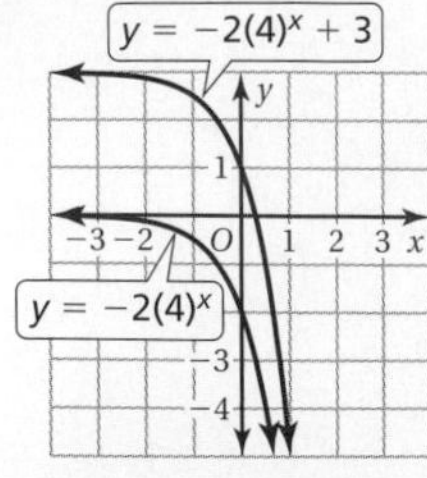

domain: all real numbers; range: all real numbers less than 3; vertical translation 3 units up of $y = -2(4)^x$

12. $y = 3(2)^x$

13. $y = 2\left(\frac{1}{2}\right)^x$

14. $x = 3$

15. no solution

16. $x = 12$

17. a. $y = 22(1.03)^t$

    b. $27.87 per hour

Review of Common Errors (continued)

- **Exercises 12 and 13** Students may write incorrect equations. Remind them that in an exponential equation of the form $y = ab^x$, a is the value of y when $x = 0$, and b is the factor that y changes by for each 1-unit increase in x.
- **Exercise 16** Students may fail to apply the Distributive Property correctly when applying the Power of a Power Property. Encourage them to first rewrite the expression using the product of the exponents, so they will see that the Distributive Property applies.
- **Exercises 18 and 19** Students may have difficulty determining the factor used to increase or decrease the dependent variable. Encourage them to look for patterns.
- **Exercise 20** Students may forget to change the percents (rates of decay) to decimals before subtracting them from 1 to find the decay factor.
- **Exercises 21 and 22** Students may label the y-axis using terms of the sequence. Remind students that they must use equal intervals.
- **Exercise 24** Students may attach the negative sign to a_1 instead of r. Explain that the terms of a geometric sequence alternate in sign when the common ratio r is negative.
- **Exercises 25 and 26** Students may not include the first term a_1. Explain that using a recursive rule to find a term a_n in these sequences depends on knowing the previous term, so the first term needs to be given.

Exercises

11. Graph $y = -2(4)^x + 3$. Describe the domain and range. Compare the graph to the graph of $y = -2(4)^x$.

Write an exponential function represented by the graph or table.

12.

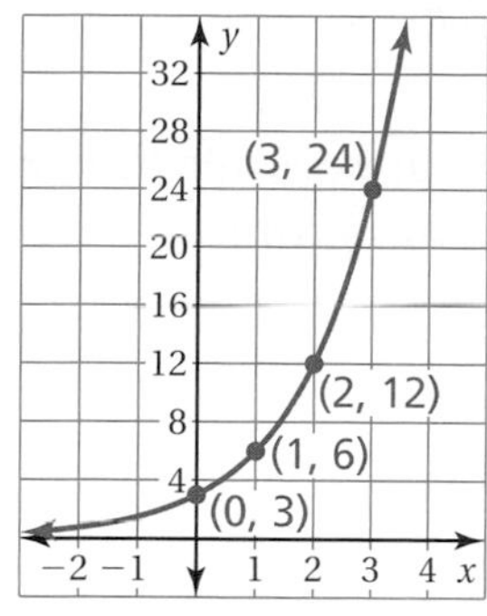

13.

x	0	1	2	3
y	2	1	0.5	0.25

Solve the equation. Check your solution, if possible.

14. $3^x = 27$

15. $5^x = 5^{x-2}$

16. $2^{5x} = 8^{2x-4}$

6.5 Exponential Growth *(pp. 294–299)*

The enrollment at a high school increases by 4% each year. In 2010, there were 800 students enrolled at the school.

a. Write a function that represents the enrollment *y* of the high school after *t* years.

$y = a(1 + r)^t$ — Write exponential growth function.

$y = 800(1 + 0.04)^t$ — Substitute 800 for *a* and 0.04 for *r*.

$y = 800(1.04)^t$ — Simplify.

b. How many students will be enrolled at the high school in 2020?

The value $t = 10$ represents 2020.

$y = 800(1.04)^t$ — Write exponential growth function.

$y = 800(1.04)^{10}$ — Substitute 10 for *t*.

≈ 1184 — Use a calculator.

Exercises

17. PLUMBER A plumber charges $22 per hour. The hourly rate increases by 3% each year.

a. Write a function that represents the plumber's hourly rate *y* (in dollars) after *t* years.

b. What is the plumber's hourly rate after 8 years?

6.6 Exponential Decay (pp. 300–305)

The table shows the value of a boat over time.

Year, t	0	1	2	3
Value, y	\$6000	\$4800	\$3840	\$3072

a. Determine whether the table represents an *exponential growth function*, an *exponential decay function*, or *neither*.

+1 +1 +1

Year, t	0	1	2	3
Value, y	\$6000	\$4800	\$3840	\$3072

× 0.8 × 0.8 × 0.8

∴ As x increases by 1, y is multiplied by 0.8. So, the table represents an exponential decay function.

b. The boat loses 20% of its value every year. Write a function that represents the value y (in dollars) of the boat after t years.

$y = a(1 - r)^t$ Write exponential decay function.

$y = 6000(1 - 0.2)^t$ Substitute 6000 for a and 0.2 for r.

$y = 6000(0.8)^t$ Simplify.

c. Graph the function from part (b). Use the graph to estimate the value of the boat after 8 years.

From the graph, you can see that the y-value is about 1000 when $t = 8$.

∴ So, the value of the boat is about \$1000 after 8 years.

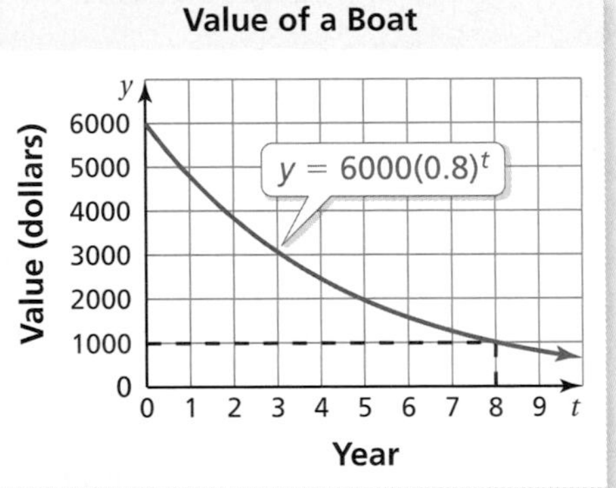

Exercises

Determine whether the table represents an *exponential growth function*, an *exponential decay function*, or *neither*.

18.

x	0	1	2	3
y	3	6	12	24

19.

x	1	2	3	4
y	162	108	72	48

20. DISCOUNT The price of a TV is \$1500. The price decreases by 6% each month. Write and graph a function that represents the price y (in dollars) of the TV after t months. Use the graph to estimate the price of the TV after 1 year.

Review Game

Using Properties of Exponents

Materials

- pencil
- paper

Directions

- Divide the class into two groups.
- Each group works together to write a set of problems (with answers) of the types below, using numbers in place of the variables (except for x).

  Simplify.

  - $\sqrt{a}$
  - $\sqrt{\frac{a}{b}}$
  - $\sqrt[n]{a}$
  - $a^{b/c}$

  Simplify. Write your answer using only positive exponents.

  - $a^b \cdot a^c$
  - $(a^b)^c$
  - $a^{b/c}$
  - $\frac{a^b}{a^c}$
  - $\left(\frac{a}{b}\right)^c$
  - Solve $b^a = b^{cx+d}$.
  - An amount a grows at a rate of r% per year. Find the amount after t years.
  - An amount a decays at a rate of r% per year. Find the amount after t years.

- Each team sends a student to the board to write the problems they created in random order. Students should leave room for the solutions.
- Playing the game: Each team sends one student at a time to the board to solve one problem written by the *other* team. When that student finishes, it is another student's turn.
- When all problems are solved, allow the students on each team to change answers they feel are wrong (one change per student).
- **Scoring:** Points are counted at the end of the game. Each team earns 1 point for each problem answered correctly.

Who wins?

The team with the most points wins.

For the Student
Additional Practice

- Lesson Tutorials
- Study Help (textbook)
- Student Website
  - Multi-Language Glossary
  - Practice Assessments

Answers

18. exponential growth function

19. exponential decay function

20. $y = 1500(0.94)^t$

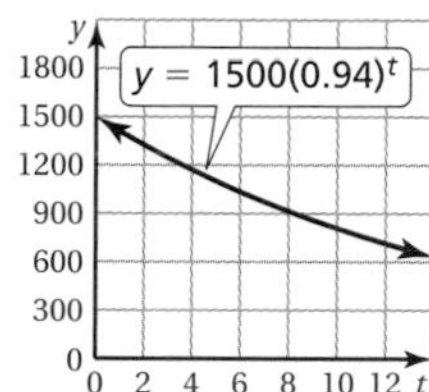

After 1 year, the price is about $714.

21. $-243, 729, -2187$

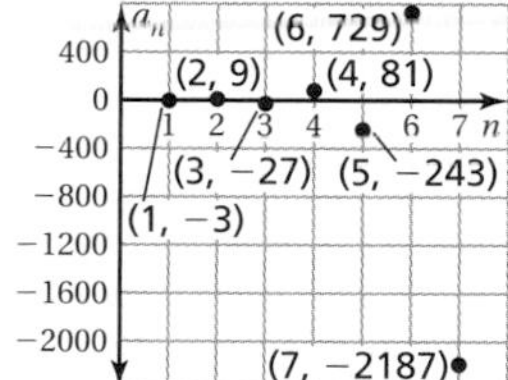

22. $\frac{3}{16}, \frac{3}{64}, \frac{3}{256}$

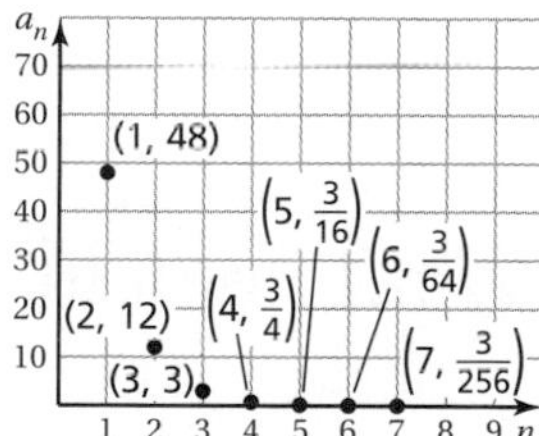

23. $a_n = 4^{n-1}$

24. $a_n = 5(-2)^{n-1}$

25. $a_1 = 3, a_n = a_{n-1} + 5$

26. $a_1 = 3, a_n = 2a_{n-1}$

My Thoughts on the Chapter

Teacher Tip

Not allowed to write in your teaching edition? Use sticky notes to record your thoughts.

What worked. . .

What did not work. . .

What I would do differently. . .

6.7 Geometric Sequences (pp. 306–315)

a. Write the next three terms of the geometric sequence 2, 6, 18, 54, ...

Use a table to organize the terms and extend the pattern.

Position	1	2	3	4	5	6	7
Term	2	6	18	54	162	486	1458

× 3 × 3 × 3 × 3 × 3 × 3

Each term is 3 times the previous term. So, the common ratio is 3.

Multiply a term by 3 to find the next term.

∴ The next three terms are 162, 486, and 1458.

b. Graph the geometric sequence 24, 12, 6, 3, 1.5, ... What do you notice?

Make a table. Then plot the ordered pairs (n, a_n).

Position, n	1	2	3	4	5
Term, a_n	24	12	6	3	1.5

∴ The points of the graph appear to lie on an exponential curve.

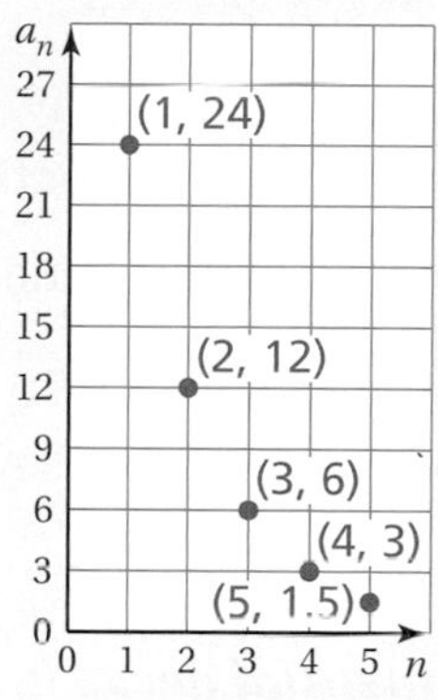

Exercises

Write the next three terms of the geometric sequence. Then graph the sequence.

21. $-3, 9, -27, 81, \ldots$

22. $48, 12, 3, \frac{3}{4}, \ldots$

Write an equation for the *n*th term of the geometric sequence.

23.

n	1	2	3	4
a_n	1	4	16	64

24.

n	1	2	3	4
a_n	5	-10	20	-40

Write a recursive rule for the sequence.

25. 3, 8, 13, 18, 23, . . .

26. 3, 6, 12, 24, 48, . . .

6 Chapter Test

Simplify the expression.

1. $\sqrt{98}$

2. $\sqrt{\frac{19}{25}}$

3. $\frac{6 - \sqrt{48}}{2}$

Simplify. Write your answer using only positive exponents.

4. $z^{-2} \cdot z^{4}$

5. $\frac{b^{-5}}{b^{-8}}$

6. $\left(\frac{2c^{4}}{5}\right)^{-3}$

Simplify the expression.

7. $\sqrt[4]{16}$

8. $729^{1/6}$

9. $32^{7/5}$

10. Graph $y = 7^x + 1$. Describe the domain and range. Compare the graph to the graph of $y = 7^x$.

Write an exponential function represented by the table.

11.

x	0	1	2	3
y	−1	−2	−4	−8

12.

x	0	1	2	3
y	3	−12	48	−192

Solve the equation. Check your solution, if possible.

13. $2^x = 128$

14. $256^{x+2} = 16^{3x-1}$

Write and graph a function that represents the situation.

15. Your $42,500 annual salary increases by 3% each year.

16. You deposit $500 in an account that earns 6.5% annual interest compounded yearly.

Determine whether the table represents an *exponential growth function*, an *exponential decay function*, or *neither*.

17.

x	0	1	2	3
y	15	30	60	120

18.

x	0	1	2	3
y	400	100	25	6.25

19. **TRAINING** You follow the training schedule from your coach.

    a. Write an equation for the nth term of the geometric sequence.

    b. Write a recursive rule for the explicit equation in part (a).

    c. On what day do you run approximately 3 kilometers?

Training On Your Own

Day 1: Run 1 km.

Each day after Day 1: Run 20% farther than the previous day.

Test Item References

Chapter Test Questions	Section to Review	Common Core State Standards
1–3	6.1	N.RN.3
4–6	6.2	N.RN.2
7–9	6.3	N.RN.1, N.RN.2
10–14	6.4	A.REI.3, A.REI.11, F.BF.3, F.IF.7e, F.LE.1a, F.LE.2
15, 16	6.5	A.SSE.1a, A.SSE.1b, F.IF.7e
17, 18	6.6	A.SSE.1a, A.SSE.1b, F.IF.7e
19	6.7	F.BF.2, F.IF.3, F.LE.2

Test-Taking Strategies

Remind students to quickly look over the entire test before they start so that they can budget their time. Have students use the **Stop** and **Think** strategy before they answer each question.

Common Assessment Errors

- **Exercises 1–3** Students may leave a perfect square factor in the radicand, or they may move the wrong factor outside the radical. Remind students to factor the radicand using its greatest perfect square factor, and bring the square root of that factor outside the radical.
- **Exercise 5** Students may divide the exponents instead of subtracting the exponents. Remind them of the Quotient of Powers Property.
- **Exercise 6** Students may forget to apply the exponent to the denominator. Remind them of the Power of a Quotient Property.
- **Exercises 11 and 12** Students may write incorrect equations. Remind them that in an exponential equation of the form $y = ab^x$, a is the value of y when $x = 0$, and b is the factor that y changes by for each 1-unit increase in x.
- **Exercise 14** Students may fail to apply the Distributive Property correctly when applying the Power of a Power Property. Encourage them to first rewrite the expression using the product of the exponents, so they will see that the Distributive Property applies.

Reteaching and Enrichment Strategies

If students need help. . .	If students got it. . .
Resources by Chapter • Practice A and Practice B • Puzzle Time Record and Practice Journal Practice Differentiating the Lesson Lesson Tutorials Practice from the Test Generator Skills Review Handbook	Resources by Chapter • Enrichment and Extension • School-to-Work • Financial Literacy • Technology Connection • Life Connections Game Closet at *BigIdeasMath.com* Start Standardized Test Practice

Answers

1. $7\sqrt{2}$
2. $\frac{\sqrt{19}}{5}$
3. $3 - 2\sqrt{3}$
4. z^2
5. b^3
6. $\frac{125}{8c^{12}}$
7. 2
8. 3
9. 128
10. 

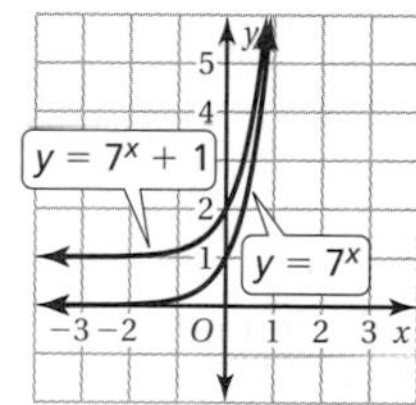

domain: all real numbers; range: all real numbers greater than 1; translation 1 unit up of graph of $y = 7^x$

11. $y = -2^x$
12. $y = 3(-4)^x$
13. $x = 7$
14. $x = 5$
15–16. See Additional Answers.
17. exponential growth function
18. exponential decay function
19. **a.** $a_n = 1.2^{n-1}$

    b. $a_1 = 1, a_n = 1.2a_{n-1}$

    c. day 7

Assessment Book

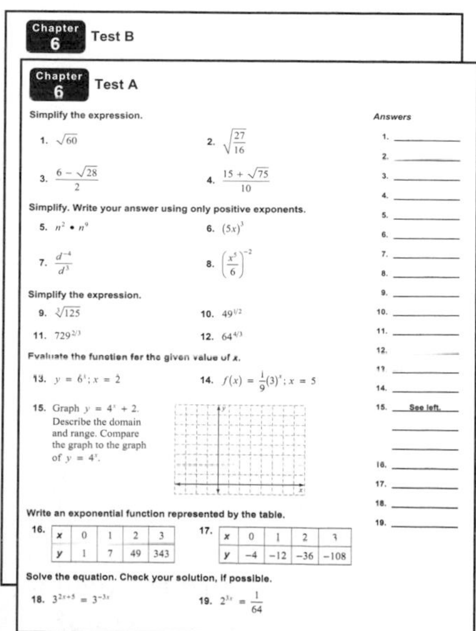

Chapter 6 Test B

Chapter 6 Test A

Simplify the expression.

1. $\sqrt{60}$
2. $\sqrt{\frac{27}{16}}$
3. $\frac{6 - \sqrt{28}}{2}$
4. $\frac{15 + \sqrt{75}}{10}$

Simplify. Write your answer using only positive exponents.

5. $n^2 \cdot n^6$
6. $(5x)^3$
7. $\frac{d^{-4}}{d^3}$
8. $\left(\frac{x^3}{6}\right)^{-2}$

Simplify the expression.

9. $\sqrt[3]{125}$
10. $49^{1/2}$
11. $729^{2/3}$
12. $64^{4/3}$

Evaluate the function for the given value of x.

13. $y = 6^x; x = 2$
14. $f(x) = \frac{1}{9}(3)^x; x = 5$
15. Graph $y = 4^x + 2$. Describe the domain and range. Compare the graph to the graph of $y = 4^x$.

Write an exponential function represented by the table.

16.

x	0	1	2	3
y	1	7	49	343

17.

x	0	1	2	3
y	−4	−12	−36	−108

Solve the equation. Check your solution, if possible.

18. $3^{2x+5} = 3^{-3x}$
19. $2^{3x} = \frac{1}{64}$

Answers

1.–19. 15. See left.

Test Taking Strategies

Available at *BigIdeasMath.com*

After Answering Easy Questions, Relax
Answer Easy Questions First
Estimate the Answer
Read All Choices before Answering
Read Question before Answering
Solve Directly or Eliminate Choices
Solve Problems before Looking at Choices
Use Intelligent Guessing
Work Backwards

About this Strategy

When taking a multiple choice test, be sure to read each question carefully and thoroughly. Sometimes you don't know the answer. So... guess intelligently! Look at the choices and choose the ones that are possible answers.

Answers

1. B
2. F
3. A
4. 81
5. I

Technology For the Teacher

Big Ideas Test Generator

Item Analysis

1. **A.** The point satisfies the first inequality but not the second.
   B. Correct answer
   C. The point satisfies the first inequality but not the second.
   D. The point satisfies the second inequality but not the first.

2. **F.** Correct answer
   G. The student divides the numerator by 5, but not the denominator.
   H. The student mishandles the negative signs.
   I. The student multiplies -10 by -3 instead of 5^{-3}.

3. **A.** Correct answer
   B. The student incorrectly chooses strict inequality.
   C. The student subtracts 0.3 from 1.9 instead of adding.
   D. The student subtracts 0.3 from 1.9 instead of adding and incorrectly chooses strict inequality.

4. **Gridded response:** Correct answer: 81
   Common Error: The student incorrectly rewrites the expression as $\frac{27^4}{27^3}$ and gets 27 for an answer.

5. **F.** The student mishandles a sign in applying properties of equality.
   G. The student mishandles signs in applying properties of equality.
   H. The student mishandles a sign in applying properties of equality.
   I. Correct answer

6 Standardized Test Practice

Test-Taking Strategy
Use Intelligent Guessing

Cats were first tamed $3 \cdot 2^{10}$ years ago in Egypt. How long ago was that?

Ⓐ 3000 Ⓑ 3072 Ⓒ 5000 Ⓓ 40

Who says I am tame? Growl. Hiss.

"It can't be 40 or 5000 because they aren't divisible by 3. So, you can intelligently guess between 3000 and 3072."

1. Which point is a solution of the system of inequalities shown below? *(A.REI.12)*

$$y > 4x - 3$$
$$3x - 2y < 4$$

A. $(-2, -7)$ **C.** $(-4, -8)$

B. $(1, 1)$ **D.** $(4, 5)$

2. What is the value of the function $y = -10(5)^x$ when $x = -3$? *(F.IF.7e)*

F. $-\frac{2}{25}$ **H.** $\frac{2}{25}$

G. $-\frac{2}{125}$ **I.** 30

3. Which graph shows the solution of $x - 1.9 \geq 0.3$? *(A.CED.1)*

A. 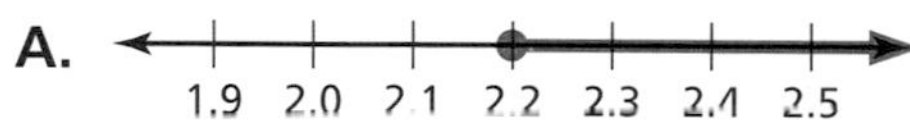

C. 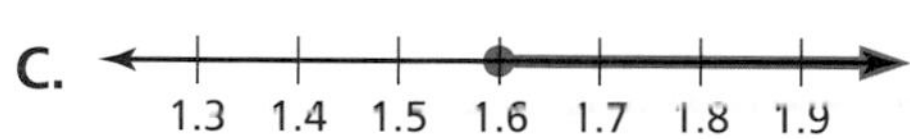

B. 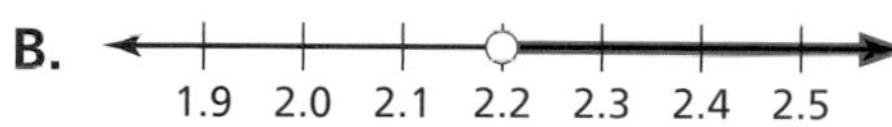

D. 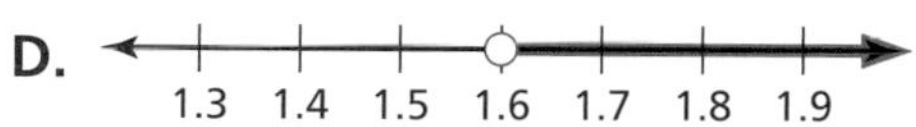

4. What is the value of $27^{4/3}$? *(N.RN.2)*

5. Which graph represents the equation $-5x - 5y = 25$? *(A.REI.10)*

F. 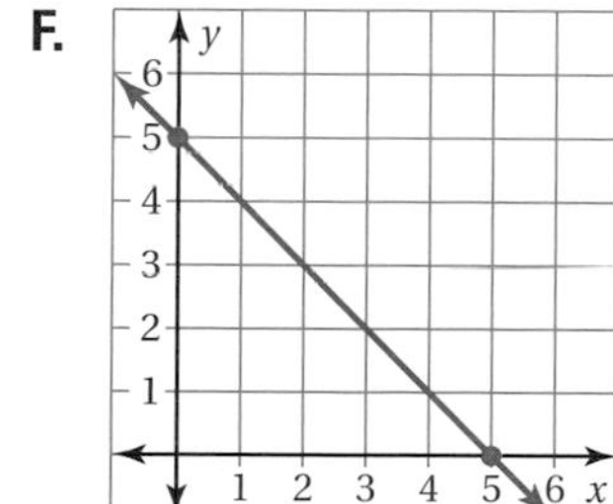

H. 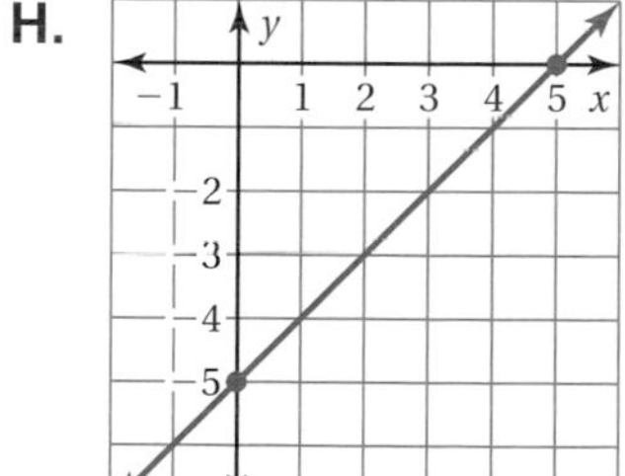

G. 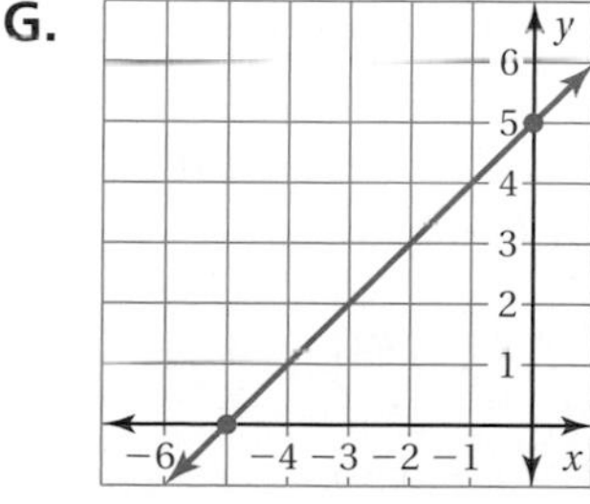

I. 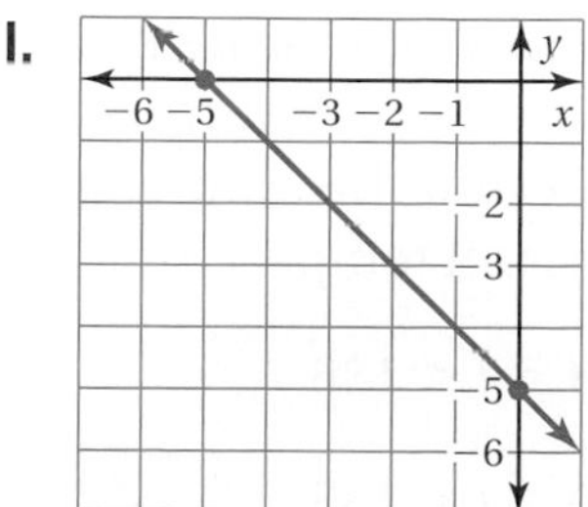

6. A system of two linear equations has infinitely many solutions. What can you conclude about the graphs of the two equations? *(8.EE.8b)*

A. The lines have the same slope and the same y-intercept.

B. The lines have the same slope and different y-intercepts.

C. The lines have different slopes and the same y-intercept.

D. The lines have different slopes and different y-intercepts.

7. The domain of the function $y = -5x + 19$ is 0, 2, 4, and 6. What is the range of the function? *(F.IF.1)*

F. $-19, -9, 1, 11$

G. $-11, -1, 9, 19$

H. $6, 4, 2, 0$

I. $-19, -11, -9, 1$

8. What is the 50th term of the sequence $20, 9, -2, -13, \ldots$? *(F.LE.2)*

9. Which graph shows an exponential decay function? *(F.IF.7E)*

A.

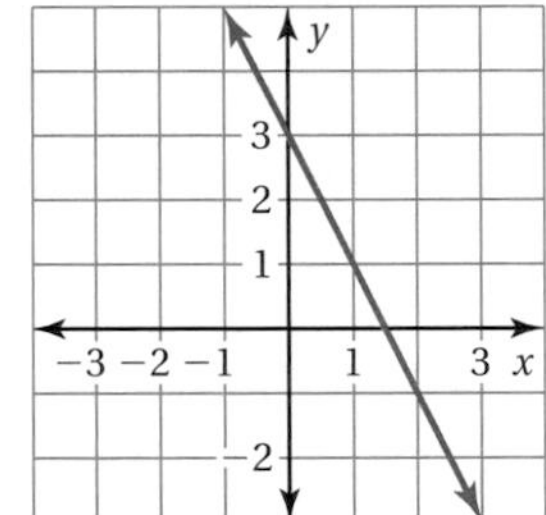

C.

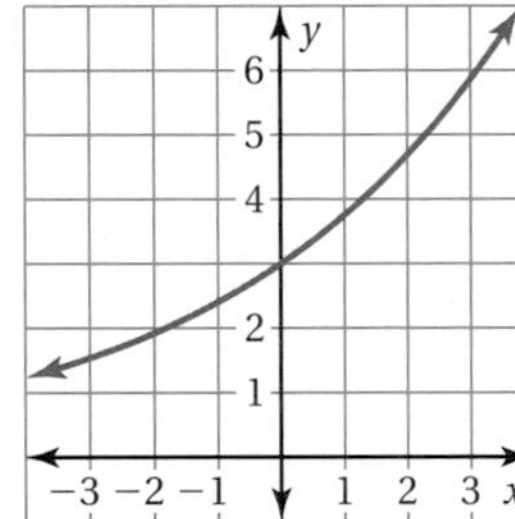

B.

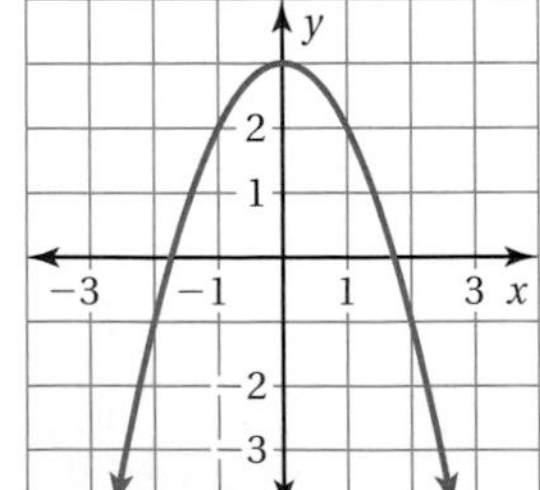

D.

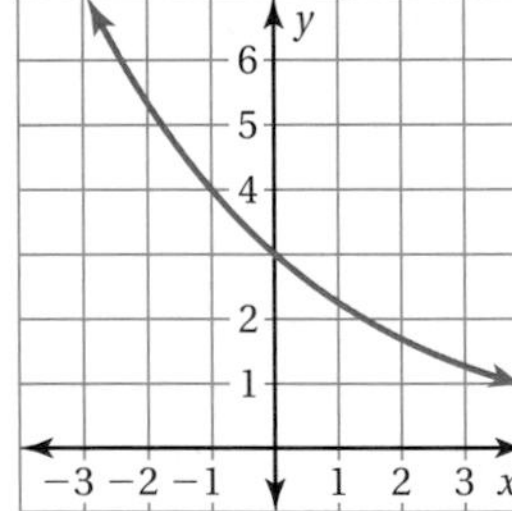

10. The lowest temperature ever recorded on Earth is $-129°$ Fahrenheit. The highest temperature ever recorded on Earth is $136°$ Fahrenheit. Let t represent the temperature, in degrees Fahrenheit. Which inequality represents all temperatures ever recorded on Earth? *(A.CED.1)*

F. $-129 < t < 136$

G. $-129 \leq t < 136$

H. $-129 \leq t \leq 136$

I. $-129 < t \leq 136$

Item Analysis (continued)

6. **A.** Correct answer
   B. The student confuses a system with infinitely many solutions and a system with no solution.
   C. The student thinks the lines must have different slopes to have infinitely many solutions.
   D. The student misinterprets "infinitely many solutions."

7. **F.** The student uses incorrect signs when listing the range.
   G. Correct answer
   H. The student thinks the range is the reverse order of the domain.
   I. The student uses incorrect signs when listing the range.

8. **Gridded response:** Correct answer: -519

   Common error: The student uses a common difference of 11 instead of -11 and gets an answer of 559.

9. **A.** The student thinks a downward slope indicates "decay."
   B. The student thinks downward concavity indicates "decay."
   C. The student confuses the characteristics of exponential growth and decay graphs.
   D. Correct answer

10. **F.** The student incorrectly uses strict inequality symbols.
    G. The student incorrectly uses one strict inequality symbol.
    H. Correct answer
    I. The student incorrectly uses one strict inequality symbol.

11. **A.** The student confuses the terms perpendicular and parallel.
    B. Correct answer
    C. The student incorrectly divides the change in x by the change in y to find the slope between the two points.
    D. The student makes a sign error in using the two points to calculate the slope.

12. **F.** The student switches the slope and the y-intercept in interpreting the slope-intercept form.
    G. The student confuses the characteristics of slope-intercept form.
    H. The student misinterprets the y-intercept as $-b$.
    I. Correct answer

Answers

6. A
7. G
8. -519
9. D
10. H

Answers

11. B

12. I

13. *Part A:* $y = 256(0.5)^x$

Part B: exponential decay; As x increases by 1, y is multiplied by $\frac{1}{2}$.

Part C:

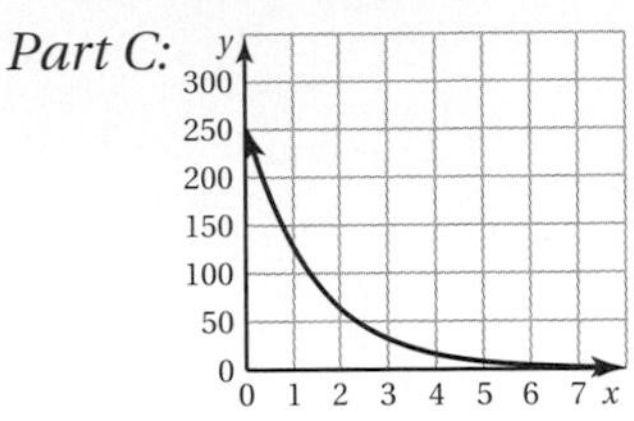

Part D: 8; When $x = 8$, $y = 1$. So, after 8 rounds there is one remaining player—the winner.

14. D

15. G

16. C

Answer for Extra Example

1. **A.** The student rewrites $\frac{4^8}{4^4}$ as 2 (by dividing out the bases of 4).

   B. Correct answer

   C. The student multiplies and divides exponents, instead of adding and subtracting.

   D. The student multiplies bases in the numerator (and adds exponents).

Item Analysis (continued)

13. 4 points The student demonstrates a thorough understanding of exponential decay functions, writing and graphing a correct function, describing why it represents exponential decay, and how it can be applied to the situation.

3 points The student demonstrates an essential but less thorough understanding of exponential decay functions, writing and graphing a correct function, but with either an incomplete description or an incorrect answer in Part D.

2 points The student understands that the situation involves exponential decay, but more than one part is wrong, incomplete, or unanswered.

1 point The student demonstrates limited understanding of exponential decay, answering only one part of the question correctly.

0 points The student provides no response, a completely incorrect or incomprehensible response, or a response that demonstrates insufficient understanding of exponential decay functions.

14. **A.** The student does not recognize the equal factors of 50.

B. The student does not recognize the equal factors of -1.

C. The student does not recognize the common difference of -10.

D. Correct answer

15. **F.** The student chooses the value -7 from the question.

G. Correct answer

H. The student adds 3(1) to 10 instead of $-3(1)$.

I. The student substitutes -7 for x.

16. **A.** The student makes an error in removing the greatest perfect square factor from the radical.

B. The student leaves the square root of the greatest perfect square factor inside the radical and removes the factor 2.

C. Correct answer

D. The student miscalculates $\sqrt{200}$ as 20.

Extra Example

1. Which of the following is equivalent to $\frac{4^8 4^5}{4^4}$? *(N.RN.2)*

A. $2 \cdot 4^5$ **C.** 4^{10}

B. 4^9 **D.** $\frac{16^{13}}{4^4}$

11. The graph of which equation is perpendicular to the line that passes through the points $(-3, -6)$ and $(5, -2)$? *(F.IF.6)*

A. $y = \frac{1}{2}x + 3$

B. $y = -2x + 7$

C. $y = -\frac{1}{2}x - 3$

D. $y = 2x + 1$

12. Which of the following is true about the graph of the linear equation $y = -7x + 5$? *(F.IF.4)*

F. The slope is 5 and the y-intercept is -7.

G. The slope is -5 and the y-intercept is -7.

H. The slope is -7 and the y-intercept is -5.

I. The slope is -7 and the y-intercept is 5.

13. At the beginning of a tennis tournament, there are 256 players. After each round, one-half of the remaining players are eliminated. *(F.IF.7e)*

Part A Write a function that represents the number of players left in the tournament after each round.

Part B Does the function in part (a) represent exponential growth or exponential decay? Explain your reasoning.

Part C Graph the function in part (a).

Part D How many tennis matches does a player have to win to win the tournament? Explain your reasoning.

14. Which sequence is neither arithmetic nor geometric? *(F.IF.3)*

A. 1, 50, 2500, 125,000, . . .

B. 10, 0, −10, −20, . . .

C. 4, −4, 4, −4, . . .

D. 0, 1, 3, 6, . . .

15. For $f(x) = -3x - 10$, what value of x makes $f(x) = -7$? *(F.IF.2)*

F. -7

G. -1

H. 1

I. 11

16. Which expression is equivalent to $20\sqrt{200}$? *(N.RN.3)*

A. $40\sqrt{2}$

B. $40\sqrt{10}$

C. $200\sqrt{2}$

D. 400

7 Polynomial Equations and Factoring

"Here's how it goes, Descartes."

"The friends of my friends are my friends. The friends of my enemies are my enemies."

"The enemies of my friends are my enemies. The enemies of my enemies are my friends."

"Descartes, which one is the monomial and which one is the polynomial?"

"Remember that poly means many and mono means one."

Connections to Previous Learning

- Evaluate algebraic expressions at specific values of their variables.
- Solve one-step linear equations.

- Simplify algebraic expressions.
- Solve multi-step problems involving linear equations.

- Perform arithmetic operations on polynomials.
- Solve problems involving polynomial equations by factoring.

Math in History

The algebraic notation used today took more than 3000 years to develop. The development occurred in three stages: the rhetorical stage, the syncopated stage, and the symbolic stage. Ancient cultures talked about problems using words without symbols during the rhetorical stage. Mathematicians started using some abbreviated words during the syncopated stage. The symbolic stage has produced the notation for writing algebraic expressions and equations over the last several centuries.

★ Ancient Babylonians (c. 2000 B.C.) used words without symbols to solve many types of problems that are solved using quadratic equations today.

★ In *Introduction to the Analytic Art* (1591), François Viète used consonants to represent arbitrary constants and vowels to represent variables, paving the way to think of quadratic equations in a general form $BA^2 + CA + D = 0$.

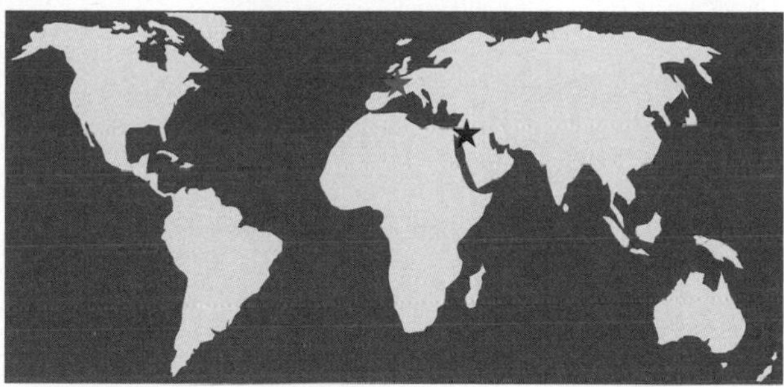

Pacing Guide for Chapter 7

Chapter Opener	1 Day
Section 1	1 Day
Section 2	2 Days
Section 3	2 Days
Section 4	2 Days
Study Help / Quiz	1 Day
Section 5	1 Day
Section 6	1 Day
Section 7	1 Day
Section 8	1 Day
Section 9	3 Days
Chapter Review/ Chapter Tests	2 Days
Total Chapter 7	18 Days
Year-to-Date	98 Days

Check Your Resources

- Record and Practice Journal
- Resources by Chapter
- Skills Review Handbook
- Assessment Book
- Worked-Out Solutions

The Dynamic Planning Tool
Editable Teacher's Resources at
BigIdeasMath.com

Common Core State Standards

7.EE.1 Apply properties of operations as strategies to add, subtract, factor, and expand linear expressions with rational coefficients.

6.NS.4 Find the greatest common factor of two whole numbers less than or equal to 100 . . ..

Additional Topics for Review

- Exponents
- Order of Operations
- Solving linear equations
- Associative Property of Addition

Try It Yourself

1. $7x - 8$
2. $-3t + 3$
3. $-z + 8$
4. $3w + 1$
5. $2g - 12$
6. $-n + 15$
7. 4
8. 12
9. 3

Record and Practice Journal

1. $-4y + 6$
2. $h + 7$
3. $5a - 4$
4. $-2m - 9$
5. $3d - 22$
6. $7q - 11$
7. $2x + 10$
8. $14x - 6$
9. 3
10. 5
11. 6
12. 16
13. 2
14. 15
15. 8
16. 21 cm

Math Background Notes

Vocabulary Review

- Algebraic Expression
- Distributive Property
- Commutative Property of Addition
- Greatest Common Factor

Simplifying Algebraic Expressions

- Students should know how to simplify algebraic expressions.
- Remind students that subtracting a term is the same as adding the opposite of the term.
- Point out how the Distributive Property is used to add or subtract like terms in Examples 1 and 2.

Finding the Greatest Common Factor

- Students should know how to find the greatest common factor of two whole numbers.
- When looking for the prime factors of a whole number, remind students that it is often easiest to start with the smallest possible factor and work toward the largest factor.
- **Teaching Tip:** Discuss how the difference of two whole numbers can sometimes be helpful in finding the greatest common factor. The GCF must be a factor of the difference, and the GCF can be no greater than the difference. For instance, in Example 4, the difference of 42 and 30 is 12, so you know the GCF is a factor of 12.

Reteaching and Enrichment Strategies

If students need help...	If students got it...
Record and Practice Journal • Fair Game Review Skills Review Handbook Lesson Tutorials	Game Closet at *BigIdeasMath.com* Start the next section

What You Learned Before

"Dear Editor, I disagree with your claim that the sum of two binomials is always a binomial."

Simplifying Algebraic Expressions (7.EE.1)

Example 1 **Simplify $5x + 7 - 2x - 3$.**

$5x + 7 - 2x - 3 = 5x - 2x + 7 - 3$	Commutative Property of Addition
$= (5 - 2)x + 7 - 3$	Distributive Property
$= 3x + 4$	Simplify.

Example 2 **Simplify $-7(y - 2) + 3y$.**

$-7(y - 2) + 3y = -7(y) - (-7)(2) + 3y$	Distributive Property
$= -7y + 14 + 3y$	Multiply.
$= -7y + 3y + 14$	Commutative Property of Addition
$= (-7 + 3)y + 14$	Distributive Property
$= -4y + 14$	Add coefficients.

Try It Yourself

Simplify the expression.

1. $3x - 8 + 4x$

2. $3t - 4 - 6t + 7$

3. $-7z + 3 + 2z + 4z + 5$

4. $3(w + 2) - 5$

5. $4g - 2(g + 6)$

6. $3(n + 1) - 4(n - 3)$

Finding the Greatest Common Factor (6.NS.4)

Example 3 **Find the greatest common factor of 50 and 75.**

$50 = 2 \cdot 5 \cdot 5$

$75 = 3 \cdot 5 \cdot 5$

The GCF is $5 \cdot 5 = 25$.

Example 4 **Find the greatest common factor of 30 and 42.**

$30 = 2 \cdot 3 \cdot 5$

$42 = 2 \cdot 3 \cdot 7$

The GCF is $2 \cdot 3 = 6$.

Try It Yourself

Find the greatest common factor.

7. 28, 64

8. 60, 72

9. 24, 27

7.1 Polynomials

COMMON CORE STATE STANDARDS
A.SSE.1a

Essential Question How can you use algebra tiles to model and classify polynomials?

1 ACTIVITY: Meaning of Prefixes

Work with a partner. Think of a word that uses one of the prefixes with one of the base words. Then define the word and write a sentence that uses the word.

Prefix	Base Word
Mono	Dactyl
Bi	Cycle
Tri	Ped
Poly	Syllabic

2 ACTIVITY: Classifying Polynomials Using Algebra Tiles

Work with a partner. Six different algebra tiles are shown at the right.

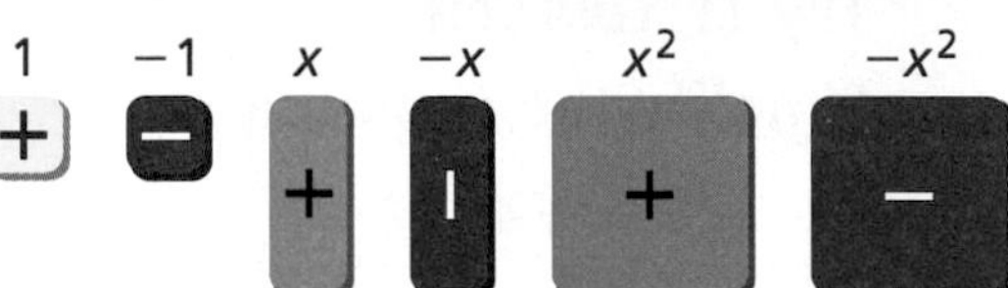

Write the polynomial that is modeled by the algebra tiles. Then classify the polynomial as a monomial, binomial, or trinomial. Explain your reasoning.

a.

b.

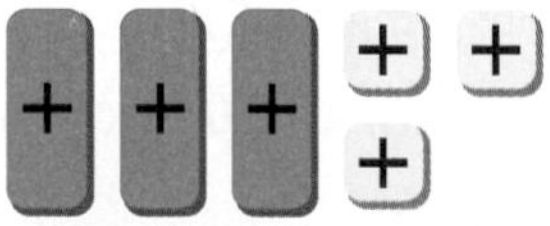

c.

d.

e.

f.

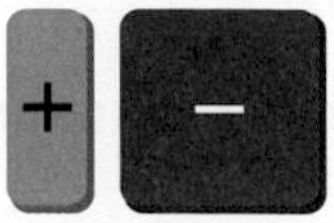

Laurie's Notes

Introduction

Standards for Mathematical Practice

- **MP5 Use Appropriate Tools Strategically:** Algebra tiles can be used to model algebraic expressions, and in particular, second-degree polynomials. As students begin working with polynomials, the tiles are an appropriate tool for developing conceptual understanding. For instance, working with algebra tiles will help students understand why $x + x = 2x$, and not x^2.

Motivate

- The warm-up today is to familiarize students with all of the algebra tile pieces. Begin by placing an x^2-tile next to an x-tile. The square tile has the same length as the x-tile, so the dimension of the square tile is x by x with an area of x^2. As with the x-tile, the color is used to distinguish between x^2 and $-x^2$.
- Show students a collection of algebra tiles and ask them what the collection represents.

? "Can the collection be simplified? (Can you remove zero pairs?)" Answers will vary based on the collection of algebra tiles.

? "What is the expression represented by the collection?" Answers will vary based on the collection of algebra tiles.

Activity Notes

Activity 1

? "What words can you think of that begin with the prefixes *deca-* or *deci-*?" Listen for decade, decathlon, decimal, and so on.

- All of these words refer to 10 of something.
- In this first activity, students match a prefix with a base word. Answers will vary.

Activity 2

- Students may be unfamiliar with the terms monomial, binomial and trinomial. The first activity, however, should give students sufficient insight to do this activity.
- It is not uncommon for students to write $3x^2 + (-x) + 1$ for part (d). Ask students for an equivalent expression so that they write $3x^2 - x + 1$.
- Discuss how the students classified the polynomials. Students who think that the polynomial in part (a) is a trinomial because there are 3 tiles should quickly change their reasoning once they look at part (b) with 6 tiles.

Common Core State Standards

A.SSE.1a Interpret parts of an expression, such as terms, factors, and coefficients.

Previous Learning

Students should know how to identify the terms and like terms in an algebraic expression.

Activity Materials	
Introduction	**Textbook**
• algebra tiles	• algebra tiles

Start Thinking! and Warm Up

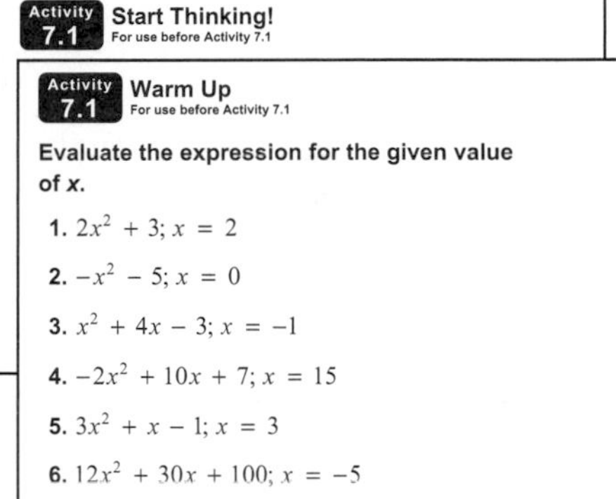

Activity 7.1 Start Thinking! For use before Activity 7.1

Activity 7.1 Warm Up For use before Activity 7.1

Evaluate the expression for the given value of x.

1. $2x^2 + 3; x = 2$
2. $-x^2 - 5; x = 0$
3. $x^2 + 4x - 3; x = -1$
4. $-2x^2 + 10x + 7; x = 15$
5. $3x^2 + x - 1; x = 3$
6. $12x^2 + 30x + 100; x = -5$

7.1 Record and Practice Journal

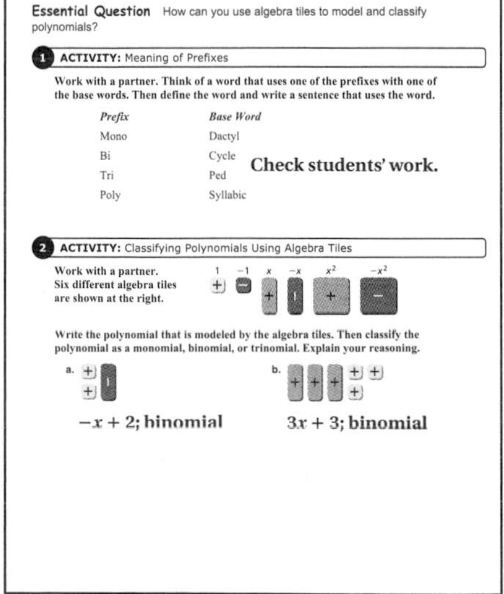

Essential Question How can you use algebra tiles to model and classify polynomials?

1 ACTIVITY: Meaning of Prefixes

Work with a partner. Think of a word that uses one of the prefixes with one of the base words. Then define the word and write a sentence that uses the word.

Prefix	Base Word
Mono	Dactyl
Bi	Cycle
Tri	Ped
Poly	Syllabic

Check students' work.

2 ACTIVITY: Classifying Polynomials Using Algebra Tiles

Work with a partner. Six different algebra tiles are shown at the right.

Write the polynomial that is modeled by the algebra tiles. Then classify the polynomial as a monomial, binomial, or trinomial. Explain your reasoning.

a. $-x + 2$; binomial

b. $3x + 3$; binomial

English Language Learners

Vocabulary

There are many new words in this chapter. To help English language learners, create a "Wall of Key Words." Using the list of key words from the *Chapter Review,* create a poster with the word, its definition, and examples for students to reference while working through the chapter.

7.1 Record and Practice Journal

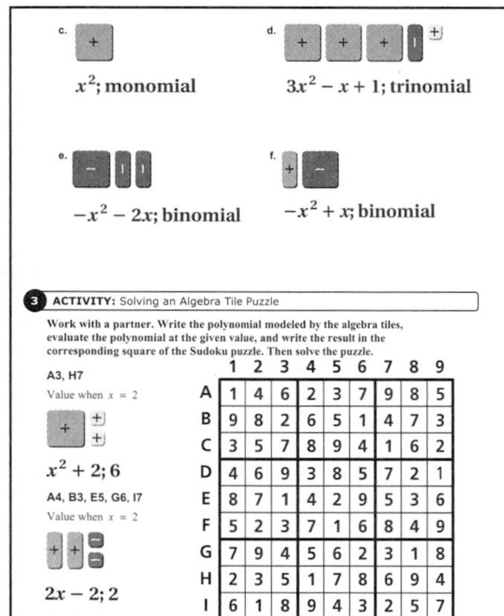

c. x^2; monomial

d. $3x^2 - x + 1$; trinomial

e. $-x^2 - 2x$; binomial

f. $-x^2 + x$; binomial

3 ACTIVITY: Solving an Algebra Tile Puzzle

Work with a partner. Write the polynomial modeled by the algebra tiles, evaluate the polynomial at the given value, and write the result in the corresponding square of the Sudoku puzzle. Then solve the puzzle.

A3, H7
Value when $x = 2$
$x^2 + 2$; 6

A4, B3, E5, G6, I7
Value when $x = 2$
$2x - 2$; 2

	1	2	3	4	5	6	7	8	9
A	1	4	6	2	3	7	9	8	5
B	9	8	2	6	5	1	4	7	3
C	3	5	7	8	9	4	1	6	2
D	4	6	9	3	8	5	7	2	1
E	8	7	1	4	2	9	5	3	6
F	5	2	3	7	1	6	8	4	9
G	7	9	4	5	6	2	3	1	8
H	2	3	5	1	7	8	6	9	4
I	6	1	8	9	4	3	2	5	7

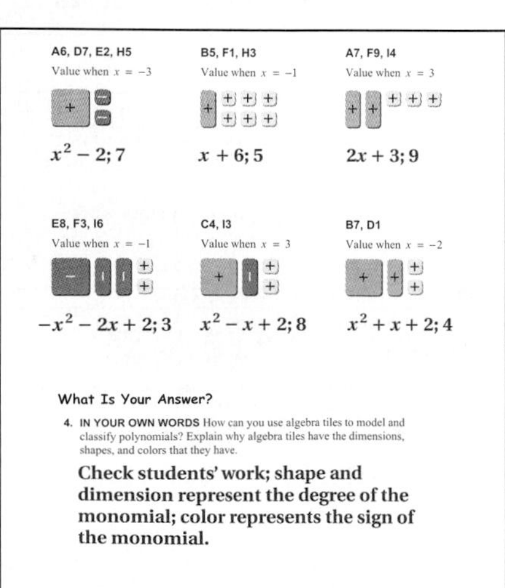

A6, D7, E2, H5
Value when $x = -3$
$x^2 - 2$; 7

B5, F1, H3
Value when $x = -1$
$x + 6$; 5

A7, F9, I4
Value when $x = 3$
$2x + 3$; 9

E8, F3, I6
Value when $x = -1$
$-x^2 - 2x + 2$; 3

C4, I3
Value when $x = 3$
$x^2 - x + 2$; 8

B7, D1
Value when $x = -2$
$x^2 + x + 2$; 4

What Is Your Answer?

4. IN YOUR OWN WORDS How can you use algebra tiles to model and classify polynomials? Explain why algebra tiles have the dimensions, shapes, and colors that they have.

Check students' work; shape and dimension represent the degree of the monomial; color represents the sign of the monomial.

Laurie's Notes

Activity 3

? "How many of you know what a Sudoku puzzle is? Have you ever tried to solve one?" Answers will vary.

- In a Sudoku puzzle, you fill in the boxes so that the nine rows, nine columns, and the nine 3×3 sections all contain every digit from 1 to 9.
- Students need to write the polynomial expression first, then evaluate it for the *x*-value given.
- Caution students to be careful where there is a $-x^2$-tile or a $-x$-tile as in the sixth and seventh expressions. The sixth expression is $-x^2 + (-2x) + 2$. When $x = -1$ the expression equals $-(-1)^2 + [-2(-1)] + 2 = -1 + 2 + 2 = 3$.
- You may want to check that students evaluated the expressions correctly before they try to complete the puzzle.
- Once the identified squares have been filled in, the puzzle has sufficient clues to finish the puzzle. For instance, D3 must be a 9 because column 3 is missing a 1, 4, 7, and 9. Row D contains the 1, 4, and 7, so D3 must be 9.

What Is Your Answer?

- Discuss students' responses to the question.

Closure

- Sketch a model of $2x^2 + 3x - 1$.

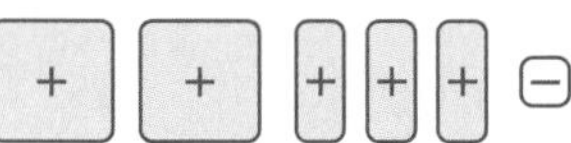

The Dynamic Planning Tool
Editable Teacher's Resources at *BigIdeasMath.com*

3 ACTIVITY: Solving an Algebra Tile Puzzle

Work with a partner. Write the polynomial modeled by the algebra tiles, evaluate the polynomial at the given value, and write the result in the corresponding square of the Sudoku puzzle. Then solve the puzzle.

A3, H7

Value when $x = 2$

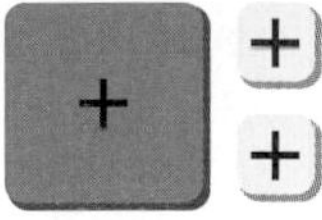

A4, B3, E5, G6, I7

Value when $x = 2$

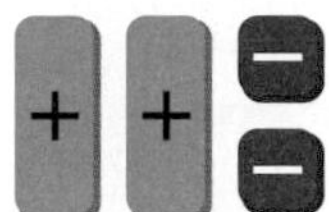

	1	2	3	4	5	6	7	8	9
A									
B									
C									
D									1
E									
F									
G									
H									
I									

A6, D7, E2, H5

Value when $x = -3$

B5, F1, H3

Value when $x = -1$

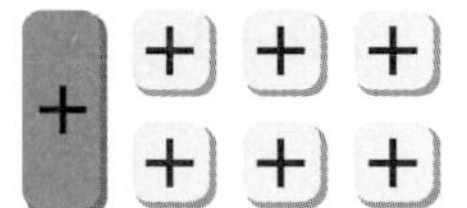

A7, F9, I4

Value when $x = 3$

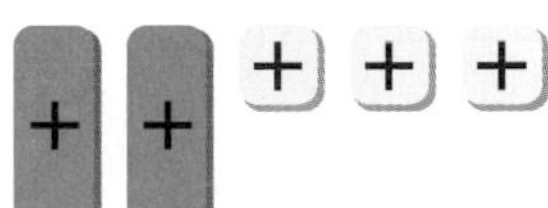

E8, F3, I6

Value when $x = -1$

C4, I3

Value when $x = 3$

B7, D1

Value when $x = -2$

What Is Your Answer?

4. **IN YOUR OWN WORDS** How can you use algebra tiles to model and classify polynomials? Explain why algebra tiles have the dimensions, shapes, and colors that they have.

Use what you learned about modeling polynomials to complete Exercises 5 and 6 on page 332.

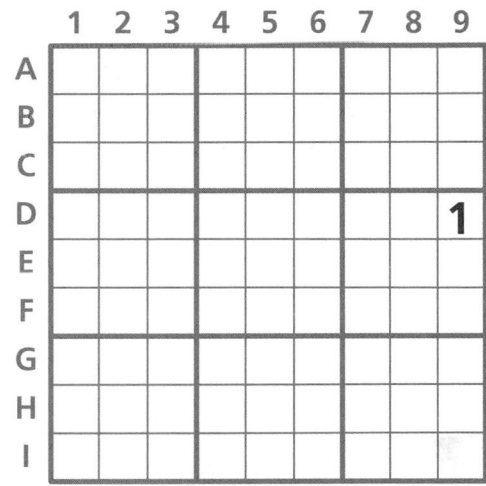

7.1 Lesson

Key Vocabulary
monomial, *p. 330*
degree of a monomial, *p. 330*
polynomial, *p. 331*
binomial, *p. 331*
trinomial, *p. 331*
degree of a polynomial, *p. 331*

A **monomial** is a number, a variable, or a product of a number and one or more variables with whole number exponents.

Monomials
-4
$\frac{1}{2}y^2$
$2.5x^2y$

Not monomials	Reason
$x^{1.5}$	Monomials must have whole number exponents.
$-\frac{2}{z}$	Monomials cannot have variables in the denominator.
7^y	Monomials cannot have variable exponents.

The **degree of a monomial** is the sum of the exponents of the variables in the monomial.

EXAMPLE 1 Finding the Degrees of Monomials

Find the degree of each monomial.

a. $5x^2$

The exponent of x is 2.

So, the degree of the monomial is 2.

b. $-\frac{1}{2}xy^3$

The exponent of x is 1 and the exponent of y is 3.

The sum of the exponents is $1 + 3 = 4$.

So, the degree of the monomial is 4.

c. -3

You can rewrite -3 as $-3x^0$.

The exponent of x is 0.

So, the degree of the monomial is 0.

Remember

For any nonzero number a, $a^0 = 1$.

On Your Own

Now You're Ready
Exercises 7–14

Find the degree of the monomial.

1. $-3x^4$

2. $7c^3d^2$

3. $\frac{5}{3}y$

4. -20.5

Laurie's Notes

Introduction

Connect

- **Yesterday:** Students used algebra tiles to classify polynomials. (MP5)
- **Today:** Students will classify polynomials.

Motivate

- Ask students to make lists of words that begin with the given prefixes.
- Examples:
  - *uni-*: unicycle, unicorn, unification
  - *bi-*: bicycle, bicoastal, bilateral, bifocal
  - *tri-*: tricycle, triangle, triad, triangulate, triathlon
  - *poly-*: polygon, polyhedron, polygraph
- **Extension:** Make a bulletin board display of numeric prefixes and words that use these prefixes (*mono-* or *uni-*, *bi-*, *tri-*, *quad-*, *penta-* or *quint-*, *hex-*, *octo-*, *deka-* or *deca-*).

Discuss

- Discuss with students that today they will use prefixes to describe mathematical expressions. Emphasize the importance of learning the vocabulary.

Lesson Notes

Discuss

- **MP6 Attend to Precision:** Write the two lists shown. Point to the monomials and say, "These are monomials." Point to the other list and say, "These are not monomials." Then ask students to define monomials. Guide them as they refine their definition.
- Write the definition of a monomial and the degree of a monomial. Refer back to the list of monomials and state the degree of each.

Example 1

? "What is the degree of $4x$? Explain." 1; The exponent of x is 1.

? "What does $4x^0$ equal? Explain." 4; $4x^0 = 4(1) = 4$

- Work through the three parts using the correct vocabulary. Repeating the vocabulary (monomial, exponent, degree) will help students become comfortable using these terms.

On Your Own

- **Neighbor Check:** Have students work independently and then have their neighbor check their work. Have students discuss any discrepancies.
- **Extension:** Have students identify the coefficient in the first three questions.

Goal Today's lesson is identifying **monomials** and classifying **polynomials**.

Start Thinking! and Warm Up

Lesson 7.1 Warm Up
For use before Lesson 7.1

Lesson 7.1 Start Thinking!
For use before Lesson 7.1

What are some benefits of using algebra tiles to model polynomials?

Extra Example 1

Find the degree of each monomial.

a. $-\frac{1}{5}y^3$ 3

b. $2s^2t^2$ 4

c. 4.7 0

On Your Own

1. 4
2. 5
3. 1
4. 0

Extra Example 2

Write each polynomial in standard form. Identify the degree and classify each polynomial by the number of terms.

a. $-3n^2 + 5n^3$ $5n^3 - 3n^2$; 3; binomial

b. $8w^5$ $8w^5$; 5; monomial

c. $-1 + 9d^2 + 4d$
$9d^2 + 4d - 1$; 2; trinomial

Extra Example 3

In Example 3, the initial vertical velocity of the baseball is 40 feet per second and the initial height of the baseball is 5 feet.

a. Write a polynomial that represents the height of the baseball.
$-16t^2 + 40t + 5$

b. What is the height of the baseball after 2 seconds? 21 ft

On Your Own

5. $-9z + 4$; 1; binomial
6. $-t^3 + t^2 - 10t$; 3; trinomial
7. $x^3 + 2.8x$; 3; binomial
8. 1 ft

Differentiated Instruction

Visual

Help students become comfortable with the vocabulary in this chapter. Have students make index cards with one of each of the following terms or operations on a card.

4, 7, x, $5x^2$, $6x^3$, $2x^4$, +, +, −, −

Give the name of a polynomial and ask students to make the polynomial from their cards. For instance, if you say, "third-degree binomial," students may make the following expression.

$6x^3$	−	7

Laurie's Notes

Discuss

- Students can become overwhelmed with vocabulary at this point. It is necessary, however, to have words to describe the expressions with which they are working.
- Refer back to the motivate portion of the lesson. Students are familiar with unicycles, bicycles and tricycles. They are all types of bikes, and the prefixes specify the number of wheels. Monomials, binomials and trinomials are all polynomials, and the prefixes specify the number of terms.
- Write several examples and discuss the vocabulary: monomial, binomial and trinomial. Define degree of a polynomial and standard form.

Example 2

? "Why is $3 + 5x$ equivalent to $5x + 3$?" Addition is commutative.

? "Explain how to rewrite $3 - 5x$ so that the x term is first." Rewrite the expression as $3 + (-5x)$ and then use the Commutative Property of Addition to rewrite it as $-5x + 3$.

- As you work through the examples, ask questions about how the expressions are rewritten, what the exponents are, and what the coefficients are.
- Students are sometimes distracted by the use of different variables. Remind students that polynomials can be written using any variable.

Example 3

- The *vertical motion model* used in this example is a polynomial with more than one variable. The v_0 and s_0 are the parameters for a particular problem. Students must use the picture to define the parameters.

? "What is the initial vertical velocity (v_0) of the baseball?" 30 feet per sec

? "What is the initial height (s_0) of the baseball?" 4 feet

- Point out that when the initial vertical velocity and initial height are substituted, the result is a polynomial in one variable (t) and it is in standard form.

On Your Own

- Check that the signs of the coefficients are correct when the polynomial is written in standard form.

Closure

- Write an example of each of the following.
  - **a.** a monomial of degree 2 with a negative coefficient *Sample answer:* $-4x^2$
  - **b.** a binomial of degree 4 *Sample answer:* $y^4 + 3y$
  - **c.** a trinomial of degree 3 *Sample answer:* $2z^3 + z^2 - 5z$

Technology For the Teacher

The Dynamic Planning Tool
Editable Teacher's Resources at *BigIdeasMath.com*

A **polynomial** is a monomial or a sum of monomials. Each monomial is called a *term* of the polynomial.

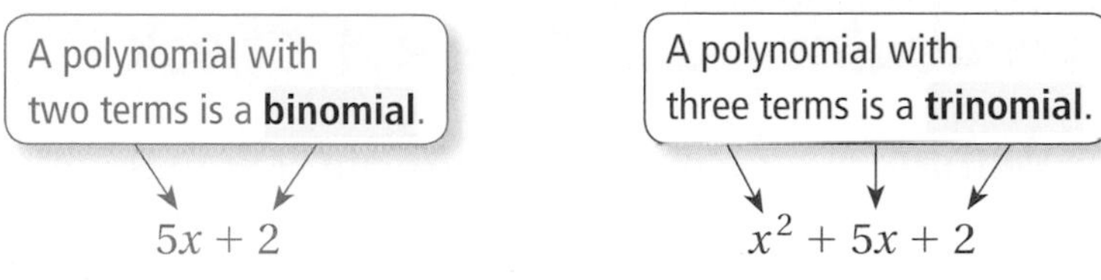

A polynomial with two terms is a **binomial**.
$5x + 2$
A polynomial with three terms is a **trinomial**.
$x^2 + 5x + 2$

The **degree of a polynomial** is the greatest degree of its terms. A polynomial in one variable is in *standard form* when the exponents of the terms decrease from left to right.

EXAMPLE 2 Classifying Polynomials

Write each polynomial in standard form. Identify the degree and classify each polynomial by the number of terms.

	Polynomial	*Standard Form*	*Degree*	*Type of Polynomial*
a.	$-3z^4$	$-3z^4$	4	monomial
b.	$4 + 5x^2 - x$	$5x^2 - x + 4$	2	trinomial
c.	$8q + q^5$	$q^5 + 8q$	5	binomial

EXAMPLE 3 Real-Life Application

The polynomial $-16t^2 + v_0t + s_0$ represents the height (in feet) of an object, where v_0 is the initial vertical velocity (in feet per second), s_0 is the initial height of the object (in feet), and t is the time (in seconds).

a. Write a polynomial that represents the height of the baseball.

$-16t^2 + v_0t + s_0 = -16t^2 + 30t + 4$ — Substitute 30 for v_0 and 4 for s_0.

b. What is the height of the baseball after 1 second?

$-16t^2 + 30t + 4 = -16(1)^2 + 30(1) + 4$ — Substitute 1 for t.

$= -16 + 30 + 4$ — Simplify.

$= 18$ — Add.

The height of the baseball after 1 second is 18 feet.

On Your Own

Now You're Ready
Exercises 15–23

Write the polynomial in standard form. Identify the degree and classify the polynomial by the number of terms.

5. $4 - 9z$ **6.** $t^2 - t^3 - 10t$ **7.** $2.8x + x^3$

8. In Example 3, the initial height is 5 feet. What is the height of the baseball after 2 seconds?

7.1 Exercises

Vocabulary and Concept Check

1. **WRITING** Is $-\frac{\pi}{3}$ a monomial? Explain your reasoning.
2. **VOCABULARY** When is a polynomial in one variable in standard form?
3. **OPEN-ENDED** Write a trinomial of degree 5 in standard form.
4. **WHICH ONE DOESN'T BELONG?** Which expression does *not* belong with the other three? Explain your reasoning.

$a^3 + 4a$	8^x	$b - 2^{-1}$	$-6y^8z$

Practice and Problem Solving

Use algebra tiles to represent the polynomial.

5. $x^2 + 2x - 4$
6. $2x^2 - x + 3$

Find the degree of the monomial.

① 7. $4g$
8. $23x^4$
9. s^8t
10. $-\frac{4}{9}$
11. $1.75k^2$
12. $\frac{1}{8}m^2n^4$
13. 2π
14. $-3q^4rs^6$

Write the polynomial in standard form. Identify the degree and classify the polynomial by the number of terms.

② 15. $7 + 3p^2$
16. $2w^6$
17. $8d - 2 - 4d^3$
18. $6.5c^2 + 1.2c^4 - c$
19. $4v^{11} - v^{12}$
20. $-\frac{1}{4}y - \frac{3}{8}y^2$
21. $7.4z^5$
22. $\sqrt{3}n^7 - 19 + \sqrt{2}n^3$
23. $\pi r^2 - \frac{5}{7}r^8 + 2r^5$

24. **ERROR ANALYSIS** Describe and correct the error in writing the polynomial in standard form.

polynomial: $3m^2 - 5m^5 + m^4$

standard form: $-5m^5 + 3m^2 + m^4$

25. **SPHERE** The expression $\frac{4}{3}\pi r^3$ represents the volume of a sphere with radius r. Why is this expression a monomial? What is its degree?

Assignment Guide and Homework Check

Level	Assignment	Homework Check
Average	1–4, 7–25 odd, 24, 26–32 even, 34–37	9, 17, 25, 28, 32
Advanced	1–4, 8–24 even, 27–33 odd, 30, 32–36	22, 27, 30, 31, 33

Common Errors

- **Exercises 7–14** Students may forget to add the exponents of each of the variables when finding the degree and choose the greatest exponent as the degree. Remind them that the degree is the sum of the exponents of all of the variables.
- **Exercises 15–23** Students may look at the first term of the given polynomial to identify the degree. Remind them that the degree of a polynomial is the greatest degree of its terms.
- **Exercises 31 and 32** Students may forget to find the height of the object after 1 second. Remind them to read the directions carefully.

7.1 Record and Practice Journal

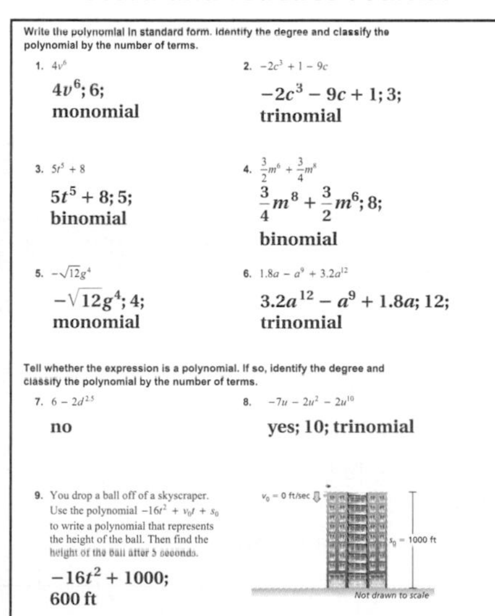

Write the polynomial in standard form. Identify the degree and classify the polynomial by the number of terms.

1. $4v^6$

   $4v^6$; 6; monomial

2. $-2c^3 + 1 - 9c$

   $-2c^3 - 9c + 1$; 3; trinomial

3. $5t^5 + 8$

   $5t^5 + 8$; 5; binomial

4. $\frac{3}{2}m^6 + \frac{3}{4}m^8$

   $\frac{3}{4}m^8 + \frac{3}{2}m^6$; 8; binomial

5. $-\sqrt{12}g^4$

   $-\sqrt{12}g^4$; 4; monomial

6. $1.8a - a^9 + 3.2a^{12}$

   $3.2a^{12} - a^9 + 1.8a$; 12; trinomial

Tell whether the expression is a polynomial. If so, identify the degree and classify the polynomial by the number of terms.

7. $6 - 2d^{2.5}$

   no

8. $-7u - 2u^2 - 2u^{10}$

   yes; 10; trinomial

9. You drop a ball off of a skyscraper. Use the polynomial $-16t^2 + v_0t + s_0$ to write a polynomial that represents the height of the ball. Then find the height of the ball after 5 seconds.

   $-16t^2 + 1000$; 600 ft

$v_0 = 0$ ft/sec
$s_0 = 1000$ ft
Not drawn to scale

Answer Presentation Tool

Vocabulary and Concept Check

1. yes; It is a number.
2. A polynomial in one variable is in standard form when the exponents of the terms decrease from left to right.
3. *Sample answer:* $4x^5 - 2x^3 + 5$
4. 8^x; It is not a polynomial.

Practice and Problem Solving

5. 
6. 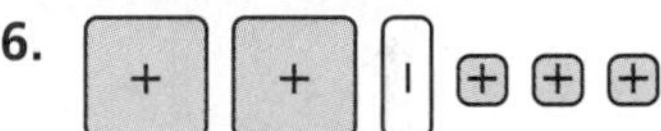
7. 1
8. 4
9. 9
10. 0
11. 2
12. 6
13. 0
14. 11
15. $3p^2 + 7$; 2; binomial
16. $2w^6$; 6; monomial
17. $-4d^3 + 8d - 2$; 3; trinomial
18. $1.2c^4 + 6.5c^2 - c$; 4; trinomial
19. $-v^{12} + 4v^{11}$; 12; binomial
20. $-\frac{3}{8}y^2 - \frac{1}{4}y$; 2; binomial
21. $7.4z^5$; 5; monomial
22. $\sqrt{3}n^7 + \sqrt{2}n^3 - 19$; 7; trinomial
23. $-\frac{5}{7}r^8 + 2r^5 + \pi r^2$; 8; trinomial
24. The exponents of the terms should decrease from left to right; $-5m^5 + m^4 + 3m^2$
25. It is the product of two numbers and a variable with a whole number exponent; 3

Practice and Problem Solving

26. polynomial; 3; monomial
27. *not* a polynomial
28. *not* a polynomial
29. polynomial; 5; trinomial
30.

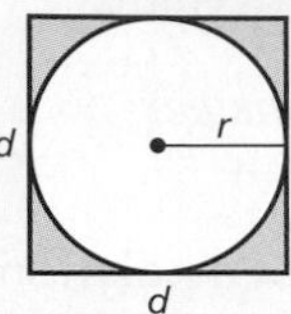

31. $-16t^2 - 45t + 200$; 139 ft
32. $-16t^2 + 16t + 3$; 3 ft
33. See *Taking Math Deeper.*

Fair Game Review

34. $5x + 4y + 13$
35. $3x - 2y$
36. $3x - 11$
37. D

Mini-Assessment

Find the degree of the monomial.

1. $1.4z^5$ 5
2. $-3x^2y$ 3

Write the polynomial in standard form. Identify the degree and classify the polynomial by the number of terms.

3. $-8p + p^3$ $p^3 - 8p$; 3; binomial
4. $9m + 2m^4 - 6m^3$ $2m^4 - 6m^3 + 9m$; 4; trinomial
5. Use the polynomial $-16t^2 + v_0t + s_0$ to write a polynomial that represents the height of a ball when $v_0 = 25$ feet per second and $s_0 = 5$ feet. Then find the height of the ball after 1 second. $-16t^2 + 25t + 5$; 14 ft

Taking Math Deeper

Exercise 33

This problem can be organized with a table.

Using the Commutative Property of Addition and the Distributive Property, you can rewrite the expression $-w^2 + 28w$ as $w(28 - w)$. Because $A = \ell w$ and the width of the rectangle is w, the length of the rectangle is $28 - w$. Try several values of w to see which gives the maximum area.

w	$28 - w$	Area $= w(28 - w)$
10	18	$10(18) = 180$ ft^2
11	17	$11(17) = 187$ ft^2
12	16	$12(16) = 192$ ft^2
13	15	$13(15) = 195$ ft^2
14	14	$14(14) = 196$ ft^2
15	13	$15(13) = 195$ ft^2
16	12	$16(12) = 192$ ft^2
17	11	$17(11) = 187$ ft^2
18	10	$18(10) = 180$ ft^2

a. The maximum area is 196 square feet when the width is 14 feet.

b. Find the perimeter of the garden.

$$P = 2\ell + 2w$$
$$= 2(14) + 2(14)$$
$$= 28 + 28$$
$$= 56 \text{ ft}$$

c. Because each packet covers 7 square feet, you need $196 \div 7 = 28$ packets.

Reteaching and Enrichment Strategies

If students need help. . .	If students got it. . .
Resources by Chapter • Practice A and Practice B • Puzzle Time Record and Practice Journal Practice Differentiating the Lesson Lesson Tutorials Skills Review Handbook	Resources by Chapter • Enrichment and Extension Start the next section

Tell whether the expression is a polynomial. If so, identify the degree and classify the polynomial by the number of terms.

26. $-g^3$

27. $7^x - 2x^2$

28. $y^{-3} + 1.5$

29. $8k^5 + 4k^3 - k$

30. LOGIC The polynomial $d^2 - \pi r^2$ represents the area of a region, where d is the diameter of a circle and r is the radius of the circle. How can this happen? Justify your answer with a diagram.

Use the polynomial $-16t^2 + v_0 t + s_0$ to write a polynomial that represents the height of the object. Then find the height of the object after 1 second.

3 **31. WATER BALLOON** You throw a water balloon from a building.

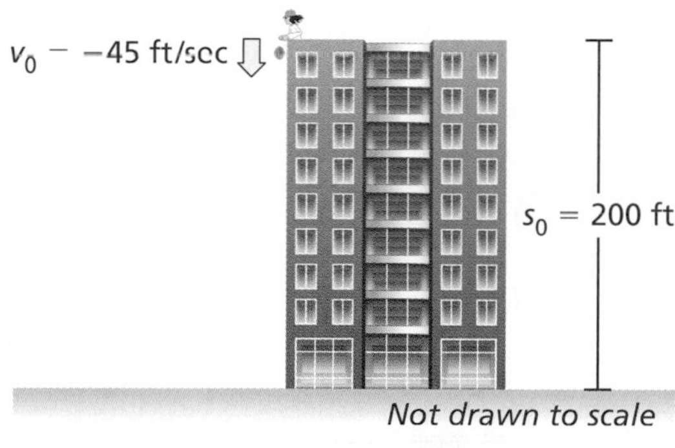

32. TENNIS You bounce a tennis ball on a racket.

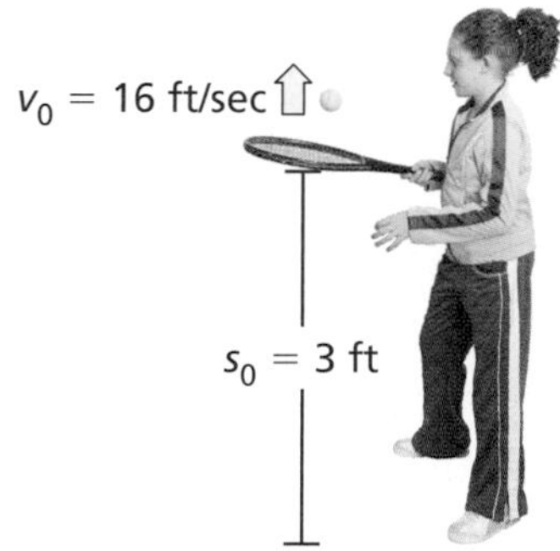

33. Number Sense The polynomial $-w^2 + 28w$ represents the area of a rectangular garden with a width of w feet.

a. Use guess, check, and revise to find the width of the garden with the maximum area. (*Hint:* The width is between 10 feet and 18 feet.)

b. What is the perimeter of the garden?

c. How many seed packets do you need for the garden?

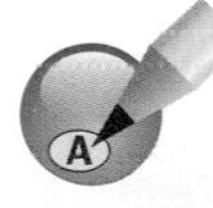

Fair Game Review What you learned in previous grades & lessons

Simplify the expression. (*Skills Review Handbook*)

34. $2x + 4y + 3x + 13$

35. $4x - x + 5y - 7y$

36. $-11 + 5x - 3x + x$

37. MULTIPLE CHOICE What is the surface area of the prism? (*Skills Review Handbook*)

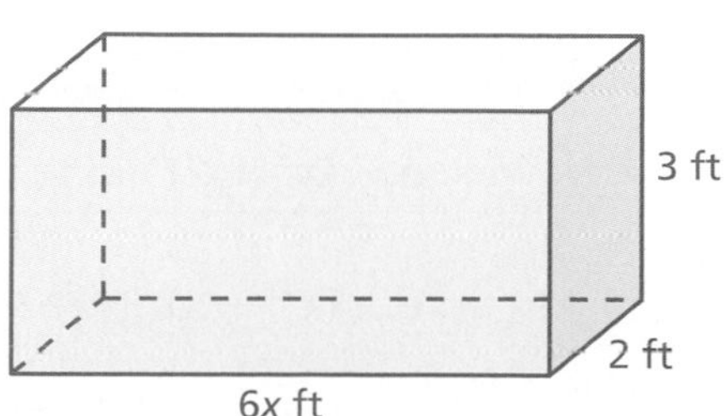

Ⓐ $11x$ ft Ⓑ $(24x + 36)$ ft

Ⓒ $(36x + 24)$ ft Ⓓ $(60x + 12)$ ft

7.2 Adding and Subtracting Polynomials

COMMON CORE STATE STANDARDS
A.APR.1

Essential Question How can you add polynomials? How can you subtract polynomials?

1 EXAMPLE: Adding Polynomials Using Algebra Tiles

Work with a partner. Six different algebra tiles are shown at the right.

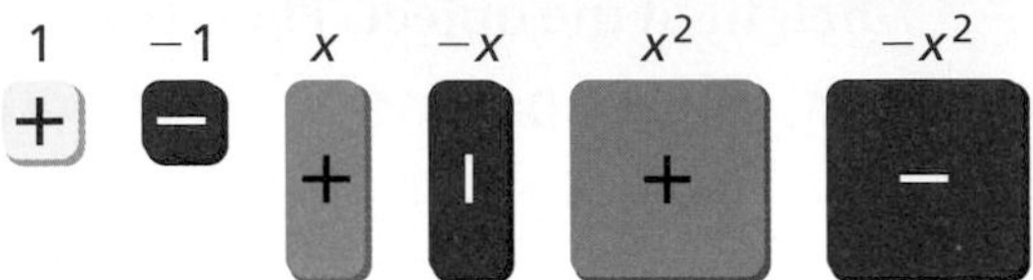

Write the polynomial addition steps shown by the algebra tiles.

Step 1: 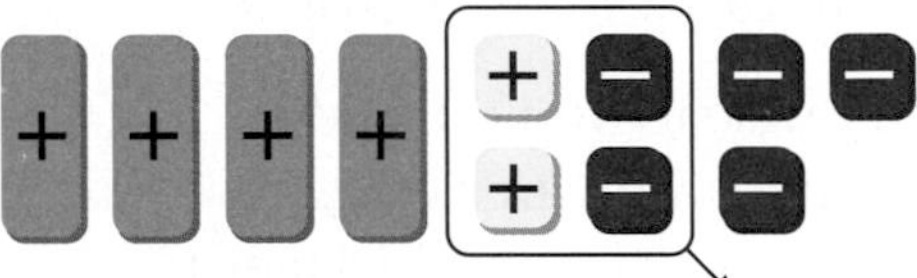 Group like tiles.

Step 2: Remove zero pairs.

Step 3: 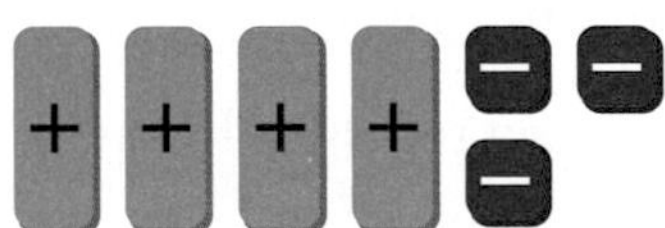 Simplify.

2 ACTIVITY: Adding Polynomials Using Algebra Tiles

Use algebra tiles to find the sum of the polynomials.

a. $(x^2 + 2x - 1) + (2x^2 - 2x + 1)$

b. $(4x + 3) + (x - 2)$

c. $(x^2 + 2) + (3x^2 + 2x + 5)$

d. $(2x^2 - 3x) + (x^2 - 2x + 4)$

e. $(x^2 - 3x + 2) + (x^2 + 4x - 1)$

f. $(4x - 3) + (2x + 1) + (-3x + 2)$

g. $(-x^2 + 3x) + (2x^2 - 2x)$

h. $(x^2 + 2x - 5) + (-x^2 - 2x + 5)$

Laurie's Notes

Introduction

Standards for Mathematical Practice

- **MP7 Look for and Make Use of Structure:** The shapes of algebra tiles are a visual clue that when you add $2x^2$ and $3x$, you do not get $5x^3$. The shapes are different so you cannot combine them by addition or subtraction. Mathematically proficient students see that $2x^2 + 3x$ means you have 2 of something (x^2-term) being added to 3 of something different (x-term).

Motivate

- Share a bit of "Did You Know?" with students today.
- The modern addition and subtraction symbols + and − were first used in Germany during the fifteenth century. They have been in use for less than 600 years.
- Ancient Greeks, such as Aristotle, used letters to represent numbers. So, variables were being used long before addition and subtraction symbols.

Activity Notes

Example 1

- Students should be comfortable with integer operations and with using algebra tiles to represent polynomials. You may want to review the addition and subtraction of integers with algebra tiles.
- Once the expressions are added (pushed together), like terms (like tiles) are grouped and zero pairs are removed.
- **MP6 Attend to Precision:** Do not allow students to use the phrase, "The positive and negative cancel each other, or cancel out." Encourage them to refer to this as the Additive Inverse Property, $a + (-a) = 0$.
- Write the problem and solution: $(3x + 2) + (x - 5) = 4x - 3$.

? "Do you think the sum of two binomials is always a binomial?" Students should be able to give examples where the sum is not a binomial.

Activity 2

- Students may be tempted to not use algebra tiles in this activity.
- As you walk around the room, ask questions about each expression using vocabulary from the previous lesson. How many terms are there? What is the coefficient of the x-term? Is the sum a trinomial?
- Students should have a record of their work and the solution.
- By manipulating the tiles, students are developing a strong sense of combining like terms and adding opposites.

Common Core State Standards

A.APR.1 Understand that polynomials form a system analogous to the integers, namely, they are closed under the operations of addition, subtraction, and multiplication; add, subtract, and multiply polynomials.

Previous Learning

Students should know how to identify the like terms in an algebraic expression.

Activity Materials
Textbook
• algebra tiles

Start Thinking! and Warm Up

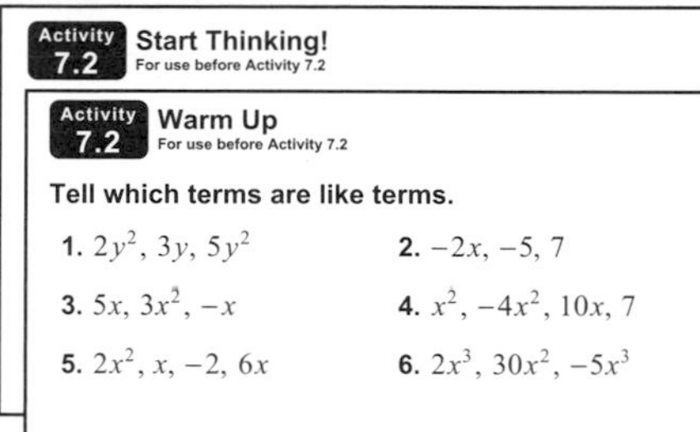
Activity 7.2 Start Thinking!
For use before Activity 7.2

Activity 7.2 Warm Up
For use before Activity 7.2

Tell which terms are like terms.

1. $2y^2, 3y, 5y^2$
2. $-2x, -5, 7$
3. $5x, 3x^2, -x$
4. $x^2, -4x^2, 10x, 7$
5. $2x^2, x, -2, 6x$
6. $2x^3, 30x^2, -5x^3$

7.2 Record and Practice Journal

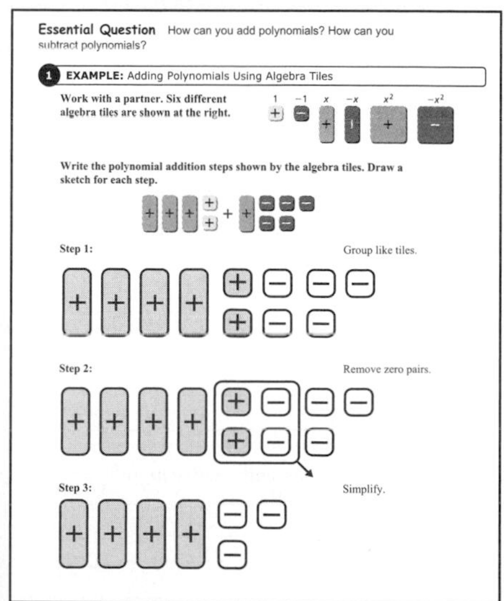
Essential Question How can you add polynomials? How can you subtract polynomials?

1 EXAMPLE: Adding Polynomials Using Algebra Tiles

Work with a partner. Six different algebra tiles are shown at the right.

Write the polynomial addition steps shown by the algebra tiles. Draw a sketch for each step.

Step 1: Group like tiles.

Step 2: Remove zero pairs.

Step 3: Simplify.

Differentiated Instruction

Organization

When subtracting polynomials, students may not subtract each term of the subtrahend. Have students break down the problem into mini-problems.

$(x^2 + 2x - 1) - (2x^2 - 2x + 1)$

$x^2 - 2x^2 = -x^2$

$2x - (-2x) = 2x + 2x = 4x$

$-1 - 1 = -2$

So, $(x^2 + 2x - 1) - (2x^2 - 2x + 1)$
$= -x^2 + 4x - 2.$

7.2 Record and Practice Journal

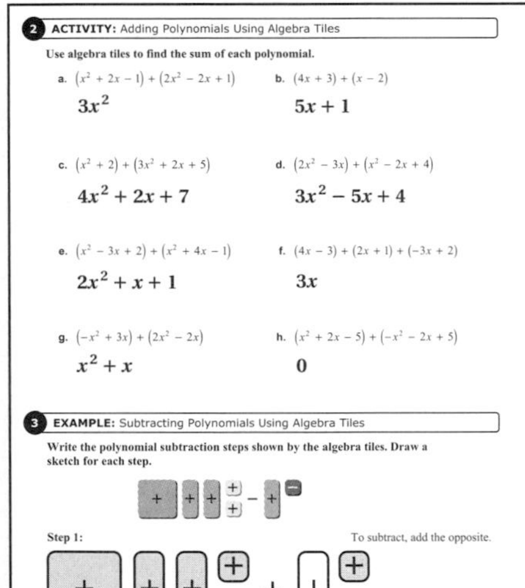

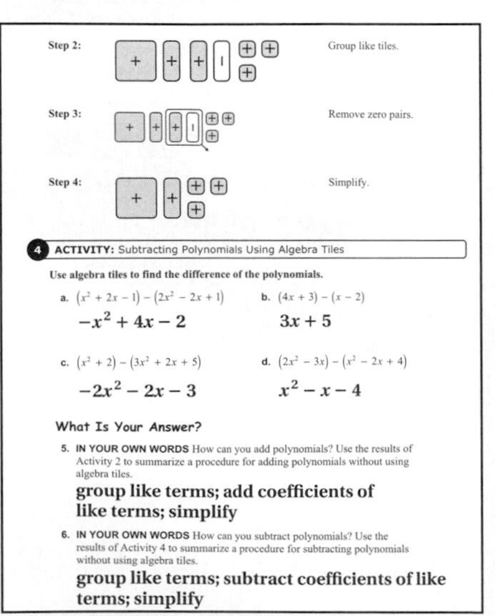

Laurie's Notes

Example 3

? "By using algebra tiles to subtract integers, we discovered that subtraction was the same as what?" adding the opposite

- Explain that Step 2 shows how to add the opposite. Change the subtraction symbol to an addition symbol and flip over each of the tiles being subtracted (or swap them out if the tiles are not two-sided).
- Write the problem and solution: $(x^2 + 2x + 2) - (x - 1) = x^2 + x + 3$.

Activity 4

- Students may be tempted to not use algebra tiles in this activity.
- As you walk around the room, ask questions about each expression using vocabulary from the previous lesson. When you subtracted the trinomial, what happened to the coefficient of each term?
- Students should have a record of their work and the solution.
- Without tiles, students often say that part (c) cannot be done because you cannot subtract $3x^2$ from nothing. But, by modeling the expressions and "adding the opposite," students will see that you obtain $-3x^2$.
- **Big Idea:** After students finish modeling these subtraction problems, they should understand that the sign of each term being subtracted changes.

What Is Your Answer?

- For Question 6, listen for the idea that you add the opposite of the polynomial being subtracted.

Closure

- Write an example to show the following:
  a. The sum of two binomials can be a trinomial.
  Sample answer: $(x^2 + 1) + (x + 1) = x^2 + x + 2$
  b. The difference of two binomials can be a monomial.
  Sample answer: $(2x + 1) - (x + 1) = x$

Technology For the Teacher

The Dynamic Planning Tool
Editable Teacher's Resources at *BigIdeasMath.com*

3 EXAMPLE: Subtracting Polynomials Using Algebra Tiles

Write the polynomial subtraction steps shown by the algebra tiles.

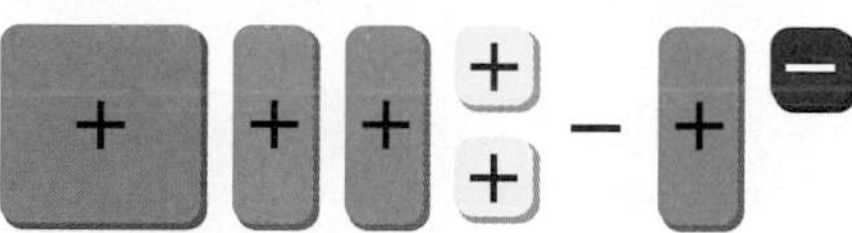

Step 1: 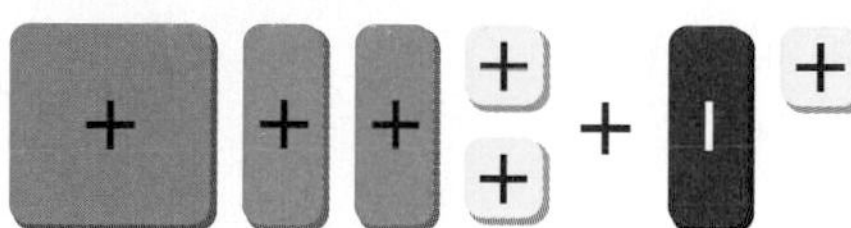 To subtract, add the opposite.

Step 2: 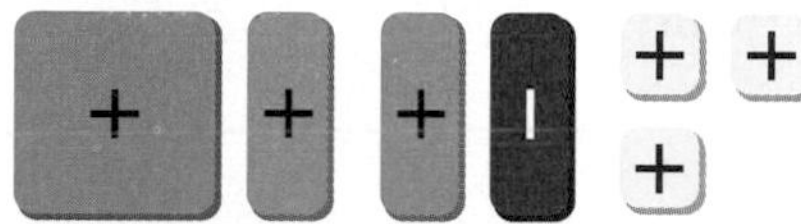 Group like tiles.

Step 3: 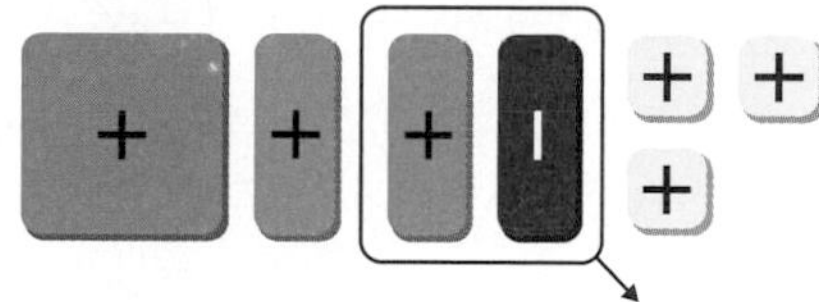 Remove zero pairs.

Step 4: 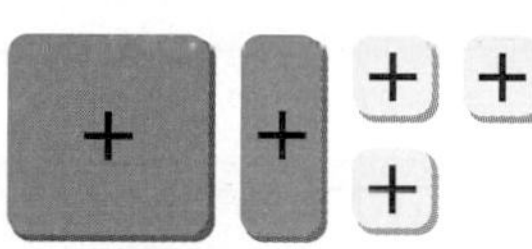 Simplify.

4 ACTIVITY: Subtracting Polynomials Using Algebra Tiles

Use algebra tiles to find the difference of the polynomials.

a. $(x^2 + 2x - 1) - (2x^2 - 2x + 1)$

b. $(4x + 3) - (x - 2)$

c. $(x^2 + 2) - (3x^2 + 2x + 5)$

d. $(2x^2 - 3x) - (x^2 - 2x + 4)$

What Is Your Answer?

5. **IN YOUR OWN WORDS** How can you add polynomials? Use the results of Activity 2 to summarize a procedure for adding polynomials without using algebra tiles.

6. **IN YOUR OWN WORDS** How can you subtract polynomials? Use the results of Activity 4 to summarize a procedure for subtracting polynomials without using algebra tiles.

Practice Use what you learned about adding and subtracting polynomials to complete Exercises 3 and 4 on page 338.

7.2 Lesson

You can add polynomials using a vertical or horizontal method to combine like terms.

EXAMPLE 1 Adding Polynomials

Find each sum.

a. $(3a^2 + 8) + (5a - 1)$ **b.** $(-x^2 + 5x + 4) + (3x^2 - 8x + 9)$

a. Vertical method: Align like terms vertically and add.

$$\begin{array}{r} 3a^2 \qquad\quad + 8 \\ + \qquad\quad 5a - 1 \\ \hline 3a^2 + 5a + 7 \end{array}$$

Leave a space for the missing term.

b. Horizontal method: Group like terms and simplify.

$$(-x^2 + 5x + 4) + (3x^2 - 8x + 9) = (-x^2 + 3x^2) + [5x + (-8x)] + (4 + 9)$$
$$= 2x^2 - 3x + 13$$

To subtract one polynomial from another polynomial, add the opposite.

EXAMPLE 2 Subtracting Polynomials

Find each difference.

a. $(y^2 + 4y + 2) - (2y^2 - 5y - 3)$ **b.** $(5x^2 + 4x - 1) - (2x^2 - 6)$

a. Use the vertical method.

$$\begin{array}{r} (y^2 + 4y + 2) \\ -\ (2y^2 - 5y - 3) \\ \hline \end{array} \quad \xrightarrow{\text{Add the opposite.}} \quad \begin{array}{r} y^2 + 4y + 2 \\ +\ -2y^2 + 5y + 3 \\ \hline -y^2 + 9y + 5 \end{array}$$

Study Tip

You can think of finding the opposite of a polynomial as finding the opposite of each term's coefficient.

b. Use the horizontal method.

$$(5x^2 + 4x - 1) - (2x^2 - 6) = (5x^2 + 4x - 1) + (-2x^2 + 6)$$
$$= [5x^2 + (-2x^2)] + 4x + (-1 + 6)$$
$$= 3x^2 + 4x + 5$$

On Your Own

Now You're Ready
Exercises 5–10 and 12–17

Find the sum or difference.

1. $(b - 10) + (4b - 3)$
2. $(x^2 - x - 2) + (7x^2 - x)$
3. $(p^2 + p + 3) - (-4p^2 - p + 3)$
4. $(-k + 5) - (3k^2 - 6)$

Laurie's Notes

Introduction

Connect

- **Yesterday:** Students used algebra tiles to add and subtract polynomials. (MP6, MP7)
- **Today:** Students will add and subtract polynomials.

Motivate

? "What is the sum of 8.2 and 0.47?" 8.67

? "Why is $8.2 + 0.47 \neq 12.9$?" You line up the decimal points before you add the decimals.

- **MP7 Look for and Make Use of Structure:** When you add and subtract whole numbers or decimals, you add and subtract the same (like) place values. Discuss with students that when you add or subtract polynomials, you add and subtract like terms.

Lesson Notes

Example 1

- Explain that when adding and subtracting polynomials, a vertical or horizontal format can be used, similar to adding and subtracting whole numbers.

? "How do you decide whether to write the addition of two whole numbers horizontally or vertically?" Answers will vary.

? "What is the degree of the binomial $3a^2 + 8$?" 2 "What is the degree of $5a - 1$?" 1

? "Can you add $3a^2$ and $5a$? Explain." no; They are not like terms.

- In part (a), point out why the space is left when writing $3a^2 + 8$. It is similar to writing the 0 in 308 when you add 308 and 51.
- In part (b), some students may be comfortable grouping the like terms mentally and simply recording the sum.

Example 2

- Subtracting polynomials is generally more challenging for students because they need to be careful with the signs of the terms.
- Remind students of the rule for subtracting integers: $a - b = a + (-b)$, and relate this back to changing signs of terms when subtracting polynomials in the activity.

? "What polynomial is being subtracted in part (a)?" $2y^2 - 5y - 3$

- In part (a), point out that adding the opposite is similar to distributing the negative.
- **Teaching Tip:** Before beginning part (b), ask students to identify like terms.

On Your Own

- Have students work in pairs to answer the questions. Ask volunteers to share their work at the board.

Goal Today's lesson is adding and subtracting polynomials.

Start Thinking! and Warm Up

Lesson 7.2 Warm Up
For use before Lesson 7.2

Lesson 7.2 Start Thinking!
For use before Lesson 7.2

When adding or subtracting polynomials, do you think it is easier to use algebra tiles or a pencil and paper? Explain.

Extra Example 1

Find each sum.

a. $(-x^2 + 4) + (3x - 9)$ $-x^2 + 3x - 5$

b. $(-3q^2 - 2q + 7) + (5q^2 + q - 4)$
$2q^2 - q + 3$

Extra Example 2

Find each difference.

a. $(2w^2 + 3w - 5) - (w^2 - 8w + 2)$
$w^2 + 11w - 7$

b. $(7d^2 - 8d + 6) - (3d^2 + 4)$
$4d^2 - 8d + 2$

On Your Own

1. $5b - 13$
2. $8x^2 - 2x - 2$
3. $5p^2 + 2p$
4. $-3k^2 - k + 11$

Extra Example 3

Find the sum of $s^2 + 5st - t^2$ and $s^2 - 2st + t^2$. $2s^2 + 3st$

Extra Example 4

In Example 4, the polynomial $-16t^2 - 30t + 200$ represents the height of the penny after *t* seconds and the polynomial $-16t^2 - 5t + 100$ represents the height of the paintbrush after *t* seconds.

a. Write a polynomial that represents the distance between the penny and the paintbrush after *t* seconds. $-25t + 100$

b. What is the distance between the objects after 1 second? 75 ft

On Your Own

5. B
6. 75 ft

English Language Learners

Visual

English language learners may find it easier to grasp the concept of addition and subtraction of like terms if terms of the polynomials are color-coded. This will be especially helpful in problems where a term does not appear in one of the polynomials.

Laurie's Notes

Example 3

- Students will find this example challenging due to the use of two variables.
- ? "What operation is being performed on the polynomials?" addition
- Ask students to identify all of the like terms before grouping them.
- If time permits, show the solution of this example using the vertical format.
- **Extension:** Remind students what it means for a set of numbers to be *closed* under an operation. Ask students if polynomials are closed under addition and subtraction.

Example 4

- **FYI:** This example uses the *vertical motion model* (the penny) and the *falling object model* (the paintbrush). These two models are commonly used in application questions in algebra.
- ? "What is the distance between the penny and the paintbrush before they are released?" 100 feet
- ? "Do you think the distance between the two objects will stay the same or change? Explain." Answers will vary.
- Interpret the meaning of the answer for part (a). The model for the distance between the two objects is $-40t + 100$. This is a linear model. The initial difference is 100 feet and the difference decreases at a rate of 40 feet per second.
- **Extension:** Make a table of values to evaluate the difference in the distances at 0, 0.5, 1, 1.5, and 2 seconds. Plot the points. The graph is linear with a slope of -40 and a *y*-intercept of 100.

On Your Own

- In Question 6, the penny now has an initial vertical velocity of -25 feet per second instead of -40 feet per second. Discuss with students how changing the initial vertical velocity affects the solution.

Closure

- Describe how to use the vertical and horizontal methods to add and subtract polynomials. Vertical method: Line up like terms vertically. Then add or subtract; Horizontal method: Group like terms. Then add or subtract.

Technology For the Teacher

The Dynamic Planning Tool
Editable Teacher's Resources at *BigIdeasMath.com*

EXAMPLE 3 Standardized Test Practice

Which polynomial represents the sum of $x^2 - 2xy - y^2$ and $x^2 + xy + y^2$?

Ⓐ $-3xy$ Ⓑ $-3xy - 2y^2$ Ⓒ $2x^2 - xy$ Ⓓ $2x^2 + 3xy + 2y^2$

Use the horizontal method to find the sum.

$$(x^2 - 2xy - y^2) + (x^2 + xy + y^2) = (x^2 + x^2) + (-2xy + xy) + (-y^2 + y^2)$$
$$= 2x^2 - xy$$

∴ The correct answer is Ⓒ.

EXAMPLE 4 Real-Life Application

A penny is thrown straight downward from a height of 200 feet. At the same time, a paintbrush falls from a height of 100 feet. The polynomials represent the heights (in feet) of the objects after t seconds.

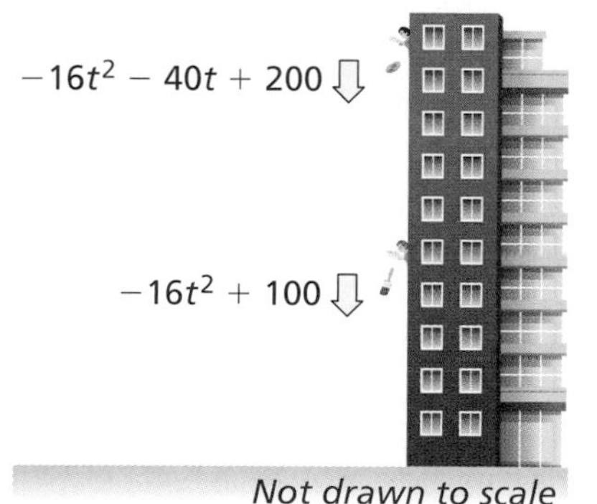

Not drawn to scale

a. Write a polynomial that represents the distance between the penny and the paintbrush after t seconds.

To find the distance between the objects after t seconds, subtract the polynomials.

Penny	***Paintbrush***

$$(-16t^2 - 40t + 200) - (-16t^2 + 100)$$
$$= (-16t^2 - 40t + 200) + (16t^2 - 100)$$
$$= (-16t^2 + 16t^2) - 40t + [200 + (-100)]$$
$$= -40t + 100$$

∴ The polynomial $-40t + 100$ represents the distance between the objects after t seconds.

b. What is the distance between the objects after 2 seconds?

Find the value of $-40t + 100$ when $t = 2$.

$-40t + 100 = -40(2) + 100$ Substitute 2 for t.

$= 20$ Simplify.

∴ After 2 seconds, the distance between the objects is 20 feet.

Study Tip

To check your answer, substitute 2 into the original polynomials and verify that the difference of the heights is 20.

On Your Own

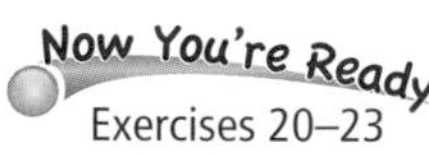

5. In Example 3, which polynomial represents the difference of the two polynomials?
6. In Example 4, the polynomial $-16t^2 - 25t + 200$ represents the height of the penny after t seconds. What is the distance between the objects after 1 second?

7.2 Exercises

Vocabulary and Concept Check

1. **WRITING** How do you add $(4x^2 - 3 + 2y^3)$ and $(-6x^2 - 15)$ using a vertical method? a horizontal method?
2. **REASONING** Describe how subtracting polynomials is similar to subtracting integers.

Practice and Problem Solving

Use algebra tiles to find the sum or difference of the polynomials.

3. $(x^2 - 3x + 2) + (x^2 + 4x - 1)$
4. $(x^2 + 2x - 5) - (-x^2 - 2x + 5)$

Find the sum.

1 5. $(5y + 4) + (-2y + 6)$
6. $(3g^2 - g) + (3g^2 - 8g + 4)$
7. $(2n^2 - 5n - 6) + (-n^2 - 3n + 11)$
8. $(-3p^2 + 5p - 2) + (-p^2 - 8p - 15)$
9. $(-a^3 + 4a - 3) + (5a^3 - a)$
10. $\left(-s^2 - \frac{2}{9}s + 1\right) + \left(-\frac{5}{9}s - 4\right)$

11. **ERROR ANALYSIS** Describe and correct the error in finding the sum of the polynomials.

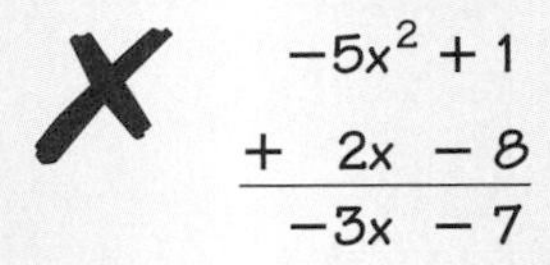

$$\begin{array}{r} -5x^2 \qquad + 1 \\ + \quad 2x \; - 8 \\ \hline -3x \; - 7 \end{array}$$

Find the difference.

2 12. $(d^2 - 9) - (3d - 1)$
13. $(k^2 - 7k + 2) - (k^2 - 12)$
14. $(x^2 - 4x + 9) - (3x^2 - 6x - 7)$
15. $(-r - 10) - (-4r^2 + r + 7)$
16. $(t^4 - t^2 + t) - (-9t^2 + 7t - 12)$
17. $\left(\frac{1}{6}q^2 + \frac{2}{3}\right) - \left(\frac{1}{12}q^2 - \frac{1}{3}\right)$

18. **ERROR ANALYSIS** Describe and correct the error in finding the difference of the polynomials.

$$\begin{aligned} (x^2 - 5x) - (-3x^2 + 2x) &= (x^2 - 5x) + (3x^2 + 2x) \\ &= (x^2 + 3x^2) + (-5x + 2x) \\ &= 4x^2 - 3x \end{aligned}$$

19. **COST** The cost (in dollars) of making *b* bracelets is represented by $4 + 5b$. The cost (in dollars) of making *b* necklaces is $8b + 6$. Write a polynomial that represents how much more it costs to make *b* necklaces than *b* bracelets.

Assignment Guide and Homework Check

Level	Assignment	Homework Check
Average	1, 2, 5–19 odd, 18, 20–24 even, 26–29	7, 13, 19, 22, 24
Advanced	1, 2, 6–18 even, 11, 21–25 odd, 24, 26–29	8, 16, 21, 24, 25

Common Errors

- **Exercises 5–10** When using the vertical method to add polynomials, students may forget to leave a space for missing terms. Remind them that they cannot combine variable terms with different exponents.
- **Exercises 12–17** Students may add the opposite of only the first term of the second polynomial. Remind them to add the opposite of all of the terms of the second polynomial.
- **Exercises 20–23** Students may have difficulty grouping like terms. Encourage them to look at the variable(s) and exponent(s) in each term.

7.2 Record and Practice Journal

Find the sum.

1. $(2d + 3) + (4d - 6)$
   $6d - 3$
2. $(5m^2 - m + 2) + (3m^2 + 10)$
   $8m^2 - m + 12$
3. $(2t^2 - 6t - 3) + (-9t^2 + 9t - 5)$
   $-7t^2 + 3t - 8$
4. $(4c^2 - 8c + 7) + (c^4 + 11c - 3)$
   $c^4 + 4c^2 + 3c + 4$

Find the difference.

5. $(3s + 4) - (6s^2 - 2s)$
   $-6s^2 + 5s + 4$
6. $(9w^2 - 5) - (4w^2 + 9w + 7)$
   $5w^2 - 9w - 12$
7. $(y^2 - 6y + 12) - (-3y^2 - 6y + 10)$
   $4y^2 + 2$
8. $(8z^3 + 6z^2 - 9) - (4z^2 - 7z - 4)$
   $8z^3 + 2z^2 + 7z - 5$
9. You are installing a swimming pool. Write a polynomial that represents the area of the walkway.
   $20x^2 + 12x + 32$

8 ft
(2x + 6) ft
(6x − 4) ft
10x ft

Answer Presentation Tool

Vocabulary and Concept Check

1. Align like terms vertically and add; Group like terms and simplify.
2. When subtracting polynomials or integers, you add the opposite.

Practice and Problem Solving

3. $2x^2 + x + 1$
4. $2x^2 + 4x - 10$
5. $3y + 10$
6. $6g^2 - 9g + 4$
7. $n^2 - 8n + 5$
8. $-4p^2 - 3p - 17$
9. $4a^3 + 3a - 3$
10. $-s^2 - \frac{7}{9}s - 3$
11. Like terms are not aligned.
$$\begin{array}{r} -5x^2 \qquad\quad + 1 \\ + \qquad\quad 2x - 8 \\ \hline -5x^2 + 2x - 7 \end{array}$$
12. $d^2 - 3d - 8$
13. $-7k + 14$
14. $-2x^2 + 2x + 16$
15. $4r^2 - 2r - 17$
16. $t^4 + 8t^2 - 6t + 12$
17. $\frac{1}{12}q^2 + 1$
18. $2x$ was added instead of the opposite of $2x$.
$$(x^2 - 5x) - (-3x^2 + 2x)$$
$$= (x^2 + 3x^2) + (-5x - 2x)$$
$$= 4x^2 - 7x$$
19. $3b + 2$

Practice and Problem Solving

20. $2c^2 - 2cd - 4d^2$

21. $-2x^2 + 3xy + 8y^2$

22. $s^2 - 12st$

23. $-3a^2 + 2ab + b^2$

24. **a.** $x^2 - 2x;\ x^2 - 12x$

b. $2x^2 - 14x$

c. 520 ft^2

d. 1.3 gallons;

$$520\ \text{ft}^2 \times \frac{1\ \text{gal}}{400\ \text{ft}^2} = 1.3\ \text{gal}$$

25. See *Taking Math Deeper.*

Fair Game Review

26. $5x + 4$

27. $2y + 7$

28. $2w - 11$

29. B

Mini-Assessment

Find the sum or difference.

1. $(a^2 - 7) + (2a^2 - a + 8)$
$3a^2 - a + 1$

2. $(8p + 2) - (6p - 4)$ $2p + 6$

3. $(4t^2 - 9t + 3) - (2t^2 - 5t - 4)$
$2t^2 - 4t + 7$

4. $(6n^2 + 3n - 9) + (-5n^2 - 4n + 1)$
$n^2 - n - 8$

5.
$$\begin{array}{r} (2x^2 + 5xy + 3y^2) \\ -\ (x^2 + 3xy + y^2) \\ \hline x^2 + 2xy + 2y^2 \end{array}$$

Taking Math Deeper

Exercise 25

This is a typical vertical motion problem.

a. Write a polynomial that represents the difference between the two heights.

$$(-16t^2 + 98) - (-16t^2 + 46t + 6)$$
$$= -16t^2 + 98 + 16t^2 - 46t - 6$$
$$= -46t + 92$$

b. Find the difference between the two heights after 1.5 seconds.

$$-46t + 92 = -46(1.5) + 92$$
$$= -69 + 92$$
$$= 23 \text{ ft}$$

c. The balls are at the same height when the difference between their heights is 0 feet. Set $-46t + 92$ equal to 0 and solve for t.

$$-46t + 92 = 0$$
$$-46t = -92$$
$$t = \frac{-92}{-46}$$
$$t = 2 \text{ sec}$$

To find the distance from the ground, substitute 2 for t in either equation.

$$-16t^2 + 98 = -16(2)^2 + 98$$
$$= 34$$

The balls are 34 feet from the ground when they are at the same height.

Reteaching and Enrichment Strategies

If students need help. . .	If students got it. . .
Resources by Chapter • Practice A and Practice B • Puzzle Time Record and Practice Journal Practice Differentiating the Lesson Lesson Tutorials Skills Review Handbook	Resources by Chapter • Enrichment and Extension Start the next section

Find the sum or difference.

3 **20.** $(c^2 - 6d^2) + (c^2 - 2cd + 2d^2)$

21. $(-x^2 + 9xy) - (x^2 + 6xy - 8y^2)$

22. $(2s^2 - 5st - t^2) - (s^2 + 7st - t^2)$

23. $(a^2 - 3ab + 2b^2) + (-4a^2 + 5ab - b^2)$

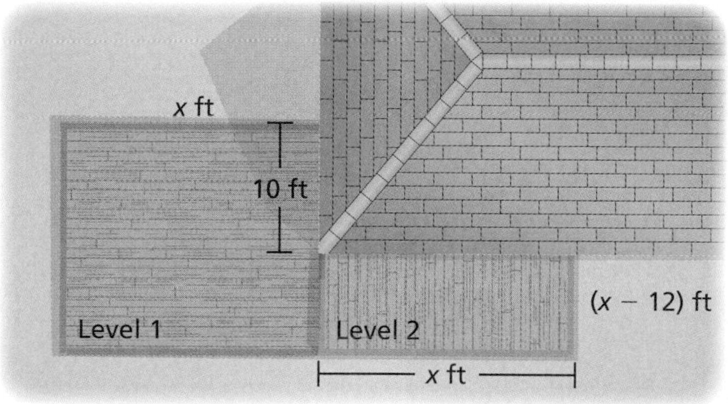

24. MODELING You are building a multi-level deck.

a. Write a polynomial that represents the area of each level.

b. Write a polynomial that represents the total area of the deck.

c. What is the total area of the deck when $x = 20$?

d. A gallon of deck sealant covers 400 square feet. How many gallons of sealant do you need to cover the deck once? Explain.

25. Problem Solving You drop a ball from a height of 98 feet. At the same time, your friend throws a ball upward. The polynomials represent the heights (in feet) of the balls after t seconds.

a. Write a polynomial that represents the distance between your ball and your friend's ball after t seconds.

b. What is the distance between the balls after 1.5 seconds?

c. After how many seconds are the balls at the same height? How far are they from the ground? Explain your reasoning.

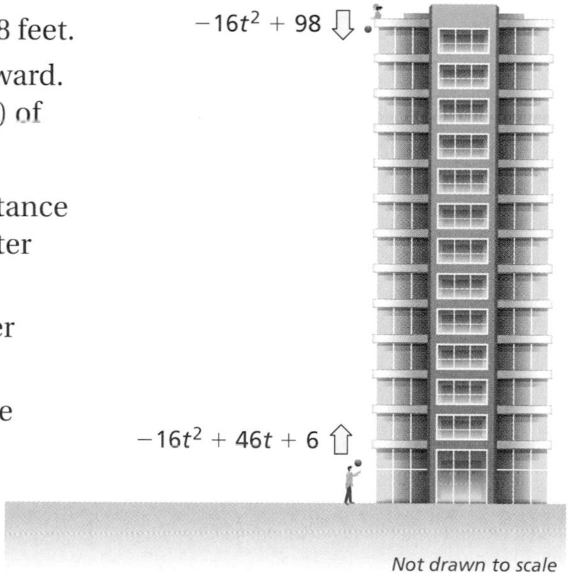

Not drawn to scale

Fair Game Review What you learned in previous grades & lessons

Simplify the expression. *(Skills Review Handbook)*

26. $2(x - 1) + 3(x + 2)$

27. $(4y - 3) - 2(y - 5)$

28. $-5(2w + 1) - 3(-4w + 2)$

29. MULTIPLE CHOICE Which inequality is represented by the graph? *(Section 3.1)*

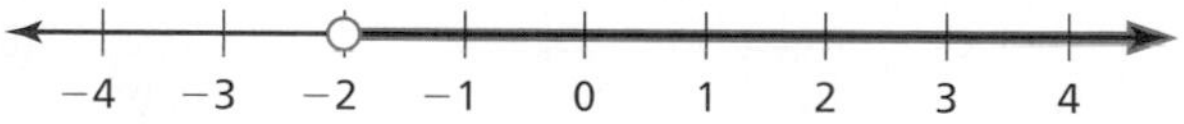

Ⓐ $x < -2$ Ⓑ $x > -2$ Ⓒ $x \leq -2$ Ⓓ $x \geq -2$

7.3 Multiplying Polynomials

Essential Question How can you multiply two binomials?

COMMON CORE STATE STANDARDS
A.APR.1

1 ACTIVITY: Multiplying Binomials Using Algebra Tiles

Work with a partner. Six different algebra tiles are shown below.

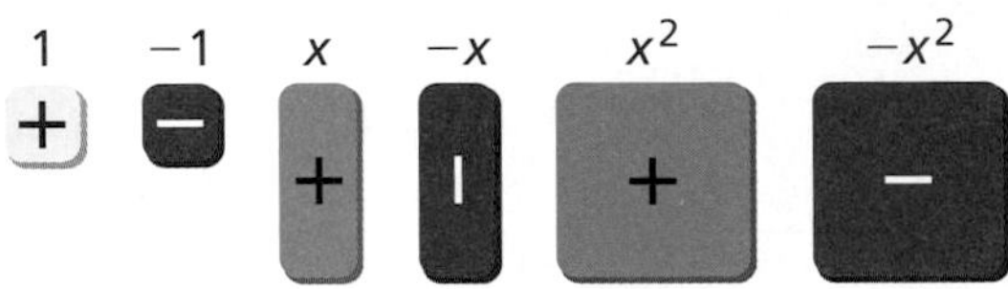

Write the product of the two binomials shown by the algebra tiles.

a. $(x + 3)(x - 2) =$ ______

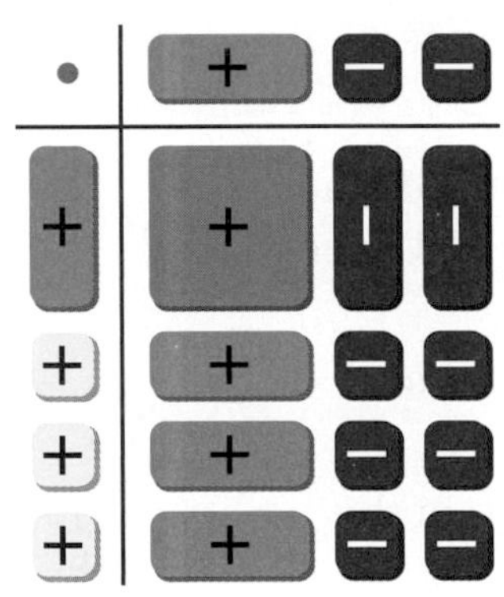

b. $(2x - 1)(2x + 1) =$ ______

c. $(x + 2)(2x - 1) =$ ______

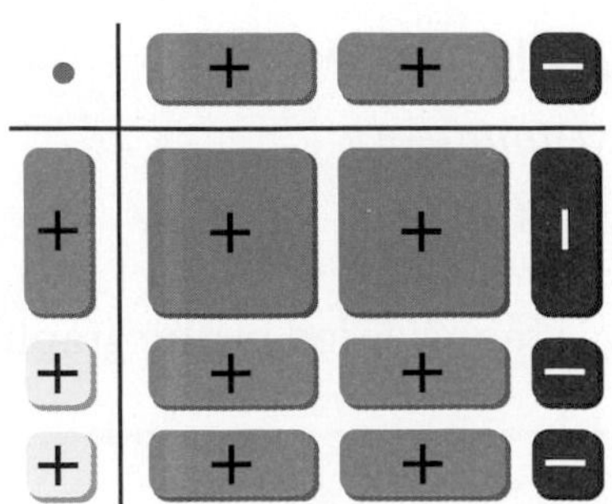

d. $(-x - 2)(x - 3) =$ ______

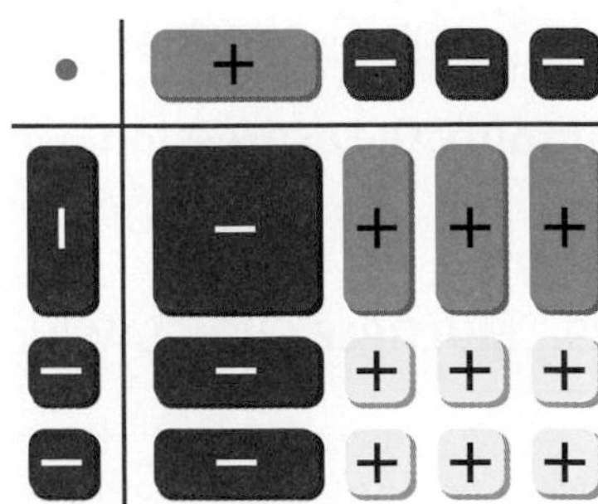

Laurie's Notes

Introduction

Standards for Mathematical Practice

- **MP1a Make Sense of Problems** and **MP7 Look for and Make Use of Structure:** Working with algebra tiles in this activity should be reminiscent of using base 10 pieces when performing multi-digit multiplication. The connection to whole number multiplication, and how it is represented, should make sense to students. The underlying structure of the problem is the same.

Motivate

- Tell students that you want to take them back a few years. Draw the following and ask them what the model represents.

	20	4
10		
6		

- Students will hopefully recall that this is the area model for multiplying 24 by 16. Ask students to fill in the box.

	20	4
10	200	40
6	120	24

$24 \times 16 = 200 + 120 + 40 + 24 = 384$

? "To multiply the two 2-digit numbers, how many multiplications do you perform?" 4; Students think of it as one multiplication even though four multiplications take place.

Activity Notes

Activity 1

- In this activity, the models for each factor and the product are shown. It is important to have a conversation about the product in part (a) before students begin.
- Make the connection to two-digit multiplication. Each binomial has two terms. There are fours multiplications to perform and the products are represented by the tiles in the lower right quadrant of the *t*-diagram.

? "What are the factors in this problem?" $(x + 3)$ and $(x - 2)$

? "What is the product of x and x?" x^2 "What is the product of x and (-2)?" $-2x$ "What is the product of 3 and x?" $3x$ "What is the product of 3 and (-2)?" -6

- When writing a product, remind students to simplify the expressions.

Common Core State Standards

A.APR.1 Understand that polynomials form a system analogous to the integers, namely, they are closed under the operations of addition, subtraction, and multiplication; add, subtract, and multiply polynomials.

Previous Learning

Students should know how to add and subtract polynomials.

Activity Materials
Textbook
• algebra tiles

Start Thinking! and Warm Up

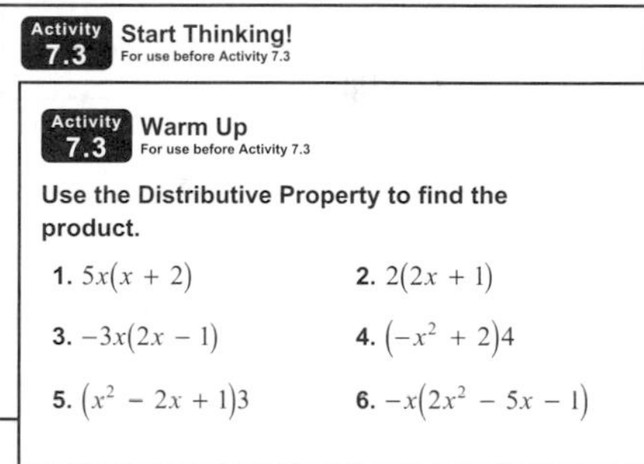

Activity 7.3 Start Thinking! For use before Activity 7.3

Activity 7.3 Warm Up For use before Activity 7.3

Use the Distributive Property to find the product.

1. $5x(x + 2)$
2. $2(2x + 1)$
3. $-3x(2x - 1)$
4. $(-x^2 + 2)4$
5. $(x^2 - 2x + 1)3$
6. $-x(2x^2 - 5x - 1)$

7.3 Record and Practice Journal

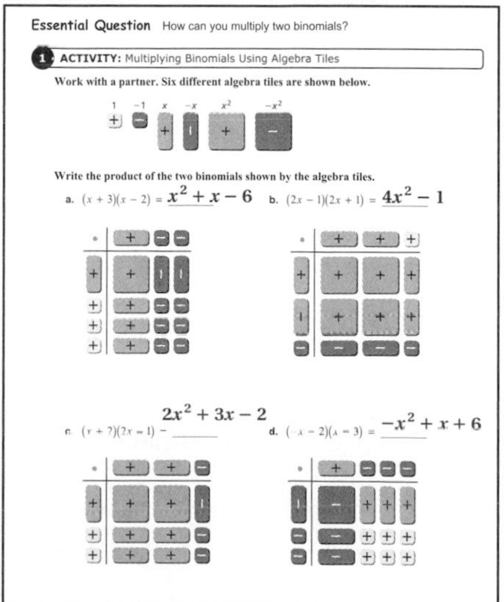

Essential Question How can you multiply two binomials?

1 ACTIVITY: Multiplying Binomials Using Algebra Tiles

Work with a partner. Six different algebra tiles are shown below.

Write the product of the two binomials shown by the algebra tiles.

a. $(x + 3)(x - 2) = x^2 + x - 6$

b. $(2x - 1)(2x + 1) = 4x^2 - 1$

c. $(x + 2)(2x - 1) = 2x^2 + 3x - 2$

d. $(-x - 2)(x - 3) = -x^2 + x + 6$

Differentiated Instruction

Organization

Students may become confused with the different methods (horizontal, vertical, and table of products) for multiplying two polynomials. Encourage students to decide which method they prefer and encourage them to use that method each time.

7.3 Record and Practice Journal

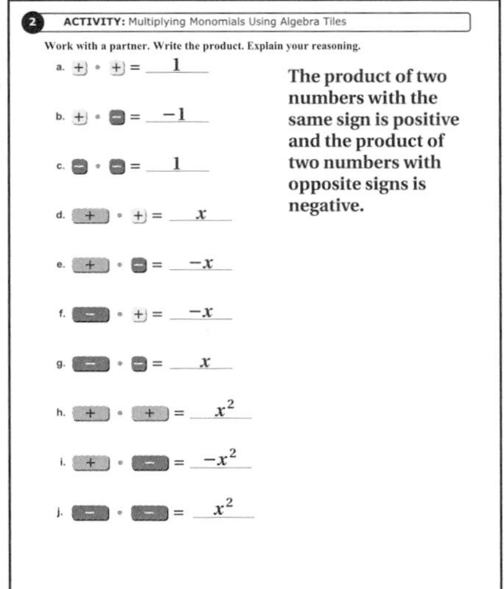
2 ACTIVITY: Multiplying Monomials Using Algebra Tiles

Work with a partner. Write the product. Explain your reasoning.

a. 1

b. -1

c. 1

d. x

e. $-x$

f. $-x$

g. x

h. x^2

i. $-x^2$

j. x^2

The product of two numbers with the same sign is positive and the product of two numbers with opposite signs is negative.

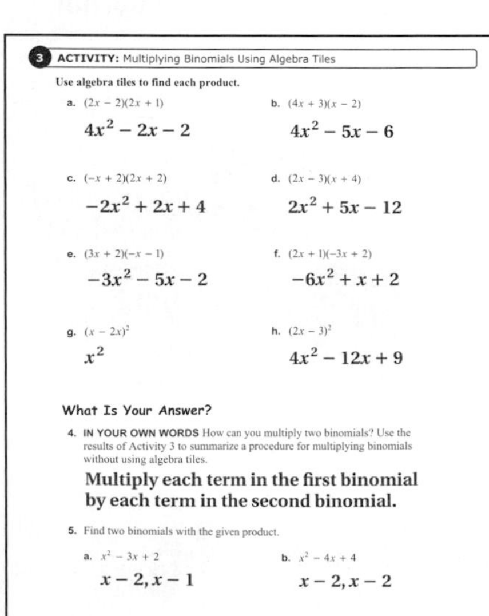
3 ACTIVITY: Multiplying Binomials Using Algebra Tiles

Use algebra tiles to find each product.

a. $(2x - 2)(2x + 1)$ $4x^2 - 2x - 2$

b. $(4x + 3)(x - 2)$ $4x^2 - 5x - 6$

c. $(-x + 2)(2x + 2)$ $-2x^2 + 2x + 4$

d. $(2x - 3)(x + 4)$ $2x^2 + 5x - 12$

e. $(3x + 2)(-x - 1)$ $-3x^2 - 5x - 2$

f. $(2x + 1)(-3x + 2)$ $-6x^2 + x + 2$

g. $(x - 2x)^2$ x^2

h. $(2x - 3)^2$ $4x^2 - 12x + 9$

What Is Your Answer?

4. IN YOUR OWN WORDS How can you multiply two binomials? Use the results of Activity 3 to summarize a procedure for multiplying binomials without using algebra tiles.

Multiply each term in the first binomial by each term in the second binomial.

5. Find two binomials with the given product.

a. $x^2 - 3x + 2$ $x - 2, x - 1$

b. $x^2 - 4x + 4$ $x - 2, x - 2$

Laurie's Notes

Activity 2

- Students multiplied monomials in Activity 1.
- In explaining their reasoning, you should expect students to write the problem using symbols, and then explain their answer. For instance, part (e) would be $(x)(-1) = -x$. The explanation might be that multiplying by -1 is the same as taking the opposite of the number.

Activity 3

- **Teaching Tip:** Make sure students have a template for setting up the multiplication problems. Simply draw two perpendicular lines intersecting as shown in Activity 1. Then place a multiplication dot in the upper left quadrant of the *t*-diagram.
- Students should keep the product looking like a rectangular array. For instance, in part (a), when multiplying $2x \cdot 2x$, the four x^2-tiles should be in a 2×2 array, not a 4×1 array. The dimensions of the array should be the same as the coefficients of the factors.
- The repetition of manipulating the tiles helps to develop the concept of performing four multiplications when two binomials are multiplied. This can also be thought of as the Distributive Property being performed twice.
- Check to see that parts (g) and (h) are done correctly.

What Is Your Answer?

- For Question 4, have students share their thoughts with the whole class.
- **Neighbor Check:** For Question 5, have students work independently and then have their neighbor check their work. Have students discuss any discrepancies.

Closure

- **Writing Prompt:** To multiply $(2x + 1)(x - 3)$ using algebra tiles, you . . .

Technology For the Teacher

The Dynamic Planning Tool

Editable Teacher's Resources at *BigIdeasMath.com*

2 ACTIVITY: Multiplying Monomials Using Algebra Tiles

Work with a partner. Write each product. Explain your reasoning.

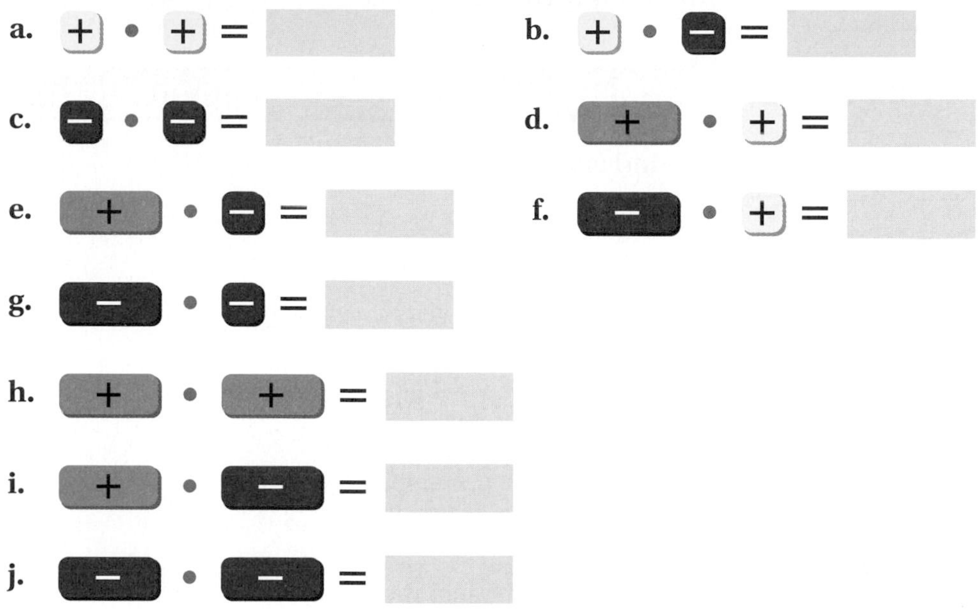

a. + • + =

b. + • − =

c. − • − =

d. + • + =

e. + • − =

f. − • + =

g. − • − =

h. + • + =

i. + • − =

j. − • − =

3 ACTIVITY: Multiplying Binomials Using Algebra Tiles

Use algebra tiles to find each product.

a. $(2x - 2)(2x + 1)$

b. $(4x + 3)(x - 2)$

c. $(-x + 2)(2x + 2)$

d. $(2x - 3)(x + 4)$

e. $(3x + 2)(-x - 1)$

f. $(2x + 1)(-3x + 2)$

g. $(x - 2)^2$

h. $(2x - 3)^2$

What Is Your Answer?

4. **IN YOUR OWN WORDS** How can you multiply two binomials? Use the results of Activity 3 to summarize a procedure for multiplying binomials without using algebra tiles.

5. Find two binomials with the given product.

   a. $x^2 - 3x + 2$

   b. $x^2 - 4x + 4$

Practice Use what you learned about multiplying binomials to complete Exercises 3 and 4 on page 345.

7.3 Lesson

Key Vocabulary
FOIL Method, *p. 343*

In Section 1.2, you used the Distributive Property to multiply a binomial by a monomial. You can also use the Distributive Property to multiply two binomials.

EXAMPLE 1 Multiplying Binomials Using the Distributive Property

Find each product.

a. $(x + 2)(x + 5)$

Use the horizontal method.

$(x + 2)(x + 5) = x(x + 5) + 2(x + 5)$	Distribute $(x + 5)$ to each term of $(x + 2)$.
$= x(x) + x(5) + 2(x) + 2(5)$	Distributive Property
$= x^2 + 5x + 2x + 10$	Multiply.
$= x^2 + 7x + 10$	Combine like terms.

b. $(x + 3)(x - 4)$

Use the vertical method.

	$x + 3$	
	$\times \quad x - 4$	Align like terms vertically.
Multiply $-4(x + 3)$.	$-4x - 12$	Distributive Property
Multiply $x(x + 3)$.	$x^2 + 3x$	Distributive Property
	$x^2 - x - 12$	Combine like terms.

The product is $x^2 - x - 12$.

EXAMPLE 2 Multiplying Binomials Using a Table

Find $(2x - 3)(x + 5)$.

Step 1: Write each binomial as a sum of terms.

$$(2x - 3)(x + 5) = [2x + (-3)](x + 5)$$

Step 2: Make a table of products.

	2x	**−3**
x	$2x^2$	$-3x$
5	$10x$	-15

The product is $2x^2 - 3x + 10x - 15$, or $2x^2 + 7x - 15$.

On Your Own

Now You're Ready
Exercises 5–13 and 16–21

Use the Distributive Property to find the product.

1. $(y + 4)(y + 1)$ **2.** $(z - 2)(z + 6)$

Use a table to find the product.

3. $(p + 3)(p - 8)$ **4.** $(r - 5)(2r - 1)$

 Multi-Language Glossary at BigIdeasMath.com.

Laurie's Notes

Introduction

Connect

- **Yesterday:** Students used algebra tiles to multiply binomials. (MP1a, MP7)
- **Today:** Students will multiply polynomials.

Motivate

- **Mental Math Challenge:** Ask students to find the following products using mental math. They should record only the answer.

  8×32 $\quad$ 9×48 $\quad$ 12×43
- Have students reveal each answer and ask how they performed the computation. Listen for strategies that likely make use of the Distributive Property.

  $8 \times 32 = 8(30 + 2) = 240 + 16 = 256$

  $9 \times 48 = 9(50 - 2) = 450 - 18 = 432$

  $12 \times 43 = 12(40 + 3) = 480 + 36 = 516$

  OR $12 \times 43 = (10 + 2)(40 + 3)$

  $= 10 \times 40 + 10 \times 3 + 2 \times 40 + 2 \times 3 = 516$
- Make sure students recognize the use of the Distributive Property, and that 12×43 can be done two different ways using the Distributive Property.

Lesson Notes

Example 1

- **Connection:** When you multiply two binomials, four multiplications are performed just like when you multiply two 2-digit numbers.
- Point out that multiplying $(x + 2)$ by $(x + 5)$ means that both terms of $(x + 2)$ must be multiplied by $(x + 5)$.
- The order in which the multiplications are performed is not important because of the Commutative Property of Addition. However, in order to make the connection to the FOIL Method, perform the operations in the order shown.
- Some students prefer to do the multiplication using a vertical format, similar to how two 2-digit multiplication is done. You need to pay attention to place value when multiplying numbers. You need to pay attention to like terms when multiplying binomials.

Example 2

- The box or table model can be used to multiply binomials. The top row and left column are labeled with the binomial factors. Each box within the table represents the product.
- **Common Error:** Students often lose the negative sign on the constant 3. Rewriting the expression as $2x + (-3)$ helps.
- Take time to interchange the rows and columns to show that the product is still the same.

On Your Own

- Ask for volunteers to share their work at the board for each of the questions.

Today's lesson is multiplying polynomials.

Start Thinking! and Warm Up

Lesson 7.3 Warm Up For use before Lesson 7.3

Lesson 7.3 Start Thinking! For use before Lesson 7.3

When you multiply binomials $(x + a)(x + b)$, you get a trinomial.

$(x + a)(x + b) = x^2 + \_\_\_\_\_x + \_\_\_\_\_$

What is the coefficient of the x-term, in terms of a and b? What is the constant, in terms of a and b?

Extra Example 1

Use the Distributive Property to find each product.

a. $(n + 9)(n + 4)$ $\quad n^2 + 13n + 36$

b. $(p + 3)(p - 7)$ $\quad p^2 - 4p - 21$

Extra Example 2

Use a table to find $(3x + 1)(x - 4)$. $3x^2 - 11x - 4$

On Your Own

1. $y^2 + 5y + 4$
2. $z^2 + 4z - 12$
3. $p^2 - 5p - 24$
4. $2r^2 - 11r + 5$

Extra Example 3

Use the FOIL Method to find each product.

a. $(k + 9)(k - 3)$ $k^2 + 6k - 27$

b. $(4n - 1)(2n + 5)$ $8n^2 + 18n - 5$

On Your Own

5. $m^2 - m - 30$

6. $x^2 - 2x - 8$

7. $6k^2 + 33k + 15$

8. $2u^2 - \frac{5}{2}u - \frac{3}{4}$

English Language Learners

Pair Activity

Have students work in pairs and assign a group of problems. One student completes the first step in multiplying the polynomials and gives the paper to the other student. The second student completes the next step and returns it to the first student. Students continue to alternate the writing of steps until the expression is written in simplest form. This interaction allows students to develop an understanding of the concept and allows the teacher the opportunity to provide one-on-one assistance.

Laurie's Notes

Key Idea

- Explain to students that the FOIL Method is simply an acronym to help them remember the four multiplications that need to be performed.
- Point out the FOIL pattern in Examples 1 and 2.

Example 3

- Some students may be comfortable with saying the process aloud and simply writing the product, skipping the first step.
- **Teaching Tip:** Use your index fingers to point to the two terms being multiplied as you say, "first terms, outer terms, inner terms, and last terms."
- In part (b), students need to pay attention to the coefficient of the x-term.

On Your Own

- Some students will prefer to do Question 8 using decimals. The results will be equivalent.

The **FOIL Method** is a shortcut for multiplying two binomials.

FOIL Method

To multiply two binomials using the FOIL Method, find the sum of the products of the

First terms,	$(x + 1)(x + 2)$	➡	$x(x) = x^2$
Outer terms,	$(x + 1)(x + 2)$	➡	$x(2) = 2x$
Inner terms, and	$(x + 1)(x + 2)$	➡	$1(x) = x$
Last terms.	$(x + 1)(x + 2)$	➡	$1(2) = 2$

$$(x + 1)(x + 2) = x^2 + 2x + x + 2 = x^2 + 3x + 2$$

EXAMPLE 3 Multiplying Binomials Using the FOIL Method

Find each product.

a. $(x - 3)(x - 6)$

First Outer Inner Last

$(x - 3)(x - 6) = x(x) + x(-6) + (-3)(x) + (-3)(-6)$ Use the FOIL Method.

$= x^2 + (-6x) + (-3x) + 18$ Multiply.

$= x^2 - 9x + 18$ Combine like terms.

b. $(2x + 1)(3x - 5)$

First Outer Inner Last

$(2x + 1)(3x - 5) = 2x(3x) + 2x(-5) + 1(3x) + 1(-5)$ Use the FOIL Method.

$= 6x^2 + (-10x) + 3x + (-5)$ Multiply.

$= 6x^2 - 7x - 5$ Combine like terms.

On Your Own

Now You're Ready
Exercises 22–30

Use the FOIL Method to find the product.

5. $(m + 5)(m - 6)$

6. $(x - 4)(x + 2)$

7. $(k + 5)(6k + 3)$

8. $\left(2u + \frac{1}{2}\right)\left(u - \frac{3}{2}\right)$

EXAMPLE 4 Multiplying a Binomial and a Trinomial

Find $(x + 5)(x^2 - 3x - 2)$.

	$x^2 - 3x - 2$	
	$\times \quad x + 5$	Align like terms vertically.
Multiply $5(x^2 - 3x - 2)$.	$5x^2 - 15x - 10$	Distributive Property
Multiply $x(x^2 - 3x - 2)$.	$x^3 - 3x^2 - 2x$	Distributive Property
	$x^3 + 2x^2 - 17x - 10$	Combine like terms.

∴ The product is $x^3 + 2x^2 - 17x - 10$.

EXAMPLE 5 Real-Life Application

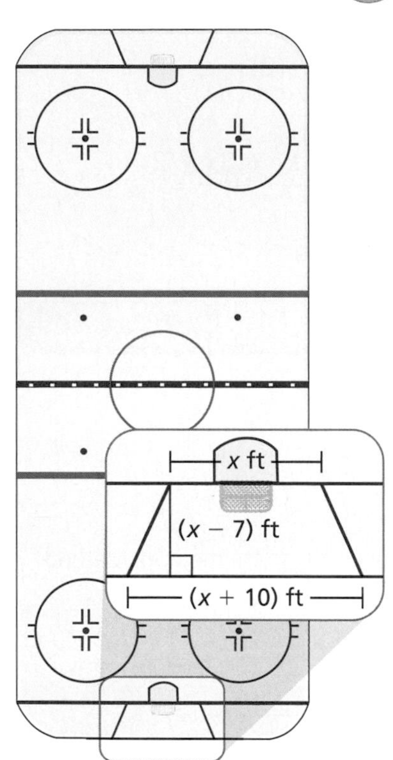

In hockey, a goalie behind the goal line can only play a puck in a trapezoidal region.

a. Write a polynomial that represents the area of the trapezoidal region.

$\frac{1}{2}h(b_1 + b_2) = \frac{1}{2}(x - 7)[x + (x + 10)]$ Substitute.

$= \frac{1}{2}(x - 7)(2x + 10)$ Combine like terms.

F O I L

$= \frac{1}{2}[2x^2 + 10x + (-14x) + (-70)]$ Use the FOIL Method.

$= \frac{1}{2}(2x^2 - 4x - 70)$ Combine like terms.

$= x^2 - 2x - 35$ Distributive Property

b. Find the area of the trapezoidal region when the shorter base is 18 feet.

Find the value of $x^2 - 2x - 35$ when $x = 18$.

$x^2 - 2x - 35 = 18^2 - 2(18) - 35$ Substitute 18 for x.

$= 324 - 36 - 35$ Simplify.

$= 253$ Subtract.

∴ The area of the trapezoidal region is 253 square feet.

On Your Own

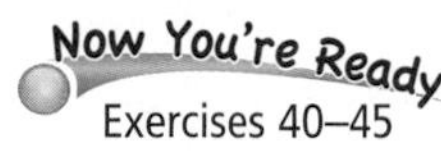

Find the product.

9. $(x + 1)(x^2 + 5x + 8)$

10. $(n - 3)(n^2 - 2n + 4)$

11. **WHAT IF?** How does the polynomial in Example 5 change if the longer base is extended by 1 foot? Explain.

Laurie's Notes

Example 4

? "How would you describe this multiplication problem?" binomial times a trinomial

- **MP7 Look for and Make Use of Structure:** This is similar to multiplying a 2-digit number and a 3-digit number.

? "How many multiplications will be performed?" 6

- Solve the problem as shown, making note of the similar terms and that they need to be lined up. The problem could also be set up and solved in a horizontal format, or it could be done using a table format (with 6 cells).
- Summarize the problem using appropriate vocabulary. The product of the binomial and trinomial is a third-degree polynomial with 4 terms.

? **Extension:** "Are polynomials closed under multiplication?" yes

Example 5

? "How do you find the area of a trapezoid?" $A = \frac{1}{2}h(b_1 + b_2)$

- Spend time identifying the dimensions in the diagram.

? "What is the height of the trapezoid?" $(x - 7)$

? "What are the lengths of the two bases?" x and $(x + 10)$

- Remind students that the $\frac{1}{2}$ is a factor that doesn't need to be multiplied until the last step. They should multiply the two binomials first to make the calculations easier.
- **Extension:** After working through part (b), ask students to find the dimensions of the trapezoid and then use the formula to find the area.

$b_1 = 18$ ft; $b_2 = 28$ ft; $h = 11$ ft; $A = \frac{1}{2}(11)(18 + 28) = 253 \text{ ft}^2$

On Your Own

- **Think-Pair-Share:** Students should read each question independently and then work with a partner to answer the questions. When they have answered the questions, the pair should compare their answers with another group and discuss any discrepancies.

Closure

- **Matching Activity:** Match the product with the correct polynomial.

1. $(x + 4)(x + 4)$ D	**A.** $x^2 - x - 12$
2. $(x + 2)(x + 8)$ B	**B.** $x^2 + 10x + 16$
3. $(x - 4)(x + 3)$ A	**C.** $x^2 + 4x - 12$
4. $(x + 6)(x - 2)$ C	**D.** $x^2 + 8x + 16$

Technology For the Teacher

The Dynamic Planning Tool
Editable Teacher's Resources at *BigIdeasMath.com*

Extra Example 4

Find $(b - 3)(b^2 + 2b - 7)$.
$b^3 - b^2 - 13b + 21$

Extra Example 5

The dimensions of a trapezoid are shown.

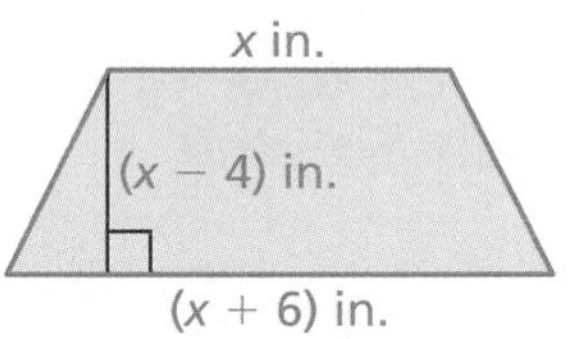

a. Write a polynomial that represents the area of the trapezoid.
$x^2 - x - 12$

b. Find the area of the trapezoid when the shorter base is 10 inches. 78 in.^{2}

On Your Own

9. $x^3 + 6x^2 + 13x + 8$

10. $n^3 - 5n^2 + 10n - 12$

11. $x^2 - 1.5x - 38.5$; The longer base becomes $(x + 11)$.

Vocabulary and Concept Check

1. *Sample answer:* Use the Distributive Property or the FOIL Method.
2. F indicates to multiply the *first* terms. O indicates to multiply the *outer* terms. I indicates to multiply the *inner* terms. L indicates to multiply the *last* terms.

Practice and Problem Solving

3. $x^2 - 4$
4. $-2x^2 + 7x - 3$
5. $x^2 + 4x + 3$
6. $y^2 + 10y + 24$
7. $z^2 - 2z - 15$
8. $a^2 + 5a - 24$
9. $g^2 - 9g + 14$
10. $n^2 - 10n + 24$
11. $3m^2 + 28m + 9$
12. $6p^2 - 8p - 8$
13. $12 - 16s + 5s^2$
14. The second binomial was not distributed to the first term of the first binomial.
$$\begin{aligned}&(t-2)(t+5)\\&= t(t+5) - 2(t+5)\\&= t^2 + 5t - 2t - 10\\&= t^2 + 3t - 10\end{aligned}$$
15. $(18x^2 + 12x + 2)$ in.^{2}
16. $x^2 + 4x + 3$
17. $y^2 + 5y - 50$
18. $h^2 - 17h + 72$
19. $8j^2 - 26j + 21$
20. $30c^2 + 61c + 30$
21. $15d^2 - 71d + 84$
22. $b^2 + 10b + 21$

Assignment Guide and Homework Check

Level	Assignment	Homework Check
Average	1, 2, 6–30 even, 33–47 odd, 50, 52–55	12, 16, 28, 33, 43, 50
Advanced	1, 2, 6–30 even, 36–46 even, 48–55	12, 16, 28, 44, 50, 51

Common Errors

- **Exercises 5–13** Students may forget to distribute the second factor to both terms of the first factor. Remind them of this process.

7.3 Record and Practice Journal

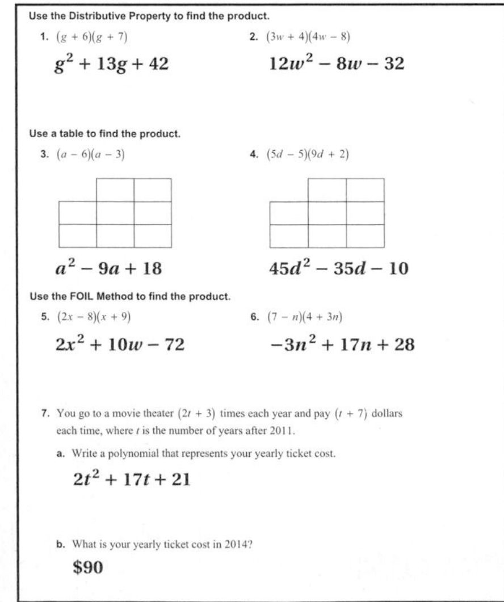

Use the Distributive Property to find the product.

1. $(g + 6)(g + 7)$
$g^2 + 13g + 42$

2. $(3w + 4)(4w - 8)$
$12w^2 - 8w - 32$

Use a table to find the product.

3. $(a - 6)(a - 3)$
$a^2 - 9a + 18$

4. $(5d - 5)(9d + 2)$
$45d^2 - 35d - 10$

Use the FOIL Method to find the product.

5. $(2x - 8)(x + 9)$
$2x^2 + 10w - 72$

6. $(7 - n)(4 + 3n)$
$-3n^2 + 17n + 28$

7. You go to a movie theater $(2t + 3)$ times each year and pay $(t + 7)$ dollars each time, where t is the number of years after 2011.

a. Write a polynomial that represents your yearly ticket cost.
$2t^2 + 17t + 21$

b. What is your yearly ticket cost in 2014?
$90

Technology For the Teacher
Answer Presentation Tool

7.3 Exercises

Vocabulary and Concept Check

1. **VOCABULARY** Describe two ways to find the product of two binomials.
2. **WRITING** Explain how the letters of the word FOIL can help you remember how to multiply two binomials.

Practice and Problem Solving

Write the product of the two binomials shown by the algebra tiles.

3. $(x - 2)(x + 2) =$ ______
4. $(-x + 3)(2x - 1) =$ ______

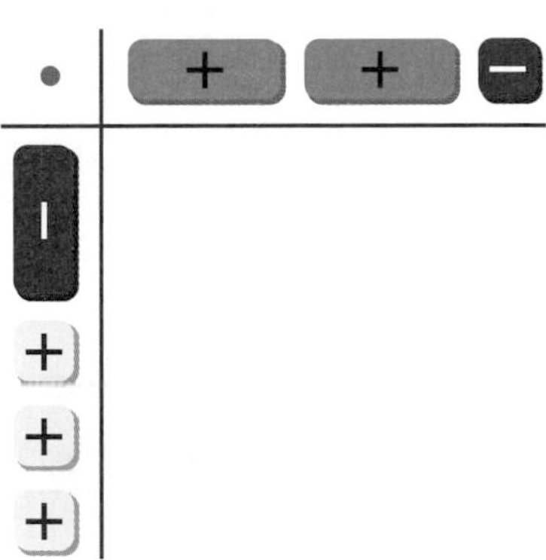

Use the Distributive Property to find the product.

1 5. $(x + 1)(x + 3)$
6. $(y + 6)(y + 4)$
7. $(z - 5)(z + 3)$
8. $(a + 8)(a - 3)$
9. $(g - 7)(g - 2)$
10. $(n - 6)(n - 4)$
11. $(3m + 1)(m + 9)$
12. $(2p - 4)(3p + 2)$
13. $(6 - 5s)(2 - s)$

14. **ERROR ANALYSIS** Describe and correct the error in finding the product.

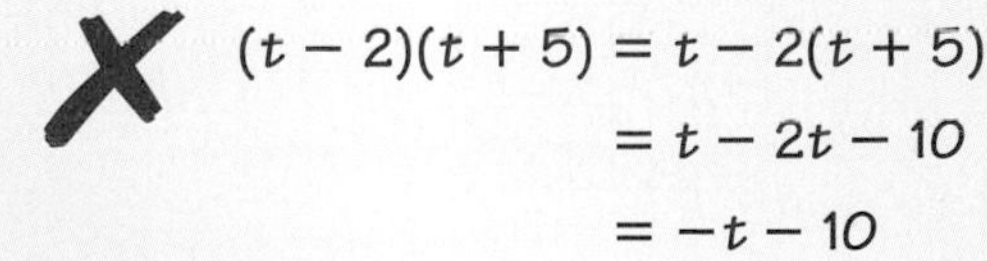

15. **CALCULATOR** The width of a calculator can be represented by $(3x + 1)$ inches. The length of the calculator is twice the width. Write a polynomial that represents the area of the calculator.

Use a table to find the product.

2 16. $(x + 3)(x + 1)$
17. $(y + 10)(y - 5)$
18. $(h - 8)(h - 9)$
19. $(-3 + 2j)(4j - 7)$
20. $(5c + 6)(6c + 5)$
21. $(5d - 12)(-7 + 3d)$

Use the FOIL Method to find the product.

3 22. $(b + 3)(b + 7)$
23. $(w + 9)(w + 6)$
24. $(k + 5)(k - 1)$
25. $(x - 4)(x + 8)$
26. $(q - 3)(q - 4)$
27. $(z - 5)(z - 9)$
28. $(t + 2)(2t + 1)$
29. $(5v - 3)(2v + 4)$
30. $(9 - r)(2 - 3r)$

31. **ERROR ANALYSIS** Describe and correct the error in finding the product.

✗ $(r + 6)(r - 7) = r(r) + r(7) + 6(r) + 6(7)$
$= r^2 + 7r + 6r + 42$
$= r^2 + 13r + 42$

32. **OPEN-ENDED** Write two binomials whose product includes the term 12.

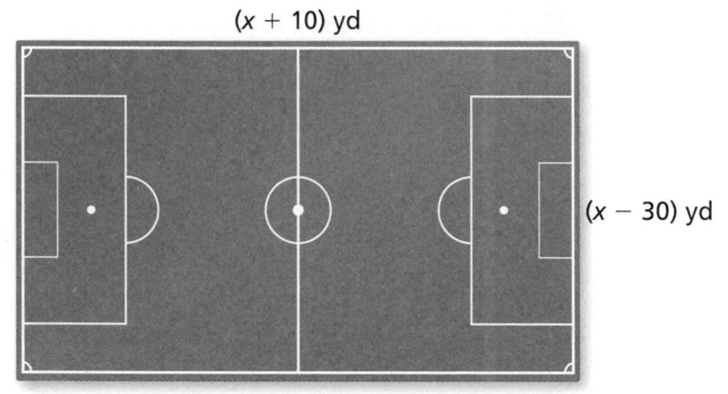

33. **SOCCER** The soccer field is rectangular.

a. Write a polynomial that represents the area of the soccer field.

b. Use the polynomial in part (a) to find the area of the field when $x = 90$.

c. A groundskeeper mows 200 square yards in 3 minutes. How long does it take the groundskeeper to mow the field?

Write a polynomial that represents the area of the shaded region.

34.
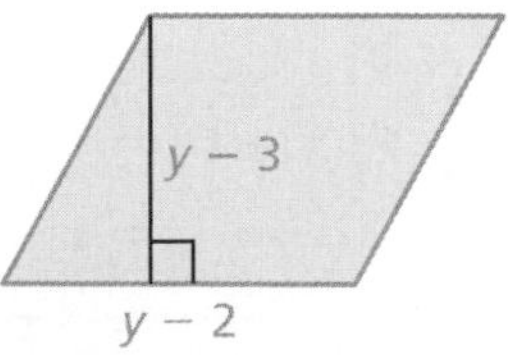

35.
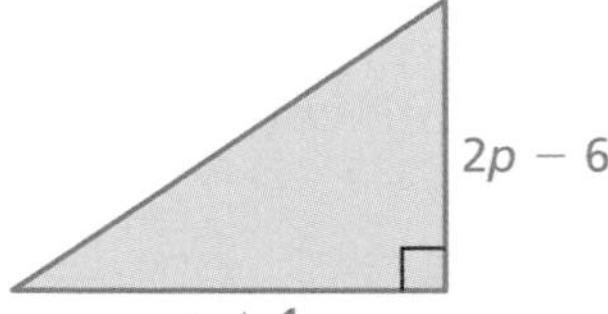

36.
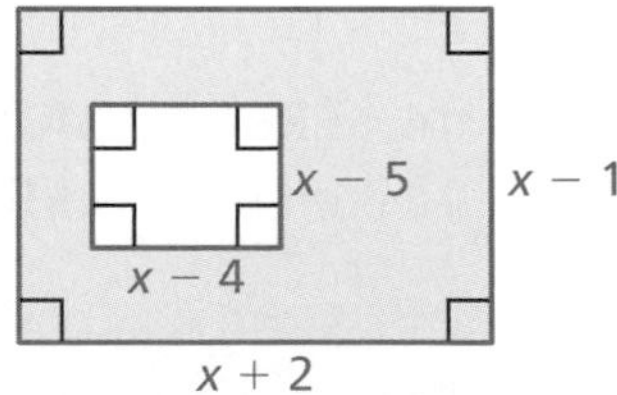

Find the product.

37. $(n + 3)(2n^2 + 1)$
38. $(x + y)(2x - y)$
39. $(2r + s)(r - 3s)$
4 40. $(x - 4)(x^2 - 3x + 2)$
41. $(f^2 + 4f - 8)(f - 1)$
42. $(3 + i)(i^2 + 8i - 2)$
43. $(t^2 - 5t + 1)(-3 + t)$
44. $(b - 4)(5b^2 - 5b + 4)$
45. $(3e^2 - 5e + 7)(6e + 1)$

46. **REASONING** Can you use the FOIL method to multiply a binomial by a trinomial? a trinomial by a trinomial? Explain your reasoning.

Common Errors

- **Exercises 16–21** Students may forget to write each binomial as a sum of terms thereby writing a negative term as a positive term in the table. Encourage students to take the time to carry out all the steps.
- **Exercises 22–30** Students may forget to perform one or more of the steps in the FOIL Method. Encourage them to draw arrows for each part of the method.
- **Exercises 22–30** Students may not include negative signs when multiplying. Encourage them to write subtraction expressions as equivalent addition expressions before multiplying.
- **Exercise 36** Students may only find the area of the larger rectangle. Remind them that the directions ask for the area of the shaded region, so they need to also find the area of the smaller rectangle and then subtract.
- **Exercises 40–45** Students may forget to align like terms when using a vertical method to multiply polynomials. Remind them that you cannot combine terms that are not alike.

Differentiated Instruction

Kinesthetic

When multiplying a binomial and a trinomial, students may leave out a partial product. Have them write the partial products and draw an arrow from the monomial back to the corresponding term in the trinomial.

$$(x + 5)(x^2 - 3x - 2)$$

$$= (x + 5)(x^2) + (x + 5)(-3x) + (x + 5)(-2)$$

Practice and Problem Solving

23. $w^2 + 15w + 54$
24. $k^2 + 4k - 5$
25. $x^2 + 4x - 32$
26. $q^2 - 7q + 12$
27. $z^2 - 14z + 45$
28. $2t^2 + 5t + 2$
29. $10v^2 + 14v - 12$
30. $18 - 29r + 3r^2$
31. See Additional Answers.
32. *Sample answer:* $2x - 6$, $x - 2$
33. **a.** $x^2 - 20x - 300$

    b. 6000 yd^2

    c. 90 minutes
34. $y^2 - 5y + 6$
35. $p^2 - 2p - 3$
36. $10x - 22$
37. $2n^3 + 6n^2 + n + 3$
38. $2x^2 + xy - y^2$
39. $2r^2 - 5rs - 3s^2$
40. $x^3 - 7x^2 + 14x - 8$
41. $f^3 + 3f^2 - 12f + 8$
42. $i^3 + 11i^2 + 22i - 6$
43. $t^3 - 8t^2 + 16t - 3$
44. $5b^3 - 25b^2 + 24b - 16$
45. $18e^3 - 27e^2 + 37e + 7$

Practice and Problem Solving

46. no; A trinomial has more than just a first term and a last term as well as more than an outer term and an inner term.

47. **a.** $x^2 + 41x + 40$

b. \$126

48. yes; *Sample answer:* The Commutative Property of Multiplication states that changing the order of factors does not change the product.

49. **a.** m is negative and n is positive; or m is positive and n is negative

b. m is positive and n is positive; or m is negative and n is negative

50. See *Taking Math Deeper.*

51. $4x^3 + 9x^2 - x - 6$

Fair Game Review

52. $-x^3 - 5x^2 + 2x$; 3; trinomial

53. $z^2 - \frac{5}{7}z$; 2; binomial

54. $-15y^7$; 7; monomial

55. B

Mini-Assessment

Find the product.

1. $(a + 2)(a + 3)$ $a^2 + 5a + 6$
2. $(t + 7)(t - 4)$ $t^2 + 3t - 28$
3. $(m - 6)(m - 5)$ $m^2 - 11m + 30$
4. $(d - 3)(2d + 5)$ $2d^2 - d - 15$
5. $(n + 4)(n^2 - 6n + 3)$
   $n^3 - 2n^2 - 21n + 12$

Taking Math Deeper

Exercise 50

This problem uses geometric reasoning and polynomial operations.

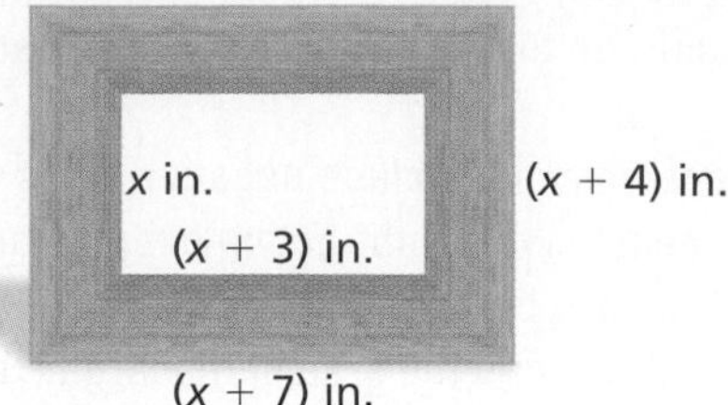

a. To find the area of wood you paint, subtract the area of the smaller rectangle from the area of the larger rectangle.

$$\begin{aligned} A &= (x + 4)(x + 7) - x(x + 3) \\ &= x^2 + 7x + 4x + 28 - (x^2 + 3x) \\ &= x^2 + 11x + 28 - x^2 - 3x \\ &= 8x + 28 \end{aligned}$$

b. If the frame displays a 5-inch by 8-inch photograph, you can conclude that $x = 5$ inches. So, the area of the frame is

$$\begin{aligned} A &= 8x + 28 \\ &= 8(5) + 28 \\ &= 40 + 28 \\ &= 68 \text{ in.}^2. \end{aligned}$$

Notice that the picture frame is made up of 4 trapezoids. Ask your students to find the area of each trapezoid when $x = 5$ inches. The answers are 14 square inches (small trapezoid) and 20 square inches (large trapezoid).

Reteaching and Enrichment Strategies

If students need help. . .	If students got it. . .
Resources by Chapter • Practice A and Practice B • Puzzle Time Record and Practice Journal Practice Differentiating the Lesson Lesson Tutorials Skills Review Handbook	Resources by Chapter • Enrichment and Extension • School-to-Work Start the next section

47. AMUSEMENT PARK You go to an amusement park $(x + 1)$ times each year and pay $(x + 40)$ dollars each time, where x is the number of years after 2011.

a. Write a polynomial that represents your yearly admission cost.

b. What is your yearly admission cost in 2013?

48. PRECISION You use the Distributive Property to multiply $(x + 3)(x - 5)$. Your friend uses the FOIL Method to multiply $(x - 5)(x + 3)$. Should your answers be equivalent? Justify your answer.

49. REASONING The product of $(x + m)(x + n)$ is $x^2 + bx + c$.

a. What do you know about m and n when $c < 0$?

b. What do you know about m and n when $c > 0$?

50. PICTURE You design the wooden picture frame and paint the front surface.

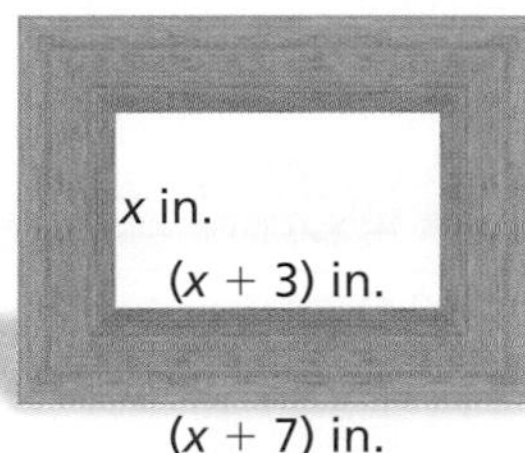

a. Write a polynomial that represents the area of wood you paint.

b. You design the picture frame to display a 5-inch by 8-inch photograph. How much wood do you paint?

51. 

The shipping container is a rectangular prism. Write a polynomial that represents the volume of the container.

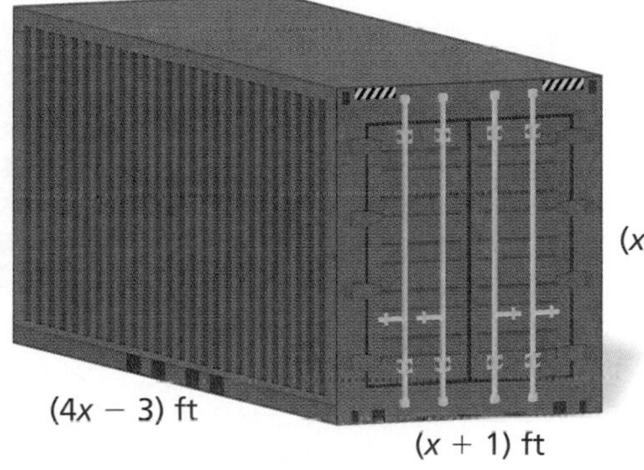

Fair Game Review What you learned in previous grades & lessons

Write the polynomial in standard form. Identify the degree and classify the polynomial by the number of terms. *(Section 7.1)*

52. $2x - 5x^2 - x^3$

53. $z^2 - \frac{5}{7}z$

54. $-15y^7$

55. MULTIPLE CHOICE Which system of linear equations does the graph represent? *(Section 4.1)*

Ⓐ $y = 3x + 4$
$y = -2x - 6$

Ⓑ $y = 2x + 1$
$y = -x - 2$

Ⓒ $y = -x + 7$
$y = 4x - 8$

Ⓓ $y = x + 10$
$y = -3x + 2$

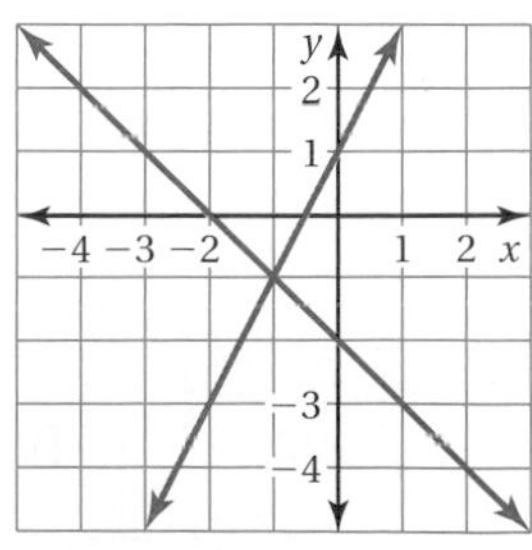

7.4 Special Products of Polynomials

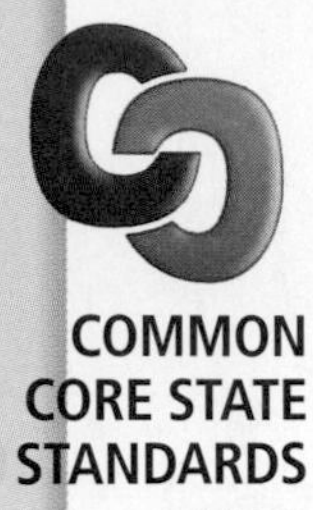

Essential Question What are the patterns in the special products $(a + b)(a - b)$, $(a + b)^2$, and $(a - b)^2$?

COMMON CORE STATE STANDARDS
A.APR.1

1 ACTIVITY: Finding a Sum and Difference Pattern

Work with a partner. Six different algebra tiles are shown below.

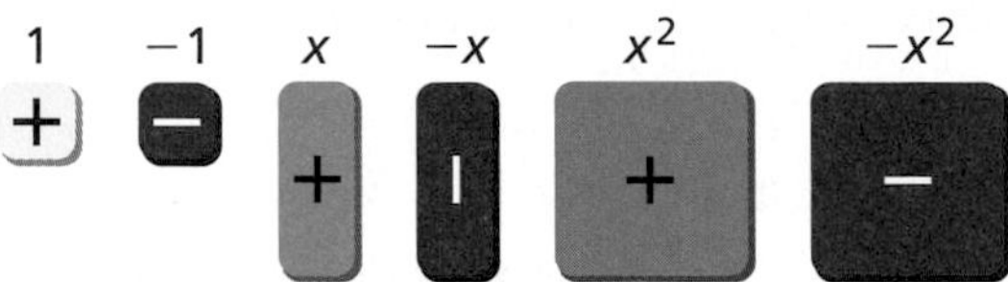

Write the product of the two binomials shown by the algebra tiles.

a. $(x + 2)(x - 2) =$ ______

b. $(2x - 1)(2x + 1) =$ ______

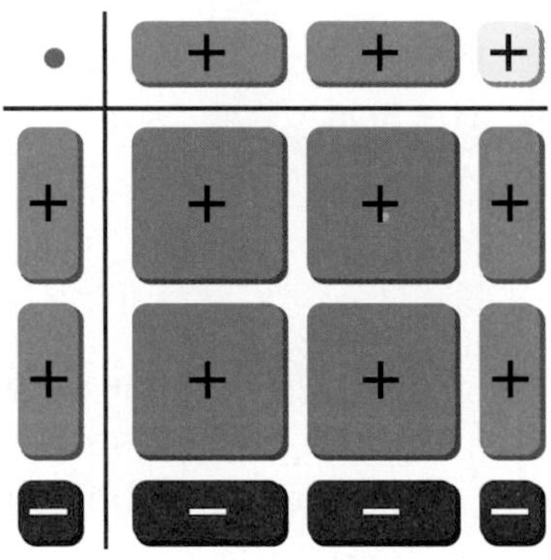

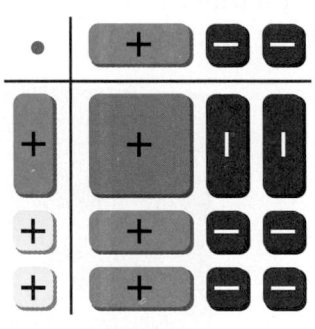

2 ACTIVITY: Describing a Sum and Difference Pattern

Work with a partner.

a. Describe the pattern for the special product: $(a + b)(a - b)$.

b. Use the pattern you described to find each product. Check your answers using algebra tiles.

i. $(x + 3)(x - 3)$ **ii.** $(x - 4)(x + 4)$ **iii.** $(3x + 1)(3x - 1)$

iv. $(3y + 4)(3y - 4)$ **v.** $(2x - 5)(2x + 5)$ **vi.** $(z + 1)(z - 1)$

Laurie's Notes

Introduction

Standards for Mathematical Practice

- **MP1a Make Sense of Problems** and **MP5 Use Appropriate Tools Strategically:** Working with algebra tiles in this activity is a visual and tactile experience that should help students make sense of, and recognize, the patterns that appear when finding special products.

Motivate

- Start writing the sequence of squares on the board and tell students to stop you if they see a pattern in the products.

$15^2 = 225 \quad 25^2 = 625 \quad 35^2 = 1225 \quad 45^2 = 2025 \quad 55^2 = 3025$
$65^2 = 4225 \quad 75^2 = 5625 \quad 85^2 = 7225 \quad 95^2 = 9025$, etc.

- For the squared numbers, the ones digit is always 5. Let the tens digit be n.
- Students should observe that the last two digits of the product are always 25 and the first digits are the product $n(n + 1)$. Example: $65^2 = 4225$

? "Does the pattern work when squaring 3-digit numbers ending in 5?" yes; $125^2 = \mathbf{15{,}6}25$

- Explain to students that today they will be investigating special products of binomials.

Activity Notes

Activity 1

- In this activity, the models for each factor and the product are shown.

? Ask the following questions.
  - "What are the factors in part (a)?" $(x + 2)$ and $(x - 2)$
  - "What is the product?" $x^2 - 4$
  - "What happened to the x-terms?" They added to zero.
  - "Did this also happen in part (b)?" yes
  - "How are the binomial factors different in each part?" The operation is different.

Activity 2

- Students may have difficulty working with two variables, a and b, versus a variable and a number. If so, have them complete part (b) with algebra tiles and then come back to part (a).
- Students may be confused by the two questions that use a variable other than x. Remind them that the rectangular tile may be named "y" and if it is, the large square is "y^2" instead of x^2.

Common Core State Standards

A.APR.1 Understand that polynomials form a system analogous to the integers, namely, they are closed under the operations of addition, subtraction, and multiplication; add, subtract, and multiply polynomials.

Previous Learning

Students should know how to multiply binomials.

Activity Materials
Textbook
• algebra tiles

Start Thinking! and Warm Up

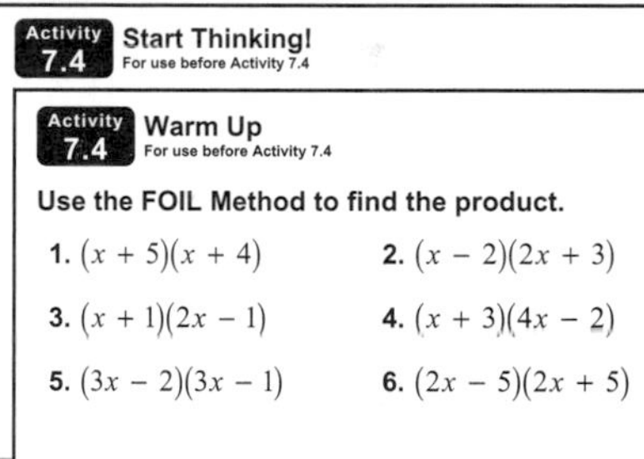

Activity 7.4 Start Thinking! For use before Activity 7.4

Activity 7.4 Warm Up For use before Activity 7.4

Use the FOIL Method to find the product.

1. $(x + 5)(x + 4)$
2. $(x - 2)(2x + 3)$
3. $(x + 1)(2x - 1)$
4. $(x + 3)(4x - 2)$
5. $(3x - 2)(3x - 1)$
6. $(2x - 5)(2x + 5)$

7.4 Record and Practice Journal

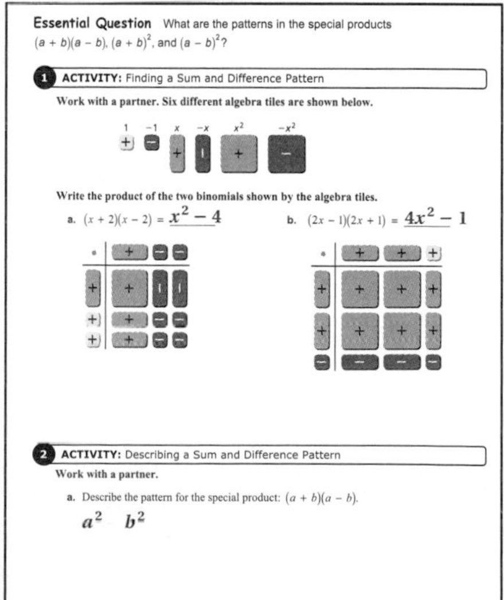

Essential Question What are the patterns in the special products $(a + b)(a - b)$, $(a + b)^2$, and $(a - b)^2$?

1 ACTIVITY: Finding a Sum and Difference Pattern

Work with a partner. Six different algebra tiles are shown below.

1   −1 x $-x$ x^2 $-x^2$

Write the product of the two binomials shown by the algebra tiles.

a. $(x + 2)(x - 2) = x^2 - 4$

b. $(2x - 1)(2x + 1) = 4x^2 - 1$

2 ACTIVITY: Describing a Sum and Difference Pattern

Work with a partner.

a. Describe the pattern for the special product: $(a + b)(a - b)$.

$a^2 \quad b^2$

Differentiated Instruction

Visual

Use product tables to help students understand the sum and difference pattern.

	x	7
x	x^2	$7x$
-7	$-7x$	-49

	$3x$	-1
$3x$	$9x^2$	$-3x$
1	$3x$	-1

Students should see that the upper right and lower left cells are always opposites and have a sum of 0.

7.4 Record and Practice Journal

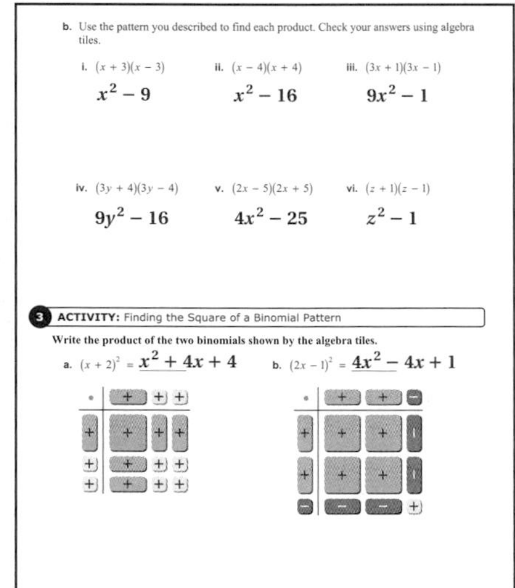

b. Use the pattern you described to find each product. Check your answers using algebra tiles.

i. $(x+3)(x-3)$ $x^2 - 9$ ii. $(x-4)(x+4)$ $x^2 - 16$ iii. $(3x+1)(3x-1)$ $9x^2 - 1$

iv. $(3y+4)(3y-4)$ $9y^2 - 16$ v. $(2x-5)(2x+5)$ $4x^2 - 25$ vi. $(z+1)(z-1)$ $z^2 - 1$

3 ACTIVITY: Finding the Square of a Binomial Pattern

Write the product of the two binomials shown by the algebra tiles.

a. $(x+2)^2 = x^2 + 4x + 4$ b. $(2x-1)^2 = 4x^2 - 4x + 1$

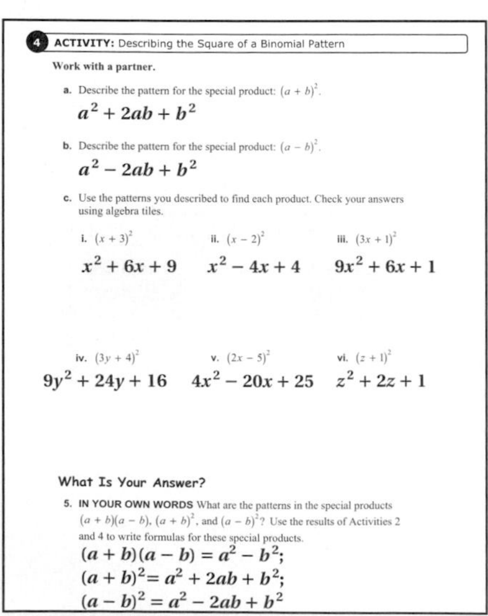

4 ACTIVITY: Describing the Square of a Binomial Pattern

Work with a partner.

a. Describe the pattern for the special product: $(a+b)^2$.
$a^2 + 2ab + b^2$

b. Describe the pattern for the special product: $(a-b)^2$.
$a^2 - 2ab + b^2$

c. Use the patterns you described to find each product. Check your answers using algebra tiles.

i. $(x+3)^2$ $x^2 + 6x + 9$ ii. $(x-2)^2$ $x^2 - 4x + 4$ iii. $(3x+1)^2$ $9x^2 + 6x + 1$

iv. $(3y+4)^2$ $9y^2 + 24y + 16$ v. $(2x-5)^2$ $4x^2 - 20x + 25$ vi. $(z+1)^2$ $z^2 + 2z + 1$

What Is Your Answer?

5. IN YOUR OWN WORDS What are the patterns in the special products $(a+b)(a-b)$, $(a+b)^2$, and $(a-b)^2$? Use the results of Activities 2 and 4 to write formulas for these special products.
$(a+b)(a-b) = a^2 - b^2$;
$(a+b)^2 = a^2 + 2ab + b^2$;
$(a-b)^2 = a^2 - 2ab + b^2$

Laurie's Notes

Activity 3

- In this activity, the models for each factor and the product are shown.
- Part (a) is read, "x plus 2 quantity squared" not "x plus 2 squared."
- Students should recognize that there are two factors that are the same.
- ? Ask the following questions.
  - "What are the factors in part (a)?" $(x + 2)$ and $(x + 2)$
  - "Are the factors the same in each part?" yes
  - "What is the product in part (b)?" $4x^2 - 4x + 1$
  - "Do you see a pattern in the x-terms of the products?" Students may not notice a pattern at this stage although they may comment on the symmetry of the product along the diagonal.

Activity 4

- Students may have difficulty working with two variables, a and b, versus a variable and a number. If so, have them complete part (c) with algebra tiles and then come back to parts (a) and (b).
- Students are more likely to see a pattern when the leading coefficient is not 1. Squaring a binomial such as $(2x - 5)$ has enough tiles that students often see a pattern even if they have difficulty stating it.

What Is Your Answer?

- **Neighbor Check:** Have students work independently and then have their neighbor check their work. Have students discuss any discrepancies.

Closure

- How could you rewrite 35^2 so that it is related to the products in Activity 3? *Sample answer:* $(30 + 5)^2$

Technology For the Teacher

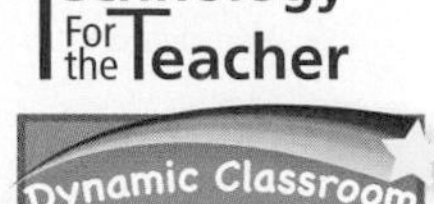

The Dynamic Planning Tool
Editable Teacher's Resources at *BigIdeasMath.com*

3 ACTIVITY: Finding the Square of a Binomial Pattern

Write the product of the two binomials shown by the algebra tiles.

a. $(x + 2)^2 =$

b. $(2x - 1)^2 =$

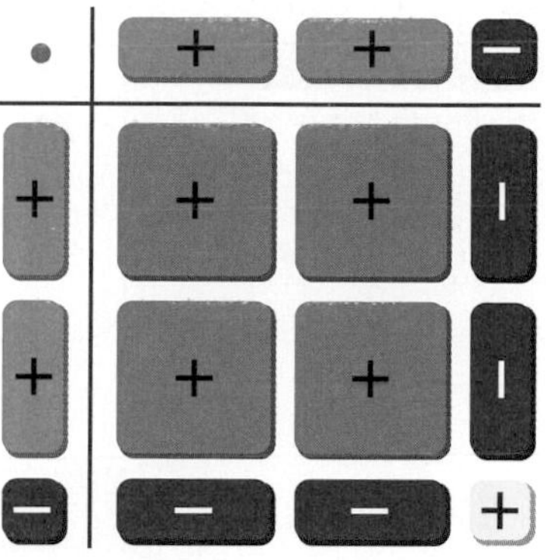

4 ACTIVITY: Describing the Square of a Binomial Pattern

Work with a partner.

a. Describe the pattern for the special product: $(a + b)^2$.

b. Describe the pattern for the special product: $(a - b)^2$.

c. Use the patterns you described to find each product. Check your answers using algebra tiles.

i. $(x + 3)^2$ **ii.** $(x - 2)^2$ **iii.** $(3x + 1)^2$

iv. $(3y + 4)^2$ **v.** $(2x - 5)^2$ **vi.** $(z + 1)^2$

What Is Your Answer?

5. **IN YOUR OWN WORDS** What are the patterns in the special products $(a + b)(a - b)$, $(a + b)^2$, and $(a - b)^2$? Use the results of Activities 2 and 4 to write formulas for these special products.

Practice Use what you learned about the patterns in special products to complete Exercises 3–5 on page 352.

7.4 Lesson

Some pairs of binomials show patterns when multiplied. You can use these patterns to multiply other similar pairs of binomials.

Key Idea

Study Tip

Because multiplication is commutative, the pattern also applies to $(a - b)(a + b)$.

Sum and Difference Pattern

Algebra

$(a + b)(a - b) = a^2 - b^2$

Example

$$(x + 3)(x - 3) = x^2 - 3^2$$
$$= x^2 - 9$$

EXAMPLE 1 Using the Sum and Difference Pattern

Find each product.

a. $(x + 7)(x - 7)$

$(a + b)(a - b) = a^2 - b^2$	Sum and Difference Pattern
$(x + 7)(x - 7) = x^2 - 7^2$	Use pattern.
$= x^2 - 49$	Simplify.

b. $(3x - 1)(3x + 1)$

$(a - b)(a + b) = a^2 - b^2$	Sum and Difference Pattern
$(3x - 1)(3x + 1) = (3x)^2 - 1^2$	Use pattern.
$= 9x^2 - 1$	Simplify.

Check

Use the FOIL Method.

$(3x - 1)(3x + 1)$

$= 9x^2 + 3x - 3x - 1$

$= 9x^2 - 1$ ✓

On Your Own

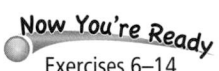

Now You're Ready Exercises 6–14

Find the product.

1. $(x - 4)(x + 4)$ **2.** $(b + 10)(b - 10)$ **3.** $(2g + 5)(2g - 5)$

Key Idea

Square of a Binomial Pattern

Algebra

$(a + b)^2 = a^2 + 2ab + b^2$

$(a - b)^2 = a^2 - 2ab + b^2$

Example

$$(x + 3)^2 = x^2 + 2(x)(3) + 3^2$$
$$= x^2 + 6x + 9$$

$$(x - 3)^2 = x^2 - 2(x)(3) + 3^2$$
$$= x^2 - 6x + 9$$

Laurie's Notes

Introduction

Connect

- **Yesterday:** Students used algebra tiles to find patterns in products of binomials. (MP1a, MP5)
- **Today:** Students will use special product patterns to multiply binomials.

Motivate

- Challenge your students to a multiplication race! Have the students choose a two-digit number. Then you choose a "compatible" two-digit number so that you can use special product patterns to multiply.
- Example: Students choose 42, you choose 38 or 58.
  $42 \times 38 = (40 + 2)(40 - 2) = 40^2 - 2^2 = 1600 - 4 = 1596$
  $42 \times 58 = (50 - 8)(50 + 8) = 50^2 - 8^2 = 2500 - 64 = 2436$
- **FYI:** Note that the number of hundreds is 1 less than a perfect square and the last two digits are the difference between 100 and a perfect square.
- After several problems, help students to see the patterns that you are using.

Lesson Notes

Key Idea

- Write the Sum and Difference Pattern. This is a special case of multiplying binomials where the product is a binomial instead of a trinomial.
- It is important to show students why the product is a binomial.
  $(x + 3)(x - 3) = x^2 - 3x + 3x - 9 = x^2 - 9$

Example 1

- **Common Error:** When substituting for a in part (b), students may write $3x^2$ instead of $(3x)^2$. Remind them to use parentheses when the coefficient is not 1.

On Your Own

- **Neighbor Check:** Have students work independently and then have their neighbor check their work. Have students discuss any discrepancies.

Key Idea

- Write the Square of a Binomial Pattern.
- Emphasize that $(x + 3)^2 \neq x^2 + 9$. Students can check their answers by rewriting the expression as $(x + 3)(x + 3)$ and using the FOIL Method.
- Point out that they can find the product $(x + 3)^2$ without expanding it. The product has a middle term which is twice the product of x and 3.

Goal Today's lesson is using special product patterns to multiply binomials.

Start Thinking! and Warm Up

Lesson 7.4 Warm Up
For use before Lesson 7.4

Lesson 7.4 Start Thinking!
For use before Lesson 7.4

Use the FOIL Method to find the products below.

$(5x + 6)(5x - 6)$

$(5x + 6)(5x + 6)$

$(5x - 6)(5x - 6)$

Use special product patterns to find the products below.

$(4x + 7)(4x - 7)$

$(4x + 7)(4x + 7)$

$(4x - 7)(4x - 7)$

Is it easier to use the FOIL Method or special product patterns to find the products?

Extra Example 1

Find each product.

a. $(k + 9)(k - 9)$ $k^2 - 81$

b. $(4q - 3)(4q + 3)$ $16q^2 - 9$

On Your Own

1. $x^2 - 16$
2. $b^2 - 100$
3. $4g^2 - 25$

Extra Example 2

Find each product.

a. $(x - 4)^2$ $x^2 - 8x + 16$

b. $(5w + 1)^2$ $25w^2 + 10w + 1$

Extra Example 3

Each of two roses has one red gene (R) and one white gene (W). The diagram shows the possible gene combinations of an offspring and the resulting colors.

	R	W
R	RR red	RW pink
W	RW pink	WW white

a. What percent of the possible gene combinations result in pink? 50%

b. The genetic makeup of an offspring can be modeled by $(0.5R + 0.5W)^2$. Use the square of a binomial pattern to model the possible gene combinations of an offspring. $0.25R^2 + 0.5RW + 0.25W^2$

On Your Own

4. $w^2 + 4w + 4$
5. $x^2 - 14x + 49$
6. $9y^2 - 6y + 1$
7. $25z^2 + 40z + 16$

English Language Learners

Vocabulary

Have students write a description of the patterns in their own words. For $(a + b)(a - b)$ a student might write, "The product of the sum and difference of two quantities is the square of the first quantity minus the square of the second quantity." Check students' descriptions.

Laurie's Notes

Example 2

- Work through the problems as shown. Note that the color-coding helps focus attention on the fact that the middle term of the product is twice the product of *ab*.
- **MP1 Make Sense of Problems and Persevere in Solving Them:** Discuss the Check. Some students will expand the square of a binomial and continue to use the FOIL Method. While this is not as efficient, it yields a correct solution and makes sense to the student.

Example 3

- Ask students if they have studied genetics in science class. If they have, they will be familiar with the Punnett square, although they may not know it by that name.
- **Common Error:** Students may forget to square the decimal coefficient.

On Your Own

- **Think-Pair-Share:** Students should read each question independently and then work with a partner to answer the questions. When they have answered the questions, the pair should compare their answers with another group and discuss any discrepancies.

Closure

- Explain how to use mental math to simplify each expression.
  - **a.** 67×73 Rewrite as $(70 - 3)(70 + 3)$ and use the sum and difference pattern to get 4891.
  - **b.** 52^2 Rewrite as $(50 + 2)^2$ and use the square of a binomial pattern to get 2704.

Technology For the Teacher

The Dynamic Planning Tool
Editable Teacher's Resources at *BigIdeasMath.com*

EXAMPLE 2 Using the Square of a Binomial Pattern

Find each product.

a. $(y + 1)^2$

$$(a + b)^2 = a^2 + 2ab + b^2$$ Square of a Binomial Pattern

$$(y + 1)^2 = y^2 + 2(y)(1) + 1^2$$ Use pattern.

$$= y^2 + 2y + 1$$ Simplify.

Check

Use the FOIL Method.

$$(2z - 3)^2 = (2z - 3)(2z - 3)$$

$$= 4z^2 - 6z - 6z + 9$$

$$= 4z^2 - 12z + 9 \checkmark$$

b. $(2z - 3)^2$

$$(a - b)^2 = a^2 - 2ab + b^2$$ Square of a Binomial Pattern

$$(2z - 3)^2 = (2z)^2 - 2(2z)(3) + 3^2$$ Use pattern.

$$= 4z^2 - 12z + 9$$ Simplify.

EXAMPLE 3 Real-Life Application

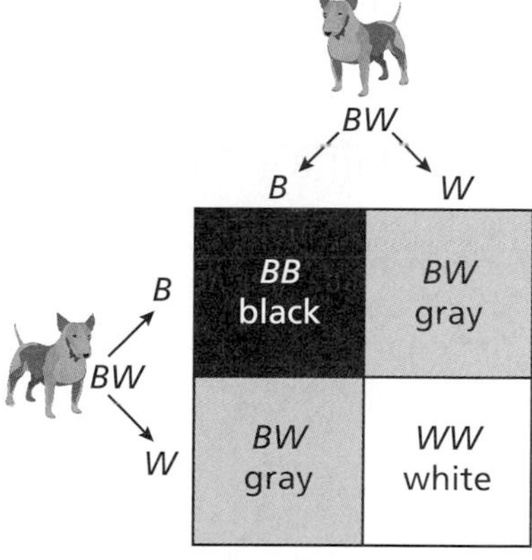

A diagram that models possible gene combinations in offspring is called a Punnett square.

Each of two dogs has one black gene (B) and one white gene (W). The diagram shows the possible gene combinations of an offspring and the resulting colors.

a. What percent of the possible gene combinations result in black?

Use the diagram. One of the four possible gene combinations results in black.

∴ So, $\frac{1}{4}$ or 25% of the possible gene combinations result in black.

b. The genetic makeup of an offspring can be modeled by $(0.5B + 0.5W)^2$. Use the square of a binomial pattern to model the possible gene combinations of an offspring.

$$(a + b)^2 = a^2 + 2ab + b^2$$ Square of a Binomial Pattern

$$(0.5B + 0.5W)^2 = (0.5B)^2 + 2(0.5B)(0.5W) + (0.5W)^2$$ Use pattern.

$$= 0.25B^2 + 0.5BW + 0.25W^2$$ Simplify.

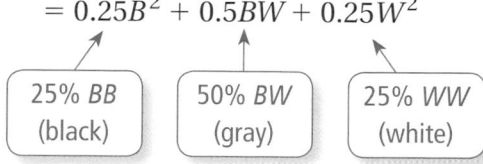

25% *BB* (black) | 50% *BW* (gray) | 25% *WW* (white)

On Your Own

Now You're Ready
Exercises 16–24

Find the product.

4. $(w + 2)^2$
5. $(x - 7)^2$
6. $(3y - 1)^2$
7. $(5z + 4)^2$

7.4 Exercises

Vocabulary and Concept Check

1. **OPEN-ENDED** Write two binomials whose product can be found using the sum and difference pattern.

2. **WHICH ONE DOESN'T BELONG?** Which expression does *not* belong with the other three? Explain your reasoning.

$(x+1)(x-1)$	$(3x+2)(3x-2)$	$(x+2)(x-3)$	$(2x+5)(2x-5)$

Practice and Problem Solving

Use algebra tiles to find the product.

3. $(x+6)(x-6)$
4. $(3y-2)(3y+2)$
5. $(2z+2)^2$

Find the product.

1 6. $(x+2)(x-2)$
7. $(g-5)(g+5)$
8. $(z-8)(z+8)$
9. $(b+12)(b-12)$
10. $(2x+1)(2x-1)$
11. $(3x-4)(3x+4)$
12. $(6x+7)(6x-7)$
13. $(9-c)(9+c)$
14. $(8-3m)(8+3m)$

15. **REASONING** Write two binomials whose product is x^2-16. Explain how you found your answer.

Find the product.

2 16. $(b-2)^2$
17. $(y+8)^2$
18. $(n+6)^2$
19. $(d-10)^2$
20. $(2f-1)^2$
21. $(5p+2)^2$
22. $(4b-5)^2$
23. $(12-x)^2$
24. $(4+7t)^2$

ERROR ANALYSIS Describe and correct the error in finding the product.

25.

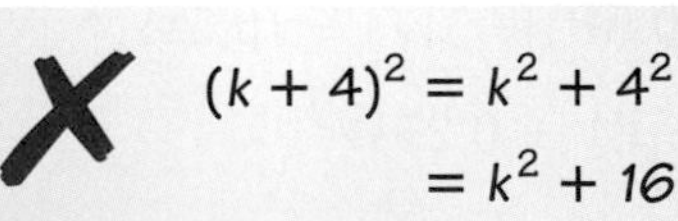

26.

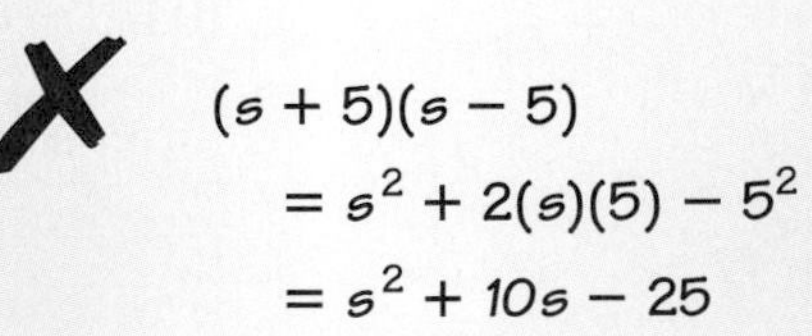

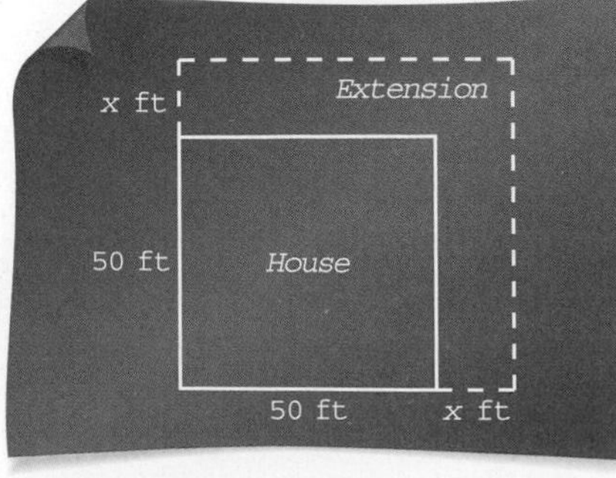

27. **CONSTRUCTION** A contractor extends a house on two sides.

    a. The area of the first level of the house after the renovation is represented by $(x+50)^2$. Find this product.

    b. Use the polynomial in part (a) to find the area of the first level when $x=15$. What is the area of the extension?

Assignment Guide and Homework Check

Level	Assignment	Homework Check
Average	1, 2, 7–27 odd, 26, 28, 30, 33, 34, 36–39	11, 19, 27, 33, 34
Advanced	1, 2, 6–26 even, 29, 31, 33–35, 36–39	14, 22, 33, 34, 35

Common Errors

- **Exercises 6–14** Students may use an addition sign in the product instead of a subtraction sign. Remind them that the sum and difference pattern is the difference of two squares not the sum. Demonstrate this by using the FOIL Method to multiply.
- **Exercises 16–24** Students may use the sum and difference pattern instead of the square of a binomial pattern when multiplying. Review the patterns with them. Encourage them to check their answers using the FOIL Method.

7.4 Record and Practice Journal

Find the product.

1. $(m - 7)(m + 7)$
   $m^2 - 49$
2. $(p + 10)(p - 10)$
   $p^2 - 100$
3. $(4s + 8)(4s - 8)$
   $16s^2 - 64$
4. $(9d - 6)(9d + 6)$
   $81d^2 - 36$
5. $(a + 5)^2$
   $a^2 + 10a + 25$
6. $(2k - 4)^2$
   $4k^2 - 16k + 16$
7. $(5 - 3r)^2$
   $9r^2 - 30r + 25$
8. $(2 + 12f)^2$
   $144f^2 + 48f + 4$
9. A garden is extended on two sides.
   a. The area of the garden after the extension is represented by $(x + 11)^2$. Find this product.
   $x^2 + 22x + 121$
   b. Use the polynomial in part (a) to find the area of the garden when $x = 4$. What is the area of the extension?
   $225 \text{ ft}^2; 104 \text{ ft}^2$

Answer Presentation Tool

Vocabulary and Concept Check

1. *Sample answer:* $x + 7, x - 7$
2. $(x + 2)(x - 3)$ does not belong because it cannot be simplified using the sum and difference pattern.

Practice and Problem Solving

3. $x^2 - 36$
4. $9y^2 - 4$
5. $4z^2 + 8z + 4$
6. $x^2 - 4$
7. $g^2 - 25$
8. $z^2 - 64$
9. $b^2 - 144$
10. $4x^2 - 1$
11. $9x^2 - 16$
12. $36x^2 - 49$
13. $81 - c^2$
14. $64 - 9m^2$
15. $x - 4$ and $x + 4$; Rewrite the expression as $x^2 - 4^2$ and apply the sum and difference pattern.
16. $b^2 - 4b + 4$
17. $y^2 + 16y + 64$
18. $n^2 + 12n + 36$
19. $d^2 - 20d + 100$
20. $4f^2 - 4f + 1$
21. $25p^2 + 20p + 4$
22. $16b^2 - 40b + 25$
23. $144 - 24x + x^2$
24. $16 + 56t + 49t^2$
25. The product should have a middle term.
    $$(k + 4)^2 = k^2 + 2(k)(4) + 4^2 = k^2 + 8k + 16$$
26. The product should not have a middle term.
    $$(s + 5)(s - 5) = s^2 - 5^2 = s^2 - 25$$

Practice and Problem Solving

27. **a.** $x^2 + 100x + 2500$

b. 4225 ft^2; 1725 ft^2

28. $x^2 + 8x + 16$

29. $4x^2 + 28x + 49$

30. $x^4 - 1$

31. $x^2 - y^2$

32. $4x^2 - 4xy + y^2$

33. **a.** 75%

b. $0.25N^2 + 0.5Na + 0.25a^2$

34. See *Taking Math Deeper.*

35. $(x + 1)^3 = x^3 + 3x^2 + 3x + 1$
$(x + 2)^3 = x^3 + 6x^2 + 12x + 8$
$(a + b)^3$
$= a^3 + 3a^2b + 3ab^2 + b^3$

Fair Game Review

36. $x^2 + 13x + 36$

37. $y^2 - 4y - 21$

38. $z^2 - 11z + 10$

39. D

Mini-Assessment

Find the product.

1. $(x - 6)(x + 6)$ $x^2 - 36$
2. $(2n + 7)(2n - 7)$ $4n^2 - 49$
3. $(s - 5)^2$ $s^2 - 10s + 25$
4. $(6d + 4)^2$ $36d^2 + 48d + 16$
5. The side length of a square can be represented by the expression $2x - 5$. Write a polynomial that represents the area of the square. $4x^2 - 20x + 25$

Taking Math Deeper

Exercise 34

Be sure students understand that the "width" of the iris is x. It is not the radius or the diameter of the eye.

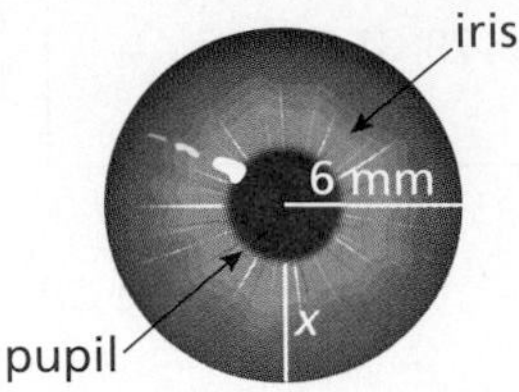

a. To find the area of the pupil, you must first find the radius of the pupil. Then use the formula for the area of a circle to find the area of the pupil.

Radius of pupil = (radius of eye) − (width of iris)
$= (6 - x)$ mm

So, the area of the pupil is represented by

$$A = \pi r^2$$
$$= \pi(6 - x)^2.$$

b. Before entering the dark room, the area is

$$A = \pi(6 - x)^2$$
$$= \pi(6 - 4)^2$$
$$= 4\pi.$$

After entering the dark room, the area is

$$A = \pi(6 - x)^2$$
$$= \pi(6 - 2)^2$$
$$= 16\pi.$$

So, the area is $\frac{16\pi}{4\pi} = 4$ times greater after entering the dark room. The reason for this is that there isn't much light in the dark room and your pupil enlarges to let in as much of the light as possible.

Reteaching and Enrichment Strategies

If students need help...	If students got it...
Resources by Chapter • Practice A and Practice B • Puzzle Time Record and Practice Journal Practice Differentiating the Lesson Lesson Tutorials Skills Review Handbook	Resources by Chapter • Enrichment and Extension • School-to-Work Start the next section

Write a polynomial that represents the area of the figure.

28.

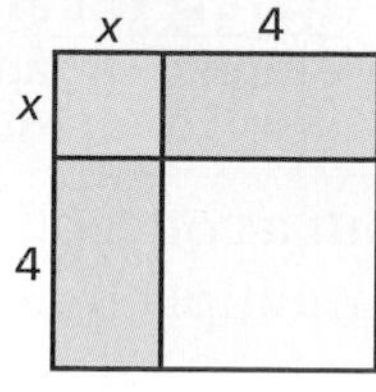

29.

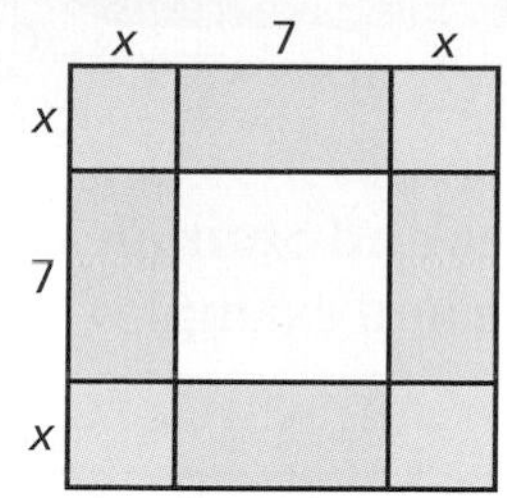

Find the product.

30. $(x^2 + 1)(x^2 - 1)$

31. $(x + y)(x - y)$

32. $(2x - y)^2$

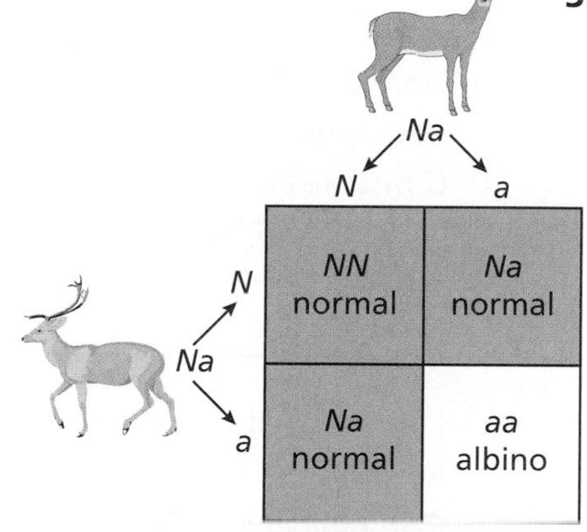

33. GENETICS In deer, the gene N is for normal coloring and the gene a is for no coloring, or albino. Any gene combination with an N results in normal coloring. The diagram shows the possible gene combinations of an offspring and the resulting colors from parents that both have the gene combination Na.

a. What percent of the possible gene combinations result in normal coloring?

b. The genetic makeup of an offspring can be modeled by $(0.5N + 0.5a)^2$. Use the square of a binomial pattern to model the possible gene combinations of an offspring.

34. VISION Your iris controls the amount of light that enters your eye by changing the size of your pupil.

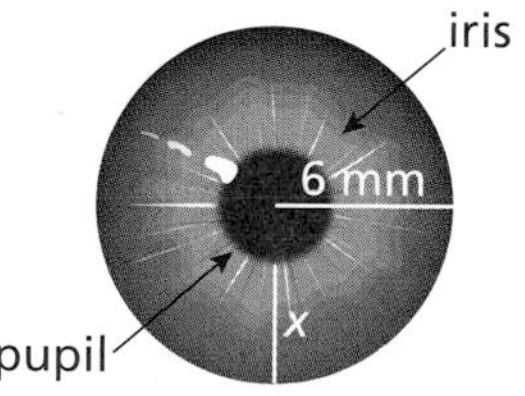

a. Write a polynomial that represents the area of your pupil. Write your answer in terms of π.

b. The width x of your iris decreases from 4 millimeters to 2 millimeters when you enter a dark room. How many times greater is the area of your pupil after entering the room than before entering the room? Explain.

35. Repeated Reasoning Find $(x + 1)^3$ and $(x + 2)^3$. Find a pattern in the terms and use it to write a pattern for the cube of a binomial $(a + b)^3$.

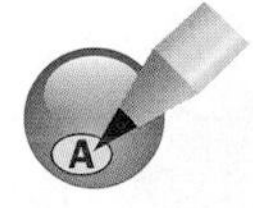

Fair Game Review What you learned in previous grades & lessons

Find the product. *(Section 7.3)*

36. $(x + 4)(x + 9)$

37. $(y - 7)(y + 3)$

38. $(z - 10)(z - 1)$

39. MULTIPLE CHOICE What is the solution of the linear system? *(Section 4.2)*

$y = 2x - 5$
$3x - 8y = 1$

Ⓐ $(-3, -1)$ Ⓑ $(-3, 1)$

Ⓒ $(3, -1)$ Ⓓ $(3, 1)$

7 Study Help

You can use an **idea and examples chart** to organize information about a concept. Here is an example of an idea and examples chart for using the FOIL Method to multiply binomials.

FOIL Method: To multiply two binomials using the FOIL Method, find the sum of the products of the First terms, Outer terms, Inner terms, and Last terms.

Example

First, Outer, Inner, Last

$(x-2)(x+3) = x(x) + x(3) + (-2)(x) + (-2)(3)$ — Use the FOIL Method.

$= x^2 + (3x) + (-2x) + (-6)$ — Multiply.

$= x^2 + x - 6$ — Combine like terms.

Example

First, Outer, Inner, Last

$(3x-1)(2x-2) = 3x(2x) + 3x(-2) + (-1)(2x) + (-1)(-2)$ — Use the FOIL Method.

$= 6x^2 + (-6x) + (-2x) + 2$ — Multiply.

$= 6x^2 - 8x + 2$ — Combine like terms.

On Your Own

Make idea and examples charts to help you study these topics.

1. degree of a polynomial
2. adding and subtracting polynomials
3. special products of polynomials

After you complete this chapter, make idea and examples charts for the following topics.

4. factored form of a polynomial
5. factoring polynomials using the GCF
6. factoring polynomials of the form $x^2 + bx + c$
7. factoring polynomials of the form $ax^2 + bx + c$

"I made an idea and examples chart to give my owner ideas for my birthday next week."

Sample Answers

1.

Degree of a Polynomial
The degree of a polynomial is the greatest degree of its terms. The degree of a polynomial in standard form is the exponent of the first term.

Example

$2x^5 + 3x^2 - 5x + 7$	The polynomial is in standard form. The degree of the polynomial is 5.

Example

$4x - 7 + 6x^3$	Standard form: $6x^3 + 4x - 7$ The degree of the polynomial is 3.

2.

Adding and Subtracting Polynomials
Vertical Method: Align like terms vertically and add or subtract.
Horizontal Method: Group like terms and simplify.

Example

Vertical Method:

$$\begin{array}{r} 3x^2 - x \qquad \\ +\ 8x^2 \qquad + 7 \\ \hline 11x^2 - x + 7 \end{array}$$

Horizontal Method:

$(3x^2 - x) + (8x^2 + 7)$
$= (3x^2 + 8x^2) + (-x) + 7$
$= 11x^2 - x + 7$

Example

Vertical Method:

$$\begin{array}{r} (6x^2 - 2) \\ -\ (4x^2 - 1) \\ \hline \end{array} \Rightarrow \begin{array}{r} 6x^2 - 2 \\ +\ -4x^2 + 1 \\ \hline 2x^2 - 1 \end{array}$$

Horizontal Method:

$(6x^2 - 2) - (4x^2 - 1)$
$= (6x^2 - 2) + (-4x^2 + 1)$
$= [6x^2 + (-4x^2)] + (-2 + 1)$
$= 2x^2 - 1$

3. Available at *BigIdeasMath.com.*

List of Organizers

Available at *BigIdeasMath.com*

Comparison Chart
Concept Circle
Example and Non-Example Chart
Formula Triangle
Four Square
Idea (Definition) and Examples Chart
Information Frame
Information Wheel
Notetaking Organizer
Process Diagram
Summary Triangle
Word Magnet
Y Chart

About this Organizer

An **Idea and Examples Chart** can be used to organize information about a concept. Students fill in the top rectangle with a term and its definition or description. Students fill in the rectangles that follow with examples to illustrate the term. Idea and examples charts are useful for concepts that can be illustrated with more than one type of example.

Technology For the Teacher
Vocabulary Puzzle Builder

Answers

1. $-8q^3$; 3; monomial
2. $d^2 - 3d - 9$; 2; trinomial
3. $-\frac{5}{6}m^6 + \frac{2}{3}m^4$; 6; binomial
4. $2z^4 + 7.4z^2 - 1.3z$; 4; trinomial
5. $x^2 + 9$
6. $-5n^2 + n - 7$
7. $-2p^2 + 7p - 15$
8. $-2ab + 2b^2$
9. $w^2 + 13w + 42$
10. $y^2 + 6y - 27$
11. $d^2 - 7d + 10$
12. $6z^2 + z - 15$
13. $h^2 - 1$
14. $p^2 - 81$
15. $t^2 + 10t + 25$
16. $q^2 - 4q + 4$
17. **a.** $x^2 - 1$
    b. 8 ft^2
18. **a.** $100r^2 + 200r + 100$
    b. $125.44
    c. $24.56

Assessment Book

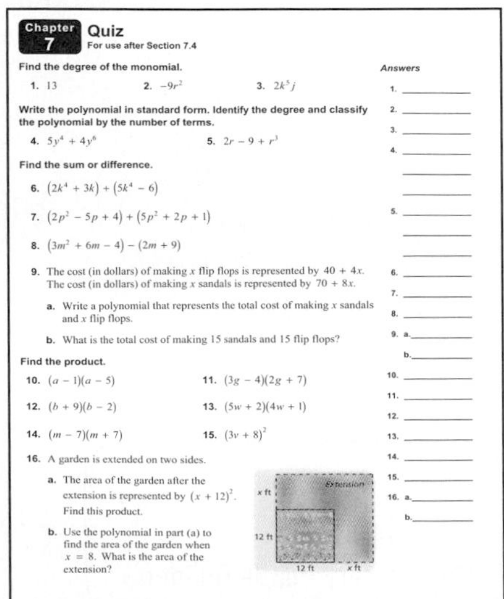

Chapter 7 Quiz
For use after Section 7.4

Find the degree of the monomial.

1. 13 2. $-9r^2$ 3. $2k^5j$

Write the polynomial in standard form. Identify the degree and classify the polynomial by the number of terms.

4. $5y^4 + 4y^6$ 5. $2r - 9 + r^3$

Find the sum or difference.

6. $(2k^4 + 3k) + (5k^4 - 6)$
7. $(2p^2 - 5p + 4) + (5p^2 + 2p + 1)$
8. $(3m^2 + 6m - 4) - (2m + 9)$
9. The cost (in dollars) of making x flip flops is represented by $40 + 4x$. The cost (in dollars) of making x sandals is represented by $70 + 8x$.
   a. Write a polynomial that represents the total cost of making x sandals and x flip flops.
   b. What is the total cost of making 15 sandals and 15 flip flops?

Find the product.

10. $(a - 1)(a - 5)$ 11. $(3g - 4)(2g + 7)$
12. $(b + 9)(b - 2)$ 13. $(5w + 2)(4w + 1)$
14. $(m - 7)(m + 7)$ 15. $(3v + 8)^2$
16. A garden is extended on two sides.
   a. The area of the garden after the extension is represented by $(x + 12)^2$. Find this product.
   b. Use the polynomial in part (a) to find the area of the garden when $x = 8$. What is the area of the extension?

Answers
1. ____
2. ____
3. ____
4. ____
5. ____
6. ____
7. ____
8. ____
9. a. ____
   b. ____
10. ____
11. ____
12. ____
13. ____
14. ____
15. ____
16. a. ____
    b. ____

Alternative Quiz Ideas

100% Quiz	Math Log
Error Notebook	Notebook Quiz
Group Quiz	Partner Quiz
Homework Quiz	**Pass the Paper**

Pass the Paper

- Work in groups of four. The first student copies the problem and completes the first step, explaining his or her work.
- The paper is passed and the second student works through the next step, also explaining his or her work.
- This process continues until the problem is completed.
- The second member of the group starts the next problem. Students should be allowed to question and debate as they are working through the quiz.
- Student groups can be selected by the teacher, by students, through a random process, or any way that works for your class.
- The teacher walks around the classroom listening to the groups and asks questions to ensure understanding.

Reteaching and Enrichment Strategies

If students need help. . .	If students got it. . .
Resources by Chapter • Study Help • Practice A and Practice B • Puzzle Time Lesson Tutorials *BigIdeasMath.com* Practice Quiz Practice from the Test Generator	Resources by Chapter • Enrichment and Extension • School-to-Work Game Closet at *BigIdeasMath.com* Start the next section

Technology For the Teacher
Answer Presentation Tool
Big Ideas Test Generator

7.1–7.4 Quiz

Write the polynomial in standard form. Identify the degree and classify the polynomial by the number of terms. *(Section 7.1)*

1. $-8q^3$
2. $-9 + d^2 - 3d$
3. $\frac{2}{3}m^4 - \frac{5}{6}m^6$
4. $-1.3z + 2z^4 + 7.4z^2$

Find the sum or difference. *(Section 7.2)*

5. $(2x^2 + 5) + (-x^2 + 4)$
6. $(-3n^2 + n) - (2n^2 + 7)$
7. $(-p^2 + 4p) - (p^2 - 3p + 15)$
8. $(a^2 - 3ab + b^2) + (-a^2 + ab + b^2)$

Find the product. *(Section 7.3 and Section 7.4)*

9. $(w + 6)(w + 7)$
10. $(y + 9)(y - 3)$
11. $(d - 2)(d - 5)$
12. $(2z - 3)(3z + 5)$
13. $(h - 1)(h + 1)$
14. $(p + 9)(p - 9)$
15. $(t + 5)^2$
16. $(q - 2)^2$

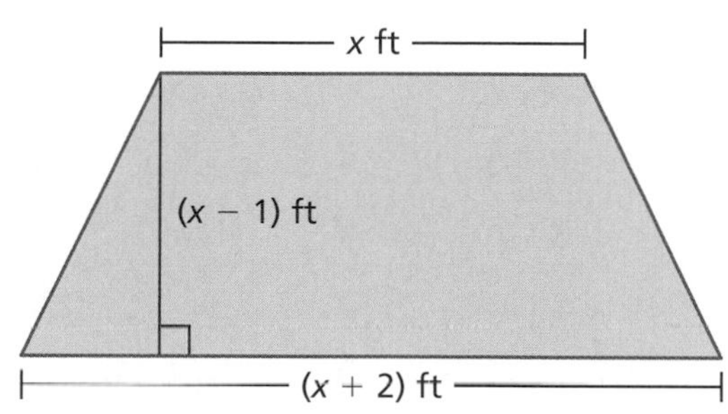

17. **WINDOW SEAT** A window seat is in the shape of a trapezoid. *(Section 7.3)*

    a. Write a polynomial that represents the area of the window seat.

    b. What is the area of the window seat when $x = 3$?

18. **COMPOUND INTEREST** You are saving for a guitar. You deposit \$100 in an account that earns interest compounded annually. The expression $100(1 + r)^2$ represents the balance after 2 years, where r is the annual interest rate in decimal form. *(Section 7.4)*

    a. Write a polynomial that represents the balance of your account.

    b. What is the balance of your account when the interest rate is 12%?

    c. How much more money do you need to save to buy the guitar?

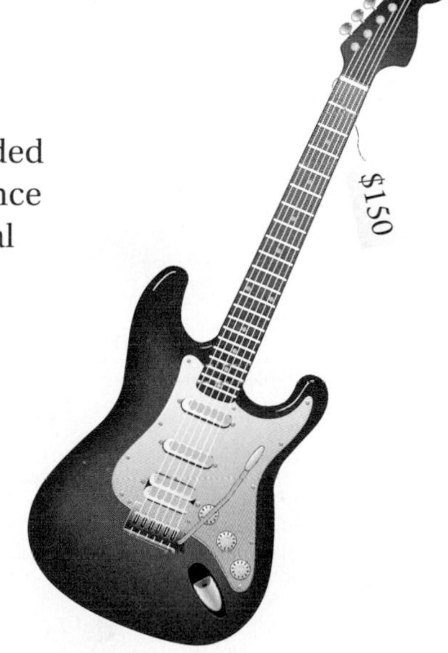

7.5 Solving Polynomial Equations in Factored Form

COMMON CORE STATE STANDARDS
A.REI.4b

Essential Question How can you solve a polynomial equation that is written in factored form?

Two polynomial equations are equivalent when they have the same solutions. For instance, the following equations are equivalent because the only solutions of each equation are $x = 1$ and $x = 2$.

Factored Form	Standard Form	Nonstandard Form
$(x - 1)(x - 2) = 0$	$x^2 - 3x + 2 = 0$	$x^2 - 3x = -2$

✓ Check this by substituting 1 and 2 for x in each equation.

1 ACTIVITY: Matching Equivalent Forms of an Equation

Work with a partner. Match each factored form of the equation with two other forms of equivalent equations. Notice that an equation is considered to be in factored form only when the product of the factors is equal to 0.

	Factored Form		Standard Form		Nonstandard Form
a.	$(x - 1)(x - 3) = 0$	**A.**	$x^2 - x - 2 = 0$	**1.**	$x^2 - 5x = -6$
b.	$(x - 2)(x - 3) = 0$	**B.**	$x^2 + x - 2 = 0$	**2.**	$(x - 1)^2 = 4$
c.	$(x + 1)(x - 2) = 0$	**C.**	$x^2 - 4x + 3 = 0$	**3.**	$x^2 - x = 2$
d.	$(x - 1)(x + 2) = 0$	**D.**	$x^2 - 5x + 6 = 0$	**4.**	$x(x + 1) = 2$
e.	$(x + 1)(x - 3) = 0$	**E.**	$x^2 - 2x - 3 = 0$	**5.**	$x^2 - 4x = -3$

2 ACTIVITY: Writing a Conjecture

Work with a partner. Substitute

1, 2, 3, 4, 5, and 6 for x

in each equation. Write a conjecture describing what you discovered.

a. $(x - 1)(x - 2) = 0$ **b.** $(x - 2)(x - 3) = 0$ **c.** $(x - 3)(x - 4) = 0$

d. $(x - 4)(x - 5) = 0$ **e.** $(x - 5)(x - 6) = 0$ **f.** $(x - 6)(x - 1) = 0$

Laurie's Notes

Introduction

Standards for Mathematical Practice

- **MP1a Make Sense of Problems** and **MP7 Look for and Make Use of Structure:** Students will solve polynomial equations by writing them in factored form and using the Zero-Product Property. The mathematically proficient student is able to transform an expression into factored form, understanding the reason to have a product that can be set equal to 0.

Motivate

- Write the following equations on the board and ask which one doesn't belong. Have students explain their reasoning.

  A. $3(x - 4) = 12$ **B.** $\frac{3}{4}x = 6$

  C. $4 - 8x = 3$ **D.** $6x - 7 = 9 + 4x$

- Students should solve the equations and reason that C does not belong. The other three equations all have the same solution, $x = 8$. These three equations are equivalent.

Activity Notes

Discuss

- Explain what equivalent polynomial equations are, connecting to the Motivate activity.
- Review the language associated with polynomials: factored form, standard form, and nonstandard form. Check the solutions in each form.

Activity 1

- Students may approach the activity in one of two ways. If they recognize the solutions from the factored form, then they may look at the other two forms and determine which forms are equivalent by substituting the solutions from the factored form. The alternate method would be to expand the factored form and see which expressions are equivalent in standard form.
- Give students time to work through the problems. Then ask for volunteers to share their findings and their method of solving.

? "How did you determine which equations are equivalent?" Expect different methods.

Activity 2

- Colors are used as a visual aid in this activity. For each problem, students will substitute the given values and evaluate the polynomial.
- **Big Idea:** Each polynomial has two solutions. When either or both of the solutions are substituted in the factored expression, the product is 0. If two numbers are substituted and neither is a solution, then the product is not equal to 0.

Common Core State Standards

A.REI.4b Solve quadratic equations by inspection, taking square roots, completing the square, the quadratic formula and factoring, as appropriate to the initial form of the equation . . ..

Previous Learning

Students should know how to solve linear equations.

Start Thinking! and Warm Up

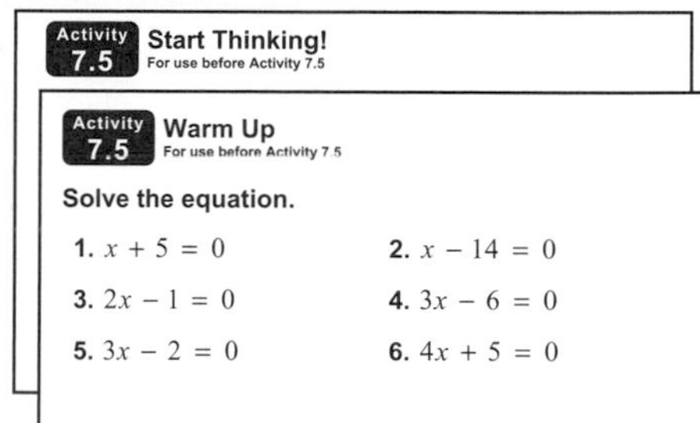

Activity 7.5 Start Thinking! For use before Activity 7.5

Activity 7.5 Warm Up For use before Activity 7.5

Solve the equation.

1. $x + 5 = 0$
2. $x - 14 = 0$
3. $2x - 1 = 0$
4. $3x - 6 = 0$
5. $3x - 2 = 0$
6. $4x + 5 = 0$

7.5 Record and Practice Journal

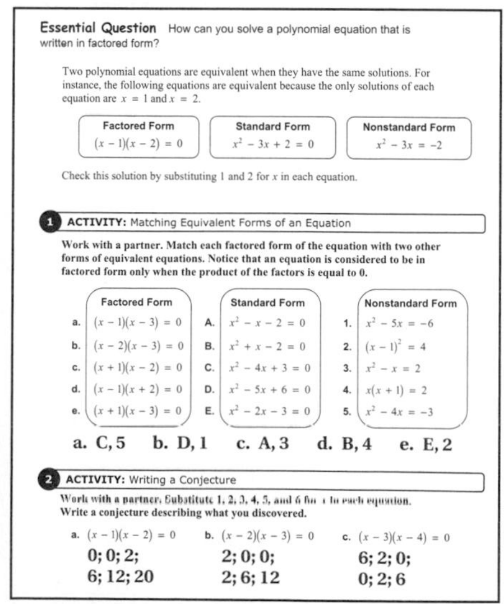

Essential Question How can you solve a polynomial equation that is written in factored form?

Two polynomial equations are equivalent when they have the same solutions. For instance, the following equations are equivalent because the only solutions of each equation are $x = 1$ and $x = 2$.

Factored Form	Standard Form	Nonstandard Form
$(x-1)(x-2)=0$	$x^2-3x+2=0$	$x^2-3x=-2$

Check this solution by substituting 1 and 2 for x in each equation.

1 ACTIVITY: Matching Equivalent Forms of an Equation

Work with a partner. Match each factored form of the equation with two other forms of equivalent equations. Notice that an equation is considered to be in factored form only when the product of the factors is equal to 0.

Factored Form	Standard Form	Nonstandard Form
a. $(x-1)(x-3)=0$	A. $x^2-x-2=0$	1. $x^2-5x=-6$
b. $(x-2)(x-3)=0$	B. $x^2+x-2=0$	2. $(x-1)^2=4$
c. $(x+1)(x-2)=0$	C. $x^2-4x+3=0$	3. $x^2-x=2$
d. $(x-1)(x+2)=0$	D. $x^2-5x+6=0$	4. $x(x+1)=2$
e. $(x+1)(x-3)=0$	E. $x^2-2x-3=0$	5. $x^2-4x=-3$

a. C, 5   b. D, 1   c. A, 3   d. B, 4   e. E, 2

2 ACTIVITY: Writing a Conjecture

Work with a partner. Substitute 1, 2, 3, 4, 5, and 6 for x in each equation. Write a conjecture describing what you discovered.

a. $(x-1)(x-2)=0$ — 0; 0; 2; 6; 12; 20

b. $(x-2)(x-3)=0$ — 2; 0; 0; 2; 6; 12

c. $(x-3)(x-4)=0$ — 6; 2; 0; 0; 2; 6

English Language Learners

Notebook Development

Have English language learners write the Multiplication Property of Zero using words and algebra in their notebooks.

Words: The product of any number and 0 is 0.

Algebra: $a \cdot 0 = 0$

Then have them do the same with the Zero-Product Property. Students should see the connection between the two properties.

7.5 Record and Practice Journal

d. $(x-4)(x-5)=0$ **12; 6; 2; 0; 0; 2**

e. $(x-5)(x-6)=0$ **20; 12; 6; 2; 0; 0**

f. $(x-6)(x-1)=0$ **0; −4; −6; −6; −4; 0**

If $(x-a)$ is a factor of an equation, then $x=a$ is a solution.

3 ACTIVITY: Special Properties of 0 and 1

Work with a partner. The numbers 0 and 1 have special properties that are shared by no other numbers. For each of the following, decide whether the property is true for 0, 1, both, or neither. Explain your reasoning.

a. If you add **0** to a number n, you get n.

b. If the product of two numbers is **0**, then one or both numbers are 0.

c. The square of **both** is equal to itself.

d. If you multiply a number n by **1**, you get n.

e. If you multiply a number n by **0**, you get 0.

f. The opposite of **neither** is equal to itself.

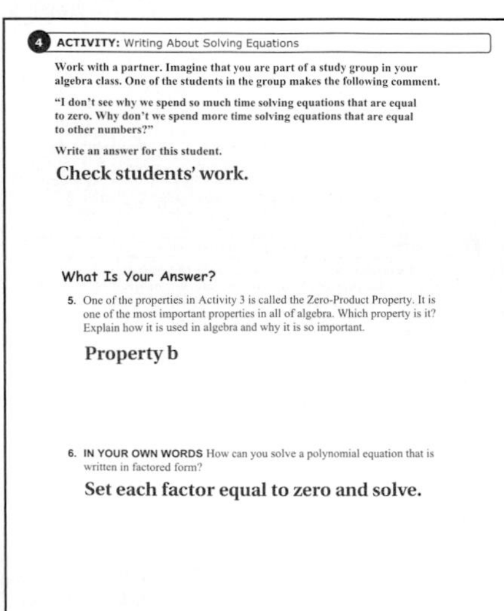

4 ACTIVITY: Writing About Solving Equations

Work with a partner. Imagine that you are part of a study group in your algebra class. One of the students in the group makes the following comment.

"I don't see why we spend so much time solving equations that are equal to zero. Why don't we spend more time solving equations that are equal to other numbers?"

Write an answer for this student.

Check students' work.

What Is Your Answer?

5. One of the properties in Activity 3 is called the Zero-Product Property. It is one of the most important properties in all of algebra. Which property is it? Explain how it is used in algebra and why it is so important.

Property b

6. IN YOUR OWN WORDS How can you solve a polynomial equation that is written in factored form?

Set each factor equal to zero and solve.

Laurie's Notes

Activity 3

- The numbers 0 and 1 have special properties. Students may not recall the formal names of the properties, but they should recall the essence of the properties.
- Remind students that they should check both 0 and 1 in each statement to determine whether the statement is true.
- When students are finished, ask for volunteers to share their answers and their reasoning.
- **Extension:** Ask if any of the statements are true when $n = -1$.

Activity 4

- Students should discuss their thoughts about the statement, and then each should write a response.
- The Zero-Product Property will be used in the lesson to solve polynomials. It is important for students to recognize the simplicity of this process, and the structure of the equation that is needed. Students will sometimes incorrectly write $x^2 - 4x = 0$ and set each term equal to zero. They have incorrectly set two terms equal to zero versus two factors equal to zero.
- Students understanding of the property will be evident in what they write.

What Is Your Answer?

- For Question 5, have students discuss which property they chose and why they chose it.
- **Neighbor Check:** For Question 6, have students work independently and then have their neighbor check their work. Have students discuss any discrepancies.

Closure

- **Exit Ticket:** What are the solutions of $(x-4)(x+2)=0$? $x=-2, x=4$

Technology For the Teacher

The Dynamic Planning Tool

Editable Teacher's Resources at *BigIdeasMath.com*

3 ACTIVITY: Special Properties of 0 and 1

Work with a partner. The numbers 0 and 1 have special properties that are shared by no other numbers. For each of the following, decide whether the property is true for 0, 1, both, or neither. Explain your reasoning.

a. If you add ▢ to a number n, you get n.

b. If the product of two numbers is ▢, then one or both numbers are 0.

c. The square of ▢ is equal to itself.

d. If you multiply a number n by ▢, you get n.

e. If you multiply a number n by ▢, you get 0.

f. The opposite of ▢ is equal to itself.

4 ACTIVITY: Writing About Solving Equations

Work with a partner. Imagine that you are part of a study group in your algebra class. One of the students in the group makes the following comment.

"I don't see why we spend so much time solving equations that are equal to zero. Why don't we spend more time solving equations that are equal to other numbers?"

Write an answer for this student.

What Is Your Answer?

5. One of the properties in Activity 3 is called the Zero-Product Property. It is one of the most important properties in all of algebra. Which property is it? Explain how it is used in algebra and why it is so important.

6. **IN YOUR OWN WORDS** How can you solve a polynomial equation that is written in factored form?

Practice Use what you learned about solving polynomial equations to complete Exercises 4–6 on page 360.

7.5 Lesson

Key Vocabulary
factored form, *p. 358*
Zero-Product Property, *p. 358*
root, *p. 358*

A polynomial is in **factored form** when it is written as a product of factors.

Standard form	*Factored form*
$x^2 + 2x$	$x(x + 2)$
$x^2 + 5x - 24$	$(x - 3)(x + 8)$

When one side of an equation is a polynomial in factored form and the other side is 0, use the **Zero-Product Property** to solve the polynomial equation. The solutions of a polynomial equation are also called **roots.**

Key Idea

Zero-Product Property

Words If the product of two real numbers is 0, then at least one of the numbers is 0.

Algebra If a and b are real numbers and $ab = 0$, then $a = 0$ or $b = 0$.

EXAMPLE 1 Solving Polynomial Equations

Solve each equation.

a. $x(x + 8) = 0$

$x(x + 8) = 0$ Write equation.

$x = 0$ *or* $x + 8 = 0$ Use Zero-Product Property.

$x = -8$ Solve for x.

The roots are $x = 0$ and $x = -8$.

Check
Substitute each solution in the original equation.

$0(0 + 8) \stackrel{?}{=} 0$

$0(8) \stackrel{?}{=} 0$

$0 = 0$ ✓

$-8(-8 + 8) \stackrel{?}{=} 0$

$-8(0) \stackrel{?}{=} 0$

$0 = 0$ ✓

b. $(x + 6)(x - 5) = 0$

$(x + 6)(x - 5) = 0$ Write equation.

$x + 6 = 0$ *or* $x - 5 = 0$ Use Zero-Product Property.

$x = -6$ *or* $x = 5$ Solve for x.

The roots are $x = -6$ and $x = 5$.

On Your Own

Now You're Ready
Exercises 4–9

Solve the equation.

1. $x(x - 1) = 0$

2. $3t(t + 2) = 0$

3. $(z - 4)(z - 6) = 0$

4. $(b + 7)^2 = 0$

Multi-Language Glossary at BigIdeasMath.com.

Laurie's Notes

Introduction

Connect

- **Yesterday:** Students explored equivalent forms of equations. (MP1a, MP7)
- **Today:** Students will solve polynomial equations using the Zero-Product Property.

Motivate

- **MP2 Reason Abstractly and Quantitatively:** Write the following multiplication problem on the board:

$$\begin{array}{r} 4y4y \\ \times \quad 6 \\ \hline 2424y \end{array}$$

? "What does y equal in this problem? Explain." $y = 0$; Explanations will vary; Listen for the fact that the product of 0 and any number is 0.

Lesson Notes

Discuss

- Write the two examples of standard form and factored form on the board.

? "What property is used to write the factored form in standard form?" Distributive Property

- In this chapter, students will work with problems that are written in standard form and the goal is to write them in factored form.
- Explain to students that the solutions of a polynomial equation are also called *roots*. Use "roots" and "solutions" interchangeably so students become comfortable with the vocabulary.

Key Idea

- Write the Zero-Product Property and connect it to the problem from the beginning of class. Use the vocabulary *factor* and *product* when discussing the Zero-Product Property and the examples.

Example 1

- Work through the examples as shown.
- **Teaching Tip:** Underline each of the factors and ask, "What are the factors that have a product of 0?" x and $(x + 8)$
- Take time to check the roots (solutions) as shown.

? Preview On Your Own Question 4 by asking, "If the product of two factors is 0, will there always be two roots? Explain." no; If the factors are the same, the roots are the same, so there is only one solution.

On Your Own

- Have volunteers solve the problems at the board.
- **Extension:** "Is $(b + 7)^2 = 0$ equivalent to $b^2 + 49 = 0$?" no; $(b + 7)^2 = 0$ is equivalent to $(b + 7)(b + 7) = 0$ or $b^2 + 14b + 49 = 0$.

Goal Today's lesson is solving polynomial equations using the **Zero-Product Property**.

Start Thinking! and Warm Up

Lesson 7.5 Warm Up
For use before Lesson 7.5

Lesson 7.5 Start Thinking!
For use before Lesson 7.5

In which form is it easier to tell that the solutions are $x = 7$ or $x = -1$?

Standard form: $x^2 - 6x - 7 = 0$

Factored form: $(x - 7)(x + 1) = 0$

Write an equation in standard form that has the solutions −2 and 9.

Write an equation in factored form that has the solutions −2 and 9.

Extra Example 1

Solve each equation.

a. $2y(y - 4) = 0$ $y = 0, y = 4$

b. $(w - 9)(w + 5) = 0$ $w = 9, w = -5$

On Your Own

1. $x = 0, x = 1$
2. $t = 0, t = -2$
3. $z = 4, z = 6$
4. $b = -7$

Extra Example 2

Solve $(3n + 8)(3n - 8) = 0$.

$n = -\frac{8}{3}, n = \frac{8}{3}$

Extra Example 3

The entrance of a cave can be modeled by $y = -\frac{1}{2}(x - 6)(x + 6)$ where x and y are measured in feet. The x-axis represents the ground. Find the width of the entrance at ground level. 12 ft

On Your Own

5. $p = -\frac{5}{3}, p = \frac{5}{3}$
6. $x = 2$
7. 8 ft

Differentiated Instruction

Organization

When solving equations such as $4m(2m - 3) = 0$, students may set the factor in parentheses equal to 0, but forget to set $4m$ equal to 0. Encourage them to write all the factors in parentheses, as shown.

$(4m)(2m - 3) = 0$

This will help identify all the factors involved.

Laurie's Notes

Example 2

? "How is this equation different from part (b) in Example 1?" The leading coefficients are not 1.

- Work through the example as shown.
- Ask students to check the two solutions for a quick review of multiplying fractions.

Example 3

- Draw the arch of the fireplace.

? "What dimensions can be used to describe the arch?" Students will likely talk about the height of the arch at the middle of the fireplace and the width of the arch at floor level.

- Now draw coordinate axes and identify the dimensions discussed.
- Write the model for the arch and solve as shown.
- **Common Error:** When multiplying each side by -9, students may think that all three factors on the right side must be multiplied by -9. To help students with this, ask them to find the product 3(4)(5). 60 Then ask them to find the product 2(3)(4)(5). 120
- **Extension:** Ask students to find the height of the arch at the middle. When $x = 0$, $y = 36$ inches.

On Your Own

- Question 6 can be challenging for students. Encourage them to write the power as the product of two binomials: $(12 - 6x)(12 - 6x) = 0$. When solving, some students may subtract 12 from each side and then divide by -6. Others may add $6x$ to each side and then divide by 6. The result is the same either way.
- Some students may find it helpful to draw a picture for Question 7.
- **Extension:** Find the height of the mine entrance in Question 7. 8 feet

Closure

- **Writing Prompt:** If $abc = 0$, then . . . $a = 0$, $b = 0$, or $c = 0$. If $(x - 1)(x + 2)(x - 3)(x + 4) = 0$, then... the solutions are $x = 1$, $x = -2$, $x = 3$, and $x = -4$.

Technology For the Teacher

The Dynamic Planning Tool
Editable Teacher's Resources at *BigIdeasMath.com*

EXAMPLE 2 Standardized Test Practice

What are the solutions of $(2a+7)(2a-7)=0$?

Ⓐ -7 and 7

Ⓑ $-\frac{7}{2}$ and $\frac{7}{2}$

Ⓒ -2 and 2

Ⓓ $-\frac{2}{7}$ and $\frac{2}{7}$

$(2a+7)(2a-7)=0$			Write equation.
$2a+7=0$	*or*	$2a-7=0$	Use Zero-Product Property.
$a=-\frac{7}{2}$	*or*	$a=\frac{7}{2}$	Solve for *a*.

∴ The correct answer is Ⓑ.

EXAMPLE 3 Real-Life Application

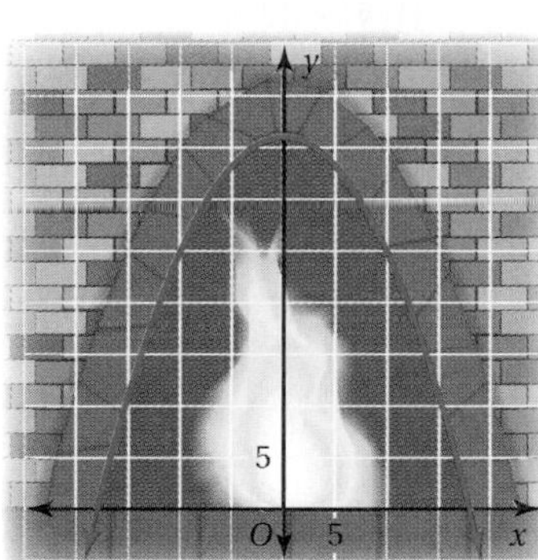

The arch of a fireplace can be modeled by $y=-\frac{1}{9}(x+18)(x-18)$, where x and y are measured in inches. The x-axis represents the floor. Find the width of the arch at floor level.

Use the x-coordinates at floor level to find the width. At floor level, $y=0$. So, substitute 0 for y and solve for x.

$y=-\frac{1}{9}(x+18)(x-18)$			Write equation.
$0=-\frac{1}{9}(x+18)(x-18)$			Substitute 0 for *y*.
$0=(x+18)(x-18)$			Multiply each side by -9.
$x+18=0$	*or*	$x-18=0$	Use Zero-Product Property.
$x=-18$	*or*	$x=18$	Solve for *x*.

The width is the distance between the x-coordinates, -18 and 18.

∴ So, the width of the arch at floor level is $18-(-18)=36$ inches.

On Your Own

Now You're Ready
Exercises 10–15

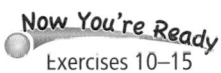

Solve the equation.

5. $(3p+5)(3p-5)=0$
6. $(12-6x)^2=0$
7. The entrance to a mine shaft can be modeled by $y=-\frac{1}{2}(x+4)(x-4)$, where x and y are measured in feet. The x-axis represents the ground. Find the width of the entrance at ground level.

7.5 Exercises

Vocabulary and Concept Check

1. **REASONING** Is $x = 3$ a solution of $(x - 3)(x + 6) = 0$? Explain.
2. **WRITING** Describe how to solve $(x - 2)(x + 1) = 0$ using the Zero-Product Property.
3. **WHICH ONE DOESN'T BELONG?** Which statement does *not* belong with the other three? Explain your reasoning.

$(n - 9)(n + 3)$	$(2k + 5)(k - 3)$
$(g + 2)^2$	$x^2 + 4x$

Practice and Problem Solving

Solve the equation.

4. $x(x + 7) = 0$
5. $12t(t - 5) = 0$
6. $(s - 9)(s - 1) = 0$
7. $(q + 3)(q - 2) = 0$
8. $(h - 8)^2 = 0$
9. $(m + 4)^2 = 0$
10. $(5 - k)(5 + k) = 0$
11. $(3 - g)(7 - g) = 0$
12. $(3p + 6)^2 = 0$
13. $(4z - 12)^2 = 0$
14. $\left(\frac{1}{2}y + 4\right)(y - 8) = 0$
15. $\left(\frac{1}{3}d - 2\right)\left(\frac{1}{3}d + 2\right) = 0$

16. **ERROR ANALYSIS** Describe and correct the error in solving the equation.

$6x(x + 5) = 0$

$x + 5 = 0$

$x = -5$

The root is $x = -5$.

Find the x-coordinates of the points where the graph crosses the x-axis.

17.

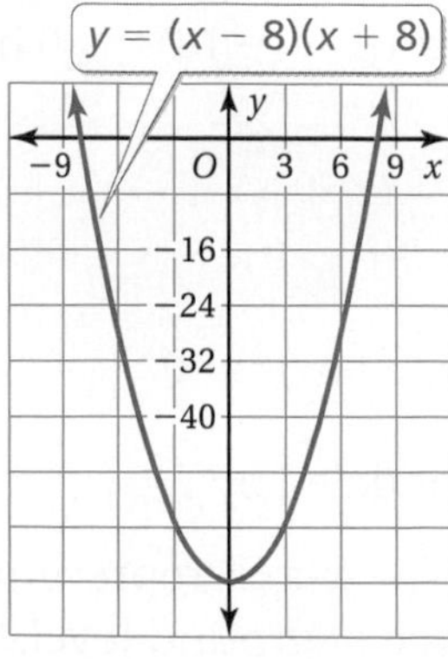

18.

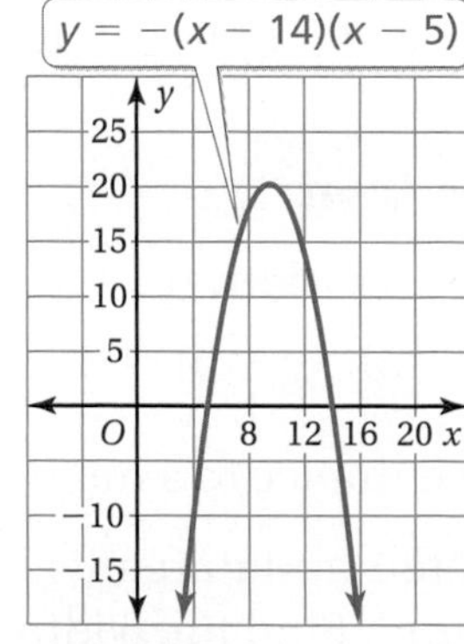

19.

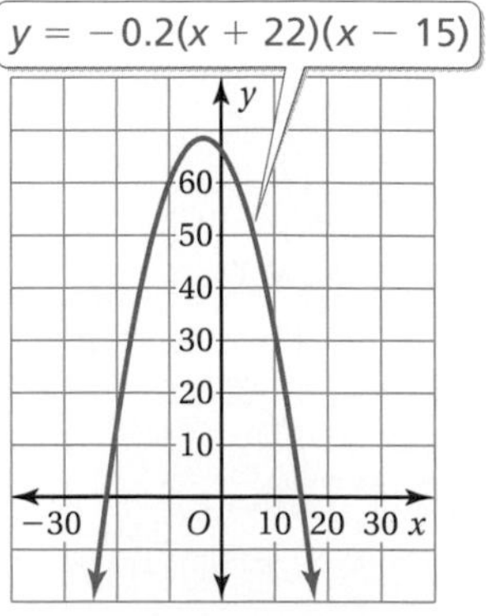

Assignment Guide and Homework Check

Level	Assignment	Homework Check
Average	1–3, 5–19 odd, 16, 20, 23, 25, 27–30	11, 13, 19, 20, 25
Advanced	1–3, 4–18 even, 22–26 even, 25, 27–30	14, 18, 24, 25, 26

Common Errors

- **Exercises 4–15** Students may set only one of the factors of the polynomial equal to zero. Remind them that they need to set both factors equal to zero when using the Zero-Product Property.
- **Exercises 17–19** Students may estimate the x-coordinates by looking at the graph. Point out that the equation is provided so that they can find the exact value of the x-coordinates.

7.5 Record and Practice Journal

Solve the equation.

1. $b(b-4)=0$
   $b=0,4$
2. $-8k(k+3)=0$
   $k=0,-3$
3. $(n-6)(n+6)=0$
   $n=6,-6$
4. $(v+11)(v+2)=0$
   $v=-11,-2$
5. $(h-9)=0$
   $h=9$
6. $(5+x)(7-x)=0$
   $x=-5,7$
7. $(3r-9)(2r+2)=0$
   $r=-1,3$
8. $\left(\frac{1}{2}p-8\right)\left(\frac{1}{4}p-1\right)=0$
   $p=4,16$
9. The arch of a bridge can be modeled by $y=-\frac{1}{170}(x-225)(x+225)$, where x and y are measured in feet. The x-axis represents the ground. Find the width of the arch of the bridge at ground level.
   450 ft

Technology For the Teacher
Answer Presentation Tool

Vocabulary and Concept Check

1. yes; When $x=3$, the factor $(x-3)$ equals 0.
2. The product $(x-2)(x+1)$ is 0, so at least one of the factors is 0. Set each factor equal to 0 and solve for x.
3. x^2+4x does not belong because it is not in factored form.

Practice and Problem Solving

4. $x=0, x=-7$
5. $t=0, t=5$
6. $s=9, s=1$
7. $q=-3, q=2$
8. $h=8$
9. $m=-4$
10. $k=5, k=-5$
11. $g=3, g=7$
12. $p=-2$
13. $z=3$
14. $y=-8, y=8$
15. $d=6, d=-6$
16. Both factors need to be set equal to 0.

    $6x(x+5)=0$

    $6x=0 \quad or \quad x+5=0$

    $x=0 \quad or \quad x=-5$
17. $x=-8, x=8$
18. $x=14, x=5$
19. $x=-22, x=15$

Practice and Problem Solving

20. 20 ft

21. $z = 0, z = -2, z = 1$

22. $w = 0, w = 6$

23. $r = 4, r = -4, r = -8$

24. $p = -\frac{3}{2}, p = \frac{3}{2}, p = -7$

25. See *Taking Math Deeper.*

26. **a.** $x = -y, x = \frac{y}{2}$

b. $x = y, x = -y, x = -4y$

Fair Game Review

27. 21

28. 3

29. 15

30. C

Mini-Assessment

Solve the equation.

1. $4p(p - 6) = 0$ $p = 0, p = 6$
2. $(n + 7)(n - 2) = 0$ $n = -7, n = 2$
3. $(k + 3)^2 = 0$ $k = -3$
4. $(2y - 8)(3y - 2) = 0$ $y = 4, y = \frac{2}{3}$
5. $8x(4x - 20)(x + 9) = 0$ $x = 0, x = 5, x = -9$

Taking Math Deeper

Exercise 25

Technically, the St. Louis Arch is a catenary, not a parabola. However, for the purpose of this exercise, you can see that the arch can be roughly modeled by a quadratic function (parabola).

a. When $y = 0$, $x = 315$ and $x = -315$. So the width of the arch at ground level is

$$\begin{aligned}\text{Width} &= 315 - (-315) \\ &= 315 + 315 \\ &= 630 \text{ ft.}\end{aligned}$$

b. The tallest point on the arch occurs when $x = 0$.

$$\begin{aligned}y &= -\frac{2}{315}(x + 315)(x - 315) \\ &= -\frac{2}{315}(0 + 315)(0 - 315) \\ &= 630 \text{ ft}\end{aligned}$$

Notice that the width and height of the arch are the same.

The St. Louis Arch is the tallest national monument in the United States. Construction began on February 12, 1963 and finished on October 28, 1965.

Reteaching and Enrichment Strategies

If students need help. . .	If students got it. . .
Resources by Chapter • Practice A and Practice B • Puzzle Time Record and Practice Journal Practice Differentiating the Lesson Lesson Tutorials Skills Review Handbook	Resources by Chapter • Enrichment and Extension • School-to-Work • Financial Literacy Start the next section

20. **CHOOSE TOOLS** The entrance of a tunnel can be modeled by $y = -\frac{11}{50}(x - 4)(x - 24)$, where x and y are measured in feet. The x-axis represents the ground. Find the width of the tunnel at ground level.

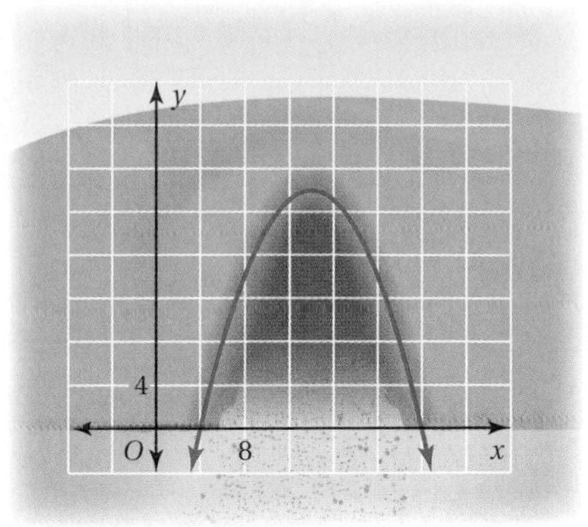

Solve the equation.

21. $5z(z + 2)(z - 1) = 0$

22. $w(w - 6)^2 = 0$

23. $(r - 4)(r + 4)(r + 8) = 0$

24. $(2p + 3)(2p - 3)(p + 7) = 0$

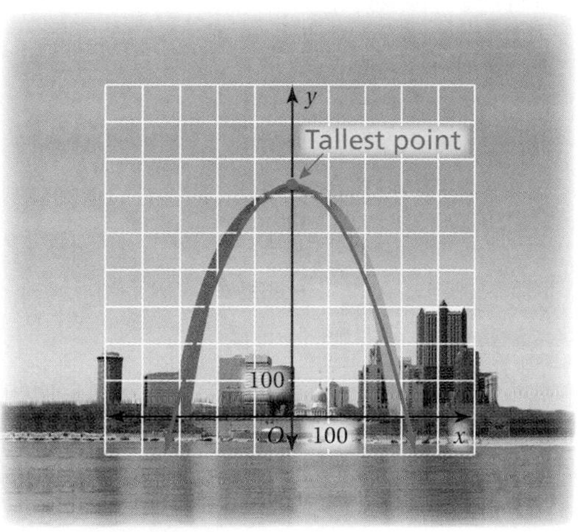

25. **GATEWAY ARCH** The Gateway Arch in St. Louis can be modeled by $y = -\frac{2}{315}(x + 315)(x - 315)$, where x and y are measured in feet. The x-axis represents the ground.

 a. Find the width of the arch at ground level.

 b. How tall is the arch?

26. **Algebra** Find the values of x in terms of y that are solutions of the equation.

 a. $(x + y)(2x - y) = 0$

 b. $(x^2 - y^2)(4x + 16y) = 0$

Fair Game Review What you learned in previous grades & lessons

Find the greatest common factor of the numbers. *(Skills Review Handbook)*

27. 21 and 63

28. 12 and 27

29. 30, 75, and 90

30. **MULTIPLE CHOICE** What is the slope of the line? *(Section 2.2)*

 Ⓐ -3 Ⓑ $-\frac{1}{3}$

 Ⓒ $\frac{1}{3}$ Ⓓ 3

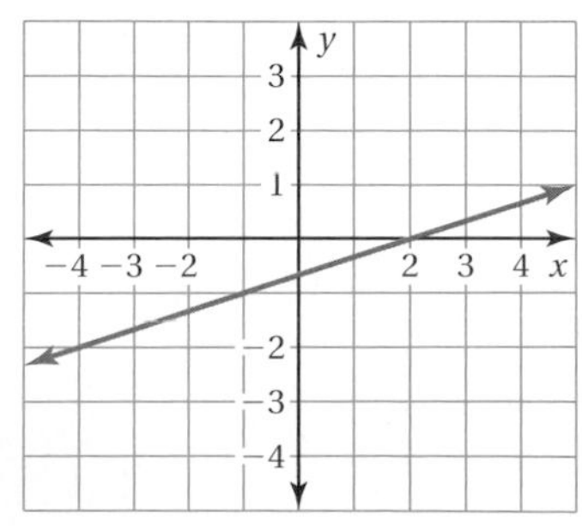

7.6 Factoring Polynomials Using the GCF

COMMON CORE STATE STANDARDS
A.REI.4b
A.SSE.3a

Essential Question How can you use common factors to write a polynomial in factored form?

1 ACTIVITY: Finding Monomial Factors

Work with a partner. Six different algebra tiles are shown below.

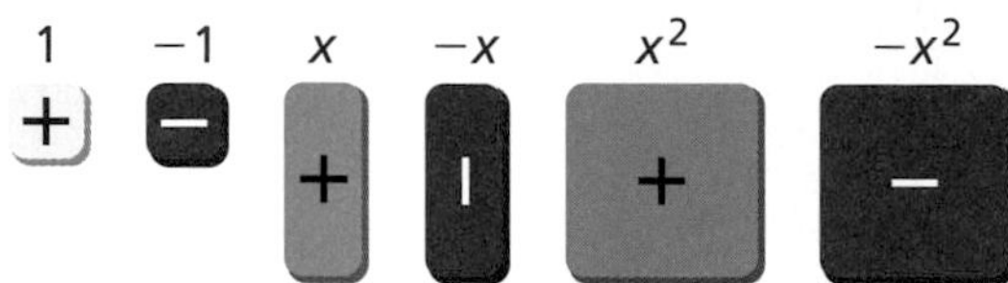

Sample:

Step 1: Look at the rectangular array for $x^2 + 3x$.

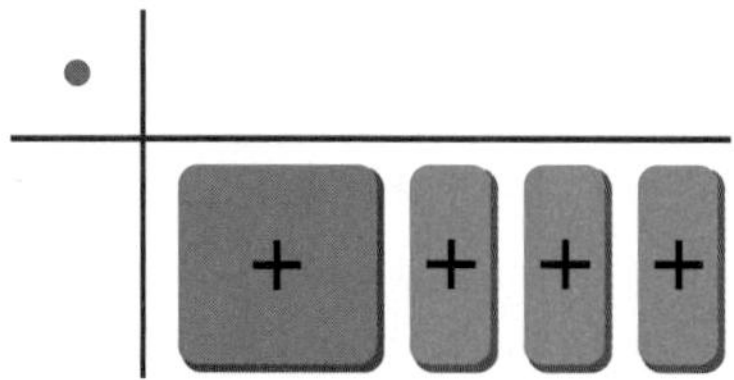

Step 2: Use algebra tiles to label the dimensions of the rectangle.

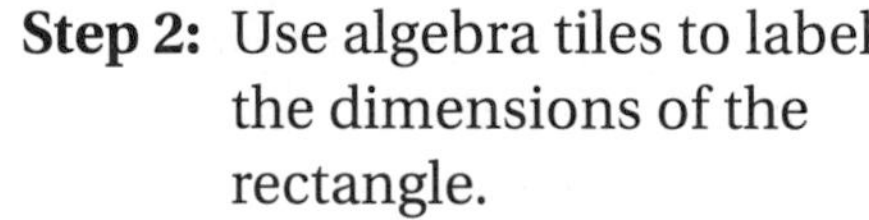

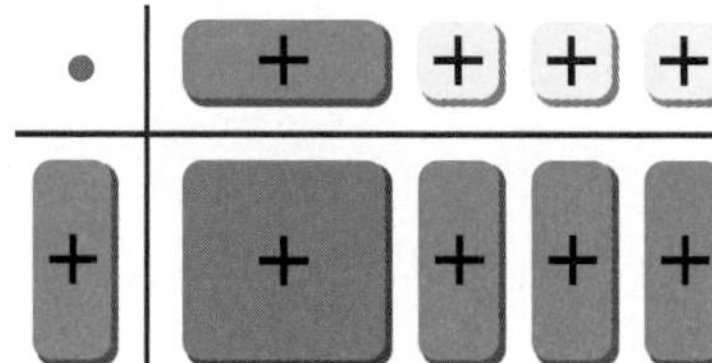

Step 3: Write the polynomial in factored form by finding the dimensions of the rectangle.

width → length

$$\text{Area} = x^2 + 3x = x(x + 3)$$

Use algebra tiles to write each polynomial in factored form.

a.

b.

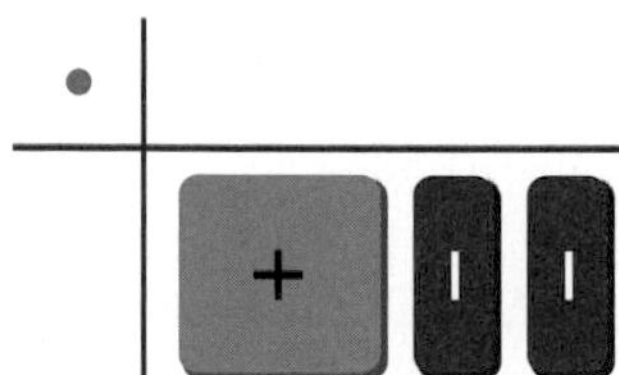

c.

d.

Laurie's Notes

Introduction

Standards for Mathematical Practice

- **MP1a Make Sense of Problems** and **MP5 Use Appropriate Tools Strategically:** Working with algebra tiles in this activity is a visual and tactile experience that should help students make sense of how to factor a binomial.

Motivate

- Play a quick game of P-F (product-factor).
- **Directions:** State a term listed below. Students write two factors with a product that equals that term. Note that the terms *change* in type.
- **Terms:** 8; 13; 24; -35; $3x$; x^2; $-3x^2$; $12x^2$
- ? "Was there a term for which everyone had the same factors?" likely 13
- ? "Which term had the most different answers?" possibly 24 or $12x^2$
- Explain that today they will be finding factors of binomials.

Activity Notes

For Your Information

- You can think of today's activity as being about the Distributive Property.
- The Distributive Property can also be written as $ab + ac = a(b + c)$, meaning that the common factor a is factored out of each term.
- Students have looked for common factors when simplifying fractions. In this activity, they look for common factors in each term of a binomial.

Activity 1

- Model the sample problem with students. Ask students to model $x^2 + 3x$.
- ? "What can you multiply to get x^2?" A common answer is x times x. Correct answers also include 1 times x^2, $0.5x$ times $2x$, and so on.
- ? "What can you multiply to get $3x$?" A common answer is 3 times x. Correct answers also include 1 times $3x$, 0.5 times $6x$, and so on.
- Use $x^2 = x \cdot x$ and $3x = 3 \cdot x$.
- Say, "There is a factor of x in each term. If we write $x^2 + 3x$ with the x factored out, we are left with $x + 3$."
- Model this using algebra tiles in the array diagram as shown.
- Have students share their work at a document camera or overhead projector if possible.
- **Teaching Tip:** When a term is negative, remind students that one of its factors must be negative.
- Show students two ways to label the dimensions of the rectangle in part (c) so that the results are $-2x^2 + x = -x(2x - 1)$ and $-2x^2 + x = x(-2x + 1)$.

Common Core State Standards

A.REI.4b Solve quadratic equations by inspection, taking square roots, completing the square, the quadratic formula and factoring, as appropriate to the initial form of the equation . . ..
A.SSE.3a Factor a quadratic expression to reveal the zeros of the function it defines.

Previous Learning

Students should know how to use the Distributive Property.

Activity Materials
Textbook
• algebra tiles

Start Thinking! and Warm Up

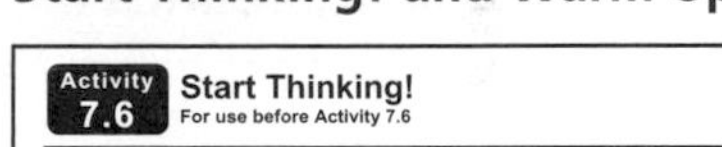

Activity 7.6 Warm Up
For use before Activity 7.6

Find the missing polynomial.

1. $3x + 6 = 3(\underline{\ ?\ })$
2. $14y^2 + 21y + 70 = 7(\underline{\ ?\ })$
3. $6z^2 + 12z + 15 = 3(\underline{\ ?\ })$
4. $15x^2 - 20x = 5x(\underline{\ ?\ })$
5. $x^2 + 6x + 5 = (x + 1)(\underline{\ ?\ })$
6. $x^2 - 7x + 12 = (x - 3)(\underline{\ ?\ })$

7.6 Record and Practice Journal

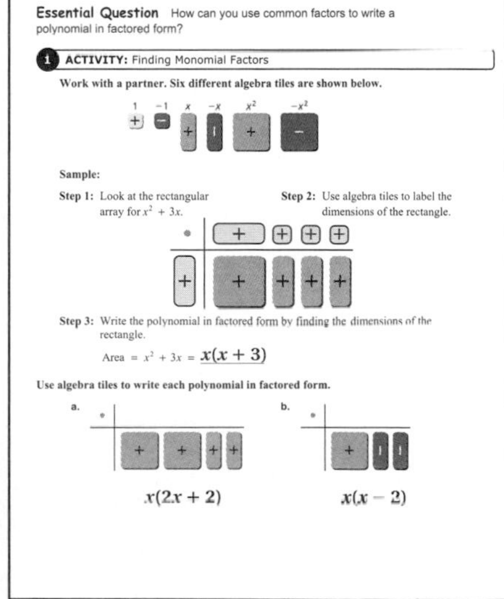

Essential Question How can you use common factors to write a polynomial in factored form?

1 ACTIVITY: Finding Monomial Factors

Work with a partner. Six different algebra tiles are shown below.

Sample:

Step 1: Look at the rectangular array for $x^2 + 3x$.

Step 2: Use algebra tiles to label the dimensions of the rectangle.

Step 3: Write the polynomial in factored form by finding the dimensions of the rectangle.

Area $= x^2 + 3x = x(x + 3)$

Use algebra tiles to write each polynomial in factored form.

a. $x(2x + 2)$

b. $x(x - 2)$

Differentiated Instruction

Kinesthetic

Have students write the factors of each term and the plus or minus sign of the polynomial on pieces of paper. Then arrange the pieces on their desks to form the expression. For example, the expression $4x^2 + 2x^3$ would be

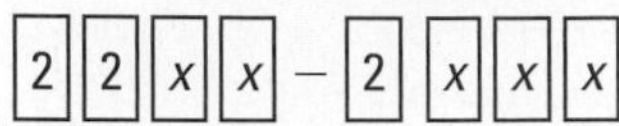

Next, students remove the pieces common to both terms, stack identical pieces in piles of two, and write the expression represented by these pieces of paper. This is one of the factors. The remaining pieces of paper are the second factor of the expression.

7.6 Record and Practice Journal

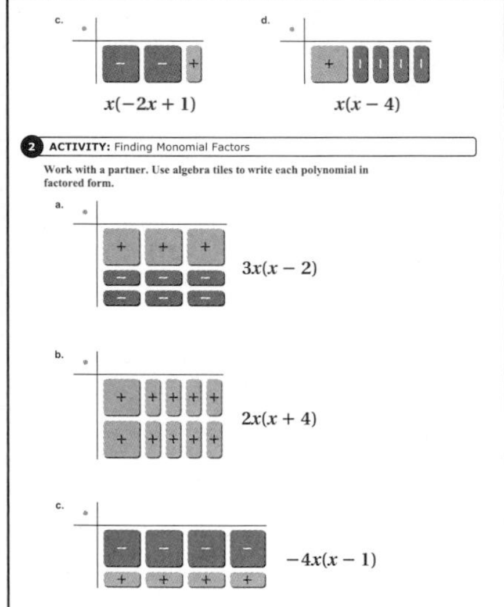

c. $x(-2x + 1)$

d. $x(x - 4)$

2 ACTIVITY: Finding Monomial Factors

Work with a partner. Use algebra tiles to write each polynomial in factored form.

a. $3x(x - 2)$

b. $2x(x + 4)$

c. $-4x(x - 1)$

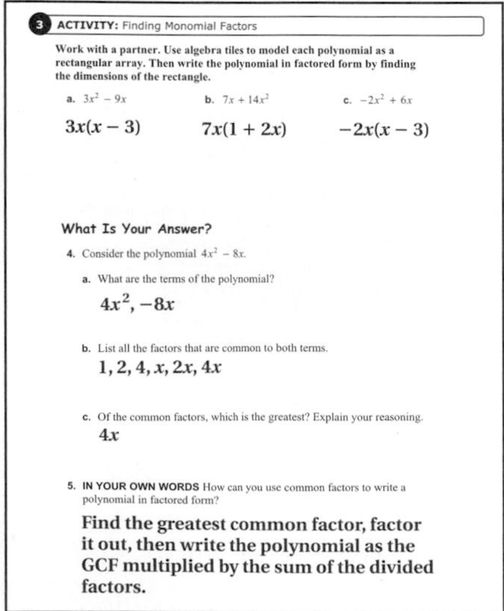

3 ACTIVITY: Finding Monomial Factors

Work with a partner. Use algebra tiles to model each polynomial as a rectangular array. Then write the polynomial in factored form by finding the dimensions of the rectangle.

a. $3x^2 - 9x$ $3x(x - 3)$

b. $7x + 14x^2$ $7x(1 + 2x)$

c. $-2x^2 + 6x$ $-2x(x - 3)$

What Is Your Answer?

4. Consider the polynomial $4x^2 - 8x$.

a. What are the terms of the polynomial?

$4x^2, -8x$

b. List all the factors that are common to both terms.

$1, 2, 4, x, 2x, 4x$

c. Of the common factors, which is the greatest? Explain your reasoning.

$4x$

5. **IN YOUR OWN WORDS** How can you use common factors to write a polynomial in factored form?

Find the greatest common factor, factor it out, then write the polynomial as the GCF multiplied by the sum of the divided factors.

Laurie's Notes

Activity 2

- **FYI:** The rectangular arrays in this activity are arranged so that students factor out a common factor that consists of a variable *and* a coefficient that is not 1.
- Show the students another way to factor Activity 1 part (a). Supply them with a new array in which the x^2-tiles are arranged vertically, and the x-tiles are arranged vertically to the right. This will lead them to write $2x^2 + 2x = 2x(x + 1)$.
- As students work the three parts with their partners, encourage them to think about what was multiplied to get the x^2-term, what was multiplied to get the x-term, and what these factors have in common.
- Have students share their work at a document camera or overhead projector if possible.

Activity 3

? "What factors do $3x^2$ and $-9x$ have in common?" Listen for x and 3.

- **Connection:** Students should see the connection to the Distributive Property in writing their solutions and in checking their work.
- Ask volunteers to share their solutions.

What Is Your Answer?

- Question 4 walks students through the process for a particular binomial. Students should then be able to describe the process in general in Question 5.

Closure

- Write each polynomial in factored form.
  a. $3x^2 + 6$ $3(x^2 + 2)$
  b. $2x^2 - 4x$ $2x(x - 2)$

Technology For the Teacher

The Dynamic Planning Tool
Editable Teacher's Resources at *BigIdeasMath.com*

2 ACTIVITY: Finding Monomial Factors

Work with a partner. Use algebra tiles to write each polynomial in factored form.

a.

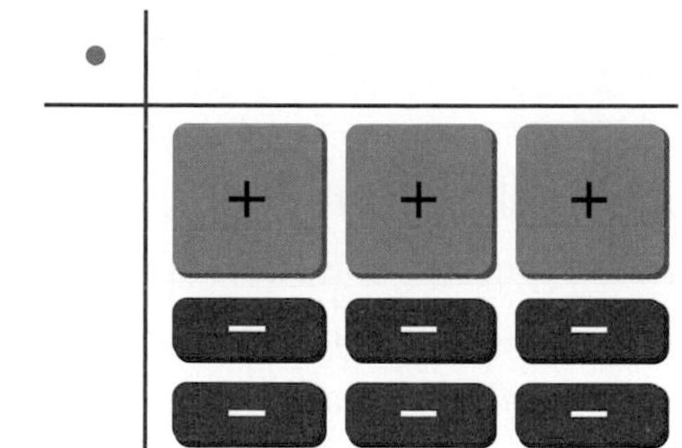

b.

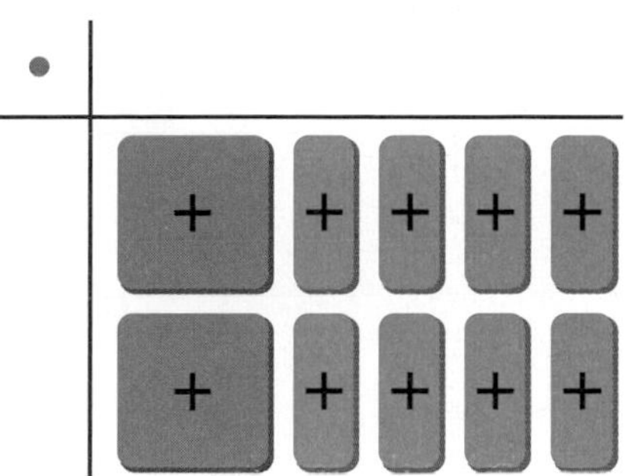

c.

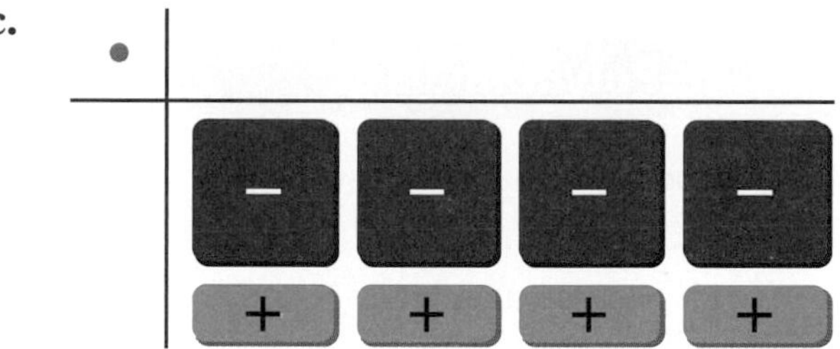

3 ACTIVITY: Finding Monomial Factors

Work with a partner. Use algebra tiles to model each polynomial as a rectangular array. Then write the polynomial in factored form by finding the dimensions of the rectangle.

a. $3x^2 - 9x$ **b.** $7x + 14x^2$ **c.** $-2x^2 + 6x$

What Is Your Answer?

4. Consider the polynomial $4x^2 + 8x$.
   - **a.** What are the terms of the polynomial?
   - **b.** List all the factors that are common to both terms.
   - **c.** Of the common factors, which is the greatest? Explain your reasoning.
5. **IN YOUR OWN WORDS** How can you use common factors to write a polynomial in factored form?

Practice Use what you learned about factoring polynomials to complete Exercises 3–5 on page 366.

7.6 Lesson

Writing a polynomial as a product of factors is called *factoring*. When the terms of a polynomial have a common factor, you can factor the polynomial as shown below.

Factoring Polynomials Using the GCF

Step 1: Find the greatest common factor (GCF) of the terms.

Step 2: Use the Distributive Property to write the polynomial as a product of the GCF and its remaining factors.

EXAMPLE 1 Factoring Polynomials

Factor each polynomial.

a. $2x^2 + 18$

Step 1: Find the GCF of the terms.

$$2x^2 = 2 \cdot x \cdot x$$
$$18 = 2 \cdot 3 \cdot 3$$

The GCF is 2.

Study Tip

When you factor a polynomial, you *undo* the multiplication of its factors.

Step 2: Write the polynomial as a product of the GCF and its remaining factors.

$$2x^2 + 18 = 2(x^2) + 2(9) \quad \text{Factor out GCF.}$$
$$= 2(x^2 + 9) \quad \text{Distributive Property}$$

b. $15y^3 + 10y^2$

Step 1: Find the GCF of the terms.

$$15y^3 = 3 \cdot 5 \cdot y \cdot y \cdot y$$
$$10y^2 = 2 \cdot 5 \cdot y \cdot y$$

The GCF is $5 \cdot y \cdot y = 5y^2$.

Step 2: Write the polynomial as a product of the GCF and its remaining factors.

$$15y^3 + 10y^2 = 5y^2(3y) + 5y^2(2) \quad \text{Factor out GCF.}$$
$$= 5y^2(3y + 2) \quad \text{Distributive Property}$$

On Your Own

Now You're Ready
Exercises 6–11

Factor the polynomial.

1. $5z^2 + 30$
2. $3x^2 + 14x$
3. $8y^2 - 24y$

Laurie's Notes

Introduction

Connect

- **Yesterday:** Students used algebra tiles to write polynomials in factored form. (MP1a, MP5)
- **Today:** Students will factor polynomials using the greatest common factor.

Motivate

? "What is the greatest common factor of 60 and 168?" 12

- Show students how to find the GCF using prime factorization.
  - Review the vocabulary: greatest common factor and prime factorization.
  - To find the GCF, first find the prime factorization of each number. Then circle the factors that the two numbers have in common. The GCF is the product of the common prime factors. For example:

$$60 = 2 \cdot 2 \cdot 3 \cdot 5$$
$$168 = 2 \cdot 2 \cdot 2 \cdot 3 \cdot 7$$
$$\text{GCF} = 2 \cdot 2 \cdot 3 = 12$$

- Discuss with students that they can also use this method when finding the GCF of two monomials, such as $8x^2$ and $12x$.

Lesson Notes

Key Idea

- Write the Key Idea.
- Note that the GCF of two numbers can be one of the numbers. Also, the GCF of two terms can be one of the terms.

Example 1

- Before working through each part, ask students to identify the terms of the expression. $2x^2$ and 18; $15y^3$ and $10y^2$
- **Common Misconception:** Students think of the Distributive Property in one direction, $2(x^2 + 9) = 2x^2 + 18$. It may be difficult for them to think of it in the other direction, $2x^2 + 18 = 2(x^2 + 9)$.
- Discuss the Study Tip.

On Your Own

- Remind students to check their answers by distributing.

Goal Today's lesson is factoring polynomials using the greatest common factor.

Start Thinking! and Warm Up

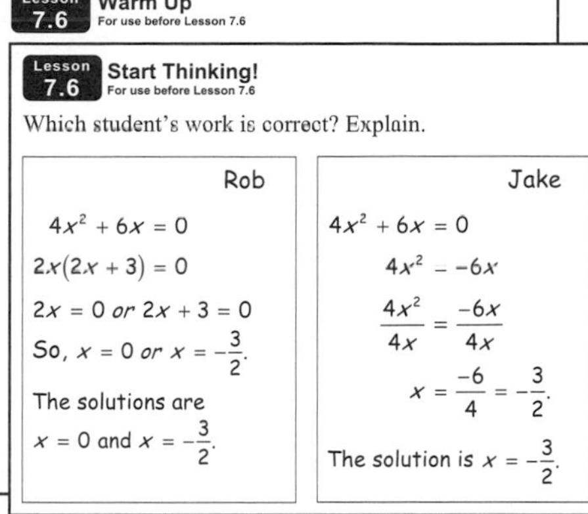

Lesson 7.6 Warm Up For use before Lesson 7.6

Lesson 7.6 Start Thinking! For use before Lesson 7.6

Which student's work is correct? Explain.

Rob	Jake
$4x^2 + 6x = 0$	$4x^2 + 6x = 0$
$2x(2x + 3) = 0$	$4x^2 = -6x$
$2x = 0$ or $2x + 3 = 0$	$\frac{4x^2}{4x} = \frac{-6x}{4x}$
So, $x = 0$ or $x = -\frac{3}{2}$.	$x = \frac{-6}{4} = -\frac{3}{2}$.
The solutions are $x = 0$ and $x = -\frac{3}{2}$.	The solution is $x = -\frac{3}{2}$.

Extra Example 1

Factor each polynomial.

a. $4x^2 + 12$ $4(x^2 + 3)$

b. $21n^3 - 18n^2$ $3n^2(7n - 6)$

On Your Own

1. $5(z^2 + 6)$
2. $x(3x + 14)$
3. $8y(y - 3)$

Extra Example 2

Solve $2w^2 = -18w$. $w = 0$, $w = -9$

On Your Own

4. $x = 0, x = -7$
5. $z = 0, z = 1$
6. $y = 0, y = 3$

Extra Example 3

A child jumps straight into the air on a trampoline. The child's height y (in feet) above the trampoline after t seconds can be modeled by $y = -16t^2 + 18t$. How many seconds is the child in the air? 1.125 sec

On Your Own

7. 0.875 sec

English Language Learners

Group Activity

Pair English language learners with English speakers. One student explains the steps while solving an equation. The other student explains the steps while checking the solutions in the equation. Students switch roles to solve and check another equation.

Laurie's Notes

Example 2

- Work through the problem as shown.
- **Common Question:** Students may ask if they can divide each side of the equation by g in the first step, leaving $4g = -6$. Mathematically, this is correct. By doing this, however, one of the solutions is lost, namely $g = 0$.

On Your Own

- **Common Error:** Question 5 will be challenging for students. When factoring $5z^2 - 5z$, students may correctly recognize that $5z$ is the GCF. However, they may write $5z(z)$ for the factored form. It is not obvious to all students that the correct answer is $5z(z - 1)$. Use the Distributive Property to show that $5z(z - 1)$ is equivalent to $5z^2 - 5z$.

Example 3

- **MP4 Model with Mathematics:** Read through the problem and write the model on the board.
- Factoring this polynomial may be a bit more challenging for students because the leading coefficient is negative.

? **Extension:** "What do the ordered pairs (0, 0) and (0.5, 2) represent?" At 0 seconds her height is 0 feet and at 0.5 second her height is 2 feet.

On Your Own

- **Think-Pair-Share:** Students should read the question independently and then work with a partner to answer the question. When they have answered the question, the pair should compare their answer with another group and discuss any discrepancies.

Closure

- **Exit Ticket:** Solve each equation.
  a. $2x^2 + 6x = 0$ $x = 0$ and $x = -3$
  b. $12x^2 - 4x = 0$ $x = 0$ and $x = \frac{1}{3}$

Technology For the Teacher

The Dynamic Planning Tool
Editable Teacher's Resources at *BigIdeasMath.com*

To solve an equation using the Zero-Product Property, you may need to first collect the terms on one side of the equation and then factor.

EXAMPLE 2 Solving an Equation by Factoring

Solve $4g^2 = -6g$.

$4g^2 = -6g$	Write equation.
$4g^2 + 6g = 0$	Add $6g$ to each side.
$2g(2g + 3) = 0$	Factor the polynomial.
$2g = 0$ *or* $2g + 3 = 0$	Use Zero-Product Property.
$g = 0$ *or* $g = -\frac{3}{2}$	Solve for g.

∴ The solutions are $g = 0$ and $g = -\frac{3}{2}$.

On Your Own

Now You're Ready
Exercises 14–22

Solve the equation.

4. $3x^2 + 21x = 0$ **5.** $5z^2 = 5z$ **6.** $18y = 6y^2$

EXAMPLE 3 Real-Life Application

A female athlete tests her vertical jump by jumping straight into the air. Her height y (in feet) after t seconds can be modeled by $y = -16t^2 + 12t$. How many seconds is she in the air?

She is on the ground when $y = 0$. So, substitute 0 for y and solve for t.

$y = -16t^2 + 12t$	Write equation.
$0 = -16t^2 + 12t$	Substitute 0 for y.
$0 = 4t(-4t + 3)$	Factor the polynomial.
$4t = 0$ *or* $-4t + 3 = 0$	Use Zero-Product Property.
$t = 0$ *or* $t = 0.75$	Solve for t.

She starts the jump at $t = 0$ and lands when $t = 0.75$.

∴ So, she is in the air for 0.75 second.

On Your Own

7. **WHAT IF?** The height of a male athlete testing his vertical jump can be modeled by $y = -16t^2 + 14t$. How many seconds is he in the air?

7.6 Exercises

Vocabulary and Concept Check

1. **REASONING** What is the greatest common factor of $12y$ and $30y^2$?
2. **WRITING** Describe how to factor a polynomial using the greatest common factor.

Practice and Problem Solving

Use algebra tiles to factor the polynomial.

3. $4x + 8$
4. $2x^2 + 4x$
5. $x^2 - 4x$

Factor the polynomial.

1 6. $5z^2 + 45z$
7. $8m^2 + 4m$
8. $3y^3 - 9y^2$
9. $20x^3 + 30x^2$
10. $4w^3 - 8w + 12$
11. $5t^2 + 20t + 50$

12. **ERROR ANALYSIS** Describe and correct the error in factoring the polynomial.

$$2x^2 + 2x = 2(x^2) + 2(x)$$
$$= 2(x^2 + x)$$

13. **INTEREST** You deposit \$100 in a savings account that earns simple interest. The balance of the account can be represented by $100 + 100rt$, where r is the annual interest rate and t is the time in years. Factor the polynomial.

Solve the equation.

2 14. $2q + 10 = 0$
15. $10x + 15 = 0$
16. $4p^2 - p = 0$
17. $6m^2 + 12m = 0$
18. $3n^2 = 9n$
19. $4r^2 = -28r$
20. $4a^3 = 44a^2$
21. $6k^3 + 39k^2 = 0$
22. $2y^2 = 2\pi y$

23. **ERROR ANALYSIS** Describe and correct the error in solving the equation.

$$3x^2 = 15x$$
$$3x^2 - 15x = 0$$
$$3x(x - 15) = 0$$
$$3x = 0 \quad or \quad x - 15 = 0$$
$$x = 0 \quad or \quad x = 15$$
The roots are $x = 0$ and $x = 15$.

24. **AGES** Your brother is y years old. Your older cousins are $2y^2$ and $6y$ years old. The difference between your cousins' ages is zero. Your brother is older than 1 year old. How old is he?

Assignment Guide and Homework Check

Level	Assignment	Homework Check
Average	1, 2, 7–31 odd, 12, 24, 33–36	9, 13, 19, 24, 31
Advanced	1, 2, 6–32 even, 23, 31, 33–36	24, 26, 30, 31, 32

For Your Information

- Students solved equations like those in Exercises 14 and 15 in Section 1.2. Here they will solve using a different method.

Common Errors

- **Exercises 6–11** Students may factor out only one or two of the common factors instead of the greatest common factor. Encourage them to rewrite each term using its prime factorization. This will make it easier to identify the greatest common factor of the terms.
- **Exercises 18–20, 22** Students may divide each side of the equation by the greatest common factor. Remind them that by doing this, one of the solutions is lost. Instead, they should collect the terms on one side of the equation and then factor.
- **Exercises 25–28** Students may factor before combining like terms and then say that a solution cannot be found. Remind them to look for like terms that can be combined before factoring.

7.6 Record and Practice Journal

Factor the polynomial.

1. $5n^2 - 15n$

   $5n(n - 3)$

2. $6t^3 + 12t^2 - 4t$

   $2t(3t^2 + 6t - 2)$

Solve the equation.

3. $4a - 16 = 0$

   $a = 4$

4. $14r^2 + 7r = 0$

   $r = 0, -\frac{1}{2}$

5. $-6w^2 = 18w$

   $w = -3, 0$

6. $14z^2 = 42z$

   $z = 0, 3$

7. $4x^3 + 36x^2 = 0$

   $x = -9, 0$

8. $-2p^2 = 9p^3 - 5p^2$

   $p = 0, \frac{1}{3}$

9. The area (in square feet) of the billboard can be represented by $18x^3 + 12x^2$.

   $(3x + 2)$ ft

   a. Write an expression that represents the length of the billboard.

   $6x^2$

   b. Find the area of the billboard when $x = 2$.

   192 ft^2

Technology For the Teacher

Answer Presentation Tool

Vocabulary and Concept Check

1. $6y$
2. Find the greatest common factor (GCF) of the terms. Use the Distributive Property to write the polynomial as a product of the GCF and its remaining factors.

Practice and Problem Solving

3. $4(x + 2)$
4. $2x(x + 2)$
5. $x(x - 4)$
6. $5z(z + 9)$
7. $4m(2m + 1)$
8. $3y^2(y - 3)$
9. $10x^2(2x + 3)$
10. $4(w^3 - 2w + 3)$
11. $5(t^2 + 4t + 10)$
12. The GCF is $2x$, not 2.

    $$2x^2 + 2x = 2x(x) + 2x(1) = 2x(x + 1)$$

13. $100(1 + rt)$
14. $q = -5$
15. $x = -\frac{3}{2}$
16. $p = 0, p = \frac{1}{4}$
17. $m = 0, m = -2$
18. $n = 0, n = 3$
19. $r = 0, r = -7$
20. $a = 0, a = 11$
21. $k = 0, k = -\frac{13}{2}$
22. $y = 0, y = \pi$
23. See Additional Answers.
24. 3 years

Practice and Problem Solving

25. $b = 0, b = 5$
26. $n = 0, n = -7$
27. $s = 0, s = -6$
28. $g = 0, g = -\frac{1}{2}$
29. *Sample answer:* $6x^2 + 3\pi x$
30. **a.** $5x - 2$

    b. Substitute 2 for x in the expression for the area, $15x^2 - 6x$. Or substitute 2 for x in the expressions for the dimensions and then multiply to find the area.
31. See *Taking Math Deeper.*
32. **a.** $x(2 + x) = x(6 - x)$

    b. 2

    c. 16 ft^2

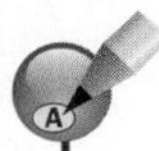

Fair Game Review

33. $y^2 + 10y + 24$
34. $m^2 - 11m + 18$
35. $4k^2 - 4k - 3$
36. D

Mini-Assessment

Factor the polynomial.

1. $7y^2 - 42y$ $7y(y - 6)$
2. $6g^2 + 15g - 24$ $3(2g^2 + 5g - 8)$

Solve the equation.

3. $4k^2 + 24k = 0$ $k = 0, k = -6$
4. $5z^3 = 25z^2$ $z = 0, z = 5$
5. A rocket is launched straight into the air. The rocket's height y (in feet) after t seconds can be modeled by $y = -16t^2 + 80t$. How many seconds is the rocket in the air? 5 sec

Taking Math Deeper

Exercise 31

This is a typical vertical motion problem.

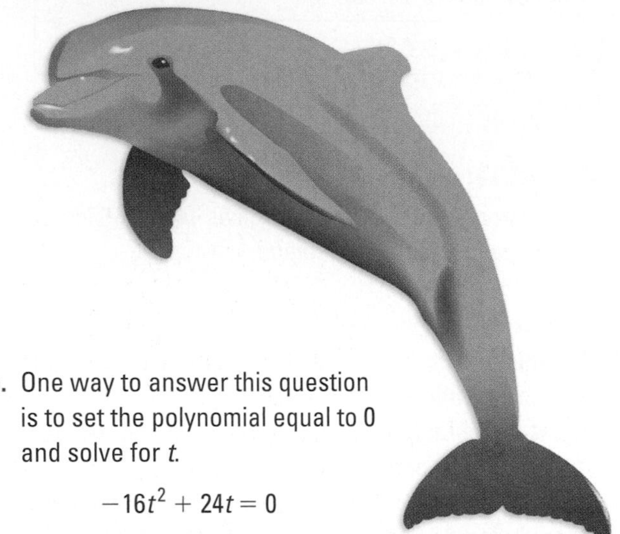

(1) **a.** One way to answer this question is to set the polynomial equal to 0 and solve for t.

$$-16t^2 + 24t = 0$$

$$t(-16t + 24) = 0$$

$$t = 0 \quad or \quad t = 1.5$$

The dolphin starts the jump at $t = 0$ and lands when $t = 1.5$. So, the dolphin is in the air for 1.5 seconds.

(2) Another way to answer this question is to make a table.

Time, t (sec)	0	0.25	0.5	0.75	1.0	1.25	1.5
Height, y (ft)	0	5	8	9	8	5	0

From the table, you can see that the dolphin is in the air for 1.5 seconds.

(3) **b.** From the table, you can see that the height of the dolphin is 9 feet (above water level) at 0.75 second.

Reteaching and Enrichment Strategies

If students need help. . .	If students got it. . .
Resources by Chapter • Practice A and Practice B • Puzzle Time Record and Practice Journal Practice Differentiating the Lesson Lesson Tutorials Skills Review Handbook	Resources by Chapter • Enrichment and Extension • School-to-Work • Financial Literacy • Technology Connection Start the next section

Solve the equation.

25. $5b^2 - 20b = b^2$

26. $5n^2 + 40n = 5n$

27. $2s^3 + 15s^2 = 3s^2$

28. $8g^3 - 2g^2 = 2g^3 - 5g^2$

29. OPEN-ENDED Write a binomial whose terms have a GCF of $3x$.

30. SCHOOL SIGN The area (in square feet) of the school sign can be represented by $15x^2 - 6x$.

a. Write an expression that represents the length of the sign.

b. Describe two ways to find the area of the sign when $x = 2$.

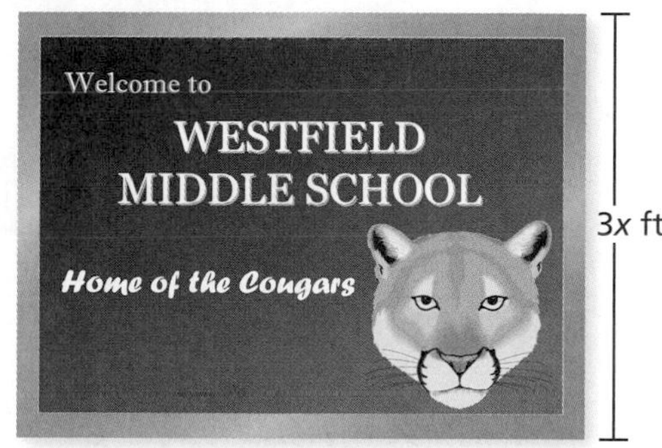

31. DOLPHIN A dolphin jumps straight into the air during a performance. The dolphin's height y (in feet) after t seconds can be modeled by $y = -16t^2 + 24t$.

a. How many seconds is the dolphin in the air?

b. The dolphin reaches its maximum height after 0.75 second. What is the maximum height of the jump?

32. **Modeling** Your teacher's work station is made up of two identical desks arranged as shown.

a. Write an equation in terms of x that relates the area of Desk 1 to the area of Desk 2.

b. What is the value of x?

c. Find the area of the top of your teacher's work station.

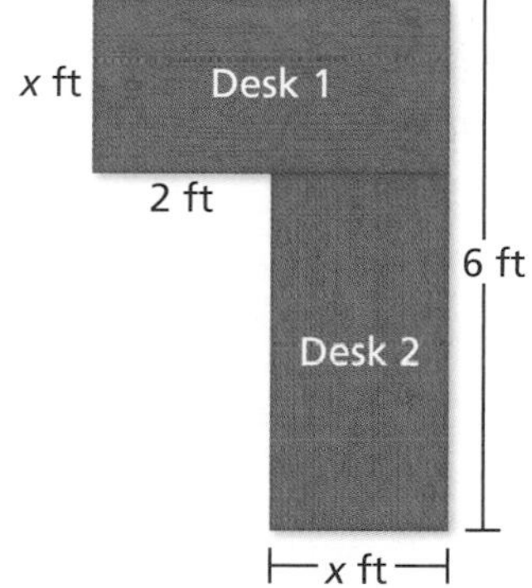

Fair Game Review What you learned in previous grades & lessons

Find the product. *(Section 7.3)*

33. $(y + 4)(y + 6)$

34. $(m - 2)(m - 9)$

35. $(2k + 1)(2k - 3)$

36. MULTIPLE CHOICE An African elephant weighs 5,200,000 grams. Write this number in scientific notation. *(Skills Review Handbook)*

Ⓐ 0.52×10^{-7} g

Ⓑ 5.2×10^{-6} g

Ⓒ 52×10^5 g

Ⓓ 5.2×10^6 g

7.7 Factoring $x^2 + bx + c$

COMMON CORE STATE STANDARDS
A.REI.4b
A.SSE.3a

Essential Question How can you factor the trinomial $x^2 + bx + c$ into the product of two binomials?

1 ACTIVITY: Finding Binomial Factors

Work with a partner. Six different algebra tiles are shown below.

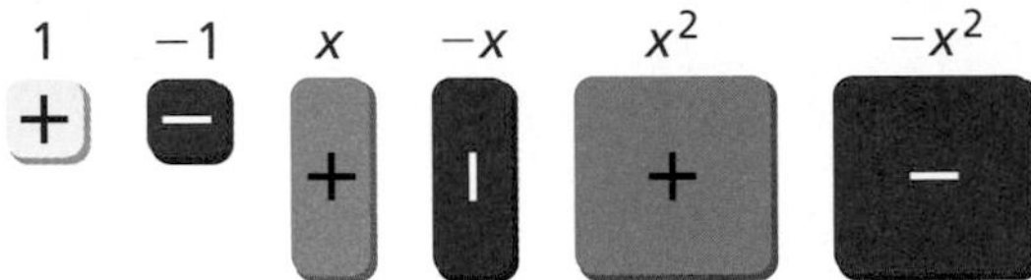

Sample:

Step 1: Arrange the algebra tiles into a rectangular array to model $x^2 + 5x + 6$.

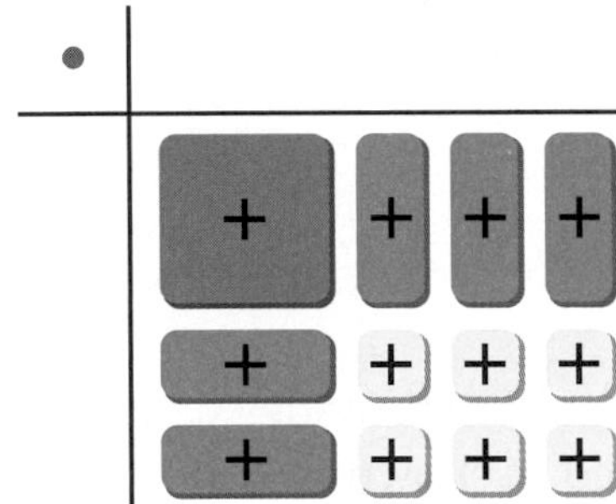

Step 2: Use algebra tiles to label the dimensions of the rectangle.

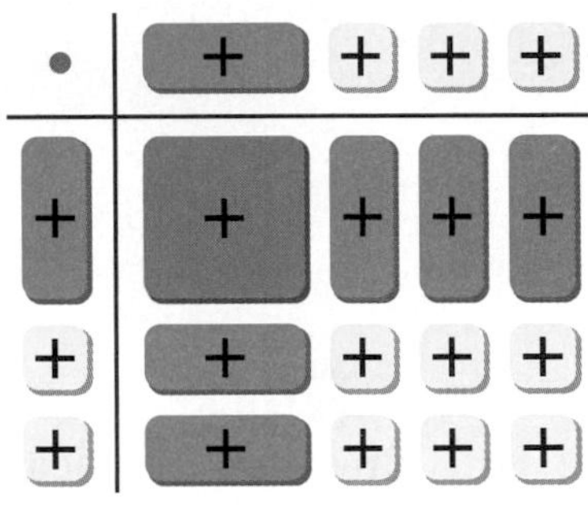

Step 3: Write the polynomial in factored form by finding the dimensions of the rectangle.

width
length

$$\text{Area} = x^2 + 5x + 6 = (x + 2)(x + 3)$$

Use algebra tiles to write each polynomial as the product of two binomials. Check your answer by multiplying.

a.

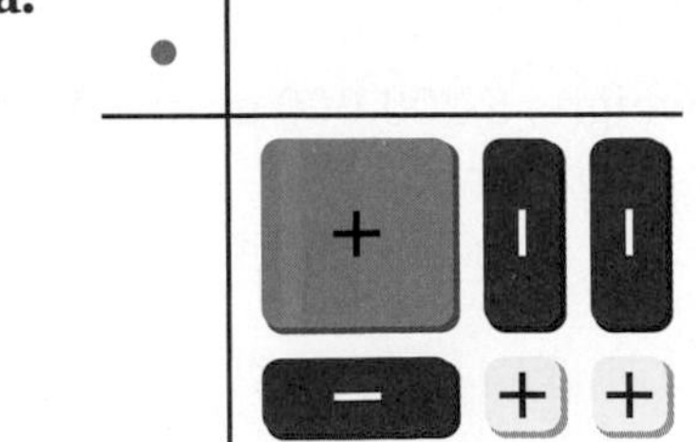

b.

Laurie's Notes

Introduction

Standards for Mathematical Practice

- **MP1a Make Sense of Problems** and **MP5 Use Appropriate Tools Strategically:** Working with algebra tiles in this activity is a visual and tactile experience that should help students make sense of how to factor a trinomial.
- In a previous lesson, students multiplied binomials that resulted in a polynomial. They are now undoing that process. The visual model is a helpful learning tool.

Motivate

- Model $x^2 + 4x + 3$ with tiles on the overhead. Have students do the same at their desks.

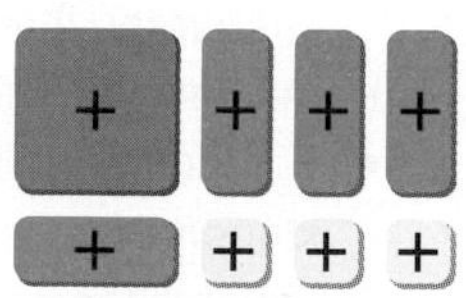

? "Can you arrange the algebra tiles so that they form a rectangle with no holes or overlaps?" yes Ask a volunteer to do so.

? "What are the dimensions of the rectangle?" $(x + 1)$ and $(x + 3)$.

- Label the dimensions. Write and say, "So $x^2 + 4x + 3 = (x + 1)(x + 3)$."
- **MP5:** This is similar to the visual model used when whole numbers are multiplied using base 10 pieces.

Activity Notes

Activity 1

- Model the sample problem with students. Ask students to model $x^2 + 5x + 6$ so that they form a rectangular array with the algebra tiles. There are two options for the six unit tiles: 1×6 or 2×3. If students try the 1×6, they will realize that they do not have enough x-tiles.

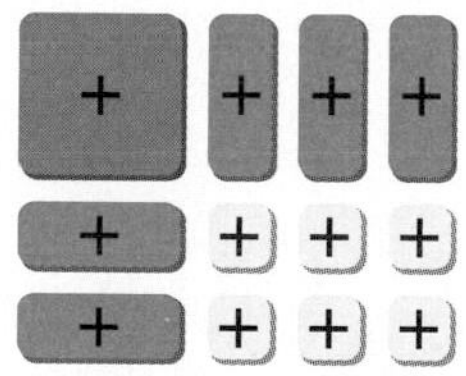

? "What are the dimensions of the rectangle?" $(x + 2)$ and $(x + 3)$

- Write and say, "So $x^2 + 5x + 6 = (x + 2)(x + 3)$." Discuss that $x^2 + 5x + 6$ is the product, or area of the rectangle, and $(x + 2)$ and $(x + 3)$ are the factors, or dimensions of the rectangle.
- As students work the two problems with their partners, encourage them to think about the dimensions of the rectangle. Students also need to pay attention to the signs of their pieces.
- Have students share their work at a document camera or overhead projector if possible.

Common Core State Standards

A.REI.4b Solve quadratic equations by inspection, taking square roots, completing the square, the quadratic formula and factoring, as appropriate to the initial form of the equation . . ..

A.SSE.3a Factor a quadratic expression to reveal the zeros of the function it defines.

Previous Learning

Students should know how to factor polynomials using the greatest common factor.

Activity Materials	
Introduction	**Textbook**
• algebra tiles	• algebra tiles

Start Thinking! and Warm Up

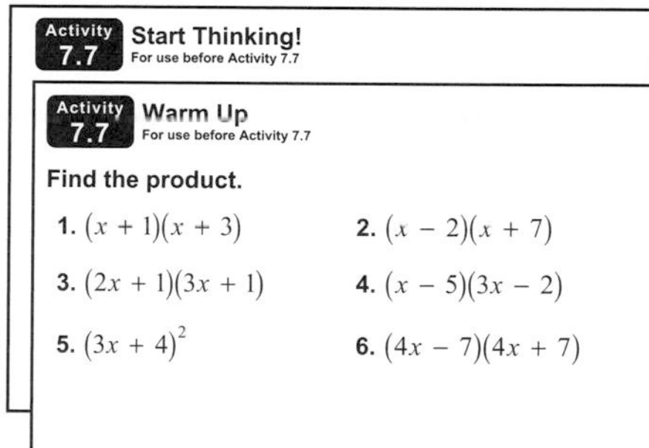

Activity 7.7 Start Thinking! For use before Activity 7.7

Activity 7.7 Warm Up For use before Activity 7.7

Find the product.

1. $(x + 1)(x + 3)$
2. $(x - 2)(x + 7)$
3. $(2x + 1)(3x + 1)$
4. $(x - 5)(3x - 2)$
5. $(3x + 4)^2$
6. $(4x - 7)(4x + 7)$

7.7 Record and Practice Journal

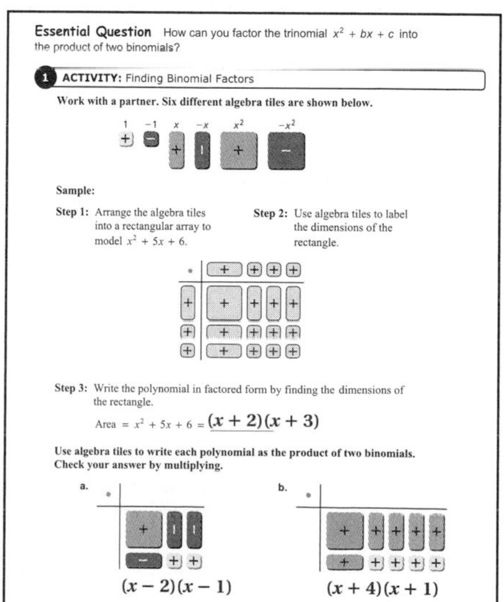

Essential Question How can you factor the trinomial $x^2 + bx + c$ into the product of two binomials?

1 ACTIVITY: Finding Binomial Factors

Work with a partner. Six different algebra tiles are shown below.

Sample:

Step 1: Arrange the algebra tiles into a rectangular array to model $x^2 + 5x + 6$.

Step 2: Use algebra tiles to label the dimensions of the rectangle.

Step 3: Write the polynomial in factored form by finding the dimensions of the rectangle.

Area $= x^2 + 5x + 6 = (x + 2)(x + 3)$

Use algebra tiles to write each polynomial as the product of two binomials. Check your answer by multiplying.

a. $(x - 2)(x - 1)$

b. $(x + 4)(x + 1)$

Differentiated Instruction

Advanced

Let students explore factoring the expressions $x^2 + 5x - 4$ and $x^2 + 3x + 6$ using algebra tiles. Have them write a brief description of their results and answer the question, "What can you conclude about these trinomials?" The trinomials can not be factored in the form $(x + p)(x + q)$.

7.7 Record and Practice Journal

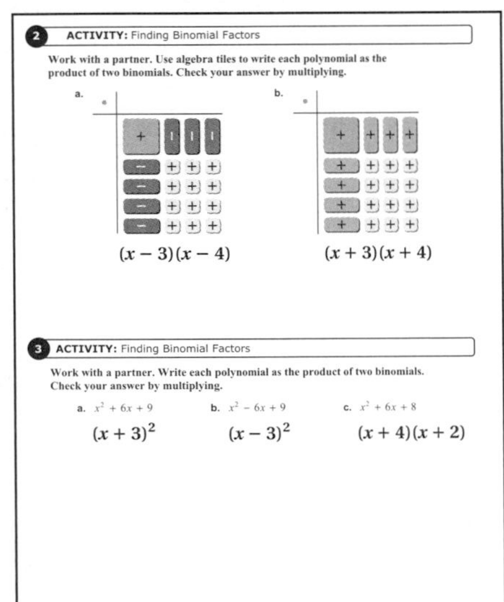

2 ACTIVITY: Finding Binomial Factors

Work with a partner. Use algebra tiles to write each polynomial as the product of two binomials. Check your answer by multiplying.

a. $(x - 3)(x - 4)$

b. $(x + 3)(x + 4)$

3 ACTIVITY: Finding Binomial Factors

Work with a partner. Write each polynomial as the product of two binomials. Check your answer by multiplying.

a. $x^2 + 6x + 9$ $(x + 3)^2$

b. $x^2 - 6x + 9$ $(x - 3)^2$

c. $x^2 + 6x + 8$ $(x + 4)(x + 2)$

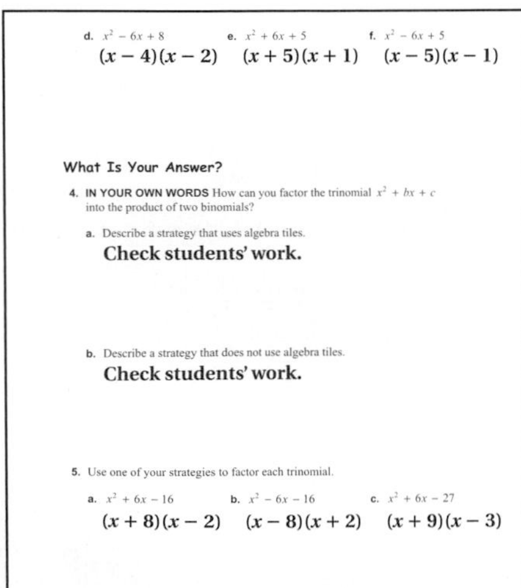

d. $x^2 - 6x + 8$ $(x - 4)(x - 2)$

e. $x^2 + 6x + 5$ $(x + 5)(x + 1)$

f. $x^2 - 6x + 5$ $(x - 5)(x - 1)$

What Is Your Answer?

4. **IN YOUR OWN WORDS** How can you factor the trinomial $x^2 + bx + c$ into the product of two binomials?

a. Describe a strategy that uses algebra tiles.
Check students' work.

b. Describe a strategy that does not use algebra tiles.
Check students' work.

5. Use one of your strategies to factor each trinomial.

a. $x^2 + 6x - 16$ $(x + 8)(x - 2)$

b. $x^2 - 6x - 16$ $(x - 8)(x + 2)$

c. $x^2 + 6x - 27$ $(x + 9)(x - 3)$

Laurie's Notes

Activity 2

- Students need to pay attention to signs.
- "How are these two problems alike?" They both have an x^2-tile, the same number of x-tiles, and 12 positive unit tiles. They both are arranged in the same size rectangular array. "How are they different?" The x-tiles have opposite signs.
- Write and say, "$x^2 - 7x + 12 = (x - 4)(x - 3)$ and $x^2 + 7x + 12 = (x + 4)(x + 3)$."

Activity 3

- Students should start by selecting the algebra tiles that represent the trinomial.
- As students work with their partners to form rectangular arrays, encourage them to think about the dimensions of the arrays.
- Remind students to pay attention to the signs of their tiles.
- Students should notice that all of the trinomials involve $\pm 6x$. Some students may start to recognize that the factors of the constant have a sum equal to the coefficient of x.
- Ask volunteers to share their solutions.

What Is Your Answer?

- Question 5 presents additional polynomials with $\pm 6x$.

Closure

- Complete the factorization. Check by multiplying the binomials.
  $x^2 + 5x + 6 = (x + \_\_)(x + \_\_)$ $(x + 2)(x + 3)$
  $x^2 - 5x + 6 = (x - \_\_)(x - \_\_)$ $(x - 2)(x - 3)$

Technology For the Teacher

The Dynamic Planning Tool
Editable Teacher's Resources at *BigIdeasMath.com*

2 ACTIVITY: Finding Binomial Factors

Work with a partner. Use algebra tiles to write each polynomial as the product of two binomials. Check your answer by multiplying.

a.

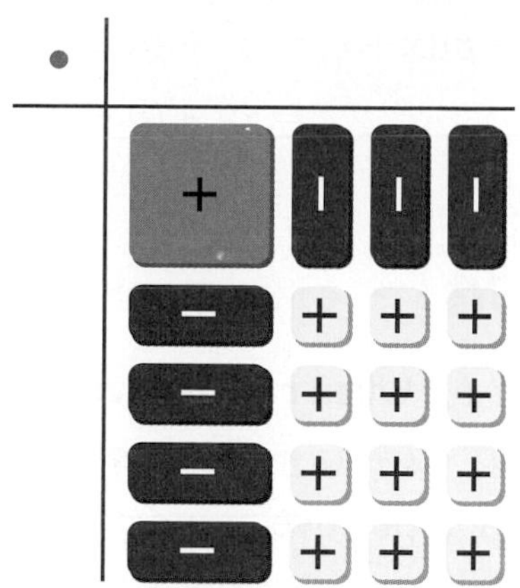

b.

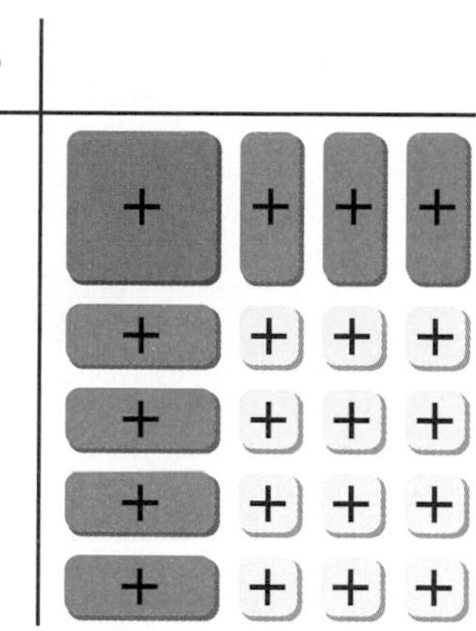

3 ACTIVITY: Finding Binomial Factors

Work with a partner. Write each polynomial as the product of two binomials. Check your answer by multiplying.

a. $x^2 + 6x + 9$ **b.** $x^2 - 6x + 9$ **c.** $x^2 + 6x + 8$

d. $x^2 - 6x + 8$ **e.** $x^2 + 6x + 5$ **f.** $x^2 - 6x + 5$

What Is Your Answer?

4. **IN YOUR OWN WORDS** How can you factor the trinomial $x^2 + bx + c$ into the product of two binomials?

   a. Describe a strategy that uses algebra tiles.

   b. Describe a strategy that does not use algebra tiles.

5. Use one of your strategies to factor each trinomial.

   a. $x^2 + 6x - 16$ **b.** $x^2 - 6x - 16$ **c.** $x^2 + 6x \quad 27$

Practice Use what you learned about factoring trinomials to complete Exercises 3–5 on page 373.

7.7 Lesson

Consider the polynomial $x^2 + bx + c$, where b and c are integers. To factor this polynomial as $(x + p)(x + q)$, you need to find integers p and q such that $p + q = b$ and $pq = c$.

$$(x + p)(x + q) = x^2 + px + qx + pq$$
$$= x^2 + (p + q)x + pq$$

Key Idea

Factoring $x^2 + bx + c$ When c Is Positive

Algebra $x^2 + bx + c = (x + p)(x + q)$ when $p + q = b$ and $pq = c$.

When c is positive, p and q have the same sign as b.

Examples $x^2 + 6x + 5 = (x + 1)(x + 5)$

$x^2 - 6x + 5 = (x - 1)(x - 5)$

EXAMPLE 1 Factoring $x^2 + bx + c$ When b and c Are Positive

Factor $x^2 + 10x + 16$.

Notice that $b = 10$ and $c = 16$.

- Because c is positive, the factors p and q must have the same sign so that pq is positive.
- Because b is also positive, p and q must each be positive so that $p + q$ is positive.

Find two positive integer factors of 16 whose sum is 10.

Factors of 16	Sum of Factors
1, 16	17
2, 8	10
4, 4	8

The values of p and q are 2 and 8.

Check
Use the FOIL Method.
$(x + 2)(x + 8)$
$= x^2 + 8x + 2x + 16$
$= x^2 + 10x + 16$ ✓

So, $x^2 + 10x + 16 = (x + 2)(x + 8)$.

On Your Own

Now You're Ready
Exercises 3–8

Factor the polynomial.

1. $x^2 + 2x + 1$
2. $x^2 + 9x + 8$
3. $y^2 + 6y + 8$
4. $z^2 + 11z + 24$

Laurie's Notes

Introduction

Connect

- **Yesterday:** Students used algebra tiles to factor $x^2 + bx + c$ into the product of two binomial factors. (MP1a, MP5)
- **Today:** Students will factor $x^2 + bx + c$.

Motivate

- Have students match the factored form with the correct standard form.

1. $(x + 2)(x + 6)$ B	**A.** $x^2 + 7x + 12$
2. $(x + 12)(x + 1)$ C	**B.** $x^2 + 8x + 12$
3. $(x + 3)(x + 4)$ A	**C.** $x^2 + 13x + 12$

? "How are the problems alike? How are they different?" Answers will vary.

- Circle the constant terms in the binomials (2, 6, 12, 1, 3, and 4), the coefficients of the linear terms (7, 8, and 13) and the constant terms (12, 12, and 12).

? "How are these numbers related?" The sum of each pair is the coefficient of the x-term and the product of each pair is the constant term.

Lesson Notes

Key Idea

- Write the Key Idea.
- **Teaching Tip:** When factoring the trinomial $x^2 + 6x + 5$, say "You need two numbers whose product is 5 and whose sum is 6."

Example 1

- Write the problem and say, "You need two numbers whose product is 16 and whose sum is 10."

? "What are the factors of 16?" 1 and 16, 2 and 8, 4 and 4

- Finish working through the problem as shown.
- Some students may find it tedious to list the factor pairs. Share with them that in future problems the signs may not both be positive and the leading coefficient may be a number other than 1.
- Point out that students should check their answers using the FOIL Method.

On Your Own

- Students should set up each problem before they factor. For example, $x^2 + 2x + 1 = (x + \_\_)(x + \_\_)$. Then they should say, "I need two numbers whose product is 1 and whose sum is 2."
- **Neighbor Check:** Have students work independently and then have their neighbor check their work. Have students discuss any discrepancies.

Goal Today's lesson is factoring $x^2 + bx + c$.

Start Thinking! and Warm Up

Lesson 7.7 Start Thinking!
For use before Lesson 7.7

How are factoring and the FOIL Method related? Give examples in your explanation.

Extra Example 1

Factor $w^2 + 11w + 28$. $(w + 4)(w + 7)$

On Your Own

1. $(x + 1)(x + 1)$
2. $(x + 8)(x + 1)$
3. $(y + 4)(y + 2)$
4. $(z + 8)(z + 3)$

Laurie's Notes

Extra Example 2

Factor $y^2 - 7y + 10$. $(y - 2)(y - 5)$

On Your Own

5. $(w - 1)(w - 3)$
6. $(n - 5)(n - 7)$
7. $(x - 2)(x - 12)$

Extra Example 3

Factor $w^2 + 8w - 33$. $(w - 3)(w + 11)$

English Language Learners

Notebook Development

For application examples, provide English language learners a handout with the detailed solution. This will allow the students to focus on clues within the text as you work through the problem. Students should highlight and make diagrams on the handout as they follow along. This can be added to their notebook and referred to as they work the exercises.

Example 2

- Make a small change to the three matching problems from the beginning of class.

1. $(x - 2)(x - 6)$	B	**A.**	$x^2 - 7x + 12$
2. $(x - 12)(x - 1)$	C	**B.**	$x^2 - 8x + 12$
3. $(x - 3)(x - 4)$	A	**C.**	$x^2 - 13x + 12$

? "If both of the binomials involve subtraction, how does the trinomial change?" The coefficient of the x-term will be negative.

- Write the problem and say, "You need two numbers whose product is 12 and whose sum is -8."
- Write the possible factors and work through the example as shown.
- Remind students to check their answers.

On Your Own

- Students should set up each problem before they factor. For example, $w^2 - 4w + 3 = (w - \_\_)(w - \_\_)$. Then they should say, "I need two numbers whose product is 3 and whose sum is -4."
- **Think-Pair-Share:** Students should read each question independently and then work with a partner to answer the questions. When they have answered the questions, the pair should compare their answers with another group and discuss any discrepancies.

Key Idea

- Write the Key Idea.
- **Common Misconception:** Students may think that in the trinomial $x^2 + bx + c$, c is positive because of the addition symbol. Remind students that the c is an unknown and it may be positive or negative. In this specific case, however, we do know that c is negative.

Example 3

- Write the problem and say, "You need two numbers whose product is -21 and whose sum is 4."

? "What are the factors of 21?" 1 and 21, 3 and 7

? "Knowing that one of the factors must be negative, is there a combination that will sum to 4?" yes; -3 and 7

- Finish working through the problem as shown.
- **Teaching Tip:** When the leading coefficient is 1 and the constant term is negative, encourage students to begin by writing $(x - \_\_)(x + \_\_)$ and then think about the pairs of factors.
- Discuss with students that when c is negative, the factor that has the greater absolute value has the same sign as b.

EXAMPLE 2 Factoring $x^2 + bx + c$ When b Is Negative and c Is Positive

Factor $x^2 - 8x + 12$.

Notice that $b = -8$ and $c = 12$.

- Because c is positive, the factors p and q must have the same sign so that pq is positive.
- Because b is negative, p and q must each be negative so that $p + q$ is negative.

Find two negative integer factors of 12 whose sum is -8.

Factors of 12	$-1, -12$	$-2, -6$	$-3, -4$
Sum of Factors	-13	-8	-7

The values of p and q are -2 and -6.

So, $x^2 - 8x + 12 = (x - 2)(x - 6)$.

Check

Use the FOIL Method.

$(x - 2)(x - 6)$

$= x^2 - 6x - 2x + 12$

$= x^2 - 8x + 12$ ✓

On Your Own

Factor the polynomial.

5. $w^2 - 4w + 3$
6. $n^2 - 12n + 35$
7. $x^2 - 14x + 24$

Key Idea

Factoring $x^2 + bx + c$ When c Is Negative

Algebra $x^2 + bx + c = (x + p)(x + q)$ when $p + q = b$ and $pq = c$.

When c is negative, p and q have different signs.

Example $x^2 - 4x - 5 = (x + 1)(x - 5)$

EXAMPLE 3 Factoring $x^2 + bx + c$ When c Is Negative

Factor $x^2 + 4x - 21$.

Notice that $b = 4$ and $c = -21$. Because c is negative, the factors p and q must have different signs so that pq is negative.

Find two integer factors of -21 whose sum is 4.

Factors of -21	$-21, 1$	$-1, 21$	$-7, 3$	$-3, 7$
Sum of Factors	-20	20	-4	4

The values of p and q are -3 and 7.

So, $x^2 + 4x - 21 = (x - 3)(x + 7)$.

EXAMPLE 4 Real-Life Application

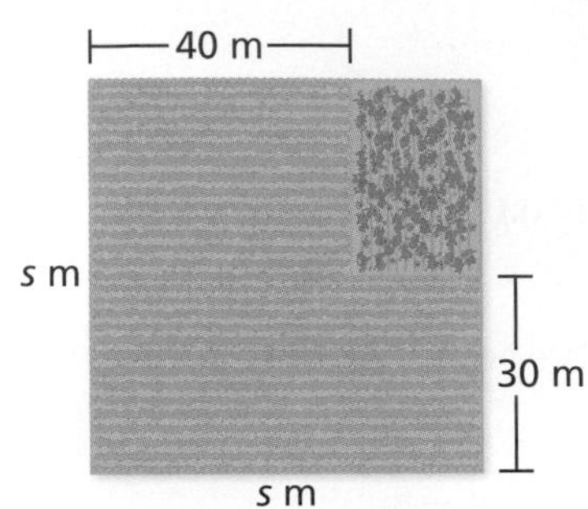

A farmer plants a rectangular pumpkin patch in the northeast corner of the square plot of land. The area of the pumpkin patch is 600 square meters. What is the area of the square plot of land?

The length of the pumpkin patch is $(s - 30)$ meters and the width is $(s - 40)$ meters. Write and solve an equation for its area.

$600 = (s - 30)(s - 40)$	Write an equation.
$600 = s^2 - 70s + 1200$	Multiply.
$0 = s^2 - 70s + 600$	Subtract 600 from each side.
$0 = (s - 10)(s - 60)$	Factor the polynomial.
$s - 10 = 0$ *or* $s - 60 = 0$	Use Zero-Product Property.
$s = 10$ *or* $s = 60$	Solve for s.

The diagram shows that the side length is at least 30 meters, so 10 meters does not make sense in this situation. The width is 60 meters.

So, the area of the square plot of land is $60(60) = 3600$ square meters.

On Your Own

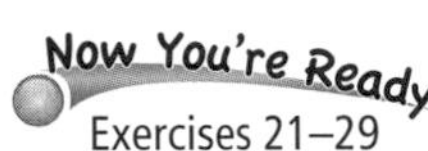

Factor the polynomial.

8. $x^2 + 2x - 15$ **9.** $y^2 + 13y - 30$ **10.** $v^2 + v - 20$

11. $z^2 - z - 12$ **12.** $m^2 - 11m - 26$ **13.** $x^2 - 3x - 40$

14. **WHAT IF?** In Example 4, the area of the pumpkin patch is 200 square meters. What is the area of the square plot of land?

Summary

Factoring $x^2 + bx + c$ as $(x + p)(x + q)$

The diagram shows the relationships between the signs of b and c and the signs of p and q.

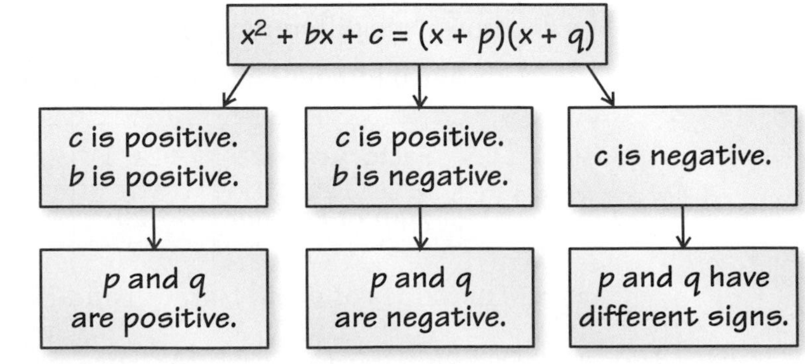

Laurie's Notes

Example 4

- Ask a volunteer to read the problem as you draw the diagram on the board.
- ? "What are the dimensions of the pumpkin patch?" $(s - 30)$ and $(s - 40)$
- Work through the problem. Explain that the area formula ($A = \ell w$) is used to write the original equation.
- **MP2 Reason Abstractly and Quantitatively:** The equation has two solutions, 10 and 60, but one of the solutions can be eliminated because of the context of the problem.

On Your Own

- Students should set up each problem before they factor. For example, $x^2 + 2x - 15 = (x - \_\_)(x + \_\_)$. Then they should say, "I need two numbers whose product is –15 and whose sum is 2."
- If time permits, have volunteers share their answers at the board.

Summary

- Draw the diagram on the board and review the techniques for factoring $x^2 + bx + c$.

Closure

- Explain your strategy for factoring the following polynomials.
  1. $x^2 - 2x - 15$ $(x - 5)(x + 3)$
  2. $x^2 + 2x - 15$ $(x - 3)(x + 5)$
  3. $x^2 + 8x + 15$ $(x + 5)(x + 3)$
  4. $x^2 - 8x + 15$ $(x - 5)(x - 3)$

Technology For the Teacher

Dynamic Classroom

The Dynamic Planning Tool
Editable Teacher's Resources at *BigIdeasMath.com*

Extra Example 4

A city builds a rectangular sandbox in the corner of a square playground. The area of the sandbox is 6 square yards. What is the area of the playground?
225 yd^2

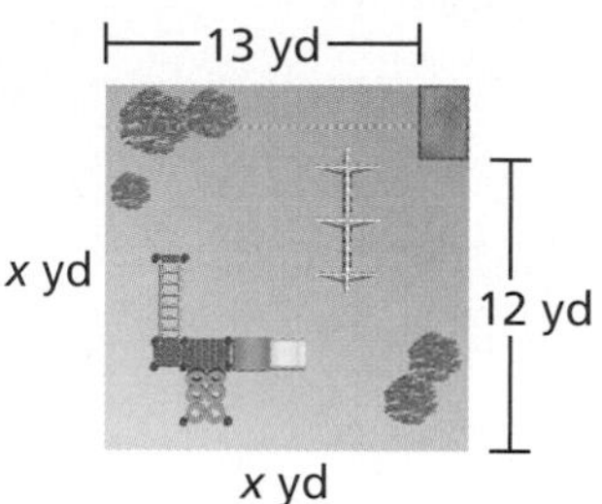

On Your Own

8. $(x - 3)(x + 5)$
9. $(y - 2)(y + 15)$
10. $(v - 4)(v + 5)$
11. $(z - 4)(z + 3)$
12. $(m - 13)(m + 2)$
13. $(x - 8)(x + 5)$
14. $2500\ m^2$

Vocabulary and Concept Check

1. Because c is negative, p and q have different signs. Because b is positive, the factor with the greater absolute value is positive.
2. *Sample answer:* $x^2 + 8x + 15$

Practice and Problem Solving

3. $(x + 1)(x + 7)$
4. $(z + 3)(z + 4)$
5. $(n + 2)(n + 6)$
6. $(s + 5)(s + 6)$
7. $(h + 2)(h + 9)$
8. $(y + 5)(y + 8)$
9. The factors of 48 (4 and 12) do not add up to 14.
   $t^2 + 14t + 48 = (t + 6)(t + 8)$
10. $(v - 1)(v - 4)$
11. $(x - 4)(x - 5)$
12. $(d - 2)(d - 3)$
13. $(k - 4)(k - 6)$
14. $(w - 8)(w - 9)$
15. $(j - 6)(j - 7)$
16. $m = -1, m = -2$
17. $x = -4, x = -7$
18. $n = 3, n = 6$
19. year 1 and year 5
20. **a.** $x - 5$

    b. 16 ft

Assignment Guide and Homework Check

Level	Assignment	Homework Check
Average	1, 2, 3–37 odd, 20, 34, 39, 43–46	7, 15, 19, 27, 34, 39
Advanced	1, 2, 4–38 even, 39–42, 43–46	8, 14, 26, 39, 40, 42

Common Errors

- **Exercises 3–8** Students may forget to check that the factors of c add up to b. Remind them that there are two steps to factoring polynomials, factoring c and checking that those factors add up to b. Encourage students to use tables as shown in the examples.
- **Exercises 10–15** Students may use the wrong sign when writing the factors of the polynomial. Remind them that when b is negative and c is positive both p and q will be negative.
- **Exercise 19** Students may solve $x^2 - 6x + 8 = 0$ instead of $x^2 - 6x + 8 = 3$. Remind them to read each question carefully.

7.7 Record and Practice Journal

Factor the polynomial.

1. $w^2 + 8x + 15$ — $(w + 3)(w + 5)$
2. $b^2 + 12b + 27$ — $(b + 9)(b + 3)$
3. $y^2 - 9y + 18$ — $(y - 6)(y - 3)$
4. $h^2 - 15h + 26$ — $(h - 2)(h - 13)$
5. $n^2 + n - 42$ — $(n + 7)(n - 6)$
6. $k^2 - 5k - 14$ — $(k - 7)(k + 2)$

Solve the equation.

7. $t^2 - 14t + 33 = 0$ — $t = 3, 11$
8. $d^2 - 3d = 54$ — $d = -6, 9$
9. The area (in square meters) covered by a building can be represented by $x^2 + 7x - 30$.
   a. Write binomials that represent the length and width of the building. — $x - 3, x + 10$
   b. Find the perimeter of the building when $x = 15$ meters. — 74 m

Technology For the Teacher
Answer Presentation Tool

7.7 Exercises

Vocabulary and Concept Check

1. **WRITING** You are factoring $x^2 + 11x - 26$. What do the signs of the terms tell you about the factors? Explain.

2. **OPEN-ENDED** Write a trinomial that can be factored as $(x + p)(x + q)$ where p and q are positive.

Practice and Problem Solving

Factor the polynomial.

1 3. $x^2 + 8x + 7$ 4. $z^2 + 7z + 12$ 5. $n^2 + 8n + 12$

6. $s^2 + 11s + 30$ 7. $h^2 + 11h + 18$ 8. $y^2 + 13y + 40$

9. **ERROR ANALYSIS** Describe and correct the error in factoring the polynomial.

$t^2 + 14t + 48 = (t + 4)(t + 12)$

Factor the polynomial.

2 10. $v^2 - 5v + 4$ 11. $x^2 - 9x + 20$ 12. $d^2 - 5d + 6$

13. $k^2 - 10k + 24$ 14. $w^2 - 17w + 72$ 15. $j^2 - 13j + 42$

Solve the equation.

16. $m^2 + 3m + 2 = 0$ 17. $x^2 + 11x + 28 = 0$ 18. $n^2 - 9n + 18 = 0$

19. **PROFIT** A company's profit (in millions of dollars) can be represented by $x^2 - 6x + 8$, where x is the number of years since the company started. When did the company have a profit of \$3 million?

20. **PROJECTION** A projector displays an image on a wall. The area (in square feet) of the rectangular projection can be represented by $x^2 - 8x + 15$.

    a. Write a binomial that represents the height of the projection.

    b. Find the perimeter of the projection when the height of the wall is 8 feet.

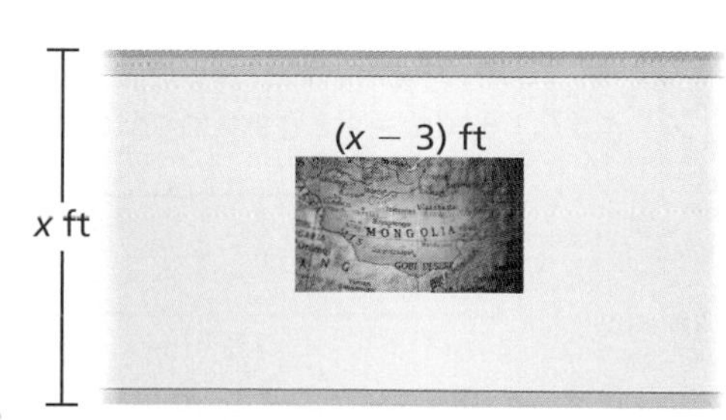

Factor the polynomial.

3 **21.** $x^2 + 3x - 4$ **22.** $z^2 + 7z - 18$ **23.** $n^2 + 4n - 12$

24. $s^2 + 3s - 40$ **25.** $h^2 + 6h - 27$ **26.** $y^2 + 2y - 48$

27. $m^2 - 6m - 7$ **28.** $x^2 - x - 20$ **29.** $t^2 - 6t - 16$

Solve the equation.

30. $v^2 + 3v - 4 = 0$ **31.** $x^2 + 5x - 14 = 0$ **32.** $n^2 - 5n = 24$

33. ERROR ANALYSIS Describe and correct the error in solving the equation.

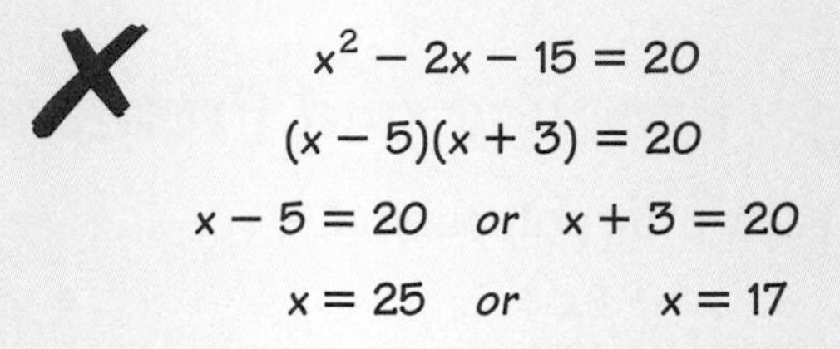

$$x^2 - 2x - 15 = 20$$
$$(x - 5)(x + 3) = 20$$
$$x - 5 = 20 \quad or \quad x + 3 = 20$$
$$x = 25 \quad or \quad x = 17$$

34. DENTIST A dentist's office and parking lot are on a rectangular piece of land. The area (in square meters) of the land can be represented by $x^2 + x - 30$.

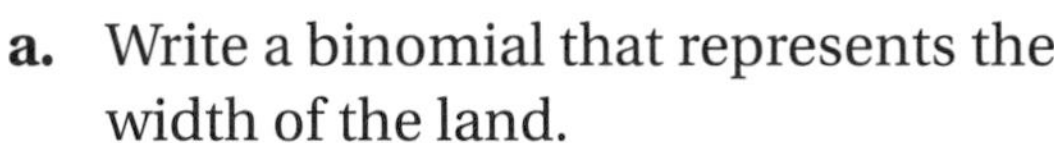

a. Write a binomial that represents the width of the land.

b. Write an expression that represents the area of the parking lot.

c. Evaluate the expressions in parts (a) and (b) when $x = 20$.

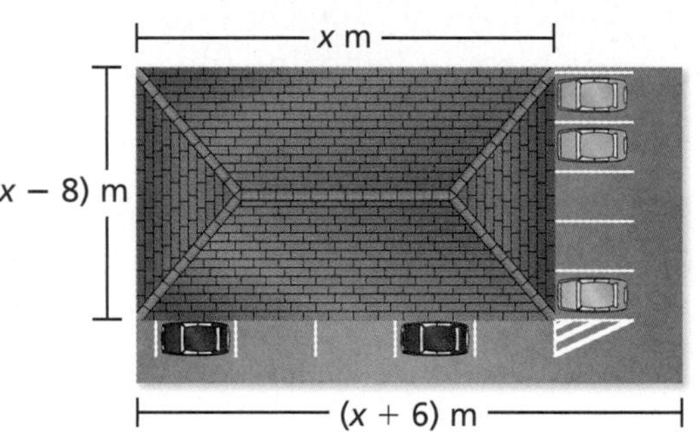

Find the dimensions of the polygon with the given area.

35. Area = 44 square feet

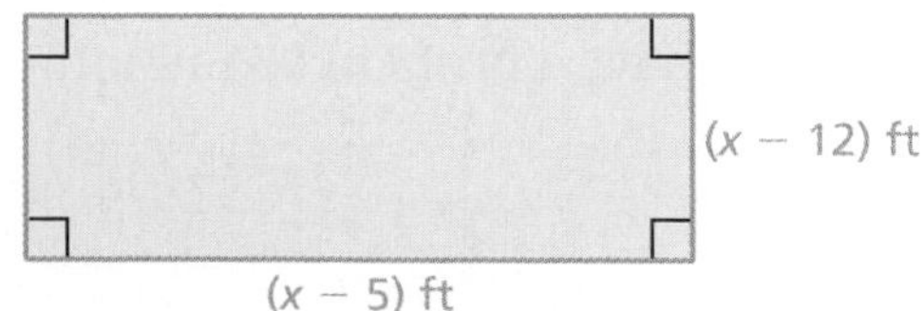

36. Area = 35 square centimeters

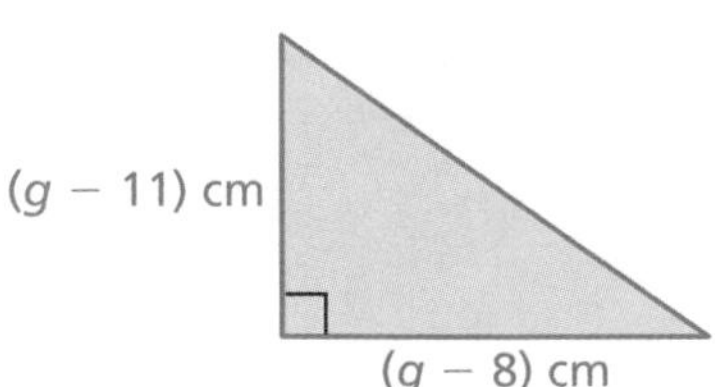

37. Area = 120 square feet

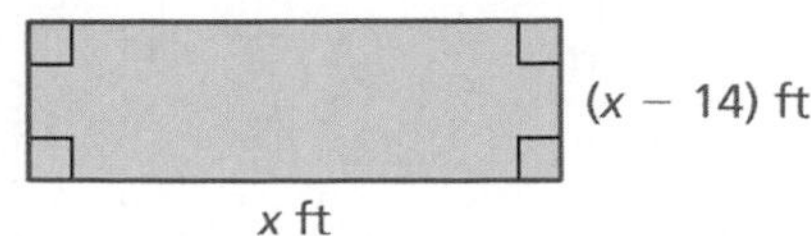

38. Area = 75 square centimeters

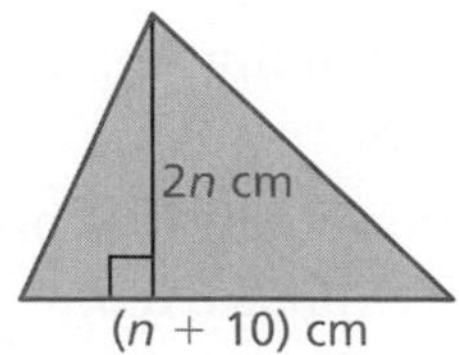

Common Errors

- **Exercises 21–29** Students may use the wrong sign when writing the factors of the polynomial. Remind them to be careful with the signs and encourage them to check their answers using the FOIL Method.
- **Exercise 32** Students may not collect all of the terms on one side of the equation before factoring. Remind them that to solve an equation by factoring, the polynomial must be equal to zero.

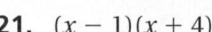

Practice and Problem Solving

21. $(x-1)(x+4)$
22. $(z-2)(z+9)$
23. $(n-2)(n+6)$
24. $(s-5)(s+8)$
25. $(h-3)(h+9)$
26. $(y-6)(y+8)$
27. $(m-7)(m+1)$
28. $(x-5)(x+4)$
29. $(t-8)(t+2)$
30. $v=1, v=-4$
31. $x=2, x=-7$
32. $n=8, n=-3$
33. All of the terms were not collected on one side of the equation before factoring.

    $x^2-2x-15=20$

    $x^2-2x-35=0$

    $(x-7)(x+5)=0$

    $x-7=0 \text{ or } x+5=0$

    $x=7 \text{ or } x=-5$
34. **a.** $x-5$

    b. $9x-30$

    c. 15 m; 150 m^2

Differentiated Instruction

Inclusion

Some students may have difficulty using the tables to factor a trinomial. In this case, partially write the solution and let students fill in the blanks.

$x^2-7x+12=(x-\_\_)(x-\_\_)$

Ask students for factors of 12 and use the FOIL method to generate the middle term. Point out that $(x-4)(x-3)$ is correct because $x(-3)+(-4)x=-7x$.

Practice and Problem Solving

35. length: 11 ft, width: 4 ft

36. base: 10 cm, height: 7 cm

37. length: 20 ft, width: 6 ft

38. base: 15 cm, height: 10 cm

39. **a.** 6

b. length: 13 in., width: 8 in.

40. **a.** $-x^2 + 38x$

b. $-x^2 + 38x = 280$; $x = 10$

c. 28 meters is not reasonable because it would make the dimensions in the diagram impossible.

41. See *Taking Math Deeper*.

42. $\pm 1, \pm 4, \pm 11$

Fair Game Review

43. $2(y - 9)$

44. $n(7n + 23)$

45. $4z^2(2z + 7)$

46. C

Mini-Assessment

Factor the polynomial.

1. $w^2 + 6w - 72$ $(w - 6)(w + 12)$
2. $k^2 - 6k - 16$ $(k - 8)(k + 2)$

Solve the equation.

3. $z^2 - 12 = z$ $z = 4, z = -3$
4. $y^2 + 8 = 52 - 7y$ $y = 4, y = -11$
5. The area of the rectangle is 80 square meters. What are the dimensions of the rectangle?
   16 m by 5 m

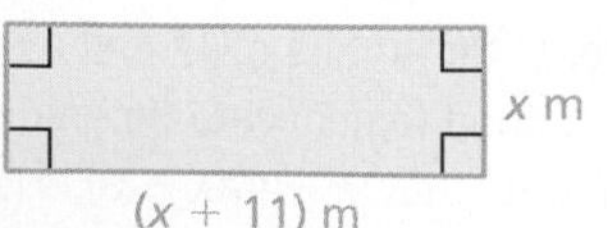

Taking Math Deeper

Exercise 41

This problem may look difficult to students. Remind them to not panic and answer one part at a time.

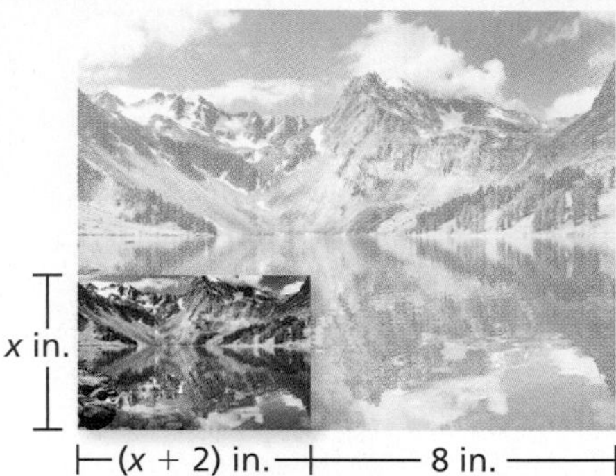

a. The area of the enlarged photograph is represented by $x^2 + 17x + 70 = (x + 7)(x + 10)$. Because $A = \ell w$, the length of the enlarged photograph is $x + 10$ and the width is $x + 7$. Notice that the diagram also shows that the length is $(x + 2) + 8 = x + 10$.

b. The length of the enlarged photograph is $(x + 10) - (x + 7) = 3$ inches greater than its width.

c. To find the dimensions of each photograph, set the given polynomial equal to 154 and solve for x.

$$x + 17x + 70 = 154$$

$$x + 17x - 84 = 0$$

$$(x - 4)(x + 21) = 0$$

$$x = 4 \text{ or } x = -21$$

The negative solution does not make sense, so $x = 4$.

Small photograph: x by $(x + 2)$ *or* 4 in. by 6 in.

Large photograph: $(x + 7)$ by $(x + 10)$ *or* 11 in. by 14 in.

Reteaching and Enrichment Strategies

If students need help. . .	If students got it. . .
Resources by Chapter • Practice A and Practice B • Puzzle Time Record and Practice Journal Practice Differentiating the Lesson Lesson Tutorials Skills Review Handbook	Resources by Chapter • Enrichment and Extension • School-to-Work • Financial Literacy • Technology Connection • Life Connections Start the next section

39. COMPUTER A web browser is open on your computer screen.

a. The area of the browser is 24 square inches. Find the value of x.

b. The browser covers $\frac{3}{13}$ of the screen. What are the dimensions of the screen?

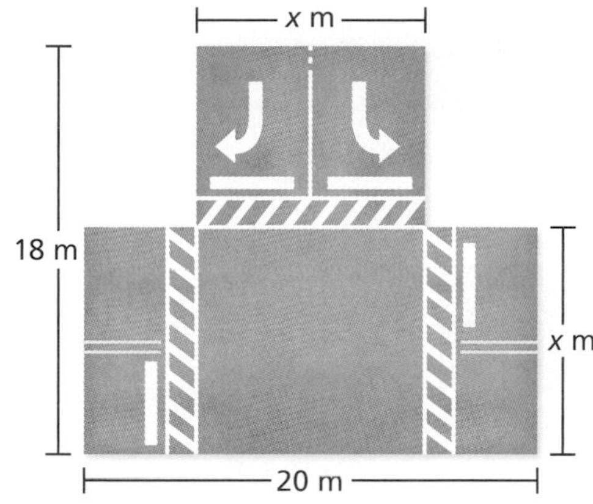

40. LOGIC Road construction workers are paving the area shown.

a. Write an expression that represents the area being paved.

b. The area being paved is 280 square meters. Write and solve an equation to find x.

c. The equation in part (b) has two solutions. Explain why one of the solutions is not reasonable.

41. PHOTOGRAPHY You enlarge a photograph on a computer. The area (in square inches) of the enlarged photograph can be represented by $x^2 + 17x + 70$.

a. Write binomials that represent the length and width of the enlarged photograph.

b. How many inches greater is the length of the enlarged photograph than the width? Explain.

c. The area of the enlarged photograph is 154 square inches. Find the dimensions of each photograph.

42. Number Sense Find all of the integer values of b for which the trinomial $x^2 + bx - 12$ has two binomial factors of the form $(x + p)$ and $(x + q)$.

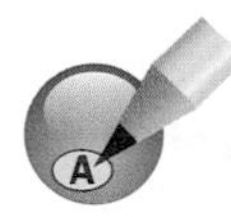

Fair Game Review What you learned in previous grades & lessons

Factor the polynomial. *(Section 7.6)*

43. $2y - 18$ **44.** $7n^2 + 23n$ **45.** $8z^3 + 28z^2$

46. MULTIPLE CHOICE Which expression is *not* equivalent to $\sqrt{\frac{9}{4}}$? *(Section 6.1)*

Ⓐ $\frac{3}{2}$ Ⓑ $\sqrt{2.25}$ Ⓒ $2\sqrt{3}$ Ⓓ $3\sqrt{\frac{1}{4}}$

7.8 Factoring $ax^2 + bx + c$

COMMON CORE STATE STANDARDS
A.REI.4b
A.SSE.3a

Essential Question How can you factor the trinomial $ax^2 + bx + c$ into the product of two binomials?

1 ACTIVITY: Finding Binomial Factors

Work with a partner. Six different algebra tiles are shown below.

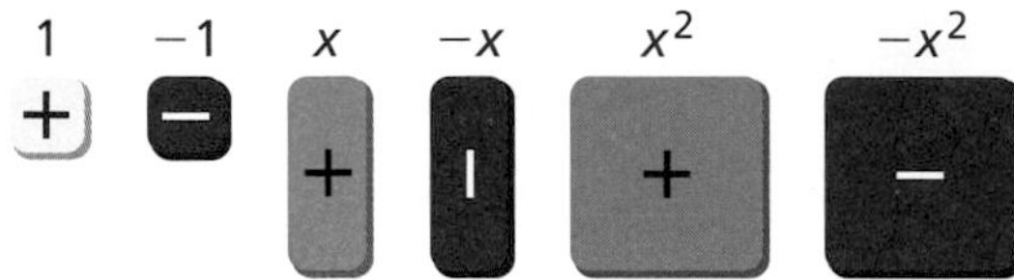

Sample:

Step 1: Arrange the algebra tiles into a rectangular array to model $2x^2 + 5x + 2$.

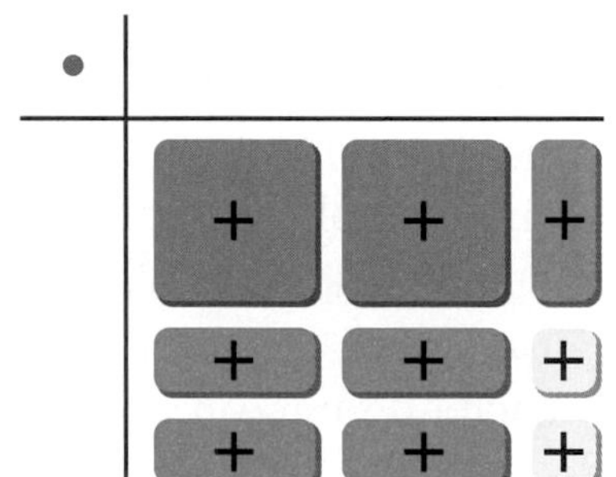

Step 2: Use algebra tiles to label the dimensions of the rectangle.

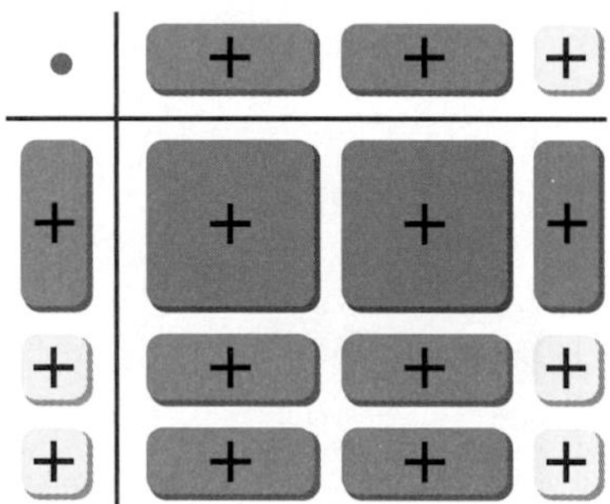

Step 3: Write the polynomial in factored form by finding the dimensions of the rectangle.

$$\text{Area} = 2x^2 + 5x + 2 = \underbrace{(2x + 1)}_{\text{length}}\underbrace{(x + 2)}_{\text{width}}$$

Use algebra tiles to write the polynomial as the product of two binomials. Check your answer by multiplying.

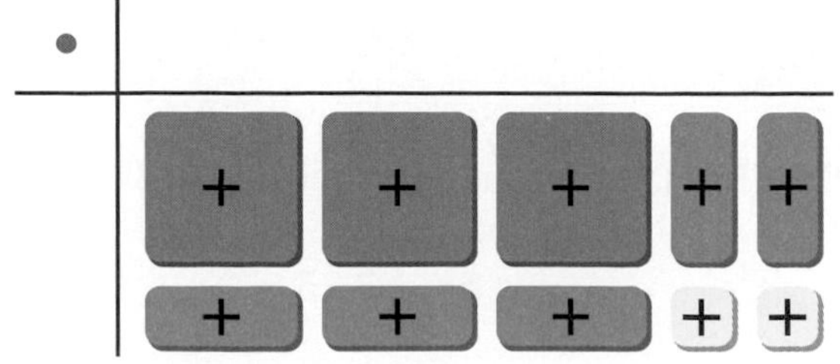

Laurie's Notes

Introduction

Standards for Mathematical Practice

- **MP1a Make Sense of Problems** and **MP5 Use Appropriate Tools Strategically:** Algebra tiles provide a visual and tactile experience that should help students make sense of how to factor a trinomial with a leading coefficient that is not 1.
- Rectangular arrays help students make sense of the factors as the dimensions and the trinomial as the area.

Motivate

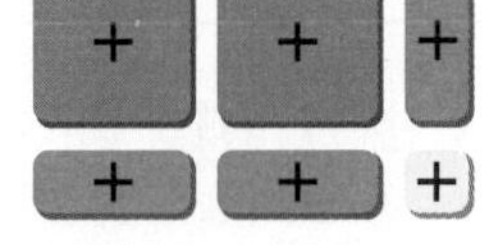

- Model $2x^2 + 3x + 1$ with tiles on the overhead. Have students do the same at their desks.
- **?** "Can you arrange the algebra tiles so they form a rectangle with no holes or overlaps?" yes
- If students try to place all three x-tiles vertically, there is no place for the unit tile to fit. Model this if students don't make the attempt.
- **?** "What are the dimensions of the rectangle?" $(x + 1)$ and $(2x + 1)$
- Label the dimensions. Explain that $2x^2 + 3x + 1 = (x + 1)(2x + 1)$.
- **MP5:** This is similar to the visual model used when whole numbers are multiplied using base 10 pieces.

Activity Notes

Activity 1

- Ask students to form a rectangular array to model $2x^2 + 5x + 2$. The two x^2-tiles need to be placed side by side, either vertically or horizontally.
- Show the following three arrangements and explain that we use the one on the left because there is space for the two remaining unit tiles.

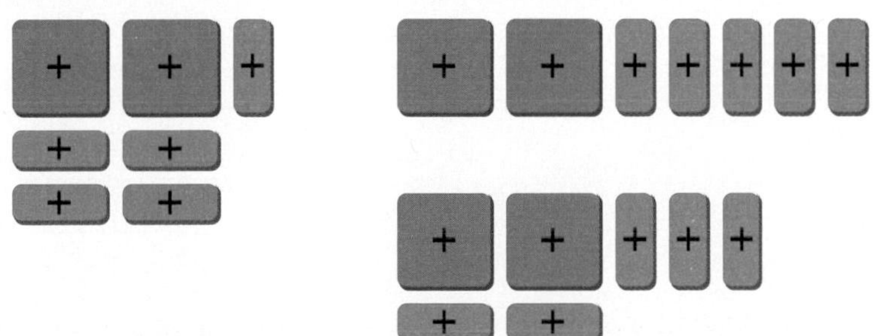

- **?** "What are the dimensions of the rectangle?" $(x + 2)$ and $(2x + 1)$
- Label the dimensions. Explain that $2x^2 + 5x + 2 = (x + 2)(2x + 1)$.
- As students work the next problem with their partners, encourage them to think about the dimensions of the rectangle.

Common Core State Standards

A.REI.4b Solve quadratic equations by inspection, taking square roots, completing the square, the quadratic formula and factoring, as appropriate to the initial form of the equation . . ..

A.SSE.3a Factor a quadratic expression to reveal the zeros of the function it defines.

Previous Learning

Students should know how to factor $x^2 + bx + c$.

Activity Materials	
Introduction	**Textbook**
• algebra tiles	• algebra tiles

Start Thinking! and Warm Up

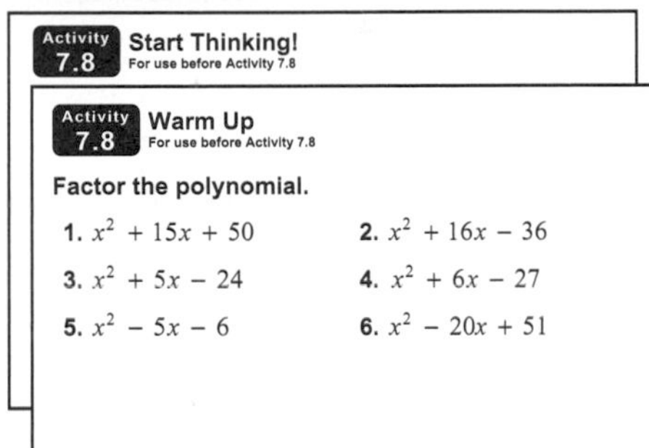

Activity 7.8 Start Thinking! For use before Activity 7.8

Activity 7.8 Warm Up For use before Activity 7.8

Factor the polynomial.

1. $x^2 + 15x + 50$
2. $x^2 + 16x - 36$
3. $x^2 + 5x - 24$
4. $x^2 + 6x - 27$
5. $x^2 - 5x - 6$
6. $x^2 - 20x + 51$

7.8 Record and Practice Journal

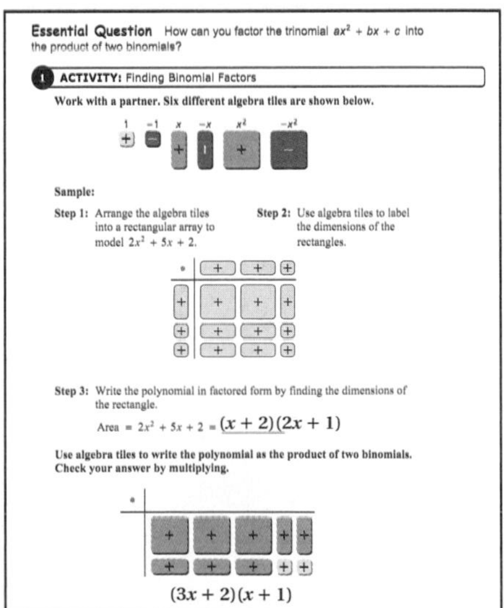

Essential Question How can you factor the trinomial $ax^2 + bx + c$ into the product of two binomials?

1 ACTIVITY: Finding Binomial Factors

Work with a partner. Six different algebra tiles are shown below.

Sample:

Step 1: Arrange the algebra tiles into a rectangular array to model $2x^2 + 5x + 2$.

Step 2: Use algebra tiles to label the dimensions of the rectangles.

Step 3: Write the polynomial in factored form by finding the dimensions of the rectangle.

Area $= 2x^2 + 5x + 2 = (x + 2)(2x + 1)$

Use algebra tiles to write the polynomial as the product of two binomials. Check your answer by multiplying.

$(3x + 2)(x + 1)$

English Language Learners

Group Activity

Provide English language learners with the opportunity to talk with their peers about the concepts in this section. Display several trinomials on the overhead for the students to factor. For the first one, show a partial solution. For example,

$3v^2 - 2v - 5 = (3v - 5)(v + ?)$. 1

7.8 Record and Practice Journal

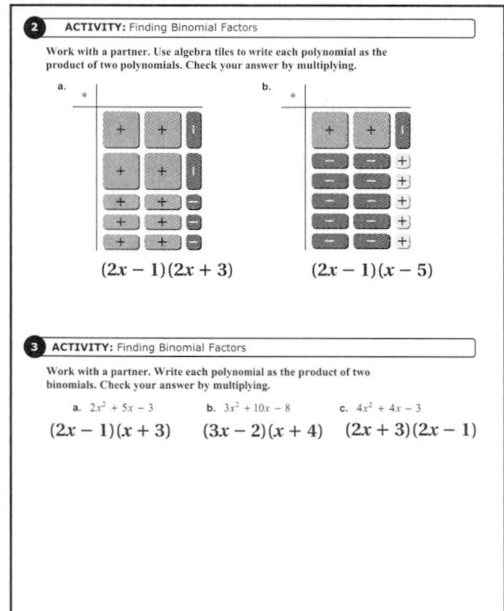

2 ACTIVITY: Finding Binomial Factors

Work with a partner. Use algebra tiles to write each polynomial as the product of two polynomials. Check your answer by multiplying.

a. $(2x - 1)(2x + 3)$

b. $(2x - 1)(x - 5)$

3 ACTIVITY: Finding Binomial Factors

Work with a partner. Write each polynomial as the product of two binomials. Check your answer by multiplying.

a. $2x^2 + 5x - 3$ $(2x - 1)(x + 3)$

b. $3x^2 + 10x - 8$ $(3x - 2)(x + 4)$

c. $4x^2 + 4x - 3$ $(2x + 3)(2x - 1)$

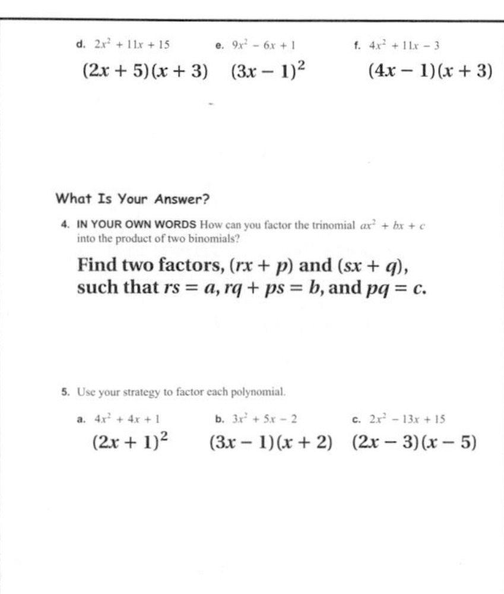

d. $2x^2 + 11x + 15$ $(2x + 5)(x + 3)$

e. $9x^2 - 6x + 1$ $(3x - 1)^2$

f. $4x^2 + 11x - 3$ $(4x - 1)(x + 3)$

What Is Your Answer?

4. **IN YOUR OWN WORDS** How can you factor the trinomial $ax^2 + bx + c$ into the product of two binomials?

Find two factors, $(rx + p)$ and $(sx + q)$, such that $rs = a$, $rq + ps = b$, and $pq = c$.

5. Use your strategy to factor each polynomial.

a. $4x^2 + 4x + 1$ $(2x + 1)^2$

b. $3x^2 + 5x - 2$ $(3x - 1)(x + 2)$

c. $2x^2 - 13x + 15$ $(2x - 3)(x - 5)$

Laurie's Notes

Activity 2

- Students need to pay attention to signs as they work these two problems.
- ? "How are these two problems alike?" They both have a factor (dimension) of $(2x - 1)$.
- Write and say, "$4x^2 + 4x - 3 = (2x - 1)(2x + 3)$ and $2x^2 - 11x + 5 = (2x - 1)(x - 5)$."
- **Big Idea:** When you multiply $(2x - 1)(2x + 3)$, the result is $4x^2 + 6x + (-2x) + (-3)$, which you can simplify as $4x^2 + 4x - 3$. This means that if you begin with the tiles for this trinomial, you need to add zero pairs to form a rectangular array.
- Students need to understand this before they begin the next activity. All of the problems in Activity 3 need zero pairs of x-tiles added except for parts (d) and (e).

Activity 3

- Students should start by selecting the algebra tiles that represent the trinomial, remembering that zero pairs may be needed to complete the rectangular array.
- Point out that when zero pairs are added, one is added to each dimension. So, one is placed vertically and one is placed horizontally.
- Ask volunteers to share their solutions.

What Is Your Answer?

- **Neighbor Check:** Have students work independently and then have their neighbor check their work. Have students discuss any discrepancies.

Closure

- Complete the factorization. Check by multiplying the binomials.

$2x^2 + 11x + 5 = (2x + \_\_\_)(x + \_\_\_)$ $(2x + 1)(x + 5)$

$2x^2 + 7x + 5 = (2x + \_\_\_)(x + \_\_\_)$ $(2x + 5)(x + 1)$

Technology For the Teacher

The Dynamic Planning Tool

Editable Teacher's Resources at *BigIdeasMath.com*

2 ACTIVITY: Finding Binomial Factors

Work with a partner. Use algebra tiles to write each polynomial as the product of two binomials. Check your answer by multiplying.

a.

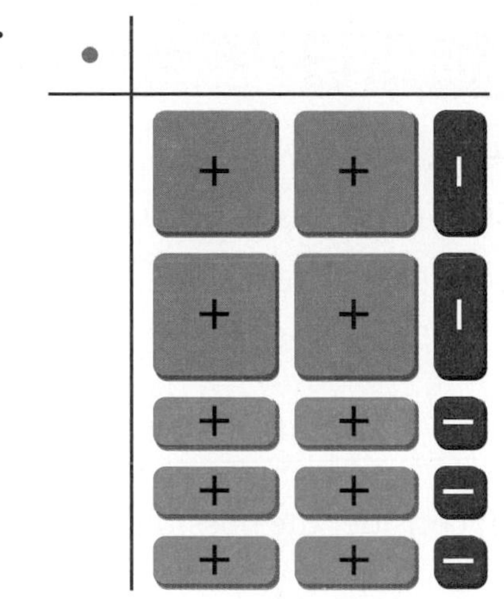

b.

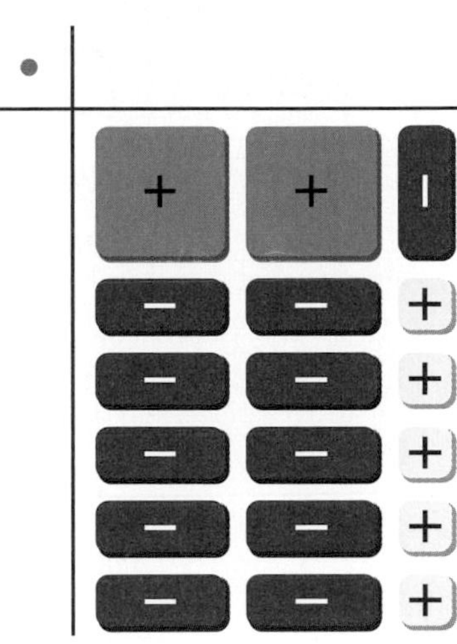

3 ACTIVITY: Finding Binomial Factors

Work with a partner. Write each polynomial as the product of two binomials. Check your answer by multiplying.

a. $2x^2 + 5x - 3$ **b.** $3x^2 + 10x - 8$ **c.** $4x^2 + 4x - 3$

d. $2x^2 + 11x + 15$ **e.** $9x^2 - 6x + 1$ **f.** $4x^2 + 11x - 3$

What Is Your Answer?

4. **IN YOUR OWN WORDS** How can you factor the trinomial $ax^2 + bx + c$ into the product of two binomials?
5. Use your strategy to factor each trinomial.

a. $4x^2 + 4x + 1$ **b.** $3x^2 + 5x - 2$ **c.** $2x^2 - 13x + 15$

Practice Use what you learned about factoring trinomials to complete Exercises 3–5 on page 380.

7.8 Lesson

In Section 7.7, you factored polynomials of the form $ax^2 + bx + c$, where $a = 1$. To factor polynomials of the form $ax^2 + bx + c$, where $a \neq 1$, first look for the GCF of the terms of the polynomial.

EXAMPLE 1 Factoring Out the GCF

Factor $5x^2 + 15x + 10$.

Notice that the GCF of the terms $5x^2$, $15x$, and 10 is 5.

$$5x^2 + 15x + 10 = 5(x^2 + 3x + 2) \quad \text{Factor out GCF.}$$
$$= 5(x + 1)(x + 2) \quad \text{Factor } x^2 + 3x + 2.$$

∴ So, $5x^2 + 15x + 10 = 5(x + 1)(x + 2)$.

When there is no GCF, consider the possible factors of a and c.

EXAMPLE 2 Factoring $ax^2 + bx + c$ When ac Is Positive

a. Factor $4x^2 + 13x + 3$.

Consider the possible factors of $a = 4$ and $c = 3$.

Factors are 1, 2, and 4. → $4x^2 + 13x + 3$ ← Factors are 1 and 3.

These factors lead to the following possible products.

$$(1x + 1)(4x + 3) \qquad (1x + 3)(4x + 1) \qquad (2x + 1)(2x + 3)$$

Multiply to find the product that is equal to the original polynomial.

$(x + 1)(4x + 3) = 4x^2 + 7x + 3$ ✗ $(2x + 1)(2x + 3) = 4x^2 + 8x + 3$ ✗

$(x + 3)(4x + 1) = 4x^2 + 13x + 3$ ✓

∴ So, $4x^2 + 13x + 3 = (x + 3)(4x + 1)$.

b. Factor $3x^2 - 7x + 2$.

Consider the possible factors of $a = 3$ and $c = 2$. Because b is negative and c is positive, both factors of c must be negative.

Factors are 1 and 3. → $3x^2 - 7x + 2$ ← Factors are −2 and −1.

These factors lead to the following possible products.

$$(1x - 1)(3x - 2) \qquad (1x - 2)(3x - 1)$$

Multiply to find the product that is equal to the original polynomial.

$(x - 1)(3x - 2) = 3x^2 - 5x + 2$ ✗ $(x - 2)(3x - 1) = 3x^2 - 7x + 2$ ✓

∴ So, $3x^2 - 7x + 2 = (x - 2)(3x - 1)$.

Study Tip

When ac is positive, the sign of b determines whether the factors of c are positive or negative.

Laurie's Notes

Introduction

Connect

- **Yesterday:** Students used algebra tiles to factor $ax^2 + bx + c$. (MP1a, MP5)
- **Today:** Students will factor $ax^2 + bx + c$, when $a \neq 1$.

Motivate

- Write the following problem on the board: $\frac{2}{3} \bullet \frac{3}{4} \bullet \frac{4}{5} \bullet \cdots \bullet \frac{49}{50} = ?$
- Students should quickly notice that the answer is $\frac{2}{50} = \frac{1}{25}$.

? "How did you solve this problem so quickly?" divided out the common factors and simplified

- Explain that in today's lesson, noticing common factors may help simplify the problem.

Lesson Notes

Example 1

- At this stage, students understand that the ability to factor a trinomial is related to the values of *a*, *b*, and *c*.
- Write the first example.

? "Do you notice anything about *a*, *b*, and *c*?" All have a common factor of 5.

? "What are we left with when 5 is factored out?" $x^2 + 3x + 2$

- Students can now proceed as in the previous section. Stress, however, that the factor of 5 is not lost. It is part of the factored solution.

? "What two numbers have a product of 2 and a sum of 3?" 1 and 2

- Write the factored solution.

Example 2

- Write part (a). Note that $a \neq 1$ and *a*, *b*, and *c* are relatively prime.

? Ask the following questions.

  - "What do you know about the *x*-terms of the two binomial factors we are trying to find?" They have a product of $4x^2$.
  - "What are the possibilities?" $4x$ and x or $2x$ and $2x$
  - "What do you know about the constant terms of the two binomial factors we are trying to find?" They have a product of 3.
  - "What are the possibilities?" 3 and 1
- **MP1b Persevere in Solving Problems:** Consider the factors of *a* and the factors of *c*. Try different factorizations to determine when the middle term is $13x$. Note that because $b > 0$ and $c > 0$, both factors of *c* are positive.

Factors of 4	Factors of 3	Possible Factorization	Middle Term
1, 4	1, 3	$(x + 1)(4x + 3)$	$3x + 4x = 7x$
1, 4	**3, 1**	$\mathbf{(x + 3)(4x + 1)}$	$\mathbf{x + 12x = 13x}$
2, 2	1, 3	$(2x + 1)(2x + 3)$	$6x + 2x = 8x$

- Use the same process in solving part (b).

Goal Today's lesson is factoring $ax^2 + bx + c$ when $a \neq 1$.

Start Thinking! and Warm Up

Lesson 7.8 Warm Up
For use before Lesson 7.8

Lesson 7.8 Start Thinking!
For use before Lesson 7.8

Factor the following polynomials.

$2x^2 + 5x + 2$

$6x^2 + 29x + 28$

$x^2 + 8x + 15$

Which polynomial is the easiest to factor? Which is the most difficult? Explain.

Extra Example 1

Factor $3x^2 - 6x - 24$. $3(x - 4)(x + 2)$

Extra Example 2

a. Factor $3x^2 + 7x + 2$. $(3x + 1)(x + 2)$

b. Factor $2x^2 - 7x + 3$. $(2x - 1)(x - 3)$

On Your Own

1. $8(x - 1)(x - 6)$
2. $(2x + 1)(x + 5)$
3. $(2x - 5)(x - 1)$
4. $(3x - 2)(x - 4)$

Extra Example 3

Factor $4x^2 + 8x - 5$.
$(2x - 1)(2x + 5)$

Extra Example 4

The length of a rectangular room is 6 feet less than three times the width. The area of the room is 360 square feet. How wide is the room? 12 ft

On Your Own

5. $(3x - 4)(2x + 3)$
6. $(4x + 1)(x - 5)$
7. 8 miles

Differentiated Instruction

Auditory

Students may have difficulty processing the information in a word problem. Have them read the problem aloud and repeat any phrases that describe a measurement. Students should describe the relationships between quantities and state the question that the problem asks in their own words.

Laurie's Notes

On Your Own

- **Neighbor Check:** Have students work independently and then have their neighbor check their work. Have students discuss any discrepancies.

Example 3

- Write the trinomial.
- ? "What information does the -7 give about the two binomials?" The constant terms in the binomials have different signs.
- The factors of a are 1 and 2. The factors of c are ± 1 and ± 7.
- **Teaching Tip:** Suggest to students that they first set up the possible binomials without using operations: $(2x \quad 1)(x \quad 7)$ or $(2x \quad 7)(x \quad 1)$.
- Work through the possibilities using mental math, remembering that the signs must be different.
- Check the solution by multiplying the binomials and simplifying.

Example 4

- Ask a student to read the problem.
- Draw the rectangular game reserve. Translate the words and label the dimensions using w and $(2w + 1)$.
- Use area = length $\times$ width to write the equation.
- ? "How do you solve a trinomial set equal to 0?" Factor and use the Zero-Product Property.
- To factor $2w^2 + w - 55$, consider the factors of 2 and factors of -55.
- ? "To get a sum of $1w$ for the middle term, what factors of -55 might be reasonable to try first?" 5 and 11, where the 5 is multiplied by the $2w$.
- Factor the trinomial and set each factor equal to 0. Although the equation has two solutions, only one makes sense in the context of the problem. The width must be a positive value.

On Your Own

- **Think-Pair-Share:** Students should read each question independently and then work with a partner to answer the questions. When they have answered the questions, the pair should compare their answers with another group and discuss any discrepancies.

Closure

- **Exit Ticket:** Factor $2x^2 - 7x + 3$. $(2x - 1)(x - 3)$

Technology For the Teacher

The Dynamic Planning Tool
Editable Teacher's Resources at *BigIdeasMath.com*

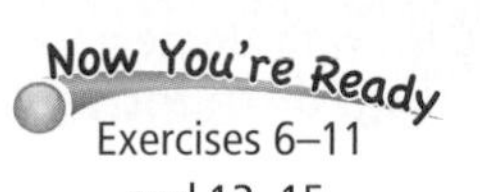

Exercises 6–11 and 13–15

On Your Own

Factor the polynomial.

1. $8x^2 - 56x + 48$
2. $2x^2 + 11x + 5$
3. $2x^2 - 7x + 5$
4. $3x^2 - 14x + 8$

EXAMPLE 3 Factoring $ax^2 + bx + c$ When ac Is Negative

Factor $2x^2 - 5x - 7$.

Consider the possible factors of $a = 2$ and $c = -7$. Because b and c are both negative, the factors of c must have different signs.

Factors are 1 and 2. → $2x^2 - 5x - 7$ ← Factors are ± 1 and ± 7.

These factors lead to the following possible products.

$(x + 1)(2x - 7)$ $(x + 7)(2x - 1)$ $(x - 1)(2x + 7)$ $(x - 7)(2x + 1)$

Multiply to find the product that is equal to the original polynomial.

$(x + 1)(2x - 7) = 2x^2 - 5x - 7$ ✓ $(x - 1)(2x + 7) = 2x^2 + 5x - 7$ ✗

$(x + 7)(2x - 1) = 2x^2 + 13x - 7$ ✗ $(x - 7)(2x + 1) = 2x^2 - 13x - 7$ ✗

So, $2x^2 - 5x - 7 = (x + 1)(2x - 7)$.

Study Tip

For polynomials of the form $ax^2 + bx + c$, where a is negative, factor out -1 first to make factoring easier. Just be sure to put -1 back in your final answer.

EXAMPLE 4 Standardized Test Practice

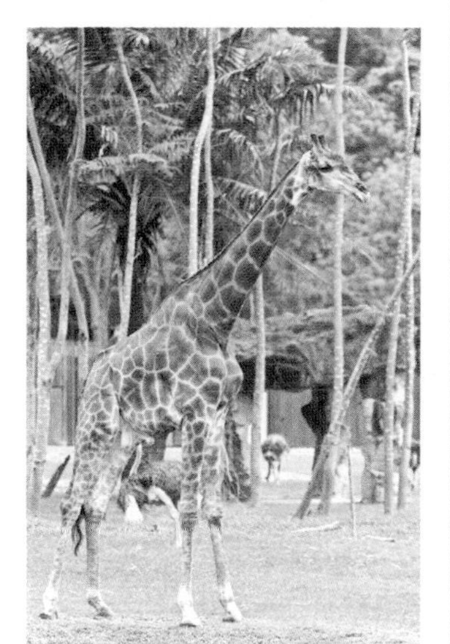

The length of a rectangular game reserve is 1 mile longer than twice the width. The area of the reserve is 55 square miles. How wide is the reserve?

(A) 2 mi (B) 2.5 mi (C) 5 mi (D) 5.5 mi

Write an equation that represents the area of the reserve. Then solve by factoring. Let w represent the width. Then $2w + 1$ represents the length.

$w(2w + 1) = 55$ Area of the reserve

$2w^2 + w - 55 = 0$ Multiply. Then subtract 55 from each side.

$(w - 5)(2w + 11) = 0$ Factor left side of the equation.

$w - 5 = 0$ *or* $2w + 11 = 0$ Use Zero-Product Property.

$w = 5$ *or* $w = -\frac{11}{2}$ Solve for w. Use the positive solution.

The correct answer is (C).

On Your Own

Now You're Ready Exercises 16–21

Factor the polynomial.

5. $6x^2 + x - 12$
6. $4x^2 - 19x - 5$
7. **WHAT IF?** In Example 4, the area of the reserve is 136 square miles. How wide is the reserve?

7.8 Exercises

Vocabulary and Concept Check

1. **WRITING** Describe how to factor polynomials of the form $ax^2 + bx + c$.
2. **WHICH ONE DOESN'T BELONG?** Which factored polynomial does *not* belong with the other three? Explain your reasoning.

$(2x - 3)(x + 2)$	$x(2x - 3) + 2(2x - 3)$	$(2x + 3)(x - 2)$	$2x(x + 2) - 3(x + 2)$

Practice and Problem Solving

Use algebra tiles to write the polynomial as the product of two binomials.

3. $2x^2 - 3x + 1$
4. $3x^2 + x - 2$
5. $4x^2 + 11x + 6$

Factor the polynomial.

1 6. $3x^2 + 3x - 6$
7. $8v^2 + 8v - 48$
8. $4k^2 + 28k + 48$
9. $6y^2 - 24y + 18$
10. $9r^2 - 36r - 45$
11. $7d^2 - 63d + 140$

12. **ERROR ANALYSIS** Describe and correct the error in factoring the polynomial.

$$2x^2 + 2x - 4 = 2x(x + 1 - 2)$$
$$= 2x(x - 1)$$

Factor the polynomial.

2 3 13. $3h^2 + 11h + 6$
14. $6x^2 - 5x + 1$
15. $8m^2 + 30m + 7$
16. $18v^2 - 15v - 18$
17. $2n^2 - 5n - 3$
18. $4z^2 - 4z - 3$
19. $8g^2 - 10g - 12$
20. $10w^2 + 19w - 15$
21. $14d^2 + 3d - 2$

22. **ERROR ANALYSIS** Describe and correct the error in factoring the polynomial.

$$6x^2 - 7x - 3 = (3x - 3)(2x + 1)$$

23. **DANCE FLOOR** The area (in square feet) of a rectangular lighted dance floor can be represented by $8x^2 + 22x + 5$. Write the expressions that represent the dimensions of the dance floor.

Assignment Guide and Homework Check

Level	Assignment	Homework Check
Average	1, 2, 7–23 odd, 12, 24, 27, 30, 31, 39–42	7, 12, 19, 23, 30
Advanced	1, 2, 6–28 even, 31–36, 39–42	20, 28, 32, 33, 36

Common Errors

- **Exercises 6–11** Some students will factor out the GCF and think they are done. Explain that factoring out the GCF is often just the first step, and emphasize the concept of factoring completely.
- **Exercises 13–21** Some students will choose factors of *a* and *c* that do not add up to *b* when the binomial factors are multiplied. Remind students that the product of the binomials must equal the original polynomial.
- **Exercises 27–29** Remind students to factor out *a* −1 before factoring polynomials with a negative leading coefficient, and to include it after they have factored.

7.8 Record and Practice Journal

Factor the polynomial

1. $5n^2 + 15n + 10$
   $5(n + 2)(n + 1)$
2. $4h^2 - 20h - 56$
   $4(h + 2)(h - 7)$
3. $2j^2 + 13j - 45$
   $(2j - 5)(j + 9)$
4. $9p^2 + 6p - 8$
   $(3p + 4)(3p - 2)$
5. $6b^2 - 7b - 24$
   $(3b - 8)(2b + 3)$
6. $12x^2 - 33x + 18$
   $3(x - 2)(4x - 3)$

Solve the equation.

7. $4y^2 + 8y + 3 = 0$
   $y = -\frac{1}{2}, -\frac{2}{3}$
8. $8d^2 - 4d = 60$
   $d = -2.5, 3$
9. The area of the surface of the trampoline is equal to twice its perimeter. Find the dimensions of the trampoline.
   8 ft by 8 ft

4x ft

(x + 6) ft

Technology For the Teacher

Answer Presentation Tool

Vocabulary and Concept Check

1. First, factor out the GCF of the terms, if possible. Then, consider the possible factors of *a* and *c*, list the possible products, and multiply.
2. $(2x + 3)(x - 2)$; $(2x + 3)(x - 2) = 2x^2 - x - 6$; The other three factored polynomials equal $2x^2 + x - 6$.

Practice and Problem Solving

3. $(2x - 1)(x - 1)$
4. $(3x - 2)(x + 1)$
5. $(4x + 3)(x + 2)$
6. $3(x - 1)(x + 2)$
7. $8(v - 2)(v + 3)$
8. $4(k + 3)(k + 4)$
9. $6(y - 3)(y - 1)$
10. $9(r + 1)(r - 5)$
11. $7(d - 5)(d - 4)$
12. The GCF is 2, not $2x$.
    $2x^2 + 2x - 4$
    $= 2(x + 2)(x - 1)$
13. $(3h + 2)(h + 3)$
14. $(2x - 1)(3x - 1)$
15. $(4m + 1)(2m + 7)$
16. $3(2v - 3)(3v + 2)$
17. $(2n + 1)(n - 3)$
18. $(2z - 3)(2z + 1)$
19. $2(g - 2)(4g + 3)$
20. $(2w + 5)(5w - 3)$
21. $(7d - 2)(2d + 1)$
22. The factorization should be $(3x + 1)(2x - 3)$ instead of $(3x - 3)(2x + 1)$.
23. $(4x + 1)$ ft by $(2x + 5)$ ft

Practice and Problem Solving

24. $x = 3$ or $x = -2$
25. $k = -2$ or $k = \frac{9}{2}$
26. $m = \frac{3}{4}$ or $m = -\frac{5}{3}$
27. $-(3w - 4)(w + 2)$
28. $-3(2x + 1)(2x - 9)$
29. $-5(4n - 1)(2n - 3)$
30. 2.5 seconds
31. See *Taking Math Deeper*.
32. yes; The dimensions of the invitation are 3 inches by 5 inches.
33. 4 ft
34. when none of the sums of the products of the factors of a and c equal b; Example: $3x^2 - 20x + 5$
35. See Additional Answers.
36. $2k(4k - 1)(5k + 2)$
37. $(2x - y)(3x + 4y)$
38. $3m(3m - n)(2m + 5n)$

Fair Game Review

39. $4x^2 - 49$
40. $k^2 + 10k + 25$
41. $9b^2 - 24b + 16$
42. D

Mini-Assessment

Factor the polynomial.

1. $5m^2 + 10m - 40$ $5(m - 2)(m + 4)$
2. $3d^2 + 10d + 7$ $(3d + 7)(d + 1)$
3. $8w^2 + 6w - 9$ $(4w - 3)(2w + 3)$
4. $12x^2 - 12x - 9$ $3(2x - 3)(2x + 1)$

Taking Math Deeper

Exercise 31

A common error in this exercise is thinking that t must be positive. Challenge your students by asking them whether t is positive or negative, then explain that "$+ tx$" does not mean that t is positive. Once that is understood, proceed by treating $2x^2 + tx + 10$ the same as other trinomials in this section.

Consider the possible factors of $a = 2$ and $c = 10$.
The possible factors of a are 1 and 2.
Because t can be positive or negative, and c is positive, the possible factors of c must have the same sign. So, the possible factors of c are ±1, ±2, ±5, and ±10.

Use these factors to form eight possible binomials products.

$(2x - 10)(x - 1)$	$(2x + 10)(x + 1)$
$(2x - 1)(x - 10)$	$(2x + 1)(x + 10)$
$(2x - 5)(x - 2)$	$(2x + 5)(x + 2)$
$(2x - 2)(x - 5)$	$(2x + 2)(x + 5)$

Now find the middle term of each product.

$-2x - 10x = -12x$	$2x + 10x = 12x$
$-20x - 1x = -21x$	$20x + 1x = 21x$
$-4x - 5x = -9x$	$4x + 5x = 9x$
$-10x - 2x = -12x$	$10x + 2x = 12x$

t can be positive or negative!

So, $2x^2 + tx + 10$ can be written as the product of two binomials when t is -21, -12, -9, 9, 12, or 21.

Reteaching and Enrichment Strategies

If students need help. . .	If students got it. . .
Resources by Chapter • Practice A and Practice B • Puzzle Time Record and Practice Journal Practice Differentiating the Lesson Lesson Tutorials Skills Review Handbook	Resources by Chapter • Enrichment and Extension • School-to-Work • Financial Literacy • Technology Connection • Life Connections • Stories in History Start the next section

Solve the equation.

24. $5x^2 - 5x - 30 = 0$ **25.** $2k^2 - 5k - 18 = 0$ **26.** $12m^2 + 11m = 15$

Factor the polynomial.

27. $-3w^2 - 2w + 8$ **28.** $-12x^2 + 48x + 27$ **29.** $-40n^2 + 70n - 15$

30. CLIFF DIVING The height h (in feet) above the water of a cliff diver is modeled by $h = -16t^2 + 8t + 80$, where t is the time (in seconds). How long is the diver in the air?

31. REASONING For what values of t can $2x^2 + tx + 10$ be written as the product of two binomials?

32. INVITATION The length of a rectangular birthday party invitation is 1 inch less than twice its width. The area of the invitation is 15 square inches. Will the invitation fit in a $3\frac{5}{8}$-inch by $5\frac{1}{8}$-inch envelope without being folded? Explain your reasoning.

33. SWIMMING POOL A rectangular swimming pool is bordered by a concrete patio. The width of the patio is the same on every side. The surface area of the pool is equal to the area of the patio border. What is the width of the patio border?

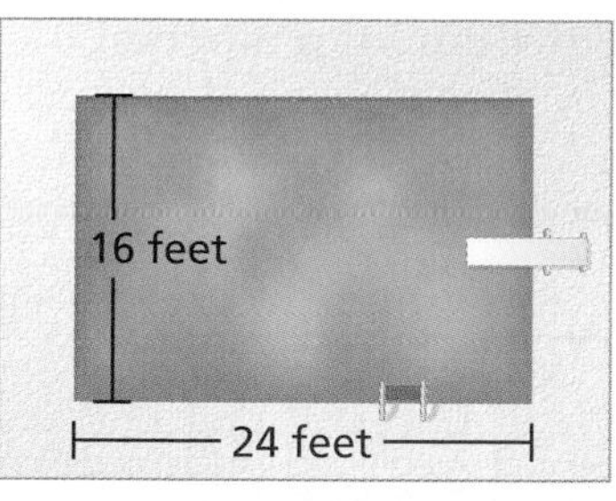

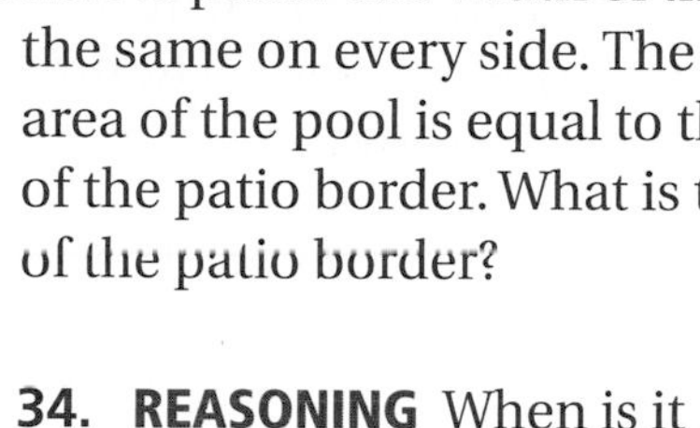

34. REASONING When is it *not* possible to factor $ax^2 + bx + c$, where $a \neq 1$? Give an example.

35. CHOOSE TOOLS A vendor can sell 50 bobbleheads per day when the price is \$40 each. For every \$2 decrease in price, 5 more bobbleheads are sold each day.

a. The revenue from yesterday was \$2160. What was the price per bobblehead? (*Note:* revenue = units sold × unit price)

b. How much should the vendor charge per bobblehead to maximize the daily revenue? Explain how you found your answer.

Structure **Factor the polynomial.**

36. $40k^3 + 6k^2 - 4k$ **37.** $6x^2 + 5xy - 4y^2$ **38.** $18m^3 + 39m^2n - 15mn^2$

Fair Game Review What you learned in previous grades & lessons

Find the product. *(Section 7.4)*

39. $(2x - 7)(2x + 7)$ **40.** $(k + 5)^2$ **41.** $(3b - 4)^2$

42. MULTIPLE CHOICE Two angles are supplementary. The measure of one of the angles is 58°. What is the measure of the other angle? *(Skills Review Handbook)*

Ⓐ 22° Ⓑ 32° Ⓒ 58° Ⓓ 122°

7.9 Factoring Special Products

COMMON CORE STATE STANDARDS
A.REI.4b
A.SSE.2
A.SSE.3a

Essential Question How can you recognize and factor special products?

1 ACTIVITY: Factoring Special Products

Work with a partner. Six different algebra tiles are shown below.

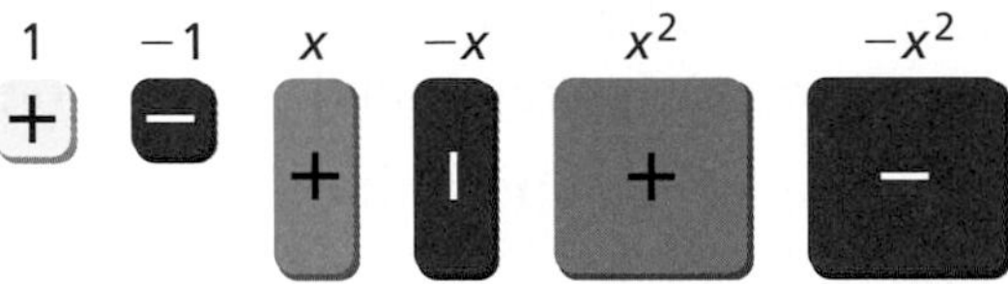

Use algebra tiles to write each polynomial as the product of two binomials. Check your answer by multiplying. State whether the product is a "special product" that you studied in Lesson 7.4.

a.

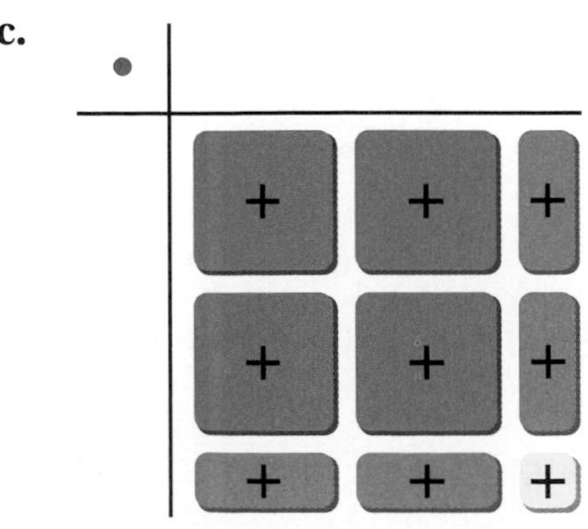

b.

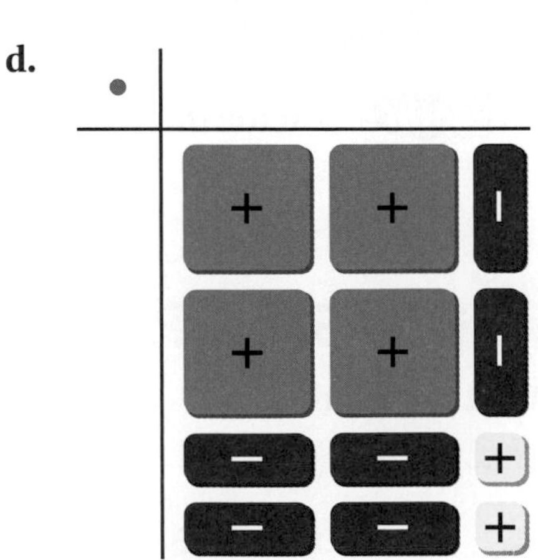

c.

d.

Laurie's Notes

Introduction

Standards for Mathematical Practice

- **MP1a Make Sense of Problems** and **MP8 Look for and Express Regularity in Repeated Reasoning:** In this activity and lesson, students will recognize patterns associated with special products from an earlier lesson. Students should be very comfortable using algebra tiles. The special products form a square array that should make sense to students.

Motivate

- Copy the four equations for students. If you have sets of numeral tiles (0 – 9), they can be used for this activity, or students can write the digits.
- **Directions:** Use the digits 0–9 to make the following statements true. Use each digit only once.

  $(x + \_\_)^2 = x^2 + 6x + \_\_$ 3; 9

  $(\_\_x - \_\_)(2x + 1) = 4x^2 - 1$ 2; 1

  $(x + \_\_)^2 = x^2 + 16x + \_\_\,\_\_$ 8; 64

  $(\_\_x - \_\_)^2 = 25x^2 - 7\_\_x + 49$ 5; 7; 0

- Completing these problems will help students review the special products from Lesson 7.4.

Activity Notes

Activity 1

- Say, "In this activity you are to determine the factors of the polynomial shown in the rectangular array. Pay attention to the signs."
- As students work through the problems with their partners, remind them to write the polynomial in both standard form and factored form.

? "Are all of the products one of the 'special products' from Lesson 7.4?" no; Part (d) is not.

Common Core State Standards

A.REI.4b Solve quadratic equations by inspection, taking square roots, completing the square, the quadratic formula and factoring, as appropriate to the initial form of the equation . . ..
A.SSE.2 Use the structure of an expression to identify ways to rewrite it.
A.SSE.3a Factor a quadratic expression to reveal the zeros of the function it defines.

Previous Learning

Students should know how to factor polynomials using the greatest common factor and how to factor trinomials of the forms $x^2 + bx + c$ and $ax^2 + bx + c$.

Activity Materials	
Introduction	**Textbook**
• numeral tiles (optional)	• algebra tiles

Start Thinking! and Warm Up

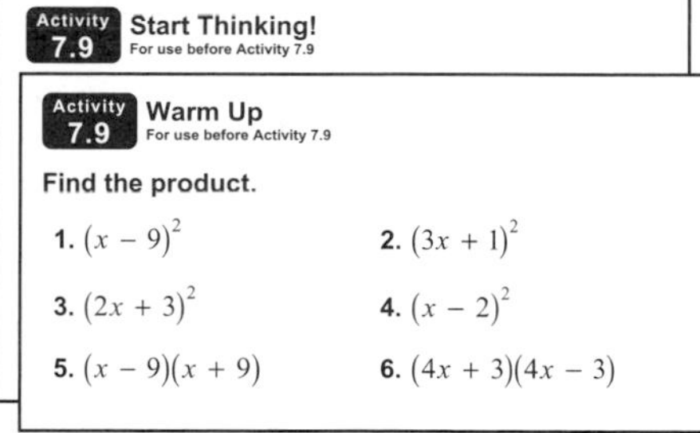

Activity 7.9 Start Thinking! For use before Activity 7.9

Activity 7.9 Warm Up For use before Activity 7.9

Find the product.

1. $(x - 9)^2$
2. $(3x + 1)^2$
3. $(2x + 3)^2$
4. $(x - 2)^2$
5. $(x - 9)(x + 9)$
6. $(4x + 3)(4x - 3)$

7.9 Record and Practice Journal

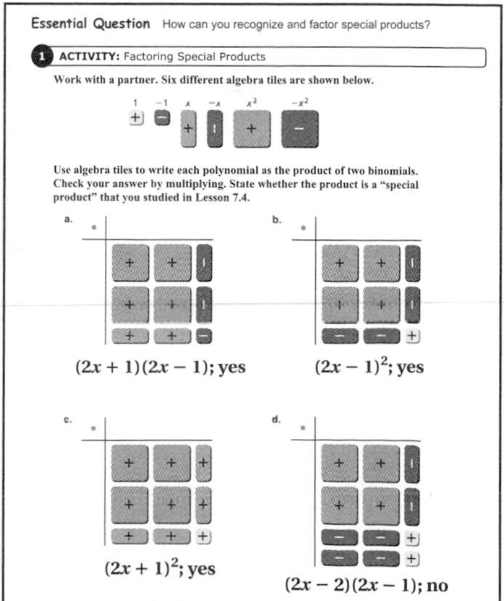

Essential Question How can you recognize and factor special products?

1 ACTIVITY: Factoring Special Products

Work with a partner. Six different algebra tiles are shown below.

Use algebra tiles to write each polynomial as the product of two binomials. Check your answer by multiplying. State whether the product is a "special product" that you studied in Lesson 7.4.

a. $(2x + 1)(2x - 1)$; yes

b. $(2x - 1)^2$; yes

c. $(2x + 1)^2$; yes

d. $(2x - 2)(2x - 1)$; no

Differentiated Instruction

Inclusion

Students may find the number of formulas and their similarities confusing. Have them use the FOIL method on the expressions $(a - b)(a + b)$, $(a + b)(a + b)$ and $(a - b)(a - b)$ to learn the patterns.

7.9 Record and Practice Journal

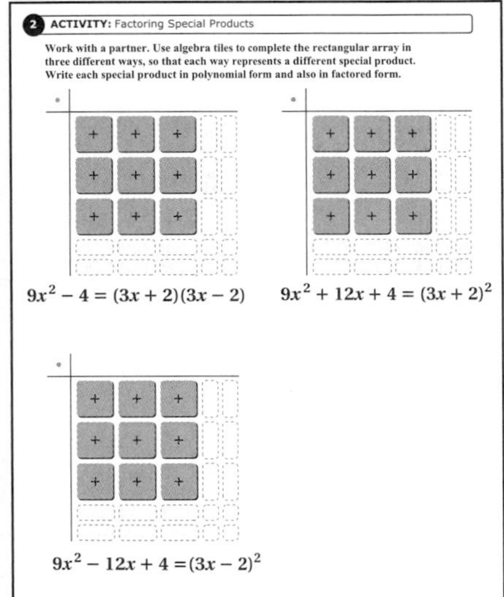

2 ACTIVITY: Factoring Special Products

Work with a partner. Use algebra tiles to complete the rectangular array in three different ways, so that each way represents a different special product. Write each special product in polynomial form and also in factored form.

$9x^2 - 4 = (3x + 2)(3x - 2)$

$9x^2 + 12x + 4 = (3x + 2)^2$

$9x^2 - 12x + 4 = (3x - 2)^2$

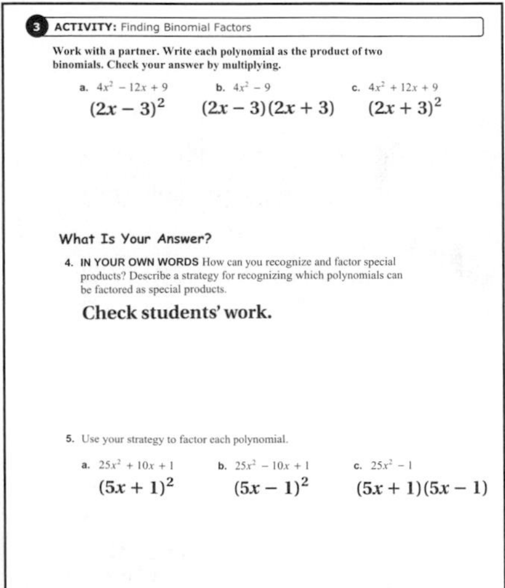

3 ACTIVITY: Finding Binomial Factors

Work with a partner. Write each polynomial as the product of two binomials. Check your answer by multiplying.

a. $4x^2 - 12x + 9$ $(2x - 3)^2$

b. $4x^2 - 9$ $(2x - 3)(2x + 3)$

c. $4x^2 + 12x + 9$ $(2x + 3)^2$

What Is Your Answer?

4. **IN YOUR OWN WORDS** How can you recognize and factor special products? Describe a strategy for recognizing which polynomials can be factored as special products.

Check students' work.

5. Use your strategy to factor each polynomial.

a. $25x^2 + 10x + 1$ $(5x + 1)^2$

b. $25x^2 - 10x + 1$ $(5x - 1)^2$

c. $25x^2 - 1$ $(5x + 1)(5x - 1)$

Laurie's Notes

Activity 2

- Before students begin, you may wish to write the three special products from Lesson 7.4. Ask students to help you.
- Note that all of the tiles will be used. Students need to determine the sign of the missing pieces.

? "What special shape is each of the rectangular arrays?" square

Activity 3

? "How are these three problems alike?" In each problem, $a = 4$ and $c = \pm 9$; 4 and 9 are both perfect squares. "How are they different?" (a) and (c) are trinomials and (b) is a binomial.

- Ask volunteers to share their solutions.

What Is Your Answer?

- For Question 4, have students share their thoughts with the whole class.

Closure

- **Exit Ticket:** Factor $9x^2 - 1$. $(3x + 1)(3x - 1)$

Technology For the Teacher

The Dynamic Planning Tool

Editable Teacher's Resources at *BigIdeasMath.com*

2 ACTIVITY: Factoring Special Products

Work with a partner. Use algebra tiles to complete the rectangular array in three different ways, so that each way represents a different special product. Write each special product in polynomial form and also in factored form.

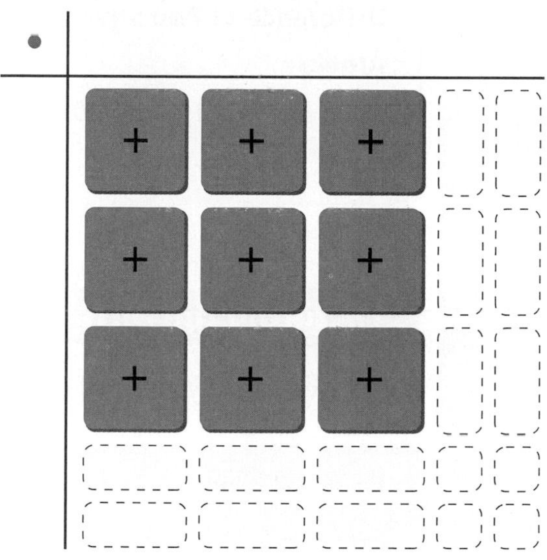

3 ACTIVITY: Finding Binomial Factors

Work with a partner. Write each polynomial as the product of two binomials. Check your answer by multiplying.

a. $4x^2 - 12x + 9$ **b.** $4x^2 - 9$ **c.** $4x^2 + 12x + 9$

What Is Your Answer?

4. **IN YOUR OWN WORDS** How can you recognize and factor special products? Describe a strategy for recognizing which polynomials can be factored as special products.

5. Use your strategy to factor each polynomial.

   a. $25x^2 + 10x + 1$ **b.** $25x^2 - 10x + 1$ **c.** $25x^2 - 1$

Practice Use what you learned about factoring polynomials as special products to complete Exercises 4–6 on page 386.

7.9 Lesson

You can use special product patterns to factor polynomials.

Difference of Two Squares Pattern

Algebra

$a^2 - b^2 = (a + b)(a - b)$

Example

$x^2 - 9 = x^2 - 3^2$

$= (x + 3)(x - 3)$

EXAMPLE 1 Factoring the Difference of Two Squares

Factor each polynomial.

a. $x^2 - 25$

$x^2 - 25 = x^2 - 5^2$ Write as $a^2 - b^2$.

$= (x + 5)(x - 5)$ Difference of Two Squares Pattern

b. $64 - y^2$

$64 - y^2 = 8^2 - y^2$ Write as $a^2 - b^2$.

$= (8 + y)(8 - y)$ Difference of Two Squares Pattern

c. $4z^2 - 1$

$4z^2 - 1 = (2z)^2 - 1^2$ Write as $a^2 - b^2$.

$= (2z + 1)(2z - 1)$ Difference of Two Squares Pattern

Remember

You can check your answers using the FOIL Method.

On Your Own

Exercises 4–8

Factor the polynomial.

1. $x^2 - 36$
2. $100 - m^2$
3. $9n^2 - 16$
4. $16h^2 - 49$

Key Idea

Perfect Square Trinomial Pattern

Algebra

$a^2 + 2ab + b^2 = (a + b)^2$

$a^2 - 2ab + b^2 = (a - b)^2$

Example

$x^2 + 6x + 9 = x^2 + 2(x)(3) + 3^2$

$= (x + 3)^2$

$x^2 - 6x + 9 = x^2 - 2(x)(3) + 3^2$

$= (x - 3)^2$

Laurie's Notes

Introduction

Connect

- **Yesterday:** Students used algebra tiles to recognize and factor special products. (MP1a, MP8)
- **Today:** Students will factor special products.

Motivate

- Play *Guess My Rule.* You write the first 4 terms of a sequence and students guess the rule.
  Examples: 4, 7, 10, 13, . . . Add 3.
  1, 3, 7, 15, . . . Add 2, 4, 8, . . . ; $2^n - 1$
  1, 4, 9, 16, . . . Add 3, 5, 7, . . . ; n^2
- Make sure that students recognize the last sequence as a list of perfect squares.
- Share with students that they will be factoring polynomials that have perfect squares.

Lesson Notes

Key Idea

- Write the Key Idea.
- Point out to students that they used the reverse of this pattern to multiply binomials earlier in this chapter.

Example 1

- Write the expression in part (a). Point out that x^2 and 25 are both perfect squares.
- Part (b) is more challenging for students because the polynomial is not written in standard form.
- **?** In part (c) ask, "What do you notice about the numbers 4 and 1, and the expression z^2?" They are all perfect squares.

On Your Own

- **Neighbor Check:** Have students work independently and then have their neighbor check their work. Have students discuss any discrepancies.

Key Idea

- Write the Key Idea. Review the meaning of a perfect square trinomial.
- Again, point out to students that they used the reverse of this pattern to square binomials earlier in this chapter.

Goal Today's lesson is factoring special products.

Start Thinking! and Warm Up

Lesson 7.9 Warm Up
For use before Lesson 7.9

Lesson 7.9 Start Thinking!
For use before Lesson 7.9

What is a perfect square trinomial?

Without factoring, tell whether each trinomial is a perfect square trinomial. Explain your reasoning.

$6x^2 + 25x + 4$

$x^2 + 24x + 8$

$16x^2 + 24x + 9$

Extra Example 1

Factor each polynomial.

a. $n^2 - 16$ $(n + 4)(n - 4)$

b. $121 - w^2$ $(11 + w)(11 - w)$

c. $25t^2 - 36$ $(5t + 6)(5t - 6)$

On Your Own

1. $(x + 6)(x - 6)$
2. $(10 + m)(10 - m)$
3. $(3n + 4)(3n - 4)$
4. $(4h + 7)(4h - 7)$

Extra Example 2

Factor each polynomial.

a. $y^2 + 16y + 64$ $(y + 8)^2$

b. $s^2 - 22s + 121$ $(s - 11)^2$

On Your Own

5. $(m - 1)^2$

6. $(d - 5)^2$

7. $(z + 10)^2$

Extra Example 3

The function $y = 64 - 16t^2$ represents the height y (in feet) of a ball t seconds after it is dropped. After how many seconds does the ball hit the ground? 2 sec

On Your Own

8. 2.25 sec

English Language Learners

Organization

Use this lesson to review all the types of factoring taught in this chapter. Have students work together to complete a table with all the essential information.

Factoring Type	Clue/ Indicator	Example

Laurie's Notes

Example 2

- Write the expression in part (a).
- ? "What sign will be used in the answer? Explain." The sign will be positive because the middle and last terms of the trinomial are positive.
- ? "Is 16 a perfect square?" yes; $16 = 4^2$
- Finish working through the problem as shown.
- Write the expression in part (b).
- ? "What sign will be used in the answer? Explain." The sign will be negative because the last term is positive but the middle term is negative.
- ? "Is 81 a perfect square?" yes; $81 = 9^2$
- Finish working through the problem as shown.

Example 3

- **FYI:** This is a *falling object model* where the polynomial is not written in standard form. The equation could have been written as $y = -16t^2 + 81$, but that does not fit the difference of two squares pattern.
- Ask a student to read the problem.
- ? "How far is the ball from the ground when it hits the tree?" 32 feet
- Work through the problem as shown.
- **MP6 Attend to Precision:** Because the coefficient of the linear term is 0, you can also solve for t using square roots. Ask students to provide reasons for each step.

$$0 = 49 - 16t^2$$
$$16t^2 = 49$$
$$t^2 = \frac{49}{16}$$
$$t = \pm\sqrt{\frac{49}{16}}$$
$$t = \frac{7}{4} \text{ or } t = -\frac{7}{4}$$

On Your Own

- Students should realize that the time must increase if the ball does not hit the tree.

Closure

- **Exit Ticket:** Factor each polynomial.
  a. $81 - x^2$ $(9 - x)(9 + x)$
  b. $x^2 - 24x + 144$ $(x - 12)^2$

Technology For the Teacher

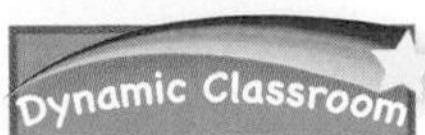

The Dynamic Planning Tool
Editable Teacher's Resources at *BigIdeasMath.com*

EXAMPLE 2 Factoring Perfect Square Trinomials

Factor each polynomial.

a. $n^2 + 8n + 16$

$$n^2 + 8n + 16 = n^2 + 2(n)(4) + 4^2$$ Write as $a^2 + 2ab + b^2$.

$$= (n + 4)^2$$ Perfect Square Trinomial Pattern

b. $x^2 - 18x + 81$

$$x^2 - 18x + 81 = x^2 - 2(x)(9) + 9^2$$ Write as $a^2 - 2ab + b^2$.

$$= (x - 9)^2$$ Perfect Square Trinomial Pattern

On Your Own

Exercises 9–12

Factor the polynomial.

5. $m^2 - 2m + 1$ **6.** $d^2 - 10d + 25$ **7.** $z^2 + 20z + 100$

EXAMPLE 3 Real-Life Application

A bird picks up a golf ball and drops it while flying. The function represents the height y (in feet) of the golf ball t seconds after it is dropped. The ball hits the top of a 32-foot tall pine tree. After how many seconds does the ball hit the tree?

Substitute 32 for y and solve for t.

$$y = 81 - 16t^2$$ Write equation.

$$32 = 81 - 16t^2$$ Substitute 32 for y.

$$0 = 49 - 16t^2$$ Subtract 32 from each side.

$$0 = 7^2 - (4t)^2$$ Write as $a^2 - b^2$.

$$0 = (7 + 4t)(7 - 4t)$$ Difference of Two Squares Pattern

$$7 + 4t = 0 \quad \textit{or} \quad 7 - 4t = 0$$ Use Zero-Product Property.

$$t = -\frac{7}{4} \quad \textit{or} \quad t = \frac{7}{4}$$ Solve for t.

A negative time does not make sense in this situation.

∴ So, the golf ball hits the tree after $\frac{7}{4}$, or 1.75 seconds.

On Your Own

8. **WHAT IF?** The golf ball does not hit the pine tree. After how many seconds does the ball hit the ground?

7.9 Exercises

Vocabulary and Concept Check

1. **WRITING** Describe two ways to show that $x^2 - 16$ is equal to $(x + 4)(x - 4)$.
2. **REASONING** Can you use the perfect square trinomial pattern to factor $y^2 + 16y + 64$? Explain.
3. **WHICH ONE DOESN'T BELONG?** Which polynomial does *not* belong with the other three? Explain your reasoning.

$n^2 - 4$	$g^2 - 6g + 9$	$r^2 + 12r + 36$	$k^2 + 25$

Practice and Problem Solving

Factor the polynomial.

① ② 4. $m^2 - 49$ 5. $9 - r^2$ 6. $4x^2 - 25$

7. $81d^2 - 64$ 8. $121 - 16t^2$ 9. $h^2 + 12h + 36$

10. $x^2 - 4x + 4$ 11. $w^2 - 14w + 49$ 12. $g^2 + 24g + 144$

13. **ERROR ANALYSIS** Describe and correct the error in factoring the polynomial.

$$n^2 - 16n + 64 = n^2 - 2(n)(8) + 8^2$$
$$= (n + 8)^2$$

Solve the equation.

14. $z^2 - 4 = 0$ 15. $s^2 + 20s + 100 = 0$ 16. $k^2 - 16k + 64 = 0$

17. $4x^2 = 49$ 18. $n^2 + 9 = -6n$ 19. $y^2 = 12y - 36$

20. **REASONING** Tell whether the polynomial can be factored. If not, change the constant term so that the polynomial can be factored using the perfect square trinomial pattern.
    a. $w^2 + 18w + 84$
    b. $y^2 - 10y + 23$
    c. $x^2 - 14x + 50$

21. **COASTER** The area (in square centimeters) of a square coaster can be represented by $d^2 + 8d + 16$. Write an expression that represents the side length of the coaster.

Assignment Guide and Homework Check

Level	Assignment	Homework Check
Average	1–3, 5–21 odd, 22–30 even, 27, 33–36	13, 21, 26, 27, 28
Advanced	1–3, 4–18 even, 13, 20, 22–32 even, 33–36	13, 20, 26, 28, 32

Common Errors

- **Exercises 4–12, 14–19** Students may use the wrong special product pattern when factoring the polynomials. Review the special product patterns and encourage students to check their work using the FOIL Method.
- **Exercises 22–25** Students may only factor out the greatest common factor. Encourage them to factor this group of polynomials using several methods.

7.9 Record and Practice Journal

Factor the polynomial.

1. $b^2 - 81$

   $(b + 9)(b - 9)$

2. $16z^2 - 36$

   $(4z + 6)(4z - 6)$

3. $k^2 - 14k + 49$

   $(k - 7)^2$

4. $f^2 + 22f + 121$

   $(f + 11)^2$

Solve the equation.

5. $x^2 - 100 = 0$

   $x = -10, 10$

6. $r^2 + 8r + 16 = 0$

   $r = -4$

7. $25a^2 - 4$

   $a = -\frac{2}{5}, \frac{2}{5}$

8. $p^2 + 169 - 26p$

   $p = 13$

9. A pinecone falls from a tree. The pinecone's height y (in feet) after t seconds can be modeled by $64 - 16t^2$. After how many seconds does the pinecone hit the ground?

   2 sec

Technology For the Teacher

Answer Presentation Tool

Vocabulary and Concept Check

1. Use the difference of two squares pattern to factor $x^2 - 16$. Find the product $(x + 4)(x - 4)$.
2. yes; $a = y$, $b = 8$, and $2ab - 16y$
3. $k^2 + 25$ does not belong because it cannot be factored.

Practice and Problem Solving

4. $(m + 7)(m - 7)$
5. $(3 + r)(3 - r)$
6. $(2x + 5)(2x - 5)$
7. $(9d + 8)(9d - 8)$
8. $(11 + 4t)(11 - 4t)$
9. $(h + 6)^2$
10. $(x - 2)^2$
11. $(w - 7)^2$
12. $(g + 12)^2$
13. The wrong sign is used.

    $n^2 - 16n + 64 = (n - 8)^2$
14. $z = -2, z = 2$
15. $s = -10$
16. $k = 8$
17. $x = -\frac{7}{2}, x = \frac{7}{2}$
18. $n = -3$
19. $y = 6$
20. **a.** no; $w^2 + 18w + 81$

    b. no; $y^2 - 10y + 25$

    c. no; $x^2 - 14x + 49$
21. $d + 4$

Practice and Problem Solving

22. $3(z + 3)(z - 3)$
23. $2m(m + 5)(m - 5)$
24. $x^2(x + 4)^2$
25. $5f(f - 2)^2$
26. a. $x - 15$
    b. $4x - 60$
27. $x = -2, x = 2$; Solve by factoring or using square roots.
28. See *Taking Math Deeper.*
29. $(2y + 1)^2$
30. $(4v - 3)^2$
31. $9(m + 2)^2$
32. a. $x^3 + 16x$
    b. $x = 3$; Solve $x^3 + 16x = 25x$. The only solution that makes sense is $x = 3$.

Fair Game Review

33. $(w + 4)(w - 3)$
34. $(x - 9)(x + 4)$
35. $(d - 10)(d + 6)$
36. C

Mini-Assessment

Factor the polynomial.

1. $25 - 4x^2$ $(5 - 2x)(5 + 2x)$
2. $n^2 + 2n + 1$ $(n + 1)^2$
3. $y^2 - 20y + 100$ $(y - 10)^2$

Solve the equation.

4. $49 = 9k^2$ $k = -\frac{7}{3}, k = \frac{7}{3}$
5. $t^2 - 16t = -64$ $t = 8$

Taking Math Deeper

Exercise 28

Artists sometimes put several pictures together to form a larger picture like the one shown.

a. The total area of the 9 picture frames is $9x^2$.

If you subtract the interiors, you obtain

$$9x^2 - 9(4^2) = 9x^2 - 144.$$

b. The area in part (a) is 81 square inches.

$$9x^2 - 144 = 81$$
$$9x^2 - 225 = 0$$
$$9(x^2 - 25) = 0$$
$$x^2 - 25 = 0$$
$$(x + 5)(x - 5) = 0$$
$$x = -5 \text{ or } x = 5$$

Choosing the positive solution, you can conclude that the side length of one of the frames is 5 inches.

Another way to solve the equation in part (b) is to use a square root.

Reteaching and Enrichment Strategies

If students need help...	If students got it...
Resources by Chapter • Practice A and Practice B • Puzzle Time Record and Practice Journal Practice Differentiating the Lesson Lesson Tutorials Skills Review Handbook	Resources by Chapter • Enrichment and Extension • School-to-Work • Financial Literacy • Technology Connection • Life Connections • Stories in History Start the next section

Factor the polynomial.

22. $3z^2 - 27$

23. $2m^3 - 50m$

24. $x^4 + 8x^3 + 16x^2$

25. $5f^3 - 20f^2 + 20f$

26. **PROBLEM SOLVING** The polynomial represents the area (in square feet) of the square playground.

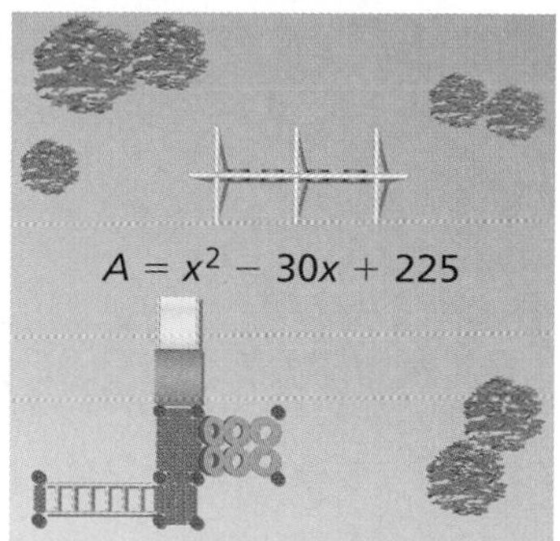

   a. Write a polynomial that represents the side length of the playground.

   b. Write an expression for the perimeter of the playground.

27. **NUMBER SENSE** Solve $28 = 64 - 9x^2$ in two ways.

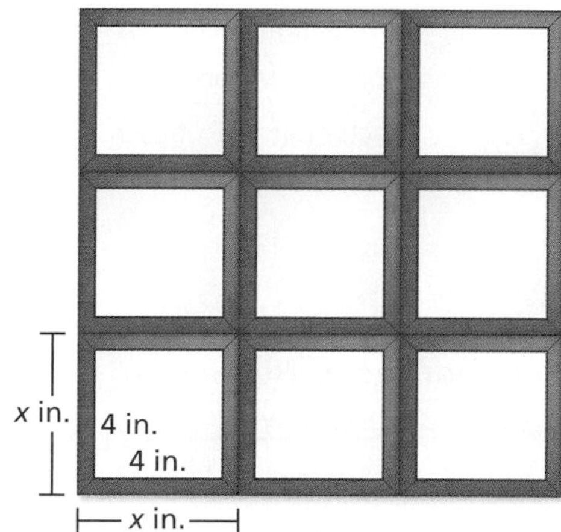

28. **INTERIOR DESIGN** You hang 9 identical square picture frames on a wall.

   a. Write a polynomial that represents the area of the picture frames, not including the pictures.

   b. The area in part (a) is 81 square inches. What is the side length of one of the picture frames? Explain your reasoning.

Factor the polynomial.

29. $4y^2 + 4y + 1$

30. $16v^2 - 24v + 9$

31. $9m^2 + 36m + 36$

32. Geometry A composite solid is made up of a cube and a rectangular prism.

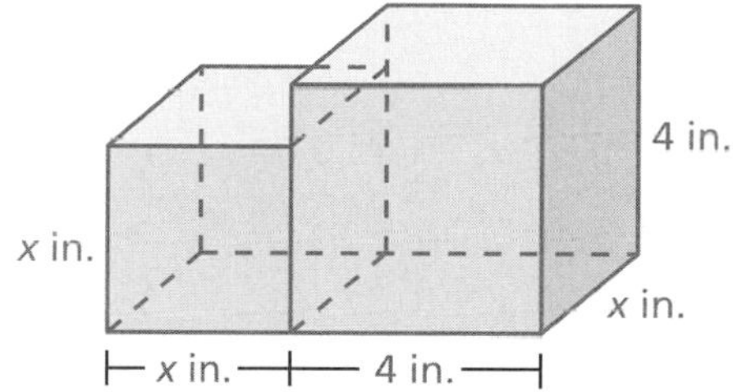

   a. Write a polynomial that represents the volume of the composite solid.

   b. The volume of the composite solid is equal to $25x$. What is the value of x? Explain your reasoning.

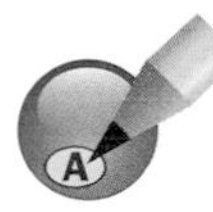

Fair Game Review What you learned in previous grades & lessons

Factor the polynomial. *(Section 7.7)*

33. $w^2 + w - 12$

34. $x^2 - 5x - 36$

35. $d^2 - 4d - 60$

36. **MULTIPLE CHOICE** You deposit $3000 in a savings account. The account earns 4% simple interest per year. What is the balance after 2 years? *(Skills Review Handbook)*

   Ⓐ $240   Ⓑ $3000   Ⓒ $3240   Ⓓ $5400

7.9b Factoring Polynomials Completely

Key Vocabulary
factoring by grouping, *p. 388*
prime polynomial, *p. 389*
factored completely, *p. 389*

To factor polynomials with four terms, group the terms into pairs, factor the GCF out of each pair of terms, and look for a common binomial factor. This process is called **factoring by grouping.**

EXAMPLE 1 Factoring by Grouping

Factor each polynomial.

a. $x^3 + 3x^2 + 2x + 6$

$x^3 + 3x^2 + 2x + 6 = (x^3 + 3x^2) + (2x + 6)$ — Group terms with common factors.

Common binomial factor is $x + 3$. → $= x^2(x + 3) + 2(x + 3)$ — Factor out GCF of each pair of terms.

$= (x + 3)(x^2 + 2)$ — Factor out $(x + 3)$.

b. $x^3 - 7 - x^2 + 7x$

The terms x^3 and -7 do not have a common factor. Rearrange the terms of the polynomial so you can group terms with common factors.

$x^3 - 7 - x^2 + 7x = x^3 - x^2 + 7x - 7$ — Rewrite polynomial.

$= (x^3 - x^2) + (7x - 7)$ — Group terms with common factors.

Common binomial factor is $x - 1$. → $= x^2(x - 1) + 7(x - 1)$ — Factor out GCF of each pair of terms.

$= (x - 1)(x^2 + 7)$ — Factor out $(x - 1)$.

c. $x^2 + y + x + xy$

$x^2 + y + x + xy = x^2 + x + xy + y$ — Rewrite polynomial.

$= (x^2 + x) + (xy + y)$ — Group terms with common factors.

$= x(x + 1) + y(x + 1)$ — Factor out GCF of each pair of terms.

$= (x + 1)(x + y)$ — Factor out $(x + 1)$.

Practice

Factor the polynomial by grouping.

1. $n^3 + 2n^2 + 5n + 10$ **2.** $p^3 - 7p^2 + 3p - 21$ **3.** $2y^3 + 8y^2 + 3y + 12$

4. $6s^3 - 16s^2 + 21s - 56$ **5.** $8v^3 + 48v - 5v^2 - 30$ **6.** $2w^3 - w^2 - 18w + 9$

7. $x^2 + xy + 3x + 3y$ **8.** $a - ab + a^2 - b$ **9.** $4xy + 20y + 3x + 15$

Multi-Language Glossary at BigIdeasMath.com.

Laurie's Notes

Introduction

Connect

- **Yesterday:** Students factored special products. (MP6)
- **Today:** Students will factor polynomials completely.

Motivate

- Today's lesson involves factoring polynomials. A quick warm-up with factoring out greatest common factors (GCF) would be helpful.
- Ask students to factor the GCF out of each of the following polynomials:
  $3x^2 - 15x = 3x(x - 5)$ $\quad -16x^3 - 8x^2 - 24x = -8x(2x^2 + x + 3)$
  $30x^2 + 5x = 5x(6x + 1)$

Lesson Notes

Discuss

- **MP7 Look for and Make Use of Structure:** All of the following expressions have the same structure. Each expression has two terms and can be simplified using the Distributive Property.
  $ab + ac \qquad a(x + 2) + b(x + 2) \qquad a(x - 4) - b(x - 4)$
- While factoring out a common binomial factor is no different from factoring out a common monomial factor, students do not always recognize the underlying structure. To help, use colors and underlining.

Example 1

? "How many terms does $x^3 + 3x^2 + 2x + 6$ have?" 4 "What is the degree and the leading coefficient?" 3; 1 "Do you think it is possible to factor this polynomial into two or more factors?" Answers will vary.

- Group (put parentheses around) terms with common factors.
- Write the next step using color and underlining: $\underline{x^2}(\underline{x + 3}) + \underline{2}(\underline{x + 3})$
- This is the point where it is important for students to view this expression as the sum of two terms that have a common factor. The common factor $(x + 3)$ can be factored out just like a monomial, leaving a new factor of $(x^2 + 2)$. The polynomial in part (b) is factored in a similar way.
- **Common Misconception:** In part (b), when factoring $x^3 - x^2$, students should recognize a common factor of x^2. When asked what is left after factoring, some may say "x − nothing." They do not immediately see $(x - 1)$. To help, refer to the Distributive Property.
- The polynomial in part (c) appears different to students because of the two variables, and because terms with common factors are not adjacent.

? "What property is used to rewrite $x^2 + y + x + xy$?" Commutative Property

- Group (put parentheses around) terms with common factors. Point out how it is similar to part (b).

Practice

- In Exercise 5, students may want to write the subtraction as "add the opposite" so the terms can be rearranged. In Exercise 6, adding 9 can also be written as subtract −9.

Goal Today's lesson is factoring polynomials completely.

Start Thinking! and Warm Up

Lesson 7.9b **Warm Up**
For use before Lesson 7.9b

Lesson 7.9b **Start Thinking!**
For use before Lesson 7.9b

Is the following polynomial *factored completely*? Explain why or why not.

$3x^2 + 9x + 6 = (3x + 3)(x + 2)$

Extra Example 1

Factor each polynomial by grouping.

a. $x^2 + 5x + 2x + 10$
$(x + 2)(x + 5)$

b. $6x^2 - 4x - 9x + 6$
$(2x - 3)(3x - 2)$

c. $dx + cx - dy - cy$
$(x - y)(d + c)$

Practice

1. $(n + 2)(n^2 + 5)$
2. $(p - 7)(p^2 + 3)$
3. $(y + 4)(2y^2 + 3)$
4. $(3s - 8)(2s^2 + 7)$
5. $(8v - 5)(v^2 + 6)$
6. $(w + 3)(w - 3)(2w - 1)$
7. $(x + 3)(x + y)$
8. $(a + 1)(a - b)$
9. $(4y + 3)(x + 5)$

Record and Practice Journal Practice
See Additional Answers.

Extra Example 2

Factor each polynomial completely, if possible.

a. $2x^3 - 2x^2 - 12x$
$2x(x + 2)(x - 3)$

b. $2x^2 - 18$
$2(x + 2)(x - 3)$

c. $x^2 - 3x + 5$
cannot be factored

Extra Example 3

Solve $3a^3 + 14a^2 = -8a$.

$a = 0, a = -\frac{2}{3}, a = -4$

Practice

10. $2x(x - 3)(x + 8)$

11. $5z^2(z + 1)(z - 1)$

12. $4c(c - 1)(c - 5)$

13. cannot be factored

14. cannot be factored

15. $3n^2(n + 4)(n - 4)$

16. $k = 0, k = 3$

17. $x = -6, x = 0, x = 4$

18. $y = -2, y = 0, y = 5$

Mini-Assessment

Factor the polynomial completely, if possible.

1. $p^3 - 6p^2 - 2p + 12$
$(p - 6)(p^2 - 2)$

2. $6t^2 - 48t + 72$
$6(t - 2)(t - 6)$

3. $5x^2 - 30x + 4$
cannot be factored

4. Solve $z^2 - 14 = -5z$.
$z = 2, z = -7$

Laurie's Notes

Discuss

- Define prime polynomial. Discuss the meaning of *factoring a polynomial completely.* Doing some examples may be helpful in terms of making sense of the definition.

Example 2

- For parts (a) and (b), review factoring polynomials into two binomials.
- **Common Error:** Students may see a factor but it won't be the greatest common factor. Always ask, "Are there any other factors?" each time common factors are suggested.
- Work through each example as shown.

Example 3

? "What is the first step in solving this equation?" Collect the terms on one side of the equal sign.

? "Are there any common factors?" yes; Some students may say x, some may say 2, and some may say the GCF, $2x$.

? "Now that the polynomial has been factored, how do you solve?" Set each factor equal to 0. "What property is this?" Zero-Product Property

- **Common Error:** There are three solutions, not two. Remember that $2x = 0$ gives the solution $x = 0$.
- **Note:** If the 2 is not factored out, then the polynomial is not factored completely. However, the solutions will still be the same. If the 2 is not factored out, then the factors will be $x(2x + 10)(x - 1)$, or $x(x + 5)(2x - 2)$. The solutions are still $x = -5$, $x = 0$, and $x = 1$.

Practice

- Students may ask if there is a way to tell if a polynomial is factorable or not. The answer is yes. Later in this book, they will learn new techniques for solving second-degree polynomials. At this stage, tell students to try all combinations suggested by the constant term and the leading coefficient.
- Students may observe that the third-degree equations have (at most) 3 solutions while the second-degree equations have (at most) 2 solutions.

Closure

- **Exit Ticket:** Solve $3x^5 - 6x^4 - 45x^3 = 0$. $x = -3, x = 0, x = 5$

Technology For the Teacher

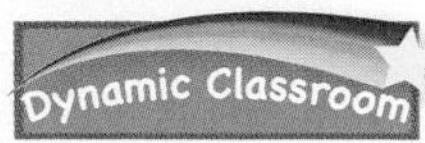

The Dynamic Planning Tool
Editable Teacher's Resources at *BigIdeasMath.com*

A **prime polynomial** is a polynomial that cannot be factored as a product of polynomials with integer coefficients. A factorable polynomial with integer coefficients is said to be **factored completely** when no more factors can be found and it is written as the product of prime factors.

EXAMPLE 2 Factoring Completely

Factor each polynomial completely.

a. $3x^3 - 18x^2 + 24x$

$3x^3 - 18x^2 + 24x = 3x(x^2 - 6x + 8)$ Factor out $3x$.

$= 3x(x - 2)(x - 4)$ Factor $x^2 - 6x + 8$.

b. $7x^4 - 28x^2$

$7x^4 - 28x^2 = 7x^2(x^2 - 4)$ Factor out $7x^2$.

$= 7x^2(x^2 - 2^2)$ Write as $a^2 - b^2$.

$= 7x^2(x + 2)(x - 2)$ Difference of Two Squares Pattern

c. $p^2 + 4p - 2$

The terms of $p^2 + 4p - 2$ have no common factors. There are no integer factors of -2 whose sum is 4. So, this polynomial is already factored completely.

EXAMPLE 3 Solving an Equation by Factoring Completely

$2x^3 + 8x^2 = 10x$ Original equation

$2x^3 + 8x^2 - 10x = 0$ Subtract $10x$ from each side.

$2x(x^2 + 4x - 5) = 0$ Factor out $2x$.

$2x(x + 5)(x - 1) = 0$ Factor $x^2 + 4x - 5$.

$2x = 0$ *or* $x + 5 = 0$ *or* $x - 1 = 0$ Use Zero-Product Property.

$x = 0$ *or* $x = -5$ *or* $x = 1$ Solve for x.

The solutions are $x = -5$, $x = 0$, and $x = 1$.

Practice

Factor the polynomial completely, if possible.

10. $2x^3 + 10x^2 - 48x$

11. $5z^4 - 5z^2$

12. $20c + 4c^3 - 24c^2$

13. $y^2 + 6y - 5$

14. $q^2 - q + 7$

15. $3n^4 - 48n^2$

Solve the equation.

16. $k^3 - 6k^2 + 9k = 0$

17. $3x^3 + 6x^2 = 72x$

18. $4y^3 - 12y^2 - 40y = 0$

7.5–7.9 Quiz

Factor the polynomial. *(Sections 7.6–7.9)*

1. $3d^2 + 11d$

2. $9z^2 - 18z$

3. $x^2 + 9x + 20$

4. $r^2 - 3r - 18$

5. $2x^2 - 3x + 1$

6. $3b^2 - 13b + 4$

7. $x^2 - 9$

8. $z^2 + 22z + 121$

Solve the equation. *(Sections 7.5–7.9)*

9. $m^2 - 11m + 18 = 0$

10. $w^3 - 9w^2 = 0$

11. $6m^2 - 5m + 1 = 0$

12. $h^2 - 8 = -3h + 10$

13. $4s^2 = 144$

14. $k^2 + 100 = 20k$

15. STORAGE The front of a storage bunker can be modeled by $y = -\frac{5}{216}(x - 72)(x + 72)$, where x and y are measured in inches. The x-axis represents the ground. Find the width of the bunker at ground level. *(Section 7.5)*

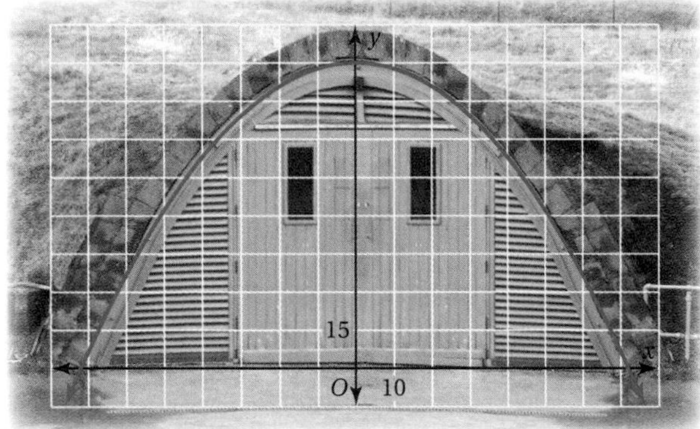

$y = 100 - 16t^2$

16. DISASTER RELIEF A helicopter drops a box of supplies after a disaster. The function represents the height y (in feet) of the box t seconds after it is dropped. After how many seconds does the box hit the ground? *(Section 7.9)*

17. MAGIC SHOW A magician's stage has a trap door. *(Section 7.7)*

a. The total area of the stage can be represented by $x^2 + 27x + 176$. Write an expression for the width of the stage.

b. The area of the trap door is 12 square feet. Find the value of x.

c. What fraction of the area of the stage is the area of the trap door?

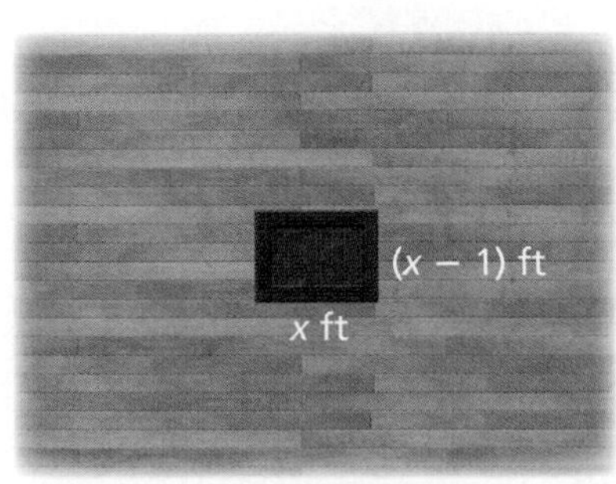

(x + 16) ft

Alternative Assessment Options

Math Chat
Student Reflective Focus Question
Structured Interview
Writing Prompt

Student Reflective Focus Question

Ask students to summarize how to factor polynomials using the different methods presented in the chapter. Be sure that they include examples. Select students at random to present to the class.

Study Help Sample Answers

Remind students to complete Graphic Organizers for the rest of the chapter.

4.

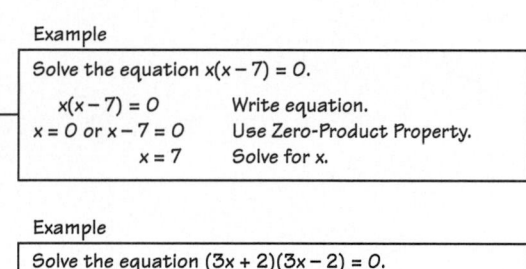

Factored Form of a Polynomial

A polynomial is in factored form when it is written as a product of factors. When one side of an equation is a polynomial in factored form and the other side is 0, you can use the *Zero-Product Property* to solve the polynomial equation.

Example

Solve the equation $x(x - 7) = 0$.

$x(x - 7) = 0$	Write equation.
$x = 0$ or $x - 7 = 0$	Use Zero-Product Property.
$x = 7$	Solve for x.

Example

Solve the equation $(3x + 2)(3x - 2) = 0$.

$(3x + 2)(3x - 2) = 0$	Write equation.
$3x + 2 = 0$ or $3x - 2 = 0$	Use Zero-Product Property.
$x = -\frac{2}{3}$ or $x = \frac{2}{3}$	Solve for x.

5–7. Available at *BigIdeasMath.com.*

Reteaching and Enrichment Strategies

If students need help. . .	If students got it. . .
Resources by Chapter • Study Help • Practice A and Practice B • Puzzle Time Lesson Tutorials *BigIdeasMath.com* Practice Quiz Practice from the Test Generator	Resources by Chapter • Enrichment and Extension • School-to-Work Game Closet at *BigIdeasMath.com* Start the Chapter Review

Answers

1. $d(3d + 11)$
2. $9z(z - 2)$
3. $(x + 4)(x + 5)$
4. $(r - 6)(r + 3)$
5. $(x - 1)(2x - 1)$
6. $(b - 4)(3b - 1)$
7. $(x + 3)(x - 3)$
8. $(z + 11)^2$
9. $m = 2, m = 9$
10. $w = 0, w = 9$
11. $m = \frac{1}{3}, m = \frac{1}{2}$
12. $h = 3, h = -6$
13. $s = -6, s = 6$
14. $k = 10$
15. 144 in.
16. 2.5 sec
17. **a.** $x + 11$
    b. 4
    c. $\frac{1}{25}$

Technology For the Teacher

Answer Presentation Tool

Assessment Book

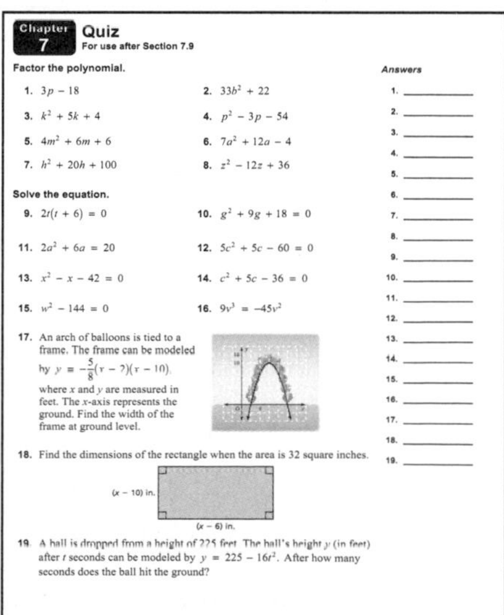

Chapter 7 **Quiz**
For use after Section 7.9

Factor the polynomial.

1. $3p - 18$
2. $33b^2 + 22$
3. $k^2 + 5k + 4$
4. $p^2 - 3p - 54$
5. $4m^2 + 6m + 6$
6. $7a^2 + 12a - 4$
7. $h^2 + 20h + 100$
8. $z^2 - 12z + 36$

Solve the equation.

9. $2t(t + 6) = 0$
10. $g^2 + 9g + 18 = 0$
11. $2a^2 + 6a = 20$
12. $5c^2 + 5c - 60 = 0$
13. $x^2 - x - 42 = 0$
14. $c^2 + 5c - 36 = 0$
15. $w^2 - 144 = 0$
16. $9v^3 = -45v^2$
17. An arch of balloons is tied to a frame. The frame can be modeled by $y = -\frac{5}{8}(x - 2)(x - 10)$, where x and y are measured in feet. The x-axis represents the ground. Find the width of the frame at ground level.
18. Find the dimensions of the rectangle when the area is 32 square inches.

19. A ball is dropped from a height of 225 feet. The ball's height y (in feet) after t seconds can be modeled by $y = 225 - 16t^2$. After how many seconds does the ball hit the ground?

Answers

1. ______
2. ______
3. ______
4. ______
5. ______
6. ______
7. ______
8. ______
9. ______
10. ______
11. ______
12. ______
13. ______
14. ______
15. ______
16. ______
17. ______
18. ______
19. ______

For the Teacher
Additional Review Options

- Big Ideas Test Generator
- Game Closet at *BigIdeasMath.com*
- Vocabulary Puzzle Builder
- Resources by Chapter
  - Puzzle Time
  - Study Help

Answers

1. $2w^3 - 4w + 3$; 3; trinomial
2. $-6y^2$; 2; monomial
3. $3t^5 - 6.2$; 5; binomial
4. $4a + 6$
5. $7x^2 + 4x + 4$
6. $-2y^2 + 6y + 4$
7. $8p^2 + p - 16$

Review of Common Errors

- **Exercises 1–3** Students may look at the first term of the given polynomial to identify the degree. Remind them that the degree of a polynomial is the greatest degree of its terms.
- **Exercises 4 and 5** When using the vertical method to add polynomials, students may forget to leave a space for missing terms. Remind them that they cannot combine variable terms with different exponents.
- **Exercises 6 and 7** Students may add the opposite of only the first term of the second polynomial. Remind them to add the opposite of all of the terms of the second polynomial.

7 Chapter Review

Check It Out
Vocabulary Help
BigIdeasMath.com

Review Key Vocabulary

monomial, *p. 330*
degree of a monomial, *p. 330*
polynomial, *p. 331*
binomial, *p. 331*
trinomial, *p. 331*
degree of a polynomial, *p. 331*
FOIL Method, *p. 343*
factored form, *p. 358*
Zero-Product Property, *p. 358*
root, *p. 358*
factoring by grouping, *p. 388*
prime polynomial, *p. 389*
factored completely, *p. 389*

Review Examples and Exercises

7.1 Polynomials *(pp. 328–333)*

a. Find the degree of $4x^2y$.

The exponent of x is 2 and the exponent of y is 1.
The sum of the exponents is $2 + 1 = 3$.

∴ So, the degree of the monomial is 3.

b. Write $x + 1 + 2x^3$ in standard form. Identify the degree and classify the polynomial by the number of terms.

Polynomial	*Standard Form*	*Degree*	*Type of Polynomial*
$x + 1 + 2x^3$	$2x^3 + x + 1$	3	trinomial

Exercises

Write the polynomial in standard form. Identify the degree and classify the polynomial by the number of terms.

1. $2w^3 + 3 - 4w$ **2.** $-6y^2$ **3.** $-6.2 + 3t^5$

7.2 Adding and Subtracting Polynomials *(pp. 334–339)*

a. $(2d^2 - 3) + (4d^2 + 2)$

$$(2d^2 - 3) + (4d^2 + 2) = (2d^2 + 4d^2) + (-3 + 2)$$
$$= 6d^2 - 1$$

b. $(c^2 + 5c + 1) - (c^2 - 2)$

$$(c^2 + 5c + 1) - (c^2 - 2) = (c^2 + 5c + 1) + (-c^2 + 2)$$
$$= [c^2 + (-c^2)] + 5c + (1 + 2) = 5c + 3$$

Exercises

Find the sum or difference.

4. $(3a + 7) + (a - 1)$ **5.** $(x^2 + 4x - 2) + (6x^2 + 6)$

6. $(-y^2 + y + 2) - (y^2 - 5y - 2)$ **7.** $(p - 9) - (-8p^2 + 7)$

7.3 Multiplying Polynomials (pp. 340–347)

Find $(x + 1)(x - 4)$.

First Outer Inner Last

$$(x + 1)(x - 4) = x(x) + x(-4) + (1)(x) + (1)(-4)$$ Use the FOIL Method.

$$= x^2 + (-4x) + (x) + (-4)$$ Multiply.

$$= x^2 - 3x - 4$$ Combine like terms.

Exercises

Find the product.

8. $(y + 4)(y - 2)$ **9.** $(q - 3)(2q + 7)$ **10.** $(-3v + 1)(v^2 - v - 2)$

7.4 Special Products of Polynomials (pp. 348–353)

Find each product.

a. $(x + 3)(x - 3)$

$$(a + b)(a - b) = a^2 - b^2$$ Sum and Difference Pattern

$$(x + 3)(x - 3) = x^2 - 3^2$$ Use pattern.

$$= x^2 - 9$$ Simplify.

b. $(y + 2)^2$

$$(a + b)^2 = a^2 + 2ab + b^2$$ Square of a Binomial Pattern

$$(y + 2)^2 = y^2 + 2(y)(2) + 2^2$$ Use pattern.

$$= y^2 + 4y + 4$$ Simplify.

Exercises

Find the product.

11. $(y + 9)(y - 9)$ **12.** $(2x + 4)(2x - 4)$

13. $(h + 4)^2$ **14.** $(-1 + 2d)^2$

7.5 Solving Polynomial Equations in Factored Form (pp. 356–361)

Solve $(x + 4)(x - 3) = 0$.

$$(x + 4)(x - 3) = 0$$ Write equation.

$$x + 4 = 0 \quad or \quad x - 3 = 0$$ Use Zero-Product Property.

$$x = -4 \quad or \quad x = 3$$ Solve for x.

The roots are $x = -4$ and $x = 3$.

Review of Common Errors (continued)

- **Exercises 8 and 9** Students may forget to distribute the second factor to both terms of the first factor. Remind them of this process.
- **Exercises 8 and 9** Students may forget to perform one or more of the steps in the FOIL Method. Encourage them to draw arrows for each part of the method.
- **Exercises 8–10** Students may not include negative signs when multiplying. Encourage them to write subtraction expressions as equivalent addition expressions before multiplying.
- **Exercises 8–10** Students may forget to align like terms when using a vertical method to multiply polynomials. Remind them that you cannot combine terms that are not alike.
- **Exercises 11 and 12** Students may use an addition sign in the product instead of a subtraction sign. Remind them that the sum and difference pattern is the difference of two squares not the sum. Demonstrate this by using the FOIL Method to multiply.
- **Exercises 13 and 14** Students may use the sum and difference pattern instead of the square of a binomial pattern when multiplying. Review the patterns with them. Encourage them to check their answers using the FOIL Method.

Answers

8. $y^2 + 2y - 8$

9. $2q^2 + q - 21$

10. $-3v^3 + 4v^2 + 5v - 2$

11. $y^2 - 81$

12. $4x^2 - 16$

13. $h^2 + 8h + 16$

14. $4d^2 - 4d + 1$

Answers

15. $x = 0, x = -2$

16. $t = 3, t = 8$

17. $a = -10$

18. $s = 0, s = -1, s = 4$

19. $6(t^2 + 6)$

20. $2x(x - 10)$

21. $3y^2(5y + 1)$

Review of Common Errors (continued)

- **Exercises 15–18** Students may set only one of the factors of the polynomial equal to zero. Remind them that they need to set all factors equal to zero when using the Zero-Product Property.
- **Exercises 19–21** Students may factor out only one or two of the common factors instead of the greatest common factor. Encourage them to rewrite each term using its prime factorization. This will make it easier to identify the greatest common factor of the terms.
- **Exercises 22–24** Students may forget to check that the factors of c add up to b. Remind them that there are two steps to factoring polynomials, factoring c and checking that those factors add up to b. Encourage students to use tables as shown in the examples.
- **Exercises 25–29** Some students will choose factors of a and c that do not add up to b when the binomial factors are multiplied. Remind students that the product of the binomials must equal the original polynomial.
- **Exercise 28** Remind students to factor out a -1 before factoring polynomials with a negative leading coefficient, and to include it after they have factored.
- **Exercises 30–33** Students may use the wrong special product pattern when factoring the polynomials. Review the special product patterns and encourage students to check their work using the FOIL Method.
- **Exercises 34 and 35** Students may only factor out the greatest common factor. Encourage them to factor this group of polynomials using several methods.

Exercises

Solve the equation.

15. $x(x + 2) = 0$

16. $(t - 3)(t - 8) = 0$

17. $(a + 10)^2 = 0$

18. $2s(s + 1)(s - 4) = 0$

7.6 Factoring Polynomials Using the GCF (pp. 362–367)

Factor $4z^2 + 32$.

Step 1: Find the GCF of the terms.

$$4z^2 = 2 \cdot 2 \cdot z \cdot z$$
$$32 = 2 \cdot 2 \cdot 2 \cdot 2 \cdot 2$$

The GCF is $2 \cdot 2 = 4$.

Step 2: Write the polynomial as a product of the GCF and its remaining factors.

$$4z^2 + 32 = 4(z^2) + 4(8) \quad \text{Factor out GCF.}$$
$$= 4(z^2 + 8) \quad \text{Distributive Property}$$

Exercises

Factor the polynomial.

19. $6t^2 + 36$

20. $2x^2 - 20x$

21. $15y^3 + 3y^2$

7.7 Factoring $x^2 + bx + c$ (pp. 368–375)

Factor $x^2 + 12x + 27$.

Notice that $b = 12$ and $c = 27$.

- Because c is positive, the factors p and q must have the same sign so that pq is positive.
- Because b is also positive, p and q must each be positive so that $p + q$ is positive.

Find two positive integer factors of 27 whose sum is 12.

Factors of 27	Sum of Factors
1, 27	28
3, 9	12

The values of p and q are 3 and 9.

So, $x^2 + 12x + 27 = (x + 3)(x + 9)$.

Exercises

Factor the polynomial.

22. $p^2 + 2p - 35$ **23.** $b^2 + 9b + 20$ **24.** $z^2 - 4z - 21$

7.8 Factoring $ax^2 + bx + c$ *(pp. 376–381)*

a. Factor $2x^2 + 13x + 15$.

Consider the possible factors of $a = 2$ and $c = 15$.

Factors are 1 and 2. → $2x^2 + 13x + 15$ ← Factors are 1, 3, 5, and 15.

These factors lead to the following possible products.

$(1x + 1)(2x + 15)$ $(1x + 3)(2x + 5)$

$(1x + 15)(2x + 1)$ $(1x + 5)(2x + 3)$

Multiply to find the product that is equal to the original polynomial.

$(x + 1)(2x + 15) = 2x^2 + 17x + 15$ ✗

$(x + 15)(2x + 1) = 2x^2 + 31x + 15$ ✗

$(x + 3)(2x + 5) = 2x^2 + 11x + 15$ ✗

$(x + 5)(2x + 3) = 2x^2 + 13x + 15$ ✓

So, $2x^2 + 13x + 15 = (x + 5)(2x + 3)$.

b. Factor $5x^2 + 4x - 9$.

Consider the possible factors of $a = 5$ and $c = -9$. Because b is positive and c is negative, the factors of c must have different signs.

Factors are 1 and 5. → $5x^2 + 4x - 9$ ← Factors are ±1, ±3, and ±9.

These factors lead to the following possible products.

$(1x + 1)(5x - 9)$ $(1x - 1)(5x + 9)$ $(1x - 3)(5x + 3)$

$(1x + 9)(5x - 1)$ $(1x - 9)(5x + 1)$ $(1x + 3)(5x - 3)$

Multiply to find the product that is equal to the original polynomial.

$(x + 1)(5x - 9) = 5x^2 - 4x - 9$ ✗

$(x + 9)(5x - 1) = 5x^2 + 44x - 9$ ✗

$(x - 1)(5x + 9) = 5x^2 + 4x - 9$ ✓

$(x - 9)(5x + 1) = 5x^2 - 44x - 9$ ✗

$(x - 3)(5x + 3) = 5x^2 - 12x - 9$ ✗

$(x + 3)(5x - 3) = 5x^2 + 12x - 9$ ✗

So, $5x^2 + 4x - 9 = (x - 1)(5x + 9)$.

Review Game

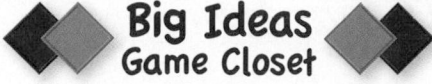

Polynomial Concentration

Big Ideas Game Closet

Materials:

- index cards

Directions:

Create playing cards by writing a polynomial on one index card and its factored form on a different index card. Make sure the number of "pairings" is an odd number so a winner can be determined. For example, there are 18 sample playing cards shown below, but only 9 pairings.

Students can play one-on-one, or team versus team. Separate the cards into two categories: "Polynomial" and "Factored Form." Lay each card face down in distinct rows and columns for its respective category. In turn, each player chooses one "Polynomial" card and one "Factored Form" card and turns them face up. If the cards are equivalent, that player wins the pair and plays again. If the cards are not equivalent, they are turned face down and play passes to the other player/team. The game ends when the last pair has been picked up.

NOTE: You can extend this game to more than two players or teams by creating a large number of playing cards and passing the turn to the left using the same rules as above. There may be a tie for first place.

Sample Playing Cards:

$x^2 + 7x + 10$	$x^2 - 7x + 10$	$x^2 - 25$
$(x + 2)(x + 5)$	$(x - 2)(x - 5)$	$(x - 5)(x + 5)$
$x^2 + 7x + 12$	$x^2 - x - 12$	$x^2 + 6x + 9$
$(x + 3)(x + 4)$	$(x + 3)(x - 4)$	$(x + 3)(x + 3)$
$2x^2 - 4x - 16$	$2x^2 + 2x - 12$	$2x^2 - 2x - 40$
$2(x + 2)(x - 4)$	$2(x + 3)(x - 2)$	$2(x + 4)(x - 5)$

Who wins?

The student or group with the most pairs of playing cards at the end of the game wins.

For the Student Additional Practice

- Lesson Tutorials
- Study Help (textbook)
- Student Website
  - Multi-Language Glossary
  - Practice Assessments

Answers

22. $(p + 7)(p - 5)$

23. $(b + 4)(b + 5)$

24. $(z + 3)(z - 7)$

25. $(2a + 1)(5a + 3)$

26. $(z + 2)(4z + 3)$

27. $(x - 14)(2x + 1)$

28. $-2(p - 2)(p + 1)$

29. $(8x + 1)$ ft by $(x + 4)$ ft

30. $(x - 3)(x + 3)$

31. $(y - 10)(y + 10)$

32. $(z + 3)^2$

33. $(m + 8)^2$

34. $(x - 3)(x + 4a)$

35. $n(n + 3)(n - 3)$

My Thoughts on the Chapter

What worked. . .

Teacher Tip

Not allowed to write in your teaching edition? Use sticky notes to record your thoughts.

What did not work. . .

What I would do differently. . .

Exercises

Factor the polynomial.

25. $10a^2 + 11a + 3$

26. $4z^2 + 11z + 6$

27. $2x^2 - 27x - 14$

28. $-2p^2 + 2p + 4$

29. OUTSIDE PATIO You are installing new tile on an outside patio. The area (in square feet) of the rectangular patio can be represented by $8x^2 + 33x + 4$. Write the expressions that represent the dimensions of the patio.

7.9 Factoring Special Products *(pp. 382–389)*

Factor each polynomial.

a. $x^2 - 16$

$x^2 - 16 = x^2 - 4^2$ — Write as $a^2 - b^2$.

$= (x + 4)(x - 4)$ — Difference of Two Squares Pattern

b. $x^2 - 2x + 1$

$x^2 - 2x + 1 = x^2 - 2(x)(1) + 1^2$ — Write as $a^2 - 2ab + b^2$.

$= (x - 1)^2$ — Perfect Square Trinomial Pattern

c. $x^3 + 4x^2 + 3x + 12$

$x^3 + 4x^2 + 3x + 12 = (x^3 + 4x^2) + (3x + 12)$ — Group terms with common factors.

$= x^2(x + 4) + 3(x + 4)$ — Factor out GCF of each pair of terms.

Common binomial factor is $x + 4$.

$= (x + 4)(x^2 + 3)$ — Factor out $(x + 4)$.

d. $2x^4 - 8x^2$

$2x^4 - 8x^2 = 2x^2(x^2 - 4)$ — Factor out $2x^2$.

$= 2x^2(x^2 - 2^2)$ — Write as $a^2 - b^2$.

$= 2x^2(x + 2)(x - 2)$ — Difference of Two Squares Pattern

Exercises

Factor the polynomial.

30. $x^2 - 9$

31. $y^2 - 100$

32. $z^2 + 6z + 9$

33. $m^2 + 16m + 64$

34. $x^2 - 3x + 4ax - 12a$

35. $n^3 - 9n$

7 Chapter Test

Write the polynomial in standard form. Identify the degree and classify the polynomial by the number of terms.

1. $-2.1w^3$
2. $7k + 4 - 3k^2$
3. $-c^8 + 9c^{12}$

Find the sum or difference.

4. $(-2p + 4) - (p^2 - 6p + 8)$
5. $(4s^2 + 2st + t) + (-3s^2 + 5st - 4t)$

Find the product.

6. $(h - 5)(h - 8)$
7. $(2w - 3)(2w + 5)$
8. $(z + 11)(z - 11)$

Factor the polynomial.

9. $7x^2 - 21x$
10. $n^2 + 7n + 10$
11. $m^2 - 2m - 24$
12. $6g^2 + 23g + 7$
13. $y^2 - 100$
14. $b^3 - 2b^2 + 3b - 6$

Solve the equation.

15. $(n - 1)(n + 6) = 0$
16. $3h^2 = -12h$
17. $s^2 - 15s + 50 = 0$
18. $5k^2 + 22k - 15 = 0$
19. $d^2 + 14d + 49 = 0$
20. $6x^4 + 8x^2 = 26x^3$

21. **TIME** The expression $\pi(r - 3)^2$ represents the area covered by the hour hand on a clock in one rotation, where r is the radius of the entire clock. Write a polynomial that represents the area covered by the hour hand in one rotation.

22. **TRAMPOLINE** You are jumping on a trampoline. Your height y (in feet) above the trampoline after t seconds can be represented by $y = -16t^2 + 24t$. How many seconds are you in the air?

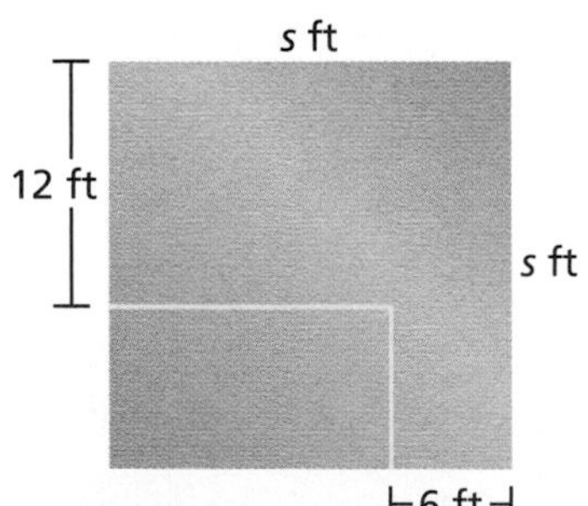

23. **CEMENT** You pour cement in a rectangular region of a square garage. The area of the rectangular region is 112 square feet.
    a. What is the area of the garage floor?
    b. You place caution tape along the two sides of the newly cemented region that are not on the wall. How many feet of caution tape do you use?

24. **ARCHERY** The area (in square inches) of the target can be represented by $\pi(x^2 + 6x + 9)$.
    a. Find the areas of the red bull's eye and the gray ring when the area of the target is 25π square inches. Write your answer in terms of π.
    b. Write a binomial that represents the radius of the target.
    c. What is the width of the gray ring? Does it change as x changes? Does its area change as x changes? Explain.

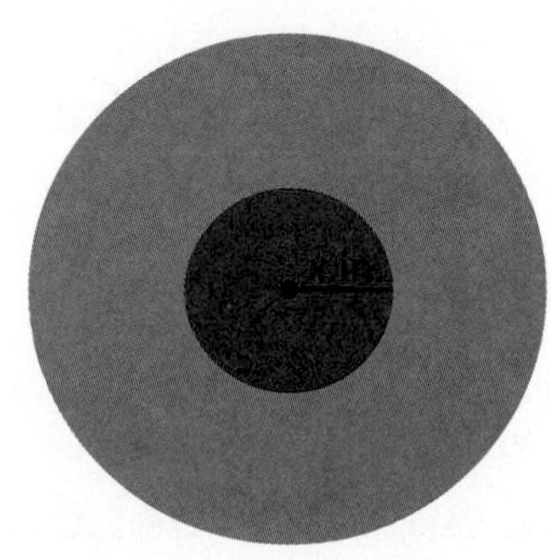

Test Item References

Chapter Test Questions	Section to Review	Common Core State Standards
1–3	7.1	A.SSE.1a
4, 5	7.2	A.APR.1
6, 7	7.3	A.APR.1
8, 21	7.4	A.APR.1
15	7.5	A.REI.4b
9, 16, 22	7.6	A.REI.4b, A.SSE.3a
10, 11, 17, 23	7.7	A.REI.4b, A.SSE.3a
12, 18	7.8	A.REI.4b, A.SSE.3a
13, 14, 19, 20, 24	7.9	A.REI.4b, A.SSE.2, A.SSE.3a

Test-Taking Strategies

Remind students to quickly look over the entire test before they start so that they can budget their time. Have students use the **Stop** and **Think** strategy before they answer each question.

Common Assessment Errors

- **Exercises 1–3** Students may look at the first term of the given polynomial to identify the degree. Remind them that the degree of a polynomial is the greatest degree of its terms.
- **Exercises 4 and 5** Students may have difficulty grouping like terms. Encourage them to look at the variable(s) and exponent(s) in each term.
- **Exercises 6 and 7** Students may forget to perform one or more of the steps in the FOIL Method. Encourage them to draw arrows for each part of the method.
- **Exercise 8** Students may use the square of a binomial pattern instead of the sum and difference pattern when multiplying.
- **Exercises 9–14** Students may use the wrong sign when writing the factors of the polynomial.

Reteaching and Enrichment Strategies

If students need help. . .	If students got it. . .
Resources by Chapter • Practice A and Practice B • Puzzle Time Record and Practice Journal Practice Differentiating the Lesson Lesson Tutorials Practice from the Test Generator Skills Review Handbook	Resources by Chapter • Enrichment and Extension • School-to-Work • Financial Literacy • Technology Connection • Life Connections • Stories in History Game Closet at *BigIdeasMath.com* Start Standardized Test Practice

Answers

1. $-2.1w^3$; 3; monomial
2. $-3k^2 + 7k + 4$; 2; trinomial
3. $9c^{12} - c^8$; 12; binomial
4. $-p^2 + 4p - 4$
5. $s^2 + 7st - 3t$
6. $h^2 - 13h + 40$
7. $4w^2 + 4w - 15$
8. $z^2 - 121$
9. $7x(x - 3)$
10. $(n + 2)(n + 5)$
11. $(m - 6)(m + 4)$
12. $(3g + 1)(2g + 7)$
13. $(y + 10)(y - 10)$
14. $(b - 2)(b^2 + 3)$
15. $n = 1, n = -6$
16. $h = 0, h = -4$
17. $s = 5, s = 10$
18. $k = \frac{3}{5}, k = -5$
19. $d = -7$

20–24. See Additional Answers.

Assessment Book

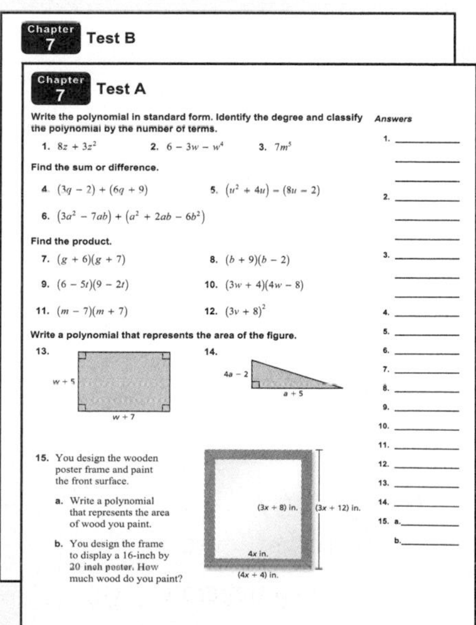
Chapter 7 Test B

Chapter 7 Test A

Write the polynomial in standard form. Identify the degree and classify the polynomial by the number of terms.

1. $8z + 3z^2$ 2. $6 - 3w - w^4$ 3. $7m^5$

Find the sum or difference.

4. $(3q - 2) + (6q + 9)$ 5. $(u^2 + 4u) - (8u - 2)$
6. $(3a^2 - 7ab) + (a^2 + 2ab - 6b^2)$

Find the product.

7. $(g + 6)(g + 7)$ 8. $(b + 9)(b - 2)$
9. $(6 - 5t)(9 - 2t)$ 10. $(3w + 4)(4w - 8)$
11. $(m - 7)(m + 7)$ 12. $(3v + 8)^2$

Write a polynomial that represents the area of the figure.

13. $w + 5$, $w + 7$ 14. $4a - 2$, $a + 5$

15. You design the wooden poster frame and paint the front surface.
 a. Write a polynomial that represents the area of wood you paint.
 b. You design the frame to display a 16-inch by 20 inch poster. How much wood do you paint?

(3x + 8) in. (3x + 12) in. 4x in. (4x + 4) in.

Answers 1. 2. 3. 4. 5. 6. 7. 8. 9. 10. 11. 12. 13. 14. 15. a. b.

Test-Taking Strategies
Available at *BigIdeasMath.com*

After Answering Easy Questions, Relax
Answer Easy Questions First
Estimate the Answer
Read All Choices before Answering
Read Question before Answering
Solve Directly or Eliminate Choices
Solve Problem before Looking at Choices
Use Intelligent Guessing
Work Backwards

About this Strategy
When taking a multiple choice test, be sure to read each question carefully and thoroughly. Before answering a question, determine exactly what is being asked, then eliminate the wrong answers and select the best choice.

Answers
1. C
2. I
3. 4
4. D
5. F

Technology For the Teacher
Big Ideas Test Generator

Item Analysis

1. **A.** The student incorrectly chooses strict inequality.
   B. The student shades the wrong half-plane and incorrectly chooses strict inequality.
   C. Correct answer
   D. The student shades the wrong half-plane.

2. **F.** The student adds exponents instead of multiplying.
   G. The student adds exponents instead of subtracting.
   H. The student adds all three exponents.
   I. Correct answer

3. **Gridded response:** Correct answer: 4

   Common error: The student assumes the polynomial is written in standard form and uses the degree 3 of the first term.

4. **A.** The student calculates the slope incorrectly.
   B. The student calculates the slope incorrectly.
   C. The student calculates the slope incorrectly.
   D. Correct answer

5. **F.** Correct answer
   G. The student forgets to divide by 5.
   H. The student thinks the roots are related to the coefficients of the factors.
   I. The student inverts the solutions.

7 Standardized Test Practice

Test-Taking Strategy

Solve Directly or Eliminate Choices

You are having x cat treats for dinner where $x^2 - x - 6 = 0$. How many is that?

Ⓐ −3 Ⓑ −2 Ⓒ 2 Ⓓ 3

Dinnertime!

"You can eliminate A and B. Then, solve directly to determine that the correct answer is D."

1. Which inequality is shown in the coordinate plane? *(A.REI.12)*

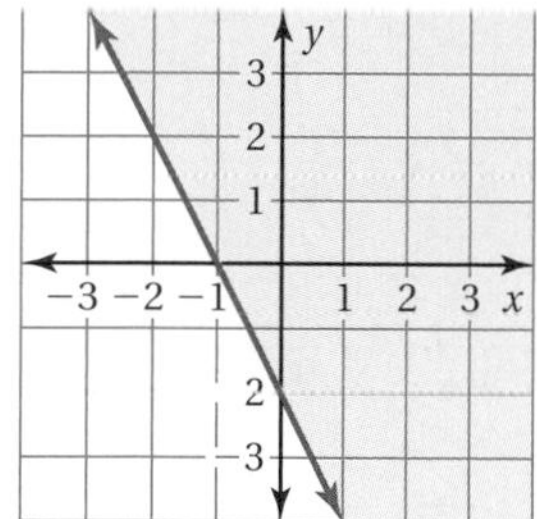

A. $y > -2x - 2$

B. $y < -2x - 2$

C. $y \geq -2x - 2$

D. $y \leq -2x - 2$

2. Which expression is equivalent to $\left(\dfrac{a^3}{a^{-2}}\right)^{-3}$? *(N.RN.2)*

F. a^2

G. $\dfrac{1}{a^3}$

H. $\dfrac{1}{a^2}$

I. $\dfrac{1}{a^{15}}$

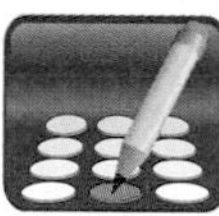

3. What is the degree of the polynomial shown below? *(A.SSE.1a)*

$$p^3 + 2p - 5p^4$$

4. Which of the following is the equation of the line that passes through the points $(-1, -6)$ and $(2, 6)$? *(A.CED.2)*

A. $y = -\dfrac{1}{4}x - \dfrac{25}{4}$

B. $y = -4x - 10$

C. $y = \dfrac{1}{4}x - \dfrac{23}{4}$

D. $y = 4x - 2$

5. What are the roots of $(5b + 3)(5b - 3) = 0$? *(A.REI.4b)*

F. $-\dfrac{3}{5}$ and $\dfrac{3}{5}$

G. -3 and 3

H. -5 and 5

I. $-\dfrac{5}{3}$ and $\dfrac{5}{3}$

6. What is the range of the function graphed in the coordinate plane below? *(F.IF.1)*

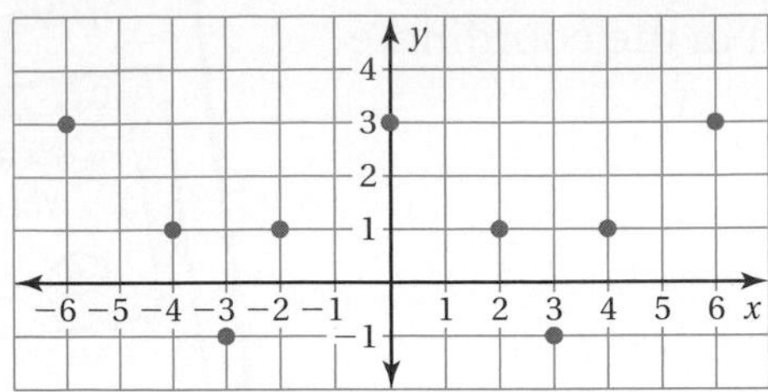

A. 1, 3

B. −1, 1, 3

C. 0, 2, 3, 4, 6

D. −6, −4, −3, −2, 0, 2, 3, 4, 6

7. Which polynomial represents the product of $2x - 4$ and $x^2 + 6x - 2$? *(A.APR.1)*

F. $2x^3 + 8x^2 - 4x + 8$

G. $2x^3 + 8$

H. $2x^3 + 8x^2 - 28x + 8$

I. $2x^3 - 24x - 2$

8. For what value of b does the system of linear equations shown below have no solution? *(8.EE.8a)*

$$y = 6x + 3$$

$$bx - 2y = -10$$

9. The graph of which equation is shown in the coordinate plane? *(F.IF.7e)*

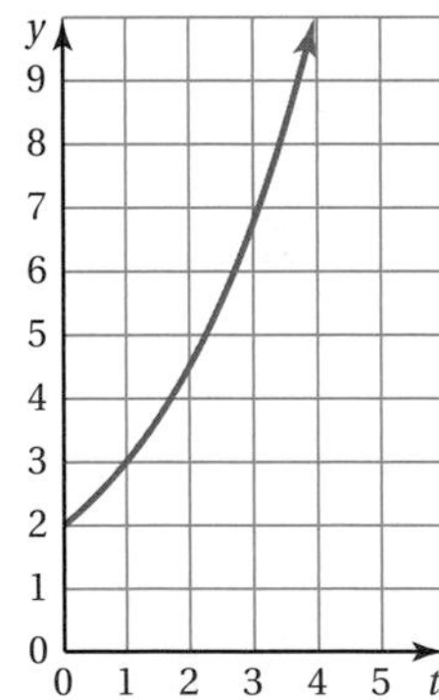

A. $y = 2(1.5)^t$

B. $y = 2^t$

C. $y = 2(0.5)^t$

D. $y = (0.5)^t$

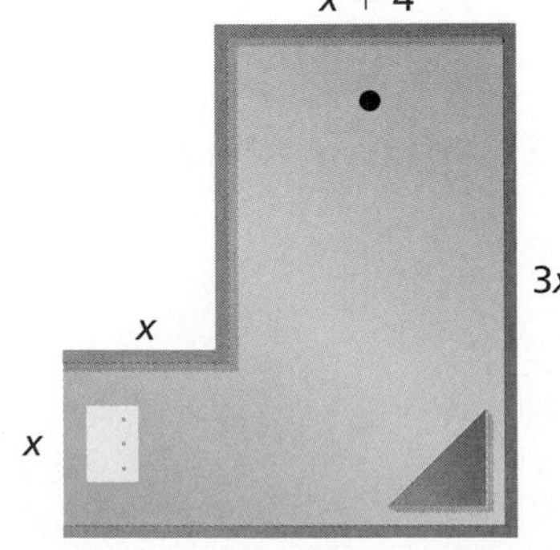

Think Solve Explain

10. The playing area of a hole on a miniature golf course is 216 square feet. What is the perimeter of the playing area? Explain. *(A.REI.4b)*

Item Analysis (continued)

6. **A.** The student lists only the positive range elements.
   B. Correct answer
   C. The student lists the nonnegative domain elements.
   D. The student lists the domain elements.

7. **F.** The student forgets to add $(-24x)$.
   G. The student multiplies the first and last terms only.
   H. Correct answer
   I. The student multiplies the first terms, the second terms, and adds the sole third term.

8. **Gridded response:** Correct answer: 12

   Common error: The student makes a sign error in solving the second equation for y, yielding an answer of -12.

9. **A.** Correct answer
   B. The student assumes that a base of 2 yields the y-intercept of 2.
   C. The student chooses an equation with the correct y-intercept, but the wrong curve.
   D. The student relates the y-intercept of 2 to the equation with a base of 0.5, the reciprocal of 2.

10. **2 points** The student demonstrates a thorough understanding of solving problems involving perimeter and area and solving a quadratic equation by factoring. The student explains fully how this process is used to find the value of x, which is used to find the perimeter.

    1 point The student's work and explanation demonstrate a lack of essential understanding. The student does not apply the factoring methods correctly to find x or fails to use the value of x to find the correct perimeter.

    0 points The student provides no response, a completely incorrect or incomprehensible response, or a response that demonstrates insufficient understanding of solving quadratic equations by factoring or solving problems involving area and perimeter.

Answers

6. B
7. H
8. 12
9. A
10. 68 feet

Answers

11. H

12. C

13. G

Item Analysis (continued)

11. **F.** The student shades the wrong half-plane for the second inequality.

G. The student shades the wrong half-plane for both inequalities.

H. Correct answer

I. The student shades the wrong half-plane for the first inequality.

12. **A.** The student makes a grouping error that changes the order of operations.

B. The student factors the difference of two squares incorrectly.

C. Correct answer

D. The student factors the difference of two squares incorrectly.

13. **F.** The student ignores the alternating signs of the terms.

G. Correct answer

H. The student ignores the alternating signs of the terms and incorrectly finds the ratio of one term to the next term.

I. The student incorrectly finds the ratio of one term to the next term.

Extra Example

1. Which expression is the completely factored form of the polynomial $9x^3 + 12x^2 - 12x$? *(A.SSE.2)*

A. $3x(3x^2 + 4x - 4)$

B. $(3x^2 + 6x)(3x - 2)$

C. $3x(3x + 2)(x - 2)$

D. $3x(3x - 2)(x + 2)$

Answer for Extra Example

1. **A.** The student stops after factoring out the GCF.

B. The student does not factor out the GCF.

C. The student factors the trinomial incorrectly.

D. Correct answer

11. Which graph shows the solution of the system of linear inequalities shown below? *(A.REI.12)*

$$2x + 3y > 6$$
$$2x - y \leq -2$$

F.

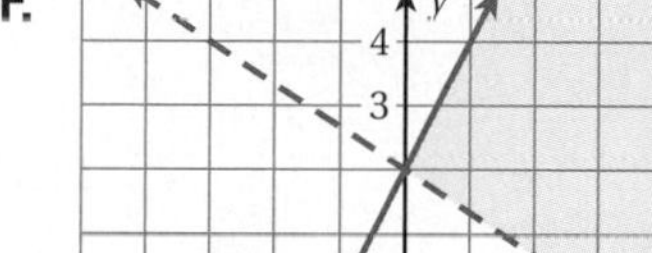

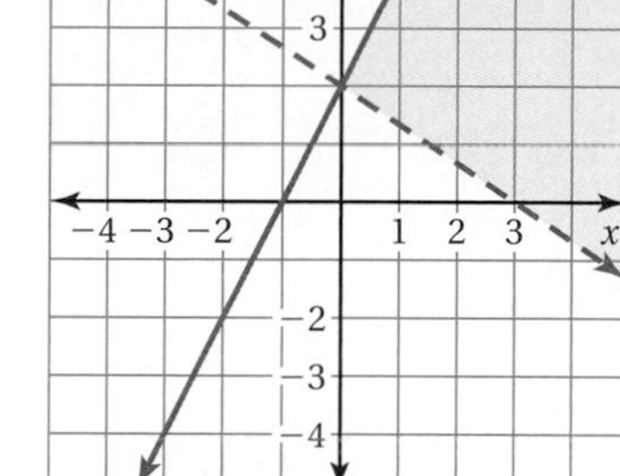

H.

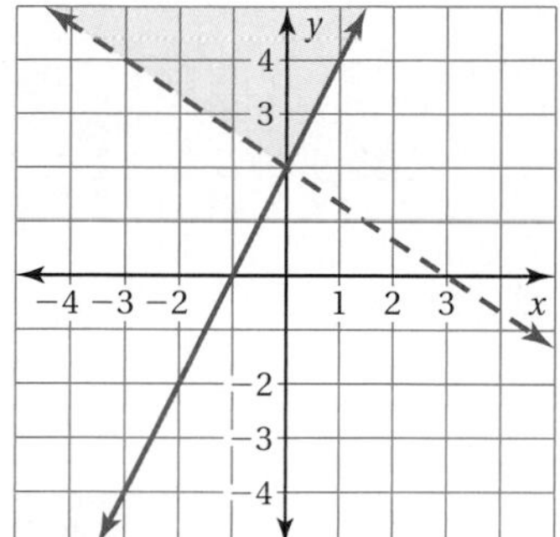

G.

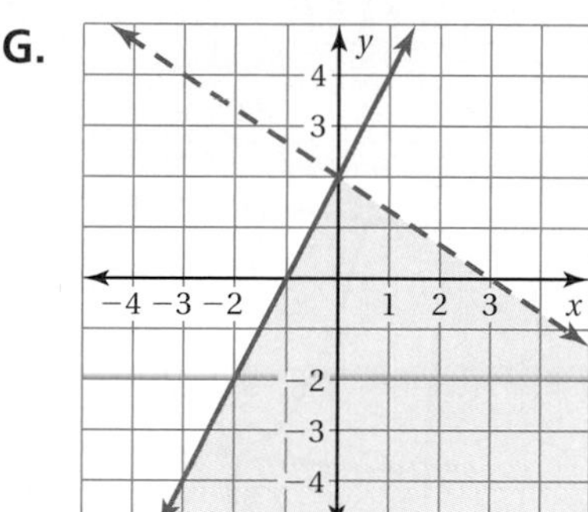

I.

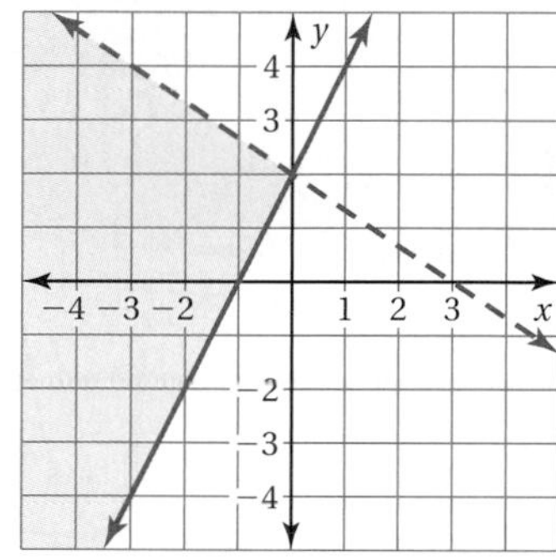

12. Andy was factoring the polynomial in the box below. *(A.SSE.2)*

$$16t^2 - 49 = 4t^2 - 7^2$$
$$= (2t + 7)(2t - 7)$$

What should Andy do to correct the error that he made?

A. Rewrite $16t^2$ as $(16t)^2$.

B. Rewrite $4t^2 - 7^2$ as $(2t - 7)^2$.

C. Rewrite $16t^2$ as $(4t)^2$.

D. Rewrite $16t^2 - 49$ as $(4t - 7)^2$.

13. What is the common ratio of the sequence 243, −81, 27, −9, . . .? *(F.LE.2)*

F. $\frac{1}{3}$

G. $-\frac{1}{3}$

H. 3

I. −3

8 Graphing Quadratic Functions

"What type of graph is this?"

"Sorry, no it's the 2nd degree."

"Let's demonstrate how changing the value of c changes the graph of $y = x^2 + c$."

"Good, now when I change $c = 0$ to $c = 3$ you jump up."

Connections to Previous Learning

- Solve unit rate problems.
- Use equations, tables, and graphs to express and analyze linear relationships between independent variables and dependent variables.

- Describe the effect of transformations on two-dimensional figures.
- Analyze proportional relationships using tables, graphs, equations, and so on, interpreting the unit rate as the slope of the graph.
- Describe the relationship between a line and its equation.

- Graph quadratic functions and interpret key features of their graphs.
- Describe how various changes in a quadratic equation affect its graph.
- Compare the rates of change of linear, exponential, and quadratic functions using graphs and tables.

Math in History

The parabola and its basic geometric properties were discovered by Greek mathematicians. Many applications of parabolas were discovered in the 16th and 17th centuries A.D.

★ Greek mathematician Menaechmus (born about 380 B.C.) discovered the parabola. He was a student of Plato and Eudoxus.

★ Italian astronomer and mathematician Galileo (born in 1564) proved that the path of a projectile is a parabola.

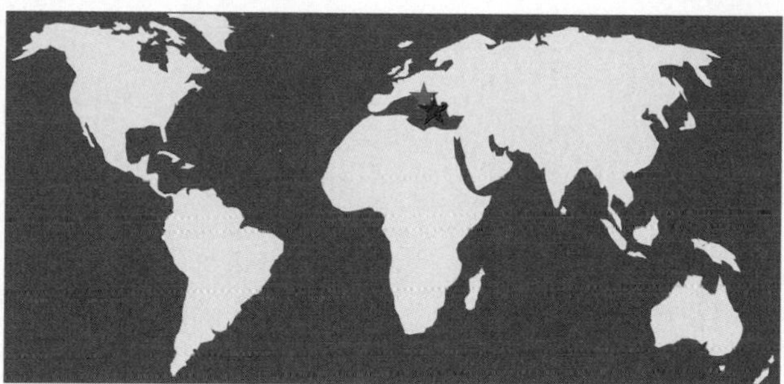

Pacing Guide for Chapter 8

Chapter Opener	1 Day
Section 1	2 Days
Section 2	1 Day
Section 3	2 Days
Study Help / Quiz	1 Day
Section 4	2 Days
Section 5	3 Days
Chapter Review / Chapter Tests	2 Days
Total Chapter 8	14 Days
Year-to-Date	112 Days

Check Your Resources

- Record and Practice Journal
- Resources by Chapter
- Skills Review Handbook
- Assessment Book
- Worked-Out Solutions

Technology For the Teacher

The Dynamic Planning Tool
Editable Teacher's Resources at
BigIdeasMath.com

Common Core State Standards

A.CED.2 . . . Graph equations on coordinate axes with labels and scales.
6.EE.2c Evaluate expressions at specific values of their variables . . .. Perform arithmetic operations, including those involving whole-number exponents, in the conventional order when there are no parentheses to specify a particular order.

Additional Topics for Review

- Domain of a function
- Range of a function
- Vertical translation
- Horizontal translation
- Common difference
- Common ratio

Try It Yourself

1. 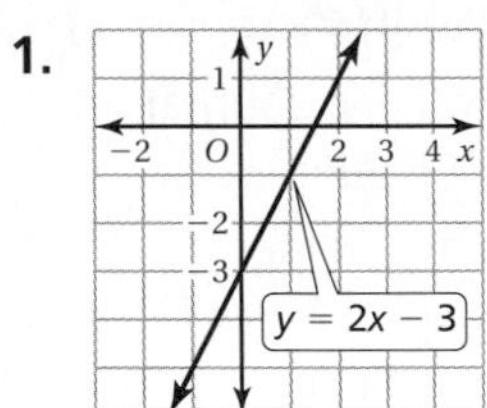

2. 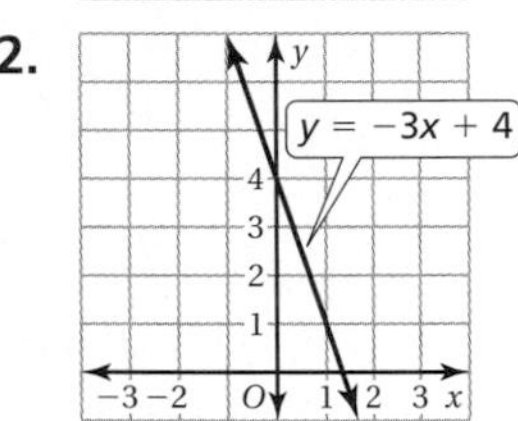

3. 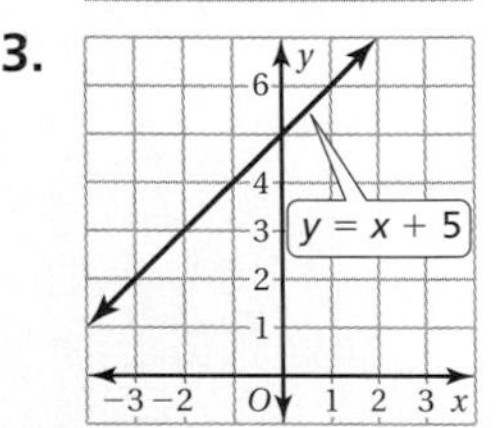

4. 5
5. 8
6. 3

Record and Practice Journal

1–6. See Additional Answers.

7. 10
8. 5
9. 7
10. −3
11. −9
12. 14
13. −18
14. 6
15. See Additional Answers.

Math Background Notes

Vocabulary Review

- Linear Equation
- Ordered Pairs
- Algebraic Expression
- Solutions of a Linear Equation

Graphing a Linear Equation

- Students should know how to graph a linear equation.
- Notice the use of color in the table. In the second column, the number in blue is the only quantity that varies (the *x*-variable). The other quantities are always the same (constant).
- **Teaching Tip:** Although any two points can be used to draw a line, using at least three or four points helps you recognize mistakes. Discuss the strategy for choosing the *x*-values.

Evaluating an Expression

- Students should know how to substitute values into a polynomial expression and evaluate the expression.
- Sometimes color-coding substitutions can help students evaluate expressions. Each time a value is substituted into an expression, substitute a colored pencil for the lead pencil.
- Suggest using parentheses where needed to make the operations clear when making substitutions.
- Remind students to use the correct order of operations in evaluating the numerical expression that remains after substituting for *x*.

Reteaching and Enrichment Strategies

If students need help. . .	If students got it. . .
Record and Practice Journal • Fair Game Review Skills Review Handbook Lesson Tutorials	Game Closet at *BigIdeasMath.com* Start the next section

What You Learned Before

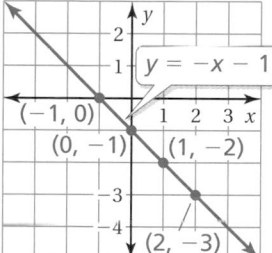

"Okay, Descartes, this will test the theory of parabolic flight paths."

Graphing a Linear Equation (A.CED.2)

Example 1 **Graph $y = -x - 1$.**

Step 1: Make a table of values.

x	$y = -x - 1$	y	(x, y)
-1	$y = -(-1) - 1$	0	$(-1, 0)$
0	$y = -(0) - 1$	-1	$(0, -1)$
1	$y = -(1) - 1$	-2	$(1, -2)$
2	$y = -(2) - 1$	-3	$(2, -3)$

Step 2: Plot the ordered pairs.

Step 3: Draw a line through the points.

Try It Yourself

Graph the linear equation.

1. $y = 2x - 3$ **2.** $y = -3x + 4$ **3.** $y = x + 5$

Evaluating an Expression (6.EE.2c)

Example 2 **Evaluate $2x^2 + 3x - 5$ when $x = -1$.**

$2x^2 + 3x - 5 = 2(-1)^2 + 3(-1) - 5$ Substitute -1 for x.

$= 2(1) + 3(-1) - 5$ Evaluate the power.

$= 2 - 3 - 5$ Multiply.

$= -6$ Subtract.

Try It Yourself

Evaluate the expression when $x = -2$.

4. $-x^2 - 4x + 1$ **5.** $3x^2 + x - 2$ **6.** $-2x^2 - 4x + 3$

8.1 Graphing $y = ax^2$

COMMON CORE STATE STANDARDS
F.BF.3

Essential Question What are the characteristics of the graph of the quadratic function $y = ax^2$? How does the value of a affect the graph of $y = ax^2$?

1 ACTIVITY: Graphing a Quadratic Function

Work with a partner.

- Complete the input-output table.
- Plot the points in the table.
- Sketch the graph by connecting the points with a smooth curve.
- What do you notice about the graphs?

a.

x	$y = x^2$
−3	
−2	
−1	
0	
1	
2	
3	

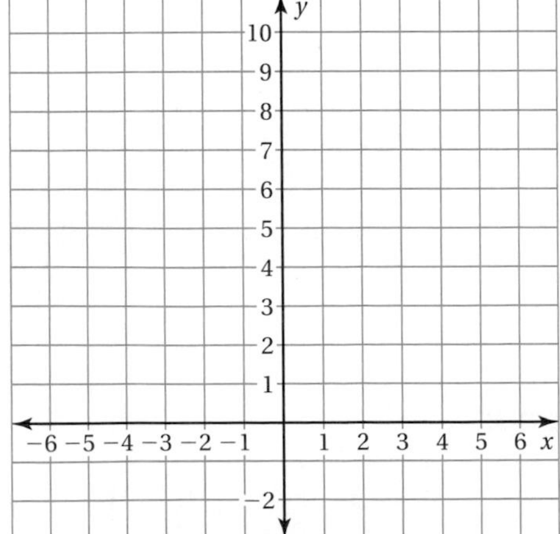

b.

x	$y = -x^2$
−3	
−2	
−1	
0	
1	
2	
3	

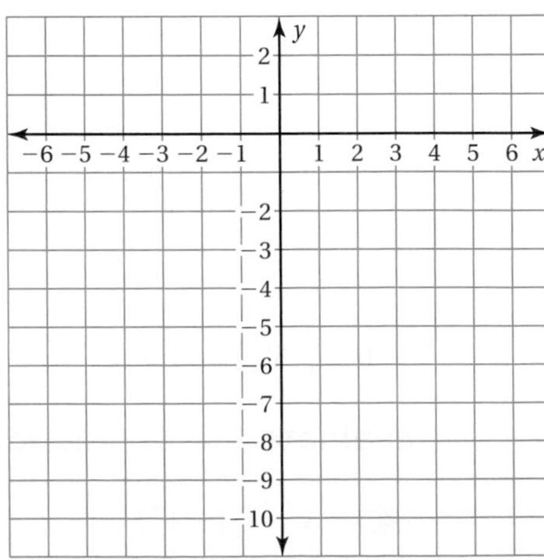

Laurie's Notes

Introduction

Standards for Mathematical Practice

- **MP1a Make Sense of Problems:** The squaring function $y = x^2$ is symmetric. When evaluating $y = x^2$, substituting $\pm x$ results in the same y-value. Mathematically proficient students will recognize this feature and understand what it means in terms of the graph of the function.

Motivate

- Fold a piece of paper in half and cut out a shape, such as a heart, pine tree, or isosceles triangle, leaving the fold intact.
- Explain that throughout this chapter they will be graphing a function that has a characteristic in common with the cutout: symmetry.

? "Have you graphed any functions that have symmetry?" yes (absolute value functions)

Activity Notes

Activity 1

- Students have completed input-output tables and plotted points. When students connect the points with a smooth curve, it should not look like the graph of $y = |x|$.
- **Common Error:** When evaluating $-x^2$, students may take the opposite of the x-value and square it instead of squaring the x-value and taking the opposite. Remind students that $-x^2 \neq (-x)^2$.
- When students finish, ask them to describe general features of the two graphs. They may use language such as "pointing up" and "pointing down." Students may also comment on the symmetry about the y-axis. The symmetry should be noted in the graph *and* the table.

Common Core State Standards

F.BF.3 Identify the effect on the graph of replacing $f(x)$ by $f(x) + k$, $kf(x)$, $f(kx)$, and $f(x + k)$ for specific values of k (both positive and negative); find the value of k given the graphs. Experiment with cases and illustrate an explanation of the effects on the graph using technology.

Previous Learning

Students should know how to graph linear functions.

Activity Materials
Introduction
• piece of paper • scissors

Start Thinking! and Warm Up

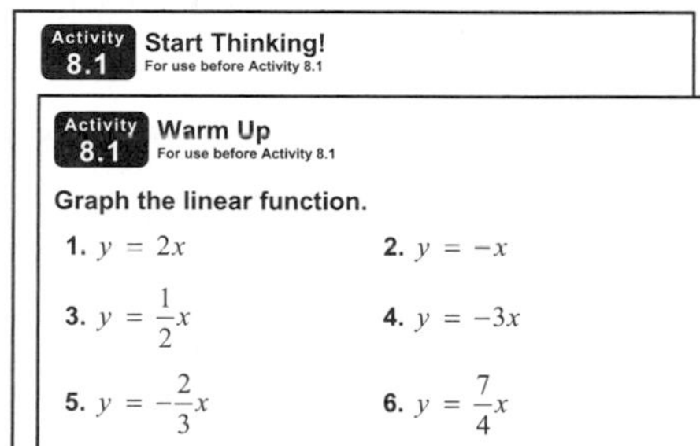

Activity 8.1 Start Thinking! For use before Activity 8.1

Activity 8.1 Warm Up For use before Activity 8.1

Graph the linear function.

1. $y = 2x$
2. $y = -x$
3. $y = \frac{1}{2}x$
4. $y = -3x$
5. $y = -\frac{2}{3}x$
6. $y = \frac{7}{4}x$

8.1 Record and Practice Journal

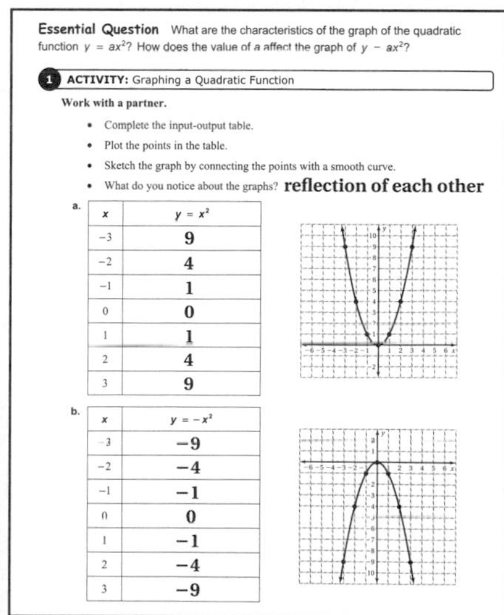

Essential Question What are the characteristics of the graph of the quadratic function $y = ax^2$? How does the value of a affect the graph of $y = ax^2$?

1 ACTIVITY: Graphing a Quadratic Function

Work with a partner.

- Complete the input-output table.
- Plot the points in the table.
- Sketch the graph by connecting the points with a smooth curve.
- What do you notice about the graphs? **reflection of each other**

a.

x	$y = x^2$
−3	9
−2	4
−1	1
0	0
1	1
2	4
3	9

b.

x	$y = -x^2$
−3	−9
−2	−4
−1	−1
0	0
1	−1
2	−4
3	−9

Differentiated Instruction

Inclusion

In the lessons on graphing quadratic functions, students are expected to make many tables and graphs. Provide a resource page with blank tables and enlarged coordinate grids for them to use.

8.1 Record and Practice Journal

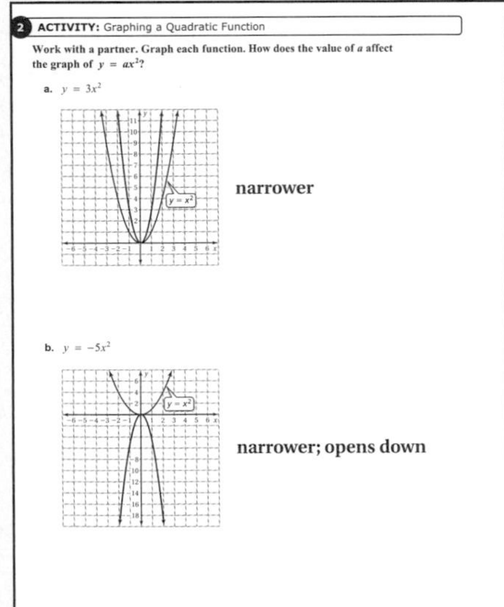

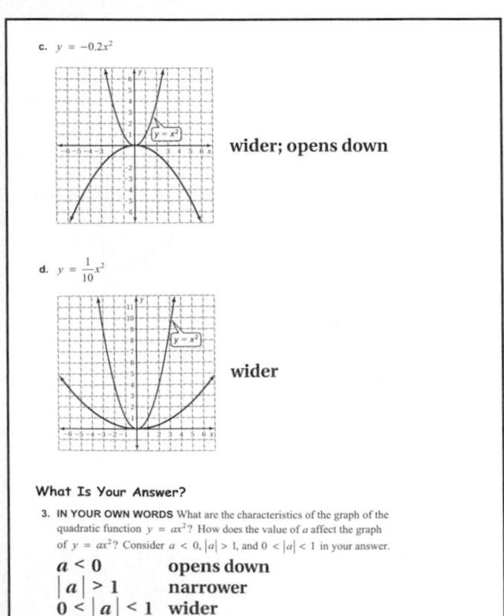

Laurie's Notes

Activity 2

- Having the graph of $y = x^2$ already on the grid may help students compare the graph of $y = ax^2$ to the graph of $y = x^2$.
- There are two features that the value of a affects, and you may need to prod student thinking as you circulate. The obvious effect is whether the graph opens up or down. The feature that is more difficult for students to explain is the horizontal stretch or shrink. Informally, this can be referred to as "more bowl-shaped" or "more narrow."
- Students often say that when $a < 1$, the graphs become more bowl-shaped. Remind students that $-5 < 1$. What they really are trying to describe are rational numbers between -1 and 1, or $|a| < 1$.
- If a student fails to notice the shape of the graph, give the student additional functions to graph.

? "Is the graph nonlinear when $a = 0$? Explain." no; When $a = 0$, the function becomes the constant function $y = 0$.

- Discuss as a class when all have finished.

? "Describe the graph when $a < 0$." open downward "Describe the graph when $|a| > 1$." narrower "Describe the graph when $|a| < 1$." more bowl-shaped

- If time allows, use a graphing calculator to check conjectures. You can also use the calculator to graph several functions and ask for reasonable equations for each graph.

What Is Your Answer?

- Have students organize their results in a table.

Closure

- **Matching:** Match the function with the color of its graph.

A. $y = 4x^2$

B. $y = 0.25x^2$

C. $y = -0.4x^2$

D. $y = -\frac{3}{2}x^2$

A. green B. red C. blue D. black

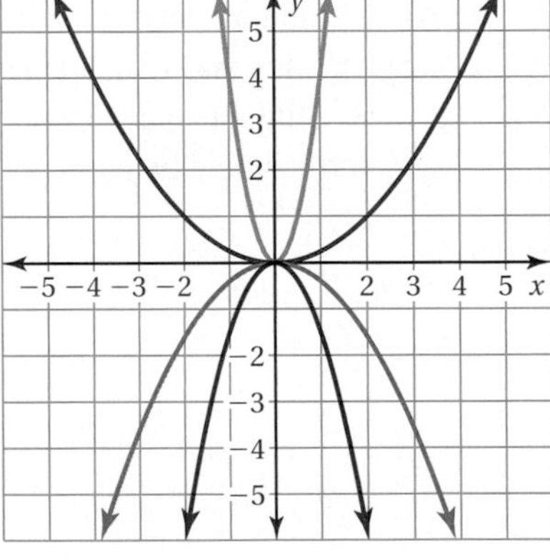

Technology For the Teacher

The Dynamic Planning Tool
Editable Teacher's Resources at *BigIdeasMath.com*

2 ACTIVITY: Graphing a Quadratic Function

Work with a partner. Graph each function. How does the value of a affect the graph of $y = ax^2$?

a. $y = 3x^2$

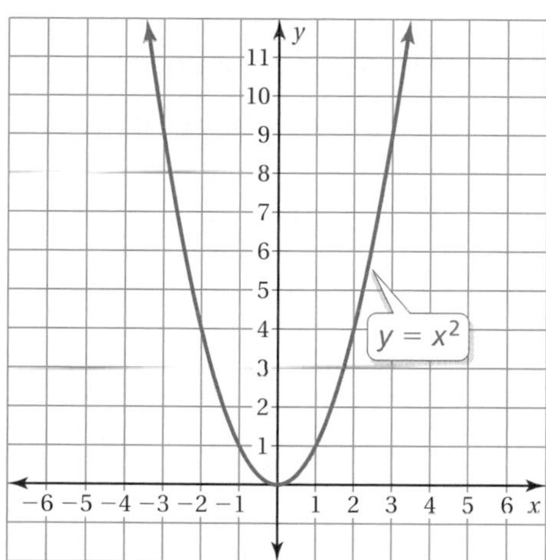

b. $y = -5x^2$

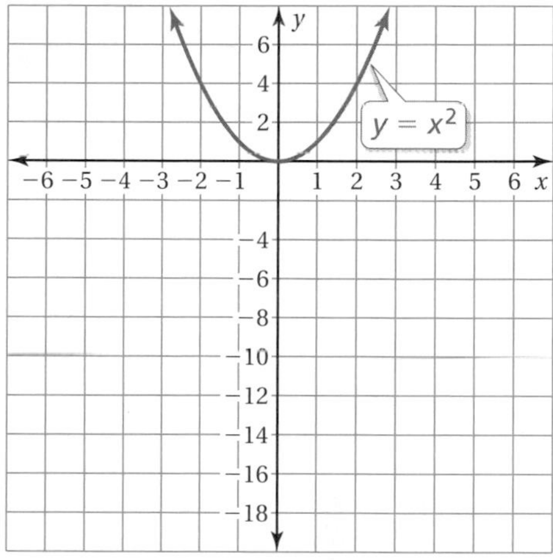

c. $y = -0.2x^2$

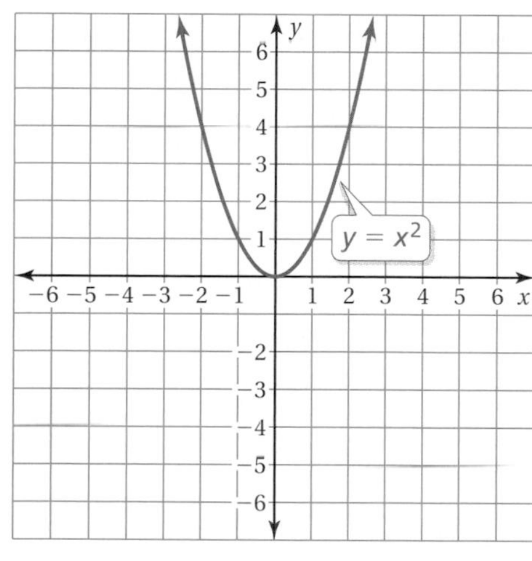

d. $y = \frac{1}{10}x^2$

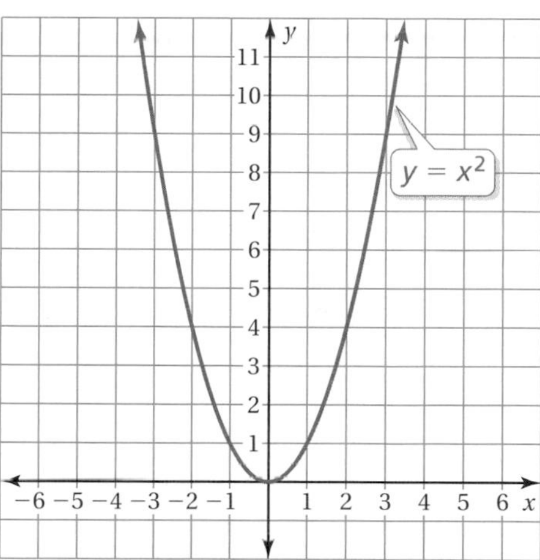

What Is Your Answer?

3. **IN YOUR OWN WORDS** What are the characteristics of the graph of the quadratic function $y = ax^2$? How does the value of a affect the graph of $y = ax^2$? Consider $a < 0$, $|a| > 1$, and $0 < |a| < 1$ in your answer.

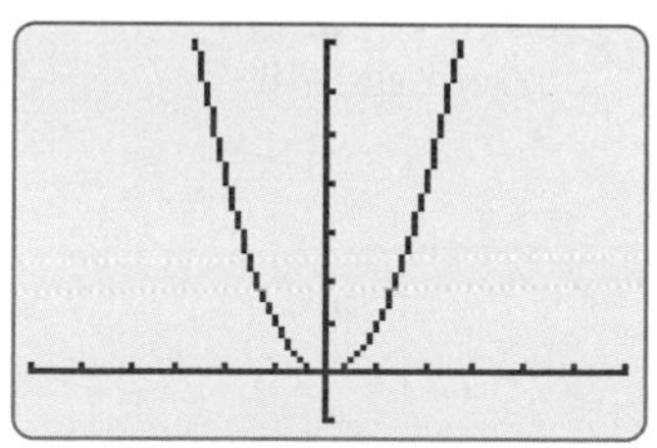

Practice Use what you learned about the graphs of quadratic functions to complete Exercises 5–7 on page 407.

8.1 Lesson

Key Vocabulary
quadratic function, *p. 404*
parabola, *p. 404*
vertex, *p. 404*
axis of symmetry, *p. 404*

A **quadratic function** is a nonlinear function that can be written in the standard form $y = ax^2 + bx + c$, where $a \neq 0$. The U-shaped graph of a quadratic function is called a **parabola.**

Characteristics of Quadratic Functions

The most basic quadratic function is $y = x^2$.

The lowest or highest point on a parabola is the **vertex.**

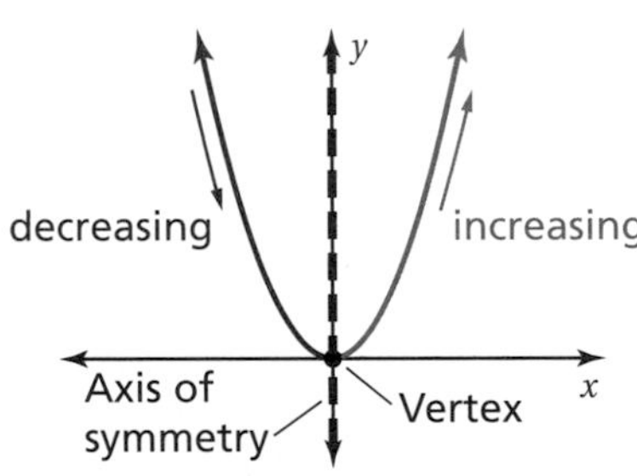

The vertical line that divides the parabola into two symmetric parts is the **axis of symmetry.** The axis of symmetry passes through the vertex.

EXAMPLE 1 Identifying Characteristics of a Quadratic Function

Consider the graph of the quadratic function.

Using the graph, you can identify the vertex, axis of symmetry, and the behavior of the graph as shown.

You can also determine the following:

- The domain is all real numbers.
- The range is all real numbers greater than or equal to -2.
- When $x < -1$, y increases as x decreases.
- When $x > -1$, y increases as x increases.

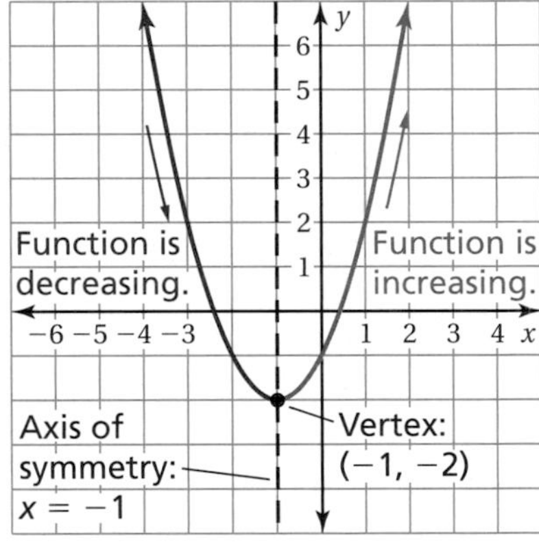

On Your Own

Now You're Ready
Exercises 8–10

Identify characteristics of the graph of the quadratic function.

1. 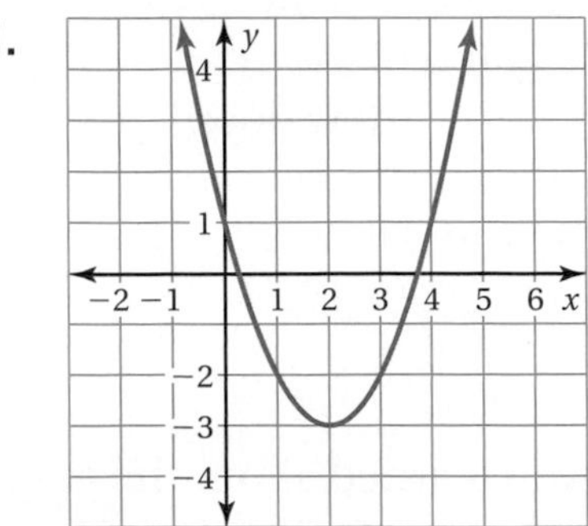

2. 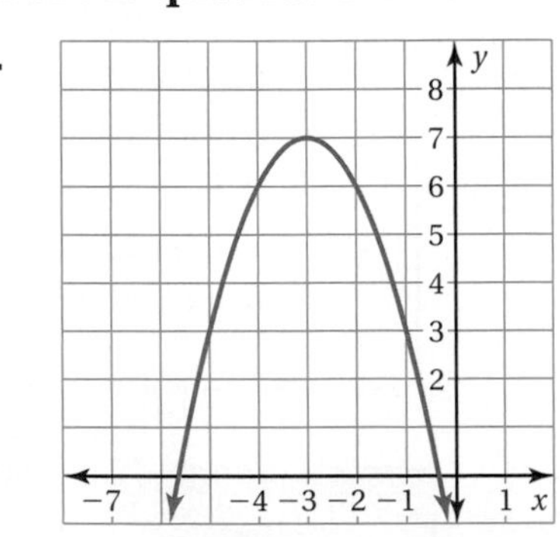

Multi-Language Glossary at BigIdeasMath.com.

Laurie's Notes

Introduction

Connect

- **Yesterday:** Students studied characteristics of the graph of the quadratic function $y = ax^2$. (MP1a)
- **Today:** Students will graph the quadratic function $y = ax^2$.

Motivate

- Use a flashlight to cast a shadow that students can describe. If you do not have a flashlight, use something that has a parabolic cross section, such as a bulb in an overhead projector. Tell students that the graphs they will be sketching represent a common shape.

Lesson Notes

Discuss

- **MP7 Look for and Make Use of Structure:** Define a quadratic function written in standard form. Explain to students that *x* is the independent variable and *y* is the dependent variable. Also explain that *a*, *b*, and *c* are parameters in the same way *m* and *b* are parameters in the linear function $y = mx + b$.
- In this section, students will graph $y = ax^2$, meaning $b, c = 0$.

Key Idea

- Write the Key Idea which describes the characteristics of quadratic functions. The *vertex* is the lowest (minimum) or highest (maximum) point on the graph. The *axis of symmetry* is where the graph could be folded onto itself.
- Discuss the red and blue portions of the graph. Graphs are read left-to-right. So when the phrase "increasing function" is used, the values of the function (*y*-values) are increasing. When the phrase "decreasing function" is used, the values of the function (*y*-values) are decreasing.

Example 1

- Sketch the graph shown with scaled units so students can identify key features or characteristics of the graph.

? "What is the vertex of the function?" $(-1, -2)$ "For what *x*-values is the function increasing?" $x > -1$ "For what *x*-values is the function decreasing?" $x < -1$ "What is the line of symmetry?" $x = -1$ "Because the graph opens up and you identified the vertex, what is the range?" $y \geq -2$ "What is the domain?" all real numbers

On Your Own

- **Question 2:** Student should recognize that when a quadratic function opens down, the function starts out as an increasing function, it reaches a maximum value, and then it becomes a decreasing function.

Today's lesson is graphing $y = ax^2$.

Activity Materials
Introduction
• flashlight • overhead projector bulb (optional)

Start Thinking! and Warm Up

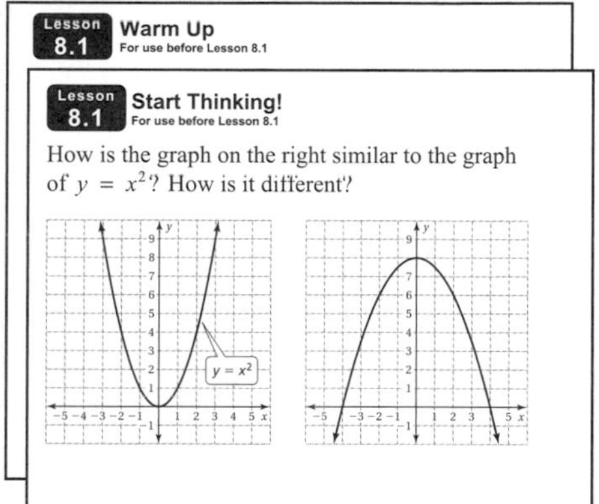

Extra Example 1

Identify characteristics of the graph of the quadratic function.

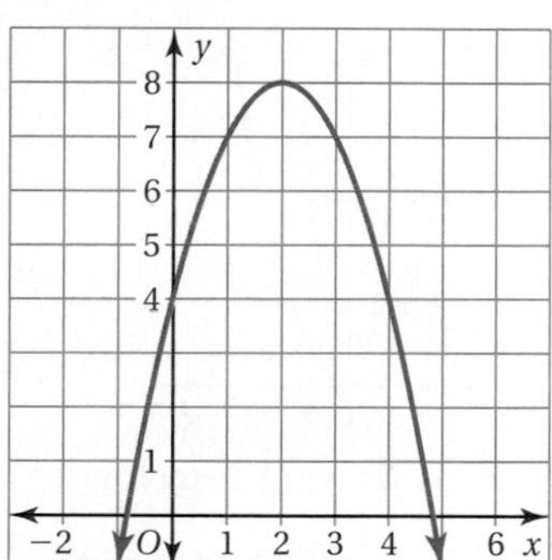

vertex: (2, 8); axis of symmetry: $x = 2$; domain: all real numbers; range: $y \leq 8$; When $x < 2$, *y* decreases as *x* decreases. When $x > 2$, *y* decreases as *x* increases.

On Your Own

1–2. See Additional Answers.

Extra Example 2

Graph $y = \frac{1}{2}x^2$. Compare the graph to the graph of $y = x^2$.

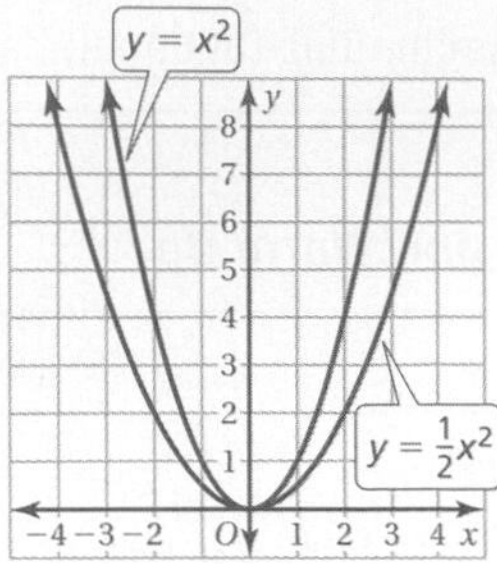

Both graphs open up and have the same vertex, (0, 0), and the same axis of symmetry, $x = 0$. The graph of $y = \frac{1}{2}x^2$ is wider than the graph of $y = x^2$.

On Your Own

3. Both graphs open up and have the same vertex, (0, 0), and the same axis of symmetry, $x = 0$. The graph of $y = 5x^2$ is narrower than the graph of $y = x^2$.

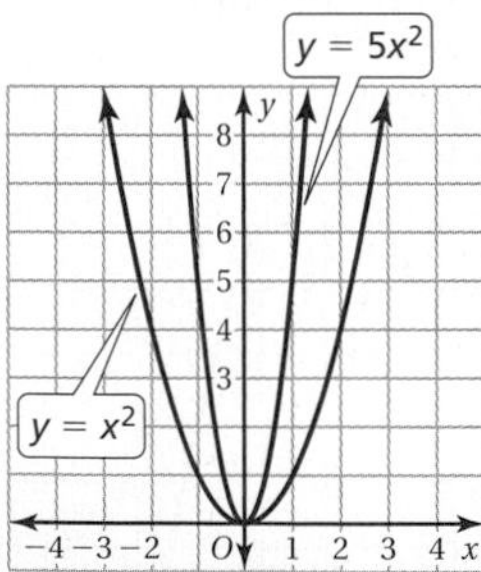

4–5. See Additional Answers.

English Language Learners

Vocabulary

To help English language learners with the vocabulary in this chapter, make a poster of the *Key Idea* from page 404. Illustrate the poster with both cases in this lesson, when $a > 0$ and when $a < 0$.

Laurie's Notes

Key Ideas

- Students should recall from the activity that the graph opened down when the leading coefficient was negative. The leading coefficient also affects the shape of the parabola.
- Write the first Key Idea. Sketch $y = x^2$ in one color. Sketch an example of $y = ax^2$ where $0 < a < 1$ in a second color. Sketch an example of $y = ax^2$ where $a > 1$ in a third color. Connect to the activity.
- Write the second Key Idea. Sketch $y = x^2$ in one color. Sketch an example of $y = ax^2$ where $-1 < a < 0$ in a second color. Sketch an example of $y = ax^2$ where $a < -1$ in a third color. Connect to the activity.

Example 2

- Write the equation. Identify the leading coefficient, $a = 2$.
- ? "When $a = 2$, what do you know?" The graph opens up because $a > 0$. It is also narrower than $y = x^2$ because $a > 1$.
- Make a table of values.
- **Common Error:** Students multiply x by 2 and then square it instead of squaring x and then multiplying by 2. Remind students about the order of operations.
- Plot the points and draw a smooth curve through the points. Ask students to identify key characteristics of the graph, including when the function is increasing and decreasing.
- **Connection:** Discuss the table of values and point out that the y-values are decreasing, then increasing. Refer to the graph and make the same observation.

On Your Own

- Check to make sure that the computations with the fractional coefficients in Questions 4 and 5 are correct.

Graphing $y = ax^2$ When $a > 0$

- When $0 < a < 1$, the graph of $y = ax^2$ opens up and is wider than the graph of $y = x^2$.
- When $a > 1$, the graph of $y = ax^2$ opens up and is narrower than the graph of $y = x^2$.

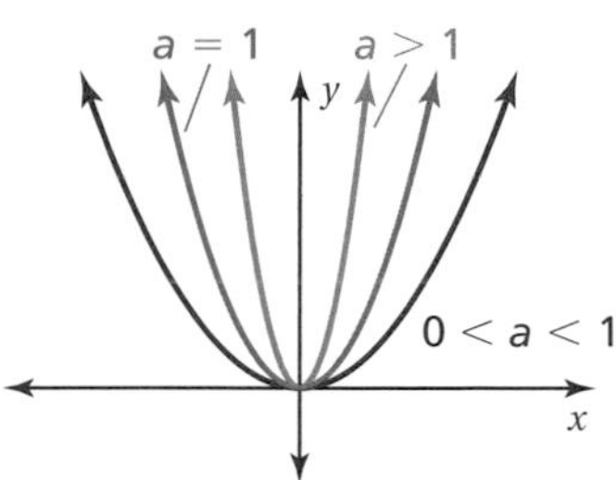

Graphing $y = ax^2$ When $a < 0$

- When $-1 < a < 0$, the graph of $y = ax^2$ opens down and is wider than the graph of $y = x^2$.
- When $a < -1$, the graph of $y = ax^2$ opens down and is narrower than the graph of $y = x^2$.

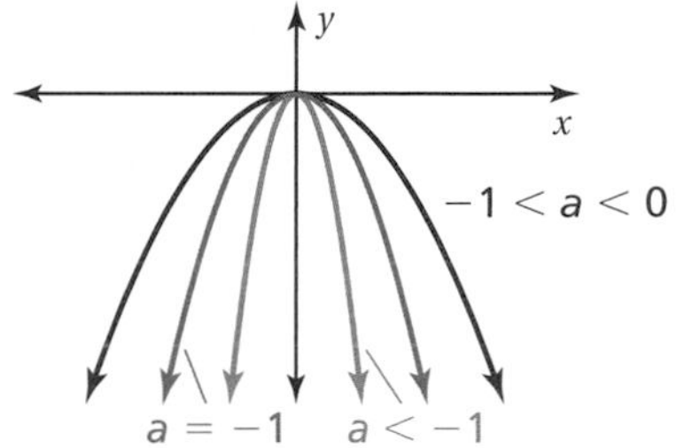

EXAMPLE 2 **Graphing $y = ax^2$ When $a > 0$**

Graph $y = 2x^2$. Compare the graph to the graph of $y = x^2$.

Step 1: Make a table of values.

x	−2	−1	0	1	2
y	8	2	0	2	8

Step 2: Plot the ordered pairs.

Step 3: Draw a smooth curve through the points.

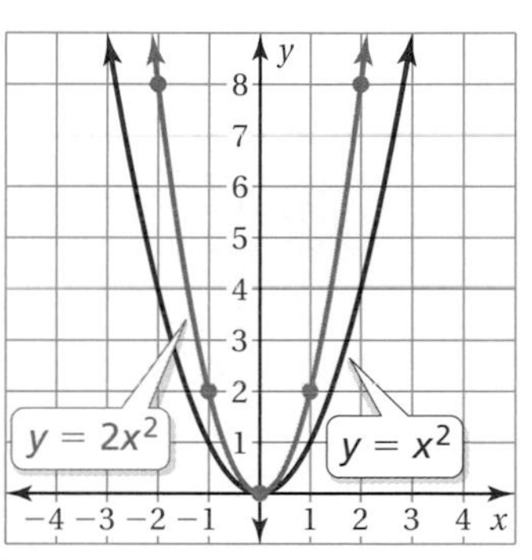

∴ Both graphs open up and have the same vertex, (0, 0), and the same axis of symmetry, $x = 0$. The graph of $y = 2x^2$ is narrower than the graph of $y = x^2$.

On Your Own

Now You're Ready
Exercises 11–16

Graph the function. Compare the graph to the graph of $y = x^2$.

3. $y = 5x^2$ **4.** $y = \frac{1}{3}x^2$ **5.** $y = \frac{3}{2}x^2$

EXAMPLE 3 Graphing $y = ax^2$ When $a < 0$

Graph $y = -\frac{1}{3}x^2$. Compare the graph to the graph of $y = x^2$.

Step 1: Make a table of values. Choose x-values that make the calculations simple.

x	−6	−3	0	3	6
y	−12	−3	0	−3	−12

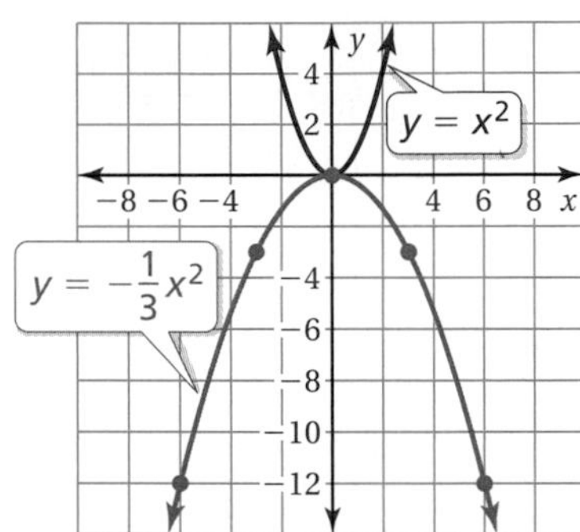

Step 2: Plot the ordered pairs.

Step 3: Draw a smooth curve through the points.

The graphs have the same vertex, (0, 0), and the same axis of symmetry, $x = 0$, but the graph of $y = -\frac{1}{3}x^2$ opens down. The graph of $y = -\frac{1}{3}x^2$ is wider than the graph of $y = x^2$.

EXAMPLE 4 Real-Life Application

The diagram shows the cross section of a satellite dish, where x and y are measured in meters. Find the width and depth of the dish.

Use the domain of the function to find the width of the dish. Use the range to find the depth.

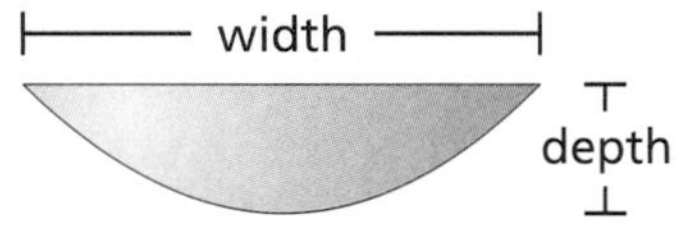

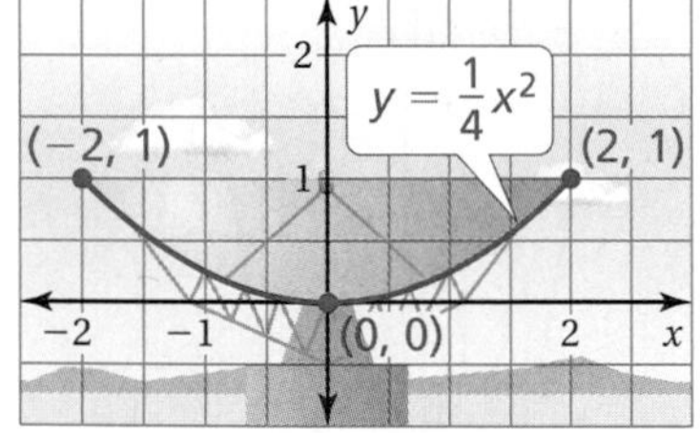

The leftmost point on the graph is (−2, 1) and the rightmost point is (2, 1). So, the domain is $-2 \le x \le 2$, which represents 4 meters.

The lowest point on the graph is (0, 0) and the highest points on the graph are (−2, 1) and (2, 1). So, the range is $0 \le y \le 1$, which represents 1 meter.

So, the satellite dish is 4 meters wide and 1 meter deep.

On Your Own

Now You're Ready
Exercises 18–23 and 34

Graph the function. Compare the graph to the graph of $y = x^2$.

6. $y = -3x^2$
7. $y = -0.1x^2$
8. $y = -\frac{1}{4}x^2$
9. The cross section of a spotlight can be modeled by the graph of $y = 0.5x^2$, where x and y are measured in inches and $-2 \le x \le 2$. Find the width and depth of the spotlight.

Laurie's Notes

Example 3

- Write the equation. Identify the leading coefficient, $a = -\frac{1}{3}$.

? "When $a = -\frac{1}{3}$, what do you know?" The graph opens down because $a < 0$. It is also wider than $y = x^2$ because $-1 < a < 0$.

- **MP6 Attend to Precision:** Make a table of values. Discuss with students that using multiples of 3 as x-values will allow for easier computations.
- Continue to work through the problem as shown.

Example 4

- The parabolic shape is one that students have likely seen. The cross section of a satellite dish is parabolic.
- Draw a sketch of the satellite dish and label the ordered pairs shown.

? "What is the domain of the function?" $-2 \leq x \leq 2$

? "What is the range of the function?" $0 \leq y \leq 1$

- Connect the domain and range to vertical and horizontal distances to finish the problem.

On Your Own

- Remind students that each graph is compared to the graph of $y = x^2$.
- Students should have a visual sense of which function is wider, Question 7 or Question 8.
- Connect Question 9 to the flashlight example in the Motivate.

Closure

- **Exit Ticket:** Describe the differences between the graphs of $y = -3x^2$ and $y = \frac{1}{3}x^2$. The graph of $y = -3x^2$ opens down and the graph of $y = \frac{1}{3}x^2$ opens up. The graph of $y = -3x^2$ is narrower than the graph of $y = \frac{1}{3}x^2$. The range of $y = -3x^2$ is $y \leq 0$. The range of $y = \frac{1}{3}x^2$ is $y \geq 0$.

Technology For the Teacher

The Dynamic Planning Tool
Editable Teacher's Resources at *BigIdeasMath.com*

Extra Example 3

Graph $y = -\frac{1}{8}x^2$. Compare the graph to the graph of $y = x^2$. See Additional Answers.

Extra Example 4

In Example 4, the function $y = \frac{1}{6}x^2$ models the cross section with a domain of $-3 \leq x \leq 3$. Find the width and depth of the dish. 6 meters wide; 1.5 meters deep

On Your Own

6. The graphs have the same vertex, (0, 0), and the same axis of symmetry, $x = 0$, but the graph of $y = -3x^2$ opens down. The graph of $y = -3x^2$ is narrower than the graph of $y = x^2$.

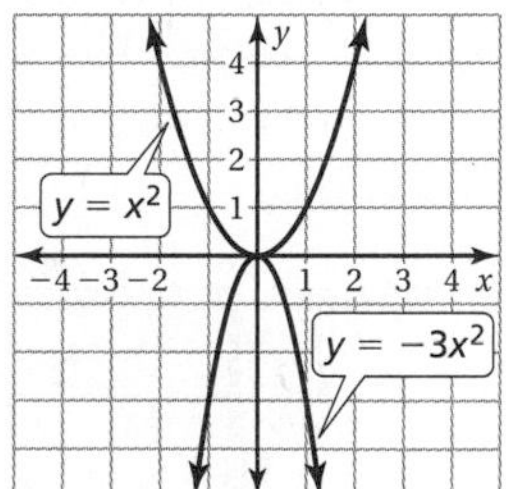

7. The graphs have the same vertex, (0, 0), and the same axis of symmetry, $x = 0$, but the graph of $y = -0.1x^2$ opens down. The graph of $y = -0.1x^2$ is wider than the graph of $y = x^2$.

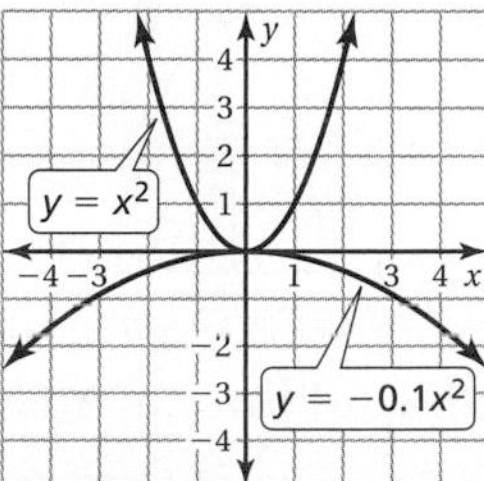

8. See Additional Answers.
9. 4 inches wide; 2 inches deep

Vocabulary and Concept Check

1. The vertex is the lowest or highest point on a parabola. The vertical line that divides the parabola into two symmetric parts is the axis of symmetry.
2. parabola
3. The graph of $y = \frac{1}{6}x^2$ is wider because when $0 < a < 1$, the graph of $y = ax^2$ is wider than the graph of $y = x^2$. When $a > 1$, the graph of $y = ax^2$ is narrower than the graph of $y = x^2$.
4. opens up when $a > 0$; opens down when $a < 0$

Practice and Problem Solving

5. Both graphs open down and have a U-shape. The graph of $y = -0.4x^2$ is wider than the graph of $y = -4x^2$.

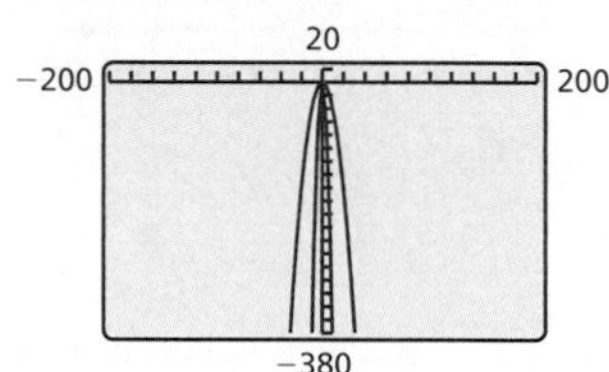

6. Both graphs open down and have a U-shape. The graph of $y = -0.04x^2$ is quite a bit wider than the graph of $y = -4x^2$.

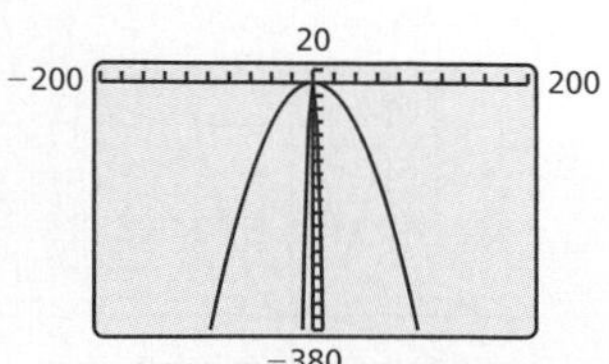

7–17. See Additional Answers.

Assignment Guide and Homework Check

Level	Assignment	Homework Check
Average	1–4, 9–25 odd, 24, 34, 41–45	9, 15, 17, 19, 34
Advanced	1–4, 17–33 odd, 24, 28, 34–40, 41–45	17, 21, 28, 34, 39

For Your Information

- **Exercise 38** Students can use a glass of water and a straw to experiment with the rotating liquid concept. What happens when they change speed? What happens when they swirl in an uneven pattern rather than a perfect circle?

Common Errors

- **Exercises 5–7** Students may not use an appropriate viewing window. Remind them that to compare the graphs, they need to see both sides of both parabolas.
- **Exercises 11–16** Students may only plot points on the right side of the y-axis, which is the axis of symmetry for quadratic functions of the form $f(x) = ax^2$. Remind students that the graphs of quadratic functions are parabolas which are U-shaped and symmetric. Students can use this symmetry to find points on both sides of the axis of symmetry.

8.1 Record and Practice Journal

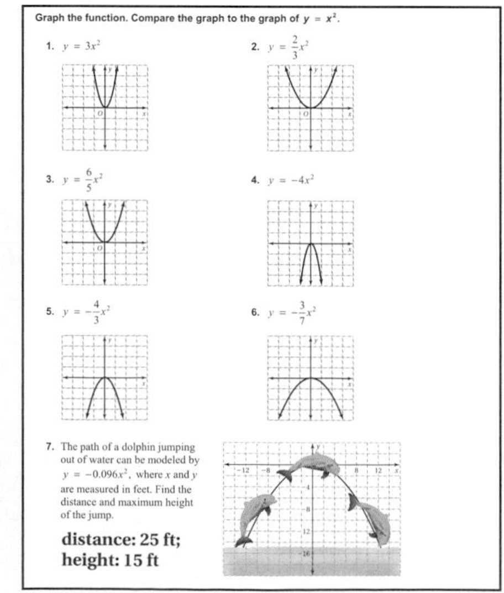

Graph the function. Compare the graph to the graph of $y = x^2$.

1. $y = 3x^2$
2. $y = \frac{2}{3}x^2$
3. $y = \frac{6}{5}x^2$
4. $y = -4x^2$
5. $y = -\frac{4}{3}x^2$
6. $y = -\frac{3}{7}x^2$
7. The path of a dolphin jumping out of water can be modeled by $y = -0.096x^2$, where x and y are measured in feet. Find the distance and maximum height of the jump.

distance: 25 ft; height: 15 ft

Technology For the Teacher
Answer Presentation Tool

8.1 Exercises

Vocabulary and Concept Check

1. **VOCABULARY** Describe the vertex and axis of symmetry of the graph of $y = ax^2$.
2. **VOCABULARY** What is the U-shaped graph of a quadratic function called?
3. **WRITING** Without graphing, which graph is wider, $y = 6x^2$ or $y = \frac{1}{6}x^2$? Explain your reasoning.
4. **WRITING** When does the graph of a quadratic function open up? open down?

Practice and Problem Solving

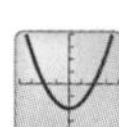

Use a graphing calculator to graph the function. Compare the graph to the graph of $y = -4x^2$.

5. $y = -0.4x^2$
6. $y = -0.04x^2$
7. $y = -0.004x^2$

Identify characteristics of the graph of the quadratic function.

8. 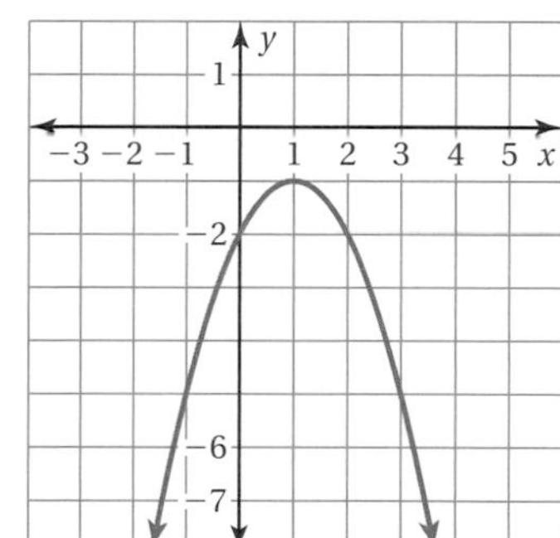

9. 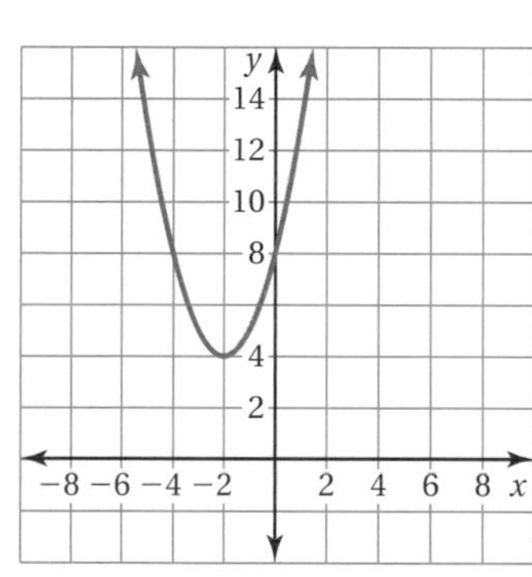

10. 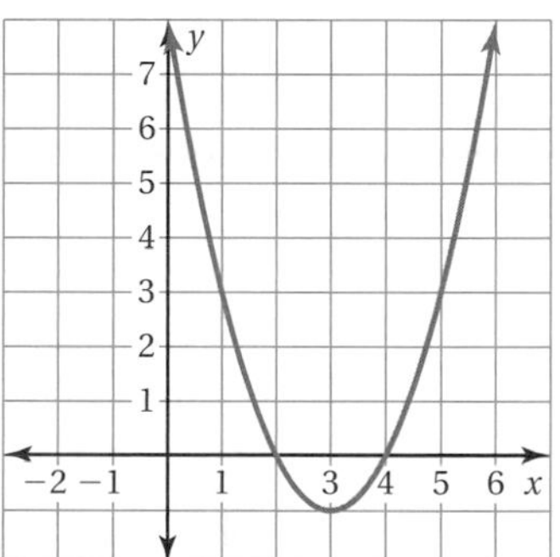

Graph the function. Compare the graph to the graph of $y = x^2$.

11. $y = 6x^2$
12. $y = 8x^2$
13. $y = \frac{1}{4}x^2$
14. $y = \frac{3}{4}x^2$
15. $y = \frac{5}{2}x^2$
16. $y = \frac{7}{5}x^2$

17. **WATERFALL** A fish swims over the waterfall. The distance y (in feet) that the fish falls is given by the function $y = 16t^2$, where t is the time (in seconds).
    a. Describe the domain and range of the function.
    b. Graph the function using the domain in part (a).
    c. Use the graph to determine when the fish lands in the water below.

Graph the function. Compare the graph to the graph of $y = x^2$.

18. $y = -2x^2$

19. $y = -7x^2$

20. $y = -\frac{1}{5}x^2$

21. $y = -\frac{5}{8}x^2$

22. $y = -\frac{5}{3}x^2$

23. $y = -\frac{9}{4}x^2$

24. **ERROR ANALYSIS** Describe and correct the error in graphing and comparing $y = x^2$ and $y = 0.5x^2$.

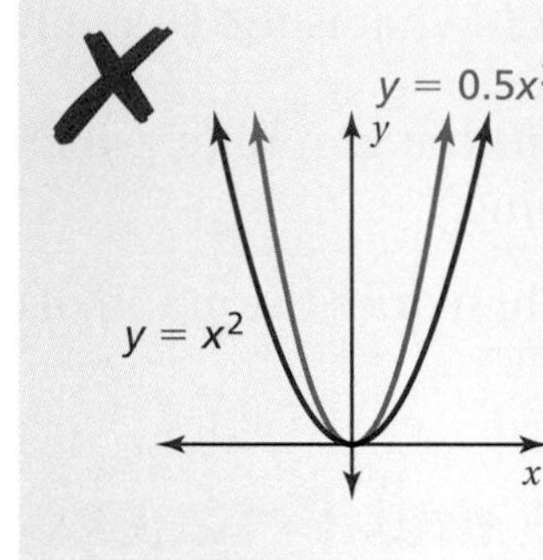

The graphs have the same vertex and the same axis of symmetry. The graph of $y = 0.5x^2$ is narrower than the graph of $y = x^2$.

Describe the possible values of *a*.

25. 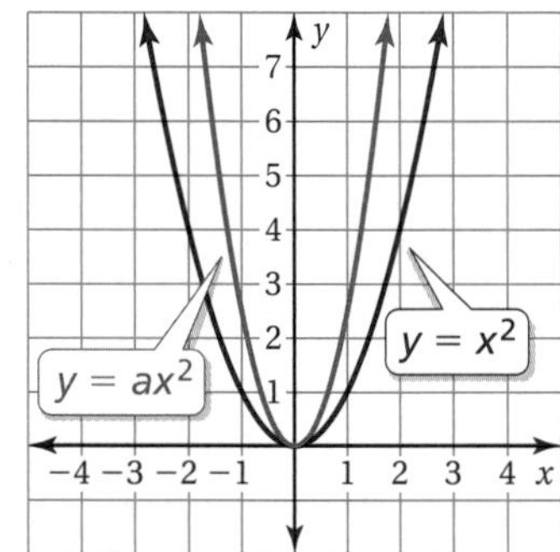

26. 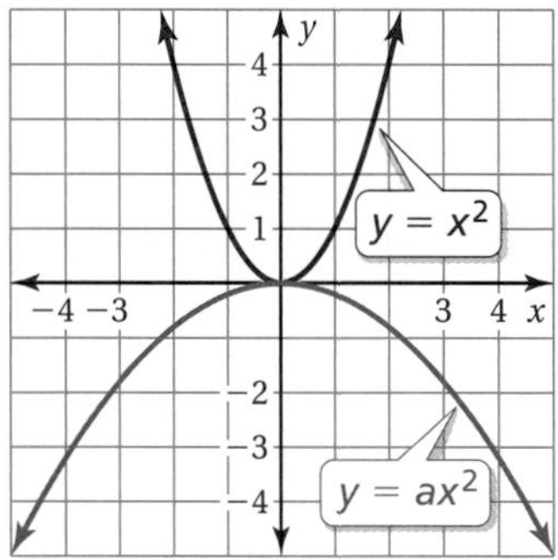

27. 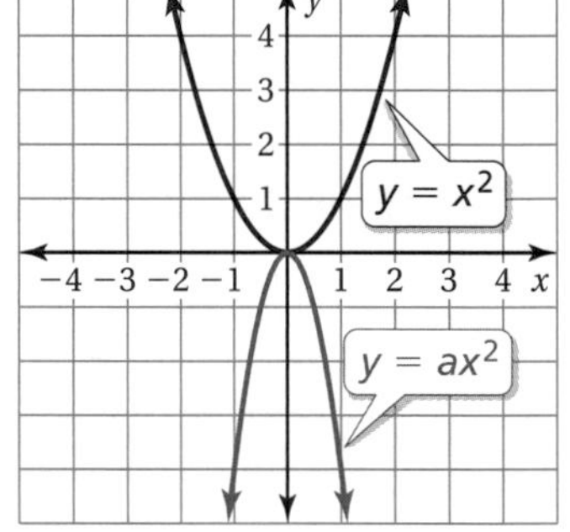

28. **REASONING** A parabola opens up and passes through $(-4, 2)$ and $(6, -3)$. How do you know that $(-4, 2)$ is not the vertex?

29. **REASONING** Describe the domain and range of the function $y = ax^2$ when (a) $a > 0$ and (b) $a < 0$.

Determine whether the statement is *always*, *sometimes*, or *never* true. Explain your reasoning.

30. The graph of $y = ax^2$ is narrower than the graph of $y = x^2$ when $a > 0$.

31. The graph of $y = ax^2$ is narrower than the graph of $y = x^2$ when $|a| > 1$.

32. The graph of $y = ax^2$ is wider than the graph of $y = x^2$ when $0 < |a| < 1$.

33. The graph of $y = ax^2$ is wider than the graph of $y = dx^2$ when $|a| > |d|$.

4 34. **BRIDGE** The arch support of a bridge can be modeled by $y = -0.0012x^2$, where x and y are measured in feet. Find the height and width of the arch.

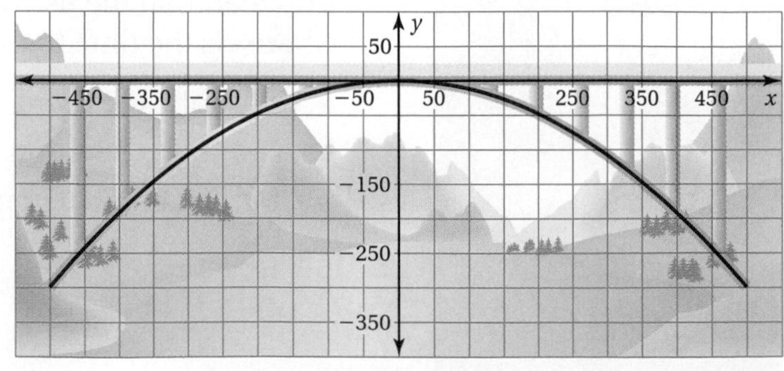

Common Errors

- **Exercises 18–23** Students may only plot points on the right side of the y-axis, which is the axis of symmetry for quadratic functions of the form $f(x) = ax^2$. Remind students that the graphs of quadratic functions are parabolas which are U-shaped and symmetric. Students can use this symmetry to find points on both sides of the axis of symmetry.

Practice and Problem Solving

18–23. See Additional Answers.

24. Because $0 < a < 1$, the graph of $y = 0.5x^2$ should be wider than the graph of $y = x^2$, not narrower.

25. $a > 1$

26. $-1 < a < 0$

27. $a < -1$

28. The graph opens up so the lowest point on the parabola is the vertex. The point $(6, -3)$ is lower than $(-4, 2)$ because $-3 < 2$. Since both points are on the graph, $(-4, 2)$ cannot be the vertex because it is not the lowest point.

29. **a.** When $a > 0$, the domain is all real numbers and the range is $y \geq 0$.

b. When $a < 0$, the domain is all real numbers and the range is $y \leq 0$.

30. See Additional Answers.

31. always; When $a > 1$ or $a < -1$, the graph of $y = ax^2$ is narrower than the graph of $y = x^2$.

32. always; When $0 < a < 1$ or $-1 < a < 0$, the graph of $y = ax^2$ is wider than the graph of $y = x^2$.

33. never; As $|a|$ increases, the graph of $y = ax^2$ gets narrower.

34. The width is 1000 feet. The height is 300 feet.

English Language Learners

Pair Activity

Pair English language learners with English speakers. Have them work together to answer Exercises 25–27. Students should not only describe the possible values of a, but be able to justify their answers.

Practice and Problem Solving

35. When $a > 0$, the graph of $y = ax^2$ is increasing from left to right when $x > 0$. When $a < 0$, the graph of $y = ax^2$ is increasing from left to right when $x < 0$.

36. When $a > 0$, the graph of $y = ax^2$ is decreasing from left to right when $x < 0$. When $a < 0$, the graph of $y = ax^2$ is decreasing from left to right when $x > 0$.

37. $\frac{3}{4}$

38–39. See Additional Answers.

40. See *Taking Math Deeper.*

Fair Game Review

41. 39 **42.** 15

43. 14 **44.** 4.5

45. B

Mini-Assessment

1. Identify the characteristics of the graph of the quadratic function.

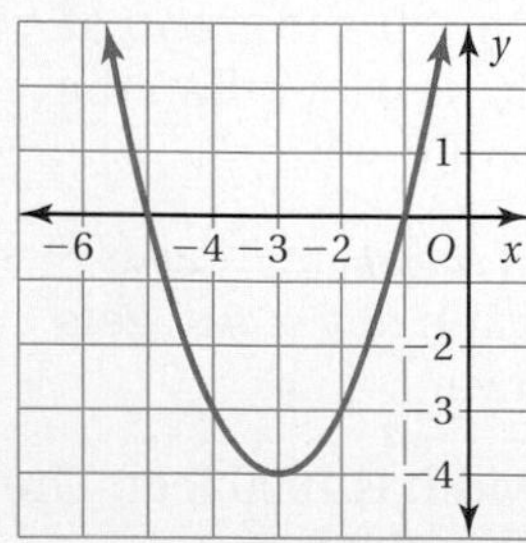

Graph the function. Compare the graph to the graph of $y = x^2$.

2. $y = 10x^2$
3. $y = \frac{1}{8}x^2$
4. $y = -4x^2$
5. $y = -\frac{3}{5}x^2$

1–5. See Additional Answers.

Taking Math Deeper

Exercise 40

Examine $y = ax^2$ when $a = 1$, $a = 2$, and $a = -\frac{1}{2}$.

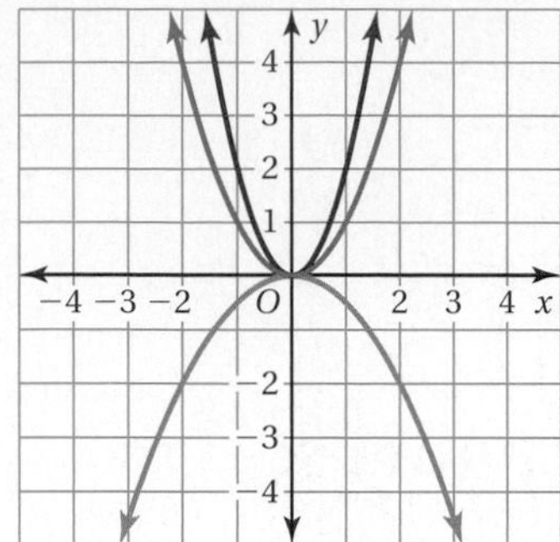

You can see that although the graphs get narrower and wider, they all have an x-intercept of 0.

You can also look at this problem algebraically. You know that the x-intercept occurs when $y = 0$. Replace y with 0 in the equation $y = ax^2$ and solve for x.

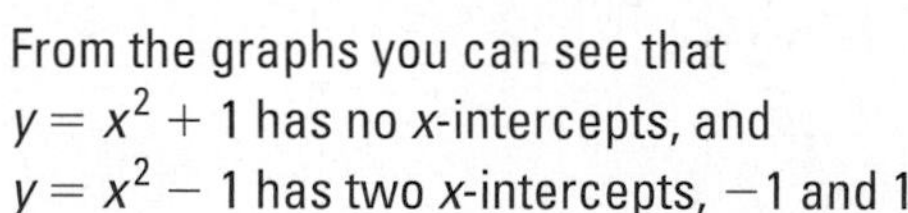

$0 = ax^2$	Substitute 0 for y.
$0 = x^2$	Divide each side by a.
$0 = x$	Solve for x.

So, the x-intercept of the graph of $y = ax^2$ is always 0.

Is there a way to change the equation so that its graph crosses the x-axis at a point other than (0, 0)? Try adding or subtracting a constant from $y = ax^2$, such as $y = x^2 + 1$ or $y = x^2 - 1$.

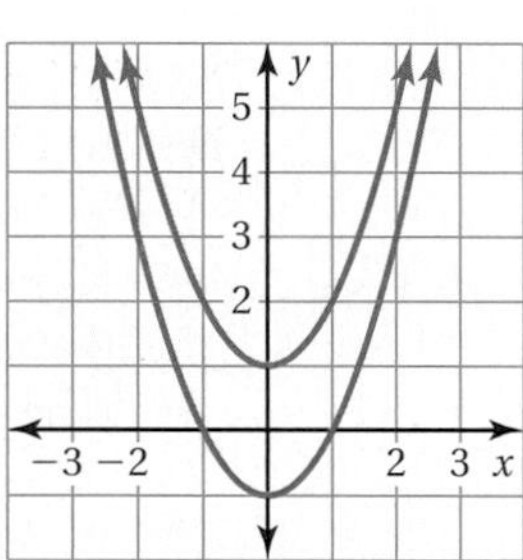

From the graphs you can see that $y = x^2 + 1$ has no x-intercepts, and $y = x^2 - 1$ has two x-intercepts, -1 and 1.

Reteaching and Enrichment Strategies

If students need help. . .	If students got it. . .
Resources by Chapter • Practice A and Practice B • Puzzle Time Record and Practice Journal Practice Differentiating the Lesson Lesson Tutorials Skills Review Handbook	Resources by Chapter • Enrichment and Extension Start the next section

In Exercises 35–37, use the graph of the function $y = ax^2$.

35. When is the function increasing?

36 When is the function decreasing?

37. Find the value of a when the graph passes through $(-2, 3)$.

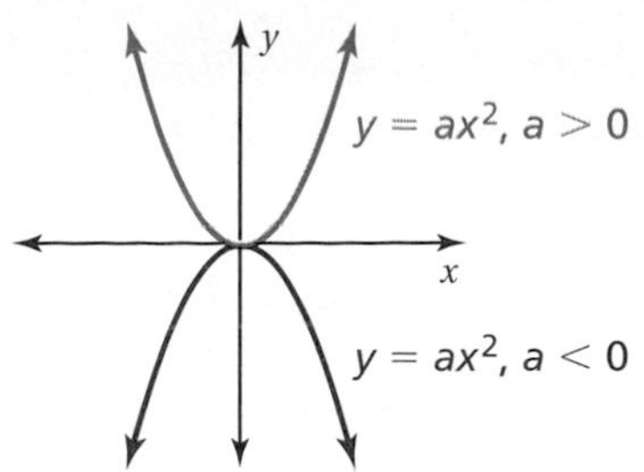

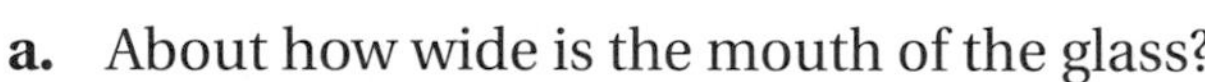

38. MODELING The diagram shows the cross section of a swirling glass of water, where x and y are measured in centimeters. The surface of the cross section of the rotating liquid is a parabola.

a. About how wide is the mouth of the glass?

b. Suppose the rotational speed of the liquid changes. The cross section can now be modeled by $y = 0.1x^2$. Did the rotational speed *increase* or *decrease*? Explain your reasoning.

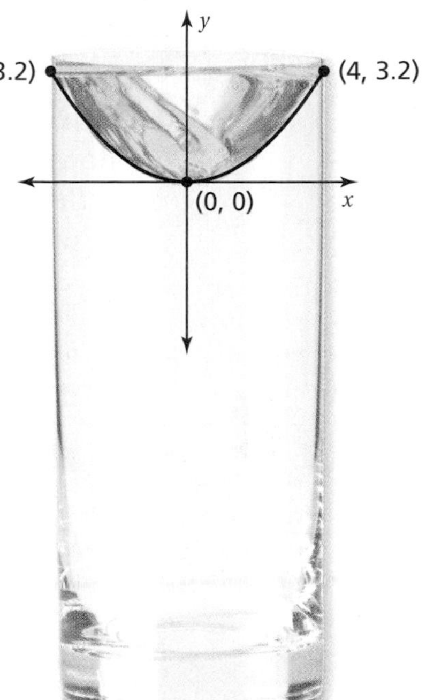

39. ASSEMBLY LINE The number y of units an assembly line can produce in 1 hour can be modeled by the function $y = 0.5x^2$, where x is the number of employees. The assembly line has a capacity of 10 employees.

a. Describe the domain of the function.

b. Graph the function using the domain in part (a).

c. Is it better for the company to run one assembly line at full capacity or two assembly lines at half capacity? Explain your reasoning.

40. **Logic** Is the x-intercept of the graph of $y = ax^2$ always 0? Justify your answer.

Fair Game Review What you learned in previous grades & lessons

Solve the proportion using the Cross Products Property. *(Skills Review Handbook)*

41. $\frac{x}{6} = \frac{13}{2}$ **42.** $\frac{5}{3} = \frac{n}{9}$ **43.** $\frac{4}{b} = \frac{6}{21}$ **44.** $\frac{14}{9} = \frac{7}{y}$

45. MULTIPLE CHOICE What is the completely factored form of $2x^5 - 8x^3$? *(Section 7.9)*

Ⓐ $2x^3(x-2)^2$ Ⓑ $2x^3(x-2)(x+2)$

Ⓒ $2x^3(x+2)^2$ Ⓓ $2x^3(x^2-4)$

8.2 Focus of a Parabola

COMMON CORE STATE STANDARDS
F.IF.4

Essential Question Why do satellite dishes and spotlight reflectors have parabolic shapes?

1 ACTIVITY: A Property of Satellite Dishes

Work with a partner. Rays are coming straight down. When they hit the parabola, they reflect off at the same angle at which they entered.

- Draw the outgoing part of each ray so that it intersects the y-axis.
- What do you notice about where the reflected rays intersect the y-axis?
- Where is the receiver for the satellite dish? Explain.

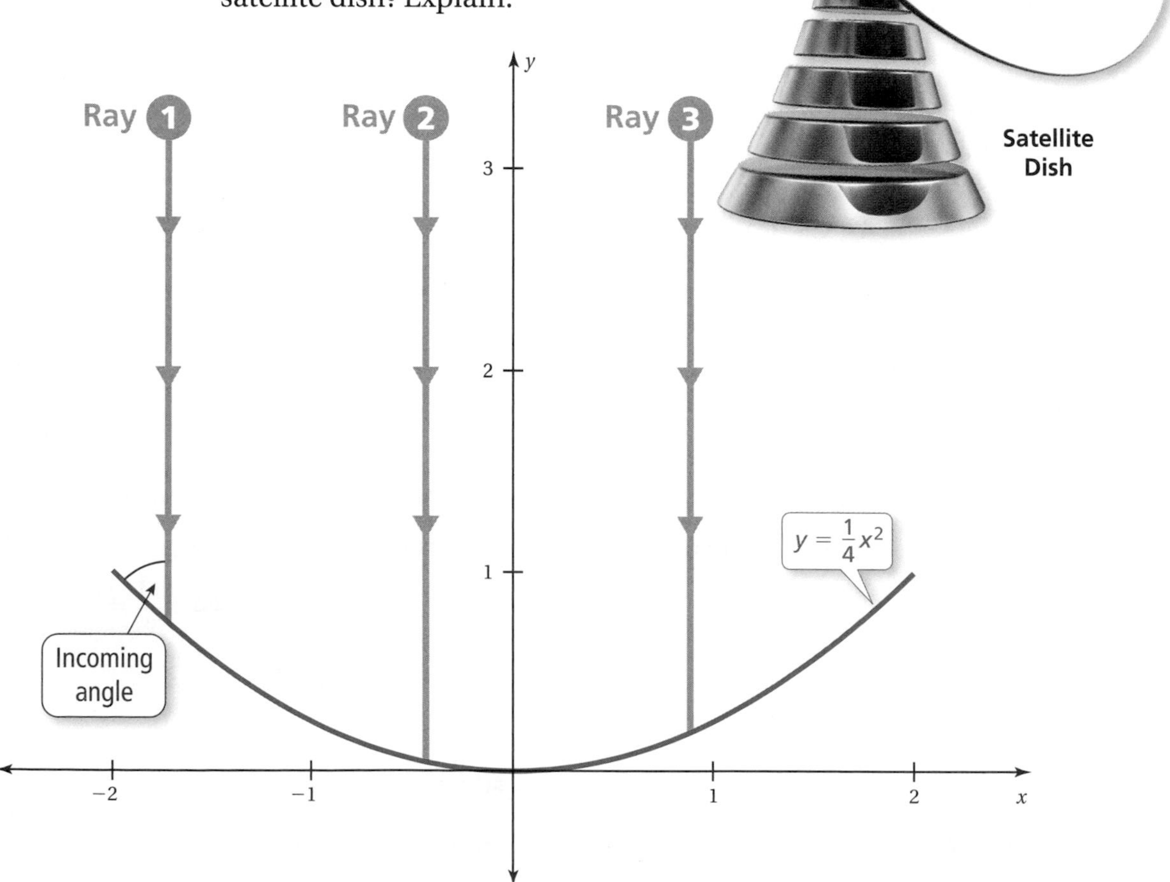

Laurie's Notes

Introduction

Standards for Mathematical Practice

- **MP4 Model with Mathematics:** The parabolic shape used in satellite dishes and projected lights (i.e. a flashlight) is very common. Students will explore many applications of the focus of a parabola in this section.

Motivate

- **Story Time:** If you have a cue stick and a billiard ball, use them as props. Tell students about a game of pool you were playing recently. You were trying to sink the 8-ball by banking it off the opposite side of the table. Sketch the following diagram.

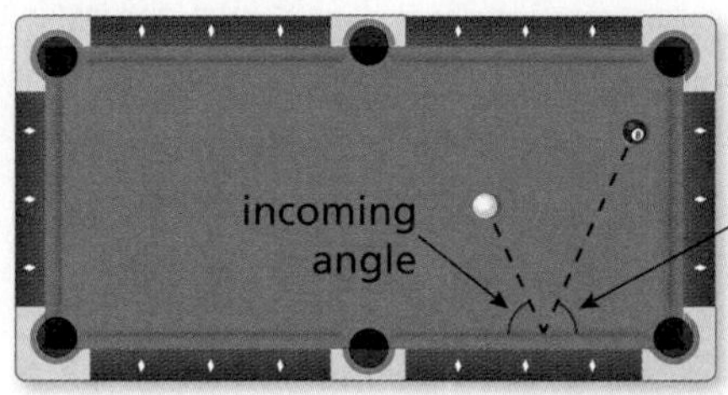

- Tell students about how carefully you measured to make sure that the angle was just right. Explain that the incoming angle is equivalent to the outgoing angle.
- **End of Story:** You sunk the 8-ball and won a math dictionary!

Activity Notes

Activity 1

- Connect this activity to the billiards scenario. The angle that the ray enters (incoming) is equivalent to the angle that it exits (outgoing).
- Students may have trouble because of the curvature of the graph of the satellite dish. Suggest to students that they use their hands to cover up part of the graph. Focus on the portion of the graph near the incoming ray.
- Students will be approximating the outgoing angle.
- If the students are fairly accurate, all of the outgoing rays should intersect the y-axis at the same point, (0, 1).
- Ask a group to share their results.

Common Core State Standards

F.IF.4 For a function that models a relationship between two quantities, interpret key features of graphs and tables in terms of the quantities, and sketch graphs showing key features given a verbal description of the relationship.

Previous Learning

Students should know how to graph $y = ax^2$.

Activity Materials
Introduction
• cue stick and billiard ball (optional)

Start Thinking! and Warm Up

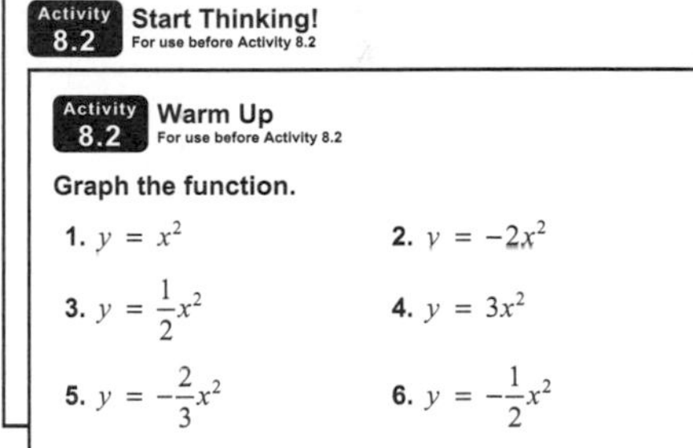

Activity 8.2 Start Thinking! For use before Activity 8.2

Activity 8.2 Warm Up For use before Activity 8.2

Graph the function.

1. $y = x^2$
2. $y = -2x^2$
3. $y = \frac{1}{2}x^2$
4. $y = 3x^2$
5. $y = -\frac{2}{3}x^2$
6. $y = -\frac{1}{2}x^2$

8.2 Record and Practice Journal

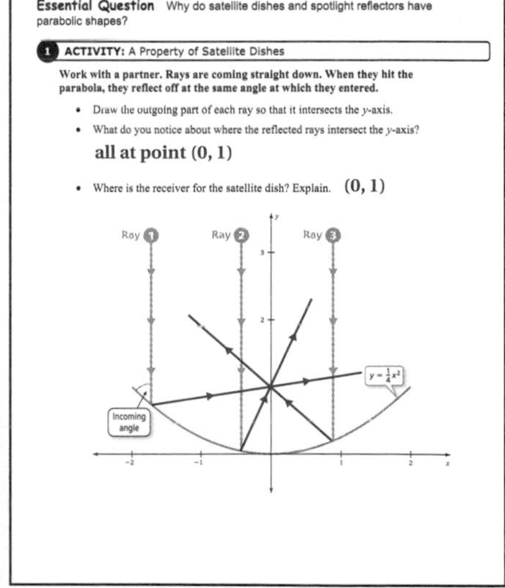

Essential Question Why do satellite dishes and spotlight reflectors have parabolic shapes?

1 ACTIVITY: A Property of Satellite Dishes

Work with a partner. Rays are coming straight down. When they hit the parabola, they reflect off at the same angle at which they entered.

- Draw the outgoing part of each ray so that it intersects the y-axis.
- What do you notice about where the reflected rays intersect the y-axis?
  all at point (0, 1)
- Where is the receiver for the satellite dish? Explain. (0, 1)

English Language Learners

Vocabulary

English language learners may relate the *focus* of a parabola with other definitions of the word, such as a center of activity or a point of concentration.

8.2 Record and Practice Journal

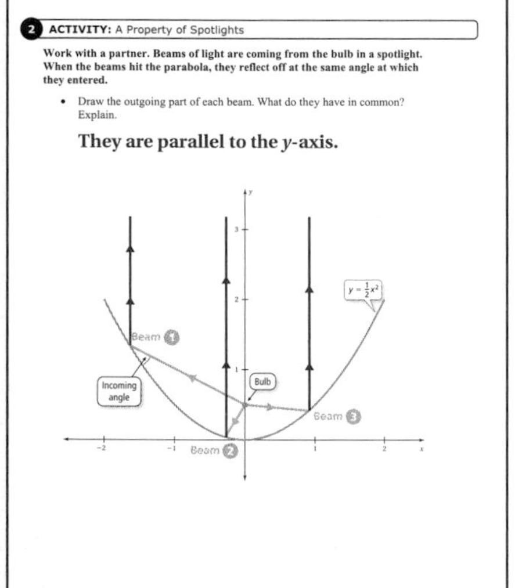

2 ACTIVITY: A Property of Spotlights

Work with a partner. Beams of light are coming from the bulb in a spotlight. When the beams hit the parabola, they reflect off at the same angle at which they entered.

- Draw the outgoing part of each beam. What do they have in common? Explain.

They are parallel to the y-axis.

What Is Your Answer?

3. **IN YOUR OWN WORDS** Why do satellite dishes and spotlight reflectors have parabolic shapes?

Satellite dishes: incoming signals are reflected to the receiver, located at the parabola's focus. Spotlight reflectors: beams of light emitted from the bulb, located at the focus, are reflected parallel to the parabola's axis of symmetry.

4. Design and draw a parabolic satellite dish. Label the dimensions of the dish. Label the receiver.

Check students' work.

Laurie's Notes

Activity 2

- This activity reverses the process used in Activity 1. The point on the y-axis is known. The light hits the reflective surface and reflects out.
- If students are fairly accurate, all of the outgoing rays should be parallel.
- If you have the flashlight or overhead projector bulb from the previous lesson, use it as a model for this activity.

What Is Your Answer?

- If time permits, let students do some research on Question 3.

Closure

- **Writing Prompt:** Other objects similar to a satellite dish and flashlight are . . .

Technology For the Teacher

The Dynamic Planning Tool
Editable Teacher's Resources at *BigIdeasMath.com*

2 ACTIVITY: A Property of Spotlights

Work with a partner. Beams of light are coming from the bulb in a spotlight. When the beams hit the parabola, they reflect off at the same angle at which they entered.

- Draw the outgoing part of each beam. What do they have in common? Explain.

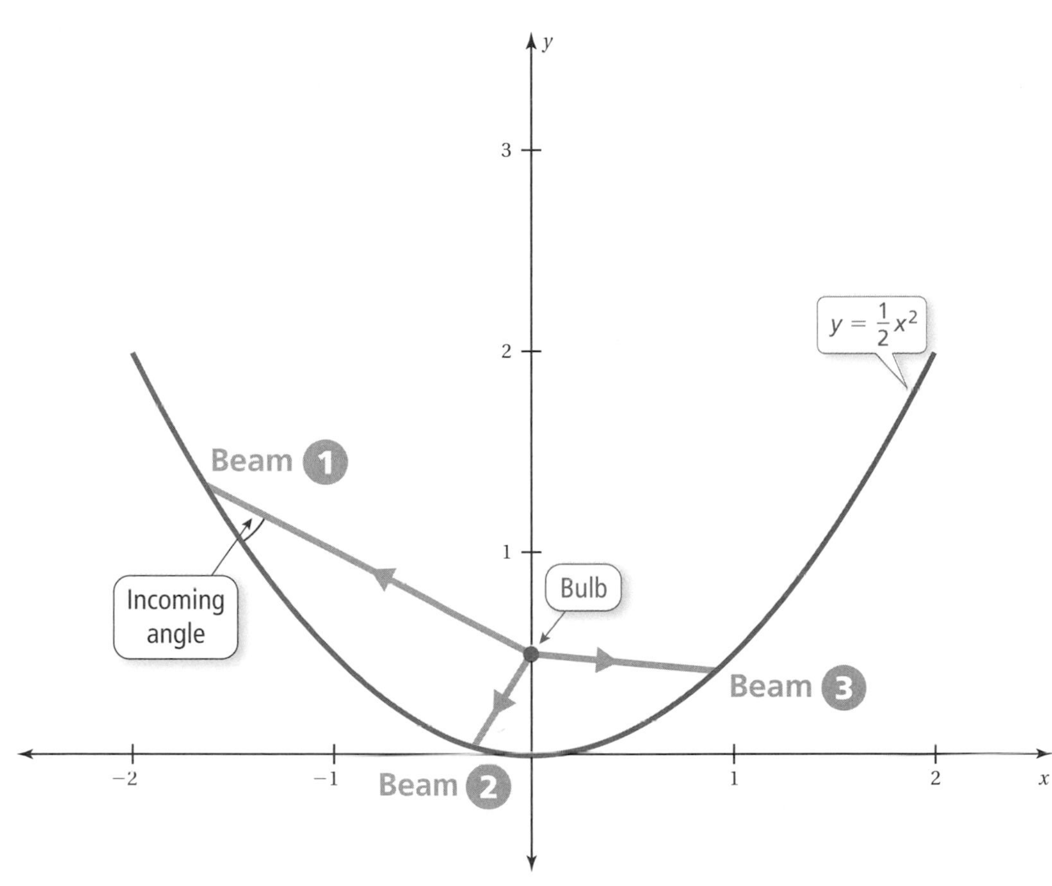

What Is Your Answer?

3. **IN YOUR OWN WORDS** Why do satellite dishes and spotlight reflectors have parabolic shapes?

4. Design and draw a parabolic satellite dish. Label the dimensions of the dish. Label the receiver.

Practice Use what you learned about parabolas to complete Exercises 4–6 on page 414.

8.2 Lesson

Key Vocabulary
focus, *p. 412*

The Focus of a Parabola

The **focus** of a parabola is a fixed point on the interior of a parabola that lies on the axis of symmetry. A parabola "wraps" around the focus.

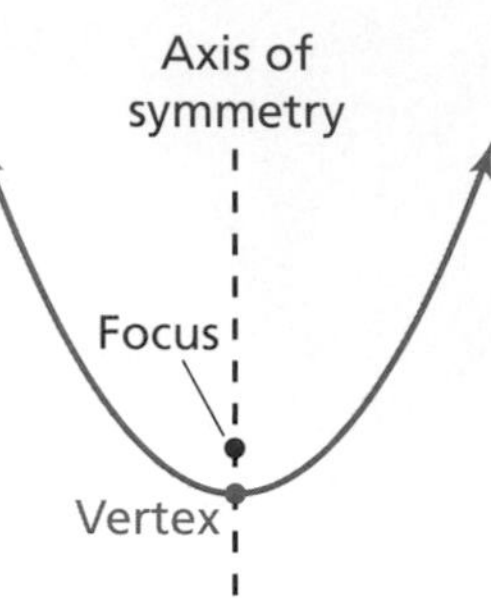

For functions of the form $y = ax^2$, the focus is $\left(0, \frac{1}{4a}\right)$.

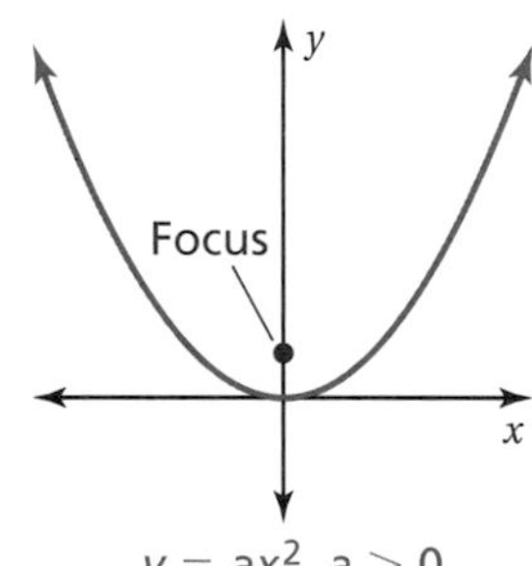

$y = ax^2, a > 0$

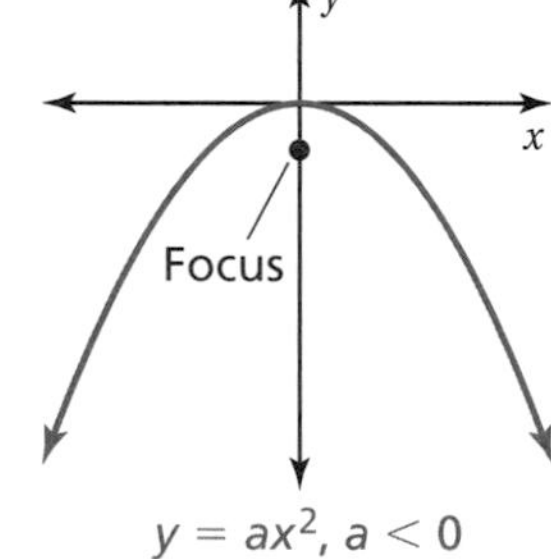

$y = ax^2, a < 0$

EXAMPLE 1 Finding the Focus of a Parabola

Graph $y = -\frac{1}{4}x^2$. Identify the focus.

Step 1: Make a table of values. Then graph.

x	−4	−2	0	2	4
y	−4	−1	0	−1	−4

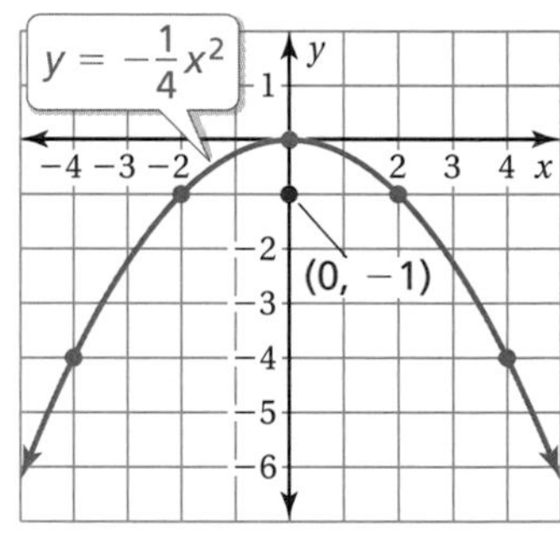

Step 2: Identify the focus. The function is of the form $y = ax^2$, so $a = -\frac{1}{4}$.

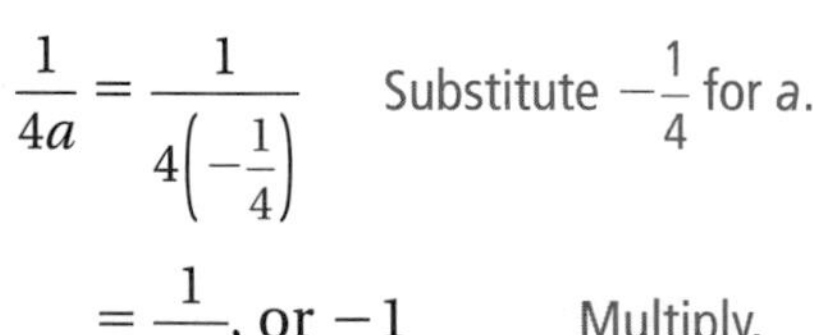

$$\frac{1}{4a} = \frac{1}{4\left(-\frac{1}{4}\right)} \quad \text{Substitute } -\frac{1}{4} \text{ for } a.$$

$$= \frac{1}{-1}, \text{ or } -1 \quad \text{Multiply.}$$

So, the focus of the function $y = -\frac{1}{4}x^2$ is $(0, -1)$.

On Your Own

Now You're Ready
Exercises 7–12

Graph the function. Identify the focus.

1. $y = 2x^2$
2. $y = \frac{1}{6}x^2$
3. $y = -3x^2$

Laurie's Notes

Introduction

Connect

- **Yesterday:** Students explored properties of parabolas. (MP4)
- **Today:** Students will find and use the focus of a parabola.

Motivate

- Show a video of how to make a parabolic solar cooker.

Lesson Notes

Key Idea

- Write the Key Idea about the focus of a parabola.
- Sketch the graphs of $y = ax^2$ for $a > 0$ and for $a < 0$. Label the focus in each graph.
- For functions of the form $y = ax^2$, the *focus* is $\left(0, \frac{1}{4a}\right)$. This will be derived in a future mathematics course.
- Explain to students that the focus is an important property of parabolas. In the activity, the focus was found by using the congruence of the incoming and outgoing angles.

Example 1

- Write the equation and identify $a = -\frac{1}{4}$.
- Make a table of values and sketch the graph of the function.
- To find the *y*-coordinate of the focus, substitute $a = -\frac{1}{4}$ into the expression $\frac{1}{4a}$. The complex fraction simplifies to -1.
- Plot the focus, $(0, -1)$.
- **Connection:** If time permits, trace rays from the focus to the parabola and then trace the outgoing rays. The outgoing rays should be parallel.

On Your Own

- These questions provide practice with graphing quadratics of the form $y = ax^2$.
- **Neighbor Check:** Have students work independently and then have their neighbor check their work. Have students discuss any discrepancies.

Goal Today's lesson is finding and using the **focus** of a parabola.

Lesson Materials
Introduction
• video

Start Thinking! and Warm Up

If you were designing the headlights of a car, would you use a parabolic reflector? Why or why not?

Extra Example 1

Graph $y = -\frac{1}{2}x^2$. Identify the focus.

focus: $\left(0, -\frac{1}{2}\right)$

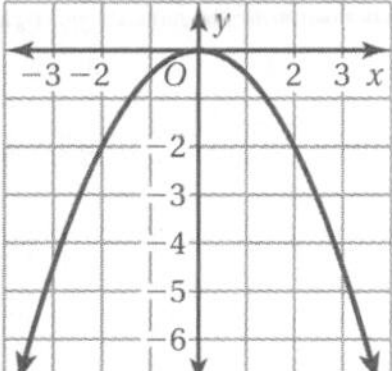

On Your Own

1. focus: $\left(0, \frac{1}{8}\right)$

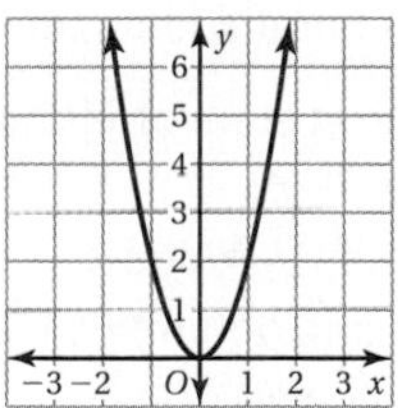

2. focus: $\left(0, \frac{3}{2}\right)$

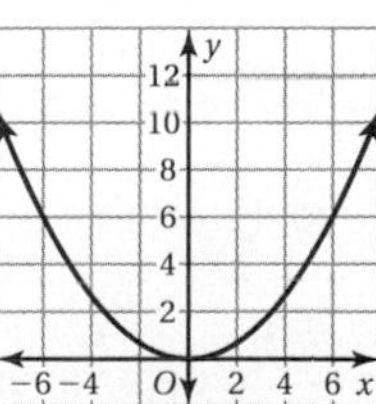

3. See Additional Answers.

Extra Example 2

Write an equation of the parabola with focus (0, 3) and vertex at the origin.

$y = \frac{1}{12}x^2$

Extra Example 3

In Example 3, the cross section of the microphone can be modeled by $y = \frac{1}{30}x^2$. What is the length of the receiver arm? 7.5 in.

On Your Own

4. $y = -\frac{1}{12}x^2$

5. 10 in.

Differentiated Instruction

Organization

In their notebooks, have students write steps for identifying features of the graph of a quadratic function.

Step 1: Write in standard form, $y = ax^2 + bx + c$.
Step 2: Identify a, b, and c.
Step 3: Use a to determine if the graph opens up or down.
Step 4: Use $|a|$ to determine if the graph is narrower or wider.

Later in the chapter, additional steps can be added.

Step 5: Use c to determine the y-intercept.

Step 6: Use $-\frac{b}{2a}$ to determine the axis of symmetry and the x-coordinate of the vertex.

Laurie's Notes

Example 2

- ? "If you know the coordinates of the focus, do you think you can write an equation of the parabola?" Students may have some thoughts about how you can find an equation.
- Plot the ordered pair (0, 4).
- ? "If (0, 4) is the focus, what do you know?" $(0, 4) = \left(0, \frac{1}{4a}\right)$
- ? "Can you find a?" Yes, set the two y-coordinates equal and solve.
- Solve the equation to show that $a = \frac{1}{16}$.
- ? "How does the graph of $y = \frac{1}{16}x^2$ compare to the graph of $y = x^2$?" It is wider.

Example 3

- This problem is similar to an example from the last lesson.
- **MP4 Model with Mathematics**: Explain that the arm length is the distance from the focus to the vertex. This distance is also called the *focal length*.
- ? "What is a in this example?" $a = \frac{1}{24}$
- Substitute to find the y-coordinate of the focus. This results in a complex fraction, which could be rewritten as $1 \div \frac{1}{6}$.
- ? "If the focus is (0, 6), what is the length of the receiver arm?" 6 inches

On Your Own

- **Think-Pair-Share:** Students should read each question independently and then work with a partner to answer the questions. When they have answered the questions, the pair should compare their answers with another group and discuss any discrepancies.

Closure

- Find the focus of $y = 6x^2$. $\left(0, \frac{1}{24}\right)$

Technology For the Teacher

The Dynamic Planning Tool
Editable Teacher's Resources at *BigIdeasMath.com*

EXAMPLE 2 Writing an Equation of a Parabola

Write an equation of the parabola with focus (0, 4) and vertex at the origin.

For $y = ax^2$, the focus is $\left(0, \frac{1}{4a}\right)$. Use the given focus, (0, 4), to write an equation to find a.

$\frac{1}{4a} = 4$ Equate the y-coordinates.

$1 = 16a$ Multiply each side by $4a$.

$\frac{1}{16} = a$ Divide each side by 16.

∴ An equation of the parabola is $y = \frac{1}{16}x^2$.

EXAMPLE 3 Real-Life Application

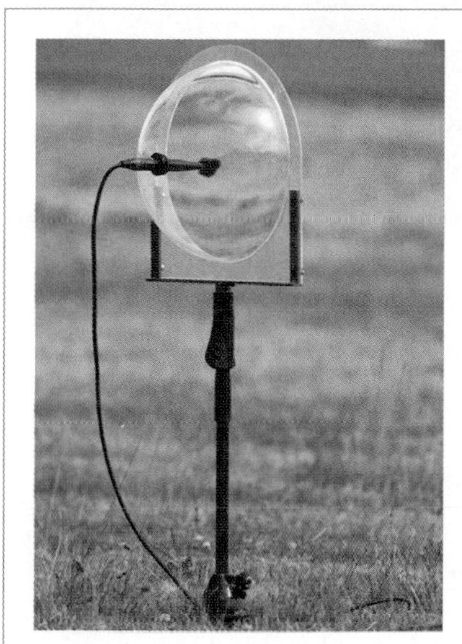

A birdwatcher uses a parabolic microphone to collect and record bird sounds. The cross section of the microphone can be modeled by $y = \frac{1}{24}x^2$, where x and y are measured in inches. The focus is located at the end of the receiver arm. What is the length of the receiver arm?

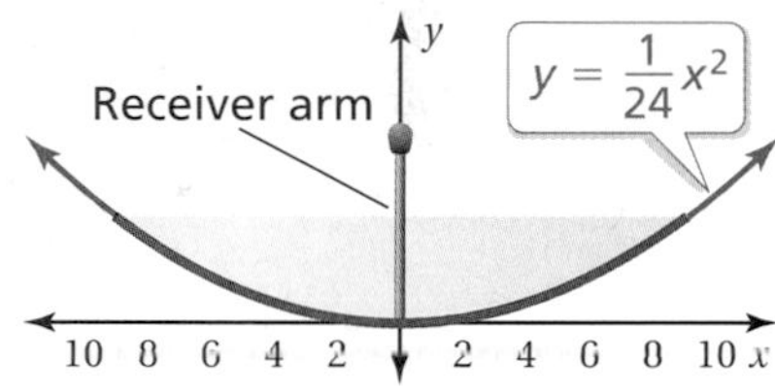

The arm length is the distance from the focus to the vertex.

Identify the focus. For the function $y = \frac{1}{24}x^2$, $a = \frac{1}{24}$.

$\frac{1}{4a} = \frac{1}{4\left(\frac{1}{24}\right)}$ Substitute $\frac{1}{24}$ for a.

$= \frac{1}{\frac{1}{6}}$ Multiply.

$= 6$ Divide.

The focus is (0, 6). The vertex is (0, 0). The distance from (0, 0) to (0, 6) is 6 units.

∴ So, the length of the receiver arm is 6 inches.

On Your Own

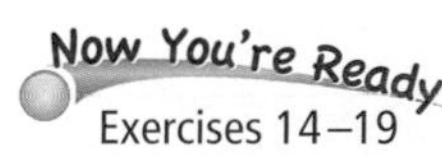

4. Write an equation of the parabola with focus (0, −3) and vertex at the origin.
5. **WHAT IF?** In Example 3, the cross section of the microphone can be modeled by $y = \frac{1}{40}x^2$. What is the length of the receiver arm?

8.2 Exercises

Vocabulary and Concept Check

1. **VOCABULARY** What is the relationship between the focus and the axis of symmetry of a parabola?
2. **WRITING** When the focus of a parabola lies below the vertex, does the parabola open up or down? Is $a > 0$ or $a < 0$? Explain.
3. **OPEN-ENDED** Write an equation of a parabola whose focus is below the x-axis.

Practice and Problem Solving

Determine whether the shape is parabolic.

4. 

5. 

6. 

Graph the function. Identify the focus.

1 7. $y = x^2$

8. $y = 4x^2$

9. $y = -12x^2$

10. $y = \frac{1}{4}x^2$

11. $y = \frac{1}{2}x^2$

12. $y = -0.75x^2$

13. **ERROR ANALYSIS** Describe and correct the error in identifying the focus.

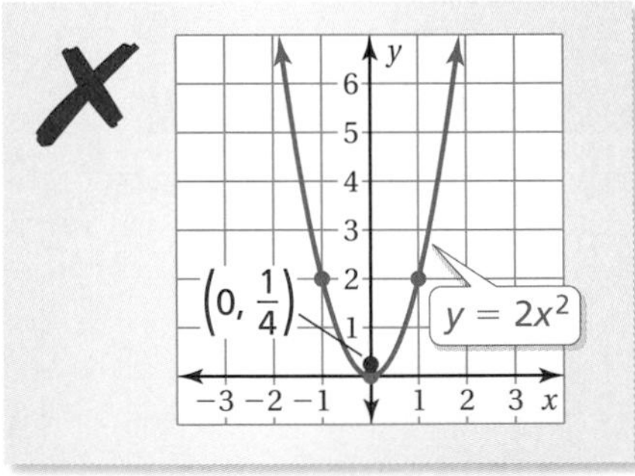

Write an equation of the parabola with a vertex at the origin and the given focus.

14. $(0, 1)$

15. $(0, -2)$

16. $(0, 5)$

17. $\left(0, -\frac{1}{4}\right)$

18. $(0, -1)$

19. $(0, 0.5)$

20. **COMET** A comet travels along a parabolic path around the Sun. The Sun is the focus of the path. When the comet is at the vertex of the path, it is 60,000,000 kilometers from the Sun. Write an equation that represents the path of the comet. Assume the focus is on the positive y-axis and the vertex is $(0, 0)$.

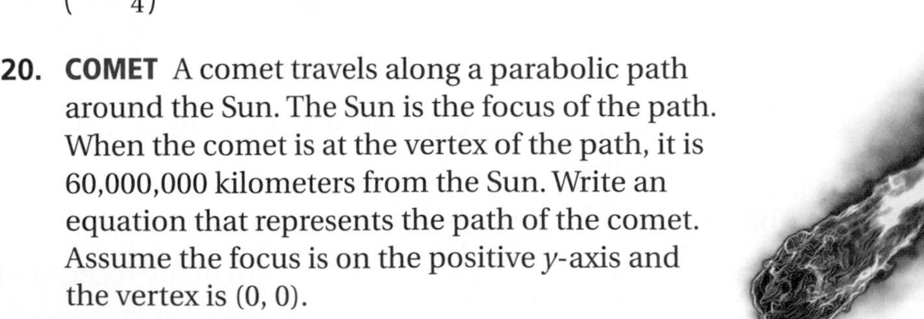

Assignment Guide and Homework Check

Level	Assignment	Homework Check
Average	1–3, 7–19 odd, 20, 24–28, 31–34	11, 13, 15, 20, 28
Advanced	1–3, 8–20 even, 13, 27–30, 31–34	13, 18, 20, 28, 30

Common Errors

- **Exercises 4–6** Students may believe all three shapes are parabolic. Point out that the skating ramp (half pipe) in Exercise 6 does not have a parabolic shape.
- **Exercises 14–19** Students may have difficulty finding the value of *a*, especially because this requires working with fractions. Encourage them to take their time.
- **Exercises 21–23** Students may have difficulty matching the equation with its graph because the foci are very close to one another. Remind students how to determine when a fraction is greater or lesser in value than another fraction.
- **Exercise 29** Students may not know how to find the equation of the parabolic reflective surface of the solar cooker. Encourage them to use a point (x, y) on the graph and the equation $y = ax^2$ to find *a*.

8.2 Record and Practice Journal

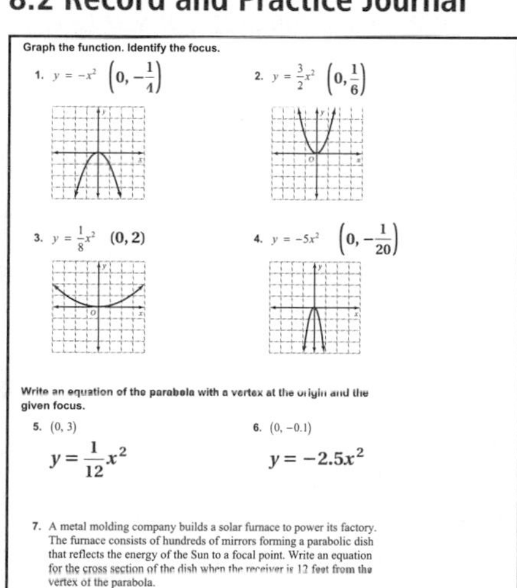

Graph the function. Identify the focus.

1. $y = -x^2$ $\left(0, -\frac{1}{4}\right)$
2. $y = \frac{3}{2}x^2$ $\left(0, \frac{1}{6}\right)$
3. $y = \frac{1}{8}x^2$ $(0, 2)$
4. $y = -5x^2$ $\left(0, -\frac{1}{20}\right)$

Write an equation of the parabola with a vertex at the origin and the given focus.

5. $(0, 3)$ $y = \frac{1}{12}x^2$
6. $(0, -0.1)$ $y = -2.5x^2$

7. A metal molding company builds a solar furnace to power its factory. The furnace consists of hundreds of mirrors forming a parabolic dish that reflects the energy of the Sun to a focal point. Write an equation for the cross section of the dish when the receiver is 12 feet from the vertex of the parabola.

$y = \frac{1}{48}x^2$

Technology For the Teacher

Answer Presentation Tool

Vocabulary and Concept Check

1. The focus is a coordinate point which lies on the axis of symmetry. If the axis of symmetry is $x = k$, then the x-coordinate of the focus is k.
2. down; $a < 0$
3. *Sample answer:* $y = -4x^2$

Practice and Problem Solving

4. parabolic
5. parabolic
6. not parabolic

7–12. See Additional Answers.

13. The student did not multiply 2 by 4. The focus is $\left(0, \frac{1}{8}\right)$.
14. $y = \frac{1}{4}x^2$
15. $y = -\frac{1}{8}x^2$
16. $y = \frac{1}{20}x^2$
17. $y = -x^2$
18. $y = -\frac{1}{4}x^2$
19. $y = \frac{1}{2}x^2$
20. $y = \frac{1}{240{,}000{,}000}x^2$

Practice and Problem Solving

21. B **22.** A

23. C

24. never; The focus of a parabola is a fixed point on the interior of a parabola.

25. always; The vertex of the graph of a function of the form $y = ax^2$ is always at the origin. So if the parabola opens down, then the focus will always lie below the x-axis.

26. sometimes; This is true only when the parabola opens up.

27. As the distance between the vertex and focus increases, the graph of $y = ax^2$ gets wider.

28. *Sample answer:* $y = \frac{1}{12}x^2$

29. 1.5 inches

30. See *Taking Math Deeper.*

Fair Game Review

31. 16 **32.** 10

33. 1 **34.** C

Mini-Assessment

Write an equation of the parabola with the given focus and vertex at the origin.

1. $(0, 6)$ $y = \frac{1}{24}x^2$

2. $(0, -3)$ $y = -\frac{1}{12}x^2$

3. $\left(0, \frac{1}{3}\right)$ $y = \frac{3}{4}x^2$

Taking Math Deeper

Exercise 30

In this exercise, it is essential to recognize that $y = (ax)^2$ is a quadratic function. Use the Power of a Product Property to rewrite the function as $y = a^2x^2$.

1 Begin by finding the focus of each parabola.

Focus of $y = ax^2$: $\left(0, \frac{1}{4a}\right)$ Focus of $y = a^2x^2$: $\left(0, \frac{1}{4a^2}\right)$

Because a can be positive or negative, the *absolute value* of the y-coordinate represents the distance from the focus to the vertex.

So, you need to find the values of a for which $\left|\frac{1}{4a}\right| < \left|\frac{1}{4a^2}\right|$.

2 This inequality can be simplified. By removing the quantities that are always positive from the absolute value, you are left with

$$\frac{1}{4|a|} < \frac{1}{4a^2}.$$

As the denominator of a fraction increases, the value of the fraction decreases. So, find the values of a for which $4|a| > 4a^2$, or $|a| > a^2$.

3 Essentially, you are looking for numbers with absolute values that are greater than their squares. Challenge students to think of positive numbers that are greater than their squares.

$1^2 = 1$ ✗ $2^2 = 4$ ✗ $\left(\frac{1}{10}\right)^2 = \frac{1}{100}$ ✓ $\left(\frac{2}{3}\right)^2 = \frac{4}{9}$ ✓

Several examples should convince students that $|a| > a^2$ when $0 < a < 1$. Because you are looking for the *absolute value* of a number that is greater than its square, this also works for $-1 < a < 0$.

So, when $0 < |a| < 1$, the distance between the focus and the vertex of the graph of $y = ax^2$ is less than the distance between the focus and the vertex of the graph of $y = (ax)^2$.

Reteaching and Enrichment Strategies

If students need help...	If students got it...
Resources by Chapter • Practice A and Practice B • Puzzle Time Record and Practice Journal Practice Differentiating the Lesson Lesson Tutorials Skills Review Handbook	Resources by Chapter • Enrichment and Extension Start the next section

Match the equation with its graph.

21. $y = 2x^2$

22. $y = 3x^2$

23. $y = 2.5x^2$

A. 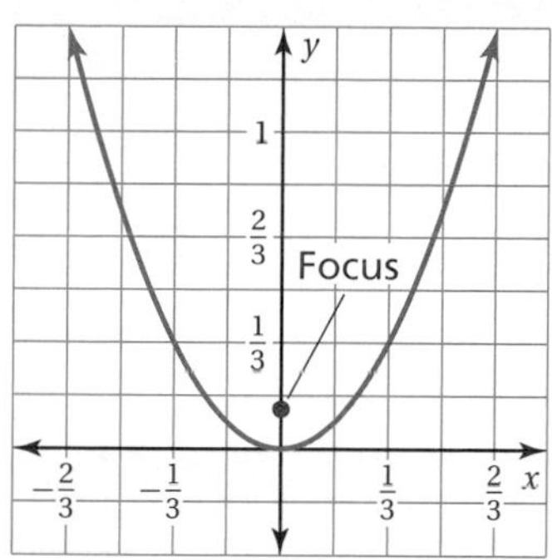

B. 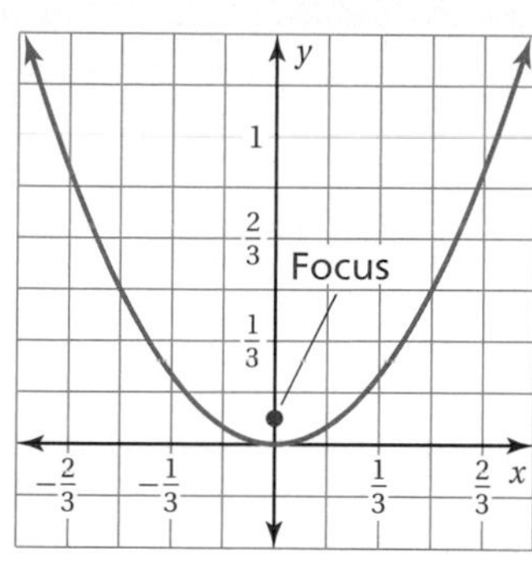

C. 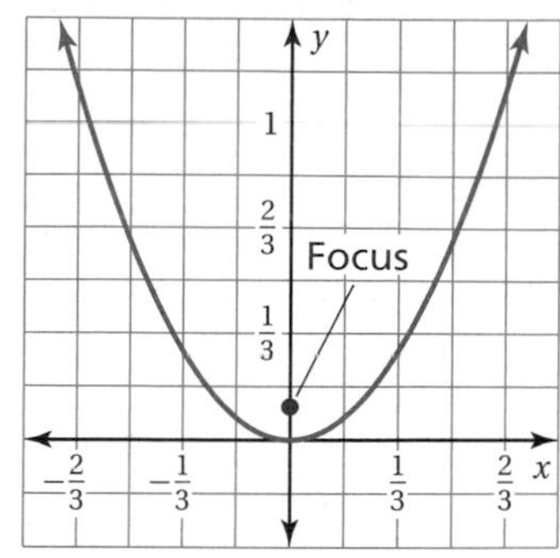

Determine whether the statement is *sometimes, always,* or *never* true for the function $y = ax^2$. Explain your reasoning.

24. The vertex and focus of a parabola can occur at the same point.

25. If a parabola opens down, then the focus lies below the x-axis.

26. The y-coordinate of the focus of a parabola is greater than the y-coordinate of the vertex.

27. **REASONING** Describe how the graph of $y = ax^2$ changes as the distance between the vertex and focus increases.

28. **WHISPER DISH** Whisper dishes are parabolic sound reflectors that transmit and receive sound waves from opposite ends of a room. For the best sound reception, you place your ear at the focus of the dish. Write an equation for the cross section of a dish when your ear is 3 feet from the vertex.

29. **SOLAR COOKING** You make a solar cooker using a parabolic reflective surface. You suspend a sausage with wire through the focus of each end piece of the cooker. How far from the bottom should you place the wire?

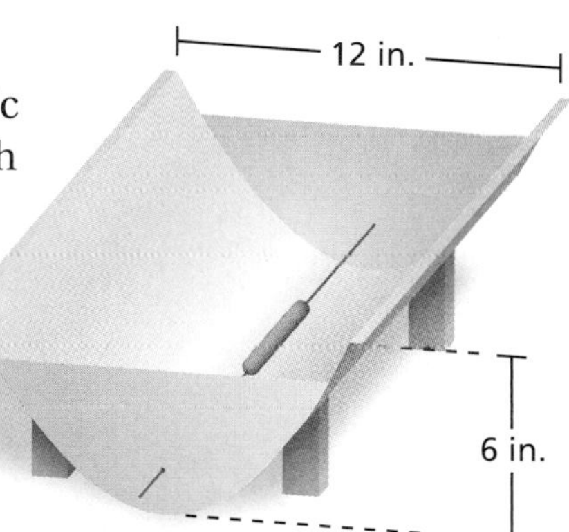

30. **Structure** For what values of a will the distance between the focus and the vertex of the graph of $y = ax^2$ be less than the distance between the focus and the vertex of the graph of $y = (ax)^2$?

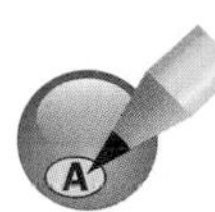

Fair Game Review What you learned in previous grades & lessons

Evaluate the expression when $x = -3$ and $n = 2$. *(Skills Review Handbook)*

31. $x^2 + 7$

32. $3n^2 - 2$

33. $x + n^2$

34. **MULTIPLE CHOICE** Which number is equivalent to $32^{3/5}$? *(Section 6.3)*

Ⓐ 2 Ⓑ 4 Ⓒ 8 Ⓓ 10

8.3 Graphing $y = ax^2 + c$

COMMON CORE STATE STANDARDS

F.BF.3

Essential Question How does the value of c affect the graph of $y = ax^2 + c$?

1 ACTIVITY: Graphing $y = ax^2 + c$

Work with a partner. Sketch the graphs of both functions in the same coordinate plane. How does the value of c affect the graph of $y = ax^2 + c$?

a. $y = x^2$ and $y = x^2 + 2$

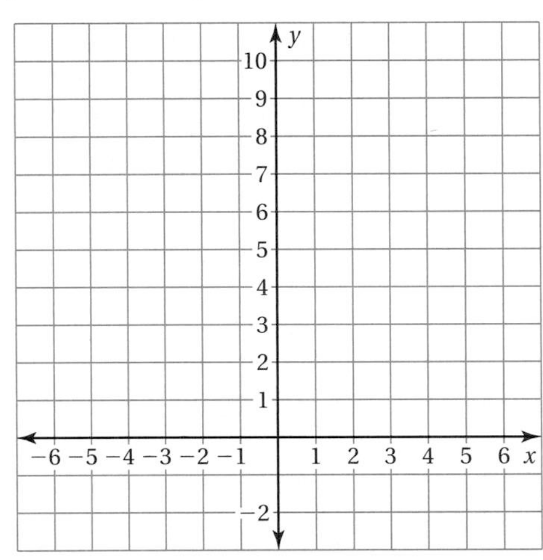

b. $y = 2x^2$ and $y = 2x^2 - 2$

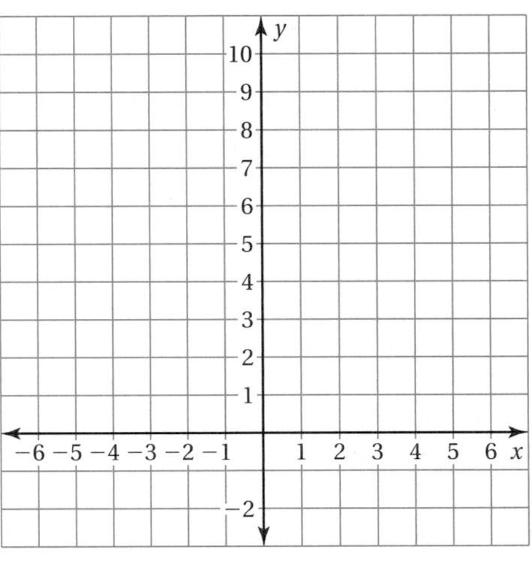

c. $y = -x^2 + 4$ and $y = -x^2 + 9$

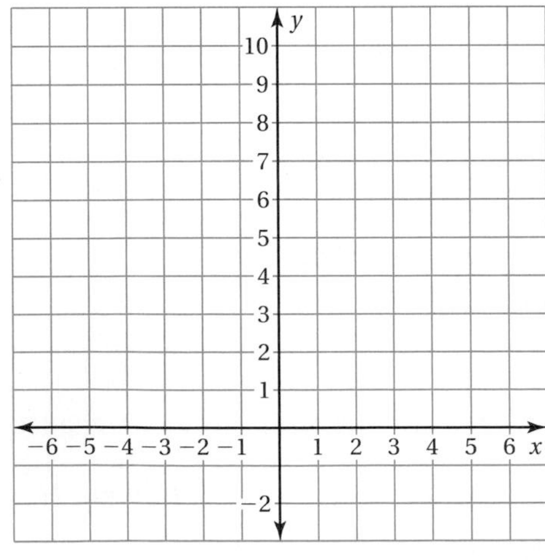

d. $y = \frac{1}{2}x^2$ and $y = \frac{1}{2}x^2 - 8$

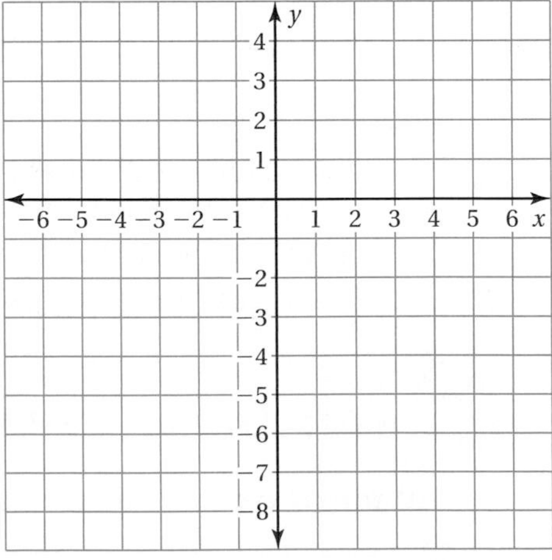

Laurie's Notes

Introduction

Standards for Mathematical Practice

- **MP1a Make Sense of Problems:** Students should be comfortable with graphing $y = x^2$ and $y = ax^2$ and explaining the effect of a on the graph of $y = ax^2$. Today, they will make sense of another parameter when they graph $y = ax^2 + c$. Using tabular and graphical representations, they make sense of the effect of c.

Motivate

- ? "How are the graphs of $y = 3x$ and $y = 3x - 4$ related?" They have the same slope, so they are parallel.
- ? "What is the equation of the line parallel to $y = -2x$ shifted up 4 units?" $y = -2x + 4$
- Explain that today they will graph pairs of functions and try to determine the relationship between the graphs.

Activity Notes

Activity 1

- Students should be comfortable graphing functions of the form $y = ax^2$ by making a table of values, plotting the ordered pairs, and drawing a smooth curve through the points.
- Students are graphing a pair of functions in the same coordinate plane.
- Discuss the characteristics of each graph, such as opens up or opens down, and wider or narrower than $y = x^2$.
- Students may only notice that the vertex is no longer at (0, 0). They may not think of "c" as a vertical shift.
- **Connection:** Discuss the idea of the vertex being the y-intercept of the graph.
- ? "Could there be more than one y-intercept? Explain." no; It would not be a function.
- **Common Error:** Students do not always see that the pairs of graphs are similar, with one graph being a translation of the other. This is sometimes the case because different parts of each graph may be plotted. This may happen when using a graphing calculator with an inappropriate viewing window.

Common Core State Standards

F.BF.3 Identify the effect on the graph of replacing $f(x)$ by $f(x) + k$, $kf(x)$, $f(kx)$, and $f(x + k)$ for specific values of k (both positive and negative); find the value of k given the graphs. Experiment with cases and illustrate an explanation of the effects on the graph using technology.

Previous Learning

Students should know how to graph $y = ax^2$.

Start Thinking! and Warm Up

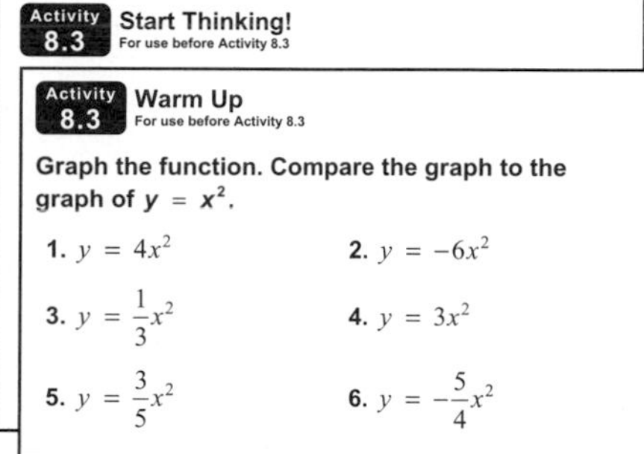

Activity 8.3 Start Thinking! For use before Activity 8.3

Activity 8.3 Warm Up For use before Activity 8.3

Graph the function. Compare the graph to the graph of $y = x^2$.

1. $y = 4x^2$
2. $y = -6x^2$
3. $y = \frac{1}{3}x^2$
4. $y = 3x^2$
5. $y = \frac{3}{5}x^2$
6. $y = -\frac{5}{4}x^2$

8.3 Record and Practice Journal

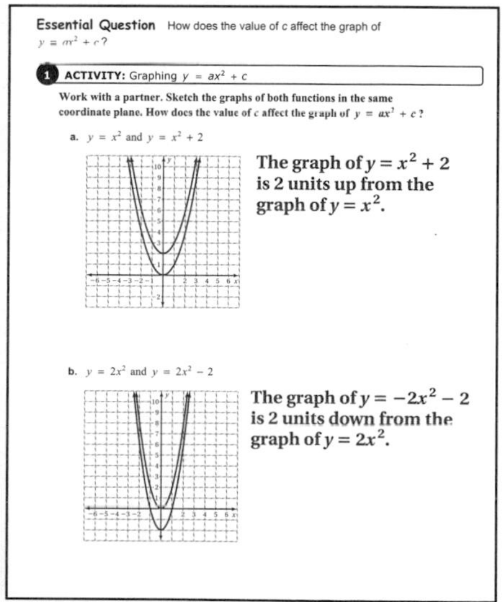

Essential Question How does the value of c affect the graph of $y = ax^2 + c$?

1 ACTIVITY: Graphing $y = ax^2 + c$

Work with a partner. Sketch the graphs of both functions in the same coordinate plane. How does the value of c affect the graph of $y = ax^2 + c$?

a. $y = x^2$ and $y = x^2 + 2$

The graph of $y = x^2 + 2$ is 2 units up from the graph of $y = x^2$.

b. $y = 2x^2$ and $y = 2x^2 - 2$

The graph of $y = -2x^2 - 2$ is 2 units down from the graph of $y = 2x^2$.

English Language Learners

Connection

The function $y = x^2$ is the most basic quadratic function just like $y = x$ is the most basic linear function. English language learners can make the connection that for both $y = 2x^2 + 4$ and $y = 2x + 4$, the 4 represents a translation of 4 units up.

8.3 Record and Practice Journal

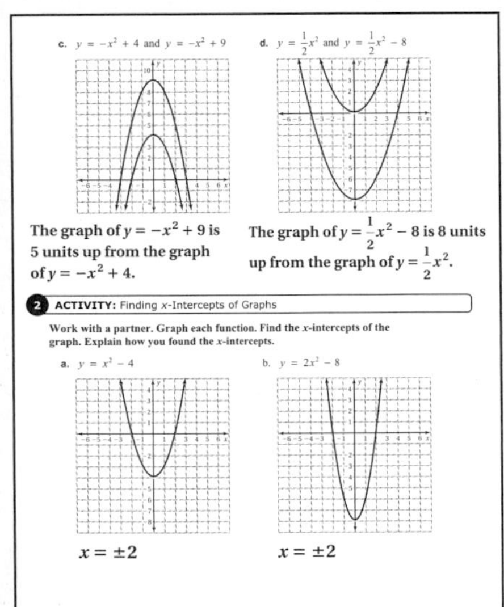

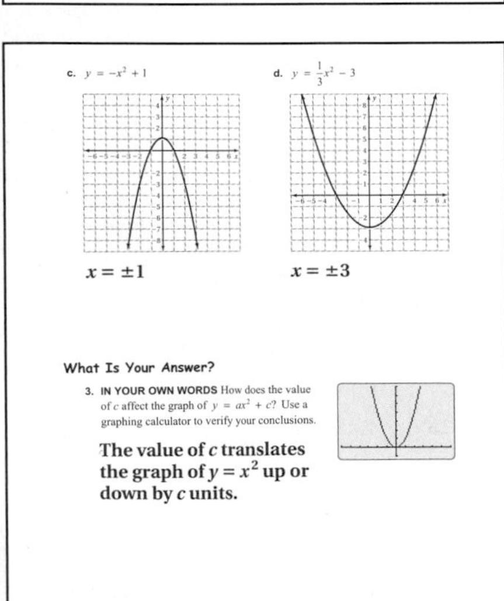

Laurie's Notes

Activity 2

- ? "What is an x-intercept?" It is where the graph crosses the x-axis.
- ? "Can a parabola have more than one x-intercept? Explain." yes; A parabola can open up and have a vertex below the x-axis, giving it two x-intercepts.
- Explain that in this activity they will be looking for the x-intercept(s) of the graph.
- Again, students are graphing by first making a table of values. When making the table, students may identify an x-intercept.
- When students have finished all four parts of the activity, ask for volunteers to share their graphs and explain how they found the x-intercepts(s). If the intercepts were not found when making the table of values, then it is possible that students may have guessed, or identified the x-intercepts from the graph.
- Discuss the line of symmetry for each graph. Because each graph is symmetric about the y-axis, the x-intercepts are opposites of one another.

What Is Your Answer?

- Make sure students are using appropriate viewing windows when verifying their conclusions.

Closure

- **Exit Ticket:** Graph $y = x^2$ and $y = x^2 - 4$ in the same coordinate plane. How are the graphs related? The graph of $y = x^2 - 4$ is a translation 4 units down of the graph of $y = x^2$.

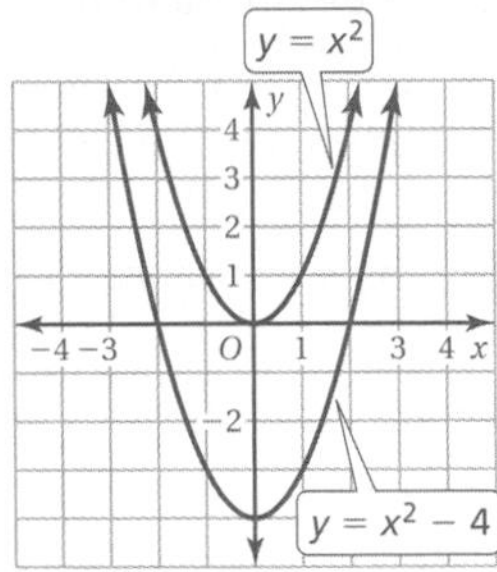

Technology For the Teacher

The Dynamic Planning Tool
Editable Teacher's Resources at *BigIdeasMath.com*

2 ACTIVITY: Finding x-Intercepts of Graphs

Work with a partner. Graph each function. Find the x-intercepts of the graph. Explain how you found the x-intercepts.

a. $y = x^2 - 4$

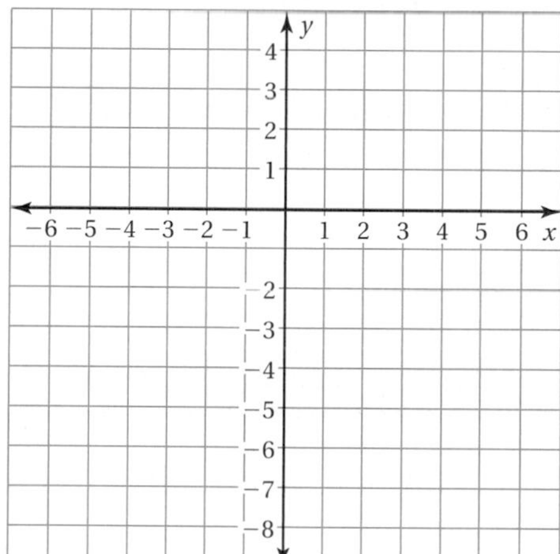

b. $y = 2x^2 - 8$

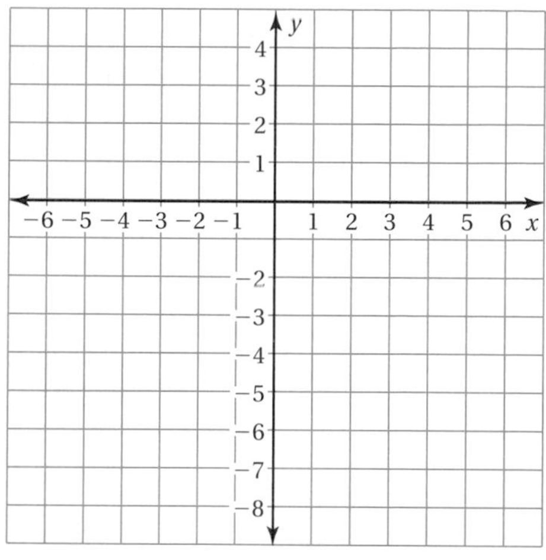

c. $y = -x^2 + 1$

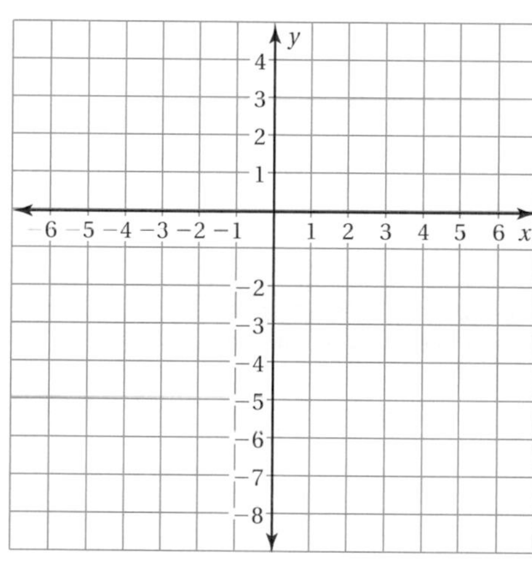

d. $y = \frac{1}{3}x^2 - 3$

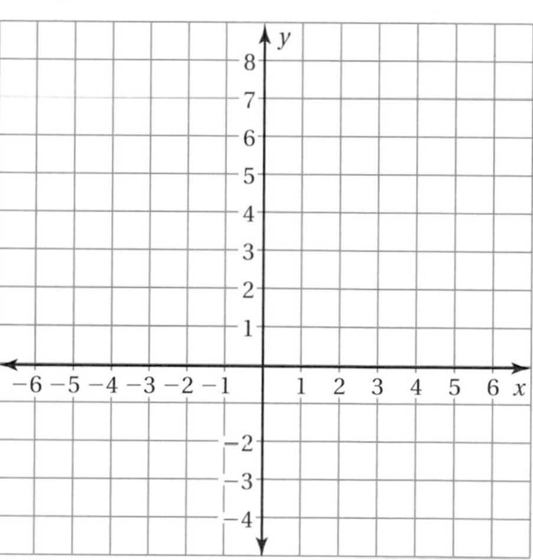

What Is Your Answer?

3. **IN YOUR OWN WORDS** How does the value of c affect the graph of $y = ax^2 + c$? Use a graphing calculator to verify your conclusions.

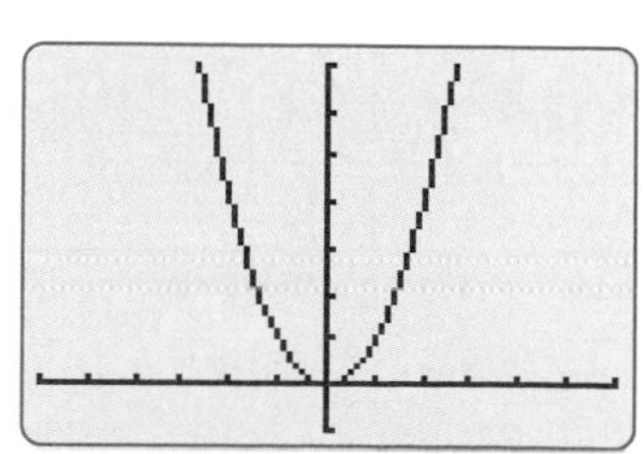

Practice Use what you learned about the graphs of quadratic functions to complete Exercises 7–9 on page 420.

Key Vocabulary
zero, *p. 419*

Graphing $y = x^2 + c$

- When $c > 0$, the graph of $y = x^2 + c$ is a vertical translation c units up of the graph of $y = x^2$.
- When $c < 0$, the graph of $y = x^2 + c$ is a vertical translation $|c|$ units down of the graph of $y = x^2$.

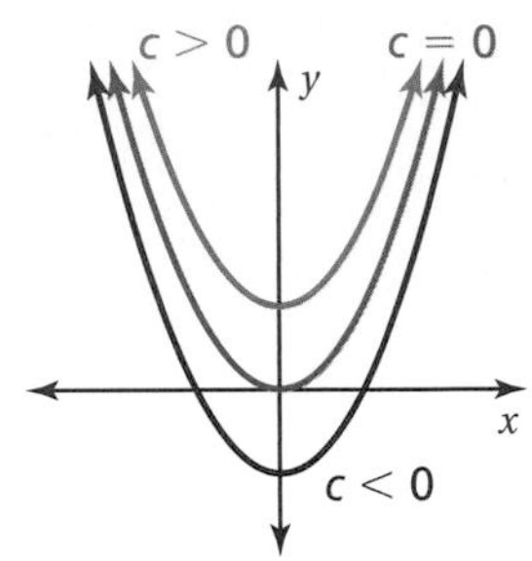

EXAMPLE 1 Graphing $y = x^2 + c$

Graph $y = x^2 - 2$. Compare the graph to the graph of $y = x^2$.

Step 1: Make a table of values.

x	−2	−1	0	1	2
y	2	−1	−2	−1	2

Step 2: Plot the ordered pairs.

Step 3: Draw a smooth curve through the points.

Both graphs open up and have the same axis of symmetry, $x = 0$. The graph of $y = x^2 - 2$ is a translation 2 units down of the graph of $y = x^2$.

EXAMPLE 2 Graphing $y = ax^2 + c$

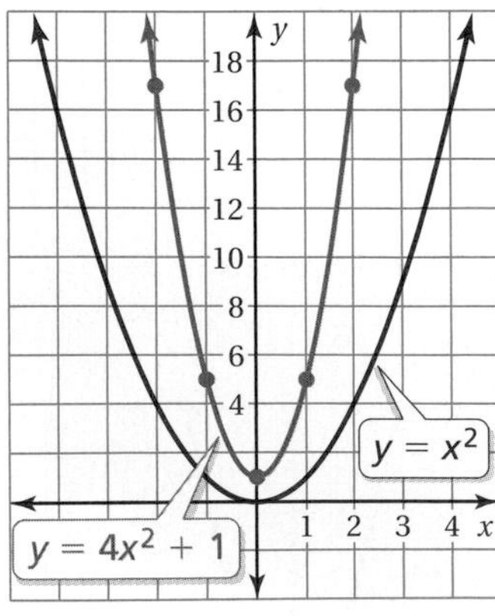

Graph $y = 4x^2 + 1$. Compare the graph to the graph of $y = x^2$.

Step 1: Make a table of values.

x	−2	−1	0	1	2
y	17	5	1	5	17

Step 2: Plot the ordered pairs.

Step 3: Draw a smooth curve through the points.

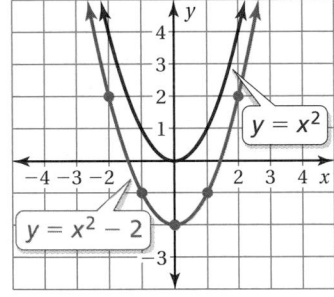

Both graphs open up and have the same axis of symmetry, $x = 0$. The graph of $y = 4x^2 + 1$ is narrower than the graph of $y = x^2$. The vertex of the graph of $y = 4x^2 + 1$ is a translation 1 unit up of the vertex of the graph of $y = x^2$.

On Your Own

Now You're Ready
Exercises 7–15

Graph the function. Compare the graph to the graph of $y = x^2$.

1. $y = x^2 + 3$
2. $y = 2x^2 - 5$
3. $y = -\frac{1}{2}x^2 + 4$

Multi-Language Glossary at BigIdeasMath.com.

Laurie's Notes

Introduction

Connect

- **Yesterday:** Students studied how the value of c affects the graph of $y = ax^2 + c$. (MP1a)
- **Today:** Students will graph the quadratic function $y = ax^2 + c$.

Motivate

? "How many y-intercepts does a parabola have?" Students may say 1 or 0. A parabola always has one y-intercept even if it cannot be seen in the standard viewing window of a graphing calculator.

? "How many x-intercepts does a parabola have?" Students will likely say 0, 1, or 2. Have them quickly sketch a parabola for each case.

Lesson Notes

Discuss

- In this section, students will graph quadratic functions that have been shifted vertically on the y-axis.

Key Idea

- Write the Key Idea.
- Sketch the function $y = x^2$ in one color. Sketch an upward translation in a second color and a downward translation in a third color. Connect this to the activity.
- Explain that the shape of the graph is not changing. The graph is shifted up or down a certain number of units.

Example 1

- Write the functions and make a table of values for each. Sketch the graph of each function.

? "How are the graphs alike?" They both open up, have the same shape, and the same line of symmetry.

? "How are the graphs different?" The vertices are different. The vertex of the graph of $y = x^2 - 2$ is 2 units below the vertex of the graph of $y = x^2$.

Example 2

- This example is similar to Example 1 except now there is a leading coefficient $a = 4$.
- Discuss the effect of $a = 4$ on the shape of the graph.
- Students should notice the symmetry of the function in the table of values and in the graph itself.
- Note that $y = 4x^2 + 1$ has no x-intercepts.
- Comparing the graph of $y = 4x^2 + 1$ to $y = 4x^2$, students should identify a vertical shift of 1 unit.

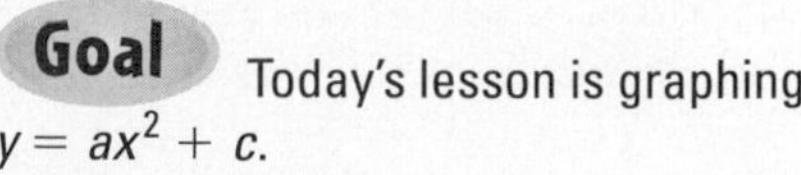

Goal Today's lesson is graphing $y = ax^2 + c$.

Lesson Materials
Textbook
• graphing calculators

Start Thinking! and Warm Up

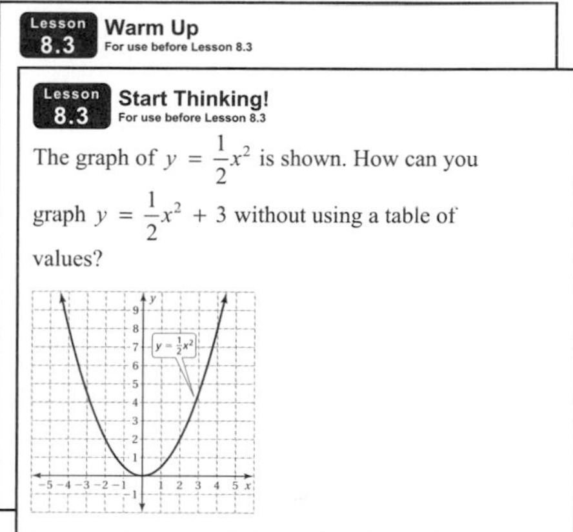

Lesson 8.3 Warm Up
For use before Lesson 8.3

Lesson 8.3 Start Thinking!
For use before Lesson 8.3

The graph of $y = \frac{1}{2}x^2$ is shown. How can you graph $y = \frac{1}{2}x^2 + 3$ without using a table of values?

Extra Example 1

Graph $y = x^2 + 6$. Compare the graph to the graph of $y = x^2$. See Additional Answers.

Extra Example 2

Graph $y = -\frac{1}{5}x^2 - 1$. Compare the graph to the graph of $y = x^2$. See Additional Answers.

On Your Own

1. Both graphs open up and have the same axis of symmetry, $x = 0$. The graph of $y = x^2 + 3$ is a translation 3 units up of the graph of $y = x^2$.

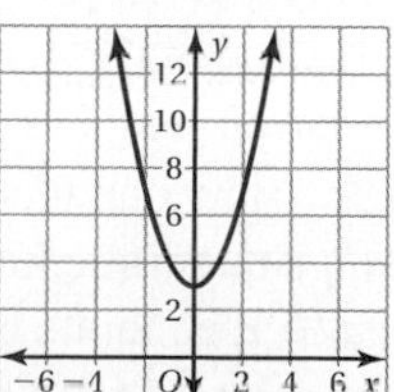

2–3. See Additional Answers.

Laurie's Notes

Example 2 (continued)

- **MP5 Use Appropriate Tools Strategically:** Graph $y = 4x^2$ and $y = 4x^2 + 1$ using a graphing calculator and look at the table of values. Students should notice the symmetry about the y-axis and that y_2 is always one more than y_1.

X	Y1	Y2
-3	36	37
-2	16	17
-1	4	5
0	0	1
1	4	5
2	16	17
3	36	37

X= -3

On Your Own

- **Neighbor Check:** Have students work independently and then have their neighbor check their work. Have students discuss any discrepancies.

Example 3

- Discuss how each graph is related to the graph of $y = x^2$.
- ? "What is the vertex of each function?" (0, −5) and (0, 2)
- ? "How far apart are the vertices?" 7 units

Example 4

- Write the falling object model, $f(t) = -16t^2 + s_0$. Discuss the independent and dependent variables. Instead of y being a function of x, the height is represented by the function $f(t)$. The initial height is s_0. Explain to students that the model ignores air resistance, wind, and other factors.
- ? "Should the table of values include t-values less than 0?" no; Time cannot be negative in the context of this problem.
- Note that the function is only graphed in the first quadrant based on the restrictions of the domain.
- From the graph and the table of values, students can determine when the egg hits the ground.
- Have students compare the graphs of $y = -16t^2 + 64$ and $y = t^2$.
- **Extension:** Look at the graph of $y = -16t^2 + 64$ on a graphing calculator.

On Your Own

- **Think-Pair-Share:** Students should read each question independently and then work with a partner to answer the questions. When they have answered the questions, the pair should compare their answers with another group and discuss any discrepancies.
- Ask a volunteer to show the work for Question 7.

Closure

- **Writing Prompt:** The graph of $y = -4x^2 + 12$ is . . .

Technology For the Teacher

The Dynamic Planning Tool
Editable Teacher's Resources at *BigIdeasMath.com*

Extra Example 3

Describe how to translate the graph of $y = x^2 + 1$ to the graph of $y = x^2 - 3$. translate 4 units down

Extra Example 4

In Example 4, an egg is dropped from a height of 256 feet. When does the egg hit the ground? 4 seconds after it is dropped

On Your Own

4. translate 2 units down
5. time cannot be negative
6. 2.5 seconds after it is dropped

Differentiated Instruction

Kinesthetic

Have students graph a quadratic function on graph paper. Then have them fold it along the axis of symmetry. Ask, "What does this tell you about the shape of a parabola?" Half of the parabola is a mirror image of the other half. Then ask, "How can this help when graphing a quadratic function by plotting points?" Plot points on one side of the axis of symmetry, then plot the mirror image of the points, and finally draw a smooth curve through the points.

EXAMPLE 3 Standardized Test Practice

Which of the following is true when you translate the graph of $y = x^2 - 5$ to the graph of $y = x^2 + 2$?

Ⓐ The graph shifts 3 units up.   Ⓑ The graph shifts 7 units up.

Ⓒ The graph shifts 7 units down.   Ⓓ The graph shifts 3 units down.

Both graphs open up and have the same axis of symmetry, $x = 0$. The vertex of $y = x^2 - 5$ is $(0, -5)$. The vertex of $y = x^2 + 2$ is $(0, 2)$. To move the vertex from $(0, -5)$ to $(0, 2)$, you must translate the graph 7 units up.

∴ The correct answer is Ⓑ.

A **zero** of a function $f(x)$ is an x-value for which $f(x) = 0$. A zero is located at the x-intercept of the graph of the function.

EXAMPLE 4 Real-Life Application

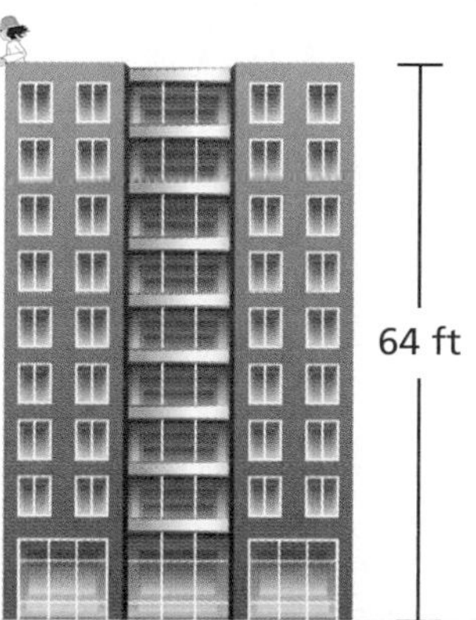

The function $f(t) = -16t^2 + s_0$ gives the approximate height (in feet) of a falling object t seconds after it is dropped from an initial height s_0 (in feet). An egg is dropped from a height of 64 feet. When does the egg hit the ground?

The initial height is 64 feet. So, the function $f(t) = -16t^2 + 64$ gives the height of the egg after t seconds. It hits the ground when $f(t) = 0$.

Step 1: Make a table of values and sketch the graph.

t	0	1	2
$f(t)$	64	48	0

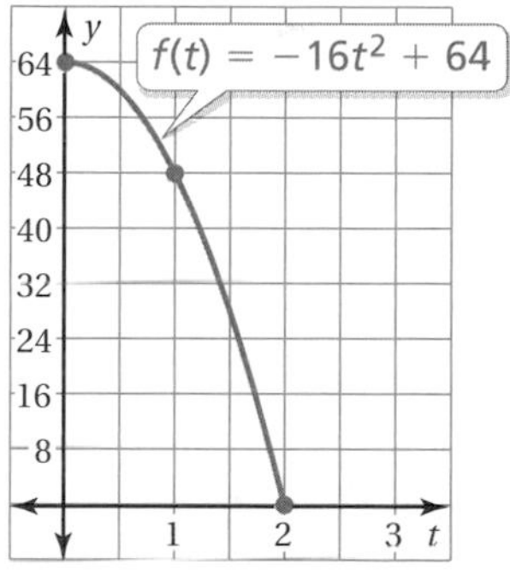

Step 2: Find the zero of the function. When $t = 2$, $f(t) = 0$. So, the zero is 2.

∴ The egg hits the ground 2 seconds after it is dropped.

Common Error

The graph in Example 4 shows the height of the object over time, not the path of the object.

On Your Own

Now You're Ready
Exercises 16–18, 21

4. The graph of $y = 2x^2 + 1$ is shifted to $y = 2x^2 - 1$. Describe the translation.
5. **REASONING** Explain why only nonnegative values of t are used in Example 4.
6. **WHAT IF?** In Example 4, the egg is dropped from a height of 100 feet. When does the egg hit the ground?

8.3 Exercises

Vocabulary and Concept Check

1. **VOCABULARY** Describe the vertex and axis of symmetry of the graph of $y = ax^2 + c$. How is the value of c related to the vertex of the graph?
2. **NUMBER SENSE** Without graphing, which graph has the greater y-intercept, $y = x^2 + 4$ or $y = x^2 - 4$? Explain your reasoning.
3. **WRITING** How does the graph of $y = ax^2 + c$ compare to the graph of $y = ax^2$?

Practice and Problem Solving

Match each function with its graph.

4. $y = 2x^2 + 3$
5. $y = -2x^2 + 3$
6. $y = 2x^2 - 3$

A. 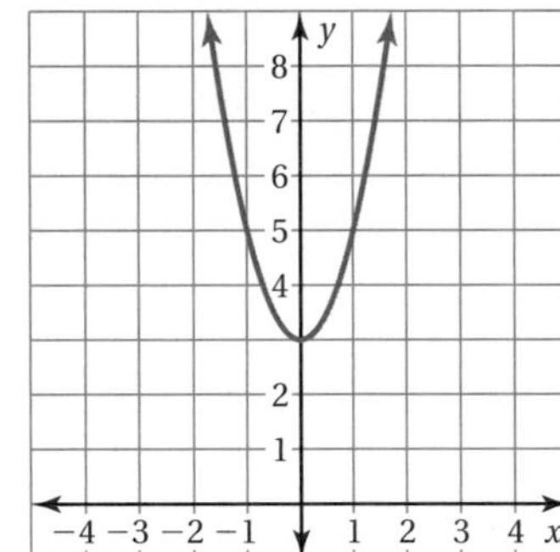

B. 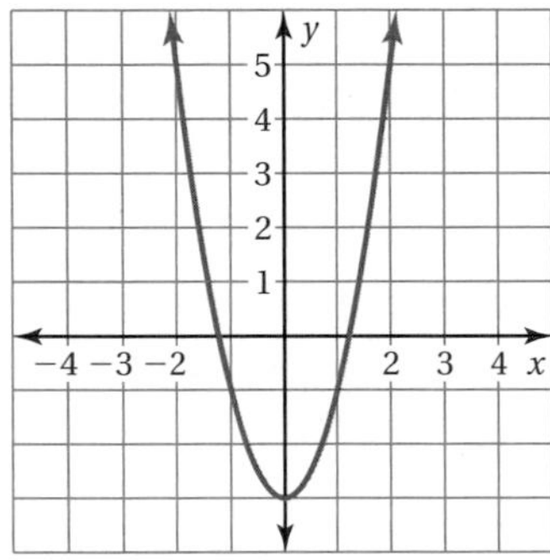

C. 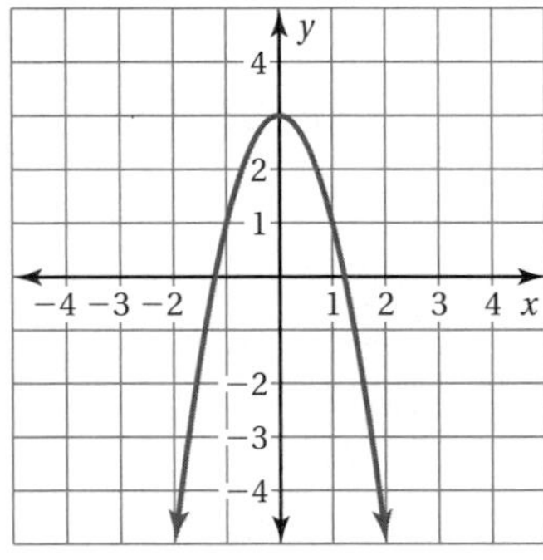

Graph the function. Compare the graph to the graph of $y = x^2$.

7. $y = x^2 + 4$
8. $y = x^2 - 3$
9. $y = -x^2 + 5$
10. $y = -x^2 - 9$
11. $y = 2x^2 - 4$
12. $y = \frac{1}{2}x^2 + 2$

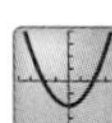

Use a graphing calculator to graph the function. Compare the graph to the graph of $y = x^2$.

13. $y = 3x^2 + 4$
14. $y = -2x^2 - 1$
15. $y = -\frac{1}{4}x^2 - \frac{1}{2}$

Describe how to translate the graph of $y = x^2 + 2$ to the graph of the given function.

(3) 16. $y = x^2 + 4$

17. $y = x^2 - 1$

18. $y = x^2 - 4.5$

19. **ERROR ANALYSIS** Describe and correct the error in comparing the graphs.

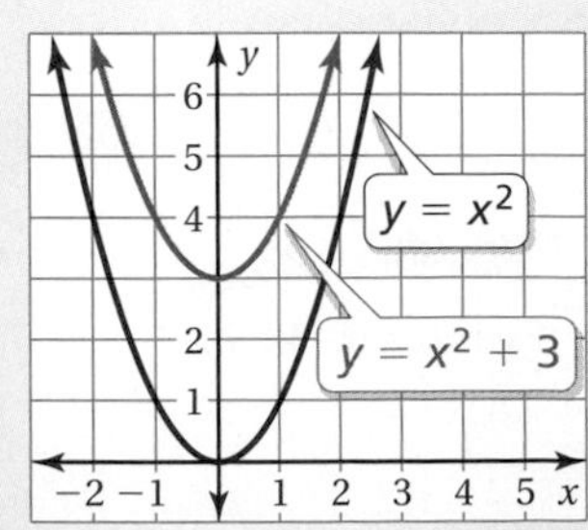

The graph of $y = x^2 + 3$ is a translation 3 units down of the graph of $y = x^2$.

20. **REASONING** The domain of $y = ax^2 + c$ is all real numbers. Describe the range when (a) $a > 0$ and (b) $a < 0$.

(4) 21. **WATER BALLOON** A water balloon is dropped from a height of 16 feet. The function $h = -16x^2 + 16$ gives the height h of the balloon after x seconds. When does it hit the ground?

Assignment Guide and Homework Check

Level	Assignment	Homework Check
Average	1–3, 8–24 even, 19, 32, 34–37	10, 19, 22, 24, 32
Advanced	1–3, 8, 12, 19, 21, 22, 30–33, 34–37	12, 19, 21, 32, 33

Common Errors

- **Exercises 7–12** Students may only plot points on the right side of the y-axis, which is the axis of symmetry for quadratic functions of the form $f(x) = ax^2 + c$. Remind students that the graphs of quadratic functions are parabolas which are U-shaped and symmetric. Students can use this symmetry to find points on both sides of the axis of symmetry.
- **Exercises 13–15** Students may not use an appropriate viewing window. Remind them that to compare the graphs, they need to see both sides of both parabolas.
- **Exercises 16–18** Students may use the difference between the absolute values of the constants to define the translation. For example, in Exercise 17, they might say $|2| - |-1| = 1$, so the translation is 1 unit down. Remind them that the translation is defined as the distance between the vertices of the graphs of the functions.

8.3 Record and Practice Journal

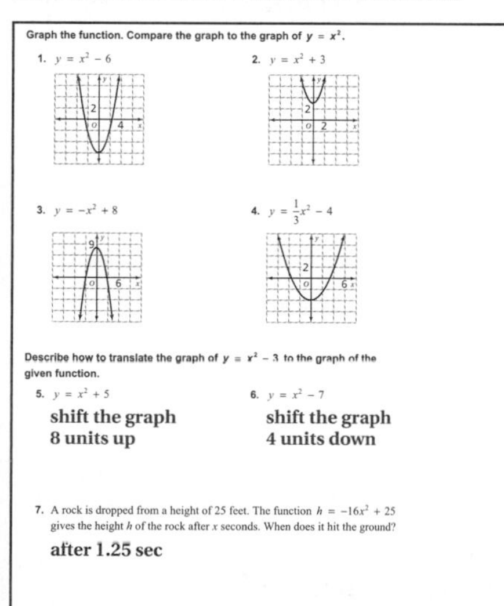
Graph the function. Compare the graph to the graph of $y = x^2$.

1. $y = x^2 - 6$
2. $y = x^2 + 3$
3. $y = -x^2 + 8$
4. $y = \frac{1}{3}x^2 - 4$

Describe how to translate the graph of $y = x^2 - 3$ to the graph of the given function.

5. $y = x^2 + 5$

   shift the graph 8 units up

6. $y = x^2 - 7$

   shift the graph 4 units down

7. A rock is dropped from a height of 25 feet. The function $h = -16x^2 + 25$ gives the height h of the rock after x seconds. When does it hit the ground?

   after 1.25 sec

Technology For the Teacher

Answer Presentation Tool

Vocabulary and Concept Check

1. The vertex is on the y-axis. The y-axis is the axis of symmetry. The value of c is the y-coordinate of the vertex.
2. $y = x^2 + 4$; The graph of $y = x^2 + 4$ has a y-intercept of (0, 4) whereas $y = x^2 - 4$ has a y-intercept of (0, −4).
3. It is a translation c units up or down of the graph of $y = ax^2$.

Practice and Problem Solving

4. A **5.** C

6. B

7–15. See Additional Answers.

16. translate 2 units up

17. translate 3 units down

18. translate 6.5 units down

19. The student incorrectly states that the graph of $y = x^2 + 3$ is a translation 3 units down of the graph of $y = x^2$. The graph is actually a translation 3 units up.

20. **a.** $y \geq c$

b. $y \leq c$

21. after 1 second

Practice and Problem Solving

22. y-intercept: (0, 36); The apple is dropped from a height of 36 feet. x-intercept: (1.5, 0); The apple hits the ground after 1.5 seconds.

23. $-1, 1$

24. $-2, 2$

25. $-3, 3$

26–29. See Additional Answers.

30. factoring and finding the square root

31. yes; $y = x^2 - 1$ has a vertex at $(0, -1)$ and a focus at $\left(0, \frac{1}{4}\right)$.

32. See *Taking Math Deeper*.

33. See Additional Answers.

Fair Game Review

34. $(x + 2)(x - 4)$

35. $(2x - 1)(x - 3)$

36. $(x - 5)(x + 7)$

37. B

Mini-Assessment

1. Graph $y = x^2 - 8$. Compare the graph to the graph of $y = x^2$. See Additional Answers.
2. Graph $y = -3x^2 + 5$. Compare the graph to the graph of $y = x^2$. See Additional Answers.
3. Describe how to translate the graph of $y = x^2 - 10$ to the graph of $y = x^2 - 4$. translate 6 units up
4. A pebble is dropped from a bridge 100 feet above the water. The function $h = -16x^2 + 100$ gives the height h of the pebble after x seconds. When does it hit the water? 2.5 seconds after it is dropped

Taking Math Deeper

Exercise 32

It is important to recognize the difference between a waterfall function in this exercise and a function that models a falling object as in Example 4. The function in Example 4 gives the height of the object over time. The functions in this exercise model the actual paths of water.

a. Use a graphing calculator to graph the functions in the same viewing window. Because distance cannot be negative, look at the graphs in the first quadrant only.

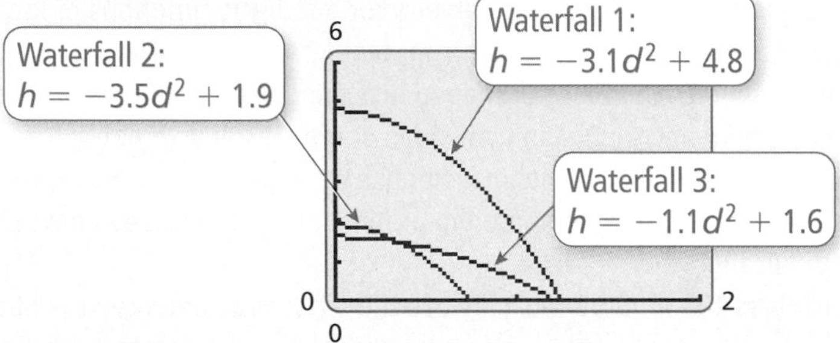

From the graph, you can see that the y-intercept of Waterfall 1 is greater than the y-intercepts of the other waterfalls. So, Waterfall 1 drops water from the highest point.

b. From the graph, you can see that the x-intercept of Waterfall 2 is to the left of the x-intercepts of the other waterfalls. So, Waterfall 2 produces the narrowest path.

c. You can also see from the graph that the x-intercept of Waterfall 1 is to the right of the x-intercepts of the other waterfalls. So, Waterfall 1 sends water the farthest.

Challenge students to examine other characteristics of the graph above. For instance, even though Waterfall 2 has a greater y-intercept than Waterfall 3, Waterfall 3 sends water farther. Students could also use a graphing calculator to identify and interpret the point of intersection of Waterfalls 2 and 3.

Reteaching and Enrichment Strategies

If students need help. . .	If students got it. . .
Resources by Chapter • Practice A and Practice B • Puzzle Time Record and Practice Journal Practice Differentiating the Lesson Lesson Tutorials Skills Review Handbook	Resources by Chapter • Enrichment and Extension • School-to-Work Start the next section

22. **APPLE** The function $y = -16x^2 + 36$ gives the height y (in feet) of an apple after falling x seconds. Find and interpret the x- and y-intercepts.

Find the zeros of the function.

23. $y = x^2 - 1$
24. $y = x^2 - 4$
25. $y = -x^2 + 9$

Sketch a quadratic function with the given characteristics.

26. The parabola opens up and the vertex is (0, 3).
27. The parabola opens down, the vertex is (0, 4), and one of the x-intercepts is 2.
28. The function is increasing when $x < 0$ and the x-intercepts of the parabola are -1 and 1.
29. The graph is below the x-axis and the highest point on the parabola is $(0, -5)$.
30. **REASONING** Describe two algebraic methods you can use to find the zeros of the function $f(t) = -16t^2 + 64$.
31. **REASONING** Can the focus and the vertex of a parabola lie on opposite sides of the x-axis? Explain your reasoning.

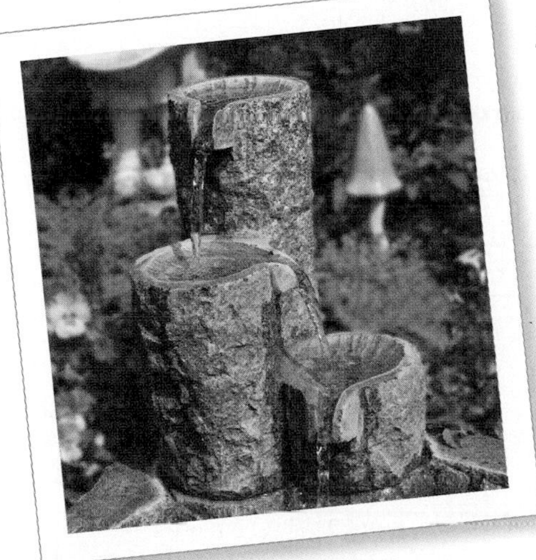

32. **PROBLEM SOLVING** The paths of water from three different garden waterfalls are given below. Each function gives the height h (in feet) and the horizontal distance d (in feet) of the water.

    Waterfall 1: $h = -3.1d^2 + 4.8$

    Waterfall 2: $h = -3.5d^2 + 1.9$

    Waterfall 3: $h = -1.1d^2 + 1.6$

    a. Which waterfall drops water from the highest point?

    b. Which waterfall follows the narrowest path?

    c. Which waterfall sends water the farthest?

33. **Logic** Let $f(x)$ be a quadratic function of the form $f(x) = ax^2 + c$.

    a. How does the graph of $f(x) + k$ compare to the graph of $f(x)$ when $k < 0$? when $k > 0$?

    b. Let k be a real number not equal to 0 or 1. How does the graph of $k \cdot f(x)$ compare to the graph of $f(x)$?

Fair Game Review What you learned in previous grades & lessons

Factor the polynomial. *(Section 7.7 and Section 7.8)*

34. $x^2 - 2x - 8$
35. $2x^2 - 7x + 3$
36. $x^2 + 2x - 35$
37. **MULTIPLE CHOICE** What is the product of $(x - 2)$ and $(x - 4)$? *(Section 7.3)*

    (A) $x^2 - 2$ (B) $x^2 - 6x + 8$ (C) $x^2 - 2x - 6$ (D) $x^2 - 4x - 8$

8 Study Help

You can use a **summary triangle** to explain a topic. Here is an example of a summary triangle for graphing a quadratic function of the form $y = ax^2$ when $a > 0$.

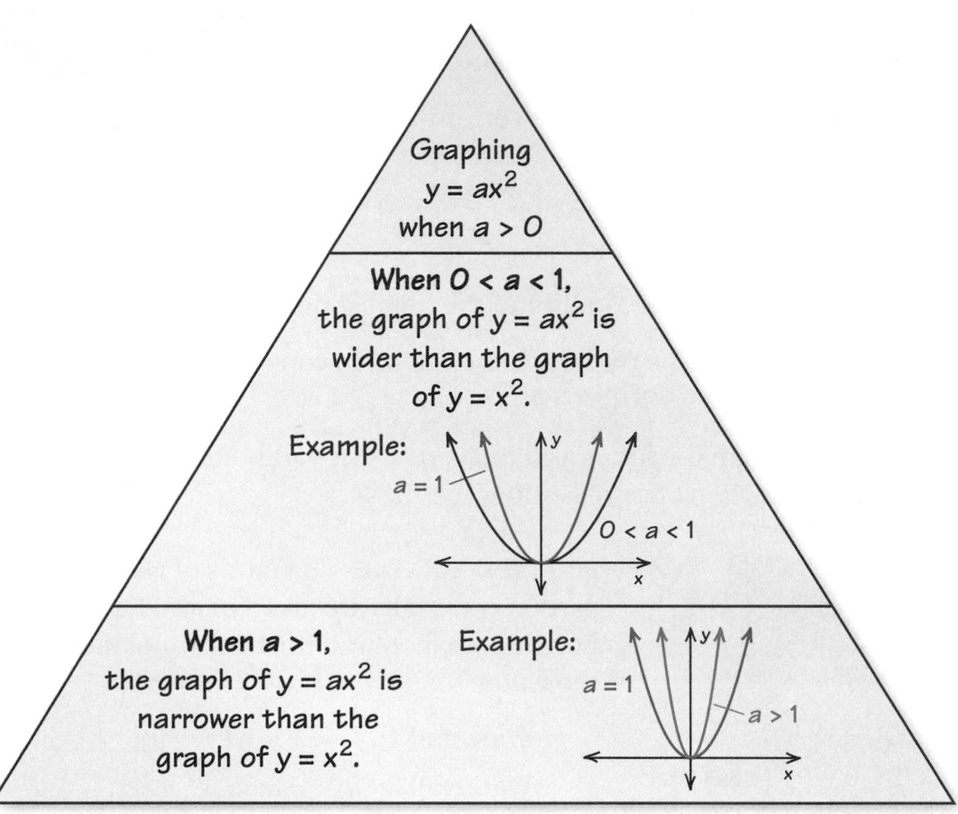

Graphing
y = ax²
when a > 0
When 0 < a < 1,
the graph of y = ax² is
wider than the graph
of y = x².
Example:
a = 1
0 < a < 1
y
x
When a > 1,
the graph of y = ax² is
narrower than the graph
graph of y = x².
Example:
a = 1
a > 1
y
x

On Your Own

Make summary triangles to help you study these topics.

1. graphing $y = ax^2$ when $a < 0$
2. identifying the focus of a parabola
3. graphing $y = ax^2 + c$

After you complete this chapter, make summary triangles for the following topics.

4. graphing $y = ax^2 + bx + c$
5. graphing $y = a(x - h)^2 + k$

"What do you call a cheese summary triangle that isn't yours?"

Sample Answers

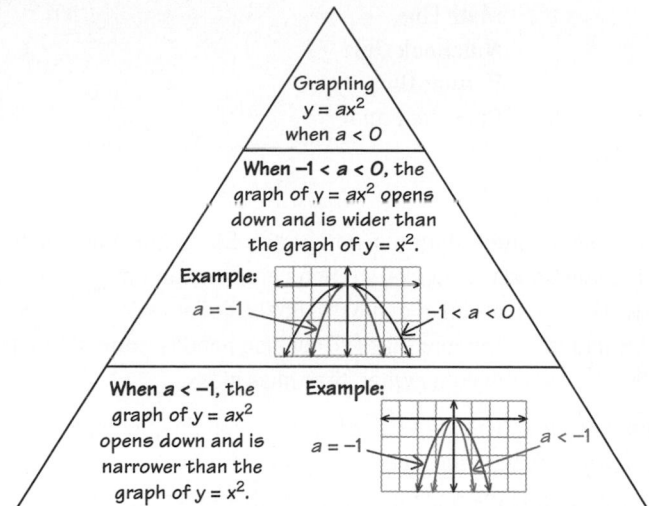

1.

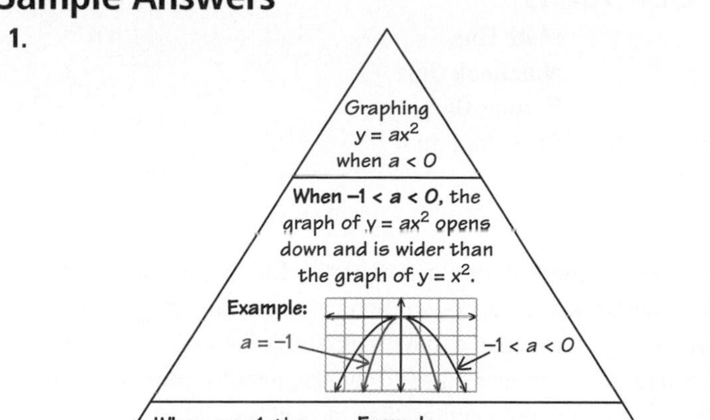

Graphing
y = ax²
when a < 0
When −1 < a < 0, the graph of y = ax² opens down and is wider than the graph of y = x².
Example:
a = −1
−1 < a < 0
When a < −1, the graph of y = ax² opens down and is narrower than the graph of y = x².
Example:
a = −1
a < −1

2.

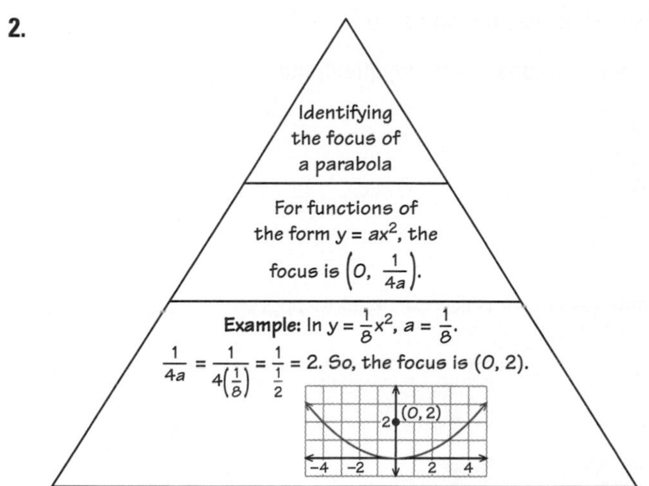

Identifying the focus of a parabola
For functions of the form y = ax², the focus is (0, 1/(4a)).
Example: In y = (1/8)x², a = 1/8.
1/(4a) = 1/(4(1/8)) = 1/(1/2) = 2. So, the focus is (0, 2).
(0, 2)

3. Available at *BigIdeasMath.com*.

List of Organizers

Available at *BigIdeasMath.com*

Comparison Chart
Concept Circle
Definition (Idea) and Example Chart
Example and Non-Example Chart
Formula Triangle
Four Square
Information Frame
Information Wheel
Notetaking Organizer
Process Diagram
Summary Triangle
Word Magnet
Y Chart

About this Organizer

A **Summary Triangle** can be used to explain a concept. Typically, the summary triangle is divided into 3 or 4 parts. In the top part, students write the concept being explained. In the middle part(s), students write any procedure, explanation, description, definition, theorem, and/or formula(s). In the bottom part, students write an example to illustrate the concept. A summary triangle can be used as an assessment tool, in which blanks are left for students to complete. Also, students can place their summary triangles on note cards to use as quick study reference.

Technology For the Teacher

Vocabulary Puzzle Builder

Answers

1–4. See Additional Answers.

5–10. See Additional Answers for graphs.

5. The graphs have the same vertex, (0, 0), and the same axis of symmetry, $x = 0$, but the graph of $y = -x^2$ opens down.

6. Both graphs open up and have the same vertex, (0, 0), and the same axis of symmetry, $x = 0$. The graph of $y = 4x^2$ is narrower than the graph of $y = x^2$.

7. Both graphs open up and have the same vertex, (0, 0), and the same axis of symmetry, $x = 0$. The graph of $y = \frac{2}{5}x^2$ is wider than the graph of $y = x^2$.

8. $\left(0, \frac{1}{20}\right)$ **9.** $\left(0, -\frac{1}{24}\right)$

10. $\left(0, \frac{3}{4}\right)$ **11.** $y = -\frac{1}{16}x^2$

12. $y = \frac{1}{8}x^2$ **13.** $y = \frac{5}{4}x^2$

14–16. See Additional Answers.

17. $\sqrt{2}$ sec **18.** 10 h

Assessment Book

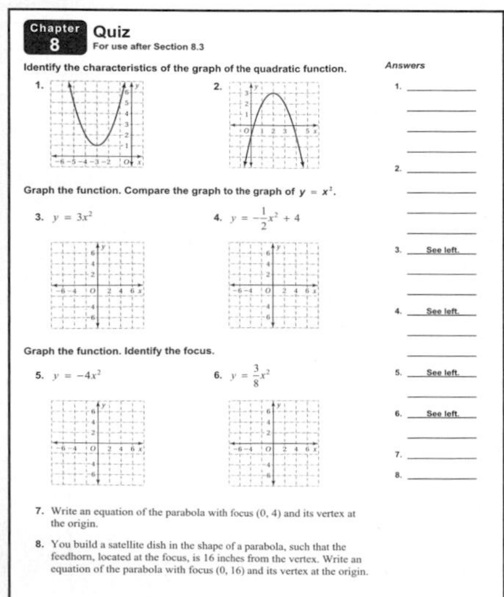

Chapter 8 **Quiz**
For use after Section 8.3

Identify the characteristics of the graph of the quadratic function.

1. 2.

Graph the function. Compare the graph to the graph of $y = x^2$.

3. $y = 3x^2$ 4. $y = -\frac{1}{2}x^2 + 4$

Graph the function. Identify the focus.

5. $y = -4x^2$ 6. $y = \frac{3}{8}x^2$

7. Write an equation of the parabola with focus (0, 4) and its vertex at the origin.

8. You build a satellite dish in the shape of a parabola, such that the feedhorn, located at the focus, is 16 inches from the vertex. Write an equation of the parabola with focus (0, 16) and its vertex at the origin.

Answers

1. ______
2. ______
3. See left.
4. See left.
5. See left.
6. See left.
7. ______
8. ______

Alternative Quiz Ideas

100% Quiz
Error Notebook
Group Quiz
Homework Quiz
Math Log
Notebook Quiz
Partner Quiz
Pass the Paper

Notebook Quiz

A notebook quiz is used to check students' notebooks. Students should be told at the beginning of the course what the expectations are for their notebooks: notes, classwork, homework, dates, problem numbers, goals, definitions, or anything else that you feel is important for your class. They also need to know that it is their responsibility to obtain the notes when they miss class.

1. On a certain day, what was the answer to the warm up question?
2. On a certain day, how was this vocabulary term defined?
3. For Section 8.2, what is the answer to On Your Own Question 5?
4. For Section 8.3, what is the answer to the Essential Question?
5. On a certain day, what was the homework assignment?

Give the students 5 minutes to answer these questions.

Reteaching and Enrichment Strategies

If students need help. . .	If students got it. . .
Resources by Chapter • Study Help • Practice A and Practice B • Puzzle Time Lesson Tutorials *BigIdeasMath.com* Practice Quiz Practice from the Test Generator	Resources by Chapter • Enrichment and Extension • School-to-Work Game Closet at *BigIdeasMath.com* Start the next section

Technology For the Teacher

Answer Presentation Tool
Big Ideas Test Generator

8.1–8.3 Quiz

Identify characteristics of the graph of the quadratic function. *(Section 8.1)*

1.

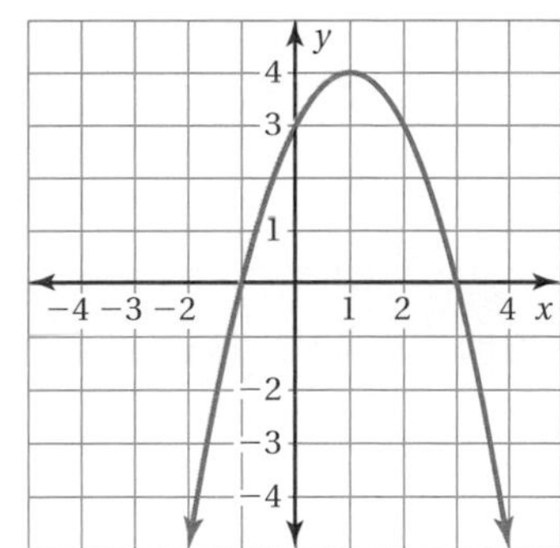

2.

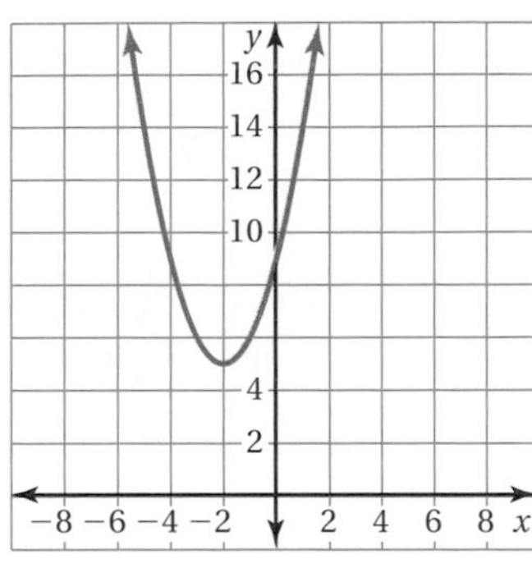

3.

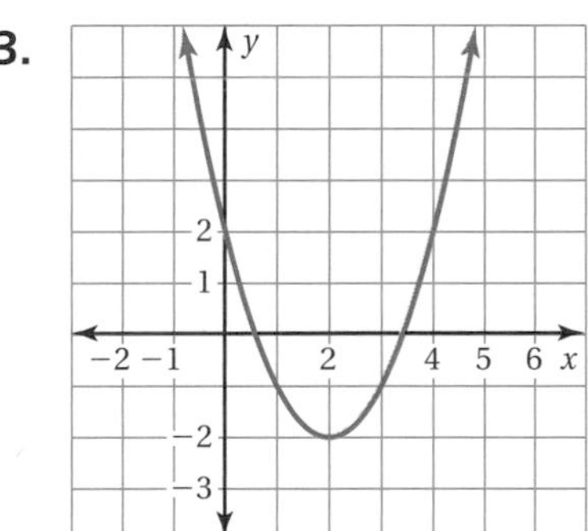

4.

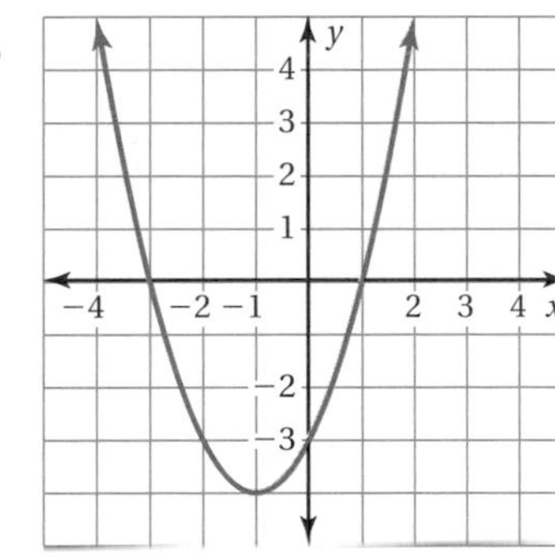

Graph the function. Compare the graph to the graph of $y = x^2$. *(Section 8.1)*

5. $y = -x^2$

6. $y = 4x^2$

7. $y = \frac{2}{5}x^2$

Graph the function. Identify the focus. *(Section 8.2)*

8. $y = 5x^2$

9. $y = -6x^2$

10. $y = \frac{1}{3}x^2$

Write an equation of the parabola with a vertex at the origin and the given focus. *(Section 8.2)*

11. $(0, -4)$

12. $(0, 2)$

13. $\left(0, \frac{1}{5}\right)$

Graph the function. Compare the graph to the graph of $y = x^2$. *(Section 8.3)*

14. $y = x^2 + 5$

15. $y = 2x^2 - 2$

16. $y = -x^2 + 3$

17. **PINEAPPLE** The distance y (in feet) that a pineapple falls is given by the function $y = 16t^2$, where t is the time (in seconds). Use a graph to determine how many seconds it takes for the pineapple to fall 32 feet. *(Section 8.1)*

18. **SMARTPHONE** A new smartphone application is available for download. The number y of downloads can be modeled by the function $y = 6.3x^2 + 3000$, where x is the number of hours since the new application was released. How many hours does it take for the number of downloads to reach 3630? *(Section 8.3)*

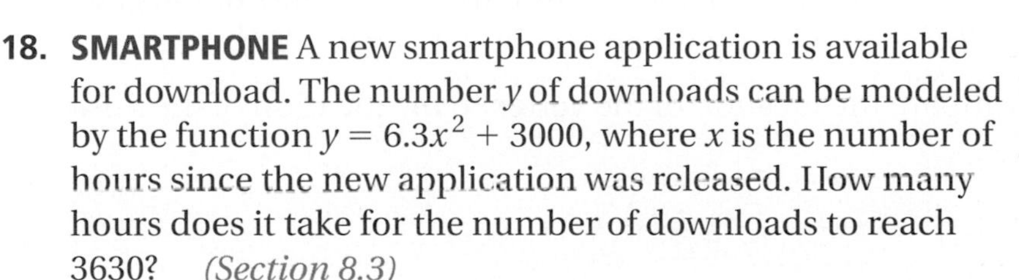

8.4 Graphing $y = ax^2 + bx + c$

Essential Question How can you find the vertex of the graph of $y = ax^2 + bx + c$?

COMMON CORE STATE STANDARDS
F.BF.3
F.IF.4
F.IF.7a

1 ACTIVITY: Comparing Two Graphs

Work with a partner.

- Sketch the graphs of $y = 2x^2 - 8x$ and $y = 2x^2 - 8x + 6$.
- What do you notice about the x-value of the vertex of each graph?

$y = 2x^2 - 8x$ $\qquad$ $y = 2x^2 - 8x + 6$

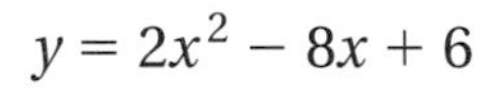

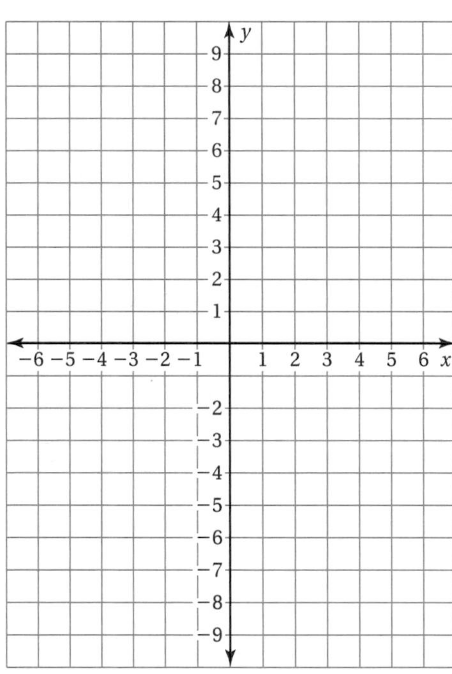

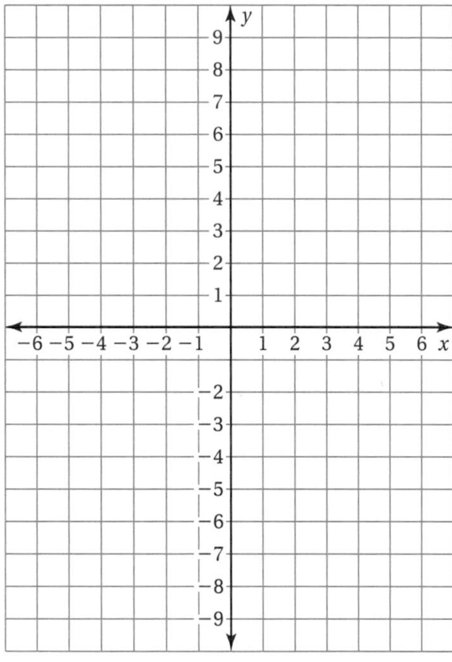

2 ACTIVITY: Comparing x-Intercepts with the Vertex

Work with a partner.

- Use the graph in Activity 1 to find the x-intercepts of the graph of $y = 2x^2 - 8x$. Verify your answer by solving $0 = 2x^2 - 8x$.
- Compare the location of the vertex to the location of the x-intercepts.

Laurie's Notes

Introduction

Standards for Mathematical Practice

- **MP7 Look for and Make Use of Structure:** All of the quadratic functions that students have graphed have been symmetric about the *y*-axis. In the equation $y = ax^2 + c$, the values of *a* and *c* do not affect the graph in a horizontal direction. In this section, quadratic functions contain an *x*-term, which shifts the axis of symmetry away from the *y*-axis.

Motivate

- Discuss with students that, in a future course, they will study conic sections: ellipses, parabolas, and hyperbolas. A form of each of those words is a part of spoken English.
- Have students find the words ellipsis, hyperbole, and parable in a dictionary. Discuss the definitions.
- In rhetoric, "hyperbolic" speech goes beyond the facts, "elliptic" speech falls short of the facts, and a "parable" is a story that fits the facts exactly.

Activity Notes

Activity 1

- Students should use a table of values to graph the functions.
- In both functions, if students use domain values of −2, −1, 0, 1, and 2, the *y*-values are decreasing.
- When they evaluate either function for $x = 3$, the *y*-values start to increase. You want students to interpret this change. The vertex occurs when $x = 2$.

? "How do you know that the vertex occurs when $x = 2$?" Students should reference the change from decreasing to increasing, and that there is symmetry of *y*-values about the line $x = 2$.

? "What do you notice about the vertex of each graph?" They both have an *x*-coordinate of 2 and the *y*-coordinates differ by 6.

Activity 2

- The *x*-intercepts may have been found when generating the table of values. If not, students should substitute 0 for *y* and solve.

? "How can you solve $0 = 2x^2 - 8x$?" Factor out $2x$ to obtain $0 = 2x(x - 4)$, which gives the solutions 0 and 4.

- **Common Error:** When solving $0 = 2x^2 - 8x$, students may add $8x$ to each side and then divide each side by $2x$ to obtain $x = 4$. Dividing by $2x$ caused the solution $x = 0$ to be lost.
- Discuss the connections between the line of symmetry, vertex, and *x*-intercepts.

Common Core State Standards

F.BF.3 Identify the effect on the graph of replacing $f(x)$ by $f(x) + k$, $kf(x)$, $f(kx)$, and $f(x + k)$ for specific values of k (both positive and negative); find the value of k given the graphs. Experiment with cases and illustrate an explanation of the effects on the graph using technology.

F.IF.7a Graph linear and quadratic functions and show intercepts, maxima, and minima.

Also **F.IF.4**

Previous Learning

Students should know how to graph $y = ax^2 + c$.

Start Thinking! and Warm Up

Start Thinking!
For use before Activity 8.4

Warm Up
For use before Activity 8.4

Evaluate the expression for the given value of *x*.

1. $4x^2 - x + 8;\ x = -1$
2. $-\frac{2}{3}x^2 + 2x + 4;\ x = 3$
3. $-x^2 - 3x + 9;\ x = 2$
4. $2x^2 + 12x - 3;\ x = -5$

8.4 Record and Practice Journal

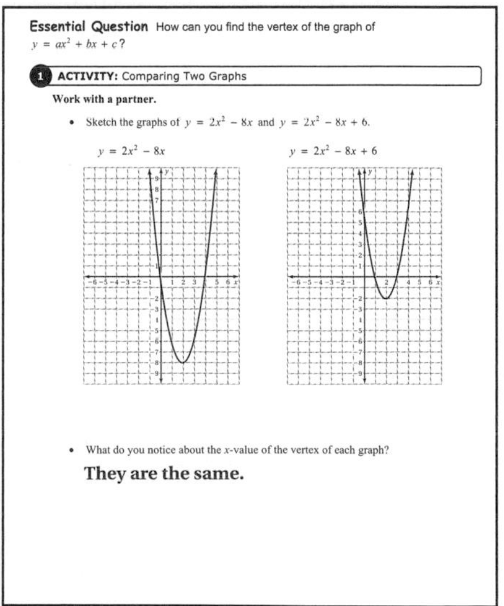

Essential Question How can you find the vertex of the graph of $y = ax^2 + bx + c$?

1 ACTIVITY: Comparing Two Graphs

Work with a partner.

- Sketch the graphs of $y = 2x^2 - 8x$ and $y = 2x^2 - 8x + 6$.

$y = 2x^2 - 8x$ $\qquad$ $y = 2x^2 - 8x + 6$

- What do you notice about the *x*-value of the vertex of each graph?

They are the same.

Differentiated Instruction

Patterns

Working in pairs, students should graph quadratic functions where $a = 1$. Next, they should fill in the table, identifying the axis of symmetry and the y-intercept.

Equation	Axis of Symmetry	y-intercept

Students should see that the axis of symmetry is $-\frac{b}{2}$ (because $a = 1$) and the y-intercept is c. Extend the activity by having students graph equations where $a > 1$.

8.4 Record and Practice Journal

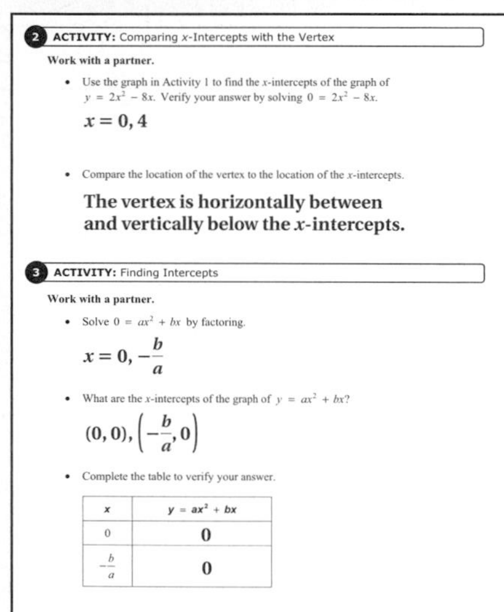

2 ACTIVITY: Comparing x-Intercepts with the Vertex

Work with a partner.

- Use the graph in Activity 1 to find the x-intercepts of the graph of $y = 2x^2 - 8x$. Verify your answer by solving $0 = 2x^2 - 8x$.

  $x = 0, 4$

- Compare the location of the vertex to the location of the x-intercepts.

  The vertex is horizontally between and vertically below the x-intercepts.

3 ACTIVITY: Finding Intercepts

Work with a partner.

- Solve $0 = ax^2 + bx$ by factoring.

  $x = 0, -\frac{b}{a}$

- What are the x-intercepts of the graph of $y = ax^2 + bx$?

  $(0, 0), \left(-\frac{b}{a}, 0\right)$

- Complete the table to verify your answer.

x	$y = ax^2 + bx$
0	0
$-\frac{b}{a}$	0

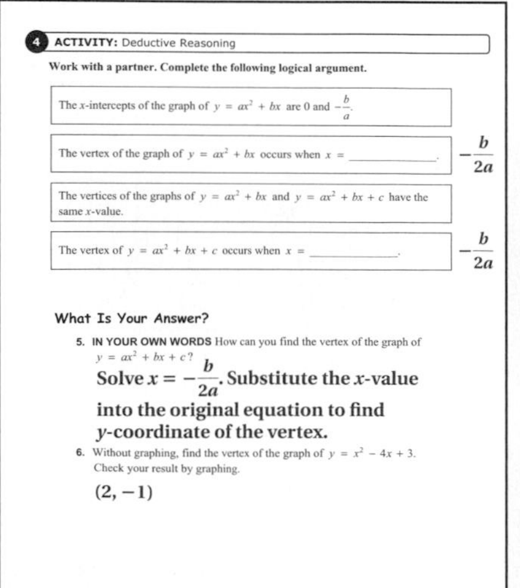

4 ACTIVITY: Deductive Reasoning

Work with a partner. Complete the following logical argument.

The x-intercepts of the graph of $y = ax^2 + bx$ are 0 and $-\frac{b}{a}$.

The vertex of the graph of $y = ax^2 + bx$ occurs when $x =$ ______. $-\frac{b}{2a}$

The vertices of the graphs of $y = ax^2 + bx$ and $y = ax^2 + bx + c$ have the same x-value.

The vertex of $y = ax^2 + bx + c$ occurs when $x =$ ______. $-\frac{b}{2a}$

What Is Your Answer?

5. **IN YOUR OWN WORDS** How can you find the vertex of the graph of $y = ax^2 + bx + c$?

   Solve $x = -\frac{b}{2a}$. Substitute the x-value into the original equation to find y-coordinate of the vertex.

6. Without graphing, find the vertex of the graph of $y = x^2 - 4x + 3$. Check your result by graphing.

   $(2, -1)$

Laurie's Notes

Activity 3

- Students may need guidance to factor $y = ax^2 + bx$. They might see 4 variables rather than 2 variables with coefficients a and b. Suggest that they treat a and b as constants, and focus on solving for x.
- When students complete the table, they are substituting the x-intercepts, 0 and $-\frac{b}{a}$, into the equation and solving. Both results should be $y = 0$.

Activity 4

- Students need to reflect back on the three activities they completed today.
- Students may describe the answer in words if they have difficulty finding an expression for x. For instance, in the first answer blank they may say the vertex occurs halfway between 0 and $-\frac{b}{a}$, not realizing that they can express this as $x = -\frac{b}{2a}$. You may need to give examples and pose questions such as, "What is halfway between 4 and 6? Halfway between 0 and a?"
- If students are having difficulty with the second answer blank, refer them back to Activity 1.

What Is Your Answer?

- Have students work in pairs. Review answers as a class.

Closure

- Given the equation $y = x^2 - 2x - 3$, complete the table of values. Identify the x-intercepts and the vertex.

x	-1	0	1	2	3	4
y	0	-3	-4	-3	0	5

$-1, 3, (1, -4)$

The Dynamic Planning Tool

Editable Teacher's Resources at *BigIdeasMath.com*

3 ACTIVITY: Finding Intercepts

Work with a partner.

- Solve $0 = ax^2 + bx$ by factoring.
- What are the x-intercepts of the graph of $y = ax^2 + bx$?
- Copy and complete the table to verify your answer.

x	$y = ax^2 + bx$
0	
$-\frac{b}{a}$	

4 ACTIVITY: Deductive Reasoning

Work with a partner. Complete the following logical argument.

The x-intercepts of the graph of $y = ax^2 + bx$ are 0 and $-\frac{b}{a}$.

The vertex of the graph of $y = ax^2 + bx$ occurs when $x =$ ______.

The vertices of the graphs of $y = ax^2 + bx$ and $y = ax^2 + bx + c$ have the same x-value.

The vertex of $y = ax^2 + bx + c$ occurs when $x =$ ______.

What Is Your Answer?

5. **IN YOUR OWN WORDS** How can you find the vertex of the graph of $y = ax^2 + bx + c$?
6. Without graphing, find the vertex of the graph of $y = x^2 - 4x + 3$. Check your result by graphing.

Practice Use what you learned about the vertices of the graphs of quadratic functions to complete Exercises 6–8 on page 429.

Key Vocabulary
maximum value, p. 427
minimum value, p. 427

Properties of the Graph of $y = ax^2 + bx + c$

- The graph opens up when $a > 0$ and the graph opens down when $a < 0$.
- The y-intercept is c.
- The x-coordinate of the vertex is $-\frac{b}{2a}$.
- The axis of symmetry is $x = -\frac{b}{2a}$.

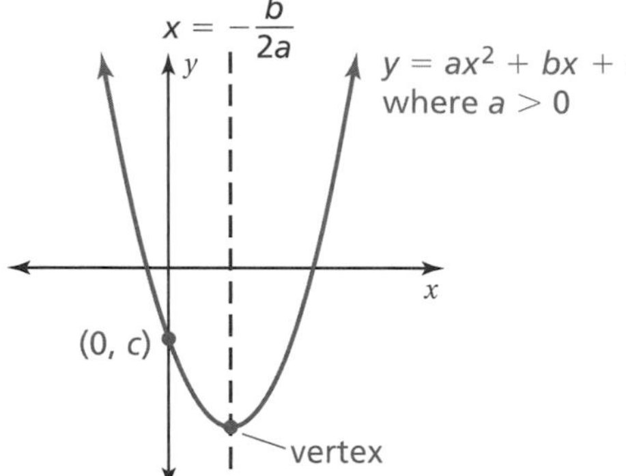

EXAMPLE 1 Finding the Axis of Symmetry and the Vertex of a Graph

Find (a) the axis of symmetry and (b) the vertex of the graph of $y = 2x^2 + 8x - 1$.

a. Find the axis of symmetry when $a = 2$ and $b = 8$.

$x = -\frac{b}{2a}$ Write the equation for the axis of symmetry.

$x = -\frac{8}{2(2)}$ Substitute 2 for a and 8 for b.

$x = -2$ Simplify.

∴ The axis of symmetry is $x = -2$.

b. The axis of symmetry is $x = -2$, so the x-coordinate of the vertex is -2. Use the function to find the y-coordinate of the vertex.

$y = 2x^2 + 8x - 1$ Write the function.

$= 2(-2)^2 + 8(-2) - 1$ Substitute -2 for x.

$= -9$ Simplify.

∴ The vertex is $(-2, -9)$.

On Your Own

Now You're Ready
Exercises 6–11

Find (a) the axis of symmetry and (b) the vertex of the graph of the function.

1. $y = 3x^2 - 2x$
2. $y = x^2 + 6x + 5$
3. $y = -\frac{1}{2}x^2 + 7x - 4$

Multi-Language Glossary at BigIdeasMath.com.

Laurie's Notes

Introduction

Connect

- **Yesterday:** Students found the vertex of a quadratic function. (MP7)
- **Today:** Students will graph $y = ax^2 + bx + c$.

Motivate

- Use a graphing calculator to graph $y = x^2 + 2x - 8$ using a viewing window of $-6 \le x \le 3$ and $-12 \le y \le 2$.

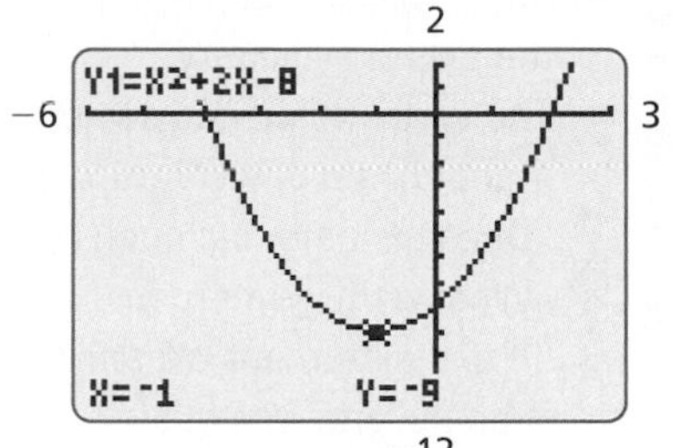

? "What are some of the features that you can identify from the graph?" *x*-intercepts, *y*-intercept, axis of symmetry, vertex

- Explain that in today's lesson, information about the graph of a quadratic function will be determined using its equation.

Lesson Notes

Key Idea

- **MP7 Look for and Make Use of Structure:** Write $y = ax^2 + bx + c$ on the board. Students should know that the coefficients a and b, and the constant c give information about the graph of the quadratic function.

? "What does the leading coefficient tell us about the graph of the function?" whether it opens up or down and the shape compared to $y = x^2$

- **Connection:** In the linear function $y = mx + b$, the constant b is the y-intercept of its graph. The constant term of a quadratic function is also the y-intercept of its graph.
- Write the remaining information about the vertex and line of symmetry from the Key Idea.

? "How can you find the y-coordinate of the vertex?" Substitute the x-coordinate of the vertex into the equation and simplify.

Example 1

- Write the equation and have students identify a, b, and c.

? "How can you find the axis of symmetry?" Listen as students describe using the formula $x = -\frac{b}{2a}$.

? "Now that you know the axis of symmetry, how do you find the vertex?" Substitute -2 for x in the equation and simplify.

On Your Own

- Check to see that students have $b = -2$ in Question 1, not $b = 2$.
- Check to see that students have $a = 1$ in Question 2.

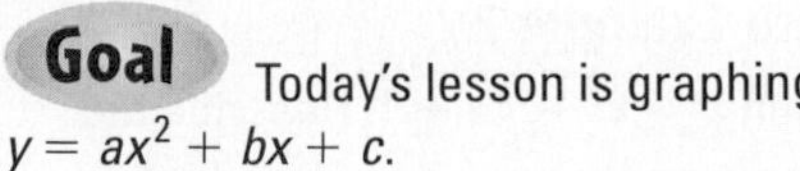

Goal Today's lesson is graphing $y = ax^2 + bx + c$.

Lesson Materials	
Introduction	**Textbook**
• graphing calculator	• graphing calculator

Start Thinking! and Warm Up

Lesson 8.4 Warm Up
For use before Lesson 8.4

Lesson 8.4 Start Thinking!
For use before Lesson 8.4

A mnemonic is a way to help you remember something. For example, a common mnemonic used in math is *Please Excuse My Dear Aunt Sally* to help students remember the order of operations (parentheses, exponents, multiplication/ division, addition/subtraction).

Can you come up with a mnemonic to help you remember that the x-coordinate of the vertex of a parabola is $-\frac{b}{2a}$?

Extra Example 1

Find (a) the axis of symmetry and (b) the vertex of the graph of $y = 3x^2 + 12x - 2$.

a. $x = -2$

b. $(-2, -14)$

On Your Own

1. **a.** $x = \frac{1}{3}$ **b.** $\left(\frac{1}{3}, -\frac{1}{3}\right)$
2. **a.** $x = -3$ **b.** $(-3, -4)$
3. **a.** $x = 7$ **b.** $\left(7, \frac{41}{2}\right)$

Extra Example 2

Graph $y = x^2 - 4x + 1$. Describe the domain and range. domain: all real numbers; range: $y \geq -3$

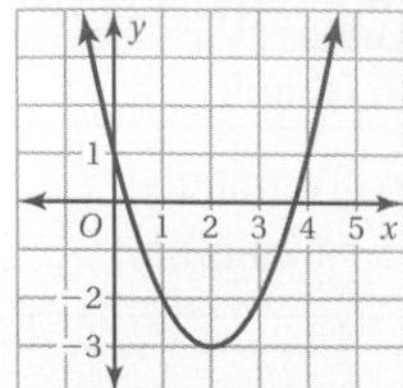

On Your Own

4. domain: all real numbers
   range: $y \geq -1$

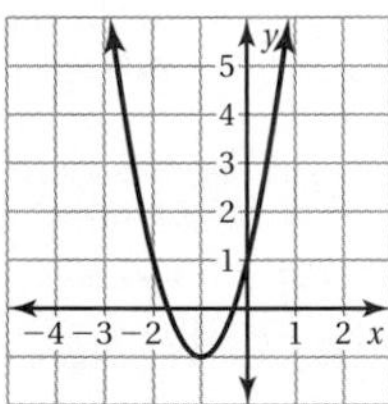

5. domain: all real numbers
   range: $y \geq -9$

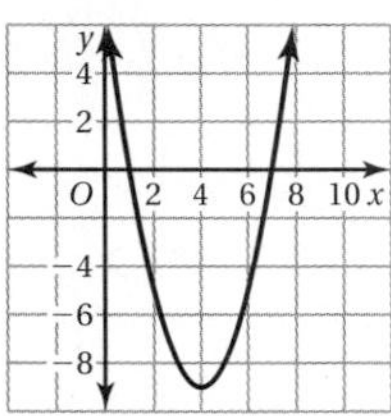

6. domain: all real numbers
   range: $y \leq 3$

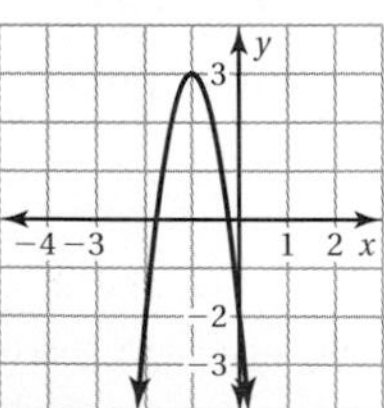

English Language Learners

Word Problems

English language learners may have difficulty in determining when a word problem is asking for a minimum or maximum value. Have them look for phrases such as *greatest height, greatest value, lowest point*, and *shortest length.*

Laurie's Notes

Example 2

- ? "What methods have you used to graph quadratic functions?" table of values and perhaps a graphing calculator
- **MP7:** Explain that another method for graphing a quadratic function is to use information found from the equation.
- Say, "The first step is to find the axis of symmetry." Use the formula to find the axis of symmetry.
- The vertex is located somewhere on the axis of symmetry. To find the y-coordinate of the vertex, substitute 1 for x in the equation and simplify.
- ? "Does the parabola open up or down? Explain." opens up because $a > 0$
- ? "What is the y-intercept? Explain." 5 because $c = 5$
- ? "If the y-intercept is 5 and the line of symmetry is $x = 1$, what other point is on the graph? Explain." (2, 5) because the parabola is symmetric about $x = 1$.
- Draw a smooth curve through the three points.
- Finish the example by finding the domain and range.

On Your Own

- Students should quickly know if the parabola opens up or down, the general shape, and the y-intercept.
- Have students share the three points they used to graph each function.

Key Ideas

- ? "If a parabola opens up and the vertex is (1, −2), what is the range?" $y \geq -2$
- ? "If a parabola opens down and the vertex is (1, −2), what is the range?" $y \leq -2$
- Write the Key Ideas.
- **Misconception:** Students may think that a *maximum* or *minimum value* is the vertex. Emphasize that a maximum or minimum value occurs at the highest or lowest point of a graph (vertex), but that a maximum or minimum value is a number (y-coordinate of the vertex).

EXAMPLE 2 Graphing $y = ax^2 + bx + c$

Graph $y = 3x^2 - 6x + 5$. Describe the domain and range.

Step 1: Find and graph the axis of symmetry.

$$x = -\frac{b}{2a} = -\frac{(-6)}{2(3)} = 1$$

Step 2: Find and plot the vertex.

The axis of symmetry is $x = 1$, so the x-coordinate of the vertex is 1. Use the function to find the y-coordinate of the vertex.

$y = 3x^2 - 6x + 5$ — Write the function.

$= 3(1)^2 - 6(1) + 5$ — Substitute 1 for x.

$= 2$ — Simplify.

So, the vertex is (1, 2).

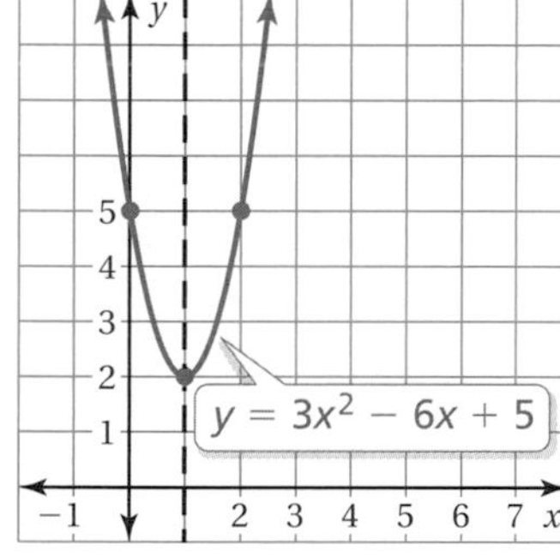

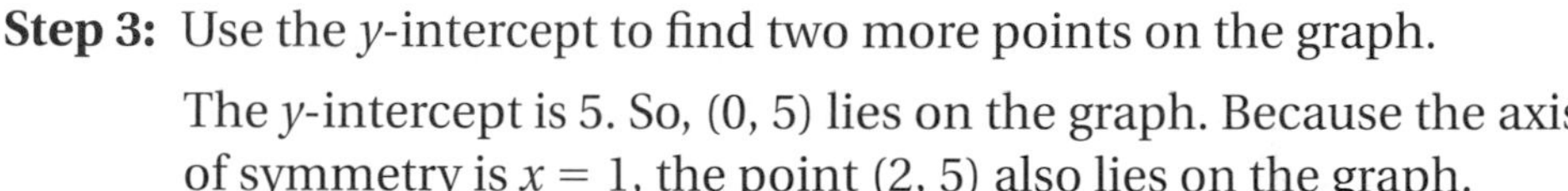

Step 3: Use the y-intercept to find two more points on the graph.

The y-intercept is 5. So, (0, 5) lies on the graph. Because the axis of symmetry is $x = 1$, the point (2, 5) also lies on the graph.

Step 4: Draw a smooth curve through the points.

∴ The domain is all real numbers. The range is $y \geq 2$.

On Your Own

Now You're Ready
Exercises 13–18

Graph the function. Describe the domain and range.

4. $y = 2x^2 + 4x + 1$
5. $y = x^2 - 8x + 7$
6. $y = -5x^2 - 10x - 2$

Key Ideas

Maximum and Minimum Values

The y-coordinate of the vertex of the graph of $y = ax^2 + bx + c$ is the **maximum value** of the function when $a < 0$ or the **minimum value** of the function when $a > 0$.

$y = ax^2 + bx + c, a < 0$

maximum

$y = ax^2 + bx + c, a > 0$

minimum

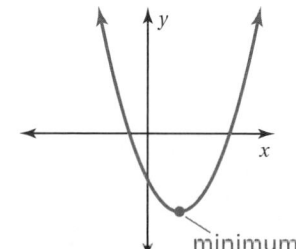

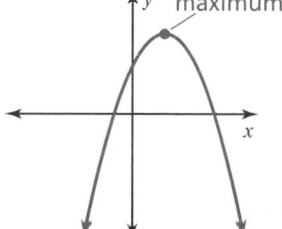

EXAMPLE 3 Finding Maximum and Minimum Values

Tell whether the function $f(x) = -4x^2 - 24x - 19$ has a minimum value or a maximum value. Then find the value.

For $f(x) = -4x^2 - 24x - 19$, $a = -4$ and $-4 < 0$. So, the parabola opens down and the function has a maximum value. To find the maximum value, find the y-coordinate of the vertex.

$$x = -\frac{b}{2a} = -\frac{(-24)}{2(-4)} = -3$$ The x-coordinate of the vertex is $-\frac{b}{2a}$.

$$f(-3) = -4(-3)^2 - 24(-3) - 19$$ Substitute −3 for x.

$$= 17$$ Simplify.

∴ The maximum value is 17.

On Your Own

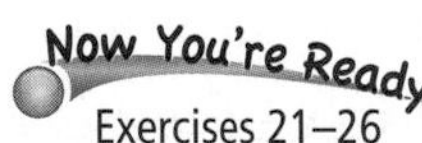

Exercises 21–26

Tell whether the function has a minimum value or a maximum value. Then find the value.

7. $g(x) = 8x^2 - 8x + 6$
8. $h(x) = -\frac{1}{4}x^2 + 3x + 1$

EXAMPLE 4 Real-Life Application

The function $f(t) = -16t^2 + 80t + 5$ gives the height (in feet) of a water balloon t seconds after it is launched. Use a graphing calculator to find the maximum height of the water balloon.

Step 1: Enter the function $f(t) = -16t^2 + 80t + 5$ into your calculator and graph it. Because time cannot be negative, use only nonnegative values of t.

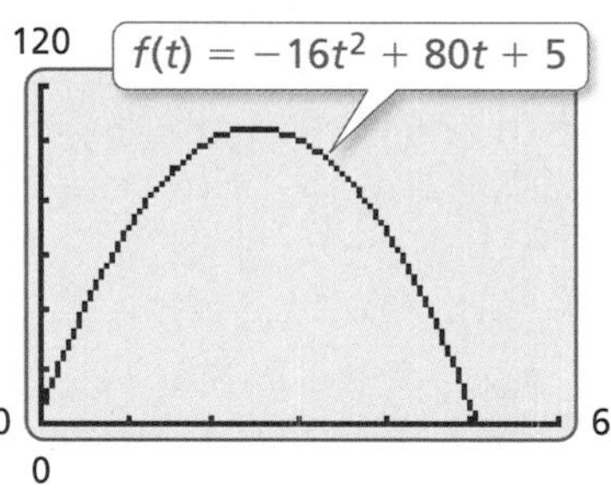

Step 2: Use the *maximum* feature to find the maximum value of the function.

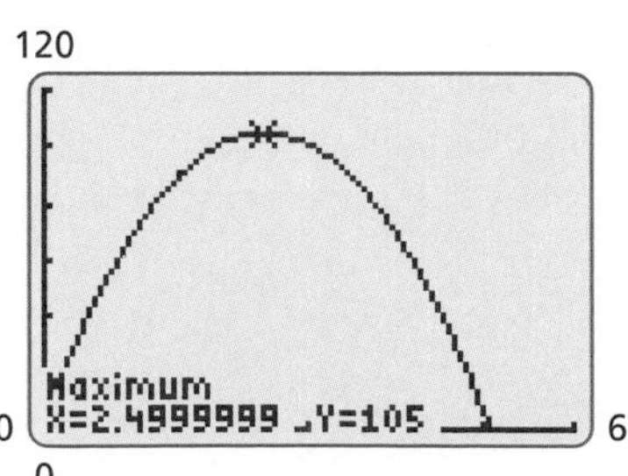

∴ The maximum height of the water balloon is 105 feet.

Study Tip

The *minimum* feature of a graphing calculator can be used for parabolas that open up.

On Your Own

9. When does the water balloon reach its maximum height?

Laurie's Notes

Example 3

- ? "What do you know about the graph of the function?" It opens down, it is narrower than $y = x^2$, and the y-intercept is -19.
- ? "Does the function have a maximum value or minimum value?" maximum
- Find the x-coordinate of the vertex and then solve for the maximum value.
- **Common Error:** Students may have the wrong signs for a, b, and c. Caution students to be careful with the signs of a, b, and c.

On Your Own

- If time permits, do a check with a graphing calculator.

Example 4

- Discuss the vertical motion model first introduced in Section 7.1. It was also used in the previous section but the initial vertical velocity was 0 so there was no t-term.
- **MP4 Model with Mathematics:** Ask a volunteer to read the problem. Remind students that the model gives the height of an object (in feet) at time t (in seconds). It is not the path of the object.
- Explain to students that in order to use a graphing calculator, the function can be thought of as $y = -16x^2 + 80x + 5$.
- **MP7:** Before using the calculator, students should be able to describe certain features of the graph: opens down, y-intercept is 5, has a maximum value, narrower than $y = x^2$.
- **MP5 Use Appropriate Tools Strategically:** If students have not used the *maximum* feature on their calculators, work through the steps necessary to find the maximum value.
- ? **Extension:** "When will the balloon hit the ground?" after about 5 seconds

On Your Own

- **Neighbor Check:** Have students work independently and then have their neighbor check their work. Have students discuss any discrepancies.

Closure

- Write an equation of a quadratic function that opens up, has a negative y-intercept, and is wider than the graph of $y = x^2$. *Sample answer:* $y = 0.25x^2 - 2$

Technology For the Teacher

The Dynamic Planning Tool
Editable Teacher's Resources at *BigIdeasMath.com*

Extra Example 3

Tell whether the function $f(x) = 3x^2 - 18x - 6$ has a minimum value or a maximum value. Then find the value. minimum; -33

On Your Own

7. minimum; 4
8. maximum; 10

Extra Example 4

The function $f(t) = -16t^2 + 40t + 6$ gives the height (in feet) of a water balloon t seconds after it is thrown. Use a graphing calculator to find the maximum height of the water balloon. 31 feet

On Your Own

9. about 2.5 seconds; the x-coordinate of the vertex

Vocabulary and Concept Check

1. For a quadratic function $y = ax^2 + bx + c$ where $a \neq 0$, if $a > 0$ then the function has a minimum value and if $a < 0$ then the function has a maximum value.
2. What is axis of symmetry of the graph of the function?; $x = 2$; 32

Practice and Problem Solving

3. vertex: $(2, -1)$;
axis of symmetry: $x = 2$;
y-intercept: 1
4. vertex: $(-3, 2)$;
axis of symmetry: $x = -3$;
y-intercept: -1
5. vertex: $(-2, -3)$;
axis of symmetry: $x = -2$;
y-intercept: 4
6. **a.** $x = 1$ **b.** $(1, -2)$
7. **a.** $x = -2$ **b.** $(-2, -12)$
8. **a.** $x = -1$ **b.** $(-1, 7)$
9. **a.** $x = 2$ **b.** $(2, 4)$
10. **a.** $x = 5$ **b.** $(5, 4)$
11. **a.** $x = 4$ **b.** $(4, -6)$
12. The equation of the axis of symmetry is $x = -\frac{b}{2a}$, not $x = \frac{b}{2a}$.

    $x = -\frac{b}{2a} = -\frac{-12}{2(3)} = 2$;

    The axis of symmetry is $x = 2$.

13–18. See Additional Answers.

Assignment Guide and Homework Check

Level	Assignment	Homework Check
Average	1, 2, 6–26 even, 19, 27, 29, 38–41	6, 16, 19, 24, 29
Advanced	1, 2, 6, 9, 13, 16, 19, 21, 24, 27–29, 33–37, 38–41	16, 19, 24, 29, 33, 36

For Your Information

- **Exercise 19** The given function represents the height of a firework, not the path of a firework.

Common Errors

- **Exercises 3–11** Students may forget to use the negative sign in $x = -\frac{b}{2a}$ when finding the axis of symmetry. Remind them of this formula.
- **Exercises 13–18** Students may not graph enough points to properly portray the shape of the graph of the function. Remind them that the graph should have a U-shape.

8.4 Record and Practice Journal

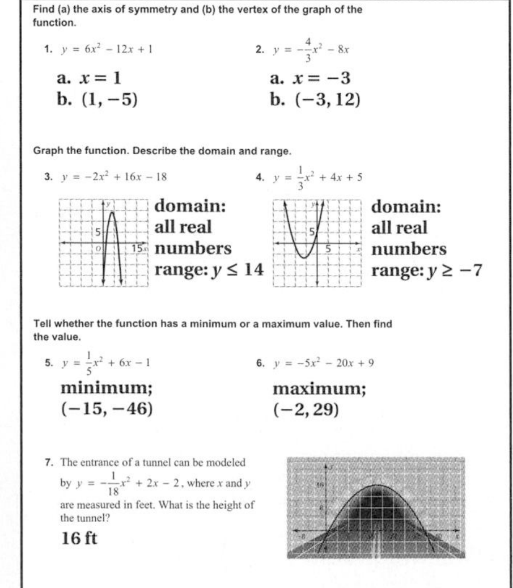

Find (a) the axis of symmetry and (b) the vertex of the graph of the function.

1. $y = 6x^2 - 12x + 1$
   a. $x = 1$
   b. $(1, -5)$
2. $y = -\frac{4}{3}x^2 - 8x$
   a. $x = -3$
   b. $(-3, 12)$

Graph the function. Describe the domain and range.

3. $y = -2x^2 + 16x - 18$
   domain: all real numbers
   range: $y \leq 14$
4. $y = \frac{1}{3}x^2 + 4x + 5$
   domain: all real numbers
   range: $y \geq -7$

Tell whether the function has a minimum or a maximum value. Then find the value.

5. $y = \frac{1}{5}x^2 + 6x - 1$
   minimum; $(-15, -46)$
6. $y = -5x^2 - 20x + 9$
   maximum; $(-2, 29)$
7. The entrance of a tunnel can be modeled by $y = -\frac{1}{18}x^2 + 2x - 2$, where x and y are measured in feet. What is the height of the tunnel?
   16 ft

Technology For the Teacher
Answer Presentation Tool

8.4 Exercises

Vocabulary and Concept Check

1. **VOCABULARY** Explain how you can tell whether a quadratic function has a maximum value or a minimum value without graphing the function.

2. **DIFFERENT WORDS, SAME QUESTION** Consider the quadratic function $y = -2x^2 + 8x + 24$. Which is different? Find "both" answers.

   What is the maximum value of the function?

   What is the y-coordinate of the vertex of the graph?

   What is the greatest number in the range of the function?

   What is the axis of symmetry of the graph of the function?

Practice and Problem Solving

Find the vertex, the axis of symmetry, and the y-intercept of the graph.

3. 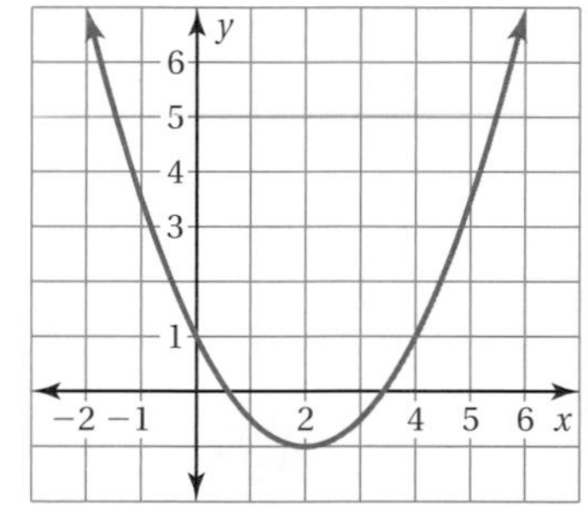

4. 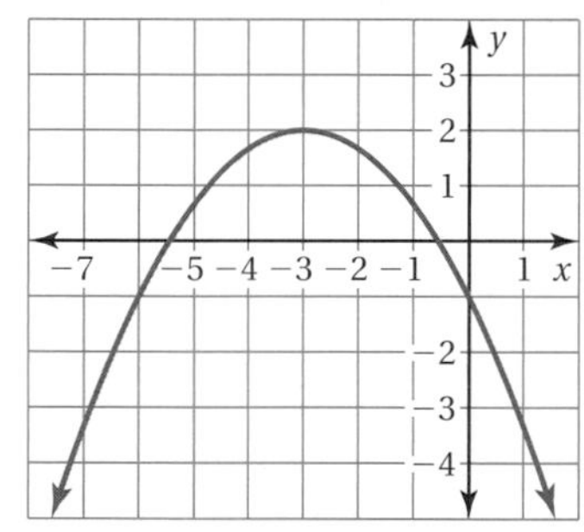

5. 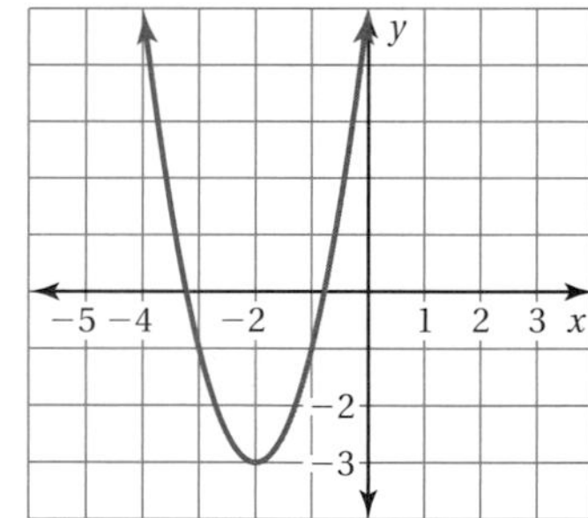

Find (a) the axis of symmetry and (b) the vertex of the graph of the function.

① 6. $y = 2x^2 - 4x$
7. $y = 3x^2 + 12x$
8. $y = -8x^2 - 16x - 1$
9. $y = -6x^2 + 24x - 20$
10. $y = \frac{2}{5}x^2 - 4x + 14$
11. $y = -\frac{3}{4}x^2 + 6x - 18$

12. **ERROR ANALYSIS** Describe and correct the error in finding the axis of symmetry of the graph of $y = 3x^2 - 12x + 11$.

$$x = -\frac{b}{2a} = \frac{-12}{2(3)} = -2$$

The axis of symmetry is $x = -2$.

Graph the function. Describe the domain and range.

② 13. $y = 2x^2 + 12x + 14$
14. $y = 4x^2 + 24x + 31$
15. $y = -8x^2 - 16x - 9$
16. $y = -5x^2 + 30x - 47$
17. $y = \frac{2}{3}x^2 - 8x + 19$
18. $y = -\frac{1}{2}x^2 - 8x - 25$

19. **FIREWORK** The function shown represents the height h (in feet) of a firework t seconds after it is launched. The firework explodes at its highest point.

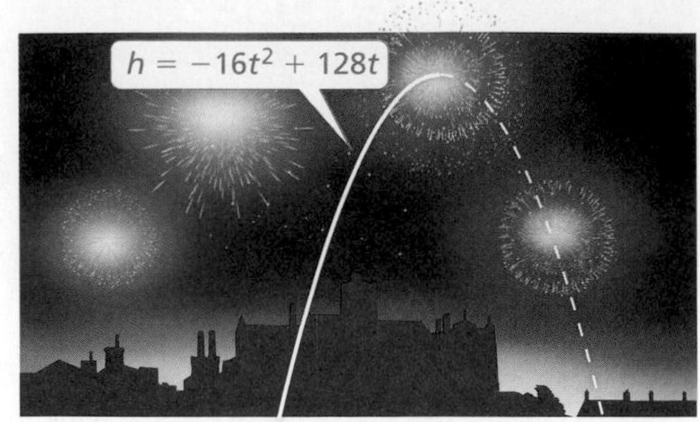

a. When does the firework explode?

b. Describe the domain and range of h.

20. **REASONING** Given the quadratic equation $y = ax^2 + bx + c$, find the axis of symmetry when $b = 0$.

Tell whether the function has a minimum value or a maximum value. Then find the value.

③ 21. $y = 3x^2 - 18x + 15$

22. $y = -5x^2 + 10x + 7$

23. $y = -4x^2 + 4x - 2$

24. $y = 2x^2 - 10x + 13$

25. $y = -\frac{1}{2}x^2 + 8x + 20$

26. $y = \frac{1}{5}x^2 - 12x + 27$

27. **PRECISION** The vertex of a graph of a quadratic function is $(3, -1)$. One point on the graph is $(6, 8)$. Find another point on the graph. Justify your answer.

28. **SUSPENSION BRIDGE** The cables between the two towers of a suspension bridge can be modeled by $y = \frac{1}{400}x^2 - x + 150$, where x and y are measured in feet. The cables are at road level midway between the towers. How high is the road above the water?

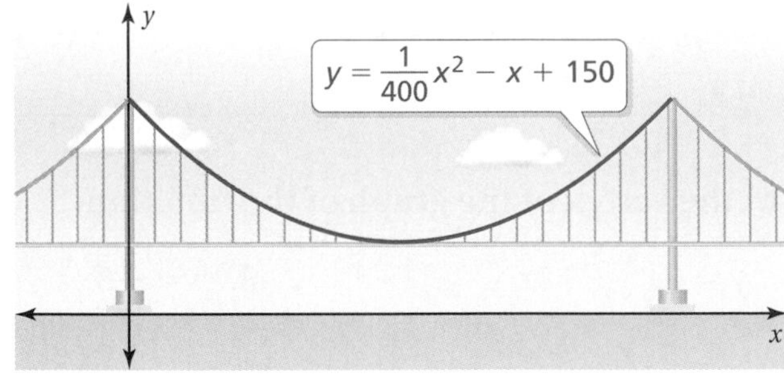

29. **STEEPLECHASE** The function $h(t) = -16t^2 + 16t$ gives the height h (in feet) of a horse t seconds after it jumps during a steeplechase.

a. When does the horse reach its maximum height?

b. Can the horse clear a fence that is 3.5 feet tall? If so, by how much?

c. How long is the horse in the air?

Common Errors

- **Exercises 30–32** Students may not properly use the minimum or maximum features of a graphing calculator. Remind them of how to use these features.
- **Exercise 35** Students may not realize that the maximum value of the function is the maximum revenue, not the price they should charge for a calculator. They must use the *x*-value of the vertex and part of the equation in part (a) to find how much should be charged for a calculator.
- **Exercise 36** Students may not realize that they need to find the intersection point of the two functions. Encourage them to stop and think about what the question is asking.

19. **a.** 4 seconds after it is launched at a height of 256 feet

b. The domain is $0 \le t \le 8$. The range is $0 \le h \le 256$.

20. $x = 0$; The *y*-axis is the axis of symmetry when $b = 0$.

21. minimum; -12

22. maximum; 12

23. maximum; -1

24. minimum; 0.5

25. maximum; 52

26. minimum; -153

27. (0, 8); Use the axis of symmetry to find the reflection of (6, 8).

28. 50 ft

29. **a.** 0.5 second after it jumps

b. yes; 0.5 ft

c. 1 sec

Differentiated Instruction

Advanced

In this lesson, students learned that the *x*-coordinate of the vertex is $-\frac{b}{2a}$. Challenge students to find the *y*-coordinate of the vertex in terms of *a*, *b*, and *c*. $y = -\frac{b^2}{4a} + c$

Practice and Problem Solving

30–34. See Additional Answers.

35. See *Taking Math Deeper.*

36. 14 ft

37. $\frac{k^2}{8}$ ft^2

Fair Game Review

38.

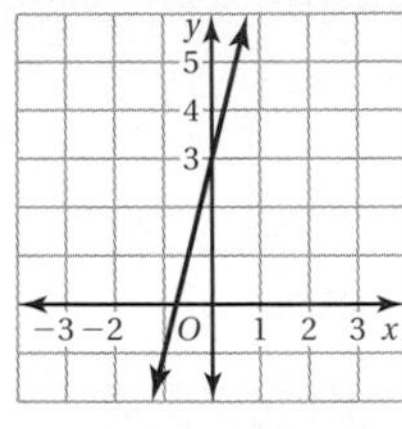

39.

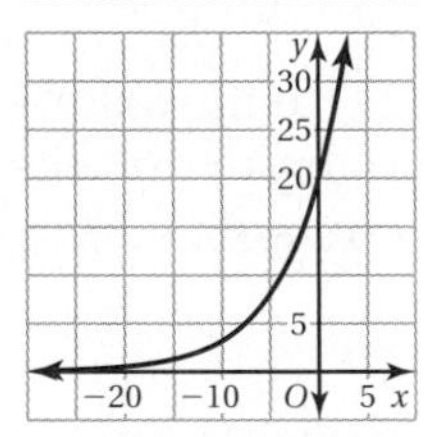

40.

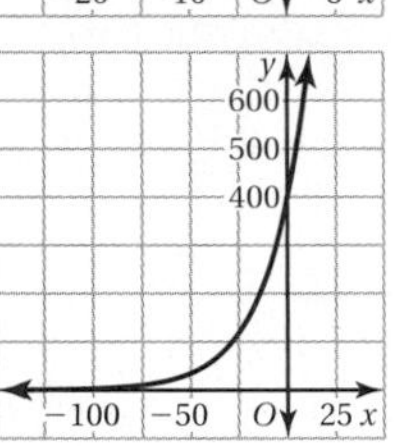

41. C

Mini-Assessment

Find (a) the axis of symmetry and (b) the vertex of the graph of the function.

1. $y = 3x^2 + 6x$

a. $x = -1$ **b.** $(-1, -3)$

2. $y = -\frac{1}{2}x^2 + 3x - 5$

a. $x = 3$ **b.** $\left(3, -\frac{1}{2}\right)$

3. Tell whether the function $y = -x^2 + 2x + 1$ has a minimum value or a maximum value. Then find the value. maximum; 2

Taking Math Deeper

Exercise 35

This exercise provides good practice for standard N.Q.2. Students that let the independent variable x represent the number of calculators sold will have difficulty writing the function.

Use the wording in the exercise as a clue for defining the independent variable.

independent variable

For each \$5 decrease in price, the store expects to sell 5 more calculators.

So, let y represent the revenue and let x represent the number of \$5 price decreases.

a. Write the function.

revenue = units sold × unit price

5 more calculators are sold for each \$5 price decrease.

$$y = (60 + 5x)(100 - 5x)$$
$$y = -25x^2 + 200x + 6000$$

b. Find the maximum value. Use the fact that the maximum value occurs on the axis of symmetry.

$$x = -\frac{b}{2a} = \frac{200}{2(-25)} = 4$$

The maximum occurs when $x = 4$. So, the store should charge $100 - 5(4) = \$80$ per calculator to maximize monthly revenue.

Challenge students to think about why the store may charge \$75 for the calculator even though an \$80 price maximizes revenue.

Project

Research the economic model called "supply and demand." How does it relate to this exercise?

Reteaching and Enrichment Strategies

If students need help. . .	If students got it. . .
Resources by Chapter • Practice A and Practice B • Puzzle Time Record and Practice Journal Practice Differentiating the Lesson Lesson Tutorials Skills Review Handbook	Resources by Chapter • Enrichment and Extension • School-to-Work Start the next section

Use the *minimum* or *maximum* feature of a graphing calculator to approximate the vertex of the graph of the function.

30. $y = -6.2x^2 + 4.8x - 1$ **31.** $y = 0.5x^2 + \sqrt{2}x - 3$ **32.** $y = \pi x^2 + 3x$

33. CHOOSE TOOLS The graph of a quadratic function passes through (4, 0), (5, 3), and (6, 4). Does the graph open up or down? Explain your reasoning.

34. REASONING For a quadratic function f, what does $f\left(-\frac{b}{2a}\right)$ represent? Explain your reasoning.

35. CALCULATORS An office supply store sells about 60 graphing calculators per month for \$100 each. For each \$5 decrease in price, the store expects to sell 5 more calculators.

a. Write a quadratic function that represents the revenue from calculator sales. (*Note:* revenue = units sold × unit price)

b. How much should the store charge per calculator to maximize monthly revenue?

36. AIR CANNON At a basketball game, an air cannon is used to launch T-shirts into the crowd. The function $y = -\frac{1}{8}x^2 + 4x$ gives the path of a T-shirt. The function $3y = 2x - 14$ gives the height of the bleachers. In both functions, y represents height (in feet) and x represents horizontal distance (in feet). At what height does the T-shirt land in the bleachers?

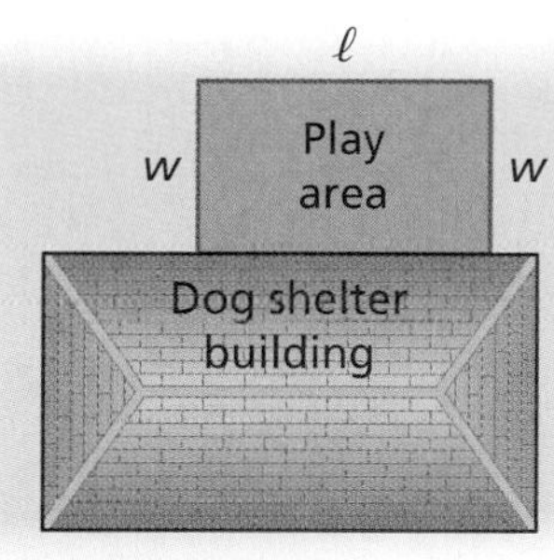

37. DOG SHELTER The owners of a dog shelter want to enclose a rectangular play area on the side of their building. They have k feet of fencing. What is the maximum area of the outside enclosure in terms of k? (*Hint:* Find the y-coordinate of the vertex of the graph of the area function.)

Fair Game Review What you learned in previous grades & lessons

Graph the function. *(Section 2.4 and Section 6.5)*

38. $-4x + y = 3$ **39.** $y = 20(1.2)^t$ **40.** $r(t) = 400(1.05)^t$

41. MULTIPLE CHOICE What is the value of $3(4)^x$ when $x = 2$? *(Section 6.4)*

(A) 6 (B) 24 (C) 48 (D) 144

8.4b Graphing $y = a(x - h)^2 + k$

Key Vocabulary
vertex form, *p. 432*

The **vertex form** of a quadratic function is $y = a(x - h)^2 + k$, where $a \neq 0$. The vertex of the parabola is (h, k).

Graphing $y = (x - h)^2$

- When $h > 0$, the graph of $y = (x - h)^2$ is a horizontal translation h units to the right of the graph of $y = x^2$.
- When $h < 0$, the graph of $y = (x - h)^2$ is a horizontal translation h units to the left of the graph of $y = x^2$.

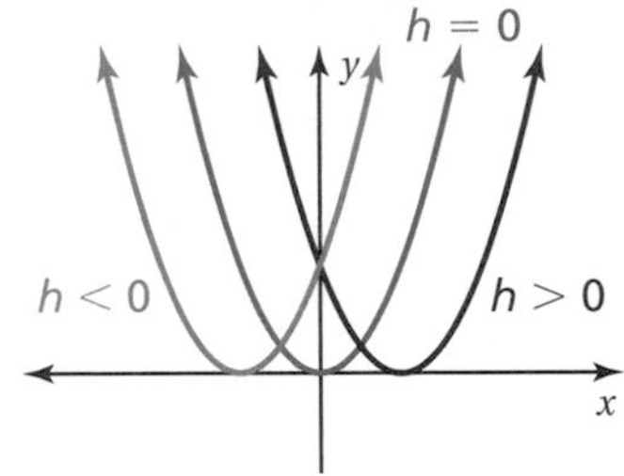

EXAMPLE 1 Graphing $y = (x - h)^2$

Graph $y = (x - 4)^2$. Compare the graph to the graph of $y = x^2$.

Step 1: Make a table of values.

x	0	2	4	6	8
y	16	4	0	4	16

Step 2: Plot the ordered pairs.

Step 3: Draw a smooth curve through the points.

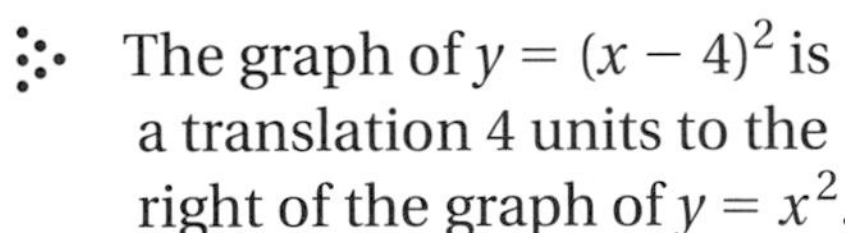

The graph of $y = (x - 4)^2$ is a translation 4 units to the right of the graph of $y = x^2$.

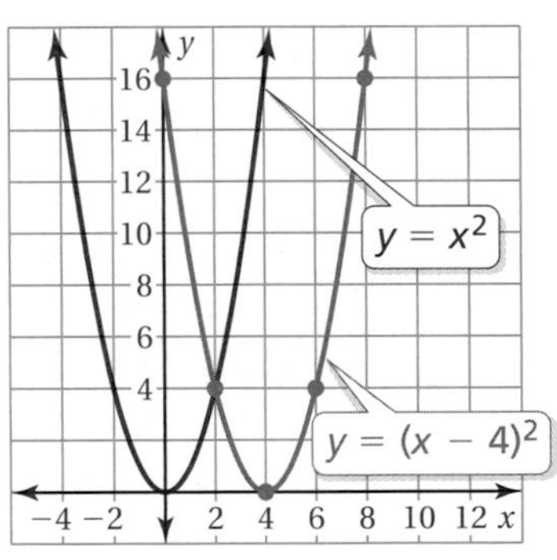

Practice

Graph the function. Compare the graph to the graph of $y = x^2$. Use a graphing calculator to check your answer.

1. $y = (x + 3)^2$
2. $y = (x - 1)^2$
3. $y = (x - 6)^2$
4. $y = (x + 10)^2$
5. $y = (x - 1.5)^2$
6. $y = \left(x + \frac{5}{2}\right)^2$
7. **REASONING** Compare the graphs of $y = x^2 + 6x + 9$ and $y = x^2$ without graphing the functions. How can factoring help you compare the parabolas? Explain.
8. **STRUCTURE** Write the function in Example 1 in the form $y = ax^2 + bx + c$. Describe advantages and disadvantages of writing the function in each form.

Laurie's Notes

Introduction

Connect
- **Yesterday:** Students graphed quadratic functions of the form $y = ax^2 + bx + c$. (MP4, MP5, MP7)
- **Today:** Students will graph quadratic functions of the form $y = a(x - h)^2 + k$.

Motivate
- Write two trinomials on the board for students to factor. Ask volunteers to share their answers.
  (a) $x^2 - 4x + 4$ $(x - 2)^2$ (b) $2x^2 + 4x + 2$ $2(x + 1)^2$
- Because $x^2 - 4x + 4 = (x - 2)^2$, the graphs of $y = x^2 - 4x + 4$ and $y = (x - 2)^2$ are the same.
- Today students will graph quadratic functions in *vertex form*.

Lesson Notes

Discuss
- **MP7 Look for and Make Use of Structure:** All of the quadratic functions that students have graphed have been in standard form $y = ax^2 + bx + c$ and b and c sometimes equal to 0. In this lesson, students will investigate the vertex form of a quadratic function. Make the connection to linear functions and the different forms in which they are written. The structure of the equation gives information about the graph of the function.
- Discuss the vertex form of a quadratic function. Write the vertex form and identify the vertex, (h, k).

Key Idea
- Write the equation $y = (x - h)^2$. Refer to $y = (x - 2)^2$ in the Motivate and explain that $h = 2$ in this equation.
- **MP7:** Note that when $h > 0$, the factor is $(x - h)$ and when $h < 0$, the factor is $(x + h)$. This can be confusing to students. Explain that when h is negative, the operation changes to addition (by adding the opposite).
- The graph of $y = (x - 2)^2$ is a translation 2 units to the right of $y = x^2$.

Example 1
? "What is the value of h?" 4
- Because $h > 0$, the graph is a translation 4 units to the right of $y = x^2$. Students can make a table of values to graph the function and verify.
- **MP5 Use Appropriate Tools Strategically:** Have students verify with a graphing calculator, understanding that the viewing window may cause the graphs to appear different.
- **Extension:** Compare the graph of $y = (x - 5)^2$ to the graph of $y = x^2$. Verify by graphing $y = x^2 - 10x + 25$.

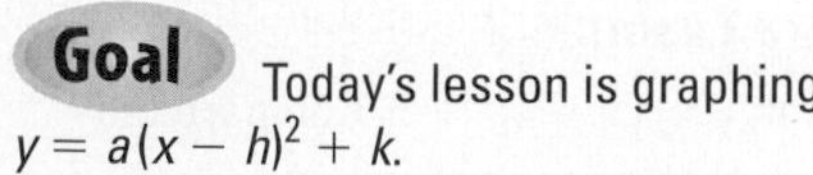

Goal Today's lesson is graphing $y = a(x - h)^2 + k$.

Lesson Materials
Textbook
• graphing calculator

Start Thinking! and Warm Up

Lesson 8.4b Warm Up
For use before Lesson 8.4b

Lesson 8.4b Start Thinking!
For use before Lesson 8.4b

How does the equation $y = 2x$ change if you want to shift the graph of $y = 2x$ to the right by 3 units? Does the equation $y = x^2$ change in the same way if you shift the graph of $y = x^2$ to the right by 3 units?

Extra Example 1
Graph $y = (x + 2)^2$. Compare the graph to the graph of $y = x^2$.

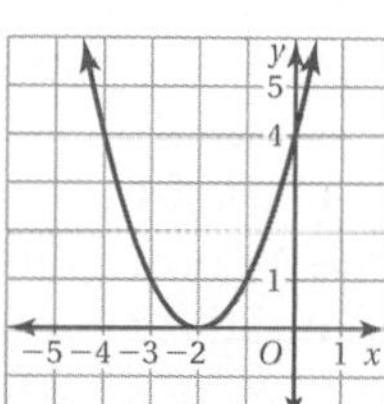

The graph of $y = (x + 2)^2$ is a translation 2 units to the left of the graph of $y = x^2$.

Practice
1–8. See Additional Answers.

Record and Practice Journal Practice
See Additional Answers.

Extra Example 2

Graph $y = (x - 4)^2 + 5$. Compare the graph to the graph of $y = x^2$.

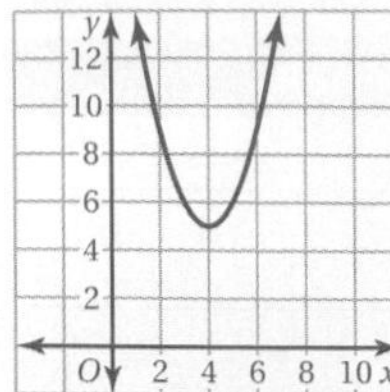

The graph of $y = (x - 4)^2 + 5$ is a translation 4 units to the right and 5 units up of the graph of $y = x^2$.

Extra Example 3

Graph $y = \frac{1}{5}(x + 5)^2 - 1$. Compare the graph to the graph of $y = x^2$.

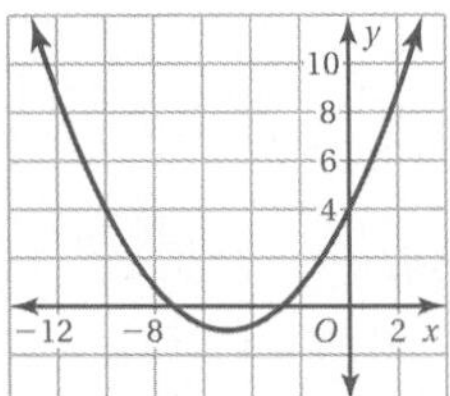

The graph of $y = \frac{1}{5}(x + 5)^2 - 1$ is wider than the graph of $y = x^2$. The vertex of the graph of $y = \frac{1}{5}(x + 5)^2 - 1$ is a translation 5 units to the left and 1 unit down of the of the vertex graph of $y = x^2$.

Practice

9–19. See Additional Answers.

20. in general, no; k represents the y-coordinate of the vertex. If $h = 0$, then k represents the y-coordinate of the vertex and the y-intercept.

Mini-Assessment

Graph the function. Compare the graph to the graph of $y = x^2$.

1. $y = (x - 7)^2$
2. $y = (x + 4)^2 + 3$
3. $y = -(x + 6)^2 - 2$

1–3. See Additional Answers.

Laurie's Notes

Practice

- Students should not have difficulty with Exercises 1–6. Discuss that the vertex is on the x-axis. The translation is horizontal only.
- Connect Exercise 7 to the Motivate.

Discuss

- Point out that k is added while h is subtracted in vertex form.

Example 2

? "What are the values of h and k?" $h = -5$ and $k = -1$

- The table of values was created so that the vertex, $(-5, -1)$, falls in the middle of the ordered pairs generated.
- Graph the function and verify that it is a translation 5 units to the left and 1 unit down of the graph of $y = x^2$.
- **MP5:** Have students verify with a graphing calculator, understanding that the viewing window may cause the graphs to appear different.

Example 3

? "What effect do you think the leading coefficient of -2 has on the graph?" A few students will suggest that $a = -2$ will make the graph open down and that it will be narrower than the graph of $y = x^2$.

- Make a table of values and plot the ordered pairs.
- Have students summarize what the values of a, h, and k tell about the graph of a quadratic function in vertex form.

Practice

- Students should work with partners to complete the exercises. Take time to discuss students' reasoning in Exercises 19 and 20.

Closure

- Consider the function $y = (x + 2)^2 + 8$.
  - **a.** Tell whether the graph opens up or down. up
  - **b.** Compare the graph to the graph of $y = x^2$. translation 2 units to the left and 8 units up

Technology For the Teacher

The Dynamic Planning Tool
Editable Teacher's Resources at *BigIdeasMath.com*

EXAMPLE 2 Graphing $y = (x - h)^2 + k$

Graph $y = (x + 5)^2 - 1$. Compare the graph to the graph of $y = x^2$.

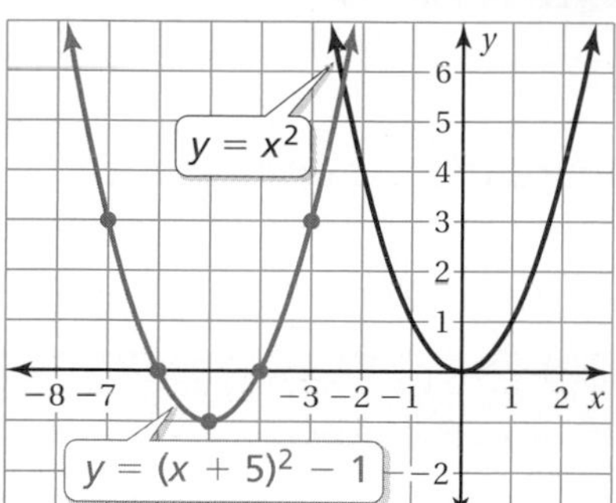

Step 1: Make a table of values.

x	−7	−6	−5	−4	−3
y	3	0	−1	0	3

Step 2: Plot the ordered pairs.

Step 3: Draw a smooth curve through the points.

∴ The graph of $y = (x + 5)^2 - 1$ is a translation 5 units to the left and 1 unit down of the graph of $y = x^2$.

EXAMPLE 3 Graphing $y = a(x - h)^2 + k$

Graph $y = -2(x + 2)^2 + 3$. Compare the graph to the graph of $y = x^2$.

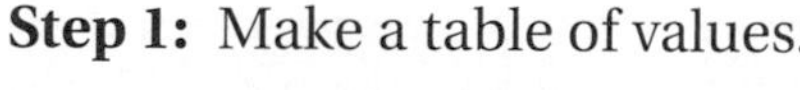

Step 1: Make a table of values.

x	−4	−3	−2	−1	0
y	−5	1	3	1	−5

Step 2: Plot the ordered pairs.

Step 3: Draw a smooth curve through the points.

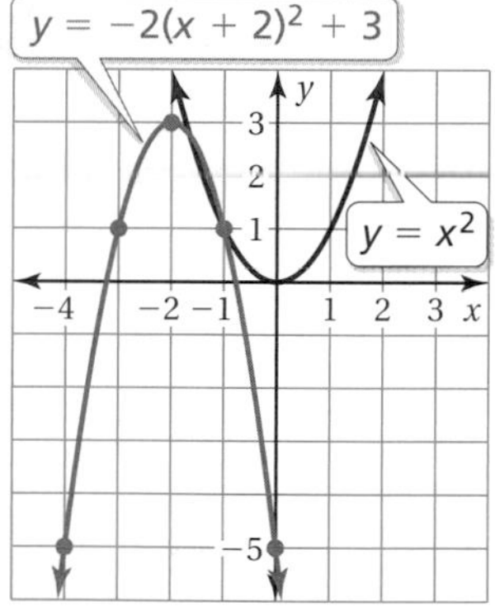

> **Study Tip**
>
> Notice what the values in vertex form represent:
>
> *a*: opens up or down, and is wider or narrower
>
> *h*: horizontal translation
>
> *k*: vertical translation
>
> (*h*, *k*): vertex

∴ The graph of $y = -2(x + 2)^2 + 3$ opens down and is narrower than the graph of $y = x^2$. The vertex of the graph of $y = -2(x + 2)^2 + 3$ is a translation 2 units to the left and 3 units up of the vertex of the graph of $y = x^2$.

Practice

Graph the function. Compare the graph to the graph of $y = x^2$. Use a graphing calculator to check your answer.

9. $y = (x - 2)^2 + 4$ **10.** $y = (x + 1)^2 - 7$ **11.** $y = (x - 8)^2 - 8$

12. $y = 3(x - 1)^2 + 6$ **13.** $y = -(x - 3)^2 - 5$ **14.** $y = \frac{1}{2}(x + 4)^2 - 2$

Describe how the graph of $g(x)$ compares to the graph of $f(x)$.

15. $g(x) = f(x) - 7$ **16.** $g(x) = f(x + 10)$ **17.** $g(x) = 5f(x)$ **18.** $g(x) = f(2x)$

19. **REASONING** The graph of $y = x^2$ is translated 2 units right and 5 units down. Write an equation for the function in vertex form and in standard form.

20. **REASONING** Does k represent the y-intercept of the graph of $y = a(x - h)^2 + k$? Explain.

8.5 Comparing Linear, Exponential, and Quadratic Functions

COMMON CORE STATE STANDARDS
F.IF.4
F.IF.6
F.IF.7a
F.LE.3

Essential Question How can you compare the growth rates of linear, exponential, and quadratic functions?

1 ACTIVITY: Comparing Speeds

Work with a partner. Three cars start traveling at the same time. The distance traveled in *t* minutes is *y* miles. Complete each table and sketch all three graphs in the same coordinate plane. Compare the speeds of the three cars. Which car has a constant speed? Which car is accelerating the most? Explain your reasoning.

t	$y = t$
0	
0.2	
0.4	
0.6	
0.8	
1.0	

t	$y = 2^t - 1$
0	
0.2	
0.4	
0.6	
0.8	
1.0	

t	$y = t^2$
0	
0.2	
0.4	
0.6	
0.8	
1.0	

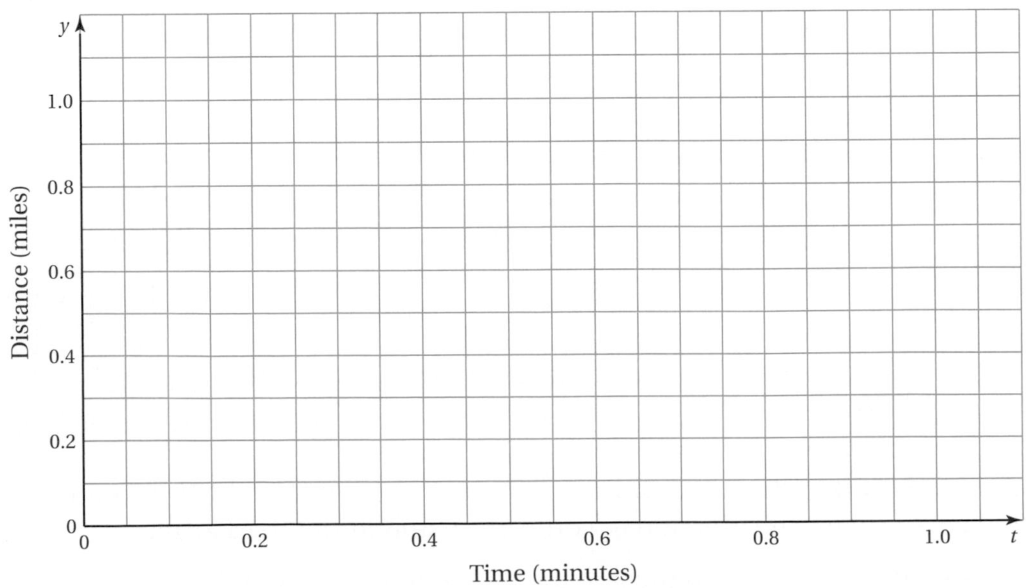

Laurie's Notes

Introduction

Standards for Mathematical Practice

- **MP4 Model with Mathematics:** Linear, exponential, and quadratic functions are common functions that can be used to model real-life applications. Mathematically proficient students can identify each of these functions from an equation, a complete graph, or a table of values. When examining a table of values, the *behavior* of the function becomes evident.

Motivate

- **Story Time:** Tell students that you test drove a sports car recently—a Lamborghini Reventón. Any student that follows cars will know that you are fibbing. The base price of a Reventón is approximately $1.6 million, a bit much for a teaching salary. So, instead of test driving, tell students you read about the car and the fact that it can go from 0 to 60 miles per hour in 3.3 seconds with a top speed of 211 miles per hour!
- A car that is a bit more affordable would be a Chevrolet Camaro, with a base price of about $37,000. It can go from 0 to 60 miles per hour in 4.9 seconds with a top speed of 155 miles per hour.
- Today students will investigate some fast cars, but no speeding!

Activity Notes

Activity 1

- This first activity reacquaints students with the three types of functions they have studied so far this year: linear, exponential, and quadratic. Students will explore the growth rates of the three functions over a small, finite domain.
- Students work with a partner to complete the table of values for each function.
- A common belief among students is that exponential functions grow most rapidly. What students discover quickly is that the functions all intersect at (0, 0) and (1, 1). It is the behavior of the functions between these two points that is of interest.
- ❓ "Were there any surprises?" Students should say yes.
- The car with a constant speed should be relatively easy to determine.
- Discuss the explanations offered for which car is accelerating the most.

Common Core State Standards

F.IF.4 For a function that models a relationship between two quantities, interpret key features of graphs and tables in terms of the quantities, and sketch graphs showing key features given a verbal description of the relationship.

F.IF.6 Calculate and interpret the average rate of change of a function (presented symbolically or as a table) over a specified interval. Estimate the rate of change from a graph.

F.LE.3 Observe using graphs and tables that a quantity increasing exponentially eventually exceeds a quantity increasing linearly, quadratically, or (more generally) as a polynomial function.

Also **F.IF.7a**

Previous Learning

Students should know how to graph linear, exponential, and quadratic functions.

Start Thinking! and Warm Up

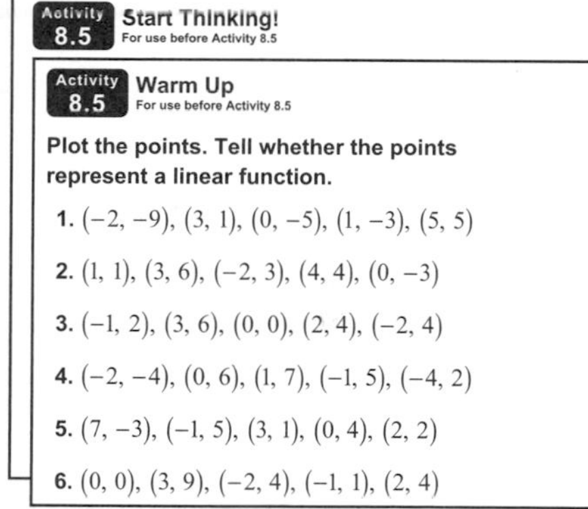

Activity 8.5 Start Thinking! For use before Activity 8.5

Activity 8.5 Warm Up For use before Activity 8.5

Plot the points. Tell whether the points represent a linear function.

1. (−2, −9), (3, 1), (0, −5), (1, −3), (5, 5)
2. (1, 1), (3, 6), (−2, 3), (4, 4), (0, −3)
3. (−1, 2), (3, 6), (0, 0), (2, 4), (−2, 4)
4. (−2, −4), (0, 6), (1, 7), (−1, 5), (−4, 2)
5. (7, −3), (−1, 5), (3, 1), (0, 4), (2, 2)
6. (0, 0), (3, 9), (−2, 4), (−1, 1), (2, 4)

8.5 Record and Practice Journal

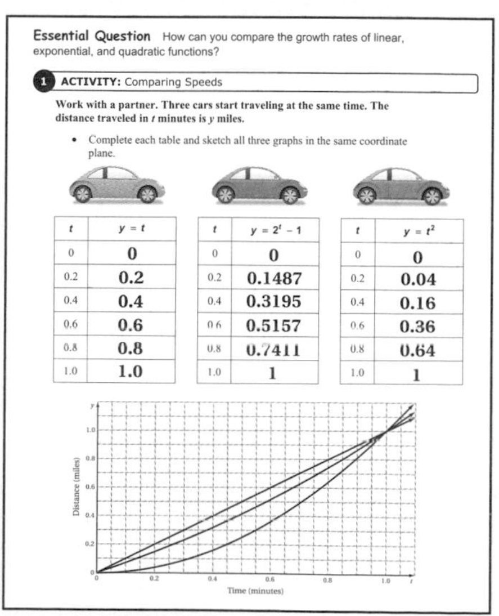

Essential Question How can you compare the growth rates of linear, exponential, and quadratic functions?

1 ACTIVITY: Comparing Speeds

Work with a partner. Three cars start traveling at the same time. The distance traveled in *t* minutes is *y* miles.

- Complete each table and sketch all three graphs in the same coordinate plane.

t	$y = t$
0	**0**
0.2	**0.2**
0.4	**0.4**
0.6	**0.6**
0.8	**0.8**
1.0	**1.0**

t	$y = 2^t - 1$
0	**0**
0.2	**0.1487**
0.4	**0.3195**
0.6	**0.5157**
0.8	**0.7411**
1.0	**1**

t	$y = t^2$
0	**0**
0.2	**0.04**
0.4	**0.16**
0.6	**0.36**
0.8	**0.64**
1.0	**1**

Differentiated Instruction

Kinesthetic

Some students benefit by having access to graphing calculators or the graphing software programs that are available on many computer operating systems. Students can enter the equations and view the graphs in the same viewing window. The viewing window can be changed to focus on different aspects of the graphs.

8.5 Record and Practice Journal

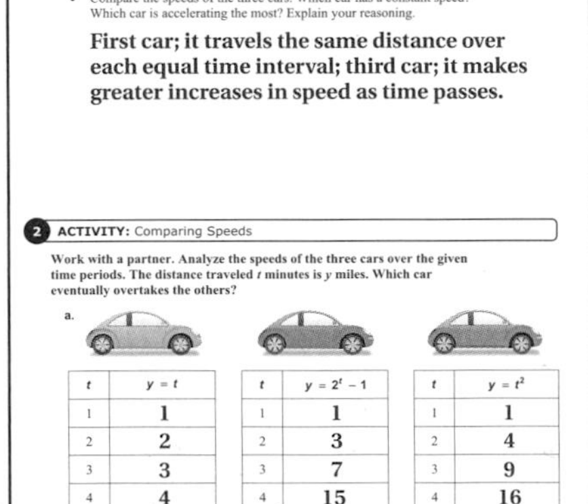

- Compare the speeds of the three cars. Which car has a constant speed? Which car is accelerating the most? Explain your reasoning.

First car; it travels the same distance over each equal time interval; third car; it makes greater increases in speed as time passes.

2 ACTIVITY: Comparing Speeds

Work with a partner. Analyze the speeds of the three cars over the given time periods. The distance traveled t minutes is y miles. Which car eventually overtakes the others?

a.

t	$y = t$
1	1
2	2
3	3
4	4

t	$y = 2^t - 1$
1	1
2	3
3	7
4	15

t	$y = t^2$
1	1
2	4
3	9
4	16

constant speed: first car; accelerating: second and third cars; third car

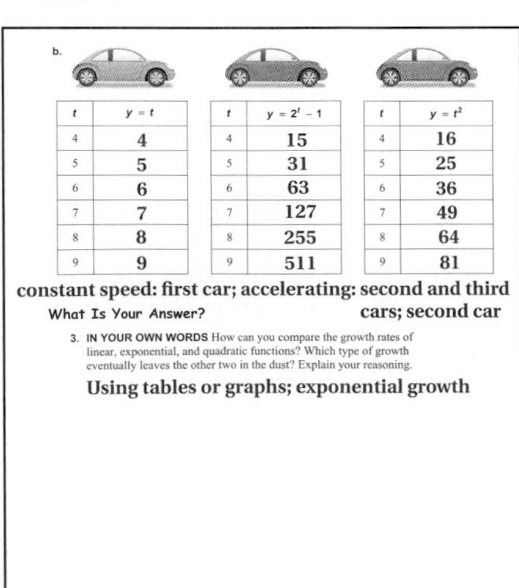

b.

t	$y = t$
4	4
5	5
6	6
7	7
8	8
9	9

t	$y = 2^t - 1$
4	15
5	31
6	63
7	127
8	255
9	511

t	$y = t^2$
4	16
5	25
6	36
7	49
8	64
9	81

constant speed: first car; accelerating: second and third cars; second car

What Is Your Answer?

3. IN YOUR OWN WORDS How can you compare the growth rates of linear, exponential, and quadratic functions? Which type of growth eventually leaves the other two in the dust? Explain your reasoning.

Using tables or graphs; exponential growth

Laurie's Notes

Activity 2

? "In part (a), how long does each car run?" 4 minutes

- Say to students that the units of time are assumed to be the same.
- The equations in part (b) are the same as part (a) but the domain is extended to include 9 minutes.
- Ask students to describe the three different speeds. Students can visually interpret the speeds in a graph, or they can calculate the speed of each car. Remember speed is distance divided by time.
- If time permits, set up a table to show the change in average speed for each car. This connects to the answer in Question 3.

Average Speeds:

t	Green (linear)	Blue (exponential)	Red (quadratic)
1	1/1 = 1	1/1 = 1	1/1 = 1
2	2/2 = 1	3/2 = 1.5	4/2 = 2
3	3/3 = 1	7/3 ≈ 2.33	9/3 = 3
4	4/4 = 1	15/4 = 3.75	16/4 = 4
5	5/5 = 1	31/5 = 6.2	25/5 = 5
6	6/6 = 1	63/6 = 10.5	36/6 = 6
7	7/7 = 1	127/7 ≈ 18.14	49/7 = 7
8	8/8 = 1	255/8 ≈ 31.88	64/8 = 8
9	9/9 = 1	511/9 ≈ 56.78	81/9 = 9

- Ask students to interpret the table. For many students, the pattern may be hard to describe. The average speed of the car does not change for the linear function, the average speed "increases at an increasing rate" for the exponential function, and the average speed increases at a constant rate for the quadratic function.

What Is Your Answer?

- **Neighbor Check:** Have students work independently and then have their neighbor check their work. Have students discuss any discrepancies.

Closure

- Write a linear, an exponential, and a quadratic equation that each have (0, 1) as a solution. Which function has a greater y-value at $x = 3$? Answers will vary depending on functions.

Technology For the Teacher

Dynamic Classroom

The Dynamic Planning Tool
Editable Teacher's Resources at *BigIdeasMath.com*

2 ACTIVITY: Comparing Speeds

Work with a partner. Analyze the speeds of the three cars over the given time periods. The distance traveled in t minutes is y miles. Which car eventually overtakes the others?

a.

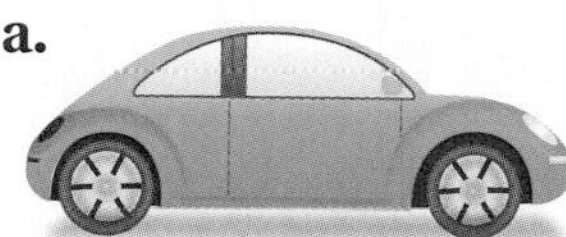

t	$y = t$
1	
2	
3	
4	

t	$y = 2^t - 1$
1	
2	
3	
4	

t	$y = t^2$
1	
2	
3	
4	

b.

t	$y = t$
4	
5	
6	
7	
8	
9	

t	$y = 2^t - 1$
4	
5	
6	
7	
8	
9	

t	$y = t^2$
4	
5	
6	
7	
8	
9	

What Is Your Answer?

3. **IN YOUR OWN WORDS** How can you compare the growth rates of linear, exponential, and quadratic functions? Which type of growth eventually leaves the other two in the dust? Explain your reasoning.

Use what you learned about comparing functions to complete Exercises 3–5 on page 439.

8.5 Lesson

Linear Function

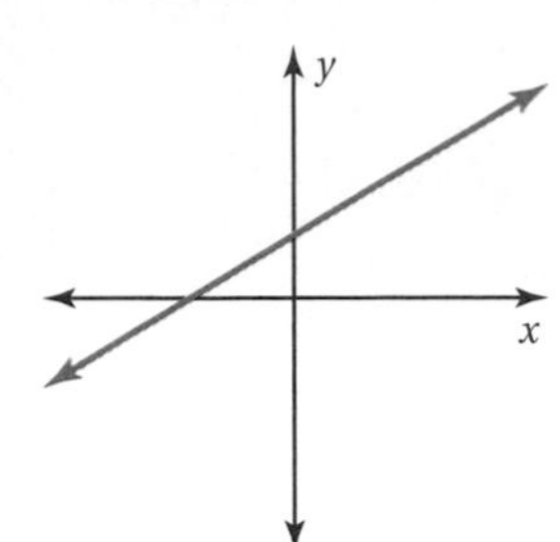

Line
$y = mx + b$

Exponential Function

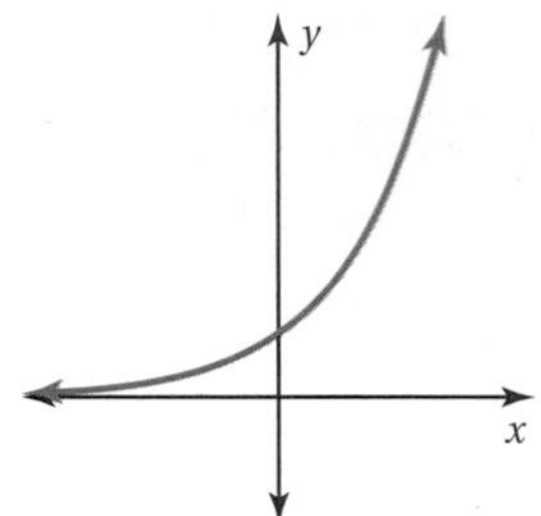

Curve
$y = ab^x$

Quadratic Function

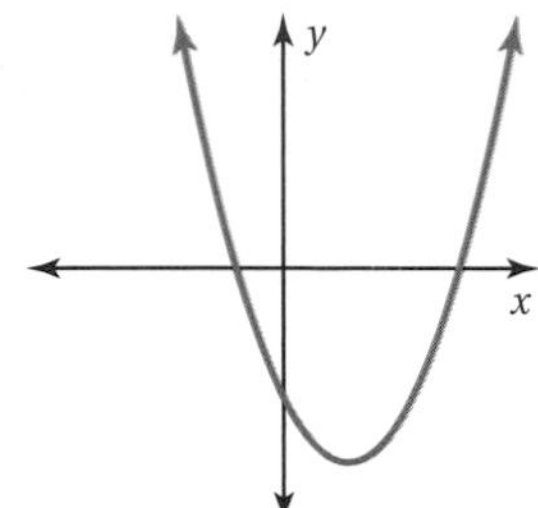

Parabola
$y = ax^2 + bx + c$

EXAMPLE 1 Identifying Functions Using Graphs

Plot the points. Tell whether the points represent a *linear*, an *exponential*, or a *quadratic* function.

a. $(4, 4), (2, 0), (0, 0), \left(1, -\frac{1}{2}\right), (-2, 4)$

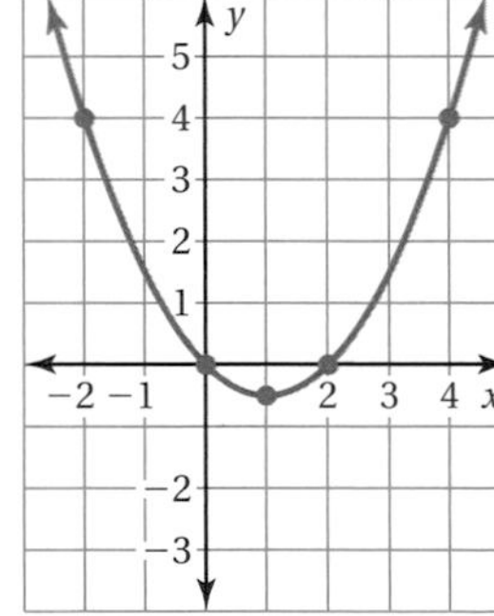

Quadratic function

b. $(0, 1), (2, 4), (4, 7), (-2, -2), (-4, -5)$

Linear function

c. $(0, 2), (2, 8), (1, 4), (-1, 1), \left(-2, \frac{1}{2}\right)$

Exponential function

On Your Own

Now You're Ready
Exercises 9–12

Plot the points. Tell whether the points represent a *linear*, an *exponential*, or a *quadratic* function.

1. $(-1, 5), (2, -1), (0, -1), (3, 5), (1, -3)$
2. $(-1, 2), (-2, 8), (-3, 32), \left(0, \frac{1}{2}\right), \left(1, \frac{1}{8}\right)$
3. $(-3, 5), (0, -1), (2, -5), (-4, 7), (1, -3)$

Laurie's Notes

Introduction

Connect

- **Yesterday:** Students compared the growth rates of linear, exponential, and quadratic functions. (MP4)
- **Today:** Students will identify linear, exponential, and quadratic functions.

Motivate

- Have students identify and describe the graphs of the three main functions that have been studied so far in this book.
- **Extension:** Make a bulletin board display of the three main functions.

Lesson Notes

Key Idea

- The Key Idea presents a summary of the three types of functions studied including a graph and a general equation.
- It would be beneficial to briefly review each equation and what the parameters tell you about the graph of each function.

Example 1

- Work through the three parts as shown.
- ? "In general, if you plot any five points, will you be able to determine which type of function the points represent?" If the points are relatively close together and do not make the features of a quadratic or exponential obvious, five points will not be sufficient.
- **MP5 Use Appropriate Tools Strategically:** Have students enter the points into their graphing calculators and plot them. If students think they know the equation, have them enter it into the calculator and graph it.

On Your Own

- **Think-Pair-Share:** Students should read each question independently and then work with a partner to answer the questions. When they have answered the questions, the pair should compare their answers with another group and discuss any discrepancies.

Goal Today's lesson is identifying linear, exponential, and quadratic functions.

Start Thinking! and Warm Up

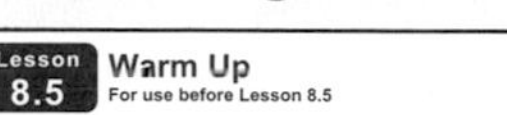

Lesson 8.5 **Start Thinking!** For use before Lesson 8.5

A function passes through the points (0, 1) and (1, 2). Can you tell whether the function is linear, quadratic, or exponential? Explain your reasoning.

Extra Example 1

Plot the points. Tell whether the points represent a *linear,* an *exponential,* or a *quadratic* function.

a. (0, 2), (3, 54), (4, 162), (2, 18), (1, 6)

b. $(-3, -9)$, $(1, -1)$, $(2, 1)$, $(-2, -7)$, $\left(\frac{1}{2}, -2\right)$

c. $(-1, 5)$, $(4, 0)$, $(2, -4)$, $(1, -3)$, $(3, -3)$

a–c. See Additional Answers.

On Your Own

1.

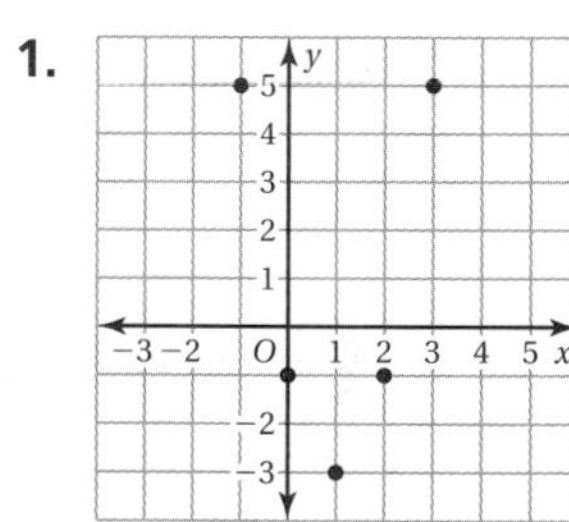

quadratic

2.

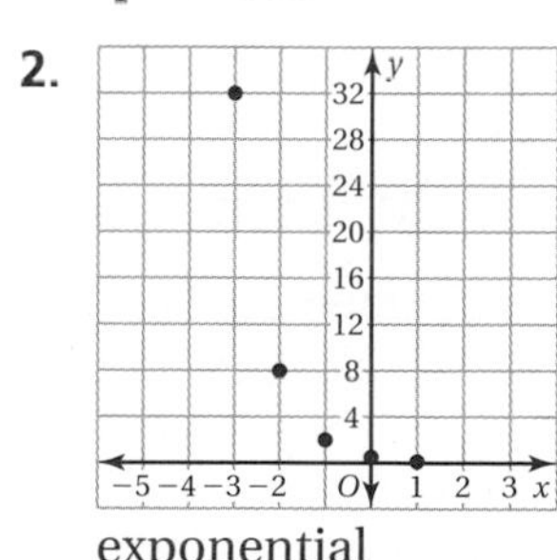

exponential

3.

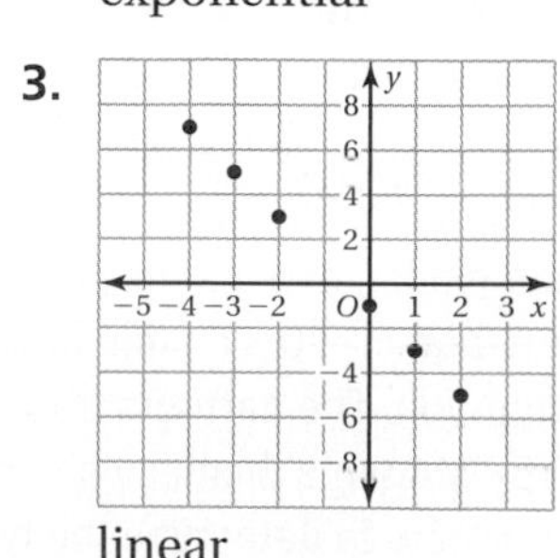

linear

Laurie's Notes

Discuss

- **Big Idea:** Take a moment to review how you can determine what type of function you have. From the first Key Idea the equation certainly tells you the type of function you have, and students should be able to distinguish between the three functions. From Example 1, when you have ordered pairs, you can graph them and generally determine the type of function, provided those ordered pairs reveal key features of the function. Another method is described in the next Key Idea.

Key Idea

- Write the Key Idea which describes a third method of determining the type of function you have given a table of values.
- Students have already used this method when working with linear and exponential functions.

? "How will you know if a table of values represents a linear function?" As x increases by a constant amount, the y-values increase by a constant amount. Note that this can also be described as a common difference.

? "How will you know if a table of values represents an exponential function?" As x increases by a constant amount, the y-values increase by a common ratio.

- **Study Tip:** The first differences are constant for a linear function.

Example 2

- Write the table of values for part (a).
- As you draw the curved arrows, ask about the pattern in both the x- and y-values.
- **Connection:** Knowing the common difference, which represents the slope, and knowing the y-intercept, ask students to write the equation of the linear function. $y = -3x + 2$
- Work through part (b) as shown.

? "Is $(0, -1)$ the vertex?" Students may quickly say yes until they recognize that the points in the table are not symmetric about $x = 0$.

- Students may ask if there is a way to determine the quadratic function. Tell them that there are several methods that they will study in future courses.

On Your Own

- Ask volunteers to share their work.

Extra Example 2

Tell whether the table of values represents a *linear*, an *exponential*, or a *quadratic* function.

x	-2	-1	0	1	2
y	3	-3	-5	-3	3

quadratic

On Your Own

4. exponential

English Language Learners

Pair Activity

Pair English language learners with English speakers. Each person chooses a function (linear, exponential, or quadratic) and creates a table of values for the function. The partners trade tables. The students then explain to each other how to determine the type of function represented by the table.

Key Idea

Differences and Ratios of Functions

Linear Function: $y = 2x + 5$

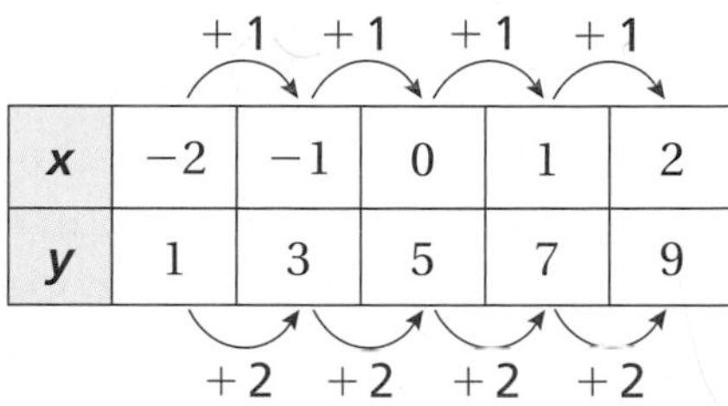

x	−2	−1	0	1	2
y	1	3	5	7	9

The y-values have a common *difference* of 2.

Exponential Function: $y = 4(2)^x$

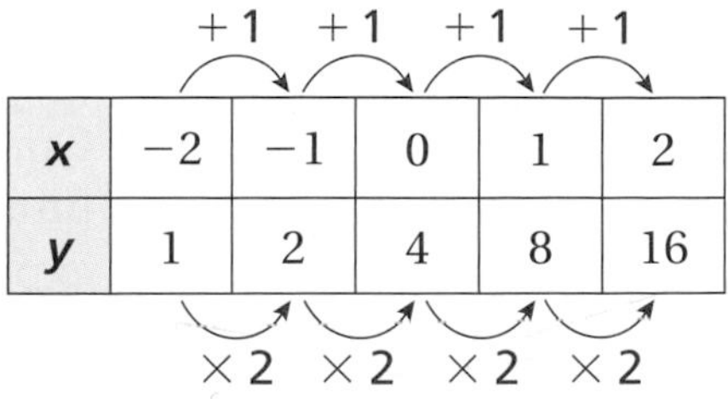

x	−2	−1	0	1	2
y	1	2	4	8	16

The y-values have a common *ratio* of 2.

Quadratic Function: $y = x^2 + 2x - 1$

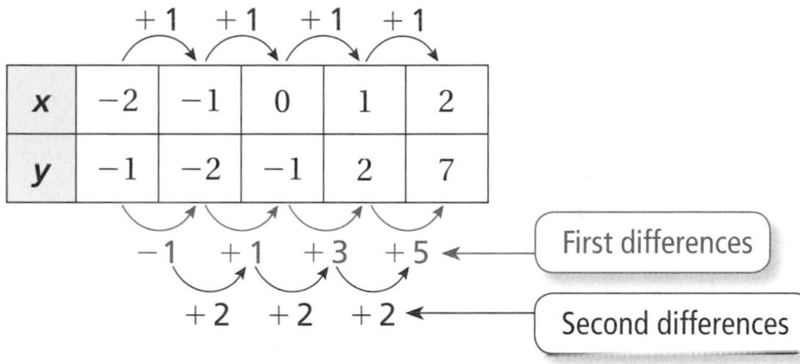

x	−2	−1	0	1	2
y	−1	−2	−1	2	7

For quadratic functions, the second differences are constant.

Study Tip

For a linear function, the first differences are constant.

EXAMPLE 2 Identifying Functions Using Differences or Ratios

Tell whether the table of values represents a *linear*, an *exponential*, or a *quadratic* function.

a.

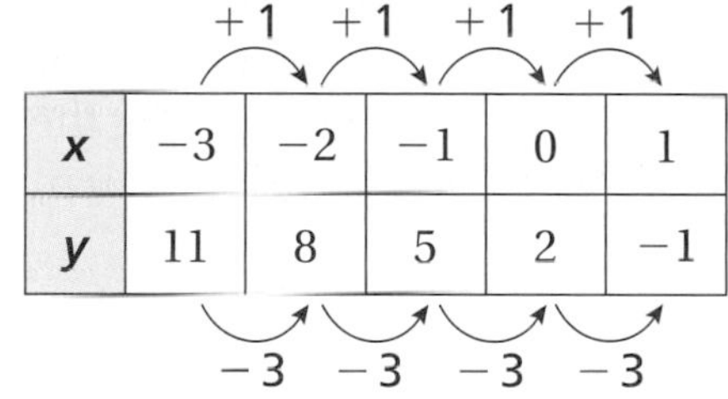

x	−3	−2	−1	0	1
y	11	8	5	2	−1

The y-values have a common difference of −3. So, the table represents a linear function.

b.

x	−1	0	1	2	3
y	0	−1	2	9	20

+1 +1 +1 +1

−1 +3 +7 +11

+4 +4 +4

The second differences are constant. So, the table represents a quadratic function.

Study Tip

For a quadratic function, the y-values will increase, then decrease, or the y-values will decrease, then increase.

On Your Own

Now You're Ready
Exercises 14–17

4. Tell whether the table of values represents a *linear*, an *exponential*, or a *quadratic* function.

x	−1	0	1	2	3
y	1	3	9	27	81

EXAMPLE 3 Identifying and Writing a Function

x	y
0	0
2	1
4	4
6	9
8	16

Tell whether the table of values represents a *linear*, an *exponential*, or a *quadratic* function. Then write an equation for the function using the form $y = mx + b$, $y = ab^x$, or $y = ax^2$.

Step 1:
Graph the data. The function appears to be exponential or quadratic.

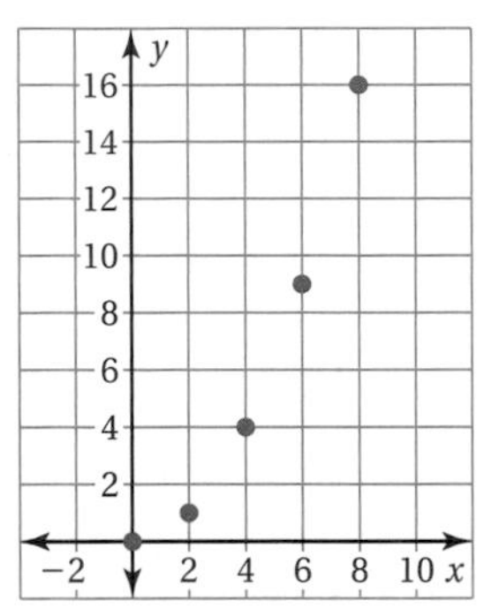

Step 2:
Check the y-values. If there is no common difference or ratio, check the second differences.

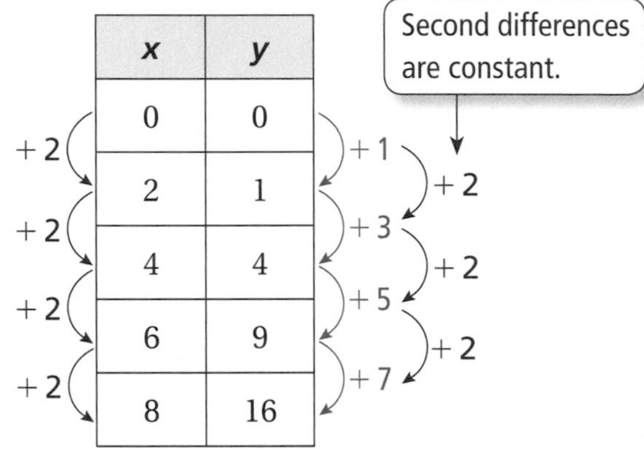

The function is quadratic.

Study Tip

To check your function in Example 3, substitute the other points from the table to see if they satisfy the function.

Step 3: Use the form $y = ax^2$.

$1 = a(2)^2$ Use the point (2, 1). Substitute 2 for x and 1 for y.

$\frac{1}{4} = a$ Solve for a.

So, an equation for the quadratic function is $y = \frac{1}{4}x^2$.

On Your Own

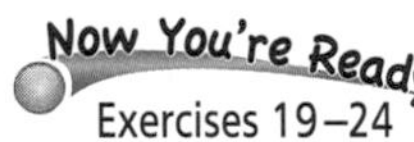

5. Tell whether the table of values represents a *linear*, an *exponential*, or a *quadratic* function. Then write an equation for the function using the form $y = mx + b$, $y = ab^x$, or $y = ax^2$.

x	-1	0	1	2	3
y	16	8	4	2	1

Summary

Linear Function

$y = mx + b$

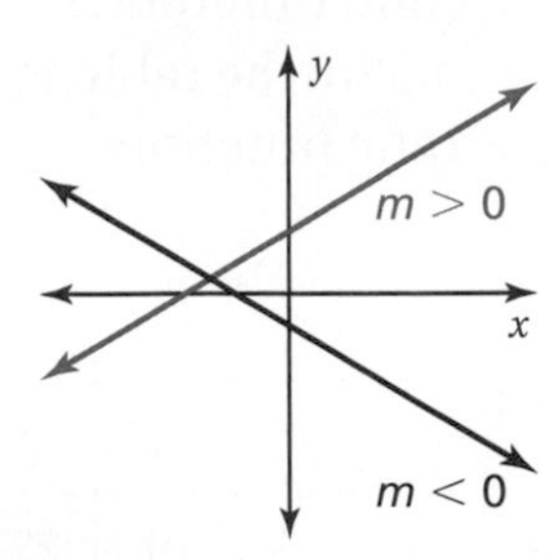

Exponential Function

$y = ab^x$, $a \neq 0$, $b \neq 1$, and $b > 0$

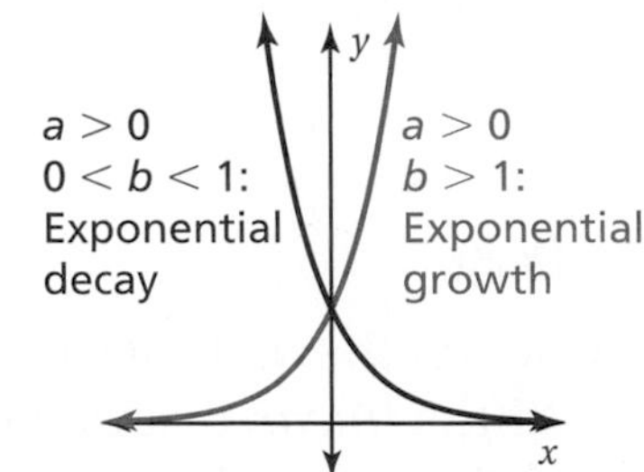

Quadratic Function

$y = ax^2 + bx + c$, $a \neq 0$

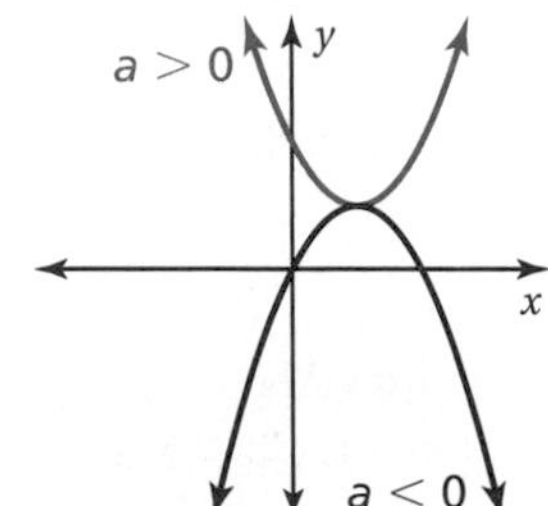

Laurie's Notes

Example 3

- "If you plot the ordered pairs, will you be able to tell what type of function it is?" Listen for student understanding that they may be able to tell if key features are present, but plotting points is not a guarantee.
- "Can you identify the type of function from the plotted points?" no; Students should know immediately that it is not linear but they cannot determine yet if it is quadratic or exponential.
- Check the table of values for a common ratio, then check the second differences to see if they are constant.
- Say, "If the quadratic function is symmetric about the y-axis with a vertex at (0, 0), then it would satisfy the equation $y = ax^2$. Let's try and verify this."
- Substitute an ordered pair other than (0, 0) for x and y and solve for a. Ask students to explain why the point (0, 0) cannot be used to find the value of a.
- Once you find the equation, the other ordered pairs in the table can be checked to see if they satisfy the equation.

On Your Own

- Have students explain their process. Anticipate that most students will recognize that the y-values are "being divided by 2," which is the same as multiplying by $\frac{1}{2}$. Because there is a common ratio, students will use the equation $y = ab^x$. Recall from an earlier lesson that the value of a is the y-intercept and the value of b is the common ratio.

Summary

- This summary extends the Key Idea at the beginning of the lesson by looking at the different cases of each type of function. Also review how to identify the y-intercept for each equation.

Closure

- Write a linear, an exponential, and a quadratic function that each have a y-intercept of 2. *Sample answer:* $y = \frac{1}{2}x + 2$, $y = 2(3)^x$, $y = x^2 - 4x + 2$

Technology For the Teacher

The Dynamic Planning Tool
Editable Teacher's Resources at *BigIdeasMath.com*

Extra Example 3

Tell whether the table of values represents a *linear*, an *exponential*, or a *quadratic* function. Then write an equation for the function using the form $y = mx + b$, $y = ab^x$, or $y = ax^2$.

x	−1	0	1	2	3
y	−11	−3	5	13	21

linear; $y = 8x - 3$

On Your Own

5. exponential; $y = 8\left(\frac{1}{2}\right)^x$

Vocabulary and Concept Check

1. linear function: y-values have a common difference; exponential function: y-values have a common ratio; quadratic function: second differences are constant
2. $n(x)$; This is an exponential function, where $f(x)$, $g(x)$, and $m(x)$ represents quadratic functions.

Practice and Problem Solving

3. $x > 1$ and $x < 0$
4. all real numbers
5. $x > -\frac{1}{2}$
6. B
7. C
8. A
9. 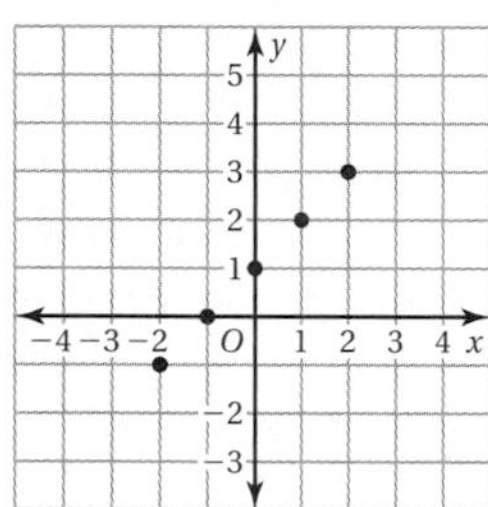

   linear
10. 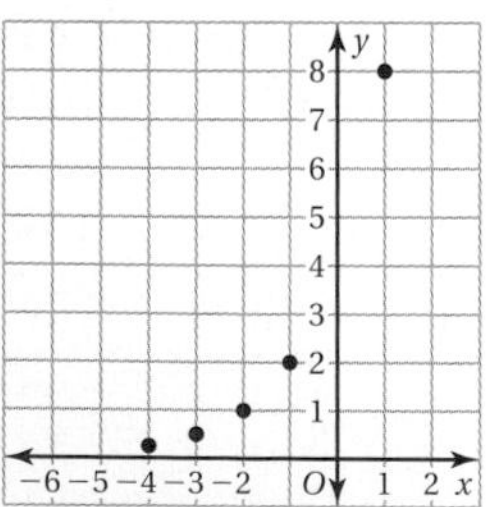

   exponential
11. 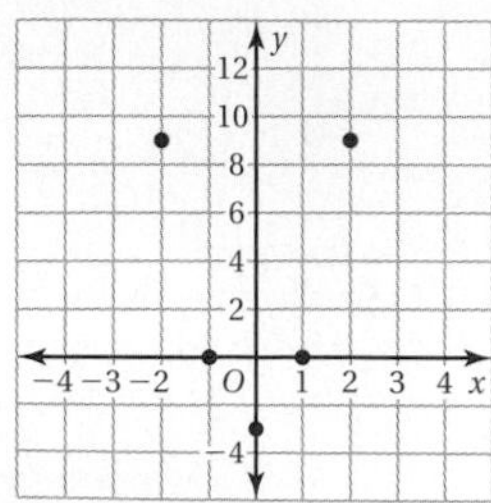

   quadratic

Assignment Guide and Homework Check

Level	Assignment	Homework Check
Average	1–5, 9–25 odd, 18, 28, 31–34	9, 13, 15, 23, 25, 28
Advanced	1, 2, 10–24 even, 13, 25–30, 31–34	10, 16, 24, 27, 28, 30

Common Errors

- **Exercises 9–12** Students may try to decipher a function type before they plot the points. This will not work because the points are not in order and the x-values may not have a constant difference. Encourage students to plot the points first and then assign a function type based on the graph.
- **Exercises 9–12** Students may plot a couple of points of a quadratic function and mistakenly classify it as exponential. Remind students to plot all the points so they can see the pattern before assigning a function type.

8.5 Record and Practice Journal

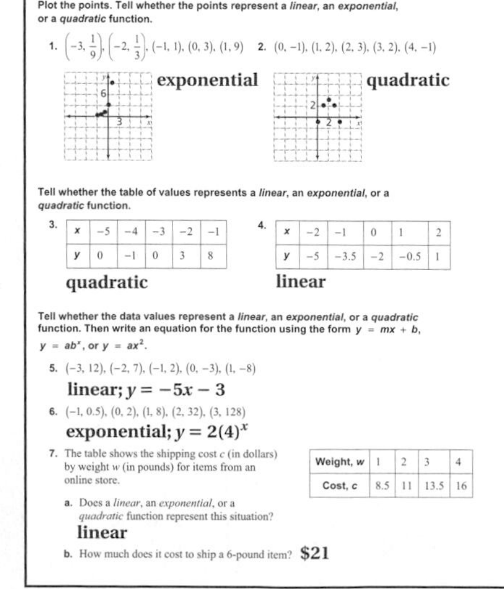

Plot the points. Tell whether the points represent a *linear*, an *exponential*, or a *quadratic* function.

1. $\left(-3, \frac{1}{9}\right), \left(-2, \frac{1}{3}\right), (-1, 1), (0, 3), (1, 9)$ **exponential**
2. $(0, -1), (1, 2), (2, 3), (3, 2), (4, -1)$ **quadratic**

Tell whether the table of values represents a *linear*, an *exponential*, or a *quadratic* function.

3.

x	−5	−4	−3	−2	−1
y	0	−1	0	3	8

quadratic

4.

x	−2	−1	0	1	2
y	−5	−3.5	−2	−0.5	1

linear

Tell whether the data values represent a *linear*, an *exponential*, or a *quadratic* function. Then write an equation for the function using the form $y = mx + b$, $y = ab^x$, or $y = ax^2$.

5. $(-3, 12), (-2, 7), (-1, 2), (0, -3), (1, -8)$

   linear; $y = -5x - 3$
6. $(-1, 0.5), (0, 2), (1, 8), (2, 32), (3, 128)$

   exponential; $y = 2(4)^x$
7. The table shows the shipping cost c (in dollars) by weight w (in pounds) for items from an online store.

Weight, w	1	2	3	4
Cost, c	8.5	11	13.5	16

   a. Does a *linear*, an *exponential*, or a *quadratic* function represent this situation? **linear**

   b. How much does it cost to ship a 6-pound item? **$21**

Technology For the Teacher

Answer Presentation Tool

8.5 Exercises

Vocabulary and Concept Check

1. **VOCABULARY** How can you decide whether to use a linear, a quadratic, or an exponential function to model a data set?

2. **WHICH ONE DOESN'T BELONG?** Which graph does *not* belong with the other three? Explain your reasoning.

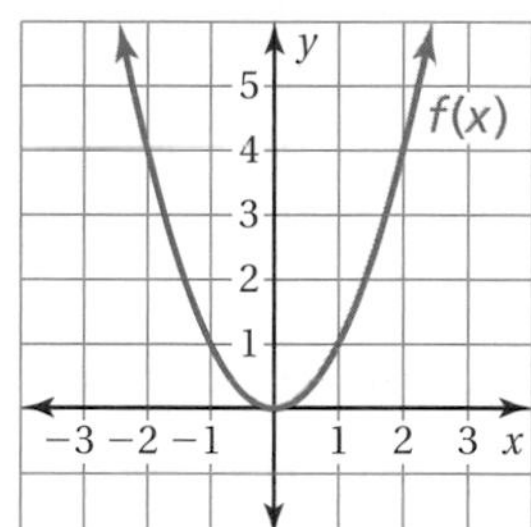

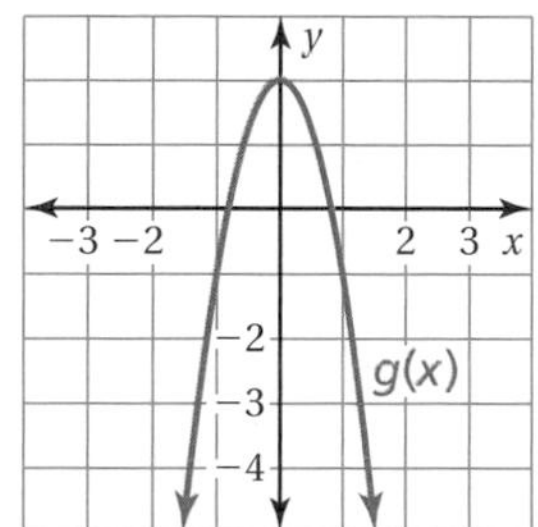

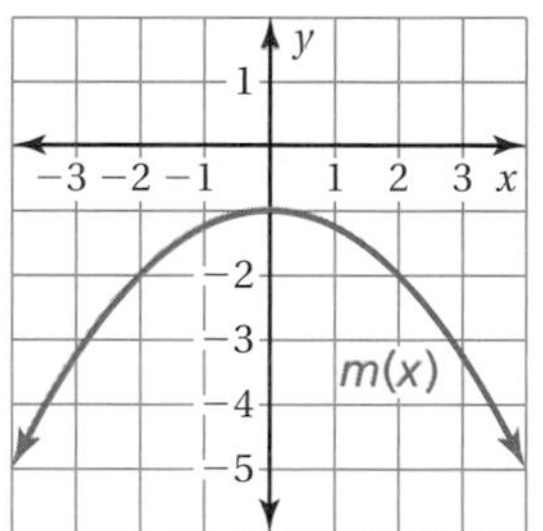

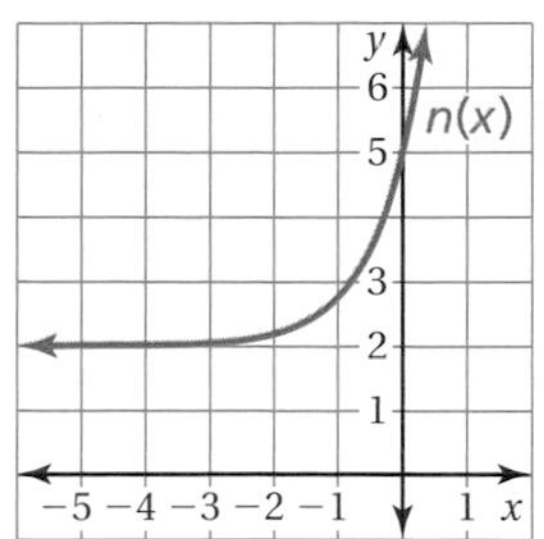

Practice and Problem Solving

Find the values of x when $f(x)$ is greater than $g(x)$.

3. $f(x) = x^2$
   $g(x) = x$

4. $f(x) = 3^x$
   $g(x) = 2x$

5. $f(x) = 4^x$
   $g(x) = 2x^2$

Match the function type with its graph.

6. Linear function

7. Exponential function

8. Quadratic function

A. 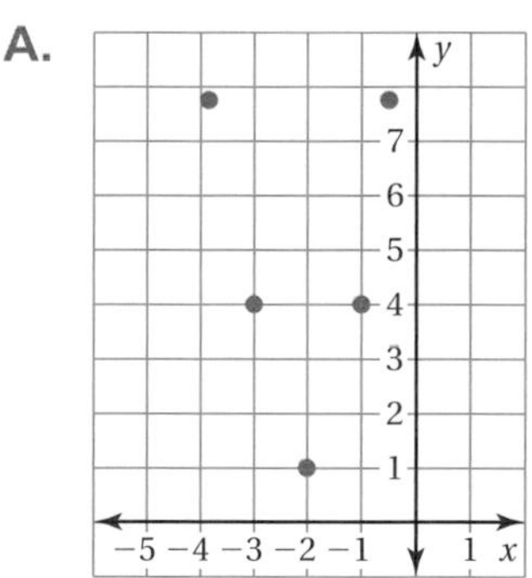

B. 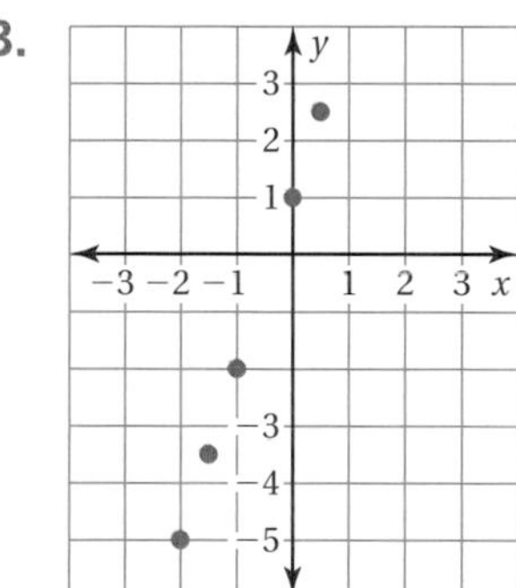

C. 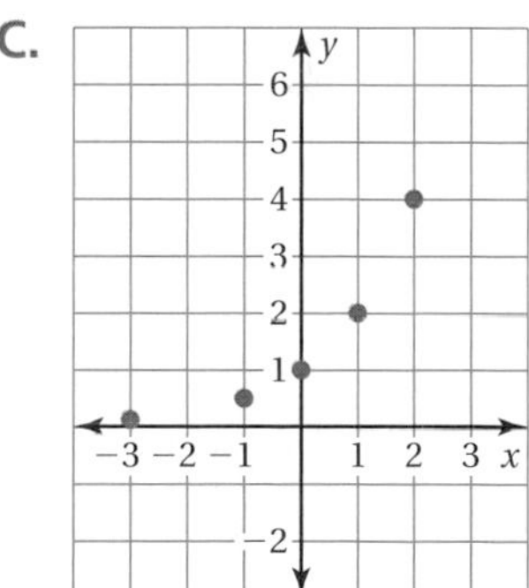

Plot the points. Tell whether the points represent a *linear*, an *exponential*, or a *quadratic* function.

1 9. $(-2, -1), (-1, 0), (1, 2), (2, 3), (0, 1)$

10. $(1, 8), \left(-4, \frac{1}{4}\right), \left(-3, \frac{1}{2}\right), (-2, 1), (-1, 2)$

11. $(0, -3), (1, 0), (2, 9), (-2, 9), (-1, 0)$

12. $(-1, -3), (-3, 5), (0, -1), (1, 5), (2, 15)$

13. **SUBWAY** A student takes a subway to a public library. The table shows the distance d (in miles) the student travels in t minutes. Tell whether the data can be modeled by a *linear*, an *exponential*, or a *quadratic* function.

Time, *t*	0.5	1	3	5
Distance, *d*	0.335	0.67	2.01	3.35

Tell whether the table of values represents a *linear*, an *exponential*, or a *quadratic* function.

(2) **14.**

x	−2	−1	0	1	2
y	0	0.5	1	1.5	2

15.

x	−1	0	1	2	3
y	0.2	1	5	25	125

16.

x	−2	−1	0	1	2
y	0.75	1.5	3	6	12

17.

x	2	3	4	5	6
y	2	4.5	8	12.5	18

18. REASONING Can the y-values of a data set have both a common difference and a common ratio? Explain your reasoning.

Tell whether the data values represent a *linear*, an *exponential*, or a *quadratic* function. Then write an equation for the function using the form $y = mx + b$, $y = ab^x$, or $y = ax^2$.

(3) **19.** (−2, 8), (−1, 2), (0, 0), (1, 2), (2, 8)

20. (−3, 8), (−2, 4), (−1, 2), (0, 1), (1, 0.5)

21.

x	−2	−1	0	1	2
y	4	1	−2	−5	−8

22.

x	−1	0	1	2	3
y	2.5	5	10	20	40

23.

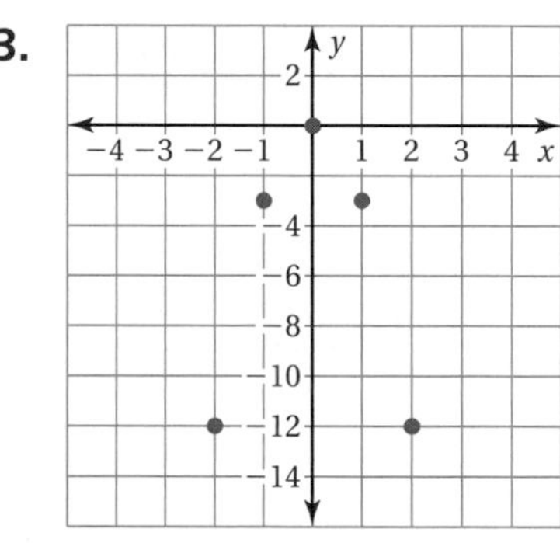

24.

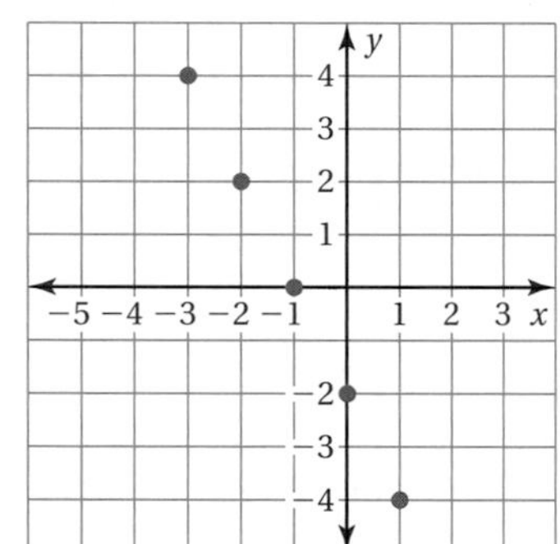

25. ERROR ANALYSIS Describe and correct the error in writing an equation for the function represented by the ordered pairs.

(−1, 4), (0, 0), (1, 4), (2, 16), (3, 36)

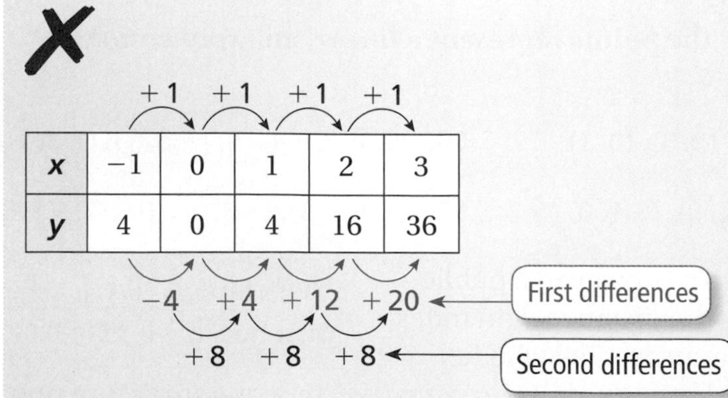

x	−1	0	1	2	3
y	4	0	4	16	36

The ordered pairs represent a quadratic function.

$y = ax^2$

$1 = a(4)^2$

$\frac{1}{16} = a$

So, an equation is $y = \frac{1}{16}x^2$.

Common Errors

- **Exercises 19 and 23** Students may incorrectly substitute the *x*-coordinate for *y* and the *y*-coordinate for *x* when writing an equation for the function. Remind them to check their equation.

Practice and Problem Solving

12.

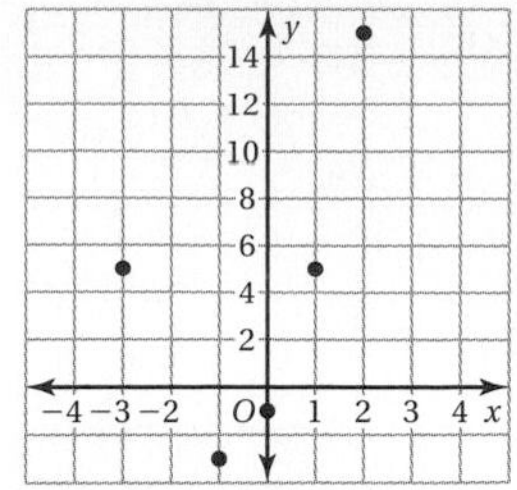

quadratic

13. linear

14. linear

15. exponential

16. exponential

17. quadratic

18. no; A function cannot be both linear and exponential.

19. quadratic; $y = 2x^2$

20. exponential; $y = 0.5^x$

21. linear; $y = -3x - 2$

22. exponential; $y = 5(2)^x$

23. quadratic; $y = -3x^2$

24. linear; $y = -2x - 2$

25. The student incorrectly substitutes the x-coordinate in for y and the y-coordinate in for x. The equation should be $y = 4x^2$.

Differentiated Instruction

Organization

Have students create a chart in their notebooks summarizing the three types of equations (linear, exponential, and quadratic) they have studied this year. The chart should include graphs of the equations.

Practice and Problem Solving

26. a. 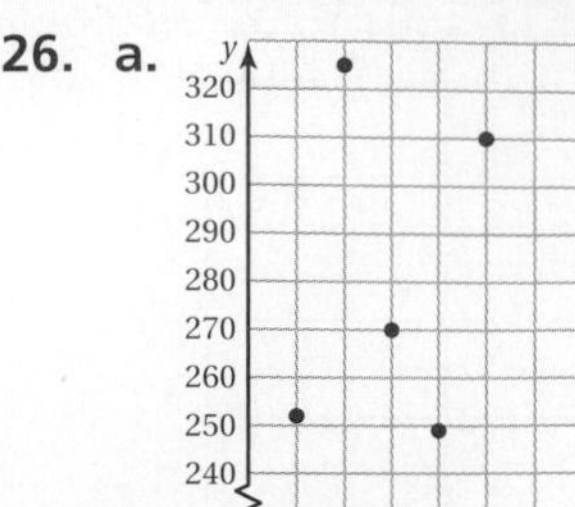

b. no; None of these types of functions fit the data.

27. in general, no; It depends on the number of points and where they are located on the graph.

28–29. See Additional Answers.

30. See *Taking Math Deeper*.

Fair Game Review

31. $-1, 1$ 32. 2

33. $-2, 2$ 34. D

Mini-Assessment

Tell whether the data values represent a *linear*, an *exponential*, or a *quadratic* function. Then write an equation for the function using the form $y = mx + b$, $y = ab^x$, or $y = ax^2$.

1. $(-3, -5), (0, 7), (-1, 3), (-2, -1), (1, 11)$ linear; $y = 4x + 7$

2.

x	−1	0	1	2	3
y	0.2	0	0.2	0.8	1.8

quadratic; $y = \frac{1}{5}x^2$

3. $\left(-1, \frac{2}{3}\right), (0, 2), (1, 6), (2, 18), (3, 54)$

exponential; $y = 2(3)^x$

Taking Math Deeper

Exercise 30

This exercise provides a great opportunity to use the problem solving strategy "work backwards."

Set up a table of values with the y-values and first differences missing, indicating that the second differences are all 3. It is a good idea to include $x = 0$ in the table so that it will show the y-intercept.

+1 +1 +1 +1

x	0	1	2	3	4
y	0	−1.5	0	4.5	12

First differences: −1.5 +1.5 +4.5 +7.5

Second differences: +3 +3 +3

Starting with an arbitrary value, fill in the first differences.

Starting with an arbitrary value, fill in the y-values.

(*Note:* Students may try adjusting the first differences to find a function that is easier to write. After a few tries, they may realize that starting with -1.5 gives the vertex in the table. Students may also realize that starting with $y = 0$ in the table may make the function easier to write.)

The ordered pairs show symmetry about the line $x = 1$. So, the vertex is $(1, -1.5)$. Use vertex form to write a function.

$$y = a(x - h)^2 + k$$

$$y = a(x - 1)^2 - 1.5$$

Substitute an ordered pair from the table to find that $a = 1.5$. So, one possible function is $y = 1.5(x - 1)^2 - 1.5$, or $y = 1.5x^2 - 3x$.

Reteaching and Enrichment Strategies

If students need help. . .	If students got it. . .
Resources by Chapter • Practice A and Practice B • Puzzle Time Record and Practice Journal Practice Differentiating the Lesson Lesson Tutorials Skills Review Handbook	Resources by Chapter • Enrichment and Extension • School-to-Work • Financial Literacy Start the next section

26. **HIGH SCHOOL FOOTBALL** The table shows the number of people attending the first five football games at a high school.

Game, g	1	2	3	4	5
Number of People, p	252	325	270	249	310

a. Plot the points.

b. Does a *linear*, an *exponential*, or a *quadratic* function represent this situation? Explain.

27. **CRITICAL THINKING** Is the graph of a set of points enough to determine whether the points represent a *linear*, an *exponential*, or a *quadratic* function? Justify your answer.

28. **RECORDING STUDIO** The table shows the amount of money (in dollars) that a musician pays for using a recording studio.

Number of Hours, h	1	2	3	4
Amount, m (dollars)	110	145	180	215

a. Plot the points. Then determine the type of function that best represents this situation.

b. Write a function that models the data.

c. How much does it cost to use the studio for 10 hours?

29. **TOURNAMENT** At the beginning of a basketball tournament, there are 64 teams. After each round, one-half of the remaining teams are eliminated.

a. Make a table showing the number of teams remaining after each round.

b. Determine the type of function that best represents this situation.

c. Write a function that models the data.

d. After which round do you know the team that won the tournament?

30. 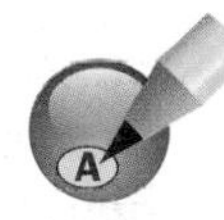

Write a function that has constant second differences of 3.

Fair Game Review What you learned in previous grades & lessons

Find the x-intercept(s) of the graph. *(Section 2.3 and Section 8.3)*

31. 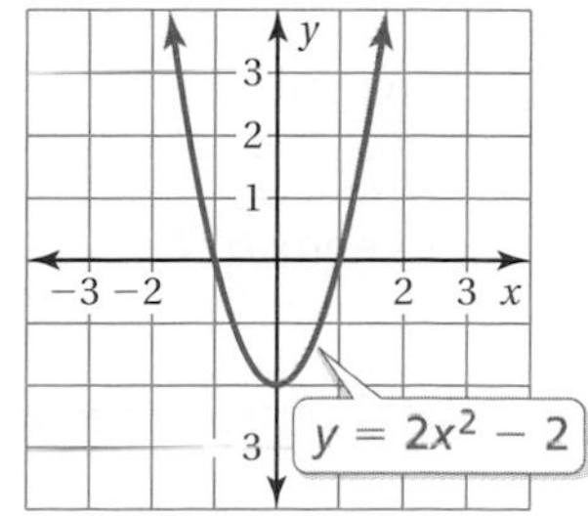

32. 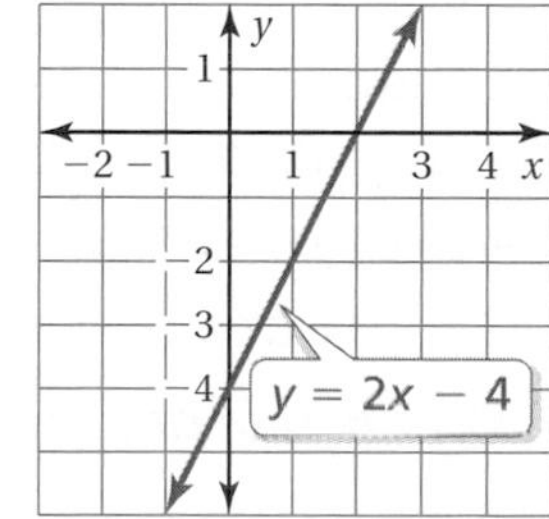

33. 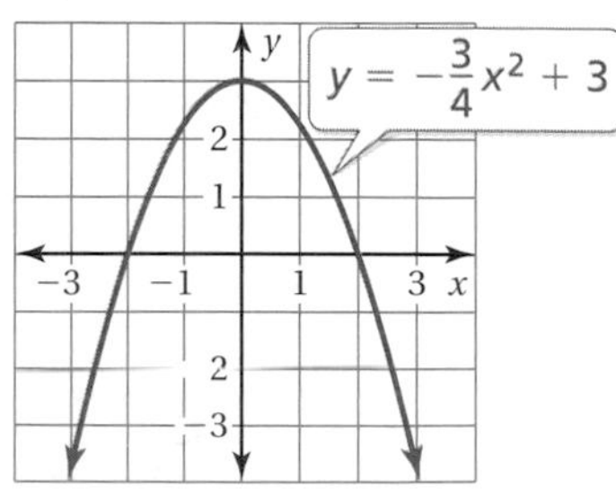

34. **MULTIPLE CHOICE** What is the factored form of $8x^3 - 18x$? *(Section 7.9)*

Ⓐ $2x(2x + 3)^2$ Ⓑ $(2x + 3)(2x - 3)$ Ⓒ $(2x - 3)^2$ Ⓓ $2x(2x + 3)(2x - 3)$

8.5b Comparing Graphs of Functions

You have already learned that the average rate of change (or slope) between any two points on a line is the change in y divided by the change in x. You can find the average rate of change between two points of a nonlinear function using the same method.

ACTIVITY 1 Rates of Change of a Quadratic Function

In Example 4 on page 428, the function $f(t) = -16t^2 + 80t + 5$ gives the height (in feet) of a water balloon t seconds after it is launched.

a. Copy and complete the table for $f(t)$.

t	0	0.5	1	1.5	2	2.5	3	3.5	4	4.5	5
$f(t)$											

b. Graph the ordered pairs from part (a). Then draw a smooth curve through the points.

c. For what values is the function increasing? decreasing?

d. Copy and complete the tables to find the average rate of change for each interval.

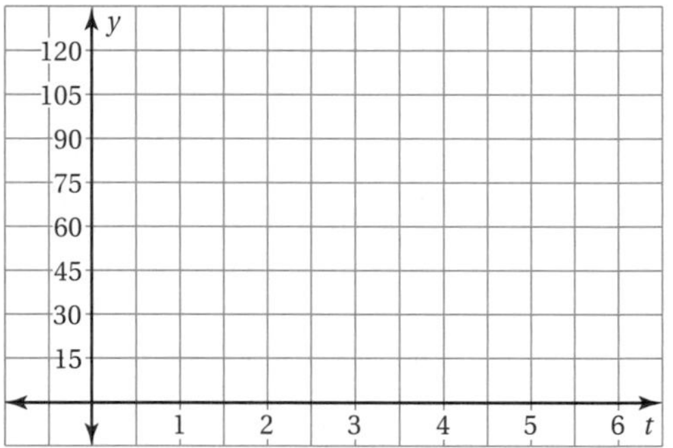

Time Interval	0 to 0.5 sec	0.5 to 1 sec	1 to 1.5 sec	1.5 to 2 sec	2 to 2.5 sec
Average Rate of Change (ft/sec)					

Time Interval	2.5 to 3 sec	3 to 3.5 sec	3.5 to 4 sec	4 to 4.5 sec	4.5 to 5 sec
Average Rate of Change (ft/sec)					

Practice

1. Compared to the average rate of change of a linear function, what do you notice about the average rate of change in part (d) of Activity 1?
2. Is the average rate of change increasing or decreasing from 0 to 2.5 seconds? How can you use the graph to justify your answer?
3. What do you notice about the average rate of change when the function is increasing and when the function is decreasing?
4. In Example 4 on page 419, the function $f(t) = -16t^2 + 64$ gives the height of an egg t seconds after it is dropped. (a) Make a table of values. Use the domain $0 \le t \le 2$ with intervals of 0.5 second. (b) Graph the ordered pairs and draw a smooth curve through the points. (c) Describe where the function is increasing and decreasing. (d) Find the average rate of change for each interval in the table. What do you notice?

Laurie's Notes

Introduction

Connect

- **Yesterday:** Students compared linear, exponential, and quadratic functions. (MP5)
- **Today:** Students will compare rates of change of linear, exponential, and quadratic functions.

Motivate

- Use a graphing calculator to display the following graphs of $y = 0.5x^2 - 2$ using the given viewing windows.

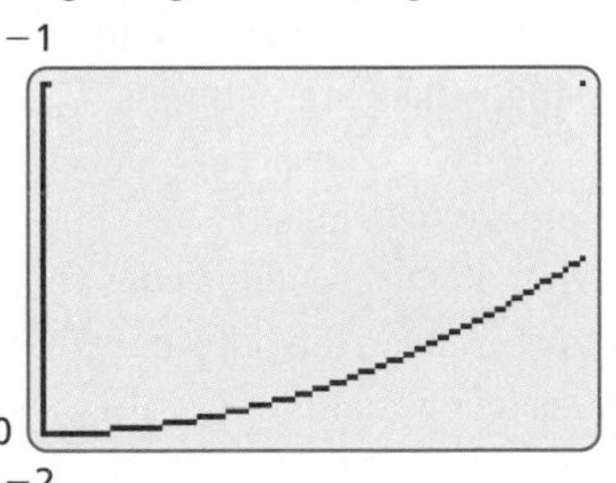

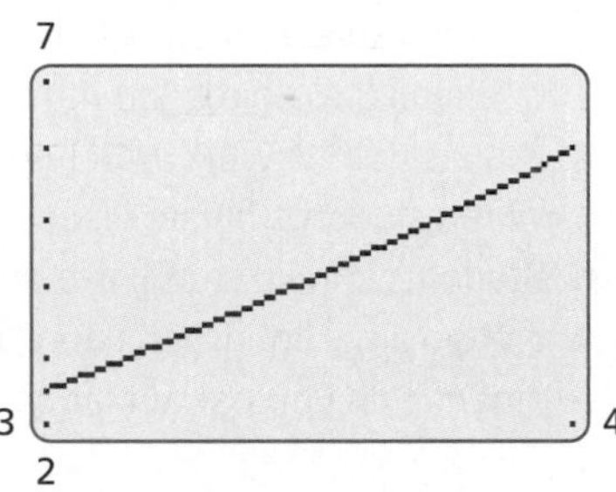

? "How would you describe the *change in x* and the *change in y* in each graph?" The change in *x* is 1. The graph on the left shows a change in *y* of about 0.5. The graph on the right shows a change in *y* of about 3.

- Now display the graph of $y = 0.5x^2 - 2$ in a standard viewing window.
- Explain that when you select any two points on the graph and connect them with a line segment, there is a difference between the curve and the segment. Depending upon where the points are selected, the slope of the segment changes. This is what students will explore today.

Lesson Notes

Discuss

- **MP5 Use Appropriate Tools Strategically:** Graphing technology can make understanding average rate of change easier. Examining a graph over different parts of the domain can give students insight into the behavior of a function. Examining a table of values can also help students develop insights. Students can use calculators in this lesson to perform quick computations. Interactive graphing software can also be used.

Activity 1

- Write the function and review the problem on page 428.
- Have students work with partners to complete parts (a)–(c).
- Discuss the average rate of change in this context. Find the difference in heights (feet) between each time interval and divide by 0.5 second.

? "How would you describe the average rate of change from the table?" Listen for understanding of positive, then negative rates that decrease.

- **Teaching Tip:** Use a pencil and model the average rate of change, showing how it increases or decreases as you move along the graph.

Goal Today's lesson is comparing average rates of change of functions.

Start Thinking! and Warm Up

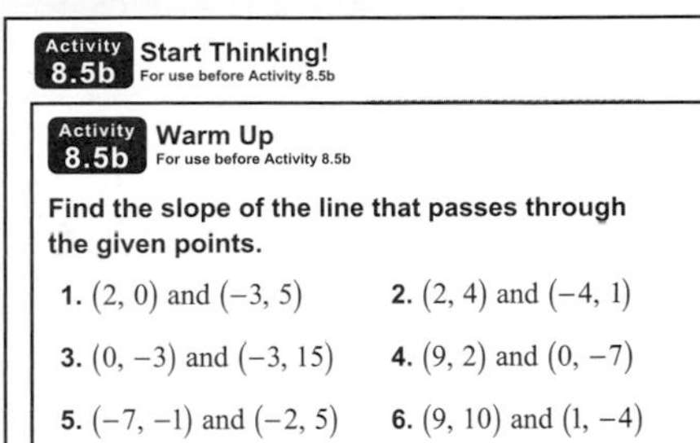

Activity 8.5b **Start Thinking!** For use before Activity 8.5b

Activity 8.5b **Warm Up** For use before Activity 8.5b

Find the slope of the line that passes through the given points.

1. (2, 0) and (−3, 5)
2. (2, 4) and (−4, 1)
3. (0, −3) and (−3, 15)
4. (9, 2) and (0, −7)
5. (−7, −1) and (−2, 5)
6. (9, 10) and (1, −4)

8.5b Record and Practice Journal

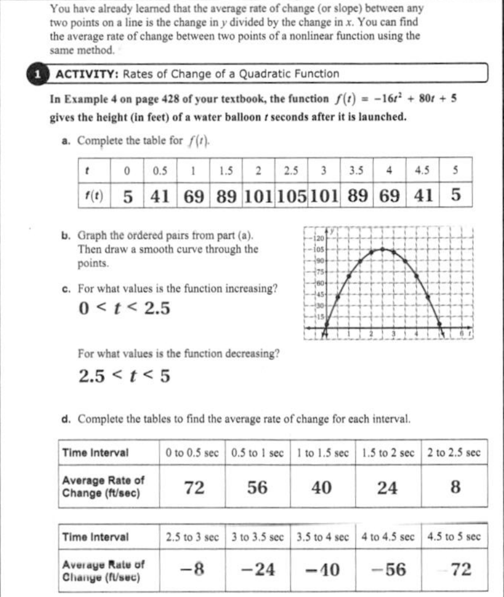

You have already learned that the average rate of change (or slope) between any two points on a line is the change in *y* divided by the change in *x*. You can find the average rate of change between two points of a nonlinear function using the same method.

1 ACTIVITY: Rates of Change of a Quadratic Function

In Example 4 on page 428 of your textbook, the function $f(t) = -16t^2 + 80t + 5$ gives the height (in feet) of a water balloon t seconds after it is launched.

a. Complete the table for $f(t)$.

t	0	0.5	1	1.5	2	2.5	3	3.5	4	4.5	5
$f(t)$	5	41	69	89	101	105	101	89	69	41	5

b. Graph the ordered pairs from part (a). Then draw a smooth curve through the points.

c. For what values is the function increasing?
$0 < t < 2.5$

For what values is the function decreasing?
$2.5 < t < 5$

d. Complete the tables to find the average rate of change for each interval.

Time Interval	0 to 0.5 sec	0.5 to 1 sec	1 to 1.5 sec	1.5 to 2 sec	2 to 2.5 sec
Average Rate of Change (ft/sec)	72	56	40	24	8

Time Interval	2.5 to 3 sec	3 to 3.5 sec	3.5 to 4 sec	4 to 4.5 sec	4.5 to 5 sec
Average Rate of Change (ft/sec)	−8	−24	−40	−56	−72

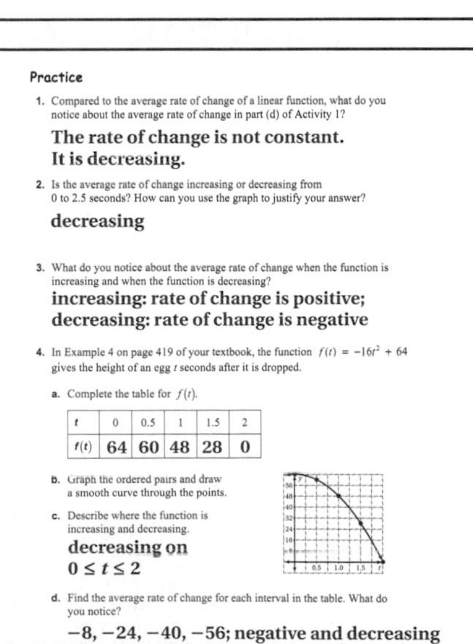

Practice

1. Compared to the average rate of change of a linear function, what do you notice about the average rate of change in part (d) of Activity 1?
   The rate of change is not constant. It is decreasing.
2. Is the average rate of change increasing or decreasing from 0 to 2.5 seconds? How can you use the graph to justify your answer?
   decreasing
3. What do you notice about the average rate of change when the function is increasing and when the function is decreasing?
   increasing: rate of change is positive; decreasing: rate of change is negative
4. In Example 4 on page 419 of your textbook, the function $f(t) = -16t^2 + 64$ gives the height of an egg t seconds after it is dropped.

   a. Complete the table for $f(t)$.

t	0	0.5	1	1.5	2
$f(t)$	64	60	48	28	0

   b. Graph the ordered pairs and draw a smooth curve through the points.

   c. Describe where the function is increasing and decreasing.
   decreasing on $0 \le t \le 2$

   d. Find the average rate of change for each interval in the table. What do you notice?
   −8, −24, −40, −56; negative and decreasing

Laurie's Notes

Practice

- Have students work with partners.
- Discuss the answers to Exercises 1–3 as a class, listening for student understanding.
- Exercise 4 follows from the worked-out activity.

Activity 2

- The equations are not given for the three functions. However, the ordered pairs provide sufficient information for the students to be able to work through the activity.

? "Is the same scale used in all three graphs?" No, they have the same scale on the x-axes but the scales on the y-axes are different.

- Have students work with partners.
- When students finish, discuss results as a whole class.
- Students should recognize that even though the steepness of segments on two different graphs may appear to be similar, the difference in scaling means that you cannot judge steepness by eyesight.

Practice

- The two questions posed are exploring the big ideas. Encourage students to be thorough with their explanations and reasoning.

Closure

- Consider the function $f(x) = 4x^2 - 4$. As x increases by 1, describe the average rate of change on the domain $-3 \le x \le 3$. Average rate of change is negative but increasing until $x = 0$, then the average rate of change becomes positive and increases.

Technology For the Teacher

The Dynamic Planning Tool
Editable Teacher's Resources at *BigIdeasMath.com*

8.5b Record and Practice Journal

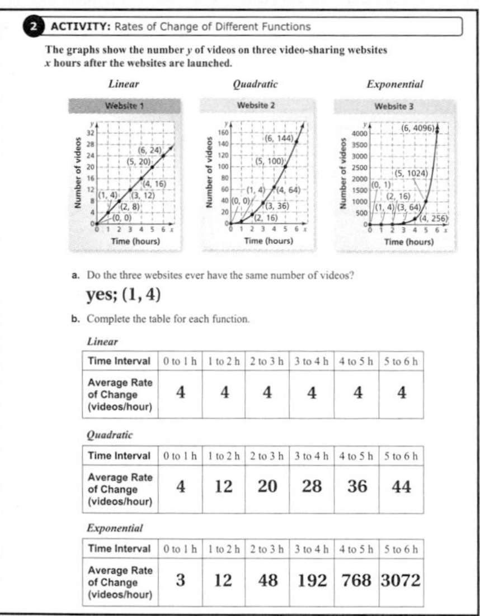

2 ACTIVITY: Rates of Change of Different Functions

The graphs show the number y of videos on three video-sharing websites x hours after the websites are launched.

a. Do the three websites ever have the same number of videos?

yes; (1, 4)

b. Complete the table for each function.

Linear

Time Interval	0 to 1 h	1 to 2 h	2 to 3 h	3 to 4 h	4 to 5 h	5 to 6 h
Average Rate of Change (videos/hour)	4	4	4	4	4	4

Quadratic

Time Interval	0 to 1 h	1 to 2 h	2 to 3 h	3 to 4 h	4 to 5 h	5 to 6 h
Average Rate of Change (videos/hour)	4	12	20	28	36	44

Exponential

Time Interval	0 to 1 h	1 to 2 h	2 to 3 h	3 to 4 h	4 to 5 h	5 to 6 h
Average Rate of Change (videos/hour)	3	12	48	192	768	3072

c. What do you notice about the average rate of change of the linear function?

It is constant.

d. What do you notice about the average rate of change of the quadratic function?

increases by a common difference of 8; arithmetic sequence

e. What do you notice about the average rate of change of the exponential function?

increases by a common ratio of 4; geometric sequence

f. Which average rate of change increases more quickly, the quadratic function or the exponential function?

the exponential function

Practice

5. REASONING How does a quantity that is increasing exponentially compare to a quantity that is increasing linearly or quadratically?

It will eventually exceed a quantity increasing linearly or quadratically.

6. REASONING Explain why the average rate of change of a linear function is constant and the average rate of change of a quadratic or exponential function is not constant.

Linear: Slope is the same between any two points. Quadratic or exponential: Function may change rapidly on some intervals, and slowly on other intervals.

ACTIVITY 2 Rates of Change of Different Functions

The graphs show the numbers y of videos on three video-sharing websites x hours after the websites are launched.

Linear	*Quadratic*	*Exponential*

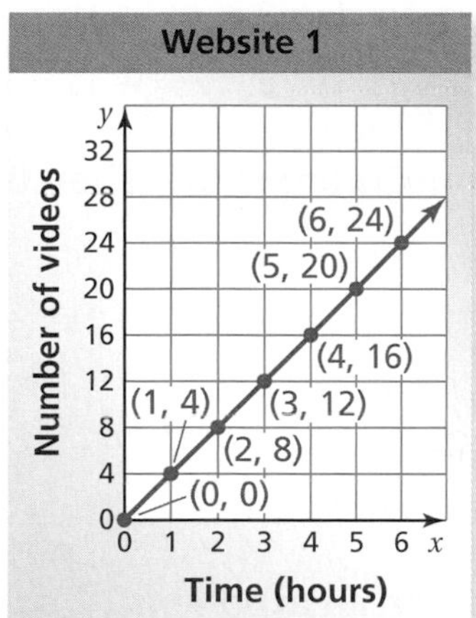

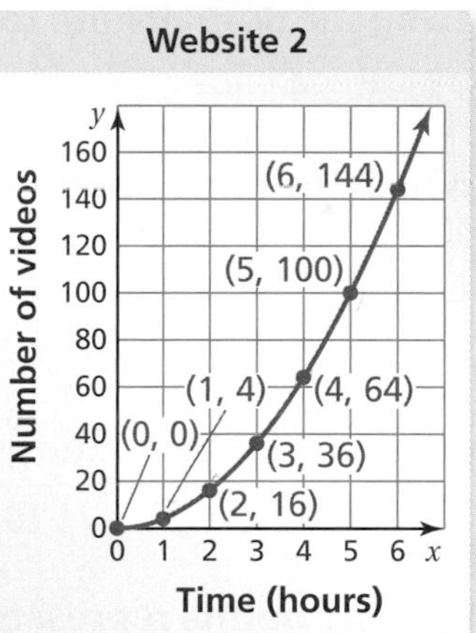

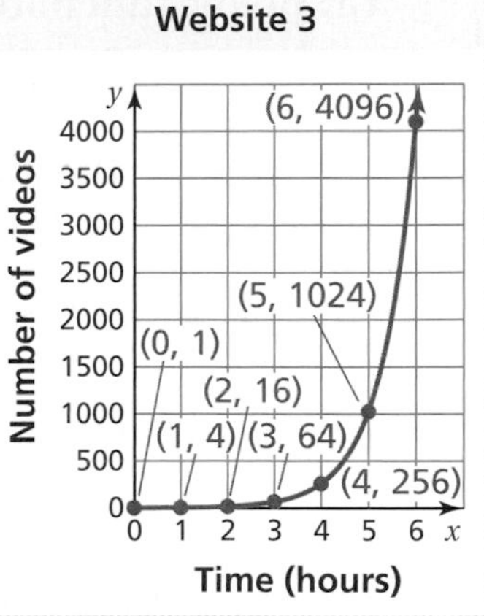

a. Do the three websites ever have the same number of videos?

b. Copy and complete the table for each function.

Time Interval	0 to 1 h	1 to 2 h	2 to 3 h	3 to 4 h	4 to 5 h	5 to 6 h
Average Rate of Change (videos/hour)						

c. What do you notice about the average rate of change of the linear function?

d. What do you notice about the average rate of change of the quadratic function?

e. What do you notice about the average rate of change of the exponential function?

f. Which average rate of change increases more quickly, the quadratic function or the exponential function?

Practice

5. **REASONING** How does a quantity that is increasing exponentially compare to a quantity that is increasing linearly or quadratically?

6. **REASONING** Explain why the average rate of change of a linear function is constant and the average rate of change of a quadratic or exponential function is not constant.

8.4–8.5 Quiz

Find (a) the axis of symmetry and (b) the vertex of the graph of the function. *(Section 8.4)*

1. $y = x^2 - 2x - 3$

2. $y = -2x^2 + 12x + 5$

Graph the function. Describe the domain and range. *(Section 8.4)*

3. $y = -4x^2 - 4x + 7$

4. $y = 2x^2 + 8x - 5$

5. $y = -4x^2 - 8x + 12$

Tell whether the function has a minimum value or a maximum value. Then find the value. *(Section 8.4)*

6. $y = 5x^2 + 10x - 3$

7. $y = -\frac{1}{2}x^2 + 2x + 16$

8. $y = -2x^2 + 8x + 3$

Graph the function. Compare the graph to the graph of $y = x^2$. *(Section 8.4)*

9. $y = (x - 5)^2$

10. $y = (x + 6)^2 - 2$

Plot the points. Tell whether the points represent a *linear*, an *exponential*, or a *quadratic* function. *(Section 8.5)*

11. (3, 6), (4, 16), (5, 30), (0, 0), (−1, 6)

12. (1, 7.5), (3, 6.5), (4, 6), (2, 7), (5, 5.5)

Tell whether the table of values represents a *linear*, an *exponential*, or a *quadratic* function. Then write an equation for the function using the form $y = mx + b$, $y = ab^x$, or $y = ax^2$. *(Section 8.5)*

13.

x	y
−1	1
0	3
1	9
2	27
3	81

14.

x	y
−3	−3
−2	−1
−1	1
0	3
1	5

15.

x	y
1	−5
2	−20
3	−45
4	−80
5	−125

16. FOOTBALL The function $h(t) = -16t^2 + 20t + 6$ gives the height (in feet) of a football t seconds after it is thrown. Describe the domain and range. Find the maximum height of the football. *(Section 8.4)*

17. DOWNLOADING MUSIC The table shows the amounts of money (in dollars) that you pay to download songs from a website. *(Section 8.5)*

Number of Songs, s	2	3	4	5
Amount, a (dollars)	2.58	3.87	5.16	6.45

a. Plot the points. Tell whether the points represent a *linear*, an *exponential*, or a *quadratic* function.

b. Write a function that models the data.

c. How much does it cost to download 15 songs?

Alternative Assessment Options

Math Chat	**Student Reflective Focus Question**
Structured Interview	Writing Prompt

Student Reflective Focus Question

Ask students to summarize the similarities and differences in graphing quadratic equations of the forms $y = ax^2$, $y = ax^2 + c$, $y = ax^2 + bx + c$, and $y = a(x - h)^2 + k$. Be sure that they include examples. Select students at random to present to the class.

Study Help Sample Answers

Remind students to complete Graphic Organizers for the rest of the chapter.

4.

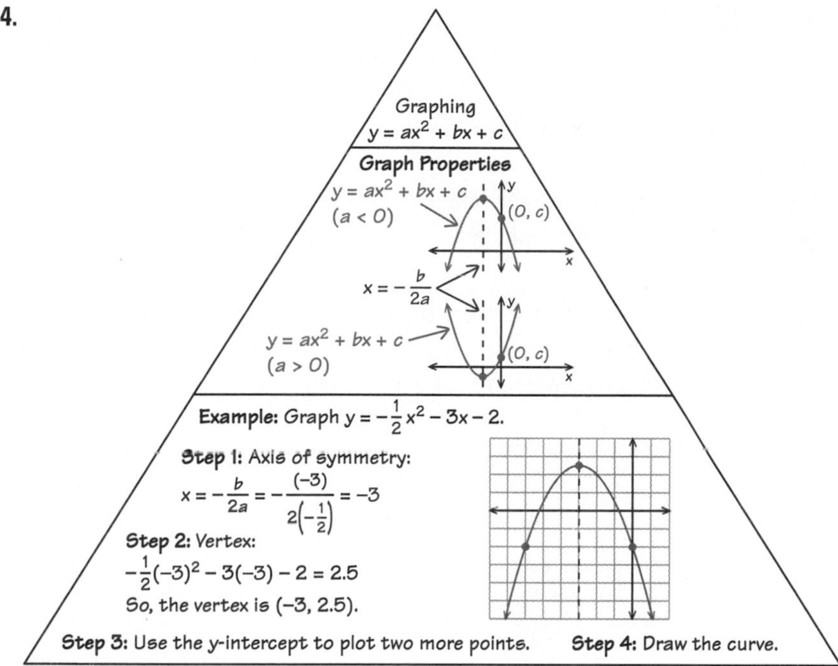

5. Available at *BigIdeasMath.com*.

Answers

1. **a.** $x = 1$ **b.** $(1, -4)$
2. **a.** $x = 3$ **b.** $(3, 23)$

3–5. See Additional Answers.

6. minimum; -8
7. maximum; 18
8. maximum; 11

9–12. See Additional Answers.

13. exponential; $y = 3(3)^x$
14. linear; $y = 2x + 3$
15. quadratic; $y = -5x^2$
16. domain: $0 \le t \le 1.5$
    range: $0 \le h \le 12.25$
    maximum height: 12.25 ft
17. **a.** See Additional Answers.
    b. $a = 1.29s$
    c. \$19.35

Technology For the Teacher

Answer Presentation Tool

Assessment Book

Chapter 8 Quiz
For use after Section 8.5

Find (a) the axis of symmetry and (b) the vertex of the graph of the function.

1. $y = 3x^2 + 12x + 5$
2. $y = -\frac{1}{2}x^2 + 4x - 7$

Graph the function. Describe the domain and range.

3. $y = x^2 + 8x + 17$
4. $y = -3x^2 - 12x - 10$

Tell whether the function has a minimum value or a maximum value. Then find the value.

5. $y = x^2 + 10x - 32$
6. $y = -\frac{1}{2}x^2 + 6x - 14$

Graph the function. Compare the graph to the graph of $y = x^2$.

7. $y = -(x - 4)^2 + 2$
8. $y = \frac{1}{3}(x + 1)^2 - 3$

9. Tell whether the table of values represents a *linear*, an *exponential*, or a *quadratic* function. Then write an equation for the function using the form $y = mx + b$, $y = ab^x$, or $y = ax^2$.

x	−3	−2	−1	0	1
y	−4.5	−2	−0.5	0	−0.5

10. The function $f(t) = -16t^2 + 32t + 42$ gives the height h (in feet) of water from a fountain t seconds after it leaves the spout. When does the water reach its maximum height? What is the maximum height?

Answers

1. a.____ b.____
2. a.____ b.____
3. See left.
4. See left.
5. ____
6. ____
7. See left.
8. See left.
9. ____
10. ____

Reteaching and Enrichment Strategies

If students need help...	If students got it...
Resources by Chapter • Study Help • Practice A and Practice B • Puzzle Time Lesson Tutorials *BigIdeasMath.com* Practice Quiz Practice from the Test Generator	Resources by Chapter • Enrichment and Extension • School-to-Work Game Closet at *BigIdeasMath.com* Start the Chapter Review

For the Teacher
Additional Review Options

- Big Ideas Test Generator
- Game Closet at *BigIdeasMath.com*
- Vocabulary Puzzle Builder
- Resources by Chapter
  - Puzzle Time
  - Study Help

Answers

1. Both graphs open up and have the same vertex, (0, 0), and the same axis of symmetry, $x = 0$. The graph of $y = 7x^2$ is narrower than the graph of $y = x^2$.

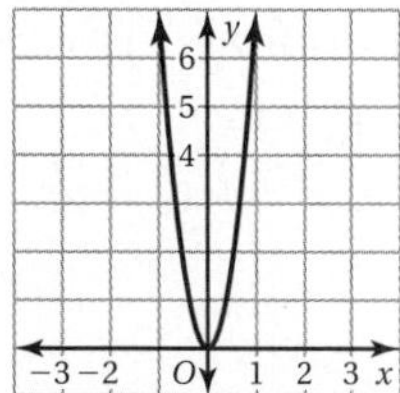

2. Both graphs open up and have the same vertex, (0, 0), and the same axis of symmetry, $x = 0$. The graph of $y = \frac{1}{2}x^2$ is wider than the graph of $y = x^2$.

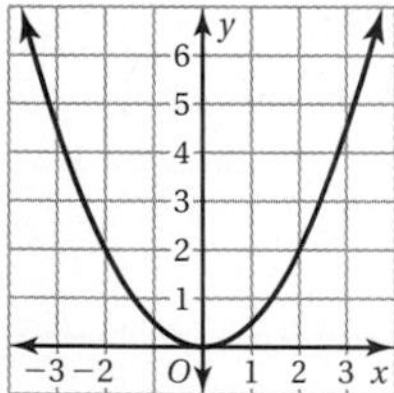

3. Both graphs have the same vertex, (0, 0), and the same axis of symmetry, $x = 0$, but the graph of $y = -\frac{3}{4}x^2$ opens down. The graph of $y = -\frac{3}{4}x^2$ is wider than the graph of $y = x^2$.

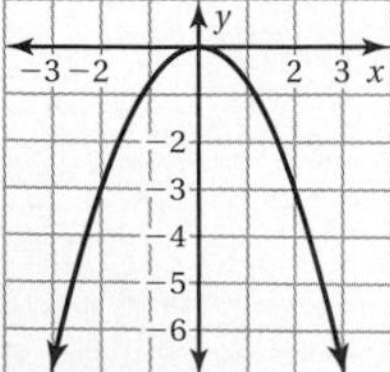

Review of Common Errors

Exercises 1–3

- Students may only plot points on the right side of the y-axis, which is the axis of symmetry for quadratic functions of the form $f(x) = ax^2$. Remind students that the graphs of quadratic functions are parabolas which are U-shaped and symmetric. Students can use this symmetry to find points on both sides of the axis of symmetry.

Exercise 5

- Students may have difficulty finding the value of a, especially because this requires working with fractions. Encourage them to take their time.

Exercises 6–8

- Students may only plot points on the right side of the y-axis, which is the axis of symmetry for quadratic functions of the form $f(x) = ax^2 + c$. Remind students that the graphs of quadratic functions are parabolas which are U-shaped and symmetric. Students can use this symmetry to find points on both sides of the axis of symmetry.

Exercises 9–11

- Students may forget to use the negative sign in $x = -\frac{b}{2a}$ when finding the axis of symmetry. Remind them of this formula.

Exercises 13–15

- Students may think that the y-intercept of a function of the form $y = a(x - h)^2 + k$ is represented by k. Remind them that k is the y-coordinate of the vertex and only represents the y-intercept when $h = 0$.

8 Chapter Review

Check It Out
Vocabulary Help
BigIdeasMath.com

Review Key Vocabulary

quadratic function, *p. 404*
parabola, *p. 404*
vertex, *p. 404*
axis of symmetry, *p. 404*
focus, *p. 412*
zero, *p. 419*
maximum value, *p. 427*
minimum value, *p. 427*
vertex form, *p. 432*

Review Examples and Exercises

8.1 Graphing $y = ax^2$ *(pp. 402–409)*

Graph $y = -4x^2$. Compare the graph to the graph of $y = x^2$.

Step 1: Make a table of values.

x	-2	-1	0	1	2
y	-16	-4	0	-4	-16

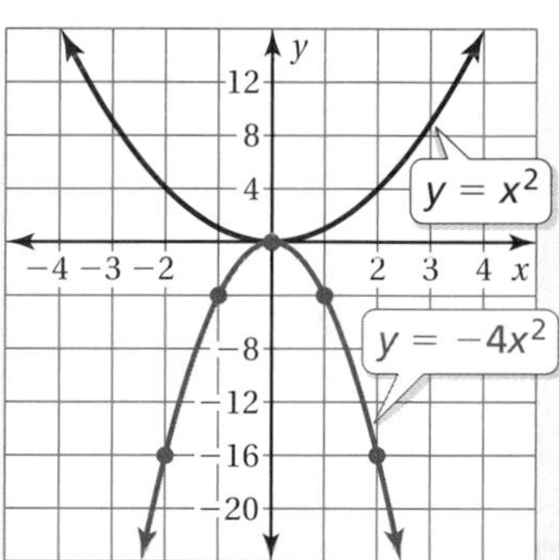

Step 2: Plot the ordered pairs.

Step 3: Draw a smooth curve through the points.

The graphs have the same vertex, (0, 0), and the same axis of symmetry, $x = 0$, but the graph of $y = -4x^2$ opens down. The graph of $y = -4x^2$ is narrower than the graph of $y = x^2$.

Exercises

Graph the function. Compare the graph to the graph of $y = x^2$.

1. $y = 7x^2$
2. $y = \frac{1}{2}x^2$
3. $y = -\frac{3}{4}x^2$

8.2 Focus of a Parabola *(pp. 410–415)*

Write an equation of the parabola with focus (0, 2) and vertex at the origin.

For $y = ax^2$, the focus is $\left(0, \frac{1}{4a}\right)$. Use the given focus, (0, 2), to write an equation to find a.

$\frac{1}{4a} = 2$ Equate the y-coordinates.

$\frac{1}{8} = a$ Solve for a.

An equation of the parabola is $y = \frac{1}{8}x^2$.

Exercises

4. Graph $y = -\frac{1}{2}x^2$. Identify the focus.

5. Write an equation of the parabola with focus (0, 10) and vertex at the origin.

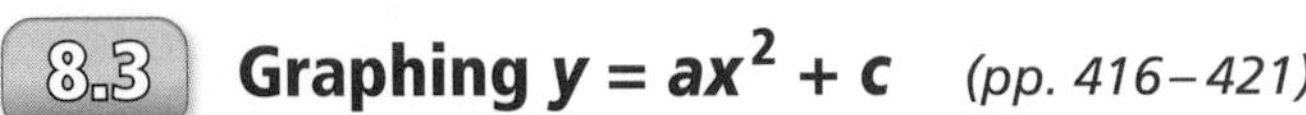

8.3 Graphing $y = ax^2 + c$ (pp. 416–421)

Graph $y = 2x^2 + 3$. Compare the graph to the graph of $y = x^2$.

Step 1: Make a table of values.

x	−2	−1	0	1	2
y	11	5	3	5	11

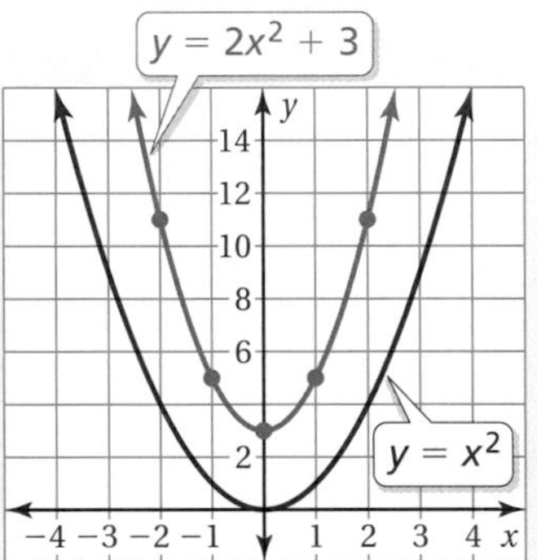

Step 2: Plot the ordered pairs.

Step 3: Draw a smooth curve through the points.

Both graphs open up and have the same axis of symmetry, $x = 0$. The graph of $y = 2x^2 + 3$ is narrower than the graph of $y = x^2$. The vertex of the graph of $y = 2x^2 + 3$ is a translation 3 units up of the vertex of the graph of $y = x^2$.

Exercises

Graph the function. Compare the graph to the graph of $y = x^2$.

6. $y = x^2 + 6$

7. $y = -x^2 - 4$

8. $y = 3x^2 - 5$

8.4 Graphing $y = ax^2 + bx + c$ (pp. 424–433)

Graph $y = 4x^2 + 8x - 1$. Describe the domain and range.

Step 1: Find and graph the axis of symmetry: $x = -\frac{b}{2a} = -\frac{8}{2(4)} = -1$.

Step 2: Find and plot the vertex. The x-coordinate of the vertex is −1. The y-coordinate is: $y = 4(-1)^2 + 8(-1) - 1 = -5$. So, the vertex is $(-1, -5)$.

Step 3: Use the y-intercept to find two more points on the graph. The y-intercept is −1. So, $(0, -1)$ lies on the graph. Because the axis of symmetry is $x = -1$, the point $(-2, -1)$ also lies on the graph.

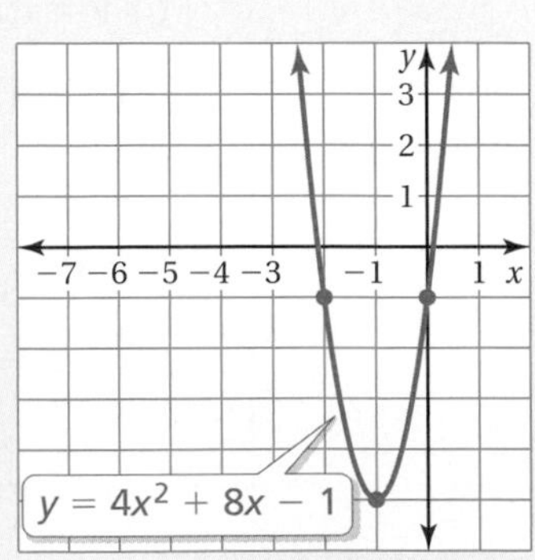

Step 4: Draw a smooth curve through the points.

The domain is all real numbers. The range is $y \geq -5$.

Review Game

Graphing Quadratic Equations

Materials per Group:

- copy of the Chapter Review from the pupil's edition
- paper
- graph paper
- pencil

Directions:
Divide the class into teams with three or four players on each team.

Setup: Each team works together to solve the Chapter Review exercises. Each team member solves a portion of the exercises. Use only one side of the paper. When all of the exercises are solved, the team members review the solutions together to make sure they are neat and correct.

Playing the game: The members of each group arrange themselves so they can see their solutions. The teacher chooses an exercise at random and gives clues about the answer. Each group uses its solutions to try to determine the exercise. The first person to raise a hand gets to name the exercise by number. If wrong, the group loses 1 point and the other teams are given an opportunity to name the exercise. The teacher may give more clues. The first team that successfully names the exercise gets 2 points.

The teacher should ask clues that require thought.

Example of clues for Exercise 15:

Clue 1: The answer includes a graph.
Clue 2: The graph is *not* narrower than the graph of $y = x^2$.
Clue 3: The graph is *not* wider than the graph of $y = x^2$.
Clue 4: The axis of symmetry is *not* $x = 0$.
Clue 5: The y-value of the vertex is *not* 0.
Clue 6: The graph has a maximum value.
Clue 7: The vertex is at (1, 1).

Who Wins?
The team with the most points at the end of the game wins.

For the Student
Additional Practice

- Lesson Tutorials
- Study Help (textbook)
- Student Website
  - Multi-Language Glossary
  - Practice Assessments

Answers

4. focus: $\left(0, -\frac{1}{2}\right)$

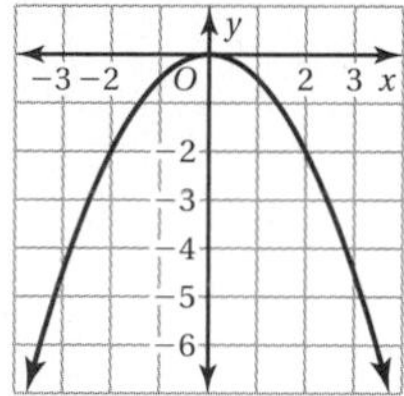

5. $y = \frac{1}{40}x^2$

6–8. See Additional Answers.

9. domain: all real numbers
range: $y \geq 6$

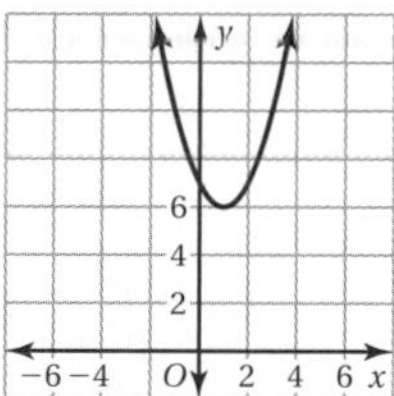

10. domain: all real numbers
range: $y \leq -\frac{13}{4}$

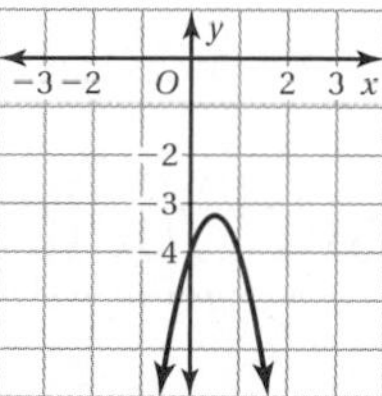

11. domain: all real numbers
range: $y \geq -8$

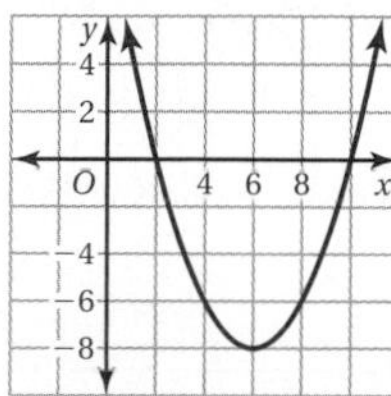

12. about 100 feet; after about 2.3 seconds

13–17. See Additional Answers.

My Thoughts on the Chapter

Teacher Tip

Not allowed to write in your teaching edition? Use sticky notes to record your thoughts.

What worked. . .

What did not work. . .

What I would do differently. . .

Exercises

Graph the function. Describe the domain and range.

9. $y = x^2 - 2x + 7$ **10.** $y = -3x^2 + 3x - 4$ **11.** $y = \frac{1}{2}x^2 - 6x + 10$

12. The function $f(t) = -16t^2 + 75t + 12$ gives the height (in feet) of a pumpkin t seconds after it is launched from a catapult. Use a graphing calculator to find the maximum height of the pumpkin. When does the pumpkin reach its maximum height?

Graph the function. Compare the graph to the graph of $y = x^2$. Use a graphing calculator to check your answer.

13. $y = (x + 5)^2$ **14.** $y = (x + 3)^2 - 2$ **15.** $y = -(x - 1)^2 + 1$

8.5 Comparing Linear, Exponential, and Quadratic Functions
(pp. 434–443)

Tell whether the data values represent a *linear*, an *exponential*, or a *quadratic* function.

a. $(-4, 1), (-3, -2), (-2, -3), (-1, -2), (0, 1)$

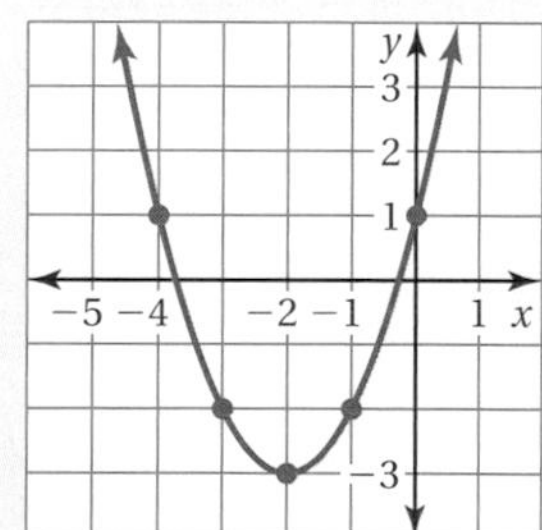

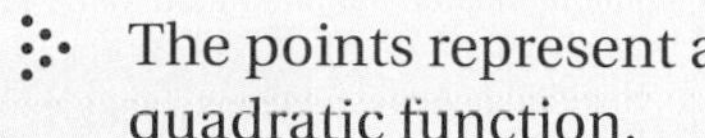

The points represent a quadratic function.

b.

x	−1	0	1	2	3
y	15	8	1	−6	−13

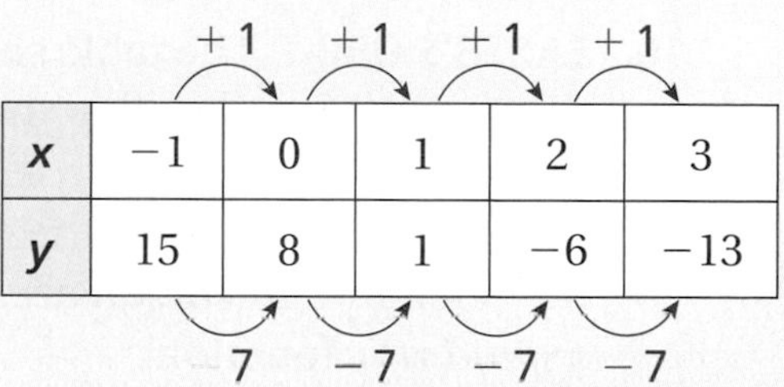

x	−1	0	1	2	3
y	15	8	1	−6	−13

The y-values have a common difference of −7. So, the table represents a linear function.

Exercises

16. Tell whether the table of values represents a *linear*, an *exponential*, or a *quadratic* function. Then write an equation for the function using the form $y = mx + b$, $y = ab^x$, or $y = ax^2$.

x	−1	0	1	2	3
y	16	8	4	2	1

17. The function $f(t) = -16t^2 + 75t + 4$ gives the height (in feet) of a baseball t seconds after it is thrown. Sketch a graph of the function. Find the average rate of change from 0 to 1 second and from 1 to 2 seconds. Is the average rate of change increasing or decreasing? How does the graph justify your answer?

8 Chapter Test

Graph the function. Compare the graph to the graph of $y = x^2$.

1. $y = 3x^2$

2. $y = 2x^2 + 2$

3. $y = -\frac{1}{2}x^2 - 1$

4. $y = (x - 3)^2$

5. $y = (x + 1)^2 - 1$

6. $y = -2(x - 5)^2$

Graph the function. Identify the focus.

7. $y = 6x^2$

8. $y = \frac{1}{5}x^2$

9. $y = -1.5x^2$

Graph the function. Describe the domain and range.

10. $y = x^2 + 2x - 1$

11. $y = -x^2 - 3x + 3$

12. $y = 2x^2 + 4x - 4$

13. Describe how the graph of $g(x) = f(x + 6)$ compares to the graph of $f(x)$.

Tell whether the table of values represents a *linear*, an *exponential*, or a *quadratic* function. Then write an equation for the function using the form $y = mx + b$, $y = ab^x$, or $y = ax^2$.

14.

x	−1	0	1	2	3
y	4	8	16	32	64

15.

x	−2	−1	0	1	2
y	−8	−2	0	−2	−8

16. EARTH'S ORBIT The table shows the distance d (in miles) that the Earth moves in its orbit around the Sun after t seconds. Tell whether the data can be modeled by a *linear*, an *exponential*, or a *quadratic* function.

Time, t	1	2	3	4	5
Distance, d	19	38	57	76	95

17. RADIO TELESCOPE An observatory uses a radio telescope to collect data from another galaxy. The cross section of the telescope's dish can be modeled by $y = \frac{1}{120}x^2$, where x and y are measured in meters. The telescope's receiver is located at the focus of the parabola. What is the distance from the vertex of the parabola to the receiver?

18. REASONING Consider the function $f(x) = x^2 + 3$. Is the average rate of change increasing or decreasing from $x = 0$ to $x = 4$? Explain.

Test Item References

Chapter Test Questions	Section to Review	Common Core State Standards
1	8.1	F.BF.3
7–9, 17	8.2	F.IF.4
2, 3	8.3	F.BF.3
4–6, 10–13	8.4	F.BF.3, F.IF.4, F.IF.7a
14–16, 18	8.5	F.IF.4, F.IF.6, F.IF.7a, F.LE.3

Test-Taking Strategies

Remind students to quickly look over the entire test before they start so that they can budget their time. Each time they complete a graph, students should look back at the equation to see if the graph has the correct properties. Encourage students to use the Stop and Think strategy before answering. **Stop** and carefully read the question, and **Think** about what the answer should look like.

Common Assessment Errors

- **Exercises 1–6** Students may only plot points on the right side of the y-axis, which is the axis of symmetry for quadratic functions of the form $f(x) = ax^2 + c$. Remind students that the graphs of quadratic functions are parabolas which are U-shaped and symmetric. Students can use this symmetry to find points on both sides of the axis of symmetry.
- **Exercises 10–12** Students may not graph enough points to properly portray the shape of the graph of the function. Remind them that the graph should have a U-shape.
- **Exercise 13** Students may say $g(x) = f(x + 6)$ is a translation 6 units to the right of the graph of $f(x)$. Explain that $g(x) = f(x + 6)$ is a translation 6 units to the left of the graph of $f(x)$.
- **Exercise 15** Students may incorrectly substitute the x-coordinate in for y and the y-coordinate in for x when writing an equation for the function. Remind them to check their equation.

Reteaching and Enrichment Strategies

If students need help. . .	If students got it. . .
Resources by Chapter • Practice A and Practice B • Puzzle Time Record and Practice Journal Practice Differentiating the Lesson Lesson Tutorials Practice from the Test Generator Skills Review Handbook	Resources by Chapter • Enrichment and Extension • School-to-Work • Financial Literacy Game Closet at *BigIdeasMath.com* Start Standardized Test Practice

Answers

1–6. See Additional Answers.

7. focus: $\left(0, \frac{1}{24}\right)$

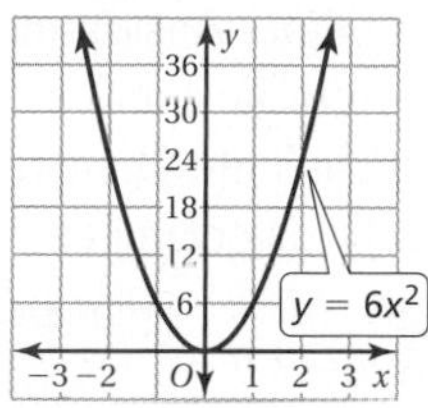

8. focus: $\left(0, \frac{5}{4}\right)$

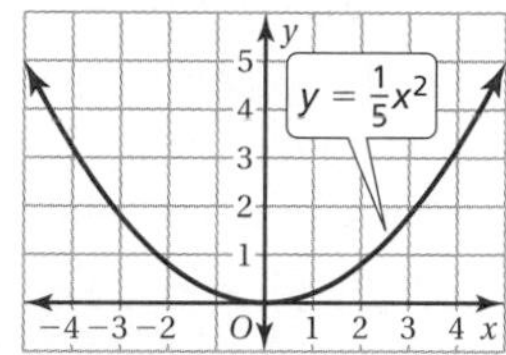

9–12. See Additional Answers.

13. The graph of $g(x)$ is a horizontal translation 6 units left of the graph of $f(x)$.

14. exponential; $y = 8(2)^x$

15. quadratic; $y = -2x^2$

16. linear

17. 30 m

18. See Additional Answers.

Assessment Book

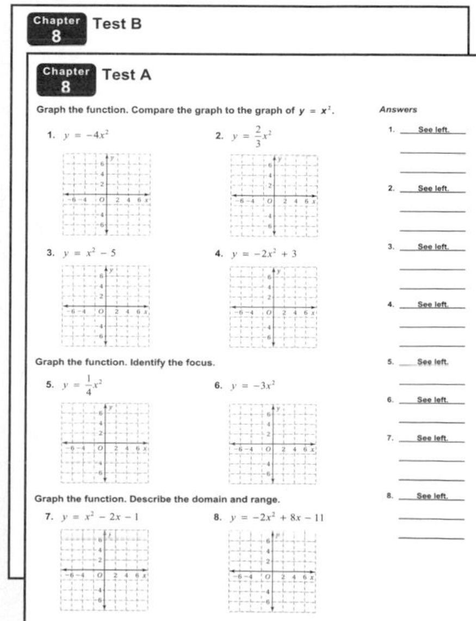

Test-Taking Strategies
Available at *BigIdeasMath.com*

After Answering Easy Questions, Relax
Answer Easy Questions First
Estimate the Answer
Read All Choices before Answering
Read Question before Answering
Solve Directly or Eliminate Choices
Solve Problem before Looking at Choices
Use Intelligent Guessing
Work Backwards

About this Strategy

When taking a multiple choice test, be sure to read each question carefully and thoroughly. When taking a timed test, it is often best to skim the test and answer the easy questions first. Be careful that you record your answer in the correct position on the answer sheet.

Answers

1. C
2. I
3. C
4. -6
5. F

Technology For the Teacher

Big Ideas Test Generator

Item Analysis

1. **A.** The student forgets the negative sign to show a reflection in the x-axis.
   B. The student forgets the negative sign to show a reflection in the x-axis and incorrectly thinks the wider graph means that $a > 1$.
   C. Correct answer
   D. The student incorrectly thinks the wider graph means that $a < -1$.

2. **F.** The students makes a sign error.
   G. The student adds the exponents.
   H. The student adds the exponents and negates the sum.
   I. Correct answer

3. **A.** The student uses the opposite of the coefficient of the x^2 term.
   B. The student makes a sign error.
   C. Correct answer
   D. The student uses the coefficient of the x^2 term.

4. **Gridded Response:** Correct answer: -6
   Common Error: The student uses the constant term 6.

5. **F.** Correct answer
   G. The student fails to divide by 4.
   H. The students uses the square root of the coefficient of the variable.
   I. The student switches the divisor and the dividend in applying the Division Property of Equality.

8 Standardized Test Practice

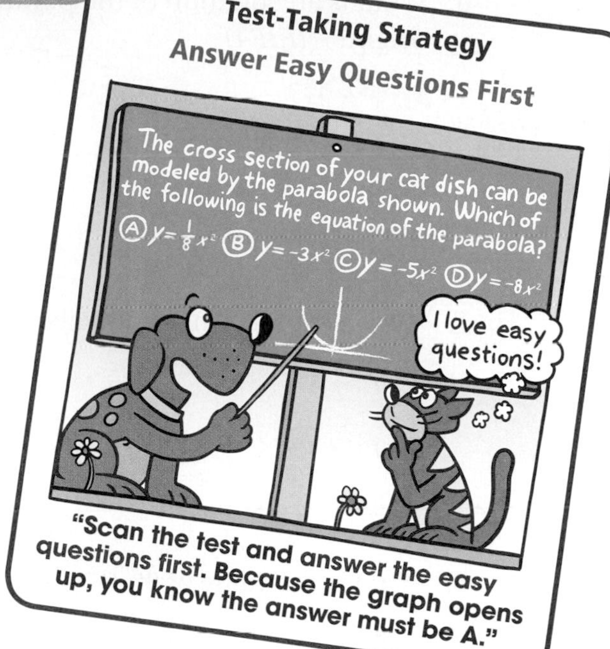

Test-Taking Strategy
Answer Easy Questions First
The cross section of your cat dish can be modeled by the parabola shown. Which of the following is the equation of the parabola?
Ⓐ $y=\frac{1}{8}x^2$ Ⓑ $y=-3x^2$ Ⓒ $y=-5x^2$ Ⓓ $y=-8x^2$
I love easy questions!
"Scan the test and answer the easy questions first. Because the graph opens up, you know the answer must be A."

1. What is the equation of the parabola shown in the graph? *(F.IF.4)*

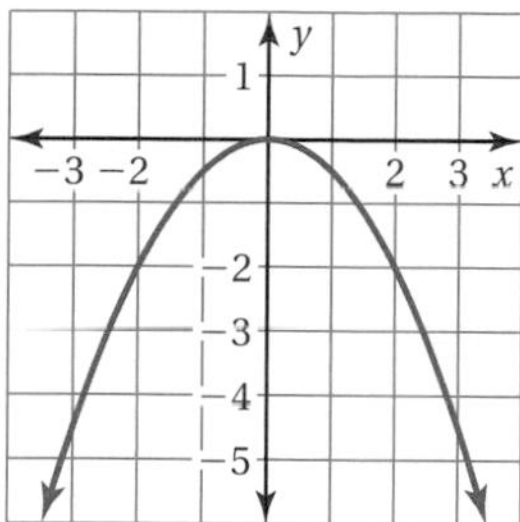

A. $y = \frac{1}{2}x^2$

B. $y = 2x^2$

C. $y = -\frac{1}{2}x^2$

D. $y = -2x^2$

2. Which expression is equivalent to $(b^{-5})^{-4}$? *(N.RN.2)*

F. b^{-20}

G. b^{-9}

H. b^{9}

I. b^{20}

3. What is the axis of symmetry of the graph of $y = 3x^2 - 6x - 14$? *(F.IF.4)*

A. $x = -3$

B. $x = -1$

C. $x = 1$

D. $x = 3$

4. What is the minimum value of the function $y = 3x^2 + 12x + 6$? *(F.IF.7a)*

5. What are the solutions of $16a^2 - 49 = 0$? *(A.REI.4b)*

F. $a = -\frac{7}{4}$ and $a = \frac{7}{4}$

G. $a = -7$ and $a = 7$

H. $a = -4$ and $a = 4$

I. $a = -\frac{4}{7}$ and $a = \frac{4}{7}$

6. What is an equation of the parabola with focus $\left(0, \frac{1}{8}\right)$ and vertex at the origin? *(F.IF.4)*

A. $y = 2x^2$

B. $y = \frac{1}{2}x^2$

C. $y = \frac{1}{8}x^2$

D. $y = \frac{1}{32}x^2$

7. Which expression is equivalent to $(t - 4)^2$? *(A.APR.1)*

F. $t^2 + 16$

G. $t^2 - 16$

H. $t^2 - 8t + 16$

I. $t^2 - 8t - 16$

8. What is the value of $8^{2/3}$? *(N.RN.2)*

9. Which of the following is the graph of $y = 4(0.5)^t$? *(F.IF.7e)*

A.

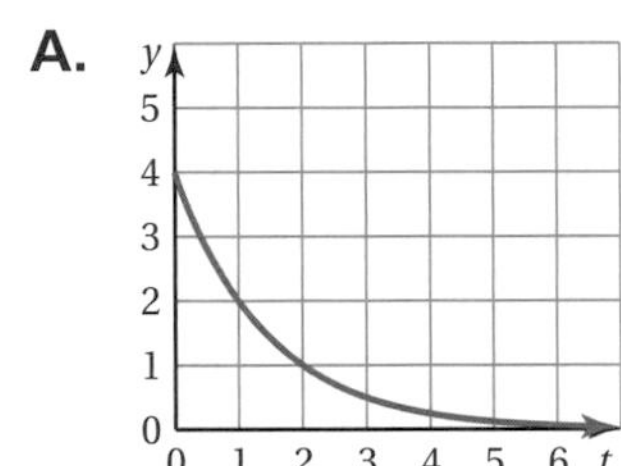

B.

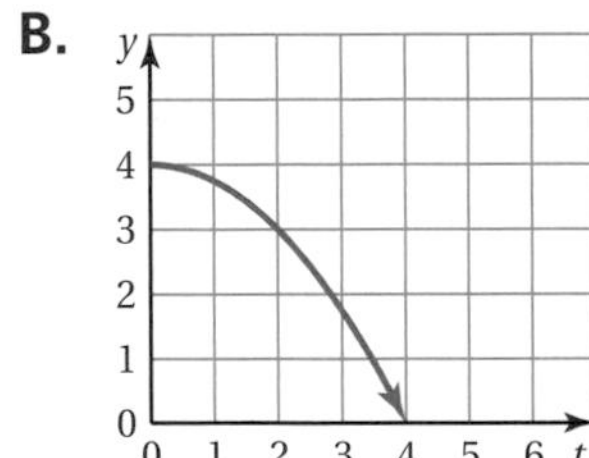

C.

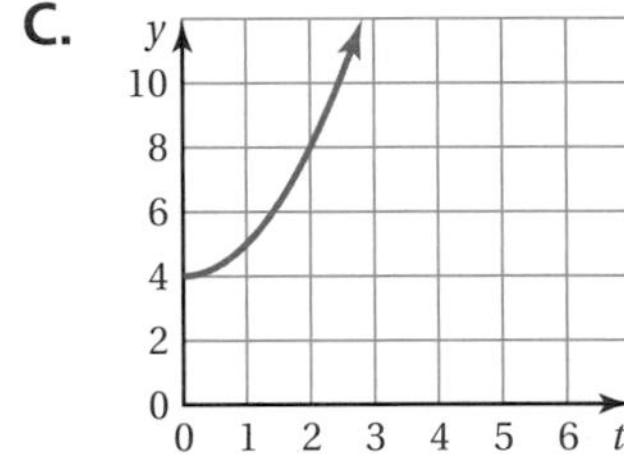

D.

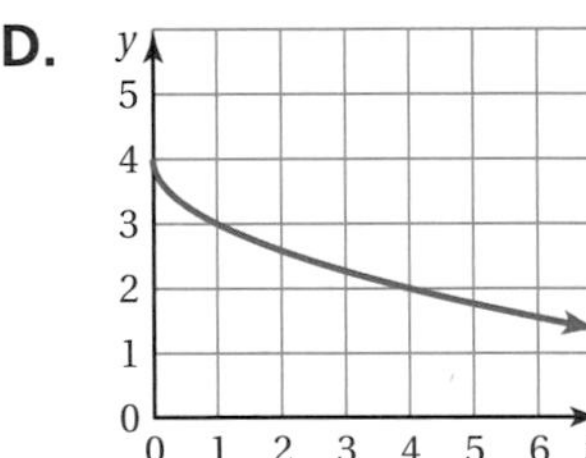

10. The function $f(t) = -16t^2 + 32t + 4$ gives the height (in feet) of a softball t seconds after it is thrown. *(F.IF.4)*

Part A Does the function have a minimum value or a maximum value? Explain your reasoning.

Part B Find the value from Part A.

Item Analysis (continued)

6. **A.** Correct answer
   B. The student divides 4 by 8 to find *a* instead of dividing 8 by 4.
   C. The student uses the *y*-value of the focus as *a*.
   D. The student calculates *a* as one-fourth of the *y*-value of the focus.

7. **F.** The student rewrites the subtraction as addition, and then squares each term instead of the whole expression.
   G. The student squares each term instead of the whole expression.
   H. Correct answer
   I. The student makes a sign error in calculating the constant 16.

8. **Gridded Response:** Correct answer: 4
   Common Error: The student divides 8^2 by 8^3 for an answer of $\frac{1}{8}$.

9. **A.** Correct answer
   B. The student thinks that the point (1, 2) is on the graph.
   C. The student thinks that because there is a positive exponent, the graph rises from left to right.
   D. The student thinks that the point (4, 2) is on the graph.

10. **2 points** The student demonstrates a thorough understanding of the graph of a quadratic function of the form $y = ax^2 + bx + c$. The student correctly explains why the curve has a maximum value and demonstrates how to find it.

    1 point The student's work and explanations demonstrate a lack of essential understanding. The student is unable to find the correct maximum value.

    0 points The student provides no response, a completely incorrect or incomprehensible response, or a response that demonstrates insufficient understanding of graphing quadratic functions of the form $y = ax^2 + bx + c$.

Answers

6. A
7. H
8. 4
9. A
10. *Part A* The function has a maximum value because $a < 0$.

    Part B 20 feet

Answers

11. I

12. C

13. F

Answer for Extra Example

1. **A.** The student incorrectly thinks that because the first differences are equal, the function is quadratic.

   B. Correct answer

   C. The student incorrectly thinks that because the first differences are increasing, the function is quadratic.

   D. The student incorrectly thinks that because the terms have a common ratio of $\frac{1}{2}$, the function is quadratic.

Item Analysis (continued)

11. **F.** The student fails to recognize that any point on the line $ax + by = c$ is a solution of the system.

G. The student fails to recognize that any point on the line $ax + by = c$ is a solution of the system.

H. The student fails to recognize that any point on the line $ax + by = c$ is a solution of the system.

I. Correct answer

12. **A.** The student fails to recognize the vertical shift.

B. The student forgets to undo the horizontal shift.

C. Correct answer

D. The student incorrectly shifts the graph down instead of up.

13. **F.** Correct answer

G. The student calculates the slope incorrectly.

H. The student calculates the slope incorrectly.

I. The student calculates the slope incorrectly.

Extra Example

1. Which table of values represents a quadratic function? *(F.IF.4)*

A.

x	0	1	2	3	4
y	-4	-1	2	5	8

B.

x	0	1	2	3	4
y	-2	0	6	16	30

C.

x	0	1	2	3	4
y	3	5	9	17	33

D.

x	0	1	2	3	4
y	-4	-2	-1	$-\frac{1}{2}$	$-\frac{1}{4}$

11. Which of the following is not a solution of the system of linear equations when $a \neq 0$, $b \neq 0$, and $c \neq 0$? *(8.EE.8a)*

$$ax + by = c$$
$$2ax + 2by = 2c$$

F. $\left(\frac{c}{2a}, \frac{c}{2b}\right)$

H. $\left(c, \frac{c - ac}{b}\right)$

G. $\left(\frac{c}{a}, 0\right)$

I. $\left(\frac{c}{b}, 0\right)$

12. Amanda was graphing $y = \frac{1}{4}x^2 + 2$. Her work is shown below. *(F.BF.3)*

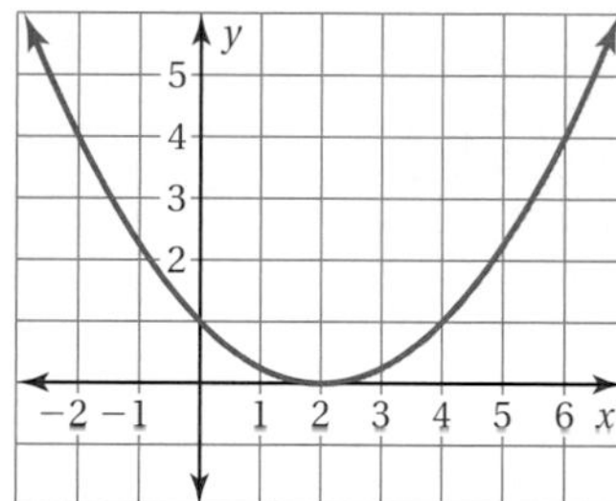

What should Amanda do to correct the error that she made?

A. Shift the graph 2 units left.

B. Shift the graph 2 units up.

C. Shift the graph 2 units left and 2 units up.

D. Shift the graph 2 units left and 2 units down.

13. Which of the following is an equation of the line that passes through the points (4, 4) and (5, 7)? *(A.CED.2)*

F. $y = 3x - 8$

H. $y = \frac{1}{3}x + \frac{8}{3}$

G. $y = \frac{1}{3}x + \frac{16}{3}$

I. $y = -3x + 16$

9 Solving Quadratic Equations

9.1 Solving Quadratic Equations by Graphing

9.2 Solving Quadratic Equations Using Square Roots

9.3 Solving Quadratic Equations by Completing the Square

9.4 Solving Quadratic Equations Using the Quadratic Formula

9.5 Solving Systems of Linear and Quadratic Equations

"Do you know why the quadratic equation $x^2 + 1 = 0$ has no real solutions?"
Imaginary that!
"It's because the graph of $y = x^2 + 1$ doesn't cross the x-axis!"

"Okay, you hold your tail straight so that there are exactly two points of intersection."

I need a firefighter to rescue me.
"That's perfect Descartes!"

Connections to Previous Learning

- Write and solve one-step linear equations in one variable.
- Evaluate expressions at specific values of their variables.

- Write and solve multi-step linear equations in one variable with one solution, no solution, or infinitely many solutions.
- Evaluate square roots of perfect squares.

- Solve quadratic equations in one variable by graphing, using square roots, completing the square, and using the quadratic formula.
- Derive the quadratic formula by completing the square.
- Solve a system of equations consisting of a linear equation and a quadratic equation in two variables algebraically and graphically.

Math in History

Around 2000 B.C., engineers in Egypt, Babylon, and China were interested in the problem of finding the side lengths that produce a given area. Many methods for solving these types of problems developed before the quadratic formula took its modern form.

★ Greek mathematician Euclid of Alexandria showed a general procedure for solving quadratic equations using strictly geometric methods in his famous text *The Elements,* written around 300 B.C.

★ Frenchman René Descartes published *La Géométrie* in 1637. In this work, Descartes used a system of representing known quantities by the first letters of the alphabet and unknown quantities by the last letters of the alphabet. This is reflected in the quadratic formula that is used today.

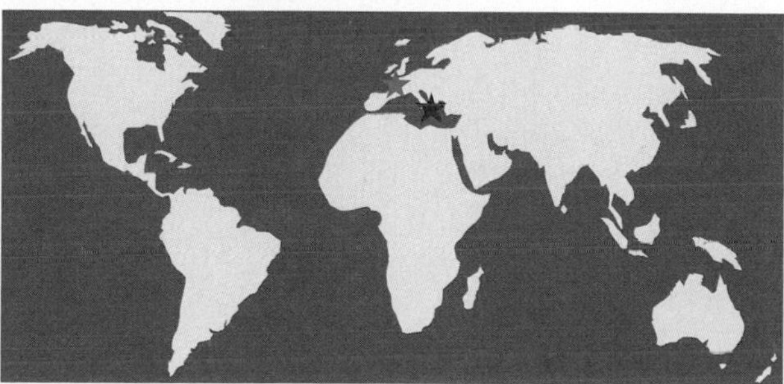

Pacing Guide for Chapter 9

Chapter Opener	1 Day
Section 1	2 Days
Section 2	1 Day
Section 3	1 Day
Study Help / Quiz	1 Day
Section 4	3 Days
Section 5	2 Days
Chapter Review / Chapter Tests	2 Days
Total Chapter 9	13 Days
Year-to-Date	125 Days

Check Your Resources

- Record and Practice Journal
- Resources by Chapter
- Skills Review Handbook
- Assessment Book
- Worked-Out Solutions

Technology For the Teacher

The Dynamic Planning Tool
Editable Teacher's Resources at
BigIdeasMath.com

Common Core State Standards

8.EE.2 . . . Evaluate square roots of small perfect squares . . ..
N.RN.2 Rewrite expressions involving radicals . . . using the properties of exponents.
A.SSE.2 Use the structure of an expression to identify ways to rewrite it.

Additional Topics for Review

- Graphing quadratic functions
- Solving polynomial equations in factored form
- Factoring polynomials
- Special products of polynomials
- Solving simple equations

Try It Yourself

1. 9 **2.** -13

3. $\pm\frac{3}{5}$ **4.** -2.5

5. $3\sqrt{6}$ **6.** $4\sqrt{5}$

7. $10\sqrt{2}$ **8.** $(x+5)^2$

9. $(m-10)^2$ **10.** $(p+6)^2$

Record and Practice Journal

1. -6 **2.** 11

3. $\frac{2}{7}$ **4.** ± 1.5

5. 3 ft **6.** 0.5 m

7. $2\sqrt{5}$ **8.** $3\sqrt{7}$

9. $6\sqrt{3}$ **10.** $12\sqrt{2}$

11. $5\sqrt{5}$ ft **12.** $8\sqrt{3}$ m

13. $(y-3)^2$ **14.** $(b+9)^2$

15. $(n+14)^2$ **16.** $(h-8)^2$

17. **a.** $(x-25)$ in.

b. $4(x-25)$ in.

Math Background Notes

Vocabulary Review

- Perfect square
- Radical sign
- Radicand
- Perfect Square Trinomial

Finding Square Roots

- Students should know how to find the square root of a perfect square.
- To find a square root of a given number, find a number that you can square to get the given number.
- **Teaching Tip:** Remind students that $\sqrt{n}$ represents the positive square root of n. Every number n has a positive square root $\sqrt{n}$ and a negative square root $-\sqrt{n}$.

Simplifying Square Roots

- Students should know how to simplify the square root of a number that is not a perfect square.
- **Teaching Tip:** The key to simplifying a square root is to factor using the greatest perfect square factor of the radicand.
- **Common Error:** Students may omit the step of using the Product Property of Square Roots and take the wrong quantity outside the radical sign. Emphasize the second and third steps in Example 3.

Factoring Perfect Square Trinomials

- Students should know how to factor a perfect square trinomial.
- Remind students of the Perfect Square Trinomial Patterns.
  $a^2 + 2ab + b^2 = (a+b)^2$
  $a^2 - 2ab + b^2 = (a-b)^2$

Reteaching and Enrichment Strategies

If students need help. . .	If students got it. . .
Record and Practice Journal • Fair Game Review Skills Review Handbook Lesson Tutorials	Game Closet at *BigIdeasMath.com* Start the next section

What You Learned Before

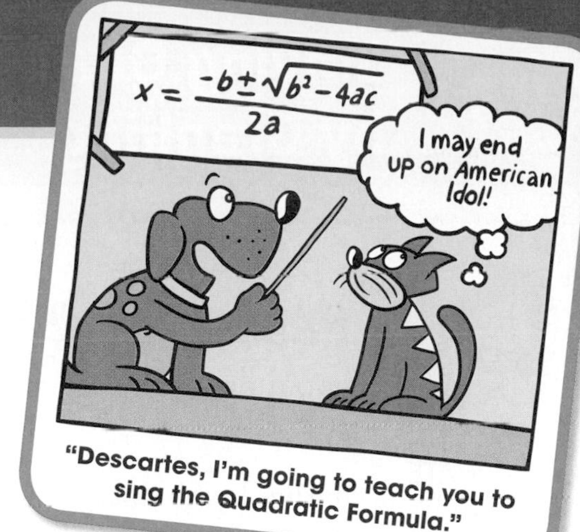

$x = \frac{-b \pm \sqrt{b^2-4ac}}{2a}$
I may end up on American Idol!

"Descartes, I'm going to teach you to sing the Quadratic Formula."

Finding Square Roots (8.EE.2)

Example 1 **Find $\sqrt{144}$.**

Because $12^2 = 144$, $\sqrt{144} = \sqrt{12^2} = 12$.

Positive square root

Example 2 **Find $-\sqrt{225}$.**

Because $15^2 = 225$, $-\sqrt{225} = -\sqrt{15^2} = -15$.

Negative square root

Try It Yourself

Find the square root(s).

1. $\sqrt{81}$ **2.** $-\sqrt{169}$ **3.** $\pm\sqrt{\frac{9}{25}}$ **4.** $-\sqrt{6.25}$

Simplifying Square Roots (N.RN.2)

Example 3 **Simplify $\sqrt{75}$.**

$\sqrt{75} = \sqrt{25 \cdot 3}$ Factor using the greatest perfect square factor.

$= \sqrt{25} \cdot \sqrt{3}$ Use the Product Property of Square Roots.

$= 5\sqrt{3}$ Simplify.

Try It Yourself

5. Simplify $\sqrt{54}$. **6.** Simplify $\sqrt{80}$. **7.** Simplify $\sqrt{200}$.

Factoring Perfect Square Trinomials (A.SSE.2)

Example 4 **Factor $x^2 + 14x + 49$.**

$x^2 + 14x + 49 = x^2 + 2(x)(7) + 7^2$ Write as $a^2 + 2ab + b^2$.

$= (x + 7)^2$ Perfect Square Trinomial Pattern

Example 5 **Factor $y^2 - 10y + 25$.**

$y^2 - 10y + 25 = y^2 - 2(y)(5) + 5^2$ Write as $a^2 - 2ab + b^2$.

$= (y - 5)^2$ Perfect Square Trinomial Pattern

Try It Yourself

Factor the trinomial.

8. $x^2 + 10x + 25$ **9.** $m^2 - 20m + 100$ **10.** $p^2 + 12p + 36$

9.1 Solving Quadratic Equations by Graphing

Essential Question How can you use a graph to solve a quadratic equation in one variable?

COMMON CORE STATE STANDARDS
A.REI.4
A.REI.11

Earlier in the book, you learned that the x-intercept of the graph of

$y = ax + b$ 2 variables

is the same as the solution of

$ax + b = 0.$ 1 variable

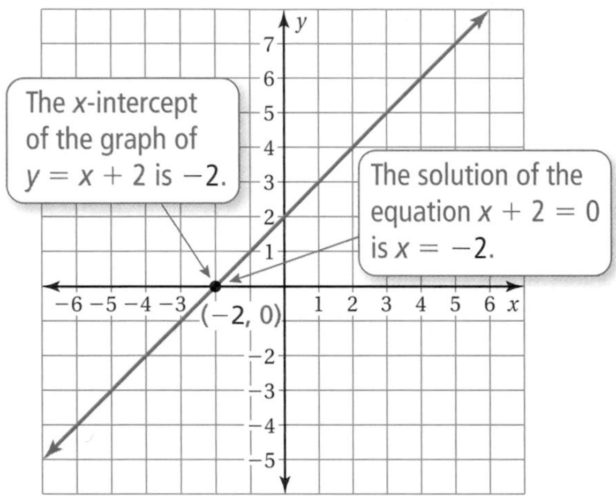

1 ACTIVITY: Solving a Quadratic Equation by Graphing

Work with a partner.

a. Sketch the graph of $y = x^2 - 2x$.

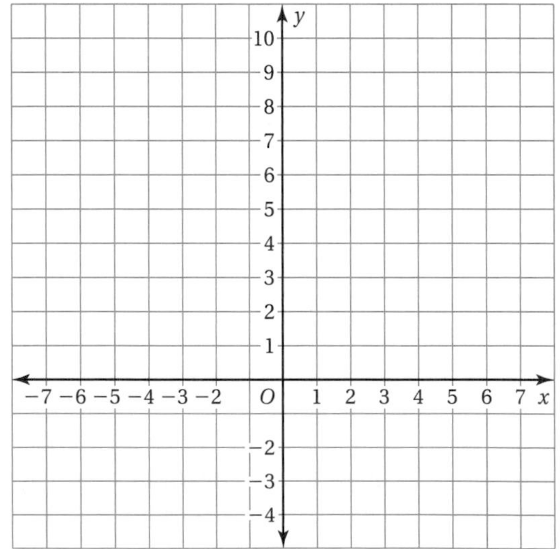

b. What is the definition of an x-intercept of a graph? How many x-intercepts does this graph have? What are they?

c. What is the definition of a solution of an equation in x? How many solutions does the equation $x^2 - 2x = 0$ have? What are they?

d. Explain how you can verify that the x-values found in part (c) are solutions of $x^2 - 2x = 0$.

Laurie's Notes

Introduction

Standards for Mathematical Practice

- **MP1a Make Sense of Problems:** When students solved quadratic equations in factored form, they used the Zero-Product Property and reasoned, "What value of x makes each factor zero?" Now they are reasoning "What values of x make the equation zero?" The values of the x-intercepts make the equation zero.

Motivate

- Show a picture of a projectile being launched from a trebuchet or a catapult.
- Ask students what they would like to know about the projectile. Hopefully a student will ask how long it takes for the projectile to land.
- Explain that in this section they will answer that type of question.

Activity Notes

Discuss

- Connect the projectile motion to the x-intercept dialogue.
- Also, connect the previous two chapters to this section by reminding students that they solved quadratic equations by factoring in Chapter 7, and they graphed quadratic functions in Chapter 8.

Activity 1

? "What do you know about the graph of $y = x^2 - 2x$?" opens up; y-intercept is 0; x-coordinate of the vertex is 1; The axis of symmetry is $x = 1$.

- Students should make an input-output table that includes x-values that show the key features of the graph.
- While students work, question different pairs of students about the vertex and the minimum value.
- **Big Idea:** In part (d), students should discuss evaluating the left side of the equation for the x-values found in part (c) and they should also discuss the x-intercept. It is not enough to solve the equation graphically because computational errors can influence the graph. An algebraic check is also important.
- **MP1a:** This approach to solving a quadratic equation should make sense to students if they think back to solving systems of equations. They can think of solving $x^2 - 2x = 0$ as solving the system $y = x^2 - 2x$ and $y = 0$. The graph of $y = 0$ is the x-axis, which intersects the graph of $y = x^2 - 2x$ at its x-intercepts.

Common Core State Standards

A.REI.4 Solve quadratic equations in one variable.

A.REI.11 Explain why the x-coordinates of the points where the graphs of the equations $y = f(x)$ and $y = g(x)$ intersect are the solutions of the equation $f(x) = g(x)$; find the solutions approximately . . ..

Previous Learning

Students should know how to graph a quadratic function and find the x-intercepts.

Start Thinking! and Warm Up

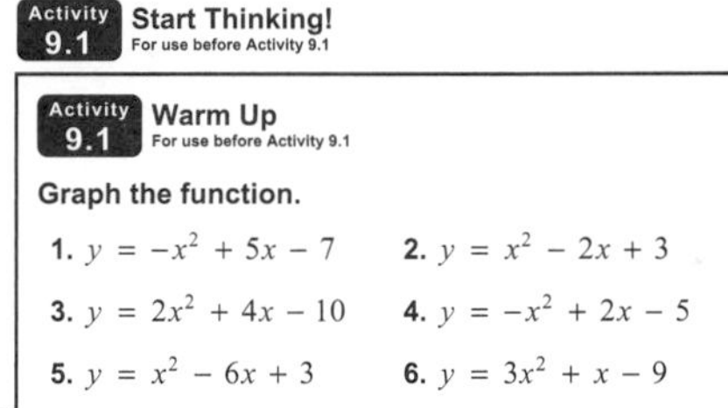

Activity 9.1 **Start Thinking!** For use before Activity 9.1

Activity 9.1 **Warm Up** For use before Activity 9.1

Graph the function.

1. $y = -x^2 + 5x - 7$
2. $y = x^2 - 2x + 3$
3. $y = 2x^2 + 4x - 10$
4. $y = -x^2 + 2x - 5$
5. $y = x^2 - 6x + 3$
6. $y = 3x^2 + x - 9$

9.1 Record and Practice Journal

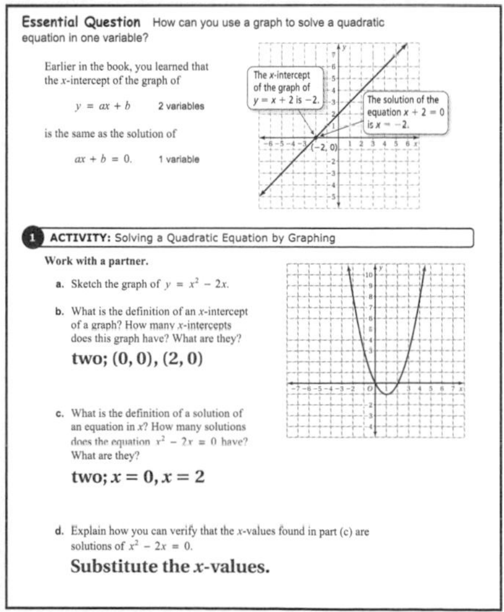

Essential Question How can you use a graph to solve a quadratic equation in one variable?

Earlier in the book, you learned that the x-intercept of the graph of

$y = ax + b$ 2 variables

is the same as the solution of

$ax + b = 0$. 1 variable

1 ACTIVITY: Solving a Quadratic Equation by Graphing

Work with a partner.

a. Sketch the graph of $y = x^2 - 2x$.

b. What is the definition of an x-intercept of a graph? How many x-intercepts does this graph have? What are they?
two; (0, 0), (2, 0)

c. What is the definition of a solution of an equation in x? How many solutions does the equation $x^2 - 2x = 0$ have? What are they?
two; $x = 0, x = 2$

d. Explain how you can verify that the x-values found in part (c) are solutions of $x^2 - 2x = 0$.
Substitute the x-values.

Differentiated Instruction

Auditory

After Activity 1, ask students to discuss the differences between an equation of the form $ax^2 + bx + c = 0$ and its related function $y = ax^2 + bx + c$. Talk about how the equation can be solved by graphing its related function and discuss the relationship between solutions and x-intercepts.

9.1 Record and Practice Journal

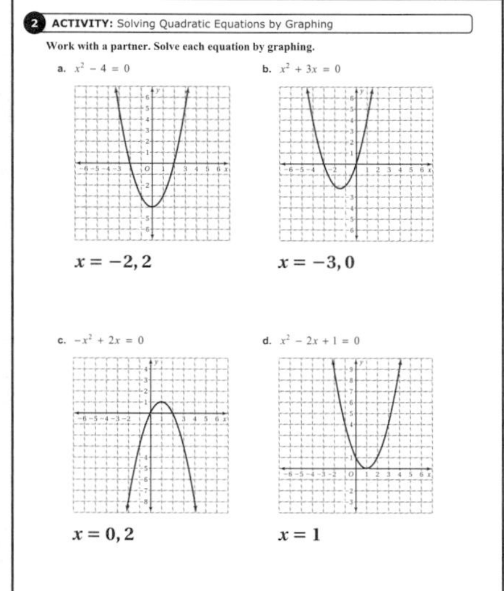

2 ACTIVITY: Solving Quadratic Equations by Graphing

Work with a partner. Solve each equation by graphing.

a. $x^2 - 4 = 0$

$x = -2, 2$

b. $x^2 + 3x = 0$

$x = -3, 0$

c. $-x^2 + 2x = 0$

$x = 0, 2$

d. $x^2 - 2x + 1 = 0$

$x = 1$

What Is Your Answer?

3. **IN YOUR OWN WORDS** How can you use a graph to solve a quadratic equation in one variable?

Get the equation equal to zero, set it equal to y, graph the resulting equation, and find its x-intercepts.

4. After you find a solution graphically, how can you check your result algebraically? Use your solutions in Activity 2 as examples.

Check that the x-values of the x-intercepts satisfy the original equation.

Laurie's Notes

Activity 2

- Explain to students that in Activity 2 they will solve four additional quadratic equations in one variable.
- As they are solving each equation by graphing, students should be checking the reasonableness of their graphs. For instance, the graph in part (c) is a parabola that opens down.
- While students work, probe different pairs of students about the vertex, the minimum or maximum value, the y-intercept, and the general shape of the graph.

? "In Activity 1, the equation had two solutions. Does each equation in Activity 2 have two solutions?" no; The equation in part (d) has only 1 solution.

? "Do you think it is possible to predict how many solutions a quadratic equation in one variable will have?" Answers will vary.

What Is Your Answer?

- **MP1a:** In Question 4, students are making sense of a problem in more than one way. It is important for students to realize that the x-intercepts correspond to the solutions of the equation, so substituting these values for x should result in a true equation.

Closure

- **Writing Prompt:** To solve a quadratic equation by graphing you . . .

Technology For the Teacher

The Dynamic Planning Tool
Editable Teacher's Resources at *BigIdeasMath.com*

2 ACTIVITY: Solving Quadratic Equations by Graphing

Work with a partner. Solve each equation by graphing.

a. $x^2 - 4 = 0$

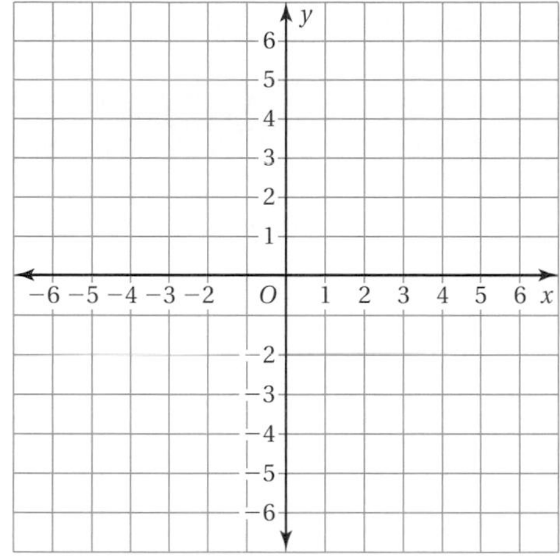

b. $x^2 + 3x = 0$

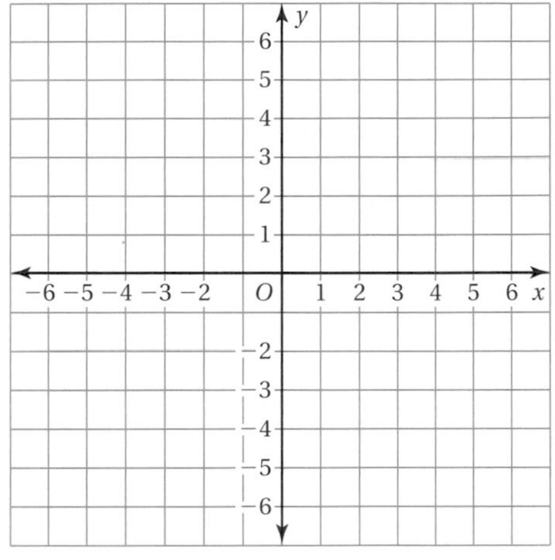

c. $-x^2 + 2x = 0$

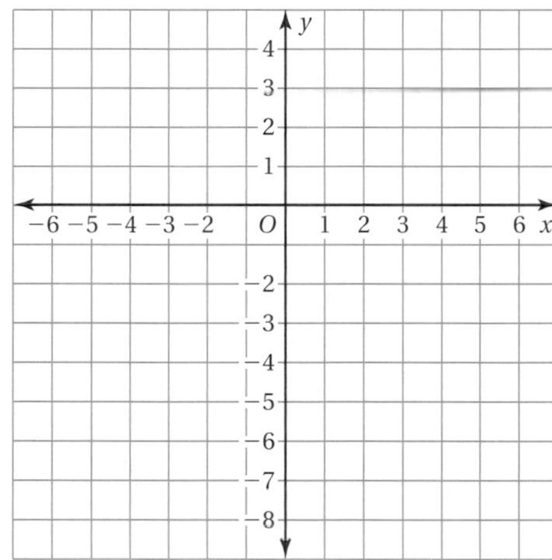

d. $x^2 - 2x + 1 = 0$

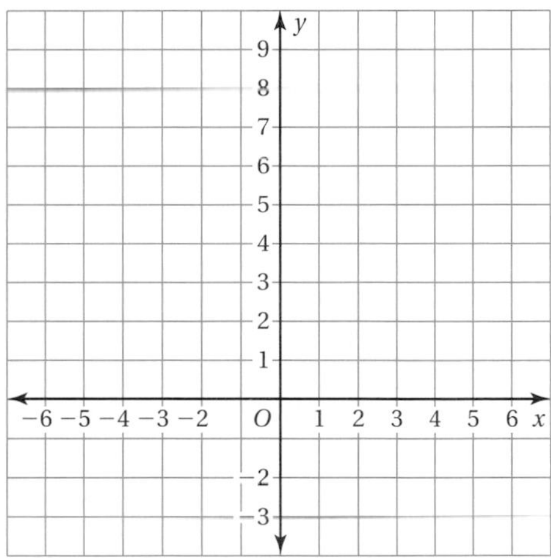

What Is Your Answer?

3. **IN YOUR OWN WORDS** How can you use a graph to solve a quadratic equation in one variable?

4. After you find a solution graphically, how can you check your result algebraically? Use your solutions in Activity 2 as examples.

Practice Use what you learned about solving quadratic equations to complete Exercises 5–7 on page 459.

9.1 Lesson

Key Vocabulary
quadratic equation, *p. 456*

A **quadratic equation** is a nonlinear equation that can be written in the standard form $ax^2 + bx + c = 0$, where $a \neq 0$.

In Chapter 7, you solved quadratic equations by factoring. You can also solve quadratic equations in standard form by finding the x-intercept(s) of the graph of the related function $y = ax^2 + bx + c$.

EXAMPLE 1 Solving a Quadratic Equation: Two Real Solutions

Remember

The solutions of a quadratic equation are also called roots.

Solve $x^2 + 2x - 3 = 0$ by graphing.

Step 1: Graph the related function $y = x^2 + 2x - 3$.

Step 2: Find the x-intercepts. They are -3 and 1.

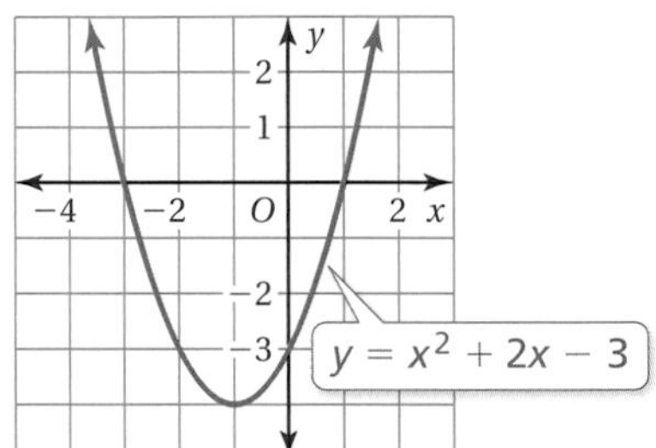

So, the solutions are $x = -3$ and $x = 1$.

Check Check each solution in the original equation.

$x^2 + 2x - 3 = 0$	Original equation	$x^2 + 2x - 3 = 0$
$(-3)^2 + 2(-3) - 3 \stackrel{?}{=} 0$	Substitute.	$1^2 + 2(1) - 3 \stackrel{?}{=} 0$
$0 = 0$ ✓	Simplify.	$0 = 0$ ✓

EXAMPLE 2 Solving a Quadratic Equation: One Real Solution

Study Tip

You can also solve the equation in Example 2 by factoring.

$x^2 - 8x + 16 = 0$

$(x - 4)(x - 4) = 0$

So, $x = 4$.

Solve $x^2 - 8x = -16$ by graphing.

Step 1: Rewrite the equation in standard form.

$x^2 - 8x = -16$ Write the equation.

$x^2 - 8x + 16 = 0$ Add 16 to each side.

Step 2: Graph the related function $y = x^2 - 8x + 16$.

Step 3: Find the x-intercept. The only x-intercept is at the vertex $(4, 0)$.

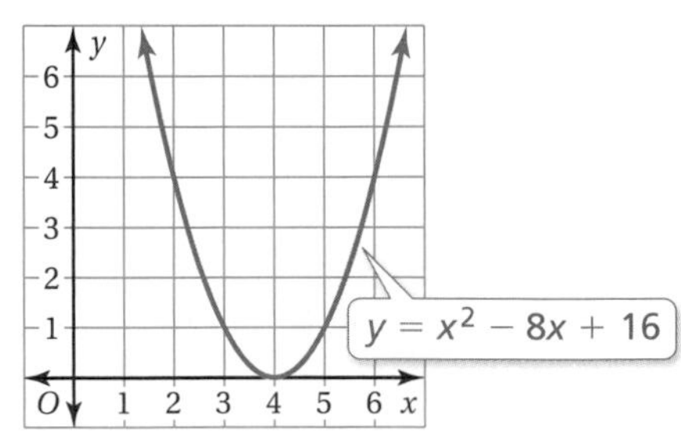

So, the solution is $x = 4$.

On Your Own

Now You're Ready
Exercises 8–10

Solve the equation by graphing. Check your solution(s).

1. $x^2 - x - 2 = 0$
2. $x^2 + 7x + 10 = 0$
3. $x^2 + x = 12$
4. $x^2 + 1 = 2x$
5. $x^2 + 4x = 0$
6. $x^2 + 10x = -25$

Multi-Language Glossary at BigIdeasMath.com.

Laurie's Notes

Introduction

Connect

- **Yesterday:** Students developed an understanding of solving quadratic equations by graphing. (MP1a)
- **Today:** Students will solve quadratic equations by graphing.

Motivate

? "What do water fountains and a kicked football have in common?" Answers will vary.
- This may prompt a lot of different responses. Listen for responses such as their paths have the same shape or they have a maximum height.

? "What questions can you answer using the equation that relates the time the football has been in the air and its height?" Answers will vary.
- Mention that today students will use a quadratic equation to determine when the football is at a specific height.

Lesson Notes

Discuss

- **MP1a Make Sense of Problems:** Throughout this lesson, discuss with students the different graphing methods they can use to solve these types of equations as well as the different checks they can use. They can check by factoring, using a graphing utility, or substituting solutions back into the original equation.
- Write the definition of the standard form of a quadratic equation.

? "How can you solve a quadratic equation such as $x^2 - 5x + 4 = 0$?" Factor the left side as $(x - 4)(x - 1) = 0$. Then solve for x.

? "Do you think all quadratic equations can be factored?" no
- Explain that today students will solve quadratic equations by writing the equation in standard form and graphing the related function $y = ax^2 + bx + c$.

Example 1

- Use an input-output table to graph the related function. You might suggest a domain from −3 to 3.

? "What values of x give an output of 0?" −3 and 1

? **MP6 Attend to Precision:** "Why is it necessary to check the solutions algebraically?" You could have made an error in graphing the equation.
- If time permits, asks students about the factored form of $x^2 + 2x - 3$.
- Remind students that the solutions of a quadratic equation are also called roots.

Example 2

? "How does $x^2 - 8x = -16$ differ from the equation in Example 1?" It is not written in standard form.
- **MP6:** "Add 16 to each side" is another way of saying "use the Addition Property of Equality."

Goal Today's lesson is solving **quadratic equations** by graphing.

Lesson Materials
Textbook
• graphing calculators

Start Thinking! and Warm Up

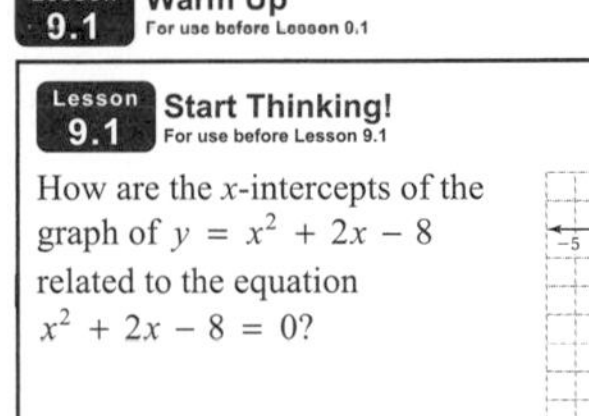
Lesson 9.1 Warm Up For use before Lesson 9.1

Lesson 9.1 Start Thinking! For use before Lesson 9.1

How are the x-intercepts of the graph of $y = x^2 + 2x - 8$ related to the equation $x^2 + 2x - 8 = 0$?

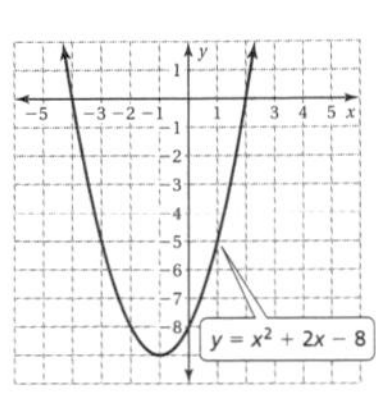

Extra Example 1

Solve $x^2 - x - 6 = 0$ by graphing.
$x = -2, x = 3$

Extra Example 2

Solve $x^2 + 9x = -20$ by graphing.
$x = -4, x = -5$

On Your Own

1. $x = -1, x = 2$
2. $x = -5, x = -2$
3. $x = -4, x = 3$
4. $x = 1$
5. $x = -4, x = 0$
6. $x = -5$

Laurie's Notes

Example 2 (continued)

- In graphing the related function, note that a domain of −3 to 3 (like in Example 1) will not include the x-intercept. Students should understand that when $x = 0$, $y = 16$ and when $x = 1$, $y = 9$. The positive leading coefficient means that the graph opens up, so the vertex must be at an x-value greater than 0. There is no need to evaluate the equation for negative values of x.
- The solution can be checked by factoring as noted in the Study Tip.

On Your Own

- If time is a concern, have students do only the even or odd exercises.

Extra Example 3

Solve $x^2 = -3x - 4$ by graphing.
no real solutions

On Your Own

7. no real solutions

8. $x = -6, x = -1$

9. no real solutions

Example 3

- Take time to work through both methods. Showing two pathways for approaching the problem helps deepen students' understanding of the problem.
- Because there are no x-intercepts when there are no solutions, it is not possible to perform a check by substituting x-values in the original equation.
- **Connection:** The second method connects to solving a system of equations (Chapter 4). In this method, each side of the equation is treated as a function and graphed.

On Your Own

- Remind students to check the reasonableness of their solution(s).

Discuss

? "What are the possible numbers of points of intersection for the graph of a linear function and the graph of a quadratic function?" 0, 1, or 2

- **Connection:** Be sure to point out the connection between this question and the Summary box.
- Note how the number of solutions is related to the position of the vertex relative to the x-axis and the direction in which the parabola opens.

? "What can you say about the vertex of a parabola that opens down when the corresponding equation has two solutions?" The vertex must be above the x-axis.

EXAMPLE 3 Solving a Quadratic Equation: No Real Solutions

Solve $-x^2 = 4x + 5$ by graphing.

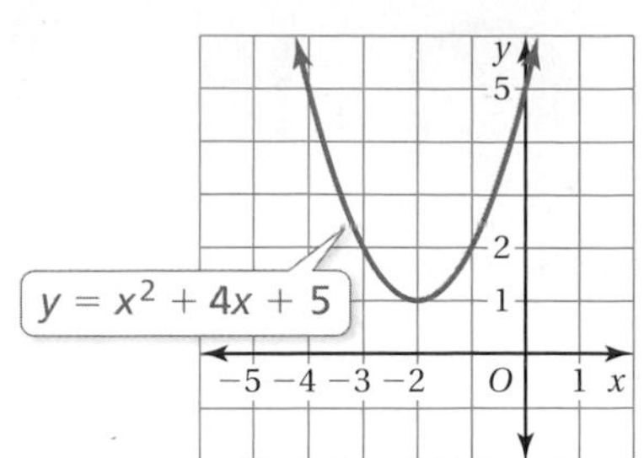

Method 1: Rewrite the equation in standard form and graph the related function $y = x^2 + 4x + 5$.

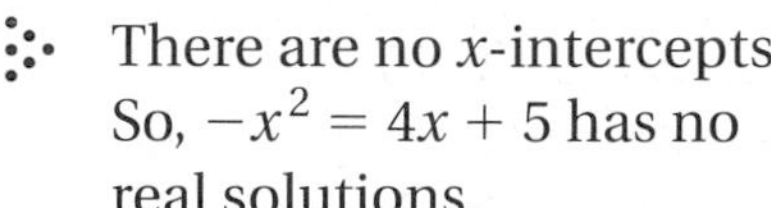

There are no x-intercepts. So, $-x^2 = 4x + 5$ has no real solutions.

Method 2: Graph each side of the equation.

$y = -x^2$ Left side

$y = 4x + 5$ Right side

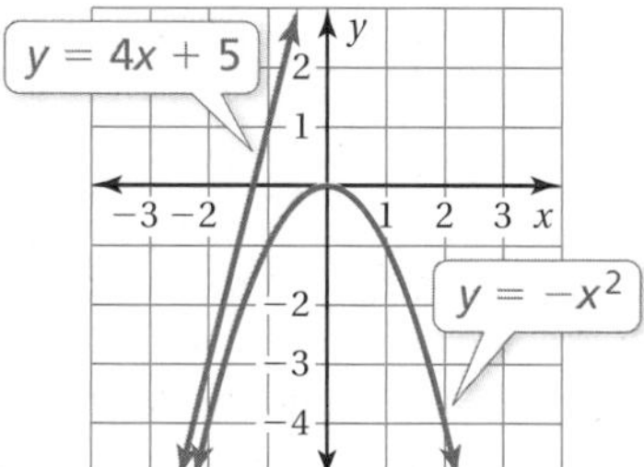

The graphs do not intersect. So, $-x^2 = 4x + 5$ has no real solutions.

On Your Own

Now You're Ready
Exercises 11–16

Solve the equation by graphing. Check your solution(s).

7. $x^2 = 3x - 3$ **8.** $x^2 + 7x = -6$ **9.** $2x + 5 = -x^2$

Summary

Quadratic equations may have two real solutions, one real solution, or no real solutions.

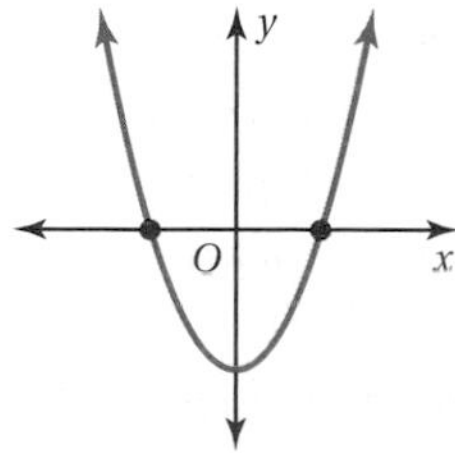

- two real solutions
- two x-intercepts

- one real solution
- one x-intercept

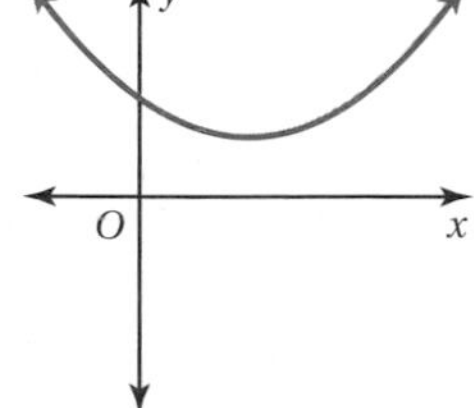

- no real solutions
- no x-intercepts

EXAMPLE 4 Real-Life Application

A football player kicks a football 2 feet above the ground with an upward velocity of 75 feet per second. The function $h = -16t^2 + 75t + 2$ gives the height h (in feet) of the football after t seconds. After how many seconds is the football 50 feet above the ground?

To determine when the football is 50 feet above the ground, find the t-values for which $h = 50$. So, solve the equation $-16t^2 + 75t + 2 = 50$.

Step 1: Rewrite the equation in standard form.

$-16t^2 + 75t + 2 = 50$ Write the equation.

$-16t^2 + 75t - 48 = 0$ Subtract 50 from each side.

Step 2: Use a graphing calculator to graph the related function $h = -16t^2 + 75t - 48$.

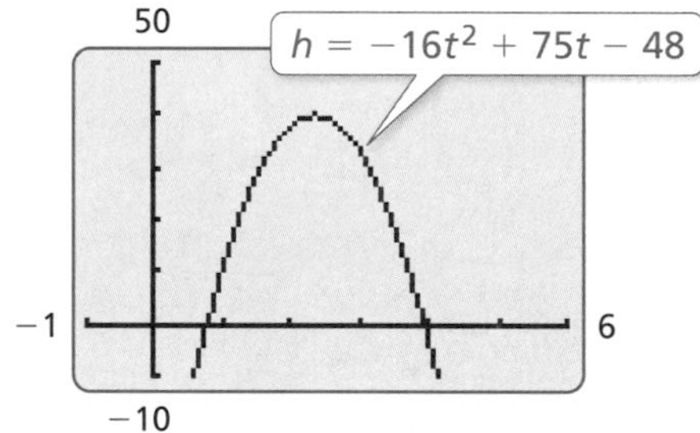

Step 3: Use the *zero* feature to find the zeros of the function.

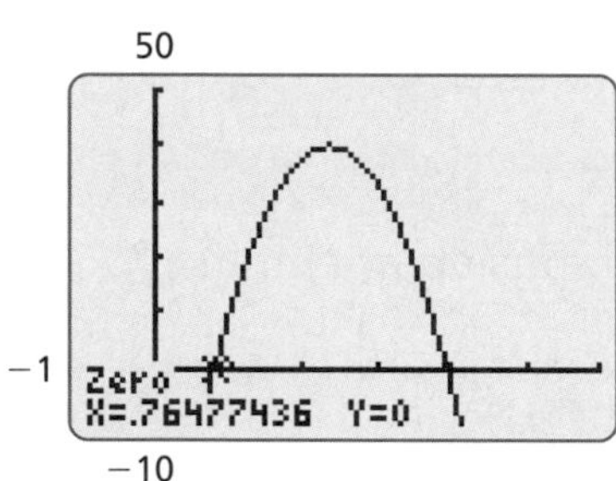

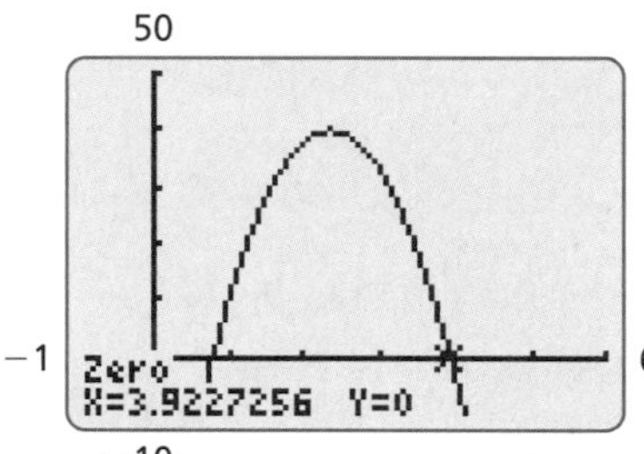

∴ The football is 50 feet above the ground after about 0.8 second and about 3.9 seconds.

Remember

A zero of a function $y = f(x)$ is an x-value for which the value of the function is zero.

On Your Own

Exercise 18

10. **WHAT IF?** After how many seconds is the football 65 feet above the ground?

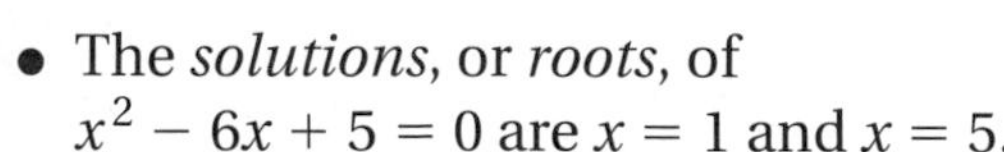

Summary

- The *solutions*, or *roots*, of $x^2 - 6x + 5 = 0$ are $x = 1$ and $x = 5$.
- The *x-intercepts* of the graph of $y = x^2 - 6x + 5$ are 1 and 5.
- The *zeros* of the function $f(x) = x^2 - 6x + 5$ are 1 and 5.

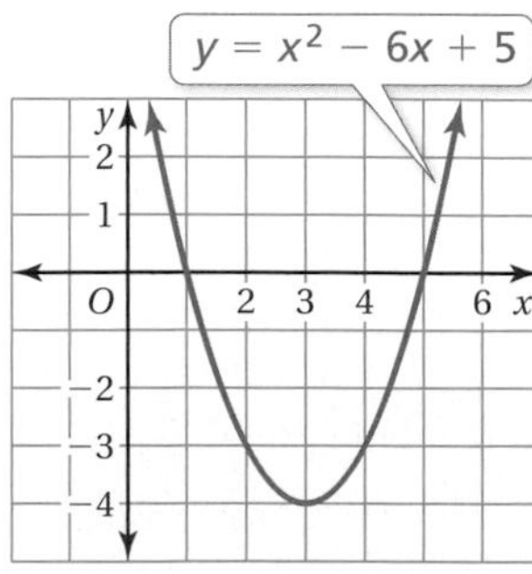

Laurie's Notes

Example 4

- **MP4 Model with Mathematics:** The position function $h = -16t^2 + v_0t + s_0$ represents the height h (in feet) of an object where v_0 is the initial upward velocity (in feet per second), s_0 is the initial height (in feet) of the object, and t is the time (in seconds). In this example, the initial upward velocity is 75 feet per second and the initial height is 2 feet.
- **MP2 Reason Abstractly and Quantitatively:** Students must reason abstractly and quantitatively to relate the model and its graph to the height of the football.
- Students are accustomed to seeing functions with ordered pairs (x, y). Remind students that each ordered pair (t, h) represents the height h of the object at time t.
- **Misconception:** The graph shows the height of the object over time, not the path of the object.
- Discuss with students what the two zeros actually represent and why the football is at the same height at two different times.
- **Connection:** Some students may ask if this problem can be solved using Method 2 in Example 3. The answer is yes. Graph each side of the equation where $y = -16t^2 + 75t + 2$ represents the left side and $y = 50$ represents the right side. The solutions are the x-values of the points of intersection of the graphs.
- This is a significant problem where many connections can be made. Do not overwhelm students with all of them until they are ready.
- **MP6:** Note that the actual zeros have been rounded to the nearest tenth of a second. This degree of precision is sufficient in this context.
- Take time to discuss the language being highlighted in the last Summary box.

Closure

- When a quadratic equation has two solutions, what do you know about the graph of its related function? It has two x-intercepts.

Technology For the Teacher

The Dynamic Planning Tool
Editable Teacher's Resources at *BigIdeasMath.com*

Extra Example 4

The function $h = -16t^2 + 32t$ gives the height h of a soccer ball t seconds after it is kicked. After how many seconds is the soccer ball 15 feet above the ground?
0.75 second and 1.25 seconds

On Your Own

10. about 1.1 seconds and about 3.6 seconds

English Language Learners

Vocabulary

To help your students understand the terminology used in this section, have them write the quadratic equation
$(x + 20)(x - 10) = 0$
in standard form and then write the related function of the form
$f(x) = ax^2 + bx + c$.
Then write the following on the board. After a volunteer reads each statement, have students write it in their notebooks.

- solutions, or roots, of $(x + 20)(x - 10) = 0$: $x = -20$, $x = 10$
- x-intercepts of the graph of f: -20 and 10
- zeros of f: -20 and 10

Vocabulary and Concept Check

1. It is a nonlinear equation that can be written in the form $ax^2 + bx + c = 0$ where $a \neq 0$.
2. $x^2 + x - 4 = 0$; It is the only equation in standard form.
3. Use the graph to find the x-intercepts.
4. The roots, or solutions, of an equation are the same as the zeros of the related function or the x-intercepts of its graph.

Practice and Problem Solving

5. $x = 4, x = 6$
6. no real solutions
7. $x = -6$
8. $x = 0, x = 4$
9. $x = 3$
10. $x = -1, x = 7$
11. no real solutions
12. $x = -2, x = 1$
13. $x = -2$
14. $x = -5, x = 3$
15. $x = 7$
16. no real solutions
17. **a.** The x-intercepts give the horizontal positions of the ball where it is struck and where it lands.

    b. 5 yards
18. about 0.6 second and about 1.3 seconds

Assignment Guide and Homework Check

Level	Assignment	Homework Check
Average	1–4, 5–29 odd, 32, 33, 46–50	5, 13, 21, 25, 33
Advanced	1–4, 14–34 even, 41–50	16, 26, 34, 41, 43

Common Errors

- **Exercise 4** Students may think these are all identical terms. Remind them of the Summary box that addresses these terms.
- **Exercise 6** Students may try using the y-intercept as a solution. Remind students that the solutions are given by the x-intercepts. When there are no x-intercepts, there are no solutions.
- **Exercises 8–16** Solutions obtained graphically may be incorrect or not exact. Make sure students use a sound graphical approach. Also, remind them to check their answers.
- **Exercise 18** Students may give only one solution. There are two solutions representing the time when the ball is 16 feet above the ground.

9.1 Record and Practice Journal

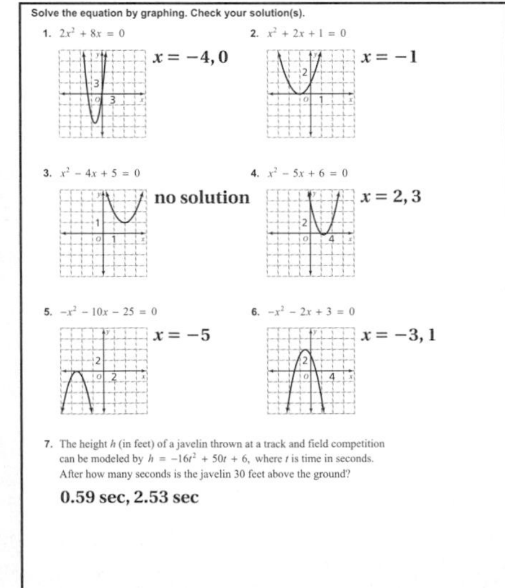

Solve the equation by graphing. Check your solution(s).

1. $2x^2 + 8x = 0$ $x = -4, 0$
2. $x^2 + 2x + 1 = 0$ $x = -1$
3. $x^2 - 4x + 5 = 0$ no solution
4. $x^2 - 5x + 6 = 0$ $x = 2, 3$
5. $-x^2 - 10x - 25 = 0$ $x = -5$
6. $-x^2 - 2x + 3 = 0$ $x = -3, 1$
7. The height h (in feet) of a javelin thrown at a track and field competition can be modeled by $h = -16t^2 + 50t + 6$, where t is time in seconds. After how many seconds is the javelin 30 feet above the ground?

   0.59 sec, 2.53 sec

Technology For the Teacher
Answer Presentation Tool

9.1 Exercises

Vocabulary and Concept Check

1. **VOCABULARY** What is a quadratic equation?

2. **WHICH ONE DOESN'T BELONG?** Which equation does *not* belong with the other three? Explain your reasoning.

   $x^2 + 5x = 20$ | $x^2 + x - 4 = 0$ | $x^2 - 6 = 4x$ | $7x + 12 = x^2$

3. **WRITING** How can you use a graph to find the number of solutions of a quadratic equation?

4. **WRITING** How are solutions, roots, x-intercepts, and zeros related?

Practice and Problem Solving

Determine the solution(s) of the equation. Check your solution(s).

5. $x^2 - 10x + 24 = 0$

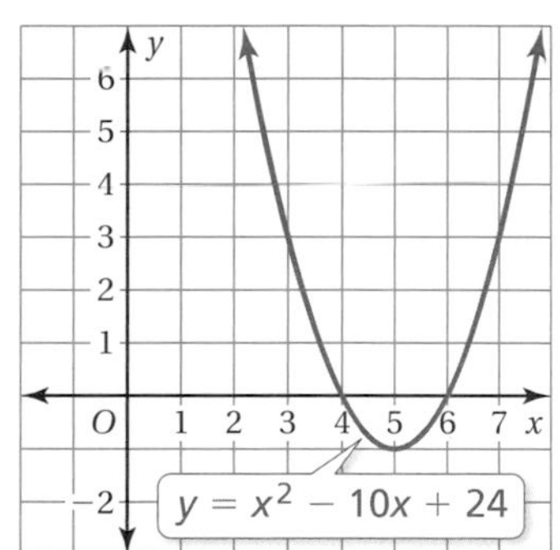

6. $-x^2 - 4x - 6 = 0$

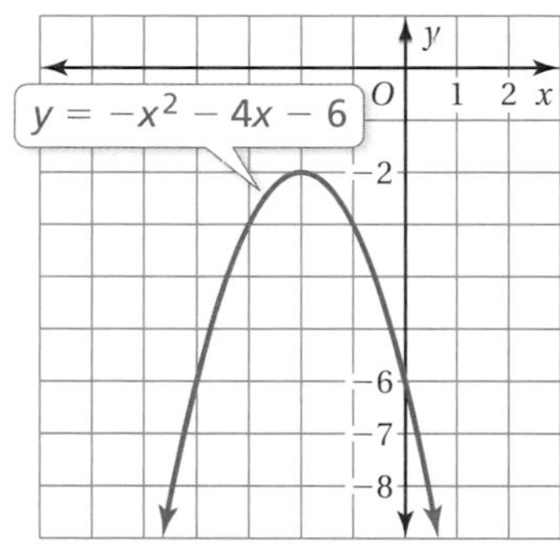

7. $x^2 + 12x + 36 = 0$

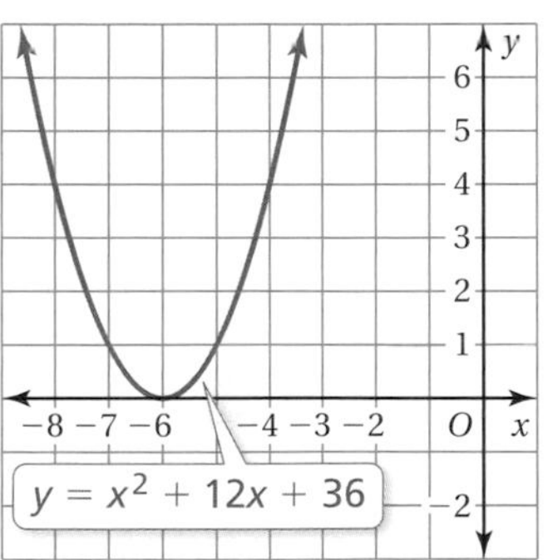

Solve the equation by graphing. Check your solution(s).

1 2 8. $x^2 - 4x = 0$

9. $x^2 - 6x + 9 = 0$

10. $x^2 - 6x - 7 = 0$

3 11. $x^2 - 2x + 5 = 0$

12. $x^2 + x - 2 = 0$

13. $x^2 + 4x + 4 = 0$

14. $-x^2 - 2x + 15 = 0$

15. $-x^2 + 14x - 49 = 0$

16. $-x^2 + 4x - 7 = 0$

17. **FLOP SHOT** The height y (in yards) of a flop shot in golf can be modeled by $y = -x^2 + 5x$, where x is the horizontal distance (in yards).

    a. Interpret the x-intercepts of the graph of the equation.

    b. How far away does the golf ball land?

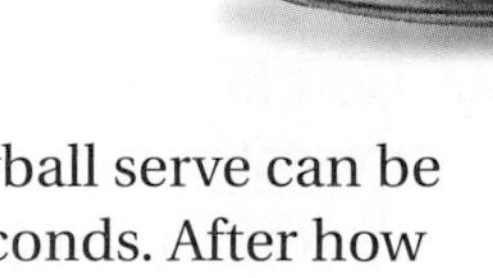

4 18. **VOLLEYBALL** The height h (in feet) of an underhand volleyball serve can be modeled by $h = -16t^2 + 30t + 4$, where t is the time in seconds. After how many seconds is the ball 16 feet above the ground?

Rewrite the equation in standard form. Then solve the equation by graphing. Check your solution(s) with a graphing calculator.

19. $x^2 = 6x - 8$ **20.** $x^2 = -1 - 2x$ **21.** $x^2 = -x - 3$

22. $x^2 = 2x - 4$ **23.** $5x - 6 = x^2$ **24.** $3x - 18 = -x^2$

Solve the equation by using Method 2 from Example 3. Check your solution(s).

25. $x^2 = 10 - 3x$ **26.** $4 - 4x = -x^2$ **27.** $5x - 7 = x^2$

28. $x^2 = 6x - 10$ **29.** $x^2 = -2x - 1$ **30.** $x^2 - 8x = 9$

31. REASONING Example 3 shows two methods for solving a quadratic equation. Which method do you prefer? Explain your reasoning.

32. ERROR ANALYSIS Describe and correct the error in solving the equation.

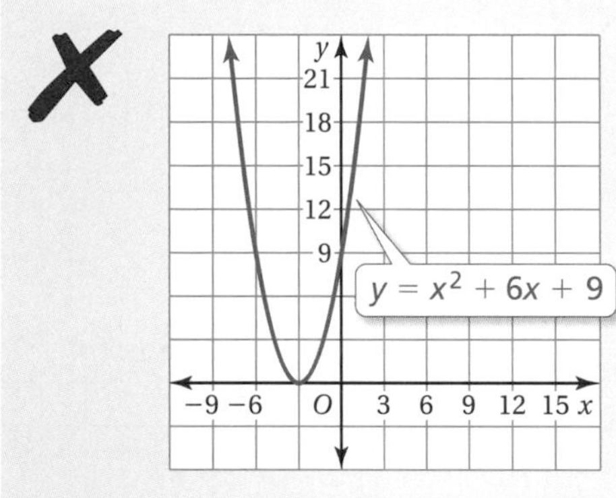

The only solution of the equation $x^2 + 6x + 9 = 0$ is $x = 9$.

33. BASEBALL A baseball player throws a baseball with an upward velocity of 24 feet per second. The release point is 6 feet above the ground. The function $h = -16t^2 + 24t + 6$ gives the height h (in feet) of the baseball after t seconds.

a. How long is the ball in the air if no one catches it?

b. How long does the ball remain above 6 feet?

34. SOFTBALL You throw a softball straight up into the air with an upward velocity of 40 feet per second. The release point is 5 feet above the ground. The function $h = -16t^2 + 40t + 5$ gives the height h (in feet) of the softball after t seconds.

a. How long is the ball in the air if you miss it?

b. How long is the ball in the air if you catch it at a height of 5 feet?

Common Errors

- **Exercises 19–24** Students may make sign errors when rewriting the equation in standard form. Remind them that they need to use the Addition or Subtraction Property of Equality to add or subtract the same quantity on each side.
- **Exercises 25–30** Students may give solutions in terms of x- and y-coordinates. Tell them that the original equation was only in x. When Method 2 is used, the solutions are the x-values of the points of intersection.
- **Exercises 35–40** Students may round their answers incorrectly. Make sure they are using the features of the graphing utility correctly and that they understand how to find solutions rounded to the nearest tenth.
- **Exercise 41** Students may have trouble setting up the equation in part (a). Explain the coefficients of the function for the height of a projectile.
- **Exercise 42** Students may not know where to start. Remind them to first write a function for the height of the keg using the initial velocity and initial height. Then write an equation to find when the keg reaches a height of 16.5 feet. When they find that there is no solution, they should realize that the keg does not clear the wall.

Practice and Problem Solving

19. $x^2 - 6x + 8 = 0; x = 2, x = 4$

20. $x^2 + 2x + 1 = 0; x = -1$

21. $x^2 + x + 3 = 0;$ no real solutions

22. $x^2 - 2x + 4 = 0;$ no real solutions

23. $x^2 - 5x + 6 = 0;$ $x = 2, x = 3$

24. $x^2 + 3x - 18 = 0;$ $x = -6, x = 3$

25. $x = -5, x = 2$

26. $x = 2$

27. no real solutions

28. no real solutions

29. $x = -1$

30. $x = -1, x = 9$

31. *Sample answer:* Method 2; You do not have to rewrite the equation.

32. The y-intercept was used instead of the x-intercept. The correct answer is $x = -3$.

33. **a.** about 1.7 seconds

b. 1.5 seconds

34. **a.** about 2.6 seconds

b. 2.5 seconds

English Language Learners

Class Activity

Form groups of 2 to 4 students with at least one English language learner and one English speaker. Select an exercise for each group and have them work together to create a poster for the exercise to present to the class. This will allow students to discuss concepts in small groups and create a visual display to aid their understanding.

Practice and Problem Solving

35. $x \approx -5.8, x \approx -0.2$
36. $x \approx -0.6, x \approx 3.6$
37. $x \approx -5.7, x \approx 0.7$
38. $x \approx -3.4, x \approx 1.4$
39. $x \approx 0.6, x \approx 3.4$
40. $x \approx 0.7, x \approx 8.3$
41. a. $h = -16t^2 + 20t + 8$
    b. about 1.6 seconds
42. See *Taking Math Deeper*.
43. sometimes; There are 2 x-intercepts when c is positive.
44. always; In each case, the parabola opens away from the x-axis.
45. never; A quadratic equation can never have more than 2 solutions.

Fair Game Review

46. 24
47. 81
48. $6\sqrt{3}$
49. $25\sqrt{2}$
50. D

Mini-Assessment

Solve the equation by graphing. Check your solution(s).

1. $x^2 + 4x + 5 = 0$ no real solutions
2. $x^2 + 7x - 8 = 0$ $x = -8, x = 1$
3. $x^2 = 10x - 25$ $x = 5$
4. $x^2 + 3x = 18$ $x = -6, x = 3$

Taking Math Deeper

Exercise 42

This exercise is good practice for using different methods to solve a problem.

Write the function that gives the height of the keg. Then write a system of equations that can be used to solve the problem.

The keg is released at a height of 5 feet with an upward velocity of 27 feet per second. So, the function is $y = -16t^2 + 27t + 5$. To determine if the keg reaches a height of 16.5 feet, use the following system.

$y = 16.5$
$y = -16t^2 + 27t + 5$

Remember 6 inches is one-half of a foot.

Graph the system. Depending on the viewing window, the graphs may look like they intersect. When a proper viewing window is used, you can see that they do not intersect.

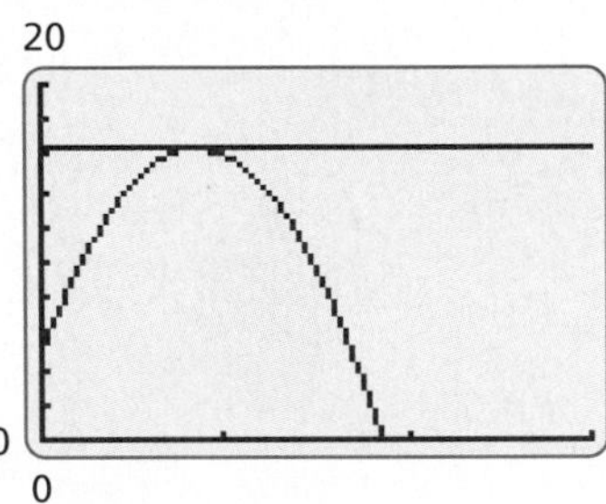

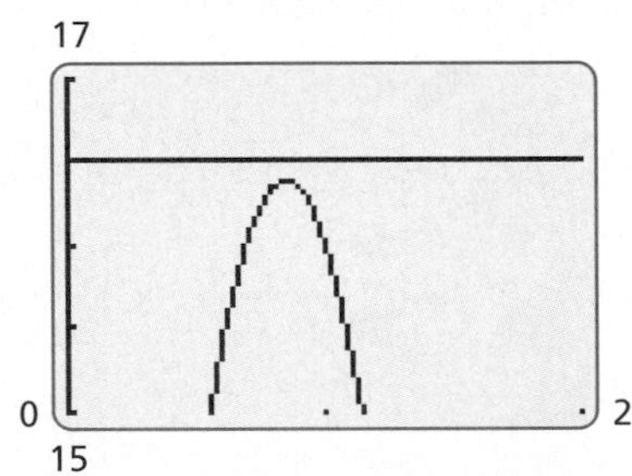

Answer the questions.

a. Because the graphs do not intersect, the throw is not high enough to clear the wall.

b. Yes, a taller competitor may release the keg at a greater height. This changes the value for the initial height of the function. Students can verify this by graphing $y = -16t^2 + 27t + 5.5$ instead of $y = -16t^2 + 27t + 5$.

Take a closer look.

Note: You may want to have students solve this problem using the *maximum* feature of their graphing calculators.

Reteaching and Enrichment Strategies

If students need help. . .	If students got it. . .
Resources by Chapter • Practice A and Practice B • Puzzle Time Record and Practice Journal Practice Differentiating the Lesson Lesson Tutorials Skills Review Handbook	Resources by Chapter • Enrichment and Extension Start the next section

Use a graphing calculator to approximate the zeros of the function to the nearest tenth.

35. $f(x) = x^2 + 6x + 1$ **36.** $f(x) = x^2 - 3x - 2$ **37.** $f(x) = x^2 + 5x - 4$

38. $f(x) = -x^2 - 2x + 5$ **39.** $f(x) = -x^2 + 4x - 2$ **40.** $f(x) = -x^2 + 9x - 6$

41. MODELING A dirt bike launches off a ramp that is 8 feet tall. The upward velocity of the dirt bike is 20 feet per second.

a. Write a function that models the height h (in feet) of the dirt bike after t seconds.

b. After how many seconds does the dirt bike land?

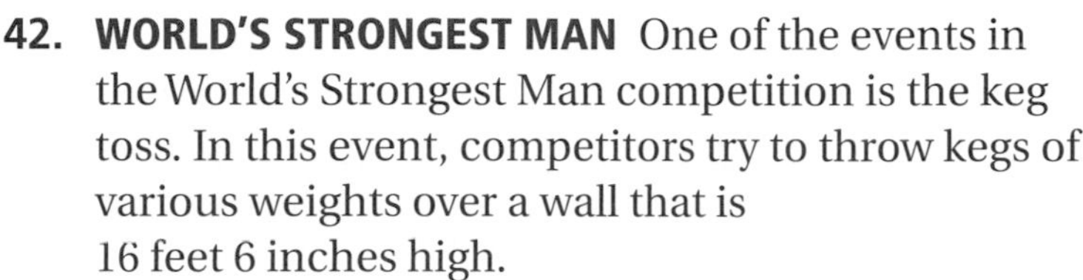

42. WORLD'S STRONGEST MAN One of the events in the World's Strongest Man competition is the keg toss. In this event, competitors try to throw kegs of various weights over a wall that is 16 feet 6 inches high.

a. A competitor releases a keg 5 feet above the ground with an upward velocity of 27 feet per second. Is this throw high enough to clear the wall? Explain your reasoning.

b. Do the heights of the competitors factor into their success at this event? Explain your reasoning.

Reasoning **Determine whether the statement is *sometimes*, *always*, or *never* true. Justify your answer.**

43. The graph of $y = ax^2 + c$ has two x-intercepts when $a = -2$.

44. The graph of $y = ax^2 + c$ has no x-intercepts when a and c have the same sign.

45. The graph of $y = ax^2 + bx + c$ has more than two zeros when $a \neq 0$.

Fair Game Review What you learned in previous grades & lessons

Simplify the expression. *(Section 6.1)*

46. $4\sqrt{36}$ **47.** $9\sqrt{81}$ **48.** $2\sqrt{27}$ **49.** $5\sqrt{50}$

50. MULTIPLE CHOICE Which expression is equivalent to $\left(\frac{2x^3}{3m^5}\right)^2$? *(Section 6.2)*

Ⓐ $\frac{2x^5}{3m^7}$ Ⓑ $\frac{2x^6}{3m^{10}}$ Ⓒ $\frac{4x^5}{9m^7}$ Ⓓ $\frac{4x^6}{9m^{10}}$

9.2 Solving Quadratic Equations Using Square Roots

COMMON CORE STATE STANDARDS

A.REI.4b

Essential Question How can you determine the number of solutions of a quadratic equation of the form $ax^2 + c = 0$?

1 ACTIVITY: The Number of Solutions of $ax^2 + c = 0$

Work with a partner. Solve each equation by graphing. Explain how the number of solutions of

$$ax^2 + c = 0 \quad \text{Quadratic equation}$$

relates to the graph of

$$y = ax^2 + c. \quad \text{Quadratic function}$$

a. $x^2 - 4 = 0$

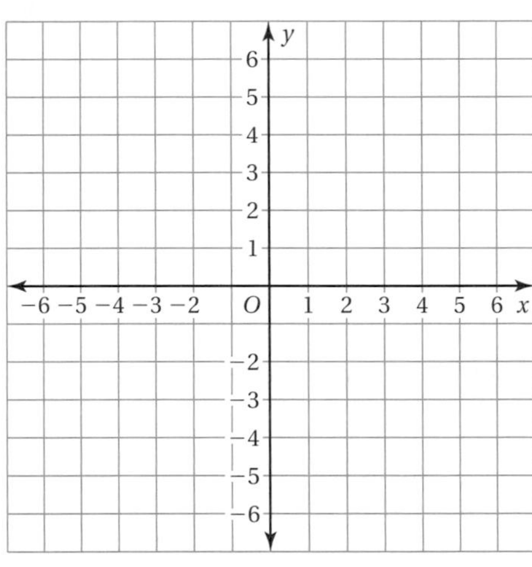

b. $2x^2 + 5 = 0$

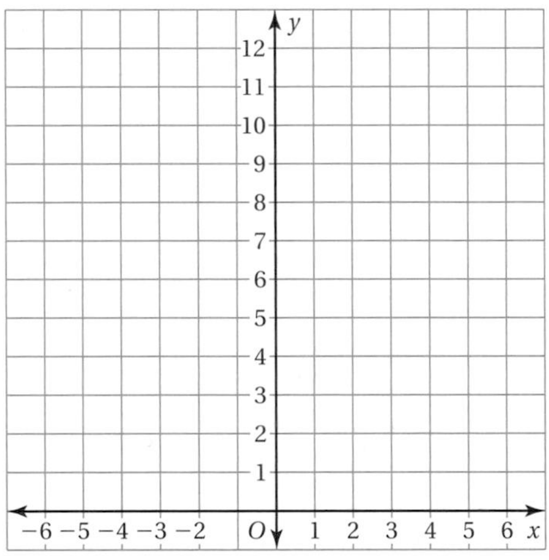

c. $x^2 = 0$

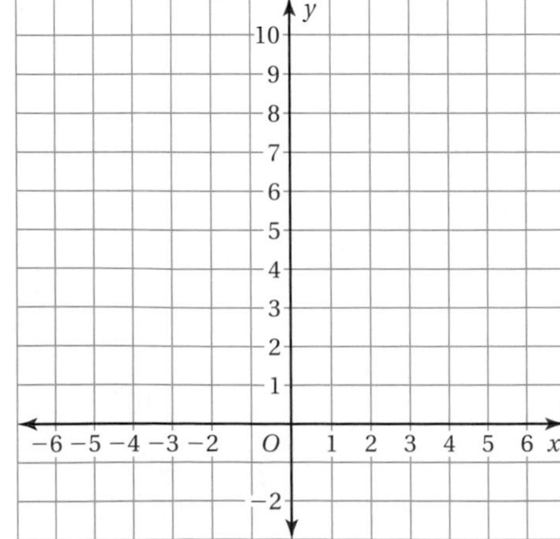

d. $x^2 - 5 = 0$

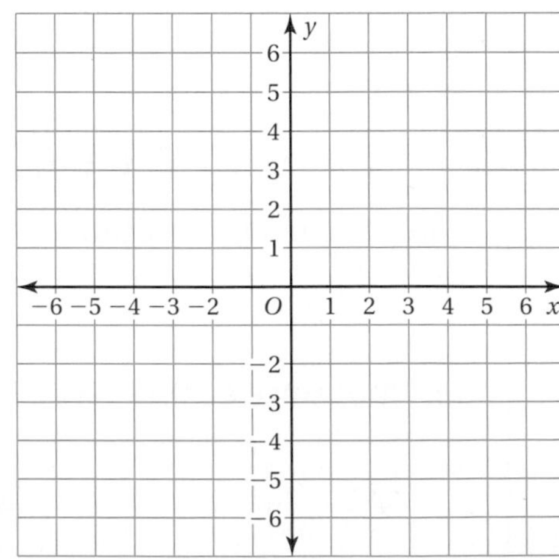

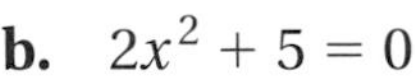

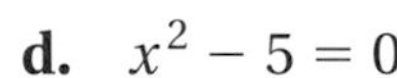

Laurie's Notes

Introduction

Standards for Mathematical Practice

- **MP3a Construct Viable Arguments** and **MP8 Look for and Express Regularity in Repeated Reasoning:** Students are asked to make a conjecture about the number of solutions of quadratic equations after graphing the related functions.

Motivate

- All of today's graphs will be symmetric about the *y*-axis.
- Draw two collections of shapes. In the first, each shape has one line of symmetry. In the second, each shape has no symmetry or more than one line of symmetry.

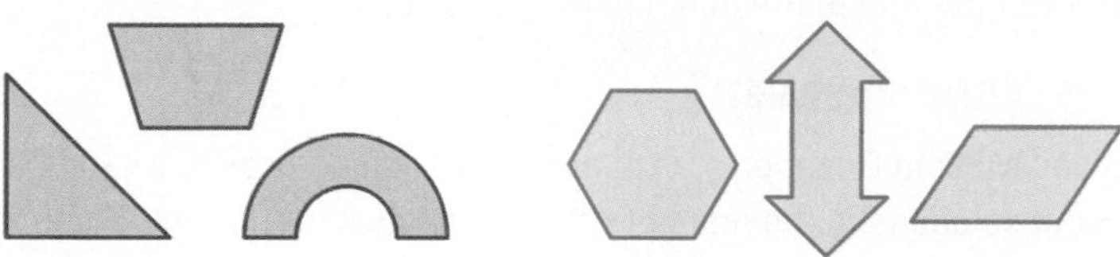

- Ask students to figure out how the groups have been sorted.

? "In which group would you place a parabola?" group on left

Activity Notes

Discuss

- Today you want students to think about the function $y = x^2$ (1 *x*-intercept) and the effect of adding a constant *c* that shifts the graph up (no *x*-intercepts) or down (2 *x*-intercepts).

Activity 1

? "What do you know about the graph of $y = x^2$?" It is a parabola with vertex (0, 0) that opens up.

- Students can make a table to graph the function. Make sure they choose a domain that displays the key features of the graph.
- Refer to the related function in part (c), $y = x^2$, as the *parent function* or *basic quadratic function.*
- **MP3a:** Conjectures will likely relate the number of solutions to the number of *x*-intercepts of the graph.
- Students might go further and say that when *c* is positive, there are no solutions, and when *c* is negative, there are two solutions. Note however that this is not true when $a < 0$.

Common Core State Standards

A.REI.4b Solve quadratic equations . . . by taking square roots . . ..

Previous Learning

Students should know how to solve quadratic equations by graphing. They should also know how to find the square roots of a number.

Activity Materials
Textbook
• calculators

Start Thinking! and Warm Up

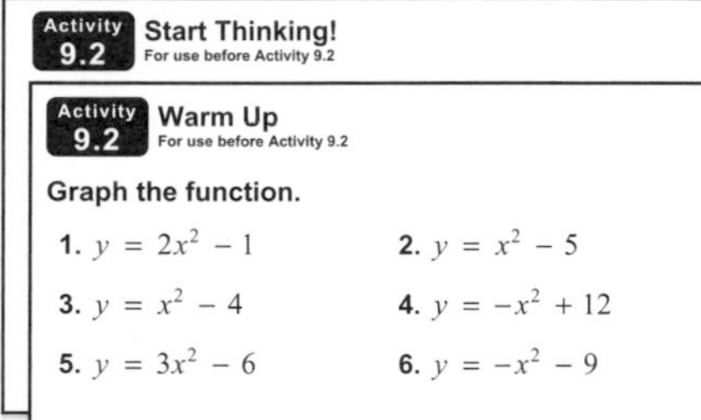

Activity 9.2 Start Thinking!
For use before Activity 9.2

Activity 9.2 Warm Up
For use before Activity 9.2

Graph the function.

1. $y = 2x^2 - 1$
2. $y = x^2 - 5$
3. $y = x^2 - 4$
4. $y = -x^2 + 12$
5. $y = 3x^2 - 6$
6. $y = -x^2 - 9$

9.2 Record and Practice Journal

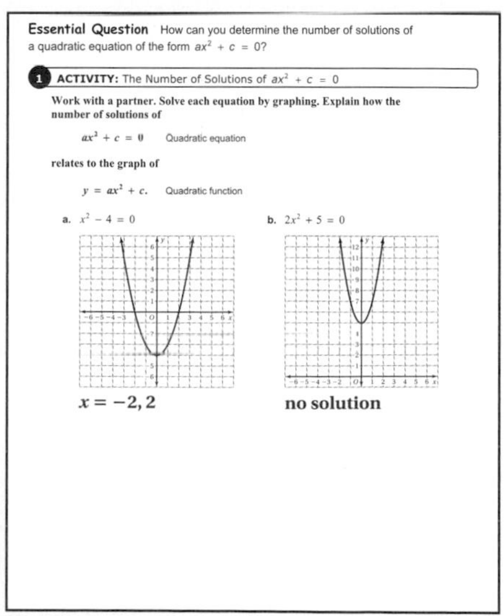

Essential Question How can you determine the number of solutions of a quadratic equation of the form $ax^2 + c = 0$?

1 ACTIVITY: The Number of Solutions of $ax^2 + c = 0$

Work with a partner. Solve each equation by graphing. Explain how the number of solutions of

$ax^2 + c = 0$ Quadratic equation

relates to the graph of

$y = ax^2 + c$. Quadratic function

a. $x^2 - 4 = 0$

$x = -2, 2$

b. $2x^2 + 5 = 0$

no solution

English Language Learners

Vocabulary

Reinforce the terms "exact solution" and "estimated solution" by asking students which term describes their solutions to $x^2 - 5 = 0$ in each activity.

9.2 Record and Practice Journal

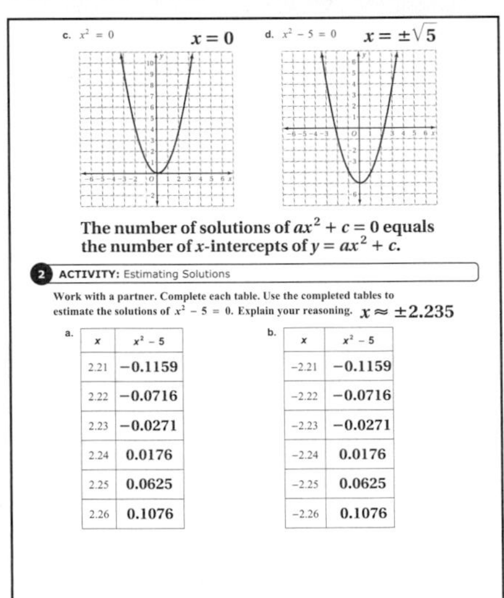

c. $x^2 = 0$ $x = 0$

d. $x^2 - 5 = 0$ $x = \pm\sqrt{5}$

The number of solutions of $ax^2 + c = 0$ equals the number of x-intercepts of $y = ax^2 + c$.

2 ACTIVITY: Estimating Solutions

Work with a partner. Complete each table. Use the completed tables to estimate the solutions of $x^2 - 5 = 0$. Explain your reasoning. $x \approx \pm 2.235$

a.

x	$x^2 - 5$
2.21	−0.1159
2.22	−0.0716
2.23	−0.0271
2.24	0.0176
2.25	0.0625
2.26	0.1076

b.

x	$x^2 - 5$
−2.21	−0.1159
−2.22	−0.0716
−2.23	−0.0271
−2.24	0.0176
−2.25	0.0625
−2.26	0.1076

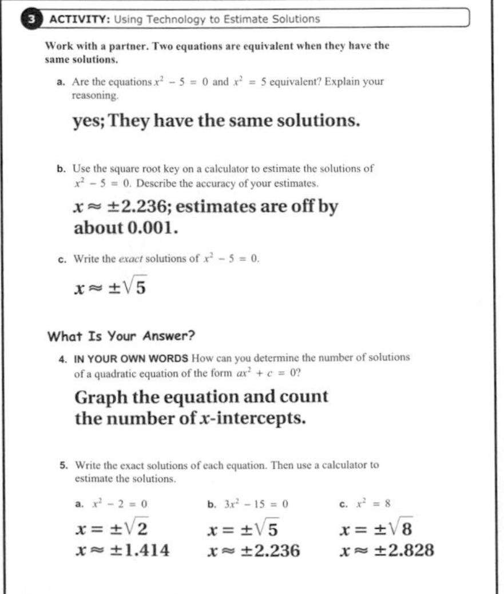

3 ACTIVITY: Using Technology to Estimate Solutions

Work with a partner. Two equations are equivalent when they have the same solutions.

a. Are the equations $x^2 - 5 = 0$ and $x^2 = 5$ equivalent? Explain your reasoning.

yes; They have the same solutions.

b. Use the square root key on a calculator to estimate the solutions of $x^2 - 5 = 0$. Describe the accuracy of your estimates.

$x \approx \pm 2.236$; estimates are off by about 0.001.

c. Write the *exact* solutions of $x^2 - 5 = 0$.

$x \approx \pm\sqrt{5}$

What Is Your Answer?

4. **IN YOUR OWN WORDS** How can you determine the number of solutions of a quadratic equation of the form $ax^2 + c = 0$?

Graph the equation and count the number of x-intercepts.

5. Write the exact solutions of each equation. Then use a calculator to estimate the solutions.

a. $x^2 - 2 = 0$ b. $3x^2 - 15 = 0$ c. $x^2 = 8$

$x = \pm\sqrt{2}$ $x = \pm\sqrt{5}$ $x = \pm\sqrt{8}$

$x \approx \pm 1.414$ $x \approx \pm 2.236$ $x \approx \pm 2.828$

Laurie's Notes

Activity 2

- A calculator is a helpful tool for this activity.
- ? "In part (a), what is happening to the x-values?" increasing by 0.01
- ? "How do the tables help you estimate the solutions of $x^2 - 5 = 0$?" The consecutive x-values at which the expression values change in sign from negative to positive indicate an approximate solution.
- **MP2 Reason Abstractly and Quantitatively:** In exploring the table of values, you are asking your students to reason quantitatively.

Activity 3

- Students are using a calculator to confirm the estimates in Activity 2.
- **MP6 Attend to Precision:** In part (b), students should have obtained approximate solutions such as ± 2.236067977. In part (c), the *exact* solutions are $\pm\sqrt{5}$. Discuss the use of exact and approximate solutions. This draws attention to the concept of precision.

What Is Your Answer?

- **MP3a** and **MP8:** In Question 4, expect student conjectures to relate the number of solutions to the number of x-intercepts of the graph. Students may also try to distinguish between whether c is a positive or negative number. If so, make sure the conjecture is correct for both $a > 0$ and $a < 0$.
- **MP6:** In Question 5, discuss whether the solutions obtained on a calculator for a quadratic equation of the form $ax^2 + c = 0$ will always be estimates.

Closure

- Besides graphing, describe how to find the solutions of $x^2 - 9 = 0$. Add 9 to each side and then take the square root of each side.

Technology For the Teacher

The Dynamic Planning Tool
Editable Teacher's Resources at *BigIdeasMath.com*

2 ACTIVITY: Estimating Solutions

Work with a partner. Complete each table. Use the completed tables to estimate the solutions of $x^2 - 5 = 0$. Explain your reasoning.

a.

x	$x^2 - 5$
2.21	
2.22	
2.23	
2.24	
2.25	
2.26	

b.

x	$x^2 - 5$
−2.21	
−2.22	
−2.23	
−2.24	
−2.25	
−2.26	

3 ACTIVITY: Using Technology to Estimate Solutions

Work with a partner. Two equations are equivalent when they have the same solutions.

a. Are the equations

$$x^2 - 5 = 0 \quad \text{and} \quad x^2 = 5$$

equivalent? Explain your reasoning.

b. Use the square root key on a calculator to estimate the solutions of $x^2 - 5 = 0$. Describe the accuracy of your estimates.

c. Write the *exact* solutions of $x^2 - 5 = 0$.

What Is Your Answer?

4. IN YOUR OWN WORDS How can you determine the number of solutions of a quadratic equation of the form $ax^2 + c = 0$?

5. Write the exact solutions of each equation. Then use a calculator to estimate the solutions.

a. $x^2 - 2 = 0$ **b.** $3x^2 - 15 = 0$ **c.** $x^2 = 8$

Practice Use what you learned about quadratic equations to complete Exercises 3–5 on page 466.

9.2 Lesson

In Section 6.1, you studied properties of square roots. Here you will use square roots to solve quadratic equations of the form $ax^2 + c = 0$.

Solving Quadratic Equations Using Square Roots

You can solve $x^2 = d$ by taking the square root of each side.

- When $d > 0$, $x^2 = d$ has two real solutions, $x = \pm\sqrt{d}$.
- When $d = 0$, $x^2 = d$ has one real solution, $x = 0$.
- When $d < 0$, $x^2 = d$ has no real solutions.

EXAMPLE 1 Solving Quadratic Equations Using Square Roots

a. Solve $3x^2 - 27 = 0$ using square roots.

$3x^2 - 27 = 0$	Write the equation.
$3x^2 = 27$	Add 27 to each side.
$x^2 = 9$	Divide each side by 3.
$x = \pm\sqrt{9}$	Take the square root of each side.
$x = \pm 3$	Simplify.

∴ The solutions are $x = 3$ and $x = -3$.

b. Solve $x^2 - 10 = -10$ using square roots.

$x^2 - 10 = -10$	Write the equation.
$x^2 = 0$	Add 10 to each side.
$x = 0$	Take the square root of each side.

∴ The only solution is $x = 0$.

c. Solve $-5x^2 + 11 = 16$ using square roots.

$-5x^2 + 11 = 16$	Write the equation.
$-5x^2 = 5$	Subtract 11 from each side.
$x^2 = -1$	Divide each side by -5.

∴ The equation has no real solutions.

Remember

The square of a real number cannot be negative. That is why the equation in part (c) has no real solutions.

On Your Own

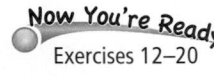

Now You're Ready
Exercises 12–20

Solve the equation using square roots.

1. $-3x^2 = -75$
2. $x^2 + 12 = 10$
3. $4x^2 - 15 = -15$

Laurie's Notes

Introduction

Connect

- **Yesterday:** Students developed an understanding of solving quadratic equations of the form $ax^2 + c = 0$. (MP2, MP3a, MP6, MP8)
- **Today:** Students will solve quadratic equations of the form $ax^2 + c = 0$ using square roots.

Motivate

- Tell students that today they will find the dimensions of a touch tank. Ask if any of them have visited an aquarium that has a touch tank.

Lesson Notes

Discuss

- **MP7 Look for and Make Use of Structure:** Help students see the similarities in solving $3x - 27 = 0$ and $3x^2 - 27 = 0$. In this lesson, students will solve the latter by performing one last step—taking the square root of each side.
- **FYI:** You can rewrite $ax^2 + c = 0$ as $x^2 = -\frac{c}{a}$. This is simplified as $x^2 = d$ in the Key Idea.

Key Idea

- Write the Key Idea on the board.
- **?** "How does this Key Idea connect to the graphing you did in the activity?" The number of solutions was shown to be 2, 1, or 0.
- Explain that today students will solve quadratic equations algebraically. Solving $ax^2 + c = 0$ is similar to solving the linear equation $ax + c = 0$.

Example 1

- Work through the three parts as shown.
- Take time to discuss with students how to solve $27 = 3x^2$. For some students, having the x^2-term on the right side of the equation makes the equation appear quite different.
- **Common Error:** Students often forget the negative square root when taking the square root of each side of an equation.

On Your Own

- **Think-Pair-Share:** Students should read each question independently and then work with a partner to answer the questions. When they have answered the questions, the pair should compare their answers with another group and discuss any discrepancies.

Goal Today's lesson is solving quadratic equations using square roots.

Lesson Materials
Textbook
• graphing calculators

Start Thinking! and Warm Up

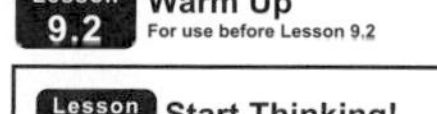

Lesson 9.2 Warm Up
For use before Lesson 9.2

Lesson 9.2 Start Thinking!
For use before Lesson 9.2

A math teacher asked her class to solve the equation $x^2 + 9 = 25$. Sophie's incorrect solution is shown below. What is Sophie's mistake?

Sophie

$$x^2 + 9 = 25$$
$$\pm\sqrt{x^2 + 9} = \pm\sqrt{25}$$
$$x + 3 = \pm 5$$
$$x = -3 \pm 5$$
$$x = -8 \text{ or } 2$$

Extra Example 1

Solve each equation using square roots.

a. $2x^2 - 32 = 0$ $x = 4, x = -4$

b. $5x^2 + 8 = 8$ $x = 0$

c. $x^2 + 8 = 5$ no real solutions

On Your Own

1. $x = 5, x = -5$
2. no real solutions
3. $x = 0$

Extra Example 2

Solve $(x - 3)^2 = 49$ using square roots. $x = -4, x = 10$

Extra Example 3

In Example 3, the volume of the tank is 450 cubic feet. Find the length and width of the tank.

width: about 7.1 ft; length: about 21.2 ft

On Your Own

4. $x = -7$
5. $x = 1.5, x = 4.5$
6. $x = \dfrac{-1 + \sqrt{35}}{2}$, $x = \dfrac{-1 - \sqrt{35}}{2}$
7. width: about 5.9 ft; length: about 17.7 ft

Differentiated Instruction

Kinesthetic

In Example 3, have students check the answer for reasonableness by using the length and width to sketch the base of the tank on graph paper (using feet as units). The drawing represents one of the three layers of unit cubes needed to fill the tank. Are the length and width reasonable?

Laurie's Notes

Example 2

- ? "How do you read $(x - 1)^2 = 25$?" Listen for "the quantity x minus 1 squared equals 25" or "x minus 1 quantity squared equals 25."
- Make sure students understand that this is about a quantity that is squared, and it is equal to 25, which is also a quantity squared.
- One technique is to place your hand or a few fingers over the expression in the parentheses and say, "Something squared is 25. Taking the square root of each side of the equation makes sense."
- Read the solution, "1 plus or minus 5." There are two solutions: 1 plus 5 and 1 minus 5.
- **Connection:** Check the solutions using a graphing calculator. The x-intercepts are -4 and 6. The vertex of the parabola occurs halfway between these numbers at $x = 1$.
- **Extension:** Discuss whether it is helpful to start the problem by first expanding $(x - 1)^2$. This expansion results in $x^2 - 2x + 1 = 25$. Have students solve this equation by factoring.

Example 3

- Discuss what the length and width represent in terms of the diagram.
- Work through the problem as shown.
- ? "What is a reasonable estimate of $\sqrt{30}$? Explain." About 5.5 because $\sqrt{25} = 5$, $\sqrt{36} = 6$, and 30 is about halfway between 25 and 36.
- Discuss the Study Tip.
- **MP6 Attend to Precision:** Point out to students that they should wait to round their answers until after they have substituted to find the length. Otherwise, they would calculate the length as $3(5.5) = 16.5$ feet.

On Your Own

- In Question 7, ask students how the new volume will affect the original dimensions in Example 3. They will increase.

Closure

- **Exit Ticket:** Match each equation with its number of solutions.

1. $2x^2 + 8 = 40$ C	**A.** 0 solutions
2. $2x^2 - 8 = -40$ A	**B.** 1 solution
3. $2x^2 = 0$ B	**C.** 2 solutions

Technology For the Teacher

The Dynamic Planning Tool
Editable Teacher's Resources at *BigIdeasMath.com*

EXAMPLE 2 Solving a Quadratic Equation Using Square Roots

Solve $(x - 1)^2 = 25$ using square roots.

$(x - 1)^2 = 25$	Write the equation.
$x - 1 = \pm 5$	Take the square root of each side.
$x = 1 \pm 5$	Add 1 to each side.

∴ So, the solutions are $x = 1 + 5 = 6$ and $x = 1 - 5 = -4$.

Check

Use a graphing calculator to check your answer. Rewrite the equation as $(x - 1)^2 - 25 = 0$. Graph the related function $y = (x - 1)^2 - 25$ and find the x-intercepts, or zeros. The zeros are -4 and 6, so the solution checks.

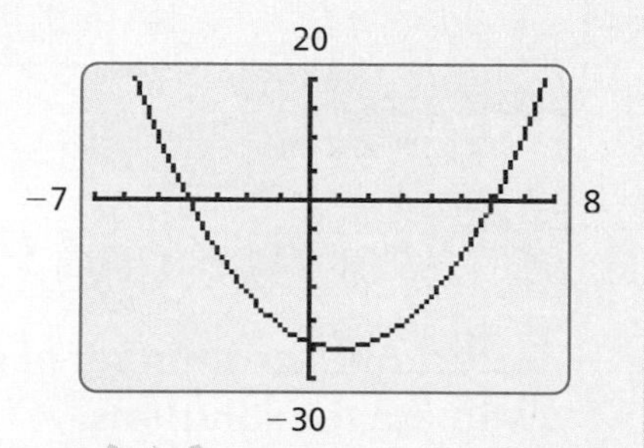

EXAMPLE 3 Real-Life Application

A touch tank has a height of 3 feet. Its length is 3 times its width. The volume of the tank is 270 cubic feet. Find the length and width of the tank.

The length ℓ is 3 times the width w, so $\ell - 3w$. Write an equation using the formula for the volume of a rectangular prism.

$V = \ell wh$	Write the formula.
$270 - 3w(w)(3)$	Substitute 270 for V, $3w$ for ℓ, and 3 for h.
$270 = 9w^2$	Multiply.
$30 = w^2$	Divide each side by 9.
$\pm\sqrt{30} = w$	Take the square root of each side.

Study Tip

Use the positive square root because negative solutions do not make sense in this context. Length and width cannot be negative.

The solutions are $\sqrt{30}$ and $-\sqrt{30}$. Use the positive solution.

∴ So, the width is $\sqrt{30} \approx 5.5$ feet and the length is $3\sqrt{30} \approx 16.4$ feet.

On Your Own

Now You're Ready
Exercises 23–28 and 32

Solve the equation using square roots.

4. $(x + 7)^2 = 0$
5. $4(x - 3)^2 = 9$
6. $(2x + 1)^2 = 35$
7. **WHAT IF?** In Example 3, the volume of the tank is 315 cubic feet. Find the length and width of the tank.

9.2 Exercises

Vocabulary and Concept Check

1. **REASONING** How many real solutions does the equation $x^2 = d$ have when d is positive? 0? negative?
2. **WHICH ONE DOESN'T BELONG?** Which equation does *not* belong with the other three? Explain your reasoning.

$x^2 = 9$ $\quad$ $x^2 = 2$ $\quad$ $x^2 = -7$ $\quad$ $x^2 = 21$

Practice and Problem Solving

Determine the number of solutions of the equation. Then use a calculator to estimate the solutions.

3. $x^2 - 11 = 0$
4. $x^2 + 10 = 0$
5. $2x^2 - 3 = 0$

Determine the number of solutions of the equation. Then solve the equation using square roots.

6. $x^2 = 25$
7. $x^2 = -36$
8. $x^2 = 8$
9. $x^2 = 21$
10. $x^2 = 0$
11. $x^2 = 169$

Solve the equation using square roots.

1 12. $x^2 - 16 = 0$
13. $x^2 + 12 = 0$
14. $x^2 + 6 = 0$
15. $x^2 - 61 = 0$
16. $2x^2 - 98 = 0$
17. $-x^2 + 9 = 9$
18. $x^2 + 13 = 7$
19. $-4x^2 - 5 = -5$
20. $-3x^2 + 8 = 8$

21. **ERROR ANALYSIS** Describe and correct the error in solving the equation.

Solve $2x^2 - 33 = 39$.

$2x^2 = 72$ Add 33 to each side.
$x^2 = 36$ Divide each side by 2.
$x = 6$ Take the square root of each side.

∴ The solution is $x = 6$.

22. **WAREHOUSE** A box falls off a warehouse shelf from a height of 16 feet. The function $h = -16x^2 + 16$ gives the height h (in feet) of the box after x seconds. When does it hit the floor?

Assignment Guide and Homework Check

Level	Assignment	Homework Check
Average	1, 2, 3–25 odd, 29, 32, 37–40	11, 13, 21, 23, 32
Advanced	1, 2, 12–28 even, 30–40	14, 20, 22, 24, 32

Common Errors

- **Exercises 3–20** Students may forget the negative square root when taking the square root of each side of the equation. Remind them to account for the negative square root when appropriate.
- **Exercises 3–20** Students may try to take the square root of a negative number. Remind them that the square of a real number cannot be negative.
- **Exercises 23–28** Students may use the Addition Property of Equality incorrectly. For example, a student may rewrite $(x - 1)^2 = 35$ as $x^2 = 36$ Remind students that the Addition Property of Equality cannot be applied to a term that is within grouping symbols.

9.2 Record and Practice Journal

Solve the equation using square roots.

1. $x^2 - 64 = 0$
   $x = \pm 8$
2. $x^2 + 18 = 0$
   no solution
3. $x^2 - 11 = -11$
   $x = 0$
4. $3x^2 - 75 = 0$
   $x = \pm 5$
5. $2x^2 + 12 = 9$
   no solution
6. $-6x^2 + 4 = 4$
   $x = 0$

Solve the equation using square roots. Use a graphing calculator to check your solution(s).

7. $(x - 4)^2 = 49$
   $x = 11$ and $x = -3$
8. $16(x + 3)^2 = 36$
   $x = -\frac{3}{2}$ and $x = -\frac{9}{2}$
9. The volume of a shed is 972 cubic feet. Find the width x of the shed.
   9 ft

8 ft
1.5x ft
x ft

Answer Presentation Tool

Vocabulary and Concept Check

1. 2; 1; 0
2. $x^2 = -7$; It is the only equation with no real solutions.

Practice and Problem Solving

3. 2; $x \approx 3.317$, $x \approx -3.317$
4. 0; no real solutions
5. 2; $x \approx 1.225$, $x \approx -1.225$
6. 2; $x = 5$, $x = -5$
7. 0; no real solutions
8. 2; $x = 2\sqrt{2}$, $x = -2\sqrt{2}$
9. 2; $x = \sqrt{21}$, $x = -\sqrt{21}$
10. 1; $x = 0$
11. 2; $x = 13$, $x = -13$
12. $x = 4$, $x = -4$
13. no real solutions
14. no real solutions
15. $x = \sqrt{61}$, $x = -\sqrt{61}$
16. $x = 7$, $x = -7$
17. $x = 0$
18. no real solutions
19. $x = 0$
20. $x = 0$
21. When taking the square root of each side, the student forgot the negative root. $x = 6$, $x = -6$
22. after 1 second
23. $x = -3$
24. $x = -1$, $x = 3$
25. $x = -4$, $x = 5$

Practice and Problem Solving

26. $x = \frac{1}{2}, x = 2$

27. $x = -\frac{7}{3}, x = \frac{1}{3}$

28. $x = -\frac{1}{2}, x = \frac{9}{2}$

29. 8 in. by 8 in.

30. $3\sqrt{13}$ cm by $2\sqrt{13}$ cm

31. 12 ft

32. length = 52.4 in., width = 26.2 in.

33. See *Taking Math Deeper.*

34. Find two integers or decimals that you know the root is between and then use a table of values.

35. a. two solutions: a is positive and c is negative, or c is positive and a is negative.

 b. one solution: $c = 0$

 c. no solutions: a and c are both positive or both negative.

36. $(-3, 9)$, $(3, 9)$; They intersect at $(\sqrt{9}, 9)$ and $(-\sqrt{9}, 9)$.

Fair Game Review

37. $x^2 + 10x + 25$

38. $w^2 - 14w + 49$

39. $4y^2 - 12y + 9$

40. B

Mini-Assessment

Solve the equation using square roots.

1. $x^2 = 100$ $x = 10, x = -10$
2. $x^2 + 10 = 0$ no real solutions
3. $5x^2 - 7 = -7$ $x = 0$
4. $2(x + 2)^2 = 72$ $x = 4, x = -8$

Taking Math Deeper

Exercise 33

You can solve this problem by recalling a property of similar figures.

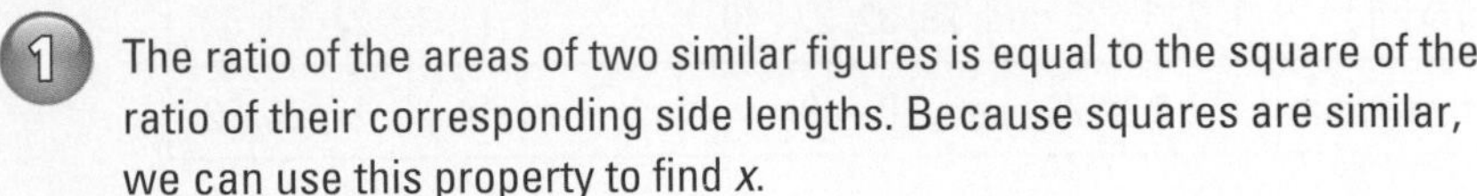

1 The ratio of the areas of two similar figures is equal to the square of the ratio of their corresponding side lengths. Because squares are similar, we can use this property to find x.

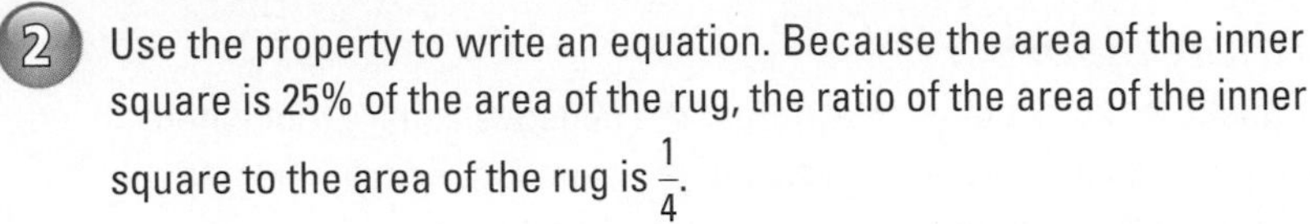

2 Use the property to write an equation. Because the area of the inner square is 25% of the area of the rug, the ratio of the area of the inner square to the area of the rug is $\frac{1}{4}$.

$$\frac{\text{Area of inner square}}{\text{Area of rug}} = \left(\frac{\text{side length of inner square}}{\text{side length of rug}}\right)^2$$

$$\frac{1}{4} = \left(\frac{x}{6}\right)^2$$

3 Solve the equation.

$$\frac{1}{4} = \left(\frac{x}{6}\right)^2$$

$$\frac{1}{4} = \frac{x^2}{36}$$

$$9 = x^2$$

$$3 = x$$

So, the side length of the inner square is 3 feet.

Project

Research Tibetan rugs and Persian carpets. Which type would you buy?

Reteaching and Enrichment Strategies

If students need help...	If students got it...
Resources by Chapter • Practice A and Practice B • Puzzle Time Record and Practice Journal Practice Differentiating the Lesson Lesson Tutorials Skills Review Handbook	Resources by Chapter • Enrichment and Extension Start the next section

Solve the equation using square roots. Use a graphing calculator to check your solution(s).

2 **23.** $(x + 3)^2 = 0$ **24.** $(x - 1)^2 = 4$ **25.** $(2x - 1)^2 = 81$

26. $(4x - 5)^2 = 9$ **27.** $9(x + 1)^2 = 16$ **28.** $4(x - 2)^2 = 25$

Use the given area A to find the dimensions of the figure.

29. $A = 64 \text{ in.}^2$

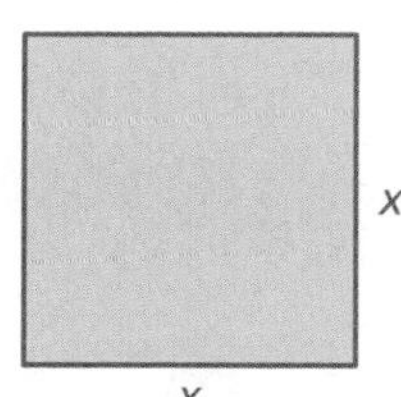

30. $A = 78 \text{ cm}^2$

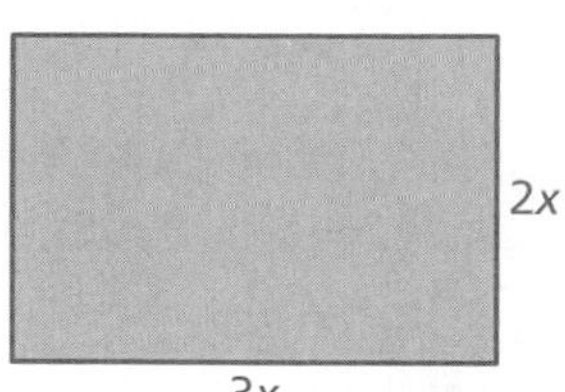

31. $A = 144\pi \text{ ft}^2$

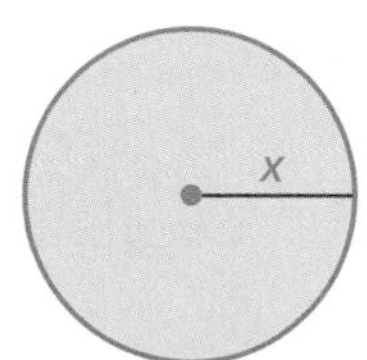

3 **32. POND** An in-ground pond has the shape of a rectangular prism. The pond has a height of 24 inches and a volume of 33,000 cubic inches. The pond's length is 2 times its width. Find the length and width of the pond.

33. AREA RUG The design of a square area rug for your living room is shown. You want the area of the inner square to be 25% of the total area of the rug. Find the side length x of the inner square.

34. WRITING How can you approximate the roots of a quadratic equation when the roots are not integers?

35. LOGIC Given the equation $ax^2 + c = 0$, describe the values of a and c so the equation has the following number of solutions.

a. two solutions **b.** one solution **c.** no solutions

36. Reasoning Without graphing, where do the graphs of $y = x^2$ and $y = 9$ intersect? Explain.

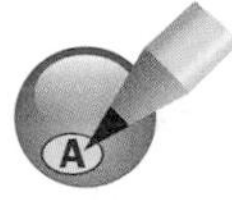

Fair Game Review What you learned in previous grades & lessons

Find the product. *(Section 7.4)*

37. $(x + 5)^2$ **38.** $(w - 7)^2$ **39.** $(2y - 3)^2$

40. MULTIPLE CHOICE What is an explicit equation for $a_1 = -3, a_n = a_{n-1} + 2$? *(Section 6.7)*

Ⓐ $a_n = 2n - 3$ Ⓑ $a_n = 2n - 5$ Ⓒ $a_n = n + 2$ Ⓓ $a_n = -3n + 2$

9.3 Solving Quadratic Equations by Completing the Square

COMMON CORE STATE STANDARDS
A.REI.4a
A.REI.4b
A.SSE.3b
F.IF.8a

Essential Question How can you use "completing the square" to solve a quadratic equation?

1 EXAMPLE: Solving by Completing the Square

Work with a partner. Five different algebra tiles are shown at the right.

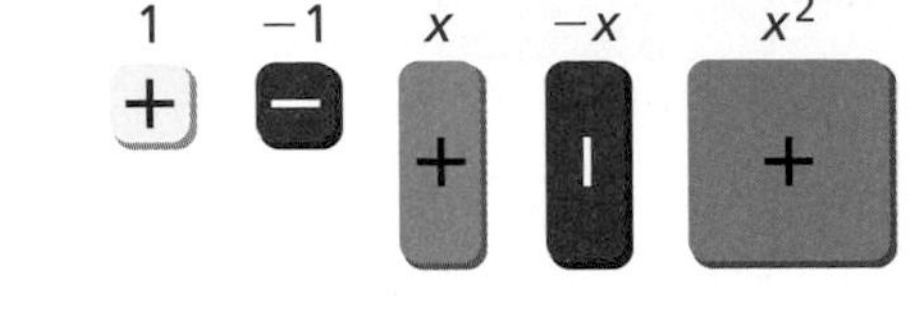

Solve $x^2 + 4x = -2$ by completing the square.

Step 1: Use algebra tiles to model the equation

$$x^2 + 4x = -2.$$

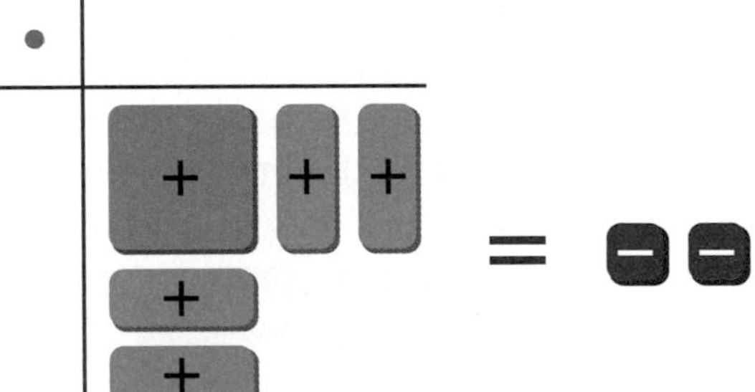

Step 2: Add four yellow tiles to the left side of the equation so that it is a perfect square. Balance the equation by also adding four yellow tiles to the right side.

$$x^2 + 4x + 4 = -2 + 4$$

$$(x + 2)^2 = 2$$

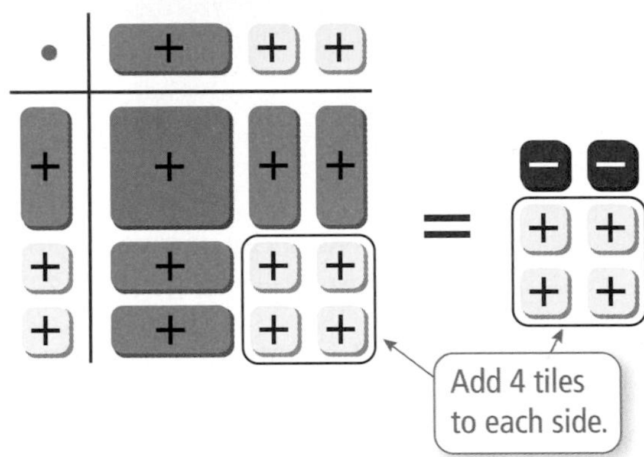

Step 3: Take the square root of each side of the equation and simplify.

$$x + 2 = \pm\sqrt{2}$$

$$x = -2 \pm \sqrt{2}$$

Check Check each solution in the original equation.

$$x^2 + 4x = -2$$

$$(-2 + \sqrt{2})^2 + 4(-2 + \sqrt{2}) \stackrel{?}{=} -2$$

$$4 - 4\sqrt{2} + 2 - 8 + 4\sqrt{2} \stackrel{?}{=} -2$$

$$4 + 2 - 8 \stackrel{?}{=} -2$$

$$-2 = -2 \checkmark$$

Now you check the other solution.

Laurie's Notes

Introduction

Standards for Mathematical Practice

- **MP5 Use Appropriate Tools Strategically:** Students are using algebra tiles to model the technique of *completing the square*.

Motivate

- Students should recall using base 10 pieces in an array to represent multiplication.
- Display the model.

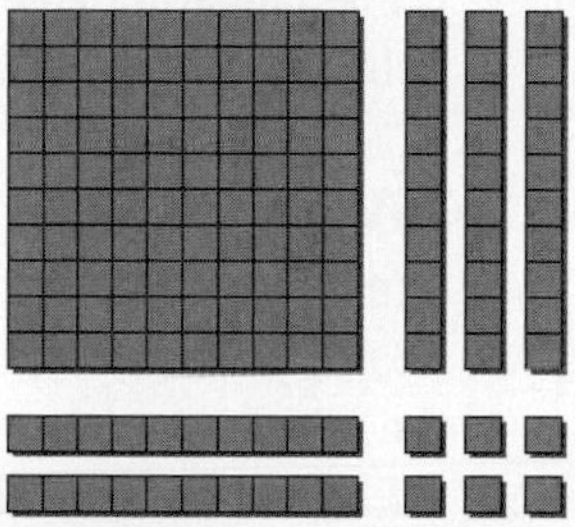

? "What multiplication problem does it represent?"
$12 \times 13 = 100 + 50 + 6 = 156$
- Discuss the value of each piece (1), the dimensions of the array ($10 + 2$ by $10 + 3$), and the answer.
- Display the model.

$15^2 =$ 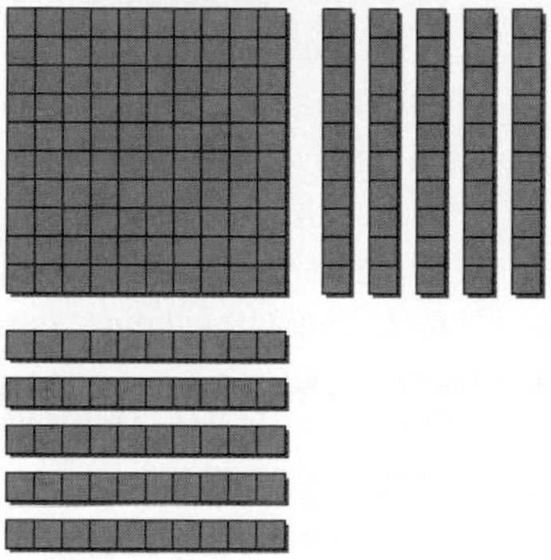

? "What is missing from the model?" 25 pieces in the bottom right corner

Activity Notes

Discuss

- **Connection:** Squaring a number (like 15^2) connects to today's work with completing the square, another technique for solving quadratic equations.

Activity 1

- Review the names and dimensions of the algebra tiles.
- Students should model the equation with the goal of adding 1-tiles to form a square array on the left side. To make this possible, start by arranging half of the x-tiles vertically and half horizontally.
- Add the number of 1-tiles needed to form a square on the left side. Add the same number of 1-tiles to the right side.
- Remind students how multiplication of binomials was modeled (rectangular array).
- Explain how to use the dimensions of the square formed on the left side of the equation to write it as the square of a binomial. The vertical and horizontal dimensions are each $x + 2$, so the left side of the equation represents $(x + 2)^2$.
- **MP2 Reason Abstractly and Quantitatively:** Students have obtained a concrete model of $(x + 2)^2 = 2$. They will now reason abstractly, using algebra to manipulate the equation.
- It is important to work through the checking process.

Common Core State Standards

A.REI.4a Use the method of completing the square to transform any quadratic equation in x into an equation of the form $(x - p)^2 = q$ that has the same solutions . . ..
A.REI.4b Solve quadratic equations by . . . completing the square . . ..
A.SSE.3b Complete the square in a quadratic expression to reveal the maximum or minimum value of the function it defines.
F.IF.8a Use the process of factoring and completing the square in a quadratic function to show zeros, extreme values, and symmetry of the graph, and interpret these in terms of a context.

Previous Learning

Students should know how to solve quadratic equations using square roots.

Start Thinking! and Warm Up

Activity 9.3 **Start Thinking!**
For use before Activity 9.3

Activity 9.3 **Warm Up**
For use before Activity 9.3

Find the square of half of the number.

1. 18	**2.** −20
3. −4	**4.** 6
5. 11	**6.** −9

9.3 Record and Practice Journal

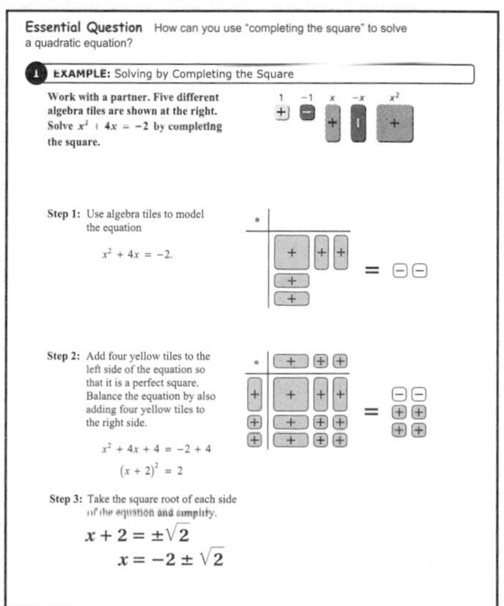

Essential Question How can you use "completing the square" to solve a quadratic equation?

1 EXAMPLE: Solving by Completing the Square

Work with a partner. Five different algebra tiles are shown at the right. Solve $x^2 + 4x = -2$ by completing the square.

Step 1: Use algebra tiles to model the equation

$x^2 + 4x = -2.$

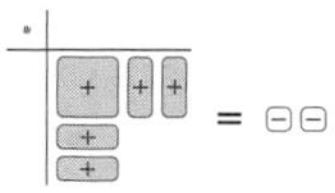

Step 2: Add four yellow tiles to the left side of the equation so that it is a perfect square. Balance the equation by also adding four yellow tiles to the right side.

$x^2 + 4x + 4 = -2 + 4$

$(x + 2)^2 = 2$

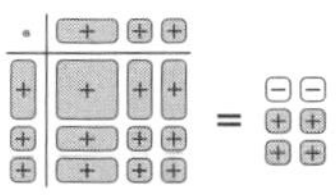

Step 3: Take the square root of each side of the equation and simplify.

$x + 2 = \pm\sqrt{2}$

$x = -2 \pm \sqrt{2}$

English Language Learners

Vocabulary

Discuss the use of the word *square* as a noun and as a verb. For example, to square (verb) the number 15, you use a model in the shape of a square (noun). The result is the square (noun) of 15. In the phrase *completing the square,* the word square is used as a noun.

9.3 Record and Practice Journal

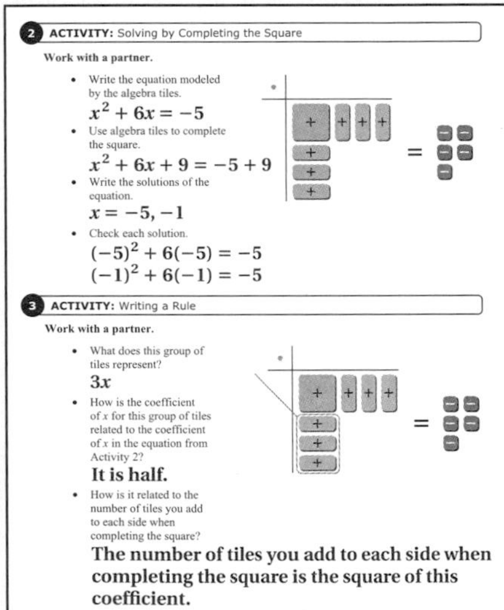

2 ACTIVITY: Solving by Completing the Square

Work with a partner.

- Write the equation modeled by the algebra tiles.
  $x^2 + 6x = -5$
- Use algebra tiles to complete the square.
  $x^2 + 6x + 9 = -5 + 9$
- Write the solutions of the equation.
  $x = -5, -1$
- Check each solution.
  $(-5)^2 + 6(-5) = -5$
  $(-1)^2 + 6(-1) = -5$

3 ACTIVITY: Writing a Rule

Work with a partner.

- What does this group of tiles represent?
  $3x$
- How is the coefficient of x for this group of tiles related to the coefficient of x in the equation from Activity 2?
  It is half.
- How is it related to the number of tiles you add to each side when completing the square?
  The number of tiles you add to each side when completing the square is the square of this coefficient.

- **WRITE A RULE** Fill in the blanks.
  To complete the square, take **half** of the coefficient of the x-term and **square** it. **Add** this number to each side of the equation.

What Is Your Answer?

4. **IN YOUR OWN WORDS** How can you use "completing the square" to solve a quadratic equation?
   Factor the left-hand side as the square of a binomial and evaluate the addition on the right-hand side of the equation. Then take the square root of each side and solve for x.
5. Solve each quadratic equation by completing the square.
   a. $x^2 - 2x = 1$ $\quad x = 1 \pm \sqrt{2}$
   b. $x^2 - 4x = -1$ $\quad x = 2 \pm \sqrt{3}$
   c. $x^2 + 4x = -3$ $\quad x = -3, -1$

Laurie's Notes

Activity 2

- Having guided students through the first activity, students should be able to work with a partner on this activity.
- ? "What equation is modeled by the algebra tiles?" $x^2 + 6x = -5$
- ? "What equation is represented by the tiles after completing the square?" $x^2 + 6x + 9 = 4$
- **MP3a Construct Viable Arguments:** Asking students to explain the thinking behind their solutions is helpful for them and for their peers.

Activity 3

- Encourage students to read through the whole activity before attempting to answer the questions. This will help them understand the focus of the activity.
- **MP2** and **MP7 Look for and Make Use of Structure:** In this activity, students compare the structure of the concrete model to the structure of the abstract equation it represents. They use these comparisons to deduce an algebraic rule for completing the square.
- The activity is intended for students to answer the first question as $3x$, realize that the coefficient 3 is one-half of the coefficient in the equation, and then realize that you square it and add that number of 1-tiles to each side. The model helps visualize this.

What Is Your Answer?

- **MP3a:** In Question 4, answers should include an understanding of writing the quadratic equation in a form that allows you to solve it using square roots.

Closure

- Ask students to repeat Activity 2 for the model shown and then compare the two problems. Procedurally it is the same (add nine 1-tiles to each side, write the equation modeled, and solve).

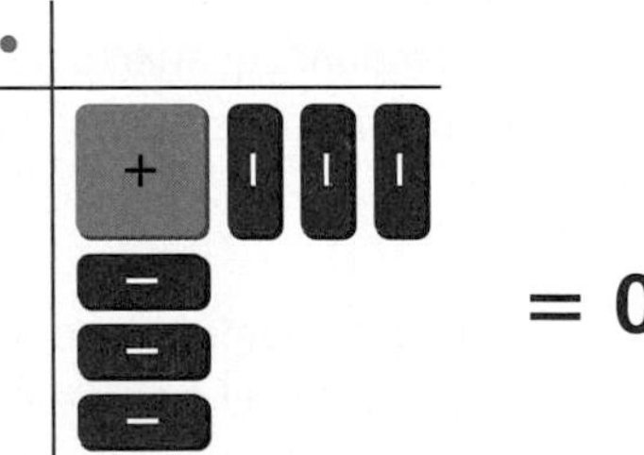

Technology For the Teacher
Dynamic Classroom
The Dynamic Planning Tool
Editable Teacher's Resources at BigIdeasMath.com

2 ACTIVITY: Solving by Completing the Square

Work with a partner.

- Write the equation modeled by the algebra tiles.
- Use algebra tiles to complete the square.
- Write the solutions of the equation.
- Check each solution.

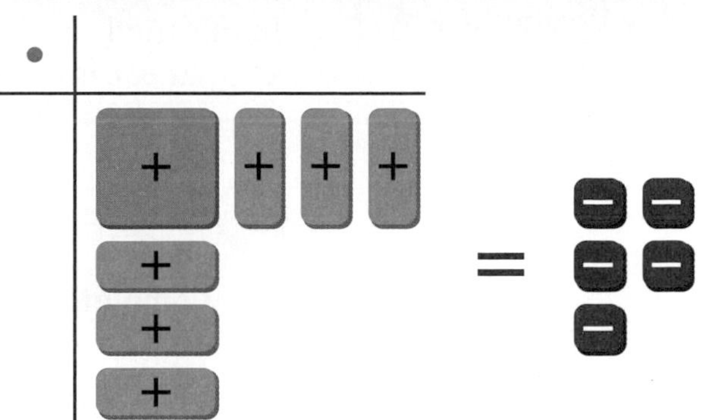

3 ACTIVITY: Writing a Rule

Work with a partner.

- What does this group of tiles represent?
- How is the coefficient of x for this group of tiles related to the coefficient of x in the equation from Activity 2? How is it related to the number of tiles you add to each side when completing the square?

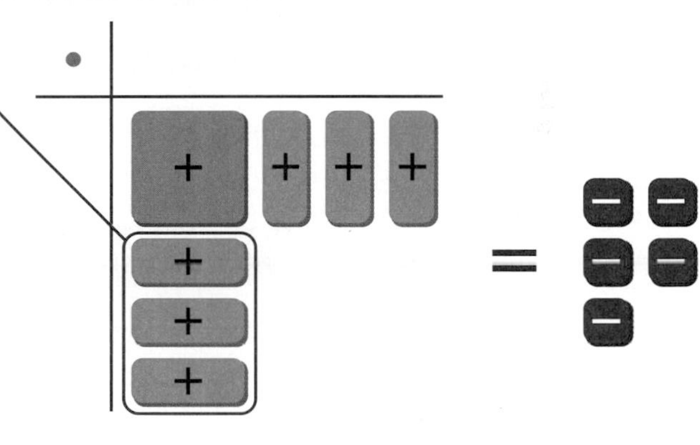

- **WRITE A RULE** Fill in the blanks.

> To complete the square, take ________ of the coefficient of the x-term and ________ it. ________ this number to each side of the equation.

What Is Your Answer?

4. **IN YOUR OWN WORDS** How can you use "completing the square" to solve a quadratic equation?
5. Solve each quadratic equation by completing the square.

   a. $x^2 - 2x = 1$ **b.** $x^2 - 4x = -1$ **c.** $x^2 + 4x = -3$

Practice Use what you learned about quadratic equations to complete Exercises 3–5 on page 472.

9.3 Lesson

Key Vocabulary
completing the square, *p. 470*

Another method for solving quadratic equations is **completing the square.** In this method, a constant c is added to the expression $x^2 + bx$ so that $x^2 + bx + c$ is a perfect square trinomial.

Key Idea

Completing the Square

Words To complete the square for an expression of the form $x^2 + bx$, follow these steps.

Step 1: Find one-half of b, the coefficient of x.
Step 2: Square the result from Step 1.
Step 3: Add the result from Step 2 to $x^2 + bx$.

Factor the resulting expression as the square of a binomial.

Algebra $x^2 + bx + \left(\frac{b}{2}\right)^2 = \left(x + \frac{b}{2}\right)^2$

EXAMPLE 1 Completing the Square

Complete the square for each expression. Then factor the trinomial.

a. $x^2 + 6x$

Step 1: Find one-half of b. $\frac{b}{2} = \frac{6}{2} = 3$

Step 2: Square the result from Step 1. $3^2 = 9$

Step 3: Add the result from Step 2 to $x^2 + bx$. $x^2 + 6x + 9$

∴ $x^2 + 6x + 9 = (x + 3)^2$

b. $x^2 - 9x$

Step 1: Find one-half of b. $\frac{b}{2} = \frac{-9}{2}$

Step 2: Square the result from Step 1. $\left(\frac{-9}{2}\right)^2 = \frac{81}{4}$

Step 3: Add the result from Step 2 to $x^2 + bx$. $x^2 - 9x + \frac{81}{4}$

∴ $x^2 - 9x + \frac{81}{4} = \left(x - \frac{9}{2}\right)^2$

On Your Own

Now You're Ready
Exercises 12–17

Complete the square for each expression. Then factor the trinomial.

1. $x^2 + 10x$
2. $x^2 - 4x$
3. $x^2 + 7x$

Multi-Language Glossary at BigIdeasMath.com.

Laurie's Notes

Introduction

Connect

- **Yesterday:** Students solved quadratic equations of the form $x^2 + bx = d$ using algebra tiles. (MP2, MP3a, MP5, MP7)
- **Today:** Students will solve quadratic equations by *completing the square.*

Motivate

- Write the sequence on the board: 1, 4, ___, 16, ___, 36, . . .
- ? "What numbers are missing and what is the pattern? Explain." 9 and 25; The terms of the sequence are perfect squares.
- ? "What is a perfect square trinomial?" a trinomial that can be factored as $(a \pm b)^2$

Lesson Notes

Discuss

- **MP7 Look for and Make Use of Structure:** In expressions of the form $x^2 + bx$, students will use the coefficient of the x-term to determine the constant that must be added to form a perfect square trinomial.
- In the previous section, students solved quadratic equations of the form $x^2 = d$. Now they will solve quadratic equations of the form $x^2 + bx = d$.

Key Idea

- Write the Key Idea on the board.
- ? "How does one-half of b in Step 1 connect to the algebra tiles activity?" Half of the x-tiles were placed on each dimension of the square.
- ? "How does Step 2 connect to the algebra tiles activity?" This gives the number of 1-tiles that were added to each side of the equation.

Example 1

- This example helps students become familiar with the technique of completing the square before they use it to solve an equation.
- It may be helpful to ask students what is missing that would make it a perfect square trinomial.

$$x^2 + 6x + \underline{\ ?\ } = (x + \underline{\ ?\ })^2$$

- Work through both parts as shown.

On Your Own

- **Neighbor Check:** Have students work independently and then have their neighbor check their work. Have students discuss any discrepancies.
- Ask a volunteer to discuss the solution of Question 3. The odd coefficient may challenge some students.

Goal Today's lesson is solving quadratic equations by **completing the square**.

Lesson Materials
Textbook
• graphing calculators

Start Thinking! and Warm Up

Lesson 9.3 Warm Up
For use before Lesson 9.3

Lesson 9.3 Start Thinking!
For use before Lesson 9.3

The steps of the solution of solving $x^2 - 6x = 5$ are shown below. Which step of the solution is completing the square? Why do you think it is called that?

1. Find the square of half of -6: $\left(\frac{-6}{2}\right)^2 = (-3)^2 = 9.$
2. $x^2 - 6x + 9 = 5 + 9$ Add 9 to each side.
3. $(x - 3)^2 = 14$ Factor.
4. $x - 3 = \pm\sqrt{14}$ Take the square root of each side.
5. $x = 3 \pm \sqrt{14}$ Add 3 to each side.

Extra Example 1

Complete the square for each expression. Then factor the trinomial.

a. $x^2 + 8x$ $x^2 + 8x + 16 = (x + 4)^2$

b. $x^2 - x$ $x^2 - x + \frac{1}{4} = \left(x - \frac{1}{2}\right)^2$

On Your Own

1. $x^2 + 10x + 25 = (x + 5)^2$
2. $x^2 - 4x + 4 = (x - 2)^2$
3. $x^2 + 7x + \frac{49}{4} = \left(x + \frac{7}{2}\right)^2$

Extra Example 2

Solve $x^2 + 4x - 5 = 7$ by completing the square. $x = -6, x = 2$

Extra Example 3

In Example 3, you throw the stone with an upward velocity of 48 feet per second. The function $h = -16t^2 + 48t + 16$ gives the height h of the stone after t seconds. When does the stone land in the water? after about 3.3 seconds

On Your Own

4. $x = -3, x = 9$
5. $x = -6 + \sqrt{34}$, $x = -6 - \sqrt{34}$
6. $x = -6, x = 4$
7. after about 4.2 seconds

Differentiated Instruction

Visual

In each step of Example 1(a), look back at the model in Activity 2 and discuss how the step is connected to the model.

Laurie's Notes

Example 2

- Write on the board: $x^2 + 8x + \underline{\ ?\ } = \underline{\ ?\ }$.
- ? "If each blank is replaced by 0, how do you complete the square?" Add 16 to each side.
- ? "If the first blank is replaced by 0 and the second blank is replaced by a nonzero number, how do you complete the square?" Add 16 to each side.
- ? "If each blank is replaced by a nonzero number, how do you complete the square?" Isolate $x^2 + 8x$ on the left side and then add 16 to each side.
- **Common Error:** Students often forget the negative square root when taking the square root of each side of the equation.
- Note that when 4 is subtracted from each side in the last step, the next step is written as $x = -4 \pm 6$ This form is preferred to $x = \pm 6 - 4$.

Example 3

- **MP4 Modeling with Mathematics:** Read through the problem. Remind students that they have used the vertical motion model previously.
- ? "What do the coefficients and the constant term have in common?" They are all divisible by 16.
- Discuss the Study Tip about having a leading coefficient of 1.
- Work through the problem as shown.
- ? "Why is -0.4 discounted as a solution?" Time cannot be negative.
- **MP5 Use Appropriate Tools Strategically:** Graph the function. The graph shows the two solutions, but only the positive solution makes sense.

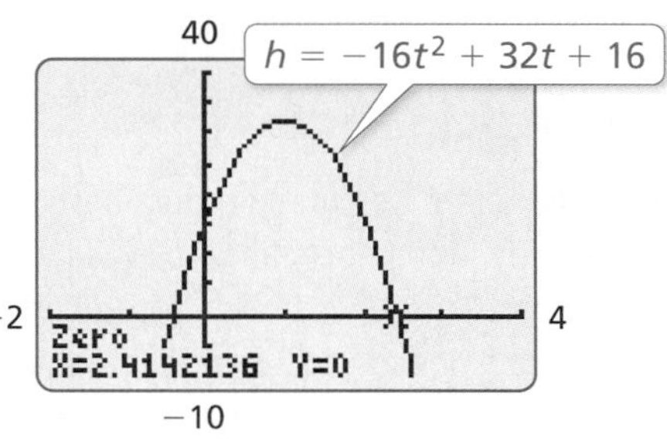

On Your Own

- In Question 7, what has changed from Example 3? the upward velocity

Closure

You have studied the methods of factoring, graphing, using square roots and completing the square to solve quadratic equations. State which method you would use to solve each equation below. (Sample answers provided.)

a. $4x^2 - 12 = 4$ square roots
b. $x^2 - 8x = 0$ completing the square
c. $x^2 + 4x + 3 = 0$ graphing or factoring

Technology For the Teacher

The Dynamic Planning Tool
Editable Teacher's Resources at *BigIdeasMath.com*

To solve a quadratic equation by completing the square, write the equation in the form $x^2 + bx = d$.

EXAMPLE 2 Solving a Quadratic Equation by Completing the Square

Solve $x^2 + 8x - 3 = 17$ by completing the square.

$x^2 + 8x - 3 = 17$	Write the equation.
$x^2 + 8x = 20$	Add 3 to each side.
Complete the square. → $x^2 + 8x + 16 = 20 + 16$	Add $\left(\frac{8}{2}\right)^2$, or 16, to each side.
$(x + 4)^2 = 36$	Factor $x^2 + 8x + 16$.
$x + 4 = \pm 6$	Take the square root of each side.
$x = -4 \pm 6$	Subtract 4 from each side.

∴ The solutions are $x = -4 + 6 = 2$ and $x = -4 - 6 = -10$.

Common Error

When completing the square, be sure to add to *both* sides of the equation.

EXAMPLE 3 Real-Life Application

You throw a stone from a height of 16 feet with an upward velocity of 32 feet per second. The function $h = -16t^2 + 32t + 16$ gives the height h of the stone after t seconds. When does the stone land in the water?

16 ft

Find the t-values for which $h = 0$. So, solve $-16t^2 + 32t + 16 = 0$.

$-16t^2 + 32t + 16 = 0$	Write the equation.
$t^2 - 2t - 1 = 0$	Divide each side by -16.
$t^2 - 2t = 1$	Add 1 to each side.
Complete the square. → $t^2 - 2t + 1 = 1 + 1$	Add $\left(\frac{-2}{2}\right)^2$, or 1, to each side.
$(t - 1)^2 = 2$	Factor $t^2 - 2t + 1$.
$t - 1 = \pm\sqrt{2}$	Take the square root of each side.
$t = 1 \pm \sqrt{2}$	Add 1 to each side.

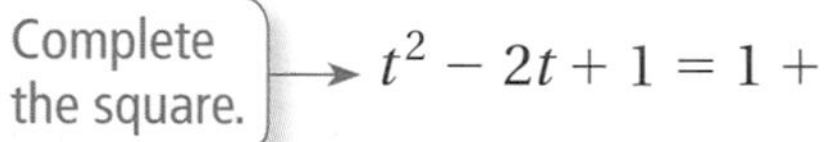

The solutions are $x = 1 + \sqrt{2} \approx 2.4$ and $x = 1 - \sqrt{2} \approx -0.4$. Use the positive solution.

∴ The stone lands in the water after about 2.4 seconds.

Study Tip

Before completing the square, make sure the leading coefficient is 1.

On Your Own

Now You're Ready
Exercises 18–23

Solve the equation by completing the square.

4. $x^2 - 6x = 27$ **5.** $x^2 + 12x + 3 = 1$ **6.** $2x^2 + 4x + 10 = 58$

7. **WHAT IF?** In Example 3, the function $h = -16t^2 + 64t + 16$ gives the height h (in feet) of the stone after t seconds. When does the stone land in the water?

9.3 Exercises

Vocabulary and Concept Check

1. **VOCABULARY** Explain how to complete the square for an expression of the form $x^2 + bx$.

2. **WRITING** For what values of b is it easier to complete the square for $x^2 + bx$? Explain.

Practice and Problem Solving

Use algebra tiles to complete the square. Then write the perfect square trinomial.

3. 

4. 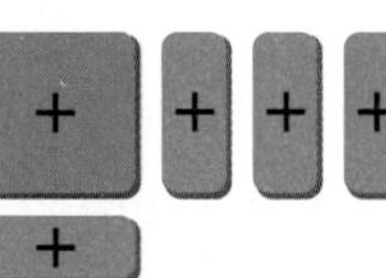

5.   

Find the value of c that completes the square.

6. $x^2 - 8x + c$
7. $x^2 + 4x + c$
8. $x^2 - 2x + c$
9. $x^2 - 14x + c$
10. $x^2 + 12x + c$
11. $x^2 + 18x + c$

Complete the square for the expression. Then factor the trinomial.

1 12. $x^2 - 10x$
13. $x^2 + 16x$
14. $x^2 + 22x$
15. $x^2 - 40x$
16. $x^2 - 3x$
17. $x^2 + 5x$

Solve the equation by completing the square.

2 18. $x^2 + 2x = 3$
19. $x^2 - 6x = 16$
20. $x^2 + 4x + 7 = -6$
21. $x^2 + 5x - 7 = -14$
22. $2x^2 - 8x = 10$
23. $2x^2 - 3x + 1 = 0$

24. **ERROR ANALYSIS** Describe and correct the error in solving the equation.

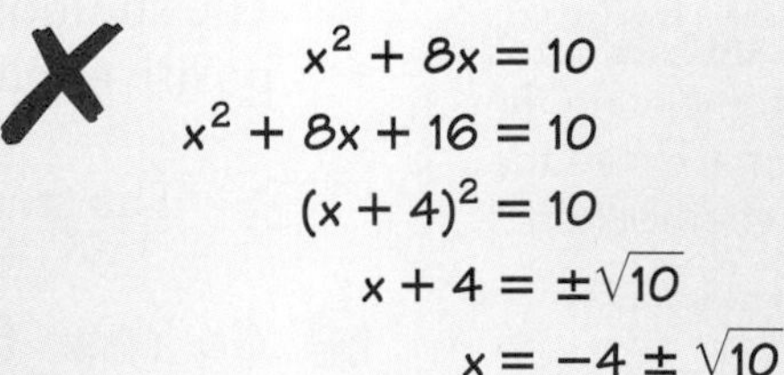

$$x^2 + 8x = 10$$
$$x^2 + 8x + 16 = 10$$
$$(x + 4)^2 = 10$$
$$x + 4 = \pm\sqrt{10}$$
$$x = -4 \pm \sqrt{10}$$

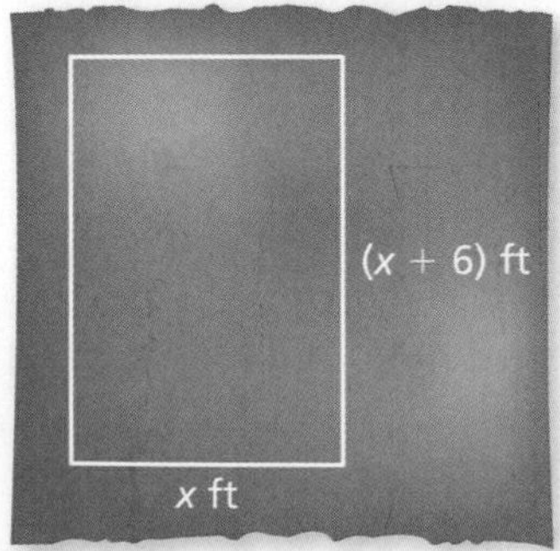

25. **PATIO** The area of the new patio is 216 square feet.
    a. Write an equation for the area of the patio.
    b. Find the dimensions of the patio by completing the square.

Assignment Guide and Homework Check

Level	Assignment	Homework Check
Average	1–4, 7–23 odd, 24, 26, 32–36	7, 15, 19, 24
Advanced	1, 2, 12–24 even, 26–36	16, 20, 24, 28

Common Errors

- **Exercises 6–23** Students may forget to divide the x-coefficient by 2 before squaring. Remind them of this process.
- **Exercises 12–23** Students may factor the trinomial as $\left[x + \left(\frac{b}{2}\right)^2\right]^2$ instead of $\left(x + \frac{b}{2}\right)^2$. Remind them that in the factored form of the trinomial, the term $\frac{b}{2}$ should not be squared.
- **Exercises 18–23** Students may not add the same value to each side of the equation when completing the square. Remind students that to form an equivalent equation, they must add the same quantity to each side.

9.3 Record and Practice Journal

Complete the square for the expression. Then factor the trinomial.

1. $x^2 + 8x$ $\quad x^2 + 8x + 16 = (x + 4)^2$
2. $x^2 - 6x$ $\quad x^2 - 6x + 9 = (x - 3)^2$
3. $x^2 - 20x$ $\quad x^2 - 20x + 100 = (x - 10)^2$
4. $x^2 + 7x$ $\quad x^2 + 7x + \frac{49}{4} = \left(x + \frac{7}{2}\right)^2$

Solve the equation by completing the square.

5. $x^2 - 2x = 8$ $\quad x = -2, 4$
6. $x^2 + 12x + 9 = -5$ $\quad x = -6 \pm \sqrt{22}$
7. $x^2 + 5x + 13 = 1$ $\quad$ no solution
8. $2x^2 - 6x + 3 = 11$ $\quad x = -1, 4$
9. Your backyard is rectangular and has an area of 760 square meters. The length of the yard is 18 meters more than the width. Find the length and width of the backyard.
   length: 38 m; width: 20 m

Answer Presentation Tool

Vocabulary and Concept Check

1. Add $\left(\frac{b}{2}\right)^2$.
2. A perfect square trinomial is a trinomial that can be factored as the square of a binomial.
   Example:
   $(x + 2)^2 = x^2 + 4x + 4$

Practice and Problem Solving

3. $x^2 - 4x + 4$

4–5. See Additional Answers.

6. 16
7. 4
8. 1
9. 49
10. 36
11. 81
12. $x^2 - 10x + 25 = (x - 5)^2$
13. $x^2 + 16x + 64 = (x + 8)^2$
14. $x^2 + 22x + 121 = (x + 11)^2$
15. $x^2 - 40x + 400 = (x - 20)^2$
16. $x^2 - 3x + \frac{9}{4} = \left(x - \frac{3}{2}\right)^2$
17. $x^2 + 5x + \frac{25}{4} = \left(x + \frac{5}{2}\right)^2$
18. $x = -3, x = 1$
19. $x = -2, x = 8$
20. no real solutions
21. no real solutions
22. $x = -1, x = 5$
23. $x = \frac{1}{2}, x = 1$
24. 16 was not added to both sides. $x = -4 \pm \sqrt{26}$
25. **a.** $216 = x(x + 6)$
    b. 12 ft by 18 ft

Practice and Problem Solving

26. 10

27. Divide each side by 3.

28. **a.** $x = -6, x = 2$

b. Evaluate $y = x^2 + 4x - 12$ when x is the mean of the solutions.

29. **a.** after about 4.4 seconds

b. 96 ft; The vertex is (2, 96). The maximum is the y-coordinate of the vertex.

30. See *Taking Math Deeper.*

31. $x(x + 1) = 42$; 6, 7

32. **a.** $y = (x + 2)^2 - 1$; The minimum value is -1.

b. $c - \frac{b^2}{4}$

Fair Game Review

33. $2\sqrt{3}$ **34.** $6\sqrt{2}$

35. $2\sqrt{5}$ **36.** B

Mini-Assessment

Solve the equation by completing the square.

1. $x^2 + 2x = 15$ $x = -5, x = 3$
2. $x^2 - 12x = -32$ $x = 4, x = 8$
3. $x^2 + 6x + 3 = -7$ no real solutions
4. $2x^2 - 11x + 4 = 10$ $x = -\frac{1}{2}, x = 6$
5. A toy rocket is launched from the top of a building. The function $h = -16t^2 + 64t + 64$ gives the height h of the rocket after t seconds. When does the rocket hit the ground? (Round your answer to the nearest tenth of a second.) after about 4.8 seconds

Taking Math Deeper

Exercise 30

The key to this exercise is that the garage forms one side of the garden. So, fencing is needed for only three of the sides.

1. Draw a diagram of the situation. Let ℓ represent the length of the garden and let w represent the width.

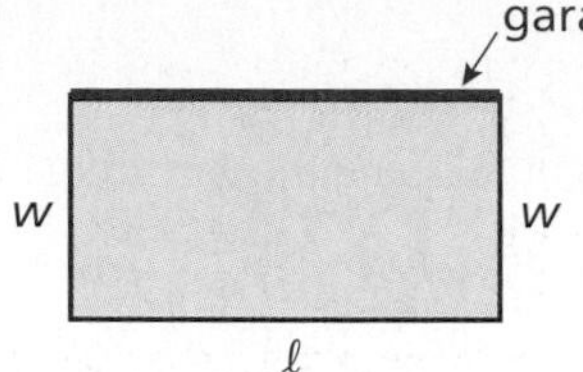

2. Use the information about the perimeter to write an equation.

$$\ell + 2w = 40, \text{ or } \ell = 40 - 2w$$

Use the information about the area to write an equation.

$$\ell w = 100, \text{ or } \ell = \frac{100}{w}$$

3. Graph the two equations and find the points of intersection. Let w be the independent variable.

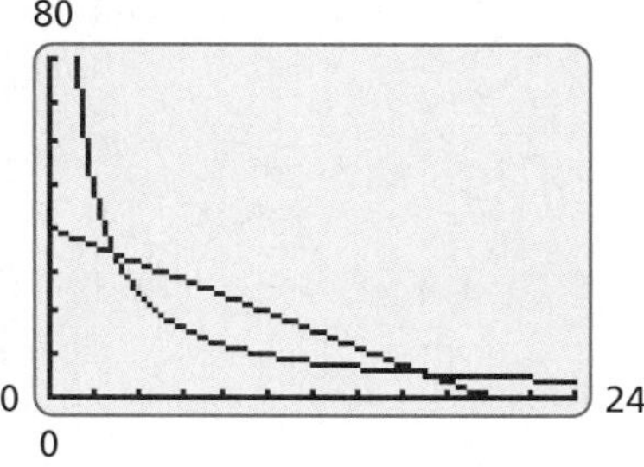

The graphs intersect at about (2.9, 34.1) and (17.1, 5.9). So, the possible dimensions of the garden are 2.9 feet by 34.1 feet and 17.1 feet by 5.9 feet.

Many students will choose the garden that is 17.1 feet by 5.9 feet because it is less narrow.

Challenge students to explain why these dimensions do not add up to a perimeter of 40 feet.

Project

Research standard garage sizes. Does this play a roll in your answer?

Reteaching and Enrichment Strategies

If students need help...	If students got it...
Resources by Chapter • Practice A and Practice B • Puzzle Time Record and Practice Journal Practice Differentiating the Lesson Lesson Tutorials Skills Review Handbook	Resources by Chapter • Enrichment and Extension • School-to-Work Start the next section

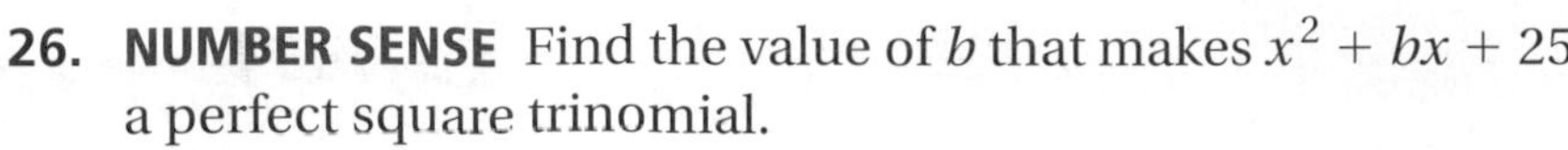

26. NUMBER SENSE Find the value of b that makes $x^2 + bx + 25$ a perfect square trinomial.

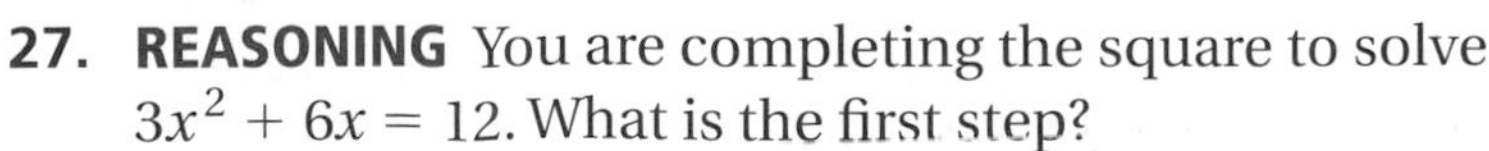

27. REASONING You are completing the square to solve $3x^2 + 6x = 12$. What is the first step?

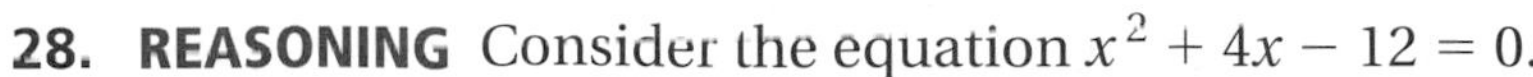

28. REASONING Consider the equation $x^2 + 4x - 12 = 0$.

a. Solve the equation by completing the square.

b. Explain how to use the solutions to find the minimum value of $y = x^2 + 4x - 12$.

29. TOY ROCKET The function $h = -16t^2 + 64t + 32$ gives the height h (in feet) of a toy rocket after t seconds.

a. When does the rocket hit the ground?

b. What is the maximum height of the rocket? Justify your answer.

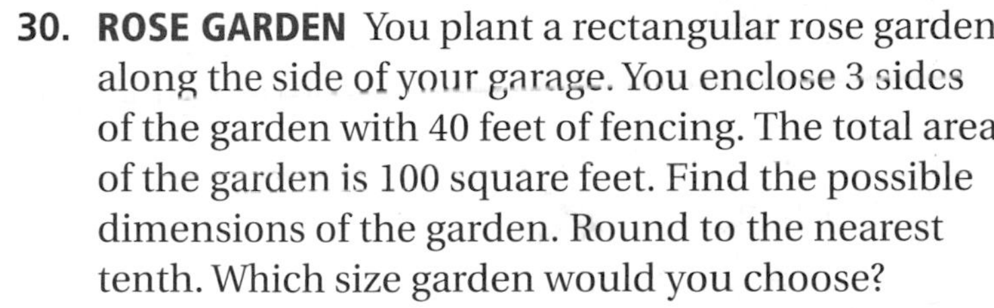

30. ROSE GARDEN You plant a rectangular rose garden along the side of your garage. You enclose 3 sides of the garden with 40 feet of fencing. The total area of the garden is 100 square feet. Find the possible dimensions of the garden. Round to the nearest tenth. Which size garden would you choose?

31. PRECISION The product of two consecutive positive integers is 42. Write and solve an equation to find the integers.

32. **Structure** Begin solving $x^2 + 4x + 3 = 0$ by completing the square. Stop when you obtain an equation of the form $(x + p)^2 = q$.

a. Write the related function in vertex form. Without graphing, determine the maximum or minimum value of the function.

b. Find the minimum value of $y = x^2 + bx + c$.

Fair Game Review What you learned in previous grades & lessons

Simplify $\sqrt{b^2 - 4ac}$ for the given values. *(Section 6.1)*

33. $a = 3, b = -6, c = 2$ **34.** $a = -2, b = 4, c = 7$ **35.** $a = 1, b = 6, c = 4$

36. MULTIPLE CHOICE What are the solutions of $x^2 - 49 = 0$? *(Section 9.2)*

Ⓐ $x = 7$ Ⓑ $x = -7, x = 7$ Ⓒ $x = 0, x = 7$ Ⓓ no solution

9 Study Help

You can use an **information wheel** to organize information about a topic. Here is an example of an information wheel for quadratic equations.

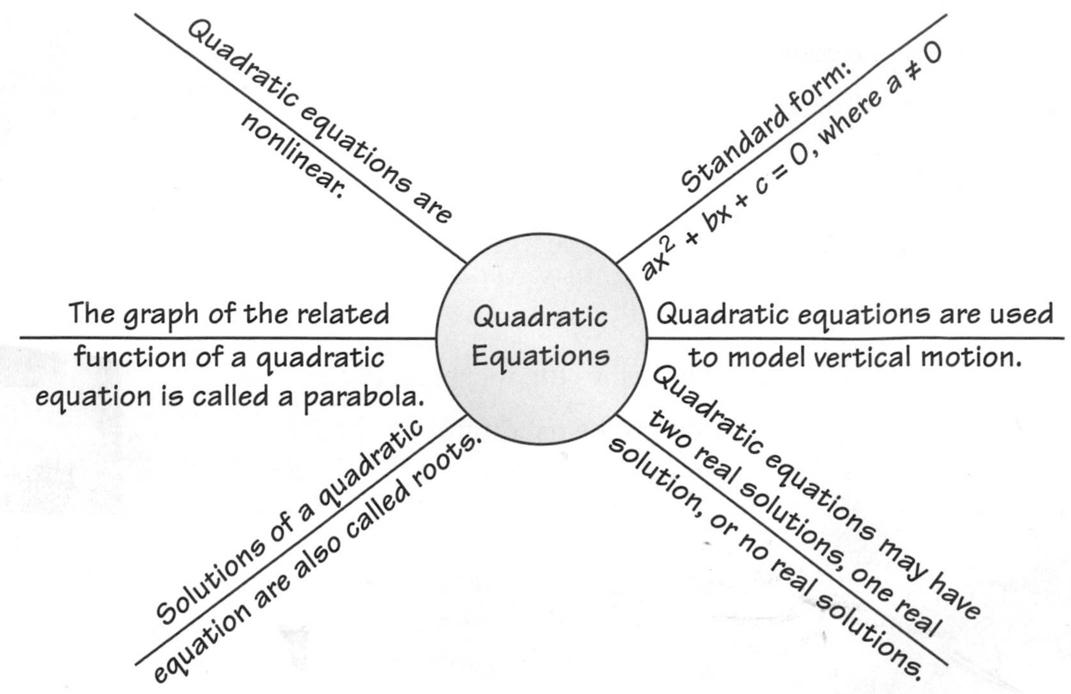

Quadratic Equations
Quadratic equations are nonlinear.
Standard form: $ax^2 + bx + c = 0$, where $a \neq 0$
The graph of the related function of a quadratic equation is called a parabola.
Quadratic equations are used to model vertical motion.
Solutions of a quadratic equation are also called roots.
Quadratic equations may have two real solutions, one real solution, or no real solutions.

On Your Own

Make information wheels to help you study these topics.

1. solving quadratic equations by graphing
2. solving quadratic equations using square roots
3. solving quadratic equations by completing the square

After you complete this chapter, make information wheels for the following topics.

4. solving quadratic equations using the quadratic formula
5. choosing a solution method for solving quadratic equations
6. solving systems of linear and quadratic equations

"My information wheel for Fluffy has matching adjectives and nouns."

Sample Answers

1.

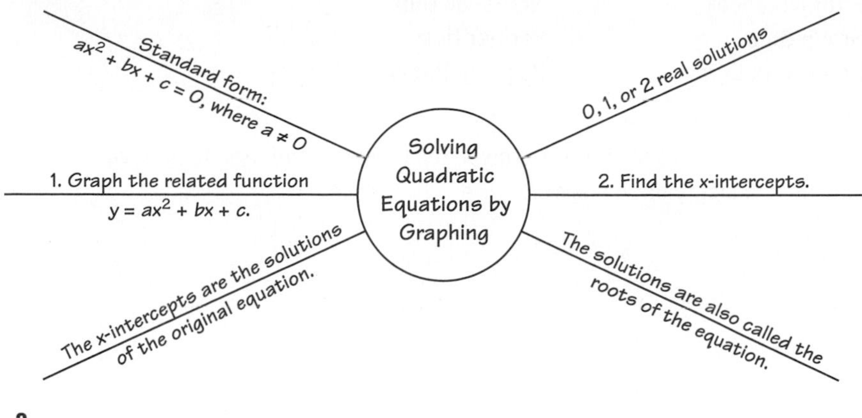

2.

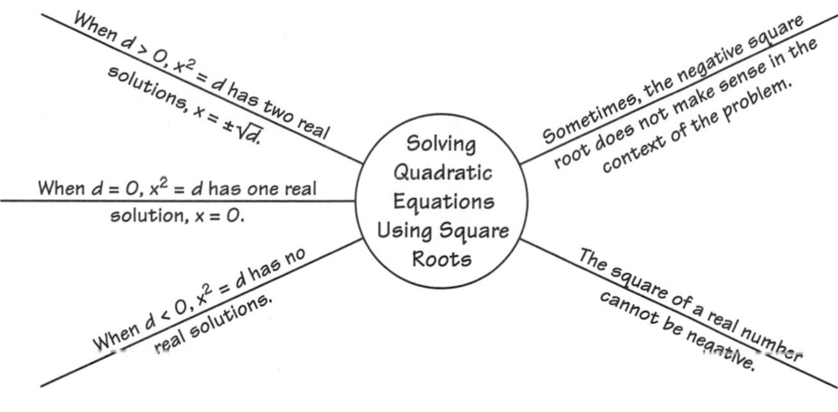

Solving Quadratic Equations Using Square Roots
When $d > 0$, $x^2 = d$ has two real solutions, $x = \pm\sqrt{d}$.
Sometimes, the negative square root does not make sense in the context of the problem.
When $d = 0$, $x^2 = d$ has one real solution, $x = 0$.
When $d < 0$, $x^2 = d$ has no real solutions.
The square of a real number cannot be negative.

3.

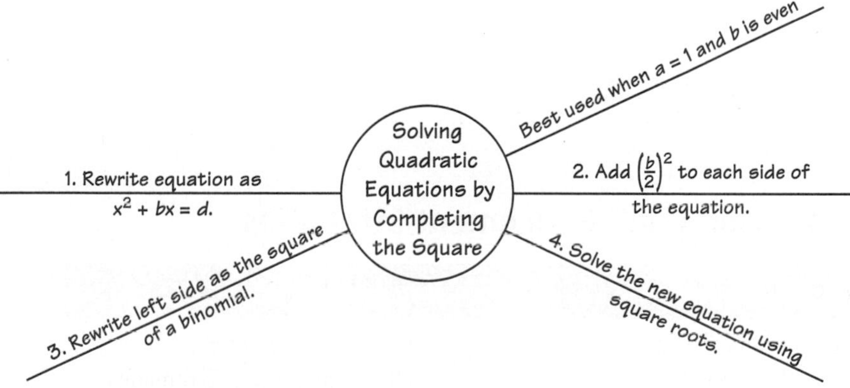

Solving Quadratic Equations by Completing the Square
Best used when $a = 1$ and b is even
1. Rewrite equation as $x^2 + bx = d$.
2. Add $\left(\frac{b}{2}\right)^2$ to each side of the equation.
3. Rewrite left side as the square of a binomial.
4. Solve the new equation using square roots.

List of Organizers

Available at *BigIdeasMath.com*

Comparison Chart
Concept Circle
Definition (Idea) and Example Chart
Example and Non-Example Chart
Formula Triangle
Four Square
Information Frame
Information Wheel
Notetaking Organizer
Process Diagram
Summary Triangle
Word Magnet
Y Chart

About this Organizer

A **Information Wheel** can be used to organize information about a concept. Students write the concept in the middle of the "wheel." Then students write information related to the concept on the "spokes" of the wheel. Related information can include, but is not limited to: vocabulary words or terms, definitions, formulas, procedures, examples, and visuals. This type of organizer serves as a good summary tool because any information related to a concept can be included.

Technology For the Teacher
Vocabulary Puzzle Builder

Answers

1. $x = -1, x = 3$
2. no real solutions
3. $x = -5$
4. $x = -7, x = -2$
5. $x = -1, x = 8$
6. no real solutions
7. $x = 4, x = -4$
8. no real solutions
9. $x = 7, x = 9$
10. $x = -9, x = 5$
11. $x = 1 + \sqrt{10}, x = 1 - \sqrt{10}$
12. $x = -7, x = 1$
13. $x = 1 + 2\sqrt{2}, x = 1 - 2\sqrt{2}$
14. Because $x^2 = 100$ has the form $x^2 = d$ with $d > 0$, there are 2 real solutions.
15. length: $4\sqrt{19} \approx 17.44$ m

    width: $\sqrt{19} \approx 4.36$ m
16. **a.** about 0.19 second and about 2.31 seconds

    b. 28 ft

Assessment Book

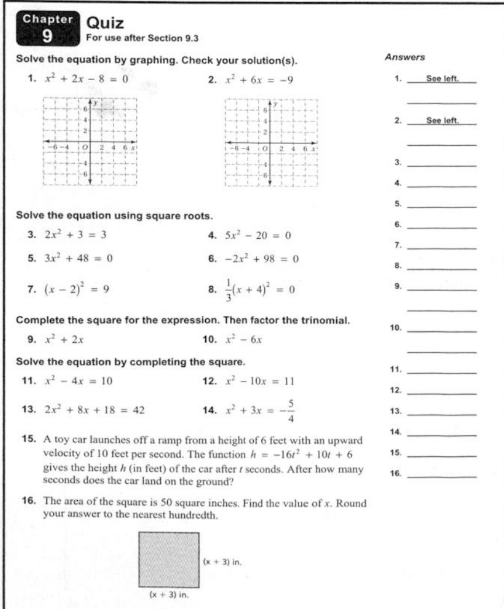

Chapter 9 **Quiz**
For use after Section 9.3

Solve the equation by graphing. Check your solution(s).

1. $x^2 + 2x - 8 = 0$
2. $x^2 + 6x = -9$

Solve the equation using square roots.

3. $2x^2 + 3 = 3$
4. $5x^2 - 20 = 0$
5. $3x^2 + 48 = 0$
6. $-2x^2 + 98 = 0$
7. $(x - 2)^2 = 9$
8. $\frac{1}{3}(x + 4)^2 = 0$

Complete the square for the expression. Then factor the trinomial.

9. $x^2 + 2x$
10. $x^2 - 6x$

Solve the equation by completing the square.

11. $x^2 - 4x = 10$
12. $x^2 - 10x = 11$
13. $2x^2 + 8x + 18 = 42$
14. $x^2 + 3x = -\frac{5}{4}$
15. A toy car launches off a ramp from a height of 6 feet with an upward velocity of 10 feet per second. The function $h = -16t^2 + 10t + 6$ gives the height h (in feet) of the car after t seconds. After how many seconds does the car land on the ground?
16. The area of the square is 50 square inches. Find the value of x. Round your answer to the nearest hundredth.

(x + 3) in.
(x + 3) in.

Answers

1. See left.
2. See left.
3.
4.
5.
6.
7.
8.
9.
10.
11.
12.
13.
14.
15.
16.

Alternative Quiz Ideas

100% Quiz
Error Notebook
Group Quiz
Homework Quiz
Math Log
Notebook Quiz
Partner Quiz
Pass the Paper

100% Quiz
This is a quiz where students are given the answers and then they have to explain and justify each answer.

Reteaching and Enrichment Strategies

If students need help...	If students got it...
Resources by Chapter • Study Help • Practice A and Practice B • Puzzle Time Lesson Tutorials *BigIdeasMath.com* Practice Quiz Practice from the Test Generator	Resources by Chapter • Enrichment and Extension • School-to-Work Game Closet at *BigIdeasMath.com* Start the next section

Technology For the Teacher
Answer Presentation Tool
Big Ideas Test Generator

9.1–9.3 Quiz

Check It Out
Progress Check
BigIdeasMath.com

Determine the solution(s) of the equation. Check your solution(s). *(Section 9.1)*

1. $x^2 - 2x - 3 = 0$

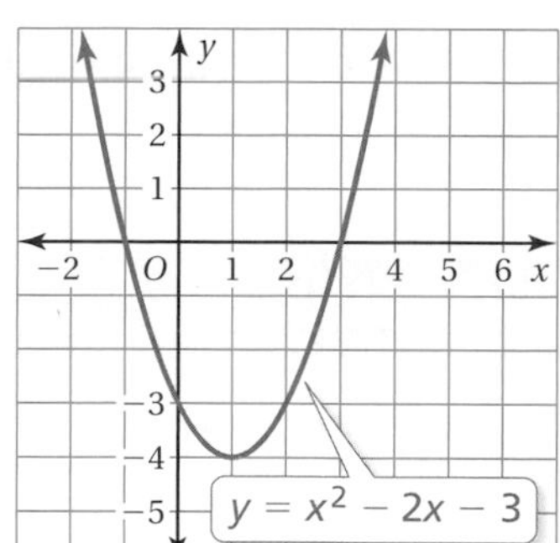

2. $x^2 - 2x + 3 = 0$

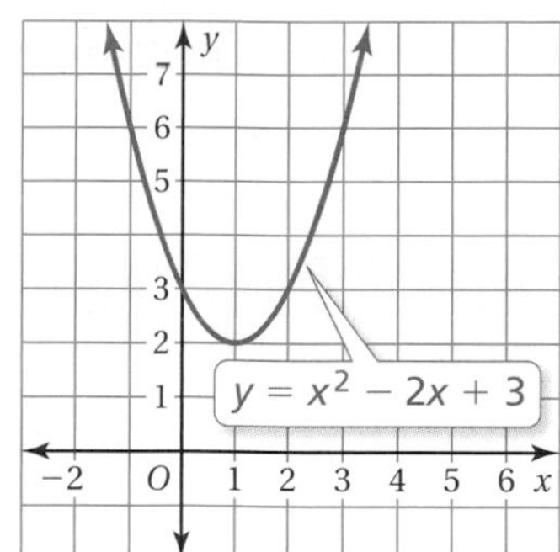

3. $x^2 + 10x + 25 = 0$

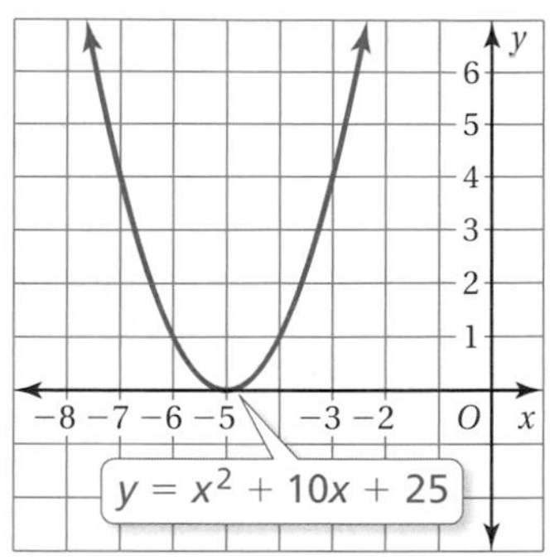

Solve the equation by graphing. Check your solution(s). *(Section 9.1)*

4. $x^2 + 9x + 14 = 0$

5. $x^2 - 7x = 8$

6. $x + 1 = -x^2$

Solve the equation using square roots. *(Section 9.2)*

7. $4x^2 = 64$

8. $-3x^2 + 6 = 10$

9. $(x - 8)^2 = 1$

Solve the equation by completing the square. *(Section 9.3)*

10. $x^2 + 4x = 45$

11. $x^2 - 2x - 1 = 8$

12. $2x^2 + 12x + 20 = 34$

13. $-4x^2 + 8x + 44 = 16$

14. REASONING Explain how to determine the number of real solutions of $x^2 = 100$ without solving. *(Section 9.2)*

15. VOLUME The length of a rectangular prism is 4 times its width. The volume of the prism is 380 cubic meters. Find the length and width of the prism. *(Section 9.2)*

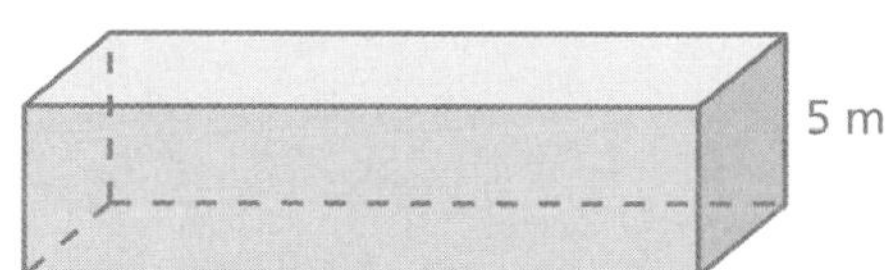

16. PROBLEM SOLVING A cannon launches a cannonball from a height of 3 feet with an upward velocity of 40 feet per second. The function $h = -16t^2 + 40t + 3$ gives the height h (in feet) of the cannonball after t seconds. *(Section 9.1 and Section 9.3)*

a. After how many seconds is the cannonball 10 feet above the ground?

b. What is the maximum height of the cannonball?

9.4 Solving Quadratic Equations Using the Quadratic Formula

COMMON CORE STATE STANDARDS
A.REI.4a
A.REI.4b

Essential Question How can you use the discriminant to determine the number of solutions of a quadratic equation?

1 ACTIVITY: Deriving the Quadratic Formula

Work with a partner. The following steps show one method of solving $ax^2 + bx + c = 0$. Explain what was done in each step.

$ax^2 + bx + c = 0$ ← 1. Write the equation.

$4a^2x^2 + 4abx + 4ac = 0$ ← 2. What was done?

$4a^2x^2 + 4abx + 4ac + b^2 = b^2$ ← 3. What was done?

$4a^2x^2 + 4abx + b^2 = b^2 - 4ac$ ← 4. What was done?

$(2ax + b)^2 = b^2 - 4ac$ ← 5. What was done?

$2ax + b = \pm\sqrt{b^2 - 4ac}$ ← 6. What was done?

$2ax = -b \pm \sqrt{b^2 - 4ac}$ ← 7. What was done?

Quadratic Formula: $x = \dfrac{-b \pm \sqrt{b^2 - 4ac}}{2a}$ ← 8. What was done?

2 ACTIVITY: Deriving the Quadratic Formula by Completing the Square

- Solve $ax^2 + bx + c = 0$ by completing the square. (*Hint:* Subtract c from each side, divide each side by a, and then proceed by completing the square.)
- Compare this method with the method in Activity 1. Explain why you think $4a$ and b^2 were chosen in Steps 2 and 3 of Activity 1.

Laurie's Notes

Introduction

Standards for Mathematical Practice

- **MP1a Make Sense of Problems:** Students read and explain the steps provided for deriving the quadratic formula. This helps deepen their understanding of the quadratic formula and the process for deriving a formula.
- **MP3b Critique the Reasoning of Others:** Students derive the quadratic formula on their own in Activity 2 by completing the square. Then they compare the two methods.

Motivate

- **Story Time:** Share with students some history about the quadratic formula.
  - Around 400 B.C., the Babylonians and Chinese use a method called "completing the square" to solve problems involving areas.
  - Around 300 B.C., the Greek mathematicians Pythagoras and Euclid use geometry to find a general procedure for solving a quadratic equation, but their methods are not considered useful.
  - Around 700 A.D., the Hindu mathematician named Brahmagupta finds the general solution for the quadratic equation. He uses irrational numbers and also recognizes the two roots in the solution.
  - Around 1100 A.D., another Hindu mathematician named Baskhara finds the complete solution. He recognizes that any positive number has two square roots.
  - In 1637, the French mathematician René Descartes publishes *La Géométrie* which presents the quadratic formula in its present form.
- Look back at Examples 2 and 3 in Section 6.1. Students will be evaluating and simplifying these types of expressions in this section.

Activity Notes

Discuss

- Tell students that today they will see one way of deriving the quadratic formula algebraically. Then they will derive the quadratic formula on their own by completing the square. They will also learn about the discriminant and discover how it relates to the number of solutions.

Activity 1

- A derivation of the quadratic formula is shown in this activity. Students justify each step.
- If students get stuck on a step, ask them what is different from the last step. Mathematically, what has changed?
- When students are finished, discuss the justifications.
- Students sometimes lose sight of what they have accomplished. Starting with the quadratic equation in general form, they solve for x algebraically to produce a formula for finding the solutions of any quadratic equation.

Common Core State Standards

A.REI.4a Use the method of completing the square to transform any quadratic equation in x into an equation of the form $(x - p)^2 = q$ that has the same solutions. Derive the quadratic formula from this form.

A.REI.4b Solve quadratic equations by . . . the quadratic formula . . ..

Previous Learning

Students should know how to solve quadratic equations by completing the square. Students should know how to find the square roots of positive numbers.

Start Thinking! and Warm Up

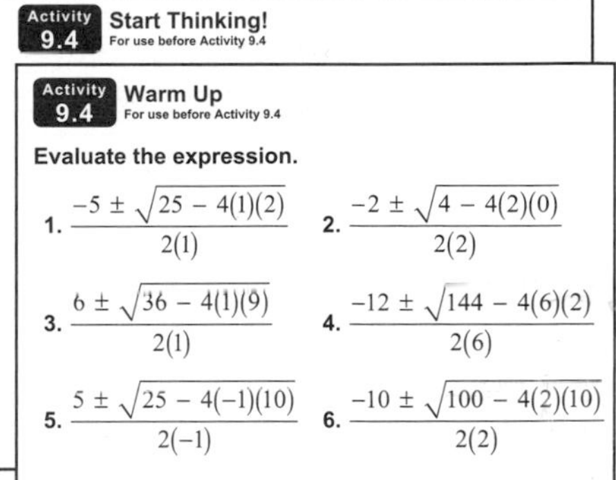

Activity 9.4 Start Thinking! For use before Activity 9.4

Activity 9.4 Warm Up For use before Activity 9.4

Evaluate the expression.

1. $\dfrac{-5 \pm \sqrt{25 - 4(1)(2)}}{2(1)}$
2. $\dfrac{-2 \pm \sqrt{4 - 4(2)(0)}}{2(2)}$
3. $\dfrac{6 \pm \sqrt{36 - 4(1)(9)}}{2(1)}$
4. $\dfrac{-12 \pm \sqrt{144 - 4(6)(2)}}{2(6)}$
5. $\dfrac{5 \pm \sqrt{25 - 4(-1)(10)}}{2(-1)}$
6. $\dfrac{-10 \pm \sqrt{100 - 4(2)(10)}}{2(2)}$

9.4 Record and Practice Journal

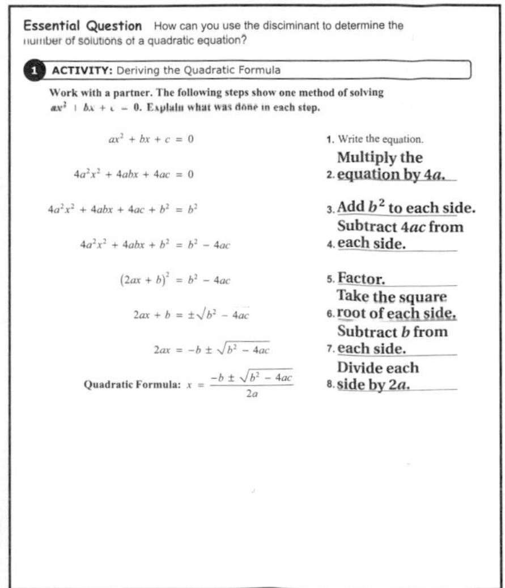

Essential Question How can you use the discriminant to determine the number of solutions of a quadratic equation?

1 ACTIVITY: Deriving the Quadratic Formula

Work with a partner. The following steps show one method of solving $ax^2 + bx + c = 0$. Explain what was done in each step.

$ax^2 + bx + c = 0$	1. Write the equation.
$4a^2x^2 + 4abx + 4ac = 0$	2. Multiply the equation by $4a$.
$4a^2x^2 + 4abx + 4ac + b^2 = b^2$	3. Add b^2 to each side.
$4a^2x^2 + 4abx + b^2 = b^2 - 4ac$	4. Subtract $4ac$ from each side.
$(2ax + b)^2 = b^2 - 4ac$	5. Factor.
$2ax + b = \pm\sqrt{b^2 - 4ac}$	6. Take the square root of each side.
$2ax = -b \pm \sqrt{b^2 - 4ac}$	7. Subtract b from each side.
Quadratic Formula: $x = \dfrac{-b \pm \sqrt{b^2 - 4ac}}{2a}$	8. Divide each side by $2a$.

Differentiated Instruction

Visual, Auditory

In Activity 3, explain that each graph represents just one example of each solution type. Show other graphs for each solution type and discuss the common characteristics of these graphs.

9.4 Record and Practice Journal

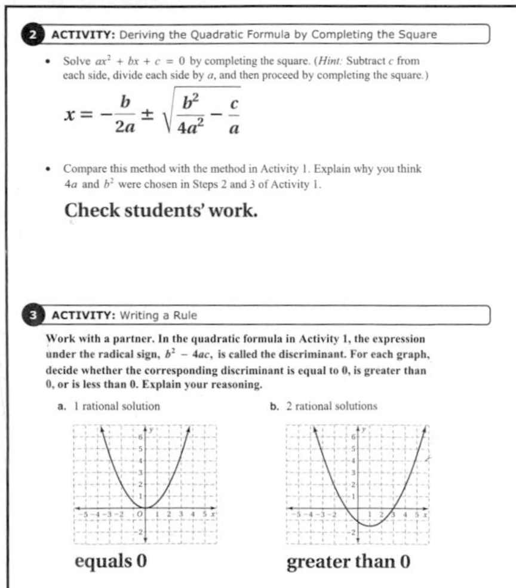

2 ACTIVITY: Deriving the Quadratic Formula by Completing the Square

- Solve $ax^2 + bx + c = 0$ by completing the square. (*Hint:* Subtract c from each side, divide each side by a, and then proceed by completing the square.)

$$x = -\frac{b}{2a} \pm \sqrt{\frac{b^2}{4a^2} - \frac{c}{a}}$$

- Compare this method with the method in Activity 1. Explain why you think $4a$ and b^2 were chosen in Steps 2 and 3 of Activity 1.

Check students' work.

3 ACTIVITY: Writing a Rule

Work with a partner. In the quadratic formula in Activity 1, the expression under the radical sign, $b^2 - 4ac$, is called the discriminant. For each graph, decide whether the corresponding discriminant is equal to 0, is greater than 0, or is less than 0. Explain your reasoning.

a. 1 rational solution

equals 0

b. 2 rational solutions

greater than 0

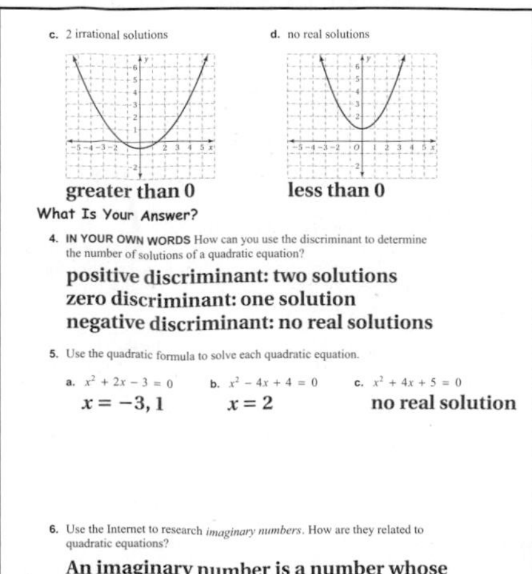

c. 2 irrational solutions

greater than 0

d. no real solutions

less than 0

What Is Your Answer?

4. **IN YOUR OWN WORDS** How can you use the discriminant to determine the number of solutions of a quadratic equation?

positive discriminant: two solutions
zero discriminant: one solution
negative discriminant: no real solutions

5. Use the quadratic formula to solve each quadratic equation.

a. $x^2 + 2x - 3 = 0$ — $x = -3, 1$

b. $x^2 - 4x + 4 = 0$ — $x = 2$

c. $x^2 + 4x + 5 = 0$ — **no real solution**

6. Use the Internet to research *imaginary numbers*. How are they related to quadratic equations?

An imaginary number is a number whose square is negative.
When the discriminant is negative.

Laurie's Notes

Activity 2

- In Section 9.3, students learned how to solve a quadratic equation by completing the square. They are using that method to derive the quadratic formula and compare it to the method used in Activity 1.
- **MP1b Persevere in Solving Problems:** Do not be too quick to rescue your students. Give them time to wrestle with the derivation. If they get stuck, have them refer back to their notes from the last section. Believe that your students can persevere in deriving the formula. Unlike many problems, they know what the formula should look like when they finish because of Activity 1.

Activity 3

- **Teaching Tip:** Begin by having students identify a, b, and c for a quadratic equation such as $3x^2 - 4x + 8 = 0$. Then have them substitute the values for a, b, and c into the quadratic formula.
- **Common Error:** Students may make mistakes when using negative signs. For $3x^2 - 4x + 8 = 0$, the value of b is -4. So, in the quadratic formula, $-b = -(4) = 4$

? "How many solutions does a quadratic equation have when the value of the discriminant is 0?" 1

What Is Your Answer?

- **MP7 Look for and Make Use of Structure:** In Question 4, students should try to list rules for the number of solutions based on the value of the discriminant.
- To solve each quadratic equation in Question 5, tell students they need to evaluate the quadratic formula for the values of a, b, and c from the equation.

Closure

- Without referring to your notes, write the quadratic formula and explain how to find a, b, and c. $x = \dfrac{-b \pm \sqrt{b^2 - 4ac}}{2a}$; The values of a, b, and c come from the standard form of the quadratic equation $ax^2 + bx + c = 0$.

Technology For the Teacher

The Dynamic Planning Tool
Editable Teacher's Resources at *BigIdeasMath.com*

3 ACTIVITY: Analyzing the Solutions of an Equation

Work with a partner. In the quadratic formula in Activity 1, the expression under the radical sign, $b^2 - 4ac$, is called the discriminant. For each graph, decide whether the corresponding discriminant is equal to 0, is greater than 0, or is less than 0. Explain your reasoning.

a. 1 rational solution

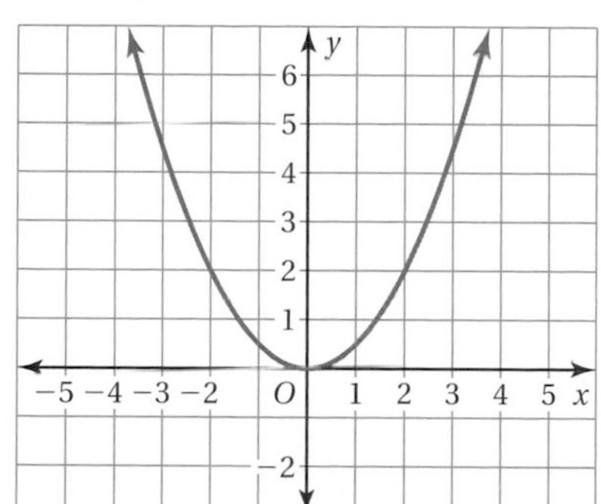

b. 2 rational solutions

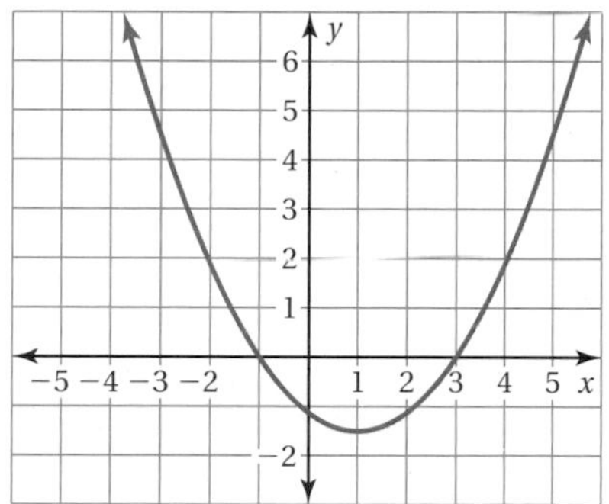

c. 2 irrational solutions

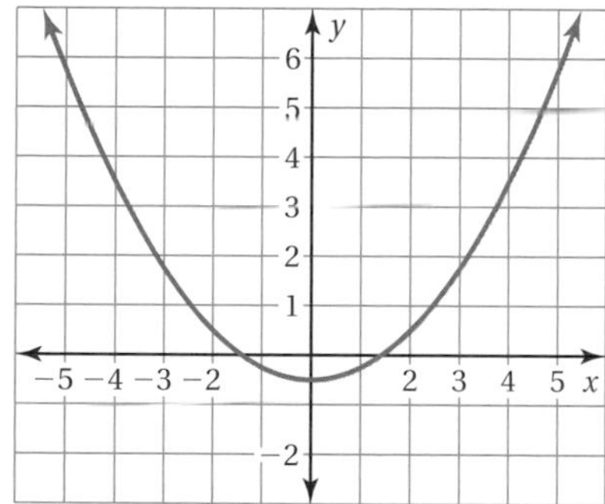

d. no real solutions

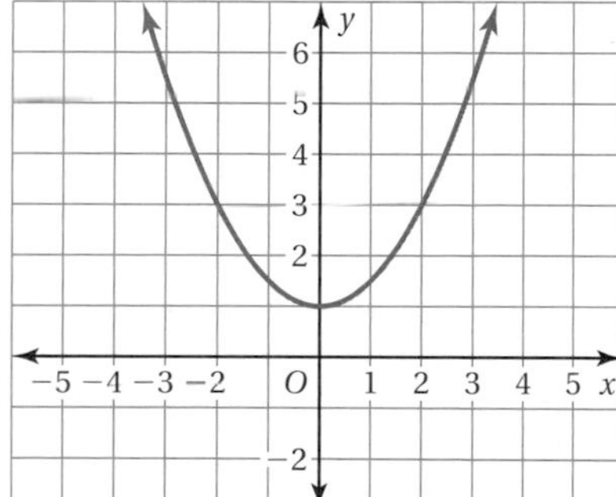

What Is Your Answer?

4. **IN YOUR OWN WORDS** How can you use the discriminant to determine the number of solutions of a quadratic equation?

5. Use the quadratic formula to solve each quadratic equation.

   a. $x^2 + 2x - 3 = 0$ **b.** $x^2 - 4x + 4 = 0$ **c.** $x^2 + 4x + 5 = 0$

6. Use the Internet to research *imaginary numbers*. How are they related to quadratic equations?

Practice Use what you learned about quadratic equations to complete Exercises 9–11 on page 481.

X e - quals neg-a-tive B Plus or mi-nus the square root of B squared mi - nus four A C all o - ver two A

9.4 Lesson

Key Vocabulary
quadratic formula, *p. 478*
discriminant, *p. 480*

Another way to solve quadratic equations is to use the *quadratic formula.*

Key Idea

Quadratic Formula

The real solutions of the quadratic equation $ax^2 + bx + c = 0$ are

$$x = \frac{-b \pm \sqrt{b^2 - 4ac}}{2a}$$

where $a \neq 0$ and $b^2 - 4ac \geq 0$. This is called the **quadratic formula.**

EXAMPLE 1 Solving a Quadratic Equation Using the Quadratic Formula

Solve $2x^2 - 5x + 3 = 0$ using the quadratic formula.

$x = \dfrac{-b \pm \sqrt{b^2 - 4ac}}{2a}$ — Quadratic Formula

$= \dfrac{-(-5) \pm \sqrt{(-5)^2 - 4(2)(3)}}{2(2)}$ — Substitute 2 for *a*, −5 for *b*, and 3 for *c*.

$= \dfrac{5 \pm \sqrt{1}}{4}$ — Simplify.

$= \dfrac{5 \pm 1}{4}$ — Evaluate the square root.

∴ So, the solutions are $x = \dfrac{5+1}{4} = \dfrac{3}{2}$ and $x = \dfrac{5-1}{4} = 1$.

Study Tip

You can use the roots of a quadratic equation to factor the related expression. In Example 1, you can use 1 and $\frac{3}{2}$ to factor $2x^2 - 5x + 3$ as $(x - 1)(2x - 3)$.

Check Check each solution in the original equation.

$2x^2 - 5x + 3 = 0$	Original equation	$2x^2 - 5x + 3 = 0$
$2\left(\frac{3}{2}\right)^2 - 5\left(\frac{3}{2}\right) + 3 \stackrel{?}{=} 0$	Substitute.	$2(1)^2 - 5(1) + 3 \stackrel{?}{=} 0$
$\frac{9}{2} - \frac{15}{2} + 3 \stackrel{?}{=} 0$	Simplify.	$2 - 5 + 3 \stackrel{?}{=} 0$
$0 = 0$ ✓	Simplify.	$0 = 0$ ✓

On Your Own

Now You're Ready
Exercises 12–14

Solve the equation using the quadratic formula.

1. $x^2 - 6x + 5 = 0$
2. $4x^2 + x - 3 = 0$
3. $-6x^2 + 7x - 2 = 0$

Laurie's Notes

Introduction

Connect

- **Yesterday:** Students derived the quadratic formula. (MP1, MP3b, MP7)
- **Today:** Students will solve quadratic equations using the quadratic formula.

Motivate

- There are many online videos of students singing the quadratic formula to the tune "Pop Goes the Weasel." Share one with students. You could return to the video at the end of the period when the students may be ready to sing along.

Lesson Notes

Discuss

- **MP2 Reason Abstractly and Quantitatively:** Students will decontextualize a quadratic model to solve for the roots and then interpret the roots in the context of the problem.
- Discuss with students the methods they have learned to solve quadratic equations. The form of the equation and the tools available often dictate the best method for solving. The quadratic formula is another method that can be used for solving quadratic equations.

Key Idea

- Write the Key Idea on the board.
- **?** "Why must one side of the quadratic equation be equal to 0?" So you can determine the values of *a, b,* and *c*.
- **?** "Why is there a restriction that $a \neq 0$?" When $a = 0$, the equation is not a quadratic equation. Also, the denominator in the quadratic formula would be 0, and you cannot divide by 0.
- **?** "Why is there a restriction that $b^2 - 4ac \geq 0$?" The expression $b^2 - 4ac$ is under a radical sign in the formula. You cannot take the square root of a negative number.

Example 1

- Write the equation on the board and ask students to identify *a, b,* and *c*. Be sure students include the sign of each number, such as $b = -5$.
- As you substitute each value into the formula, point at and read aloud the corresponding term in the quadratic equation.
- Review the order of operations as you simplify.
- **Representation:** Students may still have trouble working with the plus/minus symbol $\pm$. Read the expression slowly and translate: "5 plus or minus the square root of 1 represents the two values: 5 plus the square root of 1, and 5 minus the square root of 1."

? **Connection:** "What type of numbers are $\frac{3}{2}$ and 1?" rational

Goal Today's lesson is solving quadratic equations using the **quadratic formula**.

Lesson Materials	
Introduction	**Textbook**
• video	• calculators

Start Thinking! and Warm Up

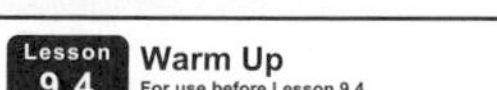

Lesson 9.4 Warm Up
For use before Lesson 9.4

Lesson 9.4 Start Thinking!
For use before Lesson 9.4

Consider the following equations:

$x^2 = 25 \qquad x^2 + 2x + 1 = 25 \qquad 3x^2 + 25x + 8 = 34$

Can all of the equations be solved using the quadratic formula? Is it the best method to solve all of the equations? Explain.

Extra Example 1

Solve $2x^2 + 7x + 3 = 0$ using the quadratic formula.

$x = -3, x = -\frac{1}{2}$

On Your Own

1. $x = 1, x = 5$
2. $x = -1, x = \frac{3}{4}$
3. $x = \frac{1}{2}, x = \frac{2}{3}$

Laurie's Notes

Example 1 (continued)

? "Recall in Chapter 7 you factored quadratic polynomials. Can $2x^2 - 5x + 3$ be factored?" yes; It factors as $(x - 1)(2x - 3)$.

- When the factored form is set equal to 0 the roots are 1 and $\frac{3}{2}$. Point out the Study Tip.
- **MP2:** By considering how the solutions of the quadratic equation are related to the factors of the related quadratic expression, students attend to the meaning of the quantities, using important reasoning that leads to a deeper understanding.

On Your Own

- Students should check their work with a neighbor after completing each question. Ask students who finish quickly to write the quadratic expressions in factored form.
- Discuss the solutions for Question 3. Rational roots are often more difficult for students to connect to the factors.

Example 2

? "Based on the discriminant, why is there only one solution?" The discriminant is 0, so adding or subtracting 0 gives the same number.

- Because there is one rational solution $x = -\frac{5}{2}$, students should understand that the expression can be factored as $(2x + 5)^2$.

Example 3

- Read through the problem statement. Ask questions to make sure that students understand the model. Interpret the *y*-intercept of the graph.

? "What is the first step in solving this equation?" Subtract 30 from each side.

- **MP2:** Students decontextualize the model to solve algebraically. Once the solutions are found, the context is considered again.
- Consider using technology to graph the function $y = 0.34x^2 + 3.0x - 21$. Use the graph to approximate the zeros. Then solve the corresponding equation by using the quadratic formula.

? "Do both solutions make sense in the context of the problem? Explain." no; Only the positive solution makes sense in the context of the problem.

? "Can the related expression be factored using only integers? Explain." no; Both solutions are irrational.

? "Do you think the trend shown by the graph will continue?" Students should recognize that while the graph increases, the number of breeding pairs cannot increase indefinitely due to the limits of nature.

On Your Own

- If time is a concern, have students do only the odd exercises.

Extra Example 2

Solve $4x^2 - 12x + 9 = 0$ using the quadratic formula.

$x = -\frac{3}{2}$

Extra Example 3

In Example 3, when were there about 55 breeding pairs?

2003

On Your Own

4. $x = \frac{1}{2}$
5. $x = -\frac{4}{5}, x = 1$
6. $x = -\frac{5}{3}, x = 1$
7. 2006

EXAMPLE 2 Solving a Quadratic Equation Using the Quadratic Formula

Solve $4x^2 + 20x + 25 = 0$ using the quadratic formula.

$$x = \frac{-b \pm \sqrt{b^2 - 4ac}}{2a}$$ Quadratic Formula

$$= \frac{-20 \pm \sqrt{20^2 - 4(4)(25)}}{2(4)}$$ Substitute 4 for *a*, 20 for *b*, and 25 for *c*.

$$= \frac{-20 \pm \sqrt{0}}{8} = -\frac{5}{2}$$ Simplify.

∴ The solution is $x = -\frac{5}{2}$.

EXAMPLE 3 Real-Life Application

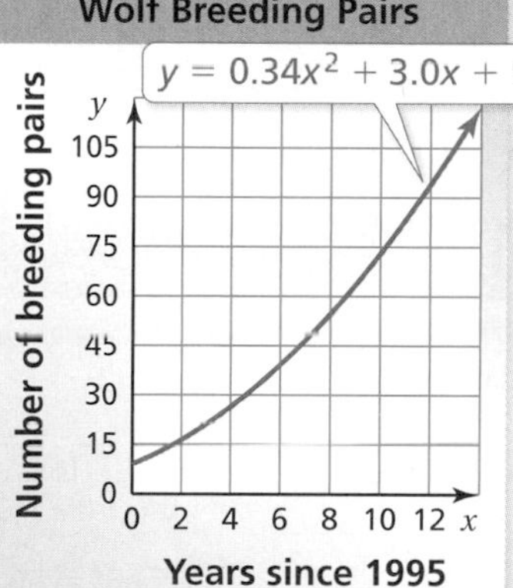

The number y of Northern Rocky Mountain wolf breeding pairs x years since 1995 can be modeled by $y = 0.34x^2 + 3.0x + 9$. When were there about 30 breeding pairs?

To determine when there were 30 breeding pairs, find the x-values for which $y = 30$. So, solve the equation $30 = 0.34x^2 + 3.0x + 9$.

$$30 = 0.34x^2 + 3.0x + 9$$ Write the equation.

$$0 = 0.34x^2 + 3.0x - 21$$ Write in standard form.

$$x = \frac{-b \pm \sqrt{b^2 - 4ac}}{2a}$$ Quadratic Formula

$$= \frac{-3.0 \pm \sqrt{3.0^2 - 4(0.34)(-21)}}{2(0.34)}$$ Substitute 0.34 for *a*, 3.0 for *b*, and -21 for *c*.

$$= \frac{-3.0 \pm \sqrt{37.56}}{0.68}$$ Simplify.

The solutions are $x = \frac{-3.0 + \sqrt{37.56}}{0.68} \approx 5$ and $x = \frac{-3.0 - \sqrt{37.56}}{0.68} \approx -13$.

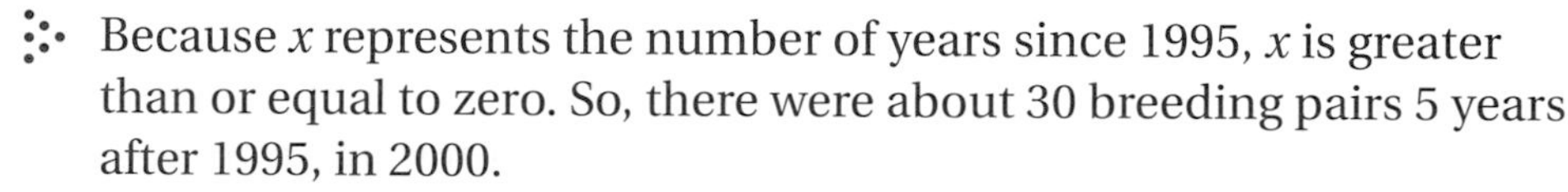

∴ Because x represents the number of years since 1995, x is greater than or equal to zero. So, there were about 30 breeding pairs 5 years after 1995, in 2000.

On Your Own

Solve the equation using the quadratic formula.

Now You're Ready
Exercises 15–23

4. $4x^2 - 4x + 1 = 0$ **5.** $-5x^2 + x = -4$ **6.** $3x^2 + 2x = 5$

7. **WHAT IF?** In Example 3, when were there about 85 breeding pairs?

The expression $b^2 - 4ac$ in the quadratic formula is the **discriminant.**

$$x = \frac{-b \pm \sqrt{b^2 - 4ac}}{2a}$$ ← discriminant

You can use the discriminant to determine the number of real solutions of a quadratic equation.

Key Idea

Interpreting the Discriminant

$b^2 - 4ac > 0$	$b^2 - 4ac = 0$	$b^2 - 4ac < 0$
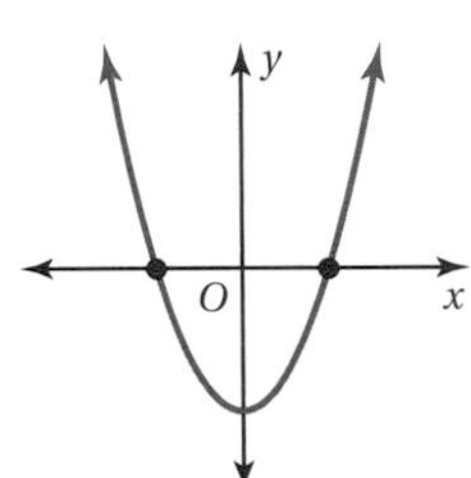	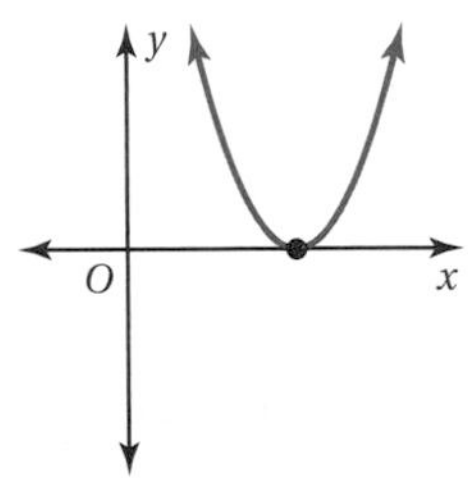	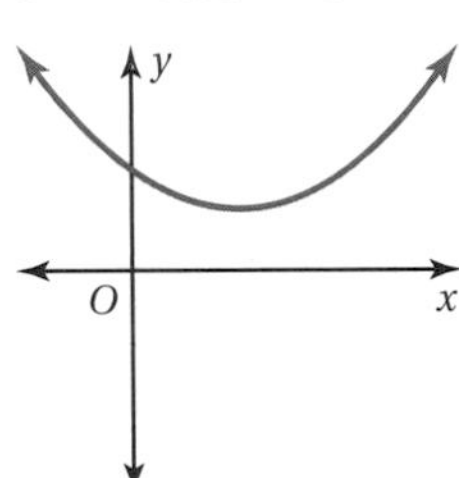
• two real solutions • two x-intercepts	• one real solution • one x-intercept	• no real solutions • no x-intercepts

Study Tip

The solutions of a quadratic equation may be real numbers or *imaginary numbers*. You will study imaginary numbers in a future course.

EXAMPLE 4 Determining the Number of Real Solutions

a. Determine the number of real solutions of $x^2 + 8x - 3 = 0$.

$b^2 - 4ac = 8^2 - 4(1)(-3)$ Substitute 1 for a, 8 for b, and -3 for c.

$= 64 + 12$ Simplify.

$= 76$ Add.

The discriminant is greater than 0, so the equation has two real solutions.

b. Determine the number of real solutions of $2x^2 + 7 = 6x$.

Write the equation in standard form: $2x^2 - 6x + 7 = 0$.

$b^2 - 4ac = (-6)^2 - 4(2)(7)$ Substitute 2 for a, -6 for b, and 7 for c.

$= 36 - 56$ Simplify.

$= -20$ Subtract.

The discriminant is less than 0, so the equation has no real solutions.

On Your Own

Now You're Ready
Exercises 27–32

Determine the number of real solutions of the equation.

8. $-x^2 + 4x - 4 = 0$ **9.** $6x^2 + 2x = -1$ **10.** $\frac{1}{2}x^2 = 7x - 1$

Laurie's Notes

Key Idea

- Discuss the discriminant and ask students to find the discriminant in each of the previous examples.
- Write the Key Idea on the board.
- **MP1a Make Sense of Problems:** This key idea helps connect previous lessons in this chapter. Ask a volunteer to discuss the ways to recognize each number of solutions when solving by graphing, factoring, and the quadratic formula.
- Discuss the Study Tip.

Example 4

- Work through each part as shown.

? After writing the equation in part (b) ask, "What are the values of *a, b,* and *c*?" Check for understanding—students need to first subtract $6x$ from each side, and then recognize that the value of b is -6, not 6.

On Your Own

- **Think-Pair-Share:** Students should read each question independently and then work with a partner to answer the questions. When they have answered the questions, the pair should compare their answers with another group and discuss any discrepancies.
- Students are often surprised that a small change in any of the three parameters (a, b, and c) can produce a big change in the graph and subsequent x-intercepts. If students finish early, have them use a graphing calculator to explore small changes in one of the parameters.

Closure

- For quadratic equations of the form $x^2 + 4x + c = 0$ determine the values of c that yield a quadratic equation with 2 roots, 1 root, and no roots. when $c < 4$, two roots; when $c = 4$, one root; when $c > 4$, no roots

Technology For the Teacher

The Dynamic Planning Tool
Editable Teacher's Resources at *BigIdeasMath.com*

Extra Example 4

a. Determine the number of real solutions of $x^2 + 2x + 5 = 0$. 0

b. Determine the number of real solutions of $4x^2 + 25 = 20x$. 1

On Your Own

8. 1 **9.** 0

10. 2

English Language Learners

Vocabulary

To check that your students can distinguish between the expressions *quadratic polynomial, quadratic equation,* and *quadratic formula*, have them state an example or explanation of each.

Discuss how the word *discriminant* relates to the word *discriminate*, which means to distinguish between things.

Vocabulary and Concept Check

1. $x = \dfrac{-b \pm \sqrt{b^2 - 4ac}}{2a}$
2. discriminant > 0: 2 real solutions; discriminant $= 0$: 1 real solution; discriminant < 0: no real solutions

Practice and Problem Solving

3. $x^2 - 7x = 0$; $a = 1, b = -7, c = 0$
4. $x^2 - 4x + 12 = 0$; $a = 1, b = -4, c = 12$
5. $-2x^2 - 5x + 1 = 0$; $a = -2, b = -5, c = 1$
6. $-4x^2 + 3x + 2 = 0$; $a = -4, b = 3, c = 2$
7. $x^2 - 6x + 4 = 0$; $a = 1, b = -6, c = 4$
8. $3x^2 + 8x + 3 = 0$; $a = 3, b = 8, c = 3$
9. $x = 6$
10. no real solutions
11. $x = -1, x = 11$
12. $x = -\frac{1}{2}, x = 1$
13. no real solutions
14. $x = \frac{1}{3}$
15. $x = \frac{2}{3}, x = \frac{3}{2}$
16. no real solutions
17. $x = \frac{1}{4}$
18. $x \approx 0.7, x \approx 4.3$
19. $x \approx -4.2, x \approx 2.2$
20. $x \approx -0.3, x \approx 1.1$

Assignment Guide and Homework Check

Level	Assignment	Homework Check
Average	1, 2, 9–21 odd, 27–29, 34, 35, 47–50	13, 15, 21, 27, 34
Advanced	1, 2, 16–22 even, 31–35, 37, 40, 43–50	16, 20, 31, 34, 44

For Your Information

- **Exercises 3–20** Note that there are two possible ways to write standard form. For example, the standard form in Exercise 6 could be written as $4x^2 - 3x - 2 = 0$ or $-4x^2 + 3x + 2 = 0$. The values of *a, b,* and *c* should come from one equation or the other.

Common Errors

- **Exercises 3–20** Students may make sign mistakes when identifying the values of *a, b,* and *c*. Emphasize how the signs are determined.
- **Exercise 23** Students may solve for h when $d = 0$. Explain that the distance from the pier is given by d, so they need to solve for d when $h = 0$.

9.4 Record and Practice Journal

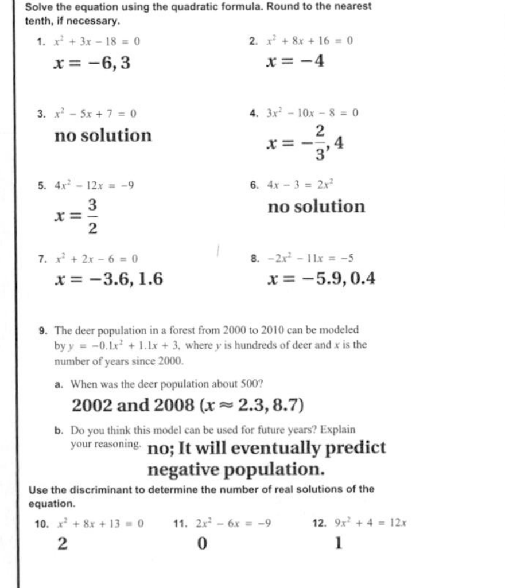

Solve the equation using the quadratic formula. Round to the nearest tenth, if necessary.

1. $x^2 + 3x - 18 = 0$ — $x = -6, 3$
2. $x^2 + 8x + 16 = 0$ — $x = -4$
3. $x^2 - 5x + 7 = 0$ — no solution
4. $3x^2 - 10x - 8 = 0$ — $x = -\frac{2}{3}, 4$
5. $4x^2 - 12x = -9$ — $x = \frac{3}{2}$
6. $4x - 3 = 2x^2$ — no solution
7. $x^2 + 2x - 6 = 0$ — $x = -3.6, 1.6$
8. $-2x^2 - 11x = -5$ — $x = -5.9, 0.4$
9. The deer population in a forest from 2000 to 2010 can be modeled by $y = -0.1x^2 + 1.1x + 3$, where y is hundreds of deer and x is the number of years since 2000.
   a. When was the deer population about 500? **2002 and 2008 ($x \approx 2.3, 8.7$)**
   b. Do you think this model can be used for future years? Explain your reasoning. **no; It will eventually predict negative population.**

Use the discriminant to determine the number of real solutions of the equation.

10. $x^2 + 8x + 13 = 0$ — 2
11. $2x^2 - 6x = -9$ — 0
12. $9x^2 + 4 = 12x$ — 1

Technology For the Teacher
Answer Presentation Tool

9.4 Exercises

Vocabulary and Concept Check

1. **VOCABULARY** Write the formula that can be used to solve any quadratic equation.
2. **VOCABULARY** What does the discriminant tell you about the number of solutions of a quadratic equation?

Practice and Problem Solving

Write the equation in standard form. Then identify the values of *a*, *b*, and *c* that you would use to solve the equation using the quadratic formula.

3. $x^2 = 7x$
4. $x^2 - 4x = -12$
5. $-2x^2 + 1 = 5x$
6. $3x + 2 = 4x^2$
7. $4 - 6x = -x^2$
8. $-8x = 3x^2 + 3$

Solve the equation using the quadratic formula. Round to the nearest tenth, if necessary.

9. $x^2 - 12x + 36 = 0$
10. $x^2 + 7x + 16 = 0$
11. $x^2 - 10x - 11 = 0$

1 12. $2x^2 - x - 1 = 0$
13. $2x^2 - 6x + 5 = 0$
14. $9x^2 - 6x + 1 = 0$

2 15. $6x^2 - 13x = -6$
16. $-3x^2 + 6x = 4$
17. $1 - 8x = -16x^2$

18. $x^2 - 5x + 3 = 0$
19. $x^2 + 2x = 9$
20. $5x^2 - 2 = 4x$

ERROR ANALYSIS **Describe and correct the error in solving the equation.**

21. $3x^2 - 7x - 6 = 0$

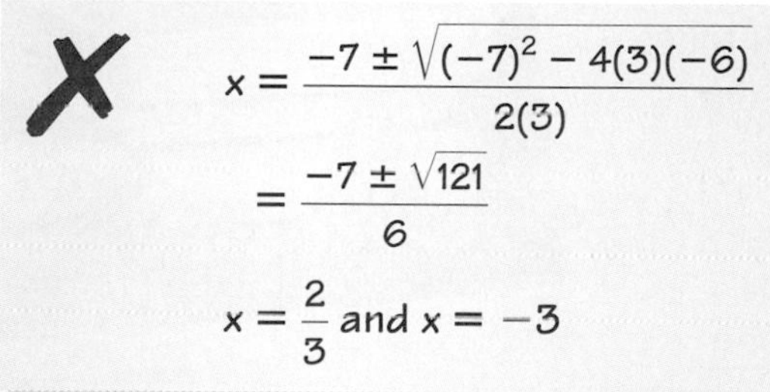

✗
$x = \dfrac{-7 \pm \sqrt{(-7)^2 - 4(3)(-6)}}{2(3)}$
$= \dfrac{-7 \pm \sqrt{121}}{6}$
$x = \dfrac{2}{3}$ and $x = -3$

22. $-2x^2 + 9x = 4$

✗

$$x = \frac{-9 \pm \sqrt{9^2 - 4(-2)(4)}}{2(-2)}$$

$$= \frac{-9 \pm \sqrt{113}}{-4}$$

$x \approx -0.41$ and $x \approx 4.91$

3 23. **PIER** A swimmer takes a running jump off a pier. The path of the swimmer can be modeled by the equation $h = -0.1d^2 + 0.1d + 3$, where h is the height (in feet) and d is the horizontal distance (in feet). How far from the pier does the swimmer enter the water?

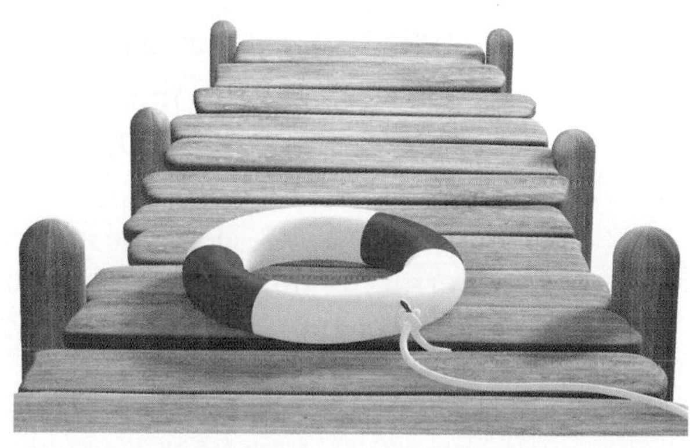

Match the discriminant with the corresponding graph.

24. $b^2 - 4ac > 0$

25. $b^2 - 4ac = 0$

26. $b^2 - 4ac < 0$

A.

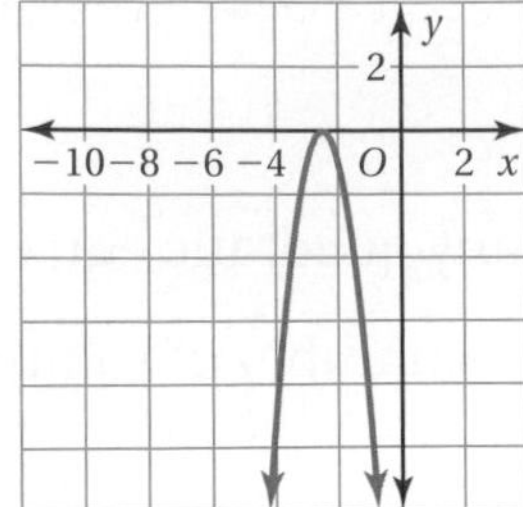

B.

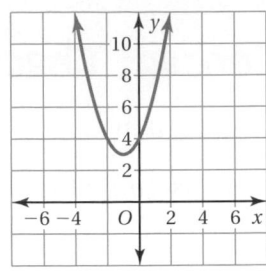

C.

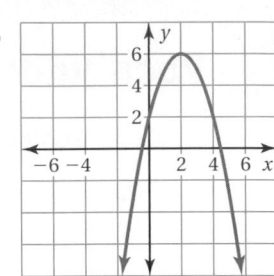

Use the discriminant to determine the number of real solutions of the equation.

27. $x^2 - 6x + 10 = 0$

28. $x^2 - 5x - 3 = 0$

29. $2x^2 - 12x = -18$

30. $4x^2 = 4x - 1$

31. $-\frac{1}{4}x^2 + 4x = -2$

32. $-5x^2 + 8x = 9$

33. REPEATED REASONING You use the quadratic formula to solve an equation.

a. You obtain solutions that are integers. Could you have used factoring to solve the equation? Explain your reasoning.

b. You obtain solutions that are fractions. Could you have used factoring to solve the equation? Explain your reasoning.

c. Make a generalization about quadratic equations with rational solutions.

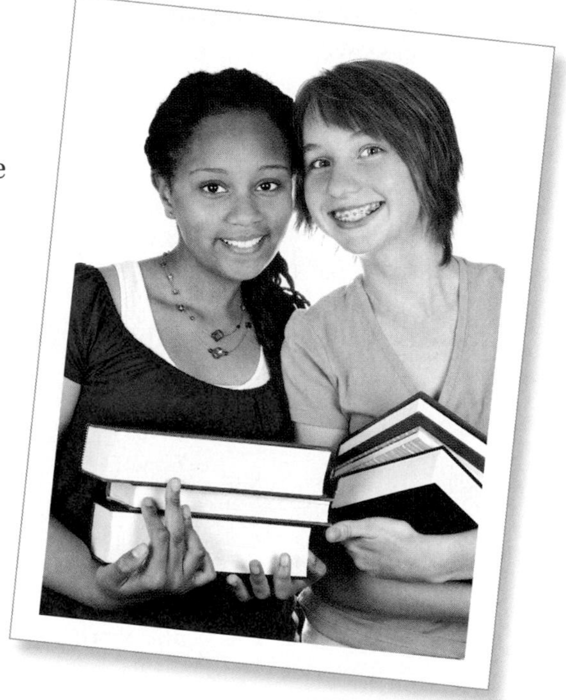

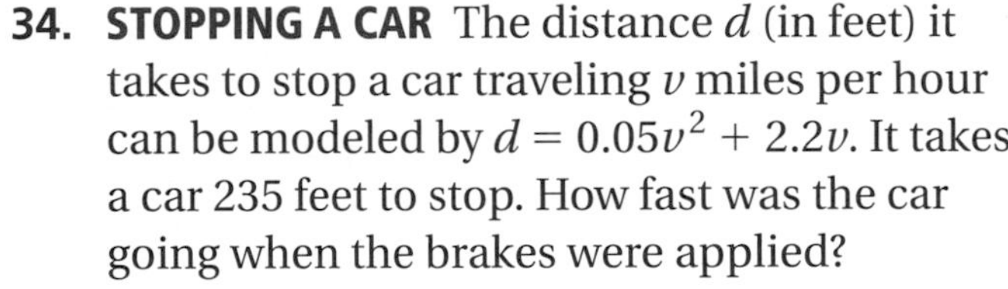

34. STOPPING A CAR The distance d (in feet) it takes to stop a car traveling v miles per hour can be modeled by $d = 0.05v^2 + 2.2v$. It takes a car 235 feet to stop. How fast was the car going when the brakes were applied?

35. FISHING The amount y of trout (in tons) caught in a lake from 1990 to 2009 can be modeled by $y = -0.08x^2 + 1.6x + 10$, where x is the number of years since 1990.

a. When were about 15 tons of trout caught in the lake?

b. Do you think this model can be used for future years? Explain your reasoning.

36. ERROR ANALYSIS Describe and correct the error in finding the number of solutions of the equation $2x^2 - 5x - 2 = -11$.

$$b^2 - 4ac = (-5)^2 - 4(2)(-2)$$
$$= 25 - (-16)$$
$$= 41$$

The equation has two solutions.

Common Errors

- **Exercises 27–32** If students use a calculator to evaluate the discriminant, they may make keystroke errors. For example, they might enter -4^2 instead of (-4^2).
- **Exercise 34** Students may solve for *d* when $v = 0$. Explain that 235 is the distance it takes the car to stop. So they need substitute 235 for *d* and solve for *v*.
- **Exercise 35** Students may neglect to give two answers. Tell them to be sure to decide whether both answers make sense in the context of the situation. In this case, they do.
- **Exercises 37–39** Students may fail to write the equation in standard form before finding the discriminant.
- **Exercise 46** Students may not know how to solve this problem. The students need to use the projectile motion model to write an expression involving *v* and set the expression equal to the height of the branch. The key is realizing that the minimum velocity is given by the value of *v* for which this equation has one solution.

Practice and Problem Solving

21. used -7 for $-b$ instead of $-(-7) = 7$; $x = -\frac{2}{3}$, $x = 3$

22. used $c = 4$ instead of $c = -4$; $x - \frac{1}{2}$, $x - 4$

23. 6 ft

24. C

25. A

26. B

27. 0

28. 2

29. 1

30. 1

31. 2

32. 0

33. **a.** yes; When the solutions m and n are integers, the standard form can be factored as $(x - m)(x - n) = 0$.

b. yes; When the solutions $\frac{m}{n}$ and $\frac{h}{k}$ are fractions, the standard form can be factored as $(nx - m)(kx - h) = 0$.

c. Any quadratic equation with rational solutions can be solved by factoring.

34. 50 miles per hour

35. **a.** 1994 and 2006

b. no; The model predicts negative numbers of fish caught after 2015.

36. Standard form was not used; $2x^2 - 5x + 9 = 0$; no real solutions

Differentiated Instruction

Kinesthetic

In their notebooks, have students create a table listing the methods they have learned for solving quadratic equations: *by graphing, using square roots, by completing the square,* and *using the quadratic formula.* A description of the method and an example with its solution should be written for each method.

Practice and Problem Solving

37. 2

38. 1

39. 0

40–42. Sample answers are given.

40. **a.** $c = 2$ **b.** $c = -5$

41. **a.** $c = 8$ **b.** $c = 2$

42. **a.** $c = -20$ **b.** $c = 4$

43. 2; When a and c have different signs, $b^2 - 4ac$ is positive.

44. rational; When the discriminant is a perfect square, the quadratic formula will have integers in the numerator which give rational solutions.

45. See *Taking Math Deeper.*

46. about 24.7 feet per second

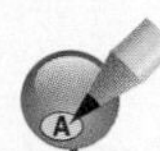

Fair Game Review

47. $(1, -1)$

48. infinitely many solutions

49. no solution

50. A

Mini-Assessment

Solve the equation using the quadratic formula. Round to the nearest tenth, if necessary.

1. $x^2 - 2x - 99 = 0$
$x = -9, x = 11$

2. $3x^2 + 16x - 35 = 0$
$x = -7, x \approx 1.7$

3. $4x^2 - 6x = -7$
no real solutions

4. $-3x^2 + 12x = 8$
$x \approx 0.8, x \approx 3.2$

Taking Math Deeper

Exercise 45

For this problem, it is important for students to read the problem carefully and list the given information before looking for an entry point to the solution.

Write an equation that represents the amount of fencing needed for both pastures and one that represents the area of each pasture.

Amount of fencing needed for both pastures:

$x + x + x + x + y + y + y = 1050$ There is 1050 feet of fencing.

$4x + 3y = 1050$ Combine like terms.

Area of each pasture:

$xy = 15{,}000$ Area = length × width

a. Solving $4x + 3y = 1050$ for y produces $y = 350 - \frac{4}{3}x$.

b. Substitute for y in the equation $xy = 15{,}000$ and solve for x.

$x\left(350 - \frac{4}{3}x\right) = 15{,}000$ Substitute for y.

$350x - \frac{4}{3}x^2 = 15{,}000$ Distributive Property

$1050x - 4x^2 = 45{,}000$ Multiply each side by 3.

$4x^2 - 1050x + 45{,}000 = 0$ Write in standard form.

a = 4,
b = -1050,
c = 45,000

Using the quadratic formula, the solutions are

$$x = \frac{1050 + \sqrt{382{,}500}}{8} \approx 208.6 \text{ and } x = \frac{1050 - \sqrt{382{,}500}}{8} \approx 53.9.$$

So, the possible lengths and widths of each section are:
$x = 208.6$ feet and $y = 71.9$ feet or $x = 53.9$ feet and $y = 278.1$ feet.

Note: You may want to challenge students by having them solve this problem using different methods.

Reteaching and Enrichment Strategies

If students need help. . .	If students got it. . .
Resources by Chapter • Practice A and Practice B • Puzzle Time Record and Practice Journal Practice Differentiating the Lesson Lesson Tutorials Skills Review Handbook	Resources by Chapter • Enrichment and Extension • School-to-Work Start the next section

Use the discriminant to determine how many times the graph of the related function intersects the x-axis.

37. $x^2 + 5x - 1 = 0$

38. $4x^2 + 4x = -1$

39. $4 - 3x = -6x^2$

Give a value for c where (a) you can factor to solve the equation and (b) you must use the quadratic formula to solve the equation.

40. $x^2 + 3x + c = 0$

41. $x^2 - 6x + c = 0$

42. $x^2 - 8x + c = 0$

43. REASONING How many solutions does $ax^2 + bx + c = 0$ have when a and c have different signs? Explain your reasoning.

44. REASONING When the discriminant is a perfect square, are the solutions of $ax^2 + bx + c = 0$ rational or irrational? Assume a, b, and c are integers. Explain your reasoning.

45. PROBLEM SOLVING A rancher constructs two rectangular horse pastures that share a side, as shown. The pastures are enclosed by 1050 feet of fencing. Each pasture has an area of 15,000 square feet.

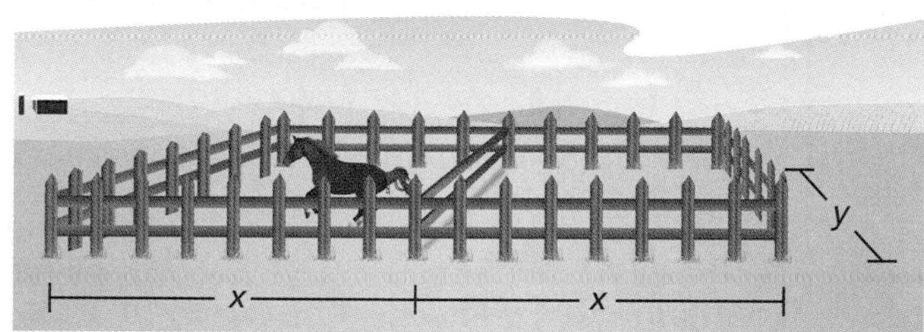

a. Show that $y = 350 - \frac{4}{3}x$.

b. Find the possible lengths and widths of each pasture.

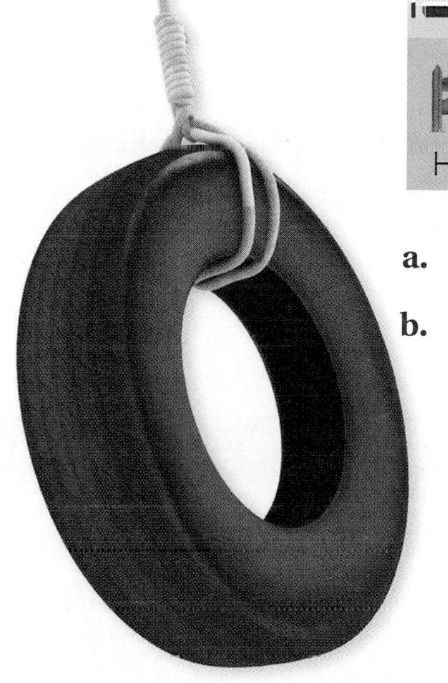

46. Critical Thinking You are trying to hang a tire swing. To get the rope over a tree branch that is 15 feet high, you tie the rope to a weight and throw it over the branch. You release the weight at a height of 5.5 feet. What is the minimum upward velocity needed to reach the branch?

Fair Game Review What you learned in previous grades & lessons

Solve the system of linear equations. *(Section 4.4)*

47. $x + y = 0$
$3x + 2y = 1$

48. $2x - 2y = 4$
$-x + y = -2$

49. $2x - 4y = -1$
$-3x + 6y = -5$

50. MULTIPLE CHOICE What is the solution of the equation $7x + 3x = 5x - 10$? *(Section 1.3)*

Ⓐ $x = -2$ Ⓑ $x = -\frac{2}{3}$ Ⓒ $x = 2$ Ⓓ $x = 4$

9.4b Choosing a Solution Method

The table shows five methods for solving quadratic equations. While there is no one correct method, some methods may be easier to use than others. Some advantages and disadvantages of each method are shown.

Key Ideas

Methods for Solving Quadratic Equations

Method	Advantages	Disadvantages
Factoring *(Lessons 7.6–7.9)*	• Straightforward when equation can be factored easily	• Some equations are not factorable.
Graphing *(Lesson 9.1)*	• Can easily see the number of solutions • Use when approximate solutions are sufficient. • Can use a graphing calculator	• May not give exact solutions
Using Square Roots *(Lesson 9.2)*	• Use to solve equations of the form $x^2 = d$.	• Can only be used for certain equations
Completing the Square *(Lesson 9.3)*	• Best used when $a = 1$ and b is even	• May involve difficult calculations
Quadratic Formula *(Lesson 9.4)*	• Can be used for *any* quadratic equation • Gives exact solutions	• Takes time to do calculations

EXAMPLE 1 Solving a Quadratic Equation Using Different Methods

Solve $x^2 + 8x + 12 = 0$ using two different methods.

Method 1: Solve by graphing. Graph the related function $y = x^2 + 8x + 12$.

The x-intercepts are -6 and -2.

$y = x^2 + 8x + 12$

∴ So, the solutions are $x = -6$ and $x = -2$.

Method 2: Solve by factoring.

$x^2 + 8x + 12 = 0$ — Write the equation.

$(x + 2)(x + 6) = 0$ — Factor left side.

$x + 2 = 0$ *or* $x + 6 = 0$ — Use Zero-Product Property.

$x = -2$ *or* $x = -6$ — Solve for x.

∴ The solutions are $x = -2$ and $x = -6$.

Study Tip

Notice that each method produces the same solutions, $x = -6$ and $x = -2$.

Laurie's Notes

Introduction

Connect

- **Yesterday:** Students solved quadratic equations using the quadratic formula. (MP1a, MP2)
- **Today:** Students will choose methods to solve quadratic equations.

Motivate

- Write three multiplication problems and three solution methods on the board.
- Tell students to choose a different solution method for each problem.

Problems	Solution methods
13×20	calculator
13×24	paper and pencil
1.3×2.42	mental math

- Explain that in today's lesson, students will choose methods to solve quadratic equations.

Lesson Notes

Key Ideas

- Write the chart on the board. Discuss the advantages and disadvantages of each method as you fill in the chart.

? "What is an example of a quadratic equation that factors easily?"
Sample answers: $x^2 - 16 = 0$, $x^2 + 2x + 1 = 0$

? "What is an example of a quadratic equation that does not factor?"
Sample answers: $x^2 + 5 = 0$, $x^2 - 4x - 2 = 0$

- Discuss with students that solutions found using a graphing utility are often approximated.

? "In a quadratic equation of the form $x^2 = d$, which term is missing?"
the x-term

- Tell students that it is possible to write a program in their calculators or in a spreadsheet that computes solutions using the quadratic formula.

Example 1

- **MP5 Use Appropriate Tools Strategically:** Write the equation on the board and ask students to strategically choose and support two methods for solving this equation.
- **MP3b Critique the Reasoning of Others:** Have students critique each method. Allow personal preference. Look for comments that are thoughtful and reasonable.
- Work through both methods as shown in the text.

Goal Today's lesson is choosing a method for solving a quadratic equation.

Start Thinking! and Warm Up

Lesson 9.4b Warm Up
For use before Lesson 9.4b

Lesson 9.4b Start Thinking!
For use before Lesson 9.4b

Name all of the methods you know for solving quadratic equations. Which is your favorite? Which is your least favorite? Why?

Extra Example 1

Solve $x^2 + x - 20 = 0$ using two different methods.
$x = -5, x = 4$

Practice

1. $x = -7 + \sqrt{41}$, $x = -7 - \sqrt{41}$
2. $x = 1, x = 9$
3. $x = 6, x = -6$

Record and Practice Journal Practice

See Additional Answers.

Extra Example 2

Solve $2x^2 - 98 = 0$ using any method. Explain your choice of method.
$x = 7, x = -7$; square roots, because no x-term

Extra Example 3

Solve $x^2 - 4x = 6$ using any method. Explain your choice of method.
$x = 2 + \sqrt{10}, x = 2 - \sqrt{10}$; The quadratic formula is most convenient.

Practice

4–12. Sample explanations are given.

4. $x = -12, x = 1$; factors easily

5. $x = 1, x = -1$; square roots, because no x-term

6. $x = \frac{1 + \sqrt{21}}{10}, x = \frac{1 - \sqrt{21}}{10}$; The quadratic formula is most convenient.

7. $x = -5, x = 8$; factors easily

8. $x = -2, x = -10$; factors easily

9. no real solutions; completing the square, because $a = 1$ and b is even

10. no real solutions; square roots, because no x-term

11. $x = -4, x = 3$; factors easily

12. $x = -7, x = 1$; factors easily

Mini-Assessment

Solve the equation using any method. Explain your choice of method.

1. $x^2 + 4x = -1$ $x = -2 + \sqrt{3}, x = -2 - \sqrt{3}$; The quadratic formula is most convenient.
2. $3x^2 = 12$ $x = 2, x = -2$; square roots, because no x-term
3. $x^2 - 5x + 6 = 0$ $x = 2, x = 3$; factors easily

Laurie's Notes

Example 2

- Write the equation as shown.
- ? "Can this equation be solved by factoring? Explain." no; $x^2 - 10x - 1$ does not factor.
- ? "Can this equation be solved using square roots? Explain." no; There is an x-term.
- ? "Can this equation be solved by completing the square? Explain." yes; $a = 1$ and b is even.
- Work through the problem as shown.
- If time permits, use the quadratic formula to solve and verify that the same solutions are found. Discuss which method the students prefer.

Example 3

- ? "Could you use either square roots or completing the square to solve this equation? Explain." Square roots will not work because there is an x-term. Completing the square is not convenient because $a \neq 1$ and b is odd.
- Consider factoring before using the quadratic formula. In this example, there are many possible products of polynomials for $a = 2$ and $c = -24$. Factoring is possible, but not easy.
- Work through the problem as shown.
- ? "How do you know this equation was factorable?" rational solutions
- ? "What are the factors?" $(x - 8)$ and $(2x + 3)$

Practice

- **MP5:** It is important that the students strategically choose the two methods they use in Exercises 1–3.
- **MP3a Construct Viable Arguments:** In explaining the methods they choose, students have to think and reason about the equations. Different students will make different choices, so it is important for students to share their thinking with the class.

Closure

- Write a quadratic equation that you would *not* solve using square roots. Look for equations that have an x-term.
- Write a quadratic equation that you would *not* solve by factoring. Look for equations that are not easily factorable.

Technology For the Teacher

The Dynamic Planning Tool
Editable Teacher's Resources at *BigIdeasMath.com*

EXAMPLE 2 Choosing a Method

Solve $x^2 - 10x = 1$ using any method. Explain your choice of method.

The coefficient of the x^2-term is 1 and the coefficient of the x-term is an even number. So, solve by completing the square.

$$x^2 - 10x = 1$$ Write the equation.

Complete the square. → $$x^2 - 10x + 25 = 1 + 25$$ Add $\left(\frac{-10}{2}\right)^2$, or 25, to each side.

$$(x - 5)^2 = 26$$ Factor $x^2 - 10x + 25$.

$$x - 5 = \pm\sqrt{26}$$ Take the square root of each side.

$$x = 5 \pm \sqrt{26}$$ Add 5 to each side.

∴ The solutions are $x = 5 + \sqrt{26} \approx 10.1$ and $x = 5 - \sqrt{26} \approx -0.1$.

EXAMPLE 3 Choosing a Method

Solve $2x^2 - 13x - 24 = 0$ using any method. Explain your choice of method.

The equation is not easily factorable and the numbers are somewhat large. So, solve using the quadratic formula.

$$x = \frac{-b \pm \sqrt{b^2 \quad 4ac}}{2a}$$ Quadratic Formula

$$= \frac{-(-13) \pm \sqrt{(-13)^2 - 4(2)(-24)}}{2(2)}$$ Substitute 2 for a, -13 for b, and -24 for c.

$$= \frac{13 \pm \sqrt{361}}{4}$$ Simplify.

$$= \frac{13 \pm 19}{4}$$ Evaluate the square root.

∴ The solutions are $x = \frac{13 + 19}{4} = 8$ and $x = \frac{13 - 19}{4} = -\frac{3}{2}$.

Practice

Solve the equation using two different methods.

1. $x^2 + 14x = -8$
2. $x^2 - 10x + 9 = 0$
3. $-4x^2 + 144 = 0$

Solve the equation using any method. Explain your choice of method.

4. $x^2 + 11x - 12 = 0$
5. $9x^2 - 5 = 4$
6. $5x^2 - x - 1 = 0$
7. $x^2 - 3x - 40 = 0$
8. $x^2 + 12x + 5 = -15$
9. $x^2 = 2x - 5$
10. $-8x^2 - 2 = 14$
11. $x^2 + x - 12 = 0$
12. $x^2 + 6x + 9 = 16$

9.5 Solving Systems of Linear and Quadratic Equations

Essential Question How can you solve a system of two equations when one is linear and the other is quadratic?

COMMON CORE STATE STANDARDS
A.REI.7

1 ACTIVITY: Solving a System of Equations

Work with a partner. Solve the system of equations using the given strategy. Which strategy do you prefer? Why?

System of Equations:

$y = x + 2$ Linear

$y = x^2 + 2x$ Quadratic

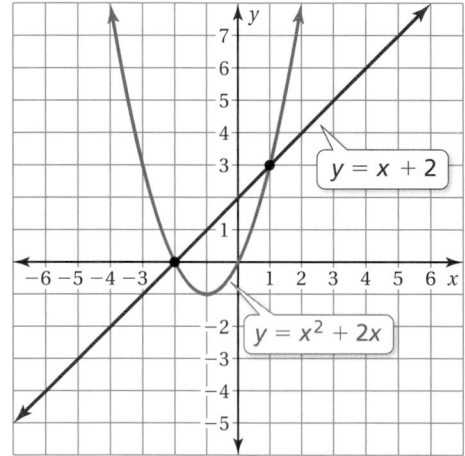

a. Solve by Graphing

Graph each equation and find the points of intersection of the line and the parabola.

$y = x + 2$

$y = x^2 + 2x$

b. Solve by Substitution

Substitute the expression for y from the quadratic equation into the linear equation to obtain

$$x^2 + 2x = x + 2.$$

Solve this equation and substitute each x-value into the linear equation $y = x + 2$ to find the corresponding y-value.

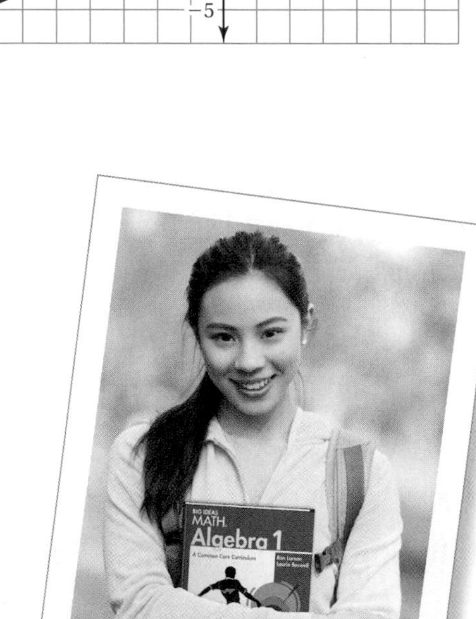

c. Solve by Elimination

Eliminate y by subtracting the linear equation from the quadratic equation to obtain

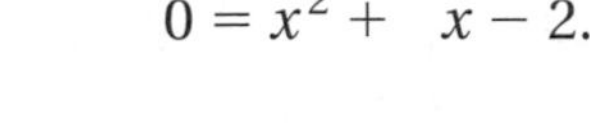

$$\begin{aligned} y &= x^2 + 2x \\ y &= \quad\quad\; x + 2 \\ \hline 0 &= x^2 + \; x - 2. \end{aligned}$$

Solve this equation and substitute each x-value into the linear equation $y = x + 2$ to find the corresponding y-value.

Laurie's Notes

Introduction

Standards for Mathematical Practice

- **MP1a Make Sense of Problems** and **MP8 Look for and Express Regularity in Repeated Reasoning:** In this activity, students extend the methods they learned for solving a system of linear equations to solving a system with a nonlinear equation. Students make sense of the problem by connecting the new process to what they know about solving linear systems and solving quadratic equations.

Motivate

- **Story Time:** Share a story about shooting clay pigeons launched into the air by a machine. In the last round, you hit the target on the way down. (Sketch the diagram shown below.)

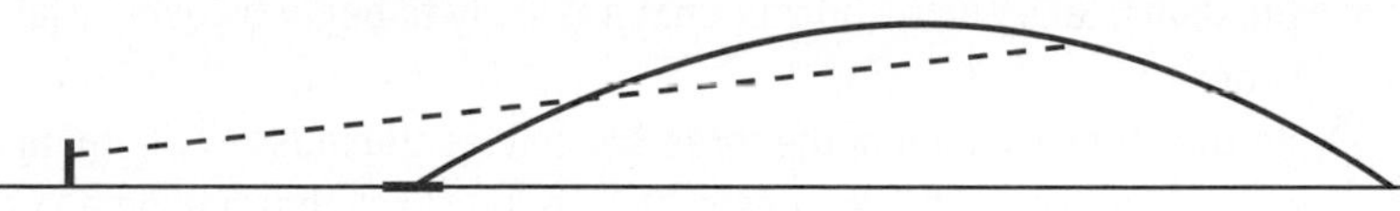

- Connect this to today's activity. Explain that you prefer to hit the target on the way up, because it is closer to you.

Activity Notes

Discuss

? "What is a system of linear equations?" a set of two or more linear equations in the same variables

? "How do you solve a system of linear equations?" graphing, substitution, or elimination

- Explain that today's activity is about systems of equations that include quadratic equations.

Activity 1

- As students graph the two equations, you should hear comments about slope, y-intercepts, the parabola opening up, and so on.
- **Big Idea:** The two equations are solved for y. So, substitution results in the expressions being set equal. The equation that results connects to solving quadratic equations by factoring, from a previous chapter.
- **Common Error:** When using elimination, students often subtract the left side of the equations but add on the right side.
- A preference for one method over another is often related to a student's comfort level with each method.
- **MP1a:** Summarize the multiple approaches used in the activity to make sense of the problem.

Common Core State Standards

A.REI.7 Solve a simple system consisting of a linear equation and a quadratic equation in two variables algebraically and graphically.

Previous Learning

Students should know how to solve systems of linear equations. They should also know how to solve quadratic equations.

Start Thinking! and Warm Up

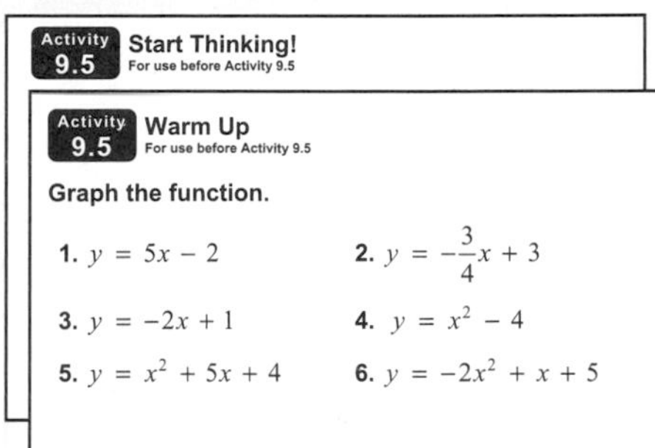

Activity 9.5 **Start Thinking!** For use before Activity 9.5

Activity 9.5 **Warm Up** For use before Activity 9.5

Graph the function.

1. $y = 5x - 2$
2. $y = -\frac{3}{4}x + 3$
3. $y = -2x + 1$
4. $y = x^2 - 4$
5. $y = x^2 + 5x + 4$
6. $y = -2x^2 + x + 5$

9.5 Record and Practice Journal

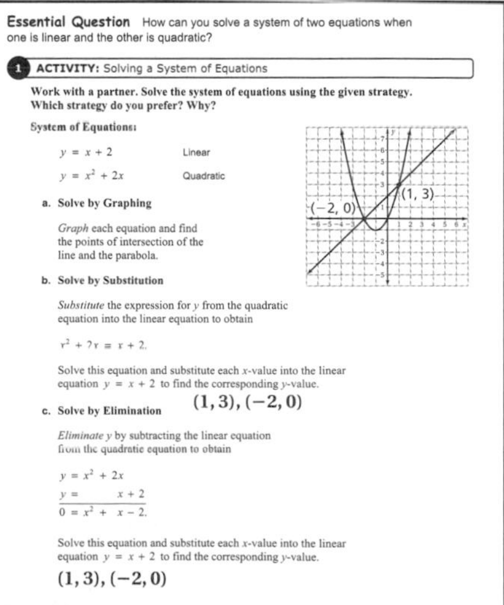

Essential Question How can you solve a system of two equations when one is linear and the other is quadratic?

1 ACTIVITY: Solving a System of Equations

Work with a partner. Solve the system of equations using the given strategy. Which strategy do you prefer? Why?

System of Equations:

$y = x + 2$ Linear

$y = x^2 + 2x$ Quadratic

a. Solve by Graphing

Graph each equation and find the points of intersection of the line and the parabola.

b. Solve by Substitution

Substitute the expression for y from the quadratic equation into the linear equation to obtain

$x^2 + 2x = x + 2.$

Solve this equation and substitute each x-value into the linear equation $y = x + 2$ to find the corresponding y-value.

(1, 3), (−2, 0)

c. Solve by Elimination

Eliminate y by subtracting the linear equation from the quadratic equation to obtain

$y = x^2 + 2x$

$y = \quad x + 2$

$0 = x^2 + x - 2.$

Solve this equation and substitute each x-value into the linear equation $y = x + 2$ to find the corresponding y-value.

(1, 3), (−2, 0)

English Language Learners

Vocabulary

To help students recall the methods of substitution and elimination, write a system of linear equations.

$y = 3x - 1$

$y = 2x + 5$

Then ask which method uses each step shown below.

a. $3x - 1 = 2x + 5$ Substitution

b. $\begin{array}{l} y = 3x - 1 \\ \underline{y = 2x + 5} \\ 0 = x - 6 \end{array}$ Elimination

9.5 Record and Practice Journal

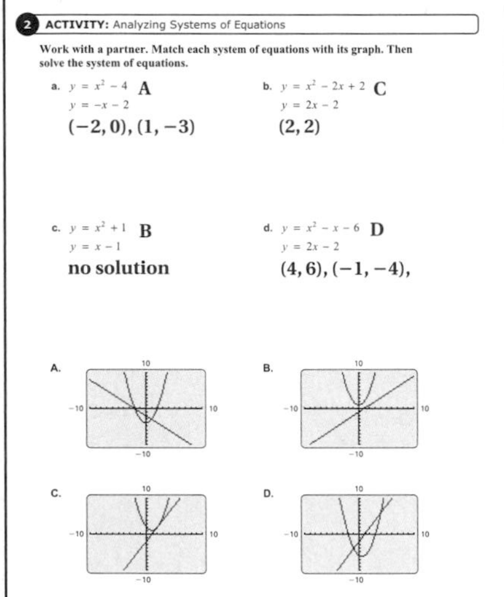

2 ACTIVITY: Analyzing Systems of Equations

Work with a partner. Match each system of equations with its graph. Then solve the system of equations.

a. $y = x^2 - 4$, $y = -x - 2$ **A** — $(-2, 0), (1, -3)$

b. $y = x^2 - 2x + 2$, $y = 2x - 2$ **C** — $(2, 2)$

c. $y = x^2 + 1$, $y = x - 1$ **B** — no solution

d. $y = x^2 - x - 6$, $y = 2x - 2$ **D** — $(4, 6), (-1, -4),$

A. B. C. D.

What Is Your Answer?

3. IN YOUR OWN WORDS How can you solve a system of two equations when one is linear and the other is quadratic?

Solve by graphing, by substitution, or by elimination.

4. Summarize your favorite strategy for solving a system of two equations when one is linear and the other is quadratic.

Check students' work.

5. Write a system of equations (one linear and one quadratic) that has the following number of solutions.

a. no solutions b. one solution c. two solutions

Your systems should be different from those in the activities.

Sample answers:

a. $y = x^2$, $y = -1$ **b.** $y = x^2$, $y = 0$ **c.** $y = x^2$, $y = 1$

Laurie's Notes

Activity 2

- In Activity 1, the focus was on the different solution methods. In Activity 2, the focus is on the number of solutions.
- The graphs visually suggest three different cases for the number of solutions. These are similar to the three cases for the number of x-intercepts of the graph of a quadratic function.
- **MP6 Attend to Precision:** Students could approximate the solution(s) for each system. Instead, make sure students find the exact solution(s).

? "What happened when you solved the system in part (c) algebraically?" found that the system has no real solutions

What Is Your Answer?

- As a follow-up to Question 4, discuss different ways to check solutions, such as solving in two different ways or substituting the solutions back into the original equations.
- Question 5 takes time and students are likely to begin by using trial and error.

? "In Question 5, which of the three cases was the most challenging and why?" Answers will vary. Many students will say that finding a system with one solution was the most challenging.

Closure

- **Writing Prompt:** Consider the graphs of $y = x^2$ and $y = -4$. As the constant function (horizontal line) is translated up, . . . the system of equations formed by the two equations goes from having no solutions, to one solution (when $y = 0$), and then to two solutions.

Technology For the Teacher

The Dynamic Planning Tool

Editable Teacher's Resources at *BigIdeasMath.com*

2 ACTIVITY: Analyzing Systems of Equations

Work with a partner. Match each system of equations with its graph. Then solve the system of equations.

a. $y = x^2 - 4$
$y = -x - 2$

b. $y = x^2 - 2x + 2$
$y = 2x - 2$

c. $y = x^2 + 1$
$y = x - 1$

d. $y = x^2 - x - 6$
$y = 2x - 2$

A.
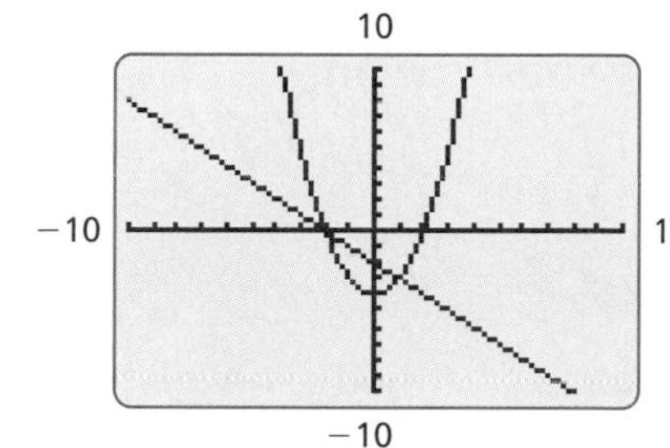

B.
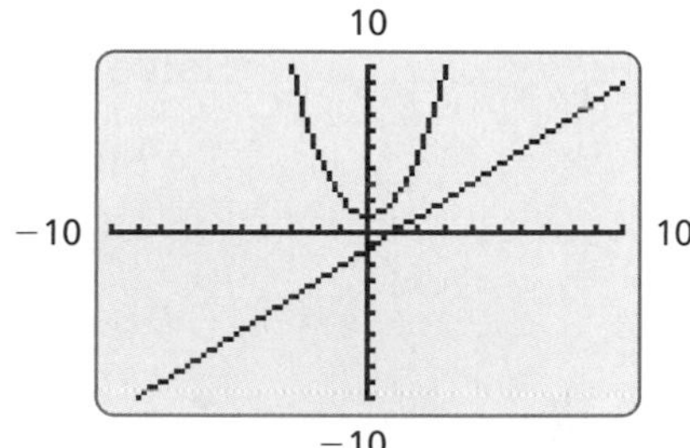

C.
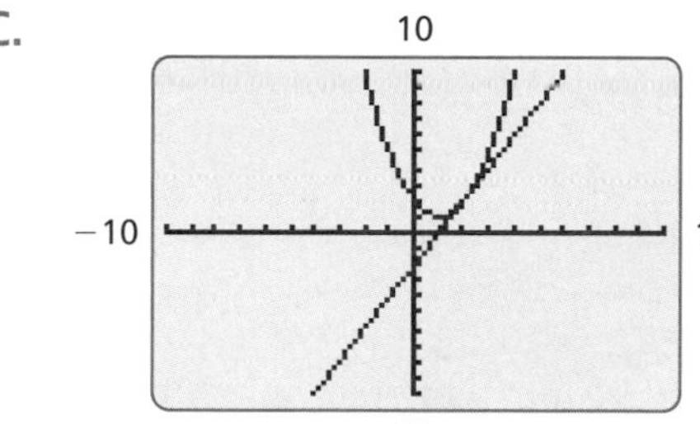

D.
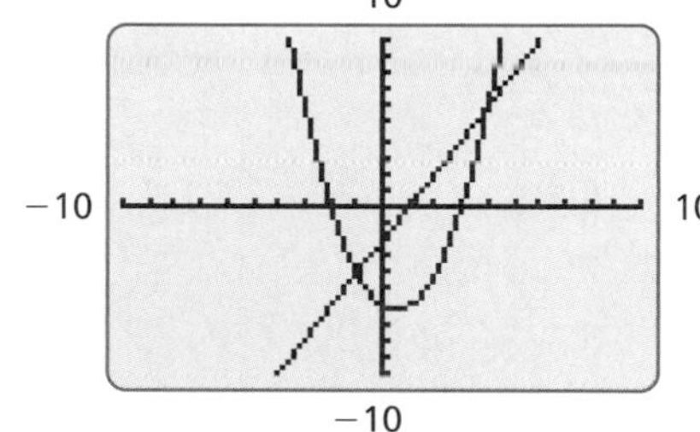

What Is Your Answer?

3. **IN YOUR OWN WORDS** How can you solve a system of two equations when one is linear and the other is quadratic?

4. Summarize your favorite strategy for solving a system of two equations when one is linear and the other is quadratic.

5. Write a system of equations (one linear and one quadratic) that has the following number of solutions.

   a. no solutions **b.** one solution **c.** two solutions

   Your systems should be different from those in the activities.

Practice Use what you learned about systems of equations to complete Exercises 3–5 on page 490.

9.5 Lesson

You learned methods for solving systems of linear equations in Chapter 4. You can use similar methods to solve systems of linear and quadratic equations.

- Solving by Graphing (Section 4.1 and Section 9.1)
- Solving by Substitution (Section 4.2)
- Solving by Elimination (Section 4.3)

EXAMPLE 1 Solving a System of Linear and Quadratic Equations

Solve the system by substitution.

$y = x^2 + x - 1$	Equation 1
$y = -2x + 3$	Equation 2

Step 1: The equations are already solved for y.

Step 2: Substitute $-2x + 3$ for y in Equation 1 and solve for x.

$y = x^2 + x - 1$	Equation 1
$-2x + 3 = x^2 + x - 1$	Substitute $-2x + 3$ for y.
$3 = x^2 + 3x - 1$	Add $2x$ to each side.
$0 = x^2 + 3x - 4$	Subtract 3 from each side.
$0 = (x + 4)(x - 1)$	Factor right side.
$x + 4 = 0$ *or* $x - 1 = 0$	Use Zero-Product Property.
$x = -4$ *or* $x = 1$	Solve for x.

Step 3: Substitute -4 and 1 for x in Equation 2 and solve for y.

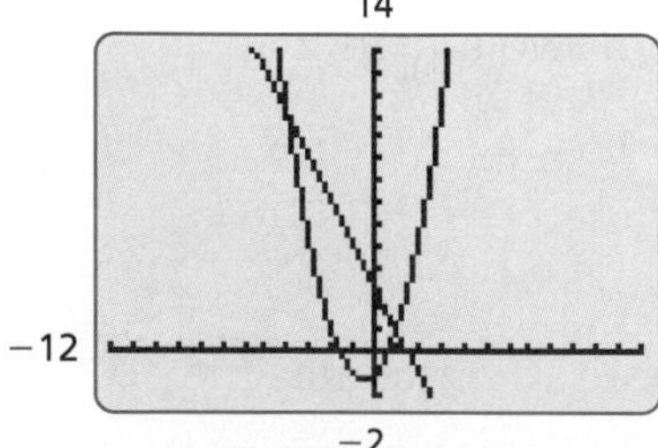

$y = -2x + 3$	Equation 2	$y = -2x + 3$
$= -2(-4) + 3$	Substitute.	$= -2(1) + 3$
$= 8 + 3$	Multiply.	$= -2 + 3$
$= 11$	Add.	$= 1$

So, the solutions are $(-4, 11)$ and $(1, 1)$.

On Your Own

Now You're Ready
Exercises 6–11

Solve the system by substitution. Check your solution(s).

1. $y = x^2 + 9$
   $y = 9$
2. $y = -5x$
   $y = x^2 - 3x - 3$
3. $y = -3x^2 + 2x + 1$
   $y = 5 - 3x$

Laurie's Notes

Introduction

Connect

- **Yesterday:** Students developed a conceptual understanding of solving a system of linear and quadratic equations. (MP1a, MP6, MP8)
- **Today:** Students will solve systems of equations consisting of one linear equation and one quadratic equation.

Motivate

- Show an illustration of the four conic sections. Discuss how a plane can intersect a double-napped cone to form a circle, a parabola, an ellipse or a hyperbola.
- Today students will look at the intersection of a line and a parabola in a plane.

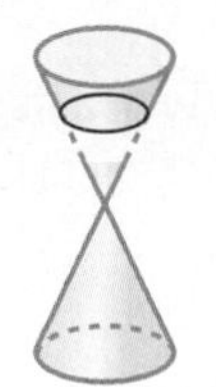

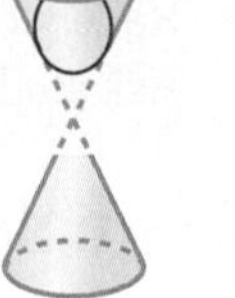

Lesson Notes

Discuss

- **MP7 Look for and Make Use of Structure:** There are no new skills in this lesson. Prior skills are applied to a different type of system. Discuss the strategies students have already learned for solving a linear system.

? "When you solved a system of linear equations, what were the possible numbers of solutions? Explain what your answers imply graphically." no solution: The lines are parallel and do not intersect; one solution: The lines intersect; infinitely many solutions: The lines are the same.

? "What are the possible numbers of solutions for a system with one linear equation and one quadratic equation? Explain." no solutions: The line and parabola do not intersect; one solution: The line intersects the parabola at one point; two solutions: The line intersects the parabola at two points.

- Some students may believe that there can be infinitely many solutions for a system of linear and quadratic equations. Explain that an entire parabola is a curve that can be intersected by a line in one or two points at most.
- If time permits, look back at Section 9.1, Example 3. Method 2 is actually solving this type of system by graphing.

Example 1

- Write the system on the board.

? "Because each equation is already solved for y, what method do you suggest using? Explain." substitution; No work is needed before setting the two expressions equal to each other.

- Collect like terms on one side, but do so in two steps, identifying the quantity added or subtracted to each side in each step.

? "So, are the solutions $x = -4$ and $x = 1$?" no; The solutions are ordered pairs, so we still need to solve for the corresponding y-values.

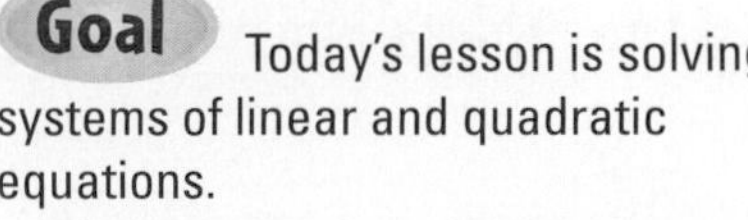

Goal Today's lesson is solving systems of linear and quadratic equations.

Lesson Materials	
Introduction	**Textbook**
• illustration	• graphing calculators

Start Thinking! and Warm Up

Warm Up
For use before Lesson 9.5

Start Thinking!
For use before Lesson 9.5

Name three methods for solving systems of linear equations.

Can all of the methods be used to solve a system of a linear equation and a quadratic equation? Explain.

Extra Example 1

Solve the system by substitution.

$y = x^2 - 4$ Equation 1

$y = -2x - 1$ Equation 2

$(-3, 5), (1, -3)$

On Your Own

1. $(0, 9)$
2. $(-3, 15), (1, -5)$
3. no real solutions

Extra Example 2

Solve the system by elimination.

$y = 2x^2 + 7x + 3$ Equation 1

$y = -x + 3$ Equation 2

(0, 3), (−4, 7)

Extra Example 3

In Example 3, how many solutions does the system have when Equation 2 is changed to $y = x - 4$? Explain.

0; The graph of $y = x - 4$ is a translation 1 unit down of the graph of $y = x - 3$, so it does not intersect the parabola.

On Your Own

4. $(-\sqrt{5}, 5 - \sqrt{5})$, $(\sqrt{5}, 5 + \sqrt{5})$
5. $\left(\frac{1}{3}, -2\frac{1}{3}\right), \left(-\frac{2}{3}, -7\frac{1}{3}\right)$
6. no real solutions
7. no; The system has 2 solutions because the graph of $y = x - 2$ is a translation 1 unit up of the graph of $y = x - 3$.

Differentiated Instruction

Kinesthetic

In Example 3, it is difficult to tell from the graph that there is exactly one solution. Have your students use an algebraic method to verify the solution.

Laurie's Notes

Example 1 (continued)

? "How can we find the y-value that corresponds to each x-value?" Substitute the x-value into one of the original equations.

- Either equation can be used to determine the y-values, though the linear equation is easier. Discuss ways of checking solutions.

On Your Own

- Have students check with a neighbor after completing each question. Ask students who finish quickly to check by graphing.

Example 2

- When solving by elimination, it is helpful to line up like terms as in Step 1.
- In Step 1, be sure that students correctly subtract each term. They are subtracting $-3x$ and -8, which means they "add the opposite."

? "How can we check the solution?" Graph each equation in the system. The graphs do not intersect. So, there are no solutions.

Example 3

- Have students graph the system in a standard viewing window of a graphing calculator.

? "Can you determine the solution(s) in a standard viewing window?" Answers will vary. Students may say no and that they need to zoom in.

- Remind students that the solution is the point of intersection. Students should be familiar with the several ways to determine the point of intersection using a graphing calculator.
- Take time to discuss the three wrong answers. These three distractors point out some common misconceptions that your students may have.

On Your Own

? "In Question 6, how does using elimination to determine that this system has no solution differ from using elimination to determine that a system of two linear equations has no solution?" Using the quadratic formula leads to an expression that contains the square root of a negative number. A linear system that has no solution leads to an equation that is never true, such as 0 = 7.

Closure

- **Exit Ticket:** Solve the system using any method. (−1, 0) and (4, 10)

$y = 2x + 2$

$y = x^2 - x - 2$

Technology For the Teacher

The Dynamic Planning Tool

Editable Teacher's Resources at *BigIdeasMath.com*

EXAMPLE 2 Solving a System of Linear and Quadratic Equations

Solve the system by elimination. $y = x^2 - 3x - 2$ Equation 1

$y = -3x - 8$ Equation 2

Step 1: Subtract.

$y = x^2 - 3x - 2$	Equation 1
$y = \quad -3x - 8$	Equation 2
$0 = x^2 \quad + 6$	Subtract the equations.

Step 2: Solve for x.

$0 = x^2 + 6$	Equation from Step 1
$-6 = x^2$	Subtract 6 from each side.

∴ The square of a real number cannot be negative. So, the system has no real solutions.

EXAMPLE 3 Standardized Test Practice

Which statement about the system is valid?

$y = 2x^2 + 5x - 1$ Equation 1

$y = x - 3$ Equation 2

Ⓐ There is one solution because the graph of $y = x - 3$ has one y-intercept.

Ⓑ There is one solution because $y = x - 3$ has one zero.

Ⓒ There is one solution because the graphs of $y = 2x^2 + 5x - 1$ and $y = x - 3$ intersect at one point.

Ⓓ There are two solutions because the graph of $y = 2x^2 + 5x - 1$ has two x-intercepts.

Use a graphing calculator to graph the system. The graphs of $y = 2x^2 + 5x - 1$ and $y = x - 3$ intersect at only one point, $(-1, -4)$.

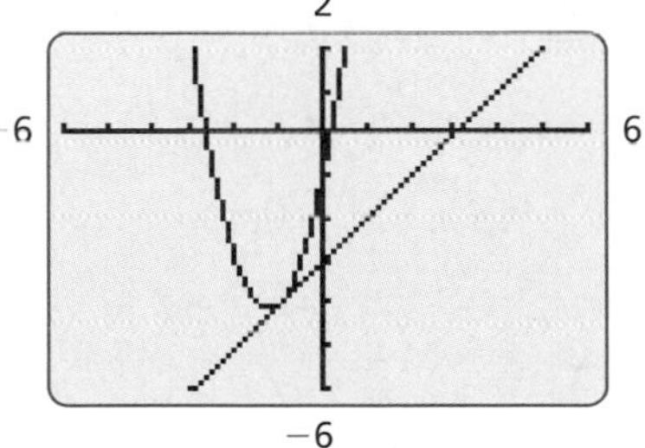

∴ So, the correct answer is Ⓒ.

On Your Own

Now You're Ready
Exercises 12–17

Solve the system by elimination. Check your solution(s).

4. $y = x^2 + x$
   $y = x + 5$

5. $y = 9x^2 + 8x - 6$
   $y = 5x - 4$

6. $y = 2x + 5$
   $y = -3x^2 + x - 4$

7. **WHAT IF?** In Example 3, does the system still have one solution when Equation 2 is changed to $y = x - 2$? Explain.

9.5 Exercises

Vocabulary and Concept Check

1. **VOCABULARY** What is a solution of a system of linear and quadratic equations?
2. **WRITING** How is solving a system of linear and quadratic equations similar to solving a system of linear equations? How is it different?

Practice and Problem Solving

Match the system of equations with its graph. Then solve the system.

3. $y = x^2 - 2x + 1$
   $y = x + 1$
4. $y = x^2 + 3x + 2$
   $y = -x - 3$
5. $y = x - 1$
   $y = -x^2 + x - 1$

A. 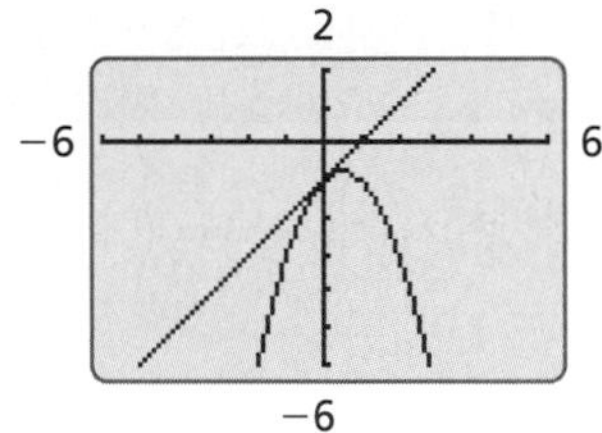

B. 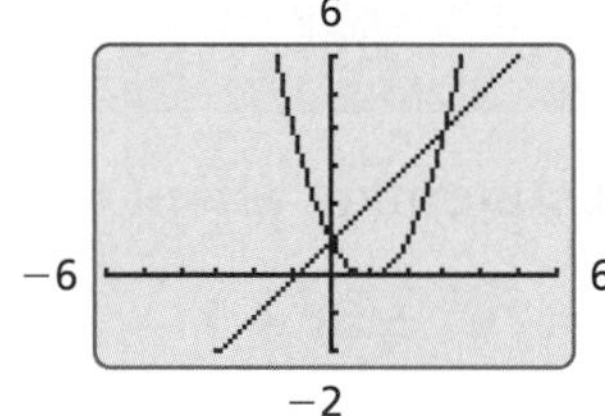

C. 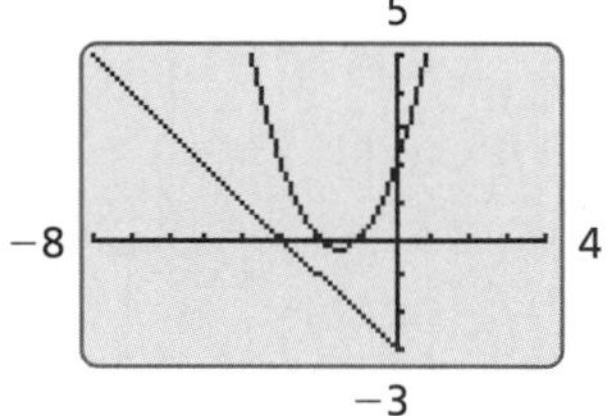

Solve the system by substitution. Check your solution(s).

1 6. $y = x - 5$
   $y = x^2 + 4x - 5$
7. $y = -2x^2$
   $y = 4x + 2$
8. $y = -x + 7$
   $y = -x^2 - 2x - 1$
9. $y = -x^2 + 7$
   $y - 2x = 4$
10. $y - 5 = -x^2$
    $y = 5$
11. $y = 2x^2 + 3x - 4$
    $y - 4x = 2$

Solve the system by elimination. Check your solution(s).

2 12. $y = -x^2 - 2x + 2$
    $y = 4x + 2$
13. $y = -2x^2 + x - 3$
    $y = 2x - 2$
14. $y = 2x - 1$
    $y = x^2$
15. $y = -2x$
    $y - x^2 = 3x$
16. $y - 1 = x^2 + x$
    $y = -x - 2$
17. $y = \frac{1}{2}x - 7$
    $y + 4x = x^2 - 2$

18. **MOVIES** The attendances y for two movies can be modeled by the following equations, where x is the number of days since the movies opened.

    $y = -x^2 + 35x + 100$ Movie A
    $y = -5x + 275$ Movie B

    When is the attendance for each movie the same?

Assignment Guide and Homework Check

Level	Assignment	Homework Check
Average	1–5, 7–17 odd, 18, 23, 27–30	7, 13, 18, 23
Advanced	1, 2, 6–18 even, 22–30	8, 16, 23, 25

Common Errors

- **Exercises 6–11** Students may give only the *x*-values instead of the coordinates of each solution. Remind them each solution is an ordered pair.
- **Exercises 15–17** Students may fail to solve for *y* before subtracting vertically. Remind students to solve both equations for *y* and to line up like terms before subtracting.
- **Exercises 19–21** Estimates obtained graphically may be incorrect or not exact. Make sure their graphical approach is correct. Also, remind them to check their answers.

9.5 Record and Practice Journal

Solve the system by substitution. Check your solution(s).

1. $y = x^2 + 5x - 4$
   $y = 3x - 1$
   $(-3, -10), (1, 2)$

2. $y = 4x + 2$
   $y = x^2 + 6$
   $(2, 10)$

3. $y = -3x^2$
   $y - 1 = 2x$
   no solution

4. $y - x = 2x^2 - 5$
   $y = x + 3$
   $(-2, 1), (2, 5)$

Solve the system by elimination. Check your solution(s).

5. $y = 4 - 2x$
   $y = -x^2 + 2x$
   $(2, 0)$

6. $y = x^2 + 5x + 8$
   $y = -2x + 2$
   $(-1, 4), (-6, 14)$

7. $y + 6x = 7$
   $y = -2x^2 + 9x$
   $\left(\frac{1}{2}, 4\right), (7, -35)$

8. $y - 4 = x^2 + 5x$
   $y = 3x - 2$
   no solution

9. The weekly profit y (in dollars) for two street vendors can be modeled by the following equations, where x is the number of weeks since the beginning of the year.

   $y = -x^2 + 9x + 100$
   $y = 5x + 103$

   When is the weekly profit for each vendor the same?
   weeks 1 and 3

Answer Presentation Tool

Vocabulary and Concept Check

1. A solution of a system of linear and quadratic equations is an ordered pair that is a solution of each equation in the system.
2. Similarities: You can solve either type of system by elimination, substitution, or graphing.

   Differences: Solving a linear system involves finding the intersection(s) of 2 lines or solving a linear equation. Solving a system of linear and quadratic equations involves finding the intersection(s) of a line and a parabola or solving a quadratic equation.

Practice and Problem Solving

3. B; (0, 1), (3, 4)
4. C; no real solutions
5. A; (0, −1)
6. (0, −5), (−3, −8)
7. (−1, −2)
8. no real solutions
9. (−3, −2), (1, 6)
10. (0, 5)
11. $\left(-\frac{3}{2}, -4\right)$, (2, 10)
12. (−6, −22), (0, 2)
13. no real solutions
14. (1, 1)
15. (0, 0), (−5, 10)
16. no real solutions
17. (2, −6), $\left(\frac{5}{2}, -\frac{23}{4}\right)$
18. after 5 days and after 35 days

Practice and Problem Solving

19. $(-1, -6), (-2, -9)$

20. no real solutions

21. $(2, 2)$

22. *Sample answer:* graphing calculator; It is easier to use a graphing calculator, especially when fractions are involved.

23. *Sample answer:* The viewing window of the graphing calculator is too small. By increasing the window you can see that $(5, 14)$ is also a solution.

24. **a.** $y = 30x + 290$

b. $(1, 320), (34, 1310)$

25. **a.** 2 **b.** 0

26. See *Taking Math Deeper*.

Fair Game Review

27. $x = -1, x = 1$

28. $x = -1, x = 5$

29. no real solutions

30. C

Mini-Assessment

Solve the system.

1. $y = x^2 + 2$
$y = 6$ $(-2, 6), (2, 6)$

2. $y = x^2 - 7x + 12$
$y = x - 4$ $(4, 0)$

3. $y = x^2 - 3x + 5$
$y + 2x = 3$ no real solutions

4. The system of equations represents the annual revenues y (in thousands of dollars) of two companies t years after 2010.

$y = 3t^2 + 32t + 10$

$y = 45t + 40$

In what year are the revenues of the two companies equal? 2016

Taking Math Deeper

Exercise 26

You can solve this problem algebraically by setting up the system using the general form of each type of equation.

Write the system.

$y = mx + n$ General form of a linear equation

$y = ax^2 + bx + c$ General form of a quadratic equation

(*Note:* m and n are used in the general form of a linear equation because a and b are used in the general form of a quadratic equation.)

Begin solving the system by subtracting the equations.

$$\begin{aligned} y &= \phantom{-ax^2 + {}} mx + n \\ y &= ax^2 + bx + c \\ \hline 0 &= -ax^2 + (m - b)x + (n - c) \end{aligned}$$

Notice that the resulting equation is a quadratic equation in one variable. The solutions of this equation represent the solutions of the system.

You learned previously that quadratic equations must have 0, 1, or 2 solutions. So, the system must have 0, 1, or 2 solutions.

Interpret the result.

This system represents ALL possible systems of linear and quadratic equations. So, a system of linear and quadratic equations cannot have an infinite number of solutions.

Project

Research cubic functions. Make a conjecture about the number of possible solutions of a system of linear and cubic equations.

Reteaching and Enrichment Strategies

If students need help...	If students got it...
Resources by Chapter • Practice A and Practice B • Puzzle Time Record and Practice Journal Practice Differentiating the Lesson Lesson Tutorials Skills Review Handbook	Resources by Chapter • Enrichment and Extension • School-to-Work • Financial Literacy Start the next section

Solve the system using a graphing calculator.

19. $y = x^2 + 6x - 1$

$y = 3x - 3$

20. $y = \frac{1}{4}x - 12$

$y = x^2 - 6x$

21. $y = \frac{1}{2}x^2$

$y = 2x - 2$

22. CHOOSE TOOLS Do you prefer to solve systems of equations by hand or using a graphing calculator? Explain your reasoning.

23. ERROR ANALYSIS Describe and correct the error in solving the system of equations.

$y = x^2 - 3x + 4$
$y = 2x + 4$

The only solution of the system of equations is (0, 4).

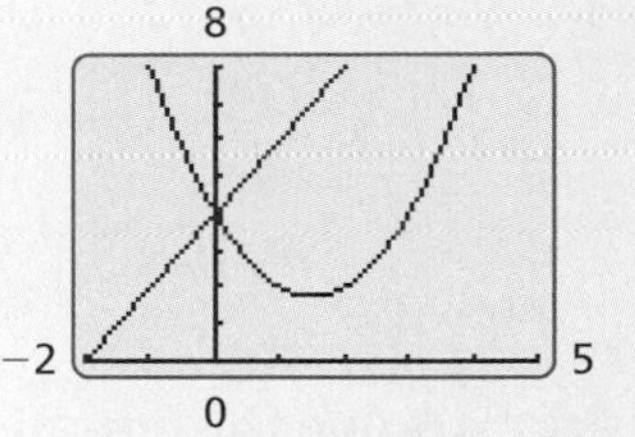

24. WEBSITES The function $y = -x^2 + 65x + 256$ models the number y of subscribers to a website, where x is the number of days since the website was launched. The number of subscribers to a competitor's website can be modeled by a linear function. The websites have the same number of subscribers on days 1 and 34.

a. Write a linear function that models the number of subscribers to the competitor's website.

b. Solve the system to verify the function from part (a).

25. REASONING The graph shows a quadratic function and the linear function $y = c$.

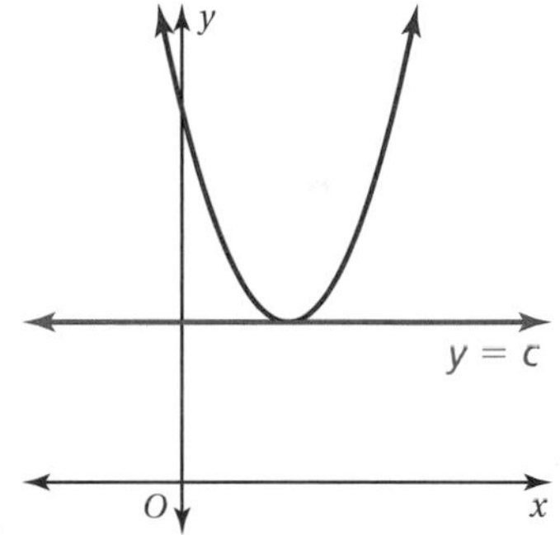

a. How many solutions will the system have when you change the linear equation to $y = c + 2$?

b. How many solutions will the system have when you change the linear equation to $y = c - 2$?

26. Writing Can a system of linear and quadratic equations have an infinite number of solutions? Explain your reasoning.

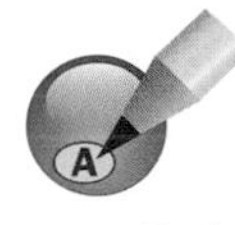

Fair Game Review What you learned in previous grades & lessons

Solve the equation by graphing. Check your solution(s). *(Section 9.1)*

27. $x^2 = 1$

28. $x^2 - 4x - 5 = 0$

29. $-x^2 = 2x + 7$

30. MULTIPLE CHOICE What is the factored form of the polynomial $x^2 - 36$? *(Section 7.9)*

Ⓐ $(x + 6)^2$ Ⓑ $(x - 6)^2$ Ⓒ $(x + 6)(x - 6)$ Ⓓ $x + 6$

9.4–9.5 Quiz

Solve the equation using the quadratic formula. *(Section 9.4)*

1. $x^2 + 8x - 20 = 0$
2. $13x = 2x^2 + 6$
3. $9 - 24x = -16x^2$

Use the discriminant to determine the number of real solutions of the equation. *(Section 9.4)*

4. $x^2 + 6x - 13 = 0$
5. $-8x^2 - x = 5$
6. $\frac{3}{4}x^2 = 3x - 3$

7. Solve $x^2 + 10x + 21 = 0$ using two different methods. *(Section 9.4)*

Solve the equation using any method. Explain your choice of method. *(Section 9.4)*

8. $x^2 + 4x - 11 = 0$
9. $-4x^2 + 1 = 0$
10. $52 = x^2 - 2x$

Solve the system. *(Section 9.5)*

11. $y = x^2 - 16$
    $y = -7$
12. $y = x^2 + 2x + 1$
    $y = 2x + 2$
13. $y = x^2 - 5x + 8$
    $y = -3x - 4$

14. **BACTERIA** The numbers y of two types of bacteria after t hours are given by the models below. *(Section 9.5)*

    $y = 3t^2 + 8t + 20$ Type 1
    $y = 27t + 60$ Type 2

    a. As t increases, which type grows more quickly? Explain.

    b. When are the numbers of Type 1 and Type 2 bacteria the same?

    c. When are there more Type 1 bacteria than Type 2? When are there more Type 2 bacteria than Type 1? Use a graph to support your answer.

15. **CELLULAR PHONE CALLS** The average monthly bill y (in dollars) for a customer's cell phone x years after 2000 can be modeled by $y = -0.2x^2 + 2x + 45$. When was the average monthly bill about \$50? *(Section 9.4)*

16. **REASONING** Do you think the model in Exercise 15 can be used for future years? Explain using a graphing calculator to support your answer. *(Section 9.4)*

Alternative Assessment Options

Math Chat	Student Reflective Focus Question
Structured Interview	Writing Prompt

Math Chat

- Have individual students work problems from the quiz on the board. The student explains the process used and justifies each step. Students in the class ask questions of the student presenting.
- The teacher explores the thought process of the student presenting, but does not teach or ask leading questions.

Study Help Sample Answers

Remind students to complete Graphic Organizers for the rest of the chapter.

4.

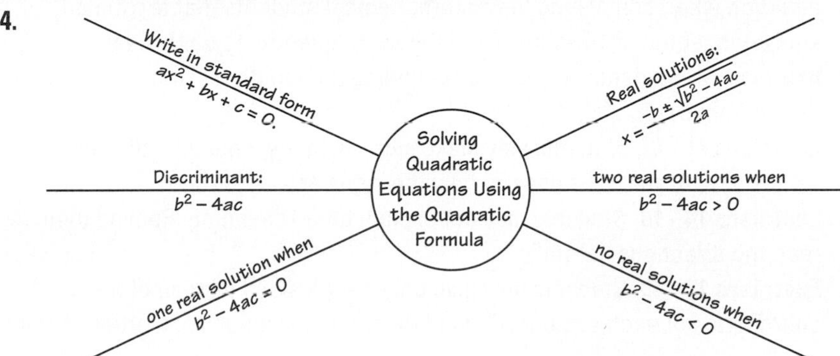

5.

Choosing a Solution Method for Solving Quadratic Equations

Factoring: Straightforward when equation can be factored easily.

Graphing: Using a graphing utility, solutions are easy to see, but not always exact.

Using square roots: Use for $x^2 = d$.

Completing the square: Best when $a = 1$, b is even.

Quadratic formula: Takes time, but always works and gives exact answers.

Always check solutions!!!

6. Available at *BigIdeasMath.com*

Reteaching and Enrichment Strategies

If students need help. . .	If students got it. . .
Resources by Chapter • Study Help • Practice A and Practice B • Puzzle Time Lesson Tutorials *BigIdeasMath.com* Practice Quiz Practice from the Test Generator	Resources by Chapter • Enrichment and Extension • School-to-Work Game Closet at *BigIdeasMath.com* Start the Chapter Review

Answers

1. $x = -10, x = 2$
2. $x = \frac{1}{2}, x = 6$
3. $x = \frac{3}{4}$
4. 2
5. 0
6. 1
7. $x = -7, x = -3$
8. $x = -2 + \sqrt{15}, x = -2 - \sqrt{15}$; *Sample answer:* completing the square, because $a = 1$ and b is even
9. $x = \frac{1}{2}, x = -\frac{1}{2}$; *Sample answer:* square roots, because no x-term
10. $x = 1 + \sqrt{53}, x = 1 - \sqrt{53}$; *Sample answer:* completing the square, because $a = 1$ and b is even
11. $(-3, -7), (3, -7)$
12. $(-1, 0), (1, 4)$
13. no real solutions

14–16. See Additional Answers.

Technology For the Teacher
Answer Presentation Tool

Assessment Book

Chapter 9 **Quiz**
For use after Section 9.5

Solve the equation using the quadratic formula.

1. $2x^2 + 5x - 3 = 0$
2. $3x^2 - x - 10 = 0$
3. $x^2 + 6x + 12 = 0$
4. $3x^2 + 4x - 2 = 0$

Use the discriminate to determine the number of real solutions of the equation.

5. $x^2 + 7x + 13 = 0$
6. $2x^2 - 8x + 8 = 0$

Solve the equation using any method. Explain your choice of method.

7. $x^2 + 8x - 9 = 0$
8. $2x^2 - 5x - 3 = 0$
9. $2x^2 + 6x - 36 = 0$
10. $x^2 + 5x = 0$

Solve the system.

11. $y = x^2 - 36$, $y = 9x$
12. $y - 3 = x^2$, $y = 2$
13. $y = x^2 - 4x + 5$, $y = x - 1$
14. $y = x^2 - 7x + 10$, $y = -5x + 10$
15. The amount of money a store earns during each hour one day can be modeled by the function $y = -8x^2 + 64x - 60$, where y is in dollars and x is the time in hours from when the store opens. During which hour will the store earn $68?
16. The numbers h of hours you work each day at two different jobs are given by the models below, where x is the number of days since you started working.
$h = -x^2 + 6x + 7$
$h = 2x + 11$
During which hour is the hourly rate the same?

Answers
1. ____ 2. ____ 3. ____ 4. ____ 5. ____ 6. ____ 7. ____ See left. 8. ____ See left. 9. ____ See left. 10. ____ See left. 11. ____ 12. ____ 13. ____ 14. ____ 15. ____ 16. ____

For the Teacher

Additional Review Options

- Big Ideas Test Generator
- Game Closet at *BigIdeasMath.com*
- Vocabulary Puzzle Builder
- Resources by Chapter
  - Puzzle Time
  - Study Help

Answers

1. $x = 3, x = 6$
2. no real solutions
3. $x = -4$
4. $x = 0$
5. no real solutions
6. $x = -10, x = 6$

Review of Common Errors

- **Exercises 1–3** Solutions obtained graphically may be incorrect or not exact. Make sure students use a sound graphical approach. Also, remind them to check their answers.
- **Exercise 5** Students may try to take the square root of a negative number. Remind them that the square of a real number cannot be negative.
- **Exercise 6** Students may forget the negative square root when taking the square root of each side of the equation. Remind them to account for the negative square root when appropriate.
- **Exercises 7–9** Students may forget to divide the *x*-coefficient by 2 before squaring. Remind them of this process.
- **Exercises 7–9** Students may not add the same value to each side of the equation when completing the square. Remind students that to form an equivalent equation, they must add the same quantity to each side.
- **Exercise 10** Students may stop after finding the length ℓ. They need to find the perimeter.
- **Exercises 11–13** Students may make sign mistakes when identifying the values of *a, b,* and *c*. Emphasize how the signs are determined.
- **Exercises 14–16** Students may not explain their reasoning. Remind them to read the directions carefully.
- **Exercises 17–19** Students may give only the *x*-values instead of the coordinates of each solution. Remind them each solution is an ordered pair.

9 Chapter Review

Check It Out
Vocabulary Help
BigIdeasMath.com

Review Key Vocabulary

quadratic equation, *p. 456*
completing the square, *p. 470*
quadratic formula, *p. 478*
discriminant, *p. 480*

Review Examples and Exercises

9.1 Solving Quadratic Equations by Graphing *(pp. 454–461)*

Solve $x^2 + 3x - 4 = 0$ by graphing.

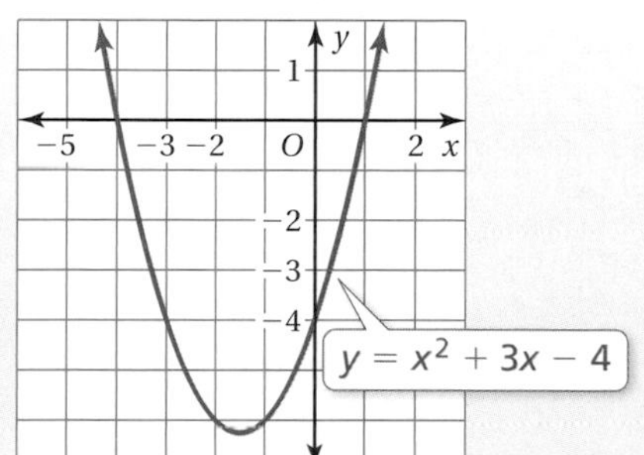

Step 1: Graph the related function $y = x^2 + 3x - 4$.

Step 2: Find the x-intercepts. They are -4 and 1.

So, the solutions are $x = -4$ and $x = 1$.

Exercises

Solve the equation by graphing. Check your solution(s).

1. $x^2 - 9x + 18 = 0$ **2.** $x^2 - 2x = -4$ **3.** $-8x - 16 = x^2$

9.2 Solving Quadratic Equations Using Square Roots *(pp. 462–467)*

A sprinkler sprays water that covers a circular region of 90π square feet. Find the diameter of the circle.

Write an equation using the formula for the area of a circle.

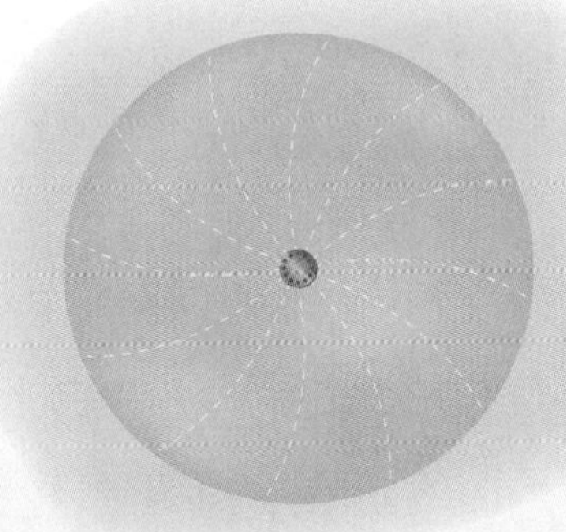

$A = \pi r^2$	Write the formula.
$90\pi = \pi r^2$	Substitute 90π for A.
$90 = r^2$	Divide each side by π.
$\pm\sqrt{90} = r$	Take the square root of each side.

A diameter cannot be negative, so use the positive square root. The diameter is twice the radius. So, the diameter is $2\sqrt{90}$.

The diameter of the circle is $2\sqrt{90} \approx 19$ feet.

Exercises

Solve the equation using square roots.

4. $x^2 - 10 = -10$ **5.** $4x^2 = -100$ **6.** $(x + 2)^2 = 64$

9.3 Solving Quadratic Equations by Completing the Square (pp. 468–473)

Solve $x^2 - 6x + 4 = 11$ by completing the square.

$x^2 - 6x + 4 = 11$	Write the equation.
$x^2 - 6x = 7$	Subtract 4 from each side.
Complete the square. → $x^2 - 6x + 9 = 7 + 9$	Add $\left(\frac{-6}{2}\right)^2$, or 9, to each side.
$(x - 3)^2 = 16$	Factor $x^2 - 6x + 9$.
$x - 3 = \pm 4$	Take the square root of each side.
$x = 3 \pm 4$	Add 3 to each side.

The solutions are $x = 3 + 4 = 7$ and $x = 3 - 4 = -1$.

Exercises

Solve the equation by completing the square.

7. $x^2 + x + 10 = 0$
8. $x^2 + 2x + 5 = 4$
9. $2x^2 - 4x = 10$
10. **CREDIT CARD** The width w of a credit card is 3 centimeters shorter than the length ℓ. The area is 46.75 square centimeters. Find the perimeter.

9.4 Solving Quadratic Equations Using the Quadratic Formula (pp. 476–485)

Solve $-3x^2 + x = -8$ using the quadratic formula.

$-3x^2 + x = -8$	Write original equation.
$-3x^2 + x + 8 = 0$	Write in standard form.
$x = \frac{-b \pm \sqrt{b^2 - 4ac}}{2a}$	Quadratic Formula
$= \frac{-1 \pm \sqrt{1^2 - 4(-3)(8)}}{2(-3)}$	Substitute −3 for *a*, 1 for *b*, and 8 for *c*.
$= \frac{-1 \pm \sqrt{97}}{-6}$	Simplify.

The solutions are $x = \frac{-1 + \sqrt{97}}{-6} \approx -1.5$ and $x = \frac{-1 - \sqrt{97}}{-6} \approx 1.8$.

Exercises

Solve the equation using the quadratic formula.

11. $x^2 + 2x - 15 = 0$
12. $2x^2 - x + 8 = 3$
13. $-5x^2 + 10x = 5$

Solve the equation using any method. Explain your choice of method.

14. $x^2 - 121 = 0$
15. $x^2 - 4x + 4 = 0$
16. $x^2 - 4x = -1$

Review Game

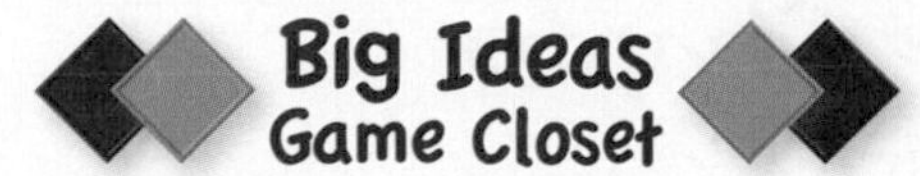

Choosing a Solution Method

Materials:

- flash cards
- paper
- pencil

Directions:

- Make flash cards ahead of time by writing quadratic equations large enough for your students to see.
- Split the class into two teams. Select a spokesperson for each team.

Playing a round:

- Lift a flash card to show a quadratic equation. The two teams race to determine:
  (1) a list of the methods that can be used to solve the equation
  (2) the solution of the equation
- The first spokesperson to raise his or her hand answers both parts.
- Next, the spokesperson from the other team either confirms or corrects the first team's answers.

Round scoring:

- The correct answer to Part (1) is a list of any of the methods (factoring, graphing, using square roots, completing the square, or quadratic formula) that can be used to solve the equation. The correct answer to Part (2) includes all solutions of the equation.
- The first team earns 2 points for each part they answer correctly. The other team earns 1 point for confirming a correct part and 3 points for correcting a wrong part.

Who wins?
The team with the greatest number of points after the last round wins.

For the Student
Additional Practice

- Lesson Tutorials
- Study Help (textbook)
- Student Website
  Multi-Language Glossary
  Practice Assessments

Answers

7. no real solutions
8. $x = -1$
9. $x = 1 + \sqrt{6}, x = 1 - \sqrt{6}$
10. 28 cm
11. $x = -5, x = 3$
12. no real solutions
13. $x = 1$
14. $x = 11, x = -11$; *Sample answer:* square roots, because no x-term
15. $x = 2$; *Sample answer:* factoring, because the left side is a perfect square trinomial
16. $x = 2 + \sqrt{3}, x = 2 - \sqrt{3}$; *Sample answer:* completing the square, because $a = 1$ and b is even
17. $(1, -5)$
18. $\left(1 - \sqrt{15}, 7 - 2\sqrt{15}\right)$, $\left(1 + \sqrt{15}, 7 + 2\sqrt{15}\right)$
19. no real solutions

My Thoughts on the Chapter

What worked. . .

Teacher Tip

Not allowed to write in your teaching edition? Use sticky notes to record your thoughts.

What did not work. . .

What I would do differently. . .

9.5 Solving Systems of Linear and Quadratic Equations *(pp. 486–491)*

Solve the system by substitution.

$$\mathbf{y = 2x^2 - 5} \qquad \text{Equation 1}$$

$$\mathbf{y = -x + 1} \qquad \text{Equation 2}$$

Step 1: The equations are already solved for y.

Step 2: Substitute $-x + 1$ for y in Equation 1 and solve for x.

$y = 2x^2 - 5$	Equation 1
$-x + 1 = 2x^2 - 5$	Substitute $-x + 1$ for y.
$1 = 2x^2 + x - 5$	Add x to each side.
$0 = 2x^2 + x - 6$	Subtract 1 from each side.
$0 = (2x - 3)(x + 2)$	Factor right side.
$2x - 3 = 0$ *or* $x + 2 = 0$	Use Zero-Product Property.
$x = \frac{3}{2}$ *or* $x = -2$	Solve for x.

Step 3: Substitute $\frac{3}{2}$ and -2 for x in Equation 2 and solve for y.

$y = -x + 1$	Equation 2	$y = -x + 1$
$= -\frac{3}{2} + 1$	Substitute.	$= -(-2) + 1$
$= -\frac{1}{2}$	Simplify.	$= 3$

So, the solutions are $\left(\frac{3}{2}, -\frac{1}{2}\right)$ and $(-2, 3)$.

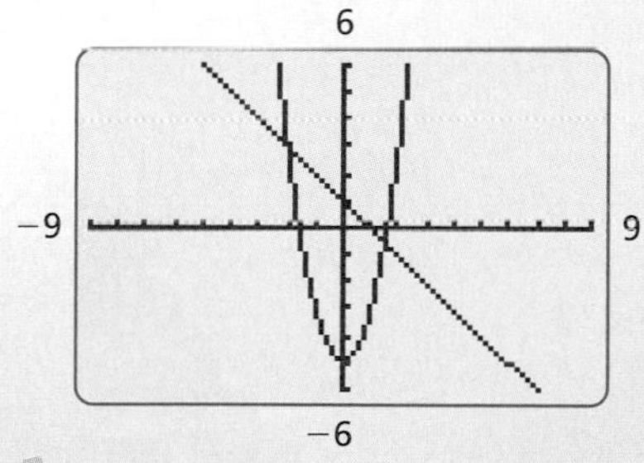

Exercises

Solve the system. Check your solution(s).

17. $y = x^2 - 2x - 4$
$y = -5$

18. $y = x^2 - 9$
$y = 2x + 5$

19. $y = 2 - 3x$
$y = -x^2 - 5x - 4$

9 Chapter Test

Check It Out
Test Practice
BigIdeasMath.com

Solve the equation by graphing.

1. $x^2 - 7x + 12 = 0$
2. $x^2 + 12x = -36$
3. $x + 1 = -x^2$

Solve the equation using square roots.

4. $14 = 2x^2$
5. $x^2 + 9 = 5$
6. $(4x + 3)^2 = 16$

Solve the equation by completing the square.

7. $x^2 - 8x + 15 = 0$
8. $x^2 - 6x = 10$
9. $x^2 - 8x = -9$
10. $16 = x^2 - 16x - 20$

Solve the equation using the quadratic formula.

11. $5x^2 + x - 4 = 0$
12. $9x^2 + 6x + 1 = 0$
13. $-2x^2 + 3x + 7 = 0$

14. **REASONING** Use the discriminant to determine how many times the graph of $y = 4x^2 - 4x + 1$ intersects the x-axis.

15. **CHOOSING A METHOD** Solve $x^2 - 9x - 10 = 0$ using any method. Explain your choice of method.

Solve the system.

16. $y = x^2 - 4x - 2$
    $y = -4x + 2$
17. $y = -5x^2 + x - 1$
    $y = -7$

18. **GEOMETRY** The area of the triangle is 35 square feet. Use a quadratic equation to find the length of the base. Round your answer to the nearest tenth.

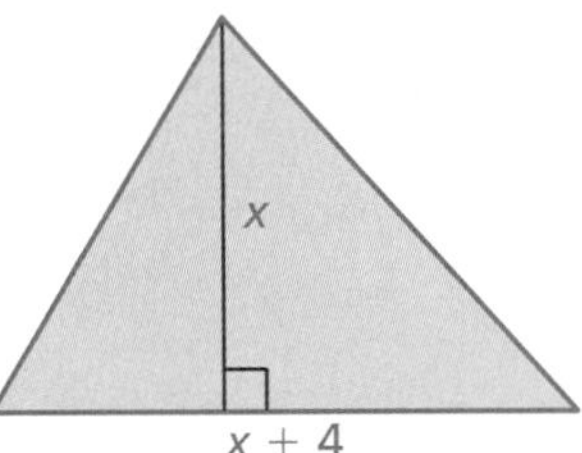

19. **SNOWBOARDING** A snowboarder leaves an 8-foot-tall ramp with an upward velocity of 28 feet per second. The function $h = -16t^2 + 28t + 8$ gives the height h (in feet) of the snowboarder after t seconds. How many points does the snowboarder earn with a perfect landing?

Criteria	Scoring
Maximum height	1 point per foot
Time in air	5 points per second
Perfect landing	25 points

Test Item References

Chapter Test Questions	Section to Review	Common Core State Standards
1–3	9.1	A.REI.4, A.REI.11
4–6	9.2	A.REI.4b
7–10	9.3	A.REI.4a, A.REI.4b, A.SSE.3b, F.IF.8a
11–14, 15, 18, 19	9.4	A.REI.4a, A.REI.4b
16, 17	9.5	A.REI.7

Test-Taking Strategies

Remind students to quickly look over the entire test before they start so that they can budget their time. Have students use the **Stop** and **Think** strategy before they answer each question.

Common Assessment Errors

- **Exercises 1–3** Solutions obtained graphically may be incorrect or not exact. Make sure students use a sound graphical approach. Also, remind them to check their answers.
- **Exercises 4–6** Students may forget the negative square root when taking the square root of each side of the equation. Remind them to account for the negative square root when appropriate.
- **Exercises 7–10** Students may not add the same value to each side of the equation when completing the square. Remind students that to form an equivalent equation, they must add the same quantity to each side.
- **Exercises 11–13** Students may make sign mistakes when identifying the values of *a, b,* and *c*. Emphasize how the signs are determined.
- **Exercises 16 and 17** Students may give only the *x*-values instead of the coordinates of each solution. Remind them each solution is an ordered pair.
- **Exercise 19** Students may have a difficult time starting the problem. Explain how to approach this problem one part at a time: Find the maximum height, the time in the air, and the points awarded.

Reteaching and Enrichment Strategies

If students need help...	If students got it...
Resources by Chapter • Practice A and Practice B • Puzzle Time Record and Practice Journal Practice Differentiating the Lesson Lesson Tutorials Practice from the Test Generator Skills Review Handbook	Resources by Chapter • Enrichment and Extension • School-to-Work • Financial Literacy Game Closet at *BigIdeasMath.com* Start Standardized Test Practice

Answers

1. $x = 3, x = 4$
2. $x = -6$
3. no real solutions
4. $x = \sqrt{7}, x = -\sqrt{7}$
5. no real solutions
6. $x = -\frac{7}{4}, x = \frac{1}{4}$
7. $x = 3, x = 5$
8. $x = 3 + \sqrt{19}, x = 3 - \sqrt{19}$
9. $x = 4 + \sqrt{7}, x = 4 - \sqrt{7}$
10. $x = -2, x = 18$
11. $x = -1, x = \frac{4}{5}$
12. $x = -\frac{1}{3}$
13. $x = \frac{3 + \sqrt{65}}{4}, x = \frac{3 - \sqrt{65}}{4}$
14. 1
15. $x = -1, x = 10$, *Sample answer:* factors easily
16. $(-2, 10), (2, -6)$
17. $(-1, -7), \left(\frac{6}{5}, -7\right)$
18. 10.6 ft
19. $55\frac{1}{4}$ points

Assessment Book

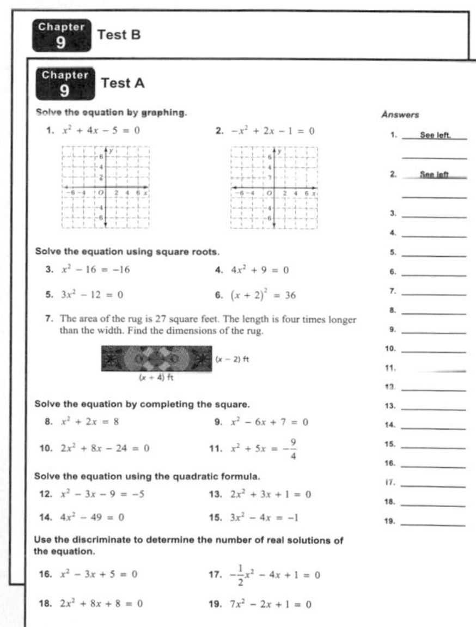

Test Taking Strategies
Available at *BigIdeasMath.com*

After Answering Easy Questions, Relax
Answer Easy Questions First
Estimate the Answer
Read All Choices before Answering
Read Question before Answering
Solve Directly or Eliminate Choices
Solve Problem before Looking at Choices
Use Intelligent Guessing
Work Backwards

About this Strategy

When taking a multiple choice test, be sure to read each question carefully and thoroughly. When taking a timed test, it is often best to skim the test and answer the easy questions first. Be careful that you record your answer in the correct position on the answer sheet.

Answers

1. B
2. I
3. D
4. -4
5. H

Technology For the Teacher

Big Ideas Test Generator

Item Analysis

1. **A.** The student confuses the patterns for the square of a binomial.

   B. Correct answer

   C. The student represents the product as a sum of two squares.

   D. The student confuses the patterns for the square of a binomial.

2. **F.** The student incorrectly rewrites the related equation as $y = (x - 25)^2 + 5$ and solves by graphing.

   G. The student is confused by the negative sign.

   H. The student incorrectly rewrites the related equation as $y = (3x + 1)^2 + 9$ and solves by graphing.

   I. Correct answer

3. **A.** The student thinks the values represent solutions, *not* zeros.

   B. The student incorrectly thinks the graph crosses the x-axis at (1, 0) and (3, 0).

   C. The student does not know that the axis of symmetry is halfway between the x-intercepts.

   D. Correct answer

4. **Gridded response:** Correct answer: -4

   Common error: The student substitutes -12 for x and evaluates as -52.

5. **F.** The student chooses an integer value of x close to where the maximum occurs.

   G. The student incorrectly estimates the value from the graph.

   H. Correct answer

   I. The student chooses an integer value of x close to where the maximum occurs.

9 Standardized Test Practice

1. Which expression represents the area of the square? *(A.APR.1)*

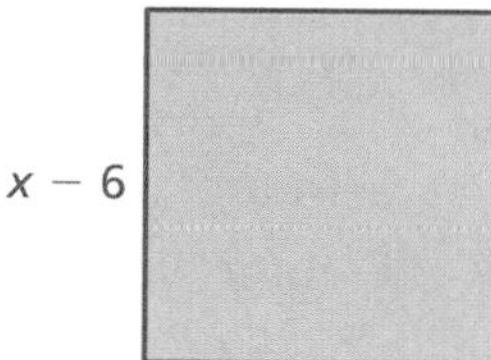

A. $x^2 + 12x + 36$ C. $x^2 + 36$

B. $x^2 - 12x + 36$ D. $x^2 - 12x - 36$

Test-Taking Strategy
Answer Easy Questions First
The function h=-16t²+45t+20 gives the height h (in feet) of a cannonball after t seconds. At what height was it launched?
Ⓐ-20 Ⓑ0 Ⓒ20 Ⓓ45
Shiver me whiskers!
"Scan the test and answer the easy questions first. You know that the constant term, 20, is the initial height. So, the answer is C.

2. Which of the following equations has *no* real solutions? *(A.REI.4b)*

F. $(x - 25)^2 = 5$ H. $(3x + 1)^2 = 9$

G. $-4x^2 = 0$ I. $2x^2 + 1 = -1$

3. Use the solution below to determine which statement about the function $y = x^2 + 4x + 3$ is false. *(A.SSE.3a)*

$$x^2 + 4x + 3 = 0$$
$$(x + 3)(x + 1) = 0$$
$$x + 3 = 0 \quad or \quad x + 1 = 0$$
$$x = -3 \qquad x = -1$$

A. The zeros are −1 and −3. C. The axis of symmetry of the graph is $x = -2$.

B. The graph crosses the x-axis at (−1, 0) and (−3, 0). D. The maximum value occurs when $x = -2$.

4. For $f(x) = 5x + 8$, what value of x makes $f(x) = -12$? *(F.IF.2)*

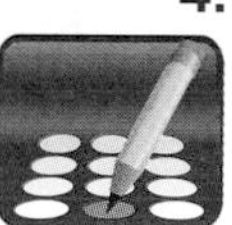

5. Which line represents the axis of symmetry of the graph of the quadratic function? *(F.IF.4)*

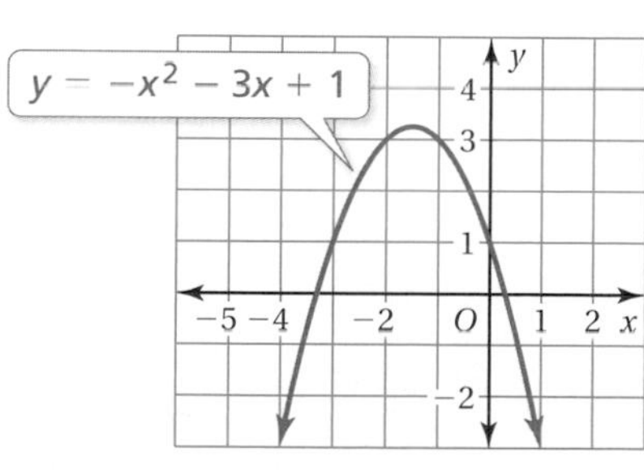

F. $x = -2$ H. $x = -\frac{3}{2}$

G. $x = -\frac{5}{3}$ I. $x = -1$

6. What are the *exact* roots of the quadratic equation $3x^2 + x - 1 = 0$? *(A.REI.4b)*

A. $-0.8, 0.4$

B. $-0.77, 0.43$

C. $\frac{-1 - \sqrt{13}}{6}, \frac{-1 + \sqrt{13}}{6}$

D. $-\frac{3}{4}, \frac{1}{2}$

7. The function $h = -16t^2 + 60t + 2$ gives the height h (in feet) of a soccer ball after t seconds. Which of the following statements is true? *(A.REI.4b)*

F. The soccer ball reaches a height of 60 feet.

G. It takes the soccer ball 2.5 seconds to reach its maximum height.

H. The soccer ball hits the ground after about 5 seconds.

I. The soccer ball is kicked from a height of 2 feet.

8. Which *best* describes the solutions of the system of equations below? *(A.REI.7)*

$y = x^2 + 2x - 8$ Equation 1

$y = 5x + 2$ Equation 2

A. Their graphs intersect at one point, $(-2, -8)$. So, there is one solution.

B. Their graphs intersect at two points, $(-2, -8)$ and $(5, 27)$. So, there are two solutions.

C. Their graphs do not intersect. So, there is no solution.

D. The graph of $y = x^2 + 2x - 8$ has two x-intercepts. So, there are two solutions.

9. Which graph shows exponential growth? *(F.LE.1c)*

F.

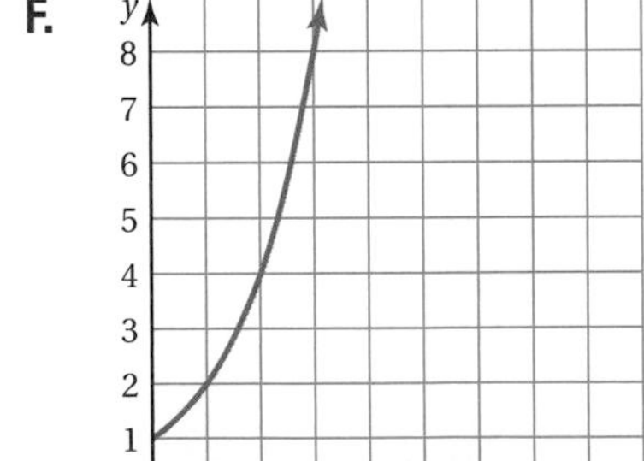

G.

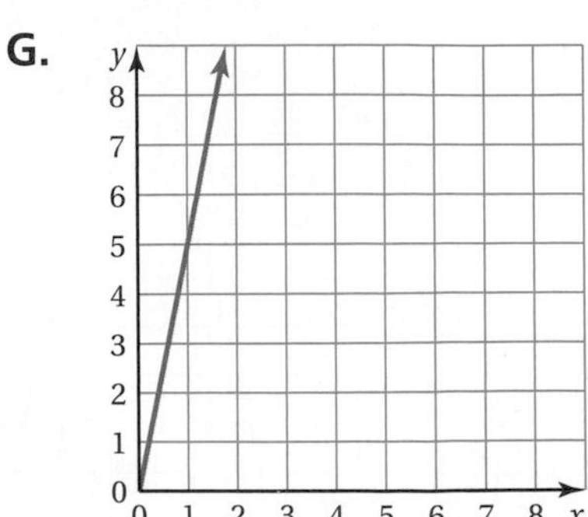

H.

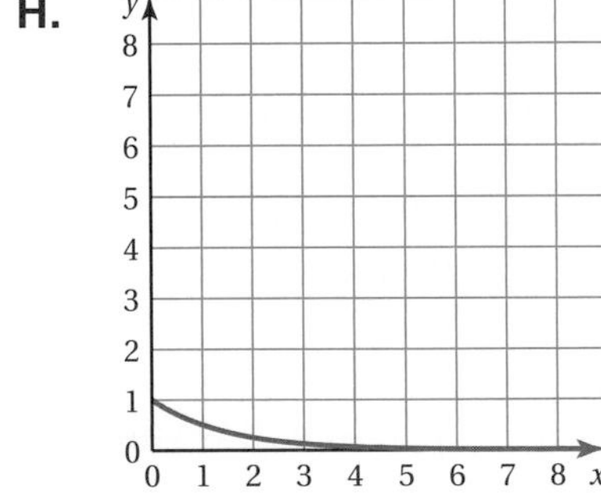

I.

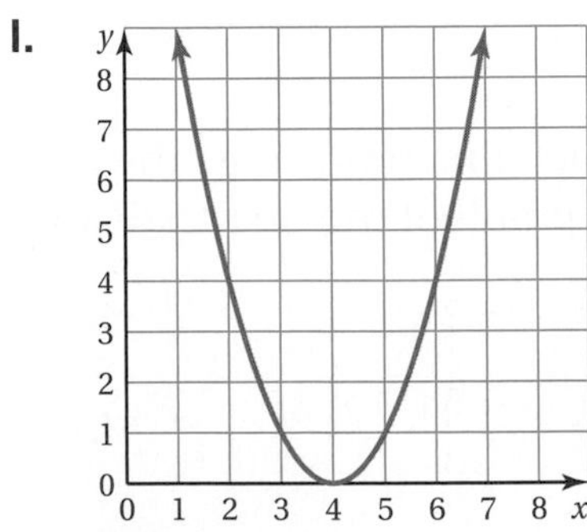

Item Analysis (continued)

6. **A.** The student rounds the solutions.
   B. The student rounds the solutions.
   C. Correct answer
   D. The student incorrectly approximates the solutions by graphing.

7. **F.** The student graphs the function and estimates that the graph reaches a height of 60 feet.
   G. The student makes a calculation error.
   H. The student makes a calculation error.
   I. Correct answer

8. **A.** Using a graphing calculator, the student graphs the system in a standard viewing window and does not see the second point of intersection.
   B. Correct answer
   C. Using a graphing calculator, the student graphs the system in a window that does not show either point of intersection.
   D. The student confuses the solutions of a system of equations with the solutions of a quadratic equation.

9. **F.** Correct answer
   G. The student confuses the graphs of exponential growth and exponential decay models.
   H. The student randomly chooses a graph that rises from left to right.
   I. The student chooses the graph of a parabola because the chapter is about solving quadratic equations.

Answers

6. C

7. I

8. B

9. F

Answers

10. *Part A:* up

Part B: (0, 4)

Part C: $x = -\frac{1}{2}$

Part D: $\left(-\frac{1}{2}, \frac{13}{4}\right)$

11. A

12. 114

13. G

Answer for Extra Example

1. **A.** The student confuses the vertical line $x = d$ for a horizontal line that does not intersect the parabola.

   B. Correct answer

   C. The student confuses the vertical line $x = d$ for a horizontal line that intersects the parabola at two points.

   D. The student confuses the vertical line $x = d$ for a horizontal line.

Item Analysis (continued)

10. **2 points** The student demonstrates a thorough understanding of how the graph of a quadratic function is related to its standard form $y = ax^2 + bx + c$. The student finds each part correctly, shows the work, and gives sound explanations.

1 point The student's work and explanation demonstrate a partial understanding. The student is unable to find one or two of the parts correctly and not all of the explanations are adequate.

0 points The student provides no response, a completely incorrect or incomprehensible response, or a response that demonstrates insufficient understanding of how the graph of a quadratic equation is related to the standard form of its equation.

11. **A.** Correct answer

B. The student incorrectly factors the perfect square trinomial.

C. The student uses $-\left(\frac{b^2}{2}\right)$ instead of $\left(\frac{b^2}{2}\right)$.

D. The student confuses this problem for the type of real-life problem in which you only use the positive root.

12. **Gridded response:** Correct answer: 114

Common error: The student forgets to write the equation in standard quadratic form and uses $c = 13$ instead of $c = -13$.

13. **F.** The student confuses the range and the domain.

G. Correct answer

H. The student thinks the zeros are the points where the graph crosses the x-axis.

I. The student confuses minimum and maximum.

Extra Example

1. Which statement best describes the number of solutions of the system, where a, b, c, and d are real numbers? *(A.REI.7)*

$y = ax^2 + bx + c$ Equation 1

$x = d$ Equation 2

A. There are no real solutions.

B. There is one solution.

C. There are two solutions.

D. There may be one, two, or no real solutions.

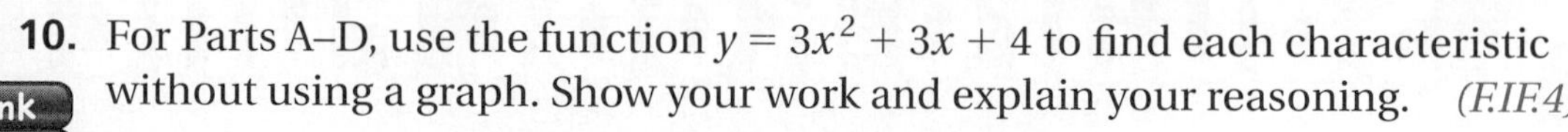

10. For Parts A–D, use the function $y = 3x^2 + 3x + 4$ to find each characteristic without using a graph. Show your work and explain your reasoning. *(F.IF.4)*

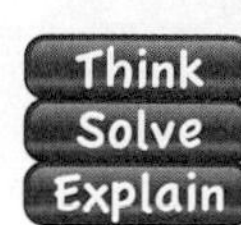

Part A direction the graph of the function opens

Part B *y*-intercept of the graph of the function

Part C axis of symmetry of the graph of the function

Part D vertex of the graph of the function

11. Jamie is solving the equation $x^2 - 14x + 7 = 18$ by completing the square.

$$x^2 - 14x + 7 = 18$$
$$x^2 - 14x = 11$$
$$x^2 - 14x + 49 = 11$$
$$(x - 7)^2 = 11$$
$$x - 7 = \pm\sqrt{11}$$
$$x = 7 \pm \sqrt{11}$$

What should Jamie do to correct the error that he made? *(A.REI.4b)*

A. Add 49 to each side of the equation.

B. Factor $x^2 - 14x + 49$ as $(x + 7)^2$.

C. Subtract 49 from each side of the equation instead of adding 49.

D. Only use the positive square root of 11.

12. What is the value of the discriminant for the quadratic equation $1.5x^2 - 6x = 13$? *(A.REI.4b)*

13. Which of the following statements is true about the quadratic function shown in the graph? *(A.REI.4b)*

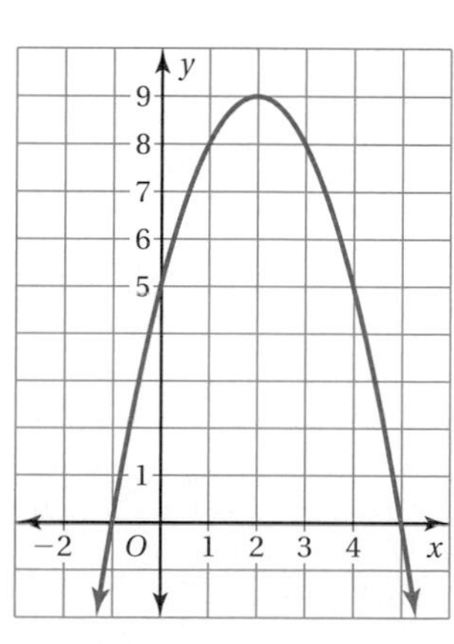

F. The range is all real numbers.

G. The domain is all real numbers.

H. The zeros are $(-1, 0)$ and $(5, 0)$.

I. A minimum occurs at the vertex.

10 Square Root Functions and Geometry

"I'm pretty sure that Pythagoras was a Greek."

"I said 'Greek', not 'Geek'."

"Let's figure out how we can measure the height of the giant hyena standing right behind you."

Connections to Previous Learning

- Write, analyze, and solve one-variable linear equations.
- Find the areas of right triangles.

- Understand the connections between proportional relationships, lines, and linear equations.
- Draw and construct triangles and describe the relationships between them.
- Evaluate square roots of small perfect squares.

- Understand and apply the Pythagorean Theorem.
- Solve square root equations.

Math in History

One form of the Pythagorean Theorem states that the area of the square on the hypotenuse of a right triangle is equal to the combined area of the squares on the other two sides. Pythagoras (c. 570–500 B.C.) may have been the first to prove this theorem, but other cultures were already aware of this property.

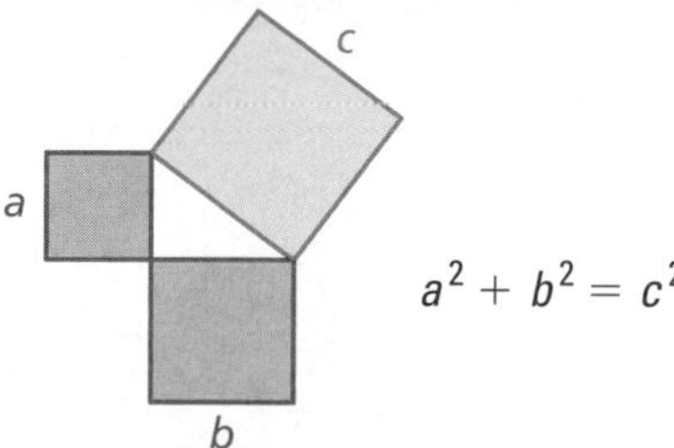

★ Sets of numbers satisfying the relationship of the side lengths of right triangles are given in an old Babylonian tablet (c.1900–1600 B.C.).

★ The relationship of the side lengths of right triangles is described in a series of writings called the *Sulbasutras* (c. 800–200 B.C.), written in India, which gives instructions for building religious altars.

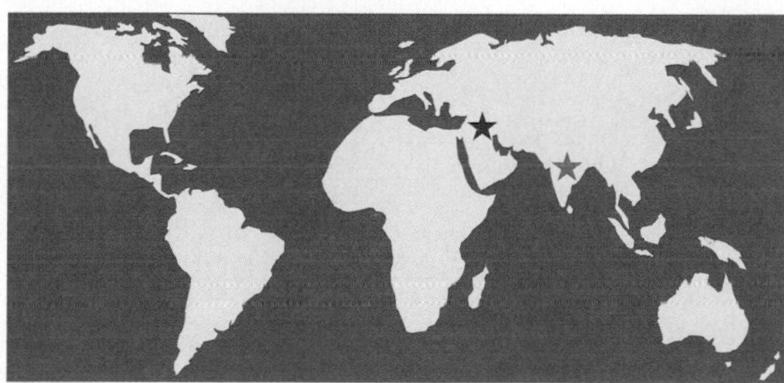

Pacing Guide for Chapter 10

Chapter Opener	1 Day
Section 1	2 Days
Section 2	2 Days
Study Help/Quiz	1 Day
Section 3	2 Days
Section 4	1 Day
Chapter Review / Chapter Tests	2 Days
Total Chapter 10	11 Days
Year-to-Date	136 Days

Check Your Resources

- Record and Practice Journal
- Resources by Chapter
- Skills Review Handbook
- Assessment Book
- Worked-Out Solutions

Technology For the Teacher

The Dynamic Planning Tool
Editable Teacher's Resources at
BigIdeasMath.com

Common Core State Standards

8.EE.2 Use square root and cube root symbols to represent solutions to equations of the form $x^2 = p$ and $x^3 = p$, where p is a positive rational number. Evaluate square roots of small perfect squares and cube roots of small perfect cubes. Know that $\sqrt{2}$ is irrational.

A.SSE.3a Factor a quadratic expression to reveal the zeros of the function it defines.

Additional Topics for Review

- Graphing functions
- Domain and range of a function
- Solving inequalities
- Solving equations
- Right triangles

Try It Yourself

1. 45
2. -10
3. -38
4. $(y + 3)(y + 9)$
5. $(n - 10)(n - 1)$
6. $(w + 6)(w - 8)$
7. $(z + 5)(z + 20)$

Record and Practice Journal

1. -2
2. 29
3. 8
4. -13
5. 20
6. -48
7. 14
8. 3
9. 16 days since launch
10. $(v - 4)(v - 8)$
11. $(d + 3)(d + 6)$
12. $(k - 7)(k + 9)$
13. $(m + 2)(m - 12)$
14. $(t - 10)(t + 9)$
15. $(f + 9)(f - 3)$
16. $(a + 5)(a + 11)$
17. $(q - 4)(q - 17)$
18. $(x + 2)$

Math Background Notes

Vocabulary Review

- Square Root
- Perfect Square
- Polynomial
- Factoring Polynomials

Evaluating an Expression Involving a Square Root

- Students should know how to apply the order of operations to evaluate an expression.
- In applying the order of operations, finding the square root of a number is equivalent to raising a number to an exponent.

Factoring $x^2 + bx + c$

- Students should know how to factor a trinomial.
- Remind students that $x^2 + bx + c = (x + p)(x + q)$ when $p + q = b$ and $pq = c$. Also remind them that when c is positive, p and q have the same sign as b, and when c is negative, p and q have different signs.
- **Common Error:** Students may forget to check that the factors of c add up to b. Remind them that there are two steps to factoring polynomials, factoring c and checking that those factors add up to b. Encourage students to use tables as shown in the example.
- Remind students to check their answers using the FOIL Method.

Reteaching and Enrichment Strategies

If students need help. . .	If students got it. . .
Record and Practice Journal • Fair Game Review Skills Review Handbook Lesson Tutorials	Game Closet at *BigIdeasMath.com* Start the next section

What You Learned Before

PyThagorean Theorem
(shorter)^2+(shorter)^2=(longest)^2
Speaking of being a square...

"I just remember that the sum of the squares of the shorter sides is equal to the square of the longest side."

Evaluating an Expression Involving a Square Root (8.EE.2)

Example 1 **Evaluate $-4(\sqrt{121} - 16)$.**

$$-4(\sqrt{121} - 16) = -4(11 - 16) \quad \text{Evaluate the square root.}$$
$$= -4(-5) \quad \text{Subtract.}$$
$$= 20 \quad \text{Multiply.}$$

Try It Yourself

Evaluate the expression.

1. $7\sqrt{25} + 10$

2. $-8 - \sqrt{\frac{64}{16}}$

3. $-2(3\sqrt{4} + 13)$

Factoring $x^2 + bx + c$ (A.SSE.3a)

Example 2 **Factor $x^2 - 3x - 28$.**

Notice that $b = -3$ and $c = -28$. Because c is negative, the factors p and q must have different signs so that pq is negative.

Find two integer factors of -28 whose sum is -3.

Factors of −28	−28, 1	−1, 28	−14, 2	−2, 14	−7, 4	−4, 7
Sum of Factors	−27	27	−12	12	−3	3

The values of p and q are -7 and 4.

So, $x^2 - 3x - 28 = (x - 7)(x + 4)$.

Try It Yourself

Factor the polynomial.

4. $y^2 + 12y + 27$

5. $n^2 - 11n + 10$

6. $w^2 - 2w - 48$

7. $z^2 + 25z + 100$

10.1 Graphing Square Root Functions

COMMON CORE STATE STANDARDS
F.IF.4
F.IF.7b

Essential Question How can you sketch the graph of a square root function?

1 ACTIVITY: Graphing Square Root Functions

Work with a partner.

- Make a table of values for the function.
- Use the table to sketch the graph of the function.
- Describe the domain of the function.
- Describe the range of the function.

a. $y = \sqrt{x}$

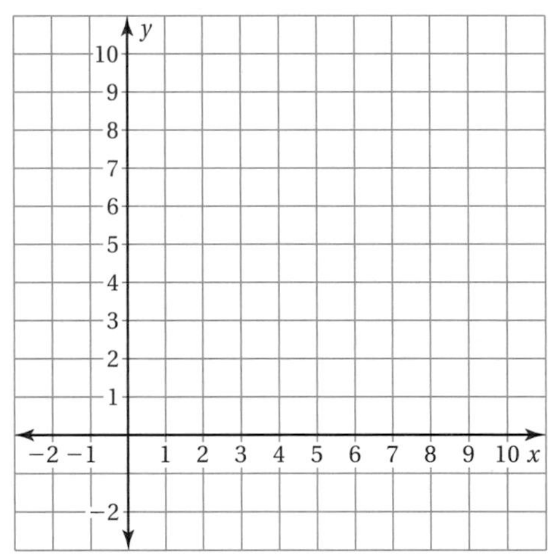

b. $y = \sqrt{x} + 2$

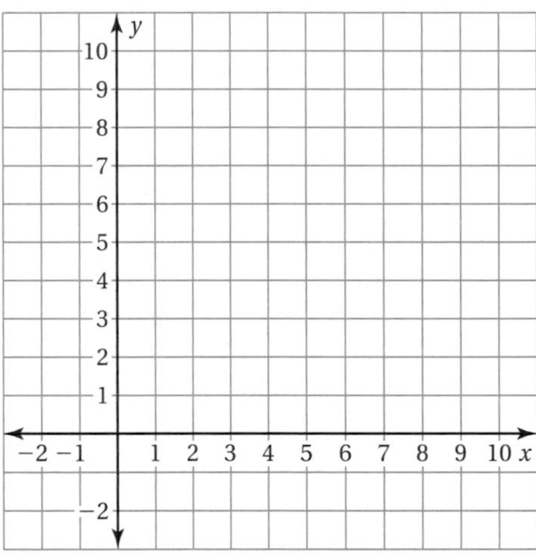

c. $y = \sqrt{x + 1}$

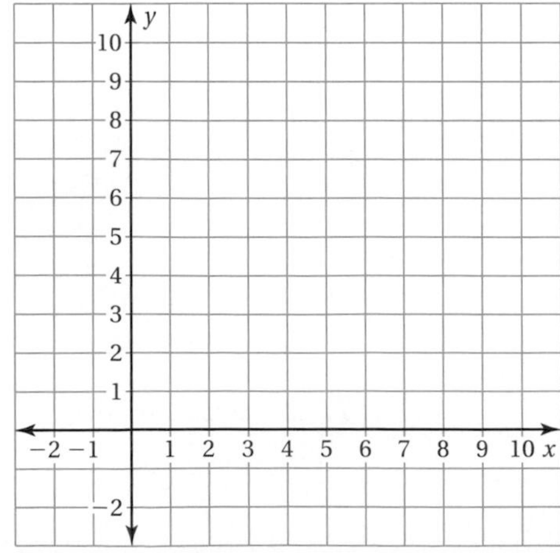

d. $y = -\sqrt{x}$

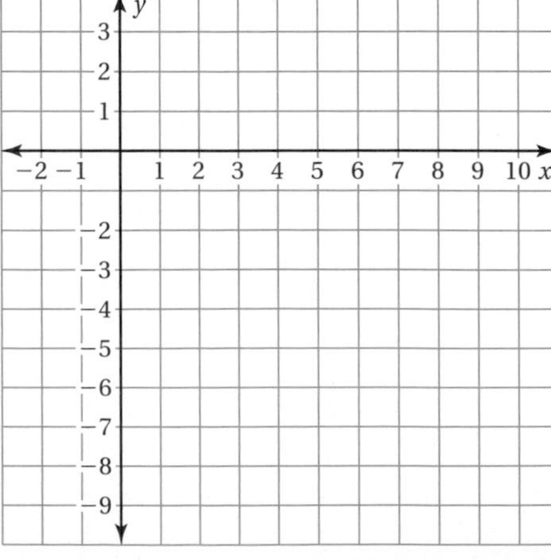

Laurie's Notes

Introduction

Standards for Mathematical Practice

- **MP5 Use Appropriate Tools Strategically:** Students will graph square root functions using a table of values. They can use a calculator to approximate square roots.

Motivate

- Play a quick round of "30-second Un-Do." In this game, students list as many actions and undoing actions as they can in 30 seconds. For example: filling a sink and draining a sink.
- Have students share a few examples after their time is up.
- Play another round, but this time the actions must involve math operations.
- Have students share a few examples. Hopefully at least one student will have an example that does not involve addition, subtraction, multiplication, or division.
- Explain today's activities using more precise language. They involve taking square roots, the inverse operation of squaring.

Activity Notes

Activity 1

- Students have graphed a variety of functions, so the directions should need no additional explanation.
- **MP6 Attend to Precision:** When using a calculator, students need to be aware of the syntax for square roots. Some calculators require the radicand to be enclosed in parentheses. Students need to recognize the difference between $\sqrt{(x)} + 2$ and $\sqrt{(x + 2)}$.

? "What is the domain of each function?" All have a domain of $x \geq 0$ except part (c), which has a domain of $x \geq \ \ 1$.

? "What is the range of each function?" Listen for correct answers.

- After students have finished all four graphs, you could have them check their work using a graphing calculator.
- You could also ask volunteers to describe the differences in the graphs. Students should be able to describe the transformations in parts (b)–(d).

Common Core State Standards

F.IF.4 For a function that models a relationship between two quantities, interpret key features of graphs and tables in terms of the quantities, and sketch graphs showing key features given a verbal description of the relationship.

F.IF.7b Graph square root, cube root, and piecewise-defined functions, including step functions and absolute value functions.

Previous Learning

Students should know how to graph a function using a table of values.

Start Thinking! and Warm Up

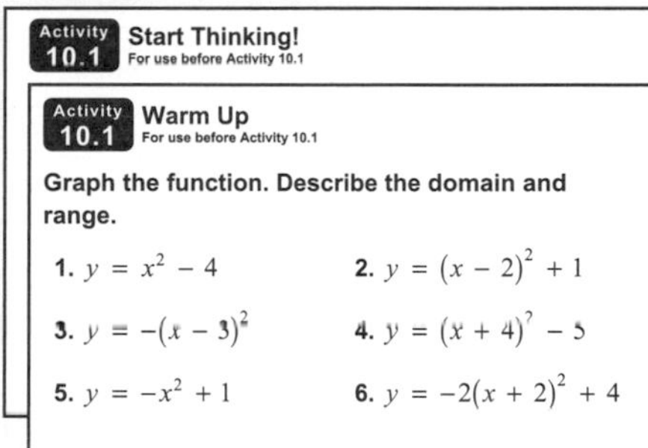

Activity 10.1 Start Thinking! For use before Activity 10.1

Activity 10.1 Warm Up For use before Activity 10.1

Graph the function. Describe the domain and range.

1. $y = x^2 - 4$
2. $y = (x - 2)^2 + 1$
3. $y = -(x - 3)^2$
4. $y = (x + 4)^2 - 5$
5. $y = -x^2 + 1$
6. $y = -2(x + 2)^2 + 4$

10.1 Record and Practice Journal

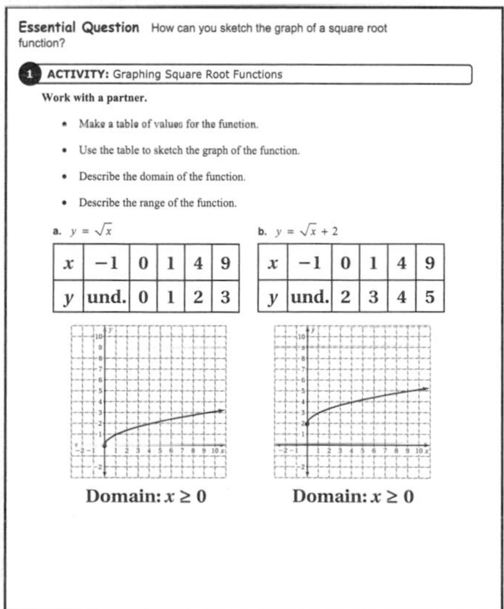

Essential Question How can you sketch the graph of a square root function?

1 ACTIVITY: Graphing Square Root Functions

Work with a partner.

- Make a table of values for the function.
- Use the table to sketch the graph of the function.
- Describe the domain of the function.
- Describe the range of the function.

a. $y = \sqrt{x}$

x	−1	0	1	4	9
y	und.	0	1	2	3

Domain: $x \geq 0$

b. $y = \sqrt{x} + 2$

x	−1	0	1	4	9
y	und.	2	3	4	5

Domain: $x \geq 0$

English Language Learners

Vocabulary

Students may think of *domain* as the set of all real numbers. Review the definition. The domain is the set of all possible input values. For a square root function, a number that makes the radicand negative is *not* a possible input value.

10.1 Record and Practice Journal

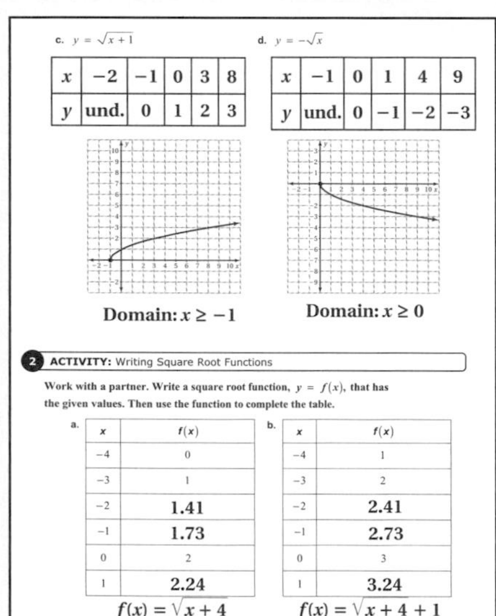

c. $y = \sqrt{x+1}$

x	-2	-1	0	3	8
y	und.	0	1	2	3

Domain: $x \geq -1$

d. $y = -\sqrt{x}$

x	-1	0	1	4	9
y	und.	0	-1	-2	-3

Domain: $x \geq 0$

2 ACTIVITY: Writing Square Root Functions

Work with a partner. Write a square root function, $y = f(x)$, that has the given values. Then use the function to complete the table.

a.

x	$f(x)$
-4	0
-3	1
-2	1.41
-1	1.73
0	2
1	2.24

$f(x) = \sqrt{x+4}$

b.

x	$f(x)$
-4	1
-3	2
-2	2.41
-1	2.73
0	3
1	3.24

$f(x) = \sqrt{x+4} + 1$

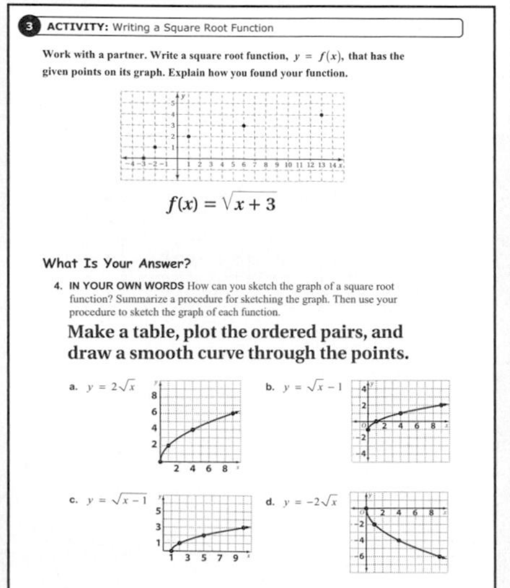

3 ACTIVITY: Writing a Square Root Function

Work with a partner. Write a square root function, $y = f(x)$, that has the given points on its graph. Explain how you found your function.

$f(x) = \sqrt{x+3}$

What Is Your Answer?

4. **IN YOUR OWN WORDS** How can you sketch the graph of a square root function? Summarize a procedure for sketching the graph. Then use your procedure to sketch the graph of each function.

Make a table, plot the ordered pairs, and draw a smooth curve through the points.

a. $y = 2\sqrt{x}$

b. $y = \sqrt{x} - 1$

c. $y = \sqrt{x-1}$

d. $y = -2\sqrt{x}$

Laurie's Notes

Activity 2

- Ordered pairs have been given so that students can determine the equation of the square root function.
- Students may need to use trial and error. For instance, students may believe part (a) is $y = \sqrt{x} + 2$ because of the y-intercept. They need to test all three ordered pairs in the equation.
- Probe students as you walk around to see if they recognize any patterns between the two parts of the activity. The y-values differ by 1. Does that influence their thinking as they start the second part?
- It is good practice to have students identify the domain and range of each function.

Activity 3

? "How is this activity like the last activity?" Ordered pairs are given, but in a graph instead of a table.

- Students have graphed several square root functions. If they have not made an observation about the general shape of the graph, ask about it now. They may describe it as half of a parabola that has been rotated.

What Is Your Answer?

- **MP7 Look for and Make Use of Structure:** Students have not graphed $y = a\sqrt{x}$ or $y = -a\sqrt{x}$. Ask them how a affects the graph of $y = \sqrt{x}$.

Closure

- Compare the graph of $y = \sqrt{x} + 2$ to the graph of $y = \sqrt{x}$. translation 2 units up

Technology For the Teacher

The Dynamic Planning Tool

Editable Teacher's Resources at *BigIdeasMath.com*

2 ACTIVITY: Writing Square Root Functions

Work with a partner. Write a square root function, $y = f(x)$, that has the given values. Then use the function to complete the table.

a.

x	$f(x)$
−4	0
−3	1
−2	
−1	
0	2
1	

b.

x	$f(x)$
−4	1
−3	2
−2	
−1	
0	3
1	

3 ACTIVITY: Writing a Square Root Function

Work with a partner. Write a square root function, $y = f(x)$, that has the given points on its graph. Explain how you found your function.

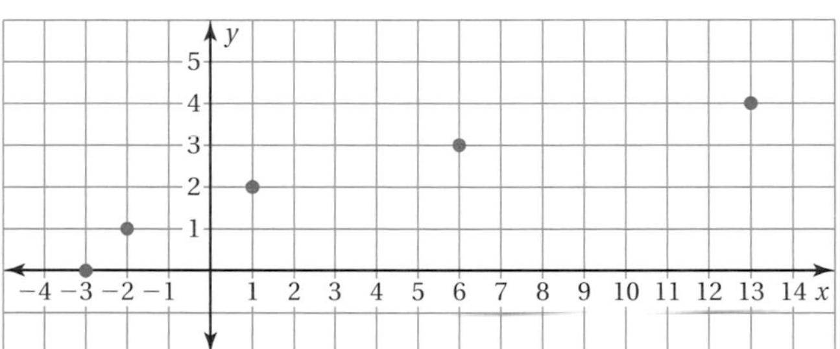

What Is Your Answer?

4. **IN YOUR OWN WORDS** How can you sketch the graph of a square root function? Summarize a procedure for sketching the graph. Then use your procedure to sketch the graph of each function.

   a. $y = 2\sqrt{x}$

   b. $y = \sqrt{x} - 1$

   c. $y = \sqrt{x - 1}$

   d. $y = -2\sqrt{x}$

Practice Use what you learned about the graphs of square root functions to complete Exercises 3–8 on page 506.

10.1 Lesson

Key Vocabulary
square root function, *p. 504*

Key Idea

Square Root Function

A **square root function** is a function that contains a square root with the independent variable in the radicand. The most basic square root function is $y = \sqrt{x}$.

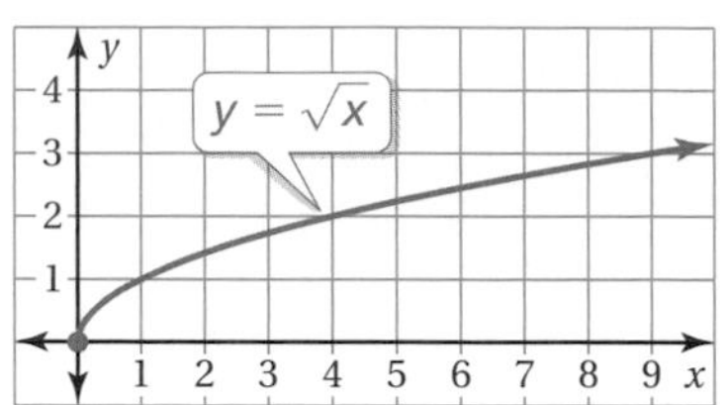

The value of the radicand in the square root function cannot be negative. So, the domain of a square root function includes x-values for which the radicand is greater than or equal to 0.

EXAMPLE 1 Finding the Domain of a Square Root Function

Find the domain of $y = 3\sqrt{x - 5}$.

The radicand cannot be negative. So, $x - 5$ is greater than or equal to 0.

$x - 5 \geq 0$ — Write an inequality for the domain.

$x \geq 5$ — Add 5 to each side.

The domain is the set of real numbers greater than or equal to 5.

On Your Own

Now You're Ready
Exercises 9–14

Find the domain of the function.

1. $y = 10\sqrt{x}$
2. $y = \sqrt{x} + 7$
3. $y = \sqrt{-x + 1}$

EXAMPLE 2 Comparing Graphs of Square Root Functions

Graph $y = \sqrt{x} + 3$. Describe the domain and range. Compare the graph to the graph of $y = \sqrt{x}$.

Step 1: Make a table of values.

x	0	1	4	9	16
y	3	4	5	6	7

Step 2: Plot the ordered pairs.

Step 3: Draw a smooth curve through the points.

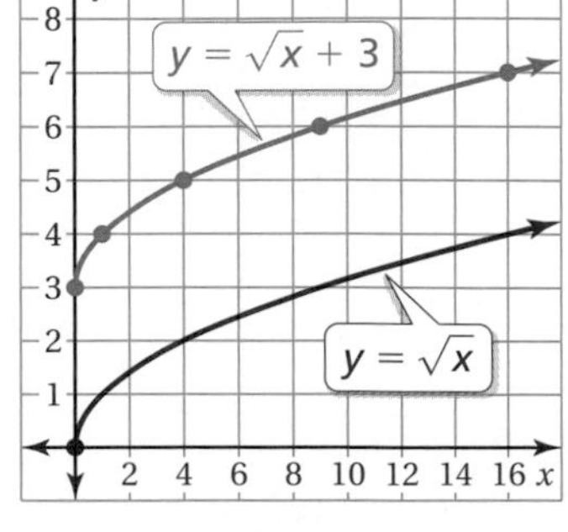

Remember
When graphing, remember $f(x) + k$ is a vertical translation of $f(x)$.

From the graph, you can see that the domain is $x \geq 0$ and the range is $y \geq 3$. The graph of $y = \sqrt{x} + 3$ is a translation 3 units up of the graph of $y = \sqrt{x}$.

Multi-Language Glossary at BigIdeasMath.com.

Laurie's Notes

Introduction

Connect

- **Yesterday:** Students explored graphs of square root functions. (MP5, MP6, MP7)
- **Today:** Students will graph square root functions.

Motivate

- Use a graphing calculator to graph $y = \sqrt{x}$, $y = \sqrt{x} - 3$, and $y = \sqrt{x - 3}$. Then ask students how the 3 affects the last two graphs.
- In today's lesson, students will graph square root functions.

Lesson Notes

Discuss

- **Big Idea:** A *square root function* is an example of a *radical function.* There are other radical functions, such as $y = \sqrt[3]{x}$. This should make sense from when students learned about rational exponents and *n*th roots.
- Point out to students that the list of functions they have studied is growing: linear, absolute value, exponential, quadratic, and radical.

Key Idea

- Write the Key Idea, which defines a square root function. The parent function is $y = \sqrt{x}$.
- Say, "You cannot take the square root of a negative number. So, the domain of a square root function includes inputs for which the radicand is nonnegative." Give examples to support this statement.

Example 1

? "Could $x = 8$?" yes "Could $x = 6$?" yes "Could $x = 4$?" no

- Continue to solve the problem as shown.

On Your Own

- **Neighbor Check:** Have students work independently and then have their neighbor check their work. Have students discuss any discrepancies.

Example 2

? "Can you use $x = 1, 2, 3, 4$, and 5 to graph the function?" Yes, but it is easier to use perfect squares so that the outputs are whole numbers.

- **MP6 Attend to Precision:** Stress that you take the square root of the input and then add three, not add three and then take the square root.
- The functions have the same domain. The range of $y = \sqrt{x} + 3$ is $y \geq 3$.

Goal Today's lesson is graphing **square root functions.**

Lesson Materials	
Introduction	**Textbook**
• graphing calculators	• graphing calculators

Start Thinking! and Warm Up

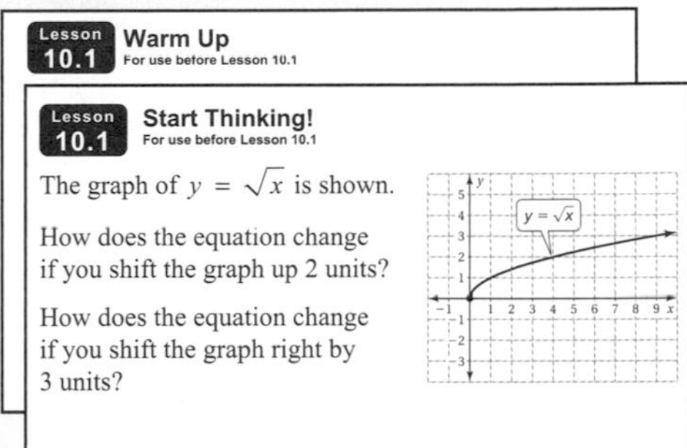
Lesson 10.1 Warm Up
For use before Lesson 10.1

Lesson 10.1 Start Thinking!
For use before Lesson 10.1

The graph of $y = \sqrt{x}$ is shown.

How does the equation change if you shift the graph up 2 units?

How does the equation change if you shift the graph right by 3 units?

Extra Example 1

Find the domain of $y = \sqrt{-x + 5}$.
$x \leq 5$

On Your Own

1. $x \geq 0$
2. $x \geq 0$
3. $x \leq 1$

Extra Example 2

Graph $y = \sqrt{x} - 6$. Describe the domain and range. Compare the graph to the graph of $y = \sqrt{x}$.

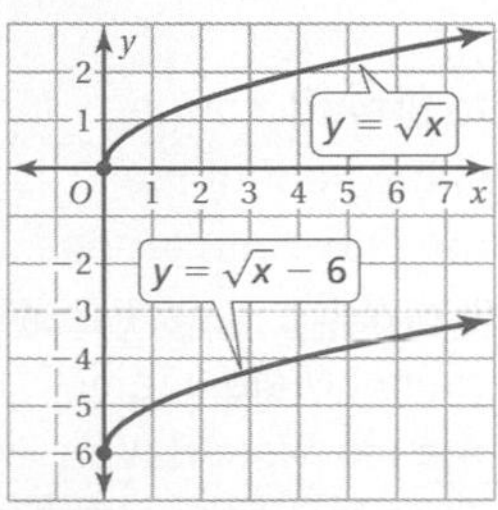

The domain is $x \geq 0$ and the range is $y \geq -6$. The graph of $y = \sqrt{x} - 6$ is a translation 6 units down of the graph of $y = \sqrt{x}$.

Laurie's Notes

Extra Example 3

Graph $y = -\sqrt{x+1} - 2$. Describe the domain and range. Compare the graph to the graph of $y = \sqrt{x}$.
See Additional Answers.

On Your Own

4–6. See Additional Answers.

Extra Example 4

In Example 4, at what depth does the velocity of the tsunami exceed 150 meters per second? about 2300 meters

On Your Own

7. domain: $x \geq 0$;
range: $y \geq 0$

8. about 1000 meters

Differentiated Instruction

Kinesthetic

Use masking tape to create a large coordinate plane on the floor of the classroom. Divide the class into groups of four. Assign each group a square root function and have them create a table of values to graph the function. Have each group model their function in the coordinate plane with string or yarn. Start with the function $y = \sqrt{x}$. Have students describe the transformation(s) as they graph the remaining functions.

Example 3

- ? "How do you think this graph will compare to the graph of $y = \sqrt{x}$?" Students may have an idea about the graph being reflected about the x-axis and translated 2 units, but they may be unsure of the direction.
- ? "What is the domain of this function? Explain." $x \geq 2$; because the radicand is negative when $x < 2$
- Make a table of values. Note that perfect squares were not selected. Two more than a perfect square would allow for easy computations.
- Work through the rest of the problem as shown.

On Your Own

- If time is short, make sure students try Question 6. Students may try to graph the function without making a table of values. Knowing where (0, 0) is transformed helps to "anchor" the graph.

Example 4

- Ask a student to read the problem.
- **MP4 Model with Mathematics:** The velocity of the tsunami is a function of the depth of the water.
- **MP5 Use Appropriate Tools Strategically:** Explain how to enter the function into a graphing calculator and set the viewing window as shown.
- Use the *trace* feature to answer the question. The units are meters per second.
- **Connection:** Have students identify a familiar distance that is about 200 meters. The tsunami travels that distance in one second.
- **Extension:** Graph $y = 200$ and use the *intersect* feature to find the solution.

On Your Own

- **Think-Pair-Share:** Students should read each question independently and then work with a partner to answer the questions. When they have answered the questions, the pair should compare their answers with another group and discuss any discrepancies.

Closure

- **Exit Ticket:** Compare the graph of $y = \sqrt{x+1} - 3$ to the graph of $y = \sqrt{x}$.
translation 1 unit to the left and 3 units down

Technology For the Teacher

The Dynamic Planning Tool
Editable Teacher's Resources at *BigIdeasMath.com*

EXAMPLE 3 Comparing Graphs of Square Root Functions

Graph $y = -\sqrt{x-2}$. Describe the domain and range. Compare the graph to the graph of $y = \sqrt{x}$.

Step 1: Make a table of values.

x	2	3	4	5	6
y	0	−1	−1.4	−1.7	−2

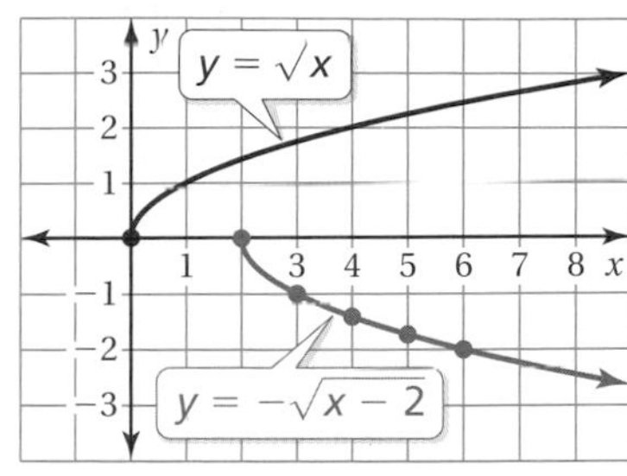

Step 2: Plot the ordered pairs.

Step 3: Draw a smooth curve through the points.

From the graph, you can see that the domain is $x \geq 2$ and the range is $y \leq 0$. The graph of $y = -\sqrt{x-2}$ is a reflection of the graph of $y = \sqrt{x}$ in the x-axis and then a translation 2 units to the right.

Study Tip

The graph of $f(x - h)$ is a horizontal translation of $f(x)$.

On Your Own

Now You're Ready Exercises 16–21

Graph the function. Describe the domain and range. Compare the graph to the graph of $y = \sqrt{x}$.

4. $y = \sqrt{x} - 4$
5. $y = \sqrt{x+5}$
6. $y = -\sqrt{x+1} + 2$

EXAMPLE 4 Real-Life Application

The velocity y (in meters per second) of a tsunami can be modeled by the function $y = \sqrt{9.8x}$, where x is the water depth (in meters). Use a graphing calculator to graph the function. At what depth does the velocity of the tsunami exceed 200 meters per second?

Step 1: Enter the function $y = \sqrt{9.8x}$ into your calculator and graph it. Because the radicand cannot be negative, use only nonnegative values of x.

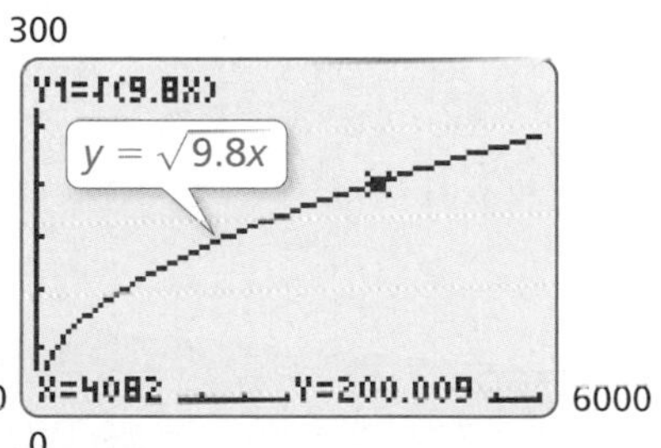

Step 2: Use the *trace* feature to find where the value of y is about 200.

The velocity exceeds 200 meters per second at a depth of about 4100 meters.

On Your Own

7. Find the domain and range of the function in Example 4.
8. **WHAT IF?** In Example 4, at what depth does the velocity of the tsunami exceed 100 meters per second?

10.1 Exercises

Vocabulary and Concept Check

1. **VOCABULARY** Is $y = 2x\sqrt{5}$ a square root function? Explain.
2. **REASONING** How do you find the domain of a square root function?

Practice and Problem Solving

Match the function with its graph.

3. $y = 8\sqrt{x}$
4. $y = \frac{5}{4}\sqrt{x}$
5. $y = -4\sqrt{x}$

A.

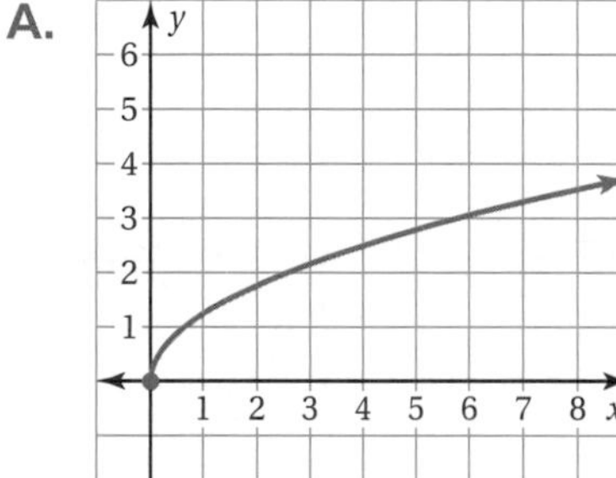

B.

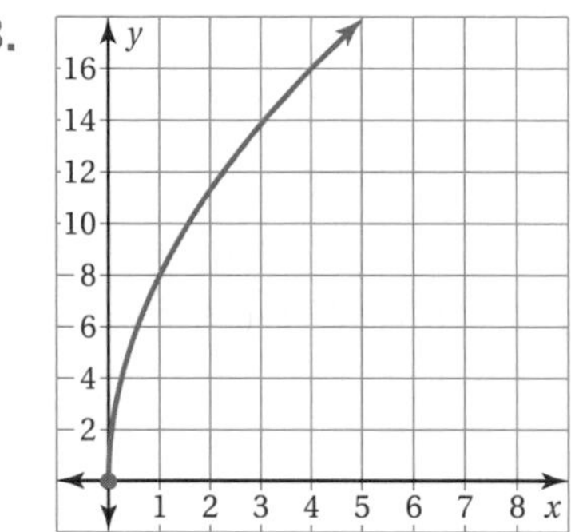

C.

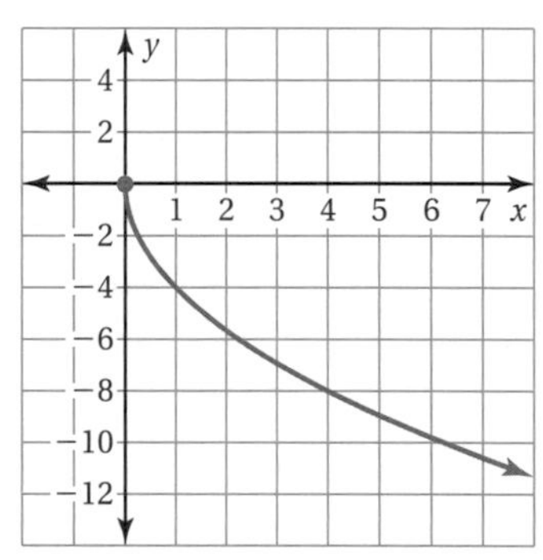

Graph the function. Describe the domain.

6. $y = 3\sqrt{x}$
7. $y = 7\sqrt{x}$
8. $y = -0.5\sqrt{x}$

Find the domain of the function.

1 9. $y = 5\sqrt{x}$
10. $y = \sqrt{x} + 1$
11. $y = \sqrt{x - 2}$
12. $y = \sqrt{-x - 1}$
13. $y = 2\sqrt{x + 4}$
14. $y = \frac{1}{2}\sqrt{-x + 2}$

15. **FIRE** The nozzle pressure of a fire hose allows firefighters to control the amount of water they spray on a fire. The flow rate f (in gallons per minute) can be modeled by the function $f = 120\sqrt{p}$, where p is the nozzle pressure (in pounds per square inch).
    a. Use a graphing calculator to graph the function.
    b. Use the *trace* feature to approximate the nozzle pressure that results in a flow rate of 300 gallons per minute.

Assignment Guide and Homework Check

Level	Assignment	Homework Check
Average	1, 2, 7–21 odd, 22, 26, 28–31	7, 13, 15, 19, 22
Advanced	1, 2, 6–20 even, 15, 23–27, 28–31	14, 15, 20, 26, 27

Common Errors

- **Exercises 12 and 14** Students may forget to reverse the direction of the inequality symbol when dividing each side by a negative number. Remind them of the Division Property of Inequality.
- **Exercises 10–14 and 16–21** Students may treat constants added to a radical expression as part of the radicand, or vice versa. Encourage them to identify the radicand before solving the exercise.

10.1 Record and Practice Journal

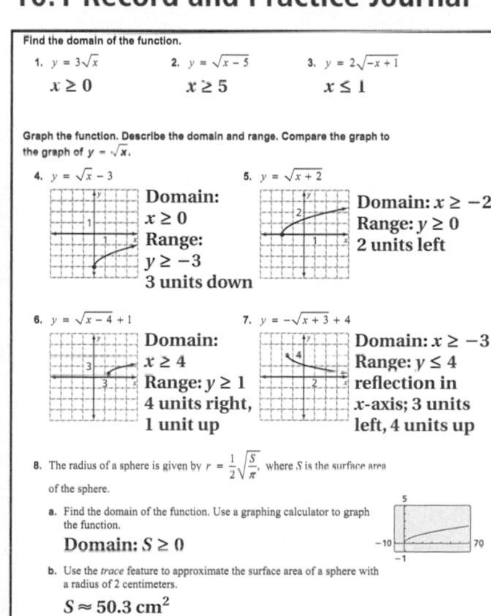
Find the domain of the function.

1. $y = 3\sqrt{x}$ $x \geq 0$
2. $y = \sqrt{x-5}$ $x \geq 5$
3. $y = 2\sqrt{-x+1}$ $x \leq 1$

Graph the function. Describe the domain and range. Compare the graph to the graph of $y = \sqrt{x}$.

4. $y = \sqrt{x} - 3$ Domain: $x \geq 0$ Range: $y \geq -3$ 3 units down
5. $y = \sqrt{x+2}$ Domain: $x \geq -2$ Range: $y \geq 0$ 2 units left
6. $y = \sqrt{x-4} + 1$ Domain: $x \geq 4$ Range: $y \geq 1$ 4 units right, 1 unit up
7. $y = -\sqrt{x+3} + 4$ Domain: $x \geq -3$ Range: $y \leq 4$ reflection in x-axis; 3 units left, 4 units up

8. The radius of a sphere is given by $r = \frac{1}{2}\sqrt{\frac{S}{\pi}}$, where S is the surface area of the sphere.
   a. Find the domain of the function. Use a graphing calculator to graph the function. Domain: $S \geq 0$
   b. Use the *trace* feature to approximate the surface area of a sphere with a radius of 2 centimeters. $S \approx 50.3\text{ cm}^2$

Technology For the Teacher
Answer Presentation Tool

Vocabulary and Concept Check

1. no; It is a linear function. A square root function contains a square root with the independent variable in the radicand.
2. The radicand cannot be negative. Write and solve an inequality with the radicand greater than or equal to zero.

Practice and Problem Solving

3. B
4. A
5. C
6. 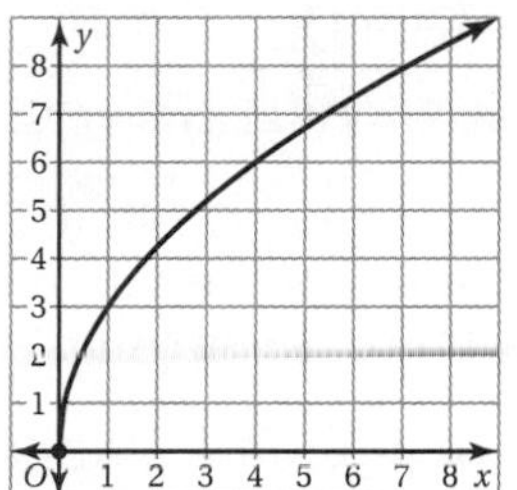

domain: $x \geq 0$
7. 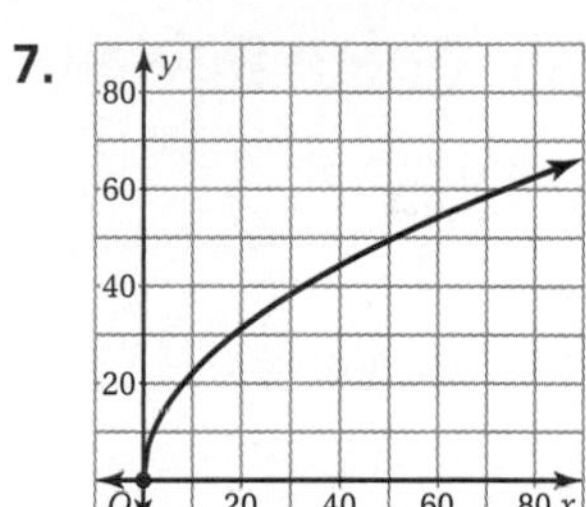

domain: $x \geq 0$
8. 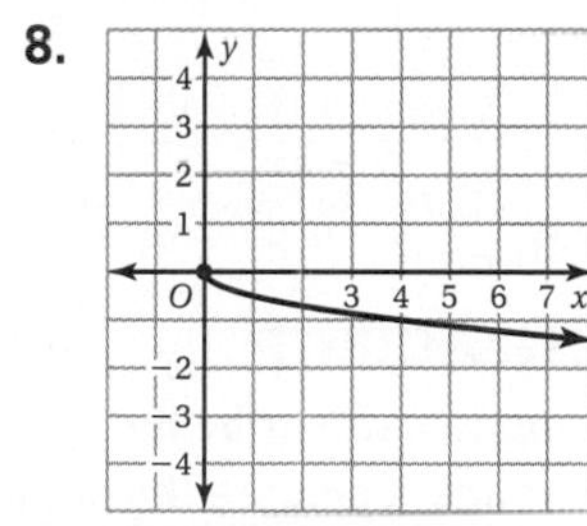

domain: $x \geq 0$
9. $x \geq 0$
10. $x \geq 0$
11. $x \geq 2$
12. $x \leq -1$
13. $x \geq -4$
14. $x \leq 2$
15. See Additional Answers.

Practice and Problem Solving

16–21. See Additional Answers.

22. The student graphed $y = \sqrt{x} + 1$ instead of $y = \sqrt{x + 1}$.

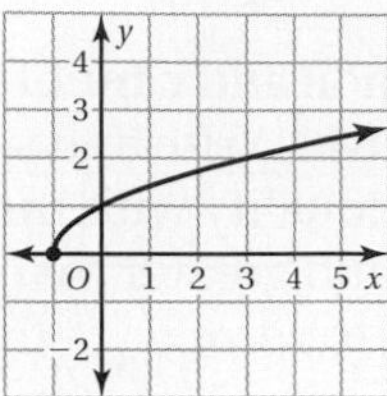

23. **a.** *Sample answer:* $y = \sqrt{x} + 1$

b. *Sample answer:* $y = -\sqrt{x}$

24. domain: yes, as long as the radicand is not negative; range: yes, the function could be a reflection or a vertical translation

25. See Additional Answers.

26. yes; Solve $40 = \sqrt{30d(0.75)}$ for d to find that the skid marks are about 71 feet long.

27. See *Taking Math Deeper.*

Fair Game Review

28. $x = 0, x = 8$

29. $x = -3$

30. $x = -2, x = 3$

31. B

Mini-Assessment

Graph the function. Describe the domain and range. Compare the graph to the graph of $y = \sqrt{x}$.

1. $y = \sqrt{x} + 5$

2. $y = \sqrt{x - 7}$

3. $y = \sqrt{x + 3} - 4$

4. $y = -\sqrt{x - 4} + 6$

1–4. See Additional Answers.

Taking Math Deeper

Exercise 27

This exercise previews the most basic cube root function, $y = \sqrt[3]{x}$, which students will study in a future course.

Make a table of values for $g(x) = \sqrt[3]{x}$.

x	−27	−8	−1	0	1	8	27
$g(x)$	−3	−2	−1	0	1	2	3

Notice that *perfect cubes* were chosen to make evaluating easier.

Graph the functions in the same coordinate plane.

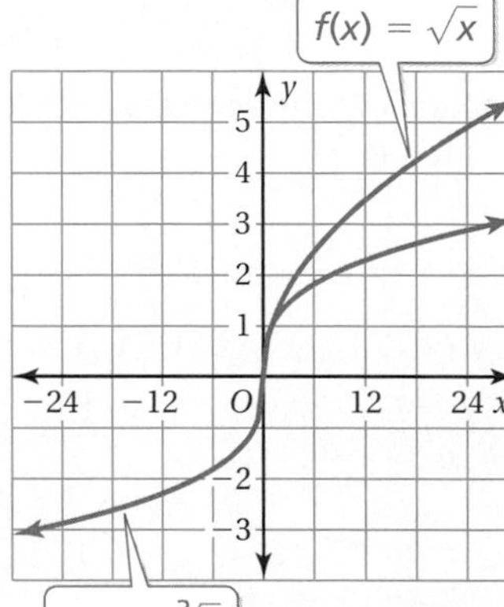

3 Compare the graphs.

Similarities:

- nonlinear functions
- pass through the origin
- *y* increases as *x* increases

Differences:

	Domain	Range	Symmetry
$f(x) = \sqrt{x}$	$x \geq 0$	$y \geq 0$	none
$g(x) = \sqrt[3]{x}$	all real numbers	all real numbers	about the origin

Project

Research *even and odd functions.* Of the functions you have studied, what types can be even functions? odd functions? Is $y = \sqrt[3]{x}$ even, odd, or neither?

Reteaching and Enrichment Strategies

If students need help. . .	If students got it. . .
Resources by Chapter • Practice A and Practice B • Puzzle Time Record and Practice Journal Practice Differentiating the Lesson Lesson Tutorials Skills Review Handbook	Resources by Chapter • Enrichment and Extension Start the next section

Graph the function. Describe the domain and range. Compare the graph to the graph of $y = \sqrt{x}$.

16. $y = \sqrt{x} - 2$
17. $y = \sqrt{x} + 4$
18. $y = \sqrt{x + 4}$
19. $y = \sqrt{x + 2} - 2$
20. $y = -\sqrt{x - 3}$
21. $y = -\sqrt{x - 1} + 3$

22. **ERROR ANALYSIS** Describe and correct the error in graphing the function $y = \sqrt{x} + 1$.

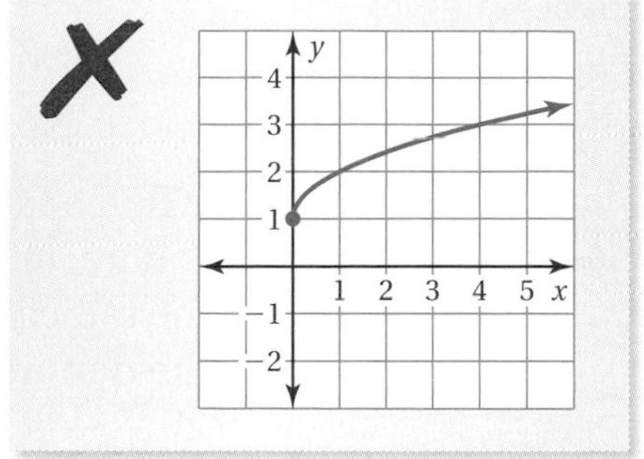

23. **OPEN-ENDED** Consider the graph of $y = \sqrt{x}$.
    a. Write a function that is a vertical translation of the graph of $y = \sqrt{x}$.
    b. Write a function that is a reflection of the graph of $y = \sqrt{x}$.

24. **REASONING** Can the domain of a square root function include negative numbers? Can the range include negative numbers? Explain your reasoning.

25. **GEOMETRY** The radius of a circle is given by $r = \sqrt{\frac{A}{\pi}}$, where A is the area of the circle.
    a. Find the domain of the function. Use a graphing calculator to graph the function.
    b. Use the *trace* feature to approximate the area of a circle with a radius of 3 inches.

26. **PROBLEM SOLVING** The speed S (in miles per hour) of a van before it skids to a stop can be modeled by the equation $S = \sqrt{30df}$, where d is the length (in feet) of the skid marks and f is the drag factor of the road surface. Suppose the drag factor is 0.75 and the speed of the van was 40 miles per hour. Is the length of the skid marks more than 65 feet long? Explain your reasoning.

27. **Precision** Compare the graphs of the functions $f(x) = \sqrt{x}$ and $g(x) = \sqrt[3]{x}$.

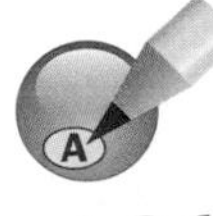

Fair Game Review What you learned in previous grades & lessons

Solve the equation. *(Section 7.5)*

28. $x(x - 8) = 0$
29. $(x + 3)^2 = 0$
30. $(x + 2)(x - 3) = 0$

31. **MULTIPLE CHOICE** What are the next three terms of the geometric sequence 240, 120, 60, 30, . . .? *(Section 6.7)*

Ⓐ 20, 10, 5 Ⓑ 15, 7.5, 3.75 Ⓒ 20, 10, 0 Ⓓ 15, 10, 5

10.1b Rationalizing the Denominator

Key Vocabulary

simplest form of a radical expression, *p. 508*

rationalizing the denominator, *p. 508*

conjugates, *p. 509*

In Section 6.1, you used properties to simplify radical expressions. A radical expression is in **simplest form** when the following are true.

- No radicands have perfect square factors other than 1.
- No radicands contain fractions.
- No radicals appear in the denominator of a fraction.

When a radicand in the denominator of a fraction is not a perfect square, multiply the fraction by an appropriate form of 1 to eliminate the radical from the denominator. This process is called **rationalizing the denominator.**

EXAMPLE 1 Simplifying a Radical Expression

Simplify $\sqrt{\frac{1}{3}}$.

$\sqrt{\frac{1}{3}} = \frac{\sqrt{1}}{\sqrt{3}}$ — Quotient Property of Square Roots

$= \frac{\sqrt{1}}{\sqrt{3}} \cdot \frac{\sqrt{3}}{\sqrt{3}}$ — Multiply by $\frac{\sqrt{3}}{\sqrt{3}}$.

$= \frac{\sqrt{1 \cdot 3}}{\sqrt{3 \cdot 3}}$ — Product Property of Square Roots

$= \frac{\sqrt{3}}{\sqrt{9}}$ — Simplify.

$= \frac{\sqrt{3}}{3}$ — Evaluate the square root.

Study Tip

Rationalizing the denominator works because you multiply the numerator and denominator by the same nonzero number a, which is the same as multiplying by $\frac{a}{a}$, or 1.

Practice

Simplify the expression.

1. $\frac{1}{\sqrt{10}}$
2. $\frac{\sqrt{2}}{\sqrt{7}}$
3. $\sqrt{\frac{9}{2}}$
4. $\sqrt{\frac{10}{21}}$
5. $\sqrt{\frac{5}{18}}$
6. $\sqrt{\frac{40}{48}}$
7. $\frac{4}{\sqrt{5}} + \frac{1}{\sqrt{5}}$
8. $\sqrt{3} - \frac{2}{\sqrt{12}}$
9. $\sqrt{\frac{16}{15}} - \frac{1}{3}$
10. **REASONING** Explain why for any number a, $\sqrt{a^2} = |a|$. Use this rule to simplify the expression $\sqrt{\frac{x^2}{2}}$.

Multi-Language Glossary at BigIdeasMath.com.

Laurie's Notes

Introduction

Connect

- **Yesterday:** Students graphed square root functions. (MP4, MP5, MP6)
- **Today:** Students will simplify square root expressions by rationalizing the denominator.

Motivate

- **Story time:** Tell a story about seeing a large ship on the horizon while at the beach. A friend says, "How far away is it?" You say, "About 3 miles. I can give you a more exact answer with a calculator."
- Tell students that by the end of the lesson, they will be able to make a similar approximation.

Lesson Notes

Discuss

- Remind students that they simplified radical expressions in Chapter 6.
- Discuss what must be true for a radical expression to be in simplest form.
- Explain the procedure called *rationalizing the denominator.*
- Point out that the name of this procedure makes sense because you rewrite the expression with a *rational* number in the *denominator.*
- Connect these previously avoided expressions with the continuous domains of square root functions. For example, when you graph $y = \sqrt{x}$, you draw the smooth curve through $\left(\frac{1}{2}, \sqrt{\frac{1}{2}}\right)$.

Example 1

- Explain that to simplify, you must "remove" the radical in the denominator.
- Remind students that whatever is done to the denominator must be done to the numerator and vice versa. This results in multiplication by 1, so it does not change the value of the expression.

? "What form of 1 is used in this example? Why is it a wise choice?" $\frac{\sqrt{3}}{\sqrt{3}}$; It results in a perfect square radicand in the denominator.

- Finish the problem. Take a moment to use a calculator to show that the original and final expressions are equivalent.

Practice

- In Exercise 5, students should recognize that multiplying by $\frac{\sqrt{2}}{\sqrt{2}}$ results in a perfect square radicand in the denominator.
- Exercises 7–9 require students to perform operations with fractions.

Goal Today's lesson is simplifying square root expressions.

Lesson Materials
Textbook
• graphing calculators

Start Thinking! and Warm Up

Lesson 10.1b Warm Up
For use before Lesson 10.1b

Lesson 10.1b Start Thinking!
For use before Lesson 10.1b

Based on the table below, make a list of criteria for a square root expression to be in simplest form.

Not in Simplest Form	In Simplest Form
$\sqrt{45}$	$\sqrt{15}$
$\sqrt{\frac{1}{3}}$	$\frac{\sqrt{3}}{2}$
$\frac{4}{\sqrt{6}}$	$\frac{\sqrt{6}}{4}$

Extra Example 1

Simplify $\sqrt{\frac{1}{6}}$. $\frac{\sqrt{6}}{6}$

Practice

1. $\frac{\sqrt{10}}{10}$
2. $\frac{\sqrt{14}}{7}$
3. $\frac{3\sqrt{2}}{2}$
4. $\frac{\sqrt{210}}{21}$
5. $\frac{\sqrt{10}}{6}$
6. $\frac{\sqrt{30}}{6}$
7. $\sqrt{5}$
8. $\frac{2\sqrt{3}}{3}$
9. $\frac{4\sqrt{15} - 5}{15}$
10. The square root of a number cannot be negative. You must include absolute value symbols for when $a < 0$; $\frac{|x|\sqrt{2}}{2}$

Record and Practice Journal Practice

See Additional Answers.

Laurie's Notes

Discuss

- ? "What is $(x-4)(x+4)$?" x^2-16
- Write $4+\sqrt{2}$ and $4-\sqrt{2}$ and say that these are called *conjugates*.
- ? "What is $(4+\sqrt{2})(4-\sqrt{2})$?" $16-\sqrt{4}=16-2=14$

Example 2

- Write the expression and say, "We need to use conjugates to simplify this expression."
- ? "How can you simplify this expression?" Multiply the numerator and denominator by the conjugate of $3+\sqrt{5}$.
- Finish the problem. Take a moment to use a calculator to show that the original and final expressions are equivalent.

Example 3

- **MP4 Model with Mathematics:** The given model for the distance you can see to the horizon is a square root function. Mathematically proficient students are able to use many function types to model real-world phenomena.
- Ask a volunteer to read the problem.
- Sketch a diagram that shows *h* and *d*. Note that the units for *h* are feet and the units for *d* are miles.
- Explain that this is how you approximated the distance to the ship in the Motivate.
- Substitute 5 for *h* and continue to solve the problem as shown.
- **MP5 Use Appropriate Tools Strategically:** Graph the function using a graphing calculator to check your answer.

Practice

- Ask volunteers to show their work at the board.

Closure

- **Exit Ticket:** Simplify $\frac{8}{\sqrt{12}}$. $\frac{4\sqrt{3}}{3}$

Technology For the Teacher

The Dynamic Planning Tool
Editable Teacher's Resources at *BigIdeasMath.com*

Extra Example 2

Simplify $\frac{2}{5+\sqrt{2}}$. $\frac{10-2\sqrt{2}}{23}$

Extra Example 3

In Example 3, how far can you see when your eye level is 6 feet above the water? 3 miles

Practice

11. $-3+3\sqrt{3}$
12. $-5\sqrt{3}-10$
13. $-2\sqrt{2}+2\sqrt{7}$
14. $\frac{\sqrt{210}}{2}\approx 7.25$ mi

Mini-Assessment

Simplify the expression.

1. $\sqrt{\frac{8}{3}}$ $\frac{2\sqrt{6}}{3}$
2. $\sqrt{2}-\frac{3}{\sqrt{5}}$ $\frac{5\sqrt{2}-3\sqrt{5}}{5}$
3. $\frac{3}{2+\sqrt{7}}$ $-2+\sqrt{7}$

The binomials $a\sqrt{b} + c\sqrt{d}$ and $a\sqrt{b} - c\sqrt{d}$ are called **conjugates.** You can use conjugates to simplify radical expressions that involve a sum or difference of radicals in the denominator.

EXAMPLE 2 Simplifying a Radical Expression

Simplify $\frac{1}{3 + \sqrt{5}}$.

Notice that the product of conjugates is a rational number.

$$\frac{1}{3+\sqrt{5}} = \frac{1}{3+\sqrt{5}} \cdot \frac{3-\sqrt{5}}{3-\sqrt{5}}$$ The conjugate of $3 + \sqrt{5}$ is $3 - \sqrt{5}$.

$$= \frac{1(3-\sqrt{5})}{3^2 - (\sqrt{5})^2}$$ Sum and Difference Pattern

$$= \frac{3-\sqrt{5}}{4}$$ Simplify.

EXAMPLE 3 Real-Life Application

The distance d (in miles) that you can see to the horizon with your eye level h feet above the water is given by $d = \sqrt{\frac{3h}{2}}$. How far can you see when your eye level is 5 feet above the water?

$$d = \sqrt{\frac{3(5)}{2}}$$ Substitute 5 for h.

$$= \frac{\sqrt{15}}{\sqrt{2}} \cdot \frac{\sqrt{2}}{\sqrt{2}}$$ Multiply by $\frac{\sqrt{2}}{\sqrt{2}}$.

$$= \frac{\sqrt{30}}{2}$$ Simplify.

Check

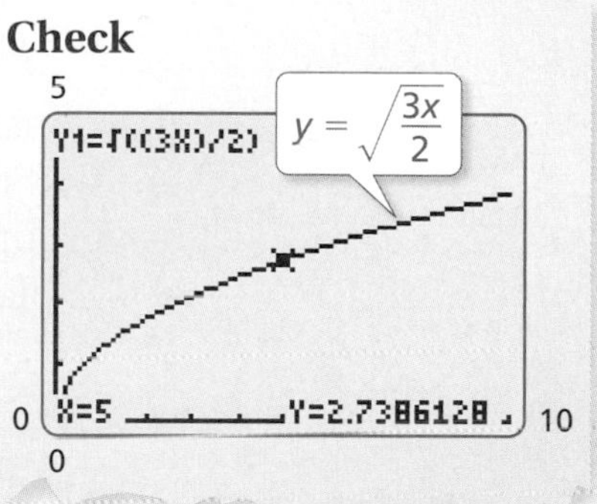

∴ You can see $\frac{\sqrt{30}}{2}$, or about 2.74 miles.

Practice

Simplify the expression.

11. $\frac{6}{1 + \sqrt{3}}$

12. $\frac{5}{\sqrt{3} - 2}$

13. $\frac{10}{\sqrt{2} + \sqrt{7}}$

14. **WHAT IF?** In Example 3, how far can you see when your eye level is 35 feet above the water?

10.2 Solving Square Root Equations

COMMON CORE STATE STANDARDS
N.RN.2

Essential Question How can you solve an equation that contains square roots?

1 ACTIVITY: Analyzing a Free-Falling Object

Work with a partner. The table shows the time t (in seconds) that it takes a free-falling object (with no air resistance) to fall d feet.

a. Sketch the graph of t as a function of d.

b. Use your graph to estimate the time it takes for a free-falling object to fall 240 feet.

c. The relationship between d and t is given by the function

$$t = \sqrt{\frac{d}{16}}.$$

Use this function to check the estimate you obtained from the graph.

d. Consider a free-falling object that takes 5 seconds to hit the ground. How far did it fall? Explain your reasoning.

d feet	t seconds
0	0.00
32	1.41
64	2.00
96	2.45
128	2.83
160	3.16
192	3.46
224	3.74
256	4.00
288	4.24
320	4.47

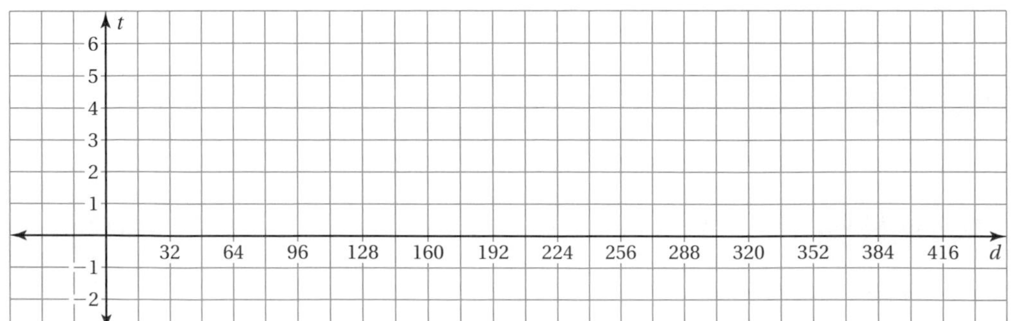

Laurie's Notes

Introduction

Standards for Mathematical Practice

- **MP6 Attend to Precision:** When working with a square root function, students recognize that small changes in the domain ($x > 1$) result in even smaller changes in the range. It is important to include a few decimal places in the result. To the nearest whole number, $y = \sqrt{x}$ rounds to 20 for values of x between 381 and 420. The context of the problem will help students determine the degree of precision needed.

Motivate

- Slide a math book off the edge of your desk or a table and let it hit the floor. This sound should get your students' attention.
- ? "How long do you think it took for the book to hit the floor from the moment it started falling?" Answers will vary.
- Drop the book again. This time, do it from your shoulder height.
- ? "How long do you think it took for the book to hit the floor from the moment it started falling?" Answers will vary. However, answers should be different from the first height.
- ? "Suppose I drop the book from the height of the ceiling. How long do you think it would take the book to hit the floor?" It would take longer than the first two heights.
- Explain that today they will investigate the time it takes for an object to fall from different heights.

Activity Notes

Activity 1

- As students begin to work on this activity make sure that they are plotting (d, t) and not (t, d).
- ? "Should the plotted points be connected with a smooth curve?" yes
- ? "What is the independent variable?" d; the distance d (in feet) that the free-falling object falls
- ? "What is the dependent variable?" t; the time (in seconds) that the free-falling object falls
- When answering part (b), students will most likely interpolate and guess a time about halfway between 3.74 and 4.00 seconds.
- Students use substitution to answer part (c).
- When answering part (d), students should substitute 5 for t and solve. To solve for d, students must square each side of the equation. Then multiply each side by 16.
- If time permits, investigate the equation on a graphing calculator. Adjusting the viewing window to see the key features of the graph is a good exercise for students. In this problem, the scale for the x-axis is much different than the scale for the y-axis.

Common Core State Standards

N.RN.2 Rewrite expressions involving radicals and rational exponents using the properties of exponents.

Previous Learning

Students should know how to solve linear and quadratic equations.

Activity Materials
Introduction
• textbook

Start Thinking! and Warm Up

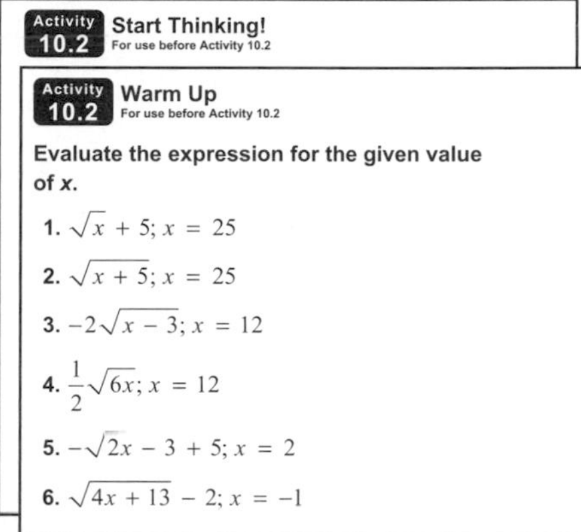

Activity 10.2 Start Thinking! For use before Activity 10.2

Activity 10.2 Warm Up For use before Activity 10.2

Evaluate the expression for the given value of x.

1. $\sqrt{x} + 5;\ x = 25$
2. $\sqrt{x + 5};\ x = 25$
3. $-2\sqrt{x - 3};\ x = 12$
4. $\frac{1}{2}\sqrt{6x};\ x = 12$
5. $-\sqrt{2x - 3} + 5;\ x = 2$
6. $\sqrt{4x + 13} - 2;\ x = -1$

10.2 Record and Practice Journal

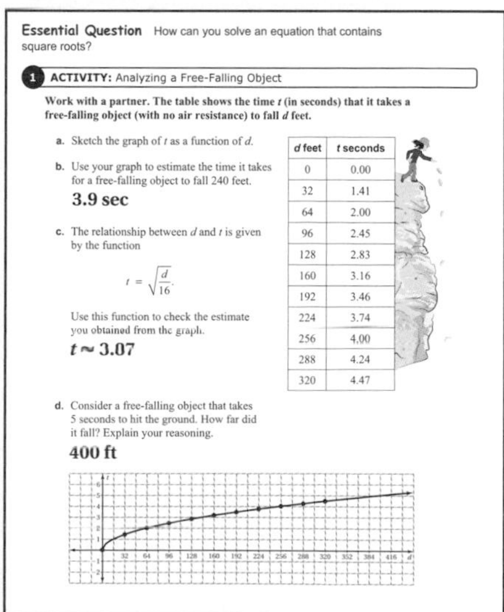

Essential Question How can you solve an equation that contains square roots?

1 ACTIVITY: Analyzing a Free-Falling Object

Work with a partner. The table shows the time t (in seconds) that it takes a free-falling object (with no air resistance) to fall d feet.

a. Sketch the graph of t as a function of d.

b. Use your graph to estimate the time it takes for a free-falling object to fall 240 feet.

3.9 sec

c. The relationship between d and t is given by the function

$$t = \sqrt{\frac{d}{16}}.$$

Use this function to check the estimate you obtained from the graph.

$t \approx 3.07$

d feet	t seconds
0	0.00
32	1.41
64	2.00
96	2.45
128	2.83
160	3.16
192	3.46
224	3.74
256	4.00
288	4.24
320	4.47

d. Consider a free-falling object that takes 5 seconds to hit the ground. How far did it fall? Explain your reasoning.

400 ft

Differentiated Instruction

Visual

Help students visualize that if two expressions are equal, then the squares of the expressions are equal. On graph paper, have students draw two horizontal line segments, 5 units in length.

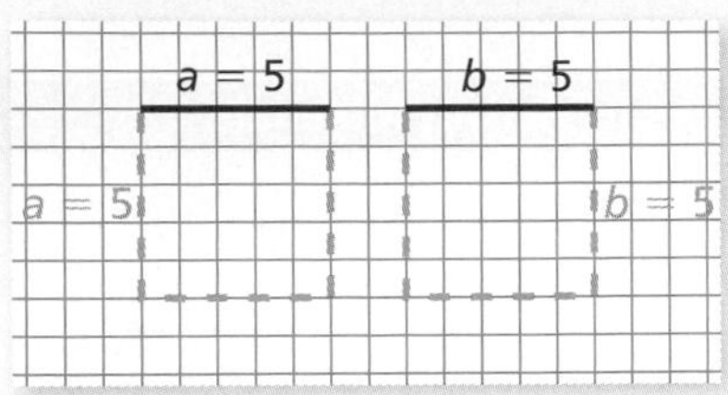

Tell them to draw two squares using the line segments and find the area of each. Point out that $a = b$ and $a^2 = b^2$.

10.2 Record and Practice Journal

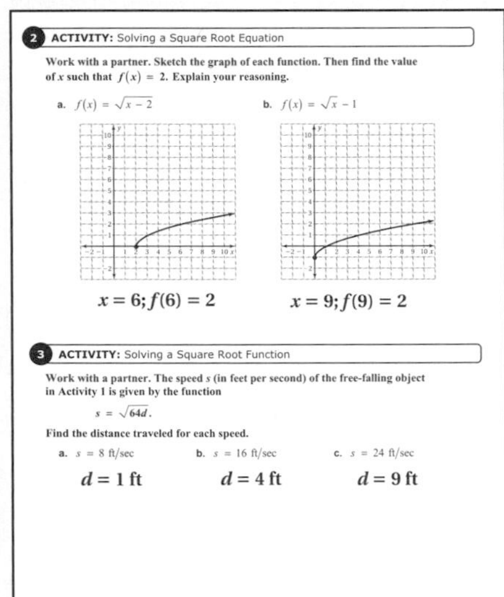

2 ACTIVITY: Solving a Square Root Equation

Work with a partner. Sketch the graph of each function. Then find the value of x such that $f(x) = 2$. Explain your reasoning.

a. $f(x) = \sqrt{x - 2}$

$x = 6; f(6) = 2$

b. $f(x) = \sqrt{x} - 1$

$x = 9; f(9) = 2$

3 ACTIVITY: Solving a Square Root Function

Work with a partner. The speed s (in feet per second) of the free-falling object in Activity 1 is given by the function

$s = \sqrt{64d}$.

Find the distance traveled for each speed.

a. $s = 8$ ft/sec $d = 1$ ft

b. $s = 16$ ft/sec $d = 4$ ft

c. $s = 24$ ft/sec $d = 9$ ft

What Is Your Answer?

4. **IN YOUR OWN WORDS** How can you solve an equation that contains square roots? Summarize a procedure for solving a square root equation. Then use your procedure to solve each equation.

Solve for the square root, then square both sides of the equation, then solve for the variable.

a. $\sqrt{x} + 2 = 3$ $x = 1$

b. $4 - \sqrt{x} = 1$ $x = 9$

c. $5 = \sqrt{x + 20}$ $x = 5$

d. $-3 = -2\sqrt{x}$ $x = \frac{9}{4}$

Laurie's Notes

Activity 2

- Students will need to make a table of values to graph the functions.

? "In part (a), what x-values allow for easy computation?" x-values that are 2 more than a perfect square

x	2	3	6	11	18	27
$f(x)$	0	1	2	3	4	5

- Students may make a comment about the scale of the axes at this point. This is very different from graphing linear, quadratic, and exponential functions. When plotting by hand, you do not want to use consecutive integer values for the domain.

? "How did you solve for the value of x that gives a function value of 2?" Listen for students describing the undoing process.

- **Extension:** Be sure to discuss the transformation of the graph $y = \sqrt{x}$ in each part. Discuss horizontal and vertical translations.

Activity 3

- Students have already solved a few square root problems. Check to be sure that students are squaring each side of the equation, then dividing by 64.

? "When the speed doubled from 8 ft/sec to 16 ft/sec, did the distance traveled double?" no; It quadrupled. "When the speed tripled from 8 ft/sec to 24 ft/sec, did the distance traveled triple?" no; It increased by a factor of 9.

What Is Your Answer?

- At this point, students have solved enough equations that they should know to isolate the expression (with the square root symbol) that contains the variable.

Closure

- Suppose the object in Activity 1 takes 3 seconds to hit the ground. How far did it fall? 144 ft

The Dynamic Planning Tool
Editable Teacher's Resources at *BigIdeasMath.com*

2 ACTIVITY: Solving a Square Root Equation

Work with a partner. Sketch the graph of each function. Then find the value of x such that $f(x) = 2$. Explain your reasoning.

a. $f(x) = \sqrt{x - 2}$

b. $f(x) = \sqrt{x} - 1$

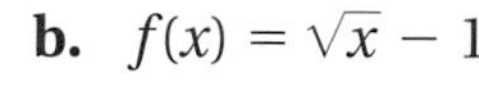

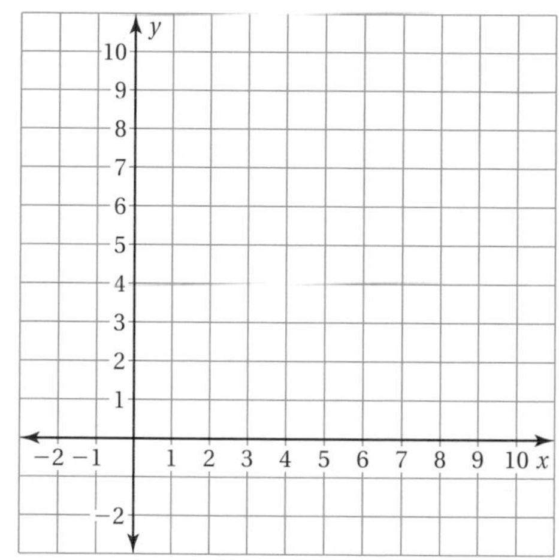

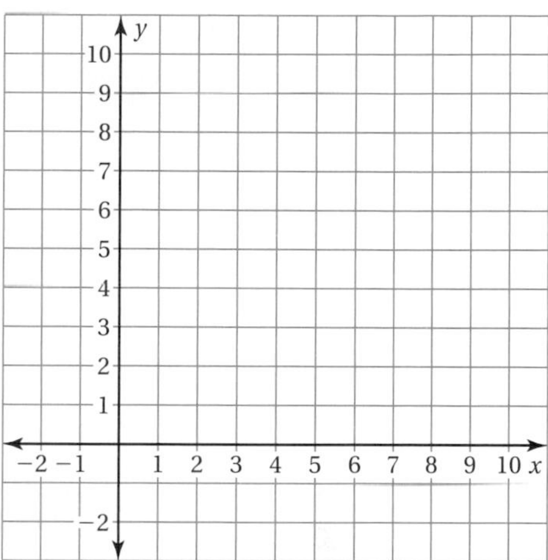

3 ACTIVITY: Solving a Square Root Equation

Work with a partner. The speed s (in feet per second) of the free-falling object in Activity 1 is given by the function

$$s = \sqrt{64d}.$$

Find the distance traveled for each speed.

a. $s = 8$ ft/sec **b.** $s = 16$ ft/sec **c.** $s = 24$ ft/sec

What Is Your Answer?

4. **IN YOUR OWN WORDS** How can you solve an equation that contains square roots? Summarize a procedure for solving a square root equation. Then use your procedure to solve each equation.

   a. $\sqrt{x} + 2 = 3$ **b.** $4 - \sqrt{x} = 1$

   c. $5 = \sqrt{x + 20}$ **d.** $-3 = -2\sqrt{x}$

Practice Use what you learned about solving square root equations to complete Exercises 3–5 on page 515.

10.2 Lesson

Key Vocabulary
square root equation *p. 512*
extraneous solution *p. 513*

A **square root equation** is an equation that contains a square root with a variable in the radicand. To solve a square root equation, use properties of equality to isolate the square root by itself on one side of the equation, then use the following property.

Key Idea

Squaring Each Side of an Equation

Words If two expressions are equal, then their squares are also equal.

Algebra If $a = b$, then $a^2 = b^2$.

EXAMPLE 1 Solving Square Root Equations

a. Solve $\sqrt{x} + 5 = 13$.

$\sqrt{x} + 5 = 13$	Write the equation.
$\sqrt{x} = 8$	Subtract 5 from each side.
$(\sqrt{x})^2 = 8^2$	Square each side of the equation.
$x = 64$	Simplify.

∴ The solution is $x = 64$.

Check

$\sqrt{x} + 5 = 13$

$\sqrt{64} + 5 \stackrel{?}{=} 13$

$8 + 5 \stackrel{?}{=} 13$

$13 = 13$ ✓

b. Solve $3 - \sqrt{x} = 0$.

$3 - \sqrt{x} = 0$	Write the equation.
$3 = \sqrt{x}$	Add $\sqrt{x}$ to each side.
$3^2 = (\sqrt{x})^2$	Square each side of the equation.
$9 = x$	Simplify.

∴ The solution is $x = 9$.

Check

On Your Own

Now You're Ready
Exercises 6–11

Solve the equation. Check your solution.

1. $\sqrt{x} = 6$
2. $\sqrt{x} - 7 = 3$
3. $\sqrt{x} + 15 = 22$
4. $1 - \sqrt{x} = -2$

Laurie's Notes

Introduction

Connect

- **Yesterday:** Students used graphs to help them solve equations involving square roots. (MP6)
- **Today:** Students will solve square root equations.

Motivate

- Read an excerpt, or find a brief online video from "The Pit and the Pendulum" by Edgar Allan Poe. Your students may have some familiarity with this classic literary work.
- This introduction provides an interesting transition to the work students will be engaged in today. Share with them that they will indeed work on a pendulum problem.

Lesson Notes

Discuss

- Discuss the square root equations students solved in the activity. They used properties of equality to isolate the square root before squaring each side of the equation.
- **MP7 Look for and Make Use of Structure:** Solving square root equations requires that students recognize the structure of the equation. The equations below all involve subtracting 5 as the first step.
  $2x + 5 = 13$ $\qquad$ $x^2 + 5 = 13$ $\qquad$ $\sqrt{x} + 5 = 13$
  In doing so, the term involving the variable x is isolated.

Key Idea

- Write the Key Idea which simply says that if you have an equation, squaring each side preserves the equality.

Example 1

? "To isolate the term involving x, what is the first step?" Subtract 5 from each side.

? "How do you undo the square root of x?" Square each side of the equation.

- In part (b), students may ask if the result is the same when you subtract 3 from each side first, then square each side. Students can verify this on their own.
- Remind students to check their solutions.

On Your Own

- **Think-Pair-Share**: Students should read each question independently and then work with a partner to answer the questions. When they have answered the questions, the pair should compare their answers with another group and discuss any discrepancies.

Goal Today's lesson is solving **square root equations**.

Lesson Materials	
Introduction	**Textbook**
• the short story "The Pit and the Pendulum"	• graphing calculators

Start Thinking! and Warm Up

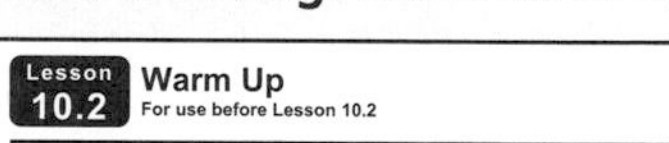

Lesson 10.2 **Start Thinking!** For use before Lesson 10.2

Can you square each side of an equation and still have a true equation?

Give an example to support your answer.

Extra Example 1

Solve each equation. Check your solution.

a. $\sqrt{x} + 1 = 4$ $\quad x = 9$

b. $7 - \sqrt{x} = -5$ $\quad x = 144$

On Your Own

1. $x = 36$
2. $x = 100$
3. $x = 49$
4. $x = 9$

Extra Example 2

Solve $3\sqrt{x+1} - 6 = 9$. Check your solution. $x = 24$

On Your Own

5. $x = 12$
6. $x = 10$
7. $x = 30$

Extra Example 3

Solve $\sqrt{2x+2} = \sqrt{3x-7}$. Check your solution. $x = 9$

On Your Own

8. $x = 8$
9. $x = \frac{1}{4}$

English Language Learners

Vocabulary

English language learners may mistake *extraneous solution* for *extra solution.* The prefix *extra-* means more than is usual or necessary. The definition of *extraneous* is not forming a necessary part. So, an extraneous solution is not a necessary part of the solution.

Laurie's Notes

Example 2

- Write the following equations on the board.

  $4(x + 2) + 3 = 19$

  $4(x + 2)^2 + 3 = 19$

  $4\sqrt{x+2} + 3 = 19$
- Discuss strategies for solving the first two equations before solving the third equation. You want students to see the underlying structure of the equation and recognize that they know the steps necessary to solve.
- **Common Error:** When squaring $\left(\sqrt{x+2}\right)^2$ students may square the x and square the 2.
- **Teaching Tip:** When squaring each side of the equation, it is a good idea to enclose the square root expression in parentheses.
- Remind students to check the solution.

On Your Own

- Ask for volunteers to share their work at the board for each question.
- If time permits, have students use graphing calculators to check their solutions. Treat each side of the equation as a function and graph each side. Then find the intersection point.

Example 3

- Students generally do not have a problem solving this type of example. Before they begin, have them consider what each side of the equation looks like graphically.

? "What is the first step?" Square each side of the equation.

? "What is the result after squaring each side?" $2x - 1 = x + 4$

- Remind students that this is an equation with variables on both sides.
- Use a graphing calculator to check the solution.

? "Will two square root functions always intersect?" no "Can they intersect at more than one point?" yes; for example, $y = \sqrt{2x}$ and $y = \sqrt{x-2} + 2$; Listen for student understanding of the shape of the square root function.

On Your Own

- **Neighbor Check:** Have students work independently and then have their neighbor check their work. Have students discuss any discrepancies.

EXAMPLE 2 Solving a Square Root Equation

Solve $4\sqrt{x+2} + 3 = 19$.

$4\sqrt{x+2} + 3 = 19$	Write the equation.
$4\sqrt{x+2} = 16$	Subtract 3 from each side.
$\sqrt{x+2} = 4$	Divide each side by 4.
$(\sqrt{x+2})^2 = 4^2$	Square each side of the equation.
$x + 2 = 16$	Simplify.
$x = 14$	Subtract 2 from each side.

The solution is $x = 14$.

Check

$4\sqrt{x+2} + 3 = 19$

$4\sqrt{14+2} + 3 \stackrel{?}{=} 19$

$4\sqrt{16} + 3 \stackrel{?}{=} 19$

$4(4) + 3 \stackrel{?}{=} 19$

$16 + 3 \stackrel{?}{=} 19$

$19 = 19$ ✓

On Your Own

Now You're Ready
Exercises 13–18

Solve the equation. Check your solution.

5. $\sqrt{x+4} + 7 = 11$
6. $8\sqrt{x-1} = 24$
7. $15 = 6 + \sqrt{3x-9}$

EXAMPLE 3 Solving an Equation with Square Roots on Both Sides

Solve $\sqrt{2x-1} = \sqrt{x+4}$.

$\sqrt{2x-1} = \sqrt{x+4}$	Write the equation.
$(\sqrt{2x-1})^2 = (\sqrt{x+4})^2$	Square each side of the equation.
$2x - 1 = x + 4$	Simplify.
$x - 1 = 4$	Subtract x from each side.
$x = 5$	Add 1 to each side.

The solution is $x = 5$.

Check

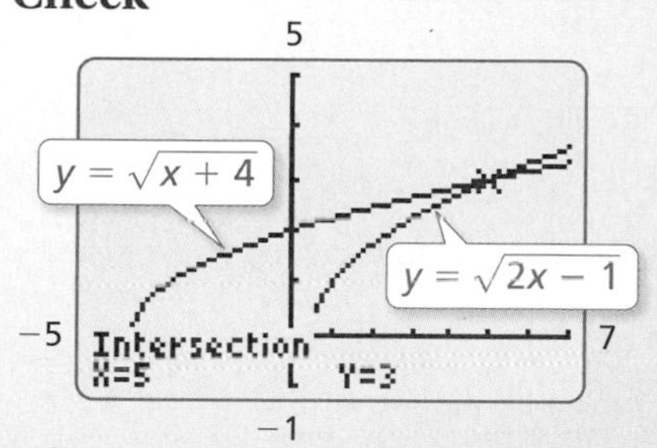

On Your Own

Now You're Ready
Exercises 22–27

Solve the equation. Check your solution.

8. $\sqrt{3x+1} = \sqrt{4x-7}$
9. $\sqrt{x} = \sqrt{5x-1}$

Squaring each side of an equation can sometimes introduce a solution that is *not* a solution of the original equation. This solution is called an **extraneous solution.** Be sure to always substitute your solutions into the original equation to check for extraneous solutions.

EXAMPLE 4 Identifying an Extraneous Solution

$x = \sqrt{x+6}$	Original equation
$x^2 = (\sqrt{x+6})^2$	Square each side of the equation.
$x^2 = x + 6$	Simplify.
$x^2 - x - 6 = 0$	Subtract *x* and 6 from each side.
$(x-3)(x+2) = 0$	Factor.
$(x-3) = 0$ *or* $(x+2) = 0$	Use Zero-Product Property.
$x = 3$ *or* $x = -2$	Solve for *x*.

Check

$3 \stackrel{?}{=} \sqrt{3+6}$	Substitute for *x*.	$-2 \stackrel{?}{=} \sqrt{-2+6}$
$3 \stackrel{?}{=} \sqrt{9}$	Simplify.	$-2 \stackrel{?}{=} \sqrt{4}$
$3 = 3$ ✓		$-2 \neq 2$ ✗

Because $x = -2$ does not check in the original equation, it is an extraneous solution. The only solution is $x = 3$.

EXAMPLE 5 Real-Life Application

Study Tip

The period of a pendulum is the amount of time it takes for the pendulum to swing back and forth.

The period *P* (in seconds) of a pendulum is given by the function $P = 2\pi\sqrt{\frac{L}{32}}$, where *L* is the pendulum length (in feet). What is the length of a pendulum that has a period of 2 seconds?

$2 = 2\pi\sqrt{\frac{L}{32}}$	Substitute 2 for *P* in the function.
$\frac{1}{\pi} = \sqrt{\frac{L}{32}}$	Divide each side by 2π and simplify.
$\frac{1}{\pi^2} = \frac{L}{32}$	Square each side and simplify.
$\frac{32}{\pi^2} = L$	Multiply both sides by 32.
$3.2 \approx L$	Use a calculator.

The length of the pendulum is about 3.2 feet.

On Your Own

Now You're Ready
Exercises 31–36

10. Solve $\sqrt{x-1} = x - 3$. Check your solution.
11. **WHAT IF?** In Example 5, what is the length of a pendulum that has a period of 4 seconds? Is your result twice the length in Example 5? Explain.

Laurie's Notes

Example 4

- Instead of telling students about extraneous solutions, work through the example and when checking the solution, emphasize that something is wrong. Then explain the possibility of introducing extraneous roots when you square each side of an equation.
- Write the original equation. Take time to discuss the graph of each side.
- ? "What are the intersection possibilities for the graphs of a linear function and a square root function?" 0, 1, or 2 points of intersection; Show students generic graphs for each case.
- Work through the problem as shown. Review how to factor a trinomial.
- **Connection:** Tell students that −2 and 3 are solutions of the quadratic equation $x^2 - x - 6 = 0$. But they are not both solutions of the original square root equation.
- **Extension:** Graph each side of the original equation on a graphing calculator. Use the *intersect* feature to find the solution.

Example 5

- It is time to return to "The Pit and the Pendulum!" Ask a student to read the problem. The Study Tip describes the period of a pendulum. Use a string with a weight on it to model the pendulum motion and to identify the period.
- Take time to identify the independent and dependent variables. Be sure to include the units of measure.
- Set the function equal to 2, which is the period of the pendulum.
- Students may want to divide by 2 and then divide by π. This can be done in one step.
- Continue to work through the problem and use a calculator to approximate the answer to the nearest tenth.
- **Extension:** Have a piece of string available that is 3.2 feet long. Place a weight on the string and model a period of 2 seconds.

On Your Own

- Discuss Question 11. This is similar to Activity 3.

Closure

- Solve each equation.

a. $2x + 1 = 10$ — $x = \frac{9}{2}$

b. $x^2 + 1 = 10$ — $x = 3, x = -3$

c. $\sqrt{x} + 1 = 10$ — $x = 81$

Technology For the Teacher

The Dynamic Planning Tool
Editable Teacher's Resources at *BigIdeasMath.com*

Extra Example 4

Solve $x = \sqrt{x + 20}$. Check your solution.
$x = 5$

Extra Example 5

In Example 5, what is the length of a pendulum that has a period of 3 seconds? about 7.3 feet

On Your Own

10. $x = 5$

11. about 13 feet; no; It is about 4 times the length of the pendulum in Example 5.

Vocabulary and Concept Check

1. no; The radicand does not contain a variable.
2. Squaring each side of a square root equation can introduce extraneous solutions, which are not solutions of the original equation.

Practice and Problem Solving

3.

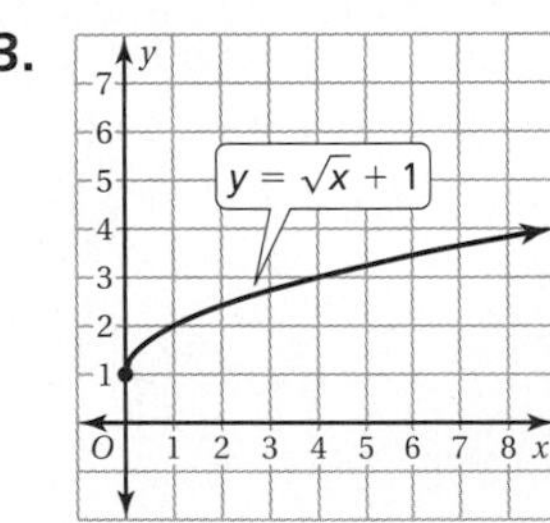

4

4.

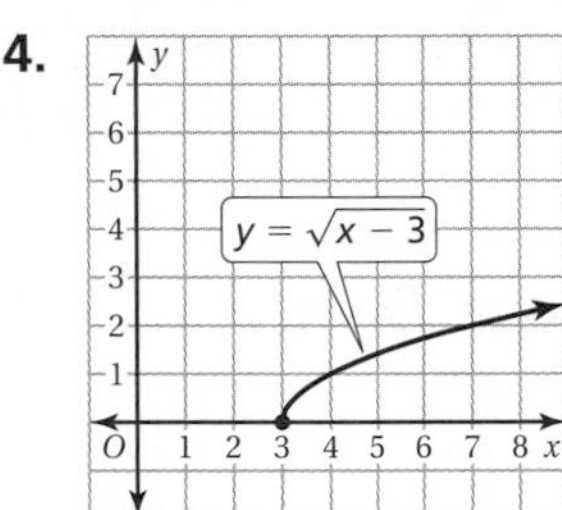

12

5.

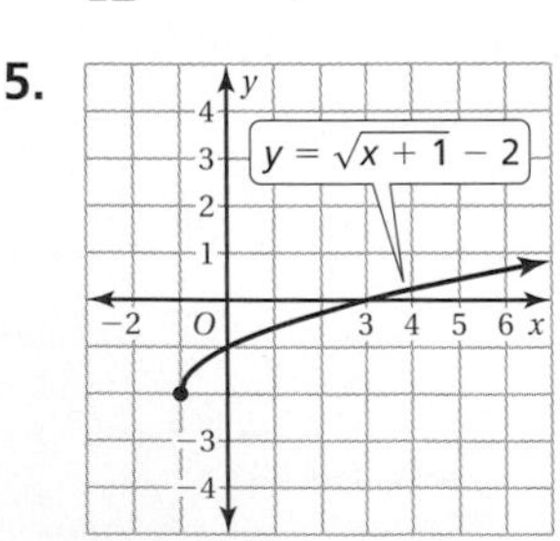

24

6. $x = 81$ 7. $x = 144$
8. $x = 16$ 9. $x = 121$
10. $x = 0$ 11. $x = 225$
12. The student did not square the negative coefficient; $x = 4$
13. $x = 19$ 14. $x = 60$
15. $x = 45$ 16. $x = 34$

Assignment Guide and Homework Check

Level	Assignment	Homework Check
Average	1, 2, 6–38 even, 19, 43, 51–54	12, 16, 20, 24, 34, 38
Advanced	1, 2, 13–37 odd, 43–50, 51–54	19, 21, 35, 45, 50

For Your Information

- **Exercise 20** BASE stands for Buildings, Antennas, Spans (bridges), and Earth (cliffs).
- **Exercise 44** Students can use graphing calculators for this exercise.

Common Errors

- **Exercises 6–11, 13–18** Students may square each side of the equation before isolating the square root on one side of the equation.

10.2 Record and Practice Journal

Solve the equation. Check your solution.

1. $\sqrt{x} + 4 = 9$
   $x = 25$
2. $-2 = 6 - \sqrt{x}$
   $x = 64$
3. $7 = 1 + 2\sqrt{x + 4}$
   $x = 5$
4. $\sqrt{5x - 11} - 3 = 5$
   $x = 15$
5. $\sqrt{4x - 3} = \sqrt{x + 6}$
   $x = 3$
6. $\sqrt{8x + 1} = \sqrt{7x + 7}$
   $x = 6$
7. $x = \sqrt{12x - 32}$
   $x = 4, 8$
8. $\sqrt{4x + 13} = x - 2$
   $x = 9$
9. The formula $\frac{S}{8} = \sqrt{df}$ relates the speed S (in feet per second), drag factor f, and distance d (in feet) it takes for a car to come to a stop after the driver applies the brakes. A car travels at 80 feet per second and the drag factor is $\frac{2}{3}$. What distance does it take for the car to stop once the driver applies the brakes?
   150 ft

Technology For the Teacher
Answer Presentation Tool

10.2 Exercises

Vocabulary and Concept Check

1. **VOCABULARY** Is $x\sqrt{3} = 4$ a square root equation? Explain your reasoning.
2. **WRITING** Why should you check every solution of a square root equation?

Practice and Problem Solving

Sketch the graph of the function. Then find the value of x such that $f(x) = 3$.

3. $f(x) = \sqrt{x} + 1$
4. $f(x) = \sqrt{x - 3}$
5. $f(x) = \sqrt{x + 1} - 2$

Solve the equation. Check your solution.

1 6. $\sqrt{x} = 9$
7. $7 = \sqrt{x} - 5$
8. $\sqrt{x} + 6 = 10$
9. $\sqrt{x} + 12 = 23$
10. $4 - \sqrt{x} = 4$
11. $-8 = 7 - \sqrt{x}$

12. **ERROR ANALYSIS** Describe and correct the error in solving the equation.

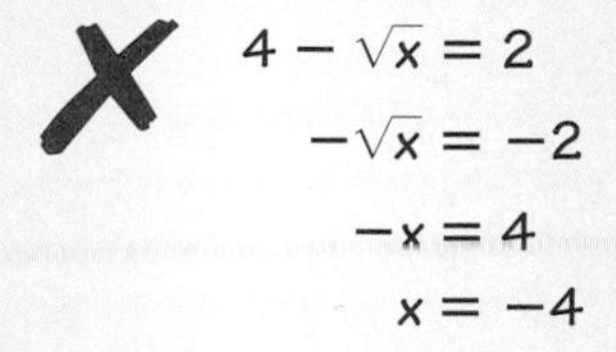

$$4 - \sqrt{x} = 2$$
$$-\sqrt{x} = -2$$
$$-x = 4$$
$$x = -4$$

Solve the equation. Check your solution.

2 13. $\sqrt{x - 3} + 5 = 9$
14. $2\sqrt{x + 4} = 16$
15. $25 = 7 + 3\sqrt{x - 9}$
16. $\sqrt{\frac{x}{2} - 1} + 14 = 18$
17. $-1 = \sqrt{5x + 1} - 7$
18. $12 = 19 - \sqrt{3x - 11}$

19. **CUBE** The formula $s = \sqrt{\frac{A}{6}}$ gives the edge length s of a cube with a surface area of A. What is the surface area of a cube with a edge length of 4 inches?

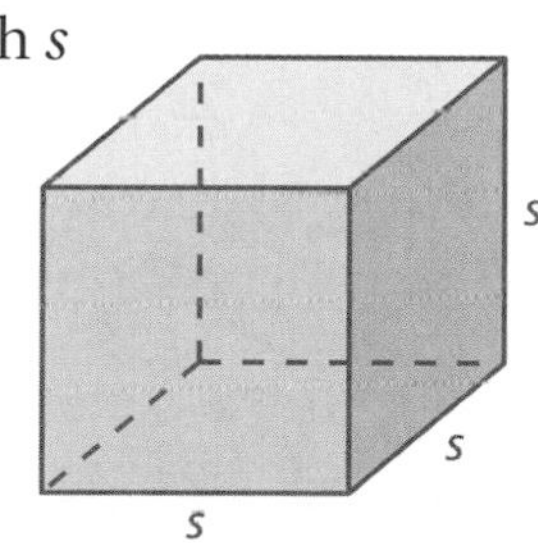

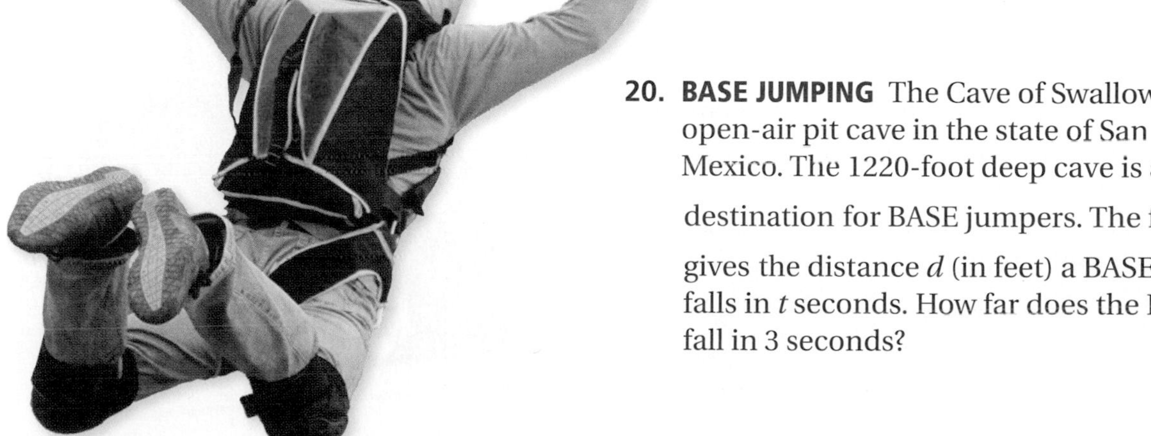

20. **BASE JUMPING** The Cave of Swallows is a natural open-air pit cave in the state of San Luis Potosi, Mexico. The 1220-foot deep cave is a popular destination for BASE jumpers. The formula $t = \sqrt{\frac{d}{16}}$ gives the distance d (in feet) a BASE jumper free falls in t seconds. How far does the BASE jumper fall in 3 seconds?

21. **WRITING** Explain how you would solve $\sqrt{m+4} - \sqrt{3m} = 0$.

Solve the equation. Check your solution.

3 22. $\sqrt{2x-9} = \sqrt{x}$

23. $\sqrt{x+1} = \sqrt{4x-8}$

24. $\sqrt{3x+1} = \sqrt{7x-19}$

25. $\sqrt{8x-7} = \sqrt{6x+7}$

26. $\sqrt{2x+1} - \sqrt{4x} = 0$

27. $\sqrt{5x} - \sqrt{8x-2} = 0$

Find the value of *x*.

28. Perimeter = 28 cm

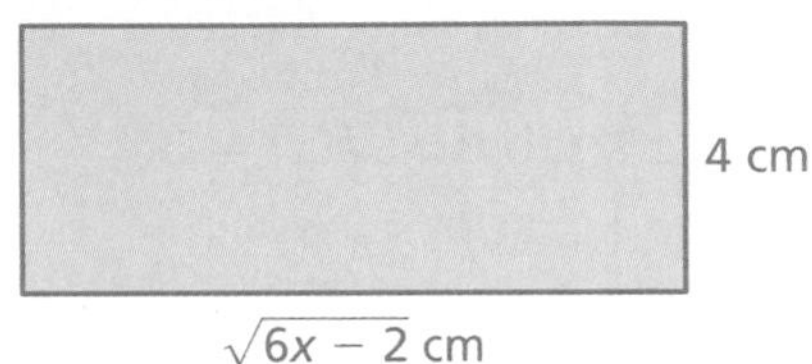

29. Area = $\sqrt{5x-4}$ ft^2

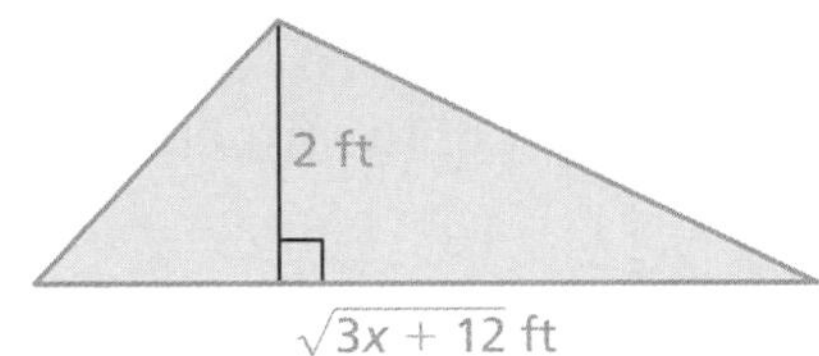

30. **OPEN-ENDED** Write a square root equation of the form $\sqrt{ax} + b = c$ that has a solution of 9.

Solve the equation. Check your solution.

4 31. $x = \sqrt{5x-4}$

32. $\sqrt{9x-14} = x$

33. $\sqrt{3x+10} = x$

34. $2x = \sqrt{6-10x}$

35. $x - 1 = \sqrt{3-x}$

36. $\sqrt{-4x-19} = x+4$

37. **ERROR ANALYSIS** Describe and correct the error in solving the equation.

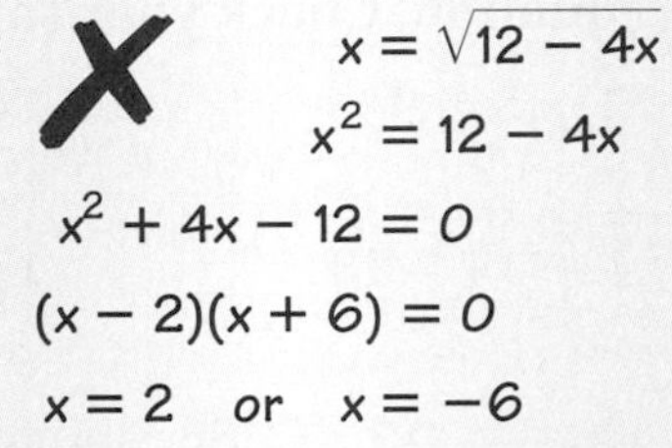

38. **REASONING** Explain how to use mental math to find the solution of $\sqrt{2x} + 5 = 1$.

Determine whether the statement is *true* or *false*.

39. If $\sqrt{a} = b$, then $(\sqrt{a})^2 = b^2$.

40. If $\sqrt{a} = \sqrt{b}$, then $a = b$.

41. If $a^2 = b^2$, then $a = b$.

42. If $a^2 = \sqrt{b}$, then $a^4 = (\sqrt{b})^2$.

43. **ELECTRICITY** The formula $V = \sqrt{PR}$ relates the voltage V (in volts), power P (in watts), and resistance R (in ohms) of an electrical circuit. What is the resistance of a 1000-watt hair dryer on a 120-volt circuit?

Common Errors

- **Exercise 37** Students may have trouble determining what the error is because the equation is solved correctly. However, the extraneous solution is not eliminated. Remind students to check their solutions.
- **Exercises 48 and 49** Students may forget that when you square an expression with two terms, the result is an expression with three terms. Remind them of this process.
- **Exercise 50** Students may forget to convert 1.5 cubic feet to cubic inches before finishing the problem.

Practice and Problem Solving

17. $x = 7$ **18.** $x = 20$

19. 96 in.^{2} **20.** 144 ft

21. Add $\sqrt{3m}$ to each side. Square each side. Subtract m from each side. Divide each side by 2. Check your solution.

22. $x = 9$ **23.** $x = 3$

24. $x = 5$ **25.** $x = 7$

26. $x = \frac{1}{2}$ **27.** $x = \frac{2}{3}$

28. $x = 17$ **29.** $x = 8$

30. *Sample answer:* $\sqrt{4x} + 10 = 16$

31. $x = 1, x = 4$

32. $x = 2, x = 7$

33. $x = 5$ **34.** $x = \frac{1}{2}$

35. $x = 2$ **36.** no solution

37. $x = -6$ does not check in the original equation, so it is extraneous. $x = 2$ is the only solution.

38. See Additional Answers.

39. true **40.** true

41. false **42.** true

43. 14.4 ohms

Differentiated Instruction

Organization

Students may have difficulty determining the order of the steps needed to solve a radical equation. Suggest they draw a box around the term containing the radical and isolate that term on one side. Then continue the process to solve for x.

$$\boxed{3\sqrt{x+1}} - 8 = 1$$
$$\boxed{3\sqrt{x+1}} = 9$$
$$\sqrt{x+1} = 3$$
$$x + 1 = 9$$
$$x = 8$$

Practice and Problem Solving

44. **a.**

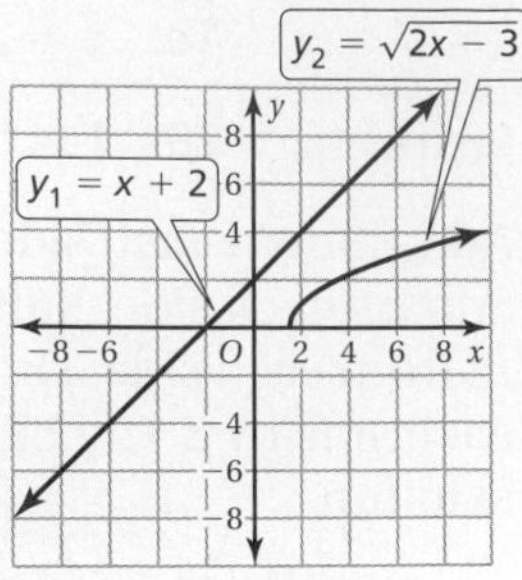

The graphs do not intersect. The equation has no solutions.

b. no solution; Solving algebraically and graphically give the same result.

c. *Sample answer:* graphing; Using a graphing calculator is convenient and accurate.

45. about 29.2 ft

46. 2; 4

47. $\sqrt{x} + 2$ has one term and $\sqrt{x + 2}$ has two terms. When you square $\sqrt{x + 2}$, the result is the radicand, $x + 2$. When you square $\sqrt{x} + 2$, the result is an expression with 3 terms that includes a radical, $x + 4\sqrt{x} + 4$.

48. $x = 5$ **49.** $x = -\frac{7}{16}$

50. See *Taking Math Deeper.*

Fair Game Review

51. 92° **52.** 80°

53. 90° **54.** B

Mini-Assessment

Solve the equation. Check your solution.

1. $\sqrt{x} - 4 = 6$ $x = 100$

2. $\sqrt{x + 7} = \sqrt{2x - 4}$ $x = 11$

3. $x = \sqrt{5x + 6}$ $x = 6$

Taking Math Deeper

Exercise 50

This problem reviews many previous skills such as converting units for volume and finding the percent of a number.

a. Substitute for A and b_2 in the formula and simplify to get $h = \sqrt{336 - 16h}$. Then solve for h.

$h^2 = 336 - 16h$	Square each side.
$h^2 + 16h - 336 = 0$	Write in standard form.
$h = 12$ *or* $h = -28$	Factor and solve for h.

The height of the speaker box is 12 inches.

b. Use the formula for the area of a trapezoid to find b_1.

$A = \frac{1}{2}h(b_1 + b_2)$	Area of a trapezoid
$168 = \frac{1}{2}(12)(b_1 + 16)$	Substitute.
$12 = b_1$	Solve for b_1.

The length of b_1 is 12 inches.

3 **c.** Convert 1.5 cubic feet to cubic inches.

$$1.5 \text{ ft}^3 \cdot \left(\frac{12 \text{ in.}}{1 \text{ ft}}\right)^3 = 1.5 \cancel{\text{ft}^3} \cdot \frac{1728 \text{ in.}^3}{1 \cancel{\text{ft}^3}} = 2592 \text{ in.}^3$$

So, the minimum volume is $0.9(2592) = 2332.8$ in.^{3} and the maximum volume is $1.1(2592) = 2851.2$ in.^{3}.

The volume of a prism is the area of the base times the height of the prism. In this problem, the height of the prism is the width w of the speaker box. Use the maximum and minimum volumes to find the range of the width.

$2332.8 = 168w$	$2851.2 = 168w$
$13.9 \approx w$	$17.0 \approx w$

To the nearest tenth, the range of the width is $13.9 \le w \le 17.0$.

Reteaching and Enrichment Strategies

If students need help. . .	If students got it. . .
Resources by Chapter • Practice A and Practice B • Puzzle Time Record and Practice Journal Practice Differentiating the Lesson Lesson Tutorials Skills Review Handbook	Resources by Chapter • Enrichment and Extension Start the next section

44. **CHOOSE TOOLS** Consider the equation $x + 2 = \sqrt{2x - 3}$.

a. Graph each side of the equation in the same coordinate plane. Solve the equation by finding points of intersection.

b. Solve the equation algebraically. How does your solution compare to the solution in part (a)?

c. Which method do you prefer? Explain your reasoning.

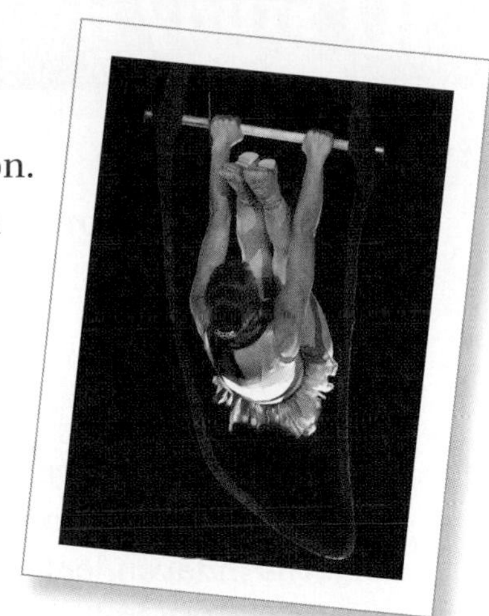

45. **TRAPEZE** The time t (in seconds) it takes a trapeze artist to swing back and forth is given by the function $t = 2\pi\sqrt{\frac{r}{32}}$, where r is the rope length (in feet). It takes 6 seconds to swing back and forth. How long is the rope? Use 3.14 for π.

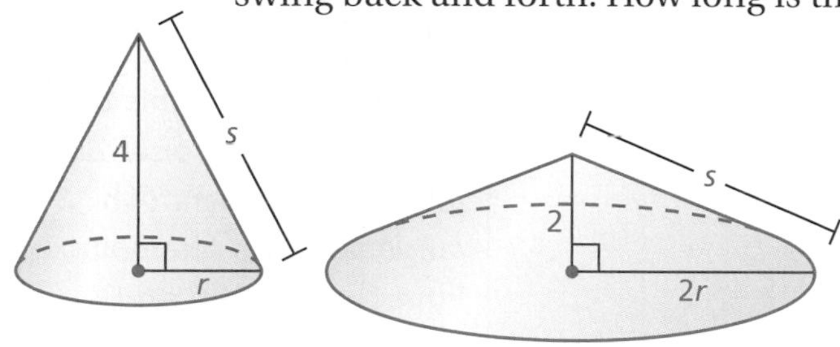

46. **GEOMETRY** The formula $s = \sqrt{r^2 + h^2}$ gives the slant height s of a cone, where r is the radius of the base, and h is the height. The slant heights of the two cones are equal. Find the radius of each cone.

47. **CRITICAL THINKING** How is squaring $\sqrt{x + 2}$ different than squaring $\sqrt{x} + 2$?

Solve the equation. Check your solution.

48. $\sqrt{x + 15} = \sqrt{x} + \sqrt{5}$

49. $2 - \sqrt{x + 1} = \sqrt{x + 2}$

50. **Modeling** The formula $h = \sqrt{2A - b_2h}$ gives the height h of the speaker box, where A is the area of one trapezoidal side, and b_2 is the length of base 2.

a. Given that $A = 168$ square inches and $b_2 = 16$ inches, find h.

b. What is the length of b_1 (base 1)?

c. Speakers work best when the volume of the speaker box is $\pm 10\%$ of the manufacturer's recommendation. Find the range of the widths w when the manufacturer recommends a volume of 1.5 cubic feet.

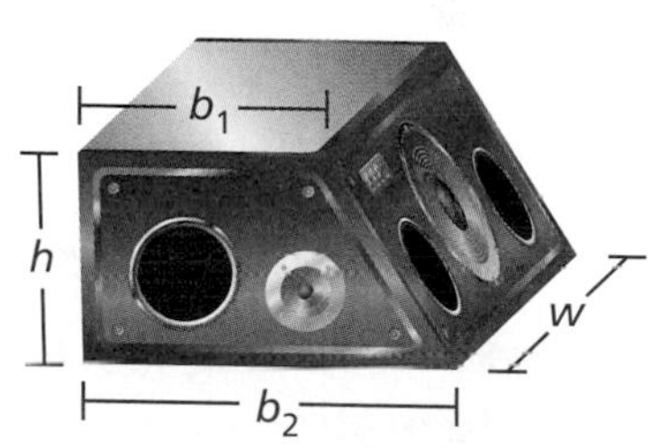

Fair Game Review What you learned in previous grades & lessons

Two angle measures of a triangle are given. Find the measure of the missing angle. *(Skills Review Handbook)*

51. 40°, 48°

52. 45°, 55°

53. 36°, 54°

54. **MULTIPLE CHOICE** Which function is represented by the ordered pairs (−1, 0.5), (0, 1), (1, 2), (2, 4), and (3, 8)? *(Section 8.5)*

Ⓐ $y = 0.5x^2$ Ⓑ $y = 2^x$ Ⓒ $y = 2x^2$ Ⓓ $y = 2x$

10 Study Help

You can use a **word magnet** to organize information associated with a vocabulary word. Here is an example of a word magnet for square root functions.

Square Root Function

Definition: A function that contains a square root with the independent variable in the radicand.

Domain: The value of the radicand cannot be negative. So, the domain is limited to x-values for which the radicand is greater than or equal to 0.

Examples:

$y = \sqrt{x} + 3$

$y = \sqrt{x - 1}$

$y = \sqrt{x + 5} - 4$

Graph: Make a table of values. Plot the ordered pairs. Draw a smooth curve through the points. Find the domain and range.

Sample Graph:

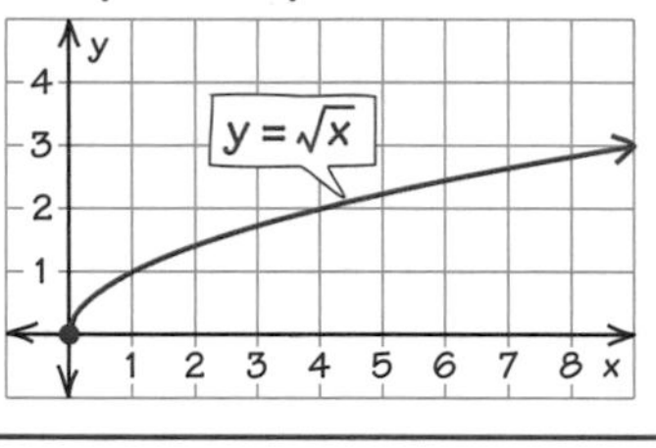

Compare: When graphing a square root function $f(x)$:

- $f(x) + k$ is a vertical translation of $f(x)$.
- $f(x + h)$ is a horizontal translation of $f(x)$.
- $-f(x)$ is a reflection of $f(x)$ in the x-axis.

On Your Own

Make word magnets to help you study these topics.

1. rationalizing the denominator
2. solving a square root equation
3. extraneous solution

After you complete this chapter, make word magnets for the following topics.

4. Pythagorean Theorem
5. converse of the Pythagorean Theorem
6. distance formula

"How do you like the word magnet I made for 'Beagle'?"

Sample Answers

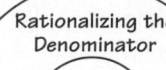

1.

Rationalizing the Denominator

Rationalizing the denominator is the process of multiplying a fraction by a form of 1 to eliminate radicals from the denominator.

Examples:

$$\frac{3}{\sqrt{5}} = \frac{3}{\sqrt{5}} \cdot \frac{\sqrt{5}}{\sqrt{5}} = \frac{3\sqrt{5}}{5}$$

$$\sqrt{\frac{2}{11}} = \frac{\sqrt{2}}{\sqrt{11}} \cdot \frac{\sqrt{11}}{\sqrt{11}} = \frac{\sqrt{22}}{11}$$

Conjugates: $a\sqrt{b} + c\sqrt{d}$ and $a\sqrt{b} - c\sqrt{d}$

The product of two conjugates does not contain a radical.

Example:

$$\frac{6}{\sqrt{7}-2} = \frac{6}{\sqrt{7}-2} \cdot \frac{\sqrt{7}+2}{\sqrt{7}+2} = \frac{6(\sqrt{7}+2)}{(\sqrt{7})^2 - 2^2} = \frac{6\sqrt{7}+12}{7-4} = 2\sqrt{7}+4$$

2.

Solving a Square Root Equation

A square root equation contains a square root with a variable in the radicand.

Squaring each side of an equation: If $a = b$, then $a^2 = b^2$.

Example:

$\sqrt{x} - 6 = 4$

$\sqrt{x} = 10$

$(\sqrt{x})^2 = 10^2$

$x = 100$

Check:

$\sqrt{100} - 6 \stackrel{?}{=} 4$

$10 - 6 = 4$ ✓

Squaring each side of an equation sometimes introduces an extraneous solution.

Example:

$x = \sqrt{x+2}$

$x^2 = (\sqrt{x+2})^2$

$x^2 = x + 2$

$x^2 - x - 2 = 0$

$(x-2)(x+1) = 0$

$x = 2$ *or* $x = -1$

Check:

$2 \stackrel{?}{=} \sqrt{2+2}$ $\quad -1 \stackrel{?}{=} \sqrt{-1+2}$

$2 = \sqrt{4}$ ✓ $\quad -1 \neq \sqrt{1}$ X

The solution is $x = 2$.

3. Available at *BigIdeasMath.com.*

List of Organizers

Available at *BigIdeasMath.com*

Comparison Chart
Concept Circle
Definition (Idea) and Example Chart
Example and Non-Example Chart
Formula Triangle
Four Square
Information Frame
Information Wheel
Notetaking Organizer
Process Diagram
Summary Triangle
Word Magnet
Y Chart

About this Organizer

A **Word Magnet** can be used to organize information associated with a vocabulary word or term. As shown, students write the word or term inside the magnet. Students write associated information on the blank lines that "radiate" from the magnet. Associated information can include, but is not limited to: other vocabulary words or terms, definitions, formulas, procedures, examples, and visuals. This type of organizer serves as a good summary tool because any information related to a topic can be included.

Technology For the Teacher

Vocabulary Puzzle Builder

Answers

1. $x \geq 0$
2. $x \geq 3$
3. $x \leq 3$

4–6. See Additional Answers.

7. $\frac{\sqrt{7}}{7}$
8. $\frac{\sqrt{3}}{3}$
9. $7\sqrt{5} - 14$
10. $x = 65$
11. $x = 4$
12. $x = 3$
13. 13
14. 4
15. a.

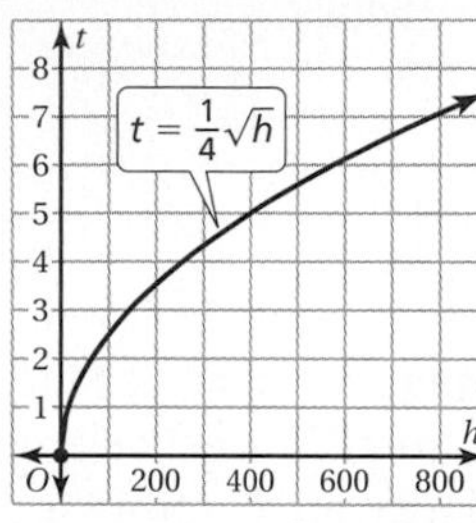

domain: $h \geq 0$; range: $t \geq 0$

b. about 876 ft

16. a. 16°C

b. 5.5 sec

Assessment Book

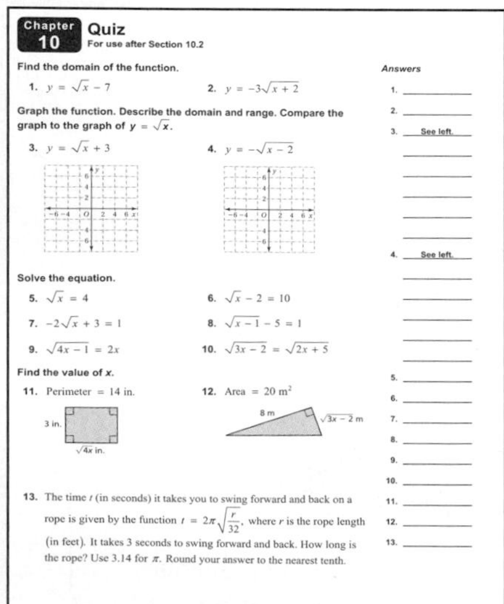

Chapter 10 Quiz
For use after Section 10.2

Find the domain of the function.

1. $y = \sqrt{x} - 7$
2. $y = -3\sqrt{x+2}$

Graph the function. Describe the domain and range. Compare the graph to the graph of $y = \sqrt{x}$.

3. $y = \sqrt{x} + 3$
4. $y = -\sqrt{x-2}$

Solve the equation.

5. $\sqrt{x} = 4$
6. $\sqrt{x} - 2 = 10$
7. $-2\sqrt{x} + 3 = 1$
8. $\sqrt{x-1} - 5 = 1$
9. $\sqrt{4x-1} = 2x$
10. $\sqrt{3x-2} = \sqrt{2x+5}$

Find the value of *x*.

11. Perimeter = 14 in. (3 in., $\sqrt{4x}$ in.)
12. Area = 20 m² (8 m, $\sqrt{3x-2}$ m)

13. The time t (in seconds) it takes you to swing forward and back on a rope is given by the function $t = 2\pi\sqrt{\frac{r}{32}}$, where r is the rope length (in feet). It takes 3 seconds to swing forward and back. How long is the rope? Use 3.14 for π. Round your answer to the nearest tenth.

Answers

1. ____
2. ____
3. See left.
4. See left.
5. ____
6. ____
7. ____
8. ____
9. ____
10. ____
11. ____
12. ____
13. ____

Alternative Quiz Ideas

100% Quiz
Error Notebook
Group Quiz
Homework Quiz
Math Log
Notebook Quiz
Partner Quiz
Pass the Paper

Math Log

Ask students to keep a math log for the chapter. Have them include diagrams, definitions, and examples. Everything should be clearly labeled. It might be helpful if they put the information in a chart. Students can add to the log as they are introduced to new topics.

Reteaching and Enrichment Strategies

If students need help. . .	If students got it. . .
Resources by Chapter • Study Help • Practice A and Practice B • Puzzle Time Lesson Tutorials *BigIdeasMath.com* Practice Quiz Practice from the Test Generator	Resources by Chapter • Enrichment and Extension • School-to-Work Game Closet at *BigIdeasMath.com* Start the next section

Technology For the Teacher
Answer Presentation Tool
Big Ideas Test Generator

10.1–10.2 Quiz

Find the domain of the function. *(Section 10.1)*

1. $y = 15\sqrt{x}$

2. $y = \sqrt{x - 3}$

3. $y = \sqrt{3 - x}$

Graph the function. Describe the domain and range. Compare the graph to the graph of $y = \sqrt{x}$. *(Section 10.1)*

4. $y = \sqrt{x} + 5$

5. $y = \sqrt{x - 4}$

6. $y = -\sqrt{x - 2} + 1$

Simplify the expression. *(Section 10.1)*

7. $\sqrt{\frac{6}{42}}$

8. $\frac{2}{\sqrt{3}} - \frac{1}{\sqrt{3}}$

9. $\frac{7}{\sqrt{5} + 2}$

Solve the equation. *(Section 10.2)*

10. $\sqrt{x - 1} + 7 = 15$

11. $\sqrt{x} = \sqrt{6x - 20}$

12. $x = \sqrt{21 - 4x}$

Find the value of x. *(Section 10.2)*

13. Perimeter = 24 mi

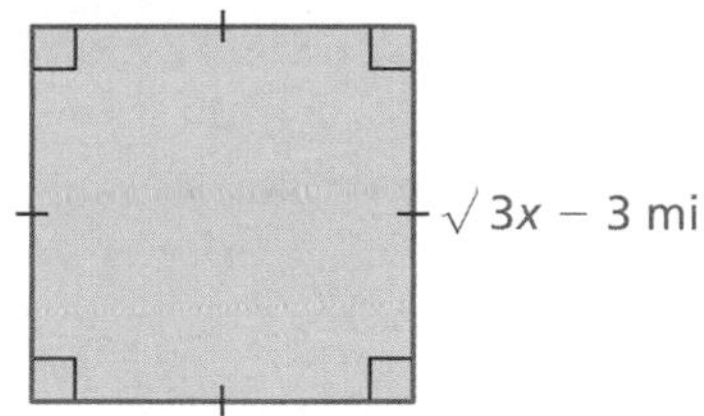

14. Area = $2\sqrt{4x - 7}$ m^2

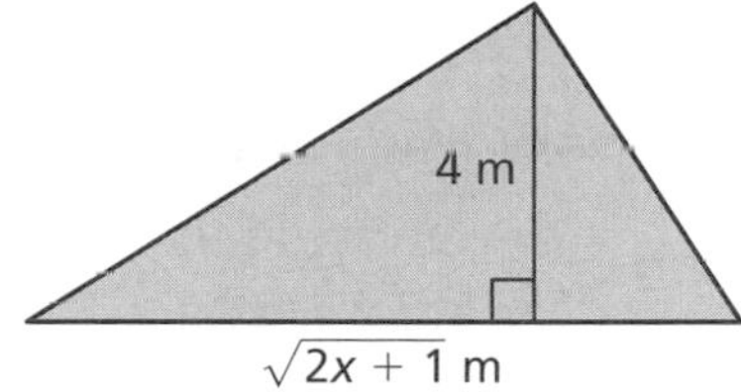

15. BRIDGE The time t (in seconds) it takes an object to drop h feet is given by $t = \frac{1}{4}\sqrt{h}$. *(Section 10.1)*

a. Graph the function. Describe the domain and range.

b. It takes about 7.4 seconds for a stone dropped from the New River Gorge Bridge in West Virginia, to reach the water below. About how high is the bridge above the New River?

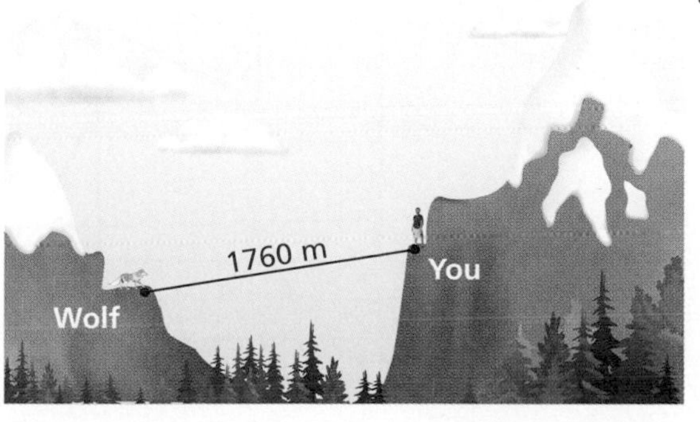

16. SPEED OF SOUND The speed of sound s (in meters per second) through air is given by $s = 20\sqrt{T + 273}$, where T is the temperature in degrees Celsius. *(Section 10.2)*

a. What is the temperature when the speed of sound is 340 meters per second?

b. How long does it take you to hear the wolf howl when the temperature is -17°C?

10.3 The Pythagorean Theorem

COMMON CORE STATE STANDARDS
8.G.6
8.G.7

Essential Question How are the lengths of the sides of a right triangle related?

Pythagoras was a Greek mathematician and philosopher who discovered one of the most famous rules in mathematics. In mathematics, a rule is called a **theorem**. So, the rule that Pythagoras discovered is called the Pythagorean Theorem.

Pythagoras
(c. 570 B.C.–c. 490 B.C.)

1 ACTIVITY: Discovering the Pythagorean Theorem

Work with a partner.

a. On grid paper, draw any right triangle. Label the lengths of the two shorter sides (the **legs**) a and b.

b. Label the length of the longest side (the **hypotenuse**) c.

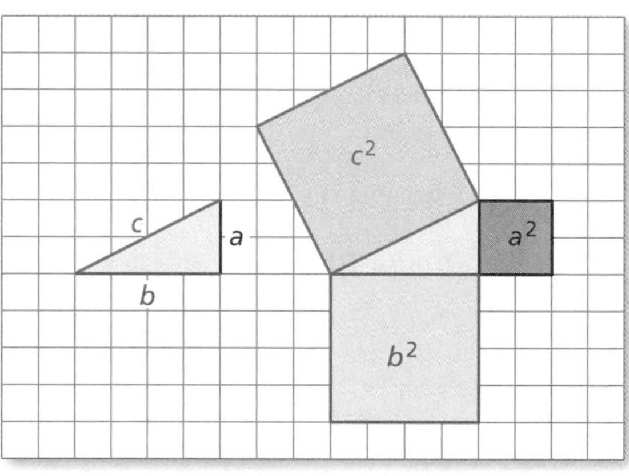

c. Draw squares along each of the three sides. Label the areas of the three squares a^2, b^2, and c^2.

d. Cut out the three squares. Make eight copies of the right triangle and cut them out. Arrange the figures to form two identical larger squares.

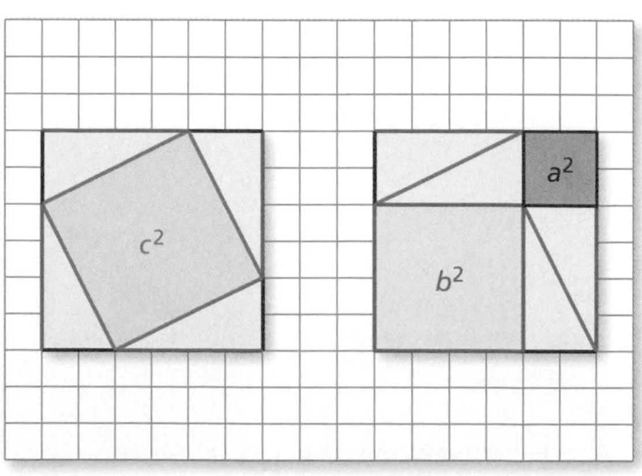

e. What does this tell you about the relationship among a^2, b^2, and c^2?

Laurie's Notes

Introduction

Standards for Mathematical Practice

- **MP4 Model with Mathematics:** Mathematically proficient students can apply what they know about the Pythagorean Theorem to model real-world phenomena. Sketching a representation of the problem, perhaps on grid paper, may be a helpful first step.

Motivate

- Share information about Pythagoras who was born in Greece in 569 B.C.
  - He is known as the *Father of Numbers*.
  - He traveled extensively in Egypt, learning math, astronomy and music.
  - Pythagoras undertook a reform of the cultural life of Cretona, urging the citizens to follow his religious, political, and philosophical goals.
  - He created a school where his followers, known as Pythagoreans, lived and worked. They observed a rule of silence called *echemythia*, the breaking of which was punishable by death. One had to remain silent for *five years* before he could contribute to the group.

Activity Notes

Activity 1

- **Suggestions:** Use centimeter grid paper for ease of manipulating the cut pieces. Suggest to students that they draw their original triangle in the upper left of the grid paper and then make a working copy of the triangle towards the middle of the paper. This gives enough room for the squares to be drawn along each side of triangle.
- Vertices of the triangle need to be on lattice points. You do not want every student in the room to use the same triangle. Suggest other leg lengths (3 and 4, 3 and 6, 2 and 4, 2 and 3, and so on).
- **MP4:** Drawing the square along the hypotenuse is the challenging step. Model one technique for accomplishing the task using a right triangle with legs 2 and 5 units.
  - Interpret the slope of the hypotenuse as "right 5 units, up 2 units."
  - Place your pencil on the upper right endpoint of the hypotenuse and rotate the paper 90° clockwise. Move your pencil right 5 units and up 2 units. Mark a point.
  - Repeat rotating and moving according to the slope of the hypotenuse until you end at the other endpoint of the original hypotenuse.
  - Use a straightedge to connect the four points (two that you marked and two on the endpoints of the hypotenuse) to form the square.
- Before students cut anything, check that they have 3 squares of the correct size.
- **MP2 Reason Abstractly and Quantitatively:** The two large squares in part (d) have equal area. Referring to areas, if $c^2 +$ (4 triangles) $=$ $a^2 + b^2 +$ (4 triangles), then $c^2 = a^2 + b^2$ by subtracting the 4 triangles from each side of the equation.

Common Core State Standards

8.G.6 Explain a proof of the Pythagorean Theorem and its converse.

8.G.7 Apply the Pythagorean Theorem to determine unknown side lengths in right triangles in real-world and mathematical problems in two and three dimensions.

Previous Learning

Students should know how to multiply fractions and decimals.

Activity Materials	
Textbook	
• scissors	• straightedge
• grid paper	• transparency grid

Start Thinking! and Warm Up

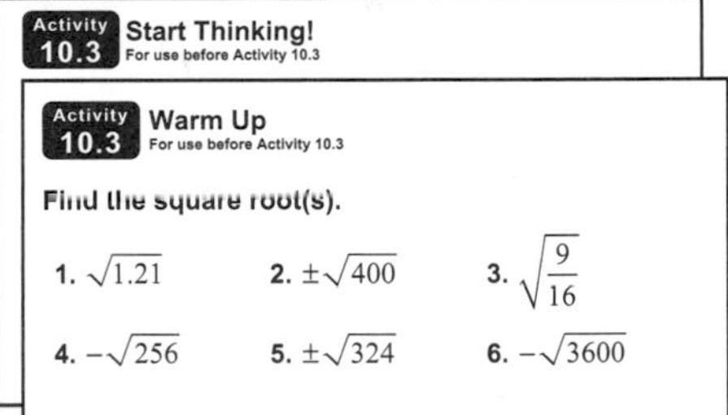

Activity 10.3 Start Thinking! For use before Activity 10.3

Activity 10.3 Warm Up For use before Activity 10.3

Find the square root(s).

1. $\sqrt{1.21}$ 2. $\pm\sqrt{400}$ 3. $\sqrt{\frac{9}{16}}$

4. $-\sqrt{256}$ 5. $\pm\sqrt{324}$ 6. $-\sqrt{3600}$

10.3 Record and Practice Journal

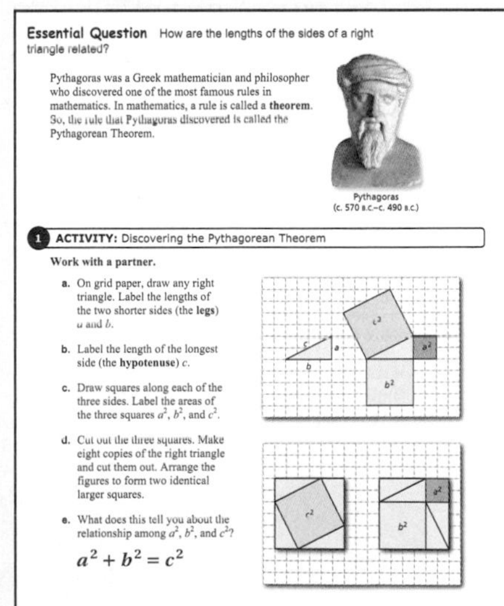

Essential Question How are the lengths of the sides of a right triangle related?

Pythagoras was a Greek mathematician and philosopher who discovered one of the most famous rules in mathematics. In mathematics, a rule is called a **theorem**. So, the rule that Pythagoras discovered is called the Pythagorean Theorem.

Pythagoras (c. 570 B.C.–c. 490 B.C.)

1 ACTIVITY: Discovering the Pythagorean Theorem

Work with a partner.

a. On grid paper, draw any right triangle. Label the lengths of the two shorter sides (the **legs**) a and b.

b. Label the length of the longest side (the **hypotenuse**) c.

c. Draw squares along each of the three sides. Label the areas of the three squares a^2, b^2, and c^2.

d. Cut out the three squares. Make eight copies of the right triangle and cut them out. Arrange the figures to form two identical larger squares.

e. What does this tell you about the relationship among a^2, b^2, and c^2?

$a^2 + b^2 = c^2$

English Language Learners

Vocabulary

Help English learners understand the meanings of the words that make up a definition. Provide students with statements containing blanks and a list of the words used to fill in the blanks.

- In any right __, the __ is the side __ the right __.
  Word list: angle, hypotenuse, opposite, triangle
  triangle, hypotenuse, opposite, angle
- In any right __, the __ are the __ sides and the __ is always the __ side.
  Word list: hypotenuse, legs, longest, shorter, triangle
  triangle, legs, shorter, hypotenuse, longest

10.3 Record and Practice Journal

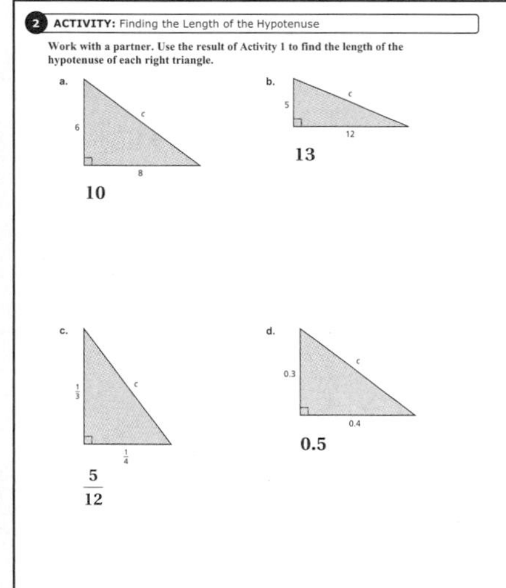

2 ACTIVITY: Finding the Length of the Hypotenuse

Work with a partner. Use the result of Activity 1 to find the length of the hypotenuse of each right triangle.

a. 10

b. 13

c. $\frac{5}{12}$

d. 0.5

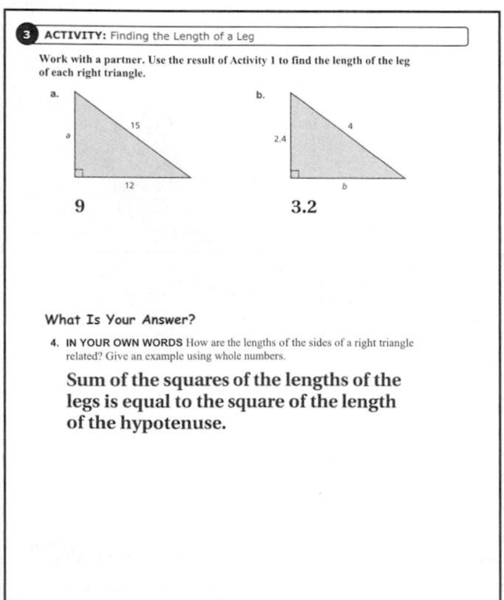

3 ACTIVITY: Finding the Length of a Leg

Work with a partner. Use the result of Activity 1 to find the length of the leg of each right triangle.

a. 9

b. 3.2

What Is Your Answer?

4. **IN YOUR OWN WORDS** How are the lengths of the sides of a right triangle related? Give an example using whole numbers.

Sum of the squares of the lengths of the legs is equal to the square of the length of the hypotenuse.

Laurie's Notes

Activity 2

- **Part (a):** This triangle is similar to the 3-4-5 right triangle. Using the property from the investigation, $6^2 + 8^2 = 36 + 64 = 100$. Students will recognize that $100 = 10^2$, so the length of the hypotenuse is 10.
- Have students share their work for each of these problems.
- **Common Error:** In part (c), when students square a fraction, they sometimes double the numerator and denominator instead of squaring each number.
  In other words, $\left(\frac{1}{3}\right)^2 \neq \frac{2}{6}$, but $\left(\frac{1}{3}\right)^2 = \frac{1}{9}$.

Activity 3

- The two triangles in this activity have a leg length missing. Building squares on the two legs of the triangle, and finding their areas gives a^2 and 12^2 for part (a). The area of the square built on the hypotenuse is 15^2. The result of Activity 1 says that $a^2 + 12^2 = 15^2$. Students should recognize this as an opportunity to solve an equation.

? "What is the first step in solving the equation $a^2 + 12^2 = 15^2$?" Evaluate 12^2 and 15^2.

? "What is the next step in solving $a^2 + 144 = 225$?" Subtract 144 from each side.

? "Finally, what positive number squared is 81?" 9

What Is Your Answer?

- **Neighbor Check:** Have students work independently and then have their neighbor check their work. Have students discuss any discrepancies.

Closure

- **Exit Ticket:** If you drew a right triangle with legs of 5 and 6 on grid paper, what would be the area of the square drawn on the hypotenuse of the triangle? 61 square units

The Dynamic Planning Tool
Editable Teacher's Resources at *BigIdeasMath.com*

2 ACTIVITY: Finding the Length of the Hypotenuse

Work with a partner. Use the result of Activity 1 to find the length of the hypotenuse of each right triangle.

a.

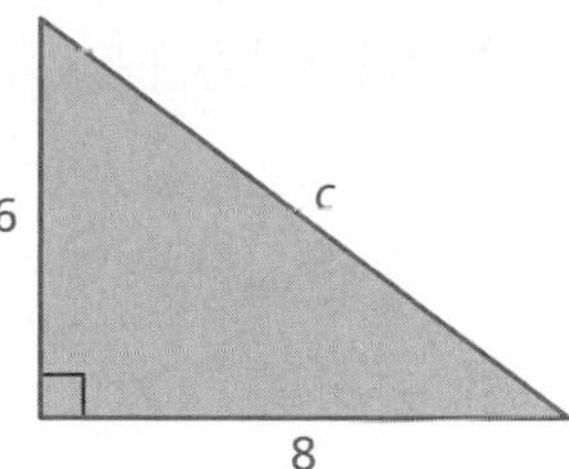

b.

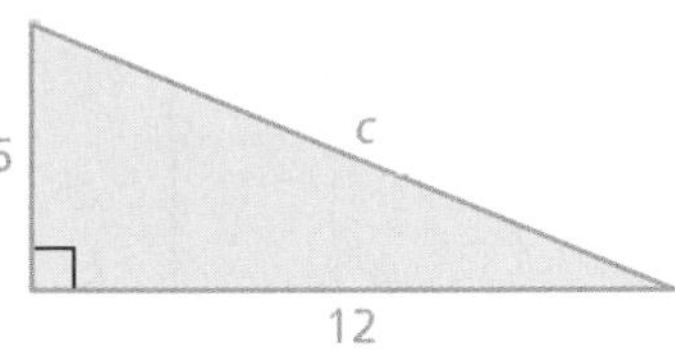

c.

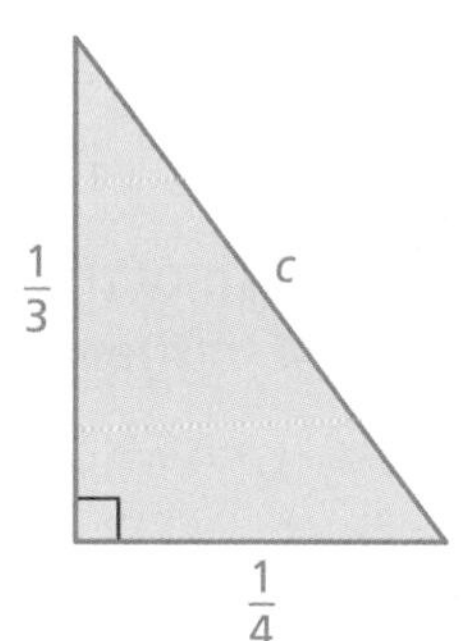

d.

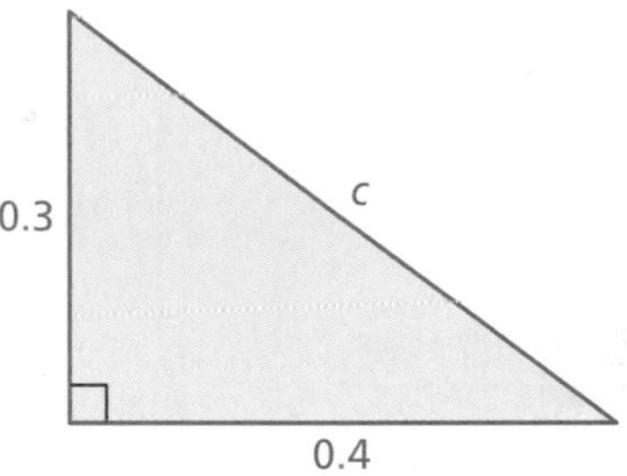

3 ACTIVITY: Finding the Length of a Leg

Work with a partner. Use the result of Activity 1 to find the length of the leg of each right triangle.

a.

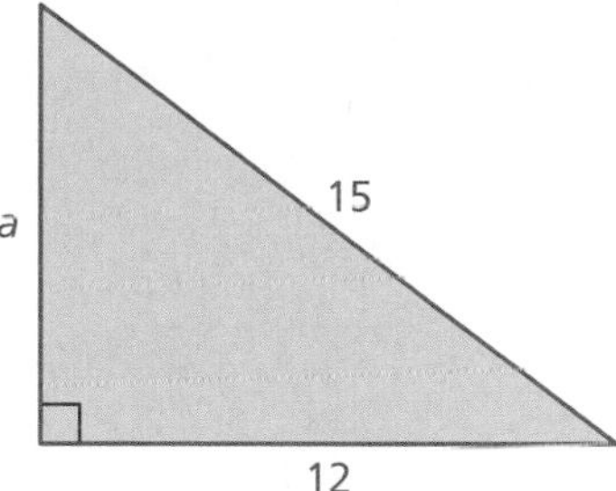

b.

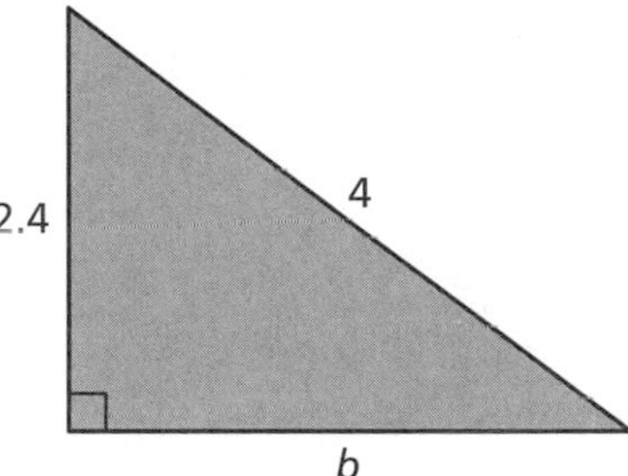

What Is Your Answer?

4. **IN YOUR OWN WORDS** How are the lengths of the sides of a right triangle related? Give an example using whole numbers.

Practice Use what you learned about the Pythagorean Theorem to complete Exercises 3–5 on page 524.

10.3 Lesson

Key Vocabulary
theorem, *p. 520*
legs, *p. 522*
hypotenuse, *p. 522*
Pythagorean Theorem, *p. 522*

Key Ideas

Sides of a Right Triangle

The sides of a right triangle have special names.

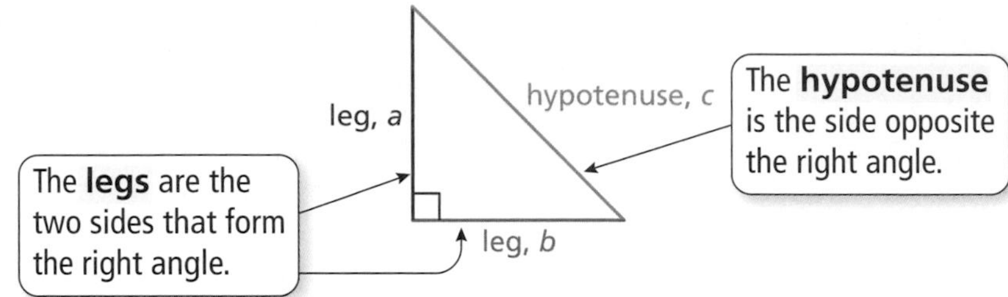

Study Tip

In a right triangle, the legs are the shorter sides and the hypotenuse is always the longest side.

The Pythagorean Theorem

Words In any right triangle, the sum of the squares of the lengths of the legs is equal to the square of the length of the hypotenuse.

Algebra $a^2 + b^2 = c^2$

EXAMPLE 1 Finding the Length of a Hypotenuse

Find the length of the hypotenuse of the triangle.

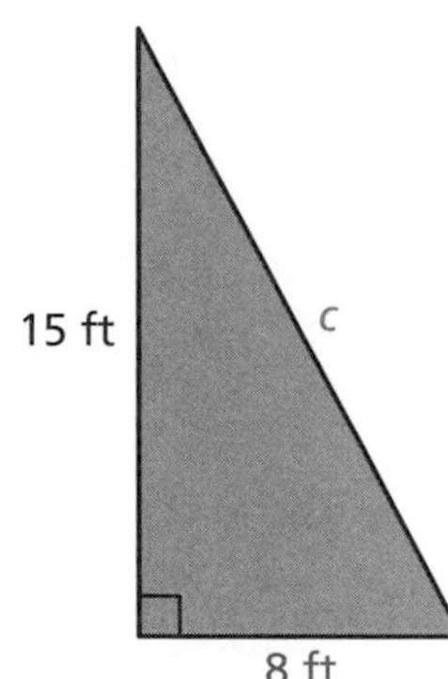

$a^2 + b^2 = c^2$	Write the Pythagorean Theorem.
$15^2 + 8^2 = c^2$	Substitute 15 for *a* and 8 for *b*.
$225 + 64 = c^2$	Evaluate powers.
$289 = c^2$	Add.
$\sqrt{289} = \sqrt{c^2}$	Take positive square root of each side.
$17 = c$	Simplify.

The length of the hypotenuse is 17 feet.

On Your Own

Find the length of the hypotenuse of the triangle.

1. **2.**

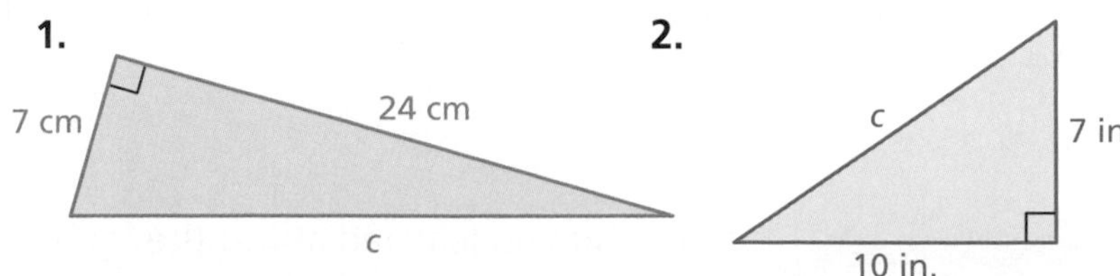

Multi-Language Glossary at BigIdeasMath.com.

Laurie's Notes

Introduction

Connect

- **Yesterday:** Students investigated a visual proof of the Pythagorean Theorem. (MP2, MP4)
- **Today:** Students will use the Pythagorean Theorem to find the missing lengths of a right triangle.

Motivate

- **Preparation:** Cut coffee stirrers (or carefully break spaghetti) so that triangles with the following side lengths can be made: 2-3-4; 3-4-5; 4-5-6.

? "What are consecutive numbers?" numbers in sequential order

- With student aid, use the coffee stirrers to make three triangles: 2-3-4; 3-4-5; and 4-5-6 on the overhead projector. If arranged carefully, all 3 will fit on the screen.
- Ask students to make observations about the 3 triangles. Students may mention that all triangles are scalene; one triangle appears to be acute, one right, one obtuse.
- They should observe that the change in the side lengths seems to have made a big change in the angle measures.

Lesson Notes

Key Ideas

- Draw a right triangle and label the *legs* and *hypotenuse.* The hypotenuse is always opposite the right angle and is the longest side of a right triangle.
- **Teaching Tip:** Try not to have all right triangles in the same orientation.
- Write the Pythagorean Theorem.
- **Common Error:** Students often forget that the Pythagorean Theorem is a relationship that is *only* true for right triangles.

Example 1

- Draw and label the triangle. Review the symbol used to show that an angle is a right angle.

? "What information is known for this triangle?" The legs are 8 ft and 15 ft.

On Your Own

- Give time for students to work the problems. Knowing their perfect squares is helpful.

Goal Today's lesson is using the **Pythagorean Theorem** to solve for lengths of a right triangle.

Lesson Materials
Introduction
• coffee stirrers • scissors

Start Thinking! and Warm Up

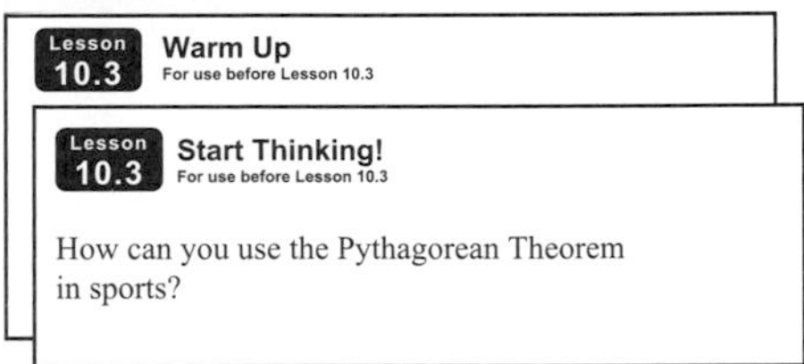

Lesson 10.3
Warm Up
For use before Lesson 10.3
Lesson 10.3
Start Thinking!
For use before Lesson 10.3
How can you use the Pythagorean Theorem in sports?

Extra Example 1

Find the length of the hypotenuse of the triangle. 5 in.

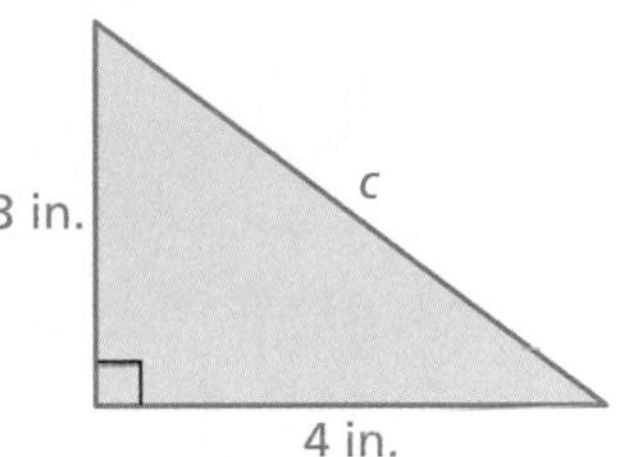

On Your Own

1. 25 cm
2. $\sqrt{149}$ in.

Extra Example 2

Find the missing length of the triangle.

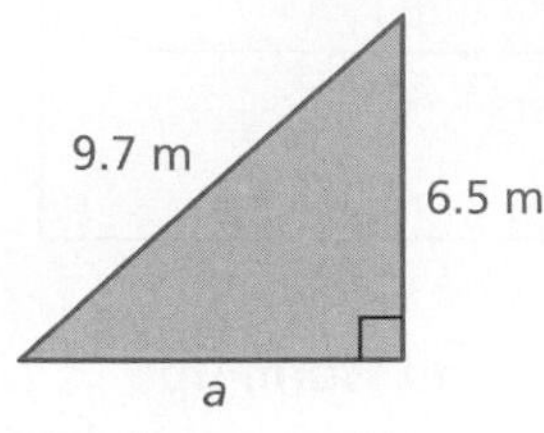

$a = 7.2$ m

Extra Example 3

Ship A is 22 kilometers north and 36 kilometers east of the port. Ship B is 12 kilometers north and 12 kilometers east of the port. How far apart are the ships? 26 km

On Your Own

3. 0.5 m

4. $2\sqrt{39}$ yd

5. 57 ft

Differentiated Instruction

Kinesthetic

Have students verify the Pythagorean Theorem by drawing right triangles with legs of a given length, measuring the hypotenuse, and then calculating the hypotenuse using the Pythagorean Theorem. Use Pythagorean triples so that students work only with whole numbers.

Leg Lengths	*Hypotenuse Length*
3, 4	5
6, 8	10
5, 12	13
8, 15	17

Laurie's Notes

Example 2

- ? "What information is known for this triangle?" One leg is 3.5 kilometers and the hypotenuse is 6.5 kilometers.
- Substitute and solve as shown.
- **Common Error:** Students need to be careful with decimal multiplication. It is very common for students to multiply the decimal by 2 instead of multiplying the decimal by itself.

Example 3

- Ask a student to read the example.
- ? "Given the compass directions stated, what is a reasonable way to represent this information?" coordinate plane
- **MP4 Model with Mathematics:** Explain that east is the positive x-direction and north is the positive y-direction. Draw the situation in a coordinate plane.
- ? "Is there enough information to use the Pythagorean Theorem? Explain." yes; The legs of the triangle can be found and then used to solve for the hypotenuse.
- **FYI:** This example previews the *distance formula,* which will be presented in the next section.

On Your Own

- **Neighbor Check:** Have students work independently and then have their neighbor check their work. Have students discuss any discrepancies.

Closure

- **Exit Ticket:** Solve for the missing length. $x = 30$ cm, $y = 1$ m

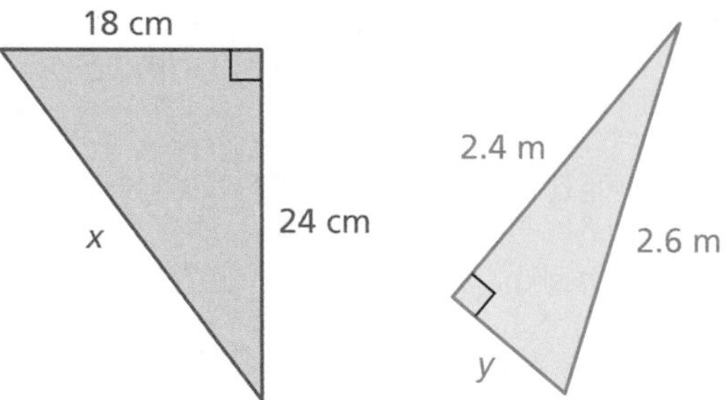

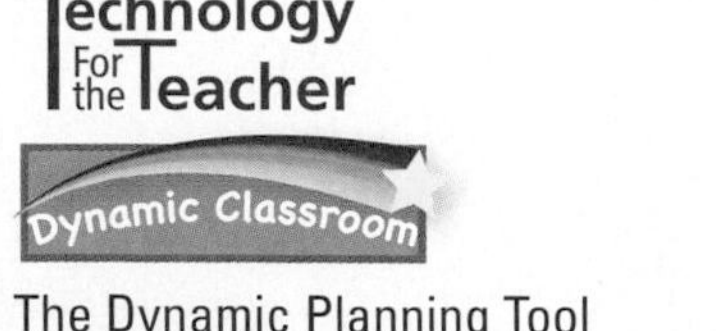

The Dynamic Planning Tool
Editable Teacher's Resources at *BigIdeasMath.com*

EXAMPLE 2 Finding the Length of a Leg

Find the missing length of the triangle.

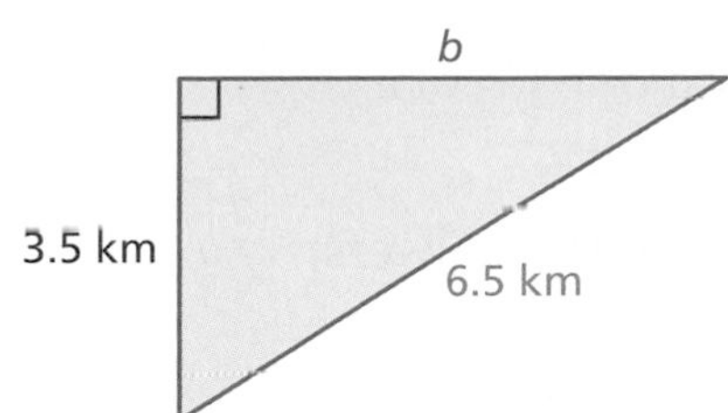

$a^2 + b^2 = c^2$	Write the Pythagorean Theorem.
$3.5^2 + b^2 = 6.5^2$	Substitute 3.5 for *a* and 6.5 for *c*.
$12.25 + b^2 = 42.25$	Evaluate powers.
$b^2 = 30$	Subtract 12.25 from each side.
$b = \sqrt{30}$	Take positive square root of each side.

The length of the leg is $\sqrt{30} \approx 5.5$ kilometers.

EXAMPLE 3 Real-Life Application

Paintball Team A is located 70 feet north and 60 feet east of the base. Team B is located 30 feet north and 30 feet east of the base. How far apart are the teams?

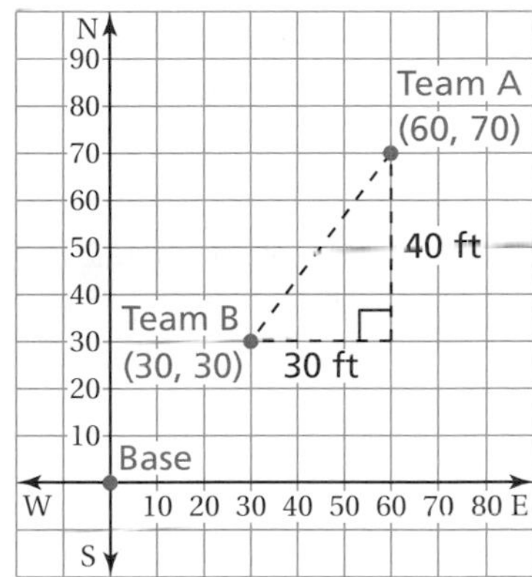

Step 1: Draw the situation in a coordinate plane. Let the base be at the origin. From the descriptions, you can plot Team A at (60, 70) and Team B at (30, 30).

Step 2: Draw a right triangle with a hypotenuse that represents the distance between the teams. The lengths of the legs are 30 feet and 40 feet.

Step 3: Use the Pythagorean Theorem to find the length of the hypotenuse.

$a^2 + b^2 = c^2$	Write the Pythagorean Theorem.
$30^2 + 40^2 = c^2$	Substitute 30 for *a* and 40 for *b*.
$900 + 1600 = c^2$	Evaluate powers.
$2500 = c^2$	Add.
$50 = c$	Take positive square root of each side.

The teams are 50 feet apart.

On Your Own

Find the missing length of the triangle.

3. 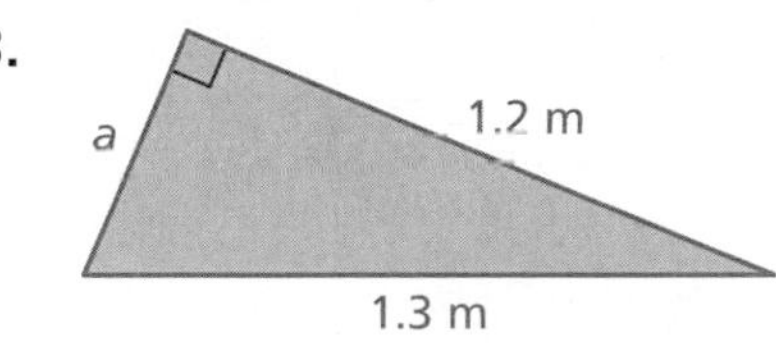

4. 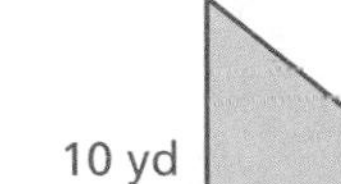
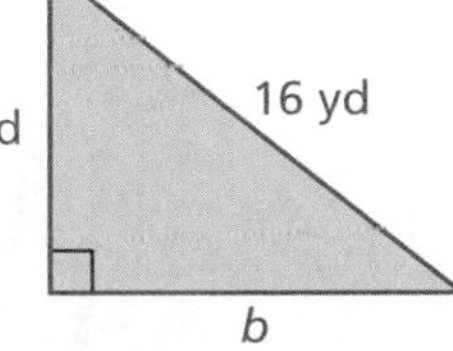

5. **WHAT IF?** In Example 3, Team B moves 10 feet to the west. How far apart are the teams to the nearest foot?

10.3 Exercises

Vocabulary and Concept Check

1. **VOCABULARY** You are given the lengths of the hypotenuse and one leg of a right triangle. Describe how you can find the length of the other leg.

2. **DIFFERENT WORDS, SAME QUESTION** Which is different? Find "both" answers.

Which side is a leg?

Which side is shortest?

Which side is longest?

Which side is part of a right angle?

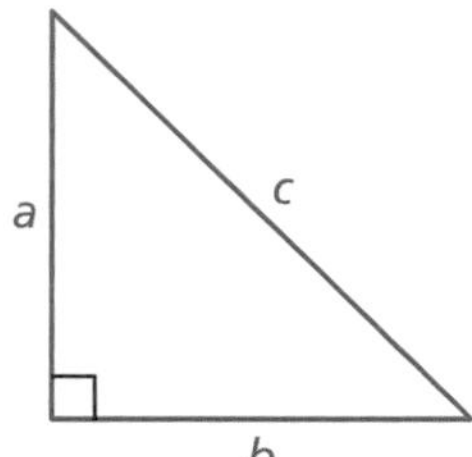

Practice and Problem Solving

Find the missing length of the triangle.

1 2 3.

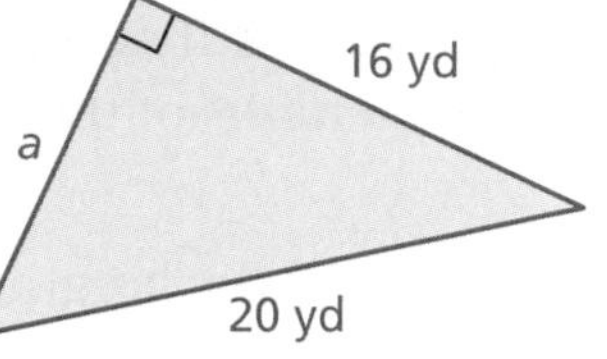

4.

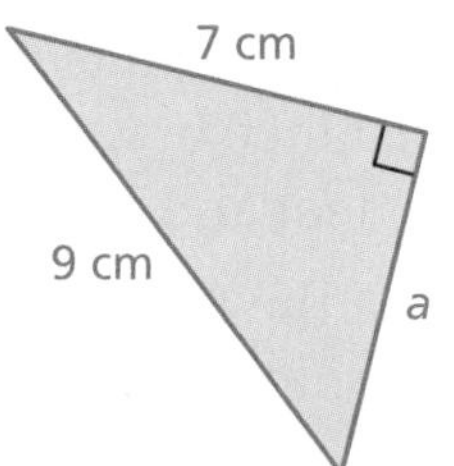

5.

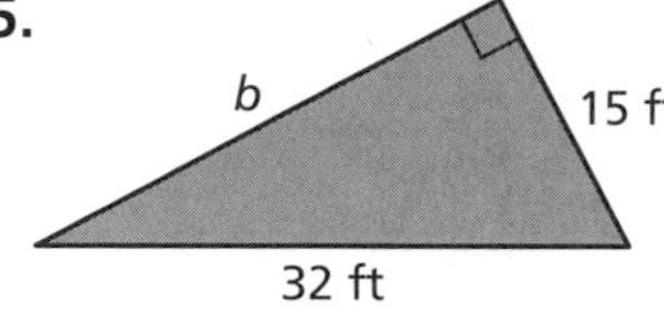

6.

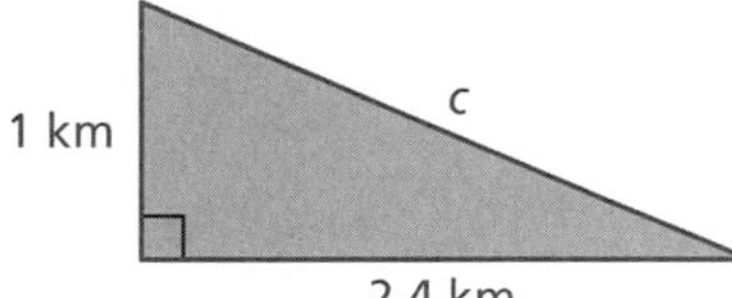

7.

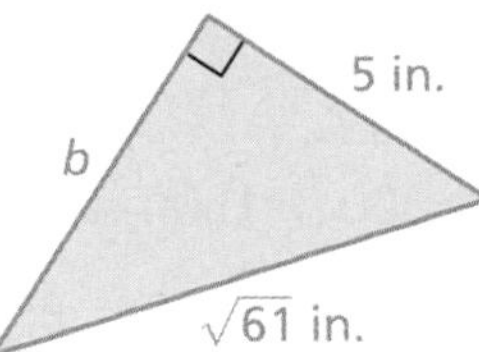

8.

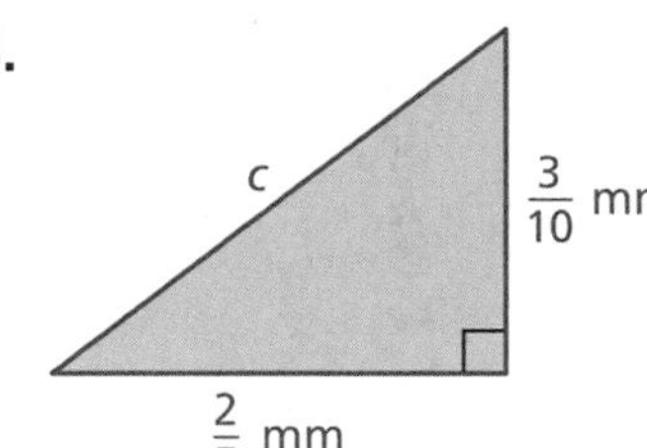

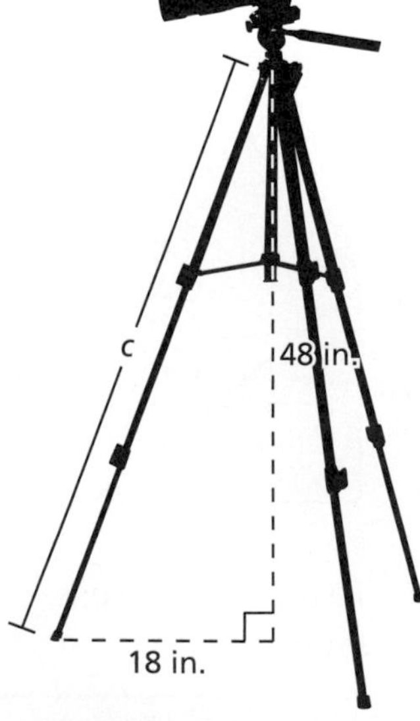

9. **ERROR ANALYSIS** Describe and correct the error in finding the missing length of the triangle.

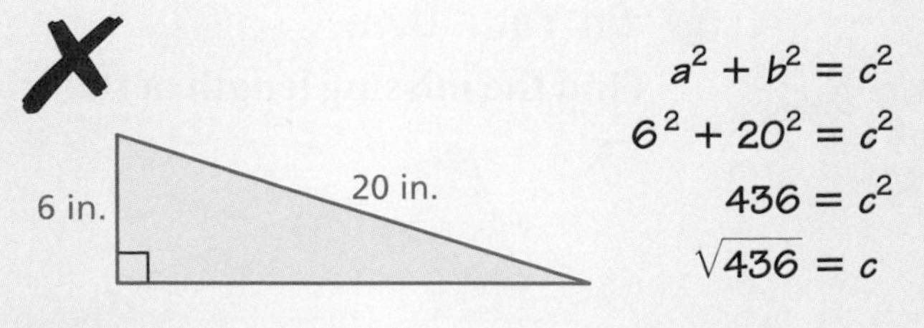

10. **TRIPOD** The center of the tripod forms a 90° angle with the ground. Find the length of the support leg to the nearest tenth of an inch.

Assignment Guide and Homework Check

Level	Assignment	Homework Check
Average	1, 2, 3–12, 14, 17–20	5, 9, 10, 14
Advanced	1, 2, 6–16, 17–20	7, 10, 14, 16

Common Errors

- **Exercises 3–8** Students may substitute the given lengths in the wrong part of the formula. For example, if they are finding one of the legs, they may write $5^2 + 13^2 = c^2$ instead of $5^2 + b^2 = 13^2$. Remind them that the side opposite the right angle is the hypotenuse c.
- **Exercises 3–8** Students may multiply each side length by two instead of squaring the side length. Remind them of the definition of exponents.

Vocabulary and Concept Check

1. Use the Pythagorean Theorem to find the missing side length.
2. Which side is longest?; c; a or b

Practice and Problem Solving

3. 12 yd
4. $4\sqrt{2}$ cm
5. $\sqrt{799}$ ft
6. 2.6 km
7. 6 in.
8. $\frac{1}{2}$ mm
9. 20 should have been substituted for c, not b. The missing length is $2\sqrt{91}$ inches.
10. about 51.3 in.

10.3 Record and Practice Journal

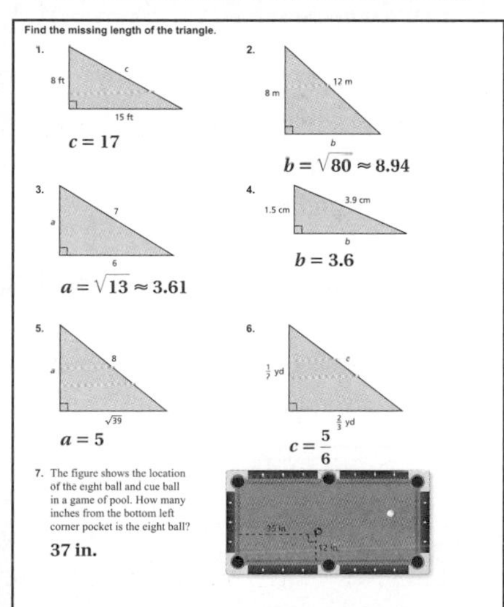
Find the missing length of the triangle.

1. $c = 17$
2. $b = \sqrt{80} \approx 8.94$
3. $a = \sqrt{13} \approx 3.61$
4. $b = 3.6$
5. $a = 5$
6. $c = \frac{5}{6}$
7. The figure shows the location of the eight ball and cue ball in a game of pool. How many inches from the bottom left corner pocket is the eight ball? **37 in.**

Technology For the Teacher
Answer Presentation Tool

Practice and Problem Solving

11. about 60 in.

12. yes; The distance between the tennis player's mouth and the referee's ear is about 41 feet.

13. See *Taking Math Deeper.*

14. $10\sqrt{2}$ ft or about 14.1 ft

15. See Additional Answers.

16. **a.** $x, x + 1, x + 2; x + 2$

b. $x^2 + (x + 1)^2 = (x + 2)^2$; integers: 3, 4, 5

Fair Game Review

17–19. See Additional Answers.

20. C

Mini-Assessment

Find the missing length of the triangle.

1. 50 ft

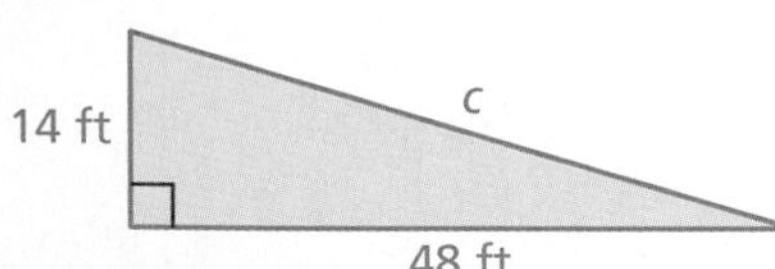

2. 24 mm

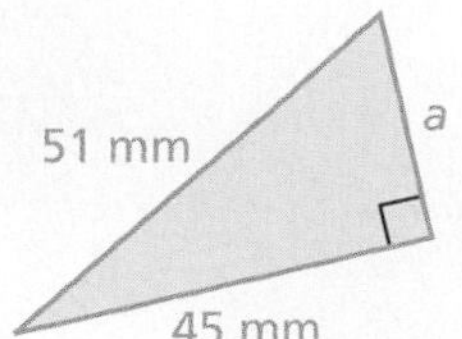

3. 12 in.

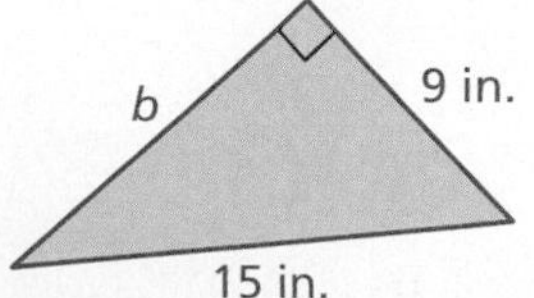

Taking Math Deeper

Exercise 13

The challenging part of this problem is realizing that the length of the hypotenuse of the right triangle is given as 145 yards at the bottom of the diagram.

Find the length of the hypotenuse of the right triangle.

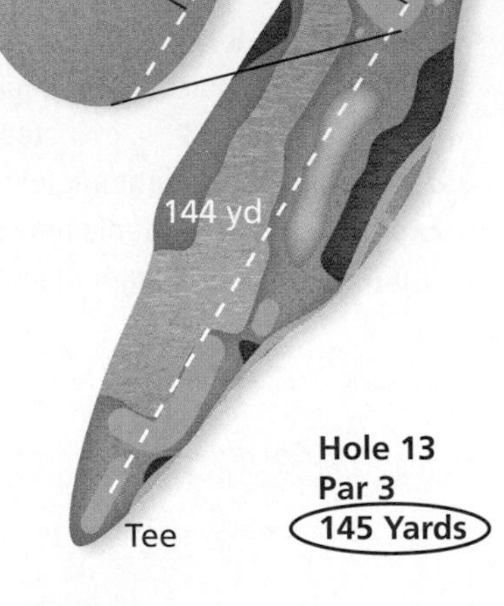

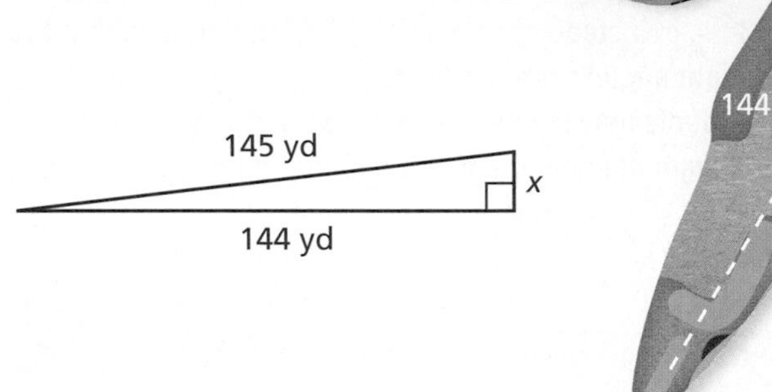

Use the Pythagorean Theorem.

$$144^2 + x^2 = 145^2$$
$$20{,}736 + x^2 = 21{,}025$$
$$x^2 = 289$$
$$x = 17$$

3 Answer the question.

The ball is 17 yards from the hole. Using the relationship of 3 feet = 1 yard, $17 \text{ yd} \times \frac{3 \text{ ft}}{1 \text{ yd}} = 51$ ft. So, the ball is 51 feet from the hole.

Reteaching and Enrichment Strategies

If students need help. . .	If students got it. . .
Resources by Chapter • Practice A and Practice B • Puzzle Time Record and Practice Journal Practice Differentiating the Lesson Lesson Tutorials Skills Review Handbook	Resources by Chapter • Enrichment and Extension • School-to-Work Start the next section

11. **TELEVISIONS** Televisions are advertised by the lengths of their diagonals. Approximate the length of the diagonal of the television to the nearest inch.

12. **TENNIS** A tennis player asks the referee a question. The sound of the player's voice only travels 50 feet. Can the referee hear the question? Explain.

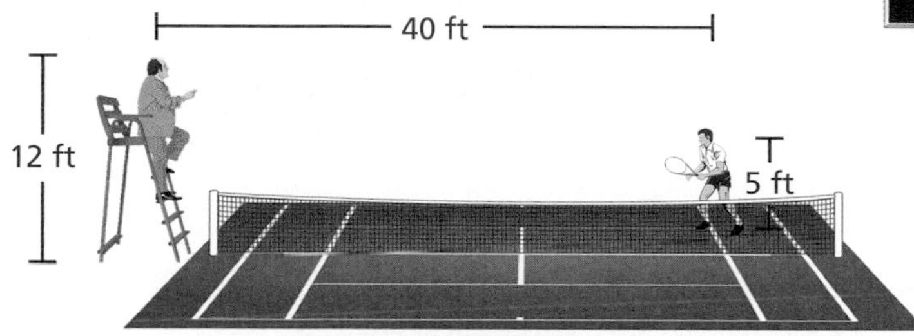

13. **GOLF** The figure shows the location of a golf ball after a tee shot. How many feet from the hole is the ball?

x
144 yd
Tee
Hole 13
Par 3
145 Yards

14. **SNOWBALLS** You and a friend throw snowballs at each other. You are 20 feet north and 15 feet east of your house. Your friend is 25 feet east and 10 feet north of your house. How far apart are you and your friend?

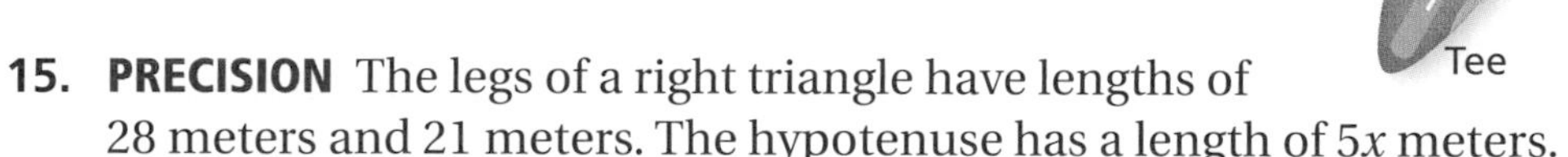

15. **PRECISION** The legs of a right triangle have lengths of 28 meters and 21 meters. The hypotenuse has a length of $5x$ meters.
   a. Write an equation to solve for x.
   b. Describe how to solve the equation by factoring and by taking a square root. Which method do you prefer? Explain.
   c. What is the value of x?

16. **Structure** The side lengths of a right triangle are three consecutive integers.
   a. Write an expression that represents each side length. Which side length represents the hypotenuse?
   b. Write and solve an equation to find the three integers.

Fair Game Review What you learned in previous grades & lessons

Graph the function. Compare the graph to the graph of $y = x^2$. *(Section 8.3)*

17. $y = -2x^2 + 4$ 18. $y = -x^2 - 6$ 19. $y = 3x^2 + 8$

20. **MULTIPLE CHOICE** Which polynomial is equivalent to $(x^2 - 3x + 1) - (-2x^2 + x - 4)$? *(Section 7.2)*

Ⓐ $3x^2 - 4x - 3$ Ⓑ $-x^2 - 2x - 5$ Ⓒ $3x^2 - 4x + 5$ Ⓓ $-x^2 + 4x + 3$

10.4 Using the Pythagorean Theorem

COMMON CORE STATE STANDARDS
8.G.6
8.G.7
8.G.8

Essential Question In what other ways can you use the Pythagorean Theorem?

The *converse* of a statement switches the hypothesis and the conclusion.

Statement:	Converse of the statement:
If p, then q.	If q, then p.

1 ACTIVITY: Analyzing Converses of Statements

Work with a partner. Write the converse of the true statement. Determine whether the converse is true or false. If it is false, give a counterexample.

a. Sample: If $a = b$, then $a^2 = b^2$.
Converse: If $a^2 = b^2$, then $a = b$.
The converse is false. A counterexample is $a = -2$ and $b = 2$.

b. If two nonvertical lines have the same slope, then the lines are parallel.

c. If a sequence has a common difference, then it is an arithmetic sequence.

d. If a and b are rational numbers, then $a + b$ is a rational number.

Is the converse of a true statement always true? always false? Explain.

2 ACTIVITY: The Converse of the Pythagorean Theorem

Work with a partner. The converse of the Pythagorean Theorem states: "If the equation $a^2 + b^2 = c^2$ is true for the side lengths of a triangle, then the triangle is a right triangle."

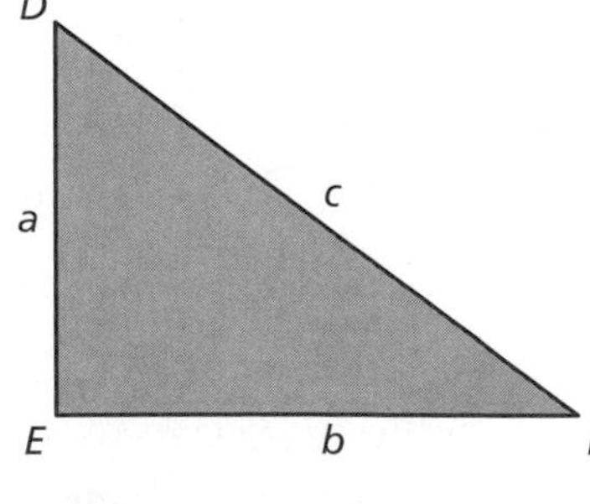

a. Do you think the converse of the Pythagorean Theorem is true or false? How could you use deductive reasoning to support your answer?

b. Consider $\triangle DEF$ with side lengths a, b, and c, such that $a^2 + b^2 = c^2$. Also consider $\triangle JKL$ with leg lengths a and b, where $\angle K = 90°$.

- What does the Pythagorean Theorem tell you about $\triangle JKL$?
- What does this tell you about c and x?
- What does this tell you about $\triangle DEF$ and $\triangle JKL$?
- What does this tell you about $\angle E$?
- What can you conclude?

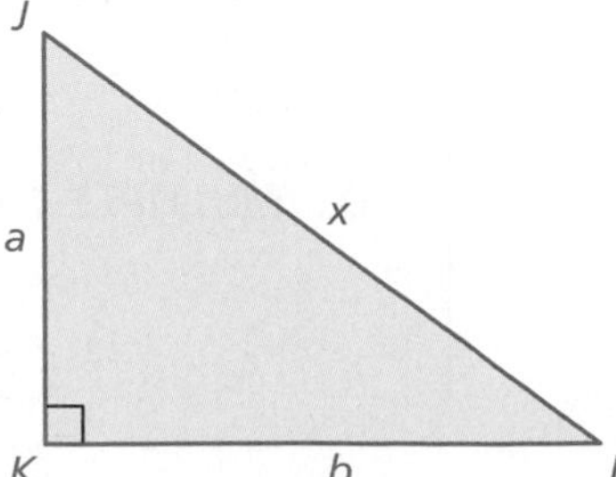

Laurie's Notes

Introduction

Standards for Mathematical Practice

- **MP3 Construct Viable Arguments and Critique the Reasoning of Others:** Students will develop a "proof" of the converse of the Pythagorean Theorem, and they will derive the distance formula. It is important that students be able to explain the steps in their work and compare it to the reasoning of their classmates.

Motivate

- Explain what the converse of a statement is.
- **Example:** If I live in Moab, then I live in Utah.
  The converse is: If I live in Utah, then I live in Moab.
  The original statement is true, but the converse is false.
- Ask students to write two if-then statements with converses that are true and two with converses that are false. The original statements must be true.
- Give students a few minutes and then share some of their examples.

Activity Notes

Activity 1

- Review the meaning of the word counterexample.
- It is important for students to understand that a statement might be false even if they cannot think of a counterexample.
- In part (d), remind students to think about irrational numbers.
- **Big Idea:** Even when a conditional statement is true, its converse does not have to be true. Students should keep this in mind for the next activity.

Activity 2

- Students often say that the Pythagorean Theorem is simply $a^2 + b^2 = c^2$. Tell them that it is actually a conditional statement. If a and b are the lengths of the legs and c is the length of the hypotenuse of a right triangle, then $a^2 + b^2 = c^2$.

? "What is the converse of the Pythagorean Theorem?" If $a^2 + b^2 = c^2$, then a triangle with side lengths a, b, and c is a right triangle.

? "Do you think the converse of the Pythagorean Theorem is true?" Students may simply guess at this point.

- **MP3:** In this activity, students use deductive reasoning to show that the converse is true. They may need guidance in linking their reasoning.
- **FYI:** Strategies for proofs are taught in later courses, so the framework of a proof is provided.
- In the second bullet, we are looking for $c^2 = x^2$ and then $c = x$. Students found this was not always true in Activity 1 part (a), but point out that it is true here because c and x must be positive.
- In Grade 7 Accelerated Additional Topic 2, students discovered that you can only construct one triangle given three side lengths. Review this concept. It will help them with the third bullet.

Common Core State Standards

8.G.6 Explain a proof of the Pythagorean Theorem and its converse.
8.G.7 Apply the Pythagorean Theorem to determine unknown side lengths in right triangles in real-world and mathematical problems in two and three dimensions.
8.G.8 Apply the Pythagorean Theorem to find the distance between two points in a coordinate system.

Previous Learning

Students should know how to use the Pythagorean Theorem.

Start Thinking! and Warm Up

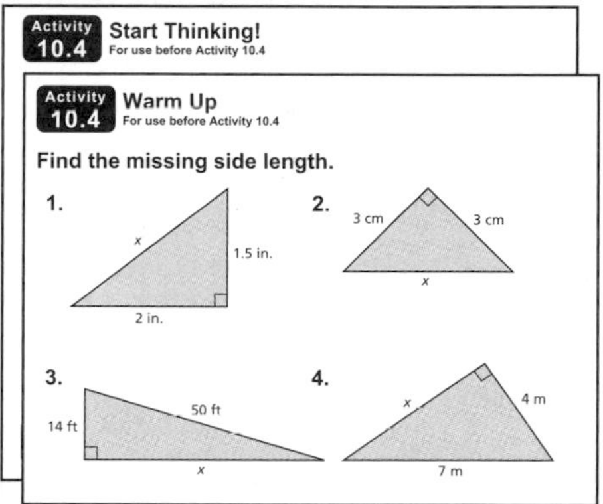

Activity 10.4 Start Thinking!
For use before Activity 10.4

Activity 10.4 Warm Up
For use before Activity 10.4

Find the missing side length.

10.4 Record and Practice Journal

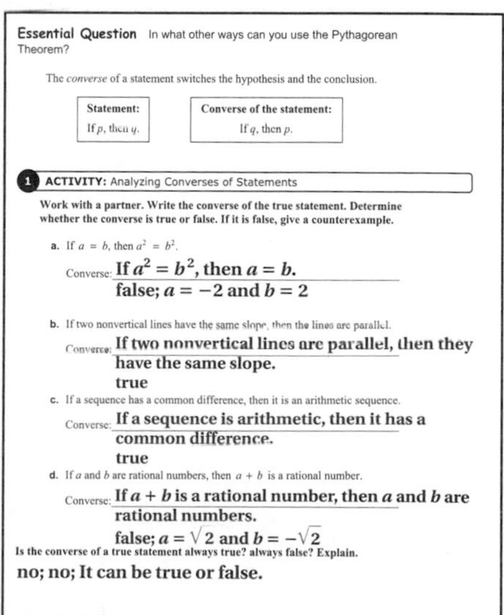

Essential Question In what other ways can you use the Pythagorean Theorem?

The *converse* of a statement switches the hypothesis and the conclusion.

Statement:	Converse of the statement:
If p, then q.	If q, then p.

1 **ACTIVITY:** Analyzing Converses of Statements

Work with a partner. Write the converse of the true statement. Determine whether the converse is true or false. If it is false, give a counterexample.

a. If $a = b$, then $a^2 = b^2$.
Converse: **If $a^2 = b^2$, then $a = b$.**
false; $a = -2$ and $b = 2$

b. If two nonvertical lines have the same slope, then the lines are parallel.
Converse: **If two nonvertical lines are parallel, then they have the same slope.**
true

c. If a sequence has a common difference, then it is an arithmetic sequence.
Converse: **If a sequence is arithmetic, then it has a common difference.**
true

d. If a and b are rational numbers, then $a + b$ is a rational number.
Converse: **If $a + b$ is a rational number, then a and b are rational numbers.**
false; $a = \sqrt{2}$ and $b = -\sqrt{2}$

Is the converse of a true statement always true? always false? Explain.
no; no; It can be true or false.

English Language Learners

Comprehension

English language learners may struggle with the concept of the converse of a statement. Some may think that the converse is always true. Give an example where the converse of a statement is not true.

Statement: *If a figure is a square, then it has four right angles.*

Converse of the Statement: *If a figure has four right angles, then the figure is a square.*

The converse of the statement is not always true. The figure could be a rectangle.

10.4 Record and Practice Journal

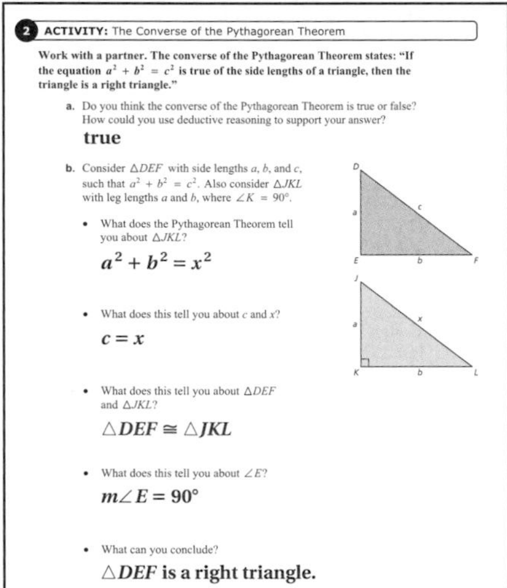

2 ACTIVITY: The Converse of the Pythagorean Theorem

Work with a partner. The converse of the Pythagorean Theorem states: "If the equation $a^2 + b^2 = c^2$ is true of the side lengths of a triangle, then the triangle is a right triangle."

a. Do you think the converse of the Pythagorean Theorem is true or false? How could you use deductive reasoning to support your answer?
true

b. Consider $\triangle DEF$ with side lengths a, b, and c, such that $a^2 + b^2 = c^2$. Also consider $\triangle JKL$ with leg lengths a and b, where $\angle K = 90°$.

- What does the Pythagorean Theorem tell you about $\triangle JKL$?
  $a^2 + b^2 = x^2$
- What does this tell you about c and x?
  $c = x$
- What does this tell you about $\triangle DEF$ and $\triangle JKL$?
  $\triangle DEF \cong \triangle JKL$
- What does this tell you about $\angle E$?
  $m\angle E = 90°$
- What can you conclude?
  $\triangle DEF$ is a right triangle.

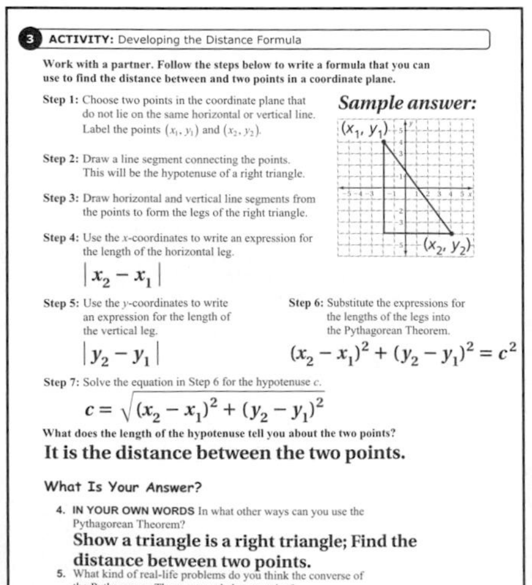

3 ACTIVITY: Developing the Distance Formula

Work with a partner. Follow the steps below to write a formula that you can use to find the distance between and two points in a coordinate plane.

Step 1: Choose two points in the coordinate plane that do not lie on the same horizontal or vertical line. Label the points (x_1, y_1) and (x_2, y_2).

Sample answer:

Step 2: Draw a line segment connecting the points. This will be the hypotenuse of a right triangle.

Step 3: Draw horizontal and vertical line segments from the points to form the legs of the right triangle.

Step 4: Use the x-coordinates to write an expression for the length of the horizontal leg.
$|x_2 - x_1|$

Step 5: Use the y-coordinates to write an expression for the length of the vertical leg.
$|y_2 - y_1|$

Step 6: Substitute the expressions for the lengths of the legs into the Pythagorean Theorem.
$(x_2 - x_1)^2 + (y_2 - y_1)^2 = c^2$

Step 7: Solve the equation in Step 6 for the hypotenuse c.
$c = \sqrt{(x_2 - x_1)^2 + (y_2 - y_1)^2}$

What does the length of the hypotenuse tell you about the two points?
It is the distance between the two points.

What Is Your Answer?

4. IN YOUR OWN WORDS In what other ways can you use the Pythagorean Theorem?
Show a triangle is a right triangle; Find the distance between two points.
5. What kind of real-life problems do you think the converse of the Pythagorean Theorem can help you solve?
Check students' work.

Laurie's Notes

Activity 3

- Visually, it will be helpful for students to select lattice points.
- In Step 3, point out that there are two distinct ways they can draw the legs. They can draw them either way because the two possible triangles are congruent.
- **MP7 Look for and Make Use of Structure:** In Steps 4 and 5, students may ask about the order in which the subtraction is performed. Tell them that in Step 6 these expressions are squared. So, the order in which the subtraction is performed does not matter.
- Give students adequate time to read carefully and work through the steps on their own. Resist the temptation to jump in and solve it for them.

What Is Your Answer?

- **Think-Pair-Share:** Students should read each question independently and then work with a partner to answer the questions. When they have answered the questions, the pair should compare their answers with another group and discuss any discrepancies.

Closure

- **Writing Prompt:** The Pythagorean Theorem can be used to . . . find missing side lengths of right triangles, find the distances between points in a coordinate plane, determine whether given side lengths form a right triangle, etc.

The Dynamic Planning Tool
Editable Teacher's Resources at *BigIdeasMath.com*

3 ACTIVITY: Developing the Distance Formula

Work with a partner. Follow the steps below to write a formula that you can use to find the distance between any two points in a coordinate plane.

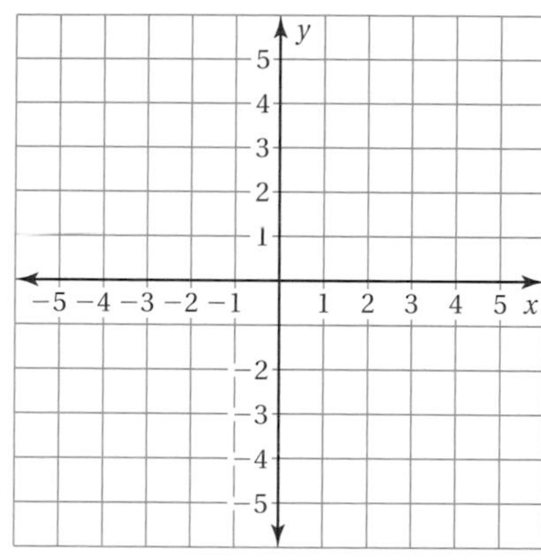

Step 1: Choose two points in the coordinate plane that do not lie on the same horizontal or vertical line. Label the points (x_1, y_1) and (x_2, y_2).

Step 2: Draw a line segment connecting the points. This will be the hypotenuse of a right triangle.

Step 3: Draw horizontal and vertical line segments from the points to form the legs of the right triangle.

Step 4: Use the x-coordinates to write an expression for the length of the horizontal leg.

Step 5: Use the y-coordinates to write an expression for the length of the vertical leg.

Step 6: Substitute the expressions for the lengths of the legs into the Pythagorean Theorem.

Step 7: Solve the equation in Step 6 for the hypotenuse c.

What does the length of the hypotenuse tell you about the two points?

What Is Your Answer?

4. **IN YOUR OWN WORDS** In what other ways can you use the Pythagorean Theorem?

5. What kind of real-life problems do you think the converse of the Pythagorean Theorem can help you solve?

Practice Use what you learned about the converse of a true statement to complete Exercises 3 and 4 on page 530.

10.4 Lesson

Key Vocabulary
distance formula, *p. 528*

Converse of the Pythagorean Theorem

If the equation $a^2 + b^2 = c^2$ is true for the side lengths of a triangle, then the triangle is a right triangle.

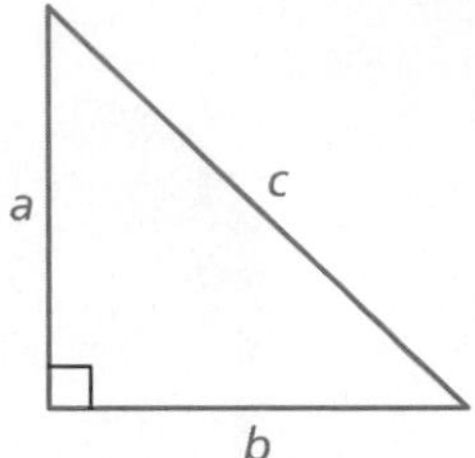

EXAMPLE 1 Identifying a Right Triangle

Study Tip
A *Pythagorean triple* is a set of three positive integers a, b, and c, where $a^2 + b^2 = c^2$.

Common Error
When using the converse of the Pythagorean Theorem, always substitute the length of the longest side for c.

Tell whether each triangle is a right triangle.

a.

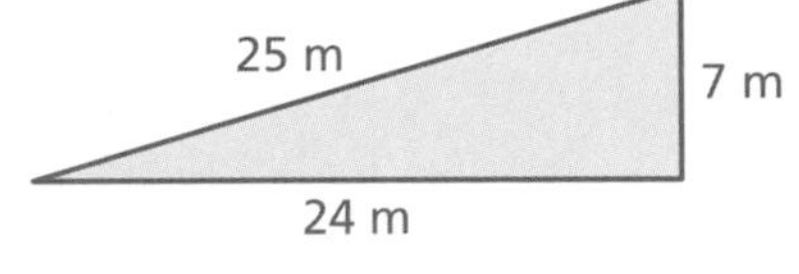

$$a^2 + b^2 = c^2$$
$$7^2 + 24^2 \stackrel{?}{=} 25^2$$
$$49 + 576 \stackrel{?}{=} 625$$
$$625 = 625 \checkmark$$

∴ It *is* a right triangle.

b.

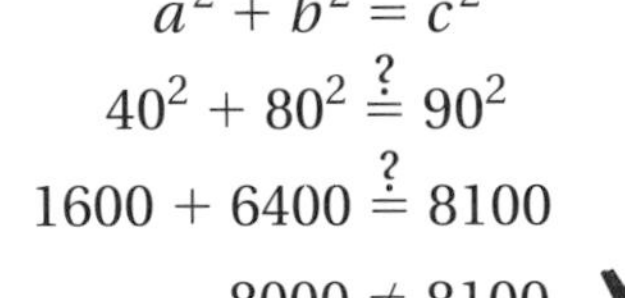

$$a^2 + b^2 = c^2$$
$$40^2 + 80^2 \stackrel{?}{=} 90^2$$
$$1600 + 6400 \stackrel{?}{=} 8100$$
$$8000 \neq 8100 \times$$

∴ It *is not* a right triangle.

On Your Own

Now You're Ready
Exercises 5–10

Tell whether the triangle with the given side lengths is a right triangle.

1. 15 cm, 10 cm, 18 cm

2. 50 yd, 40 yd, 30 yd

On page 527, you used the Pythagorean Theorem to develop the *distance formula.* You can use the **distance formula** to find the distance between any two points in a coordinate plane.

Distance Formula

The distance d between any two points (x_1, y_1) and (x_2, y_2) is given by the formula

$$d = \sqrt{(x_2 - x_1)^2 + (y_2 - y_1)^2}.$$

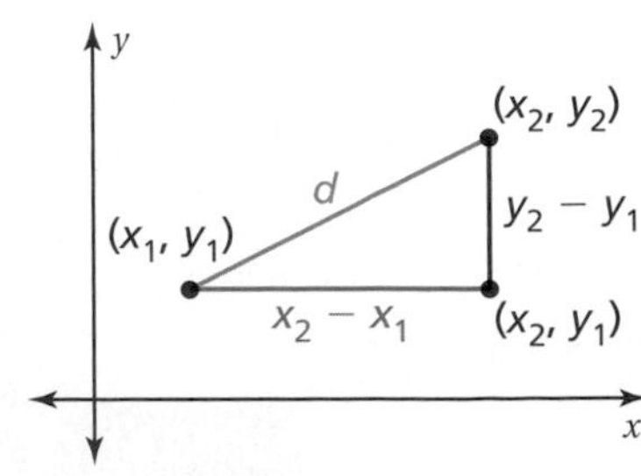

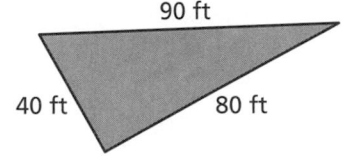

Laurie's Notes

Introduction

Connect

- **Yesterday:** Students proved the converse of the Pythagorean Theorem and developed the distance formula. (MP3, MP7)
- **Today:** Students will use the converse of the Pythagorean Theorem and the distance formula.

Motivate

- Write the following numbers on the board: 3-4-5, 5-12-13, and 8-15-17. Tell students that these are called *Pythagorean triples.* They are positive integers that satisfy the Pythagorean Theorem.
- Multiples of these examples also satisfy the Pythagorean Theorem. Have students try to name a few more Pythagorean triples.

Lesson Notes

Key Idea

- Write the Key Idea stating the converse of the Pythagorean Theorem. It is one way of determining whether given side lengths form a right triangle.

Example 1

- Draw the first triangle.
- Say, "If this is a right triangle, then the side lengths must satisfy the Pythagorean Theorem."
- "Satisfying the theorem" simply means that the sum of the squares of the two lesser numbers must equal the square of the greatest number.
- Write $a^2 + b^2 = c^2$. Substitute for each value and simplify.
- **Common Error:** Students may forget to substitute the longest side for c.
- Work through each example.
- **Extension:** If it is not a right triangle, it is possible to determine whether the triangle is acute or obtuse. Interested students can investigate this using dynamic software or by traditional research.

On Your Own

- The side lengths of the triangles are not listed from least to greatest. Make sure that students substitute correctly.

Key Idea

- Write the Key Idea that states the distance formula, derived from the Pythagorean Theorem.
- Note the use of colors in the diagram. The same colors can be used when writing the formula.
- **MP7 Look for and Make Use of Structure:** Discuss with students that the order in which the subtraction is performed is not important because the difference is squared.

Goal Today's lesson is using the converse of the Pythagorean Theorem and the **distance formula**.

Start Thinking! and Warm Up

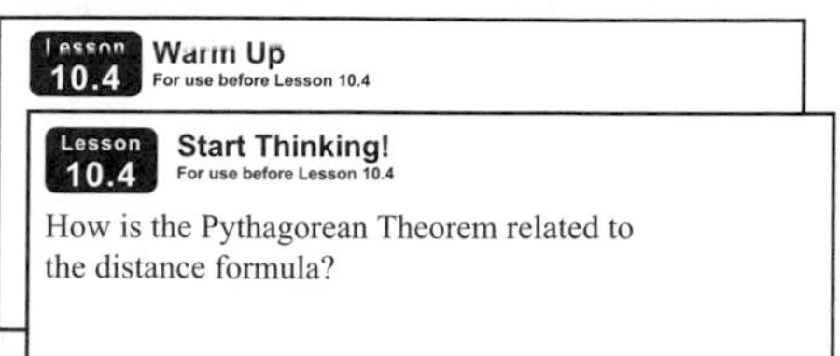

Extra Example 1

Tell whether each triangle is a right triangle.

a.

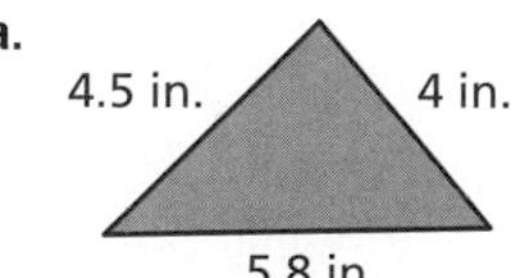

not a right triangle

b.

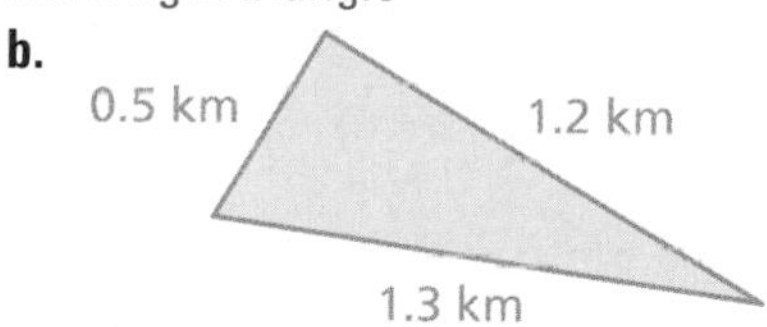

right triangle

On Your Own

1. no
2. yes

Laurie's Notes

Example 2

- Although it is not necessary, you can plot the two points as a visual aid.
- The choice of which point is (x_1, y_1) and which point is (x_2, y_2) is arbitrary. If time permits, do the problem both ways to show that the result is the same.
- Caution students to be careful with the subtraction. It is easy to make a careless calculation mistake.
- There are no units associated with the answer. In a real-life example, students would need to label the units in their answer.

Example 3

- Ask a student to read the example.
- **MP4 Modeling with Mathematics:** Sketch the coordinate plane shown with the ordered pairs identified.
- ? "How do you determine if the receiver ran the play as designed?" Check whether the triangle is a right triangle.
- ? "How do you determine if the triangle is a right triangle?" Use the converse of the Pythagorean Theorem.
- Use the distance formula to find the length of each side of the triangle.
- **MP7:** The distances found could be simplified. However, you will be using the converse and squaring again, so it is easier to leave the results as shown in the example.
- Use the converse of the Pythagorean Theorem to show that it is not a right triangle.
- Note that students could also use what they know about slopes of perpendicular lines to solve this problem.

On Your Own

- There are multiple steps required in each question. You may wish to divide the class into groups and assign a different question to each group.

Closure

- **Exit Ticket:** Find the distance between $(-4, 6)$ and $(3, -2)$. $\sqrt{113}$

Technology For the Teacher

The Dynamic Planning Tool
Editable Teacher's Resources at *BigIdeasMath.com*

Extra Example 2

Find the distance between $(1, -5)$ and $(7, 4)$. $3\sqrt{13}$

Extra Example 3

In Example 3, the receiver starts at $(40, 40)$. Did the receiver run the play as designed? yes

On Your Own

3. $\sqrt{29}$
4. $\sqrt{146}$
5. $8\sqrt{5}$
6. yes; The points $(60, 50)$, $(30, 20)$, and $(80, -30)$ form a right triangle.

Differentiated Instruction

Inclusion

Encourage students to learn the Pythagorean Theorem using the language of the triangle, $\text{leg}^2 + \text{leg}^2 = \text{hypotenuse}^2$. Have them label each side as a leg or hypotenuse before substituting the numbers into the equation.

EXAMPLE 2 Finding the Distance Between Two Points

Find the distance between (−3, 5) and (2, −1).

Let $(x_1, y_1) = (-3, 5)$ and $(x_2, y_2) = (2, -1)$.

$d = \sqrt{(x_2 \quad x_1)^2 + (y_2 - y_1)^2}$	Write the distance formula.
$= \sqrt{[2 - (-3)]^2 + (-1 - 5)^2}$	Substitute.
$= \sqrt{5^2 + (-6)^2}$	Simplify.
$= \sqrt{25 + 36}$	Evaluate powers.
$= \sqrt{61}$	Add.

EXAMPLE 3 Real-Life Application

A football coach designs a passing play in which a receiver runs down the field, makes a 90° turn, and runs to the corner of the end zone. A receiver runs the play as shown. Did the receiver run the play as designed? Each unit of the grid represents 10 feet.

Use the distance formula to find the lengths of the three sides.

$$d_1 = \sqrt{(60 - 20)^2 + (50 - 10)^2} = \sqrt{40^2 + 40^2} = \sqrt{3200} \text{ feet}$$

$$d_2 = \sqrt{(80 - 20)^2 + (-30 - 10)^2} = \sqrt{60^2 + (-40)^2} = \sqrt{5200} \text{ feet}$$

$$d_3 = \sqrt{(80 - 60)^2 + (-30 - 50)^2} = \sqrt{20^2 + (-80)^2} = \sqrt{6800} \text{ feet}$$

Use the converse of the Pythagorean Theorem to determine if the side lengths form a right triangle.

$$(\sqrt{3200})^2 + (\sqrt{5200})^2 \stackrel{?}{=} (\sqrt{6800})^2$$

$$3200 + 5200 \stackrel{?}{=} 6800$$

$$8400 \neq 6800 \quad ✗$$

It is not a right triangle. So, the receiver did not make a 90° turn.

∴ The receiver did not run the play as designed.

On Your Own

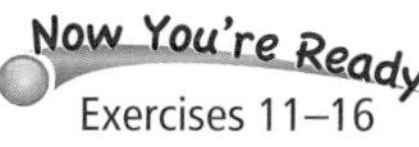
Now You're Ready
Exercises 11–16

Find the distance between the two points.

3. (0, 4), (5, 2)
4. (−1, 3), (4, −8)
5. (−10, −6), (6, 2)
6. **WHAT IF?** In Example 3, the receiver made the turn at (30, 20). Did the receiver run the play as designed? Explain.

10.4 Exercises

Vocabulary and Concept Check

1. **WRITING** Describe two ways to find the distance between two points in a coordinate plane.
2. **WHICH ONE DOESN'T BELONG?** Which set of numbers does not belong with the other three? Explain your reasoning.

3, 4, 5	8, 15, 17	18, 22, 29	9, 40, 41

Practice and Problem Solving

Write the converse of the true statement. Determine whether the converse is true or false.

3. If a is an even number, then a^2 is even.
4. If a is positive, then $|a| = a$.

Tell whether the triangle with the given side lengths is a right triangle.

1 5.

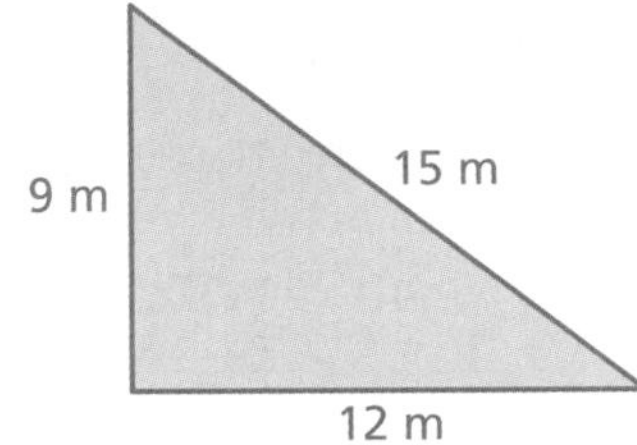

6.

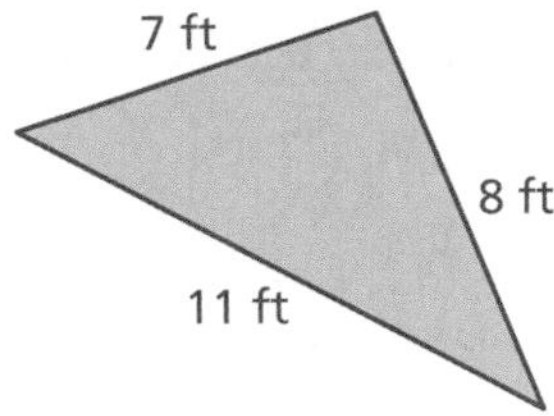

7.

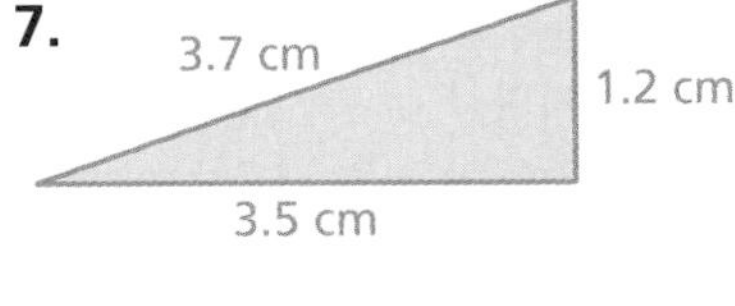

8. 16 in., 18 in., 24 in.
9. 30 yd, 22 yd, 15 yd
10. 8 mm, 15 mm, 17 mm

Find the distance between the two points.

2 11. $(4, -3), (-2, -5)$
12. $(1, 1), (4, 5)$
13. $(-7, -1), (-4, 8)$
14. $(1, 3), (7, 7)$
15. $(2, -8), (4, -1)$
16. $(-7, 4), (-3, 2)$

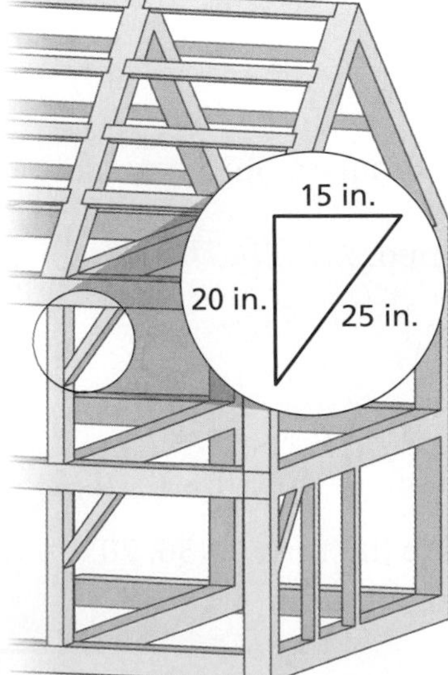

17. **ERROR ANALYSIS** Describe and correct the error in finding the distance between the points $(-4, -3)$ and $(2, -1)$.

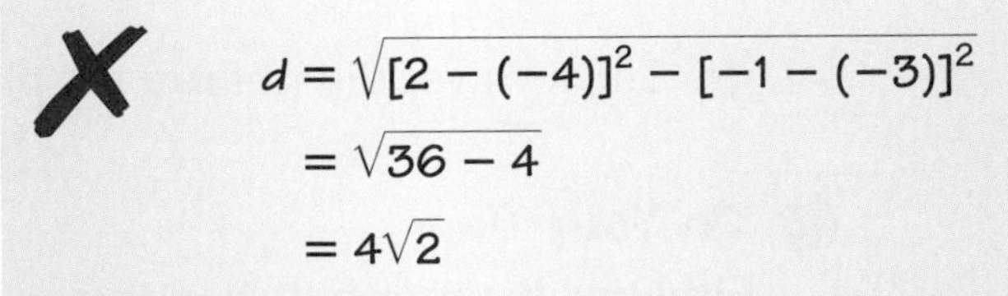

18. **CONSTRUCTION** A post and beam frame for a shed is shown in the diagram. Does the brace form a right triangle with the post and beam? Explain.

Assignment Guide and Homework Check

Level	Assignment	Homework Check
Average	1–4, 5–25 odd, 18, 27–30	9, 13, 17, 23, 25
Advanced	1–4, 6–20 even, 22–26, 27–30	8, 16, 23, 25, 26

Common Errors

- **Exercises 5–10, 18** Students may substitute the wrong value for *c* in the Pythagorean Theorem. Remind them that *c* will be the longest side, so they should substitute the greatest value for *c*.
- **Exercises 11–16** Students may mismatch the *x*-values and *y*-values when using the distance formula. This will result in students subtracting an *x* from a *y*, or vice versa. Encourage students to pair the numbers properly.
- **Exercises 11–16** Students may get careless when squaring negative numbers. Remind them that everything inside the parentheses is squared, including the minus sign, and that the square of a negative is a positive.
- **Exercise 25** Students may pick Plane A because it appears to be closer. Remind students that the drawing is not to scale. Tell them to calculate the distances before answering the question.

10.4 Record and Practice Journal

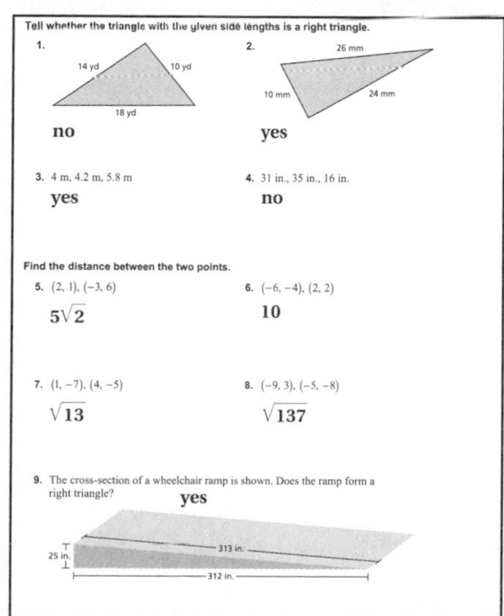
Tell whether the triangle with the given side lengths is a right triangle.

1. no

2. yes

3. 4 m, 4.2 m, 5.8 m
yes

4. 31 in., 35 in., 16 in.
no

Find the distance between the two points.

5. (2, 1), (−3, 6)
$5\sqrt{2}$

6. (−6, −4), (2, 2)
10

7. (1, −7), (4, −5)
$\sqrt{13}$

8. (−9, 3), (−5, −8)
$\sqrt{137}$

9. The cross-section of a wheelchair ramp is shown. Does the ramp form a right triangle?
yes

Technology For the Teacher
Answer Presentation Tool

Vocabulary and Concept Check

1. the Pythagorean Theorem and the distance formula
2. 18, 22, 29; This set does not form a right triangle.

Practice and Problem Solving

3. If a^2 is even, then a is an even number; true
4. If $|a| = a$, then a is positive; false (counterexample: $a = 0$)
5. yes
6. no
7. yes
8. no
9. no
10. yes
11. $2\sqrt{10}$
12. 5
13. $3\sqrt{10}$
14. $2\sqrt{13}$
15. $\sqrt{53}$
16. $2\sqrt{5}$
17. The squared quantities under the radical should be added, not subtracted; $2\sqrt{10}$
18. yes; The side lengths satisfy the converse of the Pythagorean Theorem.

Practice and Problem Solving

19. yes **20.** no

21. yes

22. yes; Use the distance formula to find the lengths of the three sides. Use the converse of the Pythagorean Theorem to show they form a right triangle.

23. See *Taking Math Deeper.*

24. yes; $\sqrt{41}$; Because you square the differences $(x_2 - x_1)$ and $(y_2 - y_1)$, it does not matter if the differences are positive or negative. The squares of opposite numbers are equivalent.

25. Plane B; Plane A is about 8.35 kilometers away and Plane B is about 8.27 kilometers away.

26. See Additional Answers.

Fair Game Review

27. $x = \frac{3}{2}, x = 1$

28. $x = -\frac{1}{2}, x = 1$

29. no real solutions

30. A

Mini-Assessment

Tell whether the triangle with the given side lengths is a right triangle.

1. 32 m, 56 m, 64 m no

2. 1.8 mi, 8 mi, 8.2 mi yes

Find the distance between the two points.

3. $(-3, -1), (6, 2)$ $3\sqrt{10}$

4. $(2, 10), (5, -4)$ $\sqrt{205}$

Taking Math Deeper

Exercise 23

You can use what you learned about slopes of perpendicular lines to solve this problem.

1 Interpret the diagram.

From the diagram, you can see that your path forms a right triangle when the angle at Container 1 is a right angle.

2 Find the slopes of the line segments that meet at Container 1.

Car to Container 1

(10, 50) to (20, −10)

$$m = \frac{y_2 - y_1}{x_2 - x_1} = \frac{-10 - 50}{20 - 10} = \frac{-60}{10} = -6$$

Container 1 to Container 2

(20, −10) to (80, 0)

$$m = \frac{y_2 - y_1}{x_2 - x_1} = \frac{0 - (-10)}{80 - 20} = \frac{10}{60} = \frac{1}{6}$$

perpendicular lines

3 Interpret the slopes.

Because $-6 \cdot \frac{1}{6} = -1$, the line segments are perpendicular. So, the angle at Container 1 is a right angle, meaning that the path formed is a right triangle.

Project

Research global positioning systems. How do they work?

Reteaching and Enrichment Strategies

If students need help...	If students got it...
Resources by Chapter • Practice A and Practice B • Puzzle Time Record and Practice Journal Practice Differentiating the Lesson Lesson Tutorials Skills Review Handbook	Resources by Chapter • Enrichment and Extension • School-to-Work Start the next section

Tell whether the set of measurements can be the side lengths of a right triangle.

19. $5\sqrt{5}$, 10, 15 **20.** 7, $3\sqrt{10}$, 6 **21.** 21, 72, 75

22. REASONING Plot the points $(-1, -2)$, $(2, 1)$, and $(-3, 6)$ in a coordinate plane. Are the points the vertices of a right triangle? Explain.

23. GEOCACHING You spend the day looking for hidden containers in a wooded area using a global positioning system (GPS). You park your car on the side of the road and then locate Container 1 and Container 2 before going back to the car. Does your path form a right triangle? Explain. Each unit of the grid represents 10 yards.

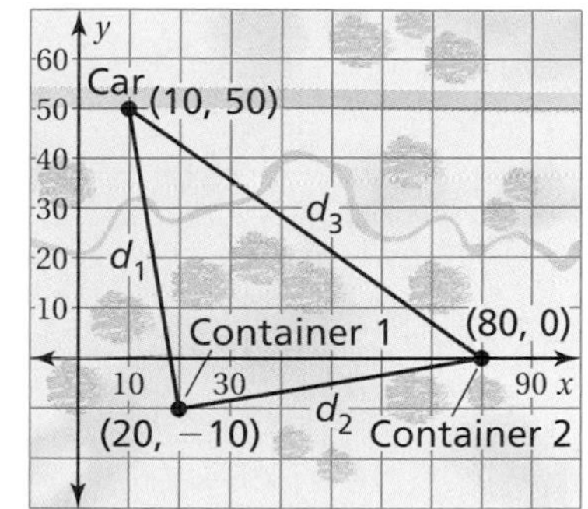

24. REASONING Your teacher wants the class to find the distance between the two points (3, 2) and (8, 6). You choose (3, 2) for (x_1, y_1) and your friend chooses (8, 6) for (x_1, y_1). Do you and your friend obtain the same answer? Justify your answer.

25. AIRPORT Which plane is closer to the base of the airport tower? Explain.

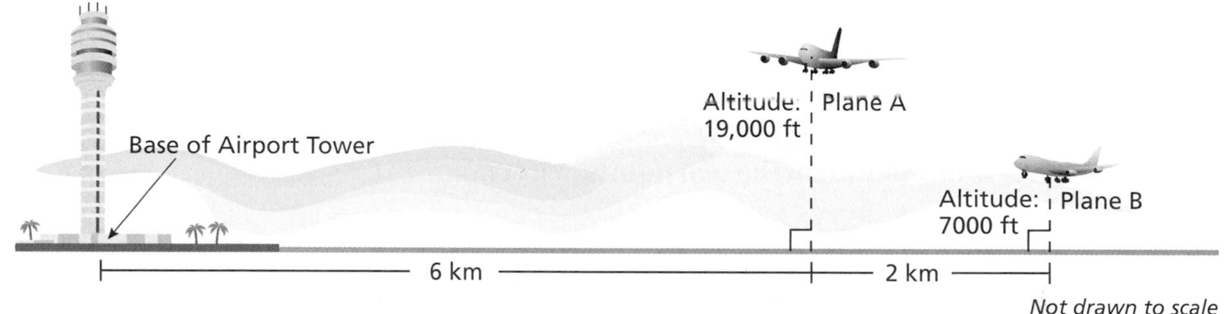

26. Structure Consider the two points (x_1, y_1) and (x_2, y_2) in the coordinate plane. How can you find the point (x_m, y_m) located in the middle of the two given points? Justify your answer using the distance formula.

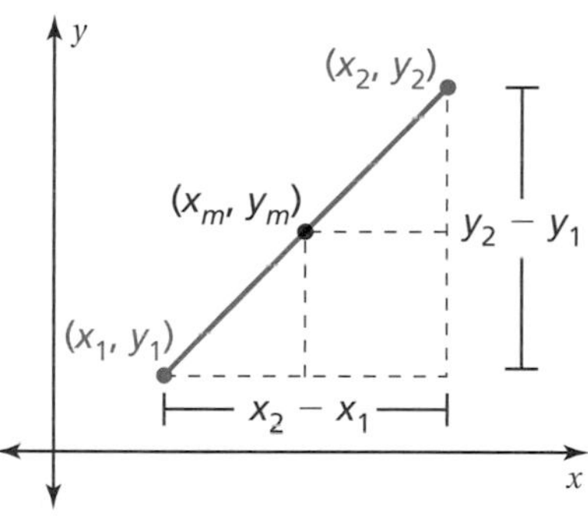

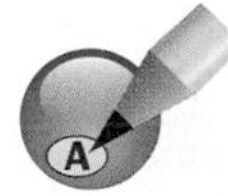

Fair Game Review What you learned in previous grades & lessons

Solve the equation using the quadratic formula. *(Section 9.4)*

27. $2x^2 - 5x + 3 = 0$ **28.** $2x^2 - x - 1 = 0$ **29.** $x^2 + 3x + 5 = 0$

30. MULTIPLE CHOICE Which point is the focus of the graph of $y = 2x^2$? *(Section 8.2)*

Ⓐ $\left(0, \frac{1}{8}\right)$ Ⓑ $\left(0, \frac{1}{4}\right)$ Ⓒ $\left(0, \frac{1}{2}\right)$ Ⓓ $\left(0, -\frac{1}{8}\right)$

10.3 – 10.4 Quiz

Find the missing length of the triangle. *(Section 10.3)*

1.

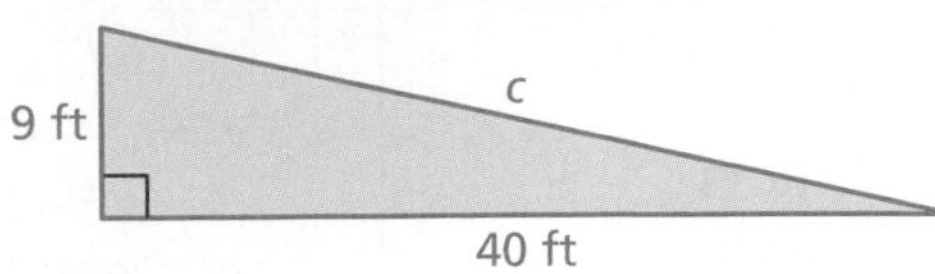

2.

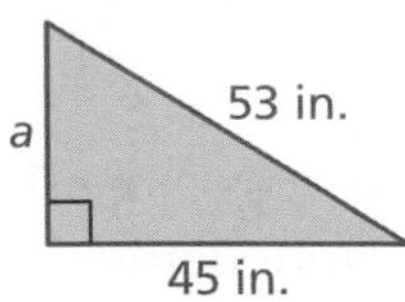

3.

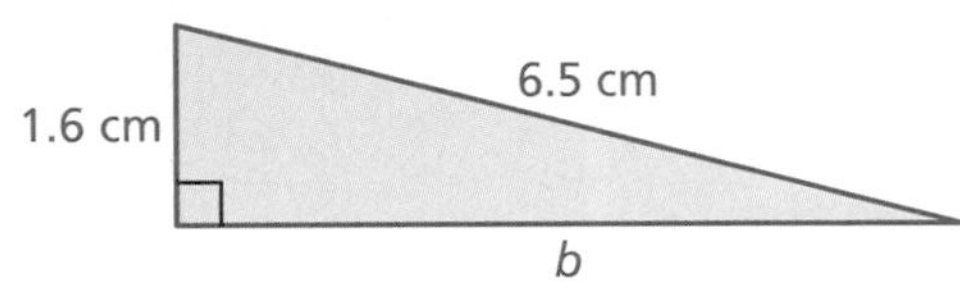

4.

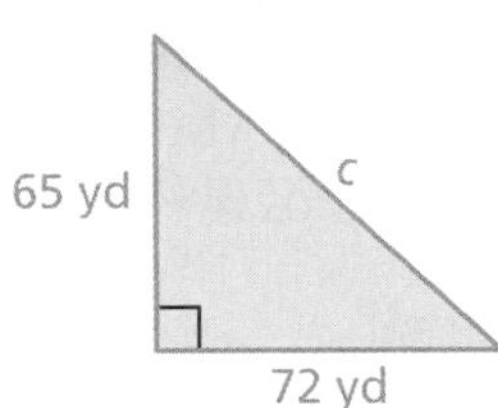

Tell whether the triangle with the given side lengths is a right triangle. *(Section 10.4)*

5.

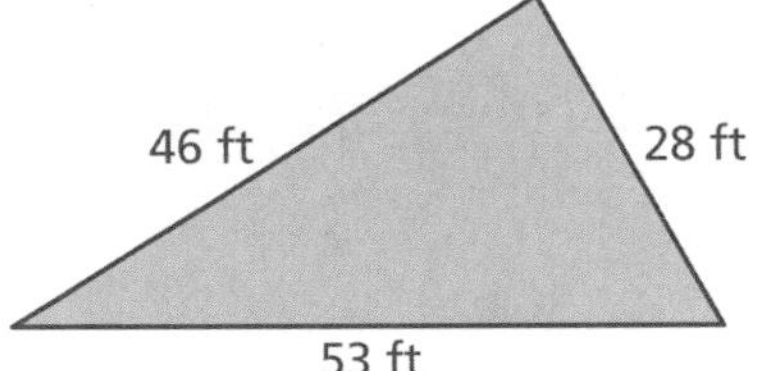

6.

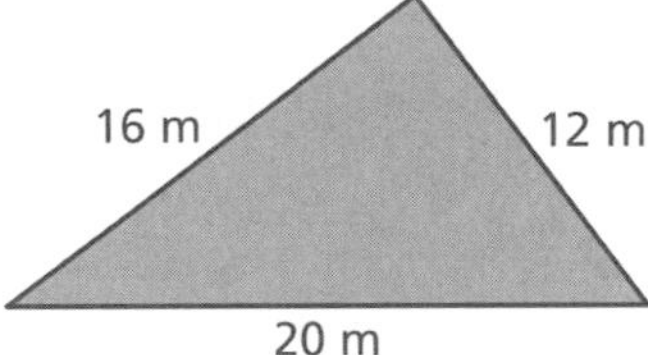

Find the distance between the two points. *(Section 10.4)*

7. $(-3, -1), (-1, -5)$

8. $(-4, 2), (5, 1)$

9. $(1, -2), (4, -5)$

10. $(-1, 1), (7, 4)$

11. $(-6, 5), (-4, -6)$

12. $(-1, 4), (1, 3)$

13. **FABRIC** You cut a rectangular piece of fabric in half along the diagonal. The fabric measures 28 inches wide and $1\frac{1}{4}$ yards long. What is the length (in inches) of the diagonal? *(Section 10.3)*

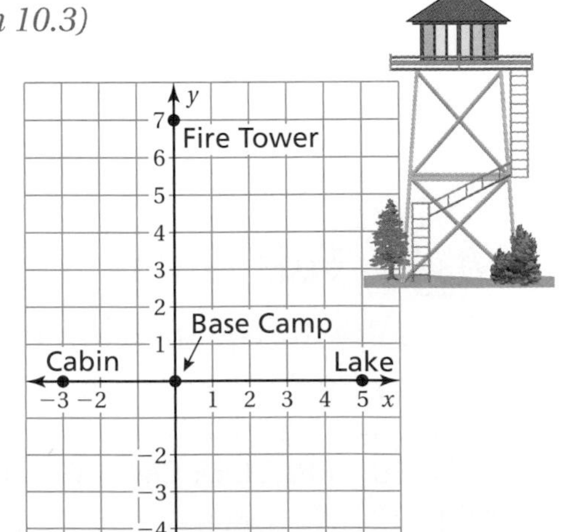

Use the figure to answer Exercises 14–17. *(Section 10.4)*

14. How far is the cabin from the peak?

15. How far is the fire tower from the lake?

16. How far is the lake from the peak?

17. You are standing at $(-5, -6)$. How far are you from the lake?

Alternative Assessment Options

Math Chat	Student Reflective Focus Question
Structured Interview	Writing Prompt

Math Chat

- Have students work in pairs. Assign Quiz Exercises 14–17 to each pair. Each student works through all four problems. After the students have worked through the problems, they take turns talking through the processes that they used to get each answer. Students analyze and evaluate the mathematical thinking and strategies used.
- The teacher should walk around the classroom listening to the pairs and ask questions to ensure understanding.

Study Help Sample Answers

Remind students to complete Graphic Organizers for the rest of the chapter.

4.

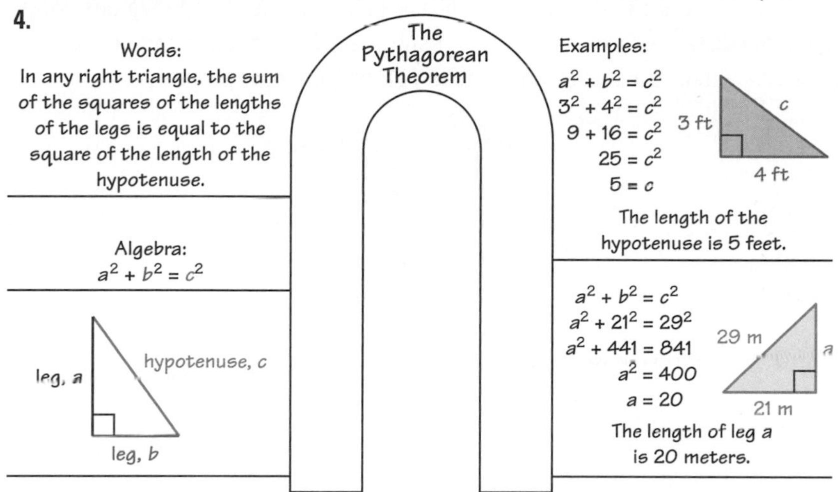

5–6. Available at *BigIdeasMath.com.*

Reteaching and Enrichment Strategies

If students need help...	If students got it...
Resources by Chapter • Study Help • Practice A and Practice B • Puzzle Time Lesson Tutorials *BigIdeasMath.com* Practice Quiz Practice from the Test Generator	Resources by Chapter • Enrichment and Extension • School-to-Work Game Closet at *BigIdeasMath.com* Start the Chapter Review

Answers

1. 41 ft	**2.** 28 in.
3. 6.3 cm	**4.** 97 yd
5. no	**6.** yes
7. $2\sqrt{5}$	**8.** $\sqrt{82}$
9. $3\sqrt{2}$	**10.** $\sqrt{73}$
11. $5\sqrt{5}$	**12.** $\sqrt{5}$
13. 53 in.	**14.** $\sqrt{34}$ km
15. $\sqrt{74}$ km	**16.** $5\sqrt{2}$ km
17. $2\sqrt{34}$ km	

Technology For the Teacher

Answer Presentation Tool

Assessment Book

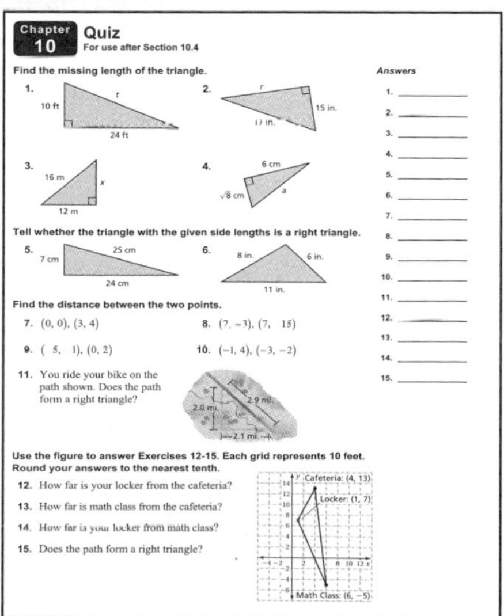

Chapter 10 **Quiz**
For use after Section 10.4

Find the missing length of the triangle.

1. 10 ft, 24 ft, t
2. r, 15 in., 17 in.
3. 16 m, x, 12 m
4. 6 cm, $\sqrt{8}$ cm, a

Tell whether the triangle with the given side lengths is a right triangle.

5. 7 cm, 25 cm, 24 cm
6. 8 in., 6 in., 11 in.

Find the distance between the two points.

7. (0, 0), (3, 4)
8. (?, −3), (7, 15)
9. (5, 1), (0, 2)
10. (−1, 4), (−3, −2)
11. You ride your bike on the path shown. Does the path form a right triangle? (2.0 mi, 2.9 mi, 2.1 mi)

Use the figure to answer Exercises 12-15. Each grid represents 10 feet. Round your answers to the nearest tenth.

12. How far is your locker from the cafeteria?
13. How far is math class from the cafeteria?
14. How far is your locker from math class?
15. Does the path form a right triangle?

Cafeteria: (4, 13)
Locker: (1, 7)
Math Class: (6, −5)

Answers
1. ____ 2. ____ 3. ____ 4. ____ 5. ____ 6. ____ 7. ____ 8. ____ 9. ____ 10. ____ 11. ____ 12. ____ 13. ____ 14. ____ 15. ____

For the Teacher
Additional Review Options

- Big Ideas Test Generator
- Game Closet at *BigIdeasMath.com*
- Vocabulary Puzzle Builder
- Resources by Chapter
    Puzzle Time
    Study Help

Answers

1. 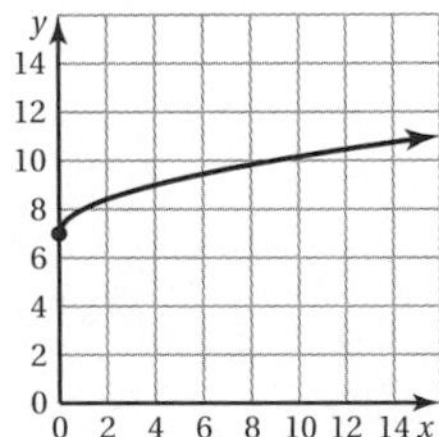

domain: $x \geq 0$; range: $y \geq 7$; The graph of $y = \sqrt{x} + 7$ is a translation 7 units up of the graph of $y = \sqrt{x}$.

2. 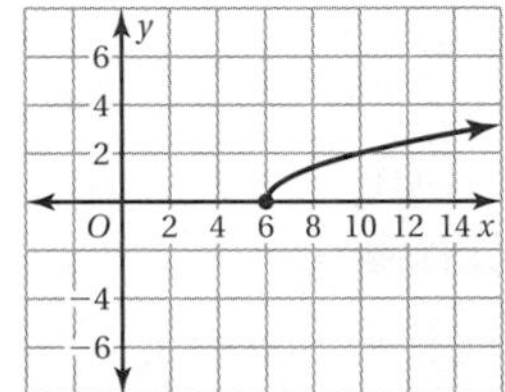

domain: $x \geq 6$; range: $y \geq 0$; The graph of $y = \sqrt{x - 6}$ is a translation 6 units to the right of the graph of $y = \sqrt{x}$.

3. 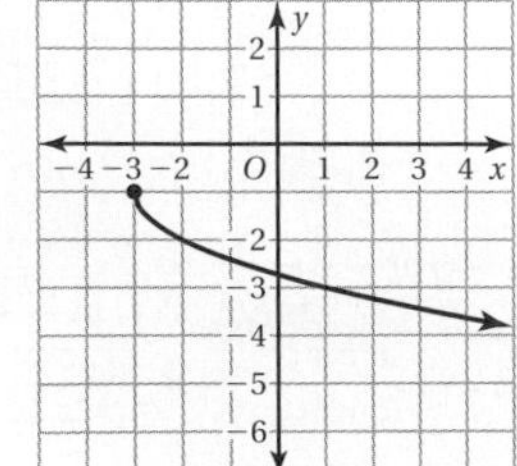

domain: $x \geq -3$; range: $y \leq -1$; The graph of $y = -\sqrt{x + 3} - 1$ is a reflection in the x-axis of the graph of $y = \sqrt{x}$ and then a translation 3 units to the left and 1 unit down.

Review of Common Errors

- **Exercises 1–3** Students may treat constants added to a radical expression as part of the radicand, or vice versa. Encourage them to identify the radicand before solving the exercise.
- **Exercises 4 and 5** Students may square each side of the equation before isolating the square root on one side of the equation.
- **Exercises 8 and 9** Students may substitute the given lengths in the wrong part of the formula. For example, if they are finding one of the legs, they may write $5^2 + 13^2 = c^2$ instead of $5^2 + b^2 = 13^2$. Remind them that the side opposite the right angle is the hypotenuse c.
- **Exercises 8 and 9** Students may multiply each side length by two instead of squaring the side length. Remind them of the definition of exponents.
- **Exercises 10 and 11** Students may substitute the wrong value for c in the Pythagorean Theorem. Remind them that c will be the longest side, so they should substitute the greatest value for c.
- **Exercises 12 and 13** Students may mismatch the x-values and y-values when using the distance formula. This will result in students subtracting an x from a y, or vice versa. Encourage students to pair the numbers properly.

Review Key Vocabulary

square root function, *p. 504*
simplest form of a radical expression, *p. 508*
rationalizing the denominator, *p. 508*
conjugates, *p. 509*
square root equation, *p. 512*
extraneous solution, *p. 513*
theorem, *p. 520*
legs, *p. 522*
hypotenuse, *p. 522*
Pythagorean Theorem, *p. 522*
distance formula, *p. 528*

Review Examples and Exercises

10.1 Graphing Square Root Functions *(pp. 502–509)*

a. Graph $y = \sqrt{x} - 1$. Describe the domain and range. Compare the graph to the graph of $y = \sqrt{x}$.

Step 1: Make a table of values.

x	0	1	4	9	16
y	−1	0	1	2	3

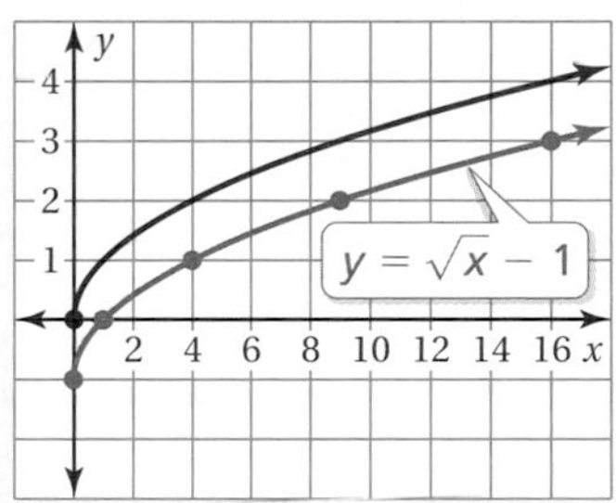

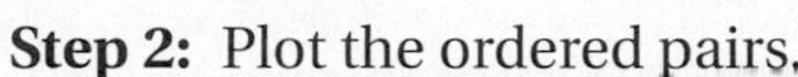

Step 2: Plot the ordered pairs.
Step 3: Draw a smooth curve through the points.

The domain is $x \geq 0$. The range is $y \geq -1$. The graph of $y = \sqrt{x} - 1$ is a translation 1 unit down of the graph of $y = \sqrt{x}$.

b. Graph $y = \sqrt{x + 2}$. Describe the domain and range. Compare the graph to the graph of $y = \sqrt{x}$.

Step 1: Make a table of values.

x	−2	−1	0	1	2
y	0	1	1.4	1.7	2

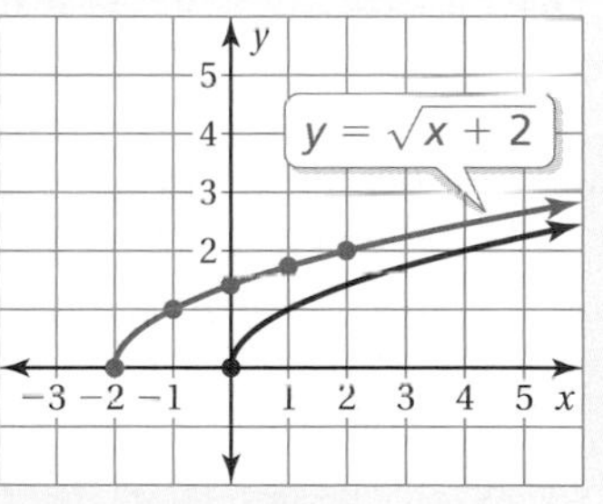

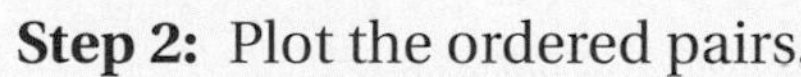

Step 2: Plot the ordered pairs.
Step 3: Draw a smooth curve through the points.

The domain is $x \geq -2$. The range is $y \geq 0$. The graph of $y = \sqrt{x + 2}$ is a translation 2 units to the left of the graph of $y = \sqrt{x}$.

Exercises

Graph the function. Describe the domain and range. Compare the graph to the graph of $y = \sqrt{x}$.

1. $y = \sqrt{x} + 7$ **2.** $y = \sqrt{x - 6}$ **3.** $y = -\sqrt{x + 3} - 1$

10.2 Solving Square Root Equations (pp. 510–517)

Solve $\sqrt{12 - x} = x$.

$\sqrt{12 - x} = x$	Write the equation.
$(\sqrt{12-x})^2 = x^2$	Square each side of the equation.
$12 - x = x^2$	Simplify.
$0 = x^2 + x - 12$	Rewrite equation.
$0 = (x - 3)(x + 4)$	Factor.
$(x - 3) = 0 \quad or \quad (x + 4) = 0$	Use Zero-Product Property.
$x = 3 \quad or \quad x = -4$	Solve for *x*.

Check	$\sqrt{12 - 3} \stackrel{?}{=} 3$	Substitute for *x*.	$\sqrt{12 - (-4)} \stackrel{?}{=} -4$
	$\sqrt{9} \stackrel{?}{=} 3$	Simplify.	$\sqrt{16} \stackrel{?}{=} -4$
	$3 = 3$ ✓		$4 \neq -4$ ✗

∴ Because $x = -4$ does not check in the original equation, it is an extraneous solution. The only solution is $x = 3$.

Exercises

Solve the equation. Check your solution.

4. $8 + \sqrt{x} = 18$
5. $\sqrt{x - 1} + 9 = 15$
6. $\sqrt{5x - 9} = \sqrt{4x}$
7. $x = \sqrt{3x + 4}$

10.3 The Pythagorean Theorem (pp. 520–525)

Find the length of the hypotenuse of the triangle.

$a^2 + b^2 = c^2$	Write the Pythagorean Theorem.
$10^2 + 24^2 = c^2$	Substitute.
$100 + 576 = c^2$	Evaluate powers.
$676 = c^2$	Add.
$\sqrt{676} = \sqrt{c^2}$	Take positive square root of each side.
$26 = c$	Simplify.

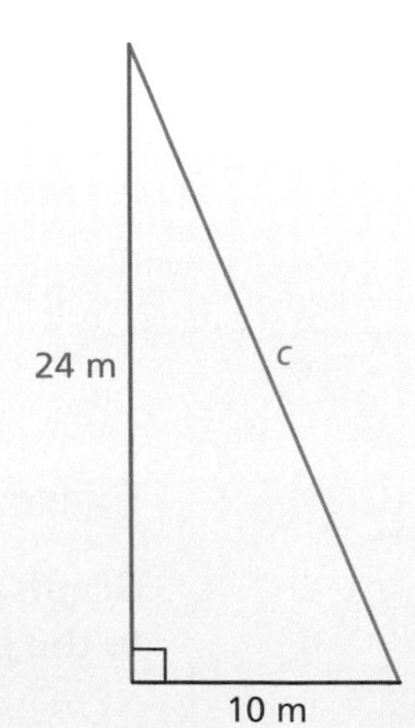

∴ The length of the hypotenuse is 26 meters.

Review Game

Pythagorean Theorem

Materials per group:

- copies of the polygons below
- pencil
- paper
- calculator

Directions

Play in pairs. Students take turns picking a polygon and finding its perimeter. (Round the perimeter to the nearest tenth, if necessary.) Play continues as students add each new perimeter to their previous perimeters, until all polygons are used. Polygons may only be used once. All measurements are in inches.

Who wins?

The student with the greatest perimeter total wins.

A.
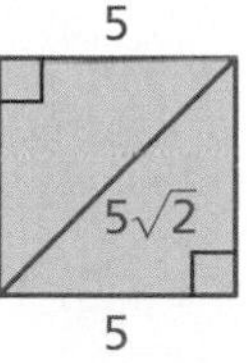

B.
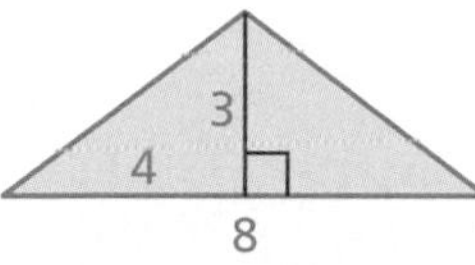

C.
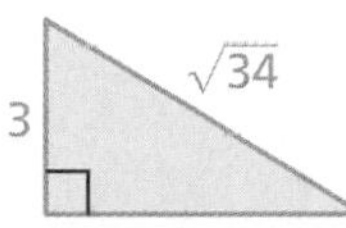

D. $\sqrt{7}$ 3

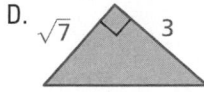

E.
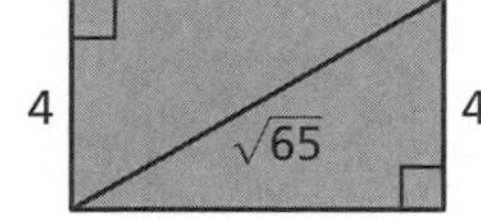

F.
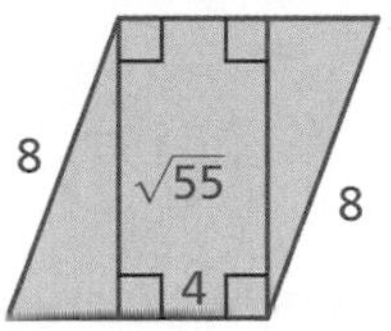

G.
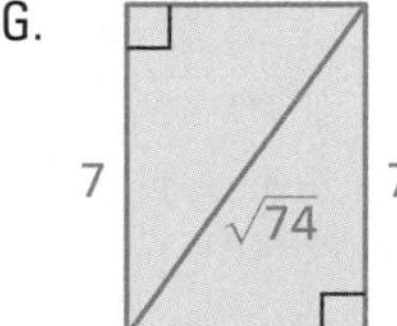

H.
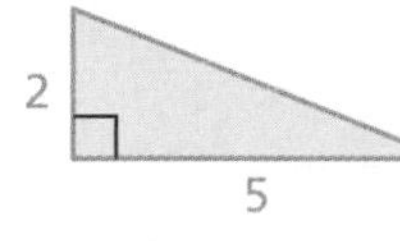

I.
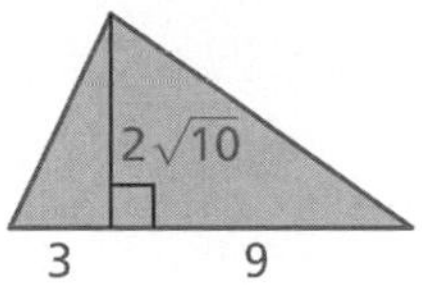

J.
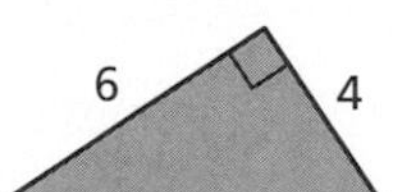

K.
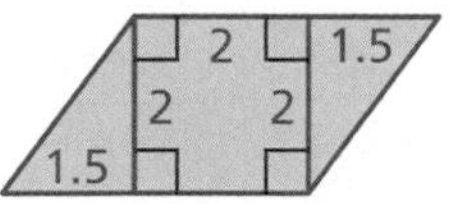

L.
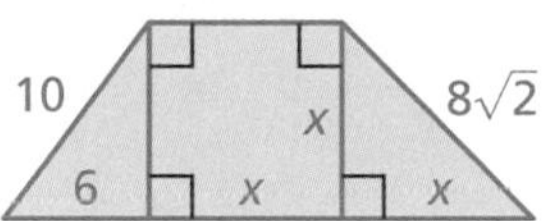

M.
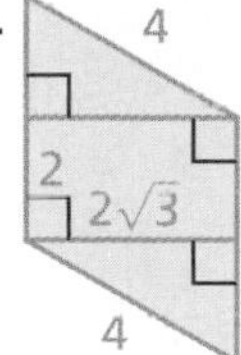

N.
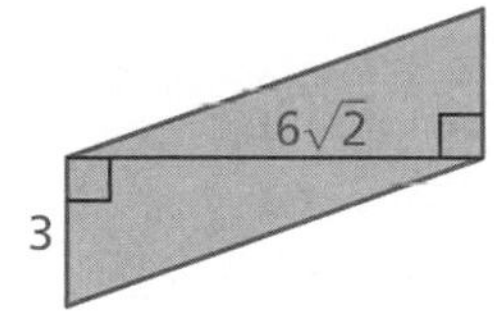

O.
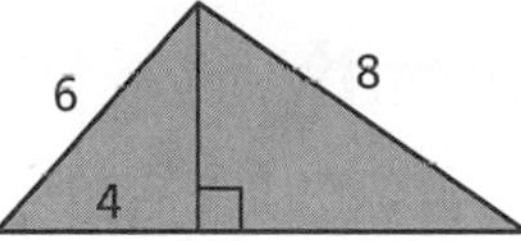

P.
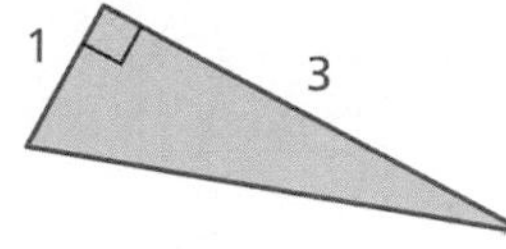

For the Student
Additional Practice

- Lesson Tutorials
- Study Help (textbook)
- Student Website
  - Multi-Language Glossary
  - Practice Assessments

Answers

4. $x = 100$
5. $x = 37$
6. $x = 9$
7. $x = 4$
8. 50 in.
9. 0.8 cm
10. yes
11. no
12. $5\sqrt{5}$
13. $\sqrt{113}$

My Thoughts on the Chapter

Teacher Tip

Not allowed to write in your teaching edition? Use sticky notes to record your thoughts.

What worked. . .

What did not work. . .

What I would do differently. . .

Exercises

Find the missing length of the triangle.

8.

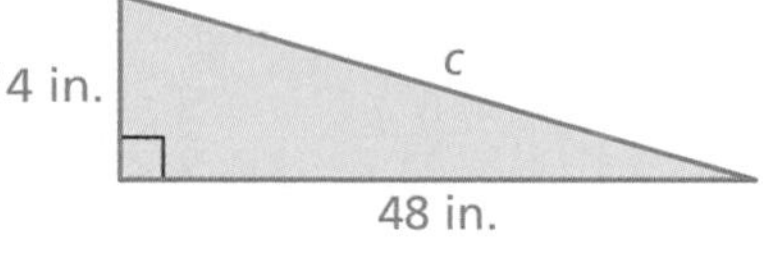

9.

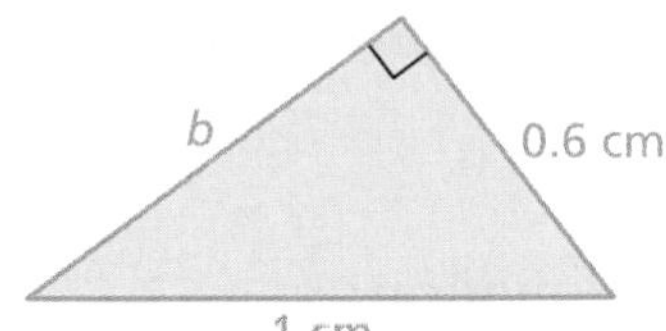

10.4 Using the Pythagorean Theorem (pp. 526–531)

a. Is the triangle formed by the rope and the tent a right triangle?

$$a^2 + b^2 = c^2$$

$$64^2 + 48^2 \stackrel{?}{=} 80^2$$

$$4096 + 2304 \stackrel{?}{=} 6400$$

$$6400 = 6400 \checkmark$$

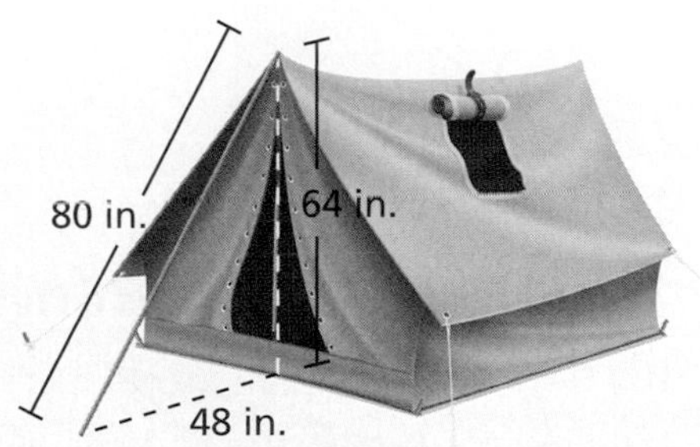

∴ It *is* a right triangle.

b. Find the distance between (−3, 1) and (4, 7).

Let $(x_1, y_1) = (-3, 1)$ and $(x_2, y_2) = (4, 7)$.

$d = \sqrt{(x_2 - x_1)^2 + (y_2 - y_1)^2}$ Write the distance formula.

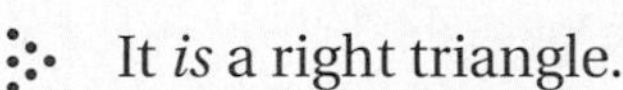

$= \sqrt{[4 - (-3)]^2 + (7 - 1)^2}$ Substitute.

$= \sqrt{7^2 + 6^2}$ Simplify.

$= \sqrt{49 + 36}$ Evaluate powers.

$= \sqrt{85}$ Add.

Exercises

Tell whether the triangle is a right triangle.

10.

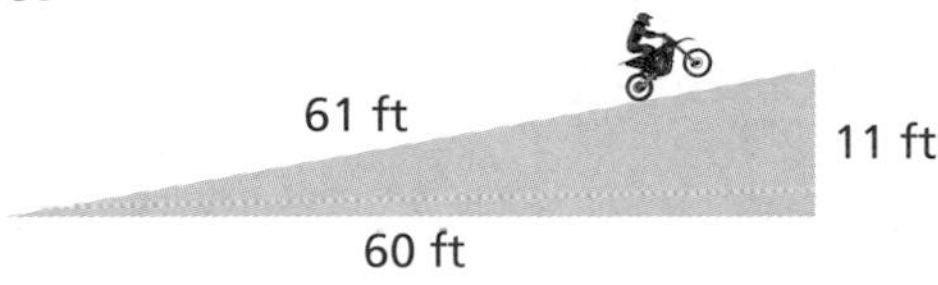

11.

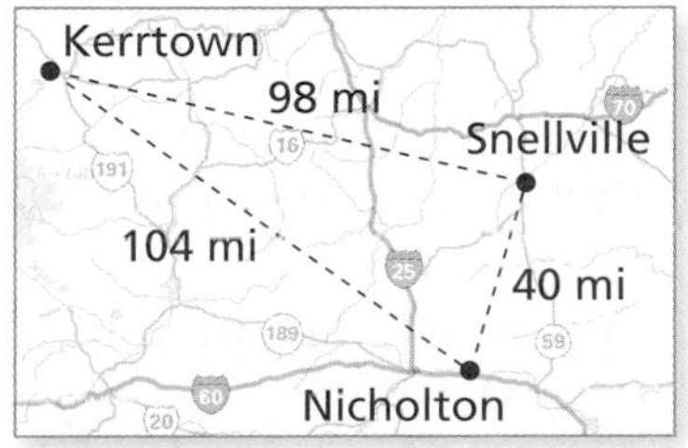

Find the distance between the two points.

12. (−2, −5), (3, 5)

13. (−4, 7), (4, 0)

10 Chapter Test

Graph the function. Describe the domain and range. Compare the graph to the graph of $y = \sqrt{x}$.

1. $y = \sqrt{x} - 6$
2. $y = \sqrt{x + 10}$
3. $y = -\sqrt{x - 2} + 3$

Solve the equation.

4. $9 - \sqrt{x} = 3$
5. $\sqrt{2x - 7} - 3 = 6$
6. $\sqrt{8x - 21} = \sqrt{18 - 5x}$
7. $x + 5 = \sqrt{7x + 53}$

Find the missing length of the triangle.

8. 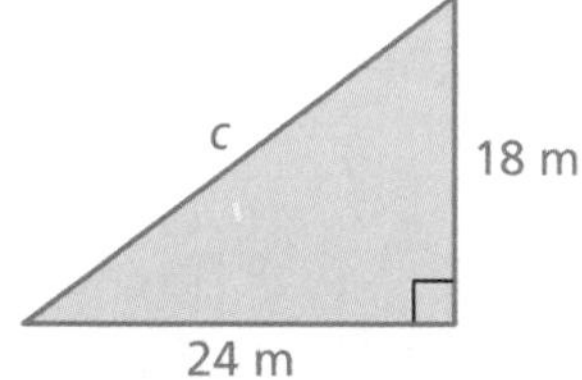

9. 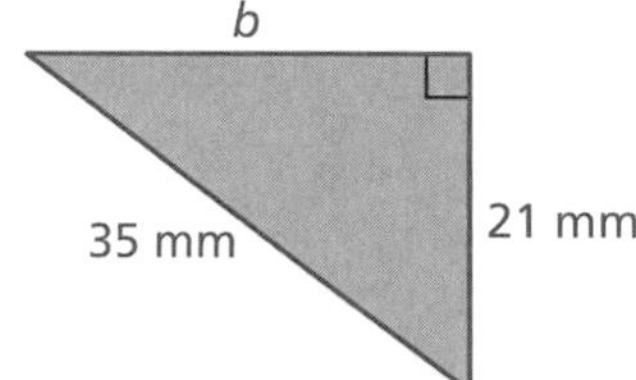

Tell whether the triangle is a right triangle.

10. 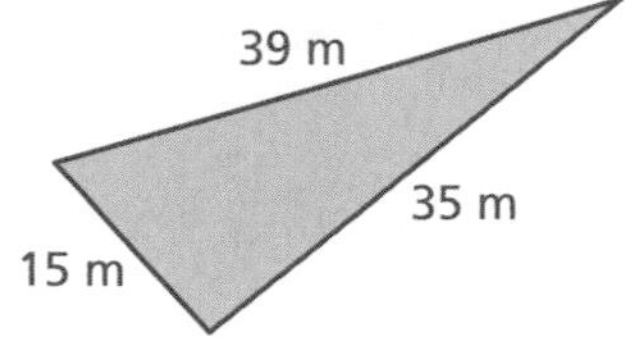

11. 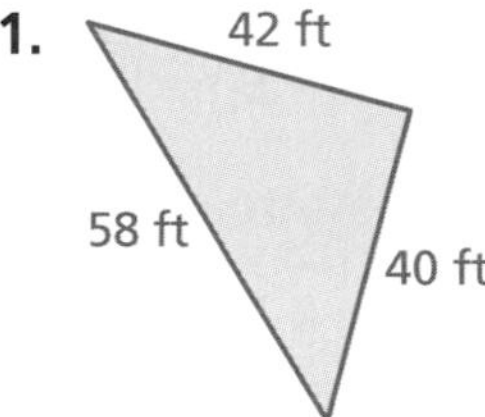

Find the distance between the two points.

12. $(-2, 3), (6, 9)$
13. $(0, -5), (4, 1)$
14. $(-3, -4), (2, -7)$

15. **ROLLER COASTER** The velocity v (in meters per second) of a roller coaster at the bottom of a hill is given by $v = \sqrt{19.6h}$, where h is the height of the hill (in meters). (a) Graph the function. Describe the domain and range. (b) How tall must the hill be for the velocity of the roller coaster at the bottom of the hill to be at least 28 meters per second?

16. **FINANCE** The average annual interest rate r (in decimal form) that an investment earns over 2 years is given by $r = \sqrt{\frac{V_2}{V_0}} - 1$, where V_0 is the initial investment and V_2 is the value of the investment after 2 years. You initially invest \$800 which earns an average annual interest of 6% over 2 years. What is the value of V_2?

17. **BUTTERFLY** Approximate the wingspan of the butterfly.

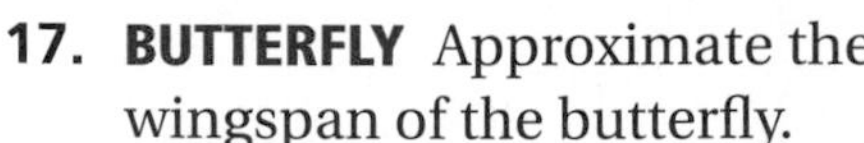

Test Item References

Chapter Test Questions	Section to Review	Common Core State Standards
1–3, 15	10.1	F.IF.4, F.IF.7b
4–7, 16	10.2	N.RN.2
8, 9, 17	10.3	8.G.6, 8.G.7
10–14	10.4	8.G.6, 8.G.7, 8.G.8

Test-Taking Strategies

Remind students to quickly look over the entire test before they start so that they can budget their time. Students should estimate and check their answers for reasonableness as they work through the test. Teach students to use the Stop and Think strategy before answering. **Stop** and carefully read the question, and **Think** about what the answer should look like.

Common Assessment Errors

- **Exercises 1–3** Students may treat constants added to a radical expression as part of the radicand, or vice versa. Encourage them to identify the radicand before solving the exercise.
- **Exercises 4 and 5** Students may square each side of the equation before isolating the square root on one side of the equation.
- **Exercises 8 and 9** Students may substitute the given lengths in the wrong part of the formula. For example, if they are finding one of the legs, they may write $5^2 + 13^2 = c^2$ instead of $5^2 + b^2 = 13^2$. Remind them that the side opposite the right angle is the hypotenuse c.
- **Exercises 8 and 9** Students may multiply each side length by two instead of squaring the side length. Remind them of the definition of exponents.
- **Exercises 10 and 11** Students may substitute the wrong value for c in the Pythagorean Theorem. Remind them that c will be the longest side, so they should substitute the greatest value for c.
- **Exercises 12–14** Students may mismatch the x-values and y-values when using the distance formula. This will result in students subtracting an x from a y, or vice versa. Encourage students to pair the numbers properly.

Reteaching and Enrichment Strategies

If students need help...	If students got it...
Resources by Chapter • Practice A and Practice B • Puzzle Time Record and Practice Journal Practice Differentiating the Lesson Lesson Tutorials Practice from the Test Generator Skills Review Handbook	Resources by Chapter • Enrichment and Extension • School-to-Work Game Closet at *BigIdeasMath.com* Start Standardized Test Practice

Answers

1–3. See Additional Answers.

4. $x = 36$ **5.** $x = 44$

6. $x = 3$ **7.** $x = 4$

8. 30 m **9.** 20 mm

10. no **11.** yes

12. 10 **13.** $2\sqrt{13}$

14. $\sqrt{34}$

15. a. $v = \sqrt{19.6h}$

domain: $h \geq 0$; range: $v \geq 0$

b. 40 meters

16. \$898.88

17. 9.6 cm

Assessment Book

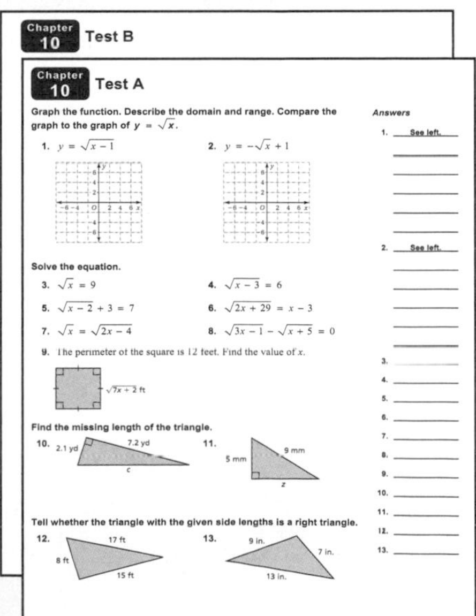
Chapter 10 Test B

Chapter 10 Test A

Graph the function. Describe the domain and range. Compare the graph to the graph of $y = \sqrt{x}$.

1. $y = \sqrt{x - 1}$
2. $y = -\sqrt{x} + 1$

Solve the equation.

3. $\sqrt{x} = 9$
4. $\sqrt{x - 3} = 6$
5. $\sqrt{x - 2} + 3 = 7$
6. $\sqrt{2x + 29} = x - 3$
7. $\sqrt{x} = \sqrt{2x - 4}$
8. $\sqrt{3x - 1} - \sqrt{x + 5} = 0$
9. The perimeter of the square is 12 feet. Find the value of x.

$\sqrt{7x + 2}$ ft

Find the missing length of the triangle.

10. 2.1 yd, 7.2 yd, c
11. 5 mm, 9 mm, z

Tell whether the triangle with the given side lengths is a right triangle.

12. 8 ft, 17 ft, 15 ft
13. 9 in., 7 in., 13 in.

Answers

1. See left.
2. See left.
3.
4.
5.
6.
7.
8.
9.
10.
11.
12.
13.

Test Taking Strategies

Available at *BigIdeasMath.com*

After Answering Easy Questions, Relax
Answer Easy Questions First
Estimate the Answer
Read All Choices before Answering
Read Question before Answering
Solve Directly or Eliminate Choices
Solve Problem before Looking at Choices
Use Intelligent Guessing
Work Backwards

About this Strategy

When taking a multiple choice test, be sure to read each question carefully and thoroughly. Sometimes it is easier to solve the problem and then look for the answer among the choices.

Answers

1. B
2. G
3. 4
4. C
5. G

Technology For the Teacher

Big Ideas Test Generator

Item Analysis

1. **A.** The student shifts the graph right instead of left.
   B. Correct answer
   C. The student confuses the vertical and horizontal shifts and shifts the graph right instead of left.
   D. The student confuses the vertical and horizontal shifts.

2. **F.** The student adds the exponents for the powers of powers.
   G. Correct answer
   H. The student rewrites the expression as $\frac{(2^3)^{10} \cdot 2^{-5}}{2^{16}}$ and then adds all of the exponents.
   I. The student rewrites $\frac{8^{10}}{16^4}$ as $\left(\frac{1}{2}\right)^{10-4} = \left(\frac{1}{2}\right)^6$.

3. **Gridded response:** Correct answer: 4

   Common error: The student fails to check the answers and chooses $x = -2$ as the answer.

4. **A.** The student makes an error either in finding the slope-intercept form of the line, or in checking the point in the equation.
   B. The student makes an error either in finding the slope-intercept form of the line, or in checking the point in the equation.
   C. Correct answer
   D. The student makes an error either in finding the slope-intercept form of the line, or in checking the point in the equation.

5. **F.** The student incorrectly determines that the solutions of $(x - 2)(x - 4) = 0$ are $x = -2$ and $x = -4$, and assumes that $x = -4$ is extraneous.
   G. Correct answer
   H. The student incorrectly assumes that $x = 2$ is extraneous.
   I. The student incorrectly determines the solutions of $(x - 2)(x - 4) = 0$.

10 Standardized Test Practice

1. Which function is shown in the graph? *(F.IF.7b)*

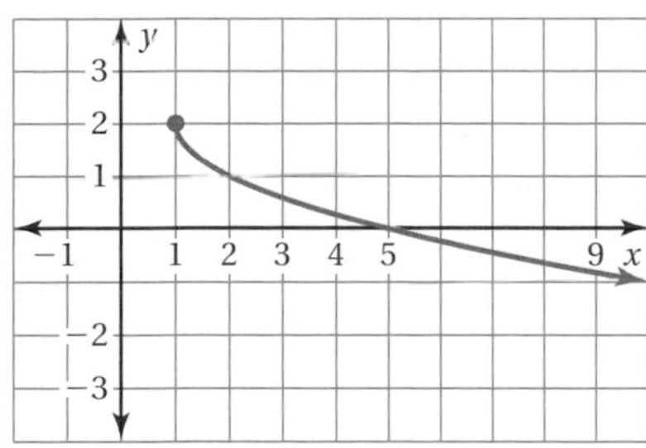

A. $y = -\sqrt{x + 1} + 2$

B. $y = -\sqrt{x - 1} + 2$

C. $y = -\sqrt{x + 2} + 1$

D. $y = -\sqrt{x - 2} + 1$

2. Which number is equivalent to $\frac{8^{10} \cdot 2^{-5}}{16^4}$? *(N.RN.2)*

F. 1

G. 2^9

H. 2^{24}

I. 2^{-11}

3. What value of x makes the equation below true? *(N.RN.2)*

$$x = \sqrt{2x + 8}$$

4. A line with a slope of -2 passes through the point $(1, -6)$. Which of the following is not a point on the line? *(A.CED.2)*

A. $(-8, 12)$

B. $(-4, 4)$

C. $(4, -4)$

D. $(8, -20)$

5. What value(s) of x make the equation below true? *(N.RN.2)*

$$\sqrt{2x - 4} = x - 2$$

F. Only -2

G. 2 and 4

H. Only 4

I. -2 and -4

6. What is the focus of the parabola? *(F.IF.4)*

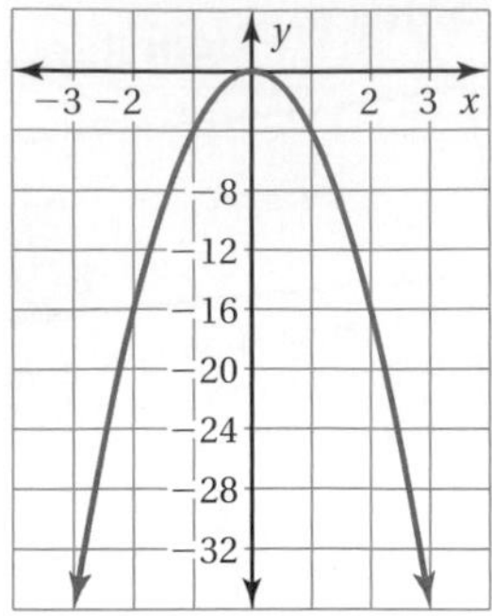

A. $\left(0, \frac{1}{16}\right)$

B. (0, 16)

C. $\left(0, -\frac{1}{16}\right)$

D. (0, −16)

7. The range of the function $y = 6x - 8$ is all real numbers from 1 to 10. What is the domain of the function? *(F.IF.1)*

F. all real numbers from 1.5 to 3

G. all real numbers from −2 to 52

H. all integers from −2 to 52

I. all real numbers

8. What is the distance between the two points in the coordinate plane? *(8.G.8)*

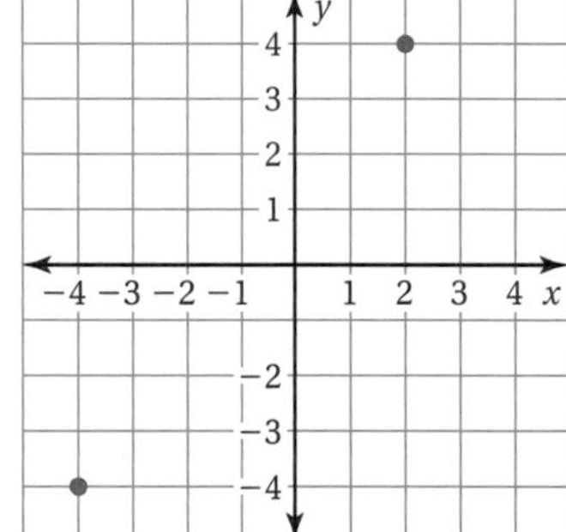

9. Which ordered pair is a solution of the system of inequalities shown in the graph? *(A.REI.12)*

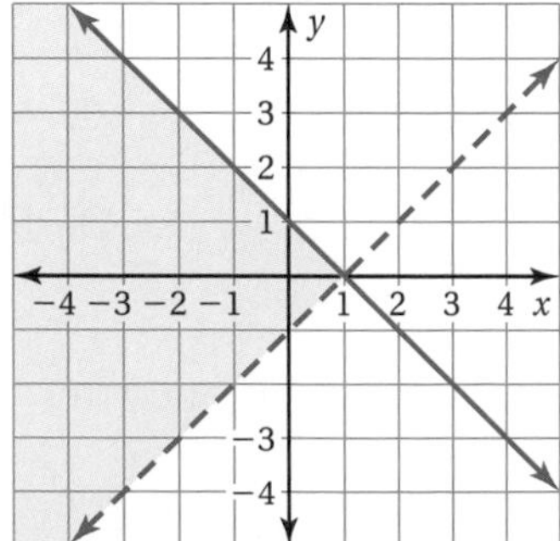

A. (1, 0)

B. (0, 1)

C. (0, −1)

D. (2, 0)

Item Analysis (continued)

6. **A.** The student forgets the negative sign at some point in the calculations.

   B. The student uses $-4a$ instead of $\frac{1}{4a}$ to find the y-value of the focus.

   C. Correct answer

   D. The student uses $4a$ instead of $\frac{1}{4a}$ to find the y-value of the focus.

7. **F.** Correct answer

   G. The student finds the range for a domain of 1 to 10 instead of the domain for a range of 1 to 10.

   H. The student specifies the integer elements of the range for a domain of 1 to 10 instead of the domain for a range of 1 to 10.

   I. The student ignores the range restriction.

8. **Gridded response:** Correct answer: 10

   Common error: The student applies the distance formula incorrectly, by adding instead of subtracting the x-terms and y-terms, and gets an answer of 2.

9. **A.** The student incorrectly thinks that a point on a dashed boundary line represents a solution of the system.

   B. Correct answer

   C. The student incorrectly thinks that a point on a dashed boundary line represents a solution of the system.

   D. The student does not realize that the solutions of the system of inequalities are in the shaded region.

10. **2 points** The student demonstrates a thorough understanding of the Pythagorean Theorem. The student correctly finds each missing distance and the total distance of each trail. The student compares the trails correctly, and provides an adequate explanation.

    1 point The student's work and explanation demonstrate an understanding of the Pythagorean Theorem, although one or more calculations are incorrect, leading to incorrect or poorly supported answers to the questions.

    0 points The student provides no response, a completely incorrect or incomprehensible response, or a response that demonstrates insufficient understanding of the Pythagorean Theorem.

Answers

6. C
7. F
8. 10
9. B

Answers

10. Trail B is longer by 10 miles; Using the Pythagorean Theorem, x and y are each 5 miles and z is 10 miles. So, Trail A is 30 miles and Trail B is 40 miles.

11. I

12. B

13. H

Answer for Extra Example

1. **A.** The student makes a computation error.

   B. The student makes a computation error.

   C. Correct answer

   D. The student makes a computation error.

Item Analysis (continued)

11. **F.** The student shades in the wrong direction and incorrectly uses an open circle.

    G. The student incorrectly uses an open circle.

    H. The student shades in the wrong direction.

    I. Correct answer

12. **A.** The student incorrectly translates the graph of $y = \sqrt{x}$ one unit to the right and two units down.

    B. Correct answer

    C. The student incorrectly translates the graph of $y = \sqrt{x}$ one unit to the left and two units up.

    D. The student incorrectly translates the graph of $y = \sqrt{x}$ one unit to the left and two units down.

13. **F.** The student makes a sign error in finding the x-coordinate of the vertex.

    G. The student makes a calculation error in finding the x-coordinate of the vertex.

    H. Correct answer

    I. The student uses $x = -\frac{b}{a}$ to find the x-coordinate of the vertex.

Extra Example

1. Which triangle is *not* a right triangle? *(8.G.7)*

A.

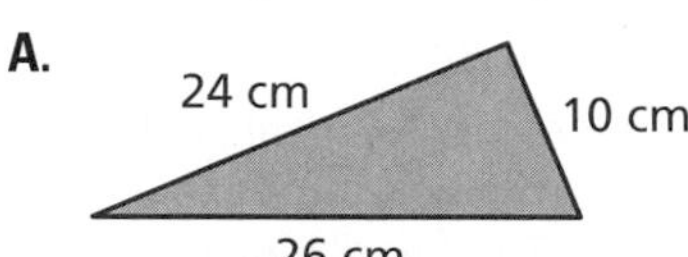

C.

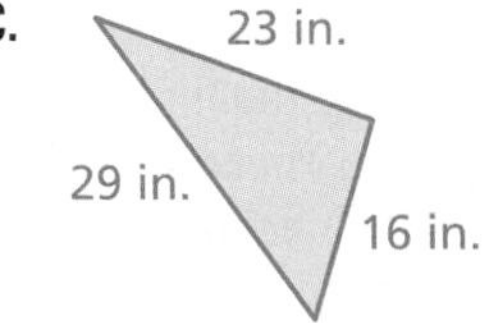

B.

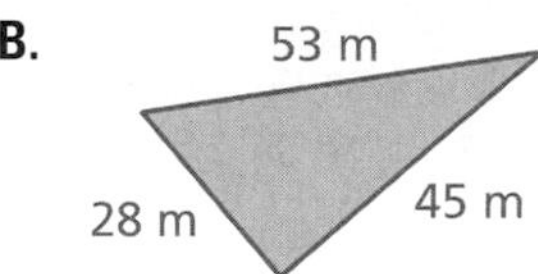

D.

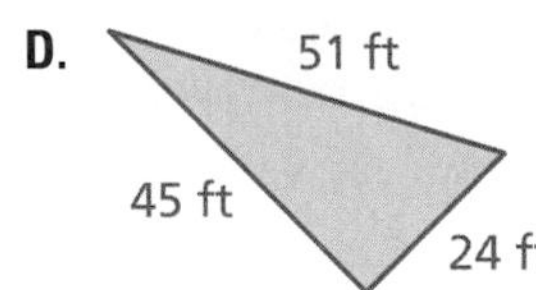

10. Two nature trails are shown below. Which trail is longer? By how much? Explain your reasoning. *(8.G.7)*

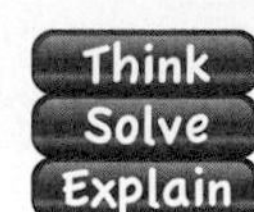

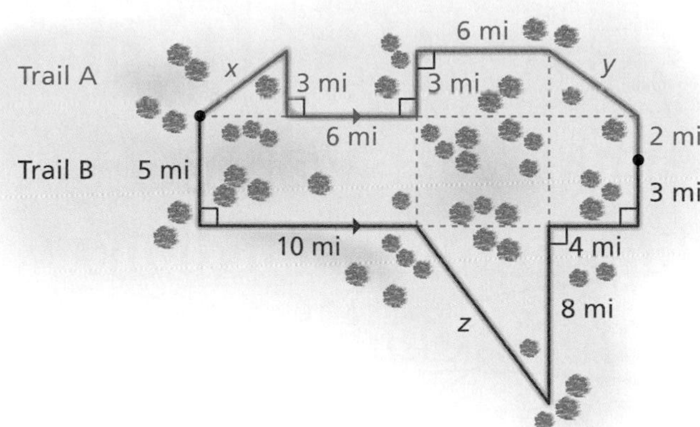

11. The solution of which inequality is shown in the graph below? *(A.REI.3)*

F. $5x - 7 \geq 3$

G. $4x + 3 \leq 11$

H. $12 - 3x < 6$

I. $10 - 2x > 6$

12. Tom was graphing $y = \sqrt{x + 2} - 1$. His work is shown below. *(F.IF.7b)*

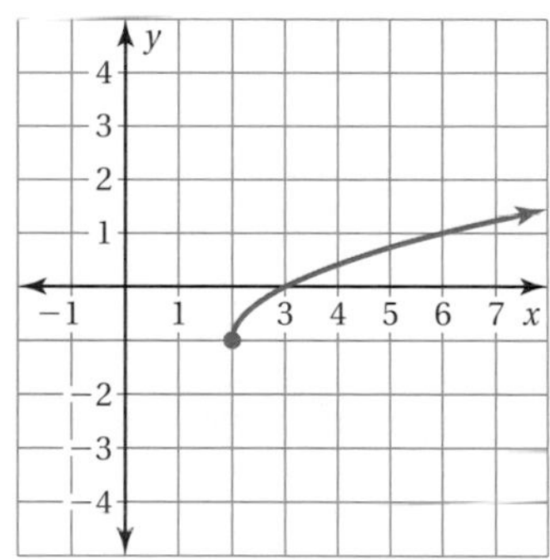

What should Tom do to correct the error that he made?

A. Shift the graph 1 unit down and 1 unit left.

B. Shift the graph 4 units left.

C. Shift the graph 3 units up and 3 units left.

D. Shift the graph 1 unit down and 3 units left.

13. What is the vertex of the graph of $y = 2x^2 - 4x + 6$? *(F.IF.4)*

F. $(-1, 12)$

G. $(-2, 22)$

H. $(1, 4)$

I. $(2, 6)$

11 Rational Equations and Functions

"Descartes, in your homework problem, you are being chased by a cat-eating hyena."

"Both of you must stay on the graph of $y = 1/x$. The safe zone is the x-axis."

"Can you ever reach the safe zone? Explain your reasoning."

"Descartes, I am keeping track of how many doggy treats my owner gives me each day."

"I am finding that my happiness is directly proportional to the day of the week."

Connections to Previous Learning

- Apply properties of operations to generate equivalent expressions.
- Solve simple one-variable equations to solve real-life and mathematical problems.
- Use variables to represent two quantities that change in relationship to one another.

- Use and simplify algebraic expressions to solve problems.
- Analyze and solve linear equations to solve real-life and mathematical problems.

- Create equations that describe numbers or relationships.
- Add, subtract, multiply, and divide rational expressions.
- Create and solve rational equations in one variable, and use them to solve problems.

Math in History

There was fierce competition for the recognition of intellectual discoveries during the 17th century. Scientists would announce new discoveries in the form of coded anagrams to avoid sharing details. This allowed them to complete their work on the discovery without the risk of sharing the credit.

★ English physicist Robert Hooke published an anagram in 1676. He published the solution to the anagram in 1678, along with the details about the direct variation formula that became known as Hooke's Law: The extension of a spring is directly proportional to the force applied to the spring.

★ An anagram was involved in the controversy between Isaac Newton (England) and Gottfried Von Leibniz (Germany) about who invented calculus. Leibniz first published his work in 1684. Although Newton did not publish any of his work until a few years later, he had sent a coded anagram of his version of the fundamental theorem of calculus to Leibniz in 1676.

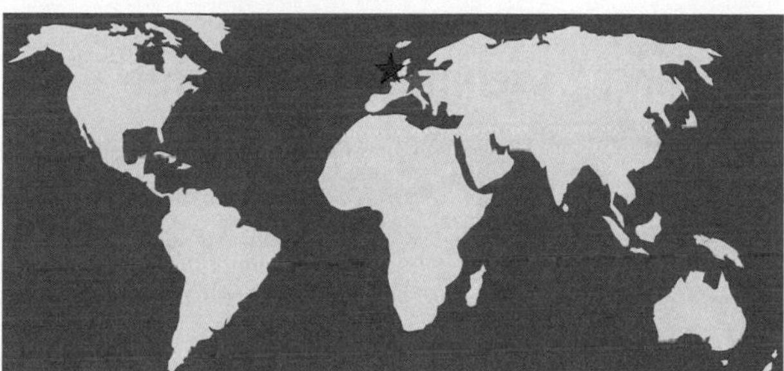

Pacing Guide for Chapter 11

Chapter Opener	1 Day
Section 1	2 Days
Section 2	3 Days
Section 3	1 Day
Study Help / Quiz	1 Day
Section 4	1 Day
Section 5	1 Day
Section 6	1 Day
Section 7	1 Day
Chapter Review / Chapter Tests	2 Days
Total Chapter 11	14 Days
Year-to-Date	150 Days

Check Your Resources

- Record and Practice Journal
- Resources by Chapter
- Skills Review Handbook
- Assessment Book
- Worked-Out Solutions

Technology For the Teacher

The Dynamic Planning Tool
Editable Teacher's Resources at *BigIdeasMath.com*

Common Core State Standards

5.NF.1 Add and subtract fractions with unlike denominators (including mixed numbers) by replacing given fractions with equivalent fractions in such a way as to produce an equivalent sum or difference of fractions with like denominators.

6.NS.1 Interpret and compute quotients of fractions, ...

7.RP.3 Use proportional relationships to solve multistep ratio . . . problems.

Additional Topics for Review

- Domain and range of a function
- Relations and functions
- Polynomials
- Long division

Try It Yourself

1. 1 **2.** $\frac{1}{2}$

3. $\frac{5}{3}$ **4.** $\frac{5}{16}$

5. 3 **6.** 2

7. 7.8

Record and Practice Journal

1. $\frac{7}{3}$ **2.** $\frac{1}{8}$

3. $-\frac{7}{10}$ **4.** $\frac{9}{4}$

5. $\frac{1}{8}$ **6.** $\frac{5}{12}$

7. $\frac{3}{5}$ **8.** 10

9. $\frac{13}{8}$ mi **10.** 6 servings

11. $d = 4$ **12.** $m = 15$

13. $a = 27$ **14.** $k = 3.2$

15. $x = 11.4$ **16.** $c = 12.75$

17. $212.50

Math Background Notes

Vocabulary Review

- Fractions
- Proportions
- Cross Products Property

Evaluating Expressions with Fractions

- Students should be able to add, subtract, multiply, and divide fractions with like and unlike denominators.
- When adding or subtracting two fractions that share a common denominator, simply add or subtract their numerators and keep the common denominator. If two fractions do not have a common denominator, students will need to find one.
- You may want to review finding the Least Common Multiple between two numbers. Remind students that finding the Least Common Multiple among the denominators will produce the least common denominator used to add or subtract fractions.
- You may want to review how to rename fractions using the least common denominator.
- Remind students that the rules for multiplying and dividing fractions are different from the rules for adding and subtracting fractions. Multiplying and dividing fractions does not require a common denominator.
- **Teaching Tip:** Most students will remember the process to divide fractions. If your students are comfortable with the process, encourage them to describe it using math vocabulary. Instead of, "change the sign and flip the second fraction," encourage "multiply by the reciprocal of the divisor."

Solving Proportions

- Students should be able to solve proportions.
- You may wish to review the Cross Products Property. Remind students that this property is unique as it can only be used in proportions.
- **Teaching Tip:** Help students to visualize the Cross Products Property by drawing the X as you work through the problem. Example:

$$\frac{2}{5} = \frac{x}{15}$$

$$30 = 5x$$

Reteaching and Enrichment Strategies

If students need help. . .	If students got it. . .
Record and Practice Journal • Fair Game Review Skills Review Handbook Lesson Tutorials	Game Closet at *BigIdeasMath.com* Start the next section

What You Learned Before

"Your namesake, René Descartes, believed in rational thinking."

Evaluating Expressions with Fractions (5.NF.1, 6.NS.1)

Example 1 Find $\frac{1}{10} + \frac{3}{5}$.

$$\frac{1}{10} + \frac{3}{5} = \frac{1}{10} + \frac{6}{10}$$
$$= \frac{1+6}{10}$$
$$= \frac{7}{10}$$

Example 2 Find $\frac{4}{3} \div \frac{5}{3}$.

$$\frac{4}{3} \div \frac{5}{3} = \frac{4}{3} \bullet \frac{3}{5}$$
$$= \frac{4 \bullet \cancel{3}}{\cancel{3} \bullet 5}$$
$$= \frac{4}{5}$$

Try It Yourself

Evaluate the expression.

1. $\frac{1}{6} + \frac{5}{6}$ **2.** $\frac{2}{3} - \frac{1}{6}$ **3.** $\frac{5}{2} \bullet \frac{2}{3}$ **4.** $\frac{5}{6} \div \frac{8}{3}$

Solving Proportions (7.RP.3)

Example 3 Solve $\frac{4}{x} = \frac{5}{12}$.

$\frac{4}{x} = \frac{5}{12}$	Write the proportion.
$4 \bullet 12 = x \bullet 5$	Use the Cross Products Property.
$48 = 5x$	Multiply.
$9.6 = x$	Divide each side by 5.

Try It Yourself

Solve the proportion.

5. $\frac{4}{6} = \frac{2}{x}$ **6.** $\frac{3}{12} = \frac{w}{8}$ **7.** $\frac{15}{y} = \frac{25}{13}$

11.1 Direct and Inverse Variation

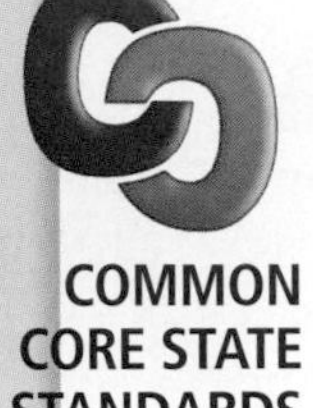

COMMON CORE STATE STANDARDS
A.REI.10

Essential Question How can you recognize when two variables vary directly? How can you recognize when they vary inversely?

1 ACTIVITY: Recognizing Direct Variation

Work with a partner. You hang different weights from the same spring.

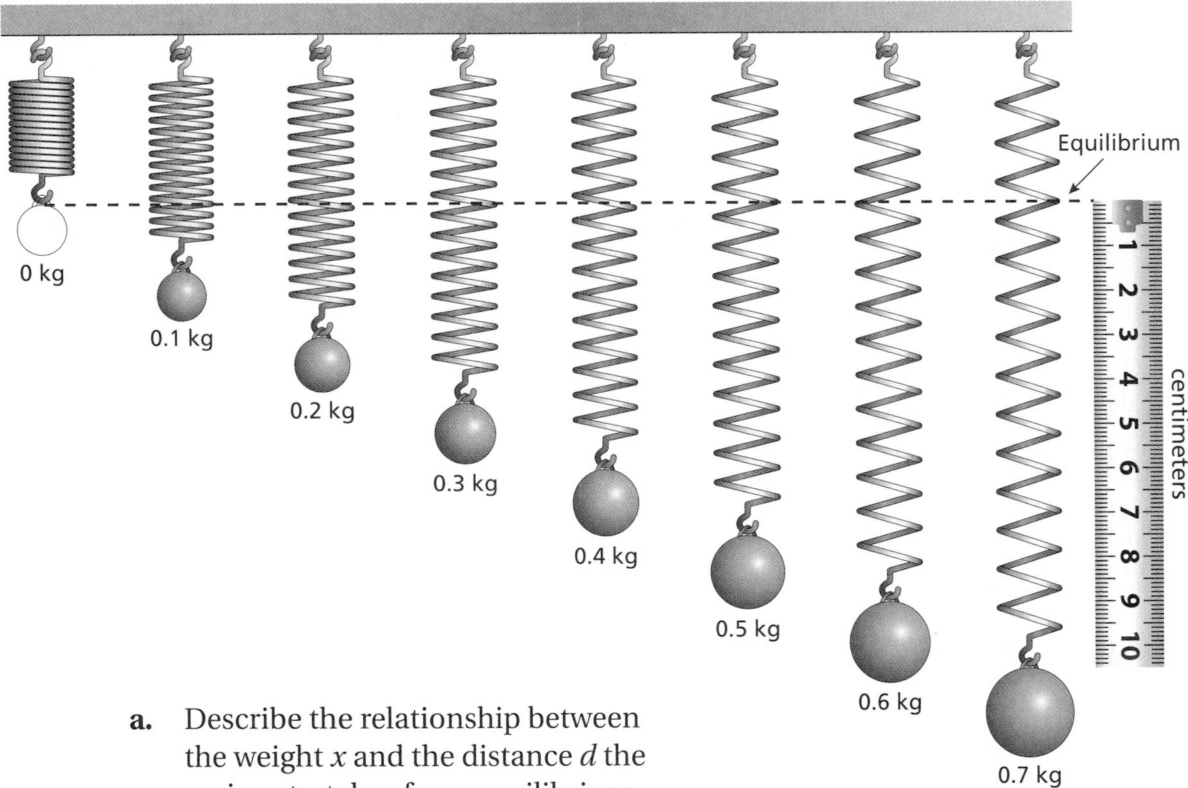

a. Describe the relationship between the weight x and the distance d the spring stretches from equilibrium. Explain why the distance is said to vary *directly* with the weight.

b. Graph the relationship between x and d. What are the characteristics of the graph?

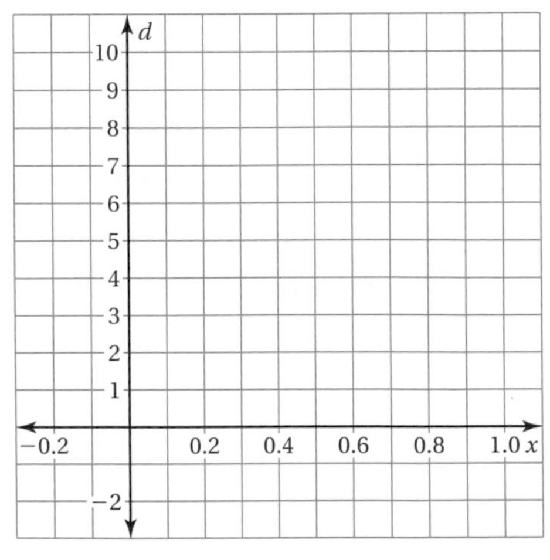

c. Write an equation that represents d as a function of x.

d. In physics, the relationship between d and x is described by Hooke's Law. How would you describe Hooke's Law?

Laurie's Notes

Introduction

Standards for Mathematical Practice

- **MP2 Reason Abstractly and Quantitatively:** Students will recognize patterns as they work with direct variation and inverse variation. It is important that students be able to describe the quantitative relationship between x and y.

Motivate

- Collect three items that have different lengths. Each item should have a length that is approximately a whole number of inches.
- Measure each of the items in inches, and again in centimeters. Record the results in a table.

Note pad	Travel mug	Paper	?
3 in.	6 in.	11 in.	
7.62 cm	15.24 cm	27.94 cm	

? "If I measure another item and tell you its length in inches, can you tell me its length in centimeters? Explain." yes, Listen for students to describe the process using the relationship 1 inch = 2.54 centimeters.

- Pretend to measure another item and say it is 4 inches long. Add it to the table and ask students for the length in centimeters.

? "If you plot the points (inches, centimeters), what type of function is represented by the graph?" linear

Activity Notes

Activity 1

- In this activity, students need to read lengths from the tape measure displayed at the right.
- **MP6 Attend to Precision:** To improve accuracy, students could use a straightedge, keeping it parallel to the dotted line.
- You could have a general discussion with students about what equilibrium means in this context.
- Each 0.1 kilogram adds approximately 1.5 centimeters to the length.
- When students have finished, ask them to share their responses in part (a). It is likely that students will mention the constant rate of change.
- When describing the graph in part (b), students should mention the slope and that the line passes through the origin.

Common Core State Standards

A.REI.10 Understand that the graph of an equation in two variables is the set of all its solutions plotted in the coordinate plane, often forming a curve (which could be a line).

Previous Learning

Students should know how to graph linear and nonlinear functions.

Start Thinking! and Warm Up

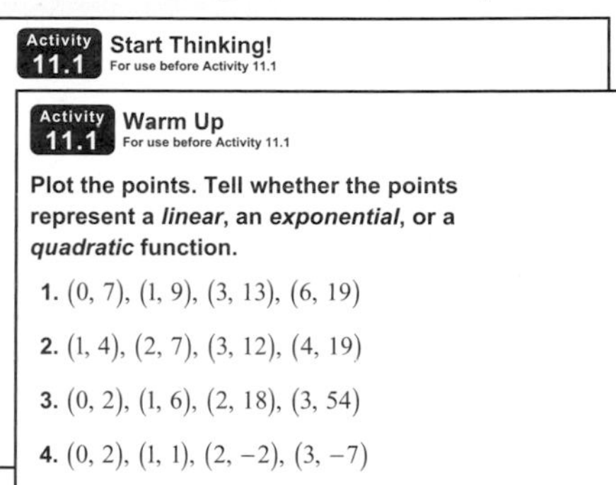
Activity 11.1 Start Thinking! For use before Activity 11.1

Activity 11.1 Warm Up For use before Activity 11.1

Plot the points. Tell whether the points represent a *linear*, an *exponential*, or a *quadratic* function.

1. (0, 7), (1, 9), (3, 13), (6, 19)
2. (1, 4), (2, 7), (3, 12), (4, 19)
3. (0, 2), (1, 6), (2, 18), (3, 54)
4. (0, 2), (1, 1), (2, −2), (3, −7)

11.1 Record and Practice Journal

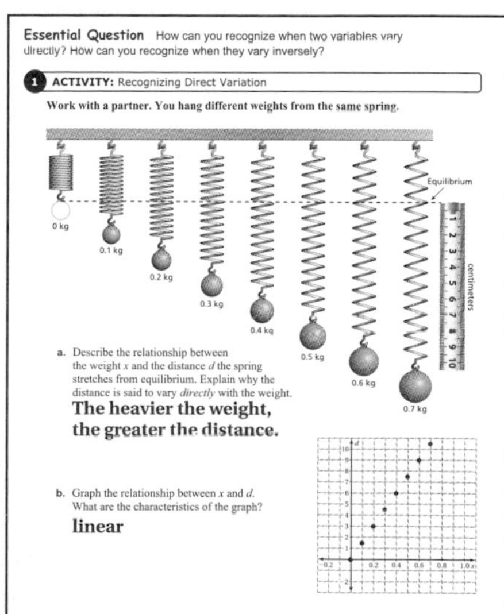
Essential Question How can you recognize when two variables vary directly? How can you recognize when they vary inversely?

1 ACTIVITY: Recognizing Direct Variation

Work with a partner. You hang different weights from the same spring.

a. Describe the relationship between the weight x and the distance d the spring stretches from equilibrium. Explain why the distance is said to vary *directly* with the weight.

The heavier the weight, the greater the distance.

b. Graph the relationship between x and d. What are the characteristics of the graph?

linear

Differentiated Instruction

Advanced

Challenge students to look for symmetry in graphs that represent inverse variation. They should see that the graph is symmetric about the line $y = x$. This occurs when exchanging x and y in the original equation produces an equivalent equation. What other equations have graphs that are symmetric about the line $y = x$?

$y = x$ and $y = -x$

11.1 Record and Practice Journal

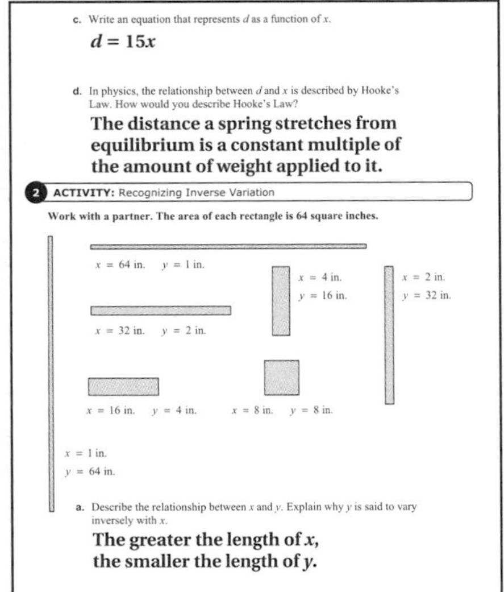

c. Write an equation that represents d as a function of x.

$d = 15x$

d. In physics, the relationship between d and x is described by Hooke's Law. How would you describe Hooke's Law?

The distance a spring stretches from equilibrium is a constant multiple of the amount of weight applied to it.

2 ACTIVITY: Recognizing Inverse Variation

Work with a partner. The area of each rectangle is 64 square inches.

a. Describe the relationship between x and y. Explain why y is said to vary inversely with x.

The greater the length of x, the smaller the length of y.

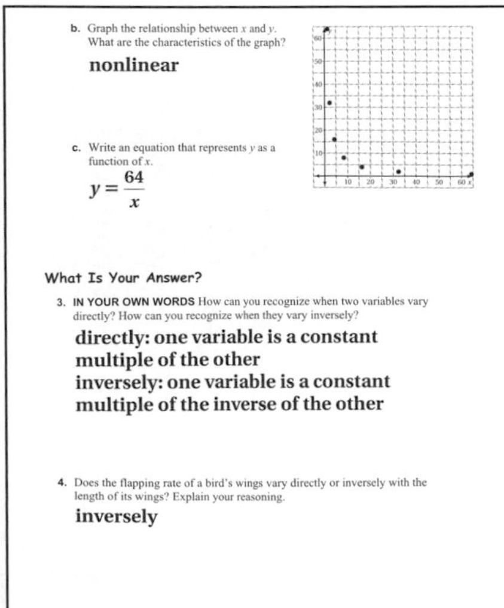

b. Graph the relationship between x and y. What are the characteristics of the graph?

nonlinear

c. Write an equation that represents y as a function of x.

$y = \frac{64}{x}$

What Is Your Answer?

3. IN YOUR OWN WORDS How can you recognize when two variables vary directly? How can you recognize when they vary inversely?

directly: one variable is a constant multiple of the other
inversely: one variable is a constant multiple of the inverse of the other

4. Does the flapping rate of a bird's wings vary directly or inversely with the length of its wings? Explain your reasoning.

inversely

Laurie's Notes

Activity 2

- Work through the activity as shown.
- If students have difficulty writing the equation in part (c), ask them what they know about the product xy. This should lead them to write $xy = 64$, which they can solve for y.
- **Common Error:** Students may believe the graph represents exponential decay. Point out that there is no y-intercept. After writing the equation in part (c), they should also recognize that x is not an exponent.

What Is Your Answer?

- You want students to recognize that with direct variation, x and y increase (or decrease) proportionally. With inverse variation, x increases as y decreases (or vice versa), and x and y are inversely proportional. Students may say that if x doubles, y is divided by 2.

Closure

- Explain why the situation in the Motivate (measuring in inches and then centimeters) involves direct variation. The measurements change proportionally.

Technology For the Teacher

The Dynamic Planning Tool
Editable Teacher's Resources at *BigIdeasMath.com*

2 ACTIVITY: Recognizing Inverse Variation

Work with a partner. The area of each rectangle is 64 square inches.

$x = 64$ in. $y = 1$ in.

$x = 4$ in.
$y = 16$ in.

$x = 2$ in.
$y = 32$ in.

$x = 32$ in. $y = 2$ in.

$x = 16$ in. $y = 4$ in.

$x = 8$ in. $y = 8$ in.

$x = 1$ in.
$y = 64$ in.

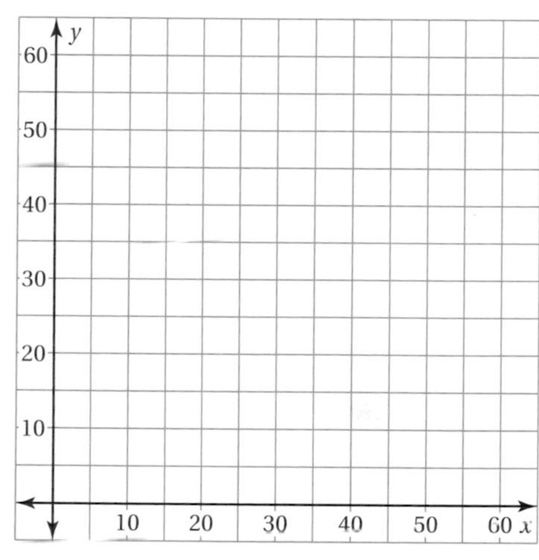

a. Describe the relationship between x and y. Explain why y is said to vary inversely with x.

b. Graph the relationship between x and y. What are the characteristics of the graph?

c. Write an equation that represents y as a function of x.

What Is Your Answer?

3. **IN YOUR OWN WORDS** How can you recognize when two variables vary directly? How can you recognize when they vary inversely?

4. Does the flapping rate of a bird's wings vary directly or inversely with the length of its wings? Explain your reasoning.

Use what you learned about direct and inverse variation to complete Exercises 3 and 4 on page 547.

11.1 Lesson

Key Vocabulary
direct variation, *p. 544*
inverse variation, *p. 544*

Direct Variation

Two quantities x and y show **direct variation** when $y = kx$, where k is a nonzero constant.

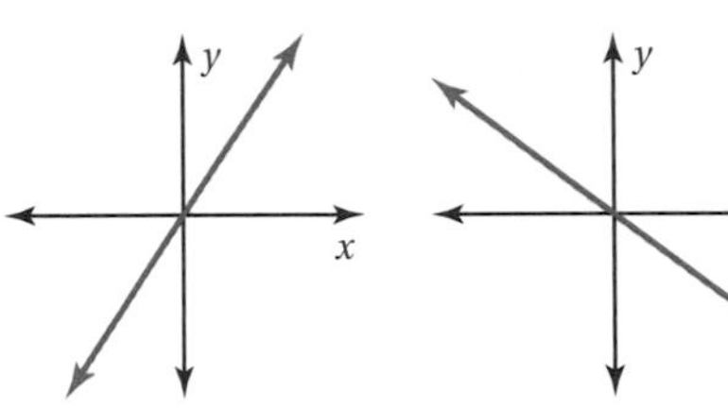

$y = kx,\ k > 0$ $\qquad$ $y = kx,\ k < 0$

The ratio $\frac{y}{x}$ is constant.

Inverse Variation

Two quantities x and y show **inverse variation** when $y = \frac{k}{x}$, where k is a nonzero constant.

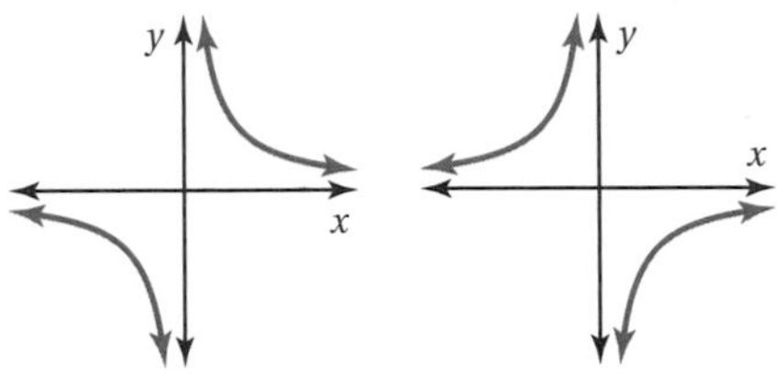

$y = \frac{k}{x},\ k > 0$ $\qquad$ $y = \frac{k}{x},\ k < 0$

The product xy is constant.

Study Tip

The constant k is called the *constant of proportionality* or the *constant of variation*.

EXAMPLE 1 Identifying Direct and Inverse Variation

Tell whether x and y show *direct variation, inverse variation,* or *neither*. Explain your reasoning.

a.

x	1	2	3	4
y	5	10	15	20

The products xy are not constant. So, the table does not show inverse variation.

Check each ratio $\frac{y}{x}$: $\quad \frac{5}{1} = 5, \quad \frac{10}{2} = 5, \quad \frac{15}{3} = 5, \quad \frac{20}{4} = 5$

∴ The ratios are constant. So, x and y show direct variation.

b. $4xy = -4$

$y = -\frac{1}{x}$ $\qquad$ Divide each side by $4x$.

∴ The equation is of the form $y = \frac{k}{x}$. So, x and y show inverse variation.

On Your Own

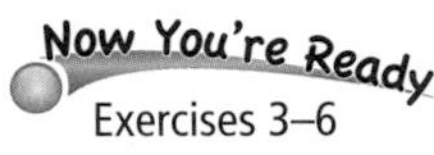

Tell whether x and y show *direct variation, inverse variation,* or *neither*. Explain your reasoning.

1.

x	1	2	3	4
y	24	12	8	6

2. $y = 3x + 1$

Multi-Language Glossary at BigIdeasMath.com.

Laurie's Notes

Introduction

Connect

- **Yesterday:** Students recognized direct and inverse variation. (MP2, MP6)
- **Today:** Students will write and graph direct and inverse variation equations.

Motivate

- **Story Time:** Tell students they have \$48 to spend at a store. They must spend all of their money and they can only purchase one type of item.
- Indicate that the purchase prices of all items are factors of 48.
- Give students 48 seconds to decide how to spend the \$48. For example, they could purchase forty-eight \$1 items or eight \$6 items.
- Have a discussion about the "purchases." Students should recognize that the more expensive the item, the fewer you can purchase. The cheaper the item, the more you can purchase.

Lesson Notes

Key Ideas

- Write the Key Ideas, defining direct variation and inverse variation.
- Discuss the constant k and the affect of $k > 0$ and $k < 0$.

? "Direct variation is a special case of what type of function?" linear

- Solve each equation for k to convince students that the ratio y to x is constant in a direct variation equation and the product of x and y is constant in an inverse variation equation.
- Discuss the vocabulary in the Study Tip. Use both "constant of variation" and "constant of proportionality" when working through examples.

Example 1

- Write the table of values in part (a).

? "How can you tell from the table whether x and y show direct variation, inverse variation, or neither?" Check for constant ratios (y to x) or constant products (xy).

- Remind students that they must check all ordered pairs, not just a few.
- Write the equation in part (b).

? MP7 Look for and Make Use of Structure: "How can you tell from the equation whether x and y show direct variation, inverse variation, or neither?" Solve for y and interpret the result.

On Your Own

- **Think-Pair-Share:** Students should read each question independently and then work with a partner to answer the questions. When they have answered the questions, the pair should compare their answers with another group and discuss any discrepancies.

Goal Today's lesson is writing and graphing **direct** and **inverse variation** equations.

Start Thinking! and Warm Up

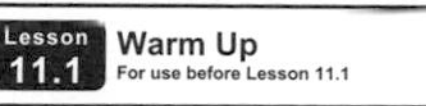

Lesson 11.1 Warm Up
For use before Lesson 11.1

Lesson 11.1 Start Thinking!
For use before Lesson 11.1

- Describe a real-life situation at a circus that represents direct variation.
- Describe a real-life situation at a circus that represents inverse variation.

Extra Example 1

Tell whether x and y show *direct variation, inverse variation,* or *neither*. Explain your reasoning.

a.

x	1	2	3	4
y	3	$\frac{3}{2}$	1	$\frac{3}{4}$

inverse variation; The products xy are constant.

b. $2y = 5x$

direct variation; The equation can be written as $y = kx$.

On Your Own

1. inverse variation; The products xy are constant.
2. neither; The equation cannot be written as $y = kx$ or $y = \frac{k}{x}$.

Extra Example 2

The variable y varies directly with x. When $x = 4$, $y = 32$. Write and graph a direct variation equation that relates x and y.

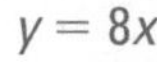

$y = 8x$

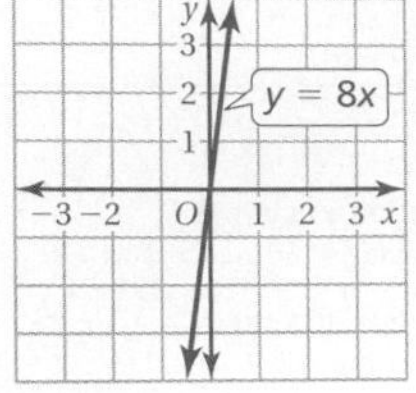

Extra Example 3

The variable y varies inversely with x. When $x = 2$, $y = -4$.

a. Write an inverse equation that relates x and y. $y = -\frac{8}{x}$

b. Graph the inverse variation equation. Describe the domain and range.

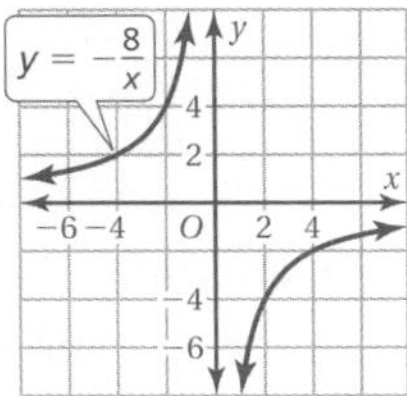

Both the domain and range are all real numbers except 0.

On Your Own

3. $y = 5x$

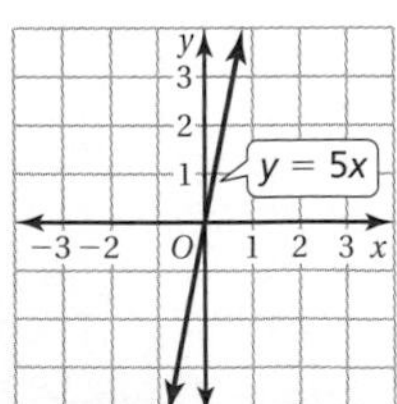

4. See Additional Answers.

English Language Learners

Vocabulary

Students may confuse the words *variation* and *variable*. A variable is a symbol, usually a letter, that represents a number that changes. Stress that variation refers to how the variable y varies in relation to the variable x.

Laurie's Notes

Example 2

- **Big Idea:** When y varies directly with x, only one ordered pair is needed to write an equation of the line. This ordered pair can be used to find k, the constant of proportionality.
- Note that the graphs of direct variation equations pass through the origin. So, you know another ordered pair, (0, 0).

? "What ordered pair satisfies the equation you are trying to write?" (12, −6)

- Write the direct variation equation. Substitute for x and y to solve for k.

? "What other ordered pairs satisfy the equation?" Answers will vary. Be sure students understand that for all ordered pairs, $\frac{y}{x} = -\frac{1}{2}$

Example 3

- **Big Idea:** When y varies inversely with x, only one ordered pair is needed to write an equation. This ordered pair can be used to find k, the constant of variation.

? "What ordered pair satisfies the equation you are trying to write?" (2, 5)

- Write the inverse variation equation. Substitute for x and y to solve for k.

? "What other ordered pairs satisfy the equation?" Answers will vary. Be sure students understand that for all ordered pairs, $xy = 10$.

- The table of values in part (b) indicates that the function is undefined when $x = 0$. Division by 0 is undefined.
- **FYI:** The graph is a *hyperbola* with vertical and horizontal *asymptotes*. Asymptotes will be explored more in the next lesson.
- **MP5 Use Appropriate Tools Strategically:** If time permits, explore the function using a graphing calculator. Use a table and small increments of x to explore what happens as x gets closer to 0. Use larger increments to explore what happens as x increases.

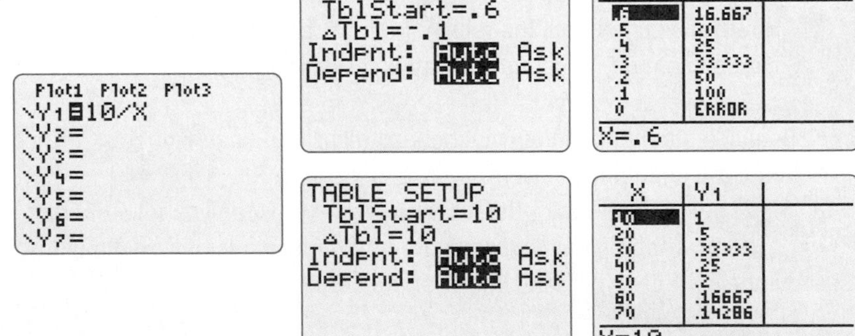

On Your Own

- **Neighbor Check:** Have students work independently and then have their neighbor check their work. Have students discuss any discrepancies.

EXAMPLE 2 Writing and Graphing a Direct Variation Equation

Reading

For direct variation equations, you can say "y varies directly with x" or "y is directly proportional to x." For inverse variation equations, you can say "y varies inversely with x" or "y is inversely proportional to x."

The variable y varies directly with x. When $x = 12$, $y = -6$. Write and graph a direct variation equation that relates x and y.

Find the value of k.

$y = kx$ — Write the direct variation equation.

$-6 = k(12)$ — Substitute 12 for x and -6 for y.

$-\frac{1}{2} = k$ — Divide each side by 12.

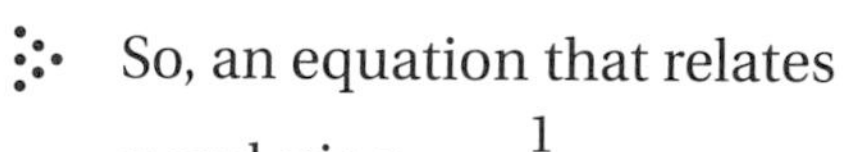

So, an equation that relates x and y is $y = -\frac{1}{2}x$.

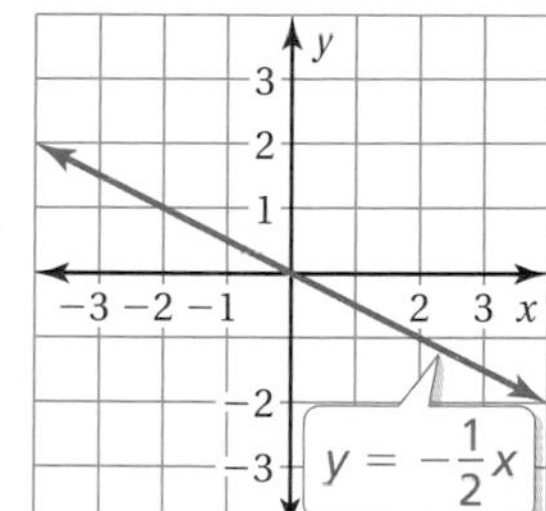

EXAMPLE  3 Writing and Graphing an Inverse Variation Equation

The variable y varies inversely with x. When $x = 2$, $y = 5$.

a. Write an inverse variation equation that relates x and y.

Find the value of k.

$y = \frac{k}{x}$ — Write the inverse variation equation.

$5 = \frac{k}{2}$ — Substitute 2 for x and 5 for y.

$10 = k$ — Multiply each side by 2.

So, an equation that relates x and y is $y = \frac{10}{x}$.

b. Graph the inverse variation equation. Describe the domain and range.

Make a table of values.

x	-10	-5	-2	0	2	5	10
y	-1	-2	-5	undef.	5	2	1

Plot the ordered pairs. Draw a smooth curve through the points in each quadrant. Both the domain and range are all real numbers except 0.

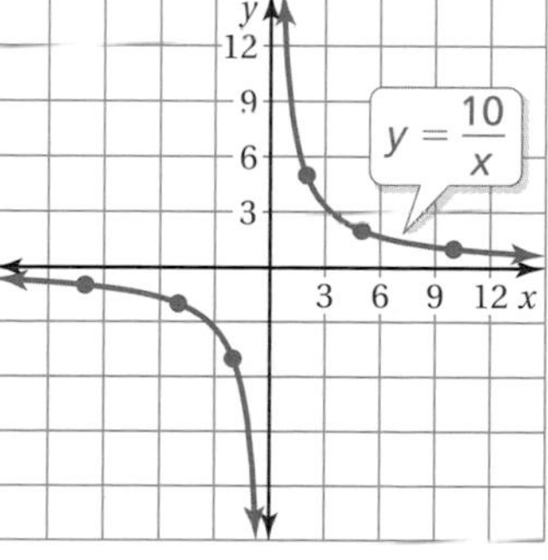

Study Tip

Notice that the equation $y = \frac{10}{x}$ is undefined when $x = 0$. So, there is no point on the graph for $x = 0$.

On Your Own

Exercises 7–12

3. The variable y varies directly with x. When $x = 3$, $y = 15$. Write and graph a direct variation equation that relates x and y.

4. The variable y varies inversely with x. When $x = 5$, $y = 4$. Write and graph an inverse variation equation that relates x and y.

EXAMPLE 4 Standardized Test Practice

Which situation represents inverse variation?

Ⓐ You buy several movie tickets for $7.50 each.

Ⓑ You earn $0.50 for each pound of aluminum cans you recycle.

Ⓒ The cost of a $600 cabin rental is shared equally by a group of friends.

Ⓓ You download several songs for $0.99 each.

Make a table of values for each situation.

Ⓐ

Number of tickets, x	1	2	3
Total cost, y	7.50	15	22.50

The ratio $\frac{y}{x}$ is constant.

Ⓑ

Number of pounds, x	1	2	3
Total earned, y	0.50	1	1.50

The ratio $\frac{y}{x}$ is constant.

Ⓒ

Number of people, x	1	2	3
Cost per person, y	600	300	200

The product xy is constant.

Ⓓ

Number of songs, x	1	2	3
Total cost, y	0.99	1.98	2.97

The ratio $\frac{y}{x}$ is constant.

∴ The correct answer is Ⓒ.

EXAMPLE 5 Real-Life Application

You bike 15 miles each morning. Your time t (in hours) to bike 15 miles is given by $t = \frac{15}{r}$, where r is your average speed (in miles per hour). Graph the function. Make a conclusion from the graph.

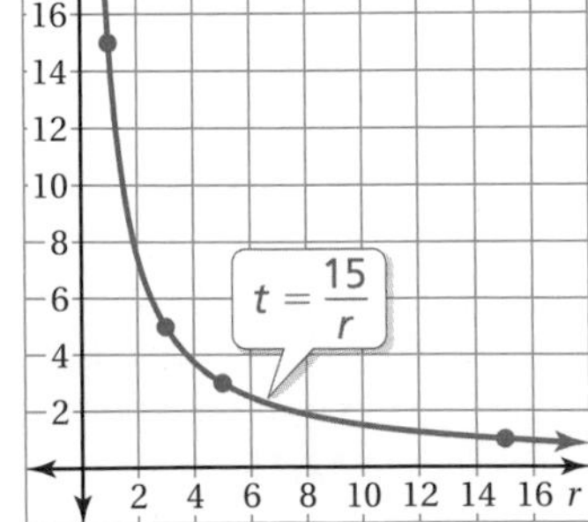

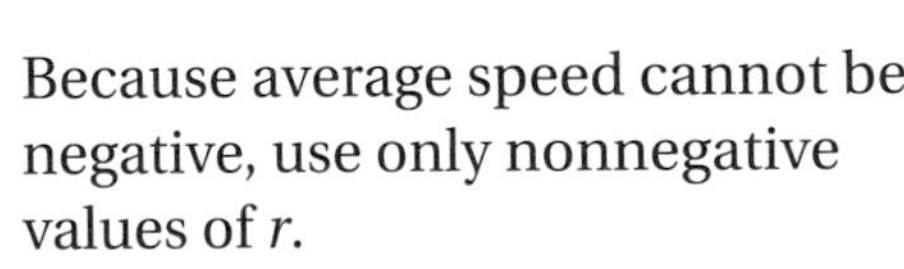

Because average speed cannot be negative, use only nonnegative values of r.

r	0	1	3	5	15
t	undef.	15	5	3	1

From the graph, you can see that as your average speed increases, the time it takes you to bike 15 miles decreases.

On Your Own

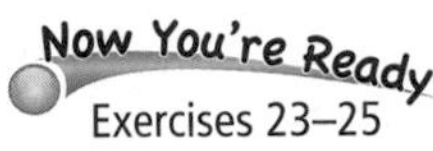

5. The cost of a taxi ride is shared equally by several friends. Does this situation represent direct variation or inverse variation? Explain.

6. **WHAT IF?** In Example 5, you bike 12 miles each morning. Write and graph a function that represents your time. Then make a conclusion from the graph.

Laurie's Notes

Example 4

- Ask a volunteer to read each of the four scenarios.
- For each choice, you want students to think about what happens to y as x increases. In three of the four choices, y increases proportionally. Only in the third choice does y decrease.
- Show the checks for constant ratios ($y : x$) and constant products (xy).

Example 5

- **MP2 Reason Abstractly and Quantitatively:** Reason through this problem with students. The faster your speed, the less time it takes to bike 15 miles. The slower your speed, the more time it takes to bike 15 miles.
- **Connection:** When the equation $y = \frac{15}{x}$ is rewritten as $xy = 15$, it should remind students of the distance, speed, and time formula $rt = d$.
- Be sure students understand that the ordered pairs are (time, speed), where time is measured in hours and speed is measured in miles per hour.
- As x (time) increases, y (speed) decreases. As y (speed) increases, x (time) decreases. These ideas could be explored using a graphing calculator, preparing students for the discussion of asymptotes.

? "What does the ordered pair (1.5, 10) mean?" It takes 1.5 hours to bike 15 miles at 10 miles per hour.

On Your Own

- In Question 5, encourage students to check their answer by assigning a value for the total cost, and then observing what happens as you divide it by greater numbers of people.

Closure

- **Exit Ticket:** Write an equation for the $48 shopping spree at the beginning of the lesson. Let y be the cost per item and let x be the number of items, $xy = 48$ or $y = \frac{48}{x}$.

Technology For the Teacher

The Dynamic Planning Tool
Editable Teacher's Resources at *BigIdeasMath.com*

Extra Example 4

The number of hours h it takes for a block of ice to melt depends on the temperature t of the room. Does this situation represent direct variation or inverse variation? Explain. inverse variation; The number of hours it takes ice to melt is inversely proportional to the temperature of the room.

Extra Example 5

The force f (in pounds) it takes to break a board is given by $f = \frac{11}{\ell}$, where ℓ is the length (in feet) of the board. Graph the function. Make a conclusion from the graph.

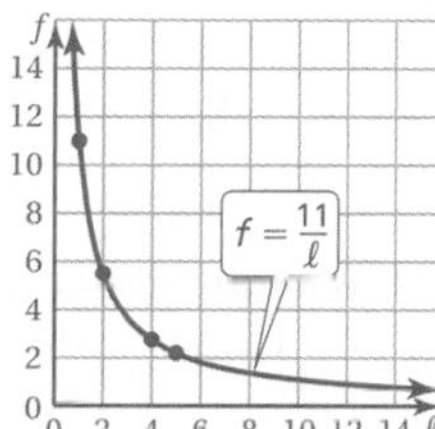

The longer the board, the less force required to break the board.

On Your Own

5. inverse variation; The number of friends is inversely proportional to the cost per person for the taxi ride.

6. The function is $y = \frac{12}{x}$. As the amount of time you take to bike 12 miles increases, your average speed decreases.

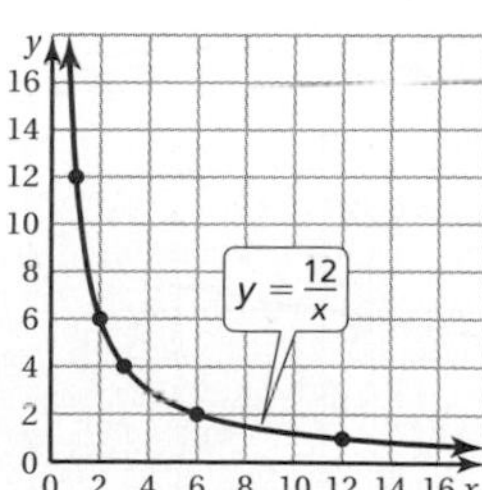

Vocabulary and Concept Check

1. In direct variation, y is the product of x and the constant k. In inverse variation, y is the quotient of the constant k and x.
2. graph of $g(x)$; It shows direct variation, whereas the rest of the graphs show inverse variation.

Practice and Problem Solving

3. direct variation; The ratios $\frac{y}{x}$ are constant.
4. inverse variation; The products xy are constant.
5. direct variation; The equation can be written as $y = kx$.
6. inverse variation; The equation can be written as $y = \frac{k}{x}$.
7. $y = 3x$

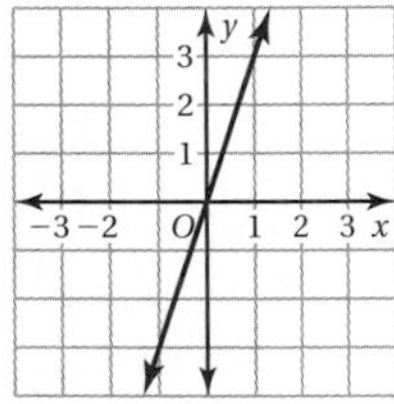

8. $y = 4x$

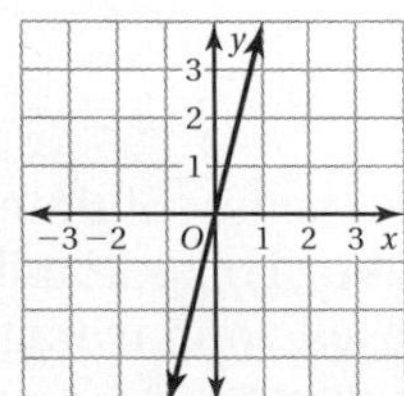

9. $y = \frac{1}{6}x$

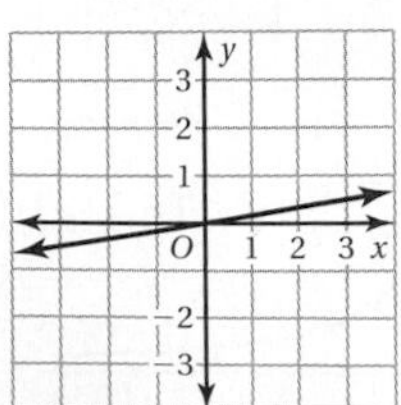

10–13. See Additional Answers.

Assignment Guide and Homework Check

Level	Assignment	Homework Check
Average	1, 2, 3–27 odd, 14, 26, 39–42	5, 11, 13, 14, 23, 26
Advanced	1, 2, 4–16 even, 23, 26–38, 39–42	6, 14, 23, 26, 29, 30

For Your Information

- **Exercises 31–37** A function is *even* if its graph is symmetric with respect to the y-axis. A function is *odd* if its graph is symmetric with respect to the origin.

Common Errors

- **Exercise 3** Students may not believe that x and y show direct variation simply because x and y are increasing by different rates. Remind them to check each product xy and each ratio $\frac{y}{x}$.
- **Exercises 5 and 6** Students may try to identify the type of variation without solving for y. Remind them of the equations $y = kx$ and $y = \frac{k}{x}$.
- **Exercises 7–12** Students may substitute the wrong values for x and y. Tell them to be careful when substituting and that k should still be in the equation after substituting for x and y.

11.1 Record and Practice Journal

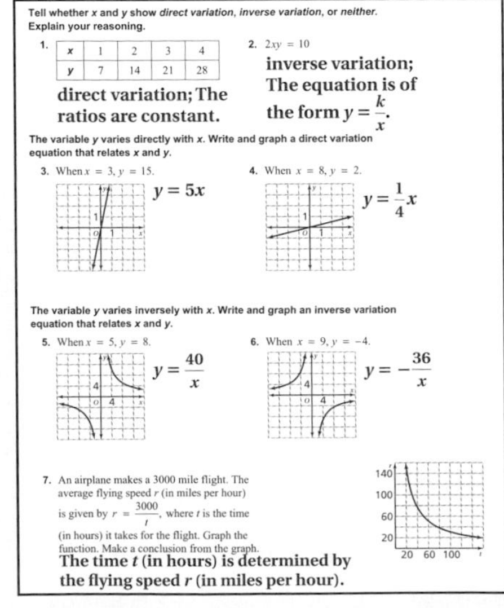

Tell whether x and y show *direct variation, inverse variation,* or *neither.* Explain your reasoning.

1.

x	1	2	3	4
y	7	14	21	28

direct variation; The ratios are constant.

2. $2xy = 10$

inverse variation; The equation is of the form $y = \frac{k}{x}$.

The variable y varies directly with x. Write and graph a direct variation equation that relates x and y.

3. When $x = 3, y = 15$. **$y = 5x$**
4. When $x = 8, y = 2$. **$y = \frac{1}{4}x$**

The variable y varies inversely with x. Write and graph an inverse variation equation that relates x and y.

5. When $x = 5, y = 8$. **$y = \frac{40}{x}$**
6. When $x = 9, y = -4$. **$y = -\frac{36}{x}$**
7. An airplane makes a 3000 mile flight. The average flying speed r (in miles per hour) is given by $r = \frac{3000}{t}$, where t is the time (in hours) it takes for the flight. Graph the function. Make a conclusion from the graph.

The time t (in hours) is determined by the flying speed r (in miles per hour).

Technology For the Teacher
Answer Presentation Tool

11.1 Exercises

 Vocabulary and Concept Check

1. **VOCABULARY** Explain how direct variation equations and inverse variation equations are different.

2. **WHICH ONE DOESN'T BELONG?** Which graph does *not* belong with the other three? Explain your reasoning.

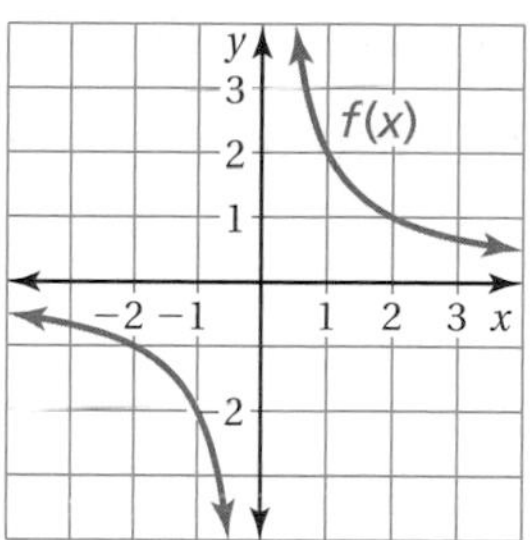

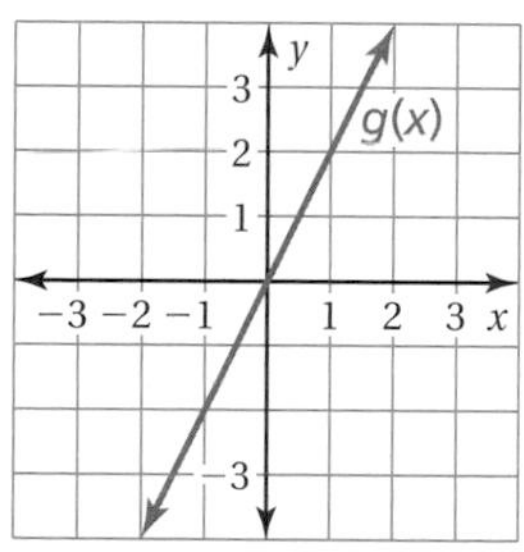

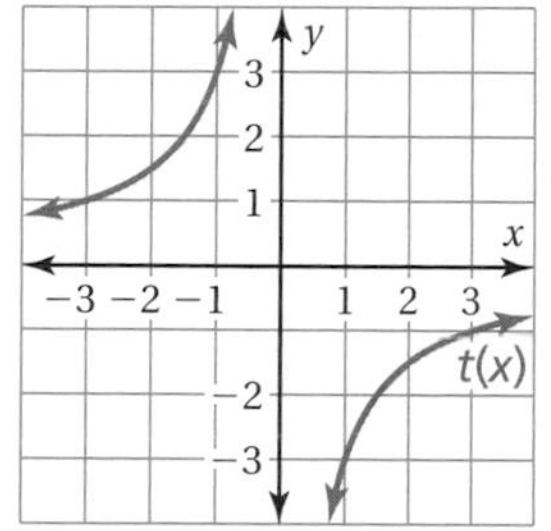

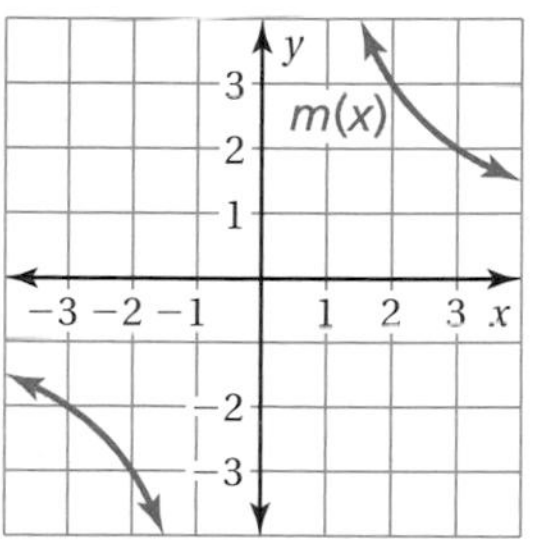

 Practice and Problem Solving

Tell whether x and y show *direct variation, inverse variation,* or *neither.* Explain your reasoning.

 3.

x	1	2	3	4
y	2	4	6	8

4.

x	1	2	3	4
y	12	6	4	3

5. $2y = x$

6. $-3xy = 6$

The variable y varies directly with x. Write and graph a direct variation equation that relates x and y.

2 7. When $x - 2$, $y = 6$.

8. When $x = 3$, $y = 12$.

9. When $x = 30$, $y = 5$.

The variable y varies inversely with x. Write and graph an inverse variation equation that relates x and y.

3 10. When $x = 3$, $y = 5$.

11. When $x = 5$, $y = 9$.

12. When $x = 5$, $y = 6$.

13. **VOLUNTEERS** You want to raise \$500 for a charity. You volunteer h hours and raise r dollars each hour. The equation $hr = 500$ represents this situation. Does this represent direct variation, inverse variation, or neither? Explain your reasoning.

14. **ERROR ANALYSIS** The variable y varies inversely with x. When $x = 8$, $y = 5$. Describe and correct the error in writing an inverse variation equation that relates x and y.

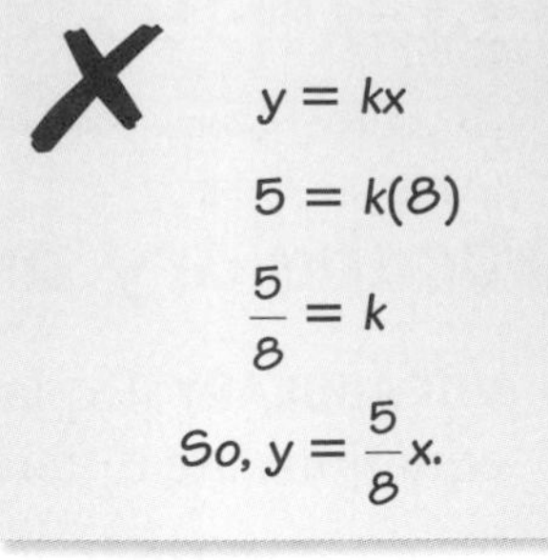

✗
$y = kx$
$5 = k(8)$
$\frac{5}{8} = k$
So, $y = \frac{5}{8}x$.

Graph the equation. Describe the domain and range.

15. $y = \frac{1}{x}$

16. $\frac{y}{x} = -\frac{1}{2}$

17. $xy = 9$

18. **REASONING** When y varies directly with x, does x vary directly with y? If so, describe the relationship between the constants of proportionality. Explain your reasoning.

The variable y varies inversely with x. Write an inverse variation equation that relates x and y. Then find the missing value of x or y.

19. When $x = 6$, $y = 2$. Find x when $y = 1$.

20. When $x = 4$, $y = 2$. Find x when $y = \frac{1}{2}$.

21. When $x = -2$, $y = -5$. Find y when $x = 4$.

22. When $x = 20$, $y = \frac{4}{5}$. Find y when $x = 8$.

Determine whether the situation represents *direct variation* or *inverse variation*. Justify your answer.

4 23. You have enough money to buy 5 hats for \$10 each or 10 hats for \$5 each.

24. Your cousin earns \$50 for mowing 2 lawns or \$75 for mowing 3 lawns.

25. The money the swim team earns from a car wash is divided evenly among the members.

5 26. **RUNNING** You race in a 200-meter dash. Your average speed r (in meters per second) is given by $r = \frac{200}{t}$, where t is the time (in seconds) it takes you to finish the race. Graph the function. Make a conclusion from the graph.

27. **VACATION** The amount v of vacation time (in hours) that an employee earns varies directly with the amount t of time (in months) she works. An employee who works 2 months earns 36 hours of vacation time.

   a. Write and graph a direct variation equation that relates v and t.

   b. How many hours of vacation time does the employee earn after working 5 months?

Common Errors

- **Exercises 6, 16 and 17** Students may struggle with multiplying or dividing by x when solving for y. Remind them of the Multiplication and Division Properties of Equality and let them know that these properties apply to variables as well as numbers.
- **Exercises 19–22** Students may substitute the wrong values for x and y. Tell them to be careful when substituting and that k should still be in the equation after substituting for x and y.

Practice and Problem Solving

14. An inverse variation equation has the form $y = \frac{k}{x}$; $y = \frac{40}{x}$

15.

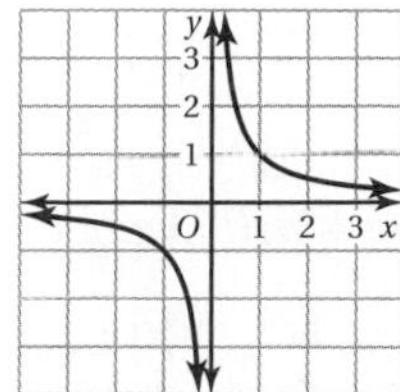

Both the domain and range are all real numbers except 0.

16.

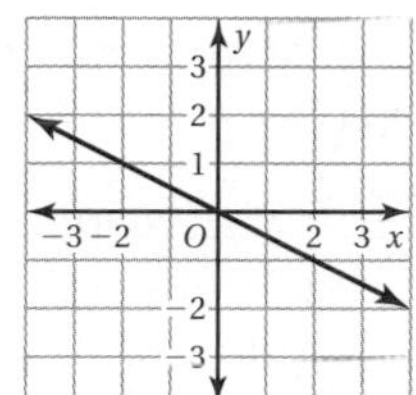

Both the domain and range are all real numbers.

17.

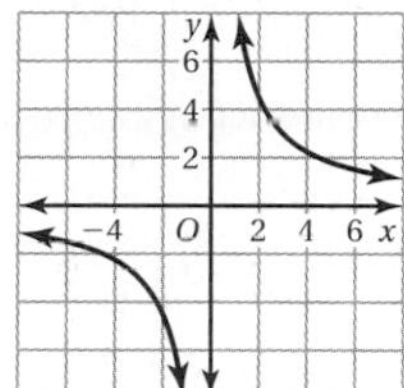

Both the domain and range are all real numbers except 0.

18. yes; The constants are reciprocals of each other.

19. $y = \frac{12}{x}$; $x = 12$

20. $y = \frac{8}{x}$; $x = 16$

21–27. See Additional Answers.

Differentiated Instruction

Connection

The direct variation equation can be written as $\frac{y}{x} = k$. Solving a direct variation problem is similar to setting up and solving a proportion. For Exercise 27(b), $k = \frac{v}{t} = \frac{36}{2}$. So, the proportion is $\frac{36}{2} = \frac{v}{5}$.

Practice and Problem Solving

28. The rate of change of v is not constant whereas the rate of change of d is constant.

29. **a.** $t = \frac{5000}{p}$

b. 200 hours

30. See *Taking Math Deeper.*

31. odd **32.** odd

33. even **34.** neither

35. even **36.** neither

37. **a.** symmetric with respect to the y-axis

b. symmetric with respect to the origin (reflection in the y-axis followed by a reflection in the x-axis)

38. yes; $f(-x) = -f(x)$ for all direct and inverse variation equations.

Fair Game Review

39–41. See Additional Answers.

42. D

Mini-Assessment

Tell whether x and y show *direct variation, inverse variation,* or *neither.*

1. $y = x + 9$ neither
2. $y = \frac{1}{6}x$ direct variation
3. $-x - y = 10$ neither
4. $\frac{y}{4} = \frac{3}{x}$ inverse variation
5. One mile is approximately equal to 1.6 kilometers. Determine whether the situation represents direct variation or inverse variation. Write an equation that relates x miles to y kilometers. direct variation; $y = 1.6x$

Taking Math Deeper

Exercise 30

A key to this exercise is realizing how to express the distance from the dog to the fulcrum.

1. Write an expression for the distance from the dog to the fulcrum.

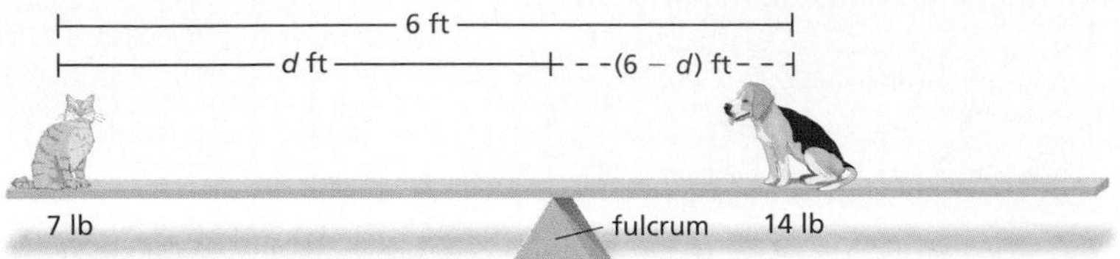

The cat and dog are 6 feet apart and the cat is d feet from the fulcrum. So, the dog is $(6 - d)$ feet from the fulcrum.

2. Let x represent the distance from the animal to the fulcrum and let y represent the weight of the animal. So, the situation can be described by the ordered pairs (distance, weight).

Use each animal to write an ordered pair.

Cat: $(d, 7)$ **Dog:** $(6 - d, 14)$

3. Because the weight varies inversely with the distance, the product xy is equivalent for each pair of points. So, given two points (x_1, y_1) and (x_2, y_2), you know that $x_1y_1 = x_2y_2$. Use this to write and solve an equation.

$$7d = 14(6 - d)$$
$$7d = 84 - 14d$$
$$d = 4$$

So, the cat is 4 feet from the fulcrum and the dog is $6 - 4 = 2$ feet from the fulcrum.

Reteaching and Enrichment Strategies

If students need help...	If students got it...
Resources by Chapter • Practice A and Practice B • Puzzle Time Record and Practice Journal Practice Differentiating the Lesson Lesson Tutorials Skills Review Handbook	Resources by Chapter • Enrichment and Extension Start the next section

28. **REASONING** Make a table using positive x-values for the inverse variation equation $v = \frac{6}{x}$ and the direct variation equation $d = 6x$. How does the rate of change of v differ from the rate of change of d?

29. **THEATER** A performing arts company is hiring actors as extras for a theater performance. The amount t of performance time (in hours per person) varies inversely with the number p of extras hired. The director estimates that he will need 20 extras performing 250 hours each.

   a. Write an inverse variation equation that relates t and p.

   b. The director decides to hire 25 extras. How much performance time will each extra receive?

30. **STRUCTURE** To balance the board in the diagram, the distance (in feet) of each animal from the center of the board must vary inversely with its weight (in pounds). What is the distance of each animal from the fulcrum?

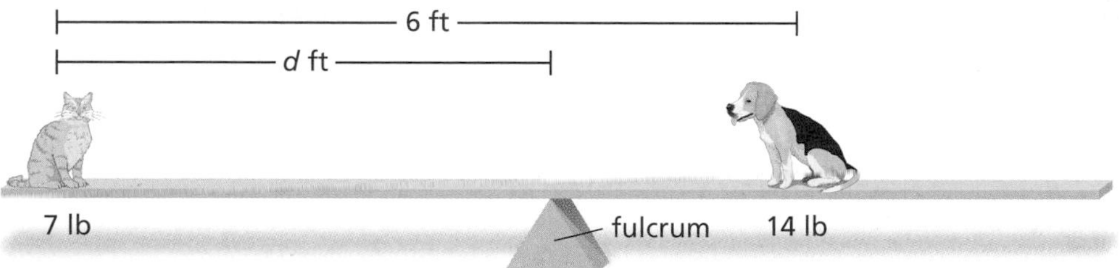

A function f is odd if $f(-x) = -f(x)$. A function f is even if $f(-x) = f(x)$. Determine whether the function is *odd*, *even*, or *neither*.

31. $f(x) = x$

32. $f(x) = \frac{1}{x}$

33. $f(x) = x^2$

34. $f(x) = \sqrt{x}$

35. $f(x) = |x|$

36. $f(x) = 2^x$

37. **REASONING** Describe the symmetry shown in the graph of (a) an even function and (b) an odd function. Justify your answers.

38. **Precision** Are all direct variation and inverse variation equations odd functions? Explain.

Fair Game Review What you learned in previous grades & lessons

Graph the function. Compare the graph to the graph of $y = x^2$. *(Section 8.1 and Section 8.2)*

39. $y = 3x^2$

40. $y = x^2 + 2$

41. $y = x^2 - 1$

42. **MULTIPLE CHOICE** What is the solution of the equation $\sqrt{x} - 5 = 4$? *(Section 10.2)*

   (A) 1 (B) 3 (C) 9 (D) 81

11.2 Graphing Rational Functions

COMMON CORE STATE STANDARDS
A.REI.10
F.BF.4a

Essential Question What are the characteristics of the graph of a rational function?

1 ACTIVITY: Graphing a Rational Function

Work with a partner. As a fundraising project, your math club is publishing an optical illusion calendar. The cost of the art, typesetting, and paper is \$850. In addition to this one-time cost, the unit cost of printing each calendar is \$3.25.

a. Let A represent the average cost of each calendar. Write a rational function that gives the average cost of printing x calendars.

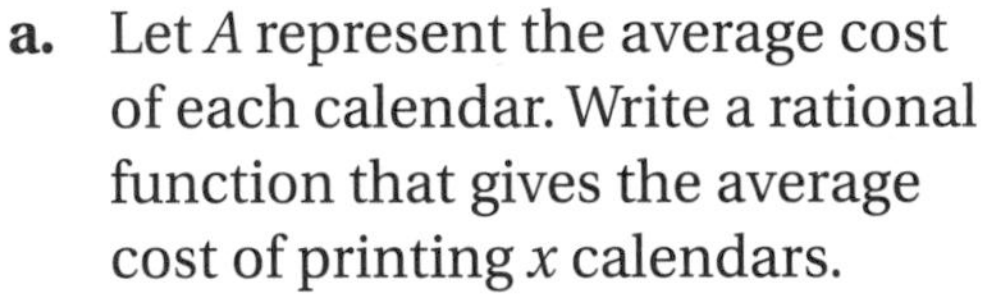

$$A = \frac{}{x}$$

b. Make a table showing the average costs for several different production amounts. Then use the table to graph the average cost function.

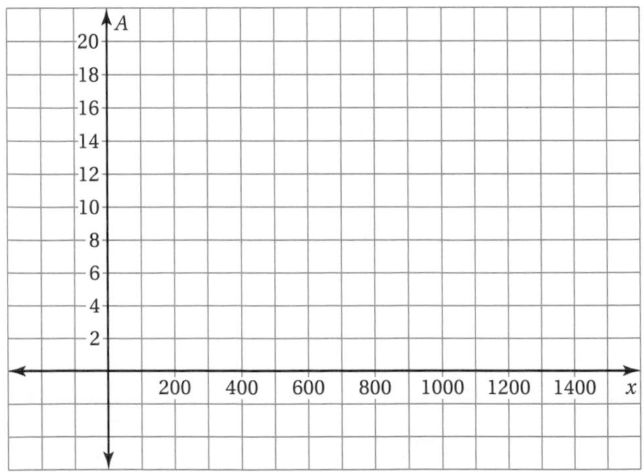

Laurie's Notes

Introduction

Standards for Mathematical Practice

- **MP2 Reason Abstractly and Quantitatively:** Vertical and horizontal asymptotes make sense to students if they have had the opportunity to investigate function values near asymptotes. Mathematically proficient students can reason quantitatively, aided by graphing technology.

Motivate

- Gather 24 like items (counters, paper clips, or wrapped mints).
- When students enter, give one student all 24 items. This will likely be met with comments about why one person got everything.
- Eventually invite a second student to share evenly in the 24 items. Each student now has 12 items.
- Repeat this multiple times, each time discussing what happens when another student joins the recipient group.
- ? "What equation describes the number y of items each person receives when 24 items are divided evenly among x people?" $y = \frac{24}{x}$
- ? "How different would the problem be if I add another item to the 24 original items each time a new person joins the recipient group?" Students may be unsure and that's okay. Today's investigation will help them explore this idea.

Activity Notes

Activity 1

- Discuss with students that there can be one-time costs in a fundraiser, such as purchasing equipment, renting, and licensing.
- **MP1a Make Sense of Problems:** In addition to these fixed costs, there can be costs that vary, such as the cost per item produced. Read through the description with students to be sure they understand the context.
- ? "What is the fixed cost in this problem?" \$850 for art, typesetting, and paper
- ? "What is the variable cost in this problem?" cost of printing each calendar
- **MP4 Model with Mathematics:** Make sure students have the correct expression in the numerator of the average cost function.
- The scales on the axes should give students a clue as to what values of x they can use to graph the function.
- **Big Idea:** As more calendars are printed, the fixed cost of \$850 is "shared" over more calendars.

Common Core State Standards

A.REI.10 Understand that the graph of an equation in two variables is the set of all its solutions plotted in the coordinate plane, often forming a curve (which could be a line).

F.BF.4a Solve an equation of the form $f(x) = c$ for a simple function f that has an inverse and write an expression for the inverse.

Previous Learning

Students should know how to graph inverse variation equations.

Start Thinking! and Warm Up

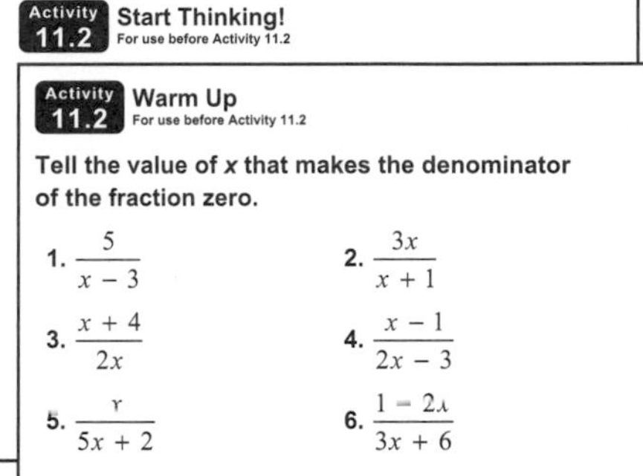

Activity 11.2 Start Thinking! For use before Activity 11.2

Activity 11.2 Warm Up For use before Activity 11.2

Tell the value of x that makes the denominator of the fraction zero.

1. $\frac{5}{x-3}$
2. $\frac{3x}{x+1}$
3. $\frac{x+4}{2x}$
4. $\frac{x-1}{2x-3}$
5. $\frac{x}{5x+2}$
6. $\frac{1-2x}{3x+6}$

11.2 Record and Practice Journal

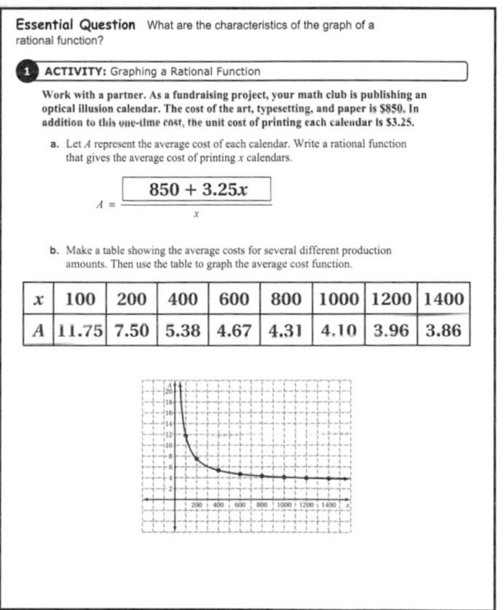

Essential Question What are the characteristics of the graph of a rational function?

1 ACTIVITY: Graphing a Rational Function

Work with a partner. As a fundraising project, your math club is publishing an optical illusion calendar. The cost of the art, typesetting, and paper is \$850. In addition to this one-time cost, the unit cost of printing each calendar is \$3.25.

a. Let A represent the average cost of each calendar. Write a rational function that gives the average cost of printing x calendars.

$A = \frac{850 + 3.25x}{x}$

b. Make a table showing the average costs for several different production amounts. Then use the table to graph the average cost function.

x	100	200	400	600	800	1000	1200	1400
A	11.75	7.50	5.38	4.67	4.31	4.10	3.96	3.86

English Language Learners

Vocabulary

English language learners should see the word *ratio* in the word *rational.* A rational number is a ratio of two integers. A rational function is the ratio of two polynomials.

11.2 Record and Practice Journal

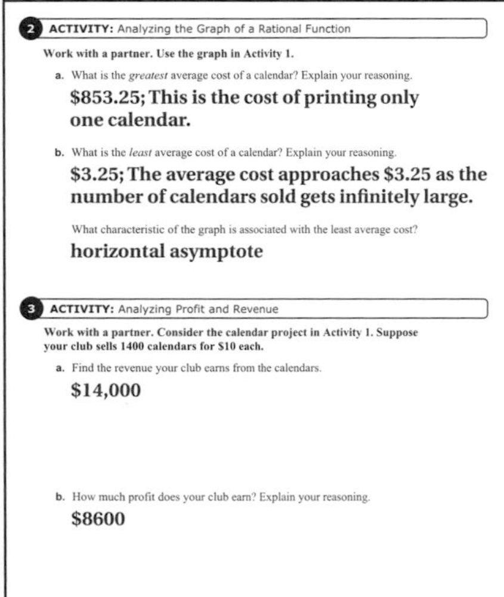

2 ACTIVITY: Analyzing the Graph of a Rational Function

Work with a partner. Use the graph in Activity 1.

a. What is the *greatest* average cost of a calendar? Explain your reasoning.

$853.25; This is the cost of printing only one calendar.

b. What is the *least* average cost of a calendar? Explain your reasoning.

$3.25; The average cost approaches $3.25 as the number of calendars sold gets infinitely large.

What characteristic of the graph is associated with the least average cost?

horizontal asymptote

3 ACTIVITY: Analyzing Profit and Revenue

Work with a partner. Consider the calendar project in Activity 1. Suppose your club sells 1400 calendars for $10 each.

a. Find the revenue your club earns from the calendars.

$14,000

b. How much profit does your club earn? Explain your reasoning.

$8600

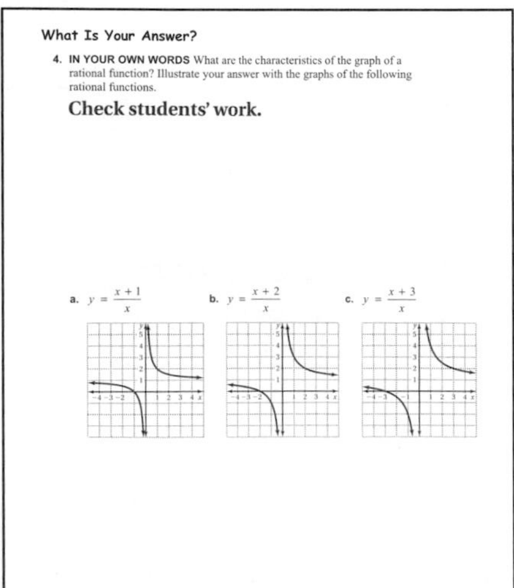

What Is Your Answer?

4. **IN YOUR OWN WORDS** What are the characteristics of the graph of a rational function? Illustrate your answer with the graphs of the following rational functions.

Check students' work.

a. $y = \frac{x+1}{x}$ b. $y = \frac{x+2}{x}$ c. $y = \frac{x+3}{x}$

Laurie's Notes

Activity 2

- Students are now asked to interpret the graph that they made in Activity 1.
- Selling just 1 calendar gives an average cost of $853.25.
- The average cost continues to decrease as the number of calendars sold increases.
- The average cost cannot go below $3.25, the unit cost of printing each calendar.
- This can be seen in the graph. As x increases, y approaches 3.25, but never reaches it.
- **Big Idea:** Students may recognize that the function can be written as $A = \frac{850}{x} + 3.25$.

Activity 3

? "How do you calculate profit?" Profit = Revenue − Expenses

- Remind students of the $850 fixed costs.
- When students have finished, have them share their reasoning.
- **MP5 Use Appropriate Tools Strategically:** Students could create a spreadsheet to explore this problem.

What Is Your Answer?

- If using graphing calculators, students should also explore the functions using tables.

Closure

- **Exit Ticket:** The variable cost increases to $3.50 per calendar. How does this change your club's profit when selling 1400 calendars for $10 each? decreases profit to $8250

Technology For the Teacher

The Dynamic Planning Tool
Editable Teacher's Resources at *BigIdeasMath.com*

2 ACTIVITY: Analyzing the Graph of a Rational Function

Work with a partner. Use the graph in Activity 1.

a. What is the *greatest* average cost of a calendar? Explain your reasoning.

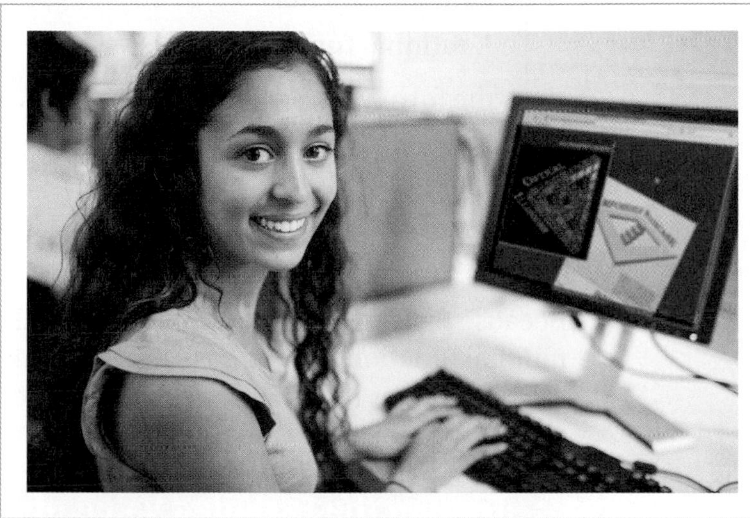

b. What is the *least* average cost of a calendar? Explain your reasoning. What characteristic of the graph is associated with the least average cost?

3 ACTIVITY: Analyzing Profit and Revenue

Work with a partner. Consider the calendar project in Activity 1. Suppose your club sells 1400 calendars for $10 each.

a. Find the revenue your club earns from the calendars.

b. How much profit does your club earn? Explain your reasoning.

What Is Your Answer?

4. **IN YOUR OWN WORDS** What are the characteristics of the graph of a rational function? Illustrate your answer with the graphs of the following rational functions.

a. $y = \dfrac{x+1}{x}$ **b.** $y = \dfrac{x+2}{x}$ **c.** $y = \dfrac{x+3}{x}$

Practice Use what you learned about the graphs of rational functions to complete Exercises 4 and 5 on page 555.

11.2 Lesson

Key Vocabulary
rational function, *p. 552*
excluded value, *p. 552*
asymptote, *p. 553*

The inverse variation equations in Section 11.1 are *rational functions.*

Rational Function

A **rational function** is a function of the form $y = \frac{\text{polynomial}}{\text{polynomial}}$, where the denominator does not equal 0. The most basic rational function is $y = \frac{1}{x}$.

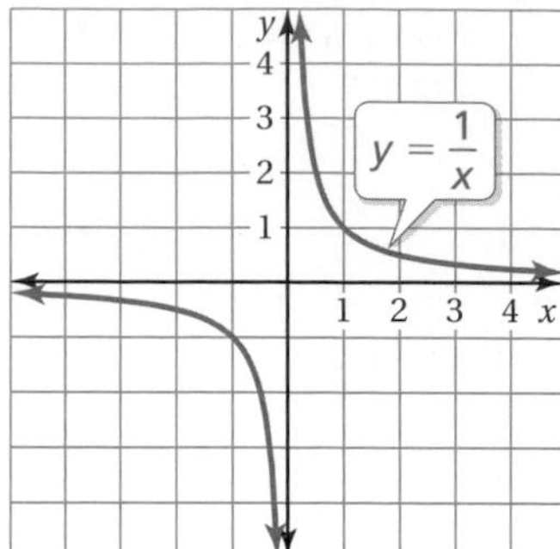

Because division by 0 is undefined, the value of the denominator of a rational function cannot be 0. So, the domain of a rational function *excludes* values that make the denominator 0. These values are called **excluded values** of the rational function.

EXAMPLE 1 Finding the Excluded Value of a Rational Function

Find the excluded value of $y = \frac{2}{x+5}$.

Find the value of x that makes the denominator 0.

$x + 5 = 0$ Use the denominator to write an equation.

$x = -5$ Subtract 5 from each side.

∴ The excluded value is $x = -5$.

EXAMPLE 2 Graphing a Rational Function

Graph $y = \frac{1}{x-1}$. Describe the domain and range.

The excluded value is $x = 1$, so choose x-values on either side of 1.

Step 1: Make a table of values.

x	−2	−1	0	0.5	1	1.5	2	3	4
y	$-\frac{1}{3}$	$-\frac{1}{2}$	−1	−2	undef.	2	1	$\frac{1}{2}$	$\frac{1}{3}$

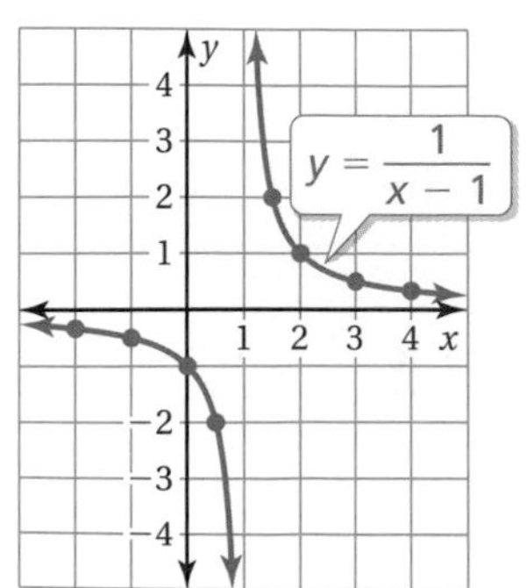

Step 2: Plot the ordered pairs.

Step 3: Draw a smooth curve through the points on each side of $x = 1$.

∴ The domain is all real numbers except 1 and the range is all real numbers except 0.

Multi-Language Glossary at BigIdeasMath.com.

Laurie's Notes

Introduction

Connect

- **Yesterday:** Students analyzed a rational function. (MP1a, MP2, MP4, MP5)
- **Today:** Students will graph rational functions.

Motivate

- Play a quick matching game to review previously studied functions and to introduce today's new function.
- Write function types (linear, exponential, quadratic, square root, rational) and sample equations on index cards. Pass them out to students and have them match each equation with a function type.
- Alternatively, you could write them on the board and work as a class. The goal is for students to recognize that there is a new function in the group.

Lesson Notes

Key Idea

- Review polynomials and different sets of numbers, including rational numbers, as a way to introduce rational functions.
- Write the Key Idea, defining a rational function. Point out that inverse variation equations are rational functions.
- You could mention that the graph is called a *hyperbola* and the two parts are called *branches*. Students may recognize the symmetry.
- Discuss the restriction on the denominator.
- Define excluded values of a rational function.

Example 1

- Write the rational function.

? "What value(s) of x will make the value of the denominator 0?" -5

Example 2

? "What value(s) of x will make the value of the denominator 0?" 1

- Because $x = 1$ is excluded from the domain, it is important to use x-values on both sides of 1 in the table of values.
- Point out to students that non-integer values are used near the excluded value. They will appreciate this technique as they graph the function.
- Complete the table of values shown.

? "What happens to the ratio of 1 divided by a positive number when the positive number is increasing?" The ratio approaches zero.

? "What happens to the ratio of 1 divided by a negative number when the negative number is decreasing?" The ratio approaches zero.

- **MP2 Reason Abstractly and Quantitatively:** Similar reasoning can be used to explore what happens to the function values as x gets closer to 1 (from either side). This prepares students for asymptotes.

Goal Today's lesson is graphing **rational functions.**

Lesson Materials	
Introduction	**Textbook**
• index cards	• graphing calculators

Start Thinking! and Warm Up

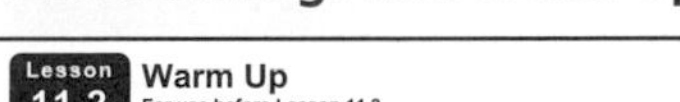

Why do you think the following function is called a rational function? Give another example of a rational function.

$$y = \frac{x}{x - 1}$$

Extra Example 1

Find the excluded value of $y = \dfrac{-3}{2x - 8}$. $x = 4$

Extra Example 2

Graph $y = \dfrac{1}{x} + 2$. Describe the domain and range.

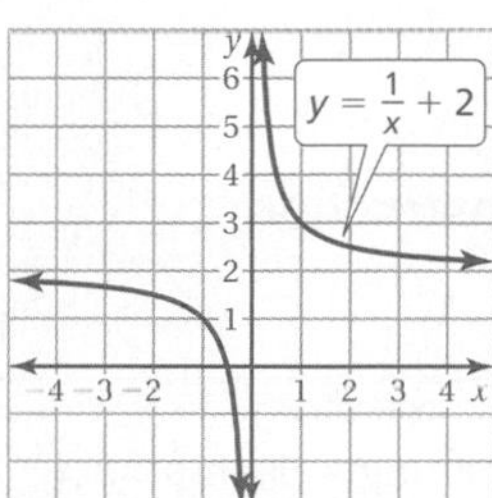

The domain is all real numbers except 0 and the range is all real numbers except 2.

On Your Own

1. $x = 0$ 2. $x = 4$

3. $x = -\frac{1}{3}$

4–6. See Additional Answers.

Extra Example 3

Identify the asymptotes of the graph of $y = \frac{1}{x+6} + 2$. Then describe the domain and range.

$x = -6$, $y = 2$; The domain is all real numbers except -6 and the range is all real numbers except 2.

On Your Own

7. $x = 0, y = 1$;
The domain is all real numbers except 0 and the range is all real numbers except 1.

8. $x = -5, y = 0$;
The domain is all real numbers except -5 and the range is all real numbers except 0.

9. $x = 3, y = -2$;
The domain is all real numbers except 3 and the range is all real numbers except -2.

Differentiated Instruction

Visual

For the graph of $y = \frac{a}{x-h} + k$, the asymptotes, $x = h$ and $y = k$, intersect at (h, k). To graph a rational function, students can locate this point and draw the asymptotes as dashed lines. Then, plot points on each side of the vertical asymptote and connect the points with a smooth curve.

Laurie's Notes

On Your Own

- Guide students in recognizing that there is a connection between the excluded value and the graph.

Discuss

- If you have taken the time to discuss what happens to the value of the function as x gets closer to the excluded value (from either side) or as x goes to infinity (in either direction), the idea of an asymptote will make sense to students.
- Describe asymptotes, referring back to previous examples.

Key Idea

- Write the Key Idea.
- The general form of a rational function may appear to contain many variables to students. Stress that x and y are the variables.
- **MP7 Look for and Make Use of Structure:** Write the general form and refer to the sample graph. The equation in the graph is of this form, where $h = 3$ and $k = 2$. The vertical asymptote is $x = 3$ and the horizontal asymptote is $y = 2$.
- Point out that dotted lines are used to represent the asymptotes.
- **Connection:** Recall that the vertex form of a quadratic function also has h and k in the equation. Students will see that these values shift the graph in the same way.

Example 3

- Write the equation, making note of the general form. Notice the use of color to identify $h = 2$ and $k = -4$.

? "What value of x is excluded from the domain?" 2

- **MP5 Use Appropriate Tools Strategically:** Take time for students to enter the equation in their calculators. Explore the function near $x = 2$ using a table of values.
- **Common Error:** Be sure that students use parentheses to enclose the denominator.
- Discuss the domain and range of the function.
- Students may notice that when using their calculators, there may be a solid vertical line that appears to be an asymptote. In fact, the calculator is just connecting the two parts of the graph.
- Some calculators have an option of "connected" and "dot" modes. The solid vertical line does not appear in dot mode.

On Your Own

- Students should work through the questions on their own, and then use a calculator to check their answers.

On Your Own

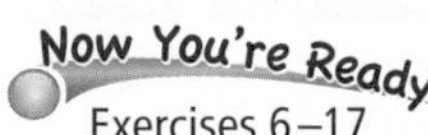

Find the excluded value of the function.

1. $y = \frac{3}{2x}$
2. $y = \frac{1}{x-4}$
3. $y = \frac{8}{3x+1}$

Graph the function. Describe the domain and range.

4. $y = -\frac{8}{x}$
5. $y = \frac{1}{x+2}$
6. $y = \frac{1}{x} - 1$

The excluded value in Example 2 is $x = 1$. Notice that the graph approaches the vertical line $x = 1$, but never intersects it. The graph also approaches the horizontal line $y = 0$, but never intersects it. These lines are called *asymptotes*. An **asymptote** is a line that a graph approaches, but never intersects.

Key Idea

Asymptotes

The graph of a rational function of the form $y = \frac{a}{x-h} + k$, where $a \neq 0$, has a vertical asymptote $x = h$ and a horizontal asymptote $y = k$.

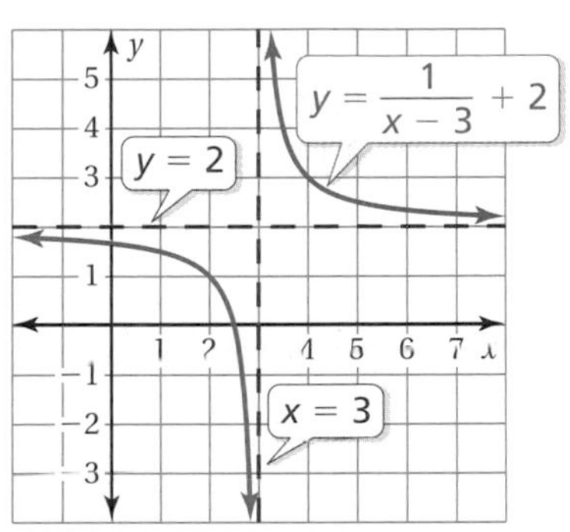

EXAMPLE 3 Identifying Asymptotes

Identify the asymptotes of the graph of $y = \frac{1}{x-2} - 4$. Then describe the domain and range.

Rewrite the function to find the asymptotes.

$$y = \frac{1}{x-2} + (-4)$$

Horizontal Asymptote: $y = -4$

Vertical Asymptote: $x = 2$

∴ The vertical asymptote is $x = 2$ and the horizontal asymptote is $y = -4$. So, the domain of the function is all real numbers except 2 and the range is all real numbers except -4.

Check

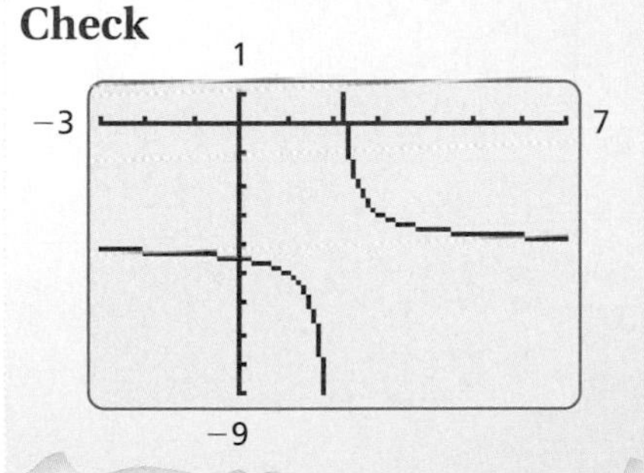

On Your Own

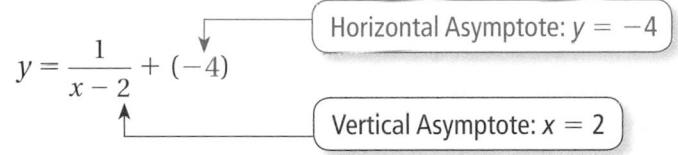

Now You're Ready
Exercises 19–24

Identify the asymptotes of the graph of the function. Then describe the domain and range.

7. $y = \frac{2}{x} + 1$
8. $y = \frac{1}{x+5}$
9. $y = \frac{8}{x-3} - 2$

EXAMPLE 4 Comparing Graphs of Rational Functions

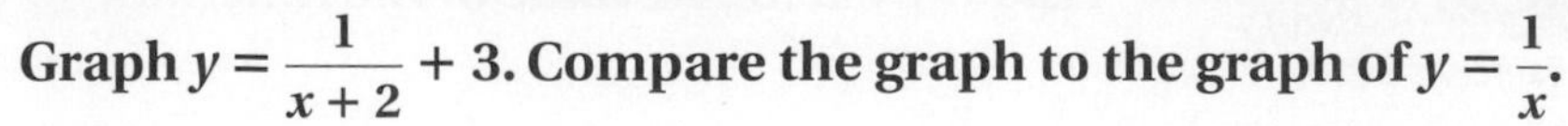

Graph $y = \dfrac{1}{x+2} + 3$. Compare the graph to the graph of $y = \dfrac{1}{x}$.

Study Tip
Use the asymptotes to help you draw the ends of the graph.

Step 1: Make a table of values. The vertical asymptote is $x = -2$, so choose x-values on either side of -2.

x	-4	-3	-2.5	-2	-1.5	-1	0
y	2.5	2	1	undef.	5	4	3.5

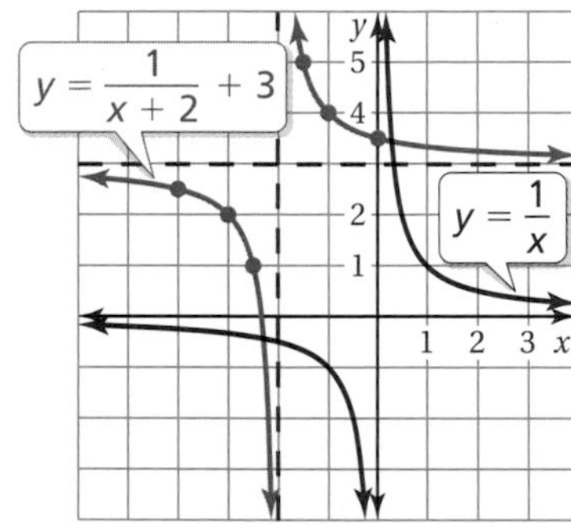

Step 2: Use dashed lines to graph the asymptotes $x = -2$ and $y = 3$. Then plot the ordered pairs.

Step 3: Draw a smooth curve through the points on each side of the vertical asymptote.

∴ The graph of $y = \dfrac{1}{x+2} + 3$ is a translation 3 units up and 2 units left of the graph of $y = \dfrac{1}{x}$.

EXAMPLE 5 Real-Life Application

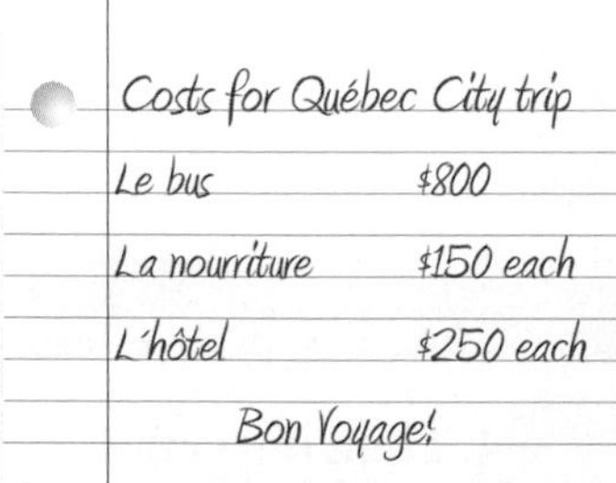

The French club is planning a trip to Québec City. The function $y = \dfrac{800}{x+2} + 400$ represents the cost y (in dollars) per student when x students and 2 chaperones go on the trip. Use a graphing calculator to graph the function. How many students must go on the trip for the cost per student to be about \$450?

Step 1: Use a graphing calculator to graph the function. Because the number of students cannot be negative, use only nonnegative values of x.

Step 2: Use the *trace* feature to find where the value of y is about 450.

850
Y1=800/(X+2)+400
X=14.042553 Y=449.86737
0 20
350

∴ About 14 students must go on the trip for the cost per student to be about \$450.

On Your Own

Now You're Ready
Exercises 28–33

Graph the function. Compare the graph to the graph of $y = \dfrac{1}{x}$.

10. $y = \dfrac{1}{x-4}$ **11.** $y = \dfrac{1}{x} - 6$ **12.** $y = -\dfrac{1}{x+3} - 3$

13. WHAT IF? In Example 5, how many students must go on the trip for the cost per student to be about \$480?

Laurie's Notes

Example 4

- ? "How does the graph of $y = (x + 2)^2 + 3$ compare to the graph of $y = x^2$?" translation 2 units to the left and 3 units up
- Say, "Let's see what happens to a rational function with similar values."
- When completing the table of values, students are reviewing rational number operations.
- Describe the translation.

Example 5

- Ask a volunteer to read the problem.
- ? "What is the domain of this function in context?" positive integers
- **MP5**: Be sure students enter the equation correctly and set a viewing window that is appropriate for the context.
- ? "What does the shape of the graph indicate about the cost per student?" As more students go on the trip, the cost per student decreases.
- ? "How do you interpret the horizontal asymptote?" The cost per student cannot be $400 or less.

On Your Own

- **Think-Pair-Share:** Students should read each question independently and then work with a partner to answer the questions. When they have answered the questions, the pair should compare their answers with another group and discuss any discrepancies.

Closure

- **Exit Ticket:** Write an equation of a rational function with a graph that has the vertical asymptote $x = -3$ and the horizontal asymptote $y = 5$.

  Sample answer: $y = \frac{1}{x+3} + 5$

Technology For the Teacher

The Dynamic Planning Tool
Editable Teacher's Resources at *BigIdeasMath.com*

Extra Example 4

Graph $y = \frac{1}{x-5} + 4$. Compare the graph to the graph of $y = \frac{1}{x}$.

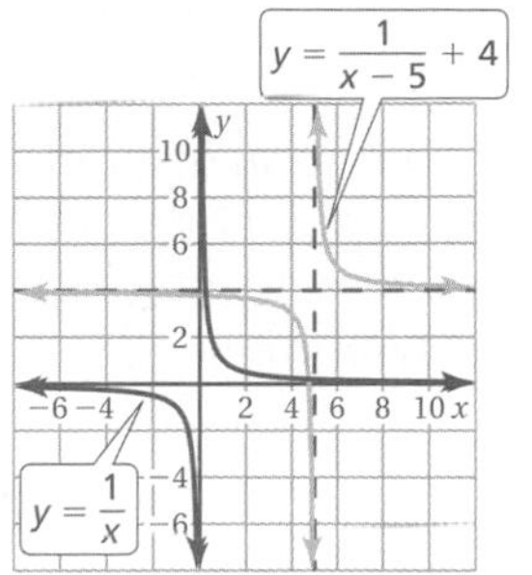

The graph of $y = \frac{1}{x-5} + 4$ is a translation 4 units up and 5 units to the right of the graph of $y = \frac{1}{x}$.

Extra Example 5

In Example 5, how many students must go on the trip for the cost per student to be about $425? 30 students

On Your Own

10.

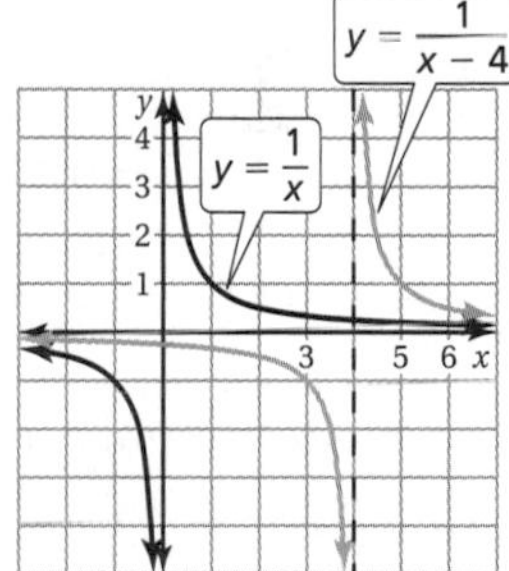

The graph of $y = \frac{1}{x-4}$ is a translation 4 units to the right of the graph of $y = \frac{1}{x}$.

11–12. See Additional Answers.

13. about 8 students

Vocabulary and Concept Check

1. no; The denominator is not a polynomial.
2. If the graph of a rational function has a vertical asymptote, it will occur at an excluded value.
3. The graph of a rational function approaches but never intersects the asymptotes.

Practice and Problem Solving

4. The graph is two smooth curves. The domain appears to be all real numbers except 0. The range appears to be all real numbers except 10. As x gets closer to 0, the graph approaches the vertical line $x = 0$. As x increases and decreases, the graph approaches the horizontal line $y = 10$.
5. See Additional Answers.

6. $x = 0$ 7. $x = -4$

8. $x = -3$ 9. $x = 9$

10. $x = 4$ 11. $x = -\frac{1}{2}$

12–17. See Additional Answers.

18. a. $x = 0$

b. 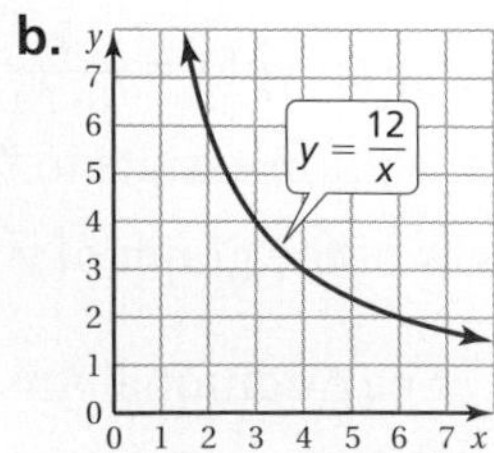

The domain is all real numbers greater than 0. The range is all real numbers greater than 0.

Assignment Guide and Homework Check

Level	Assignment	Homework Check
Average	1–3, 6–18 even, 19–35 odd, 34, 46–49	12, 18, 21, 27, 31, 34
Advanced	1–3, 10, 14, 18, 24–27, 30, 34–45, 46–49	14, 27, 30, 34, 40, 43

For Your Information

- **Exercises 42–44** Students can use a graphing calculator.

Common Errors

- **Exercises 12–17** Students may not plot ordered pairs on either side of the vertical asymptote. They also may not choose x-values that are close to the excluded value. Remind them of this process.
- **Exercise 18** Students may think the domain and range are all real numbers except 0. Negative numbers do not make sense in the context of the problem. The domain is $x > 0$ and the range is $y > 0$. Tell them that real-world problems may impose constraints on the domain and/or range.

11.2 Record and Practice Journal

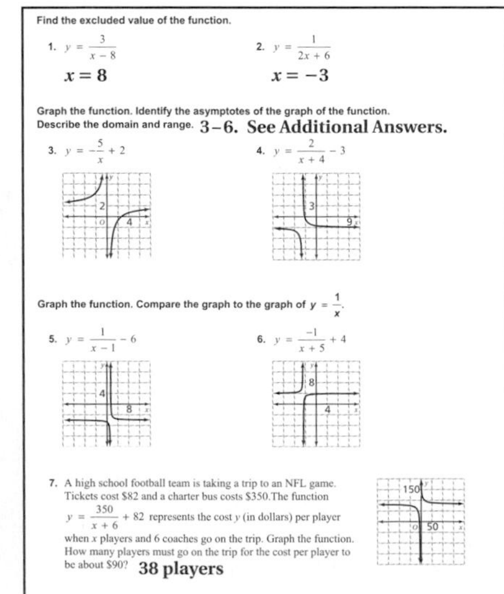

Find the excluded value of the function.

1. $y = \frac{3}{x - 8}$ **$x = 8$**

2. $y = \frac{1}{2x + 6}$ **$x = -3$**

Graph the function. Identify the asymptotes of the graph of the function. Describe the domain and range. **3–6. See Additional Answers.**

3. $y = -\frac{5}{x} + 2$

4. $y = \frac{2}{x + 4} - 3$

Graph the function. Compare the graph to the graph of $y = \frac{1}{x}$.

5. $y = \frac{1}{x - 1} - 6$

6. $y = \frac{-1}{x + 5} + 4$

7. A high school football team is taking a trip to an NFL game. Tickets cost \$82 and a charter bus costs \$350. The function $y = \frac{350}{x + 6} + 82$ represents the cost y (in dollars) per player when x players and 6 coaches go on the trip. Graph the function. How many players must go on the trip for the cost per player to be about \$90? **38 players**

Technology For the Teacher
Answer Presentation Tool

11.2 Exercises

 Vocabulary and Concept Check

1. **VOCABULARY** Is $y = \dfrac{1}{\sqrt{x} + 1}$ a rational function? Explain.
2. **VOCABULARY** How is an excluded value related to a vertical asymptote?
3. **WRITING** How can you use asymptotes to help graph a rational function?

 Practice and Problem Solving

Describe the characteristics of the graph.

4. 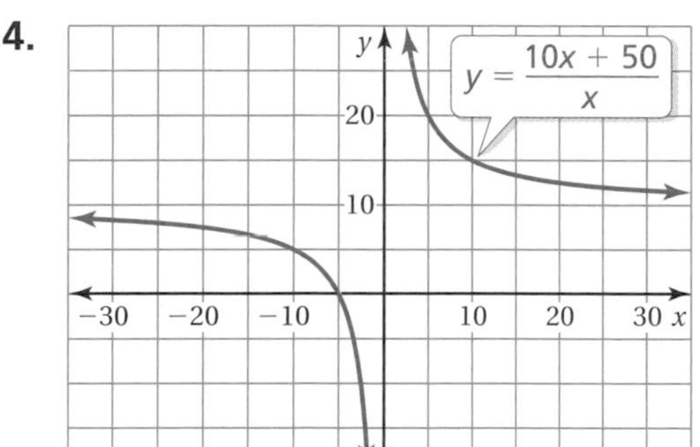

5. 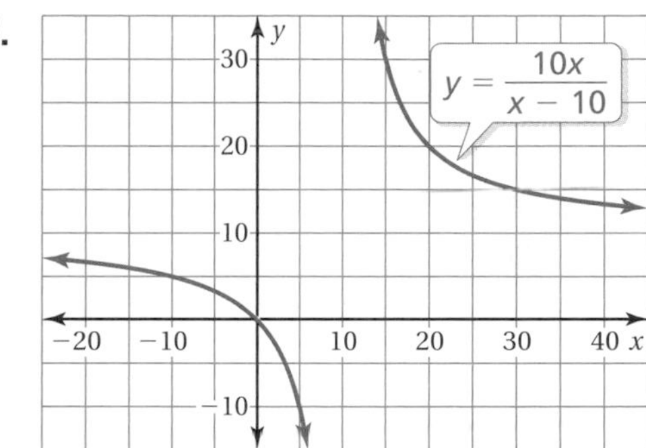

Find the excluded value of the function.

(1) 6. $y = \dfrac{3}{4x}$ 7. $y = \dfrac{2}{x + 4}$ 8. $y = \dfrac{1}{x + 3}$

9. $y = \dfrac{5}{x - 9}$ 10. $y = \dfrac{7}{8 - 2x}$ 11. $y = \dfrac{4}{3 + 6x}$

Graph the function. Describe the domain and range.

(2) 12. $y = \dfrac{5}{x}$ 13. $y = \dfrac{2}{5x}$ 14. $y = -\dfrac{3}{8x}$

15. $y = \dfrac{1}{x - 3}$ 16. $y = \dfrac{4}{x + 1}$ 17. $y = \dfrac{1}{4 - 2x}$

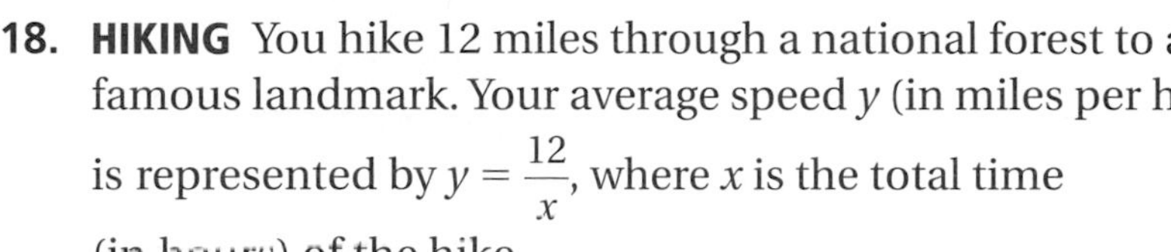

18. **HIKING** You hike 12 miles through a national forest to a famous landmark. Your average speed y (in miles per hour) is represented by $y = \dfrac{12}{x}$, where x is the total time (in hours) of the hike.

    a. Find the excluded value of the function.

    b. Graph the function. Describe the domain and range.

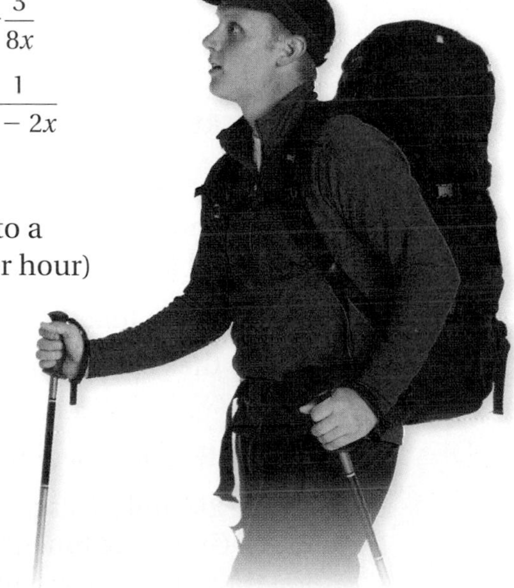

Identify the asymptotes of the graph of the function. Then describe the domain and range.

3 **19.** $y = -\frac{6}{x}$ **20.** $y = \frac{4}{x} + 8$ **21.** $y = \frac{1}{x-2} + 7$

22. $y = \frac{3}{x+4} - 4$ **23.** $y = \frac{-2}{x-5} - 2$ **24.** $y = 10 - \frac{7}{x+9}$

25. ERROR ANALYSIS Describe and correct the error in identifying the asymptotes of the graph of the function.

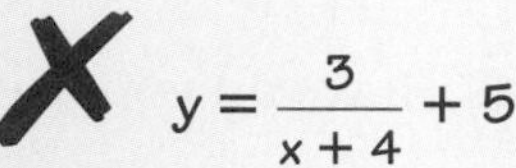

$y = \frac{3}{x+4} + 5$

The horizontal asymptote is $y = 5$.

The vertical asymptote is $x = 4$.

26. REASONING Describe the domain and range of a rational function of the form $y = \frac{a}{x-h} + k$.

27. OPEN-ENDED Write a rational function whose graph has the vertical asymptote $x = 6$ and the horizontal asymptote $y = -9$.

Graph the function. Compare the graph to the graph of $y = \frac{1}{x}$.

4 **28.** $y = \frac{1}{x} + 2$ **29.** $y = \frac{1}{x-6}$ **30.** $y = \frac{1}{x+4} - 2$

31. $y = \frac{1}{x+7} + 3$ **32.** $y = \frac{-1}{x-1} - 5$ **33.** $y = 4 - \frac{1}{x+8}$

34. SOFTBALL A softball team buys a new \$250 bat for a softball tournament. The cost of the bat is shared equally by the players on the team. Each player must also pay a \$10 registration fee. The amount y (in dollars) each player pays is represented by $y = \frac{250}{p} + 10$, where p is the number of players on the team. Graph the function. How many players must be on the team for the cost per player to be about \$28?

35. GEOMETRY The formula $h = \frac{2A}{b_1 + b_2}$ gives the height h of a trapezoid, where A is the area and b_1 and b_2 are the base lengths. Suppose $A = 60$ and $b_1 = 8$.

a. Graph the function. Describe the domain and range.

b. Use the graph to find b_2 when $h = 6$.

36. ROAD TRIP The function $t = \frac{280}{r} + 1$ models the total time t (in hours) it takes to drive 280 miles at r miles per hour. The model allows for two half-hour breaks. Graph the function. What does your average speed need to be for the total travel time to be 6 hours?

Common Errors

- **Exercises 19–24** Students may confuse the vertical and horizontal asymptotes. They may also use the wrong sign when identifying asymptotes. Remind them that in the equation $y = \frac{a}{x - h} + k$, the vertical asymptote is $x = h$ and the horizontal asymptote is $y = k$.
- **Exercises 28–33** Students may not identify the horizontal and vertical asymptotes before trying to graph the function. Encourage them to use the asymptotes to help graph the function.
- **Exercise 35** Students may think the domain is all real numbers except -8 and the range is all real numbers except 0. Negative numbers do not make sense in the context of the problem. The domain is $x > 0$ and the range is $0 < y < 15$. Tell them that real-world problems may impose constraints on the domain and/or range.
- **Exercises 37–39** Students may struggle with writing a function for the graph. Tell them to start by identifying the asymptotes. From there, they should be able to work backwards to find an equation of the form $y = \frac{1}{x - h} + k$.

Practice and Problem Solving

19. $x = 0, y = 0$; The domain is all real numbers except 0 and the range is all real numbers except 0.

20. $x = 0, y = 8$; The domain is all real numbers except 0 and the range is all real numbers except 8.

21. $x = 2, y = 7$; The domain is all real numbers except 2 and the range is all real numbers except 7.

22. $x = -4, y = -4$; The domain is all real numbers except -4 and the range is all real numbers except -4.

23. $x = 5, y = -2$; The domain is all real numbers except 5 and the range is all real numbers except -2.

24. $x = -9, y = 10$; The domain is all real numbers except -9 and the range is all real numbers except 10.

25. The wrong sign is used for the vertical asymptote; $x = -4$

26. The domain is all real numbers except h. The range is all real numbers except k.

27. *Sample answer:* $y = \frac{100}{x - 6} - 9$

28–36. See Additional Answers.

English Language Learners

Visual Aid

Create a poster of the Key Ideas on pages 552 and 553. Include new vocabulary words and any other important information. Refer to the poster in classroom discussions.

Practice and Problem Solving

37. $y = \frac{1}{x - 5}$

38. $y = \frac{1}{x + 3} + 3$

39. $y = 2 - \frac{1}{x - 1}$

40. If $|a| > 1$, the graph is wider than the graph of $y = \frac{1}{x}$.

If $0 < |a| < 1$, the graph is narrower than the graph of $y = \frac{1}{x}$.

If $a < 0$, the graph is a reflection in the x-axis of the graph of $y = \frac{1}{x}$.

41. about 23°C

42–44. See Additional Answers.

45. See *Taking Math Deeper*.

Fair Game Review

46. nonlinear; This is an inverse variation equation.

47. linear; This is a linear equation in standard form.

48. nonlinear; This is a quadratic equation.

49. D

Mini-Assessment

1. Find the excluded value of $y = \frac{5}{3x + 9}$. $x = -3$
2. Graph $y = \frac{1}{x - 8}$. Describe the domain and range.
3. Identify the asymptotes of the graph of $y = \frac{1}{x + 7} - 5$. Then describe the domain and range.
4. Graph $y = \frac{1}{x - 5} + 3$. Compare the graph to the graph of $y = \frac{1}{x}$.

2–4. See Additional Answers.

Taking Math Deeper

Exercise 45

There are a lot of numbers and information in this exercise. Students will need to read carefully and make sense of the problem before continuing.

a. Write an equation.

$$r = \frac{\text{expected monthly housing expenses}}{\text{gross monthly income}}$$

$$r = \frac{1050}{3500 + m}$$

2 Graph the equation.

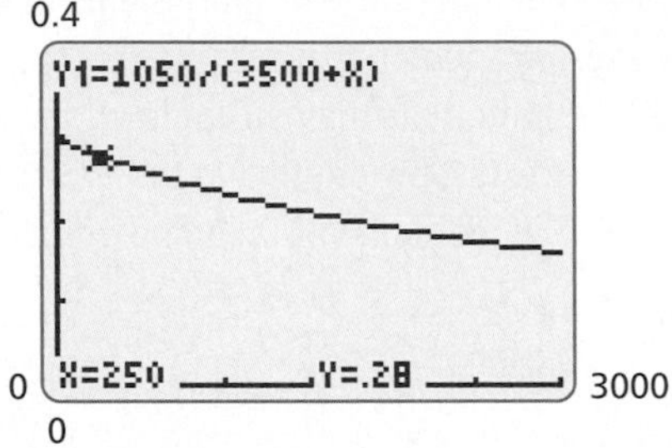

b. Substitute 0.28 for r and solve algebraically.

$$0.28 = \frac{1050}{3500 + m}$$

$$0.28(3500 + m) = 1050$$

$$980 + 0.28m = 1050$$

$$m = 250$$

The ratio decreases as m increases. So, the ratio is less than or equal to 0.28 when m is greater than or equal to \$250. You can verify this by looking at the graph.

Project

Research mortgage loans. What other criteria are taken into account before a loan is granted?

Reteaching and Enrichment Strategies

If students need help. . .	If students got it. . .
Resources by Chapter • Practice A and Practice B • Puzzle Time Record and Practice Journal Practice Differentiating the Lesson Lesson Tutorials Skills Review Handbook	Resources by Chapter • Enrichment and Extension Start the next section

Write a function for the graph.

37.

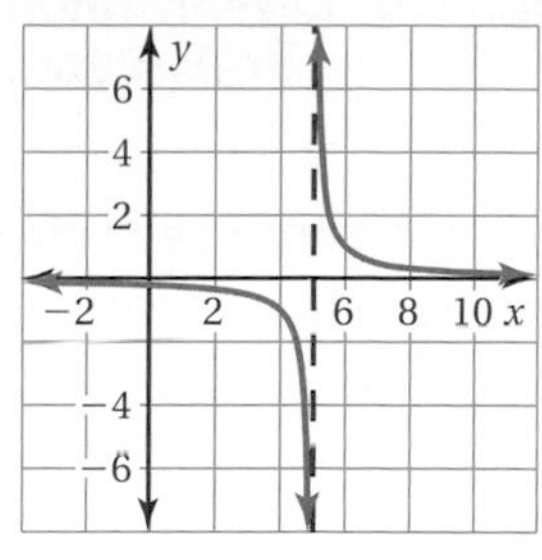

38.

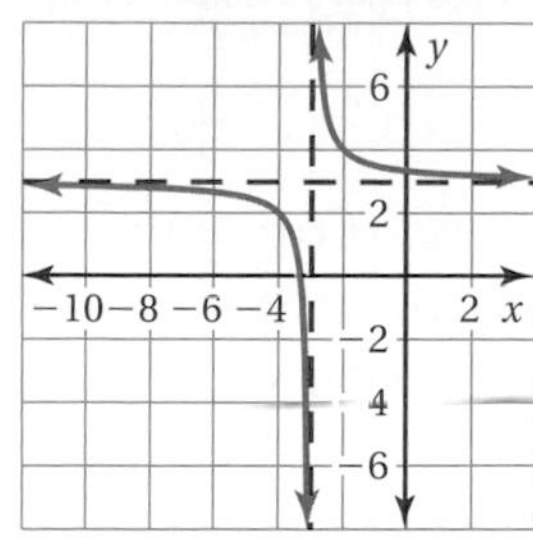

39.

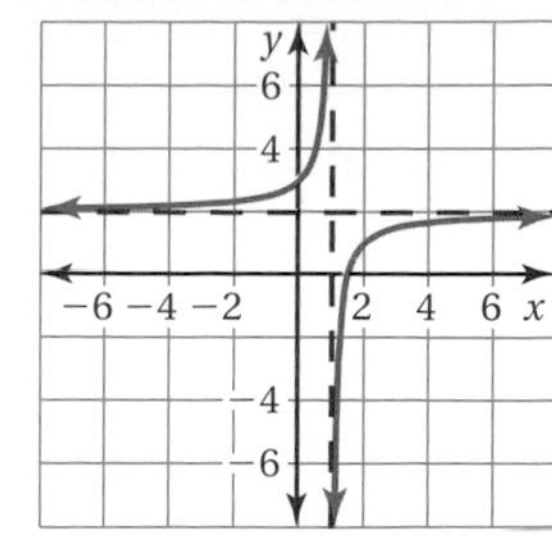

40. REPEATED REASONING Use a graphing calculator to graph the function $y = \dfrac{a}{x-1} + 2$ for several values of a. How does the value of a affect the graph? Consider $a < 0$, $|a| > 1$, and $0 < |a| < 1$ in your answer.

41. THUNDERSTORM The time t (in seconds) it takes for sound to travel 1 kilometer can be represented by $t = \dfrac{1000}{0.6T + 331}$, where T is the temperature in degrees Celsius. Use a graphing calculator to graph the function for $0 \le T \le 100$. During a thunderstorm, lightning strikes 1 kilometer away. You hear the thunder 2.9 seconds later. What is the temperature?

Graph the function. Identify the asymptotes.

42. $y = \dfrac{x}{x+1}$ **43.** $y = \dfrac{1}{x^2 - 4}$ **44.** $y = \dfrac{x+1}{x^2 - 1}$

45. **Modeling** To qualify for a mortgage, the ratio r of your expected monthly housing expenses to your gross monthly income cannot be greater than 0.28. Suppose your gross monthly income is \$3500 and you expect to pay \$1050 per month in housing expenses. You also expect to get a raise of m dollars this month.

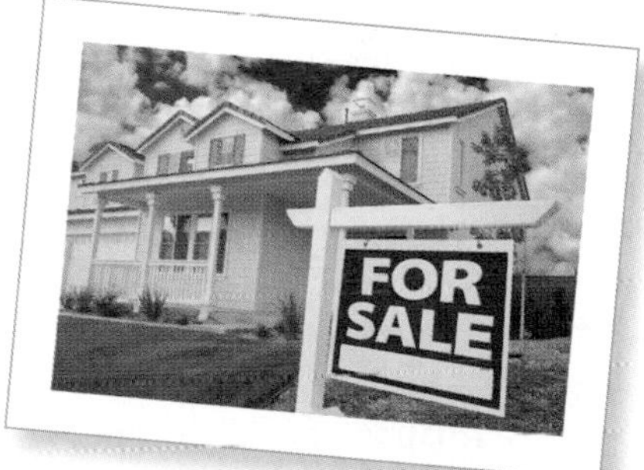

a. Write and graph an equation that gives r as a function of m.

b. How much must the raise be in order for you to qualify for a mortgage?

Fair Game Review What you learned in previous grades & lessons

Does the equation represent a *linear* or *nonlinear* function? Explain. *(Section 5.5)*

46. $3xy = 12$ **47.** $4x + 2y = 5$ **48.** $2y - x^2 = 8$

49. MULTIPLE CHOICE Which function models exponential decay? *(Section 6.6)*

Ⓐ $y = -3\left(\frac{1}{2}\right)^x$ Ⓑ $y = -\frac{1}{2}(3)^x$ Ⓒ $y = \frac{1}{2}(3)^x$ Ⓓ $y = 3\left(\frac{1}{2}\right)^x$

11.2b Inverse of a Function

Check It Out
Lesson Tutorials
BigIdeasMath.com

Key Vocabulary
inverse relation, *p. 558*
inverse function, *p. 559*

Recall that a *relation* pairs inputs with outputs. An **inverse relation** switches the input and output values of the original relation. For example, if a relation contains (a, b), then the inverse relation contains (b, a).

EXAMPLE 1 Finding Inverse Relations

Find the inverse of each relation.

a. $(-4, 7), (-2, 4), (0, 1), (2, -2), (4, -5)$

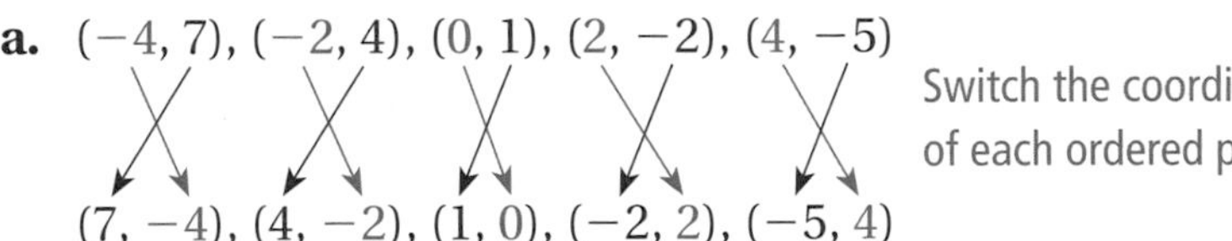
Switch the coordinates of each ordered pair.

$(7, -4), (4, -2), (1, 0), (-2, 2), (-5, 4)$

b.

Input	−1	0	1	2	3	4
Output	5	10	15	20	25	30

Switch the inputs and outputs.

Inverse relation:

Input	5	10	15	20	25	30
Output	−1	0	1	2	3	4

Practice

Find the inverse of the relation.

1. $(-5, 8), (-5, 6), (0, 0), (5, 6), (10, 8)$
2. $(-3, -4), (-2, 0), (-1, 4), (0, 8), (1, 12), (2, 16), (3, 20)$
3.

Input	−2	−1	0	1	2
Output	4	1	0	1	4

4.

Input	−2	−1	0	0	1	2
Output	3	4	5	6	7	8

5. **WRITING** How do the domain and range of a relation compare to the domain and range of its inverse relation? Explain.
6. **CRITICAL THINKING** Recall that you can use the Vertical Line Test to determine whether a graph represents a function. What kind of similar test do you think you could use to determine whether a function has an inverse that is also a function? Explain.

Laurie's Notes

Introduction

Connect

- **Yesterday:** Students graphed rational functions. (MP2, MP5, MP7)
- **Today:** Students will find inverses.

Motivate

? "What was the *average daily temperature* yesterday?" Answers will vary. Find this information in advance.

- Write a table on the board listing the date (input) and average daily temperature (output) for a specific location. Feel free to make up the data. Make sure that at least two of the temperatures are the same.

? "Does this table represent a function? Explain." yes; Every date is associated with exactly one average daily temperature.

? "If we switch the inputs and outputs, would it represent a function? Explain." no; An average daily temperature would be associated with more than one date.

Lesson Notes

Discuss

- **MP2 Reason Abstractly and Quantitatively:** Students find inverses by manipulating equations symbolically. Mathematically proficient students are able to reason abstractly, often without a context, to find inverses.
- Students should recall relations, which were introduced earlier in the book.
- Today, the inputs and outputs of a relation are switched, creating an *inverse relation.*

Example 1

- Write the set of ordered pairs in part (a), referring to them as a relation.
- Use color to draw attention to the switching of the coordinates.
- Work through each part as shown.

? "When finding an inverse relation, (x, y) becomes what?" (y, x)

Practice

- **Neighbor Check:** Have students work independently and then have their neighbor check their work. Have students discuss any discrepancies.
- Ask a student to explain the Vertical Line Test. Exercise 5 should suggest a horizontal line test in Exercise 6.

Goal Today's lesson is finding **inverse relations** and **inverse functions**.

Start Thinking! and Warm Up

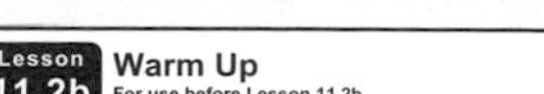

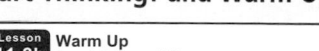

Lesson 11.2b Warm Up For use before Lesson 11.2b

Lesson 11.2b Start Thinking! For use before Lesson 11.2b

The inverse of a function "undoes" the function.

For example, $y = 2x$ and $y = \frac{1}{2}x$ are inverse functions and $y = x + 2$ and $y = x - 2$ are inverse functions.

Which function is the inverse of $y = 2x + 2$?

A. $y = 2x - 2$

B. $y = \frac{x}{2} + 2$

C. $y = \frac{x - 2}{2}$

Extra Example 1

Find the inverse of the relation: $(-6, 4), (-3, 2), (0, 0), (3, -2), (6, -4)$. $(4, -6), (2, -3), (0, 0), (-2, 3), (-4, 6)$

Practice

1. $(8, -5), (6, -5), (0, 0), (6, 5), (8, 10)$
2. $(-4, -3), (0, -2), (4, -1), (8, 0), (12, 1), (16, 2), (20, 3)$
3.

Input	4	1	0	1	4
Output	−2	−1	0	1	2

4.

Input	3	4	5	6	7	8
Output	−2	−1	0	0	1	2

5. The domain of a relation is the range of its inverse relation. The range of a relation is the domain of its inverse relation.
6. Because the domain and range switch, you could use a horizontal line test.

Record and Practice Journal Practice

See Additional Answers.

Extra Example 2

Find the inverse of each function. Graph the inverse function.

a. $f(x) = 6 - 4x$

$f^{-1}(x) = \dfrac{6 - x}{4}$

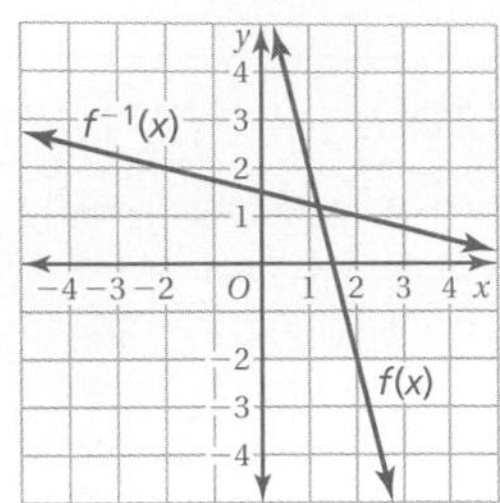

b. $f(x) = \dfrac{1}{x + 1}$

$f^{-1}(x) = \dfrac{1}{x} - 1$

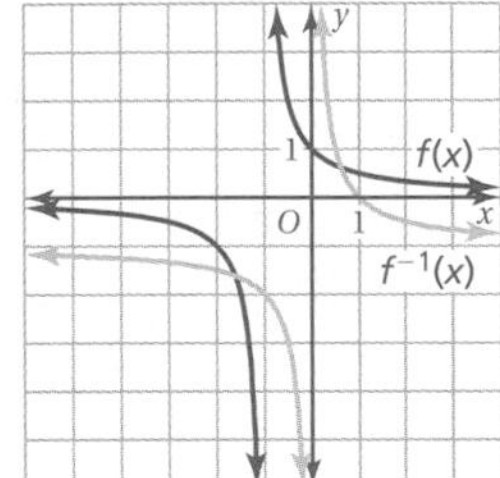

Practice

7–12. See Additional Answers.

13. -2

14–15. See Additional Answers.

Mini-Assessment

Find the inverse of the relation.

1. $(-2, 5), (-1, 3), (0, 1), (1, 1), (2, 3), (3, 5), (4, 7)$ $(5, -2), (3, -1), (1, 0), (1, 1), (3, 2), (5, 3), (7, 4)$
2. $(-27, -3), (-8, -2), (-1, -1), (0, 0), (1, 1), (8, 2), (27, 3)$ $(-3, -27), (-2, -8), (-1, -1), (0, 0), (1, 1), (2, 8), (3, 27)$

Find the inverse of the function. Graph the inverse function.

3. $f(x) = \dfrac{1}{5}x - 2$ $f^{-1}(x) = 5x + 10$
   See Additional Answers for graph.
4. $f(x) = \dfrac{1}{x} + 3$ $f^{-1}(x) = \dfrac{1}{x - 3}$
   See Additional Answers for graph.

Laurie's Notes

Discuss

- Refer back to the Motivate, reminding students that the inverse was not a function.
- Introduce the notation, $f^{-1}(x)$. Stress that the -1 is *not* an exponent.

Example 2

- Remind students that $f(x)$ can be replaced by y.
- To find the inverse, switch x and y in the equation. Relate this to switching the coordinates of the ordered pairs in Example 1.
- Solve the equation for y. The last step is to replace y with $f^{-1}(x)$.

? "If the ordered pair $(1, -3)$ satisfies the original function, what ordered pair satisfies the inverse function?" $(-3, 1)$

- **MP2:** The domain in part (b) is nonnegative, as shown in the graph. So, the *range* of the inverse must be nonnegative. If the domain of the original function had been $x \le 0$, then the inverse would be $f^{-1}(x) = -\sqrt{x}$.

Practice

- Ask volunteers to show their work at the board for Exercises 7–12.

? "What did you notice about Exercise 11?" Students may say the function did not change. This means that the function is its own inverse, which is always true with inverse variation.

- Exercise 14 helps students recognize the symmetry of functions and their inverses about the line $y = x$.
- **MP1b Persevere in Solving Problems:** You may need to help students work through Exercise 15. Without a lesson on composition of functions, students may have difficulty with the notation. Use a specific example from the lesson to discuss this problem.

Closure

- **Exit Ticket:** Explain the steps you would take to find the inverse of $f(x) = 3x - 4$. Then find the inverse. Replace $f(x)$ with y, switch x and y, solve for y, and then replace y with $f^{-1}(x)$ to obtain $f^{-1}(x) = \dfrac{1}{3}x + \dfrac{4}{3}$.

Technology For the Teacher

The Dynamic Planning Tool
Editable Teacher's Resources at *BigIdeasMath.com*

Reading

The -1 in $f^{-1}(x)$ is not an exponent.

When a relation and its inverse are functions, they are called **inverse functions**. The inverse of a function f is written as $f^{-1}(x)$. To find the inverse of a function represented by an equation, switch x and y and then solve for y.

EXAMPLE 2 Finding Inverse Functions

Find the inverse of each function. Graph the inverse function.

a. $f(x) = 2x - 5$

$y = 2x - 5$	Replace $f(x)$ with y.
$x = 2y - 5$	Switch x and y.
$x + 5 = 2y$	Add 5 to each side.
$\frac{1}{2}x + \frac{5}{2} = y$	Divide each side by 2.
$\frac{1}{2}x + \frac{5}{2} = f^{-1}(x)$	Replace y with $f^{-1}(x)$.

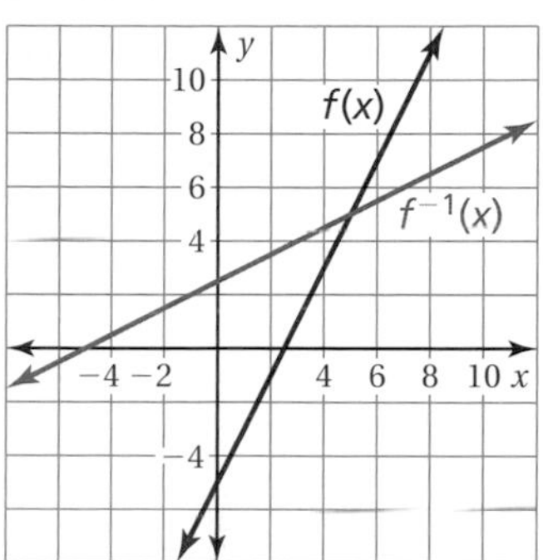

Study Tip

The domain is nonnegative in Example 2b, so the range of the inverse must be nonnegative. This is why you take only the positive square root of each side.

b. $f(x) = x^2$, where $x \geq 0$

$y = x^2$	Replace $f(x)$ with y.
$x = y^2$	Switch x and y.
$\sqrt{x} = y$	Take the positive square root of each side.
$\sqrt{x} = f^{-1}(x)$	Replace y with $f^{-1}(x)$.

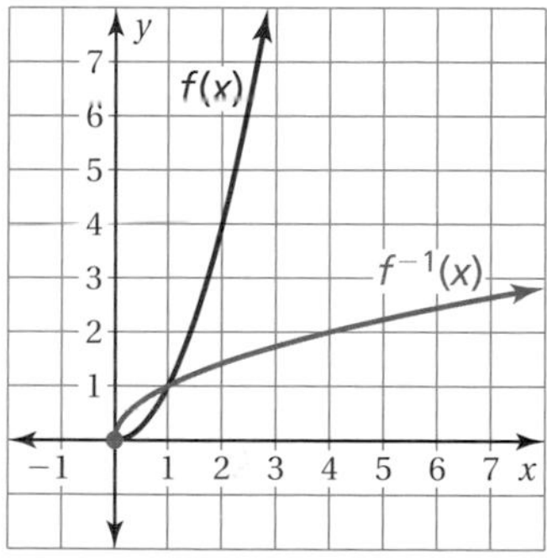

Practice

Find the inverse of the function. Graph the inverse function.

7. $f(x) = 3x - 1$

8. $f(x) = -\frac{1}{2}x + 3$

9. $f(x) = 2x^2$, where $x \geq 0$

10. $f(x) = x^2 - 3$, where $x \geq 0$

11. $f(x) = \frac{1}{x}$

12. $f(x) = \frac{1}{x - 2}$

13. **REASONING** Suppose f and f^{-1} are inverse functions and $f(-2) = 5$. What is the value of $f^{-1}(5)$?

14. **REASONING** Draw the line $y = x$ on the graph in each part of Example 2. What do you notice?

15. **LOGIC** Suppose f and g are inverse functions. What do you know about $f(g(x))$ and $g(f(x))$? Explain.

11.3 Simplifying Rational Expressions

COMMON CORE STATE STANDARDS
A.SSE.2

Essential Question How can you simplify a rational expression? What are the excluded values of a rational expression?

1 ACTIVITY: Simplifying a Rational Expression

Work with a partner.

Sample: You can see that the rational expressions

$$\frac{x^2 + 3x}{x^2} \quad \text{and} \quad \frac{x + 3}{x}$$

are equivalent by graphing the related functions

$$y = \frac{x^2 + 3x}{x^2} \quad \text{and} \quad y = \frac{x + 3}{x}.$$

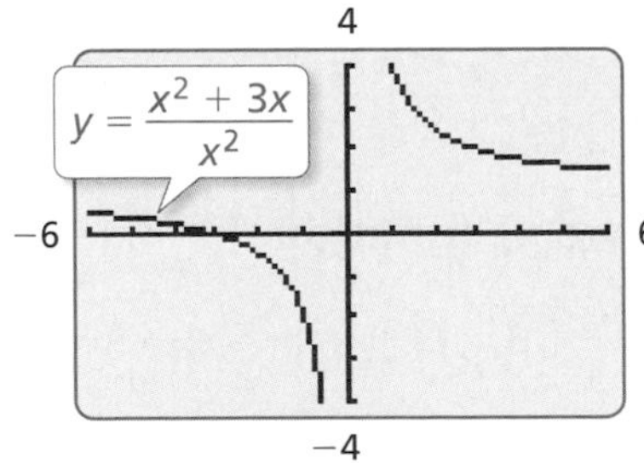

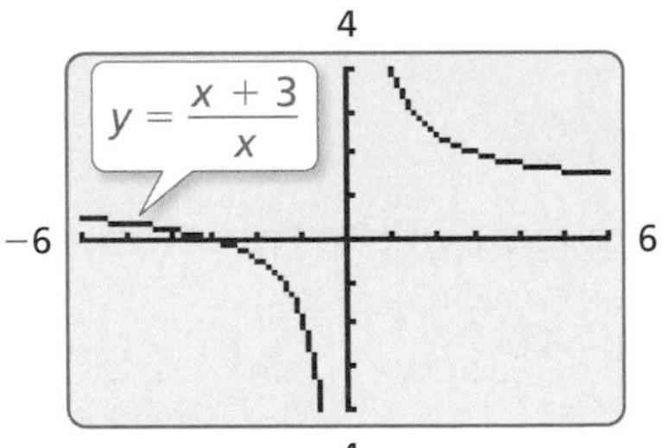

Both functions have the same graph.

Match each rational expression with its equivalent rational expression. Use a graphing calculator to check your answers.

a. $\frac{x^2 + x}{x^2}$ **b.** $\frac{x^2}{x^2 + x}$ **c.** $\frac{x + 1}{x^2 - 1}$ **d.** $\frac{x + 1}{x^2 + 2x + 1}$ **e.** $\frac{x^2 + 2x + 1}{x + 1}$

A. $\frac{1}{x + 1}$ B. $x + 1$ C. $\frac{x + 1}{x}$ D. $\frac{1}{x - 1}$ E. $\frac{x}{x + 1}$

Laurie's Notes

Introduction

Standards for Mathematical Practice

- **MP7 Look for and Make Use of Structure:** Students are simplifying rational expressions and need to remember that common factors can be divided out. Mathematically proficient students understand when an expression is in factored form and can recognize the difference between $3x + 1$ and $3(x + 1)$.

Motivate

- **Puzzle:** If 1 glob equals 8 crabs, 2 crabs equals 5 mils, and 4 mils equals 1 flake, then 1 glob is how many flakes?
- Give students time to discuss this question with a partner.
- One way to solve is shown below.

$$\frac{1 \text{ glob}}{8 \text{ crabs}} \times \frac{2 \text{ crabs}}{5 \text{ mils}} \times \frac{4 \text{ mils}}{1 \text{ flake}} = \frac{1 \text{ glob}}{5 \text{ flakes}}$$

- In today's lesson, students will work with equivalent expressions.

Activity Notes

Activity 1

? "How can you show that $\frac{x^2 + 3x}{x^2}$ is equivalent to $\frac{x + 3}{x}$?" Answers will vary. Students may suggest graphing to show that they are equivalent.

- Before graphing, ask students how to factor the numerator of the first expression. It can be factored as $x(x + 3)$. Point out that the numerator and the denominator have a common factor of x.
- Students should now enter the related functions into their graphing calculators.
- **Common Error:** When entering the functions, students may forget to put parentheses around the expression in the numerator.
- **MP1a Make Sense of Problems:** The two graphs will coincide. Students may ask if they are only seeing one of the graphs in this viewing window and the other graph is outside of the viewing window. Checking the table of values will verify that the values are the same for both functions. This helps students make sense of the problem.
- Students should spend time thinking about the rational expressions and how they might simplify an expression in order to find a match.
- **FYI:** Students will learn that the excluded values in the rational function will create a hole in the graph. The holes are sometimes not evident when using a graphing calculator.
- Have students share how they found their matches.

Common Core State Standards

A.SSE.2 Use the structure of an expression to identify ways to rewrite it.

Previous Learning

Students should know how to graph rational functions.

Activity Materials
Textbook
• graphing calculators

Start Thinking! and Warm Up

Activity 11.3 Start Thinking! For use before Activity 11.3

Activity 11.3 Warm Up For use before Activity 11.3

Factor the expression.

1. $2x^2 + x - 6$
2. $4x^2 - 49$
3. $4x^2 + 2x - 2$
4. $20x^2 - 19x - 6$
5. $x^3 + x^2 - 100x - 100$
6. $x^3 - 2x^2 - 4x + 8$

11.3 Record and Practice Journal

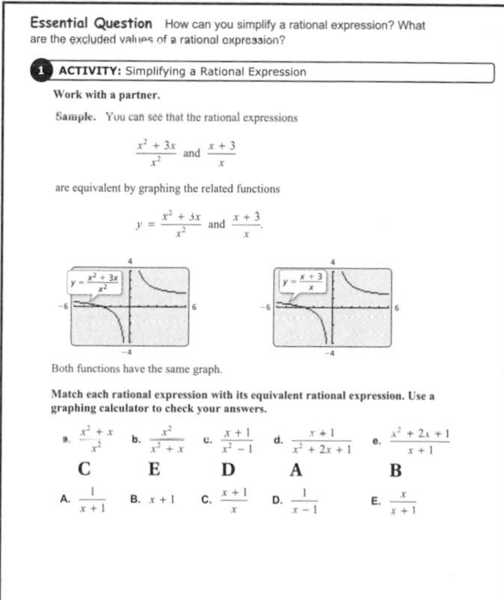

Essential Question How can you simplify a rational expression? What are the excluded values of a rational expression?

1 ACTIVITY: Simplifying a Rational Expression

Work with a partner.

Sample: You can see that the rational expressions

$\frac{x^2 + 3x}{x^2}$ and $\frac{x + 3}{x}$

are equivalent by graphing the related functions

$y = \frac{x^2 + 3x}{x^2}$ and $\frac{x + 3}{x}$.

Both functions have the same graph.

Match each rational expression with its equivalent rational expression. Use a graphing calculator to check your answers.

a. $\frac{x^2 + x}{x^2}$ C b. $\frac{x^2}{x^2 + x}$ E c. $\frac{x + 1}{x^2 - 1}$ D d. $\frac{x + 1}{x^2 + 2x + 1}$ A e. $\frac{x^2 + 2x + 1}{x + 1}$ B

A. $\frac{1}{x + 1}$ B. $x + 1$ C. $\frac{x + 1}{x}$ D. $\frac{1}{x - 1}$ E. $\frac{x}{x + 1}$

Differentiated Instruction

Kinesthetic

A common error when simplifying rational expressions is for students to divide out like terms instead of factors. For instance,

$$\frac{5x}{5(x+2)} = \frac{\cancel{5x}}{\cancel{5x}+10} = \frac{1}{10}$$

Have students write each factor of the original expression on a piece of paper and arrange the pieces as a fraction. Tell students they can only divide out factors when the pieces of paper are identical.

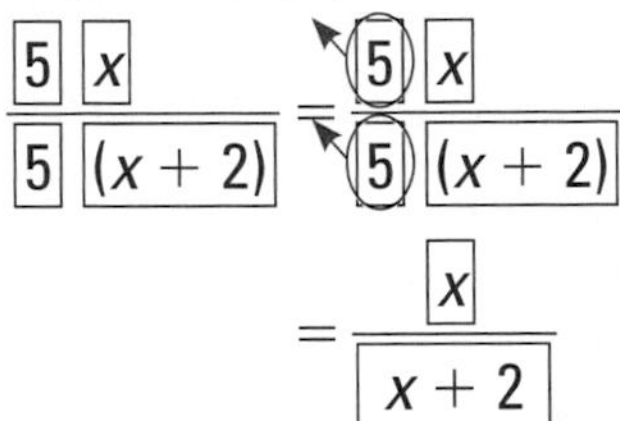

11.3 Record and Practice Journal

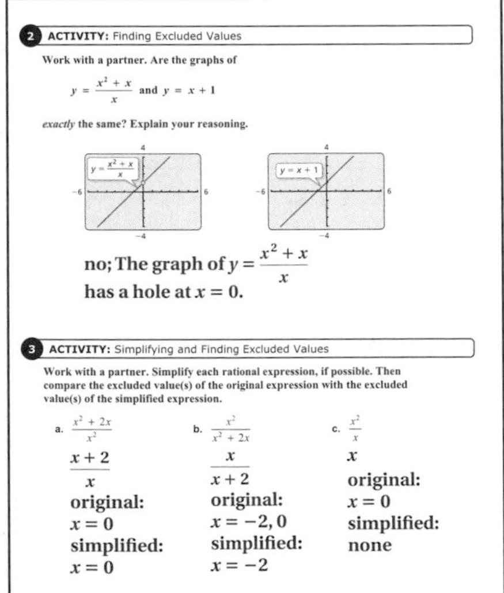

2 ACTIVITY: Finding Excluded Values

Work with a partner. Are the graphs of

$y = \frac{x^2 + x}{x}$ and $y = x + 1$

exactly the same? Explain your reasoning.

no; The graph of $y = \frac{x^2 + x}{x}$ has a hole at $x = 0$.

3 ACTIVITY: Simplifying and Finding Excluded Values

Work with a partner. Simplify each rational expression, if possible. Then compare the excluded value(s) of the original expression with the excluded value(s) of the simplified expression.

a. $\frac{x^2 + 2x}{x^2}$ — $\frac{x+2}{x}$; original: $x = 0$; simplified: $x = 0$

b. $\frac{x^2}{x^2 + 2x}$ — $\frac{x}{x+2}$; original: $x = -2, 0$; simplified: $x = -2$

c. $\frac{x^2}{x}$ — x; original: $x = 0$; simplified: none

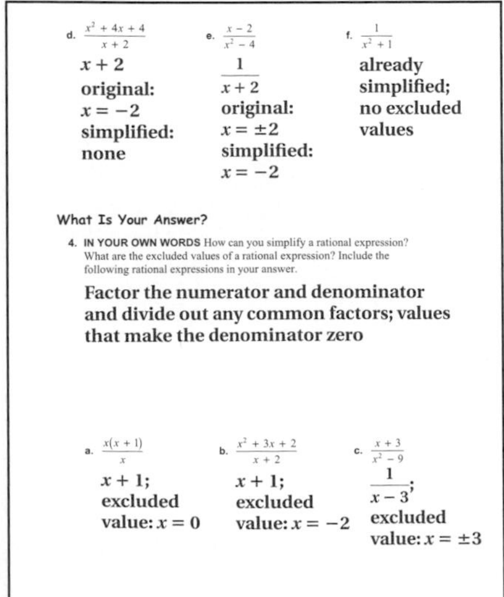

d. $\frac{x^2 + 4x + 4}{x + 2}$ — $x + 2$; original: $x = -2$; simplified: none

e. $\frac{x - 2}{x^2 - 4}$ — $\frac{1}{x+2}$; original: $x = \pm 2$; simplified: $x = -2$

f. $\frac{1}{x^2 + 1}$ — already simplified; no excluded values

What Is Your Answer?

4. IN YOUR OWN WORDS How can you simplify a rational expression? What are the excluded values of a rational expression? Include the following rational expressions in your answer.

Factor the numerator and denominator and divide out any common factors; values that make the denominator zero

a. $\frac{x(x+1)}{x}$ — $x + 1$; excluded value: $x = 0$

b. $\frac{x^2 + 3x + 2}{x + 2}$ — $x + 1$; excluded value: $x = -2$

c. $\frac{x + 3}{x^2 - 9}$ — $\frac{1}{x - 3}$; excluded value: $x = \pm 3$

Laurie's Notes

Activity 2

- The hole in the graph on the left should be evident to students.
- ? "Why do you think the graph on the left has a hole at $x = 0$?" Students may realize that $x = 0$ is an excluded value for the rational function.
- ? "Can you factor the numerator in the first function?" yes; $x(x + 1)$
- The numerator and denominator share a common factor of x that can be divided out when simplifying.

Activity 3

- Students may need to factor the numerator and/or denominator to determine if there are common factors that can divide out.
- Explain that excluded values of the original expression are still excluded values in the simplified expression even if it is a common factor that divides out.
- Working through these examples is a good review of factoring and dividing out common factors in a rational expression.
- If time permits, check solutions by graphing the related functions for the original expression and the simplified expression. Remind students to check the table of values, and not just the graphs.

What Is Your Answer?

- **Neighbor Check:** Have students work independently and then have their neighbor check their work. Have students discuss any discrepancies.

Closure

- **Writing Prompt:** Describe how the excluded values of a rational expression can differ from the excluded values of its corresponding simplified expression.

Technology For the Teacher

The Dynamic Planning Tool
Editable Teacher's Resources at *BigIdeasMath.com*

2 ACTIVITY: Finding Excluded Values

Work with a partner. Are the graphs of

$$y = \frac{x^2 + x}{x} \qquad \text{and} \qquad y = x + 1$$

***exactly* the same? Explain your reasoning.**

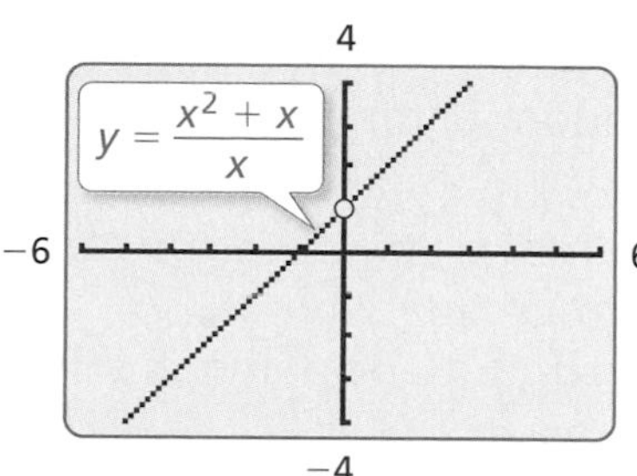

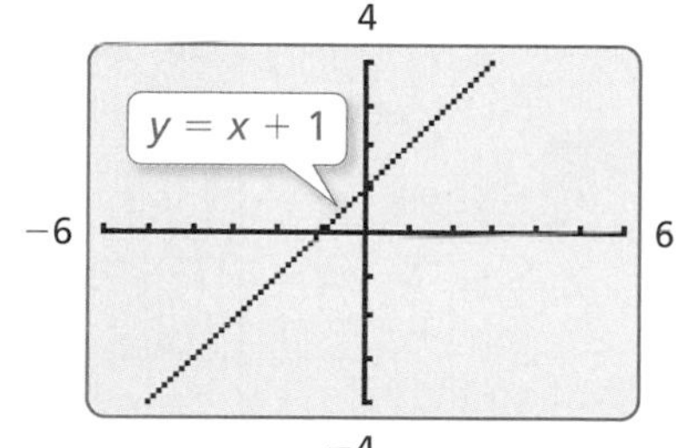

3 ACTIVITY: Simplifying and Finding Excluded Values

Work with a partner. Simplify each rational expression, if possible. Then compare the excluded value(s) of the original expression with the excluded value(s) of the simplified expression.

a. $\frac{x^2 + 2x}{x^2}$ **b.** $\frac{x^2}{x^2 + 2x}$ **c.** $\frac{x^2}{x}$

d. $\frac{x^2 + 4x + 4}{x + 2}$ **e.** $\frac{x - 2}{x^2 - 4}$ **f.** $\frac{1}{x^2 + 1}$

What Is Your Answer?

4. **IN YOUR OWN WORDS** How can you simplify a rational expression? What are the excluded values of a rational expression? Include the following rational expressions in your answer.

a. $\frac{x(x + 1)}{x}$ **b.** $\frac{x^2 + 3x + 2}{x + 2}$ **c.** $\frac{x + 3}{x^2 - 9}$

Practice Use what you learned about simplifying rational expressions to complete Exercises 3–5 on page 564.

11.3 Lesson

Key Vocabulary
rational expression, *p. 562*
simplest form of a rational expression, *p. 562*

A **rational expression** is an expression that can be written as a fraction whose numerator and denominator are polynomials. Values that make the denominator of the expression zero are *excluded values.*

Key Idea

Simplifying Rational Expressions

Words A rational expression is in **simplest form** when the numerator and denominator have no common factors except 1. To simplify a rational expression, factor the numerator and denominator and *divide out* any common factors.

Algebra Let a, b, and c be polynomials, where $b, c \neq 0$.

$$\frac{ac}{bc} = \frac{a \cdot \cancel{c}}{b \cdot \cancel{c}} = \frac{a}{b}$$

Example

$$\frac{2\cancel{(x+1)}}{5\cancel{(x+1)}} = \frac{2}{5}; x \neq -1$$

Study Tip

You can see why you can *divide out* common factors by rewriting the expression.

$$\frac{ac}{bc} = \frac{a}{b} \cdot \frac{c}{c} = \frac{a}{b} \cdot 1 = \frac{a}{b}$$

EXAMPLE 1 Simplifying Rational Expressions

Simplify each rational expression, if possible. State the excluded value(s).

a. $\dfrac{12}{2x^2} = \dfrac{\cancel{2} \cdot 2 \cdot 3}{\cancel{2} \cdot x \cdot x}$ Divide out the common factor.

$= \dfrac{6}{x^2}$ Simplify.

∴ The excluded value is $x = 0$.

b. $\dfrac{n}{n+8}$

∴ The expression is in simplest form. The excluded value is $n = -8$.

c. $\dfrac{3y^2}{6y(y-7)} = \dfrac{\cancel{3} \cdot \cancel{y} \cdot y}{2 \cdot \cancel{3} \cdot \cancel{y} \cdot (y-7)}$ Divide out the common factors.

$= \dfrac{y}{2(y-7)}$ Simplify.

∴ The excluded values are $y = 0$ and $y = 7$.

Study Tip

Make sure you find excluded values using the *original* expression.

On Your Own

Now You're Ready Exercises 3–8

Simplify the rational expression, if possible. State the excluded value(s).

1. $\dfrac{5y^3}{2y^2}$
2. $\dfrac{8x(x+1)}{12x^2}$
3. $\dfrac{m+1}{m(m+3)}$

Multi-Language Glossary at BigIdeasMath.com.

Laurie's Notes

Introduction

Connect

- **Yesterday:** Students explored how to simplify rational expressions. (MP1a, MP7)
- **Today:** Students will simplify rational expressions by factoring and dividing out common factors.

Motivate

- Write this problem on the board: $\frac{12}{17} \cdot \frac{51}{77} \cdot \frac{1}{2} \cdot \frac{11}{36} \cdot \frac{21}{13} \cdot \frac{26}{3} = ?$
- Have students work with a partner to find the product.
- Using a calculator, students can see that both the numerator and denominator are 3,675,672. So, the product is 1.

? "Without using a calculator, how can you find this product efficiently?" Divide out common factors.

- To be sure that students remember this process, rewrite the problem as $\frac{12}{17} \cdot \frac{3 \cdot 17}{7 \cdot 11} \cdot \frac{1}{2} \cdot \frac{11}{3 \cdot 12} \cdot \frac{3 \cdot 7}{13} \cdot \frac{13 \cdot 2}{3} = 1$ The numerator and denominator have the same factors that you can divide out.

Lesson Notes

Key Idea

- Write the Key Idea and discuss what it means to simplify a rational expression. Refer to the example above that students just completed.
- **MP6 Attend to Precision:** Do not use the word "cancel" in referring to the common factors. You *divide out* common factors. You can justify *dividing out* by using the Quotient of Powers Property.
- When looking at the example with the binomial factor, note that the excluded value is included in the solution. The original expression is not defined for $x = -1$, so this excluded value must be included in the solution. Students might say, "But there isn't even an x in the answer." While this is true, have students think about the related function $y = \frac{2(x + 1)}{5(x + 1)}$. The domain excludes $x = -1$.

Example 1

- When simplifying rational expressions, it is helpful to repeat the phrase, "divide out common factors" to remind students that the expression must be in factored form.
- For each part of this example, ask students to list the common factors (if any) of the numerator and denominator. Also ask them to explain the excluded values.
- **Common Error:** In part (b), students may be tempted to divide out the n. The n in the denominator is not a factor; it is being added to 8.

Goal Today's lesson is simplifying **rational expressions.**

Lesson Materials
Introduction
• calculators

Start Thinking! and Warm Up

Lesson 11.3 **Start Thinking!** For use before Lesson 11.3

Write a rational expression in which you can factor out x and $(x + 1)$ from the numerator and denominator.

Extra Example 1

Simplify each rational expression, if possible. State the excluded value(s).

a. $\frac{21x}{7x^2}$ $\frac{3}{x}$; $x - 0$

b. $\frac{3}{10 - c}$
The expression is in simplest form; $c = 10$

c. $\frac{2w^3}{8w^2(w + 5)}$ $\frac{w}{4(w + 5)}$; $w = 0$, $w = -5$

On Your Own

1. $\frac{5y}{2}$; $y = 0$
2. $\frac{2(x + 1)}{3x}$; $x = 0$
3. The expression is in simplest form; $m = -3$, $m = 0$

Extra Example 2

Simplify each rational expression, if possible. State the excluded value(s).

a. $\dfrac{d^2 - 4}{2 - d}$ $-d - 2$; $d = 2$

b. $\dfrac{s^2 - s - 6}{s^2 - 7s - 18}$ $\dfrac{s - 3}{s - 9}$; $s = -2, s = 9$

Extra Example 3

What is the surface area to volume ratio of a cube-shaped substance with side length $2x$? $\dfrac{3}{x}$

On Your Own

4. $\dfrac{2}{7}$; $b = -4$

5. $\dfrac{1}{2a}$; $a = 0, a = 3$

6. $-z - 2$; $z = 8$

7. $\dfrac{6}{x}$

Laurie's Notes

On Your Own

- **Neighbor Check:** Have students work independently and then have their neighbor check their work. Have students discuss any discrepancies.

Example 2

- **MP7 Look for and Make Use of Structure:** The rational expressions in this example require students to factor the expressions in the numerator and denominator first. Rewriting an expression in an equivalent form is understanding the structure of the expression.

? "In part (a), is the numerator factorable?" Students may not recognize that $1 - z^2$ can be factored using the Difference of Two Squares Pattern.

? "Can we divide out $(1 - z)$ and $(z - 1)$? Explain." no; They are not the same.

- Show how $(1 - z)$ can be written as $-(z - 1)$. The expressions $(1 - z)$ and $-(z - 1)$ are equivalent.
- **FYI:** Some students may recognize that $(1 - z)$ and $-(z - 1)$ are equivalent because -1 has been factored out. Most students prefer to see an explanation off to the side the first few times this technique is used.

Example 3

- Ask for a volunteer to read the problem. Discuss what the phrase "reacts faster with other substances" means. Look at the ratio *surface area* to *volume*. If this ratio increases, then the value of the numerator is increasing faster than the value of the denominator.

? "What value or values of x are excluded?" $x = 0$ and negative values of x; The context of the problem eliminates these values.

On Your Own

- Check Question 6 which requires rewriting a factor.

Closure

- Suppose that one of your friends was absent today. Write an email describing how to simplify $\dfrac{x^2 - 4}{x + 2}$.

English Language Learners

Visual

Help English language learners with visual clues when dividing out common factors. Color code the factors to reinforce the mathematical concept. For instance, in Example 2(b) color $c + 4$ red, $c - 3$ blue, and $c - 5$ green.

Technology For the Teacher

The Dynamic Planning Tool
Editable Teacher's Resources at *BigIdeasMath.com*

EXAMPLE 2 Simplifying Rational Expressions

Simplify each rational expression, if possible. State the excluded value(s).

a. $\dfrac{1 - z^2}{z - 1} = \dfrac{(1 - z)(1 + z)}{z - 1}$ — Difference of Two Squares Pattern

$= \dfrac{-(z - 1)(1 + z)}{z - 1}$ — Rewrite $1 - z$ as $-(z - 1)$.

$= \dfrac{-\cancel{(z - 1)}(1 + z)}{\cancel{z - 1}}$ — Divide out the common factor.

$= -z - 1$ — Simplify.

∴ The excluded value is $z = 1$.

b. $\dfrac{c^2 + c - 12}{c^2 - c - 20} = \dfrac{\cancel{(c + 4)}(c - 3)}{\cancel{(c + 4)}(c - 5)}$ — Factor. Divide out the common factor.

$= \dfrac{c - 3}{c - 5}$ — Simplify.

∴ The excluded values are $c = -4$ and $c = 5$.

EXAMPLE 3 Real-Life Application

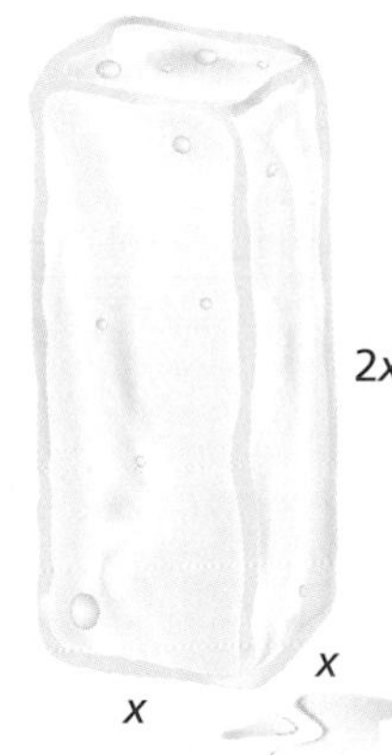

In general, as the surface area to volume ratio of a substance increases, it reacts faster with other substances. Write and simplify this ratio for a block of ice that has the shape shown.

$\dfrac{\text{Surface area}}{\text{Volume}} = \dfrac{2(x^2) + 4(2x^2)}{x(x)(2x)}$ — Write an expression.

$= \dfrac{\overset{5}{\cancel{10}}\cancel{x^2}}{\cancel{2}x^{\cancel{3}}}$ — Simplify. Divide out the common factors.

$= \dfrac{5}{x}$ — Simplify.

On Your Own

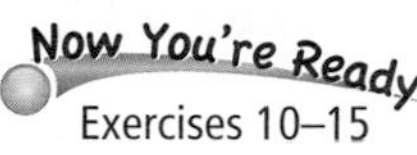

Simplify the rational expression, if possible. State the excluded value(s).

4. $\dfrac{2b + 8}{7b + 28}$

5. $\dfrac{2a - 6}{4a^2 - 12a}$

6. $\dfrac{z^2 - 6z - 16}{8 - z}$

7. What is the surface area to volume ratio of a cube-shaped substance with edge length x?

11.3 Exercises

Vocabulary and Concept Check

1. **VOCABULARY** Is $\dfrac{\sqrt{x} - 1}{x + 3}$ a rational expression? Explain.
2. **REASONING** Why is it necessary to state excluded values of a rational expression?

Practice and Problem Solving

Simplify the rational expression, if possible. State the excluded value(s).

1
3. $\dfrac{6}{18x}$
4. $\dfrac{15y^3}{5y^2}$
5. $\dfrac{n - 1}{n + 1}$
6. $\dfrac{9w^3}{12w^4}$
7. $\dfrac{4t^2}{2t(t + 11)}$
8. $\dfrac{16x^2y}{24xy^3}$

9. **ERROR ANALYSIS** Describe and correct the error in stating the excluded value(s).

$$\frac{x^3}{x^2(x - 3)} = \frac{x}{x - 3}$$

The excluded value is $x = 3$.

Simplify the rational expression. State the excluded value(s).

2
10. $\dfrac{3b + 9}{8b + 24}$
11. $\dfrac{5 - 2z}{2z - 5}$
12. $\dfrac{6a^2 + 12a}{9a^3 + 18a^2}$
13. $\dfrac{4 - y^2}{y^2 - 3y - 10}$
14. $\dfrac{n^2 + 5n + 6}{n^2 + 8n + 15}$
15. $\dfrac{3x^3 - 12x}{6x^3 - 24x^2 + 24x}$

16. **WRITING** Is $\dfrac{(x + 2)(x - 5)}{(x - 2)(5 - x)}$ in simplest form? Explain.

17. **RECYCLING** You hang recycling posters on bulletin boards at your school. Simplify the dimensions of the poster.

Assignment Guide and Homework Check

Level	Assignment	Homework Check
Average	1, 2, 3–19 odd, 16, 23, 26–28	7, 13, 17, 19, 23
Advanced	1, 2, 4–20 even, 17, 22–25, 26–28	14, 18, 23, 24, 25

Common Errors

- **Exercises 3–15 and 18–20** Students may not state the correct excluded value(s). Remind them to use the original expression to find the excluded value(s).
- **Exercise 23** Students may not remember the formulas for the volume of a cylinder and the volume of a cone. Remind them of these formulas.
- **Exercise 24** Students may not realize they must add 4 to the area of Sandbox A to make the area equivalent to the area of Sandbox B. Encourage them to identify key phrases before translating the sentence into an equation.

11.3 Record and Practice Journal

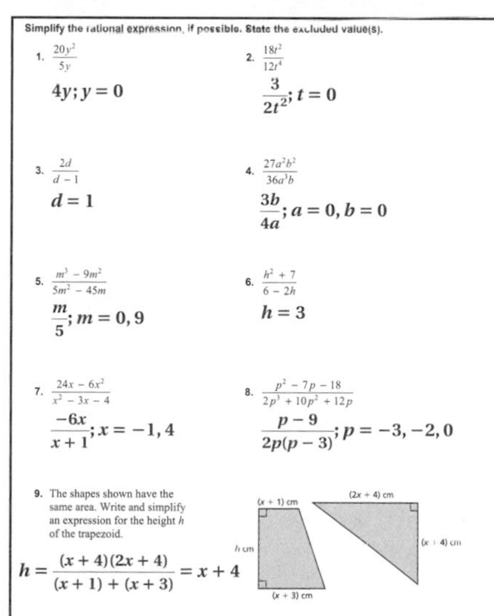

Simplify the rational expression, if possible. State the excluded value(s).

1. $\frac{20y^2}{5y}$

   $4y; y = 0$

2. $\frac{18t^2}{12t^4}$

   $\frac{3}{2t^2}; t = 0$

3. $\frac{2d}{d-1}$

   $d = 1$

4. $\frac{27a^2b^2}{36a^3b}$

   $\frac{3b}{4a}; a = 0, b = 0$

5. $\frac{m^3 - 9m^2}{5m^2 - 45m}$

   $\frac{m}{5}; m = 0, 9$

6. $\frac{h^2 + 7}{6 - 2h}$

   $h = 3$

7. $\frac{24x - 6x^2}{x^2 - 3x - 4}$

   $\frac{-6x}{x+1}; x = -1, 4$

8. $\frac{p^2 - 7p - 18}{2p^3 + 10p^2 + 12p}$

   $\frac{p-9}{2p(p-3)}; p = -3, -2, 0$

9. The shapes shown have the same area. Write and simplify an expression for the height h of the trapezoid.

   $h = \frac{(x+4)(2x+4)}{(x+1)+(x+3)} = x + 4$

Technology For the Teacher
Answer Presentation Tool

Vocabulary and Concept Check

1. no; not a ratio of two polynomials
2. Excluded values make the denominator 0. They must be stated because you cannot divide by 0.

Practice and Problem Solving

3. $\frac{1}{3x}; x = 0$
4. $3y; y = 0$
5. The expression is in simplest form; $n = -1$
6. $\frac{3}{4w}; w = 0$
7. $\frac{2t}{t+11}; t = -11, t = 0$
8. $\frac{2x}{3y^2}; x - 0, y - 0$
9. They did not list all of the excluded values; $x = 0, x = 3$
10. $\frac{3}{8}; b = -3$
11. $-1; z = \frac{5}{2}$
12. $\frac{2}{3a}; a = -2, a = 0$
13. $\frac{2-y}{y-5}; y = -2, y = 5$
14. $\frac{n+2}{n+5}; n = -5, n = -3$
15. $\frac{x+2}{2(x-2)}; x = 0, x = 2$
16. no; Factor -1 out of $(5 - x)$ to get $-(x - 5)$ and then simplify; $-\frac{x+2}{x-2}$
17. $\frac{x}{2}$ by $(x + 3)$

Practice and Problem Solving

18. $\frac{1}{x}$

19. $\frac{3(x+1)}{x(x+3)}$

20. $\frac{2x+3}{x^2+x}$

21. *Sample answer:* $\frac{1}{x^2+8x+15}$

22. The expressions are equivalent for all values of x except -2; Factor x^2-4 as $(x-2)(x+2)$ and then simplify.

23. $\left(\frac{3}{4}x+3\right)$ in.

24. $(4x+4)$ ft

25. See *Taking Math Deeper.*

Fair Game Review

26.

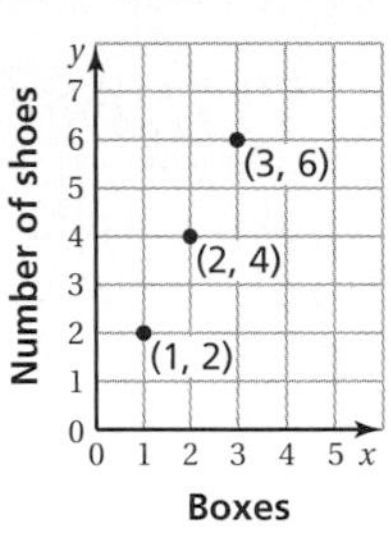

discrete

27. See Additional Answers.

28. C

Mini-Assessment

Simplify the rational expression, if possible. State the excluded value(s).

1. $\frac{2y^3-8y^2}{4y^2-16y}$ $\frac{y}{2}$; $y=0$, $y=4$

2. $\frac{3(m-1)}{5(m+1)}$ The expression is in simplest form; $m=-1$

3. $\frac{z^2-15z+56}{z^2-5z-14}$ $\frac{z-8}{z+2}$; $z=-2$, $z=7$

Taking Math Deeper

Exercise 25

For this problem, students may have difficulty finding where to begin. Students can look for an entry point to the solution by first looking at the sum $6x^2+12x$ and its factors.

The expression $6x^2+12x$ can be factored as $6x(x+2)$.
If you add the numerator and denominator of the simplified ratio $\frac{4x+1}{2x-1}$, you get

$$(4x+1)+(2x-1)=(4x+2x)+[1+(-1)]=6x.$$

So, you might conclude that the "missing factor" needed is $x+2$.

Now multiply the numerator and denominator of $\frac{4x+1}{2x-1}$ by $x+2$ because, when simplified, the binomial factors divide out.

$$\frac{4x+1}{2x-1}\cdot\frac{x+2}{x+2}=\frac{(4x+1)(x+2)}{(2x-1)(x+2)}$$
$$=\frac{4x^2+9x+2}{2x^2+3x-2}$$

You already know that the rational expression simplifies to $\frac{4x+1}{2x-1}$.

So, check to see that the sum of the two polynomials is $6x^2+12x$.

$$(4x^2+9x+2)+(2x^2+3x-2)=(4x^2+2x^2)+(9x+3x)+[2+(-2)]$$
$$=6x^2+12x\ ✓$$

So, $4x^2+9x+2$ and $2x^2+3x-2$ are the two polynomials with simplified ratio $\frac{4x+1}{2x-1}$ and with sum $6x^2+12x$.

Reteaching and Enrichment Strategies

If students need help. . .	If students got it. . .
Resources by Chapter • Practice A and Practice B • Puzzle Time Record and Practice Journal Practice Differentiating the Lesson Lesson Tutorials Skills Review Handbook	Resources by Chapter • Enrichment and Extension • School-to-Work Start the next section

Write and simplify a rational expression for the ratio of the perimeter of the figure to its area.

18.

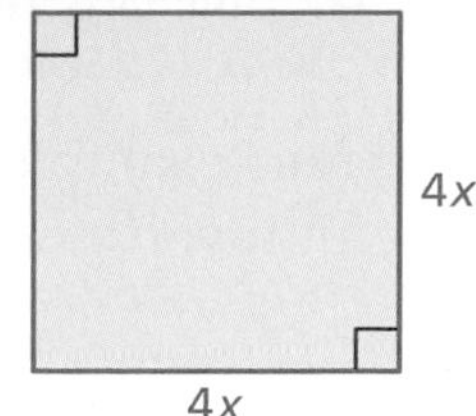

19.

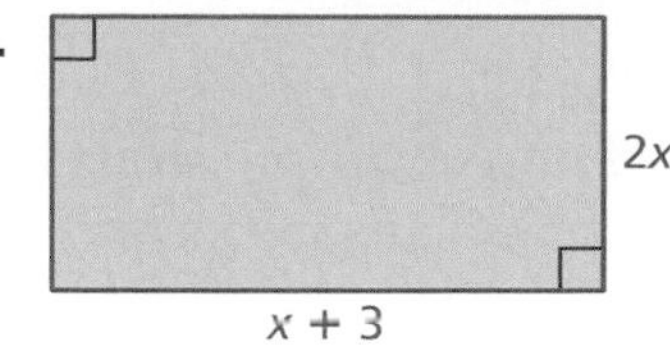

20.

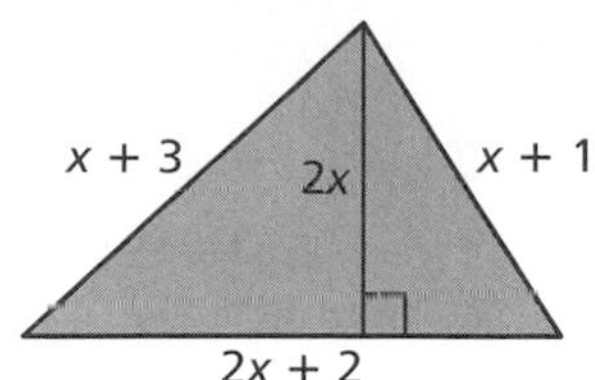

21. **OPEN-ENDED** Write a rational expression whose excluded values are -3 and -5.

22. **WRITING** Is $\dfrac{x^2 - 4}{x + 2}$ equivalent to $x - 2$? Justify your answer.

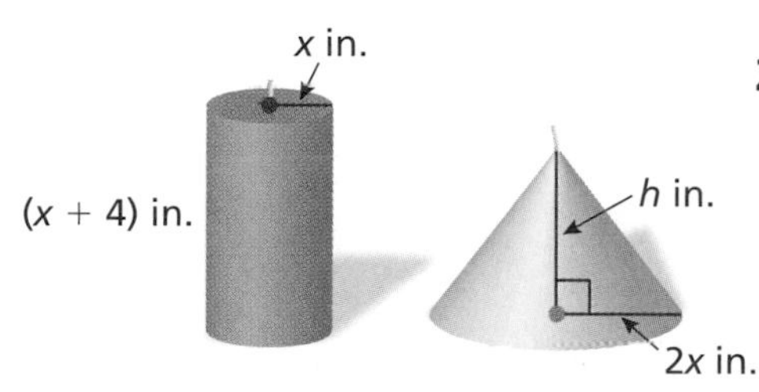

23. **PROBLEM SOLVING** The candles shown have the same volume. Write and simplify an expression for the height of the cone-shaped candle.

24. **SANDBOX** The area of Sandbox B is 4 square feet greater than the area of Sandbox A. Write and simplify an expression for the width w of Sandbox B.

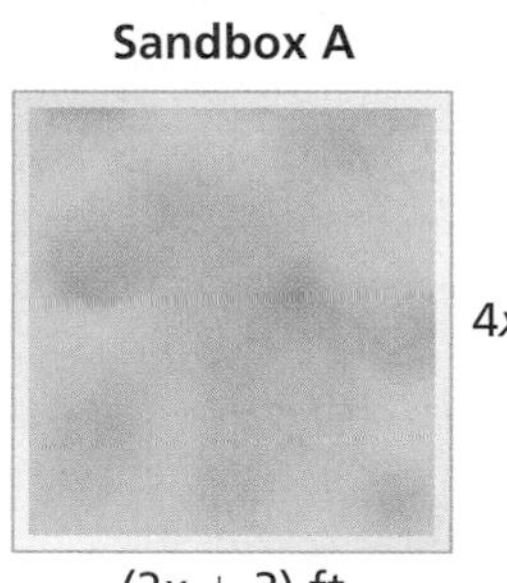

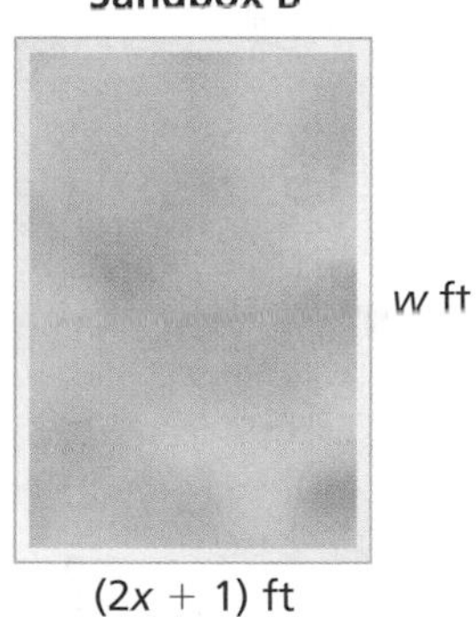

25. Critical Thinking Find two polynomials whose simplified ratio is $\dfrac{4x + 1}{2x - 1}$ and whose sum is $6x^2 + 12x$. Explain your reasoning.

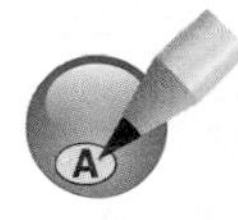

Fair Game Review What you learned in previous grades & lessons

Graph the function. Is the domain discrete or continuous? *(Section 5.2)*

26.

Input Boxes, x	Output Number of Shoes, y
1	2
2	4
3	6

27.

Input Months, x	Output Height of Plant, y (inches)
1	1.3
2	2.1
3	2.9

28. **MULTIPLE CHOICE** Consider $f(x) = 2x - 4$. What is the value of x so that $f(x) = 8$? *(Section 5.4)*

(A) 2 (B) 4 (C) 6 (D) 7

11 Study Help

You can use an **example and non-example chart** to list examples and non-examples of a vocabulary word or term. Here is an example and non-example chart for inverse variation equations.

Inverse Variation Equations

Examples	Non-Examples
$y = \frac{2}{x}$	$y = 2x$
$2 = xy$	$2 = \frac{y}{x}$
$x = \frac{2}{y}$	$y = \frac{x}{2}$
$3xy = 6$	$y = 2x + 1$

On Your Own

Make example and non-example charts to help you study these topics.

1. direct variation equations
2. rational functions
3. excluded values
4. asymptotes
5. rational expressions
6. simplest form of a rational expression

After you complete this chapter, make example and non-example charts for the following topics.

7. multiplying and dividing rational expressions
8. least common denominator of rational expressions
9. adding and subtracting rational expressions
10. rational equations

"What do you think of my example & non-example chart for popular cat toys?"

Sample Answers

1.

Direct Variation Equations

Examples	Non-Examples
$y = 3x$	$3 = xy$
$3y = x$	$3xy = 6$
$3 = \frac{y}{x}$	$x = \frac{3}{y}$
$3 = \frac{x}{y}$	$y = \frac{3}{x}$

2.

Rational Functions

Examples	Non-Examples
$y = \frac{1}{x}$	$y = x$
$y = \frac{1}{x+1}$	$y = x + 5$
$y = \frac{x+1}{x-3}$	$y = \frac{\sqrt{x+2}}{x-1}$
$y = \frac{x^2+1}{x^3-2x}$	$y = \frac{1}{\sqrt{x}}$

3.

Excluded Values

Examples	Non-Examples
$x = 0$ for $y = \frac{1}{x}$	$x = 0$ for $y = \frac{x}{x+1}$
$x = 1$ for $y = \frac{1}{x-1}$	$x = 2$ for $y = \frac{x-2}{x+4}$
$x = -1, x = 2$ for $y = \frac{x+3}{(x+1)(x-2)}$	$x = 1, x = -2$ for $y = \frac{x+3}{(x+1)(x-2)}$
$x = 0, x = 2$ for $y = \frac{x+1}{3x(x-2)}$	$x = -3, x = -2$ for $y = \frac{x+1}{3x(x-2)}$

4.

Asymptotes

Examples	Non-Examples
$x = 2, y = 3$ for $y = \frac{1}{x-2} + 3$	$x = -2, y = 0$ for $y = \frac{1}{x} - 2$
$x = 2, y = 0$ for $y = \frac{1}{x-2}$	$x = 0, y = 2$ for $y = \frac{1}{x-2}$
$x = -2, y = -1$ for $y = \frac{5}{x+2} - 1$	$x = 5, y = 1$ for $y = \frac{5}{x+2} - 1$

5–6. Available at *BigIdeasMath.com*.

List of Organizers

Available at *BigIdeasMath.com*

Comparison Chart
Concept Circle
Definition (Idea) and Example Chart
Example and Non-Example Chart
Formula Triangle
Four Square
Information Frame
Information Wheel
Notetaking Organizer
Process Diagram
Summary Triangle
Word Magnet
Y Chart

About this Organizer

An **Example and Non-Example Chart** can be used to list examples and non-examples of a vocabulary word or term. Students write examples of the word or term in the left column and non-examples in the right column. This type of organizer serves as a good tool for assessing students' knowledge of pairs of topics that have subtle but important differences, such as complementary and supplementary angles. Blank example and non-example charts can be included on tests or quizzes for this purpose.

Technology For the Teacher
Vocabulary Puzzle Builder

Answers

1. inverse variation; The products xy are constant.
2. direct variation; The ratios $\frac{y}{x}$ are constant.

3–5. See Additional Answers.

6. $x = 0$ **7.** $x = \frac{5}{4}$

8. $x = 0$; $y = -5$; The domain is all real numbers except 0 and the range is all real numbers except -5.
9. $x = 0$, $y = 0$; The domain is all real numbers except 0 and the range is all real numbers except 0.

10–12. See Additional Answers.

13. $\frac{1}{2y}$; $y = 0$
14. The expression is in simplest form; $z = -2$
15. $\frac{x-4}{2x+7}$; $x = -2$; $x = -\frac{7}{2}$
16. $2x^2$ by $(2x - 1)$
17. **a.** $c = \frac{400}{n}$

    b. $50

Assessment Book

Chapter 11 Quiz
For use after Section 11.3

Tell whether x and y show *direct variation*, *inverse variation*, or *neither*.

1.

x	1	2	3	4
y	30	15	10	7.5

2.

x	1	2	3	4
y	−4	−8	−12	−16

3. The variable y varies directly with x. When $x = 8$, $y = 6$. Write and graph a direct variation equation that relates x and y.

Identify the asymptotes of the graph of the function. Then describe the domain and range.

4. $y = \frac{3}{x-1} + 2$ 5. $y = -\frac{1}{x} - 3$

Find the inverse of the function. Graph the inverse function.

6. $f(x) = 4x + 1$ 7. $f(x) = x^2 + 2$, where $x \geq 0$

Simplify the rational expression, if possible. State the excluded value(s).

8. $\frac{3t^7}{15t^5}$ 9. $\frac{18w^2}{2w^7}$

10. $\frac{a^2 - 9}{a^2 + 2a - 3}$ 11. $\frac{b^2 + 8b + 15}{b^2 - b - 12}$

12. You have $120 to donate equally to several charities. Does this situation represent direct variation or inverse variation? Explain.
13. The side length of a square is represented by the expression $x + 1$. Write and simplify a rational expression for the ratio of the perimeter to the ratio of the area.

Answers

1. ______ 2. ______ 3. ______ See left. 4. ______ 5. ______ 6. ______ See left. 7. ______ See left. 8. ______ 9. ______ 10. ______ 11. ______ 12. ______ 13. ______

Alternative Quiz Ideas

100% Quiz	Math Log
Error Notebook	Notebook Quiz
Group Quiz	Partner Quiz
Homework Quiz	Pass the Paper

Homework Quiz

A homework notebook provides an opportunity for teachers to check that students are doing their homework regularly. Students keep their homework in a notebook. They should be told to record the page number, problem number, and copy the problem exactly in their homework notebook. Each day the teacher walks around and visually checks that homework is completed. Periodically, without advance notice, the teacher tells the students to put everything away except their homework notebook.

Questions are from students' homework.

1. What are the answers to Exercises 8 and 10 on page 547?
2. What are the answers to Exercises 23–25 on page 548?
3. What are the answers to Exercises 6 and 10 on page 555?
4. What are the answers to Exercises 19 and 21 on page 556?
5. What are the answers to Exercises 8 and 10 on page 559?
6. What are the answers to Exercises 4, 10, and 14 on page 564?

Reteaching and Enrichment Strategies

If students need help. . .	If students got it. . .
Resources by Chapter • Study Help • Practice A and Practice B • Puzzle Time Lesson Tutorials *BigIdeasMath.com* Practice Quiz Practice from the Test Generator	Resources by Chapter • Enrichment and Extension • School-to-Work Game Closet at *BigIdeasMath.com* Start the next section

Technology For the Teacher

Answer Presentation Tool
Big Ideas Test Generator

11.1–11.3 Quiz

Check It Out
Progress Check
BigIdeasMath.com

Tell whether x and y show *direct variation, inverse variation,* or *neither*. Explain your reasoning. *(Section 11.1)*

1.

x	y
1	−60
2	−30
3	−20
4	−15

2.

x	y
1	6
2	12
3	18
4	24

3.

x	y
1	−2
2	1
3	5
4	10

4. The variable y varies directly with x. When $x = 3$, $y = 15$. Write and graph a direct variation equation that relates x and y. *(Section 11.1)*

5. The variable y varies inversely with x. When $x = 2$, $y = 7$. Write and graph an inverse variation equation that relates x and y. *(Section 11.1)*

Find the excluded value of the function. *(Section 11.2)*

6. $y = \dfrac{2}{5x}$

7. $y = \dfrac{1}{4x - 5}$

Identify the asymptotes of the graph of the function. Then describe the domain and range. *(Section 11.2)*

8. $y = \dfrac{2}{x} - 5$

9. $y = -\dfrac{10}{x}$

10. $y = \dfrac{3}{x + 6} + 9$

Find the inverse of the function. Graph the inverse function. *(Section 11.2)*

11. $f(x) = 2x + 3$

12. $f(x) = x^2 + 1$, where $x \geq 0$

Simplify the rational expression, if possible. State the excluded value(s). *(Section 11.3)*

13. $\dfrac{12y^4}{24y^5}$

14. $\dfrac{2z - 1}{z + 2}$

15. $\dfrac{x^2 - 2x - 8}{2x^2 + 11x + 14}$

16. DIMENSIONS Simplify the dimensions of the computer monitor. *(Section 11.3)*

17. FISHING BOAT The cost c per person to charter a fishing boat varies inversely with the number n of people fishing. The cost to charter a boat for an entire day is $400. *(Section 11.1)*

a. Write an inverse variation equation that relates c and n.

b. How much does each person pay when 8 people fish?

11.4 Multiplying and Dividing Rational Expressions

COMMON CORE STATE STANDARDS
A.SSE.2

Essential Question How can you multiply and divide rational expressions?

1 ACTIVITY: Matching Quotients and Products

Work with a partner. Match each quotient with a product and then with a simplified expression. Explain your reasoning.

Quotient of Two Rational Expressions	*Product of Two Rational Expressions*	*Simplified Expression*
a. $\frac{2x^2}{5} \div \frac{14x}{10}$	**A.** $\frac{5x^2}{2} \cdot \frac{10}{14x}$	**1.** $\frac{25x}{14}$
b. $\frac{2x^2}{5} \div \frac{10}{14x}$	**B.** $\frac{2x^2}{5} \cdot \frac{10}{14x}$	**2.** $\frac{7x^3}{2}$
c. $\frac{5x^2}{2} \div \frac{14x}{10}$	**C.** $\frac{5x^2}{2} \cdot \frac{14x}{10}$	**3.** $\frac{14x^3}{25}$
d. $\frac{5x^2}{2} \div \frac{10}{14x}$	**D.** $\frac{2x^2}{5} \cdot \frac{14x}{10}$	**4.** $\frac{2x}{7}$

e. $\frac{x^2-1}{x+2} \div \frac{x+1}{x^2-4}$	**E.** $\frac{x^2-1}{x-2} \cdot \frac{x^2-4}{x+1}$	**5.** $x^2 - x - 2$
f. $\frac{x^2-1}{x-2} \div \frac{x+1}{x^2-4}$	**F.** $\frac{x^2-1}{x-2} \cdot \frac{x^2-4}{x-1}$	**6.** $x^2 - 3x + 2$
g. $\frac{x^2-1}{x-2} \div \frac{x-1}{x^2-4}$	**G.** $\frac{x^2-1}{x+2} \cdot \frac{x^2-4}{x-1}$	**7.** $x^2 + x - 2$
h. $\frac{x^2-1}{x+2} \div \frac{x-1}{x^2-4}$	**H.** $\frac{x^2-1}{x+2} \cdot \frac{x^2-4}{x+1}$	**8.** $x^2 + 3x + 2$

i. $\frac{x-1}{2} \div (x-1)$	**I.** $\frac{2}{x^2-1} \cdot (x-1)$	**9.** $\frac{x+1}{2}$
j. $\frac{x^2-1}{2} \div (x-1)$	**J.** $\frac{x^2-1}{2} \cdot \frac{1}{x-1}$	**10.** $\frac{1}{2}$
k. $\frac{x-1}{2} \div \frac{1}{x-1}$	**K.** $\frac{x-1}{2} \cdot \frac{1}{x-1}$	**11.** $\frac{2}{x+1}$
l. $\frac{2}{x^2-1} \div \frac{1}{x-1}$	**L.** $\frac{x-1}{2} \cdot (x-1)$	**12.** $\frac{(x-1)^2}{2}$

Laurie's Notes

Introduction

Standards for Mathematical Practice

- **MP7 Look for and Make Use of Structure:** Students must use prior skills and techniques to simplify rational expressions. They need to use the structure of an expression to divide out common factors correctly, one from the numerator and one from the denominator.

Motivate

- Write several quotients of fractions on the board.
- ? "How would you solve these problems?" Multiply by the reciprocal of the divisor.
- Ask for volunteers to share their answers.
- The purpose is to review how to divide fractions, a necessary prerequisite for this topic.

Activity Notes

Activity 1

- In this activity, students combine their knowledge of dividing fractions with simplifying rational expressions. This knowledge is generally sufficient for them to work through these problems.
- Repeat the phrase "divide out common factors" as needed while you observe students working each cluster of problems.
- Take time to listen to students explain the reasoning behind their matches. You want to be sure that they arrive at the correct answers for the correct reasons.
- Make sure students do not simply guess towards the end when there are only a few expressions left.

Common Core State Standards

A.SSE.2 Use the structure of an expression to identify ways to rewrite it.

Previous Learning

Students should know how to simplify rational expressions.

Start Thinking! and Warm Up

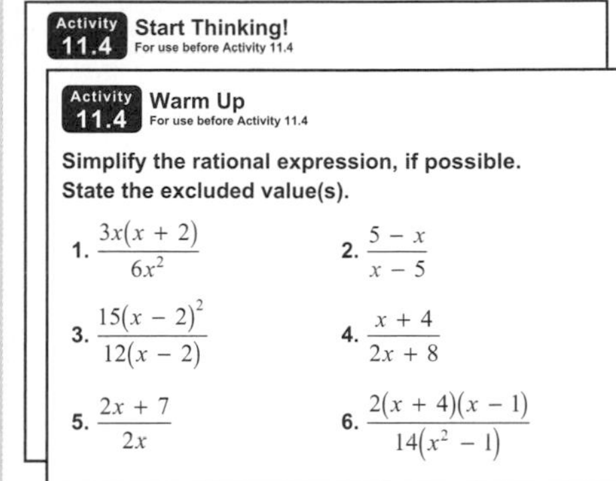

Activity 11.4 Start Thinking! For use before Activity 11.4

Activity 11.4 Warm Up For use before Activity 11.4

Simplify the rational expression, if possible. State the excluded value(s).

1. $\frac{3x(x+2)}{6x^2}$
2. $\frac{5-x}{x-5}$
3. $\frac{15(x-2)^2}{12(x-2)}$
4. $\frac{x+4}{2x+8}$
5. $\frac{2x+7}{2x}$
6. $\frac{2(x+4)(x-1)}{14(x^2-1)}$

11.4 Record and Practice Journal

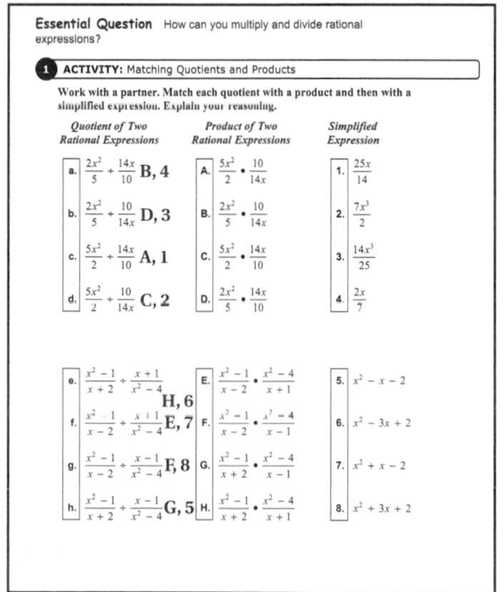

Essential Question How can you multiply and divide rational expressions?

1 ACTIVITY: Matching Quotients and Products

Work with a partner. Match each quotient with a product and then with a simplified expression. Explain your reasoning.

Quotient of Two Rational Expressions	*Product of Two Rational Expressions*	*Simplified Expression*
a. $\frac{2x^2}{5} \div \frac{14x}{10}$ **B, 4**	A. $\frac{5x^2}{2} \cdot \frac{10}{14x}$	1. $\frac{25x}{14}$
b. $\frac{2x^2}{5} \div \frac{10}{14x}$ **D, 3**	B. $\frac{2x^2}{5} \cdot \frac{10}{14x}$	2. $\frac{7x^3}{2}$
c. $\frac{5x^2}{2} \div \frac{14x}{10}$ **A, 1**	C. $\frac{5x^2}{2} \cdot \frac{14x}{10}$	3. $\frac{14x^3}{25}$
d. $\frac{5x^2}{2} \div \frac{10}{14x}$ **C, 2**	D. $\frac{2x^2}{5} \cdot \frac{14x}{10}$	4. $\frac{2x}{7}$
e. $\frac{x^2-1}{x+2} \div \frac{x+1}{x^2-4}$ **H, 6**	E. $\frac{x^2-1}{x-2} \cdot \frac{x^2-4}{x+1}$	5. x^2-x-2
f. $\frac{x^2-1}{x-2} \div \frac{x+1}{x^2-4}$ **E, 7**	F. $\frac{x^2-1}{x-2} \cdot \frac{x^2-4}{x-1}$	6. x^2-3x+2
g. $\frac{x^2-1}{x-2} \div \frac{x-1}{x^2-4}$ **F, 8**	G. $\frac{x^2-1}{x+2} \cdot \frac{x^2-4}{x-1}$	7. x^2+x-2
h. $\frac{x^2-1}{x+2} \div \frac{x-1}{x^2-4}$ **G, 5**	H. $\frac{x^2-1}{x+2} \cdot \frac{x^2-4}{x+1}$	8. x^2+3x+2

English Language Learners

Pair Activity

Pair English language learners with English speakers. Provide students with several problems of multiplying and dividing rational expressions. Have one student write the first step and explain what they did to their partner. Then have the other student write the next step and explain. The first person does the next step and the process continues until the problem is completed. Listen for the words *factor* and *divide out*.

11.4 Record and Practice Journal

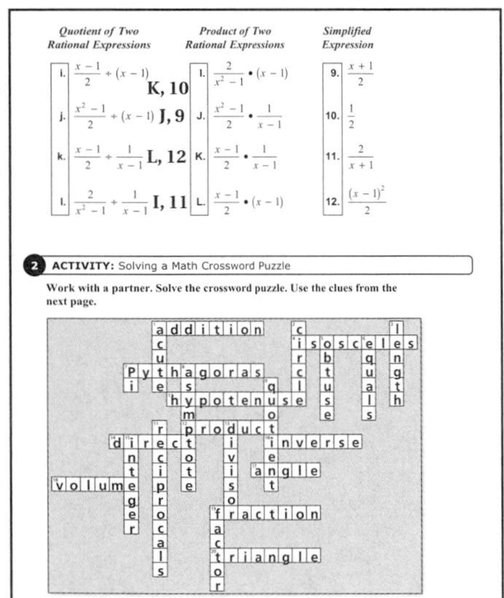

Quotient of Two Rational Expressions		*Product of Two Rational Expressions*	*Simplified Expression*
I. $\frac{x-1}{2} \div (x-1)$	**K, 10**	I. $\frac{2}{x^2-1} \cdot (x-1)$	9. $\frac{x+1}{2}$
J. $\frac{x^2-1}{2} \div (x-1)$	**J, 9**	J. $\frac{x^2-1}{2} \cdot \frac{1}{x-1}$	10. $\frac{1}{2}$
K. $\frac{x-1}{2} \div \frac{1}{x-1}$	**L, 12**	K. $\frac{x-1}{2} \cdot \frac{1}{x-1}$	11. $\frac{2}{x+1}$
L. $\frac{2}{x^2-1} \div \frac{1}{x-1}$	**I, 11**	L. $\frac{x-1}{2} \cdot (x-1)$	12. $\frac{(x-1)^2}{2}$

2 ACTIVITY: Solving a Math Crossword Puzzle

Work with a partner. Solve the crossword puzzle. Use the clues from the next page.

Across: addition, isosceles, Pythagoras, hypotenuse, product, direct, inverse, angle, volume, fraction, triangle

Down: acute, circumference, length, obtuse, equals, pi, reciprocals, integer, quotient, ratio, division, factor, numerator, integers

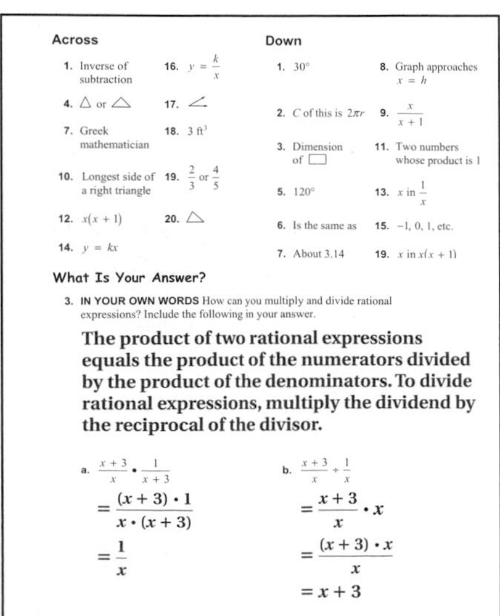

Across

1. Inverse of subtraction
4. △ or △
7. Greek mathematician
10. Longest side of a right triangle
12. $x(x+1)$
14. $y = kx$
16. $y = \frac{k}{x}$
17. ∠
18. 3 ft³
19. $\frac{2}{3}$ or $\frac{4}{5}$
20. △

Down

1. 30°
2. C of this is $2\pi r$
3. Dimension of ▭
5. 120°
6. Is the same as
7. About 3.14
8. Graph approaches $x = h$
9. $\frac{x}{x+1}$
11. Two numbers whose product is 1
13. x in $\frac{1}{x}$
15. −1, 0, 1, etc.
19. x in $x(x+1)$

What Is Your Answer?

3. **IN YOUR OWN WORDS** How can you multiply and divide rational expressions? Include the following in your answer.

The product of two rational expressions equals the product of the numerators divided by the product of the denominators. To divide rational expressions, multiply the dividend by the reciprocal of the divisor.

a. $\frac{x+3}{x} \cdot \frac{1}{x+3}$

$= \frac{(x+3) \cdot 1}{x \cdot (x+3)}$

$= \frac{1}{x}$

b. $\frac{x+3}{x} \div \frac{1}{x}$

$= \frac{x+3}{x} \cdot x$

$= \frac{(x+3) \cdot x}{x}$

$= x + 3$

Laurie's Notes

Activity 2

- Students will enjoy working on this crossword puzzle with a partner.
- If students are stuck on a word, they should come back to it later when they may have other clues to help.
- Resist the temptation to give answers or to allow pairs of students to work together.
- **MP1b Persevere in Solving Problems:** Let each pair of students work through the puzzle and persevere through their struggles.

What Is Your Answer?

- **Neighbor Check:** Have students work independently and then have their neighbor check their work. Have students discuss any discrepancies.

Closure

- **Exit Ticket:** Complete the statement.

$$\frac{x^2-1}{?} \cdot \frac{5}{?} = x + 1$$

Sample answer: $\frac{x^2-1}{5} \cdot \frac{5}{x-1} = x + 1$

Technology For the Teacher

The Dynamic Planning Tool
Editable Teacher's Resources at *BigIdeasMath.com*

2 ACTIVITY: Solving a Math Crossword Puzzle

Work with a partner. Solve the crossword puzzle.

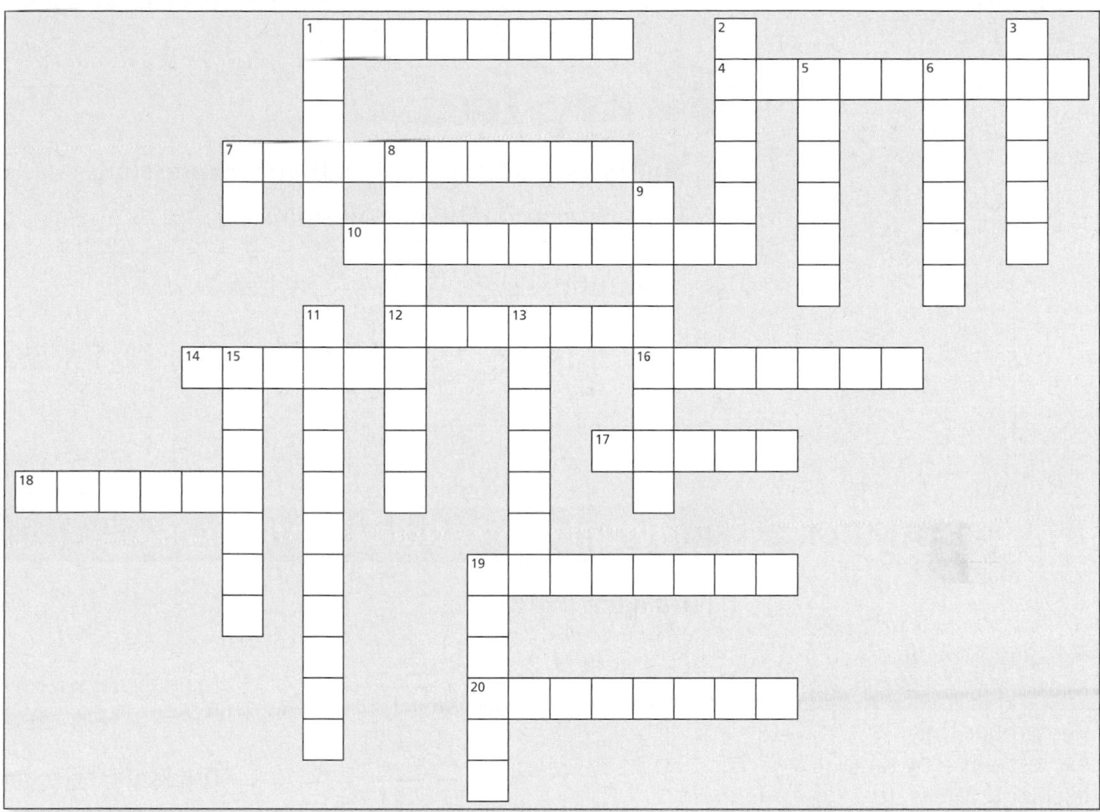

Across

1. Inverse of subtraction
4. △ or △
7. Greek mathematician
10. Longest side of a right triangle
12. $x(x + 1)$
14. $y = kx$
16. $y = \frac{k}{x}$
17. ∠
18. 3 ft^3
19. $\frac{2}{3}$ or $\frac{4}{5}$
20. △

Down

1. 30°
2. C of this is $2\pi r$
3. Dimension of ▭
5. 120°
6. Is the same as
7. About 3.14
8. Graph approaches $x = h$
9. $\frac{x}{x+1}$
11. Two numbers whose product is 1
13. x in $\frac{1}{x}$
15. $-1, 0, 1$, etc.
19. x in $x(x + 1)$

What Is Your Answer?

3. **IN YOUR OWN WORDS** How can you multiply and divide rational expressions? Include the following in your answer.

   a. $\frac{x+3}{x} \cdot \frac{1}{x+3}$

   b. $\frac{x+3}{x} \div \frac{1}{x}$

Practice Use what you learned about multiplying and dividing rational expressions to complete Exercises 4 and 10 on page 572.

11.4 Lesson

You can use the same rules that you used for multiplying and dividing fractions to multiply and divide rational expressions.

Key Idea

Multiplying and Dividing Rational Expressions

Let a, b, c, and d be polynomials.

Multiplying: $\frac{a}{b} \cdot \frac{c}{d} = \frac{ac}{bd}$, where $b, d \neq 0$

Dividing: $\frac{a}{b} \div \frac{c}{d} = \frac{a}{b} \cdot \frac{d}{c} = \frac{ad}{bc}$, where $b, c, d \neq 0$

EXAMPLE 1 Multiplying Rational Expressions

Find each product.

Remember

Remember that expressions may have excluded values. In Example 1a, the excluded values are $x = -1$ and $x = 0$.

a. $\frac{5}{2x^3} \cdot \frac{4x^3}{x+1} = \frac{5 \cdot 4x^3}{2x^3(x+1)}$ Multiply numerators and denominators.

$= \frac{5 \cdot \overset{2}{\cancel{4}}\cancel{x^3}}{\cancel{2}\cancel{x^3}(x+1)}$ Divide out the common factors.

$= \frac{10}{x+1}$ Simplify.

b. $\frac{h}{h+2} \cdot \frac{h^2+5h+6}{h^2}$

$= \frac{h}{h+2} \cdot \frac{(h+3)(h+2)}{h^2}$ Factor $h^2 + 5h + 6$.

$= \frac{h(h+3)(h+2)}{h^2(h+2)}$ Multiply numerators and denominators.

$= \frac{\cancel{h}(h+3)\cancel{(h+2)}}{\cancel{h^2}\cancel{(h+2)}}$ Divide out the common factors.

$= \frac{h+3}{h}$ Simplify.

On Your Own

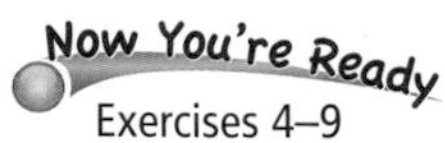

Exercises 4–9

Find the product.

1. $\frac{8y^2}{y-5} \cdot \frac{3}{4y}$ **2.** $\frac{16}{8-c} \cdot (c-8)$ **3.** $\frac{2z-4}{6} \cdot \frac{3}{z^2-7z+10}$

Laurie's Notes

Introduction

Connect

- **Yesterday:** Students recognized equivalent expressions. (MP1b, MP7)
- **Today:** Students will multiply and divide rational expressions.

Motivate

- Write the following problems on the board.

$$\frac{2}{3} \div \frac{2}{3} \bullet \frac{2}{3} \div \frac{2}{3} \bullet \frac{2}{3} = ? \qquad \frac{2}{3} \div \frac{2}{3} \bullet \frac{2}{3} \div \frac{2}{3} \bullet \frac{2}{3} \div \frac{2}{3} = ?$$

? "What are the answers?" $\frac{2}{3}$; 1

? "If this pattern were continued by adding 5 more terms (11 total fractions), what is the answer? Explain." $\frac{2}{3}$; When the number of terms is even, the answer is 1. When the number of terms is odd, the answer is $\frac{2}{3}$.

- Explain that today's lesson involves multiplying and diving rational expressions.

Lesson Notes

Key Idea

- Write the Key Idea.
- **Big Idea:** The rules for multiplying and dividing rational expressions are the same as the rules for multiplying and dividing fractions. Common factors can be divided out, and the denominator cannot be 0.

Example 1

- Write the problem in part (a). Encourage students to try factoring first and looking for common factors, rather than immediately multiplying.
- Students may need to be reminded of the properties of exponents.
- Write the problem in part (b). If students do not recognize any common factors, put parentheses around the $(h + 2)$ term. It may help them recognize that the trinomial can be factored.

? "What are the common factors?" the $(h + 2)$ terms and an h term

- Discuss the Remember box.

On Your Own

- **Teaching Tip:** Some students try to perform all of the steps mentally and may not even write the original problem. Encourage students to write each step.
- Ask volunteers to share their work and explanations with the class.
- In Question 2, students need to rewrite $(8 - c)$ as $-(c - 8)$.

Goal Today's lesson is multiplying and dividing rational expressions.

Start Thinking! and Warm Up

Warm Up
For use before Lesson 11.4

Start Thinking!
For use before Lesson 11.4

How are multiplying and dividing fractions similar to multiplying and dividing rational expressions? How are they different?

Extra Example 1

Find each product.

a. $\frac{2n-1}{4n^5} \bullet \frac{2n^3}{3}$ $\quad \frac{2n-1}{6n^2}$

b. $\frac{k-3}{k^2} \bullet \frac{k+2}{k^2-k-6}$ $\quad \frac{1}{k^2}$

On Your Own

1. $\frac{6y}{y-5}$
2. -16
3. $\frac{1}{z-5}$

Extra Example 2

Find the quotient $\frac{d+6}{d^2} \div \frac{2d+12}{d^2}$. $\frac{1}{2}$

Extra Example 3

Find the quotient

$\frac{a^2-16}{a+3} \div (a^2+7a+12)$. $\frac{a-4}{(a+3)^2}$

On Your Own

4. $2t$
5. $\frac{g-1}{g}$
6. $\frac{1}{(d-1)^2}$

Differentiated Instruction

Auditory

Guide your students in making the following list of steps used in dividing rational expressions.

1. Multiply by the reciprocal.
2. Factor each numerator and denominator.
3. Multiply the numerators and the denominators.
4. Divide out common factors.
5. Simplify.

Organize students into small groups. Assign each group an expression in the form of $\frac{a}{b} \div \frac{c}{d}$. Have each group simplify the expression and present their solution to the class. Students should read the steps from the list aloud as they explain their solution.

Laurie's Notes

Example 2

? "What is the basic procedure used to divide rational expressions?" Multiply by the reciprocal of the divisor.

- Write the original problem and solve as shown.
- **MP7 Look for and Make Use of Structure:** You may want to ask students how they could do this problem mentally. The two rational expressions have the same denominator, so you can solve by simply finding the quotient of the numerators.

Example 3

- Rewriting the divisor as a fraction will make students less likely to make a mistake when dividing.
- Write the equivalent multiplication problem.

? "Can the p^2-terms be divided out? Explain." No, the p^2-terms are not factors. These polynomials should be factored.

- Continue to work through the problem as shown.

On Your Own

- **Think-Pair-Share:** Students should read each question independently and then work with a partner to answer the questions. When they have answered the questions, the pair should compare their answers with another group and discuss any discrepancies.

Closure

- **Exit Ticket:** Find the quotient: $\frac{x^2-x-2}{5x} \div \frac{x-2}{x^2}$. $\frac{x(x+1)}{5}$

Technology For the Teacher

The Dynamic Planning Tool
Editable Teacher's Resources at *BigIdeasMath.com*

EXAMPLE 2 Standardized Test Practice

Which expression is equivalent to $\frac{8}{w-4} \div \frac{w}{w-4}$ when $w \neq 4$?

Ⓐ $\frac{8}{w}$ Ⓑ $\frac{w}{8}$ Ⓒ $\frac{8w}{(w-4)^2}$ Ⓓ $\frac{8w}{w^2-8w+16}$

$$\frac{8}{w-4} \div \frac{w}{w-4} = \frac{8}{w-4} \bullet \frac{w-4}{w}$$ Multiply by the reciprocal.

$$= \frac{8(w-4)}{w(w-4)}$$ Multiply numerators and denominators.

$$= \frac{8\cancel{(w-4)}}{w\cancel{(w-4)}}$$ Divide out the common factor.

$$= \frac{8}{w}$$ Simplify.

∴ The correct answer is Ⓐ.

EXAMPLE 3 Dividing Rational Expressions

Find the quotient $\frac{p^2-p-6}{p+1} \div (p^2-4)$.

$$\frac{p^2-p-6}{p+1} \div \frac{p^2-4}{1}$$ Write $p^2 - 4$ as a fraction.

$$= \frac{p^2-p-6}{p+1} \bullet \frac{1}{p^2-4}$$ Multiply by the reciprocal.

$$= \frac{(p-3)(p+2)}{p+1} \bullet \frac{1}{(p-2)(p+2)}$$ Factor.

$$= \frac{(p-3)(p+2)}{(p+1)(p-2)(p+2)}$$ Multiply numerators and denominators.

$$= \frac{(p-3)\cancel{(p+2)}}{(p+1)(p-2)\cancel{(p+2)}}$$ Divide out the common factor.

$$= \frac{p-3}{(p+1)(p-2)}$$ Simplify.

On Your Own

Now You're Ready
Exercises 10–15

Find the quotient.

4. $\frac{t-2}{2t} \div \frac{t-2}{4t^2}$

5. $(g+1) \div \frac{g^2+g}{g-1}$

6. $\frac{d+5}{d-1} \div (d^2+4d-5)$

11.4 Exercises

Vocabulary and Concept Check

1. **WRITING** Describe how to multiply rational expressions.
2. **WRITING** Describe how to divide rational expressions.
3. **NUMBER SENSE** Consider the expressions $\frac{x}{x-2}$ and $\frac{x+1}{x}$. For what value(s) is the product of the expressions undefined? For what value(s) is the quotient of the expressions undefined?

Practice and Problem Solving

Find the product.

1 4. $\frac{5}{3c^2} \cdot \frac{c^5}{15(c-2)}$

5. $\frac{n+3}{8n^6} \cdot \frac{4n^2}{7}$

6. $(d^2 - d) \cdot \frac{14}{1-d}$

7. $\frac{x+4}{6x} \cdot \frac{x^2}{x^2 - x - 20}$

8. $\frac{k^2 - 8k + 15}{5k^3} \cdot \frac{3k}{k-5}$

9. $\frac{-r-6}{2r^2 + 8r} \cdot \frac{4r^2 + 16r}{r^2 - 36}$

Find the quotient.

2 3 10. $\frac{2h}{h+8} \div \frac{16}{h+8}$

11. $\frac{t-5}{9t} \div \frac{t-5}{6t^2}$

12. $\frac{y+7}{7y} \div \frac{3y^2 + 21y}{14y - 5}$

13. $\frac{p^2 - 16}{p-3} \div (p-4)$

14. $\frac{g^2 - 4g - 21}{4g^2 + 12g} \div (g-7)$

15. $\frac{3z - 27}{z-6} \div (z^2 - 15z + 54)$

ERROR ANALYSIS Describe and correct the error in finding the quotient.

16.

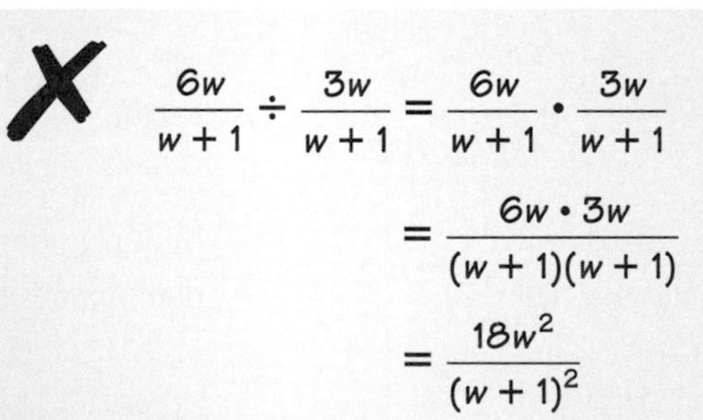

17.

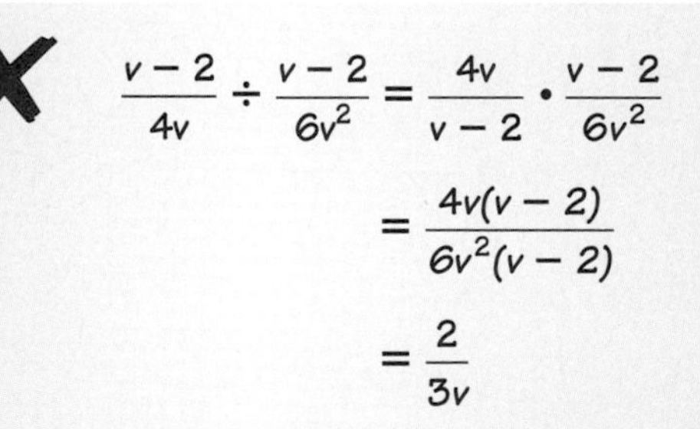

Find the total area of the red rectangle in terms of w.

18.

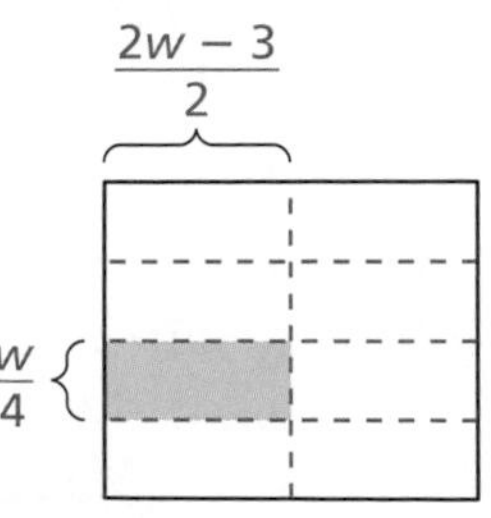

19.

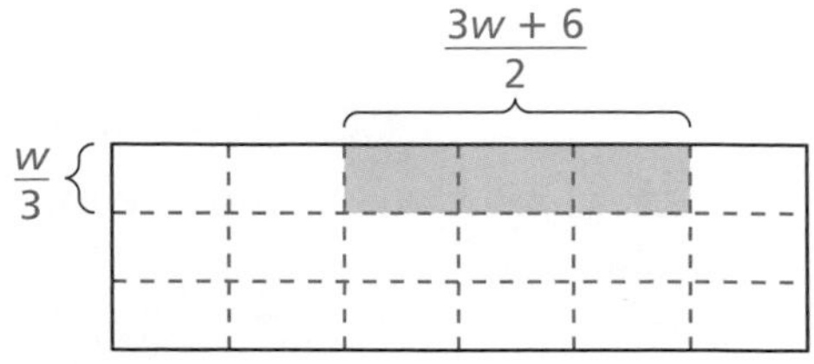

Assignment Guide and Homework Check

Level	Assignment	Homework Check
Average	1–3, 4–18 even, 20, 21, 23, 24, 27–30	4, 14, 16, 18, 23
Advanced	1–3, 5–21 odd, 22–26, 27–30	9, 17, 19, 23, 26

For Your Information

- **Exercise 26** Students may not remember the order of operations. If students are confused, tell them to think of the problem as $\left(\frac{8x^3}{x+1} \div \frac{2x-2}{3x}\right) \div \frac{16x^2}{x+1}$. The order of operations apply to rational expressions as well as numbers.

Common Errors

- **Exercises 4–15** Students may not factor completely before simplifying the rational expression. Remind students to factor completely in order to divide out common factors so that the rational expression is in simplest form.
- **Exercises 10–15** Students may not rewrite the division as multiplication. Remind students that when they divide rational expressions, they need to multiply the dividend by the reciprocal of the divisor.
- **Exercises 10–15** Students may multiply the divisor by the reciprocal of the dividend. Remind students that when they divide rational expressions, they need to multiply the dividend by the reciprocal of the divisor.

11.4 Record and Practice Journal

Find the product.

1. $\frac{8k^2}{7} \bullet \frac{k-5}{2k^4}$ **$\frac{4(k-5)}{7k^2}$**
2. $\frac{8}{w+6} \bullet \frac{3w^2+18w}{2}$ **$12w$**
3. $\frac{2b^3}{b^2-2b-15} \bullet \frac{b-5}{6b}$ **$\frac{b^2}{3b+9}$**
4. $\frac{4n^2-8n}{n^2+9n+14} \bullet \frac{n^2-4}{4n}$ **$\frac{(n-2)^2}{n+7}$**

Find the quotient.

5. $\frac{g+3}{4g^2} \div \frac{g+3}{12g}$ **$\frac{3}{g}$**
6. $\frac{y}{-5y^2-20y} \div \frac{y^2}{y^2+2y-8}$ **$\frac{y-2}{-5y^2}$**
7. $\frac{a^2-64}{a} \div (3a^3+24a^2)$ **$\frac{a-8}{3a^3}$**
8. $\frac{r^2-7r}{3} \div (r^2+r-56)$ **$\frac{r}{3r+24}$**
9. Two distinct prairie dog populations, P_1 and P_2 can be modeled by $P_1 = \frac{100x^2}{x+1}$ and $P_2 = \frac{100x^2}{x+3}$, where x is the number of years since 2000.
   a. Write a function that models the ratio of Population 1 to Population 2, that is $\frac{P_1}{P_2}$. **$\frac{x+3}{x+1}$**
   b. Find the ratio of Population 1 to Population 2 in 2004. **$\frac{7}{5}$**

Technology For the Teacher
Answer Presentation Tool

Vocabulary and Concept Check

1. (1) Factor the numerators and denominators. (2) Multiply the numerators and denominators. (3) Divide out common factors. (4) Simplify.
2. (1) Rewrite the quotient as the product of the dividend and the reciprocal of the divisor. (2) Factor the numerators and denominators. (3) Multiply the numerators and denominators. (4) Divide out common factors. (5) Simplify.
3. $x = 0, x = 2; x = -1, x = 0, x = 2$

Practice and Problem Solving

4. $\frac{c^3}{9(c-2)}$
5. $\frac{n+3}{14n^4}$
6. $-14d$
7. $\frac{x}{6(x-5)}$
8. $\frac{3(k-3)}{5k^2}$
9. $\frac{-2}{r-6}$
10. $\frac{h}{8}$
11. $\frac{2t}{3}$
12. $\frac{14y-5}{21y^2}$
13. $\frac{p+4}{p-3}$
14. $\frac{1}{4g}$
15. $\frac{3}{(z-6)^2}$
16. See Additional Answers.
17. To multiply rational expressions, you multiply by the reciprocal of the divisor, not the dividend;

$$\frac{v-2}{4v} \div \frac{v-2}{6v^2} = \frac{v-2}{4v} \bullet \frac{6v^2}{v-2} = \frac{6v^2(v-2)}{4v(v-2)} = \frac{3v}{2}$$

18. $w(2w-3)$
19. $3w(w+2)$

Practice and Problem Solving

20. $\dfrac{b+4}{2b+1}$

21. -1

22. $x = -2, x = -1, x = 3$

23. See *Taking Math Deeper*.

24. a. Their graphs coincide.

    b. The y_1 column will display ERROR for x values that are excluded.

25. a. $T = \dfrac{50 - x}{(1 - 0.05x)(0.05x^2 + 5)}$

    b. 2020; This will be 20 years after 2000 and 20 is the excluded value.

26. $\dfrac{3x^2}{4(x-1)}$

Fair Game Review

27. 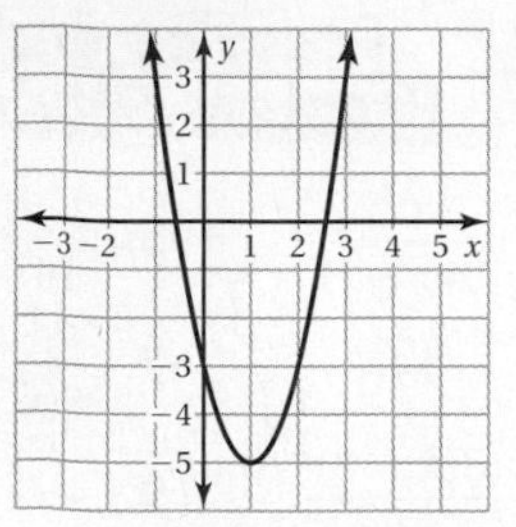

domain: all real numbers; range: $y \geq -5$

28–29. See Additional Answers.

30. D

Mini-Assessment

Find the product.

1. $\dfrac{3x^3}{x-8} \cdot \dfrac{8-x}{6x}$ $\quad -\dfrac{x^2}{2}$

2. $\dfrac{2y^2-6y}{5y^2} \cdot \dfrac{5y^2-5y-60}{y^2-7y+12}$ $\quad \dfrac{2(y+3)}{y}$

Find the quotient.

3. $\dfrac{6w}{2w-5} \div \dfrac{3}{2w-5}$ $\quad 2w$

4. $\dfrac{c^2-49}{3c} \div (2c-14)$ $\quad \dfrac{c+7}{6c}$

Taking Math Deeper

Exercise 23

Students may realize that they can find the answer using similarity.

Interpret the diagram.

Factoring the length of the campground as $2(d-2)$ shows that the shorter side length of the shaded campsites is one-half of the length of the campground.

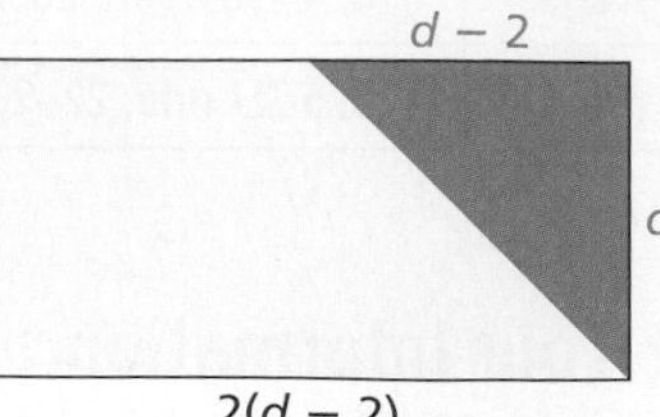

Divide the campground into two identical rectangles and label the dimensions.

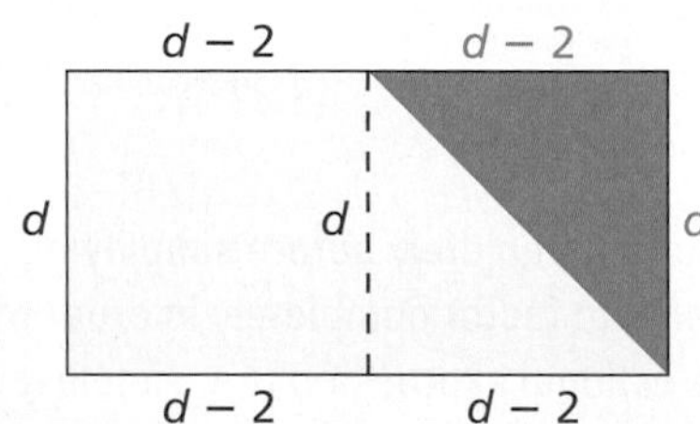

Answer the question.

From the diagram, you can see that each rectangle represents one-half of the area of the campground.

You can also see that the area of the shaded campsites is one-half of the area of each rectangle, or one-quarter of the area of the campground.

So, the probability that your campsite has shade is $\frac{1}{4}$, or 25%.

Project

Research campground regulations and fees at national parks. What regulations are shared by more than one of the parks you researched?

Reteaching and Enrichment Strategies

If students need help. . .	If students got it. . .
Resources by Chapter • Practice A and Practice B • Puzzle Time Record and Practice Journal Practice Differentiating the Lesson Lesson Tutorials Skills Review Handbook	Resources by Chapter • Enrichment and Extension • School-to-Work Start the next section

Find the product or quotient.

20. $\dfrac{2b^2 - b - 3}{b^2 - 6b - 7} \bullet \dfrac{b^2 - 3b - 28}{4b^2 - 4b - 3}$

21. $\dfrac{8y^2 + 6y - 5}{1 - 4y^2} \div \dfrac{12y^2 - y - 20}{6y^2 - 5y - 4}$

22. REASONING What are the excluded values of $\dfrac{x^2 + x - 2}{x - 3} \div \dfrac{x + 2}{x + 1}$?

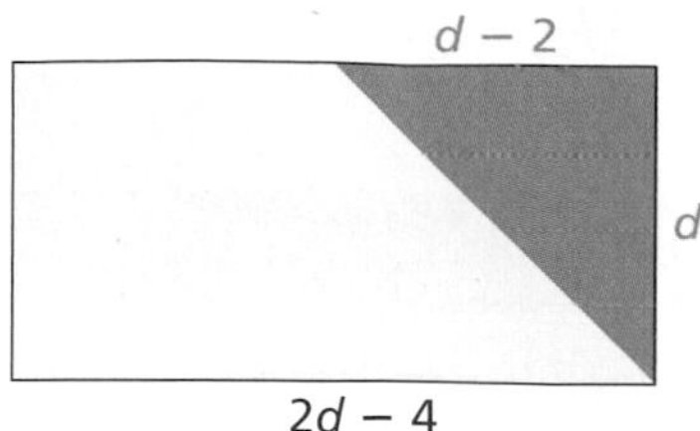

23. CAMPSITE A campsite is in the shape of a rectangle. The green region represents campsites with shade. The yellow represents campsites without shade. Your campsite is randomly assigned. What is the probability that your campsite has shade?

24. TECHNOLOGY You can use a graphing calculator to check your answers when multiplying or dividing rational expressions. For instance, graph $y = \dfrac{5}{2x^3} \bullet \dfrac{4x^3}{x + 1}$ and $y = \dfrac{10}{x + 1}$ from Example 1a in the same viewing window.

a. What do you notice about the graphs?

b. How can you use the *table* feature to find the excluded values?

25. CHARITY The revenue R (in thousands of dollars) and the average ticket price P (in dollars) for a charity event can be modeled by $R = \dfrac{50 - x}{1 - 0.05x}$ and $P = 0.05x^2 + 5$, where x is the number of years since 2000. (*Note:* revenue = tickets sold × ticket price)

a. Write an equation that models the number T of tickets sold as a function of x.

b. In what year will this model become invalid? Explain your reasoning.

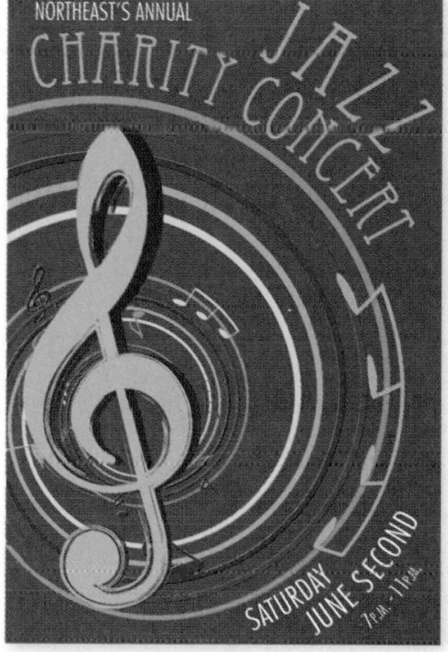

26. **Structure** Write $\dfrac{8x^3}{x + 1} \div \dfrac{2x - 2}{3x} \div \dfrac{16x^2}{x + 1}$ in simplest form.

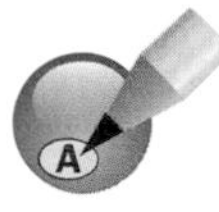

Fair Game Review What you learned in previous grades & lessons

Graph the function. Describe the domain and range. *(Section 8.4)*

27. $y = 2x^2 - 4x - 3$

28. $y = \dfrac{1}{4}x^2 - 5x + 2$

29. $y = -4x^2 + 8x + 5$

30. MULTIPLE CHOICE What is the distance between (2, 3) and (6, 5)? *(Section 10.4)*

Ⓐ $\sqrt{6}$ Ⓑ 4 Ⓒ $3\sqrt{2}$ Ⓓ $2\sqrt{5}$

11.5 Dividing Polynomials

COMMON CORE STATE STANDARDS

A.SSE.2

Essential Question How can you divide one polynomial by another polynomial?

1 ACTIVITY: Dividing Polynomials

Work with a partner. Six different algebra tiles are shown below.

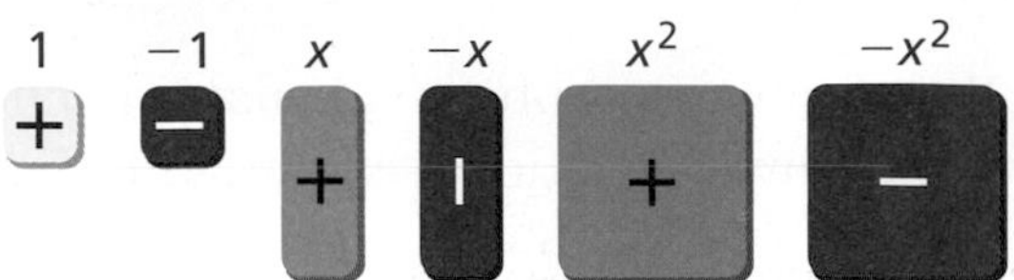

Sample:

Step 1: Arrange tiles to model

$$(x^2 + 5x + 4) \div (x + 1)$$

in a rectangular pattern.

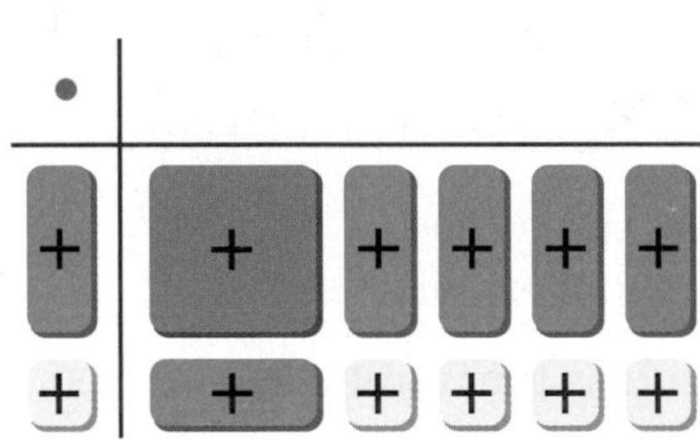

Step 2: Complete the pattern.

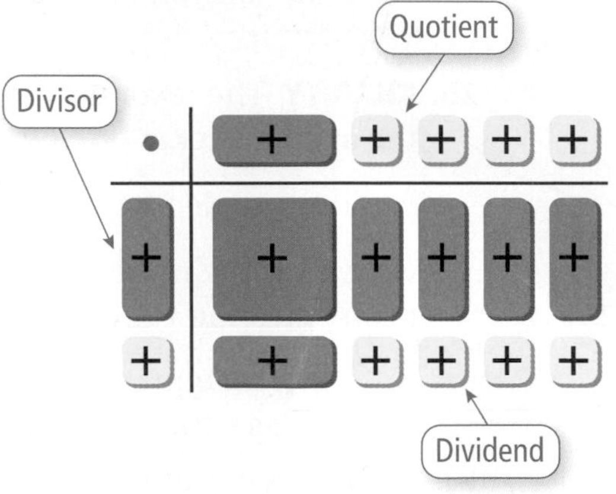

Step 3: Use the completed pattern to write

$$(x^2 + 5x + 4) \div (x + 1) = x + 4.$$

Dividend ÷ Divisor = Quotient

Complete the pattern and write the division problem.

a.

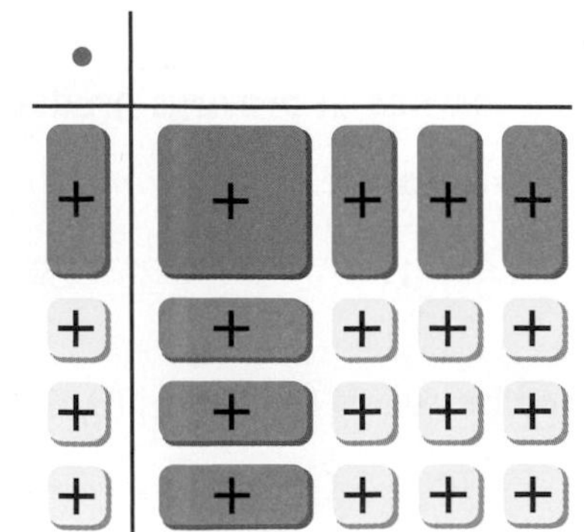

b.

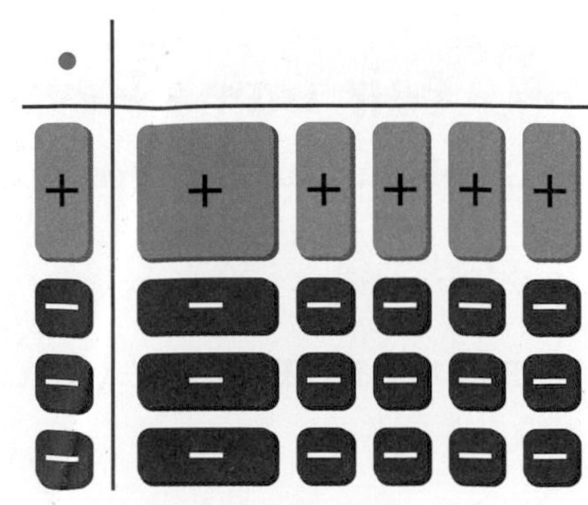

Laurie's Notes

Introduction

Standards for Mathematical Practice

- **MP7 Look for and Make Use of Structure:** Students need to understand how polynomial division and whole number division are alike. The representation and the algorithmic process are similar. Making connections to prior understanding will help students see the similarities.

Motivate

- Have students use their calculators to find the quotients.

Dividend	Divisor	Quotient
4	2	2
252	12	21
23,632	112	211
2,347,432	1112	2111
234,585,432	11,112	21,111
23,456,965,432	111,112	211,111

? "What patterns do you observe?" The dividends are palindromes; The **middle digit** of the dividend increases by 1; Digits in the quotient are the reverse of digits in the divisor; The divisor and quotient have only 1s and 2s.

- Explain that today's activity involves dividing trinomials by binomials.

Activity Notes

Activity 1

- Work through the sample with students. From previous work, students should understand that all of the tiles must be used to form a rectangle.
- **MP2 Reason Abstractly and Quantitatively:** Arrange the tiles so that one dimension is $x + 1$. This represents the division problem. The dividend is $x^2 + 5x + 4$ and the divisor is $x + 1$. The quotient is the other dimension, $x + 4$.
- **Connection:** The divisor and quotient are factors of the dividend. This problem shows that $(x + 4)(x + 1) = x^2 + 5x + 4$.
- Students may realize that this is similar to factoring a trinomial.
- Have students try the two problems on their own. In part (b), they need to pay attention to the signs.
- Students should check their answers by multiplying.

Common Core State Standards

A.SSE.2 Use the structure of an expression to identify ways to rewrite it.

Previous Learning

Students should know how to simplify rational expressions and use long division.

Activity Materials	
Introduction	**Textbook**
• calculators	• algebra tiles

Start Thinking! and Warm Up

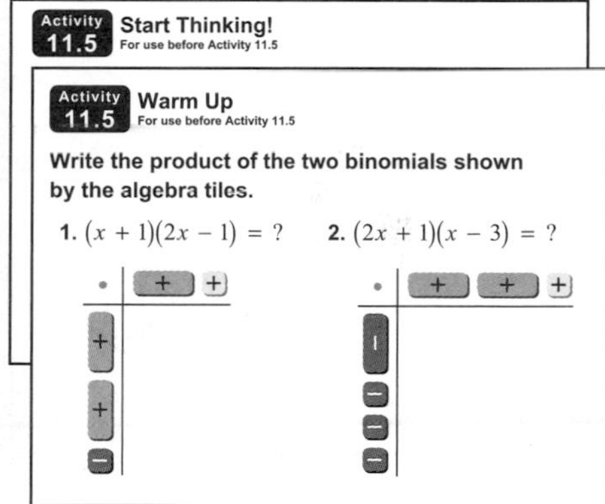

Activity 11.5 Start Thinking!
For use before Activity 11.5
Activity 11.5 Warm Up
For use before Activity 11.5
Write the product of the two binomials shown by the algebra tiles.
1. $(x + 1)(2x - 1) = ?$
2. $(2x + 1)(x - 3) = ?$

11.5 Record and Practice Journal

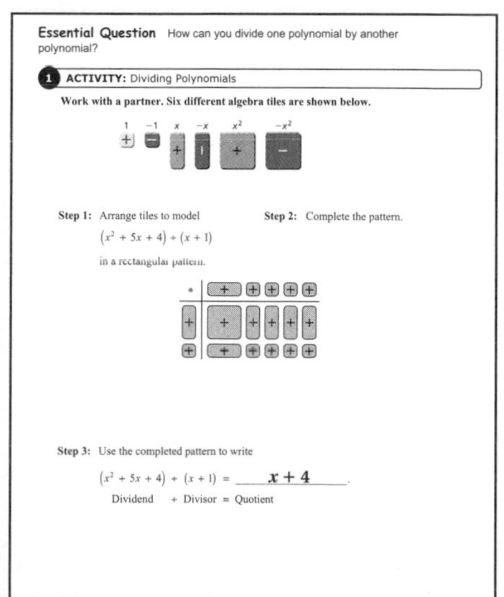

English Language Learners

Notebook Development

Provide English language learners a handout with the detailed solution of the example. This will allow the student to focus on the text as you work through the problem. Students should highlight on the handout as they follow along. This can be added to their notebook and referred to as they work the exercises.

11.5 Record and Practice Journal

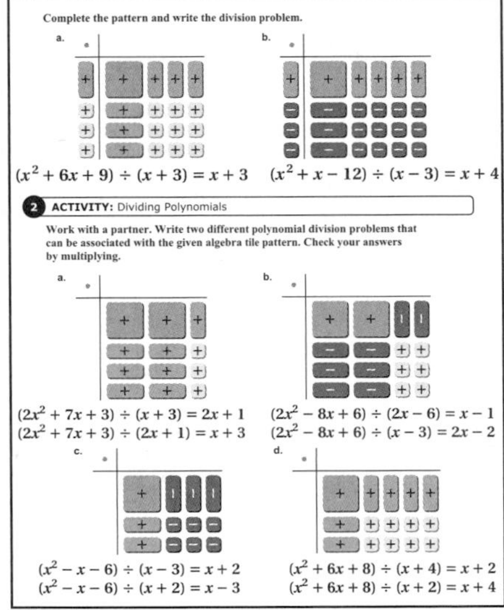

Complete the pattern and write the division problem.

a. $(x^2 + 6x + 9) \div (x + 3) = x + 3$

b. $(x^2 + x - 12) \div (x - 3) = x + 4$

2 ACTIVITY: Dividing Polynomials

Work with a partner. Write two different polynomial division problems that can be associated with the given algebra tile pattern. Check your answers by multiplying.

a. $(2x^2 + 7x + 3) \div (x + 3) = 2x + 1$
$(2x^2 + 7x + 3) \div (2x + 1) = x + 3$

b. $(2x^2 - 8x + 6) \div (2x - 6) = x - 1$
$(2x^2 - 8x + 6) \div (x - 3) = 2x - 2$

c. $(x^2 - x - 6) \div (x - 3) = x + 2$
$(x^2 - x - 6) \div (x + 2) = x - 3$

d. $(x^2 + 6x + 8) \div (x + 4) = x + 2$
$(x^2 + 6x + 8) \div (x + 2) = x + 4$

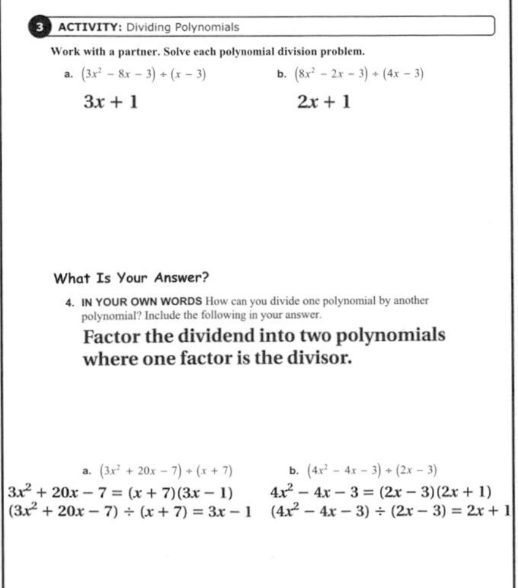

3 ACTIVITY: Dividing Polynomials

Work with a partner. Solve each polynomial division problem.

a. $(3x^2 - 8x - 3) \div (x - 3)$
$3x + 1$

b. $(8x^2 - 2x - 3) \div (4x - 3)$
$2x + 1$

What Is Your Answer?

4. IN YOUR OWN WORDS How can you divide one polynomial by another polynomial? Include the following in your answer.

Factor the dividend into two polynomials where one factor is the divisor.

a. $(3x^2 + 20x - 7) \div (x + 7)$
$3x^2 + 20x - 7 = (x + 7)(3x - 1)$
$(3x^2 + 20x - 7) \div (x + 7) = 3x - 1$

b. $(4x^2 - 4x - 3) \div (2x - 3)$
$4x^2 - 4x - 3 = (2x - 3)(2x + 1)$
$(4x^2 - 4x - 3) \div (2x - 3) = 2x + 1$

Laurie's Notes

Activity 2

- Remind students that $a \div b = c$ can also be written as $a \div c = b$. In each case, $bc = a$.
- Make sure students name the dividend correctly in each part. In part (c), the linear terms can be simplified so the dividend is $x^2 - x - 6$.
- Tell students to check their work by multiplying the quotient and the divisor to make sure the result is the dividend.

Activity 3

- In this activity, students may decide not to model the problem with algebra tiles. They may feel comfortable enough at this point to think about what the second factor (quotient) needs to be.
- Students who found factoring to be somewhat easy should not have difficulty with these problems.

? "What clues or strategies did you use in finding the quotient?" Listen for students talking about the leading coefficients and the constant terms.

What Is Your Answer?

- **Neighbor Check:** Have students work independently and then have their neighbor check their work. Have students discuss any discrepancies.

Closure

- **Exit Ticket:** What is the quotient when you divide $2x^2 + 5x + 3$ by $x + 1$? $2x + 3$

The Dynamic Planning Tool
Editable Teacher's Resources at *BigIdeasMath.com*

2 ACTIVITY: Dividing Polynomials

Work with a partner. Write two different polynomial division problems that can be associated with the given algebra tile pattern. Check your answers by multiplying.

a.

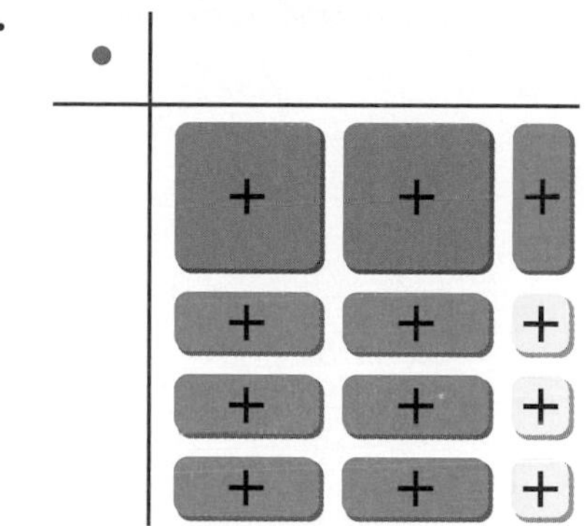

b.

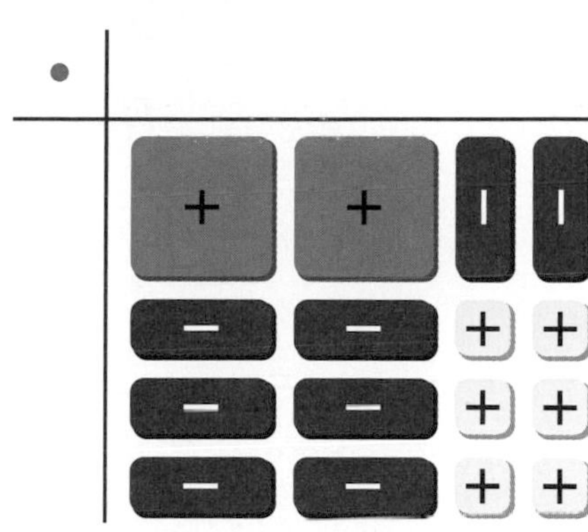

c.

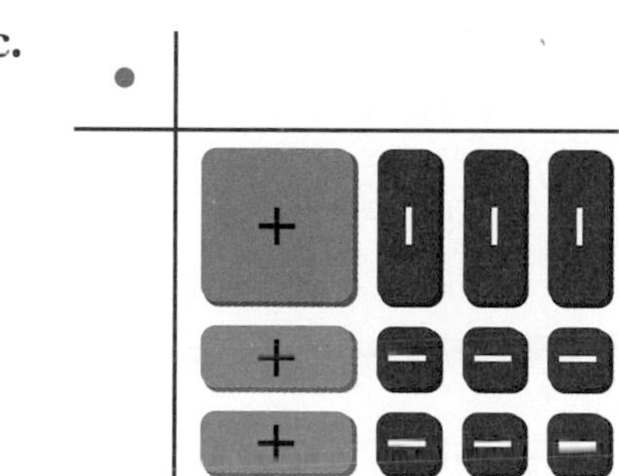

d.

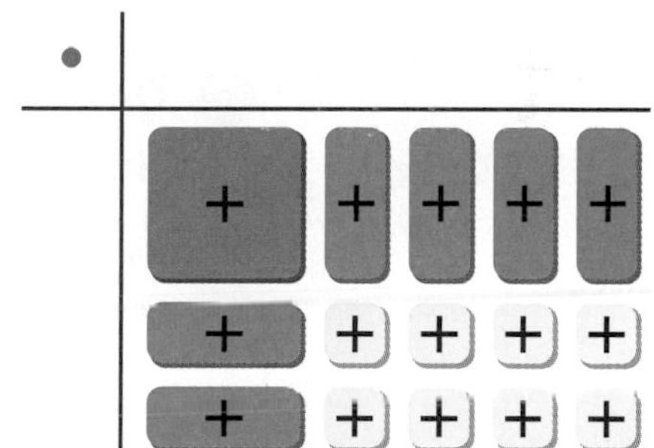

3 ACTIVITY: Dividing Polynomials

Work with a partner. Solve each polynomial division problem.

a. $(3x^2 - 8x - 3) \div (x - 3)$

b. $(8x^2 - 2x - 3) \div (4x - 3)$

What Is Your Answer?

4. **IN YOUR OWN WORDS** How can you divide one polynomial by another polynomial? Include the following in your answer.

a. $(3x^2 + 20x - 7) \div (x + 7)$

b. $(4x^2 - 4x - 3) \div (2x - 3)$

Practice Use what you learned about dividing polynomials to complete Exercises 4 and 5 on page 578.

11.5 Lesson

To divide a polynomial by a monomial, divide each term of the polynomial by the monomial.

EXAMPLE 1 Dividing a Polynomial by a Monomial

Find $(3x^2 + x - 6) \div 3x$.

$$(3x^2 + x - 6) \div 3x = \frac{3x^2 + x - 6}{3x}$$ Write as a fraction.

$$= \frac{3x^2}{3x} + \frac{x}{3x} - \frac{6}{3x}$$ Divide each term by $3x$.

$$= \frac{\cancel{3}x^{\cancel{2}}}{\cancel{3x}} + \frac{\cancel{x}}{3\cancel{x}} - \frac{\overset{2}{\cancel{6}}}{\cancel{3}x}$$ Divide out the common factors.

$$= x + \frac{1}{3} - \frac{2}{x}$$ Simplify.

On Your Own

Now You're Ready
Exercises 6 and 7

Find the quotient.

1. $(4z^2 - 18z) \div 2z$ **2.** $(n^2 - 4n + 8) \div n$ **3.** $(y^3 - 4y^2 + 9y) \div 4y$

You can use long division to divide a polynomial by a binomial.

EXAMPLE 2 Dividing a Polynomial by a Binomial: No Remainder

Find $(m^2 + 4m + 3) \div (m + 1)$.

Step 1: Divide the first term of the dividend by the first term of the divisor.

Align like terms in the quotient and dividend.

$$\begin{array}{r} m \\ m+1 \overline{) m^2 + 4m + 3} \\ \underline{m^2 + m} \\ 3m + 3 \end{array}$$

Divide: $m^2 \div m = m$.

Multiply: $m(m + 1)$.

Subtract. Bring down the 3.

Step 2: Divide the first term of $3m + 3$ by the first term of the divisor.

$$\begin{array}{r} m + 3 \\ m+1 \overline{) m^2 + 4m + 3} \\ \underline{m^2 + m} \\ 3m + 3 \\ \underline{3m + 3} \\ 0 \end{array}$$

Divide: $3m \div m = 3$.

Multiply: $3(m + 1)$.

Subtract.

So, $(m^2 + 4m + 3) \div (m + 1) = m + 3$.

Study Tip

There is no remainder in Example 2, so you could have factored the dividend and divided out a common factor.

$$\frac{m^2 + 4m + 3}{m + 1} = \frac{(m + 3)\cancel{(m + 1)}}{\cancel{m + 1}} = m + 3$$

Laurie's Notes

Introduction

Connect

- **Yesterday:** Students used algebra tiles to model polynomial division. (MP2, MP7)
- **Today:** Students will divide polynomials.

Motivate

- Ask students to describe different ways to write "48 divided by 6."
- Students should be familiar with the notations $48 \div 6$, $\frac{48}{6}$ and $6\overline{)48}$.
- In today's lesson, students will divide polynomials using these notations.

Lesson Notes

Example 1

- Write the problem as shown and then represent it as a fraction.
- ? "How do you divide a trinomial by a monomial?" Listen for students describing that you divide each term in the numerator by the denominator.
- **Big Idea:** Students should recognize how this is similar to adding and subtracting fractions with like denominators. In this problem, you start with a single fraction and then break it into three parts.
- Work through the rest of the example as shown.
- **MP7 Look for and Make Use of Structure:** Point out to students that the quotient is not a polynomial because of the $-\frac{2}{x}$ term.

On Your Own

- In Question 3, point out that $\frac{y^2}{4} = \frac{1}{4}y^2$. Ask students to explain why.

Example 2

- Say, "Now we are going to divide polynomials using long division."
- Set up the problem as shown.
- ? "How many times does m divide into m^2? m times
- Record the m in the quotient above the m-term in the dividend. Multiply each term of the divisor by m.
- **Teaching Tip:** Although the subtraction symbol is not shown, remind students that each term of $m^2 + m$ is being subtracted.
- Continue to work the problem as shown.
- The remainder is 0, meaning that the binomial $m + 1$ divides into the trinomial evenly. Refer to the Study Tip to make this connection.

Goal Today's lesson is dividing polynomials.

Start Thinking! and Warm Up

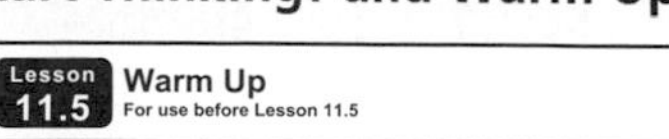

Lesson 11.5 **Start Thinking!** For use before Lesson 11.5

Complete the long division problems.

$5\overline{)2094}$ $\qquad$ $x + 1\overline{)x^2 + 5x + 9}$

Is there any difference between numerical long division and polynomial long division? Explain.

Extra Example 1

Find $(4x^2 + 6x - 8) \div 2x$. $2x + 3 - \frac{4}{x}$

On Your Own

1. $2z - 9$
2. $n - 4 + \frac{8}{n}$
3. $\frac{y^2}{4} - y + \frac{9}{4}$

Extra Example 2

Find $(c^2 - 14c + 49) \div (c - 7)$. $c - 7$

Extra Example 3

Find $(d^2 - 5d + 7) \div (d - 4)$.

$d - 1 + \dfrac{3}{d - 4}$

On Your Own

4. $s + 4$
5. $x + 2 - \dfrac{9}{x + 2}$

Extra Example 4

Find $(4y^2 - 16) \div (2y + 4)$. $2y - 4$

On Your Own

6. $z - 9 + \dfrac{87}{z + 9}$
7. $3y - 2$

Differentiated Instruction

Visual

When dividing polynomials using long division, students may find it difficult because they are not focused on the important information in each step. Students only need to focus on the first term of the divisor and the first term of the dividend. Have them use strips of paper to cover the other terms in the expressions.

$$x\square\overline{)4x^2\square\quad}$$ with quotient $4x$

Laurie's Notes

Example 3

- Not all long division problems have a remainder of 0. When there is a nonzero remainder, you add $\dfrac{\text{remainder}}{\text{divisor}}$ to the quotient. For example, the long division below shows that $49 \div 6 = 8\frac{1}{6}$.

$$\begin{array}{r} 8 \\ 6\overline{)49} \\ \underline{48} \\ 1 \end{array}$$

- Write the problem and ask how many times y divides into y^2. Caution students about the sign when subtracting: $-7y - (-3y) = -4y$.
- Continue to work through the problem as shown, writing the remainder over the divisor.

? **Extention:** "Are polynomials closed under division?" no

On Your Own

- **Think-Pair-Share:** Students should read each question independently and then work with a partner to answer the questions. When they have answered the questions, the pair should compare their answers with another group and discuss any discrepancies.

Example 4

- Perform the long division problem $1020 \div 50$. Discuss the importance of the 0s in keeping digits in the correct place values.
- **MP6 Attend to Precision:** When dividing polynomials with missing terms, insert terms with coefficients of 0 for the same purpose.
- Write the problem. Note that the q-term is missing, so it is inserted in the dividend with a coefficient of 0.

On Your Own

- The remainder is 0 in Question 7, so the divisor is a factor of the dividend.

Closure

- **Writing Prompt:** Polynomial division and whole number division are alike because . . . the notation is similar, remainders are handled the same, etc.

Technology For the Teacher

The Dynamic Planning Tool
Editable Teacher's Resources at *BigIdeasMath.com*

When you use long division to divide polynomials and you obtain a nonzero remainder, use the following rule.

$$\text{Dividend} \div \text{Divisor} = \text{Quotient} + \frac{\text{Remainder}}{\text{Divisor}}$$

EXAMPLE 3 Dividing a Polynomial by a Binomial: Remainder

Find $(2 - 7y + y^2) \div (y - 3)$.

$$\begin{array}{r l}
y - 4 & \\
y - 3\overline{)y^2 - 7y + 2} & \text{Write the dividend in standard form.} \\
\underline{y^2 - 3y} & \text{Multiply: } y(y - 3). \\
-4y + 2 & \text{Subtract. Bring down the 2.} \\
\underline{-4y + 12} & \text{Multiply: } -4(y - 3). \\
-10 & \text{Subtract.}
\end{array}$$

∴ So, $(2 - 7y + y^2) \div (y - 3) = y - 4 - \dfrac{10}{y - 3}$.

On Your Own

Now You're Ready
Exercises 8–15

Find the quotient.

4. $(s^2 - 3s - 28) \div (s - 7)$
5. $(x^2 + 4x - 5) \div (2 + x)$

EXAMPLE 4 Inserting a Missing Term

Find $(3q^2 - 8) \div (q - 2)$.

$$\begin{array}{r l}
3q + 6 & \\
q - 2\overline{)3q^2 + 0q - 8} & \text{Include a } q\text{-term with a coefficient of 0.} \\
\underline{3q^2 - 6q} & \text{Multiply: } 3q(q - 2). \\
6q - 8 & \text{Subtract. Bring down the } -8. \\
\underline{6q - 12} & \text{Multiply: } 6(q - 2). \\
4 & \text{Subtract.}
\end{array}$$

Study Tip

When dividing polynomials using long division, first write the polynomials in standard form and insert any missing terms.

∴ So, $(3q^2 - 8) \div (q - 2) = 3q + 6 + \dfrac{4}{q - 2}$.

On Your Own

Now You're Ready
Exercises 19–22

Find the quotient.

6. $(z^2 + 6) \div (z + 9)$
7. $(9y^2 - 4) \div (3y + 2)$

11.5 Exercises

Vocabulary and Concept Check

1. **WRITING** How do you divide a polynomial by a monomial? by a binomial?
2. **REASONING** How can you check your answer when dividing polynomials?
3. **NUMBER SENSE** How do you know whether a binomial is a factor of a polynomial?

Practice and Problem Solving

Use algebra tiles to find the quotient.

4. $(2x^2 + 6x - 8) \div (x + 4)$
5. $(4x^2 - 5x - 6) \div (4x + 3)$

Find the quotient.

(1) 6. $(8c^2 + 6c - 7) \div 8c$
7. $(3n^3 - 4n^2 + 12) \div 6n$

(2) (3) 8. $(m^2 - 6m - 16) \div (m + 2)$
9. $(z^2 + 10z + 21) \div (z + 3)$
10. $(5y + 8) \div (y - 4)$
11. $(3h^2 + 2h - 1) \div (1 + h)$
12. $(3 - a + 2a^2) \div (a + 5)$
13. $(2 + 8k^2 - 9k) \div (k - 1)$
14. $(6x^2 + 5 + 17x) \div (1 + 3x)$
15. $(g - 7 + 6g^2) \div (2g - 3)$

ERROR ANALYSIS **Describe and correct the error in finding the quotient.**

16. ✗

$$\begin{array}{r} 4 \\ x+1\overline{)4x+3} \\ \underline{4x+4} \\ -1 \end{array}$$

$$(4x + 3) \div (x + 1) = 4 - \frac{1}{4x + 3}$$

17.

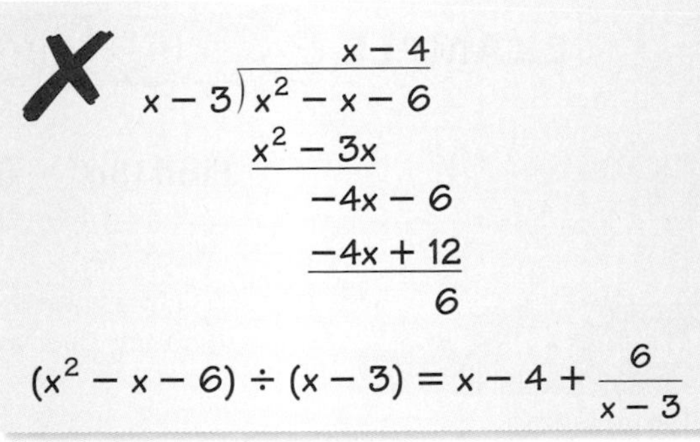

x − 4
x − 3) x² − x − 6
x² − 3x
−4x − 6
−4x + 12
6
(x² − x − 6) ÷ (x − 3) = x − 4 + 6/(x − 3)

18. **AMUSEMENT PARK** The cost of a field trip to an amusement park is represented by $35x + 300$, where x is the number of students going on the trip. The cost is shared equally by all the students except for three students whose parents are acting as chaperones. Find $(35x + 300) \div (x - 3)$ to find an expression for how much each student pays.

Assignment Guide and Homework Check

Level	Assignment	Homework Check
Average	1–3, 6–24 even, 28, 30–33	6, 12, 18, 20, 28
Advanced	1–3, 7–21 odd, 18, 26–29, 30–33	15, 18, 21, 28, 29

Common Errors

- **Exercises 12–15** Students may forget to write the polynomials in standard form before dividing the polynomials.
- **Exercises 19–22** Students may forget to insert missing terms before dividing the polynomials.
- **Exercise 22** Students may stop dividing too early. Remind them to add a constant of 0 to $10y^2 - 9y$ before dividing.
- **Exercise 27** Students may not multiply the length by the height before dividing because they have not seen an exercise where the divisor is a trinomial. Encourage them to use the process shown in the lesson for this new situation.
- **Exercise 28** Students may not be able to factor the polynomial because they do not recognize the polynomial as a difference of two squares.

11.5 Record and Practice Journal

Find the quotient.

1. $(b^2 + 11b + 18) \div (b + 2)$
   $b + 9$
2. $(n^2 - n - 30) \div (n + 5)$
   $n - 6$
3. $(4w^2 + 9w - 3) \div (w + 3)$
   $4w - 3 + \frac{6}{w + 3}$
4. $(5c^2 - 10 - 2c) \div (c - 1)$
   $5c + 3 - \frac{7}{c - 1}$
5. $(21 + 8x^2 - 26x) \div (4x - 7)$
   $2x - 3$
6. $(15h - 4 + 6h^2) \div (2h + 9)$
   $3h - 6 + \frac{50}{2h + 9}$
7. $(r^2 - 12) \div (r + 8)$
   $r - 8 + \frac{52}{r + 8}$
8. $(12y^2 + 8y) \div (3y - 4)$
   $4y + 8 + \frac{32}{3y - 4}$
9. The volume of the triangular prism is $x^3 + 6x^2 + 11x + 6$. Write an expression for the height of the prism.
   $x + 2$

$x + 3$

$2x + 2$

Technology For the Teacher
Answer Presentation Tool

Vocabulary and Concept Check

1. Monomial: Divide each term of the polynomial by the monomial.

   Binomial: You can use long division to divide a polynomial by a binomial.
2. The product of the quotient and divisor should be equal to the dividend.
3. When you divide the polynomial by the binomial, the remainder is 0.

Practice and Problem Solving

4. $2x - 2$
5. $x - 2$
6. $c + \frac{3}{4} - \frac{7}{8c}$
7. $\frac{n^2}{2} - \frac{2n}{3} + \frac{2}{n}$
8. $m - 8$
9. $z + 7$
10. $5 + \frac{28}{y - 4}$
11. $3h - 1$
12. $2a - 11 + \frac{58}{a + 5}$
13. $8k - 1 + \frac{1}{k - 1}$
14. $2x + 5$
15. $3g + 5 + \frac{8}{2g - 3}$
16. The remainder, -1, should be placed over the divisor, not the dividend;

    $(4x + 3) \div (x + 1) = 4 - \frac{1}{x + 1}$
17. See Additional Answers.
18. $35 + \frac{405}{x - 3}$

Practice and Problem Solving

19. $d - 3$ **20.** $r - 5 + \frac{35}{r + 5}$

21. $4n + 2 + \frac{5}{2n - 1}$

22. $2y - 1 - \frac{2}{5y - 2}$

23. The dividend is missing an x-term with a coefficient of 0 which resulted in like terms not being aligned;

$(2x^2 - 5) \div (x + 2) = 2x - 4 + \frac{3}{x + 2}$

24. -20

25. The sum of the degrees of the divisor and quotient are equal to the degree of the dividend.

26. **a.** The graphs coincide.

b. $x = 3, y = 3$

27. See *Taking Math Deeper.*

28. *Sample answer:* Most students will choose factoring over long division. Using long division will be messy with all the missing terms.

29. See Additional Answers.

Fair Game Review

30. $x = -1, x = 5$

31. $x = -4 + \sqrt{23}$,
$x = -4 - \sqrt{23}$

32. $x = 3 + 3\sqrt{2}, x = 3 - 3\sqrt{2}$

33. A

Mini-Assessment

Find the quotient.

1. $(12x^2 - 15x + 18) \div 3x$

$4x - 5 + \frac{6}{x}$

2. $(7 + 5y^2 - 2y) \div (3 + y)$

$5y - 17 + \frac{58}{y + 3}$

3. $(g^2 - 16) \div (g - 4)$ $g + 4$

Taking Math Deeper

Exercise 27

Students may realize that they can find the answer by observing the terms of the given expressions.

Volume of prism: $m^3 - 13m - 12$
Length of prism: $m + 3$
Height of prism: $m + 1$

1 Substitute these expressions in the formula for the volume of a rectangular prism.

$$V = \ell wh$$
$$m^3 - 13m - 12 = (m + 3)(w)(m + 1)$$

2 Determine what type of terms are in the missing expression. The expression for the volume has an m^3-term and the two known factors have an m-term. So, there must be an m-term in the missing expression.

$$m \cdot m \cdot m = m^3$$
$$m^3 - 13m - 12 = (m + 3)(m + ?)(m + 1)$$

There must be a constant term in the missing expression, otherwise there would not be a constant term in the expression for the volume.

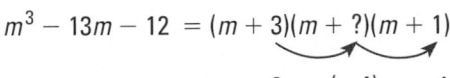

$$m^3 - 13m - 12 = (m + 3)(m + ?)(m + 1)$$
$$3 \cdot (-4) \cdot 1 = -12$$

So, an expression for the width is $m - 4$.

3 Check your solution.

$$(m + 3)(m - 4)(m + 1) = (m^2 - m - 12)(m + 1)$$
$$= m^3 + m^2 - m^2 - m - 12m - 12$$
$$= m^3 - 13m - 12$$

Reteaching and Enrichment Strategies

If students need help. . .	If students got it. . .
Resources by Chapter • Practice A and Practice B • Puzzle Time Record and Practice Journal Practice Differentiating the Lesson Lesson Tutorials Skills Review Handbook	Resources by Chapter • Enrichment and Extension • School-to-Work • Financial Literacy Start the next section

Find the quotient.

④ **19.** $(d^2 - 9) \div (d + 3)$

20. $(r^2 + 10) \div (r + 5)$

21. $(8n^2 + 3) \div (2n - 1)$

22. $(10y^2 - 9y) \div (5y - 2)$

23. ERROR ANALYSIS Describe and correct the error in finding the quotient.

$$\begin{array}{r} 2x - 9 \\ x + 2 \overline{)\, 2x^2 - 5} \\ \underline{2x^2 + 4x} \\ -9x \\ \underline{-9x - 18} \\ 18 \end{array}$$

$$(2x^2 - 5) \div (x + 2) = 2x - 9 + \frac{18}{x + 2}$$

24. REASONING Find k when $(x - 4)$ is a factor of $2x^2 - 3x + k$.

25. CRITICAL THINKING When dividing polynomials, how are the degrees of the dividend, divisor, and quotient related?

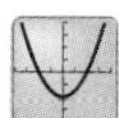

26. TECHNOLOGY Rewrite the rational function $y = \dfrac{3x - 8}{x - 3}$ in the form $y = \dfrac{a}{x - h} + k$. Graph both functions in the same viewing window of a graphing calculator.

a. What do you notice about the graphs?

b. What are the asymptotes of the graph of $y = \dfrac{3x - 8}{x - 3}$?

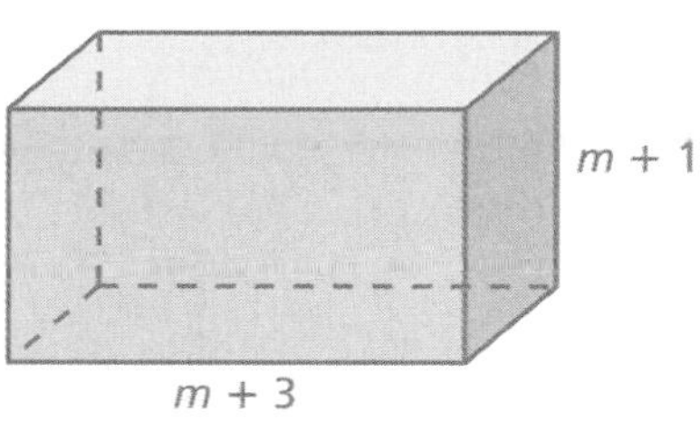

27. GEOMETRY The volume of the rectangular prism is $m^3 - 13m - 12$. Write an expression for the width of the prism.

28. CHOOSE TOOLS Would you use factoring or long division to simplify $\dfrac{x^8 - 1}{x - 1}$? Explain your reasoning.

29. Repeated Reasoning Find each quotient in the table and identify the pattern. Then predict the quotient $(x^5 - x^4 + x^3 - x^2 + x - 1) \div (x + 1)$ without calculating. Verify your prediction.

Quotient
$(x^2 - x + 1) \div (x + 1)$
$(x^3 - x^2 + x - 1) \div (x + 1)$
$(x^4 - x^3 + x^2 - x + 1) \div (x + 1)$

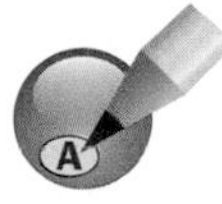

Fair Game Review What you learned in previous grades & lessons

Solve the equation by completing the square. *(Section 9.3)*

30. $x^2 - 4x = 5$

31. $x^2 + 8x - 7 = 0$

32. $2x^2 - 12x - 8 = 10$

33. MULTIPLE CHOICE What is the solution of $4^{3x} = 2^{x+1}$? *(Section 6.4)*

Ⓐ $\frac{1}{5}$ Ⓑ $\frac{1}{2}$ Ⓒ 2 Ⓓ 3

11.6 Adding and Subtracting Rational Expressions

COMMON CORE STATE STANDARDS
A.SSE.2

Essential Question How can you add and subtract rational expressions?

1 ACTIVITY: Adding Rational Expressions

Work with a partner. You and a friend have a summer job mowing lawns. Working alone it takes you 40 hours to mow all of the lawns. Working alone it takes your friend 60 hours to mow all of the lawns.

a. Write a rational expression that represents the portion of the lawns you can mow in t hours.

Portion you mow in t hours = [ ] · $\frac{[\ ]}{[\ ]}$

Time Rate

b. Write a rational expression that represents the portion of the lawns your friend can mow in t hours.

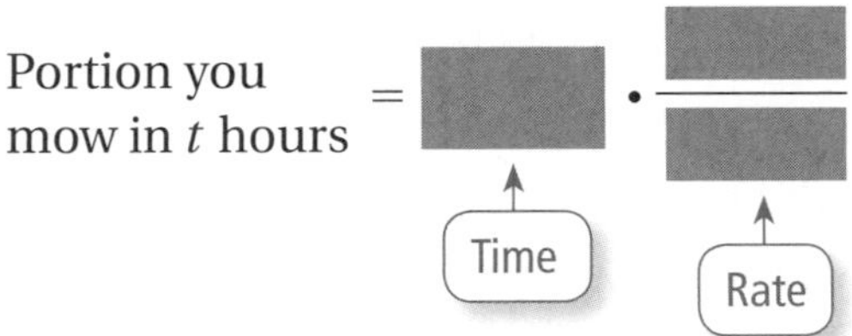

c. Add the two expressions to write a rational expression for the portion of the lawns that the two of you working together can mow in t hours.

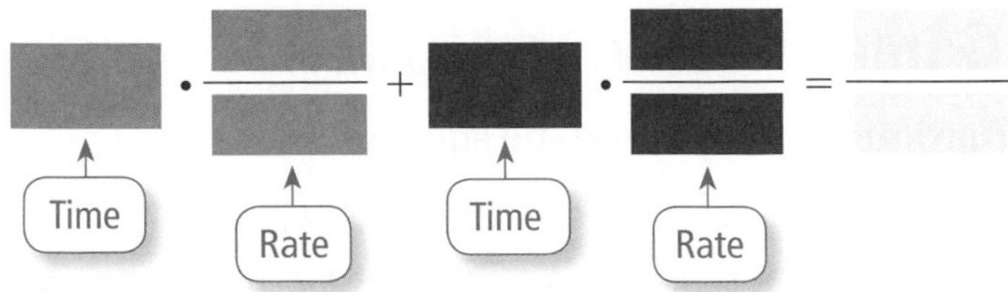

d. Use the expression in part (c) to find the total time it takes both of you working together to mow all of the lawns. Explain your reasoning.

Laurie's Notes

Introduction

Standards for Mathematical Practice

- **MP1 Make Sense of Problems and Persevere in Solving Them:** The rate problems presented in this section are common applications that students often find challenging. It is important to ease into these problems so that students build understanding and make sense of the problem.

Motivate

- **Story Time:** Tell students about a job you had collecting coins out of parking meters along a roadway at the beach. Yes, it was great to work outdoors and be near the beach, however, it took 10 hours to empty all of the meters from one end of the roadway to the other.

? "What portion of the total job is done in 5 hours? in 2 hours? in t hours?" $\frac{1}{2}, \frac{1}{5}, \frac{t}{10}$

- Write on the board:

  Portion of job completed = number of hours worked × rate
- In this problem, the rate is $\frac{1}{10}$.
- Because it took more than one 8-hour workday to finish the job, a second person is hired. This person is slower—it takes them 12 hours to empty all the meters.

? "How long do you think it takes both workers to empty the parking meters?" Students often take the average and incorrectly guess 11 hours. You are not looking for an answer, only a guess. Revisit this problem at the end of class.

Activity Notes

Activity 1

- This problem is similar to the one above so students should feel confident.
- **Teaching Tip:** Encourage students to use units in their answers.
- In part (c), students need to find a common denominator for 40 and 60, then simplify the fraction.

? "What does the rational expression $\frac{t}{24}$ represent?" the portion of the lawns mowed in t hours when you and your friend are working together

- Have students share their thoughts about part (d). It should seem reasonable that working together takes less time than either person working alone.

? "Explain why the answer should be between 20 and 30 hours when you and your friend are working together?" Students should refer to the two known rates. If they both could do the job in 40 hours, then together it takes 20 hours. If they both could do the job in 60 hours, then together it takes 30 hours. So, the answer should be between 20 and 30 hours.

Common Core State Standards

A.SSE.2 Use the structure of an expression to identify ways to rewrite it.

Previous Learning

Students should know how to multiply and divide rational expressions.

Start Thinking! and Warm Up

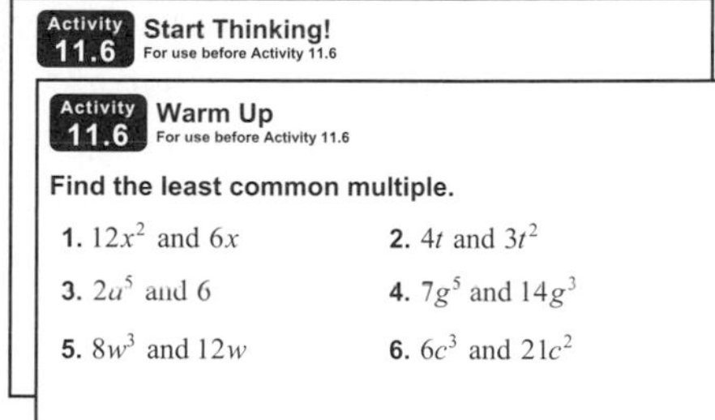

Activity 11.6 Start Thinking! For use before Activity 11.6

Activity 11.6 Warm Up For use before Activity 11.6

Find the least common multiple.

1. $12x^2$ and $6x$
2. $4t$ and $3t^2$
3. $2a^5$ and 6
4. $7g^5$ and $14g^3$
5. $8w^3$ and $12w$
6. $6c^3$ and $21c^2$

11.6 Record and Practice Journal

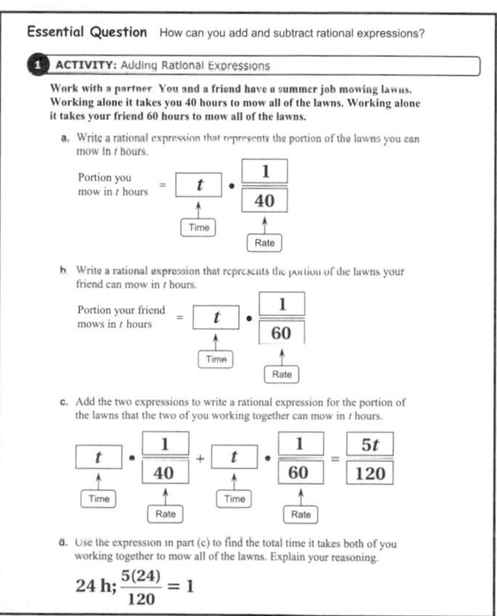

Essential Question How can you add and subtract rational expressions?

1 ACTIVITY: Adding Rational Expressions

Work with a partner. You and a friend have a summer job mowing lawns. Working alone it takes you 40 hours to mow all of the lawns. Working alone it takes your friend 60 hours to mow all of the lawns.

a. Write a rational expression that represents the portion of the lawns you can mow in t hours.

Portion you mow in t hours $= t \cdot \frac{1}{40}$ (Time, Rate)

b. Write a rational expression that represents the portion of the lawns your friend can mow in t hours.

Portion your friend mows in t hours $= t \cdot \frac{1}{60}$ (Time, Rate)

c. Add the two expressions to write a rational expression for the portion of the lawns that the two of you working together can mow in t hours.

$t \cdot \frac{1}{40} + t \cdot \frac{1}{60} = \frac{5t}{120}$ (Time, Rate, Time, Rate)

d. Use the expression in part (c) to find the total time it takes both of you working together to mow all of the lawns. Explain your reasoning.

24 h; $\frac{5(24)}{120} = 1$

Differentiated Instruction

Connection

Have students simplify the expressions $\frac{3}{4} + \frac{2}{3}$ and $\frac{7}{8} - \frac{1}{5}$ and list the steps used in the process. Make the connection between adding and subtracting rational numbers and adding and subtracting rational expressions. The process is the same in both cases. If the denominators are unlike, you must rewrite the expressions with a common denominator. Then add or subtract the numerators, and simplify.

11.6 Record and Practice Journal

2 ACTIVITY: Adding Rational Expressions

Work with a partner. You are hang gliding. For the first 10,000 feet, you travel x feet per minute. You then enter a valley in which the wind is greater, and for the next 6000 feet, you travel $2x$ feet per minute.

a. Use the formula $d = rt$ to write a rational expression that represents the time it takes you to travel the first 10,000 feet.

Time to travel first 10,000 feet $= \frac{10{,}000}{x}$ ← Distance / ← Rate

b. Use the formula $d = rt$ to write a rational expression that represents the time it takes you to travel the next 6000 feet.

Time to travel next 6000 feet $= \frac{6000}{2x}$ ← Distance / ← Rate

c. Add the two expressions to write a rational expression that represents the total time it takes you to travel 16,000 feet.

$\frac{10{,}000}{x} + \frac{6000}{2x} = \frac{13{,}000}{x}$

d. Use the expression in part (c) to find the total time it takes you to travel 16,000 feet when your rate during the first 10,000 feet is 2000 feet per minute.

6.5 min

What Is Your Answer?

3. IN YOUR OWN WORDS How can you add and subtract rational expressions? Include the following in your answer.

Rewrite (if needed) both expressions so they have a common denominator, then add or subtract the numerators over the common denominator.

a. $\frac{x}{5} + \frac{x}{10} = \frac{2x}{10} + \frac{x}{10} = \frac{3x}{10}$

b. $\frac{3}{x} + \frac{4}{x} = \frac{7}{x}$

c. $\frac{9}{x} + \frac{2}{3x} = \frac{27}{3x} + \frac{2}{3x} = \frac{29}{3x}$

d. $\frac{x}{2} - \frac{x}{4} = \frac{2x}{4} - \frac{x}{4} = \frac{x}{4}$

e. $\frac{x+1}{3} - \frac{1}{3} = \frac{x}{3}$

f. $\frac{1}{x} - \frac{1}{x^2} = \frac{x}{x^2} - \frac{1}{x^2} = \frac{x-1}{x^2}$

Laurie's Notes

Activity 2

- Before beginning this activity, review the different forms of the distance formula: $d = rt$, $r = \frac{d}{t}$, and $t = \frac{d}{r}$.
- In parts (a) and (b), students are dividing feet by $\frac{\text{feet}}{\text{minute}}$, so the answer is in minutes.
- Students should have a sense about the reasonableness of their answer. Students could find the answer in two parts: For instance, if you travel at a rate of 2000 feet per minute it takes 5 minutes to travel 10,000 feet. If your rate increases to 4000 feet per minute, it takes 1.5 minutes to travel 6000 feet. So, the time it takes to travel the entire 16,000 feet is 6.5 minutes.

What Is Your Answer?

- These six questions are practice with adding and subtracting rational expressions. Remind students that a common denominator may be needed, just as with adding and subtracting fractions.

Closure

- **Exit Ticket:** Refer back to the opening problem.
  a. Write a rational expression that represents the portion of the parking meters you emptied in t hours. $\frac{t}{10}$
  b. Write a rational expression that represents the portion of the parking meters the other person emptied in t hours. $\frac{t}{12}$
  c. Write a rational expression that represents the portion of the parking meters emptied in t hours when you both work together. $\frac{11t}{60}$

Technology For the Teacher

The Dynamic Planning Tool
Editable Teacher's Resources at *BigIdeasMath.com*

2 ACTIVITY: Adding Rational Expressions

Work with a partner. You are hang gliding. For the first 10,000 feet, you travel x feet per minute. You then enter a valley in which the wind is greater, and for the next 6000 feet, you travel $2x$ feet per minute.

a. Use the formula $d = rt$ to write a rational expression that represents the time it takes you to travel the first 10,000 feet.

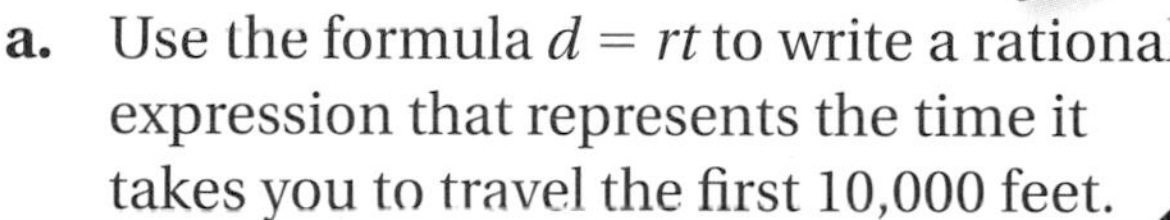

b. Use the formula $d = rt$ to write a rational expression that represents the time it takes you to travel the next 6000 feet.

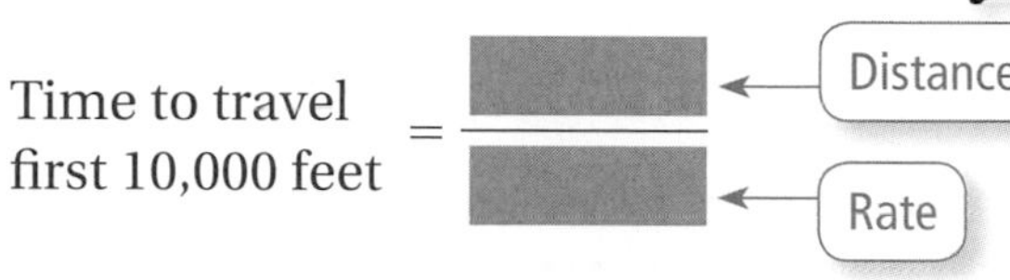

c. Add the two expressions to write a rational expression that represents the total time it takes you to travel 16,000 feet.

$$\frac{\square}{\square} + \frac{\square}{\square} = \frac{\quad}{\quad}$$

d. Use the expression in part (c) to find the total time it takes you to travel 16,000 feet when your rate during the first 10,000 feet is 2000 feet per minute.

What Is Your Answer?

3. **IN YOUR OWN WORDS** How can you add and subtract rational expressions? Include the following in your answer.

a. $\frac{x}{5} + \frac{x}{10}$ **b.** $\frac{3}{x} + \frac{4}{x}$ **c.** $\frac{9}{x} + \frac{2}{3x}$

d. $\frac{x}{2} - \frac{x}{4}$ **e.** $\frac{x+1}{3} - \frac{1}{3}$ **f.** $\frac{1}{x} - \frac{1}{x^2}$

Practice Use what you learned about adding and subtracting rational expressions to complete Exercises 3–5 on page 585.

11.6 Lesson

Key Vocabulary
least common denominator of rational expressions, *p. 583*

You can use the same rules that you used for adding and subtracting fractions to add and subtract rational expressions.

Key Idea

Adding and Subtracting Rational Expressions with Like Denominators

Let a, b, and c be polynomials, where $c \neq 0$.

Adding: $\frac{a}{c} + \frac{b}{c} = \frac{a+b}{c}$ Subtracting: $\frac{a}{c} - \frac{b}{c} = \frac{a-b}{c}$

EXAMPLE 1 Adding and Subtracting with Like Denominators

Find the sum or difference.

a. $\frac{5}{2x} + \frac{7}{2x} = \frac{5+7}{2x}$ Add the numerators.

$= \frac{12}{2x}$ Simplify.

$= \frac{\cancel{12}^{6}}{\cancel{2}x}$ Divide out the common factor.

$= \frac{6}{x}$ Simplify.

b. $\frac{3y}{y+4} - \frac{y-8}{y+4} = \frac{3y-(y-8)}{y+4}$ Subtract the numerators.

$= \frac{3y-y+8}{y+4}$ Use the Distributive Property.

$= \frac{2y+8}{y+4}$ Combine like terms.

$= \frac{2\cancel{(y+4)}}{\cancel{y+4}}$ Factor. Divide out the common factor.

$= 2$ Simplify.

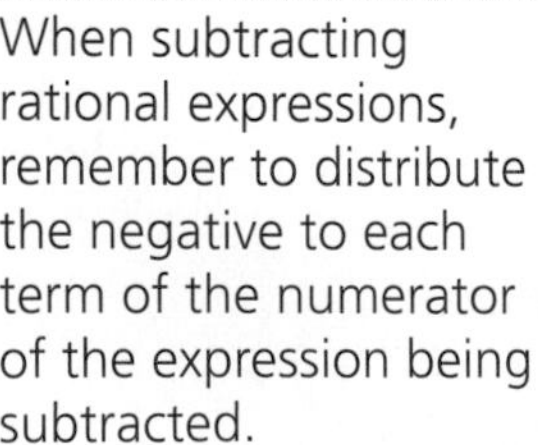

When subtracting rational expressions, remember to distribute the negative to each term of the numerator of the expression being subtracted.

On Your Own

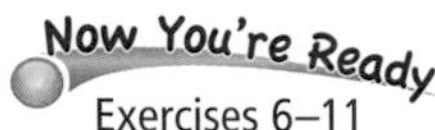

Find the sum or difference.

1. $\frac{4}{9z} - \frac{8}{9z}$
2. $\frac{3w+1}{w-1} + \frac{w}{w-1}$
3. $\frac{x+3}{x^2+x-2} - \frac{1}{x^2+x-2}$

Multi-Language Glossary at BigIdeasMath.com.

Laurie's Notes

Introduction

Connect

- **Yesterday:** Students wrote rational expressions and explored how to add and subtract them. (MP1)
- **Today:** Students will add and subtract rational expressions with like and unlike denominators.

Motivate

? "Has anyone ever kayaked or canoed before?" Answers will vary.

- The longest kayak race is the Yukon 1000, which is a 1000-mile race down the Yukon River in Canada and Alaska. It takes 7 to 12 days of 18 hours of paddling per day to finish the race. For comparison, it is about 1000 miles from New York City to Orlando, Florida.

Lesson Notes

Key Idea

- Write the Key Idea which states that adding and subtracting rational expressions is similar to adding and subtracting fractions. You need to find common denominators, and you should simplify your results.

Example 1

? "Do the fractions in part (a) have a common denominator?" yes "What is the sum of the numerators?" 12 "Can the rational expression be simplified? Explain." yes; The numerator and denominator have a common factor of 2.

- Write part (b). Make a point of the common error shown. Using parentheses when writing the difference of the two numerators will help students remember to distribute the negative to each term of the numerator being subtracted.

? "Can $2y + 8$ be rewritten?" yes; You can factor out a 2 to get $2(y + 4)$.

- Simplify the rational expression.

On Your Own

- Each question already has a common denominator.
- **Common Error:** In Question 2, the sum is $\frac{4w+1}{w-1}$. Students may try to divide out the w even though it is not a factor in the numerator or the denominator.
- Check to see that students have simplified their answers to Question 3.

Goal Today's lesson is adding and subtracting rational expressions.

Start Thinking! and Warm Up

Lesson 11.6
Warm Up
For use before Lesson 11.6

Lesson 11.6 **Start Thinking!** For use before Lesson 11.6

Solve the addition problems below. How are the steps for solving each problem similar? How are the steps different?

1. $\frac{2}{3} + \frac{4}{21}$
2. $\frac{5}{x+2} + \frac{4}{x^2+x-2}$

Extra Example 1

Find the sum or difference.

a. $\frac{6}{7h} + \frac{8}{7h}$ $\frac{2}{h}$

b. $\frac{3k+2}{k-3} - \frac{2k+5}{k-3}$ 1

On Your Own

1. $-\frac{4}{9z}$
2. $\frac{4w+1}{w-1}$
3. $\frac{1}{x-1}$

Extra Example 2

Find the LCD of $\frac{x}{x^2 - 25}$ and $\frac{x^2}{2x - 10}$.

$2(x - 5)(x + 5)$

On Your Own

4. $28g^3$
5. $n(n + 1)$
6. $(t - 2)(t + 2)$
7. $x(x - 3)(x + 2)$

Extra Example 3

Find the sum $\frac{5}{3b^3} + \frac{b - 4}{12b^2}$.

$\frac{b^2 - 4b + 20}{12b^3}$

English Language Learners

Group Activity

Organize students in groups of 2 to 4 consisting of English language learners and English speakers. Provide each group with an overhead transparency and assign them a problem from the text. Students are to solve the problem and present the solution to the class. Encourage students to include visual clues in their presentation.

Laurie's Notes

Discuss

- **?** "How do you find a common denominator for fractions with unlike denominators?" Students should be able to describe the process for finding the least common denominator.
- It is likely that students will mention finding a multiple common to both numbers. The method of using the prime factorization is less likely to be described.
- Say, "When adding or subtracting rational expressions with unlike denominators, you must first find the least common denominator (LCD). When variables are involved, the prime factorization method can be used."

Example 2

- Write the two rational expressions and say, "We need to find an expression that $10g^2$ and $12g$ both divide into."
- Write the prime factorization of each denominator. Say, "Use the greatest power of each factor that appears in either denominator to find the least common multiple (LCM) of the denominators."
- The LCM is $60g^2$.
- **?** **Extension:** "What would you multiply $10g^2$ by to get $60g^2$?" 6 "What would you multiply $12g$ by to get $60g^2$?" $5g$

On Your Own

- Questions 5–7 are different from Question 4. Question 4 involves monomials in the denominator. Questions 5–7 involve binomials and trinomials in the denominator.
- You may need to help students think about the process of prime factorization for binomials and trinomials. They need to be factored so that the denominators can be represented as products.

Example 3

- Write the problem.
- **?** "What is the LCM of 8 and 6?" 24 "What is the LCM of x and x^2?" x^2 "What is the LCM of $8x$ and $6x^2$?" $24x^2$
- **?** "Each rational expression must now be rewritten using a denominator of $24x^2$. How do you rewrite equivalent fractions?" Multiply the numerator and the denominator by the same value.
- **?** "What do you multiply $8x$ by to get $24x^2$?" $3x$ "What do you multiply $6x^2$ by to get $24x^2$?" 4
- **MP1 Make Sense of Problems and Persevere in Solving Them:** Asking each of these questions helps students make sense of the problem. Each rational expression is multiplied by 1, but how 1 is represented is different for each rational expression.
- Use colors to show how each rational expression is rewritten to have the same denominator of $24x^2$.
- Point out that $\frac{7x - 8}{24x^2}$ can be divided as shown in Section 11.5.

To add or subtract rational expressions with unlike denominators, rewrite the expressions so they have like denominators. You can do this by finding the least common multiple of the denominators, called the **least common denominator (LCD)**.

EXAMPLE 2 Finding the LCD of Two Rational Expressions

Find the LCD of $\frac{3}{10g^2}$ and $\frac{5}{12g}$.

First write the prime factorization of each denominator.

$10g^2 = 2 \cdot 5 \cdot g^2$ $\qquad$ $12g = 2^2 \cdot 3 \cdot g$

Use the greatest power of each factor that appears in either denominator to find the LCM of the denominators.

$$\text{LCM} = 2^2 \cdot 3 \cdot 5 \cdot g^2 = 60g^2$$

∴ So, the LCD of $\frac{3}{10g^2}$ and $\frac{5}{12g}$ is $60g^2$.

On Your Own

Now You're Ready
Exercises 13–18

Find the LCD of the rational expressions.

4. $\frac{2}{7g}, -\frac{15}{4g^3}$
5. $\frac{8}{n}, \frac{n}{n+1}$
6. $\frac{t}{t^2-4}, \frac{9}{t-2}$
7. $\frac{x+1}{x^2-x-6}, \frac{5}{x(x-3)}$

EXAMPLE 3 Adding with Unlike Denominators

Find the sum $\frac{1}{8x} + \frac{x-2}{6x^2}$.

Study Tip

To rewrite each expression using the LCD, multiply the numerator and denominator of each expression by the factor that makes its denominator the LCD.

Because the expressions have unlike denominators, find the LCD.

$8x = 2^3 \cdot x$ $\qquad$ $6x^2 = 2 \cdot 3 \cdot x^2$

The LCD is $2^3 \cdot 3 \cdot x^2 = 24x^2$.

$\frac{1}{8x} + \frac{x-2}{6x^2} = \frac{1(3x)}{8x(3x)} + \frac{(x-2)(4)}{6x^2(4)}$ Rewrite using the LCD, $24x^2$.

$= \frac{3x}{24x^2} + \frac{4x-8}{24x^2}$ Simplify.

$= \frac{3x+4x-8}{24x^2}$ Add the numerators.

$= \frac{7x-8}{24x^2}$ Simplify.

EXAMPLE 4 Subtracting with Unlike Denominators

Find the difference $\frac{x+3}{x^2-8x+12} - \frac{2}{x-6}$.

$$\frac{x+3}{x^2-8x+12} - \frac{2}{x-6} = \frac{x+3}{(x-6)(x-2)} - \frac{2}{x-6}$$ Factor $x^2 - 8x + 12$.

$$= \frac{x+3}{(x-6)(x-2)} - \frac{2(x-2)}{(x-6)(x-2)}$$ Rewrite using the LCD, $(x-6)(x-2)$.

$$= \frac{(x+3) - 2(x-2)}{(x-6)(x-2)}$$ Subtract the numerators.

$$= \frac{-x+7}{(x-6)(x-2)}$$ Simplify.

EXAMPLE 5 Real-Life Application

You row your kayak 5 miles downstream from your campsite to a dam, and then you row back to your campsite. You row *x* miles per hour during the entire trip, and the river current is 1 mile per hour. Write an expression for the total time of the trip.

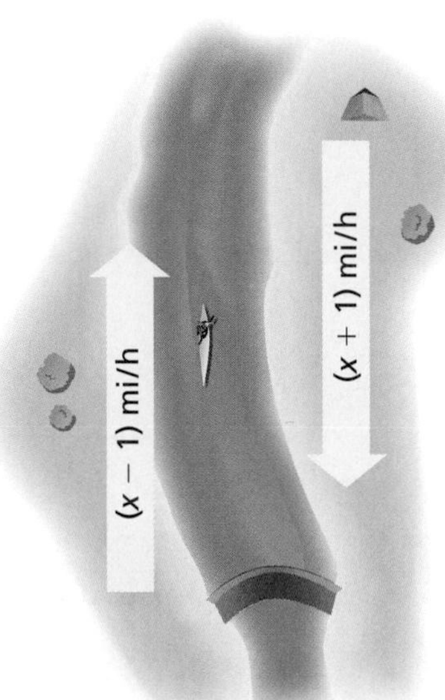

Solving the formula $d = rt$ for time t gives $t = \frac{d}{r}$. Use this to write an expression for the total time of the trip.

Time downstream				Time upstream		
Distance downstream	÷	Speed downstream	+	Distance upstream	÷	Speed upstream

$$\frac{5}{x+1} + \frac{5}{x-1} = \frac{5(x-1)}{(x+1)(x-1)} + \frac{5(x+1)}{(x-1)(x+1)}$$ Write an expression. Rewrite using the LCD.

$$= \frac{5(x-1) + 5(x+1)}{(x+1)(x-1)}$$ Add the numerators.

$$= \frac{10x}{(x+1)(x-1)}$$ Simplify.

On Your Own

Now You're Ready
Exercises 20–25

Find the sum or difference.

8. $\frac{x+5}{3x^3} - \frac{2}{9x^2}$ **9.** $\frac{2k}{k+1} + \frac{k}{k-2}$ **10.** $\frac{3y-1}{y^2-64} - \frac{3}{y+8}$

11. WHAT IF? In Example 5, the river current is 2 miles per hour. Write an expression for the total time of the trip.

Laurie's Notes

Example 4

- ? "Is $x^2 - 8x + 12$ factorable?" yes; $(x-6)(x-2)$
- Explain that the denominator of the second rational expression has one binomial factor in common with the denominator of the first rational expression.
- ? "What must the second rational expression be multiplied by to have the same denominator as the first rational expression?" $\frac{x-2}{x-2}$
- Remind students to distribute the negative to each term of the numerator being subtracted.
- The final answer can also be written so that the leading coefficient of the numerator is positive. Another form of the answer is $-\frac{x-7}{(x-6)(x-2)}$.

Example 5

- Ask for a volunteer to read the problem. Discuss the context—paddling downstream means you are paddling with the current and paddling upstream means you are paddling against the current. At a constant rate, the rate of the current is added going downstream and subtracted going upstream.
- ? "What are you trying to find an expression for?" the total time for the trip "Do you know the time for either direction?" no "In general, how do you solve for time?" Divide the distance by the rate.
- The distance is the same in each direction, but the rate changes. The rate downstream is $(x+1)$. The rate upstream is $(x-1)$.
- ? "What is the LCD for the two rational expressions?" $(x+1)(x-1)$
- Work through the problem as shown. When finished, evaluate the expression for a specific value of x, such as 3. You travel downstream at a rate of 4 miles per hour and upstream at a rate of 2 miles per hour. When $x = 3$, the total time is $\frac{30}{8}$, or 3.75 hours.

On Your Own

- If time is short, have students do Question 9.

Closure

- **Exit Ticket:** Find the difference $\frac{2}{t-4} - \frac{6}{t+4}$. $\frac{-4(t+8)}{(t+4)(t-4)}$

Technology For the Teacher

The Dynamic Planning Tool
Editable Teacher's Resources at *BigIdeasMath.com*

Extra Example 4

Find the difference $\frac{1}{x+4} - \frac{3-x}{x^2+2x-8}$. $\frac{2x-5}{(x+4)(x-2)}$

Extra Example 5

You paddle a canoe 4 miles upstream and then back downstream to your starting point. You paddle downstream 50% faster than upstream due to the current. Write an expression for the total time of the trip.
$\frac{4}{r} + \frac{4}{1.5r} = \frac{10}{1.5r} = \frac{20}{3r}$

On Your Own

8. $\frac{x+15}{9x^3}$
9. $\frac{3k(k-1)}{(k+1)(k-2)}$
10. $\frac{23}{(y+8)(y-8)}$
11. $\frac{10x}{(x+2)(x-2)}$

Vocabulary and Concept Check

1. Find the LCM of the denominators in both cases.
2. Factor $x^2 - 16$ as $(x + 4)(x - 4)$. Multiply the numerator and denominator of $\frac{1}{x+4}$ by $(x - 4)$.

Practice and Problem Solving

3. $\frac{5s}{9}$
4. $\frac{r}{16}$
5. $\frac{7}{w}$
6. $\frac{8}{3y}$
7. $\frac{3}{x+2}$
8. $\frac{2z}{z-1}$
9. $\frac{3t-4}{t-1}$
10. $\frac{1}{n-2}$
11. $\frac{1}{p-1}$
12. The denominators were added.
    $\frac{1}{x-3} + \frac{4}{x-3} = \frac{5}{x-3}$
13. $2x$
14. $36y$
15. $(m + 5)(m - 4)$
16. g
17. $(h + 3)(h - 1)$
18. $(s + 2)(s - 4)$
19. $\frac{S - 2\ell w}{2(\ell + w)}$

Assignment Guide and Homework Check

Level	Assignment	Homework Check
Average	1, 2, 3–27 odd, 26, 28, 40–43	7, 17, 19, 23, 26, 28
Advanced	1, 2, 19–39 odd, 26, 28, 38, 40–43	19, 23, 26, 29, 37, 39

Common Errors

- **Exercises 3–11** Students may add the denominators as well as the numerators. Remind them that only the numerators are added when adding rational expressions with a common denominator.
- **Exercise 9** Students may not subtract both terms in the second numerator. Remind them to distribute the negative when subtracting a binomial.
- **Exercises 13–18** Students may multiply the denominators to find the LCD. While this process will give a common denominator, it is not necessarily the LCD. Remind them to use prime factorization to find the LCD.

11.6 Record and Practice Journal

Find the sum or difference.

1. $\frac{3}{5g} + \frac{6}{5g}$ — $\frac{9}{5g}$
2. $\frac{1}{2v+3} + \frac{4}{2v+3}$ — $\frac{5}{2v+3}$
3. $\frac{11m}{4m-2} - \frac{3m+2}{4m-2}$ — $\frac{4m-1}{2m-1}$
4. $\frac{y^2}{y^2+y-6} - \frac{9}{y^2+y-6}$ — $\frac{y-3}{y-2}$
5. $\frac{3a+1}{4a} + \frac{a-2}{6a}$ — $\frac{11a-1}{12a}$
6. $\frac{k^2+8}{k-5} - k$ — $\frac{5k+8}{k-5}$
7. $\frac{4x^2-x}{3x-12} + \frac{x^2+4}{4-x}$ — $\frac{x+3}{3}$
8. $\frac{6}{d+5} - \frac{3d-7}{d^2+2d-15}$ — $\frac{3d-11}{d^2+2d-15}$
9. You drive 45 miles from home to a relative's house and 45 miles back home. Due to construction, your speed on the way back is only 60% of your speed on the way there. Let r be your speed (in miles per hour) while driving to your relative's house. Write an expression that represents the amount of time you spend driving on your trip.
   $t = \frac{120}{r}$ h

Technology For the Teacher
Answer Presentation Tool

11.6 Exercises

Vocabulary and Concept Check

1. **WRITING** Explain how finding the least common denominator of two rational expressions is similar to finding the least common denominator of two numeric fractions.

2. **REASONING** Describe how to rewrite the expressions $\dfrac{1}{x+4}$ and $\dfrac{1}{x^2-16}$ so that they have the same denominator.

Practice and Problem Solving

Find the sum or difference.

3. $\dfrac{4s}{9} + \dfrac{s}{9}$

4. $\dfrac{r}{8} - \dfrac{r}{16}$

5. $\dfrac{2}{w} + \dfrac{5}{w}$

1 6. $\dfrac{7}{3y} + \dfrac{1}{3y}$

7. $\dfrac{5}{x+2} - \dfrac{2}{x+2}$

8. $\dfrac{2z}{4(z-1)} + \dfrac{6z}{4(z-1)}$

9. $\dfrac{3t^2}{t^2-1} - \dfrac{t+4}{t^2-1}$

10. $\dfrac{2n+3}{n^2-n-2} + \dfrac{-n-2}{n^2-n-2}$

11. $\dfrac{p-2}{p^2-5p+4} - \dfrac{2}{p^2-5p+4}$

12. **ERROR ANALYSIS** Describe and correct the error in adding the rational expressions.

✗ $\dfrac{1}{x-3} + \dfrac{4}{x-3} = \dfrac{5}{2x-6}$

Find the LCD of the rational expressions.

2 13. $\dfrac{9}{2x}, \dfrac{7}{x}$

14. $\dfrac{1}{12y}, \dfrac{5}{18y}$

15. $\dfrac{m}{m+5}, \dfrac{9}{m-4}$

16. $2g, \dfrac{1}{g}$

17. $\dfrac{h}{h+3}, \dfrac{1}{h^2+2h-3}$

18. $\dfrac{s-7}{s^2-2s-8}, \dfrac{3}{s-4}$

19. **CEREAL** The height of a cereal box is given by $\dfrac{S}{2(\ell+w)} - \dfrac{2\ell w}{2(\ell+w)}$, where S is the surface area, ℓ is the length, and w is the width. Find the difference.

Find the sum or difference.

(3)(4) 20. $\frac{x+1}{2x} + \frac{2x-1}{5x}$

21. $\frac{y-3}{6y} + \frac{y+4}{8y}$

22. $3 - \frac{c-2}{c+2}$

23. $\frac{2m}{m-7} + \frac{4}{7-m}$

24. $\frac{x+2}{x^2+3x-10} + \frac{3}{2-x}$

25. $\frac{2p+3}{p^2-7p+12} - \frac{2}{p-3}$

26. **ERROR ANALYSIS** Describe and correct the error in adding the rational expressions.

$$\frac{x}{x-1} + \frac{2}{x+2} = \frac{x(x-1)+2(x+2)}{(x-1)(x+2)}$$
$$= \frac{x^2-x+2x+4}{(x-1)(x+2)}$$
$$= \frac{x^2+x+4}{(x-1)(x+2)}$$

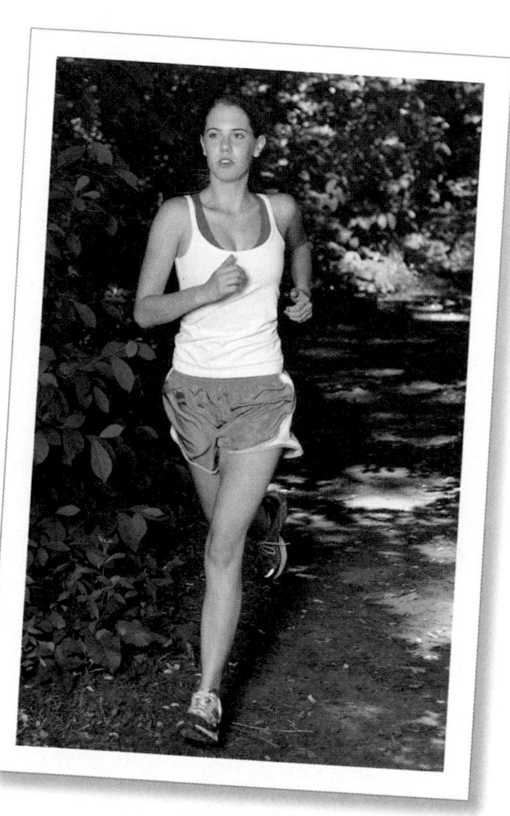

27. **REASONING** Can you find a common denominator of two rational expressions by finding the product of the denominators? Is this product always going to be the least common denominator? Justify your answers.

28. **RUNNING** You run 3 miles up a hill and 3 miles down the hill. You run 25% faster going down the hill than going up the hill. Let *r* be your speed (in miles per hour) while running up the hill. Write an expression that represents the amount of time you spend running on the hill.

29. **OPEN-ENDED** Write two rational expressions with unlike denominators.

a. Find the least common denominator of the two expressions.

b. Add the two expressions.

Write an expression for the perimeter of the figure.

30. 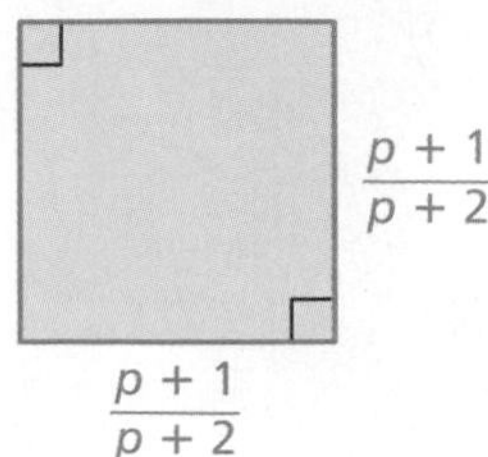

31. 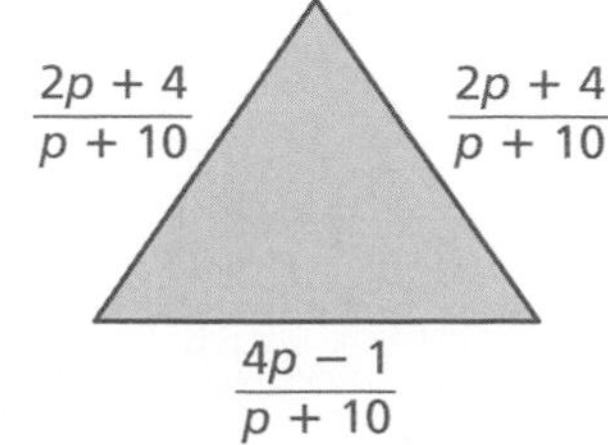

32. 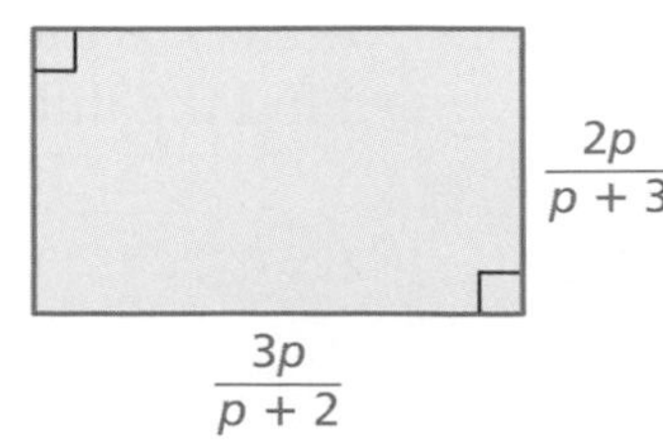

Common Errors

- **Exercises 20–25** Students may add the denominators as well as the numerators. Remind them to find the LCD and only add the numerators when adding rational expressions.
- **Exercises 20–25** Students may multiply the numerator of each rational expression by the wrong factor of the LCD. Remind them to multiply the numerator and denominator by the factor that makes the denominator the LCD.
- **Exercises 22 and 25** Students may not subtract both terms in the second numerator. Remind them to distribute the negative when subtracting a binomial.
- **Exercises 35 and 36** Students may forget about the order of operations. Remind them of this order.

Practice and Problem Solving

20. $\frac{9x+3}{10x}$ **21.** $\frac{7}{24}$

22. $\frac{2c+8}{c+2}$ **23.** $\frac{2m-4}{m-7}$

24. $\frac{-2x-13}{(x+5)(x-2)}$

25. $\frac{11}{(p-4)(p-3)}$

26. See Additional Answers.

27. yes; not always; The product of the denominators is the product of *all* factors of both denominators. The LCM of the denominators is the product of the greatest power of each factor that appears in *either* denominator.

28. $\frac{5.4}{r}$

29. *Sample answer:* $\frac{1}{x-2}, \frac{1}{x+3}$

a. $(x-2)(x+3)$

b. $\frac{2x+1}{(x-2)(x+3)}$

30. $\frac{4p+4}{p+2}$ **31.** $\frac{8p+7}{p+10}$

32. $\frac{10p^2+26p}{(p+2)(p+3)}$

Differentiated Instruction

Organization

If students have trouble finding the least common multiple (LCM) of polynomials, have them organize the factors in a chart. For example, using the expressions $2x-4$ and $6x^2+12x-48$, line up the factors in columns.

$2 \cdot \quad (x-2)$

$2 \cdot 3 \cdot (x-2) \cdot (x+4)$

$2 \cdot 3 \cdot (x-2) \cdot (x+4)$

The least common multiple is $6(x-2)(x+4)$.

Practice and Problem Solving

33. $\dfrac{2c+1}{c-1}$ **34.** $\dfrac{d+3}{d+5}$

35. $\dfrac{x(2x-1)}{(x+4)(x-5)}$

36. $\dfrac{2-(y+8)^2}{2(y-1)(y+8)}$

37. $\dfrac{33}{2b+20}$

38. **a.** $\dfrac{5x+3}{(x+3)(x-3)}$

b. about 17.7 minutes

39. *Sample answer:* $\dfrac{-3}{x+4} - \dfrac{-5}{x+3}$

Fair Game Review

40–41. See Additional Answers.

42.

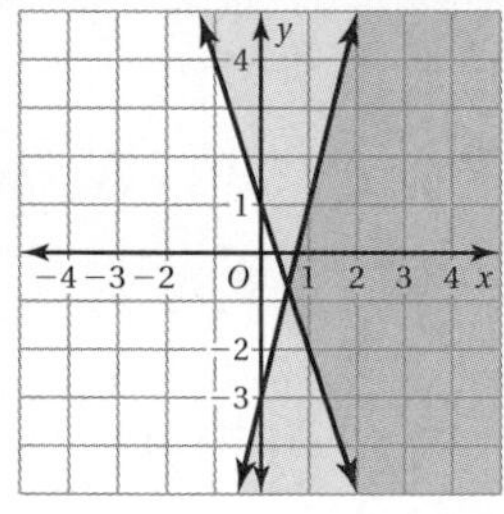

43. A

Mini-Assessment

1. Find the LCD of $\dfrac{x}{x-6}$ and $\dfrac{1}{x^2-3x-18}$. $(x-6)(x+3)$

Find the sum or difference.

2. $\dfrac{2n}{n^2-9} - \dfrac{n+3}{n^2-9}$ $\quad \dfrac{1}{n+3}$

3. $\dfrac{k}{k+4} + \dfrac{6k-36}{k^2-2k-24}$ $\quad \dfrac{k+6}{k+4}$

4. $\dfrac{w}{w-8} - \dfrac{6-3w}{w^2-10w+16}$ $\quad \dfrac{w+3}{w-8}$

Taking Math Deeper

Exercise 39

Help students understand that they are not expected to look at this problem and immediately recognize how to solve it. Just take a deep breath, relax, and begin by looking at the given information.

First, rewrite the difference $\dfrac{a}{b} - \dfrac{c}{d}$: $\dfrac{a}{b} - \dfrac{c}{d} = \dfrac{ad}{bd} - \dfrac{bc}{bd} = \dfrac{ad-bc}{bd}$.

You can then write $\dfrac{ad-bc}{bd} = \dfrac{2x+11}{(x+4)(x+3)}$.

From this equation, $b = (x+4)$ and $d = (x+3)$. Substitute these expressions.

$$\frac{a(x+3)-(x+4)c}{(x+4)(x+3)} = \frac{2x+11}{(x+4)(x+3)}$$

From this equation, you can write the following.

$a(x+3)-(x+4)c = 2x+11$	Set the numerators equal.
$ax+3a-cx-4c = 2x+11$	Distributive Property
$ax-cx+3a-4c = 2x+11$	Group like terms.
$x(a-c)+3a-4c = 2x+11$	Factor.

Write a system of equations to find a and c.

$$a - c = 2$$

$$3a - 4c = 11$$

The solution of this system is $a = -3$ and $c = -5$.

Check your solution.

$$\frac{-3}{(x+4)} - \frac{-5}{(x+3)} = \frac{-3(x+3)}{(x+4)(x+3)} - \frac{-5(x+4)}{(x+3)(x+4)}$$

$$= \frac{-3x-9+5x+20}{(x+4)(x+3)}$$

$$= \frac{2x+11}{(x+4)(x+3)}$$

Reteaching and Enrichment Strategies

If students need help. . .	If students got it. . .
Resources by Chapter • Practice A and Practice B • Puzzle Time Record and Practice Journal Practice Differentiating the Lesson Lesson Tutorials Skills Review Handbook	Resources by Chapter • Enrichment and Extension • School-to-Work • Financial Literacy • Technology Connection Start the next section

Simplify the expression.

33. $\dfrac{3c+1}{c-1} + \dfrac{c+1}{c^2-4c+3} - \dfrac{c-1}{c-3}$

34. $-\dfrac{11d-8}{d^2+d-20} + \dfrac{d}{d-4} + \dfrac{5}{d+5}$

35. $\dfrac{x}{x+4} + \dfrac{x^2}{x^2-x-20} \div \dfrac{x}{x+4}$

36. $\dfrac{1}{y^2+7y-8} - \dfrac{2}{2y-2} \bullet \dfrac{y+8}{2}$

37. **HOMEWORK** You have 20 more math exercises for homework than biology exercises. You finish 15 exercises before dinner and 18 exercises after dinner. Write an expression that represents the portion of exercises that are complete.

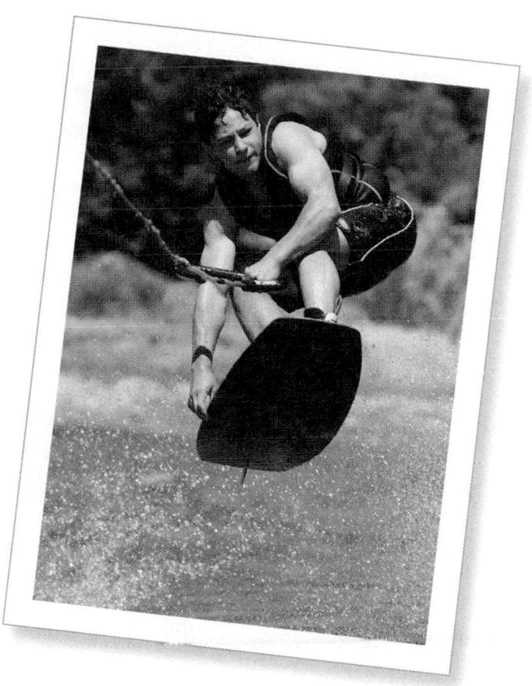

38. **WAKEBOARDING** You are wakeboarding on a river. You travel 2 miles downstream to a marina for supplies, and then you travel 3 miles upstream to a dock. The boat travels x miles per hour during the entire trip, and the river current is 3 miles per hour.

    a. Write an expression that represents the total time of the trip.

    b. How long will the trip take when the speed of the boat is 18 miles per hour?

39. 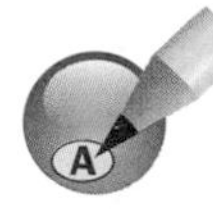

Let a, b, c, and d be polynomials. Find two rational expressions $\dfrac{a}{b}$ and $\dfrac{c}{d}$ so that $\dfrac{a}{b} - \dfrac{c}{d} = \dfrac{2x+11}{(x+4)(x+3)}$.

Fair Game Review What you learned in previous grades & lessons

Graph the system of linear inequalities. *(Section 4.5)*

40. $y > 2x + 1$
    $y < -3x + 4$

41. $y \geq -\dfrac{1}{2}x$
    $y < -x - 7$

42. $y \leq 4x - 3$
    $y \geq -3x + 1$

43. **MULTIPLE CHOICE** The graph of which function is shown at the right? *(Section 10.1)*

    Ⓐ $y = 2\sqrt{x}$ Ⓑ $y = -2\sqrt{x}$

    Ⓒ $y = \dfrac{1}{2}\sqrt{x}$ Ⓓ $y = 5\sqrt{x}$

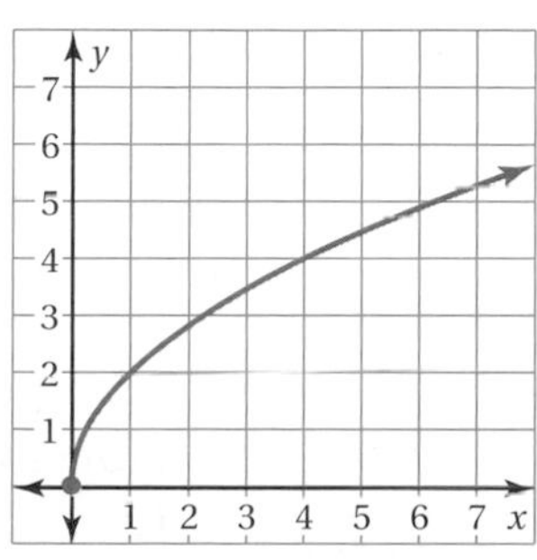

11.7 Solving Rational Equations

Essential Question How can you solve a rational equation?

COMMON CORE STATE STANDARDS
A.CED.1

1 ACTIVITY: Solving Rational Equations

Work with a partner. A hockey goalie faces 799 shots and saves 707 of them.

a. What is his save percentage?

National Hockey League goalies typically have a save percentage above .900.

b. Suppose the goalie has x additional consecutive saves. Write an expression for his new save percentage.

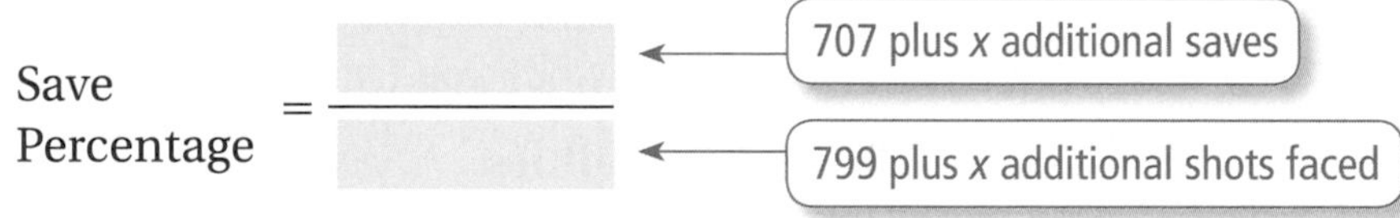

c. Complete the table showing the goalie's save percentage as x increases.

Additional Saves, x	0	20	40	60	80	100	120	140
Save Percentage								

d. The goalie wants to end the season with a save percentage of .900. How many additional consecutive saves must he have to achieve this? Justify your answer by solving an equation.

Laurie's Notes

Introduction

Standards for Mathematical Practice

- **MP7 Look for and Make Use of Structure:** Seeing the connection to proportions will help students understand why the Cross Products Property can be used to solve rational equations.

Motivate

- Crinkle up a piece of scrap paper into a ball.
- Let a student try 5 times to shoot the paper ball into a wastebasket from a distance of about 10 feet.
- Compute the success rate (shots in ÷ shots attempted).
- ? "Suppose [insert student name] took 5 more shots and made them all, what would happen to the success rate?" It would increase.
- Tell students that today they will investigate a similar type of problem for goalies and baseball players.

Activity Notes

Activity 1

- Ask a student to explain the role of a goalie in sports.
- Students should use a calculator to find the percentages in this activity.
- Make sure students write the correct expressions in part (b).
- ? "What do you observe about your answers in part (c)?" increasing
- **MP2 Reason Abstractly and Quantitatively:** Have a discussion about the value of $\frac{707 + x}{799 + x}$ as x increases.
  - The numerator and denominator are each increasing by the same amount each time. The ratio is increasing, but at a decreasing rate.
  - The ratio is getting closer to 1 but it will not reach 1. Connect this to the concept of an asymptote.
- **MP1a Make Sense of Problems:** Ask for volunteers to describe how they answered part (d). They may have:
  - used trial and error on a calculator.
  - set the ratio equal to 0.9, rewritten it as $\frac{9}{10}$, and then used the Cross Products Property.
  - set the ratio equal to 0.9, multiplied both sides of the equation by $(799 + x)$, and solved.
- **MP5 Use Appropriate Tools Strategically:** If time permits, use a graphing calculator to generate a table of values for $y = \frac{707 + x}{799 + x}$.

Common Core State Standards

A.CED.1 Create equations and inequalities in one variable and use them to solve problems.

Previous Learning

Students should know how to add, subtract, multiply, and divide rational expressions.

Activity Materials	
Introduction	**Textbook**
• paper • wastebasket	• calculators

Start Thinking! and Warm Up

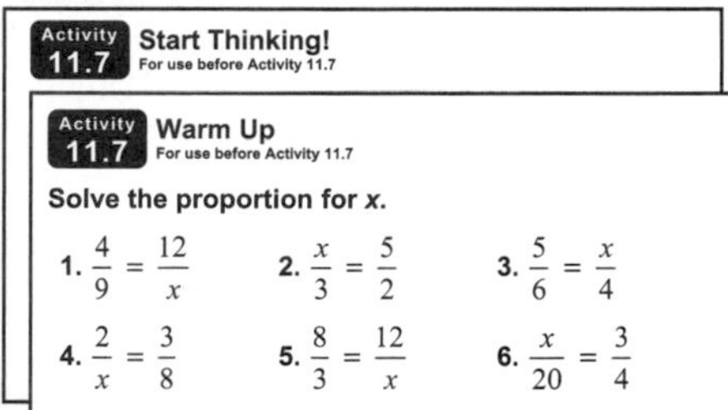

Activity 11.7 Start Thinking!
For use before Activity 11.7

Activity 11.7 Warm Up
For use before Activity 11.7

Solve the proportion for x.

1. $\frac{4}{9} = \frac{12}{x}$
2. $\frac{x}{3} = \frac{5}{2}$
3. $\frac{5}{6} = \frac{x}{4}$
4. $\frac{2}{x} = \frac{3}{8}$
5. $\frac{8}{3} = \frac{12}{x}$
6. $\frac{x}{20} = \frac{3}{4}$

11.7 Record and Practice Journal

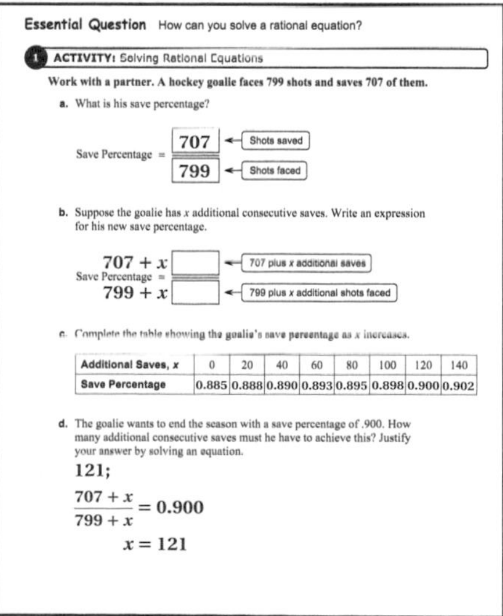

Essential Question How can you solve a rational equation?

1 ACTIVITY: Solving Rational Equations

Work with a partner. A hockey goalie faces 799 shots and saves 707 of them.

a. What is his save percentage?

Save Percentage = $\frac{707}{799}$ ← Shots saved / ← Shots faced

b. Suppose the goalie has x additional consecutive saves. Write an expression for his new save percentage.

Save Percentage = $\frac{707 + x}{799 + x}$ ← 707 plus x additional saves / ← 799 plus x additional shots faced

c. Complete the table showing the goalie's save percentage as x increases.

Additional Saves, x	0	20	40	60	80	100	120	140
Save Percentage	0.885	0.888	0.890	0.893	0.895	0.898	0.900	0.902

d. The goalie wants to end the season with a save percentage of .900. How many additional consecutive saves must he have to achieve this? Justify your answer by solving an equation.

121;

$\frac{707 + x}{799 + x} = 0.900$

$x = 121$

English Language Learners

Pair Activity

Pair English language learners with English speakers. Assign each person a problem. When they have completed their problems, they explain their solution to their partner who follows along. This will engage English language learners in conversation and help with understanding math concepts.

11.7 Record and Practice Journal

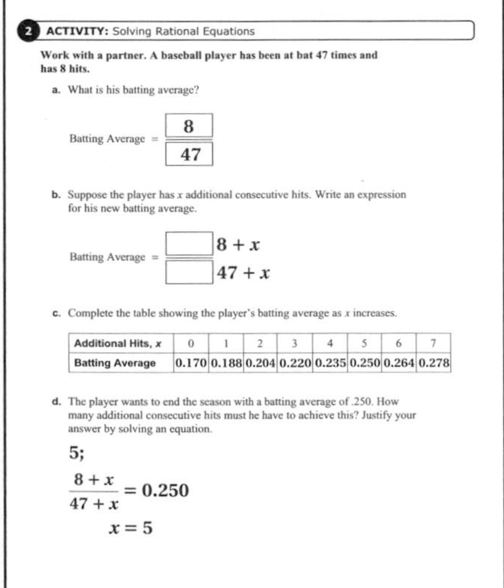

2 ACTIVITY: Solving Rational Equations

Work with a partner. A baseball player has been at bat 47 times and has 8 hits.

a. What is his batting average?

Batting Average $= \dfrac{8}{47}$

b. Suppose the player has x additional consecutive hits. Write an expression for his new batting average.

Batting Average $= \dfrac{8 + x}{47 + x}$

c. Complete the table showing the player's batting average as x increases.

Additional Hits, x	0	1	2	3	4	5	6	7
Batting Average	0.170	0.188	0.204	0.220	0.235	0.250	0.264	0.278

d. The player wants to end the season with a batting average of .250. How many additional consecutive hits must he have to achieve this? Justify your answer by solving an equation.

5;

$$\frac{8 + x}{47 + x} = 0.250$$

$$x = 5$$

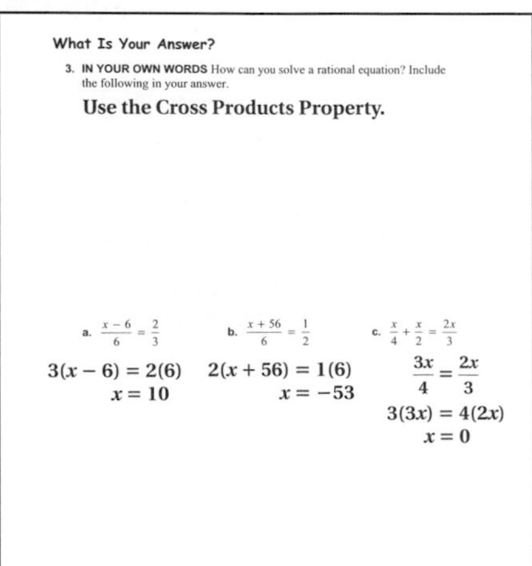

What Is Your Answer?

3. IN YOUR OWN WORDS How can you solve a rational equation? Include the following in your answer.

Use the Cross Products Property.

a. $\dfrac{x - 6}{6} = \dfrac{2}{3}$

$3(x - 6) = 2(6)$

$x = 10$

b. $\dfrac{x + 56}{6} = \dfrac{1}{2}$

$2(x + 56) = 1(6)$

$x = -53$

c. $\dfrac{x}{4} + \dfrac{x}{2} = \dfrac{2x}{3}$

$\dfrac{3x}{4} = \dfrac{2x}{3}$

$3(3x) = 4(2x)$

$x = 0$

Laurie's Notes

Activity 2

- Discuss briefly how to find a batting average. If you are not familiar with how "at bats" are defined, perhaps one of your students will explain. For example, when a batter walks, sacrifices, or gets hit by a pitch, it is not considered an "at bat."
- This activity is very similar to the first activity in that students explore ratios where the numerator and denominator are changing by the same amount each time.
- **MP2:** Have a conversation similar to the previous activity about how the ratio is changing as x increases.
- **MP1a:** Ask for volunteers to describe how they answered part (d). They may have used methods similar to those listed in Activity 1.
- **MP5:** If time permits, use a graphing calculator to generate a table of values for $y = \dfrac{8 + x}{47 + x}$.

What Is Your Answer?

- There are several ways in which students might solve the three problems shown. Having discussed a variety of methods in class will deepen student understanding.

Closure

- Ask a question related to the Motivate. For example, if the student had a success rate of 0.8, ask how many additional consecutive shots must be made to have a scoring rate of 0.9.

Technology For the Teacher

The Dynamic Planning Tool

Editable Teacher's Resources at *BigIdeasMath.com*

2 ACTIVITY: Solving Rational Equations

Work with a partner. A baseball player has been at bat 47 times and has 8 hits.

a. What is his batting average?

Batting Average $= \dfrac{\quad}{\quad}$

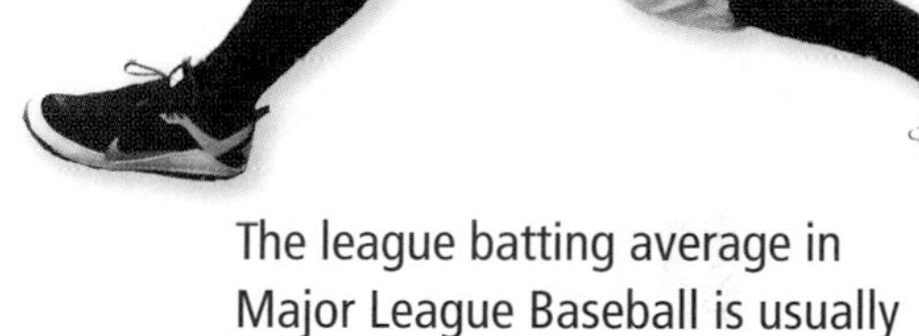

b. Suppose the player has x additional consecutive hits. Write an expression for his new batting average.

The league batting average in Major League Baseball is usually between .250 and .270.

Batting Average $=$ 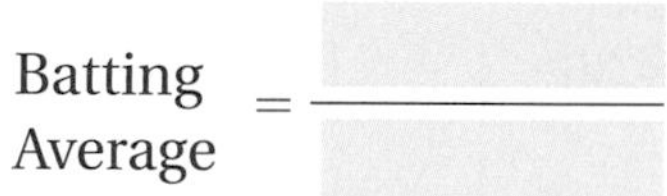

c. Complete the table showing the player's batting average as x increases.

Additional Hits, x	0	1	2	3	4	5	6	7
Batting Average								

d. The player wants to end the season with a batting average of .250. How many additional consecutive hits must he have to achieve this? Justify your answer by solving an equation.

What Is Your Answer?

3. **IN YOUR OWN WORDS** How can you solve a rational equation? Include the following in your answer.

a. $\dfrac{x-6}{6} = \dfrac{2}{3}$ **b.** $\dfrac{x+56}{6} = \dfrac{1}{2}$ **c.** $\dfrac{x}{4} + \dfrac{x}{2} = \dfrac{2x}{3}$

Practice Use what you learned about solving rational equations to complete Exercise 4 on page 592.

11.7 Lesson

Key Vocabulary
rational equation, *p. 590*

A **rational equation** is an equation that contains rational expressions. One way to solve rational equations is to use the Cross Products Property. You can use this method when each side of a rational equation consists of one rational expression.

EXAMPLE 1 Solving Rational Equations Using Cross Products

Solve each equation.

a. $\frac{5}{x+4} = \frac{4}{x-4}$

$\frac{5}{x+4} = \frac{4}{x-4}$	Write the equation.
$5(x-4) = 4(x+4)$	Cross Products Property
$5x - 20 = 4x + 16$	Distributive Property
$5x = 4x + 36$	Add 20 to each side.
$x = 36$	Subtract $4x$ from each side.

Check

$$\frac{5}{x+4} = \frac{4}{x-4}$$
$$\frac{5}{36+4} \stackrel{?}{=} \frac{4}{36-4}$$
$$\frac{1}{8} = \frac{1}{8} \checkmark$$

b. $\frac{5}{y} = \frac{y-2}{7}$

$\frac{5}{y} = \frac{y-2}{7}$	Write the equation.
$5(7) = y(y-2)$	Cross Products Property
$35 = y^2 - 2y$	Simplify.
$0 = y^2 - 2y - 35$	Subtract 35 from each side.
$0 = (y-7)(y+5)$	Factor.
$y - 7 = 0$ *or* $y + 5 = 0$	Zero-Product Property
$y = 7$ *or* $y = -5$	Solve for y.

Check

$\frac{5}{7} \stackrel{?}{=} \frac{7-2}{7}$	Substitute for y.	$\frac{5}{-5} \stackrel{?}{=} \frac{-5-2}{7}$
$\frac{5}{7} = \frac{5}{7} \checkmark$	Simplify.	$-1 = -1 \checkmark$

On Your Own

Now You're Ready
Exercises 5–10

Solve the equation. Check your solution(s).

1. $\frac{2}{x-3} = \frac{4}{x-7}$ **2.** $\frac{4}{z+4} = \frac{z}{z+1}$ **3.** $\frac{3y}{4} = \frac{6}{y+7}$

Multi-Language Glossary at BigIdeasMath.com.

Laurie's Notes

Introduction

Connect
- **Yesterday:** Students explored rational equations. (MP1a, MP2, MP5, MP7)
- **Today:** Students will solve rational equations.

Motivate
- Ask if any of your students have collectible cards (sports cards, card games, etc). These cards are often traded among collectors.
- Share some trivia with your students. For instance, you could discuss one of the rarest and most valuable baseball cards of all time—the Honus Wagner "T206" card. An owner of the Arizona Diamondbacks purchased the card for $2.8 million in 2007!
- Tell students they will solve a collectible card problem in this lesson.

Lesson Notes

Example 1
- ? "Are there any excluded values that should be noted?" yes, $x = 4$, $x = -4$
- ? "What property might be helpful in solving this equation?" Cross Products Property
- Remind students to include parentheses around each binomial when cross multiplying. The Distributive Property must now be used.
- ? "How can you check the solution?" by substitution
- **Teaching Tip**: Although the problem can be checked graphically, it is not always obvious where the graphs of two rational functions intersect. Algebraic checks are often more efficient.
- Write the equation in part (b) and ask students how it differs from the equation in part (a).
- ? "Are there any excluded values that should be noted?" yes, $y = 0$
- Check both solutions to be sure that neither is extraneous.

On Your Own
- **Think-Pair-Share:** Students should read each question independently and then work with a partner to answer the questions. When they have answered the questions, the pair should compare their answers with another group and discuss any discrepancies.

Goal Today's lesson is solving **rational equations**.

Start Thinking! and Warm Up

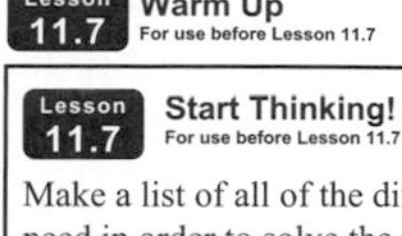

Lesson 11.7 Warm Up
For use before Lesson 11.7

Lesson 11.7 Start Thinking!
For use before Lesson 11.7

Make a list of all of the different math skills you need in order to solve the rational equation.

$$\frac{x+1}{3} = \frac{6}{x-2}$$
$$(x+1)(x-2) = 3(6)$$
$$x^2 - x - 2 = 18$$
$$x^2 - x - 20 = 0$$
$$(x-5)(x+4) = 0$$
$$x = 5 \text{ or } x = -4$$

Check $x = 5$:
$$\frac{5+1}{3} \stackrel{?}{=} \frac{6}{5-2}$$
$$2 = 2 \checkmark$$

Check $x = -4$:
$$\frac{-4+1}{3} \stackrel{?}{=} \frac{6}{-4-2}$$
$$-1 = -1 \checkmark$$

Extra Example 1
Solve each equation.

a. $\frac{3}{x-2} = \frac{6}{x+1}$ $x = 5$

b. $\frac{y+5}{2} = \frac{3}{y}$ $y = -6, y = 1$

On Your Own
1. $x = -1$
2. $z = -2, z = 2$
3. $y = -8, y = 1$

Extra Example 2

Solve $\frac{x}{x-4} - \frac{3}{4} = \frac{2}{x-4}$. $x = -4$

Extra Example 3

In Example 3, you add creature cards to the deck until it contains 45% creature cards. How many do you add? 10 cards

On Your Own

4. $p = -9$
5. $n = 3$
6. $a = -7, a = 3$
7. 5 cards

Differentiated Instruction

Connection

Show students that an equation solved using the Cross Products Property can be solved by multiplying both sides by the LCD. For instance, solving Example 1(b) would look like this.

$$\frac{5}{y} = \frac{y-2}{7}$$

$$7 \cdot \cancel{y} \cdot \frac{5}{\cancel{y}} = \cancel{7} \cdot y \cdot \frac{y-2}{\cancel{7}}$$

$$7(5) = y(y-2)$$

After multiplying by the LCD and simplifying, the resulting equation is the same as if the Cross Products Property had been used. Students who have a difficult time deciding which method to use should multiply by the LCD.

Laurie's Notes

Discuss

- **MP7 Look for and Make Use of Structure:** Rational equations can involve operations on one or both sides. Multiplying each side by the LCD of the expressions eliminates the denominators. Demonstrate this by multiplying each side of $\frac{x}{2} + \frac{3}{4} = \frac{5}{8}$ by 8.

Example 2

- This example demonstrates another way to solve rational equations.
- ? "What is the LCD of the expressions in this equation?" $3(z-2)$
- ? "Are there any excluded values that should be noted?" yes, $z = 2$
- **Big Idea:** When you multiply the left side of the equation by $3(z-2)$, you use the Distributive Property. Work through this step slowly, showing how the common factors divide out.
- **Alternate Approach:** Subtract $\frac{z}{z-2}$ from each side and simplify. Then use the Cross Products Property.
- Discuss why the rational equation has no solution.

Example 3

- **MP1a Make Sense of Problems:** Ask a volunteer to read the problem. Make sure that students understand the context, especially that only creature cards are added to deck.
- This example should remind students of the problems in the activity.
- Point out to students that using the Cross Products Property and multiplying each side by the LCD results in the same equation.
- Take time to check the solution.

On Your Own

- Have volunteers share their solutions at the board.

Closure

- **Exit Ticket:** Write a rational equation that you would solve using (a) the Cross Products Property and (b) the LCD.

  Sample answers: (a) $\frac{1}{3} = \frac{x-1}{5}$ (b) $\frac{2}{x-1} + \frac{1}{3} = \frac{8}{x-1}$

Technology For the Teacher

The Dynamic Planning Tool

Editable Teacher's Resources at *BigIdeasMath.com*

When there is more than one rational expression on one or both sides of a rational equation, multiply each side by the LCD and then solve.

EXAMPLE 2 Solving a Rational Equation Using the LCD

Solve $\dfrac{z}{z-2} - \dfrac{2}{3} = \dfrac{2}{z-2}$.

$$3(z-2) \cdot \left(\frac{z}{z-2} - \frac{2}{3}\right) = 3(z-2) \cdot \frac{2}{z-2}$$ Multiply each side by the LCD, $3(z-2)$.

$$\frac{z \cdot 3\cancel{(z-2)}}{\cancel{z-2}} - \frac{2 \cdot \cancel{3}(z-2)}{\cancel{3}} = \frac{2 \cdot 3\cancel{(z-2)}}{\cancel{z-2}}$$ Multiply. Then divide out common factors.

$$3z - 2z + 4 = 6$$ Simplify.

$$z = 2$$ Solve for z.

Because each side of the equation is undefined when $z = 2$, it is an extraneous solution.

∴ The equation has no solution.

EXAMPLE 3 Real-Life Application

Your starter deck for a collectible card game has 50 cards. The deck contains 17 creature cards. You add creature cards to the deck until it contains 50% creature cards. How many do you add?

Write an equation for the ratio of creature cards to total cards after adding x creature cards.

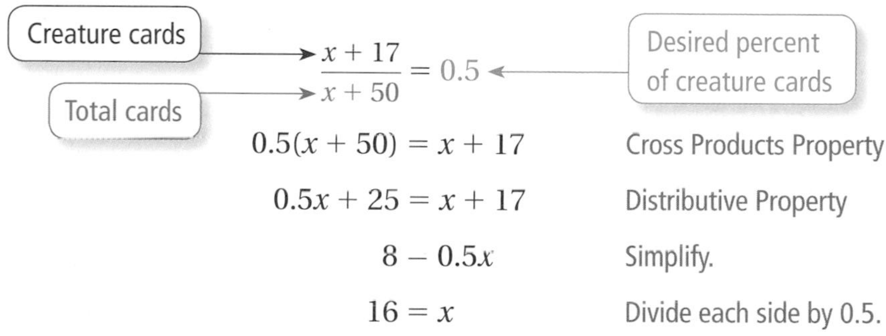

$$\frac{x+17}{x+50} = 0.5$$

$$0.5(x+50) = x + 17$$ Cross Products Property

$$0.5x + 25 = x + 17$$ Distributive Property

$$8 - 0.5x$$ Simplify.

$$16 = x$$ Divide each side by 0.5.

∴ You add 16 creature cards to the deck.

On Your Own

Now You're Ready
Exercises 13–18

Solve the equation. Check your solution(s).

4. $\dfrac{1}{p} - \dfrac{2}{3} = \dfrac{7}{p}$ **5.** $\dfrac{2}{n} + \dfrac{1}{n+3} = \dfrac{5}{n+3}$ **6.** $\dfrac{4}{a-6} + 1 = \dfrac{9}{a^2-36}$

7. WHAT IF? In Example 3, you add creature cards until the deck contains 40% creature cards. How many do you add?

11.7 Exercises

Vocabulary and Concept Check

1. **VOCABULARY** Describe two methods for solving rational equations.

2. **OPEN-ENDED** Write a rational equation that can be solved by multiplying each side by $2x(x + 1)$.

3. **WRITING** Why should you check the solutions of a rational equation?

Practice and Problem Solving

4. A basketball player attempts 64 free throws and makes 50 of them.
   a. What is her free throw percentage?
   b. Suppose the player makes x additional consecutive free throws. Write an expression for her new free throw percentage.
   c. The player wants to end the season with a free throw percentage of .800. How many additional consecutive free throws must she make to achieve this?

Solve the equation. Check your solution(s).

5. $\frac{2}{b} = \frac{6}{b+2}$

6. $\frac{2}{x-1} = \frac{3}{x+1}$

7. $\frac{4}{m-4} = \frac{m}{3}$

8. $\frac{z-1}{8} = \frac{z}{z+9}$

9. $\frac{k}{2k+5} = \frac{1}{k-2}$

10. $\frac{3w}{w+1} = \frac{w}{3-w}$

$$\frac{x}{x+1} = \frac{2}{x+1}$$
$$x(x+1) = 2(x+1)$$
$$x^2 + x = 2x + 2$$
$$x^2 - x - 2 = 0$$
$$(x-2)(x+1) = 0$$
$$x = 2 \quad or \quad x = -1$$

So, the solutions are $x = 2$ and $x = -1$.

11. **ERROR ANALYSIS** Describe and correct the error in solving the equation.

12. **WATER RESCUE** The table shows information about a water rescue team.
   a. Solve the rational equation $\frac{4}{x} = \frac{7}{x+6}$ to find the upstream speed of the rescue team.
   b. What is the downstream speed of the rescue team?

Water Rescue

Direction	Distance	Rate	Time
Upstream	4 miles	x mi/h	t hours
Downstream	7 miles	$(x + 6)$ mi/h	t hours

Assignment Guide and Homework Check

Level	Assignment	Homework Check
Average	1–3, 5–19 odd, 12, 20, 25–28	7, 11, 12, 15, 20
Advanced	1–3, 6–18 even, 19–24, 25–28	12, 16, 20, 22, 24

Common Errors

- **Exercises 5–10** Students may multiply the numerators and multiply the denominators of the rational expressions instead of finding the cross products. Remind them of the Cross Products Property.
- **Exercises 13–18** Students may not multiply each side of the equation by the LCD. Remind them of the Multiplication Property of Equality.
- **Exercises 22–24** Students may have difficulty setting up the equations for these exercises. Remind them that the sum of the portions for everyone working is the portion of the job completed in one unit of time.

11.7 Record and Practice Journal

Solve the equation. Check your solution.

1. $\frac{4}{h-3} = \frac{8}{h}$
   $h = 6$
2. $\frac{6}{q-2} = \frac{5}{q-1}$
   $q = -4$
3. $\frac{m}{m+3} = \frac{5}{m+7}$
   $m = -5, 3$
4. $\frac{c-3}{5c-6} = \frac{c}{c-3}$
   $c = \pm\frac{3}{2}$
5. $\frac{6}{z-3} - \frac{3}{z} = \frac{6}{z}$
   $z = 9$
6. $\frac{4}{k} + \frac{14}{k+3} = \frac{8}{k+5}$
   $k = -2$
7. $\frac{d}{d+3} + \frac{1}{d-1} = \frac{4}{d^2+2d-3}$
   $d = -1$
8. $\frac{t}{t-7} - \frac{5}{t-4} = \frac{3t+3}{t^2-11t+28}$
   $t = 8$
9. An academic challenge team has 24 members, 9 of which are boys. The team is required to have 50% boys and 50% girls for a competition. How many boys does the team need to add to reach this proportion?
   6 boys

Answer Presentation Tool

Vocabulary and Concept Check

1. Use the Cross Products Property or multiply each side by the LCD.
2. *Sample answer:* $\frac{1}{x} + \frac{6}{x+1} = \frac{x}{2}$
3. The solution may be extraneous.

Practice and Problem Solving

4. **a.** 0.78125, or about 0.781
   b. $\frac{50+x}{64+x}$
   c. 6 free throws
5. $b = 1$
6. $x = 5$
7. $m = -2, m = 6$
8. $z = -3, z = 3$
9. $k = -1, k = 5$
10. $w = 0, w = 2$
11. The solutions were not checked in the original equation. The solution $x = -1$ is extraneous because it is an excluded value. The only solution is $x = 2$.
12. **a.** 8 mi/h
    b. 14 mi/h

Practice and Problem Solving

13. $c = 5$
14. $y = -1$
15. no solution
16. $n = -3, n - \frac{2}{7}$
17. $a = -9, a = 6$
18. $x = -1$
19. Rewrite the left side using a common denominator of x. Then use the Cross Products Property to solve.
20. 4 pints
21. See *Taking Math Deeper.*
22. 4.8 minutes
23. 1.2 hours
24. 12 hours

Fair Game Review

25. $x = -4, x = 8$
26. $x = 2, x = 5$
27. $x = -2\frac{1}{5}, x = \frac{3}{5}$
28. B

Mini-Assessment

Sole the equation. Check your solution.

1. $\frac{7}{a} = \frac{1}{a-6}$ $a = 7$
2. $\frac{b}{b+6} = \frac{3}{b}$ $b = -3, b = 6$
3. $\frac{2c}{c+2} - 5 = \frac{7c}{c+2}$ $c = -1$

Taking Math Deeper

Exercise 21

As with many real-life problems, this one has a lot of information. Read through the problem and organize the given information.

- The club pays \$540 for a bus. So, the original bus fare per person for x club members is $\frac{540}{x}$.
- Seven hikers join the trip, so the bus fare per person is $\frac{540}{x+7}$.
- After the 7 hikers join, the bus fare per person decreases by \$7. So, the bus fare per person can also be written as $\frac{540}{x} - 7$.

Set the expressions for the bus fare per person equal and solve for x.

$$\frac{540}{x+7} = \frac{540}{x} - 7$$

$$\frac{540}{x+7} = \frac{540 - 7x}{x}$$

$$540x = (x+7)(540 - 7x)$$

$$540x = 540x - 7x^2 + 3780 - 49x$$

$$7x^2 + 49x - 3780 = 0$$

$$7(x+27)(x-20) = 0$$

$$x + 27 = 0 \quad \text{or} \quad x - 20 = 0$$

$$x = -27 \quad \text{or} \quad x = 20$$

Rappel down to the answer.

The negative solution does not make sense in this situation. So, 20 members of the rappelling club are going on the trip.

Check your solution in the original equation: $\frac{540}{20+7} = \frac{540}{20} - 7$

$$20 = 20$$

Reteaching and Enrichment Strategies

If students need help. . .	If students got it. . .
Resources by Chapter • Practice A and Practice B • Puzzle Time Record and Practice Journal Practice Differentiating the Lesson Lesson Tutorials Skills Review Handbook	Resources by Chapter • Enrichment and Extension • School-to-Work • Financial Literacy • Technology Connection • Life Connections Start the next section

Solve the equation. Check your solution(s).

2 **13.** $\frac{4}{5} - \frac{1}{c} = \frac{3}{c}$

14. $\frac{2}{y+3} - \frac{5}{y} = \frac{12}{y+3}$

15. $\frac{10}{d(d-2)} + \frac{4}{d} = \frac{5}{d-2}$

16. $\frac{n}{n-2} + \frac{2}{5} = \frac{1}{n+4}$

17. $\frac{6}{a+5} + 2 = \frac{28}{a^2 - 25}$

18. $\frac{x}{x+7} + \frac{3}{x-6} = \frac{2x+27}{x^2 + x - 42}$

19. REASONING Explain how you can use the Cross Products Property to solve $\frac{3}{x} + 1 = \frac{8}{x-3}$.

20. PAINT A department store paint mixer contains 4 pints of equal amounts of yellow and red paint. The shade of red that you want requires a paint mixture that is 75% red and 25% yellow. How many pints of red paint need to be added to the paint mixer?

21. RAPPELLING A rappelling club charters a bus for a trip to the mountains for \$540. To lower the bus fare per person, the club invites some hikers on the trip. After 7 hikers join the trip, the bus fare per person decreases by \$7. How many members of the rappelling club are going on the trip?

To solve *work problems*, find the portion of the job each person completes in 1 unit of time. The sum of these portions is the portion of the job completed in 1 unit of time.

22. You can mop a floor in 8 minutes. Your friend can mop the same floor in 12 minutes. Working together, how much time does it take to mop the floor?

23. You can mow a lawn in 3 hours. Your friend can mow the same lawn in 2 hours. Working together, how much time does it take to mow the lawn?

24. **Reasoning** A roofing contractor can shingle a roof in half the time it takes his assistant. Working together, they can shingle the roof in 8 hours. How much time does it take the roofing contractor to finish the job alone?

Fair Game Review What you learned in previous grades & lessons

Solve the equation. Check your solutions. *(Section 1.3)*

25. $|x - 2| = 6$

26. $3|2x - 7| = 9$

27. $2|5x + 4| - 1 = 13$

28. MULTIPLE CHOICE What is the solution of $-2 < -x + 5 \leq 8$? *(Section 3.4)*

Ⓐ $-7 < x \leq 3$ Ⓑ $7 > x \geq -3$ Ⓒ $x \leq -3$ and $x > 7$ Ⓓ $x < 7$ or $x \geq -3$

11.4–11.7 Quiz

Check It Out
Progress Check
BigIdeasMath.com

Find the product or quotient. *(Section 11.4)*

1. $\dfrac{c+2}{5c^3} \cdot \dfrac{4c^4}{6}$

2. $\dfrac{4ab^3 - 2b^3}{2a^3 + 4a^2} \cdot \dfrac{2a^2 + 4a}{4ab - 2b}$

3. $\dfrac{3k}{k+3} \div \dfrac{15}{k+3}$

4. $\dfrac{m^2 - 36}{m^3} \div \dfrac{m^2 + 12m + 36}{m^3}$

Find the quotient. *(Section 11.5)*

5. $(6j^3 + 12j^2 + 18j) \div 6j$

6. $(m^2 - 14m + 49) \div (m - 7)$

7. $(d^2 - 5d + 8) \div (d - 3)$

8. $(5n^2 + 7) \div (n - 1)$

Find the sum or difference. *(Section 11.6)*

9. $\dfrac{5}{v+1} - \dfrac{10}{v+1}$

10. $\dfrac{4r}{2r-3} + \dfrac{5r-1}{3-2r}$

11. $\dfrac{t^2 - 8}{6t} + \dfrac{-t^2 + 7}{4t}$

12. $\dfrac{3p + 10}{p^2 + p - 20} - \dfrac{2}{p-4}$

Solve the equation. Check your solution. *(Section 11.7)*

13. $\dfrac{3}{s-2} = \dfrac{4}{s}$

14. $2 = \dfrac{6}{2w+1}$

15. $-5 + \dfrac{2h}{h+2} = \dfrac{7h}{h+2}$

16. $\dfrac{2}{g} + \dfrac{5}{g(g+1)} = \dfrac{6}{g+1}$

17. PIGPEN You are installing a fence around a pigpen. Write an expression that represents the amount of fencing you need. *(Section 11.6)*

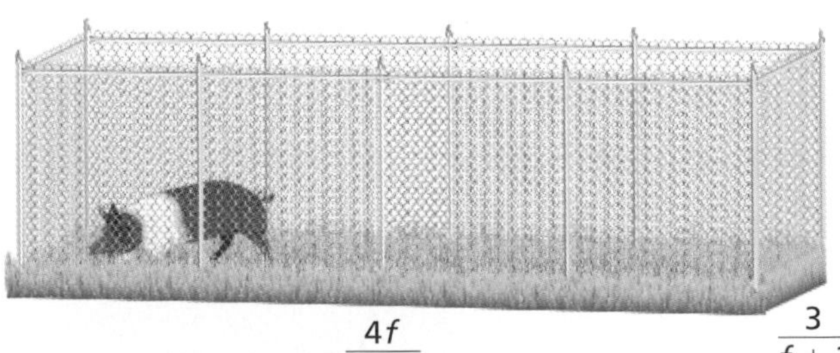

18. RAKING You can rake your front yard in 30 minutes. Your friend can rake the same yard in 50 minutes. Working together, how much time does it take to rake the yard? *(Section 11.7)*

Alternative Assessment Options

Math Chat	Student Reflective Focus Question
Structured Interview	Writing Prompt

Math Chat

- Have students work in pairs. Assign Quiz Exercises 13–16 to each pair. Each student works through all four problems. After the students have worked through the problems, they take turns talking through the processes that they used to get each answer. Students analyze and evaluate the mathematical thinking and strategies used.
- The teacher should walk around the classroom listening to the pairs and ask questions to ensure understanding.

Study Help Sample Answers

Remind students to complete Graphic Organizers for the rest of the chapter.

7.

Multiplying and Dividing Rational Expressions

Examples	Non-Examples
$\frac{2}{x} \cdot \frac{2}{x^3} = \frac{4}{x^4}$	$\frac{2}{x} \cdot \frac{2}{x^3} = 4x^4$ ✗
$\frac{2}{x} \div \frac{x^3}{2} = \frac{2}{x} \cdot \frac{2}{x^3} = \frac{4}{x^4}$	$\frac{\cancel{2}}{x} \div \frac{x^3}{\cancel{2}} = \frac{1}{x} \cdot \frac{1}{x^3} = \frac{1}{x^4}$ ✗
$\frac{x^2-1}{x+2} \cdot \frac{x+2}{x-1} = \frac{(x+1)\cancel{(x-1)}\cancel{(x+2)}}{\cancel{(x+2)}\cancel{(x-1)}}$ $= x + 1$	$\frac{x^2-1}{x+2} \cdot \frac{x+2}{x-1} = \frac{x^3+2x^2-x-2}{x^2+x-2}$ (Correct, but not simplified)
$\frac{x^2-1}{x+2} \div \frac{x-1}{x+2} = \frac{x^2-1}{x+2} \cdot \frac{x+2}{x-1}$ $= \frac{(x+1)\cancel{(x-1)}\cancel{(x+2)}}{\cancel{(x+2)}\cancel{(x-1)}}$ $= x + 1$	$\frac{x^2-1}{x+2} \div \frac{x-1}{x+2} = \frac{x+2}{x^2-1} \cdot \frac{x-1}{x+2}$ ✗ $= \frac{\cancel{(x+2)}\cancel{(x-1)}}{(x+1)\cancel{(x-1)}\cancel{(x+2)}}$ $= \frac{1}{x+1}$

8–10. Available at *BigIdeasMath.com.*

Reteaching and Enrichment Strategies

If students need help. . .	If students got it. . .
Resources by Chapter • Study Help • Practice A and Practice B • Puzzle Time Lesson Tutorials *BigIdeasMath.com* Practice Quiz Practice from the Test Generator	Resources by Chapter • Enrichment and Extension • School-to-Work Game Closet at *BigIdeasMath.com* Start the Chapter Review

Answers

1. $\frac{2c(c+2)}{15}$
2. $\frac{b^2}{a}$
3. $\frac{k}{5}$
4. $\frac{m-6}{m+6}$
5. $j^2 + 2j + 3$
6. $m - 7$
7. $d - 2 + \frac{2}{d-3}$
8. $5n + 5 + \frac{12}{n-1}$
9. $-\frac{5}{v+1}$
10. $\frac{-r+1}{2r-3}$
11. $\frac{-t^2+5}{12t}$
12. $\frac{p}{(p-4)(p+5)}$
13. $s = 8$
14. $w = 1$
15. $h = -1$
16. $g = \frac{7}{4}$
17. $\frac{8f^2 + 22f + 36}{(f+6)(f+2)}$
18. 18.75 minutes

Technology For the Teacher

Answer Presentation Tool

Assessment Book

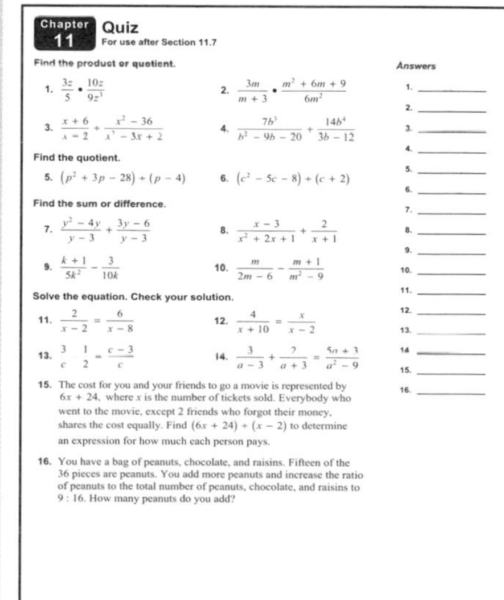

Chapter 11 **Quiz**
For use after Section 11.7

Find the product or quotient.

1. $\frac{3z}{5} \cdot \frac{10z}{9z^3}$
2. $\frac{3m}{m+3} \cdot \frac{m^2+6m+9}{6m^2}$
3. $\frac{x+6}{x-2} \div \frac{x^2-36}{x^2-3x+2}$
4. $\frac{7b^3}{b^2-9b-20} \div \frac{14b^4}{3b-12}$

Find the quotient.

5. $(p^2 + 3p - 28) \div (p - 4)$
6. $(c^2 - 5c - 8) \div (c + 2)$

Find the sum or difference.

7. $\frac{y^2-4y}{y-3} + \frac{3y-6}{y-3}$
8. $\frac{x-3}{x^2+2x+1} + \frac{2}{x+1}$
9. $\frac{k+1}{5k^2} - \frac{3}{10k}$
10. $\frac{m}{2m-6} - \frac{m+1}{m^2-9}$

Solve the equation. Check your solution.

11. $\frac{2}{x-2} = \frac{6}{x-8}$
12. $\frac{4}{x+10} = \frac{x}{x-2}$
13. $\frac{3}{c} - \frac{1}{2} = \frac{c-3}{c}$
14. $\frac{3}{a-3} + \frac{2}{a+3} = \frac{5a+3}{a^2-9}$
15. The cost for you and your friends to go a movie is represented by $6x + 24$, where x is the number of tickets sold. Everybody who went to the movie, except 2 friends who forgot their money, shares the cost equally. Find $(6x + 24) \div (x - 2)$ to determine an expression for how much each person pays.
16. You have a bag of peanuts, chocolate, and raisins. Fifteen of the 36 pieces are peanuts. You add more peanuts and increase the ratio of peanuts to the total number of peanuts, chocolate, and raisins to 9 : 16. How many peanuts do you add?

Answers

1. ______
2. ______
3. ______
4. ______
5. ______
6. ______
7. ______
8. ______
9. ______
10. ______
11. ______
12. ______
13. ______
14. ______
15. ______
16. ______

For the Teacher
Additional Review Options

- Big Ideas Test Generator
- Game Closet at *BigIdeasMath.com*
- Vocabulary Puzzle Builder
- Resources by Chapter
  - Puzzle Time
  - Study Help

Answers

1. $y = 2x$

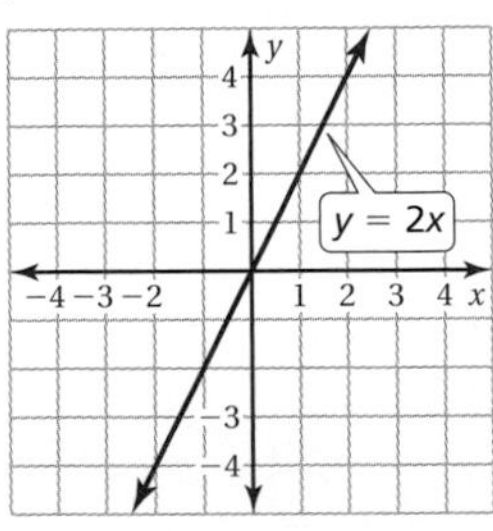

2. $y = \frac{24}{x}$

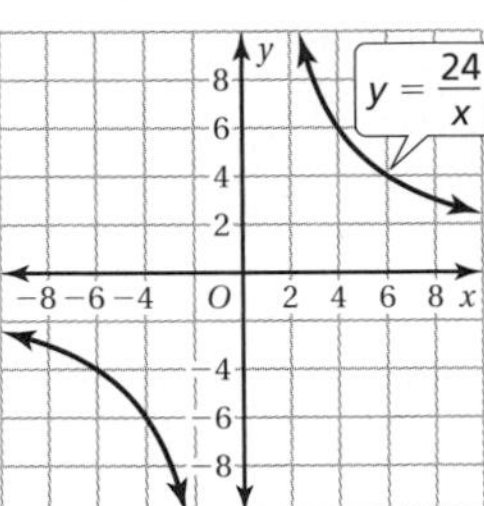

Review of Common Errors

- **Exercises 1–2** Students may substitute the wrong values for *x* and *y*. Tell them to be careful when substituting and that *k* should still be in the equation after substituting for *x* and *y*.

11 Chapter Review

Review Key Vocabulary

direct variation, *p. 544*
inverse variation, *p. 544*
rational function, *p. 552*
excluded value, *p. 552*
asymptote, *p. 553*
inverse relation, *p. 558*
inverse function, *p. 559*
rational expression, *p. 562*
simplest form of a rational expression, *p. 562*
least common denominator of rational expressions, *p. 583*
rational equation, *p. 590*

Review Examples and Exercises

11.1 Direct and Inverse Variation *(pp. 542–549)*

The variable y varies inversely with x. When $x = 3, y = 2$.

a. Write an inverse variation equation that relates x and y.

Find the value of k.

$y = \frac{k}{x}$ Write the inverse variation equation.

$2 = \frac{k}{3}$ Substitute 3 for x and 2 for y.

$6 = k$ Multiply each side by 3.

So, an equation that relates x and y is $y = \frac{6}{x}$.

b. Graph the inverse variation equation. Describe the domain and range.

Make a table of values.

x	−3	−2	−1	0	1	2	3
y	−2	−3	−6	undef.	6	3	2

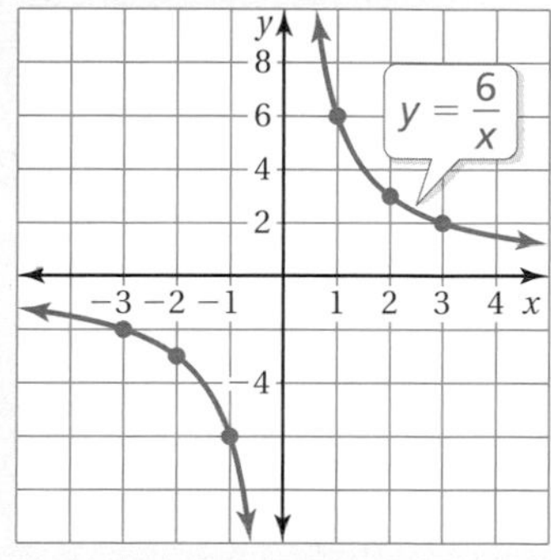

Plot the ordered pairs. Draw a smooth curve through the points in each quadrant. Both the domain and range are all real numbers except 0.

Exercises

1. The variable y varies directly with x. When $x = 6, y = 12$. Write and graph a direct variation equation that relates x and y.
2. The variable y varies inversely with x. When $x = 3, y = 8$. Write and graph an inverse variation equation that relates x and y.

11.2 Graphing Rational Functions (pp. 550–559)

Graph $y = \frac{1}{x-2} - 1$. Compare the graph to the graph of $y = \frac{1}{x}$.

Step 1: Make a table of values. The vertical asymptote is $x = 2$, so choose x-values on either side of 2.

x	0	1	1.5	2	2.5	3	4
y	−1.5	−2	−3	undef.	1	0	−0.5

Step 2: Use dashed lines to graph the asymptotes $x = 2$ and $y = -1$. Then plot the ordered pairs.

Step 3: Draw a smooth curve through the points on each side of the vertical asymptote.

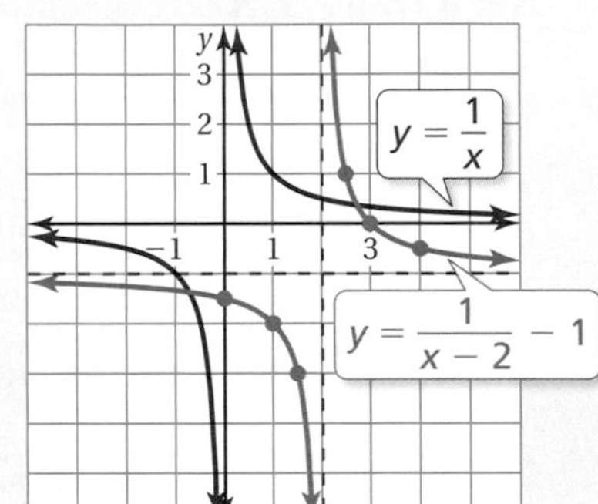

The graph of $y = \frac{1}{x-2} - 1$ is a translation 1 unit down and 2 units right of the graph of $y = \frac{1}{x}$.

Find the inverse of $f(x) = \frac{2}{x}$. Graph the inverse function.

$y = \frac{2}{x}$ Replace $f(x)$ with y.

$x = \frac{2}{y}$ Switch x and y.

$xy = 2$ Multiply each side by y.

$y = \frac{2}{x}$ Divide each side by x.

$f^{-1}(x) = \frac{2}{x}$ Replace y with $f^{-1}(x)$.

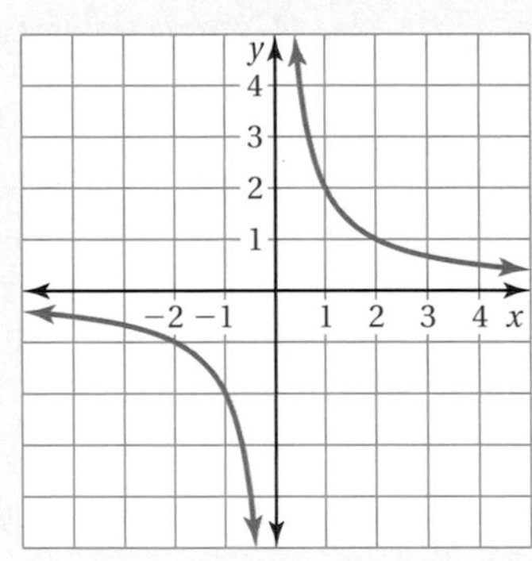

Exercises

Graph the function. Compare the graph to the graph of $y = \frac{1}{x}$.

3. $y = \frac{1}{x+5}$

4. $y = \frac{1}{x} - 4$

5. $y = \frac{1}{x-7} + 1$

Find the inverse of the function. Graph the inverse function.

6. $f(x) = x + 2$

7. $f(x) = \frac{1}{2}x - 5$

8. $f(x) = \frac{1}{x} + 7$

Review of Common Errors (continued)

- **Exercises 3–5** Students may not identify the horizontal and vertical asymptotes before trying to graph the function. Encourage them to use the asymptotes to help graph the function.
- **Exercises 6–8** Students may stop after switching x and y. Remind them to solve for y.

Answers

3. 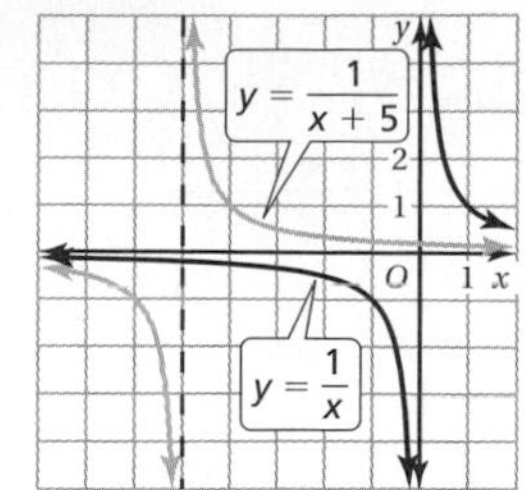

The graph of $y = \frac{1}{x+5}$ is a translation 5 units left of the graph of $y = \frac{1}{x}$.

4. 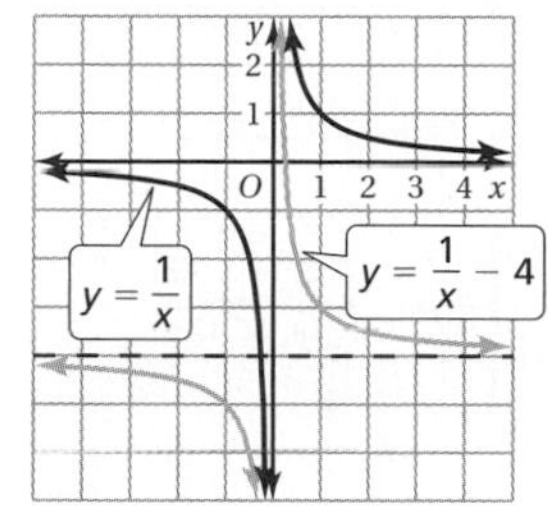

The graph of $y = \frac{1}{x} - 4$ is a translation 4 units down of the graph of $y = \frac{1}{x}$.

5. 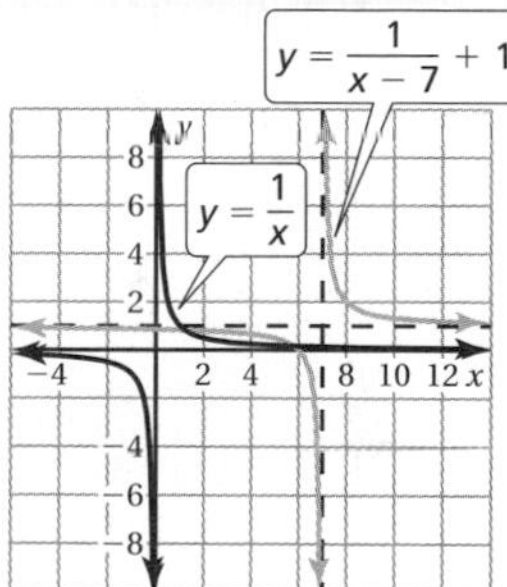

The graph of $y = \frac{1}{x-7} + 1$ is a translation 7 units right and 1 unit up of the graph of $y = \frac{1}{x}$.

6. $f^{-1}(x) = x - 2$

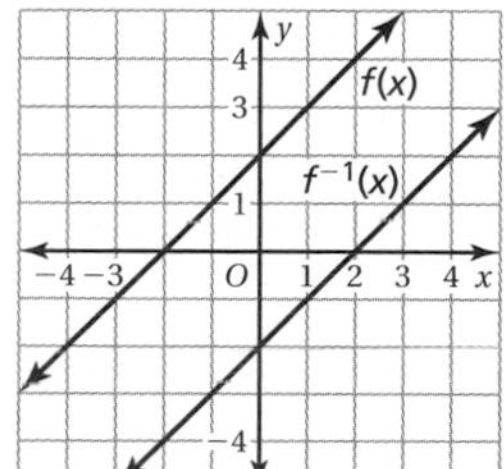

Answers

7. $f^{-1}(x) = 2x + 10$

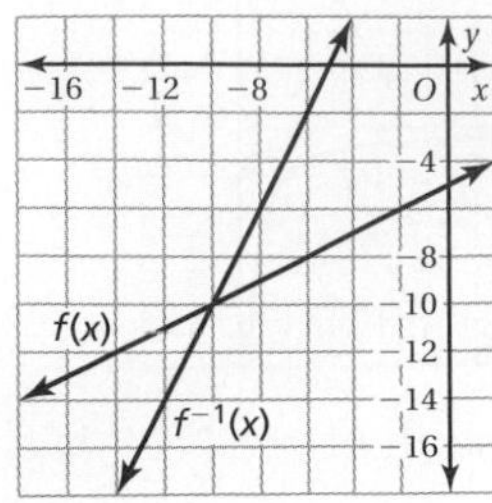

8. $f^{-1}(x) = \dfrac{1}{x - 7}$

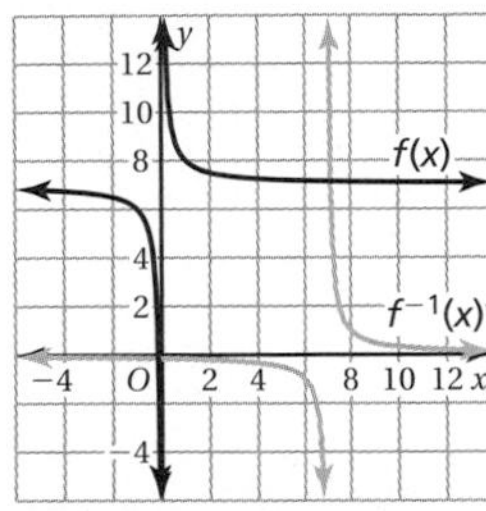

9. $\dfrac{9}{2z^2}$; $z = 0$

10. The expression is in simplest form; $n = 1$

11. $\dfrac{b + 3}{b - 5}$; $b = -6, b = 5$

Review of Common Errors (continued)

- **Exercises 9–11** Students may not state the correct excluded value(s). Remind them to use the original expression to find the excluded value(s).
- **Exercises 12–14** Students may not factor completely before simplifying the rational expression. Remind students to factor completely in order to divide out common factors so that the rational expression is in simplest form.
- **Exercises 13–14** Students may not rewrite the division as multiplication. Remind students that when they divide rational expressions, they need to multiply the dividend by the reciprocal of the divisor.
- **Exercise 16** Students may forget to insert missing terms before dividing the polynomials.
- **Exercises 19–21** Students may add the denominators as well as the numerators. Remind them that only the numerators are added when adding rational expressions with a common denominator.
- **Exercises 22–23** Students may multiply the numerators and multiply the denominators of the rational expressions instead of finding the cross products. Remind them of the Cross Products Property.
- **Exercise 24** Students may not multiply each side of the equation by the LCD. Remind them of the Multiplication Property of Equality.

11.3 Simplifying Rational Expressions (pp. 560–565)

Simplify $\frac{v^2 - 9}{v^2 - 3v}$, if possible. State the excluded value(s).

$$\frac{v^2 - 9}{v^2 - 3v} = \frac{(v - 3)(v + 3)}{v(v - 3)}$$ Factor.

$$= \frac{\cancel{(v - 3)}(v + 3)}{v\cancel{(v - 3)}}$$ Divide out the common factor.

$$= \frac{v + 3}{v}$$ Simplify.

The excluded values are $v = 0$ and $v = 3$.

Exercises

Simplify the rational expression, if possible. State the excluded value(s).

9. $\frac{18z^2}{4z^4}$

10. $\frac{n^2 + 1}{n - 1}$

11. $\frac{b^2 + 9b + 18}{b^2 + b - 30}$

11.4 Multiplying and Dividing Rational Expressions (pp. 568–573)

Find the product or quotient.

a. $$\frac{7x}{x + 4} \bullet \frac{x + 4}{x^2} = \frac{7x(x + 4)}{x^2(x + 4)}$$ Multiply numerators and denominators.

$$= \frac{7\cancel{x}\cancel{(x + 4)}}{x^{\cancel{2}}\cancel{(x + 4)}}$$ Divide out the common factors.

$$= \frac{7}{x}$$ Simplify.

b. $$\frac{t - 6}{10} \div \frac{6 - t}{12} = \frac{t - 6}{10} \bullet \frac{12}{6 - t}$$ Multiply by the reciprocal.

$$= \frac{t - 6}{10} \bullet \frac{12}{-(t - 6)}$$ Rewrite $6 - t$ as $-(t - 6)$.

$$= \frac{12(t - 6)}{-10(t - 6)}$$ Multiply numerators and denominators.

$$= \frac{{}^{6}\cancel{12}\cancel{(t - 6)}}{-\cancel{10}_{5}\cancel{(t - 6)}}$$ Divide out the common factors.

$$= -\frac{6}{5}$$ Simplify.

Exercises

Find the product or quotient.

12. $\frac{9}{10r} \cdot \frac{5r^3}{6}$

13. $\frac{k+5}{6k^2} \div \frac{5+k}{12k}$

14. $\frac{h^2+8h}{h} \div (h^2+7h-8)$

11.5 Dividing Polynomials *(pp. 574–579)*

Find $(-2x^2 + 8x + 1) \div 2x$.

$$(-2x^2+8x+1) \div 2x = \frac{-2x^2+8x+1}{2x} \qquad \text{Write as a fraction.}$$

$$= \frac{-2x^2}{2x} + \frac{8x}{2x} + \frac{1}{2x} \qquad \text{Divide each term by } 2x.$$

$$= \frac{-\cancel{2}x^{\cancel{2}}}{\cancel{2}\cancel{x}} + \frac{\overset{4}{\cancel{8}}\cancel{x}}{\cancel{2}\cancel{x}} + \frac{1}{2x} \qquad \text{Divide out the common factors.}$$

$$= -x + 4 + \frac{1}{2x} \qquad \text{Simplify.}$$

Find $(z^2 - 2z - 5) \div (z + 3)$.

Step 1: Divide the first term of the dividend by the first term of the divisor.

Align like terms in the quotient and dividend.

$$\begin{array}{r} z \\ z+3\overline{)\,z^2 - 2z - 5} \\ \underline{z^2 + 3z} \\ -5z - 5 \end{array}$$

Divide: $z^2 \div z = z$.

Multiply: $z(z+3)$.

Subtract. Bring down the -5.

Step 2: Divide the first term of $-5z - 5$ by the first term of the divisor.

$$\begin{array}{r} z - 5 \\ z+3\overline{)\,z^2 - 2z - 5} \\ \underline{z^2 + 3z} \\ -5z - 5 \\ \underline{-5z - 15} \\ 10 \end{array}$$

Divide: $-5z \div z = -5$.

Multiply: $-5(z+3)$.

Subtract.

So, $(z^2 - 2z - 5) \div (z + 3) = z - 5 + \frac{10}{z+3}$.

Exercises

Find the quotient.

15. $(8n^3 + 3n) \div 2n^2$

16. $(b^2 - 36) \div (b - 6)$

17. $(x^2 + 6x + 3) \div (x + 2)$

18. $(4c - 1) \div (c + 5)$

Review Game

Review for You

Materials per Group

- copy of the Chapter Review from the Pupil's Edition
- chalk or dry erase marker
- eraser

Directions

Divide the class into four teams. Team members gather in groups at the board.

One member of each team works the first problem from the Chapter Review. Coaching from the other team members is allowed.

Check each student's work and award 2 points to the team if the work is correct. If the work needs to be corrected, award 1 point to the team after the student makes the corrections.

Team members take turns working the problems. Repeat the process until you finish the Chapter Review or until you run out of time.

Who wins?

The team with the most points wins.

Variations

- Have students write review problems on index cards. Draw a card for students to use.
- To provide a cumulative review, write problems on index cards from anywhere in the book. Have team members select a card at random.

For the Student Additional Practice

- Lesson Tutorials
- Study Help (textbook)
- Student Website
  Multi-Language Glossary
  Practice Assessments

Answers

12. $\frac{3r^2}{4}$

13. $\frac{2}{k}$

14. $\frac{1}{h-1}$

15. $4n + \frac{3}{2n}$

16. $b + 6$

17. $x + 4 - \frac{5}{x+2}$

18. $4 - \frac{21}{c+5}$

19. $\frac{2h-9}{h-10}$

20. $\frac{13x+5}{x(x+1)}$

21. $\frac{12}{(x+8)(x-7)}$

22. no solution

23. $y = -3, y = 6$

24. $t = -\frac{4}{3}$

25. 10 first serves

My Thoughts on the Chapter

Teacher Tip

Not allowed to write in your teaching edition? Use sticky notes to record your thoughts.

What worked. . .

What did not work. . .

What I would do differently. . .

11.6 Adding and Subtracting Rational Expressions *(pp. 580–587)*

Find the difference $\frac{y+3}{7y^2} - \frac{4}{5y}$.

$$\frac{y+3}{7y^2} - \frac{4}{5y} = \frac{(y+3)(5)}{7y^2(5)} - \frac{4(7y)}{5y(7y)} \quad \text{Rewrite using the LCD, } 35y^2.$$

$$= \frac{5y+15}{35y^2} - \frac{28y}{35y^2} \quad \text{Simplify.}$$

$$= \frac{-23y+15}{35y^2} \quad \text{Subtract the numerators.}$$

Exercises

Find the sum or difference.

19. $\frac{5h-2}{h-10} - \frac{3h+7}{h-10}$ **20.** $\frac{5}{x} + \frac{8}{x+1}$ **21.** $\frac{4-x}{x^2+x-56} + \frac{1}{x-7}$

11.7 Solving Rational Equations *(pp. 588–593)*

You own a farm in a computer game. Twenty of the 120 animals on your farm are cows. You buy cows and increase the ratio of cows to total animals to 1 : 5. How many cows do you buy?

Write an equation for the ratio of cows to total animals after buying x cows.

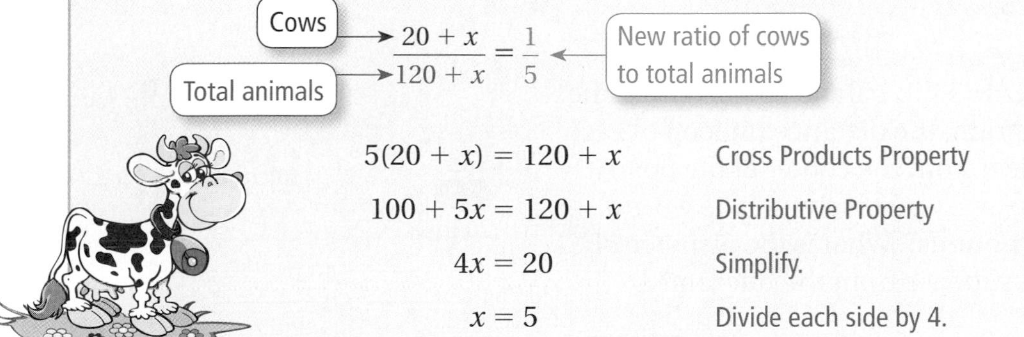

$$\frac{20+x}{120+x} = \frac{1}{5}$$

$5(20+x) = 120 + x$ Cross Products Property

$100 + 5x = 120 + x$ Distributive Property

$4x = 20$ Simplify.

$x = 5$ Divide each side by 4.

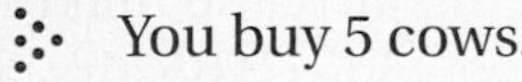

∴ You buy 5 cows.

Exercises

Solve the equation. Check your solution(s).

22. $\frac{1}{x+2} = \frac{12}{3x+6}$ **23.** $\frac{9}{y+6} = \frac{y}{y+2}$ **24.** $\frac{5}{t} - \frac{3}{4} = \frac{6}{t}$

25. TENNIS A tennis player lands 25 out of 40 first serves in bounds for a success rate of 62.5%. How many more consecutive first serves must she land in bounds to increase her success rate to 70%?

11 Chapter Test

1. The variable y varies directly with x. When $x = 3$, $y = 18$. Write and graph a direct variation equation that relates x and y.

2. The variable y varies inversely with x. When $x = 6$, $y = 4$. Write and graph an inverse variation equation that relates x and y.

Graph the function. Compare the graph to the graph of $y = \frac{1}{x}$.

3. $y = \frac{1}{x - 6}$

4. $y = \frac{1}{x} + 3$

5. $y = \frac{1}{x + 4} - 5$

Find the inverse of the function. Graph the inverse function.

6. $f(x) = x - 7$

7. $f(x) = \frac{1}{5}x - 7$

8. $f(x) = \frac{3}{x + 4}$

Simplify.

9. $\frac{y^2 + 5y - 24}{y^2 + 10y + 16}$

10. $\frac{8r^4}{5} \cdot \frac{15r}{6r^3}$

11. $\frac{x^2 - 25}{x + 5} \div (x^2 - 3x - 10)$

12. $\frac{6k + 1}{2k - 4} + \frac{2k + 3}{2k - 4}$

13. $\frac{4}{p + 6} - \frac{3}{p}$

14. $\frac{18z + 27}{z^2 + 3z - 54} + \frac{z}{z + 9}$

15. Find $(12d^3 + 8d - 6) \div 3d^2$.

16. Find $(b^2 - 4b + 10) \div (b + 3)$.

Solve the equation. Check your solution(s).

17. $\frac{1}{x - 5} = \frac{3}{2x + 7}$

18. $\frac{a}{a + 3} = \frac{4}{a + 5}$

19. $\frac{6}{n} - \frac{2}{n - 3} = \frac{5}{n - 3}$

20. **BALANCE** To balance the board in the diagram, the distance (in feet) of each object from the center of the board must vary inversely with its weight (in pounds). What is the distance of the suitcase from the fulcrum?

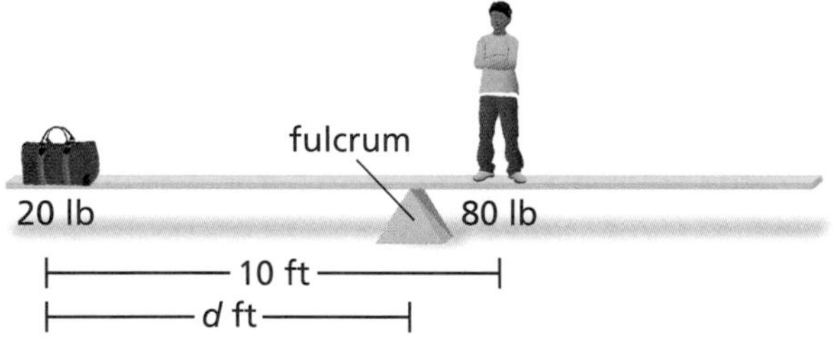

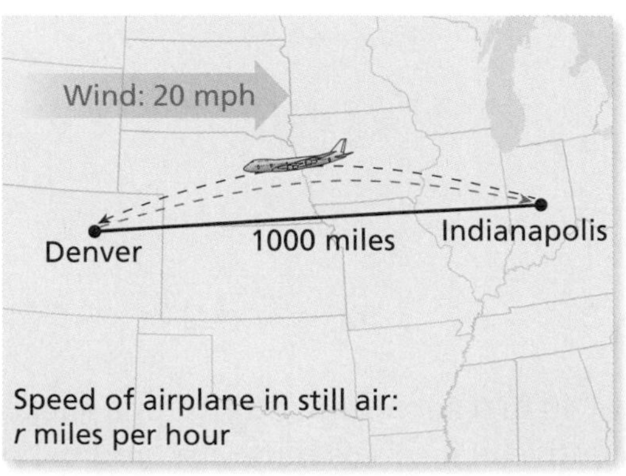

21. **AIRPLANE** An airplane makes a round trip between two cities. The airplane flies with the wind when heading east and against the wind when heading west. Write an expression for the total time of the trip.

22. **DELIVERY TRUCK** Working alone, it takes you 30 minutes, your friend 30 minutes, and your supervisor 15 minutes to unload a delivery truck. Working together, how much time does it take all three of you to unload the truck?

Test Item References

Chapter Test Questions	Section to Review	Common Core State Standards
1, 2, 21	11.1	A.REI.10
3–8	11.2	A.REI.10, F.BF.4a
9, 10	11.3	A.SSE.2
11, 12	11.4	A.SSE.2
13, 14	11.5	A.SSE.2
15–17, 22	11.6	A.SSE.2
18–20, 23	11.7	A.CED.1

Test-Taking Strategies

Remind students to quickly look over the entire test before they start so that they can budget their time. Have students use the **Stop** and **Think** strategy before they answer each question.

Common Assessment Errors

- **Exercises 1 and 2** Students may substitute the wrong values for x and y. Tell them to be careful when substituting and that k should still be in the equation after substituting for x and y.
- **Exercises 3–5** Students may not identify the horizontal and vertical asymptotes before trying to graph the function. Encourage them to use the asymptotes to help graph the function.
- **Exercises 11–12** Students may not factor completely before simplifying the rational expression. Remind students to factor completely in order to divide out common factors so that the rational expression is in simplest form.
- **Exercises 18–19** Students may multiply the numerators and multiply the denominators of the rational expressions instead of finding the cross products. Remind them of the Cross Products Property.

Reteaching and Enrichment Strategies

If students need help...	If students got it...
Resources by Chapter • Practice A and Practice B • Puzzle Time Record and Practice Journal Practice Differentiating the Lesson Lesson Tutorials Practice from the Test Generator Skills Review Handbook	Resources by Chapter • Enrichment and Extension • School-to-Work • Financial Literacy • Technology Connection • Life Connections Game Closet at *BigIdeasMath.com* Start Standardized Test Practice

Answers

1–7. See Additional Answers.

8. $f^{-1}(x) = \frac{3}{x} - 4$

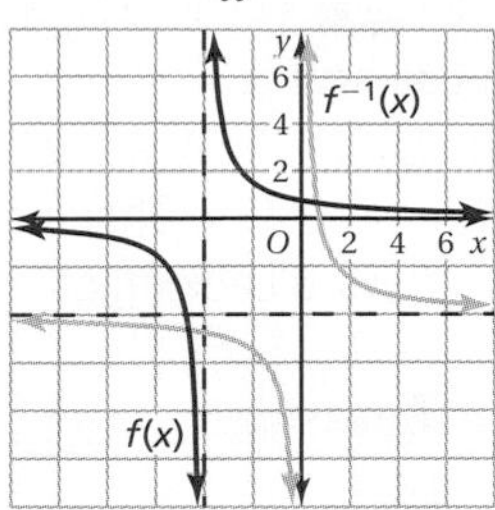

9. $\frac{y-3}{y+2}$

10. $4r^2$

11. $\frac{1}{x+2}$

12. $\frac{4k+2}{k-2}$

13. $\frac{p-18}{p(p+6)}$

14. $\frac{z+3}{z-6}$

15. $4d + \frac{8}{3d} - \frac{2}{d^2}$

16. $b - 7 + \frac{31}{b+3}$

17. $x = 22$

18. $a = -4, a = 3$

19. $n = -18$

20. 8 ft

21. $\frac{2000r}{(r+20)(r-20)}$

22. 7.5 minutes

Assessment Book

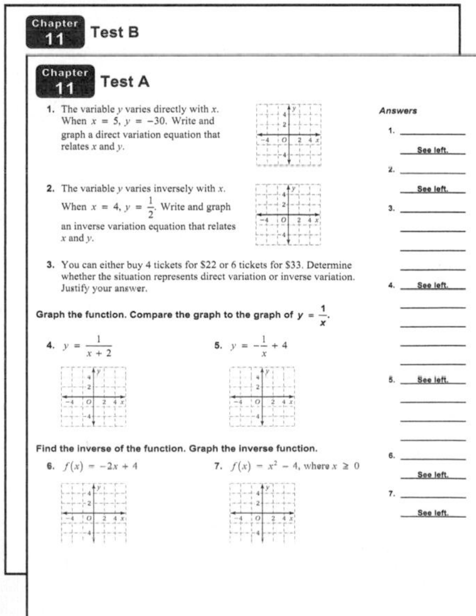

Test Taking Strategies
Available at *BigIdeasMath.com*

After Answering Easy Questions, Relax
Answer Easy Questions First
Estimate the Answer
Read All Choices before Answering
Read Question before Answering
Solve Directly or Eliminate Choices
Solve Problem before Looking at Choices
Use Intelligent Guessing
Work Backwards

About this Strategy

When taking a multiple choice test, be sure to read each question carefully and thoroughly. One way to answer the question is to work backwards. Try putting the responses into the question, one at a time and see if you get a correct solution.

Answers

1. C
2. H
3. 2
4. B

Technology For the Teacher

Big Ideas Test Generator

Item Analysis

1. **A.** The student incorrectly subtracts 3 instead of adding 3 to represent a translation 3 units up.

   B. The student switches the roles of h and k as well as their signs in writing an equation of the form $y = \frac{1}{x - h} + k$.

   C. Correct answer

   D. The student incorrectly subtracts 3 instead of adding 3 to represent a translation 3 units up, and subtracts 2 instead of adding 2 to represent a translation 2 units left.

2. **F.** The student incorrectly uses the square root of the sum of the legs.

   G. The student incorrectly uses the square root of the square of the longer leg.

   H. Correct answer

   I. The student estimates incorrectly.

3. **Gridded response:** Correct answer: 2

   Common error: The student incorrectly calculates the value of $\frac{1}{1/2}$ as $\frac{1}{2}$.

4. **A.** The student multiplies the expression by $\frac{-1}{-1}$ and incorrectly places a negative sign with the result.

   B. Correct answer

   C. The student does not recognize that because the numerator and denominator are opposites, they have a common factor.

   D. The student fails to realize that because the numerator and denominator are opposites, the result of dividing out the common factor is -1.

11 Standardized Test Practice

1. Which function is shown by the graph? *(A.REI.10)*

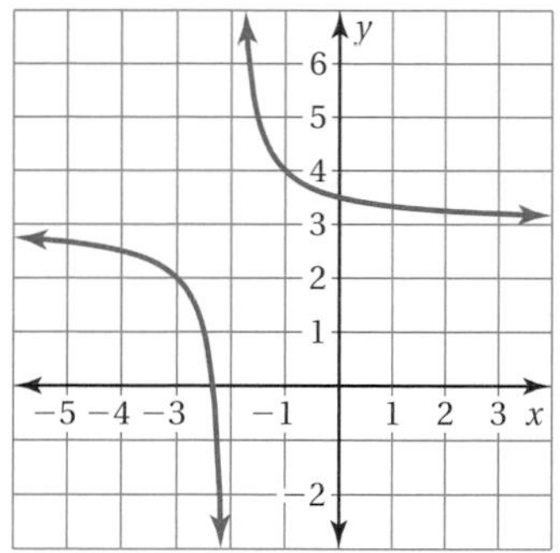

A. $y = \frac{1}{x+2} - 3$

B. $y = \frac{1}{x+3} + 2$

C. $y = \frac{1}{x+2} + 3$

D. $y = \frac{1}{x-2} - 3$

2. What is the value of c in the triangle shown? *(8.G.7)*

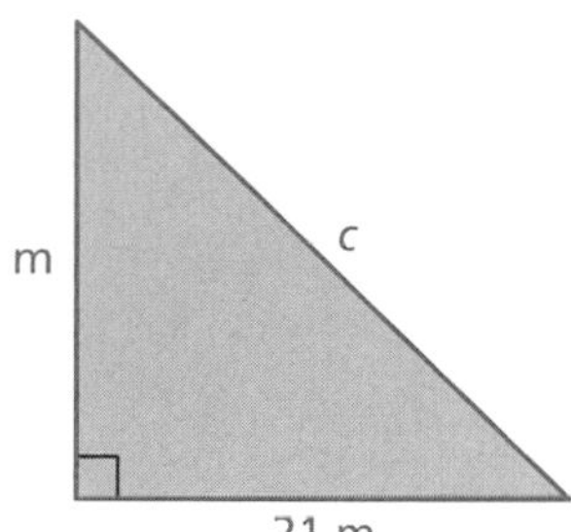

F. $\sqrt{41}$ m

G. 21 m

H. 29 m

I. 30 m

3. What is the y-coordinate of the focus of the graph of $y = \frac{1}{8}x^2$? *(F.IF.4)*

4. What is the simplest form of the rational expression $\frac{4x-3}{3-4x}$? *(A.SSE.2)*

A. $-\frac{3-4x}{4x-3}$

B. -1

C. $\frac{4x-3}{3-4x}$

D. 1

5. What is the solution of the system of equations? *(A.REI.7)*

$$y = x^2 + 2x - 7$$
$$y = 2x - 7$$

F. $(-3, -4)$

G. $(0, -7)$

H. $(-7, 0)$

I. no real solutions

6. What are the solutions of the equation? *(A.SSE.3a)*

$$x^4 - 2x^3 - 3x^2 = 0$$

A. $x = 0$

B. $x = 1, x = -3$

C. $x = 0, x = -1, x = 3$

D. $x = 0, x = 3$

7. What is the difference of the rational expressions? *(A.SSE.2)*

$$\frac{5}{3x} - \frac{3}{4x}$$

F. $-\frac{2}{x}$

G. $\frac{2}{3x}$

H. $\frac{2}{x}$

I. $\frac{11}{12x}$

8. What is the slope of the line shown in the graph? *(F.IF.6)*

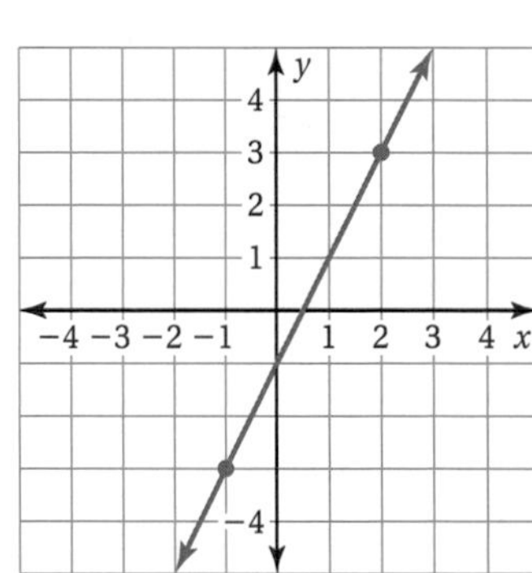

9. What is an equation for the nth term of the geometric sequence? *(F.BF.2)*

n	1	2	3	4
a_n	4	8	16	32

A. $a_n = 4(2)^{n-1}$

B. $a_n = 4^{n-1}$

C. $a_n = 2(4)^{n-1}$

D. $a_n = 2^{n-1}$

Item Analysis (continued)

5. **F.** The student uses a random solution point of the first equation.
   G. Correct answer
   H. The student reverses the coordinates of the solution.
   I. The student graphs the equations and they do not intersect due to graphing error.

6. **A.** The student notices that the equation is true when 0 is substituted for x and fails to consider other solutions.
   B. The student starts by factoring out x^2, then forgets about the x^2 after incorrectly factoring the remaining quadratic expression.
   C. Correct answer
   D. The student graphs the related function for positive values of x and does not realize that $x = -1$ is also a solution.

7. **F.** The student subtracts both the numerators and the denominators.
   G. The student makes an error in rewriting the second term using the LCD.
   H. The student subtracts both the numerators and the denominators and omits the negative sign.
   I. Correct answer

8. **Gridded response:** Correct answer: 2

   Common error: The student calculates the rise between the two points, then uses the reverse order of the two points to calculate the run, yielding a slope of -2.

9. **A.** Correct answer
   B. The student uses the form $a_n = a_1^{n-1}$ instead of $a_n = a_1r^{n-1}$.
   C. The student uses the form $a_n = ra_1^{n-1}$ instead of $a_n = a_1r^{n-1}$.
   D. The student uses the form $a_n = r_1^{n-1}$ instead of $a_n = a_1r^{n-1}$.

Answers

5. G
6. C
7. I
8. 2
9. A

Answers

10. *Part A:* Independent variable: t; dependent variable: c

Part B:

Input, t	Output, c
0	0
1	100
2	200
3	300
4	400

Part C:

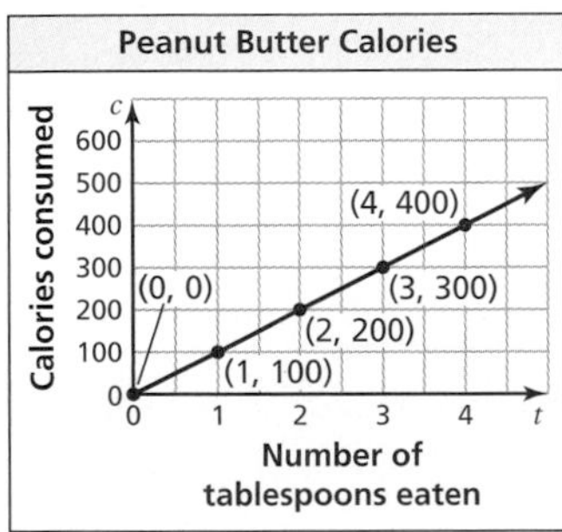

Part D: continuous; Because you can eat any part of a tablespoon of peanut butter, t can be any value greater than or equal to 0.

11. H

12. B

13. G

Answer for Extra Example

1. A. Correct answer
   B. The student makes a sign error.
   C. The student takes the reciprocal of the rational expression.
   D. The student takes the reciprocal of the rational expression and uses addition, the inverse operation of subtraction.

Item Analysis (continued)

10. **4 points** The student demonstrates a thorough understanding of discrete and continuous functions, making a correct table, graph, description of the variables, and explanation of why the domain is continuous.

    3 points The student demonstrates an essential but less than thorough understanding of discrete and continuous functions, with some part of making a table, graph, description of the variables, or the explanation incorrect or incomplete.

    2 points The student demonstrates a partial understanding of discrete and continuous functions, making a few mistakes. The student may call the function discrete, thinking that only whole number values of t are possible.

    1 point The student demonstrates limited understanding of discrete and continuous functions. The student's work is incomplete or exhibits many flaws.

    0 points The student provides no response, a completely incorrect or incomprehensible response, or a response that demonstrates insufficient understanding.

11. **F.** The student forgets to change the direction of the inequality sign when multiplying both sides by -3.
    G. The student forgets to change the direction of the inequality sign and incorrectly multiplies two negative numbers.
    H. Correct answer
    I. The student incorrectly multiplies two negative numbers.

12. **A.** The student misunderstands the process of dividing rational expressions.
    B. Correct answer
    C. The student fails to recognize that a common factor of x can be divided out of the numerator and denominator.
    D. The student misunderstands the process of dividing rational expressions.

13. **F.** The student reverses the coordinates.
    G. Correct answer
    H. The student reverses the coordinates and omits a negative sign.
    I. The student omits a negative sign.

Extra Example

1. What is the inverse of the function $f(x) = \frac{1}{x-4}$? *(F.BF.4a)*

   A. $f^{-1}(x) = \frac{1}{x} + 4$ **C.** $f^{-1}(x) = x - 4$

   B. $f^{-1}(x) = \frac{1}{x} - 4$ **D.** $f^{-1}(x) = x + 4$

10. One tablespoon of peanut butter contains 100 calories. The number c of calories consumed is a function of the number t of tablespoons of peanut butter eaten. *(F.IF.1)*

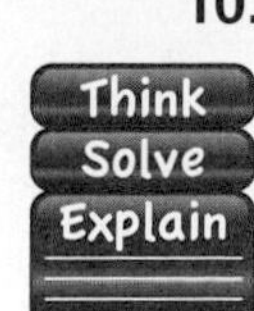

Part A Identify the independent and dependent variables.

Part B Make an input-output table.

Part C Graph the function.

Part D Is the domain discrete or continuous? Explain.

11. What is the solution of the inequality shown below? *(A.REI.3)*

$$\frac{y}{-3} - 4 > -12$$

F. $y > 24$

G. $y > -24$

H. $y < 24$

I. $y < -24$

12. John was finding the quotient of the rational expressions in the box below. *(A.SSE.2)*

$$\frac{3x}{x-4} \div \frac{2x}{4-x} = \frac{3x}{x-4} \bullet \frac{4-x}{2x}$$
$$= \frac{3x(\cancel{4-x})}{2x(\cancel{x-4})}$$
$$= \frac{3x}{2x}$$
$$= \frac{3}{2}$$

What should John do to correct the error that he made?

A. Do not multiply by the reciprocal.

B. Factor out -1 from $(4 - x)$ before dividing out common factors.

C. Divide $3x$ by $2x$ to get an answer of x.

D. Use long division to find the quotient.

13. What is the solution of the system of linear equations shown below? *(A.REI.6)*

$$y = 2x - 1$$
$$y = 3x + 5$$

F. $(-13, -6)$

G. $(-6, -13)$

H. $(-13, 6)$

I. $(-6, 13)$

12 Data Analysis and Displays

"Wow. The number of minutes I can dog paddle is growing like crazy!"

"Please hold still. I am trying to find the mean of 6, 8, and 10 by dividing their sum into three equal piles."

Connections to Previous Learning

- Determine measures of central tendency including mean, median, mode, and range.
- Select appropriate measures of central tendency to describe a data set.

- Evaluate the reasonableness of a sample to determine the appropriateness of generalizations made about the population.
- Construct and analyze histograms, stem-and-leaf plots, and circle graphs.

- Construct and analyze dot plots, histograms, box-and-whisker plots, and scatter plots.
- Use the shape of a data distribution to select appropriate measures of central tendency and dispersion, and to account for the effects of outliers in the data.
- Interpret linear models.

Math in History

Scientists such as Gregor Mendel, Charles Darwin, and Francis Galton made great strides in the fields of genetics and heredity during the 1800s.

★ English scientist Francis Galton (1822–1911) was a cousin of Charles Darwin. Galton's work in heredity challenged him to find a way to quantify the typical variation of a population. He is credited with discovering the concept of the standard deviation in the late 1860s.

★ The man credited with naming Galton's discovery was his protégé, English mathematician Karl Pearson (1857–1936). Pearson coined the term "standard deviation" during a lecture in 1893, and in a work published in 1894.

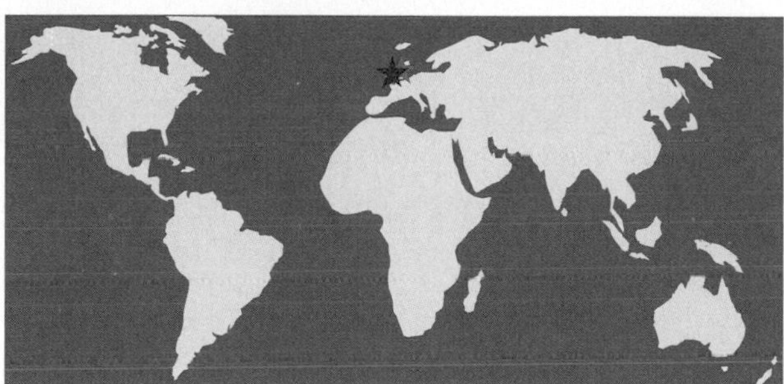

Pacing Guide for Chapter 12

Chapter Opener	1 Day
Section 1	1 Day
Section 2	1 Day
Section 3	1 Day
Section 4	1 Day
Study Help / Quiz	1 Day
Section 5	1 Day
Section 6	2 Days
Section 7	1 Day
Section 8	1 Day
Chapter Review / Chapter Tests	2 Days
Total Chapter 12	13 Days
Year-to-Date	163 Days

Check Your Resources

- Record and Practice Journal
- Resources by Chapter
- Skills Review Handbook
- Assessment Book
- Worked-Out Solutions

Technology For the Teacher

The Dynamic Planning Tool
Editable Teacher's Resources at
BigIdeasMath.com

Common Core State Standards

6.SP.5a Summarize numerical data sets in relation to their context, such as by reporting the number of observations.
6.SP.4 Display numerical data in plots on a number line, including dot plots, histograms, and box plots.

Additional Topics for Review

- Operations with real numbers
- Outliers
- Writing equations of lines
- Bar graphs
- Line graphs
- Stem-and-leaf plots

Try It Yourself

1. Answers will vary.
2. Answers will vary.

Record and Practice Journal

1. 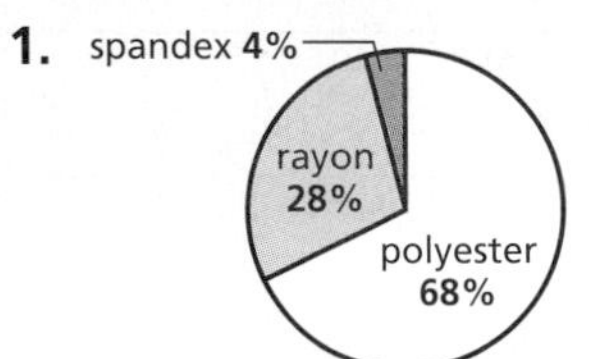

2. 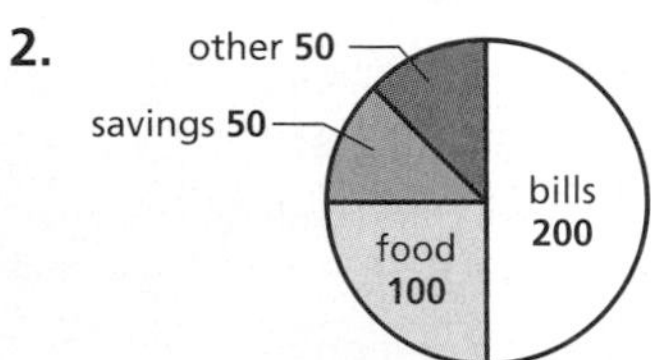

3. 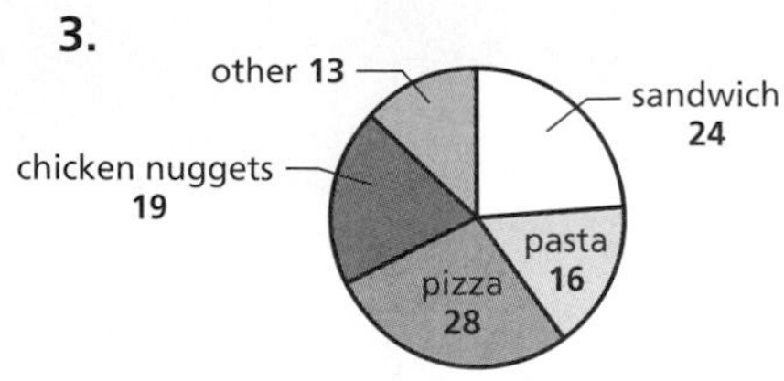

4. 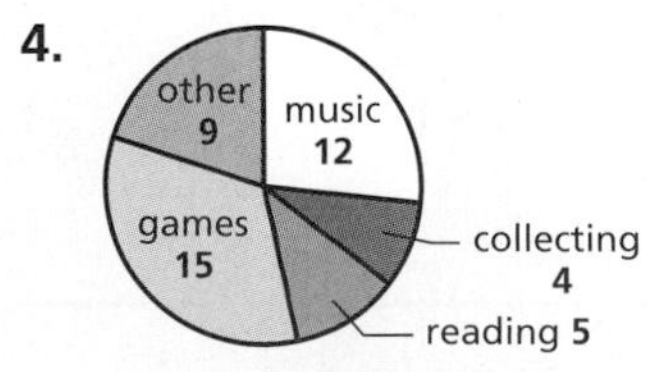

5. 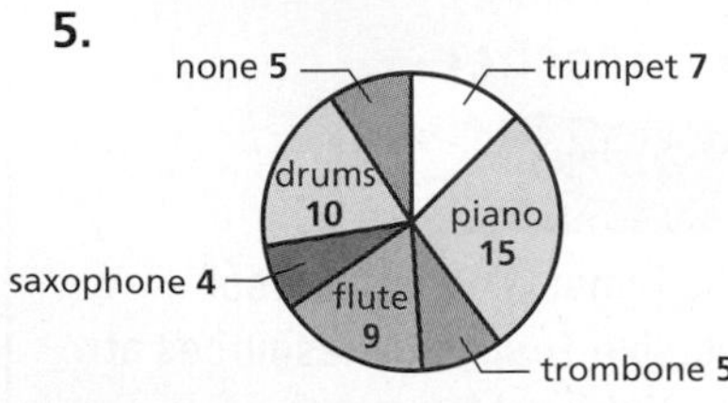

6–10. See Additional Answers.

Math Background Notes

Vocabulary Review

- Circle Graph
- Frequency Table
- Histogram

Displaying Data

- Students have collected, analyzed, and displayed data.
- **Teaching Tip:** Example 1 provides a great opportunity for review. Remind students that a circle contains 360°. This is also a good time to review using a protractor.
- Remind students that it is helpful to know the total number of people surveyed before constructing the circle graph. This number will serve as the whole (denominator of the fraction).
- **Teaching Tip:** Example 2 provides an excellent opportunity to explore students' prerequisite knowledge. Students should be familiar with bar graphs, double bar graphs, and histograms. Consider using a Venn diagram to compare and contrast the three types of displays. This will create a nice visual representation to show which characteristics go with which display.
- **Common Error:** Students might forget that a histogram uses intervals rather than individual data values. Remind students that the horizontal axis of the graph will be labeled with intervals and the vertical axis of the graph will be labeled with frequencies.
- **Common Error:** Even when an interval has a frequency of zero, it must appear on the histogram.
- **Teaching Tip:** The exercises in this set require students to take a survey to collect data and then display the data using a circle graph and a histogram. This provides a good opportunity to revisit topics such as population and sample size. What characteristics make for a fair survey? How will students ensure that the data they collect is a fair representation of the population?
- You can adapt the context of the survey to personalize it to your class.

Reteaching and Enrichment Strategies

If students need help...	If students got it...
Record and Practice Journal • Fair Game Review Skills Review Handbook Lesson Tutorials	Game Closet at *BigIdeasMath.com* Start the next section

What You Learned Before

"Okay, I have the box. But, I need your help to complete my box-and-whisker plot."

Displaying Data (6.SP.5a, 6.SP.4)

Example 1 **The table shows the results of a survey. Display the data in a circle graph.**

Class Trip Location	Water park	Museum	Zoo	Other
Students	25	11	5	4

A total of 45 students took the survey.

Water park:

$\frac{25}{45} \cdot 360^\circ = 200^\circ$

Museum:

$\frac{11}{45} \cdot 360^\circ = 88^\circ$

Zoo:

$\frac{5}{45} \cdot 360^\circ = 40^\circ$

Other:

$\frac{4}{45} \cdot 360^\circ = 32^\circ$

Class Trip Locations

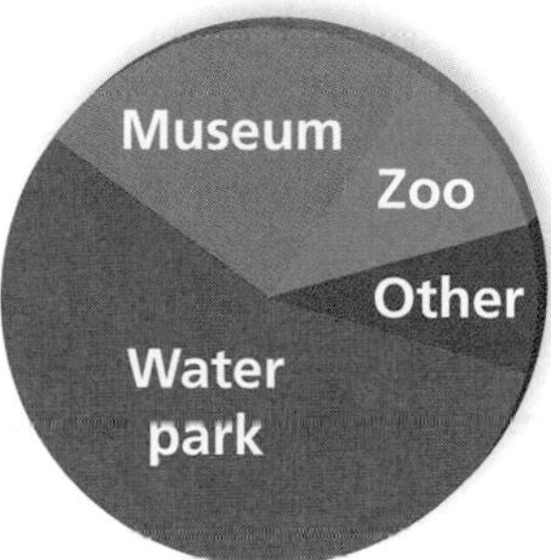

Example 2 **The frequency table shows the numbers of books that 12 people read last month. Display the data in a histogram.**

Books Read Last Month	Frequency
0–1	6
2–3	4
4–5	0
6–7	2

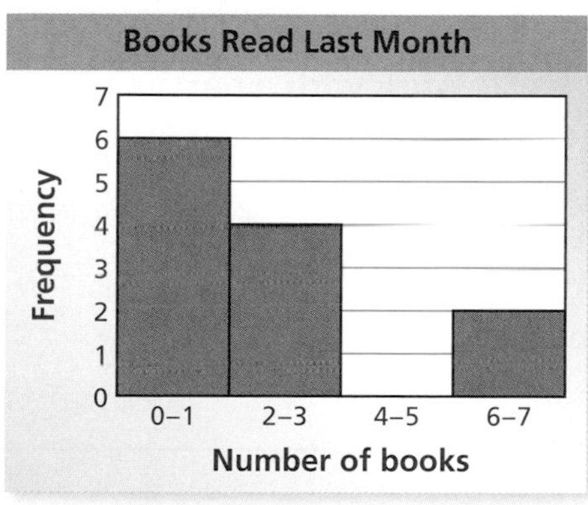

Try It Yourself

1. Conduct a survey to determine the after-school activities of students in your class. Display the results in a circle graph.

2. Conduct a survey to determine the numbers of pets owned by students in your class. Display the results in a histogram.

12.1 Measures of Central Tendency

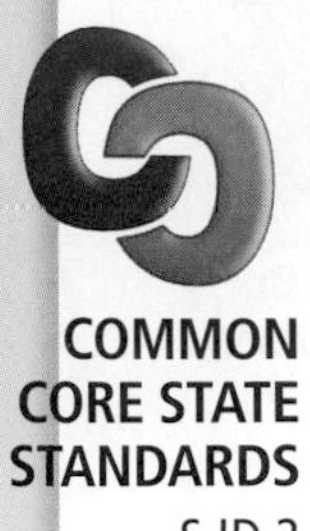

COMMON CORE STATE STANDARDS
S.ID.2
S.ID.3

Essential Question How can you use measures of central tendency to distribute an amount evenly among a group of people?

1 ACTIVITY: Exploring Mean, Median, and Mode

Work with a partner. Forty-five coins are arranged in nine stacks.

a. Record the number of coins in each stack in a table.

Stack	1	2	3	4	5	6	7	8	9
Coins									

b. Find the mean, median, and mode of the data.

c. By moving coins from one stack to another, can you change the mean? the median? the mode? Explain.

d. Is it possible to arrange the coins in stacks so that the median is 6? 8? Explain.

2 EXAMPLE: Drawing a Dot Plot

Work with a partner.

a. Draw a number line. Label the tick marks from 1 to 10.

b. Place each stack of coins in Activity 1 above the number of coins in the stack.

c. Draw a ● to represent each stack. This data display is called a *dot plot*.

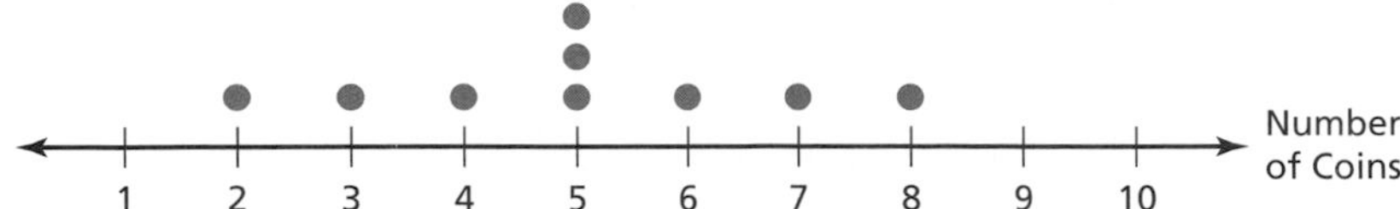

Laurie's Notes

Introduction

Standards for Mathematical Practice

- **MP2 Reason Abstractly and Quantitatively:** The physical and graphical models help students make sense of the quantities and their relationships in problem situations. In these activities, students develop an understanding of measures of central tendency.

Motivate

- Explain the work of the U.S. Census Bureau. Share the following U.S. information from the Census Bureau with students. In each case, ask students to interpret what the statistic means.
  - The mean travel time to work for workers age 16 and older is 25.1 minutes.
  - The median household income in 2009 was $50,221.
  - The *average* family size is 3.24 people.

Activity Notes

Activity 1

- For this first activity, students benefit by having some sort of manipulatives, such as coins, circular disks, square tiles, or cubes. If the manipulatives are stackable, the activity will be easier for students to follow.
- **MP2:** The concrete models help students visualize the distribution of the data, and that there could be outliers and/or gaps in the data.
- Review *mean*, *median*, and *mode*.
- Give sufficient time for students to work through each part of this activity.
- Discuss students' explanations for parts (c) and (d).
- **Big Idea:** As the median increases, the distribution becomes more skewed. One way to have a median of 8 coins is to have three stacks of 1 coin, one stack of 2 coins, and five stacks of 8 coins.

Example 2

- This example reviews dot plots.
- Reading the dot plot shown, there are three stacks that have 5 coins in them, no stacks that have 1, 9, or 10 coins in them, and all of the other possibilities have one stack each.
- This plot is reviewed so that it can be used to explore fair distributions in the next activity.

Common Core State Standards

S.ID.2 Use statistics appropriate to the shape of the data distribution to compare center (median, mean) and spread (interquartile range, standard deviation) of two or more different data sets.

S.ID.3 Interpret differences in shape, center, and spread in the context of the data sets, accounting for possible effects of extreme data points (outliers).

Previous Learning

Students should know how to find the mean, median, and mode.

Activity Materials
Textbook
• circular disks or other stackable manipulatives

Start Thinking! and Warm Up

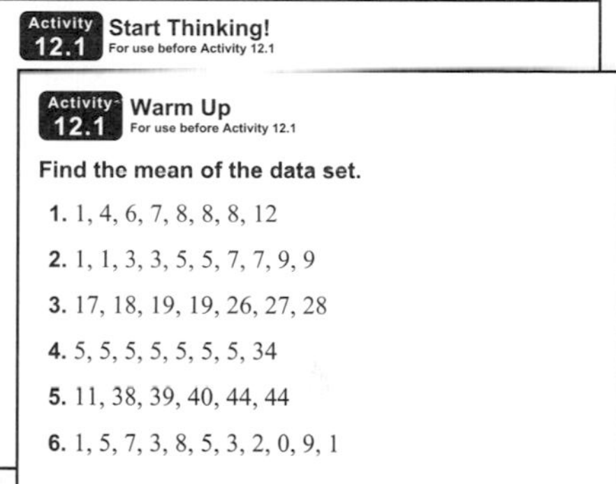

Activity 12.1 Start Thinking! For use before Activity 12.1

Activity 12.1 Warm Up For use before Activity 12.1

Find the mean of the data set.

1. 1, 4, 6, 7, 8, 8, 8, 12
2. 1, 1, 3, 3, 5, 5, 7, 7, 9, 9
3. 17, 18, 19, 19, 26, 27, 28
4. 5, 5, 5, 5, 5, 5, 5, 34
5. 11, 38, 39, 40, 44, 44
6. 1, 5, 7, 3, 8, 5, 3, 2, 0, 9, 1

12.1 Record and Practice Journal

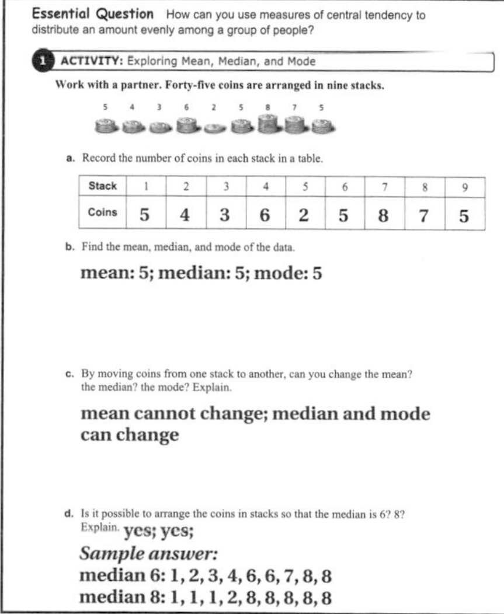

Essential Question How can you use measures of central tendency to distribute an amount evenly among a group of people?

1 ACTIVITY: Exploring Mean, Median, and Mode

Work with a partner. Forty-five coins are arranged in nine stacks.

5 4 3 6 2 5 8 7 5

a. Record the number of coins in each stack in a table.

Stack	1	2	3	4	5	6	7	8	9
Coins	5	4	3	6	2	5	8	7	5

b. Find the mean, median, and mode of the data.

mean: 5; median: 5; mode: 5

c. By moving coins from one stack to another, can you change the mean? the median? the mode? Explain.

mean cannot change; median and mode can change

d. Is it possible to arrange the coins in stacks so that the median is 6? 8? Explain. **yes; yes;**

Sample answer:
median 6: 1, 2, 3, 4, 6, 6, 7, 8, 8
median 8: 1, 1, 1, 2, 8, 8, 8, 8, 8

English Language Learners

Graphic Organizer

Provide small groups with different data sets and a copy of the organizer. Have each group find the mean, median, mode, and range of their data set and present their work to the class.

Data set:	
Mean	
Median	
Mode	
Range	

12.1 Record and Practice Journal

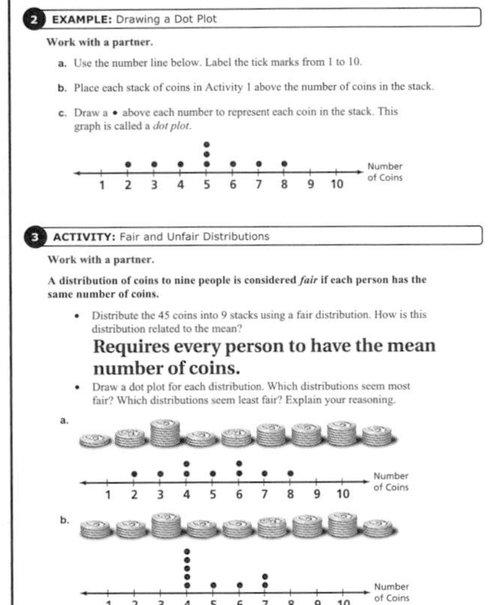

2 EXAMPLE: Drawing a Dot Plot

Work with a partner.

a. Use the number line below. Label the tick marks from 1 to 10.

b. Place each stack of coins in Activity 1 above the number of coins in the stack.

c. Draw a • above each number to represent each coin in the stack. This graph is called a *dot plot*.

3 ACTIVITY: Fair and Unfair Distributions

Work with a partner.

A distribution of coins to nine people is considered *fair* if each person has the same number of coins.

- Distribute the 45 coins into 9 stacks using a fair distribution. How is this distribution related to the mean?

  Requires every person to have the mean number of coins.

- Draw a dot plot for each distribution. Which distributions seem most fair? Which distributions seem least fair? Explain your reasoning.

a.

b.

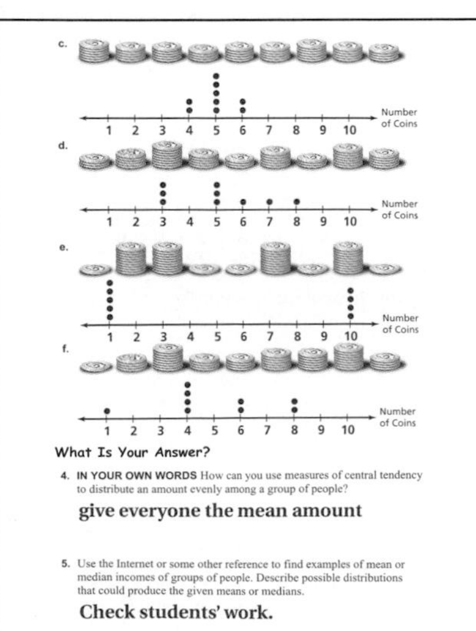

c.

d.

e.

f.

What Is Your Answer?

4. **IN YOUR OWN WORDS** How can you use measures of central tendency to distribute an amount evenly among a group of people?

   give everyone the mean amount

5. Use the Internet or some other reference to find examples of mean or median incomes of groups of people. Describe possible distributions that could produce the given means or medians.

   Check students' work.

Laurie's Notes

Activity 3

- Read the definition of a fair distribution and discuss what a fair distribution has to do with the mean.
- When students construct the dot plot for each part, they should get the sense that it is similar to constructing a bar graph. The heights of the bars in a bar graph are similar to the heights of the dots in a dot plot.
- Discuss student responses to which distributions seem most and least fair.
- **MP4 Model with Mathematics:** The coins and dot plot visually model the concept of a fair distribution. The models help students understand that mathematics can be used to model everyday concepts such as means, medians, and modes.
- **Extension:** Given the requirement that each stack has to have at least one coin, ask students to make a dot plot for the most fair distribution possible and the least fair distribution possible. The most fair would be 5 coins in each stack, so the dot plot would have nine dots above the 5 and no dots elsewhere. The least fair would likely be eight dots above the 1 and one dot above the 37, meaning the dot plot would need to be extended and there would be a big gap.
- **Big Idea:** You can think of the mean as a sharing process in which you are trying to level out the stacks. If the average amounts of money 3 students have is $10, it is possible that they all have $10 (most fair distribution) or one has $30 and the others have no money (least fair distribution). If all 3 students are known to have some money, pooling it and spreading it out into 3 stacks would level out at $10 per stack—the mean.

What Is Your Answer?

- For Question 4, the mean is the only measure of central tendency that can be used to distribute an amount evenly.

Closure

- Distribute 50 coins in 10 stacks. Make a dot plot of the distribution.

Technology For the Teacher

The Dynamic Planning Tool
Editable Teacher's Resources at *BigIdeasMath.com*

3 ACTIVITY: Fair and Unfair Distributions

Work with a partner.

A distribution of coins to nine people is considered *fair* if each person has the same number of coins.

- Distribute the 45 coins into 9 stacks using a fair distribution. How is this distribution related to the mean?
- Draw a dot plot for each distribution. Which distributions seem most fair? Which distributions seem least fair? Explain your reasoning.

a.

b.

c.

d.

e.

f.

What Is Your Answer?

4. **IN YOUR OWN WORDS** How can you use measures of central tendency to distribute an amount evenly among a group of people?
5. Use the Internet or some other reference to find examples of mean or median incomes of groups of people. Describe possible distributions that could produce the given means or medians.

Practice Use what you learned about measures of central tendency to complete Exercise 4 on page 610.

12.1 Lesson

Key Vocabulary
measure of central tendency, *p. 608*

A **measure of central tendency** is a measure that represents the center of a data set. The *mean*, *median*, and *mode* are measures of central tendency.

Key Ideas

Mean

The *mean* of a data set is the sum of the data divided by the number of data values.

Median

Order the data. For a set with an odd number of values, the *median* is the middle value. For a set with an even number of values, the *median* is the mean of the two middle values.

Mode

The *mode* of a data set is the value or values that occur most often.

Remember

Data can have one mode, more than one mode, or no mode. When each value occurs only once, there is no mode.

EXAMPLE 1 Finding the Mean, Median, and Mode

Students' Hourly Wages	
\$3.87	\$7.25
\$8.75	\$8.45
\$8.25	\$7.25
\$6.99	\$7.99

An amusement park hires students for the summer. The students' hourly wages are given in the table. Find the mean, median, and mode of the hourly wages.

Mean: $\frac{58.8}{8} = 7.35$ (sum of the data → 58.8; number of values → 8)

Median: 3.87, 6.99, 7.25, 7.25, 7.99, 8.25, 8.45, 8.75 — Order the data.

$\frac{15.24}{2} = 7.62$ — Mean of two middle values

Mode: 3.87, 6.99, 7.25, 7.25, 7.99, 8.25, 8.45, 8.75

The value 7.25 occurs most often.

∴ The mean is \$7.35, the median is \$7.62, and the mode is \$7.25.

On Your Own

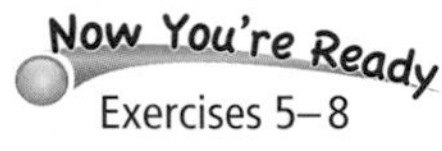

1. **WHAT IF?** In Example 1, the park hires another student at an hourly wage of \$6.99. How does this additional value affect the mean, median, and mode? Explain.

Multi-Language Glossary at BigIdeasMath.com.

Laurie's Notes

Introduction

Connect

- **Yesterday:** Students explored the connection between the mean and fair distributions. They used a dot plot to display results. (MP2, MP4)
- **Today:** Students will explore how an outlier affects the three measures of central tendency.

Motivate

- Time to play M & M's! No, it's not the candy; it's mean, median, mode time.
- To help students think about the three measures, ask a series of questions. Give the results and have students come up with the data.
  Examples:
  - Name 3 different numbers with a mean of 10.
  - Name 3 different numbers with a mean of 10 and with a median of 12.
  - Name 5 different numbers with a mean of 10 and with a median of 10.
  - Name 5 numbers with a mean and median of 10, and with a mode of 8.
- Continue to ask questions. Knowing the number of values and the mean tells you the sum of all the data. The median tells you the middle value.

Lesson Notes

Key Ideas

- Write the definition of measure of central tendency, noting that the mean, median, and mode are all measures of central tendency.

Example 1

? "Have any of you had a summer or part-time job?" Answers will vary.

- Have a general discussion of different compensation methods: hourly, hourly plus tips, salaried, and commission.

? "Are there any observations about the wages listed?" Students may mention $3.87 as an outlier. Most wages are around $7 or $8 per hour.

? "What do you need to do to compute the mean?" Add the data and divide by 8.

- You may wish to have students use calculators in this lesson.
- **Common Error:** When finding the median, students forget to sort the data first. In this example, there are an even number of data values, so you need to sort and then find the mean of the middle two values.

? "Why do you think the mean might be less than the median?" Listen for the effect of the outlier, although students may not have a sense of this yet.

On Your Own

- Give time for students to actually compute the three measures.
- Discuss the results.

Goal Today's lesson is determining the effect an outlier has on the **measures of central tendency** for a data set.

Lesson Materials
Textbook
• calculators

Start Thinking! and Warm Up

Record the number of siblings of each person in your class.

Find the mean, median, and mode of the data.

If you had instead recorded the number of children in each person's family, how would the mean, median, and mode have changed? Explain.

Extra Example 1

An amusement park hires students for the summer. The students' hourly wages are given in the table. Find the mean, median, and mode of the hourly wages.

Students' Hourly Wages	
$3.74	$7.75
$7.30	$8.43
$7.90	$7.83
$8.15	$8.50

mean: $7.45; median: $7.87; no mode

On Your Own

1. mean: $7.31, decreases; median: $7.25, decreases; Because the hourly wage of the student is less than the mean and median, both mean and median decrease. modes: $6.99 and $7.25; The data set now has two modes instead of one mode.

Laurie's Notes

Extra Example 2

Identify the outlier in Extra Example 1. How does the outlier affect the mean, median, and mode? When you remove the outlier, \$3.74, the mean increases \$7.98 − \$7.45 = \$0.53, the median increases \$7.90 − \$7.87 = \$0.03, and there still is no mode.

Extra Example 3

In Extra Example 1, the park increases each hourly wage by \$0.30. How does this increase affect the mean, median, and mode? The mean and median both increase by \$0.30, and there still is no mode.

On Your Own

2. $4\frac{1}{5}$ mi; When you remove the outlier, the mean decreases $1.85 - 1.38 = 0.47$ mile, the median decreases $1.45 - 1.4 = 0.05$ mile, and there is still no mode.
3. The mean and median both increase by $1\frac{1}{2}$ miles. There is still no mode.

Differentiated Instruction

Vocabulary

Have students add a glossary to their math notebook. Key vocabulary words should be added as they are introduced. Vocabulary words for this lesson are *measure of central tendency, outlier, mean, median,* and *mode*. Mean, median, and mode are often confusing to students. Drawing illustrations next to the vocabulary words will help in understanding and reinforcing their meanings.

Example 2

- Remind students of what an outlier is.
- Before the three measures are computed, ask students to predict what they think will happen.
- Work through the computations in the example.
- Remind students to use the correct number of values when computing the mean after removing an outlier.

? "Why did the mean increase in this example?" The data value eliminated was much less than the mean. So, the sum was not affected much and you divide by a lesser number.

? "The median increased in this example. Could it have decreased or stayed the same? Explain." yes; Depending on the middle of the data set, the median can increase or stay the same if the outlier is a low value, *or* the median can decrease or stay the same if the outlier is a high value.

- Discuss students' predictions and the actual results.

Example 3

? "What impact will there be if everyone receives a \$0.40 raise?" Answers will vary. Encourage students to reason about the problem.

- Work through the example.
- Discuss students' predictions and the actual results.
- **Big Idea:** When the same amount is added to each data value, the three measures increase by that same amount.
- **MP2 Reason Abstractly and Quantitatively:** If time permits, explore what happens when the hourly wage increases by 10%. Students need to understand the difference between an additive change to all values in the data set and a multiplicative change to all values in the data set.

On Your Own

- These questions integrate a review of fraction operations.
- For Question 3, check to see if students simply add $1\frac{1}{2}$ to each of their previous answers or if they compute the mean, median, and mode again.

Closure

- Explain the effect of an outlier on each of the three measures of central tendency.

Technology For the Teacher

The Dynamic Planning Tool
Editable Teacher's Resources at *BigIdeasMath.com*

EXAMPLE 2 Removing an Outlier

Remember

An *outlier* is a data value that is much greater than or much less than the other values.

Identify the outlier in Example 1. How does the outlier affect the mean, median, and mode?

The value \$3.87 is low compared to the other wages. It is the outlier.

Find the mean, median, and mode without the outlier.

Mean: $\frac{54.93}{7} \approx 7.85$

Median: 6.99, 7.25, 7.25, 7.99, 8.25, 8.45, 8.75 — The middle value, 7.99, is the median.

Mode: 6.99, 7.25, 7.25, 7.99, 8.25, 8.45, 8.75 — The mode is 7.25.

∴ When you remove the outlier, the mean increases \$7.85 − \$7.35 = \$0.50, the median increases \$7.99 − \$7.62 = \$0.37, and the mode is the same.

EXAMPLE 3 Changing the Values of a Data Set

In Example 1, each hourly wage increases by \$0.40. How does this increase affect the mean, median, and mode?

Students' Hourly Wages	
\$4.27	\$7.65
\$9.15	\$8.85
\$8.65	\$7.65
\$7.39	\$8.39

Make a new table by adding \$0.40 to each hourly wage.

Mean: $\frac{62}{8} = 7.75$

Median: 4.27, 7.39, 7.65, 7.65, 8.39, 8.65, 8.85, 9.15 — Order the data.

$\frac{16.04}{2} = 8.02$ — Mean of two middle values

Mode: 4.27, 7.39, 7.65, 7.65, 8.39, 8.65, 8.85, 9.15 — The mode is 7.65.

∴ When each hourly wage increases by \$0.40, the mean, median, and mode all increase by \$0.40.

On Your Own

Now You're Ready
Exercises 14–19

The figure shows the altitudes of several airplanes.

2. Identify the outlier. How does the outlier affect the mean, median, and mode? Explain.

3. Each airplane increases its altitude by $1\frac{1}{2}$ miles. How does this affect the mean, median, and mode? Explain.

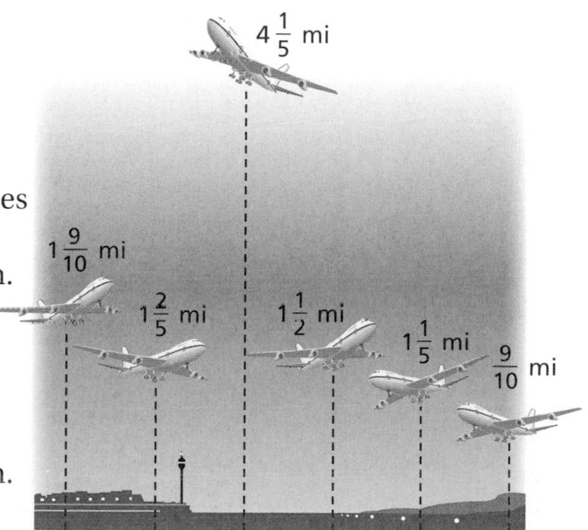

12.1 Exercises

Vocabulary and Concept Check

1. **VOCABULARY** Describe the measures of central tendency of a data set.
2. **OPEN-ENDED** Create a data set that has more than one mode.
3. **WRITING** Describe how removing an outlier from a data set affects the mean of the data set.

Practice and Problem Solving

4. Draw a dot plot of the data. Then find the mean, median, and mode of the data.

Bag	1	2	3	4	5	6	7	8	9
Strawberries	10	13	11	15	8	14	7	11	12

Find the mean, median, and mode of the data.

1 5.

Golf Scores		
3	−2	1
6	4	−1
−3	−1	2

6.

Changes in Stock Value (dollars)			
1.05	2.03	−1.78	−2.41
−2.64	0.67	4.02	1.39
0.66	−0.38	−3.01	2.20

7.

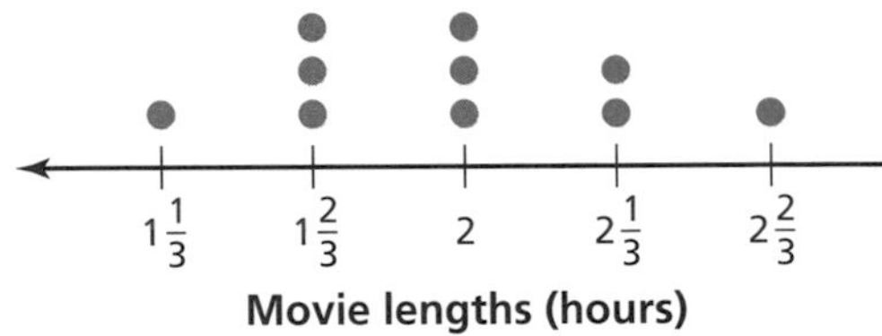

8. **Available Memory (megabytes)**

Stem	Leaf
6	5
7	0 5 5
8	0 4 5
9	4

Key: 7 | 5 = 75 megabytes

Find the value of *x*.

9. Mean is 6; 2, 8, 9, 7, 6, x
10. Mean is 0; 11.5, 12.5, −10, −7.5, x
11. Median is 14; 9, 10, 12, x, 20, 25
12. Median is 51; 30, 45, x, 100

13. **POLAR BEARS** The table shows the masses of polar bears. Find the value of x when the mean is 410 kilograms.

Masses (kilograms)			
455	262	471	358
364	553	352	x

Assignment Guide and Homework Check

Level	Assignment	Homework Check
Average	1–3, 5–19 odd, 14, 22–24	7, 13, 14, 17
Advanced	1–3, 6–20 even, 15, 21, 22–24	12, 14, 20, 21

Common Errors

- **Exercises 5–8** When finding the mean, students may forget to divide by the total number of data values and instead divide by the maximum value. Remind them that the definition of mean is an "average," so they must take into account the total number of items or numbers to get an average. Explain to students that it is as if they are dividing the total evenly among the number of groups.
- **Exercises 5–8** Students may try to identify the median without ordering the data first. Remind them that it is essential to order the data first and then find the median. This also makes finding the mode easier.
- **Exercises 9–12** Students may not know how to find the missing value. Remind them of the definition of mean and median. Encourage students to use these definitions to write an equation to find the value of x.

12.1 Record and Practice Journal

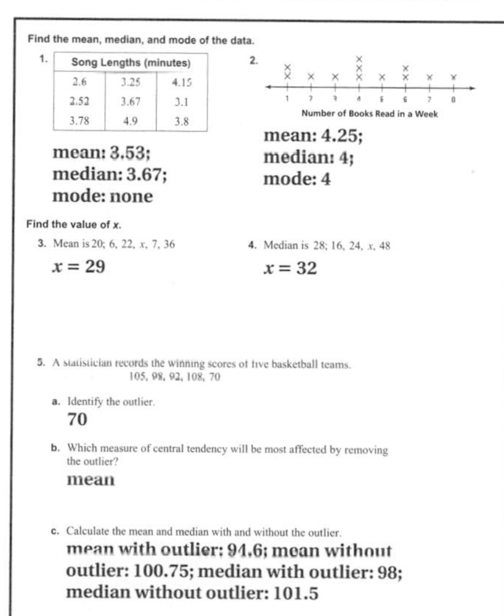

Find the mean, median, and mode of the data.

1.

Song Lengths (minutes)		
2.6	3.25	4.15
2.52	3.67	3.1
3.78	4.9	3.8

mean: 3.53; median: 3.67; mode: none

2. Number of Books Read in a Week

mean: 4.25; median: 4; mode: 4

Find the value of x.

3. Mean is 20; 6, 22, x, 7, 36

$x = 29$

4. Median is 28; 16, 24, x, 48

$x = 32$

5. A statistician records the winning scores of five basketball teams.

105, 98, 92, 108, 70

a. Identify the outlier.

70

b. Which measure of central tendency will be most affected by removing the outlier?

mean

c. Calculate the mean and median with and without the outlier.

mean with outlier: 94.6; mean without outlier: 100.75; median with outlier: 98; median without outlier: 101.5

Technology For the Teacher

Answer Presentation Tool

Vocabulary and Concept Check

1. The mean is the sum of the data divided by the number of data values. The median of an odd number of values is the middle value. The median of an even number of values is the mean of the two middle values. The mode is the value or values that occur most often.
2. *Sample answer:* 1, 2, 2, 3, 6, 8, 8, 9, 12
3. If the outlier is greater than the mean, removing it will decrease the mean. If the outlier is less than the mean, removing it will increase the mean.

Practice and Problem Solving

4. 7 8 9 10 11 12 13 14 15

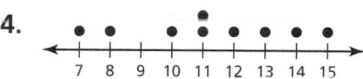

mean: $11.\overline{2}$; median: 11
mode: 11

5. mean: 1; median: 1
mode: -1

6. mean: \$0.15; median: \$0.665
mode: none

7. mean: $1\frac{29}{30}$ h
median: 2 h
modes: $1\frac{2}{3}$ h and 2 h

8. mean: 78.5 MB
median: 77.5 MB
mode: 75 MB

9. 4
10. -6.5
11. 16
12. 57
13. 465

Practice and Problem Solving

14. a. 6 in rookie season
    14 this season
    b. mean for both seasons
    c. The mean increased by about 5.3; The median increased by 5; the mode increased from 0 and 2 to 4.
15. See *Taking Math Deeper*.
16. All measures increase by 3.
17. All measures increase by k.
18. The mean and median both decrease by \$0.05 and there is still no mode.
19. All measures decrease by k.
20. a. mean: 19.37 yr
    median: 19 yr
    mode: 18 yr
    b. 37 yr; The mean decreases about $19.37 - 19.19 = 0.18$ year. The median and mode stay the same.
21. See Additional Answers.

Fair Game Review

22. $-8, -5, -3, 1, 4, 7$
23. $-4.7, -2.8, -\frac{2}{3}, 1.2, \frac{3}{2}, 5.4$
24. B

Mini-Assessment

Find the mean, median, and mode of the data.

1. 10, −4, 3, −1, 12 4, 3, no mode
2. 1.25, 3.80, −0.65, −2.40 0.5, 0.3, no mode
3. 5, 15, 8, 13, 10, 8, 6, 4, 12 9, 8, 8

Find the value of *x*.

4. Mean is 2; −4, −2, 3, x, 9 4
5. Median is 16.5; 8, 11, x, 18, 24, 26 15

Taking Math Deeper

Exercise 15

The Appalachian Trail is the longest marked trail in the U.S. at 2178 miles. The 11 shelters in this problem are in Massachusetts.

 Order the data.

0.1, 1.8, 3.3, 5.3, 6.3, 8.8, 8.8, 14, 14.3, 16.7

 Find the mean, median, and mode.

a. $\text{Mean} = \frac{79.4}{10} = 7.94$ miles

$\text{Median} = \frac{(6.3 + 8.8)}{2} = 7.55$ miles

The mode is 8.8 miles.

 A hiker begins the trail at Shelter 2 and therefore skips the 0.1-mile distance of the trail. The mean, median, and mode of the remaining distances are as follows.

1.8, 3.3, 5.3, 6.3, 8.8, 8.8, 14, 14.3, 16.7

Find the mean, median, and mode.

b. $\text{Mean} = \frac{79.3}{9} \approx 8.8$ miles

Median = 8.8 miles

The mode is 8.8 miles.

The mean increases by about 0.86 mile.
The median increases by 1.25 miles.
The mode does not change.

Project

Plan a hiking trip. List the things you need to take with you. Include the approximate amount of time you think it will take.

Reteaching and Enrichment Strategies

If students need help...	If students got it...
Resources by Chapter • Practice A and Practice B • Puzzle Time Record and Practice Journal Practice Differentiating the Lesson Lesson Tutorials Skills Review Handbook	Resources by Chapter • Enrichment and Extension Start the next section

② **14. BASEBALL** The graph shows a player's monthly home run totals in two seasons.

a. Identify the outlier in each season.

b. Which measure of central tendency is most affected by removing the outlier in each season?

c. Compare the means, medians, and modes of the home run totals in the two seasons.

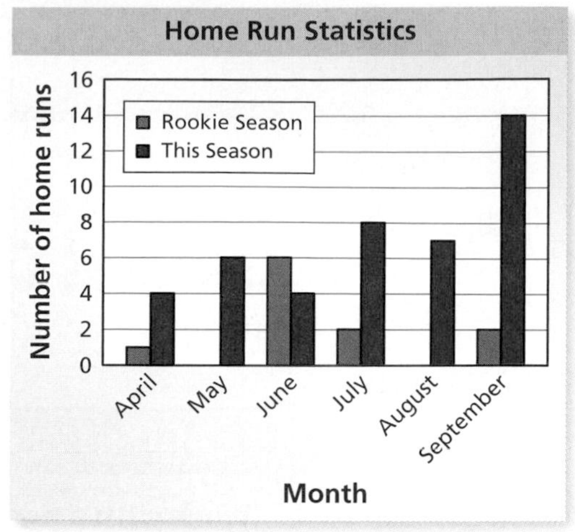

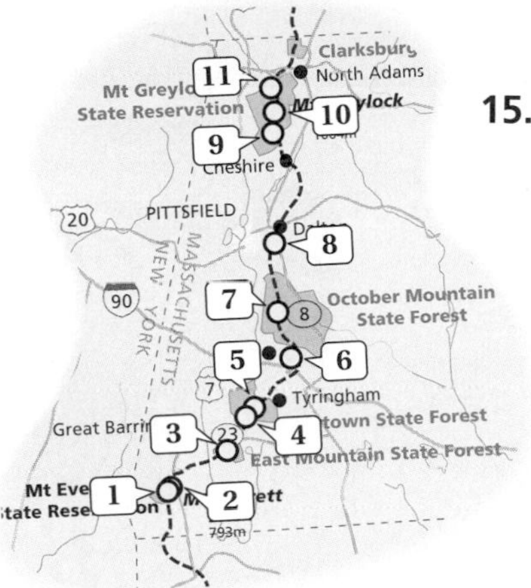

15. TRAIL The map shows the locations of 11 shelters along the Appalachian Trail. The distances (in miles) between these shelters are 0.1, 14.3, 5.3, 1.8, 14, 8.8, 8.8, 16.7, 6.3, and 3.3.

a. Find the mean, median, and mode of the distances.

b. A hiker starts at Shelter 2 and hikes to Shelter 11. How does this affect the mean, median, and mode? Explain.

In Exercises 16–19, explain how the change affects the mean, median, and mode.

③ **16.** In Exercise 4, you add 3 strawberries to each bag.

17. You add a number k to each value in a data set.

18. In Exercise 6, the value of each stock decreases by \$0.05.

19. You subtract a number k from each value in a data set.

College Student Ages

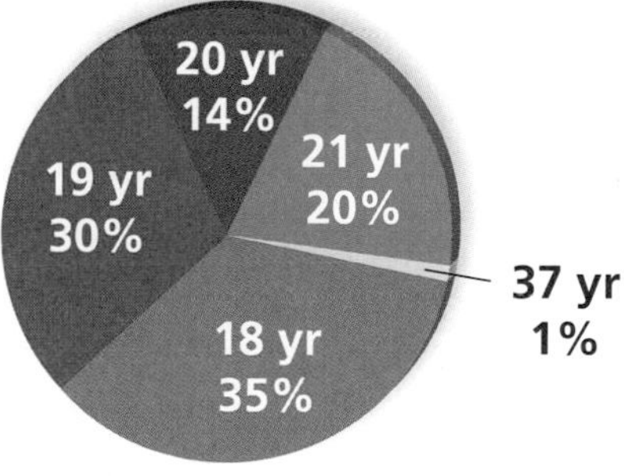

20. COLLEGE The circle graph shows the distribution of the ages of 200 students in a college psychology class.

a. Find the mean, median, and mode of the students' ages.

b. Identify the outliers. How do the outliers affect the mean, median, and mode?

21. Reasoning The mean and median hourly wage at a bagel shop is \$7.20. Hourly wages at the bagel shop increase by 10%. Where are you likely to have a greater hourly wage, at the bagel shop or at the amusement park in Example 1? Explain.

Fair Game Review What you learned in previous grades & lessons

Order the values from least to greatest. *(Skills Review Handbook)*

22. $1, -3, -8, 4, 7, -5$

23. $1.2, -2.8, \frac{3}{2}, 5.4, -4.7, -\frac{2}{3}$

24. MULTIPLE CHOICE Which equation represents a linear function? *(Section 5.5)*

Ⓐ $y = x^2$　　Ⓑ $y = 2x$　　Ⓒ $y = \frac{2}{x}$　　Ⓓ $xy = 2$

12.2 Measures of Dispersion

Essential Question How can you measure the dispersion of a data set?

COMMON CORE STATE STANDARDS
S.ID.2
S.ID.3

1 ACTIVITY: Measuring the Dispersion of Data

Work with a partner. The diagram shows the weights of 53 players on the Chicago Bears football team in 2011.

Weights of Chicago Bears

Tackles = ■ Quarterbacks = ■
Guards/Centers = ■ Running backs = □
DE/TE = ■ Wide receivers = ■
Linebackers = ■ Other players = □

155 175 195 215 235 255 275 295 315 335
Weight (pounds)

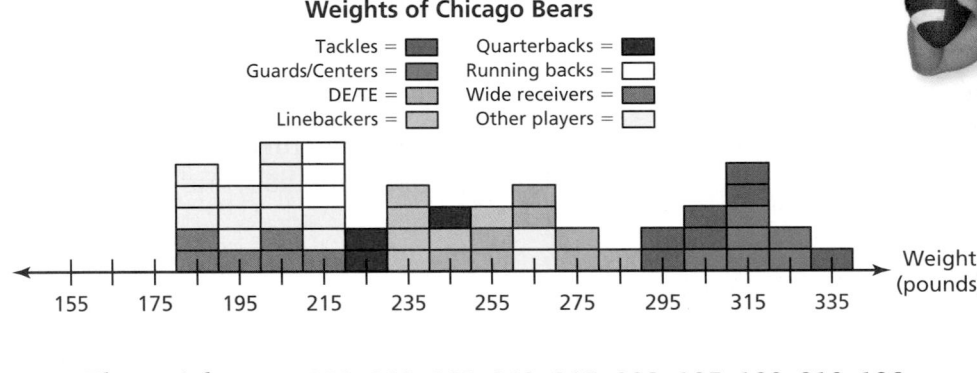

The weights are: 220, 200, 185, 240, 215, 222, 185, 180, 210, 196, 218, 190, 218, 185, 204, 180, 200, 219, 198, 196, 260, 211, 203, 239, 234, 258, 244, 230, 320, 310, 265, 309, 315, 295, 315, 275, 316, 333, 320, 308, 206, 200, 255, 267, 260, 287, 292, 300, 248, 310, 252, 238, 270.

a. Describe the data. How much are the weights dispersed from the mean weight? Explain your reasoning.

> **Definition of Dispersed:** To disperse objects means to spread them over an area. For instance, the population of Texas is much more dispersed than the population of Rhode Island.

b. Does it appear that the weight of a football player is correlated to the position that he plays? Explain your reasoning. Do you think your answer is valid for other types of professional sports, such as basketball, baseball, hockey, and soccer? Explain your reasoning.

Laurie's Notes

Introduction

Standards for Mathematical Practice

- **MP1a Make Sense of Problems** and **MP3 Construct Viable Arguments and Critique the Reasoning of Others:** Having students explain their reasoning and having whole class discussions helps all students make sense of problems. It also helps them develop an ability to form arguments and critique the reasoning of others.

Motivate

- Play a game of "Did You Know?"
  - Walter Camp is called the "Father of American Football."
  - The game originated from the game of rugby.
  - The American Professional Football Association began in 1920 at a car dealership in Canton, Ohio. This association would later become the National Football League (NFL).
  - The Green Bay Packers won the first two Super Bowls in 1967 and 1968. It took them almost 30 years to win the Super Bowl again.
  - A football is 11 inches long and weighs 14 to 15 ounces.

Activity Notes

Activity 1

- Review the differences between a histogram and a bar graph.
- Students should discuss the histogram with their partners. This histogram involves an additional element of color that is important to the problem.
- A calculator may be helpful in finding the mean weight.
- When answering part (a), students may describe only the dispersion of the two extremes values (180 and 333) from the mean. Others may talk about the clusters of numbers above and below the mean. Explain to students that they will learn different ways to measure dispersion.
- ? "In part (b), what does 'correlated' mean in the context of this problem?" Answers will vary. Students may use phrases such as "associated with" or "connected to."
- Students with knowledge of football will be able to discuss the physical attributes of different positions. Even if students have little knowledge of football, the colors used in the histogram should provide the information needed to help explain the correlation. The colors are clustered, with no symmetry.
- Students with greater knowledge about professional sports will be able to discuss the second part of this activity.

Common Core State Standards

S.ID.2 Use statistics appropriate to the shape of the data distribution to compare center (median, mean) and spread (interquartile range, standard deviation) of two or more different data sets.

S.ID.3 Interpret differences in shape, center, and spread in the context of the data sets, accounting for possible effects of extreme data points (outliers).

Previous Learning

Students should know how to find the mean, median, and mode of a data set.

Activity Materials
Textbook
• calculators

Start Thinking! and Warm Up

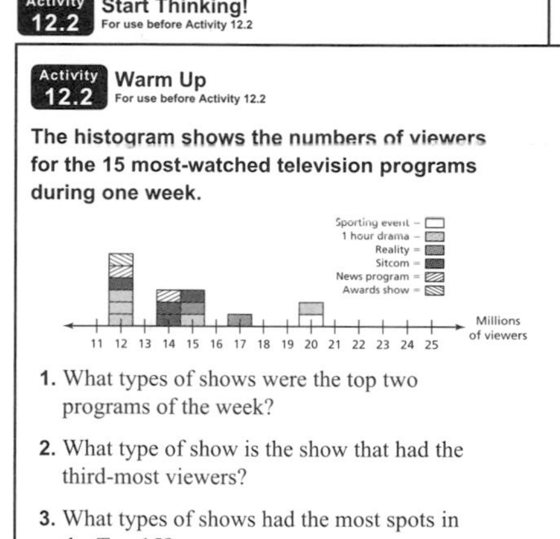

Activity 12.2 Start Thinking! For use before Activity 12.2

Activity 12.2 Warm Up For use before Activity 12.2

The histogram shows the numbers of viewers for the 15 most-watched television programs during one week.

1. What types of shows were the top two programs of the week?
2. What type of show is the show that had the third-most viewers?
3. What types of shows had the most spots in the Top 15?

12.2 Record and Practice Journal

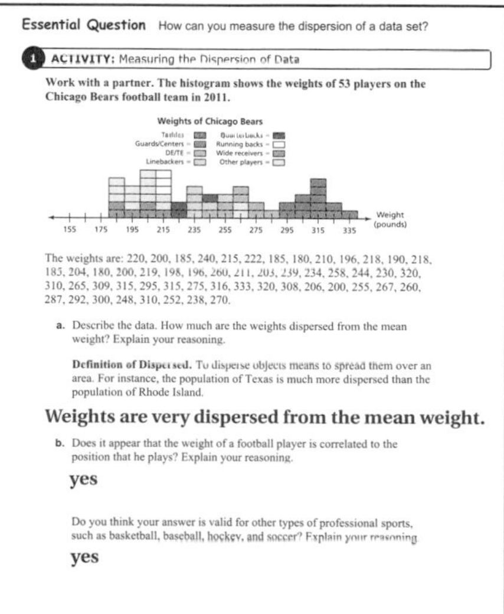

Essential Question How can you measure the dispersion of a data set?

1 ACTIVITY: Measuring the Dispersion of Data

Work with a partner. The histogram shows the weights of 53 players on the Chicago Bears football team in 2011.

The weights are: 220, 200, 185, 240, 215, 222, 185, 180, 210, 196, 218, 190, 218, 185, 204, 180, 200, 219, 198, 196, 260, 211, 203, 239, 234, 258, 244, 230, 320, 310, 265, 309, 315, 295, 315, 275, 316, 333, 320, 308, 206, 200, 255, 267, 260, 287, 292, 300, 248, 310, 252, 238, 270.

a. Describe the data. How much are the weights dispersed from the mean weight? Explain your reasoning.

Definition of Dispersed. To disperse objects means to spread them over an area. For instance, the population of Texas is much more dispersed than the population of Rhode Island.

Weights are very dispersed from the mean weight.

b. Does it appear that the weight of a football player is correlated to the position that he plays? Explain your reasoning.

yes

Do you think your answer is valid for other types of professional sports, such as basketball, baseball, hockey, and soccer? Explain your reasoning.

yes

Differentiated Instruction

Advanced

Challenge students to create two data sets with the same mean and median, but with a different range and standard deviation. Have them discuss the differences in the data sets and how measures of dispersion aid in describing a data set.

12.2 Record and Practice Journal

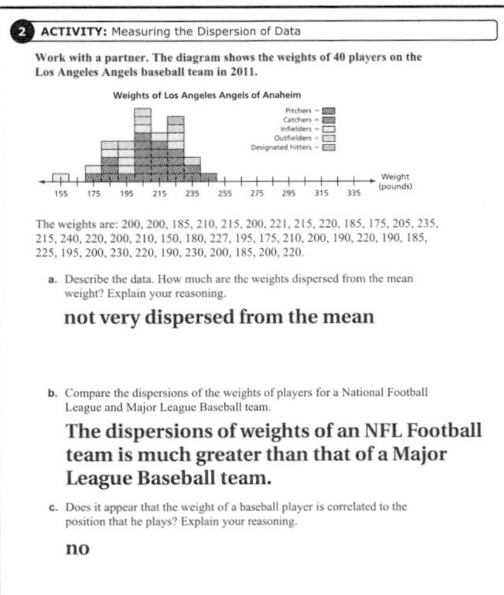

2 ACTIVITY: Measuring the Dispersion of Data

Work with a partner. The diagram shows the weights of 40 players on the Los Angeles Angels baseball team in 2011.

The weights are: 200, 200, 185, 210, 215, 200, 221, 215, 220, 185, 175, 205, 235, 215, 240, 220, 200, 210, 150, 180, 227, 195, 175, 210, 200, 190, 220, 190, 185, 225, 195, 200, 230, 220, 190, 230, 200, 185, 200, 220.

a. Describe the data. How much are the weights dispersed from the mean weight? Explain your reasoning.

not very dispersed from the mean

b. Compare the dispersions of the weights of players for a National Football League and Major League Baseball team.

The dispersions of weights of an NFL Football team is much greater than that of a Major League Baseball team.

c. Does it appear that the weight of a baseball player is correlated to the position that he plays? Explain your reasoning.

no

What Is Your Answer?

3. **IN YOUR OWN WORDS** How can you measure the dispersion of a data set? Illustrate your answer by using the positions and weights of the 15 players on the Boston Celtics basketball team in 2011.

Forward: 235; power forwards: 253, 295, 245; small forwards: 235, 235; centers: 255, 240, 325; point guards: 205, 186, 200; shooting guards: 205, 210, 180

Plot the data in a histogram and observe how spread out the data is about the mean.

Does it appear that the weight of a basketball player is correlated to the position that he plays? Explain your reasoning.

yes

Laurie's Notes

Activity 2

- This activity is similar to Activity 1 except the sport has changed. The discussion following Activity 1 helps all students to successfully complete this activity.

? "What do you notice about the colors in this histogram?" They are more symmetric, or evenly spread out.

- When comparing the dispersions of data in each activity, students should notice that the baseball data are less spread out.

What Is Your Answer?

- Provide colored pencils so students can construct a histogram of the data. If colored pencils are not available, the positions could be coded using letters.
- When comparing the dispersions of data, students should notice that the basketball data are similar to the football data in that there is a correlation between position and weight.

Closure

- Suppose you have data for the weights of players on a soccer team. When comparing the dispersions of data, which data set do you think it would be similar to? Explain your reasoning. It would most likely be similar to the basketball or baseball data.

Technology For the Teacher

The Dynamic Planning Tool

Editable Teacher's Resources at *BigIdeasMath.com*

2 ACTIVITY: Measuring the Dispersion of Data

Work with a partner. The diagram shows the weights of 40 players on the Los Angeles Angels baseball team in 2011.

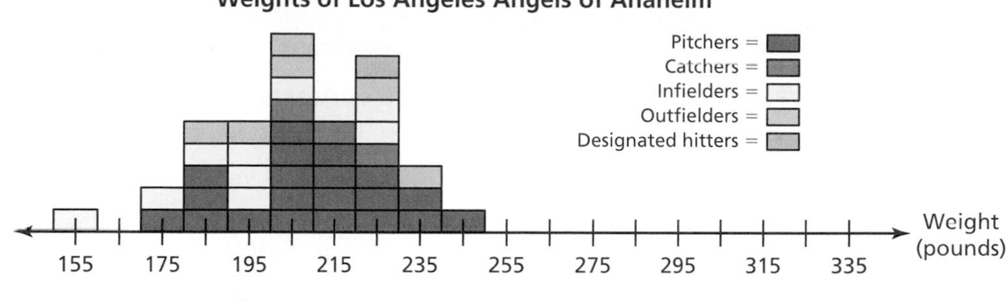

The weights are: 200, 200, 185, 210, 215, 200, 221, 215, 220, 185, 175, 205, 235, 215, 240, 220, 200, 210, 150, 180, 227, 195, 175, 210, 200, 190, 220, 190, 185, 225, 195, 200, 230, 220, 190, 230, 200, 185, 200, 220.

a. Describe the data. How much are the weights dispersed from the mean weight? Explain your reasoning.

b. Compare the dispersions of the weights of players for a National Football League team and a Major League Baseball team.

c. Does it appear that the weight of a baseball player is correlated to the position that he plays? Explain your reasoning.

What Is Your Answer?

3. **IN YOUR OWN WORDS** How can you measure the dispersion of a data set? Illustrate your answer by using the positions and weights of the 15 players on the Boston Celtics basketball team in 2011.

   Forward: 235; power forwards: 253, 295, 245; small forwards: 235, 235; centers: 255, 240, 325; point guards: 205, 186, 200; shooting guards: 205, 210, 180

   Does it appear that the weight of a basketball player is correlated to the position that he plays? Explain your reasoning.

Practice Use what you learned about measuring the dispersion of data to complete Exercises 3 and 4 on page 616.

12.2 Lesson

Key Vocabulary
measure of dispersion, *p. 614*
range, *p. 614*
standard deviation, *p. 615*

A **measure of dispersion** is a measure that describes the spread of a data set. The simplest measure of dispersion is the range. The **range** of a data set is the difference between the greatest value and the least value.

EXAMPLE 1 Finding the Range

Two reality cooking shows select 12 contestants each. The ages of the contestants are shown in the tables. Find the mean and range of the ages for each show. Compare your results.

Show A	
Ages	
20	29
19	22
25	27
27	29
30	20
21	31

Show B	
Ages	
25	19
20	27
22	25
27	22
48	21
32	24

Show A: mean $= \frac{300}{12} = 25$

Ordering the data can help you find the least and greatest ages.

19, 20, 20, 21, 22, 25, 27, 27, 29, 29, 30, 31 Order the data.

The least value is 19. The greatest value is 31.

So, the range is 31 − 19, or 12 years.

Show B: mean $= \frac{312}{12} = 26$

19, 20, 21, 22, 22, 24, 25, 25, 27, 27, 32, 48 Order the data.

The least value is 19. The greatest value is 48.

So, the range is 48 − 19, or 29 years.

The mean ages for the shows, 25 and 26, are about the same. The range of the ages for Show A is 12 years and the range for Show B is 29 years. So, the ages for Show B are more spread out.

On Your Own

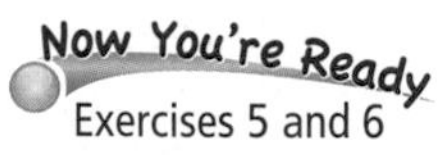

1. After the first week, the 25-year-old is voted off Show A. The 48-year-old is voted off Show B. How does this affect the mean and range of the remaining contestants on each show? Explain.

Multi-Language Glossary at BigIdeasMath.com.

Laurie's Notes

Introduction

Connect

- **Yesterday:** Students explored how to measure the dispersion of a data set. (MP1a, MP3)
- **Today:** Students will find the range and standard deviation of a data set and compare two data sets.

Motivate

- **Review:** Write two data sets on the board and ask students to find the mean, median, and mode of each data set.
  Example: Data Set 1: 85, 89, 90, 91, 95 **Data Set 2:** 0, 50, 100, 100, 200
- Discuss how the data sets are alike (5 data values in each set, mean = 90) and how they are different (median, mode, and spread).

Lesson Notes

Discuss

- Define measure of dispersion and range. Refer back to the data sets in the activity.
- **Common Error:** When finding the range, where the least value is 180 and the greatest value is 333, students may write 180 − 333, meaning 180 to 333. Stress that the range is a single number. It is the *difference* between two numbers.

Example 1

- Ask if any students watch cooking shows. Do any of your students have an interest in cooking?
- If time permits, students could begin by constructing a dot plot or line plot of the data. The shape of the distributions is important in this example.

? "How do you find the mean?" Find the sum of the values in the data set and divide by the number of data values.

- Remind students that we use the word "mean" instead of "average" because there are different types of averages—median and mode are both averages.
- Compute the mean and range for each data set.
- The means are about the same but the ranges are not. When concluding that the ages in Show B are more spread out, some students might observe that it is only one piece of data that is significantly different.

? "What is an outlier?" A data value that is significantly different than the rest of the data values.

- Students may say that if you ignore the 48-year-old contestant, the ranges would be essentially the same. The effect of the older contestant is mentioned in the On Your Own.

On Your Own

- It is important for students to understand that an outlier affects the range *and* the mean.

Goal Today's lesson is finding the **range** and **standard deviation** of a data set.

Lesson Materials
Textbook
• calculators

Start Thinking! and Warm Up

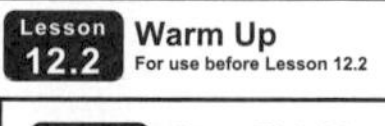

Lesson 12.2 Warm Up
For use before Lesson 12.2

Lesson 12.2 Start Thinking!
For use before Lesson 12.2

Find the age (in years) of every student in your class.

Find the number of siblings of every student in your class.

Create a dot plot for each set of data.

Find the mean and range of each set of data.

Compare and contrast the sets of data.

Extra Example 1

The bowling averages for two teams are shown in the tables. Find the mean and range of the bowling averages for each team. Compare your results.

Team A	
185	167
221	205
194	190
208	214

Team B	
182	172
168	170
175	195
190	188

The mean for Team A, 198, is greater than the mean for Team B, 180. The range for Team A, 54, is greater than the range for Team B, 27. The bowling averages for Team A are greater but more spread out than Team B.

On Your Own

1. For Show A, the mean and range remain the same. For Show B, the mean changes from 26 to 24 and the range changes from 29 to 13. The mean and range do not change for Show A because the removed data value is equal to the mean of the data set. The mean and range change for Show B because the removed data value, 48, is an outlier.

Extra Example 2

Find the standard deviation of the bowling averages for Team A in Extra Example 1. Use a table to organize your work. Interpret your result.

The standard deviation is about 16.4.This means that the typical bowling average on Team A differs from the mean by about 16.4 pins.
See Additional Answers for table.

On Your Own

2. about 7.5; This means that the typical age of a contestant on Show B differs from the mean by about 7.5 years.
3. The standard deviation for Show B is greater than that of Show A. So, the ages of the contestants on Show B are more spread out than the ages of the contestants on Show A.

English Language Learners

Visual Glossary

To help English language learners with the vocabulary and understanding the concepts, create posters for each of the words: *mean*, *median*, *mode*, *range*, and *standard deviation*. Identify whether it is a *measure of central tendency* or a *measure of dispersion*. Include the description, the formula used to find the measure, and an example.

Laurie's Notes

Key Idea

- Point out the disadvantage of using the range to describe a data set. It is easily calculated but does not adequately describe how data are spread out. The standard deviation uses all of the data values in its computation.
- Students may be overwhelmed by this formula. Remind them they can use a calculator to find a standard deviation. By looking at its calculation, students will better understand what a standard deviation represents.
- Explain the formula by saying that after the mean is calculated, it is subtracted from each data value. The difference is positive for values greater than the mean and negative for values less than the mean. These differences are squared making them all positive. The sum of all the squared differences is divided by n, finding the 'mean' of the sum of squared differences. The square root of this mean is then calculated.
- Refer to the sets in the Motivate to describe data that are clustered about the mean (Data Set 1) and data that are more spread out (Data Set 2).

Example 2

- **MP1b Persevere in Solving Problems:** Computing the standard deviation by hand requires perseverance.
- **MP5 Use Appropriate Tools Strategically:** Using a table to organize results is an appropriate tool that helps ensure accurate completion of the problem.
- Write $\bar{x} = 25$ and say, "The mean, denoted by x-bar, is 25."

? "How do you find the mean of column 4?" Add up the 12 values and divide by 12.

- Interpret the meaning of the answer by saying, "The typical age of a contestant on Show A differs from the mean by about 4.2 years."
- **Big Idea:** A standard deviation of 4.2 years must be viewed in the context of a data set where the mean is 25. A standard deviation of 4.2 years can be looked at differently if the data set has a mean of 8 or 80.

On Your Own

- Discuss the effect of the outlier for Show B.

Closure

- Refer to the sets in the Motivate. Each set has a mean of 90. The standard deviation of Data Set 1 is 3.2 and for Data Set 2, it is 66.3. Give five data values that have a mean of 90 and a standard deviation between 3.2 and 66.3. Explain why you believe you are correct. Answers will vary.

Technology For the Teacher

The Dynamic Planning Tool
Editable Teacher's Resources at *BigIdeasMath.com*

A disadvantage of using the range to describe the spread of a data set is that its calculation uses only two data values. A measure of dispersion that uses all the values of a data set is the *standard deviation*.

Key Idea

Standard Deviation

The **standard deviation** of a data set is a measure of how much a typical value in the data set differs from the mean. It is given by

$$\text{standard deviation} = \sqrt{\frac{(x_1 - \bar{x})^2 + (x_2 - \bar{x})^2 + \cdots + (x_n - \bar{x})^2}{n}}$$

where n is the number of values in the data set. The symbol $\bar{x}$ represents the mean. It is read as "x-bar."

A small standard deviation means that the data are clustered around the mean. A large standard deviation means that the data are more spread out.

Remember

The notation consisting of three dots (···) indicates that a pattern continues.

EXAMPLE 2 Finding the Standard Deviation

x	$\bar{x}$	$x - \bar{x}$	$(x - \bar{x})^2$
20	25	−5	25
29	25	4	16
19	25	−6	36
22	25	−3	9
25	25	0	0
27	25	2	4
27	25	2	4
29	25	4	16
30	25	5	25
20	25	−5	25
21	25	−4	16
31	25	6	36

Find the standard deviation of the ages for Show A in Example 1. Use a table to organize your work. Interpret your result.

Step 1: Find the mean. From Example 1, the mean is 25.

Step 2: Find the difference between each data value and the mean, $x - \bar{x}$.

Step 3: Square each difference from Step 2, $(x - \bar{x})^2$.

Step 4: Find the mean of the squares from Step 3.

$$\frac{(x_1 - \bar{x})^2 + (x_2 - \bar{x})^2 + \cdots + (x_n - \bar{x})^2}{n} = \frac{25 + 16 + \cdots + 36}{12} \approx 17.7$$

Step 5: Use a calculator to find the square root.

$$\sqrt{\frac{(x_1 - \bar{x})^2 + (x_2 - \bar{x})^2 + \cdots + (x_n - \bar{x})^2}{n}} = \sqrt{17.7} \approx 4.2$$

∴ The standard deviation is 4.2. This means that the typical age of a contestant on Show A differs from the mean by about 4.2 years.

On Your Own

Now You're Ready
Exercises 7–12

2. Find the standard deviation of the ages for Show B in Example 1. Interpret your result.
3. Compare the standard deviations for Show A and Show B. What can you conclude?

12.2 Exercises

Vocabulary and Concept Check

1. **VOCABULARY** In a data set, what does a measure of central tendency represent? What does a measure of dispersion represent?
2. **REASONING** What is an advantage of using the range to describe a data set? Why do you think the standard deviation is considered a more reliable measure of dispersion than the range?

Practice and Problem Solving

Describe the data. How much are the data dispersed from the mean? Explain your reasoning.

3. 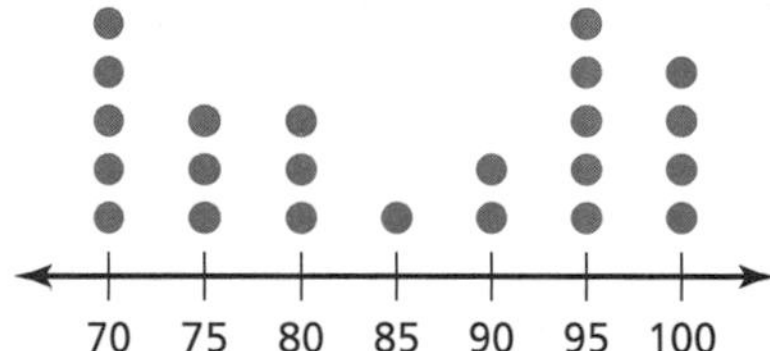

4. 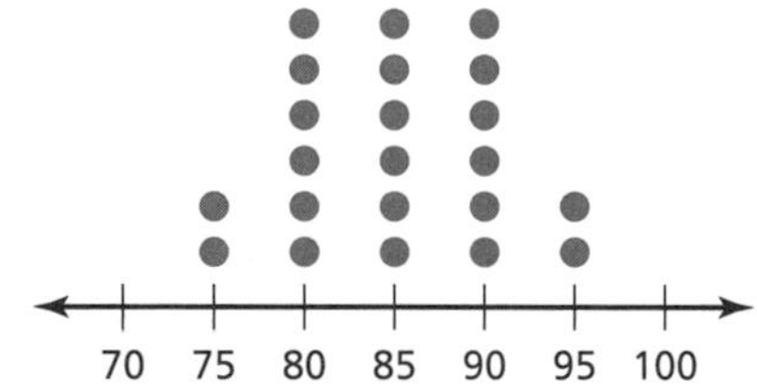

Find the mean and range of each data set. Then compare the data sets.

1 5. Heights (in inches) of two teams
Tigers: 67, 70, 65, 72, 74, 68, 67, 69
Centaurs: 74, 71, 68, 63, 75, 63, 65, 73

6. Numbers of fish caught during a week
Crew A: 120, 100, 75, 112, 135, 80, 106
Crew B: 104, 140, 159, 135, 158, 165, 140

Find the mean and standard deviation of the data.

2 7. 4, 2, 7, 3, 6, 5, 5, 8

8. 12, 4, 8, 7, 9, 13, 10

9. 18 19 20 21 22 23 24

10. 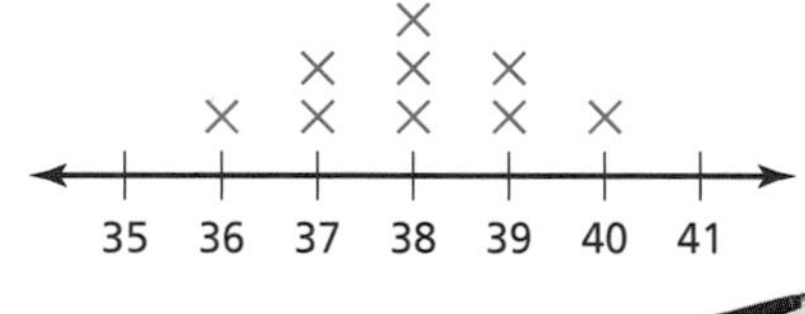

11. 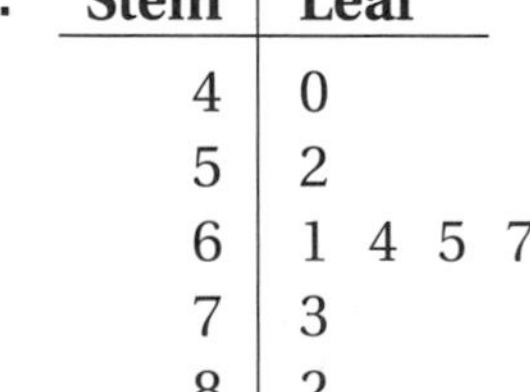

Stem	Leaf
4	0
5	2
6	1 4 5 7
7	3
8	2

Key: 6|1 = 61

12. 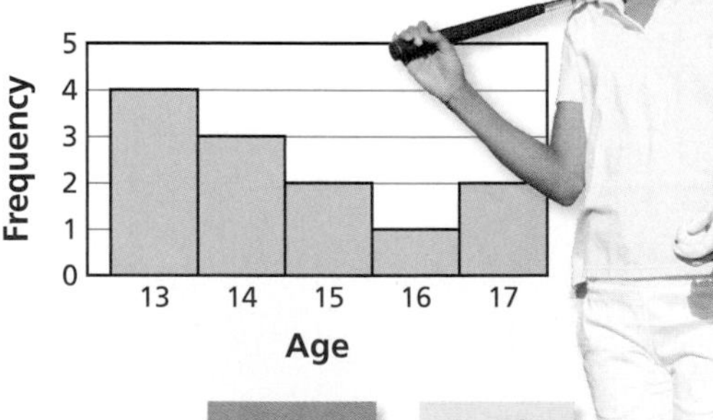

13. **GOLF** The scores for two golfers are shown.
   a. Find the mean, range, and standard deviation of the scores for each golfer. Compare your results.
   b. Which golfer do you think is more consistent? Explain.

Kirsten	
83	88
84	95
91	89
90	87
98	95

Leah	
89	87
93	95
92	94
88	91
89	92

Assignment Guide and Homework Check

Level	Assignment	Homework Check
Average	1–4, 5–17 odd, 14, 18, 22–25	5, 11, 13, 14
Advanced	1–4, 6–16 even, 17–19, 21, 22–25	6, 12, 18, 19

Common Errors

- **Exercises 5 and 6** Students may subtract the first data value from the last data value to find the range. Remind students that the range is the difference between the greatest and least data values. It helps to order the data before identifying the greatest and least data values.
- **Exercises 7–12** Students may forget to square the differences between x_i and the mean when calculating the standard deviation. The standard deviation formula is simply finding the average distance each data point is from the data set mean. Tell the students that squaring the differences is a mathematical way of making all of the distances positive, similar to the distance formula.

12.2 Record and Practice Journal

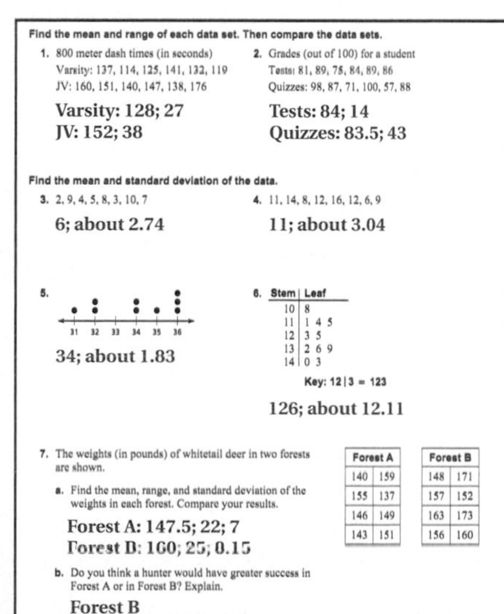

Find the mean and range of each data set. Then compare the data sets.

1. 800 meter dash times (in seconds)
   Varsity: 137, 114, 125, 141, 132, 119
   JV: 160, 151, 140, 147, 138, 176
   Varsity: 128; 27
   JV: 152; 38
2. Grades (out of 100) for a student
   Tests: 81, 89, 75, 84, 89, 86
   Quizzes: 98, 87, 71, 100, 57, 88
   Tests: 84; 14
   Quizzes: 83.5; 43

Find the mean and standard deviation of the data.

3. 2, 9, 4, 5, 8, 3, 10, 7
   6; about 2.74
4. 11, 14, 8, 12, 16, 12, 6, 9
   11; about 3.04
5. 31 32 33 34 35 36
   34; about 1.83
6. | Stem | Leaf |
   |---|---|
   | 10 | 8 |
   | 11 | 1 4 5 |
   | 12 | 3 5 |
   | 13 | 2 6 9 |
   | 14 | 0 3 |

   Key: 12 | 3 = 123
   126; about 12.11
7. The weights (in pounds) of whitetail deer in two forests are shown.

Forest A	
140	159
155	137
146	149
143	151

Forest B	
148	171
157	152
163	173
156	160

   a. Find the mean, range, and standard deviation of the weights in each forest. Compare your results.
   Forest A: 147.5; 22; 7
   Forest B: 160; 25; 8.15

   b. Do you think a hunter would have greater success in Forest A or in Forest B? Explain.
   Forest B

Technology For the Teacher
Answer Presentation Tool

Vocabulary and Concept Check

1. A measure of central tendency is a value that represents a typical value in a data set. A measure of dispersion is a value that measures how spread out a data set is.
2. The range is quick and easy to calculate, but its calculation uses only two data values from a data set and it is greatly affected by outliers. The standard deviation uses all of the data values in its calculation and is less affected by outliers compared to the range.

Practice and Problem Solving

3. The mean is 85. The majority of the data are in clusters on each side, far from the mean.
4. The mean is 85. The majority of the data are clustered around the mean.

5–6. See Additional Answers.

7. 5; about 1.9
8. 9; about 2.8
9. 21; about 1.6
10. 38; about 1.2
11. 63; about 11.9
12. 14.5; about 1.4
13. a. Kirsten: 90; 15; about 4.6; Leah: 91; 8; about 2.5; The means are about the same but Kirsten's range and standard deviation are much greater than Leah's range and standard deviation.
    b. Leah; Kirsten's scores are more spread out than Leah's scores.

Practice and Problem Solving

14. See Additional Answers.

15. 6.4; 8.2; about 3.0

16. -1; 8; about 2.7

17. See Additional Answers.

18. Data set (b) has the greatest standard deviation and data set (c) has the least standard deviation. From the dot plots, you can see that the data values of (b) are more spread out than (a) and (c). You can also see that the data values of (c) are clustered closely about the mean of 15.

19. See Additional Answers.

20. See *Taking Math Deeper.*

21. See Additional Answers.

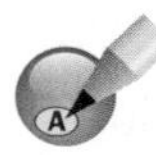

Fair Game Review

22–24. See Additional Answers.

25. B

Mini-Assessment

Find the mean, range, and standard deviation of the data sets. Then compare the data sets.

1. Number of text messages

Student A	24	21	32	16	38	29	22
Student B	20	23	15	27	14	22	19

See Additional Answers.

2. Grade point averages

Team A	3.6	3.2	3.8	3.4	3.5
Team B	3.5	3.9	2.9	4.0	3.2

See Additional Answers.

Taking Math Deeper

Exercise 20

At first glance, students may think the answer is Doctor's Office A, because the mean time is greater than Doctor's Office B. Students need to consider the standard deviation for each data set. Let's take a graphical approach.

Sketch an approximate graph of each distribution. Label the mean at the center of each distribution and the standard deviations above and below the mean. Remember the data are evenly distributed.

Interpret the graphs. From the graph for Office A, you can see that most of the waiting times are between 10 and 20 minutes. For Office B, most of the waiting times are between 3 and 25 minutes.

The standard deviation for Office B is greater, so the data are more spread out.

Waiting Times at Doctor's Office A

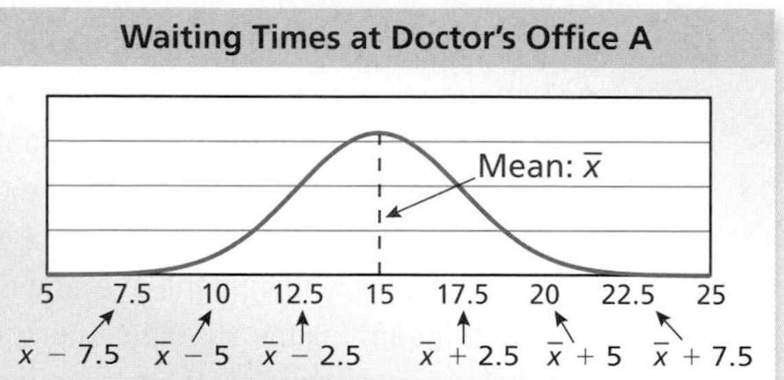

Waiting Times at Doctor's Office B

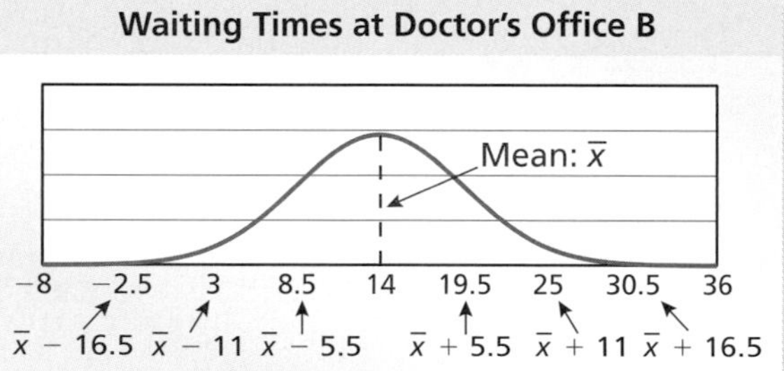

At Office A, a waiting time longer than 20 minutes is more than 2 standard deviations from the mean. At Office B, it is a little more than 1 standard deviation from the mean. So, you are more likely to wait longer than 20 minutes at Doctor's Office B.

Note: Explain to students that it is possible to spend more than 20 minutes waiting at either office, but the likelihood is greater at Office B.

Reteaching and Enrichment Strategies

If students need help...	If students got it...
Resources by Chapter • Practice A and Practice B • Puzzle Time Record and Practice Journal Practice Differentiating the Lesson Lesson Tutorials Skills Review Handbook	Resources by Chapter • Enrichment and Extension Start the next section

14. **INCLUDING A VALUE** In Exercise 13, Kirsten's score for the next round is 90, and Leah's is 80. How does each of these scores affect the mean, range, and standard deviation of each data set? Explain.

Find the mean, range, and standard deviation of the data.

15. 4.1, 2.3, 8.7, 10.5, 6.4

16. $-2, 0, 1, -5, 3, -4, 2, -3$

17. **REASONING** Two data sets have the same range. Can you assume that the standard deviations of the two data sets are about the same? Give an example to justify your answer.

18. **ADVENTURE CLUB** The dot plots show the ages of members of three different adventure clubs. Without performing calculations, which data set has the greatest standard deviation? Which has the least standard deviation? Explain your reasoning.

a.

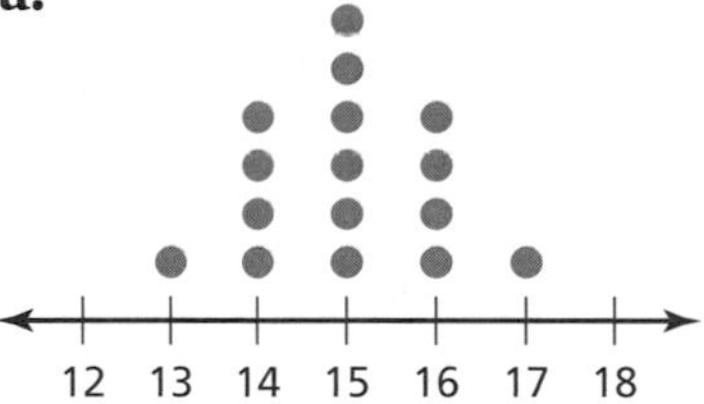

b.

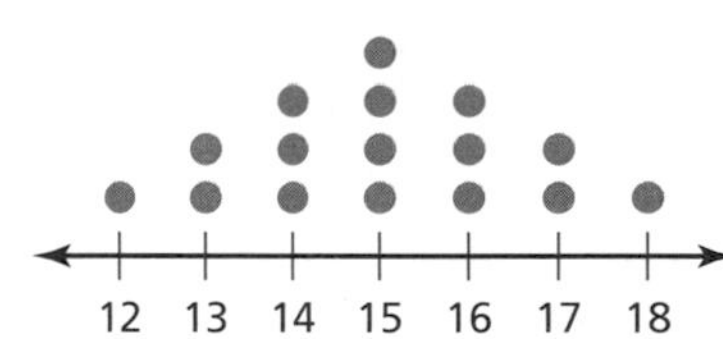

c.

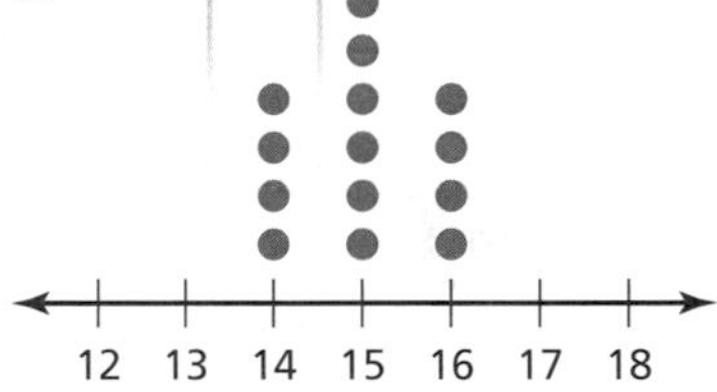

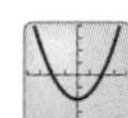

19. **PROJECT** Measure the heights (in inches) of the students in your class.

    a. Use a calculator to find the mean, range, and standard deviation of the heights.

    b. A new student who is 7 feet tall joins your class. How would you expect this person's height to affect the mean, range, and standard deviation? Verify your answer.

20. **WAITING TIMES** The waiting times at two doctors' offices are described below. At which office are you more likely to wait longer than 20 minutes? Explain. Assume the mean is at the center of each distribution and the data are evenly distributed.

    Doctor's Office A: mean = 15 minutes, standard deviation = 2.5 minutes

    Doctor's Office B: mean = 14 minutes, standard deviation = 5.5 minutes

21. Critical Thinking Can the standard deviation of a data set be 0? Can it be negative? If so, give examples to justify your answers.

Fair Game Review What you learned in previous grades & lessons

Graph the function. Describe the domain and range. *(Section 11.2)*

22. $y = -\dfrac{3}{x}$

23. $y = \dfrac{1}{x-6}$

24. $y = \dfrac{1}{x+4} - 5$

25. **MULTIPLE CHOICE** Find the quotient $(x+5) \div \dfrac{x^2+4x-5}{x+5}$. *(Section 11.4)*

Ⓐ $\dfrac{x+5}{x+1}$ Ⓑ $\dfrac{x+5}{x-1}$ Ⓒ x^2+4x-5 Ⓓ $\dfrac{x+5}{x-5}$

12.3 Box-and-Whisker Plots

Essential Question How can you use a box-and-whisker plot to describe a data set?

COMMON CORE STATE STANDARDS
S.ID.1
S.ID.2
S.ID.3

1 ACTIVITY: Drawing a Box-and-Whisker Plot

Work with a partner.

The numbers of first cousins of the students in an eighth-grade class are shown.

A box-and-whisker plot uses a number line to represent the data visually.

Numbers of First Cousins			
3	10	18	8
9	3	0	32
23	19	13	8
6	3	3	10
12	45	1	5
13	24	16	14

a. Order the data set and write it on a strip of grid paper with 24 equally spaced boxes.

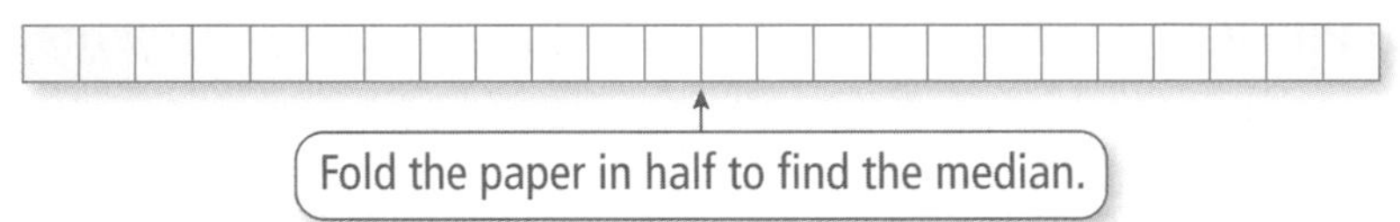

b. Fold the paper in half again to divide the data into four groups. Because there are 24 numbers in the data set, each group should have six numbers.

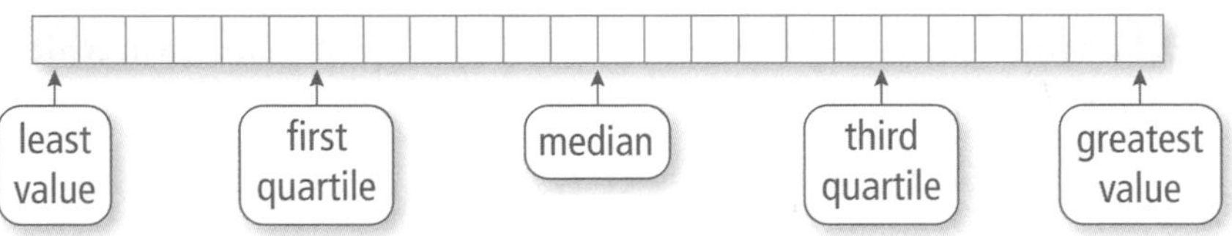

c. Draw a number line that includes the least value and the greatest value in the data set. Graph the five numbers that you found in part (b).

d. Explain how the box-and-whisker plot shown below represents the data set.

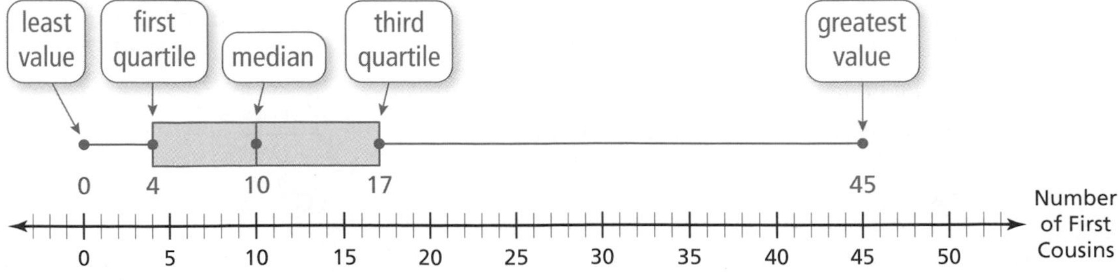

Laurie's Notes

Introduction

Standards for Mathematical Practice

- **MP3a Construct Viable Arguments:** Students create box-and-whisker plots to display data sets. Multiple box-and-whisker plots can be displayed on the same number line, therefore students are asked to construct viable arguments in comparing data sets.

Motivate

- Share a story about your commute to school today—perhaps the traffic, something you saw, or a stop you made for coffee. Conclude with how many minutes it took for your commute.
- Collect class data about the numbers of minutes it took your students to commute to school this morning, from the time they left their front doors until they walked into the school. If this is awkward data to collect, change the question. Data can be collected on slips of paper.
- Record the data on the board and leave it for later. You may want to take time to have students make comments about the data set.

Activity Notes

Activity 1

? "How many pieces of data are there?" 24

- Explain that today they are going to construct a box-and-whisker plot, a data display that is generally used for very large data sets. For instance, the results of a state test for all 8th graders could be displayed using a box-and-whisker plot.
- The box-and-whisker plot uses a number line to visually represent the data. Specific data values are *not* graphed, but characteristics of the data are still conveyed.

? "How many numbers did you graph in making the box-and-whisker plot?" 5

- **Big Idea:** The 5 numbers graphed summarize the entire data set. The least and greatest values are the boundaries. The median separates the data into two parts. The first (or lower) quartile is the median of the lower half. The third (or upper) quartile is the median of the upper half. The box encloses the middle 50% of the data.

? "What percent of the data is represented by the upper whisker?" 25%
"How many data values are in the upper whisker?" 6

? "What percent of the data is represented by the lower whisker?" 25%
"How many data values are in the lower whisker?" 6

- **MP2 Reason Abstractly and Quantitatively:** These questions provide opportunities for students to reason quantitatively.

Common Core State Standards

S.ID.1 Represent data with plots on the real number line (dot plots, histograms, and box plots).
S.ID.2 Use statistics appropriate to the shape of the data distribution to compare center (median, mean) and spread (interquartile range, standard deviation) of two or more different data sets.
S.ID.3 Interpret differences in shape, center, and spread in the context of the data sets, accounting for possible effects of extreme data points (outliers).

Previous Learning

Students should know how to find the median of a data set.

Start Thinking! and Warm Up

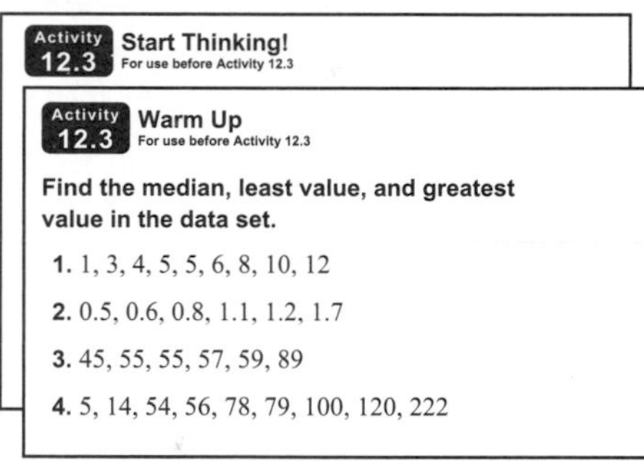

Activity 12.3 Start Thinking!
For use before Activity 12.3

Activity 12.3 Warm Up
For use before Activity 12.3

Find the median, least value, and greatest value in the data set.

1. 1, 3, 4, 5, 5, 6, 8, 10, 12
2. 0.5, 0.6, 0.8, 1.1, 1.2, 1.7
3. 45, 55, 55, 57, 59, 89
4. 5, 14, 54, 56, 78, 79, 100, 120, 222

12.3 Record and Practice Journal

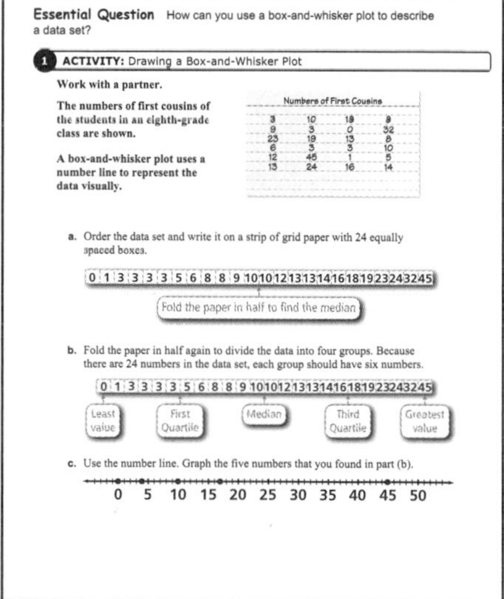

Essential Question How can you use a box-and-whisker plot to describe a data set?

1 ACTIVITY: Drawing a Box-and-Whisker Plot

Work with a partner.

The numbers of first cousins of the students in an eighth-grade class are shown.

A box-and-whisker plot uses a number line to represent the data visually.

Numbers of First Cousins			
3	10	18	8
9	3	0	32
23	19	13	8
6	3	3	10
12	45	1	5
13	24	16	14

a. Order the data set and write it on a strip of grid paper with 24 equally spaced boxes.

b. Fold the paper in half again to divide the data into four groups. Because there are 24 numbers in the data set, each group should have six numbers.

c. Use the number line. Graph the five numbers that you found in part (b).

English Language Learners

Visual

Use a diagram of a generic box-and-whisker plot on an overhead as a visual aid for English learners. Have students identify the parts of the box-and-whisker plot: *median, first quartile, third quartile, least value,* and *greatest value.* Make sure students understand that they can interpret a box-and-whisker plot that does not have a scale.

12.3 Record and Practice Journal

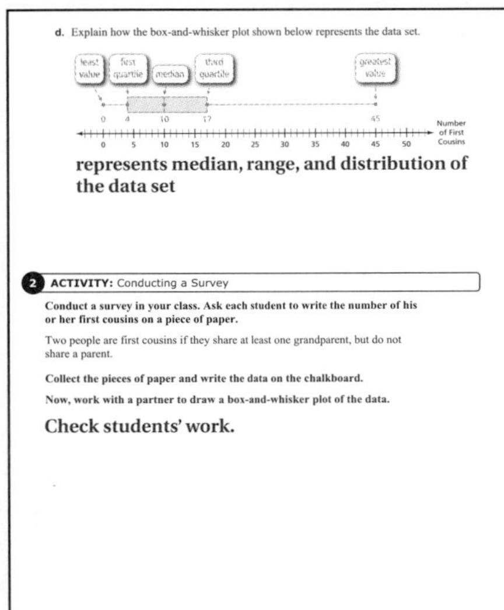

d. Explain how the box-and-whisker plot shown below represents the data set.

represents median, range, and distribution of the data set

2 ACTIVITY: Conducting a Survey

Conduct a survey in your class. Ask each student to write the number of his or her first cousins on a piece of paper.

Two people are first cousins if they share at least one grandparent, but do not share a parent.

Collect the pieces of paper and write the data on the chalkboard.

Now, work with a partner to draw a box-and-whisker plot of the data.

Check students' work.

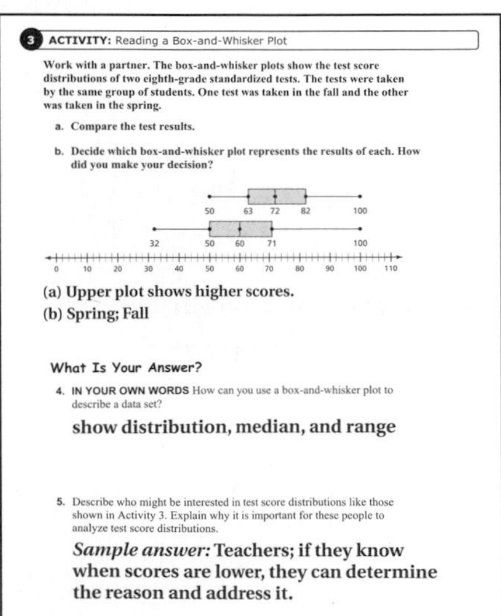

3 ACTIVITY: Reading a Box-and-Whisker Plot

Work with a partner. The box-and-whisker plots show the test score distributions of two eighth-grade standardized tests. The tests were taken by the same group of students. One test was taken in the fall and the other was taken in the spring.

a. Compare the test results.

b. Decide which box-and-whisker plot represents the results of each. How did you make your decision?

(a) Upper plot shows higher scores.
(b) Spring; Fall

What Is Your Answer?

4. **IN YOUR OWN WORDS** How can you use a box-and-whisker plot to describe a data set?

show distribution, median, and range

5. Describe who might be interested in test score distributions like those shown in Activity 3. Explain why it is important for these people to analyze test score distributions.

Sample answer: **Teachers; if they know when scores are lower, they can determine the reason and address it.**

Laurie's Notes

Activity 2

- Explain that you want to practice making a box-and-whisker plot with data collected from students. You will gather and record information about the number of first cousins for each student in your class. If this is awkward, ask a different question.
- Students should follow the steps from Activity 1 to construct this plot. If there is an odd number of students in the class, the median is the middle value of the sorted data. To find the first quartile, exclude the median and find the median of the lower half of the data. To find the third quartile, exclude the median and find the median of the upper half of the data.
- As students are making the plot, make the same plot at the overhead projector for discussion purposes.
- Ask questions about the plot: median, range, number of data values considered versus number of data values graphed.

Activity 3

- One advantage of box-and-whisker plots is that multiple plots can be displayed and analyzed using the same number line. For instance, state test scores for 5 different schools could be displayed on the same number line.
- Read the information given and analyze the two plots.
- **MP3b Critique the Reasoning of Others:** Students should be listening to the analysis offered by their partners or other classmates and critiquing their reasoning.

? "Which test is represented by which plot? Explain." Listen for students discussing the location of the median and the third quartile for each plot. The spring test is the top plot.

? "True or false: 50% of the scores in the top plot are greater than 75% of the scores in the bottom plot." true

What Is Your Answer?

- **Think-Pair-Share:** Students should read each question independently and then work with a partner to answer the questions. When they have answered the questions, the pair should compare their answers with another group and discuss any discrepancies.

Closure

- Make a box-and-whisker plot of the data collected at the beginning of class. Write one or two observations about the plot.

Technology For the Teacher

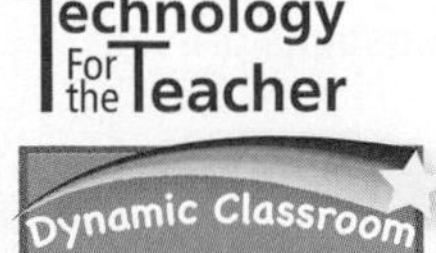

The Dynamic Planning Tool
Editable Teacher's Resources at *BigIdeasMath.com*

2 ACTIVITY: Conducting a Survey

Conduct a survey in your class. Ask each student to write the number of his or her first cousins on a piece of paper. Collect the pieces of paper and write the data on the chalkboard.

Now, work with a partner to draw a box-and-whisker plot of the data.

Two people are first cousins if they share at least one grandparent, but do not share a parent.

3 ACTIVITY: Reading a Box-and-Whisker Plot

Work with a partner. The box-and-whisker plots show the test score distributions of two eighth-grade standardized tests. The tests were taken by the same group of students. One test was taken in the fall and the other was taken in the spring.

a. Compare the test results.

b. Decide which box-and-whisker plot represents the results of each test. How did you make your decision?

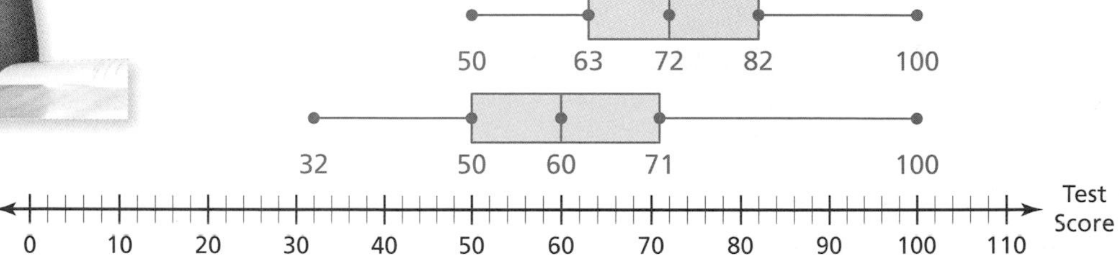

What Is Your Answer?

4. **IN YOUR OWN WORDS** How can you use a box-and-whisker plot to describe a data set?

5. Describe who might be interested in test score distributions like those shown in Activity 3. Explain why it is important for these people to analyze test score distributions.

Use what you learned about box-and-whisker plots to complete Exercise 4 on page 623.

12.3 Lesson

Key Vocabulary
box-and-whisker plot, *p. 620*
quartile, *p. 620*
five-number summary, *p. 620*
interquartile range, *p. 621*

Box-and-Whisker Plot

A **box-and-whisker plot** displays a data set along a number line using medians. **Quartiles** divide the data set into four equal parts. The median (second quartile) divides the data set into two halves. The median of the lower half is the first quartile. The median of the upper half is the third quartile.

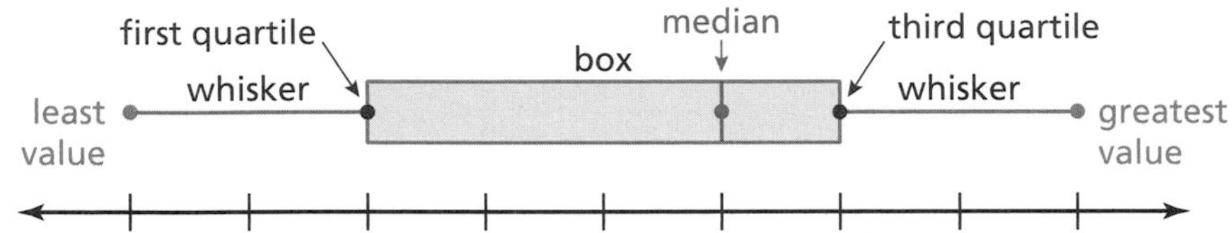

The five numbers that make up the box-and-whisker plot are called the **five-number summary** of the data set.

EXAMPLE 1 Making a Box-and-Whisker Plot

Make a box-and-whisker plot for the ages of the members of the U.S. women's wheelchair basketball team.

24, 30, 30, 22, 25, 22, 18, 25, 28, 30, 25, 27

Step 1: Order the data. Find the median and the quartiles.

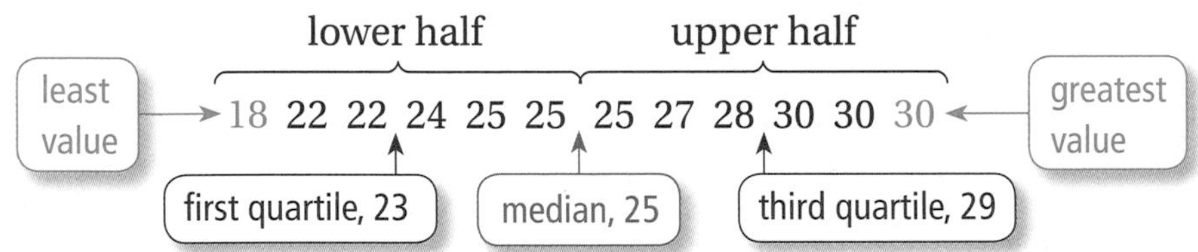

Step 2: Draw a number line that includes the least and greatest values. Graph points above the number line for the five-number summary.

Step 3: Draw a box using the quartiles. Draw a line through the median. Draw whiskers from the box to the least and greatest values.

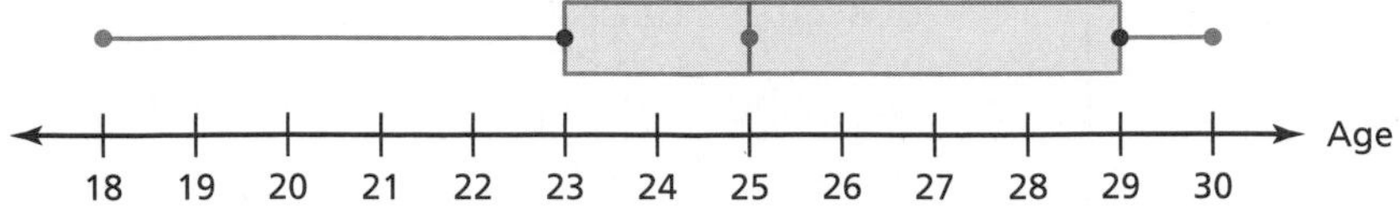

Study Tip

A box-and-whisker plot shows the *variability* of a data set.

On Your Own

Now You're Ready
Exercises 5–7

1. A basketball player scores 14, 16, 20, 5, 22, 30, 16, and 28 points during a tournament. Make a box-and-whisker plot for the points scored by the player.

Laurie's Notes

Introduction

Connect

- **Yesterday:** Students gained a general understanding of how a box-and-whisker plot is constructed. (MP2, MP3)
- **Today:** Students will construct and analyze box-and-whisker plots.

Motivate

- The physical involvement of making a human box-and-whisker plot makes a lasting impression on students.
- **Preparation:** Give each student an index card with a number written on it. Include an outlier or two on one end of the data.

? "What is the first step in making a box-and-whisker plot?" sort the data

- Students should stand up and sort themselves. Have the median, the first and third quartiles, and the least and greatest data values take one step forward. If there is an even number of data values, the middle two students must figure out how to represent the mean and take a step forward.
- Make a number line on the floor or on the board. Position the 5 key values. If the plot is done on the floor, use string to form the whiskers. Students will have to visualize the box.
- **MP2 Reason Abstractly and Quantitatively:** Discuss features of the plot. If the plot includes an outlier, the length of the string becomes an instant topic of conversation. Students recognize that the same number of data values (25% of the class) is being represented by each whisker, yet the lengths of string are very different.

Lesson Notes

Key Idea

- Define the box-and-whisker plot constructed by graphing the quartiles. Draw the sample plot and discuss the process and vocabulary.
- **Discuss:** The box-and-whisker plot shows the *variability* of the data. Refer to this idea in each example done today.
- Tell students that box-and-whisker plots are also called *boxplots*.
- Discuss the vocabulary term *five-number summary*.

Example 1

- Remind students that they must sort the data to find the quartiles.
- Notice that the first quartile (23) and the third quartile (29) are not data values from the set. This is fine. The five values are simply giving a marker for how the sorted data is spread out into four groups.

? "How many players are represented in each quartile?" 3

? "What percent of the players were older than 23?" 75%

? "What is the range of ages for the team?" 12

On Your Own

- This is a very small data set to make it manageable for students.

Goal Today's lesson is constructing and analyzing a **box-and-whisker plot**.

Lesson Materials
Introduction
• index cards • string or yarn

Start Thinking! and Warm Up

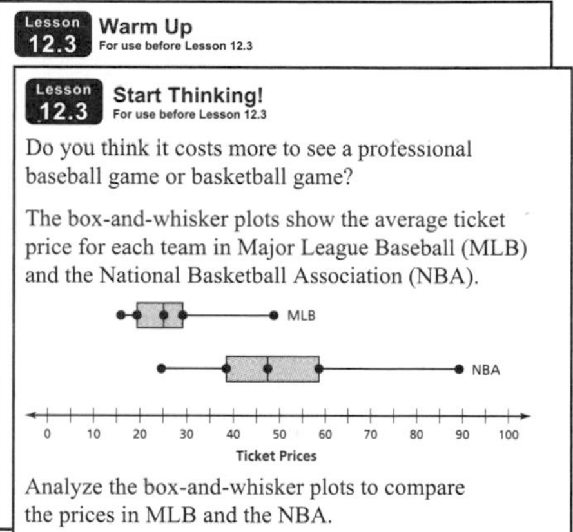

Lesson 12.3 Warm Up
For use before Lesson 12.3

Lesson 12.3 Start Thinking!
For use before Lesson 12.3

Do you think it costs more to see a professional baseball game or basketball game?

The box-and-whisker plots show the average ticket price for each team in Major League Baseball (MLB) and the National Basketball Association (NBA).

Analyze the box-and-whisker plots to compare the prices in MLB and the NBA.

Extra Example 1

Make a box-and-whisker plot for the ages of the members of a women's basketball team.

25, 22, 18, 23, 27, 20, 18, 25, 28, 17, 23, 18

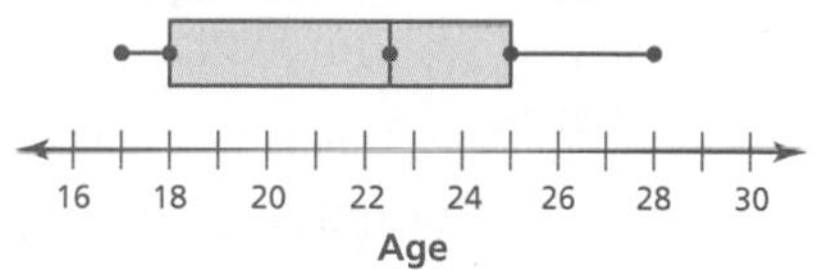

On Your Own

1.

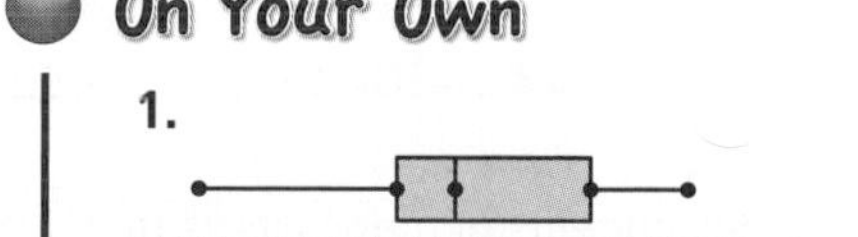
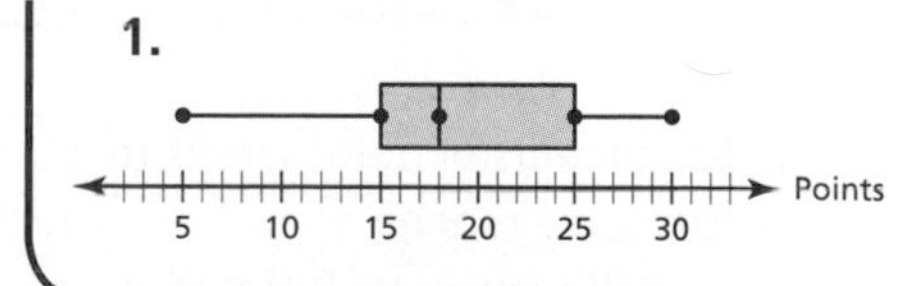

Extra Example 2

You have written several fiction stories about your favorite television character and posted them on the Internet. The box-and-whisker plot represents the number of reader reviews each of your stories received.

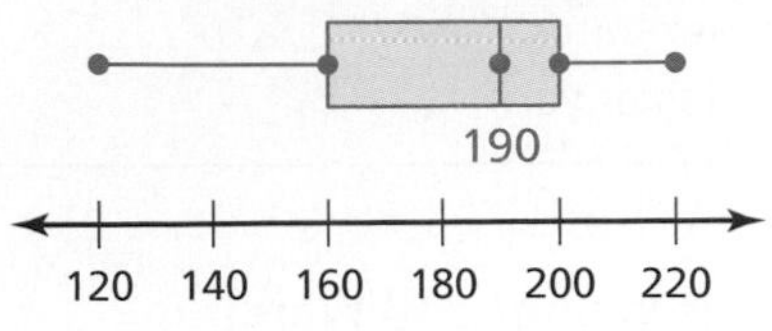

a. Find and interpret the range of the data. range: 100; This means that the number of reviews of each story varies by no more than 100.

b. Describe the distribution of the data. One-quarter of the stories received 160 reviews or less. One half of the stories received between 160 and 200 reviews. One-quarter of the stories received 200 or more reviews.

c. Find and interpret the interquartile range of the data. interquartile range: 40; This means that the middle half of the number of reviews vary by no more than 40.

On Your Own

2. range: 12 yr; This means that the ages vary by no more than 12 years; interquartile range: 6 yr; This means that the middle half of the ages vary by no more than 6 years.
3. See Additional Answers.

Differentiated Instruction

Auditory

Remind students of other words that have the same root as *quartile:* quarter and quart, for example. Define the words as *four* or *fourths*. Mention that one-fourth is the same as 25%.

Laurie's Notes

Discuss

- The box portion of the plot contains 50% of the data. Its length is called the interquartile range, IQR.
- ? "What portion of the data is in the box and to the left of the median?" $\frac{1}{4}$
- ? "What portion of the data is in the box and to the right of the median?" $\frac{1}{4}$
- In the sample shown, the lengths of the two parts of the box indicate that the data in the box are less spread out on the right side of the median than on the left side of the median.
- Similar statements about the dispersion of the data can be made by looking at the size (length) of each of the four parts of the plot.

Example 2

- ? "Do you know from the boxplot how many songs the band played at the concert?" no
- It is important for students to remember that the number of data points cannot be determined solely from a boxplot. Only the five summary points are known, and only two of those (least value and greatest value) *must* be values in the data set.
- ? "What is the range of the data and what does it mean in the context of the problem?" The range is $300 - 160 = 140$. This means that the difference between the shortest song and longest song was 140 seconds, or 2 minutes 20 seconds.
- ? "What does the length of each part of the boxplot tell you about the distribution of the data?" *Sample answers:* 50% of the songs were between 260 seconds and 300 seconds; 25% of the songs were 220 seconds or less.
- In part (c), the students are focusing on the middle 50% of the data. The IQR is 60 seconds. One way to interpret this result is that the middle half of the song lengths vary by no more than 60 seconds.

On Your Own

- **MP3a Construct Viable Arguments:** Ask different students to explain their answers. Check for correct language and valid arguments.

The figure shows how data are distributed in a box-and-whisker plot.

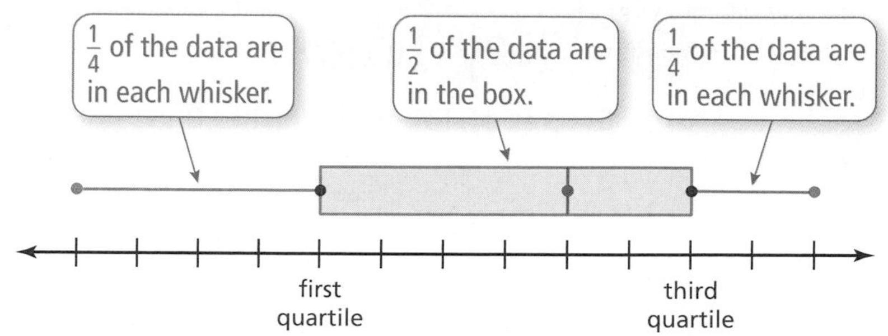

$\frac{1}{4}$ of the data are in each whisker.
$\frac{1}{2}$ of the data are in the box.
$\frac{1}{4}$ of the data are in each whisker.
first quartile
third quartile

Another measure of dispersion for a data set is the **interquartile range,** which is the difference of the third quartile and the first quartile. It represents the range of the middle half of the data.

EXAMPLE 2 Interpreting a Box-and-Whisker Plot

The box-and-whisker plot represents the lengths of songs (in seconds) played by a rock band at a concert.

140 160 180 200 220 240 260 280 300 320 Song Length (seconds)

a. Find and interpret the range of the data.

The least value is 160. The greatest value is 300.

So, the range is $300 - 160 = 140$ seconds. This means that the song lengths vary by no more than 140 seconds.

b. Describe the distribution of the data.

- 25% of the song lengths are between 160 and 220 seconds.
- 50% of the song lengths are between 220 and 280 seconds.
- 25% of the song lengths are between 280 and 300 seconds.

c. Find and interpret the interquartile range of the data.

$$\begin{aligned} \text{interquartile range} &= \text{third quartile} - \text{first quartile} \\ &= 280 - 220 \\ &= 60 \end{aligned}$$

So, the interquartile range is 60 seconds. This means that the middle half of the song lengths vary by no more than 60 seconds.

On Your Own

Now You're Ready
Exercises 10 and 11

Use the box-and-whisker plot in Example 1.

2. Find and interpret the range and interquartile range of the data.
3. Describe the distribution of the data.

A box-and-whisker plot shows the shape of a distribution.

Key Ideas

Shapes of Box-and-Whisker Plots

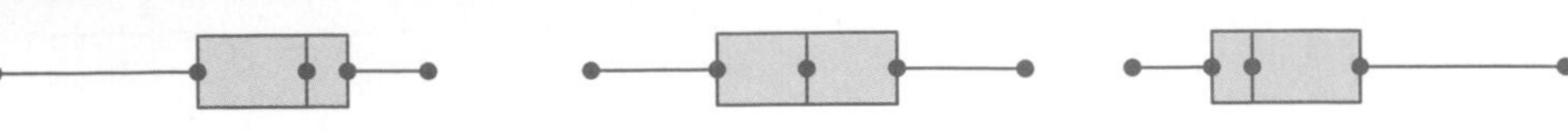

Skewed left	*Symmetric*	*Skewed right*
• Left whisker longer than right whisker	• Whiskers about same length	• Right whisker longer than left whisker
• Most data on the right	• Median in the middle of the data	• Most data on the left

Study Tip

If you can draw a line through the median of a box-and-whisker plot, and each side is a mirror image of the other, then the distribution is symmetric.

EXAMPLE 3 Comparing Box-and-Whisker Plots

The double box-and-whisker plot represents the test scores for your class and your friend's class.

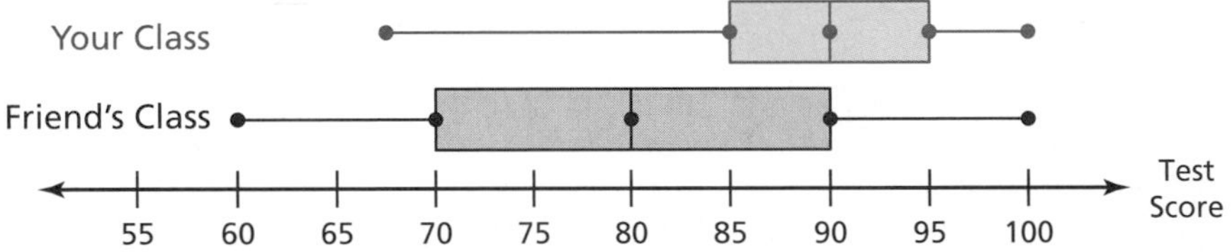

a. Identify the shape of each distribution.

For your class, the left whisker is longer than the right whisker, and most of the data are on the right side of the display. So, the distribution is skewed left.

For your friend's class, the whisker lengths are equal. The median is in the middle of the data. The data appear to be evenly distributed on both sides of the median. So, the distribution is symmetric.

b. Which test scores are more spread out?

The range of the test scores in your friend's class is greater than the range in your class. Also, because the box for your friend's class is longer than the box for your class, the interquartile range is also greater. So, the test scores in your friend's class are more spread out.

On Your Own

Now You're Ready
Exercise 20

4. The double box-and-whisker plot represents the surfboard prices at Shop A and Shop B. Identify the shape of each distribution. Which shop's prices are more spread out? Explain.

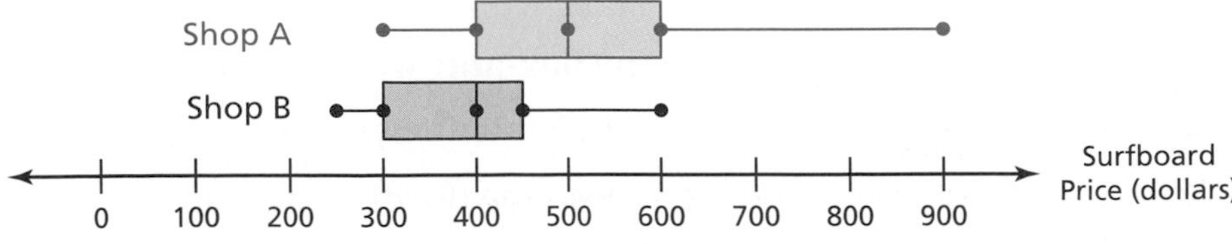

Laurie's Notes

Key Ideas

- Sketch each of the box-and-whisker plots. Discuss what the shape of the plot implies about the distribution of the data.
- ? "If the box-and-whisker plots represent the costs of jeans at three different stores and the range is the same for all three plots, what does the shape tell you about the distribution of the costs at all three stores?" Listen for student understanding of the spread of the data, referencing percentages and relative cost.

Example 3

- One advantage of a box-and-whisker plot is that multiple data sets can be shown in the same display, as with double bar graphs and double line graphs. The data sets can also have different numbers of data values.
- Part (a) asks students to identify the shape. A student might comment that a low test score in "your class" could be the reason for it being skewed. Without the outlier it could be a symmetric graph.
- **MP2:** Supplement the discussion in part (b) with quantitative comparisons, such as 50% of your class scored as well as the top 25% of your friend's class.

On Your Own

- **Neighbor Check:** Have students work independently and then have their neighbor check their work. Have students discuss any discrepancies.
- Ask other quantitative questions abut the two boxplots.

Closure

- **Exit Ticket:**
  - What are the 5 key values that are graphed in a box-and-whisker plot? least value, first quartile, median, third quartile, greatest value
  - How does an outlier affect a box-and-whisker plot? *Sample answer:* increases the length of one of the whiskers
  - Explain why two data sets of different sizes can be graphed on the same number line. Box-and-whisker plots show the distribution of the data, not individual data points.

Technology For the Teacher

The Dynamic Planning Tool
Editable Teacher's Resources at *BigIdeasMath.com*

Extra Example 3

The double box-and-whisker plot represents the length of time (in minutes) it takes students in your gym class and your friend's gym class to run 1 mile.

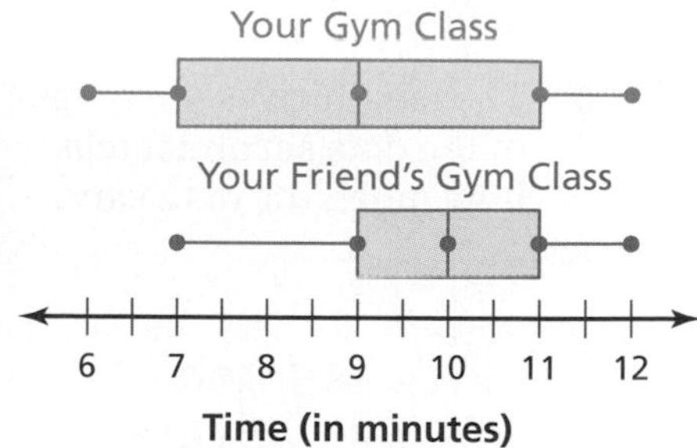

a. Identify the shape of each distribution. For your class, the whisker lengths are equal. The median is in the middle of the data. The data appear to be evenly distributed on both sides of the median. So, the distribution is symmetric. For your friend's class, the left whisker is longer than the right whisker, and most of the data are on the right side of the display. So, the distribution is skewed left.

b. Which lengths of time are more spread out? The range of the times in your class is greater than the range in your friend's class. Also, because the box for your class is longer than the box for your friend's class, the interquartile range is also greater. So, the times in your class are more spread out.

On Your Own

4. See Additional Answers.

Vocabulary and Concept Check

1. 25%; 50%
2. Find the median of the lower half of an ordered data set.
3. The length gives the range of the data set and it tells how much the data vary.

Practice and Problem Solving

4. *Sample answer:* Both sales reps have the same median number of cars sold, but Sales Rep B's numbers are more spread out than Sales Rep A's numbers.

5.

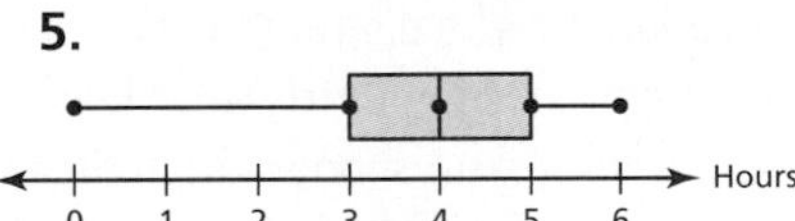

6.

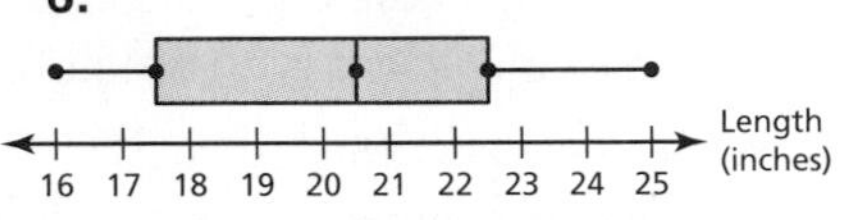

7.

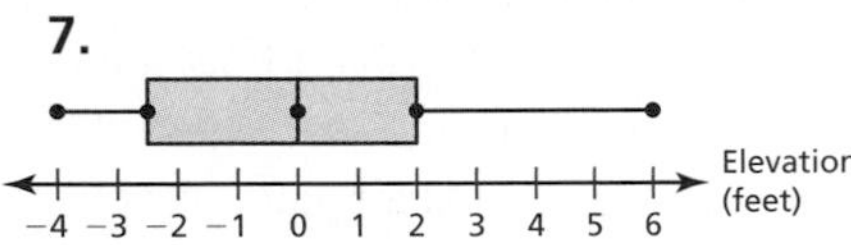

8. The median should be 5.5.

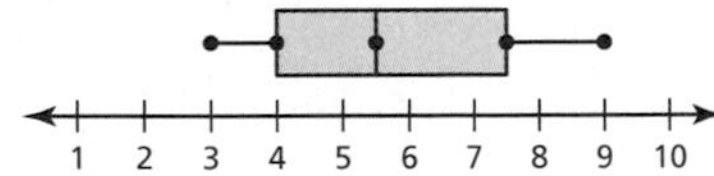

9. **a.** 11 in.

   b. about 8.5 in.; about 13.5 in.

   c.

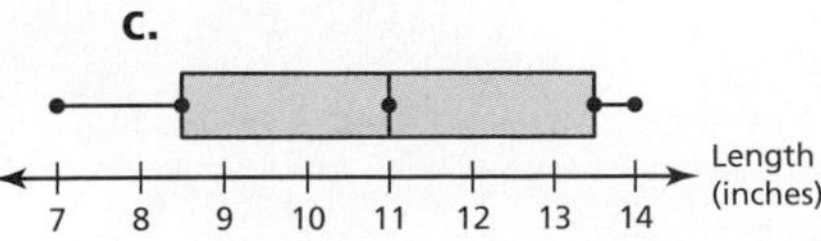

Assignment Guide and Homework Check

Level	Assignment	Homework Check
Average	1–3, 4, 5, 8–10, 12–18 even, 25–29	4, 9, 10, 14
Advanced	1–3, 6–18 even, 19–22, 25–29	10, 14,19, 21

Common Errors

- **Exercise 1** Students may confuse or forget what percent of the data from a data set is represented by each whisker and the box. Remind them of these percents.
- **Exercises 5–9** Students may have difficulty creating the box-and-whisker plots. Remind them of the five-number summary of a data set.

12.3 Record and Practice Journal

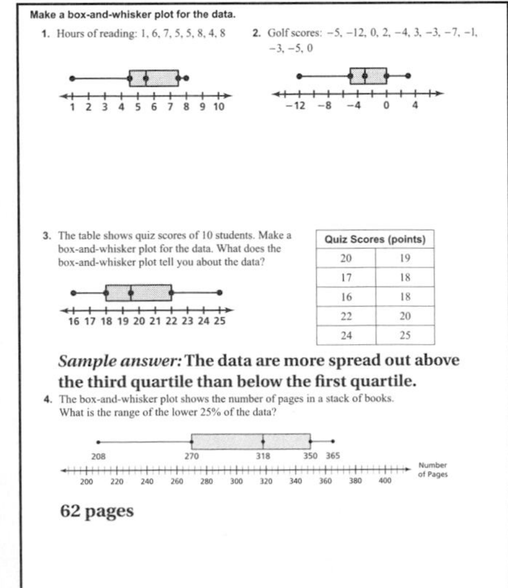

Make a box-and-whisker plot for the data.

1. Hours of reading: 1, 6, 7, 5, 5, 8, 4, 8

2. Golf scores: −5, −12, 0, 2, −4, 3, −3, −7, −1, −3, −5, 0

3. The table shows quiz scores of 10 students. Make a box-and-whisker plot for the data. What does the box-and-whisker plot tell you about the data?

Quiz Scores (points)	
20	19
17	18
16	18
22	20
24	25

***Sample answer:* The data are more spread out above the third quartile than below the first quartile.**

4. The box-and-whisker plot shows the number of pages in a stack of books. What is the range of the lower 25% of the data?

208 270 318 350 365

Number of Pages

62 pages

Technology For the Teacher

Answer Presentation Tool

12.3 Exercises

Vocabulary and Concept Check

1. **VOCABULARY** In a box-and-whisker plot, what percent of the data is represented by each whisker? by the box?
2. **WRITING** Describe how to find the first quartile of a data set.
3. **NUMBER SENSE** What does the length of a box-and-whisker plot tell you about the data?

Practice and Problem Solving

4. The box-and-whisker plots show the monthly car sales for a year for two sales representatives. Compare the sales for the two representatives.

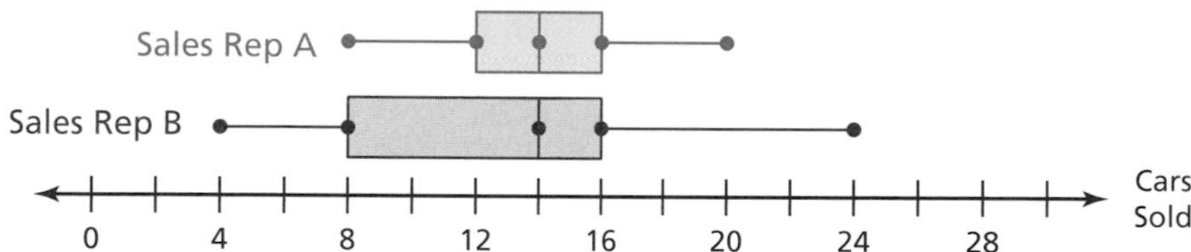

Make a box-and-whisker plot for the data.

5. Hours of television watched: 0, 3, 4, 5, 3, 4, 6, 5
6. Lengths (in inches) of cats: 16, 18, 20, 25, 17, 22, 23, 21
7. Elevations (in feet): −2, 0, 5, −4, 1, −3, 2, 0, 2, −3, 6, −1

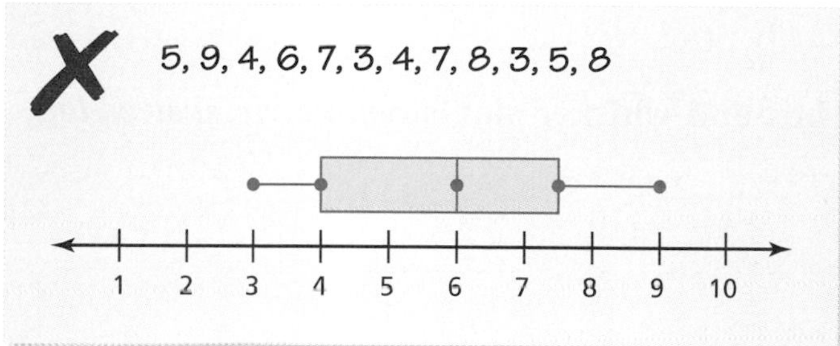

8. **ERROR ANALYSIS** Describe and correct the error in making a box-and-whisker plot for the data.

9. **FISH** The lengths (in inches) of the fish caught on a fishing trip are 9, 10, 12, 8, 13, 10, 12, 14, 7, 14, 8, and 14.
   a. What is the median of the data set?
   b. What are the first and third quartiles of the data set?
   c. Make a box-and-whisker plot for the data.

2 **10. INCHWORM** The table shows the lengths of 12 inchworms.

a. Make a box-and-whisker plot for the data.

b. Find and interpret the range of the data.

c. Describe the distribution of the data.

d. Find and interpret the interquartile range of the data.

Length (cm)	2.5	2.4	2.3	2.5	2.7	2.1	2.8	2.6	2.1	2.6	2.9	2.0

Entrée Prices (dollars)			
14.00	17.00	12.50	10.00
11.00	18.25	9.00	8.50
14.75	15.00	14.00	12.00

11. ENTRÉE The table shows the prices of entrées at a restaurant.

a. Make a box-and-whisker plot for the data.

b. Find and interpret the interquartile range of the data.

c. Describe the distribution of the data.

d. Find the standard deviation. Interpret your result.

12. WRITING Given the numbers 36 and 12, identify which number is the range, and which number is the interquartile range, of a set of data. Explain your reasoning.

Determine whether the shape of the box-and-whisker plot is *symmetric, skewed left,* or *skewed right.* Explain.

13. 

14. 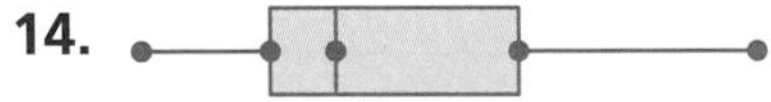

15. 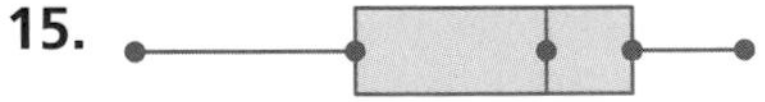

16. 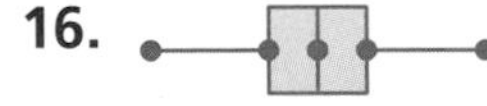

17. ERROR ANALYSIS Describe and correct the error in describing the box-and-whisker plot.

The shape of the distribution is skewed right. So, there are more data values to the right of the median than to the left of the median.

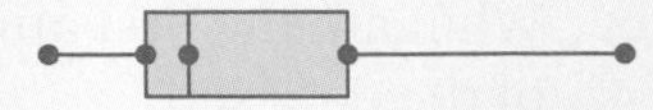

18. LOGIC Give examples of real-life data that are symmetric and real-life data that are not symmetric. Justify your answer.

Common Errors

- **Exercises 10–12, and 20** Students may confuse the range and interquartile range of a data set. Remind them of these definitions.
- **Exercises 13–16** Students may confuse the meanings of skewed left and skewed right. Tell them that the shape is determined by the longer whisker.
- **Exercise 18** Students may have difficulty coming up with examples of real-life data that are symmetric and real-life data that are not symmetric. Have them use this textbook or the Internet to search for examples.

Practice and Problem Solving

10. **a.**

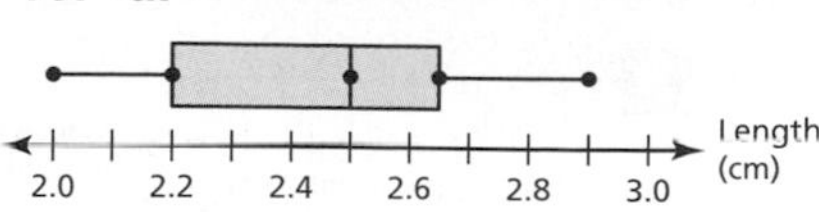

b. Inchworm lengths vary no more than 0.9 centimeter.

c. 25% of the lengths are between 2.0 and 2.2 centimeters; 50% of the lengths are between 2.2 and 2.65 centimeters; 25% of the lengths are between 2.65 and 2.9 centimeters.

d. The interquartile range is 0.45 centimeter. This means the middle half of the lengths vary by no more than 0.45 centimeter.

11. See Additional Answers.

12. range: 36; The range gives you the amount all the data vary by; interquartile range: 12; The interquartile range gives you the amount the middle half of the data vary by.

13. symmetric; The whiskers are about the same length and the median is in the middle of the data.

14–16. See Additional Answers.

17. The number of data values on each side of the median is the same.

18. *Sample Answer:* symmetric: height of students in a class; not symmetric: exam scores

Differentiated Instruction

Kinesthetic

To make ordering data easier, have students write each data value on a sticky note. Using their desk or a piece of notebook paper, students can move the notes to order the data and divide them into quartiles.

Practice and Problem Solving

19. a.

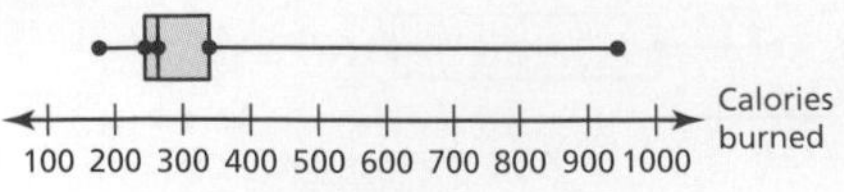

b. 944 calories

c.

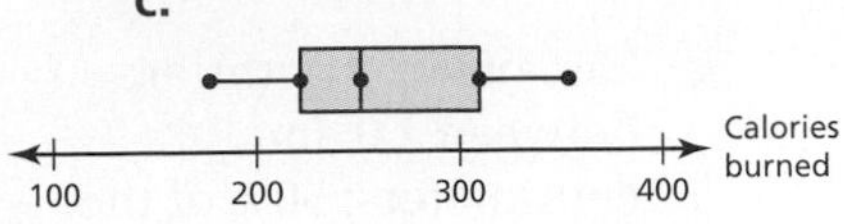

d. The outlier makes the right whisker longer, increases the length of the box, increases the third quartile, and increases the median. In this case, the first quartile and the left whisker were not affected.

20. See Additional Answers.

21. *Sample answer:* 0, 5, 10, 10, 10, 15, 20

22. See *Taking Math Deeper.*

23. *Sample answer:* 1, 7, 9, 10, 11, 11, 12

24. *Sample answer:* 1, 2, 4, 5, 10, 10, 10

Fair Game Review

25. $y = 3x + 2$

26. $y = -\frac{4}{3}x - 1$

27. $y = -\frac{1}{4}x$

28. $y = \frac{1}{2}x + 4$

29. B

Mini-Assessment

Make a box-and-whisker plot for the number of DVDs your class owns.

25, 31, 27, 36, 19, 22, 20, 24, 30, 32, 29, 27

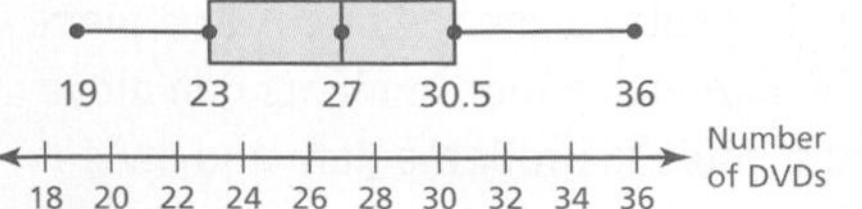

Taking Math Deeper

Exercise 22

This is a great opportunity for students to *work backwards* and demonstrate their understanding of the *five-number summary.*

Interpret the given characteristic:

Both whiskers are the same length as the box.

Express these lengths using the extreme values and the quartiles.

Length of left whisker = first quartile − least value

Length of box = third quartile − first quartile

Length of right whisker = greatest value − third quartile

Work backwards to write a data set.

The problem does not specify the length of each whisker or the box. So, assign an arbitrary value for each length, such as 10.

Assign an arbitrary value for any one of the extreme values or the quartiles (except the median). Let the greatest value be 50.

Using these values and the equations above, you can determine that the third quartile is 40, the first quartile is 30, and the least value is 20.

Choose arbitrary values to complete the data set. Make sure to add an appropriate number of values so that the extreme values and quartiles do not change. Use a box-and-whisker plot to help.

This is the tricky part.

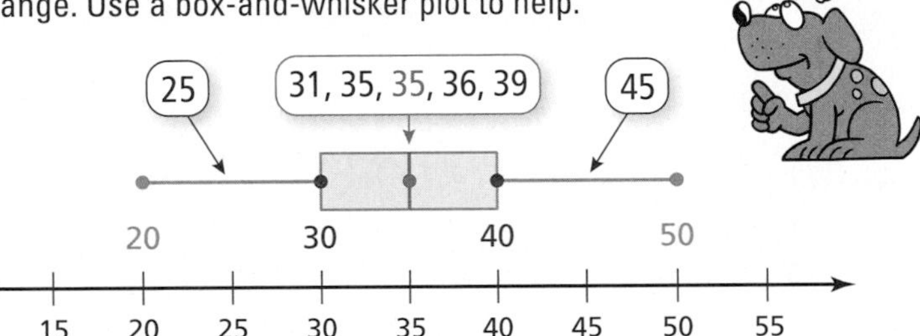

So, one possible data set is 20, 25, 30, 31, 35, 35, 36, 39, 40, 45, 50.

Reteaching and Enrichment Strategies

If students need help. . .	If students got it. . .
Resources by Chapter • Practice A and Practice B • Puzzle Time Record and Practice Journal Practice Differentiating the Lesson Lesson Tutorials Skills Review Handbook	Resources by Chapter • Enrichment and Extension • School-to-Work Start the next section

19. CALORIES The table shows the numbers of calories burned per hour for nine activities.

a. Make a box-and-whisker plot for the data.

b. Identify the outlier.

c. Make another box-and-whisker plot without the outlier.

d. WRITING Describe how the outlier affects the whiskers, the box, and the quartiles of the box-and-whisker plot.

Calories Burned per Hour	
Fishing	207
Mowing the lawn	325
Canoeing	236
Bowling	177
Hunting	295
Fencing	354
Bike racing	944
Horseback riding	236
Dancing	266

20. CELL PHONES The double box-and-whisker plot compares the battery lives (in hours) of two brands of cell phones.

a. Identify the shape of each distribution.

b. What is the range of the upper 75% of each brand?

c. Compare the interquartile ranges of the two data sets.

d. Which brand do you think has the greater standard deviation? Explain.

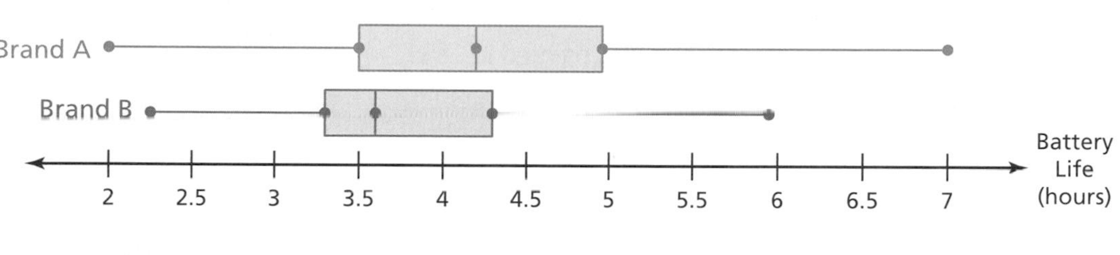

Modeling **Create a set of data values for the box-and-whisker plot that has the given characteristic(s).**

21. The least value, greatest value, quartiles, and median are all equally spaced.

22. Both whiskers are the same length as the box.

23. The box between the median and the first quartile is three times as long as the box between the median and the third quartile.

24. There is no right whisker.

Fair Game Review What you learned in previous grades & lessons

Write an equation of the line that passes through the points. *(Section 2.6)*

25. $(-4, -10), (2, 8)$

26. $(-3, 3), (0, -1)$

27. $(-4, 1), (4, -1)$

28. $(6, 7), (8, 8)$

29. MULTIPLE CHOICE What is the quotient of $(2z^2 - 13z + 21)$ and $(z - 3)$? *(Section 11.5)*

Ⓐ $2z + 7$ Ⓑ $2z - 7$ Ⓒ $z + 6$ Ⓓ $z - 7$

12.4 Shapes of Distributions

COMMON CORE STATE STANDARDS
S.ID.2
S.ID.3

Essential Question How can you use a histogram to characterize the basic shape of a distribution?

1 ACTIVITY: Analyzing a Famous Symmetric Distribution

A famous data set was collected in Scotland in the mid-1800s. It contains the chest sizes, measured in inches, of 5738 men in the Scottish Militia.

The Thin Red Line is a painting by Robert Gibb. It was painted in 1881. Only the left portion of the painting is shown in the photo at the right.

Chest Size	Number of Men
33	3
34	18
35	81
36	185
37	420
38	749
39	1073
40	1079
41	934
42	658
43	370
44	92
45	50
46	21
47	4
48	1

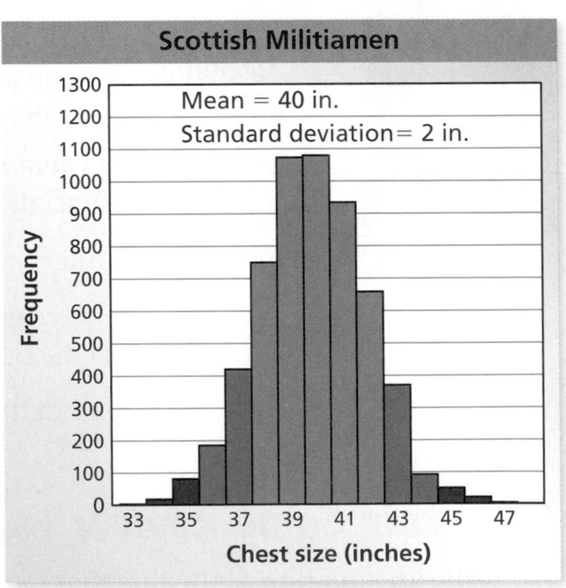

Work with a partner. What percent of the chest sizes lie within (a) 1 standard deviation, (b) 2 standard deviations, and (c) 3 standard deviations of the mean? Explain your reasoning.

Laurie's Notes

Introduction

Standards for Mathematical Practice

- **MP2 Reason Abstractly and Quantitatively:** Students will describe the shape of a distribution and choose appropriate measures of central tendency and spread. The shape of a histogram (visual display) will help students to reason abstractly about the center and spread of the data.

Motivate

- Ask students if they know what a militia is. You could give them the hint that during the American Revolutionary War, the Minutemen were members of a militia. A militia refers to a military force made up of citizens that protect their community/town in times of emergency.
- Tell students about the Scottish Militia Bill, which was passed by the Houses of Parliament in Great Britain in the early 1700s. The purpose of this bill was to help create uniformity in governing the militias across Great Britain. This bill was not passed into law by Queen Anne, because she feared the militia in Scotland would not be loyal. This would mark the last time a British ruler would veto a bill passed by both Houses of Parliament.
- Data about the Scottish Militia is the subject of today's activity.

Activity Notes

Activity 1

- The bar graph in this activity is color-coded according to the frequency table. Discuss the color-coding before students begin the activity with their partner. Students may even note the symmetry of the data about the mean, which can be recognized numerically and graphically.
- ? "In this activity, what does it mean to lie within 1 standard deviation of the mean?" It means a chest size is within 2 inches on either side of the mean of 40 inches. So, it is between 38 inches and 42 inches.
- ? "In this activity, what does it mean to lie within 2 standard deviations of the mean?" It means a chest size is within 4 inches on either side of the mean of 40 inches. So, it is between 36 inches and 44 inches.
- ? "In this activity, what does it mean to lie within 3 standard deviations of the mean?" It means a chest size is within 6 inches on either side of the mean of 40 inches. So, it is between 34 inches and 46 inches.
- A calculator is helpful in calculating these percents. Students who read carefully know that there are 5738 men represented in the frequency table and the graph. The percents are cumulative. So, the men who are within 1 standard deviation of the mean are also within 2 standard deviations of the mean, and so on.
- When students have finished, discuss their results for each part of the activity.
- **Big Idea:** More than 99% of the data are within 3 standard deviations of the mean. In this activity, 0.14% of the men have chest sizes that are more than 3 standard deviations from the mean.

Common Core State Standards

S.ID.2 Use statistics appropriate to the shape of the data distribution to compare center (median, mean) and spread (interquartile range, standard deviation) of two or more different data sets.

S.ID.3 Interpret differences in shape, center, and spread in the context of the data sets, accounting for possible effects of extreme data points (outliers).

Previous Learning

Students should know how to describe the shape of a box-and-whisker plot.

Activity Materials
Textbook
• calculators

Start Thinking! and Warm Up

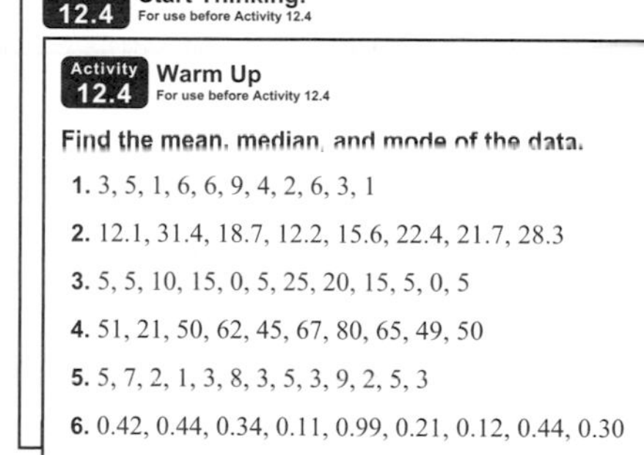

Activity 12.4 Start Thinking! For use before Activity 12.4

Activity 12.4 Warm Up For use before Activity 12.4

Find the mean, median, and mode of the data.

1. 3, 5, 1, 6, 6, 9, 4, 2, 6, 3, 1
2. 12.1, 31.4, 18.7, 12.2, 15.6, 22.4, 21.7, 28.3
3. 5, 5, 10, 15, 0, 5, 25, 20, 15, 5, 0, 5
4. 51, 21, 50, 62, 45, 67, 80, 65, 49, 50
5. 5, 7, 2, 1, 3, 8, 3, 5, 3, 9, 2, 5, 3
6. 0.42, 0.44, 0.34, 0.11, 0.99, 0.21, 0.12, 0.44, 0.30

12.4 Record and Practice Journal

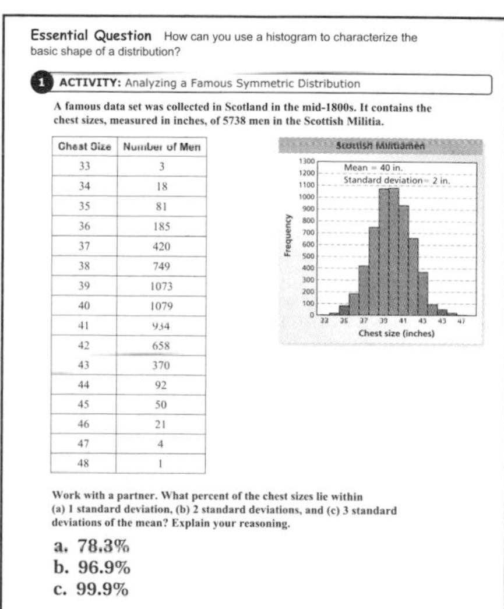

Essential Question How can you use a histogram to characterize the basic shape of a distribution?

1 ACTIVITY: Analyzing a Famous Symmetric Distribution

A famous data set was collected in Scotland in the mid-1800s. It contains the chest sizes, measured in inches, of 5738 men in the Scottish Militia.

Chest Size	Number of Men
33	3
34	18
35	81
36	185
37	420
38	749
39	1073
40	1079
41	934
42	658
43	370
44	92
45	50
46	21
47	4
48	1

Work with a partner. What percent of the chest sizes lie within (a) 1 standard deviation, (b) 2 standard deviations, and (c) 3 standard deviations of the mean? Explain your reasoning.

a. 78.3%
b. 96.9%
c. 99.9%

Differentiated Instruction

Inclusion

In this section, students are expected to make many histograms. Provide a resource page of blank axes with horizontal lines and vertical lines. Students add the vertical and horizontal scales, and labels, and color the bars to represent the data.

12.4 Record and Practice Journal

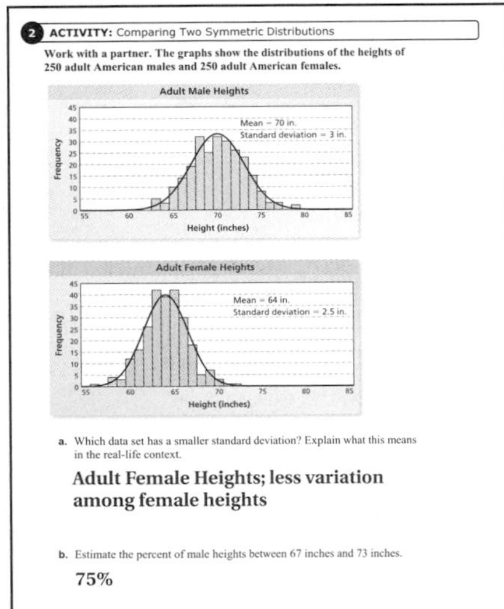

2 ACTIVITY: Comparing Two Symmetric Distributions

Work with a partner. The graphs show the distributions of the heights of 250 adult American males and 250 adult American females.

a. Which data set has a smaller standard deviation? Explain what this means in the real-life context.

Adult Female Heights; less variation among female heights

b. Estimate the percent of male heights between 67 inches and 73 inches.

75%

What Is Your Answer?

3. **IN YOUR OWN WORDS** How can you use a histogram to characterize the basic shape of a distribution?

You can determine whether the distribution is symmetric or skewed to one side, where the data is centered, and how spread out it is.

4. All three distributions in Activities 1 and 2 are roughly symmetric distributions. The histograms are called "bell-shaped."

a. What are the characteristics of a symmetric distribution?

the mean equals the median; not skewed

b. Why is a symmetric distribution called "bell-shaped"?

Histogram is shaped like a bell.

c. Give two other real-life examples of symmetric distributions.

Check students' work.

Laurie's Notes

Activity 2

- Discuss the two graphs with the class before they begin the activity. Students may ask about the curve that has been plotted on top of the histogram. They may have heard of a "bell-shaped curve." It is fine to suggest that the curve is tracing the basic shape of the distribution of heights. It is smooth versus the line segment outline of the actual graph. The bell-shaped curve will be referred to in Question 4.
- The frequencies on the vertical axis will help students estimate the size of the populations plotted.
- When students have finished, have them share their reasoning for part (a). Students should discuss the symmetry of the distributions and what that means in terms of the distribution of the data.
- Discuss the shape of the distribution and what it means in the real-life context. The mean and standard deviation for males is greater than the mean and standard deviation for females.
- **Connection:** In part (b), the range of the heights is one standard deviation about the mean.
- **MP1a Make Sense of Problems** and **MP3 Construct Viable Arguments and Critique the Reasoning of Others:** Having students explain their reasoning and having a whole class discussion, helps all students make sense of the problem and develop an ability to form arguments and critique the reasoning of others.

What Is Your Answer?

- Make sure students understand the connection between a bell shape and the distribution of the data. In the lesson that follows, students will see that not all distributions are symmetric or bell-shaped.

? "Can a distribution be symmetric and not be bell-shaped?" yes; The distribution could be U-shaped or uniform.

Closure

- Give an example of a real-life data set that may have a symmetric distribution. *Sample answers:* The pulse rates and shoe sizes of students in your class would most likely have symmetric distributions.

Technology For the Teacher

The Dynamic Planning Tool
Editable Teacher's Resources at *BigIdeasMath.com*

2 ACTIVITY: Comparing Two Symmetric Distributions

Work with a partner. The graphs show the distributions of the heights of 250 adult American males and 250 adult American females.

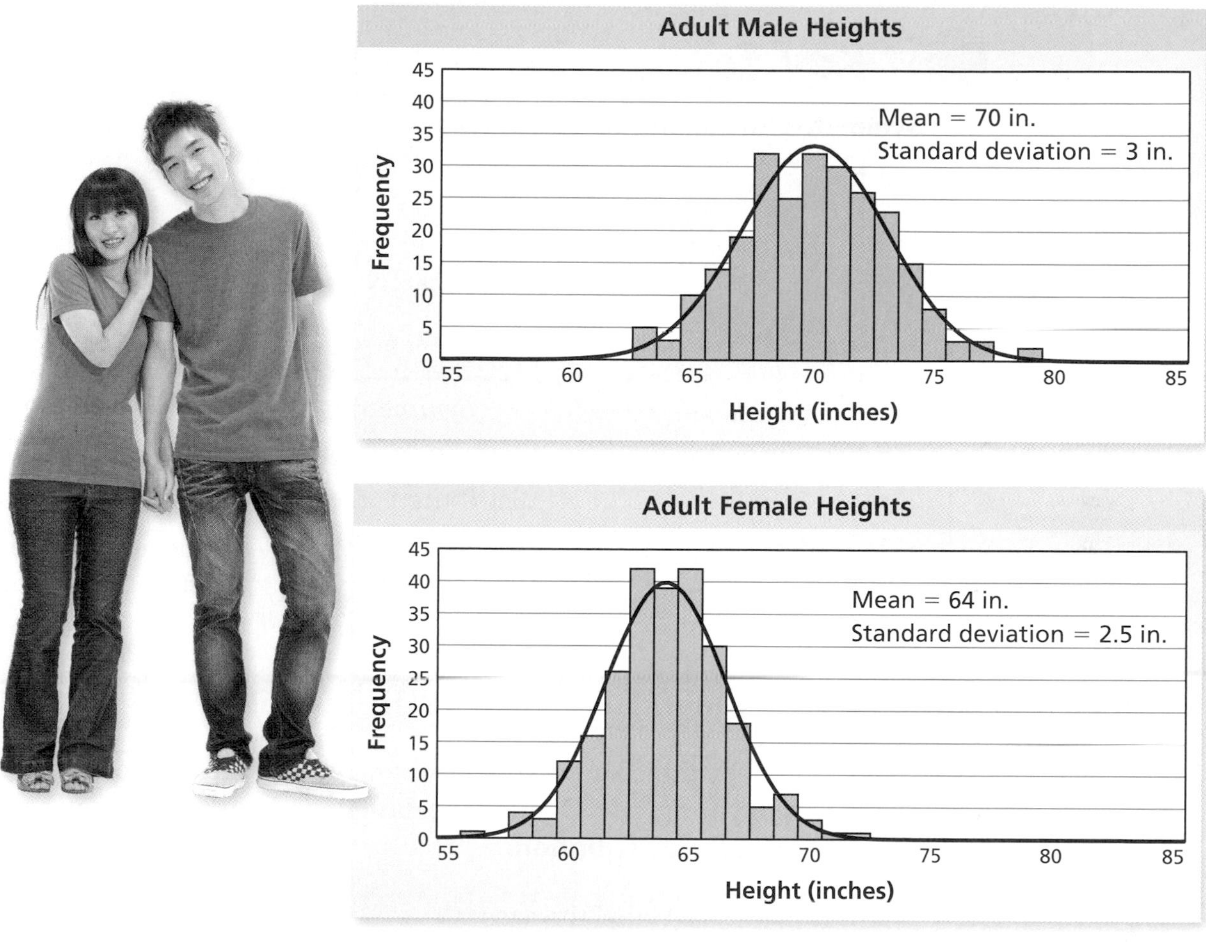

a. Which data set has a smaller standard deviation? Explain what this means in the real-life context.

b. Estimate the percent of male heights between 67 inches and 73 inches.

What Is Your Answer?

3. **IN YOUR OWN WORDS** How can you use a histogram to characterize the basic shape of a distribution?

4. All three distributions in Activities 1 and 2 are roughly symmetric distributions. The histograms are called "bell-shaped."

   a. What are the characteristics of a symmetric distribution?

   b. Why is a symmetric distribution called "bell-shaped"?

   c. Give two other real-life examples of symmetric distributions.

Practice Use what you learned about the shapes of distributions to complete Exercises 3 and 4 on page 631.

12.4 Lesson

Recall that a histogram is a bar graph that shows the frequency of data values in intervals of the same size. A histogram is another useful data display that shows the shape of a distribution.

Key Ideas

Symmetric and Skewed Distributions

Remember

If all the bars of a histogram are about the same height, then the distribution is a *flat*, or *uniform*, distribution. A uniform distribution is also symmetric.

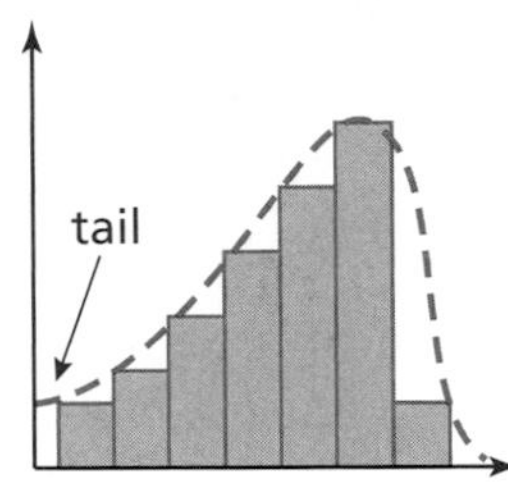

Skewed left

- The "tail" of the graph extends to the left.
- Most data are on the right.

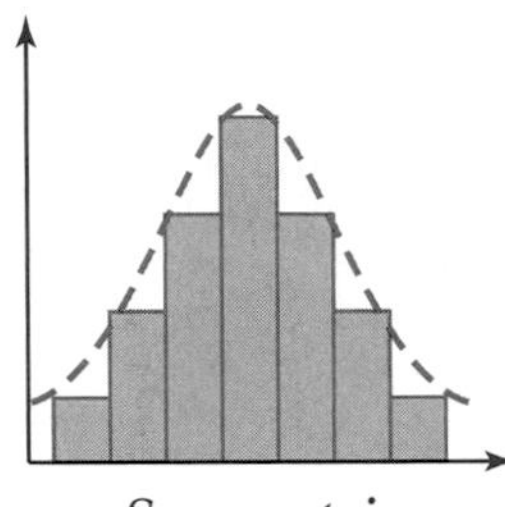

Symmetric

- The data are evenly distributed on each side of the highest bar.

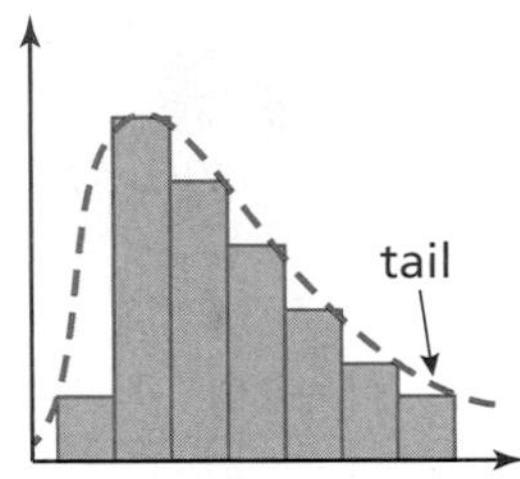

Skewed right

- The "tail" of the graph extends to the right.
- Most data are on the left.

EXAMPLE 1 Describing the Shape of a Distribution

Number of Tickets Sold	Frequency
1–8	5
9–16	9
17–24	16
25–32	25
33–40	20
41–48	8
49–56	7

The frequency table shows the numbers of raffle tickets sold by students in your grade. Display the data in a histogram. Describe the shape of the distribution.

Step 1: Draw and label the axes.

Step 2: Draw a bar to represent the frequency of each interval.

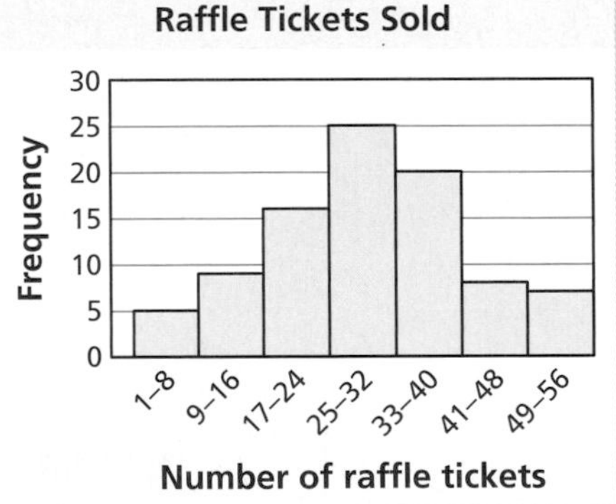

The graph is high in the middle, and the data are about evenly distributed on each side of the highest bar.

∴ So, the distribution is symmetric.

On Your Own

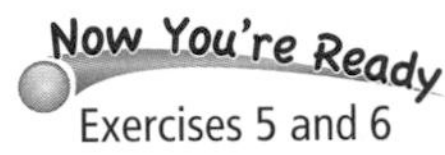

1. The frequency table shows the numbers of pounds of aluminum cans collected by students for a fundraiser. Display the data in a histogram. Describe the shape of the distribution.

Number of Pounds	Frequency
1–10	7
11–20	8
21–30	10
31–40	16
41–50	34
51–60	15

Laurie's Notes

Introduction

Connect

- **Yesterday:** Students used a histogram to describe the basic shape of a distribution. (MP1a, MP2, MP3)
- **Today:** Students will describe the shape of a distribution and choose measures of central tendency and spread to describe the data.

Motivate

- **Story time:** Tell the students that yesterday in the teacher's lounge, three teachers were describing the results of a recent test they gave.
  - **Teacher A:** Lots of low scores; Ds and Fs, some Cs, a few Bs, and one A
  - **Teacher B:** Lots of high scores; As and Bs, some Cs, a few Ds, and one F
  - **Teacher C:** Lots of average scores; Cs, some Bs and Ds, and a few Fs and As
- Ask students to quickly sketch a histogram for each of the three sets of test scores. They should sketch something similar to the three distributions in the Key Ideas.
- Discuss the attributes of a histogram: similar to a bar graph, intervals are all the same size, and frequencies of data values are displayed in the vertical direction by the bar heights.

Lesson Notes

Key Ideas

- Draw a sketch of each type of distribution and label it: skewed left, symmetric, and skewed right.
- Connect the distributions to the test score descriptions above.
- Explain that a dotted line can be drawn to help identify the greatest frequency and to help highlight the basic shape of a distribution.

? "Do all distributions fall into one of these three categories?" no; There are many different types of distributions. A distribution can be flat or uniform. It can have clusters of data or there could be gaps between data values. In short, there are distributions that have no discernible shape at all.

Example 1

- This is a good review of drawing a histogram. The intervals have already been determined, which is often the difficult part for students.
- There is no space between the bars of a histogram. When one of the frequencies is 0, the bar height for that interval is 0.
- **Extension:** Even though the actual data values are not known, have students estimate the mean of the distribution. You could also ask about the maximum range ($56 - 1 = 55$).

Goal Today's lesson is describing the shape of a distribution and determining the best way to represent it.

Start Thinking! and Warm Up

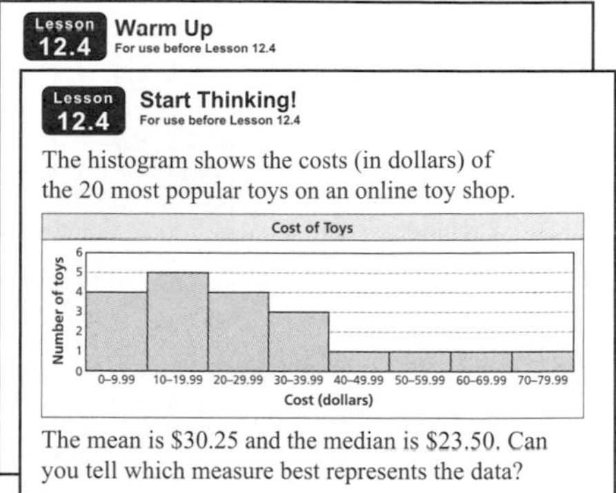
Lesson 12.4 Warm Up
For use before Lesson 12.4

Lesson 12.4 Start Thinking!
For use before Lesson 12.4

The histogram shows the costs (in dollars) of the 20 most popular toys on an online toy shop.

The mean is $30.25 and the median is $23.50. Can you tell which measure best represents the data?

Extra Example 1

The frequency table shows the numbers of pets owned by a group of students. Display the data in a histogram. Describe the shape of the distribution.

Number of Pets	Frequency
0–1	10
2–3	16
4–5	7
6–7	5
8–9	1

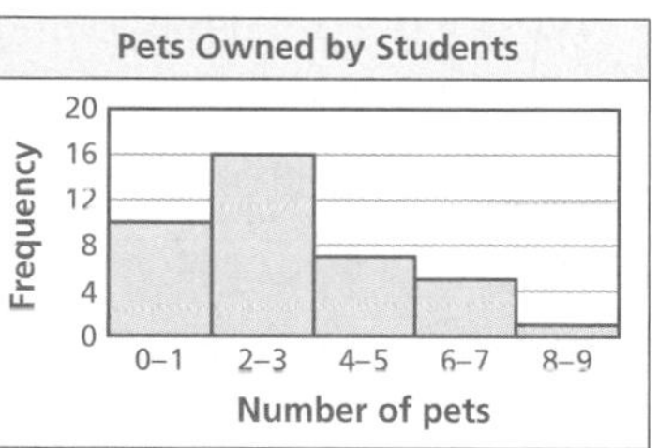

skewed right

On Your Own

1. See Additional Answers.

Extra Example 2

Which measure of central tendency best represents the data? Explain your reasoning.

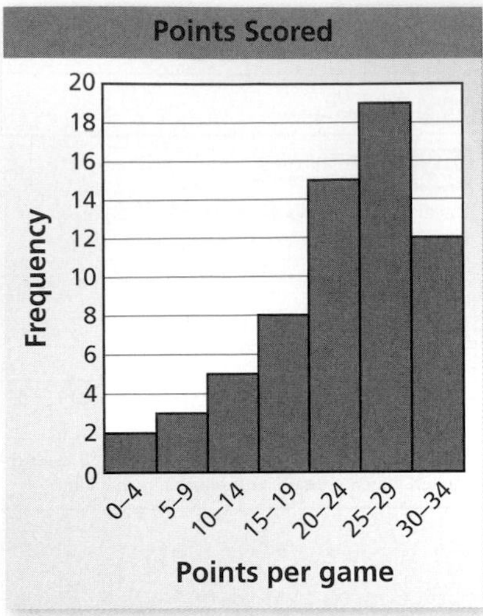

Because the distribution is high on the right and the tail of the graph extends left, the distribution is skewed left. So, the median best represents the data.

On Your Own

2. Because the distribution is high on the right and the tail of the graph extends left, the distribution is skewed left. So, the median best represents the data.

English Language Learners

Notebook Development

Provide English language learners with a handout of a simplified version of the notes from this lesson. This will allow the student to focus on the concepts being taught. Students should highlight information and add additional notes to the handout during the classroom discussion. They can refer to the page as they work through the exercises.

Laurie's Notes

On Your Own

- Students may want to answer the question without drawing the histogram. Certainly students could reason about the shape of the distribution from the frequencies given in the table, but without making a histogram, the degree of skewness might be difficult to judge.
- **Extension:** A graphing calculator can be used to construct the histogram.

Discuss

- Explain that the shape of a distribution can help you decide the most appropriate measure of central tendency when describing a data set.

? "When a distribution is symmetric, what do you know about the mean and median?" They are about the same and at the center of the distribution.

? "When a distribution is skewed left, what do you know about the mean and median?" The median is greater than the mean.

? "When a distribution is skewed right, what do you know about the mean and median?" The median is less than the mean.

? "When there is an outlier, which is affected more, the mean or the median?" mean

Example 2

? "What type of distribution is given in part (a)?" skewed right

? "Which measure of central tendency would best represent the data and why?" median; The mean is in the direction the distribution is skewed, making it greater than the median and much of the data.

- **MP3 Construct Viable Arguments and Critique the Reasoning of Others:** The explanations that students offer provide additional practice in constructing viable arguments.
- Discuss the histogram in part (b) in a similar manner.

The shape of a distribution can be used to choose the most appropriate measure of central tendency that describes a data set.

For a symmetric distribution, the mean and median are about the same, although the mean should be used to describe the center.

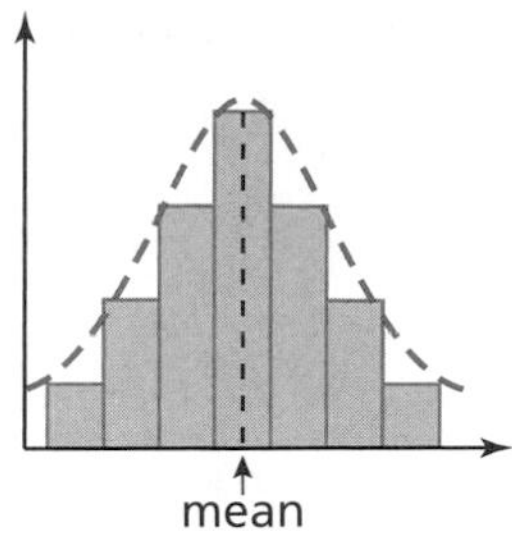

Remember

Outliers can affect the mean of a data set more than they affect the median.

When the distribution is skewed, the mean will be in the direction in which the distribution is skewed while the median will be less affected. So, when the data are skewed, use the median to describe the center.

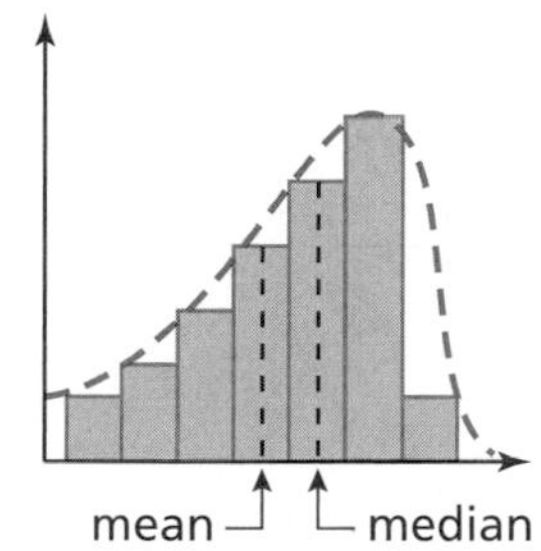

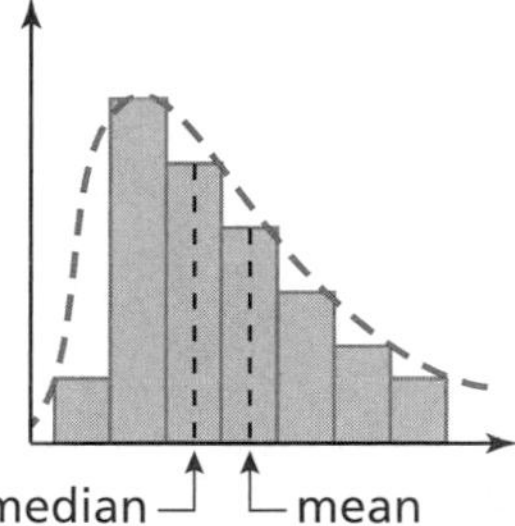

EXAMPLE 2 Choosing an Appropriate Measure of Central Tendency

Which measure of central tendency best represents the data? Explain your reasoning.

a.

b.

a. Because the distribution is high on the left and the tail of the graph extends to the right, the distribution is skewed right. So, the median best represents the data.

b. Because the distribution is high in the middle and the data are about evenly distributed on both sides, the distribution is symmetric. So, the mean best represents the data.

On Your Own

2. Which measure of central tendency best represents the data in On Your Own Question 1? Explain your reasoning.

When a distribution is symmetric, use the standard deviation to describe the spread of the data set. When a distribution is skewed, use the five-number summary to describe the spread of the data set.

EXAMPLE 3 Choosing Appropriate Measures

Speeds (mi/h)		
32	44	39
53	38	48
56	41	42
50	50	55
55	45	49
51	53	52
54	60	55
52	50	52
55	40	60
45	58	47

A police officer measures the speeds (in miles per hour) of 30 motorists. The results are shown in the table at the left.

a. Display the data in a histogram using six intervals beginning with 31–35.

Make a frequency table using the described intervals. Then use the frequency table to make a histogram.

Speed (mi/h)	Frequency
31–35	1
36–40	3
41–45	5
46–50	6
51–55	11
56–60	4

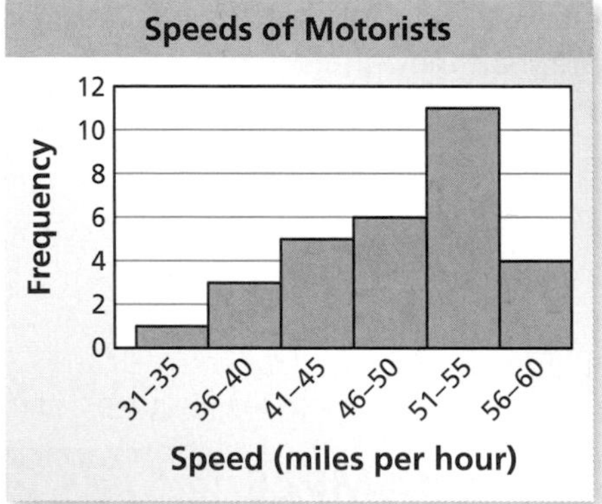

b. Which measures of central tendency and dispersion best represent the data?

Because the distribution is high on the right and the tail of the graph extends to the left, the distribution is skewed left. So, use the median to describe the center and the five-number summary to describe the spread.

c. The speed limit is 45 miles per hour. How would you interpret these results?

Because the distribution is skewed left, most of the speeds are more than 45 miles per hour. This shows that most of the motorists were speeding.

On Your Own

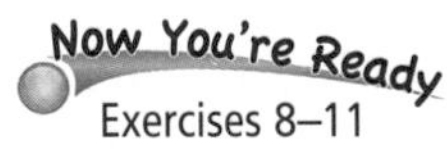

3. You record the numbers of email attachments sent by 30 employees of a company in one week. Your results are shown in the table.

Email Attachments Sent				
74	105	98	68	64
85	75	60	48	51
65	55	58	45	38
64	52	65	30	70
72	5	45	77	83
42	25	95	16	120

   a. Display the data in a histogram using six intervals beginning with 1–20.

   b. Which measures of central tendency and dispersion best represent the data? Why?

Laurie's Notes

Discuss

- When choosing appropriate measures of center and spread, tell students that the standard deviation is a measure of dispersion that is related to the mean because the mean is used in its formula. The quartiles of a five-number summary are measures of dispersion that are related to the median.

Example 3

- Ask for a volunteer to read the problem as you write the frequency table. Have students describe the process for making the histogram.

? "How would you describe the shape of the distribution?" The distribution is skewed left.

- **Connection:** When students constructed box-and-whisker plots using the five-number summary, it gave a visual representation of the spread of the data.
- **Teaching Tip:** In part (c), it may be helpful to draw a dotted vertical line along the right side of the 41–45 bar. This helps students see that much of the data is to the right of the dotted line, so most of the motorists were speeding.

On Your Own

- This problem takes additional time because students must make the frequency table before making the histogram. Circulate to check that students are using the correct intervals.
- Ask a few students to do their work on transparencies if a document camera is not available.

Closure

- Refer to the three data sets in the Motivate. Ask students questions about the most appropriate measures of central tendency and spread for each data set.

Technology For the Teacher

The Dynamic Planning Tool
Editable Teacher's Resources at *BigIdeasMath.com*

Extra Example 3

A personal trainer records the numbers of minutes that 30 clients run on a treadmill. The results are shown in the table.

Minutes on Treadmill		
25	20	35
33	50	22
28	15	34
60	50	8
25	20	37
6	39	24
30	45	15
40	25	45
18	60	30
40	35	40

a. Display the data in a histogram using six intervals beginning with 1–10.

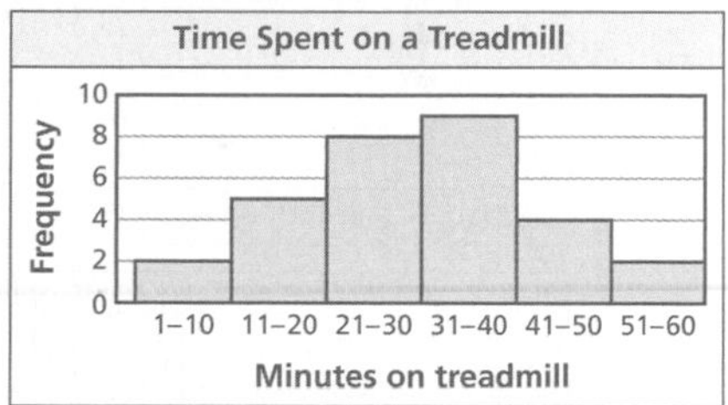

b. Which measures of central tendency and dispersion best represent the data? mean and standard deviation

On Your Own

3. a. See Additional Answers.

b. The distribution is symmetric. So, use the mean to describe the center and the standard deviation to describe the spread.

Vocabulary and Concept Check

1. The shape of a skewed distribution will have a tail on one side. The shape of a symmetric distribution is even, or symmetrical, with respect to the mean.

2. The mean should be used to describe the center of a symmetric distribution. The median should be used to describe the center of a skewed distribution.

Practice and Problem Solving

3. About 95%

4. About 95%

5.

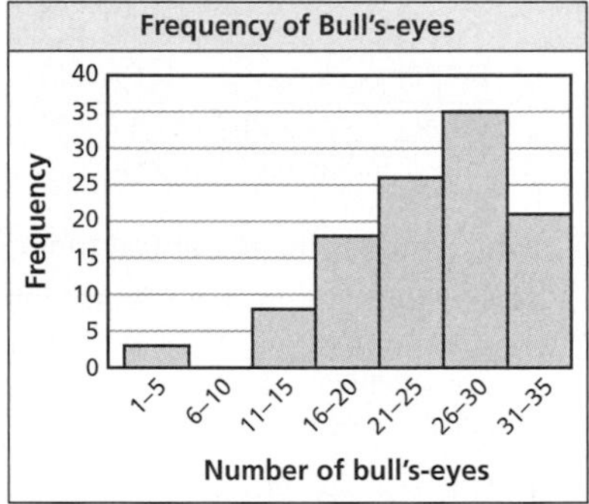

Most of the data are on the right. So, the distribution is skewed left.

6.

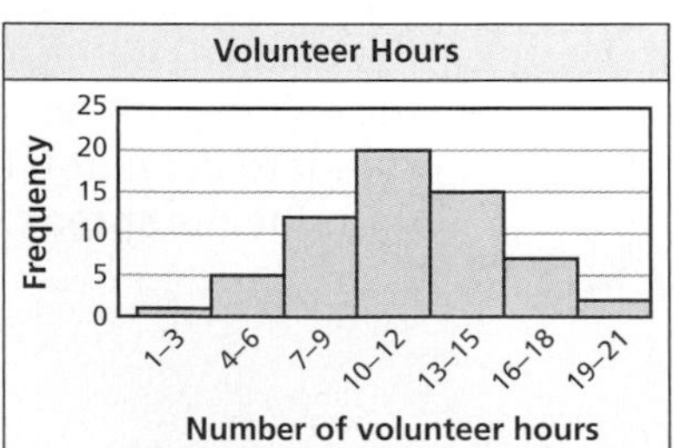

Most of the data are in the middle and evenly distributed. So, the distribution is symmetric.

7. See Additional Answers.

Assignment Guide and Homework Check

Level	Assignment	Homework Check
Average	1, 2, 3–11 odd, 12–15, 17, 19–22	7, 11, 15, 17
Advanced	1, 2, 4–10 even, 16–18, 19–22	6, 10, 16, 18

For Your Information

- **Exercise 16** Note in part (c) that as students add more IQ scores to the data set, the number of intervals will increase to the left. Students should visualize the distribution shape gradually changing from skewed right to symmetric.

Common Errors

- **Exercises 5–7** Students may create histograms with gaps between the bars. Remind students that the bars in a histogram should be touching to show that they represent an entire range of values.

12.4 Record and Practice Journal

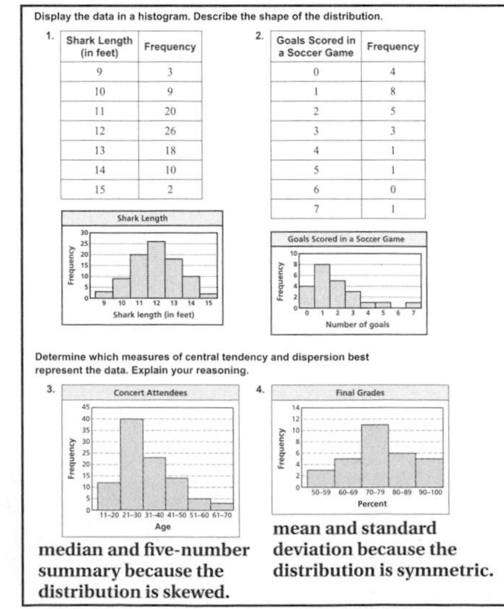
Display the data in a histogram. Describe the shape of the distribution.

1.

Shark Length (in feet)	Frequency
9	3
10	9
11	20
12	26
13	18
14	10
15	2

2.

Goals Scored in a Soccer Game	Frequency
0	4
1	8
2	5
3	3
4	1
5	1
6	0
7	1

Determine which measures of central tendency and dispersion best represent the data. Explain your reasoning.

3.

median and five-number summary because the distribution is skewed.

4.

mean and standard deviation because the distribution is symmetric.

Technology For the Teacher
Answer Presentation Tool

12.4 Exercises

Vocabulary and Concept Check

1. **VOCABULARY** How does the shape of a symmetric distribution differ from the shape of a skewed distribution?

2. **WRITING** How does the shape of a distribution help you decide which measure of central tendency best describes the data?

Practice and Problem Solving

Estimate the percent of data within 2 standard deviations of the mean.

3.

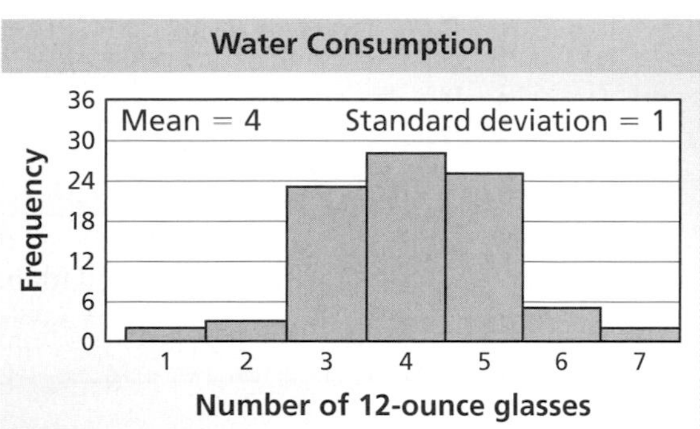

4.

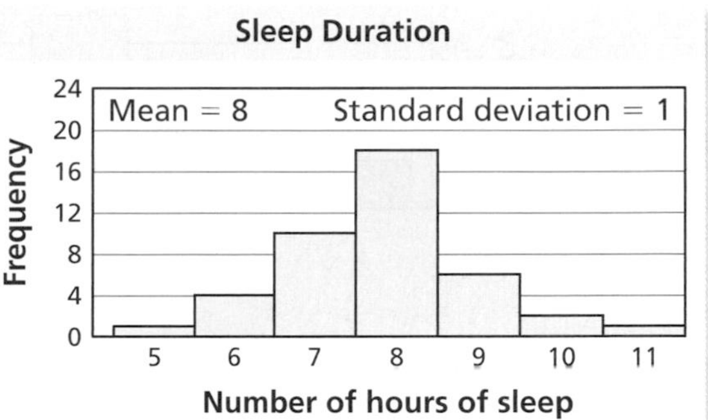

Display the data in a histogram. Describe the shape of the distribution.

1 5.

Number of Bull's-eyes	Frequency
1–5	3
6–10	0
11–15	8
16–20	18
21–25	26
26–30	35
31–35	21

6.

Number of Volunteer Hours	Frequency
1–3	1
4–6	5
7–9	12
10–12	20
13–15	15
16–18	7
19–21	2

7. **ONLINE** A survey asks people how many hours they spend online per day. The results are shown in the table. Display the data in a histogram. Describe the shape of the distribution.

Hours Online	0–2	3–5	6–8	9–11	12–14
Frequency	33	45	12	4	2

Determine which measures of central tendency and dispersion best represent the data. Explain your reasoning.

8.

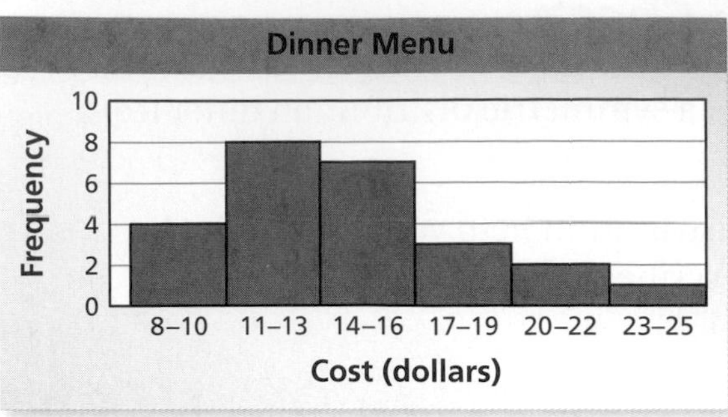

9.

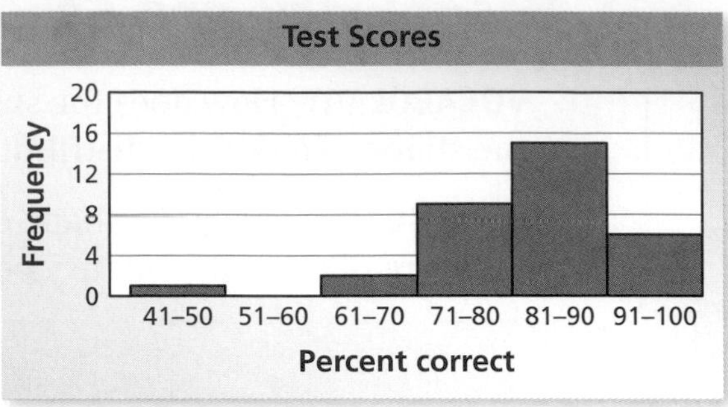

10.

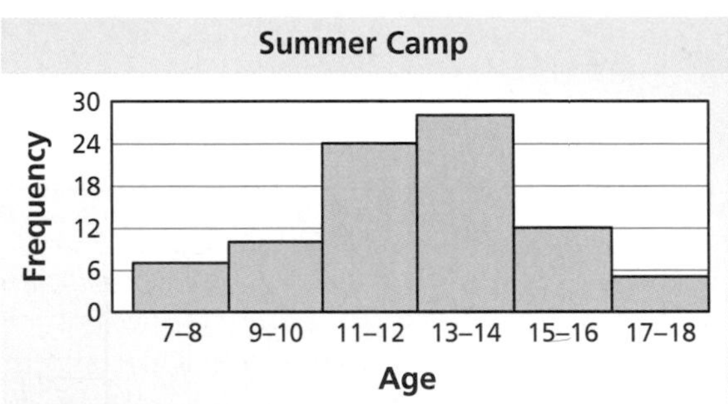

11.

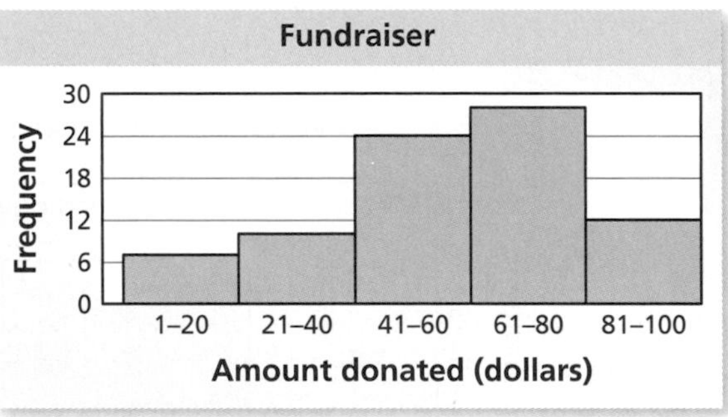

MATCHING **Match the distribution with the corresponding box-and-whisker plot.**

12.

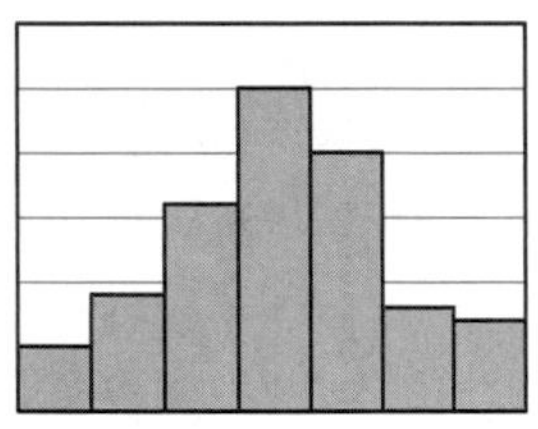

13.

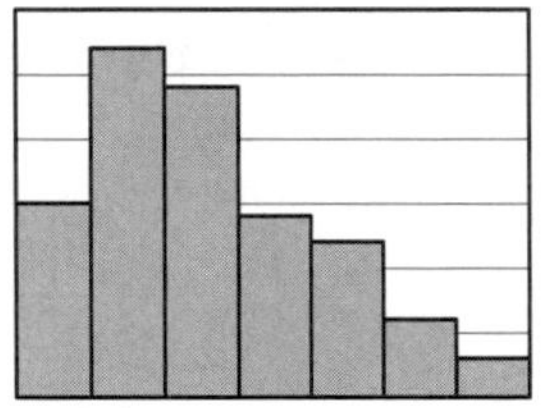

14.

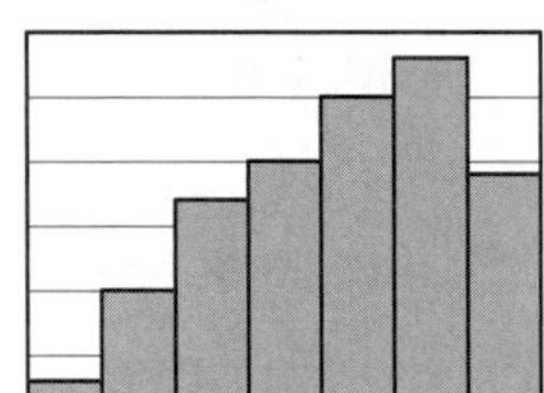

A. 

B. 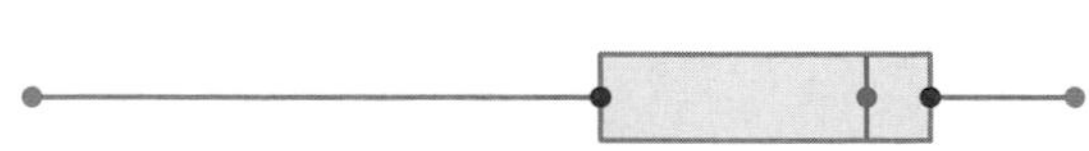

C. 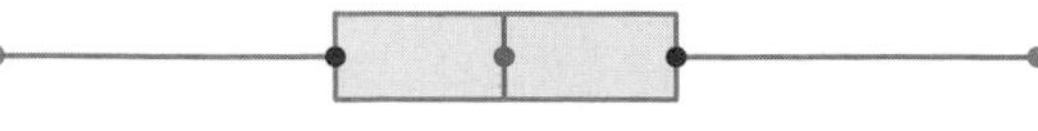

15. **CHOOSE TOOLS** A stem-and-leaf plot is another data display that shows the distribution of data. For a large data set, would you use a stem-and-leaf plot or a histogram to show the distribution of the data? Explain.

Common Errors

- **Exercises 16–18** Students may try to create a histogram without constructing a frequency table. Remind students to organize the data in a frequency table before attempting to create a histogram.
- **Exercises 16–18** Students may create histograms with gaps between the bars. Remind students that the bars in a histogram should be touching to show that they represent an entire range of values.

Practice and Problem Solving

8. Because the distribution is high on the left and the tail of the graph extends to the right, the distribution is skewed right. So, the median best represents the center of the data and the five-number summary best represents the spread of the data.
9. Because the distribution is high on the right and the tail of the graph extends to the left, the distribution is skewed left. So, the median best represents the center of the data and the five-number summary best represents the spread of the data.
10. Because the distribution is high in the middle and the data is about evenly distributed on both sides, the distribution is symmetric. So, the mean best represents the center of the data and the standard deviation best represents the spread of the data.
11. See Additional Answers.
12. C
13. A
14. B

15–16. See Additional Answers.

English Language Learners

Pair Activity

Pair English language learners with English speakers. Assign each person a problem. When they have completed their problems, they explain their solution to their partner who follows along. This will engage English language learners in conversation and help with understanding math concepts.

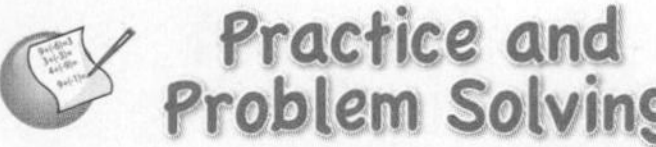

17. See *Taking Math Deeper.*

18. See Additional Answers.

Fair Game Review

19. $y = -x + 4$

20. $y = -\frac{1}{3}x + 3$

21. $y = 2x - 7$ 22. C

Mini-Assessment

1. The frequency table shows the family sizes of students in a classroom. Display the data in a histogram. Describe the shape of the distribution.

Family Size	Frequency
2–3	5
4–5	13
6–7	7
8–9	2
10–11	1

2. You record the numbers of hours that 30 students spent on homework during the past week. The results are shown in the table.

Hours of Homework		
18	22	8
20	23	20
24	19	25
13	5	21
21	24	12
27	15	4
16	17	29
18	21	14
14	25	8
23	10	18

a. Display the data in a histogram using six intervals beginning with 1–5.

b. Which measures of central tendency and dispersion best represent the data? Explain.

1–2. See Additional Answers.

Taking Math Deeper

Exercise 17

The challenge in this problem is finding the number of intervals for the histogram and the width of each interval.

1 Examine the data.

- There are 20 data values.
- All data values are multiples of 5 and 10.
- Largest withdrawal is \$100, smallest is \$10. So, the range is \$90.

2 Find the number of intervals and the width of each interval.

Using too few or too many intervals may make it difficult to determine the shape of a distribution or find a pattern. You should have at least 5 intervals and because there are only 20 data values, you want to have no more than 7 or 8 intervals. So let's use 7 intervals.

Divide the range by the number of intervals to find each interval width. You get $90 \div 7 \approx 13$.

3 Make a frequency table and draw the histogram. Use the least value, 10, as the first number in the first interval. Then add 13 to find the first number of the second interval and so on.

Withdrawal Amount	Frequency
10–22	8
23–35	5
36–48	3
49–61	3
62–74	0
75–87	0
88–100	1

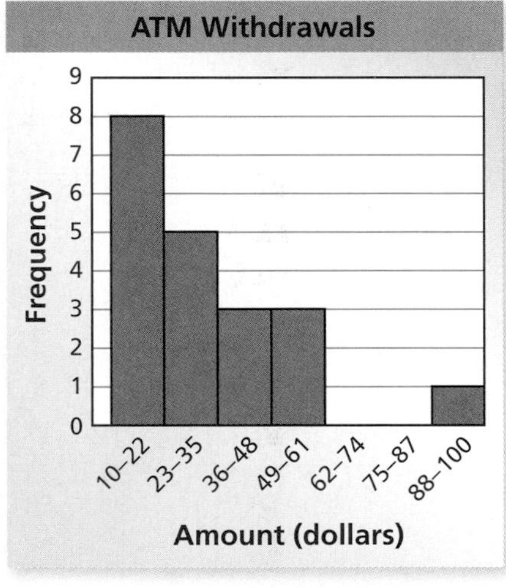

Reteaching and Enrichment Strategies

If students need help . . .	If students got it . . .
Resources by Chapter • Practice A and Practice B • Puzzle Time Record and Practice Journal Practice Differentiating the Lesson Lesson Tutorials Skills Review Handbook	Resources by Chapter • Enrichment and Extension • School-to-Work Start the next section

16. **MODELING** Measuring an IQ is an inexact science. However, IQ scores have been around for years in an attempt to measure human intelligence. The greatest known IQ scores are shown in the table.

IQ Scores		
170	190	180
160	180	210
154	170	180
195	230	160
170	186	180
225	190	170

a. Display the data in a histogram using five intervals beginning with 151–166.

b. Which measures of central tendency and dispersion best represent the data?

c. The distribution of IQ scores for the human population is symmetric. What happens to the shape of the distribution in part (a) as you include more and more IQ scores from the population in the data set?

17. **ATM** The table shows your last 20 ATM withdrawals. What intervals would you use to display the data in a histogram? Explain your reasoning. Then display the data in a histogram.

ATM Withdrawals (dollars)									
20	25	30	10	60	10	45	20	50	25
50	20	45	100	20	10	30	25	40	20

18. 

You record the following waiting times at a restaurant.

Waiting Times (minutes)														
26	38	15	8	22	42	25	10	17	26	58	35	24	31	12
29	25	0	34	44	32	20	18	7	40	42	19	32	13	21

a. Display the data in a histogram using six intervals beginning with 0–9.

b. Display the data in a histogram using twelve intervals beginning with 0–4.

c. What happens when the number of intervals is increased?

d. Which histogram best represents the data? Explain your reasoning.

Fair Game Review What you learned in previous grades & lessons

Write an equation of the line that passes through the given point and is perpendicular to the given line. *(Section 2.6)*

19. $(2, 2); y = x + 3$

20. $(-3, 4); y = 3x - 1$

21. $(1, -5); y = -\frac{1}{2}x + 4$

22. **MULTIPLE CHOICE** Which equation represents the line that passes through (0, 0) and is parallel to the line passing through (5, −2) and (1, −3)? *(Section 2.6)*

Ⓐ $y = -4x$ Ⓑ $y = -\frac{1}{4}x$ Ⓒ $y = \frac{1}{4}x$ Ⓓ $y = 4x$

12 Study Help

You can use a **concept circle** to organize information about a concept. Here is an example of a concept circle for measures of central tendency.

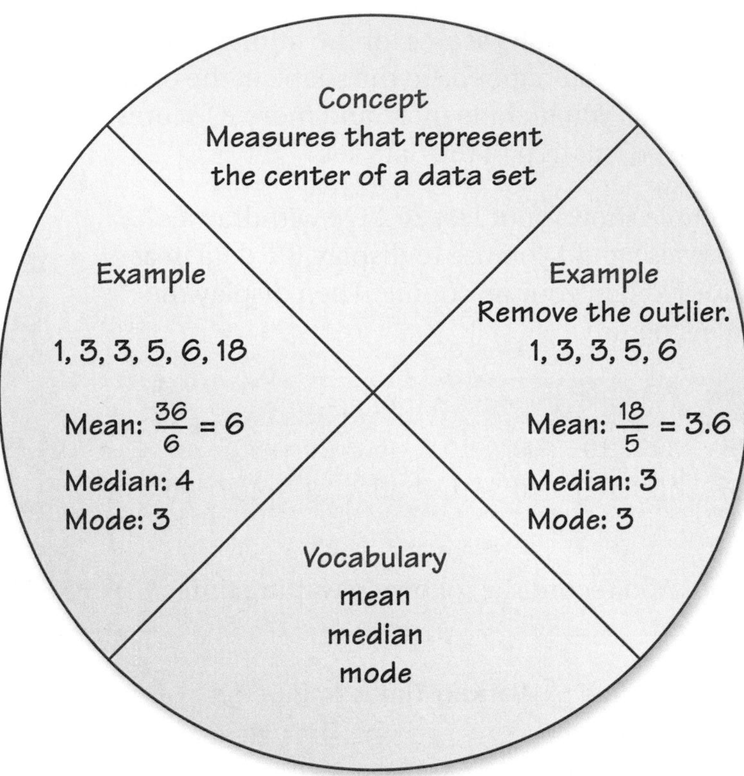

On Your Own

Make concept circles to help you study these topics.

1. measures of dispersion
2. box-and-whisker plots
3. shapes of distributions

After you complete this chapter, make concept circles for the following topics.

4. scatter plots
5. lines of fit
6. two-way tables
7. choosing a data display

"Do you think this concept circle will help my owner understand that 'Speak' and 'Sit' need motivation?"

Sample Answers

1.

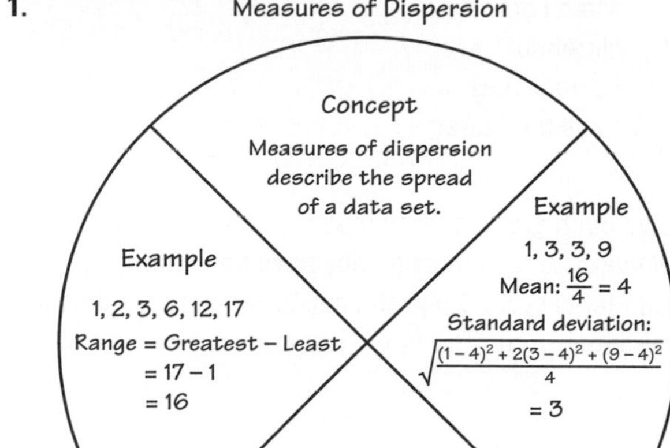

2.

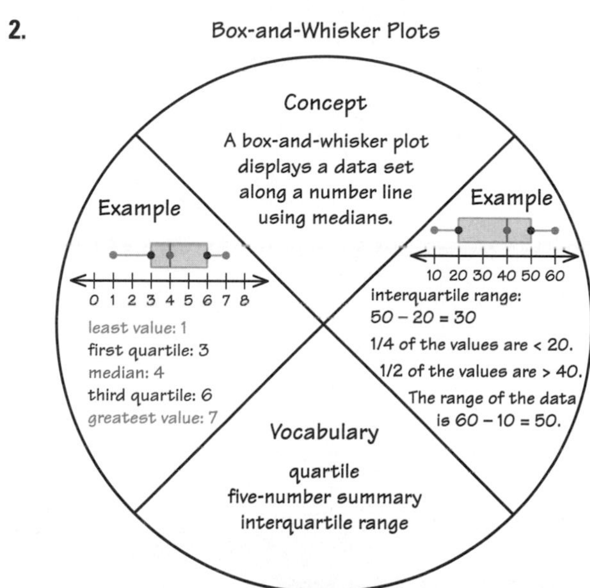

Box-and-Whisker Plots
Concept
A box-and-whisker plot displays a data set along a number line using medians.
Example
0 1 2 3 4 5 6 7 8
least value: 1
first quartile: 3
median: 4
third quartile: 6
greatest value: 7
Example
10 20 30 40 50 60
interquartile range:
50 − 20 = 30
1/4 of the values are < 20.
1/2 of the values are > 40.
The range of the data is 60 − 10 = 50.
Vocabulary
quartile
five-number summary
interquartile range

3. Available at *BigIdeasMath.com*.

List of Organizers

Available at *BigIdeasMath.com*

Comparison Chart
Concept Circle
Definition (Idea) and Example Chart
Example and Non-Example Chart
Formula Triangle
Four Square
Information Frame
Information Wheel
Notetaking Organizer
Process Diagram
Summary Triangle
Word Magnet
Y Chart

About this Organizer

A **Concept Circle** can be used to organize information about a concept. Students write the concept above the circle. Then students write associated information in the sectors of the circle. Associated information can include (an explanation of the) *Concept, Apply, Solve, Check, Example,* and *Justify*. Concept circles can have more or fewer than four sectors. Students can place their concept circles on note cards to use as a quick study reference.

Technology For the Teacher
Vocabulary Puzzle Builder

Answers

1. mean: 20
   median: 30
   mode: 40

2. mean: 3

   median: $3\frac{1}{4}$

   mode: $3\frac{1}{2}$

3. Girls: 3.8; 4; about 1.3
   Boys: 5.8; 6; about 1.9
   There tend to be 2 more boys absent each day than girls, and the number tends to differ from the mean slightly more for boys than for girls.

4. Juniors: 16.5; 11; about 3.4
   Seniors: 24; 17; about 5.7
   The seniors tend to score about 8 points per game more than the juniors and the point total tends to differ from the mean slightly more for seniors than for juniors.

5.

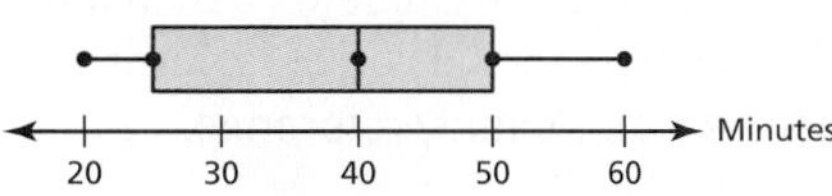

6.

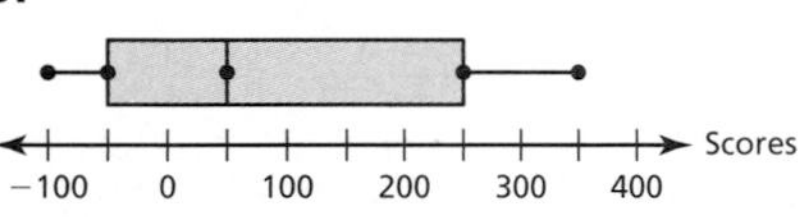

7–9. See Additional Answers.

Assessment Book

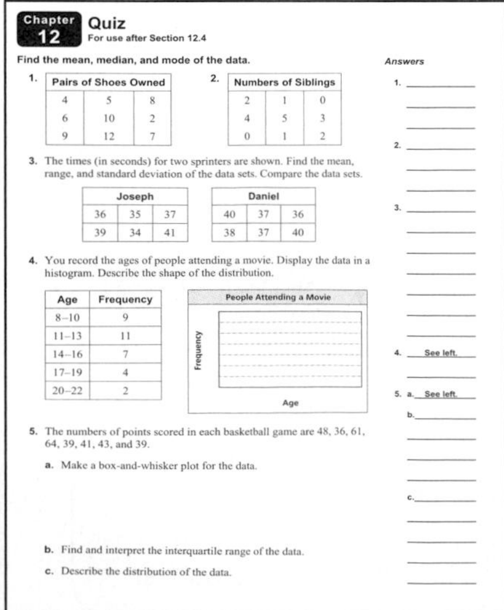

Chapter 12 **Quiz**
For use after Section 12.4

Find the mean, median, and mode of the data.

1.

Pairs of Shoes Owned		
4	5	8
6	10	2
9	12	7

2.

Numbers of Siblings		
2	1	0
4	5	3
0	1	2

3. The times (in seconds) for two sprinters are shown. Find the mean, range, and standard deviation of the data sets. Compare the data sets.

Joseph		
36	35	37
39	34	41

Daniel		
40	37	36
38	37	40

4. You record the ages of people attending a movie. Display the data in a histogram. Describe the shape of the distribution.

Age	Frequency
8–10	9
11–13	11
14–16	7
17–19	4
20–22	2

People Attending a Movie
Frequency
Age

5. The numbers of points scored in each basketball game are 48, 36, 61, 64, 39, 41, 43, and 39.

   a. Make a box-and-whisker plot for the data.

   b. Find and interpret the interquartile range of the data.

   c. Describe the distribution of the data.

Answers
1. ______
2. ______
3. ______
4. See left.
5. a. See left.
b. ______
c. ______

Alternative Quiz Ideas

100% Quiz	Math Log
Error Notebook	Notebook Quiz
Group Quiz	Partner Quiz
Homework Quiz	Pass the Paper

Group Quiz

Students work in groups. Give each group a large index card. Each group writes five questions that they feel evaluate the material they have been studying. On a separate piece of paper, students solve the problems. When they are finished, they exchange cards with another group. The new groups work through the questions on the card.

Reteaching and Enrichment Strategies

If students need help...	If students got it...
Resources by Chapter • Study Help • Practice A and Practice B • Puzzle Time Lesson Tutorials *BigIdeasMath.com* Practice Quiz Practice from the Test Generator	Resources by Chapter • Enrichment and Extension • School-to-Work Game Closet at *BigIdeasMath.com* Start the next section

Technology For the Teacher

Answer Presentation Tool
Big Ideas Test Generator

12.1–12.4 Quiz

Find the mean, median, and mode of the data. *(Section 12.1)*

1.

Checkbook Balances (dollars)		
40	10	−20
0	−10	40
30	40	50

2.

Hours Spent on Project		
$3\frac{1}{2}$	5	$2\frac{1}{2}$
3	$3\frac{1}{2}$	$\frac{1}{2}$

Find the mean, range, and standard deviation of each data set. Then compare the data sets. *(Section 12.2)*

3. Absent students during a week
Girls: 6, 2, 4, 3, 4
Boys: 5, 3, 6, 6, 9

4. Numbers of points scored
Juniors: 19, 15, 20, 10, 14, 21, 18, 15
Seniors: 22, 19, 29, 32, 15, 26, 30, 19

Make a box-and-whisker plot for the data. *(Section 12.3)*

5. Minutes of violin practice: 20, 50, 60, 40, 40, 30, 60, 40, 50, 20, 20, 35

6. Players' scores at end of first round: 200, −100, 100, 350, −50, 0, −50, 300

7. Display the data in a histogram. Describe the shape of the distribution. *(Section 12.4)*

Bowling Scores	51–100	101–150	151–200	201–250	251–300
Frequency	12	21	9	4	2

8. ANOLES The table shows the lengths of 12 green anoles. *(Section 12.1 and Section 12.3)*

a. Find the mean, median, and mode of the data.

b. Make a box-and-whisker plot for the data.

c. Find and interpret the interquartile range of the data.

d. Describe the distribution of the data.

e. How does including 8.0 in the data set affect the mean, median, and mode?

Length (cm)	17.5	17.3	16.5	16.8	17.0	16.5	17.0	16.7	16.5	17.0	17.4	17.1

9. PRESENTATIONS The times of 20 presentations are shown in the table. *(Section 12.2 and Section 12.4)*

a. Display the data in a histogram using five intervals beginning with 3–5.

b. Determine and calculate the measures of central tendency and dispersion that best represent the data.

c. The presentations are supposed to be 10 minutes long. How would you interpret these results?

Time (minutes)			
9	7	10	12
10	11	8	10
10	17	11	5
9	10	4	12
6	14	8	10

12.5 Scatter Plots and Lines of Fit

Essential Question How can you use data to predict an event?

COMMON CORE STATE STANDARDS
8.SP.1
S.ID.6a
S.ID.6c

1 ACTIVITY: Representing Data by a Linear Equation

Work with a partner. You have been working on a science project for 8 months. Each month, you have measured the length of a baby alligator.

The table shows your measurements.

	September						April
Month, x	0	1	2	3	4	5	6	7
Length (in.), y	22.0	22.5	23.5	25.0	26.0	27.5	28.5	29.5

Use the following steps to predict the baby alligator's length next September.

a. Graph the data in the table.

b. Draw the straight line that you think best approximates the points.

c. Write an equation of the line you drew.

d. Use the equation to predict the baby alligator's length next September.

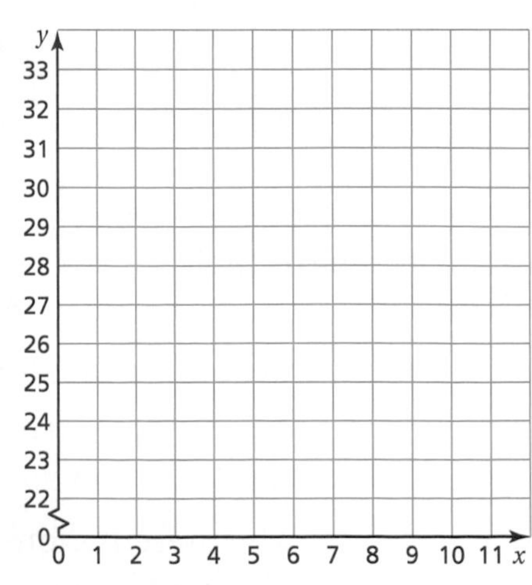

Laurie's Notes

Introduction

Standards for Mathematical Practice

- **MP1a Make Sense of Problems:** A scatter plot shows the relationship between two data sets. Students will find a line of fit for data and interpret what the equation means in the context of the problem.

Motivate

- Solicit information about what students know about alligators. Share alligator facts with students as a warm-up. (See the next page.)
- That should be enough information to set the context for this first activity.

Activity Notes

Activity 1

- ? "Look at the table of values. What do the ordered pairs represent?" (month, length of alligator)
- ? "Do the data represent the first 7 months of growth of a baby alligator? Explain." No, it does not suggest that this is from birth to age 7 months.
- ? "Are there any observations about the data in the table?" Months are increasing by 1. Lengths are increasing by about one-half to an inch each month.
- **MP6 Attend to Precision:** Students will ask what drawing a line "that best approximates the points" means. You should explain that it is a line that passes as closely as possible to all the points. Use a straightedge to lightly draw the line.
- ? "What does the jagged symbol at the bottom of the y-axis mean?" broken axis
- ? "Do you think everyone in class drew the exact same line? Explain." no; They will be close, but they do not have to be exactly the same.
- ? "How did you write the equation for the line?" Listen for an approximation of the slope (rise over run) and the y-intercept (close to 22). Write the equation in slope-intercept form.
- ? "Does everyone have the same slope?" no; They should be relatively close, however, and should match the observations made about the data when looking at the table.
- **MP1a:** Have students interpret the slope and y-intercept in the context of the problem.
- ? "How does the equation help you answer part (d)?" Substitute 12 for x, and find y.
- ? "Without the equation, can you predict the length of the alligator next September?" yes; You need to extend the graph and use eyesight to approximate the ordered pair.

Common Core State Standards

8.SP.1 Construct and interpret scatter plots for bivariate measurement data to investigate patterns of association between two quantities. Describe patterns such as clustering, outliers, positive or negative association, linear association, and nonlinear association.
S.ID.6a Fit a function to the data; use functions fitted to data to solve problems in the context of the data.
S.ID.6c Fit a linear function for a scatter plot that suggests a linear association.

Previous Learning

Students should know how to plot ordered pairs and write equations in slope-intercept form.

Start Thinking! and Warm Up

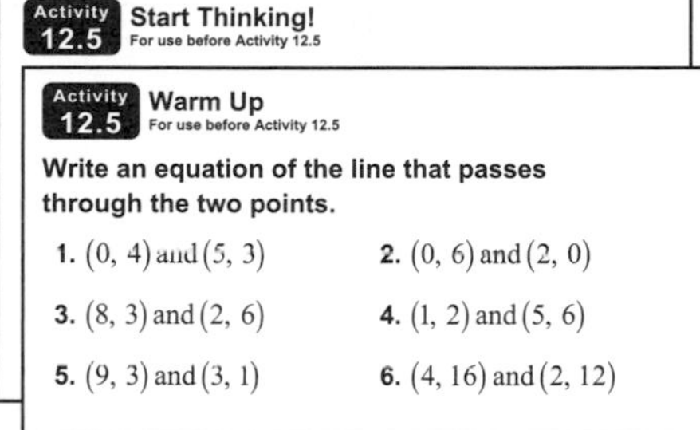

Activity 12.5 Start Thinking!
For use before Activity 12.5

Activity 12.5 Warm Up
For use before Activity 12.5

Write an equation of the line that passes through the two points.

1. (0, 4) and (5, 3)
2. (0, 6) and (2, 0)
3. (8, 3) and (2, 6)
4. (1, 2) and (5, 6)
5. (9, 3) and (3, 1)
6. (4, 16) and (2, 12)

12.5 Record and Practice Journal

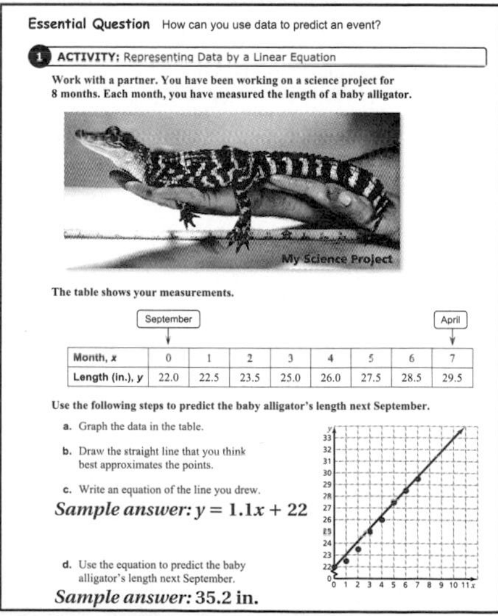

Essential Question How can you use data to predict an event?

1 ACTIVITY: Representing Data by a Linear Equation

Work with a partner. You have been working on a science project for 8 months. Each month, you have measured the length of a baby alligator.

My Science Project

The table shows your measurements.

September → 0; April → 7

Month, x	0	1	2	3	4	5	6	7
Length (in.), y	22.0	22.5	23.5	25.0	26.0	27.5	28.5	29.5

Use the following steps to predict the baby alligator's length next September.

a. Graph the data in the table.

b. Draw the straight line that you think best approximates the points.

c. Write an equation of the line you drew.
Sample answer: $y = 1.1x + 22$

d. Use the equation to predict the baby alligator's length next September.
Sample answer: 35.2 in.

English Language Learners

Class Activity

Provide English learners with an opportunity to interact while learning the concept. Draw a coordinate plane on poster board. Label the horizontal axis *shoe size* and the vertical axis *height*. Have students place a sticker on the ordered pair that represents their shoe size and height. Then have the class fit a line to the data and write an equation of the line.

12.5 Record and Practice Journal

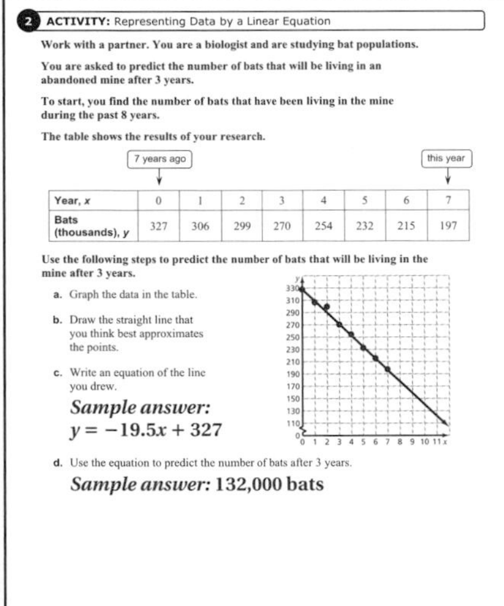

2 ACTIVITY: Representing Data by a Linear Equation

Work with a partner. You are a biologist and are studying bat populations.

You are asked to predict the number of bats that will be living in an abandoned mine after 3 years.

To start, you find the number of bats that have been living in the mine during the past 8 years.

The table shows the results of your research.

7 years ago → 0; this year → 7

Year, x	0	1	2	3	4	5	6	7
Bats (thousands), y	327	306	299	270	254	232	215	197

Use the following steps to predict the number of bats that will be living in the mine after 3 years.

a. Graph the data in the table.

b. Draw the straight line that you think best approximates the points.

c. Write an equation of the line you drew.

Sample answer: $y = -19.5x + 327$

d. Use the equation to predict the number of bats after 3 years.

Sample answer: **132,000 bats**

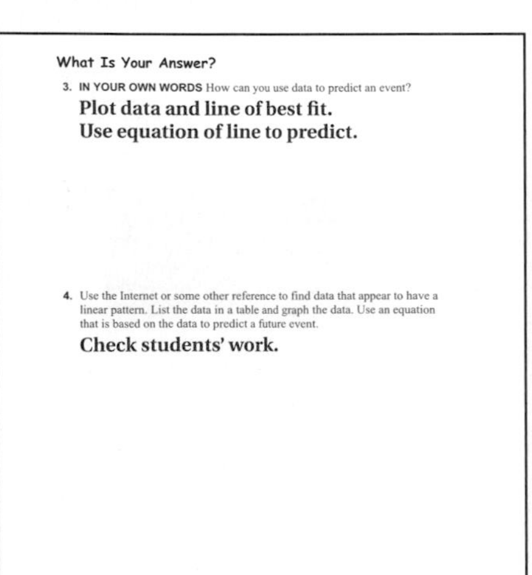

What Is Your Answer?

3. IN YOUR OWN WORDS How can you use data to predict an event?

Plot data and line of best fit.
Use equation of line to predict.

4. Use the Internet or some other reference to find data that appear to have a linear pattern. List the data in a table and graph the data. Use an equation that is based on the data to predict a future event.

Check students' work.

Laurie's Notes

Activity 2

- In this activity, the data have not been collected from an experiment. The data have been collected from a documented source and recorded in the table.
- Read the introduction. The purpose of making a scatter plot is stated. You want to make a prediction about the future by examining known data.
- ? "Are there any observations about the data in the table?" Students may recognize that as the years increase, the number of bats is decreasing by about 15–20 (in thousands) per year.
- Discuss equations written by students. Record students' results on the board. There will likely be a bit more variation of results than in the first activity.
- **MP1a:** Have students interpret what the slope and y-intercept mean in the context of the problem.
- **MP4 Model with Mathematics:** Discuss how the equation allows us to make predictions about the future.

What Is Your Answer?

- Question 4 can become a project due at the conclusion of the chapter.

Closure

- **Exit Ticket:** Describe the difference in the source of data for Activity 1 versus Activity 2. The data in Activity 1 are the result of gathering actual data from an experiment. The data in Activity 2 have been collected from a documented source and recorded in a table.

More about Alligators

- The American alligator (Alligator mississippiensis) is the largest reptile in North America. The first reptiles appeared 300 million years ago. Ancestors of the American alligator appeared 200 million years ago.
- The name alligator comes from early Spanish explorers who called them "El legarto" or "big lizard" when they first saw these giant reptiles.
- Louisiana and Florida have the most alligators. There are over one million wild alligators in each state with over a quarter million more on alligator farms.
- Alligators are about 10–12 inches in length when they are hatched from eggs. Growth rates vary from 2 inches per year to 12 inches per year, depending on the habitat, sex, size, and age of the alligator.
- Females can grow to about 9 feet in length and over 200 pounds. Males can grow to about 13 feet in length and over 500 pounds.
- The largest alligator was taken in Louisiana and measured 19 feet 2 inches.
- Alligators live about as long as humans, an average of 70 years.

Technology For the Teacher

The Dynamic Planning Tool
Editable Teacher's Resources at *BigIdeasMath.com*

2 ACTIVITY: Representing Data by a Linear Equation

Work with a partner. You are a biologist and are studying bat populations.

You are asked to predict the number of bats that will be living in an abandoned mine after 3 years.

To start, you find the number of bats that have been living in the mine during the past 8 years.

The table shows the results of your research.

	7 years ago							this year
Year, x	0	1	2	3	4	5	6	7
Bats (thousands), y	327	306	299	270	254	232	215	197

Use the following steps to predict the number of bats that will be living in the mine after 3 years.

a. Graph the data in the table.

b. Draw the straight line that you think best approximates the points.

c. Write an equation of the line you drew.

d. Use the equation to predict the number of bats after 3 years.

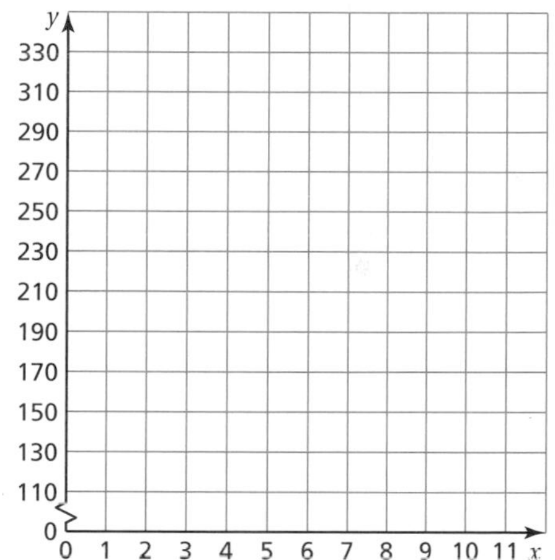

What Is Your Answer?

3. **IN YOUR OWN WORDS** How can you use data to predict an event?

4. Use the Internet or some other reference to find data that appear to have a linear pattern. List the data in a table and graph the data. Use an equation that is based on the data to predict a future event.

Practice Use what you learned about scatter plots and lines of fit to complete Exercise 3 on page 641.

12.5 Lesson

Key Vocabulary
scatter plot, *p. 638*
line of fit, *p. 640*

Scatter Plot

A **scatter plot** is a graph that shows the relationship between two data sets. The two sets of data are graphed as ordered pairs in a coordinate plane.

EXAMPLE 1 Interpreting a Scatter Plot

Restaurant Sandwiches

Calories: 0, 300, 350, 400, 450, 500, 550, 600, 650, 700, 750, 800 (y)

Fat (grams): 0, 5, 10, 15, 20, 25, 30, 35, 40, 45 (x)

The scatter plot at the left shows the amounts of fat (in grams) and the numbers of calories in 12 restaurant sandwiches.

a. How many calories are in the sandwich that contains 17 grams of fat?

Draw a horizontal line from the point that has an x-value of 17. It crosses the y-axis at 400.

So, the sandwich has 400 calories.

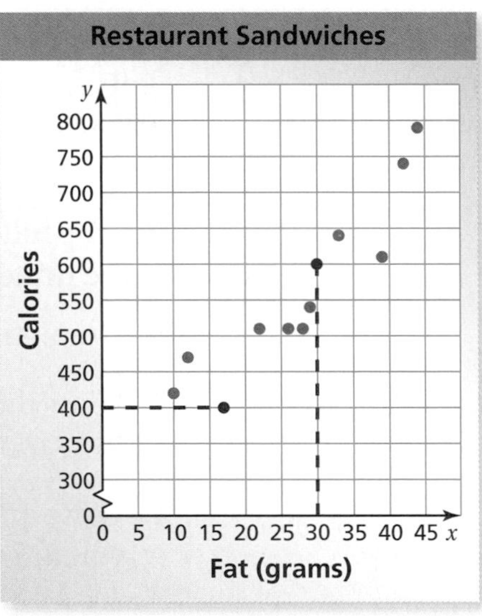

b. How many grams of fat are in the sandwich that contains 600 calories?

Draw a vertical line from the point that has a y-value of 600. It crosses the x-axis at 30.

So, the sandwich has 30 grams of fat.

c. What tends to happen to the number of calories as the number of grams of fat increases?

Looking at the graph, the plotted points go up from left to right.

So, as the number of grams of fat increases, the number of calories increases.

On Your Own

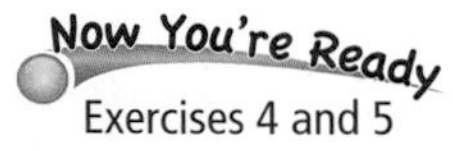

1. **WHAT IF?** A sandwich has 650 calories. Based on the scatter plot in Example 1, how many grams of fat would you expect the sandwich to have? Explain your reasoning.

Laurie's Notes

Introduction

Connect

- **Yesterday:** Students gained an intuitive understanding of how to construct scatter plots and write an equation a line of fit. (MP1a, MP4, MP6)
- **Today:** Students will construct scatter plots, draw the line of fit, and analyze the equation.

Motivate

- **Preparation:** Stop by any fast food restaurant to pick up a pamphlet, or go online to find nutritional information about the menu items.
- ? "Do you think there is a relationship between the grams of fat and number of calories in the sandwich?" yes
- Share the information about a few of the sandwiches from your pamphlet or printout to confirm students' opinions.

Lesson Notes

Key Idea

- Explain that the plot they are going to make today displays the relationship, if any, between two variables, such as grams of fat and calories.
- Define scatter plot.
- Discuss the two scatter plots made in the activity. In Activity 1, the two sets of data were months and alligator length. In Activity 2, the two sets of data were years and number of bats.
- Point out that a scatter plot differs from previous data displays in that it is bivariate (paired data).

Example 1

- **MP1a Make Sense of Problems:** This example helps students understand how a scatter plot is read and interpreted. Discuss the labels on the axes and what an ordered pair represents: (grams of fat, number of calories). There are 12 different sandwiches that are represented.
- To read information from the plot, move horizontally to the *x*-value, find the ordered pair, and then move to the *y*-axis to read the *y*-value. It is helpful to use your hands to demonstrate the motion.
- A scatter plot allows you to see trends in the data. You read a scatter plot from left to right. As the *x*-coordinate increases, is the *y*-coordinate increasing, decreasing, staying the same, *or* is there no pattern?

On Your Own

- This question implies that because you can see a particular trend in the data, you are able to make estimates about points which are not part of the data set but would fall within the trend in the data. Although it is possible that the 650 calorie sandwich has 10 grams of fat, you would not predict it based upon this scatter plot.

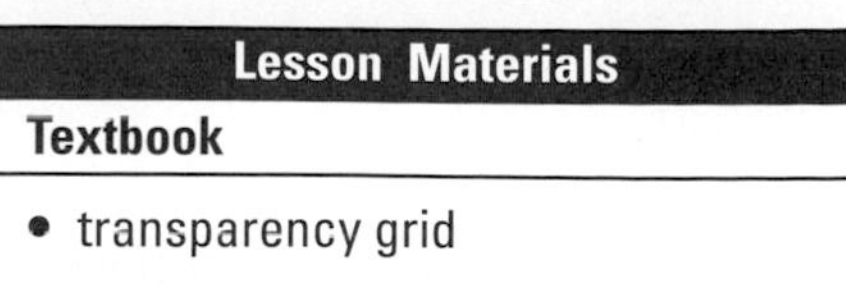

Today's lesson is making a **scatter plot** and identifying a **line of fit**.

Lesson Materials
Textbook
• transparency grid

Start Thinking! and Warm Up

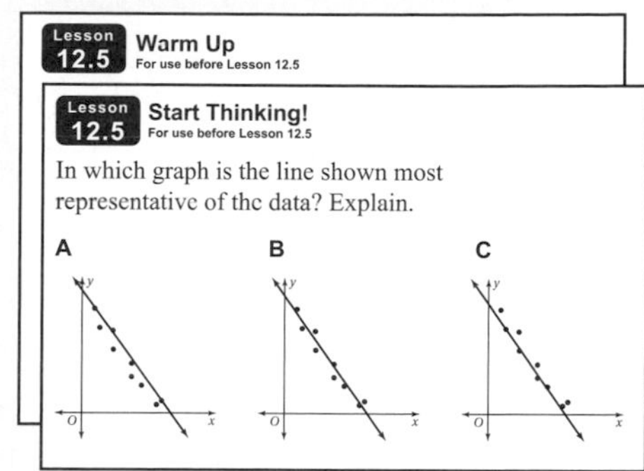

Extra Example 1

Use the scatter plot in Example 1.

a. How many grams of fat are in a sandwich that contains 740 calories? about 42 g

b. How many calories are in a sandwich that contains 33 grams of fat? about 640 calories

On Your Own

1. about 35 g; The point just below $y = 650$ has an x-value just below $x = 35$.

Extra Example 2

Tell whether the data show a *positive*, a *negative*, or *no* relationship.

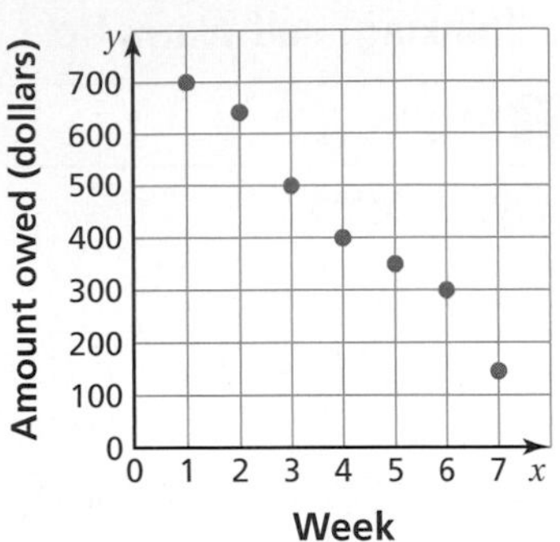

negative relationship

On Your Own

2. See Additional Answers.

3. 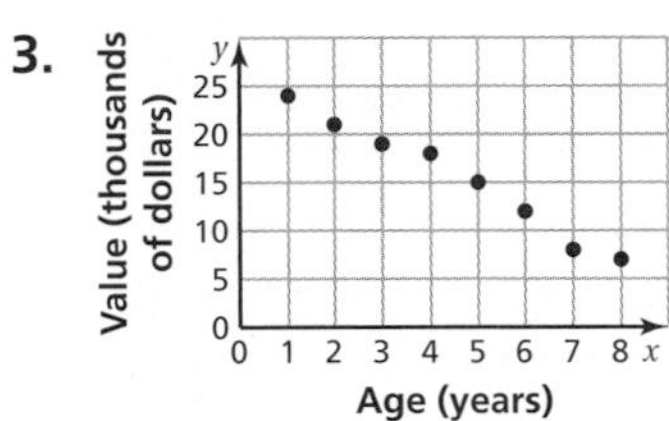

negative relationship

Differentiated Instruction

Visual

Some students may find it easier to draw a line of fit before determining if the data have a positive relationship, a negative relationship, or no relationship. A line with a *positive* slope means the data have a *positive* relationship. A line with a *negative* slope means the data have a *negative* relationship. If a line cannot be drawn, the data have *no* relationship.

Laurie's Notes

Discuss

- There are three general cases that describe the relationship between two data sets. Draw a quick example of each case.
- The alligator data was an example of a positive relationship and the bat data was an example of a negative relationship.
- Discuss the study tip and provide examples to show how these features can be seen in a scatter plot.

Example 2

- Have students review the two scatter plots shown.
- Ask students to complete this sentence. As the size of the television increases, the price increases. This is an example of a positive relationship.
- **Connection:** By this point, some students have made the connection between the slope of a line of fit from the activities and the relationship between the two data sets. A positive relationship is related to a positive slope.

? "Should there be a relationship between a person's age and the number of pets they own?" no

- Part (b) makes sense to students. There should be no trend in the data.

On Your Own

- Give time for students to complete these two scatter plots.
- **MP6 Attend to Precision:** A common difficulty for students is deciding how to scale the axes. Students should look at the range of numbers that need to be displayed, and then decide if it is necessary to start their axes at 0 or if another starting point (broken axes) makes sense.
- Have transparency grids available so that results can be shared quickly as a class.

Study Tip

Scatter plots can also show unusual features of a data set, such as outliers, or gaps and clusters in the data.

A scatter plot can show that a relationship exists between two data sets.

Positive Relationship

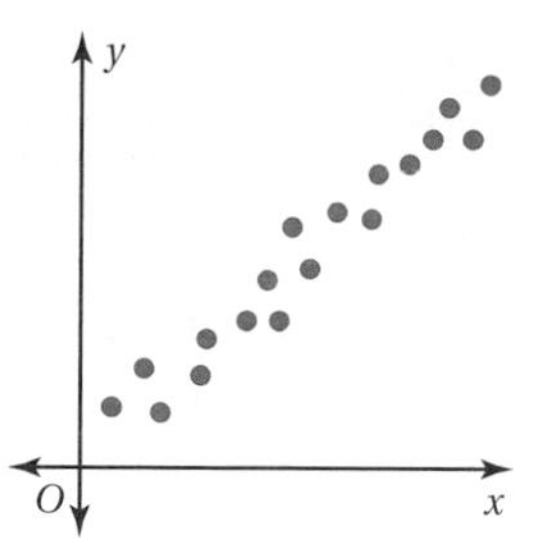

As x increases, y increases.

Negative Relationship

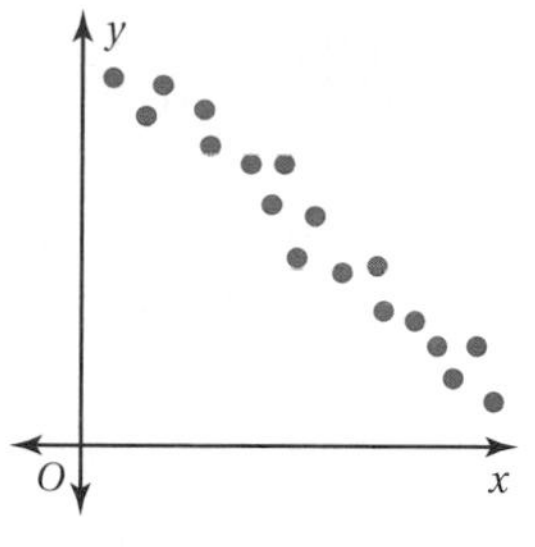

As x increases, y decreases.

No Relationship

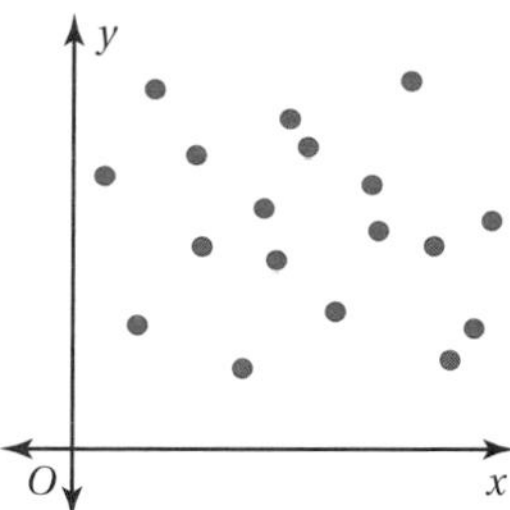

The points show no pattern.

EXAMPLE 2 Identifying a Relationship

Tell whether the data show a *positive*, a *negative*, or *no* relationship.

a. Television size and price

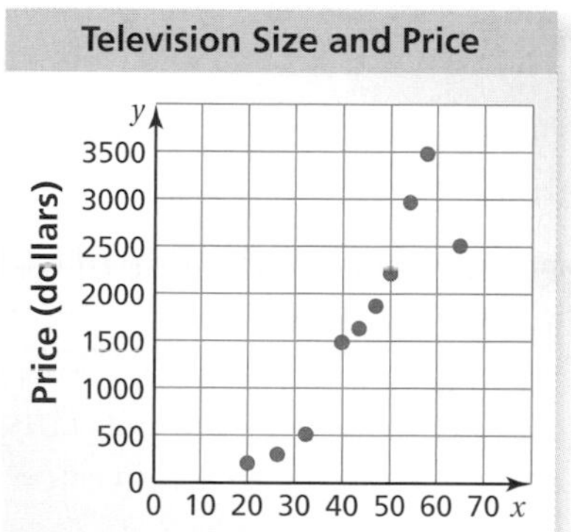

As the size of the television increases, the price increases.

So, the scatter plot shows a positive relationship.

b. Age and number of pets owned

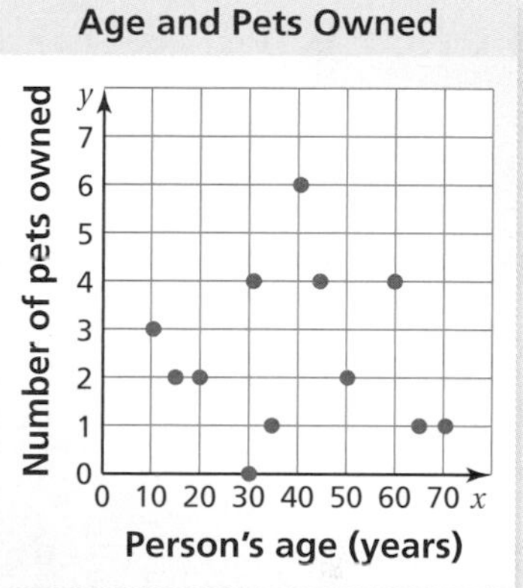

The number of pets owned does not depend on a person's age.

So, the scatter plot shows no relationship.

On Your Own

Exercises 6–8

Make a scatter plot of the data. Tell whether the data show a *positive*, a *negative*, or *no* relationship.

2.

Study Time (min), x	30	20	60	90	45	10	30	75	120	80
Test Score, y	87	74	92	97	85	62	83	90	95	91

3.

Age of a Car (years), x	1	2	3	4	5	6	7	8
Value (thousands), y	\$24	\$21	\$19	\$18	\$15	\$12	\$8	\$7

A **line of fit** is a line drawn on a scatter plot close to most of the data points. It can be used to estimate data on a graph.

EXAMPLE 3 Finding a Line of Fit

Week, x	Sales (millions), y
1	\$19
2	\$15
3	\$13
4	\$11
5	\$10
6	\$8
7	\$7
8	\$5

The table shows the weekly sales of a DVD and the number of weeks since its release. (a) Make a scatter plot of the data and draw a line of fit. (b) Write an equation of the line of fit. (c) Interpret the slope of the line of fit. (d) Predict the sales in week 9.

a. Plot the points in a coordinate plane. The scatter plot shows a negative relationship. Draw a line that is close to the data points. Try to have as many points above the line as below it.

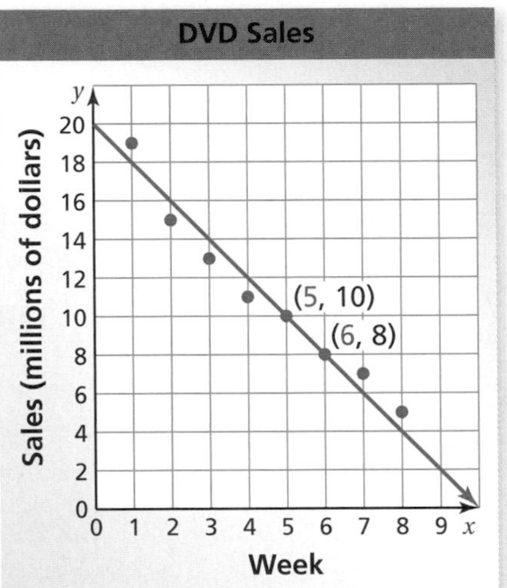

b. The line passes through (5, 10) and (6, 8).

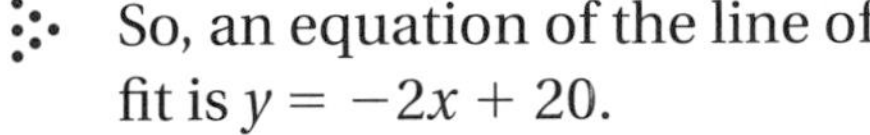

$$\text{slope} = \frac{\text{rise}}{\text{run}} = \frac{-2}{1} = -2$$

Because the line crosses the y-axis at (0, 20), the y-intercept is 20.

∴ So, an equation of the line of fit is $y = -2x + 20$.

c. The slope of the line of fit is -2. This means that the sales are decreasing by about \$2 million each week.

d. To predict the sales in week 9, substitute 9 for x in the equation of the line of fit.

$$y = -2x + 20 = -2(9) + 20 = 2$$

∴ The sales in week 9 should be about \$2 million.

Study Tip

A line of fit does not need to pass through any of the data points.

On Your Own

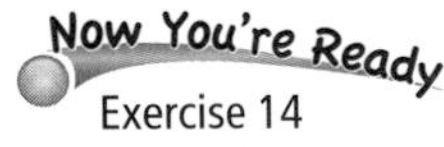

4. The table shows the numbers of people who have attended a neighborhood festival over an 8-year period.

Year, x	1	2	3	4	5	6	7	8
Attendance, y	420	500	650	900	1100	1500	1750	2400

a. Make a scatter plot of the data and draw a line of fit.

b. Write an equation of the line of fit.

c. Interpret the slope of the line of fit.

d. Predict the number of people who will attend the festival in year 10.

Laurie's Notes

Discuss

- **MP5 Use Appropriate Tools Strategically:** Define and discuss a line of fit. It is helpful to model this with a piece of spaghetti. Use a scatter plot from the previous page that was completed on the transparency. Model how the spaghetti can approximate the trend of the data.
- Move the spaghetti so that it does *not* represent the data, and then move the spaghetti so that it does. You will use your eyesight when judging where to draw the line.

Example 3

? "What observations can you make about the sales of the DVD as the weeks go on?" Sales are decreasing.

- Carefully plot the ordered pairs on a transparency grid.
- When drawing a line of fit, try to put as many points above the line as below it.
- Students may draw different lines of fit and still get a reasonable answer.
- In this example, the line passes through two actual data points, (5, 10) and (6, 8). As noted in the *Study Tip*, a line of fit does not need to pass through any of the data points.
- Finish working the problem as shown.
- **MP4 Model with Mathematics:** The purpose of writing the equation of the line of fit is to make predictions. The equation becomes a model for the data, describing its behavior.

On Your Own

- This is a nice summary problem. Students should quickly observe the positive relationship just from the table of values.
- Share results of this problem as a whole class.

Closure

- **Exit Ticket:** In Example 3, interpret the y-intercept in the context of the problem. Does it make sense in this problem? The y-intercept, 20, represents the sales in millions of dollars for week 0. It does not make sense in this problem, because there would not have been any sales before the first week.

Technology For the Teacher

The Dynamic Planning Tool
Editable Teacher's Resources at *BigIdeasMath.com*

Extra Example 3

The table shows the weekly sales of a DVD and the number of weeks since its release.

Week, x	Sales (millions), y
1	\$15
2	\$13
3	\$12
4	\$9
5	\$6
6	\$4
7	\$3

a. Make a scatter plot of the data and draw a line of fit.

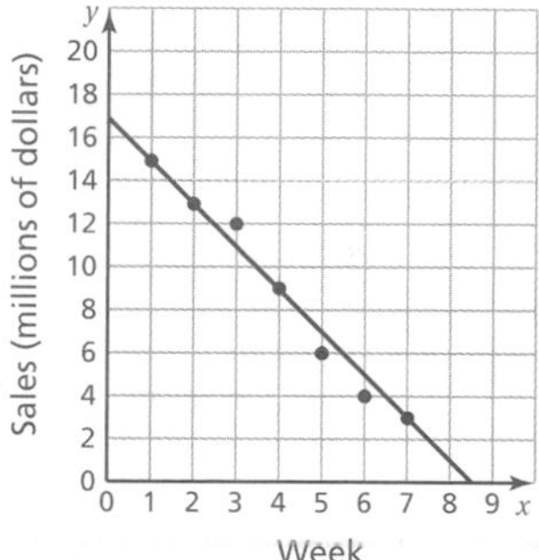

b. Write an equation of the line of fit.
$y = -2x + 17$

c. Interpret the slope of the line of fit.
The slope of the line of fit is -2. This means the sales are decreasing by about \$2 million each week.

d. Predict the sales in week 8.
about \$1 million

On Your Own

4. a. See Additional Answers.

   b. *Sample answer:* $y = 270x$

   c. *Sample answer:* The slope of the line of fit is 270. This means the number of people attending is increasing by about 270 people each year.

   d. *Sample answer:* about 2700 people

Vocabulary and Concept Check

1. They must be ordered pairs so there are equal amounts of x- and y-values.
2. You can estimate and predict values.

Practice and Problem Solving

3. a–b.

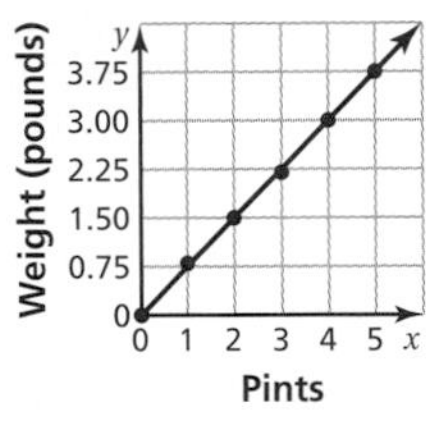

   c. *Sample answer:* $y = 0.75x$

   d. *Sample answer:* 7.5 lb

   e. *Sample answer:* \$16.88

4. a. 2007

   b. about 875 SUVs

   c. There is a negative relationship between year and number of SUVs sold.

5. a. 3.5 h

   b. \$85

   c. There is a positive relationship between hours worked and earnings.

6. negative relationship
7. positive relationship
8. no relationship
9. positive relationship
10. *Sample answer:* bank account balance after a shopping spree
11. *Sample answer:* not a good representation; Too many points in the data set lie below the line.

Assignment Guide and Homework Check

Level	Assignment	Homework Check
Average	1, 2, 4–18 even, 15, 17, 23–26	4, 6, 12, 14, 17
Advanced	1, 2, 4–22 even, 17, 23–26	4, 6, 14, 17, 18

Common Errors

- **Exercise 3** Students may use inconsistent increments or forget to label their graphs. Students should use consistent increments to represent the data. Remind them to label the axes so that information can be read from the graph.
- **Exercises 4 and 5** When finding values from the graph, students may accidentally shift over or up too far and get an answer that is off by an increment. Encourage them to start at the given value and trace the graph to where the point or line of fit is, and then trace down or left to the other axis for the answer.

12.5 Record and Practice Journal

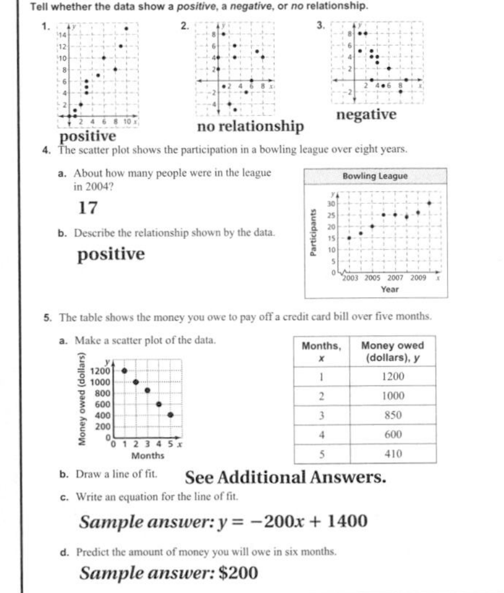

Tell whether the data show a *positive*, a *negative*, or *no* relationship.

1. **positive**
2. **no relationship**
3. **negative**

4. The scatter plot shows the participation in a bowling league over eight years.

   a. About how many people were in the league in 2004?
   17

   b. Describe the relationship shown by the data.
   positive

   Bowling League (Participants vs. Year: 2003, 2005, 2007, 2009)

5. The table shows the money you owe to pay off a credit card bill over five months.

Months, x	Money owed (dollars), y
1	1200
2	1000
3	850
4	600
5	410

   a. Make a scatter plot of the data.

   b. Draw a line of fit. **See Additional Answers.**

   c. Write an equation for the line of fit.
   Sample answer: $y = -200x + 1400$

   d. Predict the amount of money you will owe in six months.
   ***Sample answer:* \$200**

12.5 Exercises

Vocabulary and Concept Check

1. **VOCABULARY** What type of data are needed to make a scatter plot? Explain.
2. **WRITING** Explain why a line of fit is helpful when analyzing data.

Practice and Problem Solving

3. **BLUEBERRIES** The table shows the weights y of x pints of blueberries.

Number of Pints, x	0	1	2	3	4	5
Weight (pounds), y	0	0.8	1.50	2.20	3.0	3.75

a. Graph the data in the table.
b. Draw the straight line that you think best approximates the points.
c. Write an equation of the line you drew.
d. Use the equation to predict the weight of 10 pints of blueberries.
e. Blueberries cost $2.25 per pound. How much do 10 pints of blueberries cost?

1 4. **SUVS** The scatter plot shows the numbers of sport utility vehicles sold in a city from 2005 to 2010.

a. In what year were 1000 SUVs sold?
b. About how many SUVs were sold in 2009?
c. Describe the relationship shown by the data.

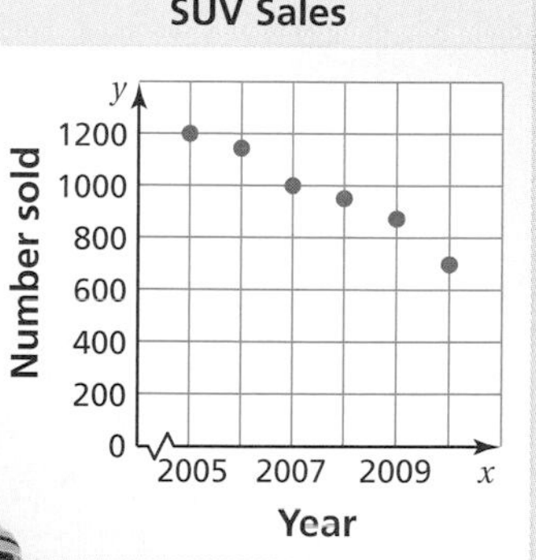

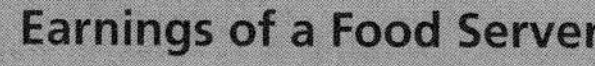

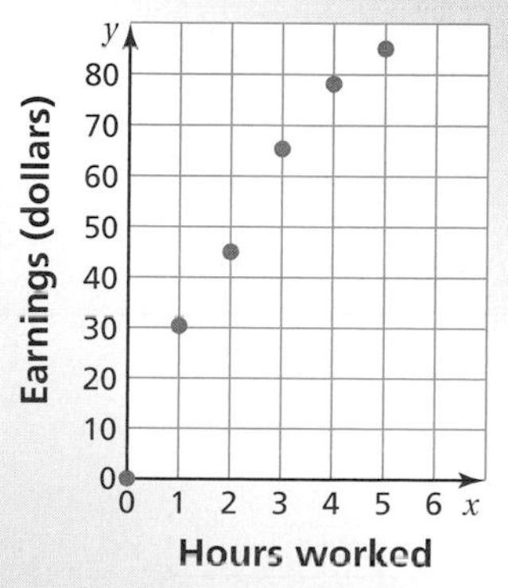

5. **EARNINGS** The scatter plot shows the total earnings (wages and tips) of a food server during 1 day.

a. About how many hours must the server work to earn $70?
b. About how much did the server earn for 5 hours of work?
c. Describe the relationship shown by the data.

Tell whether the data show a *positive*, a *negative*, or *no* relationship.

2 **6.**

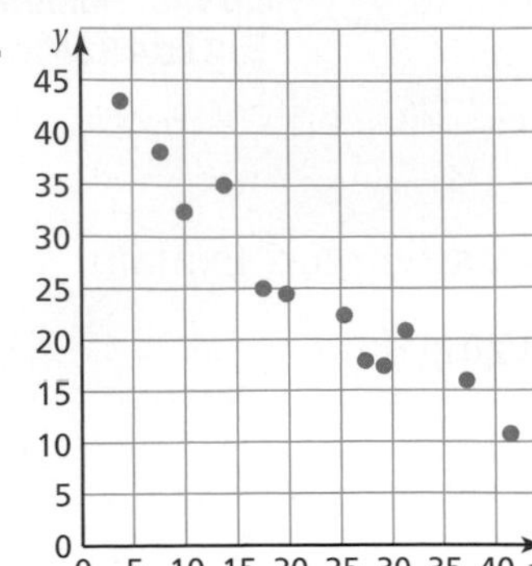

7.

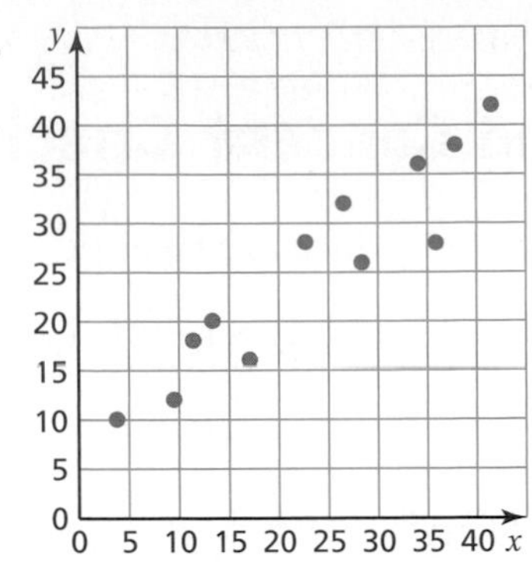

8.

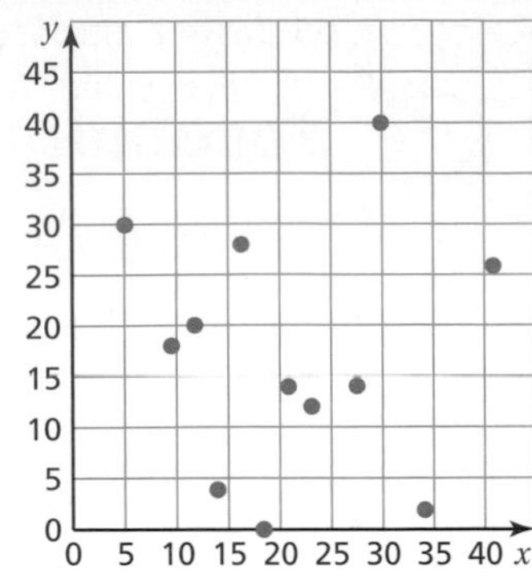

9. HONEY The table shows the average price per pound for honey in the United States from 2007 to 2010. What type of relationship do the data show?

Year, x	2007	2008	2009	2010
Average Price per Pound, y	\$1.08	\$1.42	\$1.47	\$1.60

10. OPEN-ENDED Describe a set of real-life data that has a negative relationship.

Tell whether the line drawn on the graph is a good fit for the data. Explain your reasoning.

11.

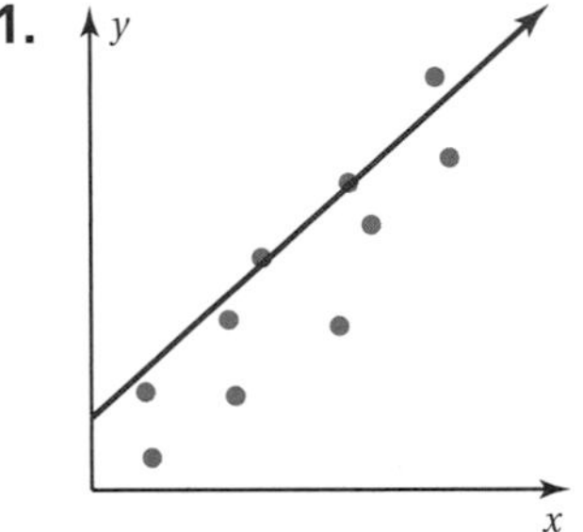

12.

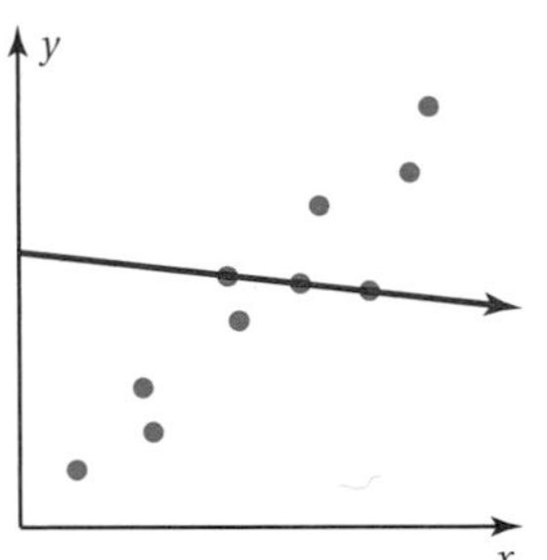

13.

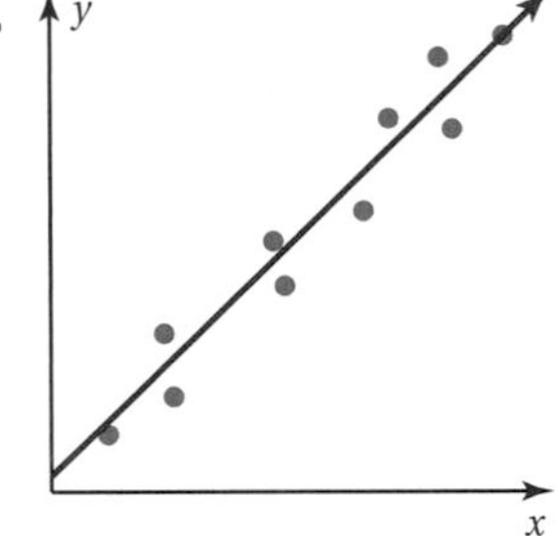

3 **14. VACATION** The table shows the distance you travel over a 6-hour period.

a. Make a scatter plot of the data and draw a line of fit.

b. Write an equation of the line of fit.

c. Interpret the slope of the line of fit.

d. Predict the distance you will travel in 7 hours.

Hours, x	Distance (miles), y
1	62
2	123
3	188
4	228
5	280
6	344

15. TEST SCORES The scatter plot shows the relationship between numbers of minutes spent studying and test scores for a science class.

a. What type of relationship do the data show?

b. Interpret the relationship.

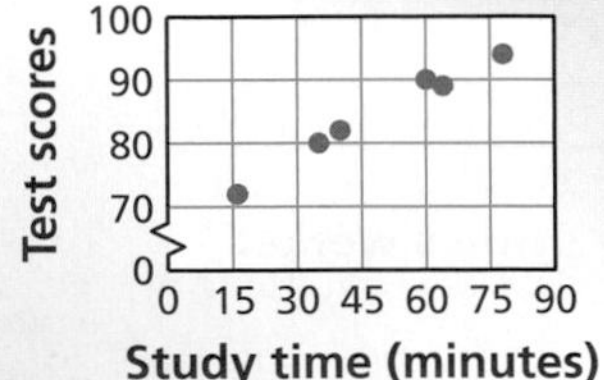

Common Errors

- **Exercises 6–8** Students may mix up positive and negative relationships. Remind them about slope. The slope is positive when the line rises from left to right and negative when it falls from left to right. The same is true for relationships in a scatter plot. If the data rises from left to right, it is a positive relationship. If it falls from left to right, it is a negative relationship.
- **Exercise 14** Students may draw a line of fit that does not accurately reflect the data trend. Remind them that the line does not have to go through any of the data points. Also remind them that the line should go through the middle of the data so that about half of the data points are above the line and half are below. One strategy is to draw an oval around the data and then draw a line through the middle of the oval. For example:

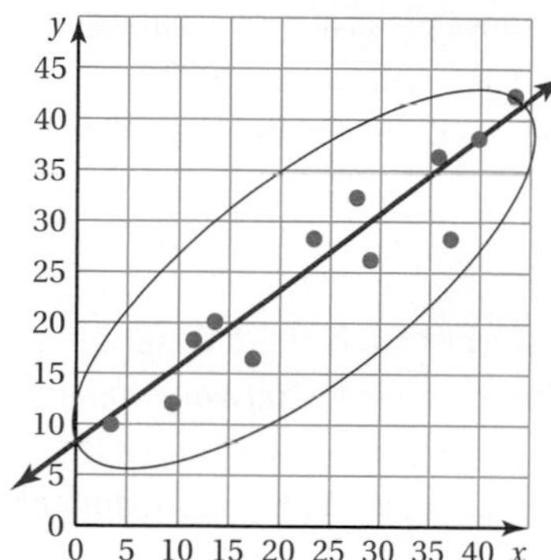

- **Exercise 14** Students may struggle writing an equation for a line of fit. When drawing the line, encourage them to try to make the line go through a lattice point. Also, students can use lattice points that are very close to the line to help them find the slope.

Practice and Problem Solving

12. *Sample answer:* not a good representation; Even though the line passes through several points, it does not indicate the overall relationship of the data set.

13. *Sample answer:* good representation; The same number of points in the data set lie above and below the line.

14. **a.**

y
350
300
250
200
150
100
50
0
0 1 2 3 4 5 6 x
Distance (miles)
Hours

b. *Sample answer:* $y = 55x + 15$

c. *Sample answer:* The slope of the line of fit is 55. This means the number of miles driven each hour is about 55.

d. *Sample answer:* 400 mi

15. **a.** positive relationship

b. The more time spent studying, the better the test score.

Differentiated Instruction

Kinesthetic

Form groups of 8 to 10 students who will create life-size models of a positive relationship, a negative relationship, and no relationship. Give two pairs of students 10-foot lengths of string and have them form the *x*- and *y*-axes of a coordinate plane. The remaining students will be the data points. Have these students represent a *positive relationship* in the coordinate plane. After students have had a few minutes, check their positions. Continue by having students represent a *negative relationship* and *no relationship*. Extend the activity by having two students hold a third string to show a line of fit.

Practice and Problem Solving

16. no; There is no line that lies close to most of the points.

17. See *Taking Math Deeper*.

18. a data point that is far removed from the other points in a data set

19–22. See Additional Answers.

Fair Game Review

23. 2 **24.** 8

25. −4 **26.** B

Mini-Assessment

The table shows the distance you travel over a 6-hour period.

Hours, x	Distance (miles), y
1	60
2	130
3	186
4	244
5	300
6	378

a. Make a scatter plot of the data and draw a line of fit.

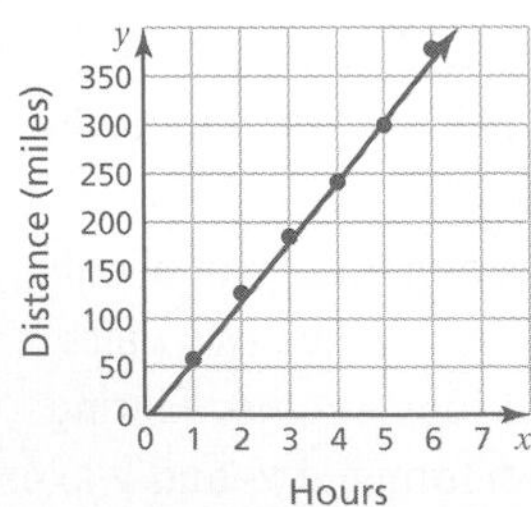

b. Write an equation of the line of fit. *Sample answer:* $y = 60x$

c. Interpret the slope of the line of fit. The slope of the line of fit is 60. This means the number of miles traveled each hour is about 60.

d. Predict the distance traveled after 7 hours. *Sample answer:* about 420 mi

Taking Math Deeper

Exercise 17

The project in the student text is described so that it can be assigned as homework. Another way to assign the project is to ask students to do the project in class.

1. Gather the data by having students measure each other's height and arm span.

2. Plot the data for the entire class in a coordinate plane. Scale the x-axis and y-axis so that the measurements of your students (in inches) will fit.

3. 

Ask your students to describe the relationship between height and arm span. The slope of the line of fit should be approximately equal to 1.

Project

Research Leonardo da Vinci's drawing of the *Vitruvian Man*. Explain the concept behind the drawing.

Reteaching and Enrichment Strategies

If students need help. . .	If students got it. . .
Resources by Chapter • Practice A and Practice B • Puzzle Time Record and Practice Journal Practice Differentiating the Lesson Lesson Tutorials Skills Review Handbook	Resources by Chapter • Enrichment and Extension • School-to-Work • Financial Literacy Start the next section

16. **REASONING** A data set has no relationship. Is it possible to find a line of fit for the data? Explain.

17. **CHOOSE TOOLS** Use a ruler or a yardstick to find the heights and arm spans of three people.

    a. Make a scatter plot using the data you collected. Then draw a line of fit for the data.

    b. Use your height and the line of fit to predict your arm span.

    c. Measure your arm span. Compare the result with your prediction in part (b).

    d. Is there a relationship between a person's height x and arm span y? Explain.

18. **REASONING** How can an outlier be identified in a scatter plot?

Describe the scatter plot and any relationship between the variables.

19. 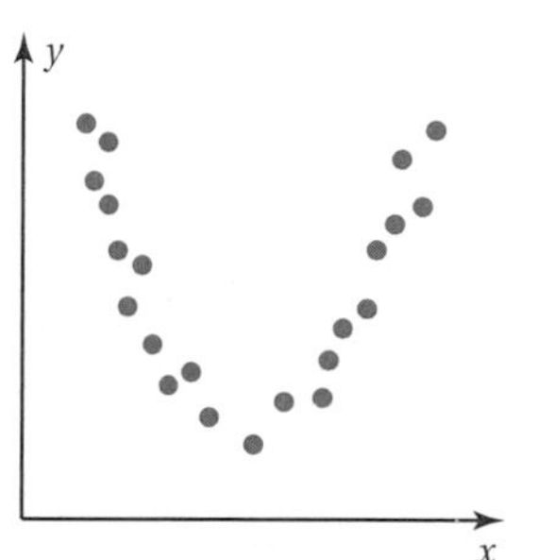

20. 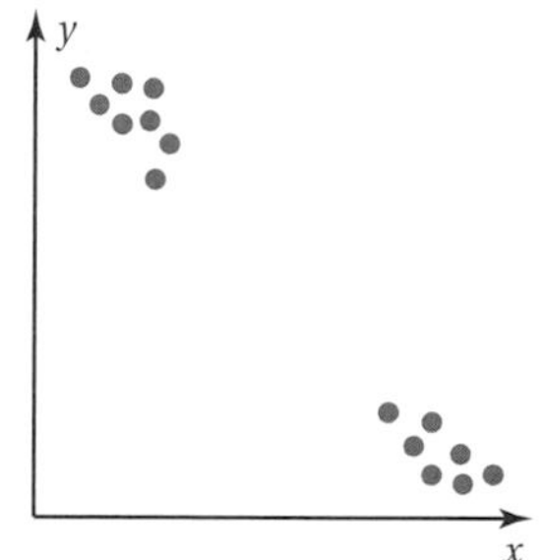

21. 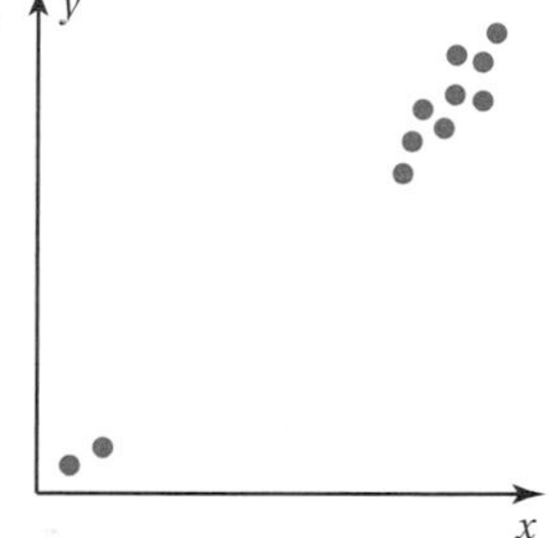

Price of Admission (dollars), x	Yearly Attendance, y
19.50	50,000
21.95	48,000
23.95	47,500
24.00	40,000
24.50	45,000
25.00	43,500

22. Critical Thinking The table shows the prices of admission to a local theater and the attendances for several years.

    a. Identify the outlier.

    b. How does the outlier affect the line of fit? Explain.

    c. Make a scatter plot of the data and draw the line of fit.

    d. Use the line of fit to predict the attendance when the admission cost is $27.

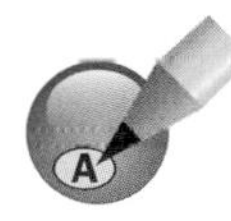

Fair Game Review What you learned in previous grades & lessons

Use a graph to solve the equation. Check your solution. *(Section 4.4)*

23. $5x = 2x + 6$

24. $7x + 3 = 9x - 13$

25. $\frac{2}{3}x = -\frac{1}{3}x - 4$

26. **MULTIPLE CHOICE** The circle graph shows the super powers chosen by a class. What percent of the students want strength as their super power? *(Skills Review Handbook)*

    (A) 10.5% (B) 12.5%

    (C) 15% (D) 25%

Super Powers

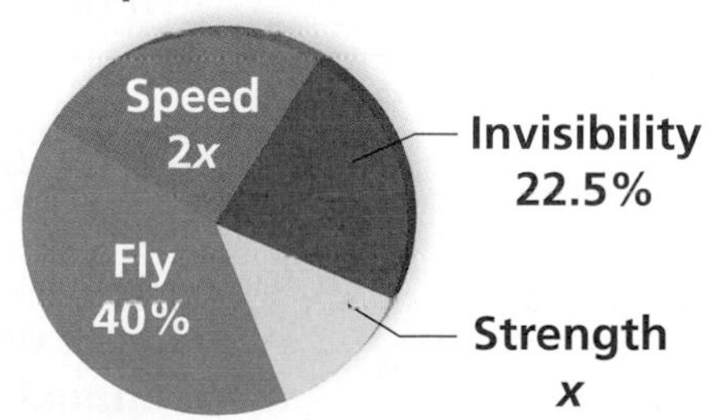

12.6 Analyzing Lines of Fit

Essential Question How can you find a line that best models a data set?

COMMON CORE STATE STANDARDS
S.ID.6b
S.ID.8
S.ID.9

1 ACTIVITY: Comparing Lines of Fit

Work with a partner. You are researching the prices of liquid crystal display (LCD) televisions. The tables show the sizes and prices of several LCD televisions.

TV Size (in.), *x*	19	19	22	24	32
Price (dollars), *y*	170	180	170	250	320

TV Size (in.), *x*	32	37	40	40	46
Price (dollars), *y*	300	400	480	500	600

TV Size (in.), *x*	46	47	52	55	55
Price (dollars), *y*	850	800	950	1000	1150

a. Make a scatter plot of the data. Describe the pattern.

b. Draw a line of fit. Then have your partner draw a different line of fit.

c. Write an equation for each line of fit.

d. Compare your line of fit with your partner's line of fit. Are they similar? Which line of fit seems to model the data better? Why?

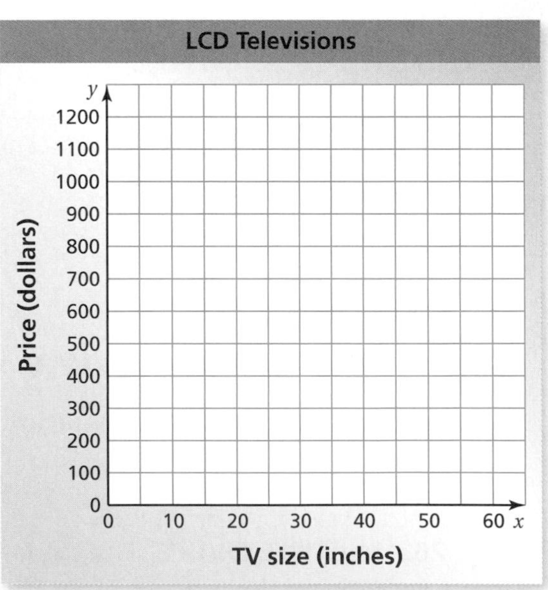

2 ACTIVITY: Choosing a Line of Fit

Compare your line of fit with the lines of fit of the other students in your class. Which line of fit do you think best models the data? What criteria did you use when choosing the line of fit? Explain your reasoning.

Laurie's Notes

Introduction

Standards for Mathematical Practice

- **MP5 Use Appropriate Tools Strategically:** To understand lines of best fit, it is important for students to fit a line by hand first. Once this is done, the regression capabilities of a graphing calculator can be used to find the line of best fit. It is important to explore the correlation coefficient to gauge how closely the equation models the data.

Motivate

- Write the following table on the board.

Number of weeks	1	2	3	4	6	7
Cumulative snowfall (in.)	11	18	32	52	73	82

- Use the data in the table to review the process of making a scatter plot using a graphing calculator. The important skills are entering the data, selecting the scatter plot option, setting an appropriate viewing window, and making the plot.

? "Do you think you could draw a line that models the data?" yes

? "Could your line help me estimate the cumulative snowfall in week 5? Explain." yes; Use the equation to find the y-value when $x = 5$.

Activity Notes

Activity 1

- Non-contextual scatter plots are often quick and easy to graph because they have *nice* domains and ranges. This contextual problem has a wide range. Students need to be as accurate as possible when plotting points.
- **Teaching Tip:** Use spaghetti to visualize lines of fit.
- A line of fit does not have to pass through two of the data points, however, students at this level may find it easier if it does.
- Discuss the results when students have finished the activity.

? "What does the slope mean in the context of the problem?" It is the average increase in price for each 1-inch increase in TV size.

Activity 2

- To help facilitate this activity, write the equations from many different students on the board.
- Give ample time for students to view all of the equations.
- While students are discussing, enter several of the equations in a graphing calculator, along with the ordered pairs.

? "Which line do you think best models the data?" Answers will vary.

? "What criteria did you use?" Student explanations should include a discussion of points being above and below the line.

Common Core State Standards

S.ID.6b Informally assess the fit of a function by plotting and analyzing residuals.

S.ID.8 Compute (using technology) and interpret the correlation coefficient of a linear fit.

S.ID.9 Distinguish between correlation and causation.

Previous Learning

Students should know how to draw and write equations of lines of fit.

Activity Materials	
Introduction	**Textbook**
• graphing calculators	• graphing calculators

Start Thinking! and Warm Up

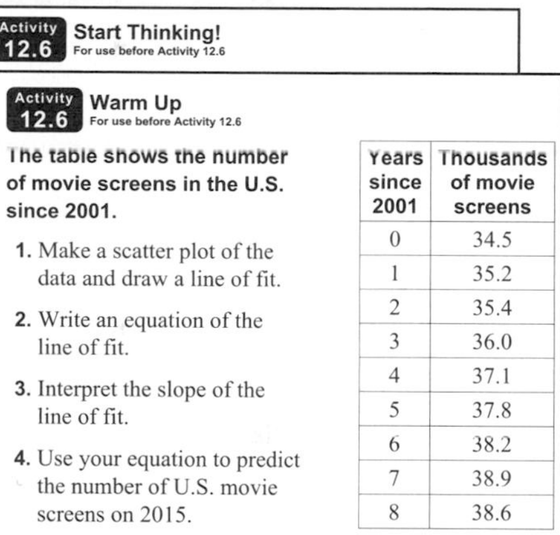

Activity 12.6 **Start Thinking!** For use before Activity 12.6

Activity 12.6 **Warm Up** For use before Activity 12.6

The table shows the number of movie screens in the U.S. since 2001.

1. Make a scatter plot of the data and draw a line of fit.
2. Write an equation of the line of fit.
3. Interpret the slope of the line of fit.
4. Use your equation to predict the number of U.S. movie screens on 2015.

Years since 2001	Thousands of movie screens
0	34.5
1	35.2
2	35.4
3	36.0
4	37.1
5	37.8
6	38.2
7	38.9
8	38.6

12.6 Record and Practice Journal

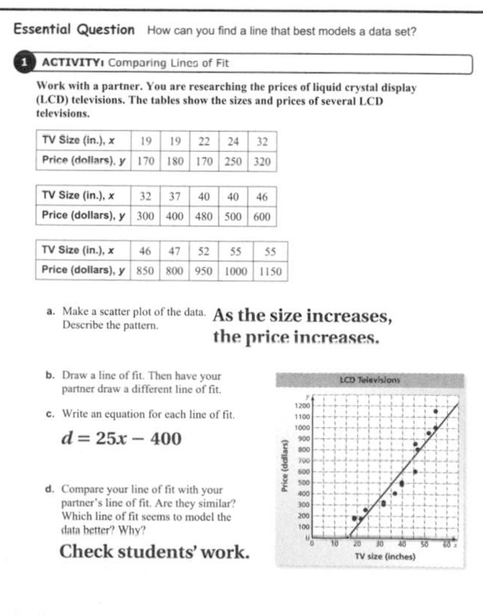

Essential Question How can you find a line that best models a data set?

1 ACTIVITY: Comparing Lines of Fit

Work with a partner. You are researching the prices of liquid crystal display (LCD) televisions. The tables show the sizes and prices of several LCD televisions.

TV Size (in.), x	19	19	22	24	32
Price (dollars), y	170	180	170	250	320

TV Size (in.), x	32	37	40	40	46
Price (dollars), y	300	400	480	500	600

TV Size (in.), x	46	47	52	55	55
Price (dollars), y	850	800	950	1000	1150

a. Make a scatter plot of the data. Describe the pattern. **As the size increases, the price increases.**

b. Draw a line of fit. Then have your partner draw a different line of fit.

c. Write an equation for each line of fit.
$d = 25x - 400$

d. Compare your line of fit with your partner's line of fit. Are they similar? Which line of fit seems to model the data better? Why?
Check students' work.

Differentiated Instruction

Kinesthetic

Give students a piece of a transparency sheet. Have them use a ruler and permanent marker to draw a straight line about 4 inches long. Use a hole punch to make a hole at each end of the line. Students can move the transparency over a scatter plot until they find a line of fit, make pencil marks in the holes, and use the marks to draw the line on the scatter plot.

12.6 Record and Practice Journal

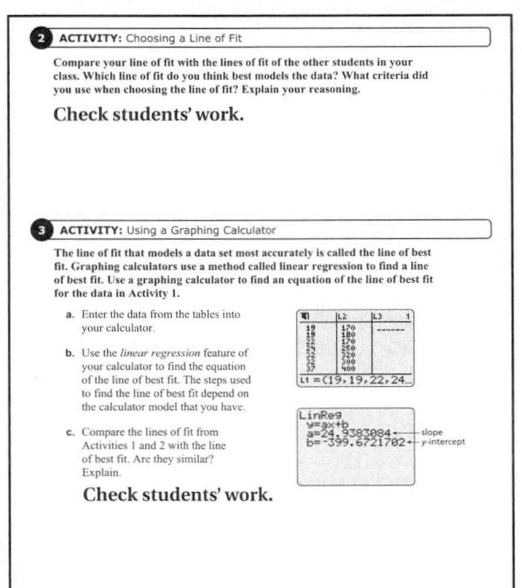

2 ACTIVITY: Choosing a Line of Fit

Compare your line of fit with the lines of fit of the other students in your class. Which line of fit do you think best models the data? What criteria did you use when choosing the line of fit? Explain your reasoning.

Check students' work.

3 ACTIVITY: Using a Graphing Calculator

The line of fit that models a data set most accurately is called the line of best fit. Graphing calculators use a method called linear regression to find a line of best fit. Use a graphing calculator to find an equation of the line of best fit for the data in Activity 1.

a. Enter the data from the tables into your calculator.

b. Use the *linear regression* feature of your calculator to find the equation of the line of best fit. The steps used to find the line of best fit depend on the calculator model that you have.

c. Compare the lines of fit from Activities 1 and 2 with the line of best fit. Are they similar? Explain.

Check students' work.

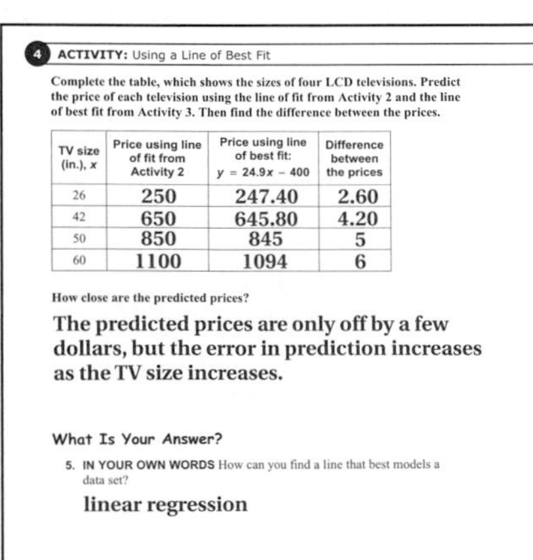

4 ACTIVITY: Using a Line of Best Fit

Complete the table, which shows the sizes of four LCD televisions. Predict the price of each television using the line of fit from Activity 2 and the line of best fit from Activity 3. Then find the difference between the prices.

TV size (in.), x	Price using line of fit from Activity 2	Price using line of best fit: $y = 24.9x - 400$	Difference between the prices
26	250	247.40	2.60
42	650	645.80	4.20
50	850	845	5
60	1100	1094	6

How close are the predicted prices?

The predicted prices are only off by a few dollars, but the error in prediction increases as the TV size increases.

What Is Your Answer?

5. **IN YOUR OWN WORDS** How can you find a line that best models a data set?

linear regression

Laurie's Notes

Activity 3

- **MP5:** Explain that the calculator has a built-in feature that gives an equation for the *line of best fit.*
- The mathematics of how the calculator actually does the calculation may be studied in a future course. What students need to understand is that the process takes into account all of the data points.
- Different calculators have different keystrokes. It would be helpful to have students bring calculator manuals.
- Circulate around the room to answer questions as they arise. Pairing students with similar calculators will help.
- When students have finished, have a discussion about the equation generated by the calculator.
- **MP6 Attend to Precision:** The calculator will give a slope and y-intercept, typically with many decimal places. Ask whether students think it is necessary to carry all of the digits to the right of the decimal point.
- **Common Error:** As students observe the equation computed by the calculator, it is likely that someone will say, "That's not what my calculator says." If this happens, then the student did not enter the data correctly.

Activity 4

- **Big Idea:** One of the reasons we write lines of best fit is to make predictions. The equation can be used to interpolate or extrapolate unknown data values.
- **Note:** The equation for the line of best fit from Activity 3 is given in the table to avoid rounding errors and discrepancies in answers.
- When students finish, discuss the differences between the prices using their lines of fit and the calculator's line of best fit. The differences should be small.
- **Extension:** If time permits, add a new data value that is an outlier, such as (36, 950). Ask how it affects the calculator's line of best fit.

What Is Your Answer?

- Ask volunteers to share their answers.

Closure

- **Exit Ticket:** Find the line of best fit for the data in the Motivate. Use the line to estimate the cumulative snowfall in week 5. $y = 12.5x - 3$; 59.5 in.

Technology For the Teacher

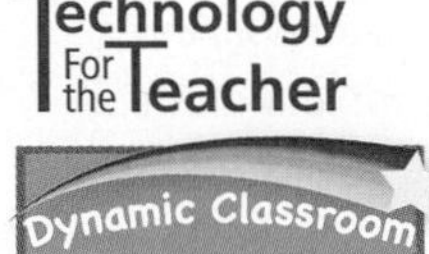

The Dynamic Planning Tool
Editable Teacher's Resources at *BigIdeasMath.com*

3 ACTIVITY: Using a Graphing Calculator

The line of fit that models a data set most accurately is called the line of best fit. Graphing calculators use a method called linear regression to find a line of best fit. Use a graphing calculator to find an equation of the line of best fit for the data in Activity 1.

a. Enter the data from the tables into your calculator.

L1	L2	L3 1
19	170	------
19	180	
22	170	
24	250	
32	320	
32	300	
37	400	

L1 ={19,19,22,24...

b. Use the *linear regression* feature of your calculator to find the equation of the line of best fit. The steps used to find the line of best fit depend on the calculator model that you have.

```
LinReg
 y=ax+b
 a=24.9383084  ← slope
 b=-399.6721702  ← y-intercept
```

c. Compare the lines of fit from Activities 1 and 2 with the line of best fit. Are they similar? Explain.

4 ACTIVITY: Using a Line of Best Fit

Copy and complete the table, which shows the sizes of four LCD televisions. Predict the price of each television using the line of fit from Activity 2 and the line of best fit from Activity 3. Then find the difference between the prices.

TV Size (in.), x	Price Using Line of Fit from Activity 2	Price Using Line of Best Fit: $y = 24.9x - 400$	Difference Between the Prices
26			
42			
50			
60			

How close are the predicted prices?

What Is Your Answer?

5. **IN YOUR OWN WORDS** How can you find a line that best models a data set?

Practice Use what you learned about analyzing lines of fit to complete Exercise 4 on page 649.

12.6 Lesson

Key Vocabulary
residual, *p. 646*
linear regression, *p. 647*
line of best fit, *p. 647*
correlation coefficient, *p. 647*
causation, *p. 648*

One way to determine how well a line of fit models a data set is to analyze *residuals.*

Residuals

A **residual** is the difference between the y-value of a data point and the corresponding y-value found using the line of fit. A residual can be positive, negative, or zero.

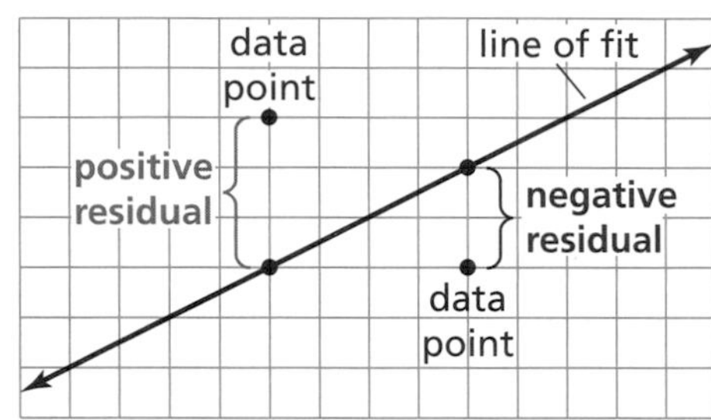

A scatter plot of the residuals shows how well a model fits a data set. If the model is a good fit, then the residual points will be randomly dispersed about the horizontal axis. If the model is not a good fit, then the residual points will form some type of pattern.

EXAMPLE 1 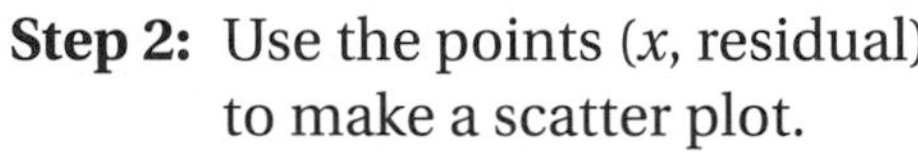

Week, x	Sales (millions), y
1	\$19
2	\$15
3	\$13
4	\$11
5	\$10
6	\$8
7	\$7
8	\$5

In Example 3 in Section 12.5, the equation $y = -2x + 20$ models the data in the table at the left. Is the model a good fit?

Step 1: Calculate the residuals and organize your results in a table.

x	y	y-Value from Model	Residual
1	19	18	$19 - 18 = 1$
2	15	16	$15 - 16 = -1$
3	13	14	$13 - 14 = -1$
4	11	12	$11 - 12 = -1$
5	10	10	$10 - 10 = 0$
6	8	8	$8 - 8 = 0$
7	7	6	$7 - 6 = 1$
8	5	4	$5 - 4 = 1$

Step 2: Use the points (x, residual) to make a scatter plot.

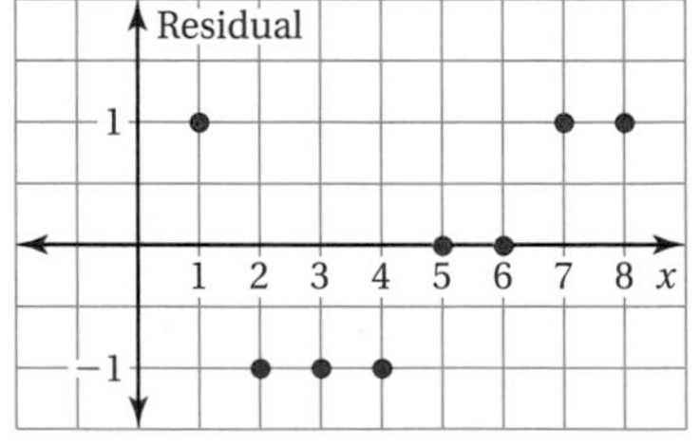

∴ The points are randomly dispersed about the horizontal axis. So, the equation $y = -2x + 20$ is a good fit.

Laurie's Notes

Introduction

Connect

- **Yesterday:** Students compared a line of fit to the line of best fit. (MP5, MP6)
- **Today:** Students will use residuals and find lines of best fit.

Motivate

- Display a scatter plot and ask students to draw a line of fit.
- Anticipate that it doesn't pass through any of the points.

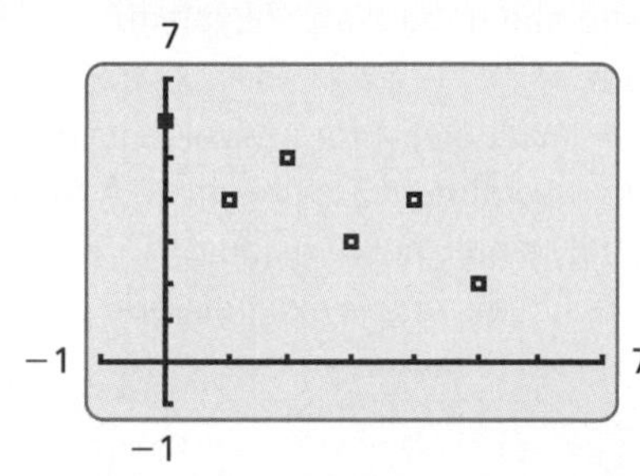

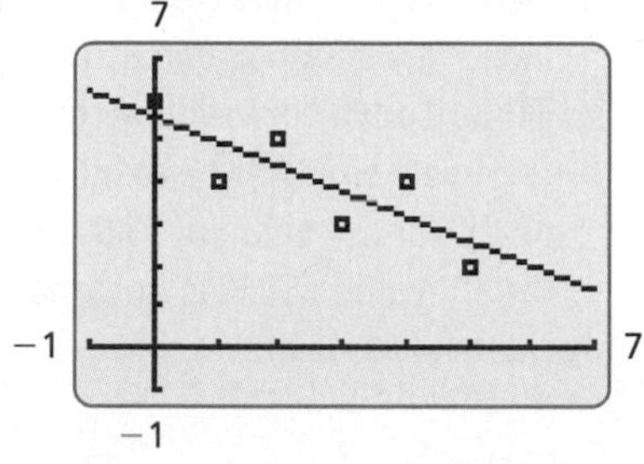

- Ask for observations about the line. Hopefully they will say that about half of the points are above the line and about half are below the line.
- You want students to focus on the difference between the y-values of the line and the actual y-values.

Lesson Notes

Key Idea

- Define residual. Use a sketch to demonstrate residuals that are positive, negative, and zero.
- Explain that each x-value has a residual.
- When a model is a good fit, the residual points form a random pattern. That is, some of the residuals will be above and some will be below the horizontal axis. They show no discernible pattern.
- When a model is not a good fit, the residuals form a pattern, such as a linear or U-shaped pattern.
- I like to draw a scatter plot with a U-shape. When you try to fit a line to the curved pattern, the left side and right side of the points will be above the line, and the middle points will be below the line.
- **Note:** You may wish to tell students that calculators use residuals to calculate lines of best fit.

Example 1

- **MP4 Model with Mathematics:** Review Example 3 from Section 12.5. The problem involved sales of DVDs over a period of 8 weeks.
- Say, "We want to compare the actual y-values and the y-values from the model."
- Begin by listing the actual ordered pairs in a table.

Goal Today's lesson is analyzing lines of fit.

Lesson Materials
Textbook
• graphing calculator

Start Thinking! and Warm Up

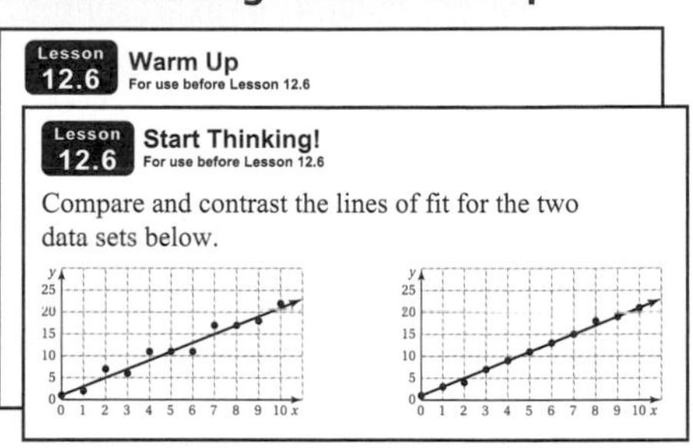
Lesson 12.6 Warm Up
For use before Lesson 12.6

Lesson 12.6 Start Thinking!
For use before Lesson 12.6

Compare and contrast the lines of fit for the two data sets below.

Extra Example 1

The equation $y = x - 1$ models the data in the table. Is the model a good fit?

x	y
2	1
3	2
4	4
5	2
6	4
7	5
8	8
9	8

The points (x, residual) are randomly dispersed about the horizontal axis. So, the equation $y = x - 1$ is a good fit.

Laurie's Notes

Example 1 (continued)

- Then find the y-values from the model. Use the language, "actual y-value" and "y-value from the model" to help students differentiate between the two.
- Finally, compute the residuals.
- The table of residuals is often plotted on a horizontal line and called a residual plot. Each x-value has a residual, even if it is 0. If time permits, explore the residual plot feature on a graphing calculator.

? "Are the residuals randomly dispersed about the x-axis? Explain." yes; They show no discernible pattern.

- **MP3a Construct Viable Arguments:** A residual of 1 for a y-value of 19 is relatively small, but a residual of 1 for a y-value of 2 is not small. A residual of 100 for a y-value of 35,000 is relatively small, but a residual of 100 for a y-value of 250 is not small. The residuals in this example are relatively small.

Example 2

- Read the problem and then create a table as in Example 1. Make a residual plot.

? "Are the residuals both positive and negative?" yes "Is there a pattern or are the points randomly dispersed?" There is a pattern; The residuals are negative at the beginning and end of the data set, and positive in the middle.

- Conclude that because the residual plot forms a $\cap$ -shape, the model does not fit the data well.

On Your Own

- Students may want to make a scatter plot and graph the model provided using a graphing calculator.

Discuss

- Remind students of the activity in which they used calculators to determine the line of best fit. It was called linear regression.
- Explain that in addition to calculating the line of best fit, the calculator also gives a value called the correlation coefficient. This is a measure of how well the line fits the data.
- Describe positive, negative, and no correlation.
- Say, "The correlation coefficient is a number between -1 and 1." Draw and label the graphic shown.
- **Common Misconception:** Students may think that a strong negative correlation is *bad*. Stress that the "negative" merely refers to the downward trend of the data, or slope.

Extra Example 2

The table shows the ounces y in a bottle of water after x minutes of exercising. The equation $y = -0.2x + 16$ models the data. Is the model a good fit?

Minutes, x	Ounces, y
5	18
10	15
15	15
20	14
25	12
30	9
35	6
40	2

The points (x, residual) form a linear pattern with a negative slope. So, the equation $y = -0.2x + 16$ does not model the data well.

On Your Own

1. The points (x, residual) are randomly dispersed about the horizontal axis. So, the model $y = -9.8x + 850$ is a good fit.

English Language Learners

Vocabulary

Help English language learners with the meaning of the key vocabulary word *correlation coefficient*. The word *correlate* can be broken into two parts, *co-* and *relate*. The prefix *co-* means "with" or "together" as in *copilot*. The word *relate* means "to show or make a connection."

EXAMPLE 2 Using Residuals

Age, x	Salary, y
35	42
37	44
41	47
43	50
45	52
47	51
53	49
55	45

The table at the left shows the ages x and salaries y (in thousands of dollars) of eight employees at a company. The equation $y = 0.2x + 38$ models the data. Is the model a good fit?

Step 1: Calculate the residuals and organize your results in a table.

x	y	y-Value from Model	Residual
35	42	45.0	$42 - 45.0 = -3.0$
37	44	45.4	$44 - 45.4 = -1.4$
41	47	46.2	$47 - 46.2 = 0.8$
43	50	46.6	$50 - 46.6 = 3.4$
45	52	47.0	$52 - 47.0 = 5.0$
47	51	47.4	$51 - 47.4 = 3.6$
53	49	48.6	$49 - 48.6 = 0.4$
55	45	49.0	$45 - 49.0 = -4.0$

Step 2: Use the points (x, residual) to make a scatter plot.

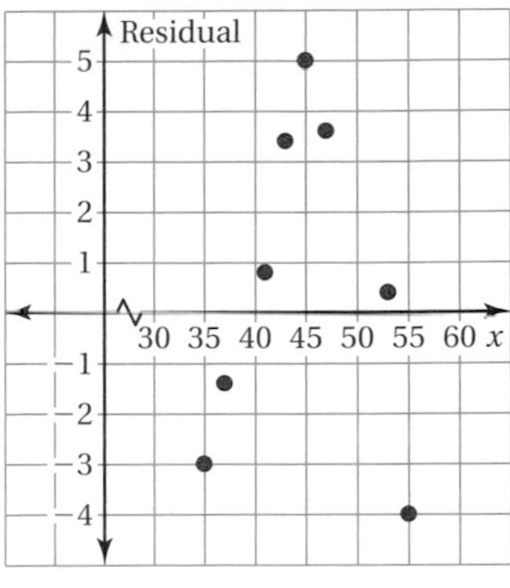

The points form a ∩-shaped pattern. So, the equation $y = 0.2x + 38$ does not model the data well.

On Your Own

1. The table shows the attendance y (in thousands) at an amusement park from 2000 to 2009, where $x = 0$ represents the year 2000. The equation $y = -9.8x + 850$ models the data. Is the model a good fit?

Year, x	0	1	2	3	4	5	6	7	8	9
Attendance, y	850	845	828	798	800	792	785	781	775	760

Study Tip

You know how to use two points to find an equation of a line of fit. When finding an equation of the line of best fit, every point in the data set is used.

Graphing calculators use a method called **linear regression** to find a precise line of fit called a **line of best fit.** This line best models a set of data. A calculator often gives a value r called the **correlation coefficient.** This value tells whether the correlation is positive or negative, and how closely the equation models the data. Values of r range from -1 to 1. When r is close to 1 or -1, there is a strong correlation between the variables. As r gets closer to 0, the correlation becomes weaker.

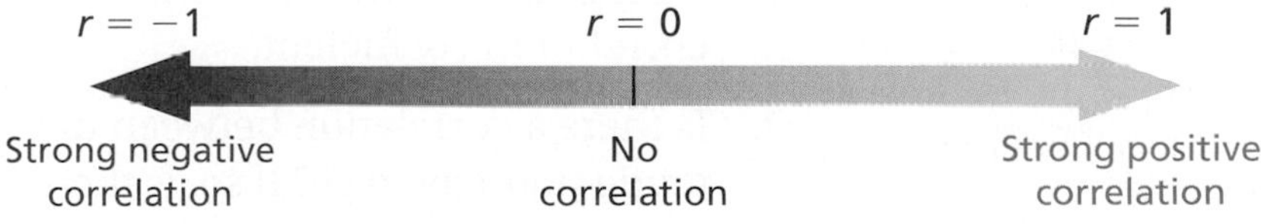

EXAMPLE 3 Finding a Line of Best Fit Using Technology

The table shows the worldwide movie ticket sales y (in billions of dollars) from 2000 to 2010, where $x = 0$ represents the year 2000. Use a graphing calculator to find an equation of the line of best fit. Identify and interpret the correlation coefficient.

Year, x	0	1	2	3	4	5	6	7	8	9	10
Ticket Sales, y	16	17	20	20	25	23	26	26	28	29	32

Step 1: Enter the data from the table into your calculator.

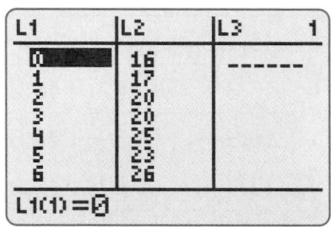

L1	L2	L3
0	16	------
1	17	
2	20	
3	20	
4	25	
5	23	
6	26	

L1(1)=0

Step 2: Use the *linear regression* feature.

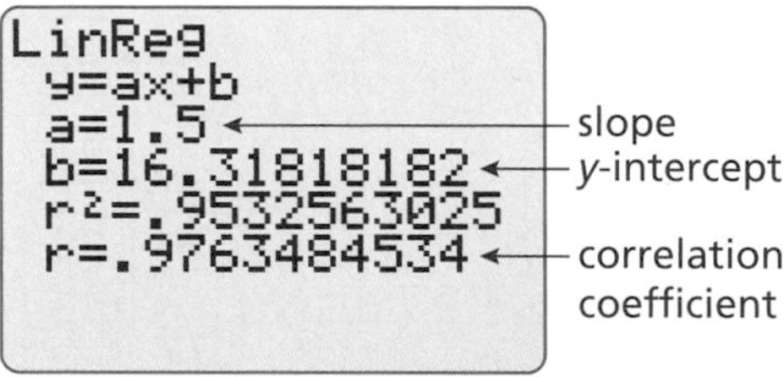

Study Tip

The slope of 1.5 indicates that sales are increasing by about $1.5 billion each year. The y-intercept of 16 represents the ticket sales of $16 billion for 2000.

An equation of the line of best fit is $y = 1.5x + 16$. The correlation coefficient is about 0.976. This means that the relationship between the years and ticket sales is a strong positive correlation and the equation closely models the data.

When a change in one variable x results in a change in another variable y, it is called **causation.** Causation produces a strong correlation between the two variables. The converse of the statement is not true. In other words, correlation does not imply causation.

EXAMPLE 4 Identifying Correlation and Causation

Tell whether a correlation is likely in the situation. If so, tell whether there is a causal relationship. Explain your reasoning.

A causal relationship exists when one variable causes a change in another variable.

a. time spent exercising and the number of calories burned

There is a positive correlation and a causal relationship because the more time you spend exercising, the more calories you burn.

b. the number of banks and the population of a city

There may be a positive correlation but no causal relationship. Building more banks will not cause the population to increase.

On Your Own

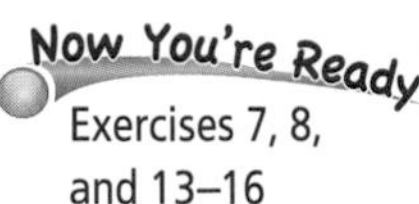

Exercises 7, 8, and 13–16

2. Use a graphing calculator to find an equation of the line of best fit for the data in On Your Own Question 1. Identify and interpret the correlation coefficient.
3. Is there a correlation between time spent playing video games and grade point average? If so, is there a causal relationship? Explain.

Laurie's Notes

Example 3

- **MP1a Make Sense of Problems:** Ask a student to read the problem. Make sure students understand that $x = 0$ represents the year 2000.
- On many calculators, the correlation coefficient is a feature that can be turned on and off. Instruct students how to turn this feature on so that it is displayed when they perform the regression. An r^2-value may also be displayed, which students may learn about in a future course.

? "What is the line of best fit?" $y = 1.5x + 16$

? "What do the slope and y-intercept mean?" A slope of 1.5 means that ticket sales are increasing by about $1.5 billion each year. A y-intercept of 16 means that in 2000, the ticket sales were about $16 billion.

? "What is the correlation coefficient and what does it mean?" $r \approx 0.976$; It implies a strong positive correlation between years and ticket sales.

Discuss

? "In Example 3, would the year increasing to 2012 *cause* the sales to increase also?" Students may be unsure of this, however, they should recognize that it is not the year that causes an increase in sales. Many other factors are the reasons behind the increases in ticket sales.

- Define causation. It is important for students to understand the difference between correlation and causation. Correlation between two variables does not imply that one causes the other.

Example 4

- Read each statement in parts (a) and (b). Discuss the explanation of causation for each.

On Your Own

- **Think-Pair-Share:** Students should read each question independently and then work with a partner to answer the questions. When they have answered the questions, the pair should compare their answers with another group and discuss any discrepancies.

Closure

- **Exit Ticket:** Give an example of a situation in which there is correlation between the variables, but no causation. *Sample answer:* The number of teachers in a school and the number of students in the school

Technology For the Teacher

The Dynamic Planning Tool
Editable Teacher's Resources at *BigIdeasMath.com*

Extra Example 3

Use a graphing calculator to find an equation of the line of best fit for the data in Extra Example 2. Identify and interpret the correlation coefficient.
$y = -0.4x + 21$; The correlation coefficient is about -0.966. This means that the relationship between minutes and ounces is a strong negative correlation and the equation closely models the data.

Extra Example 4

Is there a correlation between the number of doctors at a hospital and the number of patients in the hospital? If so, is there a causal relationship? Explain.
There may be a positive correlation but there is no causal relationship. Increasing the number of doctors will not cause the number of patients to increase.

On Your Own

2. $y = -9.6x + 845$; The correlation coefficient is about -0.964. The relationship between the years and attendance have a strong negative correlation and the equation closely models the data.
3. There may be a negative correlation. A causal relationship is possible because the more time you spend playing video games, the less time you spend studying.

Vocabulary and Concept Check

1. when actual value > value from model; when value from model > actual value
2. Plot the points (x, residual). If the points form a pattern, the line does not fit the data well.
3. -0.98, because it is closer to -1 than 0.91 is to 1. $(|-0.98| > |0.91|)$

Practice and Problem Solving

4. a. *Sample answer:* Using (1, 6) and (4, 3.9), $y = -0.7x + 6.7$.

   b. $y = -0.7x + 6.8$

   c. *Sample answer:* 2.5 in.; 2.6 in.
5. The points (x, residual) are all above the horizontal axis. So, the equation does not model the data well.
6. The points (x, residual) are randomly dispersed about the horizontal axis. So, the equation is a good fit.
7. $y = 3.5$; $r = 0$; There is no correlation between x and y. The equation does not fit the data.
8. $y = -1.2x + 7$; $r \approx -0.883$; The relationship between x and y is a fairly strong negative correlation and the equation fits the data fairly well.
9. $y = 357.5x - 495$

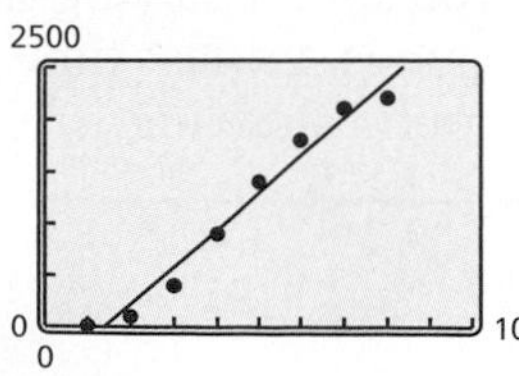

10. B

11. A

Assignment Guide and Homework Check

Level	Assignment	Homework Check
Average	1–3, 5, 7, 9, 13–19 odd, 22–24	5, 13, 17, 19
Advanced	1–3, 6, 8, 9, 14–20 even, 19, 22–24	6, 14, 19, 20

For Your Information

- **Exercise 16** This will be a good exercise for discussion. Overall, there is no correlation. There are many different types and sizes of dogs with varying tail lengths. However, an argument could be made that there is a correlation depending on the circumstance. For example, consider a growing puppy.

Common Errors

- **Exercises 5 and 6** Students may have trouble calculating the residuals. Remind them that a residual is the y-value from the data minus the y-value from the model.
- **Exercises 7–9** Students may enter the x-values into the y list of the graphing calculator, and vice versa. Encourage them to double check their data entries and understand which lists represent the x- and y-values.

12.6 Record and Practice Journal

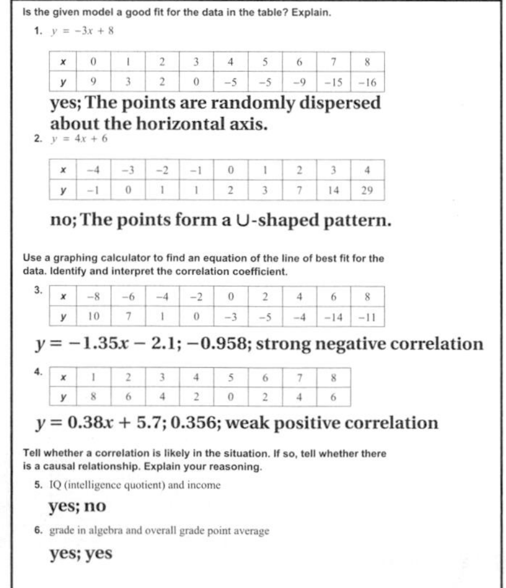

Is the given model a good fit for the data in the table? Explain.

1. $y = -3x + 8$

x	0	1	2	3	4	5	6	7	8
y	9	3	2	0	−5	−5	−9	−15	−16

yes; The points are randomly dispersed about the horizontal axis.

2. $y = 4x + 6$

x	−4	−3	−2	−1	0	1	2	3	4
y	−1	0	1	1	2	3	7	14	29

no; The points form a U-shaped pattern.

Use a graphing calculator to find an equation of the line of best fit for the data. Identify and interpret the correlation coefficient.

3.

x	−8	−6	−4	−2	0	2	4	6	8
y	10	7	1	0	−3	−5	−4	−14	−11

$y = -1.35x - 2.1$; -0.958; strong negative correlation

4.

x	1	2	3	4	5	6	7	8
y	8	6	4	2	0	2	4	6

$y = 0.38x + 5.7$; 0.356; weak positive correlation

Tell whether a correlation is likely in the situation. If so, tell whether there is a causal relationship. Explain your reasoning.

5. IQ (intelligence quotient) and income

yes; no

6. grade in algebra and overall grade point average

yes; yes

Technology For the Teacher

Answer Presentation Tool

12.6 Exercises

Vocabulary and Concept Check

1. **VOCABULARY** When is a residual positive? When is it negative?
2. **WRITING** Explain how you can use residuals to determine how well a line of fit models a data set.
3. **NUMBER SENSE** Which correlation coefficient indicates a stronger relationship, -0.98 or 0.91? Explain.

Practice and Problem Solving

4. **ANTLERS** The table shows the weekly growth y (in inches) of an elk's antlers.

Week, x	1	2	3	4	5
Growth, y	6.0	5.5	4.7	3.9	3.3

   a. Find a line of fit for the data.
   b. Use a graphing calculator to find an equation of the line of best fit.
   c. Use each model to predict the antler growth in week 6.

Is the given model a good fit for the data in the table? Explain.

5. $y - 4x - 5$

x	-4	-3	-2	-1	0	1	2	3	4
y	-18	-13	-10	-7	-2	0	6	10	15

6. $y = -1.3x + 1$

x	-8	-6	-4	-2	0	2	4	6	8
y	9	10	5	8	-1	1	-4	-12	-7

Use a graphing calculator to find an equation of the line of best fit for the data. Identify and interpret the correlation coefficient.

7.

x	0	1	2	3	4	5	6	7
y	8	5	2	-1	-1	2	5	8

8.

x	-8	-6	-4	-2	0	2	4	6	8	10
y	20	8	17	7	8	1	5	-2	2	-8

9. **EARTHQUAKE** The table shows the total number y of people reporting an earthquake x minutes after it ended. Use a graphing calculator to find an equation of the line of best fit. In the same viewing window, graph the line and plot the data.

Minutes, x	1	2	3	4	5	6	7	8
People, y	10	100	400	900	1400	1800	2100	2200

MATCHING **Match the graph of the data with its corresponding linear regression screen.**

10. 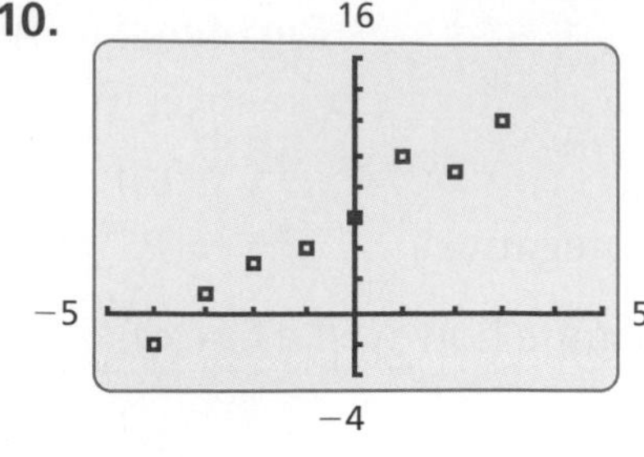

11. 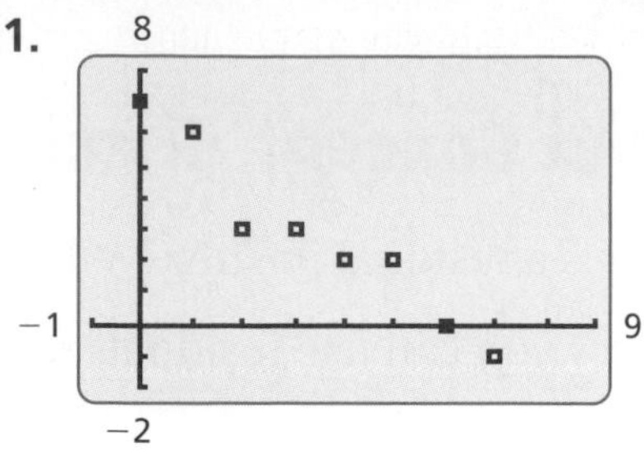

12. 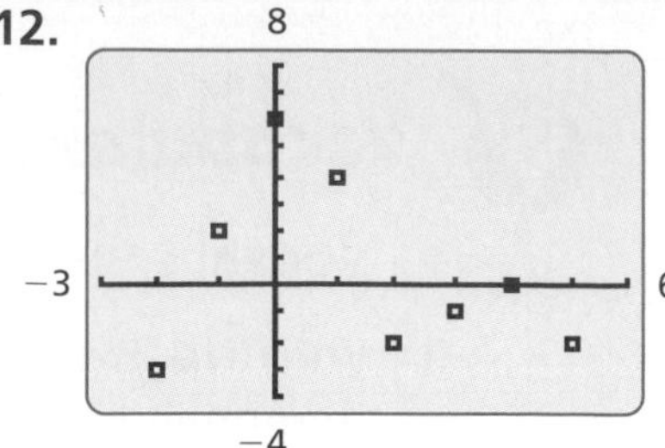

A. 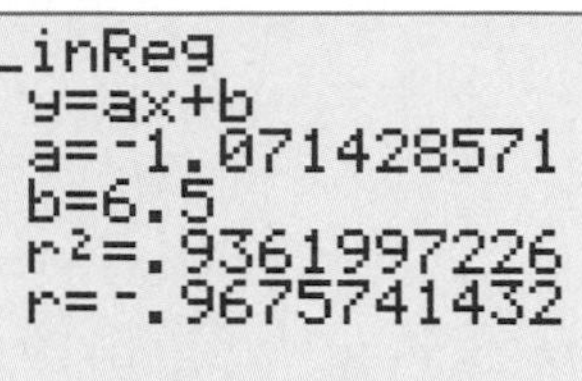

B. 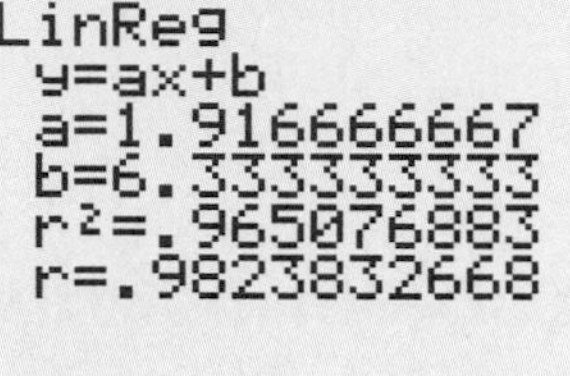

C. 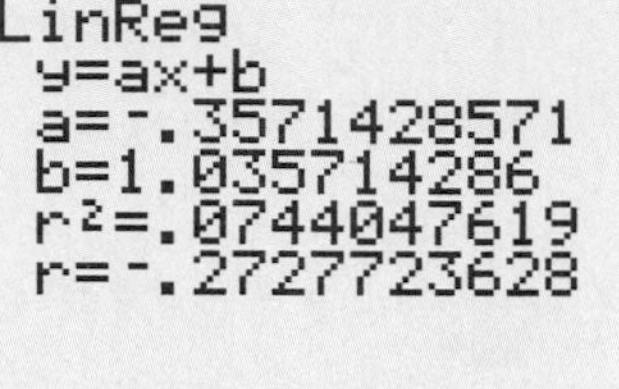

Tell whether a correlation is likely in the situation. If so, tell whether there is a causal relationship. Explain your reasoning.

(4) 13. the amount of time spent talking on a cell phone and the remaining battery life

14. the height of a toddler and the size of the toddler's vocabulary

15. the number of hats you own and the size of your head

16. the weight of a dog and the length of its tail

17. **FUEL MILEAGE** The table shows the prices x (in thousands of dollars) and fuel economies y (in miles per gallon) of several automobiles.

Price (thousands of dollars), x	24	32	30	28	35	20	22	26
Fuel Economy (miles per gallon), y	30	30	34	35	28	25	28	36

a. Use a graphing calculator to find an equation of the line of best fit. Identify and interpret the correlation coefficient.

b. Calculate the residuals. Then make a scatter plot of the residuals and interpret the results.

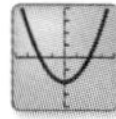

18. **TEXTING** The table shows the numbers y (in billions) of text messages sent from 2006 to 2011, where $x = 6$ represents the year 2006.

a. Use a graphing calculator to find an equation of the line of best fit. Identify and interpret the correlation coefficient.

b. Interpret the slope of the line of best fit.

c. Calculate the residuals. Then make a scatter plot of the residuals and interpret the results.

d. Predict the number of text messages sent in 2015.

Year, x	Text Messages (billions), y
6	113
7	241
8	601
9	1360
10	1806
11	2206

Common Errors

- **Exercises 13–16** Students may have a difficult time determining causation. Remind them that a correlation does not imply causation.
- **Exercises 17–21** Students may enter the x-values into the y list of the graphing calculator, and vice versa. Encourage them to double check their data entries and understand which lists represent the x- and y-values.

Practice and Problem Solving

12. C

13. yes; yes; Talking longer causes the battery life to decrease.

14. yes; no; Taller toddlers are likely older and likely to know more words, but being taller does not cause an increase in vocabulary.

15. no **16.** no

17. a. $y = 0.2x + 25$; $r \approx 0.283$; The relationship between x and y is a weak positive correlation and the equation does not fit the data well.

b.

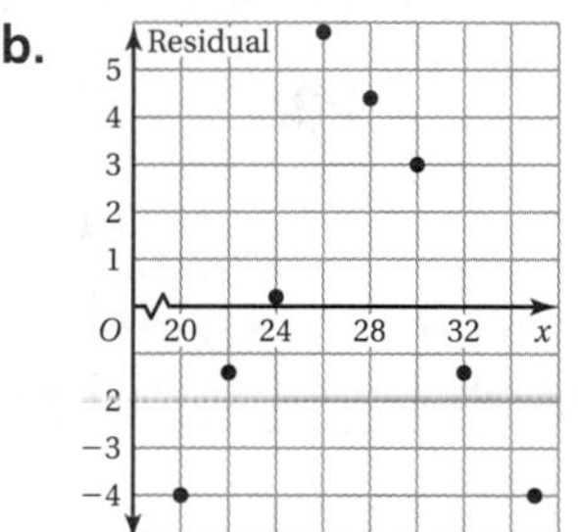

The points (x, residual) form a ∩-shaped pattern. So, the equation does not model the data well.

18–19. See Additional Answers.

English Language Learners

Pair Activity

Pair English language learners with English speakers to solve problems using a graphing calculator. English language learners will benefit by conversing in English and all students will benefit by learning how to operate the graphing calculator.

Practice and Problem Solving

20. The correlation coefficient changes from about −0.965 to about −0.667 which weakens the correlation. This happens because the new data point is an outlier.

21. See *Taking Math Deeper.*

Fair Game Review

22. 67; 19; about 7.0

23. 15; 8; 2

24. C

Mini-Assessment

1. The equation $y = 2x + 1$ models the data in the table. Is the model a good fit?

x	1	2	3	4	5	6
y	4	6	10	11	14	14

The points (x, residual) are all above the horizontal axis. So, the equation $y = 2x + 1$ does not model the data well.

2. The table shows the forecasted temperature x (in °F) and the actual temperature y (in °F) during one week. Use a graphing calculator to find an equation of the line of best fit. Identify and interpret the correlation coefficient.

x (°F)	72	75	73	82	83	78	77
y (°F)	68	74	79	85	83	81	76

$y = 1.1x - 10$; $r \approx 0.826$; The relationship between x and y is a fairly strong positive correlation and the equation fits the data fairly well.

3. In Exercise 2, is there a causal relationship between x and y? Explain. no; While there is a correlation, forecasting a temperature does not cause the actual temperature to change. The forecast is just a prediction.

Taking Math Deeper

Exercise 21

This exercise challenges students to combine previous knowledge about quadratic functions with knowledge about regression to solve a problem they have not been taught how to solve.

a. Copy and complete the table.
Each entry in the y-row must show the number of *new* responses each minute.

x	1	2	3	4	5	6	7	8
y	10	90	300	500	500	400	300	100

b. As x increases, the y-values increase, and then decrease.

A linear function will not fit the data well because the data do not show an approximately constant rate of change.

Because the y-values increase and then decrease, a quadratic function may fit the data well.

c. For linear functions, you used the *linear regression* feature. So, find and use the *quadratic regression* feature.

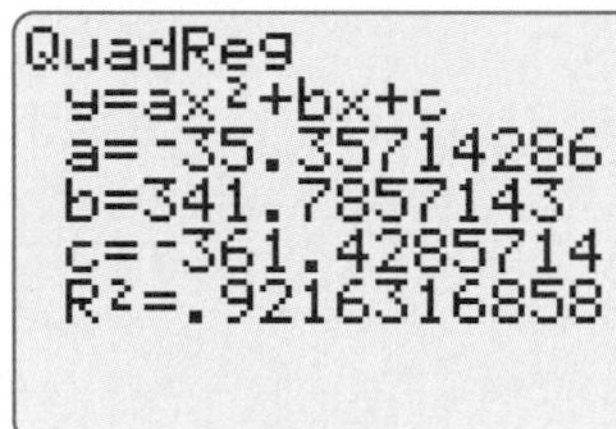

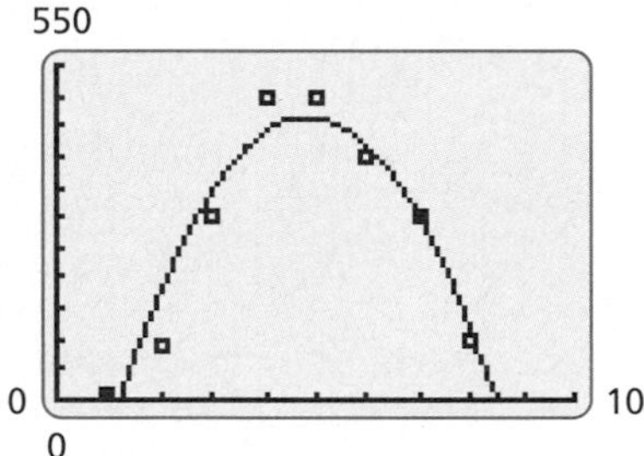

The data can be modeled by $y = -35.36x^2 + 341.8x - 361$. Challenge students to make a conjecture about the meaning of the R^2-value.

Reteaching and Enrichment Strategies

If students need help . . .	If students got it . . .
Resources by Chapter • Practice A and Practice B • Puzzle Time Record and Practice Journal Practice Differentiating the Lesson Lesson Tutorials Skills Review Handbook	Resources by Chapter • Enrichment and Extension • School-to-Work • Financial Literacy • Technology Connection Start the next section

19. GRADES The table shows the numbers x of hours spent watching television each week, and the grade point averages y of several students.

Hours, x	Grade Point Average, y
10	3.0
5	3.4
3	3.5
12	2.7
20	2.1
15	2.8
8	3.0
4	3.7
16	2.5

a. Use a graphing calculator to find an equation of the line of best fit. Identify and interpret the correlation coefficient.

b. Interpret the slope and y-intercept of the line of best fit.

c. Another student watches about 14 hours of television each week. Predict the student's grade point average.

d. Do you think watching more television each week may cause a lower grade point average? Explain.

20. REASONING A student spends 2 hours each week watching television and has a grade point average of 2.4. Include this information in the data set in Exercise 19. How does including this value affect the correlation coefficient? Explain.

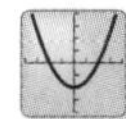

21. **Modeling** Consider the earthquake data in Exercise 9.

a. Copy and complete the table to show the number y of people reporting the earthquake in the xth minute after the earthquake ended.

x	1	2	3	4	5	6	7	8
y	10	90	300					

b. Describe how the y-values change as x increases. Do you think a linear function will fit the data well? If not, what type of function do you think will fit the data well? Explain.

c. Use a graphing calculator to find the model in part (b).

Fair Game Review What you learned in previous grades & lessons

Find the mean, range, and standard deviation of the data. *(Section 12.2)*

22. 59, 70, 62, 68, 75, 77, 58

23. 15, 14, 11, 15, 16, 19, 14, 16, 15

24. MULTIPLE CHOICE What is the interquartile range of the box-and-whisker plot? *(Section 12.3)*

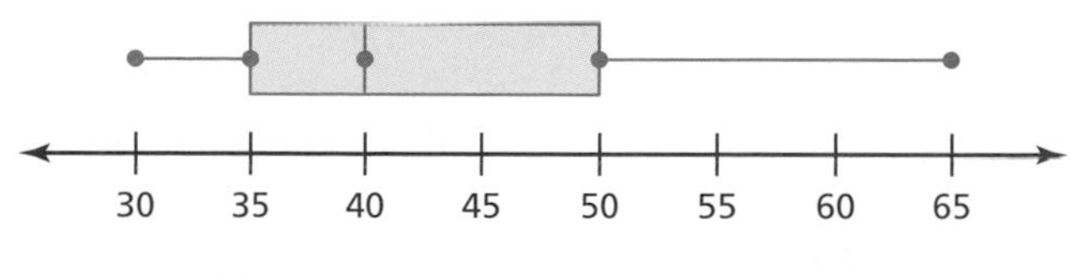

(A) 5 (B) 10 (C) 15 (D) 35

12.7 Two-Way Tables

COMMON CORE STATE STANDARDS
8.SP.4
S.ID.5

Essential Question How can you read and make a two-way table?

Two categories of data can be displayed in a **two-way table.**

1 ACTIVITY: Reading a Two-Way Table

Work with a partner. You are the manager of a sports shop. The two-way table shows the numbers of soccer T-shirts that your shop has left in stock at the end of the season.

		T-Shirt Size					
		S	M	L	XL	XXL	Total
Color	Blue/White	5	4	1	0	2	
	Blue/Gold	3	6	5	2	0	
	Red/White	4	2	4	1	3	
	Black/White	3	4	1	2	1	
	Black/Gold	5	2	3	0	2	
	Total						65

a. Complete the totals for the rows and columns.

b. Are there any black and gold XL T-shirts in stock? Justify your answer.

c. The numbers of T-shirts you ordered at the beginning of the season are shown below. Complete the two-way table.

		T-Shirt Size					
		S	M	L	XL	XXL	Total
Color	Blue/White	5	6	7	6	5	
	Blue/Gold	5	6	7	6	5	
	Red/White	5	6	7	6	5	
	Black/White	5	6	7	6	5	
	Black/Gold	5	6	7	6	5	
	Total						

d. How would you alter the numbers of T-shirts you order for next season? Explain your reasoning.

Laurie's Notes

Introduction

Standards for Mathematical Practice

- **MP2 Reason Abstractly and Quantitatively:** In this section, students are translating information into an organized table to make sense of the problem and to make observations and reason about the information. The goal is not to simply construct a table or read information from the table, but to reason about any relationships that may exist between categories in the table.

Motivate

- **Story Time:** Tell students about a few of the sessions and workshops you attended at a 3-day math conference. When you returned, you submitted your expenses.

	Day 1	Day 2	Day 3	Totals
Meals				A
Lodging				
Taxi				
Totals		B		C

- The school district does not want to share your expenses publicly so they are blacked out.
- ? "What do the numbers in A, B, and C represent?" A is the total amount spent on meals for 3 days; B is the total expenses for day 2; C is the total expenses for all 3 days.
- ? "How do you find C?" Find the sum of the last column or the sum of the last row.
- If students do not know the vocabulary—column and row, be sure to clarify.

Activity Notes

Activity 1

- Students should find that reading a two-way table is relatively easy. The term "two-way table" is new to students, yet they do not need a formal definition to make sense of the problem and what it is asking. In fact, some students will jump in and start adding the entries in the rows and columns without reading the introduction.
- When students have finished, discuss their responses to part (d).
- ? "What size(s) of shirts sold well?" XL and perhaps XXL "What size(s) of shirts did not sell well?" S and M
- ? "What color(s) of shirts sold well?" Black/White "What color(s) of shirts did not sell well?" Blue/Gold
- ? "Is there a way to quantify or rank the popular sizes and colors?" yes; You can compute the percent of each size or color that was sold.
- ? "Do you think merchants keep track of inventory in this manner?" Answers will vary. Successful merchants do track inventory to see what is selling.

Common Core State Standards

8.SP.4 Understand that patterns of association can also be seen in bivariate categorical data by displaying frequencies and relative frequencies in a two-way table. Construct and interpret a two-way table . . .. Use relative frequencies calculated for rows or columns to describe possible association between the two variables.

S.ID.5 Summarize categorical data for two categories in two-way frequency tables. Interpret relative frequencies in the context of the data . . .. Recognize possible associations and trends in the data.

Previous Learning

Students should know how to display data using different types of displays, such as histograms, box-and-whisker plots, and scatter plots.

Start Thinking! and Warm Up

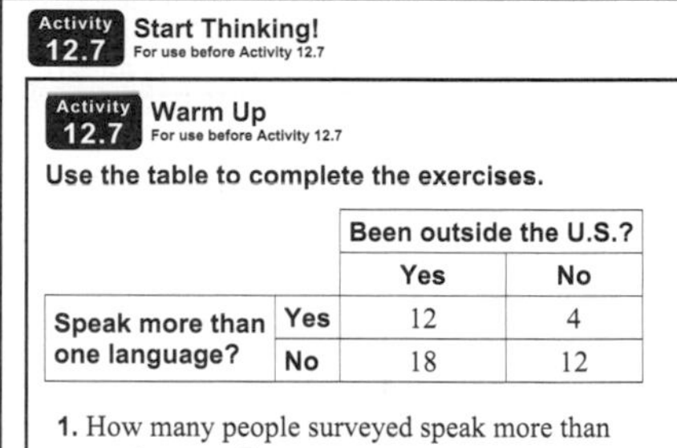

Activity 12.7 Start Thinking! For use before Activity 12.7

Activity 12.7 Warm Up For use before Activity 12.7

Use the table to complete the exercises.

		Been outside the U.S.?	
		Yes	No
Speak more than one language?	Yes	12	4
	No	18	12

1. How many people surveyed speak more than one language?
2. How many people surveyed have never traveled outside of the U.S.?
3. How many people only speak one language and have traveled outside of the U.S.?

12.7 Record and Practice Journal

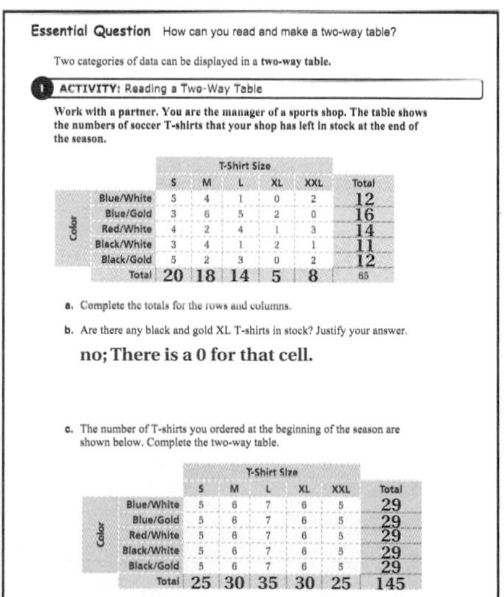

Essential Question How can you read and make a two-way table?

Two categories of data can be displayed in a **two-way table.**

1 ACTIVITY: Reading a Two-Way Table

Work with a partner. You are the manager of a sports shop. The table shows the numbers of soccer T-shirts that your shop has left in stock at the end of the season.

		T-Shirt Size					
		S	M	L	XL	XXL	Total
Color	Blue/White	5	4	1	0	2	12
	Blue/Gold	3	6	5	2	0	16
	Red/White	4	2	4	1	3	14
	Black/White	3	4	1	2	1	11
	Black/Gold	5	2	3	0	2	12
	Total	20	18	14	5	8	65

a. Complete the totals for the rows and columns.

b. Are there any black and gold XL T-shirts in stock? Justify your answer.

no; There is a 0 for that cell.

c. The number of T-shirts you ordered at the beginning of the season are shown below. Complete the two-way table.

		T-Shirt Size					
		S	M	L	XL	XXL	Total
Color	Blue/White	5	6	7	6	5	29
	Blue/Gold	5	6	7	6	5	29
	Red/White	5	6	7	6	5	29
	Black/White	5	6	7	6	5	29
	Black/Gold	5	6	7	6	5	29
	Total	25	30	35	30	25	145

Differentiated Instruction

Advanced

Have students enter the data from a two-way table into a spreadsheet. Use the *chart* feature to create a 3-D column chart (as shown on page 653), a clustered column chart, a stacked column chart, and a 100% stacked column chart. Students should describe the relationship between the data values in each chart and give an example of what information could be gathered from reading the chart.

12.7 Record and Practice Journal

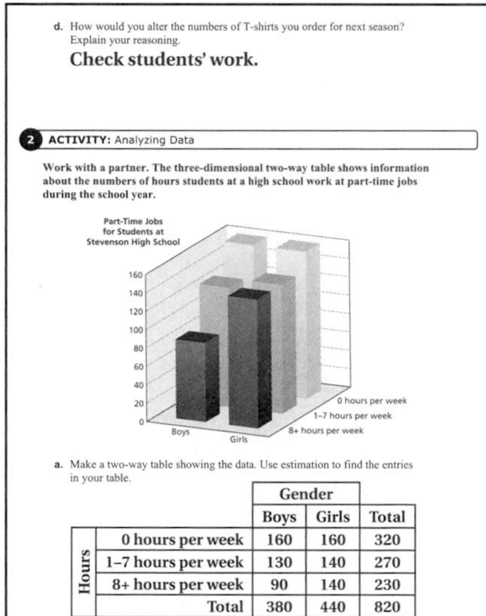

d. How would you alter the numbers of T-shirts you order for next season? Explain your reasoning.

Check students' work.

2 ACTIVITY: Analyzing Data

Work with a partner. The three-dimensional two-way table shows information about the numbers of hours students at a high school work at part-time jobs during the school year.

a. Make a two-way table showing the data. Use estimation to find the entries in your table.

		Gender		
		Boys	Girls	Total
Hours	0 hours per week	160	160	320
	1–7 hours per week	130	140	270
	8+ hours per week	90	140	230
	Total	380	440	820

b. Write two observations you can make that summarize the data in your table.

Check students' work.

c. A newspaper article claims that more boys than girls drop out of high school to work full-time. Do the data support this claim? Explain your reasoning.

no; The data is not about students dropping out or working full-time jobs.

What Is Your Answer?

3. IN YOUR OWN WORDS How can you read and make a two-way table?

Check students' work.

4. Find a real-life data set that can be represented by a two-way table. Then make a two-way table for the data set.

Check students' work.

Laurie's Notes

Activity 2

? "Have you ever seen a three-dimensional two-way table?" Students who are familiar with spreadsheets may recognize this type of display. Students may have seen them on the Internet, or in newspapers or magazines.

- Ask students to explain what one of the prisms represents.

? "What information is represented by the taller red prism?" It represents the number of girls that work 8 or more hours per week at a part-time job.

? "How many girls are in that category?" about 120

- Check that students have the correct labels for the rows and columns of the two-way table: gender (boys, girls) and hours (3 intervals shown).
- Ask a volunteer to display the two-way table.

? "Can you determine how many students attend Stevenson High School? Explain." yes; Find the sum of the last column or the last row in the table. They should be equal.

- **MP2:** In part (b), discuss student observations and listen for evidence of their statements. For instance, instead of saying that more girls have part-time jobs than boys, it is a stronger comparison to say that about 40 more girls have part-time jobs than boys. Encourage students to give quantitative evidence in their reasoning.
- Discuss part (c). Students may mention that the school should have about the same number of boys and girls. Because there are fewer boys represented in the table, you might infer that more boys have dropped out. Students may offer a viable argument as to *why* more boys might have dropped out of high school than girls, and it may not be for the purpose of working full-time.
- **MP2** and **MP3a Construct Viable Arguments:** This type of discussion is important in developing reasoning habits and constructing a logical argument.

What Is Your Answer?

- Question 4 can become a project due at the conclusion of the chapter.

Closure

- Refer back to the information about your expenses at the math conference. What would an entry in the blacked-out area represent? type of expense and day of expense How do you know if your total amount of expenses is correct? The sum of the last column should be equal to the sum of the last row.

Technology For the Teacher

Dynamic Classroom

The Dynamic Planning Tool
Editable Teacher's Resources at *BigIdeasMath.com*

2 ACTIVITY: Analyzing Data

Work with a partner. The three-dimensional two-way table shows information about the numbers of hours students at a high school work at part-time jobs during the school year.

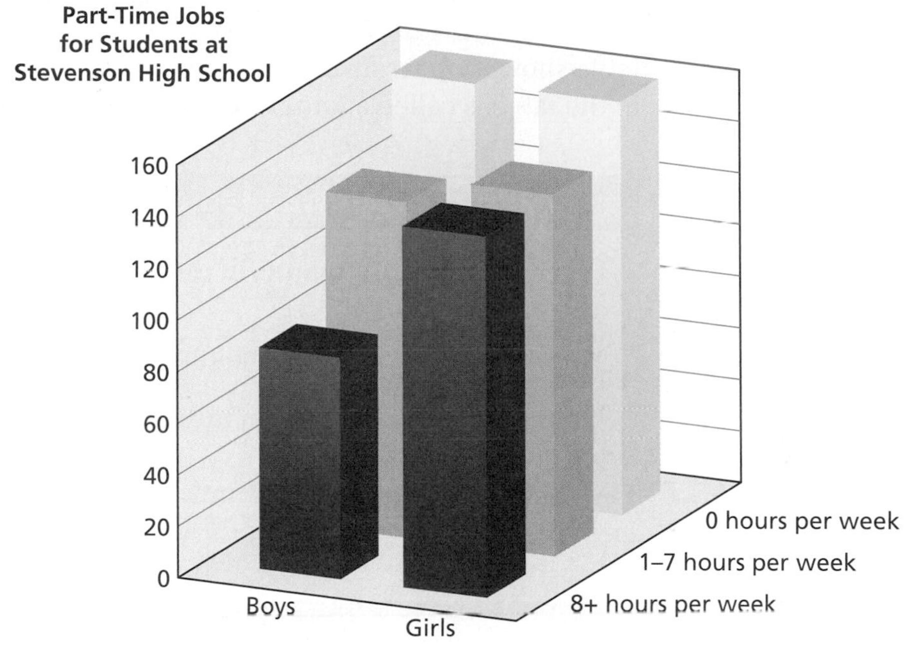

a. Make a two-way table showing the data. Use estimation to find the entries in your table.

b. Write two observations you can make that summarize the data in your table.

c. A newspaper article claims that more boys than girls drop out of high school to work full-time. Do the data support this claim? Explain your reasoning.

What Is Your Answer?

3. **IN YOUR OWN WORDS** How can you read and make a two-way table?

4. Find a real-life data set that can be represented by a two-way table. Then make a two-way table for the data set.

Practice

Use what you learned about two-way tables to complete Exercises 5 and 6 on page 656.

12.7 Lesson

Key Vocabulary
two-way table, *p. 654*
joint frequency, *p. 654*
marginal frequency, *p. 654*

A **two-way table** displays two categories of data collected from the same source.

You randomly survey students in your school about their grades on the last test and whether they studied for the test. The two-way table shows your results. Each entry in the table is called a **joint frequency.**

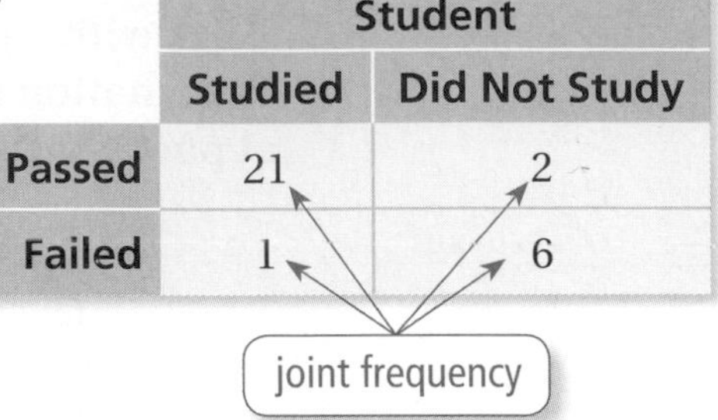

		Student	
		Studied	Did Not Study
Grade	Passed	21	2
	Failed	1	6

joint frequency

EXAMPLE 1 Reading a Two-Way Table

How many of the students in the survey above studied for the test and passed?

The entry in the "Studied" column and "Passed" row is 21.

So, 21 of the students in the survey studied for the test and passed.

The sums of the rows and columns in a two-way table are called **marginal frequencies.**

EXAMPLE 2 Finding Marginal Frequencies

Find and interpret the marginal frequencies for the survey above.

Create a new column and row for the sums. Then add the entries.

		Student		
		Studied	Did Not Study	Total
Grade	Passed	21	2	23
	Failed	1	6	7
	Total	22	8	30

23 students passed.
7 students failed.
30 students were surveyed.
22 students studied.
8 students did not study.

On Your Own

Now You're Ready
Exercises 3–6

1. You randomly survey students in a cafeteria about their plans for a football game and a school dance. The two-way table shows your results.
   a. How many students will attend the dance but not the football game?
   b. Find and interpret the marginal frequencies for the survey.

		Football Game	
		Attend	Not Attend
Dance	Attend	35	5
	Not Attend	16	20

Multi-Language Glossary at BigIdeasMath.com.

Laurie's Notes

Introduction

Connect

- **Yesterday:** Students read two-way tables. (MP2, MP3a)
- **Today:** Students will construct two-way tables and identify relationships between categories of a two-way table.

Motivate

- Tell students about data you have collected from the faculty. You asked your coworkers who own both a computer and a cell phone if they use a Mac or a PC, and if they use a smartphone or a basic cell phone.

? "How can you represent this data?" Answers will vary.

- Explain that today they will study a way in which this data can be displayed and analyzed.

Lesson Notes

Discuss

- Define two-way table. Emphasize that information is known about two categories from the same source. The focus in this lesson is drawing conclusions from the data in a two-way table.
- Refer to examples from yesterday in explaining "information about two categories from the same source," such as soccer shirts–size and color.
- Define joint-frequency. Each entry in the two-way table is a frequency for two categories, hence the name joint-frequency.

Example 1

? "What category do the rows represent?" test grade: passed or failed

? "What category do the columns represent?" preparation of the student: studied or did not study

Example 2

- Define marginal frequencies. The sums of the rows and columns appear on the *margins* of the two-way table.
- Expand the two-way table. Label the new row and new column "Total."
- Add the rows and columns. Identify the sums using the labels shown.
- Ask students general questions about the row and column totals.
- Make sure students understand that 30 students were surveyed, not 60. Because each student is tallied twice, once for each category, you do not add $22 + 8 + 23 + 7$ to find the number surveyed. The sum of the rows and the sum of the columns should be equal.

? "What can you conclude about the data?" *Sample answer:* Of the 30 students, all but one of those who studied for the test passed.

On Your Own

- **Extension:** Ask students percent questions such as what percent of the students in the survey are not planning to attend either event.

Goal Today's lesson is constructing and analyzing **two-way tables**.

Start Thinking! and Warm Up

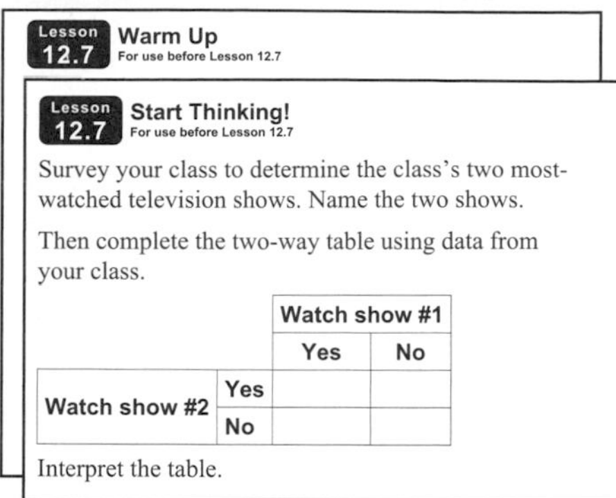

Lesson 12.7 Warm Up
For use before Lesson 12.7

Lesson 12.7 Start Thinking!
For use before Lesson 12.7

Survey your class to determine the class's two most-watched television shows. Name the two shows.

Then complete the two-way table using data from your class.

		Watch show #1	
		Yes	No
Watch show #2	Yes		
	No		

Interpret the table.

Extra Example 1

You randomly survey students about whether they like orange juice. The two-way table shows your results. How many female students in the survey like orange juice? 29

		Gender	
		Male	Female
Orange juice	No	12	22
	Yes	37	29

Extra Example 2

Find and interpret the marginal frequencies for the survey above.

A total of 34 students do not like orange juice. A total of 66 students like orange juice. A total of 49 male students participated in the survey. A total of 51 female students participated in the survey.

On Your Own

1. a. 5 students

   b. A total of 51 students will attend the game. A total of 25 students will not attend the game. A total of 40 students will attend the dance. A total of 36 students will not attend the dance.

Extra Example 3

You randomly survey students in 6th, 7th, and 8th grade about whether they are going to try to join student council. The results are shown in the tally sheets. Make a two-way table that includes the marginal frequencies. See Additional Answers.

Join

Grade	Tally
6	卌 卌 |
7	卌 卌 |||
8	卌 ||

Not Join

Grade	Tally
6	卌 卌
7	卌 卌 ||||
8	卌 卌 卌

Extra Example 4

Use the two-way table in Extra Example 3.

a. For each grade, what percent of the students in the survey are going to try to join student council? not try to join student council? Organize the results in a two-way table. Explain what one of the entries represents. See Additional Answers.

b. Does the table in part (a) show a relationship between grade and whether students are going to try to join student council? Explain. See Additional Answers.

On Your Own

2. See Additional Answers.

English Language Learners

Class Activity

Form groups of 2 to 4 students with at least one English language learner and one English speaker. Have them work together to create and conduct a survey that includes two categories. The results of the survey should be presented to the class in a two-way table that includes the marginal frequencies.

Laurie's Notes

Example 3

- Guide students through the construction of the table and ask them to explain what several of the values represent.
- The amount of data in the two-way table may be overwhelming to some students. Make sure to talk through the problem, giving students time to stop and think about what each entry represents.
- **Teaching Tip:** Use two colors in the table, one for the joint frequencies and one for the marginal frequencies. This makes it easier to read.

? "What is an advantage of the two-way table over the tally marks on the sheets of paper?" Answers will vary. Students might say the table is more organized, easier to read, and more condensed.

Example 4

- Have a student read the problem. Explain that the problem is asking about percents within each *age group*—the data represented in the columns.

? "How many 12- to 13-year-olds ride the bus?" 24 "How many 12- to 13-year-olds are in the survey?" 40 "What percent of the 12- to 13-year-olds ride the bus?" 24/40 = 60%

- Guide students through the construction of the two-way table. Ask them to explain what each entry represents.
- **MP2 Reason Abstractly and Quantitatively:** Ask students to make an observation about the percents. As age increases, students are less likely to ride the bus to school. Ask students why this might be the case, encouraging them to reason about data.

? "In part (b), the sums of the columns are each 100%. Why don't the rows add up to 100%?" The base used to compute the percents referred to each of the age groups, not whether the student rides the bus.

? "Can percents in the table be found using the row totals?" yes "What would be the first entry in the table and what would it represent?" 24/50 = 48%; 48% of the students who ride the bus are 12–13 years old.

On Your Own

- **Common Error:** Students may find the percent of students who pack a lunch out of the total number of students who pack a lunch instead of the percent for each grade level.

Closure

- In the first example, is it likely that if you study for a test you will pass? Explain. yes; The table shows that the majority of students who studied for the test passed and the majority of students who did not study for the test failed.

Technology For the Teacher

The Dynamic Planning Tool
Editable Teacher's Resources at *BigIdeasMath.com*

EXAMPLE 3 Making a Two-Way Table

Rides bus

Age	Tally
12-13	卌 卌 卌 卌 ||||
14-15	卌 卌 ||
16-17	卌 卌 ||||

Does not ride bus

Age	Tally
12-13	卌 卌 卌 |
14-15	卌 卌 |||
16-17	卌 卌 卌 卌 |

You randomly survey students between the ages of 12 and 17 about whether they ride the bus to school. The results are shown in the tally sheets. Make a two-way table that includes the marginal frequencies.

The two categories for the table are the ages and whether or not they ride the bus. Use the tally sheets to calculate each joint frequency. Then add to find each marginal frequency.

		Age			
		12–13	14–15	16–17	Total
Student	Rides Bus	24	12	14	50
	Does Not Ride Bus	16	13	21	50
	Total	40	25	35	100

EXAMPLE 4 Finding a Relationship in a Two-Way Table

Use the two-way table in Example 3.

a. For each age group, what percent of the students in the survey ride the bus to school? do not ride the bus to school? Organize the results in a two-way table. Explain what one of the entries represents.

		Age		
		12–13	14–15	16–17
Student	Rides Bus	60%	48%	40%
	Does Not Ride Bus	40%	52%	60%

$\frac{14}{35} = 0.4$

So, 40% of the 16- and 17-year-old students in the survey ride the bus to school.

b. Does the table in part (a) show a relationship between age and whether students ride the bus to school? Explain.

The table shows that as age increases, students are less likely to ride the bus to school.

On Your Own

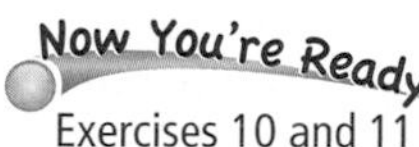

Exercises 10 and 11

Grade 6 students
11 pack lunch, 9 buy school lunch
Grade 7 students
23 pack lunch, 27 buy school lunch
Grade 8 students
16 pack lunch, 14 buy school lunch

2. You randomly survey students in a school about whether they buy a school lunch or pack a lunch. Your results are shown.
   a. Make a two-way table that includes the marginal frequencies.
   b. For each grade level, what percent of the students in the survey pack a lunch? buy a school lunch? Organize the results in a two-way table. Explain what one of the entries represents.
   c. Does the table in part (b) show a relationship between grade level and lunch choice? Explain.

12.7 Exercises

Vocabulary and Concept Check

1. **VOCABULARY** Explain the relationship between joint frequencies and marginal frequencies.
2. **OPEN-ENDED** Describe how you can use a two-way table to organize data you collect from a survey.

Practice and Problem Solving

You randomly survey students about participating in their class's yearly fundraiser. You display the two categories of data in the two-way table.

3. Find the total of each row.
4. Find the total of each column.

① 5. How many female students will be participating in the fundraiser?

6. How many male students will *not* be participating in the fundraiser?

		Fundraiser	
		No	Yes
Gender	Female	22	51
	Male	30	29

Find and interpret the marginal frequencies.

② 7.

		School Play	
		Attend	Not Attend
Class	Junior	41	30
	Senior	52	23

8.

		Cell Phone Minutes	
		Limited	Unlimited
Text Plan	Limited	78	0
	Unlimited	175	15

9. **GOALS** You randomly survey students in your school. You ask whether grades, popularity, or sports is most important to them. You display your results in the two-way table.

   a. How many 10th graders chose sports? How many 11th graders chose grades?

   b. Find and interpret the marginal frequencies for the survey.

   c. What percent of students in the survey are 9th graders who chose popularity?

		Goal		
		Grades	Popularity	Sports
Grade	9th	31	18	23
	10th	39	16	19
	11th	42	6	17

Assignment Guide and Homework Check

Level	Assignment	Homework Check
Average	1, 2, 3–7, 9, 11, 14–17	2, 5, 9, 11
Advanced	1, 2, 8–13, 14–17	2, 9, 11, 12

Common Errors

- **Exercises 1, 7–11** Students may incorrectly identify joint frequencies as marginal frequencies or marginal frequencies as joint frequencies. Remind them of these definitions.
- **Exercise 11c** Students may find the percents based on all students surveyed, not just the students from each eye color group. Encourage them to read carefully.

12.7 Record and Practice Journal

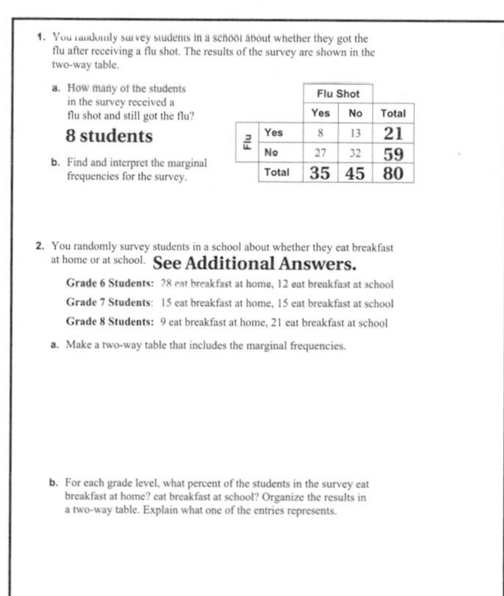

1. You randomly survey students in a school about whether they got the flu after receiving a flu shot. The results of the survey are shown in the two-way table.

   a. How many of the students in the survey received a flu shot and still got the flu?

   8 students

   b. Find and interpret the marginal frequencies for the survey.

Flu		Flu Shot: Yes	Flu Shot: No	Total
	Yes	8	13	21
	No	27	32	59
	Total	35	45	80

2. You randomly survey students in a school about whether they eat breakfast at home or at school. **See Additional Answers.**

   Grade 6 Students: 28 eat breakfast at home, 12 eat breakfast at school

   Grade 7 Students: 15 eat breakfast at home, 15 eat breakfast at school

   Grade 8 Students: 9 eat breakfast at home, 21 eat breakfast at school

   a. Make a two-way table that includes the marginal frequencies.

   b. For each grade level, what percent of the students in the survey eat breakfast at home? eat breakfast at school? Organize the results in a two-way table. Explain what one of the entries represents.

Technology For the Teacher
Answer Presentation Tool

Vocabulary and Concept Check

1. The joint frequencies are the entries in the two-way table that differentiate the two categories of data collected; The marginal frequencies are the sums of the rows and columns of the two-way table.
2. displays two categories of data collected from the same source

Practice and Problem Solving

3. total of females surveyed: 73; total of males surveyed: 59
4. total of "no" participants: 52; total of "yes" participants: 80
5. 51
6. 30
7. 71 students are juniors; 75 students are seniors; 93 students are attending the school play; 53 students are not attending the school play.
8. 78 people have limited cell phone texting plans; 190 people have unlimited cell phone texting plans; 253 people have limited cell phone minutes; 15 people have unlimited cell phone minutes.
9. a. 19; 42

   b. number of students surveyed: 9th grade: 72 10th grade: 74 11th grade: 65; 112 students chose grades, 40 students chose popularity, 59 students chose sports.

   c. about 8.5%

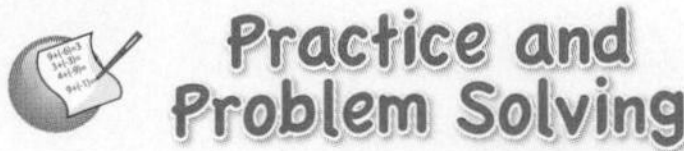

10–12. See Additional Answers.

13. See *Taking Math Deeper*.

Fair Game Review

14. $-1; x = 1$

15. $\frac{x+2}{x+4}; x = 3$ and $x = -4$

16. $\frac{5x-2}{7x+1}; x = 0$ and $x = -\frac{1}{7}$

17. B

Mini-Assessment

1. You randomly survey students about whether they are involved in school sports.

   Grade 5: 12 involved, 26 not involved
   Grade 8: 23 involved, 19 not involved

   a. Make a two-way table that includes the marginal frequencies See Additional Answers.

   b. For each grade level, what percent of the students in the survey are involved in school sports? are not involved in school sports? Organize the results in a two-way table. Explain what one of the entries represents. See Additional Answers.

   c. Does the table in part (b) show a relationship between grade level and involvement in school sports? Explain. See Additional Answers.

Taking Math Deeper

Exercise 13

This problem can help students see the benefits of a two-way table and the many different questions that can be asked regarding the entries. Encourage students to pay close attention to the wording of a question.

When finding the percent of students that are either female or have green eyes, students may mistakenly count the number of females with green eyes twice.

		Eye Color			
		Green	Blue	Brown	Total
Gender	Male	5	16	27	48
	Female	3	19	18	40
	Total	8	35	45	88

$$\frac{5}{88} + \frac{3}{88} + \frac{19}{88} + \frac{18}{88} = \frac{45}{88} \approx 51.1\%$$

To find the percent of students that are males that do not have green eyes, divide the sum of the remaining two joint frequencies by the number of students in the survey.

$$\frac{16}{88} + \frac{27}{88} = \frac{43}{88} \approx 48.9\%$$

3 The sum of the percents is $51.1\% + 48.9\% = 100\%$.

These two percents account for everyone in the survey.

Note: You may want to challenge students by asking what type of data display could be used to display the information in a two-way table, such as a double bar graph.

Reteaching and Enrichment Strategies

If students need help. . .	If students got it. . .
Resources by Chapter • Practice A and Practice B • Puzzle Time Record and Practice Journal Practice Differentiating the Lesson Lesson Tutorials Skills Review Handbook	Resources by Chapter • Enrichment and Extension • School-to-Work • Financial Literacy • Technology Connection • Life Connections Start the next section

(3) **10. PETS** You randomly survey students in your school about whether they own a pet. The results are shown in the tally sheets. Make a two-way table that includes the marginal frequencies.

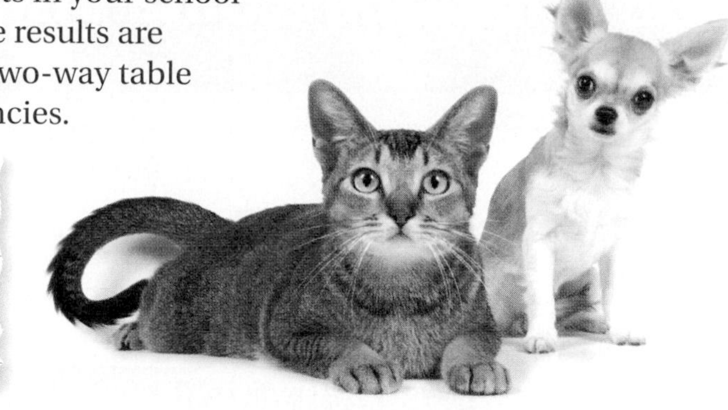

Own a Pet

Owner	Tally
Male	卌 卌 卌 |||
Female	卌 |

Don't Own a Pet

Owner	Tally
Male	卌 卌 卌 卌 卌
Female	卌 卌 ||||

(4) **11. EYE COLOR** You randomly survey students in your school about the color of their eyes. The results are shown in the tables.

a. Make a two-way table.

b. Find and interpret the marginal frequencies for the survey.

c. For each eye color, what percent of the students in the survey are male? female? Organize the results in a two-way table. Explain what two of the entries represent.

Eye Color of Males Surveyed

Green	Blue	Brown
5	16	27

Eye Color of Females Surveyed

Green	Blue	Brown
3	19	18

12. REASONING Use the information from Exercise 11. For each gender, what percent of the students in the survey have green eyes? blue eyes? brown eyes? Organize the results in a two-way table. Explain what two of the entries represent.

13. Precision What percent of students in the survey in Exercise 11 are either female or have green eyes? What percent of students in the survey are males that do not have green eyes? Find and explain the sum of these two percents.

Fair Game Review What you learned in previous grades & lessons

Simplify the rational expression, if possible. State the excluded value(s). *(Section 11.3)*

14. $\dfrac{1-x}{x-1}$

15. $\dfrac{x^2-x-6}{x^2+x-12}$

16. $\dfrac{15x^3-6x^2}{21x^3+3x^2}$

17. MULTIPLE CHOICE What is the solution of $\dfrac{1}{x}=\dfrac{3}{x+4}$? *(Section 11.7)*

(A) $x=-2$ (B) $x=2$ (C) $x=3$ (D) $x=4$

12.8 Choosing a Data Display

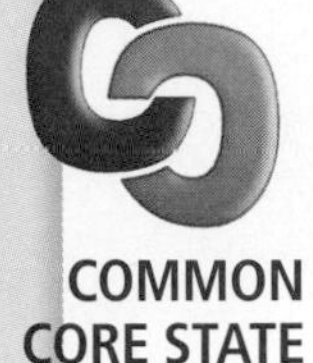

COMMON CORE STATE STANDARDS
S.ID.1

Essential Question How can you display data in a way that helps you make decisions?

1 ACTIVITY: Displaying Data

Work with a partner. Analyze and display each data set in a way that best describes the data. Explain your choice of display.

MOOSE
XING

a. ROAD KILL A group of schools in New England participated in a 2-month study and reported 3962 dead animals.

Birds 307 — Mammals 2746
Amphibians 145 — Reptiles 75
Unknown 689

b. BLACK BEAR ROAD KILL The data below show the numbers of black bears killed on a state's roads from 1993 to 2012.

Year	Number	Year	Number	Year	Number
1993	30	2000	47	2007	99
1994	37	2001	49	2008	129
1995	46	2002	61	2009	111
1996	33	2003	74	2010	127
1997	43	2004	88	2011	141
1998	35	2005	82	2012	135
1999	43	2006	109		

c. RACCOON ROAD KILL A 1-week study along a 4-mile section of road found the following weights (in pounds) of raccoons that had been killed by vehicles.

13.4	14.8	17.0	12.9
21.3	21.5	16.8	14.8
15.2	18.7	18.6	17.2
18.5	9.4	19.4	15.7
14.5	9.5	25.4	21.5
17.3	19.1	11.0	12.4
20.4	13.6	17.5	18.5
21.5	14.0	13.9	19.0

d. What do you think can be done to minimize the number of animals killed by vehicles?

Laurie's Notes

Introduction

Standards for Mathematical Practice

- **MP3 Construct Viable Arguments and Critique the Reasoning of Others:** In this section, students make decisions about how to display data. They will need to explain their reasoning for selecting a particular display. If two students select different data displays, it is important that they discuss the reasoning behind their choices.

Motivate

- The theme for the first activity is road kill. While students may giggle at the thought, automobile accidents involving large animals can be serious. I had my first and only accident with a deer 5 years ago. I was 2 miles from home and I was traveling 40 miles per hour. The deer was killed, my daughter and I were not injured, and repairs to my car were about $1400.
- Allow time for students to share personal stories.
- Use the Internet to research and share vehicular data with students, such as the number of miles of roads in the U.S., the number of registered vehicles, the number of accidents, and animal related accidents.

Activity Notes

Discuss

- Discuss the data displays with which students are familiar: pictograph, bar graph, line graph, circle graph, stem-and-leaf plot, histogram, dot plot, box-and-whisker plot, and scatter plot. Have students describe the feature(s) of each display.
- Discuss the different numerical tools they have for describing data: mean, median, mode, range, standard deviation, quartile, and interquartile range.

Activity 1

- Students need to decide what display makes sense for the type of data that they have. There may be more than one appropriate answer.
- **MP3:** Discuss students' choices and their explanations.
- Possible data displays:
  - Part (a): a circle graph (what part of the whole set is each animal) or a bar graph (compare the different categories, although there is a large difference in bar heights: 75 to 2746)
  - Part (b): a scatter plot and line of best fit (pair data, show trend over time, and make predictions for the future) or a line graph
  - Part (c): a stem-and-leaf plot (spread of data), along with calculating the mean (about 16.7) and median (17.1)
  - Part (d): As a class, discuss students' ideas for minimizing the number of animals killed by vehicles.

Common Core State Standards

S.ID.1 Represent data with plots on the real number line (dot plots, histograms, and box plots).

Previous Learning

Students should know how to construct a variety of data displays from this year and past years.

Start Thinking! and Warm Up

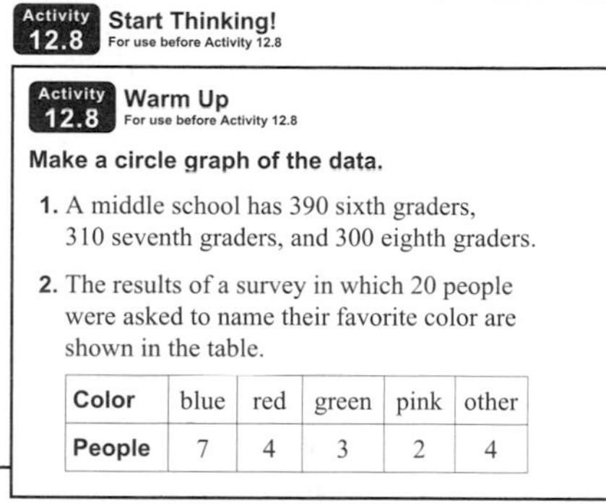

Activity 12.8 Start Thinking!
For use before Activity 12.8

Activity 12.8 Warm Up
For use before Activity 12.8

Make a circle graph of the data.

1. A middle school has 390 sixth graders, 310 seventh graders, and 300 eighth graders.
2. The results of a survey in which 20 people were asked to name their favorite color are shown in the table.

Color	blue	red	green	pink	other
People	7	4	3	2	4

12.8 Record and Practice Journal

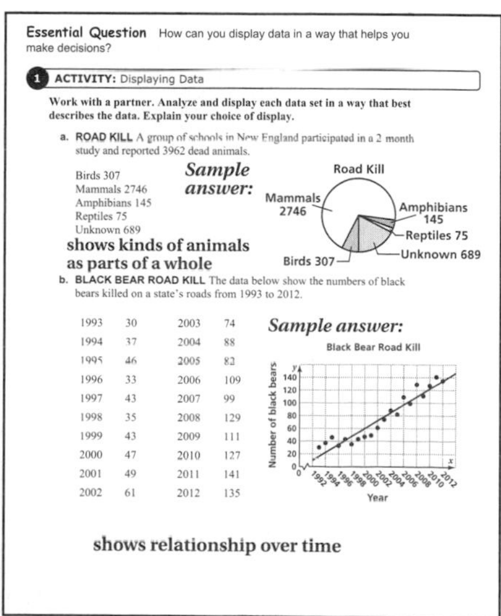

Essential Question How can you display data in a way that helps you make decisions?

1 ACTIVITY: Displaying Data

Work with a partner. Analyze and display each data set in a way that best describes the data. Explain your choice of display.

a. **ROAD KILL** A group of schools in New England participated in a 2 month study and reported 3962 dead animals.

Birds 307
Mammals 2746
Amphibians 145
Reptiles 75
Unknown 689

Sample answer: shows kinds of animals as parts of a whole

b. **BLACK BEAR ROAD KILL** The data below show the numbers of black bears killed on a state's roads from 1993 to 2012.

Year	Bears	Year	Bears
1993	30	2003	74
1994	37	2004	88
1995	46	2005	82
1996	33	2006	109
1997	43	2007	99
1998	35	2008	129
1999	43	2009	111
2000	47	2010	127
2001	49	2011	141
2002	61	2012	135

Sample answer:

shows relationship over time

English Language Learners

Vocabulary

English learners may need help understanding the word *scale*. There are several meanings in the English language. Some of the common meanings are:

a series of musical notes,
the covering of a reptile,
a device for weighing,
a ratio,
to climb.

In bar graphs, the scale is a series of markings used for measuring. Most scales start at 0 and go to (at least) the greatest value of the data.

12.8 Record and Practice Journal

c. **RACCOON ROAD KILL** A 1-week study along a 4-mile section of road found the following weights (in pounds) of raccoons that had been killed by vehicles.

13.4	14.8	17.0	12.9	21.3	21.5	16.8	14.8
15.2	18.7	18.6	17.2	18.5	9.4	19.4	15.7
14.5	9.5	25.4	21.5	17.3	19.1	11.0	12.4
20.4	13.6	17.5	18.5	21.5	14.0	13.9	19.0

See Additional Answers.

d. What do you think can be done to minimize the number of animals killed by vehicles?

Check students' work.

2 ACTIVITY: Statistics Project

ENDANGERED SPECIES PROJECT Use the Internet or some other reference to write a report about an animal species that is (or has been) endangered. Include graphical displays of the data you have gathered.

Sample: Florida Key Deer
In 1939, Florida banned the hunting of Key deer. The numbers of Key deer fell to about 100 in the 1940s.

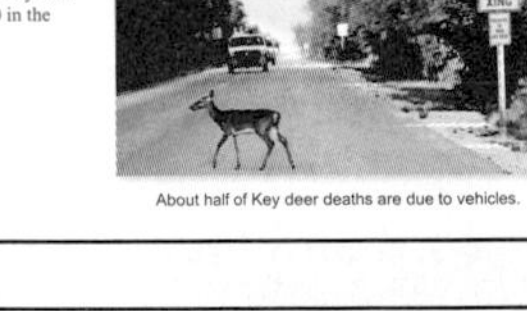

About half of Key deer deaths are due to vehicles.

In 1947, public sentiment was stirred by 11-year-old Glenn Allen from Miami. Allen organized Boy Scouts and others in a letter-writing campaign that led to the establishment of the National Key Deer Refuge in 1957. The approximately 8600-acre refuge includes 2280 acres of designated wilderness.

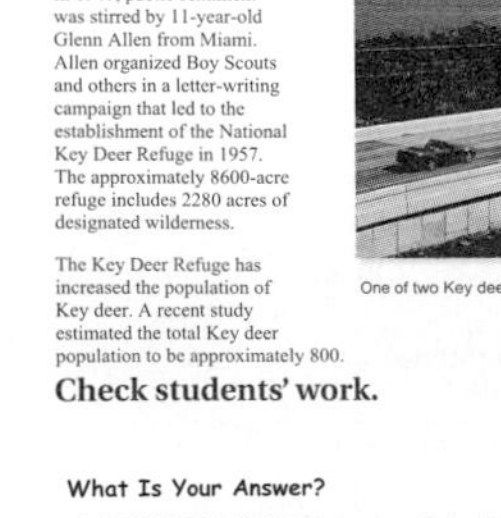

One of two Key deer wildlife underpasses on Big Pine Key.

The Key Deer Refuge has increased the population of Key deer. A recent study estimated the total Key deer population to be approximately 800.

Check students' work.

What Is Your Answer?

3. **IN YOUR OWN WORDS** How can you display data in a way that helps you make decisions? Use the Internet or some other reference to find examples of the following types of data displays.

- Bar graph
- Circle graph
- Scatter plot
- Stem-and-leaf plot
- Box-and-whisker plot

Data displays make it easy to interpret data and make conclusions.

Laurie's Notes

Activity 2

- Ask a volunteer to read the information presented about Key deer. Discuss how the actions of one person can often make a big difference.
- It would be ideal if the library or computer room is available. If not, you or your students could bring in newspapers and magazines that contain graphical displays.
- If you assign this project, students will need several days.

What Is Your Answer?

- **MP4 Model with Mathematics:** Many students can make the displays, if they are told which display to use. It is equally important that students be able to select the display based upon the data and the question you hope to answer from making the display.
- The information gathered by students can be made into classroom posters.

Closure

- **Class Discussion:** Have students present their answers to Question 3. Then have students discuss features of each display, and what types of data lend itself to each data display.

Technology For the Teacher

The Dynamic Planning Tool
Editable Teacher's Resources at *BigIdeasMath.com*

2 ACTIVITY: Statistics Project

ENDANGERED SPECIES PROJECT Use the Internet or some other reference to write a report about an animal species that is (or has been) endangered. Include graphical displays of the data you have gathered.

Sample: Florida Key Deer

In 1939, Florida banned the hunting of Key deer. The numbers of Key deer fell to about 100 in the 1940s.

About half of Key deer deaths are due to vehicles.

In 1947, public sentiment was stirred by 11-year-old Glenn Allen from Miami. Allen organized Boy Scouts and others in a letter-writing campaign that led to the establishment of the National Key Deer Refuge in 1957. The approximately 8600-acre refuge includes 2280 acres of designated wilderness.

The Key Deer Refuge has increased the population of Key deer. A recent study estimated the total Key deer population to be approximately 800.

One of two Key deer wildlife underpasses on Big Pine Key

What Is Your Answer?

3. **IN YOUR OWN WORDS** How can you display data in a way that helps you make decisions? Use the Internet or some other reference to find examples of the following types of data displays.
   - Bar graph
   - Circle graph
   - Scatter plot
   - Stem-and-leaf plot
   - Box-and-whisker plot

Practice Use what you learned about choosing data displays to complete Exercise 3 on page 662.

12.8 Lesson

Data Display	What does it do?
Pictograph	shows data using pictures
Bar Graph	shows data in specific categories
Circle Graph	shows data as parts of a whole
Line Graph	shows how data change over time
Histogram	shows frequencies of data values in intervals of the same size
Stem-and-Leaf Plot	orders numerical data and shows how they are distributed
Box-and-Whisker Plot	shows the variability of a data set using quartiles
Dot Plot	shows the number of times each value occurs in a data set
Scatter Plot	shows the relationship between two data sets using ordered pairs in a coordinate plane

EXAMPLE 1 Choosing an Appropriate Data Display

Choose an appropriate data display for the situation. Explain your reasoning.

a. the number of students in a marching band each year

∴ A line graph shows change over time. So, a line graph is an appropriate data display.

b. comparison of people's shoe sizes and their heights

∴ You want to compare two different data sets. So, a scatter plot is an appropriate data display.

On Your Own

Choose an appropriate data display for the situation. Explain your reasoning.

1. the population of the United States divided into age groups
2. the percents of students in your school who speak Spanish, French, or Haitian Creole

Laurie's Notes

Introduction

Connect

- **Yesterday:** Students reviewed data displays. (MP3, MP4)
- **Today:** Students will choose and construct an appropriate data display.

Motivate

- Make a quick sketch of the two bar graphs shown and ask students to comment on each.

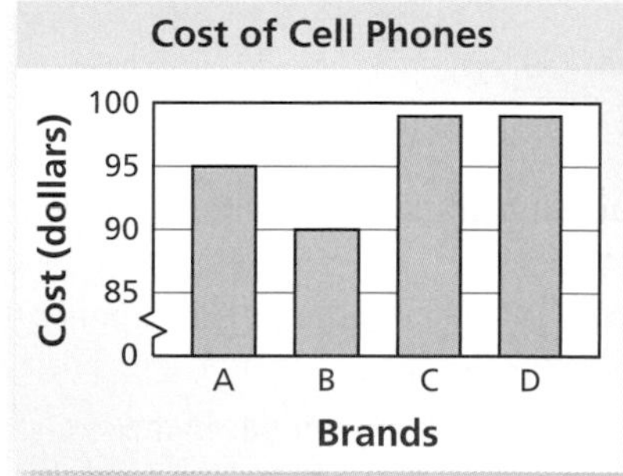

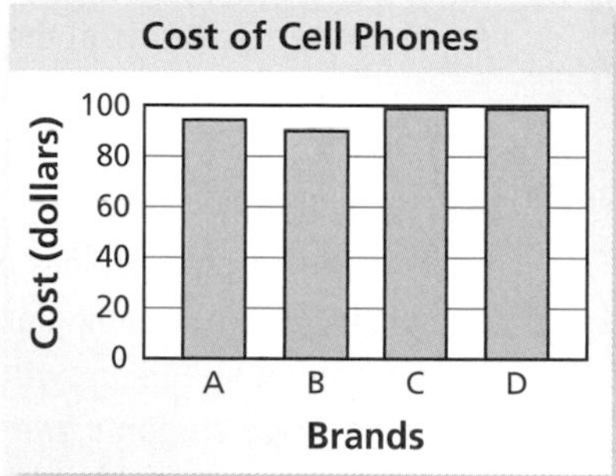

Lesson Notes

Key Idea

- Write the Key Idea. This is a terrific summary of data displays that students have learned to make.
- Emphasize that *choosing an appropriate display* is more poetry than science. On the other hand, it is clearly possible to use any of the graphs in misleading ways. This is science.
- **MP3 Construct Viable Arguments and Critique the Reasoning of Others:** Students should be able to state their reasons for selecting a particular data display, *and* why they did not select a different data display. If another student selected a different data display, students should compare their reasoning.
- There may be examples of each of these displays around your room.

Example 1

- Read each problem. Students should not have difficulty determining the appropriate data display for each problem.

On Your Own

- **Think-Pair-Share:** Students should read each question independently and then work with a partner to answer the questions. When they have answered the questions, the pair should compare their answers with another group and discuss any discrepancies.

Goal Today's lesson is choosing and constructing an appropriate data display.

Start Thinking! and Warm Up

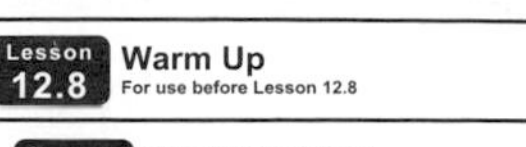

Lesson 12.8 Warm Up
For use before Lesson 12.8

Lesson 12.8 Start Thinking!
For use before Lesson 12.8

How are a bar graph and a histogram similar? How are they different?

How are a line graph and a line plot similar? How are they different?

How are a stem-and-leaf plot and box-and-whisker plot similar? How are they different?

Differentiated Instruction

Auditory

Ask students what data display would best represent the given data.

- the number of baseball cards each boy in the class has
  box-and-whisker plot
- the number of hours studying for a test and the test scores of students in a class scatter plot

Extra Example 1

You conduct a survey at your school about insects that students fear the most. Choose an appropriate data display. Explain your reasoning.
Sample answers: Circle graph: shows data as parts of a whole; Bar graph: shows data in specific categories; Pictograph: shows data using pictures.

On Your Own

1. *Sample answer:* histogram; Shows frequencies of ages (data values) in intervals of the same size.
2. *Sample answer:* bar graph; Shows data in specific categories.

Extra Example 2

Which line graph is misleading? Explain.

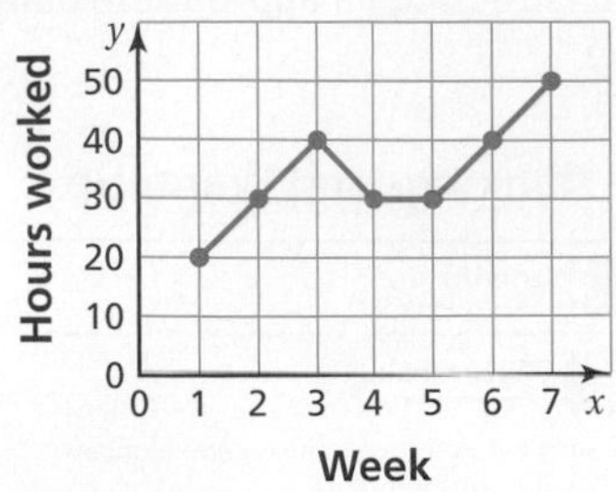

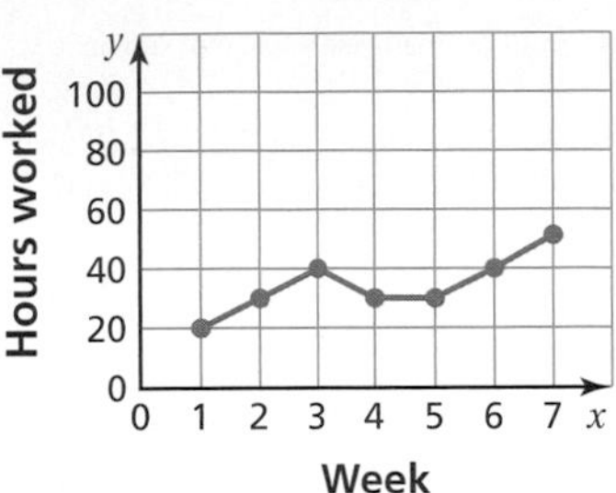

the second graph; The *y*-scale makes the change from week to week appear smaller.

Extra Example 3

Explain why the data display is misleading.

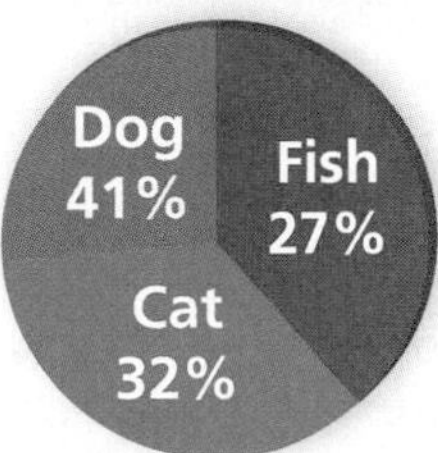

The size of each part of the circle is not proportional to the percent each choice represents.

On Your Own

3. The tickets vary in width and the break in the vertical axis makes the difference in ticket prices appear to be greater.
4. The bars become wider as the years progress, making the increase in profit appear greater.

Laurie's Notes

Discuss

- I have a collection of misleading data displays. When you find a data display in the newspaper or magazine that is misleading, cut it out and save it for later use. Ask colleagues in your school to do the same.
- **MP6 Attend to Precision:** Often what makes a graph misleading is the scale selected for one, or both, of the axes. By spreading out the scale, or condensing it, the graph becomes misleading.
- As I always tell my students, the person who makes the data display influences how we will view it. They control the extent to which we can see, or not see, features of the data.

Example 2

? "The same data are displayed in each line graph. How do the graphs differ?" The vertical scale is different.

? "Which graph is misleading and why?" first graph; It makes it appear that there has been a rapid growth in box office receipts.

- **Extension:** Have students pretend that both graphs appear in the newspaper with an article, and ask them what they would use for a headline for each article. What story does the author want readers to see when they look at each graph?

Example 3

- Have students "read" the pictograph and ask them to summarize what information it describes.
- Many students will conclude that the amount of cans and the amount of boxes is about the same due to the horizontal distance each set of icons takes up. They are mistakenly reading it more like a bar graph.

? "Approximately how many cans of food and boxes of food have been donated?" 11 cans $\times$ 20 $=$ 220 cans; 6 boxes $\times$ 20 $=$ 120 boxes

- Almost twice as many cans of food have been donated as boxes, so this is misleading. The box icon is too large. It should be the same width as the can.

On Your Own

- **Think-Pair-Share:** Students should read each question independently and then work with a partner to answer the questions. When they have answered the questions, the pair should compare their answers with another group and discuss any discrepancies.

Closure

- **Exit Ticket:** Make a pictograph for the data in Example 3 that would not be misleading.

Technology For the Teacher

The Dynamic Planning Tool
Editable Teacher's Resources at *BigIdeasMath.com*

EXAMPLE 2 Identifying a Misleading Data Display

Which line graph is misleading? Explain.

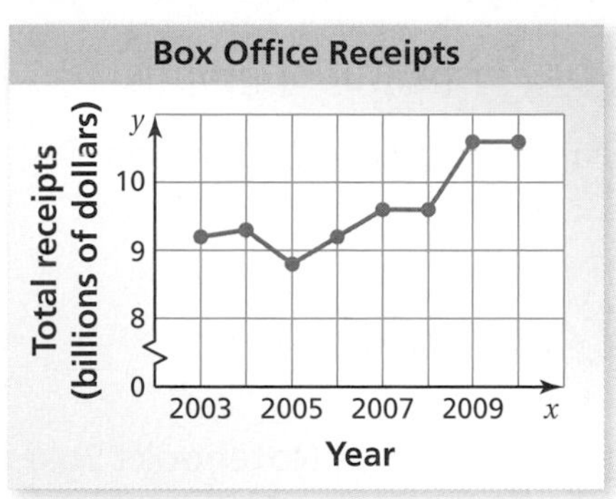

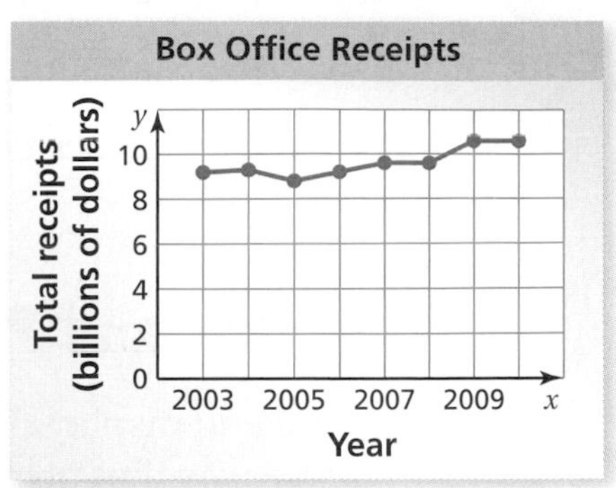

The vertical axis of the line graph on the left has a break (⚡) and begins at 8. This graph makes it appear that the total receipts increased rapidly from 2005 to 2010. The graph on the right has an unbroken axis. It is more honest and shows that the total receipts increased slowly.

∴ So, the graph on the left is misleading.

EXAMPLE 3 Analyzing a Misleading Data Display

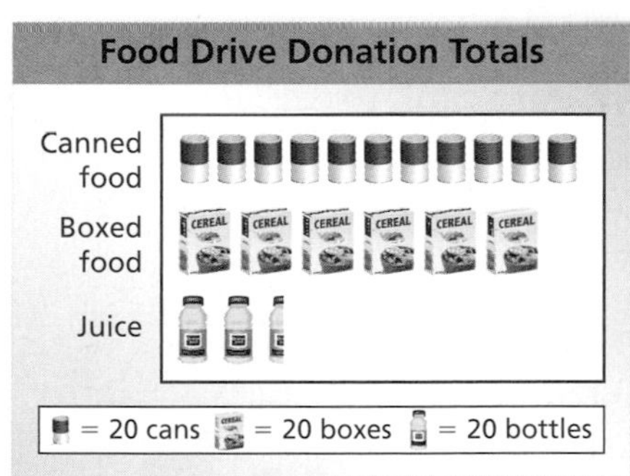

A volunteer concludes that the numbers of cans of food and boxes of food donated were about the same. Is this conclusion accurate? Explain.

Each icon represents the same number of items. Because the box icon is larger than the can icon, it looks like the number of boxes is about the same as the number of cans, but the number of boxes is actually about half of the number of cans.

∴ So, the conclusion is not accurate.

On Your Own

Now You're Ready
Exercises 9–12

Explain why the data display is misleading.

3.

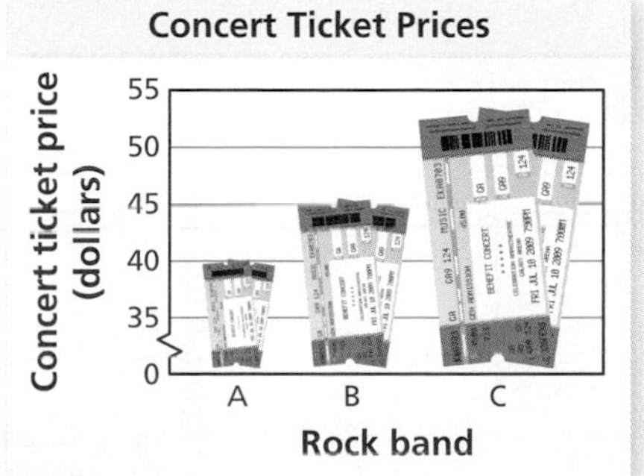

4.

Company Profits

Profit (billions of dollars)

25
20
15
10
5
0

1 2 3 4 5

Year

12.8 Exercises

Vocabulary and Concept Check

1. **REASONING** Can more than one display be appropriate for a data set? Explain.
2. **OPEN-ENDED** Describe how a histogram can be misleading.

Practice and Problem Solving

3. Analyze and display the data in a way that best describes the data. Explain your choice of display.

Notebooks Sold in One Week				
192 red	170 green	203 black	183 pink	230 blue
165 yellow	210 purple	250 orange	179 white	218 other

Choose an appropriate data display for the situation. Explain your reasoning.

1 4. a student's test scores and how the scores are spread out
5. the distance a person drives each month
6. the outcome of rolling a number cube
7. homework problems assigned each day

8. **WRITING** When would you choose a histogram instead of a bar graph to display data?

Explain why the data display is misleading.

2 3 9.

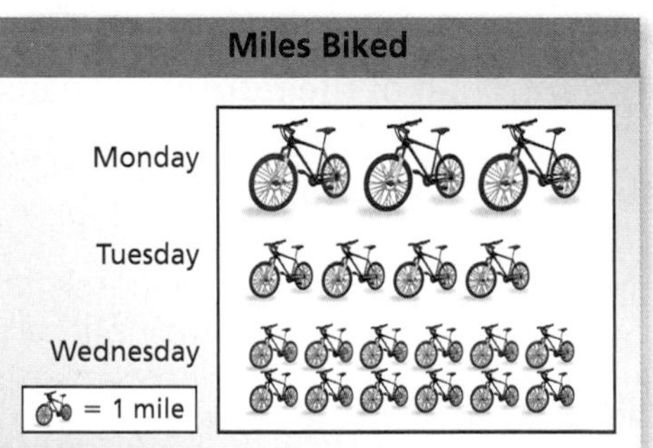

10.

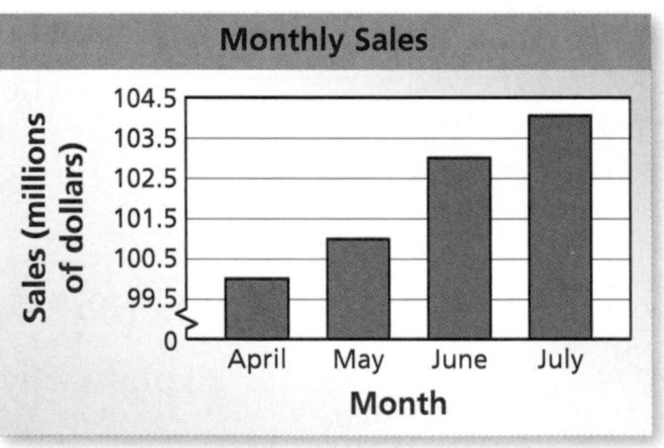

11.

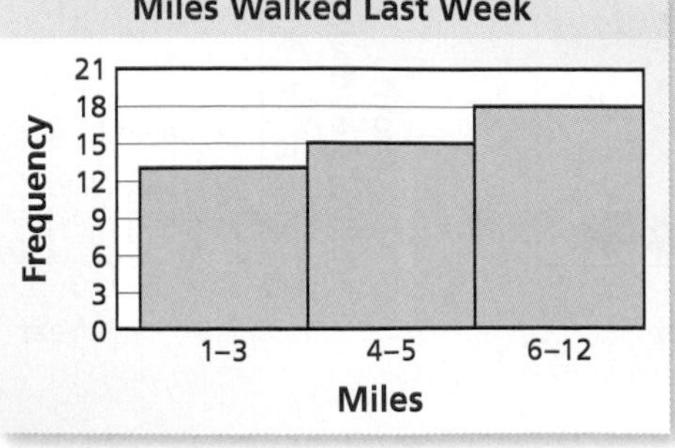

12.

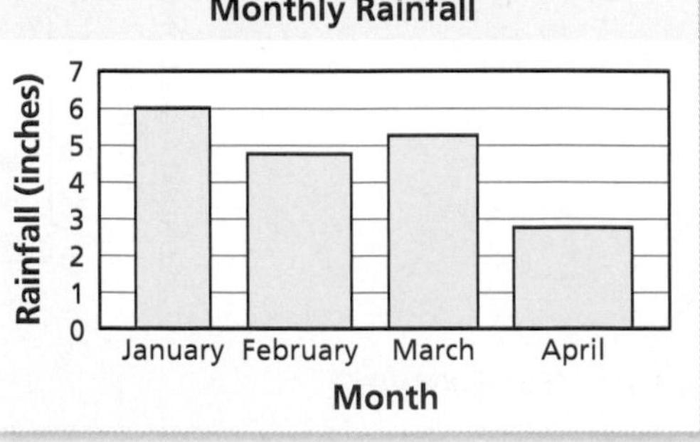

Assignment Guide and Homework Check

Level	Assignment	Homework Check
Average	1, 2, 4–14, 18–20	4, 8, 10, 14
Advanced	1, 2, 4–12 even, 13–17, 18–20	8, 10, 14, 16

Common Errors

- **Exercises 9–12** Students may not be able to recognize why the data display is misleading. As a class, make a list of things to examine when analyzing a data display. For example, check the increments or intervals for the axes.
- **Exercise 15** Students may say that the best data display for showing the mode is a stem-and-leaf plot because the leaves that have more repeated data will be wider. This display, however, could have other data in the leaf. Remind them that a dot plot isolates each data value and shows the frequency of each individual number, so this is the best data display.

12.8 Record and Practice Journal

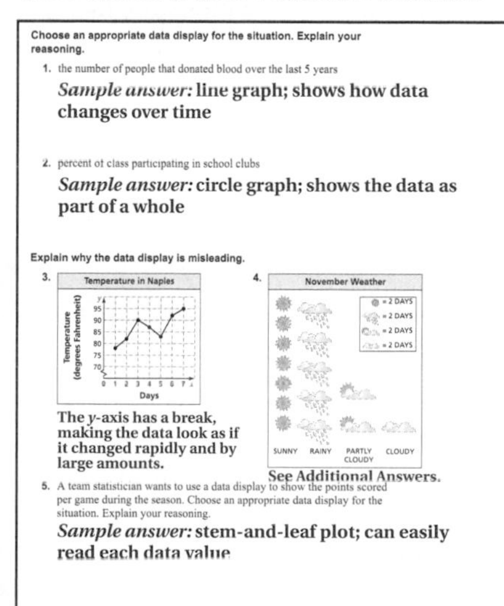
Choose an appropriate data display for the situation. Explain your reasoning.

1. the number of people that donated blood over the last 5 years

   Sample answer: **line graph; shows how data changes over time**

2. percent of class participating in school clubs

   Sample answer: **circle graph; shows the data as part of a whole**

Explain why the data display is misleading.

3. Temperature in Naples

   The y-axis has a break, making the data look as if it changed rapidly and by large amounts.

4. November Weather

   See Additional Answers.

5. A team statistician wants to use a data display to show the points scored per game during the season. Choose an appropriate data display for the situation. Explain your reasoning.

   Sample answer: **stem-and-leaf plot; can easily read each data value**

Technology For the Teacher
Answer Presentation Tool

Vocabulary and Concept Check

1. yes; Different displays may show different aspects of the data.
2. *Sample answer:* The scale of the vertical axis could be too small or too large.

Practice and Problem Solving

3. *Sample answer:*

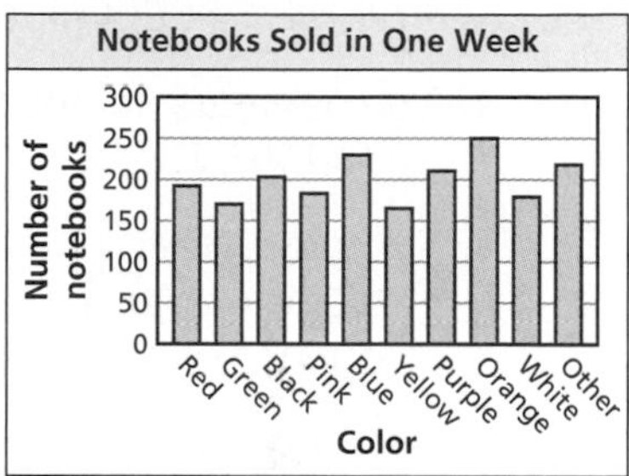

A bar graph shows the data in different color categories.

4. *Sample answer:* Stem-and-leaf plot: shows how data is distributed.
5. *Sample answer:* Dot graph: shows changes over time.
6. *Sample answer:* Line plot: shows the number of times each outcome occurs.
7. *Sample answer:* Line graph: shows changes over time.
8. when the data is in terms of intervals of one category, as opposed to multiple categories
9. The pictures of the bikes are larger on Monday, which makes it seem like the distance is the same each day.
10. The break in the scale for the vertical axis makes it appear as though there is a greater difference in sales between months.
11. The intervals are not the same size.

Practice and Problem Solving

12. The width of the bars are different, so it looks like some months have more rainfall.

13. *Sample answer:* bar graph; Each bar can represent a different vegetable.

14. yes; The vertical axis has a scale that increases by powers of 10, which makes the data appear to have a linear relationship.

15. *Sample answer:* dot plot

16. **a.** The percents do not add up to 100%.

b. *Sample answer:* bar graph; It would show the frequency of each sport.

17. See *Taking Math Deeper*.

Fair Game Review

18. $x + 3 = 5$

19. $8x = 24$

20. A

Mini-Assessment

Choose an appropriate data display for the situation. Explain your reasoning.

1. the outcome of flipping a coin
Sample answers: Pictograph: shows number of times heads or tails appears using picture of coins; Bar graph: shows number of times you get heads or tails; Dot plot: shows number of times you get a heads or tails.

2. comparison of student's test scores and how long students studied
Sample answer: Scatter plot: you want to compare two data sets.

3. the number of students participating in after-school sports each year
Sample answer: Line graph: shows how data change over time.

Taking Math Deeper

Exercise 17

This exercise introduces students to an amazing property of the number pi. Pi is an irrational number and therefore its decimal representation is not repeating. Even so, the ten digits from 0 to 9 each occur about 10 percent of the time, when one considers thousands of digits.

Display the data in a bar graph.

a.

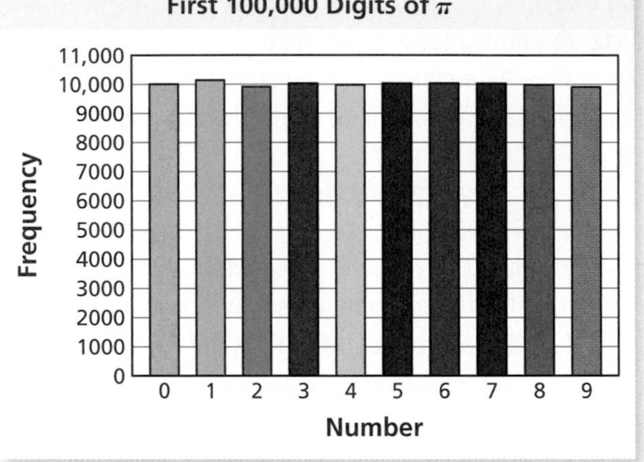

Display the data in a circle graph.

b.

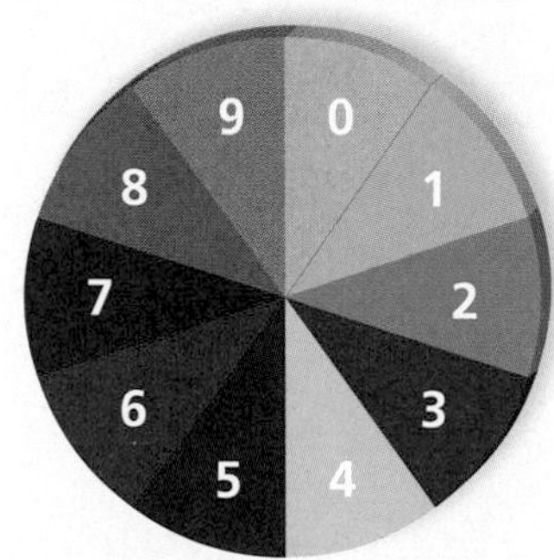

c. and d. Compare the two displays.

Both graphs show that each digit occurs about 10% of the time. The bar graph has a slight advantage because it shows that some digits occur slightly more than others.

Reteaching and Enrichment Strategies

If students need help. . .	If students got it. . .
Resources by Chapter • Practice A and Practice B • Puzzle Time Record and Practice Journal Practice Differentiating the Lesson Lesson Tutorials Skills Review Handbook	Resources by Chapter • Enrichment and Extension • School-to-Work • Financial Literacy • Technology Connection • Life Connections • Stories in History Start the next section

13. VEGETABLES A nutritionist wants to use a data display to show the favorite vegetables of the students at a school. Choose an appropriate data display for the situation. Explain your reasoning.

14. CHEMICALS A scientist gathers data about a decaying chemical compound. The results are shown in the scatter plot. Is the data display misleading? Explain.

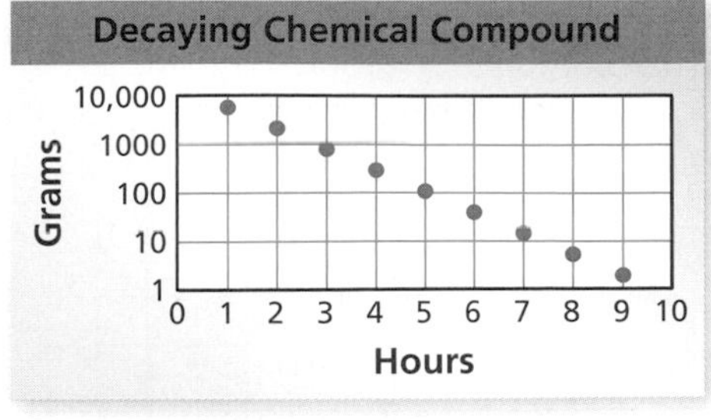

15. REASONING What type of data display is appropriate for showing the mode of a data set?

16. SPORTS A survey asked 100 students to choose their favorite sports. The results are shown in the circle graph.

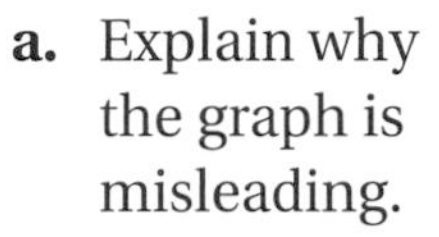

a. Explain why the graph is misleading.

b. What type of data display would be more appropriate for the data? Explain.

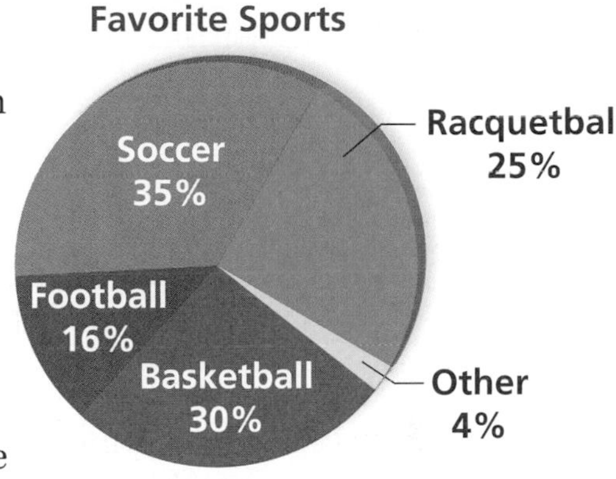

17. **Structure** With the help of computers, mathematicians have computed and analyzed billions of digits of the irrational number π. One of the things they analyze is the frequency of each of the numbers 0 through 9. The table shows the frequency of each number in the first 100,000 digits of π.

a. Display the data in a bar graph.

b. Display the data in a circle graph.

c. Which data display is more appropriate? Explain.

d. Describe the distribution.

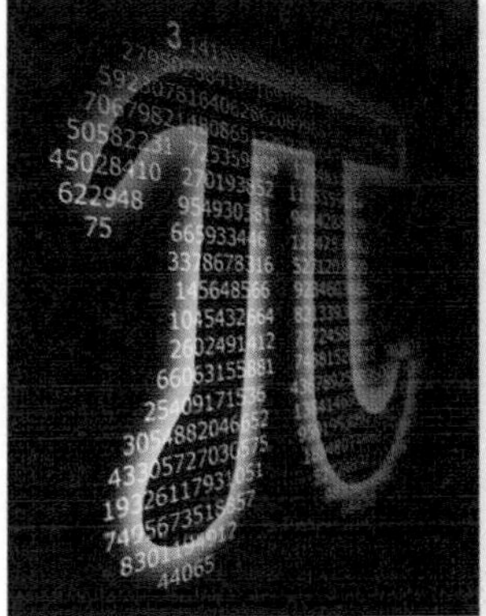

Number	0	1	2	3	4	5	6	7	8	9
Frequency	9999	10,137	9908	10,025	9971	10,026	10,029	10,025	9978	9902

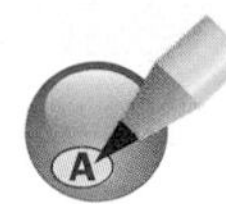

Fair Game Review What you learned in previous grades & lessons

Write the verbal statement as an equation. *(Skills Review Handbook)*

18. A number plus 3 is 5.

19. 8 times a number is 24.

20. MULTIPLE CHOICE What is 20% of 25% of 400? *(Skills Review Handbook)*

Ⓐ 20 Ⓑ 200 Ⓒ 240 Ⓓ 380

12.5–12.8 Quiz

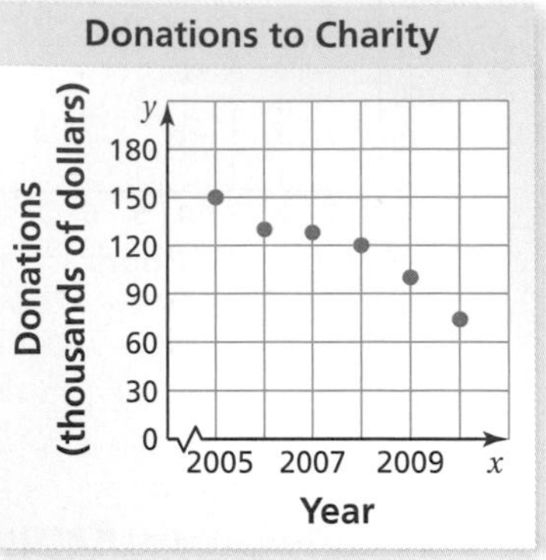

1. The scatter plot shows the amounts of money donated to a charity from 2005 to 2010. *(Section 12.5)*

   a. In what year did the charity receive $150,000?

   b. How much did the charity receive in 2008?

   c. Describe the relationship shown by the data.

2. Use a graphing calculator to find the equation of the line of best fit for the data in the table below. Identify and interpret the correlation coefficient. Make a scatter plot of the residuals and interpret the results. *(Section 12.6)*

x	0	1	2	3	4	5	6	7
y	12	16	15	14	18	22	20	25

3. The results of a recycling survey are shown in the two-way table. Find and interpret the marginal frequencies. *(Section 12.7)*

		Recycle	
		Yes	No
Gender	Female	28	9
	Male	24	14

Choose an appropriate data display for the situation. Explain your reasoning. *(Section 12.8)*

4. percent of band students in each section of instruments

5. company's profit for each week

6. **CATS** The table shows the number of cats adopted from an animal shelter each month. *(Section 12.5)*

Month	1	2	3	4	5	6	7	8	9
Cats	3	6	7	11	13	14	15	18	19

   a. Make a scatter plot of the data and draw a line of fit.

   b. Write an equation of the line that fits the data.

   c. Interpret the slope of the line.

   d. Predict how many cats will be adopted in month 10.

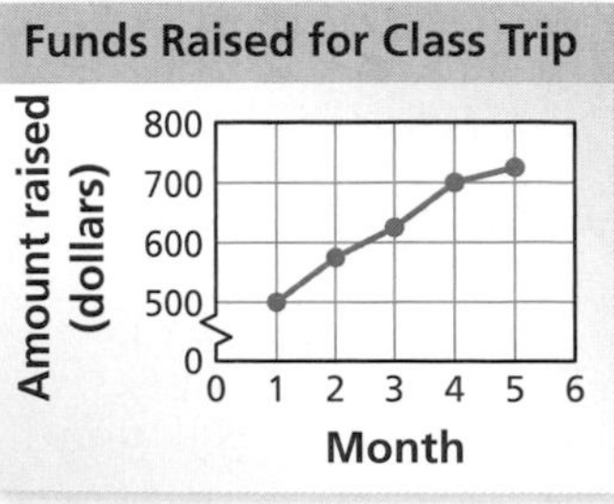

7. **FUNDRAISER** The line graph shows the amount of money that the eighth-grade students at a school raised each month to pay for a class trip. Is the graph misleading? Explain. *(Section 12.8)*

Alternative Assessment Options

Math Chat | Student Reflective Focus Question
Structured Interview | Writing Prompt

Math Chat

Ask students to use their own words to summarize how they would choose an appropriate data display. Be sure that they include examples. Select students at random to present to the class.

Study Help Sample Answers

Remind students to complete Graphic Organizers for the rest of the chapter.

4.

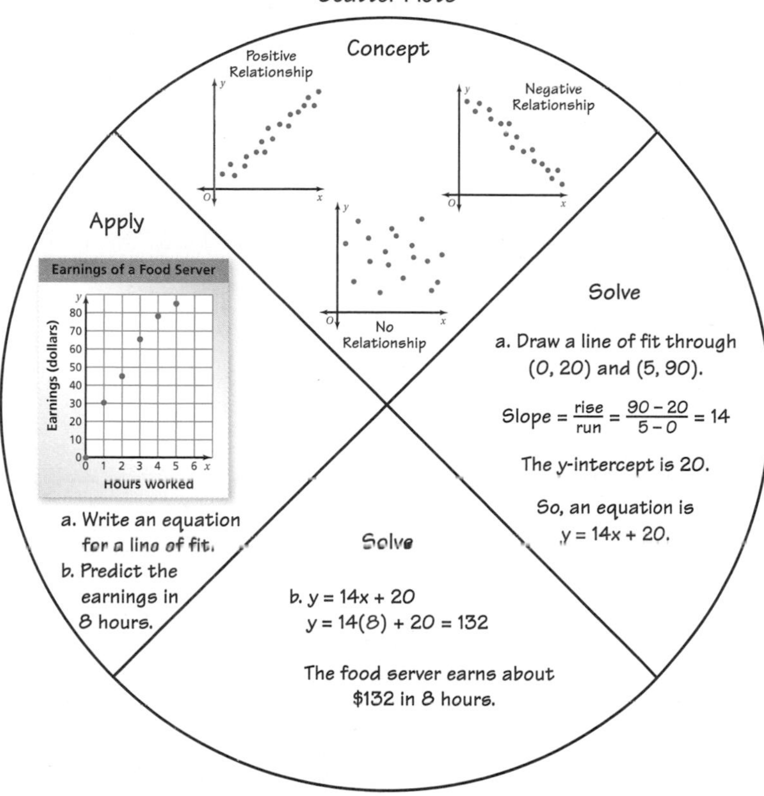

5–7. Available at *BigIdeasMath.com.*

Reteaching and Enrichment Strategies

If students need help...	If students got it...
Resources by Chapter • Study Help • Practice A and Practice B • Puzzle Time Lesson Tutorials *BigIdeasMath.com* Practice Quiz Practice from the Test Generator	Resources by Chapter • Enrichment and Extension • School-to-Work Game Closet at *BigIdeasMath.com* Start the Chapter Review

Answers

1. a. 2005
   b. $120,000
   c. There is a negative relationship between year and amount of donations.
2. $y = 1.6x + 12$; $r \approx 0.908$; This means that the relationship between x and y is a fairly strong positive correlation and the equation fits the data fairly well; The residuals are random values from -2.8 to 2.4, so the equation is a good fit.
3. 24
4. *Sample answer:* Circle graph: shows data as parts of a whole.
5. *Sample answer:* Line graph: shows changes over time.
6. See Additional Answers.
7. yes; The break in the vertical axis makes it appear that the amount of money raised increased very rapidly from month to month.

Answer Presentation Tool

Assessment Book

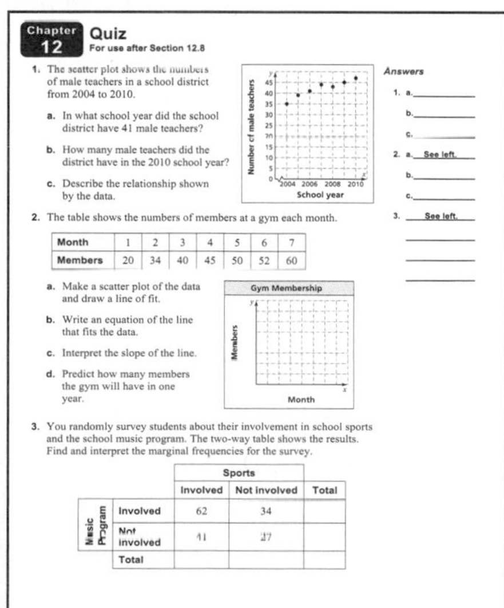

Chapter 12 **Quiz**
For use after Section 12.8

1. The scatter plot shows the numbers of male teachers in a school district from 2004 to 2010.
   a. In what school year did the school district have 41 male teachers?
   b. How many male teachers did the district have in the 2010 school year?
   c. Describe the relationship shown by the data.

2. The table shows the numbers of members at a gym each month.

Month	1	2	3	4	5	6	7
Members	20	34	40	45	50	52	60

   a. Make a scatter plot of the data and draw a line of fit.
   b. Write an equation of the line that fits the data.
   c. Interpret the slope of the line.
   d. Predict how many members the gym will have in one year.

3. You randomly survey students about their involvement in school sports and the school music program. The two-way table shows the results. Find and interpret the marginal frequencies for the survey.

		Sports: Involved	Sports: Not involved	Total
Music Program	Involved	62	34	
	Not involved	41	27	
	Total			

Answers
1. a. ____ b. ____ c. ____
2. a. See left. b. ____ c. ____
3. See left.

Review of Common Errors

- **Exercises 1–3** When finding the mean, students may forget to divide by the total number of data values and instead divide by the maximum value. Remind them that the definition of mean is an "average," so they must take into account the total number of items or numbers to get an average. Explain to students that it is as if they are dividing the total evenly among the number of groups.
- **Exercises 1–3** Students may try to identify the median without ordering the data first. Remind them that it is essential to order the data first and then find the median. This also makes finding the mode easier.

For the Teacher

Additional Review Options

- Big Ideas Test Generator
- Game Closet at *BigIdeasMath.com*
- Vocabulary Puzzle Builder
- Resources by Chapter
  Puzzle Time
  Study Help

Answers

1. The mean stays the same, the median decreases $4.2 - 4.1 = 0.1$ kilometer, and the mode changes to 4.0 and 4.3.
2. mean: 1.7
   median: 1
   mode: 1
3. mean: 4
   median: 3
   mode: 10

12 Chapter Review

Check It Out
Vocabulary Help
BigIdeasMath.com

Review Key Vocabulary

measure of central tendency, *p. 608*
measure of dispersion, *p. 614*
range, *p. 614*
standard deviation, *p. 615*
box-and-whisker plot, *p. 620*
quartile, *p. 620*
five-number summary, *p. 620*
interquartile range, *p. 621*
scatter plot, *p. 638*
line of fit, *p. 640*
residual, *p. 646*
linear regression, *p. 647*
line of best fit, *p. 647*
correlation coefficient, *p. 647*
causation, *p. 648*
two-way table, *p. 654*
joint frequency, *p. 654*
marginal frequency, *p. 654*

Review Examples and Exercises

12.1 Measures of Central Tendency *(pp. 606–611)*

The table shows the number of kilometers you ran each day for the past 10 days. Find the mean, median, and mode of the distances.

Kilometers Run	
3.5	4.1
4.0	4.3
4.4	4.5
3.9	2.0
4.3	5.0

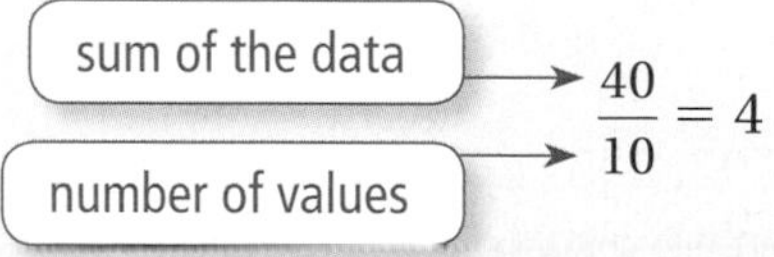

Mean: $\frac{40}{10} = 4$

Median: 2.0, 3.5, 3.9, 4.0, 4.1, 4.3, 4.3, 4.4, 4.5, 5.0 Order the data.

$\frac{8.4}{2} = 4.2$ Mean of two middle values

Mode: 2.0, 3.5, 3.9, 4.0, 4.1, 4.3, 4.3, 4.4, 4.5, 5.0

The value 4.3 occurs most often.

∴ The mean is 4 kilometers, the median is 4.2 kilometers, and the mode is 4.3 kilometers.

Exercises

1. Use the data in the example above. You run 4.0 kilometers on day 11. How does this additional value affect the mean, median, and mode? Explain.

Find the mean, median, and mode of the data.

2. 

3.

Ski Resort Temperatures (°F)		
11	3	3
0	−9	−2
10	10	10

12.2 Measures of Dispersion *(pp. 612–617)*

Find the mean, range, and standard deviation of the bowling scores for each person. Then compare the data sets.

Ryan	
205	190
185	200
210	219
174	203
194	230

Emma	
228	205
172	181
154	240
235	235
168	192

Ryan: mean $= \dfrac{2010}{10} = 201$

174, 185, 190, 194, 200, 203, 205, 210, 219, 230 Order the data.

The range is $230 - 174 = 56$.

$$\sqrt{\frac{(205-201)^2 + (185-201)^2 + \cdots + (230-201)^2}{10}} = \sqrt{242.2} \approx 15.6$$

The standard deviation is 15.6.

Emma: mean $= \dfrac{2010}{10} = 201$

154, 168, 172, 181, 192, 205, 228, 235, 235, 240 Order the data.

The range is $240 - 154 = 86$.

$$\sqrt{\frac{(228-201)^2 + (172-201)^2 + \cdots + (192-201)^2}{10}} = \sqrt{919.8} \approx 30.3$$

The standard deviation is 30.3.

∴ The mean, 201, is the same for each data set. The range for Ryan's scores is 56 and the standard deviation is 15.6. The range for Emma's scores is 86 and the standard deviation is 30.3. So, Emma's scores are more spread out than Ryan's scores.

Exercises

4. Find the mean, range, and standard deviation of the prices (in dollars) of portable keyboards at each store. Then compare the data sets.

Store A	
130	180
200	250
150	190
250	160

Store B	
225	310
260	190
200	285
210	230

Review of Common Errors (continued)

- **Exercise 4** Students may subtract the first data value from the last data value to find the range. Remind students that the range is the difference between the greatest and least data values. It helps to order the data before identifying the greatest and least data values.

Answers

4. Store A: 188.75; 120; about 41.1
Store B: 238.75; 120; about 39.7
The mean at Store B is greater, the ranges are the same, and the standard deviations are about the same. So, the portable keyboard's prices at Store B tend to be higher and they differ from the mean about as much as at Store A.

Answers

5.

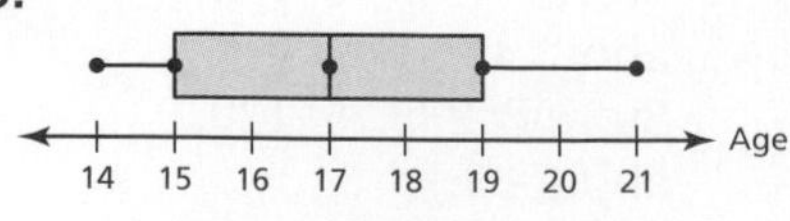

The distribution is skewed right.

6.

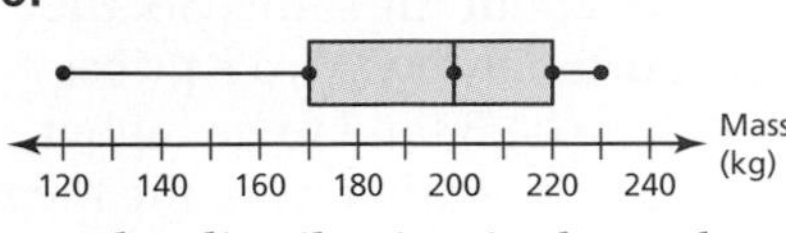

The distribution is skewed left.

7.

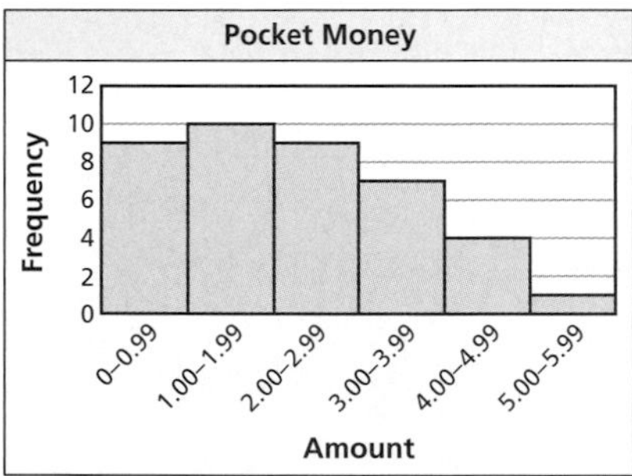

The distribution is skewed right.

8. Use the median to describe the center and the five-number summary to describe the spread.

9. *Sample answer:* $y = -0.02x + 8$; \$2.50

Review of Common Errors (continued)

- **Exercises 5–6** Students may have difficulty creating the box-and-whisker plots. Remind them of the five-number summary of a data set.
- **Exercises 5–6** Students may confuse the meanings of skewed left and skewed right. Remind them that the shape is determined by the longer whisker.
- **Exercise 7** Students may create a histogram with gaps between the bars. Remind students that the bars in a histogram should be touching to show that they represent an entire range of values.
- **Exercise 9** Students may draw a line of best fit that does not accurately reflect the data trend. Remind them that the line does not have to go through any of the data points. Also remind them that the line should go through the middle of the data so that about half of the data points are above the line and half are below. One strategy is to draw an oval around the data and then draw a line through the middle of the oval. For example:

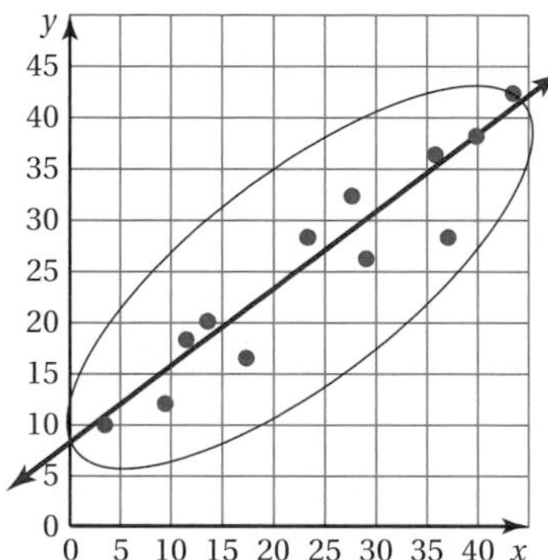

- **Exercise 9** Students may struggle writing an equation for a line of fit. When drawing the line, encourage them to try to make the line go through a lattice point. Also, students can use lattice points that are very close to the line to help them find the slope.
- **Exercise 10** Students may enter the x-values into the y list of the graphing calculator, and vice versa. Encourage them to double check their data entries and understand which lists represent the x- and y-values.
- **Exercise 11** Students may incorrectly identify joint frequencies as marginal frequencies or marginal frequencies as joint frequencies.

12.3 Box-and-Whisker Plots (pp. 618–625)

Make a box-and-whisker plot for the weights (in pounds) of pumpkins sold at a market. Identify the shape of the distribution.

16, 20, 11, 15, 10, 8, 8, 19, 11, 9, 9, 16

Step 1: Order the data. Find the median and the quartiles.

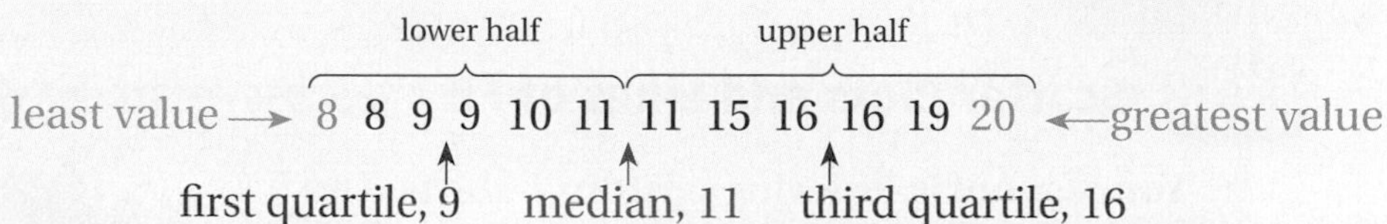

Step 2: Draw a number line that includes the least and greatest values. Graph points above the number line for the five-number summary.

Step 3: Draw a box using the quartiles. Draw a line through the median. Draw whiskers from the box to the least and greatest values.

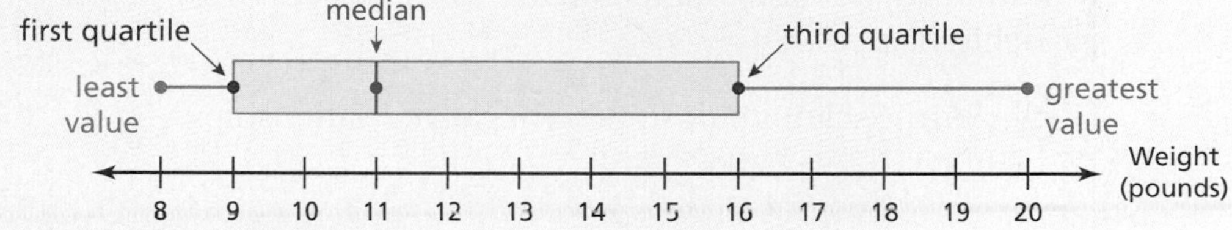

The right whisker is longer than the left whisker, and most of the data are on the left side of the display. So, the distribution is skewed right.

Exercises

Make a box-and-whisker plot for the data. Identify the shape of the distribution.

5. Ages of volunteers at a hospital: 14, 17, 20, 16, 17, 14, 21, 18
6. Masses (in kilograms) of lions: 120, 230, 180, 210, 200, 200, 230, 160

12.4 Shapes of Distributions (pp. 626–633)

The histogram shows the numbers of words spelled correctly by students at a spelling bee. Describe the shape of the distribution. Which measures of central tendency and dispersion would best represent the data?

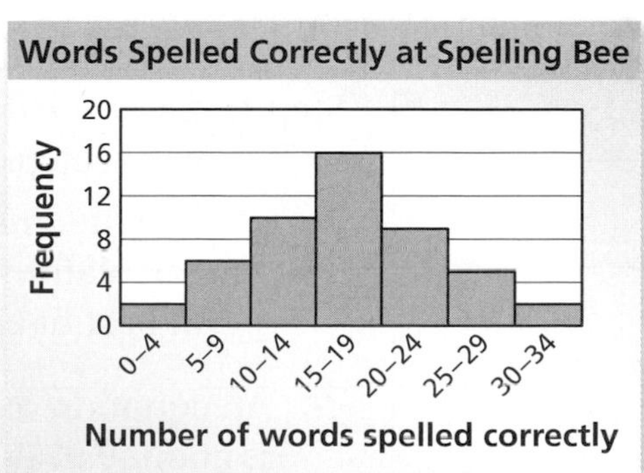

The distribution is symmetric. So, use the mean to describe the center and the standard deviation to describe the spread.

Exercises

The frequency table shows the amounts of money the students in a class have in their pockets.

Amount	Frequency
0–0.99	9
1.00–1.99	10
2.00–2.99	9
3.00–3.99	7
4.00–4.99	4
5–5.99	1

7. Display the data in a histogram. Describe the shape of the distribution.
8. Which measures of central tendency and dispersion best represent the data?

12.5 Scatter Plots and Lines of Fit *(pp. 636–643)*

Your school is ordering custom T-shirts. The scatter plot shows the costs per T-shirt for various numbers of T-shirts ordered. What type of relationship do the data show?

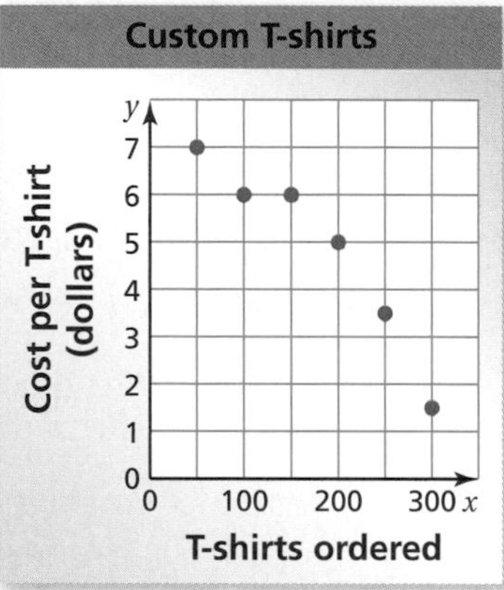

The plotted points go down from left to right. As the number of T-shirts ordered increases, the cost per T-shirt decreases.

∴ So, the scatter plot shows a negative relationship.

Exercises

9. Use the scatter plot above. Write an equation of a line of fit. Predict the cost per T-shirt when you order 275 T-shirts.

12.6 Analyzing Lines of Fit *(pp. 644–651)*

The table shows the heights x (in inches) and shoe sizes y of several students. Use a graphing calculator to find an equation of the line of best fit. Identify and interpret the correlation coefficient.

x	65	62	70	72	68	67	70	67	64	63
y	8.5	7	10.5	12	10	10.5	11	10	8	7.5

Step 1: Enter the data from the table into your calculator.

Step 2: Use the *linear regression* feature of your calculator to find an equation of the line of best fit and the correlation coefficient.

```
LinReg
 y=ax+b
 a=.4866803279
 b=-23.0102459
 r²=.943563901
 r=.9713721743
```

∴ An equation of the line of best fit is $y = 0.49x - 23.0$. The correlation coefficient is about 0.971. This means that the relationship between the heights and shoe sizes is a strong positive correlation and the equation closely models the data.

Review Game

Rolling for Data

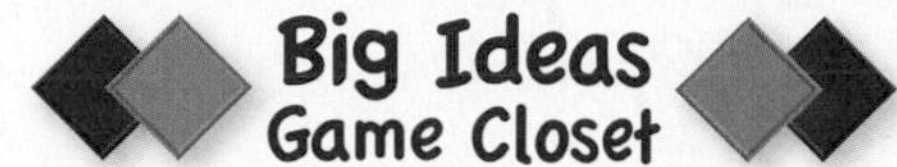

Materials per Pair

- two number cubes
- paper
- pencil

Directions

Students should work in pairs. Students in each pair take turns rolling the two number cubes one at a time. The first number they roll represents the tens digit and the second number represents the ones digit of a whole number. For example, If a 1 is rolled and then a 6, the whole number is 16. Students record the whole numbers in a stem-and-leaf plot and keep rolling until they have 10 leaves for any one stem. Once a pair acquires the 10 leaves, they race to find the mean, median, and mode of all their whole numbers and make a box-and-whisker plot to display the data.

Who Wins?

The first pair to finish all tasks wins 10 points, the second 9 points, the third 8 points, and so on. The game can be repeated as many times as desired. The pair with the most points after a predetermined number of rounds or amount of time wins.

For the Student
Additional Practice

- Lesson Tutorials
- Study Help (textbook)
- Student Website
  - Multi-Language Glossary
  - Practice Assessments

Answers

10. $y = 0.49x - 23.2$; yes; The correlation coefficient changed from about 0.971 to about 0.954. There is still a strong positive correlation.

11.

		Food Court		
		Like	Dislike	Total
Age	Adults	21	79	100
	Students	96	4	100
	Total	117	83	200

79%

12. *Sample answer:* Bar graph: shows data in specific categories.

My Thoughts on the Chapter

What worked. . .

What did not work. . .

What I would do differently. . .

Teacher Tip

Not allowed to write in your teaching edition? Use sticky notes to record your thoughts.

Exercises

10. Use the data in the example. You take height and shoe size measurements of three more students: (64, 7), (65, 9), and (71, 11). Find a new equation of the line of best fit. Did the correlation coefficient change? Explain.

12.7 Two-Way Tables *(pp. 652–657)*

You randomly survey students in your school about whether they liked a recent school play. The results are shown. Make a two-way table that includes the marginal frequencies. What percent of the students surveyed liked the play?

Male students
48 likes, 12 dislikes
Female students
56 likes, 14 dislikes

		Student		
		Liked the Play	Did Not Like the Play	Total
Gender	Male	48	12	60
	Female	56	14	70
	Total	104	26	130

Of the 130 students surveyed, 104 students liked the play.

∴ Because $\frac{104}{130} = 0.8$, 80% of the students in the survey liked the play.

Exercises

11. You randomly survey people at a mall about whether they like the new food court. The results are shown. Make a two-way table that includes the marginal frequencies. What percent of the adults surveyed dislike the new food court?

Adults
21 likes, 79 dislikes
Teenagers
96 likes, 4 dislikes

12.8 Choosing a Data Display *(pp. 658–663)*

Choose an appropriate data display for the situation. Explain your reasoning.

a. the percent of votes that each candidate received in an election

∴ A circle graph shows data as parts of a whole. So, a circle graph is an appropriate data display.

b. the distribution of the ages of U.S. presidents at their inauguration(s)

∴ A stem-and-leaf plot orders numerical data and shows how they are distributed. So, a stem-and-leaf plot is an appropriate data display.

Exercises

12. A principal wants to use a data display to compare the number of cans of food donated by each eighth-grade class. Choose an appropriate data display for the situation. Explain your reasoning.

12 Chapter Test

Check It Out
Test Practice
BigIdeasMath.com

Find the mean, median, and mode of the data.

1.

Distances (feet) Above or Below Water Level in Pool		
−3	0	−3
3	10	0
11	−6	−3

2. Cooking Times (minutes)

Stem	Leaf
3	5 8
4	0 1 8
5	0 4 4 4 5 9
6	0

Key: 4 | 1 = 41 minutes

3. TURTLES The tables show the weights (in pounds) of turtles caught in two ponds. Find the mean, range, and standard deviation of the weights of the turtles in each pond. Then compare the data sets.

Pond A			
12	13	15	6
7	8	12	7

Pond B			
9	12	5	8
12	15	16	19

4. Which type of data display would you use for the information in Exercise 3? Explain.

5. SWIMMING The table shows the numbers of hours you swam for several weeks.

Hours Swimming			
7	3.5	8	3.5
6	7	7	2
5.5	7.5	7.5	7.5

a. Make a box-and-whisker plot for the data.

b. Find the range and interquartile range.

c. Which measures of central tendency and dispersion best represent the data?

6. NEWBORNS The table shows the lengths and weights of several newborn babies.

Length (inches)	Weight (pounds)
19	6
19.5	7
20	7.75
20.25	8.5
20.5	8.5
22.5	11

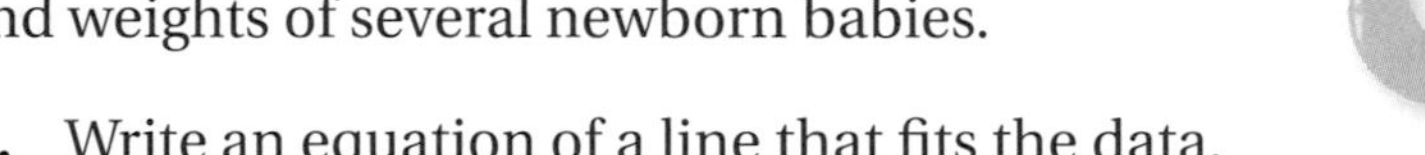

a. Write an equation of a line that fits the data.

b. Use a graphing calculator to find an equation of the line of best fit. Identify and interpret the correlation coefficient.

c. Predict the weight of a newborn that is 21 inches long using the equations from parts (a) and (b). Compare the results.

7. SAT The table shows the numbers y of students (in thousands) who took the SAT from 2003 to 2010, where $x = 3$ represents the year 2003. Use a graphing calculator to find an equation of the line of best fit. Then make a scatter plot of the residuals to tell whether the line of best fit models the data well.

x	3	4	5	6	7	8	9	10
y	1406	1419	1476	1466	1495	1519	1530	1548

8. RECYCLING You randomly survey shoppers at a supermarket about whether they use reusable bags. Of 60 male shoppers, 15 use reusable bags. Of 110 female shoppers, 60 use reusable bags. Organize your results in a two-way table. Include the marginal frequencies.

Test Item References

Chapter Test Questions	Section to Review	Common Core State Standards
1, 2	12.1	S.ID.2, S.ID.3
3	12.2	S.ID.2, S.ID.3
5a, 5b	12.3	S.ID.1, S.ID.2, S.ID.3
5c	12.4	S.ID.2, S.ID.3
6a, 6c	12.5	8.SP.1, S.ID.6a, S.ID.6c
6b, 6c, 7	12.6	S.ID.6b, S.ID.8, S.ID.9
8	12.7	8.SP.4, S.ID.5
4	12.8	S.ID.1

Test-Taking Strategies

Remind students to quickly look over the entire test before they start so that they can budget their time. When they receive their test, students should list the different types of data displays. Have students use the **Stop** and **Think** strategy before they answer each question.

Common Assessment Errors

- **Exercises 1–2** When finding the mean, students may forget to divide by the total number of data values and instead divide by the maximum value. Remind them that the definition of mean is an "average," so they must take into account the total number of items or numbers to get an average. Explain to students that it is as if they are dividing the total evenly among the number of groups.
- **Exercise 3** Students may subtract the first data value from the last data value to find the range. Remind students that the range is the difference between the greatest and least data values.
- **Exercise 5** Students may have difficulty creating the box-and-whisker plot. Remind them of the five-number summary of a data set.

Reteaching and Enrichment Strategies

If students need help. . .	If students got it. . .
Resources by Chapter • Practice A and Practice B • Puzzle Time Record and Practice Journal Practice Differentiating the Lesson Lesson Tutorials Practice from the Test Generator Skills Review Handbook	Resources by Chapter • Enrichment and Extension • School-to-Work • Financial Literacy • Technology Connection • Life Connections • Stories in History Game Closet at *BigIdeasMath.com* Start Standardized Test Practice

Answers

1. mean: 1; median: 0; mode: -3
2. mean: 49; median: 52; mode: 54
3. Pond A: 10; 9; about 3.2
   Pond B: 12; 14; about 4.3
   The turtles in Pond B tend to be heavier and their weights tend to differ more from the mean than the turtles in Pond A.
4. *Sample answer:*
   Box-and whisker plot: shows the mean and variability of the data.
5. **a.**

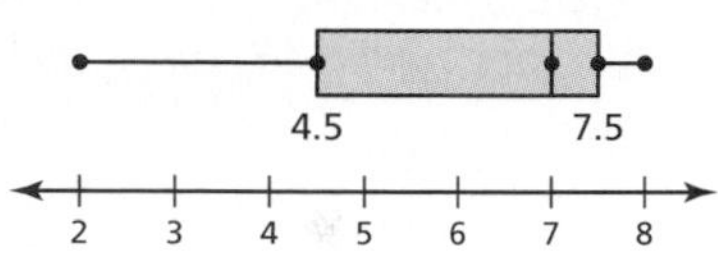

   b. 6; 3

   c. median for central tendency, five-number summary for spread
6. See Additional Answers.
7. $y = 20.3x + 1350$
8. See Additional Answers.

Assessment Book

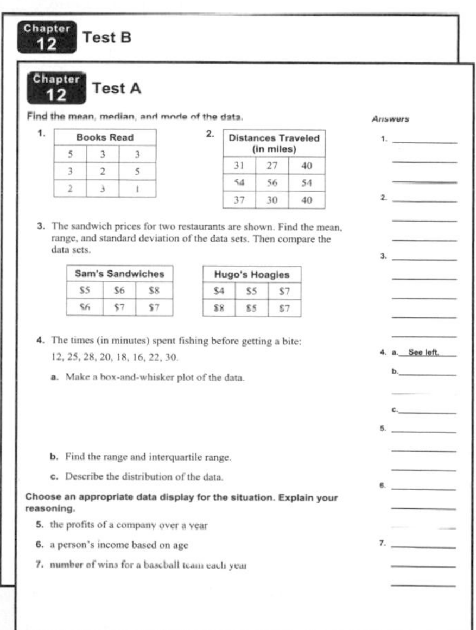

Test-Taking Strategies
Available at *BigIdeasMath.com*

After Answering Easy Questions, Relax
Answer Easy Questions First
Estimate the Answer
Read All Choices before Answering
Read Question before Answering
Solve Directly or Eliminate Choices
Solve Problem before Looking at Choices
Use Intelligent Guessing
Work Backwards

About this Strategy

When taking a multiple choice test, be sure to read each question carefully and thoroughly. Look closely for words that change the meaning of the question, such as *not*, *never*, *all*, *every*, and *always*.

Answers

1. B
2. G
3. 16
4. C

Technology For the Teacher

Big Ideas Test Generator

Item Analysis

1. **A.** The student assumes the correct answer is the least of the answer alternatives, because 4 of the 7 scores in the table are greater than 45.5.

   B. Correct answer

   C. The student uses the value of the mean given in the problem statement.

   D. The student finds the mean of the seven given scores.

2. **F.** The student confuses the direction of the vertical translation and the direction in which the parabola opens.

   G. Correct answer

   H. The student confuses the direction of the vertical translation.

   I. The student confuses the direction in which the parabola opens.

3. **Gridded Response:** Correct answer: 16

   Common Error: Instead of squaring 4, the student takes the square root of 4, yielding an answer of 2.

4. **A.** The student multiplies coefficients to get the coefficient of z and adds the constant terms.

   B. The student adds $4z$ instead of subtracting $4z$.

   C. Correct answer

   D. The student forgets the product of the outside terms.

12 Standardized Test Practice

Test-Taking Strategy
Read Question Before Answering

Of 2048 cats, how many of them will NOT answer to "Here, kitty kitty"?

Ⓐ 100% Ⓑ 1 Ⓒ $\frac{4098}{2}$ Ⓓ 2048

Hey, it means "free food."

"Be sure to read the question before choosing your answer. You may find a word that changes the meaning."

1. What is the value of x when the mean of the video game scores is 45.5? *(S.ID.2)*

Video Game Scores			
36	28	x	48
42	57	63	52

A. 35

B. 38

C. 45.5

D. 46.57

2. Which graph represents $y = 3x^2 - 1$? *(F.BF.3)*

F.

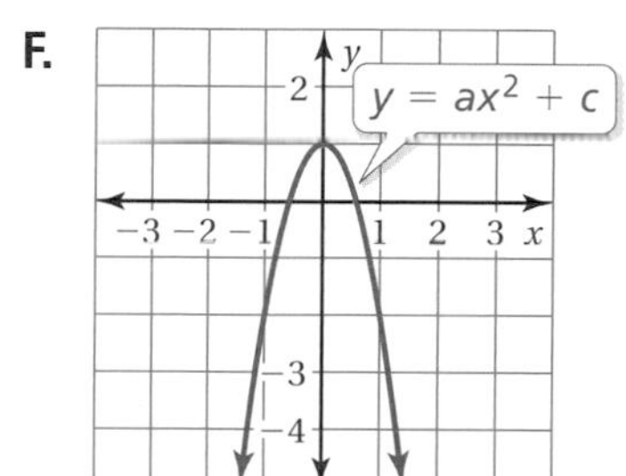

H.

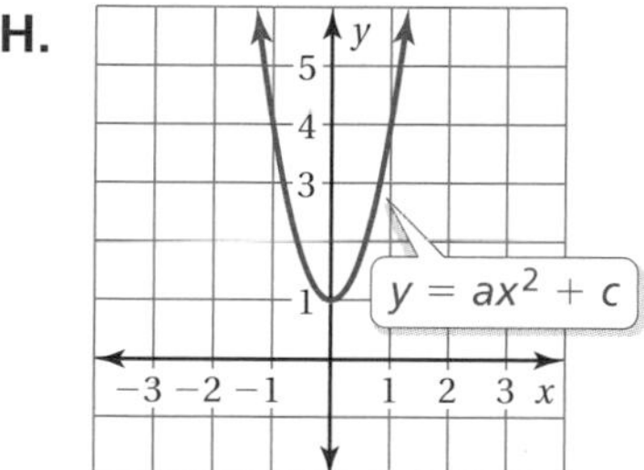

G.

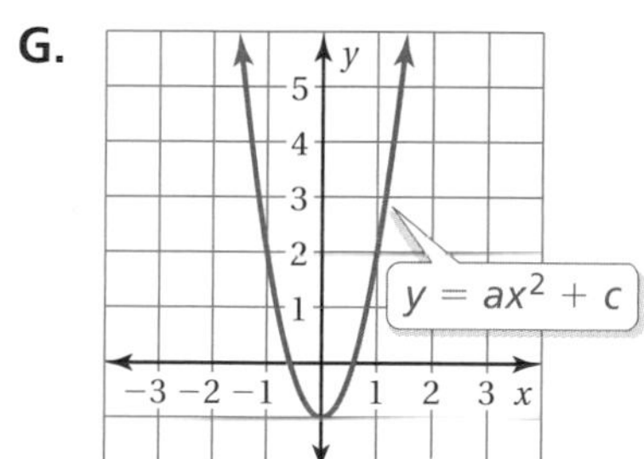

I.

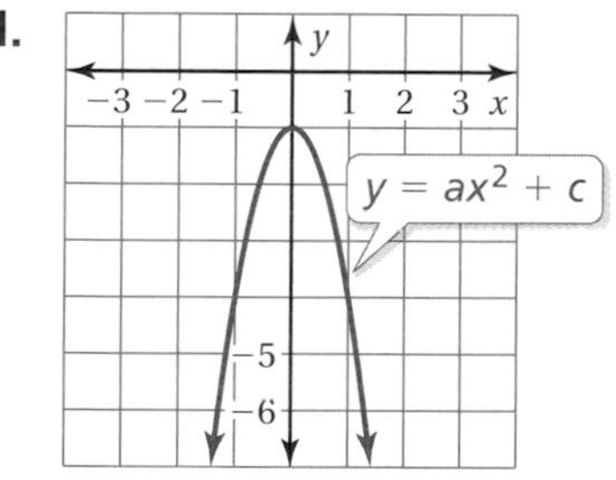

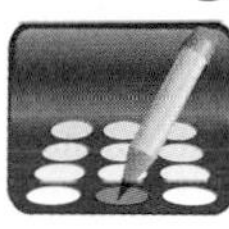

3. What is the solution of the equation $-6 + \sqrt{x} = -2$? *(N.RN.2)*

4. What is the product $(3z - 2)(2z + 4)$? *(A.APR.1)*

A. $6z + 2$

B. $6z^2 + 16z + 8$

C. $6z^2 + 8z - 8$

D. $6z^2 - 4z - 8$

5. The box-and-whisker plot represents the lengths of project presentations (in minutes) at a science fair. Find the interquartile range of the data. What does this represent in the context of the situation? *(S.ID.2)*

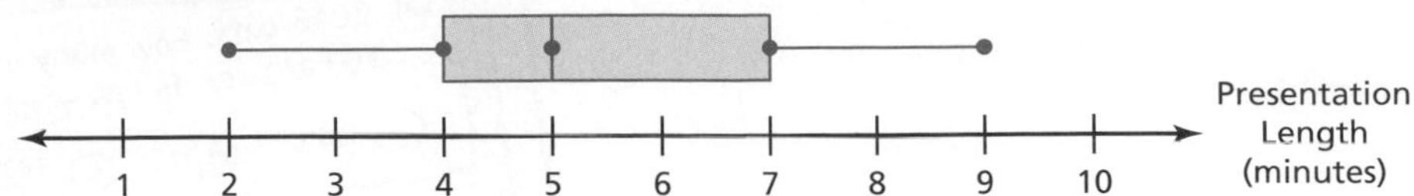

F. 7; The middle half of the presentation lengths vary by no more than 7 minutes.

G. 3; The presentation lengths vary by no more than 3 minutes.

H. 3; The middle half of the presentation lengths vary by no more than 3 minutes.

I. 7; The presentation lengths vary by no more than 7 minutes.

6. What is the simplified form of the expression? *(N.RN.2)*

$$\left(\frac{4}{5x}\right)^{-2}$$

A. $\dfrac{25}{16x^2}$

B. $\dfrac{25x^2}{16}$

C. $\dfrac{16}{25x^2}$

D. $\dfrac{16x^2}{25}$

7. Which equation shows inverse variation? *(A.REI.10)*

F. $y = -3x + 7$

G. $2y = x$

H. $y = \dfrac{1}{5}x$

I. $y = \dfrac{4}{x}$

8. You randomly survey students in your school. You ask whether they have jobs. You display your results in the two-way table. How many male students do *not* have a job? *(S.ID.5)*

		Job	
		Yes	No
Gender	Male	27	12
	Female	31	17

Item Analysis (continued)

5. **F.** The student finds the range instead of the interquartile range.

   G. The student finds the correct interquartile range, but confuses it with the range when describing what it represents.

   H. Correct answer

   I. The student confuses the range with the interquartile range.

6. **A.** The student applies the definition of negative exponent correctly to the numerical values but not to the variable.

   B. Correct answer

   C. The student applies an exponent of 2 instead of -2.

   D. The student applies the definition of negative exponent correctly to the variable, but not to the numerical values.

7. **F.** The student confuses the concept of negative slope with the concept of inverse variation.

   G. The student confuses $2y = x$ with the inverse variation form $k = xy$.

   H. The student confuses $y = \frac{1}{5}x$ with the inverse variation form $y = \frac{k}{x}$.

   I. Correct answer

8. **Gridded Response:** Correct answer: 12

   Common Error: The student adds the entries in the "No" column for an answer of 29.

Answers

5. H
6. B
7. I
8. 12

Answers

9. D

10.

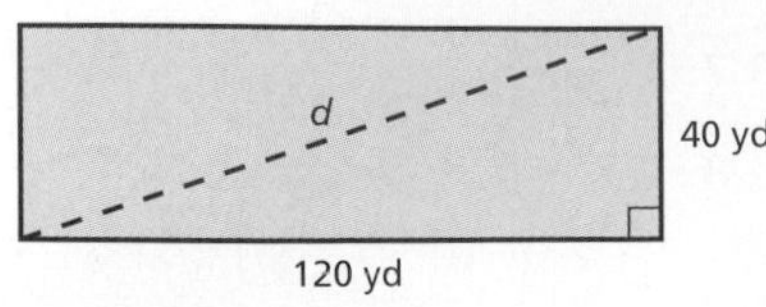

$$d^2 = 40^2 + 120^2$$
$$d = \sqrt{40^2 + 120^2}$$
$$= \sqrt{1600 + 14{,}400}$$
$$= \sqrt{16{,}000}$$
$$\approx 126.5 \text{ yd}$$

The distance between opposite corners is the length of the hypotenuse of a right triangle, so you can use the Pythagorean Theorem to find the distance.

11. I

12. C

Answer for Extra Example

1. **A.** Correct answer

B. The student confuses the words "quarterly" and "quartiles."

C. The student assumes quarterly profits can be shown as a circle graph divided into four quarters.

D. The student misunderstands the use of a dot plot.

Item Analysis (continued)

9. **A.** The student uses the incorrect form $a_n = a_1 n + d$.

B. The student uses the incorrect form $a_n = a_1 n - (a_1 - d)$.

C. The student uses the incorrect form $a_n = a_1 + (n + 1)d$.

D. Correct answer

10. **2 points** The student demonstrates a thorough understanding of the Pythagorean Theorem and its application to this problem. The student correctly calculates the distance between the opposite corners of the field and explains how the Pythagorean Theorem is used.

1 point The students demonstrates a partial understanding of the Pythagorean Theorem. The student understands that the Pythagorean Theorem applies and attempts to use it, but fails to calculate the distance correctly.

0 points The student provides no response, a completely incorrect or incomprehensible response, or a response that demonstrates insufficient understanding of the Pythagorean Theorem and its application to this problem.

11. **F.** The student confuses the characteristics that show positive and negative relationships in scatter plots.

G. The student confuses the characteristics that show no relationship and a negative relationship in scatter plots.

H. The student confuses the characteristics that show constant and negative relationships in scatter plots.

I. Correct answer

12. **A.** The student makes an error placing a decimal point.

B. The student makes an error applying the Addition Property of Equality and makes an error placing a decimal point.

C. Correct answer

D. The student makes an error applying the Addition Property of Equality.

Extra Example

1. Which data display is the most appropriate to show the trend of quarterly profits for a company? *(S.ID.1)*

A. a line graph

B. a box-and-whisker plot

C. a circle graph

D. a dot plot

9. What is an equation for the nth term of the arithmetic sequence? *(F.LE.2)*

$$-\frac{3}{4}, -\frac{1}{4}, \frac{1}{4}, \frac{3}{4}, \ldots$$

A. $a_n = -\frac{3}{4}n + \frac{1}{2}$

B. $a_n = -\frac{3}{4}n + \frac{5}{4}$

C. $a_n = \frac{1}{2}n - \frac{1}{4}$

D. $a_n = \frac{1}{2}n - \frac{5}{4}$

10. A football field is 40 yards wide and 120 yards long. Find the distance between opposite corners of the football field. Show your work and explain your reasoning. *(8.G.7)*

Think Solve Explain

11. Which scatter plot shows a negative relationship between x and y? *(8.SP.1)*

F.

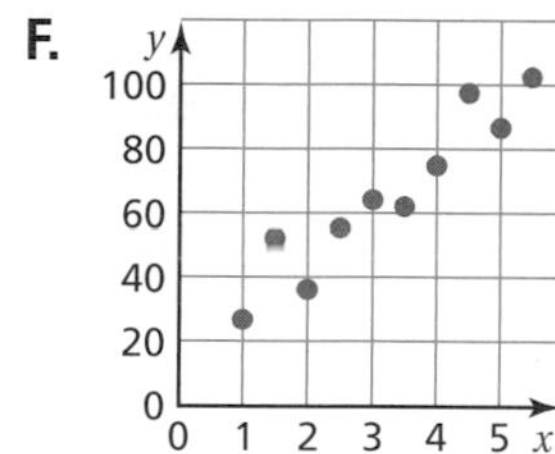

H.

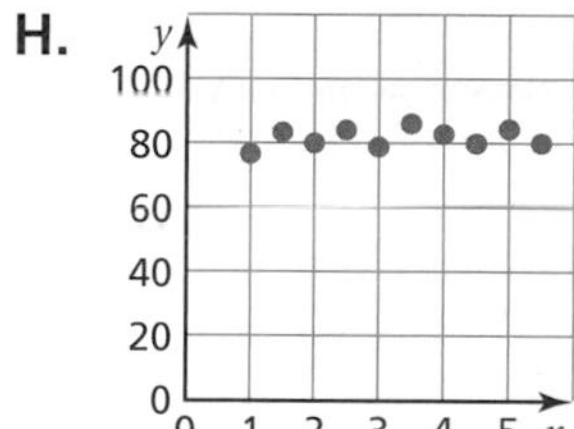

G.

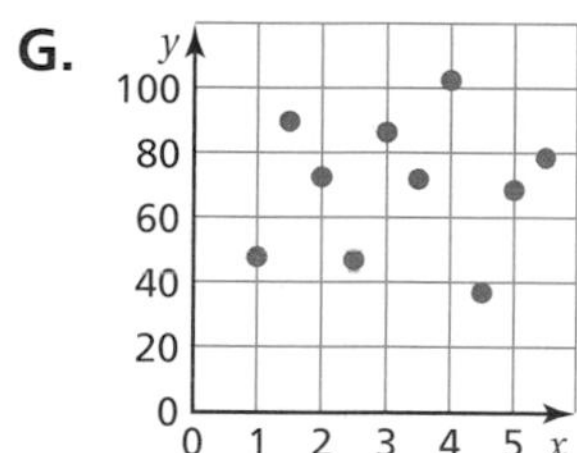

I.

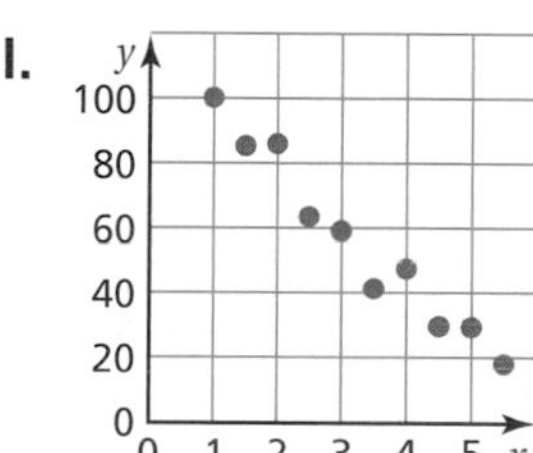

12. What is the solution of the equation? *(A.REI.3)*

$$0.22(x + 6) = 0.2x + 1.8$$

A. $x = 2.4$

B. $x = 15.6$

C. $x = 24$

D. $x = 156$

Appendix A
My Big Ideas Projects

A.1 Literature Project The Jungle Book

A.2 History Project Mathematics in Medieval Islam

About the Appendix

- The interdisciplinary projects can be used anytime throughout the year.
- The projects offer students an opportunity to build on prior knowledge, to take mathematics to a deeper level, and to develop organizational skills.
- Students will use the Essential Questions to help them form "need to knows" to focus their research.

Essential Question

- **Literature Project**
  How does the knowledge of mathematics help to visualize life in the jungle?
- **History Project**
  How did the work of al-Khowarizmi and other Arabic and Islamic mathematicians influence modern-day mathematics?

Additional Resources

BigIdeasMath.com

Additional Resources

BigIdeasMath.com

Essential Question

- **Art Project**
  How do the symmetry of the subject and the symmetry of the photograph affect a person's preference for a photograph?
- **Science Project**
  Is there a relationship, that can be described by a mathematical model, between the occurrence of tornadoes and the month of the year?

My Big Ideas Projects

A.3 Art Project
Symmetry in Photographic Art

A.4 Science Project
Tornadoes

A.1 Literature Project

The Jungle Book

1 Getting Started

The Jungle Book is a collection of stories about an orphaned boy, Mowgli, who is raised by jungle animals in India. Written by Rudyard Kipling, the book was first published in 1894.

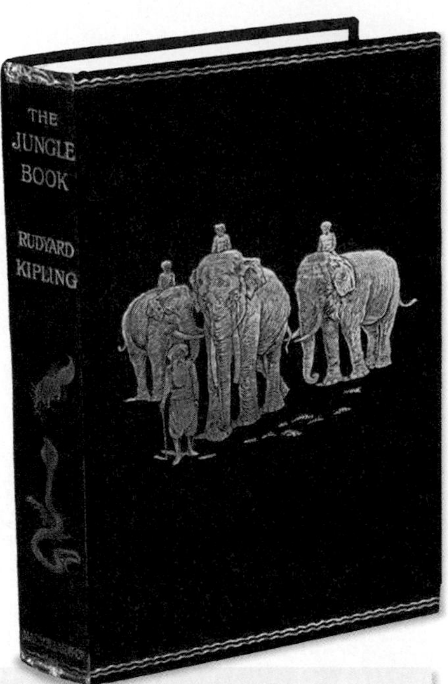

Essential Question How does the knowledge of mathematics help to visualize life in the jungle?

Read *The Jungle Book*. Look for descriptions of objects and places in the stories. Use the descriptions and mathematical formulas to estimate their size or distance.

Sample: In the first story, Mowgli almost walks into a trap, described as a square box with a drop gate. Estimate the possible volume of the trap using inequalities.

The formula for the volume of a box is $V = \ell wh$. By letting $\ell = w = h$, you can rewrite the formula as $V = h \cdot h \cdot h = h^3$.

Mowgli is between 1 and 11 years old in the story. The median height of a 1-year-old is about 2.5 feet. The median height of an 11-year-old is about 4.5 feet. So, the inequality $2.5 \le h \le 4.5$ describes Mowgli's height during this time.

If the trap is as tall as Mowgli, the possible volume of the trap would be $2.5^3 \le h^3 \le 4.5^3$, or $15.625 \le V \le 91.125$ cubic feet.

A trap this size could ensnare a large animal, such as a wolf or a bear.

Project Notes

Introduction

For the Teacher

- **Goal:** Students will read *The Jungle Book* by Rudyard Kipling and write a report about the mathematics they find in the story. Samples of things that can be included in students' reports are discussed below.
- **Management Tip:** You may want to have students work together in groups.

Essential Question

- How does the knowledge of mathematics help to visualize life in the jungle?

Things to Think About

Summary of *The Jungle Book*

- After Shere Khan, the tiger, chased the little boy's parents away, the man cub arrived at the den of Father and Mother Wolf. Mother Wolf named him Mowgli, for Mowgli the Frog. Mowgli was accepted into the pack, and as a pack member, the Law of the Jungle states that no other wolf could kill him until he, himself, had killed a wolf.
- Baloo, the sleepy brown bear, offered to teach Mowgli, as he does the other wolf cubs, the Law of the Jungle.
- Bagheera, the black panther, offered to pay a bull in order that Mowgli be accepted into the pack.
- When it was time for Akela, the Leader of the Pack, to step down from leadership, Mowgli's life in the jungle was threatened. Bagheera suggested that he go to the village and get some Red Flower (fire), of which the wolves are afraid. The result of the Council meeting was that Mowgli will leave the jungle.
- During Mowgli's time in the jungle, Baloo taught him the Master Words of the Jungle, specific for each people (the Hunting-People, the Snake-People, etc.). Thus, Mowgli was safe against accidents in the jungle.
- At seven years of age, Mowgli was kidnapped by the Bandar-logs (monkeys) and was taken among the tree tops to an abandoned ruin of a King's palace.
- Bagheera and Baloo elicited the help of Kaa, The Python, to save Mowgli.
- Mowgli went to a village over 20 miles away. There was a woman there who thought that Mowgli was her son who was taken by a tiger as a child.
- Mowgli tried to adjust to living like a human. As a herdsman, he would meet with Brother Wolf, who would alert him of Shere Khan's whereabouts.
- Shere Khan came to kill Mowgli, but with the help of Brother Wolf and Akela, Mowgli was able to orchestrate the stampeding of Shere Khan by the cattle and bulls. Afterwards, the village people turned Mowgli away. He went back to the wolves.

References

Go to *BigIdeasMath.com* to access links related to this project.

Cross-Curricular Instruction

Meet with a reading or language arts teacher and review curriculum maps to identify whether students have already or have yet to read *The Jungle Book*. If the book has already been read, you may want to discuss the work they have completed and review the book with the students. If the book has not been read, perhaps you can both work simultaneously and share notes. Or, you may want to explore activities that the reading or language arts teacher has done in the past to support student learning in this particular area.

Project Notes

Mathematics Used in the Story

- In a jungle, it is important to know your surroundings and be able to describe your environment. The Jungle Book uses phrases such as "within ten miles." This would describe a circular area encompassing all of the land and the animals ten miles or closer to them. This would be a total area of approximately 314 miles.
- Mowgli did not speak in human tongue, and, thus, could not count.
- Kaa, the Rock Snake, is thirty feet long and can twist into curves and knots.
- The Bandar-logs (monkeys) kidnapped Mowgli and brought him to the Cold Lairs. This was an old and ruined King's village. There were red sandstone reservoirs and houses with domed roofs.
- The monkeys sat in concentric circles around Mowgli, fifty and sixty deep.
- Kaa strikes with his head, backed by the strength of his body. The force is similar to that of a half ton hammer. Kaa was able to use his head and body to break the wall inside which Mowgli was held captive by the monkeys.
- Kaa could twist his body to make figure eights, squares, pentagons, and coiled mounds.

Closure

- **Rubric** An editable rubric for this project is available at *BigIdeasMath.com.*
- Students may present their reports to the class or school as a television report or public information broadcast.

2 Things to Include

- Choose an object or place from each story. Write descriptions of each.
- State a formula you can use to estimate the size of each object or place.
- Estimate the value of each variable in the formula and state the units of measure.
- Calculate the measurement.
- Summarize your findings. Consider revising or refining your estimates based on the other estimates you made.

3 Things to Remember

- You can download each part of the book at *BigIdeasMath.com*.
- Add your own illustrations to your project. Label the dimensions of the objects and places in your illustrations.
- Include as many different formulas as possible.
- Organize your report in a folder, and think of a title for your report.

A.2 History Project

Mathematics in Medieval Islam

Getting Started

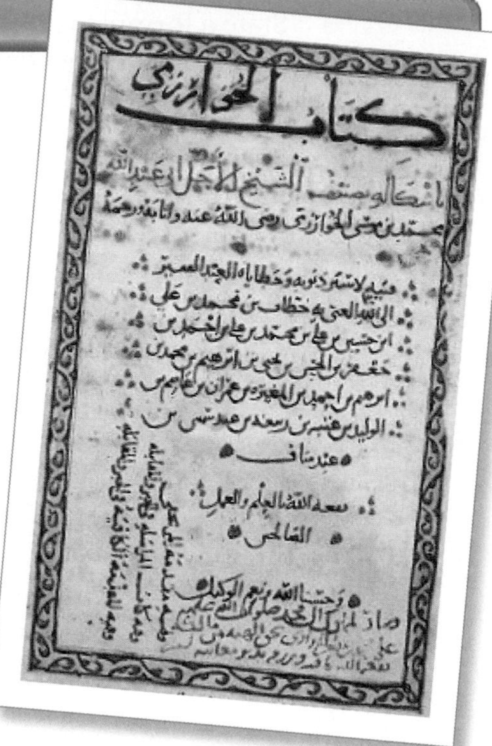

Algebraic techniques date back thousands of years. In fact, they were discovered, applied, and debated long before they received the name *algebra*.

Mohammed ibn-Musa al-Khowarizmi, an Arab mathematician, wrote a book titled, "Al-jabr wa'l muqubalah." The book was written about 825 A.D. in Baghdad.

Essential Question How did the work of al-Khowarizmi and other Arabic and Islamic mathematicians influence modern-day mathematics?

Sample: The word *algebra* is derived from the Arabic word *al-jabr*.

Al-Khowarizmi used the word *jabr* to describe moving a subtracted term to the other side of the equation.

$$x - 3 = 5$$
$$x = 8$$

Project Notes

Introduction

For the Teacher

- **Goal:** Students will discover how Arabic and Islamic people used mathematics.
- **Management Tip:** Students can work in groups to research the required topics and generate a report.

Essential Question

- How has the work of al-Khowarizmi and other Arabic and Islamic mathematicians influenced modern-day mathematics?

Things to Think About

? To what part of history does mathematics in medieval Islam refer?

- Alternative names are Islamic mathematics and Arabic mathematics.
- The title refers to the mathematics discovered or developed by the Islamic civilization between about 622 and 1600.
- The Islamic civilization covered the area east of the Iberian Peninsula, west of the Indus, and north of the Almoravid Dynasty and the Mali Empire.
- There are many unearthed Arabic manuscripts still to be studied, and many still to be found.

? To what part of mathematical history does mathematics in medieval Islam refer?

- They began work on the relationship between algebra and geometry, by combining Indian, Babylonian, and Greek mathematical discoveries.
- Islamic mathematicians played a major role in the translation of Greek mathematical writings, leading to our ability to read Greek manuscripts.
- Abu Kamil Shuja ibn Aslam was one of the Islamic mathematicians who integrated the Greek discovery of geometry-based irrational numbers into algebraic irrational numbers.
- In about 1000, al-Karaji introduced an implicit proof for arithmetic sequences.
- Omar Khayyám and Sharaf al-Din al-Tusi made advances in cubic equations.
- Medieval Islamic mathematicians translated and expanded Greek and Indian works in the area of trigonometry. By the 10th century, they were using all six trigonometric functions in applications to geometry. This was driven by the demands of both navigation and maps.

References

Go to *BigIdeasMath.com* to access links related to this project.

Cross-Curricular Instruction

Meet with a history teacher and review curriculum maps to identify whether students have already covered or have yet to discuss Arabic and Islamic people and their history. If the topic has already been covered, you may want to discuss the work they have completed and review prior knowledge with the students. If the history teacher has not discussed these concepts, perhaps you can both work simultaneously on these concepts and share notes. Or, you may want to explore activities that the history teacher has done in the past to support student learning in this particular area.

Project Notes

? What were the contributions by al-Khowarizmi?

- The formation of the digits 0, 1, 2, 3, 4, 5, 6, 7, 8, 9 that we use today.
- The presentation of the first systematic solution of linear and quadratic equations in Arabic.
- The Latin form of his name is Algoritmi, from which the words *algorithm* and *algorism* were formed.
- A method of solving linear and quadratic equations using methods of reduction (al-jabr) and balancing (al-muqãbala)
- Reduction involved transposing subtracted terms to the other side of an equation.
- Balancing involved cancelling like terms on opposite sides of an equation.
- The algebraic notation that we are familiar with had not been invented at this time, so al-Khowarizmi wrote the steps out using words rather than symbols.

Closure

- **Rubric** An editable rubric for this project is available at *BigIdeasMath.com.*
- You may hold a class debate where students can compare, defend, and discuss their findings with another student or group of students.

2 Things to Include

- In addition to *jabr,* al-Khowarizmi used the word *muqubalah*. State the definitions and explain how al-Khowarizmi used *jabr* and *muqubalah* in solving equations.
- Give an example of how al-Khowarizmi solved an equation. Then demonstrate how you would solve the same equation.
- Discuss how Greek and Hindu civilizations influenced the development of algebra in medieval Islam.
- Islamic mathematicians and scientists contributed to another branch of mathematics called *trigonometry*. Discuss the development of trigonometry.

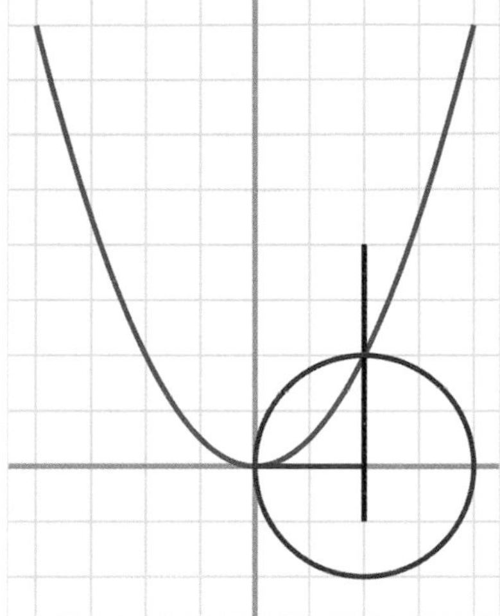

Omar Khayyám's geometric solution to a cubic equation

Medieval Islamic mosaics were produced using geometry that mathematicians did not fully understand until the 1970s.

3 Things to Remember

- Add your own illustrations to your project.
- Include as many different math concepts as possible.
- Organize your math stories in a folder, and think of a title for your report.

A.3 Art Project

Symmetry in Photographic Art

1 Getting Started

Symmetry is all around us, in nature and in manmade structures. A photographer chooses a perspective that either emphasizes the symmetry of the subject or focuses the attention on a specific part of the subject.

Essential Question How do the symmetry of the subject and the symmetry of the photograph affect a person's preference for a photograph?

Find two photographs with different perspectives of a subject having line and/or rotational symmetry. Discuss the symmetry of the subject and of the photograph. Conduct a survey asking which photograph is preferred.

Sample: In the first photograph, the bridge appears to have a vertical line of symmetry in the center of the bridge. In the second photograph, the bridge appears to have a vertical line of symmetry in the center of the bridge. There may be vertical lines of symmetry about the two vertical structures where the suspension lines are attached.

The first photograph emphasizes the symmetry of the bridge. The second photograph focuses on one of the vertical structures. The bridge in the second photograph may be symmetrical, but the photographer chose an angle that does not show the symmetry.

Of ten people surveyed, nine preferred the second photograph.

Project Notes

Introduction

For the Teacher

- **Goal:** Students will discover how a person's preference for a photograph is influenced by symmetry.
- **Management Tip:** Students may work in groups to find subjects and photographs. They can discuss the symmetry of the subjects and of the photographs as a group.

Essential Question

- How do the symmetry of the subject and the symmetry of the photograph affect a person's preference for a photograph?

Things to Think About

? Which flower do you prefer?

- Daisy: A typical daisy appears to have many lines of symmetry, following opposite petals.
- Rose: A typical rose does not appear to have any lines of symmetry.

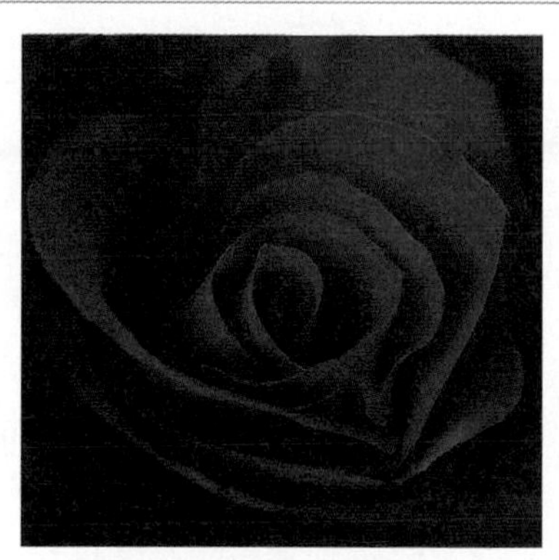

- Daisy: The straightforward photograph beautifully depicts the lines of symmetry.
- Rose: The angle of the photograph shows a possible line of symmetry, splitting the outer petals of the rose. A different angle could possibly have shown no lines of symmetry.
- Of ten people surveyed, seven preferred the photograph of the rose.

References

Go to *BigIdeasMath.com* to access links related to this project.

Cross-Curricular Instruction

Meet with an art teacher and review curriculum maps to identify whether students have already covered or have yet to discuss symmetry. If the topic has already been covered, you may want to discuss the work they have completed and review prior knowledge with the students. If the art teacher has not discussed these concepts, perhaps you can both work simultaneously on these concepts and share notes. Or, you may want to explore activities that the art teacher has done in the past to support student learning in this particular area.

Project Notes

? Which house do you prefer?

- **First house:** It appears to have a vertical line of symmetry in the middle of the house. The windows also have vertical and horizontal lines of symmetry.
- **Second house:** It appears to have a vertical line of symmetry at the peak of the roof, if you ignore the garage. If the ceilings were not angled on the top floor, then there would be a horizontal symmetry between the two floors.
- The second photograph is straightforward, emphasizing the lines of symmetry. The first photograph is at a slight angle, but it does not seem to affect the lines of symmetry.
- Of ten people surveyed, six preferred the photograph of the first house.

Closure

- **Rubric** An editable rubric for this project is available at *BigIdeasMath.com.*
- Students may present their reports to a parent panel or community members.

2 Things to Include

- Find two photographs of each of the following subjects. The subjects should have different symmetries, if possible. The photographs should have different perspectives.
    - flowers
    - houses or buildings
    - paintings or drawings by the same artist
    - sea creatures
    - a subject of your choice
- Discuss the types of symmetry (line and/or rotational) of the subject in each picture.
- Discuss whether the photographer's perspective emphasizes the symmetry of the subject in each photograph.
- Survey at least 10 people as to their visual preference of each pair.

This photograph has line symmetry and rotational symmetry. Would you prefer to view this photograph as displayed, or would you prefer to rotate it 90°?

3 Things to Remember

- Add your photographs to your project.
- Include as many different types of symmetry as possible.
- Organize your report in a folder, and think of a title for your report.

A.4 Science Project

Tornadoes

1 Getting Started

Tornadoes are considered some of nature's most violent storms. Tornadoes have strong rotating winds, which can range in speed from 40 to 318 miles per hour.

Essential Question Is there a relationship, that can be described by a mathematical model, between the occurrence of tornadoes and the month of the year?

Go to the National Oceanic and Atmospheric Administration website, *noaa.gov*, and enter the search words "storm event" to find the database containing statistics by state, county, and type of storm.

Sample: The table and scatter plot show the number of tornadoes y in a given month x, for the years 1951 to 2010, in St. Lucie County, Florida.

Month, x	Number of Tornadoes, y
1	1
2	1
3	2
4	3
5	3
6	5
7	5
8	12
9	2
10	1
11	0
12	1

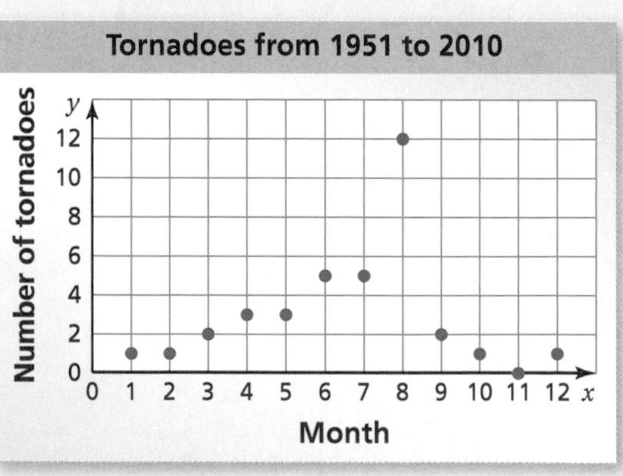

Summarize the data in the table and determine the pattern of the scatter plot (linear, quadratic, or exponential). Write an equation that roughly models the data and justify your work.

Project Notes

Introduction

For the Teacher

- **Goal:** Students will discover relationships between weather and the time of year.
- **Management Tip:** Students may work in groups to collect the data. They can discuss and compare the data as a group.

Essential Question

- Is there a relationship, that can be described by a mathematical model, between the occurrence of tornadoes and the month of the year?

Things to Think About

? How many tornadoes have occurred in St. Lucie County, Florida?

- The table and scatter plot give the numbers of tornadoes in a given month, from the years 1951 to 2010, in St. Lucie County, Florida.
- Based on the table, the month of August has the most tornado activity.
- There were a total of 37 tornadoes during this time period. About one-third of them occurred in August.
- The scatter plot shows that the summer season appears to be more active than the other seasons.

Month	Number
1	1
2	1
3	2
4	4
5	3
6	5
7	5
8	12
9	2
10	1
11	0
12	1
Total	**37**

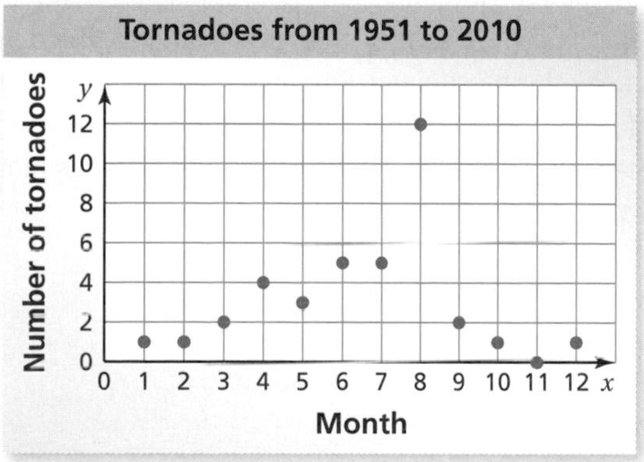

Tornadoes from 1951 to 2010

- Ignoring the month of August, which does not follow the pattern of the other months, it appears that there is a parabolic relationship between the number of tornados and the month.

References

Go to *BigIdeasMath.com* to access links related to this project.

Cross-Curricular Instruction

Meet with a science teacher and review curriculum maps to identify whether students have already covered weather. If the topic has already been covered, you may want to discuss the work students have completed and review prior knowledge with them. If the science teacher has not discussed these concepts, perhaps you can both work simultaneously on these concepts and share notes. Or, you may want to explore activities that the science teacher has done in the past to support student learning in this particular area.

Project Notes

? How many tornadoes have occurred in Nobles County, Minnesota?

- The table and scatter plot give the numbers of tornadoes in a given month, from the years 1950 to 2011, in Nobles County, Minnesota.
- Based on the table, the month of July has the most tornado activity.
- There were a total of 32 tornadoes during this time period. About 34 percent of them occurred in July.
- The scatter plot also shows that the spring and summer seasons are the most active.

Month	Number
1	0
2	0
3	5
4	3
5	6
6	3
7	11
8	2
9	0
10	2
11	0
12	0
Total	**32**

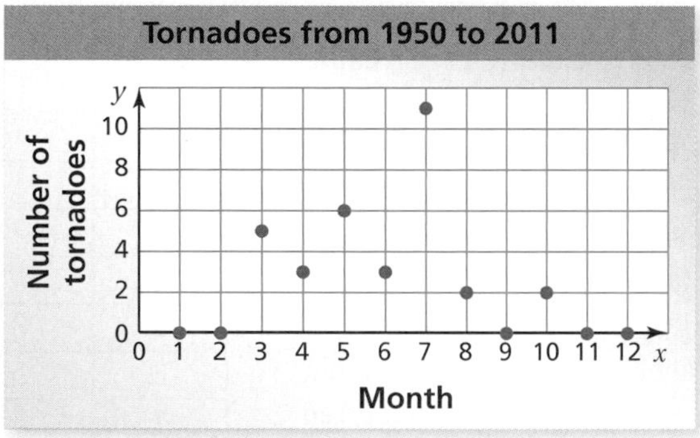

Tornadoes from 1950 to 2011

- Ignoring the month of July, which does not follow the pattern of the other months, it appears that there is a parabolic relationship between the number of tornados and the month.
- **Comparison:** Even though Florida is near the tropics and Minnesota in the cold Central Plains, the scatter plots are similar. Both counties tornado activity peak during summer. Each data set has an outlier (significantly different from the occurrences in the other months). Nobles County has more activity in the spring, but it also has more months with zero tornado activity.

Closure

- **Rubric** An editable rubric for this project is available at *BigIdeasMath.com.*
- Students may present their reports to the class or compare their report with other students' reports.

2 Things to Include

- Go to *noaa.gov* to find the dates on which tornadoes occurred in the county where you live.
- Organize the data by months in a table. Discuss the data in the table.
- Create a scatter plot of the data. Discuss the scatter plot.
- Determine if there is a linear, quadratic, or exponential relationship between the number of tornadoes and the month. Write an equation that roughly describes this relationship.
- Repeat the steps for a county in another state that has a different climate. Compare the scatter plots and equations for the two counties.

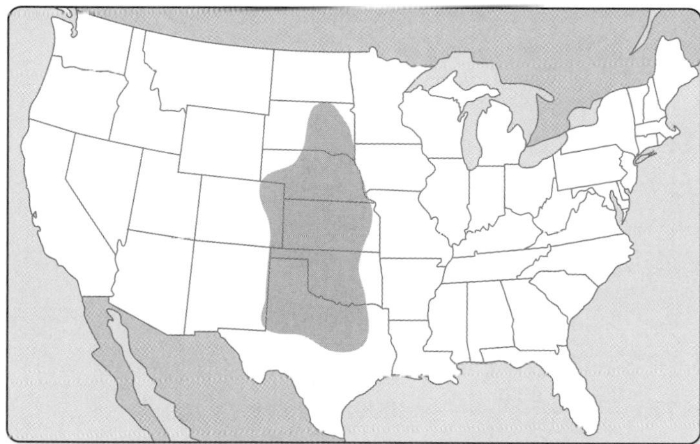

Tornado Alley in the central U.S. has a disproportionately high frequency of tornadoes during the late spring.

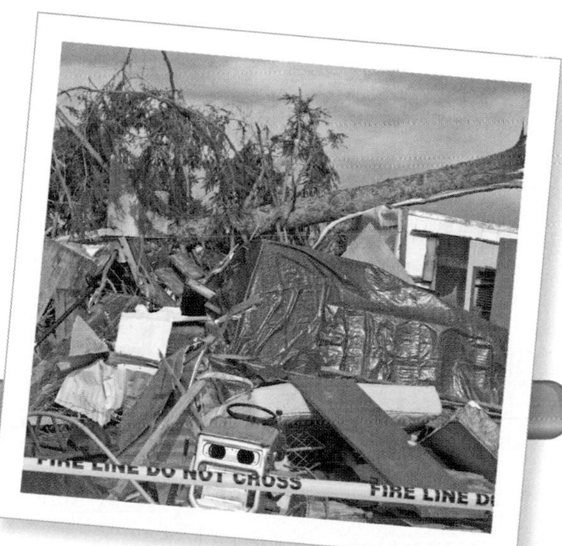

3 Things to Remember

- Add your own images to your project.
- Organize your report in a folder, and think of a title for your report.

Selected Answers

Section 1.1 Solving Simple Equations
(pages 7–9)

1. + and − are inverses. × and ÷ are inverses.

3. $x - 3 = 6$; It is the only equation that does not have $x = 6$ as a solution.

5. $x = 57$ **7.** $x = -5$ **9.** $p = 21$ **11.** $x = 9\pi$ **13.** $d = \frac{1}{2}$ **15.** $n = -4.9$

17. **a.** $105 = x + 14$; $x = 91$

b. no; Because $82 + 9 = 91$, you did not knock down the last pin with the second ball of the frame.

19. $n = -5$ **21.** $m = 7.3\pi$ **23.** $k = 1\frac{2}{3}$ **25.** $p = -2\frac{1}{3}$

27. They should have added 1.5 to each side.

$$-1.5 + k = 8.2$$
$$k = 8.2 + 1.5$$
$$k = 9.7$$

29. $6.5x = 42.25$; \$6.50 per hour

31. $420 = \frac{7}{6}b$, $b = 360$; \$60

33. $h = -7$ **35.** $q = 3.2$ **37.** $x = -1\frac{4}{9}$

39. greater than; Because a negative number divided by a negative number is a positive number.

41. 3 mg **43.** 8 in. **45.** $7x - 4$ **47.** $\frac{25}{4}g - \frac{2}{3}$

Section 1.2 Solving Multi-Step Equations
(pages 14 and 15)

1. $2 + 3x = 17$; $x = 5$ **3.** $k = 45$; 45°, 45°, 90° **5.** $b = 90$; 90°, 135°, 90°, 90°, 135°

7. $c = 0.5$ **9.** $h = -9$ **11.** $x = -\frac{2}{9}$ **13.** 20 watches

15. $4(b + 3) = 24$; 3 in.

17. $\frac{2580 + 2920 + x}{3} = 3000$; 3500 people

19. $x = \frac{-7}{b}$ **21.** $x = \frac{3b}{2c}$ **23.** $x = \frac{c - b}{a}$

25. $<$ **27.** $>$

Section 1.3 Solving Equations with Variables on Both Sides
(pages 22 and 23)

1. no; When 3 is substituted for x, the left side simplifies to 4 and the right side simplifies to 3.

3. $x = 13.2$ in. **5.** $x = 7.5$ in. **7.** $k = -0.75$

9. $p = -48$ **11.** no solution **13.** $x = -4$

15. The 4 should have been added to the right side.

$$\begin{aligned} 3x - 4 &= 2x + 1 \\ 3x - 2x - 4 &= 2x + 1 - 2x \\ x - 4 &= 1 \\ x - 4 + 4 &= 1 + 4 \\ x &= 5 \end{aligned}$$

17. $15 + 0.5m = 25 + 0.25m$; 40 mi

19. 7.5 units

21. *Sample answer:* **a.** $5x = 5x - 3$ **b.** $x + 4 = 2x - x + 4$

23. fractions; Because $\frac{1}{3}$ is hard to perform operations with when written as a decimal.

25. 25 grams

27. 15.75 cm^3

29. about 153.86 ft^3

Lesson 1.3b Solving Absolute Value Equations
(pages 24 and 25)

1. $x = 10$ or $x = -10$

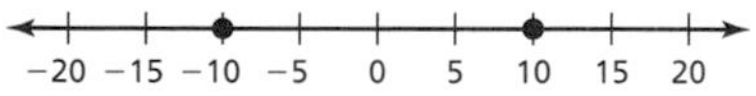

3. no solution

5. $x = 6$ or $x = -2$

7. $x = 1$ or $x = -\frac{3}{5}$

9. *Sample answer:* $|x - 24| = 8$

Section 1.4 Rewriting Equations and Formulas
(pages 30 and 31)

1. no; The equation only contains one variable.

3. **a.** $A = \frac{1}{2}bh$ **b.** $h = \frac{2A}{h}$ **c.** $b = 12$ mm

5. $y = 4 - \frac{1}{3}x$

7. $y = \frac{2}{3} - \frac{4}{9}x$

9. $y = 3x - 1.5$

11. The y should have a negative sign in front of it.

$$\begin{aligned} 2x - y &= 5 \\ -y &= -2x + 5 \\ y &= 2x - 5 \end{aligned}$$

13. **a.** $t = \frac{I}{Pr}$

b. $t = 3$ yr

15. $m = \frac{e}{c^2}$

17. $\ell = \frac{A - \frac{1}{2}\pi w^2}{2w}$

19. $w = 6g - 40$

21. **a.** $F = 32 + \frac{9}{5}(K - 273.15)$

b. 32°F

c. liquid nitrogen

23. $r^3 = \frac{3V}{4\pi}$; $r = 4.5$ in.

25. $6\frac{2}{5}$

27. $1\frac{1}{4}$

Section 2.1 Graphing Linear Equations
(pages 46 and 47)

1. a line

3. *Sample answer:*

x	0	1
$y = 3x - 1$	-1	2

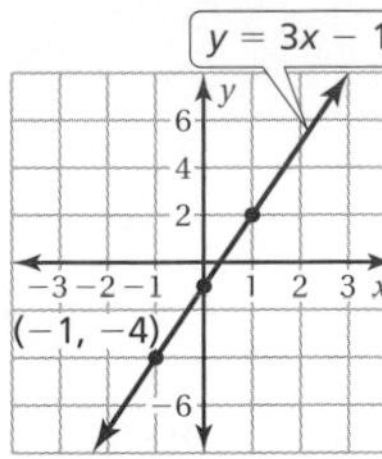

5.

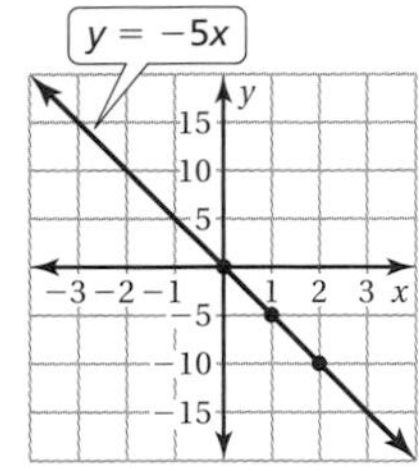

7.

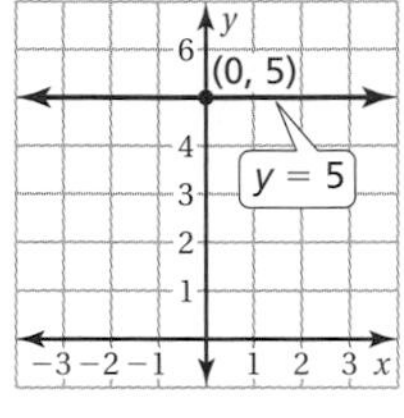

9.

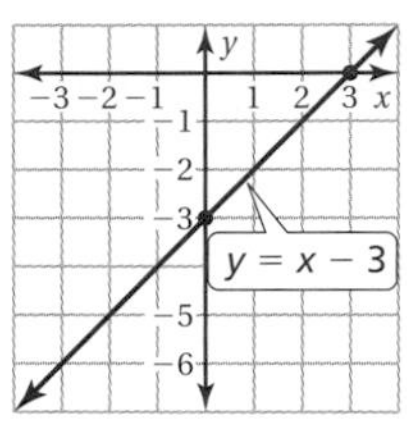

11.

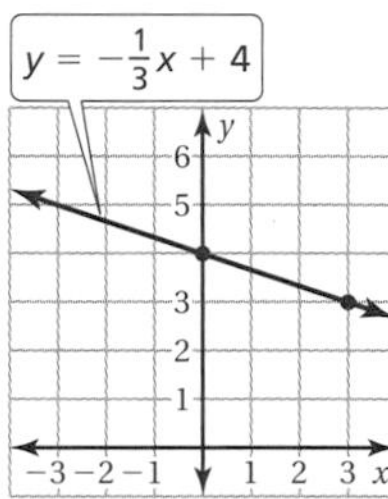

13.

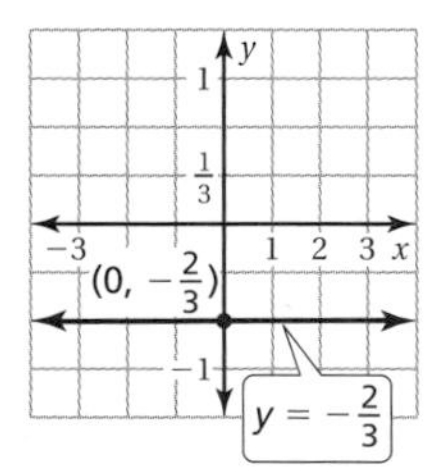

15.

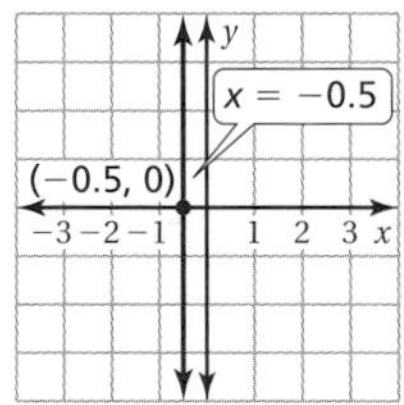

17. The equation $x = 4$ is graphed, not $y = 4$.

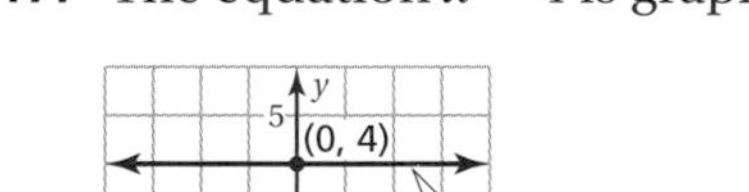

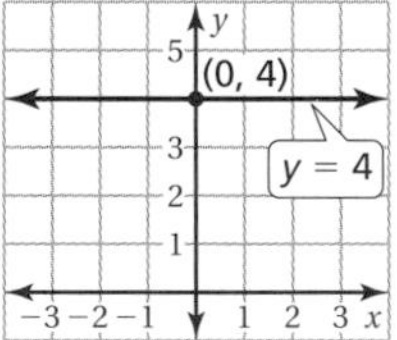

19. a.

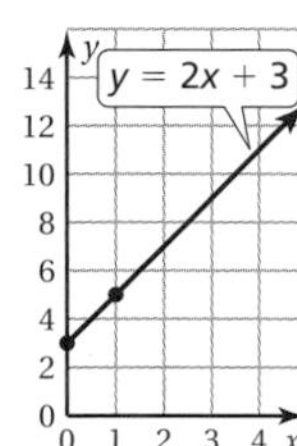

b. about \$5

c. \$5.25

21. $y = -\frac{5}{2}x + 2$

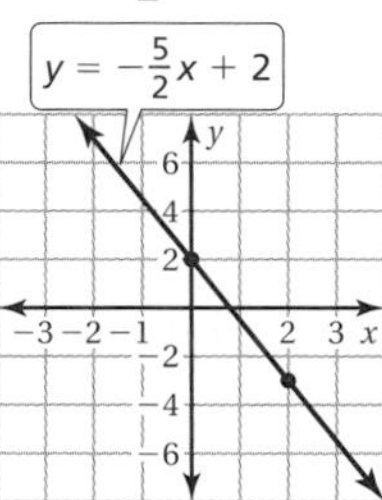

23. $y = -2x + 3$

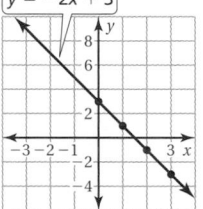

25. Begin this exercise by listing all of the given information.

Hint

27. a. *Sample answer:*

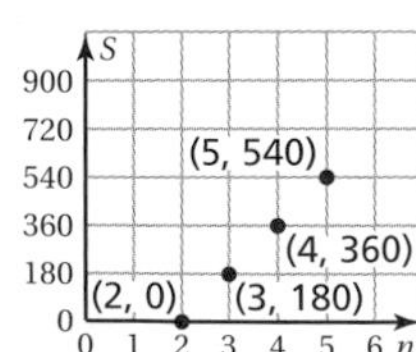

Yes, the points lie on a line.

b. No, $n = 3.5$ does not make sense because a polygon cannot have half a side.

29. $(-6, 6)$

31. $(-4, -3)$

Section 2.2 Slope of a Line
(pages 53–55)

1. **a.** B and C

   b. A

   c. no; None of the lines are vertical.

3. The line is horizontal.

5. The lines are parallel.

7. $\frac{3}{4}$

9. $-\frac{3}{5}$

11. 0

13. 0

15. undefined

17. $-\frac{11}{6}$

19. The denominator should be $2 - 4$.

    slope $= -1$

21. Choose any two points from the table and use the slope formula.

    slope $= 4$

23. Choose any two points from the table and use the slope formula.

    slope $= -\frac{3}{4}$

25. $\frac{1}{3}$

27. $k = 11$

29. $k = -5$

31. **a.** $\frac{3}{40}$

    b. The cost increases by \$3 for every 40 miles you drive, or the cost increases by \$0.075 for every mile you drive.

33. yes; The slopes are the same between the points.

35. yes; When you switch the coordinates, the differences in the numerator and denominator are the opposite of the numbers when using the slope formula. You still get the same slope.

37. 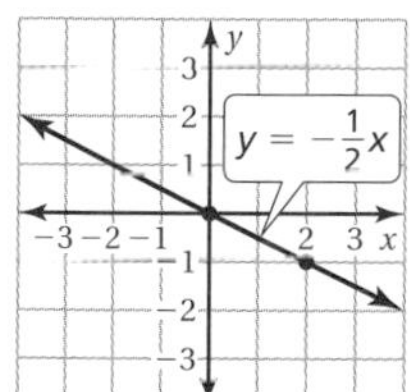

39. 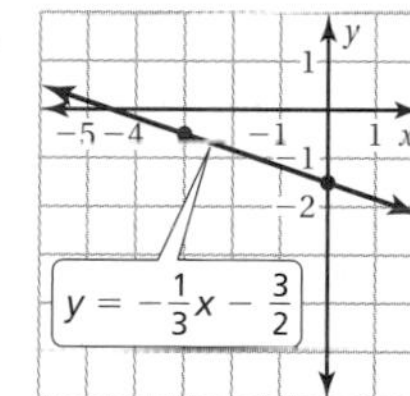

Lesson 2.2b Slopes of Parallel and Perpendicular Lines *(pages 56 and 57)*

1. blue and red; They both have a slope of −3.

3. yes; Both lines are horizontal and have a slope of 0.

5. yes; Both lines are vertical and have an undefined slope.

7. blue and green; The blue line has a slope of 6. The green line has a slope of $-\frac{1}{6}$. The product of their slopes is $6 \cdot \left(-\frac{1}{6}\right) = -1$.

9. yes; The line $x = -2$ is vertical. The line $y = 8$ is horizontal. A vertical line is perpendicular to a horizontal line.

11. yes; The line $x = 0$ is vertical. The line $y = 0$ is horizontal. A vertical line is perpendicular to a horizontal line.

Section 2.3 Graphing Linear Equations in Slope-Intercept Form *(pages 62 and 63)*

1. Find the x-coordinate of the point where the graph crosses the x-axis.

3. *Sample answer:* The amount of gasoline y (in gallons) left in your tank after you travel x miles is $y = -\frac{1}{20}x + 20$. The slope of $-\frac{1}{20}$ means the car uses 1 gallon of gas for every 20 miles driven. The y-intercept of 20 means there is originally 20 gallons of gas in the tank.

5. A; slope: $\frac{1}{3}$; y-intercept: −2

7. slope: 4; y-intercept: −5

9. slope: $-\frac{4}{5}$; y-intercept: −2

11. slope: $\frac{4}{3}$; y-intercept: −1

13. slope: −2; y-intercept: 3.5

15. slope: 1.5; y-intercept: 11

17. a.

b. The x-intercept of 300 means the skydiver lands on the ground after 300 seconds. The slope of −10 means that the skydiver falls to the ground at a rate of 10 feet per second.

19. 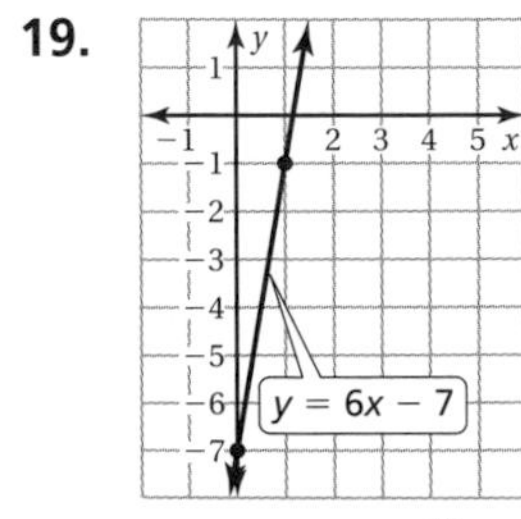

x-intercept: $\frac{7}{6}$

21. 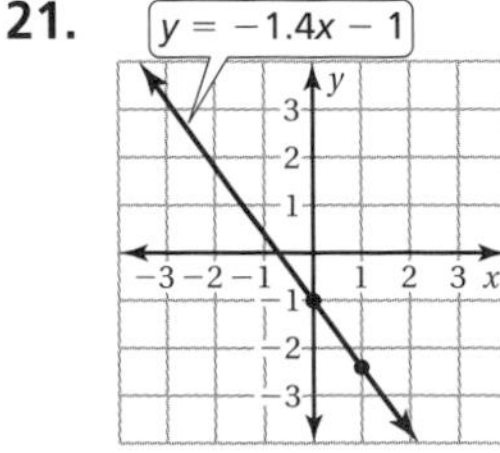

x-intercept: $-\frac{5}{7}$

23. 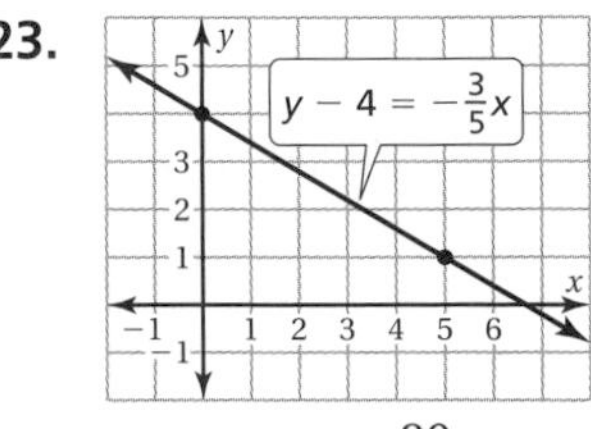

x-intercept: $\frac{20}{3}$

25. $y = 0.75x + 5$

27. $y = 0.15x + 35$

29. $y = 2x + 3$

31. $y = \frac{2}{3}x - 2$

33. B

Section 2.4 Graphing Linear Equations in Standard Form

(pages 68 and 69)

1. no; The equation is in slope-intercept form.

3. x = pounds of peaches

y = pounds of apples

$y = -\frac{4}{3}x + 10$

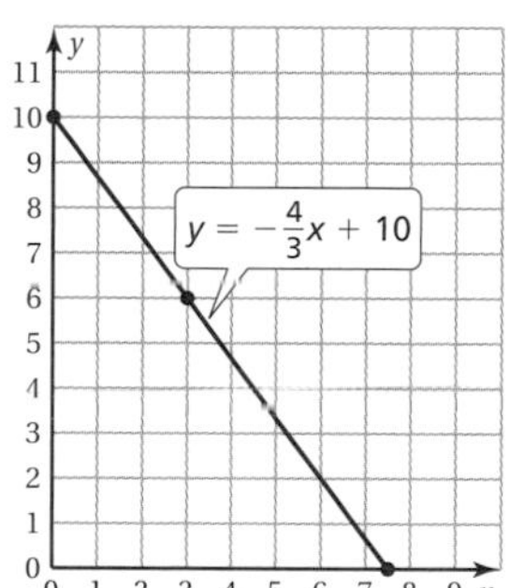

5. $y = -2x + 17$

7. $y = \frac{1}{2}x + 10$

9.

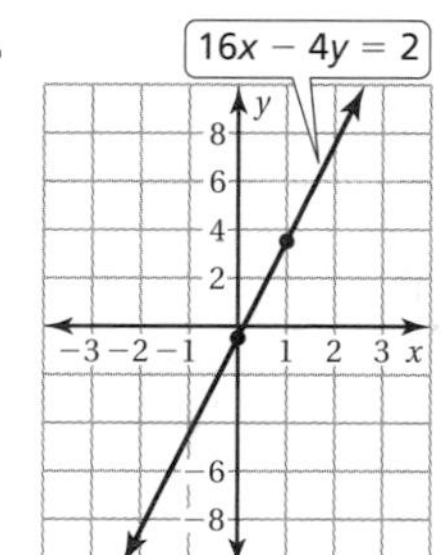

11. x-intercept: -6

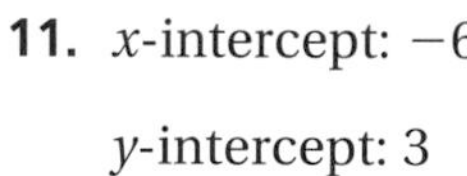

y-intercept: 3

13. x-intercept: none

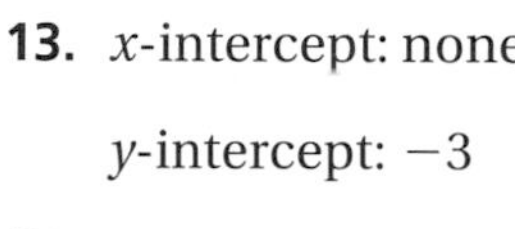

y-intercept: -3

15. a. $-25x + y = 65$

b. $390

17.

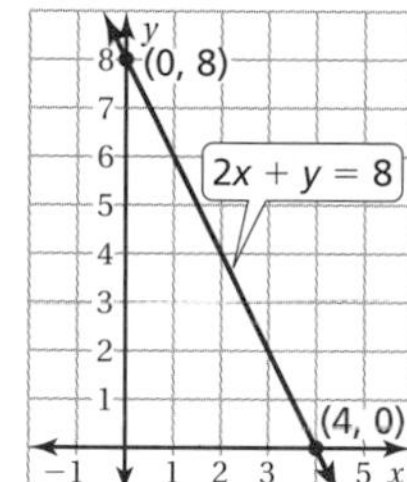

19. x-intercept: 9

y-intercept: 7

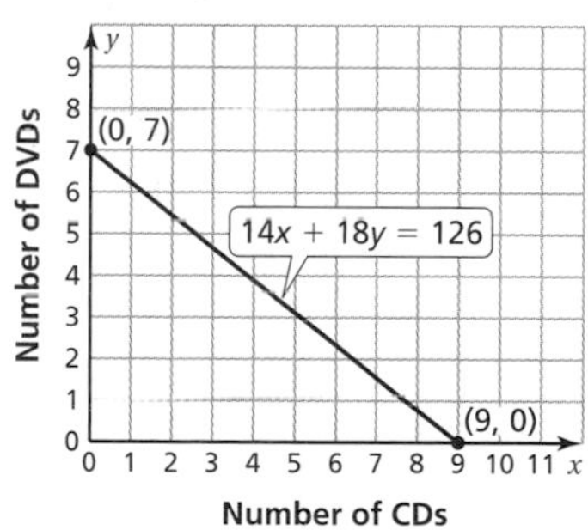

21. a. $9.45x + 7.65y = 160.65$

b.

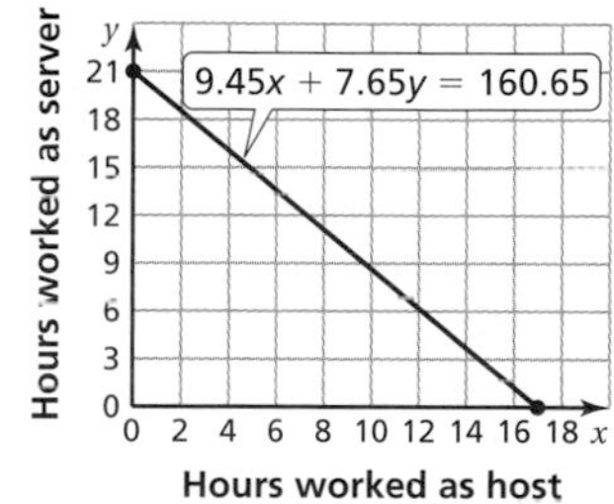

23. a. $y = 40x + 70$

b. x-intercept: $-\frac{7}{4}$; It will not be on the graph because you cannot have a negative time.

c.

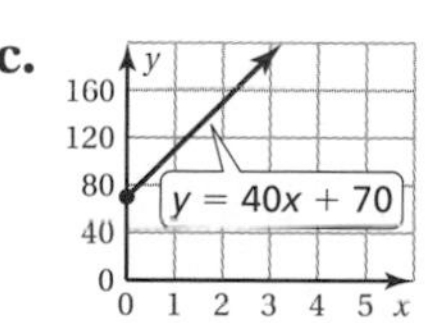

25.

x	-2	-1	0	1	2
$-5 - 3x$	1	-2	-5	-8	-11

Selected Answers

Section 2.5 Writing Equations in Slope-Intercept Form
(pages 76 and 77)

1. *Sample answer:* Find the ratio of the rise to the run between the intercepts.

3. $y = 3x + 2$; $y = 3x - 10$; $y = 5$; $y = -1$

5. $y = x + 4$

7. $y = \frac{1}{4}x + 1$

9. $y = \frac{1}{3}x - 3$

11. The x-intercept was used instead of the y-intercept. $y = \frac{1}{2}x - 2$

13. $y = 5$

15. $y = -2$

17. **a–b.**

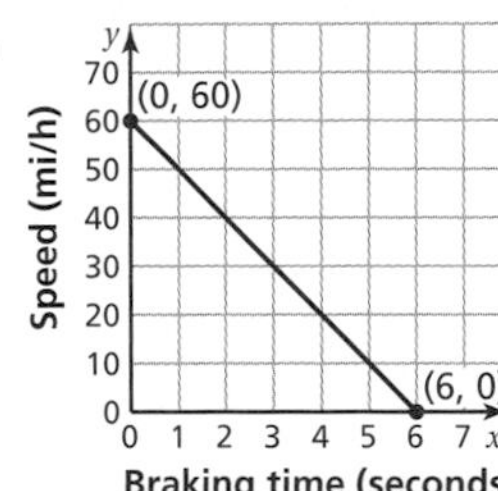

(0, 60) represents the speed of the automobile before braking. (6, 0) represents the amount of time it takes to stop. The line represents the speed y of the automobile after x seconds of braking.

c. $y = -10x + 60$

19. Be sure to check that your rate of growth will not lead to a 0-year-old tree with a negative height.

21–23.

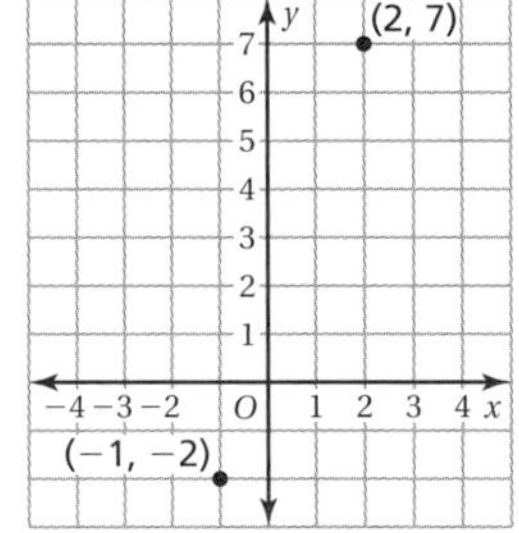

Section 2.6 Writing Equations in Point-Slope Form
(pages 82 and 83)

1. slope = −2; (−1, 3)

3. $y - 0 = \frac{1}{2}(x + 2)$

5. $y + 1 = -3(x - 3)$

7. $y - 8 = \frac{3}{4}(x - 4)$

9. $y + 5 = -\frac{1}{7}(x - 7)$

11. $y + 4 = -2(x + 1)$

13. $y = 2x$

15. $y = \frac{1}{4}x$

17. $y = x + 1$

19. **a.** $V = -4000x + 30{,}000$

    b. $30,000

21. The rate of change is 0.25 degree per chirp.

Hint

23. **a.** $y = 14x - 108.5$

    b. 4 meters

25. 175

27. D

Lesson 2.6b Writing Equations of Parallel and Perpendicular Lines (pages 84 and 85)

1. $y = 3x + 7$ **3.** $y = -4x - 21$ **5.** $y = 2x$ **7.** $y = 2x - 5$

9. $y = -x - 1$ **11.** $y = \frac{1}{5}x + \frac{19}{5}$ **13.** $y = \frac{1}{5}x - 1$

15. **Example 1**

$$y - y_1 = m(x - x_1)$$
$$y + 2 = \frac{1}{2}(x - 6)$$
$$y + 2 = \frac{1}{2}x - 3$$
$$y = \frac{1}{2}x - 5$$

Example 2

$$y = mx + b$$
$$1 = \frac{1}{4}(-3) + b$$
$$1 = -\frac{3}{4} + b$$
$$\frac{7}{4} = b$$

So, $y = \frac{1}{4}x + \frac{7}{4}$.

Sample answer: Point-slope form; The point-slope form gives you the equation of the line directly.

Section 2.7 Solving Real-Life Problems (pages 90 and 91)

1. The y-intercept is -6 because the line crosses the y-axis at the point $(0, -6)$. The x-intercept is 2 because the line crosses the x-axis at the point $(2, 0)$. You can use these two points to find the slope.

$$\text{Slope} = \frac{\text{change in } y}{\text{change in } x} = \frac{6}{2} = 3$$

3. *Sample answer:* the rate at which something is happening

5. *Sample answer:* The temperature outside is falling 3°F every hour. After 7 hours, the temperature is 0°F.

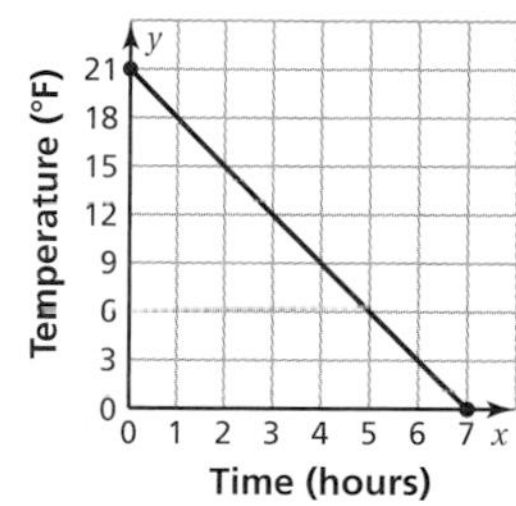

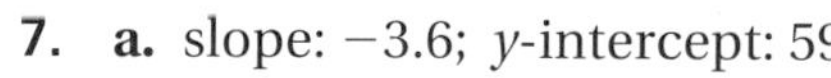

7. a. slope: -3.6; y-intercept: 59 **b.** $y = -3.6x + 59$

c. 59°F

9. a. Antananarivo: 19°S, 47°E; Denver: 39°N, 105°W; Brasilia: 16°S, 48°W; London: 51°N, 0°W; Beijing: 40°N, 116°E

b. $y = \frac{1}{221}x + \frac{8724}{221}$

c. a place that is on the prime meridian

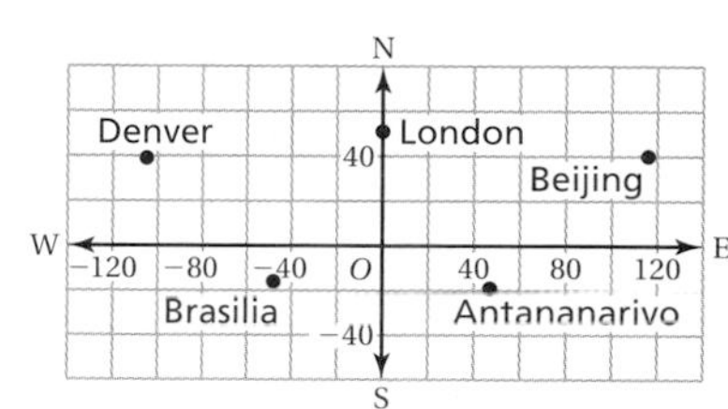

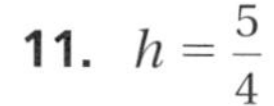

11. $h = \frac{5}{4}$ **13.** $q = -2.3$

Section 3.1 Writing and Graphing Inequalities *(pages 108 and 109)*

1. An open circle would be used because 250 is not a solution.

3. no; $x \geq -9$ is all values of x greater than or equal to -9. $-9 \geq x$ is all values of x less than or equal to -9.

5. $x < -3$; all values of x less than -3

7. $y + 5.2 < 23$

9. $k - 8.3 > 48$

11. yes

13. yes

15. no

17.

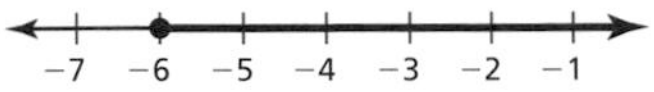

19.

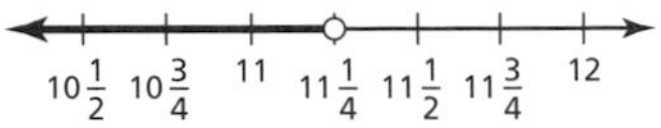

21. $x \geq 21$

23. yes

25. a. $a \geq 10$;

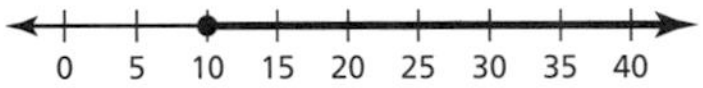

$s \geq 200$;

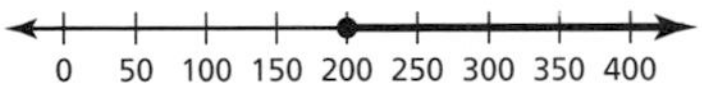

$t \geq 10$;

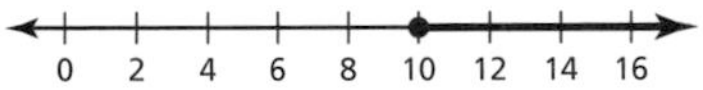

b. yes; You satisfy the swimming requirement of the course because $10(25) = 250$ and $250 \geq 200$.

27. a. $m < n$; $n \leq p$ **b.** $m < p$

c. no; Because n is no more than p and m is less than n, m cannot be equal to p.

29. -1.7

31. D

Section 3.2 Solving Inequalities Using Addition or Subtraction *(pages 114 and 115)*

1. no; The solution of $r - 5 \leq 8$ is $r \leq 13$ and the solution of $8 \leq r - 5$ is $r \geq 13$.

3. *Sample answer:* $A = 350$, $C = 275$, $Y = 3105$, $T = 50$, $N = 2$

5. *Sample answer:* $A = 400$, $C = 380$, $Y = 6510$, $T = 83$, $N = 0$

7. $t > 4$;

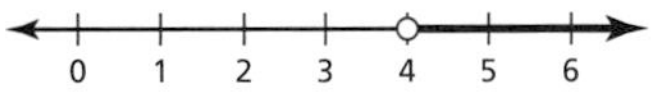

9. $a > -8$;

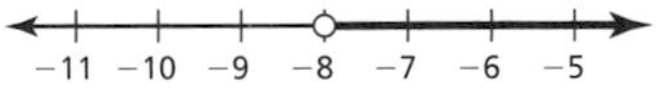

11. $-\frac{3}{5} > d$;

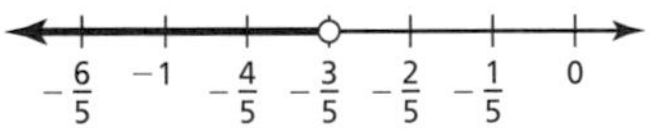

13. $m \leq 1$;

15. $h < -1.5$;

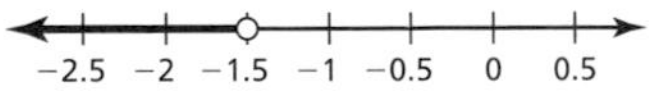

17. $9.5 \geq u$;

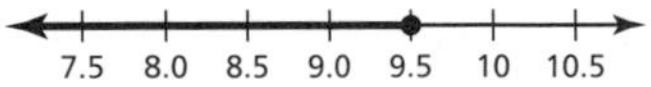

19. a. $100 + V \leq 700$; $V \leq 600$ in.^{3} **b.** $V \leq \frac{700}{3}$ in.^{3}

21. $x + 2 > 10$; $x > 8$

23. 5

25. a. $4500 + x \geq 12{,}000$; $x \geq 7500$ points

b. This changes the number added to x by 60%, so the inequality becomes $7200 + x \geq 12{,}000$. So, you need less points to advance to the next level.

27. $2\pi h + 2\pi \leq 15\pi$; $h \leq 6.5$ mm

29. 10

31. 12

33. 0.5

35. $2\sqrt{3}$

Section 3.3 Solving Inequalities Using Multiplication or Division *(pages 121–123)*

1. Multiply each side of the inequality by 6.

3. *Sample answer:* $-3x < 6$

5. $x \geq -1$

7. $x \leq -3$

9. $x < \frac{3}{2}$

11. $c \leq -36$;

13. $x < -28$;

15. $k > 2$;

17. $y \leq -4$;

19. The inequality sign should not have been reversed.

$$\frac{x}{2} < -5$$

$$2 \cdot \frac{x}{2} < 2 \cdot (-5)$$

$$x < -10$$

21. $\frac{x}{8} < -2$; $x < -16$

23. $5x > 20$; $x > 4$

25. $0.25x \leq 3.65$; $x \leq 14.6$; You can make at most 14 copies.

27. $n \geq -5$;

29. $h \leq -42$;

31. $y > \frac{11}{2}$;

33. $m > -12$;

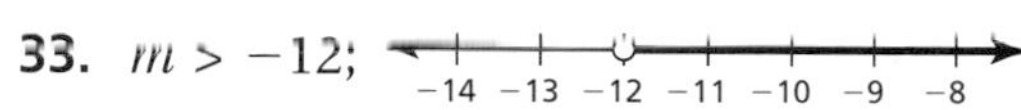

35. $b > 4$;

37. no; You need to solve the inequality for x. The solution is $x < 0$. Therefore, numbers greater than 0 are not solutions.

39. $12x \geq 102$; $x \geq 8.5$ cm

41. $\frac{x}{4} < 80$; $x < \$320$

43. *Answer should include, but is not limited to:* Using the correct number of months that the CD has been out.

45. $n \geq -6$ and $n \leq -4$;

47. $m < 20$;

49. $8\frac{1}{4}$

51. 84

Section 3.4 Solving Multi-Step Inequalities
(pages 130 and 131)

1. *Sample answer:* They use the same techniques, but when solving an inequality, you must be careful to reverse the inequality symbol when you multiply or divide by a negative number.

3. $k > 0$ and $k \leq 16$ units

5. $b \geq 1$;
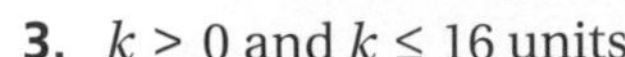

7. $m \geq -15$;

9. $p < -1$;

11. They did not perform the operations in proper order.

$$\frac{x}{4} + 6 \geq 3$$
$$\frac{x}{4} \geq -3$$
$$x \geq -12$$

13. all real numbers

15. $u < -17$

17. $z > -0.9$

19. no solutions

21. $20x + 100 \leq 320$; $x \leq 11$ \$20 bills

23. $b < 3$;

25. $500 - 20x \geq 100$; $x >$ \$0 and $x \leq$ \$20 per hour

27. **a.** $3.5x + 350 \geq 500$; $x \geq 42\frac{6}{7}$; at least 43 more cars, so at least 143 cars total

b. Because each car will pay \$1 more, fewer cars will be needed for the theater to earn \$500.

29.
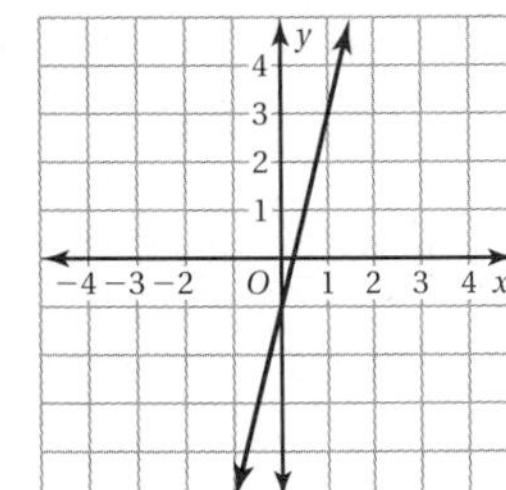

31.
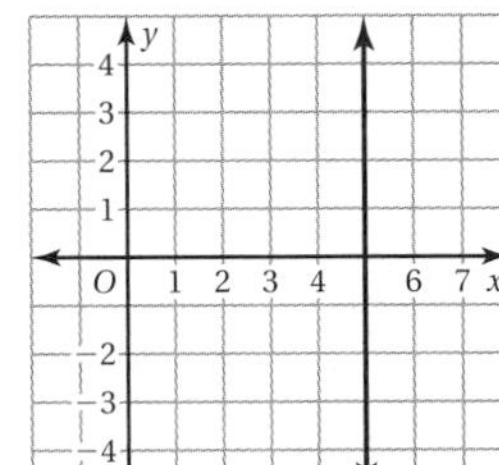

33. A

Lesson 3.4b Solving Compound Inequalities
(pages 132–135)

1. $3 < k < 9$;
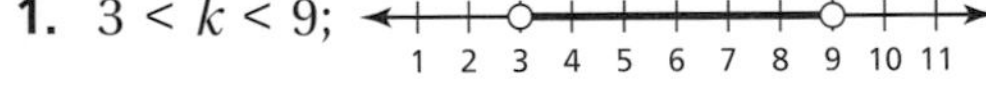

3. $w < -10$ or $w \geq -6$;

5. $-4 \leq x < -1$;
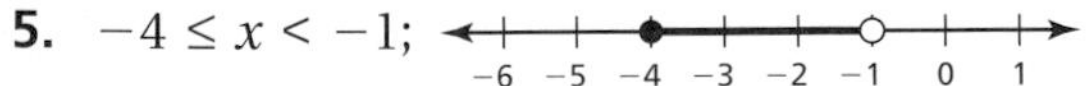

7. $9 < x < 12$;
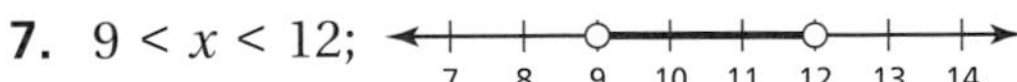

9. $-2 < x \leq 3$;

11. $x > -6$ or $x \leq -7$;
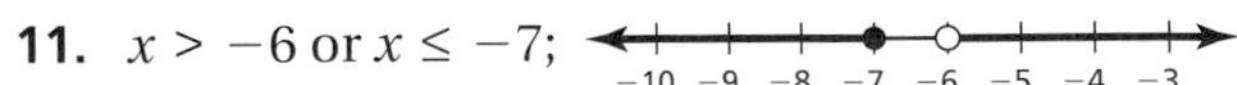

13. $x > -8$ *and* $x < -6$;

15. no solution

17. $x > 6$ *and* $x < 14$;

19. $|x - 44| \leq 3$; The least percent of voters who will vote for the new mayor is 41%. The greatest percent of voters who will vote for the new mayor is 47%.

Section 3.5 Graphing Linear Inequalities in Two Variables
(pages 141–143)

1. An ordered pair is a solution of an inequality if it makes the inequality true.

3. The graph of a linear equation in two variables will be a solid line. The graph of a linear inequality in two variables could be a solid or dashed line, and half of the coordinate plane will be shaded.

5. C

7. B

9. All the points on or above the line $y = -x + 5$.

11. yes

13. no

15. yes

17. no

19. yes

21. no

23. no

25. $2x + 3y \leq 60$; no, you cannot buy 12 yards of cotton lace and 15 yards of linen lace because (12, 15) is not a solution of the inequality.

27. The boundary line will be dashed.

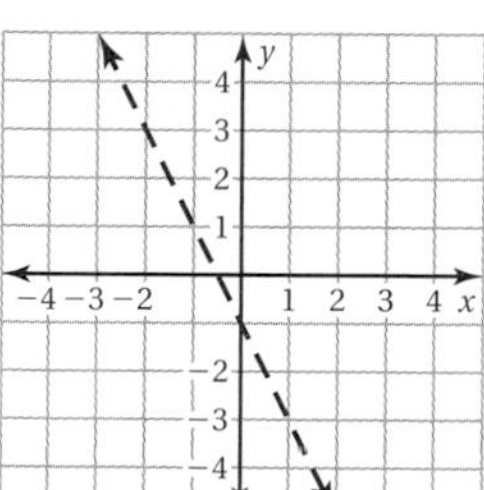

29. C

31. B

33.

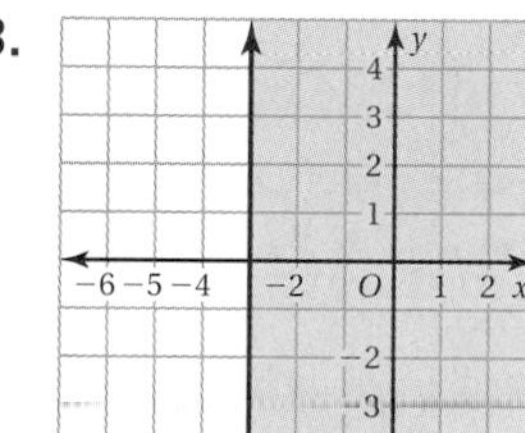

35.

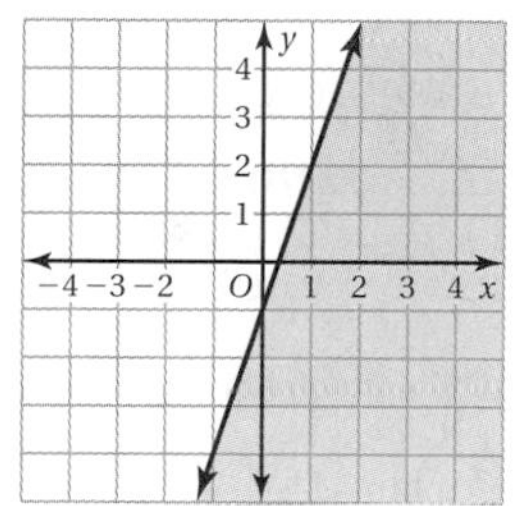

37.

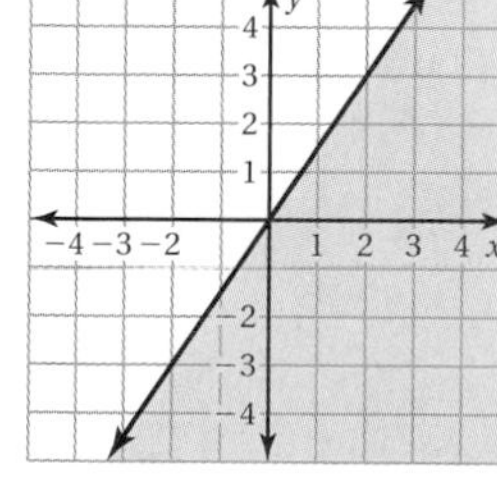

39.

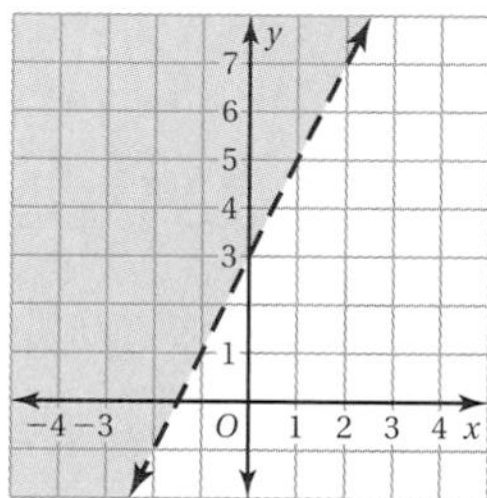

41. The boundary line should be solid instead of dashed.

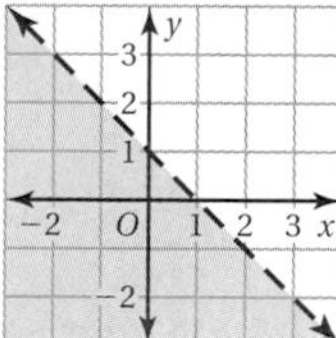

43. *Sample answer:* Choosing a test point on the boundary line will not help you determine which half-plane to shade. The test point must be in one of the half-planes.

45. yes

47. no

49. $y > 2x + 1$

51. $y \leq -\frac{1}{2}x - 2$

53. **a.** $0.75x + 2.25y \leq 9$

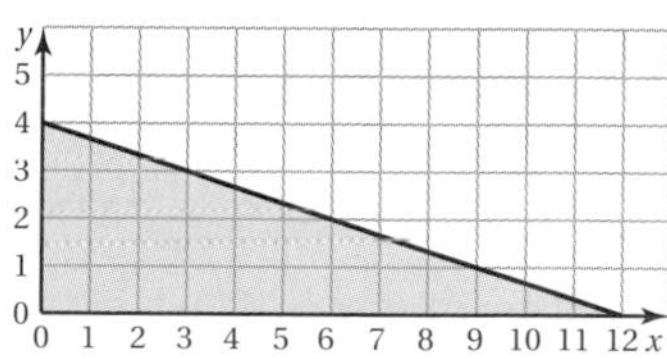

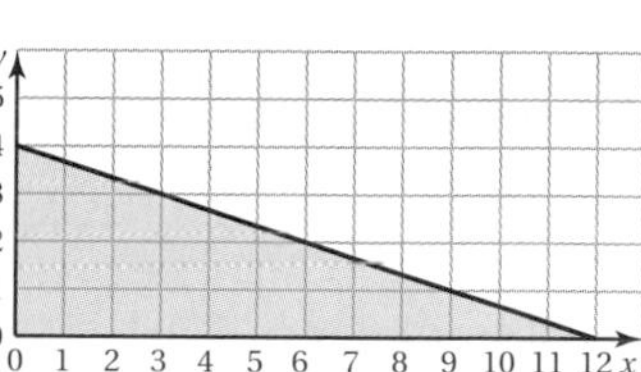

b. Two possible solutions are (1, 3) and (12, 0). So, you can play 1 arcade game and buy 3 soft drinks, or you can play 12 arcade games and buy 0 soft drinks.

55. 256

57. 243

Section 4.1 Solving Systems of Linear Equations by Graphing *(pages 158 and 159)*

1. yes; The equations are linear and in the same variables.

3. Check whether (3, 4) is a solution of each equation.

5. (4, 176)

7. B; (6, 7)

9. C; (3, −1)

11. (−5, 1)

13. (12, 15)

15. (8, 1)

17. (5, 1.5)

19. (−6, 2)

21. no; Two lines cannot intersect in exactly two points.

23. Make a table to compare your distance to your friend's distance.

25. $c = 8$

27. $x = 11$

Section 4.2 Solving Systems of Linear Equations by Substitution *(pages 164 and 165)*

1. **Step 1:** Solve one of the equations for one of the variables.

   Step 2: Substitute the expression from Step 1 into the other equation and solve.

   Step 3: Substitute the value from Step 2 into one of the original equations and solve.

3. sometimes; A solution obtained by graphing may not be exact.

5. *Sample answer:* $x + 2y = 6$
   $x - y = 3$

7. $4x - y = 3$; The coefficient of y is -1.

9. $2x + 10y = 14$; Dividing by 2 to solve for x yields integers.

11. (6, 17)

13. (4, 1)

15. $\left(\frac{1}{4}, 6\right)$

17. **a.** $x = 2y$
    $64x - 132y = 1040$

    b. adult tickets: \$8; student tickets: \$4

19. (−2, 4)

21. The expression for y was substituted back into the same equation; solution: (2, 1)

23. 30 cats, 35 dogs

25. Make a diagram to help visualize the problem.

27. $2x - 5y = -8$

29. B

Section 4.3 Solving Systems of Linear Equations by Elimination *(pages 173–175)*

1. **Step 1:** Multiply, if necessary, one or both equations by a constant so at least one pair of like terms has the same or opposite coefficients.

   Step 2: Add or subtract the equations to eliminate one of the variables.

   Step 3: Solve the resulting equation for the remaining variable.

   Step 4: Substitute the value from Step 3 into one of the original equations and solve.

3. $2x + 3y = 11$
$3x - 2y = 10$; You have to use multiplication to solve the system by elimination.

5. $(6, 2)$ 7. $(2, 1)$ 9. $(1, -3)$ 11. $(3, 2)$

13. The student added y-terms, but subtracted x-terms and constants; solution $(1, 2)$

15. a. $2x + y = 10$
$2x + 3y = 22$
    b. 6 minutes

17. $(5, -1)$ 19. $(-2, -1)$ 21. $(4, 3)$

23. a. ± 4
    b. ± 7

25. a. $23x + 10y = 86$
$28x + 5y = 76$
    b. Multiple choice: 2 points each
Short response: 4 points each

27. \$95

29. 5 grams of 90% gold alloy, 3 grams of 50% gold alloy

31. $(-1, 2, 1)$ 33. yes 35. D

Section 4.4 Solving Special Systems of Linear Equations *(pages 180 and 181)*

1. The graph of a system with no solution is two parallel lines, and the graph of a system with infinitely many solutions is one line.

3. infinitely many solutions; all points on the line $y = 4x + \frac{1}{3}$

5. no solution; The lines have the same slope and different y-intercepts.

7. infinitely many solutions; The lines are identical.

9. $(-1, -2)$

11. infinitely many solutions; all points on the line $y = -\frac{1}{6}x + 5$

13. $(-2.4, -3.5)$

15. no; because they are running at the same speed and your pig had a head start

17. When the slopes are different, there is one solution. When the slopes are the same, there is no solution if the y-intercepts are different and infinitely many solutions if the y-intercepts are the same.

19. $y = 0.99x + 10$
$y = 0.99x$
no; Because you paid \$10 before buying the same number of songs at the same price, you spend \$10 more.

21. Try using the Guess-and-Test method to help you answer this question.

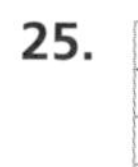

23. 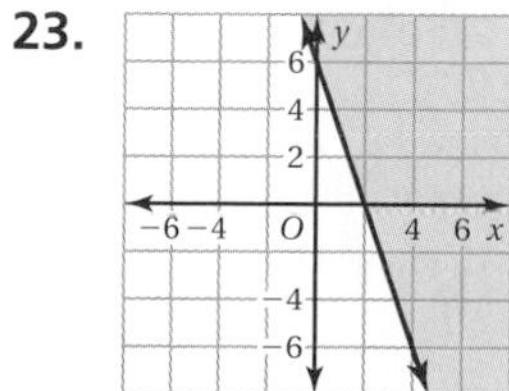

25. 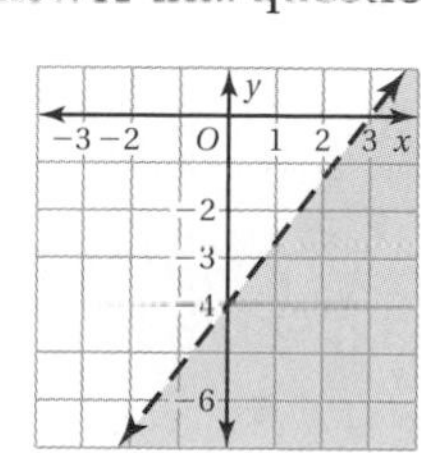

Lesson 4.4b Solving Linear Equations by Graphing

(pages 182 and 183)

1. $x = \frac{1}{2}$

3. no solution

5. $x = 2$

7. *Sample answer:* $6x - 3 = 6x$; Subtract 3 from the right side.

9. $x = \frac{21}{2}$

11. 6 mo

Section 4.5 Systems of Linear Inequalities

(pages 189–191)

1. Substitute its coordinates for x and y in each inequality of the system and simplify each side. When both resulting inequalities are true, the ordered pair is a solution. Otherwise, it is not a solution.

3. no; The point is not part of the solution of the inequality bordered by the dashed line.

5. no

7. 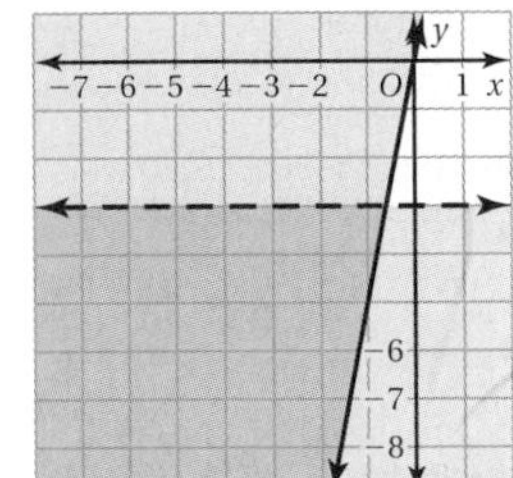

9. 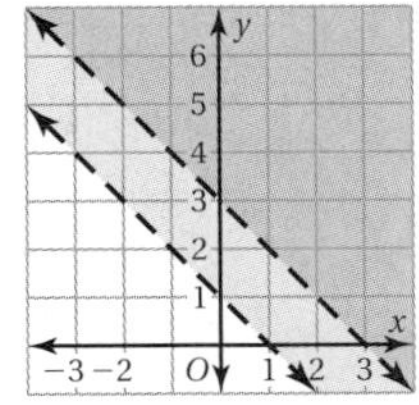

11. 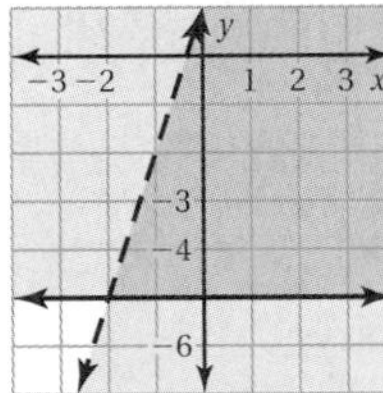

13. 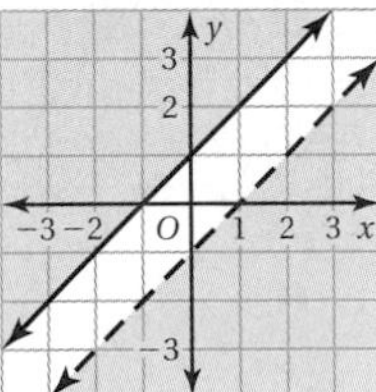

15. 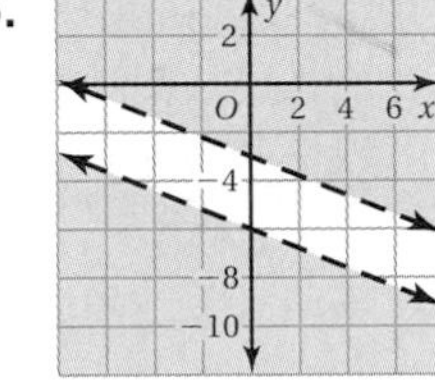

17. The solid line is incorrect for the graph of $y < -x - 2$; Change it to a dashed line.

19. B

21. C

23. a. $x + y \leq 20$

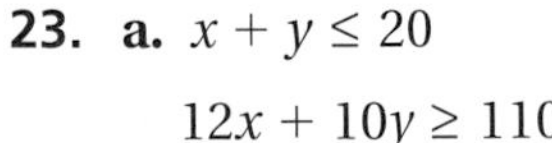

$12x + 10y \geq 110$

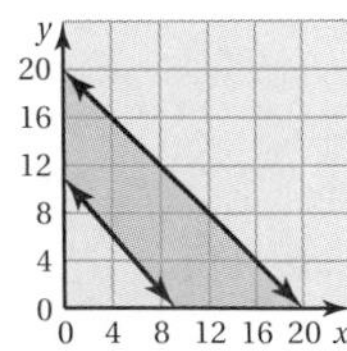

b. *Sample answer:* (10, 8); Work 10 hours at the grocery store and 8 hours as a coach.

25. $y > -2x - 1$

$y < -2x - 3$

27. 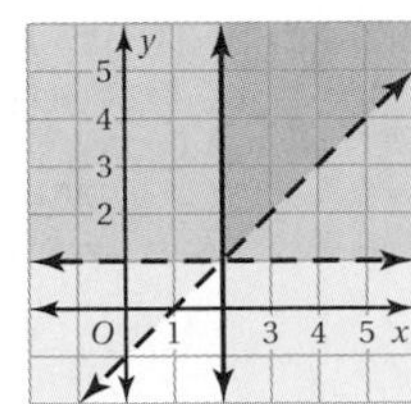

29. 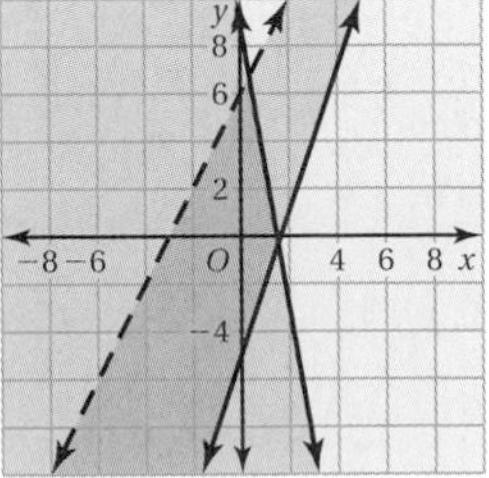

31. **a.** *Sample answer:* $y < 2x - 3$ **b.** *Sample answer:* $y > 2x - 3$

33. *Sample answer:* You drive 6 hours and your friend drives 8 hours for a total of 14 hours and a distance of 900 miles each day.

35. -5

37. 1

Section 5.1 Domain and Range of a Function (pages 206 and 207)

1. *Sample answer:* An independent variable represents an input value, and a dependent variable represents an output value.

3. **a.** $y = 6 - 2x$ **b.** domain: 0, 1, 2, 3; range: 6, 4, 2, 0

c. $x = 6$ is not in the domain because it would make y negative, and it is not possible to buy a negative number of headbands.

5. domain: $-2, -1, 0, 1, 2$; range: $-2, 0, 2$

7. The domain and range are switched. The domain is $-3, -1, 1$, and 3. The range is $-2, 0, 2$, and 4.

9.

x	−1	0	1	2
y	−4	2	8	14

domain: $-1, 0, 1, 2$

range: $-4, 2, 8, 14$

11.

x	−1	0	1	2
y	1.5	3	4.5	6

domain: $-1, 0, 1, 2$

range: 1.5, 3, 4.5, 6

13. Rewrite the percent as a fraction or decimal before writing an equation.

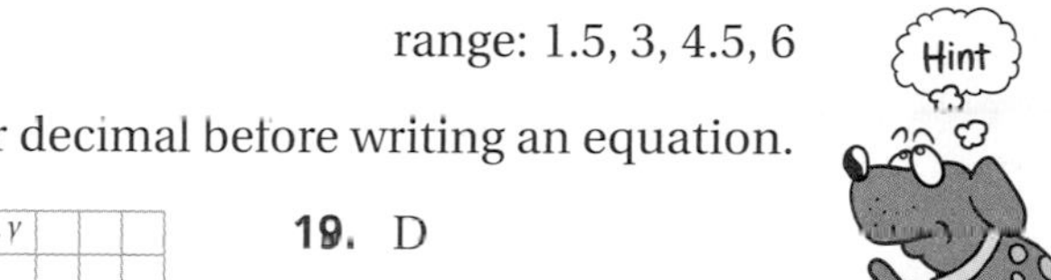

15.

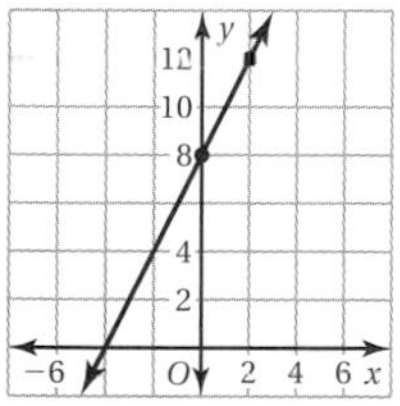

17.

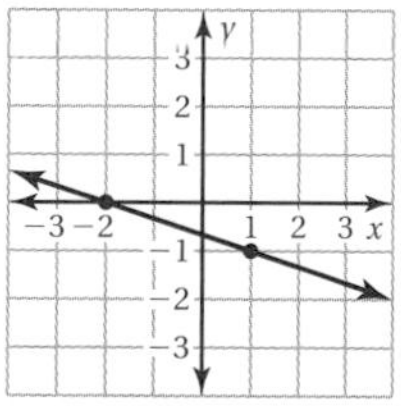

19. D

Lesson 5.1b Relations and Functions (pages 208 and 209)

1. not a function

3. not a function

5. false; A relation is not a function when an input has more than one output.

7. no; In any two games, if the winning teams had the same numbers of runs while the losing teams had different numbers of runs, then the relation is not a function.

9. function

11. no; All linear equations represent functions except for those representing vertical lines.

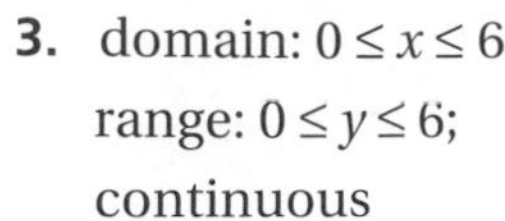

Discrete and Continuous Domains
(pages 214 and 215)

1. A discrete domain consists of only certain numbers in an interval, whereas a continuous domain consists of all numbers in an interval.

3. domain: $0 \le x \le 6$
range: $0 \le y \le 6$;
continuous

5. 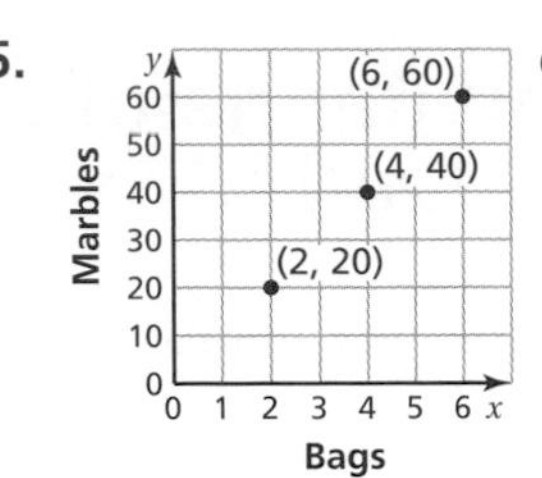

discrete

7. 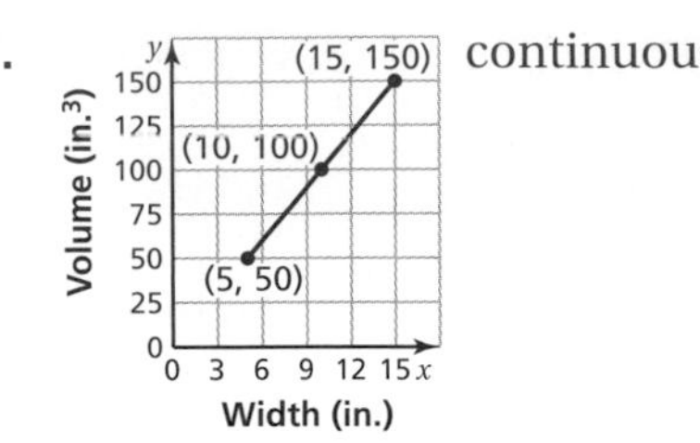

continuous

9. 2.5 is *not* in the domain, because the domain is discrete and consists only of the integers 1, 2, 3, and 4.

11. **a.** independent variable: c; dependent variable: t **b.** discrete

13. no; A height can be any positive number.

15. **a.** *Sample answer:* the elevation of a sinking ship relative to sea level

b. *Sample answer:* an overdraw leaves a negative checking account balance

17. $-\frac{5}{2}$ **19.** C

Section 5.3 Linear Function Patterns
(pages 220 and 221)

1. words, equation, table, graph

3. yes; All nonvertical lines intersect the y-axis.

5. $y = 2x$; x is the base of the triangle; y is the area of the triangle.

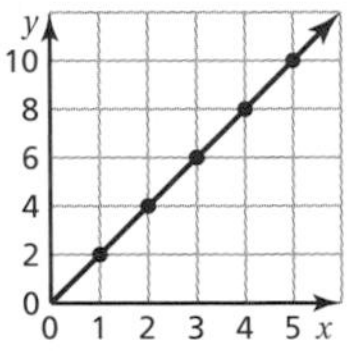

7. $y = -4x - 2$ **9.** $y = 2x$ **11.** $y = \frac{2}{3}x + 5$

13. **a.** Height (in.) vs. Weight (lb): (19, 10.2), (21, 9.8), (23, 9.4)

linear

b. $y = -0.2x + 14$

c. 9.7 in.

15. Substitute 8 for t in the equation.

17. -4; -2; 1

19. -1.25; -0.25; 1.25

Section 5.4 Function Notation
(pages 229–231)

1. Function notation assigns a name such as f to a function and $f(x)$ represents the value of the function at x. Example: $y = 3x + 1$ can be written as $f(x) = 3x + 1$.

3. a line; Changing b translates the graph vertically.

5. $g(-2) = -8$; $g(0) = -2$; $g(5) = 13$

7. $h(-2) = -5; h(0) = -7; h(5) = -12$

9. $f(-2) = 12; f(0) = 2; f(5) = -23$

11. $x = 1$

13. $x = -2$

15. $x = -6$

17. a. \$198 b. 25 hours

19. C

21. A

23. 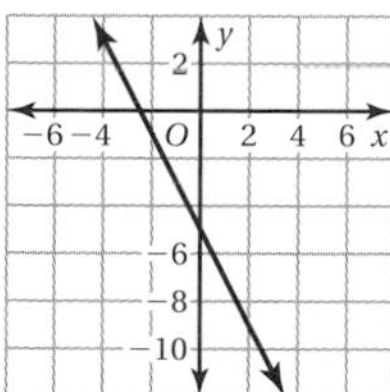

25. 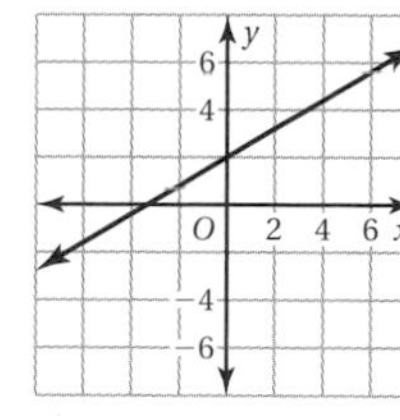

27. 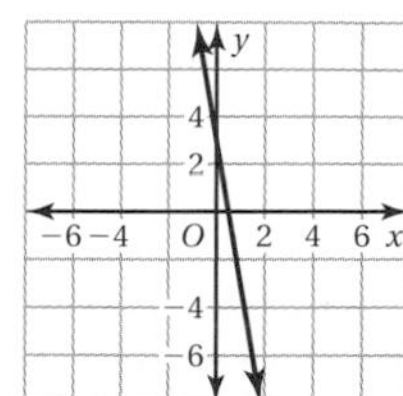

29. 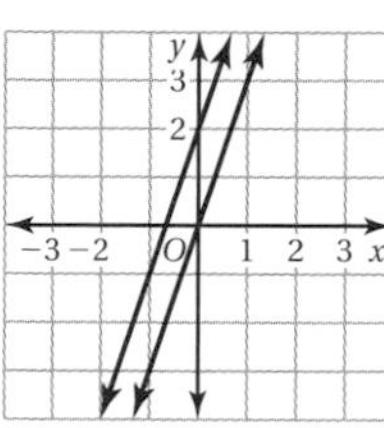

The graph of g is a translation 2 units up of the graph of f.

31. 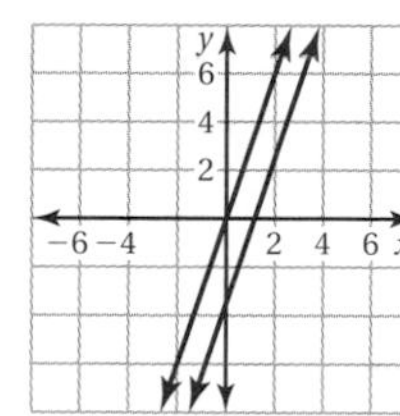

The graph of v is a translation $\frac{7}{2}$ units down of the graph of f.

33. 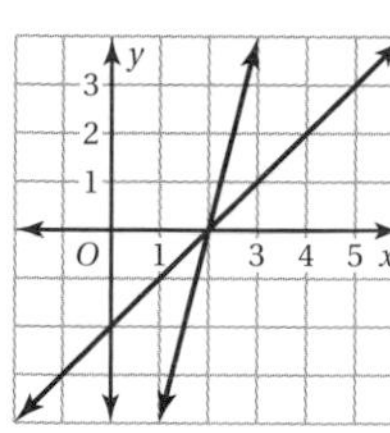

(2, 0)

35. 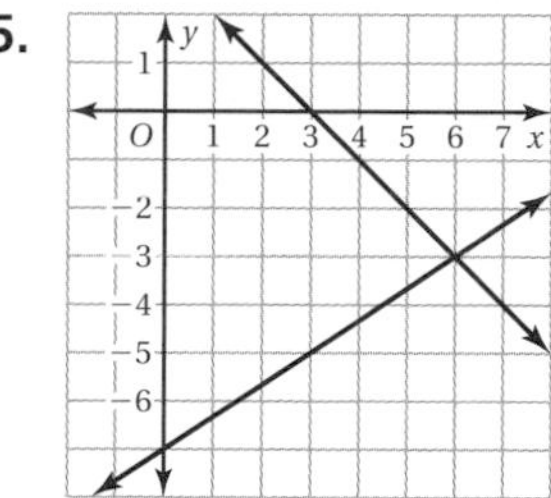

(6, −3)

37. $k = -1$

39. $k = -4$

41. Translate the graph of $y = x$ four units to the left.

43. $y = x$

45. $y = 1$

Lesson 5.4b Special Functions

(pages 232–235)

1. 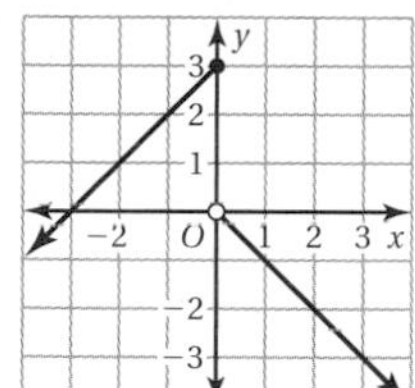

domain: all real numbers; range: $y \le 3$

3. 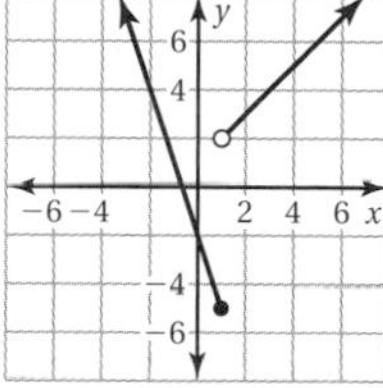

domain: all real numbers; range: $y \ge -5$

5. 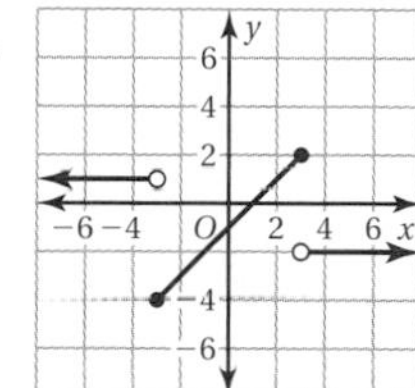

domain: all real numbers; range: $-4 \le y \le 2$

7. no; It fails the Vertical Line Test.

9. $f(x) = \begin{cases} -4, & \text{if } x < 0 \\ -x, & \text{if } x \ge 0 \end{cases}$

11. 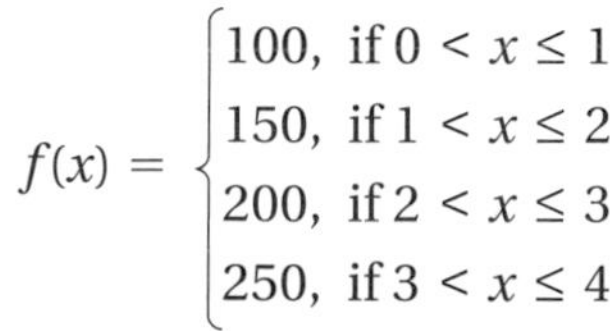

$f(x) = \begin{cases} 100, & \text{if } 0 < x \le 1 \\ 150, & \text{if } 1 < x \le 2 \\ 200, & \text{if } 2 < x \le 3 \\ 250, & \text{if } 3 < x \le 4 \end{cases}$

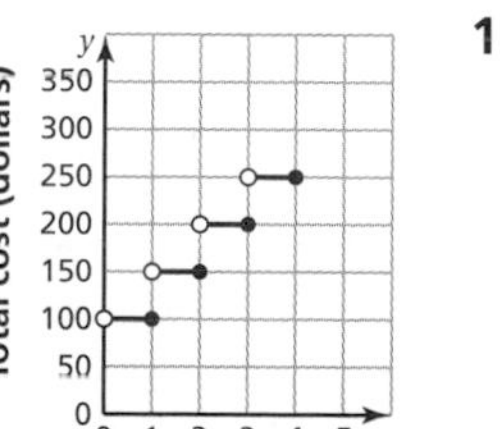

13. 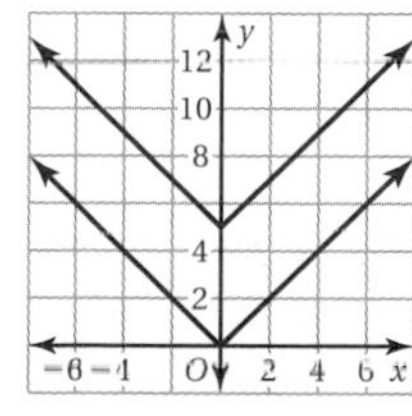

translation 5 units up; domain: all real numbers; range: $y \ge 5$

Selected Answers

Lesson 5.4b Special Functions *(continued)* *(pages 232–235)*

15. 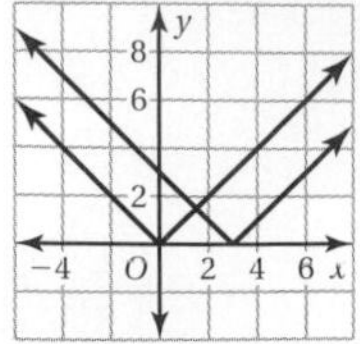

translation 3 units right; domain: all real numbers; range: $y \geq 0$

17. 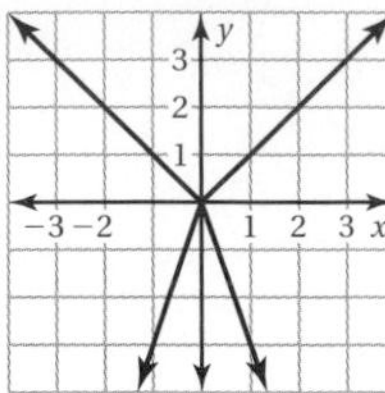

opens down and is narrower than $y = |x|$; domain: all real numbers; range: $y \leq 0$

19. 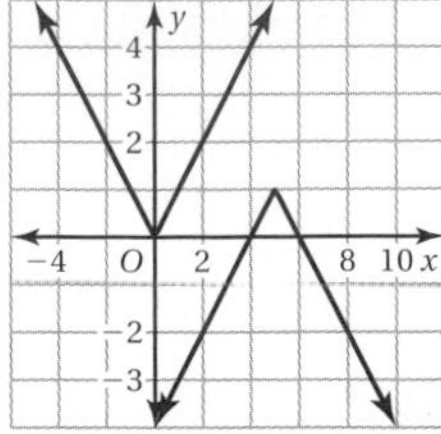

opens down and is a translation 5 units right and 1 unit up of the graph $y = |x|$; domain: all real numbers; range: $y \leq 1$

21. $y = |x| - 7$

23. $y = |x - 5| - 1$

25. **a.** positive k: translation up; negative k: translation down

b. positive h: translation to the right; negative h: translation to the left

c. positive a: Graph is narrower when $a > 1$, wider when $a < 1$; negative a: opens down, graph is narrower when $a < -1$, wider when $a > -1$.

27. $x = -7, 3$

29. *Sample answer:* $y = \begin{cases} -x - 4, & \text{if } x < -4 \\ x + 4, & \text{if } x \geq -4 \end{cases}$

Section 5.5 Comparing Linear and Nonlinear Functions *(pages 240 and 241)*

1. A linear function has a constant rate of change. A nonlinear function does not have a constant rate of change.

3. 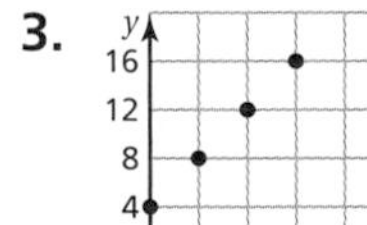 linear

5. 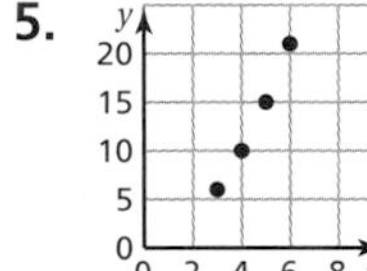 nonlinear

7. linear; The graph is a line.

9. linear; As x increases by 6, y increases by 4.

11. nonlinear; As x increases by 1, V increases by different amounts.

13. linear; You can rewrite the equation in slope-intercept form.

15. nonlinear; As x decreases by 65, y increases by different amounts.

17. **a.** linear; The equation that represents the function is $h = 1.6x$.

b. tree B; It is growing at a rate of 1.6 feet per year.

19. **a.** 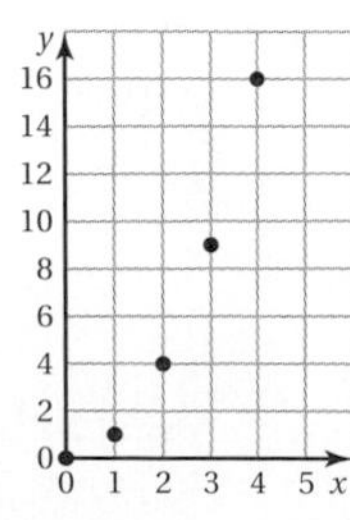 The points curve upward from left to right; nonlinear

b. $y = x^2$

21. -6

23. C

Section 5.6 Arithmetic Sequences
(pages 247–249)

1. subtract a term from the following term

3.

n	1	2	3	4
y	25	50	75	100

As n increases by 1, y increases by 25.

5. 21.5, 25, 28.5 **7.** 5 **9.** 25 **11.** −1.5

13. 22, 25, 28 **15.** 36, 41, 46 **17.** 0.1, −0.2, −0.5

19. 4, 6, 8, 10, . . . ; yes, There is a common difference of 2.

21. no **23.** yes; 6

25. 4.6, 4, 3.4

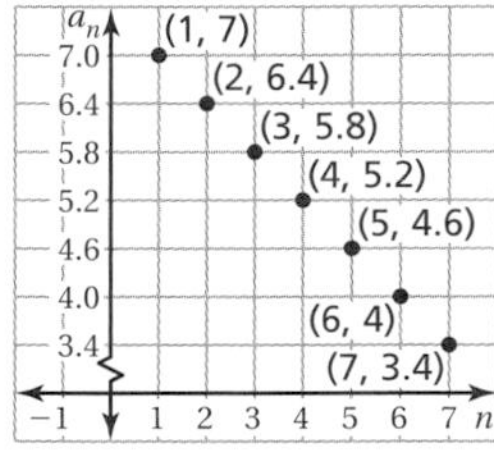

27. $1, 1\frac{1}{8}, 1\frac{1}{4}$

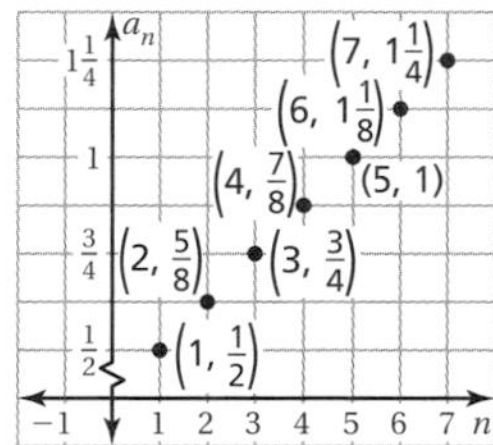

29. 7.5, 12, 16.5

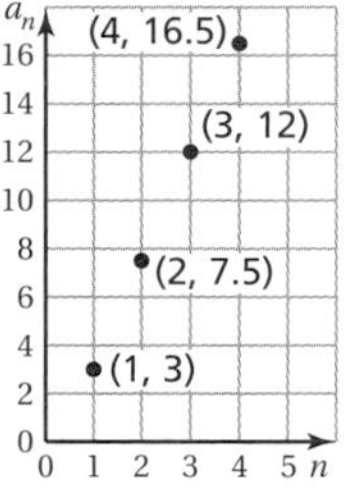

31. a.

Number of Visits in One Year	Cost
1	$8
2	$16
3	$24
4	$32

b. yes

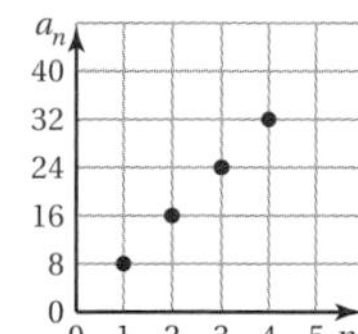

c. $48

d. 4 times; The cost for the family to visit the zoo 3 times is $120, which is less than the $130 pass. The cost for the family to visit the zoo 4 times is $160, which is more than the $130 pass.

33. $a_n = n - 6; a_{10} = 4$

35. $a_n = \frac{1}{2}n; a_{10} = 5$

37. $a_n = -10n; a_{10} = -100$

39. $a_n = -20n + 120$

41. Make a table to show your speed at each second.

43. discrete; A sequence of n terms has a domain of the integers 1 through n.

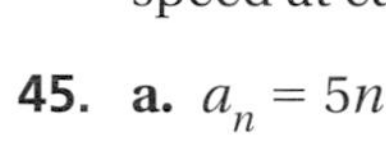

45. a. $a_n = 5n$

b. discrete

c. Substitute the number of minutes in a day, 1440, for n.

47. (4, −2) **49.** D

Section 6.1 Properties of Square Roots
(pages 264 and 265)

1. when the radicand has no perfect square factors other than 1

3. $s = 8$ ft
5. $s = \frac{3}{4}$ cm
7. $-10\sqrt{2}$
9. $4\sqrt{3}$
11. $-\frac{\sqrt{23}}{8}$
13. $\frac{3\sqrt{2}}{7}$
15. $2\sqrt{2}$
17. $2\sqrt{13}$
19. $\frac{7\sqrt{3}}{2} \approx 6.1$ amperes
21. $3 + \sqrt{11}$
23. $2 + 2\sqrt{3}$
25. $\frac{1 + \sqrt{7}}{2}$
27. $3\sqrt{10} \approx 9.5$ ft^3
29. $xy\sqrt{42}$
31. $3xy\sqrt{2xz}$
33. 243
35. 125

Section 6.2 Properties of Exponents
(pages 273–275)

1. D
3. B
5. Simplify 3^{6-3}; 3^3; 3^9
7. x^2
9. $64b^3$
11. $64a^4$
13. b^{11}
15. $\frac{1}{d^3}$
17. $\frac{1}{x^6}$
19. The exponents were added instead of multiplied; $(m^3)^4 = m^{3 \cdot 4} = m^{12}$
21. $38.44y^2$
23. $\frac{d^2}{36}$
25. $-3125x^5$
27. The exponent was not applied to the denominator.
$\left(\frac{x^3}{3}\right)^2 = \frac{(x^3)^2}{3^2} = \frac{x^6}{9}$
29. no; $(a^4)^2 = a^8$ by the Power of a Power Property.
$a^{4^2} = a^{16}$ because $4^2 = 16$.
31. $V = \frac{32\pi m^6}{3}$
33. 5.1×10^{-3}
35. 3.456×10^{-9}
37. 4×10^{-2}
39. $\frac{y^{12}}{216x^6}$
41. $\frac{5b^5c^{14}}{12}$
43. $14a^2b$ microns
45. 4.4 on the Richter Scale
47. $4\sqrt{3}$
49. $\frac{6\sqrt{5}}{11}$

Section 6.3 Radicals and Rational Exponents
(pages 280 and 281)

1. 3; Find the fourth root of 81.
3. $s = 4$ in.
5. $s = \frac{7}{8}$ ft
7. $4^{2/3}$
9. $\sqrt[3]{15}$
11. $(\sqrt[5]{78})^2$
13. 4
15. 4
17. 10
19. length: 3 ft, width: 2 ft
21. 25
23. 9
25. 2401

27. $a^{m/n}$; The mth power of the nth root of a positive number a can be written as a power with base a and an exponent of m/n.

29. always; $(x^{1/3})^3 = x^{3/3} = x^1 = x$

31. always; By definition, $x^{1/3} = \sqrt[3]{x}$.

33. always; By definition, $x^{(2/3)-(1/3)} = x^{1/3} = \sqrt[3]{x}$.

35.

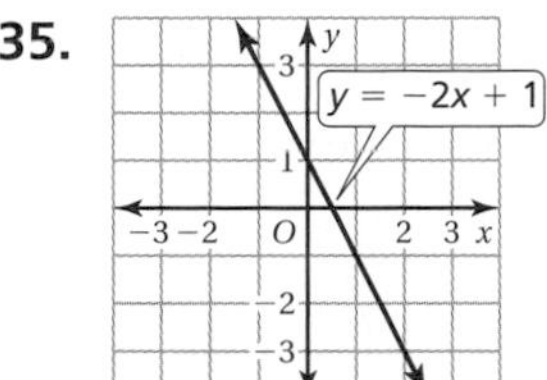

37.

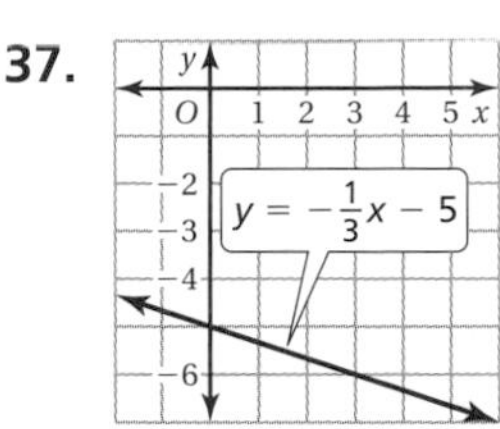

Section 6.4 Exponential Functions

(pages 289–291)

1. A linear function changes by a constant amount over equal intervals. An exponential function changes by equal factors over equal intervals.

3. $f(x) = (-3)^x$; It is not an exponential function.

5. $y = 2(2)^x$

7. exponential; As x increases by 1, y is multiplied by 2.

9. linear; As x increases by 3, y decreases by 9.

11. 1.5

13. 8

15. 2

17. 6.25%

19. A

21. $y = 9^x$

domain: all real numbers
range: all positive real numbers

23.

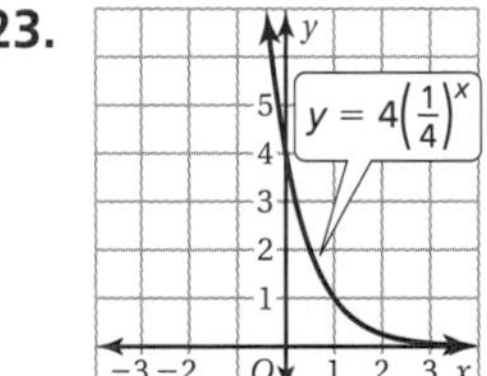

domain: all real numbers
range: all positive real numbers

25. When $b > 1$, the graph rises from left to right. When $0 < b < 1$, the graph falls from left to right.

27.

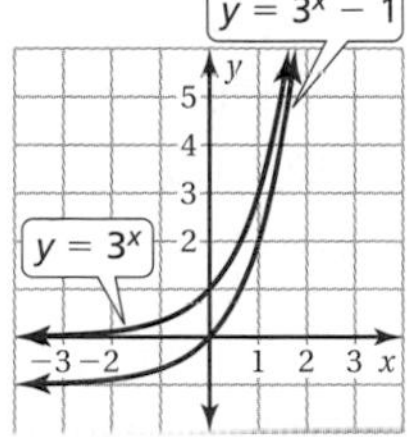

domain: all real numbers
range: all real numbers greater than -1
The graph is a translation 1 unit down of the graph of $y = 3^x$.

29.

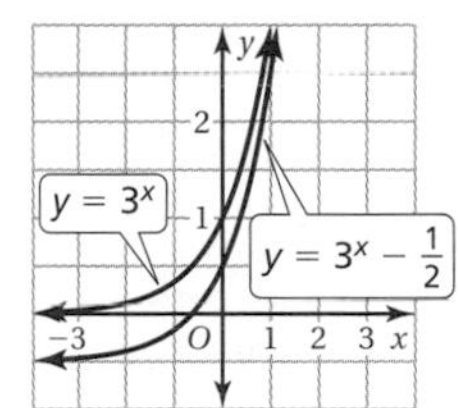

domain: all real numbers
range: all real numbers greater than $-\frac{1}{2}$
The graph is a translation $\frac{1}{2}$ unit down of the graph of $y = 3^x$.

Section 6.4 Exponential Functions *(continued)*

(pages 289–291)

31. An exponential function intersects the x-axis when it has the form $y = ab^x - c$, where a, b, and $c > 0$ or $y = ab^x + c$, where $a < 0$, $b > 0$, and $c > 0$; Example: $y = 2^x - 2$

33. $k = -1$

35. 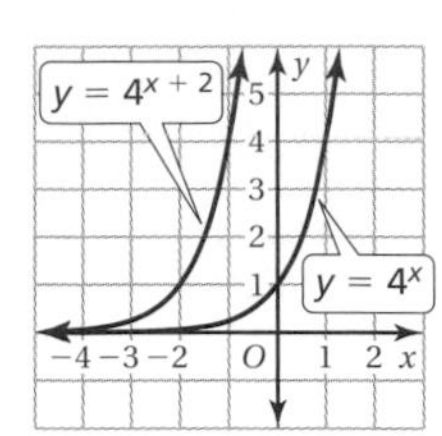

The graph is translated 2 units to the left.

37. $y = -8\left(\frac{1}{2}\right)^x$

39. $y = -3(5)^x$

41. You can begin by making a chart to show the amount of grills sold each year.

43. 0.23

45. 1.5

Lesson 6.4b Solving Exponential Equations

(pages 292 and 293)

1. $x = 4$ **3.** $x = -2$ **5.** $x = 0$ **7.** $x = \frac{7}{2}$ **9.** $x = 3$

11. *Sample answer:* Because 1 raised to any power is equal to 1, $1^x = 1^y$ is true for all real numbers x and y. The solution method fails in this case because it only gives the solution values for which $x = y$.

13. $x \approx 0.85$ **15.** $x \approx -10.00$ **17.** no solution

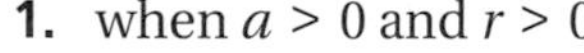

Section 6.5 Exponential Growth

(pages 298 and 299)

1. when $a > 0$ and $r > 0$

3. The terms are multiplied by a factor of about 1.2, so the pattern shows exponential growth.

5. $a = 25$, $r = 20\%$; 62.2

7. $a = 1500$, $r = 7.4\%$; 2143.4

9. $a = 6.7$, $r = 100\%$; 214.4

11. $y = 800(1.07)^t$

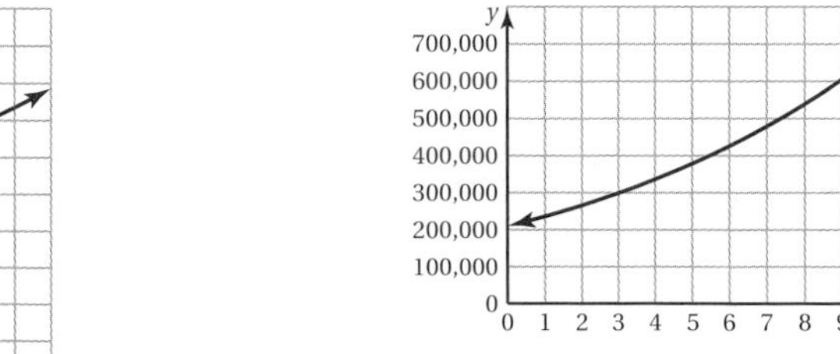

13. $y = 210{,}000(1.125)^t$

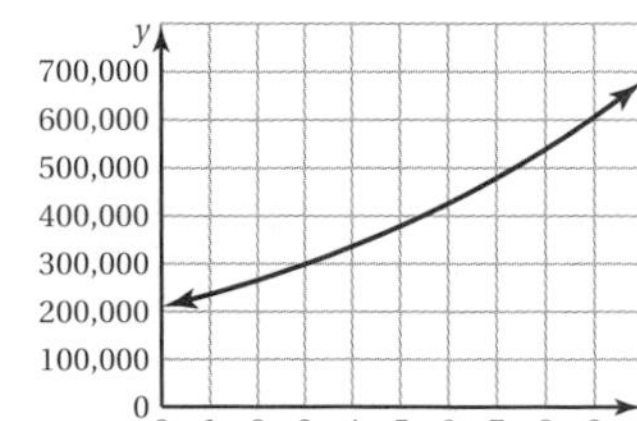

15. Because the growth rate is 150%, the growth factor should be 2.5, not 1.5.

$$b(t) = 10(1 + 1.5)^t = 10(2.5)^t$$

$$b(8) = 10(2.5)^8 \approx 15{,}258.8$$

After 8 hours, there are about 15,259 bacteria in the culture.

17. 468,000

19. $A = 6200(1.007)^{12t}$; \$7029.46

21. $C(t) = 9000(1.003)^{12t} + 480t$; $C(t)$ represents the total amount of money you have saved after t years.

23. yes; You can use the properties of exponents to transform the yearly balance function to the monthly balance function. You can use the compound interest formula with an annual interest rate of 2.4% compounded monthly to verify this.

25. $\frac{4}{9}$ **27.** $\frac{81}{625}$

Section 6.6 Exponential Decay
(pages 304 and 305)

1. When $b > 1$, the function represents exponential growth. When $0 < b < 1$, the function represents exponential decay.

3. The rate of change of the sequence is a constant -2, so it is a linear decay pattern.

5. exponential decay function **7.** exponential growth function **9.** neither

11. exponential decay function **13.** exponential growth function

15. 20% **17.** 25% **19.** A

21. *Sample answer:* Graph an exponential function by hand when the value of a is small enough to easily multiply by $1 - r$. Use a graphing calculator when the value of a is too large or when $1 - r$ has many decimal places which would make multiplying by hand more difficult.

23. The time of day is an important piece of information to consider.

25. $a_n = 3n + 6$; 51 **27.** $a_n = -4n - 3$; -63

Section 6.7 Geometric Sequences
(pages 310 and 311)

1. *Sample answer:* You add a number to extend arithmetic sequences and you multiply by a number to extend geometric sequences.

3. When the common ratio is negative, a_n will alternate between being positive and negative. These points do not lie on an exponential curve.

5. -4 **7.** 0.1 **9.** 9

11. 1250, 6250, 31,250 **13.** $1, -\frac{1}{3}, \frac{1}{9}$ **15.** $\frac{1}{36}, \frac{1}{216}, \frac{1}{1296}$

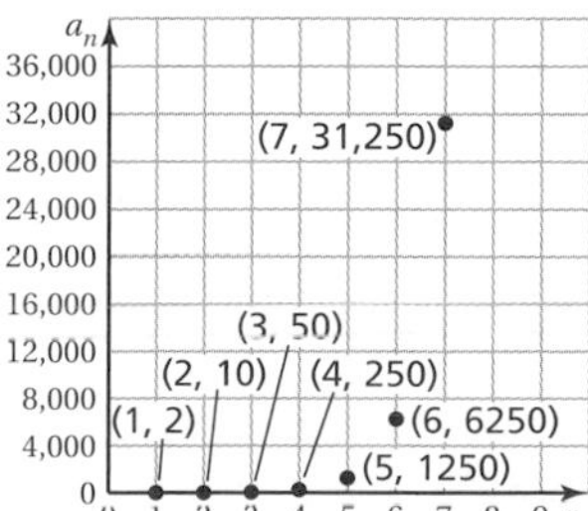

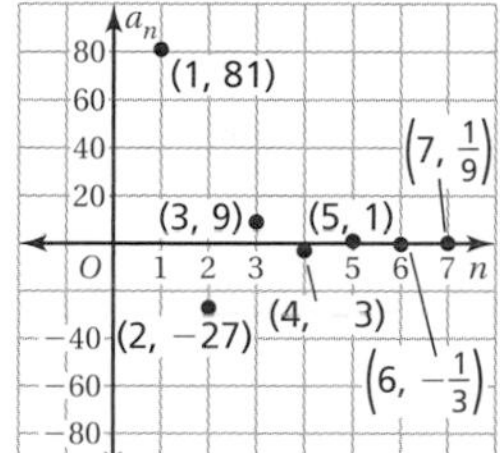

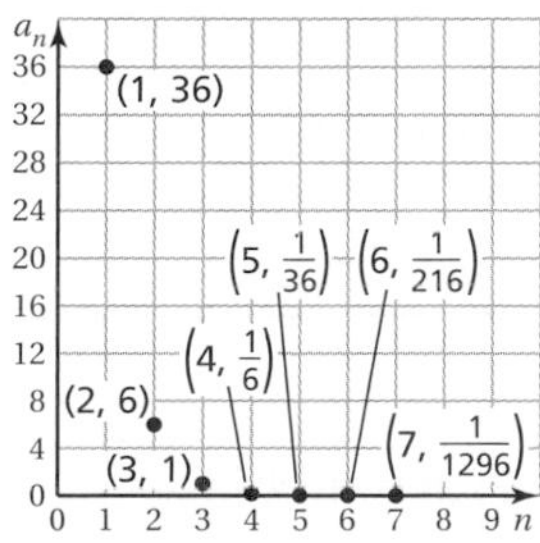

17. The ratio of the term to its previous term is $-\frac{1}{2}$, not 2. So, the next three terms are $-\frac{1}{2}$, $\frac{1}{4}$, and $-\frac{1}{8}$.

Section 6.7 Geometric Sequences *(continued)*

(pages 310 and 311)

19. arithmetic **21.** neither **23.** geometric

25. $a_n = (-5)^{n-1}$; 15,625 **27.** $a_n = 5(3)^{n-1}$; 3645

29. **a.** $a_n = (6)^{n-1}$ **b.** all positive integers; discrete

31. dependent; the terms are the output of the function; the position is the independent variable.

33. Remember that 1 gallon equals 128 fluid ounces.

35. $9x - 10$ **37.** C

Lesson 6.7b Recursively Defined Sequences

(pages 312–315)

1. 0, −8, −16, −24, −32, −40

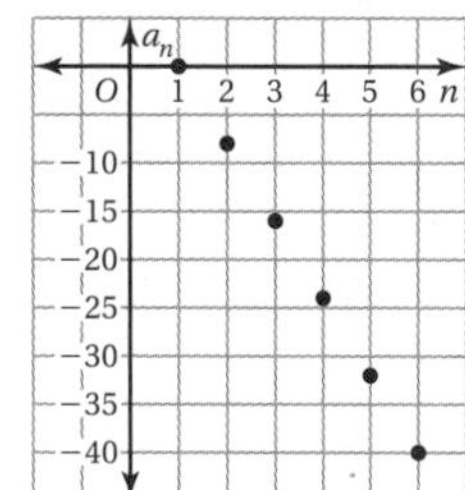

3. −36, −18, −9, −4.5, −2.25, −1.125

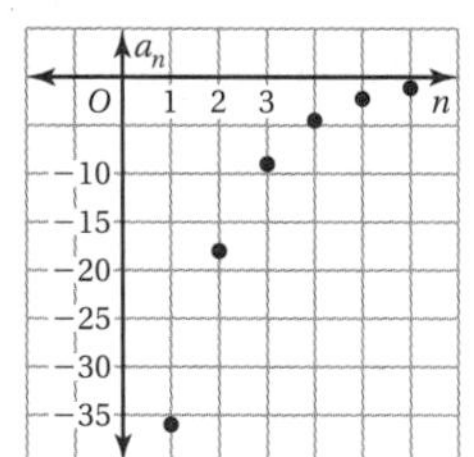

5. $a_1 = 8, a_n = a_{n-1} - 5$ **7.** $a_1 = 4, a_n = 5a_{n-1}$

9. $a_1 = 2, a_n = a_{n-1} + 1.5$ **11.** $a_n = 13(-3)^{n-1}$ **13.** $a_1 = -2.5, a_n = 2a_{n-1}$

15. $a_1 = -3, a_2 = -4,$ $a_n = a_{n-2} + a_{n-1}$; −29, −47, −76

17. $a_1 = 4, a_2 = 3,$ $a_n = a_{n-2} - a_{n-1}$; 7, −11, 18

19. $a_1 = -2, a_2 = 2.5,$ $a_n = a_{n-2} \cdot a_{n-1}$; −781.25, −48,828.125

Section 7.1 Polynomials

(pages 332 and 333)

1. yes; It is a number. **3.** *Sample answer:* $4x^5 - 2x^3 + 5$

5. [+] [+] [+] ⊖ ⊖ ⊖ ⊖

7. 1 **9.** 9 **11.** 2 **13.** 0

15. $3p^2 + 7$; 2; binomial **17.** $-4d^3 + 8d - 2$; 3; trinomial

19. $-v^{12} + 4v^{11}$; 12; binomial **21.** $7.4z^5$; 5; monomial

23. $-\frac{5}{7}r^8 + 2r^5 + \pi r^2$; 8; trinomial

25. It is the product of two numbers and a variable with a whole number exponent; 3

27. *not* a polynomial

29. polynomial; 5; trinomial

31. $-16t^2 - 45t + 200$; 139 ft

33. This problem can be organized with a table.

35. $3x - 2y$ 　 37. D

Section 7.2 Adding and Subtracting Polynomials
(pages 338 and 339)

1. Align like terms vertically and add; Group like terms and simplify.

3. $2x^2 + x + 1$ 　 5. $3y + 10$ 　 7. $n^2 - 8n + 5$ 　 9. $4a^3 + 3a - 3$

11. Like terms are not aligned.

$$\begin{array}{r} -5x^2 \qquad\quad + 1 \\ + \qquad\quad 2x - 8 \\ \hline -5x^2 + 2x - 7 \end{array}$$

13. $-7k + 14$ 　 15. $4r^2 - 2r - 17$

17. $\frac{1}{12}q^2 + 1$ 　 19. $3b + 2$

21. $-2x^2 + 3xy + 8y^2$ 　 23. $-3a^2 + 2ab + b^2$

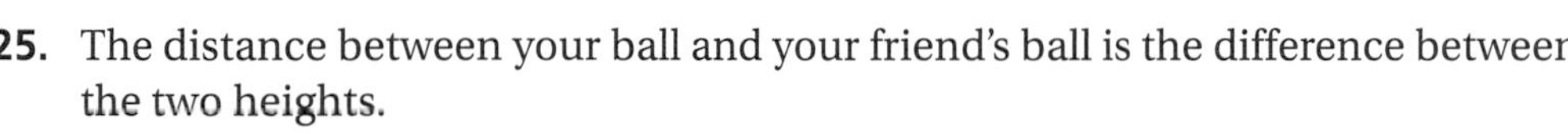

25. The distance between your ball and your friend's ball is the difference between the two heights.

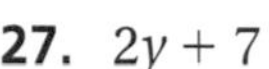

27. $2y + 7$ 　 29. B

Section 7.3 Multiplying Polynomials
(pages 345–347)

1. *Sample answer:* Use the Distributive Property or the FOIL Method.

3. $x^2 - 4$ 　 5. $x^2 + 4x + 3$ 　 7. $z^2 - 2z - 15$ 　 9. $g^2 - 9g + 14$

11. $3m^2 + 28m + 9$ 　 13. $12 - 16s + 5s^2$ 　 15. $(18x^2 + 12x + 2)$ in.^{2} 　 17. $y^2 + 5y - 50$

19. $8j^2 - 26j + 21$ 　 21. $15d^2 - 71d + 84$ 　 23. $w^2 + 15w + 54$ 　 25. $x^2 + 4x - 32$

27. $z^2 - 14z + 45$ 　 29. $10v^2 + 14v - 12$

31. The second term of the second binomial is -7, not 7.

$$\begin{aligned}(r + 6)(r - 7) &= r(r) + r(-7) + 6(r) + 6(-7) \\ &= r^2 - 7r + 6r - 42 \\ &= r^2 - r - 42\end{aligned}$$

33. **a.** $x^2 - 20x - 300$

b. 6000 yd^2

c. 90 minutes

35. $p^2 - 2p - 3$ 　 37. $2n^3 + 6n^2 + n + 3$

39. $2r^2 - 5rs - 3s^2$ 　 41. $f^3 + 3f^2 - 12f + 8$ 　 43. $t^3 - 8t^2 + 16t - 3$

Selected Answers

Section 7.3 Multiplying Polynomials *(continued)* *(pages 345–347)*

45. $18e^3 - 27e^2 + 37e + 7$

47. **a.** $x^2 + 41x + 40$ **b.** \$126

49. **a.** m is negative and n is positive; or m is positive and n is negative

b. m is positive and n is positive; or m is negative and n is negative

51. $4x^3 + 9x^2 - x - 6$

53. $z^2 - \frac{5}{7}z$; 2; binomial

55. B

Section 7.4 Special Products of Polynomials *(pages 352 and 353)*

1. *Sample answer:* $x + 7, x - 7$

3. $x^2 - 36$

5. $4z^2 + 8z + 4$

7. $g^2 - 25$

9. $b^2 - 144$

11. $9x^2 - 16$

13. $81 - c^2$

15. $x - 4$ and $x + 4$; Rewrite the expression as $x^2 - 4^2$ and apply the sum and difference pattern.

17. $y^2 + 16y + 64$

19. $d^2 - 20d + 100$

21. $25p^2 + 20p + 4$

23. $144 - 24x + x^2$

25. The product should have a middle term.

$$(k + 4)^2 = k^2 + 2(k)(4) + 4^2$$
$$= k^2 + 8k + 16$$

27. **a.** $x^2 + 100x + 2500$

b. 4225 ft^2; 1725 ft^2

29. $4x^2 + 28x + 49$

31. $x^2 - y^2$

33. **a.** 75%

b. $0.25N^2 + 0.5Na + 0.25a^2$

35. $(x + 1)^3 = x^3 + 3x^2 + 3x + 1$

$(x + 2)^3 = x^3 + 6x^2 + 12x + 8$

$(a + b)^3 = a^3 + 3a^2b + 3ab^2 + b^3$

37. $y^2 - 4y - 21$

39. D

Section 7.5 Solving Polynomial Equations in Factored Form *(pages 360 and 361)*

1. yes; When $x = 3$, the factor $(x - 3)$ equals 0.

3. $x^2 + 4x$ does not belong because it is not in factored form.

5. $t = 0, t = 5$

7. $q = -3, q = 2$

9. $m = -4$

11. $g = 3, g = 7$

13. $z = 3$

15. $d = 6, d = -6$

17. $x = -8, x = 8$

19. $x = -22, x = 15$

21. $z = 0, z = -2, z = 1$

23. $r = 4, r = -4, r = -8$

Hint

25. To find the width of the arch at ground level, find the distance between the x-intercepts.

27. 21

29. 15

Section 7.6 Factoring Polynomials Using the GCF
(pages 366 and 367)

1. $6y$ **3.** $4(x + 2)$ **5.** $x(x - 4)$ **7.** $4m(2m + 1)$

9. $10x^2(2x + 3)$ **11.** $5(t^2 + 4t + 10)$ **13.** $100(1 + rt)$ **15.** $x = -\frac{3}{2}$

17. $m = 0, m = -2$ **19.** $r = 0, r = -7$ **21.** $k = 0, k = -\frac{13}{2}$

23. 3 should be factored from 15.

$$\begin{aligned} 3x^2 &= 15x \\ 3x^2 - 15x &= 0 \\ 3x(x - 5) &= 0 \\ 3x = 0 \ \ or \ \ x - 5 &= 0 \\ x = 0 \ \ or \ \ \ \ \ \ \ x &= 5 \end{aligned}$$

25. $b = 0, b = 5$ **27.** $s = 0, s = -6$

29. *Sample answer:* $6x^2 + 3\pi x$

31. Begin by finding the time t when the height $y = 0$.

33. $y^2 + 10y + 24$

35. $4k^2 - 4k - 3$

Section 7.7 Factoring $x^2 + bx + c$
(pages 373–375)

1. Because c is negative, p and q have different signs. Because b is positive, the factor with the greater absolute value is positive.

3. $(x + 1)(x + 7)$ **5.** $(n + 2)(n + 6)$ **7.** $(h + 2)(h + 9)$

9. The factors of 48 (4 and 12) do not add up to 14.
$t^2 + 14t + 48 = (t + 6)(t + 8)$

11. $(x - 4)(x - 5)$ **13.** $(k - 4)(k - 6)$ **15.** $(j - 6)(j - 7)$ **17.** $x = -4, x = -7$

19. year 1 and year 5 **21.** $(x - 1)(x + 4)$ **23.** $(n - 2)(n + 6)$ **25.** $(h - 3)(h + 9)$

27. $(m - 7)(m + 1)$ **29.** $(t - 8)(t + 2)$ **31.** $x = 2, x = -7$

33. All of the terms were not collected on one side of the equation before factoring.

$$\begin{aligned} x^2 - 2x - 15 &= 20 \\ x^2 - 2x - 35 &= 0 \\ (x - 7)(x + 5) &= 0 \\ x - 7 = 0 \ \ or \ \ x + 5 &= 0 \\ x = 7 \ \ or \ \ \ \ \ \ \ x &= -5 \end{aligned}$$

35. length: 11 ft, width: 4 ft

37. length: 20 ft, width: 6 ft

39. **a.** 6 **b.** length: 13 in., width: 8 in.

41. Remember, area of a rectangle is length times width.

43. $2(y - 9)$

45. $4z^2(2z + 7)$

Selected Answers

Section 7.8 Factoring $ax^2 + bx + c$ (pages 380 and 381)

1. First, factor out the GCF of the terms, if possible. Then, consider the possible factors of a and c, list the possible products, and multiply.

3. $(2x - 1)(x - 1)$
5. $(4x + 3)(x + 2)$
7. $8(v - 2)(v + 3)$
9. $6(y - 3)(y - 1)$
11. $7(d - 5)(d - 4)$
13. $(3h + 2)(h + 3)$
15. $(4m + 1)(2m + 7)$
17. $(2n + 1)(n - 3)$
19. $2(g - 2)(4g + 3)$
21. $(7d - 2)(2d + 1)$
23. $(4x + 1)$ ft by $(2x + 5)$ ft
25. $k = -2$ or $k = \frac{9}{2}$
27. $-(3w - 4)(w + 2)$
29. $-5(4n - 1)(2n - 3)$
31. Remember that "$+ tx$" does not mean that t is positive.

33. 4 ft
35. **a.** \$24 or \$36

    b. \$30; *Sample answer:* By using a table, you can see that the maximum daily revenue occurs when $x = 5$. When $x = 5$, the price of the bobblehead is \$30.
37. $(2x - y)(3x + 4y)$
39. $4x^2 - 49$
41. $9b^2 - 24b + 16$

Section 7.9 Factoring Special Products (pages 386 and 387)

1. Use the difference of two squares pattern to factor $x^2 - 16$. Find the product $(x + 4)(x - 4)$.
3. $k^2 + 25$ does not belong because it cannot be factored.
5. $(3 + r)(3 - r)$
7. $(9d + 8)(9d - 8)$
9. $(h + 6)^2$
11. $(w - 7)^2$
13. The wrong sign is used.

    $n^2 - 16n + 64 = (n - 8)^2$
15. $s = -10$
17. $x = -\frac{7}{2}, x = \frac{7}{2}$
19. $y = 6$
21. $d + 4$
23. $2m(m + 5)(m - 5)$
25. $5f(f - 2)^2$
27. $x = -2, x = 2$; Solve by factoring or using square roots.
29. $(2y + 1)^2$
31. $9(m + 2)^2$
33. $(w + 4)(w - 3)$
35. $(d - 10)(d + 6)$

Lesson 7.9b Factoring Polynomials Completely (pages 388 and 389)

1. $(n + 2)(n^2 + 5)$
3. $(y + 4)(2y^2 + 3)$
5. $(8v - 5)(v^2 + 6)$
7. $(x + 3)(x + y)$
9. $(4y + 3)(x + 5)$
11. $5z^2(z + 1)(z - 1)$
13. cannot be factored
15. $3n^2(n + 4)(n - 4)$
17. $x = -6, x = 0, x = 4$

Section 8.1 Graphing $y = ax^2$

(pages 407–409)

1. The vertex is the lowest or highest point on a parabola. The vertical line that divides the parabola into two symmetric parts is the axis of symmetry.

3. The graph of $y = \frac{1}{6}x^2$ is wider because when $0 < a < 1$, the graph of $y = ax^2$ is wider than the graph of $y = x^2$. When $a > 1$, the graph of $y = ax^2$ is narrower than the graph of $y = x^2$.

5. Both graphs open down and have a U-shape. The graph of $y = -0.4x^2$ is wider than the graph of $y = -4x^2$.

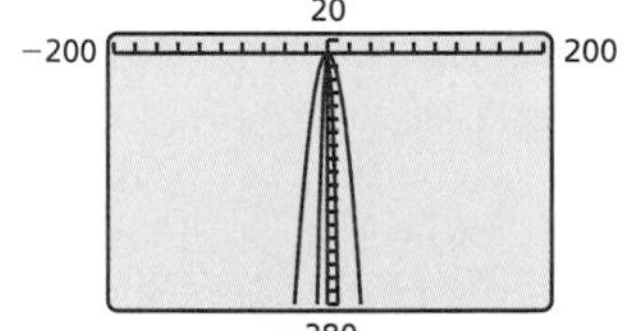

7. Both graphs open down and have a U-shape. The graph of $y = -0.004x^2$ is much wider than the graph of $y = -4x^2$.

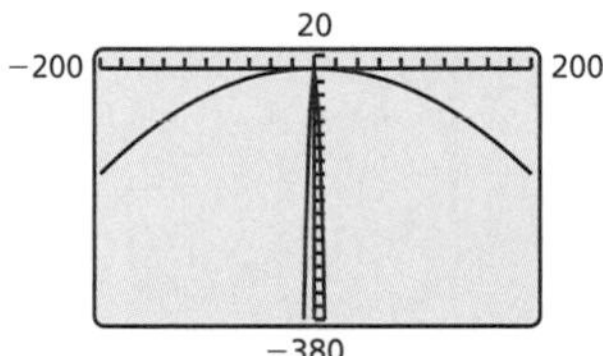

9. vertex: $(-2, 4)$; axis of symmetry: $x = -2$; domain: all real numbers; range: $y \geq 4$; When $x < -2$, y increases as x decreases. When $x > -2$, y increases as x increases.

11. Both graphs open up and have the same vertex, (0, 0), and the same axis of symmetry, $x = 0$. The graph of $y = 6x^2$ is narrower than the graph of $y = x^2$.

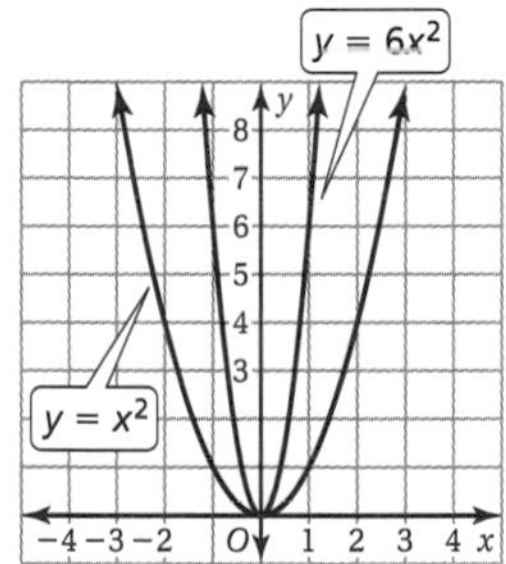

13. Both graphs open up and have the same vertex, (0, 0), and the same axis of symmetry, $x = 0$. The graph of $y = \frac{1}{4}x^2$ is wider than the graph of $y = x^2$.

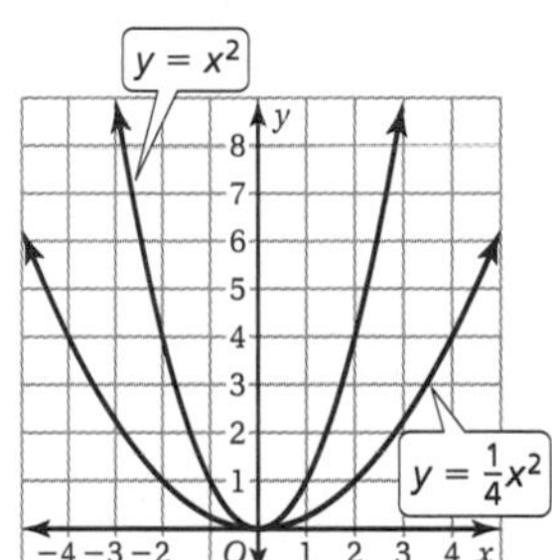

15. Both graphs open up and have the same vertex, (0, 0), and the same axis of symmetry, $x = 0$. The graph of $y = \frac{5}{2}x^2$ is narrower than the graph of $y = x^2$.

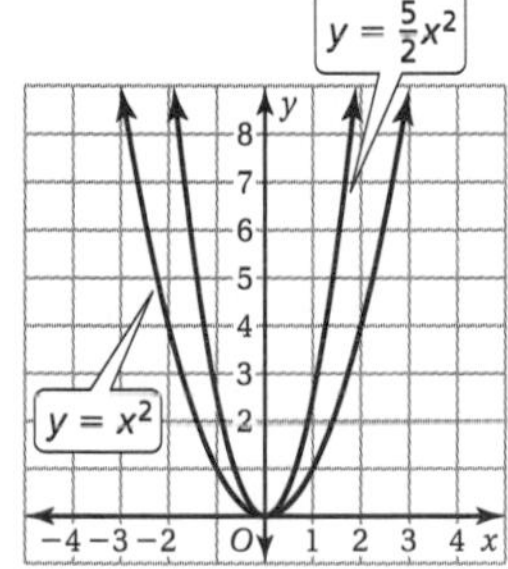

17. **a.** domain: $0 \leq t \leq 2$; range: $0 \leq y \leq 64$

b.

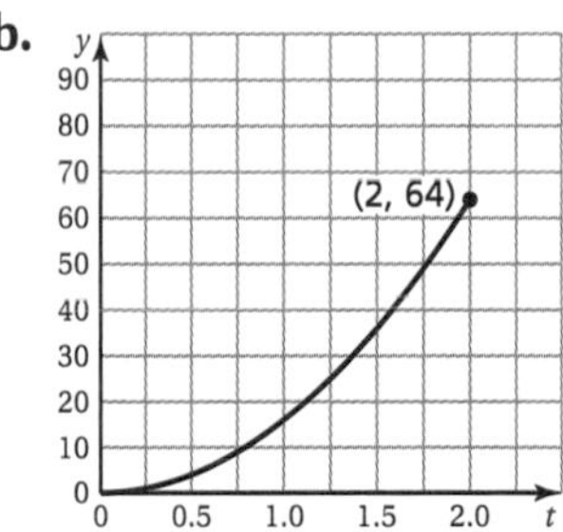

c. 2 seconds

Selected Answers

Section 8.1 Graphing $y = ax^2$ *(continued)* *(pages 407–409)*

19. Both graphs have the same vertex, (0, 0), and the same axis of symmetry, $x = 0$, but the graph of $y = -7x^2$ opens down. The graph of $y = -7x^2$ is narrower than the graph of $y = x^2$.

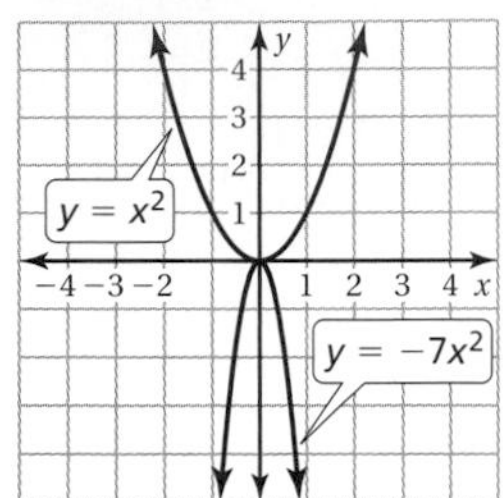

21. Both graphs have the same vertex, (0, 0), and the same axis of symmetry, $x = 0$, but the graph of $y = -\frac{5}{8}x^2$ opens down. The graph of $y = -\frac{5}{8}x^2$ is wider than the graph of $y = x^2$.

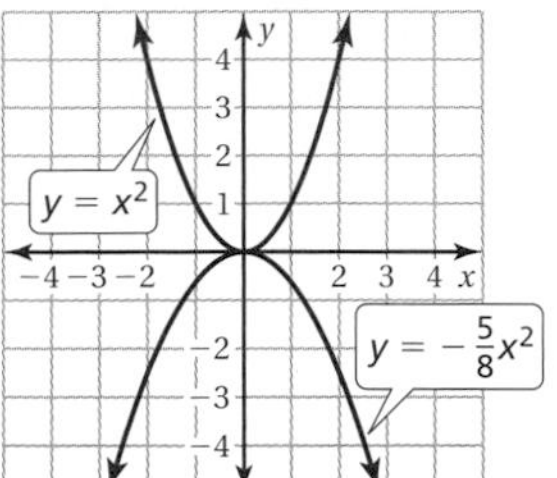

23. Both graphs have the same vertex, (0, 0), and the same axis of symmetry, $x = 0$, but the graph of $y = -\frac{9}{4}x^2$ opens down. The graph of $y = -\frac{9}{4}x^2$ is narrower than the graph of $y = x^2$.

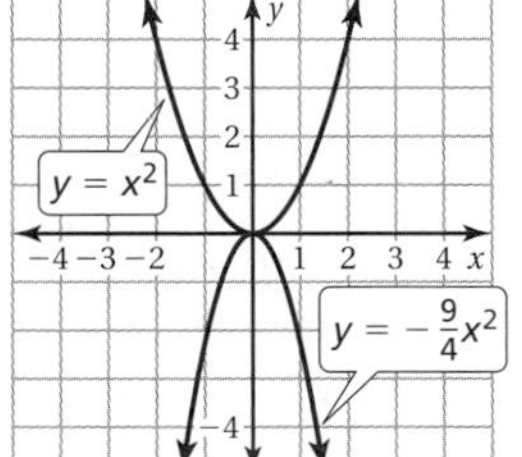

25. $a > 1$

27. $a < -1$

29. **a.** When $a > 0$, the domain is all real numbers and the range is $y \geq 0$.

b. When $a < 0$, the domain is all real numbers and the range is $y \leq 0$.

31. always; When $a > 1$ or $a < -1$, the graph of $y = ax^2$ is narrower than the graph of $y = x^2$.

33. never; As $|a|$ increases, the graph of $y = ax^2$ gets narrower.

35. When $a > 0$, the graph of $y = ax^2$ is increasing from left to right when $x > 0$. When $a < 0$, the graph of $y = ax^2$ is increasing from left to right when $x < 0$.

37. $\frac{3}{4}$

39. **a.** The domain includes the integers 0 through 10. It is discrete because you can only have a whole number of workers. It is positive because you cannot have a negative number of workers.

b.

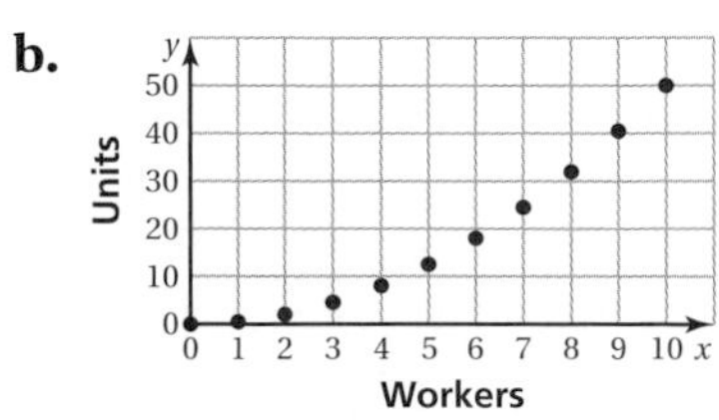

c. The company is better off running one assembly line at full capacity.

39. c. *(continued)*

Sample answer:

(1) **From a production standpoint:** Ten workers on one assembly line (full capacity) will produce 50 units per hour. Ten workers split on two assembly lines (half capacity) will produce a total of 25 units per hour. One assembly line at full capacity produces twice as many units as two assembly lines at half capacity.

(2) **From a financial standpoint:** The number of units produced is increasing at an increasing rate for each additional worker on an assembly line. This means production (i.e. revenue) is increasing faster than the labor cost when more employees work on the same assembly line. The company is getting more "bang for the buck."

41. 39 **43.** 14 **45.** B

Section 8.2 Focus of a Parabola
(pages 414 and 415)

1. The focus is a coordinate point which lies on the axis of symmetry. If the axis of symmetry is $x = k$, then the x-coordinate of the focus is k.

3. *Sample answer:* $y = -4x^2$

5. parabolic

7. focus: $\left(0, \frac{1}{4}\right)$

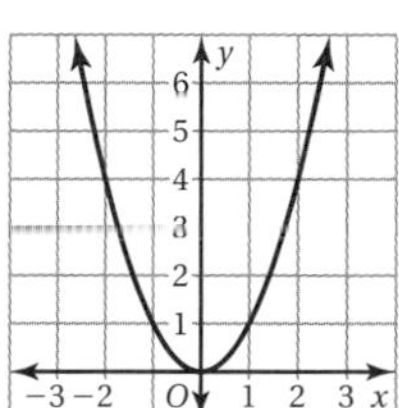

9. focus: $\left(0, -\frac{1}{48}\right)$

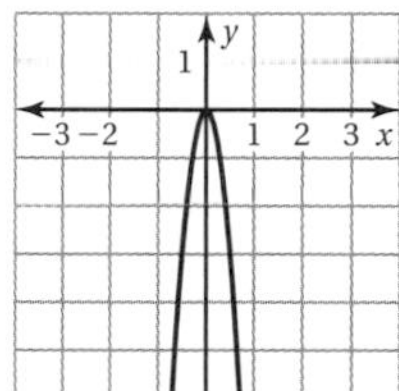

11. focus: $\left(0, \frac{1}{2}\right)$

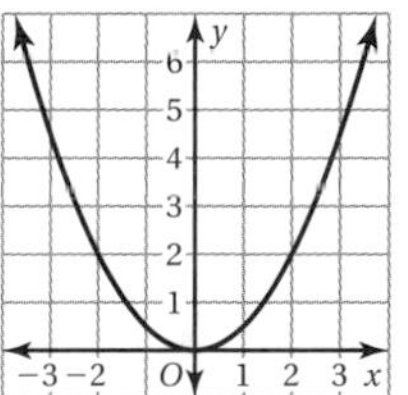

13. The student did not multiply 2 by 4. The focus is $\left(0, \frac{1}{8}\right)$.

15. $y = -\frac{1}{8}x^2$

17. $y = -x^2$

19. $y = \frac{1}{2}x^2$

21. B

23. C

25. always; The vertex of the graph of a function of the form $y = ax^2$ is always at the origin. So if the parabola opens down, then the focus will always lie below the x-axis.

27. As the distance between the vertex and focus increases, the graph of $y = ax^2$ gets wider.

29. 1.5 inches

31. 16

33. 1

Section 8.3 Graphing $y = ax^2 + c$
(pages 420 and 421)

1. The vertex is on the y-axis. The y-axis is the axis of symmetry. The value of c is the y-coordinate of the vertex.

3. It is a translation c units up or down of the graph of $y = ax^2$.

5. C

Section 8.3 Graphing $y = ax^2 + c$ (continued)
(pages 420 and 421)

7. Both graphs open up and have the same axis of symmetry, $x = 0$. The graph of $y = x^2 + 4$ is a translation 4 units up of the graph of $y = x^2$.

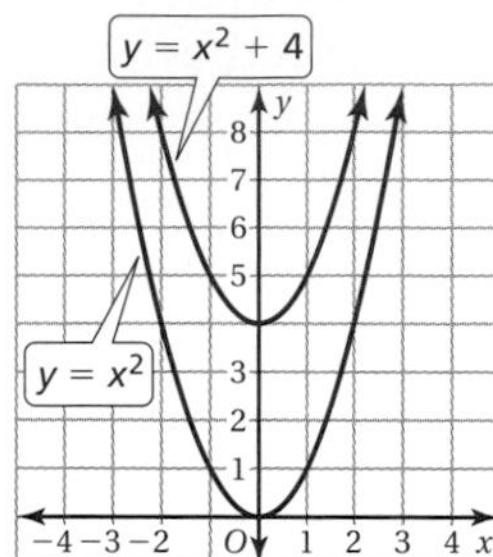

9. Both graphs have the same axis of symmetry, $x = 0$. The graph of $y = -x^2 + 5$ opens down. The vertex of the graph of $y = -x^2 + 5$ is a translation 5 units up of the vertex of the graph of $y = x^2$.

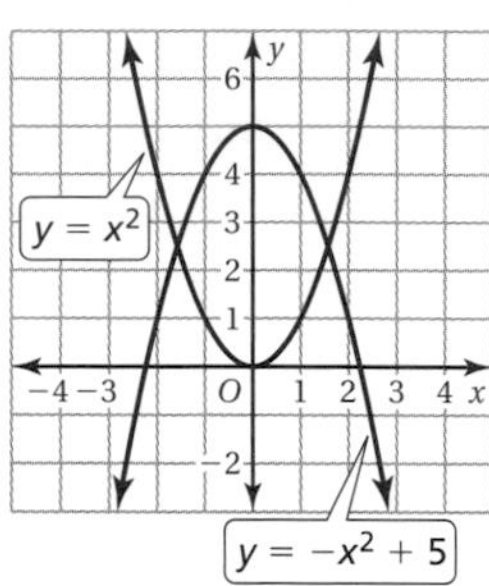

11. Both graphs open up and have the same axis of symmetry, $x = 0$. The graph of $y = 2x^2 - 4$ is narrower than the graph of $y = x^2$. The vertex of the graph of $y = 2x^2 - 4$ is a translation 4 units down of the vertex of the graph of $y = x^2$.

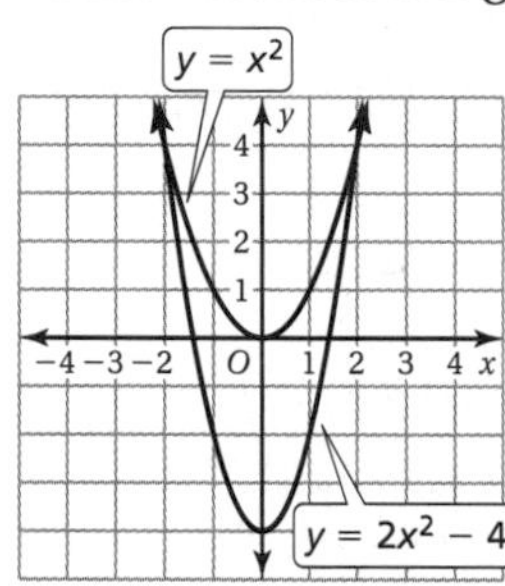

13. Both graphs open up and have the same axis of symmetry, $x = 0$. The graph of $y = 3x^2 + 4$ is narrower than the graph of $y = x^2$. The vertex of the graph of $y = 3x^2 + 4$ is a translation 4 units up of the vertex of the graph of $y = x^2$.

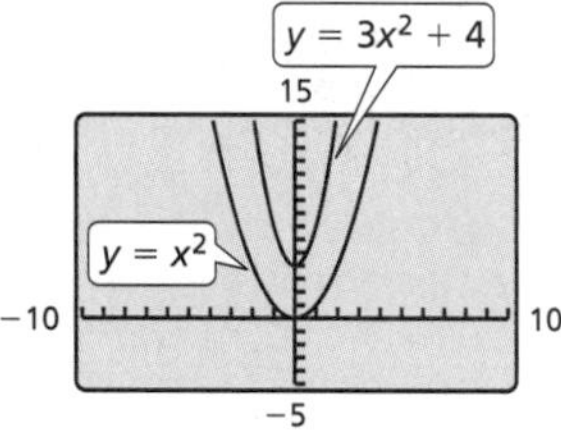

15. Both graph have the same axis of symmetry, $x = 0$. The graph of $y = -\frac{1}{4}x^2 - \frac{1}{2}$ opens down, and is wider than the graph of $y = x^2$. The vertex of the graph of $y = -\frac{1}{4}x^2 - \frac{1}{2}$ is a translation 0.5 unit down of the vertex of the graph of $y = x^2$.

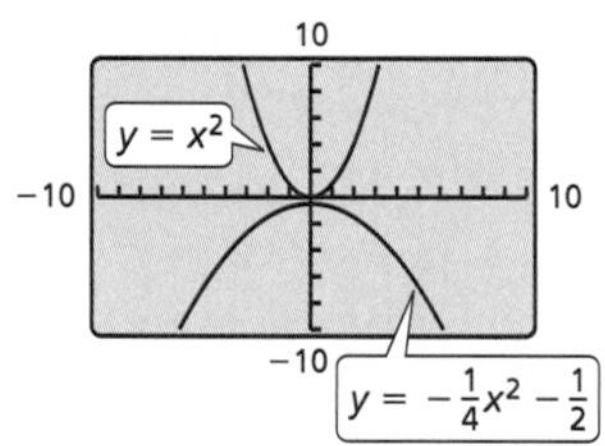

17. translate 3 units down

19. The student incorrectly states that the graph of $y = x^2 + 3$ is a translation 3 units down of the graph of $y = x^2$. The graph is actually a translation 3 units up.

21. after 1 second

23. $-1, 1$

25. $-3, 3$

27. *Sample answer:*

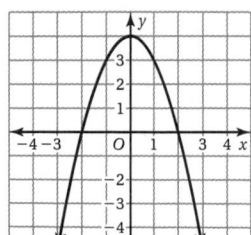

29. *Sample answer:*

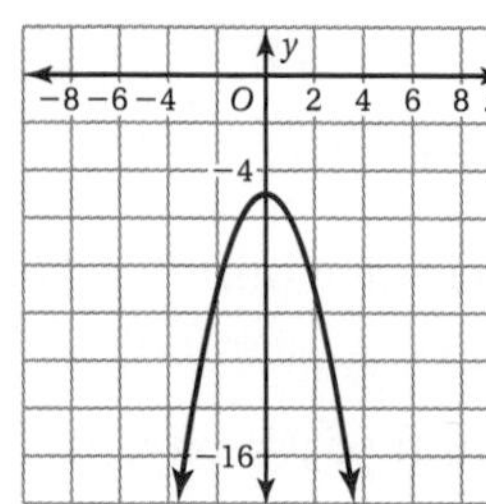

31. yes; $y = x^2 - 1$ has a vertex at $(0, -1)$ and a focus at $\left(0, \frac{1}{4}\right)$

33. **a.** When $k < 0$, the graph of $f(x) + k$ will translate the graph of $f(x)$ down. When $k > 0$, the graph of $f(x) + k$ will translate the graph of $f(x)$ up.

b. When $k > 1$, the graph of $k \cdot f(x)$ is narrower than the graph of $f(x)$. When $0 < k < 1$, the graph of $k \cdot f(x)$ is wider than the graph of $f(x)$. When $-1 < k < 0$, the graph of $k \cdot f(x)$ opens down and is wider than the graph of $f(x)$. When $k = -1$, the graph of $k \cdot f(x)$ opens down. When $k < -1$, the graph of $k \cdot f(x)$ opens down and is narrower than the graph of $f(x)$.

35. $(2x - 1)(x - 3)$

37. B

Section 8.4 Graphing $y = ax^2 + bx + c$ *(pages 429–431)*

1. For a quadratic function $y = ax^2 + bx + c$ where $a \neq 0$, if $a > 0$, then the function has a minimum value and if $a < 0$, then the function has a maximum value.

3. vertex: $(2, -1)$
axis of symmetry: $x = 2$
y-intercept: 1

5. vertex: $(-2, -3)$
axis of symmetry: $x = -2$
y-intercept: 4

7. **a.** $x = -2$
b. $(-2, -12)$

9. **a.** $x = 2$
b. $(2, 4)$

11. **a.** $x = 4$
b. $(4, -6)$

13. domain: all real numbers
range: $y \geq -4$

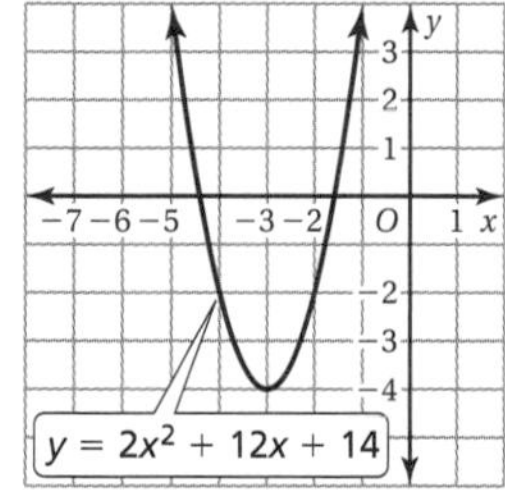

15. domain: all real numbers
range: $y \leq -1$

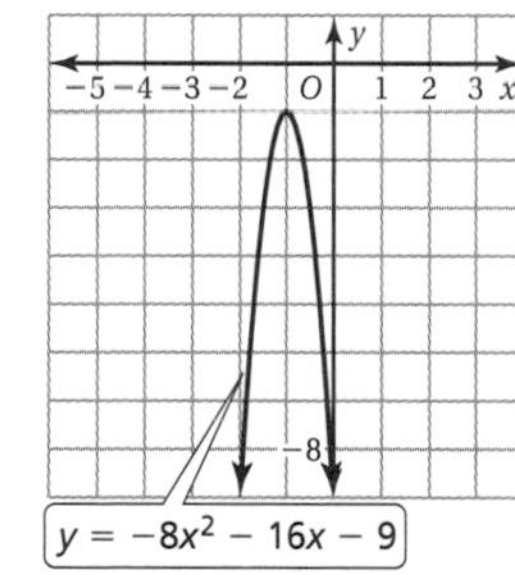

17. domain: all real numbers
range: $y \geq -5$

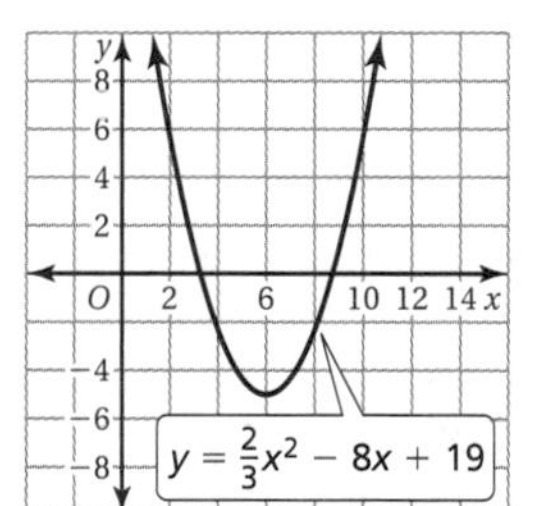

19. **a.** 4 seconds after it is launched at a height of 256 feet

b. The domain is $0 \leq t \leq 8$. The range is $0 \leq h \leq 256$.

21. minimum; -12

23. maximum; -1

25. maximum; 52

27. $(0, 8)$; Use the axis of symmetry to find the reflection of $(6, 8)$.

29. **a.** 0.5 second after it jumps

b. yes; 0.5 ft

c. 1 sec

31. vertex $\approx (-1.4, -4)$

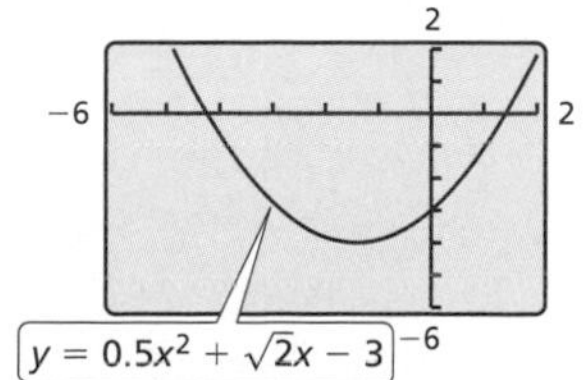

33. opens down; *Sample answer:* The y-values are increasing at a decreasing rate. This means the points are on the left side of a parabola that faces down.

Section 8.4 Graphing $y = ax^2 + bx + c$ (continued)

(pages 429–431)

35. Use the wording in the exercise as a clue for defining the independent variable.

37. $\frac{k^2}{8}$ ft^2

39. 

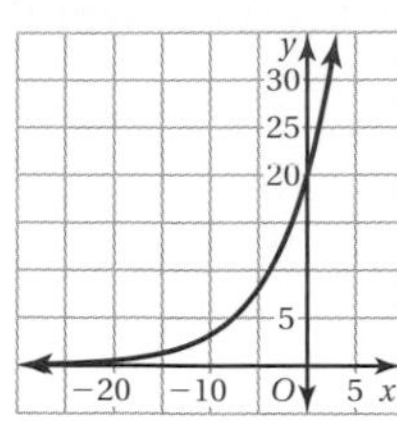

41. C

Lesson 8.4b Graphing $y = a(x - h)^2 + k$

(pages 432 and 433)

1.

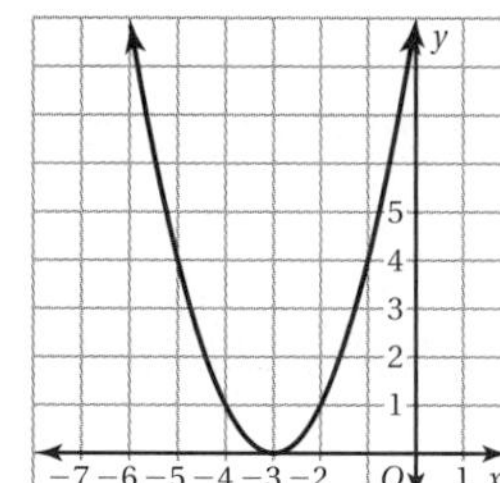

The graph of $y = (x + 3)^2$ is a translation 3 units to the left of the graph of $y = x^2$.

3.

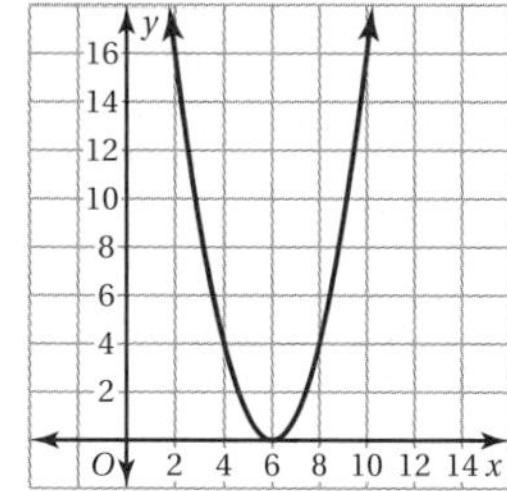

The graph of $y = (x - 6)^2$ is a translation 6 units to the right of the graph of $y = x^2$.

5.

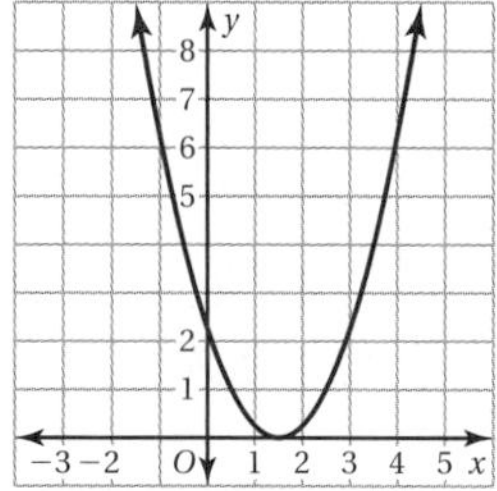

The graph of $y = (x - 1.5)^2$ is a translation 1.5 units to the right of the graph of $y = x^2$.

7. *Sample answer:* $x^2 + 6x + 9$ can be factored as $(x + 3)^2$. In this form, you can see that the graph of $y = (x + 3)^2$ is a translation 3 units to the left of the graph of $y = x^2$.

9.

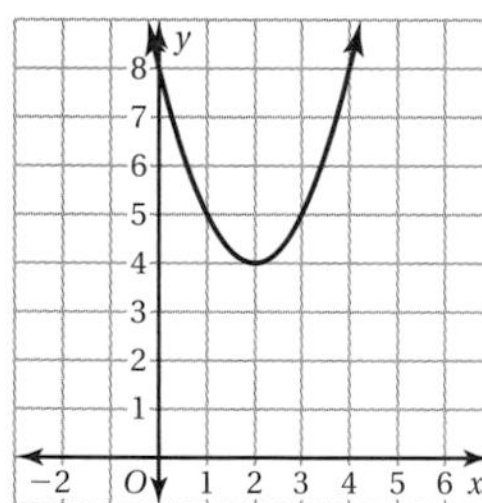

The graph of $y = (x - 2)^2 + 4$ is a translation 2 units to the right and 4 units up of the graph of $y = x^2$.

11.

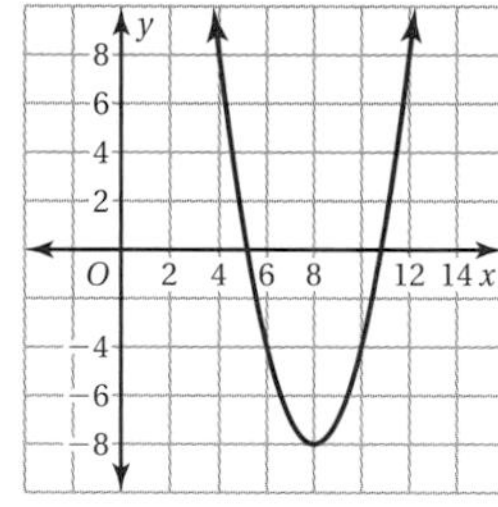

The graph of $y = (x - 8)^2 - 8$ is a translation 8 units to the right and 8 units down of the graph of $y = x^2$.

13.

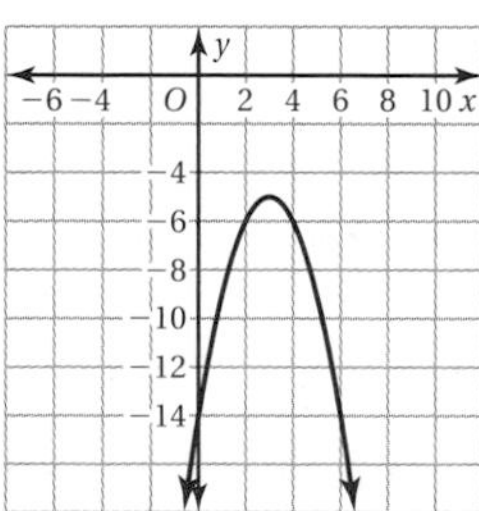

The graph of $y = -(x - 3)^2 - 5$ opens down. The vertex of the graph of $y = -(x - 3)^2 - 5$ is a translation 3 units to the right and 5 units down of the vertex of the graph of $y = x^2$.

15. $g(x)$ is a vertical translation 7 units down of $f(x)$.

17. $g(x)$ is narrower than $f(x)$.

19. $y = (x - 2)^2 - 5$; $y = x^2 - 4x - 1$

Section 8.5 Comparing Linear, Exponential, and Quadratic Functions *(pages 439–441)*

1. linear function: y-values have a common difference; exponential function: y-values have a common ratio; quadratic function: second differences are constant

3. $x > 1$ and $x < 0$

5. $x > -\frac{1}{2}$

7. C

9. 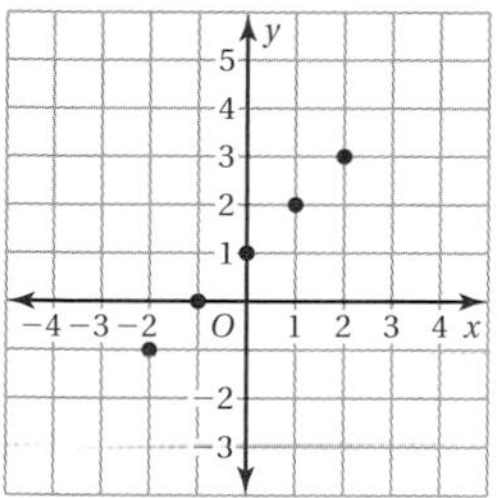 linear

11. 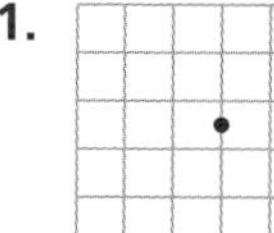 quadratic

13. linear

15. exponential

17. quadratic

19. quadratic; $y = 2x^2$

21. linear; $y = -3x - 2$

23. quadratic; $y = -3x^2$

25. The student incorrectly substitutes the x-coordinate in for y and the y-coordinate in for x. The equation should be $y = 4x^2$.

27. in general, no; It depends on the number of points and where they are located on the graph.

29. a.

Round, x	1	2	3	4	5	6
Teams Remaining, y	32	16	8	4	2	1

b. exponential

c. $y = 64(0.5)^x$

d. sixth round

31. $-1, 1$

33. $-2, 2$

Section 9.1 Solving Quadratic Equations by Graphing *(pages 459–461)*

1. It is a nonlinear equation that can be written in the form $ax^2 + bx + c = 0$, where $a \neq 0$.

3. Use the graph to find the x-intercepts.

5. $x = 4, x = 6$

7. $x = -6$

9. $x = 3$

11. no real solutions

13. $x = -2$

15. $x = 7$

17. a. The x-intercepts give the horizontal positions of the ball where it is struck and where it lands.

b. 5 yards

19. $x^2 - 6x + 8 = 0; x = 2, x = 4$

21. $x^2 + x + 3 = 0$; no real solutions

23. $x^2 - 5x + 6 = 0; x = 2, x = 3$

25. $x = -5, x = 2$

27. no real solutions

29. $x = -1$

31. *Sample answer:* Method 2; You do not have to rewrite the equation.

Section 9.1 Solving Quadratic Equations by Graphing *(continued)* *(pages 459–461)*

33. **a.** about 1.7 seconds **b.** 1.5 seconds

35. $x \approx -5.8, x \approx -0.2$

37. $x \approx -5.7, x \approx 0.7$

39. $x \approx 0.6, x \approx 3.4$

41. **a.** $h = -16t^2 + 20t + 8$
b. about 1.6 seconds

43. sometimes; There are 2 x-intercepts when c is positive.

45. never; A quadratic equation can never have more than 2 solutions.

47. 81

49. $25\sqrt{2}$

Section 9.2 Solving Quadratic Equations Using Square Roots *(pages 466 and 467)*

1. 2; 1; 0

3. 2; $x \approx 3.317, x \approx -3.317$

5. 2; $x \approx 1.225, x \approx -1.225$

7. 0; no real solutions

9. 2; $x = \sqrt{21}, x = -\sqrt{21}$

11. 2; $x = 13, x = -13$

13. no real solutions

15. $x = \sqrt{61}, x = -\sqrt{61}$

17. $x = 0$

19. $x = 0$

21. When taking the square root of each side, the student forgot the negative root; $x = 6, x = -6$

23. $x = -3$

25. $x = -4, x = 5$

27. $x = -\frac{7}{3}, x = \frac{1}{3}$

29. 8 in. by 8 in.

31. 12 ft

33. You can solve this problem by recalling a property of similar figures.

35. **a.** two solutions: a is positive and c is negative, or c is positive and a is negative.
b. one solution: $c = 0$
c. no solutions: a and c are both positive or both negative.

37. $x^2 + 10x + 25$

39. $4y^2 - 12y + 9$

Section 9.3 Solving Quadratic Equations by Completing the Square *(pages 472 and 473)*

1. Add $\left(\frac{b}{2}\right)^2$.

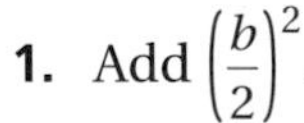

3.

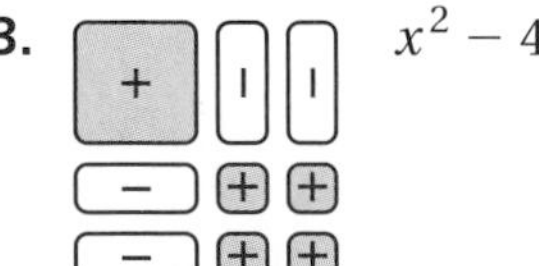

$x^2 - 4x + 4$

5.

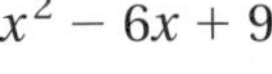

$x^2 - 6x + 9$

7. 4

9. 49

11. 81

13. $x^2 + 16x + 64 = (x + 8)^2$

15. $x^2 - 40x + 400 = (x - 20)^2$

17. $x^2 + 5x + \frac{25}{4} = \left(x + \frac{5}{2}\right)^2$

19. $x = -2, x = 8$

21. no real solutions

23. $x = \frac{1}{2}, x = 1$

25. **a.** $216 = x(x + 6)$ **b.** 12 ft by 18 ft

27. Divide each side by 3.

29. **a.** after about 4.4 seconds

b. 96 ft; The vertex is (2, 96). The maximum is the y-coordinate of the vertex.

31. $x(x + 1) = 42$; 6, 7

33. $2\sqrt{3}$

35. $2\sqrt{5}$

Section 9.4 Solving Quadratic Equations Using the Quadratic Formula *(pages 481–483)*

1. $x = \dfrac{-b \pm \sqrt{b^2 - 4ac}}{2a}$

3. $x^2 - 7x = 0$; $a = 1, b = -7, c = 0$

5. $-2x^2 - 5x + 1 = 0$; $a = -2, b = -5, c = 1$

7. $x^2 - 6x + 4 = 0$; $a = 1, b = -6, c = 4$

9. $x = 6$

11. $x = -1, x = 11$

13. no real solutions

15. $x = \dfrac{2}{3}, x = \dfrac{3}{2}$

17. $x = \dfrac{1}{4}$

19. $x \approx -4.2, x \approx 2.2$

21. Used -7 for $-b$ instead of $-(-7) = 7$; $x = -\dfrac{2}{3}, x = 3$

23. 6 ft

25. A

27. 0

29. 1

31. 2

33. **a.** yes; When the solutions m and n are integers, the standard form can be factored as $(x - m)(x - n) = 0$.

b. yes; When the solutions $\dfrac{m}{n}$ and $\dfrac{h}{k}$ are fractions, the standard form can be factored as $(nx - m)(kx - h) = 0$.

c. Any quadratic equation with rational solutions can be solved by factoring.

35. **a.** 1994 and 2006

b. no; The model predicts negative numbers of fish caught after 2015.

37. 2

39. 0

41. *Sample answer:* **a.** $c = 8$ **b.** $c = 2$

43. 2; When a and c have different signs, $b^2 - 4ac$ is positive.

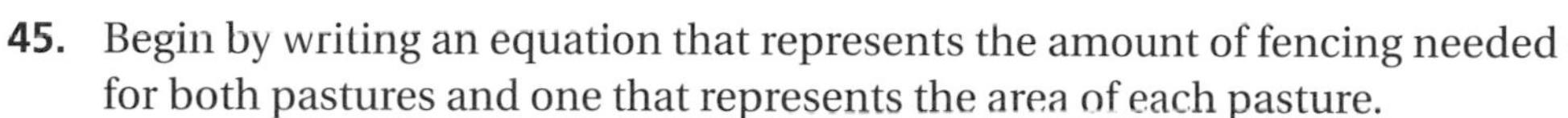

45. Begin by writing an equation that represents the amount of fencing needed for both pastures and one that represents the area of each pasture.

Hint

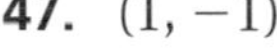

47. $(1, -1)$

49. no solution

Lesson 9.4b Choosing a Solution Method *(pages 484 and 485)*

1. $x = -7 + \sqrt{41}, x = -7 - \sqrt{41}$

3. $x = 6, x = -6$

5. $x = 1, x = -1$; square roots, because no x-term

7. $x = -5, x = 8$; factors easily

9. no real solutions; completing the square, because $a = 1$ and b is even

11. $x = -4, x = 3$; factors easily

Selected Answers

Section 9.5 Solving Systems of Linear and Quadratic Equations *(pages 490 and 491)*

1. A solution of a system of linear and quadratic equations is an ordered pair that is a solution of each equation in the system.

3. B; (0, 1), (3, 4) **5.** A; (0, −1) **7.** (−1, −2) **9.** (−3, −2), (1, 6)

11. $\left(-\frac{3}{2}, -4\right)$, (2, 10) **13.** no real solutions **15.** (0, 0), (−5, 10) **17.** (2, −6), $\left(\frac{5}{2}, -\frac{23}{4}\right)$

19. (−1, −6), (−2, −9) **21.** (2, 2)

23. *Sample answer:* The viewing window of the graphing calculator is too small. By increasing the window you can see that (5, 14) is also a solution.

25. a. 2 **b.** 0 **27.** $x = -1, x = 1$ **29.** no real solutions

Section 10.1 Graphing Square Root Functions *(pages 506 and 507)*

1. no; It is a linear function. A square root function contains a square root with the independent variable in the radicand.

3. B **5.** C

7. domain: $x \geq 0$

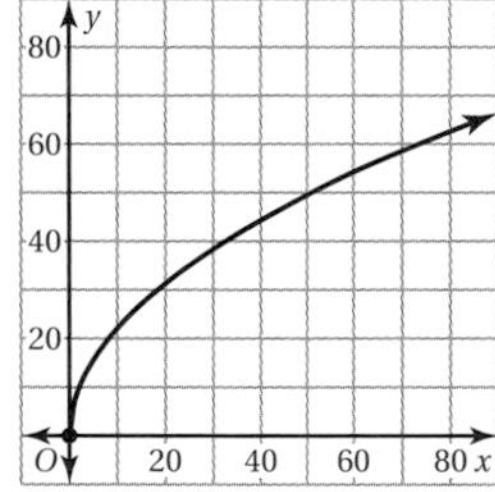

9. $x \geq 0$

11. $x \geq 2$

13. $x \geq -4$

15. a.

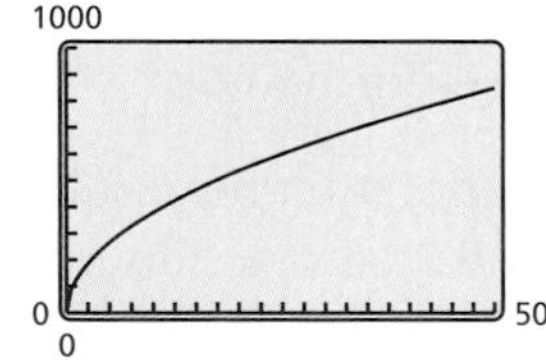

b. about 6 pounds per square inch

17. domain: $x \geq 0$; range: $y \geq 4$; The graph of $y = \sqrt{x} + 4$ is a translation 4 units up of the graph of $y = \sqrt{x}$.

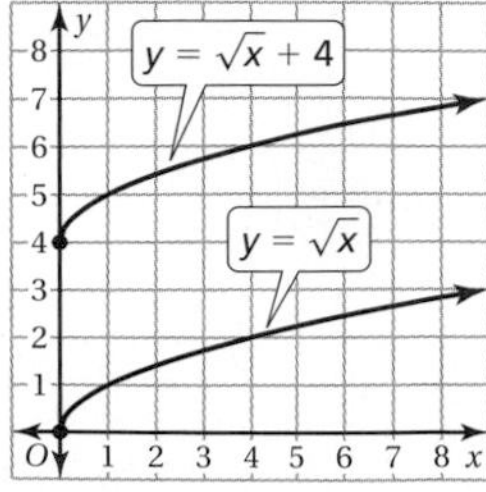

19. domain: $x \geq -2$; range: $y \geq -2$; The graph of $y = \sqrt{x+2} - 2$ is a translation 2 units to the left and 2 units down of the graph of $y = \sqrt{x}$.

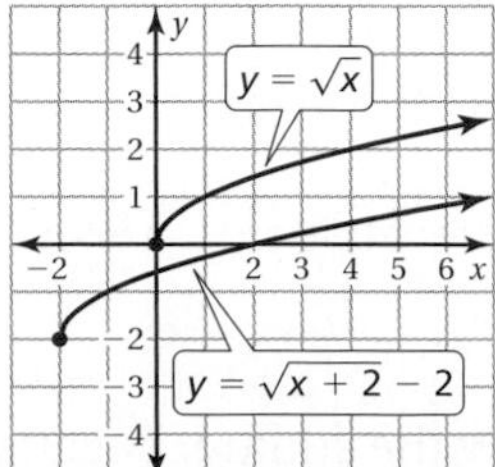

21. domain: $x \geq 1$; range: $y \leq 3$;
The graph of $y = -\sqrt{x-1} + 3$ is a reflection in the x-axis of the graph of $y = \sqrt{x}$, and then a translation 1 unit to the right, and a translation 3 units up.

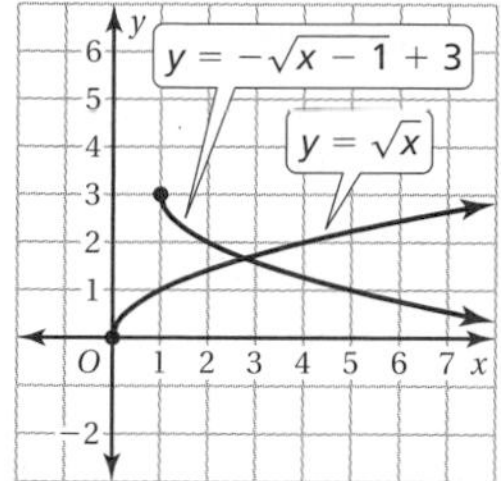

23. **a.** *Sample answer:* $y = \sqrt{x} + 1$
b. *Sample answer:* $y = -\sqrt{x}$

25. **a.** domain: $A > 0$

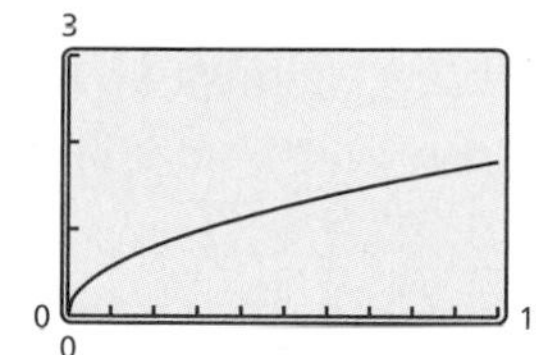

b. about 28 square inches

27. Choose perfect cubes for x to make evaluating $g(x)$ easier.

29. $x = -3$

31. B

Lesson 10.1b Rationalizing the Denominator

(pages 508 and 509)

1. $\dfrac{\sqrt{10}}{10}$

3. $\dfrac{3\sqrt{2}}{2}$

5. $\dfrac{\sqrt{10}}{6}$

7. $\sqrt{5}$

9. $\dfrac{4\sqrt{15} - 5}{15}$

11. $-3 + 3\sqrt{3}$

13. $-2\sqrt{2} + 2\sqrt{7}$

Section 10.2 Solving Square Root Equations

(pages 515–517)

1. no; The radicand does not contain a variable.

3. 4;

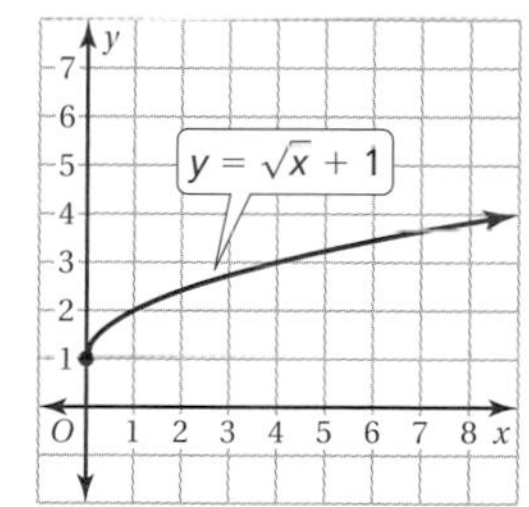

5. 24;

7. $x = 144$

9. $x = 121$

11. $x = 225$

13. $x = 19$

15. $x = 45$

17. $x = 7$

19. 96 in.^{2}

21. Add $\sqrt{3m}$ to each side. Square each side. Subtract m from each side. Divide each side by 2. Check your solution.

23. $x = 3$

25. $x = 7$

27. $x = \dfrac{2}{3}$

29. $x = 8$

31. $x = 1, x = 4$

33. $x = 5$

35. $x = 2$

Selected Answers

Section 10.2 Solving Radical Equations *(continued)* *(pages 515–517)*

37. $x = -6$ does not check in the original equation, so it is extraneous; $x = 2$ is the only solution.

39. true **41.** false **43.** 14.4 ohms **45.** about 29.2 ft

47. $\sqrt{x+2}$ has one term and $\sqrt{x} + 2$ has two terms. When you square $\sqrt{x+2}$, the result is the radicand, $x + 2$. When you square $\sqrt{x} + 2$, the result is an expression with 3 terms that includes a radical, $x + 4\sqrt{x} + 4$.

49. $x = -\frac{7}{16}$ **51.** 92° **53.** 90°

Section 10.3 The Pythagorean Theorem *(pages 524 and 525)*

1. Use the Pythagorean Theorem to find the missing side length.

3. 12 yd **5.** $\sqrt{799}$ ft **7.** 6 in.

9. 20 should have been substituted for c, not b. The missing length is $2\sqrt{91}$ inches.

11. about 60 in. **13.** The direct distance from the tee to the hole is 145 yards.

15. **a.** $28^2 + 21^2 = (5x)^2$

b. Factoring: Subtract 1225 from each side. Factor out the GCF and factor the difference of squares. Set the factors equal to zero and solve. Divide each side by 25. Then take the square root of each side. *Sample answer:* taking square roots; Taking square roots is easier than factoring.

c. 7

17. Both graphs have the same axis of symmetry, $x = 0$. The graph of $y = -2x^2 + 4$ opens down and is narrower than the graph of $y = x^2$. The vertex of the graph $y = -2x^2 + 4$ is a translation 4 units up of the vertex of the graph of $y = x^2$.

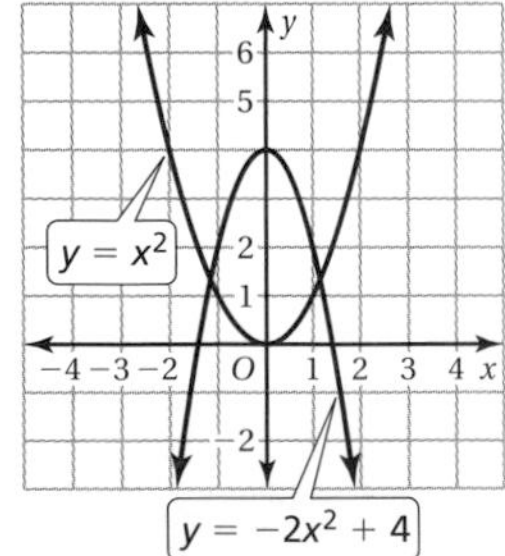

19. Both graphs open up and have the same axis of symmetry, $x = 0$. The graph of $y = 3x^2 + 8$ is narrower than the graph of $y = x^2$. The vertex of the graph of $y = 3x^2 + 8$ is a translation 8 units up of the vertex of the graph of $y = x^2$.

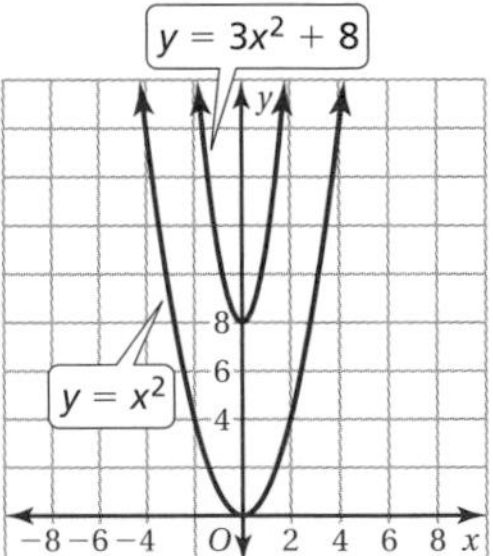

Section 10.4 Using the Pythagorean Theorem *(pages 530 and 531)*

1. the Pythagorean Theorem and the distance formula

3. If a^2 is even, then a is an even number; true

5. yes **7.** yes **9.** no **11.** $2\sqrt{10}$ **13.** $3\sqrt{10}$ **15.** $\sqrt{53}$

17. The squared quantities under the radical should be added, not subtracted; $2\sqrt{10}$

19. yes

21. yes

23. You can use what you learned about slopes of perpendicular lines to solve this problem.

Hint

25. Plane B; Plane A is about 8.35 kilometers away and Plane B is about 8.27 kilometers away.

27. $\frac{3}{2}$, 1

29. no real solutions

Section 11.1 Direct and Inverse Variation *(pages 547–549)*

1. In direct variation, y is the product of x and the constant k. In inverse variation, y is the quotient of the constant k and x.

3. direct variation; The ratio $\frac{y}{x}$ is constant.

5. direct variation; The equation can be written as $y = kx$.

7. $y = 3x$

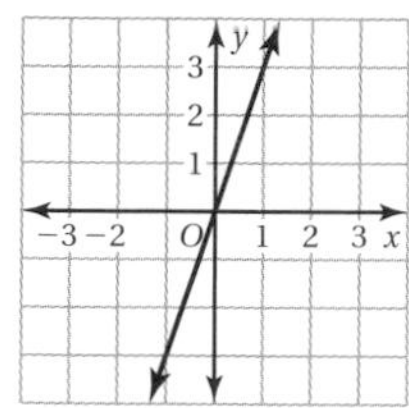

9. $y = \frac{1}{6}x$

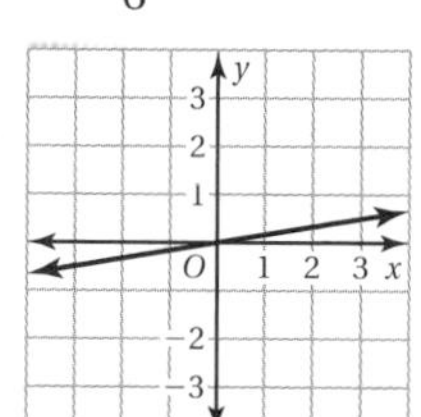

11. $y = \frac{45}{x}$

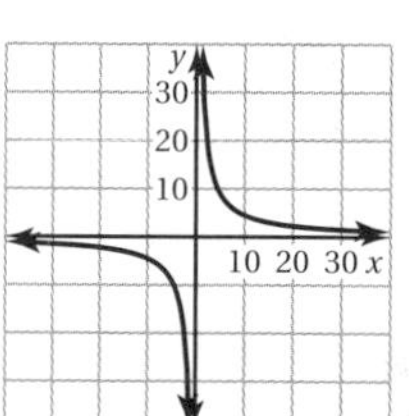

13. inverse variation; The product hr is constant.

15. Both the domain and range are all real numbers except for 0.

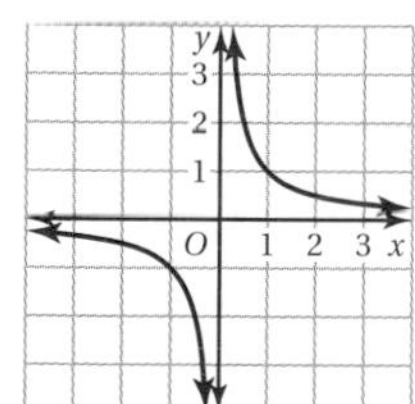

17. Both the domain and range are all real numbers except for 0.

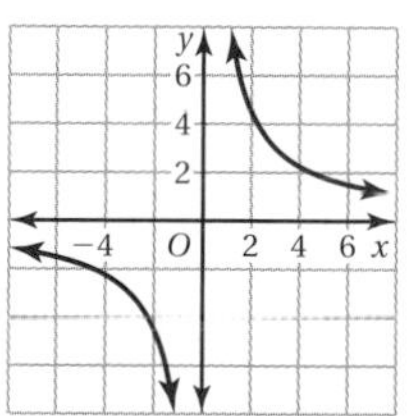

19. $y = \frac{12}{x}$; $x = 12$

21. $y = \frac{10}{x}$; $y = 2.5$

23. inverse variation;

Number of Hats, x	5	10
Cost per Hat, y	10	5

The product xy is constant.

25. inverse variation; Suppose the swim team earns \$1000.

Number of Members, x	1	5	10	20
Earnings per Member, y	1000	200	100	50

The product xy is constant.

Selected Answers

27. **a.** $v = 18t$

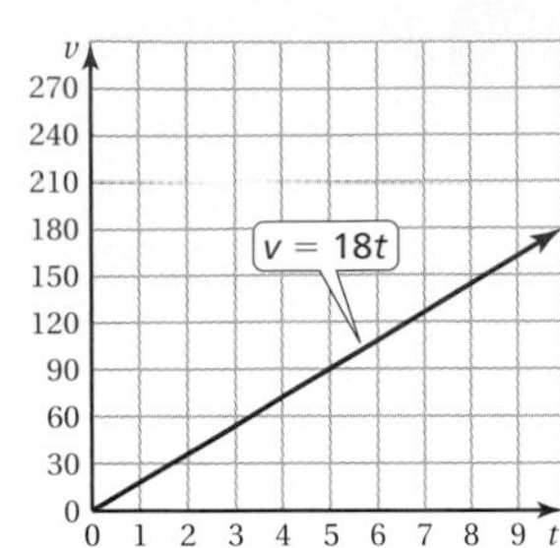

b. 90 hours

29. **a.** $t = \frac{5000}{p}$

b. 200 hours

31. odd

33. even

35. even

37. **a.** symmetric with respect to the y-axis

b. symmetric with respect to the origin (reflection in the y-axis followed by a reflection in the x-axis)

39. The graph of $y = 3x^2$ is narrower than the graph of $y = x^2$.

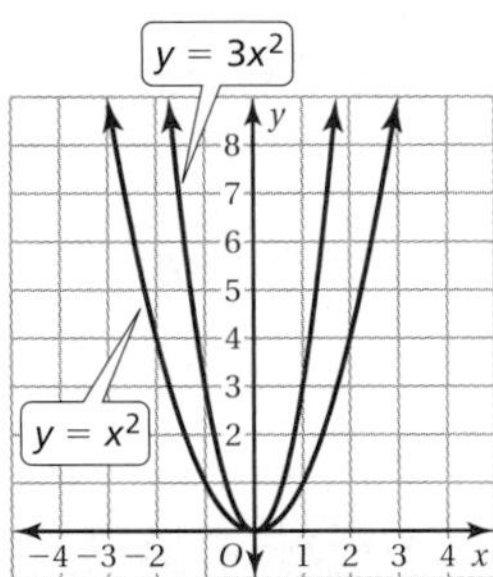

41. The graph of $y = x^2 - 1$ is a translation 1 unit down of the graph of $y = x^2$.

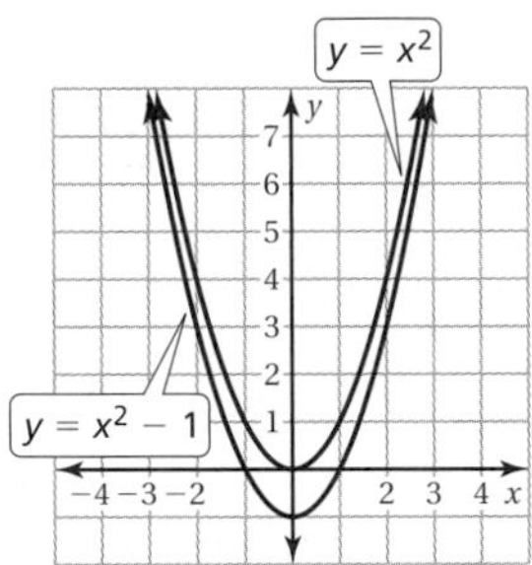

Section 11.2 Graphing Rational Functions *(pages 555–557)*

1. no; The denominator is not a polynomial.

3. The graph of a rational function approaches but never intersects the asymptotes.

5. The graph is two smooth curves. The domain appears to be all real numbers except 10. The range appears to be all real numbers except 10. As x gets closer to 10, the graph approaches the vertical line $x = 10$. As x increases and decreases, the graph approaches the horizontal line $y = 10$.

7. $x = -4$

9. $x = 9$

11. $x = -\frac{1}{2}$

13. The domain is all real numbers except 0. The range is all real numbers except 0.

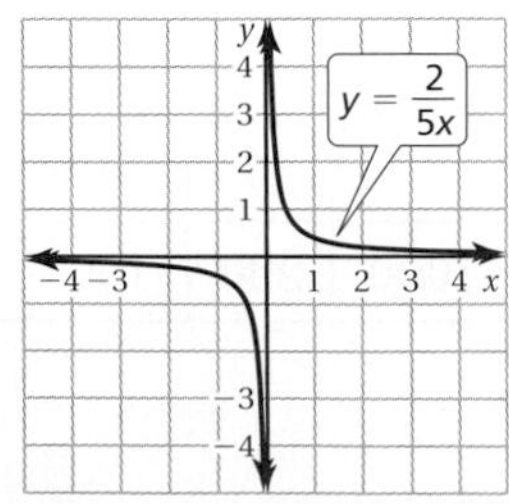

15. The domain is all real numbers except 3. The range is all real numbers except 0.

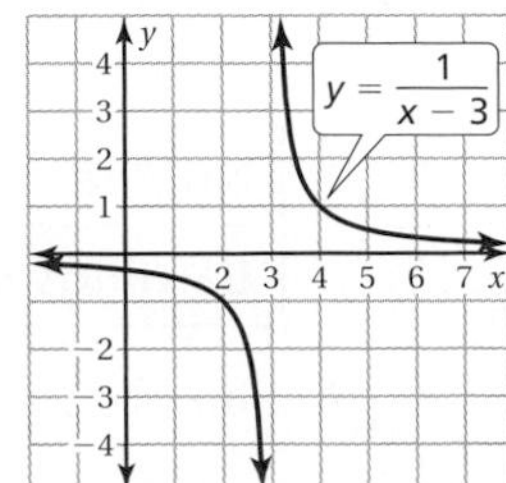

17. The domain is all real numbers except 2. The range is all real numbers except 0.

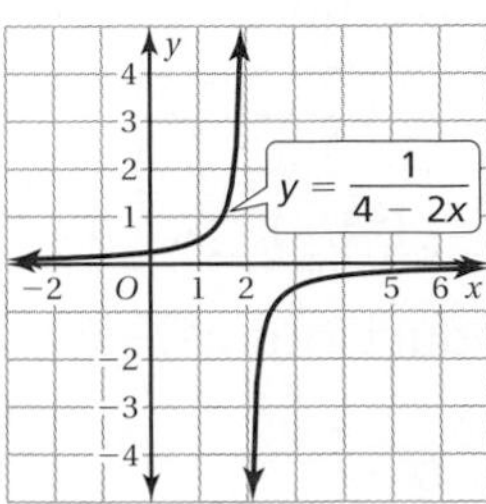

19. $x = 0$; $y = 0$; The domain is all real numbers except 0. The range is all real numbers except 0.

21. $x = 2$; $y = 7$; The domain is all real numbers except 2. The range is all real numbers except 7.

23. $x = 5$; $y = -2$; The domain is all real numbers except 5. The range is all real numbers except -2.

25. The wrong sign is used for the vertical asymptote; $x = -4$

27. *Sample answer:* $y = \dfrac{100}{x - 6} - 9$

29. The graph of $y = \dfrac{1}{x - 6}$ is a translation 6 units right of the graph of $y = \dfrac{1}{x}$.

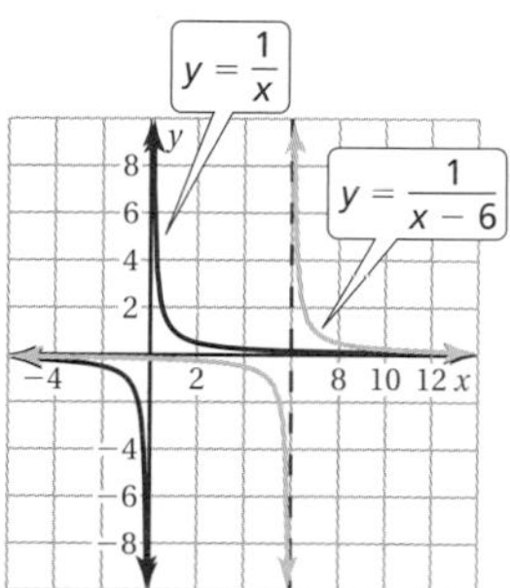

31. The graph of $y = \dfrac{1}{x + 7} + 3$ is a translation 7 units left and 3 units up of the graph of $y = \dfrac{1}{x}$.

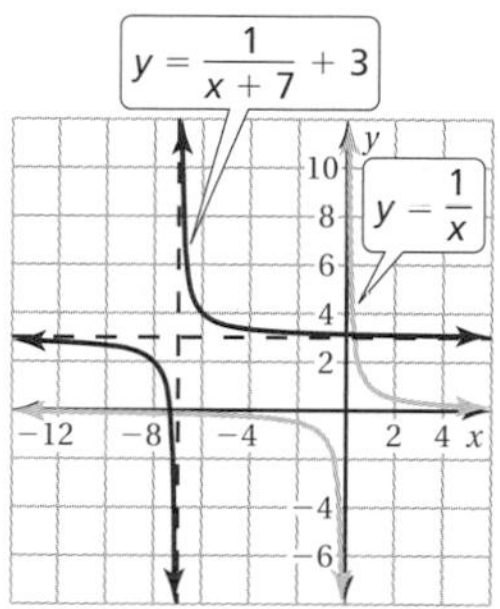

33. The graph of $y = 4 - \dfrac{1}{x + 8}$ is a reflection in the x-axis of the graph of $y = \dfrac{1}{x}$, and then a translation 8 units left and 4 units up.

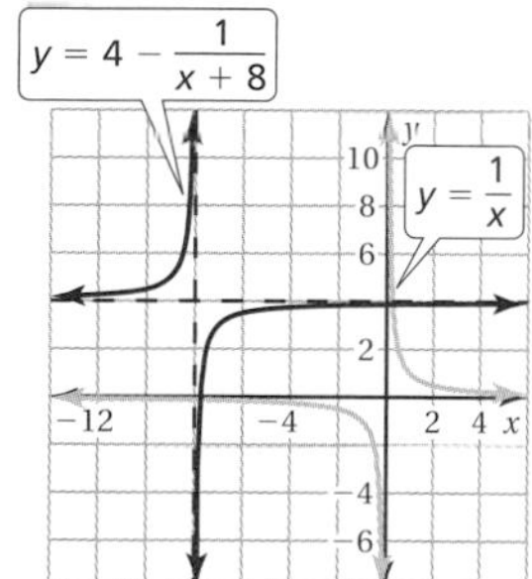

35. a. The domain is all real numbers greater than 0. The range is all real numbers greater than 0 and less than 15.

b. 12

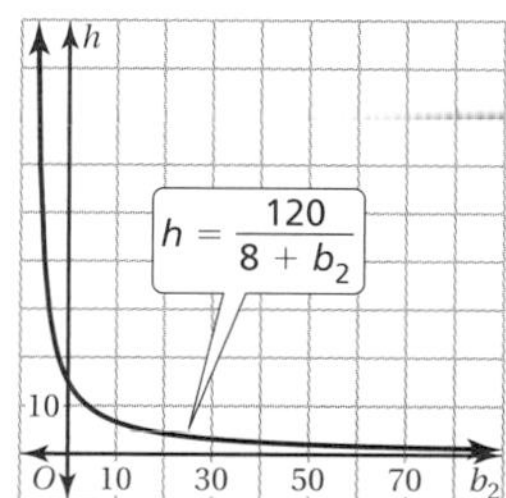

37. $y = \dfrac{1}{x - 5}$

39. $y = 2 - \dfrac{1}{x - 1}$

41. about 23°C

43. $x = -2$, $x = 2$, $y = 0$

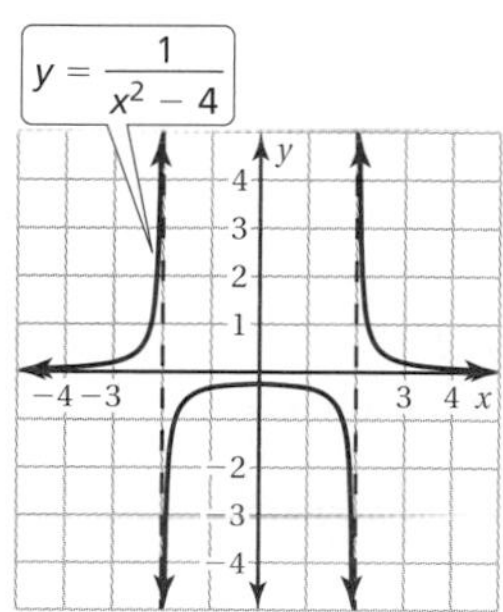

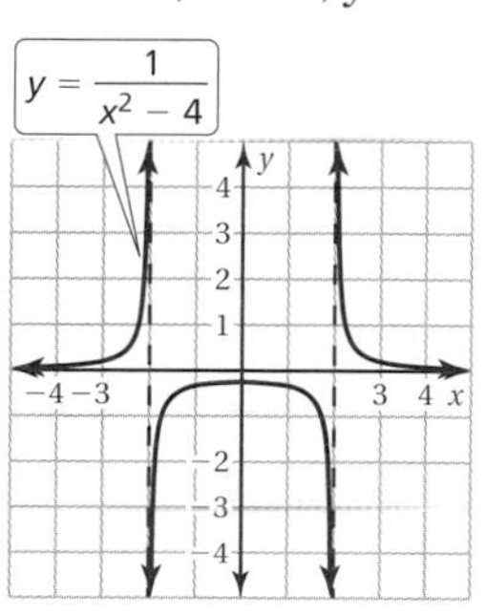

45. Read the exercise carefully before you begin.

47. linear; This is a linear equation in standard form.

49. D

Lesson 11.2b Inverse of a Function
(pages 558 and 559)

1. (8, −5), (6, −5), (0, 0), (6, 5), (8, 10)

3.

Input	4	1	0	1	4
Output	−2	−1	0	1	2

5. The domain of a relation is the range of its inverse relation. The range of a relation is the domain of its inverse relation.

7. $f^{-1}(x) = \frac{1}{3}x + \frac{1}{3}$

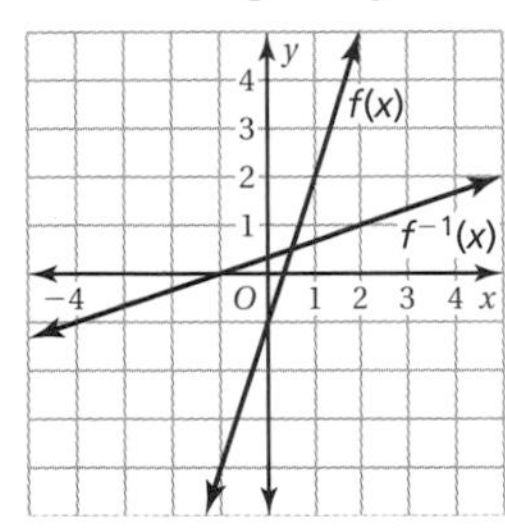

9. $f^{-1}(x) = \sqrt{0.5x}$

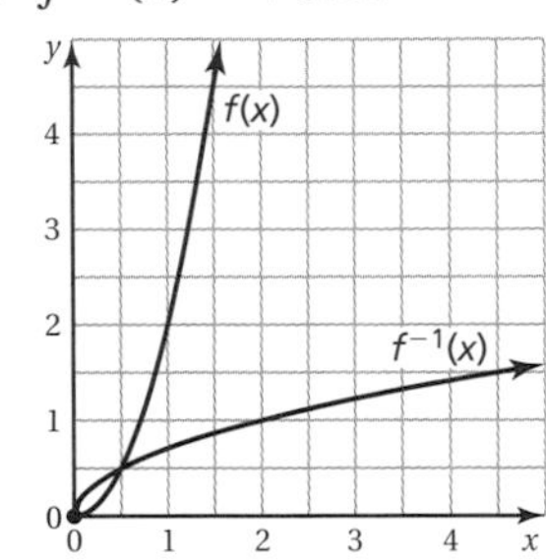

11. $f^{-1}(x) = \frac{1}{x}$

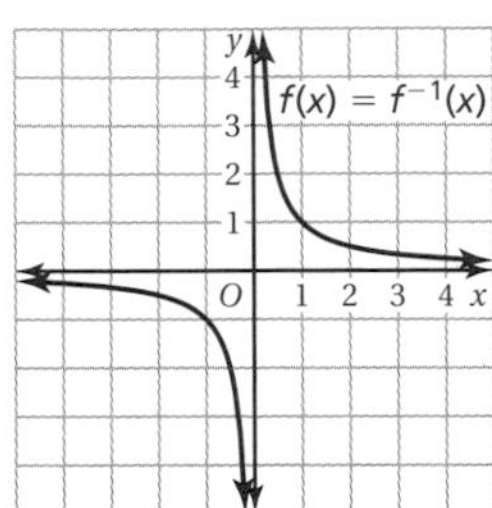

13. −2

15. Both equal x. For $f(g(x))$, $g(x)$ is the output value of x. So, then you have $f(g(x))$, which gives you back the input value, x, because they are inverses. For $g(f(x))$, $f(x)$ is the output value of x. So, then you have $g(f(x))$, which gives you back the input value, x, because they are inverses.

Section 11.3 Simplifying Rational Expressions
(pages 564 and 565)

1. no; not a ratio of two polynomials

3. $\frac{1}{3x}$; $x = 0$

5. The expression is in simplest form; $n = -1$

7. $\frac{2t}{t + 11}$; $t = -11$, $t = 0$

9. They did not list all of the excluded values; $x = 0$, $x = 3$

11. -1; $z = \frac{5}{2}$

13. $\frac{2 - y}{y - 5}$; $y = -2$, $y = 5$

15. $\frac{x + 2}{2(x - 2)}$; $x = 0$, $x = 2$

17. $\frac{x}{2}$ by $(x + 3)$

19. $\frac{3(x + 1)}{x(x + 3)}$

21. *Sample answer:* $\frac{1}{x^2 + 8x + 15}$

23. $\left(\frac{3}{4}x + 3\right)$ in.

25. Begin by looking at the sum $6x^2 + 12x$ and its factors.

Hint

27. continuous

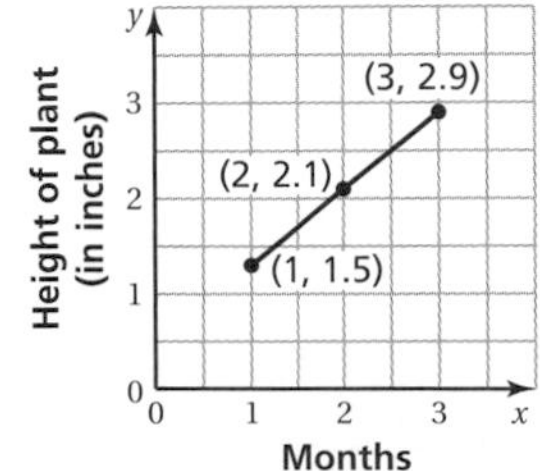

Section 11.4 Multiplying and Dividing Rational Expressions
(pages 572 and 573)

1. (1) Factor the numerators and denominators. (2) Multiply the numerators and denominators. (3) Divide out common factors. (4) Simplify.

3. $x = 0, x = 2; x = -1, x = 0, x = 2$

5. $\dfrac{n+3}{14n^4}$

7. $\dfrac{x}{6(x-5)}$

9. $\dfrac{-2}{r-6}$

11. $\dfrac{2t}{3}$

13. $\dfrac{p+4}{p-3}$

15. $\dfrac{3}{(z-6)^2}$

17. To multiply rational expressions, you multiply by the reciprocal of the divisor, not the dividend;

$$\frac{v-2}{4v} \div \frac{v-2}{6v^2} = \frac{v-2}{4v} \cdot \frac{6v^2}{v-2} = \frac{6v^2(v-2)}{4v(v-2)} = \frac{3v}{2}$$

19. $3w(w+2)$

21. -1

23. To find the probability that your campsite has shade, compare the area of the green triangle to the area of the rectangle.

25. a. $T = \dfrac{50 - x}{(1-0.05x)(0.05x^2+5)}$

b. 2020; This will be 20 years after 2000 and 20 is the excluded value.

27. domain: all real numbers
range: $y \geq -5$

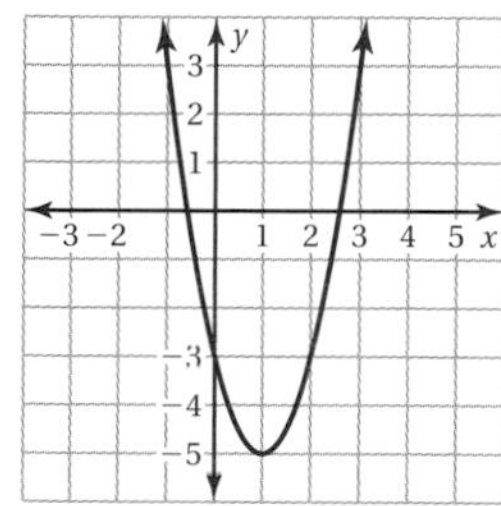

29. domain: all real numbers
range: $y \leq 9$

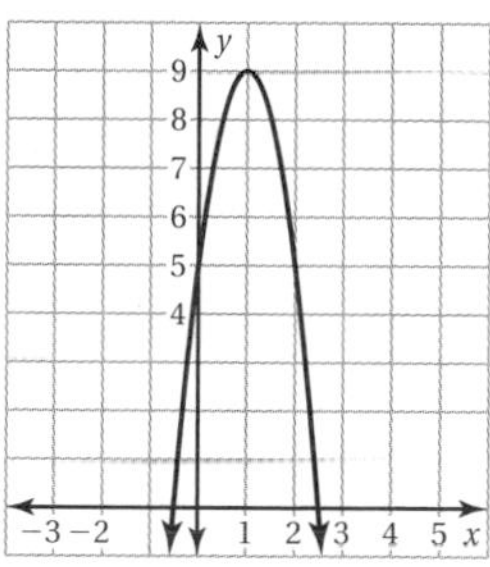

Section 11.5 Dividing Polynomials
(pages 578 and 579)

1. Monomial: Divide each term of the polynomial by the monomial.
Binomial: You can use long division to divide a polynomial by a binomial.

3. When you divide the polynomial by the binomial, the remainder is 0.

5. $x - 2$

7. $\dfrac{n^2}{2} - \dfrac{2n}{3} + \dfrac{2}{n}$

9. $z + 7$

11. $3h - 1$

13. $8k - 1 + \dfrac{1}{k-1}$

15. $3g + 5 + \dfrac{8}{2g-3}$

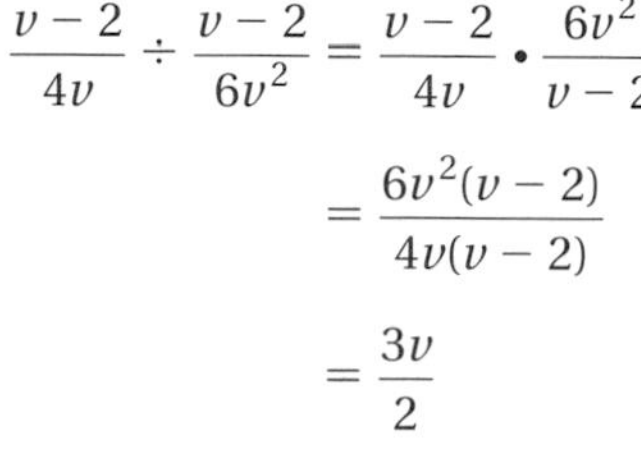

Selected Answers

Section 11.5 Dividing Polynomials *(continued)*
(pages 578 and 579)

17. When subtracting the product from the dividend, the negative sign was not distributed to the second term; $(x^2 - x - 6) \div (x - 3) = x + 2$

19. $d - 3$

21. $4n + 2 + \dfrac{5}{2n - 1}$

23. The dividend is missing an x-term with a coefficient of 0 which resulted in like terms not being aligned; $(2x^2 - 5) \div (x + 2) = 2x - 4 + \dfrac{3}{x + 2}$

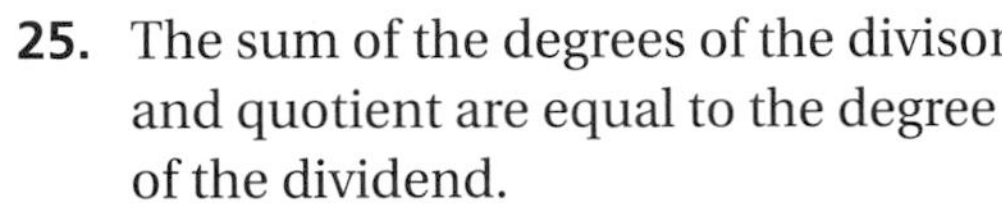

25. The sum of the degrees of the divisor and quotient are equal to the degree of the dividend.

27. The formula for the volume of a rectangular prism is $V = \ell wh$.

29.

Quotient
$(x^2 - x + 1) \div (x + 1) = x - 2 + \dfrac{3}{x + 1}$
$(x^3 - x^2 + x - 1) \div (x + 1) = x^2 - 2x + 3 - \dfrac{4}{x + 1}$
$(x^4 - x^3 + x^2 - x + 1) \div (x + 1) = x^3 - 2x^2 + 3x - 4 + \dfrac{5}{x + 1}$
$(x^5 - x^4 + x^3 - x^2 + x - 1) \div (x + 1) = x^4 - 2x^3 + 3x^2 - 4x + 5 - \dfrac{6}{x + 1}$

31. $x = -4 + \sqrt{23}, x = -4 - \sqrt{23}$

33. A

Section 11.6 Adding and Subtracting Rational Expressions
(pages 585–587)

1. Find the LCM of the denominators in both cases.

3. $\dfrac{5s}{9}$

5. $\dfrac{7}{w}$

7. $\dfrac{3}{x + 2}$

9. $\dfrac{3t - 4}{t - 1}$

11. $\dfrac{1}{p - 1}$

13. $2x$

15. $(m + 5)(m - 4)$

17. $(h + 3)(h - 1)$

19. $\dfrac{S - 2\ell w}{2(\ell + w)}$

21. $\dfrac{7}{24}$

23. $\dfrac{2m - 4}{m - 7}$

25. $\dfrac{11}{(p - 4)(p - 3)}$

27. yes; not always; The product of the denominators is the product of *all* factors of both denominators. The LCM of the denominators is the product of the greatest power of each factor that appears in *either* denominator.

29. *Sample answer:* $\dfrac{1}{x - 2}, \dfrac{1}{x + 3}$

a. $(x - 2)(x + 3)$

b. $\dfrac{2x + 1}{(x - 2)(x + 3)}$

31. $\dfrac{8p+7}{p+10}$

33. $\dfrac{2c+1}{c-1}$

35. $\dfrac{x(2x-1)}{(x+4)(x-5)}$

37. $\dfrac{33}{2b+20}$

39. Begin by subtracting the two rational expressions, $\frac{a}{b}$ and $\frac{c}{d}$.

41.

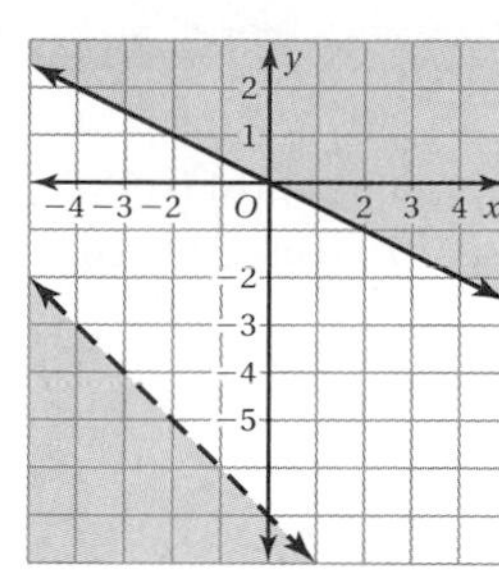

43. A

Section 11.7 Solving Rational Equations
(pages 592 and 593)

1. Use the Cross Products Property, or multiply each side by the LCD.

3. The solution may be extraneous.

5. $b = 1$

7. $m = -2, m = 6$

9. $k = -1, k = 5$

11. The solutions were not checked in the original equation. The solution $x = -1$ is extraneous because it is an excluded value. The only solution is $x = 2$.

13. $c = 5$

15. no solution

17. $a = -9, a = 6$

19. Rewrite the left side using a common denominator of x. Then use the Cross Products Property to solve.

21. Read through the problem and organize the given information.

23. 1.2 h (1 h 12 min)

25. $x = -4, x = 8$

27. $x = -2\frac{1}{5}, x = \frac{3}{5}$

Section 12.1 Measures of Central Tendency
(pages 610 and 611)

1. The mean is the sum of the data divided by the number of data values. The median of an odd number of values is the middle value. The median of an even number of values in the mean of the two middle values. The mode is the value or values that occur most often.

3. If the outlier is greater than the mean, removing it will decrease the mean. If the outlier is less than the mean, removing it will increase the mean.

5. mean: 1; median: 1; mode: -1

7. mean: $1\frac{29}{30}$ h; median: 2 h; modes: $1\frac{2}{3}$ h and 2 h

9. 4

11. 16

13. 465

15. Begin by ordering the data.

Hmmm.

17. All measures increase by k.

19. All measures decrease by k.

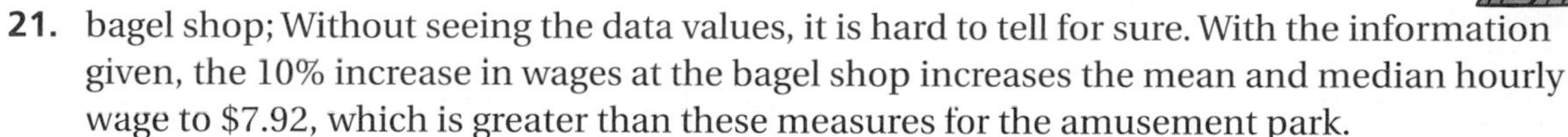

21. bagel shop; Without seeing the data values, it is hard to tell for sure. With the information given, the 10% increase in wages at the bagel shop increases the mean and median hourly wage to \$7.92, which is greater than these measures for the amusement park.

23. $-4.7, -2.8, -\frac{2}{3}, 1.2, \frac{3}{2}, 5.4$

Section 12.2 Measures of Dispersion
(pages 616 and 617)

1. A measure of central tendency is a value that represents a typical value in a data set. A measure of dispersion is a value that measures how spread out a data set is.

3. The mean is 85. The majority of the data are in clusters on each side, far from the mean.

5. Tigers: 69; 9; Centaurs: 69; 12; The means are equal but the range of the heights of the Centaurs is greater than the range of the heights of the Tigers. So, the heights of the Centaurs are more spread out.

7. 5; about 1.9

9. 21; about 1.6

11. 63; about 11.9

13. a. Kirsten: 90; 15; about 4.6
    Leah: 91; 8; about 2.5
    The means are about the same but Kirsten's range and standard deviation are much greater than Leah's range and standard deviation.
    b. Leah; Kirsten's scores are more spread out than Leah's scores.

15. 6.4; 8.2; about 3.0

17. no; Two data sets can have the same range and very different standard deviations due to outliers and/or the distribution of the data.
    Example: Data set 1: 1, 5, 6, 6, 6, 7, 11; range = 10, standard deviation = 2.7
    Data set 2: 1, 2, 2, 6, 10, 10, 11; range = 10, standard deviation = 4.0

19. a. *Sample answer:* height (in inches): 58, 58, 59, 60, 60, 61, 62, 62, 63, 65, 65, 65, 65, 66, 66, 68, 69, 71; mean = 63.5; range = 13; standard deviation $\approx$ 3.7
    b. The mean, range, and standard deviation should all increase; *Sample answer:* mean $\approx$ 64.6; range = 26; standard deviation $\approx$ 5.8

21. yes; no; The data set 4, 4, 4, 4, 4, 4, 4, 4, has a standard deviation of 0. The formula for the standard deviation of a data set involves a square root, and standard deviation is a measure of how much a typical data value differs from the mean. So, it can never be negative.

23. The domain is all real numbers except 6.
    The range is all real numbers except 0.

25. B

Section 12.3 Box-and-Whisker Plots
(pages 623–625)

1. 25%; 50%

3. The length gives the range of the data set and it tells how much the data vary.

5. 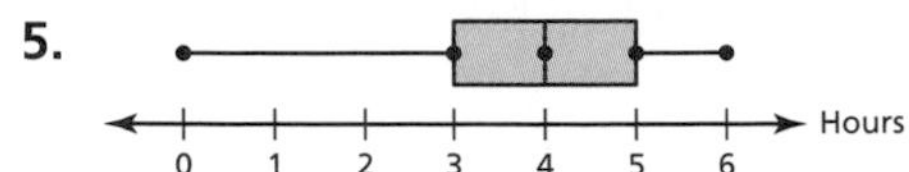

7. 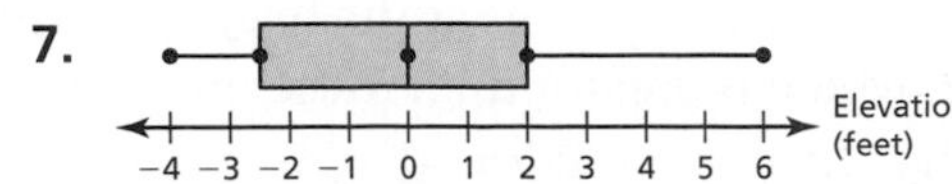

9. a. 11 in.

b. about 8.5 in.; about 13.5 in.

c.

11. a.

b. The middle half of the prices vary by no more than \$4.38.

c. One-quarter of the prices are \$10.50 or less; One-half of the prices are between \$10.50 and \$14.88; One-quarter of the prices are \$14.88 or more.

d. 2.9; The typical price differs from the mean by about \$2.90.

13. symmetric; The whiskers are about the same length and the median is in the middle of the data.

15. skewed left; The left whisker is longer than the right whisker and most of the data is on the right.

17. The number of data values on each side of the median is the same.

19. a.

b. 944 calories

c.

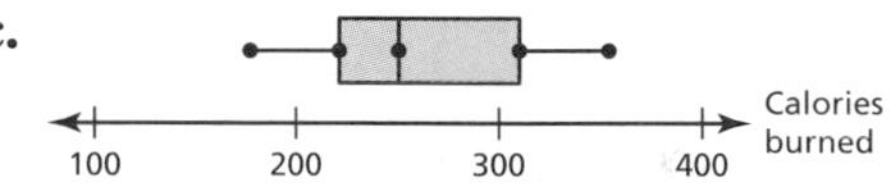

d. The outlier makes the right whisker longer, increases the length of the box, increases the third quartile, and increases the median. In this case, the first quartile and the left whisker were not affected.

21. *Sample answer:* 0, 5, 10, 10, 10, 15, 20

23. *Sample answer:* 1, 7, 9, 10, 11, 11, 12

25. $y = 3x + 2$

27. $y = -\frac{1}{4}x$

29. B

Section 12.4 Shapes of Distributions *(pages 631–633)*

1. The shape of a skewed distribution will have a tail on one side. The shape of a symmetric distribution is even, or symmetrical, with respect to the mean.

3. about 95%

5.

Most of the data are on the right. So, the distribution is skewed left.

7.

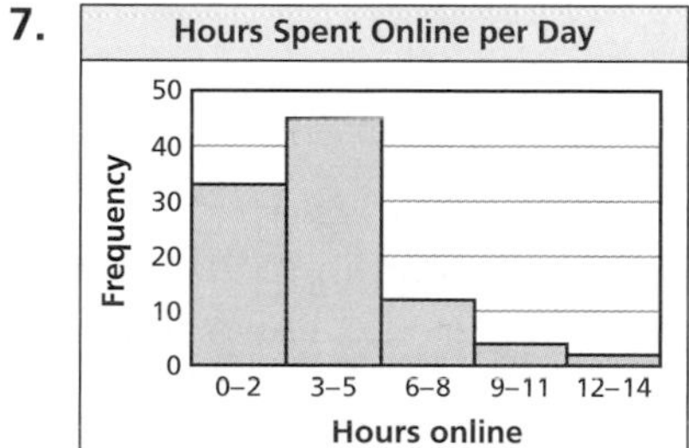

Most of the data are on the left. So, the distribution is skewed right.

Section 12.4 Shapes of Distributions *(continued)* *(pages 631–633)*

9. Because the distribution is high on the right and the tail of the graph extends to the left, the distribution is skewed left. So, the median best represents the center of the data and the five-number summary best represents the spread of the data.

11. Because the distribution is high on the right and the tail of the graph extends to the left, the distribution is skewed left. So, the median best represents the center of the data and the five-number summary best represents the spread of the data.

13. A

15. *Sample answer:* histogram; A histogram can show a variety of intervals and easily show the distribution of large data sets. A stem-and-leaf plot uses every data value and can be very tedious for large data sets. Also, stem-and-leaf plots usually only show intervals that are multiples of 10.

17. Using too few or too many intervals may make it difficult to find a pattern or determine the shape of distribution.

19. $y = -x + 4$

21. $y = 2x - 7$

Section 12.5 Scatter Plots and Lines of Fit *(pages 641–643)*

1. They must be ordered pairs so there are equal amounts of x- and y-values.

3. **a–b.**

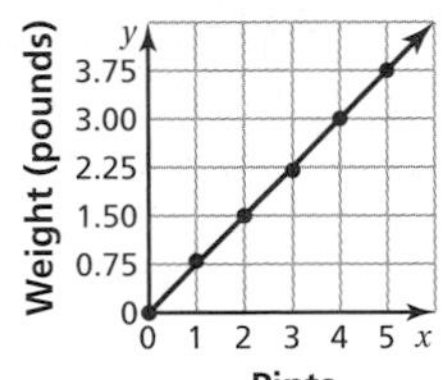

c. *Sample answer:* $y = 0.75x$

d. *Sample answer:* 7.5 lb

e. *Sample answer:* $16.88

5. **a.** 3.5 h **b.** $85

   c. There is a positive relationship between hours worked and earnings.

7. positive relationship

9. positive relationship

11. *Sample answer:* not a good representation; Too many points in the data set lie below the line.

13. *Sample answer:* good representation; The same number of points in the data set lie above and below the line.

15. **a.** positive relationship

    b. The more time spent studying, the better the test score.

17. The slope of the line of best fit should be close to 1.

19. *Sample answer:* The points follow a U-shaped pattern. There is a relationship between x and y, but it is not linear.

21. *Sample answer:* There appears to be two outliers. There appears to be a positive relationship between x and y.

23. 2

25. -4

Section 12.6 Analyzing Lines of Fit

(pages 649–651)

1. A residual is positive when the actual value is greater than value from model. A residual is negative when the actual value is less than the value from model.

3. -0.98 because it is closer to -1 than 0.91 is to 1. $(|-0.98| > |0.91|)$

5. The points (x, residual) are all above the horizontal axis. So, the equation does not model the data well.

7. $y = 3.5$; $r = 0$; There is no correlation between x and y. The equation does not fit the data.

9. $y = 357.5x - 495$

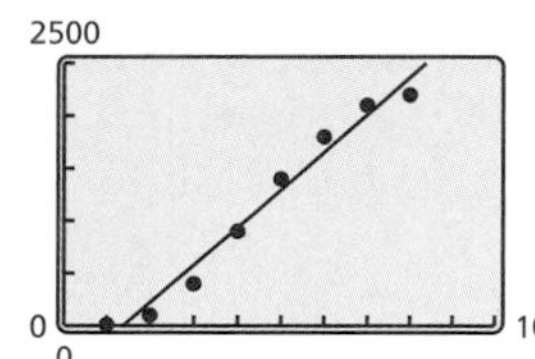

11. A

13. yes; yes; Talking longer causes life of battery to decrease.

15. no

17. **a.** $y = 0.2x + 25$; $r \approx 0.283$; The relationship between x and y is a weak positive correlation and the equation does not fit the data well.

b. The points (x, residual) form a ∩-shaped pattern. The equation does not fit the data well.

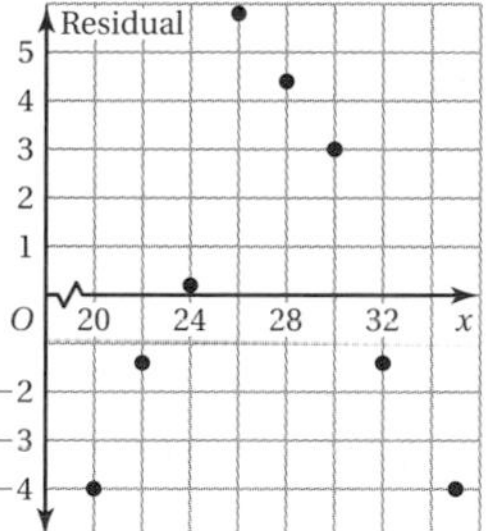

19. **a.** $y = -0.08x + 3.8$; $r \approx -0.965$; The relationship between x and y is a strong negative correlation and the equation closely models the data.

b. The slope is the change in grade point average per hour of television watched. The y-intercept is the grade point average of a student who does not watch television.

c. about 2.7

d. *Sample answer:* no; Watching television does not cause a lower grade point average, but there is a correlation. Studying less could be the cause.

21. Think back to quadratic functions and regression.

23. 15; 8; 2

Section 12.7 Two-Way Tables
(pages 656 and 657)

1. The joint frequencies are the entries in the two-way table that differentiate the two categories of data collected. The marginal frequencies are the sums of the rows and columns of the two-way table.

3. total of females surveyed: 73;
total of males surveyed: 59

5. 51

7. 71 students are juniors.
75 students are seniors.
93 students are attending the school play.
53 students are not attending the school play.

9. **a.** 19; 42

b. 72 9th-graders were surveyed.
74 10th-graders were surveyed.
65 11th-graders were surveyed.
112 students chose grades.
40 students chose popularity.
59 students chose sports.

c. about 8.5%

11. **a.**

		Eye Color			
		Green	Blue	Brown	Total
Gender	Male	5	16	27	48
	Female	3	19	18	40
	Total	8	35	45	88

b. 48 males were surveyed.
40 females were surveyed.
8 students have green eyes.
35 students have blue eyes.
45 students have brown eyes.

c.

		Eye Color		
		Green	Blue	Brown
Gender	Male	63%	46%	60%
	Female	38%	54%	40%

Sample answer: 62.5% represents that 62.5% of the students with green eyes are male. 40% represents that 40% of the students with brown eyes are female.

13. Be careful not to count the females with green eyes twice.

15. $\frac{x+2}{x+4}$; $x = 3$

17. B

Section 12.8 Choosing a Data Display
(pages 662 and 663)

1. yes; Different displays may show different aspects of the data.

3. *Sample answer:*

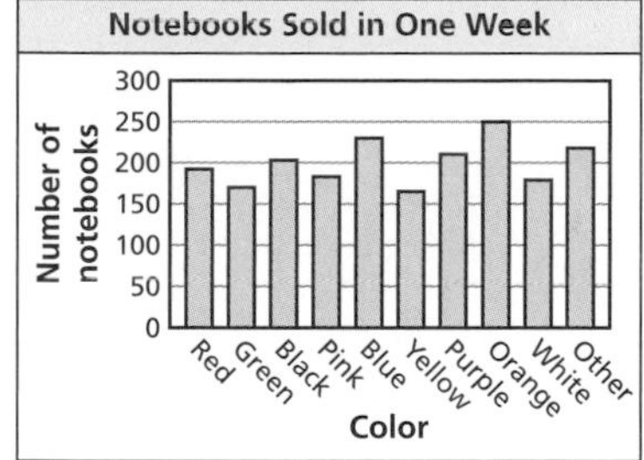

A bar graph shows the data in different color categories.

5. *Sample answer:* Dot graph: shows changes over time.

7. *Sample answer:* Line graph: shows changes changes over time.

9. The pictures of the bikes are larger on Monday, which makes it seem like the distance is the same each day.

11. The intervals are not the same size.

13. *Sample answer:* bar graph; Each bar can represent a different vegetable.

15. *Sample answer:* dot plot

17. Does one display better show the differences in digits?

19. $8x = 24$

Selected Answers

Key Vocabulary Index

Mathematical terms are best understood when you see them used and defined *in context.* This index lists where you will find key vocabulary. A full glossary is available in your Record and Practice Journal and at *BigIdeasMath.com.*

Student Index

This student-friendly index will help you find vocabulary, key ideas, and concepts. It is easily accessible and is designed to be a reference for you whether you are looking for a definition, a real-life application, or help with avoiding common errors.

A

B

C

D

E

Student Index

Student Index

H

I

J

L

M

S

T

Additional Answers

Chapter 1

Section 1.2

Record and Practice Journal

2. indigo: 45°, 45°, 90°; violet: 60°, 60°, 60°; orange: 75°, 65°, 40°; yellow: 25°, 60°, 95°; blue: 75°, 75°, 30°; green: 15°, 135°, 30°

Lesson 1.3b

Record and Practice Journal Practice

1. $x = 3$ *or* $x = -3$

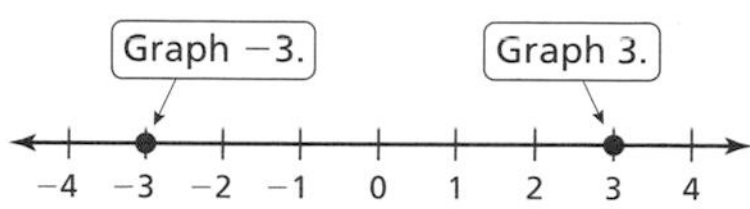

2. $x = 5$ *or* $x = 3$

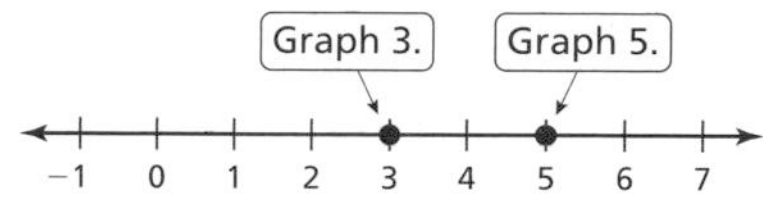

3. $x = 0$

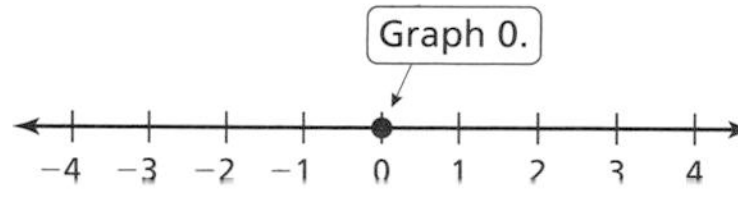

4. $x = 3$ *or* $x = -4$

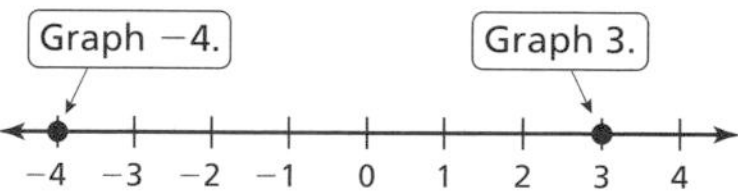

5. $x = 4$ *or* $x = -\dfrac{8}{3}$

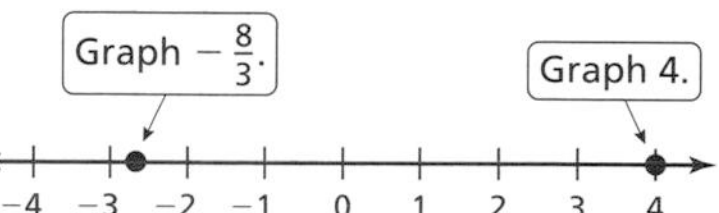

6. No solution

7. $x = -1$ *or* $x = -4$

8. $x = \dfrac{1}{2}$ *or* $x = -\dfrac{7}{2}$

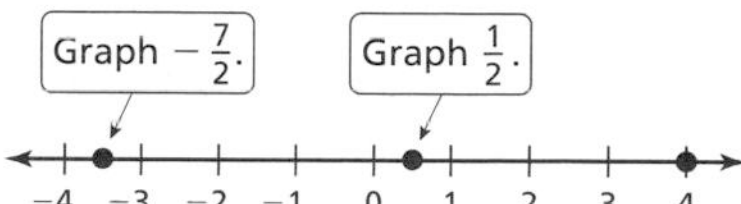

9. $x = 7$ *or* $x = 1$

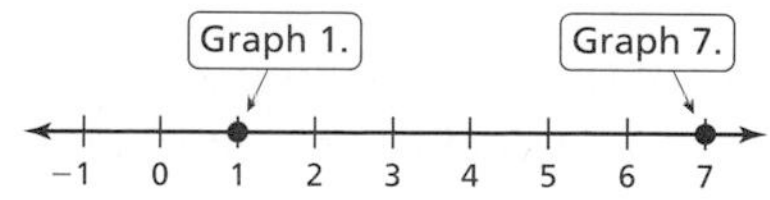

10. $x = 1$ *or* $x = -7$

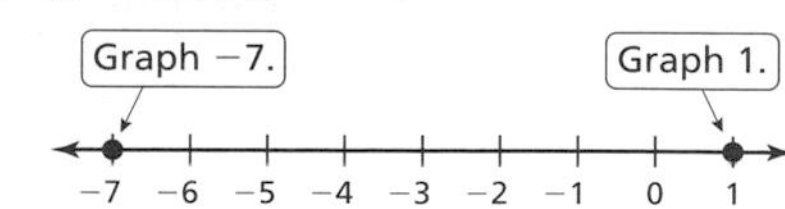

11. $|x - 4| - 1.5$

Chapter 2

Section 2.1

On Your Own

1.

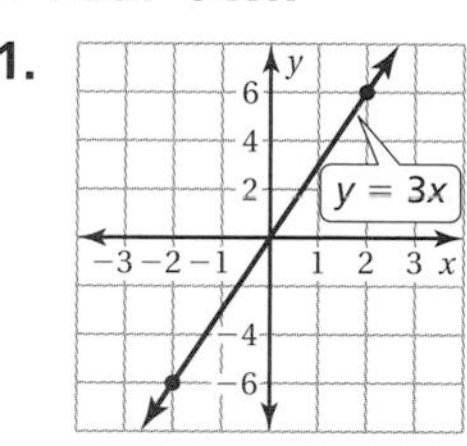

2.

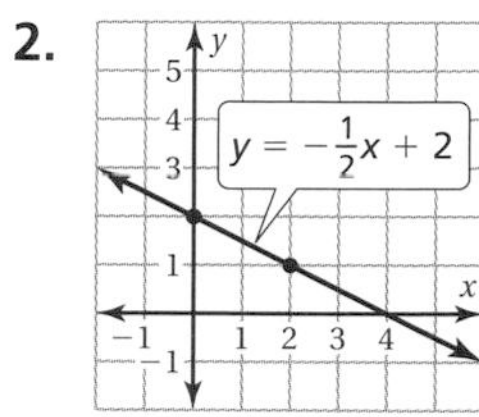

3.

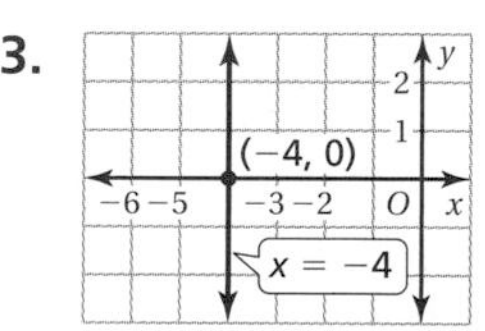

4.

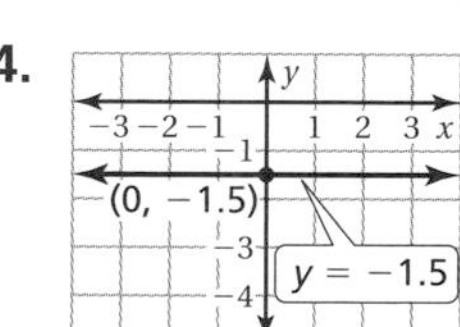

Practice and Problem Solving

7.

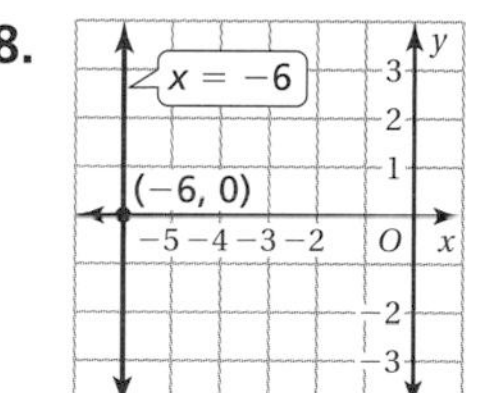

8.

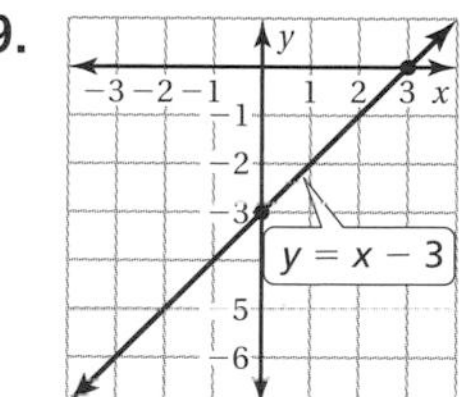

9.

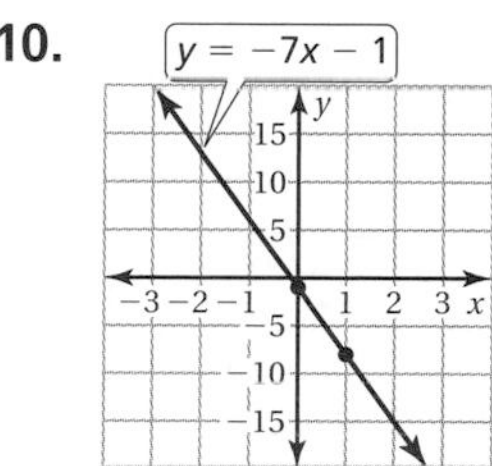

10.

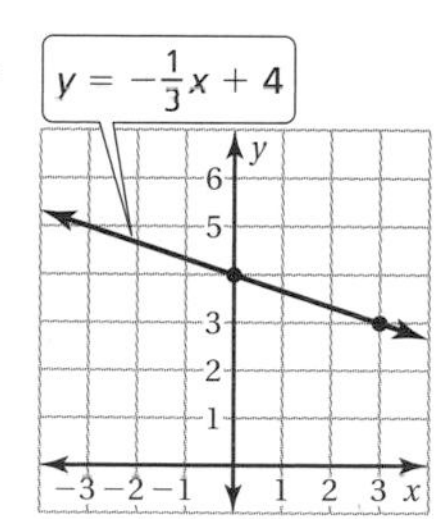

11. $y = -\frac{1}{3}x + 4$

Additional Answers

12.

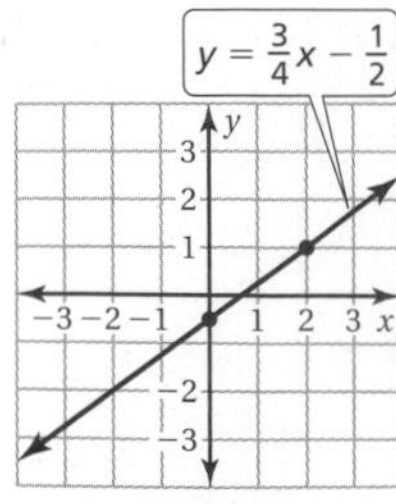

13.

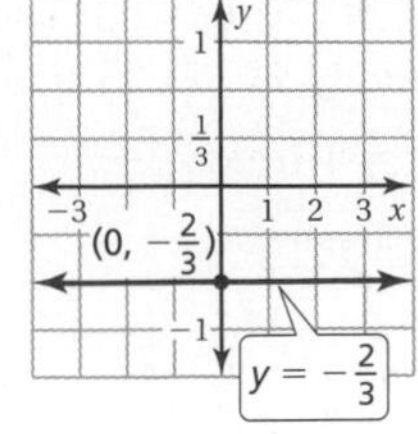

14.

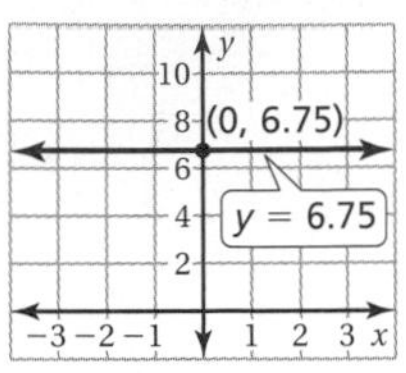

15.

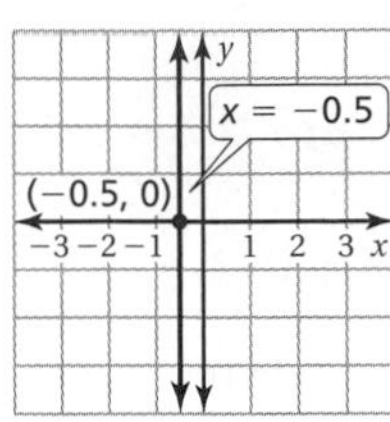

16.

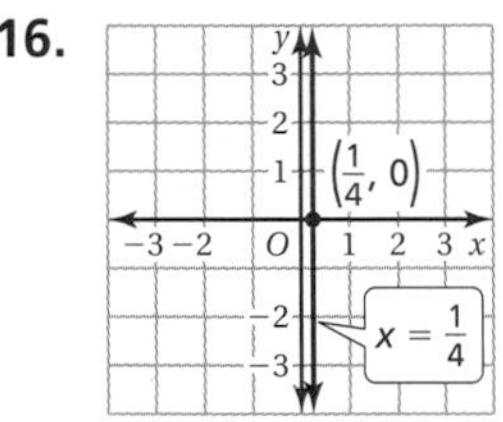

17. The equation $x = 4$ is graphed, not $y = 4$.

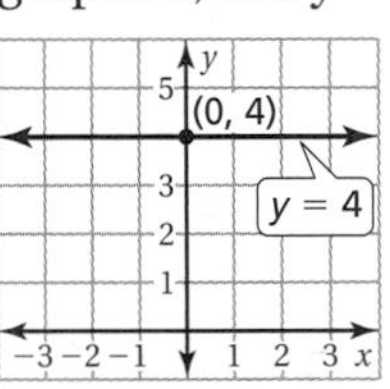

18.

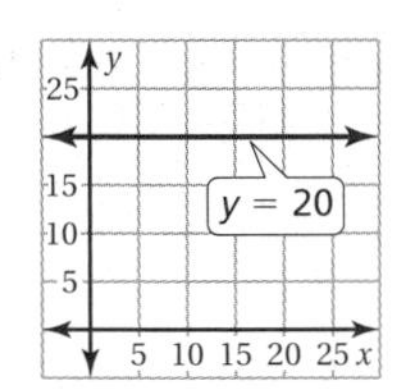

Sample answer: No matter how many text messages are sent, the cost is $20.

19. a.

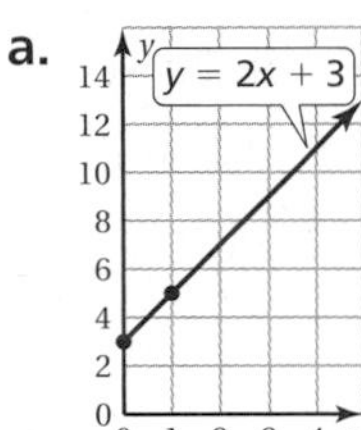

b. about $5

c. $5.25

21. $y = -\frac{5}{2}x + 2$

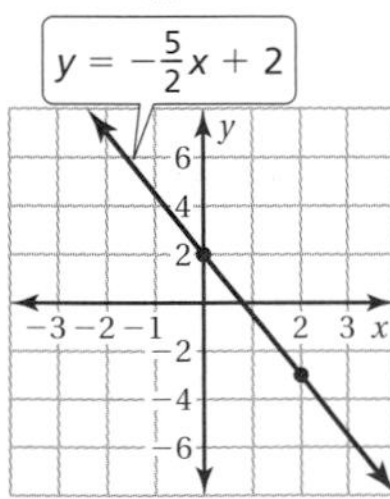

22. $y = 12x - 9$

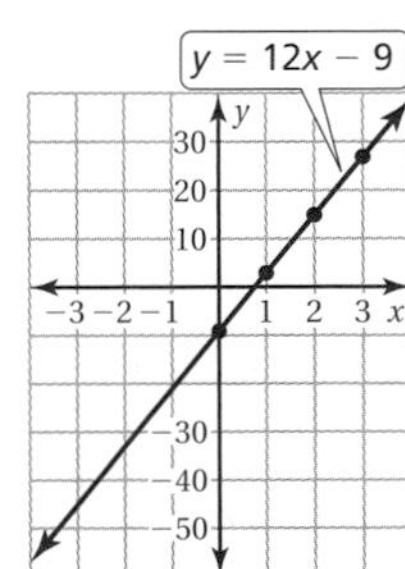

23. $y = -2x + 3$

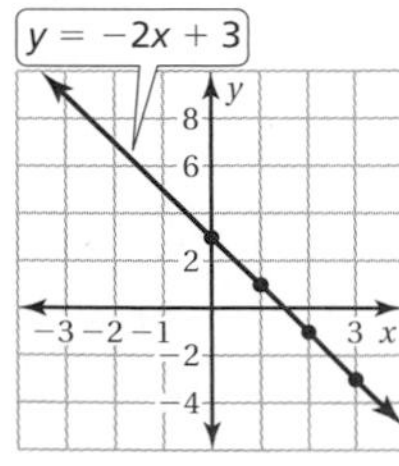

24. a. $y = 100 + 12.5x$ **b.** 6 mo

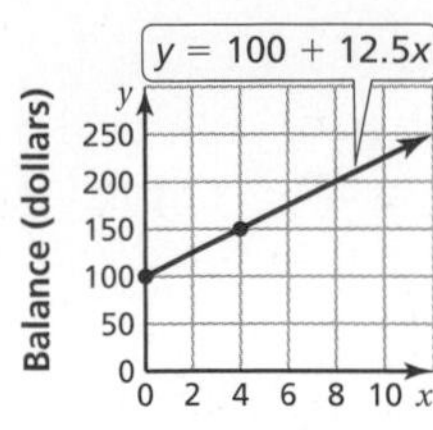

26. *Sample answer:* If you are 13 years old, the sea level has risen 26 millimeters since you were born.

$y = 2x$

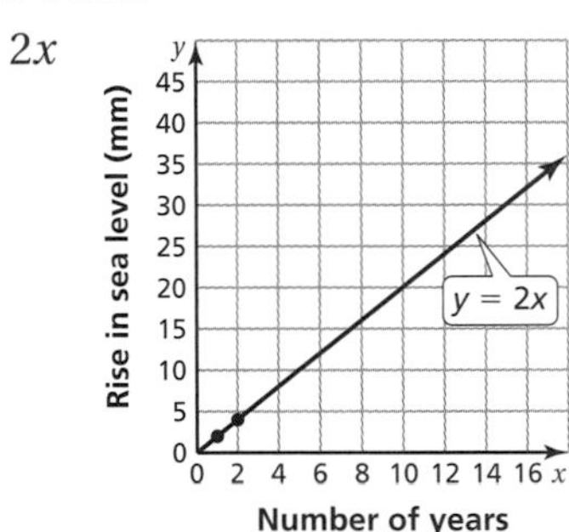

27. a. *Sample answer:*

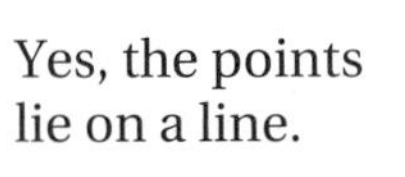

Yes, the points lie on a line.

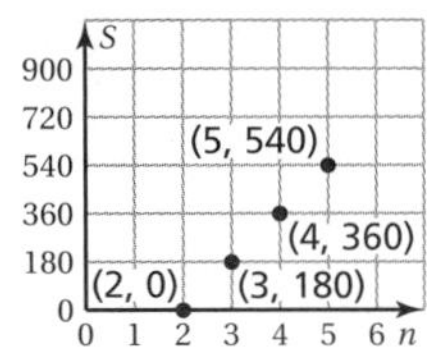

b. No, $n = 3.5$ does not make sense because a polygon cannot have half a side.

Section 2.2

Practice and Problem Solving

26. *Sample answer:*

a. Yes, it follows the guidelines.

b.

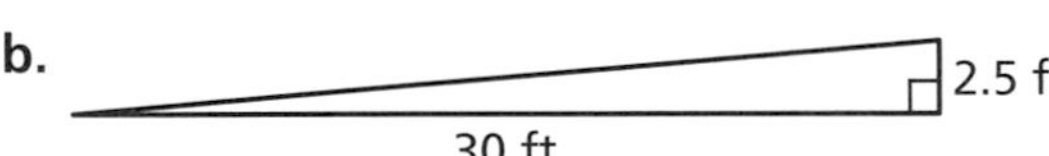

31. a. $\frac{3}{40}$

b. The cost increases by $3 for every 40 miles you drive, or the cost increases by $0.075 for every mile you drive.

32. The boat ramp, because it has a 16.67% grade.

33. yes; The slopes are the same between the points.

34. $2750 per month

35. yes; When you switch the coordinates, the differences in the numerator and denominator are the opposite of the numbers when using the slope formula. You still get the same slope.

Lesson 2.2b

Practice

6. Use the vertices of the quadrilateral to find the slope of each side. If opposite sides are parallel (have the same slope), the quadrilateral is a parallelogram.

slope of $AB = -\frac{1}{7}$; slope of $BC = -\frac{5}{2}$;

slope of $CD = -\frac{1}{6}$; slope of $DA = -\frac{5}{3}$;

No, it is not a parallelogram. Because they have different slopes, opposite sides are not parallel.

12. Use the vertices of the quadrilateral to find the slope of each side. If adjacent sides are perpendicular, then the parallelogram is a rectangle. Note that the quadrilateral is a parallelogram, so you already know that opposite sides are parallel.

slope of $JK = \frac{2}{3}$; slope of $KL = -\frac{3}{2}$

slope of $LM = \frac{2}{3}$; slope of $MJ = -\frac{3}{2}$

Yes, it is a rectangle.

JK is perpendicular to *KL* because $\frac{2}{3} \cdot \left(-\frac{3}{2}\right) = -1$.

KL is perpendicular to *LM* because $-\frac{3}{2} \cdot \frac{2}{3} = -1$.

LM is perpendicular to *MJ* because

$\frac{2}{3} \cdot \left(-\frac{3}{2}\right) = -1$.

MJ is perpendicular to *JK* because $-\frac{3}{2} \cdot \frac{2}{3} = -1$.

Record and Practice Journal Practice

1. line *B* and line *G* **2.** line *B* and line *R*

3. yes **4.** no **5.** yes

6. line *B* and line *R* **7.** line *R* and line *G*

8. yes **9.** no **10.** yes

Section 2.3

On Your Own

3.

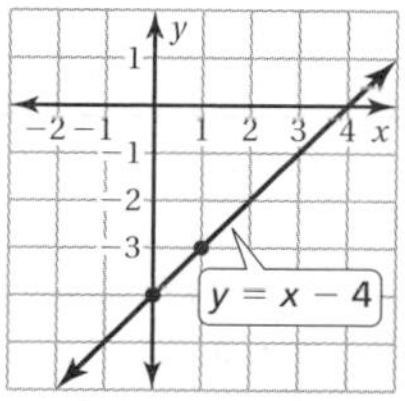

x-intercept: 4

4.

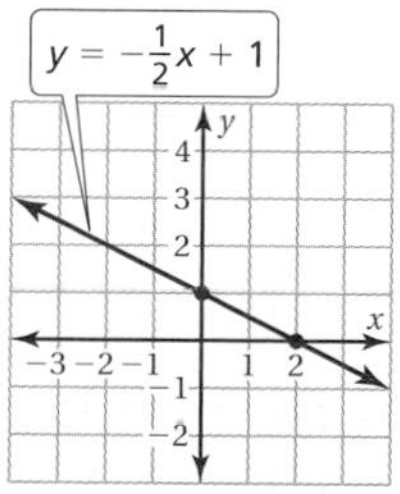

x-intercept: 2

5. The *y*-intercept means that the taxi has an initial fee of \$1.50. The slope means the taxi charges \$2 per mile.

Practice and Problem Solving

16. The *y*-intercept should be -3.
$y = 4x - 3$
The slope is 4 and the *y*-intercept is -3.

17. a.

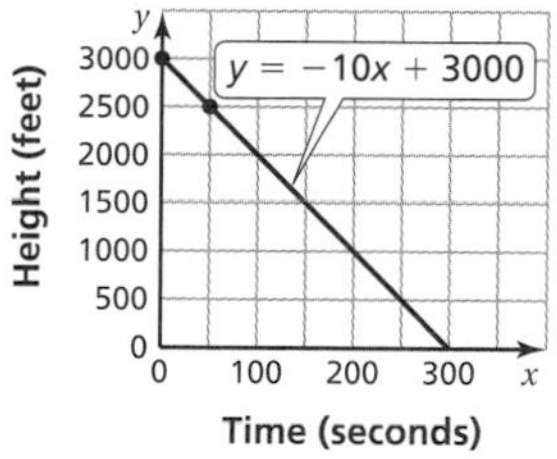

b. The *x*-intercept of 300 means the skydiver lands on the ground after 300 seconds. The slope of -10 means that the skydiver falls to the ground at a rate of 10 feet per second.

19.

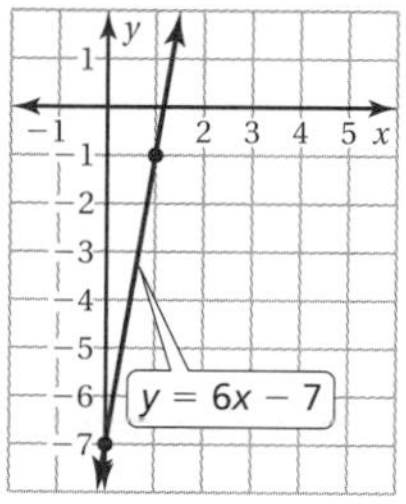

x-intercept: $\frac{7}{6}$

20.

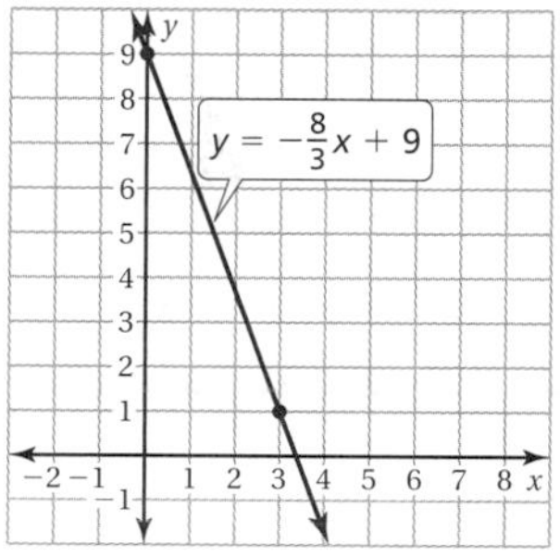

x-intercept: $\frac{27}{8}$

21.

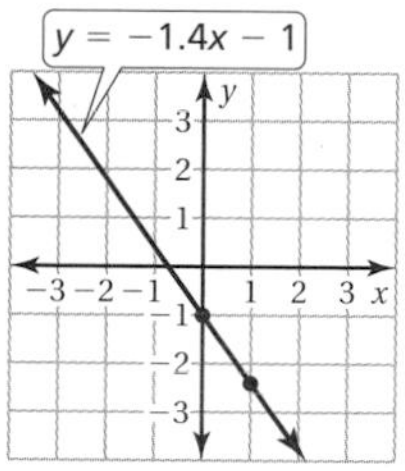

x-intercept: $-\frac{5}{7}$

22.

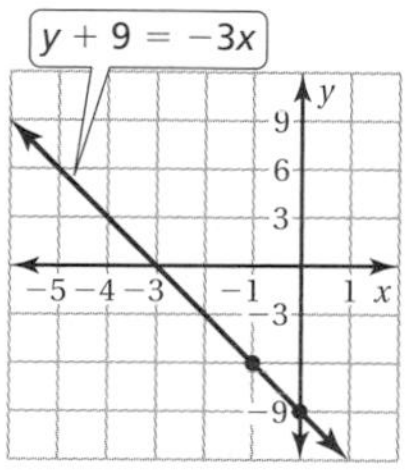

x-intercept: -3

23.

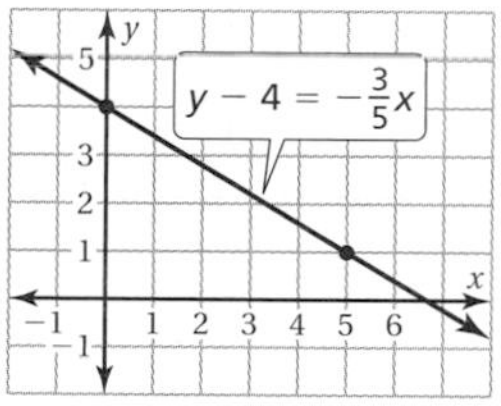

x-intercept: $\frac{20}{3}$

24. **a.**

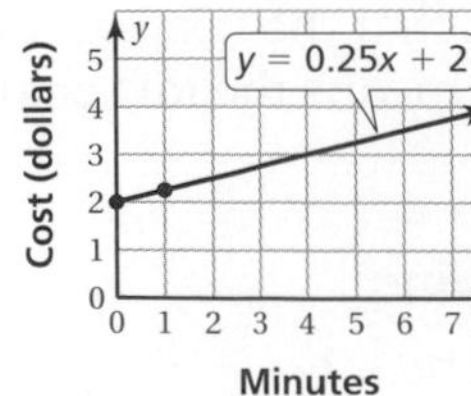

b. The slope of 0.25 means that it costs $0.25 for each minute spent making a long distance call. The y-intercept of 2 means that there is an initial fee of $2.

25. $y = 0.75x + 5$

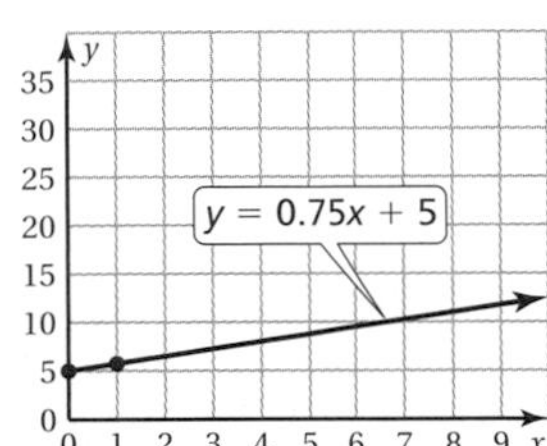

26. $y = 5x - 40$

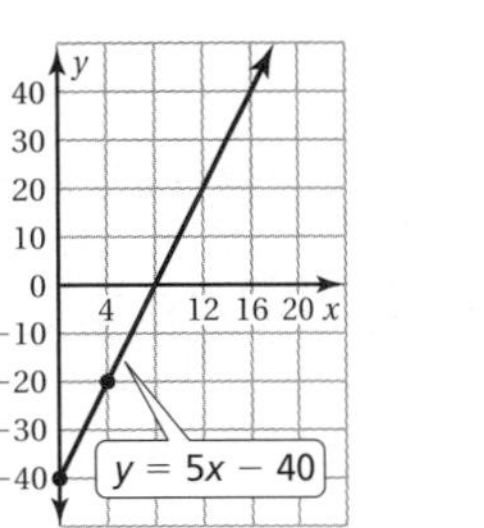

27. $y = 0.15x + 35$

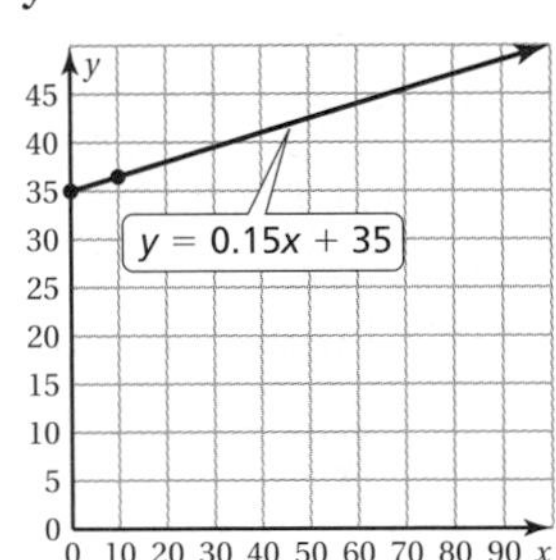

Section 2.4

On Your Own

3.

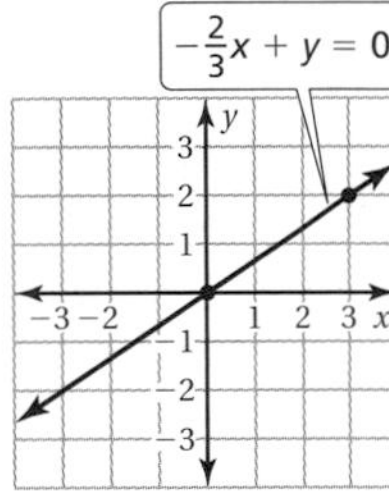

4.

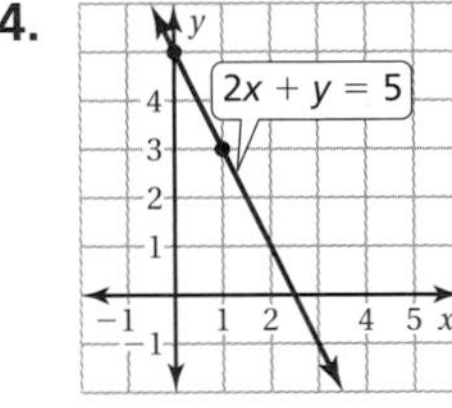

5.

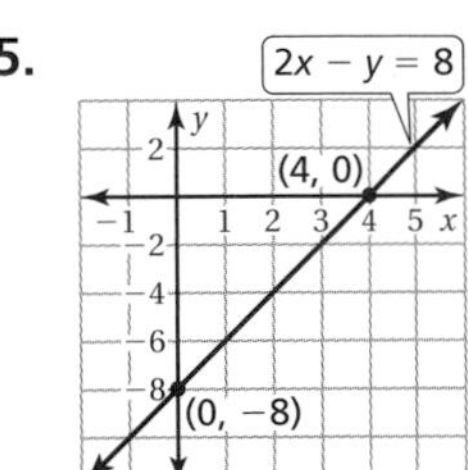

6.

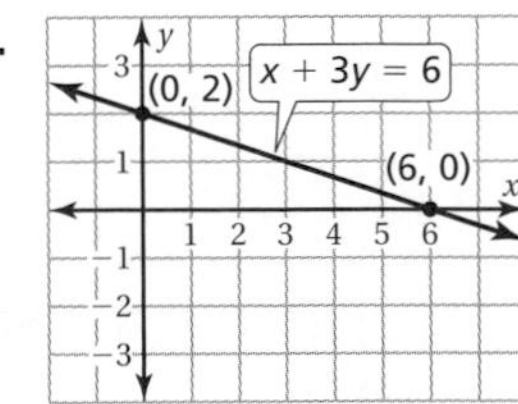

7.

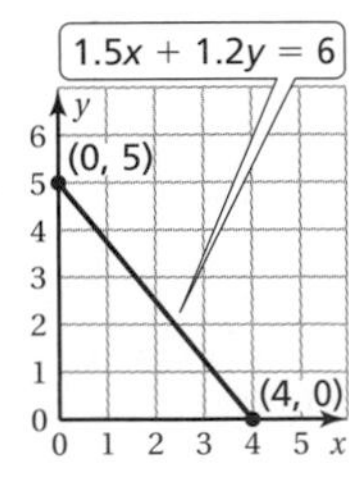

The x-intercept shows that you can buy 4 pounds of apples if you do not buy any oranges. The y-intercept shows that you can buy 5 pounds of oranges if you do not buy any apples.

Practice and Problem Solving

9.

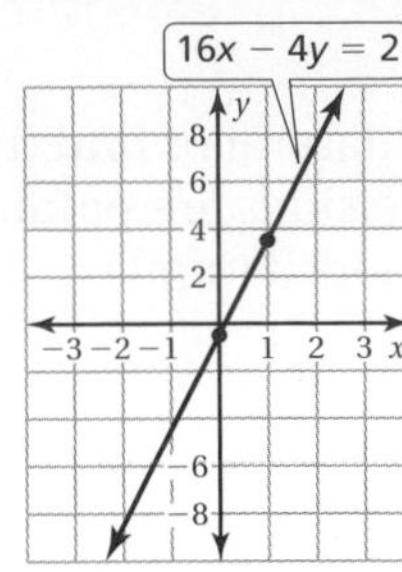

10.

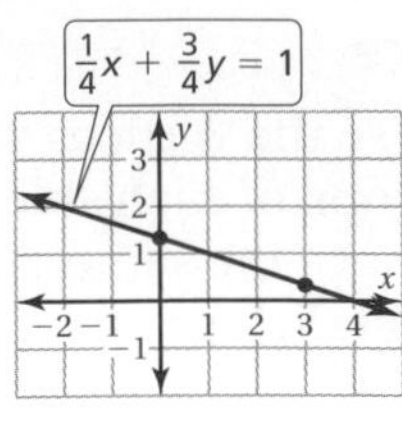

11. x-intercept: -6
y-intercept: 3

12. x-intercept: -4
y-intercept: -5

13. x-intercept: none
y-intercept: -3

14. They should have let $y = 0$, not $x = 0$.

$$-2x + 3y = 12$$
$$-2x + 3(0) = 12$$
$$-2x = 12$$
$$x = -6$$

15. **a.** $-25x + y = 65$ **b.** $390

16.

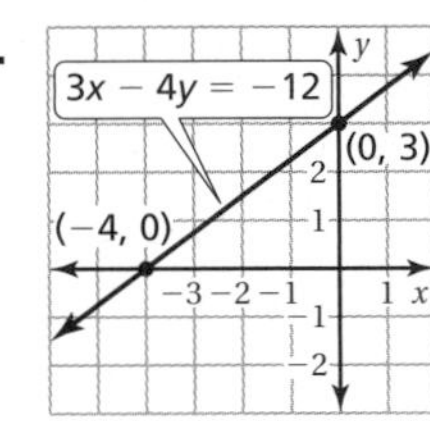

17.

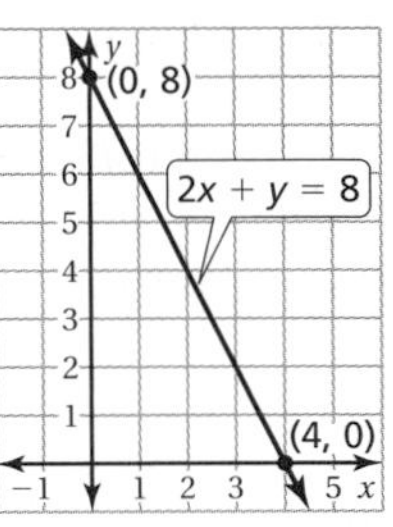

18.

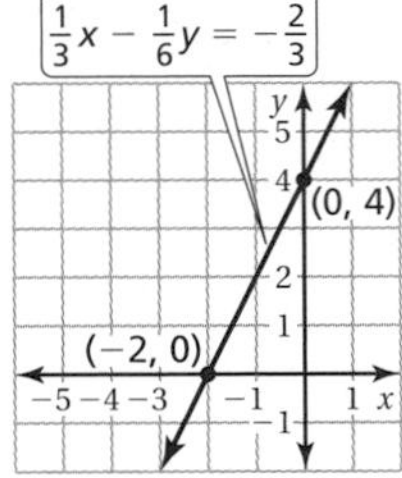

19. x-intercept: 9
y-intercept: 7

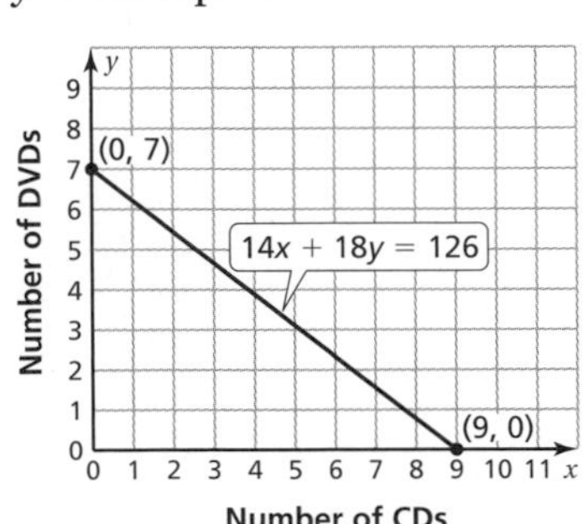

21. **a.** $9.45x + 7.65y = 160.65$

b.

9.45x + 7.65y = 160.65

Hours worked as server

Hours worked as host

22. no; For example, $y = 5$ does not have an x-intercept, neither do any horizontal lines except $y = 0$.

23. **a.** $y = 40x + 70$

b. x-intercept: $-\frac{7}{4}$; It will not be on the graph because you cannot have a negative time.

c.

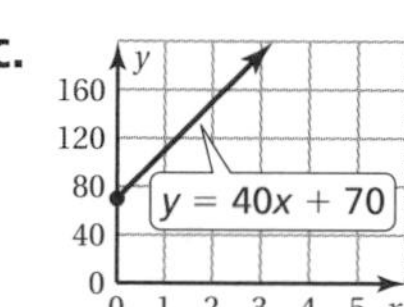

2.1–2.4 Quiz

1.

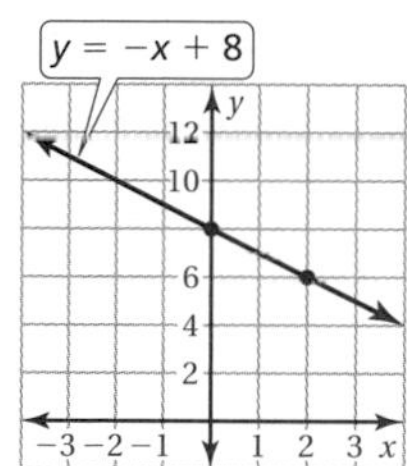

2.

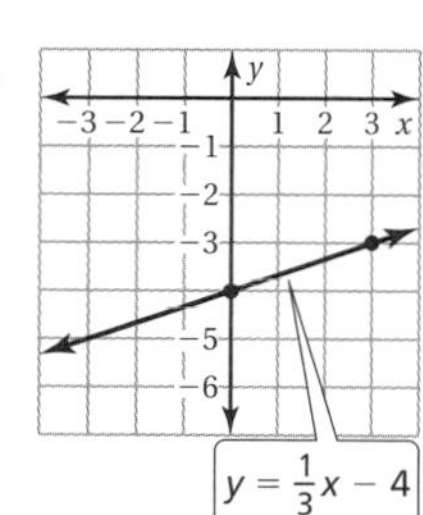

3.

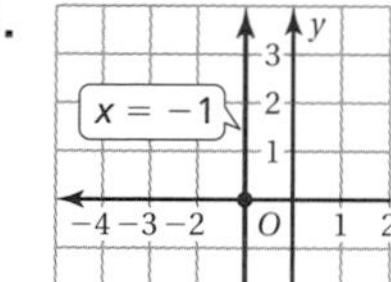

4.

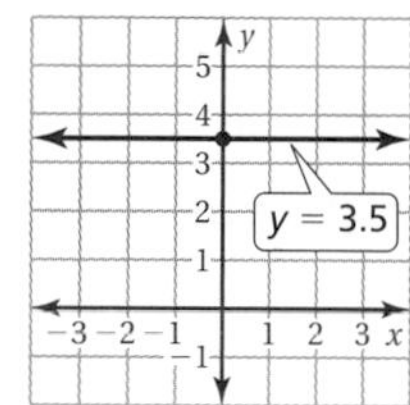

14.

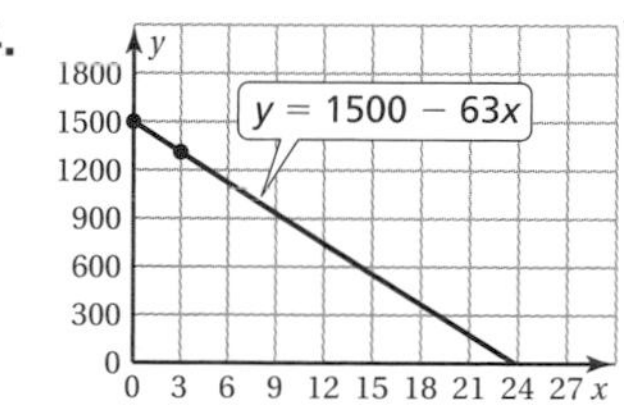

15.

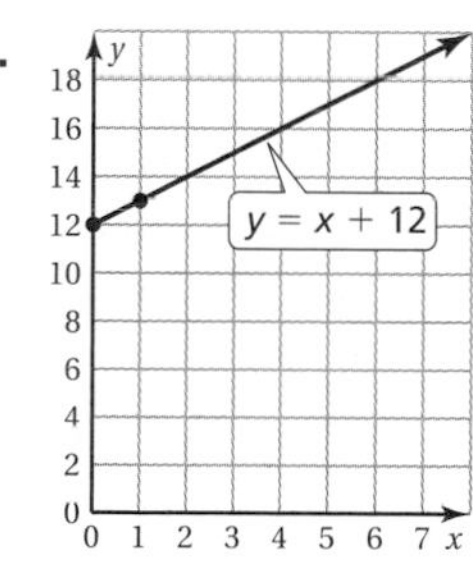

16. **a.**

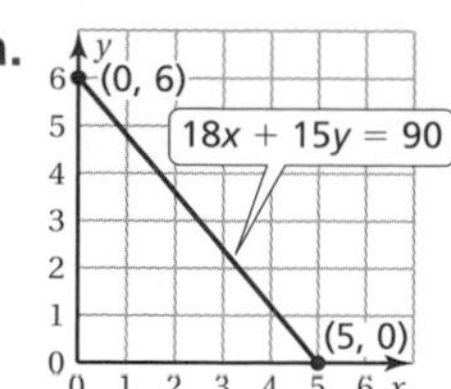

b. The x-intercept, 5, shows that you can buy 5 gallons of blue paint if you do not buy any white paint. The y-intercept, 6, shows that you can buy 6 gallons of white paint if you do not buy any blue paint.

Section 2.5

Record and Practice Journal

1.

a. top line: slope: $\frac{1}{2}$; y-intercept: 4; $y = \frac{1}{2}x + 4$

middle line: slope: $\frac{1}{2}$; y-intercept: 1; $y = \frac{1}{2}x + 1$

bottom line: slope: $\frac{1}{2}$; y-intercept: -2;

$y = \frac{1}{2}x - 2$

The lines are parallel.

b. right line: slope: -2; y-intercept: 3; $y = -2x + 3$

middle line: slope: -2; y-intercept: -1;

$y = -2x - 1$

left line: slope: -2; y-intercept: -5;

$y = -2x - 5$

The lines are parallel.

c. line passing through (3, 2): slope: $-\frac{1}{3}$;

y-intercept: 3; $y = -\frac{1}{3}x + 3$

line passing through (3, 7): slope: $\frac{4}{3}$;

y-intercept: 3; $y = \frac{4}{3}x + 3$

line passing through (6, 4): slope: $\frac{1}{6}$;

y-intercept: 3; $y = \frac{1}{6}x + 3$

The lines have the same y-intercept.

d. line passing through (1, 2): slope: 2;

y-intercept: 0; $y = 2x$

line passing through (1, -1): slope: -1;

y-intercept: 0; $y = -x$

line passing through (3, 1): slope: $\frac{1}{3}$;

y-intercept: 0; $y = \frac{1}{3}x$

The lines have the same y-intercept.

Additional Answers

2. a. $y = 4$ $\quad y = -2x - 6$
$y = -2$ $\quad y = -2x + 8$

b. $y = 5$ $\quad y = x + 1$
$y = -2$ $\quad y = x + 5$

Practice and Problem Solving

17. a–b.

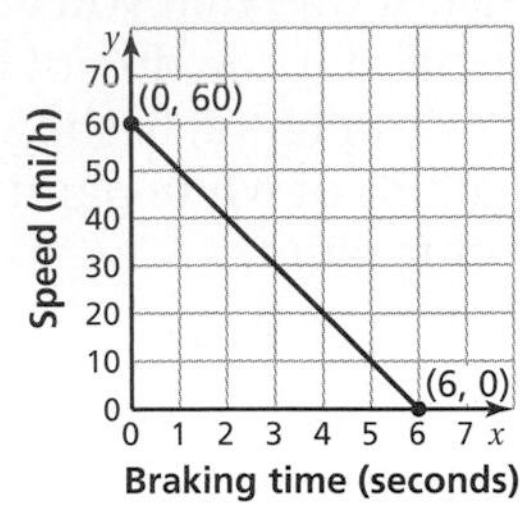

(0, 60) represents the speed of the automobile before braking. (6, 0) represents the amount of time it takes to stop. The line represents the speed y of the automobile after x seconds of braking.

c. $y = -10x + 60$

Lesson 2.6b

Practice

15. Check students' work.

Example 1	**Example 2**
$y - y_1 = m(x - x_1)$	$y = mx + b$
$y + 2 = \frac{1}{2}(x - 6)$	$1 = \frac{1}{4}(-3) + b$
$y + 2 = \frac{1}{2}x - 3$	$1 = -\frac{3}{4} + b$
$y = \frac{1}{2}x - 5$	$\frac{7}{4} = b$
	So, $y = \frac{1}{4}x + \frac{7}{4}$.

Record and Practice Journal Practice

1. $y = 2x + 1$
2. $y = -3x - 8$
3. $y = \frac{3}{5}x + 4$
4. $y = -\frac{7}{2}x + 2$
5. $y = 3x - 1$
6. $y = 3x - 5$
7. $y = 3x - 9$
8. $y = 3x + 3$
9. $y = -2x + 3$
10. $y = -\frac{1}{3}x + 6$
11. $y = \frac{1}{2}x - 7$
12. $y = \frac{7}{4}x + 24$
13. $y = -2x + 4$
14. $y = -2x - 8$
15. $y = -2x$
16. $y = -2x + 6$

Section 2.7

On Your Own

1. a.

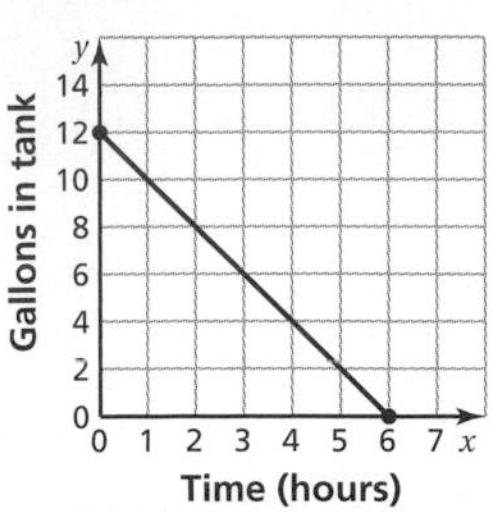

b. The x-intercept is 6. So, you can drive 6 hours before the tank is empty. The y-intercept is 12. So, there are 12 gallons in the tank before you start driving.

c. 3.5 or $3\frac{1}{2}$ h

Practice and Problem Solving

4. *Sample answer:* A gasoline tank initially has 16 gallons of gas. After 10 hours, the tank is empty.

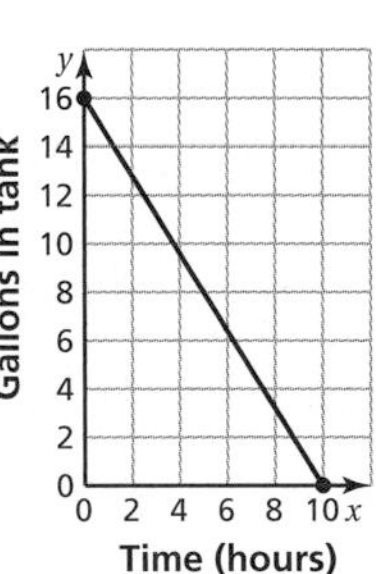

5. *Sample answer:* The temperature outside is falling 3°F every hour. After 7 hours, the temperature is 0°F.

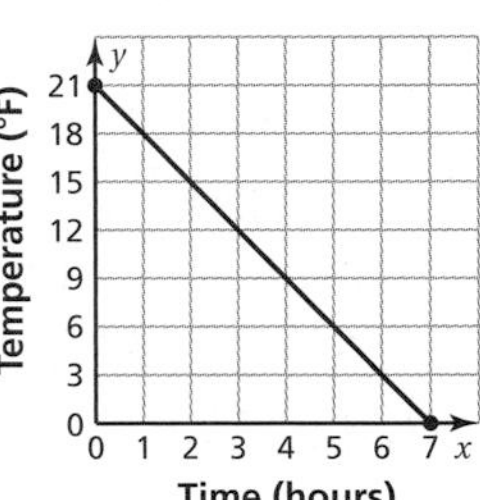

8. a. The x-intercept is 6. So, it takes 6 hours for your family to drive from Cincinnati to St. Louis. The y-intercept is 360. So, it is 360 miles from Cincinnati to St. Louis.

b. -60; Your distance from St. Louis decreases at a rate of 60 miles per hour

c. $y = -60x + 360$; Both intercepts would be less and the slope would be the same.

Record and Practice Journal Practice

1.

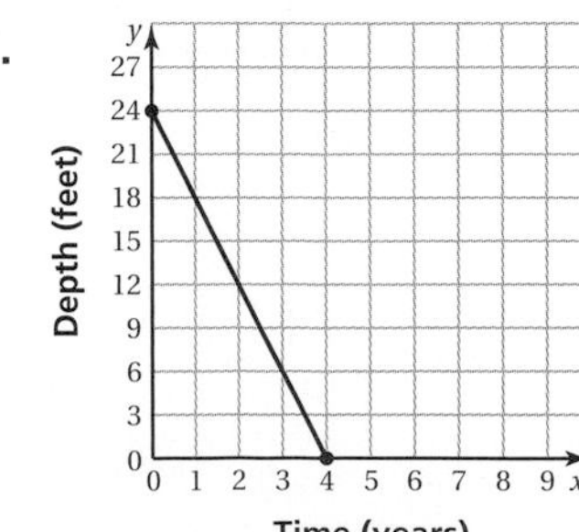

2. 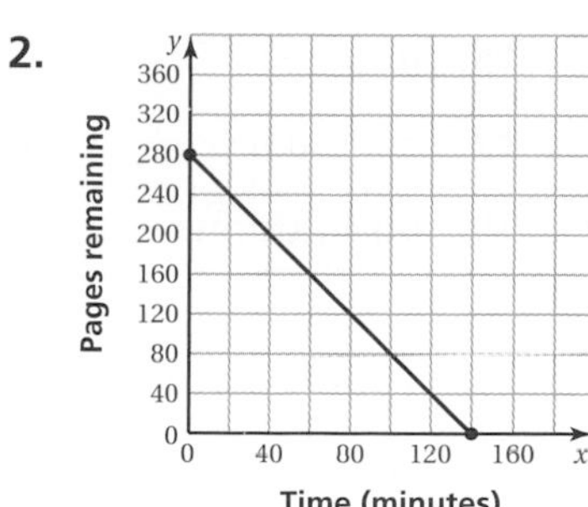

2.5–2.7 Quiz

12. a. Water remaining (liters) vs. Time (hours)

b. The x-intercept is the number of hours it takes to drain the pond. The y-intercept represents the amount of water in the pond initially.

13. a. slope: -35; y-intercept: 280

b. $y = -35x + 280$

c. 6 P.M.

Chapter 2 Test

7. 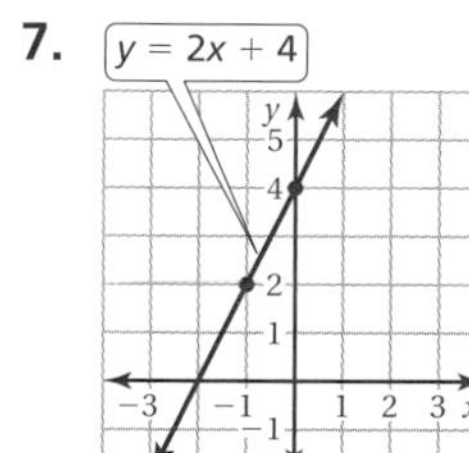

8. 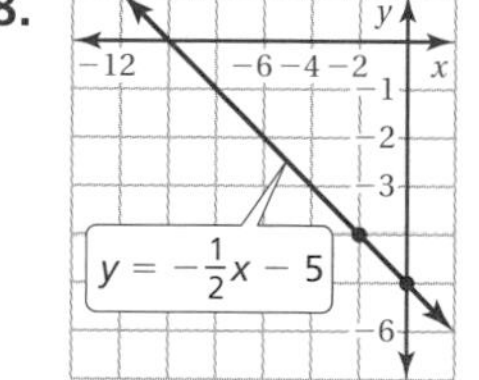

9. 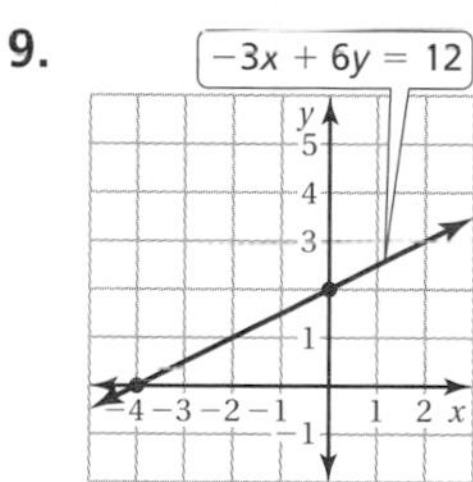

17. a. $y = 0.25x$

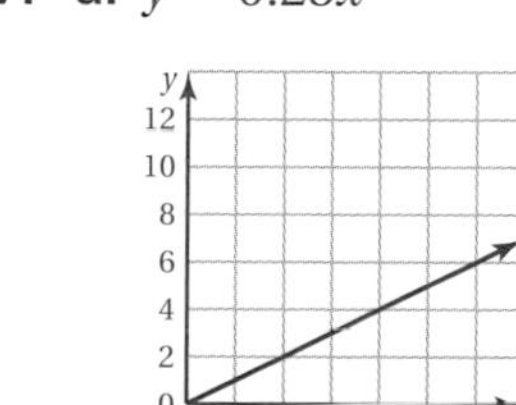

b. A person reading text can read 180 words in the same amount of time that a person reading Braille can read 45 words.

c. Because the graph has a positive slope, as x increases, y also increases.

Chapter 3

Record and Practice Journal Fair Game Review

11. 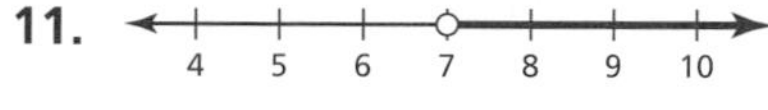

12. 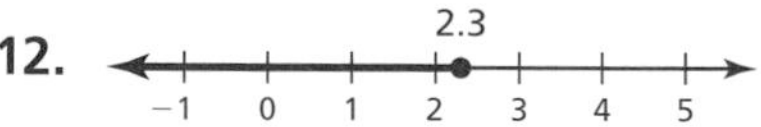

13.

14.

Section 3.1

Practice and Problem Solving

25. a. $a \geq 10$;

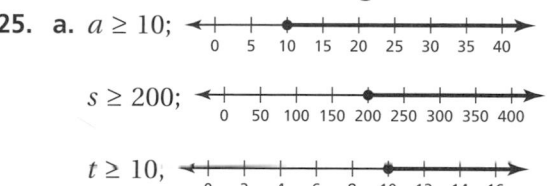

$s \geq 200$;

$t \geq 10$;

b. yes; You satisfy the swimming requirement of the course because $10(25) = 250$ and $250 \geq 200$.

Section 3.2

Practice and Problem Solving

11. $-\frac{3}{5} > d$;

12. $-\frac{1}{3} \leq g$;

13. $m \leq 1$;

14. $1.4 \leq k$;

15. $h < -1.5$;

16. $-\pi > s$;

17. $9.5 \geq u$;

Section 3.3

Practice and Problem Solving

13. $x < -28$;

Additional Answers

14. $w \geq -13$;

15. $k > 2$;

16. $x \leq -\frac{3}{8}$;

17. $y \leq -4$;

18. $b < 34.68$;

30. $x < 24$;

31. $y > \frac{11}{2}$;

32. $d \leq -8$;

33. $m > -12$;

34. $k \geq -9$;

35. $b > 4$;

42. *Sample answer:* Consider the inequality $5 > 3$. If you multiply or divide each side by -1 without reversing the direction of the inequality symbol, you obtain $-5 > -3$, which is not true. So, whenever you multiply or divide an inequality by a negative number, you must reverse the direction of the inequality symbol to obtain a true statement.

3.1–3.3 Quiz

14. $5x \leq -10$; $x \leq -2$

15. a. $s \geq 100$;

$t \geq 5$;

$u \geq 10$;

b. yes; Because 100 yards is equal to 300 feet and $350 \geq 300$.

Section 3.4

Practice and Problem Solving

11. They did not perform the operations in proper order.

$$\frac{x}{4} + 6 \geq 3$$
$$\frac{x}{4} \geq -3$$
$$x \geq -12$$

14. $h \leq -3$

15. $u < -17$

16. $n < 4.7$

17. $z > -0.9$

18. all real numbers

19. no solutions

20. no solutions

21. $20x + 100 \leq 320$; $x \leq 11$ \$20 bills

27. a. $3.5x + 350 \geq 500$; $x \geq 42\frac{6}{7}$; at least 43 more cars, so at least 143 cars total

b. Because each car will pay \$1 more, fewer cars will be needed for the theater to earn \$500.

Fair Game Review

30. 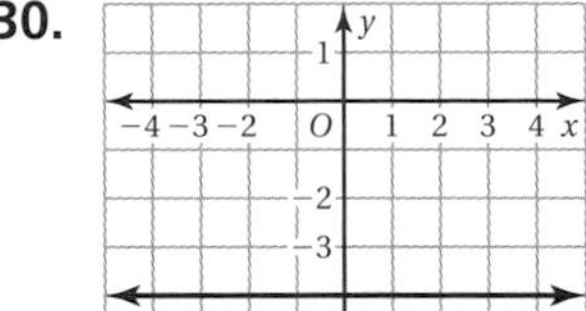

31. 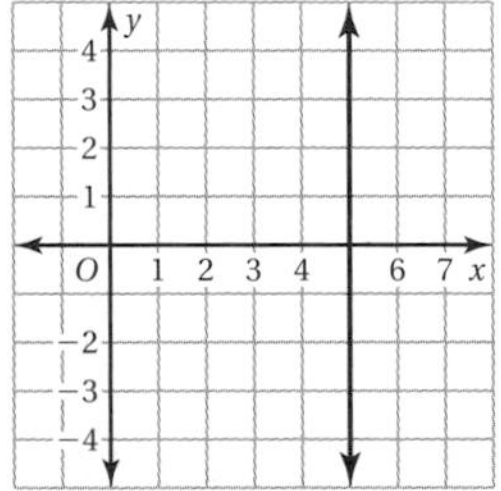

32. 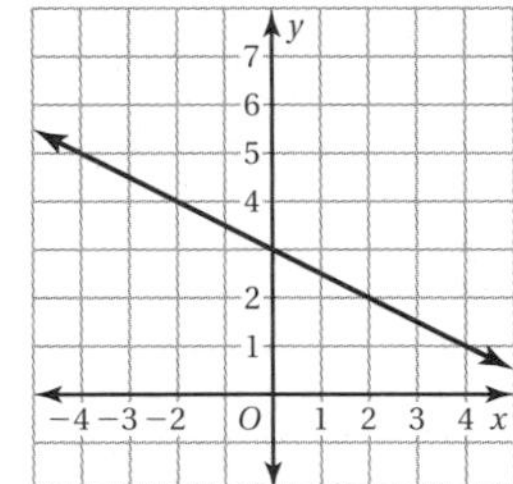

Lesson 3.4b

Practice

14. $x \geq -\frac{3}{2}$ *and* $x \leq 4$;

15. no solution

16. $x \leq -7$ *or* $x \geq 2$;

17. $x > 6$ *and* $x < 14$;

18. all real numbers; The absolute value of an expression must be greater than or equal to 0. So, $|4x - 2|$ will be greater than or equal to -6 for all values of x.

19. $|x - 44| \leq 3$; The least percent of voters who will vote for the new mayor is 41%. The greatest percent of voters who will vote for the new mayor is 47%.

Record and Practice Journal Practice

1. $4 < q < 6$;
   1 2 3 4 5 6 7 8
2. $-8 \le r < -5$;
   −10 −9 −8 −7 −6 −5 −4 −3
3. $3 \le s < 7$;
   2 3 4 5 6 7 8 9
4. $t \ge 1$ or $t < -3$;
   −4 −3 −2 −1 0 1 2 3
5. $x < -2$ or $x \ge 1$
6. $150 \le x < 200$;
   140 150 160 170 180 190 200 210
7. $1 < a < 4$;
   −1 0 1 2 3 4 5 6
8. $3 < x \le 6$;
   1 2 3 4 5 6 7 8
9. $b > 1$ or $b \le -3$;
   −4 −3 −2 −1 0 1 2 3
10. $y < 2$ or $y > 7$;
   1 2 3 4 5 6 7 8
11. $c \le 3$ or $c \ge -2$;
   −3 −2 −1 0 1 2 3 4
12. no solution
13. $|x - 3| \le 0.02$; The least weight of the coin that the country's mint will allow to be released into circulation is 2.98 grams. The greatest weight of the coin that country's mint will allow to be released into circulation is 3.02 grams.

Section 3.5

Record and Practice Journal

4. a.

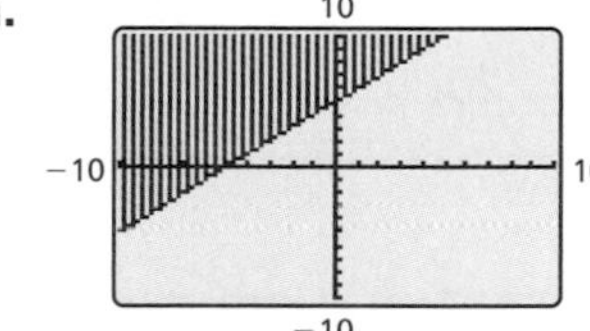

b.

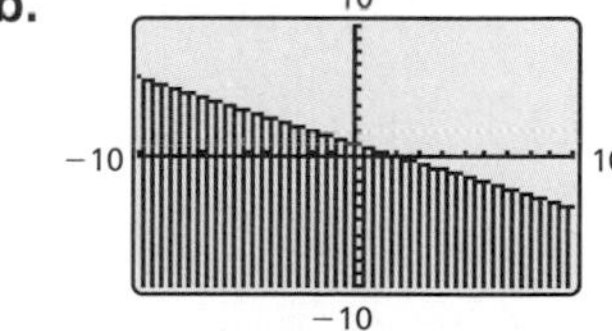

c.

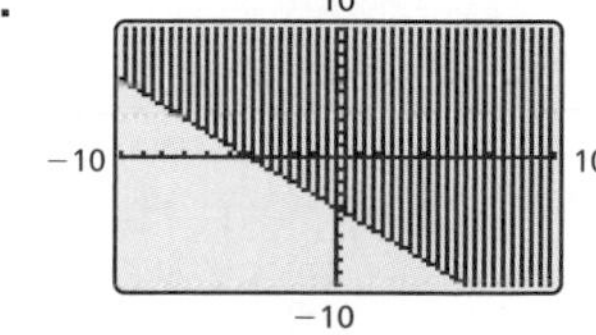

On Your Own

7.

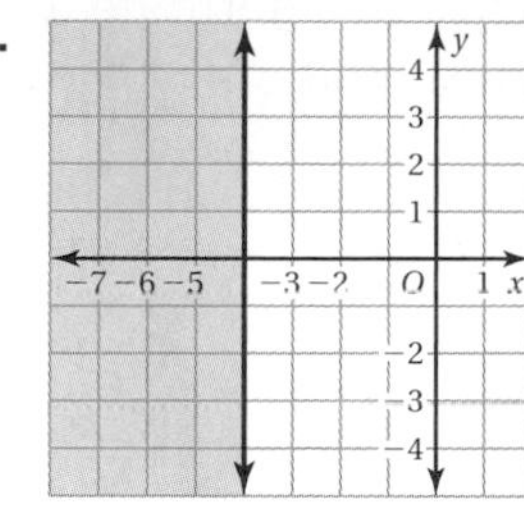

8.

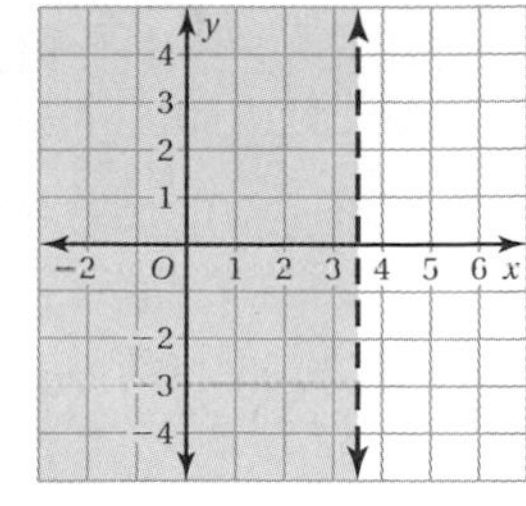

Practice and Problem Solving

33.

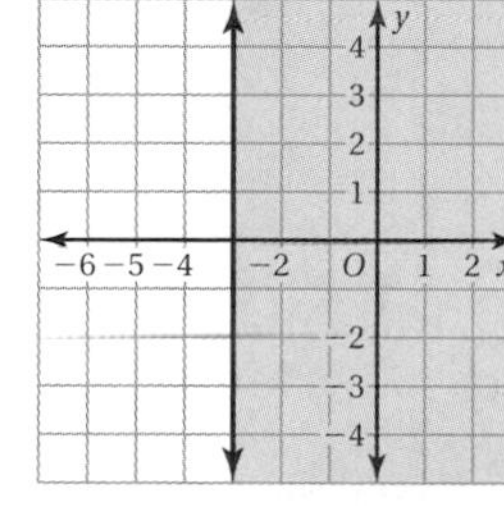

34.

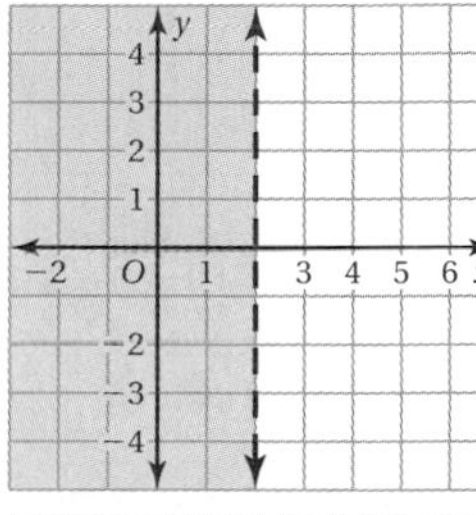

35.

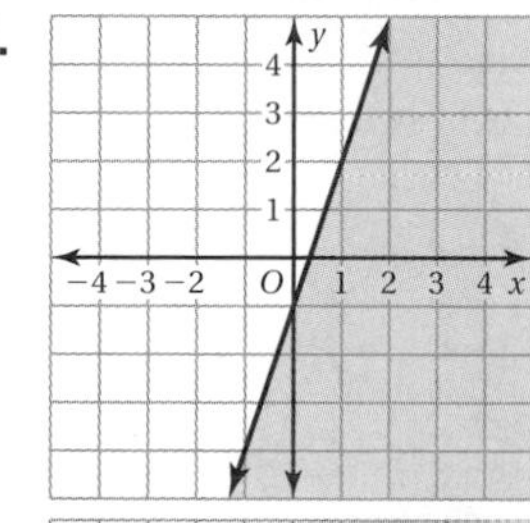

36.

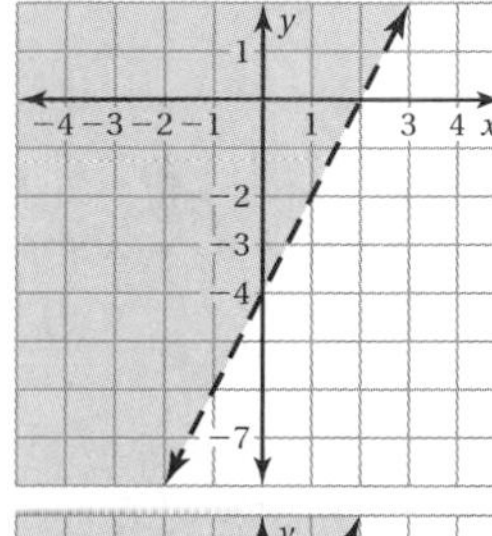

37.

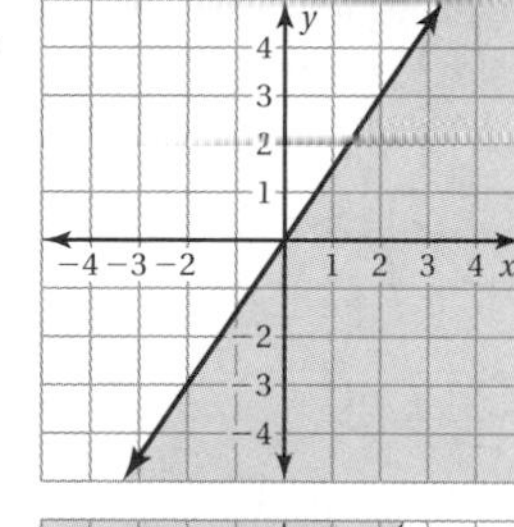

38.

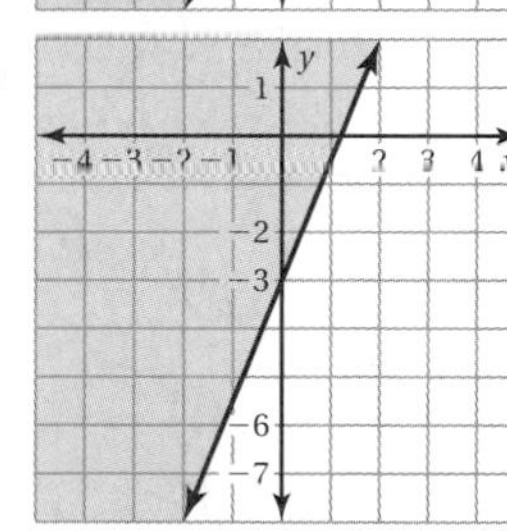

39.

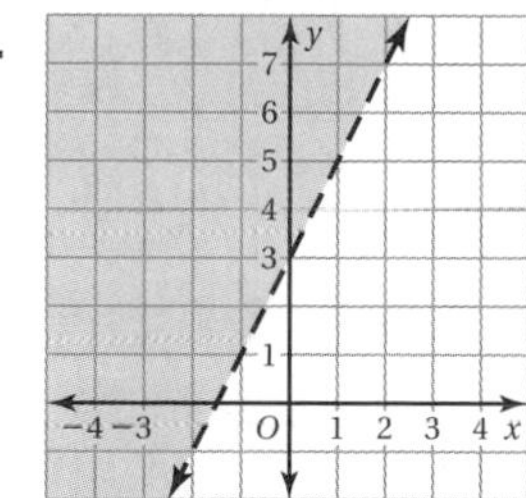

40.

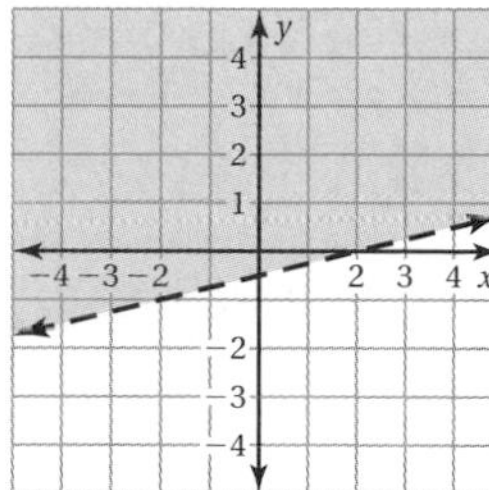

41. The boundary line should be solid instead of dashed.

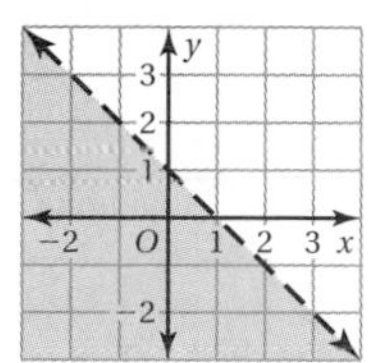

42. The wrong half-plane is shaded.

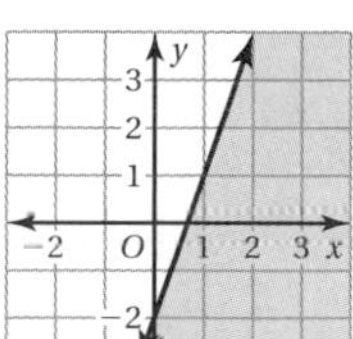

Additional Answers

43. *Sample answer:* Choosing a test point on the boundary line will not help you determine which half-plane to shade. The test point must be in one of the half-planes.

44. a. $10x + 6y \geq 1500$

b. 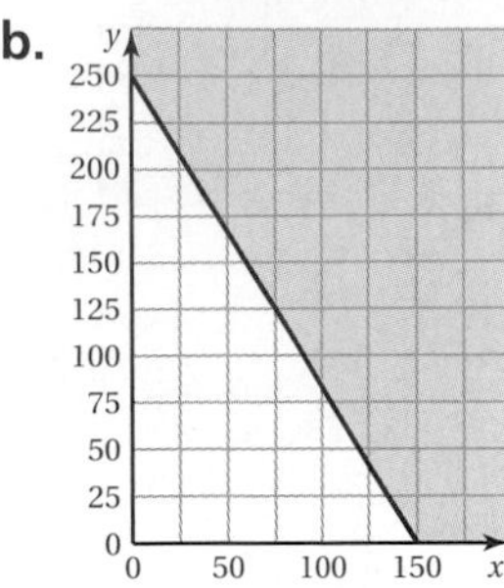

c. no; the drama club does not cover its expenses because (80, 110) is not a solution to the inequality.

53. a. $0.75x + 2.25y \leq 9$

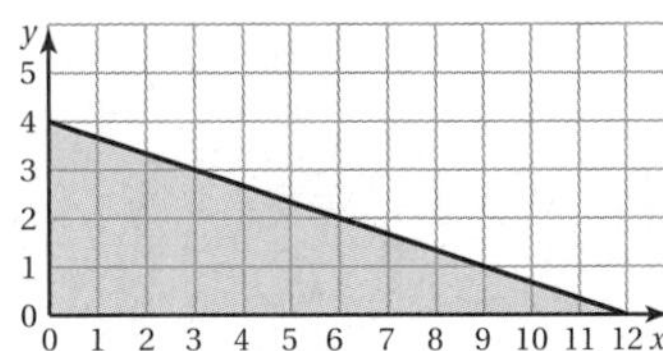

b. Two possible solutions are (1, 3) and (12, 0). So, you can play 1 arcade game and buy 3 soft drinks. or you can play 12 arcade games and buy 0 soft drinks.

3.4–3.5 Quiz

8. $z \leq -4$ or $z \geq 2$;

9. $-1 \leq b \leq 2$;

10. $r > 5$ or $r < -3$;

11.

12.

13. 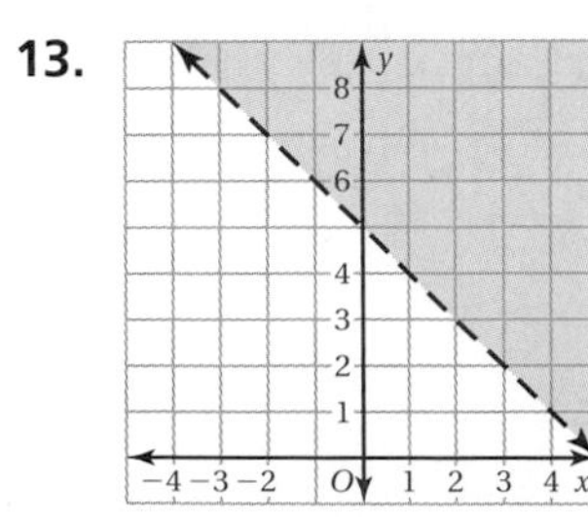

14. 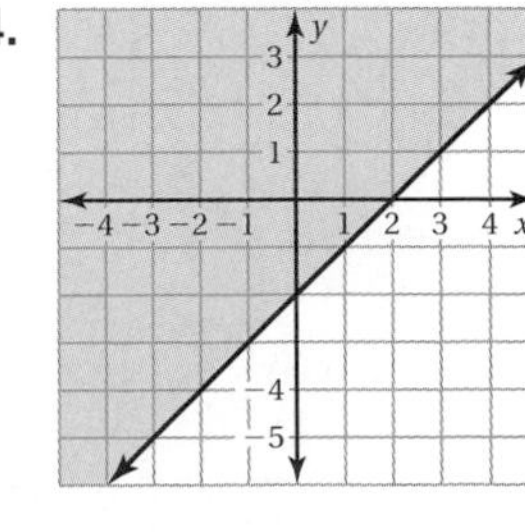

17. $0.75x + 1.5y \leq 6$; One possible solution is (2, 3). So, you can buy 2 pens and 3 notebooks.

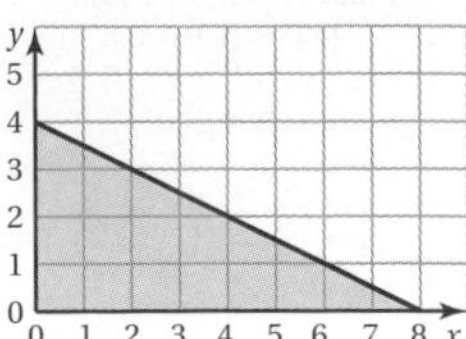

Chapter 3 Test

11. $t > 1$;

12. $-2 \leq x \leq 4$;

13. $x \leq 2$ or $x \geq 10$;

14. $-17 < x < 7$;

15.

16. 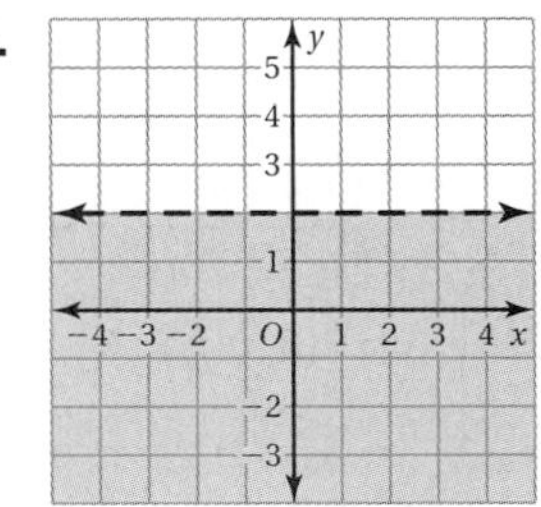

17.

18. 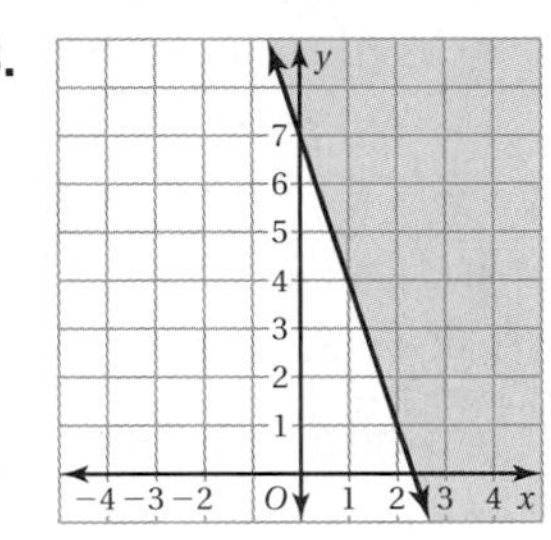

19. 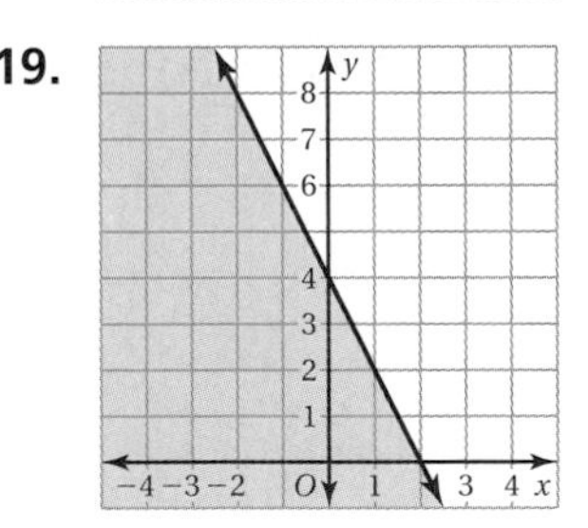

20. 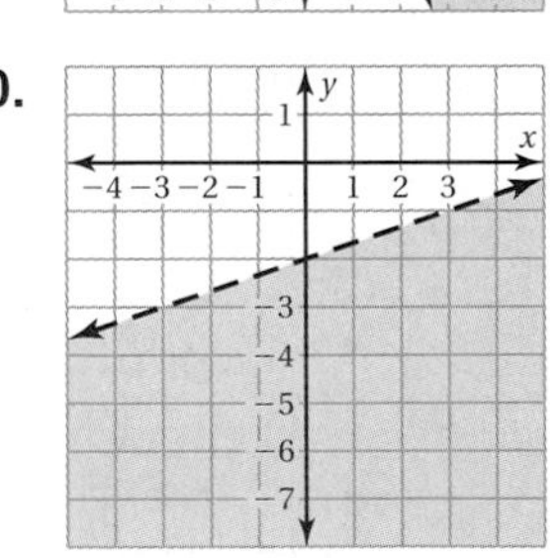

Chapter 4

Try It Yourself

7. 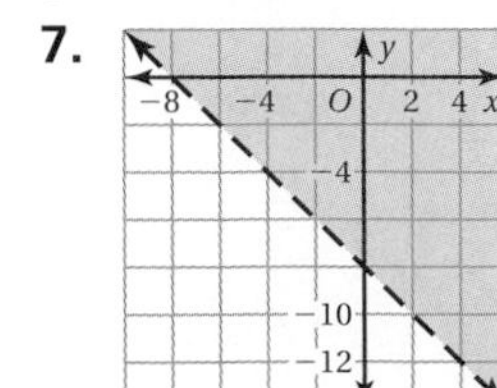

8. 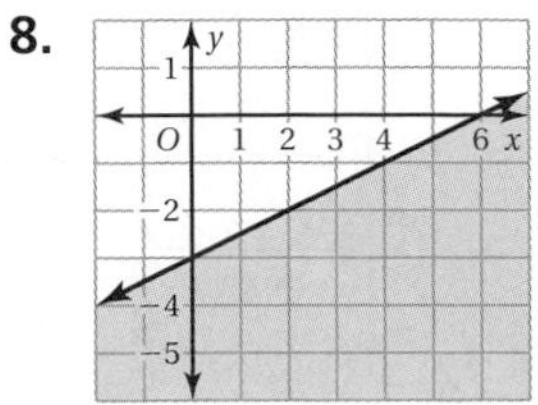

Record and Practice Journal Fair Game Review

9.

10. 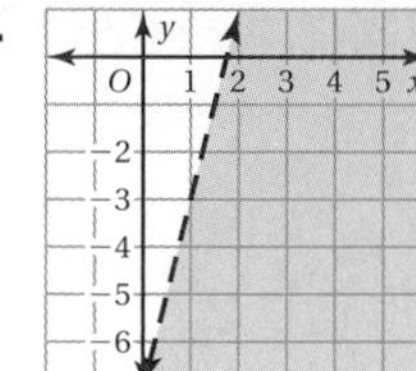

11. 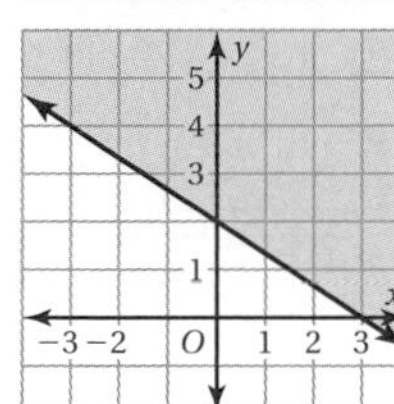

12. a. $150x + 100y \le 1200$

 b. 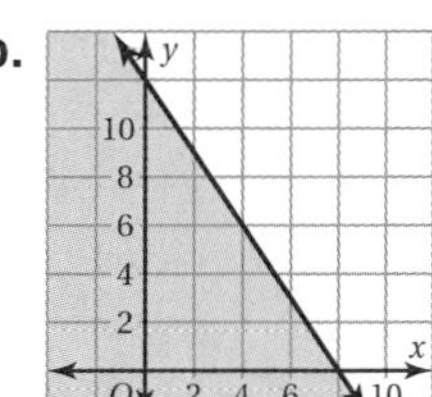

 c. yes; $4(150) + 5(100) = 1100 \le 1200$

Section 4.4

Practice and Problem Solving

17. When the slopes are different, there is one solution. When the slopes are the same, there is no solution if the y-intercepts are different and infinitely many solutions if the y-intercepts are the same.

Fair Game Review

23. 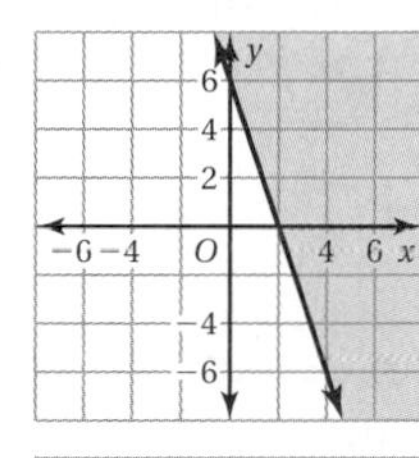

24. 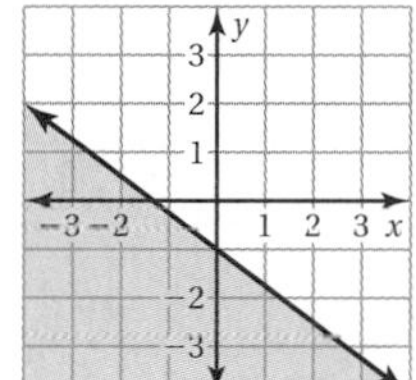

25. 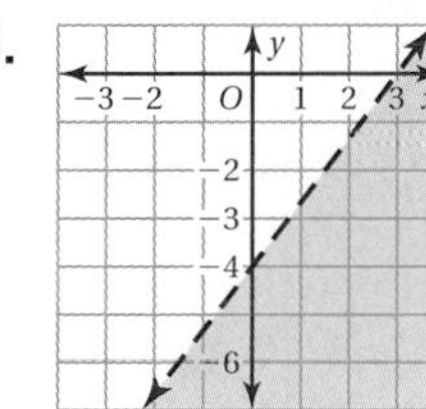

Lesson 4.4b

Record and Practice Journal Practice

1. $x = 1$ 2. $x = 3$ 3. $x = -10$

4. $x = 2$ 5. yes 6. 4 min

7. a. $25x + 500 = 15x + 750$ b. 25 years

Section 4.5

Practice and Problem Solving

7. 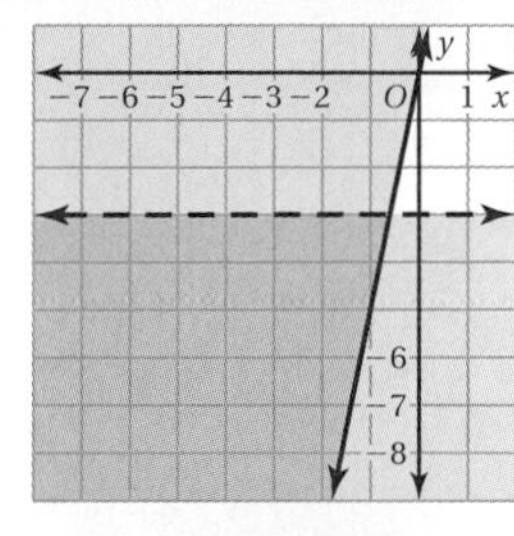

8. 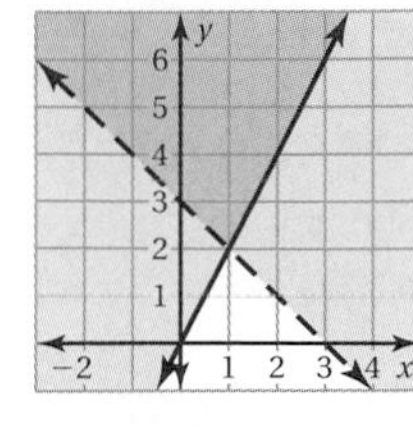

9. 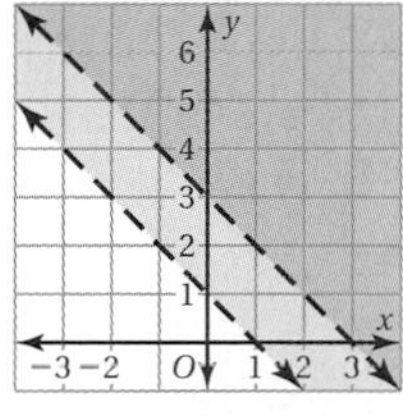

10. 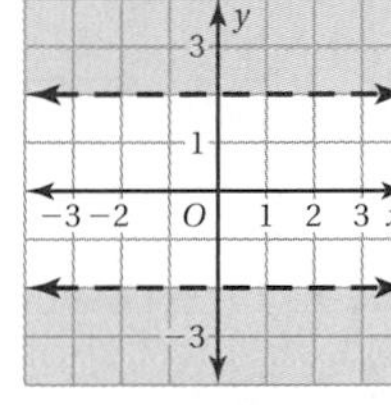

11. 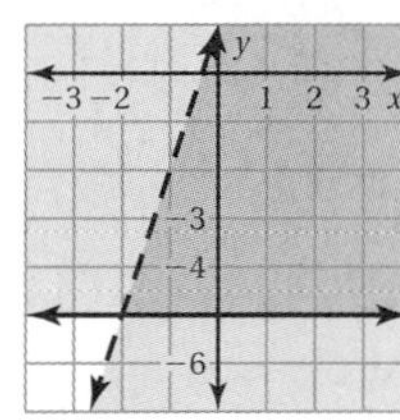

12. 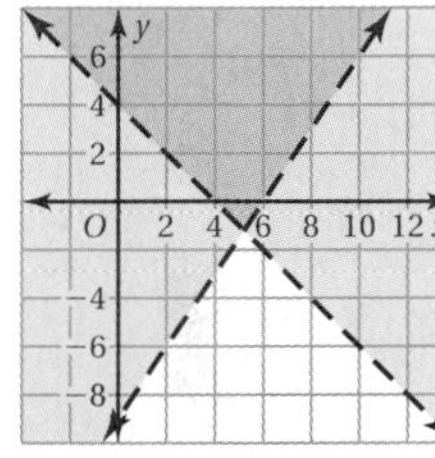

13. 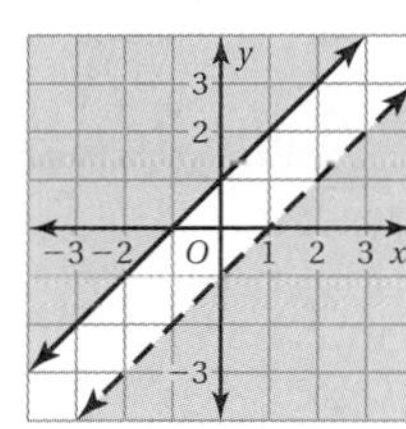

14. 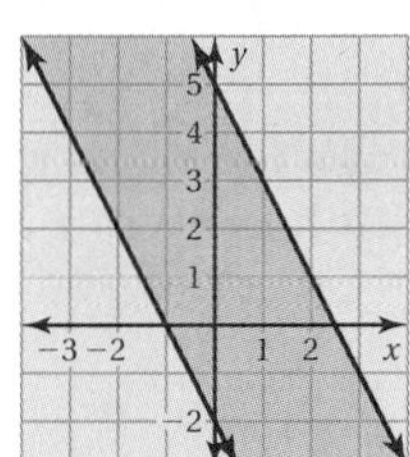

15. 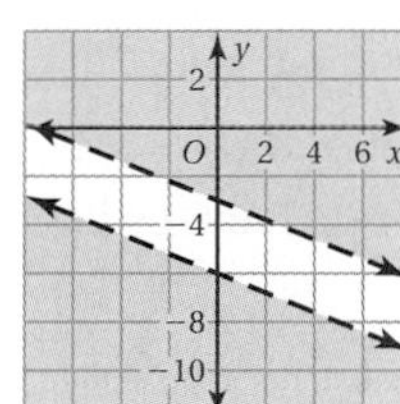

27. 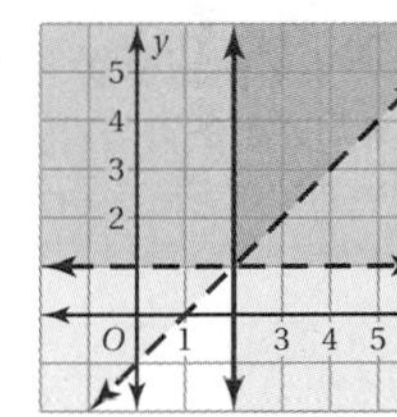

28. 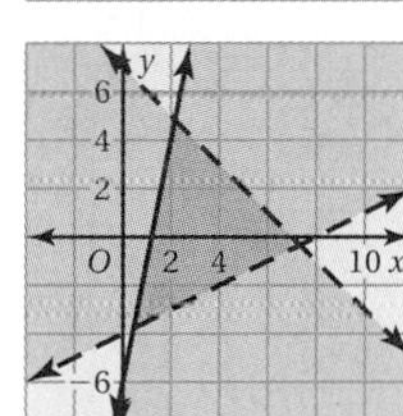

29. 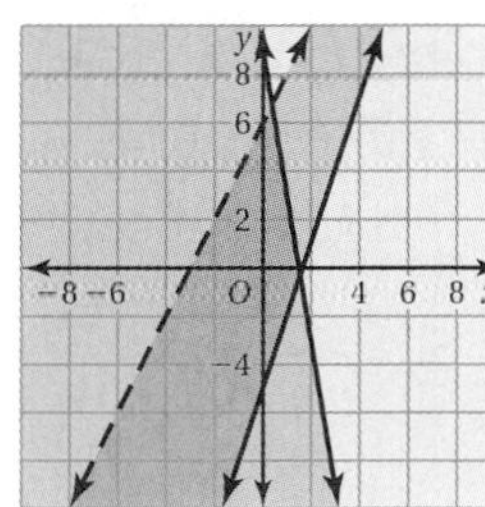

30. $y < x$

 $y > -x$

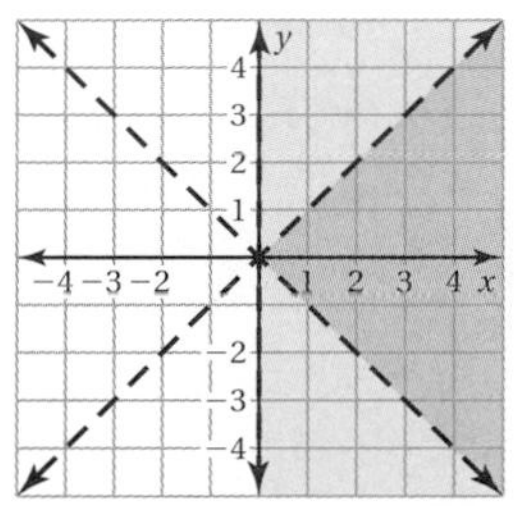

4.3–4.5 Quiz

11. 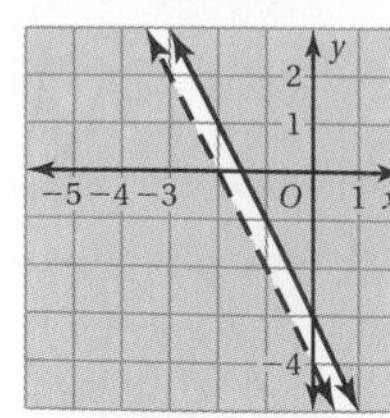

12. 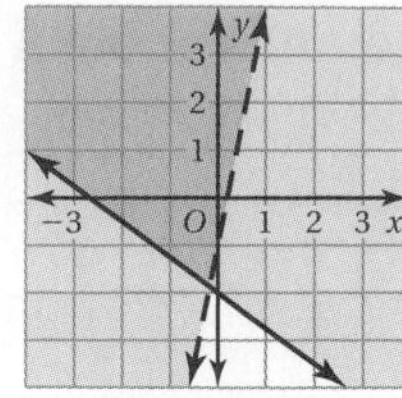

Chapter 4 Test

15. 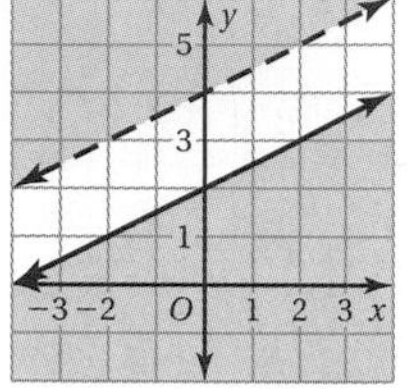

16. 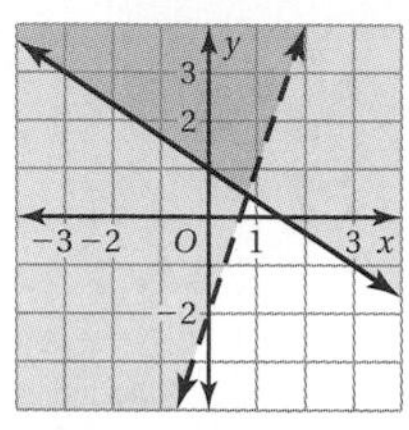

17. 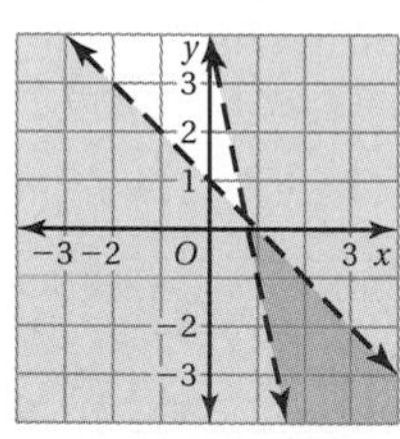

18. $x + y = 12$;
$3x + 2y = 32$;
(8, 4); 8 lilies, 4 tulips

Chapter 5

Record and Practice Journal Fair Game Review

7. Input Output

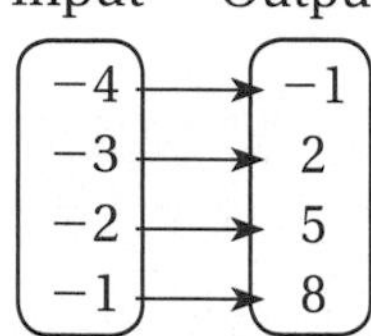

As the input increases by 1, the output increases by 3.

8. Input Output

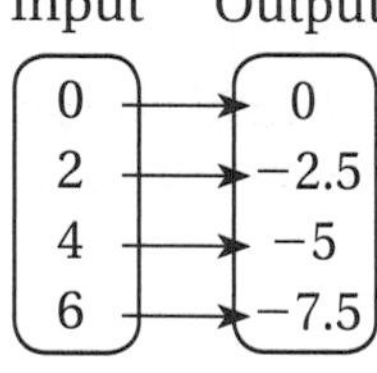

As the input increases by 2, the output decreases by 2.5.

9. Input Output

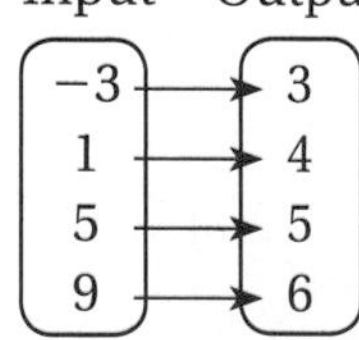

As the input increases by 4, the output increases by 1.

10. Input Output

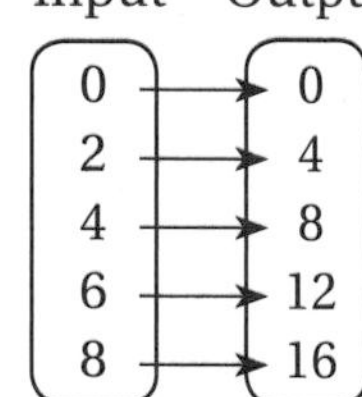

Section 5.1

Record and Practice Journal

4. b–c. women:

x (Domain)	$5\frac{1}{2}$	6	$6\frac{1}{2}$	7	$7\frac{1}{2}$	8	$8\frac{1}{2}$
y (Range)	8.8	9	9.2	9.3	9.5	9.7	9.8

x (Domain)	9	$9\frac{1}{2}$	10	$10\frac{1}{2}$	11	$11\frac{1}{2}$	12
y (Range)	10	10.2	10.3	10.5	10.7	10.8	11

4. b–c. men:

x (Domain)	$5\frac{1}{2}$	6	$6\frac{1}{2}$	7	$7\frac{1}{2}$	8	$8\frac{1}{2}$
y (Range)	9.1	9.3	9.5	9.6	9.8	10	10.1

x (Domain)	9	$9\frac{1}{2}$	10	$10\frac{1}{2}$	11	$11\frac{1}{2}$	12
y (Range)	10.3	10.5	10.6	10.8	11	11.1	11.3

Practice and Problem Solving

14. a. The domain is all real numbers because you can find the absolute value of any number. The range is all real numbers greater than or equal to 0 because the least an absolute value can be is 0.

b. The domain is all real numbers because you can find the absolute value of any number. The range is all real numbers less than or equal to 0 because the negative sign will make every y-value be 0 or negative.

c. The domain is all real numbers because you can find the absolute value of any number. The range is all real numbers greater than or equal to −6 because the least an absolute value can be is 0 and you subtract 6 from that.

d. The domain is all real numbers because you can find the absolute value of any number. The range is all real numbers less than or equal to 4 because the negative sign will make the greatest absolute value be 0 and you add 4 to that.

Fair Game Review

15. 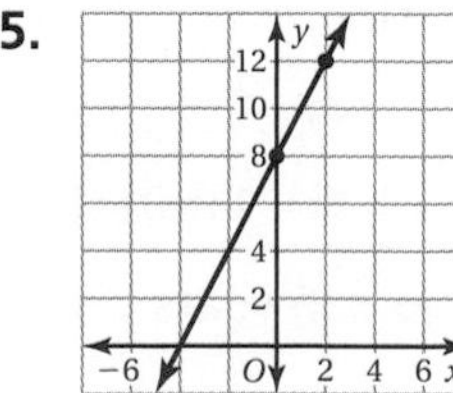

16. 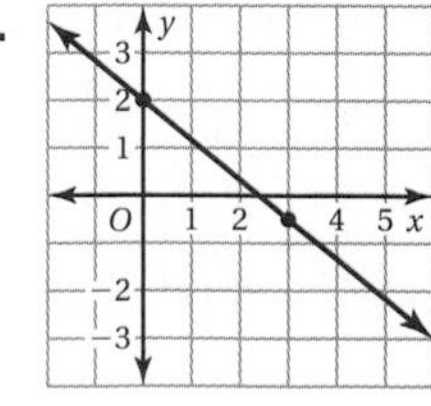

17.

18. 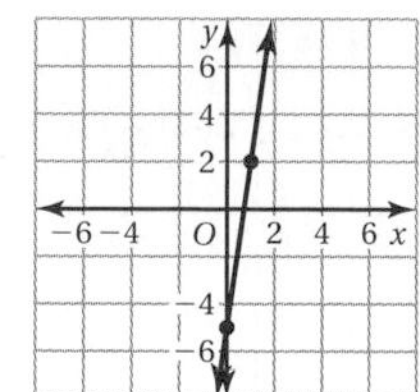

Record and Practice Journal Practice

5. b.

x	1	2	4	8	10
y	18,000	36,000	72,000	144,000	180,000

Lesson 5.1b

Record and Practice Journal Practice

1. not a function
2. function
3. function
4. not a function
5. not a function
6. function
7. function
8. not a function
9. a.

Input, x	1	2	3	4	5	6
Output, y	31	28	31	30	31	30

Input, x	7	8	9	10	11	12
Output, y	31	31	30	31	30	31

b. yes

c. no

Section 5.2

On Your Own

2. 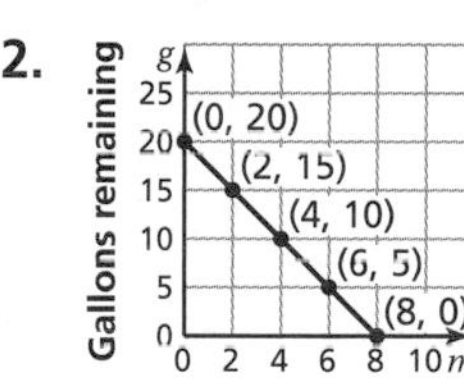
continuous

Practice and Problem Solving

6. 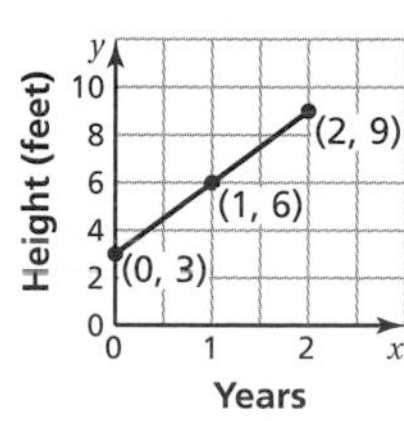
continuous

7. 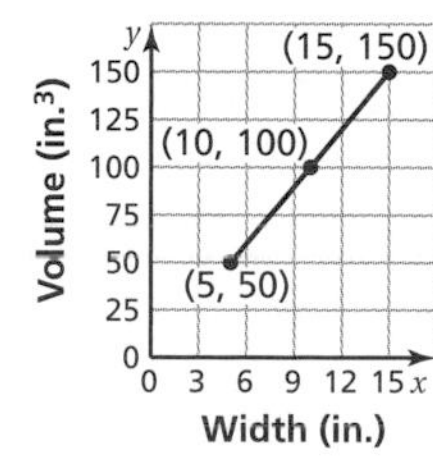
continuous

8. 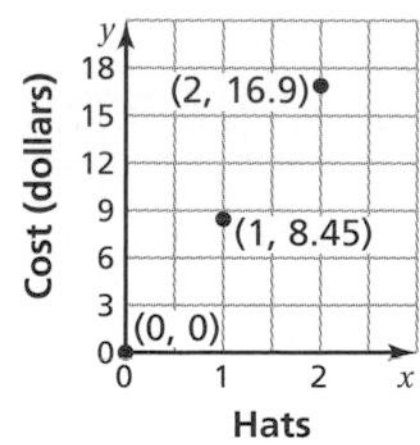
discrete

15. a. *Sample answer:* the elevation of a sinking ship relative to sea level

b. *Sample answer:* an overdraw leaves a negative checking account balance

Section 5.3

Practice and Problem Solving

16. a.

Temperature (°F), t	94	95	96	97	98
Heat Index (°F), H	122	126	130	134	138

b. independent variable: t; dependent variable: H

c. $H = 4t - 254$

d. 146°F

5.1–5.3 Quiz

8. discrete

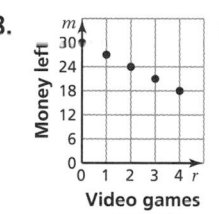

10. a. $y = 0.6x$; The independent variable is the body weight x and the dependent variable is the water weight y.

b.

x	100	120	140	160
y	60	72	84	96

c. domain: 100, 120, 140, 160
range: 60, 72, 84, 96

Lesson 5.4

Record and Practice Journal

3. a.

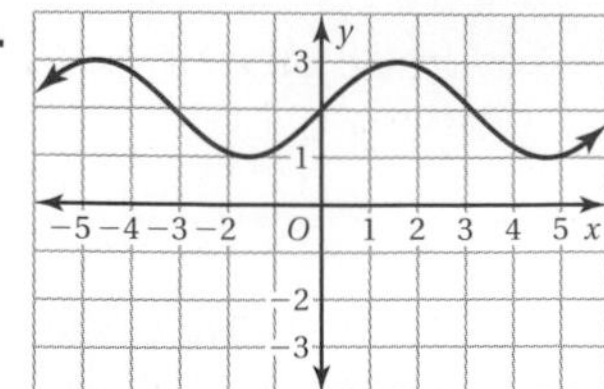

b.

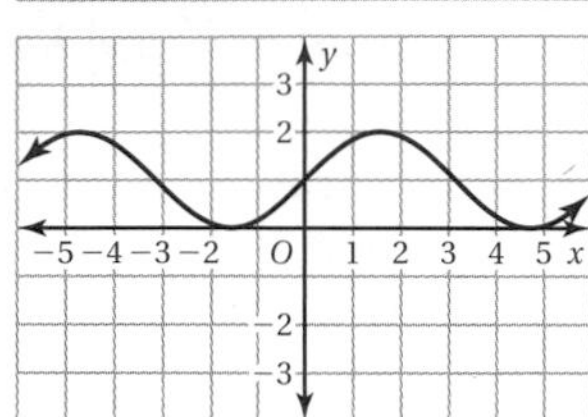

c.

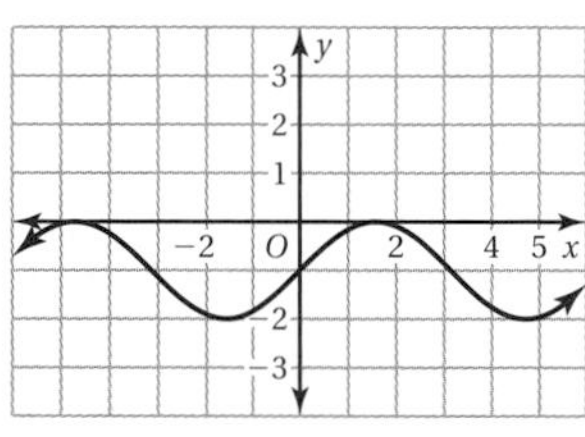

d.

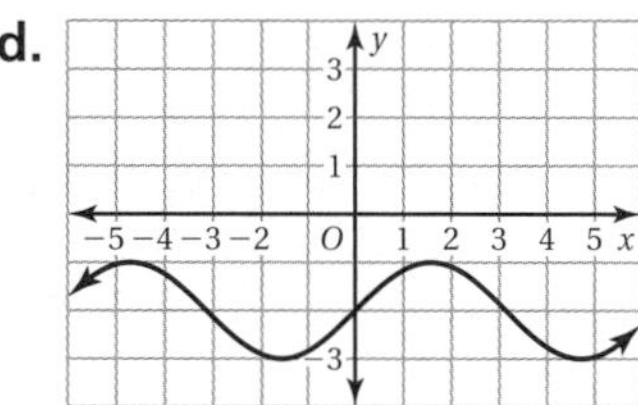

Practice and Problem Solving

25.

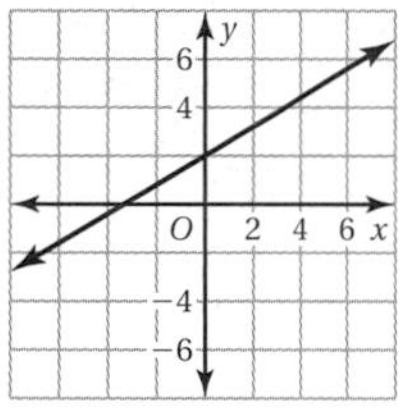

26.

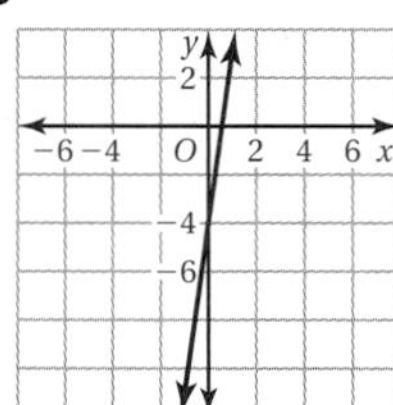

27.

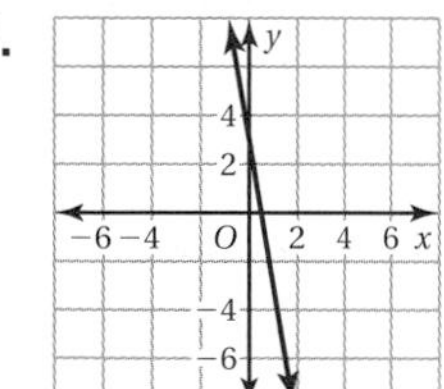

28. a.

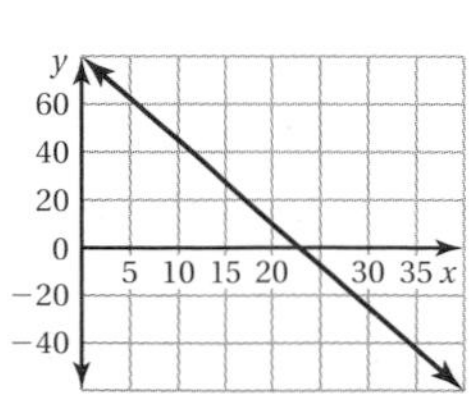

domain: $0 \le x \le 40$
range: $-60 \le t(x) \le 80$

b. $x = 30$; At an altitude of 30,000 feet, the air temperatures is $-25°$F.

29.

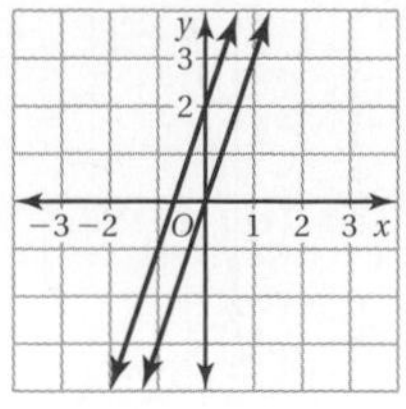

The graph of g is a translation 2 units up of the graph of f.

30.

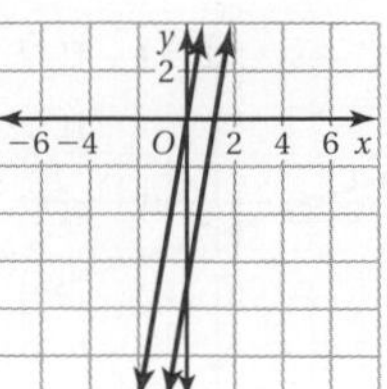

The graph of n is a translation 7 units down of the graph of f.

31.

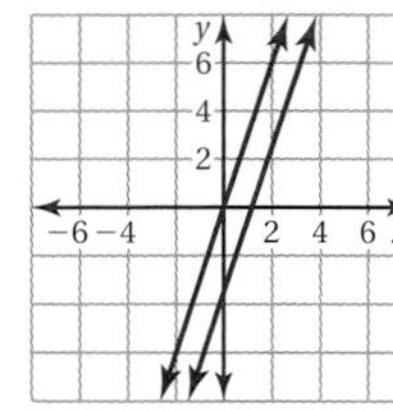

The graph of v is a translation $\frac{7}{2}$ units down of the graph of f.

Lesson 5.4b

Practice

1.

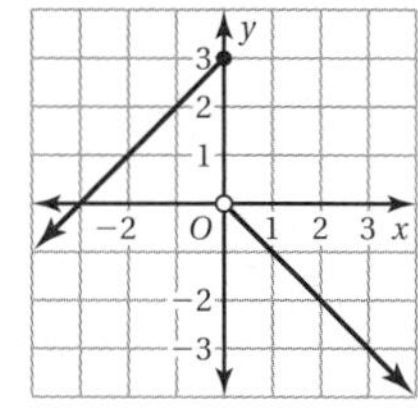

domain: all real numbers
range: $y \le 3$

2.

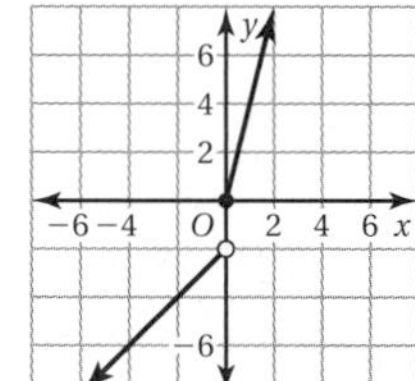

domain: all real numbers
range: $y < -2, y \ge 0$

3.

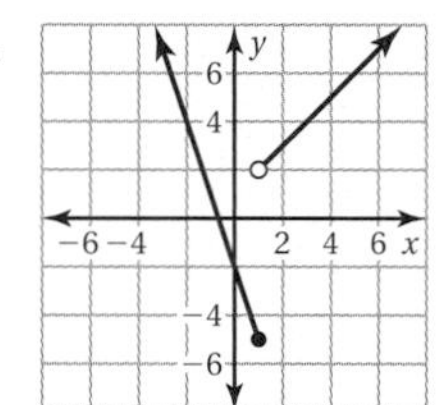

domain: all real numbers
range: $y \ge -5$

4.

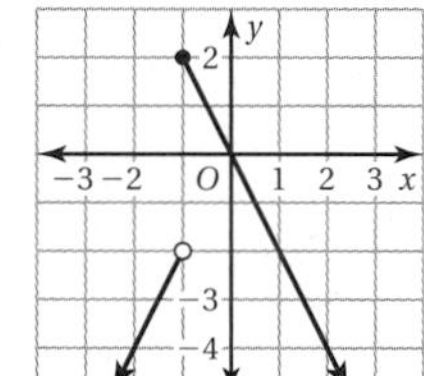

domain: all real numbers
range: $y \le 2$

5.

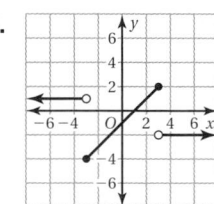

domain: all real numbers
range: $-4 \le y \le 2$

6.

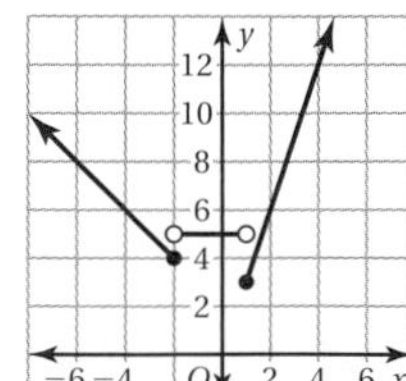

domain: all real numbers
range: $y \ge 3$

11. $f(x) = \begin{cases} 100, & \text{if } 0 < x \le 1 \\ 150, & \text{if } 1 < x \le 2 \\ 200, & \text{if } 2 < x \le 3 \\ 250, & \text{if } 3 < x \le 4 \end{cases}$

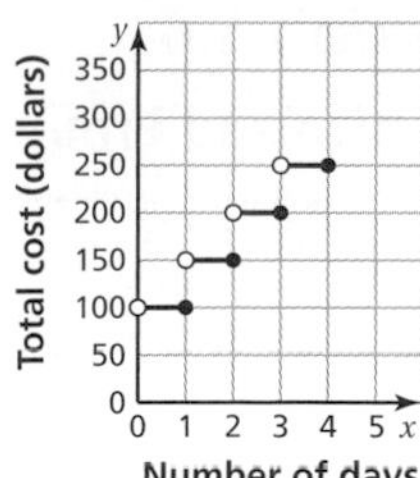

12.

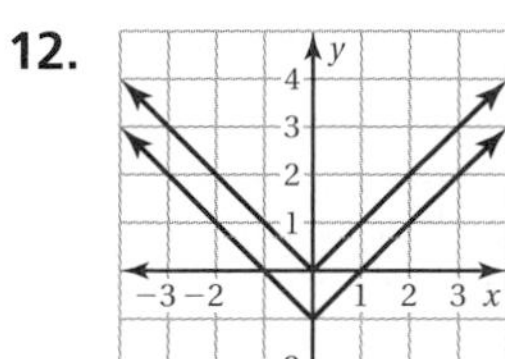

translation 1 unit down of $y = |x|$; domain: all real numbers range: $y \ge -1$

13.

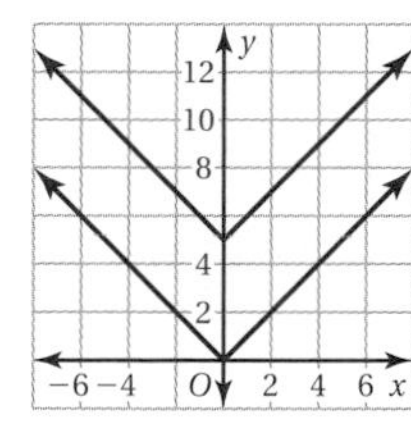

translation 5 units up of $y = |x|$; domain: all real numbers range: $y \ge 5$

14.

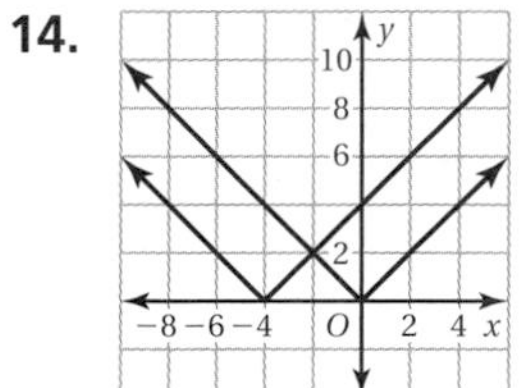

translation 4 units left of $y = |x|$; domain: all real numbers range: $y \ge 0$

15.

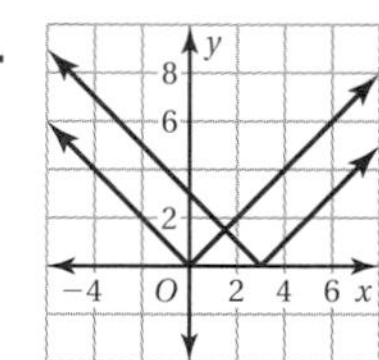

translation 3 units right of $y = |x|$; domain: all real numbers range: $y \ge 0$

16.

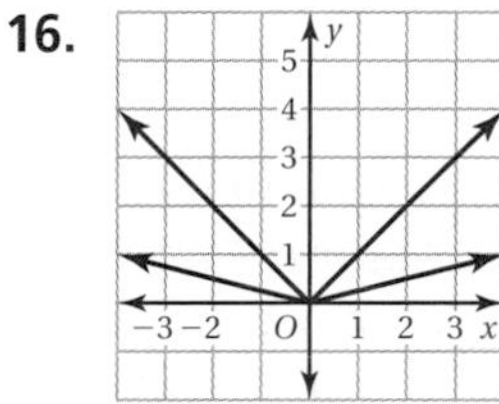

wider than $y = |x|$; domain: all real numbers range: $y \ge 0$

17.

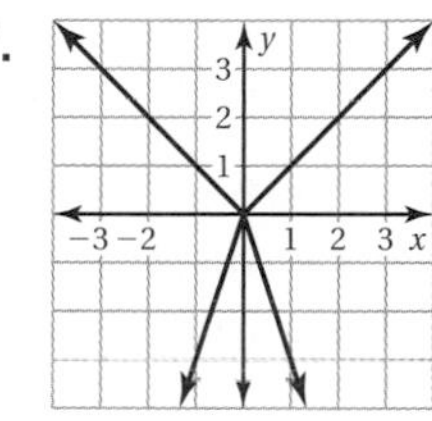

opens down and is narrower than $y = |x|$; domain: all real numbers range: $y \le 0$

18.

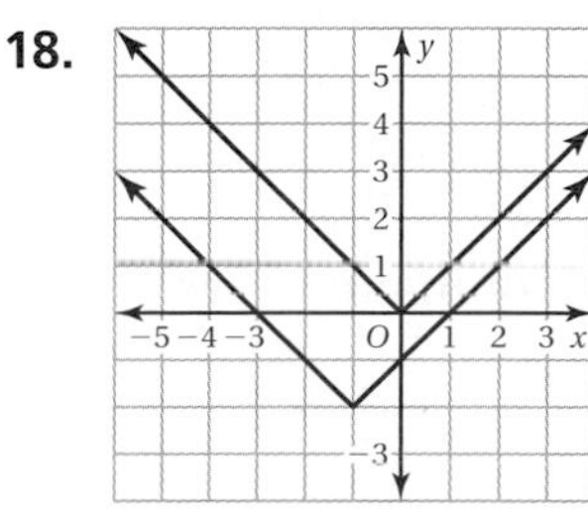

translation 1 unit left and 2 units down of $y = |x|$; domain: all real numbers range: $y \ge -2$

19.

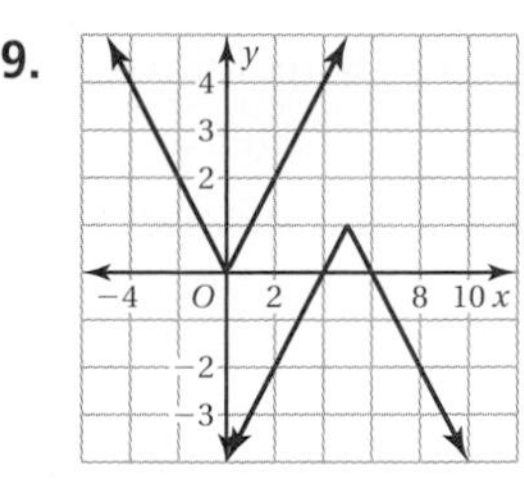

opens down, translation 5 units right and 1 unit up of $y = |x|$; domain: all real numbers range: $y \le 1$

20.

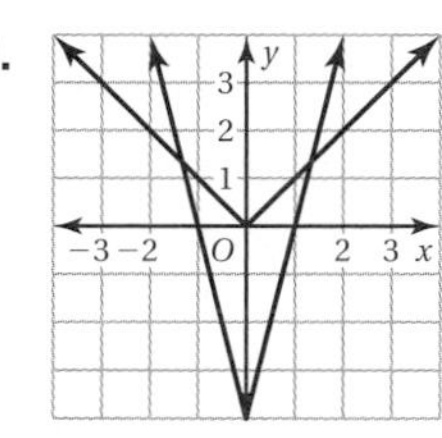

translation 4 units down and is narrower than $y = |x|$; domain: all real numbers range: $y \ge -4$

30.

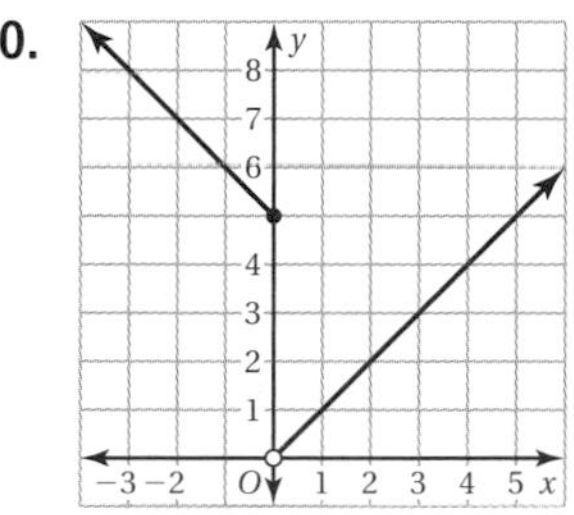

domain: all real numbers range: $y > 0$

Record and Practice Journal Practice

1.

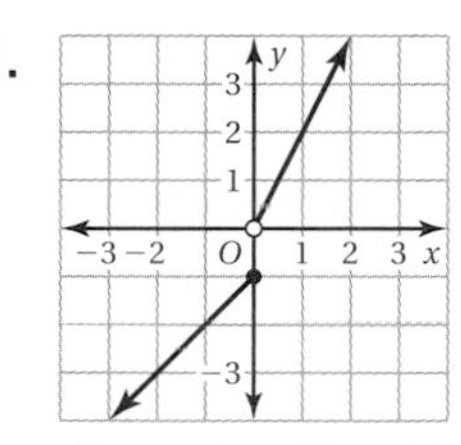

domain: all real numbers range: $y \le -1, y > 0$

2.

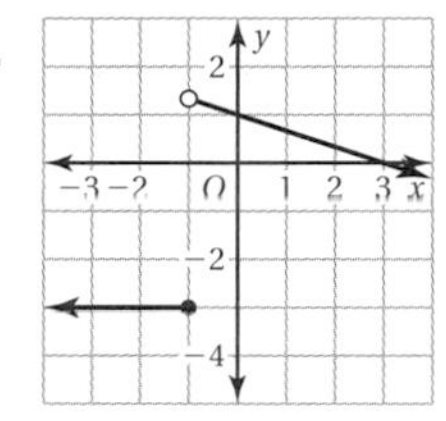

domain: all real numbers range: $y < \frac{4}{3}$

3.

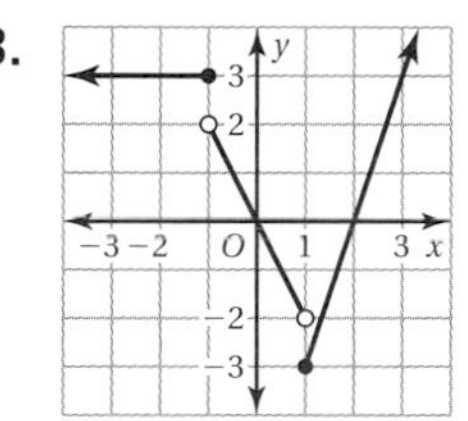

domain: all real numbers range: $y \ge -3$

4.

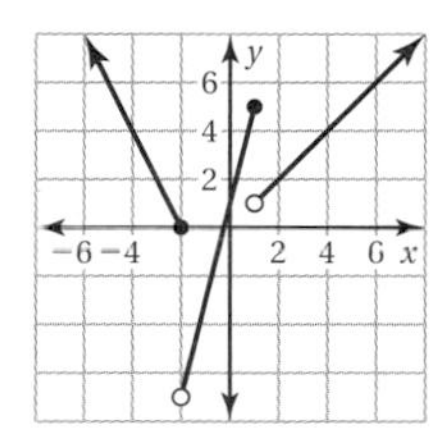

domain: all real numbers range: $y > -7$

5. $y = \begin{cases} x + 1, & \text{if } x < 1 \\ -2x + 5, & \text{if } x \ge 1 \end{cases}$

6.

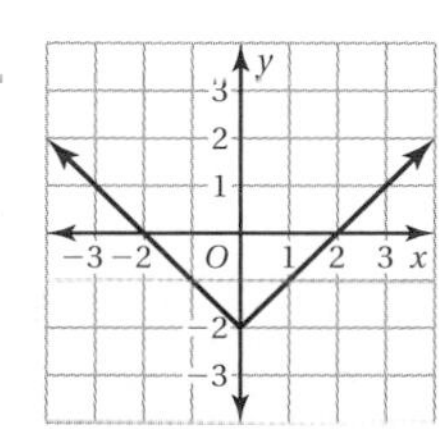

translation 2 units down of $y = |x|$; domain: all real numbers range: $y \ge -2$

Additional Answers

7. 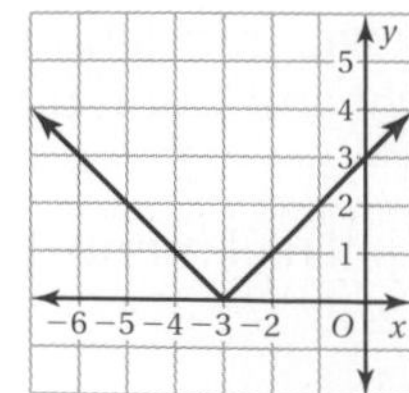

translation 2 units left of $y = |x|$;
domain: all real numbers
range: $y \geq 0$

8. 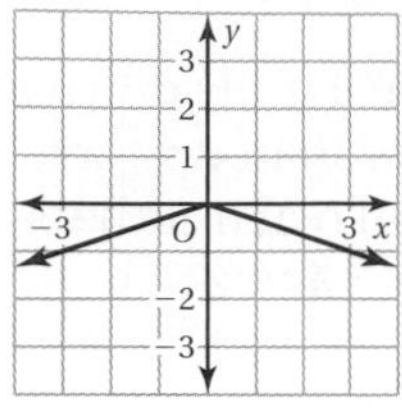

opens down and is wider than $y = |x|$;
domain: all real numbers
range: $y \leq 0$

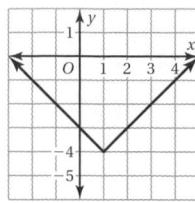

9. translation 1 unit right and 4 units down of $y = |x|$;
domain: all real numbers
range: $y \geq -4$

10. $y = |x - 3|$

11. $y = |x + 8| - 2$

Mini-Assessment

1. 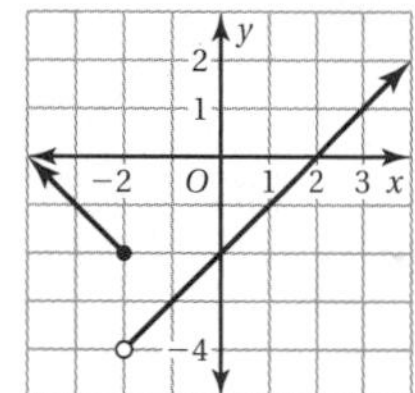

domain: all real numbers
range: $y > -4$

2. 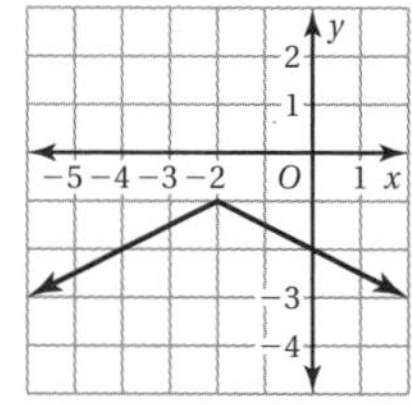

domain: all real numbers
range: $y \leq -1$

3. $f(x) = \begin{cases} 6, & \text{if } 0 < x \leq 1 \\ 8, & \text{if } 1 < x \leq 2 \\ 10, & \text{if } 2 < x \leq 3 \\ 12, & \text{if } 3 < x \leq 4 \end{cases}$

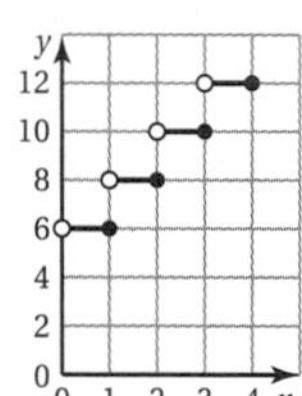

Section 5.5

Practice and Problem Solving

19. a. 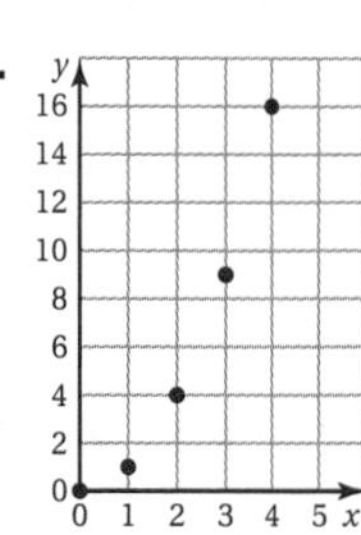

The points curve upward from left to right; nonlinear

b. $y = x^2$

Section 5.6

Practice and Problem Solving

25. 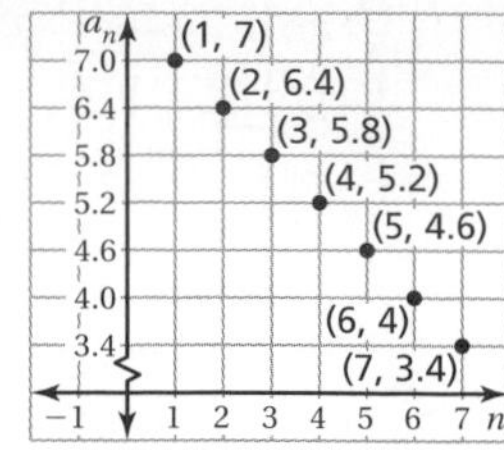

26. 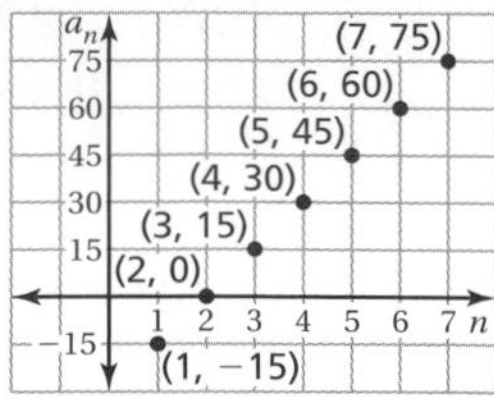

27. 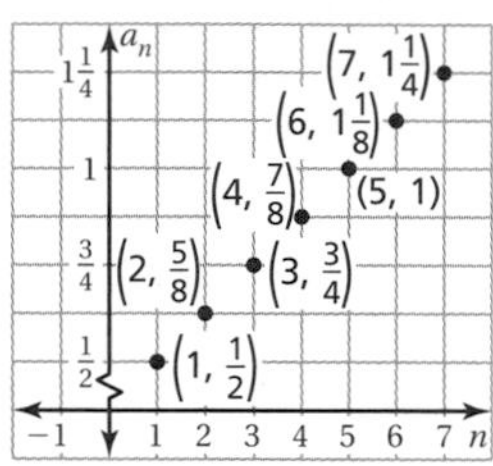

28. 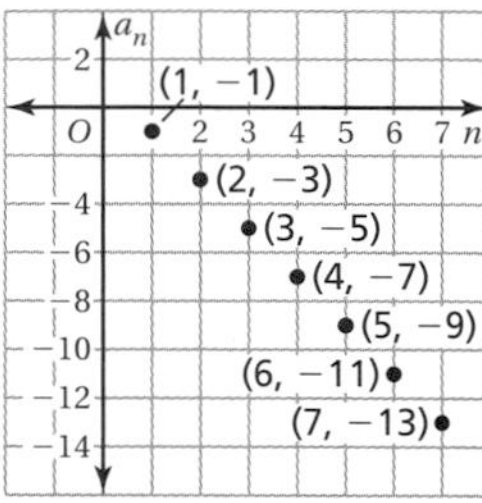

29. 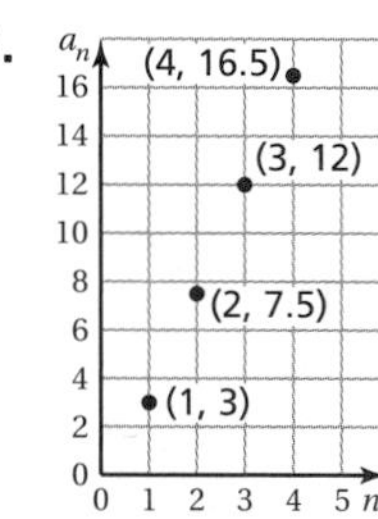

30. 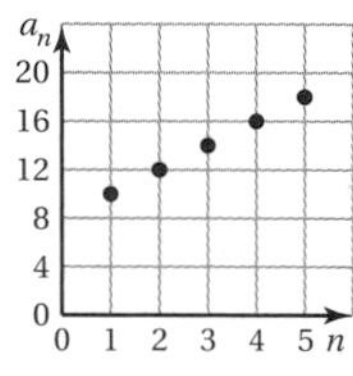

5.4–5.6 Quiz

4. 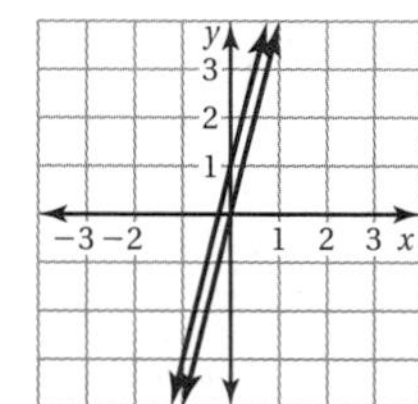

The graph of g is a translation 1 unit up of the graph of f.

5. 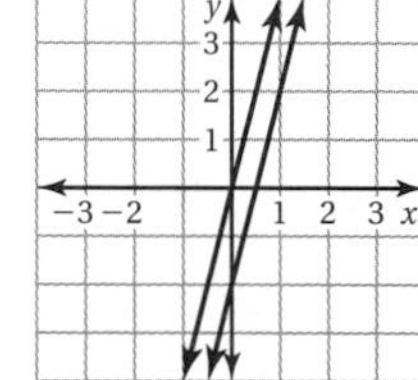

The graph of h is a translation 2 units down of the graph of f.

6. 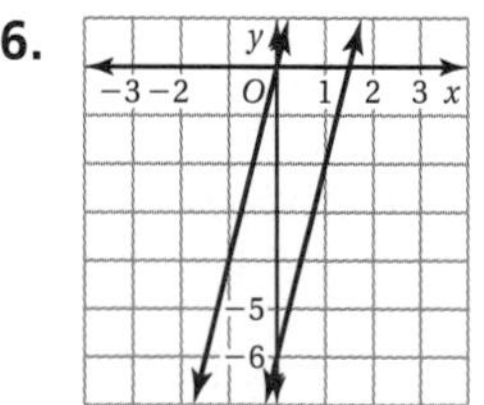

The graph of n is a translation 6 units down of the graph of f.

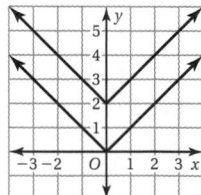

7. translation 2 units up of $y = |x|$;
domain: all real numbers
range: $y \geq 2$

8.

translation 6 units right of $y = |x|$;
domain: all real numbers
range: $y \geq 0$

9. 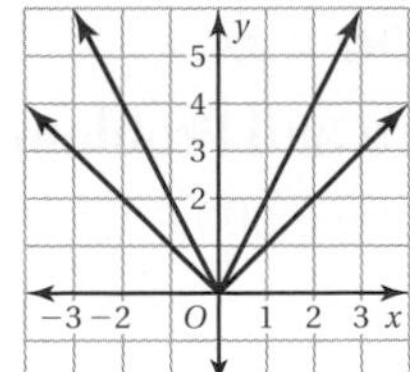

narrower than $y = |x|$;
domain: all real numbers
range: $y \geq 0$

Chapter 5 Test

8. The graph of h is a translation 2 units up of the graph of f.

9. The graph of $y = |x + 3| - 2$ is a translation 3 units left and 2 units down of the graph of $y = |x|$.

10. 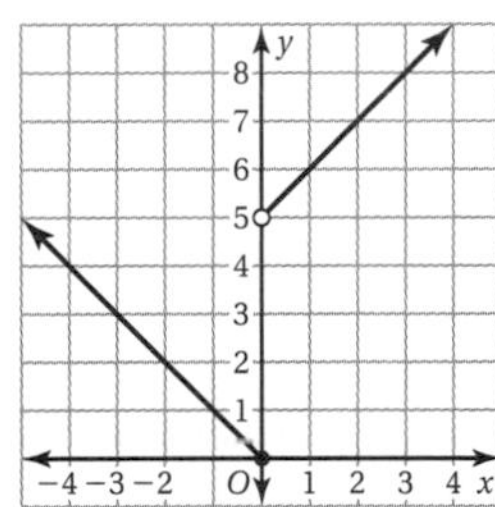

domain: all real numbers
range: $y \geq 0$

16. nonlinear; As x increases by 1, S increases by increasing amounts.

Chapter 6

Section 6.3

Fair Game Review

35. 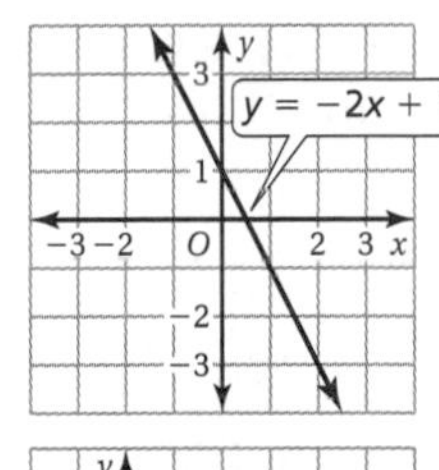

36. 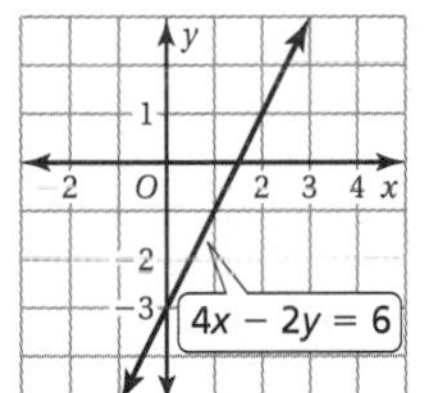

37. 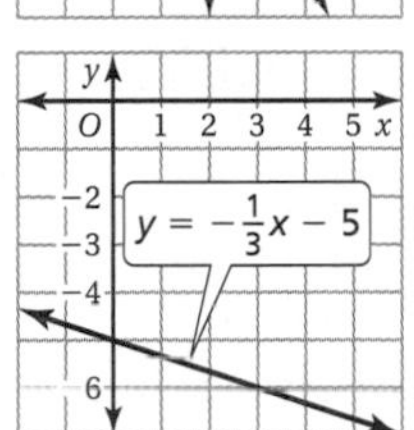

Section 6.4

On Your Own

5. 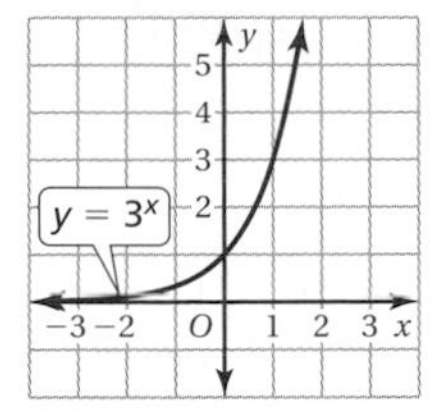

domain: all real numbers
range: all positive real numbers

6.

domain: all real numbers
range: all positive real numbers

7.

domain: all real numbers
range: all negative real numbers

8.

domain: all real numbers
range: all real numbers greater than -2;
The graph is a translation 2 units down of the graph of $y = \left(\frac{1}{2}\right)^x$.

Practice and Problem Solving

23. 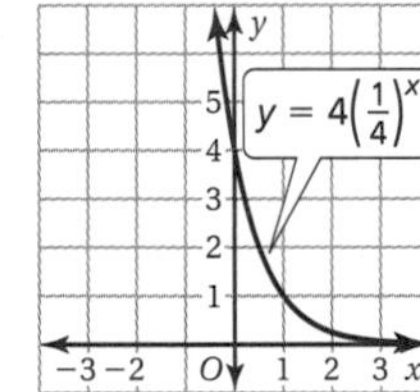

domain: all real numbers
range: all positive real numbers

26. a. 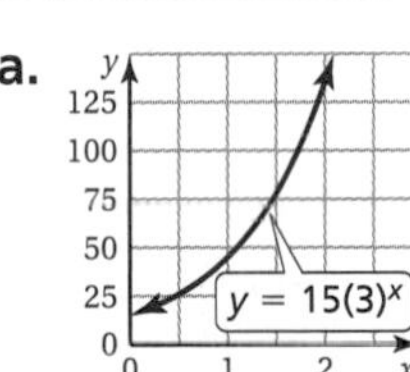

domain: all nonnegative real numbers
range: all real numbers greater than or equal to 15

b. y-intercept: 15; initial number of coyotes in the park

c. 45 coyotes

27. 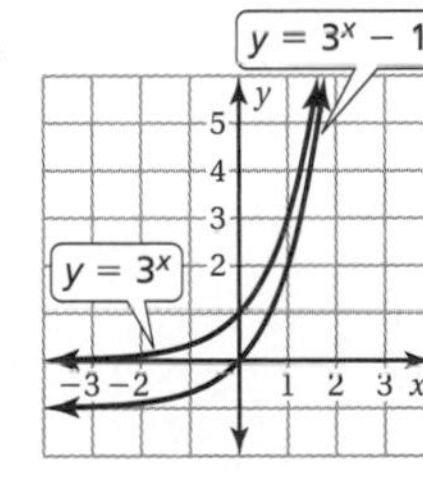

domain: all real numbers
range: all real numbers greater than -1;
The graph is a translation 1 unit down of the graph of $y = 3^x$.

Additional Answers

28. 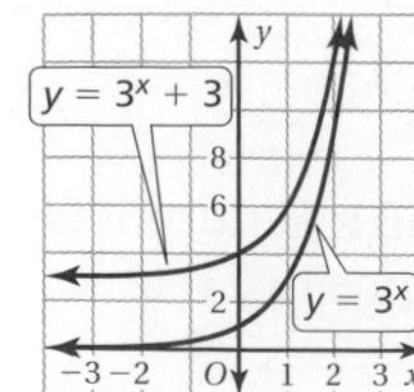

domain: all real numbers
range: all real numbers greater than 3;
The graph is a translation 3 units up of the graph of $y = 3^x$.

29. 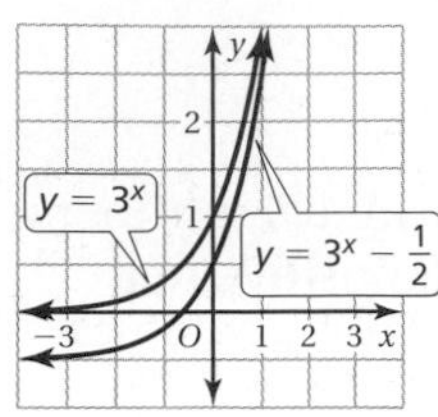

domain: all real numbers
range: all real numbers greater than $-\frac{1}{2}$;
The graph is a translation $\frac{1}{2}$ unit down of the graph of $y = 3^x$.

30. 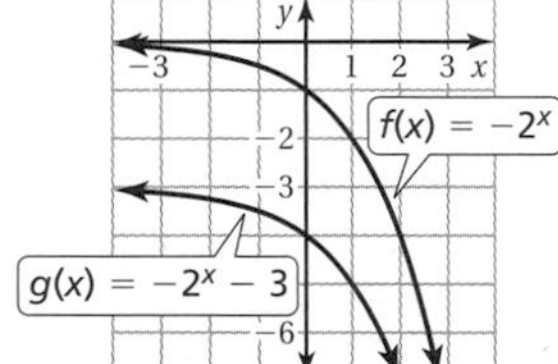

a. For f, the domain is all real numbers and the range is all negative real numbers.
For g, the domain is all real numbers and the range is all real numbers less than -3.

b. f: -1, g: -14

c. The y-intercept shifted down 3 units, the domain remained the same, and the range changed to all real numbers less than -3.

31. An exponential function intersects the x-axis when it has the form

$y = ab^x - c$, $a, b, c > 0$ or

$y = ab^x + c$, $a < 0$, $b, c > 0$;

Example: $y = 2^x - 2$

Lesson 6.4b

Record and Practice Journal Practice

1. $x = 3$
2. $x = -3$
3. $x = -1$
4. no solution
5. $x = 4$
6. $x = 6$
7. $x \approx 2.80$
8. $x \approx -0.21$
9. $x \approx 0.61$
10. $x \approx -0.57$
11. no solution
12. $x \approx -1.95$
13. a. on the third day ($x = 3$)

    b. $A = 4^3 = 64$, $B = 2^{9-3} = 64$

Section 6.5

Practice and Problem Solving

12. $y = 35{,}000(1.04)^t$

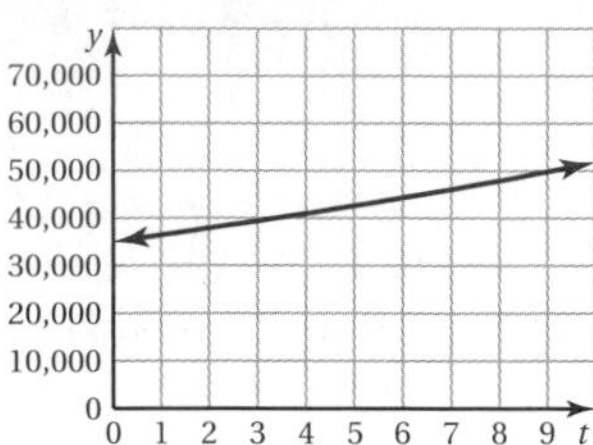

13. $y = 210{,}000(1.125)^t$

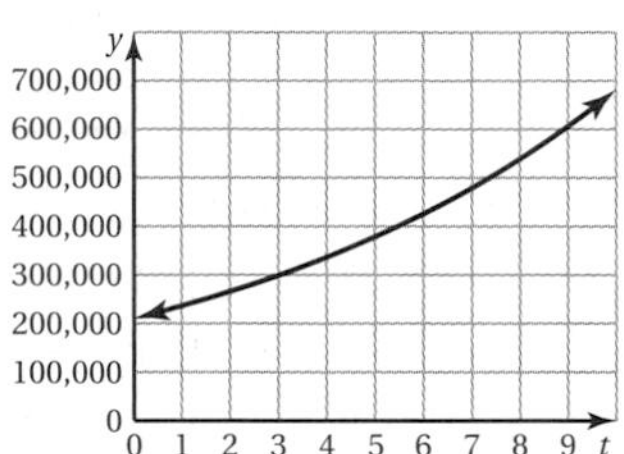

14. $y = 10{,}000(1.7)^t$

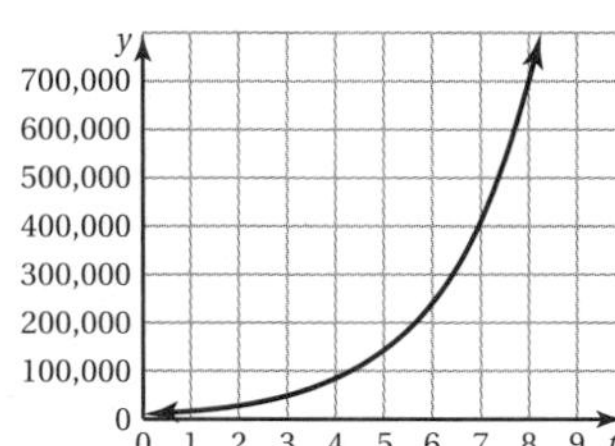

Section 6.6

Practice and Problem Solving

21. *Sample answer:* Graph an exponential function by hand when the value of a is small enough to easily multiply by $1 - r$. Use a graphing calculator when the value of a is too large or when $1 - r$ has many decimal places which would make multiplying by hand more difficult.

Section 6.7

On Your Own

2. 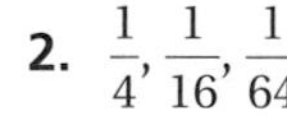 $\frac{1}{4}, \frac{1}{16}, \frac{1}{64}$

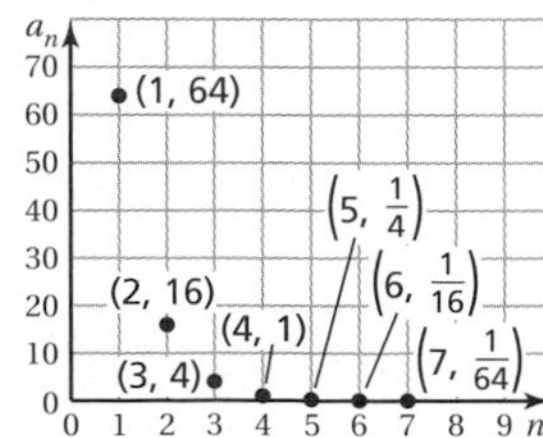

3. 5, −2.5, 1.25

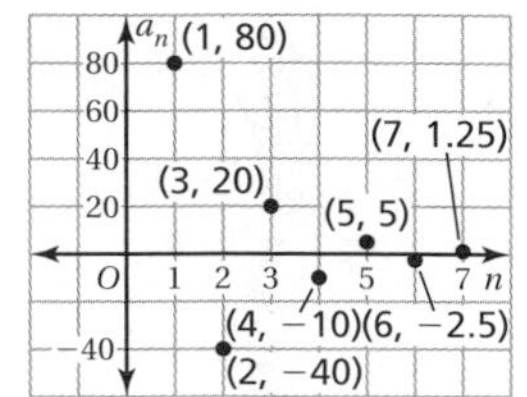

Practice and Problem Solving

11. 1250, 6250, 31,250

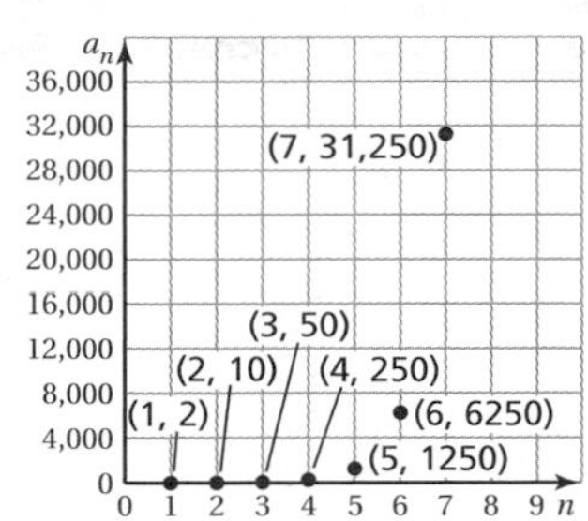

12. −112, 224, −448

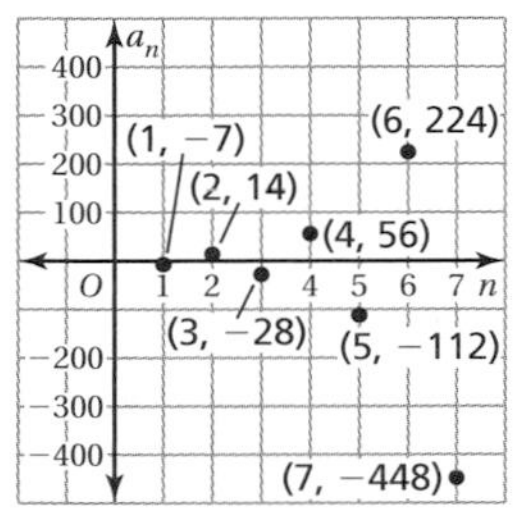

13. $1, -\frac{1}{3}, \frac{1}{9}$

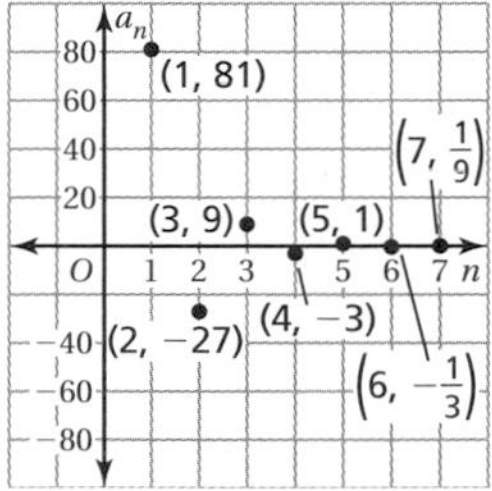

14. $-\frac{3}{5}, -\frac{3}{25}, -\frac{3}{125}$

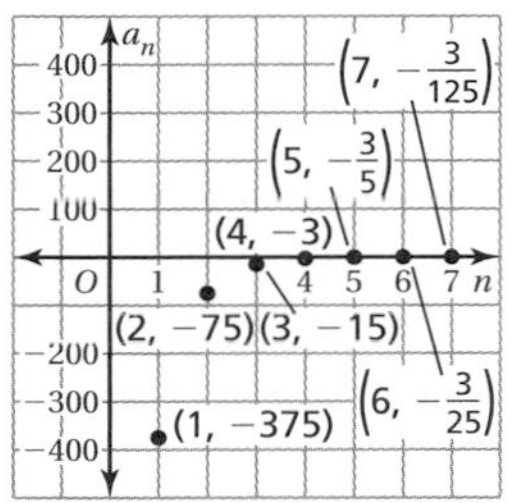

15. $\frac{1}{36}, \frac{1}{216}, \frac{1}{1296}$

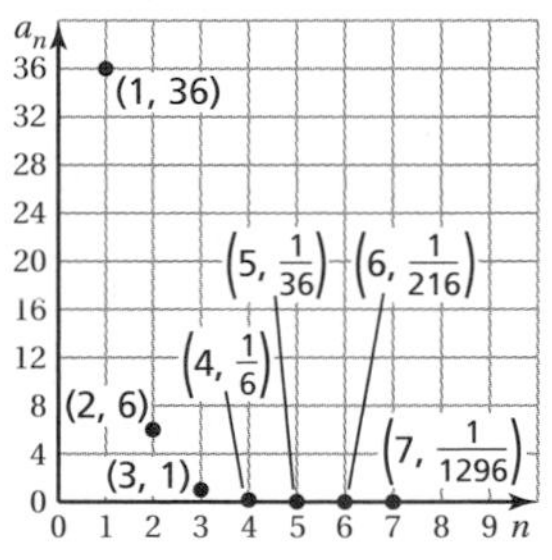

16. 49, 343, 2401

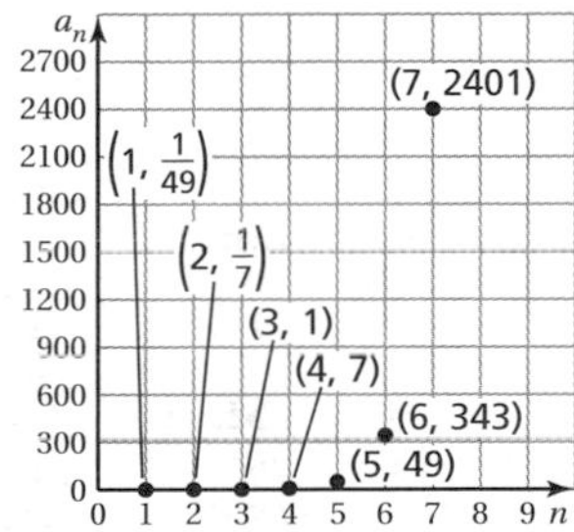

17. The ratio of the term to its previous term is $-\frac{1}{2}$, not 2. So, the next three terms are $-\frac{1}{2}, \frac{1}{4}, -\frac{1}{8}$.

18. 16 teams, 8 teams, and 4 teams

Lesson 6.7b

Practice

1. 0, −8, −16, −24, −32, −40

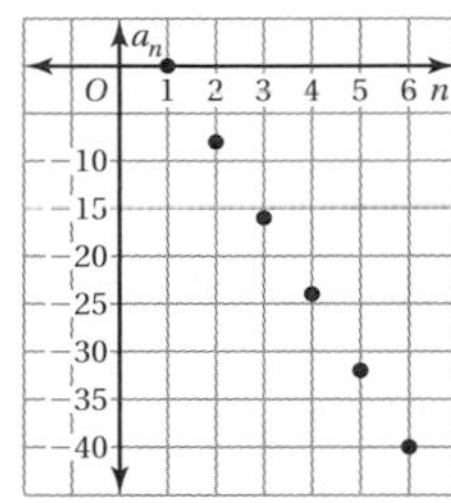

2. −7.5, −5, −2.5, 0, 2.5, 5

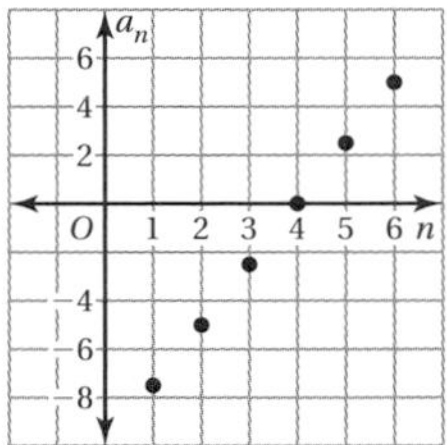

3. −36, −18, −9, −4.5, −2.25, −1.125

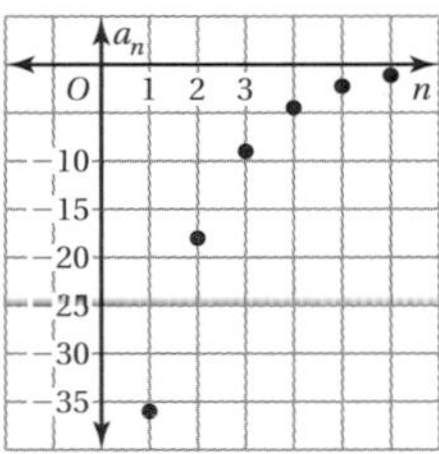

4. 0.7, 7, 70, 700, 7000, 70,000

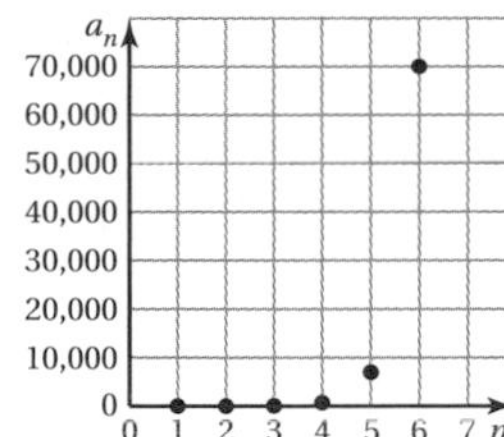

20. **a.** 1, 1, 2, 3, 5, 8 **b.** Fibonacci sequence; 13

c.

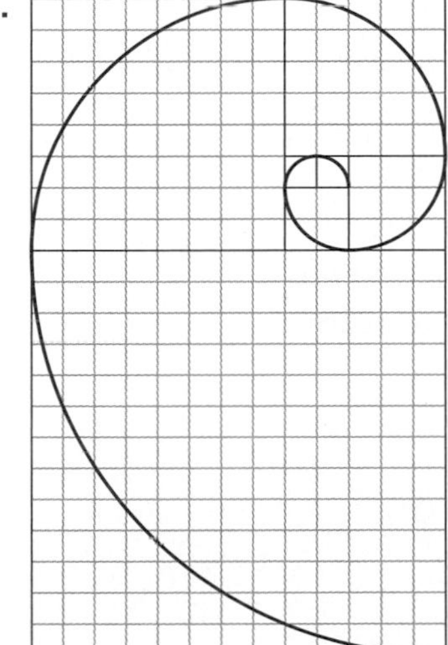

Record and Practice Journal Practice

1. 1, 7, 13, 19, 25, 31

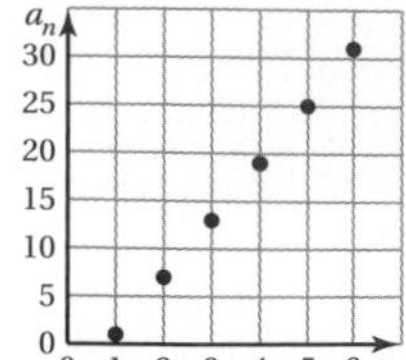

2. 64, 16, 4, 1, $\frac{1}{4}$, $\frac{1}{16}$

3. −5, −15, −45, −135, −405, −1215

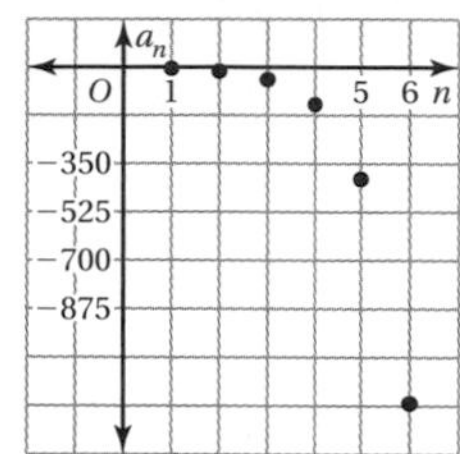

4. 20, 11, 2, −7, −16, −25

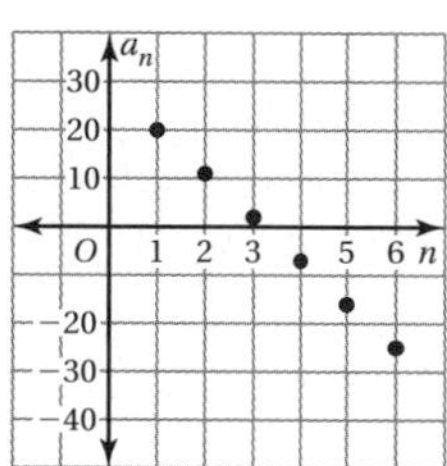

5. $a_1 = 24, a_n = a_{n-1} + 7$

6. $a_1 = -5, a_n = -6a_{n-1}$

7. $a_1 = 200, a_n = 7a_{n-1}$

8. $a_n = 54\left(\frac{1}{3}\right)^{n-1}$

9. $a_n = 12n - 12$

10. $a_1 = 60, a_n = a_{n-1} - 25$

11. $a_1 = 16, a_n = 1.5a_{n-1}$

12. $a_1 = 3, a_2 = 8, a_n = a_{n-1} + a_{n-2}$; 49, 79, 128

13. $a_1 = 1, a_2 = 2, a_n = a_{n-1} - a_{n-2}$; 1, −1, −2

14. $a_1 = 1.25, a_2 = 4, a_n = a_{n-1}a_{n-2}$; 2000, 200,000

15. $a_1 = -3, a_2 = -2, a_n = a_{n-1}a_{n-2}$; 864, −62,208

6.4–6.7 Quiz

3. domain: all real numbers;
range: all positive numbers

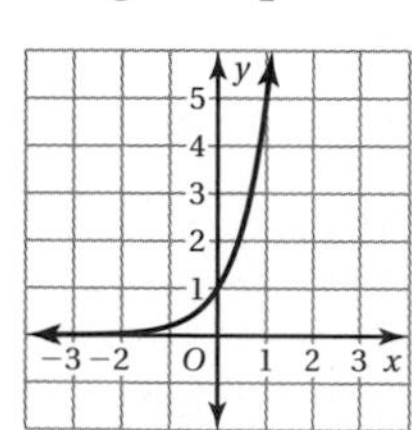

4. domain: all real numbers;
range: all negative numbers

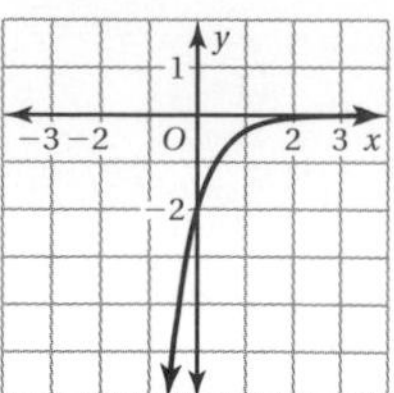

9. 1215, −3645, 10,935

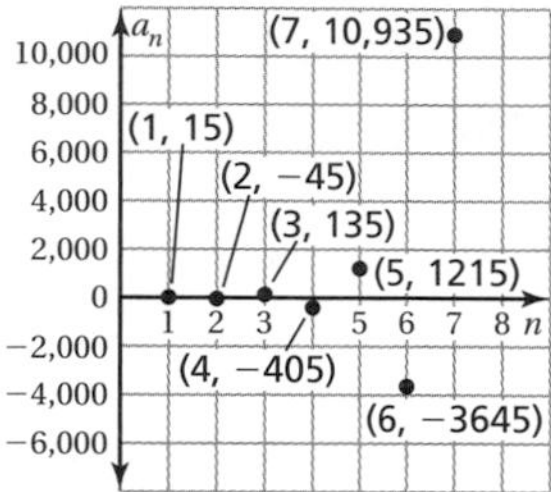

10. 3, $\frac{3}{4}$, $\frac{3}{16}$

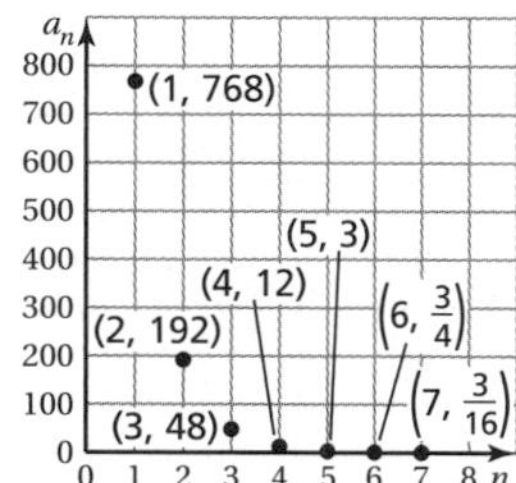

Chapter 6 Test

15. $y = 42{,}500(1.03)^t$;

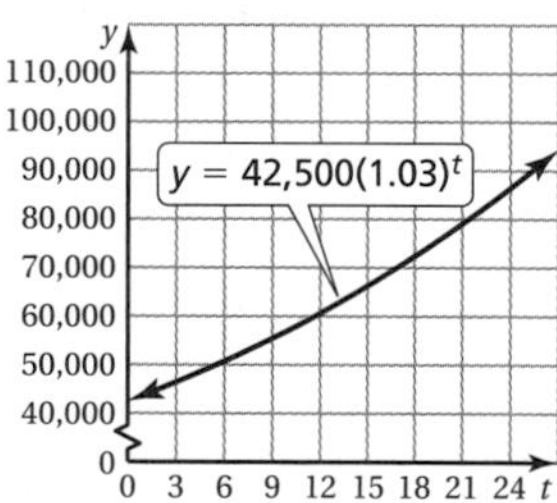

16. $y = 500(1.065)^t$;

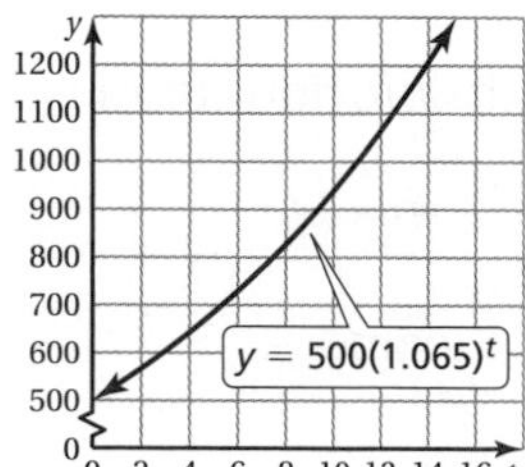

Chapter 7

Section 7.3

Practice and Problem Solving

31. The second term of the second binomial is -7, not 7.

$$\begin{aligned}(r + 6)(r - 7) &= r(r) + r(-7) + 6(r) + 6(-7)\\ &= r^2 - 7r + 6r - 42\\ &= r^2 - r - 42\end{aligned}$$

Section 7.6

Practice and Problem Solving

23. 3 should be factored from 15.

$$3x^2 = 15x$$
$$3x^2 - 15x = 0$$
$$3x(x - 5) = 0$$
$$3x = 0 \;\; or \;\; x - 5 = 0$$
$$x = 0 \;\; or \;\; x = 5$$

Section 7.8

Practice and Problem Solving

35. a. \$24 or \$36

b. \$30; *Sample answer:* By using a table, you can see that the maximum daily revenue occurs when $x = 5$. When $x = 5$, the price of the bobblehead is \$30.

Lesson 7.9b

Record and Practice Journal Practice

1. $(a^2 + 4)(a - 5)$ **2.** $(k^2 + 3)(3k + 1)$

3. $(4p^2 + 1)(2p - 7)$ **4.** $2(3t^2 - 1)(4t - 3)$

5. $(a + b)(b + 8)$ **6.** $(3x + 4)(y - 6)$

7. $4d(d + 1)(d - 9)$ **8.** $12n(n - 2)(n + 2)$

9. factored completely **10.** $6w(w + 4)^2$

11. $q = 0, 2, 4$ **12.** $r = -3, 0, 3$

13. $a = -10, 0, 3$ **14.** $f = 0, 7$

15. a. length: $x - 10$; width: $y + 8$

b. length: 110; width: 70

Chapter 7 Test

20. $x = 0, x = \frac{1}{3}, x = 4$ **21.** $\pi r^2 - 6\pi r + 9\pi$

22. 1.5 sec

23. a. 400 ft^2 **b.** 22 ft

24. a. red bull's-eye area: 4π in.^{2}; gray ring area: 21π in.^{2}

b. $x + 3$

c. 3 in.; no; yes; If x decreases, the area of the gray ring decreases. If x increases, the area of the gray ring increases.

Chapter 8

Record and Practice Journal Fair Game Review

1.

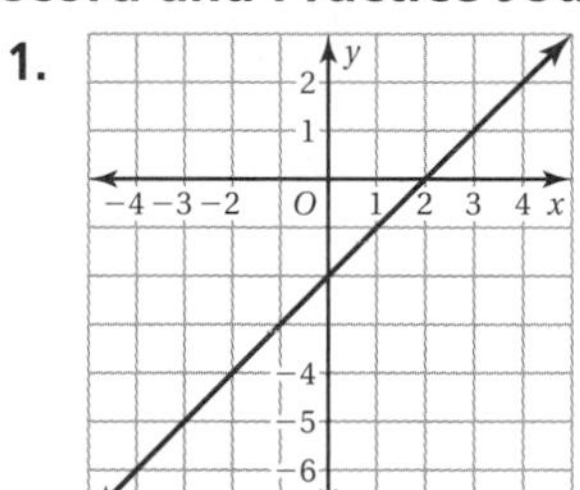

2.

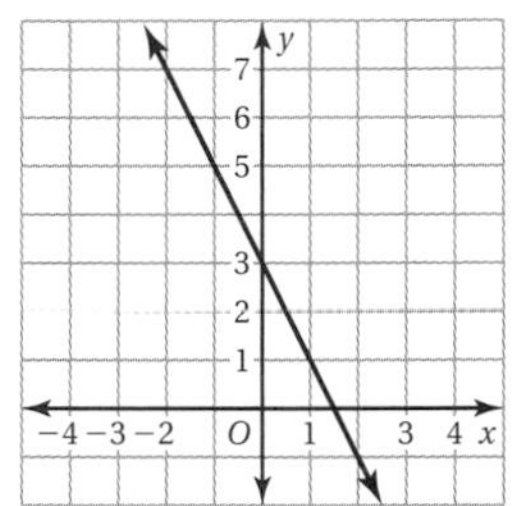

3.

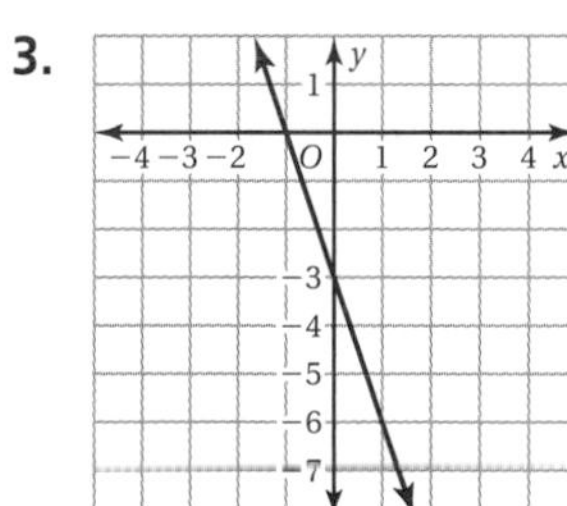

4.

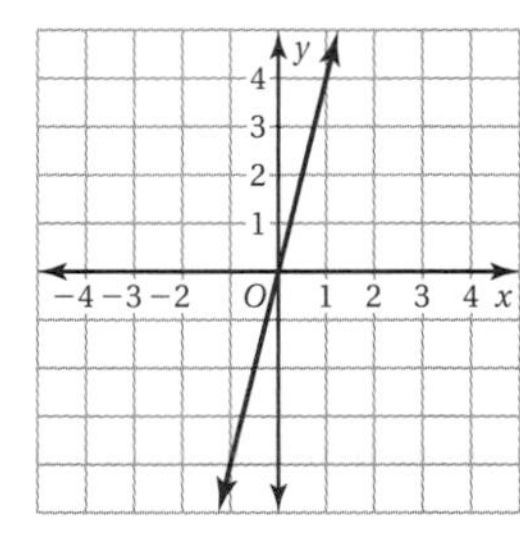

5.

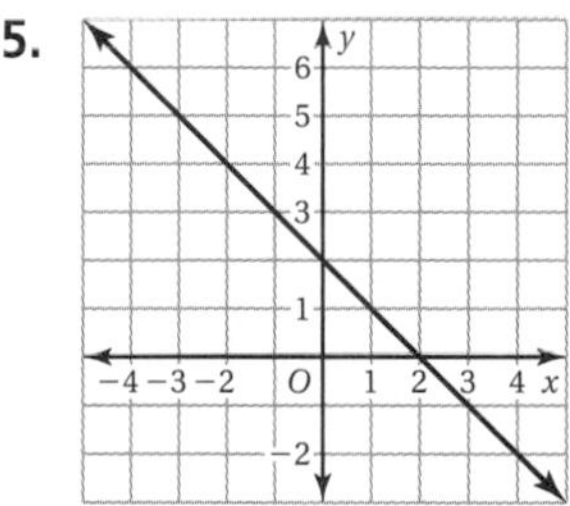

6.

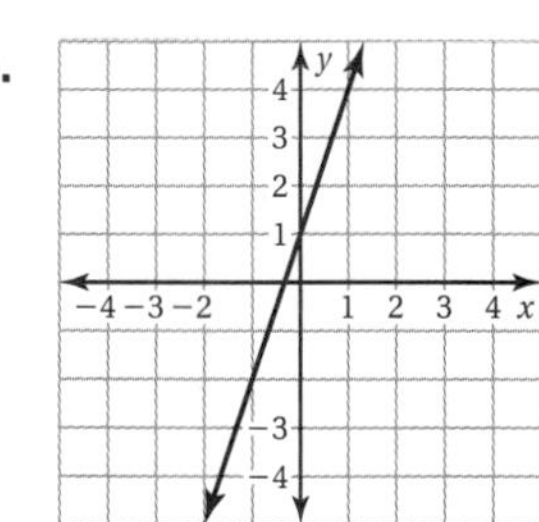

15. a. 23 ft **b.**

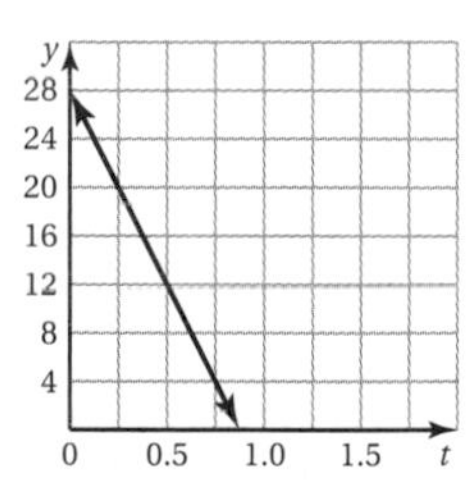

Section 8.1

On Your Own

1. vertex: $(2, -3)$; axis of symmetry: $x = 2$; domain: all real numbers; range $y \geq -3$; When $x < 2$, y increases as x decreases. When $x > 2$, y increases as x increases.

2. vertex: $(-3, 7)$; axis of symmetry: $x = -3$; domain: all real numbers; range $y \leq 7$; When $x < -3$, y decreases as x decreases. When $x > -3$, y decreases as x increases.

4. Both graphs open up and have the same vertex, (0, 0), and the same axis of symmetry, $x = 0$. The graph of $y = \frac{1}{3}x^2$ is wider than the graph of $y = x^2$.

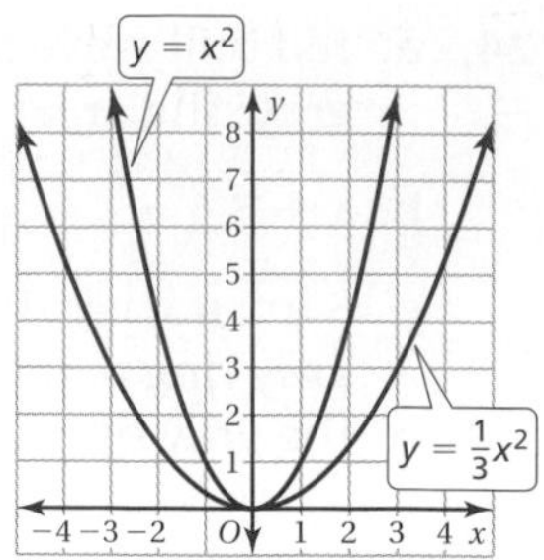

5. Both graphs open up and have the same vertex, (0, 0), and the same axis of symmetry, $x = 0$. The graph of $y = \frac{3}{2}x^2$ is narrower than the graph of $y = x^2$.

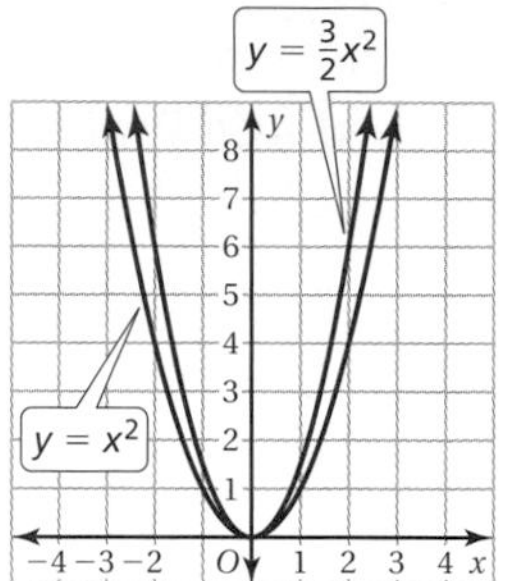

8. The graphs have the same vertex, (0, 0), and the same axis of symmetry, $x = 0$, but the graph of $y = -\frac{1}{4}x^2$ opens down. The graph of $y = -\frac{1}{4}x^2$ is wider than the graph of $y = x^2$.

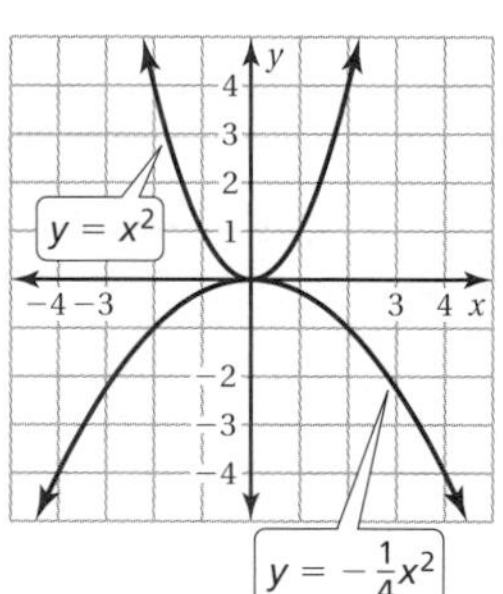

Extra Example 3

3. The graphs have the same vertex, (0, 0), and the same axis of symmetry, $x = 0$, but the graph of $y = -\frac{1}{8}x^2$ opens down. The graph of $y = -\frac{1}{8}x^2$ is wider than the graph of $y = x^2$.

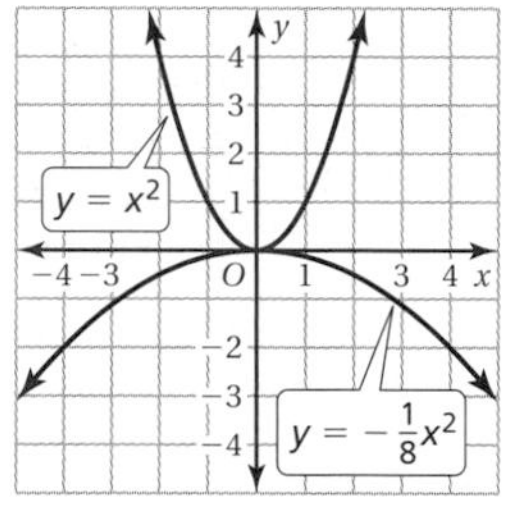

Practice and Problem Solving

7. Both graphs open down and have a U-shape. The graph of $y = -0.004x^2$ is much wider than the graph of $y = -4x^2$.

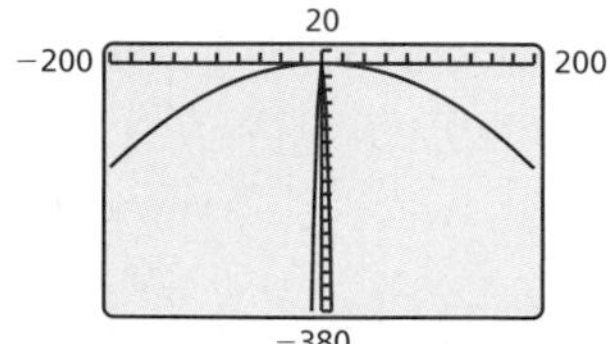

8. vertex: (1, −1); axis of symmetry: $x = 1$; domain: all real numbers; range: $y \leq -1$; When $x < 1$, y decreases as x decreases. When $x > 1$, y decreases as x increases.

9. vertex: (−2, 4); axis of symmetry: $x = -2$; domain: all real numbers; range: $y \geq 4$; When $x < -2$, y increases as x decreases. When $x > -2$, y increases as x increases.

10. vertex: (3, −1); axis of symmetry: $x = 3$; domain: all real numbers; range: $y \geq -1$; When $x < 3$, y increases as x decreases. When $x > 3$, y increases as x increases.

11. Both graphs open up and have the same vertex (0, 0), and the same axis of symmetry, $x = 0$. The graph of $y = 6x^2$ is narrower than the graph of $y = x^2$.

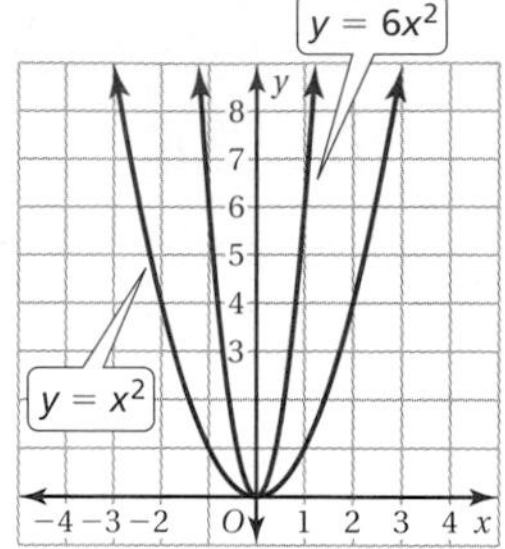

12. Both graphs open up and have the same vertex (0, 0), and the same axis of symmetry, $x = 0$. The graph of $y = 8x^2$ is narrower than the graph of $y = x^2$.

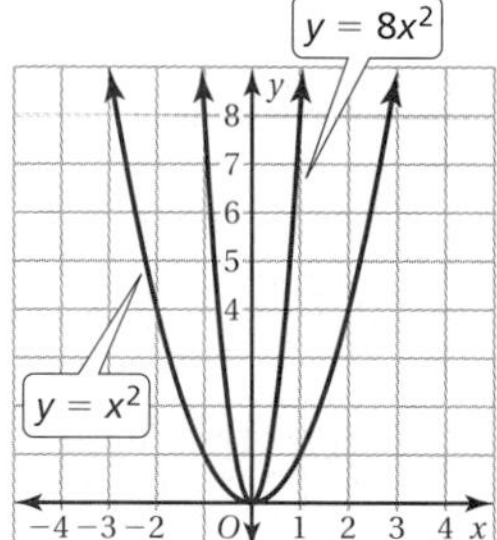

13. Both graphs open up and have the same vertex (0, 0), and the same axis of symmetry, $x = 0$. The graph of $y = \frac{1}{4}x^2$ is wider than the graph of $y = x^2$.

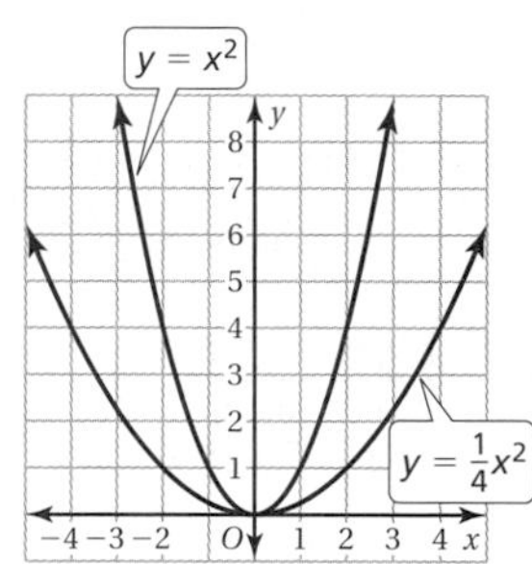

14. Both graphs open up and and have the same vertex (0, 0), and the same axis of symmetry, $x = 0$. The graph of $y = \frac{3}{4}x^2$ is wider than the graph of $y = x^2$.

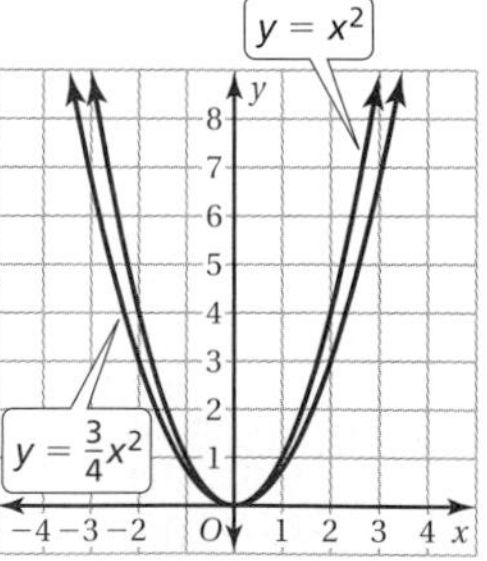

15. Both graphs open up and have the same vertex (0, 0), and the same axis of symmetry, $x = 0$. The graph of $y = \frac{5}{2}x^2$ is narrower than the graph of $y = x^2$.

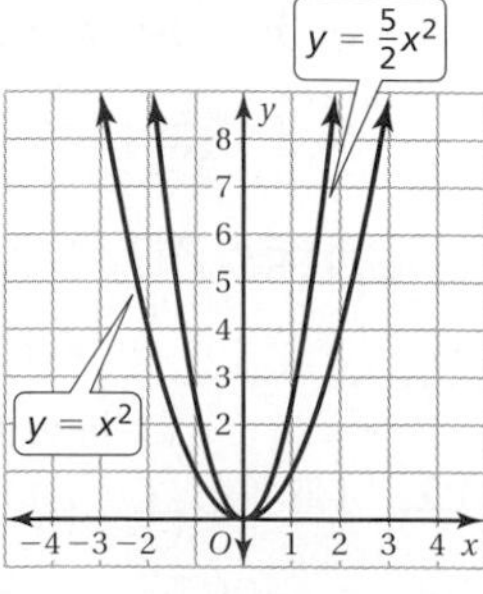

16. Both graphs open up and have the same vertex (0, 0), and the same axis of symmetry, $x = 0$. The graph of $y = \frac{7}{5}x^2$ is narrower than the graph of $y = x^2$.

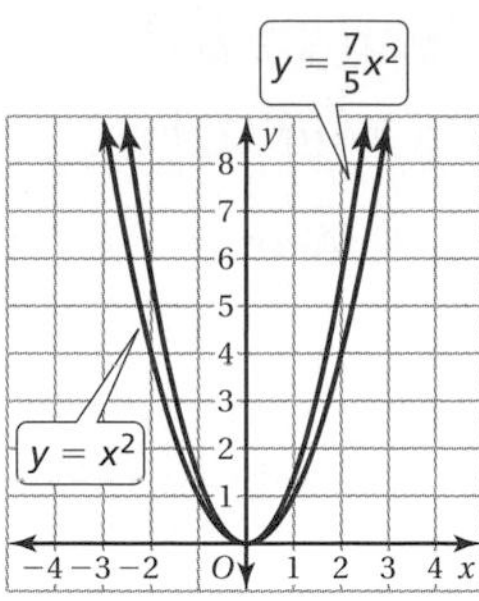

17. **a.** domain: $0 \le t \le 2$; range: $0 \le y \le 64$

b.

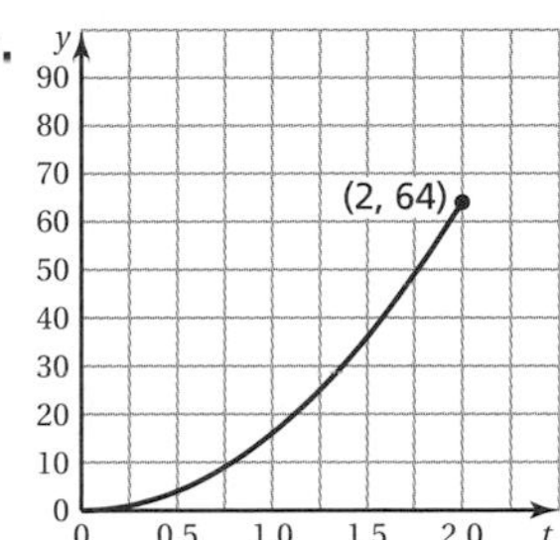

c. 2 seconds

18. Both graphs have the same same vertex (0, 0), and the same axis of symmetry, $x = 0$, but the graph of $y = -2x^2$ opens down. The graph of $y = -2x^2$ is narrower than the graph of $y = x^2$.

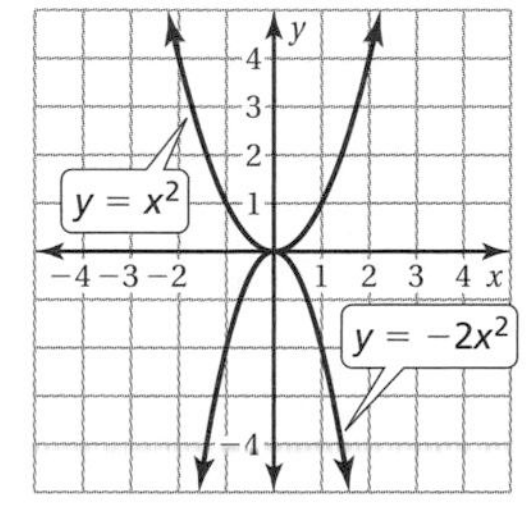

19. Both graphs have the same vertex (0, 0), and the same axis of symmetry, $x = 0$, but the graph of $y = -7x^2$ opens down. The graph of $y = -7x^2$ is narrower than the graph of $y = x^2$.

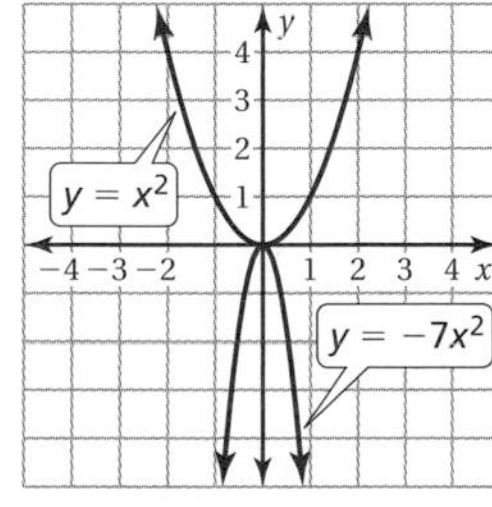

20. Both graphs have the same vertex (0, 0), and the same axis of symmetry, $x = 0$, but the graph of $y = -\frac{1}{5}x^2$ opens down. The graph of $y = -\frac{1}{5}x^2$ is wider than the graph of $y = x^2$.

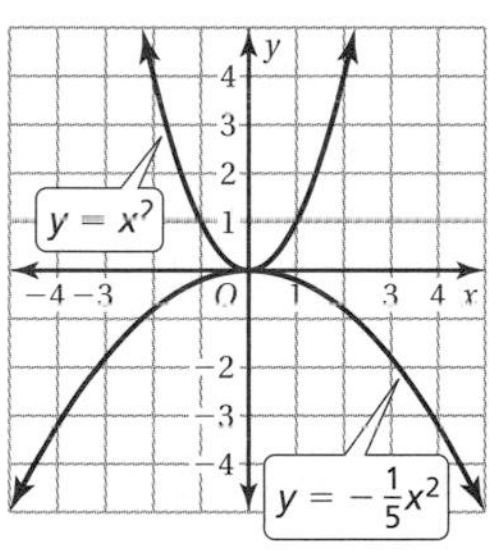

21. Both graphs have the same vertex (0, 0), and the same axis of symmetry, $x = 0$, but the graph of $y = -\frac{5}{8}x^2$ opens down. The graph of $y = -\frac{5}{8}x^2$ is wider than the graph of $y = x^2$.

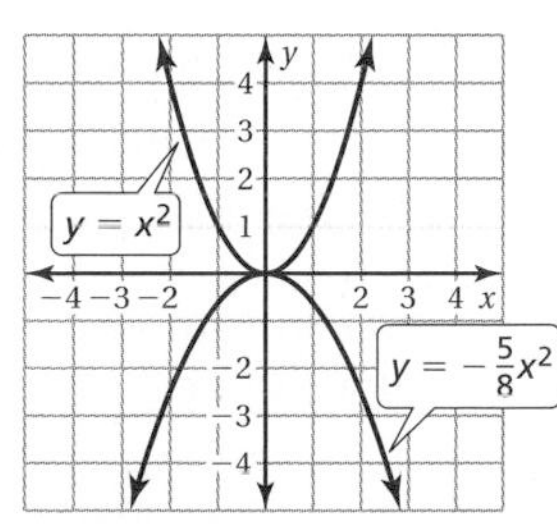

22. Both graphs have the same vertex (0, 0), and the same axis of symmetry, $x = 0$, but the graph of $y = -\frac{5}{3}x^2$ opens down. The graph of $y = -\frac{5}{3}x^2$ is narrower than the graph of $y = x^2$.

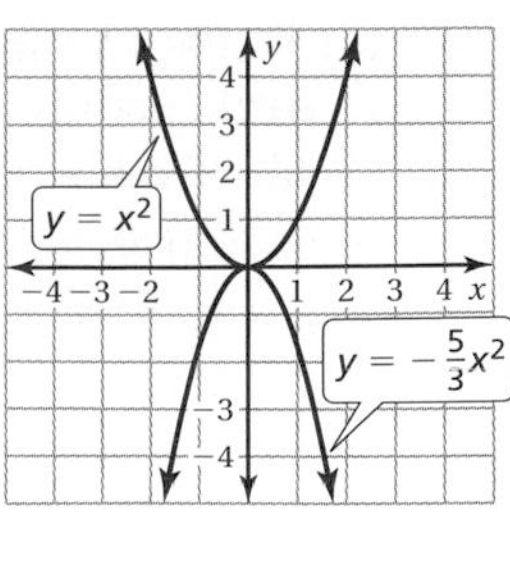

23. Both graphs have the same vertex (0, 0), and the same axis of symmetry, $x = 0$, but the graph of $y = -\frac{9}{4}x^2$ opens down. The graph of $y = -\frac{9}{4}x^2$ is narrower than the graph of $y = x^2$.

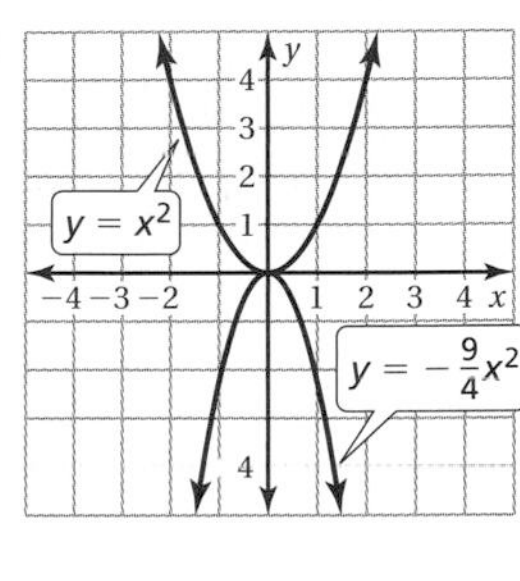

30. sometimes; When $0 < a < 1$, the graph of $y = ax^2$ is wider than the graph of $y = x^2$. When a > 1, the graph of $y = ax^2$ is narrower than the graph of $y = x^2$.

38. **a.** about 8 cm

b. decrease; the cross section as shown can be modeled by $y = 0.2x^2$. The graph of $y = 0.1x^2$ is wider than the graph of $y = 0.2x^2$. This means that the "depth" of the parabola inside the glass is less, which indicates that the rotation speed decreased.

39. **a.** The domain includes the integers 0 through 10. It is discrete because you can only have a whole number of workers. It is positive because you cannot have a negative number of workers.

b.

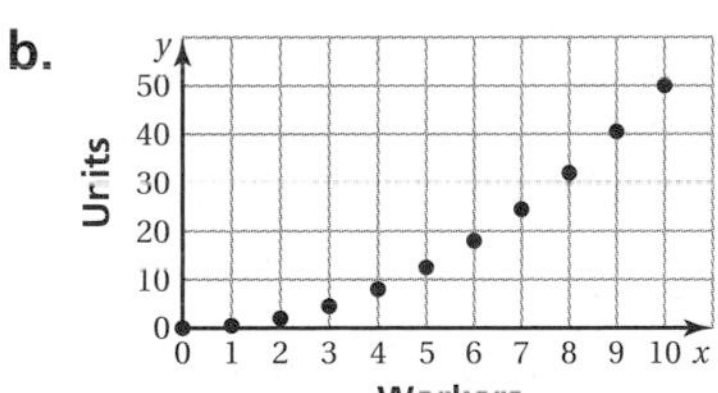

c. The company is better off running one assembly line at full capacity.

Sample answer:

(1) From a production standpoint: Ten workers on one assembly line (full capacity) will produce 50 units per hour. Ten workers split on two assembly lines (half capacity) will produce a total of 25 units per hour. One assembly line at full capacity produces twice as many units as two assembly lines at half capacity.

(2) **From a financial standpoint:** The number of units produced is increasing at an increasing rate for each additional worker on an assembly line. This means production (i.e. revenue) is increasing faster than the labor cost when more employees work on the same assembly line. The company is getting more "bang for the buck."

Mini-Assessment

1. vertex: $(-3, -4)$; axis of symmetry: $x = -3$; domain: all real numbers; range: $y \geq -4$; When $x < -3$, y increases as x decreases. When $x > -3$, y increases as x increases.

2. Both graphs open up and have the same vertex, $(0, 0)$, and the same axis of symmetry, $x = 0$. The graph of $y = 10x^2$ is narrower than the graph of $y = x^2$.

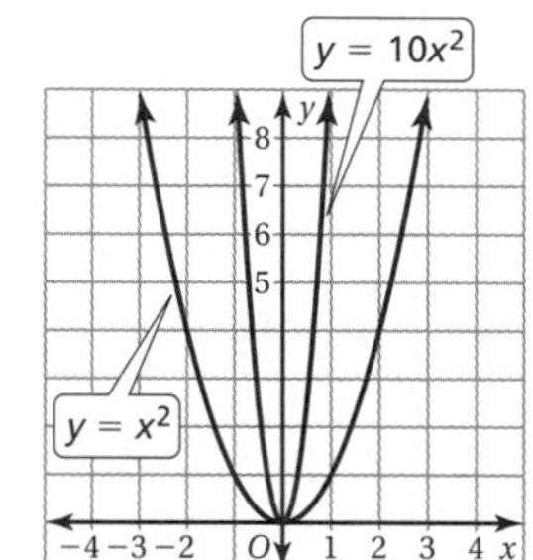

3. Both graphs open up and have the same vertex, $(0, 0)$, and the same axis of symmetry, $x = 0$. The graph of $y = \frac{1}{8}x^2$ is wider than the graph of $y = x^2$.

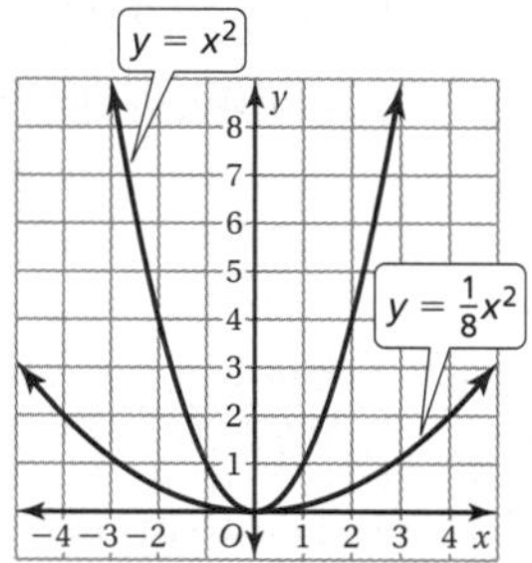

4. Both graphs have the same same vertex, $(0, 0)$, and the same axis of symmetry, $x = 0$, but the graph of $y = -4x^2$ opens down. The graph of $y = -4x^2$ is narrower than the graph of $y = x^2$.

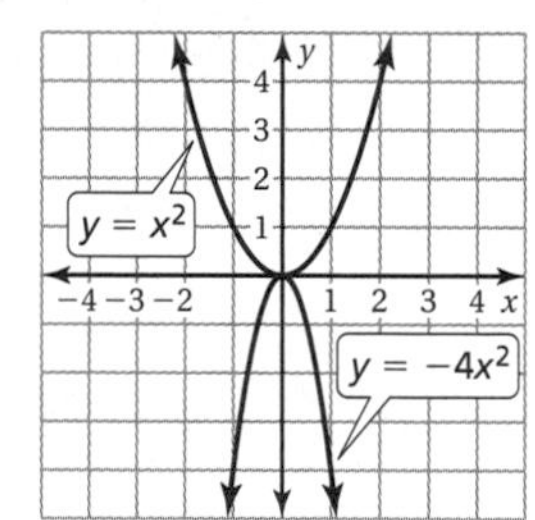

5. Both graphs have the same vertex, $(0, 0)$, and the same axis of symmetry, $x = 0$, but the graph of $y = -\frac{3}{5}x^2$ opens down. The graph of $y = -\frac{3}{5}x^2$ is wider than the graph of $y = x^2$.

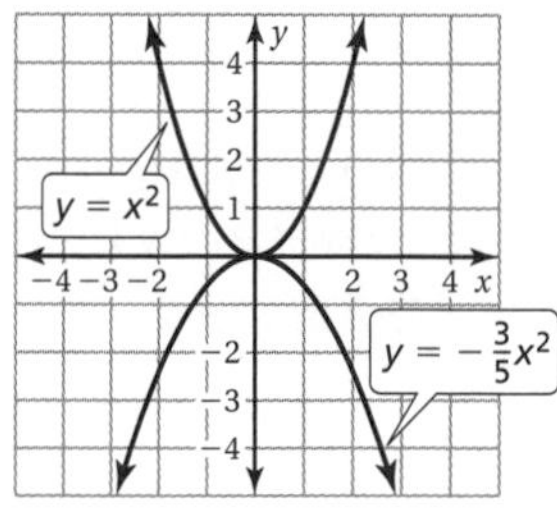

Section 8.2

On Your Own

3. focus: $\left(0, -\frac{1}{12}\right)$

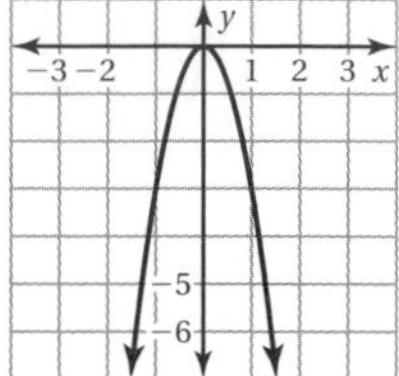

Practice and Problem Solving

7. focus: $\left(0, \frac{1}{4}\right)$

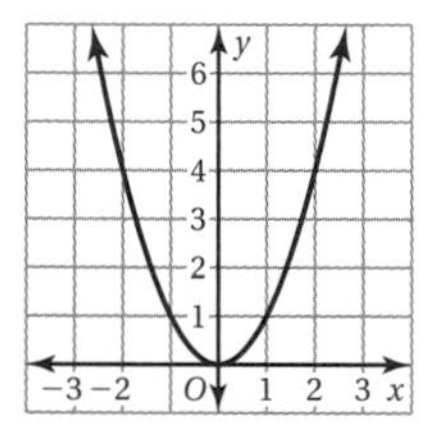

8. focus: $\left(0, \frac{1}{16}\right)$

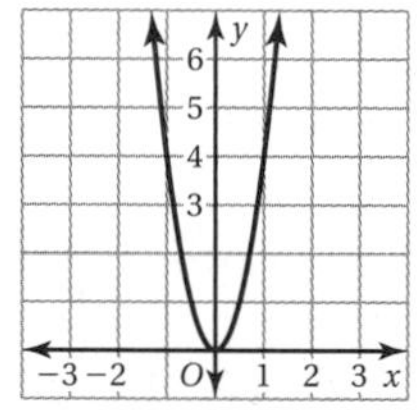

9. focus: $\left(0, -\frac{1}{48}\right)$

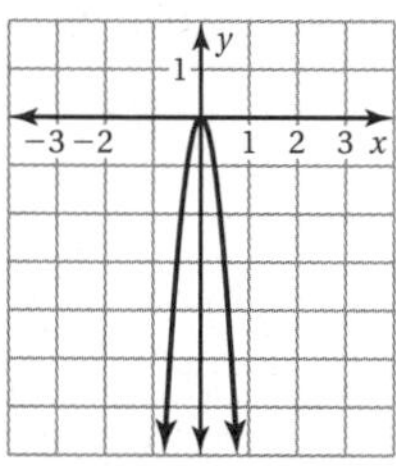

10. focus: $(0, 1)$

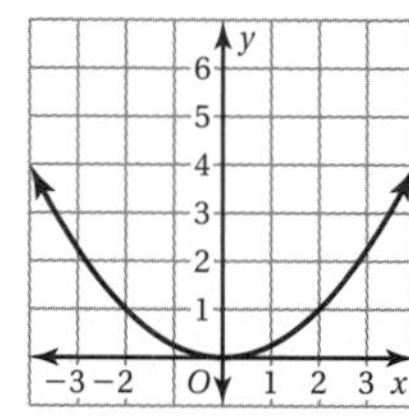

11. focus: $\left(0, \frac{1}{2}\right)$

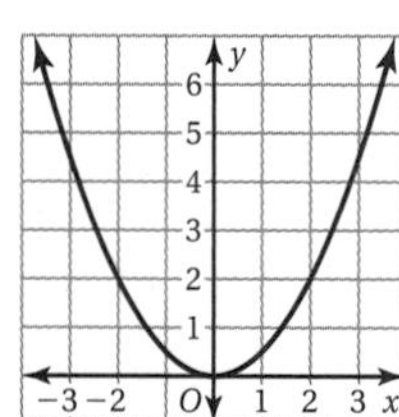

12. focus: $\left(0, -\frac{1}{3}\right)$

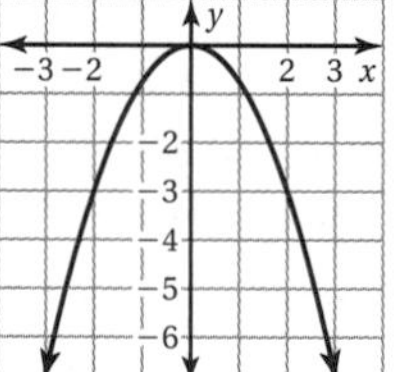

Lesson 8.3

Extra Example 1

1. Both graphs open up and have the same axis of symmetry, $x = 0$. The graph of $y = x^2 + 6$ is a translation 6 units up of the graph of $y = x^2$.

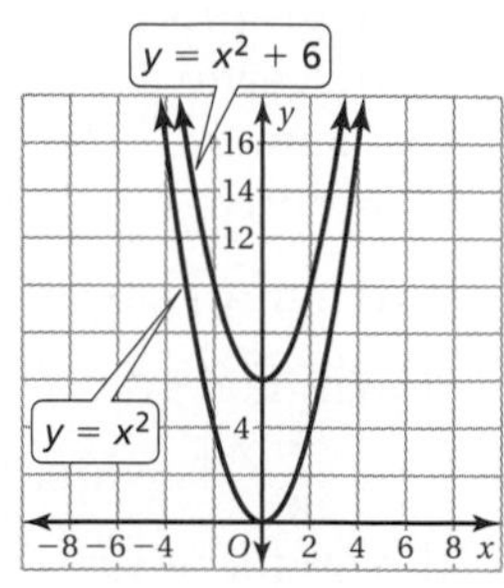

Extra Example 2

2. Both graphs have the same same axis of symmetry, $x = 0$. The graph of $y = -\frac{1}{5}x^2 - 1$ opens down and is wider than the graph of $y = x^2$. The vertex of the graph of $y = -\frac{1}{5}x^2 - 1$ is a translation 1 unit down of the vertex of the graph of $y = x^2$.

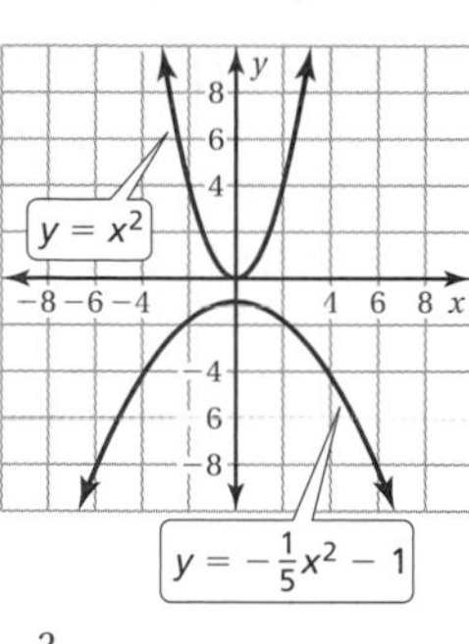

On Your Own

2. Both graphs open up and have the same axis of symmetry, $x = 0$. The graph of $y = 2x^2 - 5$ is narrower than the graph of $y = x^2$. The vertex of the graph of $y = 2x^2 - 5$ is a translation 5 units down of the vertex of the graph of $y = x^2$.

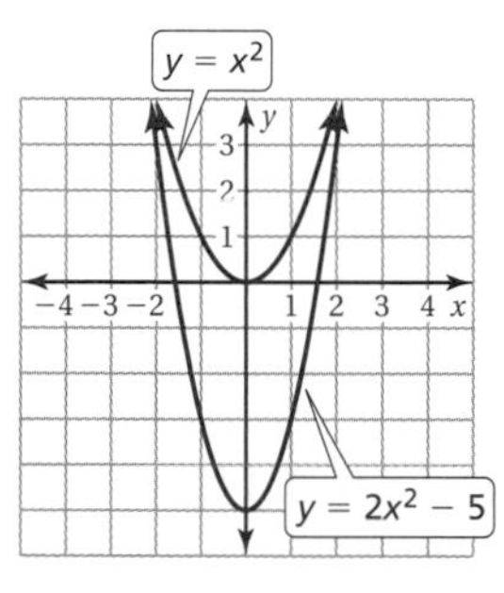

3. Both graphs have the same axis of symmetry, $x = 0$. The graph of $y = -\frac{1}{2}x^2 + 4$ opens down and is wider than the graph of $y = x^2$. The vertex of the graph of $y = -\frac{1}{2}x^2 + 4$ is a translation 4 units up of the graph of $y = x^2$.

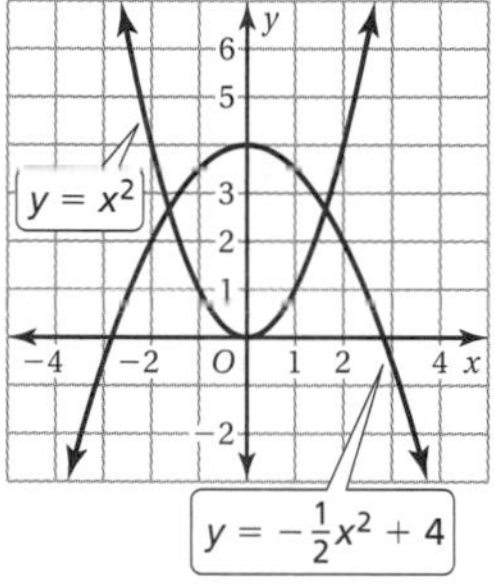

Practice and Problem Solving

7. Both graphs open up and have the same axis of symmetry, $x = 0$. The graph of $y = x^2 + 4$ is a translation 4 units up of the graph of $y = x^2$.

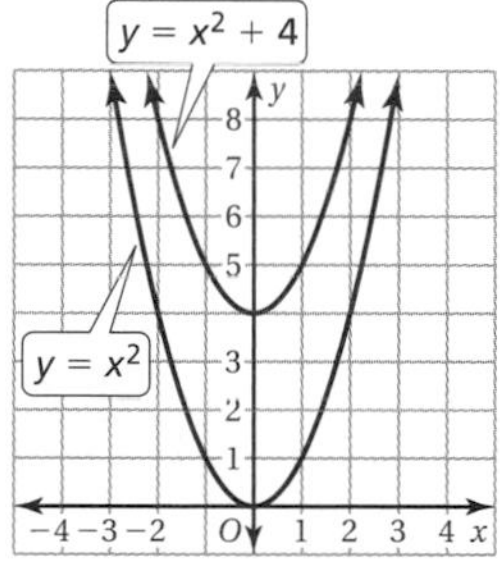

8. Both graphs open up and have the same axis of symmetry, $x = 0$. The graph of $y = x^2 - 3$ is a translation 3 units down of the graph of $y = x^2$.

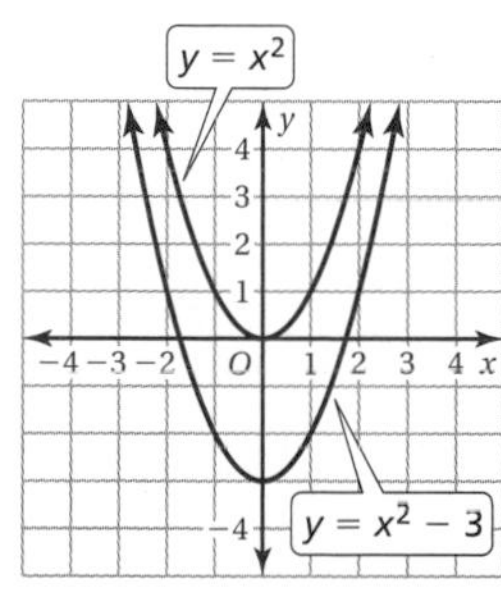

9. Both graphs have the same axis of symmetry, $x = 0$. The graph of $y = -x^2 + 5$ opens down. The vertex of the graph of $y = -x^2 + 5$ is a translation 5 units up of the vertex of the graph of $y = x^2$.

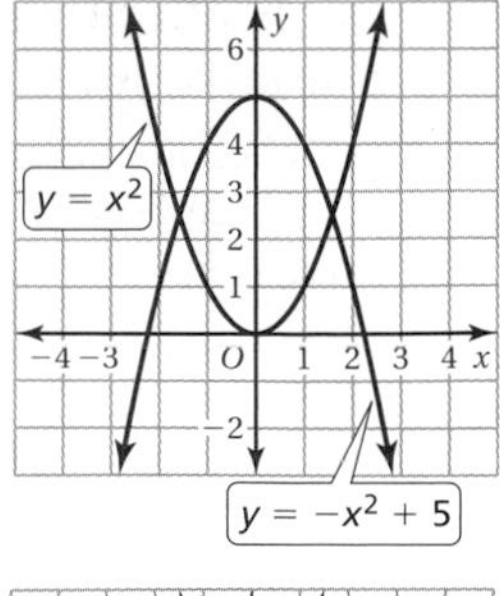

10. Both graphs have the same axis of symmetry, $x = 0$. The graph of $y = -x^2 - 9$ opens down. The vertex of the graph of $y = -x^2 - 9$ is a translation 9 units down of the vertex of the graph of $y = x^2$.

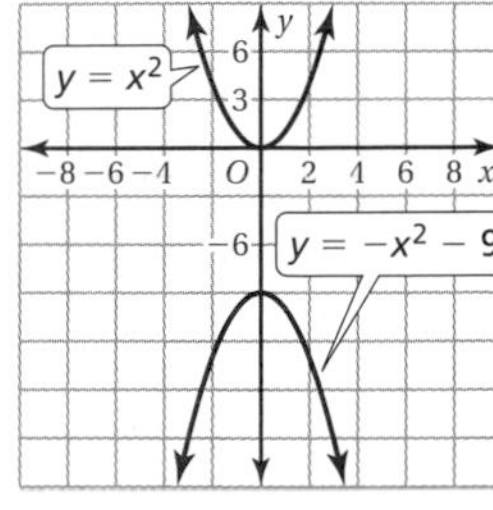

11. Both graphs open up and have the same axis of symmetry, $x = 0$. The graph of $y = 2x^2 - 4$ is narrower than the graph of $y = x^2$. The vertex of the graph of $y = 2x^2 - 4$ is a translation 4 units down of the vertex of the graph of $y = x^2$.

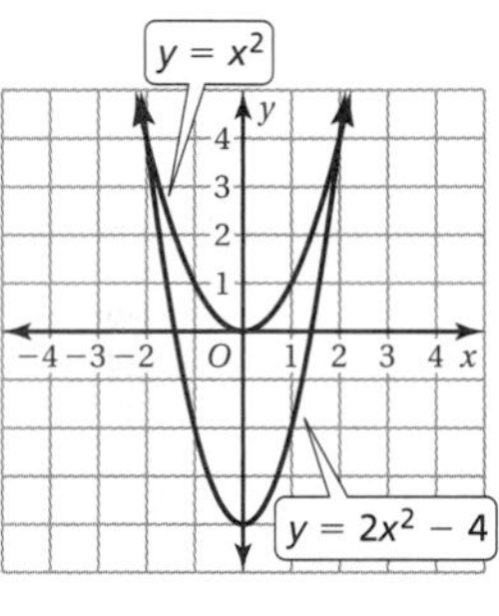

12. Both graphs open up and have the same axis of symmetry, $x = 0$. The graph of $y = \frac{1}{2}x^2 + 2$ is wider than the graph of $y = x^2$. The vertex of the graph of $y = \frac{1}{2}x^2 + 2$ is a translation 2 units up of the vertex of the graph of $y = x^2$.

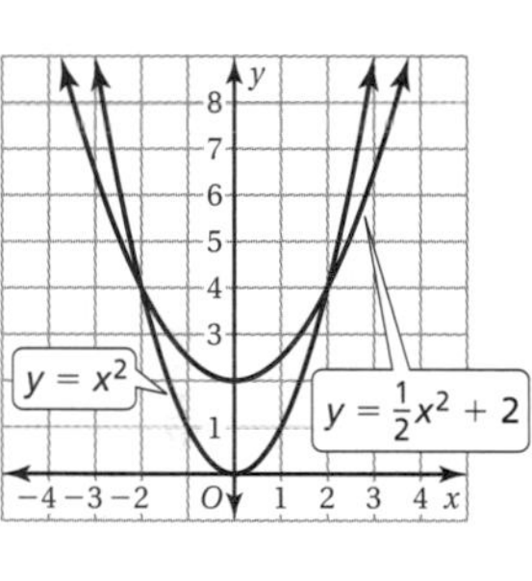

13. Both graphs open up and have the same axis of symmetry, $x = 0$. The graph of $y = 3x^2 + 4$ is narrower than the graph of $y = x^2$. The vertex of the graph of $y = 3x^2 + 4$ is a translation 4 units up of the vertex of the graph of $y = x^2$.

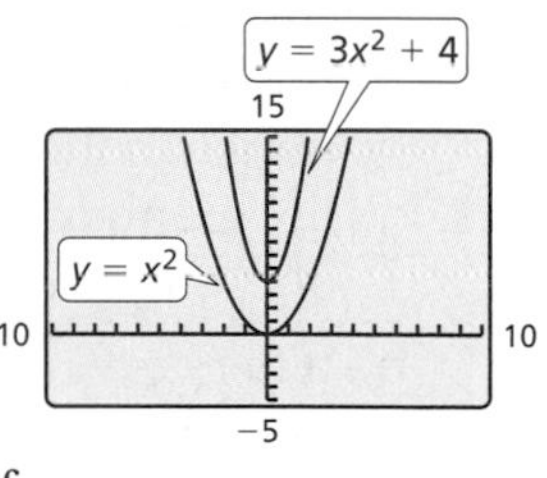

14. Both graphs have the same axis of symmetry, $x = 0$. The graph of $y = -2x^2 - 1$ opens down and is narrower than the graph of $y = x^2$. The vertex of the graph of $y = -2x^2 - 1$ is a translation 1 unit down of the vertex of the graph of $y = x^2$.

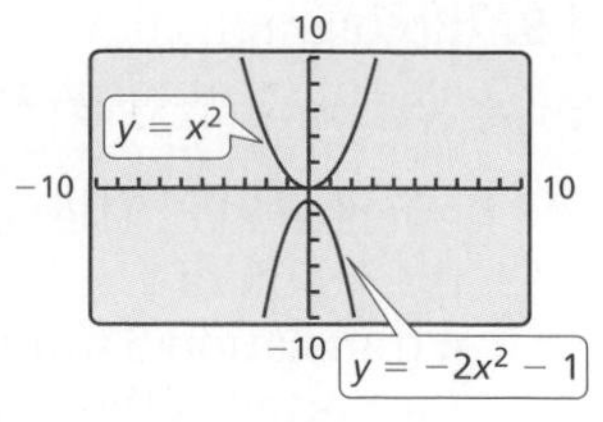

15. Both graphs have the same axis of symmetry, $x = 0$. The graph of $y = -\frac{1}{4}x^2 - \frac{1}{2}$ opens down and is wider than the graph of $y = x^2$. The vertex of the graph of $y = -\frac{1}{4}x^2 - \frac{1}{2}$ is a translation 0.5 unit down of the vertex of the graph of $y = x^2$.

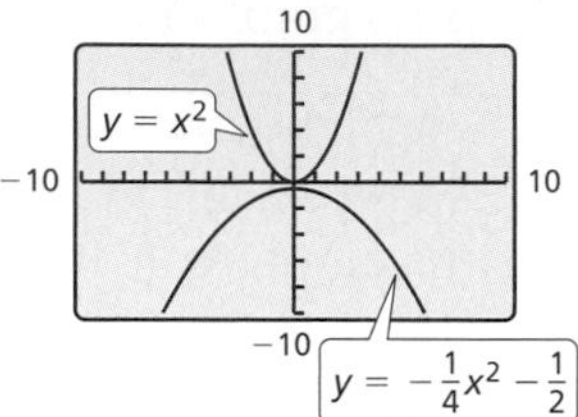

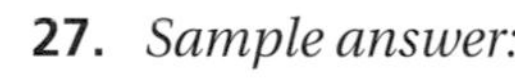

26. *Sample answer:*

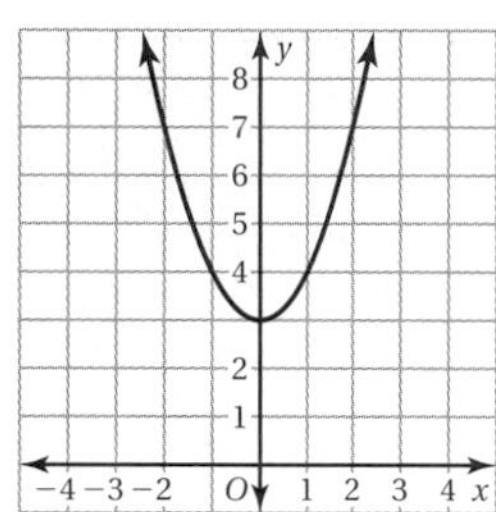

27. *Sample answer:*

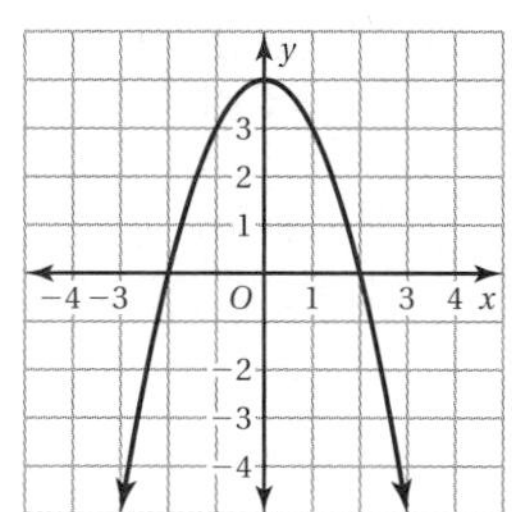

28. *Sample answer:*

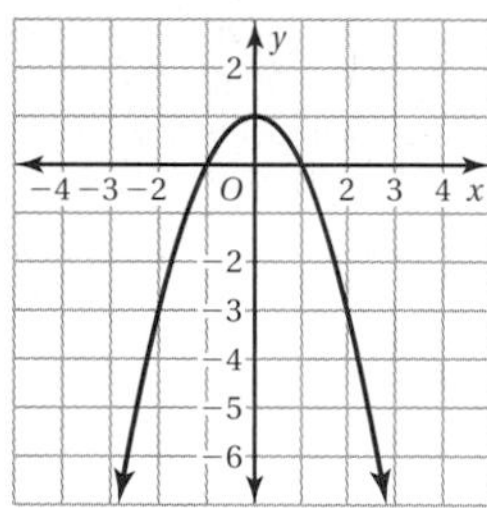

29. *Sample answer:*

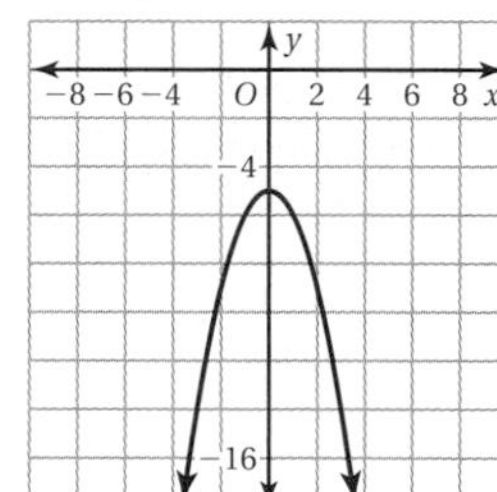

33. a. When $k < 0$, the graph of $f(x) + k$ will translate the graph of $f(x)$ down. When $k > 0$, the graph of $f(x) + k$ will translate the graph of $f(x)$ up.

b. When $k > 1$, the graph of $k \cdot f(x)$ is narrower than the graph of $f(x)$. When $0 < k < 1$, the graph of $k \cdot f(x)$ is wider than the graph of $f(x)$. When $-1 < k < 0$, the graph of $k \cdot f(x)$ opens down and is wider than the graph of $f(x)$. When $k = -1$, the graph of $k \cdot f(x)$ opens down. When $k < -1$, the graph of $k \cdot f(x)$ opens down and is narrower than the graph of $f(x)$.

Mini-Assessment

1. Both graphs open up and have the same axis of symmetry, $x = 0$. The graph of $y = x^2 - 8$ is a translation 8 units down of the graph of $y = x^2$.

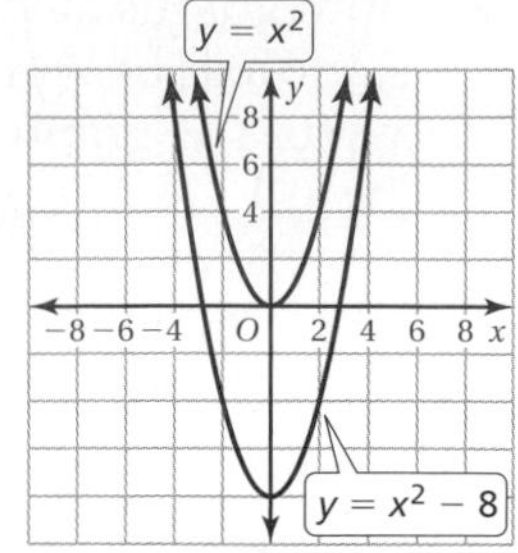

2. Both graphs have the same axis of symmetry, $x = 0$. The graph of $y = -3x^2 + 5$ opens down and is narrower than the graph of $y = x^2$. The vertex of the graph of $y = -3x^2 + 5$ is a translation 5 units up of the vertex of the graph of $y = x^2$.

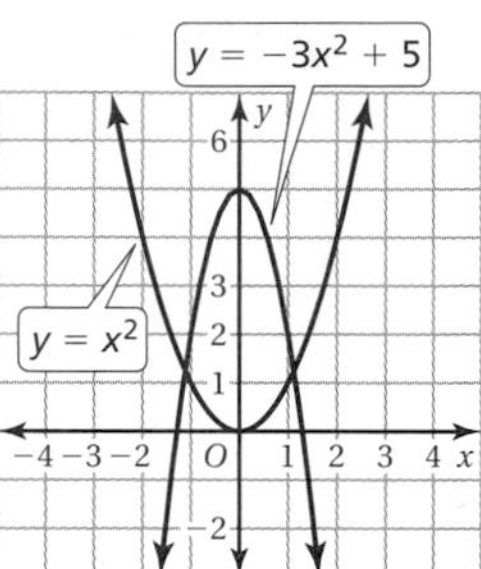

8.1–8.3 Quiz

1. vertex: (1, 4); axis of symmetry: $x = 1$; domain: all real numbers; range: $y \leq 4$; When $x < 1$, y increases as x increases. When $x > 1$, y decreases as x increases.

2. vertex: (−2, 5); axis of symmetry: $x = -2$; domain: all real numbers; range: $y \geq 5$; When $x < -2$, y decreases as x increases. When $x > -2$, y increases as x increases.

3. vertex: (2, −2); axis of symmetry: $x = 2$; domain: all real numbers; range: $y \geq -2$; When $x < 2$, y decreases as x increases. When $x > 2$, y increases as x increases.

4. vertex: (−1, −4); axis of symmetry: $x = -1$; domain: all real numbers; range: $y \geq -4$; When $x < -1$, y decreases as x increases. When $x > -1$, y increases as x increases.

5.

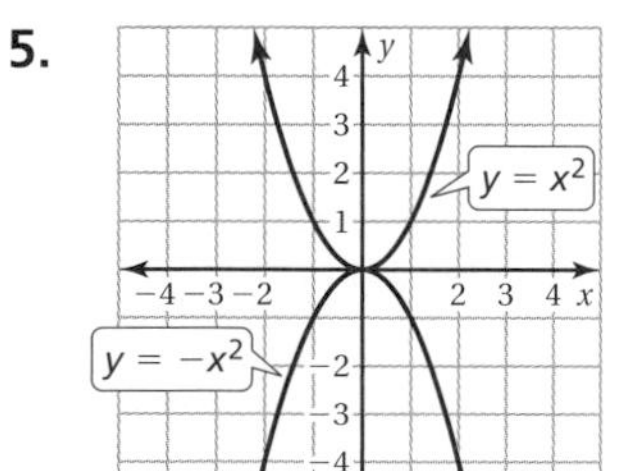

6.

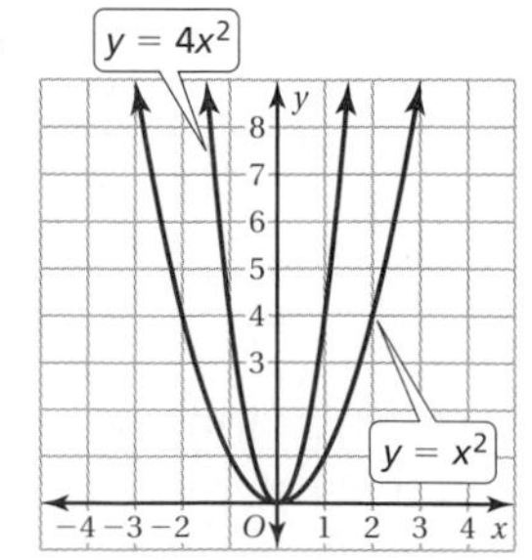

7.

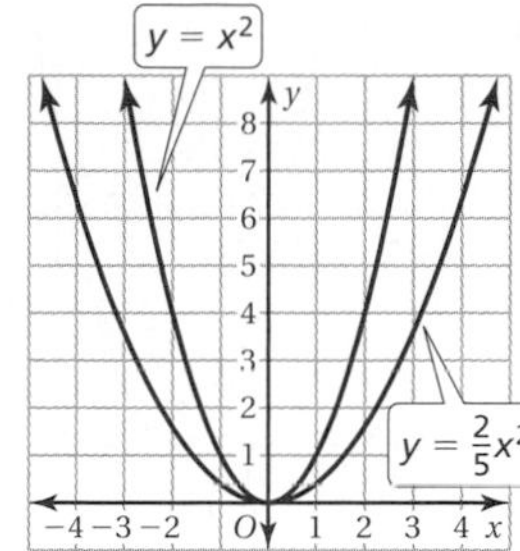

8.

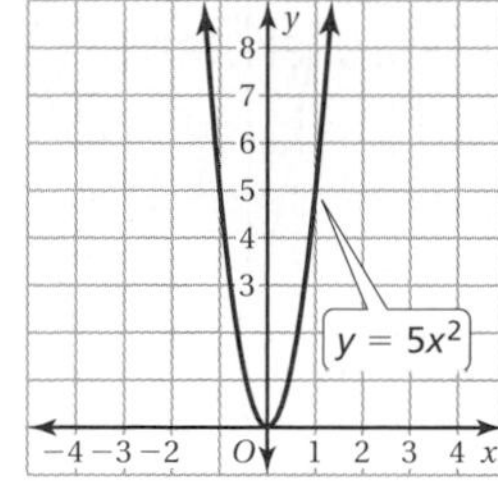

9.

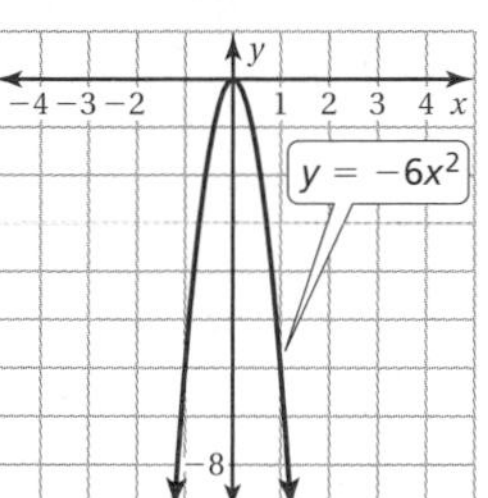

10.

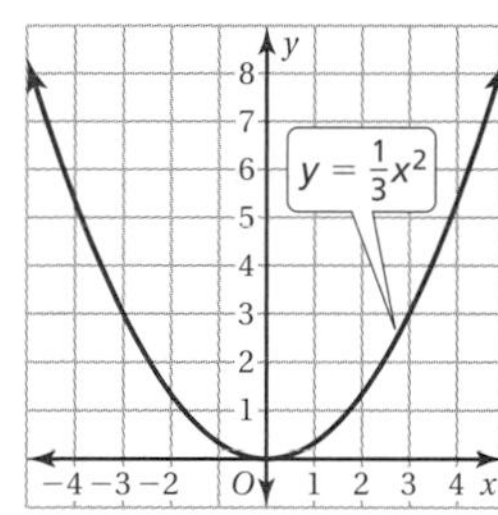

14.

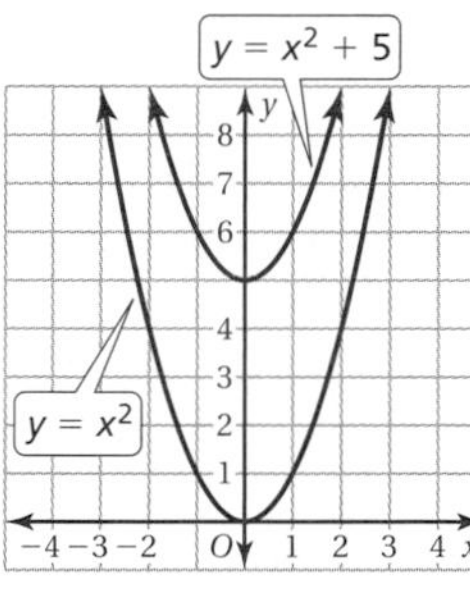

Both graphs open up and have the same axis of symmetry, $x = 0$. The graph of $y = x^2 + 5$ is a translation 5 units up of the graph of $y = x^2$.

15.

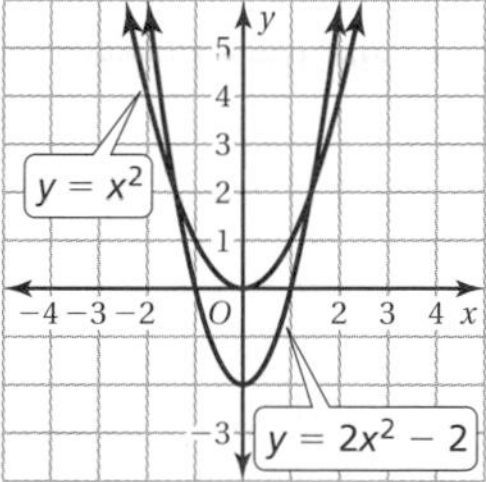

Both graphs open up and have the same axis of symmetry, $x = 0$. The graph of $y = 2x^2 - 2$ is narrower than the graph of $y = x^2$. The vertex of the graph of $y = 2x^2 - 2$ is a translation 2 units down of the vertex of the graph of $y = x^2$.

16.

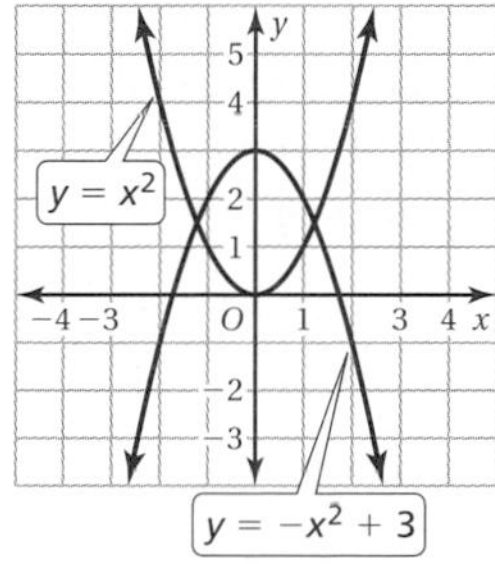

Both graphs have the same axis of symmetry, $x = 0$. The graph of $y = -x^2 + 3$ opens down. The vertex of the graph of $y = -x^2 + 3$ is a translation 3 units up of the vertex of the graph of $y = x^2$.

Lesson 8.4

Practice and Problem Solving

13. domain: all real numbers; range: $y \geq -4$

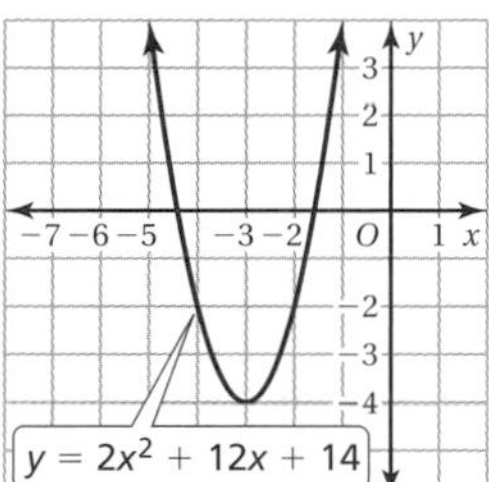

14. domain: all real numbers; range: $y \geq -5$

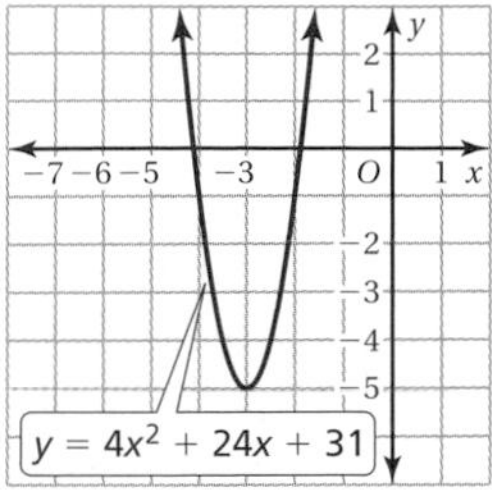

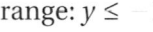

15. domain: all real numbers; range: $y \leq -1$

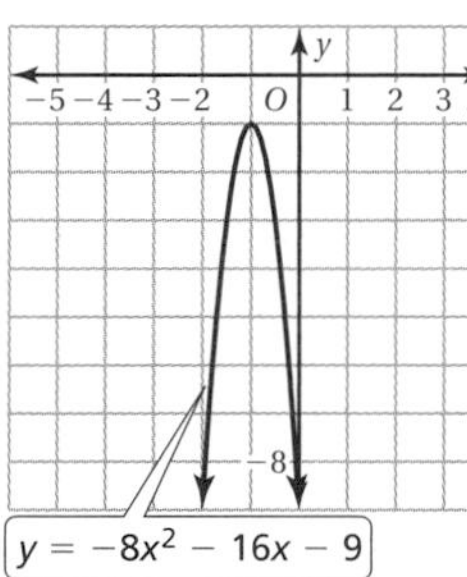

16. domain: all real numbers; range: $y \leq -2$

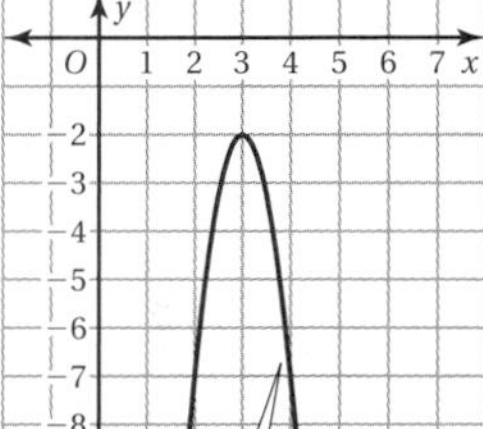

17. domain: all real numbers; range: $y \geq -5$

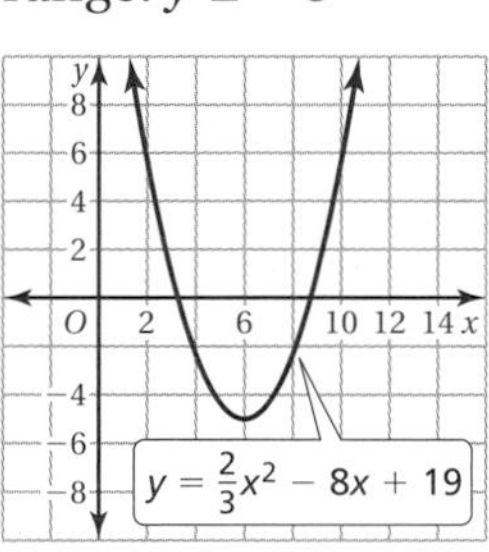

18. domain: all real numbers; range: $y \leq 7$

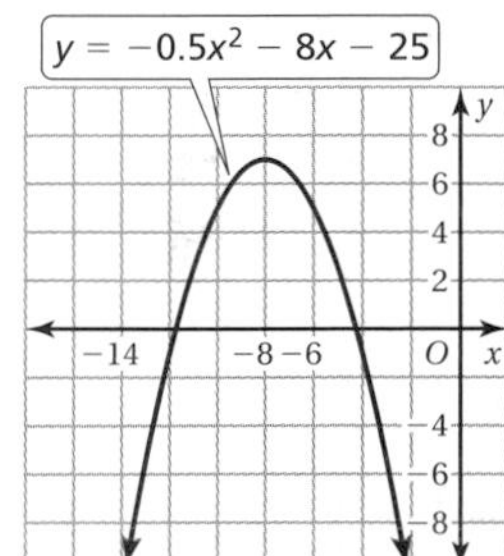

30. vertex $\approx (0.39, -0.07)$

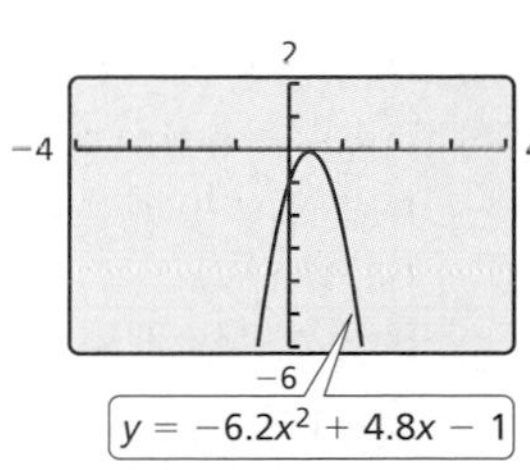

31. vertex $\approx (-1.4, -4)$

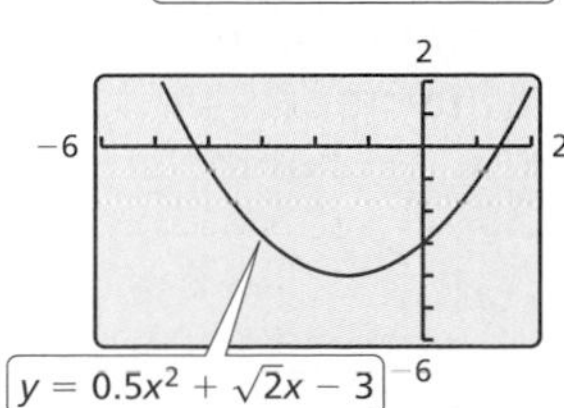

Additional Answers

32. vertex $\approx (-0.48, -0.72)$

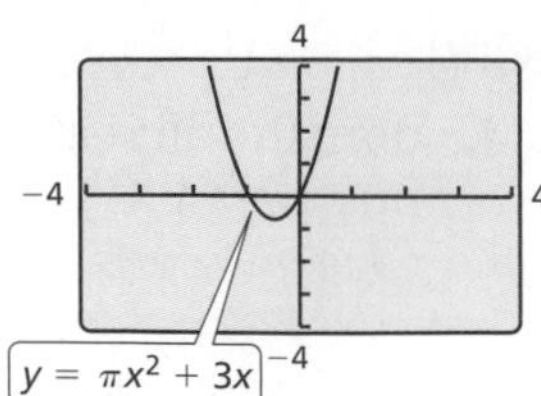

33. opens down; *Sample answer:* The y-values are increasing at a decreasing rate. This means the points are on the left side of a parabola that faces down.

34. $f\left(-\frac{b}{2a}\right)$ represents the value of the function when $x = -\frac{b}{2a}$. So, $f\left(-\frac{b}{2a}\right)$ is the y-coordinate of the vertex.

Lesson 8.4b

Practice

1. 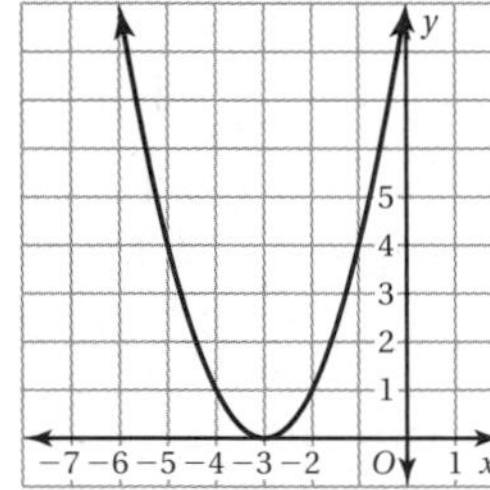

The graph of $y = (x + 3)^2$ is a translation 3 units to the left of the graph of $y = x^2$.

2. The graph of $y = (x - 1)^2$ is a translation 1 unit to the right of the graph of $y = x^2$.

3. The graph of $y = (x - 6)^2$ is a translation 6 units to the right of the graph of $y = x^2$.

4. The graph of $y = (x + 10)^2$ is a translation 10 units to the left of the graph of $y = x^2$.

5. 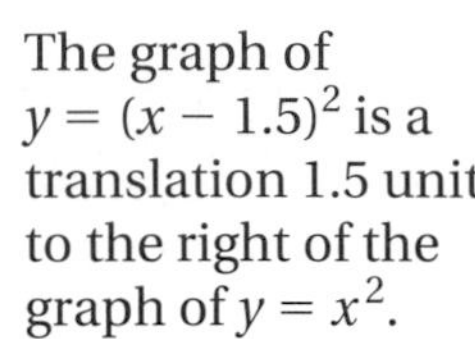

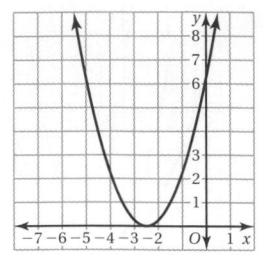

The graph of $y = (x - 1.5)^2$ is a translation 1.5 units to the right of the graph of $y = x^2$.

6. The graph of $y = \left(x + \frac{5}{2}\right)^2$ is a translation 2.5 units to the left of the graph of $y = x^2$.

7. *Sample answer:* $x^2 + 6x + 9$ can be factored as $(x + 3)^2$. In this form, you can see that the graph of $y = (x + 3)^2$ is a translation 3 units to the left of the graph of $y = x^2$.

8. $y = x^2 - 8x + 16$; *Sample answer:* Written as $y = (x - 4)^2$, it is easier to identify the vertex as $(h, k) = (4, 0)$. Written as $y = x^2 - 8x + 16$, it is easier to identify the y-intercept as $c = 16$ and that the graph opens up because $a = 1$ and $1 > 0$.

9. 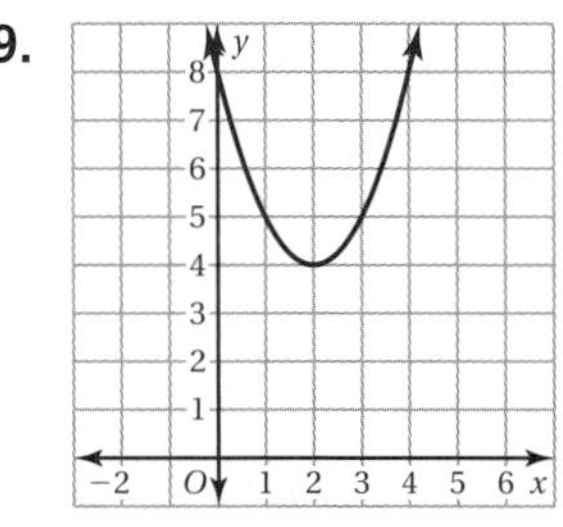

The graph of $y = (x - 2)^2 + 4$ is a translation 2 units to the right and 4 units up of the graph of $y = x^2$.

10. 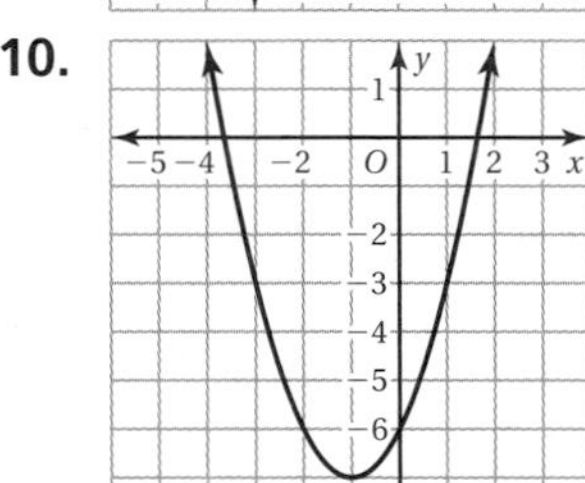

The graph of $y = (x + 1)^2 - 7$ is a translation 1 unit to the left and 7 units down of the graph of $y = x^2$.

11. 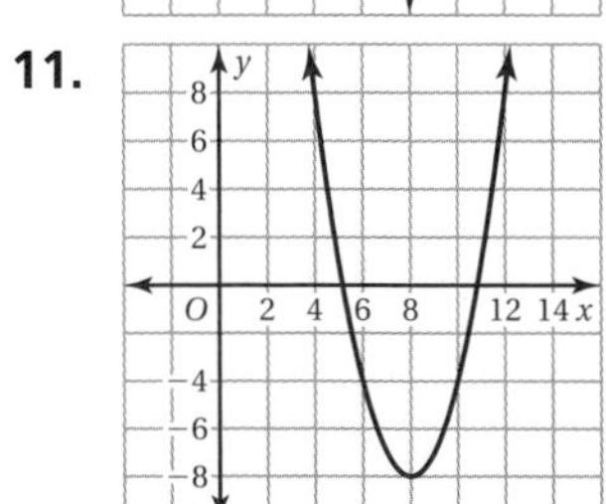

The graph of $y = (x - 8)^2 - 8$ is a translation 8 units to the right and 8 units down of the graph of $y = x^2$.

12. 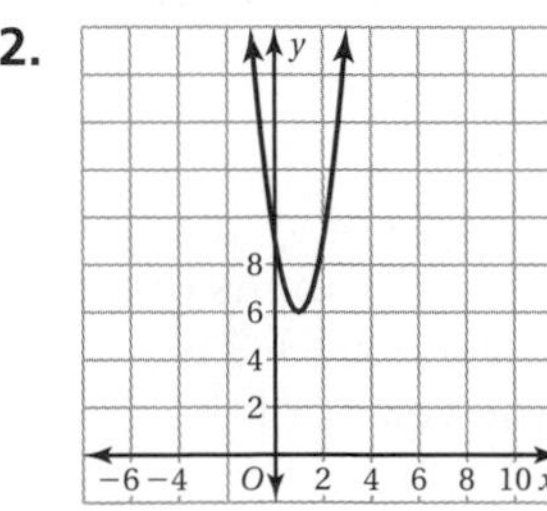

The graph of $y = 3(x - 1)^2 + 6$ is narrower than the graph of $y = x^2$. The vertex of the graph of $y = 3(x - 1)^2 + 6$ is a translation 1 unit to the right and 6 units up of the vertex of the graph of $y = x^2$.

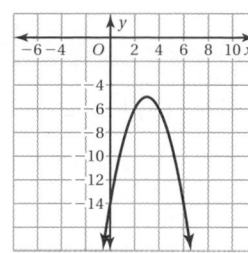

13. The graph of $y = -(x-3)^2 - 5$ opens down. The vertex of the graph of $y = -(x-3)^2 - 5$ is a translation 3 units to the right and 5 units down of the vertex of the graph of $y = x^2$.

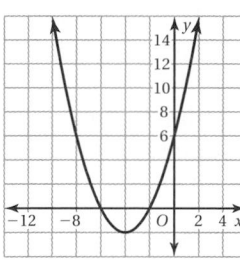

14. The graph of $y = \frac{1}{2}(x+4)^2 - 2$ is wider than the graph of $y = x^2$. The vertex of the graph of $y = \frac{1}{2}(x+4)^2 - 2$ is a translation 4 units to the left and 2 units down of the vertex of the graph of $y = x^2$.

15. $g(x)$ is a vertical translation 7 units down of $f(x)$.

16. $g(x)$ is a horizontal translation 10 units left of $f(x)$.

17. $g(x)$ is narrower than $f(x)$.

18. $g(x)$ is narrower than $f(x)$.

19. $y = (x-2)^2 - 5$; $y = x^2 - 4x - 1$

Record and Practice Journal Practice

1.

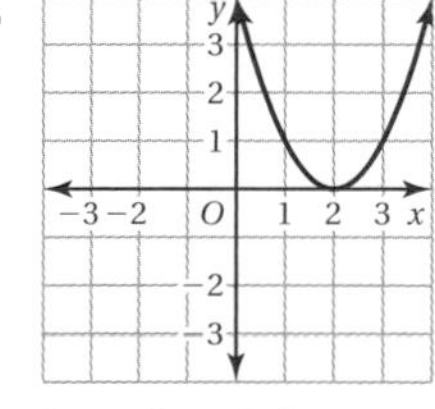

2 units right

2.

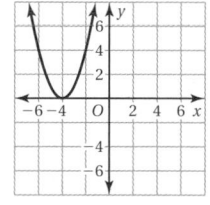

4 units left

3.

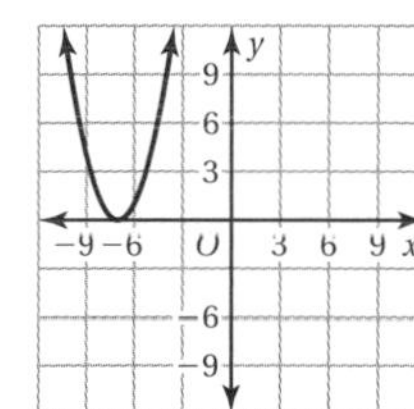

7 units left

4.

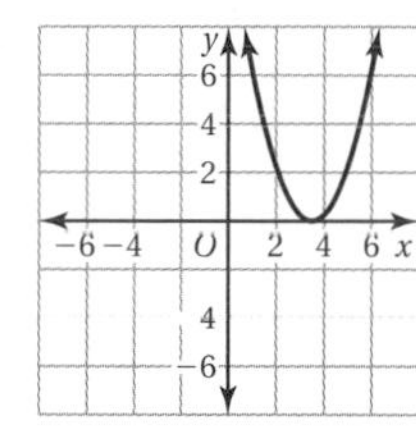

3.5 units right

5.

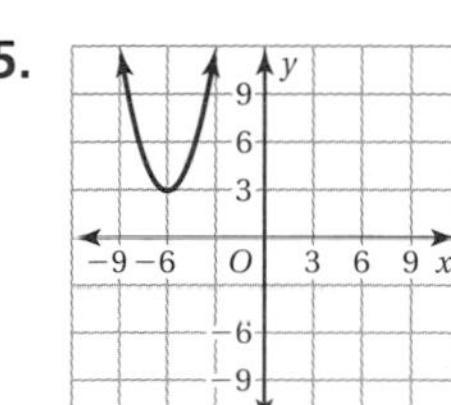

6 units left, 3 units up

6.

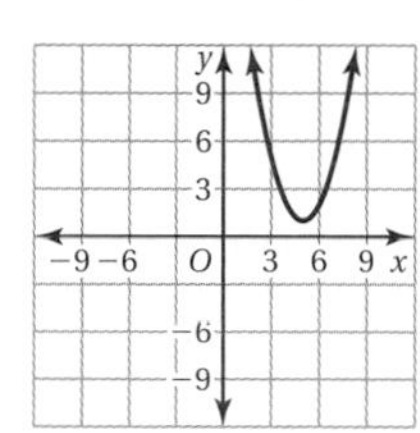

5 units right, 1 unit up

7.

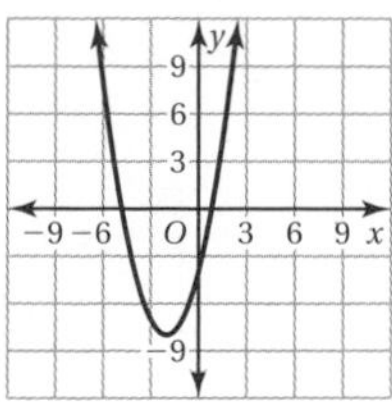

2 units left, 8 units down

8.

narrower, 1 unit right, 4 units down

9.

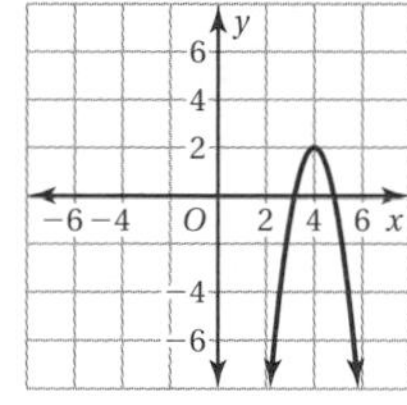

narrower, opens down, 4 units right, 2 units down

10.

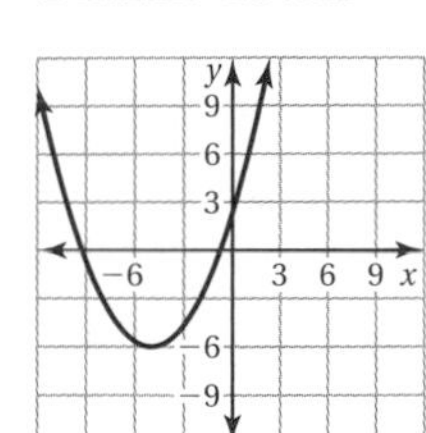

wider, 5 units left, 6 units down

11. horizontal translation 9 units right

12. narrower, vertical translation 5 units up

13. **a.** 1 million **b.** 5 years

Mini-Assessment

1.

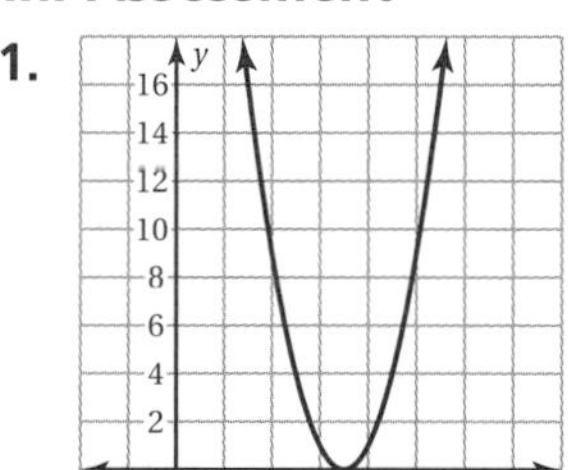

The graph of $y = (x-7)^2$ is a translation 7 units to the right of the graph of $y = x^2$.

2.

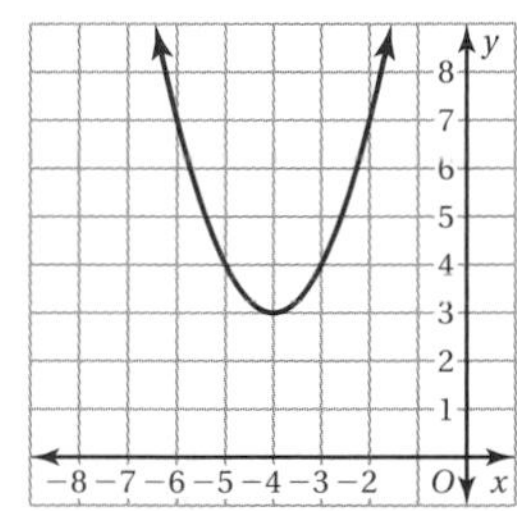

The graph of $y = (x+4)^2 + 3$ is a translation 4 units to the left and 3 units up of the graph of $y = x^2$.

3.

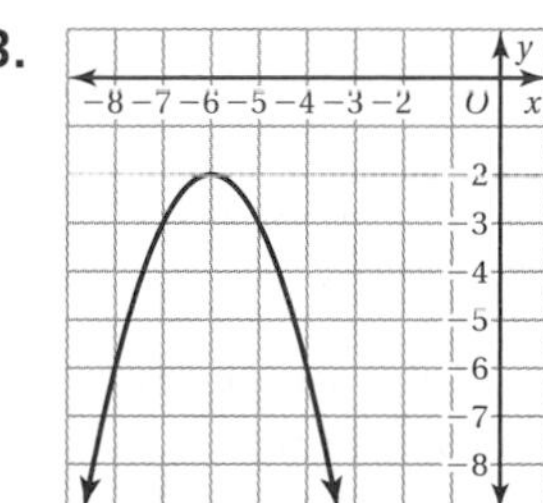

The graph of $y = -(x+6)^2 - 2$ opens down. The vertex of the graph of $y = -(x+6)^2 - 2$ is a translation 6 units to the left and 2 units down of the vertex of the graph of $y = x^2$.

Section 8.5

Extra Example 1

a.

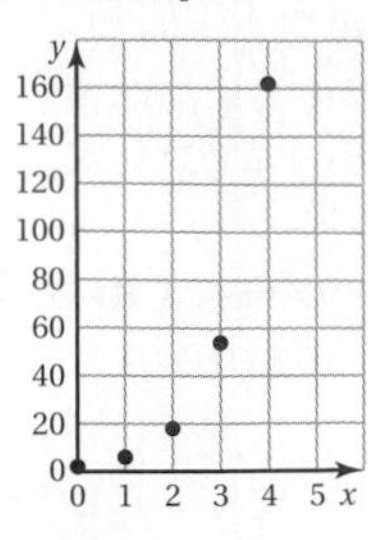

exponential

b.

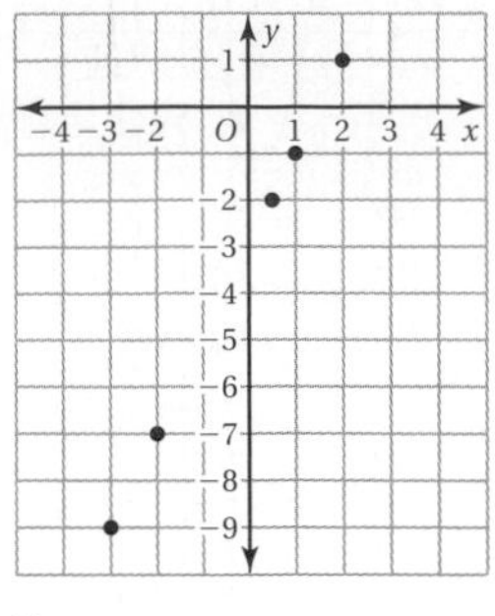

linear

c.

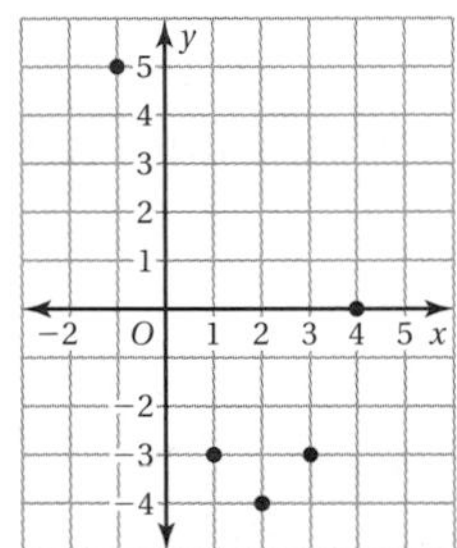

quadratic

Practice and Problem Solving

28. a.

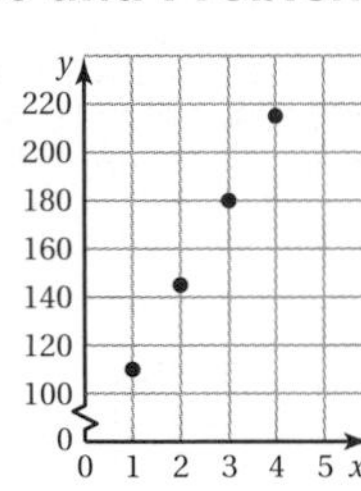

linear

b. $m = 35h + 75$

c. \$425

29. a.

Round, x	1	2	3	4	5	6
Teams Remaining, y	32	16	8	4	2	1

b. exponential

c. $y = 64(0.5)^x$

d. sixth round

8.4–8.5 Quiz

3. domain: all real numbers; range: $y \leq 8$

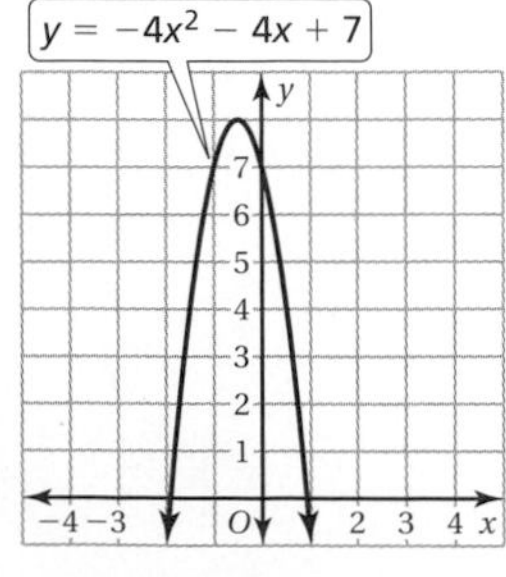

4. domain: all real numbers; range: $y \geq -13$

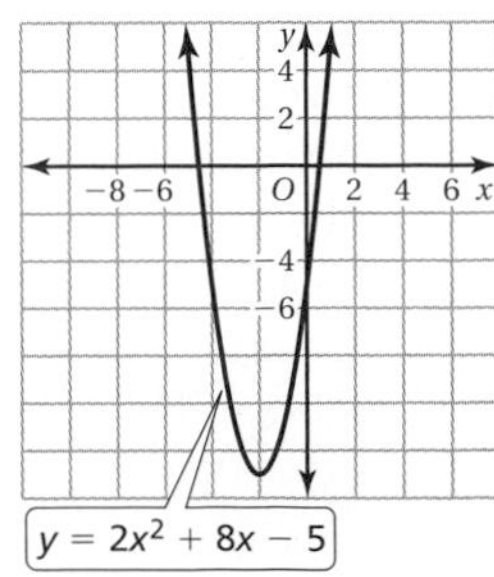

5. domain: all real numbers; range: $y \leq 16$

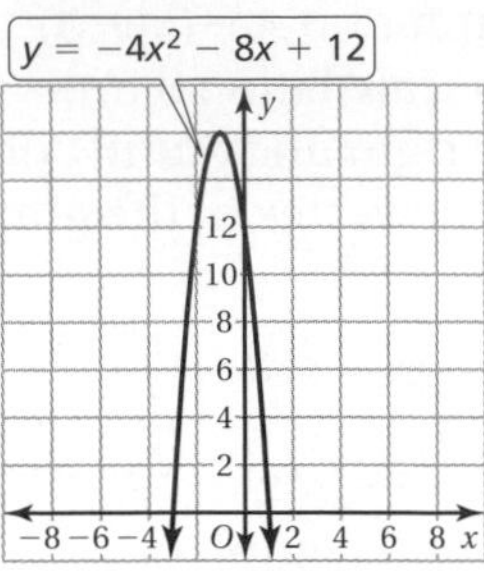

9.

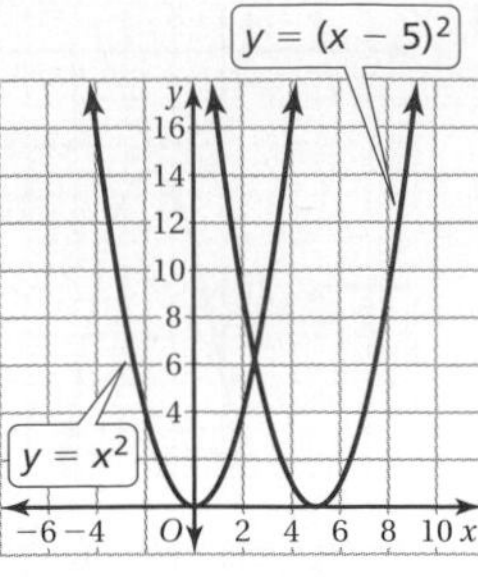

The graph of $y = (x - 5)^2$ is a translation 5 units to the right of the graph of $y = x^2$.

10.

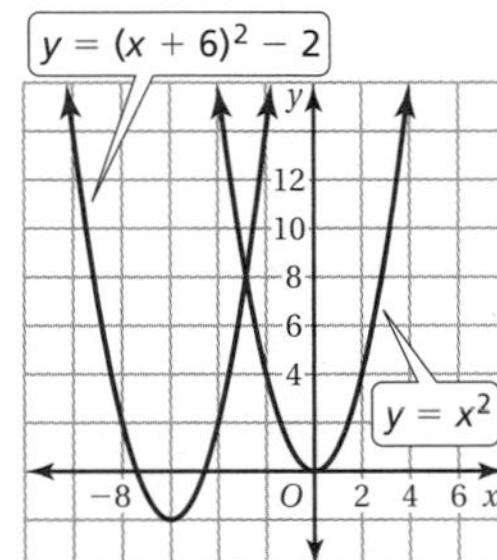

The graph of $y = (x + 6)^2 - 2$ is a translation 6 units to the left and 2 units down of the graph of $y = x^2$.

11.

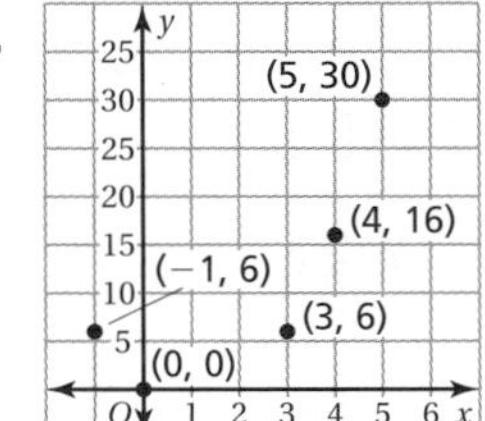

quadratic

12.

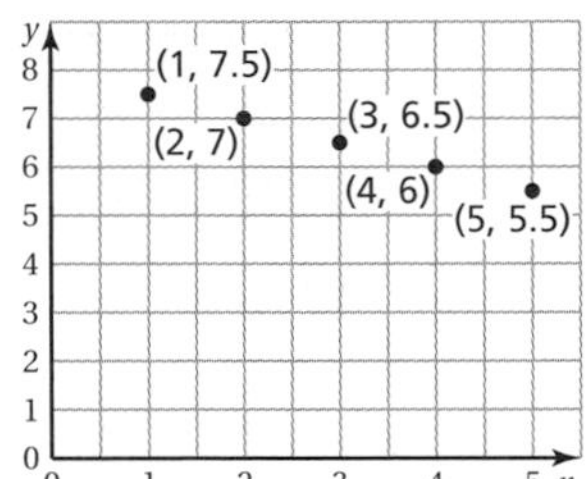

linear

17. a.

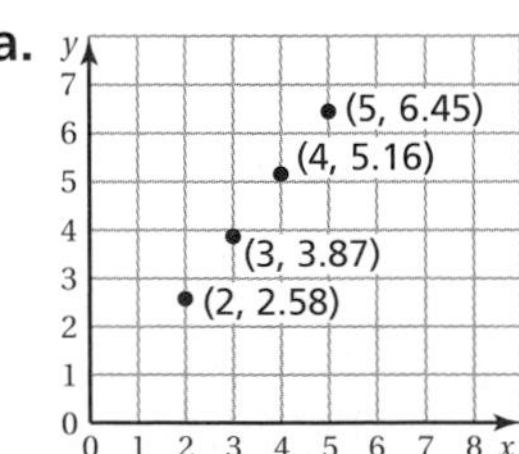

linear

Chapter 8 Review

6.

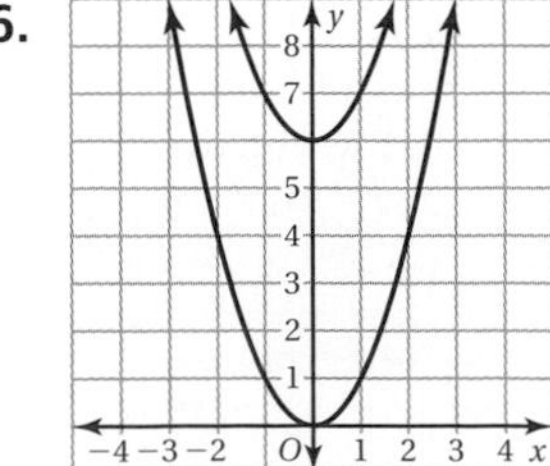

Both graphs open up and have the same axis of symmetry, $x = 0$. The graph of $y = x^2 + 6$ is a translation 6 units up of the graph of $y = x^2$.

7. 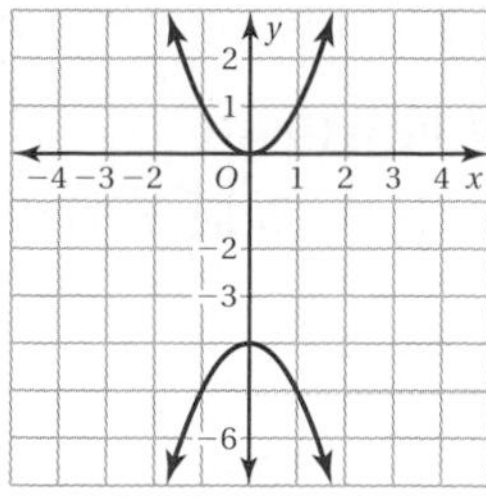

Both graphs have the same axis of symmetry, $x = 0$. The graph of $y = -x^2 - 4$ opens down. The vertex of the graph of $y = -x^2 - 4$ is a translation 4 units down of the vertex of the graph of $y = x^2$.

8. 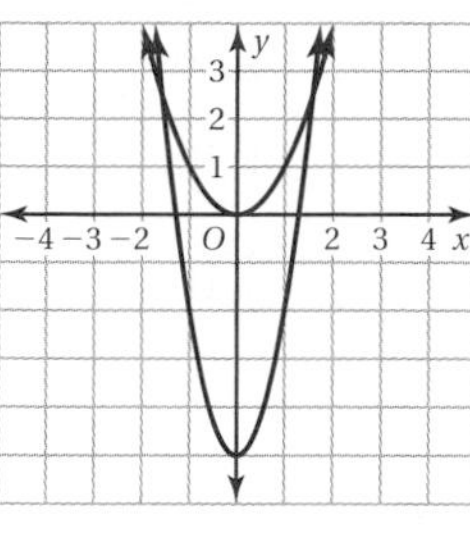

Both graphs open up and have the same axis of symmetry, $x = 0$. The graph of $y = 3x^2 - 5$ is narrower than the graph of $y = x^2$. The vertex of the graph of $y = 3x^2 - 5$ is a translation 5 units down of the vertex of the graph of $y = x^2$.

13. 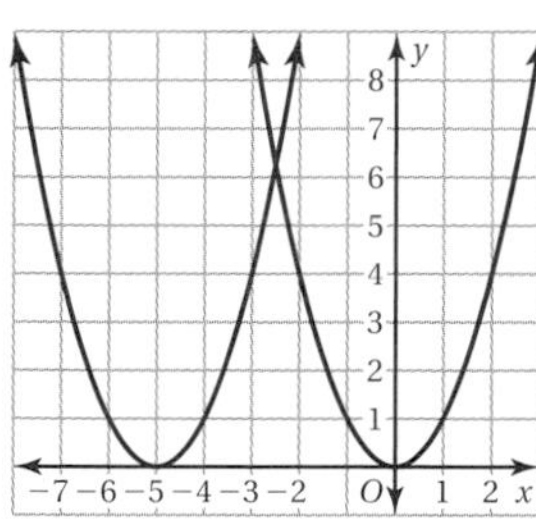

The graph of $y = (x + 5)^2$ is a translation 5 units to the left of the graph of $y = x^2$.

14. 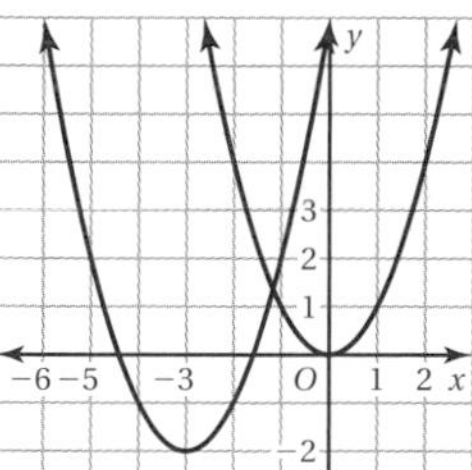

The graph of $y = (x + 3)^2 - 2$ is a translation 3 units to the left and 2 units down of the graph of $y = x^2$.

15. 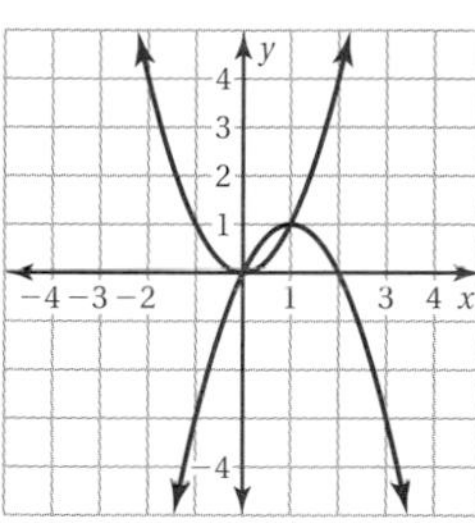

The graph of $y = -(x - 1)^2 + 1$ opens down. The vertex of the graph of $y = -(x - 1)^2 + 1$ is a translation 1 unit to the right and 1 unit up of the vertex of the graph of $y = x^2$.

16. exponential; $y = 8\left(\frac{1}{2}\right)^x$

17. 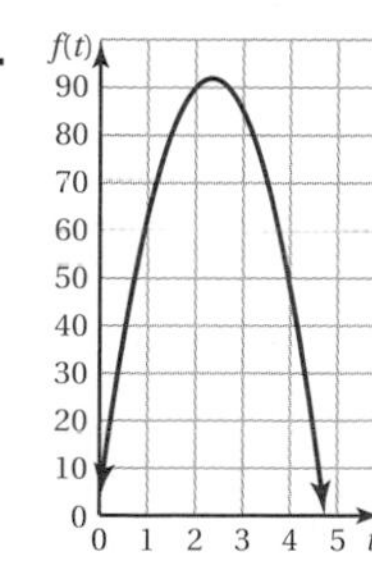

from 0 to 1 sec: 59 ft/sec; from 1 to 2 sec: 27 ft/sec; decreasing; The graph flattens as the baseball approaches its maximum height.

Chapter 8 Test

1. 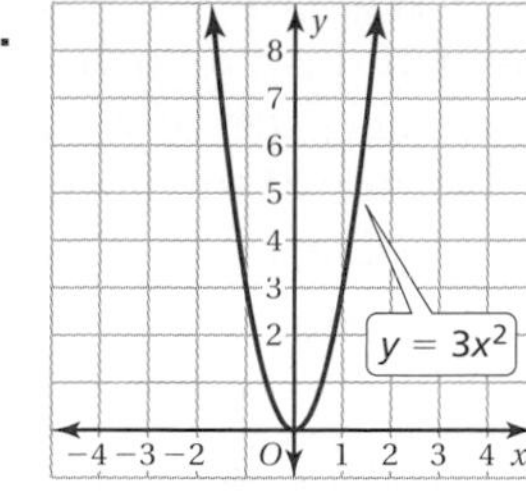

Both graphs open up and have the same vertex, (0, 0), and the same axis of symmetry, $x = 0$. The graph of $y = 3x^2$ is narrower than the graph of $y = x^2$.

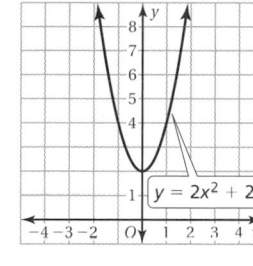

2. Both graphs open up and have the same axis of symmetry, $x = 0$. The graph of $y = 2x^2 + 2$ is narrower than the graph of $y = x^2$. The vertex of the graph of $y = 2x^2 + 2$ is a translation 2 units up of the vertex of the graph of $y = x^2$.

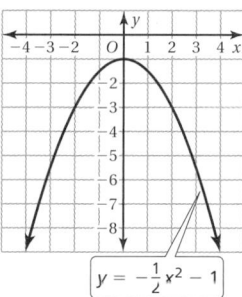

3. Both graphs have the same axis of symmetry, $x = 0$. The graph of $y = -\frac{1}{2}x^2 - 1$ opens down and is wider than the graph of $y = x^2$. The vertex of the graph of $y = -\frac{1}{2}x^2 - 1$ is a translation 1 unit down of the vertex of the graph of $y = x^2$.

4. 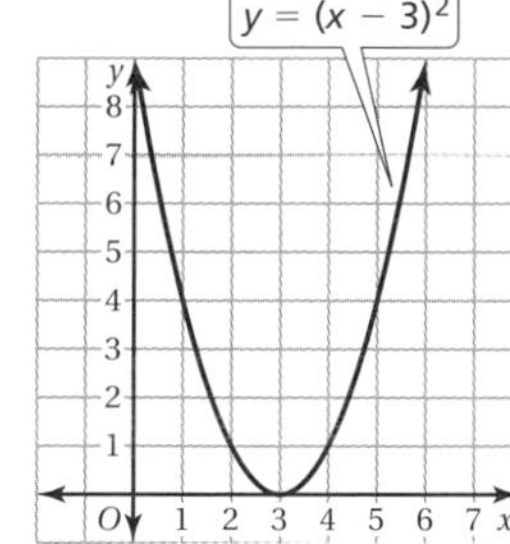

The graph of $y = (x - 3)^2$ is a translation 3 units to the right of the graph of $y = x^2$.

5. 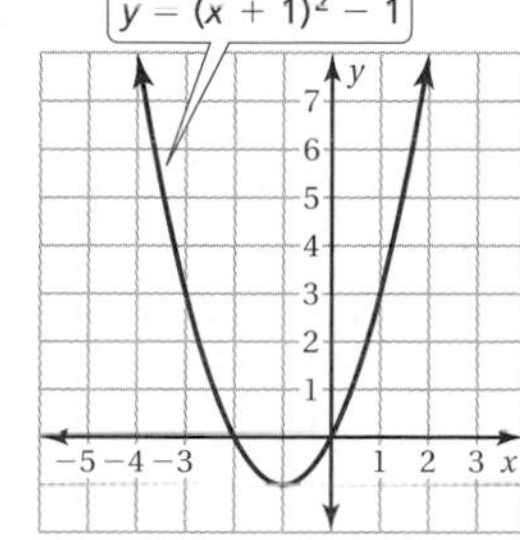

The graph of $y = (x + 1)^2 - 1$ is a translation 1 unit to the left and 1 unit down of the graph of $y = x^2$.

Additional Answers

6. 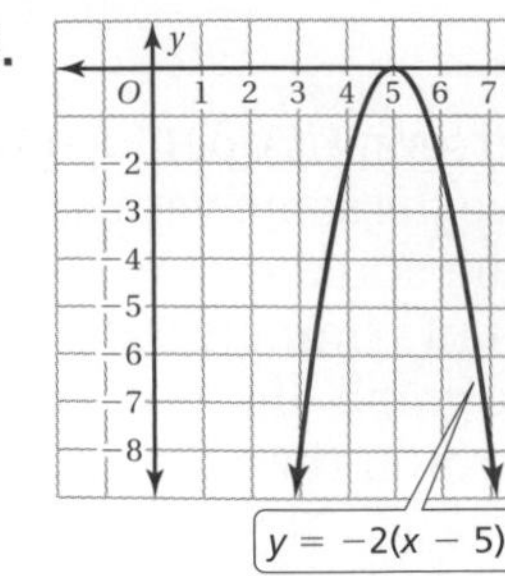

The graph of $y = -2(x-5)^2$ opens down and is narrower than the graph of $y = x^2$. The vertex of the graph of $y = -2(x-5)^2$ is a translation 5 units to the right of the vertex of the graph of $y = x^2$.

9. focus: $\left(0, -\frac{1}{6}\right)$

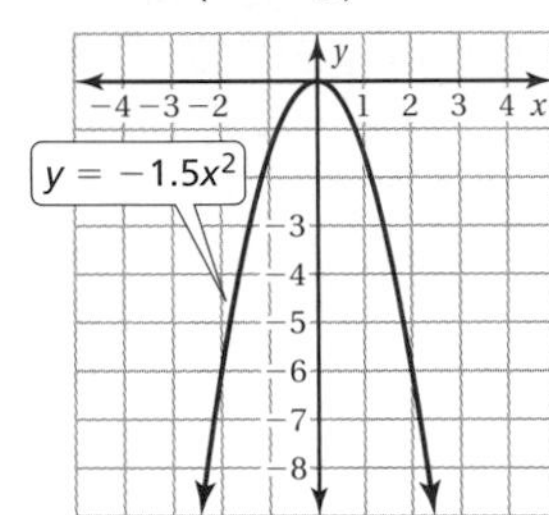

10. domain: all real numbers;
range: $y \geq -2$

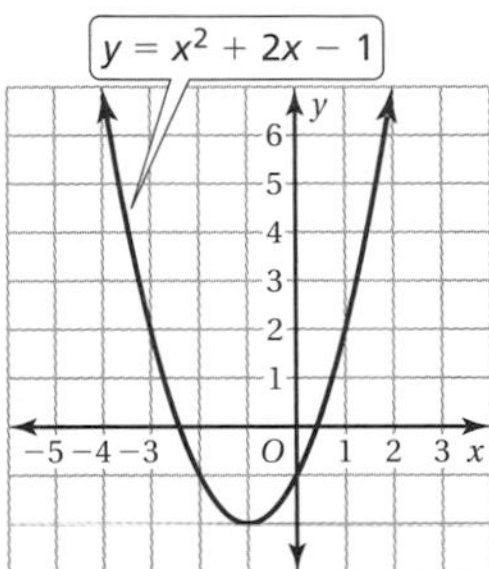

11. domain: all real numbers
range: $y \leq 5.25$

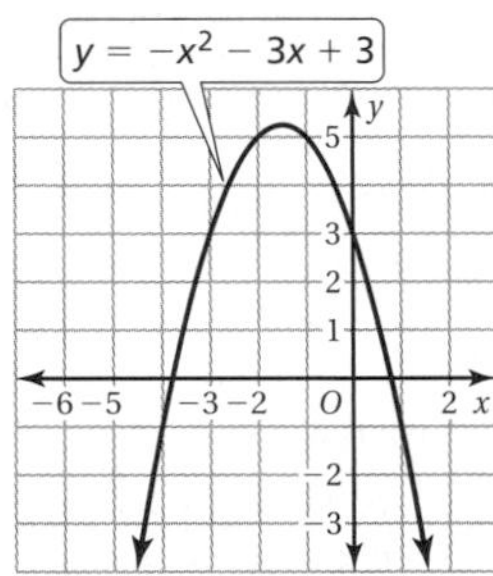

12. domain: all real numbers
range: $y \geq -6$

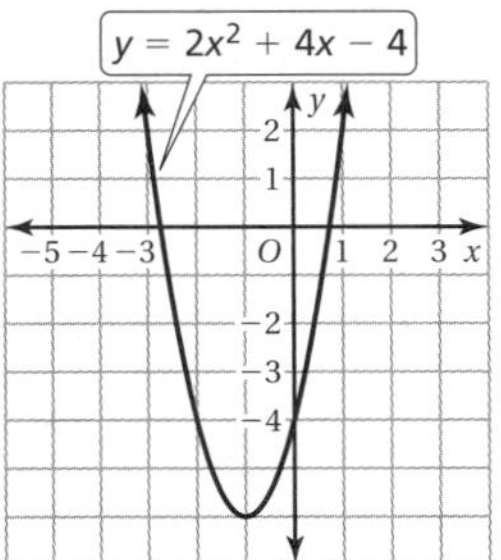

18. increasing; The average rates of change are given by the first differences for which $f(1) - f(0) < f(2) - f(1) < f(3) - f(2) < f(4) - f(3)$.

Chapter 9

Section 9.3

Practice and Problem Solving

4. 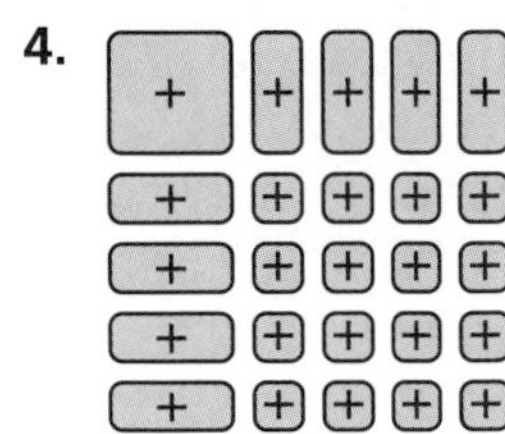

$x^2 + 8x + 16$

5. 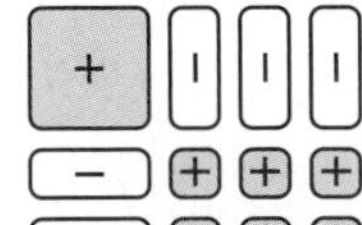

$x^2 - 6x + 9$

Lesson 9.4b

Record and Practice Journal Practice

1. $x = -2, 9$
2. $x \approx -5.54, 0.541$
3. $x = 1$
4. $x = -3, \frac{1}{4}$
5. $x = -4, 5$
6. $x = \pm 3$
7. $x = -2$
8. no solution
9. $x = 5, 7$
10. $x \approx -1.40, 1.07$
11. no solution
12. $x = -\frac{3}{2}, 6$
13. **a.** $t \approx 6.05$ sec **b.** $t = 2, 4$ sec
14. width: 19 yd; length: 13 yd

9.4–9.5 Quiz

14. **a.** Type 1 **b.** after 8 hours

c. There are more Type 1 bacteria after 8 hours and more Type 2 bacteria before 8 hours.

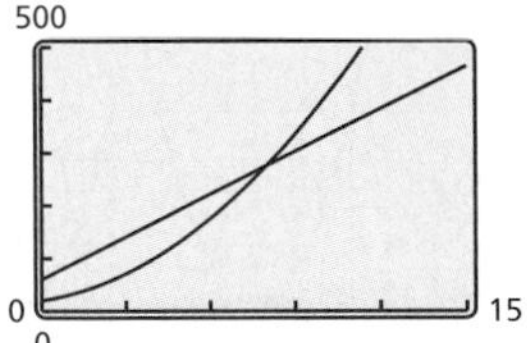

15. 2005

16. *Sample answer:* no; The model shows the average monthly bill decreasing in future years, which is unlikely.

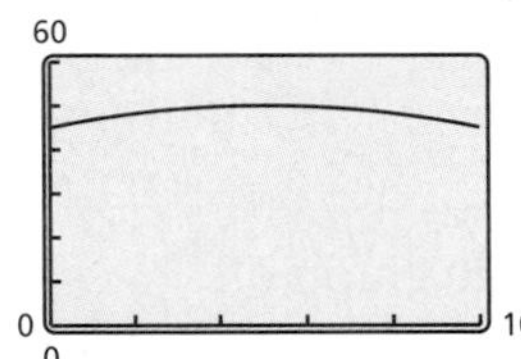

Chapter 10

Section 10.1

Extra Example 3

3. domain: $x \geq -1$; range: $y \leq -2$;

The graph of $y = -\sqrt{x+1} - 2$ is a reflection in the x-axis of the graph of $y = \sqrt{x}$, and then a translation 1 unit to the left and 2 units down.

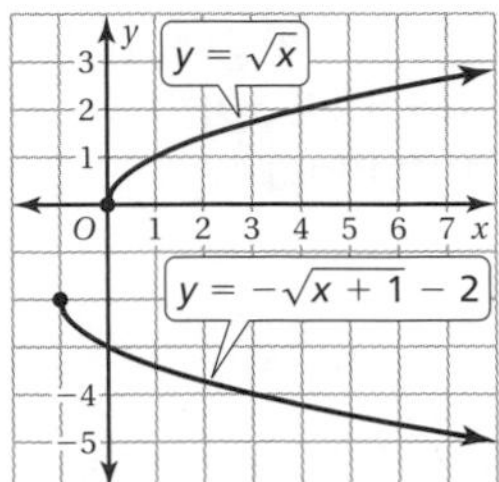

On Your Own

4. domain: $x \geq 0$; range: $y \geq -4$;
The graph of $y = \sqrt{x} - 4$ is a translation 4 units down of the graph of $y = \sqrt{x}$.

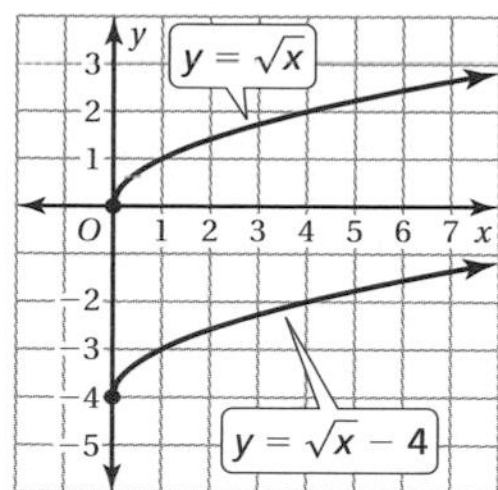

5. domain: $x \geq -5$; range: $y \geq 0$;
The graph of $y = \sqrt{x + 5}$ is a translation 5 units to the left of the graph of $y = \sqrt{x}$.

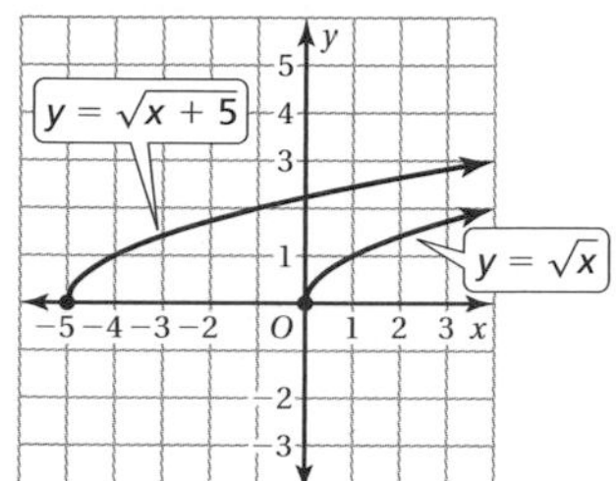

6. domain: $x \geq -1$; range: $y \leq 2$;
The graph of $y = -\sqrt{x + 1} + 2$ is a reflection in the x-axis of the graph of $y = \sqrt{x}$ and then a translation 1 unit to the left and 2 units up.

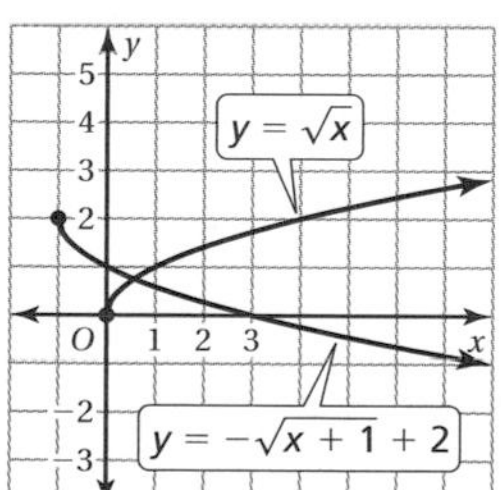

Practice and Problem Solving

15. a.

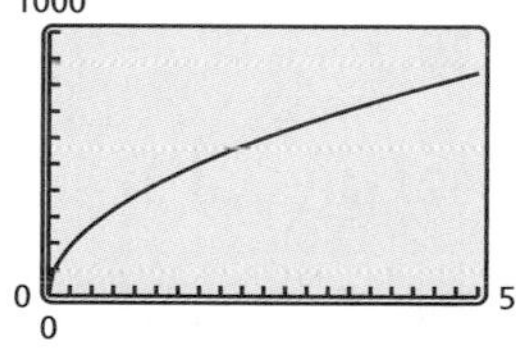

b. about 6 pounds per square inch

16. domain: $x \geq 0$; range: $y \geq -2$;
The graph of $y = \sqrt{x} - 2$ is a translation 2 units down of the graph of $y = \sqrt{x}$.

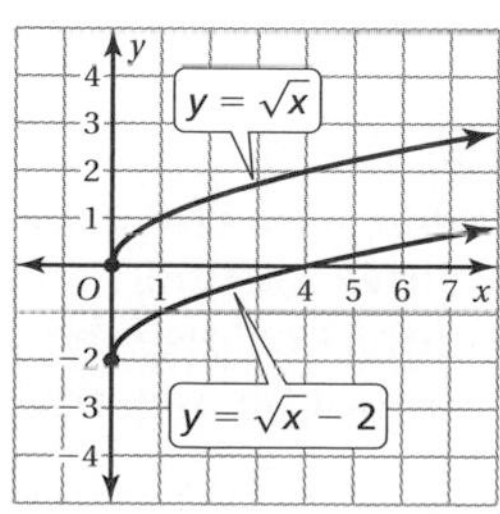

17. domain: $x \geq 0$; range: $y \geq 4$;
The graph of $y = \sqrt{x} + 4$ is a translation 4 units up of the graph of $y = \sqrt{x}$.

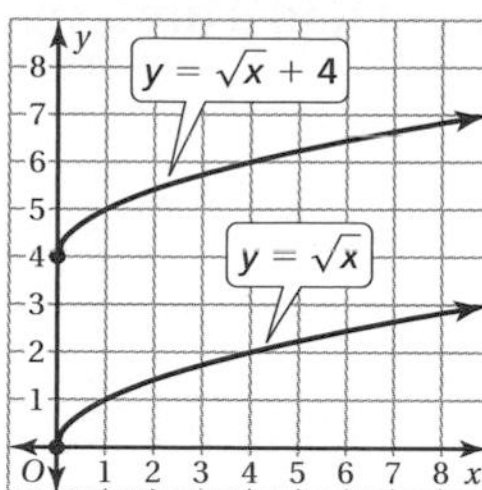

18. domain: $x \geq -4$; range: $y \geq 0$;
The graph of $y = \sqrt{x + 4}$ is a translation 4 units to the left of the graph of $y = \sqrt{x}$.

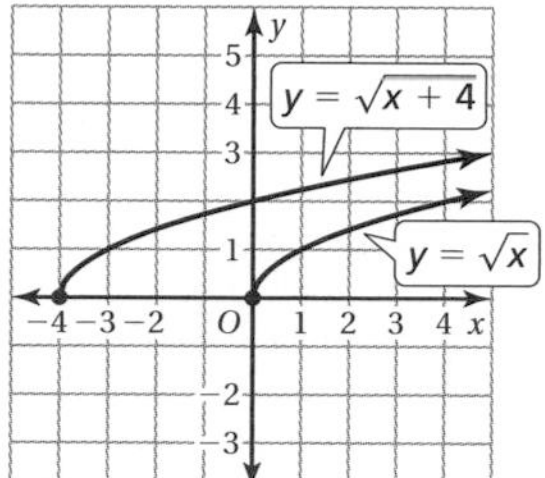

19. domain: $x \geq -2$; range: $y \geq -2$;
The graph of $y = \sqrt{x + 2} - 2$ is a translation 2 units to the left and 2 units down of the graph of $y = \sqrt{x}$.

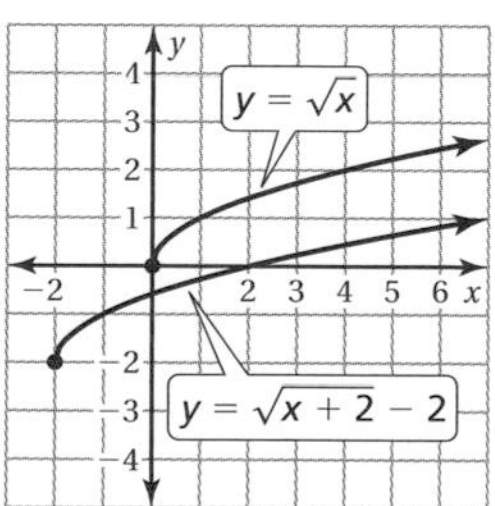

20. domain: $x \geq 3$; range: $y \leq 0$;
The graph of $y = -\sqrt{x - 3}$ is a reflection in the x-axis of the graph of $y = \sqrt{x}$ and then a translation 3 units to the right.

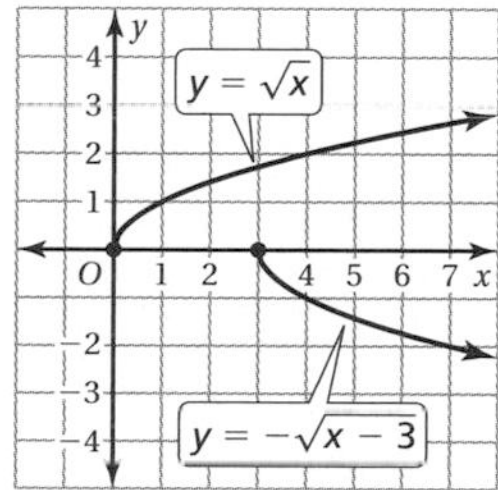

Additional Answers

21. domain: $x \geq 1$; range: $y \leq 3$;

The graph of $y = -\sqrt{x-1} + 3$ is a reflection in the x-axis of the graph of $y = \sqrt{x}$ and then a translation 1 unit to the right and 3 units up.

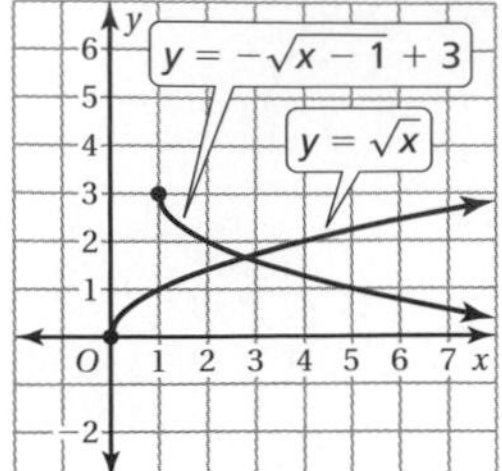

25. **a.** domain: $A > 0$

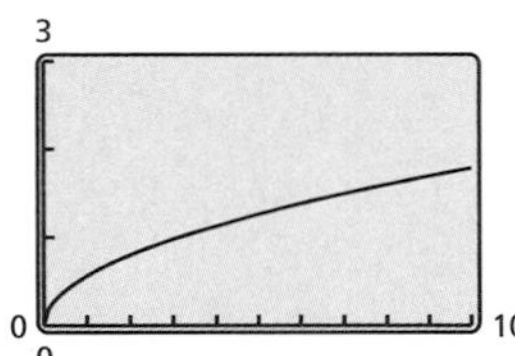

b. about 28 square inches

Mini-Assessment

1. domain: $x \geq 0$; range: $y \geq 5$;

The graph of $y = \sqrt{x} + 5$ is a translation 5 units up of the graph of $y = \sqrt{x}$.

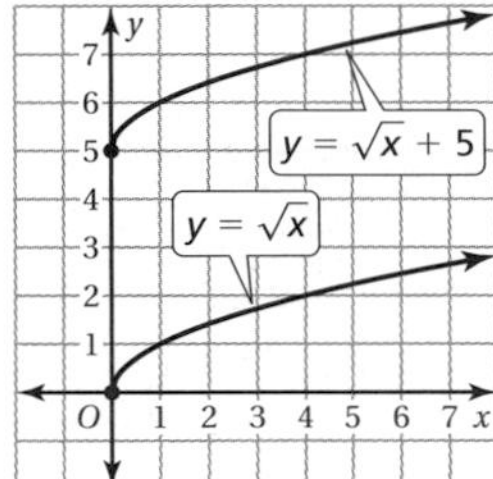

2. domain: $x \geq 7$; range: $y \geq 0$;

The graph of $y = \sqrt{x-7}$ is a translation 7 units to the right of the graph of $y = \sqrt{x}$.

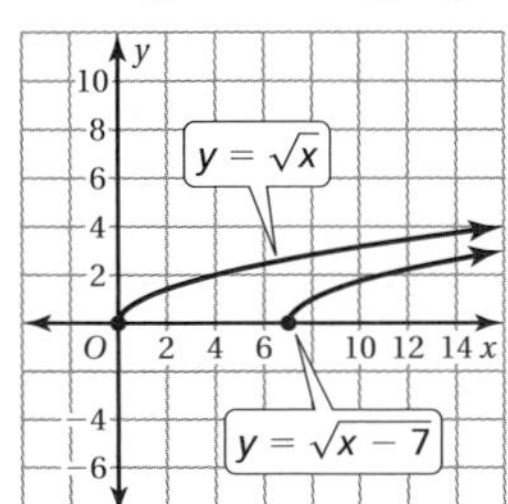

3. domain: $x \geq -3$; range: $y \geq -4$;

The graph of $y = \sqrt{x+3} - 4$ is a translation 3 units to the left and 4 units down of the graph of $y = \sqrt{x}$.

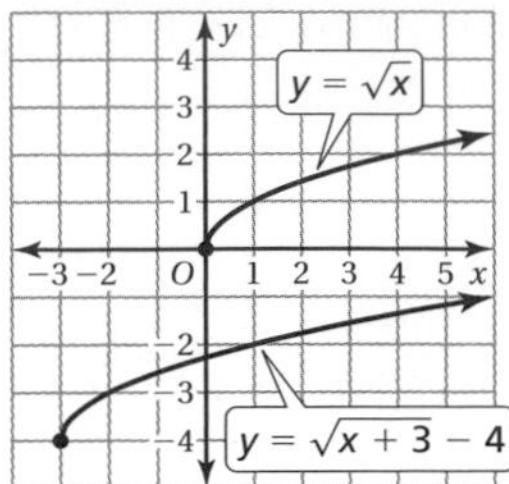

4. domain: $x \geq 4$; range: $y \leq 6$;

The graph of $y = -\sqrt{x-4} + 6$ is a reflection in the x-axis of the graph of $y = \sqrt{x}$ and then a translation 4 units to the right and 6 units up.

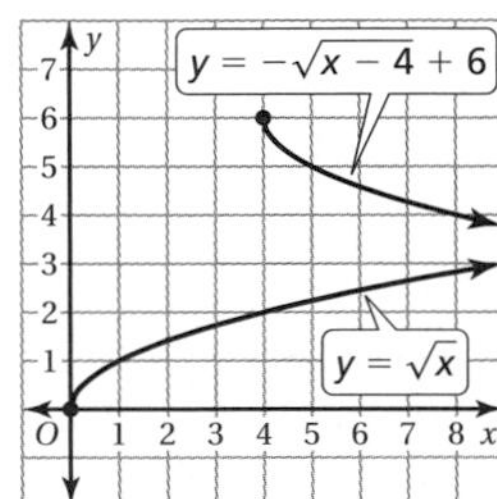

Lesson 10.1b

Record and Practice Journal Practice

1. $\dfrac{\sqrt{26}}{26}$

2. $\dfrac{\sqrt{15}}{3}$

3. $\dfrac{\sqrt{210}}{6}$

4. $\dfrac{2\sqrt{22}}{11}$

5. $\dfrac{2\sqrt{3}}{3}$

6. $2\sqrt{7}$

7. $\dfrac{3\sqrt{2}}{2}$

8. $\dfrac{60 - 3\sqrt{30}}{10}$

9. $4 + 2\sqrt{2}$

10. $\dfrac{4 - \sqrt{6}}{10}$

11. $-2\sqrt{5} - 2$

12. $10 - 5\sqrt{3}$

13. $6\sqrt{2} - 3\sqrt{5}$

14. $\dfrac{-2\sqrt{3} - \sqrt{6}}{3}$

15. $\dfrac{\sqrt{78}}{3}$ km

Section 10.2

Practice and Problem Solving

38. After subtracting 5 from each side, you get a square root equal to a negative number. The equation has no solutions because the square root of a number cannot be negative.

10.1–10.2 Quiz

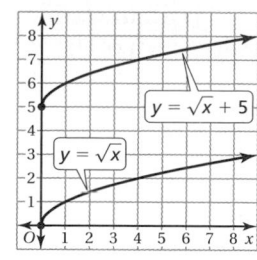

4. domain: $x \geq 0$; range: $y \geq 5$; The graph of $y = \sqrt{x} + 5$ is a translation 5 units up of the graph of $y = \sqrt{x}$.

5. 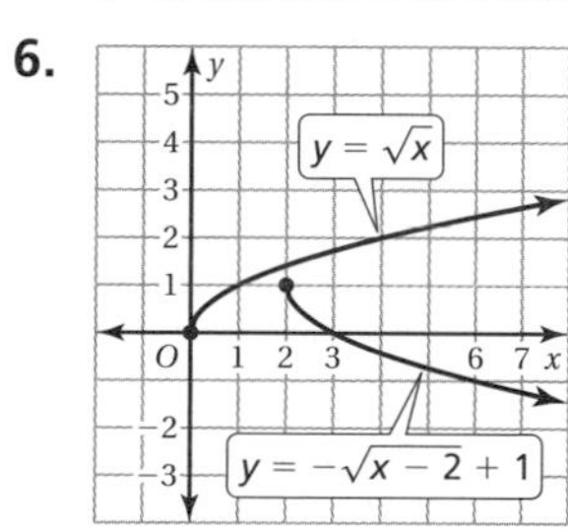

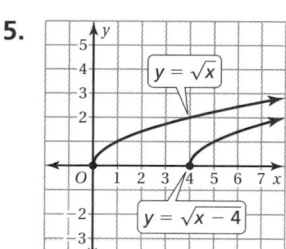

domain: $x \geq 4$; range: $y \geq 0$; The graph of $y = \sqrt{x - 4}$ is a translation 4 units to the right of the graph of $y = \sqrt{x}$.

6. domain: $x \geq 2$; range: $y \leq 1$; The graph of $y = -\sqrt{x - 2} + 1$ is a reflection in the x-axis of the graph of $y = \sqrt{x}$ and then a translation 2 units to the right and 1 unit up.

18. Both graphs have the same axis of symmetry, $x = 0$. The graph of $y = -x^2 - 6$ opens down. The vertex of the graph of $y = -x^2 - 6$ is a translation 6 units down of the vertex of the graph of $y = x^2$.

19. 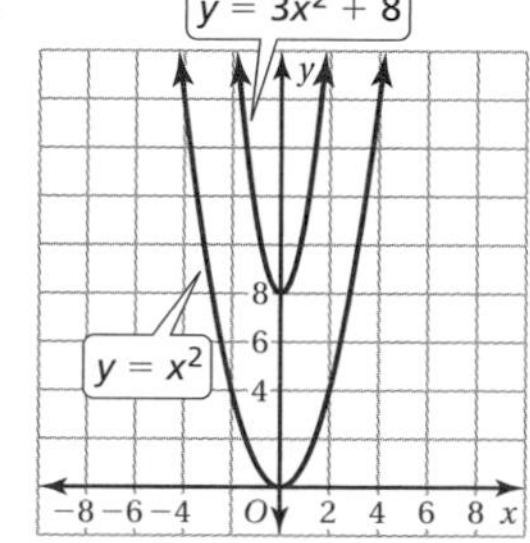

Both graphs open up and have the same axis of symmetry, $x = 0$. The graph of $y = 3x^2 + 8$ is narrower than and is a translation 8 units up of the graph of $y = x^2$.

Section 10.3

Practice and Problem Solving

15. a. $28^2 + 21^2 = (5x)^2$

b. Factoring: Subtract 1225 from each side. Factor out the GCF and factor the difference of squares. Set the factors equal to zero and solve.

Taking a square root: Divide each side by 25. Then take the square root of each side.

Sample answer: taking square roots; Taking square roots is easier than factoring.

c. 7

Fair Game Review

17. 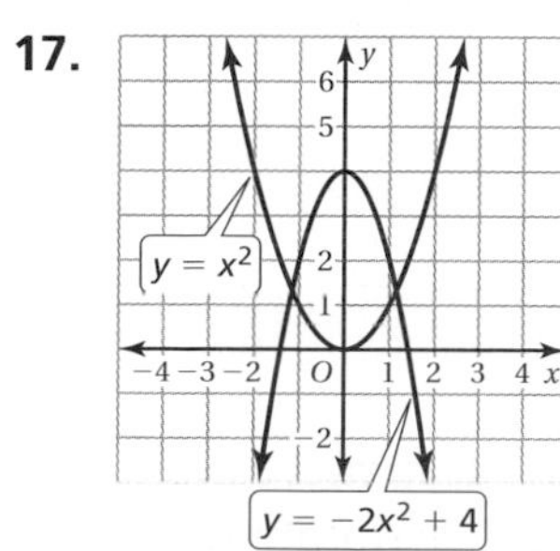

Both graphs have the same axis of symmetry, $x = 0$. The graph of $y = -2x^2 + 4$ opens down and is narrower than the graph of $y = x^2$. The vertex of the graph of $y = -2x^2 + 4$ is a translation 4 units up of the vertex of the graph of $y = x^2$.

Section 10.4

Practice and Problem Solving

26. 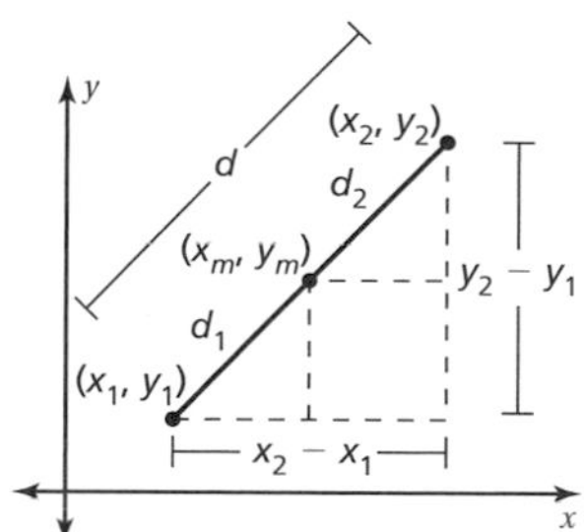

$$x_m = x_1 + \frac{1}{2}(x_2 - x_1)$$
$$= \frac{2x_1 + x_2 - x_1}{2}$$
$$= \frac{x_1 + x_2}{2}$$

Similarly, $y_m = \dfrac{y_1 + y_2}{2}$

$$d_1 = \sqrt{\left(x_1 - \frac{x_1 + x_2}{2}\right)^2 + \left(y_1 - \frac{y_1 + y_2}{2}\right)^2}$$
$$= \sqrt{\left(\frac{x_1 - x_2}{2}\right)^2 + \left(\frac{y_1 - y_2}{2}\right)^2}$$
$$= \frac{1}{2}\sqrt{(x_1 - x_2)^2 + (y_1 - y_2)^2}$$

Additional Answers

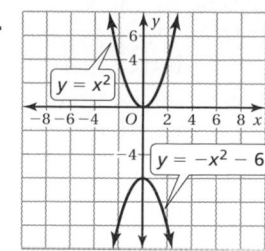

$$d_2 = \sqrt{\left(\frac{x_1 + x_2}{2} - x_2\right)^2 + \left(\frac{y_1 + y_2}{2} - y_2\right)^2}$$

$$= \sqrt{\left(\frac{x_1 - x_2}{2}\right)^2 + \left(\frac{y_1 - y_2}{2}\right)^2}$$

$$= \frac{1}{2}\sqrt{(x_1 - x_2)^2 + (y_1 - y_2)^2}$$

So, $d_1 + d_2 = \frac{1}{2}\sqrt{(x_1 - x_2)^2 + (y_1 - y_2)^2} +$

$$\frac{1}{2}\sqrt{(x_1 - x_2)^2 + (y_1 - y_2)^2}$$

$$= \sqrt{(x_1 - x_2)^2 + (y_1 - y_2)^2}$$

$$= d$$

Chapter 10 Test

1.

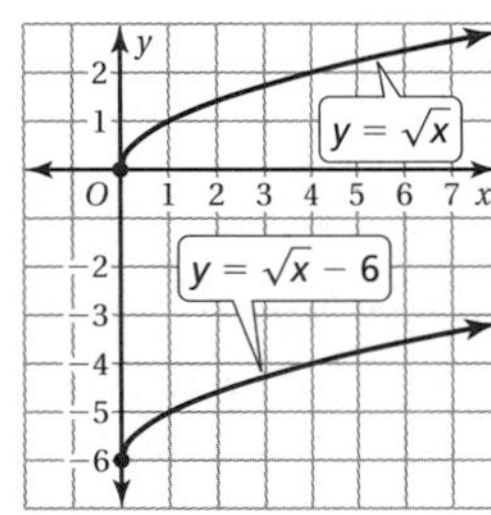

domain: $x \geq 0$; range: $y \geq -6$;
The graph of $y = \sqrt{x} - 6$ is a translation 6 units down of the graph of $y = \sqrt{x}$.

2.

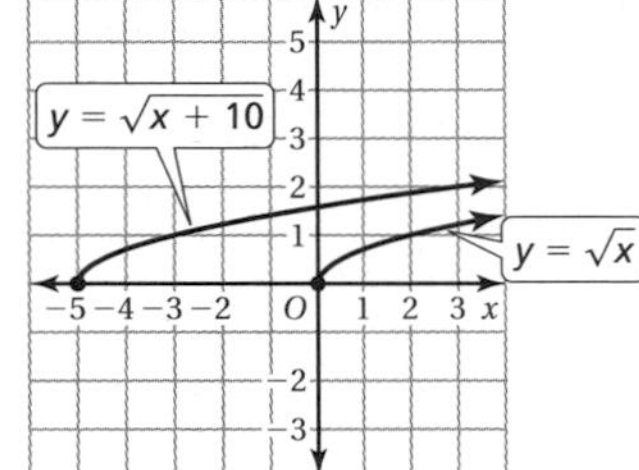

domain: $x \geq -10$; range: $y \geq 0$;
The graph of $y = \sqrt{x + 10}$ is a translation 10 units to the left of the graph of $y = \sqrt{x}$.

3.

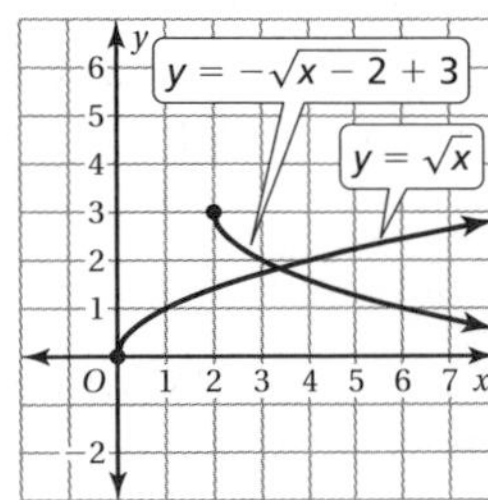

domain: $x \geq 2$; range: $y \leq 3$;
The graph of $y = -\sqrt{x - 2} + 3$ is a reflection in the x-axis of the graph of $y = \sqrt{x}$ and then a translation 2 units to the right and 3 units up.

Chapter 11

Section 11.1

On Your Own

4. $y = \frac{20}{x}$

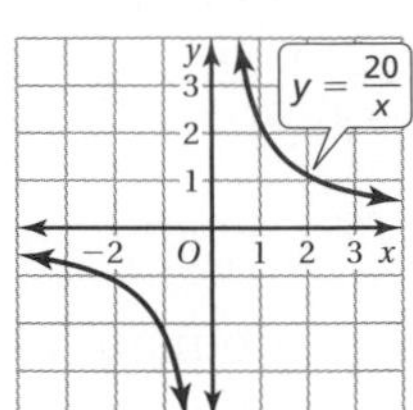

Practice and Problem Solving

10. $y = \frac{15}{x}$

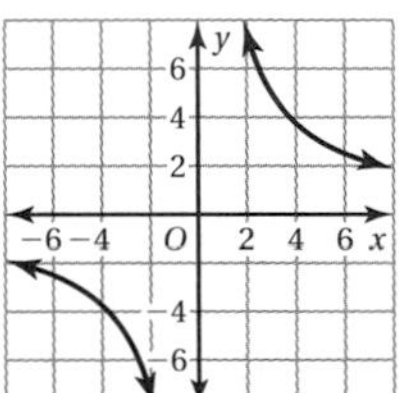

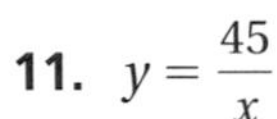

11. $y = \frac{45}{x}$

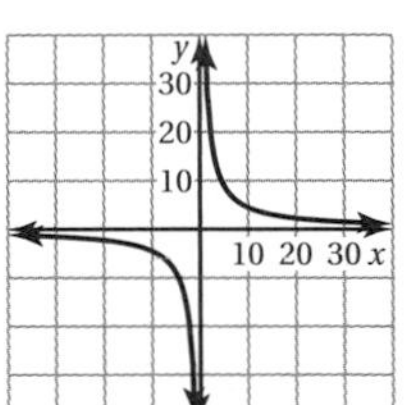

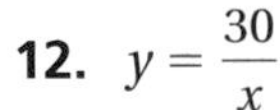

12. $y = \frac{30}{x}$

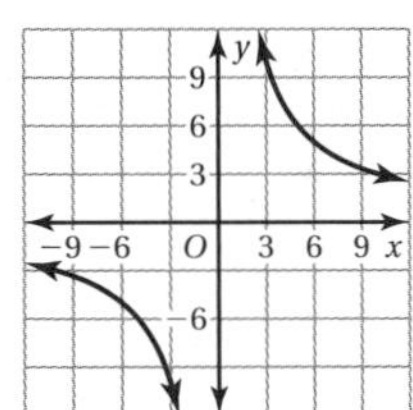

13. inverse variation; The product hr is constant.

21. $y = \frac{10}{x}$; $y = 2.5$

22. $y = \frac{16}{x}$; $y = 2$

23. inverse variation;

Number of hats, x	5	10
Cost per hat, y	10	5

The product xy is constant.

24. direct variation;

Number of lawns, x	2	3
Earnings, y	50	75

The ratio $\frac{y}{x}$ is constant.

25. inverse variation; Suppose the swim team earns $1000.

Number of members, x	1	5	10	20
Earnings per member, y	1000	200	100	50

The product xy is constant.

26. 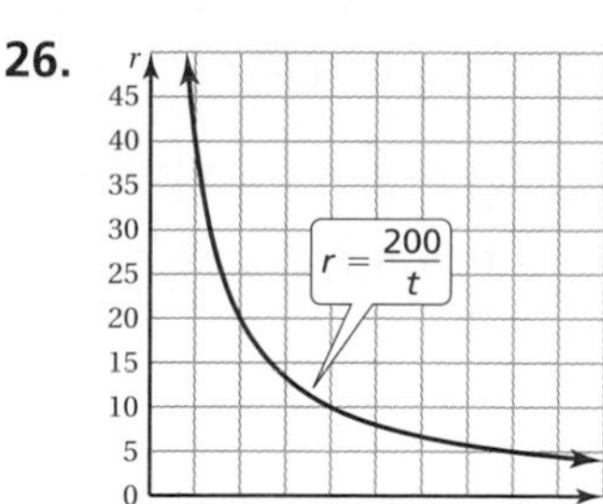

As the amount of time you take to finish the race increases, your average speed decreases.

27. a. $v = 18t$;

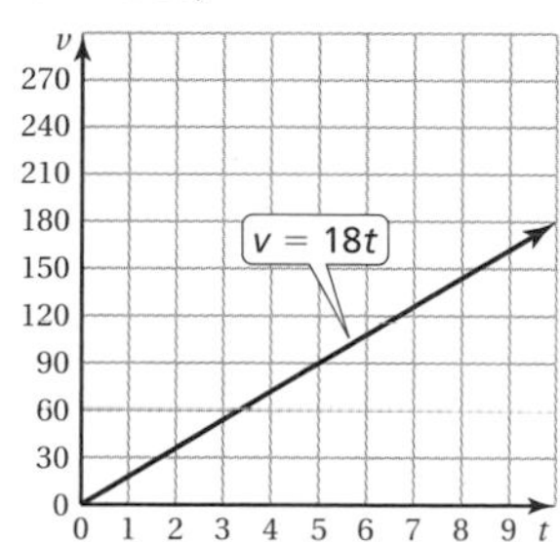

b. 90 hours

Fair Game Review

39. 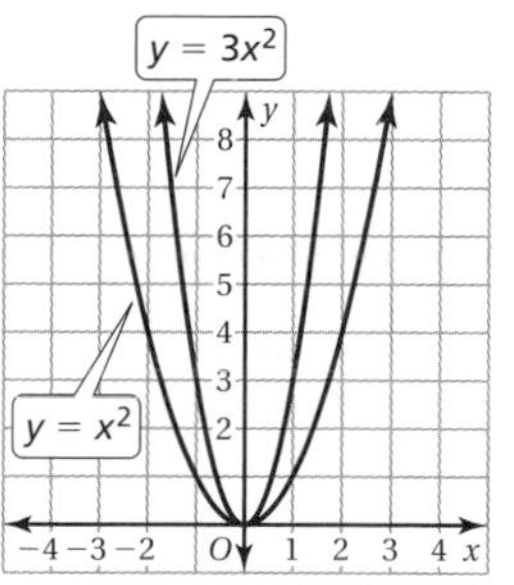

The graph of $y = 3x^2$ is narrower than the graph of $y = x^2$.

40. 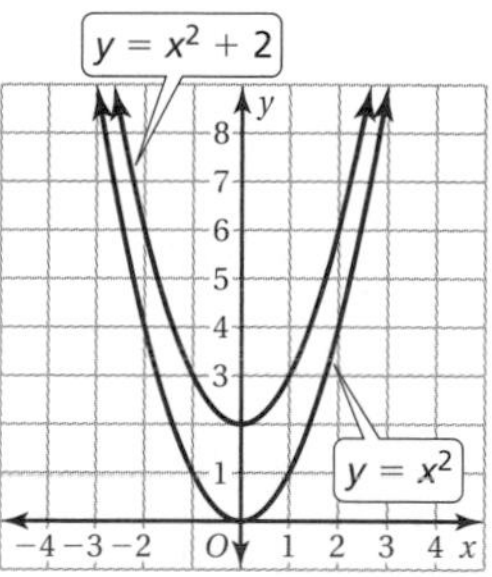

The graph of $y = x^2 + 2$ is a translation 2 units up of the graph of $y = x^2$.

41. 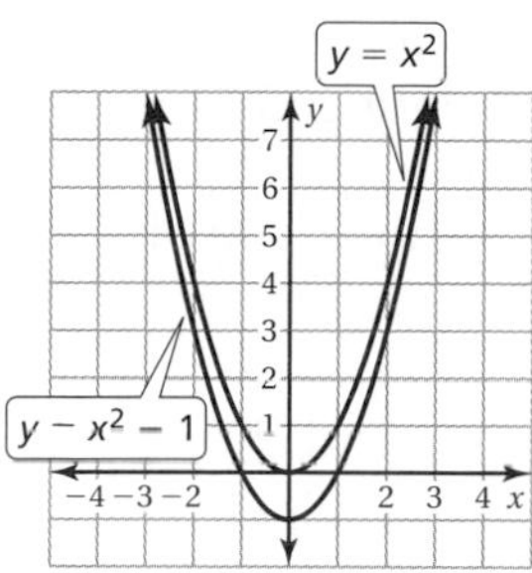

The graph of $y = x^2 - 1$ is a translation 1 unit down of the graph of $y = x^2$.

Section 11.2

On Your Own

4. 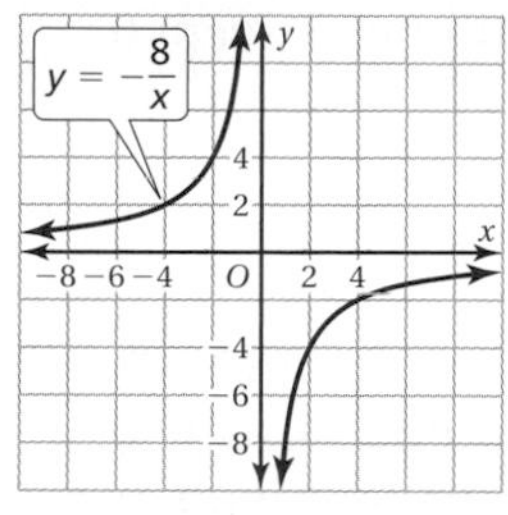

The domain is all real numbers except 0 and the range is all real numbers except 0.

5. 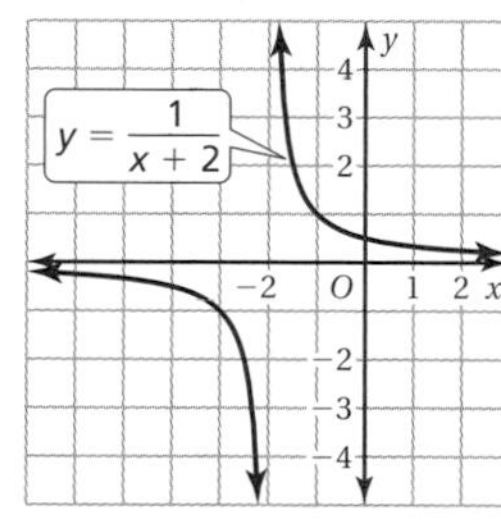

The domain is all real numbers except -2 and the range is all real numbers except 0.

6. 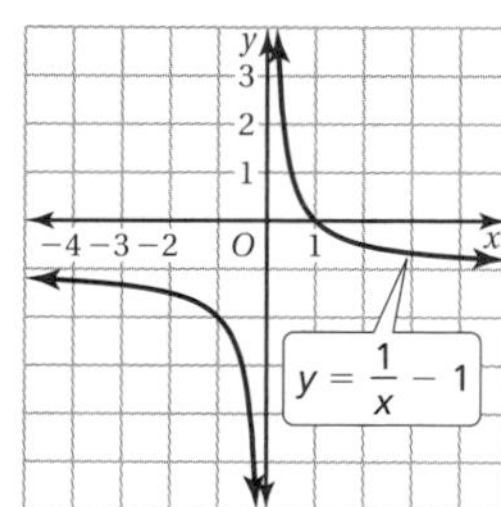

The domain is all real numbers except 0 and the range is all real numbers except -1.

11. 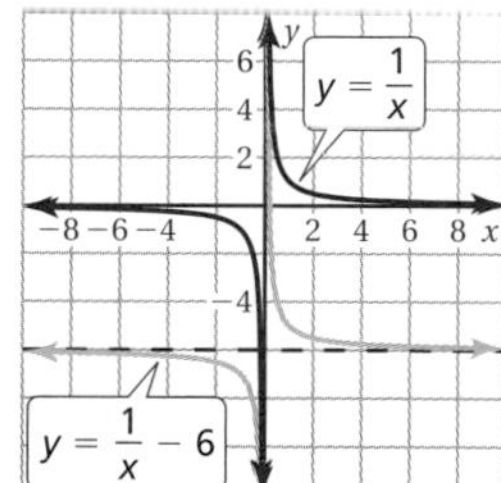

The graph of $y = \frac{1}{x} - 6$ is a translation 6 units down of the graph of $y = \frac{1}{x}$.

12. 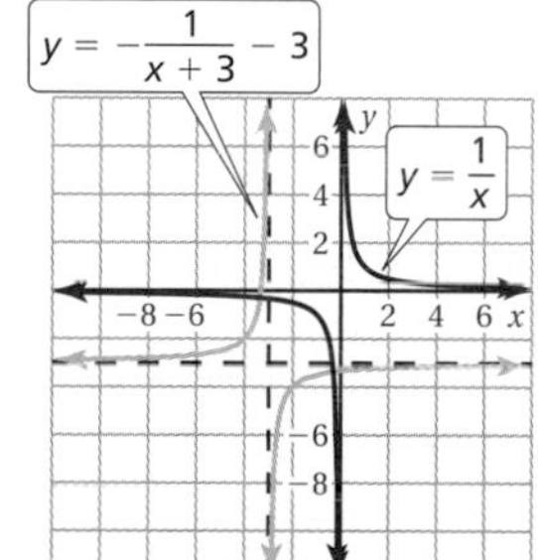

The graph of $y = -\frac{1}{x+3} - 3$ is a reflection in the x-axis of the graph of $y = \frac{1}{x}$ and then a translation 3 units to the left and 3 units down.

Practice and Problem Solving

5. The graph is two smooth curves. The domain appears to be all real numbers except 10. The range appears to be all real numbers except 10. As x gets closer to 10, the graph approaches the vertical line $x = 10$. As x increases and decreases, the graph approaches the horizontal line $y = 10$.

Additional Answers

12. 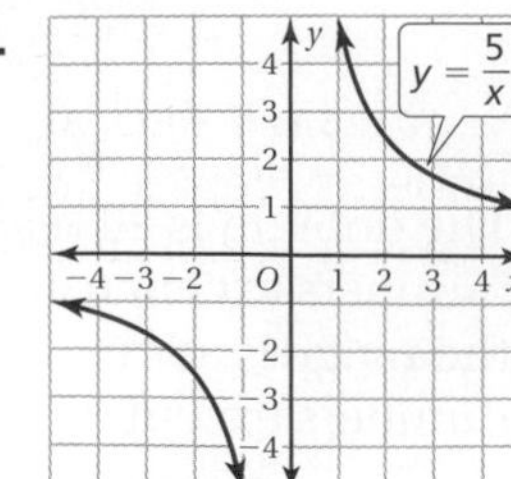

The domain is all real numbers except 0. The range is all real numbers except 0.

13. 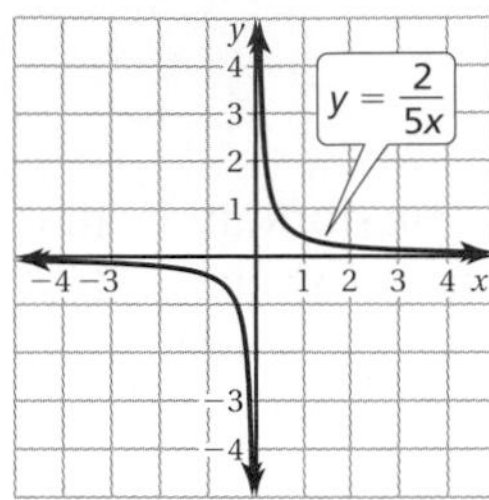

The domain is all real numbers except 0. The range is all real numbers except 0.

14. 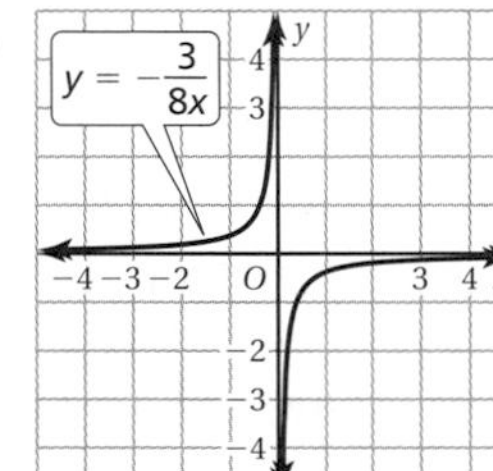

The domain is all real numbers except 0. The range is all real numbers except 0.

15. 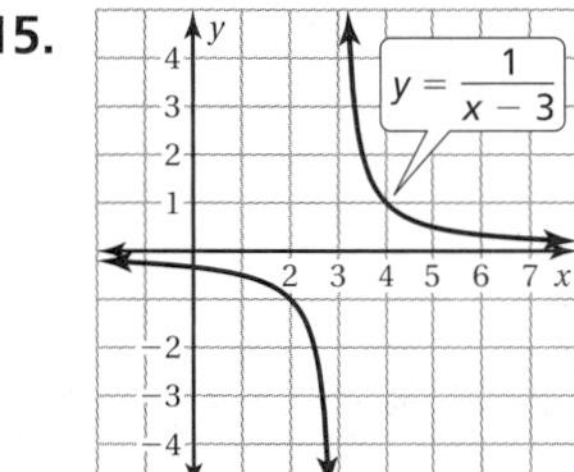

The domain is all real numbers except 3. The range is all real numbers except 0.

16. 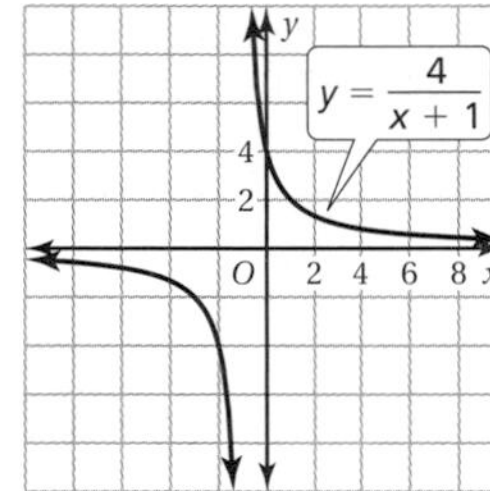

The domain is all real numbers except -1. The range is all real numbers except 0.

17. 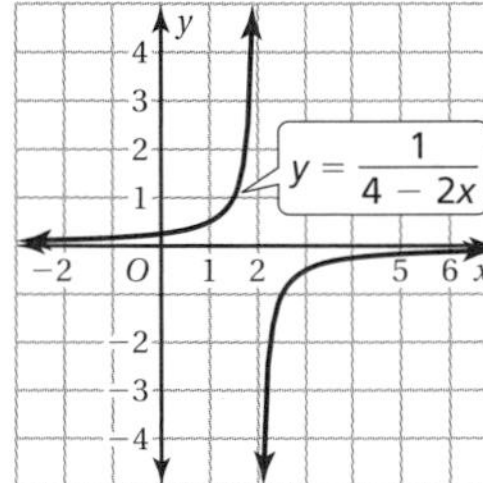

The domain is all real numbers except 2. The range is all real numbers except 0.

28. 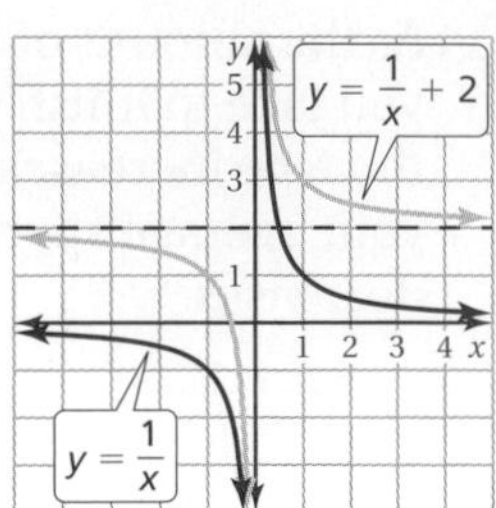

The graph of $y = \frac{1}{x} + 2$ is a translation 2 units up of the graph of $y = \frac{1}{x}$.

29. 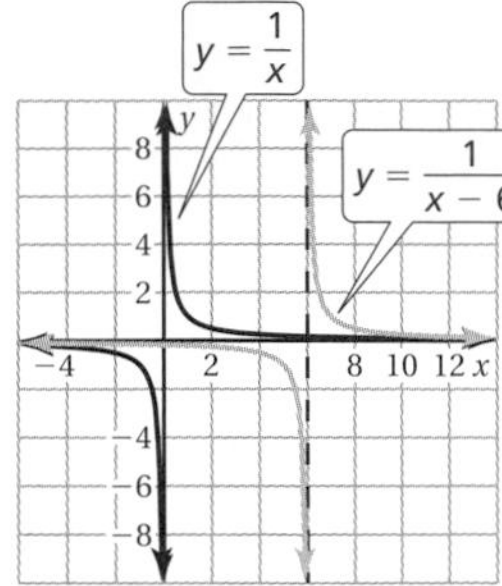

The graph of $y = \frac{1}{x-6}$ is a translation 6 units right of the graph of $y = \frac{1}{x}$.

30. 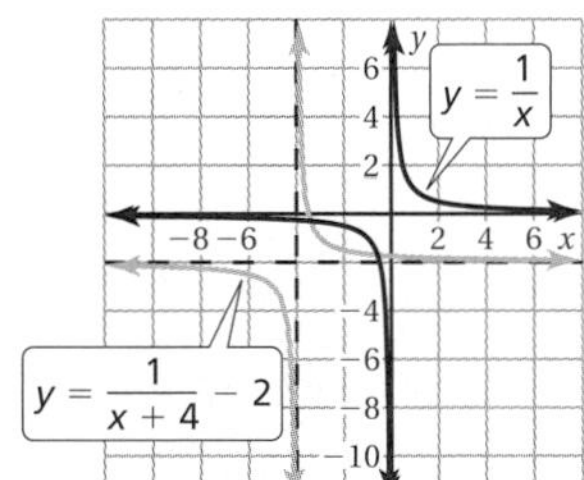

The graph of $y = \frac{1}{x+4} - 2$ is a translation 4 units left and 2 units down of the graph of $y = \frac{1}{x}$.

31. 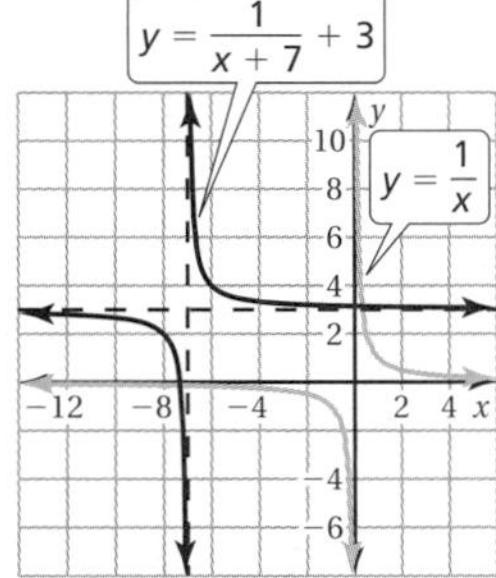

The graph of $y = \frac{1}{x+7} + 3$ is a translation 7 units left and 3 units up of the graph of $y = \frac{1}{x}$.

32. 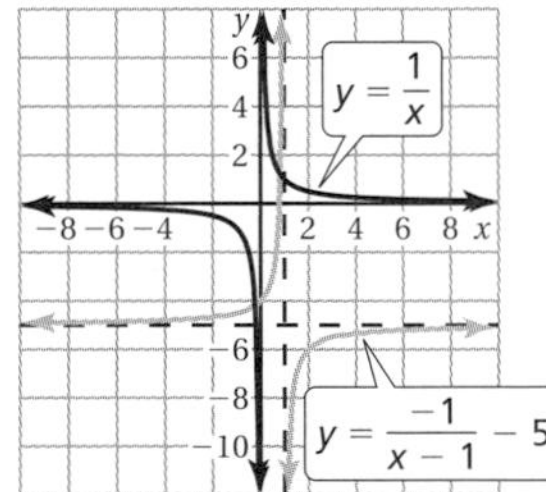

The graph of $y = \frac{-1}{x-1} - 5$ is a reflection in the x-axis of the graph of $y = \frac{1}{x}$ and then a translation 1 unit right and 5 units down.

33. 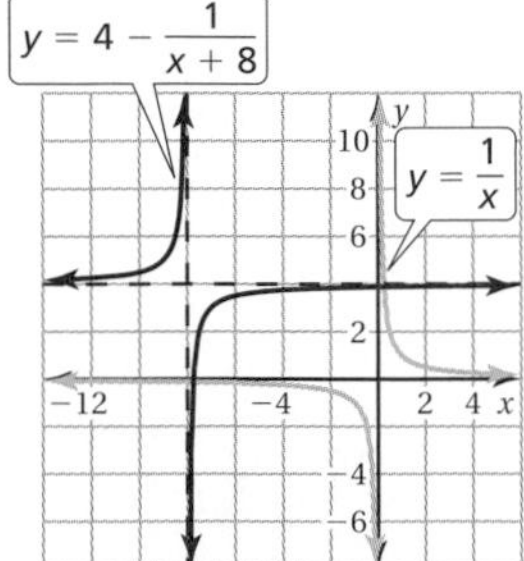

The graph of $y = 4 - \frac{1}{x+8}$ is a reflection in the x-axis of the graph of $y = \frac{1}{x}$ and then a translation 8 units left and 4 units up.

34. 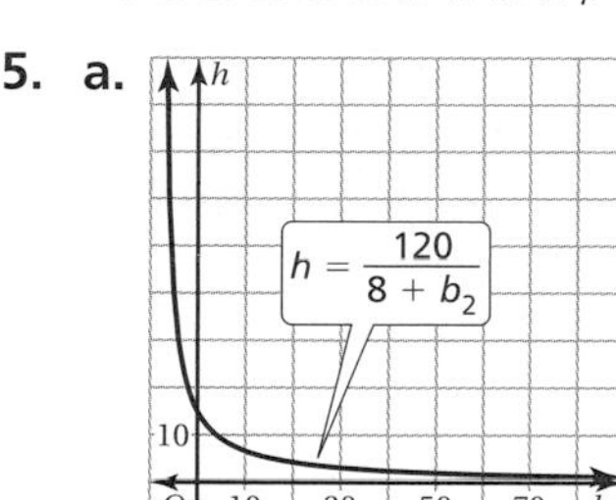

14 players

35. a.

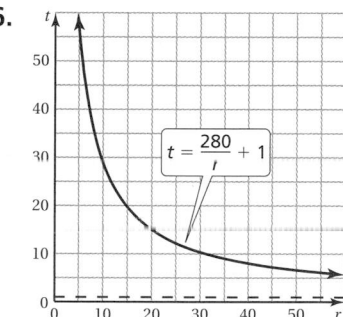

The domain is all real numbers greater than 0. The range is all real numbers greater than 0 and less than 15.

b. 12

36.

56 miles per hour

42. 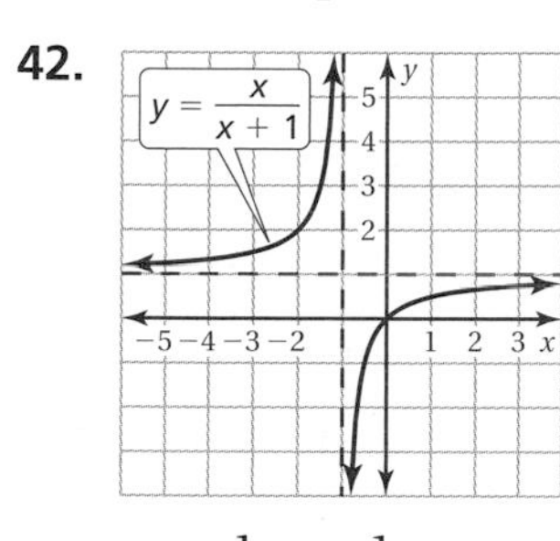

$x = -1, y = 1$

43. 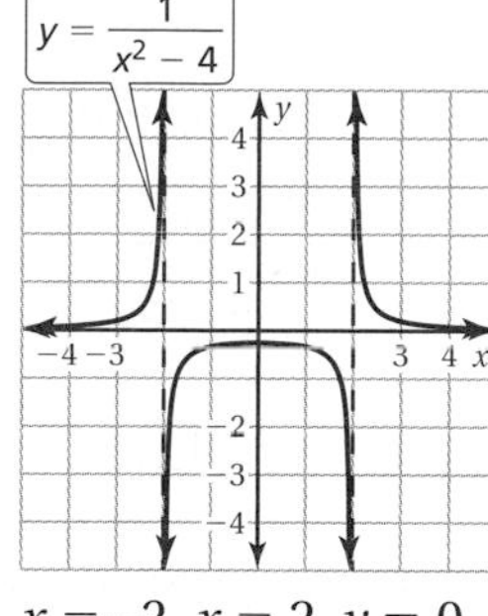

$x = -2, x = 2, y = 0$

44. 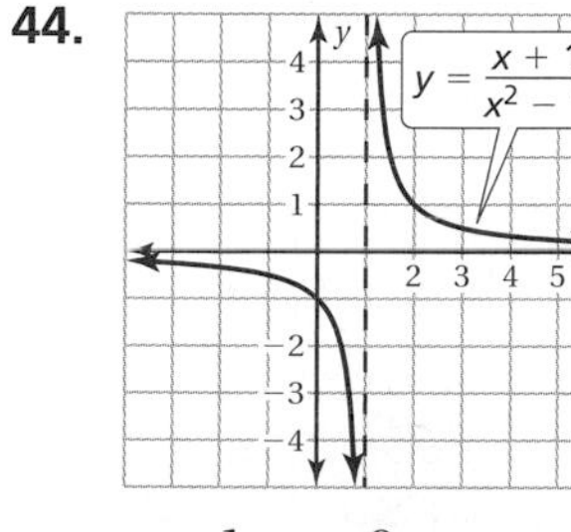

$x = 1, y = 0$

Mini-Assessment

2. 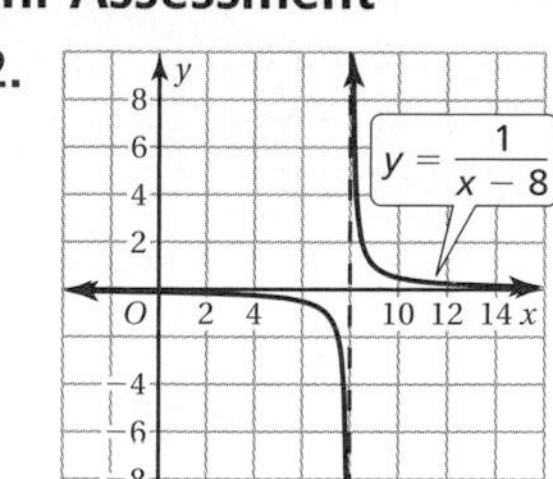

The domain is all real numbers except 8. The range is all real numbers except 0.

3. $x = -7, y = -5$;
The domain is all real numbers except -7. The range is all real numbers except -5.

4. 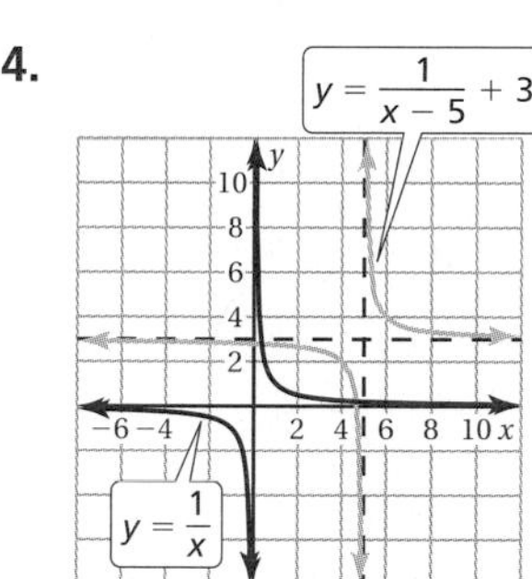

The graph of $y = \frac{1}{x-5} + 3$ is a translation 5 units right and 3 units up of the graph of $y = \frac{1}{x}$.

Record and Practice Journal Practice

3. Horizontal asymptote: $y = 2$;
Vertical asymptote: $x = 0$
Domain: All real numbers except 0
Range: All real numbers except 2

4. Horizontal asymptote: $y = -3$;
Vertical asymptote: $x = -4$
Domain: All real numbers except -4
Range: All real numbers except -3

5. The graph of $y = \frac{1}{x-1} - 6$ is a translation 6 units down and 1 unit right of the graph of $y = \frac{1}{x}$.

6. The graph of $y = \frac{-1}{x+5} + 4$ is a translation 4 units up and 5 units left and a reflection in the x-axis of the graph of $y = \frac{1}{x}$.

Lesson 11.2b

Practice

7. $f^{-1}(x) = \frac{1}{3}x + \frac{1}{3}$;

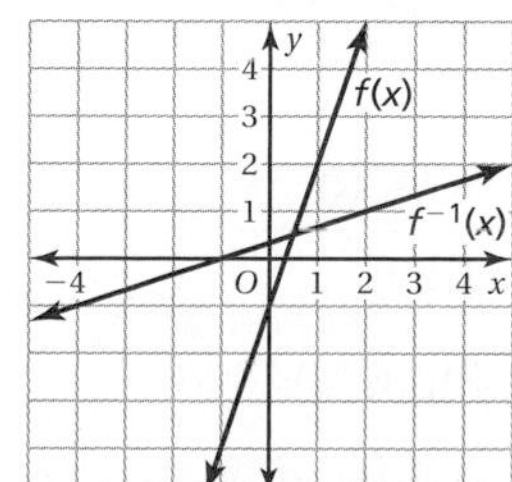

8. $f^{-1}(x) = -2x + 6$;

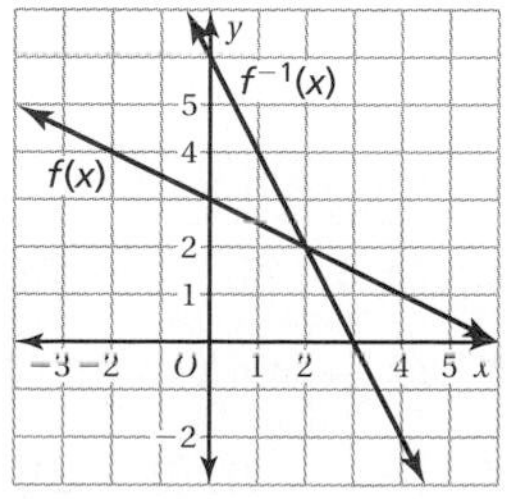

Additional Answers

9. $f^{-1}(x) = \sqrt{0.5x}$;

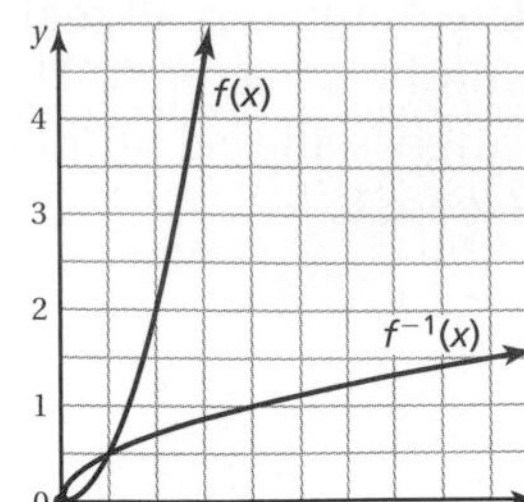

10. $f^{-1}(x) = \sqrt{x + 3}$;

11. $f^{-1}(x) = \frac{1}{x}$;

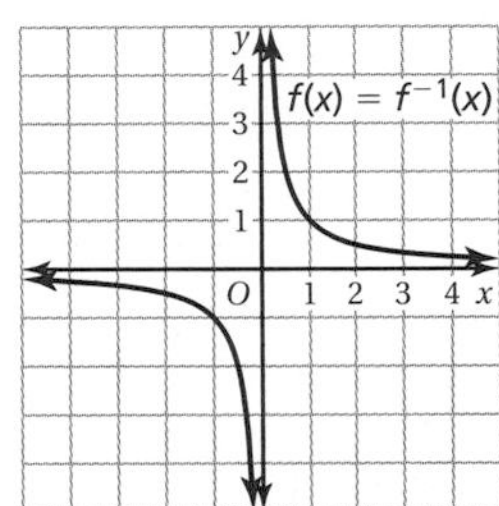

12. $f^{-1}(x) = \frac{1}{x} + 2$;

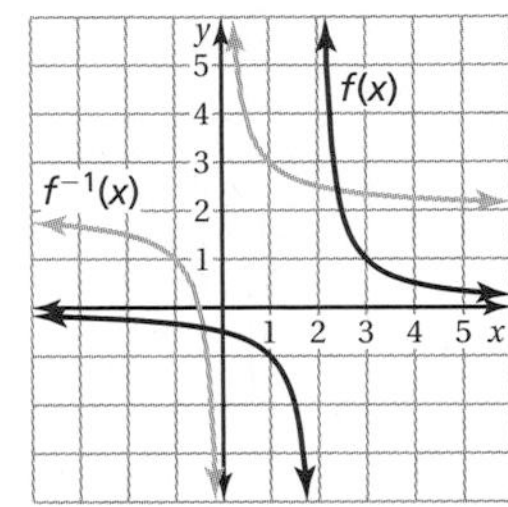

14.

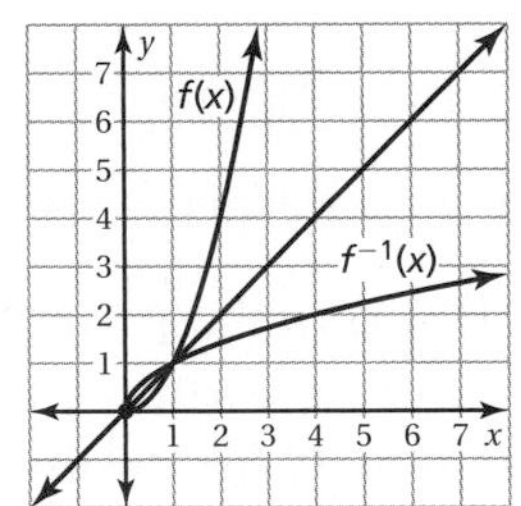

The graphs are reflections in the line $y = x$.

15. Both equal x. For $f(g(x))$, $g(x)$ is the output value of x. So, then you have $f(g(x))$ which gives you back the input value, x, because they are inverses. For $g(f(x))$, $f(x)$ is the output value of x. So, then you have $g(f(x))$ which gives you back the input value, x, because they are inverses.

Mini-Assessment

3.

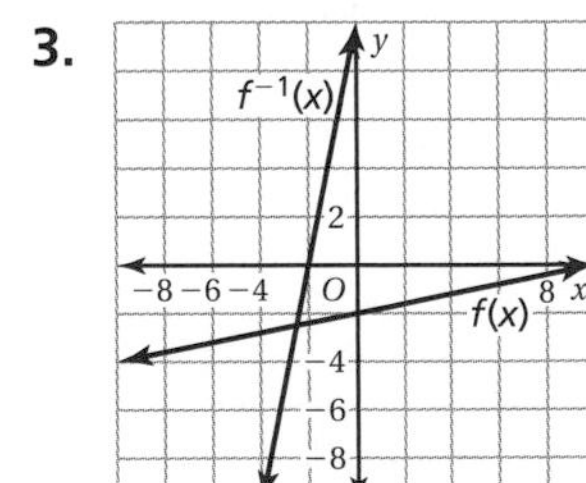

4.

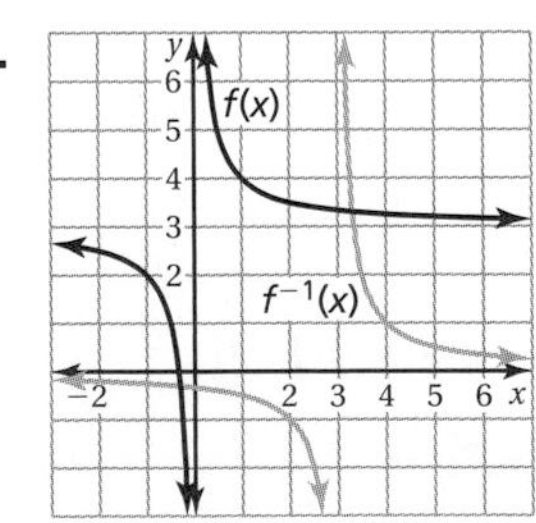

Record and Practice Journal Practice

1. (−1, −2), (2, −1), (5, 0), (8, 1), (11, 2)
2. (8, −3), (0, −1), (−1, 0), (0, 1), (8, 3)
3. (3, 1), (5, 3), (6, 4), (7, 6), (10, 6)
4. (2, −4), (3, −2), (4, 0), (5, 2), (6, 4)
5. (−1, −3), (0, −2), (−1, −1), (−4, 0), (−3, 1)
6. (12, −8), (8, −4), (4, −4), (6, 0), (4, 8)
7. $f^{-1}(x) = 2x - 2$

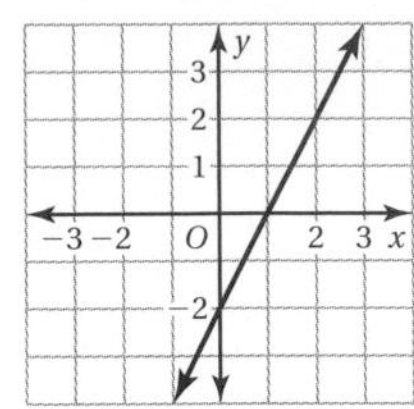

8. $f^{-1}(x) = -\frac{1}{4}x - \frac{3}{2}$

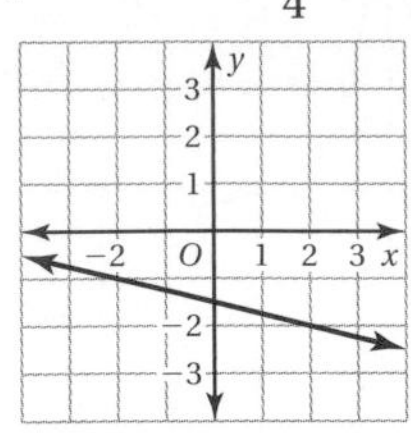

9. $f^{-1}(x) = \sqrt{x + 5}$

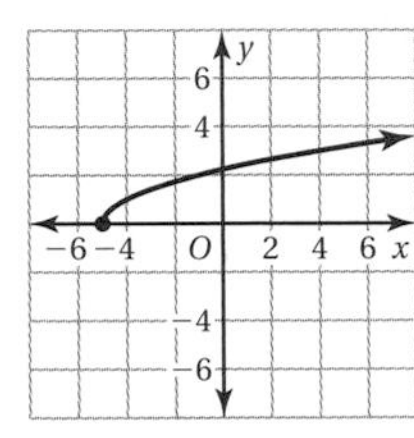

10. $f^{-1}(x) = \sqrt{3x}$

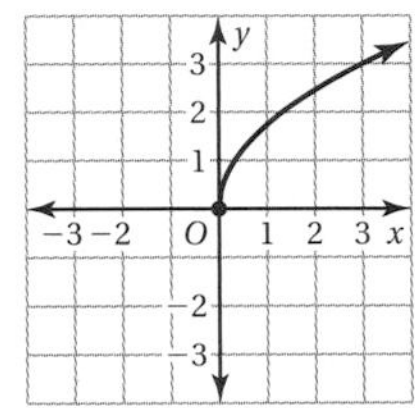

11. $f^{-1}(x) = \frac{1}{x} - 4$

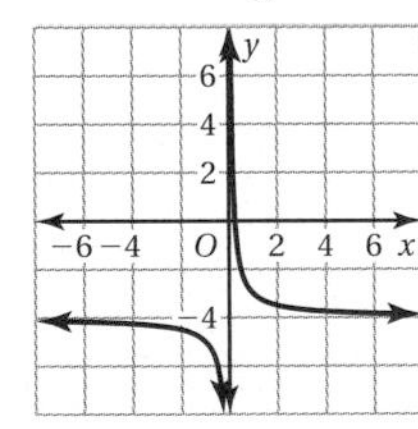

12. $f^{-1}(x) = \frac{1}{x + 3}$

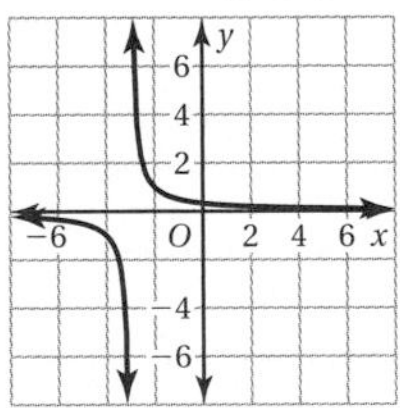

13. 8

Section 11.3

Fair Game Review

27.

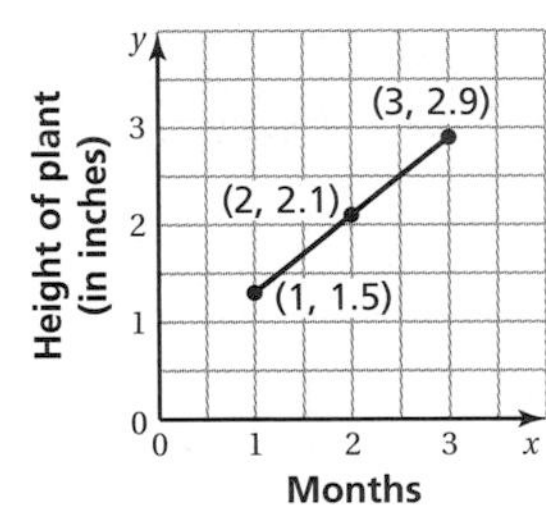

continuous

11.1–11.3 Quiz

3. neither; Neither the products xy nor the ratios $\frac{y}{x}$ are constant.

4. 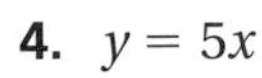
$y = 5x$

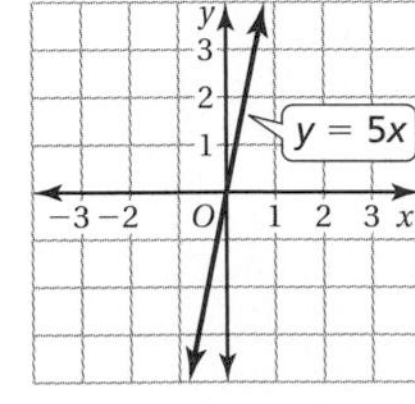

5. $y = \frac{14}{x}$

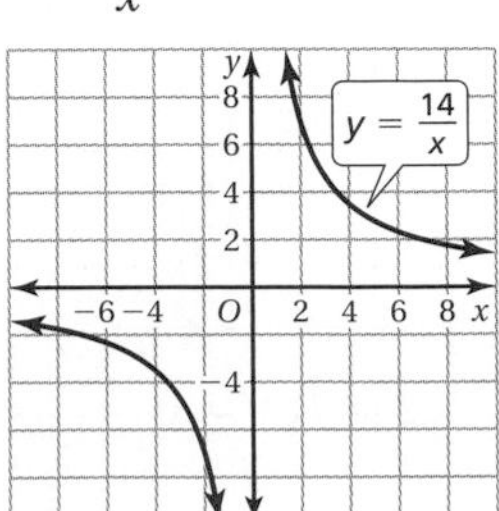

10. $x = -6, y = 9$;
The domain is all real numbers except -6 and the range is all real numbers except 9.

11. $f^{-1}(x) = \frac{1}{2}x - \frac{3}{2}$;

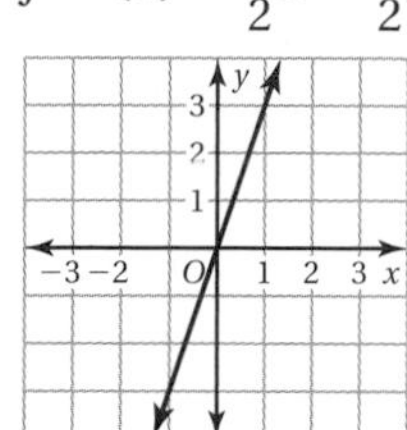

12. $f^{-1}(x) = \sqrt{x - 1}$;

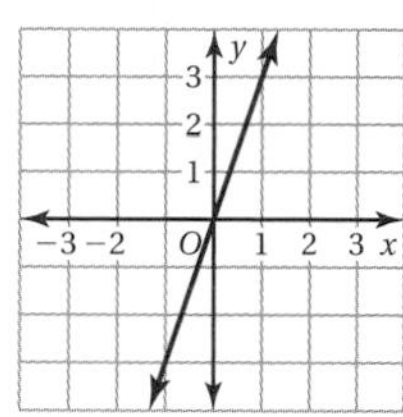

Section 11.4

Practice and Problem Solving

16. To divide by $\frac{3w}{w+1}$, you must multiply by its reciprocal, $\frac{w+1}{3w}$;

$$\frac{6w}{w+1} \div \frac{3w}{w+1} = \frac{6w}{w+1} \cdot \frac{w+1}{3w} = \frac{6w(w+1)}{3w(w+1)} = 2$$

Fair Game Review

28.

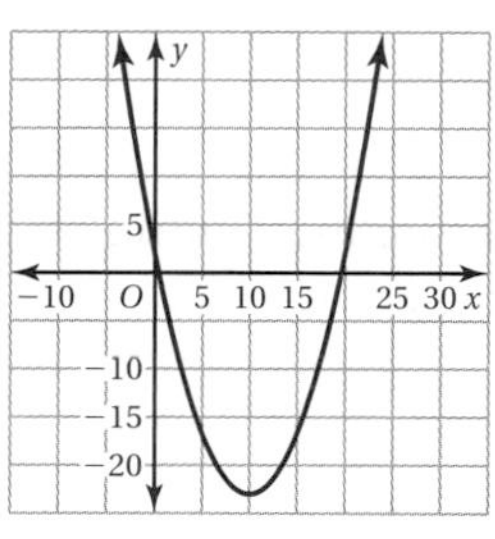

domain: all real numbers;
range: $y \geq -23$

29.

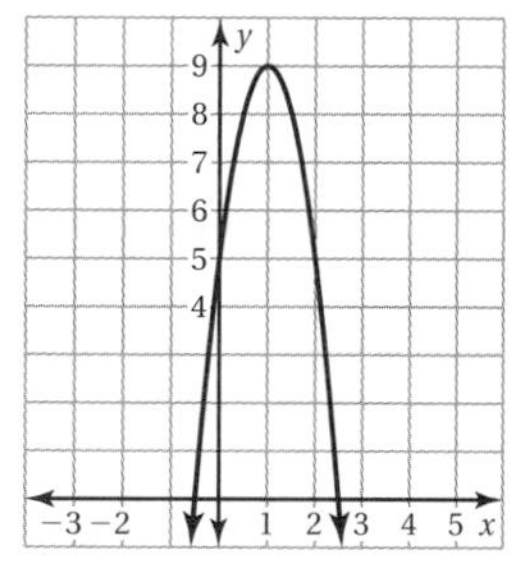

domain: all real numbers;
range: $y \leq 9$

Section 11.5

Practice and Problem Solving

17. When subtracting the product from the dividend, the negative sign was not distributed to the second term; $(x^2 - x - 6) \div (x - 3) = x + 2$

29.

Quotient
$(x^2 - x + 1) \div (x + 1) = x - 2 + \frac{3}{x+1}$
$(x^3 - x^2 + x - 1) \div (x + 1) = x^2 - 2x + 3 - \frac{4}{x+1}$
$(x^4 - x^3 + x^2 - x + 1) \div (x + 1) = x^3 - 2x^2 + 3x - 4 + \frac{5}{x+1}$
$(x^5 - x^4 + x^3 - x^2 + x - 1) \div (x + 1) = x^4 - 2x^3 + 3x^2 - 4x + 5 - \frac{6}{x+1}$

Section 11.6

Practice and Problem Solving

26. The numerator of each rational expression was multiplied by the wrong factor of the LCD.

$$\frac{x}{x-1} + \frac{2}{x+2} = \frac{x(x+2) + 2(x-1)}{(x-1)(x+2)}$$
$$= \frac{x^2 + 2x + 2x - 2}{(x-1)(x+2)}$$
$$= \frac{x^2 + 4x - 2}{(x-1)(x+2)}$$

Fair Game Review

40.

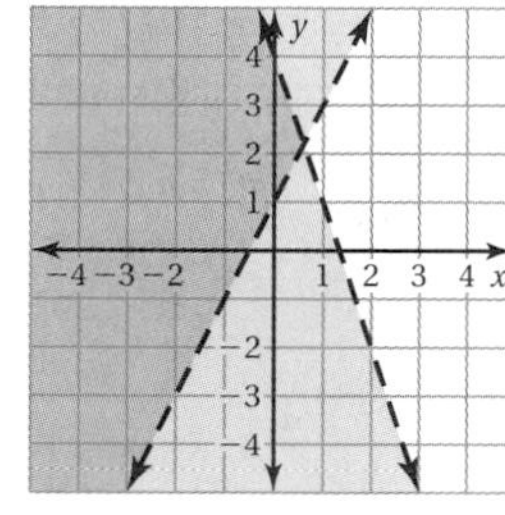

41.

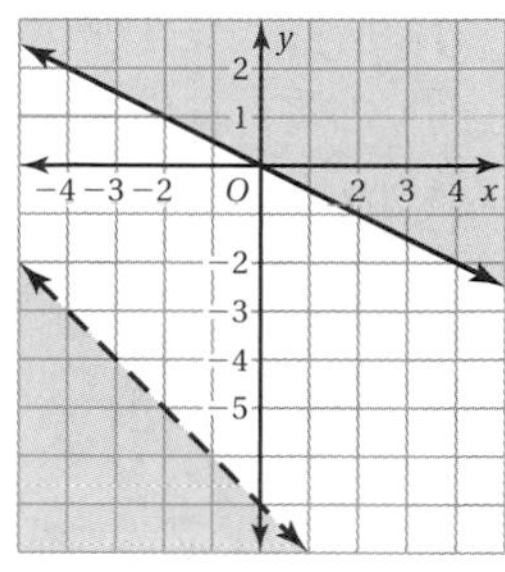

Chapter 11 Test

1. $y = 6x$

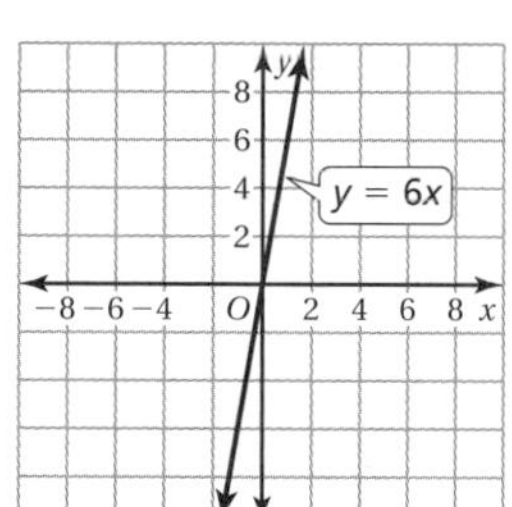

2. $y = \frac{24}{x}$

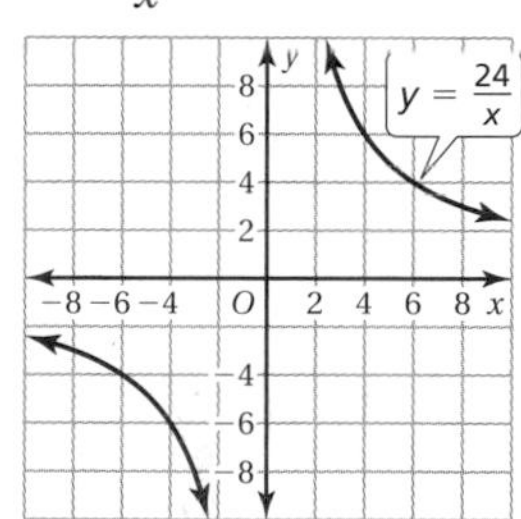

3.

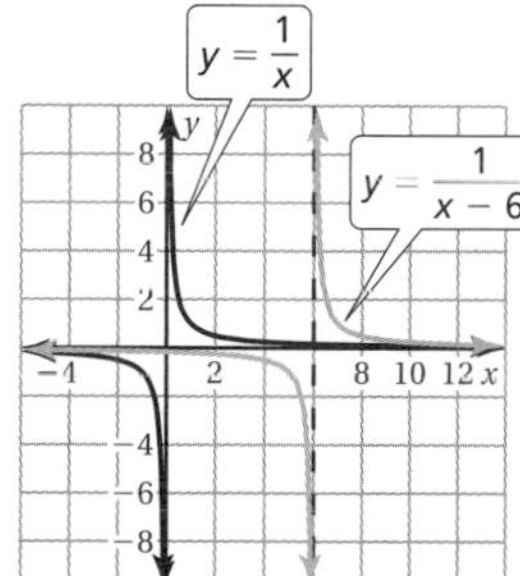

The graph of $y = \frac{1}{x-6}$ is a translation 6 units right of the graph of $y = \frac{1}{x}$.

4.

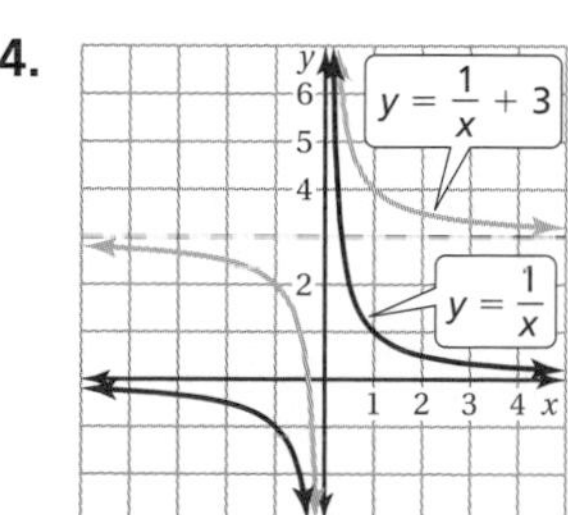

The graph of $y = \frac{1}{x} + 3$ is a translation 3 units up of the graph of $y = \frac{1}{x}$.

5. 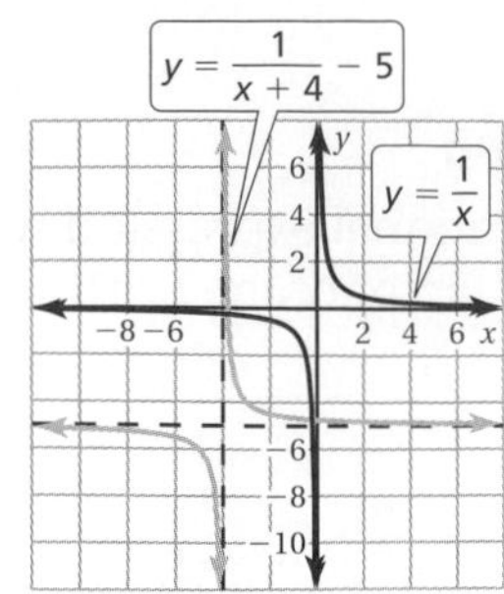

The graph of $y = \frac{1}{x+4} - 5$ is a translation 4 units left and 5 units down of the graph of $y = \frac{1}{x}$.

6. $f^{-1}(x) = x + 7$

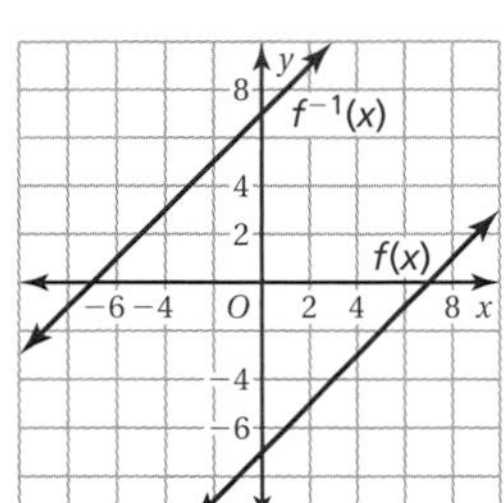

7. $f^{-1}(x) = 5x + 35$

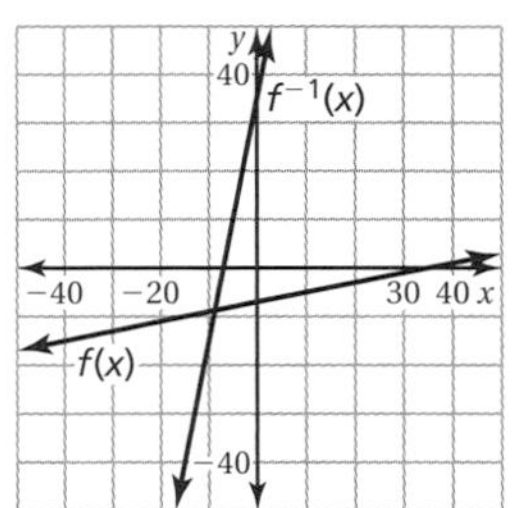

9. Money Raised

Students

Money raised

0–19, 20–39, 40–59, 60–79, 80–99

10. Siblings

Students

Number of siblings

0–1, 2–3, 4–5, 6–7, 8–9

Chapter 12

Record and Practice Journal Fair Game Review

6. Concerts

Students

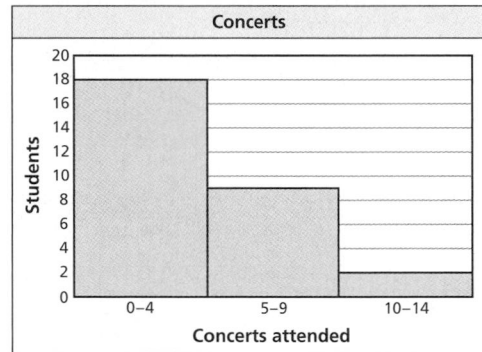

Concerts attended

0–4, 5–9, 10–14

7. 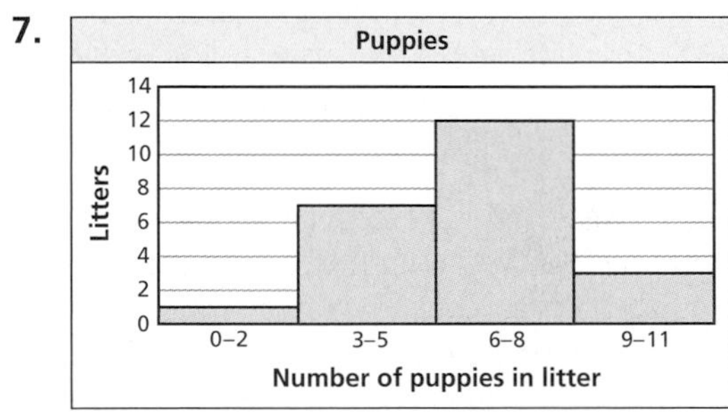

8. 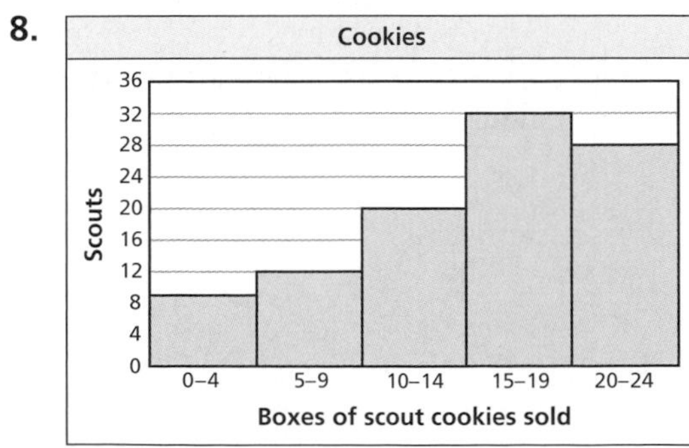

Section 12.1

Practice and Problem Solving

21. bagel shop; Without seeing the data values, it is hard to tell for sure. With the information given, the 10% increase in wages at the bagel shop increases the mean and median hourly wage to $7.92, which is greater than these measures for the amusement park.

Section 12.2

Extra Example 2

x	$\overline{x}$	$x - \overline{x}$	$(x - \overline{x})^2$
185	198	−13	169
221	198	23	529
194	198	−4	16
208	198	10	100
167	198	−31	961
205	198	7	49
190	198	−8	64
214	198	16	256

Practice and Problem Solving

5. Tigers: 69; 9; Centaurs: 69; 12; The means are equal but the range of the heights of the Centaurs is greater than the range of the heights of the Tigers. So, the heights of the Centaurs are more spread out.

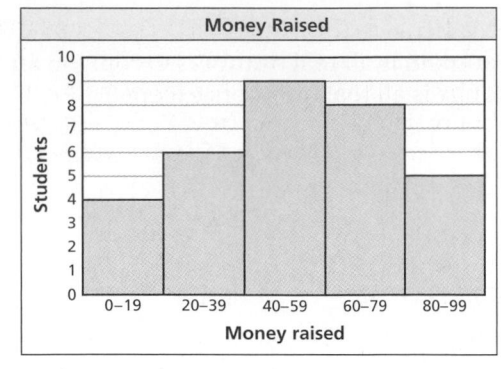

6. Crew A: 104; 60; Crew B: 143; 61; The ranges are about the same, but the mean number of fish caught by Crew A is much less than the mean number of fish caught by Crew B. So, the spreads of each data set are similar but the means are very different.

14. Kirsten: 90; 15; about 4.4; So, Kirsten's mean and range remained the same and the standard deviation decreased slightly.

    Leah: 90; 15; about 4.0; So, Leah's mean decreased, her range increased and is now the same as Kirsten's, and her standard deviation increased.

    Leah's statistics changed more than Kirsten's because her new score was very low compared to her other scores and Kirsten's new score was equal to the mean of her previous scores.

17. no; Two data sets can have the same range and very different standard deviations due to outliers and/or the distribution of the data.

    Example: Data set 1: 1, 5, 6, 6, 6, 7, 11
    range = 10, standard deviation = 2.7

    Data set 2: 1, 2, 2, 6, 10, 10, 11
    range = 10, standard deviation = 4.0

19. **a.** *Sample answer:* heights (in inches): 58, 58, 59, 60, 60, 61, 62, 62, 63, 65, 65, 65, 65, 66, 66, 68, 69, 71;
    mean = 63.5; range = 13;
    standard deviation $\approx$ 3.7

    b. The mean, range, and standard deviation should all increase. *Sample answer:* mean 64.6; range = 26; standard deviation $\approx$ 5.8

21. yes; no; The data set 4, 4, 4, 4, 4, 4, 4, 4 has a standard deviation of 0. The formula for the standard deviation of a data set involves a square root, and standard deviation is a measure of how much a typical data value differs from the mean. So, it can never be negative.

Fair Game Review

22. 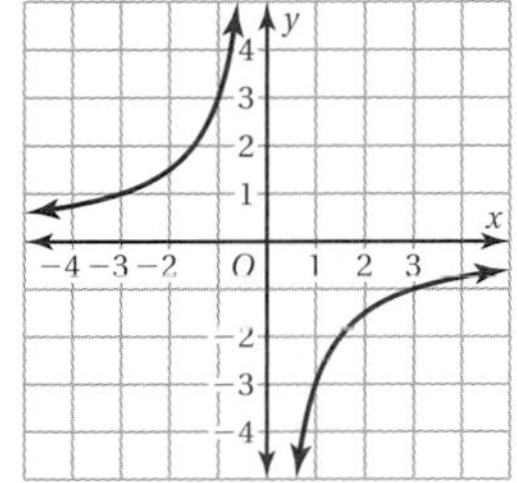

    The domain is all real numbers except 0 and the range is all real numbers except 0.

23. 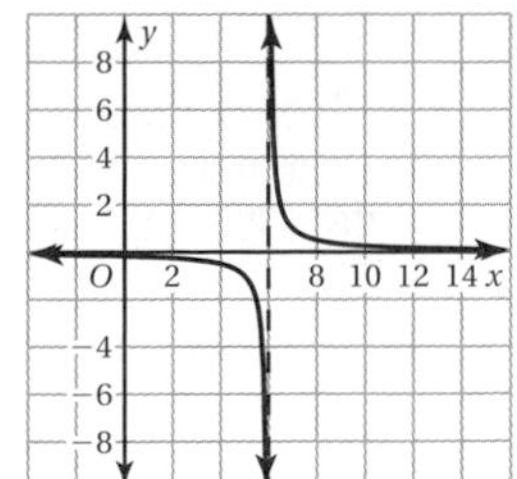

    The domain is all real numbers except 6 and the range is all real numbers except 0.

24. 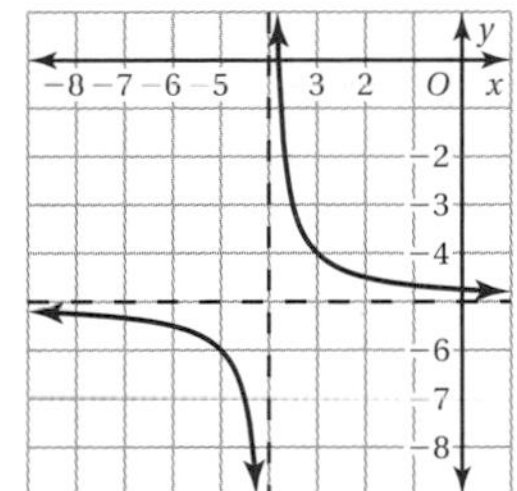

    The domain is all real numbers except −4 and the range is all real numbers except −5.

Mini-Assessment

1. Student A: 26; 22; about 6.9
   Student B: 20; 13; about 4.2
   Student A's mean, range, and standard deviation are greater than Student B's.

2. Team A: 3.5; 0.6; 0.2
   Team B: 3.5; 1.1; about 0.4
   The means are equal but Team B's range and standard deviation are greater than Team A's.

Section 12.3

On Your Own

3. 25% of the players are between 18 and 23 years old. 50% of the players are between 23 and 29 years old. 25% of the players are between 29 and 30 years old.

4. For both shops, the right whisker is longer than the left whisker and most of the data are on the left side of the display. So, both distributions are skewed right. The range of the prices of Shop A is greater than the range of Shop B. Also, because the box for Shop A is longer than the box for Shop B, the interquartile range is also greater. So, the prices of Shop A are more spread out that the prices of Shop B.

Practice and Problem Solving

11. a. 8 9 10 11 12 13 14 15 16 17 18 19 Entree prices (dollars)

b. The middle half of the prices vary by no more than \$4.38.

c. 25% of the prices are between \$8.50 and \$10.50; 50% of the prices are between \$10.50 and \$14.88; 25% of the prices are between \$14.88 and \$18.25.

d. 2.93; The typical price differs from the mean by about \$2.93.

14. skewed right; The right whisker is longer than the left whisker and most of the data is on the left.

15. skewed left; The left whisker is longer than the right whisker and most of the data is on the right.

16. symmetric; The whiskers are about the same length and the median is in the middle of the data.

20. a. Brand A: symmetric; Brand B: skewed right

b. Brand A: 3.5 hours; Brand B: 2.7 hours

c. The interquartile range of Brand A, 1.5 hours, is greater than the interquartile range of Brand B, 1 hour.

d. Brand A; The data are more spread out.

Section 12.4

On Your Own

1.

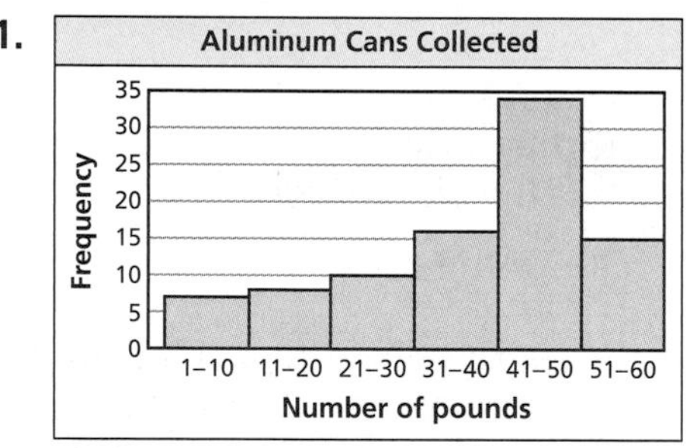

skewed left

3. a.

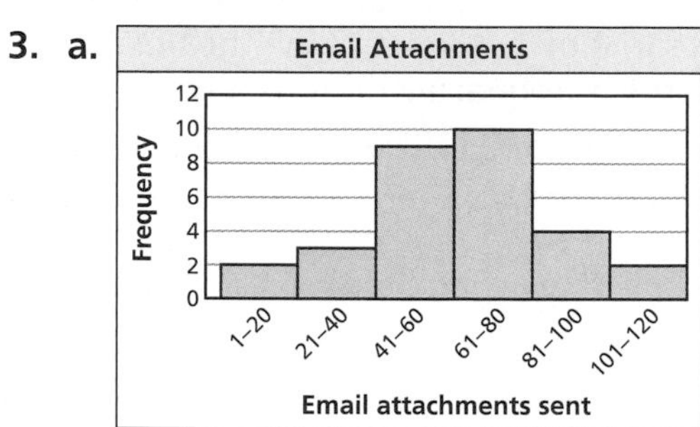

Practice and Problem Solving

7.

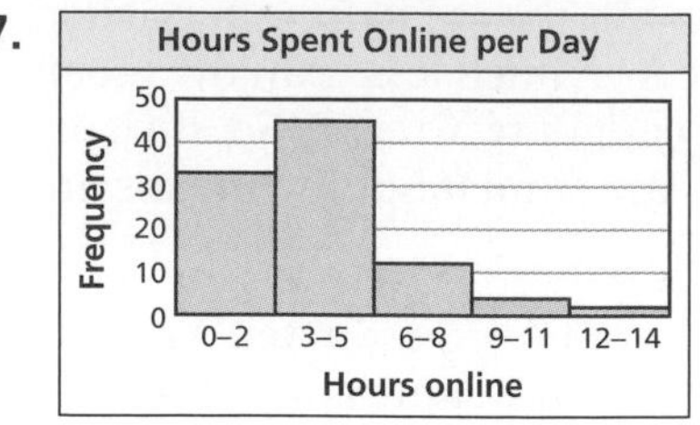

Most of the data are on the left. So, the distribution is skewed right.

11. Because the distribution is high on the right and the tail of the graph extends to the left, the distribution is skewed left. So, the median best represents the center of the data and the five-number summary best represents the spread of the data.

15. *Sample answer:* histogram; A histogram can show a variety of intervals and easily show the distribution of large data sets. A stem-and-leaf plot uses every data value and can be very tedious for large data sets. Also, stem-and-leaf plots usually only show intervals that are multiples of 10.

16. a.

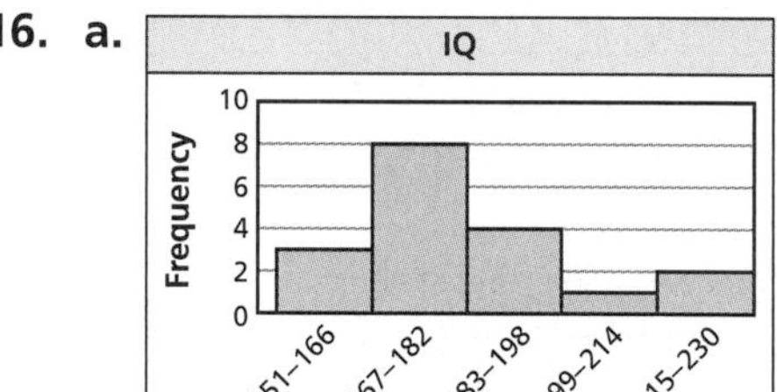

b. Because the distribution is high on the left and the tail of the graph extends to the right, the distribution is skewed right. So, the median best represents the center of the data and the five-number summary best represents the spread of the data.

c. Gradually shifts from skewed right to symmetric as you include more values.

18. a.

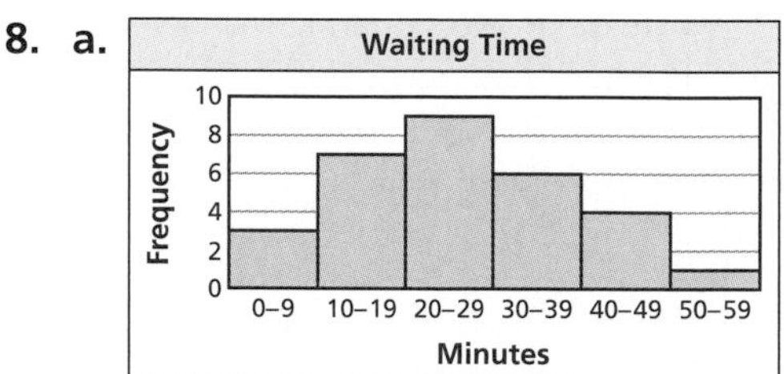

b. 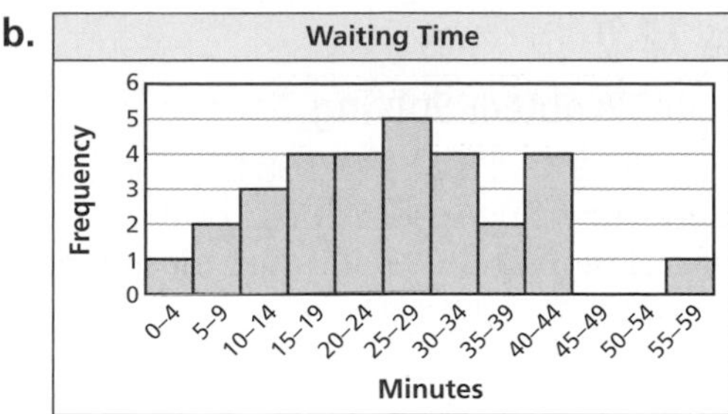

c. The distribution appears flat when the number of intervals is increased.

d. The histogram with six intervals best represents the data. It shows that the distribution is slightly skewed right. When choosing the number of intervals, an important consideration is the size of the data set. For small data sets, use a small number of intervals.

Mini-Assessment

1. 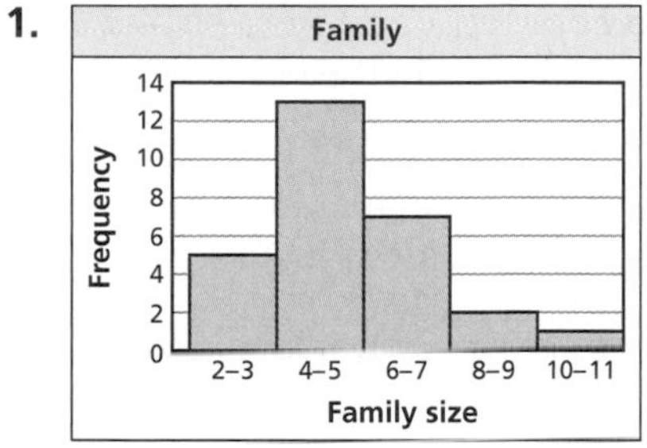

Most of the data are on the left. So, the distribution is skewed right.

2. a. 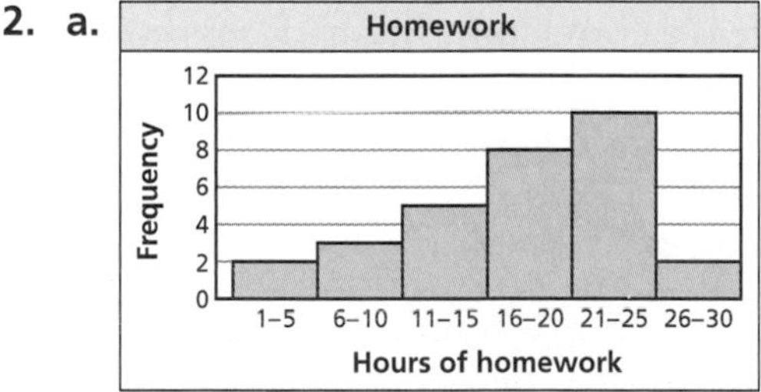

b. Because the distribution is high on the right and the tail of the graph extends to the left, the distribution is skewed left. So, the median best represents the center of the data and the five-number summary best represents the spread of the data.

12.1–12.4 Quiz

7. 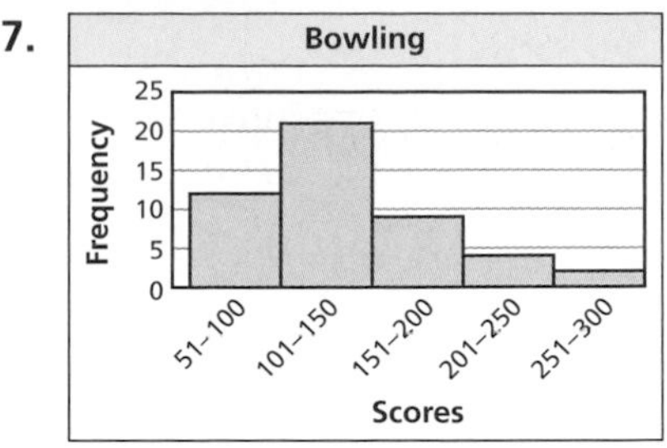

The distribution is high on the left and the tail of the graph extends to the right, so the distribution is skewed right.

8. a. mean: 16.9; median: 17.0; modes: 16.5, 17.0

b. 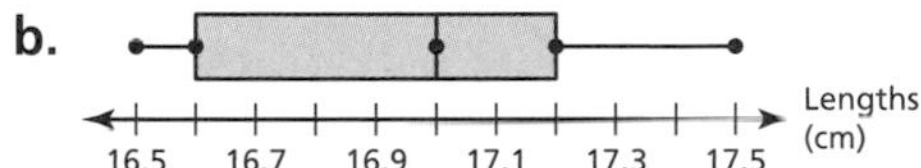

c. The interquartile range is $17.2 - 16.6 = 0.6$ centimeter. This means that the middle half of anole lengths vary by no more than 0.6 centimeter.

d One-quarter of the lengths are 16.6 centimeters or less.

One-half of the lengths are between 16.6 and 17.2 centimeters long.

One-quarter of the lengths are 17.2 centimeters or more.

e. The mean is reduced to 16.3 centimeters, but the median and mode stay the same.

9. a. 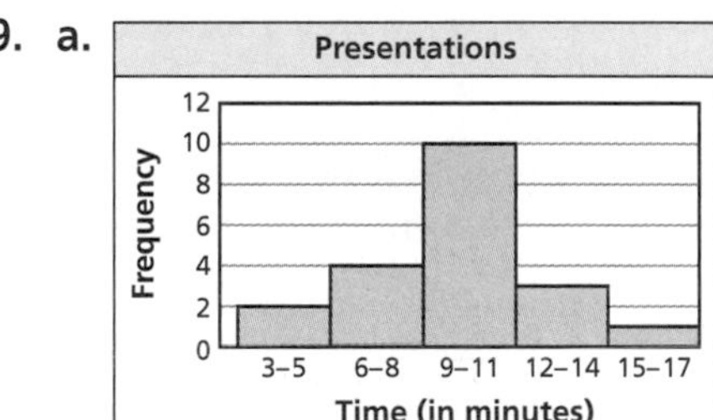

b. Because the distribution is high in the middle and evenly distributed, the distribution is symmetric. So, use the mean of 9.65 minutes to describe the center, and use the standard deviation of about 2.9 minutes to describe the spread.

c. Because the distribution is symmetric, most of the data are around the 9–11 minute interval. This shows that most of the presentations are about as long as they are supposed to be.

Section 12.5

On Your Own

2. 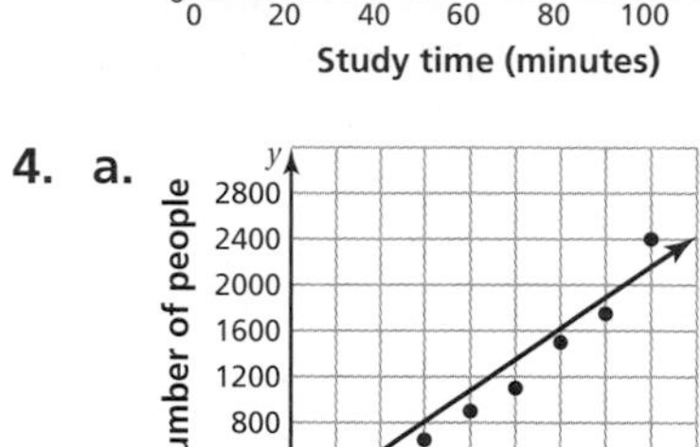

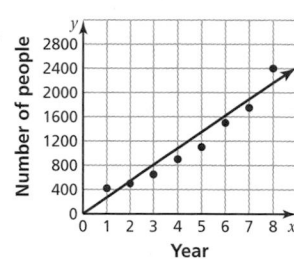

positive relationship

4. a. Number of people; Year

Practice and Problem Solving

19. *Sample answer:* The points follow a U-shaped pattern. There is a relationship between x and y, but it is not linear.

20. *Sample answer:* There are two clusters of data points. There appears to be a negative relationship between x and y.

21. *Sample answer:* There appears to be two outliers. There appears to be a positive relationship between x and y.

22. a. (24.00, 40,000)

 b. Because the outlier is below the other values, it will increase the steepness of the line of best fit.

 c. 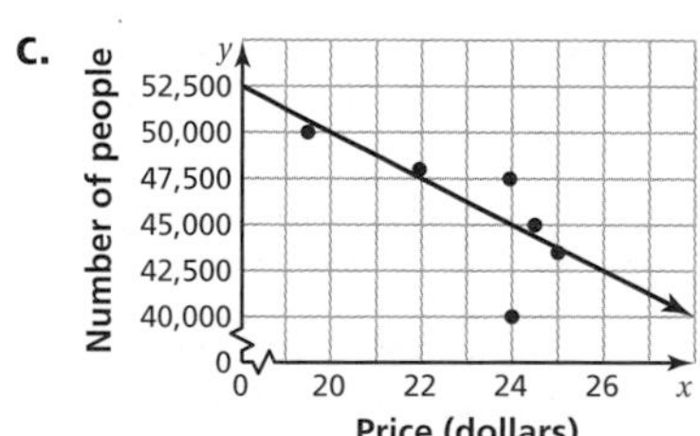

 d. *Sample answer:* 41,000 people

Record and Practice Journal Practice

5. b. 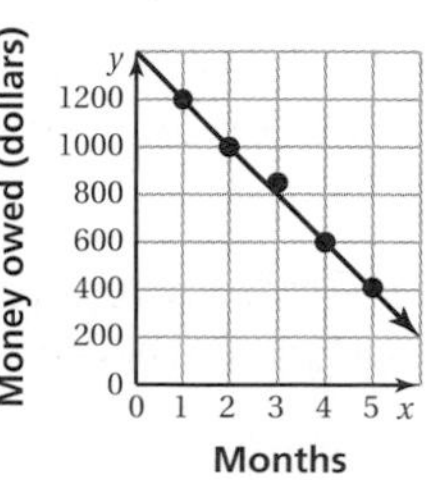

Section 12.6

Practice and Problem Solving

18. a. $y = 454.8x - 2812$; $r \approx 0.984$; The relationship between x and y is a strong positive correlation and the equation closely models the data.

 b. The slope is the average yearly increase (in billions) in texts sent.

 c. 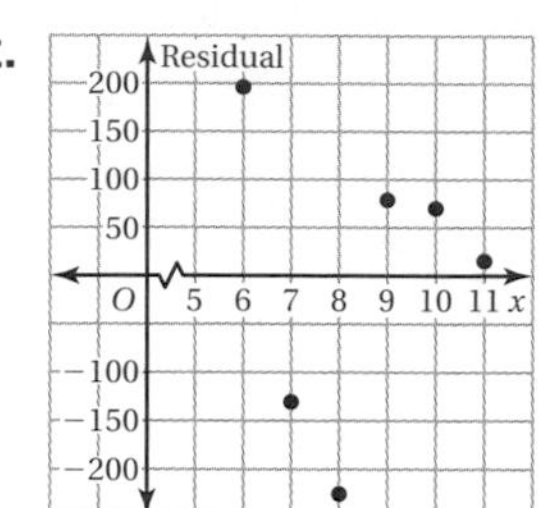

 The points (x, residual) are randomly dispersed about the horizontal axis. So, the equation is a good fit.

 d. about 4 trillion

19. a. $y = -0.08x + 3.8$; $r \approx -0.965$; The relationship between x and y is a strong negative correlation and the equation closely models the data.

 b. The slope is the change in grade point average per hour of television watched. The y-intercept is the grade point average of a student who does not watch television.

 c. about 2.7

 d. *Sample answer:* no; Watching television does not cause a lower grade point average, but there is a correlation. Studying less could be the cause.

Section 12.7

Extra Example 3

		Grade			
		6	7	8	Total
Join Student Council	Yes	11	13	7	31
	No	10	14	15	39
	Total	21	27	22	70

Extra Example 4

a.

		Grade		
		6	7	8
Join Student Council	Yes	52%	48%	32%
	No	48%	52%	68%

Sample answer: About 32% of 8th grade students in the survey are going to try to join student council.

b. The table shows that as grade level increases, students are less likely to try to join student council.

On Your Own

2. a.

		Grade			
		6	7	8	Total
Lunch	Pack	11	23	16	50
	Buy	9	27	14	50
	Total	20	50	30	100

b.

		Grade		
		6	7	8
Lunch	Pack	55%	46%	53%
	Buy	45%	54%	47%

Sample answer: 45% of the 6th grade students in the survey buy a school lunch.

c. no; About half of the students in each grade buy a school lunch.

Practice and Problem Solving

10.

		Gender		
		Male	Female	Total
Pet	Own	18	6	24
	Do not own	25	14	39
	Total	43	20	63

11. a.

		Eye Color			
		Green	Blue	Brown	Total
Gender	Male	5	16	27	48
	Female	3	19	18	40
	Total	8	35	45	88

b. 48 males were surveyed.
40 females were surveyed.
8 students have green eyes.
35 students have blue eyes.
45 students have brown eyes.

c.

		Eye Color		
		Green	Blue	Brown
Gender	Male	63%	46%	60%
	Female	38%	54%	40%

Sample answer: About 63% of the students with green eyes are male. 40% of the students with brown eyes are female.

12.

		Eye Color		
		Green	Blue	Brown
Gender	Male	10.4%	33.3%	56.3%
	Female	7.5%	47.5%	45%

Sample answers: About 10.4% of the males surveyed have green eyes. 7.5% of the females surveyed have green eyes.

Mini-Assessment

1. a.

		Grade		
		5	8	Total
School Sports	Involved	12	23	35
	Not Involved	26	19	45
	Total	38	42	80

b.

		Grade	
		5	8
School Sports	Involved	32%	55%
	Not Involved	68%	45%

Sample answer: About 55% of the 8th grade students in the survey are involved in school sports.

c. yes; Students in Grade 8 are more likely to be involved in school sports than students in Grade 5.

Record and Practice Journal Practice

2. a.

		Grade			
		6	7	8	Total
The student	Eats breakfast at home	28	15	9	52
	Eats breakfast at school	12	15	21	48
	Total	40	30	30	100

b.

		Grade		
		6	7	8
The student	Eats breakfast at home	70%	50%	30%
	Eats breakfast at school	30%	50%	70%

$\frac{9}{30} = 0.3$

So, 30% of the grade 8 students in the survey eat breakfast at home.

Section 12.8

Record and Practice Journal

1. c. *Sample answer:*

Raccoon Road Kill Weights

Stem	Leaf
9	4 5
10	
11	0
12	4 9
13	4 6 9
14	0 5 8 8
15	2 7
16	8
17	0 2 3 5
18	5 5 6 7
19	0 1 4
20	4
21	3 5 5 5
22	
23	
24	
25	4

Key: 9 | 4 = 9.4 pounds

The stem-and-leaf plot shows how the raccoon weights are distributed.

Record and Practice Journal Practice

4. Because the rain icon is larger than the sun icon, it makes it look as if there were equal amounts of sunny and rainy days when there was not.

12.5–12.8 Quiz

6. a.

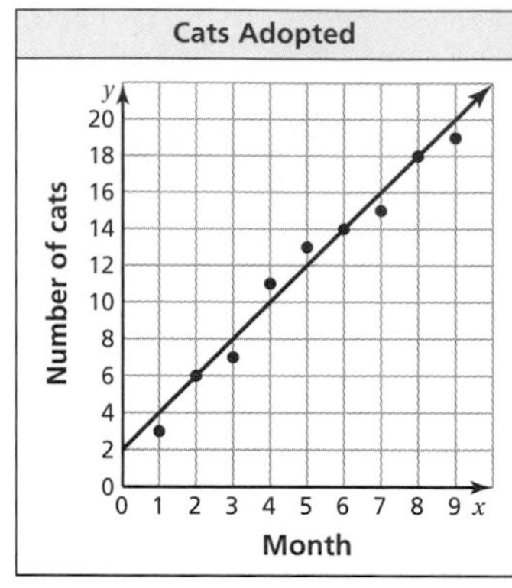

b. *Sample answer:* $y = 2x + 2$

c. The slope of the line of fit is 2. This means that the number of cats increases by 2 cats per month.

d. *Sample answer:* 22 cats

Chapter 12 Test

6. a.

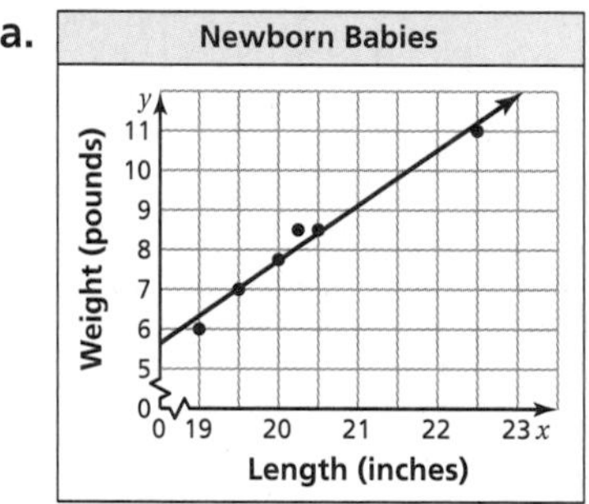

Sample answer: $y = 1.5x - 22.25$

b. $y = 1.39x - 20.1$; $r \approx 0.988$; This means that the relationship between length and weight is a strong positive correlation and the equation closely models the data.

c. 9.25 lb; 9.09 lb; The baby is slightly heavier according the equation from part (a).

8.

		Use reuseable bags?		
		Yes	No	Total
Gender	Male	15	45	60
	Female	60	50	110
	Total	75	95	170

Photo Credits

Front matter

iv Big Ideas Learning, LLC; **vi** *top* ©iStockphoto.com/Lisa Thornberg, ©iStockphoto.com/Ann Marie Kurtz; *bottom* Evok20/Shutterstock.com; **vii** *top* ©iStockphoto.com/Jonathan Larsen; *bottom* william casey/Shutterstock.com; **viii** *top* ©iStockphoto.com/ALEAIMAGE, ©iStockphoto.com/Ann Marie Kurtz; *bottom* ©iStockphoto.com/daaronj; **ix** *top* stephan kerkhofs/Shutterstock.com, Cigdem Sean Cooper/Shutterstock.com, ©iStockphoto.com/Andreas Gradin; *bottom* Kharidehal Abhirama Ashwin/Shutterstock.com; **x** *top* ©iStockphoto.com/sumnersgraphicsinc, ©iStockphoto.com/Ann Marie Kurtz; *bottom* ©iStockphoto.com/Noraznen Azit; **xi** *top* Chiyacat/Shutterstock.com, Zoom Team/Shutterstock.com; *bottom* James Flint/Shutterstock.com; **xii** *top* Varina and Jay Patel/Shutterstock.com, ©iStockphoto.com/Ann Marie Kurtz; *bottom* PETER CLOSE/Shutterstock.com; **xiii** *top* ©iStockphoto.com/Alistair Cotton; *bottom* ©iStockphoto.com/ranplett; **xiv** *top* ©iStockphoto.com/ALEAIMAGE, ©iStockphoto.com/Ann Marie Kurtz; *bottom* ©iStockphoto.com/Jane norton; **xv** *top* ©iStockphoto/Michael Flippo, ©iStockphoto.com/Ann Marie Kurtz; *bottom* wavebreakmedia ltd/Shutterstock.com; **xvi** *top* Alexander Chaikin/Shutterstock.com, ©iStockphoto.com/Ann Marie Kurtz; *bottom* Sinisa Bobic/Shutterstock.com; **xvii** Kasiap/Shutterstock.com, ©iStockphoto.com/Ann Marie Kurtz; *bottom* ©iStockphoto.com/Thomas Perkins; **xix** Gelpi/Shutterstock.com; **xxxvi** *center left* AVAVA/Shutterstock.com; *bottom right* Tomasz/Shutterstock.com, Nemida/Shutterstock.com; **xxxvii** *top left* Konstantin Chagin/Shutterstock.com; *top right* Roger Jegg - Fotodesign-Jegg.de/Shutterstock.com **xliv** Ljupco Smokovski/Shutterstock.com

Chapter 1

1 ©iStockphoto.com/Lisa Thornberg, ©iStockphoto.com/Ann Marie Kurtz; **6** ©iStockphoto.com/David Freund; **7** ©iStockphoto.com/nicolas hansen; **8** NASA; **9** ©iStockphoto.com/Ryan Lane; **12** ©iStockphoto.com/Harley McCabe; **13** ©iStockphoto.com/Jacom Stephens; **14** ©iStockphoto.com/Harry Hu; **22** ©iStockphoto.com/Andrey Krasnov, **25** Chris Curtis/Shutterstock.com; **31** *top right* ©iStockphoto.com/Alan Crawford; *center left* ©iStockphoto.com/Julio Yeste; *bottom right* ©iStockphoto.com/Mark Stay; **36** *center right* Ljupco Smokovski/Shutterstock.com; *bottom left* emel82/Shutterstock.com

Chapter 2

40 ©iStockphoto.com/Jonathan Larsen; **45** NASA; **46** ©iStockphoto.com/David Morgan; **47** *top right* NASA; *center left* ©iStockphoto.com/jsemeniuk; **54** ©iStockphoto.com/Amanda Rohde; **55** Julian Rovagnati/Shutterstock.com; **62** ©iStockphoto.com/Dreamframer; **63** *top right* Jerry Horbert/Shutterstock.com; *bottom left* ©iStockphoto.com/Chris Schmidt; **65** ©iStockphoto.com/biffspandex; **68** ©iStockphoto.com/Stephen Pothier; **69** *top left* Gina Smith/Shutterstock.com; *bottom left* Dewayne Flowers/Shutterstock.com; **71** Philip Lange/Shutterstock.com; **75** Photo courtesy of Herrenknecht AG; **76** ©iStockphoto.com/Adam Mattel; **T-77** ©iStockphoto.com/marcellus2070, ©iStockphoto.com/beetle8; **77** *top left* ©iStockphoto.com/Gene Chutka; *center right* ©iStockphoto.com/marcellus2070, ©iStockphoto.com/beetle8; **81** ©iStockphoto.com/Connie Maher; **82** ©iStockphoto.com/Jacom Stephens; **83** *top right* ©iStockphoto.com/Petr Podzemny; *center left* ©iStockphoto.com/adrian beesley; **88** ©iStockphoto.com/Ryan Putnam; **89** ©iStockphoto.com/iLexx; **90** ©iStockphoto.com/Jeremy Edwards; **91** ©iStockphoto.com/Rober Kohlhuber; **97** ©iStockphoto.com/Marcio Silva; **98** ©iStockphoto.com/Carmen Martinez Banús

Chapter 3

102 ©iStockphoto.com/ALEAIMAGE, ©iStockphoto.com/Ann Marie Kurtz; **105** ©iStockphoto.com/Floortje; **T-109** ©iStockphoto.com/Richard Goerg; **109** *top right* The ESRB rating icons are registered trademarks of the Entertainment Software Association; *center left* ©iStockphoto.com/Kevin Panizza, ©iStockphoto.com/Island Effects, ©iStockphoto.com/Nico Smit; *center right* ©iStockphoto.com/Richard Goerg; **110** ©iStockphoto.com/George Peters; **111** ©iStockphoto.com/George Peters; **114** pandapaw/Shutterstock.com; **115** ©iStockphoto.com/Daniel Van Beek; **121** Alexander Kalina/Shutterstock.com; **122** *top right* ©iStockphoto.com/Peter Firus; *bottom* Robert Pernell/Shutterstock.com; **123** ©iStockphoto.com/Trevor Fisher; **130** ©iStockphoto.com/fotoVoyager; **144** *center right* Africa Studio/Shutterstock.com; *bottom right* lanitta/Shutterstock.com

Chapter 4

152 stephan kerkhofs/Shutterstock.com, Cigdem Sean Cooper/Shutterstock.com, ©iStockphoto.com/Andreas Gradin; **154** Howard Sandler/Shutterstock.com, ©iStockphoto.com/Dori OConnell; **157** Richard Paul Kane/Shutterstock.com; **158** ©iStockphoto.com/pelicankate; **160** *top* YuriyZhuravov/Shutterstock.com; *bottom* Talvi/Shutterstock.com; **163** aguilarphoto/Shutterstock.com; **164** *Exercises 4–6* Talvi/Shutterstock.com; *bottom* Kiselev Andrey Valerevich/Shutterstock.com; **165** *center left* Susan Schmitz/Shutterstock.com; *center right* akva/Shutterstock.com; **167** Andrey Yurlov/Shutterstock.com; **168** Steve Cukrov/Shutterstock.com; **172** *top left* Le Do/Shutterstock.com; *center left* Quang Ho/Shutterstock.com, SergeyIT/Shutterstock.com, jon le-bon/Shutterstock.com; **173** Ariwasabi/Shutterstock.com; **174** Eugen Wais/Shutterstock.com; **175** *top left* Gordana Sermek/Shutterstock.com; *center right* Rashevskyi Viacheslav/Shutterstock.com; **176** ©iStockphoto.com/walik; **180** ©iStockphoto.com/inababes; **181** ©iStockphoto.com/tforgo; **183** Kateryna Larina/Shutterstock.com; **188** Sandra van der Steen/Shutterstock.com; **189** Nikola Bilic/Shutterstock.com; **190** Kristijan Zontar/Shutterstock.com; **191** Blend Images/Shutterstock.com; **192** Selena/Shutterstock.com; **196** Poznyakov/Shutterstock.com

Chapter 5

200 ©iStockphoto.com/sumnersgraphicsinc, ©iStockphoto.com/Ann Marie Kurtz; **203** *bottom left* ©iStockphoto.com/Andrew Johnson; *bottom right* Image courtesy of Zappos.Com; **205** ©iStockphoto.com/alohaspirit; **206** ©iStockphoto.com/Timur Kulgarin; **207** *top right* Primo Ponies Photography; *center left* ©iStockphoto.com/Wayne Johnson; **211** ©Fotosearch.com; **214** ©iStockphoto.com/Hannu Liivaar; **215** ©iStockphoto.com/LoopAll; **219** ©iStockphoto.com/technotr; **221** *top right* ©iStockphoto.com/Mlenny Photography; *bottom right* ©iStockphoto.com/medobear; **224** Layland Masuda/Shutterstock.com; **229** Ferenc Szelepcsenyi/Shutterstock.com; **230** Todd Hackwelder/Shutterstock.com; **233** Pinkcandy/Shutterstock.com; **237** ©iStockphoto.com/PeskyMonkey; **241** ©iStockphoto.com/Tom Buttle; **246** Africa Studio/Shutterstock.com; **247** United States coin image from United States Mint; **248** ©iStockphoto.com/Victor Maffe; **249** *top right* Rafa Irusta/Shutterstock.com; *center left* violetkaipa/Shutterstock.com; **254** A. L. Spangler/Shutterstock.com

Chapter 6

258 Chiyacat/Shutterstock.com, Zoom Team/Shutterstock.com; **265** ©iStockphoto.com/parema; **273** Vitaly Korovin/Shutterstock.com; **274** Fotana/Shutterstock.com; **275** SSSCCC/Shutterstock.com; **277** DM7/Shutterstock.com; **T-278** Jaimie Duplass/Shutterstock.com; **279** Glenda M. Powers/Shutterstock.com, siamionau pavel/Shutterstock.com; **280** ©iStockphoto.com/katkov, ©iStockphoto.com/Andrew Johnson; **283** ©iStockphoto.com/Mehrab Moghadasian; **284** *left* Kamira/Shutterstock.com; *center left* Matthew Jacques/Shutterstock.com; *center right* Rachelle Burnside/Shutterstock.com; *right* ©Norman Young; **290** Patsy Michaud/Shutterstock.com; **295** United States Dept. of the Interior, U.S. Fish and Wildlife Service, FloridaStock/Shutterstock.com; **296** locote/Shutterstock.com; **298** Viorel Sima/Shutterstock.com; **299** *top right* Elena Elisseeva/Shutterstock.com; *bottom left* Courtesy of Intel Corp. copyright ©2005; **301** photomak/Shutterstock.com; **303** risteski goce/Shutterstock.com; **305** Maen Zayyad/Shutterstock.com; **310** Feng Yu/Shutterstock.com; **311** *top left* Alexander Lukin/Shutterstock.com; *center right* Andresr/Shutterstock.com; **313** Steshkin Yevgeniy/Shutterstock.com, mayer kleinostheim/Shutterstock.com, ©iStockphoto.com/Matej Michelizza, Refat/Shutterstock.com; **315** Photo by Alvesgaspar/Joaquim Alves Gaspar, Lisboa, Portugal, modified by RDBury; **316** Faraways/Shutterstock.com

Chapter 7

326 Varina and Jay Patel/Shutterstock.com, ©iStockphoto.com/Ann Marie Kurtz; **328** ©iStockphoto.com/Derek Dammann; **331** Pinkcandy/Shutterstock.com, Dmitry Rukhlenko/Shutterstock.com; **332** ©iStockphoto.com/Mark Stay; **333** *Exercise 31* ©iStockphoto.com/edge69; *Exercise 32* Matt Antonino/Shutterstock.com; **337** ©iStockphoto.com/edge69; **338** Johanna Goodyear/Shutterstock.com; **339** ©iStockphoto.com/edge69; **344** Fejas/Shutterstock.com; **345** Rusian Ivantsov/Shutterstock.com; **347** Li Wa/Shutterstock.com; **357** ©iStockphoto.com/Aldo Murillo; **T-361** Bev Sykes; **361** Bev Sykes; **366** ©iStockphoto.com/clu; **367** ©iStockphoto.com/Chris Grissom; **373** *bottom left* mmaxer/Shutterstock.com; *bottom right* ©iStockphoto.com/RonTech2000; **T-375** Pichugin Dmitry/Shutterstock.com; **375** *top right* ©iStockphoto.com/Viktor Gmyria; *Exercise 41* Pichugin Dmitry/Shutterstock.com; **379** rattanapatphoto/Shutterstock.com; **380** Veronika Surovtseva/Shutterstock.com; **381** Martin Lehmann/Shutterstock.com; **385** G Tipene/Shutterstock.com; **386** Mark Stout Photography/Shutterstock.com; **390** ©iStockphoto.com/tirc83; **395** Elena Elisseeva/Shutterstock.com

Chapter 8

400 ©iStockphoto.com/Alistair Cotton; **407** Fred Hendriks/Shutterstock.com; **409** Wolna/Shutterstock.com, ©iStockphoto.com/PLAINVIEW; **410** Maxx-Studio/Shutterstock.com; **413** ©iStockphoto.com/Daniel Jensen; **414** *Exercise 4* Flashon Studio/Shutterstock.com; *Exercise 5* Yuri Kravchenko/Shutterstock.com; *Exercise 6* ARENA Creative/Shutterstock.com; *bottom right* Pavel Plakosh/Shutterstock.com; **415** ©iStockphoto.com/CAP53; **419** ©iStockphoto.com/edge69, L_amica/Shutterstock.com; **421** *top right* Dobrinya/Shutterstock.com; *center left* Ron Zmiri/Shutterstock.com; **423** llyashenko Oleksiy/Shutterstock.com; **430** *top right* James Thew/Shutterstock.com; *bottom right* VR Photos/Shutterstock.com; **431** Keith Lovett Photography; **447** ©iStockphoto.com/Geoffrey Holman, TyBy/Shutterstock.com; **448** cbpix/Shutterstock.com

Chapter 9

452 ©iStockphoto.com/ALEAIMAGE, ©iStockphoto.com/Ann Marie Kurtz; **458** RTimages/Shutterstock.com; **459** Ben Haslam/Haslam Photography/Shutterstock.com; **460** Golden Pixels LLC/Shutterstock.com, mikeledray/Shutterstock.com; **461** mobil11/Shutterstock.com; **463** ATurner/Shutterstock.com; **466** Digital Genetics/Shutterstock.com; **467** *center right* Zhukov Oleg/Shutterstock.com; *center left* Baker Alhashki/Shutterstock.com; **473** *top right* Maximus256/Shutterstock.com; *center left* lakov Filimonov/Shutterstock.com, Shebeko/Shutterstock.com; **475** Ivan Bondarenko/Shutterstock.com; **477** Ra Studio/Shutterstock.com; **479** Maxim Kulko/Shutterstock.com; **481** Lukiyanova Natalia/frenta/Shutterstock.com; **482** Elena Elisseeva/Shutterstock.com; **483** NMorozova/Shutterstock.com; **486** Supri Suharjoto/Shutterstock.com; **490** ©iStockphoto.com/Scott Hirko; **491** Sinisa Bobic/Shutterstock.com; **492** Fotokostic/Shutterstock.com; **496** lpatov/Shutterstock.com

Chapter 10

500 ©iStockphoto.com/Michael Flippo, ©iStockphoto.com/Ann Marie Kurtz; **505** Zacarias Pereira da Mata/Shutterstock.com; **506** Johnny Habell/Shutterstock.com; **507** Eduard Härkönen/Shutterstock.com; **509** Radovan Spurny/Shutterstock.com, Worldpics/Shutterstock.com, wong yu liang/Shutterstock.com; **514** Sergej Razvodovskij/Shutterstock.com; **515** Pavel Burchenko/Shutterstock.com; **516** Poulsons Photography/Shutterstock.com; **517** eddtoro/Shutterstock.com; **519** BrendanReals/Shutterstock.com; **520** ©oxford science archive/Heritage images/Imagestate; **524** Mikateke/Shutterstock.com; **525** kavione/Shutterstock.com, Phase4Photography/Shutterstock.com; **527** Monkey Business Images/Shutterstock.com; **535** LoopAll/Shutterstock.com; **536** ©iStockphoto.com/Cathy Keifer

Chapter 11

540 Alexander Chaikin/Shutterstock.com, ©iStockphoto.com/Ann Marie Kurtz; **543** *bottom left* KellyNelson/Shutterstock.com; *bottom right* Mircea BEZERGHEANU/Shutterstock.com; **547** mangostock/Shutterstock.com; **548** Stephen Coburn/Shutterstock.com; **549** jean schweitzer/Shutterstock.com; **550** VectoriX/Shutterstock.com, Betacam-SP/Shutterstock.com; **551** wavebreakmedia ltd/Shutterstock.com; **555** ©iStockphoto.com/4x6; **556** Mark Herreid/Shutterstock.com; **557** *center left* Rene Hartmann/Shutterstock.com; *bottom right* Andy Dean Photography/Shutterstock.com; **560** Franco Volpato/Shutterstock.com; **564** Odua Images/Shutterstock.com, Jozsef Bagota/Shutterstock.com; **567** Dmitry Rukhlenko/Shutterstock.com; **573** qushe/Shutterstock.com; **578** Racheal Grazias/Shutterstock.com; **580** Suzanne Tucker/Shutterstock.com; **581** Alexandra Lande/Shutterstock.com; **585** Supri Suharjoto/Shutterstock.com; **586** Aspen Photo/Shutterstock.com; **587** Galina Barskaya/Shutterstock.com; **588** lsantilli/Shutterstock.com; **589** Richard Paul Kane/Shutterstock.com; **591** JCElv/Shutterstock.com; **592** altug/Shutterstock.com; **593** *top left* LoopAll/Shutterstock.com; *center right* Greg Epperson/Shutterstock.com; **594** ©iStockphoto.com/Colleen Butler; **599** *center right* ©iStockphoto.com/Mark Murphy; *center left* ©iStockphoto.com/Milorad Zaric

Chapter 12

604 Kasiap/Shutterstock.com, ©iStockphoto.com/Ann Marie Kurtz; **610** ©iStockphoto.com/Jan Will; **612** ©iStockphoto.com/George Peters; **613** Steve Byland/Shutterstock.com; **614** michaeljung/Shutterstock.com; **616** Stephen Coburn/Shutterstock.com; **617** Zinin Alexei/Shutterstock.com; **619** Laurence Gough/Shutterstock.com; **620** FREDERIC J. BROWN/Staff/AFP/Getty Images; **621** ©iStockphoto.com/4x6; **623** Krasowit/Shutterstock.com; **624** *top right* Fotofermer/Shutterstock.com; *center left* Komar Maria/Shutterstock.com; **625** ©iStockphoto.com/Neustockimages; **627** East/Shutterstock.com; **631** Otna Ydur/Shutterstock.com; **633** Adam Gregor/Shutterstock.com; **635** ©iStockphoto.com/David15; **636** Gina Brockett; **637** ©iStockphoto.com/Craig Dingle; **641** ©iStockphoto.com/Jill Fromer; **642** ©iStockphoto.com/Janis Litavnieks; **644** Oleksly Mark/Shutterstock.com; **648** Sashkin/Shutterstock.com; **649** koya979/Shutterstock.com; **650** Michael Shake/Shutterstock.com; **651** Pakhnyushcha/Shutterstock.com; **652** RTimages/Shutterstock.com; **657** *top right* Suponev Vladimir/Shutterstock.com; *bottom right* Alberto Zornetta/Shutterstock.com; **658** *center left* ©iStockphoto.com/Tony Campbell; *bottom right* ©iStockphoto.com/Eric Isselée; **659** *top right* Larry Korhnak; *bottom right* Photo by Andy Newman; **663** *center left* ©iStockphoto.com/Jane norton; *bottom right* ©iStockphoto.com/Krzysztof Zmij; **664** Dwight Smith/Shutterstock.com; **666** Tiplyashin Anatoly/Shutterstock.com; **667** Nikola Bilic/Shutterstock.com; **670** *center right* Iwona Grodzka/Shutterstock.com; *bottom right* Lim Yong Hian/Shutterstock.com

Appendix A

A1 *background* ©iStockphoto.com/Björn Kindler; *panther* Clipart deSIGN; *mosaic* Pentocelo; *fractal* ravl; *tornado* EmiliaU; *daisy* tr3gin; **A2** Elnur/Shutterstock.com; **A3** *top* Clipart deSIGN/Shutterstock.com; **A5** *bottom* Pentocelo; **A6** *top right* Taras Vyshnya/Shutterstock.com; *center right* topseller/Shutterstock.com; **A7** *top right* ravl/Shutterstock.com; *center left* tr3gin/Shutterstock.com; *bottom left* Artistas/Shutterstock.com; **A8** EmiliaU/Shutterstock.com; **T-A8** *left* objectsforall/Shutterstock.com; *right* Subbotina Anna/Shutterstock.com; **T-A9** *top* ARENA Creative/Shutterstock.com; *center* clickthis/Shutterstock.com; **A9** *top right* deepspacedave/Shutterstock.com; *bottom right* Delmas Lehman/Shutterstock.com

Cartoon illustrations Tyler Stout
Cover image Lechner & Benson Design

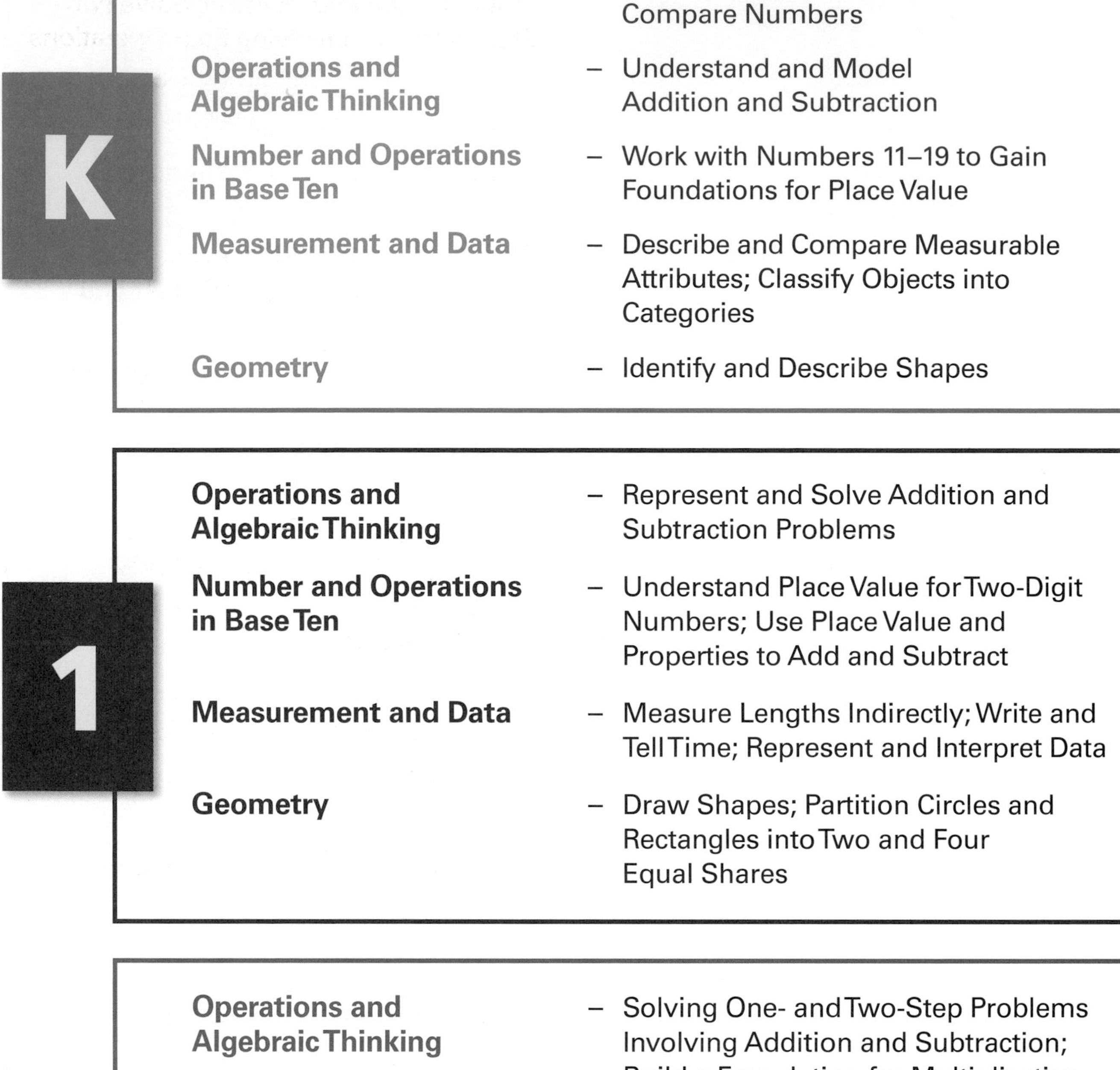

K

Domain	Cluster
Counting and Cardinality	– Count to 100 by Ones and Tens; Compare Numbers
Operations and Algebraic Thinking	– Understand and Model Addition and Subtraction
Number and Operations in Base Ten	– Work with Numbers 11–19 to Gain Foundations for Place Value
Measurement and Data	– Describe and Compare Measurable Attributes; Classify Objects into Categories
Geometry	– Identify and Describe Shapes

1

Domain	Cluster
Operations and Algebraic Thinking	– Represent and Solve Addition and Subtraction Problems
Number and Operations in Base Ten	– Understand Place Value for Two-Digit Numbers; Use Place Value and Properties to Add and Subtract
Measurement and Data	– Measure Lengths Indirectly; Write and Tell Time; Represent and Interpret Data
Geometry	– Draw Shapes; Partition Circles and Rectangles into Two and Four Equal Shares

2

Domain	Cluster
Operations and Algebraic Thinking	– Solving One- and Two-Step Problems Involving Addition and Subtraction; Build a Foundation for Multiplication
Number and Operations in Base Ten	– Understand Place Value for Three-Digit Numbers; Use Place Value and Properties to Add and Subtract
Measurement and Data	– Measure and Estimate Lengths in Standard Units; Work with Time and Money
Geometry	– Draw and Identify Shapes; Partition Circles and Rectangles into Two, Three, and Four Equal Shares

3

Domain	Cluster
Operations and Algebraic Thinking	– Represent and Solve Problems Involving Multiplication and Division; Solve Two-Step Problems Involving Four Operations
Number and Operations in Base Ten	– Round Whole Numbers; Add, Subtract, and Multiply Multi-Digit Whole Numbers
Number and Operations — Fractions	– Understand Fractions as Numbers
Measurement and Data	– Solve Time, Liquid Volume, and Mass Problems; Understand Perimeter and Area
Geometry	– Reason with Shapes and Their Attributes

4

Domain	Cluster
Operations and Algebraic Thinking	– Use the Four Operations with Whole Numbers to Solve Problems; Understand Factors and Multiples
Number and Operations in Base Ten	– Generalize Place Value Understanding; Perform Multi-Digit Arithmetic
Number and Operations — Fractions	– Build Fractions from Unit Fractions; Understand Decimal Notation for Fractions
Measurement and Data	– Convert Measurements; Understand and Measure Angles
Geometry	– Draw and Identify Lines and Angles; Classify Shapes

5

Domain	Cluster
Operations and Algebraic Thinking	– Write and Interpret Numerical Expressions
Number and Operations in Base Ten	– Perform Operations with Multi-Digit Numbers and Decimals to Hundredths
Number and Operations — Fractions	– Add, Subtract, Multiply, and Divide Fractions
Measurement and Data	– Convert Measurements within a Measurement System, Understand Volume
Geometry	– Graph Points in the First Quadrant of the Coordinate Plane; Classify Two-Dimensional Figures

Mathematics Reference Sheet

Conversions

U.S. Customary to Metric

1 inch = 2.54 centimeters
1 foot ≈ 0.30 meter
1 mile ≈ 1.61 kilometers
1 quart ≈ 0.95 liter
1 gallon ≈ 3.79 liters
1 cup ≈ 237 milliliters
1 pound ≈ 0.45 kilogram
1 ounce ≈ 28.3 grams
1 gallon ≈ 3785 cubic centimeters

Metric to U.S. Customary

1 centimeter ≈ 0.39 inch
1 meter ≈ 3.28 feet
1 kilometer ≈ 0.62 mile
1 liter ≈ 1.06 quarts
1 liter ≈ 0.26 gallon
1 kilogram ≈ 2.2 pounds
1 gram ≈ 0.035 ounce
1 cubic meter ≈ 264 gallon

Temperature

$C = \frac{5}{9}(F - 32)$

$F = \frac{9}{5}C + 32$

Number Properties

Commutative Properties of Addition and Multiplication

$a + b = b + a$
$a \cdot b = b \cdot a$

Associative Properties of Addition and Multiplication

$(a + b) + c = a + (b + c)$
$(a \cdot b) \cdot c = a \cdot (b \cdot c)$

Addition Property of Zero

$a + 0 = a$

Multiplication Properties of Zero and One

$a \cdot 0 = 0$
$a \cdot 1 = a$

Distributive Property:

$a(b + c) = ab + ac$
$a(b - c) = ab - ac$

Properties of Equality

Addition Property of Equality
If $a = b$, then $a + c = b + c$.

Subtraction Property of Equality
If $a = b$, then $a - c = b - c$.

Multiplication Property of Equality
If $a = b$, then $a \cdot c = b \cdot c$.

Division Property of Equality
If $a = b$, then $a \div c = b \div c, c \neq 0$.

Squaring both sides of an equation
If $a = b$, then $a^2 = b^2$.

Properties of Exponents

Product of Powers Property: $a^m \cdot a^n = a^{m+n}$

Quotient of Powers Property: $\frac{a^m}{a^n} = a^{m-n}, a \neq 0$

Power of a Power Property: $(a^m)^n = a^{mn}$

Power of a Product Property: $(ab)^m = a^m b^m$

Power of a Quotient Property: $\left(\frac{a}{b}\right)^m = \frac{a^m}{b^m}, b \neq 0$

Zero Exponents: $a^0 = 1, a \neq 0$

Negative Exponents: $a^{-n} = \frac{1}{a^n}, a \neq 0$

Rational Exponents: $\sqrt[n]{a} = a^{1/n}$

Properties of Square Roots

Product Property of Square Roots

$\sqrt{xy} = \sqrt{x} \cdot \sqrt{y}, x \geq 0 \text{ and } y \geq 0$

Quotient Property of Square Roots

$\sqrt{\frac{x}{y}} = \frac{\sqrt{x}}{\sqrt{y}}, x \geq 0 \text{ and } y > 0$

Slope

$$m = \frac{\text{rise}}{\text{run}}$$

$$= \frac{\text{change in } y}{\text{change in } x}$$

$$= \frac{y_2 - y_1}{x_2 - x_1}$$

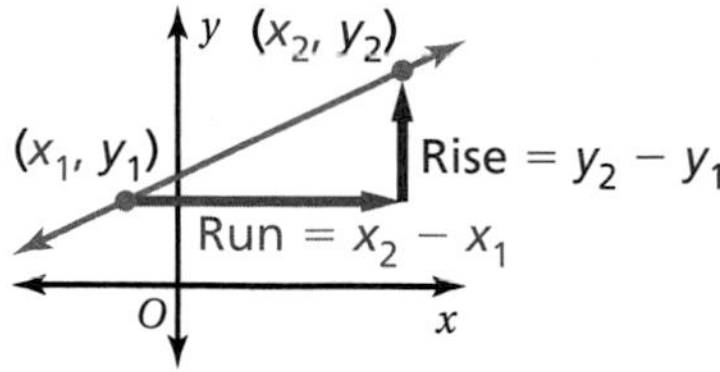

Factoring

Difference of Two Squares Pattern

$$a^2 - b^2 = (a + b)(a - b)$$

Perfect Square Trinomial Pattern

$$a^2 + 2ab + b^2 = (a + b)^2$$

$$a^2 - 2ab + b^2 = (a - b)^2$$

Pythagorean Theorem

$$a^2 + b^2 = c^2$$

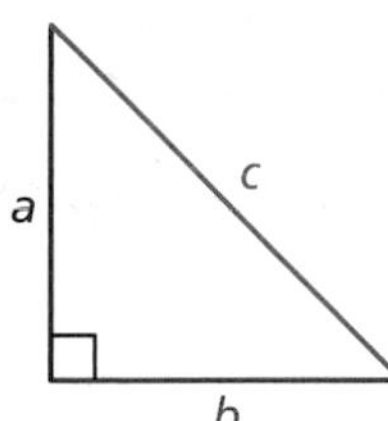

Distance Formula

$$d = \sqrt{(x_2 - x_1)^2 + (y_2 - y_1)^2}$$

Equations of Lines

Slope-intercept form

$$y = mx + b$$

Standard form

$$ax + by = c,\ a, b \neq 0$$

Point-slope form

$$y - y_1 = m(x - x_1)$$

Forms of Quadratic Functions

Standard form

$$y = ax^2 + bx + c,\ a \neq 0$$

Vertex form

$$y = a(x - h)^2 + k,\ a \neq 0$$

Quadratic Formula

$$x = \frac{-b \pm \sqrt{b^2 - 4ac}}{2a}$$ ← discriminant

Sequences

Arithmetic

$a_n = a_1 + (n - 1)d$ Explicit equation

$a_n = a_{n-1} + d$ Recursive equation

Geometric

$a_n = a_1 r^{n-1}$ Explicit equation

$a_n = r \cdot a_{n-1}$ Recursive equation

Volume

Prism

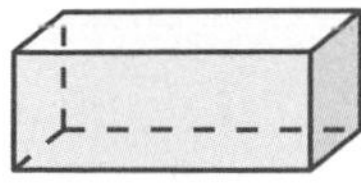

$V = Bh = \ell wh$

Cylinder

$V = Bh = \pi r^2 h$

Cone

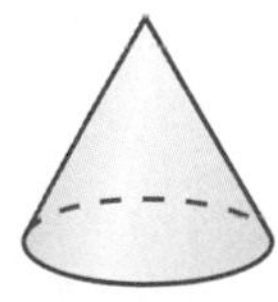

$V = \frac{1}{3}Bh = \frac{1}{3}\pi r^2 h$

Sphere

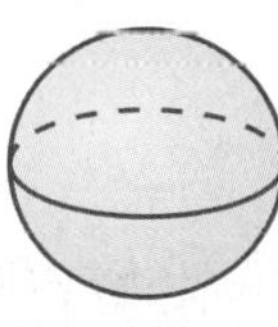

$V = \frac{4}{3}\pi r^3$